TAB. XII.

Fig. 1.

2

3

4

5

6

7

a

b

c

d

e

f

g

h

i

GENESIS cap.I.v.20.

Opus quintæ Diei.

I. Buch Mosis cap.I.v.20.

Fünfftes Tagwerck.

I.A.Corvinus sculp.

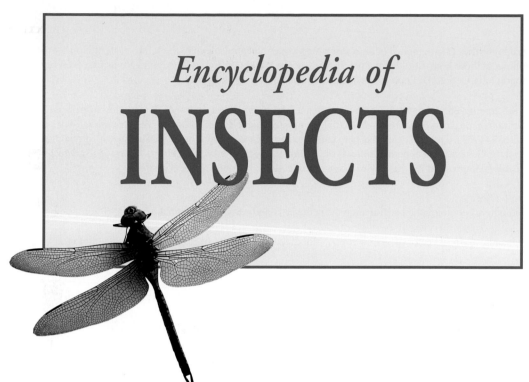

Encyclopedia of
INSECTS

Editors

VINCENT H. RESH

University of California, Berkeley

RING T. CARDÉ

University of California, Riverside

ACADEMIC PRESS

An imprint of Elsevier Science

Amsterdam   *Boston*   *London*   *New York*   *Oxford*   *Paris*   *San Diego*   *San Francisco*   *Singapore*   *Sydney*   *Tokyo*

Cover Art: Monarch butterflies photographed at an overwintering site in central Mexico. Each fall, eastern North American monarchs migrate by the millions to these high-altitude Oyamel fir forests. The weight of their dense aggregations can even cause tree limbs to break away. (Photograph by George D. Lepp, a specialist in images of the natural world.)

Frontispiece: An 18th century engraving by I. A. Covinus depicts the Biblical description of the creation of insects. The image appears in *Physique sacrée, ou histoire naturelle de la Bible,* a massive eight-volume study by the prominent Swiss naturalist Johann Jacob Scheuchzer (1672–1737). The artist Covinus here combines an accurate rendering of various insect forms with an allegorical treatment of the Book of Genesis; this approach reflects the author Scheuchzer's lifelong efforts to provide scientific evidence for the literal truth of the Bible. (Courtesy of the History of Science Collections, Cornell University Library.)

Letter-Opening Photo Credits:
R: Snakefly (Raphidioptera) photographed at Nanaimo (Vancouver Island), British Columbia, Canada. (Photograph Copyright © Jay Patterson.)
Y: *Aedes aegypti,* Uganda strain (a vector of yellow fever), bloodfeeding from the photographer's hand. (Photograph by Leonard E. Munstermann.)
Other letter-opening photographs furnished by the authors. (See relevant article for credit.)

Academic Press
An imprint of Elsevier Science
525 B Street, Suite 1900, San Diego, California 92101-4495, USA
http://www.academicpress.com

Academic Press
84 Theobald's Road, London WC1X 8RR, UK
http://www.academicpress.com

Academic Press
200 Wheeler Road, Burlington, Massachusetts 01803, USA
http://www.academicpressbooks.com

Library of Congress Catalog Card Number: 2002106355

International Standard Book Number: 0-12-586990-8

PRINTED IN CHINA
02   03   04   05   06   07   RDC   9   8   7   6   5   4   3   2   1

CONTENTS

CONTENTS BY SUBJECT AREA

INTERACTIONS WITH OTHER ORGANISMS

INTERACTIONS WITH HUMANS

HABITATS

ECOLOGY

HISTORY AND METHODOLOGY

CONTRIBUTORS

JOHN H. ACORN
University of Alberta, Canada
Teaching Resources

MICHAEL E. ADAMS
University of California, Riverside
Development, Hormonal Control of
Ecdysteroids
Juvenile Hormones

PETER H. ADLER
Clemson University
Legs

GILBERTO S. ALBUQUERQUE
*Universidade Estadual do Norte Fluminense,
Brazil*
Neuroptera

RICHARD D. ALEXANDER
University of Michigan
Crickets

ROBERT T. ALLEN
Paris, Arkansas
Diplura
Protura

MIRIAM ALTSTEIN
*Agricultural Research Organization, Volcani
Center, Israel*
Neuropeptides

SVEND O. ANDERSEN
Copenhagen University, Denmark
Cuticle
Exoskeleton
Integument

NORMAN H. ANDERSON
Oregon State University
Megaloptera

DAVID A. ANDOW
University of Minnesota, St. Paul
Genetically Modified Plants

MICHAEL F. ANTOLIN
Colorado State University
Sex Determination

LARRY G. ARLIAN
Wright State University
Chiggers and Other Disease-Causing Mites

HORST ASPÖCK
University of Vienna, Austria
Raphidioptera

ULRIKE ASPÖCK
Natural History Museum, Vienna, Austria
Raphidioptera

PETER W. ATKINSON
University of California, Riverside
Genetic Engineering

ARND BAUMANN
Forschungszentrum Jülich, Germany
Biogenic Amines

NANCY E. BECKAGE
University of California, Riverside
Immunology

PETER BELLINGER[†]
California State University, Northridge
Collembola

MAY R. BERENBAUM
University of Illinois
Movies, Insects in

MARTIN B. BERG
Loyola University Chicago
Growth, Individual

ELIZABETH A. BERNAYS
University of Arizona
Host Seeking, for Plants
Phytophagous Insects

[†]Deceased.

CHRISTER BJÖRKMAN
Swedish University of Agricultural Sciences
Body Size

SCOTT HOFFMAN BLACK
The Xerces Society, Portland, Oregon
Endangered Insects

SETH S. BLAIR
University of Wisconsin, Madison
Imaginal Discs

WOLFGANG BLENAU
Universität Potsdam, Germany
Biogenic Amines

MURRAY S. BLUM
University of Georgia
Chemical Defense

BRYONY C. BONNING
Iowa State University
Biotechnology and Insects

TIMOTHY J. BRADLEY
University of California, Irvine
Excretion

PAUL M. BRAKEFIELD
Leiden University, The Netherlands
Crypsis

JOHN E. BRITTAIN
University of Oslo, Norway
Ephemeroptera

LINCOLN P. BROWER
Sweet Briar College
Monarchs

ANDREAS BRUNE
Universität Konstanz, Germany
Symbionts Aiding Digestion

WENDELL E. BURKHOLDER
University of Wisconsin, Madison
Stored Products as Habitats

GEORGE W. BYERS
University of Kansas
Mecoptera

RING T. CARDÉ
University of California, Riverside
Insecta, Overview
Orientation
Pheromones

REGINALD F. CHAPMAN
University of Arizona
Feeding Behavior
Locusts
Mouthparts
Rostrum

LANNA CHENG
Scripps Institution of Oceanography, La Jolla
Marine Insects

KENNETH A. CHRISTIANSEN
Grinnell College, Iowa
Collembola

THOMAS M. CLARK
Indiana University, South Bend
Water and Ion Balance, Hormonal Control of

DONALD G. COCHRAN
Virginia Polytechnic Institute and State University
Blattodea

EPHRAIM COHEN
The Hebrew University of Jerusalem
Chitin
Fat Body

ANDREJ ČOKL
National Institute of Biology, Ljubljana, Slovenia
Vibrational Communication

GREGORY W. COURTNEY
Iowa State University
Diptera

CHARLES V. COVELL, JR.
University of Louisville
Collection and Preservation
Stamps, Insects and

CATHERINE L. CRAIG
Harvard University/Tufts University
Silk Production

EVA CRANE
International Bee Research Association
Apis Species
Beekeeping
Bee Products
Beeswax
Honey
Royal Jelly

PETER S. CRANSTON
University of California, Davis
Phylogeny of Insects

GREGORY A. DAHLEM
Northern Kentucky University
House Fly

DONALD L. DAHLSTEN
University of California, Berkeley
Gallmaking and Insects

GENE R. DEFOLIART
University of Wisconsin, Madison
Food, Insects as

IAN DENHOLM
Rothamsted Research, United Kingdom
Insecticide and Acaricide Resistance

DAVID L. DENLINGER
Ohio State University
Cold/Heat Protection
Diapause

GREGOR J. DEVINE
Rothamsted Research, United Kingdom
Insecticide and Acaricide Resistance

MICHAEL DICKINSON
California Institute of Technology
Flight

CHRISTOPHER H. DIETRICH
Illinois Natural History Survey
Auchenorrhyncha

HUGH DINGLE
University of California, Davis
Migration

ANGELA E. DOUGLAS
University of York, United Kingdom
Honeydew

ROBERT V. DOWELL
California Department of Food and Agriculture
Regulatory Entomology

ROBERT DUDLEY
University of California, Berkeley
Flight

JOHN D. EDMAN
University of California, Davis
Medical Entomology

BRUCE F. ELDRIDGE
University of California, Davis
Mosquitoes

JOSEPH S. ELKINTON
University of Massachusetts
Gypsy Moth
Population Ecology

MICHAEL S. ENGEL
University of Kansas
Zoraptera

JOACHIM ERBER
Technische Universität Berlin, Germany
Biogenic Amines

BRIAN A. FEDERICI
University of California, Riverside
Pathogens of Insects

LEWIS J. FELDMAN
University of California, Berkeley
Insectivorous Plants

CLÉLIA FERREIRA
University of São Paulo, Brazil
Digestive System

R. NELSON FOSTER
U.S. Department of Agriculture
Boll Weevil

GORDON W. FRANKIE
University of California, Berkeley
Pollination and Pollinators

NIGEL R. FRANKS
University of Bristol, United Kingdom
Ants

ANDREW S. FRENCH
Dalhousie University, Canada
Mechanoreception

DOUGLAS J. FUTUYMA
University of Michigan
Coevolution

ERIN C. GENTRY
University of Florida
Bioluminescence

HELEN GHIRADELLA
State University of New York, Albany
Coloration

ROSEMARY G. GILLESPIE
University of California, Berkeley
Island Biogeography and Evolution
Spiders

GONZALO GIRIBET
Harvard University
Daddy-Long-Legs

M. LEE GOFF
Chaminade University of Honolulu
Forensic Entomology

GORDON GORDH
U.S. Department of Agriculture
Anatomy: Head, Thorax, Abdomen, and Genitalia

MIODRAG GRBIĆ
University of Western Ontario, Canada
Embryogenesis

LES GREENBERG
University of California, Riverside
Fire Ants

DAVID GRIMALDI
American Museum of Natural History, New York
Fossil Record

PENNY J. GULLAN
University of California, Davis
Phylogeny of Insects
Sternorrhyncha

DARRYL T. GWYNNE
University of Toronto, Canada
Mating Behaviors

JON F. HARRISON
Arizona State University
Respiratory System
Tracheal System

MICHAEL W. HASTRITER
Brigham Young University
Siphonaptera

DAVID H. HEADRICK
California Polytechnic State University
Anatomy: Head, Thorax, Abdomen, and
Genitalia

BERND HEINRICH
University of Vermont
Thermoregulation

DAVID W. HELD
University of Kentucky
Japanese Beetle
June Beetles

RONALD A. HELLENTHAL
University of Notre Dame
Phthiraptera

JORGE HENDRICHS
FAO/IAE Division, Vienna, Austria
Sterile Insect Technique

ADAM D. HENK
Colorado State University
Sex Determination

NANCY C. HINKLE
University of Georgia
Cat Fleas
Extension Entomology

M. S. HODDLE
University of California, Riverside
Biological Control of Insect Pests

JAMES N. HOGUE
California State University, Northridge
Cultural Entomology
Folk Beliefs and Superstitions

MARILYN A. HOUCK
Texas Tech University
Phoresy

FRANCIS G. HOWARTH
B. P. Bishop Museum, Honolulu, Hawaii
Cave Insects

RON HOY
Cornell University
Hearing

LAWRENCE E. HURD
Washington and Lee University
Mantodea

MICHAEL E. IRWIN
University of Illinois
Commercial Products from Insects

RUDOLF JANDER
University of Kansas
Magnetic Sense

ROBERT L. JEANNE
University of Wisconsin, Madison
Nest Building

MATHIEU JORON
Leiden University, The Netherlands
Aposematic Coloration
Mimicry

ROBERT JOSEPHSON
University of California, Irvine
Muscle System

GAIL E. KAMPMEIER
University of Illinois
Commercial Products from Insects

KENNETH Y. KANESHIRO
University of Hawaii
Sexual Selection

MICHAEL R. KANOST
Kansas State University
Hemolymph

ALAN I. KAPLAN
*East Bay Regional Park District, Berkeley,
California*
Entomological Societies

JOE B. KEIPER
Cleveland Museum of Natural History
Diptera

GEORGE G. KENNEDY
North Carolina State University
Colorado Potato Beetle

LAWRENCE R. KIRKENDALL
University of Bergen, Norway
Parthenogenesis

KLAUS-DIETER KLASS
Zoological Museum, Dresden, Germany
Mantophasmatodea

JOHN KLOTZ
University of California, Riverside
Magnetic Sense

MARC J. KLOWDEN
University of Idaho
Oviposition Behavior
Reproduction, Male
Reproduction, Male: Hormonal Control of
Spermatheca
Spermatophore

MARCOS KOGAN
Oregon State University
Agricultural Entomology
Integrated Pest Management

VLADIMÍR KOŠT'ÁL
*Institute of Entomology, Academy of Sciences,
Czech Republic*
Temperature, Effects on Development
and Growth

ANDREAS KRUESS
University of Göttingen, Germany
Grassland Habitats

MICHAEL F. LAND
University of Sussex, United Kingdom
Eyes and Vision

ROBERT S. LANE
University of California, Berkeley
Zoonoses, Arthropod-Borne

STEPHEN G. A. LEAK
*International Trypanotolerance Centre, The
Gambia*
Tsetse Fly

RICHARD E. LEE, JR.
Miami University, Oxford, Ohio
Dormancy
Hibernation

M. J. LEHANE
University of Wales, Bangor
Blood Sucking

NORMAN C. LEPPLA
University of Florida
Rearing of Insects

RICHARD J. LESKOSKY
University of Illinois
Movies, Insects in

VERNARD R. LEWIS
University of California, Berkeley
Isoptera

JAMES K. LIEBHERR
Cornell University
Coleoptera

JAMES E. LLOYD
University of Florida
Bioluminescence

CATHERINE LOUDON
University of Kansas
Antennae

DWIGHT E. LYNN
U.S. Department of Agriculture
Cell Culture

MICHAEL E. N. MAJERUS
University of Cambridge, United Kingdom
Industrial Melanism
Ladybugs
Symbionts, Bacterial

JON H. MARTIN
The Natural History Museum, London
Sternorrhyncha

SINZO MASAKI
Hirosaki University, Japan
Aestivation

LINDA J. MASON
Purdue University
Extension Entomology

FUMIO MATSUMURA
University of California, Davis
DDT
Insecticides

JOSEPH V. MCHUGH
University of Georgia
Coleoptera

TERRI L. MEINKING
University of Miami School of Medicine
Lice, Human

RICHARD W. MERRITT
Michigan State University
Aquatic Habitats
Diptera
Growth, Individual

JOCELYN G. MILLAR
University of California, Riverside
Pheromones

THOMAS A. MILLER
University of California, Riverside
Circulatory System

NICK MILLS
University of California, Berkeley
Parasitoids

B. K. MITCHELL
University of Alberta, Canada
Chemoreception

THOMAS E. MITTLER
University of California, Berkeley
Honeydew

EDWARD L. MOCKFORD
Illinois State University
Psocoptera

MARK W. MOFFETT
University of California, Berkeley
Photography of Insects

THOMAS P. MONATH
Acambis Inc., Cambridge, Massachusetts
Yellow Fever

JOHN C. MORSE
Clemson University
Trichoptera

MAX S. MOULDS
Australian Museum, Sydney
Cicadas

LAURENCE A. MOUND
CSIRO, Canberra, Australia
Thysanoptera

BRADLEY A. MULLENS
University of California, Riverside
Veterinary Entomology

WERNER NACHTIGALL
Universität der Saarlandes, Germany
Swimming, Lake Insects

LISA NAGY
University of Arizona
Embryogenesis

OLDŘICH NEDVĚD
Institute of Entomology, Academy of Sciences,
Czech Republic
Temperature, Effects on Development
and Growth

TIM R. NEW
La Trobe University, Australia
Conservation

GORDON M. NISHIDA
University of California, Berkeley
Museums and Display Collections

BENJAMIN NORMARK
University of Massachusetts
Parthenogenesis

DAVID A. O'BROCHTA
University of Maryland Biotechnology Institute
Genetic Engineering

BARRY M. OCONNOR
University of Michigan
Mites

SEAN O'DONNELL
University of Washington
Caste
Colonies

PATRICK M. O'GRADY
American Museum of Natural History, New York
Drosophila melanogaster

DANIEL OTTE
Philadelphia Academy of Natural Sciences
Crickets

TERRY L. PAGE
Vanderbilt University
Circadian Rhythms

TIMOTHY D. PAINE
University of California, Riverside
Borers

JAMES O. PALMER
Allegheny College
Dermaptera

BERNARD PANNETON
Agriculture and Agri-Food Canada, Quebec
Physical Control of Insect Pests

DANIEL R. PAPAJ
University of Arizona
Learning

NIPAM H. PATEL
University of Chicago
Segmentation

MATS W. PETTERSSON
Swedish University of Agricultural Sciences
Body Size

JOHN D. PINTO
University of California, Riverside
Hypermetamorphosis

RUDY PLARRE
Federal German Institute for Materials Research
and Testing, Germany
Stored Products as Habitats

EDWARD G. PLATZER
University of California, Riverside
Dog Heartworm

GEORGE POINAR JR.
Oregon State University
Amber

DANIEL A. POTTER
University of Kentucky
Japanese Beetle
June Beetles

JERRY A. POWELL
University of California, Berkeley
Lepidoptera

ROGER D. PRICE
University of Minnesota
Phthiraptera

RONALD PROKOPY
University of Massachusetts
Agricultural Entomology
Integrated Pest Management

ALEXANDER H. PURCELL
University of California, Berkeley
Phytotoxemia
Plant Diseases and Insects

DONALD L. J. QUICKE
Imperial College, University of London,
United Kingdom
Hymenoptera

FRANK J. RADOVSKY
Oregon State University
Neosomy

SUSAN M. RANKIN
Allegheny College
Dermaptera

WILLIAM K. REISEN
University of California, Davis
Malaria

D. C. F. RENTZ
California Academy of Sciences, San Francisco
Grylloblattodea
Orthoptera

VINCENT H. RESH
University of California, Berkeley
Insecta, Overview
Pollution, Insect Response to
River Blindness

LYNN M. RIDDIFORD
University of Washington
Molting

JAMES RIDSDILL-SMITH
CSIRO, Canberra, Australia
Dung Beetles

ROY E. RITZMANN
Case Western Reserve University
Walking and Jumping

ALAN ROBINSON
IAEA Laboratories, Seibersdorf, Austria Sterile
Insect Technique

GENE E. ROBINSON
University of Illinois, Urbana-Champaign
Division of Labor in Insect Societies

GEORGE K. RODERICK
University of California, Berkeley
Genetic Variation
Island Biogeography and Evolution

DAVID M. ROSENBERG
Freshwater Institute, Winnipeg, Canada
Pollution, Insect Response to

EDWARD S. ROSS
California Academy of Sciences
Embiidina

MICHAEL K. RUST
University of California, Riverside
Cat Fleas
Urban Habitats

MICHEL SARTORI
Museum of Zoology, Lausanne, Switzerland
Ephemeroptera

LESLIE SAUL-GERSHENZ
*Center for Ecosystem Survival, San Francisco,
California*
Insect Zoos

CARL W. SCHAEFER
University of Connecticut
Prosorrhyncha

KATHERINE N. SCHICK
University of California, Berkeley
Gallmaking and Insects

JUSTIN O. SCHMIDT
Southwestern Biological Institute, Tucson, Arizona
Defensive Behavior
Venom
Wasps

MICHELLE PELLISSIER SCOTT
University of New Hampshire
Parental Care

THOMAS W. SCOTT
University of California, Davis
Dengue

J. MARK SCRIBER
Michigan State University
Plant–Insect Interactions

FRANTISEK SEHNAL
*Institute of Entomology, Academy of Sciences,
Czech Republic*
Temperature, Effects on Development
and Growth

IRWIN W. SHERMAN
University of California, Riverside
Bubonic Plague

RONALD A. SHERMAN
University of California, Irvine
Medicine, Insects in

DANIEL SIMBERLOFF
University of Tennessee
Introduced Insects

S. J. SIMPSON
University of Oxford, United Kingdom
Nutrition

SCOTT R. SMEDLEY
Trinity College, Connecticut
Puddling Behavior

EDWARD H. SMITH
Cornell University (Emeritus)
Asheville, North Carolina
History of Entomology

JANET R. SMITH
Asheville, North Carolina
History of Entomology

DANIEL E. SONENSHINE
Old Dominion University
Ticks

JOHN T. SORENSON
California Department of Food and Agriculture
Aphids

JOSEPH C. SPAGNA
University of California, Berkeley
Spiders

BEVERLY SPARKS
University of Georgia
Extension Entomology

FELIX A. H. SPERLING
University of Alberta, Canada
Teaching Resources

BERNHARD STATZNER
Université Lyon I, France
Swimming, Stream Insects

INGOLF STEFFAN–DEWENTER
University of Göttingen, Germany
Grassland Habitats

FREDERICK W. STEHR
Michigan State University
Caterpillars
Chrysalis
Cocoon
Larva
Metamorphosis
Ocelli and Stemmata
Pupa and Puparium

KENNETH W. STEWART
University of North Texas
Plecoptera

PETER STILING
University of South Florida
Greenhouse Gases, Global Warming, and
Insects

ANDREW J. STORER
Michigan Technological University
Forest Habitats

NIGEL E. STORK
James Cook University, Australia
Biodiversity

RICHARD STOUTHAMER
University of California, Riverside
Wolbachia

MICHAEL R. STRAND
University of Georgia
Polyembryony

NICHOLAS J. STRAUSFELD
University of Arizona
Brain and Optic Lobes

HELMUT STURM
University Hildesheim, Germany
Archaeognatha
Zygentoma

YOU NING SU
Australian National University, Canberra
Orthoptera

R. K. SUAREZ
University of California, Santa Barbara
Metabolism

DANIEL J. SULLIVAN
Fordham University
Hyperparasitism

SATOSHI TAKEDA
*National Institute of Agrobiological Sciences,
Japan*
Bombyx mori
Sericulture

CATHERINE A. TAUBER
Cornell University
Neuroptera

MAURICE J. TAUBER
Cornell University
Neuroptera

ORLEY R. TAYLOR
University of Kansas
Neotropical African Bees

WILLIAM H. TELFER
University of Pennsylvania
Vitellogenesis

K. J. TENNESSEN
Tennessee Valley Authority
Odonata

WALTER R. TERRA
University of São Paulo, Brazil
Digestion
Digestive System

CARSTEN THIES
University of Göttingen, Germany
Grassland Habitats

F. CHRISTIAN THOMPSON
U.S. Department of Agriculture
Nomenclature and Classification, Principles of

S. NELSON THOMPSON
University of California, Riverside
Metabolism
Nutrition

JAMES H. THORP
University of Kansas
Arthropoda and Related Groups

ROBBIN W. THORP
University of California, Davis
Pollination and Pollinators

ERICH H. TILGNER
University of Georgia
Phasmida

PÄIVI H. TORKKELI
Dalhousie University, Canada
Mechanoreception

JAMES F. A. TRANIELLO
Boston University
Recruitment Communication

TEJA TSCHARNTKE
University of Göttingen, Germany
Grassland Habitats

KAREN M. VAIL
University of Tennessee, Knoxville
Extension Entomology

R. G. VAN DRIESCHE
University of Massachusetts
Biological Control of Insect Pests

MACE VAUGHAN
The Xerces Society, Portland, Oregon
Endangered Insects

CHARLES VINCENT
Agriculture and Agri-Food Canada, Quebec
Physical Control of Insect Pests

META VIRANT-DOBERLET
National Institute of Biology, Ljubljana, Slovenia
Vibrational Communication

P. KIRK VISSCHER
University of California, Riverside
Dance Language
Homeostasis, Behavioral

PATRICIA J. VITTUM
University of Massachusetts
Soil Habitats

GREGORY P. WALKER
University of California, Riverside
Salivary Glands

J. BRUCE WALLACE
University of Georgia
Aquatic Habitats

GRAHAM C. WEBB
The University of Adelaide, Australia
Chromosomes

STEPHEN C. WELTER
University of California, Berkeley
Codling Moth

RONALD M. WESELOH
Connecticut Agricultural Experiment Station
Host Seeking, by Parasitoids
Predation/Predatory Insects

DIANA E. WHEELER
University of Arizona
Accessory Glands
Eggs
Egg Coverings
Ovarioles
Reproduction, Female
Reproduction, Female: Hormonal Control of

MICHAEL F. WHITING
Brigham Young University
Siphonaptera
Strepsiptera

KIPLING W. WILL
University of California, Berkeley
Research Tools, Insects as

STANLEY C. WILLIAMS
San Francisco State University
Scorpions

SHAUN L. WINTERTON
North Carolina State University
Scales and Setae

DAVID L. WOOD
University of California, Berkeley
Forest Habitats

ROBIN J. WOOTTON
University of Exeter, United Kingdom
Wings

JAYNE YACK
Carleton University, Ottawa, Canada
Hearing

JAMES E. ZABLOTNY
U.S. Department of Agriculture
Sociality

SASHA N. ZILL
Marshall University
Walking and Jumping

PETER ZWICK
Max-Planck-Institut für Limnologie, Germany
Biogeographical Patterns

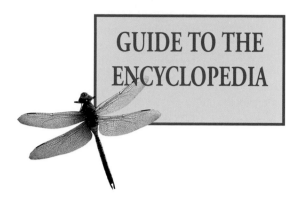

GUIDE TO THE ENCYCLOPEDIA

The *Encyclopedia of Insects* is a complete source of information on the subject of insects, contained within a single volume. Each article in the Encyclopedia provides an overview of the selected topic to inform a broad spectrum of readers, from insect biologists and scientists conducting research in related areas, to students and the interested general public.

In order that you, the reader, will derive the maximum benefit from the *Encyclopedia of Insects*, we have provided this Guide. It explains how the book is organized and how the information within its pages can be located.

SUBJECT AREAS

The *Encyclopedia of Insects* presents 271 separate articles on the entire range of entomological study. Articles in the Encyclopedia fall within twelve general subject areas, as follows:

- Anatomy
- Physiology
- Behavior
- Evolution
- Reproduction
- Development and Metamorphosis
- Major Groups and Notable Forms
- Interactions with Other Organisms
- Interactions with Humans
- Habitats
- Ecology
- History and Methodology

ORGANIZATION

The *Encyclopedia of Insects* is organized to provide the maximum ease of use for its readers. All of the articles are arranged in a single alphabetical sequence by title. An alphabetical Table of Contents for the articles can be found beginning on p. v of this introductory section.

As a reader of the Encyclopedia, you can use this alphabetical Table of Contents by itself to locate a topic. Or you can first identify the topic in the Contents by Subject Area (p. xiii) and then go to the alphabetical Table to find the page location.

So that they can be more easily identified, article titles begin with the key word or phrase indicating the topic, with any descriptive terms following this. For example, "Temperature, Effects on Development and Growth" is the title assigned to this article, rather than "Effects of Temperature on Development and Growth," because the specific term *Temperature* is the key word.

ARTICLE FORMAT

Each article in the Encyclopedia begins with an introductory paragraph that defines the topic being discussed and indicates its significance. For example, the article "Exoskeleton" begins as follows:

> The exoskeleton is noncellular material that is located on top of the epidermal cell layer and constitutes the outermost part of the integument. The local properties and appearance of the exoskeleton are highly variable, and nearly all visible features of an insect result from the exoskeleton. The exoskeleton serves as a barrier between the interior of the insect and the environment, preventing desiccation and the penetration of microorganisms. Muscles governing the movements are attached to the exoskeleton.

Major headings highlight important subtopics that are discussed in the article. For example, the article "Flight" includes these topics: "Evolution of Flight"; "Aerodynamics"; "Neural Control"; "Energetics"; "Ecology and Diversity."

CROSS-REFERENCES

The *Encyclopedia of Insects* has an extensive system of cross-referencing. References to other articles may appear either as marginal headings within the A–Z topical sequence, or as indications of related topics at the end of a particular article.

As an example of the first type of reference cited above, the following marginal entry appears in the A–Z article list between the entries "Beeswax" and "Biodiversity":

Beetle see *Coleoptera*

This reference indicates that the topic of Beetles is discussed elsewhere, under the article title "Coleoptera," which is the name of the order including this group.

An example of the second type, a cross-reference at the end of an article, can be found in the entry "DDT." This article concludes with the statement:

See Also the Following Articles
Insecticides • Integrated Pest Management • Pollution

This reference indicates that these three related articles all provide some additional information about DDT.

BIBLIOGRAPHY

The Bibliography section appears as the last element of an article, under the heading "Further Reading." This section lists recent secondary sources that will aid the reader in locating more detailed or technical information on the topic at hand. Review articles and research papers that are important to a more detailed understanding of the topic are also listed here.

The Bibliography entries in this Encyclopedia are for the benefit of the reader, to provide references for further reading or additional research on the given topic. Thus they typically consist of a limited number of entries. They are not intended to represent a complete listing of all the materials consulted by the author or authors in preparing the article. The Bibliography is in effect an extension of the article itself, and it represents the author's choice as to the best sources available for additional information.

GLOSSARY

The *Encyclopedia of Insects* presents an additional resource for the reader, following the A–Z text. A comprehensive glossary provides definitions for more than 750 specialized terms used in the articles in this Encyclopedia. The terms were identified by the contributors as helpful to the understanding of their entries, and they have been defined by these authors according to their use in the actual articles.

INDEX

The Subject Index for the *Encyclopedia of Insects* contains more than 7,000 entries. Within the entry for a given topic, references to general coverage of the topic appear first, such as a complete article on the subject. References to more specific aspects of the topic then appear below this in an indented list.

ENCYCLOPEDIA WEB SITE

The *Encyclopedia of Insects* maintains its own editorial Web page on the Internet at:

http://www.apnet.com/insects/

This site gives information about the Encyclopedia project and features links to related sites that provide information about the articles of the Encyclopedia. The site will continue to evolve as more information becomes available.

FOREWORD

I would say that creating an encyclopedia of insects was a herculean task, but I think that sells the enterprise short. After all, Hercules only had twelve labors assigned to him, and twelve years to complete them—with insects, there are over 900,000 different species and many, many more stories to tell. Twelve years from now, there will likely be even more. Why, then, would anyone undertake the seemingly impossible task of compiling an encyclopedia of insects? To an entomologist, the answer is obvious. For one thing, there's the numbers argument—over 70% of all known species are insects, so if any group merits attention in encyclopedic form, surely it's the one that happens to dominate the planet. Moreover, owing in large part to their staggering diversity, insects are in more different places in the world than virtually any other organism. There are insects in habitats ranging from the high Arctic to tropical rainforests to petroleum pools to glaciers to mines a mile below the surface to caves to sea lion nostrils and horse intestines. About the only place where insects are conspicuously absent is in the deep ocean (actually, in deep water in general), an anomaly that has frustrated more than a few entomologists who have grown accustomed to world domination. Then there's the fact that insects have been around for longer than most other high-profile life-forms. The first proto-insects date back some 400 million years; by contrast, mammals have been around only about 230 million years and humans (depending on how they're defined) a measly one million years.

Probably the best justification for an encyclopedia devoted to insects is that insects have a direct and especially economic impact on humans. In the United States alone, insects cause billions of dollars in losses to staple crops, fruit crops, truck crops, greenhouse and nursery products, forest products, livestock, stored grain and packaged food, clothing, household goods and furniture, and just about anything else people try to grow or build for sale or for their own consumption. Beyond the balance sheet, they cause incalculable losses as vectors of human pathogens. They're involved in transmission of malaria, yellow fever, typhus, plague, dengue, various forms of encephalitis, relapsing fever, river blindness, filariasis, sleeping sickness, and innumerable other debilitating or even fatal diseases, not just abroad in exotic climes but here in the United States as well. All told, insects represent a drag on the economy unequaled by any other single class of organisms, a seemingly compelling reason for keeping track of them in encyclopedic form.

In the interests of fairness, however, it should be mentioned that insects also amass economic benefits in a magnitude unequaled by most invertebrates (or even, arguably, by most vertebrates). Insect-pollinated crops in the United States exceed $9 billion in value annually, and insect products, including honey, wax, lacquer, silk, and so on, contribute millions more. Insect-based biological control of both insect and weed pests is worth additional millions in reclaimed land and crop production, and even insect disposal of dung and other waste materials, although decidedly unglamorous, is economically significant in fields, rangelands, and forests throughout the country.

So, for no reason other than economic self-interest, there's reason enough for creating an encyclopedia of insects. But what can be learned from insects that can't be learned from an encyclopedia of any other abundant group of organisms? Basically, the biology of insects is the biology of small size. Small size, which has been in large part responsible for the overwhelming success of the taxon, at the same time imposes major limits on the taxon. The range in size of living organisms, on earth at least, encompasses some 13 orders of magnitude (from a 100 metric ton blue whale to rotifers weighing less than 0.01 mg). Insects range over five orders of magnitude—from 30-g beetles to 0.03-g fairyflies—so eight orders of magnitude are missing in the class Insecta.

Problems at the upper limit involve support, transport, and overcoming inertia, issues clearly not critical for organisms, like insects, at the lower end of the range.

We humans, in the grand scheme of things, are big creatures and as a consequence we interact with the biological and physical world entirely differently. Rules that constrain human biology often are suspended for insects, which operate by a completely different set of rules. The constraints and benefits of small size are reflected in every aspect of insect biology. They hear, smell, taste, and sense the world in every other way with abilities that stagger the imagination. They are capable of physical feats that seem impossible—most fly, some glow in the dark, and others control the sex of their offspring and even occasionally engage in virgin birth, to cite a few examples. Their generation times are so short and reproductive rates so high that they can adapt and evolve at rates that continually surprise (and stymie) us. The environment is "patchier" to smaller organisms, which can divide resources more finely than can large, lumbering species. Thus, they can make a living on resources so rare or so nutrient-poor that it defies belief, such as nectar, dead bodies, and even dung.

So they're profoundly different from humans and other big animals, and the study of insects can offer many insights into life on earth that simply couldn't be gained from a study of big creatures. By the same token, though, they are cut from the same cloth—the same basic building blocks of life, same genetic code, and the like—and their utility as research organisms has provided insights into all life on the planet.

The *Encyclopedia of Insects* contains contributions from some of the greatest names in entomology today. Such a work has to be a collective effort because nobody can be an expert in everything entomological. Even writing a foreword for such a wide-ranging volume is a daunting task. To be such an expert would mean mastering every biological science from molecular biology (in which the fruit fly *Drosophila melanogaster* serves as a premier model organism) to ecosystem ecology (in which insects play an important role in rates of nutrient turnover and energy flow). But, because insects, through their ubiquity and diversity, have had a greater influence on human activities than perhaps any other class of organisms, to be the ultimate authority on insects also means mastering the minutiae of history, economics, art, literature, politics, and even popular culture. Nobody can master all of that information—and that's why this encyclopedia is such a welcome volume.

—*May R. Berenbaum*

PREFACE

Insects are ever present in human lives. They are at once awe inspiring, fascinating, beautiful, and, at the same time, a scourge of humans because of food loss and disease. Yet despite their negative effects, we depend on insects for pollination and for their products. As insects are the largest living group on earth (75% of all animal species), any understanding of ecological interactions at local or global scales depends on our knowledge about them. Given the current interest in biodiversity, and its loss, it must be remembered that insects represent the major part of existing biodiversity. Aesthetically, insect images are often with us as well: early images include Egyptian amulets of sacred scarabs; modern images include dragonfly jewelry, butterfly stationery, and children's puppets.

The idea of an *Encyclopedia of Insects* is new, but the concept of an encyclopedia is quite old. In 1745, Diderot and D'Alembert asked the best minds of their era—including Voltaire and Montesquieu—to prepare entries that would compile existing human knowledge in one place: the world's first encyclopedia. It took over 20 years to finish the first edition, which became one of the world's first best-selling books and a triumph of the Enlightenment.

What do we intend this encyclopedia to be? Our goal is to convey the exciting, dynamic story of what entomology is today. It is intended to be a concise, integrated summary of current knowledge and historical background on each of the nearly 300 entries presented. Our intention has been to make the encyclopedia scientifically uncompromising; it is to be comprehensive but not exhaustive. Cross-references point the reader to related topics, and further reading lists at the end of each article allow readers to go into topics in more detail. The presence of a certain degree of overlap is intentional, because each article is meant to be self-contained.

The *Encyclopedia of Insects* also includes organisms that are related to insects and often included in the purview of entomology. Therefore, besides the members of the class Insecta—the true insects—the biology of spiders, mites, and related arthropods is included. The core of this encyclopedia consists of the articles on the taxonomic groups—the 30 or so generally accepted orders of insects, the processes that insects depend on for their survival and success, and the range of habitats they occupy. The fact that entomology is a dynamic field is emphasized by the discovery of a new order of insects, the Mantophasmatodea, just as this encyclopedia was being completed. This is the first order of insects to be described in over 80 years, and we are pleased to be able to include it as an entry, further underscoring that there is much left to learn about insects. Some topics, especially the "poster insects"—those well-known taxa below the level of orders for which entries are presented—may not cover all that are desired by some readers. Given insect biodiversity, your indulgence is requested.

We have gathered over 260 experts worldwide to write on the entries that we have selected for inclusion. These specialists, of course, have depended on the contributions of thousands of their entomological predecessors. Because the modern study of entomology is interdisciplinary, we enlisted experts ranging from arachnologists to specialists in zoonotic diseases. Given that the two of us have spent over 25 combined years as editors of the *Annual Review of Entomology,* many of our contributors were also writers for that periodical. We thank our contributors for putting up with our compulsive editing, requests for rewrites, and seemingly endless questions.

Our intended audience is not entomological specialists but entomological generalists, whether they be students, teachers, hobbyists, or interested nonscientists. Therefore, to cover the diverse interests of this readership, we have included not just purely scientific aspects of the study of insects, but cultural (and pop-cultural) aspects as well.

We thank the staff of Academic Press for their encouragement and assistance on this project. Chuck Crumly had the original concept for this encyclopedia, convinced us of its merit, and helped us greatly in defining the format. Chris Morris provided suggestions about its development. Jocelyn Lofstrom and Joanna Dinsmore guided the book through printing. Gail Rice managed the flow of manuscripts and revisions with skill and grace, and made many valuable suggestions. Julie Todd of Iowa State University provided a crucial final edit of the completed articles. All these professionals have helped make this a rewarding and fascinating endeavor.

We dedicate our efforts in editing the *Encyclopedia of Insects* to our wives, Cheryl and Anja; their contributions to our entomological and personal lives have been indescribable.

—*Vincent H. Resh and Ring T. Cardé*

Encyclopedia of
INSECTS

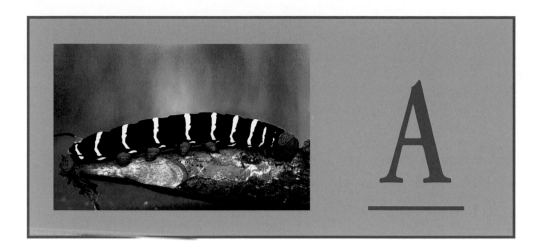

Acari

see *Mites; Ticks*

Accessory Glands

Diana E. Wheeler
University of Arizona

The accessory glands of reproductive systems in both female and male insects produce secretions that function in sperm maintenance, transport, and fertilization. In addition, accessory glands in females provide protective coatings for eggs. Accessory glands can be organs distinct from the main reproductive tract, or they can be specialized regions of the gonadal ducts (ducts leading from the ovaries or testes). Typically, glandular tissue is composed of two cell types: one that is secretory and the other that forms a duct. The interplay between male and female secretions from accessory glands is a key element in the design of diverse mating systems.

ACCESSORY GLANDS OF FEMALES

Management of Sperm and Other Male Contributions

Sperm management by females involves a wide range of processes, including liberation of sperm from a spermatophore, digestion of male secretions and sperm, transport of sperm to and from the spermatheca, maintenance of stored sperm, and fertilization.

Accessory gland secretions can have digestive functions important in sperm management. First, digestive breakdown of the spermatophore can free encapsulated sperm for fertilization and storage. Second, male contributions can provide an important nutritional benefit to their mates. Female secretions can digest the secretory components of male seminal fluid to facilitate a nutritive role. In addition, females can digest unwanted sperm to transform it into nutrients. Third, female secretions in some species are required to digest sperm coverings that inhibit fertilization.

Transfer of sperm to and from the spermatheca is generally accomplished by a combination of chemical signals and muscular contractions. Secretions of female accessory glands in some species increase sperm motility or appear to attract sperm toward the spermathecae. Transport of fluid out through the wall of the spermatheca may also create negative pressure that draws in sperm.

Sperm can be stored for some length of time in spermathecae, with the record belonging to ant queens that maintain sperm viability for a decade or more. Secretions of spermathecal glands are poorly characterized, and how sperm is maintained for such extended periods is not known. Spermathecal tissue seems to create a chemical environment that maintains sperm viability, perhaps through reduced metabolism. A nutritional function is also possible.

Transport of sperm out of storage can be facilitated by the secretions of the spermathecal gland, which presumably activate quiescent sperm to move toward the primary reproductive tract. One potential function of female accessory glands that has been explored only slightly is the production of hormone-like substances that modulate reproduction functions.

Production of Egg Coverings

Female accessory glands that produce protective coverings for eggs are termed colleterial glands. Colleterial glands have been best characterized in cockroaches, which produce an oothecal case surrounding their eggs. Interestingly, the left and right glands are anatomically different and have different products. Separation of the chemicals permits reactions to begin only at the time of mixing and ootheca formation. Other protective substances produced by glands include toxins and antibacterials.

Nourishment for Embryos or Larvae

Viviparous insects use accessory glands to provide nourishment directly to developing offspring. Tsetse flies and sheep keds are dipterans that retain single larvae within their reproductive tracts and provide them with nourishment. They give birth to mature larvae ready to pupate. The gland that produces the nourishing secretion, rich in amino acids and lipids, is known as the milk gland. The Pacific beetle roach, *Diploptera punctata*, is also viviparous and provides its developing embryos with nourishment secreted by the brood sac, an expanded portion of oviduct.

ACCESSORY GLANDS OF MALES

Accessory glands of the male reproductive tract have diverse functions related to sperm delivery and to the design of specific mating systems.

Sperm Delivery

Males of many insects use spermatophores to transfer sperm to females. A spermatophore is a bundle of sperm contained in a protective packet. Accessory glands secrete the structural proteins necessary for the spermatophore's construction. Males of the yellow mealworm, *Tenebrio molitor*, have two distinct accessory glands, one bean-shaped and the other tubular (Fig. 1). Bean-shaped accessory glands contain cells of at least seven types and produce a semisolid material that forms the wall and core of the spermatophore. Tubular accessory glands contain only one type of cell, and it produces a mix of water-soluble proteins of unknown function. Spermatophores are not absolutely required for sperm transfer in all insects. In many insects, male secretions create a fluid medium for sperm transfer.

Effects on Sperm Management and on the Female

The effects of male accessory gland secretions in the female are best known for the fruit fly, *Drosophila melanogaster*, in which the function of several gene products has been explored at the molecular level. Since insects have a diversity of mating systems, the specific functions of accessory gland secretions are likely to reflect this variation.

In *Drosophila*, the accessory glands are simple sacs consisting of a single layer of secretory cells around a central lumen (Fig. 2). Genes for more than 80 accessory gland proteins have been identified so far. These genes code for hormonelike substances and enzymes, as well as many novel

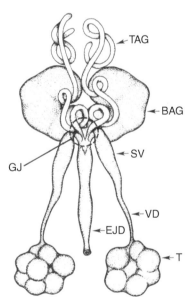

FIGURE 1 Male reproductive system of *T. molitor*, showing testes (T), ejaculatory duct (EJD), tubular accessory gland (TAG), and bean-shaped accessory gland (BAG). [From Dailey, P. D., Gadzama J. M., and Happ, G. M. (1980). Cytodifferentiation in the accessory glands of *Tenebrio molitor*. VI. A congruent map of cells and their secretions in the layered elastic product of the male bean-shaped accessory gland. *J. Morphol.* **166**, 289–322. Reprinted by permission of Wiley-Liss, Inc., a subsidiary of John Wiley & Sons, Inc.]

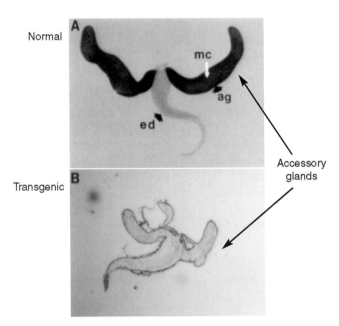

FIGURE 2 Accessory gland of *D. melanogaster*. (A) The cells in this normal accessory gland express b-galactosidase driven by a promoter of a gene for an accessory gland protein. (B) A transgenic accessory gland, cells expressing the gene have been selectively killed after eclosion. These flies were used to explore the function of accessory gland secretions. In transgenic males, accessory glands are small and translationally inert. [From Kalb, J. M., DiBenedetto, A. J., and Wolfner, M. F. (1993). Probing the function of *Drosophila melanogaster* accessory glands by directed cell ablation. *Proc. Natl. Acad. Sci. USA* **90**, 8093–8097. Copyright 1993, National Academy of Sciences, U.S.A.]

proteins. The gene products or their derivatives have diverse functions, including an increased egg-laying rate, a reduced inclination of females to mate again, increased effectiveness of sperm transfer to a female's spermatheca, and various toxic effects most likely involved in the competition of sperm from different males. A side effect of this toxicity is a shortened life span for females. Other portions of the reproductive tract contribute secretions with diverse roles. For example, the ejaculatory bulb secretes one protein that is a major constituent of the mating plug, and another that has antibacterial activity.

See Also the Following Articles
Egg Coverings • Spermatheca • Spermatophore

Further Reading
Chen, P. S. (1984). The functional morphology and biochemistry of insect male accessory glands and their secretions. *Annu. Rev. Entomol.* **29**, 233–255.
Eberhard, W. G. (1996). "Female Control: Sexual Selection by Cryptic Female Choice." Princeton University Press, Princeton, NJ.
Gillott, C. (1988). Arthropoda—Insecta. *In* "Accessory Sex Glands," (Adiyodi and Adiyodi, eds.). Vol. 3 of "Reproductive Biology of Invertebrates," pp. 319–471. Wiley, New York.
Happ, G. M. (1992). Maturation of the male reproductive system and its endocrine regulation. *Annu. Rev. Entomol.* **37**, 303–320.
Wolfner, M. F. (2001). The gifts that keep on giving: Physiological functions and evolutionary dynamics of male seminal proteins in *Drosophila*. *Heredity* **88**, 85–93.

Acoustic Behavior

see *Hearing*

Aedes Mosquito

see *Mosquitoes*

Aestivation

Sinzo Masaki
Hirosaki University

Aestivation is a dormant state for insects to pass the summer in either quiescence or diapause. Aestivating, quiescent insects may be in cryptobiosis and highly tolerant to heat and drought. Diapause for aestivation, or summer diapause, serves not only to enable the insect to tolerate the rigors of summer but also to ensure that the active phase of the life cycle occurs during the favorable time of the year.

QUIESCENCE

Quiescence for aestivation may be found in arid regions. For example, the larvae of the African chironomid midge, *Polypedilum vanderplanki,* inhabit temporary pools in hollows of rocks and become quiescent when the water evaporates. Dry larvae of this midge can "revive" when immersed in water, even after years of quiescence. The quiescent larva is in a state of cryptobiosis and tolerates the reduction of water content in its body to only 4%, surviving even brief exposure to temperatures ranging from +102°C to –270°C. Moreover, quiescent eggs of the brown locust, *Locustana pardalina,* survive in the dry soil of South Africa for several years until their water content decreases to 40%. When there is adequate rain, they absorb water, synchronously resume development, and hatch, resulting in an outburst of hopper populations. The above-mentioned examples are dramatic, but available data are so scanty that it is difficult to surmise how many species of insects can aestivate in a state of quiescence in arid tropical regions.

SUMMER DIAPAUSE

Syndrome

The external conditions that insects must tolerate differ sharply in summer and winter. Aestivating and hibernating insects may show similar diapause syndromes: cessation of growth and development, reduction of metabolic rate, accumulation of nutrients, and increased protection by body coverings (hard integument, waxy material, cocoons, etc.), which permit them to endure the long period of dormancy that probably is being mediated by the neuroendocrine system.

Migration to aestivation sites is another component of diapause syndrome found in some species of moths, butterflies, beetles, and hemipterans. In southeastern Australia, the adults of the Bogong moth, *Agrotis infusa,* emerge in late spring to migrate from the plains to the mountains, where they aestivate, forming huge aggregations in rock crevices and caves (Fig. 1).

Seasonal Cues

Summer diapause may be induced obligatorily or facultatively by such seasonal cues as daylength (nightlength) and temperature. When it occurs facultatively, the response to the cues is analogous to that for winter diapause; that is, the cues are received during the sensitive stage, which precedes the responsive (diapause) stage. The response pattern is, however, almost a mirror image of that for winter diapause (Fig. 2). Aestivating insects themselves also may be sensitive to the

FIGURE 1 Bogong moths, *Agrotis infusa,* aestivating in aggregation on the roof of a cave at Mt Gingera, A. C. T., Australia. [Photograph from Common, I. (1954). *Aust. J. Zool.* **2,** 223–263, courtesy of CSIRO Publishing.]

seasonal cues; a high temperature and a long daylength (short nightlength) decelerate, and a short daylength (long nightlength) and a low temperature accelerate the termination of diapause.

The optimal range of temperature for physiogenesis during summer diapause broadly overlaps with that for morphogenesis, or extends even to a higher range of temperature. Aestivating eggs of the brown locust, *L. pardalina,* can terminate diapause at 35°C and those of the earth mite, *Halotydeus destructor,* do this even at 70°C. The different thermal requirements for

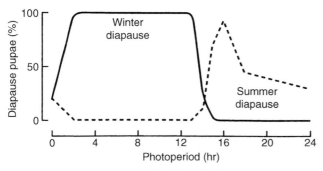

FIGURE 2 Photoperiodic response in the noctuid *M. brassicae* controlling the pupal diapause at 20°C. Note the different ranges of photoperiod for the induction of summer diapause (dashed line) and winter diapause (solid line). [From Furunishi *et al.,* 1982, reproduced with permission.]

physiogenesis clearly distinguish summer diapause from winter diapause, suggesting that despite the superficial similarity in their dormancy syndromes, the two types of diapause involve basically different physiological processes.

See Also the Following Articles
Cold/Heat Protection • Diapause • Dormancy • Migration

Further Reading

Common, I. F. B. (1954). A study of the biology of the adult Bogong moth, *Agrotis infusa* (Boisd.) (Lepidopera: Noctuidae), with special reference to its behaviour during migration and aestivation. *Austral. J. Zool.* **2,** 223–263.

Furunishi, S., Masaki, S., Hashimoto, Y., and Suzuki, M. (1982). Diapause response to photoperiod and night interruption in *Mamestra brassicae* (Lepidoptera: Noctuidae). *Appl. Entomol. Zool.* **17,** 398–409.

Hinton, H. E. (1960). Cryptobiosis in the larva of *Polypedilum vanderplanki* Hint. (Chironomidae). *J. Insect Physiol.* **5,** 286–300.

Masaki, S. (1980). Summer diapause. *Annu. Rev. Entomol.* **25,** 1–25.

Matthée, J. J. (1951). The structure and physiology of the egg of *Locustana pardalina* (Walk.). *Union S. Afr. Dept. Agric. Sci. Bull.* **316,** 1–83.

Tauber, M. J., Tauber, C. A., and Masaki, S. (1986). "Seasonal Adaptations of Insects." Oxford University Press, New York.

Africanized Bees

see *Neotropical African Bees*

Agricultural Entomology

Marcos Kogan
Oregon State University

Ronald Prokopy
University of Massachusetts

The study of all economically important insects is the object of the subdiscipline "economic entomology." Agricultural entomology, a branch of economic entomology, is dedicated to the study of insects of interest to agriculture because they help increase crop production (e.g., pollinators); help produce a commodity (e.g., honey, silk, lacquer); cause injury leading to economic losses to plants grown for food, feed, fiber, or landscaping; cause injury to farm animals; or are natural enemies of agricultural pests and, therefore, considered to be beneficial. Study of all fundamental aspects of the ecology, life history, and behavior of insects associated with agricultural crops and farm animals falls within the realm of agricultural entomology. These studies provide the foundation for the design and implementation of integrated pest management (IPM) programs (Fig. 1).

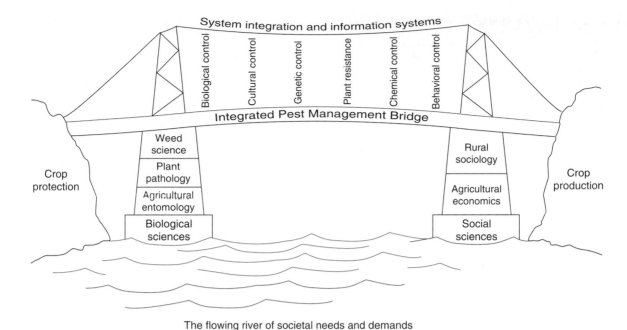

FIGURE 1 A bridge metaphor: agricultural entomology is conceived as one of the main pillars, together with plant pathology and weed science, of supporting the "integrated pest management bridge." The bridge connects two-way "traffic" between crop production and crop protection. The other pillar is provided by the social sciences of economics and sociology. The main tension cables, which are system integration and information systems, hold the vertical lines that together give stability to the bridge; these are the tactical components of IPM. Under the bridge runs the "river" of ever shifting societal needs and demands.

ECONOMIC ENTOMOLOGY

Insects are regarded by some as the main competitors of humans for dominance on the earth. Humans depend on insects for pollination of many crops, for production of honey and silk, for the decomposition of organic matter and the recycling of carbon, and for many other vital ecological roles. But it is the negative impact of insect pests that has been of greatest concern to humans. There are no reliable estimates of aggregate losses caused by insects as vectors of pathogens and parasites of humans and domestic animals, as agents causing direct damage to dwellings and other human-made structures, and as pests of crop plants and farm animals, but the amounts run to probably hundreds of billions of dollars annually. Losses caused by insects and vertebrate pests worldwide in the production of only eight principal food and cash crops (barley, coffee, cotton, maize, potato, rice, soybean, and wheat) between 1988 and 1990 have been estimated at $90.5 billion.

In the late 1800s and early 1900s, entomology became established in many academic and research institutions as a discipline equal in rank with botany and zoology. The diversity of insects and their economic importance was the justification for ranking the study of a class of animals (Insecta) as being equivalent to the study of two kingdoms of organisms (plants and animals other than insects). Through the first half of the twentieth century, there was a schism between basic and applied (or economic) entomology. Since then, common use of the expression "economic entomology" has declined, being replaced by designations of its principal branches, such as agricultural entomology, forest entomology, urban entomology, and medical and veterinary entomology. A detailed historical account is beyond the scope of this article, but Table I provides a chronology of some landmarks in the development of agricultural entomology through the ages.

The realm of agricultural entomology includes all basic studies of beneficial and pest insects associated with agricultural crops and farm animals. This article deals mainly with crops, but the general principles and concepts are equally applicable to farm animals. The starting point of such studies is a correct identification of the insect species, in accordance with the science known as biosystematics.

BIOSYSTEMATICS

Scientific nomenclature is a powerful tool for obtaining information about the basic biology of closely related species within a genus. When systematic studies have been extended beyond the naming of species (taxonomy) and contain detailed information on geographic distribution, host records, and biology of one or more species in a genus, it is often possible to extrapolate the information to other closely related species of that genus. Although details of the biology must be ascertained for each individual species, biosystematics offers a blueprint to follow when dealing with a new pest. For example, the genus *Cerotoma* (Coleoptera: Chrysomelidae) contains 10 to 12 species distributed from southern Brazil to the northeastern United States. All seem to be associated with herbaceous plants in the family Fabaceae (bean family). The biology of two of the species, *C. trifurcata* in North America and *C. arcuata* in South America (Fig. 2), has been studied extensively. Based

TABLE I   Some Landmarks in the Historical Development of Agricultural Entomology[a]

Significant events	Years ago from 2000	Date
Beginnings of agriculture	10,000	8000 B.C.E.
First records of insecticide use	4,500	2500 B.C.E.
First descriptions of insect pests	3,500	1500 B.C.E.
Soaps used to control insects in China	900	1100
Beginning of scientific nomenclature—10th edition of Linnaeus, *Systema Naturae*	242	1758
Burgeoning descriptions of insects	100–200	18th and 19th centuries
First record of plant resistance to an insect	169	1831
Charles Darwin and Alfred Wallace jointly present paper on the theory of evolution	142	1858
First successful case of biological control: the cottony cushion scale, on citrus, in California, by the vedalia beetle	112	1888
First record of widespread damage of cotton in Texas by the cotton boll weevil	106	1894
First record of an insect resistant to an insecticide	86	1914
First edition of C. L. Metcalf and W. P. Flint's *Destructive and Useful Insects*	72	1928
Discovery of DDT and beginning of the insecticide era	61	1939
First report of insect resistance to DDT	54	1946
Term "pheromone" coined by P. Karlson and P. Butenandt, who identified first such substance in the silkworm moth	45	1959
First edition of Rachel Carlson's *Silent Spring*	48	1962
Expression "integrated pest management" first appears in the press	32	1968
Rapid development of molecular biology	20	1980s
Release of Bt transgenic varieties of cotton, corn, and potato	5	1990s

[a] Based in part on Norris *et al.* (2003).

on information for these two species, it is possible to infer that the other species in the genus share at least some of the following features: eggs are laid in the soil adjacent to growing leguminous plants; larvae feed on nitrogen-fixing root nodules and pupate in soil inside pupal cases; first-generation adults emerge when seedlings emerge, and second-generation adults emerge when plants are in full vegetative growth, feeding first on foliage and, later on, switching to feeding on developing pods. The biosystematic information on the genus allows students of agricultural entomology in South, Central, or

Cerotoma eburifrons   Cerotoma atrofasciata

Cerotoma arcuata   Cerotoma trifurcata

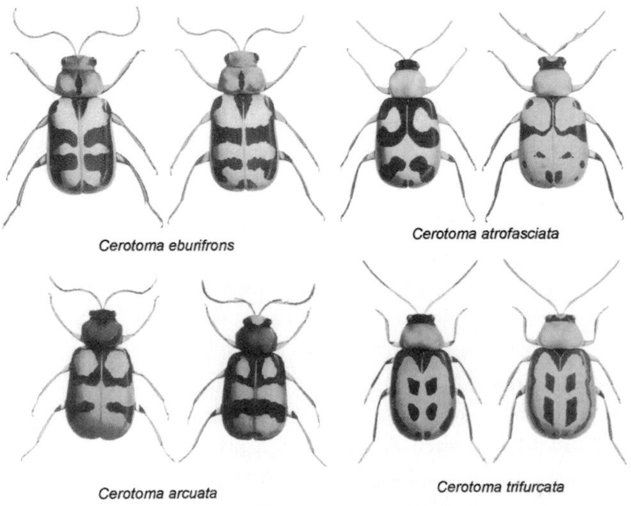

FIGURE 2 Morphological diversity and biological similarities in the genus *Cerotoma*: four of the dozen known species are illustrated by male and female specimens. The species are clearly distinguishable by morphological characters, but they have similar life histories and behaviors. (From unpublished drawings by J. Sherrod, Illinois Natural History Survey.)

North America to understand, at least in general terms, the role of any other species of *Cerotoma* within their particular agroecosystem.

The flip side of this notion is recognition that closely related and morphologically nearly undistinguishable (sibling) species may have many important biological differences. Examples of the critical need for reliable biosystematics studies are found in the biological control literature. The present account is based on studies conducted by Paul DeBach, one of the leading biological control specialists of the twentieth century. The California red scale, *Aonidella aurantii*, is a serious pest of citrus in California and other citrus-producing areas of the world. Biological control of the red scale in California had a long history of confusion and missed opportunities because of misidentification of its parasitoids. The red scale parasitoid *Aphytis chysomphali* had been known to occur in California and was not considered to be a very effective control agent. When entomologists discovered parasitized scales during foreign exploration, the parasitoids were misidentified as *A. chrysomphali* and therefore were not imported into California. It was later discovered that the parasitoids were in fact two different species, *Aphytis lingnanensis* and *A. melinus*, both more efficient natural enemies of the California red scale than *A. chrysomphali*. Now *A. lignanensis* and *A. melinus* are the principal red scale parasitoids in California. Further biosystematics studies have shown that what was once thought to be single species, *A. chrysomphali*, parasitic on the California red scale in the Orient and elsewhere, and accidentally established in California, is in fact a complex including at least seven species having different biological adaptations but nearly indistinguishable morphologically.

Knowledge of the name of a species, however, is not an indication of its true potential economic impact or pest status. A next important phase in agricultural entomology is, therefore, the assessment of benefits or losses caused by that species.

PEST IMPACT ASSESSMENT

The mere occurrence of an insect species in association with a crop or a farm animal does not necessarily mean that the species is a pest of that crop or animal. To be a pest it must cause economic losses. The assessment of economic losses from pests is the subject of studies conducted under conditions that match as closely as possible the conditions under which the crop is grown commercially or the animals are raised. Much of the methodology used in crop loss assessment has been established under the sponsorship of the Food and Agriculture Organization (FAO) of the United Nations as a means of prioritizing budget allocations and research efforts. Key data for these studies relate to the determination of the yield potential of a crop. The genetic makeup of a crop variety determines its maximum yield in the absence of adverse environmental factors. This is known as the attainable yield. To determine the attainable yield, the crop is grown under nearly ideal conditions; the actual yield is what occurs when the crop is grown under normal farming conditions. The difference between attainable and actual yields is a measure of crop loss (Fig. 3).

To assess crop losses and attribute the losses to a specific cause (e.g., the attack of a pest) requires setting up experiments to isolate the effect of the pest from all other constraints. Methodologies vary with pest category—whether the pests are insects, vertebrates, plant pathogens, or weeds, for example. The quantitative relationship between crop losses and pest population levels is the basis for computing the economic injury level for the pest. The economic injury level is a fundamental concept in IPM.

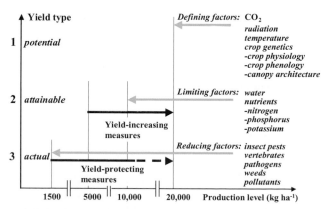

FIGURE 3 Factors impacting the yield potential of a generic crop. (Adapted from information on a Web site originated at IMI/University of Miami, Summer Institute.)

LIFE HISTORY AND HABITS

Once the identity and pest status of a species have been well established, it becomes essential to extend the informational base on the life history and habits of the species to the conditions under which the crop is grown. Economically important life history traits include information on developmental threshold temperatures and temperature-dependent developmental rates. These data are used in modeling the phenology of the pest. Other essential studies include the orientation, feeding, host selection, and sexual behavior of the species. Many of these studies provide the foundation for strategic planning in IPM and for the development of target-specific control tactics. For example, the study of sexual behavior involves the definition of the role of pheromones in mating and the identification of those pheromones. These, in turn, may be used for monitoring pest incidence and abundance or in mating disruption, both valuable components of IPM systems for many crops. The study of host selection behavior often leads to the identification of kairomones, equally important in IPM development.

PHENOLOGY

The life cycle of different insect species varies greatly, although all insects undergo the basic stages of development from egg to reproductive adult (or imago). Depending on the length of the life cycle, there is considerable variation in the number of generations per year, a phenomenon called voltinism. A univoltine species has one generation per year; a multivoltine species may have many generations per year. The range of variation in the Insecta is evident when one considers that the 17-year periodical cicada has one generation every 17 years, whereas whiteflies or mosquitoes may complete a generation in about 21 days. Under temperate climate conditions, generations often are discrete, but under warmer subtropical conditions they frequently overlap. The definition of temporal periodicity in an organism's developmental cycle is called phenology. The relationship between the phenology of the crop and the phenologies of its various pests is of interest in agricultural entomology. Figure 4 shows an example of such a relationship for soybean grown under conditions typical for the midwestern United States.

POPULATION AND COMMUNITY ECOLOGY

Population and community level studies are within the scope of insect ecology. Although the species is the focal biological entity for agricultural entomology, for management purposes it is essential to understand population and community level processes. Populations are assemblages of conspecific individuals within a defined geographical area (e.g., a crop field, a river valley, a mountain chain). Many insects have a large reproductive capacity. As calculated by Borror, Triplehorn, and Johnson, a pair of fruit flies (*Drosophila*), for

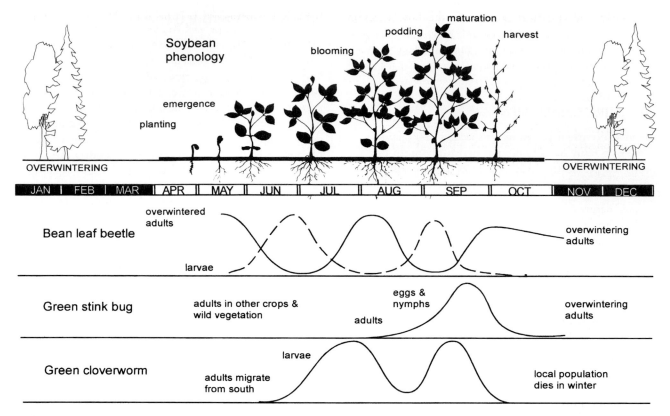

FIGURE 4 Crop phenology and pest phenology: relationship between the phenology of soybean in the midwestern United States and three of its most common insect pests, the bean leaf beetle, *C. trifurcata* (Coleoptera: Chrysomelidae); the green stink bug, *Acrosternum hilare* (Hemiptera: Pentatomidae); and the green cloverworm, *Hypena scabra* (Lepidoptera: Noctuidae).

example, produces 100 viable eggs, half of which yield females that in turn will lay 100 eggs and so on for 25 possible generations in one year; by the end of the year, the 25th generation would contain 1.192×10^{41} flies, which, if packed tightly together, 60,000 to a liter, would form a ball of flies 155 million km in diameter or a ball extending approximately from the earth to the sun. Obviously, such unlimited population growth does not occur in nature. Normally, populations are regulated by the combined actions of both physical (or abiotic) and biological (or biotic) factors of the environment. An understanding of the mortality factors that help regulate insect populations is one of the most active areas of research in agricultural entomology.

The set of species coexisting in an area and interacting to varying degrees form what is known as an ecological community. In a crop community, the crop plants and the weeds that persist within the crop field or grow along the borders are the primary producers. The animals within the crop community maintain dynamic trophic relationships: some feed on living plants, others on the decaying plants, and still others on animals. Those that feed on the plants are the herbivores, or primary consumers. Pests are primary consumers on the crop plants. Parasitoids and predators are the secondary consumers. Those that feed on the pests are beneficial natural enemies. Finally, decomposers and detritivores feed on decaying organic matter. All biotic components of the community are interconnected by "food webs." An understanding of food webs and trophic interactions in crop communities is important because it provides a basis for interpreting the nature of disturbances in crop ecosystems. Disturbances in trophic relations may lead to outbreaks of pest organisms and the need for control actions.

LINKS TO IPM SYSTEMS DEVELOPMENT

With the advent of integrated pest management and its success in the last third of the twentieth century, it has become difficult to separate agricultural entomology from IPM. In entomology, the two fields of endeavor are inextricably interconnected. A reliable database of biological information provides the means to design and develop IPM strategies. For example, there is growing interest in methods of enhancing biological control through habitat management. The technique requires information on source–sink relationships among pests and natural enemies across crop plants, neighboring crops, natural vegetation, and especially managed vegetation in the form of cover crops and field hedges. Theoretically, diversification of the crop ecosystem leads to an increase in natural enemies and to greater stability of the system. The complexity of interactions, however, makes it difficult to interpret conflicting results of experiments designed to test working hypotheses. The analysis of within-field and interfield movement, the host selection behavior of phytophagous and entomophagous insects, multitrophic interactions among community members, and the dynamics of populations, all under the scope of agricultural entomology, are only a few of the many components of the knowledge base necessary to develop advanced IPM systems.

The advent of the World Wide Web has had a major influence on accessibility to basic information on agricultural entomology. Most major agricultural research centers have developed Web pages that organize information and make it available to students worldwide. More importantly, the dynamic nature of the Web offers the opportunity to provide weather-driven modeling capabilities that greatly increase the scope and applicability of studies about the phenology and population dynamics of major pest organisms. Two sites that offer such capabilities are http://www.orst.edu/Dept/IPPC/wea/ and http://www.ipm.ucdavis.edu/PHENOLOGY/models.html.

Entomologists in the late 1800s and early 1900s studied the biology of insect pests in great detail. Articles and monographs published during that period remain valuable sources of information. These early entomologists recognized that deep knowledge of the life history of an insect and its habits could provide insights useful for the control of agricultural and other pests. The advent of organosynthetic insecticides in the mid-1940s created the illusion that pest problems now could be solved forever. Many entomologists redirected their efforts to testing new chemicals and neglected basic insect biology studies. The failure of insecticides to eradicate pests and the environmental problems engendered by the misuse of these chemicals led to the advent of IPM. For IPM to succeed, entomologists have had to return to the basics and again refocus their efforts on the study of insect biology. Agricultural entomology has come full circle as new generations of entomologists endeavor to refine knowledge of the group of animals that remain humans' most serious competitors.

See Also the Following Articles

Biological Control • History of Entomology • Integrated Pest Management • Phytophagous Insects • Plant–Insect Interactions • Population Ecology

Further Reading

Borror, D. J., Triplehorn, C. A., and Johnson, N. F. (1992). "An Introduction to the Study of Insects." Saunders College Publishers and Harcourt Brace College Publishers, Philadelphia and Fort Worth.
DeBach, P. (1964). "Biological Control of Insect Pests and Weeds." Reinhold, New York.
Evans, H. E. (1984). "Insect Biology: A Textbook of Entomology." Addison-Wesley, Reading, MA.
Huffaker, C. B. and Gutierrez, A. P. (1999). "Ecological Entomology." Wiley, New York
Jones, D. P. (1973). Agricultural entomology. In "History of Entomology." (R. F. Smith, T. E. Mittler, and C. N. Smith, eds.), pp. 307–332. Annual Reviews, Palo Alto, CA.
Norris, R. F., Caswell-Chen, E. P., and Kogan, M. (2003). "Concepts in Integrated Pest Management." Prentice Hall, Upper Saddle River, NJ.
Oerke, E. C. (1994). "Crop Production and Crop Protection: Estimated Losses in Major Food and Cash Crops." Elsevier, Amsterdam and New York.
Price, P. W. (1997). "Insect Ecology." Wiley, New York.
Schowalter, T. D. (2000). "Insect Ecology: An Ecosystem Approach." Academic Press, San Diego, CA.

Alderfly

see *Megaloptera*

Amber

George Poinar Jr.
Oregon State University

Amber is a fossilized resin ranging from several million to 300 million years of age. This material is a gold mine for the entomologist because it contains a variety of insects preserved in pristine, three-dimensional condition. Fossils in amber provide evidence of lineages dating back millions of years (Table I). External features are preserved so well that taxonomists can make detailed comparisons with living taxa to follow evolutionary development of genera and even species. Amber has a melting point between 200 and 380°C, a hardness of 2 to 3 on the Moh's scale, and a surface that is insoluble to organic solvents. Aside from providing direct evidence of an insect taxon at a particular time and place, amber insects give clues to past distributions and phylogeny, as well as indirect evidence of plants and vertebrates and the establishment of symbiotic associations, and clues for reconstructing ancient landscapes.

TABLE I Significant Amber Deposits in the World

Deposit	Location	Approximate age (million years)
Baltic	Northern Europe	40
Burmese	Burma (Myanmar)	100
Canadian	Alberta, Manitoba	70–80
Chinese	Fushun Province	40–53
Dominican	Dominican Republic	15–45
Hat Creek	British Columbia, Canada	50–55
Lebanese	Middle East	130–135
Mexican	Chiapas	22–26
New Jersey	Northeastern United States	65–95
Siberian (Taimyr)	Russian arctic	78–115
Spanish	Alava, Basque country	100–115

FIGURE 1 Origins of the honey bee lineage are provided by this primitive bee, which possesses characters of both modern *Apis* and extinct apids. Its presence in Baltic amber suggests a European origin of *Apis,* thus challenging the current view that honey bees originated in Asia.

USE OF AMBER IN TRACING INSECT LINEAGES

As a result of the excellent preservation of amber insects, specific genera (the majority, if not all, amber insects are extinct at the species level) can be recognized and compared with modern ones. In this way, lineages can be traced back tens of millions of years. An example is a small parasitic wasp of the genus *Aphelopus* (Hymenoptera: Dryinidae) trapped in Lebanese amber. This genus is still extant, and the fossil demonstrates a lineage that has survived for 130 to 135 million years.

The origin of genera can also be obtained from amber insects owing to their high degree of preservation. A recent example from Baltic amber, which contains a variety of bees, deals with the origin of the common honey bee. This fossil contains basic features characteristic of the genus *Apis* as we know it today, including pollen-collecting apparati on the hind legs and a barbed stinger (Fig. 1). This appears to be one of the most primitive *Apis* ever discovered, thus indicating a time (40 mya) and place (northern Europe) for the origin of the honey bee.

PROVIDING INDIRECT EVIDENCE OF OTHER ORGANISMS

There are size and habitat limitations to the types of organism that can be trapped in amber. For example, many plants would not normally leave flowers or leaves in the resin, and when they did, the remains would likely be difficult to identify. Vertebrates might leave hairs, feathers, or scales but these structures would also be difficult to identify. However, arthropods that are specific to certain hosts (e.g., ticks and mammals) can provide clues to other organisms that existed at that time. This use of fossils relies heavily on the principle of behavioral fixity, which asserts that, at least at the generic level, the behavior of a fossil organism would have been similar to that of its present-day descendants.

FIGURE 2 The unique morphological features (smooth, flattened body) of this fossil palm bug (*Paleodoris lattini*) in Dominican amber not only characterize it systematically but also provide clues to its lifestyle of living in confined spaces between the unopened fronds of pinnately leafed palms.

Many insects form specific associations with plants. Such associations can often be deduced by the morphological features of the insect (functional morphology). One extremely flattened hemipteran in Dominican amber (Fig. 2) that was identified as a palm bug displayed characters similar to those of an existing species in the same subfamily. The extant species lives between the closed leaves of royal palms (*Roystonea* spp.) in Cuba. This fossil provided indirect evidence that pinnately leafed palms, quite likely an extinct species of *Roystonea,* existed in the original amber forest. Other plant-specific insects, such as fig wasps and palm bruchids, provide evidence of figs and palms in the original ecosystem.

Insects that require a blood meal to complete their development can also be used as indirect evidence of a vertebrate group. Evidence of birds in the original Dominican amber forest is implied by the presence of a female *Anopheles* mosquito in amber because extant species of this subgenus normally attack birds. The presence of other vertebrate groups is implied by fleas (Siphonaptera), horseflies (Diptera: Tabanidae), biting midges (Diptera: Ceratopogonidae), and other bloodsucking arthropods such as ticks.

PROVIDING INDIRECT EVIDENCE OF SPECIFIC HABITATS

Amber insects can provide evidence of specific habitats. Diving beetles (Coleoptera: Dytiscidae), caddisflies (Trichoptera), and damsel flies (Odonata) all provide evidence of aquatic habitats. The *Anopheles* mosquito belongs to a group that normally oviposits in ground pools. Other insects can provide evidence of phytotelmata (standing water in plant parts), wood, moss, bark, and detritus.

PALEOSYMBIOSIS

Because of the sudden death of captured organisms in amber, symbiotic associations may be preserved in a manner unlikely to occur with other types of preservation. Also, the fine details of preservation may reveal morphological features characteristic

FIGURE 3 Documentation of paleophoresis is provided by a pseudoscorpion grasping the tip of the abdomen of a platypodid beetle in Dominican amber. Similar rider–carrier associations occur today, suggesting that this behavior is mandatory for survival of the pseudoscorpion.

FIGURE 4 Paleoectoparasitism is shown by two thrombidid mites attached to the mouthparts of a long-legged fly (Diptera: Dolichopodidae) in Baltic amber.

of symbiotic associations. Cases of paleosymbiosis in amber include inquilinism, commensalism, mutualism, and parasitism.

Paleoinquilinism involves two or more extinct organisms living in the same niche but neither benefiting nor harming each other. Numerous insects form inquilinistic associations under tree bark, and many pieces of amber contain flies and beetles common to this habitat.

Phoresis (one organism transported on the body of another organism) is probably the most typical type of paleocommensalism in amber. This usually involves mites and pseudoscorpions being carried by insects. The arachnid benefits by being conveyed to a new environment, where the food supply is likely to be better than the last one. The carrier generally is not harmed and only serves as a transporting agent. An example of this category in Dominican amber consists of pseudoscorpions being carried by platypodid beetles (Coleoptera: Platypodidae) (Fig. 3). The method of attachment of the pseudoscorpion to the beetle was the same then as it is today. In fact, these ancient records lead scientists to believe that such behavior is mandatory for the survival of the pseudoscorpions that live in beetle tunnels and require effective dispersal mechanisms for survival.

In paleomutualism, both organisms benefit and neither is harmed. Amber bees carrying pollen provide evidence of insect–plant mutualism in which the bee obtains a food supply and the plant is pollinated. An example of insect–insect mutualism is demonstrated by a rare fossil riodinid butterfly larva in Dominican amber. Specialized morphological features of this *Theope* caterpillar indicative of a symbiotic association are balloon setae and vibratory papillae in the neck area, and tentacle nectary organ openings on the eighth abdominal tergite. Extant caterpillars in this genus have similar features and are associated with ants. The tentacle nectary organs provide nourishment for the ants, whereas the vibratory papillae (which beat against the head capsule and make an audible sound) and balloon setae (which emit a chemical signal) are used to attract ants when the caterpillar is threatened by an

invertebrate predator or parasite. This fascinating association between butterfly larvae and ants was established at least 20 mya.

Paleoparasitism is very difficult to verify in the fossil record. There are many records of amber insects (especially wasps and flies) whose descendants today are parasitic on a wide range of organisms, but to discover an actual host–parasitic association is quite rare.

Paleoectoparasitism is the most obvious of all parasitic associations found in amber. The ectoparasite is often still attached to its host, and systematic studies can be conducted on both organisms. In amber, ectoparasites are usually parasitic mites, such as the larvae of Thrombidiidae attached to the mouthparts of a fly in Baltic amber (Fig. 4). These larval mites were feeding on the host's hemolymph, and their mouthparts are still in place. After molting to the nymphal stage, the parasites would leave the fly and become free-living predators. Large infestations could kill the host. These mites are not to be confused with phoretic ones, which are simply carried around by insects.

Paleoendoparasitism is extremely difficult to verify because internal parasites are rarely preserved as fossils. However, some parasites attempt to leave their hosts when they encounter resin. Mermithid nematodes (Mermithidae: Nematoda) and hairworms (Nematomorpha) that have nearly completed their development and are almost ready to emerge from their host will often reveal their presence (Fig. 5). Under normal conditions, they would enter soil or water and initiate a free-living existence.

BIOGEOGRAPHICAL STUDIES

In many instances amber insects provide evidence of a more extensive distribution in the past for various insect genera and families as well as indicating a warmer climatic regime in many parts of the world. Perhaps the most spectacular examples of this phenomenon are insects discovered in amber sites located far from their descendants' current habitat. Examples from

FIGURE 5 Paleoendoparasitism in amber is exemplified by a mermithid nematode (Nematoda: Mermithidae) emerging from the body of a planthopper (Homoptera: Fulgoroidea) in Baltic amber. Such records set minimum dates for the establishment of host–parasite associations.

Dominican amber include *Mastotermes* termites (Isoptera: Mastotermitidae) and *Leptomyrmex* ants (Hymenoptera: Formicidae) that obviously were part of the insect fauna some millions of years ago in the Caribbean but occur nowhere in the New World today. Both genera are represented today by a single relict species in the North and East Australian Region.

A North American example is the presence of the tropical arboreal ants of the genus *Technomyrmex* in Hat Creek amber in British Columbia, Canada, living 50 mya, hundreds of kilometers north of their present-day range. These tropical ants in Eocene Hat Creek amber provide evidence that the climate in that region of the world shifted from tropical to temperate. Other examples of past distributions involve the palm bug shown in Fig. 2, which has no present-day descendants in the Dominican Republic, with only a single living Cuban species in the subfamily. Similarly, there are no members of the genus *Theope* in the Dominican Republic or the Greater Antilles today, all living representatives being restricted to Mexico, and Central and South America. Further evidence of climatic shifts over time are clear with many of the Baltic amber insects, many of whose descendants occur in the Old World tropics today. The primitive honey bee shown in Fig. 1 evolved under subtropical conditions that characterized most of northern Europe in the Eocene. Thus, it is not surprising that most of the species and varieties of the genus *Apis* live only under tropical conditions today.

RECONSTRUCTING ANCIENT LANDSCAPES

Every amber fossil tells a story and is a piece of a jigsaw puzzle that can be used to reconstruct the natural environment at the time the amber was being produced. The challenges are to identify the inclusions, determine their biology and ecology by researching the habits of their extant descendants, and then make inferences regarding the original environment. There will always be gaps in the puzzle because there are many life-forms that are too large to become entrapped in amber or have a lifestyle that does not normally bring them into contact with the sticky resin. However, the habitat that existed in that ancient world can, in large part, be reconstructed by studying select insects that can be typified as phytophagous, soil-loving, bark inhabitants, or parasites, and identifying the associated predators, vertebrates, and special habitats.

See Also the Following Articles
Biogeographical Patterns • Fossil Record

Futher Reading
Boucot, A. J. (1990). "Evolutionary Paleobiology of Behavior and Coevolution." Elsevier, Amsterdam.
Poinar, G. O. Jr., (1992). "Life in Amber." Stanford University Press, Stanford, CA.
Poinar, G. O., Jr., and Milki, R. (2001). "Lebanese Amber. The Oldest Insect Ecosystem in Fossilized Resin." Oregon State University Press, Corvallis.
Poinar, G. O., Jr., and Poinar, R. (1994). "The Quest for Life in Amber." Addison-Wesley, Reading, MA.
Poinar, G. O., Jr., and Poinar, R. (1999). "The Amber Forest." Princeton University Press, Princeton, NJ.
Ross, A. (1998). "Amber." Harvard University Press, Cambridge, MA.
Weitschat, W., and Wichard, W. (2002). "Atlas of Plants and Animals in Baltic Amber." Pfeil, Munich.

Anatomy: Head, Thorax, Abdomen, and Genitalia

David H. Headrick
California Polytechnic State University

Gordon Gordh
U.S. Department of Agriculture

Anatomy is a subdiscipline of morphology concerned with naming and describing the structure of organisms based on gross observation, dissection, and microscopical examination. Morphology and anatomy are not synonyms. Morphology is concerned with the form and function of anatomical structure; because anatomy is an expression of organic evolution, morphology seeks to investigate possible explanations for organic diversification observed in nature. Before 1940 insect morphology focused on naming and describing anatomical structure. The need for this activity has not diminished, as much about insect anatomy remains to be revealed, described, and understood. This article focuses on the anatomical structures of the three major tagmata of the insect body: head, thorax, and abdomen, and on the external genitalia. A hypothetical ground plan for major structures is given, followed by themes in anatomical variation based on adaptation observed in the Insecta.

CONTEXT OF ANATOMICAL STUDY

Terms of Orientation and Conventions

Terms to describe orientation are not intuitive for insects. Most orientation terms are derived from the study of the human body—a body that stands upright—and their application to insects causes confusion. Some standard terms used with insects include anterior (in front), posterior (behind), dorsal (above), ventral (below), medial (middle), and lateral (side). Anatomical description usually follows in the same order, hence, we begin our discussion with the head, move on to the thorax and then the abdomen, and finish with the genitalia. Description of the relative placement of anatomical features can be cumbersome, but they are critical elements in the study of anatomical structure because relative position is one of the three basic tenets of homology, including size and shape, and embryology.

Measures of Success

The design of the insect body can be described as successful for many reasons: there are millions of species, they range in size over four orders of magnitude, their extensiveness of terrestrial and aquatic habitat exploitation (the diversity of resources), and once a successful form has been developed, there appears to be relatively little change over evolutionary time (Fig. 1). The basic insect design allows for adaptation to a variety of environmental requirements. The success of the design is rooted in the nature of the main material used for its construction.

The Building Material

When we look at an insect, it is the integument that we see. Structurally, the integument is a multiple-layered, composite organ that defines body shape, size, and color. The ultrastructure of the integument is composed of living cells and the secretory products of those cells. Each layer is of a different thickness and chemical composition, and each displays physical properties different from those of the surrounding

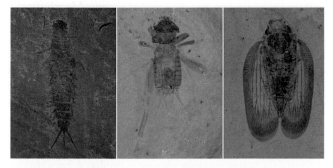

FIGURE 1 Fossil insects are easily recognizable today, indicating an early establishment of a successful design. Left to right: *Heplagenes* (Late Jurassic 150 mya, Liaoning, China); cricket (Eocene, 50 mya, Green River formation, Utah); fulgorid (Eocene, 50 mya, Green River formation, Utah).

layers. Perhaps more importantly, the integument also is the organ with the greatest diversity of structure and function.

There are two common misconceptions about the integument. First, some believe chitin is responsible for integument hardness. Actually, there is proportionally more chitin found in the soft and flexible membranous parts of the integument than in the hard, sclerotized plates. Integument hardness is attributed to an increased number of cross-linkages between protein chains contained in the integument layers. Second, some believe that the integument is rigid and that growth is incremental and limited to expansion during molting; yet some endopterygote insects are able to grow continuously between molts.

The integument determines the shape of the insect body and its appendages. One of the most captivating features of insects is their seemingly infinite variation in body shapes—everything from a simple bag (Hymenoptera grub) to a mimic of orchid flowers (Mantidae). Similarly, appendage shape is exceedingly plastic. Terms such as "pectinate," "flabbate," and "filiform" are among more than 30 terms taxonomists have proposed to describe antennal shapes. Leg shapes are similarly highly variable and express functional modifications. Among these shapes are "cursorial," "gressorial," "raptorial," "fossorial," and "scansorial." Again, these modifications of shape reflect the function of structure. Finally, wing shapes are highly variable among insects and are determined by body size and shape as well as by aerodynamic considerations.

Tagmata

Most people recognize the three tagmata—head, thorax, and abdomen—as characteristic of insects. The way they appear is rooted in a division of responsibilities. The head is for orientation, ingestion, and cognitive process; the thorax for locomotion; and the abdomen for digestion and reproduction. But even casual observations reveal further divisions of these body regions.

Segmentation of Tagmata

Two types of segmentation are evident among arthropods, primary and secondary. Primary segmentation is characteristic of soft-bodied organisms such as larval holometabolans. The body wall in these organisms is punctuated by grooves or rings that surround the anterior and posterior margin of each somite. These rings represent intersegmental lines of the body wall and define the limits of each somite. Internally, the grooves coincide with the lines of attachment of the primary longitudinal muscles. From a functional standpoint, this intrasegmental, longitudinal musculature permits flexibility and enables the body to move from side to side.

More complex plans of body organization exhibit structural modifications. Secondary segmentation is characteristic of hard-bodied arthropods, including adult and nymphal insects. Secondary body segmentation is an evolutionarily

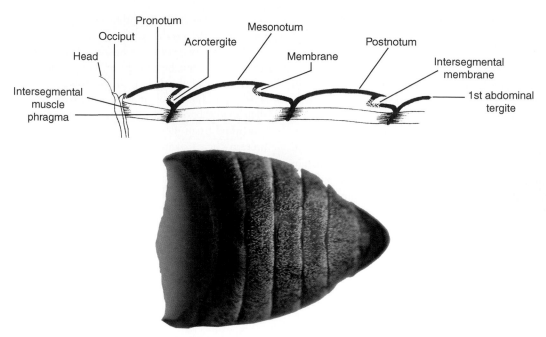

FIGURE 2 Secondary segmentation. Top: diagram of sagittal section of dorsal sclerites of thorax. Bottom: ventral view of abdominal sternites showing overlap due to secondary segmentation (Coleoptera: Scarabaeidae).

derived anatomical feature. The musculature we see in secondary segmentation is *inter*segmental, or between segments (Fig. 2). The acquisition of secondary segmentation represents a major evolutionary step in the development of the Arthropoda. The soft-bodied arthropod has primary segmentation and muscles that are *intra*segmental, or within each segment. Movement of the body and its parts is relatively simple because the body wall is flexible. However, when the body wall becomes hardened, flexibility is restricted to the articulation between hardened parts or the extension provided by intersegmental membranes. The arthropod is, in a metaphorical sense, clad in a suit of armor; most movement is possible only if soft and flexible membranes are positioned between inflexible (hardened) body parts. Exceptions may be seen in the indirect flight mechanism of pterygote insects.

In all probability secondary segmentation evolved many times, and it probably continues to evolve in response to specific problems confronting insects today. Secondary segmentation is most evident and most readily appreciated in the insect abdomen. It is less apparent in the thorax and almost totally obscured in the head.

Sclerites

The hardening of the body wall contributes significantly to the external features observed in insects. Sclerites are hardened areas of the insect body wall that are consequences of the process of sclerotization. Sclerites, also called "plates," are variable in size and shape. Sclerites do not define anatomical areas and do not reflect a common plan of segmentation. Sclerites develop as *de novo* hardening of membranous areas

of the body wall, as *de novo* separations from larger sclerotized areas of the body, and in other ways.

The hardened insect body displays many superficial and internal features that are a consequence of hardening. Understanding the distinction between these conditions and the terms applied to them is critical in understanding insect anatomy and its application in taxonomic identification. These features are of three types. First, sutures (Latin, *sutura* = seam), in the traditional sense of vertebrate anatomists, provide seams that are produced by the union of adjacent sclerotized parts of the body wall. On the insect body, sutures appear as etchings on the surface of the body and form lines of contact between sclerites. Second, sulci (Latin, *sulcus* = furrow) represent any externally visible line formed by the inflection of cuticle. Biomechanically, a sulcus forms a strengthening ridge. In contrast, lines of weakness are cuticular features that are used at molting. Lines of weakness are frequently named as if they were sutures, but they should not be viewed as such. For instance, the ecdysial cleavage line is a line of weakness that is sometimes considered to be synonymous with the epicranial suture. The two features are similar in position and appearance, but structurally they may have been derived from different conditions. Finally, apodemes (Greek, *apo* = away; *demas* = body) are hardened cuticular inflections of the body wall that are usually marked externally by a groove or pit. Structures called apophyses (Greek, *apo* = away; *phyein* = to bring forth) are armlike apodemes. Apodemes have been defined as a hollow invagination or inflection of the cuticle and an apophysis as a solid invagination. Functionally, apodemes strengthen the body wall and serve as a surface for muscle attachment.

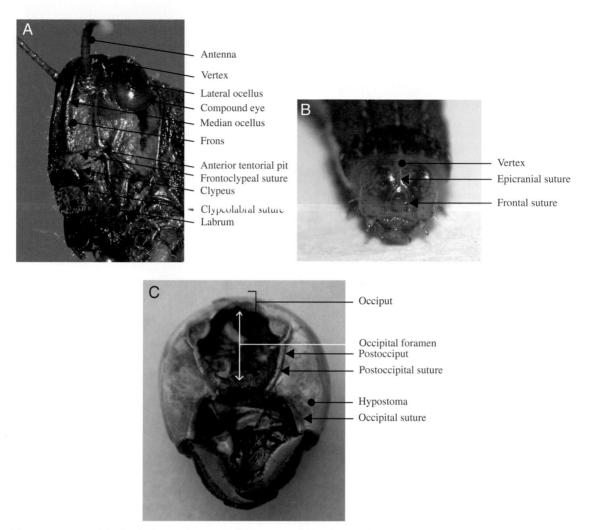

FIGURE 3 (A) Anterior view of the head of a grasshopper (Orthoptera: Acrididae). (B) Larval pterygote head showing epicranial and frontal sutures (Lepidoptera: Noctuidae). (C) Posterior aspect of the head (Orthoptera: Stenopelmatidae).

Sclerites receive different names depending upon the region of the body they are located. Tergites (Latin, *tergum* = back) are sclerites that form a subdivision of the dorsal part of the body wall (tergum). Latrotergites are sclerites that form as a subdivision of the lateral portion of the tergum. Sternites (Latin, *sternum* = breast bone) are sclerites that form as a subdivision of the ventral part of the body wall (sternum), or any of the sclerotic components of the definitive sternum. Pleurites (Greek, *pleura* = side) are sclerites in the pleural region of the body wall that are derived from limb bases.

HEAD

The head is a controversial area for anatomical nomenclature, but it provides some of the best examples of evolutionary trends in anatomy. Most insect morphologists believe that the head of modern insects represents the fusion of several segments that were present in an ancestral condition. However, the number of segments included in the ground plan of the insect head has been a contentious issue among morphologists for more than a century. Any argument that attempts to explain head segmentation must take into account comparative anatomical, embryological, and paleontological evidence, and must examine modern forms of ancestral insects.

Ground Plan of the Pterygote Head

Given the difficulty in homologizing anatomical features of the head, we describe regions associated with landmarks of a ground plan or an idealized hypognathous insect head. In terms of modern insects, the Orthoptera probably come closest to displaying all the important landmark sutures and sclerites that form the head (Fig. 3A).

The vertex (Latin, *vertex* = top; pl., vertices) is the apex or dorsal region of the head between the compound eyes for insects with a hypognathous or opisthognathous head. This definition does not apply to prognathous heads because the primary axis of the head has rotated 90° to become parallel to the primary axis of the body. The vertex is the area in which ocelli are usually located. In some insects this region has become modified or assumes different names.

The ecdysial suture (coronal suture + frontal suture, epicranial suture, ecdysial line, cleavage line) is variably developed among insects. The suture is longitudinal on the vertex and separates epicranial halves of the head (Fig. 3B). Depending on the insect, the ecdysial suture may be shaped like a Y, a U, or a V. The arms of the ecdysial suture that diverge anteroventrally, called the frontal sutures (frontogenal sutures), are not present in all insects (Fig. 3B). Some of these complexes of sutures are used by insects to emerge from the old integument during molting.

The frons is that part of the head immediately ventrad of the vertex (Fig. 3A). The frons varies in size, and its borders are sometimes difficult to establish. In most insects the frons is limited ventrally by the frontoclypeal suture (epistomal suture), a transverse suture located below the antennal sockets. As its name implies, the suture separates the dorsal frons from the ventral clypeus (Fig. 3A).

The face is a generalized term used to describe the anteromedial portion of the head bounded dorsally by the insertion of the antennae, laterally by the medial margins of the compound eyes, and ventrally by the frontoclypeal suture. In some insects the area termed the face is coincident with some, most, or all of the frons.

The clypeus (Latin, shield) is a sclerite between the face and labrum (Fig. 3A). Dorsally, the clypeus is separated from the face or frons by the frontoclypeal suture in primitive insects. Laterally, the clypeogenal suture demarcates the clypeus. Ventrally, the clypeus is separated from the labrum by the clypeolabral suture (Fig. 3A). The clypeus is highly variable in size and shape. Among insects with sucking mouthparts the clypeus is large.

The gena (Latin, cheek; pl., genae) forms the cheek or sclerotized area on each side of the head below the compound eye and extending to the gular suture (Fig. 3). The size of the gena varies considerably, and its boundaries also often are difficult to establish. In Odonata the gena is the area between compound eye, clypeus, and mouthparts. The postgena (Latin, *post* = after; *gena* = cheek; pl., postgenae) is the portion of the head immediately posteriad of the gena of pterygote insects and forms the lateral and ventral parts of the occipital arch (*sensu* Snodgrass) (Fig. 3). The subgenual area is usually narrow, located above the gnathal appendages (mandible and maxillae), and includes the hypostoma (Figs. 3 and 4) and the pleurostoma. The pleurostoma is the sclerotized area between the anterior attachment of the mandible and the ventral portion of the compound eye. The hypostoma is posteriad of the pleurostoma between the posterior attachment of the mandible and the occipital foramen. The subgenal suture forms a lateral, submarginal groove or sulcus on the head, just above the bases of the gnathal appendages (Fig. 4). The subgenal suture is continuous anteriorly with the frontoclypeal suture in the generalized pterygote head. Internally, the subgenal suture forms a subgenal

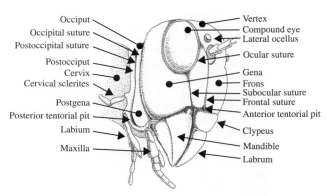

FIGURE 4 Generalized view of an insect head. [This and other line drawings after Snodgrass, R. E. (1935). "Principles of Insect Morphology," McGraw-Hill Co.].

ridge that presumably provides structural support for the head above the mandible and maxillae. In some instances, the subgenal suture is descriptively divided in two. The part of the suture that borders the proximal attachment of the mandible to the head (Fig. 4) is called the pleurostomal suture (the ventral border of the pleurostoma). The posterior part of the subgenal suture from the mandible to the occipital foramen is called the hypostomal suture (the ventral border of the hypostoma).

Head Size and Shape

The size and shape of the head and its appendages reflect functional adaptations that can be used to explain biological details of the insect—the realm of morphology as opposed to anatomy.

SIZE   Upon casual observation, the size of any given insect's head appears to be in proportion to the size of its body. A head that is disproportionately small or large relative to body size suggests that some adaptation has taken place that serves a functional need. Proportional head size varies considerably in the Insecta. Some fly families have very tiny heads in relation to their body size (e.g., Diptera: Acroceridae). Among Orthoptera, grass-feeding species typically have larger heads than herbaceous-feeding species. The large head is filled with powerful adductor muscles because grasses (monocots) are more difficult to chew than dicotyledonous plants. Furthermore, the postseedling stages of grasses are nutrient poor, meaning that more grass must be bitten, chopped, or ground to provide adequate nutrition.

SHAPE   Head shape varies considerably among insects. Many unusual shapes seem to be influenced by behavior and may be used to illustrate examples of structural form and function. The functional importance of head shape may be difficult to determine in preserved specimens. A few hours of observation with living insects can provide considerable insight into the importance of shape. Globular heads are seen in some insects, including the burrowing crickets (e.g., stenopelmatines and gryllids). This form of head is adapted for pushing soil. Hypercephalic heads are seen in the males of some Diptera (Sepsidae, Diopsidae, Drosophilidae, and Tephritoidea) and Hymenoptera (Pteromalidae and Eurytomidae); the broad heads of the males are featured in various aspects of courtship behaviors.

Topographical Features of the Head

Morphologists experience considerable difficulty in defining regions and determining homologies of structure on the insect head. We cannot unambiguously characterize topographical features of the insect head because more than a million species are involved in the definition, and they show incredible diversity in head anatomy. Shape alone is not adequate or suitable because there are many head shapes, and often a head shape can be derived independently in several unrelated lineages. Some head shapes are influenced by behavior.

AXIAL POSITION   The posture or orientation of the head in its resting position relative to the long axis of the body can be important in providing definitions of the anatomical features of the head. Axial position in insects typically falls into three basic categories: hypognathous, prognathous, and opisthognathous.

In general zoological usage, the word "hypognathous" (Greek, *hypo* = under; *gnathos* = jaw) serves to designate animals whose lower jaw is slightly longer than the upper jaw. In entomological usage, "hypognathous" refers to insects with the head vertically oriented and the mouth directed ventrad. Most insects with a hypognathous condition display an occipital foramen near the center of the posterior surface of the head. The hypognathous condition is considered by most insect morphologists to represent the primitive or generalized condition. The hypognathous position is evident in most major groups of insects and can be seen in the grasshopper, house fly, and honey bee. Other conditions are probably derived from ancestors with a hypognathous head.

In general zoological usage "prognathous" (Greek, *pro* = forward; *gnathos* = jaw) refers to animals with prominent or projecting jaws. In entomological usage, the prognathous condition is characterized by an occipital foramen near the vertexal margin with mandibles directed anteriad and positioned at the anterior margin of the head. When viewed in lateral aspect, the primary axis of the head is horizontal. Some predaceous insects, such as carabid beetles and earwigs, display the prognathous condition. In other insects, such as cucujid beetles and bethylid wasps, the prognathous position may reveal a solution to problems associated with living in concealed situations such as between bark and wood or similar confined habitats.

In general zoological usage, "opisthognathous" (Greek, *opisthos* = behind; *gnathos* = jaw) refers to animals with retreating jaws. In entomological usage, the opisthognathous condition is characterized by posteroventral position of the mouthparts resulting from a deflection of the facial region. The opisthognathous condition is displayed in many fluid-feeding Homoptera, including leafhoppers, whiteflies, and aphids.

SUTURES OF THE HEAD   Head sutures are sometimes used to delimit specific areas of the head, but there are problems. Establishing homology of sutures between families and orders is difficult. From a practical viewpoint, standards have not been developed for naming sutures among insect groups. Some names are based on the areas delimited (e.g., frontoclypeal suture); other sutures are named for the areas in which the suture is found (e.g., coronal suture). Sutures frequently have more than one name (e.g., frontoclypeal suture and epistomal suture are synonymous).

The compound eye is an important landmark on the insect head. An ocular suture surrounds the compound eye and forms an inflection or an internal ridge of the integument (Figs. 3, 4). The ocular suture is not present in all insects and is difficult to see in some insects unless the head is chemically processed for microscopic examination. When present, the ocular suture probably provides strength and prevents deformation of the compound eye.

A subocular suture extends from the lower margin of the compound eye toward the subgenal suture. In some species the subocular suture (Fig. 4). may extend to the subgenal suture; in other species it may terminate before reaching another landmark. This suture is straight and commonly found in the Hymenoptera, where it may provide additional strength for the head.

POSTERIOR ASPECT OF THE HEAD   The entire posterior surface of the head is termed the postcranium (Fig. 3). The surface may be flat, concave, or convex, depending on the group of insects. The occiput (Latin, back of head) of pterygote insects is the posterior portion of the head between the vertex and cervix (Latin, neck). The occiput is rarely present as a distinct sclerite or clearly demarcated by "benchmark" sutures. When present, the occiput signifies a primitive head segment. In some Diptera the occiput forms the entire posterior surface of the head. In other insects it forms a narrow, horseshoe-shaped sclerite.

The occipital suture (hypostomal suture *sensu* MacGillivray) is well developed in orthopteroids, but it is not present in many other groups of pterygote insects (Fig. 3C). When present, the occipital suture forms an arched, horseshoe-shaped groove on the back of the head that ends ventrally, anterior to the posterior articulation of each mandible. Internally, the occipital suture develops into a ridge, providing strength for the head.

The postoccipital suture is a landmark on the posterior surface of the head and is typically near the occipital foramen (Fig. 3C). The postoccipital suture forms a posterior submarginal groove of the head with posterior tentorial pits marking its lower ends on either side of the head. Some morphologists regard this suture as an intersegmental boundary (labium) between the first and second maxillae.

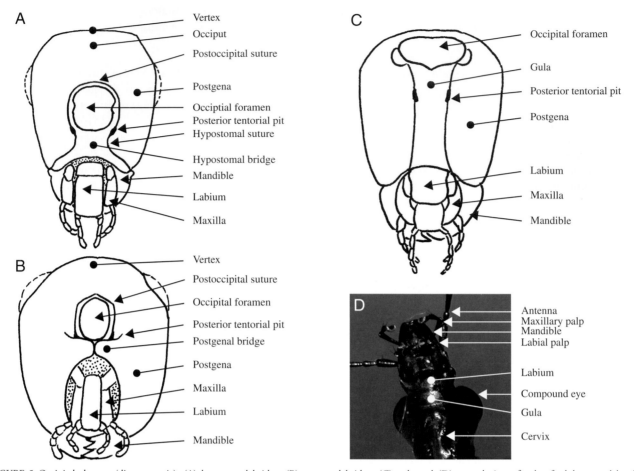

FIGURE 5 Occipital closures (diagrammatic): (A) hypostomal bridge, (B) postgenal bridge, (C) gula and (D) ventral view of gula of adult ground beetle (Coleoptera: Carabidae).

Internally, the postoccipital suture forms the postoccipital ridge that serves as an attachment for the dorsal prothoracic and cervical muscles of the head. The absence of the postoccipital suture in pterygote insects is a derived condition.

The postocciput of pterygotes forms the extreme posterior, often U-shaped sclerite that forms the rim of the head behind the postoccipital suture. The postocciput is interpreted as a sclerotic remnant of the labial somite in ancestral insects.

In pterygotes such as Orthoptera the occipital foramen and the mouth are not separated. More highly evolved insects have developed sclerotized separations between the mouthparts and the occipital foramen. At least three types of closure have been identified (Fig. 5): the hypostomal bridge, the postgenal bridge, and the gula. An understanding of these structures provides insight into the operation of the head and suggests evolutionary trends in feeding strategies.

The hypostomal bridge is usually developed in adult heads displaying a hypognathous axial orientation. The bridge is formed by medial extension and fusion of hypostomal lobes (hypostoma) (Fig. 5A). The hypostomal bridge is the ground plan condition of closure for the posterior aspect of the head, but it is not restricted to primitive insects. The hypostomal bridge is found in highly developed members of the Heteroptera, Diptera, and Hymenoptera. In Diptera the hypostomal bridge also has been called the pseudogula.

The postgenal bridge is a derived condition from the hypostomal ground plan and is developed in adults of higher Diptera and aculeate Hymenoptera. The bridge is characterized by medial extension and fusion of the postgenae, following a union of the hypostoma (Fig. 5B). The posterior tentorial pits retain their placement in the postoccipital suture.

The gula (Latin, gullet; pl., gulae) is developed in some Coleoptera, Neuroptera, and Isoptera. Typically, the gula is developed in heads displaying a prognathous axial orientation and in which posterior tentorial pits are located anteriad of the occipital foramen. (Fig. 5C, D). The median sclerite (the gula) on the ventral part of a prognathous head apparently forms *de novo* in the membranous neck region between the lateral extensions of the postocciput. The gula is a derived condition that is found in some but not all prognathous heads.

Endoskeletal Head Framework

Although the hardened integument of the head forms a structurally rigid capsule, this design is insufficient to solve the problems associated with muscle attachment and maintaining structural integrity during chewing. Thus, insects have evolved a tentorium (Latin, tent; pl., tentoria): a complex network of internal, hardened, cuticular struts that serve to reinforce the head. The tentorium forms as an invagination of four apodemal arms from the integument in most pterygote insects. The tentorium strengthens the head for chewing, provides attachment points for muscles, and also supports and protects the brain and foregut.

Anatomically, the tentorium consists of anterior and posterior arms. In most insects, the anterior arms arise from facial inflections located just above the anterior articulations of the mandibles. Externally, the arms are marked by anterior tentorial pits positioned on the frontoclypeal or subgenal (pleurostomal) suture (Fig. 4). Internally, the anterior region may form a frontal plate. Posterior arms originate at the ventral ends of the postoccipital inflection. They are marked externally by posterior tentorial pits (Fig. 4). The posterior arms usually unite to form a transverse bridge or corpus tentorii (internally) across the back of the head. Dorsal arms (rami), found in many insects, arise from the anterior arms. They attach to the inner wall of the head near antennal sockets. The dorsal arms are apparently not an invagination of cuticle, because pits do not mark them externally.

Mandible Articulation and Musculature

The hypothetical ancestor of insects is thought to possess a mandible with one point of articulation. Later, insects acquired a second point of articulation. The basis of this assumption comes from a survey of the Hexapoda in that the modern Apterygota have a monocondylic mandible and the Pterygota have a dicondylic mandible.

The term condyle (Greek, *kondylos* = knuckle) refers specifically to a knoblike process. For the mandible, the condyle is the point of articulation with the head. On the head itself is an acetabulum (Latin, *acetabulum* = vinegar cup), a concave surface or cavity for the reception and articulation of the condyle (Fig. 6).

The dicondylic mandible is the derived condition and is found in the Lepismatidae and Pterygota. The dicondylic mandible has secondarily acquired an articulation point anterior to the first point in the monocondylic mandible. These attachments form a plane of attachment. In the monocondylic mandible there is no plane of attachment, and the mandibles move forward or rearward when the muscles contract. The two points of articulation create a plane of movement that restricts the direction of mandible movement.

THORAX

The thorax represents the second tagma of the insect body. The thorax evolved early in the phylogenetic history of insects. In most Paleozoic insects the thorax is well developed and differentiated from the head and abdomen, and the three distinct tagmata probably developed during the Devonian.

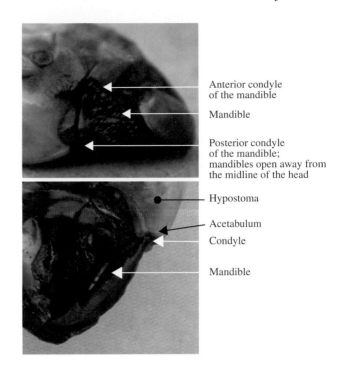

FIGURE 6 Mandible articulation. Top: lateral view; bottom: posterior view (Orthoptera: Stenoplematidae).

In terms of insect phylogeny, the thorax of Apterygota is strikingly different in shape compared with the head or abdomen. Of modern apterygotes only the Collembola display taxa in which thoracic tagmatization and segmentation are not obvious.

Apparently, the primary, functional role of the thorax has always been locomotion, since the primary modifications of the thorax have been for locomotion (first walking, and then flight). Modification for locomotion probably developed before other morphological adaptations, such as metamorphosis. Diverse independent and interdependent mechanisms for locomotion have evolved throughout the Insecta, including walking, flight, and jumping. Active participation in flight by insects is unique among invertebrates.

Anatomy of the Thorax

The cervix is the connection between the head (occipital foramen) and the anteriormost part of the thorax (pronotum) (Fig. 2). Typically, the area between the head and pronotum is membranous. The ground plan for the insect cervix contains two cervical sclerites on each side of the head that articulate with an occipital condyle of the head and the prothoracic episternum. Musculature attached to these sclerites increases or decreases the angle between the sclerites, and creates limited mobility of the head.

The thorax of modern insects consists of three segments termed the prothorax, mesothorax, and metathorax. The last two collectively are called the pterothorax (Greek, *ptero* = wing or feather) because extant insects bear wings on these segments only. The individual dorsal sclerites or terga of the thoracic segments are also known as nota (Greek, *notos* = back; sing, notum). The nota of Apterygota and many immature insects are similar to the terga of the abdomen with typical secondary segmentation. The nota of each thoracic segment are serially distinguished as the pronotum, mesonotum, and metanotum.

The size and shape of the prothorax are highly variable. The prothorax may be a large plate as in Orthoptera, Hemiptera, and Coleoptera, or reduced in size forming a narrow band between the head and mesothorax as in Hymenoptera. The prothorax is usually separated or free from the mesothorax. The sclerites are separated by a membrane that may be large and conspicuous in more primitive holometabolous insects such as Neuroptera and Coleoptera, or reduced in size in more highly evolved holometabolous insects such as Diptera and Hymenoptera.

The pterothorax includes the thoracic segments immediately posteriad of the prothorax. In winged insects the relationship between thoracic segments involved in flight can be complicated. In contrast, the thorax of larval insects and most wingless insects is relatively simple. The mesothorax and metathorax of these insects are separated by membrane. Adult winged insects show a mesothorax and metathorax that are consolidated (i.e., more or less united) to form a functional unit modified for flight.

The development of the pterothoracic segments varies among winged insects. When both pairs of wings participate equally in flight, the two thoracic segments are about the same size. This condition is seen in Odonata, some Lepidoptera, and some Neuroptera. When one pair of wings is dominant in flight, the corresponding thoracic segment is commensurately larger and modified for flight, whereas the other thoracic segment is reduced in size. This condition is seen in Diptera and Hymenoptera, where the forewing is large and dominant in flight. The reverse condition is seen in the Coleoptera, where the hind wing is large and dominant in flight.

In more closely related insect groups, such as families within an order, that are primitively wingless or in which wings have been secondarily lost in modern or extant species, many modifications to the thorax occur. Many wingless forms can be attributed to environmental factors that promote or maintain flightlessness. For instance, island-dwelling insects are commonly short winged (brachypterous), or wingless, whereas their continental relatives are winged, presumably because for island species, flight increases the likelihood of being carried aloft, moved out to sea, and subsequently lost to the reproductive effort of the population. The anatomical consequences of flightlessness can be predictable; in the Hymenoptera, short wings bring a disproportionate enlargement of the pronotum and reduction in size of the mesonotum and metanotum.

Sutures and Sclerites of Wing-Bearing Segments

The wing-bearing segments of the thorax are subdivided into a myriad of sclerites that are bounded by sutures and

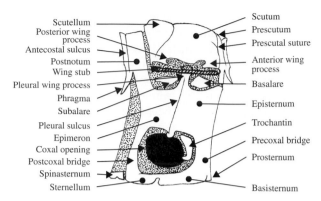

FIGURE 7 Diagram of the pterygote pterothorax.

membranous areas. These sutures and sclerites are the product of repeated modification of the thorax in response to various demands placed on the insect body by the environment. Similar modifications have occurred independently in many groups of insects; some modifications are unique. Generalizations are difficult to make, given the large number of sutures and sclerites, coupled with the number of insects that there are to consider.

Dorsal Aspect

The nota of the pterothorax are further subdivided into the prescutum, scutum, and scutellum; again, serially distinguished as mesoscutum and mesoscutellum, and metascutum and metascutellum (Fig. 7). Additionally, there are sclerites anterior and posterior to the notum, as discussed shortly.

The prescutum is the anterior portion of the scutum, laterally bearing prealar bridges separated by the prescutal suture from the mesoscutum. The scutum is the largest dorsal sclerite of the notum and is bounded posteriorly by the scutoscutellar suture, which divides the notum into the scutum and scutellum. The scutellum is generally smaller than the scutum. In Heteroptera it is a small triangular sclerite between the bases of the hemelytra. In Coleoptera the scutellum is the small triangular sclerite between the bases of the elytra. In Diptera and Hymenoptera the scutellum is relatively large, forming a subhemispherical sclerite, sometimes projecting posteriad. The posteriormost sclerite of the notum is the postnotum, separated from the notum by secondary segmentation. In some insects there is a postscutellum (metanotal acrotergite) that forms the posteriormost thoracic sclerite of the metanotum, or the posteriormost sclerite of the thorax. In Diptera the postscutellum appears as a transverse bulge below the scutellum.

The acrotergite and postnotum deserve further explanation. Again, the anteriormost sclerite is an acrotergite, the anterior precostal part of the notal plate. The postnotum is an intersegmental sclerite associated with the notum of the preceding segment. The postnotum bears the antecosta, a marginal ridge on the inner surface of the notal sclerite corresponding to the primary intersegmental fold. The postnotum also usually bears a pair of internal projecting phragmata. The antecostal suture divides the acrotergite from the antecosta, the internal ridge marking the original intersegmental boundary. Thus, when the antecosta and acrotergite are developed into larger plates and are associated with the notum anterior to them, they are referred to as a postnotum. The final structure associated with the dorsal aspect of the pterothorax is the alinotum (Greek, *ala* = wing; *notos* = back; pl., alinota). The alinotum is the wing-bearing sclerite of the pterothorax.

Wing Articulation

The thoracic components necessary for wing movement include the prealar bridge, anterior notal wing process, and posterior notal wing process. The components of the wing itself that articulate with the thoracic components are the humeral and axillary sclerites; they form the part of the wing closest to the body and are not treated in this article.

The prealar bridge is a heavily sclerotized and rigid supporting sclerite between the unsclerotized membrane of the pterothorax and the pleuron; it supports the notum above the thoracic pleura. The prealar bridge is composed of cuticular extensions from the anterior part of the prescutum and antecosta. The anterior notal wing process is the anterior lobe of the lateral margin of the alinotum supporting the first axillary sclerite (Fig. 7). The posterior notal wing process is a posterior lobe of the lateral margin of the alinotum that supports the third axillary sclerite of the wing base (Fig. 7).

Lateral Aspect

The pleuron (Greek = side; pl., pleura) is a general term associated with the lateral aspect of the thorax. Adults, nymphs, and active larvae all display extensive sclerotization of the pleural area. Sclerites forming this part of the body wall are derived from the precoxa, subscoxa, or supracoxal arch of the subcoxa.

PLEURAL REGIONS OF THE THORAX

Apterygota and Immature Plecoptera　The anapleurite is the sclerotized area above the coxa (supracoxal area) (Fig. 8). The coxopleurite is a sclerotized plate situated between the coxa and the anapleurite (Fig. 8). It bears the dorsal coxal articulation, the anterior part of which becomes the definitive trochantin. The sternopleurite, or coxosternite, is the definitive sternal sclerite that includes the areas of the limb bases and is situated beneath the coxa (Fig. 8).

Pterygota　The basalare is a sclerite near the base of the wing and anterior to the pleural wing process (Fig. 9). The basalare serves as a place of insertion for the anterior pleural muscle of the wing. The subalare is posterior to the basalare and the pleural wing process (Fig. 9). It too serves as a place for insertion of the wing's posterior pleural muscle. The tegula

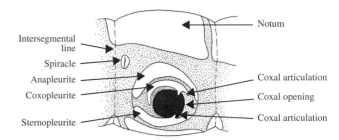

FIGURE 8 Pleural aspect of the apterygote thorax: diagrammatic.

is the anterior most independent sclerite associated with the wing base. The tegula is typically scalelike, articulates with the humeral sclerite, and protects the wing base from physical damage. The tegula is absent from Coleoptera and from the metathorax of most orders. The pleural wing process is located at the dorsal end of the pleural ridge and serves as a fulcrum for the movement of the wing (Fig. 9). The parapteron is a small sclerite, articulated on the dorsal extremity of the episternum just below the wings (Fig. 7).

The pleural suture is an easily visible landmark on the pterothoracic pleura (Fig. 9). It extends from the base of the wing to the base of the coxa. The pleural ridge is formed internally by the pleural suture and braces the pleuron above the leg. The episternum is a pleural sclerite anterior to the pleural suture and sometimes adjacent to the coxa (Fig. 9); the episternum is typically the largest lateral thoracic sclerite between the sternum and the notum. The epimeron is the posterior division of a thoracic pleuron adjacent to the coxa and posterior to the pleural suture (Fig. 9); it is typically smaller than the episternum and narrow or triangular. The episternum and the epimeron of many insects have become subdivided into several secondary sclerites bounded by sutures. The simplest condition shows the episternum divided into a dorsal anepisternum and a ventral katepisternum (Fig. 9). Similarly, the epimeron is divided into an anepimeron and katepimeron. The trochantin is a small sclerite at the base of the insect leg of some insects (Figs. 7, 9). Some workers theorize that the trochantin may have developed into the pleural wall. The trochantin is often fused to the episternum or absent.

The precoxal bridge is anterior to the trochantin, usually continuous with the episternum, frequently united with the

FIGURE 9 Lateral aspect of the pterygote thorax (Orthoptera: Acrididae).

basisternum, but also occurs as a distinct sclerite (Fig. 9). The postcoxal bridge is the postcoxal part of the pleuron, often united with the sternum behind the coxa (Fig. 9). The sclerite extends behind the coxa and connects the epimeron with the furcasternum. The meron is a lateral, postarticular basal area of the coxa and is sometimes found disassociated from the coxa and incorporated into the pleuron. The meron is typically large and conspicuous in panorpid and neuropteran insects. In Diptera the meron forms a separate sclerite in the thoracic pleuron.

Ventral Aspect

The ground plan of the sternum (Greek, *sternon* = chest; pl., sterna) consists of four sclerites, including an intersternite (spinasternite), two laterosternites (coxosternites), and a mediosternite (Fig. 10). The mediosternite and the laterosternite meet and join, and the line of union is called the laterosternal sulcus (pleurosternal suture) (Fig. 10).

Paired furcal pits are found in the laterosternal sulcus (Fig. 10). A transverse sternacostal sulcus bisects the ventral plate and thereby forms an anterior basisternite and posterior furcasternite (Fig. 10). The basisternite (basisternum) is the primary sclerite of the sternum (Fig. 10). It is positioned anterior to the sternal apophyses or sternacostal suture and laterally connected with the pleural region of the precoxal bridge. The furcasternite (furcasternum) is a distinct part of the sternum in some insects bearing the furca (Fig. 10). The spinasternum is a "spine-bearing" intersegmental sclerite of the thoracic venter, associated or united with the preceding sternum. The spinasternum may become part of the definitive prosternum or mesosternum, but not of the metasternum. The sternellum is the second sclerite of the ventral part of each thoracic segment, frequently divided into longitudinal parts that may be widely separated (Figs. 7, 10).

ABDOMEN

The abdomen is more conspicuously segmented than either the head or the thorax. Superficially, the abdomen is the least specialized of the body tagma, but there are notable exceptions such as the scale insects. The abdomen characteristically lacks appendages except cerci, reproductive organs, and pregenital appendages in adult Apterygota and larval Pterygota.

Ground Plan of the Abdomen

The ground plan abdomen of an adult insect typically consists of 11 to 12 segments and is less strongly sclerotized than the head or thorax (Fig. 11). Each segment of the abdomen is represented by a sclerotized tergum, sternum, and perhaps a pleurite. Terga are separated from each other and from the adjacent sterna or pleura by a membrane. Spiracles are located in the pleural area. Modification of this ground plan includes the fusion of terga or terga and sterna to form continuous dorsal or ventral shields or a conical tube. Some insects bear a sclerite in the pleural area called a laterotergite. Ventral sclerites are sometimes called laterosternites. The spiracles are often situated in the definitive tergum, sternum, laterotergite, or laterosternite.

During the embryonic stage of many insects and the postembryonic stage of primitive insects, 11 abdominal segments are present. In modern insects there is a tendency toward reduction in the number of the abdominal segments, but the primitive number of 11 is maintained during embryogenesis. Variation in abdominal segment number is considerable. If the Apterygota are considered to be indicative of the ground plan for pterygotes, confusion reigns: adult Protura have 12 segments, Collembola have six. The orthopteran family Acrididae has 11 segments, and a fossil specimen of Zoraptera has a 10-segmented abdomen.

Anamorphosis is present among some primitive ancestral hexapods such as the Protura—they emerge from the egg with eight abdominal segments and a terminal telson. Subsequently, three segments are added between the telson and the last abdominal segment with each molt. In contrast, most insects undergo epimorphosis in which the definitive number of segments is present at eclosion. Given the extent of variation in abdominal segmentation, morphologists conventionally discuss the abdomen in terms of pregenital, genital, and postgenital segmentation.

Abdominal Anatomy

Typically, the abdominal terga show secondary segmentation with the posterior part of a segment overlapping the anterior

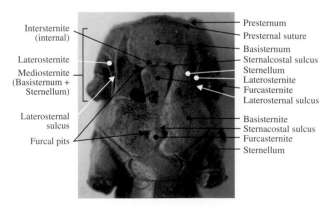

FIGURE 10 Ventral aspect of the thorax (Orthoptera: Acrididae).

FIGURE 11 Insect abdomen (Orthoptera: Acrididae).

part of the segment behind it (Fig. 11). Such overlap prevents damage or injury to the animal while it moves through the environment, particularly in confined spaces.

The pregenital segments in male insects are numbered 1 through 8; the pregenital segments in female insects are numbered 1 through 7 (Fig. 11). Among the Apterygota, male genitalia in Collembola are positioned between segments 5 and 6 and in Protura between segments 11 and the paraproct. Genital segments of Pterygota include segment 9 in males and segments 8 and 9 in females. Postgenital segments of pterygote insects are 10 and 11 in females and 9 and 10 in males.

In general there is little modification of the pregenital sclerites. A notable exception is found in the Odonata. Male Odonata do not have an intromittent organ on segment 9. Instead, the male moves the abdominal apex forward and deposits sperm in a reservoir along the anterior margin of the third abdominal sternum. Other modifications of the pregenital sclerites are not related to sexual behavior. Some of these modifications are glandular.

Modification of the genital sclerites from the ground plan is frequently observed among insects. Adult Pterygota are characterized by a well-developed reproductive system, including organs of copulation and oviposition. This duality of function has resulted in considerable differentiation of associated segments and contributed to difference of opinion regarding homology of genitalic parts. Among pterygote insects the male genitalia are generally positioned on segment 9. The ninth sternum is called a hypandrium (Greek, *hypo* = beneath; *aner* = male; Latin, *-ium* = diminutive) in many insects, including Psocoptera. In Ephemeroptera, the tenth sternum is called a hypandrium. Fusion of segments 9 and 10 in Psocoptera results in a structure called the clunium (Latin, *clunais* = buttock).

The gonopore (Greek, *gone* = seed; *poros* = channel) of the female reproductive system serves as the aperture through which the egg passes during oviposition. The gonopore usually is located on segment 8 or 9. Enlargement of sternum 8 in some female insects is called a subgenital plate.

Modification of postgenital sclerites is frequently observed and seems to be a functional response to adaptations associated with copulation and oviposition. Some modifications include fusion of the tergum, pleuron, and sternum to form a continuous sclerotized ring. The phenomenon is notable in apterygota and pterygote insects.

The eleventh abdominal segment forms the last true somite of the insect body. Frequently, this segment is found in embryonic stages of primitive insects even when it cannot be observed in postemergent stages. When the eleventh segment is present, it forms a conical endpiece that bears an anus at the apex, flanked laterally by cerci (Greek, *kerkos* = tail) (Fig. 12). The dorsal surface of the eleventh segment is called an epiproct (Greek, *epi* = upon; *proktos* = anus); the ventrolateral surface is called a paraproct (Greek, *para* = beside; *proktos* = anus) (Fig. 12). A longitudinal, medial, membranous area connects

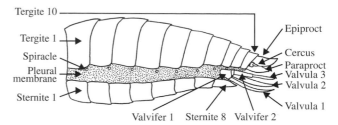

FIGURE 12 Abdominal segmentation: diagrammatic.

the paraprocts ventrally. Primitive groups of extant insects such as Thysanura and Ephemeroptera, and some fossil groups such as Paleodictyoptera, display a conspicuous, long, median filament that apparently projects from the apex of the epiproct. This is called the appendix dorsalis or caudal style. The appendage appears annulated and similar in shape to the lateral cerci, but the function of the appendix is unknown. The twelfth abdominal segment is called the periproct in Crustacea, and it forms a telson in some embryonic insects. The periproct appears in adult Protura and naiadal Odonata.

Abdominal Appendages

Presumably, the hypothetical ancestor of the Insecta was a myriapod with one pair of appendages for each body segment. Among contemporary insects the head appendages are represented by the antennae, mandibles, and the first and second maxillae. Thorax appendages are represented by legs, whereas the wings are considered to be secondary in origin. In most Apterygota, paired abdominal appendages are apparent. In most true insects embryological appendages are formed and lost before eclosion. The appendages found in embryos apparently represent ancestral conditions that are not expressed in postembryonic stages of modern insects. In modern insects, most pairs of appendages have been lost, and the irregular distribution of the remaining appendages makes a summary evaluation difficult. Abdominal appendages do not resemble the structure of thoracic legs of any insect.

Appendages are common among some entognathous hexapods, and some ancestral forms display unique abdominal appendages. Collembola are highly specialized entognathous Hexapoda. The abdomen of Collembola bear saltatorial appendages, which gives the group its common name of springtail, and a ventral tube, the collophore, which is the basis of the ordinal name.

The collophore (Greek, *kolla* = glue; *pherein* = to bear) is found on the first abdominal segment of Collembola. The collophore forms a ventromedial tube that is eversible with hydrostatic pressure and is drawn inward with retractor muscles. Some morphologists believe the collophore represents the fusion of paired, lateral appendages of an ancestor. An early explanation of the collophore function noted it was an organ of adhesion. The collophore also is used as a grooming organ in some Collembola. The collophore is connected to secretory glands in the head, and the median longitudinal channel on the venter of the thorax extends from the head to the base of the collophore.

OTHER APPENDAGES   Protura maintain short, cylindrical appendages on each of the first three abdominal segments. Each of these arises from membranous areas between the posterolateral angles of the terga and sterna. The position suggests a pleural origin.

APPENDAGES OF PTERYGOTA   The aquatic neuropteran larva *Sialis* has long, tapering, six-segmented appendages on each of the first seven body segments. These appendages articulate to pleural coxopodites. Similar appendages are found on the abdomen of some aquatic coleopteran larvae.

The tenth abdominal segment is present in most larval and adult Holometabola. As noted earlier, it is sometimes fused with segment 11. Segment 10 displays paired appendicular processes called pygopodia in Trichoptera, Coleoptera, and Lepidoptera. Pygopods form terminal eversible appendages in some beetle larvae. Pygopodia are bilaterally symmetrical, with eight podia, or feet, per side. Control of the podia is apparent because they are not always everted or inverted. Podia are withdrawn into the segment and have a common or median stalk. Each podium has several rows of equally spaced acanthae that apparently serve as holdfasts. Functionally, the acanthae enable the larvae to attach to and move on different substrates. When the larva walks on a flat substrate, the pygopodia are retracted into the body. When the larva walks on the edge of a leaf, the pygopodia are everted and used as holdfasts.

The larval prolegs of terrestrial Lepidoptera and Symphyta are not well developed, but they are adapted to grasping substrates. These structures are considered to be serially homologous with legs, but they also are referred to by some as adaptive structures with no relation to legs.

The adult pterygote abdomen has appendages that are not generally observed. These appendages are grouped for discussion based on the segments of the abdomen on which they are found.

Pregenital appendages are rare among insects. Adult white-flies have a curious structure on sternum 8 that propels honey-dew away from the body. Genitalia are segmental appendages and are treated in the next section. Postgenital appendages include cerci (Latin, circle), which are thought to represent primitive appendages because they are found in the Apterygota (except Protura) and many Pterygota. Cerci originate on abdominal segment 11 in a membranous area between the epiproct and the paraproct (Fig. 12). In insects that have lost segment 11, the cerci appear to originate on segment 10. Cerci occur in all orders among the Hemimetabola except for hemipteroids; among the Holometabola, they are found only in the Mecoptera and Symphyta.

Cerci are highly variable in size and shape and function. They are longer than the body in Thysanura, and in some Orthoptera cerci may be indistinct. Cerci resemble forceps in Japygidae and are annulated in Dictyoptera. In Dictyoptera they detect air currents, are sensitive to sound, and may be chemoreceptive. Some Ephemeroptera use cerci to propel themselves through water. Japygidae and Dermaptera probably use cerci to subdue prey. In some groups such as Embioptera and Orthoptera, cerci are sexually dimorphic and may serve a role in copulation.

There are some features on the insect body that appear as appendages but are not. Urogomphi (Greek, *oura* = tail; *gomphos* = nail; sing., urogomphus) are fixed or mobile cuticular processes on the apical abdominal segment of some coleopteran larvae. They may or may not be homologous with cerci, or other true appendages.

GENITALIA

The examination of the reproductive anatomy of different insect orders helps to develop an appreciation for the evolutionary trends in the formation of the external genitalia. The male genitalia are derived from the ninth abdominal segment. The female genitalia are derived from the eighth and ninth abdominal segments. In the female, the aperture through which the egg passes is called a gonopore. The gonopore serves as a boundary between the external and internal genitalia and is usually independent of the anus. Exceptions include some flies, such as the Tephritidae, where a common lumen termed a cloaca serves for excretion, copulation, and oviposition.

There is usually a single, medially located gonopore. The Dermaptera and Ephemeroptera are ancient groups of hemimetabolous insects. Both orders display a condition in which the lateral oviducts do not combine to form a median oviduct. Instead, the lateral oviducts independently connect with paired gonopores on the conjuctival membrane along the posterior margin of the seventh abdominal segment.

Many female insects with a genitalic opening on the posterior margin of the eighth abdominal segment display an appendicular ovipositor (Fig. 13). The ovipositor is a structure that develops from modified abdominal appendages or segments. It functions in the precise placement of eggs. It is commonly assumed that insects that do not show an ovipositor have ancestors that had an ovipositor. Thus, the structure has been lost during the course of evolutionary adaptation to a particular lifestyle.

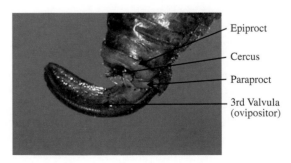

FIGURE 13 Appendicular ovipositor (Orthoptera: Tettigoniidae).

Female insects with a genitalic opening on the posterior margin of the ninth abdominal segment typically display a rudimentary or suppressed appendicular ovipositor. These insects lack special provisions for egg placement, but sometimes they reveal other abdominal modifications intended to facilitate oviposition.

Female Genitalia

Morphologists often use the Thysanura as a starting point for developing a generalized model to explain the evolution of the external reproductive system of pterygote insects. The thysanuran abdomen has basal sclerotized plates called coxopodites on which styli are attached. These plates are serially homologous along the abdomen, and the pregenital plates are regarded as identical with the genital plates. The plates located on segments 8 and 9 are considered to be genital plates. The styli associated with these segments are called gonapophyses. There are four gonapophyses on segments 8 and 9 (i.e., a pair of styli on each segment). The gonapophyses are medially concave and directed rearward. The basal sclerite is called a gonocoxa, and in some Thysanura it may be fused with the style.

The primitive pterygote with a gonopore on segment 8 has an appendicular ovipositor that consists of three components. A basal apparatus corresponds to the basal plate or primitive gonocoxite of the thysanuran abdominal appendage. The second part is the first valvifers (on the eighth sternum), and second valvifers (on the ninth sternum) are responsible for providing support and points of articulation for the tube through which the egg passes (Fig. 14). Interpolated between the first and second valvifers is a small sclerite called a gonangulum, which articulates with the second gonocoxite and tergum 9. The gonangulum is present in Odonata and Grylloblatoidea. It apparently is fused with the first valvifer in Dictyoptera and Orthoptera. In the remaining orders these structures are highly variable.

The shaft of the ovipositor consists of two pairs of elongate, closely appressed sclerites called the first and second valvulae (Fig. 14). The first pair of valvulae is positioned on the eighth abdominal sternum. The second pair of valvulae is located on the ninth abdominal sternum and is dorsal in position. Third valvulae are positioned on the posterior end of the second valvifers. These valvulae usually serve as a sheath for the shaft of the ovipositor (Figs. 13, 14).

Male Genitalia

The primary function of the male genitalia in insects is insemination of the female. Methods of achieving insemination that involve special functions of the external genitalia include clasping and holding the female, retaining the connection with the female gonopore, the construction of spermatophores, and the deposition of spermatophores or semen into the female genital tract; in some insects the injection of semen takes place directly into the female body (traumatic insemination of some Hemiptera). Other functions of the male genitalia include excretion and various sensory functions.

The genitalia of male insects exhibit such an enormous variety of shapes and constituent parts, often further complicated by structural rotation or inversion of all or some of the parts, that determination of a ground plan is virtually impossible. Examination of ancient orders shows highly variable and specialized conditions. In general, the coxites of the eighth segment in most apterygotes are reduced and without gonapophyses, and they are absent altogether in the Pterygota. Thus, the male external genitalia are derived from the ninth abdominal coxites.

Again, the Thysanura have genitalia that closely resemble that of the pterygote orders: a median intromittent organ or phallus, and paired lateral accessories (the periphallus of Snodgrass). The phallus is a conical, tubular structure of variable complexity (Fig. 15). Primitive insects may not display differentiated parts, and the entire structure may be long, sclerotized, and tapering apicad. In a ground plan

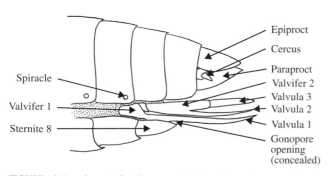

FIGURE 14 Female genitalia (diagrammatic), based on orthopteran female.

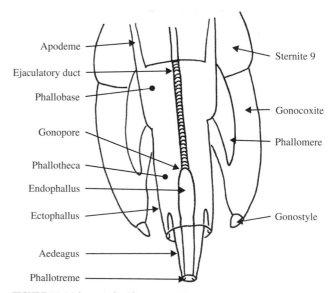

FIGURE 15 Male genitalia (diagrammatic).

condition for pterygote insects, there is a sclerotized basal portion termed the phallobase and a distal sclerotized portion called the aedeagus (Fig. 15). The phallobase in insects is characterized by highly variable development: sometimes sclerotized and supporting the aedeagus, sometimes forming a sheath for the aedeagus. The phallobase often contains an apodeme, which may provide support or a point for muscle attachment. The phallobase and aedeagus are joined by a membranous phallotheca (Fig. 15). The external walls of the phallobase and aedeagus are called the ectophallus (Fig. 15). The gonopore is positioned at the apex of the ejaculatory duct and is concealed within the phallobase. The gonopore is connected to the apex of the aedeagus via a membranous tube called the endophallus (Fig. 15). In some insects the endophallus may be everted through the aedeagus. The circular aperture at the apex of the aedeagus is called the phallotreme (Fig. 15). In some insects the endophallus and the gonopore may be everted through the phallotreme and into the female's bursa copulatrix. Genital lobes referred to as phallomeres form at the sides of the gonopore in the ontogeny of some insects. Usually the phallomeres unite to form the phallus.

See Also the Following Articles

Body Size • Integument • Legs • Mouthparts • Muscle System • Segmentation

Further Reading

Chapman, R. F. (1982). "The Insects. Structure and Function." 3rd ed. Hodder & Stoughton, London.
DuPorte, E. M. (1957). The comparative morphology of the insect head. *Annu. Rev. Entomol.* **2**, 55–77.
Gordh, G., and Headrick, D. H. (2001). "A Dictionary of Entomology." CAB Internation, Wallingford, Oxon, U.K.
Hinton, H. E. (1981). "The Biology of Insect Eggs." 3 vols., Pergamon Press, Oxford, U.K.
Matsuda, R. (1969). "Morphology and Evolution of the Insect Abdomen with Special Reference to Developmental Patterns and Their Bearings upon Systematics." Pergamon Press, Oxford, U.K.
Snodgrass, R. E. (1935). "Principles of Insect Morphology." McGraw-Hill, New York and London.
Tuxen, S. L. (1970). "Taxonomist's Glossary of Genitalia in Insects." 2nd ed. Munksgaard, Copenhagen.

Anopheles Mosquito

see *Mosquitoes*

Anoplura

see *Phthiraptera*

Antennae

Catherine Loudon
University of Kansas

Antennae are segmented appendages that function primarily as chemosensory and mechanosensory structures. An insect typically has a single pair of antennae located on its head. Antennae in juvenile insects are often very different in morphology from antennae in adult insects, typically being larger or more elaborate in the adult stage. Adult antennae may be sexually dimorphic, appearing very different in the males and females. Antennae are absent in the wingless insects belonging to the order Protura and may be extremely reduced in size in some holometabolous larvae.

STRUCTURE

The overall shape of most insect antennae is elongate and cylindrical (Fig. 1, top), although elaborations into plumose, lamellate, or pectinate forms have arisen many times in different insect lineages (Fig. 1, bottom). An elongate, cylindrical morphology, probably the ancestral condition for insect antennae, is found in fossil insects and many other arthropods. There are three parts to an insect antenna: the scape, the pedicel, and the flagellum. The scape is the first segment (most proximal) of the antenna, and it is attached to the head by a rim of flexible, intersegmental cuticle. Thus, the scape (and the rest of the antenna) can move with respect to the head. All the antennal segments are similarly joined to each other by thin, flexible cuticle.

The movements of an antenna are controlled in part by one or two pairs of muscles that attach inside the head (such

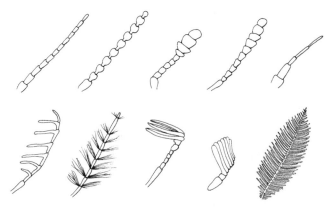

FIGURE 1 Insect antennae exhibit a variety of shapes including elongate morphologies (top) and those with lateral elaborations (bottom). [After Romoser, W. S., and Stoffolano, J. G., Jr. (1998). "The Science of Entomology," WCB/McGraw-Hill, Boston, and Loudon, C., *et al.* (1994). *J. Exp. Biol.* **193**, 233–254, published by McGraw-Hill, with permission of the McGraw-Hill Companies.]

as on the tentorium) with the other end attached inside the scape. An additional pair of muscles runs from the scape to the next segment of the antenna, the pedicel. The combined action of these two sets of muscles is capable of moving an antenna in almost any direction. The final (most distal) segment of the antenna, the flagellum, is the most variable in morphology among insects. The only insects that have intrinsic muscles in the flagellum (joining adjacent segments) are members of the wingless orders Collembola and Diplura. In all other insects (the majority), there are no muscles in the flagellum. Many specialists prefer "annulus" or "subsegment" to "segment" for an individual part of a flagellum in this latter group of insects, because "segment" is reserved for parts with their own musculature. Movements of an annulated flagellum without intrinsic musculature may still occur, such as the spreading and closing of the lamellae or lateral extensions in an antenna (Fig. 1, bottom), but these movements are driven by changes in the pressure of the hemolymph (blood) inside the antenna and thus are hydraulic rather than muscular.

In most insects, circulation of hemolymph through an antenna is facilitated by muscular pumping by an accessory heart located in the head near the base of the antenna. This antennal heart pumps the hemolymph into a blood vessel that discharges the hemolymph at the distal end of the antenna. The return flow of the hemolymph back to the head (and the general open circulatory system of the insect) is not inside a blood vessel. The lumen of an antenna also contains tracheae and nerves, which branch into any lateral extensions of the flagellum. Sensory neurons that send action potentials in response to chemical or physical stimuli sensed by the antennae terminate in the deutocerebrum of the brain. The deutocerebrum is also the site of origin for the motor neurons that stimulate the muscles associated with the antennae.

GROWTH AND DEVELOPMENT

Antennal growth and development in holometabolous insects (those that undergo complete metamorphosis) differs greatly from that in other insects. In holometabolous insects, adult antennae form from imaginal disks, which are clumps of undifferentiated cells that will develop into adult structures. The antennal imaginal disks may appear in the embryonic (fly) or late larval (moth) stage of the immature insect. Properties of the antennal imaginal disks determine to a large extent the chemical stimuli to which an adult will respond, as is seen from experiments in which antennal imaginal disks were cross-transplanted between larvae, which were then reared to adulthood and assayed.

In hemimetabolous and apterygote (wingless) insects, the nymphs are very similar in overall form and habit to the adults, and their antennae resemble smaller, shorter versions of the adult antennae. As with all external structures that are replaced at each molt, a new antenna is formed inside the old antenna. The primary morphological change that occurs at each molt is that the flagellum lengthens with the addition of

more segments or annuli, either at the distal end (orders Collembola and Diplura), the proximal end (most other insects), or along the length of the flagellum (some members of the orders Orthoptera and Odonata).

Antennae are serially homologous to mouthparts and legs, reflecting the ancestral condition of a single pair of appendages per body segment shared by arthropods and related groups. Common developmental features between legs and antennae can be seen, for example, in the action of the homeotic gene called *Antennapedia,* which results in the substitution of leglike appendages for antennae on the head when expressed ectopically in mutant *Drosophila.* Leglike appendages appearing in the antennal location in adult insects have also been observed after regeneration of antennae following injury during the larval stage (Fig. 2).

FUNCTION

The primary function of antennae is the assessment of the chemical and physical characteristics of the environment. Detection is made with innervated chemosensory and mechanosensory organs that are arrayed on the antennae. A single antenna usually has sensory organs of several types, with different properties. Most of the chemosensory organs are located on the flagellum and often take the form of microscopic chemosensory hairs (sensilla) each only 1 or 2 μm in diameter. Some antennae, such as the feathery pectinate antennae of silkworms (*Bombyx mori*), have tens of thousands of sensilla, which are capable of very thoroughly sampling the air that passes in the small spaces between them. A cockroach antenna may have hundreds of thousands of sensilla. The chemicals that may be detected by chemoreceptors on the

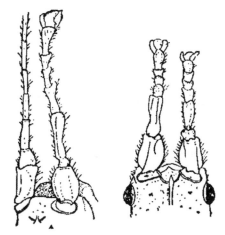

FIGURE 2 Left: head of an adult Indian stick insect (*Carausius morosus*) with a normal antenna on the left and a regenerated antenna with leglike morphology on the right. Right: head of adult *C. morosus* with two regenerated antennae with leglike morphology. [After Fig. 78 in Wigglesworth, V. B. (1971). "The Principles of Insect Physiology." Chapman & Hall, London, with the kind permission of Kluwer Academic Publishers.]

antennae are usually biological in origin and airborne (volatiles), although (depending on the insect species) the sampled chemical compounds are sometimes in a liquid or associated with a solid surface. The chemicals intercepted by antennae may alert the insect to the presence of prospective mates, food, suitable places to lay eggs, or predators.

The physical stimuli detected by mechanoreceptors on the antennae may be used by the insect to indicate air speed during flight, to detect vibrations of the air, or to detect solid boundaries in its environment by touch. While a single mechanosensory hair will send information to the brain about the local physical conditions existing at its microscopic location, an antenna also has mechanosensory organs that evaluate the physical forces acting on the antenna as a whole. These mechanosensory organs, located near the base of the antenna, include Johnston's organ, Böhm bristles, hair plates (groups of mechanosensory hairs), and campaniform sensilla (thin flexible patches of cuticle that are innervated). Johnston's organ is located in the pedicel and responds to changes of location or vibrations of the whole antenna. In contrast, the Böhm bristles, located near the scape–pedicel boundary, send information to the brain about the antennal position, rather than its movements. The variety of mechanosensory organs associated with the first two segments of the antennae are believed to act together to inform a flying insect about its air speed, because greater flying speed will cause greater deflection of the antennae by the air rushing past. Contact chemosensory hairs, so called because the chemical compounds are usually detected when the insect is touching a liquid or solid surface with the antennae, often have mechanosensory capabilities as well and are usually located near the distal ends of antennae.

The function of the antennal sensory organs will be affected by their arrangement on the antennae. For example, sensory organs on the distal tip of a very long antenna will permit chemical or physical sampling of the environment far from the body of the insect. Close packing of sensory hairs will decrease the airflow in their vicinity, and hence will modify both the chemical and physical sampling of the environment by those hairs. The function of the antennae will also be dependent on the behaviors of the insect that will affect the airflow around the antennae, such as flying, wing fanning, postural changes, or oscillating the antennae. A structure projecting into the environment is liable to collect debris that might interfere with its sensory function; both antennal grooming behaviors and modifications of leg parts against which an antenna is scraped are common in insects (Fig. 3). In some insects, antennae are modified for nonsensory functions such as clasping mates during copulation (fleas and collembolans), holding prey items (beetle larvae), or forming a temporary physical connection between an underwater air reservoir and the atmosphere (aquatic beetles).

See Also the Following Articles

Chemoreception • Imaginal Discs • Mechanoreception • Pheromones

Further Reading

Hansson, B. S., and Anton, S. (2000). Function and morphology of the antennal lobe: New developments. *Annu. Rev. Entomol.* **45**, 203–231.

Heinzel, H., and Gewecke, M. (1987). Aerodynamic and mechanical properties of the antennae as air-current sense organs in *Locusta migratoria*. II. Dynamic characteristics. *J. Comp. Physiol. A* **161**, 671–680.

Kaissling, K. E. (1971). Insect olfaction. *In* Handbook of Sensory Physiology (L. M. Beidler, ed.), Vol. IV of "Chemical Senses," Part 1, "Olfaction," pp. 351–431. Springer, Verlag, Berlin.

Keil, T. A. (1999). Morphology and development of the peripheral olfactory organs. *In* "Insect Olfaction" (B. S. Hansson, ed.), pp. 5–48. Springer-Verlag, Berlin.

Loudon, C. and Koehl, M. A. R. (2000). Sniffing by a silkworm moth: Wing fanning enhances air penetration through and pheromone interception by antennae. *J. Exp. Biol.* **203**, 2977–2990.

Pass, G. (2000). Accessory pulsatile organs: Evolutionary innovations in insects. *Annu. Rev. Entomol.* **45**, 495–518.

Schneider, D. (1964). Insect antennae. *Annu. Rev. Entomol.* **9**, 103–122.

Schneiderman, A. M., Hildebrand, J. G., Brennan, M. M., and Tumlinson, J. H. (1986). Transsexually grafted antennae alter pheromone-directed behaviour in a moth. *Nature* **323**, 801–803.

Steinbrecht, R. A. (1987). Functional morphology of pheromone-sensitive sensilla. *In* "Pheromone Biochemistry" (G. D. Prestwich and G. J. Blomquist, eds.), pp. 353–384. Academic Press, London.

Zacharuk, R. Y. (1985). Antennae and sensilla. *In* "Comprehensive Insect Physiology Biochemistry and Pharmacology" (G. A. Kerkut and L. I. Gilbert eds.), Vol. 6, pp. 1–69. Pergamon Press, New York.

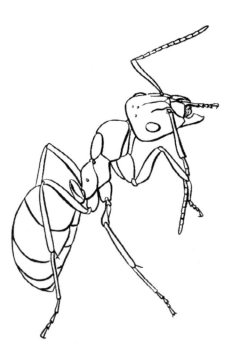

FIGURE 3 A worker ant (*Formica polyctena*) cleans one of its antennae by dragging it across the specialized comb of right foreleg. (Reprinted by permission of the publisher from THE INSECT SOCIETIES by Edward O. Wilson, Cambridge, MA.: The Belknap Press of Harvard University Press, Copyright 1971 by the President and Fellows of Harvard College. Original drawing by Turid Hölldobler.)

Ants

Nigel R. Franks
University of Bristol

The ants comprise a single family, the Formicidae, within the superfamily Vespoidea and the order Hymenoptera. There are 16 extant subfamilies of ants with a total of 296 extant genera. Some 9000 to 10,000 species of ants have been described, and it is estimated that there may be 15,000 species of ants alive in the world today. The earliest known fossil ants are from the Cretaceous (ca 105–110 mya), but ants probably did not become common until the Eocene (ca 45 mya).

EVOLUTION AND ECOLOGICAL SUCCESS

Ants are now extremely successful ecologically. There may even be an equal biomass of ants and humanity in the world today. They dominate, at their size scale, many terrestrial ecosystems from latitudes north of the boreal tree line to such southern climes as Tierra del Fuego, Chile. In certain tropical forests the contribution of ants to the biomass is spectacular. In Brazilian rain forests, for example, the biomass of ants has been estimated as approximately four times greater than the biomass of all of the vertebrates combined.

One of the reasons ants are so successful is that their colonies have extremely efficient divisions of labor: they evolved factories millions of years before we reinvented them. Another reason is that they can modify their immediate environment to suit themselves, much as we do. Leafcutter ants (*Atta*), for example, evolved agriculture tens of millions of years before humanity developed agronomy. Furthermore, leafcutter ants also use antibiotics and symbiotic bacteria to protect the crop of fungi they grow on the leaves they collect. By contrast, weaver ants (*Oecophylla*) fashion homes from living leaves by sowing them into envelopes, using their larvae as living shuttles and the silken thread they produce as glue. Ants can also dominate areas by mobilizing large numbers of well coordinated foragers; indeed, an ant colony's foragers can be so numerous and well organized that they give the impression of being everywhere at once.

Ants can also be important as seed distributors and as seed harvesters, in the turnover of soils, and in the regulation of aphid numbers and the minimization of outbreaks of defoliating insects. Economically important pest species include the imported fire ant (*Solenopsis invicta*) in North America and leafcutter ants (such as *Atta*) in the neotropics. There are also many ecologically destructive "tramp" ants or invasive species that have been distributed to alien habitats by human commerce.

Ants and plants often have closely coupled ecological relationships. Certain plants even encourage ants by producing rewards such as energy-rich elaiosomes on their seeds to encourage seed dispersal, nutritious Beltian bodies and extrafloral nectaries to entice ants to visit their leaves and shoots (hence to remove the plant's natural enemies while there), or even by supplying preformed homes (domatia) to invite ants directly to inhabit and thus better protect them. Although many ants are hunter-gatherers, very many species tend aphids for the excess honeydew they excrete. By "milking" aphids in this way, ants can in effect become primary consumers of plant products and by thus operating at a lower trophic level they can build up a larger biomass than obligate carnivores would be able to do. Yet most ants mix their diet by also consuming animal protein; for example, they will devour their own aphid milk cows if the latter become sufficiently abundant.

Arguably, the best evidence of the ecological success of ants is that their worst enemies are other ants.

EUSOCIALITY, SOCIAL ORGANIZATION, AND SOCIAL DIVERSITY

Except for a few species that have secondarily lost the worker caste, all ants are eusocial: they have an overlap of adult generations, cooperative brood care, and reproduction dominated by a minority of the colony's members. Typically, an established ant colony consists of one or more queens (each of which may have mated with one or more winged males on a nuptial flight), an all-female set of wingless workers, and the colony's brood of eggs, larvae, and pupae. The majority of queens mate only before they establish a colony. Thereafter, they store the sperm they have received.

All ants have haplodiploid sex determination. This property probably had a major role in the evolution of their eusociality through kin selection. Males are haploid, having only a single set of chromosomes, and thus the sperm that individual males produce is genetically homogeneous. Hence, the (diploid) daughters of the same mother and father are unusually closely related to one another, a circumstance likely to have favored the evolution of female workers. Nevertheless, there can be continuing conflicts within colonies between the workers and the queen (or queens) over the sex ratios they produce and which colony members produce the males. Queens can choose to produce either unfertilized (haploid) eggs destined to become males or fertilized (diploid) eggs. The latter may develop into workers or potential new queens (gynes) generally depending on how much food they receive as larvae. The workers may or may not be sterile. Fertile workers produce viable (unfertilized) haploid eggs that can develop into males. Hence, there can be conflict both among the workers and between the workers and the queen over whose sons the colony produces. Indeed, in many species of ants with only small numbers of workers in their mature colonies, there are dominance hierarchies among the workers, who fight one another over egg production. Sometimes the queen moves with active aggression against the most dominant worker to curtail its production of sons in favor of her own. In addition, even when workers are sterile and serve one, singly mated queen, they may prefer to raise more of the queen's daughters, to whom they are more closely related, than the queen's sons. For all these reasons, the study of ants has had a major impact in recent pioneering evolutionary biology because these insects provide test cases by which the evolutionary resolution of the tension between cooperation and conflict can be explored. It is clear, though, that the apparent social cohesion of ant colonies is often partly an illusion.

Among ants, there is a diversity of mating systems and social organizations. So even though it is tempting to think of the typical ant colony as having a single, singly mated queen and occupying a single nest site, the diversity of social systems among the ants is in fact huge. For example, many ant species consist of facultatively multiqueened (polygynous) colonies. Indeed, roughly half of European ant species exhibit polygyny, and there seems to be no reason to regard this as an unusual proportion. Some ant colonies are founded by solitary queens; some by groups of unrelated queens that may later fight over who will be the one to succeed. Other colonies simultaneously occupy multiple nests (polydomy), a habit often associated with polygyny, while others exhibit colony fission, with both daughter colonies usually being monogynous. Most persistent polygyny is associated with the secondary adoption of queens. Unusual social systems include queenless ants, workerless ants (inquilines), and slave-making ants. In certain queenless species, the workerlike females produce other diploid females through a parthenogenetic process called thelytoky. By contrast, certain inquilines have dispensed with the worker caste, and queens infiltrate and exploit established colonies of other species. Slave making may occur both intraspecifically and interspecifically. Interspecific slave making is also associated with nonindependent colony foundation in which slave-maker queens infiltrate established colonies of their host species, kill the host queen or queens, and produce workers that are reared by currently available host workers. The slave-maker workers raid other neighboring host colonies to capture large larvae and pupae. Such raids thus replenish the stocks of slave workers, which do all the foraging and brood rearing for the slave makers. There are also ant species in which there are polymorphic queens, others in which there are polymorphic males, and many in which there are polymorphic workers.

One of the outcomes of eusociality is that established colonies can be well defended by the workers against enemies. Thus, ant colonies are relatively K-selected; that is, they are selected to hold onto resources and to persist for long periods rather than being ephemeral, here-today-gone-tomorrow, r-strategists. Associated with this trait is the extreme longevity of ant queens. It is estimated that they can live 100 times longer than other solitary insects of a similar size. Worker populations in mature, well-established monogynous colonies range from a few tens of millions to 20 million, and certain so-called supercolonies consist of a huge network of linked nests each with many queens. One supercolony of

Formica yessensis in Japan may have as many as 300 million workers. Given such longevities and densities, it is clear that ants may also prove to be important model systems for understanding the spread of disease or the evolution of mechanisms to minimize the spread of disease among viscous populations of close kin. It is even possible that polygyny and multiple mating (polyandry) have evolved, at least in part, to promote genetic heterogeneity within colonies and thus help to minimize disease risks.

DIVISION OF LABOR

The relatively large biomass of ants in many ecosystems can be attributed not just to the way in which the ants interact with other organisms but to the way in which they interact with their nestmates in general and, in particular, to efficiencies that accrue from divisions of labor. One of the most dramatic traits associated with the division of labor among the workers is physical polymorphism, which is the presence of different physical worker forms within the same colony. In the African army ant, *Dorylus wilverthi,* for example, the smallest workers at 0.12 mg dry weight are only 1% of the dry weight of the largest workers (soldiers), and this relatively great size range is exceeded in certain other species (e.g., in *Pheidologeton diversus,* the smallest workers have a dry weight that is about 0.2% that of the largest majors). It is not just the size range that is impressive in such species but also the degree of polymorphism among the workers. Darwin, writing in *The Origin of Species,* seemed well aware not only of the phenomenon but also of its implications. Indeed, one of Darwin's most penetrating insights in his 1859 masterpiece was his suggestion that sterile forms evolved in social insects because they are "profitable to the community" and that "selection may be applied to the family, as well as to the individual." He further suggested that once such colony-level selection had begun, the sterile forms could be molded into distinct castes "Thus in [the army ant] *Eciton,* there are working and soldier neuters, with jaws and instincts extraordinarily different" (Fig. 1a, b).

Such worker polymorphism is now known to be associated with the differential growth rates of different putative tissues and body parts during the preadult stages. Indeed, the study of ants made a major contribution to the development of the concept of allometric growth (Fig. 1c, d). Notably polymorphic genera include the army ants *Eciton* and *Dorylus,* leafcutter ants (*Atta*), carpenter ants (*Camponotus*), and members of the genera *Pheidole* and *Pheidologeton.* Indeed, *Camponotus* and *Pheidole* are the two most species-rich ant genera.

However, genera with polymorphic workers are in the minority. Approximately 80% of ant genera consist entirely of species with monomorphic workers, most of the remaining genera consist of species in which there are at most only two easily recognizable worker morphs, and only about 1% of genera have species in which three or more worker morphs can be relatively easily recognized within colonies.

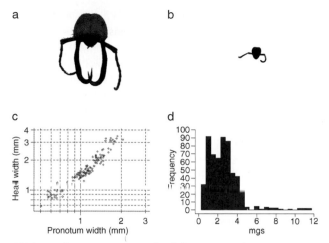

FIGURE 1 The army ant, *Eciton burchelli*. (a) Head of major worker. (b) Head of minor worker. (c) Head width vs ponotum width allometry for workers. (d) Frequency–dry weight histogram for a large sample of workers. The allometrical relationship has a slope greater than 1, so larger workers (such as majors) have disproportionately large heads. The size frequency distribution is skewed to the right so relatively few of these very large majors are produced. (Drawings © Nigel R. Franks.)

Polymorphism among the workers is mostly associated with extreme physical specialization. Thus, *Eciton* majors have ice-tong-like mandibles and are specialist defenders of the colony against would-be vertebrate predators or thieves (Fig. 1a). It has been shown that colonies of *Pheidole pallidula* can produce more defensive majors in response to stresses induced by conspecific competitors. Majors are not always for defense: large-headed majors in *Pheidole* and *Messor* serve as specialist grinders of harvested seeds. Even among such polymorphic species, however, the majority of workers belong to castes of generalists, which give their colonies an ability to respond rapidly to changes in the environment. Such generalists show behavioral flexibility not possible with the extreme morphological specialization of certain physical castes. Nevertheless, divisions of labor also occur within the majority generalist caste. Such workers typically specialize in different tasks at different times during their lives. This is known as temporal polyethism, in contrast to physical polyethism.

The sophisticated divisions of labor in monomorphic ants are being investigated. In *Leptothorax albipennis*, the workers show very little size variation, and colonies consist of, at most, a few hundred such workers living in flat crevices between rocks. Such crevices can be only 2 or 3 cm wide and deep and may have an internal cavity height of only 1 or 2 mm. Individual workers could easily roam all around such nests within a minute, but instead they have spatial fidelity zones; that is, they remain faithful to certain parts of the nest and the segregated tasks within such areas for months on end. The workers can even reconstruct their own spatial fidelity zones relative to one another if, and when, their colony is forced to emigrate to a new nest site because of the destruction of the old site. In this (and many if not all) ant

species, younger workers tend to work deep within the nest at its safe center, tending the queen and the eggs. As they get older, workers tend to move progressively out from the center of the nest, and toward the end of their lives they eventually engage in the most dangerous task of foraging in the outside world, where they are likely to meet predators and other hazards. However, the correlation between age and task is often very weak, and in an increasing number of species it has been shown that the division of labor among monomorphic workers is extremely flexible. Workers can respond to the removal of other workers by reverting to tasks that they did earlier in their lives or, if need be, they may begin foraging even when they are very young. Thus, though age may influence what workers do, it is unlikely to be the organizing principle of the division of labor in many species. Rather, it seems that workers are continuously monitoring their workloads and the delays they experience while waiting to interact with their nestmates and will flexibly change their tasks accordingly to maximize their productivity.

COMMUNICATION AND PHEROMONES

Ants have diverse systems of communication, but by far the most important medium for signaling involves the chemicals known as pheromones. Ants can deposit chemical trails to recruit nestmates to discoveries of food. Many ants can also produce highly volatile chemicals to signal alarm when they encounter dangerous predators or other hazards. Different ants in different subfamilies use a remarkable diversity of glandular structures even just to produce recruitment pheromones. These may be produced from cloacal glands, Dufour's glands, the hindgut, poison glands, pygidial glands, rectal glands, sternal glands, or even tibial glands on the back legs. Furthermore, many pheromones appear to be complex mixtures of many chemical compounds.

Pheromones can be effective in minute quantities; it has been estimated that one milligram of the trail substance of the leafcutting ant, *Atta texana,* if laid out with maximum efficiency, would be sufficient to lead a colony three times around the world.

Nestmate recognition is another important aspect of communication in ants. A pleasing metaphor for the ant colony is a factory inside a fortress. Ant colonies are dedicated to the production of more ants; but workers need to "know" that they are working for their natal colony, and colonies also need to be well defended against other ants and against infiltration by other arthropods, which might tap into their resources. Ant colonies employ colony-specific recognition cues as one of their defense systems. These are often in the form of cuticular hydrocarbons that can be spread throughout the colony both by grooming and trophallaxis (the latter is usually associated with liquid food exchange). Slave-making ants circumvent the recognition cues of their slaves by capturing them as larvae and pupae—these captives are not yet imprinted on their natal colony odor but later become imprinted on the odor of the

FIGURE 2 Scanning electron micrograph of a worker of *Lasius flavus* with a kleptoparasitic mite, *Antennophorus grandis,* gripping on its head. The mite steals food when two workers exchange nutritious liquids during trophallaxis. (Photomicrograph © Nigel R. Franks.)

and back more quickly than the ants that happen to take the longer path. All the ants lay attractive trail pheromones, and such pheromones are reinforced more rapidly on the shorter path simply because that path is shorter and quicker. In such cases, individual ants do not directly compare the lengths of the two paths, but the colony is able to choose the shorter one. Sometimes the shorter path is used exclusively, while at other times a small amount of traffic may continue to use the longer path. Having some traffic that continues to use the longer path is likely to be costly in the short term, but it may represent a beneficial insurance policy if the shorter path becomes blocked or dangerous. Self-organization also has a major role in such phenomena as brood sorting, rhythms of activity within nests, and building behaviors. This new approach may help to answer, at least in part, the age-old challenge of how ant colonies are organized.

See Also the Following Articles
Caste • Colonies • Nest Building • Pheromones • Sex Determination • Sociality

Further Reading
Bolton, B. (1994). "Identification Guide to the Ant Genera of the World." Harvard University Press, Cambridge, MA.
Bourke, A. F. G., and Franks, N. R. (1995). "Social Evolution in Ants." Monographs in Behavioral Ecology. Princeton University Press, Princeton, NJ.
Camazine, S., Deneubourg, J.-L., Franks, N. R., Sneyd, J., Theraulaz, G., and Bonabeau, E. (2001). "Self-Organization in Biology." Princeton University Press, Princeton, NJ.
Detrain, C., Deneubourg, J.-L., and Pasteels, J. (1999) "Information Processing in Social Insects." Birkhäuser Verlag, Basel, Switzerland.
Grimaldi, D., and Agosti, D. (2000). A formicine in New Jersey Cretaceous amber (Hymenoptera: Formicidae) and the early evolution of the ants. *Proc. Natl. Acad. Sci. USA* **97,** 13678–13683.
Hölldobler, B., and Wilson, E. O. (1990). "The Ants." Belknap Press, Cambridge, MA.
Keller, L., and Genoud, M. (1997). Extraordinary lifespans in ants: A test of evolutionary theories of ageing. *Nature* **389,** 958–960.
Passera, L., Roncin, E., Kauffmann, B., and Keller, L. (1996). Increased soldier production in ant colonies exposed to intraspecific competition. *Nature* **379,** 630–631.
Schmidt-Hempel, P., and Crozier, R. H. (1999). Polyandry versus polygyny versus parasites. *Philos. Trans. R. Soc. (Lond) (B)* **354,** 507–515.
Sendova-Franks, A. B., and Franks, N. R. (1994). Social resilience in individual worker ants and its role in division of labour. *Proc. R. Soc. (Lond) (B)* **256,** 305–309.

colony that kidnapped them after they have metamorphosed into adult workers. Sometimes colony-specific odors also can be influenced by chemicals picked up from the colony's environment. Nevertheless, countless species of arthropods from mites to beetles have infiltrated ant colonies. For example, more than 200 species of rove beetle (Staphilinidae) are associated with New World army ants alone, and other groups such as mites are probably even more species rich. Often these infiltrators are called "guests" simply because their relationships with their host ant colony and to its resources are unknown (Fig. 2).

SELF-ORGANIZATION, COLLECTIVE INTELLIGENCE, AND DECISION MAKING

A rapidly developing approach to the study of ants and other social insects is the application of self-organization theories. Here self-organization can be defined as a mechanism for building spatial structures and temporal patterns of activity at a global (collective or colony) level by means of multiple interactions among components at the individual (e.g., worker) level. The components interact through local, often simple, rules that do not directly or explicitly code for the global structures. The importance of studies of such self-organization is that they can show how very sophisticated structures can be produced at the colony level with a fully decentralized system of control in which the workers have no overview of the problems they are working to solve.

A simple and very intuitive example of how ants use self-organization is found in their ability to select short cuts. Certain ants can select the shortest paths to food sources. Indeed, where there is a short and a long path to the same food source, the decision-making mechanism can be surprisingly simple. The ants that happen to take the shorter path get there

Aphids

John T. Sorensen
California Department of Food and Agriculture

Aphids are remarkable, evolutionarily exquisite creatures, and among the most successful insects. Aphid evolution has been shaped through nutrient-driven selection and by the host plants on which they feed, and aphids have responded by developing intricate life cycles and complex polymorphisms. These sap-feeding hemipterans have coped with a hostile world through developing an exceptionally high reproductive rate and passive wind-borne dispersal, a strategy in which individuals are quite expendable, but survival and prosperity of their genes are guaranteed. Because of their intriguing evolutionary adaptations, aphids were among our most worthy competitors as humans entered the agricultural era.

MAJOR GROUPS AND HOST AFFILIATIONS

Aphids, as the superfamily Aphidoidea, belong to the Sternorrhyncha within the Hemiptera, a group they share with Aleyrodoidea (whiteflies), Psylloidea (jumping plant lice), and Coccoidea (scale insects and mealybugs). Aphidoidea has three families: Adelgidae (adelgids), Phylloxoridae (phylloxorids), and Aphididae (aphids), although some workers place the Adelgidae and Phylloxoridae in a separate superfamily, Phylloxoroidea. Adelgids and phylloxorids are primitive "aphids" and older groups, each with about 50 species. They differ from Aphididae by having an ovipositor and by reproducing by means of ovipary. Adelgids are restricted to conifers (Pinaceae), and some form characteristic galls (e.g., *Adelges piceae*, balsam woolly adelgid). Phylloxorids, which may also form galls, occur on plants of the Salicaceae (willow family), Fagaceae (oak family), Juglandaceae (walnut family), and Rosaceae (rose family). An exceptional species, *Daktulosphaira vitifolae*, grape phylloxera, feeds on grapes (Vitaceae), damaging European grape cultivars unless they are grafted to resistant rootstocks developed from American grape species.

Aphids originally evolved on woody plants in the Northern Hemisphere and are functionally replaced by whiteflies and psyllids in the Southern Hemisphere. As a group, they evolved and began their diversification with angiosperms, over 140 mya during the lower Cretaceous. While most fossil aphid groups became extinct during the Cretaceous–Tertiary boundary, most modern aphid groups radiated during the Miocene. Aphids have siphunculi, which vary by group from being mere pores on the abdominal surface to being very elongate tubes. They also have a cauda, which varies by group from rounded and hardly noticeable to knobbed or long and fingerlike. Aphids lack an ovipositor and are viviparous, bearing young parthenogenetically.

Aphid taxonomy is difficult; their subfamily classification has been argued and confused with nearly as many classifications as aphid taxonomists. Remaudiere and Remaudiere's 1997 classification, followed here, recognizes about 25 aphid subfamilies, with tribal groupings for about 600 genera and 4700 species of aphids. Many aphid lineages coevolved with, and radiated among, their host plant groups. Often during their phylogenetic history, however, aphid groups opportunistically switched to radically unrelated host groupings, driven by developmental requirements but tempered by evolutionary constraints.

Many aphid subfamilies are small, but several are larger and important: Chaitophorinae (e.g., *Sipha flava*, yellow sugarcane aphid), on Salicaceae and Gramineae (grass family); the closely related Myzocallidinae (e.g., *Therioaphis trifolii* f. *maculata*, spotted alfalfa aphid), Drepanosiphinae (e.g., *Drepanaphis acerifoliae*, painted maple aphid), and Phyllaphidinae (e.g., *Phyllaphis fagii*, beech aphid), often considered to be one subfamily and usually on dicotygledonous trees, but also Fabaceae (legume family) and bamboo; Lachninae (e.g., *Essigella californica*, Monterey pine aphid), mostly on Pinaceae, but also Fagaceae, Rosaceae, and Asteraceae (composite family) roots; and Pemphiginae (e.g., *Pemphigus bursarius*, lettuce root aphid), often on roots and host alternating to dicotyledonous trees forming galls. Other noteworthy subfamilies include Pterocommatinae, on Salicaceae; Greenideinae, on Fagaceae; Mindarinae, on Pinaceae; and the host-alternating Anoeciinae and Hormaphidinae, the latter causing galls.

The largest and most evolutionarily recent subfamily, Aphidinae, has two large, diverse, and agriculturally important tribes. The first tribe, Macrosiphini (e.g., *Aulacorthum solani*, foxglove aphid), is diverse in genera, which often lack attendance by ants but may alternate hosts. The second tribe, Aphidini, is diverse in species but less so in genera; these are often attended by ants. Tribe Aphidini has two important subtribes. Subtribe Rhopalosiphina (e.g., *Rhopalosiphum padi*, bird cherry–oat aphid) host alternates between Rosaceae to Gramineae or Cyperaceae (reed family). Subtribe Aphidina (e.g., *Aphis fabae*, bean aphid) host alternates mostly among Rosidae and Asteridae and is home to genus *Aphis*, which alone contains well over 1000 species.

NUTRITION-DRIVEN EVOLUTION: LIFE CYCLES AND POLYMORPHISM

Aphid life cycles are complex and may be either monoecious or dioecious, involving holocycly or anholocycly. Because of this, aphids have evolved many specialized morphs; a multitude of confusing, often synonymous names have risen among aphid biologists, but these are minimized here.

In the simple and generalized monoecious holocyclic aphid life cycle (Fig. 1A), a single host plant species is used throughout the year and sexual morphs are produced in the fall, usually in response to decreasing daylength. The males and oviparae mate, producing genetically recombinant eggs that overwinter on the host plant and often experience high mortality. In the spring, the fundatrix emerges from the egg, matures parthenogenetically, and gives live birth to nymphs that become viviparae and continue in that reproductive mode through the summer. If the aphid group produces plant galls, the fundatrix is responsible for their production.

The viviparae may be apterae or alatae (Fig. 2), but in some groups (e.g., Drepanosiphini, some Myzocallidinae) all viviparae are alatae. The parthenogenetic reproduction of

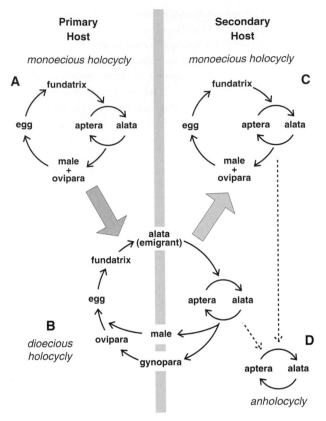

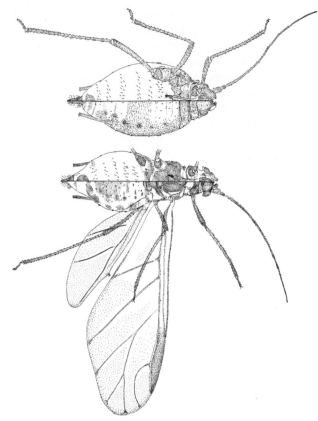

FIGURE 1 Evolutionary development of generalized aphid life cycles. Initially, aphids developed monoecious holocycly (A) on an ancestral woody primary host, where aestivation occurred because sap amino acids were unavailable during summer growth cessation. Next, multiple subfamilies independently evolved dioecious holocycly (B), where viviparae moved to summer-growing herbaceous secondary hosts but returned to their ancestral host in autumn. In some aphids, secondarily monoecious holocycly (C) developed on the secondary host when the primary host was lost. Often in warm areas, where selection for an overwintering egg is not imposed, some populations of dioecious and secondarily monoecious holocyclic aphids may lapse into facultative anholocycly (D) on their secondary hosts; this condition may become obligate anholocycly if the ability to produce sexuals is lost.

FIGURE 2 Aptera (top) and alata (bottom) viviparae of *M. persicae*. Shown in split images with ventral (upper half) and dorsal (lower half) aspects with heads to the right. [Drawings by Tokuwo Kono, modified from Kono, T., and Papp, C. S. (1977). "Handbook of Agricultural Pests, Aphids, Thrips, Mites, Snails, and Slugs." California Department of Food and Agriculture, Sacramento.]

viviparae allows very rapid buildup of numbers and collapse of generation time. When a viviparous nymph is born, it has the embryos of both its daughters and granddaughters within it, creating a "telescoping" of generations. Apterae have lost their wings and associated musculature to optimize reproduction. They produce more offspring per female than do alatae, which must invest resources in their flight apparatus. However, alatae produce progeny earlier in life than do apterae, giving their relatively reduced number of offspring a better generational turnaround time than apterae mothers can. Apterae are selectively produced when the host plant is a good source of nutrients. Once an aphid population has built, either inducing a crowding effect among apterae or stressing its host to the level of impacting nutrient levels, the population usually switches to produce alatae, which migrate to better situations. However, the risks of successful migration

are great, especially for monophagous aphids that feed on uncommon hosts, because the flight of alatae is wind-borne and relatively passive. Alatae can be blown over 1600 km, often across an ocean, and survive the trip. Upon successfully alighting on their proper host and feeding for a short time, alatae begin autolysis of their flight musculature, precluding further flight but self-cannibalistically providing nutrients for their offspring. The production of viviparae continues until fall conditions trigger production of the sexuals.

A second, more complicated dioecious life cycle (Fig. 1B) has independently evolved among several different aphid groups that show seasonal alternation between differing hosts. This dioecious cycle probably evolved in response to the seasonally inadequate supply of nitrogen-based nutrients, especially amino acids, on their primary host. The phloem sap that aphids feed on has limited nitrogen availability, and nitrogen is the limiting nutrient in aphid development. Woody deciduous plants normally translocate amino acids in quantity only during the spring, when they are foliating, and in the fall, when leaf senescence breaks down leaf protein and nitrogen is translocated to the roots for overwinter storage. Aphids groups evolving on and restricted to such plants face a nitrogen deficit during the summer, when active plant growth ceases and phloem sap is low or devoid of nitrogen. Such groups (e.g., *Periphyllus* spp.) may develop an aestivating nymph that halts growth until fall. Other aphid groups (e.g., Aphidinae) whose ancestors originated on deciduous woody plants, have evolved to leave those primary hosts during the late spring, after the nitrogen flush associated with foliation has ceased. In doing so, their spring alatae, as emigrants, migrate to herbaceous secondary hosts that actively grow and transport nitrogen during the summer. In the fall, however, as their secondary hosts die back, the aphids return to their woody primary host by producing migrating males and gynoparae. There, the aphid's sexuals, its males and oviparae, capture that host's fall nitrogen flush and mate to lay their overwintering eggs in anticipation of the spring nitrogen flush. Depending on the aphid or its group, the secondary hosts may vary from quite specific to a broad number of botanical groups; but the primary hosts are often specific to a plant genus. Most aphid lineages have adapted specific types of secondary hosts, such as grasses (e.g., *Metopolophium dirhodum*, rose-grain aphid), roots (e.g., *Smynthurodes betae*, bean root aphid), other woody plants (e.g., *Hormaphis hamamelidus*), or herbs (e.g., *Macrosiphum rosae*, rose aphid). Some aphids specialize on secondary hosts of a particular environmental ecotype; for example, *Rhopalosiphum nymphae*, waterlily aphid, uses aquatic plants in many plant families.

Some aphid lineages (e.g., *Schizaphis graminum*, greenbug) have evolved beyond dioecious holocycly, entirely leaving their primary host to remain on their secondary host, in secondarily monoecious holocycly (Fig. 1C). However, an important form of year-round residence on the secondary host occurs in warmer climates, where populations do not require an egg for overwintering survival. Under such conditions, otherwise holocyclic dioecious or monoecious populations may lapse facultatively into anholocycly on their secondary hosts (Fig. 1D). If such populations remain anholocyclic long enough, they may eventually evolve into obligate anholocycly by losing the ability to produce sexual morphs, despite undergoing environmental conditions that normally trigger their production. Depending on the aphid lineage and its adaptation to its host(s) or their alternation, nearly all aphid morphs may be winged or wingless, but the morph's wing condition is specific to the aphid group.

APHID BEHAVIOR

Aphids feed by inserting their rostrum-borne stylets into a plant and ratcheting them between plant cells, seldom penetrating any until they enter the phloem sieve tubes and extract sap. Stylet advancement is lubricated by saliva containing a pectinase that loosens the bonding between plant cells. The saliva forms a stylet sheath that is left in the plant when the stylets are withdrawn. To cope with a sap diet, aphid guts have specialized groups of cells, mycetomes, containing rickettsia-like symbiotic bacteria, mycetocytes,

which aid in synthesis of nutrients. These bacteria, which are passed from mother to embryonic daughter, have coevolved with aphid lineages, differing among them.

Whereas aphids largely rely on a high reproductive rate and great dispersive ability to maximize survival in a hostile environment, morph-specific behaviors exist to promote genetic survival of the individual or its clone. Behavior of alatae optimizes dispersion and finding a successful host. When alatae initially take to flight, they are attracted to the short wavelengths of light that predominate in a clear or cloudy sky, and fly up toward them. After flying a while, however, they come to prefer the longer light wavelengths reflected from plants, and they descend, moving to them. In some species, alatae have shown a preference for colors characteristic of their host plant's leaves. Generally alatae are attracted to yellow, a predominant hue in growing or senescent plants, which are better nitrogen sources. Upon alighting on a plant, they briefly probe below the epidermis with the rostrum to locate specialized secondary plant compounds that are of no nutritional value but are specific to the aphids' given host. If these feeding triggers are not found, the alatae move on.

In contrast, apterae usually move only when necessary to procure a better feeding site or if a predator or parasite molests them. Ants tend apterae in many aphid groups in a form of facultative mutualism; in some relationships, ants actively "farm" their aphid "cattle" by moving them among locations. Generally, however, aphid groups with elongate siphunculi are less likely to be tended by ants. In exchange for the aphid's sugary honeydew waste, the ants protect them from predators, such as coccinellid, lacewing, and syrphid fly larvae, or specialized aphid parasites, such as chalcidoid and braconid wasps. When stroked by the ant's antennae, the aphid will raise the tip of its abdomen, extruding a honeydew drop, which may be retracted if not accepted by the ant. If an ant does not accept honeydew after a while, the aphid will revert to its normal behavior of flicking the honeydew drop away with its hind leg or cauda, to prevent an accumulation of honeydew from fouling the aphid colony.

Aphids communicate by chemicals and sound. Parasites and predators are often foiled by the use of an aphid alarm pheromone, such as *trans*-β-farnesene. When molested, aphids exude microdroplets of alarm pheromone from their siphuncular pores, and in response adjacent aphids quickly withdraw their stylets from their host and drop to the ground. Shortly thereafter, the fallen aphids visually orient to vertical lines or structures and move toward them in an attempt to climb the plant stem. Aphid sexual pheromones are also used as male attractants by oviparae, being released from specialized pores on their hind tibiae. Sound communication is used by *Toxoptera* spp., which have a stridulatory mechanism consisting of a row of short pegs on the legs, which are rubbed against filelike ridges on the lower epidermis of the abdomen, just below the siphunculi. When disturbed, colonies of *T. aurantii* emit an audible high piercing stridulatory sound, to which their apterae respond.

Fundatrices of gall-forming species use species-specific patterns of feeding or probing behavior to induce characteristically shaped galls on their specialized hosts, in which their progeny can safely develop. The fundatrix of *P. bursarius* climbs the developing leaf petiole on *Populus nigra* about halfway and probes its rostrum around the petiole to create an array of punctures oriented perpendicularly to the petiole shaft. This induces a swollen globular gall with a slit oriented perpendicularly to the petiole shaft. In contrast, on the same host, a *Pemphigus spyrothecae* fundatrix probes the petiole shaft in an upward spiral array of punctures, yielding a corkscrew-shaped petiole gall. Not only do plant galls provide a protective encasement for aphid development, but aphids of even nongalling species do better on galled tissue, probably because of a local increase in plant nutrients in that tissue.

Many aphid species have some lower degree of sociality, especially among apterae, which is expressed as a gregariousness within colonies and probably confers better protection or response to attacks by natural enemies. Alatae of *Drepanosiphum platanoides,* sycamore aphid, are more likely to be distributed in a clustered manner among sycamore leaves, in groups in which the tips of their antennae and legs touch among the aphids. Some aphids have evolved a higher degree of sociality, however. The tribe Cerataphidini of the Hormaphidinae has genera in which species produce a soldier morph with enlarged forelegs, which defend their relative clones differentially. Soldiers discriminate between soldiers and nonsoldiers but do not attack soldiers of their own species. The investment in soldier production by the colony is related to areas needing defense, such as a gall's surface.

AGRICULTURAL IMPORTANCE

While aphids are among the most serious agricultural problem insects, only about 250 species are considered to be agricultural pests. Pest aphids may affect only a very specific host (e.g., *Brachycorynella asparagi*, asparagus aphid), or group of related hosts, such as crucifers (Brassicaceae) (e.g., *Brevicoryne brassicae*, cabbage aphid). Some, however, are quite polyphagous (e.g., *Aphis gossypii*, cotton aphid; *Myzus persicae*, green peach aphid), with an extremely wide host range. Some common polyphagous pest aphids represent sibling species complexes that are morphologically identical but differ in karyotype. They comprise anholocyclic clones, or biotypes, that differ in host preferences, ability to transmit diseases, or resistance to pesticides.

Aphids cause damage in several ways. They can build to high population densities and damage plants directly, by removing enough sap to cause withering and eventual plant death. If not washed off, aphid excrement, or honeydew, can build up enough on plants to serve as a medium for the growth of sooty molds, impairing photosynthesis and plant development, and eventually promoting other fungal diseases. Salivary secretions of some aphids are phytotoxic, causing stunting, leaf deformation, and gall formation. Even if the feeding effects of aphids are not apparent, they may affect plant hormone balances, changing host metabolism to their advantage, thus essentially hijacking the plant's physiological functions.

The aphid vectoring of stylet-borne and circulative plant viruses is the most serious problem to agriculture posed by aphids. Stylet-borne viruses occur on the aphid's epidermis and are not aphid specific. These viruses are acquired quickly and transmitted during the aphid's probing of the plant's epidermis. They are nonpersistent, however, and the aphid's infectiousness is lost upon molting. Circulative viruses, in contrast, live internally in the aphid's gut. The aphid must feed for a while to acquire these viruses, which require an incubation period before they can be successfully transmitted. They are persistent, however, and once infected, the aphid remains a vector throughout its life. The virus–aphid–plant linkage is fairly specific for circulative viruses, and a given virus is transmitted by only one or few aphid species. Virus-infected plants often show an aphid-attractive yellowing and have increased free amino acids, so aphids benefit by virus transmission.

APHID CONTROL IN AGRICULTURAL CROPS AND HOME GARDENS

Agricultural control of aphids best uses an integrated pest management (IPM) strategy, where species are identified and tactics reflect the allowable tolerance level on a crop. Within fields, aphids may be monitored by means of yellow water pans or sticky traps, which attract them. In some agricultural regions, especially seed-growing areas with plant virus sensitivities (e.g., the Netherlands, Idaho), specialized agencies run aerial trapping networks in which large suction traps are used to detect alates and forecast population levels. Proper aphid IPM emphasizes sustainable control, maximizing organically compatible methods to minimize effects on nontarget species, such as biological control agents, or vertebrates. IPM tactics include cultural control methods, such as minimizing weed or ant populations that promote aphids, using ultraviolet-light-reflecting or colored films near plants to repel alates, or interplanting pollen and nectar plants among crop rows to promote aphid natural enemies. Biological control agents include small wasps (e.g., *Aphidius* sp.) that parasitize aphids and disperse well within populations. Predators, which as immatures voraciously consume aphids, can be released. These include lacewings (e.g., *Chrysopa* spp.), aphid midges (e.g., *Aphidoletes* spp.), and ladybird beetle larvae (e.g., *Hippodamia convergens*). Predators may, however, disseminate when released as adults. One can apply entomopathic fungi (e.g., *Beauveria bassiana*), whose spores attach to the aphid's exoskeleton, penetrate it, and kill the aphid. Insect growth regulators applied by spray act through various means to prevent maturation of aphids. These may act in conjunction with biological control agents if the latter fail to provide adequate control. Use of chemical poisons in aphid IPM should be minimized because of the effect on nontarget species. While poison use may sometimes be necessary, heavy usage promotes insecticidal resistance in aphids, as well as secondary resurgence of aphid populations, once biological control agents have been hampered. Chemical poisons range from less toxic pyrethroids to more toxic organophosphates. They may be applied directly as contact insecticidal sprays or dusts, or indirectly as plant systemic insecticides that are ingested with the plant's sap. Cultivation of aphid-resistant crop varieties is also important.

In home gardens and yards, nontoxic controls should be emphasized. Aphid detection involves inspection of buds, stems, fruits, and the underside of leaves, where the insects are most likely to congregate. Effective control can simply involve frequently hosing aphids off plants with water, being careful to hit the leaf undersides. Spray applications of a mixture of garlic and water may repel aphids. Sprays of cuticle-disrupting insecticidal soaps, which cause fatal desiccation, often give control. Under overhanging trees, problems from aphid sooty molds on driveways, patios, and walkways are best controlled by hosing the surfaces. Control for aphid galls or leaf distortion on deciduous trees can be problematic, sometimes requiring the winter application of a dormant oil to kill overwintering eggs. Ultimately, elimination of the tree to may be required to solve problem, so tree species in yards should be carefully selected and placed, in view of their potential aphid pests.

See Also the Following Articles

Ants • Biological Control • Rostrum • Sternorrhyncha

Further Reading

Blackman, R. L., and Eastop, V. F. (1984). "Aphids on the World's Crops: An Identification Guide." Wiley, New York.

Blackman, R. L., and Eastop, V. F. (1994). "Aphids on the World's Trees: An Identification and Information Guide." CAB International, Wallingford, U.K.

Dixon, A. F. G. (1985). "Aphid Ecology." Blackie & Son, Glasgow, U.K.

Minks, A. K., and Harrewijn, P. (1987). "Aphids, Their Biology, Natural Enemies and Control," Vols. A, B , C. Elsevier, Amsterdam.

Moran, N. (1992). The evolution of aphid life cycles. *Annu. Rev. Entomol.* **37,** 321–348.

Remaudiere, G., and Remaudiere, M. (1997). "Catalogue of the World's Aphididae." INRA Editions, Paris.

Stern, D. L. (1995). Phylogenetic evidence that aphids, rather than plants, determine gall morphology. *Proc. R. Soc. Lond. B.* **260,** 85–89.

Stern, D. L., and Foster, W. A. (1996). The evolution of soldiers in aphids. *Biol. Rev. Cambridge Philos. Soc.* **71,** 27–79.

Apis Species
(Honey Bees)

Eva Crane
International Bee Research Association

Honey bees (genus *Apis*) are social insects in the family Apidae, order Hymenoptera; they are among the Aculeata (i.e., those having stingers). They evolved after the separation of the Americas and Australia from Eurasia/Africa and are native only in the Old World. The genus *Apis* probably first appeared in the Eocene, about 55 mya. Tropical species *A. dorsata* and *A. florea* existed by the end of the Oligocene 25 mya, and cavity-nesting *A. mellifera* and *A. cerana,* which can also live outside the tropics, were separate species by the end of the Pliocene about 2 mya. Therefore, the highly advanced cavity-nesting species have existed only perhaps a tenth as long as the open-nesting species, which were confined to the warmer tropics. The most important species to humans is *A. mellifera,* which has been introduced all over the world for use in beekeeping.

THE GENUS *APIS*

Known Species

The genus *Apis* contains 11 known species. *A. mellifera* (Fig. 1) is the source of most of the world's honey. It is native throughout Africa, the Middle East, and Europe except for the far north regions. All other *Apis* species are native to Asia. *A. cerana,* which is kept in hives in the temperate zone as well as the tropics, is smaller than *A. mellifera,* and it makes smaller colonies. Other Asian species that build a multiple-comb nest in a cavity are *A. koschevnikovi* and *A. nuluensis* reported in Borneo, and *A. nigrocincta* in Sulawesi.

Other *Apis* species native in parts of the Asian tropics build a single-comb nest in the open. The most important to humans is *A. dorsata,* a bee much larger than *A. cerana. A. laboriosa,* which is even larger, lives in parts of the Himalayas too high for *A. dorsata.* Much smaller than *A. cerana, A. florea* is widespread below around 500 m and can live in drier areas than *A. dorsata.*

Mating, and How Reproductive Isolation Is Achieved

Honey bees mate in flight; the process has been studied in detail in *A. mellifera,* and involves three stages. A queen flies out when only a few days old, and drones that are flying in the area, attracted by the pheromones she produces, follow her. If a drone succeeds in clasping the queen with his legs,

FIGURE 1 Worker honey bees *(Apis mellifera)* on honeycomb. (Photograph courtesy of P. Kirk Visscher.)

his endophallus is everted and mating occurs. When they separate, part of his genitalia remains in the queen, and he falls away and dies. She may mate more than once (usually on the same flight), and the semen she receives is stored in her spermatheca for use throughout her egg-laying life.

The main component of the pheromone attracting drones to the queen seems to be the same for all *Apis* species (9-oxo-*trans*-decenoic acid). In an area with more than one species, reproductive isolation can be achieved if the drones of different species fly at different times of day.

APIS MELLIFERA

Colony Life

The reproduction of individual bees takes place in the colony, and each colony normally contains a single mated female (the queen), many nonreproductive females (workers) and, during the reproductive season, a smaller number of reproductive males (drones). Colonies reproduce by swarming during a season when much food is available. The workers rear several young queens, each in a special queen cell. The old queen and perhaps half the workers of the colony leave as a swarm, which finds a new nest site. One of the young queens mates and heads the parent colony; the others are killed.

Many aspects of the beekeeping cycle and social behavior of honey bee colonies have been studied in detail (see Further Reading).

In the tropics, temperatures are never too low for plants to flower or for bees to fly, and colony activity is governed by rainfall rather than temperature. There are two seasonal cycles in the year, so colonies do not grow as large, or store as much honey, as they do in temperate zones. If the stores of a colony of *A. mellifera* become low in a dearth period, the colony may leave its hive and fly to a nearby area where plants are starting to bloom, rebuilding its combs in a nest site there. Such movements are referred to as absconding or migration, and preventing them is an important part of beekeeping in tropical Africa.

Subspecies and Their Distribution

During the Ice Ages, geographical features in Europe such as mountains confined *A. mellifera* to several separate areas, where they diversified into a number of subspecies or races. The most important in world beekeeping, and their native areas, are *A. mellifera ligustica* (Italian) in northwestern Italy south of the Alps, *A. mellifera carnica* (Carnolian) in the eastern Alps and parts of the Balkans, *A. mellifera caucasica* (Caucasian) in Georgia and the Caucasus mountains between the Black Sea and the Caspian Sea, and also *A. mellifera mellifera* north of the Alps.

The first introductions of *A. mellifera* from Europe to new continents, after 1600, enabled future beekeeping industries to build up and flourish in many countries. Some of the subsequent introductions of *A. mellifera* carried diseases or parasites

not previously present, and these caused much damage. From the late 1800s, after the movable-frame hive was devised, there was great interest in breeding more productive honey bees, and colonies of many races were transported from the Old World to other continents. Italian bees, especially, could store much honey in warm regions with consistently good nectar flows. During the 1900s, scientists introduced exotic species and races of honey bees into Europe for experimental purposes, but none is known to have survived in the wild.

Moving honey bees to new areas in tropical or subtropical environments can have wide-reaching effects. In 1956 a number of honey bee queens were transported from southern Africa to Brazil in an attempt to improve the beekeeping in that South American country, where bees of European origin performed poorly. Through an accident, a few of the African queens escaped with swarms, and this led to hydridization with bees of European origin. The consequent "Africanized" bees had characteristics that enabled them to become dominant over the "European" bees already in the American tropics, and they spread rapidly, reaching the Amazon by 1971, the north coast by 1977, Mexico by 1986, and then several southern U.S. states.

In warm regions, many native plants may be pollinated by small bees (Apoidea) whose populations are reduced if colonies of the larger *A. mellifera* are introduced, which in turn can endanger the reproduction of such native plants. This problem has been reported in Australia and Brazil.

A. mellifera is now used in beekeeping in almost every country in the world.

APIS CERANA AND RELATED SPECIES

Of the subspecies of *A. cerana,* the Asian hive bee, *A. cerana indica* is present from Yunnan in China through India to the Philippines. *A. cerana cerana* is in much of China, also the Himalayas, Afghanistan, and the Russian Far East, and *A. cerana japonica* in Tsushima Island and Japan.

After *A. mellifera* was introduced in eastern Asia, *A. cerana* became restricted to areas with native flora. Then in 1985–1986 colonies of *A. cerana* were taken from one of the Indonesian islands to Irian Jaya, also Indonesian but part of New Guinea. The bees reached Papua New Guinea by 1987 and islands in the Torres Strait by 1993. By 2000 they were found (and killed) in Brisbane, Australia, and strenuous efforts are being taken to prevent any further entry and spread of this bee in Australia because it would probably carry the varroa mite, a pest that is of serious economic importance to beekeepers.

APIS DORSATA AND RELATED SPECIES

The large single-comb nests of the giant honey bee, *A. dorsata,* built in the open, are still the most important source of honey in India and some other tropical Asian countries. *A. dorsata* is present in most of the Indo-Malayan region, from the Indus River in the west to the eastern end of the

Indonesian chain of islands, and from the Himalayas to Java in the south. *A. breviligula* is in the Philippines, and *A. binghami* in the Celebes. In the high Himalayas, *A. laboriosa*—a species even larger than *A. dorsata*—nests up to 3000 m, whereas *A. dorsata* rarely nests above 1250 m.

APIS FLOREA AND RELATED SPECIES

The area of the smaller *A. florea* extends as far northwest as Iran. It has also been reported around the Persian Gulf in Iran and Iraq, and in the Arabian peninsula. It reached parts of this last area, and also Sudan, by the aid of humans, and people may also have facilitated its spread along the coast west of the Indus Valley. It is characteristically found in hot dry areas at altitudes below 500 m; in some localities it is the only honey bee that could survive. Its eastern range does not extend as far as that of *A. dorsata,* possibly because *A. florea* could not cross wide sea channels.

A. andreniformis, rather similar to *A. florea,* occurs in southern China, Myanmar (Burma), Palawan in the Philippines, Thailand, Indonesia, Laos, and Vietnam. It is likely that some early statements about *A. florea* in these areas refer instead to *A. andreniformis.*

See Also the Following Articles

Beekeeping • Caste • Hymenoptera • Neotropical African Bees • Pollination and Pollinators

Further Reading

Crane, E. (1990). "Bees and Beekeeping: Science, Practice and World Resources." Heinemann Newnes, Oxford, U.K.

Crane, E. (1991). "The World History of Beekeeping and Honey Hunting." Duckworth, London.

Graham, J. M. (ed.) (1992). "The Hive and the Honey Bee." rev. ed. Dadant and Sons, Hamilton, IL.

Hepburn, H. R., and Radloff, S. E. (1998). "Honeybees of Africa." Springer-Verlag, Berlin.

Michener, C. D. (2000). "The Bees of the World." John Hopkins University Press, Baltimore, MD.

Otis, G. W. (1991). A review of the diversity of species within *Apis. In* "Diversity in the Genus *Apis.*" (D. R. Smith, ed.), Ch. 2. Westview Press, Boulder, CO.

Rinderer, T. E. (ed.) (1986). "Bee Genetics and Breeding." Academic Press, New York.

Ruttner, F. (1988). "Biogeography and Taxonomy of Honeybees." Springer-Verlag, Berlin.

Aposematic Coloration

Mathieu Joron

Leiden University

Insects attract collectors' attention because they are extremely diverse and often bear spectacular colors. To biologists,

FIGURE 1 *Pseudosphinx tetrio* hawk moth caterpillar from the Peruvian Amazon, showing a combination of red and black, classical colors used by aposematic insects. These larvae feed on toxic latex-sapped trees in the Apocynaceae. Length 14 cm. (Photograph © M. Joron, 1999.)

however, bright coloration has been a constantly renewed puzzle because it makes an insect a highly conspicuous prey to prospective predators. Charles Darwin understood that bright colors or exaggerated morphologies could evolve via sexual selection. However, he felt sexual selection could not account for the conspicuous color pattern of nonreproductive larvae in, for example, *Pseudosphinx* hawk moth caterpillars (Fig. 1). In a reply to Darwin about this puzzle, Alfred R. Wallace proposed that bright colors could advertise the unpalatability of the caterpillars to experienced predators. Indeed, prey that are not edible to predators are predicted to gain by exhibiting conspicuous and very recognizable colors; experienced predators can then correctly identify and subsequently avoid attacking such prey. E. B. Poulton later developed this idea, expanded it to other warning signals (i.e., sounds or smells), and coined the term "aposematism" to describe this phenomenon (from the Greek "away" and "sign").

Aposematic color patterns are found everywhere throughout the insects, from black- and yellow-striped stinging wasps to black and red, bitter-tasting lady beetles, or brightly colored, poisonous tropical butterflies. Although warning coloration has involved fascination, empirical and theoretical studies for some time, the puzzle of aposematism still motivates much debate today. First, although there is little doubt that bright coloration is often an antipredatory strategy, how aposematism evolves is far from clear. This is because brightly colored mutants in a population of cryptic (camouflaged) prey are more exposed to predators. How can a warning coloration evolve in a prey if the very first mutants exhibiting such coloration in the population are selected against? Second, the reasons for the brightness and conspicuousness of warning colors are not always clear and may be multiple. Are aposematic colors "road signs" that help predators learn better to differentiate inedible from edible prey, or are bright colors more easily memorized and associated to bad taste by predators? Did yellow and red colors, often borne by poisonous insects, evolve because of innate biases against these colors in the predators' brains, or are more complex cognitive, behavioral, frequency-dependent, or coevolutionary mechanisms involved in the evolution of warning patterns?

Finally, why are warning patterns highly diverse in the insect world, whereas all toxic prey would gain by bearing the same color, thus reducing the probability of being sampled by a naïve predator?

WHAT TO ADVERTISE

"Aposematism is quite simply the correlation between conspicuous signals, such as bright coloration, and prey unprofitability," Candy Rowe wrote in 2001. But why should some prey become unprofitable in the first place, while others do not?

Unprofitability is difficult to define, and even more difficult to measure. It is certainly contextually defined, because the propensity of an animal to eat something is highly dependent on its level of hunger and its ability to use the prey for energy once eaten. Palatability (i.e., the predator's perception of prey profitability) greatly determines whether the predator will or will not eat the prey. Predator–prey coevolution led predators in part to rely on proximal perception to gauge the prey profitability. In particular, taste sensitivity may well have evolved in predators as an assessment of food toxicity: indeed, predators usually consider toxic chemicals to be distasteful. Some insects have external defenses such as horns, or spines, many of which cause irritation. Such physical defenses may be coupled to venom, as with the irritant hairs of many caterpillars or hymenopteran stingers. These insects may be otherwise perfectly profitable, and some predators evolve ways around the physical defenses, such as bee-eaters that are able to remove a bee's stinger and venom sac. Other insects have passive chemical defenses that predators discover upon consumption, such as chemicals in the hemolymph or sequestration glands of lubber grasshoppers or monarch butterflies. Such insects usually develop extra signals such as powerful smells, at least when handled, to advertise their toxicity before being consumed.

Toxicity is not the only way an insect can be unprofitable to predators. Difficulty in capturing prey (due to fast escape, erratic flight, breakable wings, etc.), or difficulty in handling prey (due to toughness or a hard cuticle) are other ways that insects can bring no net reward to the predators that spend energy chasing them, even if the chase results in the prey being seized. However, multiple unprofitability traits might be important in the evolution of warning signals.

Predators can have three kinds of response to a prey depending on their perception of prey profitability. If consuming a prey leads the predator to be more likely to attack similar prey in the future, perhaps even using the prey's appearance as a search image, the prey is called palatable. In feeding experiments, this usually leads birds to attack nearly 100% of the palatable prey offered. Of course, the predator may satiate after consuming a number of prey, and consequently the propensity to attack may decrease at high prey densities. In contrast, if experienced predators are less likely to attack similar prey, the prey is called unpalatable. Of course, predators' memorizing capacity, and the strength of the prey unpalatability, may all influence how fast information regarding prey is acquired

and how long it is retained. However, a distasteful prey will inevitably lower the predators' instantaneous propensity to attack this prey further, an effect analogous to an immediate satiation. Finally, eating the prey may have no effect on the predator's subsequent behavior, which means that the prey is effectively neutral. This category is mainly derived from theory; there is little evidence that it exists in nature.

Variations in unpalatability among prey species, along what is called the "(un)palatability spectrum," affect the rate at which predators modify their behavior with experience. Predator's perceived toxicity is likely to be a sigmoid function of actual toxin concentration per unit prey mass, meaning that little of the palatability spectrum may fall into intermediate perceptions between "unpalatable" and "fully palatable." Although how predators learn is still under debate, experiments and theory suggest that they respond to a large extent to the (perceived) concentration of nasty chemicals they can tolerate per unit time.

The distastefulness of insects is generally linked to the host plants they utilize. Indeed, many distasteful or defended insects are herbivorous; most defended nonherbivorous insects are Hymenoptera. Some plant families, like the Solanaceae and the Passifloraceae, which are hosts to many chemically defended insects, contain alkaloids and cyanogens, respectively, as secondary metabolites. Some insects, like monarch butterflies (*Danaus plexippus*) that feed on *Asclepias* plants (milkweeds, Asclepiadaceae), sequester the compounds of such plants and store them; these insects thus avoid the toxic effects of the toxic compounds altogether. In soft-bodied insects (e.g., larvae), toxins are usually stored near the teguments or in special glands, ready to release their contents upon handling. The toxicity of insects that extract and sequester plant chemical compounds is dependent on the concentration of these compounds in the host plant. Sawfly larvae (Hymenoptera: Tenthredinidae), for example, reflex-bleed drops of hemolymph when touched; the unpalatability of such larvae is shown to be directly dependent on the glucosinolate concentration of their crucifer host plant over 24 h before "bleeding."

Other insects, however, synthesize their toxins *de novo*, like many chrysomelid beetles; they probably use the same enzymatic machinery that serves (or has served, in their ancestors) to detoxify the plant's secondary compounds. Although many of these species still use precursors derived from their food plant, these insects are usually less dependent on the plant's toxicity to develop their own noxious compounds. Some groups like ithomiine or heliconiine butterflies also get toxin precursors in their adult diet.

Whatever route to distastefulness is taken, we observe a general correlation between clades of distasteful insects and toxicity in host plant families. In butterflies, the distasteful Troidinae (Papilionidae) tend to feed on Aristolochiaceae, monarchs (Nymphalidae: Danainae) usually feed on milkweeds (Asclepiadaceae), longwing butterflies (Nymphalidae: Heliconiinae) feed on Passifloraceae, and clearwings (Nymphalidae: Ithomiinae) mainly on Solanaceae and Apocynaceae. In contrast, butterfly clades feeding on chemical-free monocotyledonous plants, like browns (Nymphalidae: Satyrinae) on grasses, or owl-butterflies (Nymphalidae: Brassolinae) on palms or Marantaceae, did not evolve distastefulness. Thus toxicity in insects may frequently have evolved as a mere by-product of adaptation to utilize new kinds of food, particularly toxic plants. The costs of detoxification or toxin production could be covered by the benefits of invading competition-free hosts, perhaps assisted by the increased survival afforded by chemical protection.

DISGUSTING, BRIGHT, SIMPLE, AND CONTRASTED: WHY AND HOW TO ADVERTISE

Why should unprofitable prey advertise? Instead of parading with gaudy colors, why should all prey not try to escape predators' detection altogether through camouflage? Although the initial steps to aposematism are not obvious, the advantage of aposematic signals once established is clear. Indeed, numerous studies have shown that most predators are able to learn and recognize, and subsequently avoid, prey they associate with a bad experience.

Some distasteful prey, such as the transparent ithomiine butterflies found in the forest understory in tropical America, are not particularly conspicuous. The rampant mimicry found in this group of inconspicuous butterflies demonstrates that predators are able to learn and avoid such prey (although other stimuli, e.g., motion, might also be important). Still, most distasteful insects are brightly colored. Why should aposematic signals usually be conspicuous, and use simple color patterns of red, yellow, or black? Most of the answer is likely to be found in the cognitive behavior of the predators that selected for such colors. Several hypotheses have been put forward to explain the correlation between bright colors and unprofitability in insects. Bright contrasted colors are thought to be (1) easier to learn, (2) more difficult to forget, and (3) as different as possible from edible prey, thereby facilitating the avoidance of recognition errors. All these mechanisms are supported by experimental data to some extent (e.g., Fig. 2). Because both predators and aposematic prey benefit from correct identification, aposematic colors are believed to take advantage of any bias in the predator's cognition system. Likewise, predators in turn gain in being biased in the same direction as that taken by the prey. Therefore, prey signaling and predators' cognition are likely to have coevolved, which, incidentally, makes experimental evidence for any of the foregoing hypotheses generally hard to establish independently. Hypothesis 3 is the most likely to involve interactions between perception and cognition in the predators, leading to fast evolution of the prey's signals.

Many aposematic insects simultaneously send signals of different kinds, and some argue that such "multimodal" warning signals may reveal unconditioned biases that are absent when each sensory modality is examined alone. Assuming that predators would rely solely on color and not behavior, motion,

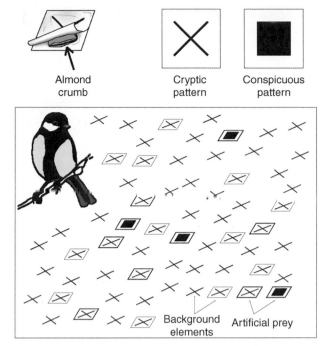

FIGURE 2 Schematic outline of the "novel world" experimental design developed by R. Alatalo and J. Mappes (University of Jyväskylä, Finland) to study the mechanisms of predator's selection on conspicuous prey. Great tits, *Parus major,* are trained to forage in a room covered with small black symbols (e.g., crosses) on a white background, some of which are actual prey. Almond crumbs are placed between two 1-cm^2 pieces of paper glued together that bear a symbol on the outside. Black-squared prey items stand out conspicuously on the black-crossed background and represent potential warning signals, whereas black-crossed prey items are cryptic. The novelty of all symbols ensures that innate or previously learned prey recognition does not interfere with the predator's response during the experiment. This setup also partly resolves one drawback of garden experiments, where the local food abundance for predators is artificially increased, making the searching costs, search images, and other predatory behavior unrealistic. By playing on the palatability of the prey items, it is possible to monitor how the birds learn to avoid the conspicuous signal. Mimics can also be incorporated in the environment at varying frequencies to study the dynamics of Batesian mimicry.

or sounds is perhaps simplistic, and it is sometimes argued that multiple signals could even be a prerequisite for the evolution of warning coloration. In fact, the reason for the apparent importance of multimodality probably lies again in the coevolutionary history of predators and their prey, which shapes innate biases. Predators are generally good entomologists for the potential prey they encounter often, and predators integrate various sensory modalities to make decisions regarding a particular action. Most aposematic insects are mimicked by edible species (Batesian mimics) that parasitize the warning function of the signal. The presence of these Batesian mimics reduces the reliability of the warning signal and means lost prey for the predators. Model species may therefore escape being mimicked by evolving new dimensions for signaling, in addition to the established one (i.e., in different sensory dimensions).

Many warningly colored insects live in groups, which enhances the warning function of their signal for three reasons (Fig. 3). First, predators tend to associate and retain noxiousness

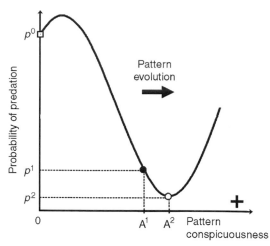

FIGURE 4 Peak shift mechanism applied to prey conspicuousness. The curve describes a fitness function that relates pattern conspicuousness to the probability of suffering predation. Predators are assumed to have knowledge of prey ranging from crypsis (open square: zero conspicuousness, tastefulness, high probability of predation) to aposematism (solid circle: high conspicuousness, A^1; distastefulness, low probability of predation, p^1). From this knowledge, predators extrapolate the palatability of prey with higher conspicuousness they have not yet experienced, hence the curve beyond the solid circle. In particular, slightly more conspicuous prey (open circle) elicit a stronger rejection response than the known aposematic prey (solid circle), and therefore suffer an even lower probability of predation, p^2. Consequently, selection should cause the conspicuousness of the distasteful prey to shift gradually from A^1 to A^2. In contrast, cryptic distasteful prey cannot drift away from crypsis (open square) toward higher levels of conspicuousness because they are more likely to be preyed upon when they become more conspicuous. An initial phenotypic jump is necessary to get to levels of conspicuousness where birds can categorize the prey as warningly colored, avoiding confusion with the normal staple of cryptic tasteful prey.

FIGURE 3 Examples of gregarious warningly colored insects. (A) Gregarious *Chromacris* lubber grasshoppers nymphs (Orthoptera: Romaleidae) feeding on a toxic *Solanum mite*. Although it is not a bright color, black is often used as a warning color by insects, presumably because it increases conspicuousness by contrast against green foliage. (B) Gregarious *Morpho* sp. caterpillars spending the day in a dense cluster. They disperse at night for feeding. Clusters of aposematic prey often create an emergent, enhanced pattern presumably perceived as a supernormal stimulus by the predators and therefore better memorized. Caterpillars about 10 cm long. (Photographs from the Peruvian Amazon, © M. Joron, 1997.)

and a warning signal more quickly when presented with multiple copies of the same signal. Second, all unpalatable prey get an individual advantage in living in groups by the immediate avoidance, similar to the immediate satiation discussed earlier, caused in the individual predator that avoids the group altogether. Third, at the population level, clustering spatially also reduces the number of individual predators the population of prey is exposed to and has to educate, which again enhances the benefit of signaling. In short, it pays to be sitting right next to the toxic individual that is being sampled by a predator, because this is where the probability of predation is lowest, hence the advantage to living in groups. However, many solitary prey also exhibit warning coloration, and gregariousness does not necessarily evolve prior to aposematism.

One common mechanism leading to fast divergence in signals throughout the animals, known as peak shift, hinges here on the coupling of prey coloration, predator experience, and predator innate aversion. Prey can be placed on a conspicuousness axis (Fig. 4), with cryptic edible prey at one end (close to 0 conspicuousness) and incipient aposematic prey or new color pattern mutants at some distance down the axis. More conspicuous prey usually elicit stronger aversion in educated predators, which extrapolate the idea that stronger conspicuousness should mean stronger noxiousness. Therefore, stronger signals (away from edible prey appearance) lead to a supernormal response in the predators that thereby select for increased conspicuousness in the prey. This mechanism is a special case of a runaway process and could be an important route to the evolution of aposematic prey that are bright and contrasted. It is thought that the coupling of such cognitive biases with the ability to learn leads to the selection of more strongly exaggerated warning colors and patterns in noxious prey than is expected in purely nonlearning predators.

EVOLUTIONARY ROUTES TO APOSEMATIC COLORATION

The Problems

There are obvious benefits to bearing warning colors in a population of warning-colored prey. As noted by early naturalists like A. R. Wallace and later E. B. Poulton, experienced predators avoid warningly colored prey, and presumably the number of prey killed during the predators' education is lower than in the absence of signaling. These benefits are clear at the group level but are not so clear at the individual level, because the first warningly colored individuals in a population of cryptic (and noxious) prey suffer strongly increased predation. Indeed, novel warningly colored prey not only suffer increased detection by prospective predators, but also elicit *no* avoidance in the predators. Consequently, there is strong positive frequency dependence, putting novel rare warning signals at a disproportionate disadvantage against an established strategy (crypsis, or another already established warning signal).

How could warning signals evolve at all if the first mutants using this strategy are killed? Laboratory experiments using the "novel world" design (Fig. 2) show rather unequivocally that aposematic patterns cannot evolve gradually in unpalatable prey. Indeed, small increases in visibility in cryptic prey increased attack rates without enhancing learning. Similarly, deviant phenotypes in established warning patterns suffered stronger predation. Finally, rare conspicuous prey suffered disproportionate predation, even when presented in groups. Therefore, a gradual increase in conspicuousness towards aposematism seems unlikely. This means that the evolving population must undergo a sudden jump, both in phenotype (to get a pattern that predators categorize as a different item) and in numbers beyond a threshold frequency (to allow the local predators to learn about the new pattern). Once the new pattern has achieved the minimum frequency and phenotypic thresholds, positive frequency dependence helps the new mutant to spread in the population. Peak shift or other processes can then occur, increasing the conspicuousness or adding other components to the signal. How can these evolutionary leaps be achieved—or circumvented—by an incipient aposematic prey?

Deterministic Evolution via Immediate Benefits

NEOPHOBIA   A new aposematic form could in theory escape the disadvantage of being rare and novel by causing immediate avoidance without having to be tasted at all by the predator. Indeed, predators are somewhat reluctant to sample novel-looking prey, particularly if novelty is associated with bright colors. This phenomenon is called "neophobia," a kind of diet conservatism in predators. Neophobia could arise from various foraging biases, such as the formation of search images in the brains of predators as they search for edible-looking prey and ignore other prey, or via cultural inheritance, as with nestlings that tend to prey upon what they were fed by their parents. Neophobia is sometimes presented as a potential route toward aposematism. However, it does not really resolve the frequency dependence problem, because it is essentially a transient phenomenon involving no information acquisition by predators. Therefore as soon as numbers grow, however slightly, neophobia tends to vanish. Neophobia should best be classified as a predator's bias, like other innate biases against colors, smells, or sound, evolved by predators in response to their prey environment. Such biases are likely to channel the ultimate form taken by the aposematic signal (to the benefit of both preys and predators), but it is unlikely that they cause its evolution in the first place.

INDIVIDUAL ADVANTAGE   One obvious way around initial obstacles is not to be killed by predators' attacks. Then, prey could both educate the predators and be avoided in subsequent encounters. Indeed, most birds taste-test their prey before ingesting them, and many aposematic prey have noxious compounds in their outer parts, making it possible to be tasted but not injured by predators. For instance, ithomiine and danaine butterflies concentrate alkaloid in their wings. Day-flying pericopine moths let a voluminous and bitter hemolymph froth out of their body, likely tasted (or smelled) by a predator before it has profoundly injured the moth. Moreover, most unpalatable butterflies have very elastic bodies, which resist crushing. Strong smells that predators take as a warning for bad taste or toxicity, like those of stinkbugs, are another way by which prey can gain immediate advantage without having to be effectively tasted by the predators.

PREY ALREADY CONSPICUOUS   Another way by which prey can overcome the difficulty of evolving conspicuous color is not to suffer any cost (i.e., avoid the necessity of a phenotypic leap) as a result of increased conspicuousness. Indeed, most flying insects are rather conspicuous in flight and rely on their difficulty of capture to escape predation. They may not suffer any cost to bearing conspicuous colors, and indeed many butterflies, if not most, irrespective of their palatability, display bright patches of colors on the upper side of their wings, visible in flight, while having cryptic underwings making them inconspicuous when sitting. Such bright dorsal colors might initially evolve as sexual signals in male–male or male–female interactions long before unpalatability evolves. Once noxiousness has evolved, predators can learn an already conspicuous pattern without making recognition errors because of the resemblance to the palatable prey they have as search images. In a way, conspicuous flying insects can be said to be "preadapted" to evolve warning colors. But such patterns can then also change or drift according to predators' biases. In particular, already bright color patterns can be enhanced toward brighter coloration through processes like peak shift, as described earlier. According to James Mallet, examples of this mechanism are the unpalatable *Taenaris* and *Hyantis* (Nymphalidae: Morphinae), which have evolved strikingly conspicuous warning spots via the enhancement of some of the less conspicuous eyespots that are still found on the undersides of their palatable relatives, the well-known blue *Morpho* butterflies.

MÜLLERIAN MIMICRY   The easiest way to avoid the cost of rarity and conspicuousness altogether is to jump to an aposematic pattern already present in the habitat and known by the local predators. The shared appearance between several defended prey is called Müllerian mimicry, and it is likely that most aposematic species evolved via this route. Indeed, mimicry rings usually include a large number of Müllerian species (all of which are noxious). Of these, only one evolved the pattern first, followed by the other species that colonized an already protected pattern. This pattern of evolution is detectable by examining the biogeography and phylogeny of the species in question. For example, *Heliconius erato* and *H. melpomene* are Müllerian mimics throughout their distribution range. However, the *H. melpomene* was shown to have much younger color pattern races, with a clearly distinct genealogy, than *H. erato*, suggesting that *H. melpomene* is a Müllerian mimic that adopted the established color patterns of *H. erato*.

Population Processes: Kin Selection, Drift, and Shifting Balance

Because many unpalatable prey are indeed gregarious, it is easy to conclude that gregariousness allows the evolution of aposematism. The evolution of aposematism through gregariousness relies on the predator rejecting the whole group after sampling only one or few individuals. This extrapolation from one prey to the whole group is analogous to a superfast learning in the predator, which can be enhanced by conspicuousness. However, it also pays for aposematic preys to live in groups, thereby increasing their apparent density to the local predators. It is therefore not clear whether gregariousness or aposematism should evolve first to trigger the evolution of the other. Groups of gregarious larvae (Fig. 3) are usually family groups, suggesting that kinship might allow a new mutation quickly to get to a locally high frequency in such little-dispersing insects through kin selection. However, one should be aware that relatedness per se is not what favors the local rise in frequency of the gene here, but simply the local founding event by one or few family groups.

In fact, many adult aposematic adult insects are either not gregarious at all or do not aggregate in family groups. Besides, some of the most gregarious insect larvae come from the joint oviposition of several unrelated females. Although these examples could have arisen after the initial evolution of warning color through kin selection, it is more parsimonious to infer that non-kin-selection arguments can also explain the evolution

of aposematism. Drift alone, particularly, followed by positive frequency dependence, is a good candidate mechanism (and in fact kin founding is only a special case of genetic drift). Indeed, when the ratio of predators to prey decreases in a locality, selection for antipredatory strategies is greatly diminished, allowing the exploration of other color pattern possibilities by the local population. Using release–recapture techniques of different warningly colored forms of *H. cydno* in Ecuador, D. Kapan showed that selection was relaxed when the butterflies were released in larger numbers. Therefore, the prey population could move via genetic drift above the required threshold, after which the new warning color invades the population. Positive frequency dependence has the interesting property that although it hinders the initial evolution of new patterns, it hinders the removal of any pattern once it has been established. If genetic drift in prey populations matches the fluctuations of selection pressures in time and space, new local aposematic patterns can be established frequently in different locations. These are essentially the first and second steps of the shifting balance theory of S. Wright. Competition between geographically adjacent warning color types then allows one pattern to spread to neighboring populations, like the traveling waves of color races documented in South America for *H. erato* or *H. melpomene*.

CONCLUSIONS

Despite the advantages of bearing a warning coloration established in the locality, the evolution of aposematism is not straightforward because proximal mechanisms seem to represent obstacles to its initial evolution. However, aposematic patterns are extremely diverse at all geographical and taxonomic levels, and this major discrepancy between theory and nature clearly suggests that positive frequency-dependent arguments are not as restrictive against the rise of novel warning colors. Similarly, predator generalization, which should not allow gradual shift of cryptic prey toward bright warning colors, does not seem to be efficient in restricting the rise of new conspicuous patterns.

In fact, both population dynamics and psychological arguments might well explain such spectacular diversification. First, positive frequency dependence would allow new local forms to be established through drift, relayed by other processes involving predator's cognitive biases. Second, the initial steps toward warning color are determined largely by which cognitive biases in the predators are exploited. That is, the initial pathway taken toward the evolution of warning coloration probably profoundly affects the aposematic phenotype that eventually evolves. Similarly, positive frequency dependence prevents deviations from the evolutionary pathway that is taken. In short, although aposematism is not expected predictably to evolve via Fisherian selection, it is such a powerful strategy once evolved that it is possibly inevitable in a contingent and varying world, where the nature and the height of the initial obstacles to its evolution fluctuate. It may thus follow a ratchetlike pattern of evolution, where more routes may lead toward aposematism than routes away from it.

See Also the Following Articles
Chemical Defense • Crypsis • Mimicry • Monarchs

Further Reading
Alatalo, R. V., and Mappes, J. (1996). Tracking the evolution of warning signals. *Nature* **382**, 708–710.
Edmunds, M. (1974). "Defence in Animals. A Survey of Anti-predator Defences." Longman, New York.
Endler, J. A. (1988). Frequency-dependent predation, crypsis, and aposematic coloration. *Philos. Trans. R. Soc. Lond. B* **319**, 505–524.
Guilford, T. (1988). The evolution of conspicuous coloration. *Am. Nat.* **131**, S7–S21.
Lindström, L., Alatalo, R. V., Lyytinen, A., and Mappes, J. (2001). Strong antiapostatic selection against novel rare aposematic prey. *Proc. Nat. Acad. Sci. U.S.A.* **98**, 9181–9184.
Mallet, J., and Joron, M. (1999). Evolution of diversity in warning color and mimicry: Polymorphisms, shifting balance and speciation. *Annu. Rev. Ecolo. System.* **30**, 201–233.
Mallet, J., and Singer, M. C. (1987). Individual selection, kin selection, and the shifting balance in the evolution of warning colors: The evidence from butterflies. *Biol. J. Linn. Soc.* **32**, 337–350.
Poulton, E. B. (1890). "The Colours of Animals." Trübner, London.
Rowe, C. (ed.). (2001). Warning signals and mimicry. Special issue of *Evolutionary Ecology* [1999, vol. 13, no 7/8]. Kluwer, Dordrecht, The Netherlands.
Sillén-Tullberg, B. (1988). Evolution of gregariousness in aposematic butterfly larvae: A phylogenetic analysis. *Evolution* **42**, 293–305.
Sword, G. A., Simpson, S. J., El Hadi, O. T. M., and Wilps, H. (2000). Density dependent aposematism in the desert locust. *Proc. R. Soc. Lond. B Biol. Sci.* **267**, 63–68.
Wallace, A. R. (1879). The protective colours of animals. *In* "Science for All" (R. Brown, ed.), pp. 128–137. Cassell, Petter, Galpin., London.

Apterygota

A pterygota is a subclass of the class Insecta in the phylum Arthropoda. It contains two orders, the Archaeognatha and the Thysanura.

Aquatic Habitats

Richard W. Merritt
Michigan State University

J. Bruce Wallace
University of Georgia

L ess than 3% of the world's total water occurs on land, and most of this is frozen in polar ice caps. Streams and rivers

are one of the more conspicuous features of the landscape; however, their total area is about 0.1% of the land surface, whereas lakes represent about 1.8% of total land surface. Some authors have questioned whether insects have been successful in water because aquatic species represent only a small portion of the total hexapod fauna. However, 13 orders of insects contain species with aquatic or semiaquatic stages, and in five of these (Ephemeroptera, Odonata, Plecoptera, Megaloptera, and Trichoptera) all species are aquatic with few exceptions (Table I). Few aquatic insects spend all of their life in water; generally any insect that lives in water during a portion of its development is considered to be "aquatic." Usually, but not always, for most "aquatic" species, it is the larval stage that develops in aquatic habitats, and the adults are terrestrial (Table I). The pupae of some taxa undergoing complete metamorphosis (i.e., holometabolous) remain within the aquatic habitat; in others the last larval instar moves onto land to pupate, providing the transition stage from the aquatic larva to the terrestrial adult.

The success of insects in freshwater environments is demonstrated by their diversity and abundance, broad distribution, and their ability to exploit most types of aquatic habitat. Some species have adapted to very restricted habitats and often have life cycles, morphological, and physiological adaptations that allow them to cope with the challenges presented by aquatic habitats. One aquatic environment in which insects have not been very successful is saltwater habitats, although some 14 orders and 1400 species of insects occur in brackish and marine habitats; only one group occurs in the open ocean. One of the most widely accepted attempts to explain why more insects do not live in marine environments is that successful resident marine invertebrates evolved long before aquatic insects and occupy many of the same niches inhabited by freshwater insects. Thus, marine invertebrates, such as crustaceans, have barred many insects from

TABLE I Occurrence of Life Stages in Major Habitat Types for Aquatic and Semiaquatic Representatives of Insect Orders (A, adult; L, larvae; P, pupae)

Order	Terrestrial	Freshwater
Collembola	A, L	A, L
Ephemeroptera	A	L
Odonata	A	L
Heteroptera	A	A, L
Orthoptera	A, L	A, L
Plecoptera	A	L
Coleoptera	A, L, P	A, L
Diptera	A, P	L, P
Hymenoptera	A	A, L, P
Lepidoptera	A	L, P
Megaloptera	A, P	L
Neuroptera	A, P	L
Trichoptera	A	L, P

Modified from Ward, J. V. (1992). "Aquatic Insect Ecology," Vol. 1, "Biology and Habitat." Wiley, New York.

marine habitats by competitive exclusion. Problems with osmoregulation have been given as another reason for the paucity of saltwater species; however, one of the two multicellular animals found in the Great Salt Lake is a member of the order Diptera (see later: Unusual Habitats), providing evidence that some insects display a strong ability to osmoregulate.

The first aquatic insects are believed to have inhabited flowing water as early as the Permian and Triassic. It was not until the late Triassic and early Jurassic that evidence of abundant lentic, or still-water, fauna arose, accompanied by rapid diversification of water beetles, aquatic bugs (Heteroptera), and primitive Diptera. On the basis of several lines of evidence including osmoregulation, fossil evidence, secondary invasions to water of many taxa, and great variation in gill structure among and within orders, some authors have suggested that the first insects may have lived in water rather than in terrestrial habitats. However, the general consensus is that an aquatic origin for insects seems unlikely and that aquatic insects may not have shown up until 60 to 70 million years later than their terrestrial counterparts.

Freshwater systems are often divided into standing (lentic) and flowing (lotic) waters. Although such a division is useful for indicating physical and biological differences, habitat diversity can vary tremendously within these two broad categories, and some of the same taxa may be found in both lentic and lotic habitats, depending on the physiological constraints of a given habit. Many factors influence successful colonization of aquatic insects to a given habitat; however, most of these would fall under four broad categories: (1) physiological constraints (e.g., oxygen demands, respiration, osmoregulation, temperature effects), (2) trophic considerations (e.g., food acquisition), (3) physical constraints (e.g., coping with harsh habitats), and (4) biotic interactions (e.g., predation, competition). However, these categories are so interrelated that detailed analysis of each factor separately is very difficult.

HABITAT, HABIT, AND TROPHIC CLASSIFICATION SYSTEM

The classification system used here for lotic and lentic habitats stresses the basic distinction between flowing water (i.e., streams, rivers) and standing water (i.e., ponds, lakes, swamps, marshes) habitats (Table II). This separation is generally useful in describing the specific microhabitats (e.g., sediments, vascular hydrophytes, detritus) in which aquatic insects may be found. Both stream/river currents and lake shoreline waves often create erosional (riffle-type) habitats and may resemble each other in their physical characteristics, whereas river floodplain pools and stream/river backwaters create depositional (pool-type) habitats that may resemble lake habitats as well (Table II). Within a given habitat, the modes by which individuals maintain their location (e.g., clingers on surfaces in fast-flowing water, sprawlers on sand

or on surfaces of floating leaves, climbers on stem-type surfaces, burrowers in soft sediments) or move about (e.g., swimmers, divers, surface skaters) have been categorized (Table III). The distribution pattern resulting from habitat selection by a given aquatic insect species reflects the optimal overlap between habit and physical environmental conditions that comprise the habitat, such as bottom type, flow, and turbulence. Because food in aquatic habitats is almost always distributed in a patchy fashion, the match between habitat and habit is maximized in certain locations. This combination will often result in the maximum occurrence of a particular species.

In view of the complex physical environment of streams, it is not surprising that benthic invertebrates have evolved a diverse array of morphological adaptations and behavioral mechanisms for exploiting foods. Throughout this article we will follow the functional classification system originally described by K. W. Cummins in 1973, which is based on the mechanisms used by invertebrates to acquire foods (Table IV). These functional groups are as follows:

- **Shredders,** which are insects and other animals that feed directly on large pieces of organic matter (e.g., decomposing leaves and fragments of wood >1 mm in size) and their associated fungi and bacteria, and convert them into fine particulate organic matter (FPOM) through maceration, defecation, and physical degradation;
- **Collector-filterers,** which have specialized anatomical structures (e.g., setae, mouth brushes, fans, etc.) or silk and silklike secretions that act as sieves to remove fine particulate matter less than 1 mm in diameter from the water column;
- **Collector-gatherers,** which gather food, primarily FPOM, that is deposited within streams or lakes;
- **Scrapers,** which have mouthparts adapted to graze or scrape materials (e.g., periphyton, or attached algae, and the associated microbes) from rock surfaces and organic substrates;
- **Predators,** which feed primarily on other animals by either engulfing their prey or piercing prey and sucking body contents.

These functional feeding groups refer primarily to modes of feeding or the means by which the food is acquired, and the food type per se (Table IV). For example, shredders may select leaves that have been colonized by fungi and bacteria; however, they also ingest attached algal cells, protozoans, and various other components of the fauna along with the leaves.

LOTIC HABITATS

Streams vary greatly in gradient, current velocity, width, depth, flow, sinuosity, cross-sectional area, and substrate type, depending on their position in the landscape with respect to geology, climate, and the basin area they drain. Anyone who has spent much time around or wading in streams is aware that these can be extremely diverse habitats, often manifesting great

TABLE II Aquatic Habitat Classification System

General category	Specific category	Description
Lotic–erosional (running-water riffles)	Sediments	Coarse sediments (cobbles, pebbles, gravel) typical of stream riffles.
	Vascular hydrophytes	Vascular plants growing on (e.g., moss, *Fontipalis*) or among (e.g., pondweed, *Potamogeton pectinatus*) coarse sediments in riffles.
	Detritus	Leaf packs (accumulations of leaf litter and other coarse particulate detritus at leading edge or behind obstructions such as logs or large cobbles and boulders) and debris (e.g., logs, branches) in riffles.
Lotic–depositional (running-water pools and margins)	Sediments	Fine sediments (sand and silt) typical of stream pools and margins.
	Vascular hydrophytes	Vascular plants growing in fine sediments (e.g., *Elodea,* broad-leaved species of *Potamogeton, Ranunculus*).
	Detritus	Leaf litter and other particulate detritus in pools and alcoves (backwaters).
Lentic–limnetic (standing water)	Open water	On the surface or in the water column of lakes, bogs, ponds.
Lentic–littoral (standing water, shallow-water area)	Erosional	Wave-swept shore area of coarse (cobbles, pebbles, gravel) sediments.
	Vascular hydrophytes	Rooted or floating (e.g., duckweed, *Lemna*) aquatic vascular plants (usually with associated macroscopic filamentous algae).
	Emergent zone	Plants of the immediate shore area (e.g., *Typha*, cattail), with most of the leaves above water.
	Floating zone	Rooted plants with large floating leaves (e.g., *Nymphaea,* pond lily), and nonrooted plants (e.g., *Lemna*).
	Submerged zone	Rooted plants with most leaves beneath the surface.
	Sediments	Fine sediments (sand and silt) of the vascular plant beds.
	Sediments	Fine sediments (fine sand, silt, and clay) mixed with organic matter of the deeper basins of lakes. (This is the only category of "lentic–profundal.")
Lentic–profundal (standing water, basin)	Freshwater lakes	Moist sand beach areas of large lakes.
Beach zone	Marine intertidal	Rocks, sand, and mud flats of the intertidal zone.

After Merritt, R. W., and Cummins, K. W. (1996). "An Introduction to the Aquatic Insects of North America." Kendall/Hunt, Dubuque, IA.

TABLE III Categorization of Aquatic Insect Habits: That Is, Mode of Existence

Category	Description
Skaters	Adapted for "skating" on the surface where they feed as scavengers on organisms trapped in the surface film (e.g., Heteroptera: Gerridae, water striders).
Planktonic	Inhabiting the open-water limnetic zone of standing waters (lentic; lakes, bogs, ponds). Representatives may float and swim about in the open water but usually exhibit a diurnal vertical migration pattern (e.g., Diptera: Chaoboridae, phantom midges) or float at the surface to obtain oxygen and food, diving when alarmed (e.g., Diptera: Culicidae, mosquitoes).
Divers	Adapted for swimming by "rowing" with the hind legs in lentic habitats and lotic pools. Representatives come to the surface to obtain oxygen, dive and swim when feeding or alarmed; may cling to or crawl on submerged objects such as vascular plants (e.g., Heteroptera: Corixidae, water boatman; Coleoptera: adult Dytiscidae, predaceous diving beetles).
Swimmers	Adapted for "fishlike" swimming in lotic or lentic habitats. Individuals usually cling to submerged objects, such as rocks (lotic riffles) or vascular plants (lentic) between short bursts of swimming (e.g., Ephemeroptera: Siphlonuridae, Leptophlebiidae).
Clingers	Representatives have behavioral (e.g., fixed retreat construction) and morphological (e.g., long, curved tarsal claws, dorsoventral flattening, ventral gills arranged as a sucker) adaptations for attachment to surfaces in stream riffles and wave-swept rocky littoral zones of lakes (e.g., Ephemeroptera: Heptageniidae; Trichoptera: Hydropsychidae; Diptera: Blephariceridae).
Sprawlers	Inhabiting the surface of floating leaves of vascular hydrophytes or fine sediments, usually with modifications for staying on top of the substrate and maintaining the respiratory surfaces free of silt (e.g., Ephemeroptera: Caenidae; Odonata: Libellulidae).
Climbers	Adapted for living on vascular hydrophytes or detrital debris (e.g., overhanging branches, roots and vegetation along streams, submerged brush in lakes) with modifications for moving vertically on stem-type surfaces (e.g., Odonata: Aeshnidae).
Burrowers	Inhabiting the fine sediments of streams (pools) and lakes. Some construct discrete burrows that may have sand grain tubes extending above the surface of the substrate or the individuals may ingest their way through the sediments (e.g., Ephemeroptera: Ephemeridae, burrowing mayflies; Diptera: most Chironominae, Chironomini, bloodworm midges). Some burrow (tunnel) into plants stems, leaves, or roots (miners).

After Merritt, R. W., and Cummins, K. W. (1996). "An Introduction to the Aquatic Insects of North America." Kendall/Hunt, Dubuque, IA.

TABLE IV   General Classification Systems for Aquatic Insect Trophic Relations

Functional group[a]	Subdivision of function group		Examples of taxa	General particle size range of food (μm)
	Dominant food	**Feeding mechanism**		
Shredders	Living vascular hydrophyte plant tissue	Herbivores—chewers and miners of live macrophytes	Trichoptera: Phyrganeidae, Leptoceridae	> 10³
	Decomposing vascular plant tissue and wood—coarse particular organic matter (CPOM)		Trichoptera: Limnephilidae	
			Plecoptera: Pteronarcyidae, Nemouridae Diptera: Tipulidae, Chironomidae	
Collectors		Detritivores—filterers or suspension feeders	Trichoptera: Hydropsychidae	< 10³
	Decomposing fine particular organic matter (FPOM)		Diptera: Simuliidae	
		Detritivores—gatherers or deposit (sediment) feeders (includes surface film feeders)	Ephemeroptera: Ephemeridae, Baetidae, Ephemerellidae Diptera: Chironomidae	
Scrapers	Periphyton—attached algae and associated material	Herbivores—grazing scrapers or mineral and organic surfaces	Trichoptera: Glossomatidae	< 10³
			Coleoptera: Psephenidae Ephemeroptera: Heptageniidae	
Predators (engulfers)	Living animal tissue	Carnivores—attack prey, pierce tissues and cells, suck fluids	Heteroptera: Belostomatidae, Odonata,	> 10³
	Living animal tissue	Carnivores—ingest whole animals (or parts)	Plecoptera: Perlidae, Perlodidae Coleoptera: Dytiscidae, Megaloptera Trichoptera: Rhyacophilidae	> 10³

After Merritt, R. W., and Cummins, K. W. (1996). "An Introduction to the Aquatic Insects of North America." Kendall/Hunt, Dubuque, IA.
[a]General category based on feeding mechanism.

differences over short distances. In the upper reaches of a catchment or drainage basin, small streams often display a range of habitats characterized by areas that are shallow, with fast flow over pebbles, cobbles, and boulders. There are also areas with steep gradients, cascades, or waterfalls when the underlying substrate is bedrock. There also may be areas of slow velocity in pools of deeper water.

In many streams draining forested watersheds, pools are found. Pools are depositional areas during normal flow as organic and inorganic particles settle to the substrate, and a similar settling process often occurs in side channels or backwater areas of streams. Pools are also created upstream of large instream pieces of wood, which may form obstructions known as debris dams. Because pools are generally characterized by reduced water velocity, many of the small particles normally suspended in fast flows settle to the bottom. In many low-gradient streams, including large rivers, bottom substrate often consists of silt, sand, and gravel-sized particles that are frequently moved by the force of the flowing water. In such systems, large pieces of woody debris entering the river from bank erosion or from adjacent floodplain or upstream areas may represent an important habitat for invertebrate colonization.

Substratum characteristics are often perceived as a major contributor to the distribution of many invertebrates; however, many other factors, including water velocity, food, feeding habits, refuge, and respiratory requirements, can be associated with specific substrates. Substratum particle size is influenced by several items, including geology, physical characteristics of the rock, past and present geomorphic processes (flowing water, glaciation, slope, etc.), climate and precipitation, and length of time over which the processes occur. These in turn influence landform, which exerts a major influence on various hydrological characteristics of aquatic habitats. Unlike many lentic environments, in lotic systems the velocity of moving water is sufficient to pass the water around the body of an insect and turbulence provides reaeration; thus, dissolved oxygen is rarely limiting to stream inhabitants. Local transport and storage of inorganic and organic materials by the current may be either detrimental (e.g., scouring action) or beneficial (as a food source). For example, most aquatic insects in flowing waters are passive filter feeders and depend on the water current for delivery of their food. Scouring flows may therefore remove from the streambed large organic particles (e.g., leaves) as well as smaller ones, creating temporary reductions in food supplies. In contrast, moderately rapid flows may facilitate feeding of some scraper or grazer insects by preventing excessive sedimentation buildup on the surfaces on which they feed.

Some Insects and Their Adaptations to Erosional Habitats

Adaptations of aquatic insects to torrential or "rapid flow" habitats include the dorsoventral flattening of the body, which serves two purposes: it increases the organism's area of contact with the surface substratum, and it offers a mechanism by which animals can remain in the boundary layer when water velocity diminishes, thereby reducing drag under subsequent exposure to high velocities. However, this second idea may be an oversimplification. Indeed, some authors have suggested that the dorsoventrally flattened shape may actually generate lift in the insect. Examples of animals inhabiting stones in torrential habitats include a number of mayflies (Ephemeroptera) belonging to the families Heptageniidae (Fig. 1A) and Ephemerellidae; some Plecoptera, such as Perlidae (Fig. 1G); some Megaloptera (i.e., Corydalidae) (Fig. 1D); and caddisflies (Trichoptera), such as Leptoceridae (*Ceraclea*).

In addition to body shape, many mayflies and stoneflies have legs that project laterally from the body, thereby reducing drag and simultaneously increasing friction with the substrate. Most of these taxa are either scrapers or gatherers on surfaces of stones or predators on other aquatic insects. Undoubtedly, the diverse physical forces encountered in aquatic environments, especially streams, influence the array of morphologies found among aquatic insects.

In some caddisflies (e.g., Glossosomatidae), the shape of the case rather than the insect is modified. The larvae of Glossosomatidae in their tortoiselike cases are frequently seen grazing on the upper surfaces of stones in riffle areas. Another

FIGURE 1 Typical insects inhabiting lotic environments. (A) Ephemeroptera: Heptageniidae (*Rhithrogena*). (Photograph by H. V. Daly.) (B) Diptera: Simuliidae (*Simulium*), (C) Trichoptera: Limnephilidae (*Dicosmoecus*), (D) Megaloptera: Corydalidae (*Corydalus*), (E) Diptera: Tipulidae (*Tipula*), (F) Plecoptera: Pteronarcyidae (*Pteronarcys*), (G) Plecoptera: Perlidae, (H) Coleoptera: Psephenidae (*Psephenus*).

curious caddisfly grazer on stone surfaces is *Helicopsyche,* whose larvae construct coiled cases of sand grains shaped like snail shells. Both glossosomatids and helicopsychids reach their greatest abundances in sunny cobble riffles, where they feed on attached periphyton or algae. Another lotic insect that relies on a rather streamlined case is the limnephilid caddisfly *Dicosmoecus* (Fig. 1C).

Larvae of the dipteran family Blephariceridae are unusual in that they possess hydraulic suckers. A V-shaped notch at the anterior edge of each of the six ventral suckers works as a valve out of which water is forced when the sucker is pressed to the substrate. The sucker operates as a piston with the aid of specialized muscles. In addition, a series of small hooks and glands that secrete a sticky substance aid sucker attachment as the larvae move in a zigzag fashion, releasing the anterior three suckers, lifting the front portion of the body to a new position, and then reattaching the anterior suckers before releasing and moving the posterior ones to a new position. These larvae are commonly found on smooth stones in very rapid velocities and are usually absent from stones covered with moss and from roughened stones that interfere with normal sucker function. Several other aquatic insects have structures that simulate the action of suckers. The enlarged gills of some mayflies (e.g., *Epeorus* sp. and *Rhithrogena* sp.: Fig. 1A) function as a friction pad, and *Drunella doddsi* has a specialized abdominal structure for the same purpose. Some chironomids have "pushing prolegs" represented by circlet of small spines that function as a false sucker when pressed to the substrate. Mountain midge larvae (Deuterophlebiidae) possibly use a similar mechanism to attach their suckerlike prolegs. Most of these animals are primarily grazers on thin films of epilithon (algae, associated fine organic matter, and microbes) found on the surface of stones.

Flowing water usually carries many organic (and inorganic) particles and a number of insects exploit these suspended particles. Filter-feeding collectors (Table IV) exploit the current for gathering food with minimal energy expenditure. For example, certain filtering collectors exploit locations where flows converge over and around substrates, thus allowing the animals to occupy sites of greater food delivery. Examples include caddisfly larvae belonging to the families Hydropsychidae and Brachycentridae. Silk is used for attachment by a number of caddisflies (e.g., Hydropsychidae, Philopotamidae, and Psychomyiidae), which build fixed nets and retreats (Fig. 2A). Although the Philopotamidae are found in riffle habitats, their fine-meshed, tubelike nets are usually found in crevices or undersides of stones in low velocity microhabitats (Fig. 2C). The nets of the caddisfly, *Neureclipsis,* are limited to moderately slow (< 25 cm s^{-1}) velocities and the large (up to 20 cm long), trumpet-shaped nets (Fig. 2D) are used to capture small animals drifting downstream. *Neureclipsis* larvae are often very abundant in some lake outflow streams where drifting zooplankton are abundant.

Some case-making caddisflies (e.g., *Brachycentrus* sp.) also use silk for attaching their cases to the substrate in regions of moderately rapid flow. Many chironomid larvae construct fixed silken retreats for attachment or silken tubes that house the larvae, with a conical catchnet spun across the lumen of the tube. Periodically, the larva devours its catchnet with adhering debris that has been swept into the burrow by the water currents. Meanwhile, other chironomid larvae such as *Rheotanytarsus* spp. construct small silk cases that are attached to the stream substratum with extended hydralike arms. The arms project up in the current and are smeared with a silklike secretion to capture particles.

Larval blackflies (Simuliidae, Fig. 1B) use a combination of hooks and silk for attachment. The thoracic proleg resembles that of chironomids and deuterophlebiids, described earlier, and the last abdominal segment bears a circlet of hooks, which it uses to anchor itself to substrates. The larva moves forward, inchwormlike, spins silk over the substrate, and attaches the proleg and then the posterior circlet of hooks to the silken web. Most blackfly larvae possess well-developed cephalic fans, which are used to filter small particles from suspension. These attached larvae twist

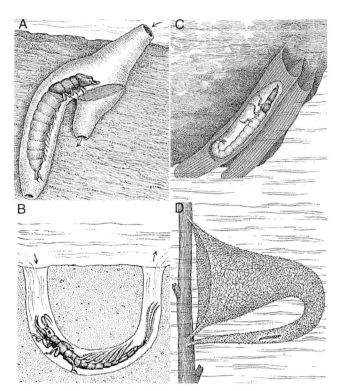

FIGURE 2 Representative lotic insects in their environment: (A) Caddisfly larva (*Macrostenum*) in its retreat grazing on materials trapped on its capture net, (B) mayfly larva of *Hexagenia* (Ephemeridae) in its U-shaped burrow, (C) tubelike nets of philopotamid caddisfly larvae (Philotamidae) on the lower surface of a stone, (D) the caddisfly larva and cornucopia-shaped net of *Neureclipsis* (Polycentropodidae). [Habitat drawings modified and taken from Wallace, J. B., and Merritt, R. W. (1980). Filter-feeding ecology of aquatic insects. *Annu. Rev. Entomol.* **25,** 103–132, (B); Merritt, R. W., and Wallace, J. B. (1981). Filter-feeding insects. *Sci. Am.* **244,** 131–144 (A, C, D).]

their bodies longitudinally from 90° to 180° with the ventral surface of the head and fans facing into the current. The fusiform body shape of blackfly larvae reduces turbulence and drag around their bodies, which are often located in regions of relatively rapid flow. Blackfly pupae are housed in silken cases that are attached to the substrate.

Although unidirectional current is the basic feature of streams, most lotic insects have not adapted to strong currents, but instead have developed behavior patterns to avoid current. Very few lotic insects are strong swimmers, probably because of the energy expenditure required to swim against a current. Downstream transport or drift requires only a movement off the substrate to enter the current. Streamlined forms, such as the mayflies *Baetis* spp., *Centroptilum, Isonychia* spp., and *Ameletus* spp., are capable of short rapid bursts of swimming, but most lotic insects move by crawling or passive displacement. One characteristic of these latter mayflies is the possession of a fusiform, or streamlined, body shape: examples include several Ephemeroptera such as *Baetis, Centroptilum,* and *Isonychia,* as well as a number of beetle (Coleoptera) larvae. A fusiform body shape reduces resistance in fluids, and within the mayflies the shape is often associated with excellent swimming abilities.

The benthic fauna in streams often can be found in cracks and crevices, between or under rocks and gravel, within the boundary layer on surfaces, or in other slack-water regions. Another method of avoiding fast currents is living in debris accumulations consisting of leaf packs and small woody debris. This debris offers both a food resource and a refuge for insects and contains a diverse array of aquatic insects including stoneflies such as Peltoperlidae and Pteronarcyidae (Fig. 1F), caddisflies such as Lepidostomatidae and some Limnephilidae, as well as dipterans such as chironomids and tipulid crane flies (Fig. 1E).

In some streams with unstable sandy or silt substrates, woody debris can represent a "hot spot" of invertebrate activity. Wood debris provides a significant portion of the stable habitat for insects in streams when the power of the flowing water is insufficient to transport the wood out of the channel. In addition to the insect component using wood primarily as a substrate, there is often a characteristic xylophilous fauna associated with particular stages of wood degradation. These include chironomid midges and scraping mayflies (*Cinygma* spp. and *Ironodes* spp.) as early colonizers, and larvae and adults of elmid beetles. In western North America, an elmid (*Lara avara*) and a caddisfly (*Heteroplectron*) are gougers of firm waterlogged wood, chironomids are tunnelers, and the tipulids, *Lipsothrix* spp., are found in wood in the latest stages of decomposition. Woody debris is most abundant in small, forested watersheds, but it is also an important habitat in larger streams with unstable beds. In the southeastern coastal plain of the United States and in low gradient mid- and southwestern streams and rivers with unstable bottom substrate, woody debris or "snags" often represent the major habitat for aquatic insect abundance and biomass. High populations and biomass of filter-feeding animals such as net-spinning caddisflies (*Hydropsyche* spp., *Cheumatopsyche* spp., and *Macrostenum*) (Fig. 2A), and blackflies occur in these streams and rivers. In addition to filter feeders, other groups such as odonates, mayflies, stoneflies, elmid beetles, nonfiltering caddisflies, and dipteran larvae can be locally abundant on large pieces of woody debris. Invertebrate shredders and scrapers promote decomposition of outer wood surfaces by scraping, gouging, and tunneling through wood. In fact, wood gouging habits of some net-spinning caddisflies have been blamed for the failure of submerged timber pilings that had been supporting a bridge!

Sand and silt substrates of rivers and streams are generally considered to be poor habitats because the shifting streambed affords unsuitable attachment sites and poor food conditions. An extreme example of this instability is the Amazon River, where strong currents move the bedload downstream, resulting in dunes of coarse sand up to 8 m in height and 180 m in length, thus largely preventing the establishment of a riverbed fauna. However, sandy substrates do not always result in poor habitat for all aquatic insects: some sandy streams are quite

productive. Blackwater (i.e., high tannic acid concentrations from leaf decomposition) streams of the southeastern United States have extensive areas of sand, with some of insects, such as small Chironomidae (< 3 mm in length), exceeding 18,000/m^{-2} in abundance. Their food is derived from fine organic matter, microbes, and algae trapped in the sandy substrate. Numerically, the inhabitants of sandy or silty areas are mostly sprawlers or burrowers, with morphological adaptations to maintain position and to keep respiratory surfaces in contact with oxygenated water. At least one insect, the mayfly *Ametropus,* is adapted for filter feeding in sand and silt substrates of large rivers. *Ametropus* uses the head, mouthparts, and forelegs to create a shallow pit in the substrate, which initiates a unique vortex (flow field in which fluid particles move in concentric paths) in front of the head and results in resuspension of fine organic matter as well as occasional sand grains. Some of these resuspended fine particles are then trapped by fine setae on the mouthparts and forelegs. Many predaceous gomphid (Odonata) larvae actually burrow into the sediments by using the flattened, wedge-shaped head and fossorial (adapted for digging) legs. The predaceous mayflies *Pseudiron* spp. and *Analetris* spp. have long, posterior-projecting legs and claws that aid in anchoring the larvae as they face upstream. Some mayflies (e.g., Caenidae and Baetiscidae) have various structures for covering and protecting gills, and others (e.g., Ephemeridae, Behningiidae) have legs and mouthparts adapted for digging. The predaceous mayfly *Dolania* spp. burrow rapidly in sandy substrates and have dense setae located at the anterior–lateral corners of the body as well as several other locations. The larva uses its hairy body and legs to form a cavity underneath the body where the ventral abdominal gills are in contact with oxygenated water.

Dense setae also are found in burrowing mayflies belonging to the family Ephemeridae that are common inhabitants of sand and silt substrates. They construct shallow U-shaped burrows and use their dorsal gills to generate water currents through the burrow (Fig. 2B), while using their hairy mouthparts and legs to filter particles from the moving water. Hairy bodies seem to be a characteristic of many animals dwelling on silt substrates, which include other collector mayflies such as *Caenis, Anepeorus,* and some Ephemerellidae.

Many dragonflies (e.g., *Cordulegaster* spp., *Hagenius* spp., and Macromiidae) have flattened bodies and long legs for sprawling on sandy and silty substrates. Some caddisflies, such as *Molanna,* have elongate slender bodies but have adapted to sand and silt substrates by constructing a flanged, flat case. They are camouflaged by dull color patterns and hairy integuments that accumulate a coating of silt. The eyes, which cap the anterolateral corners of the head, are elevated over the surrounding debris. The genus *Aphylla* (Gomphidae) is somewhat unusual in that the last abdominal segment is upturned and elongate, allowing the larvae to respire through rectal gills while buried fairly deep in mucky substrate. Some insects burrow within the upper few centimeters of the substratum in depositional areas of streams. This practice is found among some dragonflies and a number of caddisflies, including *Molanna,* and various genera of the families Sericostomatidae and Odontoceridae.

Specialized Flowing Water Habitats

The hyporheic region is the area below the bed of a stream where interstitial water moves by percolation. In gravelly substrates or glacial outwash areas, it may also extend laterally from the banks. In some situations an extensive fauna occurs down to one meter in such substrates. Most orders are represented, especially taxa with slender flexible bodies or small organisms with hard protective exoskeletons. Some stoneflies in the Flathead River of Montana spend most of their larval period in this extensive subterranean region of flow adjacent to the river. Stonefly larvae have been collected in wells over 4 m deep, located many meters from the river. Rivers draining glaciated regions where there are large boulders and cobble appear to have an exceptionally well-developed hyphoreic fauna.

Other specialized flowing water habitats include the madicolous (or hygropetric) habitats, which are areas in which thin sheets of water flow over rock. These often approach vertical conditions (e.g., in waterfalls) and have a characteristic fauna. Among common animals in these habitats are caddisflies, including several microcaddisflies (Hydroptilidae), Lepidostomatidae, beetles such as Psephenidae, and a number of Diptera larvae belonging to the Chironomidae, Ceratopogoniidae, Thaumaleidae, Tipulidae, Psychodidae, and some Stratiomyiidae.

Thermal (hot) springs often have a characteristic fauna, which is fueled by algae and bacteria adapted to high temperatures. The common inhabitants include a number of dipteran families such as Chironomidae, Stratiomyiidae, Dolochopodidae, and Ephydridae, as well as some coleopterans. A number of these survive within rather narrow zones between the thermal spring and cooler downstream areas.

LENTIC HABITATS

Lentic or standing-water habitats range from temporary pools to large deep lakes and include marshes and swamps, as well as natural (i.e., tree holes, pitcher plants) and artificial (i.e., old tires, rain barrels) containers. The available habitats and communities for insects in a pond or lake were defined in Table II. These habitats include the littoral zone, which comprises the shallow areas along the shore with light penetration to the bottom and normally contains macrophytes (rooted vascular plants). The limnetic zone is the open-water area devoid of rooted plants, whereas the deeper profundal zone is the area below which light penetration is inadequate for plant growth, water movement is minimal, and temperature may vary only slightly between summer and winter. The aquatic and semiaquatic insect communities inhabiting these zones are known as the pleuston (organisms

associated with the surface film), plankton and nekton (organisms that reside in the open water), and benthos (organisms associated with the bottom, or solid–water interface). Nektonic forms are distinguished from plankton by their directional mobility, and the latter are poorly represented in lentic waters by insects; the majority of insects found in standing-water habitats belong to the benthos. Their composition and relative abundance is dependent on a variety of factors, some of which are integrated along depth profiles. The overall taxonomic richness of benthic insect communities generally declines with increasing depth.

Among the aquatic communities of lentic habitats, the following orders of aquatic and semiaquatic insects are commonly found within the littoral, limnetic, and profundal zones: the springtails (Collembola), mayflies (Ephemeroptera), true bugs (Heteroptera), caddisflies (Trichoptera), dragonflies (Anisoptera) and damselflies (Zygoptera), true flies (flies, gnats, mosquitoes, and midges) (Diptera), moths (Lepidoptera), alderflies (Megaloptera), and beetles (Coleoptera). Not all these groups occur in lakes, and many are associated with ponds or marshes; examples of typical lentic insects are shown in Figs. 3 and 4.

The Pleuston Community

The unique properties of the water surface or air–water interface constitute the environment of the pleuston community. The Collembola, or springtails, are small in size, have a springing organ (furcula), and a water-repelling cuticle that enables them to be supported by and move across water surfaces. Among the true bugs, the Gerridae (water striders) and related families, the Veliidae (broad-shouldered water striders) and Hydrometridae (water measurers), are able to skate across the water. Adaptations for this habit include retractable preapical claws to assist in swimming, elongate legs

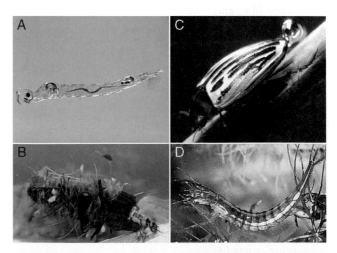

FIGURE 3 Typical insects inhabiting lentic environments. (A) Diptera: Chaoboridae (*Chaoborus*), (B) Trichoptera: Limnephilidae (*Limnephilus*), (C) Coleoptera: Dytiscidae (*Agabus*), (D) Coleoptera: Dytiscidae. (Photographs in A, B, and C by M. Higgins.)

and Coleoptera, to practically all functional feeding modes by different mosquito larvae, including collecting-filtering and gathering, scraping, and shredding (Table IV).

The Nekton and Plankton Communities

The nekton are swimmers able to navigate at will (e.g., Coleopera, Hemiptera, some Ephemeroptera), whereas plankton are floating organisms whose horizontal movements are largely dependent on water currents. The phantom midge *Chaoborus* sp. (Chaoboridae) (Fig. 3A) is normally regarded as the only planktonic insect and is abundant in many eutrophic (nutrient-rich) ponds and lakes. The tracheal system in these larvae is reduced to kidney-shaped air sacs that function solely as hydrostatic organs, and the larvae slowly descend or rise by adjusting the volume of the air sacs. *Chaoborus* remains in benthic regions during the day but moves vertically into the water column at night. These journeys are dependent on light and oxygen concentrations of the water. The larvae avoid predation by being almost transparent, and they have prehensile antennae that are used as accessory mouthparts to impale zooplankton (Fig. 3A). The only other group of insects that may be considered to be planktonic are the early chironmid instars, which have been reported in the open water column.

Among the Heteroptera, nektonic species are in the Notonectidae (back swimmers), Corixidae (water boatman), and Belostomatidae (giant water bugs), all of which are strong swimmers. Many of these rise to the water surface unless continously swimming or clinging to underwater plants. Notonectids have backs formed like the bottom of a boat and navigate upside down. They hang head downward from the surface or dive swiftly, using their long hind legs as oars. On the underside of the body, they carry a silvery film of air, which can be renewed at regular intervals, for breathing while submerged. Two genera of backswimmers (*Anisops* and *Buenoa*) use hemoglobin for buoyancy control, and this adaptation has enabled these insects to exploit the limnetic zone of fishless lentic waters, where they prey on small arthropods. They have been considered for use as biological control agents for mosquito larvae in some areas of North America. In contrast to notonectids, corixids always swim with the back up, using their elongate, flattened oarlike legs. Although some water boatmen are predators, they are the only group of semiaquatic Heteroptera that have members that are collectors, feeding on detritus and associated small plant material. The Belostomatidae are strong swimmers, but probably spend most of their time clinging to vegetation while awaiting prey, rather than actively pursuing their food in the open water. They are masters of their environment and capture and feed on a variety of insects, tadpoles, fish, and even small birds. The eggs of many belostomatids are glued to the backs of the males by the females and carried in this position until nymphs emerge, a remarkable adaptation for protection of the eggs.

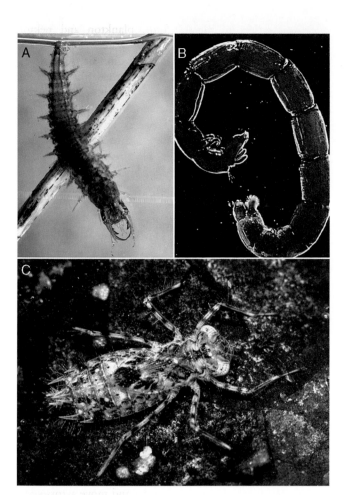

FIGURE 4 Typical insects inhabiting lentic environments (A) Coleoptera: Hydrophilidae (*Hydrochara*). (Photograph by M. Higgins.) (B) Diptera: Chironomidae (*Chironomus*), (C) Odonata: Libellulidae (*Pantala*). (Photograph by M. Higgins.)

and body to distribute the insect's weight over a large area of the surface film, and hydrofuge (nonwettable) hairpiles for support on the surface. Some gerrids also are capable of detecting surface vibrations caused by potential prey. Adult whirligig beetles (Gyrinidae) live half in and half out of water with each eye divided into upper and lower halves, permitting vision simultaneously in both the air and the water; glands keep the upper portion of the body greased to repel water. The middle and hind legs of adult gyrinids are paddle shaped, enabling them to be one of the most effective swimming invertebrates. Among the Diptera, only the mosquitoes (Culicidae) may be considered to be permanent members of the pleuston of lentic waters. The larvae and pupae of most species use the underside of the surface film for support. Larval *Anopheles* lie horizontally immediately beneath the air–water interface, supported by tufts of float hairs on each. Larvae of most other genera (*Aedes, Culex, Culiseta*) hang upside down, with an elongated terminal respiratory siphon penetrating the surface film. Feeding adaptations associated with pleuston specialization include predation by the Hemiptera

Although most aquatic beetles (Coleoptera) are associated with the substrate, members of the Dytiscidae (predaceous diving beetles) and the Hydrophilidae (water scavenger beetles) are often found swimming in the water column and together constitute the majority of all species of water beetles. The dytiscids are mainly predators in both the adult and larval stage (Fig. 3C, D), while adult hydrophilids are omnivorous, consuming both living and dead materials. The larvae of hydrophilids are predaceous. (Fig. 4A). To respire, hydrophilid adults, having their largest spiracles on the thorax, break the surface film with their antennae; dytiscids, having their largest spiracles on the abdomen, come up tail-end first, as do the larvae of both families. Overall, there are actually few truly nektonic insects, and most of them pass through the limnetic zone when surfacing for emergence. This may be, partly, because with no resting supports in the limnetic zone, maintaining position requires continuous swimming or neutral buoyancy. The vast majority of lentic insects occur in shallow water with emergent plants and are considered to be part of the benthos.

The Benthos Community

Benthos, derived from the Greek word for bottom, refers to the fauna associated with the solid–water interface and includes insects residing on the bottom or associated with plant surfaces, logs, rocks, and other solid substrates. In lentic habitats, many insects fall into this category as mentioned earlier, particularly the Chironomidae, which often represent over 90% of the fauna in the profundal (deep-water) zone of lakes and ponds. These inhabitants are mostly burrowers that feed on suspended or sedimented organic materials and are capable of tolerating low dissolved oxygen or even anaerobic conditions. Chironomid larvae build U- or J-shaped tubes with both openings at the mud–water interface. Body undulations cause a current of water, providing conditions under which oxygen and particulate food can be drawn through the tube. Some midge larvae found in sediments (mainly *Chironomus* sp.) are bright red and are known as bloodworms (Fig. 4B). The red color is caused by the respiratory pigment hemoglobin, which enables a larva to recover rapidly from anaerobic periods because the pigment takes up oxygen and passes it to the tissues more quickly than is possible by diffusion alone.

Other members of the benthos of deeper waters include the mayfly, *Hexagenia* (Ephemeridae), which inhabits the silt and mud of nearshore lake bottoms and has legs modified for digging to construct U-shaped burrows (Fig. 2B). Mayfly numbers have been increasing because of improved water quality standards for lakes and streams. Exceptions to the main constituents of the profundal zone are some immature mayflies, stoneflies, and caddisflies that have been collected at depths from 30 to 100 m in Lake Superior, Michigan. Also, a stonefly, *Utacapnia lacustra* (Capniidae), occurs at depths of 80 m in Lake Tahoe, Calfornia–Nevada, and completes its entire life cycle at this depth, never needing to surface.

Several orders of aquatic insects reach their greatest abundance and diversity in the shallow littoral zone of ponds and lakes as benthos typically associated with macrophytes (macroalgae and rooted vascular plants). The occupants are burrowers, climbers, sprawlers, clingers, swimmers, and divers (Table III) and include the Ephemeroptera, Heteroptera, Odonata, Trichoptera, Megaloptera, Lepidoptera, Coleoptera, and Diptera. The same groups occupy marshes and some swamps, which generally tend to be shallow, with an associated plant zone across the entire surface. Mayflies belonging to the families Baetidae and Siphlonuridae are generally swimmers, clingers, and climbers in vegetated ponds and marshes and mainly feed by means of collecting-filterering or -gatherering (Table IV). Heteroptera include the water scorpions (Nepidae), which have long slender respiratory filaments and are well concealed by detritus and tangled plant growth because of their sticklike appearance. These sit-and-wait predators capture organisms that frequent their place of concealment. Other families of Heteroptera adapted for moving through vegetation in ponds are the Pleidae or pygmy back-swimmers and creeping water bugs, the Naucoridae.

The Odonata, particularly the Gomphidae, are all predators and usually conceal themselves by either burrowing in substrate, sprawling among fine sediment and detritus, or climbing on vascular plants. Sprawlers are more active hunters and include the Libellulidae (Fig. 4C) and Corduliidae. Numerous setae give them a hairy appearance to help camouflage the larvae, and color is protective in patterns of mottled greens and browns. Most Zygoptera (damselflies) and the dragonfly (Anisoptera) family Aeshnidae are mainly climbers or clingers, lurking in vegetation or resting on stems of aquatic plants. The larvae stalk their prey, and both dragonfly and damselfly larvae have a unique lower lip (the labium) armed with hooks, spines, teeth, and raptorial setae that can extend to seize prey and then bring it back into the mouth, holding the food while it is being eaten. The food of larval odonates consists of other aquatic insects such as midges, semiaquatic bugs, and beetles, as well as small fish. Predators of larval odonates include aquatic birds, fish, and large predaceous insects.

In the order Megaloptera, which includes the hellgrammites or dobsonfly larvae of streams, only the predaceous larvae of the alderfly (*Sialis*) is common in ponds and lakes. They are generally found in sand or mud along the margins, but occasionally in deeper water, and they prey on insect larvae and other small animals. The only aquatic family in the related order Neuroptera is the Sisyridae (the spongilla flies), and these are found feeding on freshwater sponges that occur in some streams and the littoral zones of lakes and ponds. The larvae, which occur on the surface or in the cavities of the host, pierce the sponge cells and suck the fluids with their elongated mouthparts.

Although most caddisflies are observed living in lotic waters, several families of caddisflies are either associated with temporary ponds in the spring, aquatic vegetation in permanent ponds, lakes and marshes, or wave-swept shore lines of lakes. The Hydropsychidae (net spinners), Helicopsychidae (snail case makers), Molannidae, and Leptoceridae are often found along wave-swept shorelines of lakes, and their feeding habits range from those of scrapers and collector-filterers to predators. The Phryganeidae and several genera within the Limnephilidae are climbers, clingers, and/or sprawlers among vegetation in temporary and permanent ponds and marshes; generally, they are shredders of vascular hydrophytes and other decaying plants. The cases of lentic caddisfly families vary with the environment they are found in. Some cases consist of narrow strips of leaves put together in spiral form around a cylinder (Phryganeidae: *Phryganea* sp.), others consist of plant materials such as leaves and bark arranged transversely to produce a bulky cyclindrical case (Limnephilidae: *Limnephilus*) (Fig. 3B).

Both aquatic and semiaquatic moths (Lepidoptera) occur in lentic habitats, and several genera form close associations with vascular hydrophytes. Larvae of the family Pyralidae (*Parapoynx* sp.) spend the first two instars on the bottom and feed on submerged leaves of water lilies, whereas older larvae generally become surface feeders. Silk spun by the caterpillars is often used to build protective retreats, and pupation usually takes place in silken cocoons or silk-lined retreats. Larval habits of aquatic and semiaquatic moths include leaf mining, stem or root boring, foliage feeding, and feeding on flower or seed structures. One semiaquatic lepidopteran called the yellow water lily borer (the noctuid *Bellura gortynoides*), mines the leaves as a young caterpillar and then bores into the petioles of lilies as an older caterpillar. Within the petiole, larvae are submerged in water and must periodically back out to expose the posterior spiracles to the air before submerging again. The larvae swim to shore by undulating their bodies and overwinter under leaf litter in protected areas.

In addition to the water scavenger and predaceous diving beetles that may occur as nekton swimming through the water column, larvae and adults of other beetles are considered to be part of the benthos of ponds and marshes. These include the Haliplidae (crawling water beetles), which are clingers and climbers in vegetation, and the Staphylinidae (rove beetles), which are generally found along shorelines and beaches, as well as in the marine intertidal zone. The Scirtidae (marsh beetles) are generally found associated with vascular hydrophytes but also are a prominent inhabitant of tree holes. The aquatic Chrysomelidae (leaf beetles) occur commonly on emergent vegetation in ponds, especially floating water lily leaves. The larvae of one genus, *Donacia,* obtain air from their host plant by inserting the sharp terminal modified spiracles into the plant tissue at the base of the plant. Water lilies can be heavily consumed by larvae and adults of the chrysomelid beetle, *Galerucella* sp., and some of the aquatic herbivorous beetles belonging to the family Curculionidae (weevils) include pests of economic importance such as the rice water weevil (the curculionid *Lissorhoptrus*).

The Diptera is clearly one of the most diverse aquatic insect orders, inhabiting nearly all lentic habitats and representing all functional feeding groups and modes of existence. Although the benthic Chironomidae may reach their highest densities in the profundal zone of eutrophic lakes and ponds, they also are largely represented in the littoral zone associated with submergent and emergent plants, where they often graze on the algae attached to leaf surfaces or are vascular plant miners. Other dipteran families that occur in the littoral or limnetic zone, along with their specific habitat, habit (mode of locomotion, attachment, or concealment), and functional feeding mode are summarized in Table V. Among these, a few are of particular interest because of their high diversity and/or abundance in these habitats, namely the crane flies (Tipulidae), the shore and brine flies (Ephydridae), and the marsh flies (Sciomyzidae). The Tipulidae, the largest family of Diptera, are found along the margins of ponds and lakes, freshwater and brackish marshes, and standing waters in tree holes. A few littoral species inhabit the marine intertidal zone. To these are added the large numbers of species that are semiaquatic, spending their larval life in saturated plant debris, mud, or sand near the water's edge or in wet to saturated mosses and submerged, decayed wood. Ephydridae larvae have aquatic and semiaquatic members and occupy several different lentic habitats ranging from salt water or alkaline pools, springs, and lakes to burrowers and miners of a variety of aquatic plants in the littoral margins of these freshwater lentic habitats. All larvae utilize a variety of food, but algae and diatoms are of particular importance in their diet. The Sciomyzidae share some of the same habitat with the shore and brine flies, particularly fresh- and saltwater marshes, and along margins of ponds and lakes among vegetation and debris. The unique aspect of their larval life is that they are predators on snails, snail eggs, slugs, and fingernail clams. The aquatic predators float below the surface film and maintain buoyancy by frequently surfacing and swallowing an air bubble. Prey may be killed immediately or over a few days.

MARINE HABITATS

As noted earlier, insects have been largely unsuccessful in colonizing the open ocean, except for some members of the heteropteran family Gerridae. Most marine insects live in the intertidal zone (i.e., between high and low tide marks), especially on rocky shores or associated with decaying seaweed on sandy beaches (Table I). Although several orders have representatives in the intertidal zone, only a few orders, notably the Diptera, Coleoptera, and Collembola, have colonized these habitats in any numbers. The harsh physical environment of this area has forced these groups to occur buried in sand or mud and to hide in rock crevices or under seaweed.

UNUSUAL HABITATS

Because of adaptive radiation over evolutionary time, insects have colonized virtually every aquatic habitat on earth. Therefore, it is not surprising that these organisms are found

TABLE V   Summary of Ecological Data for Benthic Aquatic and Semiaquatic Diptera Larvae Inhabiting Lentic Habitats

Family	Habitat	Habitat	Functional feeding mode
Ceratopogonidae (biting midges, "no-see-ums")	Littoral zone (including tree holes and small temporary ponds and pools)	Generally sprawlers, burrowers or planktonic (swimmers)	Generally predators some collector-gatherers
Chironomidae (nonbiting midges)	All lentic habitats including marine, springs, tree holes	Generally burrowers, sprawlers (most are tube builders); some climber-clingers	Generally collector-gatherers, collector-filterers; some shredders and scrapers
Corethrellidae	Limnetic and littoral margins	Sprawlers	Predators
Psychodidae (moth flies)	Littoral detritus (including tree holes)	Burrowers	Collector-gatherers
Ptycopteridae (phantom crane flies)	Vascular hydrophytes (emergent zone), bogs	Burrowers	Collector-gatherers
Tipulidae (crane flies)	Littoral margins, floodplains (organic sediment)	Burrowers and sprawlers	Generally shredders, collector-gatherers
Dolichopodidae	Littoral margins, estuaries, beach zones	Sprawlers, burrowers	Predators
Stratiomyidae (soldier flies)	Littoral vascular hydrophytes; beaches (saline pools, margins)	Sprawlers	Collector-gatherers
Tabanidae (horseflies, deerflies)	Littoral (margins, sediments and detritus); beaches, marine and estuary	Sprawlers, burrowers	Predators
Canacidae (beach flies)	Beaches—marine intertidal	Burrowers	Scrapers
Ephydridae (shore and brine flies)	Littoral (margins and vascular hydrophytes)	Burrowers, sprawlers	Collector-gatherers, shredders, herbivores (miners), scrapers, predators
Muscidae	Littoral	Sprawlers	Predators
Scathophagidae (dung flies)	Vascular hydrophytes (emergent zone)	Burrower-miners (in plant stems), sprawlers	Shredders
Sciomyzidae (marsh flies)	Littoral—vascular hydrophytes (emergent zone)	Burrowers, inside snails	Predators or parasites
Syrphidae (flower flies)	Littoral (sediments and detritus), tree holes	Burrowers	Collector-gatherers

in the most unusual of aquatic habitats. The title of most versatile aquatic insect must be shared among members of the dipteran family Ephydridae, or shore flies. Shore flies can breed in pools of crude petroleum and waste oil, where the larva feed on insects that become trapped on the surface film. Other species of this family (*Ephydra cinera*), known as brine flies, occur in the Great Salt Lake, Utah, which has a salinity six times greater than that of seawater. Larva maintain water and salt balance by drinking the saline medium and excreting rectal fluid that is more than 20% salt. Another related family of flies, the Syrphidae, or "rat-tailed maggots," occur in sewage treatment lagoons and on moist substrates of trickling filter treatment facilities. Both families have larvae with breathing tubes on the terminal end, which permits the larvae to maintain contact with the air while in their environment. Some Stratiomyiidae, or soldier flies, live in the thermal hot springs of Yellowstone National Park with temperatures as high as 47°C! Other members of this family inhabit the semiaquatic medium of cow dung and dead corpses. A few species of insects have invaded caves and associated subterranean habitats, as mentioned earlier (see Lotic Habitats).

Another unsual aquatic habitat that several insect orders occupy is referred to phytotelmata or natural container habitats and include tree holes, pitcher plants, bromeliads, inflorescences, and bamboo stems. Synthetic container habitats, such as old tires, cemetery urns, rain gutters, and similar natural habitats such as hoofprints also harbor similar insects. Some of these habitats are extremely small and hold water only temporarily, but nevertheless can be quite diverse. The most common order found in these habitats is the Diptera with more than 20 families reported. Over 400 species of mosquitoes in 15 genera alone inhabit these bodies of water and some of these species are important vectors of disease agents.

Insect communities inhabiting pitcher plants (*Sarracenia purpurea*) in North America are exemplified by a sarcophagid or flesh fly (*Blaesoxipha fletcheri*), a mosquito (*Wyeomyia smithii*), and a midge (*Metriocnemus knabi*). The relative abundance of these pitcher plant inhabitants is related to the age, inasmuch as each of the three species consumes insect remains that are in different stages of decomposition. Specifically, the larvae of the flesh fly feed on freshly caught prey floating on the pitcher fluid surface. The mosquito larvae filter feed on the decomposed material in the water column, and the midge larvae feed on the remains that collect on the bottom of the pitcher chamber. Temporary habitats are important because they are populated by a variety of species, often with unique morphological, behavioral, and physiological properties.

See Also the Following Articles

Cave Insects • Marine Insects • Mosquitoes • Soil Habitats • Swimming

Further Reading

Allan, J. D. (1995). "Stream Ecology." Kluwer, Dordrecht, the Netherlands.

Brönmark, C., and Hansson, L.-A. (1998). "The Biology of Lakes and Ponds." Oxford University Press, Oxford, U.K.

Cummins, K. W. (1973). Trophic relations of aquatic insects. *Annu. Rev. Entomol.* **18**, 183–206.

Cushing, C. E., and Allan, J. D. (2001). "Streams: Their Ecology and Life." Academic Press, San Diego, CA.

Hynes, H. B. N. (1970). "The Ecology of Running Waters." University of Toronto Press, Toronto, Ont., Canada.

McCafferty, W. P. (1981). "Aquatic Entomology." Jones & Bartlett, Boston.

Merritt, R. W., and Cummins, K. W. (eds.). (1996). "An Introduction to the Aquatic Insects of North America." Kendall/Hunt, Dubuque, IA.

Merritt, R. W., and Wallace, J. B. (1981). Filter-feeding in aquatic insects. *Sci. Am.* **244**, 131–144.

Resh, V. H., and Rosenberg, D. M. (1984). "The Ecology of Aquatic Insects." Praeger Scientific, New York.

Ward, J. V. (1996). "Aquatic Insect Ecology," Vol. 1, "Biology and Habitat." Wiley, New York.

Williams, D. D. (2002). "The Ecology of Temporary Waters." Blackburn Press, Caldwell, NJ.

Williams, D. D., and Feltmate, B. W. (1992). "Aquatic Insects." CAB International, Wallingford, Oxon, U.K.

Arachnida

see *Scorpions; Spiders*

Archaeognatha
(Bristletails)

Helmut Sturm
University Hildesheim, Germany

The Archaeognatha (Microcoryphia; part of the subdivided order Thysanura) are apterygote insects with a body size between 6 and 25 mm and a cylindrical shape (Fig. 1). The eyes are large and contiguous, and there are two lateral and

FIGURE 1 A male archaeognathan (*Machilis germanica*), body length ca. 12 mm, lateral view; for details see Fig. 2.

one median ocelli (small single eyes with a single beadlike lens). The flagellate (whiplike) antennae extend one-half to three times the length of the body. The mouthparts are ectognathous (freely visible) and the mandibles are linked with the head by a monocondylic joint (i.e., one point of attachment). Some authors believe that this feature distinguishes the Archaeognatha from all other ectognathous Insecta. The seven-segmented maxillary palps are longer than the legs. The thoracic tergites II + III are in lateral view strongly arched, and the two or three tarsal segments of the legs are rigidly united. Some taxa have additional scopulae (dense brushes of specialized hairs) on the distal end of the third tarsal segment. There are mostly pairs or double pairs of eversible vesicles on the coxites of the abdomen (Fig. 3). On each of the abdominal coxites II to IX, styli (pointed, nonarticulated processes) are present (Fig. 3).

Females have two long gonapophyses on each of the abdominal segments VIII + IX, forming the ovipositor. The penis of the males on abdominal segment IX varies in length, and in Machilidae it is fitted with paired parameres on abdominal segments IX or VIII + IX. The three filiform and scaled caudal appendages (one long filum terminale and two laterally inserted cerci) are directed backward. Tergites, cerci, and coxites are always scaled. The molts continue in adult stages. Many species are petrophilous (living on and under stones). The order comprises about 500 species in two families (Machilidae and Meinertellidae).

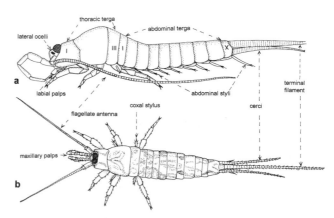

FIGURE 2 General structure of Archaeognatha, semidiagrammatic. (a) Lateral view. (b) Dorsal view, color pattern of dorsal scales intimated. (Reprinted from *Deutsche Entomologische Zeitschrift* **48**, p. 4, © 2001 by Wiley-VCH, with permission.)

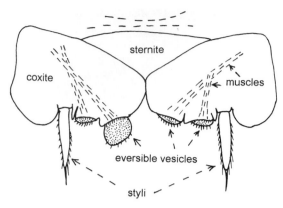

FIGURE 3 Abdominal coxites III of *M. germanica,* ventral view. The eversible vesicles can be exserted by increasing the inner pressure and retracted by muscles.

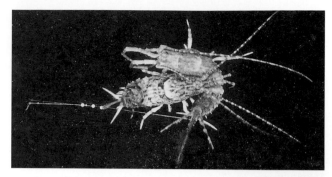

FIGURE 4 Mating position of *M. germanica,* dorsal view. The male has drawn out a secreted thread, deposited three sperm droplets on the thread, and taken up a U form. The ovipositor of the female is touching one of the sperm droplets.

FOSSIL RECORD, SYSTEMATICS, AND BIOGEOGRAPHY

The fossil record of Paleozoic and Mesozoic apterygotes is poor, and many of the fossils of ectognathous representatives cannot be clearly assigned to extant orders. From the Mesozoic, the only archaeognathan fossil is *Cretaceomachilis libanensis* from the lower Cretaceous of Lebanon. For the Cenozoic period, there are many fossils of Archaeognatha, most being amber inclusions. For example, from Baltic amber (ca. 35 mya) seven species of Machilidae are known. All extant forms and the fossils from the Cretaceous and the Tertiary can be included in the superfamily Machiloidea. This group includes two families, the more primitive Machilidae (46 genera and some 325 species), with three subfamilies (Machilinae, Petrobiinae, Petrobiellinae), and the more derived Meinertellidae (19 genera and some 170 species).

The Machiloidea are distributed worldwide. Only the Meinertellidae occur in South America, the Caribbean, South Africa, Australia, and Melanesia. Both Machilidae and Meinertellidae occur in the United States.

BEHAVIOR AND ECOLOGY

The mating behavior of archaeognathans is unique. There are three different modes of sperm transfer. In the most widespread and unique mating behavior, a carrier thread is used. In *Machilis germanica,* for example, the male approaches the female and drums on her with his long maxillary palps. The female shows "willingness" to mate by moving toward the male and bending up the tip of her abdomen. The male then attaches a secreted thread to the ground with his parameres. While the thread is being drawn out, the male secretes three to five sperm droplets onto the thread (Fig. 4). The male curves simultaneously around the female, preventing her from moving forward. Finally, the male places the sperm droplets onto the ovipositor of the female. The indirect transmission of sperm droplets, which are deposited on a thread stretched between parameres and the ground, is unique within the

animal kingdom. At least two other possibilities of indirect sperm transfer are known. In *Petrobius* (Machilidae), the sperm are moved directly from the penis onto the ovipositor of the female and in all Meinertellidae sperm are transferred by stalked spermatophores deposited on the ground.

Archaeognatha are found in habitats with very different climates. Representatives of the genus *Allopsontus* (Machilidae) are found up to 5000 m in the Himalayan region. In contrast, two meinertellid species live in the Amazonian forest. Some genera (e.g., *Petrobius*) are found near the seacoast. In tropical forests, meinertellid genera are found on the leaves of bushes and trees. Most Archaeognatha feed on green algae, lichens, and dead leaves. Several species of spiders probably are their principal predators.

Their protection against enemies is probably provided by three main factors: (1) the presence of long appendages with sensilla (filum terminale, cerci, antennae); (2) a dense scale cover on the relatively thin and flexible tergites and coxites; and (3) their ability to jump, which is fully developed in all free-living stages and in all recent representatives, and probably is their most effective defense.

PHYSIOLOGY

The cuticle of the Archaeognatha is generally thin and flexible and bears a multitude of sensory setae and other sensory organs. The hypopharynx (a mouthpart between maxillae and labium) is well developed. The coxal vesicles absorb fluids; their number varies from none to two within a coxite. In all machilid males, coxite IX bears the penis (length from $\frac{3}{4}$ of the coxite to $1\frac{1}{2}$ in Machilidae, in Meinertellidae ca. $\frac{1}{2}$ of the coxite length). In females, the paired gonapophyses on coxites VIII and IX form the ovipositor. It is of different length and has generally a specific chaetotaxy in each species.

DEVELOPMENT

The eggs (diameter 0.7–1.3 mm) are deposited into crevices of rocks or bark, and this stage lasts 60 to 380 days. Developmental stage I (first free-living larva) has a distinct prognathy, a strong forward projection of parts of the maxillae (laciniae), which aids in emergence from the egg, and rod-like setae on the head and the terga. These features disappear after the first molt. From developmental stage III onward, a scale cover is present in all terga and on abdominal coxites and caudal appendages. The molts continue during adult life.

See Also the Following Article
Zygentoma

Further Reading

Bitsch, C., and Bitsch, J. (1998). Internal anatomy and phylogenetic relationships among apterygote insect clades (Hexapoda). *Ann. Soc. Entomol. France (N.S.)* **34**, 339–363.

Mendes, L. F. (1990). An annotated list of generic and specific names of Machilidae (Microcoryphia, Insecta) with identification keys for the genera and geographical notes. *In* "Estudios, Ensaios e Documentos." *Centro Zool. Inst. Invest. Tropicall,* Lisbon.

Sturm, H., and Machida, R. (2001). Archaeognatha. *In* "Handbook of Zoology," *Vol. IV,* Part 37. Walter de Gruyter, Inc., Berlin/New York.

Wygodzinsky, P. W., and Schmidt, K. (1980). Survey of the Microcoryphia (Insecta) of the northeastern United States and adjacent provinces of Canada. *Ann. Mus. Novitates* **2701**, 1–17.

Arthropoda and Related Groups

James H. Thorp
University of Kansas

More than 75% of all described species in terrestrial, freshwater, and marine ecosystems are in the phylum Arthropoda. No other invertebrate phylum, with the possible exception of the nematodes, approaches their economic and ecological importance. This article briefly reviews all subphyla and classes of Arthropoda, with its nearly one million described species, and provides more details about major arthropod taxa not covered in other entries in the Encyclopedia. Aspects of arthropod evolutionary relationships, diversity, anatomy, physiology, and ecology are discussed.

OVERVIEW OF THE PHYLUM ARTHROPODA

The Arthropoda is a phylum more diverse than any other living or extinct animal taxon. Counted among this immense assemblage are beetles, butterflies, silverfish, centipedes, scorpions, mites, sea spiders, crabs, sow bugs, and barnacles, and many other common names too numerous to mention. Arthropods are the numerically dominant metazoan on land and rank among the most prominent benthic (bottom-dwelling)

and planktonic members of freshwater and marine ecosystems. They colonize virtually every conceivable habitat—from the equator to the poles, from high mountains to deep ocean trenches, and from rain forests to deserts and hot springs—and fill all trophic niches above the level of primary producer. Parasitism, especially ectoparasitism, is common in some groups, but most species are free-living. They range in size from tiny gall mites (80 μm) to Japanese spider crabs with leg spans of 3.6 m. While some arthropods are vectors for human disease-causing organisms and others are major agricultural competitors with humans, they are also vital to the functioning of most ecosystems and a boon to humans in many ways. In addition to deriving nutrition from some arthropods (e.g., directly or indirectly from bees, crabs, lobsters, and shrimp), humans probably could not survive without arthropods.

The name "Arthropoda" is from the Greek, meaning "jointed foot." The presence of jointed appendages is the primary feature distinguishing arthropods from other phyla. Advantages provided by these appendages, a metameric or segmented body, and a hard skeleton are the three most important reasons for the phylum's success. Arthropods are segmented like annelid worms, but the evolutionary trend has been to fuse several metameres into body regions (tagmata) with specialized functions. Spiders have two tagmata, insects have three, and many crustaceans have two; however myriapods (millipedes and centipedes) lack tagmata. Arthropods have chitinous and proteinaceous exoskeletons that are frequently strengthened with calcium salts. A modest, nonchitinized endoskeleton of inwardly projecting apodemes aids muscular attachment. To allow for continued somatic growth, the exoskeleton is shed periodically during ecdysis, a relatively strenuous and often dangerous process. Modifications of the exoskeleton have permitted arthropods to fly, swim, run, and burrow effectively.

Except for the molluscan cephalopods (e.g., the octopus), arthropods surpass all invertebrates in internal organ complexity. Although they are a phylum with a coelom, this structure no longer serves as a hydrostatic skeleton (as in annelids) but persists only as a cavity surrounding reproductive and/or excretory organs. The principal body cavity is instead the hemocoel, which is derived from the circulatory system. The open circulatory system consists of a dorsal heart, blood sinuses, and one or more discrete vessels. Hemocyanin and hemoglobin are the principal oxygen-carrying blood pigments. Respiration is achieved through the skin surface in some small species, but with gills in most aquatic organisms and tracheae and/or book lungs in terrestrial species. Excretory and osmoregulatory organs vary in type in accordance with the typical environmental moisture and salt content, as do the primary excretory products (ammonia in water and usually either uric acid or guanine on land). Cilia are absent externally and internally. The neural system is highly developed, with brain centers and complex sensory organs; indeed, next to vertebrates and cephalopods, the arthropod brain is the most complex on earth.

Most species reproduce sexually and are primarily dioecious (i.e., with an individual being a single gender), although parthenogenesis occurs in many taxa. Courtship and brood care are uncommon but are found in some members of all subphyla. Rather than possessing the spiral cleavage typical of many other protostomates, arthropods usually develop by superficial cleavage of a cytoplasmic layer above a yolky sphere. Larvae or discrete juvenile stages are common in terrestrial and aquatic taxa, but aquatic larvae never resemble the trochophore larvae that characterize related phyla.

EVOLUTIONARY RELATIONSHIPS WITH OTHER PHYLA

Arthropods were traditionally linked with the phylum Annelida in the phylogenetic clade Articulata because both are metameric phyla, but more recent molecular analyses provide no support for a particularly close relationship. Instead, arthropods seem to be linked with other phyla that must shed their cuticle during ecdysis to grow. This clade of "Ecdysozoa" also includes the phyla Tardigrada, Onychophora, Nematoda, and Nematomorpha, along with the more distantly related Priapulida and Kinorhyncha. At a greater phylogenetic distance from Ecdysozoa is the other major protostomate clade consisting of the phyla Rotifera, Annelida, Mollusca, Bryozoa (Ectoprocta), Brachiopoda, and Phoronida. Pentastomida is listed by some authors as a separate phylum with links to Arthropoda through the clade Ecdysozoa, but here it is included as a class within the arthropod subphylum Crustacea.

Within Ecdysozoa, the three closest phyla are Arthropoda, Tardigrada, and Onychophora. This conclusion is based in part on molecular studies using 18*S* rRNA. Fossil evidence from the mid-Cambrian (~520 mya) suggests that onychophoran-like limbs developed in aquatic invertebrates and may have served as a preadaptation for terrestrial life. Similarities in morphology and physiology also seem to link these phyla. For example, tardigrades possess striated muscles, paired ventral nerve cords, and a large hemocoel. In addition to these characteristics, onychophorans have the following arthropod-like features: a tracheal respiratory system, mandible-like mouth appendages, cardiac ostia, an excretory system comparable to the green gland of crustaceans, one pair of antennae, and similar defensive secretions produced by repugnatorial glands. Both Onychophora and Tardigrada, however, have some decidedly non-arthropod-like characteristics (e.g., nonjointed legs). Onychophora, Myriapoda, and Hexapoda are grouped by some systematists into Uniramia, a single phylum of arthropod-like animals having a single branch (ramus) of body appendages. According to this theory Uniramia is phylogenetically isolated from Crustacea and Chelicerata, but all are in the superphylum Arthropoda.

TAXONOMIC DIVERSITY AND INTRAPHYLETIC AFFILIATIONS

Arthropoda is treated here as a monophyletic clade of genetically diverse but evolutionarily linked species. Some zoologists, however, maintain that this alleged phylum is actually an artificial, polyphyletic grouping of similar taxa evolving multiple times from different prearthropod ancestors. Much of this debate has centered on evolutionary relationships between the phyla Arthropoda and Onychophora. Classified within Arthropoda are one extinct subphylum (sometimes called super class), the Trilobitomorpha (trilobites), and four living subphyla: Chelicerata (spiders, mites, horseshoe crabs, and sea spiders), Myriapoda (millipedes and centipedes), Hexapoda (springtails, bristletails, beetles, flies, true bugs, etc.), and Crustacea (crayfish, barnacles, water fleas, pill bugs, etc.). Sometimes the number of extant subphyla is reduced to three (Chelicerata, Uniramia, and Crustacea) or even two groups (Chelicerata and Mandibulata).

Molecular studies of arthropod phylogeny present a reasonably clear picture of relationships among three of the four living subphyla. Chelicerates are evolutionarily distinct from insects and crustaceans, and they differ from all other living arthropods in lacking a tagma for either a "head and trunk" or a "head, thorax, and abdomen." Instead, they possess an anterior prosoma without a distinct head and a posterior opisthosoma. Another major clade evident from gene sequences is Mandibulata, composed of the other three extant subphyla. Morphological observations of appendages would seem then to link Myriapoda and Hexapoda into a group (Uniramia) of taxa with only one branch to each appendage and distinct from the biramous Crustacea, but molecular evidence is inconclusive on this point. In some gene sequence trees, myriapods are tightly linked with insects, while other molecular analyses show the millipedes and centipedes as deeply entangled within other genetic branches.

Accurate estimates of both relative and absolute diversities of arthropods are often problematic because of the enormous species richness, large number of unexplored habitats, greater emphasis on studies of economically important taxa, and increasingly serious lack of qualified taxonomists. For those reasons, the literature is replete with divergent estimates of the total number of species in most groups, especially the insects and mites. Table I lists the classes of Arthropoda and includes estimates of taxonomic diversity.

SUBPHYLUM TRILOBITOMORPHA

Trilobites probably played a crucial role in the evolution of living arthropods. Members of the now extinct subphylum Trilobitomorpha began roaming primeval seas in the Precambrian, reached their zenith in the late Cambrian with 4000–10,000 species, and then slowly went extinct around 230 to 275 mya. They were flattened, bilaterally symmetrical arthropods with bodies divided by longitudinal and transverse grooves unlike living arthropods (Fig. 1). Most adults were 2 to 7 cm in length, but giants of 50 cm are known. Their organ systems probably resembled those in modern arthropods. Smaller species were probably planktonic suspension feeders, whereas most species and all larger species were probably benthic deposit feeders or facultative predators. Unlike most other arthropods, trilobites

TABLE I   Estimates of Arthropod Diversity, with Comments on Certain Arthropod Taxa and Two Related Phyla[a]

Taxon	Estimated number of species (families)	Biological features
Phylum Arthropoda	1–5 million	
Subphylum Trilobitomorpha	4,000–10,000	Extinct marine trilobites
Subphylum Chelicerata	77,000–1 million	Originally marine, but subsequent evolution has primarily been in terrestrial mites
Class Merostomata	5	Marine horseshoe crabs and extinct sea scorpions
Class Arachnida	~76,000 (550)	Spiders, scorpions, and mites
Class Pycnogonida	1,000 (8+)	Sea spiders
Subphylum Myriapoda	13,000 (140+)	Terrestrial millipedes, centipedes, and others
Class Chilopoda	3,000 (20)	Predaceous centipedes
Class Symphyla	160 (2)	Small (1–8 mm), mostly herbivorous; live in forest litter
Class Diplopoda	10,000 (120)	Millipedes
Class Pauropoda	500 (5)	Minute (< 1.5 mm) dwellers in forest litter
Subphylum Hexapoda	1–4 million	Insects, springtails, bristletails, etc.
Class Ellipura	6,000+	Wingless, entognathous (orders Protura and Collembola, or springtails)
Class Diplura	800 (9)	Blind, wingless inhabitants of forest litter; entognaths
Class Insecta	1–4 million species	Winged and wingless insects; all adults with six pairs of legs
Subclass Apterygota	600 (5)	Primitive, wingless insects (order Thysanura with bristletails and silverfish)
Subclass Pterygota	~1 million+ (906)	Mostly winged insects (grasshoppers, true bugs, beetles, flies, butterflies, ants, etc.)
Subphylum Crustacea	~50,000	Shrimp, crabs, waterfleas, barnacles, copepods, etc.
Class Cephalocarida	9 (2)	Primitive; live in soft marine sediments
Class Malacostraca	29,000 (103)	Crabs, water scuds, isopods, mantis shrimp, etc.
Class Branchiopoda	1,000 (29)	(= Phyllopoda) water fleas and brine, clam, pea, and tadpole shrimp
Class Ostracoda	6,650 (46+)	Seed shrimp enclosed in a bivalved chitinous carapace
Class Mystacocarida	11	Interstitial species living in shallow water or intertidally
Class Copepoda	8,000 (97)	Dominant crustaceans in zooplankton; a few parasites of marine fish and invertebrates
Class Branchiura	125 (1)	Fish lice (ectoparasites)
Class Pentastomida	100 (7)	Highly modified parasites of tetrapod vertebrates
Class Tantulocarida	10+ (9)	Deep-water parasites of crustaceans; some sources estimate diversity up to 1200 species
Class Remipedia	9+	Ancient, vermiform crustaceans found in marine caves; some estimate diversity as high 1200 species
Class Cirripedia	1,000 (20)	True barnacles and small groups of parasitic taxa
Phylum Onychophora	70 (2)	Velvet worms; most confined to tropical habitats
Phylum Tardigrada	800 (17)	Water bears in aquatic and moist terrestrial habitats

[a] Estimates of species richness are for living taxa only, except for the subphylum Trilobitomorpha. The reliability of these estimates varies widely among taxa.

FIGURE 1 *Asaphiscus wheeleri*, an extinct trilobite in the subphylum Trilobitomorpha. (Photograph courtesy of Sam M. Gon, III.)

proceeded gradually through a life with three larval stages, 14 or more juvenile steps in the first year of life, and multiple adult stages lasting a maximum of 3 more years.

SUBPHYLUM CHELICERATA

Few invertebrates on land or sea are so often miscast in a sinister role as the spider, which along with mites, sea spiders, and horseshoe crabs comprises the 77,000 described species in the subphylum Chelicerata. Indeed, most arachnids and all other chelicerates either are harmless to humans or are actually quite helpful in their roles as predators and parasites of insects or as decomposers of terrestrial litter. Although the subphylum evolved in the sea, the ocean now supports only the five species of horseshoe crabs (class Merostomata), a thousand or so species of sea spiders (class Pycnogonida), and a few mites.

Class Merostomata: Horseshoe Crabs

During the full or new moons of late spring and early summer when tides are the highest, vast numbers of horseshoe crabs

FIGURE 2 Representative of the subphylum Chelicerata: ventral view of an adult horseshoe crab, *Limulus polyphemus* (class Merostomata).

FIGURE 3 Desert hair scorpion, *Hadrurus arizonensis*. (Photograph by Jim Kalisch, courtesy of University of Nebraska Department of Entomology.)

(all *Limulus polyphemus;* Fig. 2) come ashore in eastern North America to breed in bays and estuaries. The five Asian and North American species of this class are remnants of this strictly marine class. Although formidable looking and up to 60 cm in length, horseshoe crabs are harmless to humans. When not breeding, these chelicerates reside on or in soft bottoms in shallow water. They are scavengers and predators on clams and worms. Although they can swim weakly by flapping their book gills (modified abdominal appendages), their primary locomotion is walking. Extinct merostomates (order Eurypterida) may have been the largest arthropods ever to have evolved (nearly 3 m long) and seem to have given rise to terrestrial arachnids. Despite their name, horseshoe crabs are not closely related to true crabs (subphylum Crustacea).

Class Arachnida

During the middle Paleozoic, chelicerates made the rigorous transition from water to land; only later did over 5000 species of arachnid mites adopt a secondary aquatic existence. Arachnids then rapidly radiated in form and species richness in association with their predaceous and parasitic exploitation of insects. One crucial factor in this success has been the diverse uses of silk by spiders, pseudoscorpions, and some mites. Although some degree of "arachnophobia" afflicts many people, relatively few of the 76,000 described species are directly harmful to humans because of their venom, link with diseases and allergies, or competition for plant resources. More than balancing their negative attributes is the substantial role in biocontrol of insect pests.

SCORPIONS, SPIDERS, AND HARVESTMEN

True Scorpions   The 1500 to 2000 species of true scorpions (order Scorpiones) are elders of the arachnid clan (Fig. 3). In addition to the true scorpions, several other arachnid orders are called "scorpions": false scorpions (Pseudoscorpiones with 2000 species), wind scorpions (Solifugae with 900 species), whip scorpions (Uropygi with 85 species), and tailless whip scorpions (Amblypygi with 70 species). All are much smaller than true scorpions, but are also typically carnivorous.

True Spiders (Order Araneae)   Most arachnids lack biting mouthparts and must, therefore, partially digest prey tissue before sucking it into their bodies. Prey are subdued with poison injected by fangs present on each chelicera. Arachnids reproduce with indirect fertilization (without a penis), often after elaborate courtship rituals. Their leglike pedipalps are used by males to transfer spermatophores. Eggs are wrapped in a protective silken cocoon, and brood care is common. Silk is produced normally by caudal spinnerets and by a small platelike organ (cribellum) in cribellate spiders only. Uses for this silk include cocoons, egg sacs, linings of retreats, and capture webs. Locomotion is typically by walking or jumping, but aerial dispersal through the process of "ballooning" with long silken threads is common in most spiderlings and adults of some smaller taxa. Most spiders are terrestrial and are found anywhere insects are located. All are carnivorous, and ecological divergence in prey type and capture method has led to the wide evolutionary radiation. In addition to insects, spiders attack other spiders, small arachnids, and a few other prey taxa including small vertebrates. Several spiders are poisonous to humans, such as the black widow and the brown recluse.

Harvestmen   The order Opiliones includes arachnids known as "daddy longlegs," a name reflecting its enormously long walking legs. They are also called "harvestmen" because some species undergo a seasonal population explosion each autumn around the farm harvest. They have "repugnatorial glands" that produce an acrid secretion to repel predators. The 5000 species are more closely related to mites than to true spiders. Most are tropical, but taxa are known from colder subarctic and alpine zones. Opilionids frequent humid forest floors, being less arboreal than true spiders. Although carnivory on small arthropods and worms is common, harvestmen are notable as the only arachnids other than mites that consume vegetation.

MITES AND TICKS
At least 30,000 species of arachnid mites and ticks have been described in the order (or subclass) Acari. The major habitat of mites is on land, where they are

either free-living or parasites of plants and animals, but lakes, streams, and even hot springs support 5000 taxa, with the marine fauna being less diverse.

Many acarines are ectoparasitic in larval and/or adult stages. Animal parasites attack mammals (including humans and domestic animals), birds, reptiles, amphibians, fish, aquatic and terrestrial insects, other arachnids, and some other invertebrates, including echinoderms, mollusks, and crustaceans. They are vectors for several human diseases, and some (e.g., chiggers) have annoying bites. Many people develop allergies to mites living on household dust. On the other hand, microscopic mites commonly consume dead tissue and oily secretions on human faces, and they are used to control harmful insects and mites. Their feeding habits and role in spreading viruses make them severe pests of natural and agricultural plants. A great diversity of mites, however, are free-living, mostly in forest and grassland litter, where they feed directly on litter or on microorganisms decomposing detritus. Many mites prey on other mites, nematodes, and small insects.

Class Pycnogonida: Sea Spiders

The body shape and gangling legs of sea spiders call to mind their terrestrial namesakes. Most of the thousand species live in shallow benthic zones at higher latitudes. They are predominantly predators of hydroids, bryozoans, and polychaetes, but some consume microorganisms, algae, and even detritus. Food is either macerated with chelae or externally predigested and then sucked into the digestive tract with a proboscis.

Sea spiders have a barely perceptible head and a body comprising four pairs of long legs joined by a narrow, segmented trunk. Most are small (1–10 mm), but some deepwater behemoths reach 6 cm in body length with a 75-cm leg span. Their eyes are mounted on an tubercle to give them a 360° arc of vision.

SUBPHYLUM MYRIAPODA

Myriapoda ("many feet") is a subphylum of elongate arthropods with bodies divided into a head and trunk with numerous segments, most of which have uniramous appendages; no pronounced tagmatization is evident. Myriapods range in length from 0.5 to 300 mm and are primarily terrestrial. Most live in humid environments, commonly in caves. Some have invaded arid habitats, but few are aquatic. Four classes are recognized: Diplopoda (millipedes), Chilopoda (centipedes), Pauropoda, and Symphyla, with 10,000, 3000, 500, and 160 species, respectively. The last two are minute dwellers of the forest floor that consume living or decaying vegetation. Symphylans look somewhat like centipedes but the adults have 14 trunk segments and 12 pairs of limbs; the posterior end of the trunk has two conical cerci and spinning glands. Members of the class Pauropoda are soft-bodied, blind myriapods with 9 to 11 leg-bearing trunk segments and branched antennae.

Class Chilopoda: Centipedes

Centipedes are dorsoventrally flattened with 15 to 173 segments, each with one pair of legs (Fig. 4A). Poisonous forcipules (fangs) enable centipedes to kill and consume insects, other centipedes, annelids, mollusks, and sometimes small vertebrates; under most circumstances, the poison is not lethal to people. The body is partially hung beneath the legs to increase stability and to allow hind legs to step over front ones, which allows the insects to run swiftly in search of prey or to escape predators. Centipedes are found in most terrestrial environments including the desert fringe; the latter is surprising given their chitinous, noncalcified exoskeleton, which is relatively permeable to water.

Class Diplopoda: Millipedes

Millipedes have a somewhat cylindrical body with 11 to 90 segments (which are really fused "diplosegments") and two pairs of legs per segment. Segmental plates are constructed to prevent "telescoping" as the body bulldozes through forest litter, while still allowing the animal to roll up or coil when threatened (Fig. 4B). Millipedes are slow moving and herbivorous by nature, eating decaying leaves and wood. They lack poisonous fangs and instead repel predators with volatile poison produced by repugnatorial glands. Millipedes are relatively long-lived, with some surviving 7 years.

SUBPHYLUM HEXAPODA

Hexapoda ("six feet") includes a tremendous diversity of winged insects (class Insecta, subclass Pterygota) and many fewer wingless insect (subclass Apterygota) and noninsect classes (Diplura and Ellipura, orders Collembola and Protura).

FIGURE 4 Members of the subphylum Myriapoda. (A) Centipede (class Chilopoda). (Photograph by Jim Kalisch, courtesy of University of Nebraska Department of Entomology.) (B) Millipede (class Diplopoda), coiled in a defensive posture. (Photograph by D. R. Parks.)

Hexapods have three major body regions (head, thorax, abdomen) and six thoracic legs.

Entognathous Hexapods: Collembola, Protura, and Diplura

Entognathous hexapods include two small taxa (class Diplura and Ellipura, order Protura) living in moist forest litter and a large group of springtails (class Ellipura, order Collembola) with at least 4000 species in terrestrial and semiaquatic environments. Most springtails live in moist terrestrial environments, but some colonize the surface film of quiet fresh and marine waters. They occur at densities much higher than almost any other invertebrate in soil litter. Unlike insects, springtails have only six abdominal segments, and cleavage of their eggs is total. Their name is derived from their ability to spring forward several centimeters when a forked structure (the furcula) flexed under the abdomen is rapidly uncocked. They have indirect fertilization, the young closely resemble adult Collembola, and adults continue molting throughout their lives (2–50+ molts). Springtails feed on decomposing organic matter or on microorganisms at the water surface. Proturans are completely terrestrial, their antennae have nearly atrophied away, and their front legs function somewhat like antennae. Diplurans are primitive hexapods whose ancestors may have given rise to both Protura and Collembola, and they are more closely related to insects than are ellipurans. Diplurans are blind and have two prominent abdominal cerci.

Class Insecta: Winged and Wingless Insects

The million or so species in the subclass Pterygota include all winged invertebrates and some insect species that have secondarily lost wings during evolution. They include two orders of ancient winged insects (Ephemeroptera and Odonata) and some 25 to 30 (depending on the classification system) orders of modern folding-wing insects. Most have 11 abdominal segments. The head features two antennae and compound eyes. Respiration is generally with internal tracheae, but aquatic species may use external, tracheate gills or other means to obtain sufficient oxygen. Fertilization is usually direct, distinct developmental stages are common, and molting generally stops with attainment of reproductive maturity. Their most prominent features are two pairs of wings, but a great many insects (e.g., fleas) lack wings or have dispensed with either the hind (e.g., flies) or fore pair (beetles). Among their beneficial attributes are pollination of most flowering plants, production of honey and silk, predation on harmful insects, decomposition of animal wastes and carcasses, and facilitation of ecological processes at all trophic levels above primary producer. Negative attributes include transmission of diseases, annoying bites, and damage to crops, stored food, ornamental plants, forests, and wooden structures.

All insect species that did not evolve from a winged hexapod and whose adults all lack wings are in the subclass Apterygota, order Thysanura. This small group of 600 or so primitive species includes bristletails, silverfish, and rock jumpers. These are small to medium-sized insects (5–25 mm) without compound eyes. They have an 11-segmented abdomen with a prominent caudal filament between two terminal cerci. Fertilization is indirect, and molting continues after the reproductive state has been reached (unlike insects). No pronounced metamorphosis is evident from subadult to adult stages. Thysanurans are swift, agile runners (probably to avoid predators) and are omnivorous scavengers of animal and plant matter. Most live in litter of forests and grasslands, but silverfish also infest houses, where they can extensively damage clothing and books.

SUBPHYLUM CRUSTACEA

Crustaceans surpass all other invertebrates in their direct contribution to human diets (from crabs, shrimp, lobsters, and crayfish) and are vitally important to many ecosystems, especially planktonic food webs. Unfortunately, they also foul boat hulls (barnacles) and destroy wooden piers in coastal waters (burrowing isopods).

Distinguishing characteristics of adults include the following: five-pairs of cephalic appendages (two mandibles, four maxillae, and two antennae), two to three tagmata, a chitinous cuticle often elaborated as a shieldlike carapace, more than 11 abdominal segments, and jointed, biramous appendages. Evolutionary trends involved specialization of mouthparts, body segments, and appendages for locomotion, sensory reception, and reproduction. Respiration is typically with gills, and hemocyanin is the principal respiratory pigment. Excretion of ammonia generally occurs through modified nephridia. Sexes are mostly separate, but hermaphroditism is common. Development always includes triangular nauplius larvae (with six appendages and a median eye), which are commonly planktonic. Many crustaceans have a relatively sophisticated behavioral repertoire and communicate visually, tactilely, and chemically.

Most of the roughly 50,000 species are marine, but crustaceans are ubiquitous in freshwater habitats and a few species have colonized saline lakes and terrestrial environments. Crustaceans are most often scavenging predators or have a generally omnivorous diet. They range in length from minute to truly gigantic (0.25 mm to 360 cm).

Class Malacostraca: Shrimp, Crayfish, and Others

About 60% of crustacean species are malacostracans, including all species consumed by humans (mostly decapods) and many important benthic crustaceans in marine and freshwater ecosystems (e.g., amphipods and isopods).

Most orders and the 29,000 species in this class are dominated by crustaceans that live near the bottom. Ecologically important exceptions include planktonic krill (order Euphausiacea), a vital prey of baleen whales. Parasitism is relatively rare and is confined mostly to ectoparasites of fish, crustaceans (both mostly in the large order Isopoda), and marine mammals (small component of the large order Amphipoda).

Class Branchiopoda: Water Fleas, Brine Shrimp, and Others

Nearly 25% of all freshwater crustaceans are branchiopods ("gill feet"), and almost all crustaceans in saline lakes are in this class; only 3% of the species occur in oceans. The class contains about 800 species of water fleas (in four orders of cladocera), 200 species of fairy and brine shrimp (order Anostraca), and a few taxa of clam (order Spinicaudata), pea (Laevicaudata), and tadpole shrimp (Notostraca). Most taxa other than cladocerans are restricted to ephemeral pools, and most branchiopods are suspension feeders. Branchiopods are noted for producing dormant embryos resistant to adverse environments.

Class Ostracoda: Seed Shrimp

The 6650 species of seed or mussel shrimp are minute crustaceans characterized by a protective, dorsal, bivalved carapace of chitin heavily impregnated with calcium carbonate. Freshwater species rarely exceed 3 mm in length, but marine taxa can reach 30 mm. Ostracodes are especially prevalent in freshwater habitats (particularly benthic areas) but are common in marine environments. A few genera have adapted to damp humus habitats of the forest floor. Ostracods are typically suspension feeders on benthic and pelagic detritus, and almost all are free-living.

Class Copepoda

In relatively permanent freshwater and marine environments, the 8000 species of copepods are the most important planktonic crustaceans, and other species contribute to the benthic fauna. Herbivory on microalgae prevails, but raptorial feeding on other zooplankton is common. Although freeliving copepods predominate, bizarre forms, barely recognizable as copepods, have evolved as parasites of fish and invertebrates.

Class Cirripedia: Barnacles

The name "barnacle" evokes a rugged image of the sea in the minds of many people, but few recognize this taxon's kinship with familiar crustaceans like shrimp and crabs because the barnacle's body is hidden inside calcareous plates and free-living species are sedentary. The 1000 species in the class Cirripedia include free-living species that live directly on a hard surface or are raised on a stalk (peduncle). This surface may be an inanimate object (e.g., rocks, floating refuse, ship hulls) or the outer layer of a living whale, turtle, invertebrate, or other larger organism. Barnacles usually cement the head to hard surfaces and employ setose legs to capture plankton. Other species, barely recognizable as barnacles, are ecto- and endoparasites of echinoderms, corals, and other crustaceans, especially crabs.

Pentastomida and Other Crustaceans

The remaining 264 to 1500 species of crustaceans are divided among six classes. Of these, only the ectoparasitic fish lice (class Branchiura), with 125 species, were recognized more than a few decades ago. Three classes closely resemble copepods—Cephalocarida, Mystacocarida, Tantulocarida—with the first two living in marine sediments and the last being ectoparasites of deep-water crustaceans. The class Tantulocarida is usually listed with 10 to 20 species, but some scientists believe that the true diversity is greater than 1000. Members of the class Remipedia are presently restricted to tropical underwater caves. Their long bodies with abundant lateral appendages call to mind segmented polychaete worms.

CLOSELY RELATED PHYLA: TARDIGRADA AND ONYCHOPHORA

Tardigrada: Water Bears

Tardigrades are called "water bears" because of their slow lumbering gait and relatively massive claws on lobopodous legs. The permeability of the cuticle limits tardigrades to aquatic habitats (often interstitial), the surface film of terrestrial mosses, and damp soil. About 800 species have been described, but many more undoubtedly exist in unexplored habitats. Faced with inhospitable microhabitats from environmental changes, both terrestrial and aquatic species may undergo cryptobiosis, where the body becomes dehydrated and metabolism is greatly reduced until favorable conditions return. Some water bears have been "resuscitated" from this state after decades! These normally dioecious organisms can also reproduce by parthenogenesis. Tardigrades typically feed on plants cells pierced by a pair of mouth stylets and sucked into the gut, but a few species are carnivorous.

Onychophora: Velvet Worms

Velvet worms are giants compared with tardigrades, for some individuals reach a length of 15 cm, but they share many characteristics with this sister phylum of Arthropoda. They are generally nocturnal and move by extending their legs, with hydrostatic pressure generated by muscular contraction within the legs.

Though most are herbivores or omnivores, many species consume small arthropods in a rather unique manner. They attack their prey and also defend themselves from predators by expelling a sticky, proteinaceous substance that entangles the target. This glue is produced by slime glands within oral papillae. The onychophoran then bites and secretes toxins to kill and partially liquefy the victim.

See Also the Following Articles
Biodiversity • Fossil Record • Insecta, Overview

Further Reading
Anderson, D. T. (ed.) (1998). "Invertebrate Zoology." Oxford University Press, Melbourne, Australia. (See especially Chaps. 10–14, 18.)
Bliss, D. E. (ed.-in-chief). (1982–1985). "The Biology of Crustacea," Vols. 1–10. Academic Press, New York.
Parker, S. P. (ed.). (1982). "Synopsis and Classification of Living Organisms," Vol. 2, McGraw-Hill, New York.
Pechenik, J. A. (2000). "Biology of the Invertebrates." 4th ed. McGraw-Hill, New York (Particularly Chaps. 14 and 15.)
Thorp, J. H., and Covich, A. P. (eds.) (2001). "Ecology and Classification of North American Freshwater Invertebrates." 2nd ed. Academic Press, San Diego, CA. (See especially Chaps. 16–23.)

Auchenorrhyncha
(Cicadas, Spittlebugs, Leafhoppers, Treehoppers, and Planthoppers)

C. H. Dietrich
Illinois Natural History Survey

The hemipteran suborder Auchenorrhyncha is the group of sapsucking insects comprising the modern superfamilies, Cercopoidea (spittlebugs, Fig. 1), Cicadoidea (cicadas, Fig. 2), Membracoidea (leafhoppers and treehoppers, Fig. 3), and Fulgoroidea (Fig. 4) Together, these groups include over 40,000 described species. Morphologically, Auchenorrhyncha differ from other Hemiptera in having the antennal flagellum

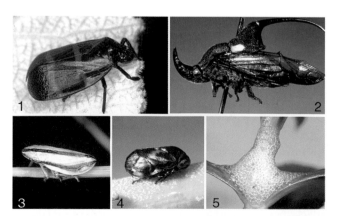

FIGURE 1 Cercopoidea: spittlebugs and froghoppers: (1) *Tomaspis* sp. (Cercopidae), Mexico, (2) *Machaerota* sp. (Machaerotidae), Vietnam, (3) *Paraphilaenus parallelus* (Aphrophoridae), Kyrgyzstan, (4) *Clastoptera obtusa* (Clastopteridae), Illinois, U.S.A., (5) spittle mass of *P. spumarius* nymph, Illinois, U.S.A.

FIGURE 2 Cicadoidea: cicadas: (6) a hairy cicada, *Tettigarcta crinita* (Tettigarctidae), Australia, (7) *Melampsalta calliope* (Cicadidae), Illinois, U.S.A., (8) a periodical cicada, *Magicicada cassini*, with a 13-year life cycle, Illinois, U.S.A., (9) a dog day cicada, *Tibicen* sp., molting into the adult stage, Illinois, U.S.A.

hairlike (aristoid), the rostrum (modified, beaklike labium) arising from the posteroventral surface of the head, a complex sound-producing tymbal apparatus, and the wing-coupling apparatus consisting of a long, downturned fold on the

FIGURE 3 Membracoidea: leafhoppers and treehoppers: (10) a brachypterous, grass-feeding leafhopper, *Doraturopsis heros*, Kyrgyzstan, (11) *Pagaronia triunata* (Cicadellidae), California, U.S.A., (12) *Eurymeloides* sp. (Cicadellidae), Australia, (13) fifth instar of *Neotartessus flavipes* (Cicadellidae), Australia, (14) female *Aetalion reticulatum* (Aetalionidae) guarding egg mass, Peru, (15) ant-attended aggregation of treehopper adults and nymphs (Membracidae: *Notogonia* sp.), Guyana.

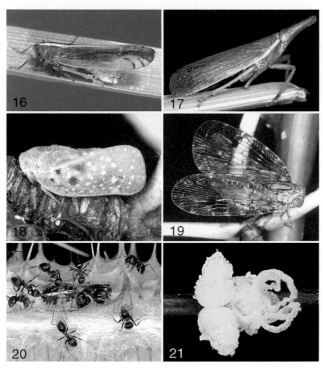

TABLE I Classification of the Hemipteran Suborder Auchenorrhyncha (synonyms and common names in parentheses) Excluding Extinct Taxa

Auchenorrhyncha (Cicadinea)
  Infraorder Cicadomorpha (Clypeorrhyncha, Clypeata)
    Superfamily Cercopoidea (spittlebugs, froghoppers)
      Aphrophoridae
      Cercopidae
      Clastopteridae
      Machaerotidae
    Superfamily Cicadoidea (cicadas)
      Cicadidae (Platypediidae, Plautillidae, Tettigadidae, Tibicinidae)
      Tettigarctidae (hairy cicadas)
    Superfamily Membracoidea (Cicadelloidea)
      Aetalionidae (Biturritiidae)
      Cicadellidae (Eurymelidae, Hylicidae, Ledridae, Ulopidae, leafhoppers)
      Melizoderidae
      Membracidae (Nicomiidae, treehoppers)
      Myerslopiidae (Cicadellidae, in part)
  Infraorder Fulgoromorpha (Archaeorrhyncha, planthoppers)
    Superfamily Fulgoroidea
      Acanaloniidae
      Achilidae
      Achilixiidae
      Cixiidae
      Delphacidae
      Derbidae
      Dictyopharidae
      Eurybrachidae
      Flatidae
      Fulgoridae (lanternflies)
      Gengidae
      Hypochthonellidae
      Issidae
      Kinnaridae
      Lophopidae
      Meenoplidae
      Nogodinidae
      Ricaniidae
      Tettigometridae
      Tropiduchidae

FIGURE 4 Fulgoroidea: planthoppers: (16) female *Stenocranus* sp. (Delphacidae) covering oviposition site with wax, Illinois, U.S.A., (17) *Chanithus scolopax* (Dictyopharidae), Kyrgyzstan, (18) *Metcalfa pruinosa* (Flatidae), Maryland, U.S.A., (19) *Biolleyana* sp. (Nogodinidae), Mexico, (20) *Tettigometra* sp. (Tettigometridae) nymphs tended by ants, Greece, (21) unidentified planthopper nymph completely covered with wax filaments, Guyana.

forewing and a short, upturned lobe on the hind wing. Auchenorrhyncha are abundant and ubiquitous insects, distributed worldwide in nearly all terrestrial habitats that support their host plants, but they are particularly diverse and speciose in the tropics. Some are important agricultural pests, injuring plants either directly through feeding and oviposition, or indirectly through the transmission of plant pathogens.

PHYLOGENY AND CLASSIFICATION

Nomenclature

The monophyly of the four existing superfamilies of Auchenorrhyncha has long been accepted, but controversy persists regarding the relationships of these lineages to each other and to various other fossil and extant hemipteran lineages. Consequently, no single classification scheme has gained universal acceptance, and the nomenclature of the various groups is presently unstable. Traditionally, Auchenorrhyncha were treated as one of three suborders of the order Homoptera. Fossil evidence, as well as phylogenetic analyses based on DNA sequences of extant taxa, suggest that Heteroptera (true bugs; Hemiptera, *sensu stricto*) arose from within Homoptera and, possibly, from within Auchenorrhyncha. Thus, many recent workers have combined Homoptera and Heteroptera into a single order. This order is usually referred to as Hemiptera (*sensu lato*), but some entomologists advocate using the ordinal name Rhynchota to avoid confusion with the more restricted definition of Hemiptera (Heteroptera) widely used in the literature. Some recent workers have further proposed dividing the Auchenorrhyncha into two suborders: Clypeorrhyncha for the lineage comprising Cicadoidea, Cercopoidea, and Membracoidea, and Archaeorrhyncha for Fulgoroidea. The older names Cicadomorpha and Fulgoromorpha, respectively (usually treated as infraorders within suborder Auchenorrhyncha), are more commonly used for these two groups. For convenience, and because the phylogenetic status of the group has not been elucidated satisfactorily, Auchenorrhyncha is retained here as the subordinal name with the caveat that this group may represent a paraphyletic assemblage rather than a monophyletic group. The current classification of families is presented in Table I.

Fossil Record

Auchenorrhyncha arose in the Paleozoic, first appearing in the fossil record in the Lower Permian (280 mya) and, judging from the abundance of forms described from Permian strata, they diversified explosively. These early auchenorrhynchans had adults with well-developed jumping abilities and somewhat resembled modern leafhoppers and spittlebugs, but nymphs (juveniles) associated with these insects were bizarrely flattened or biscuitlike, with short legs, foliaceous lobes on the head, thorax, and abdomen (similar to those of some modern Psyllidae) and elongate mouthparts, suggesting a sessile, cryptic lifestyle. The fulgoromorphan and cicadomorphan lineages (Table I) apparently diverged by the middle Permian. By the late Permian, Fulgoroidea appeared and Cicadomorpha (*sensu lato*) had diverged into the Pereboreoidea, comprising three extinct families of large cicada-like insects, and the smaller Prosboloidea, from which the three modern cicadomorphan superfamilies apparently arose. Cicadomorphans with a greatly inflated frontoclypeus (Clypeata in the paleontological literature = Clypeorrhyncha) did not appear until the Mesozoic. Prior to that, the head of Cicadomorpha resembled that of modern Psyllidae in having the frontovertex extended ventrad on the face to the antennal ledges and the lateral ocelli situated close to the eyes. This change in head structure is thought to have been associated with a shift from phloem to xylem feeding. Xylem feeding was apparently the predominant feeding strategy of the group throughout the Mesozoic, but in the late Cretaceous or early Tertiary the major lineages of phloem-feeding leafhoppers and treehoppers, which predominate in the recent fauna, arose. In these insects, the frontoclypeus became more flattened, probably because of the reduction in size of the cibarial dilator muscles. This was presumably in response to a shift from feeding on xylem, which is under negative pressure, to phloem, which is under positive pressure. Cicadoidea and Cercopoidea first appeared in the Triassic, and Membracoidea in the Jurassic. With the exception of Tettigarctidae, which arose in the late Triassic and is now confined to Australia, extant families of these groups do not appear in the fossil record until the Cretaceous or early Tertiary. Most Auchenorrhyncha from Baltic and Dominican amber of the Tertiary age are virtually indistinguishable from modern forms.

LIFE HISTORY

Courtship

Adult male and female Auchenorrhyncha locate each other by means of species-specific acoustic courtship signals. These signals are produced by specialized organs at the base of the abdomen called tymbals, present in both sexes (except female cicadas). A few cicadas and planthoppers are also able to use the stridulatory surfaces of their wings to produce sound. The loud, sometimes deafening, calls of many male cicadas are well known. In noncicadoids, the courtship calls are usually inaudible, being transmitted through the substrate, and distinct tympana are absent. The calls of some leafhoppers and planthoppers, audible only with special amplifying equipment, are among the most complex and beautiful of any produced by insects. Males move from plant to plant, signaling until they receive a response from a female. In addition to intensification of the vibrational signals, precopulatory behavior in some species may involve the male buzzing or flapping the wings, tapping the female with the legs, or repeatedly walking around or over the female. Copulation involves insertion of the male aedeagus into the female vulva at the base of the ovipositor and may last from a second or less to several hours, depending on the species. Females of most species seem to mate only once, while males often mate several times.

Oviposition and Nymphal Development

Females lay eggs singly or in batches, usually either by inserting them into plant tissue or by depositing them on plant surfaces [Figs. 3(14), 3(15), and 4(16)]. In some groups, eggs are deposited in the soil or in litter. Egg batches may be covered with plant debris, wax filaments, or secretions produced by various internal glands. Eggs may or may not undergo diapause depending on the species and climate. After hatching, the juveniles [nymphs, Figs. 2(9), 3(13), 3(15), 4(20), and 4(21)] undergo five molts prior to reaching the adult stage. In most species the nymphs feed on aboveground parts of host plant, but in cicadas, Cercopidae, a few fulgoroid families, and a few leafhopper genera, the nymphs are subterranean root feeders. Formation of galls, common among aphids and psyllids, is known in only one Auchenorrhyncha species (a leafhopper). Nymphal development requires from a few weeks to several years (in cicadas), depending on the species. Some species exhibit parental care behavior (see later).

BEHAVIOR AND ECOLOGY

Feeding and Digestion

Adult and nymphal Auchenorrhyncha feed by inserting the two pairs of feeding stylets (modified mandibles and maxillae) into the host plant tissue, injecting saliva, and ingesting fluid. Unlike Sternorrhyncha, in which the stylets pass between the cells of the host tissue (intercellular feeding), Auchenorrhyncha stylets usually pierce the cells (intracellular feeding). After selecting an appropriate feeding site based on visual and chemical cues, the insect presses the tip of the labium onto the plant surface and inserts the feeding stylets. Just prior to, and during probing of the plant tissue with the stylets, the insect secretes sheath saliva that hardens on contact with air or fluid to form an impervious salivary sheath surrounding the stylets. The sheath forms an airtight seal that prevents leakage of air or fluid during feeding. Stylet probing continues until a suitable tissue is found (xylem, phloem, or mesophyll, depending on the species), after which feeding can commence. During feeding, watery saliva is injected into the plant to aid digestion and to prevent clogging of the stylet opening. This is also the mechanism by which the insect may infect the plant with pathogens (see later). Feeding may last from a few seconds to many hours at a time, depending on the auchenorrhynchan species and the quality of the plant tissue. During feeding, droplets of liquid excretion are ejected from the anus, several droplets per second in some xylem feeders.

Plant sap is a nutritionally imbalanced food source; phloem is high in sugar and xylem is, in general, nutrient poor and extremely dilute. Auchenorrhyncha have acquired various adaptations that enable them to convert the contents of plant sap into usable nutrients. Most Cicadomorpha have part of the midgut modified into a filter chamber that facilitates rapid removal of excess water. Fulgoroidea lack a distinct filter chamber but have the midgut tightly coiled and partially or completely enclosed in a sheath of specialized cells that apparently absorb solutes from the gut contents. A broad array of transovarially transmitted (i.e., from the mother through her eggs to her offspring) prokaryotic endosymbionts have also been identified in various Auchenorrhyncha species. The roles of these endosymbionts have not been fully elucidated, but presumably they function in the conversion of the nutritionally poor plant sap on which the insects feed into essential vitamins, amino acids, and sterols. The symbionts are housed either intracellularly in specialized fat body cells called mycetocytes, intracellularly in the fat body, or in the gut epithelium. Several distinct mycetomes, consisting of groups of mycetocytes, are often present. In Cicadomorpha, each mycetome may house up to six different kinds of endosymbiont. In Fulgoroidea, only a single kind of endosymbiont is housed in each mycetome.

Host Associations

Nearly all Auchenorrhyncha are plant feeders; the few known exceptions (e.g., Fulgoroidea: Achilidae and Derbidae) feed on fungi as nymphs. Auchenorrhynchans use a wide variety of plants including mosses, horsetails, ferns, cycads, conifers, and angiosperms, but the vast majority of species feed on flowering plants. Most species appear to be restricted to a single genus or species of plants. Many species, particularly among the xylem-feeding groups, normally use a few or a single plant species but are capable of feeding and developing on a variety of alternate hosts if the preferred host is not available. A few xylem-feeding species have extremely broad host ranges. For example, the meadow spittlebug, *Philaenus spumarius,* with over 500 documented food plants, has the broadest known host range of any herbivorous insect. Phloem- and mesophyll-feeding species, comprising the majority of Auchenorrhyncha, tend to have narrower host ranges than xylem feeders, and many species appear to use a single plant family, genus, or species. Host associations appear to be conservative in some auchenorrhynchan

lineages. Delphacidae and Cicadellidae (Deltocephalinae) include large numbers of grass- and sedge-specialist species and are among the dominant herbivores in grasslands. Most of the major lineages of Auchenorrhyncha do not exhibit a distinct preference for any particular plant taxon and usually include both host-generalist and host-specialist species. Some species alternate hosts during different stages in the life cycle or in different seasons. For example, nymphs of many leafhoppers and treehoppers develop on herbs, but the adult females oviposit on a woody host.

Migration

Most species of Auchenorrhyncha are relatively sedentary, completing their life cycle within a small area. Although most species have well-developed wings and are strong fliers, few seem to move more than a kilometer from their birthplace. Many species, particularly those inhabiting grasslands and deserts, are submacropterous or brachypterous [short winged, Fig. 3(10)] and, thus, incapable of sustained flight. Some of these species occasionally produce macropterous (long-winged) females that move to new patches of suitable habitat. Other species produce both short- and long-winged forms either simultaneously or in alternate generations. The proportion of macropterous to brachypterous forms often varies in response to population density. Some Auchenorrhyncha species undergo annual migrations that may cover hundreds of kilometers. Not coincidentally, many of these accomplished migrants are important agricultural pests. Among the best studied of these are the brown planthopper (*Nilaparvata lugens)* and the potato leafhopper (*Empoasca fabae).* Neither of these species can normally overwinter in high latitudes. Populations build up in the tropical or subtropical parts of their range and migrate to higher latitudes each spring. They are assisted in their migratory flights by convection and favorable winds, and the initiation of migratory behavior is apparently triggered by favorable atmospheric conditions. Sporadic incidents of very-long-range migrations have also been documented. In one such incident in 1976, swarms of *Balclutha pauxilla* (Cicadellidae), probably originating from a source population in Angola, descended on Ascension Island, 2700 km away in the mid-Atlantic.

Thermoregulation

Most Auchenorrhyncha species appear to regulate their body temperature behaviorally, by seeking out microhabitats in which the ambient temperature remains within a narrow range and moving among alternate microhabitats as conditions change. In some cicadas, physiological mechanisms are also involved. Some species are facultatively endothermic, producing metabolic heat to facilitate calling, courtship, and other activities. This is usually accomplished by vibrating the flight or tymbal muscles until the body temperature rises to an optimal level. Some desert cicadas cool themselves by evaporation of excess water released through pores on the thorax and abdomen. In this way they are able to remain active at ambient temperatures that would kill other insects.

Defense and Escape

Because they are among the most abundant phytophagous insects in many habitats, Auchenorrhyncha are an important food source for numerous vertebrate and invertebrate predators (see next section: Natural Enemies). Species of Auchenorrhyncha exhibit myriad strategies for avoiding predation. These range from relatively simple behaviors, such as dodging around to the opposite side of a leaf or branch as a predator approaches, or hiding under a leaf sheath, to complex mutualistic associations and mimicry. Adults of many species are strong flyers and nearly all (except cicadas) are also excellent jumpers. Juvenile (nymphal) cicadas, spittlebugs, treehoppers, and some planthoppers are incapable of jumping and have adopted other strategies for avoiding predators. All cicada nymphs and many spittlebug and planthopper nymphs are subterranean; thus, their exposure to most predators is minimal. Spittlebug nymphs live within masses of froth and machaerotid nymphs live in calcareous tubes cemented to the host plant. The free-living nymphs of most other auchenorrhynchans appear to rely on cryptic coloration and body forms to escape detection by visual predators such as birds. For example, many treehopper nymphs are strongly flattened with the ventral surfaces of the body concave, enabling them to lie flat against the bark or leaf surfaces of their host plant. Others resemble plant parts such as bud scales or leaflets. Many planthopper nymphs secrete copious quantities of wax [Fig. 4(21)], with which they coat themselves and, often, surrounding parts of their host plants. The wax may prevent parasites and predators from grasping the nymphs, allowing them to leap away. Adults of some species mimic various venomous arthropods such as ants, wasps, robber flies, assassin bugs, and spiders. Some bear horns or spines on the pronotum [Membracidae, Fig. 3(15)] or scutellum [Machaerotidae, Fig. 1(2)] that make them physically difficult for some vertebrate predators to swallow. Many adult cercopids and membracids have conspicuous (aposematic) color patterns, presumably indicating that they are unpalatable. Others have the forewing apices marked with false eyespots, and a few (e.g., Fulgoroidea: Eurybrachidae) have prolongations resembling antennae; the head and thorax of such species often bear transverse lines resembling abdominal segmentation. Adults of various planthopper species mimic lizards, flowers, and lichens. Another strategy involves complex mutualistic associations with ants and other social hymenopterans. Ant mutualism has been documented in numerous lineages of Fulgoroidea and Membracoidea and occurs universally in some groups [e.g., tettigometrid planthoppers Fig. 4(20) and eurymeline leafhoppers]. In such groups, the nymphs usually form aggregations that are tended by ants. The aggressive worker

ants drive off predators and receive gifts of honeydew, a sugary excretion, from the nymphs. Ant mutualism may have facilitated the development of subsocial behavior in some groups (see Membracoidea section under Diversity).

Natural Enemies

Auchenorrhyncha are preyed upon by insectivorous vertebrates such as birds and lizards, as well as by invertebrate predators such as spiders, ants, assassin bugs, wasps, and robber flies. Auchenorrhyncha are also attacked by various parasitoids such as dryinid and chalcidoid wasps, epipyropid moths, pipunculid flies, strepsipterans, and nematodes. Because they feed on plant sap, cicadomorphans are not usually susceptible to infection by viral, bacterial, or protozoan pathogens. Thus, entomopathogenic fungi, which do not need to be ingested to infect insects, are the most important pathogens of Auchenorrhyncha.

Economic Importance

Although the vast majority of species of Auchenorrhyncha are benign, the group contains some of the most destructive pests of agriculture. Among the most important are the brown planthopper, sugarcane planthopper (*Perkinsiella saccharicida*), corn planthopper (*Peregrinus maidis*), meadow spittlebug, beet leafhopper (*Neoaliturus tenellus*), potato leafhopper, corn leafhopper (*Dalbulus* spp.), African maize leafhopper (*Cicadulina* spp.), green rice leafhopper (*Nephotettix* spp.), and various grape leafhoppers (*Arboridia* and *Erythroneura* spp.).

Auchenorrhyncha injure plants directly through feeding or oviposition or, more often, indirectly through the transmission of plant pathogens. Economic injury to plants involving cicadas, which occurs rarely, is mainly due to oviposition, although some species occasionally inflict feeding damage (e.g., on sugarcane). Spittlebugs injure plants primarily through feeding and through transmission of xylem-limited bacterial pathogens. Species of Cercopidae are the most significant pests of forage grasses in pastures in Latin America and are also destructive of sugarcane. Interestingly, much if not most of the economic damage done by spittlebugs is due to native spittlebug species colonizing nonnative hosts (e.g., introduced forage grasses, clovers, etc.). Presumably, such plants lack natural resistance to spittlebugs and are more susceptible to injury.

Leafhoppers and planthoppers are among the most significant groups of vectors of plant pathogens, transmitting viruses, bacteria, and mycoplasma-like organisms. Over 150 species are known vectors of economically important plant pathogens. The insects usually acquire the pathogen by feeding on an infected plant, but some pathogens may be transmitted transovarially from mother to offspring. Phloem-limited viral and mycoplasma-like pathogens typically multiply within the vector and enter the plant when the insect injects saliva during feeding. Some xylem-limited bacterial pathogens (e.g., *Xylella*) are apparently unable to travel from the gut to the salivary glands and require regurgitation from the foregut during vector feeding to infect the plant. Annual losses to maize, rice, and sugarcane attributed to pathogens spread by leafhoppers and planthoppers are estimated in the hundreds of millions of dollars. Xylem-feeding cicadelline leafhoppers are also the main vectors of *Xylella fastidiosa*, which causes X diseases of stone fruits (*Prunus* spp.), Pierce's disease of grape, citrus variegated chlorosis, and alfalfa dwarf.

Some Auchenorrhyncha species are considered to be beneficial. Cicadas are used as food by several human cultures. The use of Auchenorrhyncha in biocontrol of weeds has also begun to be explored. For example, a Neotropical tree hopper species (*Aconophora compressa*) has been introduced into Australia for control of *Lantana* (Verbenaceae).

Control

Control of auchenorrhynchan pests has traditionally involved the use of conventional contact insecticides, but overuse of chemical insecticides has led to the development of resistance in many pest species and has suppressed populations of their natural enemies. Modern integrated pest management has promoted greater use of resistant plant varieties, cultural control (e.g., removal of litter to reduce numbers of overwintering individuals), and biological control by means of parasitoids and pathogens, as well as more judicious use of pesticides.

CAPTURE AND PRESERVATION

Auchenorrhyncha are most commonly collected by sweeping vegetation with a heavy canvas net. Many species are also attracted to lights. Vacuum collecting is effective for collecting from dense grassy vegetation where many species reside. A gasoline-powered leaf blower fitted with a vacuum attachment can be used to suck the insects from dense vegetation. A fine-mesh insect net bag taped to the end of the intake nozzle will capture the specimens. Other effective collecting methods include Malaise trapping and insecticidal fogging of forest canopy. Auchenorrhycha may be killed in a standard insect killing jar containing potassium cyanide or ethyl acetate, or by freezing.

Specimens for morphological study are usually mounted dry on pins or point mounts. Point mounts should be glued to the right side of the thorax. To identify the species of a specimen, it is often necessary to examine the male genitalia. To do this, the abdomen is removed and soaked in 10% potassium hydroxide solution for several hours (or boiled in the same solution for a few minutes) to clear the pigment. The abdomen is then rinsed in clean water containing a small amount of glacial acetic acid, rinsed again in pure water, and immersed in glycerine. After examination, the cleared abdomen is stored in a glass or plastic microvial pinned beneath the rest of the specimen. Auchenorrhyncha may also be preserved indefinitely in 80 to 95% ethanol, but this causes some green pigments to fade to yellow.

DIVERSITY

Cercopoidea

Cercopoidea (froghoppers and spittlebugs, Fig. 1) are characterized by the following combination of morphological characters: head with frontoclypeus inflated; median ocellus absent; ocelli on crown distant from margin; pronotum extended to scutellar suture; body clothed with fine setae; hind coxae conical, tibia without rows of setae but often with one or more conspicuous spines; male subgenital plate present. The superfamily comprises four families Aphrophoridae, Cercopidae, Clastopteridae, and Macherotidae. The first Cercopoidea (Procercopidae) appear in the fossil record during the Lower Jurassic. These insects retained a median ocellus and apparently lacked the dense setal covering of modern cercopoids. Aphrophoridae and Cercopidae did not appear until the middle Cretaceous; Clastopteridae and Machaerotidae apparently arose during the Tertiary.

Approximately 2500 species and 330 genera of Cercopoidea have been described. The classification has not been revised in over 50 years, and the phylogenetic status of most cercopoid genera and higher taxa remains unknown. Cercopidae [Fig. 1(1)], the largest family, differs from Aphrophoridae [Fig. 1(3)], the next largest, in having the eyes slightly longer than wide and the posterior margin of the pronotum straight (instead of emarginate). The small families Machaerotidae and Clastopteridae differ from other Cercopoidea in having a well-developed appendix (distal membrane) on the forewing. Machaerotidae [Fig. 1(2)] differ from Clastopteridae [Fig. 1(4)] in having two or more r-m crossveins in the forewing and in lacking an outer fork on the radial vein of the hind wing.

Production of "spittle" is a unique characteristic of Cercopoidea [Fig. 1(5)]. Nymphs of Machaerotidae produce the froth during molts, while in other families nymphs live permanently surrounded by the froth. The lateral parts of nymphal abdominal segments are extended ventrally into lobes, which form an open or closed (in machaerotids) ventral cavity, filled with air. The nymphs introduce bubbles of air into their liquid excretion by bellowslike contractions of this device; periodically the tip of the abdomen is extended through the surface of spittle mass to channel air into the cavity. The same air supply is used for breathing via spiracles that open into the ventral cavity. The froth is stabilized by the action of the secretory products manufactured in the highly specialized Malpighian tubules of the nymphs and mixed into the main watery excreta. Wax secreted by plates of epidermal glands on the sixth through eighth abdominal terga (Batelli glands) may also help stabilize the froth.

The function of the spittle mass is not completely understood. It is usually assumed that it protects the insect from predators and desiccation. Cercopoid nymphs are sessile and live within the spittle mass (or, in Machaerotidae, inside fluid-filled tubes). In some species, nymphs tend to aggregate, forming large spittle masses containing hundreds of individual nymphs. Nymphs of Cercopidae apparently feed on roots, whereas aphrophorid and clastopterid nymphs occur on aboveground parts of their host plants. Nymphs of the Machaerotidae live immersed in liquid inside tubes cemented to the twigs of their host plants. The tubes are constructed from calcium carbonate and other salts secreted by the midgut and an organic matrix secreted by the Malpighian tubules. Adult cercopoids do not produce spittle and are free living. They cannot run, and often use only the front and middle legs to walk, dragging the extended hind legs. Consequently, they rely mostly on their strong jumping and flying abilities for movement.

Species of Cercopoidea are often restricted to particular habitats, but many if not most seem to be capable of utilizing a variety of host plants. Many species seem to prefer actinorrhizal and other nitrogen-fixing hosts, presumably because the xylem sap of such plants contains more amino acids and is more nutritious. Cercopoidea is a predominantly tropical group, occurring mostly in wet and mesic habitats. Nevertheless, the genus *Clastoptera,* has radiated extensively in north temperate North America, and *Aphrophora* comprises numerous arboreal species throughout the Holarctic. Cercopidae are primarily grassland insects, feeding on grasses and other herbs. The family Aphrophoridae includes both grass-feeding and arboreal species. Machaerotidae and Clastopteridae are primarily arboreal.

Members of the superfamily Cercopoidea occur worldwide. Cercopidae and Aphrophoridae are pantropical in distribution, with relatively depauperate faunas in the Holarctic. Machaerotidae are restricted to the Oriental and Australian regions. Clastopteridae are mostly New World animals, but one small genus, possibly misclassified, occurs in the oriental region. Most tribes are restricted to either the New or the Old World, and phyletic diversity seems to be highest in the oriental region. A few genera (e.g., *Philaenus* and *Aphrophora*) are widespread, partly as a result of human activities, but most are restricted to a single biogeographic realm.

Cicadoidea

Cicadoidea (cicadas, Fig. 2) are distinguished from other extant Auchenorrhyncha in having fossorial front legs (in nymphs) and three ocelli grouped in a triangle on the crown of the head; in addition they lack the ability to jump. They are conspicuous insects because of their large size (1.5–11 cm) and the loud courtship calls of the males. Most authorities recognize two families: Cicadidae and Tettigarctidae. Tettigarctidae [Cicadoprosbolidae in the paleontological literature, Fig. 2(6)], which differ from Cicadidae in having the pronotum extended to the scutellum and lacking distinct tympana, are a relict group with two extant species in southern Australia and Tasmania and several fossil taxa dating to the Lower Jurassic. Cicadidae [Fig. 2(7–9)], which do not appear in the fossil record until the Paleocene, comprise two main (possibly polyphyletic) groups, those with the

tymbals (sound-producing organs) concealed and those with exposed tymbals. These two groups are sometimes given status as separate families, Cicadidae (*sensu stricto)* and Tibicinidae, respectively. Together these groups comprise approximately 1300 extant species. Phylogenetic analyses of the major lineages are in progress and it is likely that the classification of the superfamily will be substantially revised in the near future.

Although cicadas almost always lay eggs on aboveground parts of their host plant, the nymphs drop to the ground soon after hatching and use modified (fossorial) front legs to burrow into the soil, excavating a subterranean feeding chamber adjacent to a root. They feed on the xylem of the roots of perennial plants, coating themselves and lining their burrows with "anal liquid" that appears to be similar to that produced by cercopoid nymphs. Development in most species requires from 2 to 6 years (13 or 17 years in the periodical cicadas of temperate North America). Larger nymphs of some species inhabiting wet habitats construct towers of mud that facilitate aeration of the burrow. Mature nymphs emerge from the ground and climb onto a vertical surface prior to molting into the adult stage [Fig. 2(9)]. As far as is known, all cicadas feed on xylem sap; hence the frontoclypeus is strongly inflated owing to the presence of strong cibarial dilator muscles. Like the Cercopoidea, cicadas do not walk or run well; instead they rely on flight to move over distances greater than a few centimeters. In some cicada species, males are sedentary, often forming aggregations and calling loudly in choruses to attract females. In others, the male calls are less audible, and males fly frequently from place to place in search of females. Male and female cicadas have auditory organs (tympana) at the base of the abdomen. Unlike other Auchenorrhyncha, female cicadas (except *Tettigarcta*) do not produce acoustic signals. Tettigarctidae differ from other cicadas in producing only substrate-borne signals (in males and females).

Cicadoidea are the most ecologically uniform of the Auchenorrhyncha superfamilies. Nymphs of all species are subterranean root feeders, and adults feed on the aboveground parts of their host plants. Most cicada species tend to be associated with particular habitats, and many seem to be host plant specific. Sympatric species often call at different times of day or mature during different seasons, thus temporally partitioning their habitat. The cicada faunas of deserts and savannas are particularly rich in genera and species, but tropical rain forests also harbor a great diversity of species.

Cicadoidea occur worldwide but, like the other two cicadomorphan superfamilies, are largely a tropical group. A few genera (e.g., *Cicada, Cicadetta*), occur on several continents, but most are restricted to a single biogeographic realm. Most species appear to have fairly narrow geographic ranges. The high degree of endemism in many groups has proven useful in studies of biogeography, particularly in the geologically complex island areas of the oriental and Australian regions.

Membracoidea

Membracoidea (leafhoppers and treehoppers, Fig. 3), by far the most speciose of the auchenorrhynchan superfamilies, are characterized morphologically by the narrow costal space of the forewing, the large, transversely articulated metathoracic coxae, the elongate hind femora, the longitudinal rows of enlarged setae on the hind tibiae, and the presence of scutellar apodemes. The superfamily includes Cicadellidae (leafhoppers), a paraphyletic taxon that apparently gave rise to a lineage comprising the three currently recognized families of treehoppers (Melizoderidae, Actalionidae, and Membracidae). A fifth family, Myerslopiidae, consists of two genera of small, flightless, litter-dwelling insects found only in New Zealand and Chile and thought to represent a distinct, relatively primitive lineage. Together, these groups comprise nearly 25,000 described species, currently grouped into about 3500 genera.

Membracoidea first appeared in the Jurassic, represented by the extinct family Karajassidae. These early membracoids were leafhopperlike insects with inflated faces (indicative of xylem feeding), and they retained a median ocellus and more primitive wing venation (forewing with CuA_1 free distally), but nevertheless had acquired the rows of enlarged setae on the hind tibia characteristic of modern leafhoppers. The first Cicadellidae appeared in the Lower Cretaceous. Treehoppers (Aetalionidae and Membracidae) make their first appearance in Tertiary age Mexican and Dominican amber.

The largest family, Cicadellidae [Fig. 3(10–13)], is characterized by the presence of four rows of enlarged, spine-like setae on the hind tibia, a peg-and-socket joint between the hind coxae, and the production of brochosomes. Membracidae [Fig. 3(15)], the next largest family, differ from Cicadellidae in having three or fewer rows of enlarged setae on the hind tibia, the male genital capsule with a lateral plate, and the pronotum enlarged, usually extended posteriorly over the scutellum and frequently bearing spines, horns, or other ornamentation. Like Membracidae, Aetalionidae [Fig. 3(14)]. have three or fewer setal rows on the hind tibia but differ in having the front femur fused to the trochanter, in having the scutellum completely exposed, and in having digitiform processes on the female genital capsule. Melizoderidae also resemble Membracidae but differ in having parapsidal clefts on the mesonotum. Myerslopiidae, thought to be the most primitive membracoid family, are bizarre, flightless insects with elytra-like forewings, vestigial ocelli, and a triangular mesocoxal meron resembling that of Cercopoidea. The phylogenetic status and relationships among the major lineages are only beginning to be understood.

Cicadellidae are unique among insects in producing brochosomes, which are minute proteinaceous granules synthesized in a specialized segment of the Malpighian tubules. After each molt, leafhoppers spread brochosomes over external surfaces of the body in an act known as anointing. Rows of modified setae on the legs of leafhoppers are used to distribute the brochosomes during anointing and subsequent acts of grooming. The brochosome coating of nymphal and adult leafhoppers makes the integument extremely hydrophobic and protects leafhoppers from becoming entrapped in drops of water and their own often copious excreta.

Ant mutualism and parental care behavior are widespread among treehoppers [Membracidae and Aetalionidae, Fig. 3(15)]. Females of many species guard their eggs [Fig. 3(14)] and sometimes remain with the nymphs throughout their development. In the treehopper tribes Hoplophorionini and Aconophorini, ant mutualism was lost but parental care was retained. In these groups, females are often able to drive off invertebrate predators by buzzing the wings and/or using the hind legs to kick the intruder off the plant. Acoustic alarm signals produced by the nymphs trigger the mother's defensive response. Female *Aconophora* coat the stem of the host plant on either side of their egg masses with a sticky secretion that traps predators and parasitoids.

Most species of Membracoidea seem to have fairly narrow host and habitat requirements, and this has probably contributed to their remarkable diversity. Particularly notable are the large leafhopper faunas of temperate and tropical grasslands, where they are, by far, the most speciose component of the grass-feeding herbivore fauna. Many leafhopper species in deserts and dry grasslands are flightless or only occasionally produce winged individuals. This trait has presumably reduced gene flow among populations and facilitated speciation in some lineages. In temperate forests of the Northern Hemisphere, the leafhopper subfamily Typhlocybinae has diversified extensively through specialization on individual tree genera and species. In tropical forest canopies, the treehopper family Membracidae and the leafhopper subfamilies Idiocerinae and Typhlocybinae are particularly diverse. In Australia, the endemic fauna has radiated extensively on *Eucalyptus*. The North American treehopper tribe Smiliini has radiated extensively on oak (*Quercus* spp.).

Membracoidea are distributed worldwide. Among the five currently recognized families, Cicadellidae and Membracidae occur on all continents except Antarctica. Aetalionidae have a disjunct neotropical/oriental distribution, Melizoderidae are restricted to South America, and Myerslopiidae occur only in New Zealand and Chile. Most species and genera are restricted to a single continent; many tribes and subfamilies are also restricted to particular continents.

Fulgoroidea

Fulgoroidea (planthoppers, Fig. 4) differ from other Auchenorrhyncha in having the frons occupying most of the facial part of the head and usually with distinct longitudinal carinae, tegulae usually present at the base of the forewings, the second segment (pedicel) of the antenna enlarged and (usually) bearing conspicuous placoid sensilla, the forewing anal veins confluent basad of the claval margin, and longitudinal carinae usually present on the head, pronotum, scutellum, and legs. Most have two ocelli dorsolaterally on the head, anterad of the compound eyes, but some Cixiidae also have a medial ocellus on the face. Fulgoroidea first appear in the fossil record in the middle Permian, and Cixiidae appear in the Jurassic. Other modern fulgoroid families apparently arose during the Cretaceous or early Tertiary. Twenty families are currently recognized, comprising approximately 1400 genera and 12,000 species. Fulgoroid families are distinguished from each other based mainly on the shape of the head, the spination of the hind tarsi, and the venation of the forewing. Fulgoroidea are the most morphologically variable of all auchenorrhynchan superfamilies, ranging from 1 mm to over 9 cm in length and exhibiting extensive variation in head shape, wing venation, and genital morphology.

Unlike Cicadomorphans, nymphs of Fulgoroidea apparently do not coat themselves with specialized Malpighian tubule secretions. Instead, they produce wax from specialized glands on the abdominal terga and other parts of the body. The wax forms a hydrophobic coating and may conceal some insects from predators. Adult females of many fulgoroid families also produce wax, with which they coat their eggs [Fig. 4(16)]. In certain tropical fulgoroid species, adults of both sexes produce strands of wax up to 75 cm in length. Aggregation behavior with or without ant mutualism has been documented for nymphs and adults in a few fulgoroid families, but egg guarding is known only in Tettigometridae.

In contrast to the ecologically similar Cicadoidea, the Fulgoroidea are the most ecologically diverse superfamily of Auchenorrhyncha. Nymphs of Derbidae and Achilidae live under bark or in litter, feeding on fungi, while nymphs of Cixiidae, Hypochthonellidae, and Kinnaridae are subterranean root feeders. At least four families include cavernicolous (cave-dwelling) species. Ant mutualism has been documented in several fulgoroid families and seems to occur universally among Tettigometridae [Fig. 4(20)], nymphs of which usually inhabit ant nests. Nymphs of most remaining families and nearly all adults feed on the aboveground parts of vascular plants and most seem to be host specialists. Planthopper species usually feed on woody dicotyledonous plants, but most Delphacidae are grass or sedge specialists. Several species of Delphacidae feed on emergent plants in marshes and are capable of walking on the surface of the water. Delphacidae primarily inhabit temperate and tropical grasslands, and diverse faunas of Issidae, Dictyopharidae (Orgeriinae), and Tettigometridae occur in deserts.

Fulgoroidea occur throughout the temperate and tropical regions of the world but are most diverse in the tropics. The Old World tropics harbor the greatest numbers of described families, genera, and species, but the neotropical fauna is less well studied and may be comparable in diversity. The holarctic fauna is rich in Delphacidae and Issidae, but most other families are poorly represented or absent. Tettigometridae,

Ricaniidae, Gengidae, Hypochthonellidae, and Meenoplidae are apparently restricted to the Old World. Some genera, particularly in Cixiidae and Delphacidae, are also cosmopolitan in distribution, but most appear to be restricted to a single biogeographic realm.

See Also the Following Articles
Host Seeking, for Plants • Phytophagous Insects • Phytotoxemia • Plant Diseases and Insects • Prosorrhyncha • Sternorrhyncha

Further Reading
DeLong, D. M. (1971). The bionomics of leafhoppers. *Annu. Rev. Entomol.* **16**, 179–210.
Denno, R. T., and Perfect, T. J. (eds.) (1994). "Planthoppers: Their Ecology and Management." Chapman & Hall, New York.
Hamilton, K. G. A. (1982). "The Insects and Arachnids of Canada," Part 10, "The Spittlebugs of Canada, Homoptera: Cercopidae." Agriculture Canada, Ottawa, Ontario.
Marmarosch, K., and Harris, K. F. (eds.) (1979). "Leafhopper Vectors and Plant Disease Agents." Academic Press, New York.
Metcalf, Z. P. (1960). "A Bibliography of the Cercopoidea (Homoptera: Auchenorhyncha [sic])." Waverly Press, Baltimore.
Metcalf, Z. P. (1960–1962). "General Catalogue of the Homoptera," fascicle VII, "Cercopoidea." North Carolina State College, Raleigh.
Metcalf, Z. P., and Wade, V. (1966). "A Catalogue of the Fossil Homoptera (Homoptera: Auchenorhyncha [sic])." Waverly Press, Baltimore.
Moulds, M. (1990). "Australian Cicadas." New South Wales University Press, Kensington, NSW, Australia.
Nault, L. R., and Rodriguez, J. G. (1985). "The Leafhoppers and Planthoppers." Wiley, New York.
Oman, P. W., Knight, W. J., and Nielson, M. W. (1990). "Leafhoppers (Cicadellidae): A Bibliography, Generic Check-list and Index to the World Literature 1956–1985." CAB International Institute of Entomology, Wallingford, Oxon, U.K.
Ossiannilsson, F. (1949). Insect drummers. *Opusc. Entomol. Suppl.* **10**, 1–145.
Shcherbakov, D. E. (1996). Origin and evolution of Auchenorrhyncha as shown by the fossil record. *In* "Studies on Hemiptera Phylogeny" (C. W. Schaefer, ed.). Entomological Society of America, Lanham, MD.
Sogawa, K. (1982). The rice brown planthopper: Feeding–physiology and host–plant interactions. *Annu. Rev. Entomol.* **27**, 49–73.
Wood, T. K. (1993). Diversity in the New World Membracidae. *Annu. Rev. Entomol.* **38**, 409–435.

Autohemorrhage

Many insects voluntarily discharge blood in response to a threat. This behavior, autohemorrhaging, may serve as a physical deterrent (e.g., by enveloping a potential predator or by exposing the predator to noxious substances). Many beetles in the Chrysomelidae, Meloidae, and Lampyridae are well known for this behavior, which is also called reflex bleeding. For example, the blood of blister (meloid) beetles exposes potential predators to the noxious substance cantharidin. Species apparently release blood through an increase in hydrostatic pressure. When the hydrostatic pressure returns to normal levels, much of this blood is withdrawn into the insect's hemocoel, and so little is actually lost from the insect.

Autotomy

A utotomy is a defensive response to attack involving the amputation or active breaking of a body part along a breakage plane and usually involves loss of a leg. Many invertebrates (e.g., crayfish, daddy-long-legs), including insects such as crickets, grasshoppers, and walkingsticks, and many vertebrates (e.g., salamanders) exhibit this ability. For example, walkingsticks (Phasmida) have weakened areas at the trochanter that break under stress, such as when an appendage is grasped by a predator. If the insect is not an adult, regeneration occurs at the next molt. The amputated leg of the walkingstick twitches after being detached, which may divert the predator's atten-

tion away from the attacked insect. A grasshopper, when held by a hind leg, can voluntarily discard that limb by intense muscular contraction and rupture it at the trochanter–femur junction; autotomy can also be induced by mechanical pressure or electrical shock. The individual may benefit from the loss of limb by surviving the potential predator but loss of balance, reduced ability to forage for food, and reduced ability to escape from the next predator result as well.

Sting autotomy, the self-amputation of the stinger and its glands, occurs in many social Hymenoptera as part of colony defense, especially against vertebrates, and may occur because of the size and shape of the sting barbs. Chemical cues released by the detached venom apparatus may enable other attacking individuals to orient themselves to the predator.

See Also the Following Article
Defensive Behavior

Further Reading
Personius, K. E., and Chapman, R. F. (2002). Control of muscle degeneration following autotomy of a hindleg in the grasshopper, *Barytettix humphreysii*. *J. Insect Physiol.* **48,** 91–102.

Bee

see *Apis Species*

Beekeeping

Eva Crane
International Bee Research Association

Beekeeping is the establishment and tending of colonies of social bees of any species, an activity from which the beekeeper obtains a harvest or reward. This reward is usually honey, but it may be some other bee product, or bees themselves (e.g., queens, or colonies for pollination). In beekeeping, each colony is usually in a hive, but some beekeeping is done with honey bees that build their nests in the open. Beekeeping is also done with certain nonsocial bees that are reared for pollinating crops.

TECHNIQUES OF MODERN MOVABLE-FRAME HIVE BEEKEEPING WITH *APIS MELLIFERA*

Most of the world's beekeeping is done with *A. mellifera*. In past centuries, these bees were kept primarily for the production of honey and beeswax. Beekeeping is still done mainly to produce honey, but there are also other specialized types of operation. These include the rearing of queens or package bees for other beekeepers who are producing honey. Another type of beekeeping provides colonies of bees to pollinate crops, since in many areas of large-scale agriculture

the native pollinators have been destroyed. Since the 1950s specialized beekeeping has also been developed for the production of royal jelly, pollen, and bee venom.

Each type of beekeeping requires the management of colonies to stimulate the bees to do what the beekeeper wants—for instance, to rear more young house bees to produce royal jelly, or more foragers to pollinate crops. During the 1900s, effective methods were developed for the commercial production of substances other than honey: bee brood, bee venom, beeswax, pollen, propolis, and royal jelly.

Honey Production

A colony of honey-storing bees collects nectar from which it makes honey. Nectar is not available continuously, and to store much honey a colony of bees needs many foraging bees (over, say, 10 days old) whenever a nectar "flow" is available within their flight range. Bees may fly 2 km if necessary, but the greater the distance, the more energy they expend in flight, and the more nectar or honey they consume. Thus, it is often cost-effective for the beekeeper to move hives to several nectar flows in turn during the active season.

Figure 1 shows a movable-frame hive with two "deep" boxes. The hive in Fig. 2 also has two deep boxes for brood (i.e., immature bees, eggs, larvae, and pupae), and a shallow box for honey that is less heavy to lift. Any number of honey boxes (also called supers) may be added to a hive, but these are always separated from the brood boxes by a queen excluder, to keep the honey free from brood. Some empty combs in these supers may stimulate honey storage, but supers are not added far in advance of their likely use by the bees.

It is essential that hives and frames have standard dimensions and that an accessory (spacer) be used to ensure that frames are always exactly the correct distance apart.

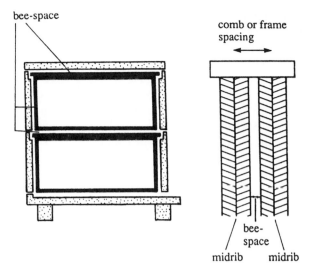

FIGURE 1 Vertical section through a movable-frame hive, showing a brood comb in each box, and the bee spaces. [After Crane, E. (1990). "Bees and Beekeeping: Science, Practice and World Resources." Heinemann, London.]

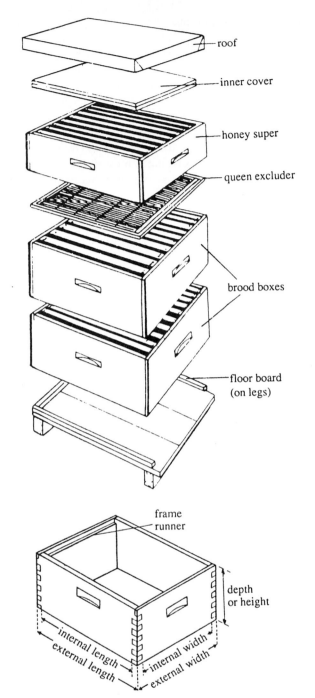

FIGURE 2 Top: exploded view of a movable-frame hive showing the component parts. Bottom: empty hive box showing one of the frame runners. [After Crane, E. (1990). "Bees and Beekeeping: Science, Practice and World Resources." Heinemann, London.]

Queen Production

Large-scale operations are done in five steps, which provide specific conditions for the successive developmental stages of the immature bees that will develop into mated and laying queens.

1. The larvae from which queens will be reared are taken from worker cells of a colony that is headed by a breeder queen selected for chosen genetic characters.

2. Very young larvae are transferred into cell cups mounted mouth down on wooden bars in a "cell-starter colony" that has been queenless for 2 to 4 h. This colony is made up of many (young) nurse bees, and little or no other brood for them to rear. Its bees build the cell cups further and feed the young larvae, and the colony can "start" the rearing of 45 to 90 queen larvae a day.

3. As the larvae grow larger, they receive more food and are better cared for if the number of nurse bees per larva is high. So it is usual to put about 15 cells in each of a number of colonies, where they are separated from the colony's queen by a queen excluder.

4. When the bees have finished feeding the larvae, they seal each immature queen in her cell. The only requirements of an immature queen during the next 7 days are appropriate conditions of temperature and humidity, and these are provided in an incubator. Each queen must emerge from her cell as an adult in a separate cage, for protection from attacks by other queens already emerged.

5. Finally each queen is placed in a "mating hive" containing a few hundred or more workers but no other queen. These hives are taken to a mating apiary, which contains a few strong colonies that include many drones (i.e., males) of a selected strain of honey bees. The apiary is located as far as possible from hives that might contain other drones; a distance of 15 km is likely to be safe, but it varies according to the terrain. When the queen is a few days old, she flies out and mates with drones, and a few days later starts to lay eggs. This shows that she is ready to head a colony.

Package Bee Production

The term "package bees" is used for a number of young worker bees (usually approximately 1 kg) hived with a newly reared and mated queen; these bees together have the potential to develop into a honey-producing colony.

Package bees are normally produced at relatively low latitudes where spring comes early, and are sold at higher latitudes where it is difficult to keep colonies over the winter; many northern beekeepers find it more cost-effective to kill some or all of their colonies when they harvest the season's honey, and to buy package bees next spring. (If they overwinter no colonies, they can follow another occupation for 6 months or more; at least one beekeeper spends the Canadian winter beekeeping in New Zealand, where it is then summer.) The site where the packages are produced should be earlier weatherwise, by 2 months or more, than the site where the bees are used. A package bee industry is most likely to be viable where a single country stretches over a sufficient north–south distance (at least 1000 km, and up to 2000 or even 2500 km). But in New Zealand, package bees are produced at the end of the bees' active season and sent by air to Canada, where the season is just starting.

Package bees are prepared as follows. First, all the bees are shaken off the combs of three or four colonies into a specially designed box, taking care that the queens are left behind. The bees are then poured through the "spout" of the box into package boxes, each standing on a weighing machine, until their weight is either 1 or 1.5 kg, as required. Each box is given a young mated queen in a cage, and a can of syrup with feeding holes. (Enough bees are flying around to return to their hives and keep the colonies functional.) For transport, the package boxes are fixed by battens in groups of three or four, slightly separated; they may travel 2400 km, and the truck needs special ventilation. Air transport, though possible, presents various difficulties.

Crop Pollination

Colonies taken to pollinate crops should be strong, with many foraging bees, and also much unsealed brood (to stimulate the bees to forage for pollen), and space for the queen to lay more eggs. Hives should not be taken to the crop before it comes into bloom, or the bees may start foraging on other plants and continue to do so when the crop flowers. If the hives are in a greenhouse, four to eight frames of bees in each may be sufficient, but the beekeeper must check regularly that the bees have enough food; alternatively, each hive may be provided with two flight entrances, one into the greenhouse and one outside. Beekeepers who hire out hives of bees for crop pollination need to have a sound legal contract with the crop grower; they should also be aware of the risks of their bees being poisoned by insecticides.

In addition to honey bees, certain native bees are especially efficient in pollinating one or more crop species, and several species are managed commercially for pollination. The following are quite widely used for the crops indicated: *Andrena* spp. for sarson and berseem in Egypt and India; *Bombus* spp. for tomato and red clover in Finland and Poland; *Megachile* spp. for alfalfa in Chile, India, South Africa, and the United States; *Nomia melanderi* for alfalfa in the United States; *Osmia* spp. for alfalfa in France; and *Xenoglossa* spp. for apple in Japan, Poland, and Spain, also cotton and curcurbits in the United States.

Special Features of Beekeeping in the Subtropics and Tropics

The subtropics (between 23.5° and 34°N, and 23.5° and 34°S) include some of the most valuable world regions for honey production. Like the temperate zones, they have an annual cycle with a distinct seasonal rhythm and a well-marked summer and winter; however, the climate is warmer and the winters are mild, so the bees can fly year-round. All the major honey-exporting countries include a belt within these subtropical latitudes: China, Mexico, Argentina, and Australia.

Between the Tropics of Cancer and Capricorn (23.5°N and S), the situation is different. The seasons (and honey bee colonies) undergo two cycles in the year because the noonday sun is overhead twice a year. So colonies do not generally grow as large as at higher latitudes, nor do they store as much honey. When forage becomes scarce, a colony may cease brood rearing, then fly as a unit to a nearby area where plants are coming into bloom; this flight is referred to as absconding or migration. So one beekeeper may lose colonies, while beekeepers in the other area put out bait hives to receive the swarms.

Beekeeping in the tropics using traditional hives has been well studied, and many development programs have been carried out to introduce more advanced methods. Francis Smith pioneered successful movable-frame hive beekeeping in tropical Africa.

In the tropics, bee diseases are of less importance than at higher latitudes, but bees in torrid zones may be subject to attack by more enemies, certain birds, mammals, and insects. Tropical honey bees therefore defend their nests more vigorously than temperate-zone honey bees. For instance, tropical African honey bees (*A. mellifera*) are easily alerted to sting and, as a result of rapid pheromone communication between individuals, they may attack *en masse*. People in tropical Africa have grown up with the bees and are accustomed to them. But after 1957, when some escaped following introduction to the South American tropics, they spread into areas where the inhabitants had known only the more gentle European bees, and those from tropical Africa were given the name "killer bees." But once beekeepers in South America had learned how to handle the new bees, they obtained much higher honey yields than from the European bees used earlier.

OTHER ASPECTS OF MODERN HIVE BEEKEEPING

World Spread of *A. mellifera*

In the early 1600s the bees were taken by sailing ship across the Atlantic from England to North America. They would have been in skeps (inverted baskets made of coiled straw), which were then used as hives. The first hives were probably landed in Virginia. The bees flourished and spread by swarming, and other colonies were taken later. By 1800 there were colonies in some 25 of the areas that are now U.S. states, and by 1850 in a further 7. The bees were kept in fixed-comb hives (skeps, logs, boxes).

The bees may possibly have been taken from Spain to Mexico in the late 1500s, but they reached other countries later: e.g., St. Kitts-Nevis in 1720, Canada in 1776, Australia in 1822, and New Zealand and South America in 1839. They were taken later to Hawaii (1857) and Greenland (1950).

In Asian countries where *A. cerana* was used for beekeeping, *A. mellifera* was introduced at the same time as movable-frame hives. Some probable dates of introduction were 1875–1876 in Japan, 1880s in India, 1896 in China, and 1908 in Vietnam.

Between 1850 and 1900 there was widespread activity among beekeepers in testing the suitability of different races of *A. mellifera* for hive beekeeping. The most favored race was Italian *(A. m. ligustica)*, named from Liguria on the west coast of Italy, south of Genoa.

Origination and World Spread of Movable-Frame Beekeeping

The production of a movable-frame hive divided the history of hive beekeeping into two distinct phases. This new hive type was invented in 1851 by Reverend Lorenzo Lorraine Langstroth in Philadelphia. He was familiar with the Greek movable-comb hive (discussed later under Traditional Movable-Comb Hive Beekeeping) and with some rectangular hives devised in Europe that contained wooden frames for the bees to build their combs in. These hives, however, had only a very small gap between the frames and the hive walls, and the bees built wax to close it. In 1853 Langstroth described how he had often pondered ways in which he "could get rid of the disagreeable necessity of cutting the attachments of the combs from the walls of the hives." He continued, "The almost self-evident idea of using the same bee-space [as between the centerlines of combs in the frames] in the shallow [honey] chambers came into my mind, and in a moment the suspended movable frames, kept at a suitable distance from each other *and from the case containing them,* came into being" (author's italics). Framed honey combs were harvested from an upper box, and the brood was in the box below. A queen excluder between the boxes prevented the queen from laying eggs in the honey chamber.

The use of hives based on Langstroth's design spread rapidly around the world, dimensions often being somewhat smaller in countries where honey yields were low. Some dates for their first known introduction are 1861, United Kingdom; 1870, Australia; 1878, South Africa; 1880s, India; and 1896, China.

Beekeeping with *A. cerana* in Movable-Frame Hives

Bees of most races of *A. cerana* are smaller than *A. mellifera;* they also build smaller colonies and are less productive for the beekeeper. Unlike *A. mellifera, A. cerana* does not collect or use propolis. *A. cerana* was the only hive bee in Asia until *A. mellifera* was introduced in the late 1800s; it had been kept in traditional hives (logs, boxes, barrels, baskets, pottery) since the first or second century A.D. in China and probably from the 300s B.C. in the upper Indus basin, now in Pakistan.

The movable-frame hives used for *A. cerana* are like a scaled-down version of those for *A. mellifera.* Colony management is similar, except that the beekeeper needs to take steps to minimize absconding by the colonies. In India 30 to 75% of colonies may abscond each year. To prevent this, a colony must always have sufficient stores of both pollen and honey or syrup, and preferably a young queen. Special care is needed to prevent robbing when syrup is fed. Colonies must also be protected against ants and wasps.

The bees at higher latitudes are larger, and in Kashmir (altitude 1500 m, and above) *A. cerana* is almost as large as *A. mellifera* and fairly similar to it in other characteristics; for instance, the colonies do not abscond.

Honey Bee Diseases, Parasites, Predators, and Poisoning

The main brood diseases of *A. mellifera,* with their causative organisms, are American foulbrood (AFB), *Paenibacillus larvae;* European foulbrood (EFB), *Melissococcus pluton;* sacbrood, sacbrood virus (Thai sacbrood virus in *A. cerana*); and chalkbrood, *Ascosphaera apis.* Diseases of adult bees are nosema disease, *Nosema apis;* amoeba disease, *Malpighamoeba mellificae;* and virus diseases. Parasites are tracheal mite, *Acarapis woodi;* varroa mites, *Varroa jacobsoni,* and *V. destructor;* the mite *Tropilaelaps clareae;* bee louse (Diptera); *Braula* spp.; and the small hive beetle, *Aethina tumida.*

Disease or parasitization debilitates the colonies, and diagnosis and treatment require time, skill, and extra expense. Most of the diseases and infestations just listed can be treated if colonies are in movable-frame hives, and in many countries bee disease inspectors provide help and advice. Colonies in fixed-comb hives and wild colonies cannot be inspected in the same way, and they can be a long-term focus of diseases. But by far the most common source of contagion is the transport into an area of bees from elsewhere.

The parasitic *Varroa* mite provides an example. It parasitized *A. cerana* in Asia, where the mite and this bee coexisted. In the Russian Far East, it transferred to introduced *A. mellifera,* whose developmental period is slightly longer, allowing more mites to be reared. Because colonies could then die from the infestation, the effects were disastrous. In the mid-1900s, some infested *A. mellifera* colonies were transported to Moscow; from there, mites were unwittingly sent with bees to other parts of Europe, and they have now reached most countries in the world.

Since the 1950s it has been increasingly easy to move honey bees (queens with attendant workers, and then packages of bees) from one country or continent to another. One result has been that diseases and parasites of the bees have been transmitted to a great many new areas, and to species or races of honey bee that had little or no resistance to them.

The development of large-scale agriculture has involved the use of insecticides, many of which are toxic to bees and can kill those taken to pollinate crops. In California alone, insecticides killed 82,000 colonies in 1962; in 1973 the number was reduced to 36,000, but in 1981 it had risen again, to 56,000. More attention is now paid to the use of practices that protect the bees, including selecting pesticides less toxic to beneficial insects, using pesticides in the forms least toxic to honey bees (e.g., granular instead of dust), spraying at night when bees are not flying, spraying only when the crop is not in flower, and using systemic insecticides and biological pest control. Possible actions by the beekeeper are less satisfactory: moving hives away from areas to be treated, or confining the bees during spraying by placing a protective cover over each hive and keeping it wet to reduce the temperature.

By 1990, legislation designed to protect bees from pesticide injury had been enacted in 38 countries, and a further 7 had established a code of practice or similar recommendations.

TRADITIONAL FIXED-COMB HIVE BEEKEEPING

A. mellifera in the Middle East, Europe, and Africa

Humans have obtained honey and wax from bees' nests in the Middle East, Europe, and Africa since very early times. Beekeeping with *A. mellifera* was probably initiated in an area when the human population increased so much that it needed more honey or wax than was available at existing nest sites, or when some change occurred that reduced the number of nest sites—for instance, when trees were felled to clear land for agriculture.

In the Middle East, population increase was linked with the development of civilizations. The earliest known hive beekeeping was done in ancient Egypt, and similar traditional beekeeping is still carried out in Egypt. In Abu Ghorab, near Cairo, an Old Kingdom bas-relief from around 2400 B.C. shows a kneeling beekeeper working at one end of hives built into a stack; smoke is used to pacify the bees, and honey is being transferred into large storage pots. Over time, the use of horizontal cylindrical hives spread throughout the Mediterranean region and Middle East, and also to tropical Africa, where hollow log hives were often fixed in trees, out of reach of predators.

In the forests of northern Europe, where honey bees nested in tree cavities, early humans obtained honey and wax from the nests. When trees were felled to clear the land, logs containing nests were stood upright on the ground as hives. As a result, later traditional hives in northern Europe were also set upright. In early types such as a log or skep, a swarm of bees built its nest by attaching parallel beeswax combs to the underside of the hive top. If the base of the hive was open as in a skep, the beekeeper harvested honey from it. Otherwise harvesting was done from the top if there was a removable cover, or through a hole previously cut in the side.

Skeps used in northwestern Europe were made small so that colonies in them swarmed early in the active season; each swarm was housed in another skep, and stored some honey. At the end of the season, bees in some skeps were killed with sulfur smoke and all their honey harvested; bees in the other skeps were overwintered, and their honey was left as food during the winter.

A. cerana in Asia

In eastern Asia the cavity-nesting honey bee was *A. cerana,* and it was kept in logs and boxes of various kinds from A.D. 200 or earlier. But farther west in the upper Indus basin horizontal hives rather similar to those of ancient Greece are used, and it has been suggested that hive beekeeping was started in the 300s B.C. by some of the soldiers of the army of Alexander the Great, who settled there after having invaded the area.

Stingless Bees (Meliponinae) in the Tropics

In the Old World tropics, much more honey could be obtained from honey bees than from stingless bees, and the latter were seldom used for beekeeping. But in the Americas, where there were no honey bees, hive beekeeping was developed especially with the stingless bee, *Melipona beecheii,* a fairly large species well suited for the purpose. It builds a horizontal nest with brood in the center and irregular cells at the extremities, where honey and pollen are stored. The Maya people in the Yucatan peninsula in Mexico still do much beekeeping with this bee. The hive is made from a hollowed wooden log, its ends being closed by a wooden or stone disk. To harvest honey, one of the disks is removed to provide access to honey cells; these are broken off with a blunt object, and a basket is placed underneath the opening to strain the honey into a receptacle below. Many similar stone disks from the 300s B.C. and later were excavated from Yucatan and from the island of Cozumel, suggesting that the practice existed in Mexico at least from that time.

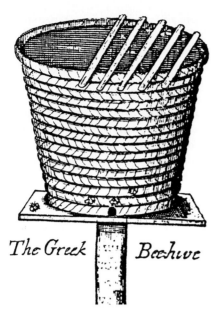

FIGURE 3 Sir George Wheler's drawing of a Greek top-bar hive. [After Wheler, G. (1682). "A Journey into Greece." W. Cademan and others, London.]

Nogueira-Neto in Brazil developed a more rational form of beekeeping with stingless bees. In Australia the native peoples did not do hive beekeeping with stingless bees, but this has recently been started.

TRADITIONAL MOVABLE-COMB HIVE BEEKEEPING

Movable-comb hive beekeeping was a crucial intermediate step between fixed-comb beekeeping, which had been done in many parts of the Old World, and the movable-frame beekeeping used today.

In a book published in 1682 in England, Sir George Wheler recounted his journeys in Greece and provided details of the hives he saw there (Fig. 3). He described the wooden bars shown lying across the top of the hive as "broad, flat sticks" and said that the bees built a comb down from each top-bar, which "may be taken out whole, without the least bruising, and with the greatest ease imaginable." So it was a movable-comb hive. The Greek beekeepers must have placed the bars at the bees' natural spacing of their combs. They made a new colony by putting half the bars and combs from a hive into an empty one; the queen would be in one of the hives, and the bees in the other would rear a new queen.

In the mountain range that separates Vietnam from China, some of the native peoples use a movable-comb hive for *A. cerana*; it is not known how old this method of beekeeping is. The bars are fitted across the top of a log hive at the correct spacing for *A. cerana*. This bee builds small combs without attaching them to the hive sides, and the combs can be lifted out by their bars. There seems to have been no development of a movable-frame hive from this movable-comb hive for *A. cerana*.

TRADITIONAL BEEKEEPING WITHOUT HIVES

A. dorsata

In tropical Asia, a nest of the giant honey bee, *A. dorsata*, which is migratory, can yield much more honey than a hive of *A. cerana*. In a form of beekeeping with *A. dorsata* practiced in a few areas, people use horizontal supports called "rafters" instead of hives. (A "rafter" is a strong pole, secured at a height convenient for the beekeeper by a wooden support, or part of a tree, at each end.) At the appropriate season, beekeepers erect rafters in a known nesting area for migratory swarms of the bees. Sheltered sites with an open space round one end are chosen, which the bees are likely to accept for nesting. After swarms have arrived and built combs from the rafters, the beekeeper harvests honey every few weeks by cutting away part of the comb containing honey but leaving the brood comb intact. When plants in the area no longer produce nectar, brood rearing ceases and the bees migrate to another site.

A. florea

The small honey bee, *A. florea*, builds a single brood comb perhaps 20 cm high, supported from the thin branch of a tree or bush. It constructs deeper cells round the supporting branch and stores honey in them. The whole comb can easily be removed by cutting through the branch at each side, and in some regions combs are then taken to an apiary where the two ends of each branch are supported on a pile of stones or some other structure. This is done, for instance, in the Indus basin near Peshawar in Pakistan, and on the north coast of Oman.

RESOURCES FOR BEEKEEPERS

There are various sources of information and help for beekeepers. Many countries publish one or more beekeeping journals, and have a beekeepers' or apiculturists' association with regional and local branches. Apimondia in Rome, Italy (http://www.apimondia.org) is the international federation of national beekeepers' associations.

In many countries, the ministry of agriculture or a similar body maintains a bee department that inspects colonies for bee diseases and often also provides an advisory service for beekeepers. Research on bees and/or beekeeping may be carried out under this ministry or by other bodies.

The International Bee Research Association in Cardiff, U.K. serves as a world center for scientific information on bees and beekeeping, and publishes international journals, including *Apicultural Abstracts,* which contains summaries of recent publications worldwide. Information about access to the Association's data banks can be obtained from its Web site (http://www.ibra.org.uk), which is linked to Ingenta.

See Also the Following Articles

Apis Species • Honey • Rearing of Insects • Royal Jelly

Further Reading

Connor, L. J., Rinderer, T., Sylvester, H. A., and Wongsiri, S. (1993). "Asian Apiculture." Wicwas Press, Cheshire, CT.

Crane, E., (ed.) (1976). "Apiculture in Tropical Climates." International Bee Research Association, London.

Crane, E. (1978). "Bibliography of Tropical Apiculture." International Bee Research Association, London.

Crane, E. (1990). "Bees and Beekeeping: Science, Practice and World Resources." Heinemann Newnes, Oxford, U.K.

Crane, E. (1999). "The World History of Beekeeping and Honey Hunting." Duckworth, London.

Crane, E., and Walker, P. (1983). "The Impact of Pest Management on Bees and Pollination." Tropical Development and Research Institute, London.

Crane, E., and Walker, P. (1984). "Pollination Directory for World Crops." International Bee Research Association, London.

Delaplane, K. S., and Mayer, D. (2000). "Crop Pollination by Bees." CAB International, Wallingford, U.K.

FAO. (1986). "Tropical and Sub-tropical Apiculture." Food and Agriculture Organisation of the United Nations, Rome.

Graham, J. M. (ed.) (1992). "The Hive and the Honey Bee." rev. ed. Dadant and Sons, Hamilton, IL.

Langstroth, L. L. (1853). "Langstroth on the Hive and the Honey-bee, a Beekeeper's Manual." Hopkins, Bridgeman, Northampton, MA.

McGregor, S. E. (1976). "Insect Pollination of Cultivated Crop Plants." U.S. Department of Agriculture, Washington, DC.

Morse, R. A., and Nowogrodzki, R. (1990). "Honey Bee Pests, Predators and Diseases." 2nd ed. Cornell University Press, Ithaca, NY.

Nogueira-Neto, P. (1997). "Vida e Criação de Abelhas Indígenas sem Ferrão." Edição Nogueirapis, São Paulo, Brazil.

Smith, F. G. (1960). "Beekeeping in the Tropics." Longmans, London.

Webster, T. C., and Delaplane, K. S. (2001). "Mites of the Honey Bee." Dadant & Sons, Hamilton, IL.

Bee Products

Eva Crane
International Bee Research Association

Honey and beeswax are the main bee products used by humans. Bee brood has been eaten by humans since ancient times in some Asian countries, but until the 1900s only honey and beeswax were produced commercially. Then in the 1950s the price of honey on the world market was depressed by surplus production, and beekeepers in certain technologically advanced countries, seeking ways of diversifying the sources of income from their bees, explored the commercialization of royal jelly, bee venom, pollen, and propolis.

BEE VENOM

Bee venom is a secretion from the venom glands of the worker or queen of a species of honey bee *(Apis)*; it is not produced by stingless bees (Meliponinae). The main components of commercial freeze-dried venom from *A. mellifera* worker bees include 15 to 17% enzymes, including phospholipase and hyaluronidase; 48 to 58% small proteins, including especially mellitin; 3% physiologically active amines, including histamine; 0.8 to 1.0% amino acids, and numerous minor components. Queen venom differs somewhat from worker venom in its composition and its pattern of change with the age of the bee. A few studies have been made on the venom of other *Apis* species; for instance, toxicity has been reported to be similar in venoms from *A. mellifera* and *A. dorsata,* less in *A. florea* venom but twice as high in *A. cerana* venom.

Bee venom is by far the most pharmacologically active product from honey bees. The general mechanism of its action on humans who are not hypersensitive is as follows. Hyaluronidase breaks down hyaluronic acid polymers that serve as intercellular cement, and the venom spreads through the tissue. (Protective antibodies that develop in the serum of most beekeepers can effectively neutralize hyaluronidase, preventing the spread of the venom.) A protease inhibitor prevents enzymatic destruction of the hyaluronidase. Simultaneously, the mast cell degranulating peptide penetrates the membrane of the mast cells, creating pores. This releases histamine, which (in combination with some small molecules of the venom) contributes to the swelling and flare, and the local itching and burning sensation. As venom penetrates blood vessels and enters the circulatory system, phospholipase A and mellitin (as a micelle, a colloidal-sized aggregate of molecules) act synergistically to rupture blood cells.

When only a few stings are received, the action just described is mostly localized, and actual toxic effects are insignificant. After massive stinging (or injection of venom directly into the circulatory system), the action may become widespread and toxic effects severe, particularly when significant amounts of venom enter the circulatory system. Apamine acts as a poison to the central nervous system, and both mellitin and phospholipase A are highly toxic. Large concentrations of histamine are produced and contribute to overall toxicity. The role of other components is unknown.

Only a very small number of people are allergic (hypersensitive) to insect venom, between 0.35 and 0.40% of the total population in one U.S. survey. In a person allergic to bee venom, the hyaluronidase may participate immediately in an antigen–antibody reaction, triggering an allergic response; both mellitin and phospholipase A can also produce allergic reactions. There may be antigen–antibody reactions to any or all of the components mentioned. Severe reactions can result in death from anaphylactic shock.

Antihistamines can give some protection to a moderately hypersensitive person if taken before exposure to stings. Systemic reactions following a sting should be treated immediately with adrenaline; extremely prompt medical treatment is essential for acute anaphylaxis.

Some allergy clinics provide carefully regulated courses of venom injection, which can decrease sensitivity to the venom; various types of immunotherapy (desensitization) have been used, involving the application of a series of graded doses of pure venom, and these can be effective in 95% of cases. If a beekeeper or another member of the household develops serious hypersensitivity to bee stings, an allergy specialist may be able to recommend a course of desensitization that will allow the beekeeper to continue.

Germany was probably the first country to produce bee venom commercially. Between 1930 and 1937, girls stationed in front of hives would pick up one worker bee at a time and press it so that it stung into a fabric tissue that absorbed the venom; the venom was extracted from the fabric with a solvent (distilled water), which later was removed by freeze-drying, leaving the venom as a crystalline powder.

A more recent method is to use a bare wire stretched to and fro across a thin membrane mounted on a horizontal frame placed directly in front of a hive entrance. When a low voltage is applied to the ends of the wire, a few "guard" bees are shocked; they sting into the membrane and also release alarm pheromone that quickly alerts other bees to sting into the membrane as well. The bees can withdraw their stings and are unharmed, and the drops of venom released are removed from the underside of the membrane; in hot weather they dry and can be scraped off.

BEE BROOD

Bee brood (immature bees) was probably a useful source of protein to hunter-gatherers in many parts of Asia and Africa, and honey bee larvae have now produced commercially, and marketed either raw or cooked. Mature *A. mellifera* larvae have been found to contain about 60% as much protein as beef and about 30% more fat (fresh weight). Pupae contain somewhat more protein and less fat. Both larvae and pupae contain vitamins A and D. Such bee brood is eaten in parts of Asia (e.g., Korea, China, Japan, Laos, Malaysia, Thailand, and Vietnam) but not in India, Pakistan, or Bangladesh. Some eastern Mediterranean religions forbade the eating of certain insects because these were regarded as unclean. One of the Dead Sea scrolls, from about 200 to 100 B.C., had the prohibition: "Let no man defile his soul with any living being or creeping thing by eating of them, from the larvae of bees [in honey] to all the living things that creep in water." (The digestive system of any animal was considered to be unclean, and it was impractical to remove these organs from individual bees.)

POLLEN

Protein is required by young adult honey bees, and it is an important component of the food they give to larvae. It is obtained from pollen (microspores of seed plants) that older bees collect from flowers and store in the nest. In one study on *A. mellifera* in the United States, bee-collected (air-dried) pollens contained 7 to 30% crude protein and 19 to 41% carbohydrates (mostly sugars from honey that bees mixed with the pollen). Pollen also contains minerals (it has an ash content of 1–6%), vitamins, enzymes, free amino acids, organic acids, flavonoids, and growth regulators.

When a worker honey bee moves past the anthers of flowers, pollen becomes trapped by her body hairs. She leaves the flowers and, with special movements of her legs, passes the pollen backward to bristles on the tibiae of her hind legs. She packs it into a "pollen load" on each of these legs, moistening it with a little nectar or honey in the process. The pollen loads carried by a foraging bee have a variety of colors, which provide clues to the plant sources.

It is relatively easy for a beekeeper to collect the pollen being brought into hives by bees: a pollen trap, fixed over the hive entrance, incorporates a grid (or two grids) through which incoming bees must push, and while they do this most pollen loads are knocked from their hind legs and drop into a tray below, although some bees get through the trap with their pollen loads. The beekeeper needs to ensure that the colony always has enough pollen to rear sufficient brood to maintain its population. (A colony can be made to collect more pollen by giving it extra combs of young brood to rear.) In 1990 pollen was known to be produced commercially in Europe (seven countries), the Americas (five), Asia (four), and Africa (one), and also Australia, where Western Australia alone produced 60 to 130 tonnes a year.

Pollen is used as a dietary supplement for humans and domestic animals, as well as for feeding to a honey bee colony to increase its brood production. Pollen from specific plant species (or cultivars) is also used for fruit pollination, in plant breeding programs, and in the study and treatment of allergic conditions such as hay fever.

PROPOLIS

Propolis is the material that honey bees and some other bees can collect from living plants, which they use alone or with beeswax in the construction and adaptation of their nests. Most of the plant sources are trees and bushes. The material collected may be a wound exudate (resin and latex) or a secretion (lipophilic substances, mucilage, and gum). Propolis thus has a much more varied origin than any other material collected by honey bees. Analyses of various samples (mostly of unknown plant origin) have shown the presence of over 100 compounds, including especially flavonoids.

A bee that collects propolis carries it back to the nest on her hind legs. She goes to a place in the hive where propolis is being used and remains there until her load is taken from her by bees using it. The propolis is mainly collected in the morning and used in the hive in the afternoon.

Where propolis is available, *A. mellifera* uses it for stopping up cracks, restricting the dimensions of its flight entrance, and other minor building works. Observations on both tropical and temperate-zone *A. cerana* indicate that this species does not collect or use propolis, even in a region where *A. mellifera* does, but uses beeswax instead. Propolis is sometimes used by *A. dorsata* to strengthen the attachment of the comb to its supporting branch. It is probably essential to *A. florea* for protecting its nest from ants. These bees build two rings of sticky propolis round the branch that supports the nest, one at each end of the comb attachment, and may "freshen" the propolis surface so that it remains sticky and ants cannot cross it.

To collect propolis from a hive, the beekeeper inserts a contrivance, such as a flat horizontal grid having slits 2–3 mm wide that will stimulate the bees to close up the gaps with propolis. On removal from the hive, the contrivance is cooled in a freezer. The propolis then becomes brittle, and a sharp blow fractures it off in pieces, which can be stored for up to a year in a plastic bag.

The total commercial world production of propolis may be between 100 and 200 tons a year. China produces more propolis (from hives of introduced *A. mellifera*) than any other country; some South American countries are next in importance. Most importing countries are in Europe.

Propolis has various pharmacological properties, partly from its flavonoid content. It is used in cosmetic and healing creams, throat pastilles, and chewing gum. A few people (in the United Kingdom about one beekeeper in 2000), are allergic (hypersensitive) to propolis, and contact with it leads to dermatitis.

Stingless bees mix much propolis with the wax they secrete before they use it in nest construction; the mixture is called cerumen.

See Also the Following Articles
Honey • Royal Jelly • Venom

Futher Reading

Crane, E. (1990). "Bees and Beekeeping: Science, Practice and World Resources," Chap. 14. Heinemann Newnes, Oxford, U.K.

Crane, E. (1999). "The World History of Beekeeping and Honey Hunting," Chap. 51. Duckworth, London.

Ghisalberti, E. L. (1979). Propolis: A review. *Bee World* **60**, 59–84.

Hocking, B., and Matsumura, F. (1960). Bee brood as food. *Bee World* **41**, 113–120.

Riches, H. R. C. (2000). "Medical Aspects of Beekeeping." HR Books, Northwood, U.K.

Stanley, R. G., and Linskens, H. F. (1974). "Pollen: Biology, Biochemistry, Management." Springer-Verlag, Berlin.

Beeswax

Eva Crane
International Bee Research Association

Beeswax is secreted by workers of most Apidae, which use it to build combs of cells in their nests, for rearing brood, and for storing food. Workers are female members of a colony of bees, active in foraging or nesting, but laying no eggs or only a few compared with a queen. The term "beeswax" is commonly used for the wax from honey bees (*Apis*),

especially that from *A. mellifera*, which is the basis of the world's beeswax industry.

PRODUCTION, SECRETION, AND USE OF BEESWAX BY *A. MELLIFERA*

Beeswax is secreted by four pairs of wax glands situated on the anterior part of the worker's last four normal sternal plates (i.e., the ventral portion); the secreted wax hardens into thin scales. In *A. mellifera* workers, the glands increase in secretory activity during the first 9 days or so after the adult bee has emerged from her cell. They usually start to regress at 17 days of age, but may be regenerated later if the colony needs new comb. Honey bees construct their combs of beeswax and also use this substance with propolis to seal small cracks in their nest structure or hive. The requirements of the colony largely determine the amount of wax secreted by its bees. Calculations have shown that an *A. mellifera* worker is likely to have the potential to secrete about half her body weight in wax during her lifetime.

COMPOSITION AND PROPERTIES OF BEESWAX

The major components of *A. mellifera* beeswax include monoesters, diesters, hydrocarbons, and free acids, which together make up more than half the total weight. Over 200 minor components have also been identified. Of the physical properties of beeswax, its thermal properties are of special practical importance, particularly the wide temperature range between its becoming plastic (32°C) and melting (61–66°C). Its relative density at 15 to 25°C is 0.96 and its refractive index at 75°C is 1.44.

Many pesticides used to control mites in the hive can contaminate beeswax.

HARVESTING AND PROCESSING

In the hive, the purest beeswax is that which has recently been secreted: in "cappings" with which cells have been sealed, and in recently built comb. Wax scraped from hive walls or frame bars may be mixed with propolis. Old, dark combs in which brood has been reared are of least value.

When a beekeeper harvests combs of honey from the hives, the honey is first extracted from the combs. Then the wax is melted and the liquid wax separated from any contaminants. On a small scale, clean wax from hives may be melted and strained through cloth, or a "solar wax extractor" may be used, in which the wax pieces are spread out on a sloping metal base in a shallow container with a double glass top, to be melted by radiation from the sun. The liquid wax flows into a container; any contaminants settle at the bottom, and clear wax flows out through an outlet near the top.

In some commercial wax extractors the wax is heated with water, floats to the top, and flows out through an appropriately placed opening. More efficient devices use a steam press. The percentage of beeswax extracted from the initial material varies according to the source of the wax and the method of extraction.

USES

Beeswax has a very rich history, with a far wider range of uses than any other bee product. In the past, beeswax was especially valued for candles, because it has a higher melting point than many other waxes, and so the candles remain upright in hot weather. Beeswax was also used for modeling and for casting. Some of the world's finest bronze statues and gold ornaments have been made by the lost-wax process, in which a beeswax model is made and encased in mud or plaster that is allowed to dry; the whole is then heated, the molten wax allowed to escape, and molten metal poured in. The metal solidifies in the exact shape of the original beeswax cast, and the casing material is then broken away.

In the batik method of dyeing cloth, and in etching on a glass or metal surface, beeswax can be used as a "resist," applied to certain areas of a surface to protect them from reaction during a subsequent process.

One of the most important current uses of beeswax is in ointments, emollient skin creams, and lotions. It also is still used in polishes and other protective coatings, and as a lubricant in the armament and other industries. Its dielectric properties have led to its use in electrical engineering.

WORLD PRODUCTION AND TRADE

Beekeeping with modern movable-frame hives aims to maximize honey production, and wax production is suppressed by providing the bees with sheets of ready-built wax comb foundation in frames. In experiments in Egypt, wax production in modern hives was only 0.4 to 0.6% of honey production, whereas in traditional hives it was 9 to 11%.

Bees secrete beeswax more readily in hot than in cold climates, and most surplus beeswax is produced in those tropical regions where traditional hives are still used. According to export figures published in 1990, relating to the preceding decade, the three regions producing most beeswax annually were Asia, Central America, and Africa (15.9, 10.5, and 8.7×10^3 tonnes, respectively). Major importing countries (in 1984) were France, German Federal Republic, United States, and Japan.

WAX FROM OTHER BEES

Because the waxes of different species of social bees differ slightly, if *A. mellifera* wax is mixed with that of other bees, its characteristics are altered. Melting points have been reported as follows for wax from other species of honey bees: *A. dorsata,* 60°C; *A. florea,* 63°C; *A. cerana,* 65°C; stingless bees, Meliponinae: *Trigona* spp. (India), 66.5°C; *T. beccarii* (Africa), 64.6°C; *T. denoiti* (Africa), 64.4°C; and bumble

bees, *Bombus,* 34–35°C. The temperature in a bumble bee nest is much lower than that in a honey bee nest.

See Also the Following Articles
Beekeeping • Commercial Products from Insects

Further Reading
Coggshall, W. L., and Morse, R. A. (1984). "Beeswax: Production, Harvesting, Processing and Products." Wicwas Press, Ithaca, NY.
Crane, E. (1990). "Bees and Beekeeping: Science, Practice and World Resources," Chap. 13. Heinemann Newnes, Oxford, U.K.
Crane, E. (1999). "The World History of Beekeeping and Honey Hunting." Duckworth, London.
Hepburn, H. R. (1986). "Honeybees and Wax." Springer-Verlag, Berlin.
Michener, C. D. (1974). "The Social Behavior of the Bees: A Comparative Study." Belknap Press, Cambridge, MA.

Beetle

see *Coleoptera*

Biodiversity

Nigel E. Stork
Cooperative Research Centre for Tropical Rainforest Ecology and Management at James Cook University, Australia

Biodiversity is a term created in the mid-1980s to represent the variety of life. It is a contraction of "biological diversity" and came into common usage following the signing in 1992 of the Convention on Biological Diversity at the United Nations Conference on Environment and Development, in Rio de Janeiro. The convention defines biodiversity as "the variability among living organisms from all sources including terrestrial, marine and other aquatic ecosystems and ecological complexes of which they are a part: this includes diversity within species, between species and of ecosystems."

However, biodiversity encompasses not just hierarchies of taxonomic and ecological scale but also other scales such as temporal and geographical scales and scaling in the body size of organisms. Biodiversity represents different things to different people. To those working in museums and herbaria it perhaps represents a new thrust for efforts to describe Earth's fauna and flora. To ecologists it may represent a growing concern about the balance of nature and how well ecosystems can function as biological diversity decreases. To economists and politicians it may represent a new and largely untapped source of needed income for developing nations.

To entomologists biodiversity *is* insects because more than half of all described species on Earth are insects.

Biodiversity is crucial to the planet's survival because as a result of it people have food, construction material, raw material for industry, and medicine, as well as the basis for all improvements to domesticated plants and animals. Biodiversity helps maintain ecosystem functions and evolutionary processes, and stores and cycles nutrients essential for life, such as carbon, nitrogen, and oxygen. Biodiversity absorbs and breaks down pollutants, including organic wastes, pesticides, and heavy metals. It also recharges groundwater, protects catchments, and buffers extreme water conditions.

The "ownership" of biodiversity and who should pay for its conservation are emotive subjects particularly in developing countries. These and other issues that relate to the sustainable utilization of biological and nonbiological resources and the maintenance of well-nurtured populations of humans throughout the world, are extremely complex.

GENETIC DIVERSITY

The individuals that make up a population are rarely identical. Such variation in the outward appearance of individuals (i.e., in their phenotype) results from the interaction of their individual inherited genetic makeup (genotype) with their surrounding environment. Most natural populations maintain a high level of such genetic diversity. This inherited genetic variation is the basis upon which evolution operates, and without it adaptation and speciation cannot occur. Genetic diversity fundamentally occurs in the form of nucleotide variation within the genome, which originates by mutation (changes in the nucleotide composition of genes, in the position of genes on chromosomes, and in the chromosome complement of individuals) and is maintained both by natural selection and by genetic drift. Other forms of genetic diversity include the amount of DNA per cell and chromosome structure and number. It is estimated that there are 10^9 genes in the world, although some of the genes for key processes vary little across organisms.

The long-term survival and success of a species depends to a large extent upon the genetic diversity within species, which makes possible both a degree of evolutionary flexibility in response to long-term climatic and other environmental change and a dynamic ecological community. The long-term aim of any conservation effort must be to maintain a self-sustaining dynamic ecological community, with the minimum of human intervention. This objective cannot be attained without recognition of the genetic diversity of the member species of the community.

SPECIES DIVERSITY

In spite of immense efforts by 19th- and 20th-century taxonomists to describe the world's fauna and flora, the true

TABLE I Comparison of the Estimated Number of Species for Vertebrates and the Four Most Species-Rich Orders of Insects, Description Rates, and Publication Effort

Order	Described species	Average description rate (species year^{-1})[a]	Publication effort[b]
Invertebrates			
Coleoptera	300,000–400,000	2308	0.01
Lepidoptera	110,000–120,000	642	0.03
Diptera	90,000–150,000	1048	0.04
Hymenoptera	100,000–125,000	1196	0.02
Vertebrates			
Birds	9,000	5	1
Mammals	4,500	26	1.8
Amphibians and reptiles	6,800	105	0.44
Fish	19,000	231	0.37

[a] Average for 1977–1988.
[b] Number of papers per number of species per year.

dimensions of species diversity remain uncertain. Understanding is hampered by lack of a consensus about the total number of species that have been named and described, with estimates ranging from 1.4 to 1.8 million species. This probably represents less than 20% of all species on Earth, and with only about 20,000 new species of all organisms being described each year, it seems that most species will remain undescribed for many years unless there is a rapid increase in species descriptions (but see http://www.all-species.org).

About 850,000 to 1,000,000 of all described species are insects. Of the 30 or so orders of insects, four dominate in terms of numbers of described species, with an estimated 600,000 to 795,000 species: Coleoptera, Diptera, Hymenoptera, and Lepidoptera (Table I). There are almost as many named species of beetle as there are of all other insects added together, or all other noninsects (plants and animals).

There is no complete catalog of names for all organisms, and for many groups it is often difficult to know what has or has not been named and described. It can sometimes be difficult for taxonomists to determine whether a series of individuals constitutes one or several species, or whether a new individual is the same species as others that have been described. On the other hand, a species may be described more than once. A taxonomist in one part of the world may not realize that a given species has already been described from elsewhere. Some species are so variable that they are described many times. For example, the ladybeetle, *Adalia decempunctata* has more than 40 synonyms. This species has many color morphs, and at various times during the last 200 years different taxonomists have given names to the color morphs without realizing that they were all one species. The level of such synonymy in some groups of organisms may be extremely high: (e.g., 80 and 35% synonymy for Papilionidae and Aphididae, respectively).

The question of how many species in total there are on Earth, including undescribed species, also remains a mystery. In 1833 the British natural historian John Westwood estimated that there might be some 20,000 species of insects worldwide. Today it is recognized that there are about this number of insect species in Britain alone. Estimates for how many species there are on Earth have continued to rise, and still it seems that the answer cannot be provided to within a factor of 100. Groups such as birds, large mammals, and some woody plants are well known, and estimates of their global numbers of species can be made with a fair degree of confidence. However, the scientific rationale for almost all estimates of global numbers of species for the remainder taxa, including insects, is surprisingly thin. Although estimates for global numbers of all species, from bacteria to vertebrates, vary from as low as 2 million to more than 100 million, much evidence seems to support estimates on the lower end of this scale: 5 to 15 million species.

Much of the recent literature on global species estimates has focused on insects and in particular on tropical forest insects. Until the 1980s most entomologists thought that there might be about 2 to 5 million insect species on Earth. However, Terry Erwin of the Smithsonian Institution in 1982 calculated that there are 30 million species of tropical arthropods alone, based on his knockdown insecticide fogging samples of beetles from the canopy of Central American tropical forests. He sampled 1200 species of beetles from the canopy of a single species of tree in Panama and suggested that 13.5% of these (162) must be specific to that tree. He arrived at his total of 30 million by suggesting that (1) all 50,000 species of tropical tree had the same level of insect host specificity, (2) beetles represented 40% of canopy arthropods, and (3) the canopy is twice as rich in arthropods as the ground. Others have since criticized all the steps in Erwin's calculation, suggesting that he overestimated the relative proportion of ground to canopy species, the relative proportion of beetle species to other groups of insects and, perhaps most important of all, the number of species that are host specific to a given species of tree. Another argument Stork and others have proposed is based on well-known insect faunas such as those for Britain and for butterflies. There are some 22,000 insect species in Britain and 67 of these are butterflies. It is also estimated that there are 15,000 to 20,000 species of butterflies in the world. Therefore, if the ratio found in Britain of butterfly species to all other insect species is the same for the whole world, this would indicate that there are 4.9 to 6.6 million species of insects on Earth. These and other analyses indicate that lower estimates (5–10 million insect species worldwide) may be realistic.

One of the reasons so few species have been described is that there are few taxonomists, and most of these are in the developed world. For example, 80% of insect taxonomists are found in North America and Europe. Another critical factor is that most of the type specimens on which species names depend are found in European and (to a lesser extent) North American museums.

It may seem that a great deal is known about the biology, distribution, and threatened or nonthreatened status of insects. In practice, this is far from the truth. For well-known insect faunas, such as those of Britain and other areas of Europe, virtually all species (but, surprisingly, not all) have been described. Even so, distribution maps for these species are often extremely poor, and the data used are often based on records more than 50 years old. For other parts of the world, particularly tropical regions, knowledge of the biota is largely nonexistent. Rarely are there even species lists for some of the better known groups, let alone taxonomic keys and field guides to identify these and other less well-known insects.

Much of the information on the distribution and biology of species is housed in the museums, herbaria, and libraries of developed countries. Some of this information is on index cards. There is now a growing effort to place information associated with specimens in the collection into electronic databases and to make this information readily available. Similarly, the biology and conservation status of the vast majority of insect species remain unknown. For this reason the International Union for the Conservation of Nature's (IUCN) Red Data Books on the threatened status of organisms are mostly limited to groups of large vertebrates and higher plants.

THE EVOLUTION AND EXTINCTION OF BIODIVERSITY

Evolution, simply speaking, is change through time. In genetic terms, evolution is an alteration in the frequency with which different genes are represented in a population, and it results primarily from the processes of natural selection and random drift. Natural selection operates through differential survival and reproductive success of individuals in a population, which determines their contribution to the genetic composition of the next generation. Natural selection acts on individual phenotypes best suited to the environment.

There has been life on Earth for at least 3.5 billion of the 4.6 billion years that the planet has existed. Multicellular plants and animals have evolved in just the last 1.4 billion years. The earliest fossil insect, or insect relative, is a hexapod, the collembollan *Rhyniella praecursor*, from the Lower Devonian (about 380 mya) from Scotland. It is unlikely that insects existed before the Devonian, and there was extensive radiation during the Carboniferous. There are fossils from 300 mya of several nonextinct groups, such as Paleodictyoptera, Meganisoptera, Megasecoptera, and Diaphanopterodea. The only extant orders represented by Carboniferous fossils are Ephemeroptera, Blattodea, and Orthoptera. Orders of modern insects, except Hymenoptera and Lepidoptera, appear to have been established by the Triassic (225 mya), and some of the early groups had disappeared by the late Permian. The massive explosion of insect diversity appears to coincide with that of the flowering plants (angiosperms) in the Cretaceous (135–65 mya).

Numerous studies have shown that there have been periods of rapid evolution of biodiversity and even more dramatic periods of extinction. Four of the five big episodes of extinction in the last 500 million years of the fossil record saw the removal of approximately 65 to 85% of the animal species in the ocean that are preserved as fossils, and the fifth resulted in the loss of 95% or more. In spite of these huge losses, it is now estimated that through subsequent rapid evolution, the present-day diversity of organisms, at both the species level and higher taxonomic levels, is greater than at any other time. Some suggest that present-day diversity may represent roughly 1% of all the species that have ever existed.

There have been many attempts to estimate the life span of species in the fossil record and these range from 0.5 to 13 million, although a few species present today appear to be unchanged in the fossil record for up to 50 million. Some data suggest that the average life span for species is 4 to 5 million.

The extinction of species, just like the evolution of species, is a natural process, and thus the extinction of existing species should occur at the same time as the evolution of new ones. The current list of all plants and animals that are recognized as having become extinct in the last few hundred years is relatively short. In total this amounts to just 600 plant and 491 animal species, and of these only 72 are insects. It is not surprising therefore that the fate of many thousands of threatened species of insects, other invertebrates, and fungi is almost completely overlooked. The death of the last passenger pigeon, "Martha," in 1914 is well known to many conservation biologists, yet the coextinction of two species of lice (*Columbicola extinctus* and *Campanulotes defectus*) that were host specific to this bird went unheralded. Some estimated extinction rates would indicate that most insect species are more likely to become extinct than to be named by taxonomists

Of the 72 species of insects listed on the IUCN's Red Data List as extinct, more than 40 are from Hawaii, and many of the others are from other islands. Proving that a species as small as an insect has become extinct can be very difficult, and indeed one of the largest species of insects that was thought to be extinct, the 15-cm-long Lord Howe Island stick insect (Phasmatodea), was discovered surviving in a remote part of this small island 80 years after its extinction had been declared. Of the insect species that no longer exist, most were driven to extinction by the introduction of other animals such as rats or invasive insects, whereas the demise of most extinct species of birds and mammals resulted from overhunting or loss of habitat.

It seems that there is a genetic or population threshold below which the survival of a species diminishes rapidly. For some species this "minimum viable population" may be 10 individuals and for others, hundreds or thousands. Such species with numbers of individuals below this threshold, the "living dead," although not presently extinct, appear to be doomed to extinction in the near future. A critical factor in the long-term survival of a single species or group of species is the maintenance of the intricate web of interacting species that are important in some way or other for each other's survival. For example, the Brazil nut tree, *Bertholletia excelsa,* relies on euglossine bees for pollination and seed setting, whereas the bees rely on the availability of other resources in the forest to complete their life cycle. Loss of these resources through forest fragmentation or disturbance could lead to the loss of the bees. The Brazil nut tree, however, might survive for many years before becoming extinct. This is just one example from the continuum of cosurvival of species, from those that are entirely dependent on the existence of one other species to those that are only in part dependent on one or a number of species. In this way, the survival or extinction of species or groups of species is linked to the survival of whole habitats or ecosystems.

THE DISTRIBUTION OF BIODIVERSITY

Life-forms of one kind or another are to be found in almost all parts of the surface of Earth, and insects are known to exist in most of these environments except the marine ecosystem. Clearly there is a strong latitudinal gradient in biodiversity, with few species occurring in higher latitudes and most species occurring in the tropics, peaking in tropical rain forests and coral reefs. Freshwater systems occupy a very small part of Earth's surface. Only 2.5% of all water on Earth is nonmarine, and most of this is unavailable to life; 69% of all fresh water exists as ice, principally in the polar regions, and another 30% is present underground. Just 0.3% of Earth's fresh water is freely available in rivers, streams, lakes, and freshwater wetlands, taking up only about 1% of the planet's surface! Although occupying a tiny percentage of Earth's surface, freshwater ecosystems support a rich and varied insect fauna. For some groups, the number of freshwater inhabitants is seemingly out of proportion to the representation of existing freshwater systems.

Of the world's open forest and shrubland, 75 and 42%, respectively, lie within tropical boundaries. At least two-thirds of all plant species are tropical, and thus, 6 to 7% of Earth's surface may contain 50 to 90% of all species of plants and animals. The high species richness of tropical forests is illustrated by La Selva forest of Costa Rica, 13.7 km^2 of which harbors 1500 species of plants, more than the total in the 243,500 km^2 of Great Britain. This Central American area also contains 388 species of birds, 63 of bats, and 42 of fish, as well as 122 reptile species and 143 butterfly species. A single site in southeastern Peru has yielded more than 1200 species of butterflies—almost a quarter of the 5000 species thought to be found in South America.

Two strata in forests are particularly noteworthy, both for their important roles in the functioning of animal and plant communities and for their high insect species richness: the canopy and the soil. The canopy of trees has been called by some the "last biotic frontier" because of the immense diversity of insects, plants, and fungi found there. Forest canopies came to the attention of biologists largely through the work of entomologists using knockdown insecticides to collect insects from the tops of trees. In 1982 Stork used knockdown insecticides released by a fogging machine hoisted in to the canopy of a 75-m-high rain forest tree in Borneo to collect canopy insects. When the collection had been sorted by taxonomists at the Natural History Museum in London, there were more than 1000 species, and yet the area of collecting sheets on the ground was only 20 m². In total, 4000 to 5000 species of insects were collected and sorted in a similar way from just 10 Bornean trees. For one group, the Chalcidoidea wasps, 1455 individuals were collected, but after sorting it was found that this represents 739 species. Because fewer than 100 chalcid species had been recorded before from Borneo, this indicates how little is known about the diversity of insects in some ecosystems.

Elsewhere, 43 species of ants were collected by canopy fogging from a single tropical tree in Peru, a number approximately equal to the ant fauna of the British isles. Tropical forests may cover only a small percentage of Earth's surface, but they are vital for the global cycling of energy, water, and nutrients. Most terrestrial life is found in temperate and tropical forests and grasslands. Some other vegetation types, such as the fynbos of South Africa, are also extremely species rich. This system supports more plant species per square meter than any other place on Earth, with more than 8500 species in total, 68% of which are endemic.

Perhaps less attention has been paid to the diversity of life in soils and associated leaf litter and dead wood. It is probable that there are at least as many species of insects specific to the soil as to the canopy. The diversity of soil organism assemblages and their importance in ecosystem functioning is just beginning to be understood. Relatively obscure groups such as fungi, springtails (Collembola), mites, and nematodes are all rich in species in the soil and are extremely important in ensuring that organic material is broken down and the resulting nutrients made available for the growth of plants. Earthworms in temperate regions and termites in tropical regions are critical for the production, turnover, and enrichment of the soil. They also help to aerate the soil and increase the through flow of water, hence reducing water runoff and soil erosion.

THE ROLE OF SYSTEMATICS IN BIODIVERSITY ASSESSMENT

Systematics is the part of comparative biology that tracks the diversity of organisms with regard to specified relationships among those organisms. It is the branch of biology responsible for recognizing, comparing, classifying, and naming the millions of different sorts of organism that exist. Taxonomy is the theory and practice of describing the diversity of organisms and the arrangement of these organisms into classifications.

Widely accepted as the most basic of natural taxa is the *species*. However, there is still some argument over what

FIGURE 1 Insecticide fogging being released from a knockdown insecticide fogging machine in a hardwood plantation in Cameroon. Note the circular catchment trays suspended above the ground to catch the falling insects released by the insecticide. (Photograph by N. Stork.)

exactly a species is. A major problem stems from variation observed among individual organisms, and the species question is largely one of how biologists attempt to classify individual organisms, all of which differ to a greater or lesser extent when compared with one another, into discrete groups or taxa. There is a range of definitions that largely reflects the various theories of the origin of diversity. When biological classification was first developed, organisms were considered each to have a fundamental design and the task of the taxonomist was to discover the essential features of these "types." Even after the publication of Darwin's theory of biological evolution, this concept did not change.

It was only with the emergence of a reliable theory of inheritance, and the development of the disciplines of genetics and population biology, that biologists began to develop rational explanations for the origin of diversity and then apply this knowledge to the species concept. The initial step forward was the recognition of geographical variation, first as "varieties," then as subspecies. This led to the concept of the species as a group of populations that reflected both common ancestry and adaptation to local conditions. In turn, this view was developed into the biological species concept, which defined the species as "groups of interbreeding natural populations that are isolated from other such groups." This species concept is perhaps the most widely accepted today, but it applies only to sexually reproducing species.

After the recognition of species, the next step in taxonomy is to classify the relationship of these species. A number of methods have been developed by which phylogenetic relationships can be estimated. Of these, cladistic analysis is now widely acknowledged as the best. Cladistic analysis rests upon three basic assumptions: features shared by organisms (termed homologies or apomorphies) form a hierarchic pattern; this pattern can be expressed as a branching diagram (cladogram); and each branching point symbolizes the features held in common by all the species arising from that node. Cladograms are the most efficient method for representing information about organisms, hence are the most predictive of unknown properties of those organisms.

Once a cladogram of taxa has been established, the next stage is to formally recognize and name the species and higher taxa. Names are assigned to these taxa according to a system based upon that first developed by the Swedish naturalist Linnaeus in the mid-18th century. Species are grouped into genera, and these in turn are grouped into families, orders, classes, phyla, and kingdoms. The ultimate goal of this nomenclature is to produce a universal system of unambiguous names for all recognized taxa. Animals, plants, and bacteria each have a separate set of rules or codes, which are applied voluntarily by taxonomists and are designed to promote stability and consistency in taxonomic nomenclature, and thus to biological science in general. Traditionally, life-form have been grouped into two kingdoms, Animalia (including the insects) and Plantae, but in the last few decades this view has been questioned by experts, with other kingdoms being recognized. Recent work using analyses of ribosomal RNA sequences has shown that the total genetic diversity of the traditionally well-known groups such as fungi, plants, and animals is only a tiny proportion of the genetic diversity shown by microorganisms. The term "domain" has now replaced "kingdom," with the higher organisms (fungi, plants, and animals) being grouped in the domain Eukarya and a variety of microorganisms being included in two further domains, Archaea and Bacteria.

THE ECOLOGICAL CONTEXT

Ecology is the study of the relations between organisms and the totality of the physical and biological factors affecting them or influenced by them, or more simply, as the study of patterns in nature. Ecologists investigate the biology of organisms, looking for consistent patterns in their behavior, structure, and organization. Although a relatively new field in comparison to systematics, ecology has already provided considerable insights into the organization of taxa.

Ecosystem function refers to the sum total of processes operating at the ecosystem level, such as the cycling of matter, energy, and nutrients. The species in a community influence its productivity, nutrient cycling, and fluxes of carbon, water, and energy. Ultimately, species may be responsible for such factors as the maintenance of atmospheric composition, the

dispersal and breakdown of waste material, the amelioration of weather patterns, the hydrological cycle, the development of fertile soils, and even the protection of many coastal areas.

Biogeochemical cycling is the movement of materials including carbon, nitrogen, phosphorus, and calcium through an ecosystem as individuals of different trophic levels are consumed by others at higher trophic levels. These nutrients are returned eventually to the abiotic "nutrient pool," where they are again available to primary producers.

Some of the important roles played by different species in biochemical cycling can be outlined briefly.

By their photosynthetic activity, plants play a fundamental role in the carbon cycle, introducing carbon into the food web. Microorganisms are also crucial. It is estimated that algae and cyanobacteria are responsible for 40% of the carbon fixed by photosynthesis on Earth. At the other end of the process, wood-decaying fungi release approximately 85 billion metric tons of carbon into the atmosphere each year as carbon dioxide. Termites also play an important role in global carbon cycling (hence, potentially, in global climate change) through their production of methane. Earth's nitrogen cycle is dependent on bacteria for nitrogen fixation and the release of nitrogen by denitrification. The microbial community thus controls the amount of nitrogen available to an ecosystem, determining ecosystem productivity in areas where nitrogen is limiting. By absorbing water from soils or other surrounding media, plants have a fundamental effect on the water cycle.

There is an ongoing debate between those who believe that all species in a given ecosystem are important and those who say that some are "functionally redundant." That is, if a species is removed from an ecosystem, can other species fulfill the same role? Two factors influence the importance of a species in ecosystem functioning: the number of ecologically similar species in the community and the extent to which a species has qualitative or quantitative effects on the ecosystem.

SPECIES INTRODUCTIONS

The introduction of exotic species has been responsible for great perturbations in many ecosystems. The arrival of predators, competitors, pests, and pathogens has caused decreases in populations of native species in many areas. Native or endemic species often occupy narrow ranges, have small population sizes, and lack defenses, all of which make them vulnerable to species introductions. The arrival of alien species is generally a more serious problem on islands, especially remote islands than in continental areas. For example, dramatic changes have occurred on the Hawaiian islands as a result of species introductions since human colonization. Although introductions may increase local diversity, most colonizers are cosmopolitan and are not endangered, whereas many endemic species are potentially threatened. Ultimately, many local ecosystem types may be lost worldwide, leading to a more homogenous global biota.

See Also the Following Articles

*Conservation • Endangered Insects • Genetic Variation •
Introduced Insects • Island Biogeography and Evolution*

Further Reading

Ehrlich, P. R., and Ehrlich, A. (1981). "Extinction. The Causes of the Disappearance of Species." Random House, New York.

Groombridge, B., and Jenkins, M. D. (2000). "Global Biodiversity: Earth's Living Resources in the 21st Century." World Conservation Press, Cambridge, U.K.

Gullan, P. J., and Cranston, P. S. (1994). The insects. *In* "An Outline of Entomology." Chapman & Hall, London.

Hillis, D. M., and Moritz, C. (1990). "Molecular Systematics." Sinauer, Sunderland, MA.

Lawton, J. H., and May, R. M. (eds.). (1995). "Extinction Rates." Oxford University Press, Oxford, U.K.

McNeely, J. A. (1988). "The Economics of Biological Diversity." International Union for Conservation of Nature, Gland, Switzerland.

Mawdsley, N. A., and Stork, N. E. (1995). Species extinctions in insects: Ecological and biogeographical considerations. *In* "Insects in the Changing Environment." (R. Harrington and N. E. Stork, eds.), pp. 321–369. Academic Press, London.

May, R. M. (1992). How many species inhabit the earth? *Sci. Am.* October, 18–24.

Stork, N. E. (1988). Insect diversity: Facts, fiction and speculation. *Biol. J. Linn. Soc.* **35,** 321–337.

Stork, N. E. (1995). The magnitude of global biodiversity and its decline. *In* "The Living Planet in Crisis: Biodiversity Science and Policy" (J. Cracraft and F. T. Grifo, eds.), pp. 3–32. Columbia University Press, New York.

United Nations Environmental Programme. (1995). "Global Biodiversity Assessment." Cambridge University Press, Cambridge, U.K.

World Conservation Monitoring Centre. (1992). "Global Biodiversity. Status of the Earth's Living Resources." Chapman & Hall, London.

Wilson, E. O. (1993). "The Diversity of Life." Allen Lane, Penguin Press, London.

Wilson, E. O., and Peter, F. M. (eds.). (1988). "Biodiversity," pp. 3–18. National Academy Press, Washington, DC.

Biogenic Amines

Arnd Baumann
Forschungszentrum Jülich, Germany

Wolfgang Blenau
Universität Potsdam, Germany

Joachim Erber
Technische Universität Berlin, Germany

Biogenic amines are important messenger substances and regulators of cell functions. In insects, these small organic compounds act as neurotransmitters, neuromodulators, and neurohormones. Biogenic amines control endocrine and exocrine secretion, the contraction properties of muscles, the activity of neurons, and the generation of motor patterns. In addition, certain biogenic amines are involved in learning and the formation of memory. Biogenic amines mediate these diverse cellular and physiological effects by binding to specific membrane proteins that primarily belong to the superfamily of G-protein-coupled receptors.

Specialized Terms

Ca^{2+} signaling Change in the intracellular Ca^{2+} concentration, through the release of Ca^{2+} ions from intracellular stores by the opening of ligand-gated ion channels or the entry of Ca^{2+} ions into the cell through different types of Ca^{2+}-selective channels located in the plasma membrane, that plays a role in the regulation of various cellular processes, including cell metabolism, gene expression, cytoskeletal dynamics, and neurotransmission.

cyclic AMP Cyclic derivative of adenosine monophosphate that is synthesized from ATP by adenylyl cyclase. Intracellular second messenger involved in the regulation or modulation of ion channels, protein kinase activity, and gene expression.

G-protein-coupled receptors Integral membrane proteins that constitute a large family of neurotransmitter, hormone, or olfactory receptors. Characterized by seven transmembrane regions. When agonists bind to these receptors trimeric GTP-binding (G) proteins are activated that then regulate the activity of intracellular secondary effectors, which change intracellular concentrations of second messengers or ion channel activity.

neurohormone Small organic or peptidergic substance that is produced in neurosecretory cells. Released into the hemolymph at special regions called neurohemal organs. Transported to target tissues with the hemolymph.

neuromodulator Neuroactive substance that is released by synaptic terminals. Simultaneously acts on large numbers of cells in the proximity of the releasing cell and modifies the properties of synaptic transmission and the properties of target cells.

neurotransmitter Chemical substance that is released from the presynaptic endings of a neuron. Transmits information across the synaptic cleft to specific receptors located on the surface of postsynaptic cells.

phosphorylation Transient, reversible posttranslational modification of proteins in which the terminal phosphate group of ATP is transferred to specific residues of a polypeptide by kinases and often alters the properties of the protein.

second messenger Intracellular substance, such as Ca^{2+}, cyclic AMP, inositol-1,4,5-trisphosphate, that modifies or modulates cellular responses. Concentration changes in response to activation of G-protein-coupled receptors.

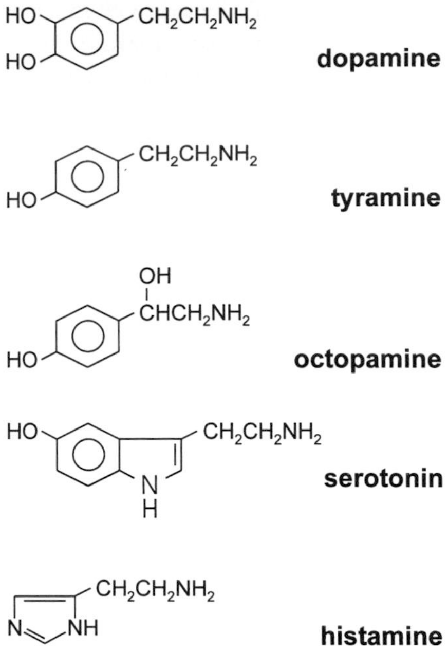

HO—⟨benzene ring⟩—CH₂CH₂NH₂   **dopamine**

HO—⟨benzene ring⟩—CH₂CH₂NH₂   **tyramine**

HO—⟨benzene ring⟩—CHCH₂NH₂ (with OH)   **octopamine**

HO—⟨indole ring⟩—CH₂CH₂NH₂   **serotonin**

⟨imidazole ring⟩—CH₂CH₂NH₂   **histamine**

FIGURE 1 Biogenic amines of invertebrates. In insects five substances have been identified as biogenic amines: dopamine, tyramine, octopamine, serotonin, and histamine.

BIOSYNTHESIS OF BIOGENIC AMINES

Biogenic amines are involved in a variety of regulatory functions. Five primary amines are considered biogenic amines in invertebrates: histamine (HA), serotonin (5-HT), dopamine (DA), tyramine (TA), and octopamine (OA) (see Fig. 1). These small organic compounds are synthesized from three different amino acids by single to multistep enzymatic reactions.

LOCALIZATION OF BIOGENIC AMINES

Aminergic systems in insects and vertebrates are quite different. In insects, OA and TA are present in relatively high concentrations, whereas they appear to have only minor significance in vertebrates. In contrast, the catecholamines norepinephrine and epinephrine are important chemical messengers in vertebrates, whereas in the insect nervous system they are detected only in very low concentrations if at all. Several additional catecholamines are involved in the process of cuticle tanning, hardening, and sclerotization in insects. These catecholamines are cross-linking reagents for cuticle proteins and chitin.

Considerable physiological, biochemical, and histochemical evidence suggests that HA, 5-HT, DA, OA, and TA act as transmitters or modulators in the central and peripheral nervous systems of insects. Antisera to HA, 5-HT, DA, and OA often label interneurons that have wide branching patterns within the central nervous system, sometimes innervating neuropils bilaterally. The dorsal and ventral unpaired median neurons, which can contain OA, are well-known examples of such large-field cells. Amine-containing

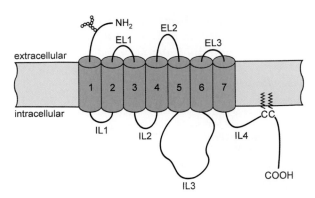

FIGURE 2 Transmembrane topography of G-protein-coupled receptors. The polypeptide spans the membrane seven times. The transmembrane regions (TM 1–7) are depicted as cylinders. The N-terminus (NH₂) is located extracellularly and often contains glycosylated residues (o). The C-terminus (COOH) is located intracellularly. The membrane-spanning regions are linked by three extracellular loops (EL1–EL3) that alternate with three intracellular loops (IL1–IL3). Posttranslational palmitoylation of cysteine residues (C) in the cytoplasmic tail creates a fourth intracellular loop (IL4).

neurons with large arborizations are well suited to act on large groups of other neurons simultaneously. In addition to these large-field cells there are small-field aminergic neurons, especially in the central complex and in the optic lobes. The neuroanatomy of these cells suggests that they communicate with a limited number of target cells.

BIOGENIC AMINE RECEPTORS

Biogenic amines bind to specific integral membrane receptors belonging predominantly to the superfamily of G-protein-coupled receptors. Physicochemical, biochemical, and immunochemical investigations show that these polypeptides share the common motif of seven transmembrane (TM) segments (Fig. 2). The N-terminus is located extracellularly, whereas the C-terminus is located intracellularly. The N-terminus is the target of a common posttranslational modification. In this part of the polypeptide consensus sequence motifs are often glycosylated. The membrane-spanning regions are linked by three extracellular loops (EL) that alternate with three intracellular loops (IL). Cysteine residues in the C-terminus of the polypeptides are the target of posttranslational palmitoylation. This modification creates a fourth intracellular loop.

A receptor is activated after binding of the specific biogenic amine in a binding pocket formed by the TM regions in the plane of the membrane. Individual residues in TM3, TM5, and TM6 were shown to participate in ligand binding. Once the ligand is bound, the receptor changes its conformation. This structural alteration usually is registered by intracellular trimeric GTP-binding proteins (G proteins). Residues that reside in close proximity to the plasma membrane in IL2, IL3, and IL4 of the receptor proteins determine the specificity and efficacy of the interaction between receptor and G protein.

GENERAL FUNCTIONS OF BIOGENIC AMINES

Biogenic amines have diverse functions controlling all phases of the life cycle of an insect. They are important chemical messengers during embryonic and larval development and they participate in the synaptic organization of the brain in the adult. As neuroactive substances they act on sensory receptors, inter- and motoneurons, and muscles and other peripheral organs (fat body, firefly lantern, salivary glands, corpora allata and corpora cardiaca, oviduct, etc.). Biogenic amines can initiate or modulate different types of behavior and they are involved in learning and the formation of memory in insects.

The effects of biogenic amines in the insect central nervous system are studied with the techniques of electrophysiological recordings, primary cell cultures, microinjections of amines and receptor ligands, and behavioral assays. Often the physiological responses to biogenic amines last for many minutes, which suggests that they can also act as neuromodulators. Biogenic amines modulate neuronal activity and the efficacy of synaptic transmission in all parts of the nervous system. The huge projection fields of many aminergic neurons support the idea of parallel modulation of entire neuronal circuits by just a few aminergic cells. In addition to synaptic neurotransmission, some aminergic neurons release the amine into the hemolymph. The substances are transported throughout the body and may thus have hormonal functions in specific target tissues.

The physiological role of OA at different levels of the organism is well documented. As a stress hormone in the periphery and in the central nervous system OA prepares the animal for energy-demanding behaviors. This monoamine stimulates glycogenolysis, modifies muscle contraction, supports long-term flight, and regulates "arousal" in the central nervous system. OA and OA agonists can enhance behavioral responses, like escape or aggressive behavior in crickets and sucrose responsiveness in honey bees. Injection of OA can elicit flight motor behavior in locusts, even in isolated thoracic ganglia. It is assumed that in insects OA has functions similar to those of the adrenergic system in vertebrates.

Both OA and 5-HT can modulate sensory receptors and receptor organs in insects. In many cases the sensitivities of the receptors are enhanced. Different funcions of OA and 5-HT at the sensory periphery are not very well understood, because the two amines often differ only in the degree of modulation. The increased sensitivity of sensory receptors due to the action of OA can modify behavior and is part of the "fight or flight" function. Studies on the *Drosophila* tyramine receptor mutant *hono* suggest that TA can also modulate the sensitivity of olfactory receptor cells, thus modulating behavioral responses to olfactory repellents.

The modulation of interneurons or effector neurons by biogenic amines is another level of modifying signal processing. OA and 5-HT can have functional antagonistic effects in a number of different systems. In these systems OA usually enhances the sensitivity or activity of single neurons and 5-HT usually has the opposite action. These effects, which can be measured at both the behavioral and the single-cell level, are dependent on the state of the insect. OA can induce a state of "arousal" in inactive animals and has only minor effects on very active animals, whereas 5-HT shows the largest effects in active animals.

In addition to modulatory functions during the adult life of an insect, DA and 5-HT have important functions during development. In *Drosophila*, high DA concentrations coincide with larval and pupal molts. Reduced levels of DA during larval stages lead to developmental retardation and decreased fertility in adults. 5-HT similarly acts as a chemical signal during larval development in *Drosophila*. Impaired 5-HT synthesis can lead to abnormal gastrulation movements, cuticular defects, and even embryonic death.

The neurotransmitter HA is released from photoreceptors in the compound eyes and ocelli in response to illumination. HA has also been detected in mechanosensory cells in *Drosophila*.

FUNCTIONS IN LEARNING AND MEMORY

Biogenic amines are involved in different forms of learning and memory formation in *Drosophila* and honey bees. However, it has not been unequivocally proven that the same biogenic amines serve identical functions in both species. Research on the neuronal and molecular bases of learning and memory over the past two decades in insects has focused on the mushroom bodies and antennal lobes of the brain. These two structures are involved primarily in processing of olfactory stimuli. Experimental evidence suggests that DA signals the presence of reinforcers and modulates intrinsic mushroom body neurons during conditioning in *Drosophila*. Thus DA could trigger signaling cascades that affect the storage of information about the conditioned stimulus.

In the honey bee, OA appears to be the modulatory transmitter which conveys information about rewarding sucrose stimuli and induces medium- to long-term modifications in interneurons during associative olfactory learning. Electrical stimulation of an identified octopaminergic cell, the ventral unpaired median VUM$_{mx1}$ neuron, can substitute for the sucrose reward during olfactory conditioning. This neuron has extensive arborizations in different brain regions, including the antennal lobes and the mushroom bodies. Microinjections of OA into these two neuropiles of the bee brain confirmed that OA in fact induces associative learning.

See Also the Following Articles
Brain and Optic Lobes • Chemoreception • Learning

Further Reading
Blenau, W., and Baumann, A. (2001). Molecular and pharmacological properties of insect biogenic amine receptors: Lessons from *Drosophila melanogaster* and *Apis mellifera*. *Arch. Insect. Biochem. Physiol.* **48**, 13–38.

Davis, R. L. (1996). Physiology and biochemistry of *Drosophila* learning mutants. *Physiol. Rev.* **76**, 299–317.

Erber, J., Kloppenburg, P., and Scheidler, A. (1993). Neuromodulation by serotonin and octopamine in the honeybee: Behavior, neuroanatomy and electrophysiology. *Experientia* **49**, 1073–1083.

Hammer, M. (1997). The neural basis of associative reward learning in honeybees. *Trends Neurosci.* **20**, 245–252.

Homberg, U. (1994). Distribution of neurotransmitters in the insect brain. In "Progress in Zoology" (W. Rathmayer, ed.), Vol. 40, VCH, Stuttgart.

Kutsukake, M., Komatsu, A., Yamamoto, D., and Ishiwa-Chigusa, S. (2000). A tyramine receptor gene mutation causes a defective olfactory behavior in *Drosophila melanogaster*. *Gene* **245**, 31–42.

Monastirioti, M. (1999). Biogenic amine systems in the fruit fly *Drosophila melanogaster*. *Microsc. Res. Tech.* **45**, 106–121.

Nässel, D. R. (1999). Histamine in the brain of insects: A review. *Microsc. Res. Tech.* **44**, 121–136.

Osborne, R. H. (1996). Insect neurotransmission: Neurotransmitters and their receptors. *Pharmacol. Ther.* **69**, 117–142.

Roeder, T. (1999). Octopamine in invertebrates. *Prog. Neurobiol.* **59**, 533–561.

Stevenson, P. A., and Spörhase-Eichmann, U. (1995). Localization of octopaminergic neurons in insects. *Comp. Biochem. Physiol.* **110A**, 203–215.

Vanden Broeck, J. J. M. (1996). G-protein-coupled receptors in insect cells. *Int. Rev. Cytol.* **164**, 189–268.

Wright, T. R. F. (1987). The genetics of biogenic amine metabolism, sclerotization, and melanization in *Drosophila melanogaster*. *Adv. Genet.* **24**, 127–222.

Biogeographical Patterns

Peter Zwick

Max-Planck-Institut für Limnologie

Biogeography, which deals with the description and interpretation of plant and animal distributions, is linked with other sciences, especially ecology and (paleo-) geography; zoogeography is the branch addressing animal distribution.

Most animal species inhabit restricted ranges, and only relatively few are cosmopolitan. A comparison of the areas inhabited by different species reveals common distributional patterns that are complex reflections of the ecology and of active and passive animal dispersal, but also of the evolutionary history of both the species and the earth's surface. Zoogeography was sometimes divided into different disciplines, descriptive as opposed to causal zoogeography; the latter was then subdivided into ecological and historical zoogeography. Although studies may differ in their emphasis, the interrelations among these disciplines are too close for a formal division. This article describes the major zoogeographical patterns and uses selected examples from among the insects to highlight the significance of some of the factors just mentioned.

Insects are of great geological age, and most orders existed and were diverse when familiar vertebrates were only begin-

ning to appear. Therefore, the distribution of most insect orders dates back much further than the distributions of many birds and mammals.

Insects are generally absent from some habitats. For example, with the exception of a few littoral specialists for unknown reasons, the only insects in the sea are some high ocean surface skaters among the bugs. Therefore, marine distribution patterns need not be considered here. The salt content of seawater is not the cause of this absence; insects are well represented in epicontinental waters of all kinds: fresh, brackish, and even hypersaline. Aquatic insects played an important role in the development of modern insect zoogeography. Because of their specific habitat ties, aquatic insects are easily collected, and the distributions of many are exceptionally well documented. Their distributions resemble those of terrestrial insects, in part because most aquatic insects have terrestrial adults that disperse over land.

The early explorers were struck by overall differences between the faunas of the lands they visited. The recognition of distinct faunal regions on a global scale thus has a long tradition and is briefly presented as an introduction. In addition to landmass topography, ecological conditions provide the basic setting for animal distributions; a brief outline of the major bioregions with similar overall ecology is therefore also presented.

Reproductively isolated species are the only naturally defined animal taxa; subspecific taxa can interbreed, whereas supraspecific taxa such as genera or families are human abstractions that change with conventions. It is convenient to use extant species to explain some concepts related to ranges and to discuss insect dispersal. Next, distribution patterns shaped by Pleistocene events are used to illustrate the importance of ecological change. The final focus is on disjunct (discontinuous, divided) distributions of monophyletic taxa that can best be explained by much older events, particularly continental drift.

ZOOGEOGRAPHICAL REGIONS

The major faunal regions (or realms) of the world only partly coincide with major landmasses (Fig. 1). Each region has a characteristic fauna distinguished by the particular combination of endemic taxa that exist in only this one region and those occurring also elsewhere. This early descriptive approach has long dominated zoogeography.

The Holarctic region is the largest region and is composed of the Palearctic and Nearctic regions, with many animals distributed over all the entire Holarctic region. Although a narrow land bridge (i.e., Central America) connects the Nearctic with the Neotropical region, the faunal change is pronounced. This land bridge is recent and was available only intermittently in the past. The Sahara Desert separates the Palearctic region from the Ethiopian (or Afrotropical) region, which includes the Arabian peninsula; Madagascar is now recognized as a distinct subregion. In Southeast Asia, climatic

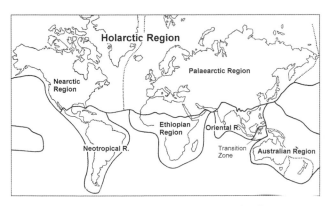

FIGURE 1 Terrestrial zoogeographical regions. [Modified from DeLattin, G. (1967). "Grundriss der Zoogeographie." Gustav Fischer Verlag © Spektrum Akademischer Verlag, Heidelberg.]

and other ecological differences cause a rather abrupt change of the biota south of the Himalayas, where the Palearctic and Oriental regions meet. In the southeast, the Oriental region is in contact with the Australian region, which includes New Guinea, New Zealand, New Caledonia, and the Oceanic subregion. The Australian region is most distinct, but the change toward the Oriental region is nevertheless not abrupt. Depending on the animal group studied, different variants of a border line (named Wallace's line after Alfred Russel Wallace, the earliest observer) were proposed in the past. The Oriental–Australian transition zone is sometimes called Wallacea.

There are puzzling resemblances among the faunas of different zoogeographical regions that are not in physical contact, and related animals may live on widely separate continents. Examples are provided by similarities between the faunas of eastern South America and West Central Africa, between Madagascar and India, or between Andean South America and the Australian region. Also, the fauna of eastern North America has resemblances to the European fauna, and the Far East Asian fauna to that of northern and western North America, despite the intervening oceans. On the other hand, insects in western North America are more distinct from those in the east, and those in Europe differ more from those in Asia, than one would expect in view of the continuous landmasses. These inconsistencies cannot be explained from present geography or ecology but reflect histories of ancient landmasses.

BIOREGIONS OR BIOMES

Seashores, glaciers, high mountains, and deserts pose obvious physical limits to animal distribution. Even in the absence of physical barriers, however, most species inhabit only part of a major landmass, because of ecological constraints. It is rare that a single ecological factor, or a precise combination of factors, limits an insect's distribution. However, most ranges

can readily be assigned to a particular biome or bioregion, that is, a large landscape with characteristic overall ecological conditions. Biomes can conveniently be described by general landscape physiognomy, mainly by reference to plant cover, which, among other things, determines the microclimate the insects experience. Biomes do not coincide with zoogeographic regions and each biome comprises separate areas on different continents. A particular biome may harbor animals that look similar or behave similarly but are not necessarily closely related. Instead, they may be characteristic life-forms exhibiting certain traits evolved independently, in response to similar ecological conditions; desert beetles offer examples. Authors differentiate and subdivide biomes to different degrees; some clearly distinct and almost universally recognized biomes are briefly discussed here (Fig. 2).

Arboreal Bioregions

The arboreal biome includes the areas supporting forests, as opposed to only individual trees. Patches of meadows, rocky outcrops, or swamps may occur because of local ecological conditions, and although they are mostly treeless, they still form part of the arboreal biome. Temperature and humidity mainly determine the type of forest occuring in an area. Only the large zonal types are briefly characterized; most are more or less disjunct today.

HYLAEA A name orginally proposed for the Amazonian rain forest, Hylaea is now widely used to designate all tropical evergreen rain forests—dense, multi-storied forests, with little light reaching the forest floor. Animals and plants are adapted to favorable conditions such as temperatures, precipitation, and air humidity that are continuously high. Biodiversity is very high, probably partly because of the presumedly continuous existence of tropical rain forests over exceptionally long periods of time. Processing of shed plant material is fast, and little detritus accumulates on the forest floor. Recent studies using fumigation techniques have shown that most insects inhabit the tree crowns. From an amazingly large number of undescribed species discovered by this method, the total number of existing insect species would be 35 million; more broadly based estimates range from 10 to 30 million species of insects.

The climate supporting the Hylaea is basically nonseasonal. However, seasonal snowmelt in the Andean headwaters of the Amazon results in a seasonal discharge regime that leads to months-long seasonal flooding of vast rain forest areas and drastic seasonal changes of conditions for all life. Similar situations may occur elsewhere. Evergreen tropical rain forests exist in parts of South America, in Central America, in a discontinuous belt across equatorial Africa, and in parts of Southeast Asia, from whence they extend into tropical northeastern Australia, where only small remnants remain.

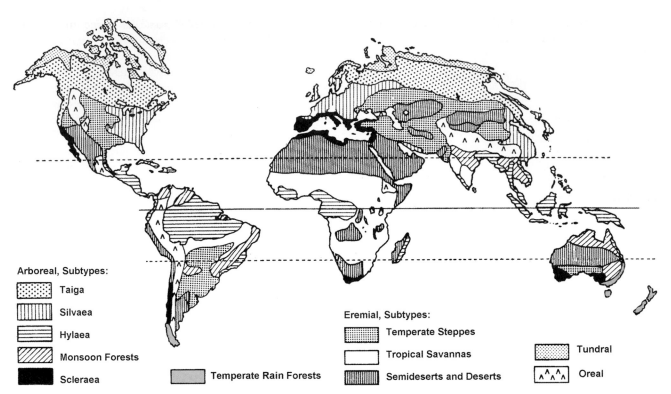

FIGURE 2 Main biomes. [Modified from a map of vegetation zones in Cox, B. C., and Moore, P. D. (1985). "Einführung in die Biogeographie." UTB 1408. Gustav Fischer Verlag © Spektrum Akademischer Verlag, Heidelberg.]

SILVAEA The term "Silvaea" refers to summer green deciduous broad-leafed forests, which, like the Hylaea, were much more widespread in the Tertiary than they are today. They occur in oceanic-to-suboceanic subtropical-to-temperate areas, mainly in eastern North America, in central Europe and the northern portion of southern Europe and Asia Minor, and in eastern China, Korea, and Japan. Adequate humidity is permanently available, and the species-rich vegetation offers protection against wind and radiation. During the vegetation period, the biota experiences favorable temperatures. In autumn, insects withdraw and most are inactive during winter, which may be frosty, although the soil does not freeze to a great depth. Strictly seasonal leaf shedding provides enormous amounts of dead plant material. Because low temperatures reduce production less than decomposition, much detritus accumulates, which provides habitat and food for many specialists among the diverse arthropod and insect fauna.

The evergreen temperate rain forests are located in restricted areas of the Southern Hemisphere, especially in Patagonia, southeastern Australia (including Tasmania), and New Zealand. They are in many ways similar to the two previously mentioned types but geographically disjunct, except along the Australian east coast where tropical and temperate rain forests meet and intergrade. Large tropical and subtropical areas with monsoon climate support forests that are only seasonally green (Fig. 2).

SCLERAEA Hard-leafed trees and shrubs dominate in the Scleraea in subtropical areas with rainy, mild winters and hot, dry summers. This evergreen forest type occurs not only along the western borders of the large landmasses, mainly in California and the European Mediterranean region, but also in middle Chile, the southwestern Cape of Africa, and southwestern Australia. Thick bark, as well as hard, reflectant, and often wax-covered leaf surfaces, or feltlike, often rolled-in leaf undersides and other modifications protect the deeply rooted plants against summer heat and dryness; winters tend to be wet and cool, but frost is rare. Life cycles are fitted to this pronouncedly seasonal regime, and some specialized insects also move to protected underground habitats where there are minute, blind, soil-inhabiting ground beetles and rove beetles (Carabidae and Staphylinidae). Plant cover is diverse and sufficiently dense to provide food and protection so that overall conditions for insect life are good; insect diversity is therefore very high.

TAIGA A small number of conifer species form a belt of northern evergreen forests known as taiga. The taiga ranges from North America through northern Asia to Scandinavia. Summer temperatures during the short vegetation period may be high, but the duration and severity of winters, and also the relative monotony of vegetation, limit the number of insect taxa. During the Pleistocene, the taiga was displaced southward; the more southern montane conifer forests in the

Northern Hemisphere date back to this period. Biodiversity in the taiga is generally low; a few species dominate and may inhabit vast areas.

Eremial Bioregions

The large generally treeless arid areas on earth, mainly steppes, semideserts, and deserts, are collectively called "eremial." An almost continuous belt extends through North Africa and Asia, from Mauretania to eastern Mongolia. The western portion, including the Indian Thar Desert, is hot and dry; the more northeasterly areas experience extreme winter cold. The Ethiopian eremial center and the Kalahari were originally separate but became more or less connected to Arabia and the northern Palearctic eremial belt via dry savannas in central and East Africa, after the Pleistocene. This permitted some exchange of eremial fauna, but the eremial centers on other continents remain isolated from each other. Therefore, the faunas are phylogenetically different, but all must be tolerant of lack of cover, as well as dryness, strong solar irradiation, extreme diel temperature changes, and often strong winds. Compared with other biomes, insect biodiversity is low, but even the most extreme, vegetation-free types of desert are not totally void of insects; several darkling beetles (Tenebrionidae), for example, survive on wind-transported organic material.

Oreal and Tundral Bioregions

High mountains above the tree line constitute the oreal biome, and treeless areas close to the poles form the tundral biome. The oreal and the tundral biomes are ecologically similar and often are considered together as oreotundral. This classification is particularly justified in view of contacts between the two realms during the Pleistocene. Treelessness is caused by cold temperatures and the short vegetation period, and sometimes also by exposure to wind. The Arctic tundra is a large zonal biome, and large areas have permafrost soil. During the short summer, the deeper soil never thaws; thus meltwater remains on the surface, leading to the establishment of extensive swamps and bogs. Low temperatures impede the rotting of dead plants, which are largely mosses, and peat formation is therefore common. Tundral areas in the Southern Hemisphere have no permafrost soil; they are highly fragmented and essentially restricted to the subantarctic islands. Insect biodiversity is low, especially in the tundral, and less so in the oreal.

Dinodal Bioregions

The overall agreement of the distribution of freshwater insects with terrestrial fauna is in line with the universal experience that freshwater bodies in many ways mirror the conditions in their catchments. This applies particularly to running waters that are largely allotrophic (i.e., depend on inputs of organic material from the environment), most obviously in that members of the shredder functional feeding group depend on dead leaves and other coarse organic material that only terrestrial inputs make available in streams. However, not all trophic groups are equally dependent on the terrestrial environment. Based on the presence of several endemic stream caddisflies in formerly glaciated or severely impacted areas in central Europe, swiftly and vehemently flowing streams (usually assigned to the arboreal biome) are believed to constitute a separate new biome, the dinodal, which would be largely independent of the biome in the catchment. (See work by Malicky, 1983.)

INSECT RANGES

The geographical area in which a species regularly occurs and maintains itself through natural reproduction is called its distribution area or range. Species ranges differ in size from individual small islands (or some other island habitats, like an isolated mountaintop, a particular lake, or some individual cave) to entire continents to the entire Northern or Southern Hemisphere or even to almost global distributions. The term "range" is also used to describe the distribution of geographical races (subspecies) and supraspecific taxa, such as genera and families.

Within their ranges, insects are not randomly or evenly distributed. Specimens are usually clumped and restricted to habitats fulfilling the species' particular ecological requirements. Abiotic factors, or the presence of particular food or host plants, but also the absence of predators, parasites, competitors, and others, may be important in determining their occurrence. Where suitable habitats are at some distance, more or less isolated subpopulations, which only occasionally interbreed, result. However, as long as gene flow is not completely disrupted there is a single, continuous range. Within-range aspects, such as fine-scale patterns of distribution and clinal or discrete variation in morphology, physiology, or other characters across ranges, occur but are not considered here. Similarly, seasonal or diurnal differences of specimen distribution, habitat changes between different life stages, and other intrinsic details occur but are not dealt with here.

Species that occur in several spatially separate, reproductively isolated populations are called disjunct. Disjunctions arise because ranges experience extensions and restrictions that are often induced by a combination of factors. Active and passive animal dispersal, changing ecological conditions, and changes of the earth's surface—for example, by sea transgressions (e.g., level changes), orogenesis (e.g., mountain building), or continental drift—may be involved. Because of this complexity, size and shape of ranges are not generally related to insect size and mobility. To provide some examples from butterflies that are strong flyers, the birdwing, *Ornithoptera aesacus,* is endemic to small Obi Island in the Maluku Islands, whereas the peacock, *Inachis io,* is endemic to Eurasia.

Dispersal

ACTIVE DISPERSAL   Random movement of individuals in a growing population leads by itself to some peripheral range extension until eventually the entire inhabitable space is occupied. Small-scale ecological change and normal insect activity lead to range extensions, or to restrictions, if conditions deteriorate. Although some active dispersal is involved in all range extensions, the term is most often applied to long-distance movements. These are often observed in migratory species, but only the area in which a species regularly reproduces is called its range. The area in which it appears only during migration is separately recorded. Long-distance migrations may be by single specimens or by large numbers and may or may not lead to temporary or lasting range extensions.

Under favorable conditions, the population density of some insects such as in the notorious migratory locusts, can become high enough to induce emigration of large numbers. Similar situations occur in dragonflies, for example, the European fourspotted chaser *(Libellula quadrimaculata),* and in other species. However, migrants usually move within the general range of the species, attaining only temporary and marginal range extensions. In Europe, some butterflies—for example, the painted lady and the red admiral *(Vanessa cardui* and *V. atalanta)*—regularly migrate to north of the Alps, and some Mediterranean moths (e.g., the death's head hawk moth, *Acherontia atropos,* the convolvulus hawk moth, *Agrius convolvuli,* and *Daphnis nerii)* do the same in warm summers. However, photoperiodic cues or winter temperatures do not permit lasting establishment in central Europe. Seasonal mass migrations are performed by the monarch butterfly, *Danaus plexippus,* and long-distance dispersal of individual butterflies is often observed. *Danaus* established itself in New Zealand, Australia, and elsewhere but only after humans introduced milkweeds, the food plants that had originally been absent from these areas.

Active dispersal is most easily noticed in spectacular forms such as those mentioned earlier. In most insects, the numbers moving and the distances traveled remain unknown but may be important. For example, many insects, such as hoverflies (Diptera: Syrphidae) and moths (Noctuidae) but also large numbers of dragonflies, were observed migrating across some high alpine passes in Switzerland. On most days, thousands were trapped in malaise traps or light traps.

PASSIVE DISPERSAL   Passive dispersal or transport of insects occurred naturally long before the involvement of human traffic. For range extensions, passive dispersal may be equally or more important than active movements. Transport in the pelt or plumage of larger animals occasionally occurs and may be important for the colonization of, for example, isolated ponds. Flooding streams and rivers move huge amounts of riparian organic debris plus the associated fauna downstream, sometimes over large distances. In large, tropical rivers, floating trees with a diverse fauna or entire vegetation islands have been observed traveling substantial distances, eventually also over sea. There is now also widespread agreement that the post-Pleistocene (re)colonization of parts of Scandinavia, Iceland, and Greenland was through drifting ice carrying soil and associated biota from refugial areas in western Europe.

Species with limited flight capacity and strong flyers alike are exposed to air transport. Collections made on ships stationed on the open sea or from airplanes or the outfall on high mountains show that amazingly large numbers of insects and spiders travel as aerial plankton. These air-transported species are mainly small organisms and not only those spending part of their life in some resistant inactive state, such as rotifers and tardigrades. Apparently through passive aerial dispersal, some of the smallest animals have some of the largest ranges, and some very small species are even of global distribution. Air transport ("ballooning") may form part of distributional strategies, for example, in spiders and also some first-instar Lepidoptera that produce silk strands, facilitating being caught and carried by moving air. Examples can be found in the arrival and partly the subsequent establishment of a number of butterflies and probably also other insects in New Zealand during the last 150 years.

However, the importance of passive transport has sometimes been overestimated. For example, at a time when no other explanations seemed to exist, transport by westerly storms encircling the southern end of the world in the "Roaring Forties" latitudes was thought to have caused continental disjunctions that can today more convincingly be explained by continental drift.

Various quarantine measures are presently taken against unintended human transport of insects, but the problem is an old one. For example, when the Vikings came to North America, they contracted human fleas. Preserved fleas were found in Viking settlements on Greenland whence they were apparently carried to Europe, where human fleas first appeared around the year 1000. Several soil-dwelling beetles were introduced to North America with ship ballast collected in Europe at sites where the particular beetles abound. Their survival today indicates suitable ecological conditions in America, but only some of the beetles dispersed widely. Others spread easily, such as the Colorado potato beetle *(Leptinotarsa decemlineata),* which expanded its restricted natural range over much of North America and over Europe, where it was also introduced. As with the monarch butterfly, the intentional introduction of the food plant had prepared its way.

LARGE-SCALE ECOLOGICAL CHANGE: EFFECTS OF THE PLEISTOCENE

The ecological relations of extant (i.e., existing) insects are assumed to have been the same in the past as they are now; if different assumptions are made, they must be explained. The drastic climatic changes since the end of the Tertiary and especially during Pleistocene glaciations profoundly

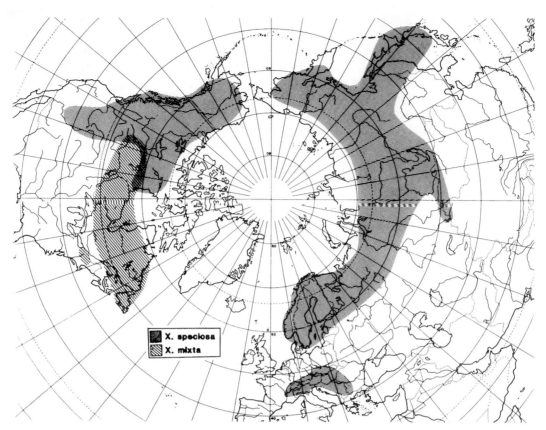

FIGURE 3 Transberingian and arctoalpine disjunctions in circumpolar Noctuidae of the genus *Xestia;* the Old World *X. speciosa* is represented in western North America by the subspecies *aklavicensis;* the origin of the eastern North American *X. mixta* is thought to predate the Wisconsin glaciation. [From Mikkola, K., Lafontaine, J. D., and Kononenko, V. S. (1991). Zoogeography of the Holarctic species of the Noctuidae (Lepidoptera): Importance of the Beringian refuge. *Entomol. Fenn.* **2,** 157–173.]

impacted insect distributions. The north polar ice expanded south, leading to a southward displacement of zonal biomes in the Holarctic. The tops of major mountains further south also acquired ice caps. The Silvaea and its fauna were driven into southern refugia, whereas cold-adapted oreotundral insects became established in their former place. The process was reversed when the ice began to retreat. The zonal biomes and their associated biota then shifted again northward; they probably continue to do so today.

The essentially north–south orientation of major mountain ranges in North America facilitated the displacements. In Europe, where major mountains run mainly from east to west, there remained an ice-free corridor between the polar ice front and the glaciated Alps, inhabited by a mixed fauna of northern and alpine origin, respectively. Many animals were driven further southwest, toward the Pyrenees, or southeast into the mountains of the Balkan peninsula. About 18,000 years ago, the ice began to retreat and cold-adapted insects followed it. There was a partial faunal exchange, and a number of fracturing of ranges, or disjunctions, resulted. Today, representatives of several insect orders exhibit boreoalpine or boreomontane (or arctoalpine and arctomontane, respectively) disjunctions (Fig. 3), but only rarely has this led to perceptibly divergent evolution, or even speciation.

Because of European topography, Pleistocene refugia of the south-retreating Silvaea and its fauna were mainly on the three large Mediterranean peninsulas. Many of the present central European species can clearly be assigned to one particular refugium because their ranges tend to coincide and occupy all of the former refugium, even though postglacial climate change and human impact strongly fragmented the Mediterranean deciduous forests. In contrast, at the northern range limits the individual species returned, variably far into once devastated areas. Postglacial recolonization by European insects was apparently fast because not all insects were affected by barriers like the English Channel, which formed about 8000 years ago, or the Baltic Sea straits separating Jutland from Scandinavia, which definitely formed some 6400 years before the present. The foregoing scenarios, initially inferred from distribution patterns, were later backed up by fossils, especially well-sclerotized and easily preserved beetles. Today, molecular genetic studies in a new line of research, phylogeography, provide support to these historical reconstructions.

The binding of much water as ice during the glacial periods lowered the sea level by about 100 meters during the last glacial period, which made important land bridges available. Tasmania was connected to Australia, which in turn established contact with New Guinea, which itself had ties with the Oriental region. The latter included a large continuous landmass, the Sunda plate, where numerous separate islands remain today. Japan was connected to the Asian, and England to the European mainland.

The Bering bridge connected East Asia and western Alaska, which were covered by tundra. Thus, there is a fair number of shared species or pairs of sibling species in East Asia and northern North America, mainly among tundral insects. Numbers of terrestrial as well as aquatic insect species are of circumpolar distribution (Fig. 3). Ecological demands seem to mostly prevent a southward spread of these northern species.

ANCIENT DISTRIBUTION PATTERNS AND CHANGING EARTH SURFACE

Changing Concepts

Widely disjunct distributions of older origin than discussed so far are revealed when taxa of higher rank, for example, families, are considered. Mainly in the Southern Hemisphere, ranges of close relatives may be separated by wide oceans. Explanations proposed for these patterns changed in accordance with the developing understanding of animal evolution and of changes occurring on the surface of the earth.

First, scientists proposed the former existence of numerous land bridges in early times to explain disjunctions. Most of the proposed land bridges never existed, but a few indeed did occur beyond those entirely caused by Pleistocene sea level fluctuations. Greenland, for example, was long connected with North America; the so-called DeGeer route connected northern Greenland with the extreme northwest of the then larger European continent. To the south, the Thule bridge connected Iceland, the British Isles, and the rest of Europe.

Later, and for as long as continents were believed to have been stable, insect dispersal was the favorite explanation for disjunctions. Routes and corridors along which animals might have moved, and bottlenecks or filters allowing only the passage of selected taxa, were discussed. Dispersal over long distances and across major hazardous obstacles along so-called sweepstake routes, where only few taxa would succeed, and largely by chance, also was considered. Actually, some dispersal can never be excluded, but the phenomenon is by itself insufficient to explain many insect distributions. This is also obvious from the studies of L. Croizat in 1958, who developed a method that he called pan-biogeography. Croizat connected ranges of related animals by lines (or tracks) and observed similar patterns in quite different groups of animals, suggesting the existence of "generalized tracks." Some of these tracks connected landmasses across oceans that all the many taxa with different dispersal capacities could not possibly have crossed; however, no general explanation of wide transoceanic disjunctions was offered.

Major advances in both animal systematics and in the earth sciences profoundly changed the situation, and this occurred only a few decades ago. On the zoological side, the work of

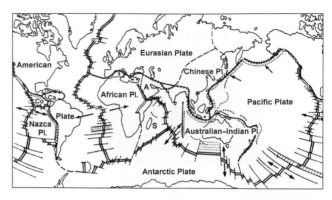

FIGURE 4 Midoceanic ridges and continental plates: A, Arabic plate; Ca, Caribic plate; Co, Cocos plate; S, Somalian plate. Arrows indicate directions of plate movements. Double lines represent midoceanic ridges; transverse lines across them are faults. Subduction and compression zones are mainly along deep-sea rift valleys (dotted lines) or mountain chains (crosses). Figures and stippling identify million years of seafloor spreading. [Modified after Thenius, E. (1979). "Die Evolution der Säugetiere." UTB 865. Gustav Fischer Verlag © Spektrum Akademischer Verlag, Heidelberg.]

Willi Hennig was instrumental in the development of modern zoogeography. Henning showed how the degree of phylogenetic relationship, or closeness of common ancestry (as opposed to some vague relatedness) between taxa can be recognized and reflected in the animal system. He also explained that postulating former land connections is logically justified only if sets of phylogenetically related taxa (i.e., branched sections from the hierarchical animal cladogram) exhibit similar disjunctions. Otherwise, relic distribution from once wider ranges or, alternatively, chance dispersals, are no less probable, even among widely disjunct individual sister taxa.

In the earth sciences, Alfred Wegener in 1912 suggested that continent positions are not stable but change over time. Evidence presented in support of continental displacements included the good fit of continental shelf lines, as well as observations of areas with particular deposits or minerals, traces of paleozoic glaciations, and particular mountain chains on separate continents. However, as long as no mechanism providing the power for movements of continents could be identified, this evidence remained unconvincing.

The situation changed a few decades ago when midoceanic ridges on the seafloors (Fig. 4) were recognized as sources of magma from the fluid interior of the earth; ridges form a network delimiting the continental plates. As magma appears at the surface, it pushes sideward, and the seafloor is spreading. Magnetic particles in the magma become uniformly oriented in the global magnetic field. This orientation is preserved when the magma cools and hardens. Bands of seafloor differing in magnetic orientation (or in paleomagnetism) extend parallel to the midoceanic ridges; evidently, the global magnetic pattern is at times reversed. Several centimeters of new seafloor is produced per year. In combination with measurements of paleomagnetism, the age of seafloors was estimated and found to increase with distance from midoceanic ridges, from contemporary at the ridge to only about 65 mya at the distant sites. Seafloors are generally young, the most ancient ones are only about 200 mya old. Seafloor spreading provides the power that shifts the continental plates, which because of overall differences in elemental composition, are of lower specific weight than seafloor. Therefore, most of the continental material remains afloat while essentially excess seafloor is subducted back into the fluid center of the earth. Such subduction zones occur in deep-sea valleys, mostly along continental edges. Subduction zones coincide with arcs of major vulcanism and earthquake activities. Floating continents may slide past each other along friction zones, continents may collide and cause upfolding of mountains, or they may merge or break up.

Overview of Continental Drift Pattern

Once the mechanism driving continental movements had been recognized, continental drift was widely accepted. Continental plates are moving, merging, and breaking up since their formation. Using all available evidence, paleogeography can describe past changes of the earth's surface in fair detail (Fig. 5). The origin of life in general and

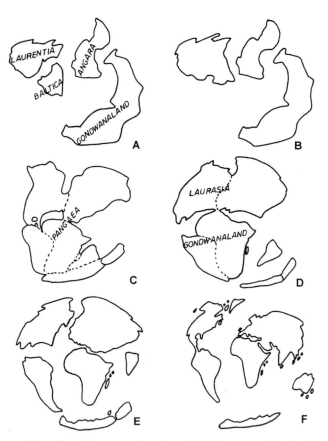

FIGURE 5 Pictorial summary of continental drift, from approximately 330 mya to present times: (A) 330 mya, (B) 300 mya, (C) a single landmass, Pangaea, 280–200 mya, (D) separation of Gondwanaland and Laurasia (180 mya), (E) 40 mya, (F) the Americas reunited, 25 mya. [From Vickery, V. R. (1989). The biogeography of Canadian Grylloptera and Orthoptera. *Can. Entomol.* **121**, 389–424.]

also of several insect orders dates back farther than the formation of a single supercontinent, Pangaea, but the methods of zoogeography cannot provide insight into earlier events. Zoogeography deals mainly with subsequent changes, in particular, the breakup of Pangaea into the northern and southern continents, Laurasia and Gondwanaland, respectively, and their further fates.

Consequences for Biogeography

Knowledge of seafloor spreading, plate tectonics, and continental drift has profoundly affected zoogeography and increased the relative importance of its alternative approaches. The so-called vicariance approach, based on evolutionary theory and on phylogenetic systematics in combination with information on continental drift, gained great explanatory power. Closely related taxa inhabiting separate ranges are called vicariant. Splitting up of populations with disruption of gene flow, which are called vicariance events, becomes the starting point of divergent evolution, eventually leading to differences in species. Over geological time, different vicariant sister clades may evolve from the ancestral species.

Today, one can understand how the breakup of supercontinents led to wide disjunctions and induced separate evolution of related taxa, on separate continents. Continental drift actually provided for means of transport, and insects can now be seen rafting on drifting continents instead of dispersing between them, across wide oceans. Vicariance biogeography seeks for congruences between the evolution of landmasses and the evolution of animals living on them and envisions the first process driving the second. There are now elaborate methodological considerations as well and they are described in works by Humphries and Parenti, and Wiley.

Ancient Disjunctions in Northern Hemisphere

Intra-American faunal differences provide evidence that contemporary insect distributions are almost always the result of a variety of causes that were effective at different times. Most disjunctions between related groups inside North America occur along a line that runs through the central plains, from northwest to southeast. This separation line results from present ecological differences between the mountains that support mainly the arboreal biome and the essentially eremial plains, from the past existence of a midcontinental seaway in the area of the present plains, and from past affiliations of the mountainous eastern and western halves of North America with other continents.

The areas adjacent to the present Bering Strait support tundras, and so did the ice-free areas on and around the former Bering bridge, but forest-dwelling insects had no access to this land bridge. Nevertheless, among the more southern arboreal insects, genera are often shared between eastern Asia and North America; species tend to differ between continents (Fig. 6). These disjunctions date back much further than the

Pleistocene. Because the Angara shield was separated from the Canadian shield only about 2 mya, by the opening of the northern Pacific Ocean, range disjunctions at the specific or generic levels between the Asian Far East and western North America are observed in many orders.

The northern continents formed through fusion of the ancient Canadian (or Laurentian), Fennoscandian (or Baltic), and Angara continental shields that were subsequently again divided and reunited, until the present pattern appeared. North America and Europe were connected until the opening of the Atlantic Ocean, through seafloor spreading, about 70 mya ago. Close phylogenetic relations between various insect groups in eastern North America and Europe are evidence of the past unity (Fig. 7). Europe was to the east long separated from Siberia by the Turgai Strait east of the Ural Mountains, which explains differences in the European and Asian faunas, despite the present continuity of land (Fig. 6).

Southern Hemisphere Case Study

The stepwise disintegration of Gondwanaland caused some of the most striking disjunctions and long unexplained "transantarctic" or "amphinotic" relations between animals living in Andean South America, Australia, and New Zealand. These landmasses are now known to have long remained connected with, or closely adjacent to, the then forested and inhabitable Antarctic continent.

The fundamental change in views, from dispersalism to continental drift, is recent; the Plecoptera (or stoneflies) provide an example. In 1961 stonefly evolution was still explained entirely by long-distance dispersal involving transgressions of the equator, two in each of the two then recognized suborders. An initial movement of primitive taxa from south to north was assumed, followed by the return of evolutionarily advanced forms to the Southern Hemisphere. In 1965 elements of cladistics and continental drift were added to this scenario. A few years later, a cladistic approach to

Plecoptera systematics led to a widely accepted revised system, suggesting that continental drift steered stonefly evolution. The breakup of Pangaea into Laurasia and Gondwanaland seems to have caused the separation into distinct Southern and Northern Hemisphere suborders, the Antarctoperlaria and Arctoperlaria, respectively. When Gondwanaland fell apart, the ranges of the suborder Antarctoperlaria and its families became disjunct, distinct representatives that evolved on each of the distant landmasses (Fig. 8).

However, continental drift alone can probably not explain all of the present Plecoptera distribution. Ecology and also dispersal remain important. Antarctoperlaria must have been present on Gondwanaland before Africa and India broke away from it. Ecological change, perhaps past dryness, is thought to have caused their disappearance from these lands. More difficult are two arctoperlarian families of which subordinate endemic groups are present also in the Southern Hemisphere. Contrary to widespread belief, not all Plecoptera are cool adapted. The Australian Gripopterygidae, Eustheniidae, the arctoperlarian Leuctridae, and the Nemouridae include many tropical species; they are most numerous in the large family Perlidae. The many *Neoperla* in the Ethiopian region are clearly of northern origin, but the origin of the diverse South American Perlidae is uncertain. Most problematic, however, is the so-called family Notonemouridae. Its monophyly is doubtful; it may represent independent early branches of the Nemouridae. Nevertheless, all notonemourids live in temperate parts of South America, Australia, New Zealand, South Africa, and Madagascar, but not in India. Dispersal seems to have contributed to these distributions that, admittedly, remain essentially unexplained. A practical test using the methodological refinements of vicariance biogeography proposed by Humphries and Parenti would require a better understanding of phylogenetic relationships among Plecoptera than is presently available.

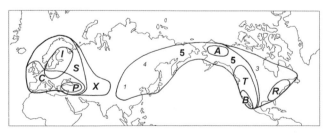

FIGURE 6 America–Asian relations, and the distinctness of the European fauna: distribution of the genera of Chloroperlinae (Plecoptera: Chloroperlidae). The ranges of five genera (*Alloperla, Haploperla, Plumiperla, Suwallia,* and *Sweltsa*) indicated by **5** in northeastern Asia and northwestern America, largely overlap. Numbers of genera of this group decline east- and westward; figures in italics are numbers present in the respective areas. The other genera are A, *Alaskaperla;* B, *Bisancora;* C, *Chloroperla;* I, *Isoptena;* P, *Pontoperla;* R, *Rasvena;* S, *Siphonoperla;* T, *Triznaka;* X, *Xanthoperla.*

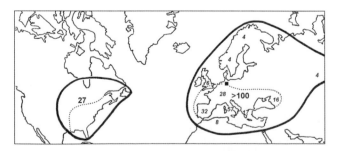

FIGURE 7 American–European disjunctions. Solid lines: range of the stonefly genus *Leuctra* (Plecoptera: Leuctridae); bold figures are total numbers of species per continent, figures in italics are regional numbers of species. A single species, *L. fusca,* occurs all over Europe and extends through Siberia to the southern portion of the Russian Far East. Broken lines: ranges of the extant ants *Ponera pennsylvanica* (America) and *P. coarctata* (Europe); *P. atavia* (black square) is an amber fossil (Hymenoptera: Formicidae). [Range information after Noonan, G. R. (1988). Faunal relationships between Eastern North America and Europe as shown by insects. *Mem. Entomol. Soc. Can.* **144,** 39–53.]

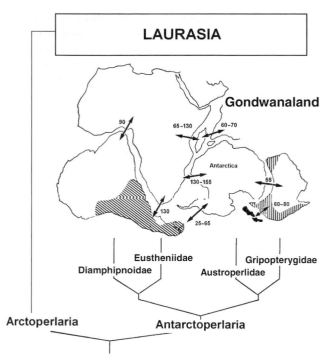

FIGURE 8 Phylogenetic system and distribution of the Plecoptera. The range of the family Gripopterygidae is shown on a map of Gondwanaland at the end of the Cretaceous, with dates of last possible faunal exchange (mya). Different shading indicates that each of the disjunct areas has an endemic fauna; no genus is shared. The Eustheniidae and Austroperlidae are distributed in the same way but have narrower ranges, the five species of the Diamphipnoidae are all South American. [Map based on Crosskey, R.W. (1990). "The Natural History of Blackflies," copyright The Natural History Museum, London.]

It is not common for the first step in the breakup of Pangaea to be clearly reflected in the phylogenetic system; otherwise, however, the Plecoptera have many parallels among other insects. Southern Hemisphere disjunctions suggesting a Gondwanian origin are widespread among aquatic (e.g., Ephemeroptera, Odonata, various dipteran midges) and terrestrial insects, for example, in the Hemiptera, Neuroptera, Mecoptera, and Coleoptera, to name a few. Some of these disjunct groups also comprise African representatives. The phylogenetic relationships within several of these groups of insects appear to reflect the proposed sequence of the disintegration of Gondwanaland. The breakup provided series of vicariance events enabling phylogenetic divergence.

See Also the Following Articles
Biodiversity • Fossil Record • Introduced Insects • Island Biogeography and Evolution • Population Ecology

Further Reading
Briden, J. C., Drewry, G. E., and Smith, A. G. (1974). Phanerozoic equal-area world maps. *J. Geol.* **82**, 555–574.
Cranston, P. S., and Naumann, I. D. (1991). Biogeography. *In* "The Insects of Australia," Vol. 1, pp. 180–197. Melbourne University Press, Melbourne.
Croizat, L. (1958). "Panbiogeography," Vols. 1, 2a, 2b. Caracas.
Downes, J. A., and Kavanaugh, D. H., eds. (1988). Origins of the North American insect fauna. *Mem. Entomol. Soc. Can.* **144**, 1–168.
Hennig, W. (1960). Die Dipteren-Fauna von Neuseeland als systematisches und tiergeographisches Problem. *Beitr. Entomol.* **10**, 221–329.
Humphries, C. J., and Parenti, L. R. (1986). "Cladistic Biogeography." Oxford Monographs on Biogeography 2. Clarendon Press, Oxford, U.K.
Malicky, H. (1983). Chorological patterns and biome types of European Trichoptera and other freshwater insects. *Arch. Hydrobiol.* **96**, 223–244.
Morgan, A. V., and Morgan, A. (1980). Faunal assemblages and distributional shifts of Coleoptera during the late Pleistocene in Canada and the northern United States. *Can. Entomol.* **112**, 1105–1128.
Morrone, J. J., and Crisci, J. V. (1995). Historical biogeography: Introduction to methods. *Annu. Rev. Ecol. Syst.* **26**, 373–401.
Platnick, N. I., and Nelson, G. (1978). A method of analysis for historical biogeography. *Syst. Zool.* **27**, 1–16.
Raven, P. H., and Axelrod, D. I. (1974). Angiosperm biogeography and past continental movements. *Ann. Missouri Botan. Garden* **61**, 539–673.
Taberlet, P., Fumagalli, L., and Wust-Saucy, G. (1998). Comparative phylogeography and postglacial colonization routes in Europe. *Mole. Ecol.* **7**, 453–464.
Tarling, D. H., and Tarling, M. P. (1975). "Continental Drift." 2nd ed. Doubleday, Garden City, NY.
Wiley, E. O. (1988). Vicariance biogeography. *Annu. Rev. Ecol. Syst.* **19**, 513–542.
Zwick, P. (2000). Phylogenetic system and zoogeography of the Plecoptera. *Annu. Rev. Entomol.* **45**, 709–746.

Biological Control of Insect Pests

R. G. Van Driesche
University of Massachusetts, Amherst

M. S. Hoddle
University of California, Riverside

Biological control is a form of pest control that uses living organisms to suppress pest densities to lower levels. It is a form of ecologically based pest management that uses one kind of organism (the "natural enemies") to control another (the pest species). Types of natural enemies vary with the type of pest. For example, populations of pest insects such as scales are often suppressed by manipulating populations of parasitoids, which are insects that develop in or on the pest insects they attack and kill. Populations of plant-feeding mites, such as the common twospotted spider mite (*Tetranychus urticae*) are often limited by predators, especially mites in the family Phytoseiidae. Populations of weeds can be suppressed by specialized herbivorous insects that feed on them. Finally, many insect populations have pathogens (e.g., bacteria, viruses, or fungi) that infect them. Such pathogens, whether they occur naturally or are applied artificially as microbial pesticides, can locally and temporarily suppress a pest's numbers.

Biological control is thus about the relative numbers of pests and their natural enemies. Predators (or pathogens, parasitoids, herbivorous insects) increase in number over time and feed on the pest, whose population then declines, because of higher mortality or lower birthrates caused by predation by natural enemies. Biological control agents are living organisms that increase in number through reproduction in response to pests that are used for nutrition. Biological control, at least in some of its forms, has the potential to be permanent in its action, through the reproduction and spread of the natural enemies as they track target pest populations.

There are four broad ways in which people have manipulated natural enemies to enhance their action: natural enemy importation, augmentation, conservation, and application of microbial pesticides. Each of these approaches has its own rationale, history, and level of past successful use. Biological control has important advantages compared with other methods of pest control. Moreover, because pest control is relatively permanent and requires no further capital input, biological control through self-sustaining forms (natural enemy importation and conservation) is often cheaper than use of pesticides. Although there may be significant initial costs (especially for projects that import new species of natural enemies from the pest's country of origin), costs drop to low or even zero levels in later years, whereas the benefits of the pest control achieved continue to accrue for years. For other forms of biological control (natural enemy augmentation and use of microbial pesticides), control is not permanent and costs recur annually, as with pesticides. For the latter two approaches, biological control may be either more or less expensive than other approaches depending on details such as the cost of natural enemy production by commercial insectaries that sell beneficial organisms, and the efficacy of other control tactics. In all four forms, biological control has the advantage of being virtually harmless to people and vertebrates, whereas pesticides must be actively managed for safe use to mitigate harm to humans and other nontarget organisms.

Biological control as a scientific endeavor has a history of about 125 years of effective use (beginning in the 1880s), over which time new information, techniques, and technologies have increased humankind's ability to use biological control agents with increasingly greater understanding and effectiveness. Before this period of active use, there were several centuries during which ideas about predators, parasitoids, pathogens, and their links to pest populations evolved. These ideas had to be developed before biological control as an applied pest management activity could be conceptualized.

HISTORICAL DEVELOPMENT

Biological control is the deliberate attempt by people to make practical use of the capacity of predation, parasitism, herbivory, and disease to restrain the growth of plant and animal populations. The ability to make practical use of these processes depends on an understanding of how pest population densities are controlled by natural enemies. Biological control also requires detailed knowledge of the biology of pests and their key natural enemies, because such knowledge often provides the means for their practical manipulation.

Predation

Predation by vertebrates on other vertebrates has long been part of human knowledge. Predation as a force affecting pest insects was recognized when people first saw individual acts of predation taking place on their crops. A close observer of an aphid colony, for example, cannot help but see the predatory action of ladybird beetles and cecidomyiid midge larvae as they devour their aphid prey. The predaceous effect of some species of ants on pest insects associated with citrus was recognized thousands of years ago by farmers in Yemen and China, who used the knowledge to suppress these pests by moving ant colonies into new orchards. In Europe during the Renaissance, the emergence of natural history as a subject worthy of observation and thought led keen-eyed naturalists to arrive at similar findings. The father of the classification of plants and animals, Caralus Linnaeus, observed in 1752 that "Every insect has its predator which follows and destroys it. Such predatory insects should be caught and used for disinfesting crop-plants." By the early 1800s, such observations led naturalists such as Erasmus Darwin and American entomologists such as Asa Fitch to suggest that predaceous insects should be used to suppress pest insects by making releases of the predators in places where they were lacking. These suggestions formed the fundamental basis for the modern use of augmentative biological control in greenhouses, vegetable production, and various outdoor crops.

Parasitism

The action of insect parasitoids on their hosts has no direct analogue among animals that people could easily observe before the invention of magnifying lenses. Consequently, the concept of parasitism took longer to become recognized. Because many parasitoids feed inside their hosts, their presence was not easily recognized, and the detection of parasitoids required that insects be either reared or dissected. By the 1600s, European naturalists were noticing the occurrence of parasitoids. Aldrovandi, for example, in 1602, reared tiny parasitoid wasps from the pupae of a nymphalid butterfly and recorded what he saw in woodcut print. Because he misunderstood the process, however, he wrongly concluded that the tiny wasps were an alternate adult form to the usual butterfly. The first person to publish a correct interpretation of insect parasitism was the English physician Martin Lister, who in 1685 noted that the ichneumon wasps seen emerging from a caterpillar were a distinct kind of insect that originated from eggs inserted into the caterpillars. No one thought of any way to make practical use of such parasitoids, however, until 1855 when Asa Fitch proposed the importation of parasitoids from Europe to America to help suppress a nonnative invasive pest of wheat, the wheat midge *Sitodiplosis mosellana*. Fitch's ideas provided a clear plan for the modern practice of biological control through natural enemy importation, but they were not acted on for nearly 30 years. The first importations of exotic species of parasitoids between continents occurred in the 1880s, when *Cotesia glomerata* was brought to the United States from Europe to suppress *Pieris rapae,* a European pest of cabbage that had invaded North America in 1860.

Insect Diseases

The study of the diseases of insects started not for purposes of killing pest insects, but rather for protecting economically important species such as silkworms and honey bees. In the mid- to late nineteenth century, microscopes made it possible to observe bacteria and microscopic fungi, and the study of these organisms as pathogens of domesticated insects initiated insect pathology. The infectious nature of insect diseases was first demonstrated by Agostino Bassi of Italy, who in 1835 studied a fungal disease of silkworm larvae caused by the fungus *Beauveria bassiana.* Louis Pasteur continued work on silkworm diseases in France in the 1860s. The first attempt to use pathogens to destroy pest insects was made in 1884 by the Russian entomologist Elie Metchnikoff, who reared *Metarhizium anisopliae,* a fungal pathogen, and attempted to suppress the sugar beet curculio, *Cleonus punctiventris*, with application of the fungal spores. In 1911 the German scientist Berliner observed a bacterial disease of larvae of the flour moth, *Anagasta kuehniella,* and by 1938 this bacterium, *Bacillus thuringiensis,* was being marketed as a microbial pesticide for control of some species of caterpillars. These early efforts established the concepts that insects were subject to infectious diseases and that the causative agents could be reared in quantity artificially. Technical methods to use reared pathogens to reliably infect insects in crops, hence to achieve biological control, came later.

Use of Insects to Suppress Weedy Plants

Although humans have known for millennia that insects damage and even kill plants, the idea that specialized herbivorous insects could be manipulated to suppress plants considered to be weeds is a relatively new concept. The first person to suggest such use was Asa Fitch. In 1855, Fitch noted that some European plants that had invaded North America, such as toadflax (*Linaria vulgaris*), had no American insects that fed on them. He suggested that importation of insects from Europe might help suppress these invasive plants. In 1863 this concept was implemented when a scale insect was moved from northern to southern India for the purpose of damaging an invasive nonnative species of cactus

(*Opuntia vulgaris*). Two related cacti, *O. stricta* and *O. inermis,* were introduced as ornamentals to Australia and became highly invasive and damaging. By 1925, these plants occurred in dense stands over approximately 20 million hectares of land. The Australian government began a survey of South America (the home of these cacti) looking for specialized insects attacking the ornamentals and a moth, *Cactoblastis cactorum,* was released into Australia in 1926. By 1932, the cacti were killed over most of the infested area and native vegetation and crops were able to reclaim the cleared ground.

These early suggestions and projects laid out the concepts that many plants are limited in number by specialized insects and that plants moved to new regions often become separated from these specialized insects because they are not moved along with the seed or nursery stock used to import the plant.

IMPLEMENTATION METHODS

The observation that nonnative insects and plants could be suppressed by importing missing specialized natural enemies from their homelands led to the first successful method for practical use of biological control. This approach is called classical biological control (because it was the first deliberate, successful application of biological control as a technology), or importation biological control, or simply natural enemy introduction.

After World War II, the chemical industry began the rapid development and marketing of chemicals to control pest insects by poisoning them. Pesticides became very popular and were used on a large scale in the second half of the twentieth century, such that the frequent application of insect-killing poisons to crops became routine. Widespread pesticide use led to a substantial reduction in the level of natural control provided by predators and parasitoids of pest insects, necessitating the further use of pesticides to suppress pest insect populations. However, many pests became resistant to one or more pesticides. This resistance sparked an interest in restoring natural control by reducing the use of insecticides in crops and making their use less damaging to natural enemies by manipulating their timing, placement, or formulation. The effort to restore natural control while making judicious use of pesticides formed the basis of the integrated pest management (IPM) movement in the late 1950s. Efforts to restore and protect natural controls by removing damaging influences such as pesticides are referred to as conservation biological control. A more recent, and less successful, mode of conservation biological control has been the attempt to increase natural enemy numbers by actively providing them with better food sources or habitats. Ideas that have been investigated include a variety of vegetation manipulations in or near crop fields, including ground covers between crop rows and unmowed field borders, where flowering plants provide nectar and pollen for natural enemies.

In the 1970s farmers in Europe producing vegetables in greenhouses were also interested in enhancing natural control of pests such as whiteflies in tomato and cucumber crops, because frequent development of pesticide resistance had rendered pesticides alone unreliable. The desire to reduce or even completely avoid the use of insecticides in tomato and cucumber crops was further stimulated in the 1980s when growers began the practice of placing colonies of bumble bees inside their greenhouses for crop pollination. Inside greenhouses, however, there were very few natural enemies because natural enemy immigration from outdoors is difficult in the indoor, sealed crop environment. Placing whitefly parasitoids, such as *Encaria formosa,* or predatory mites, such as *Phytoseiulus persimilis,* in the crop shortly after planting allowed natural control to develop and act on incipient pest populations. Once there was demand by growers for natural enemies, it became possible for specialized businesses (insectaries) to rear and sell natural enemies. This kind of pest management is called augmentative biological control, because the goal is to augment or initiate a natural enemy population.

In some crops, pests that feed directly on the edible part of the plant may not be adequately controlled by natural enemies before damage occurs. For example, cabbageworm larvae feed on cabbage heads and codling moth (*Cydia pomonella*) larvae burrow into apples. A faster acting form of biological control that can be applied when and where needed may be necessary in these situations. Microbial pesticides can be used in this way and were developed to meet immediate needs without resorting to disruptive chemical pesticides. The most successful of these products are those containing *B. thuringiensis,* a bacterium that produces toxic proteins that kill insects within a few days of ingestion. There are many subspecies of this bacterium that can be used to control the larvae of some moths, butterflies, beetles, and flies.

CLASSICAL BIOLOGICAL CONTROL

Ecological Justification

People routinely move species such as crop plants and ornamentals across natural barriers such as mountain ranges or oceans that would otherwise limit their spread. These plants may carry with them small, unrecognized infestations of pest insects. In some cases, the plants themselves may spread and become damaging. Both invasive plants and insects often escape their specialized natural enemies when they cross geographic barriers and establish in new locations. This allows these species to reach abnormally high densities and become damaging pests. Classical or importation biological control is based on the premise that the pest was originally limited to lower densities in its area of origin by specialized natural enemies, that these control agents are missing in the invaded area, and that densities of the pest in the invaded area can be reduced by importing the missing specialized natural enemies. Two recent biological control success stories from Africa, the control of cassava mealybug

(*Phenacoccus manihoti*) and water hyacinth (*Eichhornia crassipes*) illustrate these processes.

CASSAVA MEALYBUG Cassava (*Manihot esculenta*) is a tropical shrub that produces starchy tubers used much like potatoes, as a staple food source. The crop is a native of the Americas, but it is now a basic crop in all tropical countries, from Asia to Africa. In the 1970s, an unknown species of mealybug appeared on cassava in West Africa and spread rapidly throughout the cassava belt of tropical Africa. In this region, cassava was a basic food for some 200 million people. Within a few years, cassava crops began to fail as plants suffered extreme damage from high-density mealybug populations. Because the pest was clearly an exotic invader in Africa, importation biological control was seen as a means to suppress it. Furthermore, this method was chosen because it offered the possibility of providing permanent control that would not require the region's cash-poor farmers to repeatedly buy expensive pesticides and application equipment.

Cassava mealybug was believed to be from the Americas, the area of origin of the crop plant. The pest, however, was initially an unknown species. Therefore, no one knew where it could be found in South or Central America. With international funding, a cassava mealybug control project was organized. Crop protection laboratories in Africa (the International Institute of Tropical Agriculture in Benin) and South America (Centro International de Agricultura, in Colombia) worked with the Commonwealth Institute of Biological Control in Trinidad (now CABI-Bioscience, a private biological control organization in the United Kingdom) to find the pest, locate specialized natural enemies attacking it, import natural enemies to quarantine laboratories in the United Kingdom, and ship pure cultures of natural enemies on to Africa for release and evaluation in the effort to control the cassava mealybug.

Initial efforts were frustrated by an inability to find the pest in the Americas. Eventually, cassava mealybug and its parasitoid, the encyrtid wasp *Epidinocarsis lopezi,* were found in Paraguay. Upon release of this parasitoid, control was rapidly achieved. The parasitoid has spread (both naturally and from releases made by entomologists) throughout the cassava region, covering more than 26 countries. In 95% of the region, this single parasitoid has achieved stable, permanent control of this pest.

The net result of this project has been to increase food security in a region that frequently experiences food shortages. A pest has been controlled permanently (for nearly 20 years now in some areas), at no recurring cost, with no use of contaminating pesticides, and no damage to native plants or wildlife.

WATER HYACINTH *E. crassipes* is both a plant used in ornamental fish ponds and the world's worst aquatic weed. Its beautiful lavender flowers have led people to take it far from its native range in the Amazon basin of South America. Wherever water hyacinth has been introduced into subtropical or tropical climates, it has escaped into the wild, forming gigantic mats that clog rivers and cover over bays and ponds. Among the many places invaded by water hyacinth is Lake Victoria in East Africa. The pest was first recorded there in 1980 and by the mid-1990s some 12,000 ha of weed mats had clogged bays and inlets around the lake. Economic losses resulted for fisheries (the mats impede the launching of boats and the use of nets) and for waterworks and hydroelectric power plants. Ecologically, the weed threatened one of evolution's greatest products—the radiation of cichlid fishes in the lake, some 200 to 400 species of endemic fish that have evolved in the lake. These fish, often separated by mating habits based on bright colors, were threatened by hybridization among species induced by low light under weed mats, where color-based visual recognition mating systems could not be sustained.

Controls efforts recommended to the governments of the affected countries (Uganda, Kenya, and Tanzania) included applying herbicide to the mats, using harvester boats to cut the mats, and releasing specialized herbivorous insects. Two weevils, *Neochetina eichhorniae* and *N. bruchi,* known to be specialists on water hyacinth from earlier work in Florida, were chosen for release. In 1995 Uganda was first to release biological control insects against the weed, followed by the other two countries in 1997. On the Ugandan shore, weed mats began to show damage and disappear by late 1998. By 1999 some 75% of the mats had died and sunk into the lake. *Neochetina* weevils also produced dramatic results on a water hyacinth infestation in Kenya in only a few months in 1999 (Figs. 1 and 2).

Description of the Process

The following steps are typical of importation biological control projects.

1. Choice of the target pest. There should be broad social agreement that the species chosen as targets of importation biological control are pests and need to be reduced in density. Targets should be species that are strongly regulated by natural enemies in their native ranges, and these species should be missing in the areas invaded by the pest.

2. Pest identification and taxonomy. Correct identification of the target pest is essential. Mistakes at this stage cause project delays or failure. If the pest is an unknown species, its nearest relatives need to be identified, for this information can provide clues to the pest's likely native range.

3. Identification of the native range. The region in which the pest evolved needs to be identified to facilitate the search for specialized natural enemies that evolved with the pest. Several criteria can be used, including the center of the geographic range of the pest, the area where the principal host plant of the pest evolved, regions where the pest is recorded to occur but remains at low densities, and regions with the largest numbers of species closely related to the pest.

4. Surveys to collect natural enemies. Natural enemy collection, or foreign exploration, needs to be done extensively over the range of locations and habitats where the pest is found naturally, and in the proper seasons. Surveys of natural enemies in the invaded area are unlikely to locate effective natural enemies but are needed to identify any natural enemies that may already be present because of their own natural invasion of the region.

FIGURE 1 Water hyacinth infestation at a yacht club in Kisumu, Kenya, May 6, 1999. (Photograph courtesy of Mic Julien.)

FIGURE 2 Reduction of the water hyacinth infestation by *Neochetina* weevils at the yacht club in Kisumu, Kenya, December 16, 1999. (Photograph courtesy of Mic Julien.)

5. Importation to quarantine. Promising natural enemies collected in surveys need to be shipped to quarantine laboratories, where they can be colonized and maintained on the pest for further study.

6. Host specificity and biology studies. To promote selection of safe species for importation, the biology and degree of host specificity of each candidate biological control agent must be determined through a mixture of field observations in the area of origin and laboratory studies in quarantine before release into a new area is approved.

7. Release and colonization in the field. Releases need to be made at numerous locations where the target pest is present, and over extended periods, until efficient means to establish the natural enemies in the invaded area have been discovered or until it is clear that the agents are unable to establish. Once established, natural enemies are further redistributed throughout the range of the pest.

8. Evaluation of efficacy. Field experiments in the invaded area comparing pest density in plots having and lacking the introduced natural enemy are needed to measure the degree to which the natural enemy is able to reduce the density of the pest.

9. Documentation of benefits. Economic and ecological consequences of the project need to be recorded and published.

Extent of Successful Use

Following introductions of natural enemies, pest densities may be reduced, sometimes by 90 to 99% or more. This has been achieved for a variety of kinds of pest insects, including caterpillars, sawflies, aphids, scales, whiteflies, and mealybugs. Over the past 125 years, some 1200 projects of insect biological control have been attempted. Of these, 60% have resulted in a reduction of the pest's density. In 17% of projects, no further controls were needed and control was complete. Introductions of specialized herbivores have been attempted against about 133 species of invasive plants and, of these, 41 species (31%) have been completely controlled.

Economics

Importation biological control is an activity conducted by governments for the benefit of society. Funds for such work are typically provided by governments but may come from grower organizations representing particular crops in a region. Costs of projects are concentrated at the beginning of the work, as costs to search for and study new candidate natural enemies are high. Use of biological control agents of proven value in new locations (where need arises because of the continued spread of the pest into new regions) is cheaper, as much of the initial work need not be repeated and known natural enemies can quickly be introduced. Benefits of successful projects accrue indefinitely into the future, and benefit-to-cost ratios of past projects have averaged 17:1,

with some projects reaching as high as 200:1. In successful programs, control is permanent and does not require continued annual investments to sustain the benefits, in contrast to other forms of pest control (e.g., pesticide applications). This makes the method particularly attractive for the protection of natural areas and of crops in countries with resource-poor farmers. Biological control also promotes good environmental stewardship of farmlands in developed countries.

Safety of Natural Enemy Importations

Insects may be released as natural enemies of either invasive plants or invasive insects. Both biological weed control and biological insect control show a very high level of safety to human health and to the health of all other vertebrates. There are three safety issues when insects (herbivores, predators, or parasitoids) are imported to a new region: identification of unwanted contaminants, recognition of organisms damaging to other biological control agents, and potential damage to nontarget species (e.g., native insects or plants) in the area of release by natural enemies with broad host ranges.

The first two safety concerns are addressed by the use of quarantine facilities, which are designed to prevent the unintentional release of new species into the environment following importation. In quarantine, desired natural enemies are separated from miscellaneous insects that might have been accidentally included in the package by the collector, as well as from extraneous plant materials and soil inadvertently sent along.

A taxonomist then confirms the species identification of the organism and ensures that all individuals collected are the same species. Voucher specimens are deposited with an entomological museum for possible future reference. Natural enemy identification indicates either the name of the organism or, sometimes, that it is a species new to science and has not yet been described. New species can usually be placed in a known genus, for which some biological information may exist. A sample of the natural enemies is also submitted to a pathologist to determine whether they carry any microbial or nematode infections. If they do, they are either destroyed or, if possible, treated with antibiotics to cure the infection. This group of field-collected, healthy individuals is then bred in the laboratory on the target host. This series of steps eliminates any undesirable parasitoids (for herbivores attacking weeds) or hyperparasitoids (for insect agents) that might exist in the collected material and, if established, could damage the biological control project by reducing the efficacy of imported natural enemies. For insect parasitoids, rearing for one generation on the target host excludes the possibility that a hyperparasitoid has been obtained by mistake, since such agents typically do not breed on the host itself because they use the natural enemy as nutrient source.

The third safety concern—potential attack on nontarget species after release—requires that scientists estimate the host range of the natural enemy proposed for release and that this information be carefully evaluated as part of the decision whether to release the species from quarantine. For both weed and insect biological control agents, estimation of an agent's host range is based on several sources of information, including the hosts known to be attacked by the agent in the region from which it is collected, any species of interest that occur with the agent in its home range but are not attacked, and data from laboratory tests. For herbivorous insects released for weed biological control, these laboratory tests include studies of the adult's preference for where it lays its eggs, the immature feeding stages' preferences to eat various plants, and the ability of these plants to sustain normal growth of the agent's larvae to maturity. Similar tests can be applied to the study of parasitoids (i.e., both oviposition preferences and survival of the immature stages on a given host). For predators, oviposition preferences may sometimes exist; feeding preferences of both adults and larvae must be measured.

Estimation of host ranges of herbivorous insects used against weeds began in the 1920s, evolving from initial testing of local crops only to a phylogenetically based attempt to define the limits of the host range by testing first plants in the same genus as the target weed, then plants in the same tribe, and finally plants in the same or other families. This process has been highly successful in avoiding the introduction of insects whose host ranges are wider than initially thought. Attacks of introduced herbivores on nontarget plants have largely been limited to other species in the same genus. Also, some attacks were forecast by quarantine studies and judged acceptable by agencies granting permission for release, rather than being unforeseen attacks. Of 117 species introduced into North America, Hawaii, or the Caribbean for biological weed control, only one species (the lacebug *Teleonemia scrupulosa,* introduced into Hawaii in 1902 against the shrub *Lantana camara*) has attacked nontarget plants that were neither in the same genus as the target weed, nor a very closely related genus (for the lacebug, the native shrub *Myoporum sandwicense*).

Estimation of host ranges of parasitoids and predators introduced for biological control of insects began in the 1990s, in response to changing views on the ecological and conservation value of native nontarget insects. Techniques for making estimates of arthropod natural enemy safety are less well developed than those for herbivorous biological control agents. A few examples of harm from parasitoids or predaceous insects to nontarget insects have been reported. Importation of generalist species that have broad host ranges should be avoided because of such potential to harm native insects.

Laws governing biological control importations exist principally in New Zealand and Australia. Laws in the United States regulate importation of herbivorous insects used against weeds but do not currently regulate importation of parasitoids or predators.

NATURAL ENEMY CONSERVATION

Concept of Natural Control

All insects and plants, to various degrees, are attacked by natural enemies independent of any deliberate manipulations by people. Such natural control is rarely sufficient to suppress an invasive species: the local natural enemies lack specialized relationships to the invader, since by definition, the new pest is outside the evolutionary experience of the prey species. For native species, however, natural control may suppress plants and insects below pest levels. Also, for invasive species against which specialized natural enemies have been imported and established, the latter become part of the fauna, providing naturally occurring control. Thus, for all species, apart from invaders not yet subject to natural enemy importations, natural control exists and may be sufficient to suppress such pests adequately for human needs. However, in crops and other artificial landscapes, people can disrupt natural control, particularly with the application of pesticides that kill, sterilize, or repel important natural enemies. Conservation as a form of biological control aims to avoid this loss of natural control either from the use of pesticides or habitat simplification. Sometimes active intervention on behalf of natural enemies to provide them with key missing foods or hosts is necessary.

Effects of Pesticides on Natural Enemies

Before 1947, few synthetic pesticides were used in crops. Most available materials were stomach poisons based on heavy metals such as lead and arsenic, which kill only if eaten. Some botanical extracts, such as rotenone and pyrethrum, both of which quickly degrade in the environment, were also used. After World War II, a business revolution occurred when it became recognized that a variety of compounds that could be artificially synthesized in laboratories were highly effective in killing insects by mere physical contact. Beginning with DDT in 1947, many types of chemicals were marketed to kill insects. One of the undesirable consequences of this change in farming practice was the mass destruction of beneficial insects in crops, resulting in a substantial decrease in natural control. Indeed, insecticides often killed natural enemies more efficiently than they killed the target pest. This unintended consequence was due to the smaller body size, greater relative surface area, and lower levels of detoxification enzymes possessed by parasitic Hymenoptera and other natural enemies, compared with herbivorous pests.

PEST RESURGENCE Occasionally, farmers found that pests for which they applied pesticides were, within a few months, more numerous than they had been before the application of insecticide. This population rebound has been termed pest resurgence. The steps in resurgence are as follows:

1. The pest population is reduced by the insecticide.
2. The same insecticide application destroys most of the natural enemies that were partially suppressing the pest before the application.
3. Natural enemies are slower to increase in number than the pest after the pesticide residue from the application has degraded to levels unable to kill insects.
4. In the absence of the pesticide and with few remaining natural enemies, the survival and reproductive rates of the pest population increase, leading to higher densities.

In rice crops in Asia, outbreaks of a sucking insect called rice brown planthopper (*Nilaparvata lugens*) were rare before the 1960s. In the 1970s, outbreaks occurred with greater frequency and intensity, as insecticide use increased to control this pest. Research conducted at the International Rice Research Institute in the Philippines demonstrated that this was a classic example of pest resurgence and that pesticide applications were destroying spiders and other generalist predators that were otherwise usually able to suppress rice brown planthopper. As farmers used pesticides more often, outbreaks became larger and more frequent. This phenomenon led to a pesticide treadmill for rice brown planthopper control. A program of grower education was supported by the Food and Agriculture Organization of the United Nations to help rice farmers to understand pest resurgence, natural enemy recognition, and the beneficial role of natural enemies in rice paddies. This outreach program successfully reduced pesticide use on rice crops in Asia, ending a cycle of damaging pesticide use and crop loss.

SECONDARY PEST OUTBREAK   A related population process occurs when insecticides applied to suppress a primary pest induce a different species, formerly not damaging, to become a pest. This is called a secondary pest outbreak. In apple crops in the eastern United States, growers must control two serious direct pests of the fruit, apple maggot (*Rhagoletis pomonella*) and plum curculio (*Conotrachelus nenuphar*). These species are most often controlled by repeated application of insecticides to foliage with chemicals that have long periods of residual activity. These applications destroy the parasitoids of leafminers and predators associated with spider mites. Outbreaks of these two foliar pests later in the summer are a direct consequence of grower efforts to control these two key fruit pests.

SEEKING PESTICIDES COMPATIBLE WITH NATURAL ENEMIES   To reduce the destruction of natural enemy populations caused by insecticides, there are two potential solutions: using pesticides that have intrinsically selective action or using application systems that are ecologically selective.

Selective Pesticides   Three kinds of insecticide have shown the greatest compatibility with natural enemies: stomach poisons, systemic pesticides, and insect growth regulators. Stomach poisons are materials that must be ingested to kill. Materials such as the microbial pesticide. *B. thuringiensis* and some mineral compounds such as kryolite are examples. Pests eating foliage with residues of these materials are killed, but natural enemies walking on treated foliage are not affected.

Systemic pesticides are materials that enter plant tissues and are translocated through the plant. These compounds may be applied to soil and absorbed by roots, or they may move translaminarly into leaves after application to the foliage. Because residues are available only to insects that feed on the crops' tissue or sap, natural enemies resting or walking on plants are not affected.

Insect growth regulators are chemicals that mimic or disrupt insect hormones, preventing normal molting. These compounds kill only when the insect tries to molt. Such materials can be selective if only the pest is likely to be exposed in a susceptible stage. In principle, screening programs could identify specific insecticide–natural enemy combinations in which any contact pesticide might turn out to be selective relative to some particular natural enemy. However, because such materials tend to be rare and screening trials to discover them are costly, only a few are available.

Ecologically Selective Methods of Pesticide Use   Manipulation of a pesticide's formulation, timing, or method of application is another method for achieving selectivity in control. Granular formulations of pesticides that fall to the soil, for example, are unlikely to damage natural enemies that forage for hosts or prey on the foliage. Thus, a granular material may be applied at transplant into a cabbage field to protect the roots of young plants from feeding in the soil by larvae of cabbage maggot (*Delia radicum*) without injuring the braconid parasitoids that search the leaves to find and parasitize cabbage aphids (*Brevicoryne brassicae*). More complex methods of separating the pesticide from the natural enemies exist, such as monitoring the emergence of key natural enemies and applying pesticides either earlier or later than the peak activity period of the natural enemy. However, methods that require effort on the part of growers, or are at all complex, tend not to be used.

Loss of Natural Control through Simplification of Crop Fields

Natural control of pest insects and mites in crops has also been reduced by habitat simplification and physical changes in crop plants used in commercial, large-scale agriculture. To sustain their populations, parasitoids need hosts, carbohydrates, and secure places to live that are not subject to insecticide application or physical destruction by plowing, flooding, or fire. Predators need prey and can benefit from or even subsist on alternative nonprey foods such as pollen.

Natural control in crops can be maintained or improved by considering the degree to which these basic necessities of natural enemies are provided within or adjacent to the crop. A few examples illustrate the process.

ADDING POLLEN TO ENHANCE PREDATOR MITES Phytoseiid predatory mites are often important in control of pest spider mites. In some crops, numbers of such phytoseiids may be too low to provide effective control. One approach to increasing phytoseiid numbers is to provide pollen as an alternative food, especially for periods when spider mite densities are low. Levels of pollen on foliage of citrus and other orchard crops may be increased by use of species of trees in windbreaks around orchards or species of grasses as ground covers within orchards that are prolific pollen producers. Effective use of this approach has been made in South African citrus orchards for control of citrus thrips (*Scirtothrips aurantii*) with the phytoseiid *Euseius addoensis addoensis.*

KEEPING USEFUL STRUCTURES ON PLANTS Many plants, such as cotton, have sugar-secreting glands called nectaries both inside and outside of flowers. Many species of natural enemies feed on these sugars. Plant breeding has made it possible to eliminate such nectaries in some crops, and this is sometimes done to deny pests access to the carbohydrate resources. The decision to eliminate or retain nectaries needs to be based on studies of the net benefit to pest control of these structures. Plants (e.g., grapes) also often have on their leaves pits or pockets, called domatia, that provide physical refuges for phytoseiid mites. Varieties with domatia often have higher phytoseiid densities and fewer pest mites. Retention of such structures in new crop varieties may be important and should be an explicit part of plant breeding.

ENHANCING SPACES BETWEEN CROP ROWS OR AROUND CROP FIELDS AS REFUGES Natural enemies of some species remain tightly linked to the plant and are little affected by the larger environment. Parasitoids of scales on citrus trees, for example, have all their needs met on citrus trees, provided insecticides are not used and some scales are present year-round for parasitism, host feeding (feeding on host body fluids), or production of honeydew (a sticky carbohydrate waste product produced by homopterans that parasitoids use for food). Other species of natural enemies move about more, passing through the spaces between crops rows, or moving back and forth between crops and noncrop vegetation in uncultivated borders. Species such as spiders and carabid beetles are generalist predators of value in vegetable plantings. However, bare or plowed soil between rows often becomes too dry and hot to favor these predators. Reduced tillage, through greater use of herbicides, or use of cover crops between rows, can enhance populations of these predators. Plants between crop rows, however, must not compete with the cash crop for water or nutrients, or crop yield may be reduced. In cereal crops in the United Kingdom, populations of ground beetles, generalist predators that eat cereal aphids, can be increased by leaving low dikes through fields that are not plowed. These dikes produce perennial grass and herb communities that act as refuges for carabid beetles, which then forage in the cereal plots and consume aphids. Also, in crops that are sometimes treated with pesticides, nontreated patches of noncrop vegetation along crop borders can act to reinoculate crops with natural enemies.

Extent of Successful Use of Natural Enemy Conservation

Natural control is ubiquitous and contributes extensively to pest control in most settings. Conservation of natural enemies through reduction of conflicts with pesticides is a major focus of integrated pest management (IPM) philosophy and practice, and many studies have been conducted that have led to better conservation of natural enemies in crops such as citrus, avocados, apples, and greenhouse tomatoes. Because it is often associated with reductions in out-of-pocket costs, this form of conservation is particularly acceptable to growers, who often are asked to reduce or stop altogether a costly practice (such as applying a pesticide). In contrast, practices that require positive action, such as providing a resource or manipulating vegetation in or near the crop, have been adopted much less often. To be valued by growers, such measures must clearly produce pest control benefits that significantly exceed the costs of undertaking them. Practical use of these ideas presently is limited to organic growers and others who wish to produce crops with little or no use of synthetic pesticides.

Safety

Conservation biological control is universally considered to be a very safe activity. Measures to reduce insecticide use, or to convert to selective or compatible materials, both reduce risks to people working on or living near farms and minimize environmental contamination.

BIOLOGICAL CONTROL THROUGH AUGMENTATION

Pros and Cons of Augmenting Natural Enemies

Entomologists and farmers, working together, have developed methods to rear some species of predators and parasitoids that attack pest insects. This approach of deliberately rearing natural enemies and releasing them against target pests has been applied against insects and mites of both greenhouse and outdoor crops.

The use of this practice in greenhouse-grown tomatoes was begun in the 1920s with the rearing by English growers of *Encarsia formosa*, a parasitoid of the greenhouse whitefly

(*Trialeurodes vaporariorum*). This control program died out because of grower use of pesticides. Biological control was revived in the 1970s by Dutch greenhouse tomato growers because whiteflies had developed resistance to pesticides. In greenhouses that are closed up against the cold early in the crop cycle, natural enemies may be scarce or absent. Augmentative biological control was seen as a way of correcting this natural enemy absence. Natural enemy rearing for the greenhouse industry started when one grower began producing natural enemies for his own use, but soon he was selling surplus parasitoids or predators to other growers, and the operation became a separate business (an insectary). From 1970 to 2000, the number of commercial insectaries grew from just a few to several dozen firms, which collectively produce about 100 species of natural enemies for sale. A few species (mainly the parasitoid *E. formosa* and the predatory mite *Phytoseiulus persimilis*), however, make up most of the sales. Today, a variety of natural enemies are used in indoor settings that include greenhouses, plant conservatories, mushroom houses, and animal holding buildings such as dairies, hog-rearing facilities, poultry barns, and zoos.

Outdoor releases of several species of predators and parasitoids are regularly made by growers in various countries. Egg parasitoids in the genus *Trichogramma* (Hymenoptera: Trichogrammatidae) have been used extensively throughout the twentieth century to suppress pest weevils and caterpillars in cotton, corn, and sugarcane, especially in China, Russia, and tropical sugar-producing countries. Predators of mealybugs for release on citrus crops in parts of California have been reared by a growers' cooperative since 1926. One of the more common current uses of augmentative biological control on outdoor crops is the release of various species of predatory phytoseiid mites for control of pest spider mites, an approach that has been used most often with strawberries and with foliage plants grown outdoors in shade houses.

There are two different approaches to augmentative biological control. Most indoor releases of natural enemies intend only to seed the crop with a founding population of the natural enemy, which then reproduces and eventually suppresses the pest after its numbers have increased naturally in the crop. This approach is called inoculative biological control. Cost of this approach is minimized because smaller numbers of the natural enemy are needed. In contrast, with inundative biological control, an attempt is made to release enough natural enemies to control the pest immediately. Because much higher numbers are released, this approach is economical only against natural enemies with very low production costs, and use has been most successful on crops with a high cash value per hectare.

How Insectaries Turn Natural Enemies into Mass Market Products

To profitably market a natural enemy, an insectary must succeed in a series of activities.

1. Find a suitable natural enemy. Commercial augmentative biological control starts with the discovery of a natural enemy that research suggests may be effective. The natural enemy must attack an important pest efficiently, be able to be reared under mass production conditions, be easily harvested and able to survive transit stress, and be competitive in price with other forms of pest control available to growers.

2. Develop a mass rearing system. To commercially produce a natural enemy, insectaries must be able to make a financial profit on the species. Successful production systems vary. For some species, such as whitefly parasitoids, production can use natural hosts on their favored plants. *E. formosa,* for example, is reared in greenhouse whitefly produced on tobacco plants. Similarly, the important predatory mite. *P. persimilis* is grown on the spider mite *Tetranychus pacificus* on bean plants in greenhouses. In other examples, costs of production or the scale of production are improved by rearing species other than the target pest. Most *Trichogramma* wasp species are grown on the eggs of moths that feed on stored grain, rather than on eggs of the target moths themselves, because colonies of grain-feeding moths can be reared much more cheaply, allowing the production of *Trichogramma* in huge numbers at low cost.

3. Develop harvest, storage, and shipping methods to get the product to customers. Most predators and parasitoids must be used within a few days or weeks of production. For some species, induction of an arrested state called diapause can be used to store immature parasitoids inside parasitized hosts for months. Shipping to customers must use rapid transport (1–3 days) and avoid delays at international borders. Longer delays invariably result in the deaths of natural enemies due to heat, desiccation, continued development, or starvation.

4. Provide clear instructions on effective release methods and rates for customers. The final step in the effective use of natural enemies reared in insectaries is their release by the farmer at the right rate and in the correct manner. Effective rates are discovered by controlled trials in universities and government laboratories, and by ascertaining the experience of growers who have used products in accordance with advice from producers.

Extent of Successful Use

INDOOR CROPS The use of augmentative biological control has become widespread in greenhouses in northern Europe and Canada that produce vegetables, with over 5000 ha using *E. formosa* for whitefly control and over 2800 ha using *P. persimilis* for spider mite control. These amounts, however, still represent only a small percentage of the world's protected culture because these biological control agents are used much less often in southern Europe and Japan, areas with extensive greenhouse vegetable production but with differences in temperatures and open rather than closed greenhouses. Similarly, use of biological control is very limited in greenhouses producing bedding plants or floral crops, the major focus of greenhouse production in the United States.

OUTDOOR CROPS The scientific use of augmentative natural enemy releases in outdoor crops is best established in northern Europe for control of European corn borer (*Ostrinia nubilalis*) in corn. Use is greatest in Germany and France, where over 3200 ha is protected annually with *Trichogramma* releases. This fraction is, however, small compared with the total corn acreage in Europe, and use of biological control is concentrated principally where pesticide use is not allowed because of concern for health of people living near cornfields. Natural enemy releases for mite control have been successful in strawberries in California, Florida, and the northeastern United States, and in outdoor shade houses used for production of foliage plants in Florida. In Mexico, Russia, China, and other countries, large-scale releases of *Trichogramma* spp. have been made for a variety of moth and beetle pests of corn, sorghum, and cotton, but the efficacy of these releases has not been well demonstrated. Some of these activities have been state supported, and their actual economic value for pest control is not clear.

Safety

Release of parasitoids and predators replaces pesticide application and thus enhances human safety. For workers in insectaries, handling of large quantities of insects or mites constitutes an allergy risk. Where problems arise, risk can be reduced through air exchange or filtration to reduce concentrations of airborne particles and through use of gloves and long-sleeved shirts to reduce skin contact with arthropod body fragments. Risk to native species posed by releases of nonnative natural enemies can be of concern, as well. Generalist, nonnative species released in large numbers may establish outdoors and attack or suppress populations of native species, or they may reduce densities of native natural enemies through competition for resources. Consequently, some governments, such as those of Hawaii, Australia, and New Zealand, restrict importation of natural enemies used in augmentative biological control. For example, importation of North American green lacewing species (Neuroptera: Chrysopidae, *Chrysopa* spp.), used in greenhouses as predators of aphids, might lead to establishment of such species in the wild, increasing competition with the endemic native lacewings in Hawaii, which have conservation value as unique native wildlife.

MICROBIAL PESTICIDES

Using Microbes as Tools

Insects suffer from diseases caused by pathogens of several kinds, including bacteria, viruses, fungi, nematodes, and protozoa. Sometimes natural outbreaks of disease occur that locally and temporarily influence the density of pest populations. Microbial control seeks to use pathogens as tools to suppress pest insects. This process involves finding pathogens able to kill pest species of concern, followed by development of methods to rear pathogens economically. Methods must be developed to store the pathogen's infective stages without loss of viability and to apply the pathogen to the target in ways that result in high rates of infection and thus control. Details of the biology of each pathogen and the effects of environmental conditions on infectivity after application are crucial in determining whether any given pathogen can be used effectively as a microbial pesticide.

Bacteria

Many of the bacteria that infect insects are lethal only in stressed insects because the bacteria, lacking effective means of escaping from the host's gut after ingestion, are unable to enter its body cavity. Species in the genera *Bacillus* (*B. thuringiensis, B. sphaericus,* and *B. popililae*) and *Serratia* (*S. entomophila*) are the main bacteria that have been used as microbial pesticides. Of these, only *B. thuringiensis* has been widely successful. This species produces toxic crystalline proteins inside its spores. Crystals from different strains of this bacterium vary in their ability to bind to the gut membranes of different species, thus shaping the host ranges of each subspecies of the pathogen. If crystals are able to bind to the gut membranes, these tissues are degraded, allowing bacteria to penetrate the body cavity and kill the host.

Strains discovered in the 1920s infected only some species of caterpillars. Later, new strains were discovered that were able to infect mosquito larvae, chrysomelid beetle larvae (such as the Colorado potato beetle, *L. decemlineata*), and scarabs (such as the Japanese beetle, *Popillia japonica*). Commercial use of this pathogen is possible because it can be successfully mass-reared in fermentation media without any use of living hosts. This makes its production inexpensive. Applications of *B. thuringiensis* have advantages for use in forests, where residues of conventional pesticides are objectionable because of potential harm to native wildlife, and in IPM programs in crops where conservation of natural enemies is desired. *B. thuringiensis* is compatible with most natural enemies because it must be ingested to have any effect and because its toxic proteins are selective in their gut binding properties. Genes from *B. thuringiensis* that code for toxic proteins have been isolated and inserted into plants where they are expressed and produce insecticidal proteins in plant tissue and pollen. Transgenic varieties of such major crops as corn, soybeans, and cotton exist and are widely planted in the United States.

Other species of bacteria have had limited commercial use. *B. sphaericus* is formulated for use against some species of mosquito larvae. *B. popilliae* was once commercially produced for use against larvae of Japanese beetle (an important pest of turf), but this bacterium must be reared in living host larvae, which has made its production uneconomical. In New Zealand, *S. entomophila* causes an infection known as amber disease in a native turf grub (*Costelytra zealandica*), and its commercial use is being promoted. As with *B. thuringiensis,* the ability of *S. entomophila* to be reared in fermentation media apart from living hosts has been a key feature in promoting its commercial use.

Fungi

Species of Deuteromycotina fungi in several genera, including *Beauveria, Metarhizium, Verticillium,* and *Hirsutella,* infect insects and can be grown on fermentation media in solid culture. The spores of these species, when applied, adhere to the bodies of insects, and special hyphae from the spores use enzymes and mechanical pressure to penetrate through the insect's cuticle to cause infection. Infection requires spore germination, a process that is sensitive to environmental conditions. In general, many fungal strains or species require a minimum number of hours (often 12–24) of high relative humidity (often above 80%) to germinate. However, these requirements vary within species and among isolates from different locations and hosts. Spore germination requirements, if not met, can lead to control failures. Successful commercial use of entomopathogenic fungi has focused on ways to either meet these requirements by manipulating the formulations of the product applied (e.g., adding oils when used in arid climates), using these products in inherently favorable climates (e.g., greenhouses), using them in favorable habitats (e.g., soil), or finding strains or species with less stringent environmental requirements for spore germination. Commercial use of these fungi is also affected, but not prevented, by the inability of most species of fungi to produce spores under water. This prevents the use of liquid culture methods, requiring the use of solid media (like boiled rice) or a diphasic system in which mycelial growth takes place in liquid culture, followed by plating out of fungi on solid media for spore production as a second production step.

Successful use of microbial pesticides based on fungi has been achieved by an international consortium (LUBILOSA) in Africa, which has developed the fungus *Metarhizium anisopliae* var. *acridum* (Green Muscle) for control of locusts in Africa. This locust control project is highly beneficial to the environment because this selective, naturally occurring fungus replaces the use of highly toxic, often persistent, pesticides such as dieldrin. Field trials in a number of African countries have demonstrated both high levels of efficacy and costs competitive with the use of conventional pesticides (about U.S. $12 ha^{-1}). Success in this effort involved screening over 160 isolates of fungi to find the best fungus and the development of formulations for both storage (without refrigeration in hot climates) and application. Field trials demonstrated high initial levels of mortality and pathogen recycling, leading to persistence of suppression.

Similar success of fungal pesticides in general agriculture has not yet occurred. In the United States, for example, only one species, *Beauveria bassiana,* is commercially available, and its use is extremely limited.

Nematodes

Nematodes are multicellular organisms as opposed to unicellular microbes, but they are formulated and applied like microbial pesticides. Nematodes in more than 10 families infect insects, but only those in the families Steinernematidae and Heterorhabditidae have been commercially employed for insect control. These species, unlike those in other families, can be reared in fermentation media apart from living hosts. Techniques for large-scale production in liquid broths containing ingredients from dog food can be used to rear about six species in these families. Entomophagous nematodes actively penetrate insect hosts through the insect integument or natural body openings (spiracles, mouth, anus). Once inside the insect body cavity, the nematode defecates specialized bacteria that it carries symbiotically. These bacteria (in the genera *Xenorhabdus* and *Photorhabdus*) quickly kill the host with toxins. Nematodes then reproduce as saprophytes in the decaying host tissues. Entomopathogenic nematodes are sensitive to desiccation, which has limited their use in pest control. Applications made to dry foliage are ineffective because nematodes usually die before encountering hosts. Successful use of these nematodes has been limited to control of pests in moist habitats, such as fungus gnats and scarab grubs in soil, and lepidopteran borers in plant stems.

Viruses

Insects are subject to infections by viruses in a number of families. However, only those in the highly specialized Baculoviridae have been considered for use as microbial pesticides. Viruses in this family infect only insects and are very safe to people and wildlife. However, all viruses are obligate parasites of living cells, and none can be grown in fermentation media. Currently, they are produced in live host insects, which themselves must be mass-reared. This makes viral products relatively expensive, although the governments of some countries, notably Brazil, have promoted their use. A further aspect of the biology of viruses is their high level of host specificity. Extreme specificity of viruses reduces the economic value of products because they kill very few species of pests. Because of these economic factors, no virus products have been economically successful in the United States or Europe, although a few have been developed and briefly marketed.

Safety of Microbial Pesticides

In the United States and many other countries, microbial preparations (but not nematodes) sold for pest control are considered to be pesticides that require government approval and product registration before sale. Requirements for registration have been modified to reflect differences between chemical and microbial pesticides. Manufacturers are required to specify the exact identity of the microbe in their products, the production process, including controls to prevent contamination, and safety data on infection and allergenic properties of the pathogen and the product as a whole. The safety record to date suggests that risks from such products are either nonexistent or too low to detect.

Degree of Use

Except for genetically transformed plants that express the *B. thuringiensis* toxin (which are not microbial pesticides, but a related development), microbial pesticides are niche products. In no control programs have microbial pesticides widely displaced synthetic pesticides from pest control markets. *B. thuringiensis* is the most widely used organism, but *B. thuringiensis* products represent 1 to 2% of the pesticide market. These products do, however, have important value as pesticides because they are more readily incorporated into IPM programs that include natural enemies.

FUTURE USE OF BIOLOGICAL CONTROL

Biological control can be implemented through four different approaches: conservation of existing natural enemies, importation of new species for permanent establishment, temporary natural enemy augmentation, and use of microbial pesticides. The first two methods are most widely applicable and have produced the greatest benefits. Conservation biological control is the foundation of all insect control. Importation biological control is the method that is appropriate to combat exotic invasive pests (whose numbers are large and increasing). Augmentative biological control is limited by cost factors and largely restricted to high-value crops in greenhouses. Microbial pesticides are niche market tools useful in IPM programs but are limited by high production costs or the narrow host ranges of the pathogens. Biological control's greatest strengths are in public sector applications (conservation, importation) rather than private sector approaches (augmentative, microbial pesticides). Expanded use of biological control will require increased commitment of public resources and recognition that publicly supported programs are more effective for biological control implementation.

See Also the Following Articles

Agricultural Entomology • Genetically Modified Plants • Host Seeking by Parasitoids • Integrated Pest Management • Physical Control of Insect Pests • Predation

Further Reading

Barbosa, P. (ed.) (1998). "Conservation Biological Control." Academic Press, San Diego.
Bellows, T. S., and Fisher, T. W. (eds.) (1999). "Handbook of Biological Control: Principles and Applications of Biological Control." Academic Press, San Diego.
Clausen, C. P. (ed.) (1978). "Introduced Parasites and Predators of Arthropod Pests and Weeds: A World Review." Agricultural Handbook 480. U.S. Department of Agriculture, Washington, DC.
DeBach, P., and Rosen, D. (1991). "Biological Control by Natural Enemies." Cambridge University Press, Cambridge, U.K.
Follett, P. A., and Duan, J. J. (eds.) (2000). "Nontarget Effects of Biological Control." Kluwer, Boston.
Gaugler, R., and Kaya, H. K. (eds.) (1990). "Entomopathogenic Nematodes in Biological Control." CRC Press, Boca Raton, FL.
Gurr, G., and Wratten, S. (eds.). (2000). "Biological Control: Measures of Success." Kluwer, Dordrecht, The Netherlands.
Jervis, M., and Kidd, N. (eds.) (1996). "Insect Natural Enemies: Practical Approaches to Their Study and Evaluation." Chapman & Hall, London.
Julien, M. H., and Griffiths, M. W. (eds.) (1998). "Biological Control of Weeds, a World Catalogue of Agents and their Target Weeds." 4th ed. CAB International, Wallingford, U.K.
Pickett, C. H., and Bugg, R. L. (eds.) (1998). "Enhancing Biological Control: Habitat Management to Promote Natural Enemies of Agricultural Pests." University of California Press, Berkeley.
Tanada, Y., and Kaya, H. K. (1993). "Insect Pathology." Academic Press, San Diego.
Van Driesche, J., and Van Driesche, R. G. (2000). "Nature Out of Place: Biological Invasions in a Global Age." Island Press, Washington, DC.
Van Driesche, R. G., and Bellows, T. S. (1996). "Biological Control." Chapman & Hall, New York.
Van Driesche, R. G., and Hoddle, M. S. (2000). Classical arthropod biological control: Measuring success, step by step. *In* "Biological Control: Measures of Success." (G. Gurr and S. Wratten, eds.), pp. 39–75. Kluwer, Dordrecht, The Netherlands.

Bioluminescence

James E. Lloyd and Erin C. Gentry
University of Florida

Light that is produced in a chemical reaction by an organism is called bioluminescence. This "living light" is most commonly produced in tissues or organs within and shines out of the emitter's body, but luminous secretions are produced by some organisms and oozed or squirted out, even smeared on attackers. Chemiluminescence is but one of several forms of light emission collectively known as luminescence, which occurs when atoms of a substance emit photons (packets of light energy) as their electrons return to their stable state after being lifted to a higher and unstable energy level by input energy.

The best-known insect bioluminescence is that of beetles of the family Lampyridae. They are known as fireflies, lightningbugs, blinkies, and many other local and colloquial names around the world.

TAXONOMIC OCCURRENCE (PHOTIC BIODIVERSITY)

Bioluminescence occurs "everywhere" among organisms, with self-lighting species appearing in all kingdoms of a four-kingdom classification—Monera, Fungi, Plantae, and Animalia—in 11 of 29 phyla. In the phylum Arthropoda, luminescent forms also are found among the sea spiders, crustaceans, millipedes, and centipedes. Self-luminescent insects occur in the families Poduridae and Onychiuridae (Collembola), Mycetophilidae (Diptera), and coleopterans Lampyridae, Phengodidae, Elateridae, and Staphylinidae and, possibly, the Telegeusidae.

CHEMISTRY AND EVOLUTIONARY ORIGIN

Bioluminescence chemistry varies widely among organisms. Bacteria use riboflavin phosphate, the sea pansy uses diphosphoadenosine, and fireflies use adenosine triphosphate (ATP) in the oxidative decarboxylation of substrates generically known as luciferins, with enzymes termed luciferases. The present, cautious conclusion would be that bioluminescence has evolved from many separate biochemical origins.

Molecular structures and their alterations along light-producing pathways of some systems are illustrated in general references, but many systems have not been investigated. Many use oxidative mechanisms that involve two major stages: the first creates an energy-rich molecule ("large energy quantum") by combining molecules, the second then excites a luminescent molecule that unloads this energy as a photon of light when it returns to its stable state. Among insects, photons range in color from an unbelievable bright, ruby red in the headlight of the railroadworm (*Phrixothrix tiemanni*, Phengodidae, Coleoptera) to the demure blue of glowing Appalachian glowworm larvae (*Orfelia fultoni*, Mycetophilidae, Diptera). In twilight-active fireflies, longer wavelengths (orange-yellow), with appropriate filters in the eyes, may be connected with enhancing signal reception against (noisy) backgrounds of green foliage. The different colors are caused by alterations in the amino acid composition of the luciferase that shift the emission peaks.

ANATOMY OF EMISSION

The ultrastructure of the flashing lantern was first seen in the 1960s, when the electron microscope revealed that a miniature and new type of structure, the tracheal end organ, occurred throughout the flashing lantern, and that each microunit was obviously involved in controlling the photocytes associated with it. The light-emitting layer of a flashing lantern is organized into a sheet of rosettes, each with a central channel (cylinder), through which air-supply tubes and nerve trunks pass, and surrounding photocytes, which abut the photocytes of neighboring rosettes (Fig. 1).

The flashing lantern of adult fireflies does not develop (ontogenetically) from the glowing lantern of juveniles. The difference in the light output of these two lantern types is remarkable. On the one hand, larval lanterns require perhaps a full second to reach their full but much lesser intensity, and in

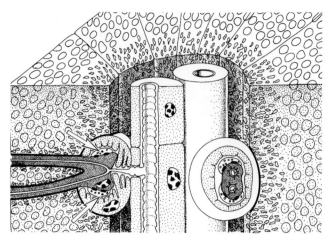

FIGURE 1 Diagram of part of a single rosette ("unit") in the flashing lantern. The central channel (cylinder) is ringed by photocytes, which are differentiated into inner and outer regions. Within the cylinder are two tracheal trunks and two tracheal end organs, one cut longitudinally, one transversely. Note within the end organ the intimate contact of the air supply, nerve ending (arrows), tracheal end cell, and tracheolar cell (the last two indicated by lightly and densely stippled structures, respectively). Original "magnification" approximately 1800×. (Reproduced, with permission, from H. Ghiradella, 1983, Permeable sites in the firefly lantern tracheal system: Use of osmium tetroxide vapor as a tracer. *J. Morphol.* **177**, 145–156. See Ghiradella 1998.)

an array of lantern types their behavior is little removed from the granules in subdermal fat cells or excretory tubules that glow continuously or in a simple circadian (daily) rhythm (*Keroplatus, Orfelia*), which is perhaps controlled by changes in hormone level. On the other hand, a flashing lantern is capable of photic finesse that can be appreciated only with electronic detector systems—the flashes of male fireflies of some species have very sharp *on* transients, and field recordings of flying males reveal that a flash can reach its bright peak in 20 ms (Figs. 2A and 2B), the flicker signal of a Florida *Photuris* species is modulated up to 45 Hz (Fig. 2C), and the four subliminal peaks of what appears to the eye to be a single flash of an Andean Mountain *Photinus* occur at 25 Hz (Fig. 2D).

The triggering of such light emission is currently thought to be connected with the release or gating of oxygen into the photocytes. This occurs in response to patterned volleys from the central nervous system, delivered by neurons that connect to or are closely associated with other key elements within the tracheal end organ (Fig. 1). A recent study suggests that nitric oxide gas plays a key role in the release of oxygen into the photocytes.

DESCRIBING/QUANTIFYING EMISSIONS

Humans have observed and written about light-emitting insects for more than 2 millennia, but early in the 20th century they began to give careful scientific attention to different colors and forms of firefly emissions, noting glows, flashes, flickers, tremulations, scintillations, and so on, and they borrowed descriptive terms from other senses, such as

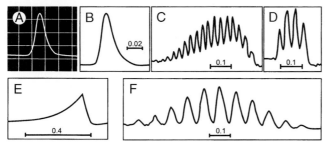

FIGURE 2 Flash patterns of fireflies as displayed for examination and measurement from electronic (frequency-modulated) field recordings, as described in the text, with standard display convention: horizontal axis, time; vertical axis, relative intensity (not total photon flux, which would be virtually impossible to measure from the entire surface of a flying light organ, but intensity change through time as detected from a single position in space, as visible to another firefly). Bars show time scale in seconds; horizonal grid in A, 0.02 s. (A) Nearly symmetrical flash of *Photuris salina*, photographed from the CRT screen of a storage oscilloscope. (B) Same flash scanned from ink tracing of a chart recorder. (C) Crescendo flash of *Photuris* unnamed species "D", with subliminal (to human eyes) modulations at the light organ. (D) Flash pattern of a Colombian *Photinus*, with subliminal modulations. (E) Crescendo flash of *Photuris cinctipennis*. (F) Visibly flickering flash pattern of a New Guinea *Luciola* species.

"crescendo" and "ramp" (Fig. 2E)—both describing a flash that begins dimly and gradually rises in intensity.

The simplest form of light emission is a glow of indeterminate length, as produced by fungus gnats, some Collembola, beetle larvae, adult fire beetles (elaters), phengodid beetle larvae and adult females, and lampyrid glowworm firefly females. A useful description of glows requires only a statement of apparent brightness (distance visible) and, cautiously, apparent color. In contrast, an adequate description of many adult lightningbug emissions often requires a chart, such as first published by Frank McDermott in 1914, with relative-intensity/time on the axes and notes of variations among flash patterns, from pattern to pattern, and at different ambient temperatures.

In the 1930s electronic technology, with photocells, string galvanometers, and ocillographs, made it possible to electronically chart firefly flashes in the laboratory. A generation later photomultiplier-tube systems permitted recording of the flashes of fireflies flying in the field. With today's miniaturized solid-state, digital circuits, detectors, and tape recorders flash detecting/recording systems can be hand-held. The flash patterns shown in Fig. 2 were all recorded in flight in the field, with a photomultiplier tube as detector, whose output was frequency modulated (fm) to encode intensity information; the fm conversions were recorded on magnetic tape. In the laboratory the recorded fm patterns were demodulated and fed into an oscilloscope (Fig. 2A) and chart recorder (Figs. 2B to 2F), for display and measurement. Video cameras add yet another dimension to observation and flashing behavior analyses.

There is no evidence that bioluminescent insects make use of color discrimination—it should be noted that the color a human observer perceives in the field often errs because of the intensity of the luminescence, its background (sky, street) lighting, and the degree of dark adaptation of the viewer's eyes. For example, dim green light may appear white because color vision apparatus (cone vision) is not stimulated, and the yellow flashes of roadside *Photinus pyralis* may appear green when under a sodium-vapor streetlight.

BEHAVIORAL ECOLOGY

Behavioral ecology is the analysis of behavioral features of the phenotype as ecological and reproductive adaptations. Insect bioluminescence offers remarkable opportunities for applying the "adaptationist's program" of behavioral ecology, observation, speculation, systematic observation, and experimentation, in the laboratory and field. The experimenter can enter these informational transactions with a penlight or computer-driven light-emitting diodes. Further, interactions often occur quickly and can be photographed, videotaped, and electronically recorded for precise analysis.

The mating signals of lightningbug fireflies are the most commonly seen example of insect luminescence, but others are easily found if sought in their habitat: prey-attracting glows of larval Appalachian glowworm flies *(O. fultoni)* in beds of impatiens at roadside springs and under overhanging mossy banks of streams along dark mountain roads; glowing *Arachnocampa luminosa* larvae hanging from ceilings of New Zealand caves, attracting midges from streams below and tourists from around the world; and prey-attracting glows of larval termitophageous click beetles *(Pyrearinus termitilluminans)*, which make termite mounds look like high-rise buildings of a metropolis seen from the air, in the dry-scrub region (open-formation cerrado) of northeastern Brazil.

The significance of many luminosities remains problematic: why do sparkling, galaxy-like arrays of flashing Collembola result when rotting forest litter under damp logs is scratched with a hand cultivator? Is each individual, when stimulated by our invasive touch, warning relatives, or a predator, say, a firefly larva? Several firefly knowns and unknowns are illustrated in Fig. 3, in which black circles, teardrops, stripes, and beads represent emissions of different forms. Coordinates on the axes guide attention to specific locations in the scene—near 3M an armadillo views a flashing firefly under its nose, perhaps retrieving memory data that flashing lights taste terrible or, previously when eaten, vomiting followed. The blood of some fireflies has been found to have cardioglycosides that can be deadly for some animals. (Recently, to the regret of pet owners, several exotic lizards died after eating North American fireflies.) This fact makes a warning (aposematic) function of luminosity a testable explanation.

The most often seen and best understood bioluminescent emissions are the mating flashes of lightningbug fireflies. Nearly all flying emissions seen afield are the mating signals of males, signaling over and over, advertising, with their species' sexual-recognition flash patterns. The male flash patterns of many species are distinctive and diagnostic in a

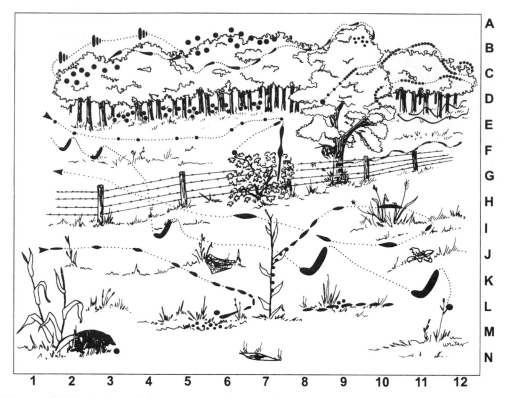

FIGURE 3 A graphic scene of firefly emissions, with alphanumeric coordinates to index text discussion. The diagram illustrates flash patterns, response flashes, a warning flash (3N), an attack flash (7E), illumination flashes, a flash pattern default (9D), some (probably?) meaningless flashes of "stressed" fireflies (6J, 7N, 9F), and a *Photuris* larval glow (8L–10L).

given location. In Fig. 3 several examples are illustrated. A male *Photinus macdermotti* flies from 2E to 7E, repeating his two-flash pattern every few seconds; a *Photinus collustrans* presents several of his low arcing flash patterns behind the fence 9F–12F; *Photuris frontalis* males fly low in the woods beyond the fence at 3E–7D; a *Photinus pyralis* repeats his J-stroked flash pattern diagonally across the view between 1F and 11K, and three or possibly four *Photuris* species cruise over the forest canopy where high-flying one-short-flash patterns are often difficult to identify without attracting the emitters to view them in the hand, but the crescendo flash at 3B–9A is that of *Photuris lucicrescens*. Low in the woods at the upper right (10E–12D) two glowing males of the glowworm firefly *Phausis reticulata* cruise low over the ground emitting their green light continuously.

In the signal system used by most North American lightningbugs, perched females flash responses to the flash patterns of males of their own species. The response signals of females are single flashes in most cases, emitted after a slight delay (<1 s) and, after a brief dialogue of flash patterns and responses, males reach and mount answering females. However, the female delay in the common *Photinus pyralis* is distinctive and varies between 2 and 4 s, depending upon temperature. A female *Photinus pyralis* is answering her male from a perch up a spike of grass near 12L.

After mating, males return to mate searching and females turn to ovipositing and/or hunting *(Photuris)*; this explains why males nearly always greatly outnumber females during

their mating-time window, although the sex ratio at egg laying and at adult eclosion is expected to be 1:1. The operational sex ratio is reversed in some species toward the end of the season. Under such circumstances males theoretically are expected to become the more discriminating (i.e., the choosier) of a courting pair. Individuals may receive information that influences their flashing and mating behavior from the number of other flash patterns they observe around them.

Among the dangers that flying, signaling males encounter in the dark are the predaceous firefly females of the genus *Photuris*. These versatile *femmes fatales* mimic the flash responses of females of other species, attract males, and eat them (Fig. 4; termed variously aggressive, Batesian–Wallacian,

FIGURE 4 A predatory female of the genus *Photuris* (Florida member of the *Photur. versicolor* complex) devouring a male *Photinus tanytoxus* she has attracted with false mating signals. When hunting females of this species are presented with simulated "hovering" male flash patterns they sometimes launch aerial attacks.

or Peckhammian mimicry); they also launch aerial attacks on flying, flashing males, aiming at their luminescent emissions. In Fig. 3, the flash pattern of the *Photinus macdermotti* male is answered (6F) but he flies on, then is attacked in the air (7E) by the *Photuris* female. Such predators may be expected to have had a strong evolutionary impact on both the signal coding (countermeasure trickery) of prey species and the flight paths and bioluminescent signals of mate-seeking males. Predaceous females obtain the defensive steroidal pyrones called lucibufagens from some firefly prey, which they incorporate into their own and their eggs' defenses.

The hawking *Photuris* female (7E) flashes just as she reaches the male, and they fall glowing into the shrub; possibly illuminating her attack so she could seize him. *Photuris* females use what appear to be illumination flashes in other situations: as they approach the ground or vegetation for landing (2J–6L); as they climb vegetation and take flight, when their flashes gradually become less frequent, then stop altogether as the females disappear into the night (7L–6I); and as they walk around in tangles of vegetation on the ground (5M–6M, 9L), perhaps seeking oviposition sites. Other flashes of unlikely if any behavioral-ecological significance are those emitted by fireflies in spider webs (6J), water puddles (7N), and tangles of Spanish moss (9F).

Males of many *Photuris* species use two or more flash patterns during mate search. Several species in one species group use a flicker with modulations timed exactly like those in the flash pattern of *Pyractomena angulata* (similar to the flicker shown in Fig. 1F), in addition to their species' own identification flash pattern. For example, *Photuris tremulans* males usually emit a single short flash every 2 s (20°C; Fig. 3, 2C–3C), but occasionally several or all males in a local population will emit the described *Py. angulata* flicker pattern (except it is green, not amber). When a short-flashing *Photuris tremulans* male is answered with a short flash, he approaches the respondent (female or penlight), maintaining a dialogue, and lands near; but, when a flickering male is answered, he switches (defaults) to his short flash pattern and then approaches as described (11C–9D).

When males of a related *Photuris* species receive an answer to their *Py. angulata* flicker (supernumerary) pattern, they default to their species' identification pattern, which is a pulsed pattern like the one seen above the trees at 2B–4B, or a variation of it. This firefly demonstrates another twist in the use of supernumerary flash patterns that will be important for understanding bioluminescent communication. For a few minutes at the onset of evening activity all males begin with the pulsing pattern but soon some, then more and more of them use their *Py. angulata* flicker pattern, until in an hour or so a peak of 50 to 80% flickering is reached; then, the proportion of males using the flicker pattern gradually decreases across the midnight hour.

Contrasting behavior of *Photuris lucicrescens* may be helpful for understanding signals of other *Photuris* species. They also use two flash patterns, a short flash and a crescendo flash (5B–7B), but they apparently do not switch patterns

FIGURE 5 Male fireflies of two *Photuris* species flashing over an oldfield in Connecticut seeking mates. One species is responsible for nearly all of the patterns seen here.

(default) during approach, and a nocturnal changeover has not yet been recognized. There is much to be learned about the behavioral ecology of *Photuris* fireflies and, in particular, the sexual selection aspects of their bioluminescent signals (Fig. 5). Note that the supernumerary patterns of several *Photuris* species are copies of flash patterns of known prey of the males' own females. In other words, pattern-copying *Photuris* males use a signal that hunting conspecific females will answer. This suggests a possible evolutionary origin of these prey-mimicking mate search patterns. Without this information the behavioral ecologist cannot understand selection pressures that have influenced the evolution of the bioluminescent signals we see today.

HUMAN APPLICATIONS OF FIREFLY CHEMISTRY

Bioluminescent insects are choice subjects for behavioral studies, but of equal or greater significance is the human-serving use that molecular and cellular biology and medicine have made of firefly bioluminescence chemistry. As one of the few known instances in which ATP is involved in light production, firefly light has been used as a research tool for understanding the machinery of cellular energy conversion and a variety of other applications, including medical diagnoses. ATP, the energy currency of life, is produced, stored, and used by living cells, and each photon of firefly light requires the energy released "from" (the terminal phosphate of) one ATP molecule. Light is easily quantified in the laboratory at extremely low levels by photo detectors (luminometers) and recorders that do not intrude into an ongoing living or chemical systems. Thus, photon monitoring (of ATP) in real time can reveal even subtle variations in reaction rates and enzyme kinetics under different experimental conditions.

Glow intensity of an extract or culture will vary directly as ATP is produced or used ("up") by a cellular reaction and can reveal the diurnal rhythmicity, quiescence or torpor, and health and vigor of living tissue, the last being of particular interest when exploring a bacterial culture's sensitivity to antibiotic substances. Firefly genes whose products are involved in this reaction have been put into the genomes of bacteria, mammals, and plants.

CULTURE AND LITERATURE

The Aztecs combined science and humanities with a firefly: "'A firefly in the night' the Nahuas called their songs: a tiny light in a great darkness, a little truth within the ignorance surrounding them." It is a culturally impoverished American who has not heard of Wah-wah-taysee, the firefly in a memorized passage from Longfellow's *Song of Hiawatha* or does not know where to find reference to a glowworm in *Hamlet,* or what Robert Frost said about fireflies, or what a glowworm did for the Mills Brothers. There are firefly books in English for children, but nothing to compare with the literature available in Japan, especially in connection with the Yokosuka City Museum and their dynamic firefly program.

See Also the Following Articles
Chemical Defense • Coleoptera • Mating Behaviors • Mimicry

Further Reading
Eisner, T., Goetz, M. A., Hill, D. E., Smedley, S. R., and Meinwald, J. (1997). Firefly "femmes fatales" acquire defensive steroids (lucibufagins) from their firefly prey. *Proc. Natl. Acad. Sci. U.S.A.* **94,** 9723–9728.

Ghiradella, H. (1998). The anatomy of light production: The fine structure of the firefly lantern. *In* "Microscopic Anatomy of Invertebrates," Vol. 11A, pp. 363–381. Wiley–Liss, New York.

Harvey, E. N. (1952). "Bioluminescence." Academic Press, New York.

Harvey, E. N. (1957). "A History of Luminescence." Am. Philos. Soc., Philadelphia.

Herring, P. (ed.) (1978). "Bioluminescence in Action." Academic Press, New York.

Lloyd, J. E. (1979). Sexual selection in luminescent beetles. *In* "Sexual Selection and Reproductive Competition in Insects" (M. Blum and A. Blum, eds.). Academic Press, New York.

Lloyd, J. E. (1983). Bioluminescence and communication in insects. *Annu. Rev. Entomol.* **28,** 131–60.

Lloyd, J. E. (1997). Firefly mating ecology, selection and evolution. *In* "Mating Systems in Insects and Arachnids" (J. C. Choe and B. J. Crespi, eds.), Chap. 10. Cambridge University Press, Cambridge, UK.

Sivinski, J. (1981). The nature and possible functions of luminescence in coleoptera larvae. *Coleoptera Bull.* **35,** 167–180.

Wickler, W. (1968). "Mimicry in Plants and Animals." McGraw–Hill, New York.

Wood, K. V. (1995). The chemical mechanism and evolutionary development of beetle bioluminescence. *Photochem. Photobiol.* **62(4),** 662–673.

Biotechnology and Insects

Bryony C. Bonning
Iowa State University

B iotechnology can be broadly defined to include all practical uses of living organisms. As such, biotechnology has been practiced since the beginning of recorded history through endeavors such as fermentation of microorganisms for production of beer, selective breeding of crops, beekeeping for the production of honey, and maintenance of silkworms for the production of silk. Laboratory techniques developed within the last 20 years that enable transfer of genes from one organism to another have resulted in tremendous scientific and commercial interest and investment in biotechnology. The word "biotechnology" is now commonly used to refer to manipulation of organisms at the molecular level. This article reviews insect-derived tools used for biotechnological research and the use of recombinant DNA technology for management of insect pests and insect-borne disease.

INSECT-DERIVED TOOLS USED FOR BIOTECHNOLOGICAL RESEARCH

Expression of Foreign Proteins in Insect Cells

Production of large amounts of a particular protein is extremely valuable for both research and industrial purposes. Baculoviruses, which are double-stranded DNA viruses that infect mainly insects, have been developed as baculovirus expression vectors (BEVs) by genetic modification to include a gene of interest. BEVs can replicate in lepidopteran cells and larvae, thereby efficiently transferring foreign genes into eukaryotic cells. The foreign gene is usually under transcriptional control of a viral promoter so that the gene is transcribed by the virus, but translated by the host cell biosynthetic machinery. The BEV system is one of the best tools for recombinant protein expression in a eukaryotic host and has been used for the production of many different proteins for research purposes. The BEV system also has potential industrial application for the production of proteins used in vaccines, therapeutic agents, and diagnostic reagents. Advantages of this protein production system include production of large quantities of foreign protein, and eukaryotic protein processing allowing production of more authentic eukaryotic proteins. The BEV expression system is only transient, however, because the baculovirus ultimately kills the host cells. Baculoviruses do not infect vertebrates and therefore provide relative safety for laboratory manipulation. The use of a baculovirus for production of a foreign protein was first demonstrated by expression of human β-interferon and *Escherichia coli* β-galactosidase.

Insect cells can also be engineered directly to express the recombinant protein, without the baculovirus expression vector intermediate. Such insect cells are stably transformed to constitutively express a foreign gene. Expression levels are usually lower than for the BEV system, but stably transformed cells produce recombinant proteins continuously and process them more efficiently than infected cells.

Insect-Derived Genes Used in Biotechnology

Reporter enzymes allow monitoring of gene expression in living tissues and cells. The gene encoding the reporter enzyme is typically inserted under control of the promoter of the gene of interest, and production of the enzyme is monitored by means of an enzyme assay. Luciferases belong to a unique group of enzymes that produce light as an end product of catalysis. The luciferases derived from the North American firefly *Photinus pyralis* (Coleoptera) and the Jamaican click beetle *Pyrophorus plagiophthalamus* (Coleoptera) have been used as genetic reporter enzymes in virtually every experimental biological system, including prokaryotic and eukaryotic cell cultures, transgenic plants and animals, and cell-free expression systems. These luciferases, which evolved for the nocturnal mating behavior of the beetles, use ATP, oxygen, and D-luciferin as substrates in the catalysis of a light-producing reaction. The ease and reliability with which luciferase can be assayed, combined with the sensitivity of the technique, has made this enzyme a highly valuable research tool.

USE OF BIOTECHNOLOGY FOR MANAGEMENT OF INSECT PESTS IN AGRICULTURE

The ability to move genes from one organism to another has enabled scientists to develop insect-resistant transgenic crops and insect pathogens with enhanced insecticidal properties. The technology also has the potential to protect beneficial insects from chemical pesticides.

Insect-Resistant Transgenic Plants

Despite the progress made in recent years, a significant proportion of the world's food supply is lost to the activities of insect pests. The deleterious impact of chemical pesticides on the environment, combined with the emergence of technologies enabling plants to be transformed with foreign genes, has driven the seed industry to develop transgenic plants as novel, environmentally benign means of pest control. Insect-protected crops were among the first products of biotechnology to have a significant impact on crop protection, and at times their use has resulted in decreased application of classical chemical pesticides.

The bacterium *Bacillus thuringiensis* (Bt) *kurstaki* has served as a microbial insecticide for many years, but widespread use was limited by its instability when exposed to ultraviolet light and its poor retention on plant surfaces in wet weather. The high toxicity of the Bt toxins to a variety of insect pests, and the ease with which the gene could be isolated from bacterial plasmids, made it an obvious choice for development of the first insect-resistant transgenic plants. The active Bt toxin binds to a receptor in cells lining the insect gut and creates a channel allowing free passage of ions. The cells lining the gut die, and very soon, the insect dies, too. Different strains of Bt contain plasmids encoding different toxins with different specificities of action against insects. A particular toxin is generally effective against only a limited range of closely related species. Bt toxins are used in a variety of transgenic crops in the United States, including cotton, for protection against various lepidopteran pests, corn (maize),

for protection against the European corn borer *Ostrinia nubilalis* (Lepidoptera), and potatoes, for protection against the Colorado potato beetle *Leptinotarsa decemlineata*.

Industry has expended enormous effort to identify new isolates of Bt. with different specificities and increased virulence for development of insect-resistant crops. Other bacteria also provide a resource for identification of insect-specific toxin genes such as those derived from *Bacillus cereus* and the entomopathogenic nematode-associated bacterium *Photorhabdus luminscens*.

Plants have a variety of strategies to avoid or survive attack by insects, and genes encoding endogenous plant defensive compounds are also candidates for enhancing the resistance of crops to insect pests. Such factors include inhibitors of digestive proteinases that disrupt digestion by phytophagous insects. However, expression of serine protease inhibitors rarely results in high mortality of the insect pest, and the levels of mortality achieved were below what is required for commercial viability. A variety of lectins that bind specifically to carbohydrate residues have also been expressed in transgenic plants for protection against insects. A gene encoding the snowdrop lectin has been engineered into transgenic plants, and it confers protection against a variety of pests, including aphids and planthoppers. As with proteinase inhibitors, the levels of protection conferred by the foreign lectins are not sufficient for commercial viability.

Transgenic Arthropod Natural Enemies

Recombinant DNA methods may be applied to produce improved strains of natural enemies such as predatory arthropods and parasitoids, but techniques are in the early stages of development. For example, the western predatory mite, *Metaseiulus occidentalis* (Acari), is among a group of mites that are mass-reared for the control of spider mites. However, pesticides applied for control of other pest species often wipe out the predatory mites. Engineering beneficial insects such as the western predatory mite with insecticide resistance genes would in theory provide protection from chemical sprays applied for control of insect pest species.

Engineered Insect Pathogens for Pest Control

Insect pathogenic bacteria, viruses, fungi, and nematodes have been used for the management of insect pests in various niche markets. However, each agent suffers from at least one major limitation, such as susceptibility to environmental stresses, temperature extremes, desiccation, or solar radiation. Most work has been done on the genetic enhancement of bacteria and viruses, in part because of the relative ease of genetic manipulation of these organisms. Genetic engineering to enhance the insecticidal properties of entomopathogenic nematodes and fungi is in its infancy.

Genetic engineering has been used to enhance the insecticidal efficacy of various strains of Bt. by increasing virulence, extending host range, and increasing field stability, and by introducing alternative toxins to facilitate resistance management. Techniques have been developed for production by genetic means of new strains of Bt. with new combinations of toxin genes.

Considerable progress has been made toward optimization of entomopathogenic viruses at the genetic level. The baculoviruses are arthropod-specific viruses that have been studied extensively both as protein expression vectors and as insect pest control agents. These viruses have been genetically engineered with genes encoding insect-specific toxins that are active within the hemocoel of the insect. Upon infection of the insect host, the toxin is produced as the virus replicates, and the infected insect dies from the effects of the toxin delivered by the virus. Baculoviruses have been engineered with genes encoding a variety of insecticidal proteins and peptides. The most effective enhancement of insecticidal properties of the virus was achieved by introduction of genes into the virus genome that encode insect-specific neurotoxins derived from scorpion venom. Recombinant baculovirus insecticides have been developed that now approach the efficacy of the classical chemical insecticides.

USE OF BIOTECHNOLOGY FOR MANAGEMENT OF INSECT PESTS AND INSECT-BORNE DISEASE

Transgenic Insects

Transposable elements are mobile segments of DNA that can move from site to site within a genome and can be used for delivery of foreign DNA into the genomes of insects. Although the vinegar fly *Drosophila melanogaster* (Diptera) was, in 1982, the first organism to be transformed, leading to tremendous advancements in genetics research, the application of this technology to other insects has been slow. Recent successes however indicate that stable transformation of insects may become more routine in the foreseeable future. Transformation using transposable elements has been achieved for relatively few species, mostly within the Lepidoptera and Diptera (Table I). Other gene transfer

systems using viruses or gene expression from transformed bacterial endosymbionts (so-called paratransgenesis) have been used for some species that are not amenable to direct transformation. The genomes of bacteria and viruses are also significantly easier to engineer than eukaryotic genomes. Bacteria and viruses have been used as vectors for both transient and stable foreign gene expression in insects. For example, the bacterial symbionts of the kissing bug, *Rhodnius prolixis,* were successfully engineered to reduce the quantity of *Trypanosoma cruzi,* the parasitic protozoan that causes Chagas disease and is carried by this vector. The bacterial endosymbionts were engineered to express an antimicrobial peptide or antibodies that specifically target the parasite. Similar methods are being developed to prevent transmission of the malaria parasite *Plasmodium* by its mosquito vectors.

GENETIC APPROACHES FOR MANAGEMENT OF INSECT PEST POPULATIONS

The sterile insect technique (SIT) relies on release of large numbers of sterile male insects that mate with wild females, thereby reducing reproductive potential or, if sufficient numbers of males are released over time, resulting in eradication of the pest population in a given area. Successful SIT programs have been conducted against the screwworm, *Cochliomyia hominivorax,* the Mediterranean fruit fly, *Ceratitis capitata,* and the tsetse fly, *Glossina* spp. One of the problems associated with SIT is that laboratory rearing and sterilization of males results in reduced fitness of the insects.

Alternative genetic control systems include use of natural sterility such as cytoplasmic incompatibility induced by infection with the bacterium *Wolbachia,* and conditional lethal traits. For a conditional lethal release, insects are engineered to carry a lethal trait that is active only under certain conditions, such as certain temperatures, or at diapause. Since the trait is not lethal immediately, it can spread in a population. Genetic techniques have also been developed that allow induction of female-specific lethality. These autocidal control strategies have been demonstrated only in the model organism *Drosophila* thus far. The ability

TABLE I   Genetic Transformation of Nondrosophilid Insects

Order	Species transformed	Common name	Pest status
Diptera	*Anopheles stephensi*		Disease vectors
	A. albimanus		
	Aedes aegypti	Yellow fever mosquito	
	Culex quinquefasciatus	Southern house mosquito	
	Musca domestica	House fly	
	Stomoxys calcitrans	Stable fly	
	Ceratitis capitata	Mediterranean fruit fly	Horticultural pests
	Bactrocera tryoni	Queensland fruit fly	
	B. dorsalis	Oriental fruit fly	
	Anastrepha suspensa	Caribbean fruit fly	
Coleoptera	*Tribolium castaneum*	Red flour beetle	Stored-product pest
Lepidoptera	*Bombyx mori*	Silkworm	None
	Pectinophora gossypiella	Pink bollworm	Cotton pest

to insert the desired genes into insect genomes will be critical to the success of these genetic approaches for management of insect pests in the future.

See Also the Following Articles

Genetically Modified Plants • *Genetic Engineering* • *Pathogens of Insects* • *Sterile Insect Technique* • *Wolbachia*

Further Reading

Atkinson, P. W., Pinkerton, A. C., and O'Brochta, D. A. (2001). Genetic transformation systems in insects. *Annu. Rev. Entomol.* **46**, 317–346.

Durvasula, R. V, Gumbs, A., Panackal, A., Kruglov, O., Aksoy, S., Merrifield, R. B., Richards, F. F., and Beard, C. B. (1997). Prevention of insect-borne disease: An approach using transgenic symbiotic bacteria. *Proc. Natl. Acad. Sci. U.S.A.,* **94**, 3274–3278.

Durvasula, R. V., Gumbs, A., Panackal, A., Kruglov, O., Taneja, J., Kang, A. S., Cordon-Rosales, C., Richards, F. F., Whitham, R. G., and Beard, C. B. (1999). Expression of a functional antibody fragment in the gut of *Rhodnius prolixis* via transgenic bacterial *Rhodococcus rhodnii*. *Med. Vet. Entomol.* **13**, 115–119.

Estruch, J. J., Warren, G. W., Mullins, M. A., Nye, G. J., Craig, J. A., and Koziel, M. G. (1996). Vip3A, a novel *Bacillus thuringiensis* vegetative insecticidal protein with a wide spectrum of activities against lepidopteran insects. *Proc. Natl. Acad. Sci. U.S.A.* **93**, 5389–5394.

ffrench-Constant, R. H., and Bowen, D. J. (2000). Novel insecticidal toxins from nematode-symbiotic bacteria. *Cell Mol. Life Sci.* **57**, 828–833.

Gatehouse, J. A., and Gatehouse, A. M. R. (2000). Genetic engineering of plants for insect resistance. *In* "Biological and Biotechnological Control of Insect Pests" (J. E. Rechcigl and N. A. Rechcigl, eds.), pp. 211–241. CRC Press, Boca Raton, FL.

Handler, A. M. (2001). A current perspective on insect gene transformation. *Insect Biochem. Mol. Biol.* **31**, 111–128.

Harrison, R. L., and Bonning, B. C. (2000). Genetic engineering of bio-control agents for insects. *In* "Biological and Biotechnological Control of Insect Pests" (J. E. Rechcigl and N. A. Rechcigl, eds.), pp. 243–280. CRC Press, Boca Raton, FL.

James, A. A., Beerntsen, B. T., Capuur, M. D., Coates, C. J., Coleman, J. Jasinskiene, N., and Krettli, A. U. (1999). Controlling malaria transmission with genetically-engineered, *Plasmodium*-resistant mosquitoes: Milestones in a model system. *Parasitologia* **41**, 461–471.

Jarvis, D. L. (1997). Baculovirus expression vectors. *In* "The Baculoviruses" (L. K. Miller, ed.), pp. 389–431. Plenum Press, New York.

Naylor, L. H. (1999). Reporter gene technology: The future looks bright. *Biochem Pharmacol.* **58**, 749–757.

Thomas, D. D., Donelly, C. A., Wood, R. J., and Alphey, L. S. (2000). Insect population control using a dominant, repressible, lethal genetic system. *Science* **287**, 2474–2476.

Blattodea
(Cockroaches)

Donald G. Cochran

Virginia Polytechnic Institute and State University

Cockroaches are an ancient and highly successful form of insect life. They were among the groups of insects that evolved during the first great radiation of insects and have been in existence for at least 350 million years, or since early Carboniferous times. They seem to have achieved an optimum body form and other features early in their evolutionary history. Fossil specimens are relatively abundant; some that are at least 250 million years old are easily recognizable as cockroaches and could pass for modern species. Among the features that allowed them to escape the extinction that claimed many of the earlier insect groups was the ability to fold their wings over the body. This allowed them to more easily hide from predators and escape other dangers. They also evolved early in their existence an ootheca that could be hidden, hence offering some measure of protection for their eggs.

Cockroaches are referred to as generalized orthopteroid insects, which classifies them with the true Orthoptera (crickets, katydids, grasshoppers, locusts), Phasmatodea (walkingsticks), Mantodea (praying mantids), Plecoptera (stoneflies), Dermaptera (earwigs), Isoptera (termites), and a few other minor groups. The phylogenetic relationships among all these groups are not firmly established, although several theories exist. The closest relatives of cockroaches are believed to be the mantids, and some modern taxonomists prefer to place these two groups, as well as termites, in the order Dictyoptera. Indications are that termites evolved out of the cockroach stem or that cockroaches and termites both evolved from a common ancestor. One family of cockroaches (Cryptocercidae) and one extant relic species of termite (*Mastotermes darwiniensis*) have certain characteristics in common. Among them are the segmental origin of specific structures in the female reproductive system and that both deposit their eggs in similar blattarian-type oothecae. They also share a system of fat body endosymbiotic bacteria that is common to all cockroaches but is unique to *Mastotermes* among the termites.

THE SPECIES OF COCKROACHES

Between 3500 and 4000 species of cockroaches have been identified, with one relatively simple classification scheme dividing this group into five families as follows:

- **Cryptocercidae** is the most primitive family and consists of one genus with fewer than 10 species. These cockroaches live as isolated family groups in decaying logs and occur in the United States, Korea, China, and Russia. They are large, reddish brown insects that are wingless at adults.
- **Blattidae** is a diverse family with many genera and hundreds of species. Those classified as *Periplaneta* and *Blatta* are widely distributed, while other genera are more regional. They are large insects that tend to live outdoors. Several species are referred to locally as palmetto bugs.
- **Blattellidae** is also a diverse family with many genera and around 1000 species. These cockroaches are widely distributed in the world but are concentrated in the tropics and subtropics. Blattellids are mostly small outdoor

cockroaches, including those called wood cockroaches. The genus *Blattella* contains the German cockroach.

• **Blaberidae** is the largest family of cockroaches, with dozens of genera and more than 2000 species. These insects are widely distributed outdoors in tropical and subtropical regions. Some members of the genus *Blaberus* are extremely large, reaching more than 80 mm in length. These are the most highly evolved cockroaches, having developed the ability to incubate their eggs internally and, in some species, to nourish them.

• **Polyphagidae** is a small family with only a few described genera and 100 to 200 species. Females of most species are wingless. These cockroaches are widely distributed in harsh environments, such as deserts and other arid climates. Some members of the genus *Arenivaga* have evolved structures that can absorb moisture from humid air.

Other, more complex, classification schemes for cockroaches also exist, indicating that the subject is not entirely settled. In addition, insect collections from formerly unexplored locations often include many undescribed cockroach species. Thus, it is likely that the total number of extant species is much higher than the figures just given. Indeed, the vast majority of cockroaches live in the tropical regions of the world, many of which have not been adequately assessed to establish the diversity of insect life, including cockroaches, occurring there.

Cockroaches, being hemimetabolous insects, have egg, nymph, and adult stages and grow through a series of molts. They vary greatly in size, ranging from a few millimeters to over 100 mm in length. Many cockroaches are dark brown, but some are black or tan, and others show a surprising amount of color variation and cuticular color patterns. Most species have four wings as adults, and some are capable of rapid, sustained flight; others are wingless or have wings that are variously reduced in size. The majority of species are either nocturnal (or are hidden from view because of where they live), but some are diurnal. Cockroaches occupy diverse habitats, such as among or under dead or decaying leaves, under stones or rubbish, under the bark of trees, under drift materials near beaches, on flowers, leaves, grass, or brush, in the canopy of tall trees, in caves or burrows, in the nests of ants, wasps, or termites, in semiaquatic environments, and burrowing in wood. Thus, the common view of cockroaches as pests is not representative of the group as a whole.

COCKROACHES AS PESTS

The most important of the several reasons for considering some cockroaches to be pests is based on the species that invade people's homes and other buildings and become very numerous. Most people find such infestations to be objectionable, in part because the important pest species also have an unpleasant odor and soil foods, fabrics, and surfaces over which they crawl. However, on a worldwide basis less

FIGURE 1 German cockroach. From left: adult male, adult female, nymph, ootheca.

than 1% of all known cockroach species interact with humans sufficiently to be considered pests. The actual number varies depending on location, because some pest species are greatly restricted in their global distribution. It is also true that more pest species are encountered in tropical locations than in the colder parts of the world. Of the 25 to 30 species that can be a problem, more than half are only occasionally of importance and should be rated as minor or even incidental pests. Of the remaining species, only four or five are of global importance as pests, with the other nine or ten species having regional significance only.

The most important pest species is the German cockroach, *Blattella germanica* (Blattellidae) (Fig. 1). It has a worldwide distribution and can survive well in association with any human habitation that provides warmth, moisture, and food. It is small, measuring 10 to 15 mm in length. Adults are yellowish-tan but nymphs are black, with a light-colored stripe up the mid-dorsum. There are two longitudinal, black, parallel bands on the promotum of both nymphs and adults. The wings cover most of the body in adults of both sexes. This is a nocturnal species that lives mainly in kitchen and bathroom areas. When a person enters the kitchen of an infested house at night and turns on a light, the cockroaches scurry out of sight—a startling experience that adds to the desire to eliminate them. There are three or four generations per year. Each egg mass (ootheca) contains from 30 to 50 eggs, and each female can produce three to six oothecae. The potential for rapid population expansion is obvious.

The oriental cockroach, *Blatta orientalis* (Blattidae) (Fig. 2), the next most important pest species, is restricted to the more temperate regions of the world. It is large, measuring 20 to 27 mm in length. All stages are dark brown to black. Females are essentially wingless, but in males the wings cover about two-thirds of the abdomen. This cockroach frequents basements and crawlspaces under buildings, where temperatures are cooler, and often lives outdoors. It is a long-lived insect and may require 1 to 2 years to complete its life cycle. The ootheca contains 16 eggs, and one female may produce eight

FIGURE 2 Oriental cockroach. From left: adult male, adult female, nymph, ootheca.

FIGURE 4 Brown-banded cockroach. From left: adult male, adult female, nymph, ootheca.

or more oothecae. Under favorable conditions *B. orientalis* can become very numerous.

There are five or six species belonging to the genus *Periplaneta* (Blattidae) (Fig. 3) that are important pests. The American cockroach, *P. americana,* is the most notorious. It measures 35 to 40 mm in length and is a chocolate-brown color in all stages. Adults of both sexes are fully winged and may undertake a weak flight. *P. americana* is widely distributed around the world but does not extend into the temperate zones as far as does the German cockroach. It requires 6 to 9 months to complete its life cycle. Among other *Periplaneta* species of importance are the Australian cockroach, *P. australasiae,* the smoky-brown cockroach, *P. fuliginosa,* and the Japanese cockroach, *P. japonica.* Each has a more restricted distribution, with *P. japonica,* for example, being found in Japan and China. They are all large cockroaches with a long life cycle but can become numerous under certain conditions. Although tending to be outdoor cockroaches, they often occupy buildings in which food is stored, prepared, or served.

The brown-banded cockroach, *Supella longipalpa* (Blattellidae) (Fig. 4), is a nearly cosmopolitan pest. It is small,

FIGURE 3 American cockroach. From left: adult male, adult female, nymph, ootheca.

measuring 10 to 14 mm in length. As its common name indicates, there are two dark, transverse stripes or bands on the dorsum. The pronotum lacks the two black bands found on the German cockroach. Nymphs are light colored. Females produce numerous oothecae and glue them in inconspicuous places. Each one has about 16 eggs. The life cycle requires approximately 3 months to complete. This cockroach occupies homes and other buildings but unlike the German cockroach is not restricted to the kitchen and bathroom.

Other pest species include the Turkistan cockroach, *Blatta lateralis* (Blattidae), two species in the genus *Polyphaga* (Polyphagidae), the Madeira cockroach, *Rhyparobia maderae* (Blaberidae), the lobster cockroach, *Nauphoeta cinerea* (Blaberidae), the Suriname cockroach, *Pycnoscelus surinamensis* (Blaberidae), the Asian cockroach, *Blattella asahinai* (Blattellidae), the harlequin cockroach, *Neostylopyga rhombifolia* (Blattidae), and the Florida cockroach, *Eurycotis floridana* (Blattidae). Most of these species are of regional concern as pests.

COCKROACHES AND HUMAN HEALTH

Cockroaches harbor many species of pathogenic bacteria and other types of harmful organism on or inside their bodies, but they do not transmit human diseases in the same manner as do mosquitoes. They acquire the harmful organisms because of their habit of feeding on almost any type of organic matter, including human and animal wastes. These cockroach-borne organisms can remain viable for a considerable period of time. If the cockroach next visits and soils food intended for human consumption, it is likely that harmful organisms will be deposited on the food. Consuming such food can lead to gastroenteritis, diarrhea, and intestinal infections and pathogenic conditions of other types.

Cockroaches have been shown to harbor pathogenic bacteria belonging to the genera *Mycobacterium, Shigella, Staphylococcus, Salmonella, Escherichia, Streptococcus,* and *Clostridium.* They also harbor pathogenic protozoa in the genera *Balantidium, Entamoeba, Giardia,* and *Toxoplasma,* and parasites in the genera *Schistosoma, Taenia, Ascaris, Ancylostoma,* and *Necator.* From these lists of organisms it is clear that cockroaches are important as potential disease vectors.

Cockroaches are also important because people can become allergic to them, especially under conditions of constant exposure. These reactions usually involve the skin and/or respiratory system. Studies have shown that people who exhibit skin or bronchial responses to cockroaches have elevated levels of cockroach-specific antibodies. These responses can be severe and may require treatment. The species most commonly involved in producing allergic reactions are the German and American cockroaches.

COCKROACH CONTROL

There are several considerations that come into play in any discussion of cockroach control. The first is understanding how infestations arise. The usual modes of entry for German and brown-banded cockroaches are through infested parcels containing food or other materials, and by movement from abutting dwellings. Most of the other, larger pest species tend to live outdoors and can move from one building to another. They can also be introduced in parcels. Thus, the next consideration is prevention of entry. All entering parcels should be inspected to be sure they do not contain cockroaches, and dwelling defects should be corrected to exclude invaders. Finally, human living space should be kept free of clutter, which can act as hiding places for cockroaches, and food left on dishes, in sinks, or on floors, which can feed a population of cockroaches, should be disposed of properly.

When infestations occur, there are two main methods of control. Nonchemical methods include trapping and vacuuming cockroaches, both of which can significantly reduce the size of an infestation. In addition, freezing, overheating, or flooding structures with a nontoxic gas can be used to kill the pests. Some of the latter procedures require specialized equipment and are best done by professional pest control operators.

The most common method of control is the use of chemical poisons. A large variety of insecticides exist that will kill cockroaches. Some of them are contact poisons that are absorbed as cockroaches walk over treated surfaces. The most common of these belong to the chemical classes called pyrethroids, organophosphates, and carbamates. They kill by disrupting the insect's nervous system, each in a specific manner. Other insecticides are administered in bait formulations that must be eaten by the cockroach. Among them are avermectin and fipronil, which also attack the nervous system, hydramethylnon, which disrupts cellular respiration, and boric acid, which destroys the cells lining the insect gut wall. Each of these materials, as well as others not mentioned, has its own chemical characteristics and must be used in accordance with label instructions.

New insecticides are regularly being introduced that can kill cockroaches, and older ones are being phased out. A critical goal is to develop safer chemicals and safer methods of applying them. For example, the older practice of applying insecticides to surfaces over which cockroaches are expected to crawl is being used less frequently and, as a consequence, the organophosphate and carbamate insecticides especially are being phased out. The practice of dispensing chemicals as baits has largely replaced the surface application method. With baits, the insecticide is more confined and the safety (of humans and pets) is thereby enhanced. The use of baits has become practical in recent years because some of the newer chemicals are highly palatable for cockroaches in bait formulations.

Cockroach control in the future will likely depend on the availability of new insecticides as well as the development of better methods of applying them. Among the approaches that are possible is searching for chemicals that act on sites not previously exploited. For example, a combination of two chemicals is known that prevents cockroaches from producing uric acid. Previous research has shown that storing and recycling the chemical constituents in uric acid is critical to the survival of cockroaches. The functioning of this system is dependent on the fat body endosymbiotic bacteria, mentioned earlier. Other points of metabolic vulnerability will also probably be found in the future.

Another reason for the need for new chemical approaches is that the most important cockroach pest, *B. germanica,* has become resistant to many of the older insecticides. When this occurs, either the effectiveness of those chemicals is greatly reduced or they become useless against resistant populations. With continued use of the newer chemicals, resistance to some of them will probably develop. A steady supply of new chemicals with new modes of action will greatly alleviate this problem and facilitate continued control.

See Also the Following Articles

Isoptera • Medical Entomology • Orthoptera • Phasmida • Urban Habitats

Further Reading

Baumholtz, M. A., Parish, L. C., Witkowski, J. A., and Nutting, W. B. (1997). The medical importance of cockroaches. *Int. J. Dermatol.* **36**, 90–96.

Bell, W. J., and Adiyodi, K. G. (1982). "The American Cockroach." Chapman & Hall, New York, NY.

Cloaric, A., Rivault, C., Fontaine, F., and LeGuyader, A. (1992). Cockroaches as carriers of bacteria in multi-family dwellings. *Epidemiol. Infect.* **109**, 483–490.

Cochran, D. G. (1999). "Cockroaches: Their Biology, Distribution, and Control." World Health Organization/CDS/CPC/WHOPES/99.3. WHO, Geneva, Switzerland.

Cornwell, P. B. (1968). "The Cockroach," Vol. 1. Hutchinson, London.

Helm, R. M., Burks, W., Williams, L. W., Milne, D. E., and Brenner, R. J. (1993). Identification of cockroach aeroallergins from living cultures of German and American cockroaches. *Int. Arch. Allergy Appl. Immunol.* **101**, 359–363.

Labandeira, C. C., and Sepkoski, J. J., Jr. (1993). Insect diversity in the fossil record. *Science* **261**, 310–315.

McKittrick, F. A. (1964). Evolutionary studies of cockroaches. Cornell University Agricultural Experimental Station Memorandum 389. Cornell University, Ithaca, NY.

Nalepa, C. A., and Lenz, M. (2000). The ootheca of *Mastotermes darwiniensis* Froggatt (Isoptera: Mastotermitidae): Homology with cockroach oothecae. *Proc. R. Soc. Lond. B* **267**, 1809–1813.

Roth, L. M. (1989). *Sliferia,* a new ovoviviparous genus (Blattellidae) and the evolution of ovoviviparity in Blattaria (Dictyoptera). *Proc. Entomol. Soc. Wash.* **91**, 441–451.

Rust, M. K., Owens, J. M., and Reierson, D. A. (eds.) (1995). "Understanding and Controlling the German Cockroach." Oxford University Press, New York.

Thorne, B. L., and Carpenter, J. M. (1992). Phylogeny of the Dictyoptera. *Syst. Entomol.* **17**, 253–268.

Blood

see *Circulatory System*

Blood Sucking

M. J. Lehane
University of Wales, Bangor

O f the 1 million described insect species, only 300 to 400 species feed on blood. The best known groups of blood-sucking insects are the lice, fleas, mosquitoes, sand flies, black flies, and bugs. But there are also several lesser known groups such as the nycteribiids and streblids, two families of cyclorrhaphous flies found exclusively on bats; the Rhagionidae or snipe flies, a little-studied group of brachyceran flies; some lepidopterans (e.g., *Calpe eustrigata*); and even some coleopterans (e.g., *Platypsyllus castoris*) that appear to have started on the evolutionary road to hematophagy.

Blood-sucking insects are of immense importance to humans, primarily because of the diseases they transmit. They also cause huge losses in animal husbandry because of disease transmission and because of direct losses linked to the pain and irritation they cause to animals. The most spectacular example of this agricultural loss is the prevention of the development of a cattle industry worth billions of dollars a year through much of sub-Saharan Africa because of tsetse fly-transmitted trypanosomiasis, although some argue that this has been Africa's savior because it has preserved wildlife and prevented desertification. Blood-sucking insects also cause serious losses in the tourist industry in areas as diparate as the French Camargue, the Scottish Highlands, and the state of Florida. We cannot ignore the sheer annoyance they can cause to us all.

EVOLUTION OF THE BLOOD-SUCKING HABIT

Although blood-sucking insects are poorly represented in the fossil record, it seems probable that they emerged along with the first nesting or communal dwelling vertebrates (reptiles) in the Mesozoic era (65–225 mya). Evolution of the blood-sucking habit probably occurred in two main ways. The first route involved the attraction of insects to vertebrates, with the attraction being either to the protection of the nest environment or for the utilization of vertebrate-associated resources such as dung. The second route involved morphological preadaptations that permitted the rapid adoption of the blood-sucking habit.

Many insects would have been drawn to vertebrate nests because of the protected environment and abundance of food there. Gradually, some would have progressed to feeding on cast skin or feathers. Phoresy also would have permitted easy travel from one nest to another. Once phoresy was adopted, the insects may have begun to feed directly on the host animal and thus established an even more permanent association with the host; mallophagan lice make a good example of this type of association. Regular accidental encounters with blood may then have led rapidly to the evolution of the blood-sucking habit because of the highly nutritious nature of blood compared to skin, fur, and feather.

Other insects are attracted to vertebrates outside the nest situation to utilize other vertebrate-associated resources, notably dung. Dung is used by a wide variety of organisms and there is strong competition to be the first to lay eggs in it. So, for example, the female horn fly *Haematobia irritans* lays its eggs in dung within 15 s of its deposition. To do this, the insect must remain permanently with the vertebrate; to do that, it must feed on the vertebrate. The high nutritional content of blood will then make hematophagy a favored evolutionary route.

Some insects also had morphological preadaptations for piercing surfaces, facilitating the relatively easy switch to blood feeding. Entomophagous insects (those that feed on other insects) and plant-feeding insects are prime candidates. For example, the Boreidae are a group of small apterous scorpion flies who are capable of jumping. They live in moss and feed on other insects by piercing them with their mouthparts. They are commonly found in nests because of the moss content and abundance of insects found there. It is easy to imagine such a lineage developing into fleas.

Insects that feed on plants may also have switched to the blood-feeding habit. An unusual example is a blood-feeding moth, *C. eustrigata*. This moth belongs to a group of noctuids having a proboscis that is hardened and modified to allow them to penetrate fruit rinds. *C. eustrigata* has used the morphological preadaptation to feed on vertebrate blood.

HOST CHOICE

The question of host choice is an extremely important one because it defines patterns of disease transmission and economic damage caused by blood-sucking insects. Blood-sucking insects in general feed on a range of different hosts, including birds, reptiles, mammals, and amphibians. Even invertebrates such as annelids, arachnids, and other insects are sometimes included in the diet. But any particular blood-sucking insect generally feeds only from a small segment of the available hosts. This segment of choice is preferred but it is not immutable. This can be clearly seen around zoos where the exotic animals are quickly incorporated into the diet of the local blood-sucking insects.

The determinants of host choice are complex, but probably one of the most important factors is simply host availability. Changes in host availability because of more intensive animal husbandry, coupled with decreasing rural populations of humans and improved, mosquito-free housing, were a major factor in the disappearance of autochthonous malaria from Northern Europe in the past century. Despite our poor understanding of the factors determining host choice, there is a direct relationship between the number of hosts that blood-sucking insects utilize and the insects' locomotory abilities (which is often reflected in the amount of time they spend with the host). Thus, ectoparasites (which have poor locomotory abilities and usually remain permanently on hosts) are often restricted to a single host species. A good example is the louse *Haematomyzus elephantis,* which is restricted to elephants. At the other extreme, those flying blood-sucking insects such as mosquitoes that make contact with the host only long enough to take a blood meal often display a very catholic host choice. For example, a sample population of the mosquito *Culex salinarius* was shown to take 45% of its blood meals from birds, 17% from equines, and 15% from canines; moreover, 13% of the meals was a mixture of blood from more than one host!

In general terms, the most common hosts chosen are large herbivores. Large, social herbivores present an abundant, easily visible food source that is reliable and predictable from season to season. Carnivores in comparison are fewer in number, often solitary, and range unpredictably over wide areas. Another reason large herbivores are chosen is that they are poor at defending themselves from attack compared to small, agile animals that will often kill and/or eat attacking blood-sucking insects.

HOST LOCATION

For lice and other blood-sucking insects that are permanently present on the host, finding a new host is simply a matter of moving from one to the other when the hosts are in bodily contact. For blood-sucking insects that are only in temporary contact with the host, finding a host is a more difficult proposition. The following host-seeking behaviors are not rigidly patterned but they probably typically follow one another in a loose sequence. For most blood-sucking insects, olfactory stimuli are the first host-related signals perceived, and visual signals from the host probably are apt to become important at a later stage in host location. Blood-sucking insects make use of this predictability by permitting the current behavioral response to lower the threshold required for the next host signal to elicit the next behavioral response in the host location sequence. The increasing strength and

diversity of host-derived signals that the blood-sucking insect receives as it moves closer to the host are thereby enhanced.

Host location is usually restricted to particular times of the day for each species of blood-sucking insect. Thus, tsetse flies tend to be crepuscular, *Anopheles gambiae* (the most important vector of malaria) is a night feeder, and the stable fly *Stomoxys calcitrans* bites during the day. As hunger increases, bouts of host location behavior intensify. For many blood-sucking insects such as the tsetse, the first behavior is often to choose a resting site where they have a good chance of encountering a host-derived signal and once there to remain motionless and wait for a host-derived signal. This strategy combines minimal energy usage with a good chance of encountering a host. Other blood-sucking insects use more active strategies. If a gentle wind is blowing from one direction the optimum strategy can be to fly across the wind so that the probability of contact with a host odor plume is enhanced.

Host-derived olfactory clues used include carbon dioxide, lactic acid, acetone, octenol, butanone, and phenolic compounds found in urine. These are probably used in combination by each insect's sensitivity to different combinations of smells. For example, we can look at the power of phenolic components found in bovine urine to draw tsetse flies to a bait. Used singly, 3-*n*-propylphenol draws roughly equal numbers of *Glossina pallidipes* and *G. m. morsitans*. In contrast, when 3-*n*-propylphenol is used in combination with 4-methylphenol catches of *G. pallidipes* increase 400%, whereas catches of *G. m. morsitans* decrease. The explanation for this may be that first, mixtures of odors are a stronger guide to the presence of a host than a single odor alone and so will minimize energy consumption from chasing false trails. Second, mixtures may help in host choice by guiding blood-sucking insects to particular host species.

Tracking the source of an odor plume while in flight is a major task. It is believed that many blood-sucking insects achieve this by using upwind optomotor anemotaxis. During flight, insects will be blown off course by any wind that is blowing. They can use this fact to determine wind direction. To do this, they observe the perceived movements of fixed objects on the ground and by comparing this to the direction in which they are trying to fly determine wind direction. The suggestion is that the insect flies across wind until an odor plume is encountered, when it turns upwind. If the odor plume is lost, it recommences flying across wind until it refinds the odor and turns upwind once more. This proceeds until the insect comes into the immediate vicinity of the host. It is believed that hosts can be detected by odor at about 90 m by tsetse flies and at 15 to 80 m by some mosquitoes.

Vision is also used in host location by the majority of blood-sucking insects and is used most extensively by day-feeding insects in open habitats. In general, blood-sucking insects can detect and discriminate between objects on the basis of color contrast, relative brightness (intensity contrast), movement, and shape. Insects are quite sensitive to movement and their color vision stretches up into the UV but not down to the red.

Night-feeding blood-sucking insects have relatively better intensity contrast than color contrast, whereas for day-biting blood-sucking insects movement perception and color contrast may be particularly important. Large individual herbivores (as opposed to herds) are thought to be detected by vision at about 50 m by tsetse flies and at 5 to 20 m by some mosquitoes.

Once the blood-sucking insect is in proximity to the host, heat and humidity become important factors in location in addition to the continuing importance of vision and odor. Temperature is probably a useful guide from about 5 cm to a meter or so from the host depending on insect species. Even when they have contacted a host, blood-sucking insects will imbibe a blood meal only if it provides the correct biochemical characteristics (i.e., taste).

THE BLOOD MEAL

Blood-sucking insects take huge meals. Temporary ectoparasites such as the tsetse fly typically ingest more than their own unfed body weight in blood. The reasons are twofold. First, taking a blood meal is a very dangerous activity and taking huge blood meals minimizes the number of times an insect must associate with the host. Second, locating the host is often difficult and huge blood meals are a way of making the most of each encounter. Mouthparts are adapted to the blood-feeding habit. Typically, they are either of the piercing kind seen in mosquitoes, bugs, lice, and fleas or the cutting kind seen in tabanids, black flies, and biting flies.

The host usually responds to feeding activity, particularly the injection of saliva, by mounting an immune response that includes pruritis (itching). Typically this begins to occur about 3 min after feeding commences. Thus, there is a selective advantage in completing the blood meal within this 3-min "safety period" after which the host will be alerted to the presence of the insect. To help achieve this, blood-sucking insects have produced a range of antihemostatic molecules in the saliva, one of the major functions of which is to minimize host contact time.

Antihemostatic molecules produced by the blood-sucking insect include anticoagulant molecules working variously, for example, on thrombin or factors VIII and X. However, platelet plugging of small wounds is probably of more importance to blood-sucking insects than blood coagulation. Consequently, they also produce anti-platelet aggregating factors such as apyrase. These are used to impede the plugging of the penetration wound in capillaries and to prevent clogging of the insect mouthparts. The insect saliva also contains powerful vasodilatory substances to increase blood flow to the wound and anti-histamines that will minimize inflammation and itching, possibly extending the "safe period." Salivary components are also important as they can facilitate the transmission of arthropod-borne pathogens. For example, the production of *Leishmania*-enhancing factor in the saliva of the sand fly *Lutzomyia longipalpis* enhances the establishment of the parasite *Leishmania major* in the vertebrate host.

It has also been shown that such effects may be limited to naïve hosts, suggesting that the history of exposure to vector saliva may influence the outcome of potentially infectious inoculations. Parasites can also manipulate the salivary glands to their own advantage. Thus, malaria sporozoites damage the salivary glands of mosquitoes, reducing antihemostatic effectiveness, and thus extend probing time and increase the chances they will be transmitted to a new host.

Some blood-sucking insects feed only on blood during their entire life. Examples include the tsetse flies, streblids, hippoboscids and nycteribiids, triatomine and cimicid bugs, and lice. Blood is deficient in certain nutrients such as the B-group vitamins and pantothenic acid, and the insect cannot make these itself. To make up for this deficiency, these obligate hematophages harbor symbiotic microorganisms that produce these extra nutrients. These symbionts are often housed in a specialized body compartment, traditionally called a mycetome or, more recently, a bacteriome. For example, the tsetse fly *Glossina* harbors three symbiotic microorganisms, including *Wigglesworthia glossinia,* which is from the γ-subdivision of the Proteobacteria, in the bacteriome of the anterior gut.

HOST–INSECT INTERACTIONS

There are several evident morphological adaptations for a blood-sucking life. Piercing or cutting mouthparts are the clearest example. In addition, many periodic and permanent ectoparasites such as fleas and lice are laterally or dorsoventrally flattened and are wingless, which are adaptations allowing them to move easily through the pelage or feathers and to avoid being groomed by permitting them to flatten themselves against the skin. Most of these ectoparasites also have cuticular extensions in the form of spines and combs. These are longer and spinier in bird-infesting forms than in those found on mammals. The combs in particular are found covering weak spots in the body such as the articulations between body segments. The spacing of the tips of the combs correlates well with the diameter of the hairs on the body of the host. This suggests that these combs have a dual function: protecting the body from abrasion and acting as an anchoring device for the ectoparasite.

The host regulates the numbers of permanent ectoparasites by grooming, usually with both the toes and the teeth. This often limits ectoparasite distribution on the host to those areas the host can groom least efficiently, such as the head and neck. The immune response mounted against these blood-sucking insects is often very localized. It makes feeding on these protected areas of the skin difficult, with the result that the insects feed less well or move to less affected areas of the body where they are more easily groomed. The result is that the host regulates ectoparasite numbers.

The host also shows behavioral defenses to temporary ectoparasites such as mosquitoes. The level of defensive behavior is usually density dependent and thus can have important consequences for disease transmission. For example, the arbovirus eastern equine Encephalitis (EEE), which is naturally found in birds, is transmitted in the United States by the mosquito *Culiseta melanura*. During spring and early summer, these mosquitoes feed almost exclusively on passerine birds, transmitting the virus among them. Later in the season, as mosquito numbers increase, bird defensive behavior increases and mosquitoes are more willing to feed on other vertebrate hosts. This is when EEE is transmitted to other vertebrates including horses and humans.

See Also the Following Articles

Medical Entomology • Mosquitoes • Phthiraptera • Siphonaptera • Tsetse Fly • Veterinary Entomology

Further Reading

Beaty, B., and Marquardt, W. (eds.) (1996). "The Biology of Disease Vectors." University of Colorado Press, Boulder.

Braks, M. A. H., Anderson, R. A., and Knols, B. G. J. (1999). Infochemicals in mosquito host selection: Human skin microflora and *Plasmodium* parasites. *Parasitol. Today* **15**, 409–413.

Charlab, R., Valenzuela, J. G., Rowton, E. D., and Ribeiro, J. M. C. (1999). Toward an understanding of the biochemical and pharmacological complexity of the saliva of a hematophagous sand fly *Lutzomyia longipalpis*. *Proc. Nat. Acad. Sci. U.S.A.* **96**, 15155–15160.

Clements, A. (1999). "The Biology of Mosquitoes." CABI Int., Oxon.

Dye, C. (1992). The analysis of parasite transmission by bloodsucking insects. *Annu. Rev. Entomol.* **37**, 1–19.

Hurd, H., Hogg, J. C., and Renshaw, M. (1995). Interactions between bloodfeeding, fecundity and infection in mosquitos. *Parasitol. Today* **11**, 411–416.

Lane, R., and Crosskey, R. (eds.) (1993) "Medical Insects and Arachnids." Chapman & Hall, London.

Lehane, M. J. (1991). "Biology of Blood-Sucking Insects." Chapman & Hall, London.

Sutcliffe, J. F. (1986). Black fly host location: A review. *Can. J. Zool.* **64**, 1041–1053.

Titus, R. G., and Ribeiro, J. M. C. (1990). The role of vector saliva in transmission of arthropod-borne disease. *Parasitol. Today* **6**, 157–160.

Vale, G. A., Hall, D. R., and Gough, A. J. E. (1988). The olfactory responses of tsetse flies, *Glossina* spp. (Diptera: Glossinidae), to phenols and urine in the field. *Bull. Entomol. Res.* **78**, 293–300.

Waage, J. K. (1979). The evolution of insect/vertebrate associations. *Biol. J. Linn. Soc.* **12**, 187–224.

Body Size

Christer Björkman and Mats W. Pettersson
Swedish University of Agricultural Sciences

B ecause of its structure, the environment offers many more niches for small organisms than for large ones. The relatively small size of insects, which is one of the reasons for their success, has therefore made them very diverse, a characteristic that has resulted in a high number of species. In spite of their generally small size, limited by their method of gas exchange, insects show as large a range in size as other groups of organisms.

SIZE VARIATION

Size varies tremendously among and within orders, families, and species of insects. The smallest extant insects known are about 0.2 mm in length and can be found among beetles of the family Ptiliidae and wasps of the family Mymaridae, which are egg parasitoids. Insects of this minute size are smaller than the largest one-celled protozoans. The largest extant insects are phasmid (walkingstick) species (up to 30 cm long), sphingid moths (wingspan of up to 30 cm), and some beetles of the genera *Megasoma, Dynastes,* or *Goliathus* (up to 100 gr). The high number of possible runners-up for the title "largest insect on earth" highlights another difficulty when dealing with size: What measure should be used?

Although insects are not as large as some other organisms, the range in size among insects is almost as large. For example, the difference in volume between the largest mammal (the blue whale) and the smallest (a minute shrew) is about $1:2 \times 10^8$. This is comparable to $1:1.5 \times 10^8$ for the two extremes among insects (ptiliid beetle and Goliath beetle).

HOW TO MEASURE INSECTS

Body length (measured from tip of head to tip of abdomen) is probably the most often used measure of size and the easiest to comprehend. Weight (or biomass) is a measure of size that interests many ecologists because it often correlates well with fecundity. Fecundity, in turn, is often a major fitness component and may be a key feature in population dynamics. For practical reasons, traits that are correlated with weight are used instead, such as hind tibia length, front wing length, elytra length, head capsule width, or body volume. The ratios between these and other morphological traits are often used in taxonomic descriptions of insect species. These ratios give descriptions of shape. The main objective in describing size is to identify traits that are easy to measure (e.g., does not have a curved shape) and are stable on prepared specimens.

SOME EXPLANATIONS FOR VARIATIONS IN SIZE

The size of an insect individual is determined by its genes and by the environment in which it grows. Temperature, crowding, food quantity, and food quality are examples of environmental factors that affect size, but insects may make up for such effects by compensatory feeding.

The size of female insects often determines their fecundity, which may be manifested in giving birth to many small offspring or a smaller number of large ones. To be able to produce many large offspring, which may be adaptive for survival in a harsh environment, the female herself must be large. The importance of size for female fecundity can often be seen in the sexual dimorphism of insect species, in which males typically are much smaller than females.

Although the primary role of male insects is to fertilize the eggs, males may benefit from being large because they contribute to the realized fecundity of females by providing resources through their ejaculate or in competing with other males to obtain mates. An example of the latter characteristic is provided by some species of digger wasps (Sphecidae), in which males compete intensely with each other for females only half their size.

Sometimes, adult or larval foods come in packages or shapes that allow only very small insects to use them. Such foods include very small items like seeds and insect eggs, or very thin items like pores of fungi. Many insect families that use these foods [e.g., bruchid beetles, mymarid parasitic wasps, and nanosellin (Ptiliidae) beetles, respectively] have been adapted to and have radiated into several species under such living conditions.

Insects smaller than 1 mm operate in a world where gravity and molecular forces are in the same order of magnitude. This can be advantageous when, for example, insects find it easier to climb vertical surfaces. However, it can also lead to problems when, for example, an insect is trapped in a drop of water by the water's surface tension.

FACTORS THAT LIMIT SIZE

The smallest insects will have difficulty making room for the internal organs that are necessary for their existence. For example, some ptiliid beetles can lay only one egg at the time because their eggs may be up to 0.7 times the size of the whole insect.

The largest size an insect can reach is limited by the tracheal system. In insects, gas exchange with air is mediated directly to the tissues by a highly branched system of chitin-lined tubes called tracheae. No cells in the insect body are more than 2 to 3 μm from a tracheole. Diffusion along a concentration gradient can supply enough oxygen for small insects, but forms that weigh more than about a gram, or are highly active, require some degree of ventilation. Most insects have ventilating mechanisms to move air in and out through the tracheal system, but the need to allow enough oxygen to reach the tissue by diffusion imposes limits on tracheal length. Most large insects present today have long slender bodies, a trait that also limits tracheal length. Furthermore, elaborations of the tracheal system could not be made without destroying the water balance in large insects. However, there are exceptions: some of the heaviest extant beetles have bulky bodies, but these insects are not (or do not have to be) very quick and do not fly.

CHANGE OVER TIME

It has been suggested that organisms increase in size over an evolutionary time scale. However, there is no evidence to support this suggestion, and perhaps natural selection acts on correlated traits that constrain the evolution of increased size. In fact, fossils reveal that some insects in the past were much larger than their extant relatives. For example, many winged Carboniferous and Permian insects, existing about 300 mya, had wingspans exceeding 45 cm; the largest was the Permian

dragonfly *Meganeuropsis schusteri,* which had a wingspan of 71 cm. These insects certainly also had long, narrow bodies, to reduce the length of the trachea. During these prehistoric times, the atmospheric oxygen concentration was much higher (up to about 35%) than the present level (20.9%), which may have allowed sufficient oxygen to reach the innermost tissues of very large insects. However, such an oxygen-rich atmosphere also would have augmented aerodynamic properties in early flying insects. It has been suggested that later appearing insects could not evolve to a large size because of competition for niches with birds and other later appearing animals.

OTHER RELATIONSHIPS

Ectotherms, including insects, in contrast to endotherms, seem to follow the converse of Bergmann's rule: that is, they are smaller toward higher latitudes and altitudes. A high degree of genetic determinism seems to underlie this pattern. A possible reason could be an evolved response to geographic patterns in season length.

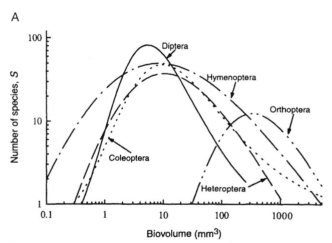

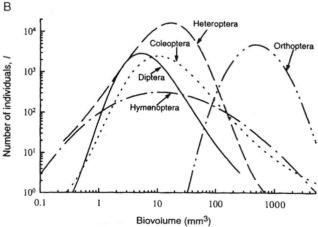

FIGURE 1 Relationship between insect size (body volume) and (A) species richness and (B) number of individuals for five insect orders. [Redrawn from Fig. 2 in Siemann, E., Tilman, D., and Haarstad, J. (1996). Insect species diversity, abundance, and body size relationships. *Nature* **380,** 704–706, with permission from *Nature.*]

Insect assemblages are thought to be structured by competition, with most of the insects found in medium-sized classes. Thus, the size of a particular insect is governed by its living habits and its feeding guild, in which competition with similar insects has forced some to evolve a larger or smaller body size. Empirical data show that species diversity in any taxonomic group of insects peaks at some intermediate body size (Fig. 1). For some authors, this implies that there may be fewer undescribed small insect species than previously suggested, which in turn, suggests that global biodiversity probably is lower than the highest estimates (30–50 million species). However, it is not clear whether such a pattern results from biological processes or from statistical or sampling properties. The size distribution of cars parked at Heathrow Airport also shows a peak in diversity at an intermediate size. Nevertheless, more knowledge about the causes behind size distribution patterns among insects and other organisms may provide key information in the effort to preserve biodiversity.

See Also the Following Articles

Biogeographical Patterns • Growth, Individual • Tracheal System

Further Reading

Conner, J., and Via, S. (1992). Natural selection on body size in *Tribolium:* Possible genetic constraints on adaptive evolution. *Heredity* **69**, 73–83.

Dudley, R. (1998). Atmospheric oxygen, giant Paleozoic insects and the evolution of aerial locomotor performance. *J. Exp. Biol.* **201**, 1043–1050.

Gaston, K. J., Blackburn, T. M., and Lawton, J. H. (1993). Comparing animals and automobiles: A vehicle for understanding body size and abundance relationships in species assemblages. *Oikos* **66**, 172–179.

Mousseau, T. A. (1997). Ectotherms follow the converse to Bergmann's rule. *Evolution* **51**, 630–632.

Price, P. W. (1984). "Insect Ecology." Wiley-Interscience, New York.

Siemann, E., Tilman, D., and Haarstad, J. (1996). Insect species diversity, abundance and body size relationships. *Nature* **380**, 704–706.

van Voorhies, W. A. (1997). On the adaptive nature of Bergmann size clines: A reply to Mousseau, Partridge and Coyne. *Evolution* **51**, 635–640.

Boll Weevil

R. Nelson Foster
United States Department of Agriculture

The boll weevil, *Authonomus grandis grandis,* is a beetle belonging to the family Curculionidae. A native of Mexico or Central America, it was first described in 1843 by the Swedish entomologist Boheman. Boll weevil is a new world pest of cultivated cotton and occurs in all Central America countries where cotton is cultivated, Mexico, the United States, Cuba, Hispaniola, Colombia, Venezuela, Brazil, and Peru. Where established, it is probably the most serious pest of cotton. Since its arrival in the United States, it is estimated that actual damage loss combined with control costs attributed to boll weevil exceed $16 billion (consumer-price-index-adjusted value of $91 billion). The boll weevil is important to examine in detail because of its destructive impact in cultivated cotton, use of aggregation pheromone, overwintering diapause, existence as three different but difficult-to-distinguish forms of the species, and status as the target of successful, area-wide eradication efforts in the United States.

MIGRATION AND DISPERSAL IN THE UNITED STATES

Boll weevil is a migratory pest, and its movement is largely dependent on wind direction and speed. It has been known to travel as far as 272 km and can hitchhike on cars, trucks, and trains. Boll weevil was first reported in the United States in the fall of 1894 from Brownsville, Texas, and may have been established as early as 1892. By 1895, the weevil had spread north to San Antonio and eastward to Wharton, Texas. The weevil reached Louisiana in 1903, Mississippi in 1907, and Georgia by 1916; by the 1920s it had infested cotton throughout the Mississippi Delta and the southeastern United States, and by 1922 had become established almost to the northern limits of cotton production. Northern and western portions of Texas became infested as a result of a sequence of expansions of the pest range between 1953 and 1966. Arizona was plagued with problems from the boll weevil beginning in the late 1970s, and in 1982 the weevil was detected in the southern desert valleys of California. The weevil became established in New Mexico in the early 1990s.

DESCRIPTION

Eggs of the boll weevil are pearly white, usually elliptical, and approximately 0.8 mm long by 0.5 mm wide. Legless white grubs hatch from eggs; they have light brown heads approximately 1 mm long. The larva, as it feeds and grows for 7 to 14 days, exhibits a ventrally curved, crescent form, with the dorsum strongly wrinkled and the venter smooth. The pupa is white at first but turns brown as it develops for 3 to 5 days. The adults are initially light in color but darken with age to colors from reddish brown to gray; depending on abundance and nature of food, they range in size from about 2.5 to 7 mm. The dark snout of this weevil is about half as long as the body (Fig. 1). Three forms (Mexican boll weevil, southeastern boll weevil, and thurberia weevil) have been separated by using characters ranging from morphological profemoral width/length ratios and spermatheca shapes to analysis systems relying on molecular biology.

LIFE HISTORY

The boll weevil belongs to a family of insects that is strictly phytophagous. This group is highly host specific and

FIGURE 1 Boll weevil on cotton leaf.

generally prefers flower buds for feeding. Cotton (*Gossypium* spp.) is the principal host plant of boll weevil, but it also develops on certain species of the related genera *Thespesia, Cienfuegosia,* and *Hampea*. The boll weevil passes the winter in diapause in the adult stage sheathed beneath brush and ground litter and in other protected locations in or around cotton fields. In arid areas, overwintering sites may be associated with increased moisture habitats such as near irrigation canals and rivers. Winter survivors emerge from overwintering sites in the spring and begin feeding on the tips of cotton seedlings and squares (i.e., the cotton flower buds). Weevils that emerge before cotton plants have begun to form squares, feed on leaf buds and growing terminals, and live for only a week or two; those that emerge later produce eggs for 3 to 6 weeks. Later generations survive the winter in a diapause state. The female deposits eggs singly in the bottom of punctures she makes in the cotton squares and later in the season in bolls. Overwintered females produce fewer than 100 eggs, but later generations produce 300 or more eggs. The average female's rate of reproduction is 5 or 6 eggs a day. Depending on the temperature, larvae hatch in 3 to 4 days. Larvae feed for 7 to 14 days and pupate. Adults emerge 3 to 6 days later.

A sex pheromone facilitates mating, after which the females begin laying eggs in 3 to 5 days. Two to seven generations can occur in a season. However, as many as 10 generations may develop under favorable conditions. Late in the season as cotton ceases to produce fruit, boll weevils move in large numbers from cotton fields to overwintering sites. Only 1 to 20% of weevils survive the winter. Reduced survival is seen after unusually cold winters, and unusually dry summers also cause some mortality through loss of moisture in overwintering sites.

CONTROL AND ERADICATION

Sound cultural practices combined with chemical control relying on numerous compounds has been the traditional practice for combating boll weevil. Coordinated eradication efforts in the United States have been quite effective. The eradication effort in the United States is based on three major activities: mapping all cotton fields, evaluating weevil presence in each field with pheromone traps, and applying control treatments. The program consists of a series of sequential expansions and usually lasts 3 to 5 years in any particular area. The program relies on intensive, carefully coordinated, ground and aerial treatments of ultra low volume malathion (almost exclusively), concentrated over one to three seasons, in response to predetermined numbers of weevils caught in pheromone traps. The program started in 1983 in the Carolinas and has expanded to parts of all of the cotton-growing states in the country. Active eradication has been completed in Virginia, North Carolina, South Carolina, Georgia, Florida, Alabama, Arizona, California, and in some parts of the other cotton-growing states. It is projected to be completed in 2005. In the southeastern states, where active eradication has been completed, a remarkable increase in cotton production has occurred. When completed, nationwide eradication will result in substantial economic and environmental benefits throughout the areas once plagued by the boll weevil.

See Also the Following Articles

Agricultural Entomology • Coleoptera • Migration • Regulatory Entomology

Further Reading

Burke, H. R. (1997). Early boll weevil fighters. *Southwest Entomol.* **22**, 248–277.

Burke, H. R., Clark, W. E., Cate, J. R., and Fryxell, P. A. (1986). Origin and dispersal of the boll weevil. *Bull. Entomol. Soc. Am.* **32**, 228–238.

Haney, P. B., (exec. ed.) (2001). "History of Boll Weevil Eradication in the United States." National Cotton Council Series. Herff-Jones, Montgomery, AL.

Pfadt, R. E. (1985). Insects pests of cotton *In* "Fundamentals of Applied Entomology. 4th ed. (R. E. Pfadt, ed.). Macmillan, New York.

Bombyx mori

Satoshi Takeda

National Institute of Agrobiological Sciences, Japan

The silkworm, *Bombyx mori,* is used for sericulture and is one of the most economically important insects in the world (Fig. 1). The species of silkworm usually raised by sericulturists is *B. mori* (Lepidoptera: Bombycidae). A closely allied species is the mulberry wild silkworm, *B. mandarina*. Its morphology does not differ markedly that of from *B. mori,* and hybrids are highly fertile.

Along with *Drosophila melanogaster, B. mori* larvae have been used as a model for various biological studies for many years. These two insect species have greatly contributed to the progress of research in several scientific fields.

FIGURE 1 Larva of the silkworm, *B. mori*. Larva at 4 days of fifth instar.

LIFE CYCLE OF *B. MORI*

Silkworms undergo complete metamorphosis (Fig. 2). Larvae feed on the leaves of the mulberry (family Moraceae, genus *Morus*); they will consume other genera in this family, but growth rate is reduced. At 23 to 25°C the five instars require 25 to 30 days to hatch. Toward the end of fifth instar, *B. mori* spin a cocoon over a 3-day period and pupate within that cocoon; the pupal stage lasts for about 10 days. After molting to the adult has occurred inside the cocoon, the moth emerges. The moth softens the cocoon by orally excreting a

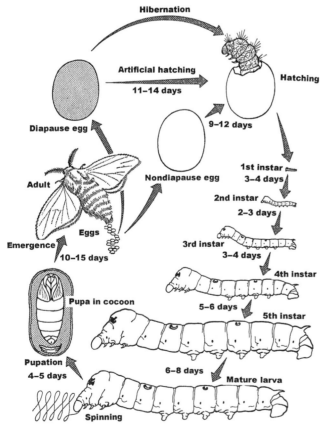

FIGURE 2 Life cycle of *B. mori* reared at 23 to 25°C. [From Mori, T. (1970). Life cycle of *Bombyx mori. In* "The Silkworm—New Experimental Tool in Biology" (T. Mori, ed.), p. 17. Sanseido Press, Tokyo.]

special enzyme, cocoonase, and then emerges from the end of the cocoon. The female moth mates the same day as emergence and begins egg deposition. One generation of *B. mori* spans 40 to 45 days.

Voltinism (i.e., the number of generations occurring in a single year) has a genetic basis, but it is also strongly affected by environmental factors. Some races are univoltine (only one generation a year); others are bivoltine (two generations a year) or polyvoltine (three or more generations a year, as seen in tropical silkworms that do not undergo diapause). Voltinism is closely tied to the geographic distribution of silkworms, which are divided into Japanese, Chinese, European, and tropical races. Typically, the silkworms distributed in the cold regions are univoltine, those distributed and adapted to warm regions are bivoltine, and those in tropical regions are polyvoltine.

EMBRYONIC DIAPAUSE OF *B. MORI*

Silkworms undergo diapause, during which the embryo within the egg stops growing. In the *B. mori* embryo, diapause is primarily determined by the temperature, light, nutrition, and other conditions of the lifetime of silkworms. Of the various factors, temperature and light during the egg (embryo) stage have the greatest influence. When the eggs of bivoltine silkworms were incubated at relatively high temperatures (25°C) with long days (> 13 h of light), all silkworms that grew from these eggs laid diapausing eggs. When the eggs were incubated at 15°C with short days (< 13 h of light), adults of all these silkworms subsequently laid nondiapausing eggs.

The subesophageal ganglion, located just below the brain, secretes a peptide hormone that induces the embryonic diapause of *B. mori*. In 1951, K. Hasegawa and S. Fukuda, in separate studies, demonstrated that the subesophageal ganglion plays role in induction of diapause. The diapause hormone is composed of 24 amino acids and is produced and secreted by six pairs of neurosecretory cells of the subesophageal ganglion.

SILK PRODUCTION BY *B. MORI*

B. mori larvae have a unique metabolic system for producing a large amount of cocoon protein and efficiently using dietary nitrogen. Both male and female silkworms digest and absorb about two-thirds of the nitrogen in the mulberry leaves they consume, and high percentages of the digested and absorbed nitrogen (66% in females and 70% in males) are utilized in the production of cocoon protein.

During the last larval stage (fifth instar), the silk gland produces the silk for the cocoon from a pair of curved glands found on the ventral side of the digestive tube. The weight of this organ accounts for about 25% of the weight of larvae in the late fifth instar.

The silk gland can produce massive amounts of fibroin and sericin, the proteins constituting silk. Sericin surrounds a fibroin core. The ratio of fibroin to sericin is approximately

3:1. Fibroin is rich in four amino acids: glycine (Gly), alanine (Ala), serine (Ser), and tyrosine (Tyr). The fibroin molecule contains repeats of a section composed of a regular arrangement of three amino acids, Gly, Ala, and Ser. Major amino acids constituting sericin are Ser, aspartic acid, glutamic acid, and glycine. A characteristic of sericin is that, unlike fibroin, it is soluble in hot water. Therefore, when cocoon threads are reeled, most of the sericin is removed, and the remaining raw silk is composed of fibroin alone.

Studies on *B. mori* greatly contributed to early discoveries in insect endocrinology and to the isolation and analysis of insect peptide hormones in 1980s and 1990s. The large size of these insects made experimental morphological studies easier, and because of the importance of this species to the sericultural industry, large quantities of materials for hormone extracts were made available.

In silkworms, larval ecdysis is induced by a molting hormone secreted by the prothoracic gland, which is located inside the first thoracic spiracle. The role of the prothoracic gland in ecdysis was discovered in 1944, and the molting hormone, ecdysone, was structurally determined in 1954 in studies that used large amounts of silkworm pupae as material. Ecdysone was the first hormone to be isolated from an insect species. In addition, the function of the corpora allata in Lepidoptera was also first discovered in silkworms in 1942. The corpora allata, which are small organs located adjacent to the brain, secrete juvenile hormone, which controls silkworm development together with molting hormone. Among the peptide hormones, the molecular structures of prothoracicotropic hormone (initially named "brain hormone"), which controls the secretion of molting hormone, and the diapause hormone, which induces silkworm egg diapause, were elucidated by using silkworms.

TRANSGENIC SILKWORMS AND INSECT FACTORIES

The National Institute of Sericultural and Entomological Science in Japan created transgenic silkworms with a jellyfish fluorescent protein gene as a marker. Prior to this, there were only a few transgenic insects in the Diptera (e.g., *Drosophila, Aedes*). This success resulted from development of a microinjector to introduce DNA into silkworm eggs, and the use of an effective transposon vector. The transgenic silkworms are expected to provide new opportunities for silk production.

A promising use of silkworms outside the clothing industry is in so-called insect factories, where silkworms are used for biological production of peptides or proteins useful for humans. Silkworms are infected with nuclear polyhedrosis virus (one species of baculovirus) to enable them to produce useful substances on a large scale.

See Also the Following Articles
Biotechnology and Insects • Diapause • Ecdysteroids • Lepidoptera • Sericulture • Silk Production

Further Reading
Maeda, S., Kawai, M., Obinata, H., Fujiwara, T., Horiuchi, T., Saeki, T., Sato, Y. and Furusawa, M. (1985). Production of human α-interferon in silkworm using a baculovirus vector. *Nature* **315**, 592–594.
Mori, T. (1970). Life cycle of *Bombyx mori. In* "The Silkworm—New Experimental Tool in Biology" (T. Mori, ed.), In Japanese. Sanseido Press. Tokyo.
Tajima, Y. (1978) "The Silkworm: An Important Laboratory Tool." Kodansha, Tokyo.

Book Louse

see *Psocoptera*

Borers

Timothy D. Paine
University of California, Riverside

Insects that are borers belong to a wide range of taxonomic groups, but they all share a common life history trait: they spend all or part of their larval life feeding within the tissues of their host plant. Some borer species deposit eggs within host plant tissues, whereas other species oviposit on the external surface and the larvae bore into the plant. Although there may be some feeding activity within the phloem and cambial tissues the larvae typically excavate feeding galleries within the woody tissues of perennial plants, within the stems of annual plants, and within the stalks or stems of grasses. Adult borers are free-living outside the host plant.

TAXONOMIC AFFILIATIONS

Insect orders that include species commonly referred to as borers include Lepidoptera, Coleoptera, Hymenoptera, and Diptera. The most varied and numerous representatives are among the moths and the beetles. The primitive wasps, which include the horntails (Siricidae), wood wasps (Xiphydriidae and Syntexidae), and sawflies (Tenthredinidae and Cephidae), are the only representatives of the Hymenoptera. The horntails and wood wasps prefer to colonize weakened hosts and the larvae construct feeding galleries in the wood. Larvae of boring sawfly species often feed within the center of tender shoots, twigs, and stems of their host plants. There are a small number of fly species in the family Agromyzidae with life history strategies that leave injury patterns that could be characterized as boring, but the larvae actually mine the cambial tissue, and the trees overgrow the galleries and the mine remains in the wood.

There are many moth families in the Lepidoptera that include species of larval borers. Among the most important families comprising only species that have a boring life history or having large numbers of boring species are the Hepialidae (ghost moths or swifts), the Sesiidae (clearwing moths), the Cossidae (carpenterworm and leopard moths), and the Tortricidae (leafroller and olethreutine). Larvae of the Hepialidae and the Cossidae tunnel extensively into the wood of their host plants and can cause substantial damage. Larvae of many species of Tortricidae bore through the twigs and tender terminals of vigorous trees and shrubs, whereas the life histories of species of Sesiidae can be highly variable and may include boring in bark, cambium, wood, roots, or gall tissues. Other families of Lepidoptera with at least some species that can be characterized as borers include the Agonoxenidae, Argyresthiidae, Gelechiidae, Momphidae, Nepticulidae, Noctuidae, Pterophoridae, Pyralidae, and Thyrididae. Larvae of one important agricultural pest, the European corn borer, *Ostrinia nubialis* (Lepidoptera: Crambidae), bore into the stems of woody host seedlings as well as the stems of grasses, grains, and herbaceous host plants.

The Coleoptera include a large number families composed exclusively of species with larvae that are boring or having very few representatives that have evolved alternative life history strategies. Although not an entirely comprehensive list, the families include Anobiidae (deathwatch and drugstore beetles), Bostrichidae (false powderpost beetles), Brentidae (brentid beetles), Buprestidae (metallic or flatheaded wood borers), Cerambycidae (longhorned or roundheaded wood borers), Lyctidae (powderpost beetles), Lymexylidae (timber beetles), Platypodidae (ambrosia beetles), and Scolytidae (bark beetles). The Curculionidae (snout beetles or weevils) is a very diverse family that includes a number of species with larvae that bore into plant tissues.

FEEDING STRATEGIES

Many different plant parts serve as sites of insect feeding activity. In general terms, borers can be distinguished from miners. Typically, larvae of miners feed within plant foliage, whereas larval borers may feed within other plant tissues, including roots, stems and twigs, meristems, fruit, conductive tissues, galls, and bark. The variety of plant tissues that are used by borers also spans an array of plant groups that range from the ferns and gymnosperms to the grasses and dicotyledonous angiosperms.

All plant tissues may be subject to borer colonization. Larvae of a number of families (e.g., Cerambycidae and Hepialidae) may construct feeding tunnels, or galleries, within the large roots of broadleaf trees and conifers, which may weaken the trees directly or provide entry points for invasion by pathogenic fungi. At the other extreme, there are many species of insects that colonize the meristematic tissues at branch terminals, tips, twigs, and canes. Some of these insects feed in the phloem tissues girdling the twigs, whereas

FIGURE 1 Adult pairs of the European elm bark beetle *S. multistriatus* excavate parental galleries in the cambium and phloem of host elm trees. Larvae eclose from eggs laid in niches cut into the margins of the parental galleries and construct feeding galleries that extend laterally into the same host tissues.

larvae of other species burrow through the growing tips and into the elongating stems. These types of larval feeding can reduce plant growth, apical dominance, and plant form.

Between the twigs or apical tips and the roots is the main stem or trunk of the tree. The woody xylem tissues, cambial layers, phloem tissues, and bark may all have different groups of specialist borers. For example, larvae of a few species of clearwing moths feed within the bark of their host plant. Scribble-barked gums are species of *Eucalyptus* in Australia that derive their common name from the twisting galleries constructed in the outer bark by lepidopteran larvae. In different feeding strategies, bark beetle larvae feed within the cambial and phloem tissues of their hosts (Fig. 1), whereas larvae of many species of longhorned and flatheaded borers feed in the outer layers of phloem and cambium but then bore deep into the wood to pupate. Alternatively, many other cerambycid and buprestid species feed almost entirely within the wood of their host trees. Larvae of cossid moths also feed entirely within woody tissues and may take several years to complete their larval development. Woody plant tissues are not as rich in nutrients as the cambial tissues and the quality deteriorates as the tissues age. Consequently, many wood borers may have prolonged larval development and long generation times.

Not only the larvae bore into woody tissues but also the adults in a number of species within a variety of families (e.g., some species of Scolytidae and Platypodidae) bore into the plant. The larvae of ambrosia beetles are found in galleries excavated within the wood, but they feed on a fungus inoculated into the tissue by the parental adults rather than on the plant itself. The fungi acquire nutrients from a large volume of plant tissue as the hyphae ramify throughout the wood.

PLANT CONDITION

Plants in a wide range of physiological conditions may be subject to colonization by borers. Although some species of borers use healthy hosts or healthy host tissues, plants that are suffering from some type of stressful condition either attract or inhibit further dispersal behavior by many other borer species. Insects that bore into tender tips and stems frequently colonize young and vigorously growing plants. Consequently, younger plants may suffer more damage than mature plants. Open wounds or stressed, damaged, or weakened plant tissues may be subject to invasion. Weakened or stressed host plants may result from chronic growing conditions (poor-quality site) or from acute detrimental changes (e.g., fire, flood, drought, lightning strikes). Infections by pathogens, particularly plant pathogenic fungi, nematodes, and parasitic plants, weaken host plants and increase their susceptibility to subsequent borer infestations. In addition, previous infestation by other insect herbivores may weaken the host plant and increase susceptibility to subsequent borer colonization. Recently killed and dying trees are particularly suitable for colonization by a range of borers. For example, there are several species of wood wasps and flatheaded wood borers that are attracted to trees that have been recently killed by fires. A wide range of borers have developed complex relationships with tree-killing pathogens and are responsible for transmitting the pathogens into the host trees. Scolytid bark beetles transmit species of pathogenic *Ophiostoma* and *Ceratocystis* fungi into a variety of hosts (e.g., *Scolytus scolytus, S. multistriatus,* and *Hylurgopinus rufipes* transmit *Ophiostoma ulmi,* the causal agent of Dutch elm disease). Cerambycids in the genus *Monochamus* are responsible for transmitting the nematode, *Bursaphelenchus xylophilus,* the pathogen causing pine wilt disease, into susceptible host pines. Females of the European woodwasp, *Sirex noctilio,* inject a phytotoxin and spores of the pathogenic fungus *Amylostereum areolatum* into susceptible host trees during oviposition. Borers that are adapted to colonize the woody tissues of dead or dying plants may also colonize trees that have been cut during commercial logging or even timber that has been milled into lumber. It is not unusual for adult borers to emerge from products or materials constructed from infested wood that has not been kiln dried or otherwise treated to kill the infesting insects.

See Also the Following Articles

Forest Habitats • Integrated Pest Management • Plant Diseases and Insects

Further Reading

Creffield, J. W. (1996). "Wood Destroying Insects: Wood Borers and Termites," 2nd ed. CSIRO Australia, Collingwood.

Furniss, R. L., and Carolin, V. M. (1980). "Western Forest Insects." U.S. Department of Agriculture Forest Service, Washington, DC. [Miscellaneous Publication 1339].

Hanks, L. M. (1999). Influence of the larval host plant on reproductive strategies of cerambycid beetles. *Annu. Rev. Entomol.* **44,** 483–505.

Johnson, W. T., and Lyon, H. H. (1988). "Insects That Feed on Trees and Shrubs." Cornell University Press, Ithaca, NY.

Paine, T. D., Raffa, K. F., and Harrington, T. C. (1997). Interactions among scolytid bark beetles, their associated fungi and live host conifers. *Annu. Rev. Entomol.* **42,** 179–206.

Solomon, J. D. (1995). "Guide to the Insect Borers of North American Broadleaf Trees and Shrubs." U.S. Department of Agriculture Forest Service, Washington, DC. [Agricultural Handbook 706]

U.S. Department of Agriculture Forest Service. (1985). "Insects of Eastern Forests." U.S. Department of Agriculture Forest Service, Washington, DC. [Miscellaneous Publication 1426]

Brain and Optic Lobes

Nicholas J. Strausfeld

University of Arizona

Authors variously use the term brain either to include all neuropils located within the head capsule or, restrictively, to refer to only those neuropils (called preoral neuropils) that lie dorsal to the esophagus. These are considered to lie anterior to the mouth. Preoral neuropils are also known as the supraesophageal ganglion, which comprises three fused ganglia: the protocerebrum, deutocerebrum, and tritocerebrum. The preoral brain of the larger Hymenoptera, such as the predatory wasp *Pepsis thisbe,* can contain well over a million neurons, with more than a third of a million neurons in each mushroom body. The extreme density of neurons packed into a small volume, and the likelihood that single nerve cells can be functionally divided into several discrete elements, suggests that the largest insect brains have impressive computational power.

The first definition of the brain includes neuropils of the subesophageal ganglion, which is composed of the fused ganglia from three postoral segmental neuromeres. These are located ventrally with respect to the digestive tract, as are ganglia of the thorax and abdomen. In most hemimetabolous insects, and in many paleopterans, the subesophageal ganglion is connected by paired circumesophageal commissures to the supraesophageal ganglion. In many crown taxa (those representing more recent evolved lineages) the subesophageal and supraesophageal ganglia are fused, as is the case in honey bees or the fruit fly *Drosophila melanogaster,* which is the taxon here used to summarize the major divisions of the brain (Figs. 1–6). A consequence of fusion is that tracts of axons that would otherwise form the circumesophageal commissures are embedded within a contiguous neuropil.

In insect embryos, the three preoral segmental neuromeres providing the proto-, deuto-, and tritocerebrum are contiguous

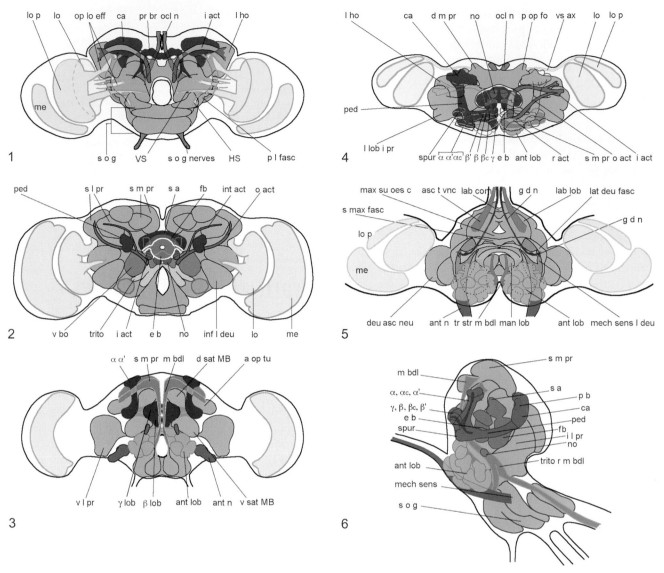

FIGURES 1–6 Summary diagrams of the brain of the fruit fly *D. melanogaster.* The reader is referred to these searchable figures in the atlas of the *Drosophila* brain at http://www.flybrain.org/Flybrain/html/atlas/schematic/index.html. The higher centers of the mushroom bodies and central complex are shown in reds and oranges. Optic lobe regions are yellow. Antennal lobes are light green and their axon projections are dark green. The median bundle is shown in light blue. Other neuropil areas are gray/pink. (1) Posterior aspect, vertical section. According to the neuraxis (see Fig. 7), up is rostral in Figs. 1–3. (2) Middle aspect, vertical section, at the level of the central body and associated regions. (3) Frontal aspect, vertical section, at the level of the antennal lobes (green) and mushroom body lobes (red). Dark green profiles in Figs. 1–3 are the antennocerebral tracts. (4) Top-down view, looking onto the mushroom bodies and central complex. One mushroom body only is shown to the left, with the antennocerebral tracts from the antennal lobes to the lateral protocerebrum shown to the right. The front of the brain is down, the rear of the brain is up. (5) Top-down view of the deutocerebrum/tritocerebrum and the root of the ventral nerve cord. (6) Side-on view of the brain, emphasizing the ascending tracts (blue) from the subesophageal ganglion to the rostral protocerebrum via the median bundle. Note the disposition of the mushroom body and central complex. Abbreviations used: α, α', αc, (β, β',βc, γ) **lobes,** subdivisions of the mushroom body medial (β, β',βc, γ) and vertical (α, α', αc) lobes; **ant n,** antennal nerve; **ant lob,** antennal lobe; **a op tu,** anterior optic tubercle (optic glomerulus); **asc t vnc,** ascending tracts of ventral nerve cord; **ca,** calyx of mushroom body; **deu asc neu,** deutocerebral neuropil receiving ascending terminals; **d m pr,** dorsal median protocerebrum; **e b,** ellipsoid body of the central complex; **fb,** fan-shaped body of the central complex; **g d n,** giant descending neuron (typifies descending pathways); **inf l deu,** inferior lateral deutocerebrum; **i act,** inner antennocerebral tract; **int act,** intermediate antennocerebral tract; **l lob i pr,** lateral lobe of the inferior protocerebrum; **lat deu fasc,** lateral deutocerebral fascicle; **lab lob,** labral lobe; **lab com,** labral commissure; **lo,** lobula; **lo p,** lobula plate; **l ho,** lateral horn; **max su oes c,** maxillary subesophageal connective; **me,** medulla; **mech sens l deu,** mechanosensory neuropil of the lateral deutocerebrum; **mech sens,** mechanosensory strand and neuropil supplied by the antennal nerve; **m bdl,** median bundle; **no,** noduli of the central complex; **op lo eff,** optic lobe efferents; **ocl n,** ocellar nerve; **o act,** outer antennocerebral tract; **pr br,** protocerebral bridge of the central complex; **p l fasc,** posterior lateral fascicle; **p op fo,** posterior optic focus (glomerulus); **r act,** root of antennocerebral tract; **s a,** superior arch of the central complex; **s o g,** subesophageal ganglion; **s l pr,** superior lateral protocerebrum; **s m pr,** superior median protocerebrum; **s o g nerves,** nerve bundles of subesophageal neuromeres; **spur,** spur of mushroom body; **trito,** tritocerebrum; **tr str m bdl,** tritocerebral strand of the median bundle; **trito r m bdl,** tritocerebral root of the median bundle; **VS, HS,** axons of giant vertical and horizontal cells (movement sensitive neurons); **vs ax,** visual interneuron axons; **v bo,** ventral body (also known as lateral accessory lobes); **v sat MB, d sat MB,** ventral and dorsal satellite neuropils of the mushroom bodies.

with the three postoral neuromeres that will give rise to the subesophageal ganglion. These neuromeres are, in turn, contiguous with fused neuromeres of the thorax and abdomen. In many species of hemimetabolous insects, such as locusts and cockroaches, the sub- and supraesophageal ganglia separate postembryonically and are connected by paired tracts. In cockroaches, each segmental ganglion is separate from the next, except for the last three abdominal ganglia, which are specialized to serve receptors of the cerci and contain the dendrites of giant ascending neurons and local networks of interneurons that mediate escape reactions. However, in many holometabolous insects there are various degrees of ganglion fusion, one of the most extreme being in certain Heteroptera such as the water strider *Gerris* sp. In *Gerris,* the supraesophageal, subesophageal, and thoracic–abdominal ganglia comprise a contiguous mass perforated by the gut. This arrangement is reminiscent of the nervous systems of another group of arthropods, the chelicerates. In adult cyclorrhaphan flies the three thoracic ganglia and all abdominal ganglia are fused into a single mass connected to the sub- and supra-esophageal ganglion by long neck connectives (this has also been achieved in the nervous systems of crabs).

The subesophageal ganglion, which comprises the mandibular, maxillary, and labial neuromeres, has a ground pattern organization comparable to that of the thoracic and abdominal ganglia. The roots of motor neurons (the exit point of motor neuron axons) are generally dorsal with respect to incoming sensory axons. This arrangement is the opposite of that in the vertebrate spinal cord.

The names of the subesophageal ganglia reflect the appendages that their motor neurons control and from which they receive sensory supply. However, this relationship is not a strict one. For example, in flies mechanosensory neuropil extending into the subesophageal ganglion also receive afferents from mechanosensilla on the head, including between facets of the compound eyes, around the margin of the eyes, the frons, between and flanking the ocelli, and at various positions on the rear of the head capsule. As on the thorax and abdomen, or on the limbs, wings, and halteres (modified wings in Diptera that are organs of balance), sensilla on the head provide receptor neuron axons to defined locations in their target ganglia. Principles underlying the development and organization of the central representation of sensilla are best known from Walthall and Murphey's 1988 studies of cricket cerci or studies on the central projections of receptors to discrete regions of the thoracic ganglia of dipterans, also by Murphey and colleagues in 1989. In flies, groups of receptors encoding different modalities at a segment supply axons to modality-specific regions within the ganglion. In such regions, the peripheral locations of receptors within a sensory field can be represented as a map of axon terminals onto the dendritic trees of postsynaptic neurons. Burrows and Newland have shown that such maps play important roles in the activation of the postsynaptic elements that participate in circuits controlling limb actions and position.

THE PREORAL BRAIN

Although there has been in the past endless debate about how many segments contribute to the head and to the brain, expression of homeobox genes now confirms three embryonic brain segments only. The main issue of contention focused on the neuromeric identity of the optic lobes, which were claimed by some to have a distinct segmental origin. Developmental studies, which are summarized by Meinertzhagen and Hanson's 1993 review, showed that the inner optic lobe neuropils in the adult insect brain develop from a lateral outgrowth of the protocerebrum. George Boyan and his colleagues at the University of Munich have provided crucial evidence supporting a three-neuromere origin of the supraesophageal ganglion from studies of the segment polarity gene *engrailed,* which is expressed in cells (including neuroblasts) in the posterior compartment of each segment. The expression of *engrailed* in the first wave of neuroblast generation shows the delineation of the tritocerebrum from the first (maxillary) subesophageal neuromere, as well as the delineation between the tritocerebrum and the deuto-cerebrum and the delineation between the deutocerebrum and the protocerebrum, the last being the most rostral segmental neuromere. Crucially, the expression of *engrailed* shows the latter to be segmentally indistinct from the developing optic lobes.

THE PROTOCEREBRUM

The ground structure of the protocerebrum suggests its ancestral affinities with segmental ganglia. In the protocerebrum, as in postoral ganglia, ascending sensory interneuron tracts enter it ventrally, whereas premotor interganglionic interneurons exit dorsally. Afferents (here the optic lobe output neurons; see later) distribute to local interneurons in a manner reminiscent of sensory afferents within postoral ganglia.

Despite its basic similarities with segmental ganglia, the protocerebrum contains neuropils that are not normally found in other segments and appear to have no counterparts in other ganglia, unless generated ectopically by genetic manipulation. Unique protocerebral neuropils comprise: (1) the central complex and (2) the mushroom bodies and some satellite neuropils belonging to both of these. A midline indentation between the two protocerebral lobes, called the pars intercerebralis, with its accompanying populations of neuromodulatory neurons, may also be unique to the protocerebrum. But without the relevant developmental studies on thoracic ganglia, it is not clear whether any of the unique clusters of neurons at their dorsal midlines are segmental counterparts of neurons at the pars intercerebralis.

The structure of the protocerebrum is best approached by understanding the basic organization of major axon tracts that extend between its two halves. Studies of *Drosophila* embryos show that major cerebral tracts appear early in development and pioneer the trajectories of interneurons linking later developing neuropil regions. Again, research by Boyan and colleagues on the development of locust embryos provides important insights into early brain development and demonstrates that neurons developmentally ascribed to the protocerebrum can actually end up distant from it in the adult. For example, studies of *engrailed* expression show that in locusts the first episode of neuroblast generation in the protocerebrum includes three neuroblasts that migrate caudally to lie beside the glomerular antennal lobes, which are structures usually ascribed to the deutocerebrum. The segmental origin of these three neuroblasts, which contribute neurons to the antennal lobe system, cautions against uncritically ascribing segmental identities to neurons in the adult brain.

Nevertheless, many of the tracts and neuropils described from the adult have been both ascribed to one of its segments and named, even though only a few are yet understood at a functional and developmental level. The reader is referred to two brain atlases, one by Strausfeld and the other an electronic publication, FLYBRAIN (www.flybrain.org), both of which focus on the adult structure of dipteran brains (the housefly *Musca domestica* and the fruit fly *D. melanogaster*). The basic divisions of the *Drosophila* brain are shown in Figs. 1–6. The following neuropils, or neuropil groups, comprise the salient regions of the adult protocerebrum.

The Central Complex

Insect and crustacean protocerebra contain unique midline neuropils and, in more advanced taxa, satellite neuropils associated with them. These structures and their inter-relationships have been described by several authors, one cardinal study being by L. Williams in 1975. This and other studies summarized here are described in some detail in a recent article by Strausfeld published in 1999.

The midline component of the central complex, called the central body, is similar to a unique midline neuropil in the brains of chilopods, branchiopod crustaceans, and archaeognathan insects. Comparative studies suggest that these neuropils have become elaborated through time. In the flightless Zygentoma (e.g., "silverfish") as well as in the Palaeoptera (e.g., mayflies and dragonflies), several paired satellite neuropils are reciprocally connected to two midline neuropils: the columnar ellipsoid body and, above it with respect to the brain's neuraxis, the fan-shaped body, which is usually recognized by its scalloped profile. Further elaboration has occurred in the Neoptera, in which a distinct stratum called the superior arch is attached to the fan-shaped body anteriorly. A bridge of neuropils, called the protocerebral bridge, connects the two protocerebral lobes and provides axons that extend into the fan-shaped body and to the ellipsoid body behind it. In many taxa (e.g., locusts, flies, wasps) the protocerebral bridge is divided into 16 discrete modules, 8 each side of the midline. These connect

to the 16 modules of the fan-shaped body and ellipsoid body. The most lateral module on one side of the bridge is linked to the most medial module of the other side. The next most lateral module is linked to the next most medial one, and so on. These connections provide an elaborate pattern of chiasmata between the bridge and the fan-shaped and ellipsoid bodies. Some of these neurons also extend to a pair of ball-like structures, called the noduli, situated caudally with respect to the fan-shaped and ellipsoid bodies. Two synaptic zones in the noduli, a core and an outer layer, receive connections from the fan-shaped body such that one-half is represented in the core of the contralateral nodulus, whereas the other half is represented in the outer layer of the ipsilateral nodulus. A recent account by Renn and colleagues (see FLYBRAIN database) uses genetic markers to dissect these various components and trace their development.

The protocerebral bridge receives a system of elongated fibers from the medial protocerebrum, which is itself supplied by terminals of ascending interganglionic interneurons that originate in thoracic ganglia. These elongated protocerebral fibers extend through dendritic trees that contribute to the modules across the bridge and are assumed to provide inputs to their dendrites, although this awaits confirmation.

The superior arch appears to be distinct from protocerebral bridge inputs and is connected heterolaterally to neuropils of the protocerebral lobes, themselves receiving terminals from the median bundle, a midline tract originating from the subesophageal ganglion and ventral cord and ascending along the midline of the ventral surface of the protocerebrum. The superior arch shares local interneurons with the fan-shaped body and ellipsoid body.

The fan-shaped body and ellipsoid body each receive fan-like terminals from axons that originate at dendritic trees in various lateral neuropils of the protocerebrum. Both the ellipsoid and the fan-shaped bodies supply outputs that extend to lateral protocerebral neuropils, particularly a ventrocaudal region called the ventral bodies, known also as the lateral accessory lobes. These lobes are invaded by the dendritic collaterals of many of the descending neurons leaving the brain for thoracic and abdominal ganglia.

The central complex is strictly a higher center that is distant from sensory inputs. Dye fills fail to demonstrate any sensory interneuron inputs to central complex neuropils nor do the antennal lobes or optic lobes provide direct connections to the central complex. The central complex has no direct connections with the mushroom bodies. Instead, various regions of the protocerebrum that are connected to the central complex are also connected to the mushroom bodies and to higher level sensory neuropils, such as the lobula of the optic lobes and antennal and vertical lobes of the deutocerebrum (see later).

As described in Nässel's 1993 review, the central complex is richly supplied by peptidergic neurons that originate from the pars intercerebralis. The pars also provides a wealth of peptidergic neurons whose axons leave the brain for the retrocerebral complex (the corpora allata and corpora cardiaca) via the corpora cardiaca nerve NCC1. Other neurosecretory cells sending axons out of the brain lie lateral and rostral with respect to the protocerebrum (NCC2) and in the lateral tritocerebrum (NCC3). An exquisitely detailed enhancer trap analysis of these systems has been published by Siegmund and Korbe. The central complex is implicated in the control of motor actions, although exactly what it controls is not yet known. Studies of motor-coordination-defective *Drosophila* show that certain behavioral mutants have midline lesions of their protocerebral bridge or the fan-shaped bodies. Roland Strauss, at the University of Würzburg, has shown that these mutant flies are incapable of adjusting step length during turning. A similar disruption across the midline occurs in nature: certain rowing Heteroptera, such as the water strider *Gerris,* have split protocerebral bridges, minute noduli, and reduced modules in their fan-shaped and ellipsoid bodies. In contrast, insects that show sophisticated asymmetric but highly coordinated limb actions, such as are employed in grooming, object manipulation, or cell construction, possess elaborately modular central complexes and complete protocerebral bridges.

Prominent connections between the fan-shaped and the ellipsoid bodies with the lateral accessory lobes of the protocerebrum are of functional interest. The lateral accessory lobes are visited by dendritic collaterals from many of the interganglionic descending interneurons that send axons from lateral protocerebral regions to neuropils of the thoracic and abdominal ganglia. One interpretation of this organization is that the central complex plays a role in gating outgoing information from the brain.

The central complex is richly supplied from protocerebral regions involved in sensory discrimination. This organization, with the elaborate arrangement of repeat units (modules and chiasmata) within and between the protocerebral bridge and midline neuropils, might suggest that the central complex assesses the context and occurrence of sensory stimuli around the animal and that this plays a crucial role in modifying descending information to motor circuits.

Mushroom Bodies

The mushroom bodies, discovered by Félix Dujardin in 1850, were the first brain centers to be recognized as distinct entities. Dujardin's suggestion that mushroom bodies supported intelligent actions was with reference to social insects, in which mushroom bodies are largest and most elaborate. Since his 1850 paper, the mushroom bodies have been considered to be centers crucial to learning and memory.

The mushroom bodies are paired lobed neuropils. Comparative studies by Strausfeld, Ito, and others have identified mushroom bodies in all groups of insects except the archaeognathan. In zygentoman and palaeopteran insects, mushroom bodies comprise two sets of lobes, one set extending medially toward the midline (medial lobes), the

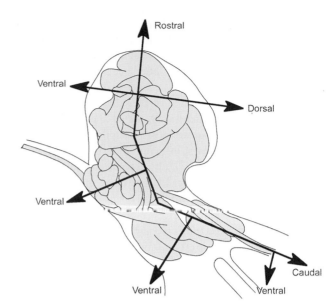

FIGURE 7 Neuraxis. Many descriptions in the literature rarely make the point that the brain's orientation is not that of the body. During postembryonic development, the brain undergoes morphogenic movements, tilting upward and back. This brings the dorsal surface of the brain to face caudally with respect to the body's axis. The front of the brain is its ventral side according to the neuraxis. The top part of the brain is rostral. Likewise, the antennal lobes are ventral, not frontal.

other extending ventrally, with respect to the neuraxis. However, because most brains are tilted upward (Fig. 7), these lobes can point forward or even upward. They are thus collectively referred to as the vertical lobes.

The mushroom body lobes comprise many thousands of approximately parallel-running processes. These originate from clusters of minute globuli cells situated dorsorostrally in the protocerebrum's cell body rind. In neopteran insects, these neurons have distal dendritic trees that contribute to rostral neuropils called the calyces. Each mushroom body has a pair of calyces, each of which is divided into two halves. A crucial study by Kei Ito and colleagues demonstrated that each half is generated by one of a quartet of embryonic neuroblasts. The four half-calyces are supplied by four lineages of globuli cells, all of which provide dendrites in the calyces and long axon-like processes in the lobes. These "intrinsic neurons" of the mushroom bodies are known as Kenyon cells, named after their discoverer. Lineage analysis of the *Drosophila* mushroom bodies has shown that each of the four neuroblasts generates the same sequence of Kenyon cells, certain types of which differentiate before others. Different types of Kenyon cells contribute to different and discrete subdivisions of the lobes. Farris's studies on the mushroom bodies of the cockroach *Periplaneta americana* and the worker honey bee *Apis mellifera* have shown that the sequence of Kenyon cell production and segregation to subdivisions is similar to that in *Drosophila* described by Lee *et al.* in 2001.

Observations of the cockroach and honey bee calyces show these neuropils as organized into nested zones, each of which is defined by the types of afferents supplying it. The most up-to-date study on the honey bee demonstrates that an outer region called the lip comprises three zones, each of which receives axon collaterals from neurons that project from glomeruli of the antennal lobes (see later) to regions of the lateral protocerebrum. A second region of the calyx, which comprises the collar, is further divided into discrete zones, each of which is defined by visual and other afferent endings, such as from gustatory neuropils of the subesophageal ganglion. However, in many other insect orders, the calyces receive sparse inputs, if any, from the visual system. A central region of the calyx called the basal ring is similarly divided into modality-specific zones.

Kenyon cells having dendrites in one of these zones send their axon-like processes into a specific stratum that extends all the way through the vertical and medial lobes. Each stratum thus represents a zone of the calyces. However, a special class of Kenyon cells that is generated earliest in development supplies axon-like processes to a separate division of the mushroom bodies, called the gamma lobe. Depending on the taxon, this lobe lies parallel to the medial (flies), vertical (honey bees), or both (cockroach) lobes. Important studies by Zars and Heisenberg on gene expression in different parts of the mushroom body of *Drosophila* have implicated the gamma lobe in supporting short-term memory.

One pervasive misconception is that the calyces are the "input region" of the mushroom bodies, whereas their lobes are their output regions. This view of the mushroom body does much to confuse and mislead theoretical considerations about how the mushroom bodies might work. Palaeopteran insects lack calyces supplied by sensory interneurons, yet their lobes both receive afferent endings from other protocerebral neuropils and provide efferents that extend back to protocerebral neuropils. In neopteran insects, the lobes likewise receive inputs and provide outputs, with the axon-like processes of Kenyon cells providing local circuits between them. However, in neopterans, Kenyon cells also supply calyces with dendrites that are visited by sensory interneurons. The role of the calyces is not fully understood. Possibly, afferents ending on Kenyon cell dendrites serve to modify the activity of local circuits in the lobes that are supplied by the processes of Kenyon cells, thereby providing sensory context dependence to computations that occur via Kenyon cell processes between inputs and outputs at the mushroom body lobes. It is also possible that inputs to the calyces provide persisting memory-like alterations of groups of Kenyon cells. Peptidergic and other modulatory neurons (e.g., octopaminergic, dopaminergic) associated with the mushroom bodies have been implicated in memory formation, and genetic disruption of vesicle recycling in a modulatory neuron of the *Drosophila* mushroom body shortens memory. It is still somewhat of a mystery why there are two sets of lobes, with most Kenyon cell processes dividing into each of them. However, as shown by Pascual and Préat, working at the CNRS in France, if the vertical lobes are absent, as in one type of *Drosophila* mutant, then long-term memory cannot be established. A role for the mushroom bodies in learning and memory has also been suggested by chemical ablation of the mushroom body neuroblasts, and a consequent lack of the mushroom body abolishes olfactory associative learning by the adult fly. However, such ablations also remove a set of local circuit neurons in the antennal lobes, complicating the interpretation of such experiments. A further complication in interpreting the mushroom body's role in memory acquisition is Dubnau's recent finding that synaptic transmission by mushroom body neurons is necessary only for memory retrieval and not for memory formation.

The Rostral Lateral Protocerebrum and Lateral Horn

The protocerebrum is composed of many discrete centers, most of which do not have obvious order and neat geometries, as in the mushroom bodies and central complex. Nevertheless, each protocerebral center is a unique entity and specific centers can be identified across different species. It is likely that studies of enhancer trap lines, as well as genetic labeling of clonally related neurons, developed by Liqun Luo and his colleagues, will in the near future reveal many new features of the cellular organization of the protocerebrum. But, so far, few studies have been done on these neuropils even though they together impart great complexity to the brain. This section focuses on just two neighboring regions, the lateral protocerebrum and lateral horn, which are now known to be second-order olfactory neuropils.

Antennal lobe projection neurons relay information from olfactory glomeruli to various areas of the brain, via three axon bundles called the inner, intermediate, and outer antennocerebral tracts. Axons of the inner antennocerebral tract provide axon collaterals to the mushroom body calyces. However, olfactory projection neurons providing input to the calyces do not terminate there but end in a region of the protocerebrum called the lateral horn and, caudally adjacent to it, the lateral protocerebrum. Axons of the intermediate and outer antennocerebral tracts also invade these neuropils, which therefore must be considered second-order olfactory processing centers of the brain. In honey bees, certain axons of the intermediate tract also target some neuropils that lie in front of and beneath the calyces as well as neuropils enwrapping the vertical lobes. Studies from Liqun Luo's laboratory at Stanford University have now shown that discrete fields of endings in the lateral horn and lateral protocerebral areas lying immediately caudal to it are supplied by specific groups of antennal lobe glomeruli, thus showing that the olfactory map that occurs among antennal lobe glomeruli is partially maintained within this lateral protocerebral area. With the exception of the calyces, neuropils targeted by antennal lobe projection neurons are second-order olfactory centers. These neuropils are not, however, unimodal olfactory centers as they also receive inputs from the optic lobes via large ascending fascicles.

The lateral horn and lateral protocerebrum give rise to systems of local interneurons as well as long-axoned interneurons, certain of which terminate in the mushroom body lobes. However, the relationship of the lateral protocerebrum with descending pathways is not yet known. A further area of ignorance is its relationship with the central complex.

OPTIC LOBES

Studies on the fly visual system dominate the literature on insect vision, and descriptions of visual regions have been mainly from dipteran insects, whose structure and physiology have been described in great detail in numerous papers.

Retinotopic Organization

The optic lobes of palaeopteran and neopteran insects consist of three retinotopic neuropils. These are the lamina, medulla, and lobula complex. In certain orders of insects (e.g., Diptera Lepidoptera, Coleoptera) the lobula complex is divided into two separate neuropils: a lenticular lobula that is mainly composed of columnar neurons and a tectum-like lobula plate that is hallmarked by wide-field tangential neurons. However, in insects with an undivided lobula, deeper layers comprise tangential neurons that probably have the same functions as tangential neurons in the lobula plate. Connections between the medulla and the lobula plate in Diptera are homologous to connections between the medulla and the deep lobula layers in honey bees. The lobula plate or its equivalent supports achromatic motion vision, whereas the lobula is thought to support object and color vision. The lobula plate sends axons to dorsal neuropil of the lateral and medial protocerebrum from which descending neurons supply neuropils belonging to the neck and flight motor systems. The lobula supplies bundles of axons to discrete glomerular-like neuropils of the ventrolateral protocerebrum, certain of which retain retinotopic organization. These optic glomeruli (also called optic foci) are invaded by processes of local interneurons and relay neurons.

The Lamina

The lamina is the first neuropil of the optic lobes and the best known with respect to its cellular organization, synaptology, and development. It consists of relatively few types of neurons whose relationships achieve surprising complexity. These have been reviewed by several authors, including Nässel and Strausfeld and Meinertzhagen and Sorra. The following summary focuses on the fly's visual system because of its preeminence in vision research.

Each ommatidium of the compound eye contains eight (in honey bees nine) receptor cells. As summarized in Hardie's 1986 review, the axons (called short visual fibers) of six blue-green sensitive photoreceptors (termed R1–R6) end in the lamina, whereas the axons (long visual fibers, termed R7 and R8) of the other two photoreceptors (blue and UV sensitive) terminate in the medulla. In insects equipped with apposition eyes or neural superposition eyes, a set of six short visual fibers terminate at each columnar subunit of the lamina, called an optic cartridge. The six endings belong to six photoreceptors that share the same optical alignment and thus "look" at the same restricted area of visual space even though, in neural superposition eyes, the optically coherent photoreceptors are distributed among six ommatidia, a discovery made by Kuno Kirschfeld in 1967 and elegantly explained from a developmental view point by Meinertzhagen in his 2000 review article. Nicholas Franceschini has termed an optically coherent set of receptors a visual sampling unit or VSU.

Each VSU is associated with several classes of interneurons, the most prominent of which are cells known as large monopolar cells or LMCs. In Diptera, LMCs include the pair of radial monopolar cells (also called L1 and L2) that are postsynaptic to all six terminals of a VSU. A third monopolar cell (the brush or unilateral monopolar cells, also called L3) is postsynaptic to all six receptors of a VSU but has about two-thirds fewer dendrites than either L1 or L2. This triplet of neurons sends axons to the medulla, alongside the axons of the R7 and R8 receptor neurons that originate from the optically relevant ommatidium. In insects with superposition eyes, the lamina is not obviously divided into cartridges and the dendrites of second-order neurons (monopolar cells) extend across many photoreceptor terminals.

In addition to the L1–L3 monopolar cells, the lamina contains two types of amacrine neurons whose processes provide local interactions between receptor terminals of several cartridges and the dendrites of other efferent neurons that leave each cartridge. These are the types L4 and L5 monopolar cells and the basket cell, referred to as the T1 efferent neuron. Amacrine cells, which are local interneurons that lack axons and provide local circuits, are postsynaptic to receptors from many VSUs and are presynaptic onto the dendritic trees of T1 and dendrites of L4 neurons. Each L4 neuron contributes axon collaterals to a rectilinear network of connections beneath the lamina. These collaterals are presynaptic to the axons of L1 and L2 and reciprocally pre- and postsynaptic to the L4 neuron collaterals of neighboring cartridges. A fifth species of efferent neuron, called the midget monopolar cell, or L5, has one or two minute tufts of dendrites that are visited by a second species of amacrine cell. A comprehensive review by Strausfeld and Nässel demonstrates that similar types of neurons have been identified in crustaceans and in honey bees.

Four distinctive types of centrifugal cells visit the lamina. Two are associated with each cartridge. One is the type C2 centrifugal cell, which is presynaptic onto the L1 and L2 monopolars at a level above their dendrites. The second is the type C3 centrifugal cell, which is presynaptic to the L1 and L2 monopolar cells at the level of their dendrites. Both C2 and C3 are GABAergic and both have dendrites at various levels in the outer and inner medulla. Two wide-field centrifugal cells, called the types 1 and 2 lamina tangential cells, have bistratified and concentric dendritic fields in the medulla and send axons back out to the lamina, where their terminals provide isomorphic plexi of presynaptic endings.

The Optic Chiasmata

Axons originating from the front of the lamina end in the back of the medulla; axons from the back of the lamina end in the front of the medulla, so reversing the horizontal order of vertical rows of VSUs. The order is rereversed in the lobula by a second chiasma beneath the medulla.

THE MEDULLA   The arrangement of retinotopic columns in the medulla is defined by pairs of long visual fiber endings, each pair accompanied by the sextet of endings of neurons leaving the optically corresponding optic cartridge (L1–L5, T1). These endings terminate at specific levels of the medulla where they coincide with the processes of amacrine cells, the dendrites of relay neurons, and dendrites and terminals of tangential neurons. The medulla is immensely complicated. Each column may contain as many as 40 relay neurons and many strata of amacrine cells intersect these. Thus, the medulla has one of the highest densities of nerve cells outside the mushroom bodies.

A broad layer of incoming and outgoing axons belonging to tangential neurons separates the outer two-thirds of the medulla from its inner third. Tangential cells have dendritic trees or terminal fields that extend across many retinotopic columns. Axons of centripetal tangential cells project centrally via the posterior optic tract, to reach ipsi- and or contralateral regions of the dorsal protocerebrum. Some tangentials connect the medullae of the two lobes. Tangential endings in the medulla derive from dendrites within the medial protocerebrum. These neurons carry processed information about motion stimuli back out to peripheral layers of the visual system.

Retinotopic columns may each consist of as many as 40 different morphological types of axonal neurons. These neurons have dendrites at characteristic levels, spreading through defined fields of neighboring columns. Each morphological cell type sends its axon to a characteristic deeper level of the optic lobe. The class of neurons called transmedullary cells (Tm cells) sends its axons to various depths of the lobula. A subset of transmedullary cells, whose dendrites are restricted to within a retinotopic column, supplies a special sheet of synaptic neuropil over the surface of the lobula where they end among quartets of bush-like dendritic trees called T5 cells. These neurons were shown by J. K. Douglass to be the first in the system to exhibit directional- and orientation-selective responses to moving visual stimuli. Their axon terminals segregate to four levels in the lobula plate where they end on layered systems of tangential neurons. Pioneering studies by Klaus Hausen, at

the Max Planck Institute in Tübingen, demonstrated that these large-field neurons respond selectively to wide-field directional motion across the retina, relaying this information to the midbrain and to the contralateral optic lobes so that signals from both eyes can be integrated.

A second class of transmedullary cells (called intrinsic transmedullary neurons) serves to link the outer layer of the medulla with its inner layer. A third class of transmedullary cells, called Tm Y cells, consists of neurons whose axons branch, one tributary reaching the lobula plate and the other reaching the lobula. Retinotopic neurons also originate from the inner layer of the medulla. These must derive their inputs from other transmedullary cells because their dendrites lie beneath afferents from the lamina. Neurons from the inner medulla send axons to the lobula or provide bifurcating axons to the lobula plate and lobula.

The lobula contains ensembles of tree-like neurons organized retinotopically. The spacing of their axons coarsens the original retinotopic mosaic so that an oval ensemble of retinotopic inputs from the medulla visits the dendritic tree of a single lobula neuron. The size of the dendritic fields of these columnar neurons varies, however, depending on the cell type. The smallest lobula neuron has a dendritic field equivalent to an approximately oval array of nine VSUs. However, it cannot be assumed that such fields are the physiologically receptive fields because the functional organization among their inputs is not known.

Columnar neurons in the lobula are likely to be tuned to highly specific visual features although few recordings have been made from medulla neurons supplying the lobula. Those that have been recorded suggest that the lobula receives information about orientation but not about wide-field motion. However, the presence in the lobulas of male flies of sex-specific neurons that respond to movement of small objects in the visual field must imply that a class of directionally selective neurons from the medulla supplies at least a part if not all of the lobula.

Each ensemble of identically shaped columnar neurons provides a coherent bundle of axons that targets a circum-scribed region of neuropil in the lateral protocerebrum. Such neuropils are called optic foci and are reminiscent of and may be functionally equivalent to olfactory glomeruli of the antennal lobe.

OUTPUTS FROM THE OPTIC LOBES   The now classic studies by Walter Gehring and his colleagues on the ectopic expression of compound eyes in *Drosophila* by genes controlling eye formation have demonstrated that super-numerary eyes are formed on limbs. If the ancestral origin of the compound eye was a limb, it would follow that the arrangements of sensory neurons leading centrally from the compound eye neuropils (comprising the optic lobe) should reflect arrangements of sensory neurons and interneurons that, in other ganglia, serve the appendages. Cell tracer studies indeed suggest that central projections from the optic lobes are reminiscent of sensory-interneuron arrangements in thoracic ganglia. Afferents from the optic lobes do not appear to terminate directly onto descending neurons, as was once assumed. Rather, optic lobe outputs end on systems of local interneurons and some appear to map the retinotopic mosaic into optic glomeruli, suggesting further high-order visual reconstruction. Certain local interneurons connect optic glomeruli. Others have long axons that connect glomeruli on both sides of the brain. Yet others project anteriorly, via a thick fascicle of axons, to end rostrally in the lateral protocerebrum where they meet the terminals of olfactory projection neurons.

Optic glomeruli are also associated with neurons that reach the fan-shaped and ellipsoid bodies of the central complex or extend to the mushroom body lobes. Thus, optic lobes supply protocerebral centers from which interneurons extend to higher centers. Descending neurons mainly receive their optic lobe inputs via intermediate local interneurons. Thus, organization in the protocerebrum between sensory inputs (that is, lobula and lobula plate outputs) and interneurons is reminiscent of sensory-to-interneuron arrangements in a segmental ganglion.

Exceptions to this general arrangement include the giant vertical motion-sensitive neurons of the lobula plate, which, with neurons from the ocelli, establish mixed electrical and chemical synapses onto descending neurons that are involved in the stabilization of roll and pitch during flight and the control of visually induced head movements.

ACCESSORY MEDULLA AND CIRCADIAN RHYTHM
The optic lobes also support pathways involved in circadian rhythms. The most important of these, described from *Drosophila* and the locust by Helfrich-Forster and others, involve systems of neurosecretory pacemaker cells associated with a small satellite neuropil in the optic lobes, called the accessory medulla.

DESCENDING OUTPUTS FROM THE BRAIN
Dye fills into the ventral nerve cord of flies reveal large numbers (>200) of cell bodies on each side of the brain. These belong to descending neurons that carry information from the brain to thoracic ganglia circuits that control flight, walking, and other motor actions. Descending neurons that have their dendrites in dorsal neuropils of protocerebrum and receive inputs from the lobula plate terminate dorsally in thoracic ganglia where they contribute to the visual and mechanosensory stabilization of flight. Descending neurons with dendrites in more ventral protocerebral neuropils terminate in ventral thoracic and abdominal regions.

Electrophysiological recordings from locusts, dragonflies, and flies have all shown that descending neurons are multimodal, carrying integrated information about visual, olfactory, mechanosensory, and acoustic stimuli. Descending neurons are activated by correlative information from different modalities. Thus, descending neurons involved in controlling the direction of flight, and which respond selectively to panoramic movement around the vertical axis of the body (yaw motion), also relay information about the corresponding displacement of head hairs that occurs when the body undergoes a yaw displacement.

Among descending neurons are systems of axons that provide extremely fast motor actions in response to defined visual stimuli. These comprise the class of "escape circuits," the best known of which is the lateral giant motion detector and descending contralateral motion detector system of the locust and the Col A–giant descending neuron system in flies. In both locusts and flies optic lobe neurons (LGMD, Col A) provide mixed chemical and electrical synapses onto large axon diameter descending neurons (DCMD, GDN). In flies the paired GDN are electrically coupled and are electrically contiguous with the tergotrochanteral motor neurons that provide sudden midleg extension. At the same time, the direct (power) muscles of the wings are activated by electrically coupled local interneurons relaying signals from the GDN terminal to axons of motor neurons supplying the longitudinal flight muscles. A comparable connection between the DCMD and the hindleg extensor muscles is found in locusts.

ASCENDING SUPPLY TO THE BRAIN
Sensory afferents supplying segmental ganglia distribute outputs to interganglionic interneurons. Many of these interneurons extend only locally, between neighboring or next-to-neighbor ganglia, and serve functions in regulating leg movements and posture. However, a large number of ascending neurons have axons that ascend through ganglia to reach the brain via ventrally disposed tracts of axons. Dye filling these axons demonstrates their terminals in ventrolateral and ventromedial protocerebral neuropils and in neuropils of the deutocerebrum, tritocerebrum, and subesophageal ganglion. Ascending axons do not, however, terminate in any of the mushroom body neuropils nor in neuropils of the central complex.

Functional studies of ascending pathways, exemplified by the recent studies of Nebeling, have mainly focused on ascending acoustic interneurons in crickets, which terminate in specific caudolateral and caudomedial protocerebral neuropils. However, the distribution of terminals in many protocerebral neuropils, including those receiving inputs from the optic lobes and antennal lobes, adds weight to the idea that a large volume of the protocerebrum is involved in multimodal integration. That such areas provide inputs to, and receive outputs from, the mushroom bodies adds credence to these paired centers being higher integration neuropils. Likewise, the relationship of the fan-shaped and ellipsoid bodies with protocerebral neuropils also suggests their crucial role as a higher integrator for the control of motor actions.

THE DEUTOCEREBRUM

The second preoral neuromere is called the deutocerebrum. It consists of sensory and motor neuropils and is constructed along the ground pattern typical of postoral ganglia.

Ventrally its neuropils comprise two paired sensory centers: the antennal lobe and, caudal to it, the vertical lobe. Three recent reviews by Vosshall, Hildebrand, and Hansson provide useful summaries of how olfactory receptors wire into the antennal lobes, how the lobes themselves are structured, and how these structures relate to the rest of the brain. Antennal lobe neuropils consist of discrete islets, called glomeruli. Most of these receive mainly olfactory receptor endings from the antenna's funiculus, although certain glomeruli receive inputs from the maxillary palps. Studies on *Drosophila* and the moth *Manduca sexta,* which are reviewed, have demonstrated that as a rule glomeruli have unique identities and positions in the antennal lobe.

The vertical lobes receive mainly mechanosensory terminals from the scapus and pedicellus. Some glomeruli situated caudally in the antennal lobe, bordering the dorsal lobe, receive inputs from antennal thermoreceptors, receptors responding to water vapor, and some mechanoreceptors. Motor neurons controlling antennal musculature originate lateral to and dorsal to these sensory regions. However, there is little information about the arrangements of interneurons and relay neurons supplying antennal lobe motor neurons.

Glomerular antennal lobes appear to be typical of neopteran insects. The antennae of primary apterygotes and palaeopterans supply columnar and layered neuropils of the vertical lobes, as do mechanosensory axons from the first two antennal segments (scapus and pedicellus) in neopterans.

Cellular arrangements in the antennal lobes are reminiscent of arrangements in malacostracan crustaceans and in vertebrates. Whether these similarities are a consequence of convergent evolution is debatable. However, in both phyla olfactory receptor neurons tuned to a specific odor molecule converge to the same address (glomerulus) in the antennal lobe. These addresses are then represented as a coarser map in the neuropils of the lateral protocerebrum by the axons of projection neurons originating in glomeruli. Antennal lobes also integrate olfactory information by virtue of complex connections provided by local inhibitory local interneurons. Sexual dimorphism also occurs in the antennal lobes, particularly in the Lepidoptera, in which receptors encoding components of the female pheromone blend send their axons to discrete glomeruli of the male-specific macroglomerular complex.

THE TRITOCEREBRUM

The tritocerebrum is the third and structurally the most discrete of the three preoral neuromeres, with sensory motor connections with the third metamere of the head. An analysis using expression of the segmentation border gene *engrailed* has demonstrated the appendage-type identity of the labrum, which supplies a major input into the tritocerebral ganglion, thus establishing its sensory supply and metameric relationships. The tritocerebrum also supplies motor neurons to labral muscles. The tritocerebrum receives a substantial supply of terminals from interganglionic interneurons with dendrites in the anterior medial protocerebrum (according to the neuraxis), which itself receives inputs from the median bundle originating in the subesophageal ganglion with some additional elements recruited from the thoracic ganglia. The tritocerebrum gives rise to descending neurons to the thoracic and abdominal ganglia, as does the deutocerebrum. However, even though Rajashekhar and Singh have described its general architecture and relationships with the protocerebrum, relatively little is known of the tritocerebrum's relationships to other brain regions or about its intrinsic connections. In all neopteran insects it is the protocerebrum that shows the greatest elaboration.

THE POSTORAL BRAIN (SUBESOPHAGEAL GANGLION)

Various functions have been ascribed to the subesophageal ganglion, including arousal prior to motor actions and sensory convergence from the brain. Generally, the three neuromeres of the subesophageal ganglion relate to the metameric identity of the mouthparts. Edgecomb and Murdock describe from flies that the labial neuromere of the subesophageal ganglion receives sensory axons from the dorsal cibarial organ, labellar sensilla, and labral sense organs and possibly some intersegmental inputs arrive from the tarsi. Stocker and Schorderet have supplied evidence from *Drosophila* that mechanosensory and gustatory chemo- (taste) receptors segregate out in the labral neuromere into discrete modality-specific zones. In honey bees, inputs from the mandibles invade the mandibular neuromere, which has been suggested to be absent in flies. However, it is unlikely that an entire neuromere has been eliminated and anatomical evidence for it is indeed present. In 1992 Shanbhag and Singh made the attractive suggestion that taste receptors segregate into chemospecific zones, because horseradish peroxidase uptake by the tips of functional species of receptors identified seven arborization areas in the subesophageal ganglion. Shanbhag and Singh suggested that these areas correspond to seven types of gustatory sensilla. These findings have not been contradicted and in larvae such receptor-specific zones appear to be substantiated by genetic markers of specific chemosensory axons to target neuropils. What is uncontroversial is that, at any neuromere of the subesophageal ganglion, neuropils are divided into a number of discrete synaptic regions. Some of these clearly belong to interneuron–motor neuron assemblages. In flies, local interneurons responding to sucrose reflect the discrete partitioning of lateral subesophageal neuropils.

See Also the Following Article
Eyes and Vision

Further Reading
Bacon, J. P., and Strausfeld, N. J. (1986). The dipteran 'giant fibre' pathway: Neurons and signals. *J. Comp. Physiol. (A)* **158,** 529–548.
Boyan, G. S., and Williams, J. L. (2001). A single cell analysis of *engrailed* expression in the early embryonic brain of the grasshopper *Schistocerca*

gregaria: Ontogeny and identity of the secondary headspot cells. *Arthropod Struct. Dev.* **30**, 207–218.

Burrows, M., and Newland, P. L. (1993). Correlations between the receptive fields of locust interneurons, their dendritic morphology, and the central projections of mechanosensory neurons. *J. Comp. Neurol.* **329**, 412–426.

Douglass, J. K., and Strausfeld, N. J. (1998). Functionally and anatomically segregated visual pathways in the lobula complex of a calliphorid fly. *J. Comp. Neurol.* **396**, 84–104.

Douglass, J. K., and Strausfeld, N. J. (1996). Visual motion-detection circuits in flies: Parallel direction- and non-direction-sensitive pathways between the medulla and lobula plate. *J. Neurosci.* **16**, 4551–4562.

Dubnau, J., Grady, L., Kitamoto, T, and Tully, T. (2001). Disruption of neurotransmission in *Drosophila* mushroom body blocks retrieval but not acquisition of memory. *Nature* **411**, 476–480.

Edgecomb, R. S., and Murdock, L. L. (1992). Central projections of axons from taste hairs on the labellum and tarsi of the blowfly, *Phormia regina* Meigen. *J. Comp. Neurol.* **315**, 431–444.

Farris, S. M., and Strausfeld, N. J. (2001). Development of laminar organization in the mushroom bodies of the cockroach: Kenyon cell proliferation, outgrowth, and maturation. *J. Comp. Neurol.* **439**, 331–351.

Franceschini, N. (1975). Sampling of the visual environment by the compound eye of the fly: Fundamentals and applications. *In* "Photoreceptor Optics" (A. W. Snyder and R. Menzel, eds.), pp. 98–125. Springer-Verlag, Berlin.

Gilbert, C., and Strausfeld, N. J. (1991). The functional organization of male-specific visual neurons in flies. *J. Comp. Physiol. [A]* **169**, 395–411.

Halder, G., Callaerts, P., and Gehring, W. J. (1995). Induction of ectopic eyes by targeted expression of the eyeless gene in *Drosophila*. *Science* **267**, 1788–1792.

Hansson, B. S. (2002). A bug's smell—Research into insect olfaction. *Trends Neurosci.* **25**, 270–274.

Hardie, R. C. (1986). The photoreceptor array of the dipteran retina. *Trends Neurosci.* **9**, 419–423.

Hausen, K. (1984). The lobula complex of the fly: Structure, function, and significance in visual behavior. *In* "Photoreception and Vision in Invertebrates" (M. A. Ali, ed.), pp. 523–599. Plenum Press, New York.

Heisenberg, M. (1983). What do the mushroom bodies do for the insect brain? An introduction. *Learn. Memory* **5**, 1–10.

Helfrich-Forster, C., Stengl, M., and Homberg, U. (1998). Organization of the circadian system in insects. *Chronobiol. Int.* **15**, 567–594.

Hildebrand, J. G. (1996). Olfactory control of behavior in moths: Central processing of odor information and the functional significance of olfactory glomeruli. *J. Comp. Physiol. [A].* **178**, 5–19.

Ito, K., Awano, W., Suzuki, K., Hiromi, Y., and Yamamoto, D. (1997). The *Drosophila* mushroom body is a quadruple structure of clonal units each of which contains a virtually identical set of neurones and glial cells. *Development* **124**, 761–771.

Lee, T., Lee, A., and Luo, L. (1999). Development of the *Drosophila* mushroom bodies: Sequential generation of three distinct types of neurons from a neuroblast. *Development* **126**, 4065–4076.

Ludwig, P., Williams, L., and Boyan, G. (2001). The pars intercerebralis of the locust brain: A developmental and comparative study. *Microsc. Res. Tech.* **56**, 174–188.

Marin, E. C., Jefferis, G. S., Komiyama, T., Zhu. H., and Luo, L. (2002). Representation of the glomerular olfactory map in the *Drosophila* brain. *Cell* **109**, 243–255.

Meinertzhagen, I. A. (2000). Wiring the fly's eye. *Neuron* **28**, 310–313.

Meinertzhagen, I. A., and Hanson, T. E. (1993). The development of the optic lobe. *In* "The Development of *Drosophila melanogaster*" (M. Bate and A. Martinez-Arias, eds.), pp. 1363–1491. Cold Spring Harbor Laboratory Press, Cold Spring Harbor, NY.

Meinertzhagen, I. A., and Sorra, K. E. (2001). Synaptic organization in the fly's optic lamina: Few cells, many synapses and divergent microcircuits. *Prog. Brain Res.* **131**, 53–69.

Mitchell, B. K., Itagaki, H., and Rivet, M. P. (1999). Peripheral and central structures involved in insect gustation. *Microsc. Res. Tech.* **47**, 401–415.

Murphey, R. K. Possidente, G., Pollack, D., and Merritt, D. J. (1989). Modality-specific axonal projections in the CNS of the flies *Phormia* and *Drosophila*. *J. Comp. Neurol.* **290**, 185–200.

Nässel, D. R. (1993). Neuropeptides in the insect brain: A review. *Cell Tissue Res.* **273**, 1–29.

Nassif, C., Noveen, A., and Hartenstein, V. (1998). Embryonic development of the *Drosophila* brain. I. Pattern of pioneer tracts. *J. Comp. Neurol.* **402**, 10–31.

Nebeling, B. (2000). Morphology and physiology of auditory and vibratory ascending interneurones in bushcrickets. *J. Exp. Zool.* **286**, 219–230.

Pascual, A., and Préat, T. (2001). Localization of long-term memory within the *Drosophila* mushroom body. *Science* **294**, 1115–1117.

Rajashekhar, K. P., and Singh, R. N. (1994). Neuroarchitecture of the tritocerebrum of *Drosophila melanogaster*. *J. Comp. Neurol.* **349**, 633–345.

Rowell, C. H., and Reichert, H. (1986). Three descending interneurons reporting deviation from course in the locust. II. Physiology. *J. Comp. Physiol. [A]* **158**, 775–794.

Scott, K., Brady, R., Jr., Cravchik, A., Morozov, P., Rzhetsky, A., Zuker, C., and Axel, R. (2001). A chemosensory gene family encoding candidate gustatory and olfactory receptors in *Drosophila*. *Cell* **104**, 661–673.

Siegmund, T., and Korge, G. (2001). Innervation of the ring gland of *Drosophila melanogaster*. *J. Comp. Neurol.* **431**, 481–491.

Shanbhag, S. R., Singh, K, and Singh, R. N. (1995). Fine structure and primary sensory projections of sensilla located in the sacculus of the antenna of *Drosophila melanogaster*. *Cell Tissue Res.* **282**, 237–249.

Stocker, R. F. (1994). The organization of the chemosensory system of *Drosophila melanogaster:* A review. *Cell Tissue Res.* **275**, 3–26.

Strausfeld, N. J. (1976). "Atlas of an Insect Brain." Springer-Verlag, Berlin.

Strausfeld, N. J. (1999). A brain region in insects that supervises walking. *Prog. Brain. Res.* **123**, 274–284.

Strausfeld, N. J. (2002). Organization of the honey bee mushroom body: Representation of the calyx within the vertical and gamma lobes. *J. Comp. Neurol.* **450**, 4–33.

Strausfeld, N. J., and Nässel, D. R. (1980). Neuroarchitectures serving compound eyes of Crustacea and insects. *In* "Handbook of Sensory Physiology VII/68" (H. Autrum, ed.), pp. 1–132. Springer-Verlag, Berlin.

Vosshall, L. B. (2001). The molecular logic of olfaction in *Drosophila*. *Chem. Senses.* **26**, 207–213.

Walthall, W. W., and Murphey, R. K. (1986). Positional information, compartments and the cercal system of crickets. *Dev. Biol.* **113**, 182–200.

Zars, T., Fischer, M., Schulz, R., and Heisenberg, M. (2000). Localization of a short-term memory in *Drosophila*. *Science* **288**, 672–675.

Bristletail

see *Archaeognatha*

Bubonic Plague

Irwin W. Sherman
University of California, Riverside

B ubonic plague, a devastating bacterial disease most commonly transmitted by fleas, has produced profound changes in human societies throughout history.

THE DISEASE IN HUMANS

During the last 2000 years, three great plague pandemics, including the Black Death of the 14th century, have resulted in social and economic upheavals that are unmatched by armed conflicts or any other infectious disease. Plague, caused by the rod-shaped, gram-negative bacterium *Yersinia pestis* (*Pasteurella pestis* was the name used before 1970), is a zoonotic infection, transmitted by any one of several species of fleas, that predominantly affects small mammals such as rodents; humans actually are accidental hosts.

At present, most human cases of the plague are of the bubonic form, which results from the bite of a flea, usually the common rat flea *Xenopsylla cheopis,* that has fed on an infected rodent. The bacteria spread to the lymph nodes (armpits and neck but frequently the area of the groin) that drain the site of the bite, and these swollen and tender lymph nodes give the classic sign of bubonic plague, the bubo (from the Greek word *boubon* meaning groin). Three days after the buboes appear, there is a high fever, the infected individual becomes delirious, and hemorrhages in the skin result in black splotches. Some contend that these dark spots on the skin gave the disease the name Black Death, whereas others believe "black" is simply a mistranslation of "pestis atra" meaning not black, but a dark or sinister disease.

The buboes continue to enlarge, sometimes reaching the size of a hen's egg, and when these buboes burst there is agonizing pain. Death can come 2 to 4 days after the onset of symptoms. Sometimes, however, the bacteria enter the bloodstream. This second form of the disease, which may occur without the development of buboes, is called septicemic plague. Septicemic plague is characterized by fever, chills, headache, malaise, massive hemorrhaging, and death. Septicemic plague has a higher mortality than bubonic plague.

In addition, the bacteria may move via the bloodstream to the alveolar spaces of the lungs, leading to a suppurating pneumonia or pneumonic plague. Pneumonic plague, the only form of the disease that allows for human-to-human transmission, is characterized by a watery and sometimes bloody sputum containing live bacteria. Coughing and spitting produce airborne droplets laden with the highly infectious bacteria, and by inhalation others may become infected. Pneumonic plague is the rapidly fatal form of the disease, and death can occur within 24 h of exposure. It is likely that this form of transmission produced the devastating Black Death. The nursery rhyme "Ring around the rosies, a pocket full of posies, Achoo! Achoo! We all fall down" refers to plague in 17th-century England: the rosies are the initial pink body rash, posies the perfumed bunches of flowers used to ward off the stench of death, "achoo" is the coughing and sneezing, and death is signified by "we all fall down."

Y. pestis is one of the most pathogenic bacteria: the lethal dose that will kill 50% of exposed mice is only a single bacterium that is injected intravenously. Typically, *Y. pestis* is spread from rodent to rodent by flea bites, but it can also survive for a few days in a decaying corpse and can persist for years in a frozen body.

THE DISEASE IN THE FLEA

The disease in fleas also has a distinctive pattern. Small mammals, such as urban and sylvatic (or wood) rats, as well as squirrels, prairie dogs, rabbits, voles, coyotes, and domestic cats, are the principal hosts for *Y. pestis*. More than 80 different species of fleas are involved as plague vectors. Fleas are bloodsucking insects, and when a flea bites a plague-infected host (at the bacteremic/septicemic stage) it ingests the rod-shaped bacteria; these multiply in the blood clot in the proventriculus (foregut) of the flea. This bacteria-laden clot obstructs the flea's bloodsucking apparatus and, as a consequence, the flea is unable to pump blood into the midgut, where normally it would be digested. The flea becomes hungrier and in this ravenous state bites the host repeatedly; with each bite, it regurgitates plague bacteria into the wound. In this way, infection is initiated. *Y. pestis* can also be pathogenic for the flea, and fleas with their foregut blocked rapidly starve to death. If the mammalian host dies, its body cools down, and fleas respond by moving off the corpse to seek another live warm-blooded host. However, if there is an extensive die-off of rodents, the fleas move on to less preferred hosts such as humans, and so an epidemic may begin.

HISTORICAL

Plague outbreaks occurred prior to the current era (i.e., 2000 years ago), but the numbers affected and the societal impact of the events remain unrecorded. During the current era, however, there have been three well-documented plague pandemics. The first, the plague of Justinian, arrived in 542 and raged intermittently until 750. It came to the Mediterranean region from an original focus in northeastern India or central Africa and was spread by infected rats hitchhiking on ships. It is estimated that a million people died. In Constantinople, the capital of the Roman Empire in the east, plague contributed to Justinian's failure to restore imperial unity because of a diminution in resources, which in turn prevented Roman and Persian forces from offering more than token resistance to the Muslim armies that swarmed out of Arabia in 634.

In the year 1346, the second pandemic began, and by the time it disappeared in 1352 the population of Europe and the Middle East had been reduced from 100 million to 80 million people. It is estimated that in cities such as Siena, Marseilles, and London, at the height of the pandemic, approximately 1500 people died each day from the plague. This devastating phenomenon, known as the Black Death, the Great Dying, or the Great Pestilence, put an end to the rise in the human population that had begun in 5000 B.C., and it took more than 150 years before the population returned to its former size.

Providing the source of the second pandemic were bacteria from the first pandemic that had moved eastward and remained endemic for seven centuries in the highly susceptible black rats *(Rattus rattus)* of the Gobi Desert. Plague-infected rats and their fleas moved westward along the Silk Road, the caravan routes between Asia and the Mediterranean; plague traveled from central Asia around the Caspian Sea to the Crimea. There the rats and their fleas boarded ships and moved from port to port and country to country, spreading plague to the human populations living in filthy rat-infested cities. Indeed, the story of the Pied Piper of Hamlin may have had its roots in the plague-ridden cities of Germany.

By 1347, there were plague outbreaks in Kaffa, Constantinople, and Genoa. By 1348 it had spread via North Africa to Spain, and it was also present in France, Germany, Switzerland, and Great Britain. In 1349 a ship from London carrying its crew, wool cargo, and infected rats landed in Bergen, Norway. In this way plague came to Scandinavia. In 1351, it was in Poland and when, in 1352, it reached Russia, the plague had completed its circuit.

Because no one in medieval times knew that microbes cause infectious diseases, any public heath measures were crude and generally ineffectual: ships were restricted in their entry into ports and sailors had to remain on board for 40 days while their vessels were tied up at the dock, a practice that gave rise to the term quarantine (from *quarant,* meaning 40 in French). But the disease continued unabated because flea-bearing rats left the ships by means of docking lines. Cordon sanitaires (i.e., quarantine zones) may have had some effect, but oftentimes infected individuals were shut up in their homes with the uninfected members of the family and the flea-infested rats, conditions that actually led to higher mortality. More effective measures included the burning of clothing and bedding, and the burying of the dead as quickly as possible.

The public, unable to identify the real source of the plague, used Jews, prostitutes, the poor, and foreigners as scapegoats. The Black Death led to societal and religious changes: feudal institutions began to break down; the laboring class became more mobile; merchants and craftsmen became more powerful; and guild structures were strengthened. There was also a decline in papal authority, and people lost faith in a Catholic Church that was powerless to stem the tide of death. The horrors of the plague during this time are depicted in Pieter Brueghel's 1562 painting *Triumph of Death* and graphically described in the introduction to Giovanni Boccaccio's classic collection of short stories, the *Decameron.* "Plague doctors" who ministered to the dying wore special costumes depicted in drawings and engravings (Fig. 1), as seen in popular movies such as *The Seventh Seal,* directed by Ingmar Bergman.

Though the Black Death was undoubtedly the most dramatic outbreak of plague ever visited upon Europe, it did not disappear altogether after 1352. Between 1347 and

FIGURE 1 During the plague physicians wore protective clothing in an attempt to avoid acquiring the disease from their patients. The beaklike mask was supplied with aromatic substances and perfumes to ward off the stench of death, the stick was used to touch the afflicted. To prevent the disease vapors from entering the body of the "plague doctors," a hat was worn, as well as a coat impregnated with a waxy material. (Illustration from *Der Pestarzt Dr. François Chicoyneau.* © Germanisches National Museum, Munich. Reproduced with permission.)

1722, plague epidemics struck Europe at infrequent intervals and occurred without the introduction from caravans from Asia. In England, the epidemics occurred at 2- to 5-year intervals between 1361 and 1480. In 1656–1657, 60% of the population of Genoa died, half of Milan in 1630, and 30% in Marseilles in 1720. In the Great Plague of 1665, which was described in the diary of Samuel Pepys (and fictionalized in Daniel Defoe's *Journal of the Plague Year*), at least 68,000 Londoners died.

The third and current pandemic began in the 1860s in the war-torn Yunnan region of China. Troop movements from the war in that area allowed the disease to spread to the southern coast of China. Plague-infected rodents, now assisted in their travels by modern steamships and railways, quickly spread the disease to the rest of the world. By 1894, plague had arrived in Hong Kong, and there Alexander Yersin (1863–1943) and Shibasaburo Kitasato (1852–1931), by taking material from buboes, independently discovered the occurrence of the bacillus in humans. Yersin also isolated the same bacterium from dead rats, thus demonstrating the importance of these rodents in transmission. Four years later, during the phase of the epidemic that swept over India, Paul-

Louis Simond and Masanori Ogata independently determined that the flea was the vector of plague.

THE BACTERIUM, *Y. PESTIS*

Y. pestis has been subdivided into three phenotypic biovarieties—Antiqua, Medievalis, and Orientalis. Based on epidemiological and historical records, it has been hypothesized that Antiqua, presently resident in Africa, is descended from bacteria that caused the first pandemic, whereas Medievalis, resident in central Asia, is descended from bacteria that caused the second pandemic; those of the third pandemic, and currently widespread, are all Orientalis. It is believed that *Y. pestis* probably evolved during the last 1500 to 2000 years because of changes in social and economic factors that were themselves the result of a dramatic increase in the size of the human population, which was coincident with the development of agriculture.

This increase in food supply for humans allowed rodent populations to expand as well. Increased numbers of rodents coupled with changes in behavior (i.e., living in and around sylvatic rodents and human habitation) triggered the evolution of virulent *Y. pestis* from the enteric, food-borne, avirulent pathogen *Y. pseudotuberculosis*. This occurred by means of several genetic changes. For example, development of a gene whose product is involved in the storage of hemin resulted in the ability of the bacteria to block the flea proventriculus, enhancing flea-mediated transmission. Other gene products (phospholipase D and plasminogen activator) facilitated blood dissemination in the mammalian body and allowed for the infection of a variety of hosts by fleas.

DIAGNOSIS, TREATMENT, AND VACCINE

Mortality and morbidity from plague were significantly reduced in the 20th century. However, the disease has not been eradicated. Plague remains endemic in regions of Africa, Asia, and North and South America. From 1983 to 1997, there were 28,570 cases with 2331 deaths in 24 countries reported to the World Health Organization (WHO). In 1997 the total number of cases reported by 14 countries to the WHO was 5419, of which 274 were fatal. Epidemics occurred in Madagascar in 1991 and 1997, in Malawi, Zimbabwe, and India in 1994, and in Zambia and China in 1996. In contrast, there were four cases and only one death in the United States in 1997.

Though human disease is rare, a feverish patient who has been exposed to rodents or flea bites in plague endemic areas should be considered to be a possible plague victim.

Diagnosis can be made by Gram stain and culture of bubo aspirates or sputum. The bacteria grow aerobically and form small colonies on blood and MacConkey agar.

Unless specific treatment is given, the condition of a plague-infected individual deteriorates rapidly and death can occur in 3 to 5 days. Untreated plague has a mortality of more than 50%. A variety of antibiotics including streptomycin, sulfonamide, and tetracycline are effective against bubonic plague. Tetracycline can be used prophylactically, and chloramphenicol is used to treat plague meningitis. No antibiotic resistance has been reported.

Two plague vaccines have been approved for use in humans. One is a formaldehyde-killed, whole-cell vaccine first used in 1942, and the other is a live vaccine used in the former Soviet Union since 1939. A new subunit vaccine that uses the bacterial capsular antigens F1 and V for immunization is under development.

See Also the Following Articles
Blood Sucking • Medical Entomology • Siphonaptera • Zoonoses

Further Reading
Achtman, M., *et al.* (1999). *Yersinia pestis,* the cause of plague is a recently emerged clone of *Yersinia pseudotuberculosis. Proc. Natl. Acad, Sci. U.S.A.* **96,** 14043–14048.
Carniel, E. (2000). Plague. *In* "Encyclopedia of Microbiology," (J. Lederberg, ed.), Vol. 3, pp. 654–661. Academic Press, San Diego, CA.
Herlihy, D. (1997). "The Black Death and the Transformation of the West." Harvard University Press, Cambridge, MA.
Hinnebusch, B. J. (1997). Bubonic plague: A molecular genetic case history of the emergence of an infectious disease. *J. Mol. Med.* **75,** 645–652.
McNeill, W. (1998). "Plagues and People." Anchor Books, New York.
Perry, R., and Fetherston, J. (1997). *Yersinia pestis*—etiologic agent of plague. *Clin. Microbiol. Rev.* **10,** 35–66.
Zeigler, P. (1969). "The Black Death." Harper & Row, New York.

Bumblebee

see *Hymenoptera*

Butterfly

see *Lepidoptera*

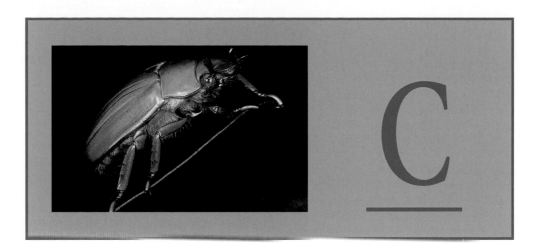

Caddisfly

see *Trichoptera*

Caste

Sean O'Donnell
University of Washington, Seattle

The term "caste" refers to long-term, stable differences among insect colony members that affect the roles played by individuals in their social group. It was the existence of distinct morphological castes in insect colonies that led Charles Darwin to identify social insects as a major challenge to his theory of evolution by natural selection. Few topics are more central to the study of social insect biology than caste. It is ironic, then, that few topics have generated more controversy and debate among social insect biologists. Beyond the deceptively simple definition just offered, there is little agreement on how the term should be defined, or on what characteristics should be used to identify the castes of individual insects. In part, the controversy exists because published definitions of caste are often not operational. For example, some definitions do not specify whether caste differences must be developmentally fixed and permanent or, alternatively, whether individuals can exhibit caste flexibility. As a result, "caste" has been applied to a wide array of physiological and behavioral phenomena.

The diversity of caste systems between and within evolutionary lineages of social insects may preclude a simple, universal definition of caste. Rather than advocate a single definitional point of view, this article explores the diversity of ways in which the concept has been used and the array of important phenomena caste encompasses in different insect societies.

REPRODUCTIVE CASTES

Differences in Reproductive Function

A fully social or eusocial group is generally understood to exhibit reproductive division of labor. This means that eusocial groups must include some individuals that forgo direct reproduction and instead aid the rearing of the offspring of others in their group. In eusocial insects, the helpers comprise the worker caste and reproductive females are referred to as queens. Termite colonies possess long-lived royal couples (a queen and a king), whereas in eusocial Hymenoptera, males are sometimes referred to as drones. Males in the order Hymenoptera (bees, ants, and wasps) rarely work for their colonies and typically die soon after mating. In contrast, male eusocial thrips (Thysanoptera) and termites (Isoptera) comprise part of the worker force and participate fully in colony labor.

Social insect species vary according to whether the group's members are permanently relegated to reproductive versus worker roles and in the degree of fecundity differences between reproducers and workers. There is a general evolutionary trend toward increased reproductive caste specialization as more complex, larger societies evolve from smaller, simpler ones. In some ants, workers lack reproductive organs and are permanently sterile. In most species, however, workers can achieve limited direct reproduction under some conditions.

Morphological Differences

Some species are reproductively monomorphic, and reproductives do not differ significantly in body structure from workers. Many sweat bees and bumble bees, some paper wasps, and even some primitive ants are examples of reproductively monomorphic species. Workers in monomorphic species are often smaller than reproductives, but there can be considerable overlap in body size distributions among the reproductive castes. In some cases, clear physiological differences distinguish workers from reproductives when morphology does not. For example, temperate *Polistes* paper wasp colonies produce gynes (potential female reproductives) at the end of the summer. Gynes possess enlarged, nutrient-laden fat bodies, not present in female workers, that permit them to overwinter in a quiescent state.

In contrast, consistent reproductive caste differences in body size and shape have evolved in several lineages of social insects. Most eusocial insects with wingless workers, such as ants and termites, retain a morphologically distinct reproductive caste with wings. In species with flying workers, developmental allometry can still result in the production of distinct, nonoverlapping body forms for reproductives and workers. Morphologically discrete reproductive castes are found among honey bees, stingless bees, and some paper wasps. Reproductives are often larger than workers, but also differ in body proportions (hence shape) in ways that suggest specialization in egg laying, such as relatively enlarged abdomens. The degree of morphological differentiation between reproductive castes probably evolves in response to a complex array of natural selection pressures. For example, the degree to which the colony occupies a defensible, long-lasting nest site may in part determine the whether queens can afford to adopt relatively immobile body forms.

Caste Determination: Immature Development and Adult Interactions

Other than an interesting exception in the ant *Harpagoxenus sublaevis,* there are no well-documented cases of genetic differences that affect reproductive caste differentiation. Often caste differentiation must depend in part on differential patterns of gene expression during development, particularly in species with distinct caste morphology.

Differences in environmental conditions during immature development can have strong effects on an individual's caste. Nutritional effects on reproductive caste have been documented in numerous taxa and appear to be widespread, if not universal, among eusocial insects. Differences in the amount of food provided to larvae may underlie many of the differences between reproductives and workers, especially in species exhibiting the common pattern of larger body size for reproductives. However, differences in food quality, possibly including the addition of glandular secretions and pheromones, cannot be ruled out. Especially interesting in this regard are those eusocial wasps whose reproductives are smaller than workers (genus *Apoica*) or identical in size but different in shape (genus *Pseudopolybia*).

Social interactions among adults may also influence reproductive caste, particularly in species without apparent morphological caste differences. For example, dominance interactions among paper wasp (*Polistes*) females, which often cooperate to start new colonies, determine which female acts as the sole reproductive. Subordinate *Polistes* females function as workers.

CASTES IN THE WORKER FORCE

Morphological Castes

DISCRETE WORKER MORPHOLOGY   In all termites and in approximately 10% of ant species, workers exhibit developmental allometry resulting in body shape variation within the worker caste. Interestingly, this type of morphological caste has not been documented in social insects with flying workers, such as bees and wasps. There is typically some association between a worker's body form and the tasks that she performs. One of the most common types of morphological specialization is the assignment of large workers, called soldiers, to the special role of colony defense. When the colony is threatened by an animal, the soldiers advance and attack, while other workers flee. Often the soldiers uniquely possess heavily armored exoskeletons and some type of weaponry, including enlarged muscular heads, long, piercing mouthparts, or glands that produce defensive chemicals. In other cases, worker body shape variation affects the performance of more mundane tasks such as food collection. In army ants (*Eciton* spp.), longer-legged workers select larger food items to carry back to their colonies. In leafcutter ants (*Atta* spp.), the largest workers are soldiers, the medium-sized workers cut and transport leaves, and the smaller workers usually remain in the nest to tend the colony's fungus garden. An ant worker's body size and shape are fixed upon adult emergence; further growth is not possible. In contrast, some termite workers (*Zootermopsis* spp.) exhibit considerable caste plasticity, potentially molting among different body forms, and even switching from soldier to nonwinged reproductive castes under certain conditions.

BODY SIZE EFFECTS   Even in monomorphic species, body size differences can influence the tasks that workers perform. In some species, larger-bodied workers dominate their smaller nestmates (*Polistes metricus, P. fuscatus,* and *P. dominulus*), and dominance status in turn affects the tasks a worker performs. In some bumble bees (*Bombus* spp.), however, larger workers are more likely to perform certain tasks such as foraging to collect food for the colony, independently of obvious worker aggression.

Behavioral Castes

Workers can be assigned to behavioral castes when they specialize on a subset of the tasks that the colony needs. In some eusocial insect species such as *Bombus* and *Polistes*, workers exhibit a great deal of flexibility, switching among tasks often, and behavioral castes are weakly defined. In honey bees (*Apis mellifera*) and swarm-founding paper wasps (*Polybia* spp.), on the other hand, workers specialize more consistently.

AGE OR TEMPORAL POLYETHISM Changes in task specialization as workers age are among the best-studied factors that influence workers' behavioral caste. "Age" or "temporal polyethism" refers to an ordered, predictable sequence of task specializations through which an adult worker passes as it ages. Typically, species with temporal polyethism exhibit centripetal development: workers begin by working deep inside the nest, close to the queen(s) and brood; they later perform tasks at the nest periphery; and they finally move further out to perform risky tasks such as foraging and nest defense (Fig. 1). This centripetal pattern of development is remarkably similar among the diverse eusocial insect species that exhibit well-developed temporal polyethism. Workers usually follow the same sequence of task specializations, but individuals vary in their rate of passage through the sequence. Changes in hormone titers, such as juvenile hormone, have been implicated in determining the rate of temporal polyethism in *Apis* and *Polybia*.

Workers' relative age influences social status and task performance in some species. In the paper wasps *P. exclamans,* the first-emerging (and consequently the oldest) workers in the nest tend to socially dominate their nestmates, a pattern referred to as gerontocracy, which is independent of body size variation. In this case, age influences workers' behavioral caste in a static way, rather than in a dynamic way as in temporal polyethism.

INDIVIDUAL DIFFERENCES AND SPECIALIZATION Superimposed on broader patterns of division of labor, such as body size or age effects, workers sometimes exhibit idiosyncratic specialization on tasks. For example, *Apis* and *Polybia* foragers often specialize by collecting one of the several materials their colony needs to function. Such specialization may be benefit the colony by increasing the efficiency or reliability of task performance.

Genetic Effects Genetic relatedness among the offspring in the worker force is highest when the workers are born to a single reproductive female, which has mated with a single male. Some social insect species exhibit mating behavior or social structure that decreases the genetic relatedness among the offspring workers within colonies. When queens mate with several different, unrelated males (polyandry), or when several reproductive females are present in the colony (polygyny), workers can find themselves sharing a nest with a combination of more closely and more distantly related individuals. In a number of polyandrous and polygynous species, including *Apis* spp. and stingless bees (genus *Partamona*), several species of ants, and *Polybia* spp., workers that are more genetically similar have been found to specialize on similar tasks.

Experience and Learning A predicted benefit of task specialization is that workers can improve performance as they gain experience. There is evidence that some insect workers learn to perform tasks more effectively with experience. Bumble bees collecting nectar and pollen from complex flowers learn to do so more rapidly after repeated attempts to handle a given type of flower. *Polybia* foragers are less likely to return from foraging trips empty-handed as they gain foraging experience.

COLONY-LEVEL INVESTMENT

Investment in Growth and Maintenance versus Reproduction

One of the major challenges that faces growing organisms is the developmental decision of how many resources to invest in growth and how many to invest in reproduction. Insect colonies can be treated as organisms in this sense, since each colony must decide how much it will invest in different castes (i.e., in workers vs reproductives). To the extent that colonies

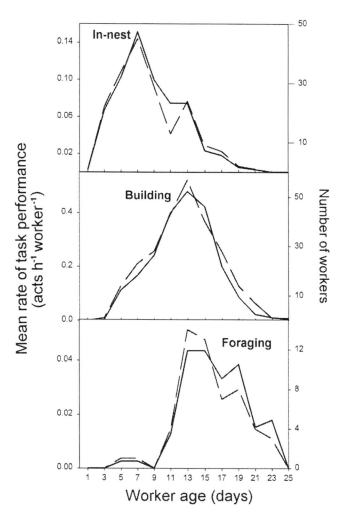

FIGURE 1 Temporal or age polyethism in the paper wasp *Polybia aequatorialis.* Data were collected on 130 individually marked, known-age workers. Two measures of worker activity at three task sets are plotted against worker age (solid lines: mean rate of task performance; dashed lines: number of workers performing the task). Note the typical centripetal developmental sequence: in-nest tasks (mostly nest cleaning) are followed by building on the nest exterior, and later by foraging (leaving the nest and returning with food and building materials).

are reproductive units, optimality theory predicts that natural selection will favor colonies that allocate their limited resources efficiently into different castes. Many insect societies segregate the production of workers (early in colony development) from the production of new queens and males (later in colony development).

Worker Caste Ratios

Insect colonies appear to behave in an adaptive manner by adjusting their worker caste ratios to meet current colony needs. Production of different worker castes reflects a trade-off between the costs and benefits of producing and maintaining workers of different kinds. As ant colonies with morphologically specialized workers grow in size, their amount of investment in large-bodied workers increases, and many eusocial insects produce tiny nanitic workers early in colony development. Colonies of the ant *Pheidole pallidula* increase their rate of production of soldiers when exposed to potential competitors. Similar colony flexibility is apparent in age–caste distributions. In honey bees and paper wasps, if the level-of-colony need for foragers changes, some workers accelerate or reverse their behavioral development, performing the age-atypical tasks that are in greatest demand. Identifying the mechanisms that link individuals' developmental plasticity with the level of colony need remains as a central challenge in the study of caste.

See Also the Following Articles
Colonies • Division of Labor • Hymenoptera • Isoptera • Juvenile Hormone • Sociality

Further Reading
Evans, J. D., and Wheeler, D. E. (2001). Gene expression and the evolution of insect polyphenisms. *Bioessays* **23**, 62–68.
Hölldobler, B., and Wilson, E. O. (1990). "The Ants." Harvard University Press, Cambridge, MA.
Hunt, J. H. (1994). Nourishment and social evolution in wasps sensu lato. *In* "Nourishment and Evolution in Insect Societies" (J. H. Hunt and C. A. Nalepa, eds.), pp. 221–254. Westview Press, Boulder, CO.
O'Donnell, S. (1998). Reproductive caste determination in eusocial wasps (Hymenoptera: Vespidae). *Annu. Rev. Entomol.* **43**, 323–346.
Oster, G. W., and Wilson, E. O. (1978). "Caste and Ecology in the Social Insects." Princeton University Press, Princeton, NJ.
Peters, C., and Ito, F. (2001). Colony dispersal and the evolution of queen morphology in social Hymenoptera. *Annu. Rev. Entomol.* **46**, 601–630.

Caterpillars

Frederick W. Stehr
Michigan State University

The larvae of butterflies, skippers, and moths of the order Lepidoptera are generally known as caterpillars. Caterpillars

FIGURE 1 Caterpillar of the polyphemus moth, *Antheraea polyphemus,* showing the five pairs of prolegs bearing crochets (hooks). (Photograph by Joseph L. Spencer, Illinois Natural History Survey.)

come in a diversity of sizes, shapes, and colors. The most common form has a conspicuous head, a thorax with three pairs of legs, and an abdomen with five pairs of prolegs that bear crochets (hooks) (Fig. 1) that enable the caterpillar to cling tightly to or wedge itself between materials. In fact, some of the giant silk moth caterpillars (Saturniidae) can cling so tightly to a twig that a proleg can be ripped from the body if they are pulled too hard. A few other orders of insects contain larvae that are caterpillar-like, but only the larvae of the leaf-feeding sawflies (Hymenoptera) are commonly encountered. They are easily mistaken for caterpillars, but they usually feed in groups (as do some caterpillars), rear up when disturbed, have more than five pairs of prolegs on the abdomen, and never have crochets on the prolegs.

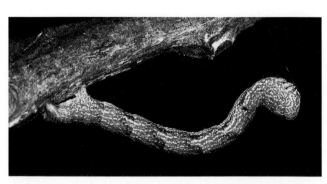

FIGURE 2 Twig-mimic inchworm caterpillar of a moth (Geometridae). (Photograph by Fred Stehr, Department of Entomology, Michigan State University.)

FIGURE 3 "Bird dropping" mimicry by the caterpillar of the orangedog, *Papilio cresphontes.* (Photograph by J. Mark Scriber, Department of Entomology, Michigan State University.)

FIGURE 4 A slug caterpillar, *Euclea delphinii* (Limacodidae), showing the lack of prolegs and the sucker-like discs. (Photograph by Jane Ruffin, Rosemont, Pennsylvania.)

BIOLOGY AND ECOLOGY

Caterpillars are commonly encountered because many are leaf-feeders and are not concealed feeders, although they may be cryptically shaped or colored. The realistic twig mimicry (Fig. 2) and behavior found in some inchworms (Geometridae) are complete with body markings that resemble bark irregularities, scars, and stipules. Another excellent mimic is the caterpillar of the orangedog, *Papilio cresphontes,* whose black and white larvae resemble bird droppings in all instars (Fig. 3). Some other swallowtail caterpillars resemble bird droppings in the early instars, whereas the later instars change to a greenish color (with black and orange, or yellow, markings) that camouflage them on their host plants. One of the most unsual diet-induced camouflage systems is that of the geometrid moth, *Nemoria arizonaria.* This moth has two generations, the first feeds on oak catkins in the spring and takes on the shape and colors of the catkins; the second generation hatches when the catkins are gone and only leaves are available for food, resulting in the caterpillar being a twig mimic.

Nearly all species of plants are fed upon by caterpillars, and many caterpillars are important pests. They also form galls; are scavengers on dead plant materials, fungi, and animal materials such as feathers, wool, or fur (clothes moths, Tineidae); and some are important pests of stored products (meal moths, Pyralidae). Some larvae of the family Pyralidae are truly aquatic and have gills, and some caterpillars of other families feed in or on aquatic plants. A few caterpillars are predators such as the lycaenid butterfly, *Feniseca tarquinius,* which feeds on and among clusters of wooly aphids. Some tropical lycaenids are predators on ant larvae in their nests. The ants benefit from this arrangement by obtaining sweet secretions from the dorsal abdominal glands of the caterpillars. For other lycaenid caterpillars that feed on plants above ground, the relationship is clearly symbiotic, with the larvae providing secretions for the ants and the ants protecting the larvae from predators and parasitoids. Probably the most unusual predators are the Hawaiian geometrids in the genus *Eupethecia,* whose larvae strike backward with their enlarged thoracic legs, seizing any prey that touches their rear end.

Most caterpillars move by a wavelike movement of the legs from rear to front. However, those with reduced numbers of prolegs or none at all proceed in a different manner. The inchworms (Fig. 2), which usually have two pairs of prolegs, and the looper cutworms of the Noctuidae, which have only three or four pairs of prolegs, move by pulling the hind end close to the thoracic legs and then looping the body forward. Caterpillars of the Limacodidae (Fig. 4), the slug caterpillars, have no prolegs and move on sucker-like discs, gliding along in a manner similar to slugs.

Caterpillars may be aposematically colored to advertise that they are distasteful or poisonous. They may also bear diverse

FIGURE 5 The ill-smelling, eversible, dorsal, prothoracic glands of a tiger swallowtail caterpillar, *Papilio glaucus.* (Photograph by J. Mark Scriber, Department of Entomology, Michigan State University.)

FIGURE 6 The snake-mimicking behavior of the spicebush swallowtail larva, *Papilio troilus.* (Photograph by J. Mark Scriber, Department of Entomology, Michigan State University.)

FIGURE 7 Tent of the western tent caterpillar, *Malacosoma californicum* (Lasiocampidae). (Photograph by Fred Stehr, Department of Entomology, Michigan State University.)

lobes, spines, horns, knobs, and urticating hairs or spines that may or may not be irritating in some way. Most caterpillars that feed in protected or concealed locations do not possess such structures or defenses. The ill-smelling secretions of the eversible, dorsal prothoracic glands (Fig. 5) of swallowtail butterfly caterpillars (Papilionidae) are well known, as are the stinging properties of saddleback caterpillars (Limacodidae). Some papilionid and sphingid caterpillars have large eye-like spots on the thorax and rear up their head and expanded thorax, thereby mimicking small snakes (Fig. 6). The distastefulness to birds of many monarch larvae and adults that feed on toxic species of milkweeds is common knowledge, but monarch larvae that have fed on the relatively nontoxic species of milkweeds are quite edible, as are their adults.

Caterpillars spin silk through a conspicuous labial spinneret. Species use silk in diverse ways, from webbing together leaves or other materials (Tortricidae, Pyralidae, and many others) to constructing webs [fall webworm (Arctiidae)], large "tents" of the tent caterpillars (Fig. 7) (Lasiocampidae: *Malacosoma*), or silken tube shelters (some Pyralidae, Tineidae). Some, such as the casebearers (Coleophoridae), the bagworms (Psychidae), and the sackbearers (Mimallonidae), make tough, complex shelters that they carry with them. The "sack" of the mimallonid caterpillars is particularly interesting because it is extremely tough, with an opening at both ends that can be blocked by the caterpillar's head or its hard anal plate, either of which fits tightly against either opening.

Many caterpillars spin cocoons in which they pupate, ranging from the tight and commercially valuable cocoons of the silkworm, *Bombyx mori,* to those of the gypsy moth, *Lymantria dispar,* which are at best a loose net of silk (even though it was originally brought to the United States because it was believed to be a good prospect for silk production). Other common caterpillars, such as those of sphinx moths,

cutworms, and some saturniids, pupate in the ground or litter where they form cells that can be silk lined.

Caterpillars are occasionally an item of commerce. Canned caterpillars (gusanos) are eaten in (and sometimes imported from) Mexico. Tequila worms, usually in the family Megathymidae (giant skippers), were formerly imported in bottles of tequila from Mexico. If the tequila worm was well preserved, presumably the alcohol concentration was satisfactory.

See Also the Following Articles
Cocoon • Larva • Lepidoptera

Further Reading
Stehr, F. W. (ed.) (1987). "Immature Insects," Vol. 1. Kendall/Hunt, Dubuque, IA.

Cat Fleas

Nancy C. Hinkle
University of Georgia

Michael K. Rust
University of California, Riverside

Fleas are small (2 mm), dark, reddish brown, wingless, bloodsucking insects. Their bodies are laterally compressed (i.e., flattened side to side) and covered with many hairs and short spines directed backward, permitting forward movement through hairs on the host's body. Their hind legs are long and well adapted for jumping. Adult flea mouthparts are equipped for sucking blood from the host.

In North America, the cat flea, *Ctenocephalides felis felis,* is the most common ectoparasite of dogs and cats. In tropical areas it is a year-round pest, whereas in temperate climates its season varies. Because they attack a range of warm-blooded hosts, including humans and pets, cat fleas are both a veterinary problem and a household pest.

SIGNIFICANCE OF FLEAS

Fleas are important to humans because of their potential as disease vectors, in addition to the annoyance they produce merely by biting. Pathogen transmission is facilitated by their habit of feeding sequentially on several hosts. The best known disease associated with fleas is bubonic plague; the plague bacterium, *Yersinia pestis,* is transmitted almost exclusively by rodent fleas. Murine typhus is another disease for which cat fleas have been implicated in the transmission cycle. The flea-borne typhus causative agent, *Rickettsia typhi,* is transmitted from its rodent reservoir by several flea species, including *C. felis.* Fleas probably play a role in maintenance and transmission of several other disease organisms such as *Bartonella henselae,* causing cat scratch disease. The cat flea is the intermediate host for the dog tapeworm, *Dipylidium caninum,* which can affect small children as well as dogs and cats.

Pets infested with fleas bite and scratch themselves repeatedly. In situations in which flea numbers are high, veterinarians occasionally see kittens and puppies near death from flea-produced anemia. Sensitized people suffer from flea bites, which can cause intense itching, with scratching opening the skin to infection.

Flea allergy dermatitis (FAD) is a severe condition found primarily in dogs, but also occasionally seen in cats. In a flea-allergic animal, flea salivary antigens initiate a cascade of symptoms, resulting in intense pruritus accompanied by scratching, biting, and self-inflicted trauma. An affected animal typically displays obsessive grooming behavior, with accompanying depilation, leaving the skin with weeping sores, often resulting in secondary infection. FAD is treated with corticosteroids, which possess undesirable side effects, especially when continuous use is required as in chronic FAD cases. Until development of FAD immunotherapy, successful treatment involves flea elimination from the animal's environment and flea bite prevention.

Fleas and their associated diseases can constitute over half a veterinary practice's caseload in some areas of the country. More energy and money are spent battling these insects than any other problem in veterinary medicine.

LIFE HISTORY

The cat flea is a cosmopolitan, eclectic species, having been recorded from more than three dozen species, including opossums, raccoons, kangaroos, and even birds. This wide host range explains this flea's ability to repopulate domestic animals after suppression efforts. Because it lacks host specificity and tends to feed on humans, the cat flea is a pest of both companion animals and humans with whom they share their abode.

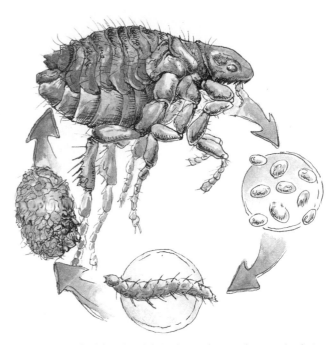

FIGURE 1 Cat flea life cycle: adult (top), eggs, larva, and cocoon (enclosing pupa). (Illustration by Cara J. Mitten.)

Adults

Once adult cat fleas locate a host, they tend to remain on that animal unless dislodged. They feed readily and mate on the host. Female fleas lay eggs while on the host and because the eggs are not sticky, they readily fall off into the host's environment, with large numbers accumulating in areas frequented by the animal. Each female flea can produce more than two dozen eggs per day. Adult fleas are about 1 to 4 mm in length and are strongly flattened from side to side. They are equipped with relatively long legs armed with strong outwardly projecting spines. Cat fleas have a collar of spines (ctenidium) on the back and another row of spines above the mouth. These characteristics allow for rapid movement through the host's hairs and also serve to resist removal from the fur.

Once the adult flea finds a host, it begins to feed. Typically the female mates and begins oviposition within a couple of days. On the host, a female flea averages about one egg per hour and, as a female flea can live on the host for several weeks, potential production can amount to hundreds of eggs in her lifetime. Only the adult stage is parasitic; all other life stages develop off the host (Fig. 1).

Eggs

Cat flea eggs are approximately 1 mm in length, with little surface structure other than aeropyles (permitting gaseous exchange for the developing embryo) and micropyles (for sperm entrance during fertilization). Typically, the larvae hatch within 24 to 48 h after oviposition, with more rapid hatching at warm temperatures.

Larvae

From the eggs emerge small, white, eyeless, legless larvae with chewing mouthparts. Because they seldom travel far from where they hatch, cat flea larvae are usually found in furniture, carpeting, or outside in areas frequented by flea hosts. Flea larvae have three instars that, under favorable conditions, can be completed in as little as 10 days. Larvae will develop only in protected microhabitats in which the relative humidity exceeds 75%. Cool temperatures, food shortages, or other unsuitable environmental conditions may extend larval developmental time to several weeks or a month. The third instar voids its gut approximately 24 h before initiation of cocoon construction. The white prepupa wanders until it locates an appropriate site for pupation and then begins to spin a silk cocoon. Frequently, environmental debris is incorporated into the cocoon, adhering to the sticky silk fibers, so that the cocoon may appear as a small dirt clod or lint ball.

Pupae

Within its cocoon, the prepupa molts to the pupa and continues metamorphosis to the adult flea within about 4 days, under favorable conditions. Length of the preemerged adult stadium is the most variable in the flea life cycle, ranging from less than a day to several months (or perhaps over a year). The mechanisms are not completely understood, but it appears that some individuals are programmed to delay emergence. Likely this is an evolutionary strategy whereby offspring emerge over an extended interval, ensuring that some successfully achieve hosts. Stimuli such as pressure, carbon dioxide, and warmth (triggers associated with mammalian hosts) serve as releasers, causing the adult flea to emerge from the cocoon. Upon emergence, if the flea does not locate a host immediately, it can survive for approximately 7 to 10 days (or longer under high-humidity and low-temperature conditions).

FLEA SUPPRESSION

Because fleas must have blood from a mammalian host to survive, treating host animals is the most efficient and effective means of suppressing fleas. There are several on-animal products that are effective for flea control. Many contain pyrethrins, which are safe, effective products but kill only fleas on the animal at the time of treatment and do not provide residual control. Other over-the-counter compounds include spot-on permethrin products, which are limited to canine use because they can be lethal to cats.

Veterinarians can recommend products that provide several weeks of control with a single application. Products prescribed by veterinarians for on-animal flea control are applied in a small volume (a few milliliters) on the back of the animal's neck. The material distributes over the body surface in skin oils. In addition to spot-on formulations, some products are available as sprays. These adulticides kill fleas on the animal within a few hours, then provide residual flea suppression for several weeks.

To forestall flea infestations, pets can be started on flea developmental inhibitors early in the season. Products containing insect development inhibitors can be applied topically, given orally (once monthly as a pill for dogs or a liquid added to a cat's food), or given as a 6-month injectable formulation for cats. Female fleas that feed on blood of treated animals subsequently are unable to reproduce.

Once pets have been treated, it will take a while for fleas in the environment to die off. Meanwhile, as they emerge, fleas will hop onto the animal; the host will continue to "harvest" fleas from the surrounding environment until they have been killed and no more are emerging. Insect growth regulators can be used to break the flea life cycle. Although these compounds do not kill adult fleas, they do prevent eggs and larvae from completing their development, ensuring that any fleas brought into the area will not establish a sustaining population.

Sanitation is an important flea suppression tactic; by eliminating larval developmental sites and destroying immature stages before they develop to the pestiferous adult stage, pets and people can be protected from fleas. Areas frequented by pets accumulate flea eggs and larval food, so these microhabitats should be vacuumed and treated to prevent flea infestations. These might include areas under furniture, animal bedding and sleeping quarters, and utility rooms or other areas where the pet spends time.

See Also the Following Articles
Bubonic Plague • Medical Entomology • Siphonaptera • Veterinary Entomology

Further Reading
Dryden, M. W. (1997). Fleas. *In* "Mallis Handbook of Pest Control," 8th ed., Chap. 16, pp. 747–770. Franzak & Foster, Cleveland.
Dryden, M. W., and Rust, M. K. (1994). The cat flea: Biology, ecology and control. *Vet. Parasitol.* **52,** 1–19.
Hinkle, N. C., Rust, M. K., and Reierson, D. A. (1997). Biorational approaches to flea (Siphonaptera: Pulicidae) suppression: Present and future. *J. Agric. Entomol.* **14,** 309–321.
Rust, M. K., and Dryden, M. W. (1997). The biology, ecology and management of the cat flea. *Annu. Rev. Entomol.* **42,** 451–473.
Taylor, M. A. (2001). Recent developments in ectoparasiticides. *Vet. J.* **161,** 253–268.

Cave Insects

Francis G. Howarth
Hawaii Biological Survey, B. P. Bishop Museum

Caves and associated subterranean voids harbor extraordinary ecosystems inhabited by equally remarkable animals. Insects and arachnids dominate terrestrial habitats, whereas crustaceans dominate aquatic systems. This article describes the subterranean biome, highlighting terrestrial systems and the insects that are obligately adapted to live permanently in underground voids.

DISCOVERY AND CHARACTERIZATION OF CAVE ARTHROPODS

Why an animal would abandon the lighted world and lose such adaptive characters as eyes, pigment, and dispersal ability to live permanently in perpetually damp, dark, barren caves has long fascinated both biologists and laymen. In fact, it is these pale, blind obligate cave species that one usually envisages under the rubric of cave animal, and it is this group that is featured in this article. However, numerous other animals live all or part of their life cycles in caves or are regular visitors. Although some cave insects were known in ancient times, the first scientific writings on this topic began in northern Italy in the mid-16th century with the discovery of blind aquatic crustaceans in cave streams. The science of cave biology (biospeleology) was founded in the mid- to late 19th century with studies of limestone caves in southern Europe by Schinner and continued into the early 20th century by Racovitza and colleagues. They devised the currently used classification scheme for cavernicoles, based on the degree of association with caves. Also, in the mid-1800s, obligate cave animals were discovered in Mammoth Cave, Kentucky, and in a few other North American limestone caves, but the study of North American cave faunas generally lagged behind Europe for the next century. The loss of eyes and other apparently adaptive characters led to a revival, circa the turn of the 20th century, of Lamarckian (i.e., acquired rather than inherited characteristics) theories to explain their evolution. During this period several major expeditions went to tropical regions to search for obligate cave faunas, but for a variety of reasons none were found or recognized. The apparent absence of tropical troglobites (obligate cave species) and the relictual nature of temperate cave animals led to the development and general acceptance of the theory that these animals evolved only after populations were stranded in caves by changing climates that extinguished their surface relatives.

Troglobitic Adaptations

The most conspicuous aspect displayed by obligate cave arthropods is the reduction of structures normally considered adaptive (e.g., eyes, pigment, wings, and cuticle thickness). Compare the closely related surface and cave insects shown in Figs. 1A and 1B. Cave species also often lack a circadian rhythm and have relatively low metabolic and reproductive rates. A few characters are often enhanced, including modified structures such as increased hairiness, enlarged sensory organs, longer appendages, and specialized tarsi. These morphological, physiological, and behavioral changes allow the animals to

FIGURE 1 Cave and surface cixiid planthoppers. (A) Rain forest *Oliarus* species from Maui Island; note large eyes, dark color, and functional wings. (B) Adult female cave-adapted *Oliarus polyphemus* from Hawaii Island; note absence of eyes, enlarged antennae, and reduced wings and pigment. (Photographs by W. P. Mull, used with permission.)

maintain water balance, breathe unusual gas mixtures, disperse, reproduce, and locate food and other resources in their environment. The remarkable convergent evolution of troglomorphy (adaptations to caves) among unrelated cave species in different regions of the world indicates that selective pressures must be similar in all such environments.

Taxonomic Overview of Troglobites

TERRESTRIAL CAVE ARTHROPODS Insects, arachnids, and millipedes are the dominant terrestrial groups living in caves. Not all orders are represented, however. Among the Hexapoda, the orders Collembola, Orthoptera, Hemiptera, Coleoptera, and Diptera predominate. The springtails (Collembola) are represented by many troglophilic (facultative cave residents) and troglobitic species and are important scavengers in many caves. Most cavernicolous orthopterans are troglophilic or trogloxenic (roosting in caves), with the cave crickets (Rhiphidophoridae) being the best known. As more tropical caves are studied, many new species of troglobitic true crickets (Gryllidae) are being described. Among

Hemiptera, both suborders occur in caves: true bugs (Heteroptera) and planthoppers and allies (Auchenorrhyncha or Homoptera of some classification systems). Thread-legged bugs (Reduviidae) are common troglophiles in warmer caves, and a few troglobitic forms are known from the tropics. Most species are cryptic, and many new cave species await discovery. Several other hemipteran families contain troglophilic species. Planthoppers, especially Cixiidae (Fig. 1B), are common in tropical caves. Ongoing surveys for cixiids indicate that each isolated cave system may harbor one or more cave-adapted species, and the group may be among the most speciose families in caves, rivaling even the carabid ground beetles in temperate caves.

Beetles (Coleoptera), especially the families Carabidae, Leiodidae, and Staphylinidae, are especially well represented in the temperate caves. For example, the endemic North American ground beetle genus *Pseudanophthalmus* contains at least 250 species, which, with one exception, are found only in caves. Flies (Diptera) are dominant troglophiles in both tropical and temperate caves, but only a few blind, flightless troglobitic species are known.

Troglobitic species are also found in the orders Diplura, Thysanura, Blattodea, Dermaptera, Grylloblattodea, Psocoptera, and Lepidoptera. Troglobitic bristletails (Diplura) occur mainly in temperate caves. Cockroaches (Blattodea) are well represented in tropical caves, and many new species await description. Only a few cave-adapted earwigs are known, and most are from oceanic islands. Grylloblattids are restricted to glaciated mountains in northwestern North America and eastern Asia. They characteristically inhabit caves and crevices; however, most species also venture outside to feed in damp surface habitats. Many moths habitually roost in caves, and some are troglophilic scavengers or root feeders. A few are blind and flightless troglobites.

The arachnids are second only to the insects in numbers of terrestrial cave species. The spiders are common denizens of caves, with numerous troglobitic forms known from temperate and tropical caves. In many tropical caves, spiders instead of ground beetles are the top predators. Pseudoscorpions are also well represented in temperate and tropical caves, and over 300 cave-adapted species representing most families are known. Harvestmen (Opilionida) are more restricted in distribution, but most of the 26 families contain troglobitic species. Some surface species roost in caves in huge numbers. Mites (Acari) are often abundant and diverse in caves, especially species associated with guano. Most terrestrial cave species are troglophilic, but a few families, such as the Rhagidiidae, contain many troglobites. Cavernicolous species are also known among the palpigrades, schizomids, amblypygids, scorpions, and ricinuleids.

Myriapods are also well represented in caves. The millipedes are the third major group of cavernicolous arthropods, especially in temperate caves, where they are often the dominant scavengers in the ecosystem. The orders containing the most cave species are Julida, with numerous troglobites in Europe and North America; Chordeumatida and Polydesmida, with

troglobites in Europe, North America and Japan; and Callipodida, with troglobites in Europe and the Near East. Four other orders (Polyxenida, Glomerida, Spirobolida, and Spirostreptida) each have a few cave-adapted species. Cave millipedes from the tropics are still poorly known, and many new species undoubtedly await discovery. Many ground-inhabiting centipedes regularly enter caves. Whether they can live and reproduce underground is unknown for most species, but a few are troglophilic or troglobitic. The rock centipedes (Lithobiomorpha) are widespread and include several troglobitic species. A few troglobitic giant centipedes (Scolopendromorpha) are known from the tropics. An undescribed 8-cm-long Scutigeromorpha from North Queensland, Australia, is one of the largest terrestrial troglobites known.

Two groups of terrestrial Crustacea are found in caves. Isopods in the suborder Oniscidea have adapted to caves many times, especially in the Mediterranean region and in the tropics. Fourteen of the 34 recognized families contain cave species. In contrast, only a few terrestrial amphipods (Talitridae) are found in caves, and most are from islands.

AQUATIC CAVE ARTHROPODS Aquatic subterranean habitats include underground lakes and streams, perched pools of water, water films, and water-filled phreatic aquifers. These aquatic habitats support diverse faunas of troglobitic (or stygobitic) arthropods. By far the dominant group is the crustaceans, with about 2700 cave-adapted species known worldwide. Water mites (Acari) are also well represented, especially in smaller interstitial habitats. Few insects have invaded subterranean aquatic habitats. The most successful group is the dytiscid diving beetles, several species of which are known from aquifers in Africa, Europe, North America, and Japan. Two troglobitic water bugs are known: a blind water scorpion (Nepidae) from a cave in Romania and a terrestrial water treader (Mesoveliidae) from Hawaii.

Zoogeography of Cave Arthropods

Until recently, obligate cave species were thought to occur mainly in temperate limestone caves, and the cave faunas of temperate Europe and North America are well characterized. Diverse cave faunas are also known from Japan, Tasmania, and New Zealand. However, in the past few decades discoveries of significant cave faunas in tropical caves, lava tubes, and even fractured rock layers have revolutionized our understanding of cave biology. These findings suggest that troglobites have evolved wherever suitable subterranean voids are available for sufficient time. They are now known from most regions that have been appropriately investigated. Thus rather than being exceptional, cave adaptation must be a general and predictable process among animals adapting to exploit underground resources.

In hindsight, the early expeditions to the tropics missed troglobites for three main reasons. (1) The environment of caves: Troglobites are restricted to deeper, constantly moist passages. Because cave temperatures are usually near the mean annual surface temperature (MAST) over the cave and, in the tropics, the surface temperature rises above and falls below MAST almost every day, most tropical caves are subjected to drying winds created by the sinking cold nighttime air. (2) Accessibility: The higher solution rate of limestone in the tropics creates large open cave systems, exacerbating the effects of the daily drying winds and making the deeper moist cave passages, where the troglobites are found, beyond the limits of safe exploration using the equipment available at that time. In addition, the caves found and explored were often bird and bat roosts, and the biologists could fill their containers with new species without going deeper. (3) Systematics: Ironically, many troglobites were collected, but the species belonged to groups unrelated to the animals found in temperate caves, and in fact unrelated to anything the temperate-based taxonomists had seen, so their status in the cave went unrecognized. As in all fields in biology, evolutionary biology is only as good as the systematics research upon which it is based.

Each cave region is inhabited by representatives of the surface fauna currently or historically living over the caves. Only a few surface taxa within each region successfully invaded caves. In general, the surface ancestors possessed characters that facilitated their shift into underground environments; i.e., they were already adapted to live in dark, moist rocky habitats and utilized food that was relatively common in caves. The chief ancestral habitats for terrestrial cave species include rocky margins of rivers, lakes, and seashores; leaf litter and moss in wet forests; and moist rocky terrains. Each cave system harbors relatively few species of troglobites; even the most diverse known fauna—that in the Postojna–Planina System, Slovenia—totals only 84 species. In North America, Mammoth Cave supports the most species (41). Among lava tubes, Bayliss Cave (North Queensland, Australia) contains the highest number (25). Because of the restricted distribution of each species, cave habitats are often likened to islands. Despite the few species found in each cave, the overall number of troglobites is quite large since subterranean habitats are much more extensive and widespread than is often assumed. Karst landscapes cover about 15% of the earth's surface, and cavernous lava and fractured rock habitats have not been mapped but may cover another 5% or more. Submarine caves have barely been investigated, but the diverse fauna derived from marine ancestors found in anchialine systems along seacoasts indicates that caves and cave-like habitats below the seafloor may harbor diverse ecosystems at least in shallow coastal areas.

SUBTERRANEAN BIOME

Caves and Voids

Caves are subterranean voids large enough for humans to enter, but intermediate-sized voids (i.e., mesocaverns) smaller

than caves but larger than capillary spaces are also important for terrestrial cave insects. Terrestrial animals rarely exploit capillary-sized spaces underground, but water-filled pore spaces (i.e., interstitial habitats) are often inhabited by numerous tiny species of stygobites. Caves and voids can form in three ways: solution, erosion, and volcanism. The largest and best known caves are dissolved in limestone, calcium carbonate. Limestone is structurally strong yet readily dissolves in weak acid, such as the small amounts of carbonic acid normally found in groundwater. The process is slow, but over millennia large interconnected systems of caves and voids can form in limestone exposed to weathering. Caves created by solution can also form in other soluble rocks, such as gypsum (hydrated calcium sulfate) and dolomite (magnesium calcium carbonate), but the caves formed are usually less stable than those in limestone. Erosional caves form during landslides and tectonic events, as well as by groundwater removing loose material from under a cap rock. Erosional caves are usually ephemeral but, in some areas, they are re-created continuously and so remain available for colonization. Tectonic caves are common on volcanoes, but lava tubes are more familiar cave features. Lava tubes form by the roofing over of lava channels during an eruption. Because the roof insulates the flow, lava tubes become efficient transporters of lava away from the vent, and long complex caves can be built over time by long-lived eruptions. Mesocavernous habitats are more extensive than caves and can be found in rock strata not suitable for supporting cave-size passages. Mesocaverns also occur in fractured rock strata and in cobbles deposited by rivers.

Environment of Caves

The terrestrial cave environment is strongly zonal (Fig. 2). Three zones are obvious: (1) the entrance zone where the surface and underground environments meet; (2) the twilight zone between the limits of vascular plants and total darkness; and (3) the dark zone. From biological and environmental perspectives, the dark zone can be subdivided into three zones: (a) the transition zone where short-term

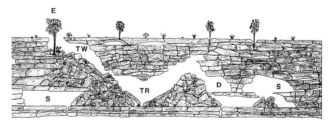

FIGURE 2 Profile view of a representative cave showing the five environmental zones. Not shown to scale; length and depth are compressed. Key: D, deep zone; E, entrance zone; S, stagnant air zone; TR, transition zone; and TW, twilight zone. [Illustration by N. C. Howarth. Reproduced, with permission, from E. C. Dudley (ed.), 1991, "The Unity of Evolutionary Biology," Dioscorides Press, an imprint of Timber Press, Portland, OR.]

climatic events on the surface are still felt; (b) the deep cave zone where the atmosphere remains saturated with water vapor; and (c) the stagnant air zone where decomposition gases, especially carbon dioxide, can accumulate. The boundary between each zone is often dynamic and is determined by size, shape, orientation, and location of entrances in relation to the surface environment and size and shape of the cave passages, as well as to the climate on the surface and availability of water. Because air exchange is reduced in smaller spaces, the environment within most mesocaverns probably remains in the stagnant air zone. Each zone often harbors a different community of organisms, with the obligate cave species found only in the inner two zones. The deep cave and stagnant air zones contain a harsh environment for most surface-dwelling organisms. It is a perpetually dark, wet, three-dimensional maze without many of the cues used by surface species and with often abnormally high concentrations of carbon dioxide. In many caves in temperate regions, the transition zone is evident only in winter when the outside temperature is below cave temperature.

Energy Sources and Nutrient Cycling in Caves

Unlike capillary spaces typical of soils, which act as filters capturing water and nutrients near the surface, caves and mesocaverns act as conduits for water and nutrients. In cavernous regions, a significant amount of organic material sinks or is carried into deeper underground voids where it is inaccessible to most species adapted to surface habitats. The principal mechanisms that transport material underground are sinking streams, percolating rainwater, trogloxenes, animals blundering into caves, and deeply penetrating plant roots. A few cave communities are known to rely on food energy created underground without the aid of sunlight by chemoautotrophic microbes. Sinking streams are more important in transporting food into limestone caves than in lava and other caves, because streams are important in creating and maintaining solution caves. Plants growing on barren rocky substrates such as lava and limestone often must send their roots deep into crevices and caves to obtain water and nutrients. Because higher temperatures result in higher rates of water loss from leaves and higher rates of leaching of tropical soils, and because there is a continuous growing season without a spring recharge of water, plant roots must penetrate deeper underground (sometimes in excess of 100 m) and are, therefore, generally more important in tropical caves than in temperate caves.

Most troglobites are detritivores or scavengers feeding on decaying organic matter and the associated microbes. Living tree roots provide food directly for several obligate cave insects. A relatively large percentage of troglobites are predators, attesting to the role of lost surface animals in bringing in food. It is these available food resources that enable the evolution of troglobites, which are highly specialized to exploit resources within medium-sized subterranean voids. They colonize or temporarily exploit cave-sized passages only where the physical environment is suitable. Most caves appear barren and therefore often are believed to be food-poor environments. However, food can be locally abundant, and exploiting such a patchy resource in a harsh, maze-like environment is probably more critical than paucity per se.

In addition to troglobites many other organisms enter caves. Many arthropods seek out caves for estivation or hibernation sites during periods of harsh weather. Some, such as agrotine moths and cave crickets, use caves for daytime retreats and sometimes oviposition sites and emerge at night to forage in the neighboring forest. Troglophilic arthropods enter to feed on guano and other organic material deposited or brought in by roosting bats, birds, crickets, and other trogloxenes. Parasites and other associates of trogloxenes also live in caves, and some of these, such as nycteribiid and streblid flies on bats, show some troglomorphies. Many leaf-litter and soil arthropods living in caves feed on accumulations of organic material left by sinking streams. These resources are usually more abundant near entrances and in the transition zone. Only a portion of the surface-inhabiting species in each region can cope with the environment and exploit these food resources. Some troglophiles apparently leave caves only to disperse to new sites, but most show no morphological adaptations to living in caves.

CONSERVATION OF CAVE LIFE

The fantastic adaptations displayed by obligate cave animals have long intrigued biologists. Their often narrow environmental tolerances, coupled with their island-like habitats, have reinforced the view that these animals are fragile, lead an endangered existence, and are in need of conservation. However, development of conservation programs is hampered by a severe lack of data about the species present and their status. Discoveries in the past few decades of cave ecosystems in a variety of cavernous rocks in diverse regions have revolutionized our understanding of cave life. We now believe that cave colonization and adaptation are general phenomena and occur wherever there are suitable underground voids available for evolutionary time. Most cave species remain undiscovered; in fact, the cave faunas of large areas containing caves, especially in the tropics, remain unsurveyed and unknown. Unfortunately, many cave systems are being destroyed before their faunas become known. The major anthropogenic threats to cave faunas include (1) mining of the surrounding rock, (2) changes in land use over subterranean habitats such as deforestation and urbanization, (3) alteration of groundwater flow patterns, (4) waste disposal and pollution, (5) invasion by nonindigenous species, (6) disruption of food inputs, and (7) direct human disturbance during visitation. Biological surveys are urgently needed. Also, recent systematic studies reveal that cave arthropod faunas are far more diverse than previously thought, indicating that priority should be focused on recognizing and protecting each distinct population rather than protecting a single population of each conventional species.

Conservation efforts must mitigate threats affecting the system, as well as recognize emerging threats. Generally, species extinctions result from novel perturbations, e.g., new stresses with which a species has had little experience during its evolution. Ecological studies are needed that improve our understanding of the functioning ecosystem, as well as understanding of natural successional processes. However, experimental ecological studies in caves are problematic because in few other habitats are humans so dramatically intruders as in caves. Not only do researchers affect the environment of the passages they study, but also they cannot sample the medium-sized voids where the major activity usually occurs. Caves are a fragile window through which we can see and study the fauna living within cavernous rock. Protected areas must include a sustainable portion of the ecosystem as well as suitable source areas for food and water resources. This usually represents an area larger than the footprint of the known cave.

RESEARCH OPPORTUNITIES

The bizarre adaptations displayed by troglobites make them excellent animals for evolutionary research. Recent advances in phylogenetic methods and molecular techniques provide important new tools for deciphering relationships among cave animals and their surface relatives. The discovery that close surface relatives are still extant for many tropical and island troglobites allows more appropriate comparative studies between species pairs adapted to wildly different environments. These studies should provide more critical understanding of how certain adaptations correlate with environmental parameters, as well as a better understanding of evolution in general. Some of these studies are in progress, for example, the work of Culver and colleagues on *Gammarus minus* in springs and caves in the eastern United States. Individual species of troglobites frequently have restricted distributions even within a given area of caves. Usually such a limited distribution indicates the existence of a barrier to subterranean dispersal, but not always. Critical morphological and behavioral studies, corroborated by modern molecular techniques, are showing that some troglobites thought to be widespread actually are composed of several more or less reproductively isolated populations. It has been assumed that cave adaptation was a dead end and that each of these populations evolved separately from the same or closely related surface ancestors that independently invaded caves. However, recent research by Hoch and colleagues on Hawaiian cixiid planthoppers suggests that some troglobites can disperse to new caves through underground voids and diverge into new species.

Caves are island-like habitats that support distinct ecosystems composed of communities of highly specialized organisms. Because the environment is discrete, rigorous, and easily defined, it provides an ideal system in which to conduct ecological studies. The number of species is usually manageable. The physical environment is rigidly constrained by the geological and environmental setting, and the environmental parameters can be determined with great precision because the habitat is surrounded and moderated by thick layers of rock. However, it is a rigorous, high-stress environment and difficult for humans to access and envision because it is so foreign to human experience. Also, one cannot enter or sample the mesocaverns where perhaps most cave animals live. These disadvantages can be overcome by comparing passages differing in the parameter of interest or by designing experiments that manipulate the parameter being studied in the natural environment. Biospeleology is still in the discovery phase. Although our understanding of cave biology has progressed substantially, results of future studies on evolution and ecology will be exciting and add significantly to our fascination with caves.

See Also the Following Articles
Aquatic Habitats • Conservation

Further Reading
Barr, T. C. (1968). Cave ecology and the evolution of troglobites. *Evol. Biol.* **2,** 35–102.
Camacho, A. I. (ed.) (1992). "The Natural History of Biospeleology." Museo Nacional de Ciencias Naturales, Madrid.
Chapman, P. (1993). "Caves and Cave Life." Harper Collins, London.
Culver, D. C. (1982). "Cave Life: Evolution and Ecology." Harvard University Press, Cambridge, MA.
Culver, D. C. (ed.) (1985). Special issue, regressive evolution. *Nat. Speleol. Soc. Bull.* **47(2),** 70–162.
Culver, D. C., Kane, T. C., and Fong, D. W. (1995). "Adaptation and Natural Selection in Caves. The Evolution of *Gammarus minus*." Harvard University Press, Cambridge, MA.
Hoch, H., and Howarth, F. G. (1993). Evolutionary dynamics of behavioral divergence among populations of the Hawaiian cave-dwelling planthopper *Oliarus polyphemus* (Homoptera: Fulgoroidea). *Pac. Sci.* **47,** 303–318.
Howarth, F. G. (1983). Ecology of cave arthropods. *Annu. Rev. Entomol.* **28,** 365–389.
Humphries, W. F. (ed.) (1993). The biogeography of Cape Range, Western Australia. *Rec. West. Aust. Mus.* Suppl. 45.
Juberthie, C., and Decu, V. (1996). "Encyclopaedia Biospeologica," Vol. I. Soc. Biospeologie, Moulis, France.
Juberthie, C., and Decu, V. (1996). "Encyclopaedia Biospeologica," Vol. II. Soc. Biospeologie, Moulis, France.
Vandel, A. (1965). "Biospeleology. The Biology of Cavernicolous Animals." Pergamon, Oxford. [Translated by B. E. Freeman]
Wilkens, H., Culver, D. C., and Humphries, W. F. (eds.) (2000). "Subterranean Ecosystems." Elsevier, Amsterdam.

Cell Culture

Dwight E. Lynn
U. S. Department of Agriculture

Cell culture is the technique in which cells are removed from an organism and placed in a fluid medium. Under proper conditions, the cells can live and even grow. The

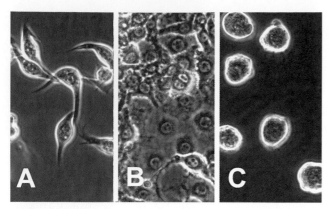

FIGURE 1 Typical appearances of insect cells in culture by phase contrast microscopy: (A) spindle shaped (fibroblast-like), (B) epithelial shaped, and (C) round.

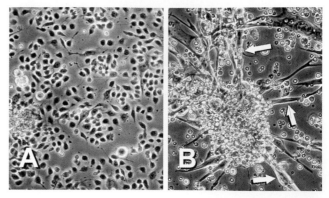

FIGURE 2 Cell shape change in response to treatment with the insect molting hormone: (A) Untreated cells. (B) Cells treated for 2 weeks with 20-hydroxyecdysone. Arrows point to cells that were contracting in the culture.

growth can be characterized by cell division (mitosis) or by other processes, such as differentiation, during which the cells can change into specific types that are capable of functions analogous to tissues or organs in the whole organism. The practice of cell culture (and its close cousins, tissue culture and organ culture) originated in a Yale University laboratory in 1907, when Ross Harrison removed nerves of a frog and maintained them in a simple salt solution for several days. Within a very few years a visiting scientist in Harrison's laboratory, Richard Goldschmidt, reported on the first cell cultures from an insect. For the next half-century, insect cell culture was used periodically in a variety of experiments, such as studying the pathogenesis of viruses, but the field received a great boost when the Australian Thomas D. C. Grace succeeded in obtaining four cell lines from the emperor gum moth, *Antheraea eucalypti*. These lines were capable of continuous growth, requiring periodic subculturing.

In the years since Grace's report, numerous other continuous insect cell lines have been developed—over 500 lines from more than 100 different insect species. Under microscopic examination, cells take on one of several distinct morphologies, including spindle shaped, epithelial, and round to oval (Fig. 1). Cell cultures are frequently used in research and biotechnology.

USES IN PHYSIOLOGY AND DEVELOPMENTAL BIOLOGY

Harrison's earliest work was designed to examine the physiology of a nerve outside the living organism, and similar uses have been made of insect cells and tissue. By removing the tissue or cells from the insect, it is possible to delineate how individual compounds affect them. Some of the most useful work has been with ecdysone, the insect molting hormone. In 1972 Anne-Marie Courgeon showed that exposing a cell line from *Drosophila melanogaster* to β-ecdysone (a particular form of ecdysone now known as 20-hydroxyedysone) caused rounded cells to change to aggregates of highly elongated cells. Lynn and

Hung found that a cell line from a small wasp can undergo a similar morphogenesis with the added feature that the elongated cells are highly contractile, like muscle cells (Fig. 2).

USES IN PATHOLOGY

Certain disease-causing organisms, such as viruses, rickettsia, and certain protozoans are obligate pathogens. Cell cultures can be extremely useful with such pathogens because this is often the only way to grow them outside a whole animal or plant and make them much easier to study. Some of the early work with insect cell culture was initiated with a group of viruses known as nucleopolyhedrovirus. These viruses cause diseases in a large number of pest insects and thus are potential biological control agents. Cell cultures grown in large volumes can be used to produce these viruses for biologically based pesticides.

Some insects are also notorious for their ability to transmit diseases (such as malaria and yellow fever) to higher animals and plants. Cell cultures from mosquitoes and other insects can also be used to study these pathogens.

USES IN MEDICINE AND PHARMACOLOGY

Decades ago, researchers showed that a gene in baculoviruses could be replaced through genetic engineering with genes for other proteins. When insect cells are infected with these modified viruses, the cultures can produce a large quantity of the protein. This technique—the baculovirus expression vector system—has facilitated a new use for insect cell cultures for the production of vaccines, growth factors, and other materials useful in medicine. Over 1800 articles have been published on the use of insect cells to produce various recombinant proteins.

See Also the Following Articles
Biotechnology and Insects • Genetic Engineering

Further Reading

Freshney, R. I. (2000). "Culture of Animal Cells: A Manual of Basic Technique." 4th ed. Wiley-Liss, New York.

King, L. A., and Possee, R. D. (1992). "The Baculovirus Expression System: A Laboratory Guide." Chapman & Hall, London.

Lynn, D. E. (1996). Development and characterization of insect cell lines. *Cytotechnology* **20,** 3–11.

Lynn, D. E., and Hung, A. C. F. (1991). Development of continuous cell lines from the egg parasitoids *Trichogramma confusum* and *T. exiguum. Arch. Insect Biochem. Physiol.* **18,** 99–104.

Maramorosch, K., and McIntosh, A. H. (eds.) (1994). "Insect Cell Biotechnology." CRC Press, Boca Raton, FL.

Maramorosch, K., and Mitsuhashi, J. (eds.) (1997). "Invertebrate Cell Culture: Novel Directions and Biotechnology Applications." Science Publishers, Enfield, NH.

Shuler, M. L., Wood, H. A., Granados, R. R., and Hammer, D. A. (eds.) (1995). "Baculovirus Expression Systems and Biopesticides." Wiley-Liss, New York.

Smith, G. E., Fraser, M. J., and Summers, M. D. (1983). Molecular engineering of the *Autographa californica* nuclear polyhedrosis virus genome: Deletion mutations within the polyhedrin gene. *J. Virol.* **46,** 584–593.

Chemical Defense

Murray S. Blum
University of Georgia

Biologists have become keenly aware that insects possess a remarkable ability to biosynthesize a large variety of compounds for use as agents of chemical defense against their omnipresent enemies. Many of these compounds are unique products (e.g., cantharidin, or Spanish fly, produced by blister beetles) with diverse modes of toxicity against a variety of vertebrate and invertebrate predators. These defensive secretions often originate from unlikely sources that appear to optimize the effectiveness of the chemical defensive systems. Ultimately, for countless species of insects, chemical defense and survival are synonymous.

ECLECTIC ORIGINS, FUNCTIONS, AND RESERVOIRS OF DEFENSIVE COMPOUNDS

It would be no exaggeration to state that the tremendous abundance of insects constitutes the primary food source for diverse vertebrate and invertebrate predators. For insects in a variety of orders, blunting the attacks of their omnipresent predators is identified either with the production of defensive compounds in exocrine glands or with the acquisition of these compounds from external sources. These deterrent allomones sometimes represent novel natural products that have a very limited distribution in the Insecta. In short, exocrine compounds, characteristic of species in orders or genera, have evolved to function as versatile agents of chemical defense.

It has been generally assumed that *de novo* biosynthesis characterizes the origins of insect defensive compounds. However, recent investigations suggest that novel insect defensive allomones, including the complex amide pederin from staphylinid beetles (*Paederus* spp.) and unique steroids from dytiscid beetles, are biosynthesized by endosymbiotes. These results raise the question of whether other novel insect allomones, including cantharidin and steroids in chrysomelids and lampyrids, may have microbial origins.

Often, however, the deterrent allomones constitute ingested allelochemicals such as cardenolides (milkweeds) and toxic pyrrolizidine alkaloids (asters, heliotrope). Furthermore, some of these plant natural products have been metabolized after ingestion into products that are suitable for sequestration and use as deterrents, as for ingested steroids from milkweeds by the monarch butterfly, *Danaus plexippus.* These compounds are also transferred to eggs to function as effective predator deterrents. In addition, these allelochemicals may be added to the secretions of exocrine glands, further increasing the deterrent properties of these exudates. The dependence on ingested plant natural products of some insect species is further emphasized by the utilization of "stolen" defensive exudates that essentially represent mixtures of pure plant allelochemicals that have been appropriated, unchanged, from the host plants.

In some species, ingested allelochemicals are sexually transmitted by the male as a copulatory "bonus" for the female. For example, the sperm-rich spermatophore of ithomiine butterflies is accompanied by pyrrolizidine alkaloids that provide protection for the female and her eggs. Importantly, this very adaptive system is functional because the spermatozoa are resistant to the well-known toxic effect of these alkaloids.

Some allelochemicals also possess great selective value for insects as antibiotic agents. Alkaloids such as α-tomatine, a constituent of tomatoes, reduce the infectivity of bacteria and fungi for lepidopterous larvae. Other compounds reduce the activity of viruses and in some cases are highly toxic to insect parasitoids.

Insects have adapted for defensive functions a variety of glands not identified as defensive organs. For example, salivary glands have been converted into defensive structures that deliver deterrent compounds biosynthesized in these glands. Even respiratory structures have assumed the role of deterrent organs as further testimony to the insect emphasis on defensive adaptations. For a variety of insect species, chemical defense is clearly identified with survival.

The defensive value of insect allomones has been further enhanced by the ability of these arthropods to adapt a variety of these natural products to subserve a surprising variety of multiple functions. This phenomenon, semiochemical parsimony, has been particularly emphasized by insect species such as fire ants, whose alkaloidal venoms possess a dazzling variety of pharmacological activities. The same may be said of cantharidin, the potent vesicant from blister beetles (Spanish fly).

Things are seldom what they seem. The sting-associated glands of bees and wasps are obvious candidates for the production of compounds with considerable deterrent activities. These glands have evolved as biosynthetic centers clearly dedicated to the biogenesis of pharmacologically active compounds that can be delivered by the sting in an unambiguous act of defense. On the other hand, some glands clearly identified with nondefensive functions have been adapted by a variety of insect species to function as defensive organs with varied functions. Furthermore, the deterrent efficiency of these secretions may be considerably enhanced by adding repellent plant natural products to the exudate. And insects have not neglected adapting enteric products to discourage their omnipresent predators. If all else fails, many insects eject blood, sometimes fortified with toxic allomones, at their adversaries with startling results. It is no exaggeration to state that for these species, bleeding has often provided an extraordinary means of deterring a variety of aggressive predators.

VARIETY OF SALIVARY DEFENSIVE FUNCTIONS

Salivary Venoms

The spitting cobra, *Naja nigricollis,* has an insect parallel, both in terms of the general chemistry of the saliva and the ability to accurately "fire" the venom at a moving target. For example, *Platymeris rhadamanthus* is a black and orange assassin bug (Reduviidae) that is very conspicuous because of its aposematic (warning) coloration. This insect can eject its saliva for a distance up to 30 cm, and if this enzyme-rich solution (proteases, hyaluronidase, phospholipase) strikes the nose or eye membranes of a vertebrate, intense pain, edema, and considerable vasodilation may follow. The saliva of *P. rhadamanthus* is admirably suited to deter vertebrate predators, including birds and reptiles. This salivary venom has clearly been evolved for predation on invertebrates, and rather than a specific site of action, it is reported to attack many organs simultaneously. Its speed of paralytic action is very pronounced: an American cockroach (*Periplaneta americana*) can be totally immobilized in about 4 s.

Entspannungsschwimmen (Chemically Induced Aquatic Propulsion)

The proteinaceous saliva of the hemipteran *Velia capraii* has been adapted to promote escape from potential predators in aquatic environments. This aquatic true bug will discharge its saliva onto the water surface, a reaction that results in lowering the surface tension of the water behind the bug. Under these circumstances, *V. capraii* is rapidly propelled across the water surface, putting considerable distance between itself and the source of the disturbance. Discharge of saliva posteriorly from the rostrum may project the bug 10 to 25 cm on the contracting water surface on which it is riding.

Allomonal Pheromones

Male bumble bees (*Bombus* spp.) scent-mark territorial sites with cephalic products that are very odoriferous. The secretions, which originate in the cephalic lobes of the salivary glands, are dominated by terpenes, some of which are well-known defensive compounds. This appears to be an excellent example of semiochemical parsimony, with the males utilizing the compounds both as territorial pheromones and as defensive allomones.

Salivary "Glues"

Termite workers in both primitive and highly evolved genera secrete defensive exudates that are rapidly converted to rubberlike or resinous products that can rapidly entangle small predators such as ants. This conversion frequently reflects the polymerization of salivary proteins that have reacted with *p*-benzoquinone, a highly reactive salivary defensive product. Similar systems for generating entangling salivas have been detected in a diversity of termite genera, including *Mastotermes, Microtermes, Hypotermes,* and *Odontotermes.*

Termites in other genera discharge cephalic exudates that are fortified with toxic terpenes. Species of *Nasutitermes* and *Tenuirostritermes* secrete mixtures of compounds that rapidly form a resin that entangles ants and other small predators. The presence of monoterpene hydrocarbons is probably responsible for killing ants and, in addition, may function as an alarm pheromone for recruiting termite soldiers.

NONSALIVARY ENTANGLING SECRETIONS

The posterior abdominal tergites and cerci of cockroaches in a variety of genera are covered with a viscous secretion that can act as an entangling glue for small predators. Species in genera as diverse as *Blatta* and *Pseudoderopeltis* produce proteinaceous secretions on the abdominal tergites that would be readily encountered by predators pursuing these cockroaches. After seizing the cockroaches, predatory centipedes, beetles, and ants rapidly release their prey while cleaning their mouthparts. The fleeing cockroaches generally have more than ample time to effect their escape.

Aphid species in many genera also utilize an entangling secretion as a primary means of defense. In this case the exudate is discharged in response to a confrontation, often hardening to a waxy plaque on an adversary within 30 s. This defensive behavior, which appears to be widespread in the Aphididae, uses tubular secretory organs, the cornicles, on the fifth and sixth abdominal tergites. The secretions, which are dominated by triglycerides, have been characterized in a range of genera, including *Aphis, Myzus, Acyrthosiphon,* and *Therecaphis.* The cornicular secretions are clearly more effective against generalized predators (e.g., ants) than they are against specialized predators (coccinellids, nabids). The secretions also contains alarm pheromones, *E-β*-farnesene and germacrene A, which release dispersive behavior that may cause aphids to drop off plants.

A variety of glands have been evolved by ants as sources of viscous defensive secretions. Many species in the subfamily Dolichoderinae discharge a pygidial (anal) gland secretion that is dominated by cyclopentanoid monoterpenes such as iridodial, compounds that rapidly polymerize on exposure to air. The viscous polymer effectively entangles small predators such as ants. Myrmicine species in the genus *Pheidole* also use the pygidial glands as a source of an entangling glue and in addition, an alarm pheromone. In contrast, a myrmicine species in the genus *Crematogaster* secretes a potent viscous deterrent from the hypertrophied metapleural glands. On the other hand, minor workers of a *Camponotus* sp. (Formicinae) produce a secretory "glue" in the capacious mandibular glands that extend through the entire body. Mechanical disturbance of the workers results in contraction of the gaster and eventual altruistic rupture, liberating the mandibular gland contents, which are very sticky and readily immobilize attacking ants.

DEFENSIVE FROTHS FROM DIVERSE GLANDS

A surprising diversity of defensive secretions has been converted to froths that may literally bathe small adversaries with compounds that seem to adversely stimulate the olfactory and gustatory receptors of their predators. The independent evolution of deterrent froths by moths, grasshoppers, and ants demonstrates that this form of defensive discharge can be highly efficacious in adverserial contexts.

Species in moth genera in the families Arctiidae (aposematic tiger moths), Hypsidae, and Zygaenidae secrete froths, the production of which is often accompanied by a hissing sound and a pungent odor. The aposematism of these moths is enhanced by secretions discharged from brightly colored areas on or near the prothorax. These secretions do not seem to contain plant natural products but rather, toxic *de novo* synthesized compounds such as pharmacologically active choline esters. Some arctiid froths contain blood, but its importance is unknown.

Frothing is highly adaptive in the ant genus *Crematogaster.* Workers in this very successful myrmicine genus do not possess a hypodermic penetrating sting, but rather, a spatulate sting that is enlarged at the tip. Venom accumulates at the tip and can be smeared onto small adversaries such as ants as if with a paintbrush. This mode of administration of venom is obviously identified with a topical toxicant that can penetrate the insect cuticle much as an insecticide does. There is no indication that tracheal air is added to the venom to generate the discharged froth.

Two grasshopper species produce froths that are derived from a mixture of tracheal air and glandular secretion. Both species are eminently aposematic, and this warning coloration is enhanced by a powerful odor emanating from the froths of the pyrgomorphid *Poekilocerus bufonius* and the acridid *Romalea guttata.*

P. bufonius, a specialized milkweed feeder, is brilliantly colored, exhibiting a dark bluish gray background with contrasting yellow spots and orange hind wings. From a bilobed gland opening between the first two abdominal tergites, disturbed grasshoppers discharge a viscous secretion that is converted to a froth when it mixes with air while passing over the second abdominal spiracle. The froth enhances the aposematism of *P. bufonius* by appearing to be rainbow tinted in contrast to the dark background. This grasshopper is well protected from predation because its exudate contains *de novo* synthesized toxins and sequestered plant natural products that are strongly emetic.

In contrast to *P. bufonius, R. guttata* is a generalist feeder but as with *P. bufonius,* its defensive exudate is discharged as a froth that contains plant natural products as well as compounds synthesized by the grasshopper. If *R. guttata* temporarily specializes on a plant species rich in allelochemicals (*Allium* spp.), its defensive froth can be highly repellent. The aposematism of this acridid rivals that of the pyrgomorphid, and the warning coloration of the former is considerably enhanced by a loud hissing that accompanies the very odoriferous secretory froth.

EXTERNALIZING ALLOMONES BY REFLEX BLEEDING

Many insect species, particularly beetles, externalize their distinctive defensive compounds in a blood carrier rather than discharging them as components in an exocrine secretion.

Cantharidin, the terpenoid anhydride synthesized by adult beetles in the families Meloidae and Oedemeridae, is externalized in blood discharged reflexively from the femorotibial joints. The repellent properties of cantharidin were established more than 100 years ago, and the ability of amphibians to feed on these beetles with impunity has been long known, as well. Cantharidin possesses a wide spectrum of activities, including inducing priapism in the human male, and it has been reported to cause remission of epidermal cancer in mammals. Although its role as a repellent and lesion producer certainly documents its efficacy as a predator deterrent, its potent antifungal activity may be of particularly great adaptiveness in protecting developing meloid embryos from entomopathogenic fungi present in their moist environment.

Autohemorrhage, from the femorotibial joints, is widespread in many species of ladybird beetles (Coccinellidae), most of which are aposematic. The blood is generally fortified with novel alkaloids that are outstanding repellents and emetics (i.e., inducers of vomiting) as well.

Adult fireflies (*Photinus* spp.) produce novel steroids (lucibufagins) that are effective repellents and inducers of emesis in invertebrates and vertebrates. Reflex bleeding from specialized weak spots in the cuticle along the elytra and antennal sockets externalizes these steroids.

Sometimes, rapidly coagulating blood, free of allomones, is used defensively.

BLOOD AS PART OF A GLANDULAR SECRETION

Often the secretions of defensive glands are fortified with blood. The mechanism by which blood penetrates the defensive gland preparatory to being discharged remains to be determined, and indeed, the exact function of the blood itself is not known.

Arctiid moths (e.g., *Arctia caja*) discharge odoriferous froths from prothoracic glands, and these exudates contain pharmacologically active choline esters that are accompanied by blood. A similar system characterizes the pyrgomorphid grasshopper *P. bufonius*. Nymphs of this species possess abdominal defensive glands that sequester two of the six compounds ingested from their milkweed hosts and, in addition, synthesize high concentrations of histamine. These compounds are accompanied by blood as a normal component of the secretion.

NONGLANDULAR DISCHARGES OF PLANT ORIGIN

Certain insects have evolved storage reservoirs for plant natural products that can be discharged in response to traumatic stimuli. This evolutionary development reflects the insect's appropriation of plant allelochemicals (defensive compounds) for subsequent utilization as defensive allomones. In essence, the insects have sequestered the plant's defenses and stored them in reservoirs, where they are available as defensive agents. This defensive system does not require the evolution of any biosynthetic pathways for the storage of compounds in nonglandular reservoirs.

Adults of hemipterous species in the family Lygaeidae possess dorsolateral (reservoirs) and abdominal spaces that contain a fluid very similar to that of the proteins in the blood. This fluid sequesters steroids (cardenolides) present in the milkweeds on which these species feed. The cardenolides are about 100-fold more concentrated in the dorsolateral fluid than they are the blood, and they thus constitute a formidable deterrent system.

Sequestration of plant natural products in nonglandular reservoirs also characterizes larvae of the European sawfly, *Neodiprion sertifer*. Feeding on pine (*Pinus* spp.), these larvae sequester both mono- and sesquiterpenes in capacious diverticular pouches of the foregut. Young larvae, feeding only on pine needles, sequester only three terpenes, whereas older larvae also ingest resin acids. These acids also serve to entangle would-be predators, thus providing a dual protective function.

PLANT NATURAL PRODUCTS IN EXOCRINE SECRETIONS

Herbivorous insects may incorporate plant natural products into exocrine and nonexocrine defensive secretions. By selectively adding proven plant repellent compounds to their own deterrent secretions, insects can increase the effectiveness of their chemical deterrents. These plant-derived compounds are generally unrelated to the constituents in the defensive exudates of their herbivores. In all likelihood, these plant additives may augment the repellency of the deterrents by reacting with olfactory chemoreceptors different from those targeted by the insect-derived repellents.

The large milkweed bug, *Oncopeltus fasciatus,* in common with many species of true bugs, uses the secretion of the metathoracic scent gland as an effective defensive exudate. Nymphs of this species generate defensive secretions with midsorsal glandular fluid. The repellent secretions also contain cardenolides derived from the milkweed host plants of this species. These toxic and emetic steroids undoubtedly augment the deterrent effectiveness of the *de novo* synthesized compounds in the glandular exudates.

Similarly, *R. guttata* sequesters in the metathoracic defensive glands plant allelochemicals that can considerably augment the deterrent effectiveness of the secretion. Unlike *O. fasciatus*, *R. guttata* is a generalist that feeds on and sequesters a potpourri of plant natural products. As a consequence, the compositions of the glandular exudates can be variable, sometimes resulting in secretions that are considerably more repellent than those derived from insects that had fed on a limited number of host plant species.

REGURGITATION AND DEFECATION OF ALLELOCHEMICALS

Enteric defense may be widespread in insects as a means of using the proven repellencies of a variety of plant natural products. In a sense, the intestine is functioning as a defensive organ once repellent plant products have been ingested, and it is likely that the presence of pharmacologically active plant compounds in the intestine renders the insect distasteful or emetic. Therefore, transfer of gut contents to the outside by either regurgitation or defecation could actually constitute the externalization of the internal enteric defenses.

When tactilely stimulated, acridid grasshoppers readily regurgitate, and this discharge, fortified with plant natural products, is very repellent to ants. Similarly, larvae of the moth *Eloria noyesi* regurgitate when molested. The enteric discharge, which contains cocaine extracted from the larval food plant, is very repellent to ants.

Defecation can also serve to externalize deterrent plant natural products. The large milkweed bug defecates readily when subject to traumatic stimuli, the discharge being fortified with emetic and distasteful cardenolides (steroids) derived from the milkweed host plants. The anal discharge, containing concentrated cardenolides, is very repellent to ants.

See Also the Following Articles
Aposematic Coloration • *Autohemorrhage* • *Defensive Behavior* • *Monarchs* • *Venom*

Further Reading
Blum, M. S. (1981). "Chemical Defenses of Arthropods." Academic Press, New York.

Blum, M. S. (1996). Semiochemical parsimony in the Arthropoda. *Annu. Rev. Entomol.* **41**, 353–374.

Dettner, K. (1987). Chemosystematics and evolution of beetle chemical defenses. *Annu. Rev. Entomol.* **32**, 17–48.

Duffey, S. S. (1977). Arthropod allomones: Chemical effronteries and antagonists. *Proc. XV Int. Cong. Entomol.* Washington, DC, pp. 323–394.

Duffey, S. S. (1980). Sequestration of plant natural products by insects. *Annu. Rev. Entomol.* **25**, 447–477.

Edwards, J. S. (1961). The action and composition of the saliva of an assassin bug *Platymeris rhadamanthus* Gaerst. (Hemiptera: Reduviidae). *J. Expt. Biol.* **38**, 61–77.

Eisner, T. (1970). Chemical defense against predation in arthropods. In Chemical Ecology" (E. Sondheimer and J. B. Simeone, eds.), pp. 157–217. Academic Press, New York.

Hartmann, T. (1995). Pyrrolizidine alkaloids between plants and insects: A new chapter of an old story. *Chemoecology* **5**, 139–146.

Kellner, R. L. L. (1999). What is the basis of pederin polymorphism in *Paederus riparius* rove beetles? The endosymbiotic hypothesis. *Entomol. Exp. Appl.* **93**, 41–49.

Pasteels, J. M., and Gregoire, J.-C. (1983). The chemical ecology of defense in arthropods. *Annu. Rev. Entomol.* **28**, 263–289.

Chemoreception

B. K. Mitchell
University of Alberta

Insects are acutely aware of many aspects of their environment, as anyone knows who has tried to catch a fly perched on a slice of pizza. In the chemical realm, and depending on the chemicals and insects involved, insects are often outstandingly sensitive. The most famous and best-studied aspects of chemoreception in insects are mate recognition and finding. Like many, if not most animals, insects produce chemicals called pheromones that allow individuals of one sex in a species to recognize and find individuals of the opposite sex. Usually the female produces a mixture of chemicals to which the male responds. Other important, life-or-death decisions largely based on chemicals include choice of site for egg laying, decisions about what to eat and what to avoid, and communications about immediate danger. How insects taste and smell is therefore of great interest and, given that many insects are serious agricultural pests and vectors of disease, research in this area is both fundamental and practical.

As with most physiological systems, model animals are vitally important for scientists who explore the specific workings of what is always a complex series of interactions. For studies of insect chemoreception, adult moths and caterpillars, flies, cockroaches, and leaf beetles have provided some of the best models. Large moths such as the silkworm, *Bombyx mori,* and the tobacco hornworm, *Manduca sexta,* have been essential in studies of pheromones, whereas flies such as the black blowfly, *Phormia regina,* caterpillars such as the cabbage butterfly, *Pieris brassicae,* and *M. sexta,* and

beetles such as the Colorado potato beetle, *Leptinotarsa decemlineata,* have helped unravel the role of chemoreception in food and oviposition-related behavior.

For an insect to sense and respond appropriately to the presence of a chemical, or more often a mixture of chemicals, requires a large number of cuticular, cellular, and molecular processes. Because insects are covered in cuticle, it is appropriate to begin there. The cells involved include the sensory cells themselves and closely associated accessory cells whereas the molecules include a wide array of extracellular, intracellular, and membrane-bound proteins. The processes involved in tasting and smelling include sampling the environment, transport of stimulus molecules to receptors, reception, transduction, coding, and transmission to the higher brain centers. This article looks at both gustation (taste) and olfaction (smell), though in many areas detailed knowledge is more complete for smell than it is for taste.

ROLE OF CUTICLE IN TASTE AND SMELL

Insects, like all arthropods, are covered with a chitin–protein complex called cuticle, which in turn is covered with wax to prevent desiccation. For the creature to taste or smell anything, there must be a pathway from the outside to the sensory cells inside. On various parts of the insect body, but particularly on the antennae, mouthparts, legs, and ovipositor (egg-laying structure) insects possess a variety of cuticular elaborations in which are housed chemically sensitive cells. These cuticular structures take the form of hairs (trichoids), pegs, pegs in pits, flat surfaces, and several other shapes. Common to them all is a modified cuticular region that will provide one or more pores through which chemicals can gain entrance. For water conservation, and to keep the important sensory cells functional, these pores cannot allow direct contact of the sensory cell membrane with air. All these pores are small (in the submicrometer range), and there is always a water–protein pathway from the pore to the cell membrane. The cuticular structures plus the associated cells collectively are referred to as sensilla.

Figure 1 represents a reconstruction of a typical mouthpart gustatory sensillum in a caterpillar. All caterpillars so far investigated have this type of sensillum, and it is always important in the food selection processes. The reconstruction is based on careful observations of hundreds of images taken with the electron microscope. The cellular details shown in the drawing cannot be seen with the light microscope. Most of the parts of this drawing below the cuticle could be mistaken for those in an olfactory sensillum. This is because chemosensory cells in insect sensilla are modified cilia and the accessory cells are also basically the same in both types. This involvement of cilia is not surprising, because most of the sensory cells of animals, including light, touch, and hearing, as well as chemical sensors, are modified cilia. Only the sensory cells are modified cilia. The accessory cells are more ordinary, although still specialized, epidermal cells, and they

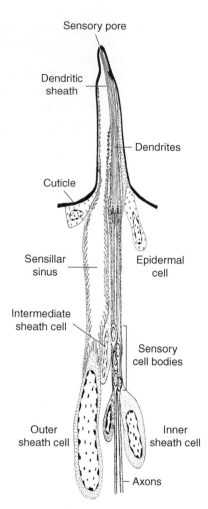

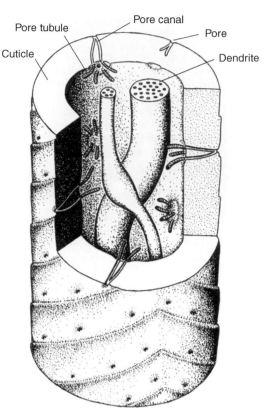

FIGURE 2 Schematic view of a section of a pheromone sensillum in a moth. The features are those revealed in an electron microscopic examination. Olfactory sensilla may have as few as two sensory dendrites, as here, or many more. The arrangement shown is typical of many moth pheromone sensilla. [Relabeled from Keil, T. A. (1999). *In* "Insect Olfaction" (B. S. Hansson, ed.), Fig, 17a, p. 39. © Springer-Verlag GmbH & Co. KG, Berlin.]

FIGURE 1 Reconstruction of a taste sensillum of the type typically found on the mouthparts of caterpillars. Associated with the maxilla there are four such sensilla, each with four gustatory cells, and it is clear that caterpillars rely heavily on the information provided by the cells to make food choices. The cuticular modification, accessory cells, and sensory cells are all necessary for the sensillum to function properly. In addition to providing the sense of taste, these sensilla are also sensitive to touch. [From Shields, V. D. C. (1994). *Can. J. Zool.* **72**, 2016–2031, as modified by Mitchell, B. K., *et al.* (1999). *Microsc. Res. Technol.* **47**, 401–415.]

have two very different functions. During the development of a sensillum (i.e., between molts) these cells are involved in secreting all the cuticular elements of the sensillum, including the base, the shaft, the cuticular pore or pores, and the dendritic sheath surrounding the dendrites (above the cillary rootlets) of the sensory cells. Once the dendritic sheath is in place, the dendrites are physically separated from the rest of the sensillum lumen, though chemicals can pass through. The dendritic sheath is much longer in taste sensilla, as depicted in Fig. 1, running all the way to the single pore in the tip. The dendritic sheath in olfactory sensilla stops nearer the base of the sensillum, and the dendrites are free in the lumen. In both types, the dendritic sheath provides mechanical stabilization for the sensory cells. When development is complete, the accessory cells provide the particular chemical

ionic mix that surrounds the dendrites (note the microvilli in the outer sheath cell). The fluid surrounding the dendrites is very different from the general body fluid (hemolymph), and its high cation concentration is critical in allowing the cell to signal its contact with an appropriate chemical stimulus. This signal is in the form of a potential change across the dendritic cell membrane that is eventually turned into normal action potentials near the sensory cell body.

The structural features discussed so far are shared by olfactory and gustatory sensilla. The major differences between the two types have to do with the way chemicals get into the system and the underlying cuticular modifications. Chemicals enter gustatory sensilla via the single pore in the tip. This pore contains a sugar–protein complex (mucopolysaccharide) that protects the dendrites from desiccation and probably limits the types of chemicals that can pass (though this latter point is in need of further study). Once past this barrier, the chemical enters the solution around the dendrites and potentially can interact with the cell. Olfactory sensilla typically have many pores, and they are different in origin from those in gustatory sensilla. Figure 2 illustrates a section of a typical olfactory sensillum from the pheromone system of a moth. To understand the nature of the numerous pores on this structure requires knowledge of insect cuticle in general. The surface of insect cuticle is in constant communication with the inside of the animal for the purpose of wax renewal. This communication is provided by numerous pore canals, microscopic and tortuous passages through the cuticle. These canals are filled with a water–wax mix. In olfactory sensilla, the pore canals are taken over for the function of providing access of stimuli to the sensory dendrites. On the inside end of some pore canals are structures called pore tubules; these delicate structures can be seen only in electron micrographs of carefully prepared tissue. It was once thought that pore tubules provided a hydrophobic route for odor molecules to pass from the outside waxy surface of the sensillum to the surface of the dendrite (which is surrounded by water and salts). Discovery of additional molecular components of this system replaced this long-standing and attractive hypothesis.

THE ODOR PATH

Substances animals taste are usually much more water soluble than those that they smell, and the sensory dendrites of both gustatory and olfactory sensilla are in an aqueous medium. Thus, the problem of getting the stimulus to the receptor has received much more attention in olfactory research. In insects, odor molecules first contact the cuticular surface, and because it is waxy, they easily dissolve. From here they move in two dimensions, and some find their way into the opening of a pore canal. Since the pore canal contains wax, passage through it is probably easy, and passage in pore tubules may be similar. Eventually, however, before it arrives at the receptor surface of a dendrite, the hydrophobic odor molecule will encounter water. Recent work, particularly with the antennae of large moths, has uncovered at least two types of protein in the extracellular spaces of sensilla. One type specifically binds chemicals that are part of the moths' pheromone mixture, and are therefore called pheromone binding proteins (PBP). The other type binds less specifically a variety of nonpheromone molecules (e.g., food odors) and are called general odorant binding proteins (GOBP).

The odorant binding proteins (OBP) act as shuttles and carry odor molecules through the aqueous medium to the surface of the dendrite. In the membrane of the sensory cell are receptors for various odors, depending on the specificity of the cell. Cells that respond to only a single pheromone would be expected to have only one type of receptor molecule. More typically, a cell that is sensitive to food odors has a variety of related receptors covering various stimuli. In either case, the odorant binding protein, now carrying the odor molecule, comes in contact with a receptor. What happens next is now under investigation, and there are two competing hypotheses. The OBP may simply deliver the stimulus, which itself then interacts with the receptor protein; or, the stimulus–OBP complex may be the actual stimulus. That is, the receptor site may be configured as to recognize only the combined stimulus and OBP; either alone

will not fit. The latter hypothesis may also explain how these systems can turn on and off so quickly: namely, because moths can follow a discontinuous (patchy) odor trail, making minute adjustments in flight pattern on a millisecond scale. This precise behavior is corroborated by electrophysiological measurements showing that the sensory cells can follow an on-and-off pattern of odor stimulation, also in the millisecond range. It is possible that the OBP–stimulus complex, when first formed, is the effective stimulus for the receptor. During the interaction with the receptor, however, the OBP–stimulus complex changes slightly, becomes inactive, and immediately leaves the receptor. Later, it is broken down by other proteins (enzymes) in the sensillum lumen. Figure 3 summarizes this complex series of events and emphasizes the second hypothesis.

CHEMICAL-TO-ELECTRICAL TRANSDUCTION

In almost all studies of animal sensory systems, the stimulus being sensed is in a different energetic form than the chemoelectrical transmission used by the nervous system. Thus, in eyes, light (photon) energy needs to be transduced into chemoelectrical energy via photo pigments. Similarly, with a chemical stimulus–receptor complex, once binding between stimulus and receptor has occurred, the event must be communicated to other parts of the sensory cell to ensure that the end result is a message, composed of action potentials, transmitted to the brain. Understanding of chemical transduction in insects is far enough along to permit the statement that the

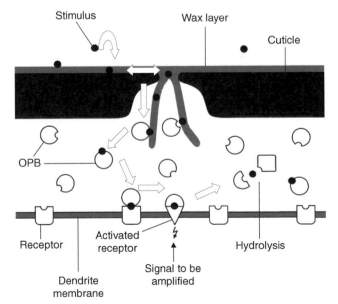

FIGURE 3 Schematic summary of the movement (arrows) of an odor molecule (solid circles) from the surface of a sensillum to the dendritic membrane. Specialized proteins (various shapes) act sequentially as carriers, receptors, and hydrolytic agents to make precise detection of the odorant possible. See text for details. [Relabeled, from Stengl *et al.* (1999). *In* "Insect Olfaction" (B. S. Hansson, ed.), Fig. 1, p. 66. © Springer-Verlag GmbH & Co. KG, Berlin.]

basic elements are probably very much like the arrangement in the vertebrates. There will be differences in detail, but these will continue to be the subjects of active research for some time. Basically, most chemotransduction requires (1) a more or less specific receptor molecule (thus the stimulus–receptor complex can be formed), (2) an amplification step (involving a series of membrane-bound and intracellular molecules) that turns a few stimulus–receptor events into a significant, momentary elevation of some chemical (often calcium) inside the cell, (3) at least one ion channel that senses the rise in calcium and opens, allowing depolarization, and (4) a braking (deactivation) system, composed of more molecular interactions, so the system can be precisely controlled.

Parts of a complete transduction system are beginning to emerge from electrophysiological (patch-clamp) studies of cultured olfactory cells, pharmocological experiments on these cells and on whole-sensillum studies of fly taste sensilla, and from genetic work with *Drosophila* fruit flies. The fruit fly work used specific searches of the now complete *D. melanogaster* genetic database to find some likely candidates for sugar receptor proteins. Carlson has used this information to make specific fluorescent probes, and some of these probes bound only with cells in gustatory sensilla. Combinations of genetic analysis, molecular biology, electrophysiology, and pharmacology will be needed to define all the necessary components.

CHEMOSENSORY CODING AT THE PERIPHERY

In the real world, animals encounter thousands of chemicals. Most of these are meaningless, in the sense that no behavioral response is required, whereas some are critical. A sensory system thus serves two opposing functions. First, the effective sensory system must act as a filter, allowing the animal to ignore most potential stimuli so that it can concentrate on the important ones. Second, the same system must be sensitive, sometimes exquisitely sensitive, to biologically relevant stimuli and must continuously transmit a "summary" report to the brain or central nervous system. The receptor proteins and associated transduction molecules provide the specificity (only some things are adequate stimuli) and sensitivity (the effectiveness of the amplification step in transduction). The nature of the summary neural message is the problem addressed in studies of peripheral sensory coding. In insect chemosensory coding, the problem can be as simple as a few highly specific receptor proteins recognizing a three- or four-component blend of pheromone molecules all housed on a pair of cells found in each of many thousand antennal sensilla. At the other extreme, a leaf beetle may be faced with a food choice of two closely related plants, each with many chemicals to which its tens of gustatory cells are capable of responding. When one is comparing these two scenarios, it is not the number of sensory cells that constitutes the relative scale of the coding problem, but the number of chemical compounds that can be sensed by these cells, and the combinations of compounds that are possible.

In the pheromone example, there are two cell types (each sensillum has one of each type). They respond differently to, for example, four pheromone molecules and not much else. Also, one or two of the pheromone molecules may be completely nonstimulatory to one of the two cells. In addition, only two of the four compounds in the blend may be sufficient to stimulate a full array of behaviors necessary for the male to find the female. The coding problem, though overly simplified to make the point, could thus be reduced to the following: cell A responds only to compound A, and cell B responds only to compound B. Both cells continuously signal to the antennal lobe the levels of compounds A and B detected in the air. If cell A is firing at twice the rate of cell B and both cells are firing at some rate, then the moth flies upwind. Thus the code is a simple comparison, and the large number of cells involved is a kind of amplifier, reflecting the overwhelming importance of the pheromone system to the animal. The two cells, A and B in this example, can be thought of as labeled lines, each sending unique information about the concentration of compound A or B. The central nervous system uses a simple hardwired rule to compare this paired input, and, accordingly, behavior is or is not released.

The beetle, potentially, has a more difficult coding problem. Many experiments have shown that gustatory cells of plant-feeding insects are affected by numerous single plant compounds. Ubiquitous compounds such as water, salts, amino acids, and sugars are sensed by some cells on the mouthparts of all such insects. Less widely distributed chemicals such as alkaloids, terpenes, glucosinolates, and other so-called secondary plant compounds, are stimuli for cells that are variously scattered throughout the class Insecta. To exemplify this coding problem, consider a Colorado potato beetle facing the choice of a potato leaf (host plant) or a tomato leaf (marginal host) (Fig. 4A). The gustatory cells in the beetle's mouthpart sensilla (on the galea), are all sensitive to different compounds. Both direct stimulation by some molecules and inhibition of one molecule by another are known, as are some injury effects in the presence of when too much glycoalkaloid (compounds in potatoes and tomatoes). Not surprisingly, the summary report such a four-cell system sends to the brain comprises two kinds of message, one for potato and one for tomato (Fig. 4B). The complex array of stimuli represented by potato actually stimulate a single cell—the others may well be inhibited. The tomato leaf juice, on the other hand, causes several cells to fire in an inconsistent pattern. The first is another example of a labeled-line type of code; while the second is an across-fiber pattern. In the latter type of code, the brain is receiving information from several physiologically distinct cells, and it is the pattern that is important. It is thought that the across-fiber code pattern prevails in many situations involving complex chemical mixtures. Progress in this area is impeded by the inherent variability of the types of recording possible in the across-fiber pattern (see, e.g., Fig. 4B).

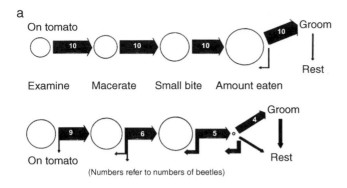

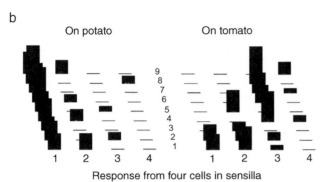

FIGURE 4 (a) Summary of the behaviors exhibited by newly emerged Colorado potato beetle adults when provided with either potato (host plant) or tomato (nonhost plant); numbers of beetles indicated inside heavy arrows. Beetles first examine the leaf, then they squeeze it between their mandibles (macerate) before taking a small bite, which they taste for only a short time. If the plant is acceptable, they very quickly move to sustained feeding. If the plant is less acceptable, few beetles will feed. The decision to not feed is made after considerable time has been spent in examining, macerating, taking small bites, and sometimes repeating one or more of these steps. [Modified from Harrison, G. D., (1987). Host–plant discrimination and evolution of feeding preferences in the Colorado potato beetle, *Leptinotarsa decemlineata. Physiol. Entomol.* **12**, 407–415.] (b) Taste sensilla are important in making the kinds of decisions shown in (a). If potato leaf juice is the stimulus, four cells in nine sensilla on the mouthparts respond by sending a clear, almost labeled-line (cell 1), message to the central nervous system. When tomato leaf juice is the stimulus, a mixed message is provided from the four cells housed in each of the nine sensilla, and this message varies considerably across the available sensilla. The result is a type of across-fiber pattern that signals "do not eat." [Modified from Haley Sperling, J. L., and Mitchell, B. K. (1991). A comparative study of host recognition and the sense of taste in *Leptinotarsa. J. Exp. Biol.* **157**, 439–459. © Company of Biologists LTD.]

CENTRAL PROCESSING OF CHEMOSENSORY INPUT

Over the past 20 years, studies of insect olfactory systems have produced a rich literature on the topic of central processing, particularly for pheromonal systems. Work on gustatory systems is far less advanced. The section on insect pheromones provides more information on olfactory processing. This section simply contrasts the gross morphology of the two systems. Both olfactory and gustatory sensory cells are primary neurons; that is, they connect the periphery (sensillum) directly with the central nervous system. Olfactory cells, on the antennae as well as on the palpi, send their axons directly to the antennal lobe, which

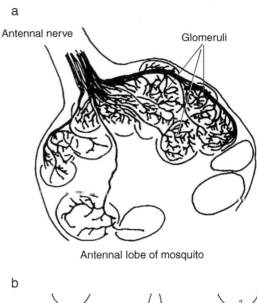

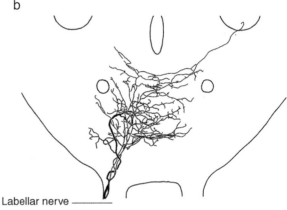

FIGURE 5 In insects, both olfactory and gustatory cells send axons (afferents) directly to the central nervous system. The first synapse (information relay point) is in a particular part of the central nervous system for each sensory modality. (a) Olfactory afferents go to the antennal lobe, where the input is organized in a manner resembling a bunch of grapes—glomerular organization. [From Anton, S., and Homberg, U. (1999). *In* "Insect Olfaction" (B. S. Hansson, ed.), Fig. 6a, p. 110. © Springer-Verlag GmbH & Co. KG, Berlin.] (b) Gustatory afferents from mouthpart sensilla go to the subesophageal ganglion, where they project into a discrete space that is not organized into glomeruli. For both (a) and (b), subsequent processing is done by first-level and higher interneurons. [From Edgecomb, R. S., and Murdock, L. L. (1992). Central projections of axons from the taste hairs on the labellum and tarsi of the blowfly, *Phormia regina* Melgeri. *J. Comp. Neurol.* **315**, 431–444. Reprinted by permission of Wiley-Liss, Inc., a subsidiary of John Wiley & Sons, Inc.]

is a part of the insect brain. Gustatory cells, for the most part, send their axons to the ganglion for the segment in which the sensory cell occurs. Figure 5A shows a typical innervation pattern for antennal and mouthpart olfactory cells in a mosquito, and Fig. 5B shows innervation from the gustatory cells in the mouthparts of a blowfly. A striking difference in the organization of the two parts of the central nervous systems receiving these imputs is repeated across many animal phyla. Olfactory systems are characterized by a glomerular arrangement (like a bunch of grapes) of the neural centers (neuropile) that receive olfactory afferents (input), but gustatory systems have no such patterned arrangement. The distribution of olfactory inputs into glomeruli suggests a strong association of structure with function, and this is most clearly seen in the macroglomeruli, which receive only pheromonal afferents in male moths. There is undoubtedly an association of structure with function in the way gustatory inputs are arranged, but the lack of a glomerular substructure makes any such system far less obvious. The two ways of organizing chemosensory input, throughout animals, may also point to important differences in coding and or evolution.

See Also the Following Articles

Antennae • Brain • Feeding Behavior • Mechanoreception • Mouthparts • Pheromones

Further Reading

Clyne, P. J., Warr, C. G., and Carlson, J. R. (2000). Candidate taste receptors in *Drosophila. Science* **287**, 1830–1834.

Hansson, B. S. (ed.) (1999). "Insect Olfaction." Springer-Verlag, Berlin, Heidelberg. (See especially Chaps. 1, 2, 3 and 4.)

Hildebrand, J. G., and Shephers, G. M. (1997). Mechanisms of olfactory discrimination: Converging evidence from common principles across phyla. *Annu. Rev. Neurosci.* **20**, 595–631.

Mitchell, B. K. (1994). The chemosensory basis of host–plant recognition in Chrysomelidae. *In* "Novel Aspects of the Biology of the Chrysomelidae" (P. H. Jolivet, M. L. Cox, and E. Petitpierre, eds.), pp. 141–151. Kluwer, Dordrecht, The Netherlands.

Mitchell, B. K., Itagaki, H., and Rivet, M.-P. (1999). Peripheral and central structures involved in insect gustation. *Microsc. Res. Technique* **47**, 401–415.

Pollack, G. S., and Balakrishnan, R. (1997). Taste sensilla of flies: Function, central neuronal projections, and development. *Microsc. Res. Technique* **39**, 532–546.

Schoonhoven, L. M., Jermy, T., and van Loon, J. J. A. (1998). "Insect–Plant Biology." Chapman Hall, London. (See especially Chaps. 5 and 6.)

Steinbrecht, R. A. (1997). Pore structures in insect olfactory sensilla: A review of data and concepts. *Int. J. Insect Morphol. Embryol.* **26**, 229–245.

Zacharuk, R. Y., and Shields, V. D. (1991). Sensilla of immature insects. *Annu. Rev. Entomol.* **36**, 331–354.

Chiggers and Other Disease-Causing Mites

Larry G. Arlian
Wright State University

The Acari (mites and ticks) represent a large array of organisms that exhibit very diverse lifestyles. This article deals with the acarines that are of importance to human health, a group that includes human parasites, natural parasites of other mammals and birds that in particular situations may bite humans, and acarines whose fecal matter, body secretions, and disintegrating bodies are sources of potent allergens.

The parasitic Acari of vertebrates are physiologically dependent on their host and must obtain nourishment from tissue fluids, blood, and cytoplasm from the host to survive, complete the life cycle, and reproduce. Thus, these are obligate parasites. Some species are temporary parasites (e.g., ticks), which visit and feed on the host intermittently. In contrast, other species of parasitic Acari (e.g., scabies and follicle mites) are permanently associated with the host and perish if they become separated from the host. For some species, only one life stage in the life cycle is a parasite (e.g., chiggers), whereas for other species each life stage must feed from a vertebrate host to complete the life cycle (e.g., scabies mites and ticks).

There is usually an intimate interrelationship between acarine parasites and their hosts. Specific host factors, such as carbon dioxide, body odor, and temperature, allow the parasite to locate a host. For example, scabies mites are attracted to the host by body odor and temperature. Permanent parasites may be directed to specific areas of the host body by factors in the skin. The host–parasite interactions for most parasitic acarines have not been well studied and thus are not well understood. This article discusses mites that bite humans, live in the skin of humans, or produce substances that induce immune and/or inflammatory reactions. Because acarine parasites can induce inflammatory and adaptive immune responses, an understanding of the relationship between these two responses is important if one is to understand the symptoms associated with bites from parasitic mites or reactions to body parts, secretions, and fecal matter.

INFLAMMATORY AND IMMUNE RESPONSES

When feeding from the host skin surface, acarine parasites inject or secrete into the host an array of immunogenic and pharmacokinetic molecules. Likewise, acarine parasites that live in the skin, hair follicles, sebaceous glands, and respiratory tree and lungs release immunogenic molecules both while living and after death, from their disintegrating bodies. Substances injected or released may induce an inflammatory (i.e., innate) and/or immune (i.e., adaptive) response by the host. Pharmacokinetic molecules can modulate specific aspects of the host immune or inflammatory responses.

Innate Immune Response

After a person has been bitten by a parasitic acarine, a red (or erythematous) swollen (i.e., edema), and irritated (i.e., painful) lesion may develop at the bite site. These symptoms may be the result of a localized innate inflammatory reaction and not an adaptive immune reaction. In an inflammatory reaction, components of the saliva and body secretions of mites that feed from the skin surface or in tissue (e.g., follicle or scabies mites) cause cells of the skin (epidermis and dermis) such as keratinocytes, fibroblasts, and antigen-presenting cells (Langerhans, macrophages, natural killer cells) to release an array of chemical mediators (cytokines, kinins, and others). These substances cause arterioles to dilate, which results in increased blood flow to the tissue. Increased blood flow to the skin where a mite has bitten or is located imparts a red appearance. In addition, the tight junctions between endothelial cells of the capillary wall become less tight, which allows fluid from the blood to leak from the capillary lumen into the surrounding tissue, causing it to swell. These cytokines also cause local endothelial cells in the capillaries and white blood cells that pass by in the capillaries to express or increase expression of adhesion molecules (i.e., the receptors) in their surface membranes. White blood cells in the blood vessels stop and adhere to the endothelial cells of the capillary. These cells (cellular infiltrate) then migrate out of the capillary space between endothelial cells to the source of the molecules that induced the reaction. The infiltrating cells may include neutrophils, eosinophils, macrophages, and lymphocytes. The molecules from damaged or stimulated cells and secreted cytokines from the infiltrating cells stimulate pain receptors in the vicinity, causing an irritating sensation. This type of a host response is referred to as innate immunity, and it is not altered with repeated exposure to a particular mite or tick. The time and intensity of the response reaction is the same each time the individual is challenged.

Adaptive Immune Response

In contrast, the molecules introduced into the body by acarine parasites may induce an adaptive immune response that is highly specific for a particular epitope (sequential or structural) on an immunogenic molecule (antigen) from the parasite. An epitope is the part of the antigen that receptors on B and T lymphocytes recognize. The adaptive immune response is stronger and quicker with successive exposures and involves T and B lymphocytes and memory cells of each type. It may be accompanied by an inflammatory reaction too that can be delayed. With the help of type 2 T-helper cells (Th2) B cells become plasma cells that produce antibody directed at the offending molecules from the mite. Activated Th1-type helper cells activate cytotoxic T cells (Tc) that perform functions that kill the parasite directly or damage it. Helper T cells release specific cytokines such as Interleukin 2 (IL-2), interferon γ (IFN-γ), and other interleukins (IL-4, IL-6, IL-10, and IL-13), which act as signals to activate Tc and B cells.

PARASITIC MITES

Family Trombiculidae

Chiggers are the parasitic larval stage of prostigmatid mites that belong to the family Trombiculidae (Fig. 1). Chiggers are also known as harvest bugs in Europe and scrub-itch mites in Asia and Australia. Trombiculid mites are prevalent in moist, warm temperate climates and in tropical climates worldwide. These mites live in moist soil covered with vegetation such as grassy and weedy areas. More than 3000

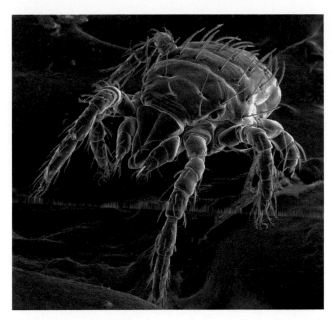

FIGURE 1 *Tromicula alfreddugesi,* the mite that causes chiggers. (Photograph © David Scharf.)

species of chiggers are known, but only about 15 species frequently bite humans and cause a cutaneous reaction.

Unlike many mites, male and female chiggers do not copulate directly. Instead, males deposit a stalked spermatophore (sperm packet) on the substrate. Females insert it into their genital pores to fertilize the eggs, which are then deposited on moist soils. Larva emerge from the eggs and complete development into an active hexapodal (six-legged) larva (chigger). The larva is parasitic and must feed from a mammal, bird, or reptile host before development can progress to the nymphal stages and the adult. The active nymphal stages and adults are predators and prey on small arthropods (insects and mites) or their eggs. The larval stage (chigger) generally feeds on rodents, mice, birds, and reptiles, and some species bite humans.

Chiggers can cause dermatitis and transmit the agent *Rickettsia tsutsugamushi,* which causes scrub typhus in humans. Scrub typhus is characterized by an ulcer at the site of the bite, high fever, and headache. Scrub typhus is present in tropical climates such as parts of India, Pakistan, Southeast Asia, Philippines, Indonesia, Korea, Japan, China, some Pacific Islands, and coastal Queensland, Australia. The principal vectors are species of the chigger genus *Leptotrombidium.* The reservoir hosts for this disease are rodents (mainly rats). In nature, the pathogen is transferred from rodent to rodent by many chiggers species. Humans become infected when they venture into an enzootic area and are bitten by infected larva. The larval stage feeds only once and acquires the pathogen from infected moles, mice, rats, and other small rodents. Therefore, chiggers can only acquire the rickettsia or, if they were already infected, transmit it, but not both. The rickettsia acquired by the larva is carried (trans-stadially) throughout the developmental stages to the adult. Rickettsia acquired by the larva multiply in the subsequent developmental life stages and infect the ovaries of the adult, from which they are passed to the egg (transovarially) and then to the larva of the next generation. The rickettsia in transovarially infected larva infect the salivary gland and are transmitted to humans when the larva feed.

The chigger feeds from the surface of the skin much like a tick. Its piercing mouth-parts (chelicerae) are inserted through the epidermis into the dermis. Saliva is introduced into the host during feeding. In humans, these salivary components induce both an innate inflammatory reaction and an adaptive immune response. These reactions are characterized by the production of circulating antibody and by cellular infiltration into the feeding lesion. Repeated exposures result in a more rapid and intense adaptive immune response. It is unclear whether chiggers induce an innate inflammatory response independent of the immune response. Clinically, however, the bite manifests as a reddish (erythematous), swollen (edema), and epidermally-thickened papular and irritating lesion. Histologically, the feeding lesion appears as a cylinder of tightly packed cells surrounding a strawlike channel that extends from the dermis to the skin surface where the chigger is located. The chigger sucks fluids from the surface of the channel until it is engorged, and then it drops off the host. Chiggers do not feed on blood; rather, they feed on extracellular fluid from the dermis.

Family Demodicidae

The prostigmatid mites of the family Demodicidae are small (approx. 100 μm in length) and have an elongated, wormlike body. The podosoma bears retractible, short, stumpy, telescoping legs. The opistosoma is transversly striated and elongate. Two species, *Demodex folliculonum* and *D. brevis,* parasitize humans and are commonly called follicle mites. Both species are most often obtained from the face, particularly along the nose, forehead, scalp, and eyelids. *D. folliculonum* lives in the hair follicle alongside the hair shaft and is positioned with its capitulum (mouthparts) down in the follicle. *D. brevis* resides in the sebaceous gland off the follicle. The entire life cycle is completed in the follicle and sebaceous gland. Generally, these mites cause little pathology in humans who practice good facial hygiene and are not immunocompromised. However, they may be associated with acne, blackheads, and acne rosacea.

Families Laelaptidae, Dermanyssidae, and Macronyssidae

The Mesostigmata contains many species of mites that are parasitic on reptiles, birds, and mammals. Included are hematophagous (blood-feeding) species in the families Laelaptidae, Dermanyssidae, and Macronyssidae. Among these are *Dermanyssus gallinae* (chicken mite), *Ornithonyssus bacoti* (tropical rat mite), *O. bursa* (tropical fowl mite), *O. sylviarum* (northern fowl mite), *Echinolaelaps echidninus* (spiny rat mite), *Liponyssus sanguineus, Haemogamasus pontiger,* and

Eulaelaps stabularis. These species are attracted to warm objects and usually live on their host or in the nest of their host. Some of these species will attack humans if their normal hosts are not available. This situation may result after roosts and nests of birds (e.g., pigeons, sparrows, starlings) and nests of rodents (mice, rats, squirrels) in homes (attics, behind shutters, etc.) are destroyed. In the absence of a natural host, the mites invade homes and attack humans. Also, species that infest poultry *(O. sylviarum, O. bursa, O. gallinae)* can be a problem for workers who handle infected chickens and turkeys. Bites of these mesostigmatid mites can cause an irritating inflammatory reaction. There may also be an allergic reaction in some individuals, but this remains to be confirmed. *Siponyssoides sanguineus* parasitizes house mice and rats and can transmit *Rickettsia abari,* which causes rickettsial pox in humans. Western equine encephalitis and St. Louis encephalitis viruses have been isolated from *D. gallinae,* but there are no documented cases of transmission of these viruses to humans.

Species in the families Rhinonyssidae, Entonyssidae, and Halarachnidae live in the nasal cavity and lungs of birds and some mammals (e.g., dogs, monkeys, seals, baboons). Human infections by these mites have not been reported.

Family Sarcoptidae

The astigmatid mites, (e.g., *Sarcoptes scabiei*) are permanent obligate parasites that live in the stratum corneum of the skin of at least 17 families of mammals. These mites cause a disease known as scabies. Scabies is a common contagious disease of humans. There is little morphological difference between the strains of *S. scabiei* that parasitize different host mammals, and at this time, the strains from different host species are not considered to be different species by most experts. However, the strains from different host species are host specific and generally cannot permanently infest an unnatural host. For example, the strain from dogs causes only temporary self-limiting infestations in humans, cats, pigs, cattle, goats, and mice, yet scabies naturally occurs on these host species. The host factors and physiological differences between mite strains that do not allow one strain to establish an infestation on strange hosts are not known.

Scabies mites are small. The male and female are 213 to 285 μm and 300 to 504 μm in length, respectively. The life cycle, consisting of egg, larvae, protonymph, tritonymph, and adult males and females, is completed in about 10 to 13 days on the host. All active stages are oval, with a characteristic tortoise like body with stout dorsal setae, cuticular spines, and cuticular striations.

When separated from the host at room temperature, scabies mites must infest a new host within 24 to 36 h to survive. Under cool (4 or 10°C) and humid conditions, females of the strain that infests humans (var. *hominus*) remain infective for at least 4 days. Therefore, fomites (i.e., clothing, bedding, and furniture that harbor dislodged mites) can be important sources of infection for humans. Body odor and temperature attract these mites to a host. Once on the host skin, females begin to burrow into the skin within minutes, and they can be completely submerged within the stratum corneum within a half-hour. Males, nymphs, and larval stages penetrate more quickly than females.

Scabies is common in nursing homes, day-care centers, and among the general population in the United States. It often mimics other skin diseases and is difficult to diagnose. Scabies is prevalent in some populations in Africa, Central America, South America, Egypt, India, and Australia. Human scabies infestations are manifested in the vicinity of the burrowing mite by itching, red, papular and vesicular lesions. There symptoms generally develop in 6 to 8 weeks after a primary (first) infestation, but they are evident within a few days of a subsequent infestation. Lesions most commonly occur on the interdigital, elbow, and chest (breast area) skin. However, other areas that may be infested are the penis, buttocks, knees, soles and insteps of the feet, wrists, waistline, and axillae.

Scabies mites induce both cell-mediated (Th1) and circulating antibody (Th2) immune responses and an associated inflammatory reaction. The cell-mediated/inflammatory response is characterized by a mixed cellular infiltrate in the skin lesion that consists of plasma cells, lymphocytes, mast cells, neutrophils, Langerhans cells, and eosinophils.

An infestation with scabies induces some immune resistance to subsequent infestations. The balance between the Th1 and Th2 responses appears to be a key aspect in protective immunity. Hosts that develop protective immunity exhibit up-regulated Th1 and weaker Th2 responses. In contrast, hosts that do not develop protective immunity exhibit strongly up-regulated Th2 response (circulating antibody) but a weaker Th1-cell-mediated response. Infected hosts produce serum antibodies to at least 12 antigens from sarcoptic mites. Some of these antigens are cross-reactive with antigens from the related house dust mites *Dermatophagoides farinae, D. pteronyssinus,* and *Euroglyphus maynei.* In some humans, antigens from *S. scabiei* can also induce an IgE-mediated allergic reaction and circulating IgE-type antibody.

Family Pyemotidae

Pyemotid mites are prostigmatids that have an elongate cigar-shaped body with the first two pair of legs widely spaced from the posterior two pair of legs. They have stylettiform (needlelike) chelicerae and are usually parasitic on the larvae of insects. Unlike other mites, pyemotid female mites retain internally the eggs from which the immatures hatch and pass through all developmental stages. As a result, the female's opisthosoma (region behind the last pair of legs) becomes enormously swollen before the offspring are born.

Pyemotes tritici (straw itch mite) and *P. ventricosus* (grain itch mite) are parasitic on the larvae of grain moths, boring and stored grain beetle larvae, and other insects. Humans may contact these species when working with grain and hay. Also, hordes of these insects may emerge from the flowers of

cattails brought into a home to make a floral arrangement. These mites will attack humans and cause red, itchy inflammatory dermatitis.

Families Tetranychidae and Eriophyidae

Many species of prostigmatid mites such as those in the families Eriophyidae and Tetranychidae parasitize plants and can become an economic problem on food crops (e.g., fruit trees; vegetable and grain crops) and yard/garden and green houseplants. Humans come into contact with these mites when working in fields, orchards, greenhouses, gardens, and yards, when handling infested food crops/produce, or by living near an area in which food crops are grown. The importance to human health of most of these pest species has yet to be determined. However, it is clearly documented that a few species are the source of allergens that induce allergic reactions in predisposed individuals. Farmers working in apple orchards and children living around citrus orchards have become sensitized and/or had allergic reactions to *Tetranychus urticae* (two-spotted spider mite) and *Panonychus ulmi* (European red mite) and *P. citrilis* (citrus red mite).

Family Phytoseiidae

Humans come into contact with predaceous mites that are used for biological control of pest species such as the tetranychids just mentioned. The predaceous mite *Phytoseilus persimilis,* which feeds on spider mites, can cause allergic reactions.

Family Hemisarcoptidae

Hemisarcoptes cooremani is an astigmatid mite that is a predator of scale insects that parasitize woody plants. The body of this mite is the source of at least two allergenic proteins. Close contact with these mites can result in production of serum IgE and allergic symptoms. Therefore, gardeners and nursery workers may become sensitized to this mite and have allergic reactions.

NONPARASITIC MITES

Family Pyroglyphidae

The family Pyroglyphidae contains mainly species of astigmatid mites that live in the nests of birds and mammals, where they feed on the epidermal detritus (skin, feathers) left by the host. Three species, *Dermatophagoides farinae, D. pteronyssinus,* and *Euroglyphus maynei,* are commonly found in homes of humans. In homes, these mites are most prevalent in high-use areas, where shed skin scales collect and serve as their food. Therefore, the greatest densities are found in carpets around sofas and easy chairs, in fabric-covered overstuffed furniture, and in mattresses. However, they may also be found in bedding, on pillows, on clothing, on automobile and train seats, and sometimes in schools and in the workplace. Each species is the source of multiple potent allergens that sensitize and trigger allergic reactions in predisposed people. These allergens cause perennial rhinitis, asthma, and atopic dermatitis.

Ambient relative humidity is the key factor that determines the prevalence and geographical distribution of these mites. This is because water vapor in humid air is the main source of water for their survival. They survive and thrive well at relative humidities above 50% but desiccate and die at relative humidities below this. Therefore, dust mites and the allergies they cause are a significant problem only for people who live in humid, tropical, and temperate geographical areas. *D. farinae* and/or *D. pteronyssinus* are prevalent in homes in the United States, Europe, South America, and Asia. Most homes are coinhabited by multiple species. However, the most prevalent species varies both between homes in a geographical area and between geographical areas. For example, in the United States, both *D. farinae* and *D. pteronyssinus* are prevalent in homes. However, in South America, *D. pteronyssinus* is prevalent in homes, whereas *D. farinae* is not.

In temperate climates, population densities of *D. farinae* and *D. pteronyssinus* exhibit pronounced seasonal fluctuations that parallel the seasonal fluctuations in indoor relative humidity. High densities occur during the humid summer and low densities during winter.

The life stages of the dust mites are egg, larva, protonymph, tritonymph, and adult male and female. Length of the life cycle is temperature dependent when relative humidity is above 60%. At 23°C the life cycle takes 34 and 36 days to complete for *D. farinae* and *D. pteronyssinus,* respectively. Females produce 2 or 3 eggs daily during the reproductive period at 23°C. *D. pteronyssinus* takes 23 and 15 days to complete development at 16 and 35°C, respectively. *D. farinae* does not develop well at 16 and 35°C.

A desiccant-resistant quiescent protonymphal stage can develop that allows survival during long periods (months) under dry (low relative humidity) conditions. When relative humidity conditions become optimal, the quiescence is broken and development continues.

Allergens from these mites are associated with fecal material, body secretions, and body anatomy. Fourteen different groups of mite allergens have been characterized. The frequency of reactivity to most of these allergens is above 40% among patients sensitive to dust mites. Sensitivity to allergens varies both within and between individuals. Allergens from one species may be species specific, or they may cross-react with allergens from another mite species. Most patients with sensitivities are allergic to multiple allergens of a species and to multiple mite species.

Families Acaridae, Glycyphagidae, Carpoglyphidae, Echimyopididae, and Chortoglyphidae

Many species of the astigmatid families Acaridae, Glycyphagidae, Carpoglyphidae, Echimyopididae, and Chortoglyphidae are medically important because they are the sources of potent allergens. Many species of these mites are often referred to as "storage mites" because they occur in stored hay, grain, and straw, in processed foods made from grain (flour, baking mixes), and in dust in grain and hay at storage, transfer, and livestock feeding facilities. Humans may be exposed to storage mites, and their allergens, occupationally and in the home. Inhalation or contact on the skin with allergens from storage mites can induce allergic reactions. These mites and their allergens can also occur in bread, pancakes, cakes, pizza, pasta, and bread made from ingredients contaminated with mites. Humans have had anaphylactic reactions after eating these mite-contaminated foods.

Species known to be the sources of allergens include *Blomia tropicalis* (Echimyopididae); *Acarus siro, Tyrophagus putrescentiae, T. longior,* and *Aleuroglyphus ovatus* (Acaridae); *Lepidoglyphus destructor* and *Glycyphagus domesticus* (Glycyphagidae); *Carpoglyphus* spp. (Carpoglyphidae); *Chortoglyphus arcuatus* (Chortoglyphidae); and *Suidasia medanensis* (Suidasiidae). *T. putrescentiae* is the source of 14 allergens, with the number recognized as allergens by individuals ranging from 5 to 11. *B. tropicalis,* which is common in house dust in tropical climates and may be more prevalent than pyroglyphid mites, has been reported in small numbers in some homes in the southern subtropical United States. Several allergens from *B. tropicalis* have been characterized and/or produced by recombinant technology. There is little cross-reactivity between storage mites and house dust mites. However, many patients are sensitive to both storage mites and the pyroglyphid house dust mites.

See Also the Following Articles
Medical Entomology • Mites • Ticks • Veterinary Entomology

Further Reading
Arlian, L. G. (1989). Biology, host relations, and epidemiology of *Sarcoptes scabiei. Annu. Rev. Entomol.* **34,** 139–161.
Arlian, L. G. (1992). Water balance and humidity requirements of house dust mites. *Exp. Appl. Acarol.* **6,** 15–35.
Arlian, L. G. (1996). Immunology of scabies. *In* "The Immunology of Host–Ectoparasitic Arthropod Relationships." (S. Wikel, ed.), pp. 232–258. CAB International, Wallingford, U.K.
Arlian, L. G. (2002). Arthropod allergens and human health. *Annu. Rev. Entomol.* **47,** 395–433.
Arlian, L. G, Bernstein, D., Bernstein, I. L., *et al.* (1992). Prevalence of dust mites in homes of people with asthma living in eight different geographic areas of the United States. *J. Allergy Clin. Immunol.* **90,** 292–300.
Arlian, L. G., Neal, J. S., Morgan, M. S., *et al.* (2001). Reducing relative humidity is a practical way to control dust mites and their allergens in homes in temperate climates. *J. Allergy Clin. Immunol.* **107,** 99–104.
Johansson, E., Johansson, S. G. O., and van Hage-Hamsten, M. (1994). Allergenic characterization of *Acarus siro* and *Tyrophagus putrescentiae* and their cross reactivity with *Lepidoglyphus destructor* and *Dermatophagoides pteronyssinus. Clin. Exp. Allergy* **24,** 743–751.
Kim, Y. K., Lee, M. H., Jee, Y. K., *et al.* (1999). Spider mite allergy in apple-cultivating farmers: European red mite (*Panonychus ulmi*) and the two-spotted spider mite (*Tetranychus urticae*) may be important allergens in the development of work-related asthma and rhinitis symptoms. *J. Allergy Clin. Immunol.* **104,** 1285–1292.

Lee, M. H., Cho, S. H., Park, H. S., *et al.* (2000). Citrus red mite (*Panonychus citrilis*) is a common sensitizing allergen among children living around citrus orchards. *Ann. Allergy Asthma Immunol.* **85,** 200–204.

Radovsky, F. J. (1994). The evolution of parasitism and the distribution of some dermanyssoid mites (Mesostigmata) on vertebrate hosts. *In* "Mites, Ecological and Evolutionary Analysis of Life-History Patterns." (M. A. Houck, ed.), pp. 186–217. Chapman & Hall, New York.

Thomas, W. R., and Smith, W. (1999). Towards defining the full spectrum of house dust mite allergens. *Clin. Exp. Allergy* **29,** 1583–1587.

Wikel, S. K., Ramachandra, R. N., and Bergman, D. K. (1996). Arthropod modulation of host immune responses. *In* "The Immunology of Host-Ectoparasitic Arthropod Relationships." (S. Wikel, ed.), pp. 107–130. CAB International, Wallingford, U.K.

Wrenn, W. (1996). Immune responses to chigger mites and chiggers. *In* "The Immunology of Host-Ectoparasitic Arthropod Relationships." (S. Wikel, ed.), pp. 259–289. CAB International, Wallingford, U.K.

Chitin

Ephraim Cohen
The Hebrew University of Jerusalem

Chitin is a globally abundant biopolymer, second only to cellulose and possibly lignin in terms of biomass. Owing to extensive hydrolytic activity mainly by soil and marine chitinolytic microorganisms, chitin is not accumulated in the biosphere, because it is similar to cellulose and unlike lignin. Chitin, which is absent from plants and vertebrates, is present to a small or large extent in most invertebrates, notably in cuticles of arthropods, in primary septum and scar buds of yeast, and in cell walls of most filamentous fungi. Chemically detectable chitin has been verified in 25-million-year-old insect fossils. Chitin is almost invariably covalently or noncovalently associated with other structural molecules in contact with the external environment; examples include carbohydrate polymers in fungi and the cuticular proteins that comprise up to 50% by weight of arthropod cuticles. The chitoprotein supramolecular matrix occurs in peritrophic membranes of insects and in the arthropod exoskeleton, where the rigid chitin microfibrils contribute greatly to its mechanical strength.

STRUCTURE

Chitin is a large water-insoluble, linear aminocarbohydrate homopolymer composed of β_{1-4}-linked *N*-acetyl-D-glucosamine units with a three-dimentional α-helix configuration (Fig. 1). Intramolecular hydrogen bondings stabilize the α-helical configuration of the macromolecule. In nature, chitin polymers coalesce extracellularly by intermolecular hydrogen bonds to form crystalline microfibrils that may appear in various polymorphs (α, β, and γ). The most abundant one in insects is the antiparallel arrangement of the α-chitin polymorph.

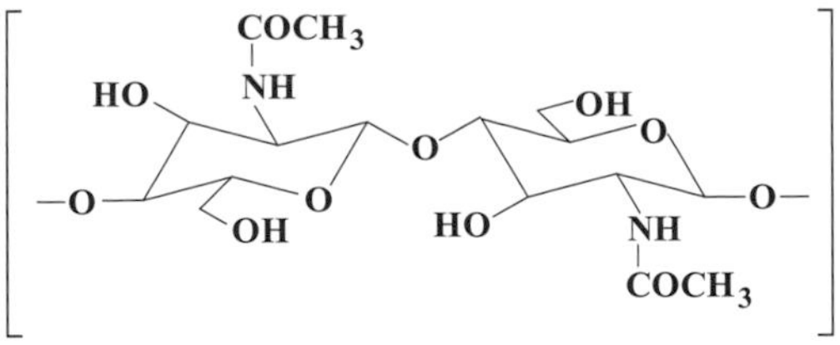

FIGURE 1 The dimer unit of chitin polymer.

CHITIN SYNTHESIS

Chitin synthesis occurs throughout the insect's life cycle and is under hormonal control of ecdysteroids. Bursts of synthetic activity that are associated with the buildup of the new cuticles occur in particular at the last phase of embryonic development, and as larvae or pupae molt. Chitin synthesis is the end result of a cascade of interconnected biochemical and biophysical events that link the mobilization of substrate molecules, polymerization by the enzyme chitin synthase, and translocation of the nascent amino polymer across the plasma membrane (Fig. 2). Individual chitin chains coalesce outside the plasma membrane, forming fibril crystallites by intramolecular hydrogen bonds. The UDP-*N*-acetyl-D-glucosamine substrate is the end point of a series of biochemical transformations that include successive steps of phosphorylation, amination, and acetylation of starting precursors such as trehalose or glucose. Chitin synthase is a relatively large membrane-bound enzyme with multiple transmembrane segments. The active site of the enzyme faces the cytoplasm, and the catalysis involves linking together dimer amino sugar substrates. The question of how chitin polymers are translocated across the cell membrane remains unresolved. Hydrophobic transmembrane segments of chitin synthase are implicated in this process.

The complete chitin synthase cDNA and deduced amino acid sequences of the insects *Drosophila melanogaster* and *Lucila cuprina,* and the nematode *Caenorhabditis elegans* have

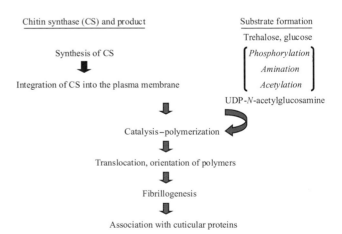

FIGURE 2 Polymer formation and deposition.

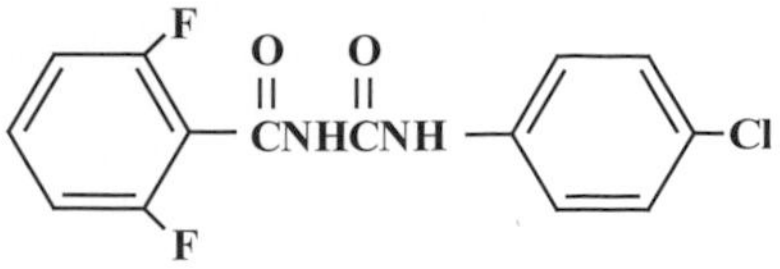

FIGURE 3 Chemical structure of the acylurea compound diflubenzuron (Dimilin).

been described. The deduced amino acid sequence revealed a protein (about 180 kDa) with a large number (15–18) of potential transmembrane segments that may be involved in the translocation of chitin polymers.

CHITIN DEGRADATION

Degradation of chitin is physiologically crucial for normal growth and development of insects. Chitin is degraded by the joint action of chitinase, which yields oligomeric fragments, and exochitinase, or β-N-acetylglucosaminidase, which hydrolyzes terminal polymers or dimers. These hydrolytic enzymes are widespread in plants, vertebrates, invertebrates, and microorganisms. During the complex molting process in arthropods, the chitin in the cuticular region (the endocuticle), which is close to the epidermal cells, is degraded. Since chitin microfibrils are tightly associated with various cuticular proteins, proteolytic activity accompanies and facilitates chitin hydrolysis. Hydrolysis of chitin does not occur in the exocuticle, where sclerotization of the cuticular protein takes place. Formation and secretion of chitinases by epidermal cells, processes that are under hormonal control, are vital for the molting process. The mono- and disaccharide degradation products are absorbed by the epithelial cells and may be recycled to serve for biosynthesis of the new chitin.

INHIBITION OF CHITIN SYNTHESIS AND DEGRADATION

Because chitin is present in invertebrates (abundantly in arthropods) and absent from vertebrates and plants, it is a logical target for selective pest control. Acylurea compounds, discovered serendipitously by Dutch scientists in 1972, inhibit chitin synthesis, resulting in deformed and weak cuticles that cause molting failure and death by desiccation. Acylureas do not inhibit the catalytic step of polymerization, and their exact biochemical lesion is unresolved. It appears that the mode of action is associated with the process of chitin translocation from site of catalysis across cell membranes to the region of deposition and fibrillogenesis. The first commercial product reaching the market was diflubenzuron (Dimilin) (Fig. 3), which was followed by a large number of structurally similar bioactive molecules. The acylurea compounds, which act as insect growth regulators, are widely used in integrated pest management (IPM) programs.

See Also the Following Articles
Cuticle • Exoskeleton • Integument • Molting

Further Reading
Cohen, E. (1987). Chitin biochemistry: Synthesis and inhibition. *Annu. Rev. Entomol.* **32**, 71–93.
Cohen, E. (1993). Chitin synthesis and degradation as targets for pesticide action. *Arch. Insect Biochem. Physiol.* **22**, 245–261.
Muzzarelli, R. A. A. (1977). "Chitin." Pergamon Press, Oxford, U.K.
Stankiewicz, B. A., Briggs, D. E. G., Evershed, R. P., Flanerry, M. B., and Wuttke, M. (1997). Preservation of chitin in 25-million-year-old fossils. *Science* **276**, 1541–1543.
Tellam, R. L., Vuocolo, T., Johnson, S. E., Jarmey, J., and Pearson, R. D. (2000). Characterization of insect chitin synthase: cDNA sequence, gene organization and expression. *Eur. J. Biochem.* **267**, 6025–6042.
Verloop, A., and Ferrell, C. D. (1977). Benzoylphenyl ureas—A new group of larvicides interfering with chitin deposition. *In* "Pesticide Chemistry in the 20th Century" (J. R. Plimmer, ed.), pp. 237–270. ACS Symposium Series 37, American Chemical Society, Washington DC.

Chromosomes

Graham C. Webb
The University of Adelaide, Australia

Chromosomes in insects display almost the whole range of variation seen in the chromosomes of higher plants and animals. In these groups the deoxyribonucleic acid (DNA), which contains the genetic code determining development and inheritance, is contained in a nucleus in each cell. At interphase, the DNA is organized into the complex linear structures that are chromosomes, which can be seen in a conveniently condensed state when the cell is dividing.

The study of insect chromosomes is less intensive now than formerly for three possible reasons: (1) the thoroughness of the early investigators, (2) the commercialization of science, which has pushed the study of chromosomes (cytogenetics) in animals toward more lucrative mammalian, and particularly human, fields, and (3) the replacement of cytogenetic with molecular methods. The third point was predicted by Michael White in the conclusion to his famous 1973 textbook, *Animal Cytology and Evolution.* Through his work, almost entirely on insects, White is widely regarded as the founder of the study of evolutionary cytogenetics in animals and one of its foremost authorities; his book remains a most comprehensive authority on most aspects of insect chromosomes.

In 1978 White was firmly convinced that evolution is essentially a cytogenetic process, and he did much to demonstrate this at the level of speciation. At a higher evolutionary level, the integrated chromosomal characteristics of the various insect orders seem to support this view. However more recently authors such as King have de-emphasized the importance of chromosomal changes in species evolution.

SOURCES AND PREPARATION OF CHROMOSOMES FROM INSECTS

Mitotic chromosomes undergoing the familiar stages of prophase, metaphase, anaphase, and telophase can be prepared from any insect somatic tissues with dividing cells. Embryos are the best sources of mitotic divisions, but they are also seen in the midgut ceca of adults and juveniles and in the follicle cells covering very early ova in females.

Insect cytogeneticists now usually use colchicine or other mitostatic agents to arrest the chromosomes at metaphase of mitosis by inhibiting the formation of the spindle fibers required for the cells to progress to anaphase. Squashing, under a coverslip, spreads the chromosomes, and for squash preparations the cells are usually prestained. Insect cytogeneticists now often use air-drying to spread the chromosomes, since this process has the advantage of making the chromosomes immediately available for modern banding and molecular cytogenetic methods.

Male meiosis is very commonly used to analyze the chromosomes of insects and to analyze sex-determining mechanisms. The structure of the insect testis is very favorable to chromosomal studies because each lobe has a single apical cell that divides by a number *(s)* of spermatogonial divisions (Fig. 1A) to yield 2^s primary spermatocytes, which then undergo synchronous first and second meiotic divisions to yield $2^s + 1$ secondary spermatocytes and $2^s + 2$ sperm.

First meiotic prophase in insects involves the usual stages (Fig. 1). Replication of the DNA is followed by the prophase stages of leptotene (strand forming), zygotene (chromosome pairing to form bivalents), pachytene (crossing over to yield recombinants), diplotene (repulsion of the homologues), diakinesis (completion of repulsion), and premetaphase (bivalents fully condensed).

Metaphase I is followed by first anaphase, which can be a very informative stage and, in contrast to mammals, is readily available in insects. Second meiotic division is also readily observed in insects (Fig. 1) and can be useful for confirming events in earlier stages.

Meiotic chromosomes in insect females are difficult to prepare and are usually studied only in special cases, such as parthenogenesis.

TYPES OF CHROMOSOME IN INSECTS

Autosomal chromosomes are usually represented as two haploid sets, one from each parent, in the diploid tissues of insects. With the addition of the sex chromosomes from each parent, the haploid set is known as "*n*" and the diploid set as "$2n$." Major exceptions to diploidy in both sexes occur, such as in almost all species in the the orders Hymenoptera (ants, bees, and wasps), Thysanoptera (thrips), and some species of Heteroptera and Coleoptera, where the females are diploid, the males being normally haploid (i.e., derived from unfertilized eggs, arrhenotoky). Arrhenotoky determines the sex of

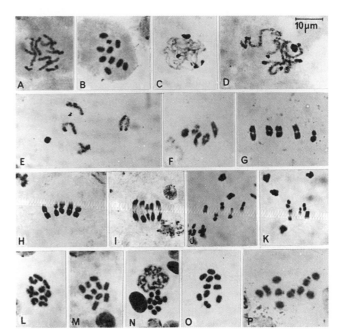

FIGURE 1 Mitotic and meiotic holocentric chromosomes in an earwig, *Labidura truncata.* Orcein-stained squash preparations, B, L, M-P colchicine-treated. (A) Spermatogonial division in prophase with the Y at bottom left and the X to the right, both more condensed than the autosomes. (B) Spermatogonial metaphase with the small Y chromosome obvious. (C) Leptotene, with the sex chromosomes at the top very condensed and the heterochromatic ends of some autosomes also condensed. Two nucleoli are visible, one at 11 o'clock and the other at 5 o'clock. (D) Zygotene–pachytene with the heterochromatic ends of the autosomes more obvious. (E) Diplotene displaying the four autosomal bivalents and the condensed sex chromosomes separately. (F) Diakinesis, one autosomal bivalent showing a chiasmata that is quite interstitial. (G, H) First metaphases with the larger X seem to be paired with the smaller Y. First anaphase with the neocentromere actively moving the chromosomes apart. (J, K) Second metaphases; J shows the X dyad, K shows the smaller Y dyad. (L–P) Female mitotic chromosomes, late and early prophase in L and N, respectively; M–P show metaphases, with O and P showing secondary constrictions. The primary constrictions of fixed centromeres do not show, and uninterrupted chromatids, characteristic of holocentric chromosomes, are particularly obvious in M. [From Giles, E. T., and Webb, G. C. (1973). The systematics and karyotype of *Labidma Truncata* Kirby, 1903 (Dermoptera: Labiduridae). *J. Aust. Entomol. Soc.* **11**, Plate 1, with permission.]

about 20% of all animal species. This mechanism has allowed one species of Australian ant, *Myrmecia croslandi,* to achieve the lowest possible chromosome number, $n = 1$, in the parthenogenically derived male.

Chromosomal imprinting has not been demonstrated in insects, so gametes from both sexes are not necessarily required. Indeed, accidental development of unfertilized eggs (thelytoky) can form a parthenogenetic insect if sufficient double-haploid cells arise in critical tissues in the $n/2n$ mosaic.

Sex chromosomes are usually involved in sex determination in insects, but by a variety of genetic mechanisms. The male is usually the heterogametic sex in insects, the exceptions being the orders Lepidoptera (butterflies and moths) and Trichoptera, in which the females are heterogametic. The mammalian system of having genes determining the sex and other male functions on the Y chromosome almost certainly does not occur in insects. In the earwigs (Dermaptera), male determination by the presence of a Y chromosome seemed to be the rule, until XO/XX mechanisms were found in two species.

As in other animals, the insect heterogametic male has half the number of X chromosomes as the female; most commonly the sexes are XO male and XX female, but multiple X-chromosome systems frequently occur. Fusions of autosomes to the X chromosome can cause the formation of XY/XX and further fusions to form $X_1X_2Y/X_1X_1X_2X_2$ systems. X_1X_2Y males are almost the rule in the mantids (Mantodea). In the most of the Hymenoptera, sex is determined by the diploid females being heterozygous, and the haploid males hemizygous, for multiple alleles at a single genetic locus on one chromosome; that might still be regarded as an X chromosome.

The karyotype is the set of chromosomes, both autosomes and sex chromosomes, in an organism. The karyotype found in 90% of the large family of short-horned grasshoppers, Acrididae, is usually given as $2n\male$ = 23 [22 + X (or XO)], the female karyotype, 23 (22 + XX), being usually inferred from the male. Some authors carefully confirm diploidy and subdifferentiate the autosomes and the sex chromosomes, for example, for the earwigs *Chaetospania brunneri* $2n \male$ = 31 (13AA + $X_1X_2X_3X_4Y$) and *Nala lividipes*, with one pair of autosomes being exceptionally long, $2n\male$ = 34 (A^LA^L + 15AA + XY). Since in insects, the karyotypic nomenclature is variable and somewhat confusing, it would seem preferable to adopt the simple karyotypic nomenclature used for mammals [e.g., human: $2n\male$ = 46,XY].

Monocentric chromosomes are the norm in most insect orders (Fig. 2), with the single centromere characterized by a primary constriction, a structure seen in many other animals and in plants. After replication of the DNA and other chromosomal constituents during interphase, the chromosomes at metaphase show two identical chromatids. Following the discovery that each chromatid must be terminated by a telomere, geneticists concluded that it is highly probable that a chromosome must always have two arms, one on each side of the centromere. If these arms are of appreciable length, the chromosomes are called metacentric (arms of about equal length), or submetacentric (arms of unequal length). If one of the arms is very short, perhaps invisible under normal microscopy, the chromosome is said to be acrocentric. It is now widely accepted that a chromosome cannot normally be telocentric (terminated by a centromere). The sequence of nucleotides repeated many times to make up the DNA of the telomeres of most insects is TTAGG, but it is not universal.

Holocentric chromosomes occur in the insect orders Heteroptera, Dermaptera (Fig. 1), Mallophaga, Anoplura, and Lepidoptera. The centromeres are elongated across much of the length of the chromosomes, although not usually extending to the telomeres. During mitotic anaphase, the spindle fibers pull equally on most of the length of the chromosome so that only the distal ends can be seen to be

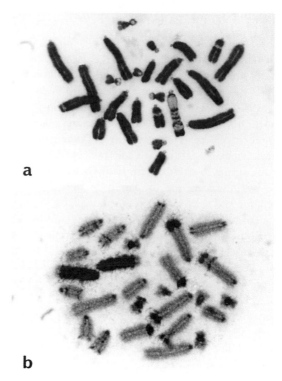

FIGURE 2 Monocentric chromosomes of the locust *Chortoicetes terminifera*, mostly acrocentric with some of the smaller ones submetacentric. (a) With one B chromosome, which is distinctively G-banded by a trypsin treatment that has produced comparatively minor effects in the A chromosomes. (b) with two B chromosomes showing positive C-banding for most of their length. The A chromosomes mostly have small centromeric C bands, but they show variable interstitial and distal C-banded segments.

trailing. Operation of an elongate centromere during first meiosis would break the crossovers, or chiasmata, which have formed between the the paired chromosomes. Apparently to preserve this chromosomal bivalent, the holocentric chromosomes develop neocentric activity at one telomeric end only. The neocentromeres behave like those of monocentric chromosomes, and they persist through second metaphase of meiosis (Fig. 1). Broken holocentric chromosomes seem to be able to retain attachment to the spindle fibers: in earwigs, each piece of a broken chromosome forms a bivalent with a neocentromere during meiosis. Breakage of holocentric chromosomes probably also explains the wide range of chromosome numbers seen in butterflies, from $2n$ = 14 to $2n$ = 446, and the extreme of $2n$ = 4 to $2n$ = 192 found by Cook in a single genus of scale insect, *Apiomorpha*. Breakage probably also accounts for the common finding of multiple X chromosomes in insects with holocentric chromosomes. The addition of telomeres to the broken ends of holocentric chromosomes might be a function of the complex enzyme telomerase.

Polytene chromosomes are large chromosomes formed by the repeated replication, without intervening division, of chromatids that remain uncondensed as in interphase (Fig. 3). Polytene chromosomes often contain thousands of chromatid strands, and the homologous chromosomes are usually closely somatically paired, so that inversions in them are accommo-

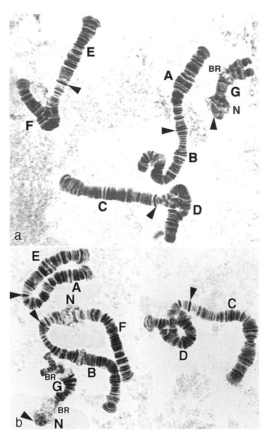

FIGURE 3 Polytene chromosomes in the salivary glands of the larvae of two species of chironomid midge. Orcein-stained squash preparations. (a) From the North American species *Chironomus decorus*, species b; (b) From the Australian species *C. oppositus*. For both species, labels A–F indicate arms of metacentric chromosomes, with arrowheads indicating the centromeres. The acrocentric chromosome G shows some breakdown of somatic pairing at the distal end in both species. Chromosomes AB and EF in *C. decorus* b have undergone whole-arm exchanges to form AE and BF chromosomes in *C. oppositus.* N and BR indicate nucleoli and Balbiani rings, respectively. Loop pairing, resulting from heterozygosity for paracentric inversions, can be seen in arms D and F in *C. decorus* b and in arm D in *C. oppositus* (Images kindly supplied by Dr. Jon Martin, University of Melbourne.)

dated by the formation of loops. Transcription of ribonucleic acid (RNA) from the DNA is accomplished at expanded regions called Balbiani rings, (BR in Fig. 3B), and the attachments of the polytene chromosomes to the nucleoli (N in Fig. 3B) by the nucleolar organizing regions are obvious. Polytene chromosomes have been most famously studied in the salivary glands and other glandular tissues in insects of the order Diptera, particularly in the fruit fly, *Drosophila melanogaster*. They display a large number of bands without any special staining, and the detail revealed is most useful for the localization of DNA sequences of various types, including single gene probes.

Supernumary or B chromosomes occur occasionally in insects of most orders. B chromosomes, when present, are in addition to the always present A chromosomes. Certainly the most variable and spectacular B chromosomes ever seen were found in the Australian plague locust, *Chortoicetes terminifera*

(Fig. 2). These B chromosomes display over 20 different banding patterns after treatment with trypsin, and this treatment allowed the harmless identification of carriers of B chromosomes using interphase cells in the hemolymph, thus facilitating breeding experiments. These breeding experiments showed that single B chromosomes in males of *C. terminifera* were distributed into the sperm with a 50% frequency, but in females single B chromosomes were driven into the egg with a frequency of 80%. This meiotic drive in females should have ensured that every individual in the population carried a B chromosome. Since, however, they were found in only 10% of individuals, the B chromosomes must have been lowering the fitness of carriers. The situation supported a "parasitic" mechanism for the maintenance of B chromosomes in the population.

SUBCHROMOSOMAL ORGANIZATION IN INSECTS

Euchromatin and heterochromatin can be distinguished in insects in various ways. Euchromatin contains the active genes, and heterochromatin, contains mainly repetitious, transcriptionally inactive DNA. Heterochromatic segments of the chromosomes can be observed in meiosis because of their high degree of condensation during first prophase (Fig. 1). Heterochromatin may also be detected by hybridization *in situ* of repetitous DNA sequences, such as satellite DNA, to the chromosomes. The DNA of heterochromatin is replicated later in the S phase of the cell cycle than the DNA of the euchromatin. Examples of DNA replication that is both late to start and late to finish has been seen in the B chromosomes of *C. terminifera,* and in the sex chromosomes of the common earwig, *Forficula auricularia.* The C-banding technique (described shortly) can also be used to stain heterochromatic segments.

In most cases the heterochromatin of insects is constitutive (i.e., in a permanent state), but in some insects with peculiar life cycles, such as the Cecidomyidae, individual chromosomes or sets of them may be made facultatively heterochromatic before being eliminated from the soma or the germ line of one of the sexes.

Chromosome banding in insects is largely limited to C-banding (Fig. 2b), originally named because the repetitious DNA proximal to the centromeres was stained. The position of the C bands is often procentric in insects, but these bands can be procentric, terminal, or interstitially distributed on the chromosome arms in different races of the same species, as in the grasshopper *Caledia captiva.*

The results of treating insect chromosomes with trypsin or other reagents that induce the narrow G bands, which are distributed all over the chromosomes of vertebrates, are disappointing in insects. The dark G bands correspond to the chromomeres, regions of the chromosomes that are contracted during meiotic prophase, and since insect chromosomes display chromomeres during meiosis, it is surprising that they do not show typical G bands. The bands revealed by trypsin treatment of the B chromosomes of *C. terminifera* seem to be exceptional, and they are a reflection of the C-banding patterns of these chromosomes (Fig. 2).

POLYPLOIDY

Polyploidy occurs when the zygote, or first cell, has more than two sets of haploid chromosomes. In insects, polyploidy is mainly restricted to parthenogenetic species and is largely limited to $3n$ and $4n$. Chromosomal sex determination is regarded as a major barrier to the formation of polyploids among bisexual species of insects because duplicated sex chromosomes, such as XXYY, would lead to uniformly XY sperm and therefore no possibility of sex determination.

Endopolyploidy is the occurrence of a multiplicity of the ploidy in the zygote in the somatic tissues of an organism. The term "endomitosis" is used if the chromosomes appear during cycles of endoreduplication but with no formation of a mitotic spindle and no cell division. Endopolyploidy, including endomitosis, is commonly seen in the somatic tissues of most insects. In a special case of endoreduplication, involving only one round of replication under a variety of treatments, the chromosomes that subsequently appear may remain closely associated at the centromeres, forming diplochromosomes, which were first seen in a locust. The formation of polytene chromosomes (Fig. 3) is also a special form of endoreduplication. Endopolyploid cells are very common in the tissues of all insects, and the phenomenon seems to reflect a tendency for insects to increase the bulk of certain tissues by increasing cellular size rather than cell number.

CHROMOSOMAL REARRANGEMENTS

Rearrangements occur in the chromosomes of insects when they occasionally break and rejoin in an irregular fashion. If any chromosomal rearrangement is maintained heterozygously in a population at a frequency greater than can be explained by recurrent chromosomal mutation, it is said to be polymorphic. There are a number of chromosomal rearrangements.

Paracentric Inversions

Paracentric inversions result when two breaks in one chromosome arm rejoin after the excised piece has inverted. These rearrangements are commonly recorded in polytene chromosomes, where the presence of them is shown by the formation of a loop allowing the homologues to be closely paired (Fig. 3). The presence of a chiasmata at meiosis within paracentrically inverted segments results in a dicentric chromosome and an acentric fragment, which cannnot be regularly transmitted. Paracentric inversions survive for long periods in many dipteran species because there is no chiasma formation in males and because the products of female meiosis are organized to ensure that a nonrecombinant for any paracentric inversion is deposited in the egg nucleus, with recombinants being placed in the unused polar bodies.

Because they have the capacity to lock up long combinations of syntenic genes, it has been assumed that inversion polymorphisms can be adaptive. For paracentric inversions, many studies with dipterans have been undertaken to link paracentric inversion polymorphism to aspects of the environment in which the particular insect exists.

Pericentric Inversions

Pericentric inversions result from breaks in each arm of a chromosome that rejoin after the excised piece containing the centromere has inverted. Pericentric inversion polymorphism was perhaps most famously studied in the morabine grasshopper, *Keyacris scurra*. White and coworkers used this rearrangement to develop adaptive topographies (defined by Sewell Wright) for various populations of *K. scurra,* that were on a saddle between adaptive peaks. The duplications and deletions that are the consequences of recombination within mutually pericentrically inverted segments seem to be largely avoided in insects bearing them at polymorphic frequencies. This is because the chromosomes are able to pair during meiosis without the inverted regions undergoing synapsis: so-called "torsion pairing."

Translocations

Translocations result from breaks in two chromosomes that allow exchange of pieces between the chromosomes. For breaks that are interstitial on the chromosome arms, a reciprocal translocation results, and in heterozygotes the translocated chromosomes synapse together at meiosis to form a quadrivalent (or a multivalent, if exchanges are more frequent). Multiple translocation heterozygosity has been observed, with resulting ring multivalents, in the cockroaches (Blattodea).

Centric Fusions

Centric fusions, or Robertsonian translocations, are special cases of translocation in which two breaks are very close to the centromeres of acrocentric chromosomes, causing the formation of a large metacentric or submetacentric chromosome and a very small remnant, which is lost. Centric fusions commonly distinguish chromosomal races or species in insects, but they are not seen to be maintained polymorphic in populations as frequently as they are in mammals. Centric fusions between sex chromosomes and autosomes results in the formation of neo-XY and $X_1 X_2 Y$ systems in insects.

Dissociation

Dissociation, the reverse of fusion, involves the formation of two acrocentrics from a metacentric chromosome. Dissociation is rare because a donor centromere, a short arm, and a telomere are required; however this rearrangement was shown to occur in the dissociation that formed the two chromosomal races of the morabine grasshopper *K. scurra.*

Whole-Arm Interchanges

Whole-arm interchanges occur when chromosomal breaks and rejoinings near the centromeres of metacentric chromosomes result in the exchange of whole chromosome arms (Fig. 3). It has been noted that such exchanges distinguish races and species more frequently than reciprocal translocations, perhaps because the former maintain a sequence of coadapted genes in the arms concerned.

Complex Rearrangement

Complex rearrangements such as insertions, involving three or more breaks, have been noted in insect chromosomes, particularly after damage induced by radiation. Such work, particularly by H. Müller in *D. melanogaster,* led to each arm of the chromosome being defined as oriented from the centromere to the telomere.

See Also the Following Articles
Genetic Engineering • Parthenogenesis • Sex Determination

Further Reading

Cook, L. G. (2000). Extraordinary and extensive karyotypic variation: A 48-fold range in chromosome number in the gall-inducing scale insect *Apiomorpha* (Hemiptera: Coccoidea: Eriococcidae). *Genome* **43,** 169–190.

Crosland, M. W. J., and Crozier, R. H. (1986). *Myrmecia pilosula,* an ant with one pair of chromosomes. *Science* **231,** 1278.

Giles, E. T., and Webb, G. C. (1972). The systematics and karyotype of *Labidura truncata* Kirby, 1903 (Dermaptera: Labiduridae). *J. Aust. Entomol. Soc.* **11,** 253–256.

Gregg, P. C., Webb, G. C., and Adena, M. A. (1984). The dynamics of B chromosomes in populations of the Australian plague-locust, *Chortoicetes terminifera* (Walker). *Can. J. Genet. Cytol.* **26,** 194–208.

Hoy, M. (1994). "Insect Molecular Genetics." Academic Press, San Diego.

John, B. (ed.). (1974–1990). "Animal Cytogenetics," Vol. 3, "Insecta." Bornträger, Berlin.

Jones, R. N., and Rees, H. (1982). "B Chromosomes." Academic Press, London.

King, M. (1993). "Species Evolution: The Role of Chromosome Change." Cambridge University Press, Cambridge, U.K.

Kipling, D. (1995). "The Telomere." Oxford University Press, Oxford, U.K.

Shaw, D. D., Webb, G. C., and Wilkinson, P. (1976). Population cytogenetics of the genus *Caledia* (Orthoptera: Acridinae). II: Variation in the pattern of C-banding. *Chromosoma* **56,** 169–190.

White, M. J. D. (1973). "Animal Cytology and Evolution," 3rd ed. Cambridge University Press, London.

White, M. J. D. (1978). "Modes of Speciation." Freeman, San Francisco.

Chrysalis

Frederick W. Stehr
Michigan State University

A chrysalis (plural chrysalids) is the pupa of a butterfly, usually belonging to the family Papilionidae, Pieridae, or

FIGURE 1 Pupa of the butterfly *Papilio cresphontes,* showing the strand of silk that holds some butterfly pupae in an upright position. (Photograph from the Teaching Collection, Department of Entomology, Michigan State University.)

Nymphalidae. It is commonly found suspended or hanging from a leaf, twig, or branch, or even a windowsill, arbor, or other suitable structure. Not all species in these families form chrysalids. For example, the parnassians in the Papilionidae and the wood nymphs (Satyrinae) in the Nymphalidae pupate in a minimal cocoon in grass, leaves, or litter. The pupae of the families Lycaenidae and Riodinidae are also not suspended and usually are concealed in leaves or litter.

In forming the chrysalis, the prepupal caterpillar has to perform the seemingly impossible maneuver of spinning the silk pad to attach its cremaster (caudal pupal hooks) while maintaining its grip; it then must molt the larval skin as it attaches to the silk pad. Members of the Pieridae and Papilionidae (except Parnassinae) secure the chrysalis in an upright position with a band of silk around the middle (Fig. 1).

Chrysalids are usually angular, with projections, tubercles, spines, and sometimes gold or silver flecks. They are often cryptically colored so that they blend into the surrounding materials but some, like the monarch chrysalis, are smooth with gold flecks. In emerging from the chrysalis the adult splits the chrysalis ventrally and dorsally at the anterior end, crawls out, and suspends itself from the pupal skin while its wings expand.

See Also the Following Article
Pupa and Puparium

Further Reading

Stehr, F. W. (ed.) (1987). "Immature Insects," Vol. 1. Kendall/Hunt, Dubuque, IA.

Stehr, F. W. (ed.) (1992). "Immature Insects," Vol. 2. Kendall/Hunt, Dubuque, IA.

Cicadas

M. S. Moulds

Australian Museum, Sydney

Cicadas form a small part of the order Homoptera, a diverse group of insects whose mouthparts comprise a jointed rostrum for piercing and sucking up liquid food. They make up the superfamily Cicadoidea, distinguished by having three ocelli, an antennal flagellum usually of five segments, and a complete tenorium (internal development of the head for attachment of muscles); nymphs burrow and develop underground. The family arrangement for cicadas remains in a state of flux but is now generally accepted as being two families, the majority falling within the Cicadidae, and just two extant species plus some fossil species in the Tettigarctidae. There are almost 2000 named species, with perhaps as many again awaiting description.

Cicadas are mostly tropical or subtropical insects, but many also inhabit temperate regions. Some are minor pests of sugarcane, rice, coffee, and fruit trees, either reducing the vigor of the plants by nymphal feeding or weakening branches by oviposition, which in turn may cause the branches to break under crop load.

STRUCTURE AND FUNCTION

Cicadas typically possess a broad head delimited by a pair of large compound eyes, a large pro- and mesothorax housing mostly wing and leg muscles, a small metathorax, an abdomen that in the male is highly modified to accommodate the organs of sound production and reception, and two pairs of membranous wings that are usually held tentlike over the body at rest.

The head is dominated by a large, noselike postclypeus that houses muscles for sucking sap through the rostrum; the three jewellike ocelli detect the direction of light sources and, if asymmetrically covered, cause erratic flight.

The foreleg femora are characteristically enlarged and swollen. On the nymph these are even more enlarged (Fig. 1), serving the nymph for subterranean tunneling.

The abdomen carries the organs of reproduction and of hearing and, in males, also sound production.

SOUND PRODUCTION AND RECEPTION

Cicadas are best known for their ability to produce loud sound. No other insect has developed such an effective and specialized means of doing so. The calls are mating songs produced only by the males. Each species has its own distinctive song and attracts only females of its own kind (Fig. 2).

The organs of sound production are the tymbals, a pair of ribbed cuticular membranes located on either side of the first abdominal tergite (Fig. 3). In many species the tymbals are partly or entirely concealed by tymbal covers, platelike anterior projections of the second abdominal tergite. Contraction of internal tymbal muscles causes the tymbals to buckle inward, and relaxation of these muscles allows the tymbals to pop back to their original position. The sound produced is amplified by the substantially hollow abdomen, which acts as a resonator.

Many species sing during the heat of the day, but some restrict their calling to semidarkness at dusk. Often the species that sing at dusk are cryptic in coloration and gain further protection from predatory birds by confining their activity to dusk. The loud noise produced by some communal day-singing species actually repels birds, probably because the noise is painful to the birds' ears and interferes with their normal communication. The American periodical cicadas

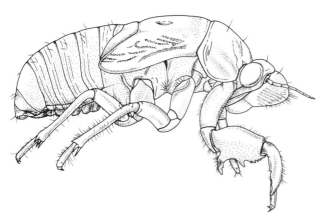

FIGURE 1 Mature nymph of *Cyclochila australasiae,* lateral view.

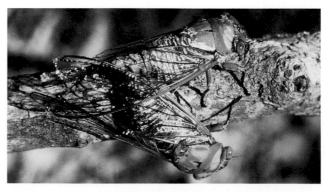

FIGURE 2 A mating pair of northern cherrynose, *Macrotristria sylvara* (family Cicadidae). This large and colorful species is found in tropical northeastern Australia. (Photograph by Max Moulds.)

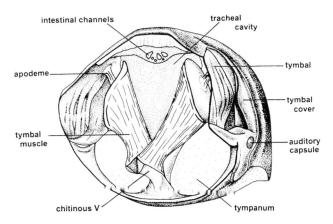

FIGURE 3 Transverse section of male abdomen of *Tamasa tristigma* at the first abdominal segment with the thorax removed. Exposed are the large tymbal muscles anchored basally to a chitinous V and attached dorsally via an apodeme to the sound-producing tymbals. Sound received by the tympana is transferred to the auditory capsules.

have mass emergences, and although their song is not sufficiently loud to repel birds, the number of individuals is so large that predatory birds soon lose their appetite for them.

Both sexes have organs for hearing. Sound is received by a pair of large, mirrorlike membranes, the tympana, which are often concealed below the opercula (Fig. 3). The tympana are connected to an auditory organ by a short slender apodeme. A singing male creases the tympana to avoid being deafened by his own song.

LIFE CYCLE

Eggs are laid in branches of trees and shrubs or in the stems of grasses (the nymphal food plants) in small slits cut into the surface by the female's spearlike ovipositor. The number of eggs laid in each slit varies between both species and individuals. Usually it is about 10 to 16, although the number laid per slit by a single female can range from 3 or fewer to more than 20.

A female makes many egg slits and often distributes her eggs at more than one site. A batch of eggs can number 300 or more. Some species, such as many *Cicadetta*, select only living tissue for oviposition, whereas others choose only dead or dying tissue. Many days, often in excess of 100, may pass before the nymphs hatch.

On hatching, the young nymphs are encased in a thin transparent skin that encloses the appendages separately but restricts their function. These pronymphs quickly wriggle their way along the egg slit to its entrance. A spine at the apex of the abdomen probably assists this exit and also in casting off the pronymphal skin. The young nymphs fall to the ground, whereupon they immediately seek shelter in the soil and later search for a root from which to feed by sucking sap.

Cicadas spend most of their life underground, slowly growing to maturity through five instars (Fig. 1). The length of life cycle is known only for a small number of species.

Some grass-feeding species mature within a year. The American periodical cicadas, *Magicicada* species, have a life cycle spanning 13 or 17 years, the longest known for any insect. Periodical cicadas are consistently regular in their life cycle length, but most other cicadas change by a year or two, and even individuals from a single egg batch can mature at different rates.

For most species, emergence from the final nymphal skin occurs during the first few hours after dark; the laborious process can last an hour or more. The adult life usually lasts 2 to 4 weeks, but some grass-dwelling species possibly live only 3 to 4 days. Some of the larger tree-inhabiting species probably live 8 or more weeks.

FAMILY CICADIDAE

The family Cicadidae includes all but two extant species. Two subfamilies are widely recognized, the Cicadinae, which have tymbal covers present, and the Tibicinae, with tymbal covers absent.

FAMILY TETTIGARCTIDAE

The family Tettigarctidae includes the other two extant species, *Tettigarcta crinita* and *T. tomentosa,* both found only in Australia. This family also includes 13 genera known from Cenozoic fossils.

The Tettigarctidae differ from other cicadas in several features. Most notable is the presence of tymbals in both sexes, but instead of producing airborne songs, they create low-level vibrations of the substrate below the adult. These substrate vibrations are detected by sensory empodia between the claws on all legs; the tympana used for hearing in other cicadas are lacking.

See Also the Following Articles

Auchenorrhyncha • Hearing • Sternorrhyncha • Vibrational Communication

Further Reading

Boer, A. J., de (1995). Islands and cicadas adrift in the west-Pacific. Biogeographic patterns related to plate tectonics. *Tijdschr. Entomol.* **138,** 169–244.

Boer, A. J., de, and Duffels, J. P. (1996). Historical biogeography of the cicadas of Wallacea, New Guinea and the West Pacific: A geotectonic explanation. *Palaeogeogr. Palaeoclimato. Palaeoecol.* **124,** 153–177.

Claridge, M. F. (1985). Acoustic signals in the Homoptera: Behaviour, taxonomy, and evolution. *Annu. Rev. Entomol.* **30,** 297–317.

Kato, M. (1956). The Biology of the Cicadas [Bulletin of the Cicadidae Museum]. Iwasaki Shoten, Jinbocho Kanda, Tokyo. [In Japanese; headings, subheadings, captions and index in English. Facsimile reprint, 1981, Scientist Inc., Japan.]

Duffels, J. P. (1988). The cicadas of the Fiji, Samoa and Tonga Islands, their taxonomy and biogeography (Homoptera, Cicadoidea). *Entomonograph* **10,** 1–108.

Duffels, J. P., and van der Laan, P. A. (1985). "Catalogue of the Cicadoidea (Homoptera, Auchenorhyncha) 1956–1980," Series Entomologica, Vol. 34. Junk, Dordrecht, Netherlands.

Metcalf, Z. P. (1963). "General Catalogue of the Homoptera," Fasc. 8, Cicadoidea. Part 1, Cicadidae. Part 2, Tibicinidae. [Species index by Virginia Wade, 1964.] University of North Carolina State College, Raleigh.

Moulds, M. S. (1990). "Australian Cicadas." New South Wales University Press, Kensington, Australia.

Moulton, J. C. (1923). Cicadas of Malaysia. *J. Fed. Malay States Mus.* **11**, 69–182, pls. 1–5.

Williams, K. S., and Simon, C. (1995). The ecology, behaviour, and evolution of periodical cicadas. *Annu. Rev. Entomol.* **40**, 269–295.

Circadian Rhythms

Terry L. Page
Vanderbilt University

Circadian rhythms are daily oscillations in physiology, metabolism, or behavior that persist (or free run) in organisms that have been isolated from periodic fluctuations in the environment. These rhythms are under the control of innate regulatory systems that are based on internal oscillators (or pacemakers) whose periods approximate those of the naturally recurring 24-h environmental cycles. The oscillators are subject to control by a limited number of these environmental cycles that synchronize or entrain the period to exactly 24 h and establish specific phase relationships between the rhythms and the external world (Fig. 1). Light cycles are virtually universally effective in the entrainment of

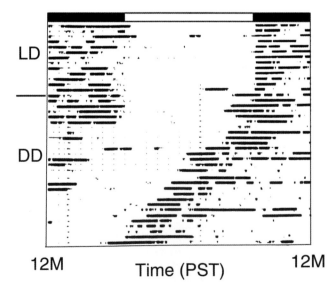

LD

DD

12M   **Time (PST)**   **12M**

FIGURE 1 Event recording of the wheel-running activity of a cockroach, *L. maderae.* Data for successive days are placed one below the other in chronological order. The bar at the top of the record indicates the light cycle to which the animal was exposed during the first 14 days of the recording. Then animal was then placed in constant darkness (DD) and its endogenously generated, free-running circadian rhythm was expressed for the remainder of the record with a period of about 23.5 h.

circadian rhythms, and in insects, daily cycles of temperature are also effective.

FUNCTION OF THE CIRCADIAN SYSTEM

In insects, the circadian system is responsible for imposing daily rhythmicity on a variety of processes, including locomotor activity, stridulation, oviposition, hatching, pupation and pupal eclosion, pheromone release, retinal sensitivity to light, and daily cuticle growth. This list is by no means exhaustive. It is generally accepted that the functional importance of this control is to restrict processes that are best undertaken at a particular phase of the environmental cycle to a particular time of day. It has also been suggested that a secondary role of the circadian system is to provide for internal temporal organization, coordinating the timing of various processes within the individual.

In addition to its role in generating daily rhythms, the circadian clock has been shown to be involved in photoperiodic time measurement for seasonal regulation of reproduction, development, and diapause in many insects. In honey bees (*Apis mellifera*), it is also involved in time measurement necessary for time-compensated sun orientation and in *Zeitgedachtnis*, which is the ability to return at the appropriate time to a food source that is available only at particular times of day. Thus the circadian system functions as a biological clock, capable of providing the individual with information on the time of day and with the ability to measure lapse of time.

PHYSIOLOGICAL BASIS OF CIRCADIAN ORGANIZATION

The study of the anatomical and physiological organization of circadian systems of insects has a long and productive history. The heuristic model generally used in these studies is illustrated in Fig. 2. There are four essential elements: (1) a pacemaker or oscillator that generates the primary timing signal, (2) photoreceptors for entrainment, and two coupling pathways, (3) one that mediates the flow of entrainment information from the photoreceptor to the pacemaker and (4) another that couples the pacemaker to the effector mechanisms that it controls. The model identifies several basic questions. Can the anatomical location of the circadian clock be identified? What are the pathways and mechanisms by which inputs to the pacemaking system regulate its phase and period? Finally, what are the neural and endocrine signals

FIGURE 2 Functionally defined model of the circadian system. An entrainment pathway that consists of a photoreceptor and coupling mechanism (input) synchronizes a self-sustaining oscillator (pacemaker) to the external light/dark cycle. The output of the pacemaker regulates the timing of various processes (e.g., activity) via coupling to the effector mechanisms.

by which the pacemaking system regulates the various processes under its control?

Circadian Oscillations Are Generated by Discrete, Localized Populations of Cells

Studies on pacemaker localization in insects have largely focused on behavioral rhythms (locomotor activity or eclosion) and their control by the nervous system. Compelling evidence that the brain is the site of generation of circadian timing signals for rhythms in behavior has been obtained in several species, with much of the early work involving studies on the locomotor activity rhythm of the cockroach.

In 1968 it was first discovered that surgical removal of both optic lobes or disconnecting them from the rest of the brain by section of the optic tracts abolished the activity rhythm of the Madeira cockroach, *Leucophaea maderae.* Results of lesion studies on other cockroach species, several species of crickets, and beetles have also suggested that the optic lobes might contain the pacemaker. Compelling evidence arose from the observation that it is possible to transplant optic lobes between cockroaches whose activity rhythms had quite different free-running periods. Animals that received transplanted optic lobes recovered rhythmicity in a few weeks with regeneration of the optic tracts, and the preoperative period of the donor and the postoperative period of the host were strongly correlated. Thus, the transplantation of the optic lobes not only restored the rhythm of locomotor activity but also, critically, imposed the period of the donor animal's rhythm on the activity of the host. Other studies involving small electrolytic lesions indicated that the cells responsible for generating the circadian signal have their somata and/or processes in the proximal half of the optic lobe, likely in a group of cells located ventrally near the medulla.

In contrast to cockroaches, crickets, and beetles, in a variety of other insects the optic lobes do not appear to be required for rhythmicity and the pacemaker appears instead to reside in the cerebral lobes (midbrain). In a classic series of experiments by James Truman and colleagues it was shown that the circadian pacemaker that controls the timing of eclosion in two silkmoth species, *Hyalophora cecropia* and *Antheraea pernyi,* is located in the cerebral lobes of the brain. The time of day at which eclosion occurs is different for the two species. When the insects are maintained in a photoperiod of 17:7 (L:D) hours, *H. cecropia* emerges shortly after lights-on while *A. pernyi* emerges just before lights-off. Removal of the brain did not prevent eclosion, but did disrupt its timing. However, if the brain was reimplanted in the abdomen, normal rhythmicity was restored under both entrained and free-running conditions. When brains were transplanted between species, individuals exhibited normal species-specific eclosion behavior, but the phase of the rhythm was characteristic of the donor and not the host. The demonstration that the transplanted brains restored rhythmicity and determined the phase of the rhythm left little

doubt that the circadian pacemaker that regulates the timing of the eclosion rhythm is located in the brains of these moths. The fact that the pacemaker was located in the cerebral lobes and not the optic lobes was demonstrated by subdividing the brain prior to transplantation. It was found that the optic lobes were unnecessary and that transplantation of the cerebral lobes alone was sufficient to restore rhythmicity.

Similarly, in a variety of dipterans, including the fruit fly, the house fly, the blow fly, and the mosquito, regions of the nervous system controlling locomotor activity rhythms have been dissected with both surgical and genetic lesions, and in each instance the pacemaking oscillation appears to be generated in the cerebral lobes. In the fruit fly, *Drosophila melanogaster,* extensive behavioral and genetic evidence demonstrates a crucial role for the *period (per)* gene in the circadian pacemaker controlling locomotor activity and eclosion rhythms (see later). The *per* gene is widely, and in some cell types rhythmically, expressed in the fly, including the head, thorax, and abdomen; thus, its spatial expression pattern in wild-type flies provided no definitive localization of the central pacemaker. However, the expression pattern has been altered by numerous genetic and molecular manipulations and it has been possible to determine the identity of the pacemaker cells in *Drosophila* by correlating *per* expression in specific cell types with the presence or absence of behavioral rhythmicity. The results suggest that only a few neurons between the lateral protocerebrum and the medulla of the optic lobes, the lateral neurons, are necessary for the generation of a circadian rhythm in locomotor activity.

The potential for further cellular identification of pacemaker neurons in insects was provided by an observation that in cockroaches and crickets optic lobe neurons that fulfilled the predicted anatomical criteria to be pacemaker cells were labeled by an antibody to crustacean pigment-dispersing hormone (PDH). When anti-PDH was applied to *Drosophila* brains, it labeled a ventral subset (LNv) of the *per*-expressing lateral neurons that were identified as pacemaker neurons in genetic studies. Taken together, the results indicated that the PDH-immunoreactive neurons are strong candidates for pacemaker neurons in insects and raise the possiblity that the insect version of crustacean PDH (called pigment-dispersing factor or PDF) may be an important temporal signaling molecule.

Interestingly, the numbers and projection patterns of PDF neurons in cockroaches and crickets are strikingly similar to those of the LNv of *Drosophila,* suggesting that they are functionally homologous. The most salient difference in the morphology of these neurons is in the locations of their somata: between the lobula and the medulla of the optic lobes in cockroaches and crickets, as opposed to between the medulla of the optic lobes and the lateral margin of the cerebral lobes in fruit flies. This difference may be sufficient to account for the fact that the lesion and transplant studies suggested different anatomical organizations for pacemaker structures in the central nervous system in different insects.

Circadian Pacemakers Are Also Found in Tissues outside the Nervous System in Insects

The localization of circadian pacemakers that regulate behavioral rhythms to the brain raised the question of whether other rhythms are controlled by the same clock. In crickets, beetles, and cockroaches studies indicate that the pacemaker regulating the daily rhythm in retinal sensitivity to light as measured by electroretinogram (ERG) amplitude is located in the optic lobe and suggest that the same pacemaker controls both the ERG amplitude and the activity rhythms. However, in other cases rhythms have been found to be regulated by pacemakers outside the nervous system. These include rhythmic secretion of cuticular layers in newly molted cockroaches, the release of sperm from the testis into the seminal ducts in gypsy moths, and the timing of ecdysteroid release from the prothoracic gland of the cynthia moth, *Samia cynthia*. In each of these examples the rhythms were shown to persist *in vitro* in the absence of neural pacemaking structures.

These results indicate that the distribution of circadian pacemaking centers may be widespread in insects. In support of this view, one recent study by Plautz and co-workers with *Drosophila*, in which the *per* promoter was coupled to the coding sequence for luciferase, indicated that rhythmic promoter activity could be detected in a variety of tissues, including the wing, leg, proboscis, and antennae maintained in isolation in tissue culture.

The fact that the circadian system in the individual may be composed of several widely distributed oscillators raises the question of whether there is communication between component oscillators. In general the answer is uncertain. Work on cockroaches has shown that the bilaterally distributed oscillators in the two optic lobes are connected to one another (mutually coupled) and suggested that the coupling was relatively strong. In contrast, both in the beetle, *Blaps gigas,* and in crickets the data indicated that coupling between optic lobe pacemakers is either absent or weak. Coupling relationships among other oscillators have not yet been systematically explored.

Photoreceptors for Entrainment

Extraretinal photoreceptors are typically involved in entrainment of behavioral rhythms. The classical example is the silkworm, in which it was shown that the photoreceptor for entrainment of the eclosion rhythm resides in the brain. Brains were removed from silkworm pupae and were either replaced in the head region or transplanted to the abdomen. The pupae were then placed in holes in a partition that separated two chambers in which the light/dark cycles were out of phase. Whether the pupae entrained to the light cycle to which the anterior end of the pupae was exposed or entrained to the light cycle at the posterior end corresponded to the location of the brain.

Additional evidence for extraretinally mediated entrainment of pacemakers that are located in the nervous system has been obtained in a variety of other insects, including other lepidopterans, dipterans, and orthopterans. In those instances in which there is evidence on the location of the photoreceptor, the brain appears to be the most likely site. However, more precise identification of the cells involved in the phototransduction has not been accomplished.

Even though the compound eyes may not be necessary for entrainment, they may nevertheless participate. In *Drosophila,* for example, genetic lesions to the eyes or the phototransduction pathway can alter the entrainment pattern. Further, there are at least two insects, the cockroach and the cricket, in which the compound eyes appear to be the exclusive photoreceptors for entrainment because sectioning the optic nerves between the eyes and the optic lobe or painting over the compound eyes eliminated entrainment of the locomotor activity rhythm to light cycles.

As noted above, there are several instances in which there is convincing evidence for circadian pacemakers outside the nervous system. In the case of the moth testis, since the rhythm measured *in vitro* responds to light, some cells in the testis–seminal duct complex must be photosensitive. Similarly, in the saturnid moth *S. cynthia,* the photoreceptor for entrainment of the pacemaker in the prothoracic glands appears to be in the gland itself.

Signals to Communicate Timing Information

Another important issue is how circadian oscillators impose periodicity on the various physiological and behavioral processes they control. *A priori,* several alternative mechanisms are plausible. Timing information within the individual could be represented by the level of a circulating hormone, impulse frequency in specific neural circuits, changes in general levels of neural excitability through neuromodulation, or, as the weight of the available evidence suggests, some combination of these mechanisms.

There are a large number of studies that suggest that secretion of a variety of insect hormones, including ecdysone, prothoracicotropic hormone, and eclosion hormone, is under the control of the circadian system during development. The experiments involving the transplantation of the silkworm brain, described above, provide the clearest demonstration of a hormonal link in the control of behavior by the circadian system. The signal for the eclosion behavior is the eclosion hormone that is produced in neurosecretory cells located in a region near the midline of the brain, the pars intercerebralis, and released via the neurohemal organs, the corpora cardiaca. The release of the hormone triggers release of two other peptide hormones, pre-ecdysis-triggering hormone (PETH) and ecdysis triggering hormone (ETH). PETH and ETH act on the central nervous system to initiate a stereotyped sequence of behavior that ultimately results in the emergence of the adult moth from the pupal case.

The role of humoral factors in the regulation of adult behaviors in insects (e.g., locomotor activity) is less clear. In cockroaches and crickets, the timing signal that originates in the optic lobe is transmitted to the brain via the optic tracts, and transmission from the brain to the activity centers in the thorax requires that the connectives of the ventral nerve cord be intact. Nerve impulse activity is rhythmic in both the optic tracts and the cervical connective.

In summary, the mechanism by which circadian phase information is transmitted to behavioral effectors in insects is generally not well understood. The emerging picture is that temporal regulation of behavior involves a modulation of excitability in the central nervous system. Axonal connections between the brain and the lower elements of the central nervous system are clearly required for the maintenance of some behavioral rhythms (e.g., cockroach locomotor activity), whereas others appear to rely heavily on hormonal mechanisms (moth eclosion). An important step in understanding how temporal information is transmitted will be the identification of the signal molecules involved.

MOLECULAR BASIS OF CIRCADIAN ORGANIZATION

There has been remarkable progress in the past 15 years in identifying the molecular basis of circadian clocks in a variety of organisms. In animals, much of this progress has resulted from pioneering work with the fruit fly. In 1971 the first clock gene, the *period* gene, was discovered in a mutagenesis screen in *D. melanogaster.* A decade later the gene was cloned, paving the way for studies of the gene's regulation. This work led to the discovery of several other genes in *D. melanogaster* that appear to be part of the clock mechanism, including those involved in entrainment.

Molecular Basis of the Clock

Four genes, the transcriptional regulators *period (per), timeless (tim), cycle (cyc),* and *clock (clk),* have been shown to be critical components for generating the basic circadian oscillation. Of the four, three, *per, tim,* and *clk,* are rhythmically expressed and circadian oscillations in both mRNA and protein levels are well documented. The fourth gene, *cyc,* is expressed at relatively constant levels throughout the day. Both CLK and CYC proteins are transcription factors that utilize basic helix-loop-helix domains to bind to E boxes, and both contain protein–protein interaction domains (PAS domains) that likely mediate the association of the two proteins with each other, thus forming heterodimers. The fundamental mechanism for generating the oscillation involves a transcription/ translation negative feedback loop. The basic loop is illustrated in Fig. 3. A heterodimer composed of CLK and CYC binds to promoters of *per* and *tim,* leading to an increase in transcription of these two genes that continues throughout the day. Levels of mRNA for the

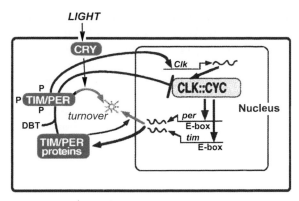

FIGURE 3 Molecular model of the circadian pacemaker of *Drosophila* showing the proposed negative feedback loop of the oscillation. CLK and CYC heterodimers bind to E boxes of nuclear DNA promoting transcription of *per* and *tim* genes. TIM and PER proteins heterodimerize and are phosphorylated by DBT. The heterodimer enters the nucleus and inhibits the positive regulation by the CLK/CYC heterodimer. Light enters the system through CRY and promotes turnover of TIM and PER (modified from Dunlap, 1999).

two genes peak in the early night. Protein products of these two genes increase as well, but peak levels of protein are delayed by several hours, peaking after the middle of the subjective night. PER and TIM themselves form a heterodimer, interacting through PAS domains. The heterodimer moves to the nucleus and functions as the negative element in the feedback loop, acting on the positive regulators CLK and CYC to suppress their activation of the *per/tim* promoters. This leads to a decline in the *per* and *tim* mRNA levels that continues throughout the night. The degradation of PER and TIM allows the cycle to start over.

The time delay between mRNA synthesis and the accumulation of PER and TIM is likely to be a critical element in the generation of the oscillation. PER is unstable in the absence of TIM. The dimerization stabilizes PER and promotes nuclear entry. In addition, both PER and TIM are phosphorylated, probably through the action of a homolog of casein kinase identified as *double-time (dbt).* This phosphorylation appears to be involved in regulation of PER turnover.

Mechanism of Entrainment

In *Drosophila,* light acts to cause a rapid decrease in the levels of TIM, and because TIM stabilizes PER, PER levels also decline. In the late day and early night when levels of these proteins are increasing, their destruction delays the progress of the oscillation, whereas in the late night and early day PER and TIM levels are decreasing, and hastening their demise advances the oscillation. Interestingly, genetic ablation of the eyes or mutations in the visual phototransduction pathway, although reducing sensitivity of the circadian clock to light, do not block its entrainment. The altered sensitivity to light observed with mutations that affect the visual system indicates that an opsin-based photoreceptor can contribute to entrainment of the circadian rhythm of locomotor activity, but the persistence of entrainment in these mutants implicates an extraretinal photoreceptor. Action spectra for entrainment have suggested a flavin-based photoreceptor. Cryptochrome (CRY), is a member of a family of flavoproteins, which includes photolyases and plant blue-light receptors. A mutant allele of the *cry* gene disrupts normal light responses of the locomotor activity rhythm, whereas flies overexpressing CRY are hypersensitive to light pulses. Further, in the periphery, CRY is required for light-dependent TIM degradation. These results suggest that CRY is a central element in the phototransduction pathway for entrainment.

The extent to which the molecular mechanisms detailed for *Drosophila* are applicable to other insects is not yet clear. However, there has been considerable progress in identifying homologous proteins in mammals, and although there are differences in detail, the basic framework of the oscillator seems to have persisted through the evolutionary process, giving confidence that the story will be broadly applicable to other insects as well.

Output of the Molecular Clock

The general supposition is that the clock ultimately regulates rhythms through the regulation of gene expression. This view is supported by the observation that there are several clock-controlled genes (CCGs) in *Drosophila*. However, at this point there are no examples in which a CCG has been linked directly to an overtly expressed physiological or behavioral rhythm, and this is an area of research that is likely to receive increased attention as researchers work to further elucidate the molecular details of the circadian system.

See Also the Following Articles

Brain and Optic Lobes • *Drosophila melanogaster*

Further Reading

Dunlap, J. C. (1999). Molecular bases for circadian clocks. *Cell* **96**, 271–290.

Giebultowicz, J. M. (2000). Molecular mechanism and cellular distribution of insect circadian clocks. *Annu. Rev. Entomol.* **45**, 769–793.

Hall, J. C. (1995). Tripping along the trail to the molecular mechanisms of biological clocks. *Trends Neurosci.* **18**, 230–240.

Page, T. L. (1985). Clocks and circadian rhythms in insects. *In* "Sensory Physiology" (G. Kerkut and L. Gilbert, eds.), Vol. VI of "Comprehensive Insect Biochemistry, Physiology, and Pharmacology," pp. 577–652. Pergamon Press, Oxford.

Page, T. L. (1990). Circadian organization in the cockroach. *In* "Cockroaches as Models for Neurobiology: Applications in Biomedical Research" (I. Huber, ed.), pp. 225–246. CRC Press, Boca Raton, FL.

Plautz, J. D., Kaneko, M., Hall, J. C., and Kay, S. A. (1997). Independent photoreceptive circadian clocks throughout *Drosophila*. *Science* **278**, 1632–1635.

Saunders, D. (1982). "Insect Clocks," 2nd ed. Pergamon Press, Oxford.

Zitnan, D., Ross, L. S., Zitnanova, I., Hermesman, J. L., Gill, S. S., and Adams, M. E. (1999). Steroid induction of a peptide hormone gene leads to orchestration of a defined behavioral sequence. *Neuron* **23**, 523–535.

Circulatory System

Thomas A. Miller
University of California, Riverside

Insects have an open circulatory system. This means that the internal organs and tissues are bathed in hemolymph, which is propelled actively to all internal surfaces by specialized pumps, pressure pulses, and body movements and is directed by vessels, tubes, and diaphragms. Without such constant bathing, tissues would die. The internal organs and tissues depend on the circulatory system for the delivery of nutrients, both to carry away excretion products and as the chemical communication pathway by which hormone messengers coordinate development and other processes. So vital is this function that it can be equated to the umbilical cord of a human fetus during development (Fig. 1).

Gas exchange in insects occurs via the tracheal system, which supplies all internal organs with tracheole tubules from spiracular openings in the body wall of terrestrial insects or from gill structures in aquatic insects. However, the hemolymph has the capacity to dissolve carbon dioxide gas in the form of bicarbonate ions. A few insects live in low oxygen environments and have a type of hemoglobin that binds oxygen at very low partial pressures, but for the most part oxygen is supplied and carbon dioxide is removed by ventilation through the tracheal system via the same system.

Besides the functions already mentioned, the circulatory system provides a medium in which battles are fought between the insect host and a myriad of invading disease microorganisms, including viruses, bacteria, fungi, and insect parasites. Principal participants in these interactions are the blood cells or hemocytes.

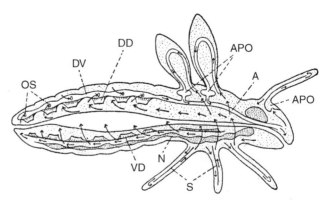

FIGURE 1 Delivery of the hemolymph to all tissues is so vital that a number of structures have evolved to ensure complete circulation including accessory pulsatile organs (APO), aorta (A), dorsal vessel (DV), dorsal diaphragm (DD), ostial valves (OS), and septa (S). Perfusion of the abdominal ventral nerve cord (N) is associated with a ventral diaphragm (VD). [Modified from Wigglesworth, V. B. (1972). "Principles of Insect Physiology." 7th ed. Chapman & Hall, New York, London, with kind permission of Kluwer Academic Publishers.]

While maintaining the body tissues, the circulatory system is the medium in which homeostasis is ensured, including the regulation of pH and inorganic ions, as well as the maintenance of proper levels of amino acids, proteins, nucleic acids, carbohydrates, and lipids. Any change in the hemolymph immediately affects all organs bathed. The time for complete mixing of the hemolymph depends on the size of the insect, but it can be up to 5 min in a resting adult cockroach weighing about a gram. Any substance injected into a healthy insect will eventually appear at the extreme ends of all appendages in a few minutes, emphasizing the efficiency of the delivery mechanisms, which can be marvels of microhydraulic engineering.

DORSAL VESSEL

The principal organ of hemolymph propulsion is the dorsal vessel, or at least it is the most visible organ associated with hemolymph movement. By tradition, the abdominal portion of the dorsal vessel is called the "heart" and the thoracic and cephalic (head) portion the "aorta." Both terms are borrowed from better-known vertebrate structures and give an inaccurate impression of the roles of those structures.

The dorsal vessel, especially in the abdomen of insects, is suspended in the pericardial sinus, which is delimited by the dorsal cuticle and the dorsal diaphragm (when present). Contractions of the dorsal vessel operate against the pull of lateral connective tissues, which are responsible for dilating or opening the vessel (in diastole) following each contractile stroke (called systole or systolic contraction). In most larvae of holometabolous insects (insects that possess a pupal stage) and in most primitive insects, the dorsal vessel is a simple tube running from the rear of the abdomen to the head, where it becomes closely associated with the top of the foregut and then travels under and opens just underneath or in front of the brain. This arrangement ensures a constant supply of nutrients and removal of waste products to and from the brain mass. In addition, the dorsal vessel is often intimately associated with the retrocerebral nervous system (including the hypocerebral ganglion, corpora cardiaca, and corpora allata complex) just behind the brain, which may deliver neurohormones and possibly other hormones into the aorta by way of specialized release sites.

The dorsal vessel has openings called ostia along the sides and ventral surfaces of each segment of the abdominal heart. The most common ostia allow the flow of hemolymph into the dorsal vessel and contain valves to prevent backflow. These are called incurrent ostia. Some insects have openings without valves through which the hemolymph moves constantly; these are called excurrent ostia and are common, for example, on the ventral side of the grasshopper heart, which also has a full complement of paired incurrent ostia associated with each segment in the abdomen.

Occasionally, insects have structures that branch out from the dorsal vessel. In the American cockroach (*Periplaneta americana*) and some other orthopteroid insects (e.g., crickets and mantids) there are paired segmental vessels diverging from the heart laterally. In the cockroach, these vessels are simple sacs of connective tissue and have no inherent musculature, thus providing a simple channel to the lateral aspects of the pericardial sinus in the middle segments of the abdomen. These specialized vessels ensure lateral perfusion of the pericardial sinus in moderate to large insects. Lateral tubes and vessels are not known in small insects.

The dorsal vessel is composed of muscle cells (collectively called the myocardium) that lie sometimes as opposed pairs and sometimes as spiral bands to form the cylinder of the dorsal vessel. The myocardium in all insects is spontaneously active usually beginning in the embryonic stages. This type of heart is termed myogenic because the electrical activity underlying contractions arises in the myocardium itself. This is in contrast to a neurogenic heart present in, for example, crustaceans such as crabs and lobsters, in which a barrage of nervous impulses drives the heartbeat from a discrete cardiac ganglion center.

In the pupae (and sometimes in resting adults) of holometabolous insects, the heartbeat exhibits reversal during which peristaltic contractile waves first push hemolymph from back to front (anterograde peristalsis) then at other times exclusively from front to back (retrograde peristalsis). Because heartbeat reversal is characteristic of even highly mobile mosquito pupae, reversal of hemolymph flow is thought to be an adaptation to an insect body that is rigid (the front end of a mosquito pupa is a rigid structure in which internal tissue and organs are undergoing drastic changes in shape to form adult structures, including the wings).

Nervous stimulation or mechanical disturbance causes the anterograde pulsations to revert to retrograde peristalsis. Because substances that block nerve impulses can cause the anterograde peristalsis to disappear, nervous signals (possibly inhibitory signals) are assumed to be responsible for alternating between the two peristalsis conditions, with retrograde being the basic condition.

Until recently, little was known about the innervation and control of heartbeat activity. Although the basal heartbeat rate of most insects is around 60 beats min^{-1} at room temperature and at rest (American cockroach and the locust, *Locusta migratoria*), the heartbeat of adult house flies (*Musca domestica*) is extremely unusual in that it fluctuates seemingly at random from over 300 beats min^{-1} to zero regardless of activity of the insect, flying or at rest.

The central nervous system of the adult house fly is composed of the brain and a thoracic ganglion mass. No ganglia are present in the abdomen. Because of this unusual anatomy, the dorsal vessel in the abdomen can be separated from all innervation from the central nervous system simply by cutting between the thorax and abdomen. After this operation, the heartbeat of the fly becomes quite regular at around 60 beats min^{-1}. This indicates that the house fly heart is innervated by both inhibitory and excitatory motor neurons from the central nervous system. Recently Ruthann

Nichols demonstrated inhibition caused by one or more neuropeptides in the fruit fly, *Drosophila melanogaster.*

VENTRAL DIAPHRAGM

The ventral diaphragm plays a prominent role in perfusing the ventral nerve cord of insects (Fig. 1). Nearly 40 years ago Glenn Richards surveyed the ventral diaphragms in insects and found that insects with a well-defined ventral nerve cord in the abdomen also had a well-developed ventral diaphragm. In contrast, insects with the ventral nerve cord condensed into a complex ganglion structure in the thorax invariably lacked a ventral diaphragm. This correlation suggests that the role of the ventral diaphragm is inexorably tied to perfusion of the ventral nerve cord in the abdomen.

The thorax of most insects is so packed with muscles involved in locomotion that other tissues are greatly reduced. Thus, the foregut is a simple tube passing through a small opening in the middle of the thorax and a well-defined ventral diaphragm (if present) is reduced. When present, the ventral diaphragm loosely defines a perineural sinus below and the perivisceral sinus above containing the gut.

In some insects, the ventral diaphragm is a strong muscular structure with a great deal of contractile activity. The activity of the ventral diaphragm is dictated by innervation from the central nervous system. In some large flying insects, the ventral diaphragm assists in hemolymph flow during thermoregulation by facilitating the removal of warm hemolymph from the hot thoracic muscles to the abdomen for cooling.

The intimate association between the ventral diaphragm in insects and perfusion of the ventral nerve cord is strengthened by considering the structure in the American and Madeira (*Leucophaea maderae*) cockroaches that takes the place of a proper diaphragm. In these two insects, four strips of muscles are attached at the back of the thorax and inserted on the ninth sternite. This structure has been called the hyperneural muscle because it does not form a true diaphragm above the ventral nerve cord and therefore is given a distinctive name. The hyperneural muscle is attached near the back of each of the abdominal ganglia, and the muscles contract slowly but not in a rhythmic order.

The hyperneural muscles are electrically inexcitable, which means that they do not contract myogenically, as the myocardium does, but instead are neurally driven by motor neurons located in the ventral ganglia. Thus each of the ventral nerve cords in these two cockroach species (*P. americana* and *L. maderae*) has its own muscle supply that pulls the ganglia back and forth along the midline of the abdomen upon demand. This entire structure is designed to increase the mixing and contact between the ganglia and the hemolymph.

DORSAL DIAPHRAGM

A cross section of the abdomen of insects reveals a pericardial sinus near the dorsal cuticle. The dorsal diaphragm can be a thin sheet of muscular tissue, or it can be fenestrated (Fig. 1). In most cases, there are muscles present in the diaphragm, which are called alary muscles because when vitally stained they give the appearance of "wings" projecting laterally from each abdominal segment of the dorsal vessel. The presence of paired alary muscles and paired ostia in each segment of the dorsal vessel in the abdomen reinforces the concept of "chambers" of the dorsal vessel in each abdominal segment.

Although mistakenly sometimes thought to play a key role in heartbeat, the alary muscles are more properly called muscles of the dorsal diaphragm. Whereas the myocardium is specialized to contract rapidly and constantly, the ultrastructure of the alary muscles is compatible with muscles that contract infrequently and slowly, having long sarcomeres and few mitochondria to provide only moderate amounts of energy.

In some insects, such as the tsetse fly and some moths, the alary muscles of the vestigial dorsal diaphragm extend from lateral cuticular attachment to join the dorsal vessel in the abdomen, turn, and travel along the dorsal vessel for some distance. Where this occurs, it is more difficult to determine the role of such alary muscles in the heartbeat.

ACCESSORY PULSATILE ORGANS

Because the circulation of hemolymph is vital to all insect tissues, several intricate structures ensure circulation of hemolymph through the appendages. Collectively, these are termed the accessory pulsatile organs (APOs), but modifications to ensure circulation in the appendages also include diaphragms and directed channels. When present, APOs occur at the bases of wings, antennae, legs, and cercal appendages at the back of the abdomen.

Early studies of the neuromusculature of the locust leg revealed a proximal bundle of muscles in the extensor tibia (jumping muscle) of the hind leg that exhibited rhythmic contraction. Amputating the hind leg of the grasshopper or locust very near the connection with the thorax, and attaching the femur to a convenient substrate with the tibia pointing straight up, demonstrates this rhythmic activity. After a delay of several minutes, the tibia will move back and forth spontaneously.

A small patch of muscles (called a "leg heart") near the coxal–trochanter–femur joint generates rhythmic pulsations thought to assist in the movement of hemolymph in the large femur of the jumping leg. To ensure hemolymph supply to the entire leg, there is a delivery route out and a collecting route back. Movement of hemolymph in an open circulatory system may be assisted by gross movement of internal organs, such as contractile activity of the Malpighian tubules and of the mid- and hindgut.

EXTRACARDIAC PULSATIONS

First described in 1971, extracardiac pulsations of insects are the simultaneous contractions of intersegmental muscles, usually of the abdomen of insects, that cause a sharp increase in the pressure in the insect body. The amount of movement accompanying each pulse is too small to be seen, but it can be readily measured as a slight shortening or telescoping of the abdomen as measured from its tip. The extracardiac pulses should not be confused with larger overt movements of the abdomen, especially in bees and bumble bees, that accompany ventilation during times or high activity or exertion such as flight.

Either the extracardiac pulsations occur in coordination with openings of certain of the spiracles, and therefore can play a role in ventilation, or they occur when all the spiracles are tightly closed, hence affecting hemolymph movement. The extracardiac pulsations become suppressed only in quiescent stages of insect development, such as during diapause, but they can be evoked immediately upon disturbance or stimulation.

The extracardiac pulsations are driven by a part of the nervous system for which Karel Slama coined the name "coelopulse nervous system." The pressures induced by extracardiac pulsations are 100 to 500 times greater than pressures caused by contractions of the dorsal vessel and are transmitted by the hemolymph throughout the entire body of the insect, influencing hemolymph movement at some distance from the dorsal vessel and APO structures.

TIDAL FLOW OF HEMOLYMPH

To keep body weight to a minimum, large flying insects decrease the amount of water in the hemolymph. The remaining amount of hemolymph is first delivered into the thorax and directed into the wing veins. To make room for the hemolymph in the wing veins, an accompanying tracheal tube collapses as the hemolymph is pumped into the vein. Thus, the space in the wing veins is first taken up by hemolymph at the expense of air in the tracheal tube and then, as the hemolymph is pumped out, the air reenters the wing vein. Lutz Wasserthal called this periodic exchange of air and hemolymph a "tidal flow" of hemolymph.

The entire circulatory system of the insect is recruited to operate the tidal flow of hemolymph in large flying insects. Thus both the dorsal vessel and the ventral diaphragm are recruited to direct the hemolymph in the proper direction during each tidal cycle. This implies a very sophisticated control mechanism that must operate from the central nervous system.

THERMOREGULATION

Before the extracardiac pulsations were reported and before the tidal flow of hemolymph had been described in insects, Bernd Heinrich wrote about the use of the hemolymph in thermoregulation of flying insects. The optimum temperature for flight muscle contraction in many insects, such as the tobacco hornworm, *Manduca sexta,* is surprisingly high, up to 45°C. Before this moth can fly, it must warm the thorax to near this temperature, which it accomplishes by means of a series of simultaneous isometric contractions of the antagonistic pairs of flight muscles that appear to the casual observer as "shivering," or vibrations of the wings (Fig. 2).

A "thermometer" in the thoracic ganglia detects the proper temperature. When the thoracic temperature is below

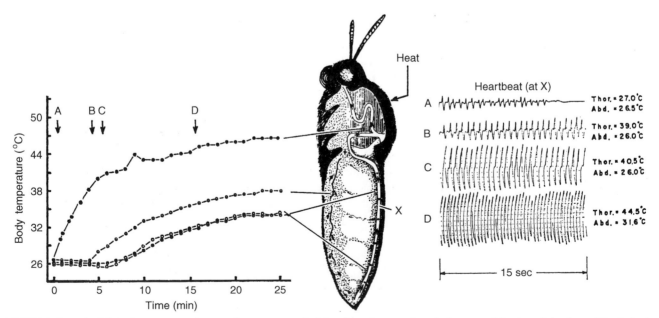

FIGURE 2 Control of thoracic temperature by central nervous control of dorsal vessel contractions during external heating of the thorax (Heat). At the optimum temperature, hemolymph is pumped at maximum frequency and amplitude through the dorsal vessel to conduct heat from the thorax to the abdomen, where is it dissipated. [From Heinrich, B. (1970). Nervous control of the heart during thoracic temperature regulation in a sphinx moth. *Science* **169,** 606–607, Copyright 1970 American Association for the Advancement of Science.]

optimum, the central nervous system signals the dorsal vessel to circulate hemolymph slowly. When the thoracic temperature rises above optimum, the central nervous system brings about maximal amplitude and rate of heartbeat to drive hemolymph through the thoracic muscles. The increased hemolymph flow pulls heat away from the flight muscles in the thorax and eventually delivers hot hemolymph to the abdomen, where the heat is dissipated. Then relatively cool hemolymph is redelivered to the thoracic muscles by the dorsal vessel, completing the thermoregulation cycle.

The warm hemolymph is then delivered to the head and percolates back past the ventral ganglia in the thorax to the abdomen, where the heat is dissipated. The cooler hemolymph is then delivered again to the thorax. The dorsal vessel and the very strong ventral diaphragm in the tobacco hornworm act together to move hemolymph in this analogy to an automobile radiator. When the thorax is too warm, both the amplitude and the frequency of heartbeat contractions are increased, and the rate of delivery of hemolymph increases. When the thorax is too cool, amplitude and frequency of contraction of the dorsal vessel are decreased. The activity of the ventral diaphragm acts in concert with that of the dorsal vessel.

Thermoregulation of the flight muscles of the tobacco hornworm implies a sophisticated nervous control. The overall nervous control can be easily demonstrated by severing the ventral nerve cord between the thorax and abdomen. When this is done, the moth can no longer thermoregulate because the feedback loop of temperature detection by the thoracic ganglia has been destroyed, and control over ventral diaphragm and dorsal vessel contractions has been lost. An extreme form of modified circulatory system to accommodate thermoregulation is shown in Fig. 3.

AUTONOMIC NERVOUS SYSTEM

The tidal flow of hemolymph, the extracardiac pulsations, heartbeat reversal, and thermoregulation all imply a very sophisticated control of circulation by the central nervous system. The central nervous system also plays a role in regulation of the respiratory system. It seems increasingly clear that the activities of circulatory and respiratory systems are coordinated by the central nervous system, perhaps to an extent not fully appreciated, but strongly implied by the tidal flow of hemolymph concept of Lutz Wasserthal.

It would be convenient and satisfying to be able to point out a particular part of the central nervous system and related peripheral nerves in insects that might comprise this regulatory system; however, outside existing evidence that the meso- and/or metathoracic ganglia play a major role in certain of these functions, entomologists know of no such discrete structure or structures, possibly because these interregulatory functions have been undertaken by different parts of the nervous system in different insects. It is known that insects have a number of regulatory mechanisms that can be recruited to achieve such control, from motor and sensory neurons to neurosecretory neurons to neurohormonal organs located all over the insect hemocoel.

See Also the Following Articles
Hemolymph • Immunology • Respiratory System • Thermoregulation

Further Reading
Ai, H., and Kuwasawa, K. (1995). Neural pathways for cardiac reflexes triggered by external mechanical stimuli in larvae of *Bombyx mori. J. Insect Physiol.* **41,** 1119–1131.

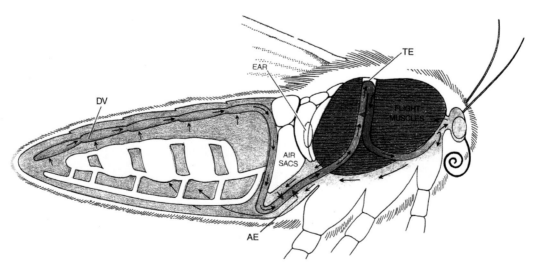

FIGURE 3 Circulation under precise nervous control of the dorsal vessel (DV) keeps the flight muscles (red) warm in the noctuid winter moth with the aid of two strategically placed countercurrent heat exchangers (TE and AE). [Modified from Heinrich, B. (1987). Thermoregulation in winter moths. *Sci. Am.* **256,** 104–111. Illustration by Patricia J. Wynne.]

Heinrich, B. (1971). Temperature regulation in the sphinx moth, *Manduca sexta*. II: Regulation of heat loss by control of blood circulation. *J. Exp. Biol.* **54,** 153–166.

Jones, J. C. (1977). "The Circulatory System of Insects." Thomas, Springfield, IL.

Miller, T. A. (1997). Control of circulation in insects. *Gen. Pharmacol.* **29,** 23–38.

Pass, G. (2000). Accessory pulsatile organs: Evolutionary innovations in insects. *Annu. Rev. Entomol.* **45,** 495–518.

Slama, K. (2000). Extracardiac versus cardiac haemocoelic pulsations in pupae of the mealworm (*Tenebrio molitor* L.). *J. Insect Physiol.* **46,** 977–992.

Wasserthal, L. T. (1996). Interaction of circulation and tracheal ventilation in holometabolous insects. *Adv. Insect Physiol.* **26,** 297–351.

Classification

see *Nomenclature and Classification*

Cockroach

see *Blattodea*

Cocoon

Frederick W. Stehr
Michigan State University

A cocoon is commonly believed to be the silken protective covering within which the caterpillars of many moths and a few butterflies pupate. Other orders of insects also spin silk and form cocoons, including Siphonaptera (fleas), Hymenoptera (ants, bees, and wasps), Neuroptera (lacewings and antlions), and Trichoptera (caddisflies).

Lepidoptera cocoons can be very thick and tough, such as that of the silkworm, *Bombyx mori,* or they can be limited to a relatively few strands of silk that keep the pupa from falling or that hold materials together to form a shelter. Cocoons such as those of the silkworm are composed of a single continuous strand of silk that is unwound in commercial silk production. Other lepidopteran cocoons may also be composed of a single strand, but the strand is usually intertwined in a way that is very difficult to unwind.

There are many kinds of lepidopteran cocoons. Some are formed from substrate materials held together by silk, others are so tough and juglike that they have a special escape lid woven at the end for the emergence of the adult (slug caterpillars, Limacodidae). Some lepidopterans such as the cecropia moth,

Hyalophora cecropia (Saturniidae), spin an elongate cocoon with a one-way escape hatch at the end. The tent caterpillars, *Malacosoma* (Lasiocampidae), spin a complete cocoon that is infused with a yellow or whitish powder that can be irritating to predators. Emergence is accomplished by secretion of a fluid that softens and dissolves part of the cocoon. Many woollybear caterpillars (Arctiidae) incorporate the caterpillar's spiny setae (which can also be irritating) into the cocoon.

Emergence from a pupa in the soil is usually by the adult crawling upward and often occurs after a rain softens the soil. Other adults emerge by cutting or forcing their way through the wall of the cocoon with sharp structures on the pupal head that are moved by the adult inside the pupal skin. Generally, if the pupa is capable of forward movement, it only partially protrudes from the cocoon, because it is held in place by forward-projecting spines near its rear that anchor it within the cocoon, enabling the adult to pull out of the pupal skin more easily.

Caterpillars that live in bags or cases attach the bag firmly with silk to the host (or some other substrate) and pupate within these structures. Winged male bagworms (Psychidae) emerge from the bag but the females of most species are wingless and remain inside the bag where they are fertilized by the males' insertion of their long abdomens into the open end of the bag. Females usually lay their eggs inside the bag and die.

Hymenopteran cocoons are highly variable in appearance, ranging from the tightly spun cocoons of some parasitic ichneumonid wasps that are suspended on a long silken thread to the more loosely spun cocoons of some braconid wasps such as those found clustered on the outside of caterpillars (Fig. 1). Ants also spin cocoons. When ant nests are opened, many ants rush off carrying larvae and cocoons (the larger, smoother objects that look like short hotdog buns and are thought to be eggs by some). Sawfly larvae spin cocoons that are similar to ant cocoons, and most bee and wasp larvae spin cocoons inside the cells provided by the adults.

Flea (Siphonaptera) larvae spin cocoons covered with debris in areas where the larvae have been feeding. Lacewing larvae

FIGURE 1 Multiple parasitic wasp cocoons (Braconidae) formed by larvae after emerging from the slug caterpillar, *Lithacodes fasciola*. (Photograph by David J. Wagner, University of Connecticut.)

(Neuroptera: Chrysopidae) spin tight, egg-shaped cocoons in some snug spot near where they have been feeding. Caddisfly larvae (Trichoptera) are aquatic; many use their cases made of silk or materials spun together with silk as cocoons by attaching them to the substrate and sealing off the ends.

See Also the Following Articles

Caterpillar • Larva • Pupa and Puparium

Codling Moth

Stephen C. Welter

University of California, Berkeley

Codling moth, *Cydia pomonella* (Fig. 1), is a key insect pest of apples, pears, and walnuts nearly worldwide. Codling moth acquired its name because of its attacks on a greenish, elongated English cooking apple referred to as a "codling" apple. The codling moth was noted to be a pest in Europe as early as 1635, well before the development of large-scale planting of apples or pears; the first report of codling moth in the United States was in 1750.

IMPORTANCE AND DAMAGE

Codling moth spread from its presumed site of origin in Eurasia, along with the cultivation of its hosts, particularly apples; other agricultural hosts include quince, apricot, plum, peach, and nectarine, although codling moth is less of a problem in these crops. Damage results from shallow feeding wounds that cause scarring of the fruit, from direct feeding damage to the fruit pulp or seeds, or from indirect contamination of the fruit by larval feces.

LIFE HISTORY

Codling moth has between one and four generations per year, depending on temperature and other climatic factors.

FIGURE 1 Codling moth *(Cydia pomonella).* (Photograph by Mark Skevington, Whetstone, Leicestershire, U. K.)

Adult codling moth females lay single eggs on the fruit or leaves of their host. Although some larvae feed on the surface of the fruit, most larvae bore directly into the fruit within 24 h, continue to feed briefly under the surface of the skin, and then move through the flesh of the fruit to feed on the seeds. There are five larval instars. Mature larvae exit the fruit and most frequently pupate under the bark. As daylength shortens with the approach of winter, mature fifth instars spin overwintering cocoons under bark, in debris, or wood fruit. The mature larvae spend the winter in a state of arrested development until spring conditions trigger development.

MANAGEMENT

Insecticides

Management of codling moth populations in orchards traditionally has relied on synthetic pesticides. Although newer, more selective pesticides provide effective control of codling moth, older pesticides have been associated with nontarget environmental and human health risks. In addition, the evolution of resistance in codling moth to many different groups of insecticides (the chlorinated hydrocarbons, organophosphates, carbamates, pyrethroids, and newer insect growth regulators) has made the long-term reliance on these compounds more problematic.

Pheromone Mating Disruption

A recent alternative to insecticides relies on the disruption of codling moth mating using sex pheromones. Artificial emitters of the female attractant interfere with the male's ability to find females. The most common dispensers are variations on different reservoir designs, which are tied or placed in orchard tree canopies. Synthetic pheromone from these emitters then permeates the orchard canopies. Although the exact mechanisms explaining this approach are unclear, program efficacy has been demonstrated in almost all growing regions of the world. However, mating disruption is often not efficacious initially in orchards with high pest densities, so that some use of conventional insecticides may be required. Mating disruption has been widely implemented in some areas such as the western United States, where up to 40 to 50% of the pear and apple acreage (e.g., in northern California) uses this technique.

Biological Control

Although management of codling moth based on control by natural enemies has proven elusive, significant reductions in population densities have been made by using both native and introduced natural enemies of codling moth. One of the more thoroughly studied natural enemies in North America, Europe, or the former Soviet Union is the *Trichogramma* egg parasitoid. Large numbers of these minute wasps are periodically released into an orchard to seek out and kill the eggs of codling moth. The eggs of *Trichogramma* are laid into the eggs of codling moth; the death of the egg occurs as the *Trichogramma* larvae develop. Other parasitoids that attack larval or pupal stages have also been introduced or accidentally released into new regions, including *Pimpla pterelas* and *Ascogaster quadridentata*. However, parasitism levels rarely reach more than 5%, except for some regions in central Asia where levels are as high as 50%. Nonspecialized parasitoids of egg, prepupal, or pupal stages comprise the majority of the natural enemies in North America; more specialized larval parasitoids are found in Europe and Central Asia.

General predators such as birds, predaceous insects, and spiders have been reported as suppressive agents of codling moth; these include woodpeckers, carabid beetles, and mirid bugs.

Although codling moth is susceptible to several diseases, a granulosis virus that can be applied in water, similar to insecticide applications, can cause significant reductions in codling moth densities. However, problems with production, formulation, and the short residual activity of the virus restrict its usage. Some reductions in codling moth populations also have been associated with applications of the bacterium *Bacillus thuringiensis*, but its efficacy is limited.

See Also the Following Articles
Agricultural Entomology • *Biological Control of Insect Pests* • *Integrated Pest Management* • *Pheromones*

Further Reading
Aliniazee, M. T., and Croft, B. A. (1999). Biological control in deciduous fruit crops. *In* "Handbook of Biological Control: Principles and Applications of Biological Control" (T. S. Bellows and T. W. Fisher, eds.), pp. 750–753. Academic Press, San Diego.
Barnes, M. M. (1991). Codling moth occurrence, host race formation, and damage. *In* "Tortricid Pests: Their Biology, Natural Enemies, and Control" (L. P. S. v. d. Geest and H. H. Evenhuis, eds.), Vol. 5, pp. 313–328. Elsevier, Amsterdam.
Calkins, C. O. (1998). Review of the codling moth areawide suppression program in the western United States. *J. Agric. Entomol.* **15**, 327–333.
Cardé, R. T., and Minks, A. K. (1995). Control of moth pests by mating disruption: Successes and constraints. *Annu. Rev. Entomol.* **40**, 559–585.
Cross, J. V., Solomon, M. G., Babandreier, D., Blommers, L., Easterbrook, M. A., Jay, C. N., Jenser, G., Jolly, R. L., Kuhlmann, U., Lilley, R., Olivella, E., Toepfer, S., and Vidal, S. (1999). Biocontrol of pests of apples and pears in northern and central Europe. 2. Parasitoids. *Biocontrol Sci. Technol.* **9**, 277–314.

Coevolution

Douglas J. Futuyma
State University of New York, Stony Brook

The term "coevolution" usually refers to the joint evolution of two or more species or genomes, owing to interactions between them. These interactions include interspecific competition, mutualism, and interactions between "consumers" and "victims" (encompassing predator/prey, herbivore/plant, and parasite/host relationships), as well as other interactions such as mimicry. Although it is often difficult to prove that true coevolution has occurred, it has probably had profound effects on the diversity of organisms and the evolution of their characteristics. Insects have figured prominently in research on coevolution.

CONCEPTS OF COEVOLUTION

Coevolution refers to several processes. One possible form of coevolution is cospeciation, the coordinated branching (speciation) of interacting species (such as host and parasite). To the extent that this has occurred, concordant (or matching) phylogenies of host and parasite clades (or evolutionary lines) would be expected. Cospeciation might be caused by the interaction between species, but it could also result from a joint history of geographic isolation, assuming that divergence and reproductive isolation evolve at similar rates in the two groups. Concordance of the two phylogenies implies a longer history of association, and of opportunity for reciprocal adaptation, than, for example, when parasites or symbionts have frequently switched from one host to another. Host switching can be inferred from certain patterns of discordance between host and symbiont phylogenies. Both cospeciation and host switching have been revealed in herbivorous insects, symbiotic bacteria, and parasites. For example, lice associated with gophers and with certain seabirds appear to have cospeciated to a considerable extent, and endosymbiotic, mutualistic bacteria *(Buchnera)* display almost complete phylogenetic concordance with their aphid hosts, from the family level down through relationships among conspecific populations.

In its most frequent usage, coevolution refers to genetic changes in the characteristics of interacting species resulting from natural selection imposed by each on the other—i.e., reciprocal adaptation of lineages to each other. Such changes are referred to as specific or pairwise coevolution if the evolutionary responses of two species to each other have no impact on their interactions with other species. Diffuse or guild coevolution occurs when the genetic change in at least one species affects its interaction with two or more other species. For example, cucumber genotypes with high levels of the chemical cucurbitacin have enhanced resistance to mites but also enhanced attractiveness to cucumber beetles; this is an instance of a negative genetic correlation in resistance. Early season attack by flea beetles makes sumac plants more susceptible to stem-boring cerambycid beetles, and so resistance to the former would also reduce the impact of the latter.

In one of the seminal papers on coevolution, Ehrlich and Raven postulated in 1964 what has since been named "escape and radiate" coevolution—a process in which evolutionary changes temporarily reduce or eliminate the ecological interactions between species. Applying this concept to plants and herbivorous insects, Ehrlich and Raven postulated that in response to selection by herbivores, a plant species may evolve new defenses that enable it to escape herbivory and to flourish so well that it gives rise to a clade of descendant species with similar defenses. At some later time, one or more species of herbivores adapt to the defenses and give rise to an adaptive radiation of species that feed on the plant clade. In this scenario, the evolutionary diversification of both herbivores and plants is enhanced by their interactions.

Despite a common misconception, coevolution need not promote stable coexistence of species, and it certainly need not enhance mutual harmony. For example, parasites may evolve to become more virulent or less, depending on their life history. The Darwinian fitness of a genotype of parasite is measured by the average reproductive success of an individual of that genotype. Extracting more resource from a host, thereby reducing its chance of survival, often enhances the parasite's reproductive success, as long as the parasite individual, or its offspring, can escape to new hosts before the current host dies. Evolution of the parasite, by individual selection, may result in such high virulence that the prey or host population is extinguished. Extinction of prey populations does not alter the relative fitnesses of individual parasite genotypes and so does not select for reduced virulence. However, group selection may favor lower virulence or proficiency. If populations of more virulent parasites suffer higher extinction rates than less virulent populations, the species as a whole might evolve lower virulence. Although individual selection is likely to be stronger than group selection in most species, the population structure of some parasites may provide an opportunity for group selection to affect their evolution.

COEVOLUTION OF COMPETING SPECIES

Darwin argued that competition is an important agent of natural selection for adaptation to different habitats or resources by different species. Indeed, a common theme in community ecology is that coexisting species differ in food or other components of their ecological niches and that such differences are ordinarily necessary for species to coexist in the long term.

Quantitative genetic models of the evolution of competitors assume that in each of two or more species, a heritable, continuously varying trait, such as an animal's body size or mouth size, determines the mean and variance of resources (e.g., size of prey) consumed. Because competition for limiting resources decreases an individual's fitness, genotypes of species 1 that use a resource different from that used by species 2 are likely to increase in frequency, so that the mean phenotype (and resource use) shifts away from that of the other species. At evolutionary equilibrium, the species will still overlap in resource use to a greater or lesser extent, depending on the abundance of different resources, but the variance in each (the breadth of resources used) is likely to be lower than in a solitary species. Three or more species may evolve differences from each other in phenotype (e.g., size) and resource use. Such coevolutionary changes should promote coexistence. However, if competition between species is asymmetrical (e.g., if larger individuals reduce the fitness of smaller ones more than the converse), a species may converge toward the other, use its resources, and "chase" it to extinction.

Considerable evidence, mostly from vertebrates, supports this coevolutionary theory. For example, closely related sympatric species of Darwin's finches, woodpeckers, and some other animals each use a narrower variety of food types or microhabitats than do species that occur singly on islands. Evidence for evolutionary response to competition is provided by some instances of character displacement—a greater difference between two species where they occur together than where each occurs alone. Some lakes left by retreating glaciers in northwestern North America are inhabited by a single species of stickleback fish (*Gasterosteus aculeatus* complex), which feeds both near the bottom and in open water. In other lakes, two coexisting species have evolved. Relative to the solitary form, the coexisting species have diverged and specialized in morphology and behavior: one feeds on benthic prey and the other on plankton. Experiments have shown that competition among similar phenotypes reduces growth of juveniles more than among dissimilar phenotypes. In one of the few cases of ecological character displacement reported for insects, sympatric populations of two species of rhinoceros beetles (Scarabaeidae: *Chalcosoma*) overlap less in altitudinal range and differ more in size than allopatric populations. However, it has not been shown that these differences stem from competition for resources.

Coevolution of competitors may explain some patterns in community structure. For example, differences in body size or trophic structures among sympatric pairs of species of bird-eating hawks, carnivorous mammals, and seed-eating Galapagos finches are greater than if the species had been assembled at random. In a remarkable example of coevolutionary consistency, ecologically and morphologically equivalent sets of species of *Anolis* lizards have evolved independently on each of the four islands of the Greater Antilles.

COEVOLUTION OF CONSUMERS AND VICTIMS

We might expect predators and prey, herbivores and plants, and parasites and their hosts to evolve in an "arms race," whereby the victim evolves ever greater resistance, defense, or evasion, and the consumer evolves ever greater proficiency in finding and attacking the victim. However, the coevolutionary dynamics may be more complex than this, because of factors such as costs of adaptation and diffuse coevolution.

Considerable evidence supports the assumption that greater elaboration of a defensive or offensive feature imposes costs resulting from the character's interfering with another function or simply from the energy required for its development.

The population dynamics and the course of character evolution depend on many parameters and are often sensitive to starting conditions. An indefinitely extended arms race or escalation of the two species' characters is unlikely, because the cost of a sufficiently elaborated character eventually exceeds its benefit. Rather, the characters of both the prey and the predator may evolve to an intermediate stable state. Perhaps counterintuitively, species may become less proficient in attack or defense; for instance, a prey species may evolve a lower level of defense if it is so well defended that the predator becomes rare and thus becomes a weaker agent of selection than the energetic cost of defense. In some models, both the population densities and the character means of both species may change indefinitely, either in stable limit cycles or chaotically, and may even result in extinction.

When the consumer feeds on multiple species of victims, or a victim is attacked by multiple consumer species, diffuse coevolution may affect the outcome. For example, if there exists a negative genetic correlation between a host's resistance to different species of parasites, then resistance to each carries a "cost," selection will vary in time and space, depending on the relative abundance of the two parasites, and resistance to each parasite is constrained. Diffuse coevolution can be very difficult to document and might often be sluggish. Because prey species have characteristics (e.g., cryptic coloration, distastefulness, speed of escape) that provide protection against many species of predators, and predators likewise have characteristics that enable them to capture and handle many prey species, changes in the relative abundance of different predators (or prey) may not greatly alter selection. During the "Mesozoic marine revolution," lineages of crustaceans and fishes capable of crushing hard shells evolved, and many groups of molluscs evolved features (e.g., thicker shells, spines) that made them more difficult to crush. Surely these changes reflect diffuse coevolution, but our inability to ascribe changes in any one species to changes in any other one species makes it hard to discern a coevolutionary process.

Predators and Prey

Geographic variation in the identity and strength of interactions among species provides some of the best evidence of coevolution. For example, the shape of the cones of lodgepole pine *(Pinus contorta)* differs among populations, depending on whether its major seed predator, the red squirrel *(Tamiasciurus hudsonicus),* is present or absent. In mountain ranges without red squirrels, red crossbills *(Loxia curvirostra* complex) are abundant seed predators. In these areas, the pine has evolved modifications of the cone that reduce seed extraction by this species of bird, and the shape and size of the crossbill's bill have evolved to enhance seed extraction.

Such evidence of coevolution, however, is rare, compared with evidence of unilateral adaptation. For example, Mediterranean populations of the braconid parasitoid *Asobara tabida* have higher "virulence" (capacity to survive host defenses) than northern European populations. Although one of its hosts, *Drosophila melanogaster,* shows a somewhat parallel geographic pattern in defense, the cline in *Asobara* appears to be most parsimoniously explained not by coevolution, but by the fact that *D. melanogaster,* its major host in the south, has stronger defenses than the major northern host, *D. subobscura.*

Parasite–Host Interactions

The evolution of interactions between hosts and parasites (including pathogenic microorganisms) can differ from predator–prey interactions in several respects. Whereas improvement in a predator or prey trait (such as size or fleetness) is likely to enhance fitness regardless of the specific genotype in the opponent species, parasite–host interactions are more likely to be affected by "gene-for-gene" interactions, in which each allele for host resistance is matched by a parasite "virulence allele" that enables the parasite to overcome resistance. Such gene-for-gene relationships have been described for several plant–fungus interactions and for the relationship between the Hessian fly (Cecidomyiidae: *Mayetiola destructor*) and resistant genotypes of wheat. Selection in gene-for-gene systems may be frequency-dependent: as a parasite allele that matches the most common host allele increases in frequency, rare host alleles acquire a selective advantage by conferring resistance against most of the parasites and so increase in frequency and initiate selection for a currently rare virulence allele. The genetic composition of local populations is likely to differ at any one time, because these oscillatory genetic dynamics may be out of phase unless the populations are connected by high gene flow. Geographic variation in genetic composition has been reported for trematodes and snails, trematodes and fish, microsporidians and *Daphnia,* and fungal parasites and plants. In most of these parasite–host pairs, populations of the parasite are best adapted to their local host populations, suggesting that the parasites adapt faster than their hosts.

The fitness of a parasite genotype may be approximately measured by the number of potential hosts it infects, compared with other genotypes. Often, the rate of transmission to new hosts is proportional to the parasite's reproductive rate, which in turn often (though not always) determines the parasite's virulence to the host. For example, the probability that progeny of a virus are transmitted by a mosquito is a function of the density of viral particles in the host's blood. However, the probability of transmission is reduced if the host dies too soon, i.e., if the parasites die before transmission. Such conflicting factors result in an evolutionary equilibrium level of virulence that is determined by several factors, especially the mode of transmission. If transmission is "vertical," i.e., only to the offspring of infected individuals, then parasite fitness is proportional to the number of surviving host offspring, and selection favors benign, relatively avirulent parasite genotypes. If transmission is "horizontal," i.e., among hosts of the same generation, the equilibrium level of virulence is likely to be higher, because (a) an individual parasite's fitness does not depend on successful reproduction of its individual host and (b) the likelihood is higher that an individual host will be infected by multiple parasite genotypes that compete for transmission to new hosts. As predicted by this theory, among species of nematodes that parasitize fig wasps (Agaonidae), those that are mostly horizontally transmitted cause a greater reduction of their hosts' fitness than those that are vertically transmitted.

Herbivores and Plants

Most of the many thousands of species of herbivorous insects are fairly or highly specialized, feeding on closely related species—sometimes just a single species—of plants. At a proximal level, this specificity is largely the result of behavioral responses to plant features, especially the many "secondary chemicals" that distinguish plant taxa. Insects often react to compounds in nonhost plants as deterrents to oviposition or feeding and to certain compounds in host plants as stimulants. Phytochemicals may not only deter feeding but also reduce insect fitness by acting as toxins or interfering with digestion. It is generally thought that chemical and other differences among plants select for host-specificity in insects, on the supposition that physiological costs impose trade-offs among adaptations to different plant characteristics; however, only a minority of genetic and physiological studies has supported this hypothesis. Other proposed advantages of host specificity include use of specific plants as rendezvous sites for mating, greater efficiency of finding hosts, and predator escape by several means, such as sequestering defensive plant compounds.

Phylogenetic studies show that associations between some insect clades and plant clades are very old, often dating to the

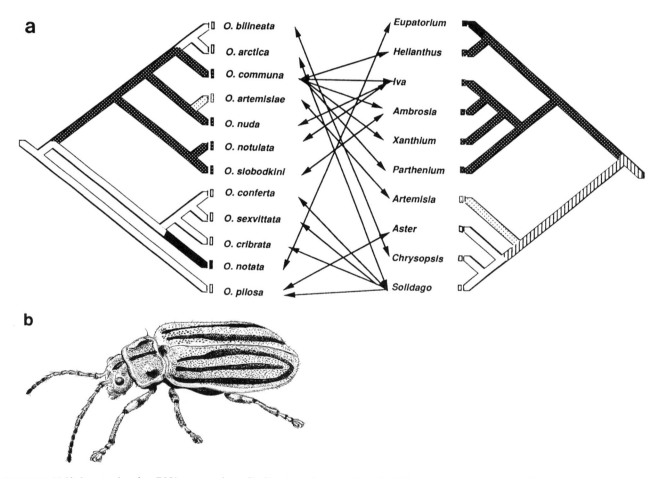

FIGURE 1 (a) Phylogenies, based on DNA sequence data, of leaf beetles in the genus *Ophraella* (left) and their host plants (right). Arrows join each beetle species to its host plant. Different shading patterns represent the four tribes of Asteraceae into which the host plants fall; the shading of branches is a parsimonious inference of the tribes with which ancestral *Ophraella* lineages were associated. Note that most host shifts associated with beetle speciation have been between plants in the same tribe. The incongruence between the phylogeny of the insects and that of their host plants is one of several indications that the beetles and plants did not cospeciate. These plant lineages represent only a few of the tribes of Asteraceae and of the genera within each tribe. [After D. J. Funk *et al.* (1995). *Evolution* **49**, 1008–1017. The Society for the Study of Evolution.] (b) The leaf beetle *O. sexvittata,* which feeds on *Solidago* species, tribe Astereae. (Original illustration by author.)

early Tertiary and in some cases to the Cretaceous or even Jurassic. Nevertheless, only a few instances of cospeciation and phylogenetic concordance have been described. In most cases, much of the speciation within an insect clade has occurred after the host plants diversified, but new species have shifted to plant species closely related to the ancestral host (Fig. 1). That these host shifts have been facilitated by chemical similarity of related plants is supported by instances in which phylogenetic relationships among insect species (e.g., *Blepharida* flea beetles, melitaeine butterflies) more closely match the hosts' chemical similarities than phylogenetic relationships. Patterns of genetic variation in the ability of host-specific *Ophraella* leaf beetles to feed and develop on nonhost plants, all within the Asteraceae, indicated greater genetic potential to adapt to those plants that were most closely related to the insect's normal host.

Although physiological, morphological, behavioral, and phenological adaptations of insects to host plants are many and obvious, demonstrating that plant characters have evolved because of selection for their defensive functions has been more difficult. Many chemical and morphological features of plants have the effect of reducing attack or damage by some or many insect species, but some authors have argued that they actually evolved for physiological reasons or as defenses against mammalian herbivores rather than insects. However, both phytochemicals (e.g., furanocoumarins) and morphological features (e.g., trichomes) have been shown to determine fitness differences among genotypes due to their effect on insect herbivores, and the distribution of many plant compounds among tissues conforms to what we should expect if they were adaptively deployed defenses. Still, there have been few demonstrations of adaptive geographic variation in plant defenses in relation to the abundance or identity of particular herbivorous insects. In one of the few examples of probable coevolution at the population level, populations of wild parsnip *(Pastinaca sativa)* have diverged in their profile of toxic furanocoumarins, and parsnip webworms *(Depressaria pastinacella)* are adapted to their local host population.

Ehrlich and Raven's escape-and-radiate model of coevolution between plants and herbivorous insects has found some support. Most lineages of plants that have independently evolved latex or resin canals (potent deterrents to most insects) are richer in species than their canalless sister groups, supporting the hypothesis that new plant defenses enhance the rate of diversification. Likewise, herbivorous clades of insects are generally more diverse than their nonherbivorous sister groups. Clades of phytophagous beetles that are thought to be primitively associated with gymnosperms have fewer species than sister taxa that have shifted to angiosperms, perhaps because the latter are so very diverse. The diversity of several moth taxa that feed on Apiaceae with presumably "advanced" chemical defenses is greater than that of those that feed on Apiaceae with "primitive" defenses, paralleling the differences in plant diversity, but phylogenetic analysis is needed to confirm that the diversification rate has been enhanced by novel plant defenses and insect counteradaptations.

EVOLUTION OF MUTUALISM

In mutualistic interactions between species, each species uses the other as a resource. That is, each exploits the other, and the degree of exploitation may determine whether an interaction is mutualistic or parasitic. Mutualisms include interactions both between free-living organisms, such as plants and pollinating animals, and between symbionts, one of which spends most of the life cycle on or in the other. Microbes are partners in many symbiotic mutualisms. Mutualists often have adaptations for encouraging the interaction or even nurturing the associate, such as foliar nectaries in plants, which attract ants that defend the plants against herbivores, or the root nodules of legumes, which house and nourish nitrogen-fixing rhizobial bacteria. In some intimate symbioses, the symbiont functions as an organ or organelle, as in the case of host-specific bacteria that reside within special cells in aphids and supply essential amino acids to their host.

For each mutualist, the interaction has both a benefit and a cost. Legumes, for example, obtain nitrogen from rhizobia, but expend energy and materials on the symbionts. Excessive growth of the rhizobia would reduce the plant's growth to the point of diminishing its fitness. Likewise, excessive proliferation of mitochondria or plastids, which originated as symbiotic bacteria, would reduce the fitness of the eukaryotic cell or organism that carries them. Thus, selection will always favor protective mechanisms to prevent overexploitation by an organism's mutualist. Whether selection on a mutualist favors restraint depends on how much an individual's fitness depends on the fitness of its individual host. When a mutualist can readily move from one host to another, as pollinating insects can from plant to plant, it does not suffer from the reproductive failure of any one host, and selfishness or overexploitation may be favored. For example, many pollinating insects "cheat." The larvae of yucca moths *(Tegeticula)* feed on developing yucca seeds in flowers that their mothers actively pollinated. However, several species of *Tegeticula* have independently lost the pollinating behavior, having evolved the habit of ovipositing in flowers that other species have already pollinated. Moreover, the pollinating species lay only a few eggs in each flower, so that the few larvae do not consume all the developing seeds. This reproductive restraint has evolved in response to a defensive tactic of the plant, which aborts developing fruits that contain more than a few eggs. However, the "cheater" species of *Tegeticula* circumvent the plant's defense by laying eggs after the developmental window for fruit abortion, and they lay so many eggs that the larvae consume most or all of the seeds. Deception and cheating has also evolved in some plants, such as orchids that provide no reward to the naive bees that visit them; other orchids mimic the female sex pheromone of an insect species, the males of which effect pollination by "copulating" with the flower.

Vertical transmission of a symbiont favors restraint and reciprocal benefit, just as it favors lower virulence in parasites, because the fitness of the individual symbiont is then proportional to its host's reproductive success. This principle can explain why internal symbionts such as aphids' bacteria or corals' zooxanthellae (or eukaryotes' mitochondria) divide at rates commensurate with their host's growth. It is conceivable that hosts may evolve mechanisms to prevent horizontal transmission (mixing) of symbionts and thus maintain conditions under which "selfishness" would be disadvantageous to the symbiont. By extension, such principles explain the conditions for the evolution of coordination versus conflict among different genes in a single genome, i.e., the evolution and maintenance of integrated organisms.

CONSEQUENCES OF COEVOLUTION

Coevolution has undoubtedly had major effects on the history and diversity of life. Many of the adaptive differences among organisms—the many thousands of toxic defensive compounds in different plants, insects, and fungi, the many forms of flowers, the diverse growth forms of plants, the sometimes astonishingly specialized diets of animals—have issued from interactions among species. The numbers of species, too, may have been augmented by coevolution, as Ehrlich and Raven proposed. Coevolution among competitors can also augment the species diversity in communities, producing suites of specialized species that finely partition resources among them. In theory, such coevolution may result in ecosystem-level effects such as higher productivity and resource consumption, but the evidence on this subject is very sparse.

See Also the Following Articles
Insectivorous Plants • Parasitoids • Plant–Insect Interactions • Predation • Symbionts

Further Reading
Abrams, P. A. (2000). The evolution of predator–prey interactions: Theory and evidence. *Annu. Rev. Ecol. Syst.* **31**, 79–105.
Benkman, C. W., Holiman, W. C., and Smith, J. W. (2001). The influence of a competitor on the geographic mosaic of coevolution between crossbills and lodgepole pine. *Evolution* **55**, 282–294.
Berenbaum, M., and Zangerl, A. (1998). Chemical phenotype matching between a plant and its insect herbivore. *Proc. Nat. Acad. Sci. USA* **95**, 13473–13478.
Ehrlich, P. R., and Raven, P. A. (1964). Butterflies and plants: A study in coevolution. *Evolution* **18**, 586–608.
Farrell, B. D. (1998). "Inordinate fondness" explained: Why are there so many beetles? *Science* **281**, 555–559.
Frank, S. A. (1996). Models of parasite virulence. *Q. Rev. Biol.* **71**, 37–78.
Fritz, R. S., and Simms, E. L. (eds.) (1992). "Plant Resistance to Herbivores and Pathogens: Ecology, Evolution, and Genetics." University of Chicago Press, Chicago.
Futuyma, D. J. (1998). "Evolutionary Biology," 3rd ed., pp. 539–560. Sinauer, Sunderland, MA.
Futuyma, D. J., and Mitter, C. (1996). Insect–plant interactions: The evolution of component communities. *Philos. Trans. R. Soc. London B* **351**, 1361–1366.
Kraaijeveld, A. R., and Godfray, H. C. J. (1999). Geographic patterns in the evolution of resistance and virulence in *Drosophila* and its parasitoids. *Am. Naturalist* **153**, S61–S74.
Lively, C. M. (1999). Migration, virulence, and the geographic mosaic of adaptation by parasites. *Am. Naturalist* **153**, S34–S47.
Schluter, D. (2000). "The Ecology of Adaptive Radiation." Oxford University Press, New York.
Thompson, J. N. (1994). "The Coevolutionary Process." University of Chicago Press, Chicago.
Thompson, J. N. (1999). Specific hypotheses on the geographic mosaic of coevolution. *Am. Naturalist* **153**, S1–S14.
Wahlberg, N. (2001). The phylogenetics and biochemistry of host–plant specialization in melitaeine butterflies. *Evolution* **55**, 522–537.

Cold/Heat Protection

David L. Denlinger
Ohio State University

As poikilotherms, insects are largely at the mercy of environmental temperatures. There are a few exceptions: some moths, bees, and flies do have the capacity to elevate their body temperature by vigorously contracting their flight muscles to generate heat, and a few species of cicadas, grasshoppers, and other desert species exploit evaporative cooling to lower their body temperature on hot days. The majority of species, however, have a limited capacity to alter their body temperature. Insects survive, perform, and reproduce across a broad temperature range, but they do so with varying levels of success at different temperatures. A thermal performance curve (Fig. 1) can be generated for nearly any quantitative trait. The curve delimits the body temperature at which a certain activity can be performed (tolerance zone). The low extreme is the critical thermal minimum, and the upper extreme is the critical thermal maximum.

Construction of such a curve will demonstrate that any activity has a temperature at which performance is optimal (optimum body temperature). Characteristically, the drop in performance outside the optimum body temperature is more precipitous at the high end of the temperature scale than at the low end. When given a choice, insects will readily select temperatures at which performance is maximized. But, the extremes of the daily temperature cycle and the dominant temperatures that prevail during major portions of the year pose significant obstacles for insect growth, development, and performance. This review describes the nature of the injury inflicted by high and low temperatures and discusses the protective mechanisms used by insects to counter these forms of injury.

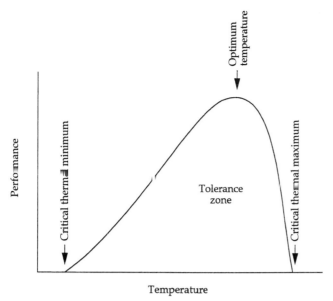

FIGURE 1 Hypothetical performance curve delineating the tolerance zone, critical thermal minimum temperature, optimum temperature, and critical thermal maximum temperature at which any quantitative trait can be performed. Note that the decline in performance above the optimal temperature is usually more precipitous at the high end of the temperature scale than at the low end.

PROTECTION AGAINST HIGH-TEMPERATURE INJURY

Heat Injury

Lethality at high temperature is a function of both temperature and time; the higher the temperature, the shorter the exposure time needed to kill the insect (Fig. 2). But, injury can manifest itself in more subtle forms at less extreme temperatures. For example, temperatures that prevent reproduction are lower than the temperatures that cause immediate mortality. At still less severe temperatures, adults are reproductively functional but emergence may be delayed or occur at the "wrong" time of day.

Heat shock can also produce developmental abnormalities known as phenocopies (developmental abnormalities resembling mutations but caused by environmental conditions), a phenomenon especially well known for the fruit fly, *Drosophila melanogaster*. Flies heat shocked during embryogenesis or metamorphosis yield interesting phenocopies with aberrant adult bristle shapes, colors, and wing formations. Which defect is observed is dependent upon the age of the fly at the time of exposure. The sensitive period for the production of each phenocopy is brief, usually less than 2 h. The various phenocopies are generated by disruption of a heat-sensitive developmental process that is specific to a particular developmental window. For example, heat shock can shut down phenol oxidase, the enzyme needed for melanin production. If heat shock is thus administered during the interval when this enzyme is needed to generate the black color normally associated with bristles, the blond-bristle phenocopy will be produced instead.

At the cellular level, a number of abnormalities are elicited in response to heat stress. These include declines in hemolymph pH; disruption of the normal pattern of protein synthesis; loss of conformational integrity of RNA, DNA, and protein; and deformation of the cellular membrane. Many cell processes are thus vulnerable to injury. Which cell process is the primary site of thermal wounding is still not clear, but two models have been proposed. One model suggests that the plasma membrane is the primary site of thermal wounding. In this model, disruption of the plasma membrane sets in motion a cascade of events involving inactivation of membrane proteins and subsequent leakage of K^+ out of the cell and movement of Ca^{2+} and Na^+ into the

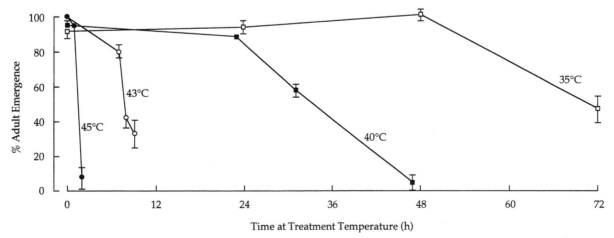

FIGURE 2 Mortality is a function of both temperature and duration of exposure, as demonstrated by the survival curves for *S. crassipalpis*. The flies were exposed to four different temperatures for various durations several days before adult emergence and survival was based on success of adult emergence. [Reproduced, with permission, from Yocum and Denlinger, (1994). Copyright Blackwell Science.]

cell. This loss of the cell's bioelectrical properties leads to a breakdown in cell metabolism, loss of homeostasis, and finally death. An alternative model also focuses on the plasma membrane but suggests that the subsequent protein denaturation is the critical cause of death. Denatured protein adheres to the chromatin and restricts enzymatic access to the DNA. The cell eventually dies as a consequence of an increase in DNA damage. But, it is also evident that an enzyme will lose its metabolic function at a fairly low level of heat stress, long before denaturation is complete. Thus, it is difficult to point to any single factor as the cause of death because high temperature adversely affects many aspects of the cell or organism's physiology simultaneously.

Thermotolerance

Thermotolerance (tolerance of high temperature) can be increased several ways. (1) Genetic adaptation: Differences in thermotolerance can be detected in diverse geographic populations, as well as in laboratory lines that have been selected for heat-shock survival. (2) Long-term acclimation: Rearing individuals for long durations at high temperatures can result in a striking increase in thermotolerance. (3) Rapid heat hardening: A brief exposure to an intermediately high temperature provides protection from injury at a more severe temperature.

Rapid heat hardening is the best studied response. Heat shock is the thermal injury caused by a sudden increase in temperature. This form of injury can be reduced dramatically if an organism is first exposed to an intermediate temperature (rapid heat hardening). For example, the flesh fly, *Sarcophaga*

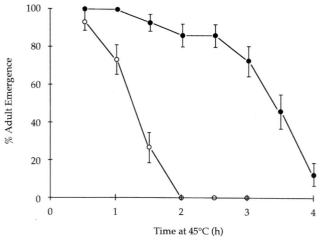

FIGURE 3 Thermotolerance can be demonstrated by the higher survival rates noted in flies that were first exposed to a moderately high temperature. In this example, based on *S. crassipalpis*, flies reared at 25°C and then transferred directly a few days before adult emergence to 45°C (open circles) survived poorly as indicated by success of adult emergence. By contrast, flies that were first exposed to 40°C (solid circles) survived exposure to 45°C much better. [Reproduced, with permission, from Chen *et al.* (1990), copyright Springer-Verlag.]

crassipalpis, can tolerate only a brief time at 45°C if it is transferred there directly from 25°C, but survival at 45°C is greatly extended (Fig. 3) if the flies are first exposed to 40°C for 2 h. The thermotolerance that protects against heat-shock injury is acquired quickly, within minutes, reaches a maximum within a few hours, and then decays rather slowly over several days.

Heat-shock proteins are the best known contributors to thermotolerance. In response to heat stress, the normal pattern of protein synthesis is suppressed, and concurrently several new proteins, the heat-shock proteins, are synthesized. These proteins are classified according to their molecular mass and in *D. melanogaster* include a high-molecular-mass protein (82 kDa), members of the 70-kDa family, and small heat-shock proteins with molecular masses of 22, 23, 26, and 27 kDa. The most highly expressed heat-shock proteins, members of the heat-shock protein 70 (Hsp70) family, are highly conserved. The gene that encodes Hsp70 is over 50% identical in bacteria and *D. melanogaster*. In response to heat stress Hsp70 levels in the cell may increase more than 1000-fold.

Though heat shock was the first stress known to elicit synthesis of these proteins, it is now evident that many other forms of stress (e.g., heavy metals, alcohols, metabolic poisons, aberrant proteins, cold shock, desiccation) can elicit synthesis of these same proteins. It is thus clear that these proteins are involved in diverse stress responses.

For years the linkage between heat-shock proteins and thermotolerance was based strictly on correlation between the presence of the proteins and the expression of thermotolerance, but more recently the linkage has been strengthened with new experimental evidence. Cultured *D. melanogaster* cells and whole flies transformed with extra copies of the Hsp70 gene acquire thermotolerance more rapidly than normal cells or flies, while cells transformed with Hsp70 antisense genes acquire thermotolerance more slowly.

How do the heat-shock proteins contribute to thermotolerance? Members of the Hsp70 family function as molecular "chaperones" that facilitate the process of protein folding and assembly. Hsp70 can reduce high-temperature damage by interacting with susceptible proteins to prevent their interactions with other reactive surfaces, thus helping to maintain the integrity of proteins present in the cell.

Although heat-shock proteins have received the most attention in studies of thermotolerance, other molecules, including sugars such as trehalose and polyols such as glycerol and sorbitol, are also suspected of contributing to the protective mechanism.

Thermosensitivity

While it is widely appreciated that previous exposure to an elevated temperature can generate tolerance to high temperature (thermotolerance, as discussed earlier), it is less well appreciated that some high temperatures can decrease an insect's ability to survive a subsequent high-temperature exposure. It is this loss of tolerance that is referred to as thermosensitivity. For example, *S. crassipalpis* appears to readily survive a 1-h exposure to 45°C, but if the fly is subjected to a second high-temperature pulse 1 day later, the effect will be lethal, even if the second pulse is considerably less severe, e.g., 35°C. Such observations suggest that some form of injury caused by the first challenge made the flies considerably more vulnerable to the second heat pulse. Without the second challenge, the initial injury can apparently be repaired, but the problem arises if the insect is challenged a second time before it has fully recovered. The temperatures that produce thermosensitivity are generally above the temperatures that generate thermotolerance.

An intriguing practical implication of thermosensitivity is that the pattern of administering a thermal stress has important consequences for an insect's survival. Two relatively modest pulses of high temperature may be just as effective in causing death as a single pulse of a higher temperature. From an economic perspective, this type of wounding may require less energy input than needed to administer a single pulse of a higher temperature.

PROTECTION AGAINST LOW-TEMPERATURE INJURY

Insects are frequently classified as being either freeze tolerant or freeze susceptible. Freeze tolerance implies that the insect can actually survive ice formation within its body. Relatively few insects have this capacity, but it is well documented in some insects such as the goldenrod gall fly, *Eurosta solidaginis.* By contrast, most insects are freeze susceptible, which means that they cannot tolerate internal ice formation. This, however, does not mean that all freeze-susceptible species can survive temperatures approaching the point at which their body will freeze. Many such species are fatally injured at temperatures well above their freezing point.

Within a single species, huge differences in cold tolerance may be evident in different stages of the life cycle. For example, in *S. crassipalpis,* the adult is the stage most susceptible to cold injury, while the pupa is least susceptible. If the pupa is in diapause, the overwintering state of dormancy, the pupa is even more cold tolerant. Characteristically, diapausing stages are highly tolerant of low temperature and are capable of withstanding far lower temperatures, and for much longer, than nondiapausing stages. In the flesh fly, diapausing pupae can tolerate temperatures of −20°C (a few degrees above their supercooling point) for many months, while nondiapausing pupae will be killed with an exposure of just a few hours to −10°C.

Supercooling and Ice Nucleation

Understanding the nature of supercooling and ice nucleation is critical for understanding the strategies used by insects to survive at subzero temperatures. One might assume that an insect will freeze when its body temperature reaches 0°C, but this does not occur. Instead, the body water supercools, a process that is enhanced in many cases by the production of cryoprotectants that dramatically reduce the freezing point and thus enable the insect to remain unfrozen at temperatures down to −20°C or lower. The temperature at which the body liquid turns to ice is called the supercooling point or the temperature of crystallization. This point is easily detected by monitoring body temperature and noting the appearance of an exotherm, the burst of heat given off by crystallization as the body water freezes. The insect body contains a number of agents that can affect the supercooling point. Cryoprotectants are capable of lowering the supercooling point, whereas ice-nucleating agents elicit the opposite response. Ice-nucleating agents act as catalysts to promote ice nucleation at higher temperatures than would occur in their absence. Formation of ice at rather high temperatures is especially common in freeze-tolerant species. In such cases it is advantageous to initiate ice formation at a rather high subzero temperature, a feature that enhances survival by slowing down the processes of ice formation. By contrast, freeze-susceptible insects exploit the use of cryoprotectants to suppress the supercooling point and thus avoid freezing.

Cold Injury

In addition to lethality, cold injury can be manifested in failure of reproduction and the appearance of developmental abnormalities. A cold shock can induce an extra molt in some species such as the greater wax moth, *Galleria mellonella.* Phenocopy defects, like those noted in *D. melanogaster* at high temperature, can also be elicited by low temperature: The incidence of aristapedia (in which antennae are transformed into legs) increases at low temperature. Sex ratios can be distorted, in some species favoring females and in others, males.

Many freeze-susceptible species are killed at temperatures well above their freezing points. The mechanism involved in this form of nonfreezing injury is poorly understood but may result from a decline in the rate of enzyme function at low temperatures or to irreversible changes in tertiary structure of critical proteins. Nonfreezing injury resulting from low temperature is frequently associated with damage to the plasma membrane. At some point chilling induces fluid to gel phase transitions in cell membranes that result in major alterations in membrane permeability and reduction in activity of membrane-bound enzymes.

Among freeze-tolerant species it is commonly assumed that survival of freezing requires that ice formation be restricted to extracellular spaces. This, however, is not always the case. Intracellular freezing does occur in some tissues such as the fat body cells of the goldenrod gall fly. Ice formation normally is initiated outside of the cell. Only water is added to the ice lattice, thus the remaining body fluids become more concentrated. This, in turn, causes osmotic removal of water from the cells. Although mechanical injury due to ice formation can be a deleterious effect, it is likely that the primary initial stress results from cell dehydration and the accumulation of excess amounts of solutes in the body fluid. The high concentrations of solutes, particularly electrolytes, can cause protein denaturation and extreme shifts in pH that result in irreversible membrane damage.

Certain systems are more vulnerable to injury than others. The neuromuscular system appears to be particularly vulnerable. As temperatures decline insects gradually lose their ability to fly, and at still lower temperatures they lose their ability to walk. Chill coma, the point at which the insect loses its ability to walk, coincides with the temperature at which the muscles and nerves lose their electrical excitability.

The reproductive system is also quite vulnerable to cold injury. Insects may appear normal but fail to reproduce following a cold shock. Both the number of eggs produced and the fertility of the eggs may be lowered by cold injury.

Cold Hardiness

The injury caused by low temperature can frequently be mitigated by prior exposure to less severe low temperatures. Like the acquisition of thermotolerance at high temperatures, cold hardening enables an insect to survive at low temperatures that would otherwise prove lethal. Cold hardening can be either a long-term process attained after weeks or months at low temperature or a very rapid process (rapid cold hardening) invoked within minutes or hours after exposure to low temperature.

The traditional view of cold hardening depicts a slow process that gradually increases the insect's tolerance to low temperature. As seasonal temperatures drop in the autumn, many insects become progressively more cold hardy. Thus, a field-collected insect from the north temperate region evaluated in January is likely to be more cold tolerant than one collected in September. In contrast, rapid cold hardening is a very fast process that allows an insect to respond to daily changes in temperature. For example, *S. crassipalpis,* when reared at 25°C cannot survive an immediate transfer to −10°C, but if the fly is first placed at 0°C for as short a time as 10 min, it can readily survive a subsequent 2-h exposure to −10°C. The capacity for rapid cold hardening appears to be common among insects and presumably functions in enabling them to track daily and other forms of rapid temperature change.

Several diverse physiological mechanisms contribute to cold hardiness. For freeze-susceptible insects, one of the most important mechanisms involves the elimination of ice nucleators. The presence of ice nucleators limits the insect's ability to supercool; thus getting rid of potential nucleators is a critical feature of cold hardiness. Food particles present in the gut are among the most powerful ice nucleators; thus it

is perhaps no surprise that many insects purge their gut prior to overwintering.

Another common cold-hardening mechanism used by freeze-susceptible insects is the synthesis and accumulation of high concentrations of low-molecular-mass polyols (glycerol, sorbitol, mannitol) and sugars (trehalose). Like a classic antifreeze, the polyols and sugars reduce the supercooling point and thus allow the insect to avoid freezing at temperatures well below 0°C. Hemolymph concentrations of polyols sometimes reach multimolar levels.

Thermal hysteresis refers to a difference between the freezing and the melting point of the body fluid. At equilibrium one would expect these two points to be nearly identical, but this relationship can be altered by thermal hysteresis proteins, also known as antifreeze proteins. Thermal hysteresis proteins depress the freezing point while leaving the melting point unchanged. This lowering of the freezing point can thus expand an insect's low-temperature tolerance. Such proteins were first discovered in cold-water, marine fish but were found more recently in several species of beetles.

Ice nucleator proteins function in a manner opposite to that of thermal hysteresis proteins. Rather than inhibiting freezing, these proteins promote freezing. Ice nucleator proteins facilitate the organization of water molecules into embryo crystals, which, in turn, seed the supercooled solution, causing freezing at relatively high temperatures. As discussed above, this is advantageous for freeze-tolerant species.

Synthesis of heat-shock proteins is a well-documented response to high temperature, but some of the same proteins are also synthesized in response to low-temperature shocks. As with heat-shock, the most prominent heat shock protein elicited by cold shock is a member of the 70-kDa family of heat-shock proteins. These stress proteins are most evident following the cold shock, thus suggesting they may play a role in the recovery process.

Insects thus have at their disposal an array of mechanisms to counter the adverse effects of low temperature. Cold hardening can entail a complex suite of responses and should not be regarded as a process driven by a single biochemical event, but species differences are likely to dictate that one particular process may be more important in one species than in another.

PRACTICAL IMPLICATIONS

Exploiting temperature for pest management is an attractive alternative to the use of pesticides. The manipulations can be safely administered and no harmful residues remain. Heat and cold treatments are emerging as the treatment of choice for quarantine treatment of fresh fruits and vegetables. Temperature treatment is especially popular in this industry because the major fumigant, methyl bromide, is being removed from the market due to its role as an ozone depleter. Soils and planting beds are being treated with heat, and both high and low temperatures are being used to treat houses and other structures. Stored grain can be effectively protected from insects with temperature treatments, and even some field crops can be protected with novel applications of heat applied directly to the plant.

Cold storage is used extensively to increase the "shelf life" of parasitic wasps and other biological control agents, as well as the hosts on which they are reared. The cryopreservation of embryos of *D. melanogaster* and other insects is a goal sought by numerous researchers. This ability could facilitate the long-term maintenance of valuable genetic stocks and reduce the care and expense required to continuously propagate insects used for research.

Insects have a wealth of behavioral and physiological responses to counter the effects of high- and low-temperature stress, and if temperatures are to be exploited for use in integrated pest management systems, these mechanisms must be either overridden or disabled. For example, the generation of thermotolerance can be prevented by applying heat stress in a nonoxygenated environment. Combination treatments that simultaneously apply both heat and anoxia or thermosensitization (application of two temporally separated treatments at moderately high temperatures) are especially attractive because they can cause mortality with less energy input. The low temperatures that prevail during winter are frequently just a few degrees above the insect's lower limit of tolerance. Attempts to further reduce the insect's body temperature by destroying the insect's protective winter habitat offer promise. Recent discoveries of ice-nucleating bacteria and fungi that are active on insects suggest new tools for manipulating the supercooling point. The diverse protective responses operating in insects suggest a similar richness of targets that could be rendered vulnerable to heat or cold injury.

See Also the Following Articles

Diapause • Hibernation • Temperature, Effects on Development and Growth • Thermoregulation

Further Reading

Bowler, K., and Fuller, B. J. (eds.) (1987). "Temperature and Animal Cells." Society for Experimental Biology Symposium 41, Cambridge, U.K.

Chen, C.-P., Lee, R. E., Jr., and Denlinger, D. L. (1990). A comparison of the responses of tropical and temperate flies (Diptera: Sarcophagidae) to cold and heat stress: *J. Comp. Physiol. B* **160**, 543–547.

Hallman, G. J., and Denlinger, D. L. (eds.) (1998). "Temperature Sensitivity in Insects and Application in Integrated Pest Management." Westview Press, Boulder, CO.

Heinrich, B. (1993). "The Hot-Blooded Insects." Harvard University Press, Cambridge, MA.

Johnston, I. A., and Bennett, A. F. (eds.) (1996). "Animals and Temperature: Phenotypic and Evolutionary Adaptation." Society for Experimental Biology Symposium 59, Cambridge, U.K.

Lee, R. E., Jr., and Denlinger, D. L. (eds.) (1991). "Insects at Low Temperature." Chapman & Hall, New York.

Somero, G. (1995). Proteins and temperature. *Annu. Rev. Physiol.* **57**, 43–68.

Yocum, G. D., and Denlinger, D. L. (1994). Anoxia blocks thermotolerance and the induction of rapid cold hardening in the flesh fly, *Sarcophaga crassipalpis. Physiol. Entomol.* **19**, 152–158.

Coleoptera
(Beetles, Weevils, Fireflies)

James K. Liebherr
Cornell University

Joseph V. McHugh
University of Georgia

Beetle diversity so characterizes Earth that instead of telling future extraterrestrial colleagues we come from the "blue planet," we might better state that we come from the "beetle planet." Beetles comprise 25% of all described animals and plants, single-handedly making them the primary contributor to earth's biodiversity. The 350,000 described beetle species are members of largest order of life on Earth, Coleoptera.

Familiar beetles are known by various names including fireflies, ladybugs, june bugs, and weevils. The vast number of beetle species is reflected by a bewildering array of anatomical and biological diversity in the order. Coleoptera are represented in nearly all biogeographic regions and nonmarine habitats. Most adult beetles can fly; when not in use, however, the delicate flight wings are usually concealed beneath protective shell-like elytra, permitting beetles to utilize diverse resources and engage in a broad range of activities that otherwise would be restricted to either winged or wingless insects. Most beetles are herbivores, fungivores, or predaceous carnivores in the larval and adult stages. Many are considered to be serious pests of our homes, forests, crops, and stored products, whereas some beneficial species are regularly employed as biological control agents. Countless curious youngsters, including Charles Darwin, Alfred Russel Wallace, and Henry Walter Bates, have started their broader studies of biology through beetle collecting, as beetle species often are consistently found in specific sorts of habitats.

The technical name, Coleoptera, was coined by Aristotle to signify the hardened, shieldlike forewings (*coleo* = shield + *ptera* = wing). Although several other insect orders possess hardened forewings, beetles are considered to be a monophyletic assemblage based on their sum of shared evolutionary derivations that include the following:

1. A holometabolous life cycle, wherein the larval stages are developmentally separated from the adult by the pupal stage.

2. Possession of hardened forewings, called elytra, that abut medially. Flight is powered predominantly by the metathoracic wings, which are folded longitudinally and usually transversely to lie under the elytra when the beetle is walking or at rest. The mesothoracic scutellum is visible as a triangle situated medially between the bases of the two elytral halves.

3. A prothorax that is distinct from, and most often freely articulating with, the following mesothorax. The meso- and metathoracic segments are fused to form the pterothorax.

4. A generally depressed body shape, whereby the legs are situated on the ventral surface of the body. The leg bases, or coxae, are recessed into cavities formed by heavily sclerotized thoracic sclerites.

5. Abdominal sternites that are much more heavily sclerotized than the tergites. These sternites may close tightly against the lateral edges of the elytra, protecting the hind body from the attentions of predators and parasitoids.

6. Antennae usually with 11 or fewer segments.

7. Terminal genitalia that are not visible when in repose; that is, the male aedeagus and the female ovipositor are retracted into the abdominal apex when not in use.

Insects in several other orders may appear superficially similar to beetles. For example, various Hemiptera in the superfamily Pentatomoidea possess an enlarged triangular scutellum and heavily sclerotized forewings. However, these bugs can be distinguished by their beaklike suctorial mouthparts, whereas beetles retain the more generalized mandibulate mouthparts seen throughout orders such as Odonata, Orthoptera, and Hymenoptera. In addition, the forewings of Hemiptera always retain an apical membranous portion, whereas beetle forewings are consistently sclerotized throughout their length. Also, Dermaptera, or earwigs, exhibit quadrate forewings, looking much like the foreshortened elytra of staphylinoid, or rove beetles. Earwigs, however, exhibit a radial wing folding mechanism versus the transverse folding system of beetles, retain the presence of abdominal cerci, represented by large tonglike forceps at their abdominal apex, and do not undergo complete metamorphosis incorporating the pupal stage.

BEETLE DIVERSITY

Although beetles share characters supporting their common evolutionary origin, remarkable variations have evolved on the beetle theme. For example, adult body size ranges from the 0.4-mm-long *Nanosella fungi* ptiliid feather-winged beetles of North America to the 200-mm-long *Titanus giganteus* cerambycid long-horned beetles of South America. A rough estimate based on maximum dimensions for adult length, breadth, and depth puts the disparity in volume at a factor of 2.8×10^7. Life cycles also can vary in extraordinary ways, depending on the larval food resources used for development. The mushroom-inhabiting aleocharine staphylinid *Phanerota fasciata* completes three instars in 3.2 days at room temperature. Even more impressive, *Anisotoma* round fungus beetles of the family Leiodidae can complete larval development on short-lived slime mold fruiting bodies in as little as 2 days, making them arguably the fastest developing beetles yet recorded. Conversely, C. V. Riley, the first entomologist of the U.S. Department of Agriculture, reported that a larva of the

dermestid carpet beetle, *Trogoderma inclusum,* survived for 3.5 years in a tight tin box. These larvae feed on the dried proteinaceous matter in animal remains, and even if Riley's larva had started with a tin full of insect specimens, the feat of solitary confinement is remarkable. *Trogoderma* larvae can even molt to a smaller size under starvation conditions, then regain size by progressively molting when food returns. Stan Beck found that mature larvae molted retrogressively eight times during a year of starvation, dropping from an initial weight of 9.24 mg to a final, svelte 1.38 mg (an 85% weight loss!).

Dramatic variation in reproductive capacity is also observed across the Coleoptera. An abundant plant pest such as the chrysomelid northern corn rootworm, *Diabrotica barberi,* can colonize cornfields and build populations quickly, since each female lays on average nine clutches of eggs, spaced 6 days apart, totaling 274 eggs over the reproductive period. At the opposite extreme we once again find the diminutive, feather-winged Ptiliidae. In eight species of *Bambara* ptiliids from Sri Lanka, the males produce spermatozoons that range in length from 220 to 600 µm; the largest size being more than two-thirds the length of the adult male producing them. After mating, these giant sperm pack the female spermatheca, with up to 28 spermatozoons recorded filling this structure. The length of the female spermathecae of various *Bambara* species is consistent within species and varies in proportion to the length of the complementary male sperm, whereas the diameter of the spermathecal duct varies in proportion to the diameter of the sperm. The female also invests heavily in her progeny, maturing one relatively giant egg in her abdomen at a time. The highly complementary male spermatozoons and female spermathecae ensure reproductive isolation because of biomechanical incompatibilities associated with any attempted interspecific matings.

Beetles are among the earliest diversifying groups of the Holometabola. Together with the orders Megaloptera, Raphidioptera, and Neuroptera, they are classified in the superorder Neuropterodea. The order Coleoptera is divisible into four major lineages, which are recognized as the suborders Archostemata, Adephaga, Myxophaga, and Polyphaga (Table I). Present-day diversity among the four coleopteran suborders is highly skewed toward the Polyphaga. Taking the numbers of beetle species estimated for Australia, John Lawrence and Everard Britton calculated that Archostemata (9 species) make up 0.03% of the Australian beetle fauna, Adephaga, with 2730 species comprise 9.6%, Myxophaga, with 2 species (0.007%), and, with 25,600 species, Polyphoga, dominates at 90.4% of the fauna. Extrapolating these figures to the estimated world total of 350,000 described beetle species suggests that Polyphaga would account for more than 300,000 species.

Consensus concerning the phylogenetic relationships among all four suborders has yet to be achieved. Recent summaries of morphological data and separate efforts using molecular sequence data reach different conclusions based on the character types and sets of taxa included. Recent studies

TABLE I   **Classification of Beetle Suborders, Series, Superfamilies, and Families of the Order Coleoptera**

Suborder Archostemata
   Cupedoidea
      1. Ommatidae
      2. Cupedidae
      3. Micromalthidae
Suborder Adephaga
   Caraboidea
      4. Gyrinidae
      5. Haliplidae
      6. Hygrobiidae
      7. Amphizoidae
      8. Dytiscidae
      9. Noteridae
     10. Trachypachidae
     11. Carabidae (incl. Rhysodini, Cicindelini)
Suborder Myxophaga
     12. Torridincolidae
     13. Cyathoceridae
     14. Hydroscaphidae
     15. Microsporidae
Suborder Polyphaga
   Staphyliniformia
     Hydrophiloidea
      16. Hydrophilidae
      17. Sphaeritidae
      18. Synteliidae
      19. Histeridae
     Staphylinoidea
      20. Hydraenidae
      21. Ptiliidae
      22. Agyrtidae
      23. Leiodidae
      24. Scydmaenidae
      25. Silphidae
      26. Staphylinidae
   Sciritiformia
     Scirtoidea
      27. Scirtidae
      28. Eucinetidae
      29. Clambidae
   Scarabaeiformia
     Scarabaeoidea
      30. Lucanidae
      31. Passalidae
      32. Trogidae
      33. Glaresidae
      34. Pleocomidae
      35. Diphyllostomatidae
      36. Geotrupidae
      37. Ochodaeidae
      38. Ceratocanthidae
      39. Hybosoridae
      40. Glaphyridae
      41. Scarabaeidae
   Elateriformia
     Dascilloidea
      42. Dascillidae
      43. Rhipiceridae
     Buprestoidea
      44. Buprestidae
     Byrrhoidea
      45. Byrrhidae

      46. Dryopidae
      47. Lutrochidae
      48. Elmidae
      49. Limnichidae
      50. Heteroceridae
      51. Psephenidae
      52. Callirhipidae
      53. Eulichadidae
      54. Ptilodactylidae
      55. Chelonariidae
      56. Cneoglossidae
   Elateroidea
      57. Artematopidae
      58. Rhinorhipidae
      59. Brachypsectridae
      60. Cerophytidae
      61. Eucnemidae
      62. Throscidae
      63. Elateridae
      64. Plastoceridae
      65. Drilidae
      66. Omalisidae
      67. Lycidae
      68. Telegeusidae
      69. Phengodidae
      70. Lampyridae
      71. Omethidae
      72. Cantharidae
   Bostrichiformia
     Derodontoidea
      73. Derodontidae
     Bostrichoidea
      74. Jacobsoniidae
      75. Nosodendridae
      76. Dermestidae
      77. Endecatomidae
      78. Bostrichidae
      79. Anobiidae
   Cucujiformia
     Lymexyloidea
      80. Lymexylidae
     Cleroidea
      81. Phloiophilidae
      82. Trogossitidae
      83. Chaetosomatidae
      84. Cleridae
      85. Acanthocnemidae
      86. Phycosecidae
      87. Melyridae
     Cucujoidea
      88. Protocucujidae
      89. Sphindidae
      90. Nitidulidae
      91. Monotomidae
      92. Boganiidae
      93. Helotidae
      94. Phloeostichidae
      95. Silvanidae
      96. Passandridae
      97. Cucujidae
      98. Laemophloeidae
      99. Propalticidae

(continues)

TABLE I (Continued)

100. Phalacridae
101. Hobartiidae
102. Cavognathidae
103. Cryptophagidae
104. Lamingtoniidae
105. Languriidae
106. Erotylidae
107. Biphyllidae
108. Byturidae
109. Bothrideridae
110. Cerylonidae
111. Discolomidae
112. Endomychidae
113. Alexiidae
114. Coccinellidae
115. Corylophidae
116. Latridiidae

Tenebrionoidea
117. Mycetophagidae
118. Archaeocrypticidae
119. Pterogeniidae
120. Ciidae
121. Tetratomidae
122. Melandryidae
123. Mordellidae
124. Rhipiphoridae
125. Colydiidae
126. Monommatidae
127. Zopheridae
128. Perimylopidae
129. Chalcodryidae
130. Trachelostenidae
131. Tenebrionidae
132. Prostomidae
133. Synchroidae
134. Oedemeridae
135. Stenotrachelidae
136. Meloidae
137. Mycteridae
138. Boridae
139. Trictenotomidae
140. Pythidae
141. Pyrochroidae
142. Salpingidae
143. Anthicidae
144. Aderidae
145. Scraptiidae

Chrysomeloidea
146. Cerambycidae
147. Chrysomelidae

Curculionoidea
148. Nemonychidae
149. Anthribidae
150. Urodontidae
151. Oxycorynidae
152. Aglycyderidae
153. Belidae
154. Attelabidae
155. Caridae
156. Ithyceridae
157. Brentidae
158. Curculionidae

Source: Modified from Lawrence, J. F., and Britton, E. B. (1994). "Australian Beetles." Melbourne University Press, Melbourne, Australia.

agree that the Archostemata are the sister group to the other three suborders. The position of Myxophaga remains ambiguous, though Beutel and Haas's comprehensive morphological analysis places them as the sister group to Polyphaga.

The burgeoning discoveries of beetle diversity throughout the course of modern scientific endeavor has begged the question, "Why?" The noted geneticist J. B. S. Haldane, in a lecture on the biological aspects of space exploration, stated that "the Creator, if he exists, has a special preference for beetles, and so we might be more likely to meet them than any other type of animal on a planet that would support life." No single answer provides the definitive biological explanation for the present-day preponderance of beetle diversity. A number of answers are consistent with the pattern of diversity, with some better supported by the comparative totals of species in the different suborders and the major families.

First, the origin of Coleoptera, relatively early in the Triassic compared with other holometabolous orders, provided ample time for diversification. Having been in existence throughout the breakup of Pangaea, which started in the Jurassic, distinct beetle biotas have evolved in place on the various continental fragments of that supercontinent.

Second, beetle diversification has been explained as the result of a successful body plan incorporating protective elytra and a flexibly articulating prothorax. Although beetles are generally not regarded as fast or agile fliers, representatives of various beetle families have routinely colonized the most remote island systems in the world. In many families, the outward appearance and function of the walking beetle has been maintained, while the metathoracic flight wings have been reduced to nonfunctional straps or vestigial flaps. This brachypterous condition eliminates the possibility of winged dispersal by individuals and is associated with increased speciation and endemism, most often in ecologically stable, geographically isolated montane, desert, or island habitats.

Third, as representatives of the Holometabola, the larval and adult beetle life stages have been morphologically decoupled via the intervening pupal stage. Larvae may exhibit morphological specializations not observed in the adult stages, and may live in particular microhabitats not primarily occupied by the adults.

Fourth, the early diversification of beetles in the Jurassic placed many lineages in prime position to exploit ecological opportunities associated with the Cretaceous diversification of flowering plants. Many of the largest families of Polyphaga (e.g., Buprestidae, Scarabaeidae, Chrysomelidae, Cerambycidae, and Curculionidae) include lineages that are intimately associated with angiosperms. These host plant associations are based on the use of various portions of the particular species or sets of species of flowering plants as larval or adult food. In addition, many other beetle groups use fungi as a food source, and fidelity to fungi of particular types is not atypical. The ability to specialize along with their larval and adult hosts has clearly been associated with extensive speciation across the Coleoptera.

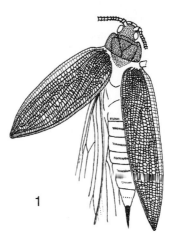
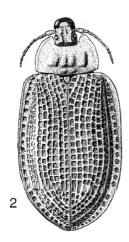

FIGURES 1–2 Fossil beetles. (1) *Moravocoleus permianus* (Tshekardocoleidae: Protocoleoptera, Permian). (© Czech Geological Survey.) (2) *Notocupes picturatus* (Cupedidae: Coleoptera, Triassic).

EVOLUTIONARY HISTORY

The earliest beetlelike insects are known from Lower Permian (280 mya) fossil deposits in Moravia, Czech Republic, and the Ural Mountains of Russia. These insects, classified in the family Tshekardocoleidae, order Protocoleoptera, resemble present-day species of the archostematan families Ommatidae and Cupedidae. They differ from true beetles in having 13-segmented antennae, elytra with more well-developed venation and more irregular longitudinal ribbing, and an abdomen and ovipositor extending beyond the apex of the elytra (Fig. 1).

We now date the origin of true Coleoptera as Triassic, about 240 mya. These fossils exhibit the coleopteran 11-segmented antennae, have more regular longitudinal ribbing on the elytra, and possess internal genitalia (Fig. 2). The earliest fossil beetle faunas have been described from Queensland in Australia, South Africa, and central Asia. The four lineages now recognized as suborders appear to have been extant at this time. The Archostemata were represented by species assignable to Ommatidae and Cupedidae, plus others belonging to families not lasting past the Mesozoic. The Adephaga included species sharing enlarged hind coxal plates such as are seen in present-day Haliplidae, plus other ground beetle-like species of Trachypachidae. Myxophagan ancestors included a variety of genera in the extinct families Catiniidae and Schizophoridae. The currently dominant suborder Polyphaga was represented in these faunas by members of the Elateroidea and Curculionoidea. These earliest beetles inhabited a world made up of early forked-leaved pteridosperms, lycopods, cycads, gingkos, and early conifers. The large animals of these communities included therapsid reptiles and dinosaurs; however, neither birds nor true mammals had yet evolved.

During the Jurassic period (210–145 mya), known family-level beetle diversity increased dramatically. Among the Adephaga, first appearances are documented for the whirlygig beetle family Gyrinidae, the ground beetle family Carabidae, and the predaceous diving beetle family

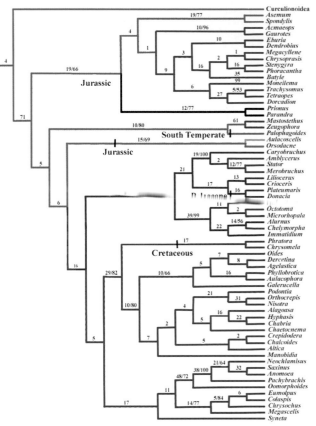

FIGURE 3 Strict-consensus estimate of the phylogeny of Chrysomeloidea and outgroups, with host groups mapped onto the cladogram. Numbers of synapomorphies/bootstrap values exceeding 50% shown along branches. Colors indicate major host group attributable to common ancestor of each group (green, Coniferae; mustard, Cycadales; red, dicotyledonous angiosperms; blue, monocotyledonous angiosperms; black, do not feed on living plants). Approximate ages of Mesozoic and early Tertiary fossils only are indicated where known, since almost all subfamily groups are present in the mid-Tertiary fossil record. [Redrawn with permission from Farrell, B. D. (1998). "Inordinate fondness" explained: Why are there so many beetles? *Science* **281**, 555–559. © 1998 American Association for the Advancement of Science.]

Dytiscidae. In all three families, the predaceous habit would be considered to be the ancestral condition. Among Polyphaga, the major families Staphylinidae, Scarabaeidae, Tenebrionidae, and Chrysomelidae are first documented. Other earliest occurrences include members of the scavenging water beetles (Hydrophilidae), carrion beetles (Silphidae), ovoid bark-gnawing beetles (Trogossitidae), tumbling flower beetles (Mordellidae), sap beetles (Nitidulidae), and false blister beetles (Oedemeridae). Of these, Scarabaeidae, Chrysomelidae, Oedemeridae, Mordellidae, plus the Triassic-aged Curculionoidea are strictly phytophagous or saprophagous. Members of the large, diverse present-day assemblage of Chrysomelidae use a broad diversity of plant hosts, ranging from cycads to conifers to angiosperms. Based on a phylogenetic hypothesis derived from extant species, the basal chrysomelid lineages are associated with primitive conifers (*Araucaria* spp.) and cycads (Fig. 3). The

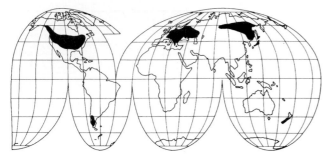

FIGURE 4 World distribution of Derodontidae. Areas supporting species include North America, Europe, Siberia, Japan, the Valdivian forest of Chile, and the South Island of New Zealand. [From Crowson, R. W. (1981). "The Biology of Coleoptera," p. 349, Fig. 2. Academic Press, London.]

Curculionoidea, the sister group to chrysomeloids, also exhibits this ancestral association with conifers and cycads. Third, the larvae of present-day Oedemeridae are borers in conifers. Thus it appears that at least several lineages of phytophagous Coleoptera were in place before the evolutionary advent of the angiosperms.

The Cretaceous witnessed initiation of the most recent round of southern landmass fragmentation, via the opening of the southern Atlantic Ocean and the isolation of New Zealand. South America and Antarctica plus Australia became progressively isolated from Africa, although they maintained contact with one another. Beetle families responded to this pattern of vicariance, with relictual distributions of several extant taxa supporting their origin during this time (Fig. 4). Continuing vicariance of the southern portions of Gondwana continued into early Tertiary, with progressive isolation of Australia, and finally the separation of Antarctica and South America at the start of the Oligocene (38 mya). This last event permitted formation of the circum-Antarctic current, helping plunge the world into a latitudinally zonated climate similar to that of today.

Preservation of beetles in amber has provided unparalleled levels of information about extinct taxa. The deposits of Baltic amber dated at 35 to 50 mya, and Dominican amber dated 15 to 40 mya, open windows onto the transition from the tropical world of the Eocene to the climatically zonated world of today. Most often, amber fossils (Fig. 5) indicate historically broader distributions for taxa presently known from only one continent (Fig. 6). This range contraction, continuing from the Eocene until the present day, suggests one explanation for the current latitudinal pattern of biodiversity. Many of the tropically adapted groups of organisms, of which beetles count significantly, have been progressively excluded from higher latitudes through the advent of cool to cold higher latitude climes, followed by the dramatic climatic perturbations associated with Pleistocene glaciation. G. Russell Coope goes so far as to argue that Pleistocene glaciation has put a halt to speciation of beetles in the temperate zones most influenced by the glaciation. His argument is based a simple fact: as he and his students studied subfossil beetle bits interred in wetland peats throughout various portions of Europe and North America, they found that all species taken from deposits younger than Pliocene could be identified as currently extant. These findings contrast starkly with those from tropical island systems, where speciation may have occurred in far younger areas. In Hawaii, for example, cave-adapted carabid beetles with reduced eyes and elongate legs have evolved from fully eyed, short-legged, epigean ancestors on the younger volcanoes of East Maui and Hawaii Island, which respectively broke the ocean surface no earlier than 750,000 and 430,000 years ago. Numerous Hawaiian beetle radiations in the Carabidae, Anobiidae, Nitidulidae, Cerambycidae, and Curculionidae demonstrate the many rapid and extensive bouts of speciation that occur in newly evolving tropical island communities.

ADULT SPECIALIZATION

It is impossible to argue for or against the proposition that possession of elytra has helped beetles' evolutionary success because possession of elytra is a defining character of "beetle-

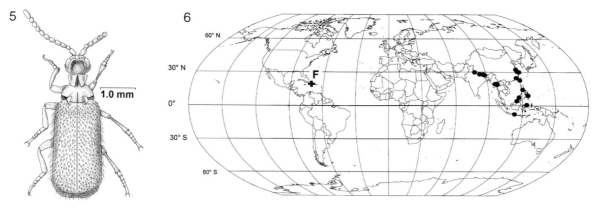

FIGURES 5–6 *Protopaussus pristinus* (Carabidae), described from Dominican amber. (5) Reconstruction of adult, dorsal view. (6) Distribution of *Protopaussus:* **F**, *P. pristinus* fossil; ● localities of extant *Protopaussus*. [From Nagel, P. (1997) New fossil paussids from Dominican amber with notes on the phylogenetic systematics of the paussine complex. *Syst. Entomol.* **22**, 345–362. © Blackwell Science Ltd.]

ness," and beetle families vary so dramatically in their diversity. For example, one of the earliest evolving beetle groups, the Cupedidae, is currently represented by only 26 species worldwide. Possession of elytra thus is only one step among many leading to successful diversification of Coleoptera. Nonetheless, functional study of the beetle body plan illustrates many instances in which "beetleness" has predisposed lineages to enter and proliferate in particular habitats.

Most generally, the organization of beetle bodies that has permitted entry into confining, laminar microhabitats involves (1) thick hard cuticle on the head and prothorax, (2) a prothorax flexibly articulating with the pterothorax, and (3) a pterothorax topped by elytra that cover folded flight wings and soft, expansible abdominal tergites. Carabid beetles utilize wedge-pushing locomotion to move through leaf litter and under loose tree bark. In these beetles, a rounded projection on the base of the hind femur impinges on the meta-trochanter, which articulates only in a horizontal plane with the immobile hind coxa. Pulling the hind leg forward pushes the apex of the femur away from the body, thereby elevating the carabid's dorsum (the wedge). This upward motion is then followed by a thrust of the hind legs, forcing the beetle body forward (the push). Using this mechanism, rhysodine carabid beetles can move through dying or even living wood without leaving a trace; the wood simply closes up behind them! Their goal in this unlikely activity is foraging on the amoeboid plasmodia of slime molds (Myxomycetes).

For beetles to both fly and move through confining spaces, their wings must be stowed under the elytra while walking or wedge pushing, yet quickly unfurled for flight. All wing folding is controlled through muscles attached to the wing base; as long as tension is applied so that the radial and cubital veins are pulled apart, the wing surface remains flattened. However, relaxation of this tension brings natural folds into play, so that, with the wing apex folding in upon itself (Fig. 7), the medius comes to lie above the radius posterior (Fig. 8).

Numerous variations on wing folding have evolved depending on elytral configuration. In the archostematan Cupedidae, the wing apex rolls up longitudinally (Fig. 9). Wing folding can proceed even given the evolutionary reduction of wing venation observed in tiny beetles such as Microsporidae or Ptiliidae (Figs. 10–11). In addition to the folding characteristics of the wings, setose binding patches occurring on the wing surface, inner elytral surface, and abdominal terga are used to manage the wing folding, thus ensuring safe stowage of the wing membranes (Fig. 12).

The generalized thickening of cuticle characteristic of archostematans, adephagans, and many polyphagans results in an adult insect that is highly constrained in internal volume. Abdominal sutures between ventrites allow these segments to move against one another so that well-fed or gravid individuals will exhibit an abdomen extending beyond the elytral apex. Nonetheless, longitudinal abdominal extension can be minimized and external structural integrity maintained if the body is allowed to expand in another

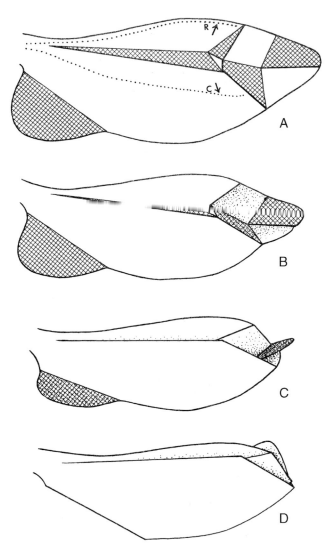

FIGURE 7 Paper model of right wing of *Cantharis* sp. arranged to demonstrate wing folding. Cross-hatched areas face ventrally in fully folded wing. In the extended wing (A) the principal veins—radius (*R*) and cubitus (*C*)—are apart by muscular action from the wing base. When this action ceases, the wing apex automatically folds (B, C) until wing is fully folded (D). [From Hammond, P. M. (1979). Wing-folding mechanisms of beetles, with special reference to investigations of adephagan phylogeny. *In* "Carabid Beetles: Their Evolution, Natural History, and Classification" (T. L. Erwin, G. E. Ball, D. R. Whitehead, and A. L. Halpern, eds.), p. 122, Fig. 1. Junk, the Hague. With kind permission of Kluwer Academic Publishers.]

direction. Beetles accomplish this increase in body volume through dorsoventral expansion of the abdomen.

In a newly eclosing beetle, the lateral reaches of the abdominal tergites lie both between and below the lateral portions of the abdominal sternites. The tergites and sternites are joined by extensive membranes, within which the spiracles are situated. These membranes may stretch, and the tergites may move dorsally relative to the stationary lateral margins of the sternites, dramatically increasing the volume of the abdomen. This volumetric expansion is accomplished without any compromise to the external armor represented

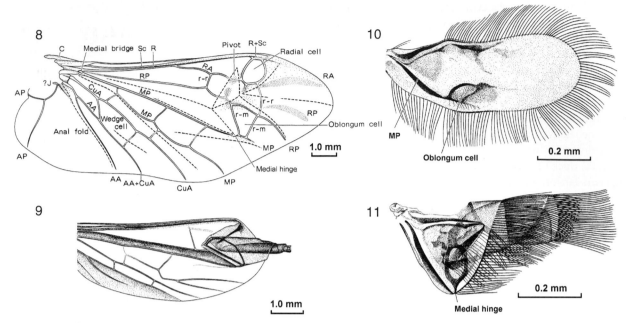

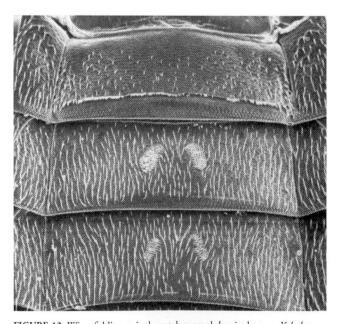

FIGURES 8–11 Hind wings of Coleoptera. (8) *Omma stanleyi* (Ommatidae): AA, anal anterior; AP, anal posterior; C, costa; CuA, cubitus anterior; J, jugal; MP, medial posterior; R, radius; RA, radius anterior; r-m, radial-median crossvein; RP, radius posterior; r-r, radial crossvein; Sc, subcosta. (9) *Adinolepis mathesoni* (Cupedidae). (10) Open hind wing of *Microsporus vensensis* (Microsporidae). (11) Folded wing of *M. ovensensis* (Images provided by copyright holder, CSIRO Entomology, Canberra, ACT, Australia.)

by the cuticle. The soft, flexible abdominal tergites are protected by the elytra, except when the beetle is flying. At this time the soft membranes and flexible tergites are vulnerable to attack by predators or parasites.

FIGURE 12 Wing-folding spicule patches on abdominal terga, *Xylodromus concinnus* (Staphylinidae) (see Fig. 7). [From Hammond, P. M. (1979). Wing-folding mechanisms of beetles, with special reference to investigations of adephagan physiology. *In* "Carabid Beetles: Their Evolution, Natural History, and Classification" (T. L. Erwin, G. E. Ball, D. R. Whitehead, and A. L. Halpern, eds.), p. 122, Fig. 1. Junk, the Hague. With kind permission of Kluwer Academic Publishers.]

In the floricolous, day-flying Buprestidae and scarab beetles of the subfamily Cetoniinae, flight is undertaken without significant separation or lifting of the elytra, with the metathoracic wings extended under the lateral elytral margins. In the buprestids, this posture allows the aposematic coloration of the elytra to be visible both in flight and at rest. In other polyphagans and the Archostemata, the elytra are held at an angle during flight, beating synchronously with the flight wings, and thereby providing some degree of aerodynamic lift.

Given the need to exchange oxygen and carbon dioxide at a liquid interface on the surfaces of the tracheolar cells, respiration represents the major activity through which an insect can lose water. This source of water loss is of particular importance for an animal of small body volume. The beetle respiratory system opens via large metathoracic spiracles and up to eight pairs of abdominal spiracles, all of which open onto the subelytral cavity. Thus, in addition to controlling gas exchange via the spiracular openings, a beetle can modulate respiration by the position of the abdominal venter relative to the elytra. Reduction of the elytra to the quadrate condition seen in Staphylinidae has resulted in secondary exposure of the abdominal spiracles.

Beetles have invaded freshwater aquatic habitats several times during their evolutionary history. In all instances, adult aquatic beetles retain the spiracular respiratory system of their terrestrial relatives, requiring that they regularly have access to atmospheric gases. The subelytral space provides the means to hold an air bubble while the beetle is active underwater. This bubble can be replenished by periodic surfacing of the beetle, during which the tip of the abdomen breaks the water surface, permitting exchange of gases.

Because of the makeup of our atmosphere (79% nitrogen, 21% oxygen), the subelytral air bubble serves as a compressible gill, permitting extended underwater sojourns. As the beetle uses oxygen, more oxygen diffuses into the bubble from the surrounding water. The carbon dioxide produced through the beetle's respiration, being highly soluble in water, quickly leaves the bubble. Because nitrogen dissolves slowly into the water, there is a gradual reduction in bubble size. The beetle can use up to eight times as much oxygen than was in the original bubble before being required to surface to replenish its air supply. Swimming beetles using these simple subelytral compressible gills include various Adephaga (e.g., Haliplidae and Dytiscidae).

In other families of the aquatic realm, oxygen is supplied to the subelytral bubble by a plastron composed of microfuge hairs or other columnar evaginations of the cuticle that are close together along their outer surface, excluding water by its surface-filming qualities. Oxygen diffuses into the plastron without any change in plastron gas volume, allowing the beetle to remain indefinitely below the water's surface. Nonetheless, plastron respiration can work only in highly oxygenated water, so beetles with plastrons are usually found in moving waters. Plastron breathers also are less active than the adephagan compressible gill breathers, because the plastron cannot provide the high levels of oxygen required for intense activity. This type of structure has evolved repeatedly in the order, being found in the Hydrophilidae, Dryopidae, Elmidae, and some Curculionidae.

LARVAL SPECIALIZATION

Among the four suborders of Coleoptera, life histories of the predaceous Adephaga most closely resemble those of the beetles' phylogenetic sister group, the neuropteran orders. Adephagan larvae are generally campodeiform, that is, elongate and slightly dorsoventrally flattened, with long thoracic legs and a posteriorly tapered, dorsally sclerotized abdomen (Fig. 13). They typically have anteriorly directed mouthparts that often include elongate, sickle-shaped mandibles with a reduced mola (Fig. 14). The legs are six-segmented (coxa, trochanter, femur, tibia, tarsus, claws) as in the Megaloptera, Raphidioptera, and Neuroptera. The ninth abdominal tergite usually bears a pair of dorsolateral appendages (urogomphi) that may be short and unsegmented, or longer and variously segmented. These are secondarily evolved structures of the Coleoptera, and not homologous with the cerci of, for example, the orthopteroid orders. Adephagan larvae usually develop through three instars before pupation.

The larvae of Archostemata deviate from this generalized configuration by representing the syndrome that has evolved repeatedly in taxa characterized by the larval wood-boring habit. In these groups, the larvae are lightly sclerotized, more or less tubular, with shortened or reduced legs, and various ampullae on the thoracic and abdominal segments (Fig. 15). The archostematan family Micromalthidae exhibits probably

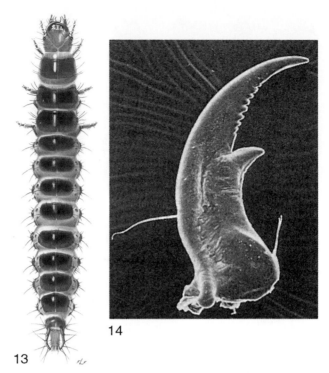

FIGURES 13–14 (13) *Abaris bigenera* mature larva (Carabidae), dorsal view (larval length, 8.3 mm). (Image © F. L. Fawcett.) (14) Right larval mandible, ventral view, *Platynus* sp. (Carabidae). Note absence of basal mola, and presence of large retinacular tooth and serrate incisor. [From Lawrence, J. F. (1991). Order Coleoptera. *In* "Immature Insects," Vol. 2 (F. W. Stehr, ed.), Fig. 34.19. Kendall/Hunt, Dubuque, IA.]

the most bizarre set of larval forms and associated life cycle seen in Insecta. The campodeiform first instar is an active triungulin. It molts to become a legless, feeding cerambycoid larva, which in turn may undergo four types of molt. It may pupate directly to become an adult diploid female. Alternatively, it may develop into one of three kinds of larviform reproductive: a thelytokous pedogenetic female that parthenogenetically produces viviparously a number of diploid triungulins; an arrhenotokous pedogenetic female that lays a single egg, from which hatches a stump-legged curculionoid larva that in turn devours the mother, pupates, and then emerges as an adult haploid male; and an amphitokous pedogenetic female, which may produce either form. The hormonal controls of this system are not known, although production of the various larval types seems to be affected by environmental conditions.

The demographic consequences of this life cycle include the ability to quickly multiply and to use available rotting wood in the production of numerous dispersive adults. The triungulin larvae (Fig. 16) can expand the infestation to adjacent portions of the rotten log or timber. The cerambycoid larvae (Fig. 17), more typical of other archostematan larvae, can efficiently feed in confined galleries in rotting wood. The pedogenetic form (Fig. 18) can itself produce many more triungulins, enhancing the rate of increase of the population. The adults (Fig. 19) are produced in massive numbers, with these winged colonists establishing new colonies. Natural

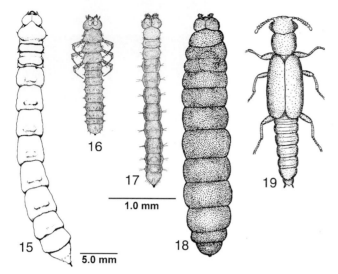

FIGURES 15–19 (15) *Tenomerga concolor* mature larva (Cupedidae), dorsal view. [From Lawrence, J. F. (1991). Order Coleoptera. *In* "Immature Insects," Vol. 2 (F. W. Stehr, ed.), Fig. 34.67a. Kendall/Hunt, Dubuque, IA. Figures 16–19, *Micromalthus debilis* (Micromalthidae), dorsal view. (16) Triungulin first instar larva. (17) Cerambycoid larva. (18) Pedogenetic larva. (19) Adult female. (Drawings, Figs. 16–19, courtesy of the copyright holder, the Royal Entomological Society, London.)

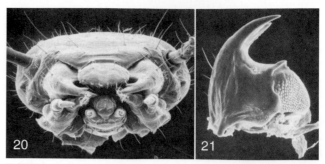

FIGURES 20–21 *Anisotoma errans* larva (Leiodidae). (20) Head capsule, anterior view. (21) Right mandible, ventral view. Note large asperate mola at base. [From Newton, A. F., Jr. (1991). Leiodidae, pp. 327–329. *In* "Immature Insects," Vol. 2 (F. W. Stehr, ed.), Figs. 34.152a and 34.154. Kendall/Hunt, Dubuque, IA.

infestations have been reported in large *Quercus* (oak) or *Castanea* (chestnut) logs across the beetles' native range in northeastern North America. Other human-associated infestations have been reported from timbers deep in a South African diamond mine, and in thick oak paneling used to line the vaults of the Federal Reserve Bank in New York City.

The small suborder Myxophaga is characterized by adults and larvae of extremely small size, with both larvae and adults living interstitially in riparian areas, where they feed on algae. As opposed to the Archostemata and Adephaga, the larval legs are five-segmented, with the tarsus and claws fused into a single segment, the tarsungulus. The abdomen may or may not bear urogomphi on the ninth tergite. Like many other beetle species that feed on small particulate matter (pollen, spores, conidia, etc.), the larval mandibles bear a basal mola. Because they are aquatic in all stages, the adults bear a plastron, and the larvae may breathe by means of a plastron that covers the spiracles or via vesicular gills (i.e., a balloonlike expansion of the spiracular peritreme with an apical opening).

It is in the order Polyphaga that divergence of larval and adult lifestyles becomes evolutionarily significant. Among basal polyphagans in the superfamilies Staphylinoidea and Hydrophiloidea, larval anatomy remains generally of the campodeiform type, although mouthparts may be specialized for feeding on fungal food through development of broadly papillate molar regions on the mandible (Figs. 20–21). As in the Myxophaga, the larval leg has five segments. Aquatic forms may bear lateral gills on the thorax or abdomen (Fig. 22). Urogomphi of various configurations also may be present.

The larvae of the superfamilies Dascilloidea (Fig. 23), Byrrhoidea, and Bostrichoidea exhibit a dorsally convex body

configuration that has evolved into the much more exaggerated C-shaped grub characteristic of the Scarabaeoidea (Fig. 24). Scarab grubs can develop in a variety of microhabitats. Primitive scarabaeoids such as stag beetles (Lucanidae) and bess beetles (Passalidae) develop as saprophagous larvae in rotting wood. Larvae of the Geotrupidae and scarab subfamilies Scarabaeinae and Onthophaginae develop in mammalian herbivore dung where they also feed on fungi. Flowering plant roots are fed on by larvae of species in the more highly derived scarab subfamilies Melolonthinae, Rutelinae, and Dynastinae. Many species in these subfamilies are of economic concern, because they feed on commodities such as corn, small grains, vegetable crops, grasses, turf, fruits, and nursery stock. The C-shaped larval configuration results in an increased abdominal capacity relative to the head and thoracic forebody. This increased capacity is directly connected to the scarab larva's penchant for feeding on large amounts of food in order to pupate at a large size. Scarabs are well represented among the largest beetles, with the impressive *Goliathus* beetles of Africa and Asia attaining the greatest body mass of any beetle known.

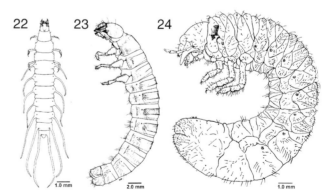

FIGURES 22–24 Beetle larvae. (22) *Berosus metalliceps* (Hydrophilidae), dorsal view. [From Spangler, J. P. (1991). Hydrophilidae, pp. 355–358. *In* "Immature Insects," Vol. 2 (F. W. Stehr, ed.), Fig. 34.296. Kendall/Hunt, Dubuque, IA.] (23) *Dascillus davidsoni* (Dascillidae), lateral view. [From Lawrence, J. F. (1991). Order Coleoptera. *In* "Immature Insects," Vol. 2 (F. W. Stehr, ed.), Fig. 34.323a. Kendall/Hunt, Dubuque, IA.] (24) *Popillia japonica* (Scarabaeidae), lateral view. (© New York Entomological Society.)

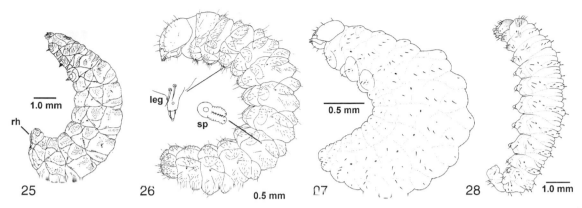

FIGURES 25–28 Larvae of phytophagous Chrysomeloidea and Curculionoidea. (25) *Donacia* sp. (Chrysomelidae), lateral view, sharp respiratory horns (rh) insert into underwater stems of water lily, providing air to the spiracles located at base of horns. (© New York Entomological Society.) (26) *Neocimberis pilosus* (Nemonychidae), lateral view (sp, spiracle). [From Anderson, W. H. (1947). *Ann. Entomol. Soc. Am.* **40**, 489–517. © Entomological Society of America.] (27) *Apion griseum* (Apionidae), lateral view. (28) *Hypera nigrirostris* (Curculionidae), lateral view.

A C-shaped larva has evolved independently in another phytophagous group with concealed larval stages; the Curculionidae. The curculionid sister group, the Chrysomeloidea, is primitively characterized by larval stages superficially similar to those of Eucinetoidea and Dascilloidea, that is, larvae of moderately convex dorsal habitus (Fig. 25). Primitive weevils retain evidence of thoracic legs (Fig. 26); however, all evidence of thoracic appendages has been evolutionarily erased in higher weevils (Fig. 27). As phytophagous weevils have specialized, taxa have moved from being internal feeders to foraging on the external surfaces of their hosts. External feeders such as the lesser clover-leaf weevil gain a foothold on their host plant through ventral abdominal ampullae (Fig. 28), analogous to the prolegs of Hymenoptera and Lepidoptera. A parallel transition from hidden feeders to exposed foliage feeders has also evolved in the weevil sister group, the Chrysomeloidea.

The Cerambycidae comprise one basal division of the chrysomeloids, with all their larvae internal feeders. The Palophaginae represent the earliest divergent lineage of Chrysomelidae, based both on late Jurassic fossils (> 145 mya), and phylogenetic analysis of living species. Larvae of this subfamily attack the male strobili of *Araucaria* (Coniferales: Araucariaceae).

The subfamily Aulacoscelinae represents another early-diverging chrysomelid lineage. Larvae of this group are internal feeders on cycads. From this syndrome of hidden feeding, leaf beetle larvae have evolved to live on open plant tissues of many of the world's angiosperms. Where plants have evolved the ability to incorporate secondary chemical compounds in their tissues, herbivorous chrysomelids have evolved to use these chemicals to recognize food and stimulate oviposition. They have also evolved the ability to sequester these broadly toxic chemicals into their tissues to gain protection from predators. Today it is commonplace to observe brightly colored larvae and adults of protected leaf beetles congregated on exposed plant surfaces, serving as a communal warning to predators regarding their unpalatability.

The wood-boring larval body plan of the Archostemata is well represented in the Polyphaga, having independently evolved in the Buprestidae (Fig. 29), Eucnemidae, and Cerambycidae (Figs. 30–31). Larvae in all these families can bore through freshly dead or dying wood by using their well-developed, anteriorly directed mandibles. Laterally expanded thoracic segments or abdominal ampullae serve to anchor these larvae in their tunnels, facilitating purchase by the

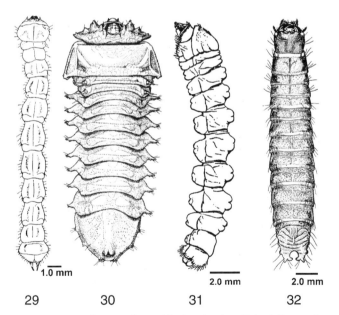

FIGURES 29–32 Larvae of wood-boring beetles. (29) *Agilus anxius* (Buprestidae), dorsal view. (30) Unidentified lepturine larva (Cerambycidae), ventral view, scale unknown. [Figs. 29, 30 from Böving, A. G., and Craighead, F. C. (1930). An illustrated synopsis of the principal forms of the order Coleoptera. *Entomol. Am.* **11**, 1–125. © New York Entomological Society.] (31) *Platyzorilespe variegata* (Cerambycidae), lateral view. [From Gardner, J. C. M. (1944). On some coleopterous larvae from India. *Ind. J. Entomol.* **6**, 111–116. © Entomological Society of India.] (32) *Hemicrepidius memnonius* (Elateridae), dorsal view. [From Dietrich, H. (1945). Cornell University Agricultural Experiment Station Memoir 269, plate IV.2. © Cornell University.]

mandibles on the wood surface. Leg reduction has proceeded during diversification of cerambycid borers, with larvae of more basally divergent subfamilies such as the Prioninae and Lepturinae having shortened thoracic legs (Fig. 30), whereas larvae of the highly derived subfamily Lamiinae (Fig. 31) are legless.

Where wood-boring beetles have gone, similarly shaped predatory beetles have followed. These tubular larvae in the Elateroidea and Cleroidea may be highly sclerotized, and they bear well-sclerotized head capsules and/or urogomphal plates (Fig. 32) that armor them appropriately for their habitats (e.g., under bark, within wood-boring beetle galleries). Elaterid larvae have diverse feeding habits, with many groups being phytophagous or saprophagous. However, all forms, regardless of food habit, imbibe their food as an extraorally predigested liquid.

Other elateroid larvae, such as fireflies (Lampyridae) and soldier beetles (Cantharidae), lack the heavy armor of the concealed gallery feeders, and prey on other arthropods among leaf and ground litter. These larvae use grooved mandibles to suck up the liquefied contents of their prey. In Elateridae, and independently in Phengodidae and Lampyridae, larval and adult stages have evolved the ability to produce light using organs composed of modified cuticular cells. The significance of larval luminescence has been variously explained. For example, night active *Pyrearinus* larvae in the elaterid subfamily Pyrophorinae use light organs to attract flying insect prey to Brazilian termite mounds where they make their home. Phengodid larvae of the genus *Phrixothrix* possess medial photic organs on the head that use red light to illuminate potential prey. But they also possess lateral abdominal light organs that emit green light. These abdominal light organs are homologous with those of Lampyridae and most likely serve to advertise that the larvae are chemically protected. Increasingly complicated light communication systems have evolved in the adult stages of various phengodid and lampyrid taxa.

Cucujoidea and Tenebrionoidea are diverse superfamilies whose larval forms blend imperceptibly into each other morphologically and biologically. It is in these groups that saprophagous and mycophagous feeding habits are associated with extensive larval diversification. Primitive larvae of both superfamilies are similar and typical of Polyphaga in many ways (e.g., five-segmented legs, urogomphi, moderate degree of sclerotization, etc.) Evolutionary trends in one often are mirrored in the other. Cucujoid and tenebrionoid larvae are usually small to moderate in size, and somewhat dorsoventrally compressed. Many are cryptozoic, occurring in leaf litter, under bark, in fungus, or in rotting wood, where they feed on fungi or on fungus-altered plant matter. Groups specialized for feeding on spores, conidia, loose hyphae, or other small particles exhibit various specializations correlated with microphagy. Most notably, these include a well-formed mandibular mola and prostheca (Fig. 33).

Extreme dorsoventral compression of the body has occurred repeatedly in response to the selective pressures of

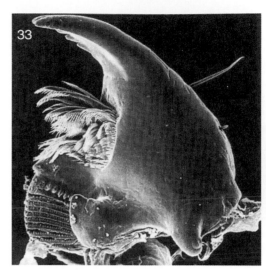

FIGURE 33 Left larval mandible, ventral view, of *Anchorius lineatus* (Biphyllidae), showing basal mola (lower left) and prostheca with comb hairs; mandible width, 0.16 mm. [From Lawrence, J. F. (1989). Mycophagy in the Coleoptera: Feeding strategies and morphological adaptations. *In* "Insect–Fungus Interactions" (N. Wilding, N. M. Collins, P. M. Hammond, and J. F. Webber, eds.), p. 6, Fig. 6. Academic Press, London.]

occupying subcortical and interstitial leaf litter habitats. Sometimes (Fig. 34), the body form is simply flattened, with a reorientation of the head to a protracted, prognathous condition and a migration of the leg articulations to more lateral positions. Flattening, however, may be accompanied by an additional transition to an onisciform (or pie-plate-shaped) body through extensive development of tergal flanges, resulting in a broadly oval body outline in some Cerylonidae, Corylophidae (Fig. 35), Discolomidae, and Nilioninae (Tenebrionidae). Larvae specialized for life under bark, in fungi, or in rotting wood typically have short, stout,

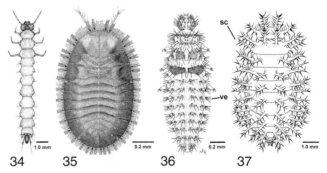

FIGURES 34–35 Flattened beetle larvae, dorsal view. (34) *Dendrophagus americanus* (Cucujidae). [From Lawrence, J. F. (1991). Order Coleoptera. *In* "Immature Insects," Vol. 2 (F. W. Stehr, ed.), Fig. 34.527. Kendall/Hunt, Dubuque, IA.] (35) Corylophidae, genus unknown [From Lawrence, J. F. (1991). In "Immature Insects," Vol. 2 (F. W. Stehr, ed.), Fig. 34.628a. Kendall/Hunt, Dubuque, IA.] Figures 36–37 Larvae of Coccinellidae, dorsal view. (36) Predaceous *Stethoris histrio*. (37) Phytophagous *Epilachna varivestis:* sc, scolus; ve, verruca. [From Le Sage, L. (1991). Coccinellidae, pp. 485–494. *In* "Immature Insects," Vol. 2 (F. W. Stehr, ed.), Fig. 34.570. Kendall/Hunt, Dubuque, IA.]

unarticulated, and unsegmented urogomphi. The apex is typically recurved to point dorsally. This form of urogomphi is thought to help larvae move about in cramped habitats.

Several tenebrionoid and cucujoid groups experienced parallel transitions to a parasitic lifestyle, including Meloidae, Rhipiphoridae, and some Cucujidae and Bothrideridae. The Rhipiphoridae provide a glimpse at parasitism involving both externally and internally feeding stages. In Rhipiphorinae, the triungulin first instar locates and attaches itself to an adult of a suitable hymenopteran host. After being carried back to the host's nest, the triungulin detaches itself and searches for a host larva. Once the host has been located, the larva burrows inside, where feeding continues (endophagy) until the parasitoid becomes greatly swollen. As the host larva reaches maturity, the parasitoid exits from its thorax, switching to feed externally (ectophagy), eventually killing it.

In Rhipidiinae, the reverse sequence of internal and external feeding occurs: the triungulin locates a cockroach as a potential host, inserts its head and thorax into a membranous region on its venter, and begins to feed while most of its body remains outside the host. Later, the larva transforms into a less mobile, legless form and moves entirely inside the host, where it begins to grow rapidly. Near the end of its development, the larva molts to a form with legs and emerges from the host to pupate.

Cucujoid and tenebrionoid taxa that are adapted for external feeding typically have a more eruciform (caterpillar-like) shape resulting from elongation of the legs, reorientation of the head to a more hypognathous position, and dorsoventral inflation to a more cylindrical shape. These external feeders also tend to exhibit defensive modifications. Aposematic coloration is common in these groups. Tergal and pleural armature, which is absent or modest in most cucujoids and tenebrionoids, becomes exaggerated in some predators (e.g., *stethoris*, Coccinellidae, Fig. 36), surface feeding herbivores (e.g., the coccinellid genus *Epilachna*, Fig. 37), and fungus feeders (e.g., the erotylid genus *Aegithus*), to form various structures such as setose, tuberculate verrucae, and complexly branched scoli.

Within Cucujoidea and Tenebrionoidea there is a recurring evolutionary transition from mycophagy/saprophagy to a lifestyle of true phytophagy as a borer in healthy herbaceous stems or wood. This entire sequence can be observed within individual families (e.g., Melandryidae), where there is a range of larval feeding that extends from boring in fungus sporophores to boring in fungus-infested wood and finally to boring in sound wood. Cucujoid and tenebrionoid wood borers tend to have fleshy bodies with conspicuous sclerotized plates usually restricted to the anterior end of the body. The head capsule tends to be prognathous and often bears a median endocarina, an internal keel on the dorsum associated with the development of especially powerful mandibular muscles.

Predatory larval forms have arisen repeatedly within Cucujoidea and Tenebrionoidea, most notably in the Coccinellidae. Accompanying this trophic transition is a suite of morphological changes to produce a campodeiform body (Fig. 36). The head typically has a more prognathous orientation. The mandibles are more prominent and lack a mola.

The beetle pupa is adecticous and usually exarate (i.e., the mandibles are fixed in position, and the head and thoracic appendages are free). Several groups have independently evolved the obtect condition; among them staphylinine Staphylinidae, Ptiliidae, and Coccinellidae, and the hispine Chrysomelidae. If the pupa rests concealed in a pupal chamber, it lies on its dorsum elevated from the substrate by numerous thoracic and abdominal setae. Pupae may be enclosed in a cocoon made of silk (aleocharine Staphylinidae, Tenebrionidae, Curculionidae), fecal material (Passalidae and some Scarabaeidae), or the larval fecal case (cryptocephaline Chrysomelidae).

Exposed pupae, as in Coccinellidae, Chrysomelidae, and Erotylinae (Erotylidae) may remain attached to their host plant or fungus via the sloughed-off last larval cuticle, which encircles the anal portion of the pupa. Such exposed pupae may be protected by defensive secretions remaining in the shed larval skin. Beetle pupae retain the ability to move the abdomen by using the flexible abdominal intersegmental membranes. Sclerotized processes on opposing margins of the abdominal segments, called gin traps, have been suggested as defensive devices used to pinch and drive off mites and other predators.

ECOLOGICAL SPECIALIZATION

One means of estimating the ground plan feeding habits of Coleoptera uses observations of extant taxa, interpreted in context of phylogenetic hypotheses for the various lineages making up the order. By this method, we would deduce that the most primitive beetles were either saprophagous wood borers as larvae, such as extant Archostemata, or that they were campodeiform predators, such as the Adephaga and the sister group to Coleoptera, the Neuroptera+Raphidioptera+ Neuroptera. Examining the fossil record of Coleoptera as well as suggestive damage to fossil plants of the Triassic and Jurassic formations containing the earliest beetle fossils provides a second means of making such an estimation. By this method, we find that archostematans and primitive weevils predate fossils of all other types, suggesting that the earliest feeding habits were either saprophagous or herbivorous. Of course, fossil evidence of predation is not likely to be preserved, nor interpretable as such if it were. These two viewpoints, phylogenetic and paleontological, represent the diversity of opinion about how the first beetles lived their lives.

The two viewpoints can be reconciled if we view fossil data drawn from the various periods in light of phylogenetic estimates based on a diversity of taxa and characters. To do this, we must assume that the lifestyles of recent taxa represent those of their related fossil relatives. By this reasoning, it is very apparent that herbivorous taxa have

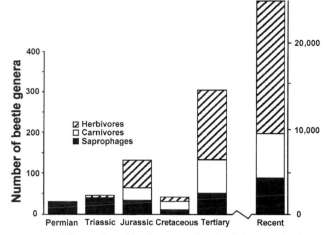

FIGURE 38 The number of beetle genera of each of three trophic levels from Permian to recent epoch. Permian genera represent Protocoleoptera. [Redrawn with permission from Farrell, B. D. (1998). "Inordinate fondness" explained: Why are there so many beetles? *Science* **281**, 555–559. © 1998 American Association for the Advancement of Science.]

constituted a majority of the major life-forms, measured by their recognition as genera, for coleopteran faunas from the Jurassic and Tertiary to recent times (Fig. 38). Even before the advent of the flowering plants, more than half the variety of beetle life-forms had evolved to focus their feeding attentions on plants. We can investigate the impact of the origin and diversification of angiosperms on beetle diversity by looking at beetle sister taxa in which one group is restricted to gymnospermous plants, whereas its sister is found on angiosperms. Brian Farrell examined lineages within the Chrysomeloidea and Curculionoidea. He found angiosperm-feeding taxa to be far more rich in species today than their gymnosperm-feeding sister groups. Clearly, angiosperm feeding has enhanced the species-level diversification of beetles living on them.

In addition to internal feeding on stem tissue, and feeding on saprophagous growth in decaying cambium, the angiosperms offer floral resources unavailable from gymnosperms. Adult beetles of many families characterized by phytophagous, saprophagous, or scavenging larvae may be found feeding in or on flowers, associated exudates, or pollen. Melolonthine and cetoniine scarab beetles, whose larvae are subterranean root feeders or rotten wood feeders, respectively, often feed on flowers. Dermestid beetle larvae scavenge dead animal matter, then move to flowers to feed on pollen after they have eclosed as adults. Once a female dermestid has fed, she becomes negatively phototactic and searches for cavities containing animal remains, where she will oviposit. Other families well represented among the pollen-feeding adults include Buprestidae, Lycidae, Nitidulidae, Mordellidae, Rhipiphoridae, Meloidae, Anthicidae, and Cerambycidae. Meloids and rhipiphorids not only feed at flowers but oviposit there, with their hatching triungulin larvae waiting in the flower to climb on passing bees and wasps, which they

parasitize. Feeding on hard pollen grains is facilitated by possession of mandibles bearing a well-developed mola. Such mandibles are also associated with fungal feeding, and families such as the Nitidulidae, Tenebrionidae, and Oedemeridae contain species representing both adult feeding habits; individual oedemerid species have been reported to feed on both fungi and pollen.

Coleopteran relationships with fungi are widespread throughout the order and diverse in form. Approximately 25 extant families of beetles are primarily mycophagous. Greatly unappreciated, however, are the less obvious trophic relationships between fungi and many beetles that are ostensibly saprophagous or phytophagous. Many beetles eat plant tissue only after it has been partially broken down by fungi. Some harbor endosymbiotic fungi that allow digestion of plant tissue or provide essential nutrients. Others are thought to ingest and acquire fungal enzymes that are essential for their existence as herbivores. John Lawrence estimated that as many as half of all beetle families either are truly mycophagous or feed on plant matter that has been altered by fungal enzymes.

Ancient Greeks believed fungi were merely homes of insects. A rich insectan fauna often dwells in larger fungi, and much of it comprises mycophagous and predaceous beetles. Through evolutionary time few fungus taxa have escaped the interest of beetles. Mycophagous families seem to be especially concentrated in the polyphagan superfamilies Cucujoidea, Tenebrionoidea, and Staphylinoidea. However, fungivory arose repeatedly in various other lineages within the order as well.

Fungi are tremendously diverse physically, chemically, behaviorally, and ecologically. Mushrooms, woody conks, puffballs, truffles, yeasts, smuts, rusts, and molds present separate special challenges as food sources. In addition, a single fungus often represents a composite of resources. For example, a single polypore shelf on a log may provide a delicate layer of spore-bearing tissue on the underside, a hard, woody context, and an area where its hyphae penetrate decaying wood. Some mycophagous beetles have a broad range of acceptable hosts; however, many are more selective, feeding only on some portions of fruiting structures from a few species at a particular stage of development or decay. Host specificity tends to be narrower for immature stages. Specialization of beetles has occurred in response to the various resources and challenges that fungi present.

Woody polypore shelves offer large, persistent sources of food for mycovores. There are many different strategies for the use of the soft spore-bearing tissue of wood polypore fungi. Species of *Ellipticus* (Erotylidae) have robust mandibles capable of gouging off chunks of hymenium and its supporting tissues (Fig. 39). Larvae of *Holopsis* (Corylophidae) have found another method of tapping this resource. They use a slender, snoutlike elongation of the head to graze on the inner surface of individual spore tubes (Fig. 40). The Nannosellinae (Ptiliidae) exhibit another evolutionary solution, namely,

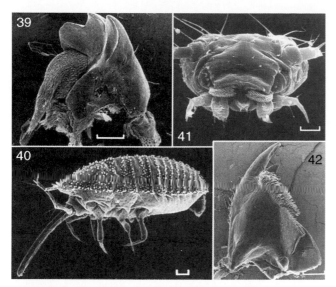

FIGURES 39–42 Feeding structures of fungus-feeding beetle larvae. (39) Larval mandible, ventral view, of *Ellipticus* sp. (Erotylidae); multidentate apex and setose lobe near base (lower left) are used to bite large chunks off fungal substrate, which are then swallowed whole (scale, 100 μm). [From Lawrence, J. F. (1989). Mycophagy in the Coleoptera: Feeding strategies and morphological adaptations. *In* "Insect–Fungus Interactions" (N. Wilding, N. M. Collins, P. M. Hammond, and J. F. Webber, eds.), p. 16, Fig. 21. Academic Press, London.] (40) *Holopsis* sp. (Corylophidae), lateral view, with long feeding rostrum bearing apical mandibles, allowing feeding inside pore tubes of sporocarp fungi (scale, 100 μm). [From Lawrence, J. F. (1989). Mycophagy in the Coleoptera: Feeding strategies and morphological adaptations. *In* "Insect–Fungus Interactions" (N. Wilding, N. M. Collins, P. M. Hammond, and J. F. Webber, eds.), p. 16, Fig. 17. Academic Press, London.] (41) Larval head, anterior view, of *Dasycerus* sp. (Staphylinidae), showing brushy mandibular apices used to remove spores or hyphae from the substrate to the mouth (scale, 50 μ). [From Lawrence, J. F. (1989). Mycophagy in the Coleoptera: Feeding strategies and morphological adaptations. *In* "Insect–Fungus Interactions" (N. Wilding, N. M. Collins, P. M. Hammond, and J. F. Webber, eds.), p. 11, Fig. 12. Academic Press, London.] (42) Larval right mandible, ventral view, of *Nosodendron unicolor* (Nosodendridae), showing food press near base (lower right) that concentrates particulate food while ejecting liquid (scale, 100 μm). [From Lawrence, J. F. (1989). Mycophagy in the Coleoptera: Feeding strategies and morphological adaptations. *In* "Insect–Fungus Interactions" (N. Wilding, N. M. Collins, P. M. Hammond, and J. F. Webber, eds.), p. 6, Fig. 6. Academic Press, London.]

miniaturization: fully grown adults, only 0.4 mm in length, crawl inside individual spore tubes to feed directly on the soft spore-bearing tissue.

Specialists on fleshy mushrooms [e.g., *Oxyporus* (Staphylinidae),] face different challenges. Unable to fly around to look for new mushrooms, larvae must complete their feeding on their ephemeral host before it decays. Many of the fungus beetles that specialize on soft mushrooms exhibit greatly accelerated development. Their mandibles are more bladelike and are capable of slicing through large chunks of soft fungal tissue.

Beetles preferring small, scattered fungal spores or conidia as food often have a suite of features related to their microphagous habits. The mouthparts tend to be brushy and capable of sweeping tiny particles from the substrate into their mouth. These modifications often involve the maxillae, but in larval *Dasycerus* (Staphylinidae) the mandibular apices are modified for this function as well (Fig. 41). The mandibular mola is also commonly modified in spore feeders. Spores are ground between opposing molar grinding surfaces on each mandible, with an action much like that of a millstone grinding wheat into flour. Nosodendridae, which feed partially on yeasts that occur in sap fluxes, use their brushy mouthparts to filter the fungal cells from the fluid (Fig. 42).

Less exploitative symbiotic relationships with fungi also are widespread and diverse within Coleoptera. The best-studied examples of mutualism with fungi are the relationships occurring in the bark and ambrosia beetles (Platypodinae and Scolytinae of the Curculionidae). Perhaps the most familiar case is that of Dutch elm disease. At the corners of this "ecological triangle" are the bark beetles (*Scolytus* spp.), the fungus (*Ceratostomella ulmi*), and the host elm trees (*Ulmus* spp.). Adult beetles nibble on tree twigs and thereby inoculate them with fungal spores. Following germination of the spores, the fungus attacks the tree and ultimately kills it. The beetles prefer to oviposit on recently killed *Ulmus* trees, many of which were recent victims of *C. ulmi*. Upon hatching, their larvae bore about, feeding on fungus-infested wood. The final link in the cycle is completed when newly emerging adults pick up fungal spores as they move around the gallery before flying off to dine on some living elm twigs.

A broad range of variants stems from the basic pattern observed in Dutch elm disease. In some cases the link between the fungus and the beetles weakens to the point of being merely incidental. In Lymexylidae and at least some Platypodinae, the relationship is a tighter, obligatory one in which the beetles farm a fungus to feed their brood. The wood of the host tree is important to the beetle only as a substrate for the fungal garden. In these evolutionarily linked relationships, the beetles often have specialized pockets called mycangia on their body to aid in the transportation of spores or conidia to new substrates (Figs. 43–44). Mycangia sometimes have associated glands that help to keep the fungal tissue viable until it is needed to start a new garden. There also is a tendency for these ambrosia fungi to be less invasive and destructive to the tree, instead staying near the galleries in which they are cultivated. Neither the fungi nor the beetles in these closer relationships can exist independently.

Another solution to digestion of plant matter is seen in some Cerambycidae and Anobiidae. Instead of using fungi to externally convert plant matter to digestible food, they rely on endosymbiotic yeasts and bacteria to accomplish the feat internally. Although yeasts and bacteria are common inhabitants of the gut in many insects, the relationship between some yeasts and beetles is one of obligatory symbiosis. Endosymbiotic yeasts may be harbored in the lumen of the gut, in diverticula (Fig. 45), or in specialized cells in the cytoplasm called mycetocytes. Clusters of

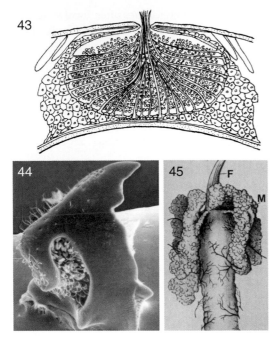

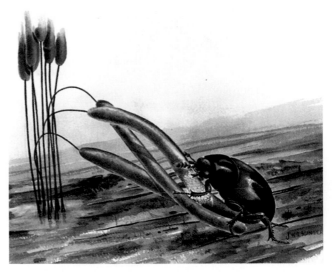

FIGURE 46 *Anisotoma basalis* (Leiodidae) feeding on a *Stemonitis* myxomycete fruiting body. [From McHugh, J. V., and Wheeler, Q. D. (1989). *Cornell Plantations Q.* **44(3),** cover figure. © Cornell Plantations.]

FIGURES 43–45 Beetle mycangia and mycetome. (43) *Scolytoplatypus* sp. (Curculionidae), transverse section of front part of adult pronotum, showing mycangial cavity filled with spores. [From Crowson, R. A. (1981). "The Biology of Coleoptera," p. 562, Fig. 286. Academic Press, London.] (44) *Eurysphindus hirtus* (Sphindidae), left adult mandible, dorsal view, showing spores of myxomycete inside dorsal cavity that is presumed to serve as a mycangium. [From McHugh, J. V. (1993). A revision of *Eurysphindus* LeConte and a review of sphindid classification and phylogeny. *Syst. Entomol.* **18,** 57–92. © Blackwell Science Ltd.] (45) Foregut (F) and anterior portion of midgut of *Lixus* sp. larva (Curculionidae), showing mycetomes (M). [From Crowson, R. A. (1981). "The Biology of Coleoptera," p. 562, Fig. 286. Academic Press, London.]

mycetocytes can form small organs called mycetomes. Yeasts may permit the breakdown of cellulose and provide various nutrients to their host. In the drugstore beetle, *Stegobium paniceum*, endosymbiotic yeasts are credited with providing riboflavin, niacin, pyridoxine, pantothenic acid, folic acid, and biotin.

To perpetuate endosymbiotic relationships, the gut of offspring must be charged with endosymbionts early in development. Yeasts are passed from adult beetles to larvae in various ways. The egg chorion may be inoculated with yeast so that the young are charged upon chewing out of egg and ingesting the chorion. In some Cucujidae, Silvanidae, Lyctidae, and Curculionidae, yeasts migrate into the egg within the female before the chorion is secreted. A third method of yeast transmission results following migration into the testes of the father. The yeast and sperm then enter the egg through the micropyle.

Formerly classified as fungi and studied by mycologists, the Myxomycetes are now recognized as protozoan animals. Despite their phylogenetic position, Myxomycetes are similar to fungi in some respects, and as a result beetle–myxomycete interactions share parallels with beetle–fungus interactions. In the plasmodial stage, Myxomycetes flow around their

environment, consuming bacteria. Rhysodine Carabidae and Cerylonidae feed, at least facultatively, on the plasmodial stage. When these colonial protozoans well up as plasmodia to form a sporocarp, they take on many funguslike features. This stage has attracted specialist beetles from no fewer than seven families: Leiodidae (Fig. 46), Staphylinidae, Clambidae, Eucinetidae, Cerylonidae, Sphindidae, and Lathridiidae. Pits in the mandibles of Sphindidae, an entirely myxomycophagous family (Fig. 44), and the venter of slime-mold-feeding latridiid species have been found to house myxomycetan spores.

Whereas other holometabolous insect orders such as the Hymenoptera and Diptera include parasitoid lineages of great diversity, the Coleoptera have not diversified to any great extent via parasitism on animal hosts. In addition to meloid and rhipiphorid hymenopteran parasites, parasitism of single host individuals has been infrequently observed. Aleocharine Staphylinidae parasitize the pupae of higher flies (order Diptera, suborder Cyclorrhapha). Within the Carabidae, the bombardier beetles, or Brachinini, parasitize the pupae of Gyrinidae, and species of the genus *Lebia* parasitize chrysomelid leaf beetles. *Lebia* beetles imitate various alticine flea beetle species with which they co-occur. The quick-jumping alticines are protected from predatory birds by their ability to disappear via a jump, suggesting that the *Lebia* have evolved a similar appearance through mimetic evolution. Coccinellid predatory larvae approach the specialization seen in some parasitoids, as some of the smaller species require only one to several homopterous prey individuals to complete larval development. Nonetheless, these species can switch prey species depending on the density of various hosts.

Platypsyllus castoris beetles of the family Leiodidae are specialists on beavers, with both the flattened, highly modified adults (Fig. 47) and the larval stages living in the animals' fur. Related leiodids in the subfamily Leptininae live

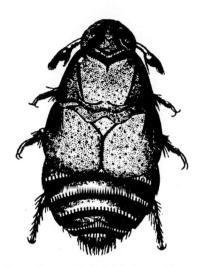

FIGURE 47 *Platypsyllus castoris* (Leiodidae), parasitic on beaver (*Castor* spp.). [From Crowson, R. A. (1981). "The Biology of Coleoptera," p. 549, Fig. 280. Academic Press, London.]

on the bodies of rodents, though they exhibit much less extreme body forms, and a lower level of host specificity, than the beaver beetles. The highly specific host relationship of *Platypsyllus* probably evolved from a more general predaceous habit. Such nest inquilines are found in a variety of lineages within the Staphylinidae, with adults and larvae variously preying on flea larvae or other nest-associated scavengers.

INTRASPECIFIC INTERACTIONS

The newly eclosed adult beetle faces the various tasks of dispersing from the pupal habitat, finding a mate, mating, finding a suitable larval habitat, ovipositing, and possibly guarding or facilitating the development of its young, all the while avoiding natural enemies. To beetles, flight may be a rare event. Many species undergo only a nuptial flight from the larval habitat to a new habitat, where mating and oviposition occur. Others may move from a breeding habitat to a drier microhabitat for overwintering, and thence back to the breeding habitat the next spring, making three flight periods in their lifetime. Others, such as the floricolous cerambycid long-horned beetles, buprestid jewel beetles, herbivorous Chrysomelidae, and Homoptera-feeding Coccinellidae may fly more or less continuously during their adult life span as they move from plant to plant. Beetle flight always requires the unfolding of the flight wings. Typically, beetles will climb some sort of prominence, use their antennae as "windsocks" and orient their body so that their initial liftoff is against the wind, and then open their elytra and unfold their flight wings prior to takeoff. During the nuptial flight, beetles are likely to be reproductively incompetent. In some scarab beetles, vast amounts of air are swallowed prior to flight, resulting in a distended gut unsuitable for feeding.

Mate finding may be facilitated by aggregation on host plants. The crushed leaves of host- and non-host-plant species are attractive to both male and female scarab beetles.

Adult emergence occurs as synchronous mass flights. Once near or on the host plants, female pheromones attract male scarabs. Ruteline scarabs utilize sex pheromones derived via fatty acid biosynthesis, whereas scarabs in the not distantly related Melolonthinae utilize amino acid derivatives and terpenoid compounds. The compounds of different classes are released from glands on different parts of the body: ruteline pheromones from epithelial cells lining inner surfaces of the abdominal apical segments, for example, or melolonthine pheromones from eversible glands on the abdominal apex. Pheromones used in the other beetle superfamilies span these scarab pheromone classes (e.g., terpenoids in the Curculionidae, fatty-acid-derived aldehydes and acetates in the Elateridae, esters in the Dermestidae).

Beetles use the other sensory modalities in mate finding, outdoing diversity observed in any other insect order. The anobiid deathwatch beetle acquired its ominous name through the predisposition of its males to bang their head capsules on host wood, telegraphically inquiring whether a receptive female is in the vicinity. This rapping was thought to foretell an imminent death. A males initially taps an average of five times, and if a female responds with a single tap, he moves a short distance and taps once. If he determines that the second female tap is fainter than the first, he turns at various angles to attempt to approach the female. Males receiving no returning female tap to their five-tap overture move greater straight-line distances between tapping bouts, searching greater expanses of wooden habitat for a responsive female (Fig. 48).

The use of light for mate finding has been evolutionarily refined in the Elateroidea, with the Lampyridae using flashing signals produced in abdominal light organs to engage in complex male–female dialogue before mating. These light organs are modified fat body cells with transparent outer surfaces, backed with highly reflective uric acid crystals. The light is highly efficiently produced via the oxidation of luciferin by the enzyme luciferase in the presence of adenosine triphosphate (ATP) and oxygen, producing oxyluciferin, carbon dioxide, and light. Male flashes are composed of species-specific series of flashes of varying duration, composition, and in some instances intensity. Males of different species fly in different patterns and at different heights, while females respond with a simpler flash that encodes species identity by the response delay to the male flashing sequence, by the flash duration, and in several species by a multiple-flash sequence. This sexual communication has been co-opted as a predation mechanism in *Photuris* fireflies. Males and females of these adult-feeding lampyrids use a typical male–female light dialogue to mate, whereupon the female's nervous system is affected so that she sends species-specific mating responses coded for sympatric, smaller *Photinus* species. *Photinus* males who venture too near the faux-*Photinus* female responses sent by the *Photuris* females are eaten.

Males and females may undertake various types of precopulatory behavior before mating. These may involve the sensing of species-specific alkene aphrodisiacs related to cuticular

A

Position of
female and end
of male path

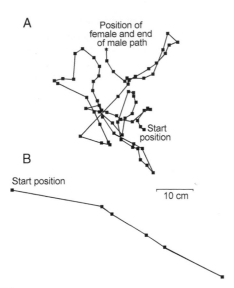

Start
position

B

Start position

10 cm

FIGURE 48 Paths taken by male deathwatch beetle, *Xestobium rufovillosum* (squares indicate positions at which male stopped to tap). (A) Female responded to male head tapping and was successfully found by male. (B) Male tapped in absence of any responding female. [Redrawn from Goulson, D., Birch, M. C., and Wyatt, T. D. (1994). Mate location in the deathwatch beetle, *Xestobium rufovillosum* DeGeer (Anobiidae): Orientation to substrate vibrations. *Anim. Behav.* **47**, 899–907.]

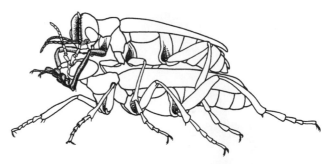

FIGURE 49 *Eupompha fissiceps* (Meloidae), male antennating female while rubbing his tarsi under her head; male antennae bring female's antennae alternately into his cephalic sulcus. [From Pinto, J. D. (1977). *Ann. Entomol. Soc. Am.* **70**, 937–952. © Entomological Society of America.]

hydrocarbons, as in aleocharine Staphylinidae. Males or females may stridulate as part of their behavioral repertoire. In Meloidae, a male will climb onto the dorsum of the female and antennate her head, palps, or antennae. In *Eupompha* meloids, the males draw the antennae of the females along a longitudinal sulcus on the male vertex (Fig. 49). Genitalic insertion by the male is successfully attempted only after antennation of the female. The passage of a nuptial gift of the highly toxic compound cantharidin has been incorporated into mating behavior in the pyrochroid fire beetles. In *Neopyrochroa flabellata*, the female samples an exudate from a transverse sulcus on the male vertex. If the exudate contains the terpenoid cantharidin (better known as the mammalian "aphrodisiac" Spanish fly), the male successfully mates, whereupon he transfers to the female, along with his sperm, about half the cantharidin stored in the accessory glands of his reproductive tract. The female translocates this cantharidin from her spermatheca to the developing eggs, which are thus chemically protected from predation by cantharidin-sensitive predators. Although meloids are known to produce cantharidin, transfer of this compound during meloid mating has not been documented. Conversely, although pyrochroids utilize this chemical in their mating behavior, they do not seem to be able to synthesize it, and the natural source of cantharidin that facilitates their behavior remains to be discovered.

Precopulatory behavior may involve more than a male and a female, especially in species in which male-specific structures have evolved in elaborate fashion. The enlarged male mandibles of stag beetles, Lucanidae, and prominent horns on the heads and pronotum of scarab beetles, are used by males to joust for advantageous mating sites with females. Dynastine and other scarabs seek out branches of shrubs and low trees upon which to mate. Males competitively maneuver for the top position on the branch, which is favored by females entering the fray for mating.

Copulation occurs with male dorsal to the female, the male grasping the female with the fore- and midlegs, and sometimes the mandibles, as in the tiger beetles or cicindeline Carabidae. The male aedeagus is inserted into the female gonopore. An aedeagal internal sac may be everted to place the male's gonopore near the entrance to the female spermatheca, and a spermatophore may be passed that encloses the sperm. Most beetles exhibit a monotrysian female reproductive tract: that is, the eggs pass out of the same structures used for copulation. However, in the dytiscid water beetles, a ditrysian configuration has evolved whereby copulation and oviposition occur via parallel, though connected, passages in the female.

The necessity for mating and copulation has been obviated in various groups of Curculionidae, Chrysomelidae, and Carabidae through thelytokous parthenogenesis. Species may be composed entirely of parthenogenetic populations, or such populations may be restricted to peripheral portions of the range. Parthenogenesis may also be associated with polyploidy, especially in weevils.

Although most mated female beetles oviposit into appropriate microhabitats where the larvae will develop, some families are characterized by eggs being laid in masses (e.g., Coccinellidae). Some tortoise shell chrysomelid females, Cassidinae, will lay eggs in a mass and then hover over the mass through hatching and the early days of the larvae. Female pterostichine carabids of the genera *Abax* and *Molops* similarly guard their eggs, although only until hatching. Females of several staphylinid species of *Oxyporus*, voracious mycovores with large sicklelike mandibles, have been reported to oviposit several eggs within a cavity in a soft mushroom, then stay with the larvae as they quickly develop to pupation over 3 to 6 days.

Ovoviviparity, or the holding of eggs until larvae hatch, has evolved several times across the Coleoptera. Typically it occurs in beetles occupying marginal environments dangerous to egg development. Chrysomelid females of montane or subarctic species hold developing eggs in the reproductive tract while basking on sun-drenched leaves to hasten egg development before larviposition. In *Pseudomorpha hubbardi* carabid beetles, females hold developing eggs until the larvae can be deposited, whereupon the larvae complete development as inquilines in an ant nest.

Male and female cooperative brood rearing has evolved repeatedly in various groups of Coleoptera. The long-known, and oft-revered dung-rolling Scarabaeidae provision nest burrows with rolled dung balls, upon which the eggs are laid. Females undertake this activity alone in some species, whereas the sexes work together in others. Burrows may be dug before dung balls are cut from mammalian dung pats, requiring navigation from the dung pat to a predetermined burrow location, or the burrow may be dug after the dung ball has been constructed.

In the Australian *Cephalodesmius armiger*, males and females pair up, with males actively foraging for decomposing leaves, flowers, fruit, and seeds, which are brought back to the female, staying in the nest. The female works the plant materials into a compressed ball, to which she adds her fecal material. The microbiological action of fungi from her feces causes fermentation in this external rumen after larval brood balls have been made from the mass. As the larvae develop, they feed on the brood ball from the inside out. When the thickness of the walls of the brood ball drops to about 2 mm, the increased volume of larval stridulations sensed by the female stimulates her to add more decaying material to the brood ball. Four to ten brood balls are made per nesting pair. When the larvae finish their development, the female seals the brood ball with a combination of larval and female feces, the larva having ejected its fecal material through cracks in the brood ball before pupation. Both parents die before adult emergence of their young. The new adult beetles feed on the walls of the brood ball, inoculating their gut with the fungi used by the mothers to produce fermentation in the external rumen.

Like the nest-building scarab, beetles of the silphid carrion beetle genus *Nicrophorus* raise their young on a concealed, highly desirable resource, a decaying carcass. Adult *Nicrophorus* actively fly long distances searching for a carcass. If a male discovers one, he emits a pheromone that attracts a female, with mating occurring on the carcass. Male and female then cooperatively bury the carcass by digging underneath it, and maneuver the corpse into a ball. Their activities isolate the corpse from competing silphids, and insulate it from microorganisms in the soil. After repeated mating, the female lays eggs in the surrounding soil. Upon hatching, the larvae crawl to the carcass, attracted by olfactory cues and adult stridulation. The adults precondition part of the carcass for larval feeding by chewing on it. They first feed the young larvae by regurgitating predigested carrion. Older larvae feed on their own, developing in 1 to 3 weeks, during which the female stays on the carcass. Upon maturation, the larvae crawl into the adjoining soil to pupate, and the female leaves

to search for a new carcass. Variations on this scenario include more than one pair of adults supported by a larger carcass, mated females raising their larvae alone when they discover a carcass without a resident male, and larger *Nicrophorus* species usurping a carcass through killing the original colonizing adults and their larvae.

True sociality, wherein more than one generation of adult lives together, and reproduction is restricted to a portion of the individuals, is likely to be rare in Coleoptera. This behavior has been reported only twice, and conclusive studies to completely document interactions among adults and larvae have not been fully documented for either family, represented by the wood-inhabiting bess beetles (Passalidae) and the ambrosia beetle *Austroplatypus incompertus* (Curculionidae: Scolytinae). There is no doubt that adults and larvae live together, and that fungi are passed from generation to generation. For this arrangement to qualify as eusociality, the existence of individuals that assist reproductives but do not themselves reproduce, at least during a portion of their life, must be documented.

INTERSPECIFIC INTERACTIONS

Beetles exhibit defensive behavior that is mostly rooted in the attributes of their cuticle. Many beetles living an exposed portion of their life cycle on vegetation will use the "drop-off" reflex if disturbed (i.e., simply close the legs and tumble off the leaf or branch and fall to the ground, where their often cryptic coloration helps protect them from visually oriented predators). The drop-off reflex can be combined with thanatosis, in which the beetle lies still with legs appressed to the body. Alternatively, the legs may be held at irregular positions by muscular tetanus (catalepsy), or the individual may roll up into a ball with the antennae, mouthparts, and legs hidden from view. More brightly colored species do not use the drop-off reflex. Chrysomelid flea beetles have enlarged hind femora containing strong tibial extensor muscles; a cuticular femoral spring releases the stored energy, catapulting them into the air.

Defensive chemical secretions that protect beetle adults from predators have evolved numerous times. Toluquinone is a defensive constituent common to several major terrestrial families (Carabidae, Staphylinidae, and Tenebrionidae), suggesting that this was one of the earliest defensive secretion types to have evolved. Since quinones are used in the tanning process of new cuticle, they would have been evolutionarily available in large quantities in well-sclerotized ancestral lineages of these families. Their tanning nature is not restricted to insect cuticle, as attested by the darkly stained fingertips of anyone who picks up an oozing *Eleodes* tenebrionid beetle.

Perhaps the most famous defensive chemical reaction in beetles is observed in the crepitating bombardier beetles of the carabid tribe Brachinini. These beetles, like other carabids, possess pygidial defensive glands that empty from the lateral edges of the intersegmental membranes between the seventh and eighth abdominal segments, Brachinine bombardier beetles plus carabid beetles of several other tribes (Metriini and Paussini) eject a combination of hydroquinones plus hydrogen peroxide held in one chamber of the gland, and catalases plus hydrogen peroxidase held in a second chamber. These chemicals combined result in an explosive ejection of hot (100°C) secretion, with liberation of the oxygen of H_2O_2, thus reducing hydroquinone to quinone, with the released O_2 propelling the spray (Fig. 50).

In addition to quinone compounds, beetles have evolved to use a variety of other defensive chemicals. The more recently evolved carabid beetle groups spray formic acid, a chemical also utilized as a defensive agent by their omnipresent antagonists, the ants, or Formicidae. Brightly colored or starkly patterned beetles are candidates for chemical protection via defensive gland secretions. The buprestid jewel beetles are often colored in black and yellow stripes to appear like the Hymenoptera with which they cohabit in various flowers. Jewel beetles are highly protected by bitter chemicals named buprestins. Not only are these chemicals distasteful to mammals (viz., organic chemists!), but ants reject sugar solutions laced with buprestins. Jewel beetles form mimetic complexes with lycid beetles, themselves protected by defensive secretions composed of various substituted parazines, reportedly among the most powerful odorous substances known (Fig. 51).

Various other beetle families regularly contribute members to lycid-based mimicry rings, including Cerambycidae, Meloidae, and Oedemeridae. Given that the meloids and oedemerids can synthesize cantharidin, it is likely that most beetles in such rings are distasteful, making Müllerian

mimicry the dominant basis for such common color patterns (Fig. 51). Other mimicry rings center on the dangerously toxic *Paederus* staphylinid beetles (Fig. 52), the cuticle of which exudes pederin. When such a beetle is scraped or crushed, contact with the pederin released results in human whiplash dermatitis (Fig. 53).

INTERACTIONS WITH HUMANS

Throughout history, humans have had diverse interactions with and perceptions of beetles. Coccinellid beetles were once perceived to have a close association with the Virgin Mary, hence their common name "ladybugs." Ancient Egyptians recognized dung beetles (Scarabaeidae) as a symbol of Ra, the sun god, because of parallels between the beetles' behavior and cosmic activities credited to the deity. Much as the scarabs rolled dung balls across the desert, Ra was thought to guide the sun across the sky each day. The symbolism of sacred scarabs has continued until today, as scarab images are still incorporated into jewelry, signifying good luck to the buyer or wearer.

The mystery and aesthetic beauty of beetles has been captured in paintings, sculptures, dances, poems, songs, and other art forms. Beetles have been used by many cultures for decoration. The brilliant metallic elytra of Buprestidae serve as natural sequins on textiles, and as biological gems in jewelry. In some cultures, beetle horns are included in jewelry because they are thought to increase sexual potency.

Live stag beetles (Lucanidae) are prized as pets in Japan, where a considerable amount of study has been given to their care in captivity. In Thailand the practice of "fighting" male

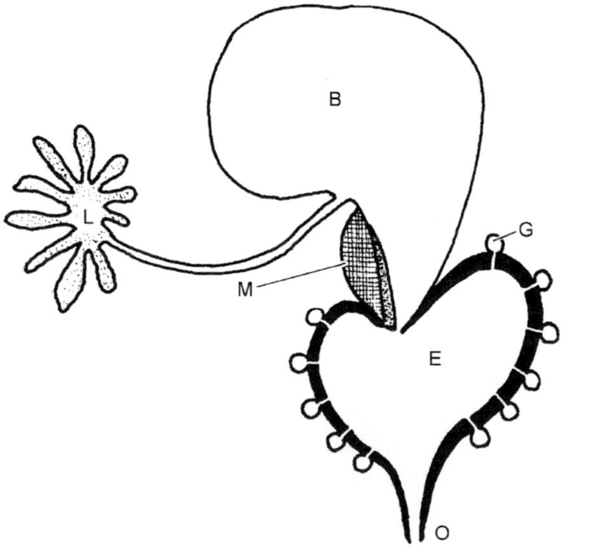

FIGURE 50 Cross section of pygidial defense gland of *Brachinus* bombardier beetle adult (Carabidae): L, secretory lobes; B, collecting vesicle; M, sphincter muscle; E explosion chamber; G, ectodermal glands that secrete catalase; O, outlet. Vesicle B contains mixture of hydroquinone and hydrogen peroxide, exploded by catalase, when it passes into E. [From Crowson, R. A. (1981). "The Biology of Coleoptera," p. 502, Fig. 265. Academic Press, London.]

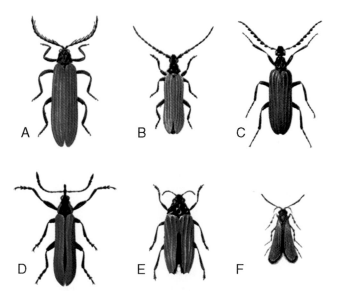

FIGURE 51 Five Australian beetles and a moth forming part of a mimicry ring: (A) *Metriorrhynchus rhipidius* (Lycidae), (B) *Eroschema poweri* (Cerambycidae), (C) *Tmesidera rufipennis* (Meloidae), (D) *Rhinotia haemoptera* (Belidae), (E) *Stigmodera nasuta* (Buprestidae), (F) *Snellenia lineata* (Lepidoptera: Oecophoridae). (Images provided by copyright holder, CSIRO Entomology, Canberra, ACT, Australia.)

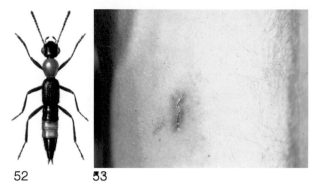

52   53

FIGURES 52–53 (52) *Paederus cruenticollis* (Staphylinidae) exhibiting warning coloration observed in many species of this genus. (Image provided by copyright holder, CSIRO Entomology, Canberra, ACT, Australia.) (53) Dermatitis linearis on human forearm at 66 h after an adult *Paederus* beetle had been crushed on volunteer's skin. (Photograph courtesy J. Howard Frank.)

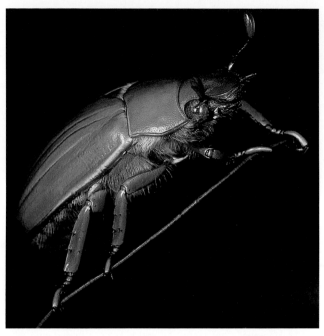

FIGURE 54 The jewel scarab, *Chrysina cusuquensis*, known only from a restricted fragment of forest in northern Guatemala. (Photograph courtesy of David Hawks.)

Hercules beetles (Scarabaeidae) is a traditional source of entertainment. With a referee controlling the action, two males are introduced into an arena. When a female is placed nearby, her mating pheromones trigger the combatants to engage each other. The match ends and a victor is declared when one male becomes exhausted or backs down from the advances of his opponent. In Central America local craftsmen blur the distinction between "pet" and "jewelry" by gluing rhinestones, glass beads, and a small chain to the dorsal surface of zopherid beetles. When the tiny chain is pinned to clothing, the tethered beetle becomes living jewelry.

Entomophagy, the eating of insects, is common in many parts of the world, and beetles often make up part of the menu. Larvae of palm weevils (Curculionidae) are considered to be a delicacy on the islands of the South Pacific. Similarly the fleshy, sausagelike larvae of various long-horned beetles (Cerambycidae) and scarabs are relished by people around the world. Mealworms, the larvae of some tenebrionid beetles, are easily reared and have become standard fare for culinary demonstrations of entomophagy.

Beetles attract the most attention when they become economic pests of agriculture, horticulture, and forestry. Two families, the snout beetles (Curculionidae) and the leaf beetles (Chrysomelidae), include many serious pest species. In the middle to late 1800s, the Colorado potato beetle, *Leptinotarsa decimlineata* (Chrysomelidae) abruptly expanded its range across North America and then colonized Europe and neighboring regions. Great efforts were made to thwart the invader each time it appeared, but ultimately the beetles succeeded. Throughout the 20th century an epic battle was waged against the notorious boll weevil, *Anthonomus grandis grandis* (Curculionidae), in the Cotton Belt of the southern United States, where it inflicted great financial losses. A sustained and coordinated effort to control this pest succeeded in eradicating the boll weevil from portions of several states by the turn of the millennium.

Predaceous ladybugs are often used in biological control to suppress populations of homopterous crop pests (i.e.,

aphids and scales). In the first successful biological control introduction, an Australian ladybug, *Rodalia cardinalis*, suppressed the cottony cushion scale (Hemiptera) on citrus crops in California. Phytophagous beetles have been employed to control weeds. In the 1960s the cattle-rearing industry in Australia faced a dilemma: because cows are not native to the continent, no natural bovine dung entomofauna was available to use their feces. Therefore cow patties persisted for months, during which time they served as breeding grounds for pestiferous horn flies. After careful study, Australian entomologists introduced South African *Onthophagus* dung beetles (Scarabaeidae). The measure was successful, and the problem quickly abated.

Perhaps the least appreciated human–beetle interactions are those in which human population pressure inflicts a negative impact on beetle populations. Coleopteran diversity is largely attributable to their specialization for particular geographic locales, microhabitats, and food. As human populations grow and people alter the Earth for their needs, destruction of spatially restricted resources is an inevitable result, leading to extinction of species associated with those resources. Ironically, a characteristic that helped Coleoptera to attain the astounding degree of diversity that it exhibits today also predisposes many beetle species to anthropogenic extinction (Fig. 54).

See Also the Following Articles
Boll Weevil • Cultural Entomology • Dung Beetles •
Hymenoptera • Japanese Beetle • June Beetles • Ladybugs

Further Reading

Arnett, R. H., Jr., and Thomas, M. C. (eds.). (2001). "American Beetles." Vol. 1. CRC Press, Boca Raton, FL.

Arnett, R. H., Jr., Thomas, M. C. Skelley, P. E., and Frank, J. H. (2002). "American Beetles." Vol. 2. CRC Press, Boca Raton, FL.

Beutel, R. G., and Haas, F. (2000). Phylogenetic relationships of the suborders of Coleoptera (Insecta). *Cladistics* **16,** 103–141.

Branham, M. A., and Wenzel, J. W. (2001). The evolution of bioluminescence in cantharoids (Coleoptera: Elateroidea). *Fla. Entomol.* **84,** 565–586.

Carpenter, F. M. (1992). Arthropoda 4. *In* "Treatise on Invertebrate Paleontology," Part R (R. L. Kaesler, ed.). Geological Society of America, Boulder, CO, and University of Kansas, Lawrence.

Choe, J. C., and Crespi, B. J. (eds.). (1997). "The Evolution of Social Behavior in Insects and Arthropods." Cambridge University Press, Cambridge, U.K.

Crowson, R. A. (1981). "The Biology of Coleoptera." Academic Press, London.

Elias, S. A. (1994). "Quaternary Insects and Their Environments." Smithsonian Institution Press, Washington, DC.

Evans, A. V., and Bellamy, C. L. (1996). "An Inordinate Fondness for Beetles." Holt, New York.

Farrell, B. D. (1998). "Inordinate fondness" explained: Why are there so many beetles? *Science* **281,** 555–559.

Lawrence, J. F. (1989). Mycophagy in the Coleoptera: Feeding strategies and morphological adaptations. *In* "Insect–Fungus Interactions" (N. Wilding, N. M. Collins, P. M. Hammond, and J. F. Webber, eds.), pp. 1–23. Academic Press, London.

Lawrence, J. F. (1991). Order Coleoptera. *In* "Immature Insects." Vol. 2 (F. W. Stehr, ed.), pp. 144–658. Kendall/Hunt, Dubuque, IA.

Lawrence, J. F., and Britton, E. B. (1994). "Australian Beetles." Melbourne University Press, Melbourne, Australia.

Lawrence, J. F., Hastings, A. M., Dallwitz, M. J., Paine, T. A., and Zurcher, E. J. (2000). Beetles of the World (CD-ROM, Windows version). CSIRO Publishing, Victoria, Australia.

Lawrence, J. F., Hastings, A. M., Dallwitz, M. J., Paine T. A., and Zurcher, E. J. (2000). Beetle Larvae of the World (CD-ROM, Windows version). CSIRO Publishing, Victoria, Australia.

McCormick, J. P., and Carrel, J. E. (1997). Cantharidin biosynthesis and function in meloid beetles. *In* "Pheromone Biochemistry" (G. D. Prestwich and G. J. Blomquist, eds.), pp. 307–350. Academic Press, London.

Collection and Preservation

Charles V. Covell, Jr.
University of Louisville

Insect collecting often begins in youth, when one discovers the love of making specimens for school, scouts, 4-H clubs, and other projects or as a fascinating pastime in its own right. The great diversity and numbers of insects, plus their rapid life cycles, usually mean insect populations can afford to give up some of their numbers and not be adversely affected by most collecting activities.

As one becomes engaged in various facets of insect biology as a researcher, the collection of specimens is important for taxonomic research, ecological studies, bioassessment and biomonitoring, and physiological and genetic studies. Because each labeled specimen is a historical record of that species' occurrence in time and place, proper methods of collecting, preparing, labeling, and storing are vital.

The general habitats, collecting equipment needs, and methods of collection and storage for the major insect orders and order groupings are presented in Table I. Below, a description of each type of equipment and its use are given. For more extensive illustrations and descriptions consult the books listed under Further Reading.

BASIC EQUIPMENT FOR COLLECTING INSECTS

1. Aerial net—A net bag made of translucent netting so one can see what's inside; it can be used as a beating net if needed. The net is used to grab insects off plants or to cover them on the ground. Since insects tend to crawl or fly upward, hold the net so they move toward its closed end once they are inside.

2. Beating (or sweeping) net—A heavy cloth bag, perhaps with small netted area at the bottom; it is used to sweep "like a broom" through vegetation many times. To use, strongly wave the net to concentrate insects in bottom of the net before placing net with insects into a killing jar until movement ceases. Then pick out what is desired and allow the rest to revive and go free.

3. Aquatic net—A heavy-duty metal hoop that can be D-shaped or round supports the netting. The former type is best for stream bottoms. The mesh and heavy cloth skirt have to be strong enough to take a beating. To use, hold the net against the bottom of the stream riffle and disturb the substrate upstream to allow insects to flow into net, or "work" the net among plants or debris to catch pond insects.

4. Malaise trap—A tent-like structure made of netting and designed to direct insects that encounter it to climb upward and follow the seams to a collecting container into which they fall. Container can be designed for live capture or killing in alcohol or by means of a dry poison such as cyanide powder.

5. Lights and light traps—A battery-powered light bulb such as a 15-W fluorescent "black light" or self-ballasted mercury vapor lamp can be hung from a tree limb or other support about one-half meter in front of a white sheet strung between two trees in the forest. The collector then picks the desired insects off the sheet. Various trap designs are available from supply houses, in which lights attract the insects that hit one of four vanes (or baffles) surrounding the bulb and above a funnel, into which the insects fall when they hit a vane. Ethyl acetate in tins with "wicks" of cloth provide a killing agent; crumpled paper also can be used in the bucket below for live capture.

6. Pitfall traps—Tin cans, jars, or pails can be placed in holes dug in the ground and filled with earth to the outside rims. One may bait with dead animal matter or other attractants. Ethylene glycol (antifreeze) is often used as a

TABLE I   Collection and Preservation of Insect Specimens for Insect Orders

Taxon	Habitat	Equipment to use	Collection method	Preparation
Protura, Diplura, and Collembola	Leaf litter, rotten logs and stumps, birds' nests, other detritus	Berlese funnel, aspirator, wet brush	Place in funnel for several days, jar of alcohol beneath, light above	70% EtOH, mount on microslides
Thysanura and Microcoryphia	Buildings (silverfish), leaf litter, logs, seashores	Forceps, Berlese funnel	Same as above	70% EtOH
Ephemeroptera	Naiads: streams, rivers, lakes Adults: fields and forests	Dip nets, grab samplers Aerial nets, light traps	Kick samples, pick off stones Pick off plants or from light sheet	70% EtOH
Odonata	Naiads: streams, lakes, ponds Adults: fields, near streams and ponds	Dip nets Aerial nets	Dredge or kick sample with net Sweep fast from behind with net	70% EtOH, place in envelope, wings folded over back, and card with collecting data; spread for display
Plecoptera	Naiads: streams Adults: along streams, at lights	Aquatic nets Light trap, aerial and sweep nets, light trapping	Kick-netting in riffles, pick off stones, sweep shore vegetation	70% EtOH
Orthoptera and other orthopteroids	Fields, forests, gardens, and other terrestrial habitats	Sweep nets, light traps, aerial nets, hand capture	Sweep and aerial netting, light trap sampling	Mount on insect pins, support body until dry
Hemiptera, Homoptera, and other hemipteroids	All terrestrial habitats	Sweep nets, beating sheet, examine plants, light traps	Sweep and aerial netting, light trap sampling	Pin large bugs, small ones on card points or store in 70% EtOH, scales on microslides
Phthiraptera	Avian and mammalian hosts	Forceps, aspirator	Scrape fur and feathers	70% EtOH, mount on microslides
Thysanoptera	Plant axils, flower parts, and other plant parts	Aspirator	Examine plants and aspirate	70% EtOH, mount on microslides
Neuroptera and Megaloptera	Larvae aquatic (mostly streams) or on plants	Aquatic nets, sweep nets and light traps	Kick sampling in riffles, sweep vegetation, examine trap samples	70% EtOH or pin
Coleoptera	All habitats	Aquatic, aerial, and sweep nets: light, malaise and pitfall traps	Bait pitfall traps with rotting animal flesh, other methods as above	Pin or mount on card points
Mecoptera	Woodland glades, understories	Sweep and aerial nets, light trap	Follow and net individuals seen, use light trap (Meropeidae)	Pin or place in 70% EtOH
Lepidoptera	All habitats, esp. fields and woods	Aerial net, sweep net, bait, malaise and light traps	Net resting butterflies, bait traps with rotting animal flesh and excrement or fermenting fruit, sweep or examine plants for larvae	Relax, and then spread on spreading boards, use 70% EtOH or special fluids for larvae
Trichoptera	Running water, esp. streams for larvae Adults may be near or far from breeding sites	Aquatic, sweeping, and aerial nets Malaise and light traps	Kick samples for some larvae, others must be picked in cases off rocks in stream Adults come to lights or can be swept from streamside vegetation	Store all stages in 70% EtOH
Diptera	All habitats; larvae most common in aquatic or moist habitats in water and land or animal hosts	All kinds of nets, dippers, light traps, malaise traps	Examine plant and animal hosts, capture in net, traps	Pin, place on card points, or store in 70% EtOH
Siphonaptera	Bodies and nests of birds and mammals	Aspirator or moistened brush, sweep net	Comb animal, break up nest over white background, sweep grassy areas around infested buildings	Place in 70% EtOH, mount on microslides
Hymenoptera	All terrestrial habitats	Nets, all trap types	Collect from flowers, sweep, extract from light, malaise, pitfall, and other traps	Mount on pins; on card points or in 70% EtOH if small

killing agent. Walls of boards can also be erected narrowing to the opening of the pitfall to direct arthropods to the pit.

7. Beating sheet—A square of bed sheet or similar white cloth placed under a bush or tree to catch insects when they are knocked off after the plants are struck with a large stick, such as an axe handle. Insects are then collected by aspirator or forceps.

8. Aspirator—A tube plugged with a rubber cork in which are inserted two tubes: one bent and used to point at tiny insects; the other connected to a rubber tube for inhaling quickly to suck the insect into the tube. The latter one has a tiny screen attached to the inside end to prevent insects from getting into one's mouth.

9. Berlese funnel—A commercial funnel of any size is needed, equipped with a screen inserted just above the narrow spout to prevent material from falling out. Leaf litter, birds' nests, and other organic matter are put into the funnel, which is mounted on a rack or ring stand. A light bulb is placed over the top to dry out the organic material, driving arthropods downward as they seek moisture. Insects then fall through the screen and into a jar of 70% alcohol placed under the spout. The Berlese funnel is left in place until the organic matter is completely dried out.

10. Relaxing box—A tight container (plastic refrigerator boxes are excellent) is chosen in a size needed. Cut or fold paper toweling to line the bottom of the box at least 1 cm deep. Moisten the paper thoroughly with water, but leave no water standing. Add a small amount of an anti-mold chemical such as paradichlorobenzene or carbolic acid (phenol). Place a piece of stiff cardboard above the wet paper as a platform for the specimens. Freshly killed insects, or dried ones you wish to pin or spread, can be softened in the box. If left in the "relaxer" too long, however, they may mold or turn mushy and disintegrate.

11. Killing jar—A glass or plastic jar of desired size can be made into a killing jar by putting about a 1-cm layer of plaster of Paris in the bottom, or use just a pad of absorbent material such as cellucotton, cotton, or soft tissue. A fluid killing agent such as ethyl acetate or fingernail polish remover containing acetone is added to be absorbed by the plaster or other material. Be sure not to have any fluid on the walls of the jar, or specimens will be spoiled. If you use cotton or other absorbent material, cut a cardboard disk to separate the insects from the pad of killing agent.

METHODS

Insects are prepared for study and storage in three basic ways: pinning, fluid storage, and mounting on microslides. Adult insects or the immature forms of hard-bodied insects such as those with incomplete metamorphosis are pinned through the thorax of the body, unless too tiny, and then they are mounted on card points (see later). Insect pins, available from supply houses, are long and very sharp. They range from tiny headless "minuten nadeln" for mounting specimens on tiny blocks of foam, which in turn are put on regular insect

pins, to pins that are numbered to match the general size of the insect. Size 000 is the smallest made and bends very easily. Most small insects that can be pinned are at least 5 mm in length, with a thorax big enough to hold the pin. Most medium and large insects are pinned on sizes 1 to 3. Sizes 4 to 7 are sometimes available for large specimens.

Preserving Insect Specimens in Fluid

Insects that are too small, or the bodies of which are too brittle or soft, should not be pinned. They should be stored in glass vials in 70% ethyl alcohol (EtOH). Other special fluids, especially those that preserve colors, can be learned from the works under Further Reading. Actually collecting in alcohol can be done using traps of any type (light, malaise, pitfall, and some bait traps). The larger insects can be dried out later and pinned. However, collecting in fluid is NOT recommended for collecting Lepidoptera (butterflies and moths) or Culicidae (mosquitoes) because they have patterns formed of colored scales and those may be ruined by the fluid.

Vials used are often of the "patent lip" type with neoprene stoppers. The author prefers to use 4-dram vials with size 0 stoppers and store them in plastic racks and cardboard boxes with partitions available from supply houses. A better alternative is the screw-cap vial, which should be equipped with "polyseal" plastic sealing inserts. One of the biggest problems with liquid-stored specimens is the drying out of the fluid. I believe the latter storage to be superior because the alcohol does not discolor with years nor does the cap change shape (stoppers swell or stick to the glass).

Many tiny insects such as lice, fleas, and thrips can be stored in EtOH until such time as they can be made into permanent microslide mounts with Euparol, Canada balsam, or some other mounting medium.

Pinning Insect Specimens

1. Be sure the insects to be pinned are soft enough so that they will not crumble when you handle them and attempt to pierce them with the pin. These can be just-caught, or they can be softened, if dry, in a relaxing box (see earlier).

2. Select the pin and pierce the high point of the thorax with the point. Push the pin straight through the thorax. Check straightness by observing from front and side to see if the pin is perfectly perpendicular to both the transverse and the longitudinal axis of the insect.

3. When the insect, such as a grasshopper, has a middorsal ridge in the thorax, pin just to the right of the ridge.

4. For beetles, insert the pin in the right elytron (front wing) close to the midline. Do not pin beetles through the prothorax.

5. Push the pin on through when you are satisfied with the position. One-third to one-fourth of the pin should be showing above the insect's thorax.

6. If the abdomen or legs are drooping, push the pinned insect into a block of foam plastic or a cardboard box to

support these parts until they are dry. Then remove the insect and label it.

7. Most museum specimens do not have legs and antennae adjusted to a life-like position when they are pinned. However, for display purposes or personal satisfaction one may move these body parts into desired positions on the foam or cardboard support and fix them temporarily with pins over or against them.

Placing Insects on Card Points

A card point is a small wedge of high quality (100% cotton content) cardstock, punched from the sheet with a special punch obtainable from a supply house. There are several different shapes, but the author prefers the ones with the wide end rounded.

Card pointing is used for tiny insects that are hardbodied enough not to lose shape when dried. Size usually ranges from 1 to 5 mm or slightly larger in length. The author normally selects from large samples of dried specimens collected in sweep samples or light traps.

1. Punch out a number of card points. Place them on top of a firm foam plastic or cardboard surface.

2. Push the point of an insect pin into the wide end a short way from the very end, and push the card point up the pin by inserting the pin with the card point into the top hole of a 3-step pinning block (wooden block with three fine holes of different depths to provide uniform heights of labels on pins) and pushing the point up until it stops. It should be about 1/3 the distance from the top of the pin.

3. Use forceps to turn the very tip of the card point downward at a right angle to form a vertical surface.

4. Put a tiny dab of glue on the vertical surface you have made with the forceps. When doing a number of specimens, put a small drop of glue on a piece of card or paper to use (although it will tend to harden on the surface after a minute or two).

5. Position the insect so that the right side of the thorax is accessible, and touch the glue-covered surface of the card point to the right side of the thorax. (The insect should appear to be "holding onto the card point with its right hand"). Use forceps to position the insect firmly against the glued surface and have it positioned so that its orientation to the ground is as it would be in life.

6. Fill out your insect label with locality, date, and collector's name. Trim it to be as small as possible (avoid large, oversized "barn door" labels). Labels should be printed on 100% cotton light card stock in permanent black (India) ink or can be done on a postscript laser printer.

7. Position the label on the pin and push it up the pin at the middle hole of the 3-step pinning block. As you read the label, the card point and insect should be projecting to the left of the pin shaft. Make sure both card point and label are not tilted or crooked.

8. Place the specimen in a temporary holding unit tray until it can be identified and put in the collection. Identification labels should be affixed below the collecting data label and in a position so that both labels can be read from the same angle. The lowest step on the 3-step pinning block is normally used for the identification label.

INSTRUCTIONS FOR SPREADING BUTTERFLIES AND MOTHS

1. Have all needed items ready: well-softened specimens (stored in relaxing box or freezer after collecting), spreading boards of proper sizes, straight strips of tracing or waxed paper or other material, setting needles or picks, insect pins for specimens, glass-headed pins for holding paper in place (insect or dressmaker's pins okay). See Fig. 1.

2. Fix paper strips along the side boards of spreading board, slightly back from the notch to allow you to work the wings into place. Use two or three pins at the top of board to hold the paper even down its length.

3. Push a proper sized insect pin straight down through the thorax of specimen, so it is not tilted in any direction. Push the pin far enough that the top of the thorax is one-third to one-fourth the distance down from the pin head (Fig. 1A).

4. Push the pin down into the soft material in the notch of the spreading board so that it is not tilted in any direction. Also, push it far enough that the wings, when out straight to the side of the insect, rest flat on the side boards of the spreading board. Be sure you do not place the insect too close to the top of the board (leave room to pull wings into proper position).

5. Push an insect pin down along the left rear of the thorax, behind the base of the left hind wing, to keep the body from swinging left as you position the wings.

6. Place paper over the wings. Hold the left-hand paper strip in the thumb and forefinger of your left hand while you now begin to position the wings.

7. Insert a sharp insect pin or setting pick behind the costa vein close to the base of the left forewing. Swing that wing upward until the inner (anal) margin is at a right angle to the plane of the body (Fig. 1B). Be sure not to let the hind wing pop out from below the forewing. Insert a glass headed pin into the paper above the costa near the base and inner margin near the anal angle to hold the wing secure.

8. Pull the left hind wing forward by inserting a setting pin or pick behind the radial vein near the wing base, and swinging it forward. Leave a small triangular space between the outer margin of the hind wing and the inner margin of the forewing. Fasten paper over the left hind wing by putting a pin below it near the wing base.

9. Repeat procedures 6 to 8 on the right side, and be sure you have produced symmetrical results (Fig. 1C).

10. Position the antennae with pins to look as shown in Fig. 1. The abdomen may need to be supported with crossed pins beneath it or held down straight with crossed pins above it.

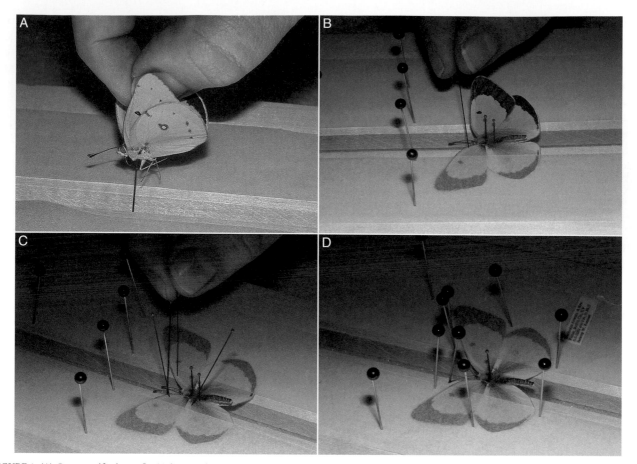

FIGURE 1 (A) Orange sulfur butterfly *(Colias eurytheme)* with insect pin inserted at proper height, ready to place in groove of spreading board. (B) Insect pin inserted behind thick costa margin and pulled forward so that inner margin of forewing is at right angle to groove. (C) After left hindwing is pulled forward and secured, right forewing and hindwing positioned to match left. (D) Glass-headed pins in proper position to hold tracing paper tight for at least one week, until the insect dries and can be removed; label ready to add.

11. Write data (where, when, and by whom collected) on the paper strip holding down the wing or make a label and tuck it under the paper strip until the specimen is taken off the board (Fig. 1D).

12. Add other specimens below, as close together as you can, if you have many specimens to spread.

13. Make a notation of the date of spreading on the paper strip to remind you of how long the specimens have been on the boards.

14. Store the board in a pest-free, dry place such as a steel or wood cabinet. Fumigation of the storage enclosure is recommended.

15. Allow specimens to dry for at least a week, longer if possible. If the abdomen is completely dry and stiff, the specimen should be ready to remove.

STORAGE OF SPECIMENS

Specimens that you would normally pin or spread after pinning can be placed in envelopes. This is known as "papering." Glassine stamp envelopes are excellent, but any kind will do. To make triangular paper envelopes, cut rectangles of paper, one side about a half-inch longer than the other. Fold into a triangle and then fold down the remaining "flaps" after putting the insects inside. Be sure butterflies and moths have wings folded over their backs for best results. They can be softened in the relaxing box at any later time. Don't forget to put collection data on the envelope.

Storage of pinned and papered specimens must be in tight containers so that museum pests such as Dermestidae (carpet beetles) and booklice (Psocoptera) cannot get to them. These can also be repelled by fumigants such as napthalene (moth flakes or moth balls), PDB (paradichlorobenzen), or dichlorvos-impregnated "strips" cut into blocks. However, the trend is away from museum fumigants because of possible health problems from exposure to them. The better method is freezing. Whole boxes can be left in a freezer for a few days on an annual basis to kill any pests that may have entered.

Drawers and boxes housing pinned specimens must have tight-fitting lids with inner flanges higher than the outer walls of the unit. Thus, a tight seal can be achieved, which usually keeps pests out. Equipment dealers offer high quality "Schmitt" boxes and standard cabinet drawers of different dimensions (Cornell, U.S. National Museum, and California Academy types are most common), as well as cabinets to house them. Homemade boxes and cigar boxes will do in a pinch; just add a foam plastic lining. However, one cannot expect such boxes to be pest-proof without fumigation.

Vials with alcohol-preserved specimens and microscope slides can be stored in special boxes or cabinets also available from dealers or built yourself.

See Also the Following Articles

Museums and Display Collections • Photography of Insects • Population Ecology

Further Reading

Anonymous (1992). "How to Make an Insect Collection." BioQuip Products, Inc., Gardena, CA 90248.

Borrer, D. J., Triplehorn, C. A., and Johnson, N. F. (1992). "Introduction to the Study of Insects," 6th ed. Harcourt Brace, New York.

Borrer, D. J., and White, R. E. (1970). "A Field Guide to the Insects." Houghton–Mifflin, Boston.

Covell, C. V., Jr. (1984). "A Field Guide to the Moths of Eastern North America." Houghton–Mifflin, Boston.

Martin, J. E. H. (1977). "The Insects and Arachnids of Canada," Part 2, "Collecting, Preparing and Preserving Insects, Mites and Spiders." Biosystematics Research Institute, Ottawa, Ontario. [Publication No. 1643]

Merritt, R. W., and Cummins, K. W. (1996). "An Introduction to the Aquatic Insects of North America," 3rd ed. Kendall/Hunt, Dubuque, IA.

Opler, P. A. (1998). "A Field Guide to the Butterflies of Eastern North America." Houghton–Mifflin, Boston.

White, R. E. (1983). "A Field Guide to the Beetles of North America." Houghton–Mifflin, Boston.

Winter, W. D. (2000). "Basic Techniques for Observing and Studying Moths and Butterflies." Lepidopterists' Society, Natural History Museum, Los Angeles, CA 90007-4057.

Collembola
(Springtails, Snow Fleas)

Kenneth A. Christiansen
Grinnell College, Grinnell, Iowa

Peter Bellinger
California State University, Northridge

Collembola or springtails comprise one of the most widespread and abundant groups of terrestrial arthropods. They are found everywhere, to the utmost reaches of multicellular animals in the Antarctic and Arctic and in all habitats except the open oceans and deep areas of large lakes. These all-wingless hexapods range in adult size from 0.4 to

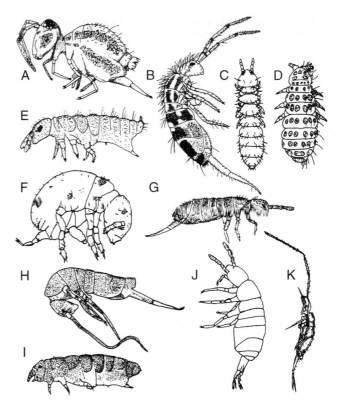

FIGURE 1 Variety of Collembola forms (not to scale). (A) Sminthuridae. (B) Entomobryidae. (C) Onychiuridae. (D) Neanuridae. (E) Hypogastruridae. (F) Neelidae. (G) Isotomidae. (H) Tomoceridae. (I) Odontellidae. (J) Oncopoduridae. (K) Paronellidae.

over 10 mm. Their small size generally results in their being overlooked, but they display an enormous range of body forms (Fig. 1), habitats, and habits. While most feed on fungi, bacteria, and decaying vegetation, some are carnivores, others are herbivores, and a number are fluid feeders. There are many commensal but no parasitic forms. They are most common in soils and leaf litter, but many species live in vegetation, littoral and neustonic habitats, caves, and ice fields or glaciers. Collembola have been classified with the insects but are now generally considered to belong to an order closely related to the Diplura and Protura. There are approximately 9000 described species belonging to about 27 families (Table I).

ANATOMY

All Collembola are primitively wingless hexapods. All have three thoracic segments and six or fewer abdominal segments, including a telson consisting of a dorsal and two ventral valves surrounding the anus.

There are typically four antennal segments, each with musculature (this distinguishes them from true insects, with three, and Diplura, with many antennal segments). Collembola vary enormously in form and somewhat in internal anatomy,

TABLE I   Families and Numbers of Species of Collembola

Family	Number of species
Suborder Arthropleona	
Hypogastruidae	800
Odontellidae	150
Brachystomellidae	130
Neanuridae	1500
Onychiuridae	800
Poduridae	1
Isotogastruridae	5
Isotomidae	1500
Coenaletidae	2
Actaletidae	9
Entomobryidae	1800
Microfalculidae	1
Paronellidae	450
Cyphoderidae	185
Oncopoduridae	50
Tomoceridae	200
Mackenziellidae	1
Suborder Symphypleona	
Sminthurididae	170
Katiannidae	350
Sturmiidae	1
Spinothecidae	8
Dicyrtomidae	220
Bourletiellidae	300
Sminthuridae	270
Suborder Neelipleona	
Neelidae	30

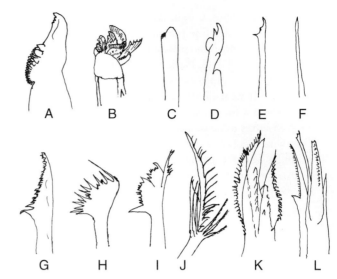

FIGURE 3   Collembolan mouthparts. (A) typical mandible and (B) maxilla; (C) reduced mandible and (D) maxilla of Cyphoderidae; (E) piercing and sucking mandible and (F) maxilla of *Neanura*. (G–I) Various mandibles of Neanuridae and (J–L) various maxillae of Neanuridae.

but all lack Malpighian tubules and most have paired labial nephridia that empty into the ventral groove at the base of the labium. One universal and unique feature is the ventral tube or collophore (Fig. 2)—a distally weakly paired projection from the first abdominal segment with membranous, sometimes eversible, distal margins. Probable functions include imbibition, excretion, respiration, and adhesion to smooth surface. Collembolan mouthparts are said to be entognathous, being concealed by the head capsule, and typically adapted for chewing. The mandible usually has apical teeth and a molar

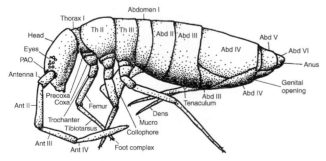

FIGURE 2   Typical Collembola anatomy.

plate, and the maxilla varies greatly and bears a number of complex lamellae.

In some Neanuridae and a few other groups, the mouthparts are simplified and the mandible may be lost (in connection with adaptation for specialized, including liquid, diets). In other Neanuridae the mandibles and maxillae show an inexplicable complexity (Fig. 3) and diversity of form equal to that seen in any other order of insects. The mouth opening is connected to the anterior surface of the ventral tube by a ventral groove through which fluids may flow.

Collembola are equipped maximally with 8 + 8 ommatidia but often have a supplementary light sensory organ between the antennae on the dorsum of the head. A few Collembola possess rudimentary trachea; however, respiration is normally through their thin cuticle and the membranous surface of the ventral tube. The reproductive system consists of paired ovaries or testes opening on the venter on the fifth abdominal segment. Collembolan legs consist of one or two apparent subcoxal segments, a coxa, femur, trochanter, fused tibiotarsus, and distal, normally four bladed, unguis. An opposable smaller lamellate unguiculus is usually present.

Most Collembola have a forked ventral jumping apparatus or furcula on the fourth abdominal segment, consisting of a single basal manubrium and paired distal dentes and mucrones. It is held in place by the latch-like tenaculum on the third abdominal segment. When the tenaculum releases, the furcula catapults the animal, as much as 10 cm. All Collembola are covered with setae but their number, size, and structure vary greatly from group to group. The cuticle of Collembola is extremely varied and often has elaborate surface structures.

FOSSIL HISTORY

The first fossil Collembola occur in the 400 million-years-old Rhynie chert deposits of the Devonian, although there are secondary fossil hints of earlier Collembola occurrence. These fossils display very modern collembolan features, including typical entognathous, chewing mandibles; ventral tube; and, probably, a furcula. The single described species—*Rhyniella praecursor*—has been placed in a variety of families, including recently Isotomidae; however, all family placement must be considered very tentative and it is likely that one or two additional species are in this deposit. A single specimen of a very probable member of the family Entomobryidae was found in Permian shale of South Africa but extensive collembolan fossils are limited to amber of the Cretaceous, Oligocene, Miocene, and Pliocene. Collembola represent only a small fraction of the hexapods found in amber, and they are absent from many amber deposits; however, there are over 70 specimens from late Cretaceous Canadian amber, over 160 from the Baltic Eocene amber, about 130 from Miocene amber of Chiapas and the Dominican Republic, and 16 from Pliocene Japanese amber. The Cretaceous material has no identifiable specimens of extant genera and most specimens can be placed in one of eight extinct genera. All the remaining amber specimens can be placed in extant genera and in a few cases in extant species. Since the Eocene, generic extinction appears to have been absent, a unique feature among hexapods well represented in Eocene deposits.

VARIETY OF BODY FORM

Although the generally considered primitive Collembola (Fig. 2) display most of the features described above, most genera differ from this. All families have some forms with reduced numbers of eyes, and Neanuridae, Hypogastruridae, and Isotomidae (Figs. 1E and 1G) often have reduced or no furcula. The Neanuridae (Fig. 1D) often have large spines on the body as well as spectacularly complex mouthparts. Indeed these are so complex and varied (Fig. 3) that species can be identified by their mouthparts alone. The Onychiuridae (Fig. 1C) all lack eyes and almost all lack pigment and a furcula. They are characterized by the presence of pseudocelli through which defensive toxic and/or repulsive fluids are secreted. These along with the Hypogastruridae, Poduridae, and Neanuridae have well developed, seta bearing, first thoracic segments; the remaining families all have greatly reduced, nonsetaceous, first thoracic segments (Fig. 2), and some families have fusion of abdominal segments. The Neelidae and Sminthuridae have the first four abdominal segments fused and more or less fused with thoracic segments. Some Entomobryidae and Sminthuridae (as well as most Tomoceridae) have antennal subsegmentation, giving the appearance of more than four antennal segments. The largest species are found in the Neanuridae, Entomobryidae, and Tomoceridae, often reaching 5 mm and occasionally over

10 mm in length, but the Neelidae and Mackenziellidae rarely reach 1 mm.

HABITATS AND HABITS

Most Collembola in temperate and arctic zones live in the soil or ground litter, but there are several groups, most notably the Sminthuridae, that largely inhabit vegetation. In tropical regions Collembola are abundant in trees and epiphytic plants. In rain forests, they are rare in soils but abundant in trees. Collembola are abundant in many caves and are frequent in marine littoral zones. They are also common in the interstitial sand regions of marine beaches and the surface of standing fresh water. In all these examples there are many species specialized for these habitats. Collembola have recently been discovered at depths up to 20 m in both fresh and salt water, but nothing is known of the habits of such forms. Many species are found in bird and mammal nests, and microcavernicole habitats are frequently exploited but such forms show no particular specializations, being also found either in litter or in soil habitats. Ant and termite nests are frequently occupied, and one family, the Cyphoderidae, consists largely of species limited to and highly adapted for life in these habitats. Some of the most striking examples of presumed commensalism occur in the genus *Axelsonia* (Isotomidae), of which one species lives in the gill chamber of land crabs, and in the family Coenaletidae, of which all species are confined to the shell of terrestrial hermit crabs.

The forms living in the different habitats often display a suite of morphological characteristics correlated to their habitat. Thus, forms that have reduced furcula, reduced or no eyes, weak pigment, and reduced pointed tenent hairs are characteristically found in soil. Forms with no eyes or pigment; well-developed furcula; elongate, slender untoothed ungues; and reduced, pointed tenent hairs (troglomorphic) are almost always cave dwelling. Almost all species with well-

FIGURE 4 Collembolans discovered in various habitats in Reading, UK. (A) *Podura aquatica* (Poduridae), from the surface of a garden pond. (B) *Kalaphorura burmeisteri* (Onychiuridae), from soil. (C) *Dicyrtoma fusca* (Dicyrtomidae), from leaf litter. (D) *Entomobrya nicoleti* (Entomobryidae), under surface debris. (Photographs by Steve Hopkin.)

marked color patterns and well-developed furcula are either litter or vegetation dwelling (Fig. 4).

REPRODUCTION AND DEVELOPMENT

Fertilization is internal; however, exchange of sperm occurs in a variety of fashions. Sexual receptivity is associated with adult molting and in some species pheromones to facilitate aggregation of sexes. The sperm is produced in a packet, often with a stalk holding it above the substrate. In some groups (most Onychiuridae) these packets are produced randomly and fertilization occurs by accidental contact of the female with the packet of sperm. In a number of species the packets are produced only in the presence of females, but the most elaborate procedures are seen in *Podura aquatica* and the Sminthuridae. Here, often, there are elaborate courtship and maneuvering associated with fertilization. This is often accompanied by modifications in male anatomy, which ensure the appropriate species response and/or positioning for sperm packet uptake. Most of these species are brightly colored and patterned, which may also be associated with species recognition. In these forms, sexual dimorphism is the rule and often extreme. This is also true of many marine littoral species, but in these, the method of sperm transfer is still unknown and the function of the dimorphic structures (usually male) is unclear.

In some members of the family Isotomidae secondary sexual characters alternate with molts, being expressed in stages in which the animals are sexually receptive and not expressed in stages in which they are not receptive. In most Collembola there is little or no sexual dimorphism and sexes can be separated only by the difference in their genital openings. Both males and females occur in most species but parthenogenesis is common, especially in some genera of the Tullbergiidae.

Development is direct, with the young generally very similar to the adults except for the absence of sexually associated features and body ratios and some aspects of the setae clothing. The main exception to this generalization is in the Tomoceridae, whose juveniles have been assigned to genera different from those of the adults. Collembola continue to molt after reaching sexual maturity and some species can molt very large numbers of times (the record is 52). They stop reproducing at some point and later molts result in reduced rather than increased size. Although some Collembola have been known to live more than 5 years in captivity, their life span in the wild is undoubtedly much shorter.

UNUSUAL FEATURES

One remarkable feature of some members of the family Onychiuridae is that some male-only specialized setae on the venter of the abdomen achieve full development only several molts after sexual maturity. Their function is unknown.

Many species of Collembola, almost entirely of the families Isotomidae and Hypogastruridae, go through a period of reduced activity, wherein they develop a unique morphology, often associated with the development of heavy abdominal spines and wrinkled surface and reduced mouthparts and digestive systems. When this is associated with particular ecological conditions (most commonly drying or elevated temperature), it is termed ecomorphosis: feeding ceases and the structural changes are usually striking. The cessation of the causal conditions results in a quick molt and return to normal anatomy and activity. When these conditions are part of a regular cycle the process is called cyclomorphosis.

A number of Collembola are also capable of anhydrobiosis, that is, they can become completely dry without dying. In some (but not all) instances these animals forms small ball-like capsules around themselves before entering this state. If wetted, the animal resumes normal activity in an hour or two. Recent studies with sand dune Collembola suggest that this capacity may be more widespread than currently established. Another unusual feature of Collembola is the ability of some species to live very long periods without food. This characteristic appears to be best developed in some cave forms, and in several instances animals reproduced after not being fed for 30 weeks. The longest survival was a specimen of *Onychiurus,* which lived over a year without food and was then accidentally killed.

ECOLOGY AND ROLES IN ECOSYSTEM

Because Collembola are found in all habitats, from the coldest to the hottest supporting multicellular life, and from treetops to the deepest soil layers supporting multicellular animals, it is clear that their responses to various abiotic conditions must vary enormously. Humidity is usually the most important factor in determining Collembola distribution. High humidity is seldom a problem for Collembola but desiccation is often serious. Collembola resist desiccation by moving into microenvironments of high humidity (under stones or into deeper soil layers) and/or limiting activity to nights and by morphological adaptations (such as cuticular thickening, ornamentation, and scales). Some species, as already discussed, change form radically and cease feeding, while others go into anhydrobiosis. Many species lay eggs that are much more resistant to drying and they survive desiccation in this stage, often accompanying this with short postembryonic life cycles.

Collembolans have vastly different temperature tolerances and preferences, ranging from a species of *Sminthurides* found in volcanic vents with temperatures as high as 48°C to an Antarctic species shown to survive temperatures below −30°C. Survival (and activity) in low temperatures has been studied extensively. Some Collembola are primarily inhabitants of glaciers and ice fields and others are dominant members of the arthropod faunas of high latitudes. Winter-active Collembola in temperate climates often build up large numbers under snow and on suitable warm days pour out onto the snow in vast numbers as snow fleas. Extreme cold tolerance always involves supercooling with the accumulation of cryoprotective substances.

Oxygen requirements of Collembola also vary enormously. The greatest tolerances discovered are in the Antarctic *Cryptopygus antarcticus,* which has a 30% survival rate after 30 days in pure nitrogen atmosphere. In many Collembola, respiration when submerged is via air films surrounding the animals as a result of their hydrophobic cuticle, but this apparently not necessary in all forms. In many forms the eggs are more resistant to immersion than in other stages.

Collembola, even in uniform soils, are never randomly distributed, but show strong clumping because of pheromones or local food abundance or simply as a result of limited dispersion after founding events and subsequent population growth.

Competition between Collembola species in cultures has in at least a few instances shown that there is no evidence for competitive exclusion, even under long-term clearly competitive conditions. In addition it has been shown that interactions between two species can be either positive or negative depending upon the nature of the interaction (airborne allomones, substrate-transmitted allomones, or direct contact).

While most soil- and litter-inhabiting Collembola feed primarily on decaying vegetation and fungi (and appear to be general feeders), experimental studies have shown that, given a choice, they may be very selective as to both the decay state and nature of the vegetation and the species of fungi. A number of Collembola are occasionally or primarily (and in a few species exclusively) carnivores, different species feeding on a variety of organisms, ranging from rotifers to other Collembola. Probably the most commonly eaten prey is nematodes. Vegetation-inhabiting Collembola eat primarily unicellular algae, pollen, and soft parts of vegetation and fungal spores. Many Collembola are coprophagic, feeding largely on arthropod feces. Some littoral species appear to feed largely on diatoms or unicellular algae, and forms with piercing–sucking mouthparts feed largely on fungal hyphae juices. Thus their primary role in the environment is that of reducer; however, another major role is that of prey. The ability to jump is the major defense mechanism of Collembola; however, many Poduromorpha, particularly those with the furcula short or absent, have body fluids that are repellent to predators, and they may release these by reflex bleeding when attacked. Most carnivorous soil organisms feed on Collembola, and many beetles, ants, and wasps are specialized for feeding on them.

HUMAN INTERACTIONS

Collembola rarely interact overtly with humans. There are few agricultural pests and, except for the introduced Lucerne flea (*Sminthurus viridis*) in Australia, which is a pest in pastures and horticultural crops, these are of little economic importance. There are no parasitic Collembola and they are not known to transmit any diseases. Mass emergences occur and may cause a temporary problem with household infestation but they are generally short lived there. The true household Collembola are unobtrusive and generally overlooked. Collembola play an important role in the development and maintenance of healthy soils, but this is not generally appreciated. Here they are usually abundant and may reach densities up to a trillion per square meter.

See Also the Following Articles
Arthropoda • Diplura • Protura

Further Reading
Christiansen, K., and Bellinger, P. (1998). "The Collembola of North America North of the Rio Grande: A Taxonomic Analysis," 2nd ed. Grinnell College, Grinnell, IA.
Fjellberg, A. (1998). The Collembola of Fennoscandia and Denmark. Part I. Poduromorpha. *Fauna Entomol. Scand.* **35,** 183.
Hopkin, S. P. (1997). "Biology of the Springtails (Insecta, Collembola)." Oxford University Press, London.
Lubbock, J. (1873). "Monograph of the Collembola and Thysanura." Ray Soc., London.
Maynard, E. A. (1951). "A Monograph of the Collembola or Springtail Insects of New York State." Comstock, Ithaca, NY.

Colonies

Sean O'Donnell
University of Washington, Seattle

Some species of insects spend much or all of their life living in organized social groups called colonies. Insect colonies have long fascinated biologists because they resemble superorganisms. Although insect societies are composed of distinct individuals, they possess group organization and coherence. Colonies exhibit emergent developmental properties, which are characteristics that cannot be explained or predicted by examining the behavior of their component parts. Insect colonies can serve as useful models of biological processes that occur in other complex living systems. One powerful analogy has been to compare the initiation, growth, and reproduction of an insect colony to the process of development of multicellular organisms.

Like individual plants and animals, insect colonies are initiated by propagules that are produced by parents (mother colonies); they then grow, reproduce, and often decline in old age. However, a wide array of developmental patterns have evolved in insect societies. Some of this variation can be explained by abiotic factors, such as the climate that prevails in the geographic range of a given species. Seasonality of temperature, daylength, and rainfall appear to have far-reaching effects on colony development. Climatic variables are not the whole story, however, since a diversity of colony cycles can be found among closely related species that live in the same area. Pressure from natural enemies, such as predators and parasites, as well as pressure from social competitors, has shaped the evolution of colony development.

MODES OF FOUNDATION: INDEPENDENT, SWARMING, AND BUDDING

Parent Colony Investment Decisions

SIZE VERSUS NUMBER OF PROPAGULES Insect colonies vary widely in the amount of investment they make in each of their offspring colonies. At the low end are independently founded colonies, wherein single inseminated females (such as eusocial thrips and aphids and some Hymenoptera) initiate new colonies alone. In these species, the colony passes through a solitary phase. Examples of independent founders include sweat bees (Halictidae), bumble bees *(Bombus),* several genera of paper wasps [most Vespinae (hornets and yellowjackets), *Parapolybia,* some *Ropalidia, Mischocyttarus,* and *Polistes*], and many ants (Formicidae). In some species, the lone foundress may be later joined by one or more conspecific cofoundresses. In other species, cofoundresses are not tolerated. In many termites, the smallest possible social group founds the new colony: a single reproductive male–female pair. At the other extreme, the relatively large colonies of some species issue discrete colony-founding swarms. Swarms are made up of reproductives and workers that migrate to a new nest site as a coordinated unit. Swarms often include a sizeable portion of the worker force, and they represent a large investment. Swarm-founding lineages include honey bees *(Apis),* swarm-founding wasps (tribe Epiponini), and army ants *(Eciton).*

There is an inherent trade-off between the size of the offspring colony propagule and the number of propagules that a given parent colony can produce. Large propagules are logically restricted to species with large colony sizes, but not all large-colony species reproduce by swarming or budding. *Vespula* paper wasps, higher termites (Termitidae), and leafcutter ants *(Atta)* achieve mature colony sizes of thousands or millions of adults, yet reproduce by issuing solitary dispersing reproductives. Production of new colonies by swarms has evolved independently in bees (honey and stingless bees), paper wasps (Neotropical *Epiponini,* some *Ropalidia, Provespa,* and *Polybioides*), and ants (*Eciton* army ants). Some species of ants produce new colonies by budding, wherein portions of the colony that occupy discrete nests gradually reduce interchange of members and eventually become independent.

SURVIVAL OF PROPAGULES: PREDATORS AND ENVIRONMENTAL EFFECTS One important set of selective pressures that may explain variation in propagule size is negative biotic interactions. These can take the form of predation, attack by other natural enemies such as parasites,

and conflict with conspecific competitors. Larger incipient colonies result from swarming and budding. These larger groups possess a defensive worker force and are more likely to resist destruction or consumption by enemies.

Abiotic challenges may also select for larger numbers of participants during incipient colony formation. Larger social groups may be better able to resist desiccation and temperature fluctuations, especially when they nest in enclosed spaces. Interesting in this regard are ant colonies that exhibit seasonal polydomy. Polydomy occurs when a single colony occupies several distinct nest cavities or structures. *Leptothorax* ants nest in small cavities in the leaf litter, such as hollow twigs. The colonies of some *Leptothorax* species divide themselves among several nests in summer when milder weather prevails, later coalescing into a single nest cavity as winter approaches.

Independent Foundation and Options for Social Cooperation

TO JOIN OR NOT TO JOIN In some species of independent-founding eusocial Hymenoptera, reproductives have the option of joining an already-initiated nest as a cofoundress, rather than starting one of their own. The degree of division of reproductive rights among the cofoundresses can be analyzed as a type of social contract. Often, the cooperating females are closely related. Differences in social status and reproductive capacity may be influenced by the degree of genetic relatedness among the cofoundresses. Dominant females can attempt to monopolize reproduction, or they can share a portion of reproduction as an incentive to stay and help on the part of subordinates. Kin selection theory predicts that the incipient society should be more equitable if the social partners are less closely related, since a greater incentive to help is required of nonrelatives. Cooperative colony founding may also represent a form of bet hedging and may be favored irrespective of genetic relatedness. If lone nest founders have little chance of succeeding, then cooperating can be favored by all individuals, even in the face of complete reproductive division of labor. In some cases, such as bull-horn *Acacia*-inhabiting *Pseudomyrmex* ants, female reproductives of different species may occupy a young plant, even though only one colony will eventually emerge to monopolize the tree.

USURPATION AND SOCIAL PARASITISM Another option for reproductives of some species is to steal or usurp a young colony from a conspecific or from another species. Social parasitism occurs when an invading reproductive uses the workers of a nest she did not construct to rear her reproductive offspring. A range of degrees of integration of social parasites into their host colonies can be observed in a diversity of insect lineages. Good examples occur in yellowjacket wasps (Vespinae), European *Polistes* paper wasps, bumble bees and their *Psythris* parasites, and ants. In the simplest cases, queens attack conspecific colonies and kill the resident reproductive, taking over the worker force. Simple heterospecific parasitism is similar to conspecific takeovers, in that the invading queen kills the resident queen. Often, females of socially parasitic species exhibit adaptations to improve their chances of winning queen vs queen combat, such as enlarged heads and mandibles. Parasitic species are often incapable of producing workers of their own, so the colony switches to producing new parasite reproductives after a takeover. In some species of ants, the socially parasitic queens are better integrated into the host society (e.g., *Teleutomyrmex* invading *Tetramorium* colonies). The parasitic queens coexist with the host queen and allow her to continue to produce a worker force, while the parasites produce reproductive offspring.

Social Groups as Founding Units

DIVISION OF LABOR   When new colonies are founded by swarms or by buds, a worker force is always present. One potential advantage to this strategy is the increased efficiency of the colony resulting from division of tasks among the group members. An important form of division of labor, which swarm-founders generally exploit, is the removal of the reproductives from the need to perform such risky and expensive tasks as food collection and nest defense. Division of labor is often weaker in independently founded colonies and is absent by definition for solitary foundresses.

DEFENSE   A group of workers can protect incipient colonies from natural enemies. New nests that are left unattended when solitary foundresses leave to forage are often attacked by parasites and predators. Survival of colony propagules increases dramatically with group size, particularly in areas where negative biotic pressures are most intense. Several studies of independent-founding paper wasps (*Polistes* and *Mischocyttarus* spp.) have shown that young colonies with cofoundresses fare dramatically better than singly founded nests.

THE NEED FOR COMMUNICATION   A special challenge facing swarm-founding species, and perhaps to a lesser extent budding species, is the need to coordinate movement from the parent nest to the offspring nest site. Special communicative mechanisms are used, such as the dance language in honey bees (*Apis* spp.), and trail pheromones in stingless bees and epiponine wasps. The need to evolve communicative mechanisms may constrain the evolution of swarming as a mode of colony foundation.

COLONY GROWTH

Social insects provide interesting and accessible models for testing life history theory because workers are roughly equivalent to the soma or body of a metazoan organism, while the reproductives can be treated as the germ or reproductive line. This analogy becomes weaker when the workers have some opportunity for direct reproduction. Nonetheless, insect colonies often develop in ways that suggest a trade-off between growth (i.e., worker production) and reproduction [i.e., production of gynes (new queens) and males].

One important decision that colonies make is the size at which to reproduce. This varies widely among even closely related species. For example, average size for mature colonies varies over at least five orders of magnitude among eusocial paper wasps (Vespidae).

Another important concern is the timing of reproduction. In some species, colonies exhibit a big-bang pattern of reproduction. The worker population of the nest increases as the growing season progresses, often exponentially. At some critical point in development the colony ceases to produce workers, switching entirely to the production of gynes and/or males. Colony decline or senescence follows reproduction. Temperate bumble bees and yellowjackets often approximate a big-bang approach to reproduction, and their life cycles resemble those of annual plants. Other species produce workers and reproductives simultaneously. In the extreme case, some males and gynes may emerge among the earliest offspring from the nest. For example, some Neotropical paper wasps (*Mischocyttarus* spp.) exhibit a great deal of overlap of worker and reproductive production. Swarm-founding species frequently undergo several bouts of reproduction, issuing reproductive swarms sequentially over a long period without undergoing parent colony decline.

For eusocial Hymenoptera, production of male offspring is potentially costly to the colony, and selection for labor efficiency may act to delay male production. Male Hymenoptera rarely work for their colonies and are often thought to represent a drain on colony resources. This cost does not accrue to termites, both sexes of which participate fully as workers.

REPRODUCTION

Timing and Synchrony

In seasonal habitats, the proper conditions for nest foundation can be constrained to a narrow window of time. This can select for a high degree of synchrony among colonies in a population in the timing of release of reproductives. In some species, reproductive offspring that depart from their natal nest must mate and either overwinter or initiate a new nest or perish. This pattern is apparently common to many ants and termites. In other species, reproductive females (honey bees, some bumble bees) and males (other bumble bees, some tropical *Mischocyttarus* wasps) can leave to find mates, but then return to the natal nest.

Sex Ratios and Sex Allocation

Beyond the germ line vs soma distinction, investment in the different sexes is an important consideration facing many insect societies. This is especially important for eusocial Hymenoptera, the males of which perform little or no labor for their colonies to offset their production and maintenance. Sex allocation theory, which attempts to predict the optimal investment an organism should make in the proportion of male compared to female offspring, has been applied to ant colonies. There is some evidence to suggest that, as predicted, colonies alter their relative amount of investment in males and gynes, depending on such environmental conditions as food availability.

COLONY SENESCENCE

Seasonal Effects versus Programmed Senescence

Colonies of many temperate eusocial insects are annual and appear to exhibit a programmed decline and senescence. Colony decline appears to be related to queen longevity and queen condition. For example, late-season colony breakdown appears to follow a decline in the queen's ability to suppress worker reproduction or the queen's death, in temperate bumble bees, *Polistes* paper wasps, and yellowjackets. The queen is not the whole story, however. Queens lost early in the season can be replaced or supplanted by reproducing workers without colony decline. The larger worker forces that are present late in colony development may be harder for the queens or their replacements to suppress. However, closely related species in less seasonal habitats do not exhibit time- or stage-determined colony decline. The plasticity of colony development exhibited in subtropical and tropical habitats by temperate invaders, such as German yellowjackets *(Vespula germanica),* may provide valuable insights into the factors that cause colony decline. German yellowjacket colonies in invaded sites (e.g., Hawaii and New Zealand) can be polygynous, accepting new queens into established nests, and are often perennial. These colonies can grow to much larger sizes than occur in temperate habitats, and the invading populations have become serious pests.

Can Colonies Be Immortal?

When abiotic forces do not terminate colonies, their longevity can be determined by the longevity of the reproductives. Queens and nests of *Atta* leafcutter ants may survive a decade or more in the wild. If colonies can replace dying queens, there is no inherent limit on colony longevity. The polygyne (multiple queen) of the imported fire ants *Solenopsis invicta* in the United States is an example of a species that accepts new, young queens into active nests. In this case, colonies may not senesce, and the observed upper limit on colony longevity will be set by the background rate of colony mortality. In other words, the chance of colony termination may be independent of colony age. Particularly interesting in this regard are some unicolonial invasive ant species, such as the Argentine ant, *Linepithema humile.* In

habitats outside their native South American range, such as the western United States, these ants fail to show internest aggression. Colony boundaries are fluid, and workers, brood, and reproductives are freely exchanged among nests. The entire population, which at present extends over a range greater than 1000 km in length, functions as a single colony. Colony longevity therefore equals the time to population extinction, and these may prove to be the longest lived insect colonies.

See Also the Following Articles
Ants • Apis Species • Division of Labor • Isoptera • Recruitment Communication • Sociality

Further Reading
Hölldobler, B., and Wilson, E. O. (1990). "The Ants." Harvard University Press, Cambridge, MA.
Jeanne, R. L. (1991). The swarm-founding Polistinae. *In* "The Social Biology of Wasps" (K. G. Ross and R. W. Matthews, eds.), pp. 191–231. Cornell University Press, Ithaca, NY.
Jeanne, R. L., and Davidson, D. W. (1984). Population regulation in social insects. *In* "Ecological Entomology" (C. B. Huffaker and R. L. Raab, eds.), pp. 559–590. Wiley, New York.
Oster, G. W., and Wilson, E. O. (1978). "Caste and Ecology in the Social Insects." Princeton University Press, Princeton, NJ.
Tschinkel, W. R. (1991). Insect sociometry: A field in search of data. *Insect Soc.* **38,** 77–82.

Colorado Potato Beetle

George G. Kennedy
North Carolina State University

The Colorado potato beetle, *Leptinotarsa decemlineata* (Coleoptera: Chrysomelidae), is the most devastating, defoliating, insect pest of potato *(Solanum tuberosum).* Uncontrolled, it is capable of causing complete crop failure. The potato beetle is important because of the damage it causes to potato and some related crops, as well as its extraordinary ability to evolve resistance to insecticides used in its control.

GEOGRAPHICAL SPREAD AND HOST RANGE

The Colorado potato beetle is native to Mexico. It was first recorded in the United States in 1811, feeding on a native plant, buffalo bur *(Solanum rostratum)* near the Iowa/Nebraska border. It was first reported as a pest on potato in Nebraska in 1859. The expansion of its host range to include potato allowed the beetle to spread rapidly eastward, moving among farm and garden plantings of potato. By 1874, it had expanded its geographic range to the East Coast of the United States. The potato beetle now occurs in North America throughout Mexico, the United States, and Canada, except

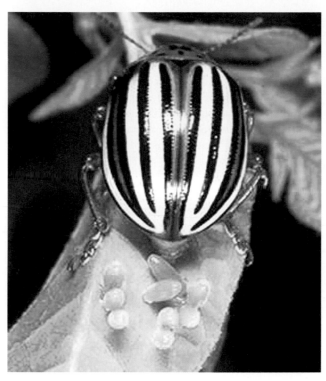

FIGURE 1 Adult female Colorado potato beetle depositing eggs on foliage of potato. Note evidence of feeding by the beetle at the leaflet tip. (Image © 2001–2003 www.arttoday.com.)

California, Nevada, and the coastal area of the Pacific Northwest, between 15 and 55° N latitude. It was accidentally introduced into France in 1922 and subsequently spread throughout Europe (except Great Britain) and the former Soviet Union; it now occurs in China, Greece, Turkey, and northern Iran.

The host range of the Colorado potato beetle is largely restricted to plants in the genus *Solanum* but includes some plants, such as tomato *(Lycopersicon)*, in related genera. Although geographically isolated potato beetle populations vary in their ability to utilize particular plant species as hosts, potato is the preferred host for most populations.

LIFE HISTORY AND CROP INJURY

Colorado potato beetles overwinter as adults in the soil within potato fields or in field margins. There are typically one to three generations per year, depending on latitude and the availability of host plants. Adult Colorado potato beetles are oval and approximately 9.5 mm in length and 6.4 mm in width. They are yellow-orange with 10 narrow, black, longitudinal stripes on their elytra (Fig. 1). Adults typically consume 130 to 1200 mm² of foliage per day and are highly fecund, depositing up to 3000 yellow eggs in clusters of 10 to 50 eggs on the lower surface of host leaves over a period of several weeks. All larvae within an egg mass hatch simultaneously, typically within 4 to 14 days, depending on temperature. There are four instars, and larvae have a

distinctive "hunchbacked" appearance, a black head capsule, and two rows of black spots on each side of the body. Instars 1 and 2 are brick red, whereas instars 3 and 4 are pink to salmon. The larvae are voracious feeders, with fourth instars consuming as much as 500 mm² of potato foliage per day. Larval development requires as little as 8 days or as long as 28 days at average temperatures of 29 and 14°C, respectively. Mature fourth instars burrow into the soil where they pupate. The pupal stage typically lasts 8 to 18 days, depending on temperature.

The Colorado potato beetle is primarily a pest of potatoes, but in some locations is also a pest of tomato *(L. esculentum)* and eggplant *(Solanum melongena)*. Damage results from defoliation by adult and larval feeding. In potato, yield reductions are related to both the amount of defoliation and the stage of plant growth during which it occurs. Yield reductions in tomato and eggplant result from feeding injury to the fruits, as well as from defoliation.

MANAGEMENT

Although a number of cultural measures, including crop rotation, isolation from previous potato crops, planting of nonpreferred and early maturing potato varieties, and use of trap crops, were recommended measures for potato beetle control, hand removal of adults, eggs, and larvae from infested plants was the primary means of control prior to the introduction of the insecticide Paris Green (copper acetoarsenite) in the late 1800s. Arsenic-based insecticides remained the primary means of control until DDT replaced them in the late 1940s. Resistance of the Colorado potato beetle to DDT was first reported in New York in the early 1950s. Resistance to other chlorinated hydrocarbon insecticides soon followed throughout much of the potato-growing region of the eastern United States. A series of insecticides was used to control the beetle during the succeeding decades and the potato beetle developed resistance to each. By the early 1980s, insecticide resistance had reached a crisis level. In many locations, potato beetle populations could not be controlled using insecticides. This stimulated a burst of research activity, which resulted in the development of more holistic pest management approaches. These involved foliar applications of the bacterial pathogen *Bacillus thuringiensis tenebrionis*, crop rotation, naturally occurring biological control, scouting and the use of economic thresholds, and the use of narrow-spectrum insecticides. By the late 1990s, several new, highly effective, narrow-spectrum insecticides had become available to control resistant potato beetle populations. Currently, potato beetle management relies on these new insecticides but heavily emphasizes their use within a pest management context, which is designed to minimize selection for insecticide resistance and negative environmental impacts.

During the mid-1990s, transgenic potato varieties were commercialized that expressed a protein from *B. thuringiensis tenebrionis,* which is highly toxic to the Colorado potato beetle. These varieties produce high-quality potatoes and are highly effective in controlling the potato beetle. Nonetheless, they have received only limited use because of their inability to compete with insecticides that controlled other insect pests (aphids and leafhoppers) in addition to the Colorado potato beetle and because of concern that consumers would not buy potato products made from transgenic potatoes. It is not clear at this time whether transgenic potato varieties will play a significant role in the future management of the Colorado potato beetle.

See Also the Following Articles

Agricultural Entomology • Coleoptera • Insecticide and Acaricide Resistance • Integrated Pest Management • Plant–Insect Interactions

Further Reading

Bishop, B. A., and Grafius, E. J. (1996). Insecticide resistance in the Colorado potato beetle. *In* "The Classification, Phylogeny and Genetics" (P. H. A. Jolivet and M. L. Cox, eds.), Vol. 1 of "Chrysomelidae Biology." SPB Academic Pub., Amsterdam.

Casagrande, R. A. (1987). The Colorado potato beetle: 125 years of mismanagement. *Bull. Entomol. Soc. Am.* **33,** 142–150.

Ferro, D. N. (2000). Success and failure of Bt products: Colorado potato beetle—A case history. *In* "Emerging Technologies for Integrated Pest Management: Concepts, Research, and Implementation" (G. G. Kennedy and T. B. Sutton, eds.). APS Press, St. Paul.

Hare, J. D. (1990). Ecology and management of the Colorado potato beetle. *Annu. Rev. Entomol.* **35,** 81–100.

Lashomb, J. H., and Casagrande, R. (eds.) (1981). "Advances in Potato Pest Management." Hutchinson Ross Pub., Stroudsburg, PA.

Zehnder, G. W., Powelson, M. L., Jansson, R. K., and Raman, K. V. (eds.) (1994). "Advances in Potato Pest Biology and Management." APS Press, St. Paul.

Coloration

Helen Ghiradella
State University of New York, Albany

Coloration is, as the word implies, the tapestry of hues with which an organism arrays the surfaces that it presents to the world. The signals thus produced may aid in species identification, camouflage, warning, and temperature regulation; all in all, they serve as a mute "language" with which an individual organism may communicate its place in the community within which it lives.

Insects are master chemists whose virtuosity is particularly evident in the design of the cuticle, the nonliving material that makes up the exoskeleton and serves as the boundary between the living animal and the outside world. Cuticle, a composite of chitin fibrils and various proteins and lipids, can be tailored for strength, rigidity, flexibility, permeability,

or elasticity, as needs dictate. It is also a technical and artistic medium with which insects, who are also master physicists and optical engineers, manipulate light to attire themselves with brilliant color on their bodies and wings. This article briefly reviews the bases of this ability. It begins, however, with an overview of the physics of color production, particularly with respect to structural colors, because only with this background can the reader really appreciate what a biological system, in its handling of light and color, can do.

TYPES OF COLOR

"Light" by definition involves wavelengths within the visible part of the electromagnetic spectrum. For humans it consists of wavelengths ranging from approximately 400 nm (violet) to approximately 725 nm (red). Many organisms, including insects, extend this range into the near ultraviolet (300–400 nm). "White" light for a particular organism consists of all wavelengths visible to that organism. Colored light has an incomplete spectrum in which only some wavelengths are represented.

Matter interacts with white light in various ways to produce color. One way is by selective absorption of particular wavelengths by a chemical, or pigment. The absorbed wavelengths (which are determined by the pigment's molecular structure) are essentially subtracted from the total spectrum, whereas the rest are reflected or transmitted to produce the visible color. Because pigments subtract colors, as additional pigments are added to a mix, additional wavelengths are absorbed and lost to view, changing the perceived color. When all wavelengths of the visible spectrum are absorbed, we call the sensation "black." (This is a somewhat simplified view: visual physiologists and psychophysicists would point out that additional processing by the visual system tempers what humans actually "see.") Pigmentary colors may be found in the cuticle or, if that be transparent, in the underlying tissues and even in the gut contents.

A second basis for color is structural, caused by the interaction of white light with minute and precise arrays on or in the material. The effects depend on the architecture, rather than the chemical makeup of the material. Light may be reflected, refracted, or scattered, but it is not absorbed, and so structural colors are "additive": if two are combined, both sets of wavelengths are represented in the final effect. If all wavelengths of the visible spectrum are reflected, we call the sensation "white." (Technically, white, even if caused by a pigment, is always a structural color, because it is the absence of any absorption of light.) Because the underlying architecture must generally be precise and stable, most structural colors are typically produced by stiff, nonliving materials, and of these insect cuticle is literally a brilliant example.

In biological systems, pigmentary colors are more common in the "warm" range—red, orange, and yellow—although green and blue pigments do exist. Biological

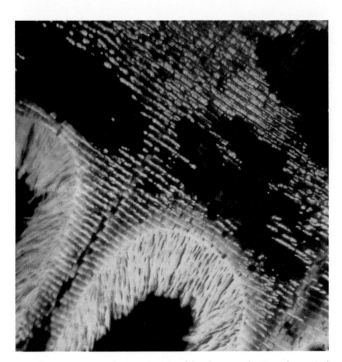

FIGURE 1 *Uranus riphaeus,* portion of hind wing, showing the typical lepidopteran investiture of shingle-like scales (on the surface) and bristles (at the edges). The scales in the black areas are colored by a pigment, probably melanin, whereas the iridescent scales and the white bristles are structurally colored.

structural colors, in contrast, are more likely to be "cool"— green, blue, violet, and ultraviolet. Figure 1 shows part of a butterfly wing: the dark colors are pigmentary, whereas the iridescent colors and the whites are structural. Many insects display both types, which are sometimes used together to produce yet additional effects. For example, a structural blue may be added to a pigmentary red to make a luminous violet, or a structural color may be "deepened" or intensified by a "backing" pigment that absorbs stray light leaking in from the "wrong" direction.

This article considers both pigmentary and structural colors. The following is a review of insect pigments, abstracted from the reviews of Chapman, Fox, and Nijhout.

INSECT PIGMENTS

Insects can make most of their pigments (some apparently from waste products that were historically simply stored or excreted), whereas others must come from their diets. Several general classes of pigments are recognized. These differ in the color ranges they generate and in the precursors used to produce them. As they share the same underlying mechanism of color production (selective absorption of some wavelengths of light), they can be reviewed with a simple list.

Melanins are black, brown, tan, or reddish brown pigments whose production and deployment involve a complex system of gene products and biochemical pathways. They are often present as granules in the exocuticle, although in lepidopteran scales they may be diffusely distributed, and they are responsible for most of the dark patterning in the body and wings. Eumelanin, the black form, commonly requires dopamine and tyrosine as precursors, while the chemistry of phaeomelanin, the brown, tan, or reddish brown form, is less well understood and may require the incorporation of additional kinds of molecules into the compound.

Pterins are white, yellow, or red pigments derived from a purine, guanosine triphosphate. Some function as cofactors of enzymes important in growth and differentiation; they may help control these processes. They are also cofactors in ommochrome (see later) production and often occur with these latter pigments, for example in the screening-pigment cells in the ommatidia of the eyes.

Ommochromes are red, yellow, or brown pigments derived from tryptophan, which they may serve to use up if it is in excess supply during times of high protein turnover (e.g., in metamorphosis). They usually occur in granules coupled with proteins and, as mentioned above, are present as screening pigments in the eyes as well as in the colors on the body. In insects displaying Tyndall blue (see later), they may serve as background pigments to absorb extraneous light.

Tetrapyrroles are pigments commonly classified into two groups. The first, the ring-shaped porphyrins, may add and incorporate iron to become hemes, which in turn may link to proteins to become (1) cytochromes, proteins important in cellular respiration in all higher organisms, or (2) hemoglobin, the protein that vertebrates and other organisms use to facilitate oxygen transport to their cells. Of necessity, all insects make cytochromes. Some that live in habitats of very low oxygen tension may make hemoglobin as well.

The other class of tetrapyrroles, the bilins, may in themselves be green or may link with proteins to make blue chromoproteins. These may in turn link with carotenoid pigments (see later) to make many insect greens.

Papiliochromes are yellow and red/brown pigments found only in butterflies of the family Papilionidae.

Quinone pigments are pigments of uncertain origin found in the Homoptera. Anthraquinones are found in members of the family Coccidae, in which they give red and sometimes yellow coloration; these include cochineal dye of historical importance. Aphins are characteristic of aphids, to whom they impart a purple or black coloration.

Carotenoids are yellow, orange, red, and, if bound to the appropriate protein, blue pigments that are made from dietary carotenes and their oxidized derivatives, xanthophylls. In combination with blue pigments (often bilins) they may produce an insect green, insectoverdin. They are also sources of retinal, a component of the photopigment of the eye.

Flavonoids are plant-derived pigments that produce cream or yellow colors, particularly in the Lepidoptera. Like the carotenoids, they cannot be synthesized but must come from the diet.

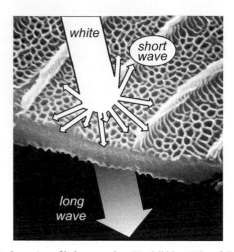

FIGURE 2 Scattering of light to produce Tyndall blue. When full-spectrum (white) light encounters structures or particles of the right dimensions, the shorter wavelengths are preferentially scattered in all directions, including toward the eye of the observer, who sees a blue color. The longer wave light passes through unscattered (and therefore bypasses the observer).

STRUCTURAL COLORS

There are many mechanisms by which structural colors can be produced. All depend directly or indirectly on the fact that a particular piece of material scatters or refracts different wavelengths of light to different degrees. This property of the material can be expressed in terms of its index of refraction, n, a measure of the degree to which a given wavelength of light entering the material is "retarded" or slowed down. For insect cuticle, n typically ranges from 1.5 for long-wave (red) light to 1.6 for short-wave (UV) light, although in special cases n less than 1.4 has been reported (for comparison, n for air is by definition 1). Structural colors described so far in biological systems fall into two general classes, scattering and interference.

Scattering

Scattering of light occurs when white light encounters a distributed cloud or array of molecules, particles, or other

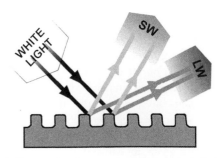

FIGURE 3 Diffraction (in this example, from a grating). Light hitting an edge or discontinuity gets bent or refracted to different degrees, depending on its wavelength. When it is then reflected, which of the component wavelengths are reinforced varies with the position of the observer, so that from one angle shorter wave light (SW) predominates, whereas from another, longer wave (LW) light predominates.

structures (Fig. 2). At least some of the component wavelengths of the beam will be reflected in random directions, including toward the observer. If the scattering agents are relatively large (700 nm or more), all visible wavelengths are scattered, and the resulting color is a matte white (the color of whole milk is an example of such scattering). If the particles are smaller (in the 400 nm range), the short wavelengths are scattered to a much greater degree than the long ones, which tend to pass on through the system and not reach the eye of the observer. The resulting color, Tyndall blue, is commonly seen in blue eyes and bluejay feathers; in insects it occurs in blue dragonflies and in some blue butterflies. Often, the blue structure is underlaid by a layer of ommochrome pigments, which, as mentioned above, deepen and intensify the color by absorbing stray light. Lacking such pigment backing, the blue is a dilute "powder" blue.

Interference

The general category of interference includes those situations in which the rays of a beam of white light are temporarily separated and then brought back together in such a manner that some have traveled a longer path than others. Depending on geometry, when the rays recombine, certain wavelengths are in phase and reinforced ("constructive interference"), whereas others are out of phase and cancel each other ("destructive interference"). The results are the brilliant, shimmering colors we call "iridescent." There are many ways of producing iridescence; this article considers only those of known importance in insects.

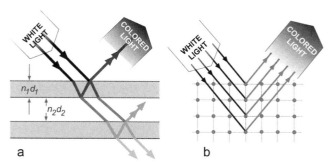

FIGURE 4 Two forms of interference from layers. (a) Thin film. A thin film can be described in terms of its optical thickness, its index of refraction, n, times its actual thickness, d. When white light encounters such a film, part of the light reflects from the top surface and part from the bottom. When these two beams recombine, those wavelengths four times the optical thickness of the film are constructively reinforced and the others not. If many films are stacked, light not reflected by the first film may be so by the others; if the films are alternated with others of equal optical thickness but of a different refractive index (so that $n_1 d_1 = n_2 d_2$), the stack reflects essentially all light of the reinforced wavelength. (b) Lattice. A lattice of points, spheres, or other structures reflects light in a manner analogous to that of a crystal. Each plane reflects part of a beam and transmits the rest (transmitted light not diagrammed here). If the planes are evenly spaced, they reflect light the wavelength of which is twice the spacing, i.e., they will form a half-wave reflector. As in the case of thin films, with enough reflective planes, essentially all the light of the reinforced wavelength will be reflected.

DIFFRACTION Diffraction occurs when light strikes the edge of a slit, groove, or ridge. Different wavelengths bend around the edge to different degrees and the spectrum fans out into its components. If many such grooves or ridges occur in a regularly spaced array (for example, a "diffraction grating" such as that in Fig. 3) light of different wavelengths is reinforced at different angles so that the colors change with the position of the viewer (e.g., consider iridescent bumper stickers and other shimmering plastic labels). Many insect cuticles have fine gratings etched into them; these and the ridge and crossrib structures (see later) of some lepidopteran scales and bristles produce diffraction colors.

THIN-FILM INTERFERENCE Thin-film interference involves, as the name implies, the interaction of light with ultrathin films of a material (e.g., iridescence from soap bubbles and oil slicks). Light reflecting from the top surface of such a film interacts with that reflecting from the bottom surface (Fig. 4a) and depending on the optical thickness of the film (its index of refraction, n, times its actual thickness, d), some wavelengths are reinforced and others not. Because the wavelengths of the reinforced light are four times the optical thickness of the film (i.e., a film of 100 nm optical thickness results in reflected light of wavelength 400 nm), such films are commonly called "quarter-wave interference reflectors" or "quarter-wave films." Because a slanted beam of light has to penetrate a greater thickness of film, thereby changing the effective optical geometry, thin-film colors shift toward the shorter wavelengths when the films are tilted with respect to the light source (e.g., the familiar blue of the morpho butterflies becomes more violet).

Of course any film thin enough to act as a quarter-wave reflector can catch and reflect only a portion of the incident light; the rest passes through. The presence of other films below the first increases the likelihood that light will be reflected, and in fact the most efficient of these reflector systems are stacks of thin films of the material in question, separated by other films with a different refractive index or by air ($n = 1$), so that the light is reflected from layers of alternating high and low n. If all the films are equivalent in nd, their optical thickness, the emerging colors are relatively pure, whereas varied spacing produces a less intense but broader range of reflection. As in all these systems, there may be behind the "mirror" a layer of pigment that intensifies the color by eliminating stray light that would otherwise interfere with the efficiency of the interference and thereby dilute the color.

LATTICES Many iridescent colors are produced, not by thin films per se but by thin-film analogues, systems that achieve similar effects without actual discrete films. One such mechanism is the "Bragg" or space lattice (Fig. 4b), a highly regular array of spheres or other units. Light entering such a lattice is reflected from the various layers, and the beams interfere in a manner analogous to that in thin-film stacks. In

FIGURE 5 This beetle shows the metallic coloration typical of many beetles and flies. The colors have at least two possible origins: they may be caused by a thin-film stack in the exocuticle (or sometimes the endocuticle) (Fig. 6d) or they may be the result of a helicoidal arrangement of chitin fibrils in the exocuticle (Fig. 6e). The latter effect is analogous to that produced by certain types of liquid crystal in common technological use. The red and black coloration in the eyes, on the other hand, is almost certainly pigmentary.

this example, the wavelength reinforced is twice that of the spacing between the layers of the lattice, which therefore acts as a half-wave reflector. The familiar brilliance of the mineral opal is an example of this type of interference, caused in this instance by a lattice of tiny silica spheres. These lattices are very common in the biological world; those described so far in insects are "reverse" lattices, consisting of spheres of air in a matrix of cuticle.

HELICOIDS The metallically colored cuticles of many beetles and flies (Fig. 5) either are thin film (Fig. 6d) or owe their iridescence to yet another mechanism, one analogous to that shown by the familiar and brightly colored liquid crystal displays in our electronic world. Cuticle is of course a composite of chitin fibrils in a complex matrix that is laid down sequentially in what can be considered a series of layers. If the fibrils in a particular layer are lined up in the same direction, the layer exhibits form birefringence, i.e., different indices of refraction parallel to and normal to the fibrils. In many cuticles, the layers precess, that is, each is laid down slightly rotated relative to the previous one (Fig. 6e). In essence, the structure can be considered a helicoid, and like all helical structures, it repeats itself with a certain spacing (called a "pitch"). As the layers precess, so does the difference in refractive index, so that viewed from a given direction a helicoidal array displays what are essentially layers of alternate high and low n, reminiscent of those in thin films. (Unlike thin films, helicoids also circularly polarize light, which insects may be able to see and which may therefore carry additional information to them.) If the spacing is regular and the pitch is appropriate, the helicoid behaves like a half-wave interference reflector, i.e., it reflects light of wavelengths twice the pitch. In the typical metallic cuticles,

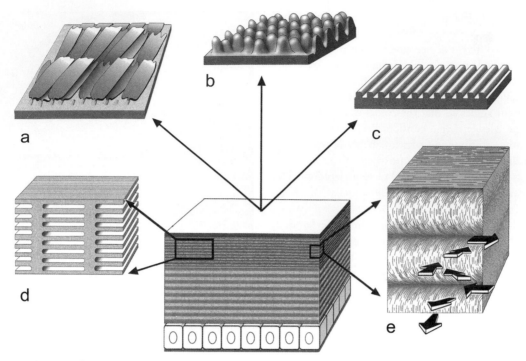

FIGURE 6 How to make an interference color. A block of hard insect cuticle (bottom center) typically consists of a relatively thin epicuticle (here represented as a featureless covering layer) and an inner procuticle, which in turn consists of a distal exocuticle and an inner endocuticle (this diagram also shows the attendant epithelial cells). The layering of the procuticle is common in most (but not all) cuticles and is the visible manifestation of the helicoidal architecture of the chitin fibrils. Such a block of cuticle may be modified in any of several ways to manipulate light: (a) The surface investiture (the scales and/or bristles) may be modified to produce scattering or iridescent colors (see Fig. 7). (b, c) The cuticle surface may be sculpted into fine protuberances that serve as an antiglare coating (see Figs. 8 and 9) or into fine parallel grooves that act as diffraction gratings. (d) Part of the procuticle may be elaborated into a quarter-wave thin-film reflector stack. (e) The chitin fibrils of the exocuticle may be arranged in a "helicoidal" array, analogous to that in a liquid crystal and producing color by a similar mechanism. (The apparent parabolic bending of the fibers is an optical illusion.)

the helicoids of the exocuticle are so tuned, and because the helicoidal arrangements of their fibrils resemble that of the molecules in one iridescent class of liquid crystals, they are often referred to as "liquid crystal analogs." Some insects intensify the effect by doping the cuticle with uric acid, which increases its birefringence.

BASES OF STRUCTURAL COLORS IN INSECTS

As mentioned above, because scattering colors (whites and Tyndall blues) can be produced by granules or droplets as well as hard structures, these may come from the epidermis and internal tissues, as well as from the integument. Interference colors, which require stable structures to produce them, are limited to the cuticle and its investiture. Figure 6 shows diagrammatically a patch of cuticle with its two basic layers, the thin outer epicuticle and the inner procuticle. The procuticle commonly shows the helicoidal arrangement described above, which results in a banded or layered appearance in section. In hard or stiffened cuticle, the procuticle is commonly further subdivided into a cross-linked, more tightly woven distal exocuticle and a basal, more loosely structured endocuticle.

The cuticular surface and the exocuticle are most likely to be modified to produce structural colors, although in some

insects the endocuticle may be as well. Several possibilities exist (Fig. 6). For example, the surface may be invested with layers of scales and/or bristles (Fig. 6a), which carry the color, especially in the Lepidoptera (see later). Alternatively, it can be sculpted into a series of nipple-like protuberances (Fig. 6b—more about this later) or into the fine grooves that characterize diffraction gratings (Fig. 6c). In the exocuticle (and sometimes the endocuticle) metallic colors can be produced by stacks of thin films of alternating refractive indices ($n = 1.58$ alternating with $n = 1.38$ has been measured in one of these systems) (Fig. 6d) or by appropriately tuned helical rotation of the chitin fibrils (Fig. 6e). (As yet another example of insect command of light, in many corneas, the helicoidal architecture of the cuticle is tailored not to produce structural colors but to control refractive index, so that incoming light is appropriately focused as it enters the ommatidia.)

Scales and bristles are particularly impressive in the variety and complexity of their architecture (Fig. 7). They commonly exist in two and sometimes three layers on the body or wing surface (Fig. 7a), and each layer may be modified in shape and color. A typical scale consists of a flattened sleeve of cuticle whose lower surface (that toward the wing) is relatively featureless, whereas the upper surface is elaborated into a reticular network of longitudinal ridges joined at intervals by

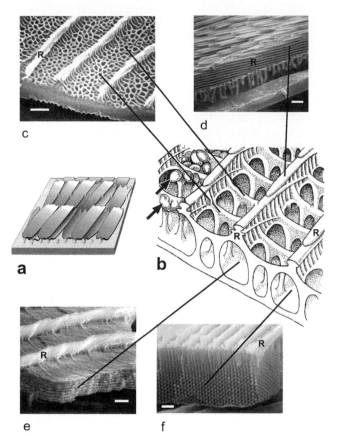

FIGURE 8 Effects of an antiglare coating. Even though the wing of this clearwing moth is somewhat wrinkled and parts of it would therefore be expected to reflect light, its matte surface (Fig. 9) allows the text to be read through it with minimal loss.

FIGURE 7 Closer look at the investiture (scales and bristles), which in the Lepidoptera typically carries the color. Scales and bristles are complex cuticular structures each elaborated by a single cell, and they are often both pigmentarily and structurally colored. (a) As in Fig. 6a, a patch of cuticle surface showing several overlapping scales and one empty socket. (b) Diagrammatic view of a small fragment of a more or less typical unspecialized scale. The scale may be thought of as a flattened sac, the two surfaces of which are joined by fine pillars. (A bristle would be cylindrical, rather than flattened, but it is essentially the same type of structure.) The upper surface is a rectangular grid made up of longitudinal ridges (R) joined at regular intervals by transverse crossribs from which, in some species, hang pigment granules (arrows) (in other insects, pigment is incorporated into the cuticle itself). Ridges and crossribs together frame a series of windows opening into the interior of the scale. Virtually any part of this basic scale may be elaborated into a reflective structure. In the following examples, scales have been fractured to show their interior structures; lines indicate which basic scale structures have been elaborated to produce each structural color. [Modified from Ghiradella, H. (1998). Hairs, bristles and scales. *In* "Insecta." (M. Locke, ed.), vol. 11A of "Microscopic Anatomy of Invertebrates" (F. W. Harrison, ed.) pp. 257–287. Wiley, New York. Copyright 1998 John Wiley & Sons. Reprinted by permission of John Wiley & Sons.] (c) *Papilio zalmoxis*, fragment of upper scale surface. The ridges are low and unornamented, but the crossribs have "filled in" the windows with a network of "alveolae" that scatter light to produce a Tyndall blue color (compare Fig. 2). Bar, 1 µm. (d) *Morpho menelaus*, fragment of deep blue iridescent scale, fractured longitudinally to show a side view of a ridge (R), together with the pillars that join it to the bottom layer of the scale. The ridge (and those behind it) has been elaborated into stacks of slanting thin films that reflect the characteristic blue of this butterfly. Bar, 1 µm. (e) *Urania riphaeus*, fractured green iridescent scale (see Fig. 1). The ridges and crossribs are not particularly elaborate, but the interior of the scale is filled with a stack of thin films that produce the color. Bar, 1 µm. (f) *Teinopalpus* sp., fractured green iridescent scale. The scale interior is filled with a space lattice that produces the color. Bar, 1 µm.

transverse crossribs (Fig. 7b). Fine flutings or microribs line the sides of the ridges and sometimes run out across the crossribs. Slender pillars join top and bottom surfaces. Pigments in some groups (typically the Pieridae) may exist in discrete granules, whereas in other insects they are laid into the scale cuticle itself.

Virtually any part of this basic scale can be elaborated to produce a structural color. The spacing of the ridges and/or crossribs may be appropriate to produce diffraction colors. The crossribs and microribs may extend to fill in the windows with a network of "alveolae" that reflects Tyndall blue (Fig. 7c). The scale ridges may bear stacks of thin films (examples known so far reflect green, blue, or ultraviolet) (Fig. 7d). The interior of the scale may be filled with stacks of thin films tuned to produce green or blue (Fig. 7e), or it may contain a space lattice that reflects iridescent green (Fig. 7f). And, as mentioned earlier, these structural colors may be combined with pigments to give yet additional colors and effects.

More detailed study of some of these systems is revealing yet more complicated and sophisticated optical effects. For example, in blue Morpho butterflies, the deep blue iridescent scales (whose color comes from thin-film iridescence on the ridges—Fig. 7d) are overlaid by a layer of "glass scales," which, though otherwise transparent, do have iridescent ridges. The apparent function of the glass scales is to broaden the effective angle of reflection (see Vukusic *et al.*, 1999). The iridescent scales of *Papilio palinurus* have stacks of internal thin films, but rather than being flat, the stacks are puckered into shallow cup-shaped depressions whose bottoms reflect yellow light, whereas the sides reflect blue, giving the human observer the sensation of green (see Vukusic *et al.*, 2000)—it is not known why these animals have developed this mechanism to produce green scales when other iridescent greens are produced by more conventional thin films or by lattices. There are other intriguing scale and bristle types whose optics are now being studied, and from these insect systems new and sophisticated insights into the effective control of light can be expected.

ANTIGLARE COATINGS

Insect handling of light does not stop with the production of colors. Figure 8 shows the clarity with which light may be transmitted through the wing of a clearwing moth. Although

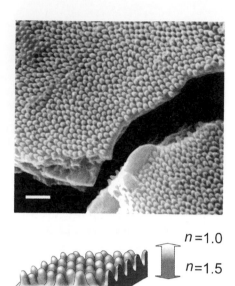

$n = 1.0$

$n = 1.5$

FIGURE 9 (Top) *Podesesia syringae,* patch of wing fractured to show its internal structure as well as the fine protuberances or nipples that form the antiglare coating. A few of those on the wing reverse show through the break at bottom center. Bar, 1 μm. (Bottom) Basis of the antiglare effect. The tapered shape of the protuberances produces a gradual change in refractive index from that of air ($n = 1$) to that of cuticle ($n = 1.5$ in this example), so that at the interface there is neither refraction nor reflection to disturb the passage of light.

the cuticle is somewhat rippled and one would expect some of its surfaces to show glare, they do not. Figure 9 shows why: the wing surface is covered by fine arrays of protuberances that are commonly found on cuticles that are engineered not to maximize the reflection of light but to minimize it (besides the wings of these clearwing moths, such arrays have been reported on the eye corneas of nocturnal moths). The arrays provide a gradual transition in refractive index from that of air ($n = 1$) to that of cuticle (typically $n = 1.5$ to 1.6) so that there is no sharp interface to refract or reflect light as it passes from one phase to the other (the basis for antiglare coatings on eyeglasses).

In summary, the complexity associated with insect colors extends, for pigments, to the sophisticated biochemistry with which insects make (and often recycle) the compounds that characterize their chemical colors and, for structural colors, to the production and control, often by single cells, of the precise cuticular architecture reviewed here. Other effects abound; for example, many insects are capable of physiological color change, by reversibly hydrating or dehydrating their cuticles to change the optical thickness of the layers or by moving pigment about.

PERSPECTIVE

An easy question is why such arrays of color? Insects share the same challenges as humans and so they use color and patterning for species and mate recognition, camouflage,

startling potential predators, and mimicry. Energy is also almost certainly a factor: dark colors absorb more heat, and butterflies, for example, may use pigments and possibly interference mechanisms to increase the absorption of infrared. It can only be speculated as to why structural colors predominate at the short end of the visible spectrum. As a biological material, cuticle is assumed to have a limited range of refractive indices and if so, only shorter wavelengths may be refracted and scattered effectively enough to produce the needed effects. It could also be that short-wave structural colors are metabolically "cheaper" (i.e., require less energy to produce) or easier to make than short-wave pigments, which do seem relatively rare in biological systems. Further study may enlighten both biologists and engineers.

How are these structures and color patterns made? On one time scale the question is developmental: how can an animal transform its genetic information into the complicated structures observed? This is the general question of pattern formation, the nested series of instructions that must be carried out by a developing organism on many levels at once. A developing butterfly must specify, for example, the general shape of its wing, the precise venation pattern, the distribution of scale and/or bristle types on both sides and on all edges of the wing, the distribution of pigment(s), and finally whether scales are to be structurally colored and if so, what type of structure they are to have. Nijhout has presented a compact and authoritative review of pattern formation in butterfly wing systems; many other researchers are currently studying the molecular and genetic mechanisms underlying pigment formation and deployment. Common themes are emerging, but much still remains to be done, especially on the role of physical forces that almost certainly work along with the biochemical ones to bring forth the final form.

The formation of the microarchitecture underlying structural color systems is less well understood. Ghiradella in 1998 reviewed what was known about development of structural colors in scale systems, and Neville in 1993 presented a comprehensive review of the formation of helicoidal and other fibrous composite systems. However, despite their value as potential models for human research and development, particularly of optical systems, very little is known about these systems. There surely are lessons to be learned here. For example, Bragg lattices are of interest to engineers seeking more efficient transmission of information along optical fibers, and scale optics is becoming of interest to the photonics research community, which is seeking to develop structures and materials that can control light for purposes of communication, paints, surface coatings, electronic displays, etc. Again, the insect systems have a lot to teach us, especially since their structures are made at room temperature and without toxic solvents.

On the longer time scale, how did these systems evolve? In some examples there are grounds for speculation. As mentioned above, many pigments may have originally been metabolic by-products that, because of actual or potential ability to absorb some wavelengths of light, were somehow co-opted for purposes of display. The helicoidal arrangement of chitin fibrils in cuticle is part of a larger structural adaptation of cuticle as a building material. As in all skeletons, fibril orientation in cuticle is tailored to local challenges. Helicoidal arrangements, with their multidirectional fibril orientation, are well equipped to to provide toughness and strength in the face of multidirectional stresses and are common in areas exposed to such stresses. Having evolved such a helicoidal arrangement to confer a particular type of strength, the animals needed only to make the pitches of the helices regular and to tune them to have fine iridescent reflectors at the same time.

The evolution of the thin-film, diffraction, and other systems is at present a very open question. They appear to be of great antiquity. Parker reported diffraction and antiglare structures in Burgess shale fossils and suggested that the emergence at the beginning of the Cambrian period of image-forming eyes (to quote Parker, "…the lights were effectively turned on…") may have produced extreme selection pressure for potential prey animals to develop rigid armor (with its inherent potential for forming structural colors) and at the same time a need for, and an opportunity to develop, camouflage, recognition patterns, and all the other common uses and expressions of biological color. While there is no question about the utility of these color mechanisms, it is still hard to imagine how so much can have been accomplished, even with millions of years of research and development in a competitive and presumably highly selective world.

To this point coloration has been considered in terms of passive and static displays on the surfaces of insects. But in living insects, the color-producing structures are situated on a moving body with moving appendages, and so the displays are modulated over time. The resulting signals are four dimensional, which adds to them a richness of information that we cannot begin to appreciate, especially because the true capabilities of the insect eye (which is much "faster" than that of the human) in its processing of either color or movement are not known.

The subject of insect mastery of light must also include bioluminescence. Lantern types and flash patterns come in a variety of forms; superficially, the mechanisms by which they are produced seem to differ radically from those already discussed. But here too, the insect displays mastery of architecture—in the design of the lantern cells themselves—and of chemistry to create light signals that can be controlled in space and time, but at those times of day when sunlight is not available to power the display. In doing so, the insects have truly made "the lights come on," replacing in their signaling the warmth of sunlight with their own cold light. As researchers continue to learn about these systems, they are exploring worlds within worlds of complexity and can only gain in appreciation of the enormous capabilities of biological systems in their communication with their environments … and with each other.

See Also the Following Articles
Aposematic Coloration • Bioluminescence • Cuticle • Eyes and Vision • Industrial Melanism • Integument

Further Reading
Carroll, S. B., Grenier, J. K., and Weatherbee, S. D. (2001). "From DNA to Diversity: Molecular Genetics and the Evolution of Animal Design." Blackwell Sci., Malden, MA.
Chapman, R. F. (1998). "The Insects: Structure and Function." Cambridge University Press, Cambridge, U.K.
Fox, D. L. (1976). "Animal Biochromes and Structural Colors." University of California Press, Berkeley.
Ghiradella, H. (1998). Hairs, bristles and scales. In "Insecta" (M. Locke, ed.), Vol. 11A of "Microscopic Anatomy of Invertebrates," (F. W. Harrison, ed.), pp. 257–287. Wiley, New York.
Kunzig, R. (2001). Trapping light. *Discover* **22**, 72–79.
Mason, C. W. (1926). Structural colors in insects. I. *J. Phys. Chem.* **30**, 383–395.
Mason, C. W. (1927). Structural colors in insects. II. *J. Phys. Chem.* **31**, 321–354.
Mason, C. W. (1927). Structural colors in insects. III. *J. Phys. Chem.* **31**, 1856–1872.
Neville, A. C. (1975). "Biology of the Arthropod Cuticle." Springer-Verlag, New York.
Neville, A. C. (1993). "Biology of Fibrous Composites." Cambridge University Press, Cambridge, U.K.
Nijhout, H. F. (1991). "The Development and Evolution of Butterfly Wing Patterns." Random House (Smithsonian Inst. Press), Washington, DC.
Parker, A. (1999). Light-reflection strategies. *Am. Sci.* **87**, 248–255.
Parker, A. (1999). The Cambrian light switch. *Biologist* **46**, 26–30.
Vukusic, P., Sambles, J. R., Lawrence, C. R., and Wootton, R. J. (1999). Quantified interference and diffraction in single *Morpho* butterfly scales. *Proc. R. Soc. London B* **266**, 1403–1411.
Vukusic, P., Sambles, J. R., and Lawrence, C. R. (2000). Colour mixing in wing scales of a butterfly. *Nature* **404**, 457.

Commercial Products from Insects

Michael E. Irwin and Gail E. Kampmeier
University of Illinois and Illinois Natural History Survey

When people contemplate how insects are marketed as consumer products, images of novelties, gimmick foods, cuddly toys, odd adornments, and cartoon images are invoked, calling forth a range of emotions from repugnance to warmth. But that is just the tip of the iceberg. Insects can be very big business. They and their products are sold for crop pollination, pharmaceuticals, health and agricultural protection, and human, pet, and livestock nutrition, as implements for conducting research, and for a host of other uses. This article focuses on commercialization of insects and their products.

FIGURE 1 Bees hired out for pollination of apple orchard. (Photograph by Eugene Killion.)

MARKETING LIVING INSECTS

Crop Pollination

Flowering plants are fertilized by several groups of insects. By far the most common pollinators are bees, and the honey bee, *Apis mellifera,* plays the dominant role in pollinating large tracts of agriculture. The domestication of the honey bee for pollinating crops had its beginnings at least 4000 years ago. Since that time, beekeeping has flourished and is now a thriving industry. In the United States alone, $15 billion worth of crops (fruits, vegetables, flowers) are pollinated by domesticated honey bees each year. Commercial apiaries lease their beehives to growers who need their crops pollinated. The keepers manage the hives, moving the bees from field to field to ensure crop pollination (Fig. 1). Although worker bees are not sold as such, their labor is. Moreover, the commercial interdependency of the honey bee industry is complex. Keepers buy high-quality queen bees from specialized suppliers, who, along with the keepers, purchase bee-tending equipment from other specialized suppliers, and the entire industry is dependent on information contained in specialized books, journals, and magazines.

A number of crops are more efficiently pollinated by bees of other kinds. Leaf-cutting bees, or mason bees, are a good example. These "solitary" bees, unlike honey bees, do not live in colonies. Solitary bees produce no honey or wax but are relatively docile and not likely to sting. One species of leaf-cutting bee, *Osmia cornifrons,* is widely used in Japan for apple pollination. It was imported to the eastern and midwestern United States for the same purpose. Another leaf-cutting bee, *O. lignaria,* a native to parts of the United States, is also widely used for orchard pollination. Pollinating a hectare of apples requires on average either 750 female hornfaced bees *(O. cornifrons),* 600 female blue orchard bees *(O. lignaria),* or 50,000 honey bee workers. Other mason bees, bred and sold to alfalfa growers in the western portions of the United States, ensure the production of high quality alfalfa seed. Not only are mason bees better pollinators of a

number of crops, they are also immune to the devastating effects of tracheal and varroa mites, which can decimate honey bee colonies.

Another crop pollinator is the bumble bee, which is less affected by extreme weather than the honey bee and is better adapted to perform under confined greenhouse conditions. By vibrating as they extract nectar and pollen, bumble bees efficiently pollinate flowers and encourage high fruit set under greenhouse conditions. Bumble bees are bred, reared, and packaged for sale to growers for pollinating vegetable crops (particularly tomatoes) grown under greenhouse and plastic tunnel conditions. Entire industries are founded on the production and sale of bumble bees, especially in the Mediterranean region, from Spain to Israel.

Agricultural and Human Protection

One who has never witnessed the devastation of a crop by insect pests would be alarmed by the rapidity with which it can occur. One of the best ways to counter the buildup and devastation caused by insect pests is to unleash on them their own natural enemies. A vibrant industry is built on supplying the natural enemies or "beneficials" needed to manage pests and pest outbreaks, both for protecting agriculture and for preserving human health. These beneficials can take the form of insect pathogens, insects that prey on the pests (predators), insects that parasitize them (parasitoids), or insects that destroy weeds. This industry is increasingly in demand as growers, horticulturists, home gardeners, and vector control organizations alike turn from chemically oriented pest suppression measures to the principles of integrated pest management (IPM) and practices more attuned to organic farming. Many companies are in the business of rearing and supplying beneficial organisms, not just for agriculture and health in their broadest senses, but also for parklands, green corridors, and home gardens. This challenging industry must take into account knowledge of the systematics of the pests and the beneficial organisms that attack them, methods to efficiently and inexpensively mass-produce the desired beneficials, ways to maintain genetically viable and aggressive beneficial organisms, procedures to efficiently transport beneficials to targeted release sites, and knowledge to ensure that the habitats of the release sites are conducive to optimal utilization by the beneficials for controlling the pest. This last point is especially critical because the beneficials can simply move from the release site and take up residence elsewhere, providing a neighbor, instead of the grower who purchased them, with pest suppression.

Although the mass rearing and marketing of beneficial insects is an expanding business, the mass irradiation and release of sterile males is of a considerably larger scale. The Mediterranean fruit fly, New World screwworm, tsetse fly, and boll weevil have been successfully controlled through inundative releases of sterile males. The technology of irradiation is so notably complex, the scale of releases so great, and the costs of mass irradiation so high that these services are almost always provided by a governmental agency. The International Atomic Energy Agency (IAEA), the Food and Agriculture Organization of the United Nations (FAO), and the U.S. Department of Agriculture (USDA) have been instrumental in pioneering sterile male irradiation and release. The equipment needed to mass-rear, irradiate, and release sterile males is costly. Industry, its major supplier, is exploiting the need for this technology by developing and marketing specialized products.

Live Insects and Human Therapy

The thought of using live insects to treat human ailments would make most pale, but the results can sometimes outperform drugs and surgery typical of more traditional Western medicines. Honey bees, fly maggots, ants, and *Plasmodium*-carrying mosquitoes have all been used in human therapy.

The venom of honey bees is used to ameliorate inflammatory and autoimmune conditions such as multiple sclerosis, arthritis, rheumatism, chronic pain, neurological diseases, asthma, and dermatological conditions. The venom can be administered by humans or injected via the sting of a bee. Venom therapy is widely used in China, Korea, Bulgaria, Romania, Russia, Brazil, and the United States. Much of the research with venom therapy in the United States focuses on treatment of multiple sclerosis and chronic pain. Of the more than 40 components identified in bee venom, 18 are considered to be active. One of these, melittin, is among the most powerful anti-inflammatory substances known. Although not a conventional form of treatment in the United States, anecdotal evidence of its efficacy is accumulating, and companies market bees and bee products for therapeutic purposes. This form of therapy should be undertaken only with qualified supervision and, since some people go into anaphylactic shock when stung by bees, should be administered with adequate precaution.

Maggot debridement therapy uses maggots of *Phaenicia sericata* to cleanse wounds of necrotic tissue without attacking healthy underlying tissues. Maggots have been used to treat abscesses, burns, cellulitis, gangrene, ulcers, osteomyelitis, and mastoiditis. Their use has lessened the need for amputations and has been especially useful where diabetes is a complicating factor. Therapy involving the cleansing effect of these maggots dates to the 16th century. Despite the pioneering work in the early part of the 20th century, the practice of debridement therapy fell to disuse with the advent of antibiotics and new surgical techniques in the mid-1940s. The increase in resistance to antibiotics in the late 1980s elicited a resurgence of interest in debridement therapy. The mechanisms underlying success of this treatment remain poorly understood to this day.

Live ants, particularly Amazonian army ants and carpenter ants of Africa, India, and the Mediterranean region, have been used to close wounds and surgical incisions. The sharp mandibles of the soldiers lock when their jaws are closed, irrevocably fastened in place even if the bodies are severed from their heads. Although these live suturing instruments are unlikely to grace surgical theaters in modern hospitals, they have long been used by native peoples.

Another use of live insects for therapy has fallen to disuse now that better alternatives are available. Before other treatments were available, it was known that the progress of syphilis could be halted when the body temperature was raised above 40°C. Because the effects of syphilis were so devastating, mosquitoes bearing a relatively mild strain of malaria were used to infect such patients. The *Plasmodium* pathogen caused high fevers that exterminated the syphilis pathogen. Although the patients were then infected with malaria, the cure was deemed worth the consequences.

Living Insects on Parade

There is little doubt that insects fascinate. Perhaps that is why they are so often featured in zoos and living museum displays, sold as pets, bred and released to celebrate special events, filmed and videotaped for movie and television productions, and ubiquitously adopted for live entertainment and education.

Why anyone would purchase live immature insects and rear them to adulthood may baffle some, but marketing immatures for that purpose is a thriving industry. Hobbyists in the United Kingdom and around the world order exotic butterfly chrysalids for the sheer joy of observing the spectacularly adorned adults emerge. Ant farms, butterfly houses with living chrysalids, a wide variety of butterfly immatures, exotic live tropical stick insects, praying mantid and cockroach oothecae, pea aphids, *Drosophila*, mealworms, and lady beetles are sold directly by suppliers or by auction. Rearing these insects is both fun and educational. For insect collectors, it offers a way to obtain exotic species.

Insect zoos, petting zoos, live museum displays, and insects in botanical gardens provide a sometimes exotic backdrop for educating the curious. Staged cricket and beetle fights are popular pastimes in Japan and elsewhere in Southeast Asia. Entomology departments at universities and museum and entomological societies in a number of countries have sponsored insect expositions (insect expos) that draw school groups and families from great distances to view these fascinating creatures firsthand. Cockroach races are regular features at insect expos. The human flea, *Pulex irritans*, was the center of attraction in American flea circuses, where their antics would attract Depression-era audiences to see a show at more than the cost of a double-feature movie.

Butterflies are a charismatic group of insects that are recognized and appreciated by almost everyone. The butterflies' spectacular, often iridescent beauty has caught the eye of naturalists and collectors alike. In Victorian times, Lord Rothschild employed more than 400 explorers to seek out and collect butterflies for what became the largest personal butterfly collection in the world. Although rearing, buying, and trading butterflies has been a popular pastime in Europe since the days of Queen Victoria, one of the first butterfly houses was inaugurated only relatively recently, in 1977, to attract tourists to Britain's island of Guernsey, whose poor weather left little to recommend it. After Guernsey's commercial tomato industry failed and the plastic growing houses were abandoned, someone thought to plant tropical gardens in those plastic houses and populate them with exotic butterflies. The idea was a success and was copied elsewhere well into the 1980s. In 1977, exotic butterfly suppliers were unknown, but the industry has since spread to Thailand, Malaysia, the Philippines, El Salvador, Costa Rica, Taiwan, Kenya, Madagascar, and the United States.

Why not release butterflies at your wedding ceremony or next celebration? Environmentally correct, butterflies can be impressive when they take flight. Forget the rice, birdseed, and confetti. Painted lady *(Vanessa cardui)* and monarch *(Danaus plexippus)* butterflies are bred and sold for such releases. One particularly poignant use of butterfly releases took place at a fund-raising event in Costa Rica, where each member of the legislature released a butterfly and simultaneously called out the name of a street child to whom the release was dedicated, lifting the children's hopes and aspirations skyward.

Who can resist the calming effects of sounds emanating from nature? Many commercial stores play bird, frog, and other nature sounds as a means of enticing customers to come in and shop. Compact discs (CDs) are recorded and sold for commercial use and for households wishing to bring that calming quality into the home. Among the insect sounds recorded and marketed are those of cicadas, grasshoppers, tree crickets, mole crickets, ground crickets, and katydids singing; June beetles flying; honey bees, bumble bees, yellowjackets, and midges swarming; and medleys of insects communicating or otherwise sounding off in nature. These recordings are sometimes played at insect expos to help bring a sense of reality to those who come to imbibe the amazing presence of the insect world.

The entertainment industry takes advantage of fascinating, educational, scary, and exciting properties of insects by featuring them in movies and on television. Insects are topics of education and wonder on various television series. The cost of producing these documentaries, largely filmed in the wild, is covered by sponsorship. Like early films with insect subjects, children's movies rarely film living insects; instead they use graphical characterizations and cartoon images. Fictional films made for more mature audiences, however, usually present the frightening or horrifying aspects of insects, and for this purpose, people are hired as "insect wranglers" to supply and manage live insects on the set. Such management demands a basic understanding of insect behavior, including knowing how to influence the insects to "act" in the way desired by the film director. Discovering that dead insects

were easier to manipulate than live ones, Wladislaw Starewicz wired dead specimens and manipulated them frame by frame to simulate desired actions in his early short, *The Fight of the Stag Beetles.* In the 1978 film, *The Swarm,* killer bee invasions were depicted by filming actual bees.

Waging War with Insects

Insect-borne diseases have taken the lives of countless soldiers throughout the ages. Millions have fallen to malaria, yellow fever, dengue, and a host of other diseases transmitted by mosquitoes. The purposeful waging of war with living insects dates to at least the 14th century when the Tartar army catapulted bodies of bubonic plague victims into Kaffa. Although knowledge that fleas spread this dread disease would not come until much later, the tactic nonetheless served its purpose.

Using insects to destroy agricultural crops seems to have emerged as a weapon of war only in modern times. Harlequin bugs, *Murgantia histrionica,* were introduced into the South, presumably in an effort to destroy the crops of the Confederacy during the American Civil War. Insects were used in both world wars as purposeful weapons. During World War II, the Japanese undertook the first large-scale use of insects as weapons of war by mass-producing an astonishing 500 million fleas bearing plague bacilli per year! In 1950, during the Cold War, the United States was accused of dropping Colorado potato beetles over East Germany. The Korean War brought to the Far East theater some 14 additional insects purportedly propagated in the United States as agricultural and medical warfare agents. The Vietnam War introduced additional entomological agents of war, especially as vectors of anticrop agents like plant viruses (e.g., beet curly top and Fiji disease), and fungi, (e.g., fire blight, cornwilt).

It was not until 1972 that insects were explicitly banned as weapons of mass destruction by the Biological Weapons Convention. Even though the mass production of these biological weapons was carried out exclusively by governmental agencies acting in secret, the trickle-down effects on local economies of producing entomological "weapons" must have been notable.

Entomological warfare does not stop with wars, where humans square off against each other. In 1990, another relatively large-scale war was waged, this time on the illicit drug trade. In fact, the U.S. government allocated $6.5 million to investigate, breed, and air-drop lepidopterous caterpillars to devour fields planted to coca in tropical Peru.

Insect Identification Services

There are so many insects in this world that it is difficult to identify them. Only by having authoritative determinations can many of the various insect-oriented industries succeed. Because of this demand, identification services have sprung up around the world. Some are geared toward the identification of agriculturally or medically related insects, but many focus on identifying insects in the context of biodiversity, especially of benthic invertebrates.

FARMING INSECTS FOR THEIR PRODUCTS AND BY-PRODUCTS

Not only living insects are marketed. Dead insects and products derived from them can also be of high commercial value. In fact, insect products and by-products probably account for the lion's share of insect commercialization.

Implements for Research

Insects provide critical basic tools for studying a great many aspects of biology. Because *Drosophila melanogaster,* a common fruit fly, is small, has a short life cycle, and is inexpensive and easy to rear, it is an extremely valuable organism for biological research, particularly in the fields of genetics and developmental biology. *Drosophila* has been used extensively and intensively as a model organism for research for almost a century, primarily to uncover the relatedness of genes to proteins and to study and map the underlying mechanisms of genetic inheritance and gene expression. More recently, the field of developmental biology, especially embryology, has relied on *Drosophila* in explorations of how a complex organism arises from a relatively simple fertilized egg. The genome of *Drosophila*, recently sequenced, maps the gene structure of that seminal organismal model. Gene products such as *Drosophila* polypeptides and transcripts, and investigative tools such as the *Drosophila* Activity Monitor for circadian rhythm research, provide highly marketable products for the scientific supply industry. Moreover, specific, even mutant strains of *Drosophila* may be purchased, as well as supplies for rearing and maintaining cultures, and specialized equipment for conducting experiments.

Insect products are also marketed for other research functions. For instance, they are used for genetic and molecular markers. The enzyme luciferase, derived from fireflies, is an excellent marker for assaying gene expression. These markers are produced and sold commercially. Indeed, specialized equipment for detecting the expressed bioluminescence is also marketed. Cell lines derived from insects is another powerful research tool. For example, protein-based human and veterinary vaccines and therapeutic proteins are produced by using baculovirus expression vector systems in insect cell lines. Human and animal protein products derived from insect cell lines are marketed for a number of purposes, including drug screening and clinical trials.

Food Products

Insects are an extremely rich source of high-quality proteins, fats, essential vitamins, and minerals. It is therefore not surprising that dead insects and products derived from them are marketed for their nutritional value. These products can take the form of human food, pet food, and livestock feed.

HONEY   One can hardly think of insects as a source of human food without envisioning honey, diligently produced by worker bees. Honey was used as a sweetener in ancient Egypt and continues to be popular today, both in cooking and for sweetening foods to be consumed immediately. Entire industries are built around honey bees both as crop pollinators and as master producers of honey. The latter industry ends with the sale of honey products on the supermarket shelf, but the intermediaries are varied and include, beyond those involved in rearing honey bees, equipment for extracting honey from combs, devices for straining and clarifying honey, and beekeeping books and magazines that keep the honey producer up to date on the latest developments in the industry.

HUMAN FOOD   One often thinks of insects as human food in a novelty context, like being dared to eat fried mealworms, crickets, or chocolate-covered ants at the county fair. But insects have been a serious source of human nutrition for a very long time. This association substantially waned as urbanization and "westernization" spread, but in the less developed corners of the globe it continues unabated. Accordingly, about 500 species in some 260 genera and 70 families of insects are used for human food somewhere in the world, especially in central and southern Africa, Asia, Australia, and Latin America. Even in the West, insect foods need not be a novelty. Where they are consumed, insects provide 5.10% of the annual animal protein of indigenous peoples. Some Native American peoples consumed saturniid moth larvae as a main part of their diets. Currently, more than 100 species of insects are sold as human food at local markets in rural Mexico, where they constitute a regular part of the local diets. In Thailand, the specialized sex pheromone gland from giant water bugs provides a flavoring to shrimp paste. Thus, marketing insect-derived foodstuffs in selected regions of the globe contributes to local economies, but repugnancy of insect foods in western cultures continues to thwart economic opportunity for mass-producing and marketing these products in the West.

PET FOOD   Birds, lizards, fish, caiman, crocodiles, turtles, and a host of other insectivorous pets survive and breed much better if supplied with protein and nutrients that are available from live or dead insects. Rearing and selling these insects to the public is a thriving business (Fig. 2) Madagascar hissing roaches are sold as reptile food, whereas crickets are marketed for consumption by a variety of pets.

LIVESTOCK FEED   Beyond pet food, insects can provide a highly nutritious food source for domestic animals and livestock. Although low in such amino acids as methionine-cysteine, arginine, and tryptophan, when supplemented by

FIGURE 2 Superworms *(Zophobus morio)* are popular large mealworms originating from South America available in pet stores for reptile and amphibian food. (Photograph by Gail Kampmeier.)

these, insect protein forms an excellent feed. Under clinical trials, white rats (the universal experimental animal for testing new medical and pharmaceutical findings) fed Mormon cricket meal demonstrated the great potential of insects as a major source of protein for rats. China recognizes the potential nutritive value of insects as feed for fish, poultry, pig, and farm-grown mink. There, experiments have demonstrated that insect-derived diets are cost-effective alternatives to more conventional fish meal diets. House fly larvae and pupae, silkworm pupae, and mealworm larvae are the major source of these insect-based diets. Fly larvae fed on poultry manure have been experimentally incorporated back into poultry feed. When this system is in place, it will take the concept of recycling to a whole new level.

Secretions and Dyes

A number of insects have the ability to secrete substances such as waxes and resins through specialized glands. Dyes too can be extracted from insect tissues. Many of these products are of high commercial value.

SERICULTURE Among fine fabrics made of natural products such as wool, cotton, linen, and leather, silk is almost always the most highly prized. Silk cloth is woven from a secretion of the silkworm, *Bombyx mori*. In the Orient, sericulture, a 4700-year-old industry, has built up around this insect and its precious secretion. The silk is a continuous-filament fiber consisting of fibroin protein, secreted from two larval salivary glands in the insect's head, and a gum called sericin, which cements the two filaments together. Silkworm larvae secrete this substance to weave cocoons within which they pupate. To obtain the fibroin protein filaments, cocoons are softened in hot water to remove the sericin. Single filaments are drawn from cocoons in water bowls and combined to form yarn, which is drawn under tension and wound onto reels, dried, packed according to quality, and sold as raw silk. It was once believed that silklike synthetic fibers would replace silk, thus decimating the silk industry, but that has not occurred. In fact, world silk production nearly doubled over the last 30 years. Together, China and Japan manufacture more than half of the world production. Other countries, like Nepal, are intensifying their silk production. The sericulture industry is complex, and many suppliers commercially produce and sell products to culture silkworms, obtain the raw silk, refine the silk, weave it, produce clothing from it, and sell the products on the market. Wild sericulture also exists: that is, fibers from cocoons other than the silkworm are used, often by native peoples, in a similar manner. This industry is less relevant to the modern world of commerce, but it fuels local industry and provides clothing and other needs of native peoples, especially in India.

SPIDER SILK Spider silk, like that of silkworms, is composed of fibroin. However, unlike silkworms, which secrete silk from salivary glands in the head, spiders secrete silk from glands at the tip of the abdomen. Depending of the type of silk that is to be made, the spider mixes the fluid from up to six different glands and regulates the speed and volume of release. Spider silk is an extraordinarily strong and elastic material. On a weight basis, it is stronger than steel; a pencil-thick strand of silk is strong enough to snare a Boeing 747 airplane in midair. DuPont advertises that the company's researchers are studying biopolymer structures of the spiderwebs. They have used recombinant DNA technology to produce analogues of spider silk in yeast and bacteria and are planning to promote this synthesized material for all manner of construction purposes.

ROYAL JELLY Royal jelly, a substance secreted by the salivary glands of worker honey bees, stimulates the growth and development of queen honey bees. It is one of the most difficult of all foods to harvest, commanding astronomical prices because of its scarcity and high demand, fueled by belief in its healing properties. What royal jelly can do for humans is controversial, but it purportedly reinvigorates the body and extends the life span. Pantothenic acid, a major ingredient, is useful in treating some bone and joint disorders. Rheumatoid arthritis symptoms may subside with the injection of this acid. When pantothenic acid is combined with royal jelly, even better results are reported. This product is sold by many health food companies.

BEESWAX Glands on the underside of young worker honey bee abdomens secrete small wax platelets, which are masticated and molded inside the hive into a comb of hexagonal cells that are then filled with honey. Additional wax is used to cap the cells for honey storage. Of all the primary products of the honey bee, wax has been, and remains, the most versatile and most widely used material. For centuries, beeswax has been regarded as the best material for making candles. An excellent wax for polishing woods and floors, it is also an ingredient in general-purpose varnishes. It has uses in packaging, processing, and preserving foods, and as a separation agent in the confectionery industry and in cigarette filters. Textiles and papers are waterproofed with products containing beeswax. Emulsions containing beeswax clean and soften leather goods.

Batik, an Asian method of coloring cloth, is based on the principle that wax (traditionally beeswax) protects areas that are not to be stained when the cloth is immersed in the dye solution. This protection feature is used for waterproofing and as an anticorrosion rust inhibitor to prevent dissolution of the metal in steel drums used to store and ship honey. Materials for embedding or electrically insulating circuits of high and ultrahigh frequency include beeswax. Beeswax is used as a binder when lubricant characteristics are desired or if mixtures are to be ingested. It is an ingredient in slow-release pellets of pyrethrum pesticides. Glass can be etched with hydrofluoric acid when areas that are not to be etched have been protected with beeswax. Various inks, pens, markers, and even carbon paper often contain small amounts of beeswax. Ancient jewelers and artisans formed delicate objects from wax and cast them later in precious metals. Colors of 2000-year-old wall paintings, as well as wrappings of Egyptian mummies, contain beeswax. Beeswax has long found use in medicines and body lotions. As a coating for pills, beeswax facilitates ingestion. Other products in which beeswax is a traditional ingredient are grafting wax, crayons, sealing wax, protective car polishes, and thread for sewing sails and shoes.

RESINS Shellac has been in use for 3200 years and is made from an insect native to India and Myanmar, the lac scale, *Laccifer lacca*. Lac females infest branches of fig trees and cover their bodies with a resinous secretion that hardens into a shield. Between 17,000 and 90,000 insects are needed to produce a pound of lac. The resins are ground to free the lac granules, which are then crushed and boiled in water. The floating lac is skimmed off, dried, and placed in burlap bags, which are stretched over a fire. As it is heated, the bags are twisted and the melted lac drips out. Before hardening, the lac is stretched like toffee. After hardening, the lac is broken into pieces and sold. Lac is the basic ingredient of a vast list of products besides shellac, including stiffening agents in the toes and soles of shoes and felt hats, shoe polishes, artificial fruits, lithographic ink, glazes in confections, phonographic records, playing card finishes, and hair dyes.

INKS Iron gall ink is arguably the most important ink in the history of the Western civilization. It is made of vitriol, gum, water, and, most notably, tannin extracted from Aleppo galls. Oaks produce Aleppo galls in response to a chemical substance secreted by larvae of the cynipid wasp, *Cynips*

gallae-tinctoriae. The gall provides both food and protection for the larva. Tannin content of the gall is highest before the wasp exits. Iron gall ink is still sold and used for many purposes. Because iron gall ink is indelible, it was the ink of choice for documentation from the late Middle Ages to the middle of the twentieth century. It was very popular with artists as a drawing ink, used with quill, reed pen, or brush. It is now used by the U.S. Treasury in the ink for printing money. The range of objects that contain iron gall ink is enormous. It was used for most manuscripts, music scores, drawings, letters, maps, and official documents such as wills, bookkeeping records, logs, and real estate transactions.

DYES Historically, adult female Mediterranean scales (*Kermes iticies* and *K. vermilio*), Oriental lac insects (*Kerria lacca*), central European scales (*Porphyrophora polonica*), and New World cochineal scales (*Dactylopius coccus*) were used in the preparation of red dye by a number of indigenous populations. Today, cochineal dye is the most important. It is obtained from an extract of the bodies of scale females found feeding on a cactus native to Mexico and Central America. The insects' bodies contain the pigment called carminic acid, which is effective in repelling potential predators such as ants. This substance is obtained by subjecting a mass of the crushed insects to steam or dry heat. Because 70,000 scale bodies are needed to produce a pound of cochineal, the dye is extremely expensive. Once commonly used as a scarlet-red mordant dye for wool and as a food coloring, cochineal has been largely replaced by synthetic products. It continues to be used as a coloring agent in cosmetics and beverages. Furthermore, the art of cochineal dying is practiced by natives in southern Mexico. The cochineal scale is still widely cultivated as a source of commercial dye in the Canary Islands and in parts of Central and South America. It is sold and chiefly used now as a biological stain.

Pharmacology

Even 3600 years ago, insects, their parts, and toxins derived from them were used to alleviate a number of human ills. Some of the remedies were less than effective (e.g., notably hirsute flies and bees used to treat baldness). Other insect-derived remedies were more credible because they have at their core a chemical property that today confirms their efficacy. For example, the hemolymph of cicadas has a high sodium ion concentration and was recommended in preparations to treat bladder and kidney dysfunction. Hemolymph is known to possess antibacterial properties and has thus been recommended in prescriptions to treat bacterial infections and sepsis. Traditional Chinese medicine includes a wealth of insects and other arthropods in its pharmacopoeia. Dried cockroaches, blister beetles, maggots, silkworm larvae, cicada exuviae, cicada nymphs and adults, and recipes using mole crickets, mantid oothecae, and silkworm frass can be purchased at traditional Chinese drugstores.

Aside from bee venom therapy described earlier, products from honey bees have long been used to promote health and as a food source (Fig. 3). Honey, royal jelly, bee pollen, and propolis are all sold to treat a variety of ailments from anorexia to insomnia to cardiovascular diseases, and to promote wound healing. More information can be obtained from the American Apitherapy Society.

Blister beetles are the major source of cantharidin, the active ingredient of "Spanish fly." This chemical has been used to topically treat warts and can be ingested for its aphrodisiac properties. Acute renal failure and death can arise from overdosing on cantharidin. These findings have prompted the removal of cantharidin from use in the United States, but Chinese researchers have discovered that beetles (e.g., *Mylabris phalerata* and *M. cichorii*), long used in traditional medicines, contain antitumor properties. Researchers are attempting to balance the potential cancer-fighting properties with undesirable side effects by testing less toxic analogues of cantharidin.

Adornments and Displays

Certain insects lend themselves or their products to the making of spectacular jewelry. Beetles are probably the most notable because of their durable, often iridescent, hardened forewings, called elytra, and interesting body shapes. They can be made into brooches or encased in plastic for key chains and paperweights; many tropical species are reared specifically for this purpose. Beetle elytra have also been woven into textiles. Insect galls and morpho butterfly and dragonfly wings have been incorporated into jewelry designs. Caddisfly larvae glue together tiny stones, grains of sand, and bits of litter to form cases that protect them in their aquatic environment. Furnished specific materials such as gold nuggets, shells, or semiprecious stones, they will incorporate these materials into their protective cases, which can then be harvested and made into earrings, necklaces, tie tacks, and pins. Insects trapped in

FIGURE 3 Honey, skin care products, and cough lozenges all make use of products from honey bees. (Photograph by Ann Coddington Rast.)

fossil amber also are sold for jewelry and displays. Although butterflies and beetles are commonly encountered in displays, a wide variety of insects are sold for those purposes, as well whether as decoration or for educational uses.

Party Favors and Pranks

For the prankster, live Madagascar hissing cockroaches are sold as party favors and "stocking stuffers" for the holidays. Honey bees embedded in plastic cubes shaped like ice can be purchased to be placed in a guest's drink. Mexican jumping beans, which are bean seeds containing larvae of a small moth *Carpocapsa saltitans*, have been popular as novelties for decades.

MARKETING INSECTS IN THE ENVIRONMENT

A more nebulous category of insect commercialization surrounds the marketing of insects in the wild. Bioprospecting, ecotourism, and conservation enhancement are modes through which insects are marketed in an environmental context. These modes frequently interact to serve the broader intent of environmental protection.

Biodiversity Prospecting

Biodiversity prospecting involves the exploration, extraction, and screening of commercially valuable genetic and biochemically active compounds of plants, arthropods, and microorganisms for pharmaceutical development and agricultural and industrial use. That some 200 pharmaceutical corporations and biotechnology companies are now stalking the wilds in search of biological riches is convincing evidence of the economic potential ascribed to bioprospecting. The vast array of insect compounds that are being discovered, reexamined, and put to new uses in disease treatments lags behind that of the botanicals currently being exploited. In combination with the tendency of many insects to sequester or change plant compounds they have ingested, there is an enormous untapped source of potential insect or insect derived compounds for medicine in the biodiversity of this planet.

With advances in molecular biology and the availability of more sophisticated diagnostic screening tools, it is increasingly cost-effective for commercial organizations in search of new pharmaceuticals to seek out natural products. This trend has resulted in a soaring market for candidate biological specimens, a market that currently tops $40 million per year for the pharmaceutical industry alone. Because of the difficulty many governments have encountered in maintaining sovereignty and control over their resources, there has been a surge of interest in legislation governing access to resources and in ensuring that host countries benefit from the commercial products fashioned from their native species.

Ecotourism

Tourism is the leading economic sector in several tropical countries. It is dependent on the lure of a warm climate, relatively low prices, and perceptions of relaxation, excitement, and even educational appeal. Ecotourism takes advantage of the attractiveness of adventure by offering the enticement and wonder of nature in an exotic setting. Insects, too, provide tourist attractions, and perhaps the best example involves the monarch butterfly, a popular insect in North America. A tropical species, it extends its range northward well into Canada during the growing season but cannot overwinter there. Individuals retreat southward for thousands of kilometers each autumn to take up residence in climes more amenable to their survival. These butterflies are attractive to ecotourism enterprises precisely because of this pattern of movement accompanying the remarkable biology of the insects. Almost anyone can view these beautiful butterflies flitting around meadows and parklands during the summer months. But as autumn approaches, they begin remarkable journeys southward and westward towards one of two destinations, depending on where they grew up. Those east of the Rocky Mountains migrate to the Monarch Butterfly Biosphere Reserve at the high-altitude oyamel fir forests of Michoacán, in central Mexico, where they overwinter in extraordinary aggregations of millions of individuals. Those born west of the Continental Divide migrate southwestward and take up residency in the monterey pines, cypress, and introduced eucalyptus trees of Natural Bridges State Park and Monarch Grove Sanctuary in Pacific Grove, on the Monterey Peninsula of California, where they too overwinter in large aggregations. These two localities are ecotourist destinations. Entire tourist industries surrounding each locality are based on this amazing insect and its habitat. Accommodation, guided tours, and, in the case of Pacific Grove, considerable emphasis on fine dining, are featured. Organizations like Friends of the Monarch in Pacific Grove promote this ecotourism.

Conservation Pursuits

Conservation efforts fold together the concepts of ecotourism and bioprospecting in an effort to protect the landscape and the biota it contains. One intent of ecotourism is to sustain the environments that attract the tourists, permitting the business to remain viable. The indigenous Ejido community of central Mexico, for example, depended on income from logging in the buffer zone of the Sierra Chincua sanctuary, the largest and most pristine monarch butterfly overwintering area in the world. Through a leasing contract, the community agreed to cease logging sanctuary forests in exchange for compensation of lost income from ecotourism profits. When agreements are made with the care of the earth as a goal, bioprospecting can also be an instrument for conservation.

Although not big business, conservation efforts can involve the production and sale of insects. Indigenous populations that use natural areas will maintain them if profitable industries, based on gathering and selling renewable resources of the system, can be developed. Jewelry made from beetle elytra and sold at local tourist markets is an example. Insects are sometimes bred and released into the wild to enhance the preservation of the species. A butterfly breeding industry has emerged in many corners of the world where pupae are sold to collectors and accumulated for release into habitats where the species is, for one reason or another, becoming rare. In Papua New Guinea, participants in a butterfly farming project sell live and preserved butterflies to collectors around the world. They earn between $2500 and $5000 per year, 50 to 100 times the average per-capita income of $50. Residents who gain from this industry have a stake in protecting the local environment where wild butterfly stocks originate. Conservation groups encourage the sale of reared butterflies because that reduces the pressure on threatened and endangered species in the wild. Furthermore, by releasing a portion of the reared specimens back into the wild, the industry encourages ecotourism, which, in turn, brings added wealth to the community. A butterfly ranching project in Barra del Colorado in northeastern Costa Rica, is an example. It provides sustainable income for its participants and assigns a portion of the stock bred from wild and captive butterflies for release back into the wild.

See Also the Following Articles

Bee Products • *Food, Insects as* • *Honey* • *Medicine, Insects in* • *Silk Production*

Further Reading

Akre, R. D., Hansen, L. D., and Zack, R. S. (1991). Insect jewelry. *Am. Entomol.* **37(2),** 91–95.

Beekeeping/Apiculture/Imkerei/Apicultura. (2002). http://www.beekeeping.org/ Last updated January 10, 2002. Accessed March 12, 2002.

Cherry, R. H. (1987). History of sericulture. *Bull. Entomol. Soc. Am.* **33(2),** 83–84.

Crane, E. (1983). "The Archaeology of Beekeeping." Cornell University Press, Ithaca, NY.

DeFoliart, G. R. (1992). Insects—An overlooked food resource. (J. Adams, ed.), *In* "Insect Potpourri: Adventures in Entomology." pp. 44–48. Sandhill Crane Press, Gainesville, FL.

Family of Nature Websites. (2000). The Butterfly Website: Public butterfly gardens and zoos. http://butterflywebsite.com/gardens/index.cfm/ Last updated June 14, 2000. Accessed March 12, 2002.

Genetics Society. (1997). FlyBase: A database of the *Drosophila* genome. http://flybase.bio.indiana.edu/ Accessed March 12, 2002.

Krell, R. (1996). Value-added products from beekeeping. FAO Agricultural Service. Bulltin 124. http://www.fao.org/docrep/w0076e/w0076e00.htm/ Accessed March 12, 2002.

Lockwood, J. A. (1987). Entomological warfare: History of the use of insects as weapons of war. *Bull. Entomol. Soc. Am.* **33,** 76–82.

Manning, G. (2000). A quick and simple introduction to *Drosophila melanogaster.* http://ceolas.org/fly/intro.html/ Last updated Oct. 23, 2000. Accessed March 12, 2002.

National Honey Board. (2001). http://www.nhb.org/ Accessed March 12, 2002.

Reid, W. V., Laird, S. A., Meyer, C. A., Gámez, R., Sittenfeld, A., Janzen, D. H., Gollin, M. A., and Juma, C. (1993). Biodiversity prospecting: Using genetic resources for sustainable development. World Resources Institute. Washington, DC.

Strickler, K. (2000). Solitary bees: An addition to honeybees. http://www.pollinatorparadise.com/Solitary_Bees/SOLITARY.HTM/ Last updated 21 Feb. 2001. Accessed 12 March 2002.

Tauber, M. J., Tauber, C. A., Daane, K. M., and Hagen, K. S. (2000). Commercialization of predators: Recent lessons from green lacewings (Neuroptera: Chrysopidae: *Chrysoperla*). *Am. Entomol.* **46(1)**, 26–38.

Thompson, J. C. (1996). "Manuscript Inks." Caber Press, Portland, OR.

Conservation

Tim R. New
La Trobe University, Melbourne, Australia

Insect conservation includes two main contexts. Insects may be conservation "targets," whereby particular species become the focus of concern because of their perceived decline in abundance or distribution, or insects may be conservation "tools," in which they are incorporated into broader aspects of conservation concern through their sensitivity to environmental changes and used as "signals" to monitor or herald changes to natural environments. This role is facilitated by their high richness and diversity in most terrestrial and freshwater environments. Both contexts reflect concern over human intervention with the natural world and the desire to sustain both components (i.e., species and equivalent entities) and processes in natural ecosystems. The major roles of insects in sustaining ecosystem services and processes acknowledge their immense richness and biomass and are reflected in E. O. Wilson's famous characterization of invertebrates as "the little things that run the world."

Nevertheless, with few exceptions, ideas of conserving insects are difficult for many people to accept. In contrast to higher vertebrates and many vascular plants, which people accept readily as objects worthy of conservation, insects have a poor image and are more commonly viewed as objects for suppression or elimination. They are regarded broadly as pests or nuisances or by some disparaging epithet such as "bird food" (however important that categorization may be in sustaining community integrity).

Insect conservation has a long history, mainly through focus on the more popular groups, such as predominantly butterflies, dragonflies, and some showy beetles. These insects are accepted widely as "worthy," simply because people like them and regard them as harmless. It is also revealing to see the commonly polarized perceptions of "a butterfly" and "a moth" despite these being artificial segregates of the same insect order. Concerns arose over decline of particular species from the mid-19th century onward. Initial concerns, and the foundations of modern insect conservation practice, were in western Europe and North America but have expanded to encompass many parts of the world. Conservation in practice includes application of biological knowledge to manage or sustain species and other higher ecological levels, which reflects the total biodiversity and linkages that occur within the complex, imposed framework of regulation and socioe-conomic needs that provides for ever-increasing human populations. "Biodiversity" encompasses both taxonomic and genetic diversity, with conservation aiming, broadly, to prevent its loss—either by the extinction of threatened or rare entities or by preventing other entities from decline to that state. As major components of biodiversity, in terms of species richness, ubiquity, and ecological variety, insects are an important and increasingly appreciated component of global conservation need.

PROBLEMS WITH INSECT CONSERVATION

Traditionally, most insects have been largely disregarded in conservation, on the premise that they may be secure under measures taken to conserve more charismatic taxa such as warm-blooded vertebrates. The latter are supposed widely to act as "umbrellas" for most or all coexisting species, but this idea is now recognized as oversimplistic, because many invertebrates are ecologically specialized and need detailed management to sustain them in the face of environmental change. However, without past emphasis on vertebrates, many habitats and sites recognized as of considerable importance for insects would surely have been lost. One attraction of basing conservation on groups such as birds or mammals is simply that they are relatively well known: their diversity is limited and tangible, most of the species are named, and many are recognizable without having to capture and kill the animals for detailed examination; their biology and habitat needs are reasonably well understood, and their distributional ranges and patterns defined; even the numbers and population sizes of many species can be evaluated reliably. Parallels with large showy butterflies and dragonflies have led to these being referred to as "birdwatchers' bugs," but they contrast dramatically with most other insect groups. Uncertainties over levels of species richness, that most species are still unnamed or even uncollected, and fragmentary or nonexistent ecological and distributional knowledge provide severe impediments to defining the patterns of diversity and distributions that may constitute the template for conservation evaluation. For many insect habitats in most parts of the world, we have little idea of insect species richness and identity. Many insect species are known solely from long-dead museum specimens and may never be seen alive. Of the world's 12 "megadiverse countries" (collectively estimated to harbor more than 70% of earth's animal and higher plant species), only for Australia can reasonably informed approximations of the extent, distribution, and ecological features of the insect fauna be deduced sufficiently to make conservation recommendations above the universal need to safeguard natural habitats. For the far more species-rich tropical countries, the paucity of resident entomologists and differing priorities render such data very approximate and their accumulation a low priority. Costa Rica, recently subject to an internationally sponsored biodiversity inventory through its national Biodiversity Institute, is an important exception. It is salutary to reflect that a decline in individual insect species in well-known (predominantly temperate-region) faunas can arouse substantial conservation interest and action, whereas tropical habitats supporting far more insect species than the total fauna of any European country disappear rapidly.

There is little reason to doubt that numerous insect species have become extinct as a direct result of human activities during the past few decades, although most have not been documented, and that the process continues. Insects are a major component of what has sometimes been referred to as "the sixth great extinction," considered likely to result in the loss of a substantial proportion of the world's species within a few decades. Efforts in insect conservation are an important avenue to increasing the understanding of human impact on natural ecosystems and of the subtle steps needed to safeguard them in the face of accelerating losses. However, the complexity of the issues involved demands a clear perspective and allocation of priorities, so that limited funding and expertise can be deployed for the greatest benefit.

Developing such perspective has involved: (a) increasing fundamental documentation of patterns of insect species richness on a variety of geographical scales, perhaps streamlining the process by concentrating on selected focal taxa because of the immense difficulty of enumerating all insect groups; (b) selecting the most deserving taxa for conservation targets, based on urgency of need to prevent extinction; (c) defining and alleviating threats to taxa and to their host environments; (d) public and administrative education to communicate the importance of insects in the natural world, and hence the need for their conservation; and (e) evaluating the contributions of insects in broader conservation activities. These parameters recognize that, despite ethical problems with any such selection, the diversity of insect species is such as to necessitate some form of "triage" in selecting the most deserving species for management and recovery action. One consequence has been a tendency to increase the scale of conservation concern; whereas single species are the most popular conservation targets, because they are defined tangible entities to which people can relate easily, their value as "flagships" or "umbrellas" for their habitats and other community members is of increasing importance in seeking wider benefits. Most fundamentally, support will never be sufficient to treat all deserving insect species individually as conservation targets needing expensive long-term recovery actions, and so any constructive shortcuts must be explored.

RANKING TAXA FOR CONSERVATION PRIORITY

The World Conservation Union (IUCN) has initiated schemes whereby species can be signaled as of conservation concern through being included on a global *Red List of Threatened Animals* and progressively allocated to a category of threat severity based on quantitative estimates of risk of extinction. The year 2000 *Red List* includes 747 insect species, including representatives of 15 orders, but is dominated by Lepidoptera (284), Hymenoptera (152), and Odonata (137 species); daunting though this number may seem, it is no more than the detected tip of the iceberg of needy insect taxa. Many of the species included have not been evaluated critically in relation to their close relatives, for example; some are listed as the result of the zeal of individual nominators; and many insect groups have no such champions to promote their welfare.

A number of regional red data books dealing with insects have established more local priorities, as have a greater number of Action Plans and similar documents arising from country- or state-based conservation legislations. In common with other taxa, the "listing" of an insect on a schedule of protected taxa often confers legal obligation to define and pursue the necessary conservation measures needed to ensure its well-being. It is important to recognize that simply being "rare" does not necessarily indicate conservation need. Many insects are known from single localities or otherwise very small areas. "Rarity" has connotations of one or more of small numbers, limited distribution, and ecological specialization, with the rarest species being ecological specialists occupying very small areas and occurring in very low numbers. However, rarity can be a stable condition. Conservation concern arises more properly from threats caused by human intervention increasing the level of rarity, such that a risk of extinction is imposed on a formerly stable balance or a trajectory of decline is accelerated. Conditions of rarity may predispose the species to stochastic effects and increase its vulnerability if the external threat spectrum increases.

Unlike many conservation assessments for vertebrates, quantitative population data on insects are rarely available, and even large numerical fluctuations between successive generations may be entirely normal. Detection of numerical decline is thereby difficult, and the quantitative thresholds for allocating a species to the IUCN categories of "critically endangered," "endangered," and "vulnerable" generally cannot be met. For most insects for which *any* biological information is available, which is a small minority, even the basic pattern of population structure is generally unclear. Many butterfly species previously assumed to have closed populations, for example, are now known to manifest a metapopulation structure, wherein discrete demographic units (nominally subpopulations) occur disjunctly in patches of habitat across a wider area, and the whole population is maintained through rolling series of extirpations and recolonizations of the suitable habitat patches in the wider environment. Thus, even loss of whole apparent populations may be entirely normal, and the practical conservation dilemma is to distinguish these from declines and loss caused by imposition of external threats.

THREATENING PROCESSES

The following are the major threats cited in decline and loss of insect species and assemblages.

Habitat Loss and Change

Many insects depend on very intricate and specialized ecological conditions for their survival, so that critical habitat and resource parameters can be very subtle. Whereas destruction of a forest, for example, is an obvious form of habitat loss, relatively small changes in vegetation composition or microclimate may lead to decline of ecologically specialized insects. Many lycaenid butterflies, for example, depend on a tripartite association whereby their caterpillars have obligate mutualistic relationships with particular species of ants, as well as specific larval food plants, so that both of these are critical resources, in addition to the need for nectar sources for the adult butterflies. On a broader scale, many insects are limited to or associated with particular vegetation types, so that any process that diminishes forests, grasslands, heathlands, alpine meadows, mangroves, and many other habitats may harm them. The area needed for many insect populations to be self-sustaining is not large—colonies of many butterfly species can thrive on areas of less than a hectare—so that the widespread pattern of habitat fragmentation through agricultural and urban conversion so damaging to many other taxa may not necessarily be harmful for insects. But, by the same token, even limited habitat destruction or change might exterminate the entire population or species.

Despite wide supposition to the contrary, many insects do not disperse readily or far. Some butterflies are reluctant to traverse even narrow bands of open ground between sheltered or shaded habitats, so even apparently unobtrusive habitat fragmentation (such as by construction of an access road) may have severe demographic and genetic consequences through promoting isolation.

Habitat loss is the paramount threat concern in insect conservation and is potentially universal. Many insect conservation programs stress the need for habitat security and management as the most important single conservation measure. The latter aspect is critical; simply that an insect is represented in a high-quality reserve such as a National Park does not in itself guarantee its well-being, because conditions may continue to change through succession or management for other priorities. For example, in Britain, several butterflies declined following changed grassland management involving removal of grazing by domestic animals or rabbits. This led to the decline of attendant ant species because of changes in ground microclimates from denser overlying vegetation; particular grazing regimes are an integral part of habitat management for such taxa. Simply "locking up" a habitat in a reserve may be a vital first step in ensuring security, but is not an end point in conservation practice.

Invasive Species

Replacement of native flora by exotic plants has characterized much human endeavor. In Australia, native grasslands in the south east are regarded as among the country's most endangered ecosystems, having been reduced to around 1% of their original extent. There is still "plenty of grassland," but most of it is composed of exotic grass species introduced to improve pasture quality for domestic stock. Many insects (including wingless morabine grasshoppers and some Lepidoptera) that depend on native grass species are now of considerable conservation concern, as representing putative remnant populations confined to small patches of their original much wider range. Introduced plants, be they agricultural or forestry crops, weeds or ornamentals, provide opportunity for exotic herbivores to establish and thrive—often as insect pests demanding control in order to protect commercial interests. Classical biological control of introduced pest weeds and arthropods has led to numerous introductions of insect consumers, be they herbivores, predators, or parasitoids. The practice has aroused concern among insect conservationists, because of the propensity of some such taxa to invade natural environments and attack native species, rather than being restricted to the (predominantly) agroecosystem environments where their impacts are needed. Protocols for screening for safety of biological control agents continue to improve, but some recent pest management practices need careful appraisal. Neoclassical biological control (whereby exotic natural enemies are introduced to combat native pest species) is a highly controversial practice, for example, as witness the recent debate over the possible side effects of exotic wasp parasitoids against innocuous native grasshoppers coexisting with the few destructive rangeland species in North America. In such situations, lack of host specificity is a prerequisite, as the agent is to attack "new species"; the agents are thereby seen as predisposed to become invasive and attack a wider host spectrum. For classical biological control, much concern has arisen from isolated island environments such as Hawaii, where there is strong suggestion that extinctions of sensitive native insects have resulted from nonspecific agents invading natural environments. A tachinid fly, *Compsilura concinnata,* introduced to combat gypsy moth (an introduced major forest pest in North America), is known to attack a wide variety of native Lepidoptera, with recent concerns for its effects on some giant silkmoths (Saturniidae).

Invasive social Hymenoptera, particularly ants such as the Argentine ant, *Linepithima humile,* and bigheaded ant, *Pheidole megacephala,* and vespoid wasps (such as *Vespula* in New Zealand), are known to outcompete native species and to disrupt the structure of natural communities in many parts of the world. As with other invasive taxa, many of these insects are extremely difficult to eradicate once they become established, and continuing spread is a major conservation concern.

Exploitation

Exploitation of insects as a threat revolves largely around issues of "overcollecting," a highly controversial and emotive theme in insect conservation. Collector demands for rare butterflies and beetles, in particular, have led to the listing of a number of species on schedules of the Convention on

International Trade in Endangered Species (CITES) as either prohibited in trade (a few species such as Queen Alexandra's birdwing butterfly, *Ornithoptera alexandrae*) or for which numbers in trade must be monitored. Much protective legislation for insects prohibits or restricts take of specimens. Individual rare insects can command sums of many thousands of dollars on the black market or more openly in dealers' catalogs. Trade in insects is the predominant aspect of exploitation and has three main components (as nominated by Collins and Morris for swallowtail butterflies): the low-value high-volume trade (mainly in common species, for the souvenir trade and general collector supply), the high-value low-volume trade (of very rare species for collectors), and the live trade (mainly of long-lived showy species for display in butterfly houses). The second of these is the major conservation concern, with potential for illicit measures to circumvent protective measures for exceedingly scarce taxa and which has led to development of butterfly farming (or butterfly ranching) activities to help satisfy demand for high-quality reared specimens. This approach was pioneered in Papua New Guinea and has major conservation benefits in helping to reduce human pressures on primary forest habitats. Rearing butterflies for sale through a centralized government-supported agency has provided income sufficient to curtail needs for continued agricultural development in places and has allowed people to recognize forests as resources on which their sustainable incomes depend.

However, and despite widespread assertion to the contrary, there is very little evidence that overcollecting is a common threat to insect species or populations. It is almost always subsidiary to changes to habitats. Very small, isolated populations of highly desirable species may indeed be "tipped over the brink" by imposition of any additional pressures and mortality, but measures to prohibit take must be seen as a responsible action. They should be balanced against the possible loss of information to be gained from hobbyists, who have contributed most of the information available on collectable insects such as butterflies.

Pesticides and Pollution

Pesticides are a special category of environmental pollutants, in that they are chemicals designed specifically to kill insects, rather than simply the by-product of industrial and other manufacturing processes. Pesticides can have nontarget effects, with the practice of greatest conservation concern being aerial spraying of insects in noncrop environments, either accidentally or purposely. Wiest's sphinx moth, *Euproserpinus weisti,* was almost exterminated when its last known site in Colorado was sprayed with malathion, for example, and aerial use of fenitrothion against massing plague locusts in Australia before they reach cropping areas remains controversial and a stimulus to develop alternative management strategies.

Other forms of chemical contamination, of both land and freshwater environments, have been documented as harmful to

insect assemblages, and the more widespread acid rains in the northern hemisphere have undoubtedly threatened insects in forest environments. Many local pollutant effects could be cited as occasional threats. Concerns have been expressed widely over effects of insecticides on dispersive pollinators such as honey bees that can forage 5 km or more from their hives.

INSECTS AS CONSERVATION TARGETS

The most familiar forms of insect conservation are associated with campaigns to conserve individual species, often "crisis-management exercises" stimulated by perception of decline or of current or impending threats from anthropogenic changes to the environment. Many are initially local exercises, but the species may also be of national or (as regional endemics) global concern. Detection and quantification of conservation need are followed, ideally, by well-designed and effectively coordinated management or recovery plans, with adequate monitoring to determine their effectiveness. Sound biological knowledge of the species underlies any such program, and in many programs an initial research phase to elucidate key ecological features must precede optimal management. However, for this to occur, the species may need interim regulatory protection, such as a moratorium on future despoiling of its habitat.

With few exceptions, decline of insects has been difficult to quantify, because of lack of knowledge of population dynamics and absence of historical data on numbers and distribution. Many declines have been inferred from loss of habitat, on the basis of persistence in small remnant habitat patches and presumed losses elsewhere. For some well-known faunas, particularly for butterflies in western Europe, disappearances have been documented more effectively through a century or more of collector intelligence and accumulation of labeled specimens and information. The most complete example is for the British butterflies, a fauna of fewer than 60 resident species for which data are sufficient to plot reliable high-resolution maps of species incidence and change over much of the past century. Such "atlases" have been produced on a 10 by 10-km^2 scale and have progressively spawned similar compilations for other insect groups and countries, together with numerous recording schemes to define current situations. In North America, the annual Fourth of July Butterfly Count developed through the Xerces Society is an important initiative helping to define the template for insect conservation needs. However, for most of the tropics, which is the most species-rich part of the world for insects, such schemes are impracticable because of the lack of sufficiently informed resident entomologists/hobbyists and the complexity of the faunas, as well as vastly different local priorities.

Most concern for species, then, arises from perceived or anticipated declines to taxa considered rare or threatened in some way. For practical conservation the need is to define the severity of the threat(s), integrating this with knowledge of the species' ecology, and to determine and pursue the measures needed to alleviate the threat and either (a) prevent decline and/or loss or (b) enable recovery from decline to occur. Development of an Action Plan or Recovery Plan is often accompanied by listing the species on some form of legal or regulatory protection schedule, a step that commonly accords formal obligation to investigate and pursue conservation needs and confers priority for limited support over nonlisted species. However, listing has sometimes been seen as an end in itself, being considered as practical conservation rather than as a facilitating mechanism for practical conservation measures. Many legislations provide for eventual delisting of species as secure, and this can come about in two ways:

1. The more intensive examination of the species that results from listing it reveals that it is more secure than supposed previously, so that continued categorization as of conservation concern is not warranted.

2. Recovery (or broader management) measures are successful and render the species, and its habitat, secure.

Either outcome is positive, but the process emphasizes the need for periodic review of all listed species to determine changes in security and the effectiveness of conservation measures. The practical steps needed, as for a variety of other taxa, are varied but may include increasing habitat security, intensifying site management (for example, by enriching it with food plants and eliminating competitive weeds), increasing insect numbers and distribution through *ex situ* measures such as captive breeding and release, and translocations to sites within the historical range from strong donor colonies. Any such program should be monitored fully and coordinated and managed effectively by a species recovery team whose membership includes informed entomologists. Because of the novelty of insect species management in this way, many such teams still tend to focus on expertise derived from vertebrates rather than reflect invertebrate expertise strongly.

In addition to species focus, assemblages or communities of insects are sometimes adopted as conservation targets, leading to a larger scale of consideration.

INSECTS AS CONSERVATION TOOLS

The high numbers of insect species and higher taxonomic and ecological categories (guilds) comprise a significant proportion of easily sampled biodiversity, with many easily categorized forms, in terrestrial and freshwater environments. They have attracted considerable attention in attempts to document communities and to measure the impacts of changes, whereby aspects of diversity, species composition, and ecological integrity can be evaluated by using insects as highly informative surrogates or indicators in various ways. A burgeoning literature on these topics reflects movements to conserve entities above single species and emphasizes the growing awareness of the key roles of insects in ecosystems and as "signals" of environmental health. Not all insects are

amenable to use in this collective way; they are simply too poorly known. The desirable features of insect groups used as indicators include their high diversity and abundance, being widespread within the target ecosystems; being taxonomically tractable and recognizable (not always to species, and genera or families can be used instead of species in some groups, but it is highly desirable that adequate handbooks and identification keys suitable for use by nonspecialists are available); being easily sampled quantitatively or semiquantitatively by simple methods; showing demonstrable changes in response to particular sets of disturbances or otherwise being ecologically responsive; and being sufficiently understood biologically that normal fluctuations in abundance, incidence, and distribution are not confounded with disturbance effects. The best documented insect groups have naturally attracted most attention, and it is sometimes difficult to distinguish the really useful insect groups from those that simply have strong advocates but less proven worth. One constructive approach is to focus on several different taxonomic groups simultaneously and so to incorporate additional ecological breadth.

Another useful approach has been to determine the incidence of "functional groups" among ecologically diverse taxa (such as ants) in which different genera, tribes, or subfamilies coexist but have different trophic habits; respond to different physical, vegetational, or climatic cues; and interact in various ways. Local faunas can thus be characterized in functional terms and changes in balance of the different guilds used to evaluate environmental changes in, often, subtle ways. In freshwater environments, groups such as chironomid flies are diverse and have likewise been used to signal wider effects of pollution or temperature changes. In such contexts, insect indicators are the equivalent of the "miner's canary," with the strong implication that their responses may be sufficiently subtle to indicate environmental changes before the effects are reflected in other changes in biota. On a broader scale, the dependence of some insects on particular microclimates may provide a basis to monitor effects of longer term climate change. Thus, in Britain many insects are on the northernmost fringe of a broader European distribution and are confined to south-facing slopes with high insolation; with change in climate, their distribution may well also change.

Insects also have values as flagship or umbrella taxa, much as with some vertebrates in the past. Selected popular insects can capture public sympathy and are of vital importance in spreading advocacy for insect conservation and the broader values of invertebrates. These need not be indicators, but species adopted as local or broader emblems for conservation have been one of the main imperatives in development of insect conservation through bodies such as the Xerces Society (United States) and the former Joint Committee for Conservation of British Insects (United Kingdom, most recently known as Invertebrate Link). Many leading entomological societies now have sections for members interested in conservation and conservation committees to help serve these wider interests. Even more broadly, characterization of community condition in terms of "representativeness" or "typicalness" or the principle of selecting nature reserves on their values as centers of diversity, evolution, or endemism can all benefit from incorporating insects in such evaluations rather than relying on low-diversity, sometimes atypical, vertebrate assessments alone.

See Also the Following Articles

Biodiversity • Biogeographical Patterns • Greenhouse Gases, Global Warming, and Insects • Endangered Insects • Insect Zoos

Further Reading

Collins, N. M., and Thomas, J. A. (eds.) (1991). "Conservation of Insects and Their Habitats." Academic Press, London.

Collins, N. M., and Morris, M. G. (1985). "Threatened Swallowtail Butterflies of the World." IUCN, Gland/Cambridge.

Gaston, K. J., New, T. R., and Samways, M. J. (eds.) (1994). "Perspectives on Insect Conservation." Intercept, Andover, MA.

Journal of Insect Conservation (various dates). Kluwer, Dordrecht.

New, T. R. (1995). "Introduction to Invertebrate Conservation Biology." Oxford University Press, Oxford.

New, T. R. (1997). "Butterfly Conservation," 2nd ed. Oxford University Press, Melbourne.

New, T. R. (1998). "Invertebrate Surveys for Conservation." Oxford University Press, Oxford.

New, T. R., Pyle, R. M., Thomas, J. A., Thomas, C. D., and Hammond, P. C. (1995). Butterfly conservation management. *Annu. Rev. Entomol.* **40,** 57–83.

Pyle, R., Bentzien, M., and Opler, P. (1981). Insect conservation. *Annu. Rev. Entomol.* **26,** 233–258.

Samways, M. J. (1994). "Insect Conservation Biology." Chapman & Hall, London.

Wells, S. M., Pyle, R. M., and Collins, N. M. (1983). "The IUCN Invertebrate Red Data Book." IUCN, Gland/Cambridge.

Crickets

Richard D. Alexander
University of Michigan

Daniel Otte
Philadelphia Academy of Natural Sciences

Crickets are insects in the order Orthoptera that comprise the ensiferan family Gryllidae. Some authors regard them as the superfamily Grylloidae with four families: Myrmecophilidae, Gryllotalpidae, Mogoplistidae, and Gryllidae. The group dates from the Triassic Period and today includes 3726 known living species and 43 extinct ones, 22 extant subfamilies and 7 extinct ones, 528 extant genera and 27 extinct ones. Most extant subfamilies are distributed worldwide.

DIAGNOSTIC FEATURES

The Orthoptera also include katydids, long-horned and meadow grasshoppers, short-horned grasshoppers and locusts, pigmy locusts, and wetas. Orthoptera are related to stick insects (order Phasmatodea), cockroaches (order Blattodea), and mantids (order Mantodea), all of which lack jumping hind legs. Phasmatodea have three tarsal segments, Blattodea and Mantodea five tarsal segments. Crickets are further classified in the suborder Ensifera, the members of which share jumping hind legs, two pairs of wings (rarely one) or none, either three or four tarsal segments, and thread-like antennae that are longer than the body except in subterranean forms.

Crickets all have long thread-like antennae, two slender tactual abdominal cerci, three tarsal segments, and some bulbous sensory setae basally on the insides of the cerci. No other insects share all these features; the last is closest to a single defining trait, shared by only certain Stenopelmatidae (Jerusalem crickets with four tarsal segments).

Some Ensifera have been called crickets mainly because they resemble Gryllidae, lack established common names, or otherwise have obscure family connections. All of these, however, have four tarsal segments and are probably related more closely to one another, and to other noncrickets, than to crickets. Examples are bush crickets (a term used in Europe for certain Tettigoniidae), sand and stone crickets (Schizodactylidae), and camel crickets, cave crickets, wetas, and their relatives (Tettigoniidae, Gryllacrididae, Rhaphidophoridae, Anostostomatidae, Stenopelmatidae, and Cooloolidae). Some caeliferan insects related to grasshoppers are called pigmy mole crickets and have two tarsal segments.

VARIATION

The smallest crickets are tiny, wingless forms comprising the subfamily Myrmecophilinae (ca. 1 mm); they apparently live and reproduce only in ant nests. The largest are the short-tailed crickets (Brachytrupinae) called bull crickets (ca. 5 cm); they excavate burrows a meter or more deep. Different cricket groups vary from having slender, fragile, whitish or greenish bodies with virtually transparent forewings (tree

FIGURE 1 Adult male *Gryllus pennsylvanicus.* (Photograph courtesy of David H. Funk.)

FIGURE 2 Adult male *Oecanthus latipennis,* with the forewings in singing position. (Photograph courtesy of David H. Funk.)

crickets: Oecanthinae, Fig. 1) to heavy-bodied, aggressive brown and black defenders of burrows and territories (field crickets: Gryllinae, Fig. 2, short-tailed crickets: Bachytrupinae; mole crickets: Gryllotalpinae). James Thurber said of one grylline, the sturdily built European burrowing field cricket *(Gryllus campestris),* that it "has the aspect of a wrecked Buick."

HABITATS

Crickets live in virtually all terrestrial habitats from treetops to a meter or more beneath the ground. Members of multiple subfamilies live in or near treetops and in bushes, grasses, and other herbaceous plants (Oecanthinae, Mogoplistinae, Eneopterinae, Podoscirtinae, Trigonidiinae) (Fig 3); on the soil surface (Nemobiinae, Gryllinae); in caves (Phalangopsinae, Pentacentrinae); and in shallow or deep burrows (Gryllotal-

FIGURE 3 Adult male *Orocharis saltator.* (Photograph courtesy of David H. Funk.)

pinae, Brachytrupinae). Some excavate burrows in logs or standing trees (Pteroplistinae). Some beach-dwelling species of Trigonidiinae run and jump readily on water.

Females of different groups lay eggs in stems or twigs, in wood, under bark, in the ground, or in burrows. Apparently all females in the widely distributed burrowing subfamilies Brachytrupinae (short-tailed crickets, 223 species) and Gryllotalpinae (mole crickets, 76 species) are parental toward their eggs and also toward their juveniles.

GEOGRAPHIC DISTRIBUTION

Crickets occur almost everywhere on the earth between the regions of taiga vegetation that start at approximately 55° north and south latitude. Excellent dispersers, they are the only orthopterans that readily colonize far-flung Pacific islands. Some colonizers fly, others tend to lay their eggs in wood, which is carried by water far and wide. Flightless soil ovipositors are poor colonizers, except via human transport. The largest numbers of cricket species audible from a single location occur in tropical regions. In Malaysia, along about 1 km of road northeast of Kuala Lumpur, the songs of 88 species can be heard, including almost every extreme of structure in cricket songs across the world. In the richest habitats of tropical Queensland (Australia), midwestern United States (southern Ohio), and some other tropical locations, up to about 25 species can be heard in single locales. Mute species, more difficult to locate, must be added to these numbers to obtain the number of cricket species in each region.

WINGS AND FLIGHT

The forewings of crickets, when present, are typically stiff and leathery; the hind wings are membranous and fold fan-like under the forewing when not being used. The hind wings can be miniature nonflying organs (microptery), longer than the forewings (macroptery), or absent. Some macropterous individuals shed their hind wings. The hind wings may also be pulled off and eaten by their bearer or by a female being courted by a macropterous male. Some macropterous crickets, such as the subtropical and tropical American species, *Gryllus assimilis,* take off, fly, and land so adeptly as to be wasp-like; others, such as mole crickets, fly in almost comically ponderous and slow manners, some with their abdomens hanging almost vertically.

SONGS AND COMMUNICATION

In most cricket species the males chirp or trill, producing clear, rhythmic, musical sounds distinctive to their family. An upturned scraper on one forewing is rubbed along a row of fewer than 10 to more than 1300 teeth, on the underside of the other forewing; tooth number, often species-specific, correlates with pulse rate and length. The dominant frequency in the sound depends upon the tooth-strike rate,

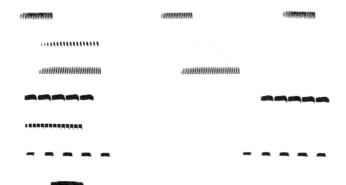

FIGURE 4 Drawings from audiospectrographs of the songs of 7 of the 17 known species of Western Australian desert crickets in the genus Eurygryllodes. Top to bottom the species are warrilla (a), warrilla (b), warrami, wirangis, yoothapina, buntinus, and diminutus. *E. warrilla* (a) and (b) have not yet been treated as different species because too little is known about them, and the available specimens have not been distinguished morphologically (from Otte and Alexander, 1983, p. 81).

ca. 1.5 to 10 kHz (Figs. 4 and 5). Sounds are pulses caused by the individual closing strokes of the wings and separated by the silent opening strokes. Pulse rates vary from one every 3 s (a Hawaiian trigonidiine) to more than 200 per second (a Malaysian gryllotalpine). The communicative significance of the songs lies in the rates and patterning of the pulses; pitch

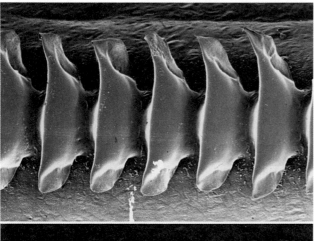

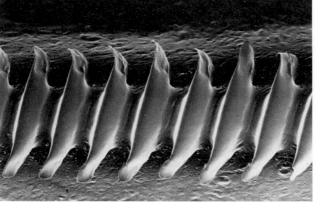

FIGURE 5 Teeth on the stridulatory files of field crickets (genus *Gryllus*) from the Galapagos Islands.

is important because the tympana on the forelegs are for each species tuned to the sounds of conspecific individuals. The calling song, best known, attracts females and challenges nearby males. The courtship or mating song induces the female to move into the copulatory position, and the fighting chirps (most obvious in surface-dwelling and subterranean species) challenge other males at close range.

Cricket sounds vary with temperature. The North American snowy tree cricket (*Oecanthus fultoni*), sometimes called the thermometer cricket, chirps at a steady countable rate, such that degrees Fahrenheit is represented by the number of chirps in 15 s plus 40. Individuals of this nocturnal species also synchronize their chirps so that a dense colony produces an intense, monotonous beat. Ralph Waldo Emerson said that if moonlight could be heard it would sound like this intense, beautiful cricket song.

Tree- and bush-living crickets tend to sing only at night, those in grasses and on the ground both day and night; presumably this situation has come about because of a history of trade-offs between obtaining mates and avoiding predators, mainly birds. Some crickets are also parasitized by flies that locate males by song. In Europe and Australia certain sounds of nightjars (Caprimulgidae) are so similar to the songs of mole crickets in the same locations that it seems likely (though unproved) that these insectivorous birds attract mole crickets, both sexes of which respond to songs in flight.

Crickets were among the first musicians on earth. The calling song of each species today—as with all so-called "singing" insects—is invariably distinct from the songs of all other species that breed in the same times and places. Species that mature and mate in different regions, or at different times in the same region, sometimes have songs so similar as to be unlikely to be distinguished by the crickets.

Cricket acoustical communication evidently evolved but once, yet has been lost many times. In Australia the stridulum (file) has been lost at least 27 times, in Africa 17 times. The auditory tympana are often retained after acoustical communication has ceased, but only in flying species, suggesting that the tympana are used to avoid bat predation (Fig. 6).

Varying shapes and locations of the auditory tympana on the fore tibiae of crickets, and variations in the structure of the stridulatory device on the forewings of males, cause the morphological devices of crickets, as well as their songs, to be important to students of phylogeny and classification. Because of their species distinctiveness, cricket songs are unusually fine tools for locating previously unknown species and for the rapid study of geographic and ecological distribution, biogeography and phylogeny, species density variations, population size, seasonal and geographic overlap of species, character displacement, and the nature of the life cycle and overwintering stage from the seasonal distribution of adults.

Biologists have analyzed cricket songs intensively for almost 70 years because they are audible, recognizable, and amenable to sophisticated acoustical, physiological, and

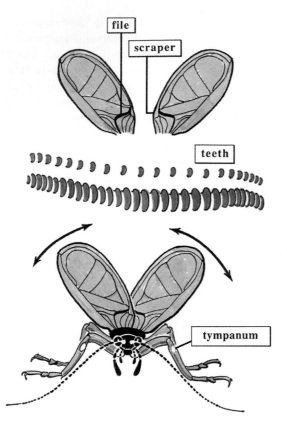

FIGURE 6 The sound-producing and auditory apparatus of a tree cricket. The stridulatory file is located on the underside of the forewing near its base. Each time the forewings close during song the teeth are rubbed against a scraper located on the underside of the other forewing. The forewings are thus caused to vibrate, more or less as a whole, as fast as the teeth are struck, producing separated pulses of a whistle-like tone. The hearing organ, or tympanum, is located on the upper portion of the front tibiae. It is tuned to the frequency of its own species' wing vibrations, causing it to transmit the rate and pattern of pulse production of all songs at that frequency to the central nervous system. The central nervous system is able to distinguish its own species' songs from those of other species with songs of similar frequencies that sing in the same times and places. In some crickets the tympanum is exposed through openings on both the inner and the outer faces of the tibia; in others there is only one opening, most often on the outer surface of the tibia, but in some (typically burrowing) species on the inner surface only.

behavioral analysis and because hybridization of different species provides insight into the genetics of song differences. As a result of the use of song to locate new species, during the past 50 years the number of known cricket species has more than doubled. With respect to genetic background, nerve and muscle physiology, and behavioral functions, the cricket acoustical system is almost certainly the most thoroughly understood of all animal communicative systems.

EGG-LAYING

Most female crickets inject their eggs into the soil or into plant stems through long, slender ovipositors. The oviposition slashes of tree crickets often seriously damage berry canes and small twigs. Females of the two subterranean subfamilies do not inject their eggs into the soil and have lost the external

ovipositor; a few Brachytrupinae retain short ovipositors and inject eggs shallowly into burrow walls. Most subterranean forms lay only a few eggs in one place in an underground chamber. In those studied, such as the North American species *Anurogryllus arboreus,* plant materials are also stored in the burrow, and a special defecation chamber is excavated; juveniles eat stored plant materials, small unfertilized eggs produced by the mother after the young hatch, and eventually the body of the dead mother. A female may dig her own burrow or ferociously take over and defend the burrow of a courting male before allowing him to mate with her at the entrance.

LIFE HISTORIES

In northern (and probably southern) latitudes most crickets overwinter as eggs and mature in late summer. A few burrowers overwinter as partly grown juveniles and mature in early summer. There are 6 to 12 nymphal molts, and the adults usually live 6 to 8 weeks. In latitudes with significant winters, life cycles vary from one generation every 2 years in a mole cricket to two generations each year. Diapause occurs in the overwintering stage. Nondiapausing crickets such as the house cricket *(Acheta domesticus)* have a generation time of a few weeks, varying with rearing temperature. Diapause also occurs during droughts in some tropical countries. Eggs or adults live through droughts, with rain causing nymphs to hatch and adults to oviposit.

FOOD

Many crickets are omnivorous. Some seem to feed almost entirely on vegetable matter, yet sometimes consume carrion and even ferociously kill and eat other insects. Several species frequent human dwellings and refuse heaps, most notably *A. domesticus* and the decorated cricket, *Gryllodes sigillatus.* Subterranean species feed mostly on roots and can be injurious when abundant in crops, gardens, lawns, golf courses, and newly reseeded forests.

HUMAN USE

House and field crickets are reared and sold in large numbers as fish bait and food for laboratory animals in many parts of the world. In the Orient, male crickets are caged for their songs, and staged battles between aggressive males in surface and subterranean species have been a favorite sport for over 1000 years. Males can be primed for serious fights—even to the death—by tickling them with brushes resembling antennae and by providing them with lairs made of small boxes that enhance their motivation to defend the site. Burrowing crickets can be extracted from their burrows by eliciting aggressive reactions to tiny pebbles rolled into the burrow followed by blades of grass used so as to imitate cricket antennae. Allowing a male to mate sets him into guarding the female for further mating, thus also priming him to fight ferociously.

MATING SYSTEM

The long-range female-attracting songs and long tactual cerci of crickets are components of a unique mating system, some aspects of which evidently trace to the earliest instances of copulation in the insect line and help explain changes leading to the current major groups of insects. Thus, none of the primitively wingless modern insects copulate, while all winged and secondarily wingless insects do, the majority with the male mounting the female and in some way holding or forcing her. In primitively wingless insects, however, a sac or bulb containing the sperm (a spermatophore) is transferred indirectly to the female without direct copulation. Like crickets, some of these particular primitively wingless insects possess prominent tactual cerci (e.g., Thysanura), used to guide the female during spermatophore transfer, as also in cockroaches and mayflies. In all insect groups of ancient origin that have prominent tactual cerci, transfer of the spermatophore is a luring act in which the female either mounts (winged and secondarily wingless forms) or stands beside the male (primitively wingless forms). In some crickets, such as the field cricket genus *Gryllus,* the copulatory act appears unique among all animals in being entirely luring, with no evidence of controlling force by the male at any stage. The female is attracted initially by the long-range calling song and then by the male's close-range courtship song and probably the fluttering touches of his antennae (Fig. 7). As in nearly all crickets, most close relatives of crickets, and most cockroaches and mayflies (the last aerially), the female mounts (or flies above) the male in the copulatory act. Apparently in correlation with the male field cricket having minimal ability to clasp the female's genital parts, the spermatophore is transferred quickly, in 15 to 90 s. The spermatophore is osmotically self-emptying, so that sperm injection occurs largely after the female dismounts from the male. In forms related to crickets, such as Tettigoniidae and Caelifera, in which males have evolved terminal claspers on the abdomen, the tactual cerci have disappeared and copulation is much lengthier. In Caelifera the mating act has evolved such that the male mounts the female, though still reaching beneath her to attach the genitalia; here, unlike Tettigoniidae, the antennae have also become much shorter. Apparently luring copulatory acts in insects have repeatedly evolved into

acts involving significant force, but the reverse does not seem to have happened. Groups of features related to the history of insect mating acts have significance for interpreting changes in diagnostic features of major groups of insects, including cerci, antennae, genitalia, wing structure, long-range communication, and modes of pair formation.

Distinctive morphological and behavioral features of crickets, especially those related to their methods of pair formation and mating behavior, make them a pivotal group in understanding insect evolution and phylogeny.

See Also the Following Articles
Cultural Entomology • Folk Beliefs and Superstitions • Hearing • Orthoptera

Further Reading
Alexander, R. D. (1961). Aggressiveness, territoriality, and sexual behaviour in field crickets (Orthoptera: Gryllidae). *Behaviour* **17,** 130–223.

Alexander, R. D. (1966) The evolution of cricket chirps. *Nat. History Mag.* **75,** 26–31.

Alexander, R. D. (1968). Life cycle origins, speciation, and related phenomena in crickets. *Q. Rev. Biol.* **43,** 1–42.

Alexander, R. D. (1969). Arthropods. *In* "Animal Communication" (R. Sebeok, ed.), 1st ed., pp. 167–215. Indiana University Press, Bloomington.

Alexander, R. D., and Otte, D. (1967). The evolution of genitalia and mating behavior in crickets (Gryllidae) and other Orthoptera. University of Michigan Museum of Zoology Miscellaneous Publication No. **133,** pp. 1–62.

Huber, F., Moore, T. E., and Loher, W. (eds.) (1989). "Cricket Behaviour and Neurobiology." Cornell University Press, Ithaca, NY.

Otte, D. (1992). Evolution of cricket songs. *J. Orthoptera Res.* **1,** 25–49.

Otte, D. (1994). The crickets of Hawaii: Origins, systematics, and evolution. *In* "Publications on Orthopteran Diversity." The Orthopterists' Society, Academy of Natural Sciences, Philadelphia.

Otte, D. (1994). Orthoptera species file 1: Crickets (Grylloidea). *In* "Publications on Orthopteran Diversity." The Orthopterists' Society, Academy of Natural Sciences, Philadelphia.

Otte, D., and Alexander, R. D. (1983). The Australian crickets (Orthoptera: Gryllidae). Academy of Natural Sciences of Philadelphia Monograph No. **22,** pp. 1–447.

Otte, D. and Naskrecki, P. (1997). Orthoptera species file online at http://viceroy.eeb.uconn.edu./Orthoptera.

Crypsis

Paul M. Brakefield
Leiden University

Organisms with bright and conspicuous color patterns tend to attract the most attention both scientifically and aesthetically. However, the majority of insects and other animals rely on camouflage or crypsis for survival from predators that hunt them by sight. Furthermore, crypsis may extend to include the other senses, namely, smell, touch, and sound. Indeed, any stimulus or signal that can alert a potential predator could be expected to become part of a

FIGURE 7 Adult female (left) and male (right) *Phyllopalpus pulchellus,* the latter with forewings in singing position. (Photograph courtesy of David H. Funk.)

FIGURE 1 Crypsis illustrated for different insects. (A) An individual of the dry season form of the evening brown, *Melanitis leda,* resting among dead leaves on the forest floor in the Shimba Hills, Kenya. The insect is at the center with head pointing to the right; forewing length is ca. 4.5 cm. (B) A small moth that resembles a dead patch on a large leaf in a forest in Costa Rica (wing span is ca. 3 cm). (C) The caterpillar of a moth of the family Lasiocampidae resting on a tree trunk in the Shimba Hills, Kenya; it is ca. 6 cm in length and is positioned horizontally, head to the right, in the center of the photo (image has been rotated 90 degrees). (D) The same larva when actively moving in the same direction along a twig. See text for further details. (Photographs by the author.)

coordinated suite of cryptic traits. A form of crypsis is also shown by some predators that disguise themselves by assuming the same color and patterns as the background on which they hunt. H. B. Cott in 1940 wrote perhaps the best known book on animal color patterns, but many of the great entomologists of the 19th century had already considered insect camouflage. It is not usual to consider insect crypsis as a subject of applied biology but there are certainly many parallels with military expertise in either the hiding of or the searching for personnel and armaments in a landscape.

COLOR MATCHING AND CRYPSIS

An insect that is perfectly camouflaged is perhaps one of the most striking exhibitions of the power of evolution by natural selection to mold and adapt organisms to fit their environment and to maximize survival and reproductive success. Wonderful examples of camouflage are presented by many species of insects, including some butterflies in tropical forests (Fig. 1A), which rest on carpets of dead brown leaves. The apparent perfection of crypsis is emphasized in many such insects by a similarity of, and matching of, the color pattern of the wings, body, and appendages to the background on which they normally rest. The color pattern of these different body parts and structures must involve

different genetic and developmental pathways, and yet evolution has led to a corresponding perfection of matching, albeit using entirely different mechanisms of pattern formation. Such an example of an underlying complexity of patterning is given by some caterpillars of the family Lasiocampidae that rest on the bark of trees and survive by resemblance to the background color pattern of the bark, including epiphytic lichens and algae (Fig. 1C). Such larvae are encircled by long hairs that are flattened around their margin when at rest. This breaks up their shape, smoothing their outline. These hairs are also patterned in a very specific way and one that is fully coordinated with the body cuticle, including the short bristles of the dorsal areas of the body segments. These elements are exposed, and the whole insect becomes highly conspicuous as soon as a larva is forced to move along a twig of fine diameter (Fig. 1D).

Furthermore, color matching in crypsis is almost always only one component of the strategy for survival; both habitat choice and, frequently, the adoption of very specific patterns of behavior and activity are required for effective crypsis. One such example is shown by some species of moths that attain crypsis by appearing to be a dead patch of tissue within a large leaf on which they rest (Fig. 1B). They achieve this not only through the generally brown color of their wings and some details of patterning, which may resemble small patches of fungal-attacked leaf tissue, but also through a precise positioning on the leaf. For example, the moth in the photograph of Fig. 1B has rolled up the leading edge of its forewing, wrapped its abdomen along the trailing edge of one hind wing, hidden its appendages, and positioned itself alongside the midrib of the leaf.

Despite the potential fascination of understanding crypsis, it is only relatively recently that scientists have begun to analyze what is meant precisely when it is stated that an organism is well camouflaged. John Endler in 1978 stated that "a color pattern is cryptic if it resembles a random sample of the background perceived by predators at the time and age, and in the microhabitat where prey is most vulnerable to visually hunting predators." There are several crucial components in this definition. First, a color pattern is cryptic only with respect to the specific environment in which the organism is potentially encountered by the predator or the guild of predators to whom the pattern is an adaptive response. What is a cryptic pattern on the resting background of that environment may be conspicuous and ineffective on any other background. Second, the effectiveness of a particular pattern is considered with respect to the normal time and lighting conditions under which crypsis is functional. Third, to be cryptic the color pattern of a prey organism must essentially reflect a random sample of the background on which it rests.

INDUSTRIAL MELANISM AND CRYPSIS

Perhaps the first analysis of crypsis and the evolution of a color pattern from the perspective of changes in camouflage involved industrial melanism in the salt-and-pepper moth, *Biston betularia*. Industrial melanism refers to an association of high frequencies of dark, melanic forms or phenotypes of a species with high levels of air pollution. The fundamental components of this classic example of the evolution of an adaptive trait also apply to numerous other species of moth and other insects that have evolved melanism as a response to environments influenced by air pollution. These components are: (1) the environment was changed by air pollution in such a way that the camouflage of the "typical" or wild type of color pattern was impaired, (2) a mutant phenotype occurred in this new environment that had a functional design or color pattern that improved survival from birds hunting the moths at rest, and (3) the dominant allele at the gene that specified this favored mutant phenotype then increased in frequency under the influence of natural selection, leading to the species exhibiting industrial melanism.

In the salt-and-pepper moth, we know from museum collections that prior to the middle of the 19th century in northern England the moths had pale-colored wings with a speckling of dark dots (the typical form). Also, up until that time in the early industrial revolution the bark of trees was predominantly pale and covered in epiphytic lichens and algae. The salt-and-pepper moth rests on bark, and females lay their eggs under foliose lichens or in cracks in the bark. The moths

are active at night and rely on background matching and crypsis for survival from birds during daylight hours. Survival enables males to mate at night and females to lay their eggs over a number of nights. The gaseous (e.g., sulfur dioxide) and particulate (soot) air pollution produced by industry both killed the epiphytic communities on the trees and blackened the resting surfaces of the moths. The typical, pale-colored moths became more conspicuous. The fully black, melanic form known as *carbonaria* was not collected until 1848, near Manchester. It may have occurred shortly before through a mutation (producing a new allele of the gene), or perhaps it had already existed for some time in that region as a rare allele. Whatever its precise origin, the *carbonaria* form rose rapidly in frequency and spread extensively through the industrial regions of Great Britain over the following decades; the adult moth as well as newly emerged larvae can move long distances. Clear geographical associations were established between the amount of air pollution and the frequency of the fully melanic *carbonaria* and also of several intermediate melanic forms known as *insularia.*

Up until the mid-20th century this remained a verbal, albeit persuasive, reasoning for the evolution of melanism as an adaptive response to a changed environment. It was only then that some classic early experiments in evolutionary biology began to add scientific rigor to this explanation. Several researchers performed a series of experiments that showed beyond doubt that, whereas the survival of the pale typical form was higher in rural, unpolluted regions of Great Britain than that of the *carbonaria* form, this relationship is reversed in the polluted industrial environments. Although there have been discussions about the precise details of some of these types of experiments, the fundamental finding of a switch in survival and relative fitnesses (reproductive success) of the pale and dark phenotypes across the extreme environments, principally the result of corresponding changes in crypsis, has been corroborated. Other differences in fitness among the phenotypes that are not directly related to the visual differences in color pattern may also be involved in determining the precise dynamics of the evolution.

There has, however, more recently been an additional finding that proves beyond any doubt the role of evolution by natural selection. Great Britain and other countries in northern Europe have over the past few decades reduced levels of air pollution from soot and gases such as sulfur dioxide. This has in turn led to declines in the frequencies of the melanic forms and the coining of the phrase "evolution in reverse." As the resting environment returns, at least in a qualitative sense, back toward the original, unpolluted state, the relative fitnesses are also reversed, leading to present-day declines in melanism. Although it has not been precisely quantified, the conclusion must be that in previously polluted regions, while the fully black melanic *(carbonaria)* has again become conspicuous and vulnerable to birds, the paler typicals have become well camouflaged on the changed background.

ANALYSIS OF CRYPSIS

This example of the salt-and-pepper moth illustrates that crypsis still needed to be scientifically measured and fully quantified. In 1984 Endler began to use early techniques of image analysis to mathematically describe how well matched in terms of color patterning were moths in a North American woodland community with respect to different potential resting environments. If crypsis is "optimal" the patterning of the insect will represent a random assemblage of the pattern elements of the background. Endler also pointed out that there will be matching with respect to different components of the color patterns of both insect and resting background, namely, size, color, shape, and brightness. In some backgrounds, such as pine needles or bark with striations, the component of orientation should also be added. Failure to match with respect to any one of these components will lead to mismatching and ineffective crypsis. Because the color vision of many predators, including birds and insects, extends into the ultraviolet part of the spectrum, when color matching in crypsis is considered it often has to include the UV. Researchers have recently begun to use computer-generated patterns, image analysis, and "visual predators" to explore more fully the potential effects of interactions among predators and their prey that lead to the evolution of cryptic color patterns.

Cryptic color patterns may also include an element of banding, which is disruptive and can serve to break up the outline of the prey. Usually, such an element also has to blend into the resting background in terms of the prey representing a random assemblage of its pattern. However, this restriction is perhaps relaxed when crypsis is used only to protect a prey from a distance, such as in the brightly colored, banded moth caterpillers, including the cinnabar, *Tyria jacobaeae,* and the strikingly striped forewings of some arctiid moths, for example, *Callimorpha quadripunctaria.*

CRYPSIS AND NATURAL SELECTION

Although testing of these ideas, at least in the context of animal color patterns and their camouflage, has not been completed, Endler has also performed experiments with guppies that dramatically illustrate the power of natural selection to lead to the evolution of effective crypsis. Male guppies can be very colorful with a patterning of bright spots and patches on their lateral flanks and fins. Laboratory experiments in which females can choose whether to mate with males of different patterns show that there is female preference for the more brightly colored males. In the wild in Trinidad, there is a correlation between the degree of color patterning on males in a population and the presence of predatory fish and invertebrates ranging from weak to strong mortality factors on guppies. Male fish are colorful and brightly patterned when either no predators or only weak predators are present, whereas they are drab and unpatterned brown fish when strong predators such as certain cichlids are present. A series of experimental pools with natural backgrounds in a greenhouse was established to examine the efficacy of natural selection on crypsis in this system. Endler showed that guppy populations with the weak predators showed no divergence over subsequent generations in their average color pattern; in contrast, in those pools to which strong predators were added the guppies showed a marked and progressive decline in the brightness and spottiness of the males. This result was highly consistent with selection favoring a more effective crypsis through a lower conspicuousness and improved background matching of the prey populations. In the absence of such strong predators, the balance of sexual selection through female choice and of natural selection by visually hunting predators favors colorful males because they survive to maturity and then achieve a higher mating success than their less colorful competitors.

Such a balance of selection on animal color patterns is probably the norm in natural populations. Thus, in animal communication, a color pattern is usually a compromise between being conspicuous to conspecifics and being poorly visible to predators (or prey). Indeed, one of the potential disadvantages of adopting crypsis as the primary means of survival is that it almost inevitably ties the organism down to a sedentary style of life at least during the hours of daylight. In contrast, when organisms are distasteful and adopt a conspicuous, aposematic lifestyle or when they evolve Batesian mimicry to resemble such warningly colored species, there is no such disadvantage associated with daytime activity.

INTERACTION OF CRYPSIS AND OTHER DEFENSES

In many insects, an organism may not rely only on crypsis for survival. There may be some secondary means of defense once crypsis has failed and the prey has been detected by a potential predator. Insects that are cryptic at a distance but conspicuous when seen close up (including the banded larvae and arctiid moths mentioned above) are often chemically protected. This type of multiple defense is also illustrated by the moth caterpillar in Fig. 1C. If the caterpillar is disturbed and begins to move it can expose a series of glands in the dorsal cuticle of several segments toward the front of the body. These are visible as a pair of partial bands in Fig. 1D, the largest immediately to the right of the largest white-colored region. These produce a pungent odor and probably provide a potential chemical defense against birds and other predators.

The effectiveness of crypsis will also show complex interactions with the visual processing abilities of the specialist predator or the guild of predators. Some insects that rely on camouflage for survival often exhibit extreme individual variation. One example is the tropical evening brown, *Melanitis leda.* This large brown butterfly is common throughout the Old World tropics. In wet–dry seasonal environments, the species shows classical seasonal polyphenism (i.e., distinct color patterns that result from phenotypic plasticity), with a wet season form having conspicuous marginal eyespots and a cryptic dry season form without such eyespots. The latter form relies on survival through crypsis on a resting background of dead brown leaves (Fig. 1A). In large numbers of the dry season form it is difficult to find two individuals with exactly the same color pattern. Dramatic variation across individuals is produced by high genetic variation in several different pattern elements across the wing (such as the contrast and brightness of particular patches and bands and the background wing color in different regions). This variation can be interpreted as an evolutionary response involving "apostatic selection" to make it more difficult for browsing predators in the leaf litter to form a specific "search image" for a particular form of dead leaf pattern corresponding to the color pattern of the prey. Although like many of the detailed ideas about the significance of crypsis and particular animal color patterns, this hypothesis remains to be tested rigorously, it does once again illustrate the fascination of crypsis.

See Also the Following Articles

Aposematic Coloration • Defensive Behavior • Eyes and Vision • Industrial Melanism • Mimicry

Further Reading

Bond, A. B., and Kamil, A. C. (2002). Visual predators select for crypticity and polymorphism in virtual prey. *Nature* **415**, 609–613.
Bradbury, J. W., and Vehrencamp, S. L. (1998). "Principles of Animal Communication." Sinauer, Sunderland, MA.
Cott, H. B. (1940). "Adaptive Coloration in Animals." Methuen, London.
Endler, J. A. (1978). A predator's view of animal colour patterns. *Evol. Biol.* **11**, 319–364.
Endler, J. A. (1984). Progressive background matching in moths, and a quantitative measure of crypsis. *Biol. J. Linn. Soc.* **22**, 187–231.
Kettlewell, H. B. D. (1973). "The Evolution of Melanism." Clarendon Press, Oxford.
Majerus, M. E. N. (1998). "Melanism: Evolution in Action." Oxford University Press, Oxford.
Sargent, T. D. (1976). "Legion of Night: The Underwing Moths." University of Massachusetts Press, Amherst.
Thery, M., and Casas, J. (2002). Predator and prey views of spider camouflage. *Nature* **415**, 133.

Cultural Entomology

James N. Hogue
California State University, Northridge

Since the dawn of humanity, the organisms that share our world have captured our imagination and influenced our thoughts, dreams, and fears. This influence is particularly true of insects, which impact nearly every facet of human activity. In addition to serving as objects of scientific inquiry, competitors for resources, carriers of disease, and food, insects have made a marked impact on the cultural aspects of human societies. Cultural entomology is the study of the role of insects in those human affairs that are practiced for the nourishment of the mind and soul, such as language and literature, music, folklore, religion, art, and recreation. These activities that pervade primitive and modern human societies are concerned primarily with life's meaning rather than its function.

Despite their extra appendages and different strategies for making a living, insects look and behave enough like humans to serve as models for friends, enemies, teachers, and entertainers. This status permits insects to act as objects on which to impart human qualities and as the source of qualities that can be incorporated into the framework of human ideology and social structure. It is not surprising then to find insects playing a host of roles in the oral and written traditions throughout human history, ranging from folk tales to the holy writings of the world's most prominent religions.

FOLKLORE, MYTHOLOGY, AND RELIGION

The derivation of stories and myths is a universal tendency of all human societies. Both myths and folk tales differ enormously in their morphology and their social function. They are used to mediate perceived contradictions in phenomena observed in the natural world, they serve as vehicles of wish fulfillment, they may embody a lesson, or they may serve to preserve a piece of a culture's history. Myth and folklore also differ from one another in their origin and purpose, but application of these distinctions is difficult to discuss here. Originally, mythology meant no more than telling stories, such as traditional tales passed from generation to generation. Later, some of these tales acquired new meaning and status and evolved more symbolic or religious functions. All tales, whether classified as folklore or myth, are not generated in isolation, but derive their inspiration, elements, and messages from the environment, including the host of other species that surround us. These tales are often used to derive commonsense explanations of natural phenomena observed in the environment. Conversely, such observations may also serve as the basis for the superstitious beliefs and tales surrounding aspects of human existence such as healing practices and other utilitarian activities such as agriculture.

Entomological mythology commonly employs transformations of beings between the insect and the human form (and combinations thereof), the acquisition of souls by insects, and ultimately the deification of insect forms. Insects are also used symbolically throughout the world's religions in a variety of roles.

Insects figure prominently in the creation myths of many cultures. The widespread recognition of insects in this role probably stems from an innate recognition of insects as ancient members of the living world that must have been present at its creation or soon thereafter. Beetles, for example, play central roles in the creation myths of two native American tribes. According to the Cherokee of the southeast, the world was originally covered by water. The first land was brought forth by the water beetle that dived under the water and brought mud to the surface.

The behavior of beetles in the genus *Eleodes* (Tenebrionidae), which raise their abdomens in the air by standing on their heads when disturbed, is explained by the role this beetle played in the creation of the universe according to the Cochiti of the American Southwest. The beetle was given the responsibility of transporting a bag of stars that would later be carefully named and placed into the sky. But the beetle's carelessness resulted in most of the stars being spilled into what is now the Milky Way. He was punished with blindness and today expresses his embarrassment at his mistake by hiding his head when approached. Any hope of this beetle regaining its sight was apparently completely lost at a later date, as this seems to be the same beetle that lost his eyes in a bet with a spider; this is how the spider got its extra sets of eyes.

Insects appear throughout Mayan codices and Aztec reliefs. The use of insects in this manner indicates an appreciation of their existence and their inclusion in cultural events. In addition to scorpions and some unknown bugs and worms, references to seven different insects are found in the Mayan book of the dawn of life, the Popul Vuh. These include lice, leafcutter ants, mosquitoes, fireflies, bees, yellowjackets, and another type of wasp. Yellowjackets were used as weapons by the Quiché against the enemy tribes during an attack on the Quiché citadel at Hacauitz. Fireflies were used by the brothers Hunahpu and Xbalanque, who later became the sun and the full moon, respectively. They placed these insects in the tips of cigars as false lights to deceive the Xibalban sentries of the underworld that watched over them during their night in the Dark House.

Observations of metamorphosis led people throughout history and from various parts of the world to equate pupation with death of the earthbound larva and the emergence of the often beautiful, winged adult with resurrection. The adult insect is additionally equated with the soul in many circumstances. The equation of souls or spirits of the afterlife with imaginal insects may be why angels are traditionally depicted bearing wings. Insect analogies in descriptions of death, resurrection, and the journey to the afterlife continue to be used today. For example, a Doris Stickney book uses a story of growth and emergence of dragonflies to explain a Christian concept of death to young children.

Insects have also been incorporated into the astrological and cosmological traditions of various societies. Aquatic insects were used as water symbols associated with the coming of rain by Chumash astrologers of southern California, who believed that rain was a gift from the sun. The guardians of the four cardinal points in Warao (Orinoco delta of Venezuela) cosmology are insects: arboreal termites, two kinds of stingless bees, and a paper wasp. There is even a constellation of the southern fly, *Musca australis.*

In Aztec culture, Xochiquetzal, represented by the swallowtail butterfly, *Papilio multicaudatus,* was the goddess of beauty, love, and flowers; patron of domestic labor and the courtesans; and the symbol of the soul and the dead. The mother deity and goddess of human sacrifice, war, and travelers, Itzpapalotl (the saturniid moth *Rothschildia*), was also the personification of the earth and moon. Images of these and other deified insects are found in many Aztec and Mayan reliefs.

The most famous deified insect is the scarab of ancient Egypt. The scarab is a symbol of the sun god Khepera (Fig. 1) and also equated with the creator god Atum. One representation of the scarab was as the agent responsible for moving the sun through the sky, in the manner that these beetles move balls of dung across the ground. Another prominent representation of the scarab was that of the soul emerging from the body, and it was commonly associated with mummies. Just as the actions of the beetles and balls of earth and dung give rise to new beetles, the buried human dead will rise again. Scarab figures are nearly always found on Egyptian mummy sarcophagi, and amulets and pendants bearing the scarab likeness were worn as jewelry by royalty

FIGURE 1 Depiction of the Egyptian scarab god, Khepera. [Illustration modified from Bodenheimer (1928).]

and included in funeral caches as symbols of new life. Another testament of the association of these beetles with life comes from Saint Ambrose, the Archbishop of Milan, who wrote of Jesus as "the good Scarabaeus, who rolled up before him the hitherto unshapen mud of our bodies."

Recent English translations of the Bible, based solely on the original texts, have shed new light on biblical references to insects, particularly with respect to the identity of the insects themselves. Of the 98 references to insects in the Revised English Version, most focus on negative aspects of their activities and as vehicles for God's wrath. Three of the 10 plagues (maggot infestation, swarming flies, locusts) visited upon Egypt prior to the Exodus were mediated by insects. Other references deal with more utilitarian or beneficial aspects of insect life. Insects are included as part of the instructions of what kinds of animals are permitted as food (Leviticus 11:22), some writings are merely observations of insects and their habits (Exodus 16:20), and other references use them as examples of virtuous characteristics (Proverbs 6:6–8 and 30:25–27). Sometimes insects are used metaphorically, as in Psalms 118:12: "They surround me like bees at the honey, they attack me, as fire attacks brushwood, but in the Lord's name I will drive them away." One or more kinds of scarab beetle may have even served as the inspiration for the prophetic visions of Ezekial.

SYMBOLISM AND REVERENCE

Throughout human existence, many insects have been admired for their ingenuity, beauty, fantastic shapes, and behaviors. In some instances, the use of insects as totemic figures that may symbolize ancestry or kinship of humans with these organisms leads to a deep sense of adoration and reverence. In other cases, the resultant admiration has developed into a reverence for their inspirational and historical nature and a medium for symbolizing a variety of aspects of human life. In these situations, characterizations of organisms, in both illustration and sculpture, act as vehicles to convey human feelings rather than as objective expressions of entomological facts. Insect symbolism is best developed in the most advanced ancient civilizations of Egypt, Greece, and especially Central America, where the people were surrounded by a multitude of insects.

Of all the insect groups, the flies (Diptera) most frequently play negative roles in human symbolism. Flies typically represent evil, pestilence, torment, disease, and all things dirty. This association is likely a result of the fact that those flies most familiar to people have a close association with filth. Beelzebub, the Lord of the Flies, is a fallen angel who presides as a leader of demons and an agent of destruction and putrefaction. In the ancient lore of Persia, the devil Ahriman created an evil counterpart for every element of good put on Earth by the creator. Many insects, particularly flies, were thus formed and they continue to be associated with evil and filth. Some flies were considered so bad that they became symbols of qualities revered by humans. The Order of the Golden Fly was a military decoration of the New Kingdom of Egypt (1550–1069 B.C.) awarded for valor. Derived from encounters with the stable fly, *Stomoxys calcitrans,* soldiers observed these flies to fiercely bite and return to bite again, even in the face of persecution.

Because of the perceived similarities between human and insect societies, social insects figure prominently in the symbolic representation of insects. Social insects such as ants, termites, and some bees represent desirable qualities such as unity, cooperation, and industriousness. For example, ants represent the benefits of teamwork and cooperation for the good of all. Many symbolic depictions feature the ancient activities of honey hunting and beekeeping. In Europe, bees and hives also are widely used in various signage and as heraldic emblems, perhaps extolling various qualities of bees upon their bearer. A fine example of the latter is found on the coat of arms of Pope Urban VIII, Maffeo Barberini, who consecrated the present church in St. Peter's Basilica in 1626. The three Barberini bees adorn various ornamentations at the church and many papal objects located in the Vatican museum, including the building itself. In the United States, honey bees are used to symbolize virtuous qualities. The designation of Utah as the "Beehive State" originates from the adoption of the beehive as a motif by the Mormon leaders in 1849 and may be based on impressions of the bees as hard-working, industrious creatures.

Some insect groups have such wide representation in the symbolism of past and present human societies that it is impossible to make general statements about their meaning. Butterflies and moths, for example, are very common elements in symbolism of societies worldwide. Within the limited scope of Western art, Ronald Gagliardi describes the use of butterflies and moths in 74 different symbolic contexts. These insects adorn the artwork of many societies, not only because of their beauty but also because they are widely used to symbolize spirits. Butterflies are often equated with the souls of the dead or sometimes of souls passing through Purgatory (Irish folklore) and are thus often used to represent life after death. The Greek goddess Psyche, who represented the soul, is typically depicted bearing butterfly wings. Moths are depicted as a symbol of the soul's quest for truth, and just as the moth is attracted to a light, so the soul is drawn to divine truth.

Butterfly images are common adornments of pottery, featherwork, and the deeply religious characters hewn in stone in ancient Mexico. The Hopi of the American Southwest have a ritual called the "Butterfly Dance" and have kachina figures that anthropomorphize the butterfly spirit. The Blackfoot Indians of North America believe that dreams are brought to sleeping people by the butterfly.

Symbolic depictions of insects also serve to bestow honor on the insects themselves. Insects have been featured on a few coins and on several thousand postage stamps worldwide (Fig. 2). One fifth century Roman coin bears a honey bee,

FIGURE 2 Postage stamps featuring a selection of moths. (Stamps from the author's collection.)

the emblem of the city of Ephesus. Their "Great Mother" was also known as the Queen Bee and her priestesses were called "melissae," from the Greek for honey bee, in analogous reference to worker bees and their servitude to the queen. There are currently 39 U.S. states that have designated an official state insect, chosen typically to represent something beautiful or inspirational from the state or merely an insect familiar to many. The honey bee is the insect of choice for 16 states and has been nominated, along with the monarch butterfly, as a candidate for the national insect of the United States.

Some insects, particularly those that symbolize aggression, have found their way onto the playing field in the form of sports team mascots. Teams often choose insects, such as wasps, that symbolize aggression (e.g., the Charlotte Hornets of the National Basketball Association and the Yellowjackets of Georgia Institute of Technology).

Other forms of insect celebration involve periodical events of recognition or appreciation for the actions or beauty of local insects. Cities and towns celebrate the beneficial industry and products of some insects, such as honey bees, or the pestiferous activities of those insects that affect the local economy. Annual festivals are held to celebrate honey bees in Illinois and Georgia, woollybear caterpillars in Kentucky and North Carolina, and monarch butterflies in California. In other places, celebrations recognizing the local impact of pestiferous insects are held, including a fire ant festival in Texas, a phylloxera festival in Spain, and a permanent tribute to a weevil in Alabama. The monument to the boll weevil in Enterprise, Alabama, is a large statue of a woman holding a larger than life weevil high over her head. It was dedicated in 1919 to honor the pest for the roll it played in the history of the town. The farmers were forced to switch from planting cotton to a diversity of other crops, particularly peanuts, and the town prospered as a result.

A very successful type of organized celebration of insects that has become common in recent years is the insect fair. These events serve to congregate people with a common interest in insects where they can participate in and enjoy a variety of insect-based fun, contests, food, and dialogue. Insect fairs also provide opportunities to see and purchase nearly anything of entomological interest.

ART, LANGUAGE, AND LITERATURE

Art draws its inspiration from the environment of the artist. It is therefore not surprising that insects have pervaded all forms and aspects of art. Images of insects are found as adornment on all types of objects from textiles and pottery to weapons and jewelry and even the tattoos on human bodies. Insects are also found, either as the primary subjects or merely as curious elements of lesser status, throughout all types of paintings.

Like the illustration of other animals, insect illustration began as a form of decoration. The earliest clearly identifiable drawing of an insect apparently dates to around 20,000 years ago. It is of a cave-dwelling rhaphidophorid cave cricket, inscribed on a piece of bison bone by Cro-Magnon people in southern France. From this humble beginning, depictions of insects have adorned everything from the walls of caves and temples of ancient societies to the paintings and textiles of modern artisans.

Insects, particularly butterflies, were used for decorative purposes in the painted illuminations of medieval manuscripts. By the 15th century, insects had become as important as birds in this respect. The rich iconographic use of insects at this time, often associated with folklore composed of a mixture of misinformation and factual representations, formed the basis on which the first scientific naturalists started their work in the 16th century. One insect painter, who was primarily an artist rather than a biologist, was Maria Sibylla Meriam. She reconciled the old aesthetic realism of medieval origin with the new tradition of practical engravings of the elaborately illustrated natural history treatises of the day and helped form the foundations of modern scientific investigations and writings on insect subjects.

In addition to paintings and textiles bearing artistic depictions of insects, their bodies, parts, and products often serve as the media for art. The metallic, brightly colored elytra of some buprestid beetles have been used as decorative cover on sculptures and textiles and as accessories in jewelry. Similarly, pieces of the colorful wings of butterflies are used in various parts of the world in collages to create artistic images. Beeswax was used to fashion figures and was the wax used to make the positive images in the "lost wax" technique for casting metal figures that originated in the third millennium B.C. in the Middle East.

One art form in which insects have been widely used as models is jewelry. Jewelry resembling insects has been used as aesthetic adornment around the world, throughout history and currently. The insects most commonly used as models for jewelry are beetles, flies, bees, butterflies, and dragonflies.

Some of these, such as flies and bees, had symbolic significance in ancient societies. Others, such as dragonflies and butterflies, are more likely used because of their beauty. A particularly interesting form of insect-based adornment is living jewelry. In Mexico, small jewels, glass beads, and metallic ornaments are set or glued to the elytra and pronotum of living ironclad beetles (Zopheridae) that are then attached to a fine chain pinned to the blouse and allowed to act as a living brooch. Some brilliantly metallic buprestids are used in a similar manner in parts of tropical Asia, and living fireflies and luminescent elaterids are used as decorations in hair or attached to clothing.

Many 15th and 16th century paintings include the motif of common-looking flies perched on various subjects, including people. The depiction of flies in this manner was done in mischievous jest or to invoke shock, perhaps to symbolize the worthiness of even the smallest objects of creation in association with the images of humans, as an expression of artistic privilege, or to indicate that the person in a portrait had died. Sometimes flies were included simply as imitation of such *musca depicta* done by previous painters.

In addition to their roles in mythology and folklore, insects and their symbolic representations have been adapted into the language and philosophy of various cultures. Symbols are used to suggest some idea or quality other than itself. One example is insect symbols in ideographic or phonetic symbols in written language. Examples are found in Assyro-Babylonian cuneiform and the ideographic writing of the Chinese and Japanese. The Greek word for mosquito, "Konops," is the source of the word for canopy, such as that made of mosquito netting. The medieval word "mead" refers to an alcoholic drink made from fermented honey and water that was used as an elixir. This word is the basis for the word "medicine" in recognition of its purported healing properties, and the word "madness" is in reference to the state of some people under the influence of mead. Insects even form the basis for geographical place names. Chapultepec, the hill of the grasshoppers, is where the castle of Aztec Emperor Montezuma stood in what is now part of Mexico City. Urubamba, which means the plain of the insect, is the sacred valley of the Incas near Cuzco in Peru. Japan was once known as Akitsushima, meaning dragonfly island.

Insects have also lent their names and attributes to a variety of descriptors of people and their personalities. People may be described as "busy as a bee," "nit picky," or "antsy." They may act "merry as a cricket" or feel as though they have "butterflies in their stomach." Connotations associated with particular insects may be used to convey similar traits in people. In many parts of the world, the reference of someone as a cockroach signifies an utter contempt for the individual and implies that their life is without value. The Spanish word for butterfly, "mariposa," is street slang for male homosexuals in Mexico. Lastly, insects enter language as metaphor. For example the self-ascribed desirable qualities of boxer Muhammed Ali are that he can "float like a butterfly and sting like a bee."

Aside from purely scientific works, insects have been represented in word and verse in a variety of contexts. In literature, insects are found as subjects of humor, as examples of aesthetically interesting natural subjects of wonder and appreciation, and as characters in fairy tales and in science fiction, mystery, and fantasy novels. Insects sometimes even serve as the storytellers themselves.

The essence of insects in literary humor typically involves the superimposition of insects into aspects of human behavior. The depiction of insects engaged in human activities is a common avenue of insect humor. This is particularly true of the role of insects in comic strips and cartoons, such as in the Far Side cartoons by Gary Larson. In other works, factual entomological information is cleverly presented in a humorous format. Such essays serve to popularize insects and their study, to educate, and, of most relevance here, to entertain.

Insects with endearing qualities, such as beautiful appearance or song, are used in fanciful stories and celebrated in poetry and verse. In Roald Dahl's *James and the Giant Peach,* a group of larger than life insects join a young boy as companions in a surreal adventure inside a monstrous fruit. A cricket and its song play a central role in Charles Dicken's fairy tale of home, *The Cricket on the Hearth.* The people of the house are gladdened and cherish the pleasant voice of the cricket as they listen to its fireside music. The melodious tune made for a happy home and served as an inspiration for those that heard it. Selections of insect poetry are typically written to convey particular feelings or to celebrate insects themselves. A contemporary example comes from the late D. K. McE. Kevan, the author of many humorous entomocentric verses, who wrote *An Embiopteran Epitaph* (reprinted from the *Bulletin of the Entomological Society of Canada* 6(1), 29, 1974).

We embiid web-spinners,
When seeking out our dinners,
Run back and forth in tunnels made of silk;
But, when we get the urge,
We occasion'ly emerge
From beneath a log, or places of that ilk.

We like our climates warm;
We're of dimorphic form;
We're soft and have a tendency to shrink.
"One does not often see 'em!"
Says the man in the museum,
But we're really not so rare as people think!

On the other hand, insects with undesirable qualities or strange traits are typically the subject of horror and mystery stories. Hundreds of science fiction and fantasy stories that use insects in a variety of prominent roles have been published. Franz Kafka's short story *The Metamorphosis* is about a young man who awakes one morning to find out he has turned into a giant insect. In Edgar Alan Poe's tale of the hunt for a pirate's treasure, *The Gold Bug,* an insect is used to find the buried loot. As per the directions on a coded map, the gold beetle, tied to the end of a string and passed through the left eye of a skull nailed high in a tall tree, indicates the spot of a landmark from which the location of the treasure can be deciphered.

The role of insects in science fiction is particularly well established in film, where various insects appear as horrific creatures. Some of these insect fear films, e.g., *The Hellstrom Chronicle* (1971), merely embellish factual information in order to prey on the entomophobic tendencies of the general populace and the potential fleetingness of the future of humans on Earth in the face of the insect hordes. Others use fantastic representations of insects with supranormal characteristics, typically the result of science and technology gone awry, to instill fear and malevolence toward the insect characters, and as a lesson of what can happen when humans arrogantly fool with nature. Ants are common subjects in these roles and appear as giant mutants invading southern California in *Them* (1954) and a housing development in Florida in *Empire of the Ants* (1977). In *The Naked Jungle* (1954) and *Phase IV* (1974), the ant attackers are of normal size, but possess supernatural intelligence and aggression. Because they are widely despised by humans, cockroaches and flies are predisposed to be good villains in these films. In *Bug* (1975), hordes of carnivorous, self-combustible cockroaches wreak havoc on the population; and in the classic insect horror film *The Fly* (1958), the bodies of scientist and insect become inextricably combined with horrific consequences.

Not all fiction films starring insects depict them in a negative light. Insects sometimes fill the role of funny or entertaining characters. For example, in *Joe's Apartment* (1996), the singing and dancing cockroaches are crudely humorous roommates. The literary or cinematic use of insects in humor or as subjects of entertainment invariably leads to the creation of bugfolk. Bugfolk are humanized insects and other related arthropods that dress or talk like humans or are little people with wings, antennae, or other insect features. Bugfolk appear in nearly every literary and art form and are favorite characters for young audiences because of their teaching and entertainment abilities.

Certainly the most familiar bugfolk to Americans and many others worldwide is Jiminy Cricket, of Walt Disney's 1940 animated film *Pinocchio.* Like many of his kind, but unlike his true insect model, Jiminy Cricket bears only four limbs and acts and appears very human. Although morphological correctness is commonly practiced in more recently derived motion picture bugfolk, four-leggedness continues to be seen particularly when a friendly character relationship is desired. Six leggedness, e.g., the evil "Hopper" and his gang of grasshoppers in Disney's animated feature *A Bug's Life,* is used perhaps to provide a farther-from-human image and invoke disdain. Many other bugfolk are featured in comics, as children's toys, and as subjects in literature and art.

FIGURE 3 Trio of bugfolk extolling the virtues of sociality. [Illustration from *Episodes of Insect Life,* by Acheta Domestica, M.E.S. (1851).]

The use of bugfolk in literature and film enables people to see and learn something about themselves through these characters, in perhaps a different light than would be achieved through a strictly human relationship. For example, insect humor often involves a comparison of human behavior and what an insect might be supposed to do in comparable situations. In this manner, insects are found dressed as humans engaged in human activities, such as attending a festive party or dance, or as subjects in amusing or thought-provoking situations (Fig. 3). Particularly creative illustrations of anthropomorphized insects enjoying themselves are found in Grandeville's *Scènes de la vie privée et publique des animaux* and in Aldridge's *Butterfly Ball and the Grasshopper's Feast.*

MUSIC, ENTERTAINMENT, AND CEREMONY

The songs, sounds, and other qualities of insects have inspired many musicians and songwriters. The sounds produced by various insects serve as songs for direct enjoyment or as the inspiration for man-made music. Singing insects have a rich social history in Asia where celebrations and festivals are routinely held. People in both Japan and China have long kept singing insects, chiefly crickets and katydids, in small cages, like birds, for the enjoyment of their songs. The inspiring influence of insect sound for human musicians is exemplified in Nicolas Rimsky-Korsakov's famous musical composition the *Flight of the Bumblebee* and in that of another piece by Korsakov's pupil Anatol Liadov, the *Dance of the Mosquito.* As subject matter in song, insects such as cockroaches are common in blues and folk songs such as the famous Mexican folk song "La Cucaracha," about the troubles of a cockroach down on his luck.

Although insect collection and observation is generally done as an educational activity, many people find great enjoyment in capturing insects for specimens, to keep as pets, and to use in a variety of entertaining tasks. This is particularly true of children living in rural areas of Japan where insects have achieved a lofty cultural status.

These activities support an entire industry devoted to providing the equipment used to capture, observe, and keep insects in captivity. Some insects, particularly large dynastine scarabs and lucanids, are even mass reared and sold in vending machines.

Insects serve as the models for games or may be active, albeit unwilling, participants in a variety of six-legged sporting events. In the children's game "Cootie," the object is to be the first player to assemble a complete insect from a set of body parts such as antennae, proboscis, and six legs. In many parts of the world children fly insects instead of kites. Large insects, such as big beetles and dragonflies, are tethered to strings and allowed to fly for the amusement of people. In places where they occur naturally, large male dynastine scarabs or lucanids are collected and made to fight each other for sport.

Bouts and games involving insects are a source of enjoyment as well as an opportunity for gambling, such as with cricket fighting in China and Thailand and water bug roulette. In the latter contest, water beetles or water bugs are released into the center of a circular container filled with water. The inside perimeter is bounded by a continuous series of marked slots into which the insect can enter. Entrance of a particular insect into a slot is analogous to the landing of the ball on a particular number on a roulette wheel and the appropriate prize is awarded. In addition to being pitted against each other in battle, insects are commonly matched in foot races. For example, caterpillar races are held in Banner Elk, North Carolina, during the Woolly-Bear Festival, and cockroach racing is popular in many parts of the world, particularly in China and India.

Other forms of insect-based entertainment for humans include flea circuses and entertaining displays of both living and dead insects. Flea circuses use tiny performing fleas that are "trained" to perform a variety of circus acts for the amusement of the audience. Living insects are displayed in venues such as butterfly houses, where they can be viewed and enjoyed flying about their enclosures by an appreciative public. Dead insects have been similarly displayed as objects of aesthetic pleasure, sometimes with added adornments such as miniature clothing. Dead fleas are dressed in tiny costumes and displayed in folk art exhibits in Mexico. In Plano, Texas, the Cockroach Hall of Fame Museum features dead roaches dressed as famous people engaged in various activities.

As is true for other organisms that are held in high regard and for those that serve some utilitarian function such as food, some insects fill symbolic roles in human ceremonies. For example, although insects are regularly eaten in many parts of the world for sustenance, the consumption of insects

FIGURE 4 Grasshoppers being carried to a feast to celebrate the Assyrian defeat of the Elamites, from a relief of Ashurbanipal at Ninevah. [Illustration modified from Bodenheimer (1928).]

was sometimes reserved for ceremonies or other special occasions (Fig. 4). The Kaua of Brazil perform a dance known as the "Dance of the Dung Beetles" that is used to drive away demons. The dancers attempt to transfer powers to themselves from the spirit world by taking on the image of the beetles. They do this by imitating the actions of beetles rolling a ball of dung. Other insects play a more active role and are used for a particular ceremonial or ritualistic purpose. Because of their powerful stings, giant hunting ants are used by indigenous peoples in Amazonia in male initiation and virility rites ceremonies. Large numbers of ants are tied to a woven mat and the mat with the now enraged ants is applied to the initiate's bare skin. Those who endure the excruciating pain without complaint, and live, are deemed worthy.

The Indians of central and southern California also made ceremonial use of ants. Male youths of the Kitanemuk, Tübatulabal, and Kawaiisu were taken by their elders for three days of fasting, after which they were given numerous live "red" or "yellow" ants to eat. The ants were consumed in order to gain power and induce a trance-like state during which spiritual insight would be gained.

HISTORICAL EVENTS MEDIATED BY INSECTS

Finally, insects have made their mark on human cultures by influencing events that shape history, such as wars, or by changing the way societies can or cannot accomplish things. The Panama Canal was built and ultimately controlled by the United States in part because the earlier effort by France was thwarted by mosquito-borne yellow fever. As vectors of African sleeping sickness, *Glossina* spp. (Diptera) have made huge pieces of land in Africa uninhabitable by humans. Bubonic plague, spread by its flea vector, helped cause drastic changes in the social and economic structure in Europe during the 14th and 15th centuries. The populations, and thus the sites, of more than one ancient eastern Mediterranean city moved because of the actions of insects, particularly flies.

In many military campaigns, the number of casualties attributed to insects has exceeded that caused by actual fighting. The activities of insects, primarily by transmitting disease to troops in battle, have determined the outcome of entire wars. Napoleon's invading army lost hundreds of thousands of men and was decimated by the louse-borne disease typhus during their eastward march across Europe in 1812 and 1813.

Insects have also served as important determinants in the fates of human societies and economies throughout human history. The survival of the Israelites during their extended journey through the Sinai Desert was apparently made possible by insects. The manna that they gathered, ate, and survived upon was most likely the excretions of scale insects. If not for the arrival and help of divinely inspired seagulls, a plague of mormon crickets in 1848 may have ruined the crops and doomed the Mormons soon after their arrival in their new home in Utah. The silk trade was central to the economy of the Chinese Empire as was cochineal to the Aztecs of central Mexico. This is also true on a smaller scale for producers of honey and shellac, and for the thriving modern-day trade in insects sold for scientific, educational, and hobbyist uses.

The action of insects even helped to revolutionize the production of one product that has greatly shaped the whole of human civilization over the past 2000 years. Since the "invention" of paper was first proclaimed to the Chinese emperor Ho Ti in 105 A.D. by Ts'ai Lun, a variety of plant fibers were used in the production of paper. As writing flourished, supplies of raw materials for making paper became in short supply. Such was the situation in 16th century Europe, where paper was made from cotton and linen. It was here that observations of paper wasps inspired the French naturalist and physicist Rene Antoine Ferchault de Réaumer to suggest the use of wood as a papermaking fiber in 1719. These wasps, which chew wood and mix the fibers with saliva to make their nests, served as the inspiration for the use of the plentiful fiber on which modern papermaking is based.

CONCLUSION

The sources of published information dealing with the roles of insects in human culture have until recently been found in a diffuse body of literature. Such information is often hidden in historical documents, anthropological works, and ethnoentomological notations in travel logs and journals. In the past 20 years, a wider aesthetic and cultural appreciation for insects has been realized. The celebration of insects and their attributes as they relate to the development of human societies is generally accepted as a worthwhile endeavor. This interest has spawned a number of review articles and books that summarize, synthesize, and sometimes popularize much of the previously diffuse literature and serve as a starting point for those interested in this fascinating subject. Some periodicals, namely, *The American Entomologist,* regularly publish cultural entomological articles, and the recently introduced periodical, *Cultural Entomology Digest* (http://www.bugbios.com/ced/), is devoted entirely to this topic.

Along with the modernization of the world, the perceived relevance of insects to human life is slowly eroded. As this happens, the various roles of insects in human cultural affairs may change or be lost. However, many people continue to carry mythological modes of thought, expression, and communication into this supposedly scientific age and others still find pleasure in observing and contemplating their six-legged companions on Earth. Therefore, the importance of insects as subjects of entertainment and aesthetic pleasure should continue to enter into the thoughts of future people and mold aspects of human culture. As some relationships between human and insect are lost, others are formed. Because of the dominant place in the function of the world's ecosystems and their influence on human existence, insects have played and will continue to play a prominent role in our perception of life and pursuit of aesthetically pleasing activities and for the enlightenment of human societies.

See Also the Following Articles
Entomological Societies • Folk Beliefs and Superstitions • History of Entomology • Movies, Insects in • Stamps, Insects and

Further Reading
Akre, R. D., Hansen, L. D., and Zack, R. S. (1991). Insect jewelry. *Am. Entomol.* **37,** 90–95.
Berenbaum, M. R. (1995). "Bugs in the System." Addison Wesley, Reading, MA.
Berenbaum, M. R. (2000). "Buzzwords." Henry Press, Washington, DC.
Bodenheimer, F. S. (1928). "Materialien zur Geschichteder Entomologie bis Linné," Vol. I. Junk, Berlin.
Cherry, R. H. (1993). Insects in the mythology of native Americans. *Am. Entomol.* **39,** 16–21.
Clausen, L. W. (1954). "Insect Fact and Folklore." MacMillan Co., New York.
Cloudsley-Thompson, J. L. (1976). "Insects and History." St. Martin's Press, New York.
Dicke, M. (2000). Insects in western art. *Am. Entomol.* **46,** 228–236.
Gagliardi, R. A. (1976). "The Butterfly and Moth as Symbols in Western Art." Southern Connecticut State College, New Haven. [Masters thesis]
Hamel, D. R. (1991). "Atlas of Insects on Stamps of the World." Tico Press, Falls Church, VA.
Hogue, C. L. (1980). Commentaries in cultural entomology. 1. Definition of cultural entomology. *Entomol. News* **91,** 33–36.
Hogue, C. L. (1985). Amazonian insect myths. *Terra* **23,** 10–15.

Hogue, C. L. (1987). Cultural entomology. *Annu. Rev. Entomol.* **32,** 181–199.

Kritzky, G., and Cherry, R. (2000). "Insect Mythology." Writers Club Press, San Jose, CA.

Laurent, E. L. (2000). Children, 'insects' and play in Japan. *In* "Companion Animals and Us" (A. L. Podberscek, E. S. Paul, and J. A. Serpell, eds.), pp. 61–89. Cambridge University Press, Cambridge, U.K.

Pearson, G. A. (1996). Insect tattoos on humans: A "demographic" study. *Am. Entomol.* **42,** 99–105.

Stickney, D. (1997). "Water Bugs and Dragonflies." Pilgrim Press, Cleveland.

Tedlock, D. (1985). "Popul Vuh." Simon & Schuster, New York.

Cuticle

Svend O. Andersen

Copenhagen University

The cuticle is an extracellular layer that covers the complete external surface of insects, as well as the surfaces of their foreguts and hindguts, and acts both as a skeleton for muscle attachment and as a protective barrier between the animal and its environment. The cuticle is an integral part of a complex dynamic tissue, the integument, which also includes the cuticle-producing epidermal cells, and various glands and sense organs.

GENERAL PROPERTIES OF CUTICLES

The cuticular layer varies in thickness from a few micrometers to a few millimeters, depending upon the insect species, developmental stage, and body region, but cuticles typically are between 100 and 300 μm thick. Cuticles are highly diverse in their mechanical properties. They can be divided into two groups: stiff and hard cuticles, and soft and pliant cuticles. Intermediate degrees of stiffness also exist, and some types of cuticle have special properties, such as rubberlike elasticity or extreme extensibility. Cuticles differ in color and in surface sculpturing, but electron microscopy shows that all types of cuticle are built according to a common plan. The details in structure and properties of the various cuticular regions are such that for each species they are optimal for the functioning of the living insect in its natural surroundings.

EPICUTICLE

The outermost layer of a cuticle is called epicuticle; it forms a continuous layer covering the complete cuticular surface. Seldom more than 2 μm thick, it is responsible for the waterproofing properties of the cuticle. Electron microscopy shows that the epicuticle can be subdivided into several layers, of which the inner epicuticle, also called the dense layer, is the thickest. It is covered by the thin, outer epicuticle, sometimes called the cuticulin layer, which is assumed to be responsible for the mechanical stiffness of the epicuticle. The inner and outer epicuticle are composed of polymerized lipids and protein, and they contain no chitin. These two layers remain poorly characterized because they are difficult to purify, dissolve, and degrade.

The outer epicuticle is covered by a waterproofing wax layer, containing complex mixtures of extractable lipids, secreted during the molting process from integumental oenocytes and epidermal cells. This layer is again covered by a protective cement layer, secreted immediately after ecdysis from glands in the integument.

The extractable lipids in the wax layer have been characterized for several insect species. They appear to be species-specific mixtures of a wide range of lipids, including normal and branched, saturated and unsaturated hydrocarbons, fatty acids, alcohols, esters, sterols, and aldehydes. Differences in lipid composition have been used to discern closely related insect species. The epicuticular lipid composition also can vary between instars and sex of the same species, and these lipids often play an essential role in recognition and communication between insects.

PROCUTICLE

The region of the cuticle, located between epicuticle and the epidermal cell layer, is called procuticle; it constitutes the main part of the total cuticle. Histologically, the sclerotized regions (sclerites) are often subdivided into layers with different staining properties: (1) the outermost layer, the exocuticle, may be dark colored because of sclerotization, but is refractory to staining; (2) the innermost, uncolored layer, the endocuticle, stains blue; and (3) in between one often observes a layer of mesocuticle, staining red with Mallory triple stain. The flexible cuticle (arthrodial membranes), which connects the sclerites, stains blue with Mallory throughout most of its thickness. Exocuticle may correspond to the part of the procuticle deposited before ecdysis, stabilized by sclerotization. Mesocuticle plus endocuticle often correspond to the post-ecdysially deposited procuticle, and if these layers are sclerotized at all, it is only slightly.

The procuticle consists mainly of chitin and proteins; water is an essential component, and other materials, such as lipids, phenolic compunds, salts, pigments, and uric acid may be present. Chitin (poly 1,4-β-*N*-acetylglucosamine) is a polysaccharide, present as long and nearly straight microfibrils, usually about 2.8 nm in diameter and of indeterminate length. The filaments tend to run parallel to the cuticular surface, but columns of chitin filaments running perpendicular to the surface have been described for some types of cuticle (lepidopteran larval cuticle). The function of such chitinous columns remains uncertain.

The chitin microfibrils are organized in various patterns, and the organization seems to have importance for the mechanical properties of the cuticle. The most commonly observed patterns are the heliocoidal pattern, where the microfibril direction changes by a small, constant angle between neighboring layers; the preferred, unidirectional orientation, where the fibrils run in the same direction in all layers, and the pseudo-orthogonal orientation, where unidirectional layers of chitin microfibrils alternate with layers running at nearly right angles to each other. In certain cuticles the pattern of chitin microfibrils depends on a daily rhythm: in locust tibiae, heliocoidal cuticle is deposited during the night and unidirectional cuticle is deposited during the day, making it possible to determine the number of days since ecdysis.

The chitin microfibrils are embedded in a protein matrix; the protein content tends to equal the chitin content in flexible cuticles and is usually three to four times higher than the chitin content in hard cuticles. The number of different proteins present in a given type of cuticle can vary from about 10 to 100. Different types of protein are present in flexible and hard cuticles; the proteins are species specific, and some of them are also specific for certain cuticular regions. A characteristic amino acid sequence region, common to a large number of cuticular proteins, is supposed to have a function in the linking of proteins to the chitin microfibrils. The proteins are often extractable immediately after deposition. In many cuticular regions, however, they are later rendered inextractable by sclerotization, whereby low molecular weight phenolic compounds are covalently incorporated into the cuticular matrix, cross-linking the proteins, and making the cuticle harder and stiffer, and more difficult to digest with enzymes.

Sclerotization may occur soon after a molt when the insect has expanded its new cuticle to a larger size, but the regions that are not enlarged may have been sclerotized in the pharate stage, which is the stage that is present before emergence from the exuvium, or old cuticle. The elastic protein, resilin, present in rubberlike cuticular regions, is cross-linked as soon as it is deposited extracellularly. The cross-linking process is different from that in sclerotized cuticle because no low molecular weight compounds are involved, but tyrosine residues in the protein chains are oxidatively coupled to each other, forming di- and trityrosine residues.

SUBCUTICLE

A narrow, histochemically distinct layer, called subcuticle, is situated between the procuticle and the epidermal cells. It stains positively for muco- and glycoproteins. It has been suggested that it serves to bind cuticle and epidermis together and that this layer is the deposition zone, where new cuticular material is assembled and added to the already existing cuticle.

See Also the Following Articles
Chitin • Exoskeleton • Integument • Molting

Further Reading

Andersen, S. O. (1985). Sclerotization and tanning of the cuticle. *In* "Comprehensive Insect Physiology, Biochemistry and Pharmacology" (G. A. Kerkut, and L. I. Gilbert, eds.), Vol. 3, Chap. 2. Pergamon Press, Oxford, U.K.

Blomquist, G. J., and Dillwith, J. W. (1985). Cuticular lipids. *In* "Comprehensive Insect Physiology, Biochemistry and Pharmacology" (G. A. Kerkut, and L. I. Gilbert, eds.), Vol. 3, Chap. 4. Pergamon Press, Oxford, U.K.

Hepburn, H. R. (ed.). (1976). "The Insect Integument." Elsevier, Amsterdam.

Hepburn, H. R. (1985). Structure of the integument. *In* "Comprehensive Insect Physiology, Biochemistry and Pharmacology" (G. A. Kerkut, and L. I. Gilbert, eds.), Vol. 3, Chap. 1. Pergamon Press, Oxford, U.K.

Kramer, K. J., Dziadik-Turner, C., and Koga, D. (1985). Chitin metabolism in insects. *In* "Comprehensive Insect Physiology, Biochemistry and Pharmacology" (G. A. Kerkut, and L. I. Gilbert, eds.), Vol. 3, Chap. 3. Pergamon Press, Oxford, U.K.

Neville, A. C. (1975). "Biology of the Arthropod Cuticle." Springer-Verlag, Berlin.

Wigglesworth, V. B. (1972). "The Principles of Insect Physiology." 7th ed. Chapman & Hall, London.

Daddy-Long-Legs
(Opiliones)

Gonzalo Giribet
Harvard University

The Opiliones, commonly known as daddy-long-legs, harvestmen, shepherd spiders, or harvest spiders (among many other names), are a very interesting group of arachnids that are well known by farmerss. Opiliones constitute the third most speciose arachnid order (after Acari and Araneae), comprising approximately 1500 genera and 5000 species in 45 families. They are the only nonmite or tick arachnids that ingest vegetable matter, but generally they prey on insects, other arachnids, snails, and worms and have the ability to ingest particulate food; this is unlike most arachnids, which ingest only liquefied substances.

Opiliones are divided into four suborders: Cyphophthalmi, Eupnoi, Dyspnoi, and Laniatores.

DESCRIPTION

Daddy-long-legs present all the typical characteristics of arachnids, with the body divided into two regions, cephalothorax and abdomen, although these two regions are not clearly differentiated, giving daddy-long-legs the aspect of "waistless spiders." The cephalothorax generally has a pair of median eyes on top of an ocular tubercle. The eyes are simple, i.e., not compound as in insects and crustaceans. The cyphophthalmids lack eyes entirely or have a pair of lateral eyes. The cephalothorax also bears a pair of chemical-secreting organs, known as repugnatorial glands.

The cephalothorax has one pair of chelicerae for manipulating the food particles, one pair of palps of either tactile or prehensile function, and four pairs of walking legs, enormously long in some Eupnoi and Laniatores species, surpassing 15 cm in some species. The palps of most Laniatores are relatively large and have two rows of spines acting as a grasping organ. The second pair of walking legs is sometimes modified and acquires a tactile function.

Another distinctive characteristic of the Opiliones is that the females have a long ovipositor with sensory organs on the tip that are used to check the soil quality where they will lay the eggs. Except for some mites, similar organs are not known for any other arachnids. The males have a penis or copulatory organ, which may be muscular or alternatively operated by hydraulic pressure. Copulatory organs are also unique among the arachnids, again with the exception of certain mites. Fertilization is thus internal and direct (unlike in spiders, which use the palps for the indirect internal fertilization).

Figure 1 is an example of a typical daddy-long-leg, *Odiellus troguloides,* from the western Mediterranean.

FIGURE 1 *O. troguloides,* one of the most typical daddy-long-legs from the western Mediterranean region, with an elongated body reaching almost 1 cm in length. This species has considerable sexual dimorphism, females being much larger and more globose than males. Juveniles of this species hatch in the spring, quickly reach maturity, and die in the fall.

LIFE HISTORY AND BIOLOGY

Opiliones are oviparous and deposit between one (in cyphophthalmids) and several hundred (in phalangiids) eggs. Life cycles and longevity are variable. Many species live 1 year, with embryonic development occurring during the winter, with hatching in the spring, and reaching maturity in the fall, after five to seven molting periods. This is the typical seasonal life history of most Northern Hemisphere phalangiids. Others have an overlap of adults and juveniles throughout their life cycles during the favorable seasons, dying in the winter. Finally, cyphophthalmids and most laniatorids live several years, with cases recorded up to 5 years.

Sexual dimorphism is evident in some species. All cyphophthalmid males have a spur on the tarsus of the fourth walking leg. This structure, named an adenostyle, possibly secretes a pheromone. The families Pettalidae and Sironidae in the Cyphophthalmi have male anal glands, and the pettalids may have extreme modifications of the male anal regions.

Opiliones are generally small to medium in size (body measuring less than 1 mm to almost 2.5 cm in the European species *Trogulus torosus*), inhabit all types of moist to wet habitats, and occur on all the continents. The Laniatores include the large (up to more than 2 cm) and the most colorful Opiliones, and their distribution reaches a peak of diversity in tropical regions and in the Southern Hemisphere. The Eupnoi and Dyspnoi are more widely distributed, but especially abundant in the Northern Hemisphere. Finally, the Cyphophthalmi are distributed more uniformly worldwide, but are the smallest (down to 1 mm) and most obscure of the Opiliones.

No Opiliones are harmful to humans, and they do not contain any type of venom or other substance. Some Opiliones are reported as highly poisonous although not having the capacity of biting humans. This myth seems to be a confusion with the highly neurotoxic venom of some spiders. These are differentiated from Opiliones by the presence of a waist that separates the prosoma from the opisthosoma, among many other characters. In fact, Opiliones are supposed to be beneficial, and they are good indicators of undisturbed environments.

See Also the Following Articles
Arthropoda and Related Groups • Spiders

Further Reading
Edgar, A. L. (1990). Opiliones (Phalangida). *In* "Soil Biology Guide" (D. L. Dindal, ed.), pp. 529–581. Wiley, New York.
Giribet, G., Edgecombe, G. D., Wheeler, W. C., and Babbitt, C. (2002). Phylogeny and systematic position of Opiliones: A combined analysis of chelicerate relationships using morphological and molecular data. *Cladistics* **18**, 5–70.
Hillyard, P. D., and Sankey, J. H. P. (1989). "Harvestmen." Brill, Leiden.
Shear, W. A. (1982). Opiliones. *In* "Synopsis and Classification of Living Organisms" (S. P. Parker, ed.), pp. 104–110. McGraw–Hill, New York.

Damselfly

see *Odonata*

Dance Language

P. Kirk Visscher
University of California, Riverside

The "dance language" of honey bees refers to patterned, repetitive movements performed by bees that serve to communicate to their nestmates the location of food sources or nest sites.

RECRUITMENT

If a saucer of honey is placed outdoors, many hours or days may go by before a bee finds it and feeds on it. Soon after this first visit, however, large numbers of bees will arrive. Interest in honey bees goes back to prehistory, because their colonies provided human ancestors' most concentrated source of sugar. At least as far back as Aristotle, people have inferred that the bees that first discover a food source must recruit their nestmates to share in the collection of the food, thus accounting for the rapid buildup once a discovery has been made. The same kind of buildup occurs at flowers, bees' natural source of their sugary food.

Recruitment to food is one of the most important adaptations of nearly all social insects, and there are many forms of recruitment among them. Being able to recruit nestmates to food sources allows colonies of insects to realize one of the advantages of living in groups: the ability to harvest food that would not be as readily available to an individual foraging alone. Such edible items might include prey bigger than an individual could subdue, food resources that are rich but so widely scattered that an individual would not be likely to discover their source and sources that are ephemeral and thus more effectively harvested by means of group foraging during the short time the source is available. Cooperative foraging also is important to social animals in overcoming one of the inherent disadvantages of group living: since members of groups generally will compete with other members of the group for local food resources, without some compensating foraging advantage, solitary individuals would have better access to food than those in groups.

SIGNIFICANCE OF THE DANCE LANGUAGE

The best known of the mechanisms of recruitment in social insects is the honey bee (*Apis* spp.) dance language, in terms of both its fame outside the realm of specialists and the depth in which it has been studied. The dance language is famous for a number of reasons. It is frequently cited as the premiere example of symbolic communication among nonhuman animals, and it is one of the first and best examples of such communication aside from human language. The discovery that mere insects could perform such a complex behavior led to a reassessment of the behavioral complexity possible among these animals with relatively small nervous systems, which had formerly been regarded as simple automatons governed by instinct and reflex. Finally, the dance language has provided a tool for studying the perceptual world and behavioral response of bees that has illuminated our understanding of their vision, olfaction, memory, orientation, learning, and social organization, and has provided a model for understanding these areas about insects in general.

DISCOVERY OF THE DANCE LANGUAGE

Observers of bees had repeatedly noted that sometimes a bee in a colony will perform repeated circular movements, closely followed by other bees, but it was Karl von Frisch who firmly established the connection between these movements and recruitment, and, in the course of a long career, discovered many aspects of communication by the dance language.

Von Frisch began his studies of the dance language in 1919, with the simple yet powerful approach of marking bees with paint as they fed at a flower he had enriched with a drop of sugar syrup (and in later experiments with a simple scented syrup feeder). He then watched their behavior when they returned to a glass-walled observation beehive. He observed his marked bees doing circular "round dances," which were followed attentively by other bees in the hive. He then observed that bees, presumably those that had followed the dances, would investigate nearby flowers of the same type as those at which the marked bee had fed but did not investigate flowers of other types as much. Von Frisch inferred that the dance stimulated recruits to look for food, and that odor in the nectar, and on the body of the dancing bee, communicated to the recruits the scent to seek. He also described a "waggle" form of the dance in which a dancing bee rapidly waggles her abdomen laterally while moving in a particular direction on the comb, then turns back more or less to the starting point, repeats the waggle on the same course, turns back the other way, and so on, describing a squat figure-eight with the waggle in the middle. The artificially small scale of his early work, in a small, walled, Munich garden, caused von Frisch to mistakenly conclude that the two kinds of dance he saw indicated different types of food. The waggle dancers often had pollen on their legs, whereas the bees he provided with nectar did not, so he concluded that waggle dances indicated pollen and the round dances nectar.

This error persisted for 25 years, but von Frisch himself discovered the full story when, during World War II, he was forced to take his studies away from the war-torn city to rural Brunnwinkel, Austria. There, in 1944 and 1945, working under conditions that more accurately reflected the natural scale of bees' foraging, he found that when bees fed at long distances from the hive they performed the waggle dances for nectar, as well. At the same time, he also made the startling discovery that the bees were communicating the direction and distance to the food source, as well as its odor.

COMMUNICATION OF DISTANCE AND DIRECTION IN THE DANCE

The waggle dance of honey bees can be thought of as a miniaturized reenactment of the flight from the hive to the food source (Fig. 1). As the flight distance to the food becomes longer, the duration of the waggle portion of the dance also becomes longer. The angle that a bee flies during the flight to the food, relative to the sun azimuth (the horizontal component of the direction toward the sun), is mirrored in the angle on the comb at which the waggle portion of the dance is performed. If the food is to be found directly toward the sun, a bee will dance straight upward. If the food is directly away from the sun, the bee will dance straight downward. If food is at 35° to the right of the sun, then the dance is performed with the waggle run at 35° to the right of vertical, and so forth. Bees make a transition from round dances for food sources near the nest to waggle dances at greater distance, with the transitional distance varying somewhat between different subspecies of *A. mellifera*.

While the bee is waggling her abdomen, she also produces bursts of buzzing sound from her wings, which are perceived by dance-following bees with the Johnston's organ at the base of the antennae. Recent work by Wolfgang Kirchner has shown that even the round dance contains directional

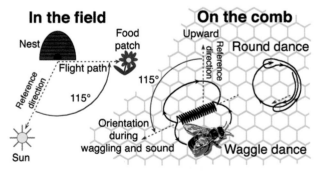

FIGURE 1 How direction to the food patch is encoded in the honey bee dance language. As a bee flies to flowers in the field (left), she learns the direction to the food patch relative to a reference direction of the sun azimuth (here the food is 115° to the left of the sun). When she dances on the vertical combs of the dark hive (right), she uses the direction upward as a reference and performs the waggling portion of the dance at the same angle, relative to this upward reference, to indicate that the food is to be found relative to the sun direction reference in the field (here, 115° to the left of upward). Dancing bees produce buzzing sounds during the waggle portion of the dance. In the round dance (far right), the dancing bee changes direction more randomly and does not waggle, but does buzz when moving in the direction that would indicate the direction to the food.

information, because these sounds are produced at the time in the round dance at which the circling bee is facing in the direction on the comb in which waggle runs would be performed for more distant food sources in the same direction. However, recruit bees seem to search the vicinity of the nest equally in all directions in response to round dances. This scatter in search area, however, is not really greater than the area searched at greater distances, though because of its proximity to the nest it includes all angles, whereas more distant searches are mostly within a restricted range of angles. Thus it is uncertain whether recruit bees can perceive the direction information in round dances.

MEASUREMENT OF DISTANCE AND DIRECTION

The ability to bees to communicate distance and direction to a food source requires that the recruiting bee and the recruits be able to measure these parameters. The study of how bees do this provides an example of how the dance language gives a readout of the perception of the bees. This in turn makes possible sophisticated analyses of the mechanisms by which bees acquire the information, analyses that are vastly more difficult to perform with insects that do not report their findings in a format entomologists have learned to decode.

Von Frisch found that wind, height differences between the feeder and hive, or adding additional weights or airfoils to bees changed the tempo of their dances. This finding indicated that something about these conditions had changed the bees' perception of distance to the food source. One aspect that was changed was the time of flight to the source, but the changes in dance tempo did not correlate well with the changes in flight time, and so this was rejected as the way the bees measured distance. Instead, it was concluded that the bees were measuring energy use, because all these conditions would affect energy use. This was consistent with observations that, on the flight to the food source, either a headwind or flying uphill would increase perceived distance, whereas either a tailwind or flying downhill would decrease it.

However, more recent work by Harald Esch and others suggests that it is not energy that is measured, but the movement of landscape objects across the visual field, or optic flow. Humans experience the apparent motion of landmarks as faster when riding in a car than when flying in an airplane. Similarly, when a bee flies close to the ground, she experiences rapid optic flow, whereas at greater altitudes the optic flow is less. In von Frisch's experiments, the changing conditions also affected the height off the ground of the bees' flight, so that energy use and optic flow were confounded. In experiments in which bees are trained to feeders at different distances from the ground, the distance that a bee perceives, as indicated by the tempo of her dances, is shorter for higher feeders, even though more energy is needed to fly to them and the length of the flight path is greater. The progress of entomologists' understanding of the mechanism by which bees measure distance provides an excellent example of how

the conclusions from an experiment may reject incorrect hypotheses, but may also accept incorrect ones, if the predictions of the latter are the same as another alternative hypothesis not considered in the design of the experiment.

Martin Lindauer described the way in which bees measure the angle of their body with respect to gravity, using groups of sensory hairs in the joints between head and thorax and thorax and abdomen. When Lindauer severed the nerves to these hairs, bees were no longer able to do oriented dances on a vertical comb. When flying in the field, bees use their compound eyes to measure their angle of flight relative to the sun, searching out the patterns of polarized light in the blue sky itself, even if the sun is not visible. The polarized light is produced by a phenomenon called Rayleigh scattering; the angles of polarization occur in a pattern that is consistent relative to the position of the sun, and this pattern moves across the sky as the earth moves relative to the sun. Rüdiger Wehner and S. Rössel discovered that the bees use a "celestial compass" to interpret the polarization patterns, which consists of the layout of ommatidia in the dorsal portion of the bees' compound eyes. Each ommatidium is selectively sensitive to a particular angle of polarization of light, and each ommatidium also gathers light exclusively from a particular region of the visual field of the bee. The layout of the ommatidia is such that when a bee is facing directly away from the sun, each ommatidium is looking at the region of the sky that contains the angle of polarized light to which it is most sensitive. Thus, as the bee rotates in flight, the summed response from these specialized ommatidia will reach a peak when the bee is aligned with the sun azimuth and fall away as she turns off it. Although the way in which a bee uses this system to hold a fixed course at a particular angle relative to the sun is not known yet, this compass provides a beautiful example of how a solution to a tremendously complex analysis can be built into the design of the sensory system, so that only relatively simple neural processing is needed to execute the behavior.

USE OF THE DANCE LANGUAGE

Honey bees are known to use the dance language to recruit nestmates in several contexts. In the context of foraging, bees dance to indicate the location of sources of nectar, pollen, water, and propolis (a resinous material collected from plants and used to seal cracks and waterproof the nest cavity). As far as is known, the dances for these different materials are the same, but this area has not been systematically investigated.

When a swarm of bees leaves its natal colony to build a new nest elsewhere, scout bees report the location of cavities they have found by means of the dance, and other bees inspect the advertised sites and may dance in turn. Over the course of hours or days the swarm as a whole makes a choice among the alternative sites discovered by different scouts and arrives at a unanimous decision on a single site. The swarm then takes off and flies to the new nest site. Only a small minority of the bees in the swarm has ever visited the chosen cavity. Therefore, although the information transferred by the dance could be important in guiding other bees to the site, there are probably other mechanisms, perhaps visual or olfactory, involved as well. The question of how swarms find their way, and the question of just how the dance language is used in the course of the swarm coming to a collective decision on a single nestsite, are still being investigated.

The sharing of information about food sources makes it possible for a honey bee colony to serve as an information center, pooling the reconnaissance of its many foragers, surveying a vast area around the nest, and focusing the bulk of its foraging force on the best sources discovered. In the 1980s a study by Kirk Visscher and Tom Seeley decoded the dances of a colony living in a deciduous forest in New York State to show the dynamics of colony food patch use that result from these interactions. Research by Seeley has shown that integration of foraging information via the dance language is quite flexible, and Seeley has worked out many of the mechanisms by which a honey bee colony responds rapidly to changes in the relative quality of food sources and colony need for food.

THE DANCE LANGUAGE CONTROVERSY

In the late 1960s Adrian Wenner, Patrick Wells, and Dennis Johnson challenged von Frisch's interpretation of the bee dances. While they did not question that the dances contain correlations of distance and direction, they pointed out that many experiments claimed by von Frisch to show that bees actually used this vector information in their searches could also be interpreted as the bees simply orienting with respect to odors. These ambiguous results were recorded when the recruiters' feeder was placed in the center of an array of scented bait stations and recruits were observed to come more frequently to stations near the center. This behavior, von Frisch's critics argued, would be predicted regardless of whether bees were using distance and direction (and odor) information or just odor information. Johnson and Wenner performed experiments at relatively short distances and with strong odors, and the results followed the expectations of recruits relying strongly on odor produced by bees feeding at the bait stations, but not the expectations of the location information in the dance.

Not all of von Frisch's experimental results were readily reinterpreted in terms of the odor-only hypothesis. For example, when a hive is turned on its side, bees are unable to use gravity as a reference for their dances and so do disoriented dances, and von Frisch showed that recruits were less well oriented under these conditions, although odor cues would not have been affected. Several lines of subsequent work have indicated that the search distribution of recruits can indeed be influenced by distance and direction information from the dance alone. The challenge in such studies is that normally odor information and dance vector information is highly correlated, so definitive experiments required means of unlinking them.

In the 1970s James Gould unlinked the location (and odor) of the food source on which dancers had foraged from the directional information in their dances. To achieve this, he shined a bright light from the side as bees danced. In this situation, recruiters or recruits will normally perform or interpret dances using the position of the light as the "sun" angle reference, rather than the direction upward. However, if a bee's ocelli are painted over with opaque paint, the bee becomes less sensitive to light, and so this shift in reference does not occur. By having recruiters with painted ocelli (and a reference of up) dancing, followed by recruits with unpainted ocelli (and reading the dances relative to a reference of the light, at some other angle), Gould was able to show that recruits could interpret a direction from the dance that was independent of the direction to the food source. The recruits then searched principally in the direction predicted by the modified dance information, rather than the true direction of the feeder, as would have been predicted by the odor-only hypothesis.

In the early 1990s Axel Michelsen, Martin Lindauer, and Wolfgang Kirchner constructed a computer-controlled robot bee that mimicked the behavior of a dancing bee. Recruits followed this robot bee and searched for food preferentially in the directions indicated by the dance angles programmed for the robot. Changes in the length of the robot bee's dances also changed the distribution of distances at which recruits were captured. The robot bee recruited rather imprecisely, with even more scatter than the rather large scatter of recruits from real bee dances. However, the demonstration that changing nothing but the computer programming was enough to cause significant shifts in the search distribution of recruits in the predicted manner was conclusive evidence that recruits were decoding distance and direction information from the dances.

CURRENT QUESTIONS

Although it is now quite clear that bees do decode the dances, odor does play a strong role in recruitment to food sources. It is appropriate to think of the dance as giving recruits a general idea of the direction and distance to the food source. Recruits then search in this area for sources matching the odors they have learned from the food carried by the dancing bee. Depending on the distribution of available food sources, the distance and direction information might be crucial in organizing a colony's food collection, or relatively unimportant. However, the relative importance of these two mechanisms in different habitats is just beginning to be investigated.

The angular scatter in the dance itself decreases with increasing distance indicated, as von Frisch reported. This change in scatter may be the result of changing duration of the waggle runs of the dance, but it also may be an adaptation to recruit bees to patches of more or less constant size at varying distances. This idea is supported by Seeley and Burmann's finding that the dances of scouts for nest sites, which are always single points rather than patches, have less scatter than those of nectar foragers. However, these results differ from those reported by Will Towne on the same issue, and this remains a question of current research.

The evolutionary origin of the honey bee dance remains incompletely discerned. All species of *Apis* perform recruitment dances, though there are interspecific variations in a number of the aspects discussed earlier. The stingless bees (Meliponini), the bumble bees (Bombini), and the orchid bees (Euglossini) are the closest relatives of *Apis,* but the phylogeny of these different taxa within the Apidae remains controversial. Stingless bees are highly social and have a variety of mechanisms of recruitment that may provide possible antecedents to the dance language, but a determinination of how the current form of the dance language might have arisen from these components must await both a greater understanding of recruitment mechanisms within the stingless bees and a more firmly established phylogeny within the family Apidae.

See Also the Following Articles

Apis Species • Feeding Behavior • Orientation • Recruitment Communication

Further Reading

Dyer, F. C. (2002). The biology of the dance language. *Annu. Rev. Entomol.* **47,** 917–949.
Frisch, K. von. (1967). "The Dance Language and Orientation of Bees." Harvard University Press, Cambridge, MA.
Moffett, M. W. (1990). Dance of the electronic bee. *Natl. Geogr.* **177(1),** 135–140.
Seeley, T. D. (1995). "The Wisdom of the Hive: The Social Physiology of Honey Bee Colonies." Harvard University Press, Cambridge, MA.

DDT

Fumio Matsumura
University of California, Davis

DDT (dichloro-diphenyl-trichloroethane) is an old insecticide that has been banned from use in most countries of the world since the 1970s. However, DDT, its metabolites, and some of its derivatives, which are mostly produced as impurities in technical insecticide preparations, still contaminate the environment. DDT residues continue to cause deleterious biological effects, most notably, environmental endocrine disruptions.

From the viewpoint of environmental toxicology and chemistry, DDT is by far the best-studied chemical. Many models of bioaccumulation, atmospheric transport, transfer mechanisms within soil compartments, and from soil to air, and soil to water are based on data generated from studies of DDT residues in the environment.

CHEMICAL CHARACTERISTICS

DDT is one of several typical chlorinated hydrocarbon insecticides discovered in the early 1940s and known for their persistent insecticidal activities, their lipophilic attributes, and their stable chemical properties. The insecticidal properties of DDT itself were discovered in 1939 by Paul H. Müller of Switzerland, who later received the Nobel Prize for his work. Since DDT was the first organic synthetic insecticide that possessed advantages such as low mammalian toxicity, wide spectrum, long-lasting properties, and low cost in comparison to arsenicals and other inorganic insecticides, most entomologists embraced its use to such an extent that more than 100 million pounds of DDT was being produced annually by the mid-1950s.

The insecticidal active ingredient of DDT preparations is *p,p'*-DDT (Fig. 1A). Its 1-dechlorination product, *p,p'*-DDD (Fig. 1B) retains reasonable levels of toxicity for some insects, but its dehydrochlorination product, *p,p'*-DDE (Fig. 1C), shows no insecticidal property, although *p,p'*-DDE could still have a toxic effect in other organisms. Other components often found in insecticidal DDT preparations are *o,p'*-DDT, *p,p'*-DDD, and *o,p'*-DDD. All these can be found as environmental residues.

DDE (dichloro-diphenyl-ethylene) one of the residues derived from DDT most frequently found in the environment, is produced mainly by metabolic activities in biological systems and is particularly prevalent in insects and in some mammalian species. Although both *p,p'*-DDE and *o,p*-DDE are found in the environment, the former is more abundant and more frequently encountered. In assessing residue levels of all DDT-derived compounds today, scientists express the entire spectrum of DDT-related (DDT-R) compounds or DDT-derived compounds as total DDT residue, or DDTs.

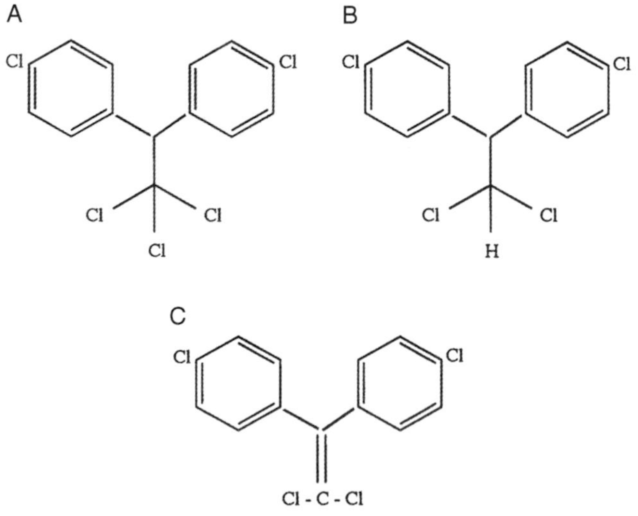

FIGURE 1 (A) 1,1,1-Trichlor-2,2-bis (*p*-chlorophenyl) ethane (*p, p'*-DDT). (B) *p, p'*-DDD. (C) *p, p'*-DDE.

EFFECTS ON INSECTS

The main action mechanism by which p,p'-DDT causes the death of insects is the destabilization of the sodium channel, the main vehicle that propagates excitation signals on the surface of neurons, so that affected neurons become easily excitable. Insects poisoned by DDT show typical hyperexcitation symptoms that lead to exhaustion and death. This phenomenon may be better understood as an electrophysiological manifestation in which neurons affected by DDT show a typical excitation pattern called "repetitive discharges." Such a neuron that has been excited by a stimulus remains in an excited state and continues to discharge for several minutes.

The most well-known use of insecticidal DDT is probably for mosquito control in malaria eradication programs. The most frequently used technique was that of "wall painting" of the interior of buildings with DDT in areas where malaria was prevalent. Because mosquitoes transmit malaria directly from human to human (i.e., without going through other hosts), this method effectively cuts off the link to continued transmission. The two key properties of DDT responsible for its effectiveness are the extreme susceptibility of mosquitoes to DDT and the long-lasting nature of DDT, particularly in indoors and dry environments.

DDT was also well known for its role in the control of cotton insect pests that posed a serious problem to cotton growers in the southern United States. The most commonly used formulation was a mixture of DDT and toxaphene. DDT was also used to control many other pests including the bark beetle vectoring Dutch elm disease, locusts, and forest pests (e.g., spruce budworm); these wider uses resulted in environmental loading of DDT-R.

ENVIRONMENTAL EFFECTS

Although p,p'-DDT is really the only component of DDT-R potent enough to be an insecticidal ingredient (as far as environmental effects are concerned), all the DDT-related compounds are presumed to be potentially toxic. Perhaps the best example of the extreme toxicity of DDE is its effects on bird reproduction. Because DDT is slowly converted into DDE in the environment over many years, environmental samples of DDT-R today are actually mostly DDE.

Another important compound is o,p'-DDT, which is known to mimic the actions of estrogen in several vertebrate biological systems. The action of o,p'-DDT can be attributed to its ability to bind to the estrogen receptor as an agonist, like estrogen itself, and to activate estrogen signals in the organism. Interestingly, p,p'-DDE acts as an antagonist to the androgen receptor in males, thereby blocking male sex hormone signaling in many vertebrate species.

Of all the effects of DDT-related compounds on wildlife, the biological damage cited most frequently is that of eggshell thinning. This phenomenon was originally reported by Ratcliffe in 1967 and verified by Anderson and Hickey in 1976 in North America. In addition to DDT, both DDE and polychlorinated biphenyls also have deleterious effects on eggshell production. Eggs affected by these chemicals crack easily and contribute to the decline of vulnerable bird species.

Eggshell thinning is not the only harmful effect for which DDT-R has been implicated. DDT-R has also been shown to contribute to the increased mortality as well as myriad reproductive problems among a broad range of wildlife including birds, fish, and other aquatic organisms. Behavioral changes are also caused by exposure to DDT-R.

A current view among scientists is to interpret many of these effects as "endocrine disruptions" caused by the hydrocarbon pollutants, with DDT-R being one of the prominent study materials. Certainly, DDT-R, particularly o,p'-DDT, acts in an estrogen-like manner, whereas p,p'-DDE acts as an anti-androgen. Deleterious effects of such endocrine disruptions by DDT-R in birds are well documented. Because disruptions of endocrine actions, including those of some vitamins, are expected to cause serious effects on reproduction, development, and nutritional balance of animals, this topic is likely attract increased attention in the scientific community.

Despite the difficulty of conducting and evaluating environmental effects studies, evidence for the harmful biological effects of DDT on wildlife and ecosystems has been overwhelming. Clearly, the decision to ban the use of DDT was sound.

See Also the Following Articles

Insecticides • Integrated Pest Management • Pollution

Further Reading

Bradley, D. J. (1998). The particular and general issue of specificity and verticality in the history of malaria control. *Parasitologia (Rome)* **40,** 5–10.

Fry, D. M. (1995). Reproductive effects in birds exposed to pesticides and industrial chemicals. *Environ. Health Perspect.* **103** (suppl. 7), 165–171.

Matsumura, F. (1985). "Toxicology of Insecticides," 2nd ed, pp. 51–55. Plenum Press, New York.

Metcalf, R. M. (1955). "Organic Insecticides," pp. 127–180. Interscience, New York.

Peakall, D. (1970). Pesticides and the reproduction of birds. *Sci. Am.* **222,** 72–78.

Peterle, T. J. (1991). "Wildlife Toxicology," pp. 157–172. Van Nostrand Reinhold, New York.

Ratcliffe, D. A. (1967). Decrease in eggshell weight in certain birds of prey. *Nature* **215,** 208–210.

Defensive Behavior

Justin O. Schmidt
Southwestern Biological Institute, Tucson

Defensive behaviors are the responses of organisms to perceived threats by potential predators. The responses can be active and obvious to an outside observer, including the predator; they can be subtle and difficult to observe; or they can be completely inapparent. Obvious responses might include escape flight or changing to a menacing posture, a subtle response might be the "freezing" of a slowly moving insect, and an inapparent behavior might be the warming of flight muscles by a large moth or beetle in anticipation of flight from a detected predator. The goal of this article is to describe the major types of defensive behaviors of insects and to illustrate how, when, and why these defenses are of survival benefit.

All animals must eat. Food choices for animals are limited to materials derived from other life forms, with flesh from animals being among the richest food sources in energy and nutrients. This sets the evolutionary stage for fierce competition among organisms to eat others, yet not to be eaten themselves. Good defenses and defensive behaviors tip the balance from mere survival of a population (or its extinction) to success and domination of a niche.

Survival and reproduction are the key elements of life. For both elements, defense is a paramount feature; without defense, survival, and therefore reproduction, is unlikely. Insects must defend against microorganisms, parasites, and predators and use different strategies against each. The defenses against these attackers differ. The ultimate defense against microorganisms, including bacteria, viruses, protozoans, and fungi, is the immune system. Parasites pose a different challenge. These multicellular organisms live in or on the insect body, sapping vital nutrients and reserves, sometimes damaging essential tissues or organs and causing death. Defenses against parasites are primarily behavioral and life history strategies, with backup from the immune system after parasite attack. Parasitoids are a curious group of attackers that share properties of both parasites and predators. They, like parasites, live in or on the body of the host insect and feed on its blood and tissues. Like parasites, they also do not immediately kill the insect. Parasitoids resemble predators, in fact some consider them predators, because their mode of delivery usually involves direct physical attack on the prey. Parasitoids differ from parasites because instead of directly killing the host, the parasitoid lays one or more eggs or larvae on or in the host and then the parasitoid larva(e) consumes and kills the host. Insect defenses against parasitoids are a combination of defenses used against predators and against parasites. Attacks by adult parasitoids are met with behavioral and morphological defenses similar to those used against predators. Deposited eggs and larvae are resisted by encapsulation by the immune system and other physiological defenses. Predators, in contrast to microorganisms and parasites, directly attack and kill or paralyze their insect prey. They also possess more complex nervous systems than parasites and use this added ability to combine enhanced sensory awareness, decision-making, and learning to challenge the limits of the insect prey to detect, respond to, and defend against the predator. Consequently insects have evolved a myriad of defenses of dazzling form and complexity against their predatory enemies.

FACTORS AFFECTING PREDATORS AND PREY

The predator–prey equation is never constant. Age and size of respective predator and prey, hunger levels, population sizes, presence of alternative prey, and behavioral factors are ever changing. Size of a potential predator relative to prey size is an obvious factor affecting defensive capabilities of an insect. For example, an ant's mandibles might be an effective defense against a small jumping spider, but likely are ineffective against an anteater. Hunger is an important, often overlooked, factor in the defensive equation. Investigators sometimes starve a potential predator for a period of time to ensure that it is hungry when presented with a potential prey. This often yields false impressions, because a starved predator is much more likely to try to attack almost anything that might be edible than would a well-fed predator. An analogy from human experience is instructive. Humans faced with starvation from war or other disasters have eaten rats or cockroaches in an effort to survive, but these same people would not consider such items when not starving. The effect of predator hunger can greatly affect the success of insect defenses. Hunger level can also affect prey insects by inducing them to forage for food during more dangerous times and for longer periods.

Population levels also influence the success of insect defenses. Cryptic (concealed) caterpillars (Figs. 1 and 2) are at more risk of failure of their defensive concealment when high populations of paper wasps *(Polistes)* are present than at times of low wasp numbers. The opposite situation, high populations of prey, can turn the defensive tables in favor of the prey. The synchronous emergence of periodic cicadas and mayflies not only serves reproductive benefit but also saturates the predators in the environment, reducing the risk to each individual cicada or mayfly. Presence or absence of

FIGURE 1 Looper caterpillar (Geometridae) with fleshy body projections whose shape and appearance closely match the vegetation of its host plant, *Polygonella* sp., providing excellent camouflage and protection from visually searching predators. (Author photograph, Florida, U.S.A.)

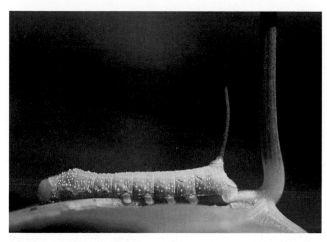

FIGURE 2 Second instar of *Eumorpha typhon* cryptically matching background grape leaf as it rests. (Author photograph, Arizona, U.S.A.)

alternative prey affects the success of various insect defenses. Some species of caterpillars exhibit several different color patterns. These color differences form the basis for "apostatic selection," which confers protective benefit on the rare color morph (form). The rarer morph is safer because birds adopt search images or searching behaviors oriented toward discovery of the common-color morph and often miss rare-color morphs that do not fit the image. Predator behavioral factors can determine the effectiveness of the defensive behavior of prey. Prey speed and flight ability often provide excellent protection from predators that actively search for prey. Flies and bees rarely fall prey to roving spiders, but often are captured by ambush sit-and-wait crab spiders. In the examples of vinegaroons (whiptail scorpions, *Mastigoproctus giganteus*) and tarantula spiders, which are classical ambush predators, fast and powerful prey such as sulpugids (wind scorpions) and centipedes are surprised by the ambush predator and fall prey. The element of surprise is crucial. Without surprise, the powerful jaws and quickness of alert sulpugids and centipedes make them formidable prey that would be difficult, if not impossible, to overcome by predators that have, at best, only equal equipment.

OVERRIDING PRINCIPLES

Predator–prey relationships are not chaotic, but are based upon time-tested principles with constant refining and generation of new approaches. Principles for prey include: (1) it is better to avoid attack than to defend against an attack; (2) the higher the cost or penalty suffered by a predator, the greater protection gained by the prey; and (3) avoid sharing time and space with as many potential predators as possible. Combinations of these principles form the basis for most successful defenses of insects. However, the story is complex: insects usually face not just one, or even a few, species of predators, but rather a whole suite of potential predators. And the biggest, fiercest predator usually is not the

one that poses the greatest risk. To be successful, insect populations must maximize their defensive success against the summation of all predator attacks encountered. This is the combination of the probability of attack by each potential type of predator and the probability of success in countering the attack. Herein lies the experimentalist's dilemma: how can we know all of the predators that have impacted the evolution of an insect's defenses, the frequency of the attacks by each predator type, the success of the attacks by each predator type, and whether "phantom" predators, predators that might have posed serious threats in the past or become threats only at very infrequent intervals, exist. In addition, to access accurately the role of predators, the effects of predators during both bottleneck and outbreak periods of the insect's population must be evaluated. A clever experimentalist is an outstanding observer of natural history and the biology of the insect in question and successfully controls as many variables as possible to resemble nature.

EVOLUTIONARY STRATEGIES AND DEFENSIVE BEHAVIORS

Crypsis

Crypsis, or avoiding detection by blending into the background, is one of the most common and successful defenses. Classical examples of crypsis include stick insects in the order Phasmida, leaf-mimicking moths, and ambush bugs (Phymatidae) that resemble the flowers in which they hide. In the first two examples, crypsis functions to avoid detection by visually hunting predators such as birds and monkeys. The crypsis of ambush bugs serves dual purposes of concealment from potential predators and from their prey, flower-visiting insects. Variations of the cryptic theme can take many forms. Many caterpillars, moths, and other insects are patterned and colored like the vegetation they eat or the twigs, bark, or other substrate upon which they rest (Figs. 1 and 2). In this way they blend toward invisibility in the eyes of all but the best predators. Other cryptic specialists resemble dead objects ranging from bird droppings, for some swallowtail butterfly caterpillars (*Papilio* spp.), to stones, for some grasshoppers *(Eremocharis insignis)* and toad bugs (Gelastocoridae). Cryptic insects match behavior to lifestyle. To maintain their concealment cryptic insects tend to move little during the day, and when they do move it is slow and deliberate to avoid notice. Exceptions are cryptic predators that must move swiftly during the act of prey capture. Cryptic insects tend to select resting backgrounds, lighting conditions, and positions to match their own appearance. How they recognize and choose matching backgrounds is unclear.

Aposematism

The opposite of crypsis is aposematism, or a warning signal to predators. Aposematic insects usually appear and behave

FIGURE 3 Aposematic venomous caterpillar, *Automeris metzli* (Saturniidae), whose painful urticating hairs can cause a long-lasting burning rash. (Michael F. Wilson photograph, Chihuahua, Mexico; used with permission.)

very different from cryptic insects. These warning-colored insects sport bright pattern and color combinations of reds, yellows, oranges, whites, and blacks that are notably conspicuous within the background setting and advertise the insect's presence (Fig. 3). Insects with warning coloration, like cryptic insects, tend to move slowly in their environment. Similarity ends there; the slow, deliberate movement of aposematically colored insects serves not to avoid detection, but to provide time for potential predators to detect and recognize the insect before acting. Warning coloration boldly signals that this insect is toxic, is bad tasting, or can sting or cause injury.

Aposematic warnings have evolved for every sensory system used by predators. Acoustic stridulations, hisses, and other sounds advertise to potential predators that the emitting insect is dangerous. Toxic tiger moths (Arctiidae) send loud return sounds to approaching insectivorous bats. These sounds warn the bat that the moth is unpalatable and potentially harmful, and may also interfere with the echolocation capabilities of bats. Often aposematic sounds share similarities: they have broad frequency ranges, have low pattern complexity, and resemble "white noise." Such signals are readily detected by a wide range of predators, differ from most nonwarning sounds produced by insects or other animals, and are poorly structured for precise intraspecific communication, e.g., courtship sounds. These are exactly the properties a good warning signal needs: they are readily detected, conspicuous, and generalizable and they indicate no form of communication other than warning. Aposematic sounds are produced by a wide range of insects, including velvet ants (Mutillidae), bees and flies, assassin bugs (Reduviidae), moths, many groups of beetles, cockroaches, and grasshoppers.

Aposematic odors and tastes serving as warnings need not be toxic, only readily apparent. Examples of warning odors include the strong, unpleasant fragrances of pyrazines produced by butterflies and a variety of other insects, ketones

produced by velvet ants, and uncharacterized compounds produced in mandibular glands of tarantula hawks (Pompilidae, *Pepsis*). Warning tastes are probably widespread among noxious insects, but few have been investigated. Predators such as lizards, birds, fish, and spiders will often "taste" an insect before actually killing and eating it. If the sampled insect has a compound with an aposematic taste, it is often released unharmed. Tiger moths externalize some of their pyrrolizidines, which, when tasted by orb weaver spiders, cause the spider to cut its web to release the unharmed moth.

Mimicry

Aposematic organisms having effective defensive systems become models for biological copycats. Some of these mimicking organisms are cheaters who do not actually possess noxious or dangerous properties like those of the models they resemble. They "trick" predators into perceiving them as noxious and unpalatable. This kind of mimicry is called Batesian mimicry, in honor of the famous naturalist Henry Bates, who first recognized the phenomenon. The pipevine swallowtail *(Battus philenor)* is a toxic butterfly that serves as the model for several Batesian mimics, including the dark morph of the palatable female (but not male) tiger swallowtail *(Papilio glaucus)*. In regions where model pipevine swallowtails are abundant, most tiger swallowtail females are mimics; but in areas where pipevine swallowtail models are rare, many female tiger swallowtails have patterns that are not mimics. Other mimicking species are "truthful" because they themselves are noxious or toxic and resemble other noxious or toxic insects. Such "Müllerian" mimicry abounds among butterflies, especially within the passion-flower-feeding butterflies in the genus *Heliconius* in South America or the queen butterflies *(Danaus* and relatives) of Africa. Between the extremes of Batesian and Müllerian mimicry exists a continuous gradation of mimicking species that possess varying degrees of noxiousness.

Allomones

Allomones are chemical defenses. Many hundreds of allomonal chemicals have been identified from insects, most being distasteful, damaging, or toxic to other animals or synergizing the activity of other active chemicals in the secretion. Beyond their common properties for defense, allomones share little chemical similarity and contain compounds of almost all imaginable types, including organic acids, alcohols, ketones, aldehydes, esters and lactones, hydrocarbons, terpenes, phenolics, quinones and hydroquinones, amines, alkaloids, sulfurous compounds, steroids, polysaccharides, peptides, and proteins. Allomones are the defensive arsenal's "backup artillery" used to blunt direct attack by predators. The effectiveness of allomones is embodied in the common names of some insects and their relatives: stink bugs, blister beetles, bombardier beetles, and vinegaroons. The aldehydic secretions of stink bugs spread over the body surface, providing a repellent odor barrier that doubles as a repugnant contact liquid when touched by a predator. Bombardier beetles (*Brachinus* spp.) and vinegaroons (*Mastigoproctus* spp.) spray corrosive quinones/hydroquinones and concentrated acetic acid solutions, respectively, at approaching predators, thereby actively extending protection to a distance and reducing the risks of physical attack by the predator. Insect venoms are effective specialized allomones typically consisting of water-soluble proteins and other components that are injected into the body of an assailant. The effectiveness of stings as defenses against large predators is evident from the names "killer bees," "fire ants," and "cow killers." Other names, including hornet and wasp, have become terms synonymous with pain, power, and fear.

DEFENSIVE PLOYS

For virtually any sensory system and behavior used by predators to detect prey, prey insects have evolved counterstrategies or defensive ploys. For predators relying mainly on vision, insects possess physical properties and behaviors either to avoid being seen or to maximize being seen. For predators that rely primarily on sound, prey have counterbehaviors to minimize sound generation, to evade sound-emitting predators, or to counter with effective sounds of their own. For predators using olfaction as their primary searching sense, prey have evolved systems to reduce their own odor, to mask it, to mimic the odors of unsuitable prey, or to blunt sensory orientation with allomones and aposematic odors. As a rule, insects supplement general defenses and behavioral strategies with multiple suites of defenses directed toward specific sensory systems.

Concealment and Hiding

Hiding from predators is a nearly universal tactic of insects. Even well-defended insects such as stinging wasps conceal their nests within dense vegetation, among roots, or in holes. Aposematic insects such as *Dasymutilla occidentalis* tend to rest in concealed places during periods of inactivity and run and hide under leaves or among vegetation when an approaching potential predator is sensed. Toxic butterflies often rest with wings folded and in among vegetation that hides them from view. Many other insects are masters of concealment and are so cryptically colored and patterned that finding them in a photograph has become an educational and entertaining challenge for children and adults alike. Some insects are concealed only during particular times. Caterpillars are commonly concealed on bark or in the ground while at rest, but are more apparent while feeding on leaves.

Concealment and hiding take many forms. Less obvious than daytime concealment against visual predators, but an equally frequent and effective defense, is use of time for concealment. Looper caterpillars (Geometridae) and others that are cryptically colored and concealed feed on leaves during daylight. At night when they are not cryptic or concealed from spiders, beetles, ants, and predators that search for prey mechanically and by olfaction and vibration, some conceal themselves by terminating feeding and hanging below the vegetation on long silken threads. An extreme of concealment is used by mayflies, which "conceal" their adulthood by reducing their adult life to only about a day, just enough time to mate, lay eggs, and die.

Escape

If an insect is capable of flying, jumping, running, crawling, or dropping to safety, escape is often the first response to detection by a predator. Nevertheless, not all insects can, or do, attempt to escape when approached by predators. Many larvae live in confined spaces and can move little or slowly, and none can fly. These individuals must rely primarily on other means of protection such as concealment, crypsis, or chemical defense. Other species rely first on their aposematism and/or noxious nature for protection and only attempt escape secondarily.

Fighting Back

When concealment and escape fail, most insects resist and fight back by biting with mandibles, kicking and struggling, stinging, and releasing allomones, with varying success. Leg spurs and spines are used effectively by some large moths, including sphinx moths in the genera *Manduca* and *Eumorpha* and cockroaches (*Archimandrita marmorata*), to defeat the grasp of even large potential predators. These sharp spines not only can painfully pierce skin but also can anchor strong kicks to free the slippery insect from grasp. Some male wasps possess either sharp genitalia or separate "pseudostings" that are jabbed into grasping predators. Jabbed predators might mistake pseudostings for actual stings of female wasps and release the male.

Pain

Pain is the early warning system to indicate that bodily damage is occurring, has occurred, or is about to occur. Bodily damage is a serious threat and risk to an organism's ability to survive, feed, and reproduce. When given a choice between a meal with accompanying pain (plus perceived bodily damage) and the loss of a meal, predators often opt for the latter. The venomous stings of wasps, bees, and ants are legendary for their abilities to cause pain and deter predation. Spiny caterpillars and an assortment of biting bugs and beetles, including assassin bugs (Reduviidae), giant water bugs (Belostomatidae), water scorpions (Nepidae), and predaceous diving beetles (Dytiscidae), also produce painful venoms. Allomones can be effective by causing immediate pain. Examples are formic acid, sprayed by ants in the subfamily Formicinae, carabid beetles, and notodontid caterpillars, and quinones, released by tenebrionid and carabid beetles. Although bites and kicks might induce pain, their overall effectiveness relative to venoms and allomones suffers from lesser ability to produce pain and from predator familiarity with them and their expected effects.

Warnings

Stereotyped warnings are used to threaten and intimidate predators. Paper wasps (*Polistes* spp.) on their nest face large adversaries with raised wings, waving front legs, abdomens curved toward the predator, and wings flipped, fluttered, or buzzed. These threats inform the predator that it is spotted and an attack will ensue if the advance continues. Hissing cockroaches *(Grophadorhina portentosa)* threaten by hissing, which resembles the defensive hiss of a snake. Many flies and harmless bees and wasps buzz loudly when grabbed. These aposematic buzzes sound similar to those of painfully stinging honey bees and wasps and often serve as effective warnings.

Surprise and Startle

Insects that perceive an approaching predator can use the elements of surprise and startle to escape. Surprise combined with rapid escape flight are often a sufficient defense. If a predator is adept at pursuit, the execution of surprise and rapid flight, followed by instantaneous concealment upon landing, becomes a powerful defense. Startle is the combination of the elements of surprise and fright. Examples of startle are dull cryptic moths, which, when detected by a bird or monkey, flash their hidden hind wings, revealing bright colors (Fig. 4) or large frightening eyespots mimicking those of an owl or large predator (Fig. 5). Desert clacker grasshoppers, *Arphia pseudoneitana,* are grand masters of surprise and startle. When approached by a large animal, these inconspicuous desert grassland insects suddenly jump into the air, fly away amidst a confusion of bright red wing flashes and loud clacking noise, and disappear into distant

vegetation as suddenly as they appeared. In addition to startle, this behavioral display provides the predator with a search image for red color, which can cause the predator to overlook the dull grasshopper.

Confusion

Individuals in schools of fish, flocks of birds, and herds of running African ungulates present difficult targets for predators. Use of confusion of predators via mass motion is little studied in insects but likely is an important defensive behavior in some situations. The constant movement and hopping of masses of migratory locusts and the seemingly erratic circling flights of flies disturbed from a fresh cow patty are likely examples of the use of confusion as a defense.

Aggregations

When many individuals aggregate in a group, each member receives protection through the presence of the others. Not only does a group present fewer locations with prey but also the individuals within a group gain protection through reduction in the chance of being the chosen prey by a discovering predator. Aggregation as defense is particularly effective if the individuals are toxic or are defended as are ladybird beetles (Coccinellidae), monarch butterflies, milkweed bugs (Lygaeidae), or social wasps on a nest. In these insects, a potential predator need sample only one or a few individuals to learn the unsuitability of the whole.

Association

Protection can be achieved by living near or associating with a defended or noxious species. The tropical paper wasp, *Mischocyttarus immarginatus,* prefers to make its small nest with few individuals near the much larger nest of stinging *Polybia occidentalis,* a common social wasp. The arrangement seems to provide protection for the *Mischocyttarus* from vertebrate predators. No potential benefit to the *Polybia* has been demonstrated. Honey bees and some species of ants are known to nest in portions of termite mounds. The exact

FIGURE 4 Underwing moth (Noctuidae) with cryptic leaf-mimicking front wings (note that the mimicry even includes "mold" spots) that normally cover the bright hind wings, which are exposed to startle predators. (Author photograph, Borneo.)

FIGURE 5 Warning eyespots resembling the eyes of an owl or mammal are displayed as a threat by the leaf-mimicking silk moth, *Automeris cecrops,* when disturbed. (Michael F. Wilson photograph, Arizona, U.S.A; used with permission.)

nature of these associations is unclear and the ants generally attack termites if given the chance. The benefits to the bees and ants are more obvious; they not only share the moderated temperature and humidity environments produced by the termites, but they also gain protection and reduced risk of discovery by being in the termite mound.

Sociality

Unlike a solitary individual that must detect and defend against predators alone, individuals of social species enjoy benefits of group defense. A group of many coordinated individuals can more readily detect predators than solitary individuals, can then recruit others via alarm pheromones or vibrational signals to the common defense, and can launch effective attacks *en masse*. Group attacks are particularly effective when individuals possess painful stings or bites and when the attackers are nonreproductive workers who can sacrifice themselves in battle with little reproductive loss to themselves or the colony as a whole. Predators confronted by a "cloud" of attackers cannot devote attention to defending against each attacker, and reduced predator vigilance enhances attacker chances of scoring an effective sting or bite. Sociality and defense are strongly synergistic and, when combined, go a long way toward explaining the success of social insects.

Overall, insects as a class have taken defensive behaviors to levels unsurpassed in number, complexity, and creative diversity within animal life.

See Also the Following Articles

Aposematic Coloration • Crypsis • Ladybugs • Mimicry • Venom • Wasps

Further Reading

Blum, M. S. (1981). "Chemical Defenses of Arthropods." Academic Press, New York.
Cott, H. B. (1940). "Adaptive Coloration in Animals." Methuen, London.
Edmunds, M. (1974). "Defence in Animals." Longman, Essex, U.K.
Evans, D. L., and Schmidt, J. O. (eds.) (1990). "Insect Defenses: Adaptive Mechanisms and Strategies of Prey and Predators." State University of New York Press, Albany.
Schmidt, J. O. (1982). Biochemistry of insect venoms. *Annu. Rev. Entomol.* **27**, 339–368.
Wickler, W. (1968). "Mimicry in Plants and Animals." McGraw–Hill, New York.

Dengue

Thomas W. Scott
University of California, Davis

Dengue is a human disease caused by a virus that is transmitted from one person to another by the bite of infected mosquitoes. Worldwide, dengue virus infections cause more human morbidity and mortality than any other arthropod-borne virus. It is estimated that 2.5 to 3.0 billion people are at risk of infection each year, and millions have been infected during recent epidemics. Dengue occurs throughout the tropics, where incidence rates have steadily increased since the 1950s. The most severely affected areas are urban centers of Southeast Asia, where dengue hemorrhagic fever and dengue shock syndrome (DHF/DSS) are among the leading causes of pediatric hospitalization. During the past 20 years dengue has emerged as a major international public health threat; during that time, changes in dengue epidemiology were most pronounced in the Americas. Epidemics in Cuba and Venezuela during the early 1980s have elevated concern that the progression of dengue outbreaks in the Western Hemisphere is following a pattern similar to that observed in Asia over the past 50 years, putting people in the New World tropics at increased risk for severe, life-threatening disease.

HISTORY

Dengue viruses are believed to have originated in tropical forested habitats, moved from there to rural environments, and finally invaded urban centers. The word dengue most likely originated from Swahili, and following a series of modifications in pronunciation and spelling the word evolved to its present form. The earliest recorded epidemics of a dengue-like illness were in China during the Chin dynasty (265–420 AD). During the 18th and 19th centuries sporadic epidemics were reported in Asia and the Americas. Following World War II, the pattern changed from one of periodic outbreaks to one of continuous transmission of multiple virus serotypes in Asian cities. It was from that situation that DHF/DSS surfaced in 1954 in the Philippines.

DISEASE

Dengue fever (DF), DHF, and DSS are caused by four closely related but antigenically distinct single-stranded RNA viruses (DEN-1, 2, 3, and 4) in the genus *Flavivirus*, family Flaviridae. All four serotypes cause a range of human disease, including asymptomatic infections; undifferentiated fever; classic DF, an acute febrile illness with headache, body aches, and rash; DHF; and DSS. Sequential infections with different serotypes are possible because infection with one serotype provides lifelong protection from a homologous infection, but is only briefly cross-protective against heterologous serotypes. Even though most infections, especially in children under 15 years of age, are asymptomatic, it is estimated that annually there are between 50 and 100 million DF cases and between 250,000 and 500,000 DHF/DSS cases worldwide. If untreated, the case fatality rate for DHF/DSS can approach 20%; however, with supportive therapy (fluid and electrolyte management and oxygen) fewer than 1% of severely ill patients die.

The etiology of serious illness is not completely understood. Risk of DHF/DSS is highest in places where two or more viral serotypes are simultaneously transmitted. People with preexisting dengue antibodies, which can be obtained actively from a previous infection or passively by infants from their mothers, are 100-fold more likely to experience severe disease following infection with a heterologous virus serotype than are people without preexisting anti-dengue antibodies. The mechanism most commonly attributed to severe dengue is immune enhancement. Preexisting nonneutralizing antibodies complex with the infecting heterologous serotype virus, which enhances phagocytosis (entry) and replication of virus in mononuclear cells, which then leads to increased vascular permeability and hemorrhaging. An alternative explanation for severe disease is that different serotypes or strains of virus vary in virulence. Severe disease and death have been reported following primary dengue infections. Recent studies in the Americas suggest that the two hypotheses are not mutually exclusive. Immune enhancement may require secondary infection by genotypes of virus from Southeast Asia.

VIRUS TRANSMISSION

Although other mosquitoes in the subgenus *Stegomyia*, including *Aedes albopictus* and *Ae. polynesiensis*, have been implicated as vectors in jungles and rural habitats, and on islands in the South Pacific, *Ae. aegypti* is the most important dengue vector. Although there is some evidence of sylvatic transmission cycles that include nonhuman primates and vertical transmission from an infected female mosquito to her progeny, the majority of dengue virus transmission is between mosquitoes and human hosts. Horizontal virus transmission begins when a mosquito imbibes viremic human blood. Virus enters and replicates in midgut epithelial cells, disseminates to the hemocoel, and infects secondary target tissues, including the salivary glands. Following replication in salivary gland acinar cells, virus is released into the salivary matrix and can be transmitted the next time the infective mosquito probes its mouthparts into a human host in an attempt to locate blood. Extrinsic incubation in the mosquito requires 10 or more days, depending on the ambient temperature. Once infective, *Ae. aegypti* can transmit virus each time it probes its mouthparts into a host or imbibes a blood meal. Incubation in the human host typically ranges from 4 to 7 days, after which the person is viremic for ~5 days. Fever subsides in concert with the inability to detect virus in the blood.

MOSQUITO VECTOR

Ae. aegypti is uniquely adapted to a close association with humans and efficient transmission of dengue virus. Immature forms develop primarily in man-made containers. Highly anthropophilic adult females rest inside houses where they feed on human blood. Unlike most other mosquito species, which engage in a feeding duality of plant carbohydrates for synthesis of nutrient reserves and blood for egg development, female *Ae. aegypti* forego sugar meals and feed almost exclusively and frequently on human blood. Relatively low concentrations of the amino acid isoleucine in human blood are believed to be responsible for the ability of *Ae. aegytpi* to use only blood to meet their energy needs and to complete vitellogenesis. Females fed only human blood have higher measures of fitness (survival and reproduction), and thus a selective advantage, over those fed sugar and blood. Because females seldom disperse beyond 100 meters and consequently food, mates, and oviposition substrates are readily available within the human habitations where they reside, rapid synthesis of glycogen from sugar substrates for extended flight is not necessary. To meet their energy and reproductive needs, females must imbibe more than one blood meal in each gontorophic cycle, something that increases contact with human hosts and opportunities to contract or transmit a viral infection. Because of their unusual propensity to make frequent and preferential contact with humans, *Ae. aegypti* is an exceedingly efficient vector of dengue virus even though compared to other mosquito species they are not especially susceptible to virus infection. Relatively low *Ae. aegypti* population densities have been associated with virus transmission. It is expected that entomological thresholds for dengue virus are quite low.

CONTROL

Dengue control is dependent on control of *Ae. aegypti* because there is no licensed vaccine for dengue, which will need to be tetravalent because of the phenomena of immune enhancement, and no clinical cure has been found. Despite a history of detailed study of *Ae. aegypti*, vector control programs for dengue control are often nonexistent or ineffective. Outdoor applications of aerosol insecticides to kill adults have in most instances not been effective because the majority of females rest indoors where they avoid contact with the insecticide. Over any considerable period of time, control of immatures using chemicals, biological control, or community-based source reduction has been effective only in authoritative systems in which negative consequences are associated with noncompliance. Disease control based on genetic manipulation of mosquito vectors, rendering them incapable of transmitting virus, is currently being investigated but this method will require extensive evaluation before it can be deployed. Of great concern is the observation that explosive dengue epidemics occur even when *Ae. aegypti* population densities are low. This apparent paradox illustrates that aspects of *Ae. aegypti* biology other than population density, such as their blood-feeding behavior, duration of extrinsic incubation, and female survival, play an important role in defining virus transmission dynamics. Control of dengue constitutes a formidable challenge for public health officials.

See Also the Following Articles

Medical Entomology • Mosquitoes • Zoonoses

Further Reading

Gubler, D. J., and Kuno, G. (1997). "Dengue and Dengue Hemorrhagic Fever," p. 462. CAB Int., Wallingford, U.K.

Halstead, S. B., Rojanasuphot, S., and Sangkawibha, N. (1983). Original antigenic sin in dengue. *Am. J. Trop. Med. Hyg.* **32**, 154–156.

Harrington, L. C., Edman, J. D., and Scott, T. W. (2001). Why do female *Aedes aegypti* (Diptera: Culicidae) feed preferentially and frequently on human blood? *J. Med. Entomol.* **38**, 411–422.

Kuno, G. (1995). Review of the factors modulating dengue transmission. *Epidemiol. Rev.* **17**, 321–335.

Monath, T. P. (1994). Dengue: The risk to developed and developing countries. *Proc. Natl. Acad. Sci. USA.* **91**, 2395–2400.

Rigau-Perez, J., Clark, G. G., Gubler, D. J., Reiter, P., Sanders, E. J., and Vordam, A. V., (1998). Dengue and dengue haemorrhagic fever. *Lancet* **352**, 971–977.

Scott, T. W., Amerasinghe, P. H., Morrison, A. C., Lorenz, L. H., Clark, G. G., Strickman, D., Kittayapong, P., and Edman, J. D., (2000). Longitudinal studies of *Aedes aegypti* (L.) (Dipterta: Culicidae) in Thailand and Puerto Rico: Blood feeding frequency. *J. Med. Entomol.* **37**, 89–101.

Watts, D. M., Porter, K. R., Putvatana, P., Vasquez, B., Calampa, C., Hayes, C. G., and Halstead, S. B. (1999). Failure of secondary infection with American genotype dengue 2 to cause dengue haemorrhagic fever. *Lancet* **354**, 1431–1434.

Dermaptera
(Earwigs)

Susan M. Rankin and James O. Palmer

Allegheny College

The Dermaptera (earwigs) comprise a small, relatively old, hemimetabolous order of insects characterized in their external anatomy by paired cerci (forceps) at the posterior end, and (in winged forms) short tegmina incompletely covering hind wings that are also unique structurally (Fig. 1). Behaviorally, earwigs are thigmotactic, nocturnal, and subsocial, in a system whereby the female parent broods, grooms, and defends eggs and young nymphs (Fig. 2). Internal anatomy is typical of orthopteroids, except that the corpora allata have undergone fusion to a single median structure, and the paired ovaries are primitively polytrophic (i.e., each follicle contains an oocyte and a single nurse cell).

Earwigs are members of the orthopteroid assemblage and have a strong sister-group relationship with the Dictyoptera; they also may be closely related to the Grylloblattodea. Four suborders are generally recognized, and of the three extant ones, the Hemimerina and Arixinina are small groups of viviparous ectoparasites of vertebrates; most species of earwigs are oviparous members of the third group, the Forficulina. Typically, classification schemes have relied on features of the

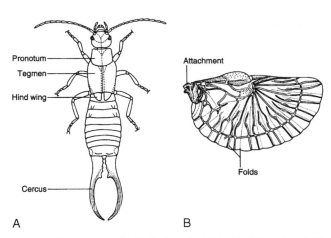

FIGURE 1 European earwig *(Forficula auricularia)*: (A) adult male and (B) his right hind wing. [Reprinted with the permission of Scribner, a Division of Simon & Schuster, from "College Entomology" by E. O. Essig (Macmillan, New York, 1942).]

male genitalia. Earwigs are not of medical importance; they do not crawl in people's ears (occasional anecdotal accounts notwithstanding), and they do not bite, although some may pinch with their forcepslike cerci. Some earwigs may be pests of gardens or households; alternatively, some species are important biocontrol agents, feeding on agricultural pests, such as aphids, armyworms, mites, and scale insects.

GENERAL CHARACTERISTICS

In 1773 DeGeer coined the term Dermaptera (but used the name for all orthopteroids). Kirby in 1815 introduced the name in its current sense, as a small insect order of about 2000 described species. The oldest known examples of dermapterans are Jurassic fossils dating from about 208 mya. These elongate, slender, hemimetabolous (incompletely metamorphic) insects have chewing mouthparts, three-segmented tarsi (in extant groups), (usually) compound eyes, and no ocelli. The presence of abdominal cerci makes them easy to distinguish from beetles. The cerci are typically forcepslike (though they are filiform in at least some parasitic forms) and are sexually dimorphic (Fig. 1). The forceps are used for a variety of purposes, including prey capture,

FIGURE 2 Female ringlegged earwig *(Euborellia annulipes)*, brooding over her clutch of eggs.

defense, fighting, and as aids in copulation and in folding of hind wings. Earwigs are diploid (i.e., they have a double set of chromosomes); males are heterogametic (i.e., they produce gametes with different sex chromosomes, e.g., X and Y).

Some earwig species are wingless as adults, but most have short front wings (tegmina) that do not cover the abdomen. The derivation of the name "dermaptera" (*derma,* skin, *ptera,* wing) refers to the thickened or "skinlike" front wings. The hind wings are unlike those of any other group of insects: they are semicircular and membranous, with radially arranged veins. They fold fanlike beneath the front pair when the insects are at rest (Fig. 1). The derivation of the common name (earwig) may be a corruption of "earwing," in reference to the hind wing resemblance to a human ear. Alternatively, it could be a reference to the ancient Anglo-Saxon legend that these insects crawl in ears of sleeping humans. Additionally, the forceps of some species look like instruments once used for piercing women's ears for earrings.

Earwigs typically display parental care of offspring (although there are almost no observations of maternal care in the viviparous, ectoparasitic forms). Eggs are typically deposited in soil (or protected whorls of monocotyledonous plants); the females "roost" on the eggs until the young hatch and then they care for them (Fig. 2). The period of maternal care appears to be a time of nonfeeding of the brooding female; physiologically, her levels of both juvenile hormone and ecdysteroids (see later) are likely low during the period of egg care.

INTERNAL ANATOMY

The major elements of the neuroendocrine system are the brain, the subesophageal ganglion, three thoracic ganglia, and six abdominal ganglia (with thick, paired connectives between the ventral ganglia). Paired neurohemal corpora cardiaca are connected to the brain and frontal ganglion by strong nervous connections; the closely associated single median corpus allatum produces and releases juvenile hormone III and is in close proximity to the neurohemal dorsal aorta. The digestive system contains the typical regions of fore-, mid-, and hindgut (though gastic caecae are lacking); the midgut–hindgut junction is characterized by the presence of numerous long, slender (excretory) Malpighian tubules.

The female reproductive system consists of paired ovaries, lateral oviducts, a median oviduct, spermatheca (for sperm storage), and genital chamber. Earwigs are unusual in that the female genital opening (gonopore) is just behind the seventh abdominal segment. The ovaries are primitively polytrophic; in some species the long ovarioles branch off the lateral oviduct, while in others, short ovarioles appear in series around the oviduct. The viviparous species are pseudoplacental, with egg maturation and embryonic development taking place in the greatly enlarged vitellarium. The male reproductive system is complex, with paired testes, paired vasa deferentia, paired or single vesicula seminalis (for sperm storage), and a paired or single common ejaculatory duct ending in the sclerotized virga (penislike structure). Anatomy of the male reproductive system has been used extensively in classification schemes of earwigs, as discussed shortly.

RELATIONSHIPS TO OTHER INSECTS

Earwigs are members of the informal orthopteroid assemblage and share a sister-group relationship with the dictyopterans (cockroaches). The Grylloblattodea (rock crawlers) may be linked to the Dictyoptera/Dermaptera. Alternatively, earwigs may be closer to the Grylloblattodea than to any other orthopteroid order.

PHYLOGENY AND DISTRIBUTION OF DERMAPTERA

Early in the 20th century, Burr established suborders of the Dermaptera recognized by most contemporary systematists. The four suborders (three of them recent) are as follows:

1. Archidermaptera, represented by 10 fossil specimens from the Jurassic; they are characterized by unsegmented cerci and tarsi having four or five segments.
2. Forficulina, the suborder containing most earwigs (i.e., 1800 described species, in 180 genera); cerci are unsegmented (except in a few primitive larvae) and forcepslike.
3. Hemimerina, composed of 10 species in one genus; they have filiform (segmented) cerci, and are wingless, blind, viviparous (pseudoplacental) ectoparasites of African rats.
4. Arixenina, composed of five species in two genera; like the Hemimerina, they are viviparous (pseudoplacental), wingless, blind, and ectoparasitic of vertebrates. The Arixenina live on bats in Malayan–Philippine region; Popham considered this group to be a sister group of the Labiidae.

Popham based his phylogeny of families on the structure of the male genitalia (Fig. 3). The Hemimerina consists only of

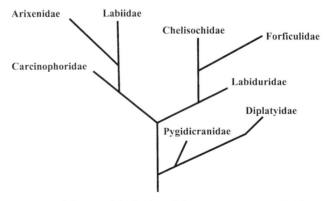

FIGURE 3 Phylogeny of the families of the Dermaptera. [Amended figure reprinted with the permission of Cambridge University Press from Popham, E. (1965). The functional morphology of reproductive organs of the common earwig *(Forficula auricularia)* and other Dermaptera with reference to the natural classification of the order. *J. Zool.* **146**, 1–43.]

the family Hemimerinidae. The Pygidicranidae are frequently regarded as the most primitive family of earwigs, in as much as the males have two functional penis lobes. The Carcinophoridae also show less specialized, ancestral traits (and are typically treated as wingless, though some are fully winged). The Arixinidae and Labiidae are treated as sister groups (though this is not generally accepted). The Chelisochidae and Forficulidae are the most specialized families (i.e., display the most derived traits). The Forficulidae are considered to be "higher earwigs" (more recently derived) because the males have a single functional penis lobe. This family is the best-represented family in North America.

The current geographical distribution of most families of earwigs was largely determined by continental drift, with two main centers of radiation before the Triassic opening of the Pacific ocean being the equatorial region of the eastern Pacific (Pygidicranidae, Carcinophoridae, Arixinidae, and Labiidae) and the Afro-Indian circumtropical center (Labiduridae, Chelosochidae, and Forficulidae).

The distribution has also been largely affected by climatic conditions: earwigs were "discouraged" from spreading northward from the tropics by mountain ranges of southern Europe and Asia; only more specialized members have become established in the Palearctic region (and none have been reported in the polar regions). Although most earwigs have wings, they seldom fly. The cosmopolitan distribution of some species can be attributed to their habit of hiding in crevices, especially in timber or other material that is transported by commerce.

NATURAL HISTORIES AND BIOTIC ASSOCIATIONS

Development

In oviparous species such as European earwig, *Forficula auricularia,* and ringlegged earwig, *Euborellia annulipes,* clutches of ovoid, creamy white eggs are laid in protected burrows; eggs can be up to 2 mm in length in larger species. Nymphs generally resemble adults but can be distinguished from them by lighter color, shorter antennae, a male-type 10-segmented abdomen (rather than the 8-segmented abdomen of the adult female), and typically female-type forceps. Sexes are not easily distinguished externally in nymphs.

Habits

Most earwigs are thigmotactic and nocturnal, inhabiting crevices of various types, bark, fallen logs, and debris. Cavernicolous (cave-dwelling) blind species have been reported in the Hawaiian islands and in South Africa. Food typically consists of a wide array of living and dead plant and animal matter. Some earwigs have scent glands opening onto the dorsal side of the third and fourth abdominal segments, and from these they can squirt a foul-smelling yellowish-brown fluid some 10 or so cm, presumably for protection.

Reproductive Strategies

Among the Forficulina, reproductive strategies range from oviparous iteroparity (many clutches, the likely ancestral condition) to semelparity (one clutch, a trait derived in colder climates). In contrast, the Hemimeridae are viviparous ectoparasites, residing in the fur of central African rats of the genera *Cricetomys* and *Beamys* and dining on the skin and body secretions of the host, while the viviparous Arixenina are ectoparasites of bats.

Natural Enemies

Some beetles, toads, snakes, birds, and bats have been reported as predators of earwigs. *F. auricularia* can be parasitized by gregarines (sporozoans, which may be harmless) in the gut, as well as nematodes, cestodes, mites, and tachinid flies (a potential biocontrol agent). A fungal parasite has also been associated with several earwig species.

MEDICAL AND ECONOMIC IMPORTANCE

Earwigs are harmless to humans: they carry no known pathogens of humans, and their mouthparts are incapable of biting humans (although some species can pinch).

Some genera (e.g., *Forficula, Labidura, Euborellia*) are repeatedly reported as a pest of homes, gardens, and orchards. Their thigmotactic nature, coupled with (known and suspected) aggregation pheromones, can lead to high densities of earwigs in and around homes. In gardens, earwigs may attack seedlings and soft fruit. Management in backyard gardens can be accomplished by persistent trapping in bamboo tubes, rolled-up newspaper, or low-sided cans filled with vegetable oil. Removing refuge sites, such as ivy and piles of leaves, is also helpful.

Some species are also of importance to commercial agriculture, being pests of ginger, maize, and of honey bee colonies. However, earwigs are also regarded as valuable biocontrol agents for crop pests, consuming armyworms, aphids (of various types), mites, scale insects, sugarcane rootstock borers, and tropical corn borers. Several dermapteran species are found in commercial egg houses and have potential as biocontrol agents for fly eggs and larvae.

See Also the Following Articles
Cave Insects • Juvenile Hormone • Orthoptera

Further Reading
Burr, M. (1911). "Genera Insectorum" (P. Wytsman, ed.), pp. 1–112. L. Desmet-Verteneuil, Bruxelles.
Crumb, S. E., Eide, P. M., and Bonn, A. E. (1941). The European earwig. U.S. Department of Agriculture Technical Bulletin 766.
DeGeer, C. (1773). "Mémoires pour servir à l'histoire des insectes," Vol. 3. Hesselberg, Stockholm. Lucas, W. J. (transl.) (1920). "A Monograph of the British Orthoptera." The Ray Society, London.
Giles, E. T. (1963). The comparative external morphology and affinities of the Dermaptera. *Trans R. Entomol. Soc. Lond.* **115**, 95–164.

Popham, E. J. (1963). The geographical distribution of the Dermaptera. *Entomologist* June, 131–144.

Popham, E. J. (1965). The functional morphology of the reproductive organs of the common earwig (*Forficula auricularia*) and other Dermaptera with reference to the natural classification of the order. *J. Zool.* **146**, 1–43.

Development, Hormonal Control of

Michael E. Adams
University of California, Riverside

The majority of insects undergo embryonic development within an egg and then advance through a series of immature larval stages culminating in metamorphosis to the adult, reproductive form. It has been suggested by Lynn Riddiford and James Truman that the stunning evolutionary success of insects is attributable to complete metamorphosis, whereby the immature larva, essentially a gut covered with cuticle, is exquisitely adapted for resource exploitation, rapid growth, and avoidance of competition with its conspecific adult, reproductive stage. Metamorphosis is a magnificent transformation of one body form into a completely different one under the control of hormones. Upon reaching the requisite body size, precisely timed hormonal signals are released, committing the animal to a postembryonic rebirth. Carroll Williams summarized this basic developmental process with the following anecdote:

> The earth-bound stages built enormous digestive tracts and hauled them around on caterpillar treads. Later in the life-history these assets could be liquidated and reinvested in the construction of an entirely new organism—a flying machine devoted to sex.

Conversion of the energy accumulated by the larval form into the adult during metamorphosis is a fascinating process of organismal remodeling. It is accomplished via programmed cell death of larval-specific cells, reprogramming of others, and postembryonic birth of new cells from imaginal disc tissues upon receipt of precisely timed and coordinated hormonal signals.

INSECT BODY PLANS AND DEVELOPMENTAL PROGRAMS

Three distinct patterns of growth are observed in insects, distinguished by body form and type of metamorphosis: ametabolous, hemimetabolous, and holometabolous. Insect orders exhibiting ametabolous development belong to the Apterygota, or wingless insects. Included in this group are the primitive orders Protura, Collembola, and Diplura, whose earliest stages are miniature adults in form, except for the absence of external genitalia. They grow continuously and lack metamorphosis. Attainment of reproductive competence occurs at an indefinite time in development, and they continue to molt even as adults. Little is known about the hormonal control of development in the Ametabola.

The majority of insects are winged and have either partial or complete metamorphosis. The Hemimetabola, such as grasshoppers and crickets, emerge from the egg formed as small, immature versions of the adult and are called nymphs. They lack wings and functional reproductive organs. After a series of molts, the number usually constant from one generation to another, nymphs pass directly to the winged, reproductive adult stage in a single step. This mode of development is referred to as incomplete metamorphosis. The more advanced insect orders, including moths, beetles, flies, and wasps, develop as vermiform (wormlike) larvae during the immature stages. The complete metamorphosis of these groups is a two-step process in which a sessile, nonfeeding pupal stage is intermediate between larva and adult. The pupal stage allows for a complete change in body form from larva to winged, hexapod adult. This total transformation of body form during complete metamorphosis requires a high degree of postembryonic cellular programming under the control of hormones. Three types of cellular processes are dictated, mostly by hormonal signaling involving ecdysteroids and juvenile hormones. First, many larval-specific structures such as body wall muscles and neurons must be eliminated through a programmed cell death known as apoptosis. Second, some cells persist to the adult stage, but are extensively remodeled to serve adult functions. Finally, new cells derived from imaginal discs are born.

The diversity of insect groups places limits on generalizations about hormonal control of development. The effects of hormones in one group may not be the same for other groups, because of different patterns of growth and cellular specification. In the Lepidoptera, for example, larval epidermal cells change their cuticle secretory program at metamorphosis and switch to production of pupal and subsequently adult cuticle. Exogenous juvenile hormone (JH) application at this time maintains the larval secretory program, resulting in supernumerary (extra) instars. Similarly, properly timed application of JH in the early pupal stage of moths causes a second pupal stage. However, the fruit fly *Drosophila* and other higher flies are largely resistant to such effects of JH. This may relate to differences in the developmental program of fly epidermal cells. The entire epidermis in the head and thorax is programmed for secretion of larval cuticle only and dies at metamorphosis. It is replaced by imaginal disc tissue, which remains undifferentiated throughout larval life in the presence of JH. Abdominal epidermal cells also die after pupation and are replaced by abdominal histoblasts. Thus the development program of epidermis in higher flies is entirely distinct from the Lepidoptera, and its lack of response to JH in the early stages may be a consequence of this very different design.

These differences in responses to hormones complicate interpretation of many findings, especially because it has become fashionable to test overall hypotheses using the moth, *Manduca,* for physiology and endocrinology experiments, and *Drosophila* for genetic manipulations. Generalizing about common mechanisms for these two evolutionarily distant groups should be done cautiously.

INSECT DEVELOPMENTAL HORMONES

Ecdysteroids

Ecdysteroids are relatively polar steroid hormones released by the prothoracic glands in immature stages and the gonads in adults. They are the chief regulators of gene expression in insect development and reproduction. The first structure to be elucidated was α-ecdysone (αE), the immediate precursor to 20-hydroxyecdysone (20HE), accepted as the main protagonist in ecdysteroid actions. While 20HE indeed is associated with the majority of ecdysteroid actions, there is evidence to support αE as a signaling molecule in certain instances. Other suspected ecdysteroids are 20,26HE and makisterones.

Juvenile Hormones

JHs, sesquiterpenoid derivatives from the sterol synthesis pathway, are released by the corpora allata. Six types are known in insects, with JH III being the predominant form. The chief action of JHs is to modulate ecdysteroid-mediated gene expression. No receptors for JH have been clearly identified, but it seems likely that the hormone interacts with intracellular receptors or proteins that modulate ecdysteroid signaling. The signature of JH action is to promote expression of the immature phenotype.

Prothoracicotropic Hormone

Prothoracicotropic hormone (PTTH) is a large peptide hormone released by brain neurosecretory cells from terminals in the corpora cardiaca or corpora allata. PTTH stimulates the prothoracic gland to synthesize and release αE or, in some instances, 3-dehydroecdysone, which are both then converted to 20HE in the hemolymph or by target tissues. PTTH is a homodimer consisting of two 12-kDa subunits joined by a disulfide bond. Release of PTTH is regulated by sensory inputs to the brain, which convey information about body size and nutritional state.

Bursicon

Bursicon is a 30- to 40-kDa peptide that accelerates sclerotization of cuticle. It is released from neurosecretory cells in the brain and ventral nerve cord after each ecdysis. Since insects are particularly vulnerable to predation during and after ecdysis, rapid hardening of the cuticle maximizes survival.

Eclosion Hormone

Eclosion hormone (EH) is a 62-amino-acid peptide hormone that mediates circadian-mediated eclosion to the adult stage and each larval ecdysis. Eclosion hormone is released into the blood from ventral median neurosecretory cells of the brain, causing release of ecdysis-triggering hormone (ETH) from Inka cells of the epitracheal endocrine system. It is also implicated in the elevation of cyclic GMP in a subset of neurons in the central nervous system that initiate ecdysis behavior.

Ecdysis-Triggering Hormones

ETHs are peptide hormones synthesized and released from Inka cells at the end of the molt. ETHs act directly on the central nervous system (CNS) to cause a behavioral sequence that leads to shedding of the cuticle. ETH release is caused by eclosion hormone, and the action of ETH on the CNS leads to release of eclosion hormone from VM neurons. It has been proposed that ETH and EH engage in a positive-feedback signaling pathway that results in depletion of ETH from Inka cells, which is necessary for the transition from preecdysis to ecdysis behaviors.

MODES OF HORMONE ACTION

Hormones are chemical messengers that travel throughout the body to effect responses in specific tissues. Targeted cells have receptors that, upon binding the hormone, transduce the signal into a cellular response. Insect hormones regulate development by activation of either intracellular receptors or receptors at the cell membrane.

Ecdysteroids and Juvenile Hormones Activate Intracellular Receptors

Ecdysteroids and juvenile hormones are relatively lipophilic signaling molecules able to easily traverse the cell membrane. Upon entry, they bind to intracellular proteins called nuclear receptors or nuclear transcription factors, which reside either in the cytoplasm or in the nucleus. Regardless of their initial location, hormone binding triggers passage to the nucleus, where the receptor forms a complex with other proteins and then binds directly to DNA, inducing or repressing gene expression.

Several factors govern diverse, stage-specific responses of target cells to ecdysteroids. First, ecdysteroid receptors occur as multiple subtypes, including EcR-A, EcR-B1, and EcR-B2. The response to ecdysteroids is governed by the subtype expressed by target cells, as well as which subtype of USP (USP-1 or USP-2), the EcR partner, is expressed. The transient availability of receptors in target cells leads to sensitive periods at specific stages of development. For example, both EcR-A and EcR-B1 are present throughout larval life, but the ratio of the two favors EcR-B1. In contrast, levels of EcR-A increase during metamorphosis. Second, the affinity of EcR subtypes can vary, as can the kinetics of a given response to an ecdysteroid peak. These patterns of receptor and dimer partner expression appear to mediate different cellular responses to ecdysteroids at different developmental times.

Peptide Hormones Activate Cell-Surface Receptors

Most peptide hormones bind to a class of integral membrane proteins on the plasma membrane of the target cell. The majority of these are G-protein-coupled receptors that trigger intracellular second messenger cascades.

EMBRYOGENESIS

Distinct patterns of embryonic development are observed in hemimetabolous and holometabolous insects. In both types, the embryo produces multiple cuticular layers, and the appearance of these coincides with pulses of ecdysteroids. Juvenile hormone levels are generally low during early embryogenesis, but climb later to program nymphal or larval cuticle formation upon appearance of an ecdysteroid peak.

In the hemimetabolous grasshopper *Locusta*, four peaks of ecdysteroids are observed, corresponding to production of serosal cuticle and three embryonic cuticle layers (Fig. 1A). JH levels are elevated immediately after oviposition, because of maternal contribution to the yolk, but rapidly decrease to

low levels. At ~20% of embryonic development and before prothoracic glands are developed, the first ecdysteroid peak consisting exclusively of αE occurs in the presence of relatively low JH levels. This first peak comes from the release of maternal ecdysteroids stored as polar conjugates, and shortly thereafter the serosal cuticle is secreted. A second αE peak occurs at ~30% development, leading to formation of the first embryonic cuticle. A third ecdysteroid peak occurs just after differentiation of the prothoracic glands, and this coincides with the first appearance of 20HE and 20,26HE. Nevertheless, levels of αE together with 20,26HE predominate at this time, leading to the secretion of the second embryonic cuticle, which Truman and Riddiford refer to as pronymphal cuticle. It is the first layer of cuticle tough enough to require shedding via ecdysis behavior. The coincidence of the early, αE peaks with cuticle secretion suggests that αE is not only a 20HE precursor, but also a biologically active hormone at certain times of development. At 70% development, the first embryonic cuticle is shed and followed quickly by a large peak of ecdysteroids. This is the first exposure of the embryo to substantial 20HE levels in the presence of JH, leading to synthesis of the first instar nymphal cuticle. This peak of ecdysteroids contains large amounts of αE, 20HE, and 20,26HE.

At hatching, the grasshopper emerges from the egg under the ground, still surrounded by the pronymphal cuticle. Despite its hexapod body plan, the animal exhibits a classic vermiform (wormlike) locomotory pattern as it escapes the egg pod and maneuvers through the substrate to the surface. In a matter of seconds to minutes, the pronymphal cuticle is shed, and the animal stretches its legs and switches abruptly to hexapod behavior. Truman and Riddiford have called attention to many similarities between the hemimetabolous pronymph and the holometabolous larva, suggesting that the latter has resulted from a hormonal shift in embryogenesis, resulting in an extended postembryonic phase of pronymph development. The ancestral pronymph undergoes an extended, multistage developmental sequence as a larva in the Holometabola.

The importance of a JH-free period during early embryogenesis of hemimetabolous insects (grasshopper—*Schistocerca*, cricket—*Acheta*) has been demonstrated by treatment of eggs with JH analogs. This results in inhibition of blastokinesis, reduction in the number of embryonic cuticle layers produced, premature appearance of nymphal cuticle and mouthparts, and reduced body size.

Embryogenesis in holometabolous Lepidoptera is somewhat simpler, with the secretion of only three cuticles, one serosal and two embryonic cuticles. Levels of both ecdysteroids and JH are undetectable early in embryogenesis, but rise earlier compared with the Hemimetabola, or at about 30% development (Fig. 1B), preceding the ecdysteroid peak. Therefore, unlike hemimetabolous embryogenesis, the first exposure to ecdysteroids occurs in the presence of JH, leading to production of the first larval cuticle. An

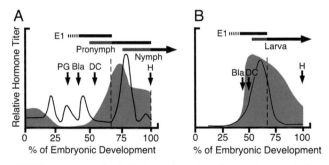

FIGURE 1 Hormone levels during embryonic molts in a hemimetabolous insect, the grasshopper *Locusta migratoria* (A), and in a holometabolous insect, the moth, *Manduca sexta* (B). During *Locusta* embryogenesis, four peaks of ecdysteroid are observed, each corresponding to secretion of a layer of cuticle. The first peak of predominantly αE at 20% development initiates secretion of the serosal cuticle (not shown). A second peak prior to prothoracic gland differentiation (PG) causes secretion of the first embryonic cuticle (E1). Just after blastokinesis (Bla), a third ecdysteroid peak leads to secretion of the pronymph cuticle. Shading of the pronymph time line at the top of the plot indicates the pharate stage, which ends with ecdysis of the E1 cuticle (vertical dotted line). The fourth ecdysteroid peak, occurring for the first time in the presence of JH, contains approximately equal amounts of αE, 20HE, and 20,26HE. This causes secretion of the first-stage nymphal cuticle. At hatching the nymph sheds the pronymph cuticle upon escaping from the substrate. Secretion of the first embryonic cuticle in *Manduca* occurs in the absence of an ecdysteroid peak. The first larval cuticle is secreted in response to elevated ecdysteroids in the presence of JH just after dorsal closure (DC). Note that JH levels in *Manduca* rise earlier in development than in *Locusta* and that ecdysteroid signaling always occurs in the presence of JH. (Adapted from Truman and Riddiford, 1999, 2002.)

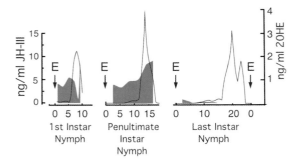

FIGURE 2 Hormonal regulation of development in postembryonic stages of the cockroach *Nauphoeta cinerea*, representing hemimetabolous development. During the immature stages of *Nauphoeta*, JH levels are elevated each time ecdysteroid levels rise to initiate a molt, resulting in secretion of nymphal cuticle. Following ecdysis to the last instar nymph, JH levels drop precipitously. Adult commitment is signaled by a biphasic ecdysteroid peak in the absence of JH. Incomplete metamorphosis occurs during the period between ecdysteroid elevation and ecdysis to the adult stage. (Adapted from Lanzrein *et al.*, 1985.)

embryonic ecdysis occurs at 70% development, and first instar larval hatching does not involve cuticle shedding. The importance of a JH-free period observed for hemimetabolous insects is not the case for embryogenesis in the Holometabola. Embryos are largely insensitive to exogenous JH treatment, suggesting perhaps the absence of receptors for these hormones until later in embryogenesis.

LARVAL DEVELOPMENT

The Intermolt

Because the exoskeleton places limits on growth, insect development occurs in stages, each ending with molting and cuticle shedding, or ecdysis. During the intermolt, which follows ecdysis, JH levels are maintained around 1 to 10 ng/ml in the blood (Fig. 2). It is presumed that these JH levels promote a high metabolic rate, active feeding behavior, synthesis of larval cuticle proteins, and continuous proliferation (but not differentiation) of imaginal discs. Growth during the immature stages is possible because the immature integument is predominantly unsclerotized procuticle, which is quite flexible compared with hard, sclerotized adult cuticle. Several mechanisms allow for larval cuticle expansion. The epidermal cells add new protein to the cuticle throughout the intermolt, increasing the surface area by intussusception. In addition, new cuticle in *Manduca* is deposited in vertical columns that are gradually reoriented during the feeding stage to allow for expansion. The increase in size during the larval stage can be quite impressive, as in fifth-instar *Manduca*, which increases its body weight from ~1 g on the first day of development to 15 g by the end of the instar, and its cuticular surface area by approximately fivefold just prior to pupation. Blood-feeding insects such as *Rhodnius* are known to release serotonin after a blood meal,

which acts as a plasticizing agent, facilitating the enormous expansion of the body wall after a blood meal.

The Molt

At some point during each immature stage, growth results in a decision by the brain to initiate the molt. In *Rhodnius,* the simplest case known, stretch-receptor input from the abdomen to the brain causes release of PTTH, which induces synthesis and secretion of ecdysteroids from the prothoracic glands. In most insects, the decision to release PTTH is more complicated and less well understood, but it has to do with body weight, nutritional state, and time spent at that stage. The immediate effects of ecdysteroid elevation include cessation of feeding and apolysis, the detachment of the old cuticle from underlying epidermal cells. Apolysis of larvae results in head capsule slip, which occurs because the new head capsule is larger than the old one. This is the most visible sign that the molt has been initiated. If elevation of ecdysteroids occurs in the presence of JH, epidermal cells maintain the secretory program for immature phenotype, and larval cuticle is secreted (Fig. 2). Through the action of molting fluid, most components of the old cuticle are broken down and recycled into the new layer.

During the period of new cuticle synthesis, ecdysteroids also orchestrate gene expression crucial to the synthesis and action of peptide hormones that control ecdysis behaviors. Ecdysis is a complex process in which the old cuticle is shed not only from the surface of the animal, but also from the lining of the foregut, the hindgut, and the inner walls of the tracheal system. Success in this process depends on completion of new cuticular synthesis, attachment of the musculature to this new cuticle, and digestion of the old cuticle. In addition, the animal prepares for a sequence of Houdini-like escape behaviors necessary to shed the old cuticle. These consist of preecdysis and ecdysis behaviors. The ability to perform these behaviors depends on orchestration of a peptide signaling cascade involving the central nervous system and the epitracheal endocrine system.

For the ecdysis signaling cascade to be functional at the appropriate time, ecdysteroids orchestrate gene expression in four ways. Genes are activated in epitracheal glands to increase production of ecdysis triggering hormones (ETHs). Release of ETHs initiates ecdysis behaviors through direct action on the CNS. Although the CNS is not sensitive to ETHs during the feeding stage, acquisition of sensitivity occurs upon elevation of ecdysteroids, specifically around the time of apolysis. Third, the nervous system becomes competent to release EH, a peptide hormone that targets Inka cells to cause release of ETH. Finally, elevated ecdysteroids exert a negative influence on the secretory competence of Inka cells. As long as ecdysteroids remain high, Inka cells are unable to secrete ETHs in response to EH exposure. This latter effect of ecdysteroids, to block release of ETHs from Inka cells,

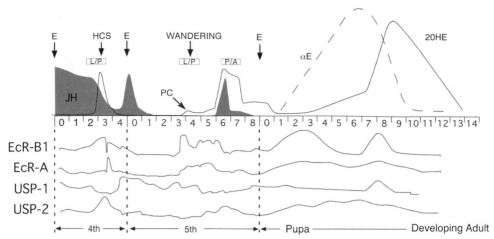

FIGURE 3 Hormonal regulation of development in postembryonic stages of the moth, *M. sexta,* exemplifying the Holometabola. Elevated JH levels occur throughout larval development in *Manduca,* in which metamorphosis is encoded by two ecdysteroid peaks. During the fifth instar, JH levels drop and an ecdysteroid pupal commitment peak (PC) signals a change in the way epidermal cells respond to the secretory program from larval to pupal phenotype. This peak also triggers cessation of feeding and wandering behavior. The next time epidermal cells are exposed to ecdysteroid and JH on day 6 leads to secretion of pupal cuticle. During the pupal stage, αE and 20HE rise in the complete absence of JH, signaling adult commitment. Sensitive periods of epidermal cell commitment are shown as L/P (larval vs pupal specification) and P/A (pupal vs adult specification of imaginal disc tissue). Fluctuating levels of ecdysteroid receptor subtypes (EcR- B1, EcR-A) and USP subtypes (USP-1, USP-2) are shown below. (Adapted from Riddiford and Truman, 2001, and Baker *et al.,* 1987.)

appears designed to ensure that ecdysis does not occur prematurely. The mechanism of block involves a crucial step in the EH-induced secretory mechanism of Inka cells. Declining ecdysteroid levels at the end of the molt provide the necessary signal permitting expression of one or more genes needed for secretory competence.

It is believed that initiation of ecdysis behaviors occurs as a result of an ongoing conversation between EH neurons and Inka cells. When Inka cells become sensitive to EH, ETH release initiates preecdysis behavior, which is thought to loosen the remaining connections between the new and the old cuticle. The transition from preecdysis to ecdysis occurs upon depletion of ETH from Inka cells. It is thought that the action of ETH activates a downstream cascade of peptide signaling within the CNS to regulate each unit of the behavioral sequence. Included in this cascade is a neuropeptide called crustacean cardioactive peptide (CCAP), named for its initial discovery and biological activity. In the context of insect ecdysis, CCAP appears to be an immediate chemical signal within the CNS for activation of peristaltic ecdysis behavior. Upon escaping the old cuticle, the animal is surrounded by a new soft cuticle and is therefore extremely vulnerable to injury. Release of the neuropeptide bursicon from neurosecretory cells of the CNS accelerates sclerotization of the cuticle.

In summary, ETH, EH, CCAP, and bursicon regulate ecdysis at all stages. This includes embryonic ecdysis in *Manduca,* and adult eclosion.

METAMORPHOSIS

The transition from immature to adult is signaled by the elevation of ecdysteroid levels in the absence of JH. This is a one-step process in hemimetabolous insects. During the last

nymphal instar of the cockroach *Nauphoeta,* JH levels fall from 5 to 10 ng/ml to less than 1 ng/ml prior to the next ecdysteroid peak (Fig. 2). Appearance of ecdysteroids at this low JH level signals a commitment to an adult gene expression pattern. Some examples of cellular responses to this adult commitment peak include mitosis in wingpad tissue, development of flight muscles, competence of gonadal accessory glands to differentiate, formation of external genitalia, and reorganization of the nervous system to accommodate these new adult structures. Also included are more subtle alterations, such as the relative proportions of body parts and addition of secondary sexual characteristics such as acoustic organs for communication.

In the Holometabola, complete metamorphosis requires an intervening pupal stage for remodeling of the larva into an adult. During the last larval instar of *Manduca,* ecdysteroids rise on two occasions, first in the absence of JH and later in its presence (Fig. 3). The first ecdysteroid pulse is a small one during days 3 to 4, referred to as the pupal commitment peak. This is the first time in the life history of the animal that a peak of 20HE occurs in the absence of JH. This triggers cessation of feeding and a wandering behavior aimed at locating a suitable site for pupation. The pupal commitment peak prepares the genome for its response to the next ecdysteroid peak. Although the second pulse of ecdysteroids occurs in the presence of JH, the commitment peak has changed the response of epidermal cells from a larval to a pupal secretory program. Similarly, imaginal discs respond to this peak by differentiating into adult tissues, something not observed in the previous larval stages.

In the Lepidoptera, the new hormonal milieu that triggers metamorphosis produces striking changes in the CNS and musculature. These tissues must be drastically altered during construction of the adult body form. Simultaneously, undifferentiated cells in imaginal discs proliferate and differentiate. For these tissues, the pupal commitment peak sets the stage for three types of cellular responses to the subsequent ecdysteroid peak: programmed cell death (apoptosis), cellular remodeling, or differentiation of imaginal discs. For example, some motoneurons that innervate larval-specific structures such as prolegs die shortly after pupal ecdysis. Others persist because of their involvement in the motor patterns involved in adult eclosion and die soon thereafter. Most larval neurons survive, but are remodeled to play roles in adult behavior.

The precise mechanisms governing cellular responses to the pupal commitment peak remain obscure, but the identification of EcR and USP subtypes has allowed monitoring of their expression during metamorphosis, for example, the response of epidermal cells of *Manduca* to the ecdysteroid peak during days 2 to 3 of the fourth instar by up-regulation EcR-B1, no change in EcR-A, and down-regulation of USP-1 (Fig. 3). However, during the fifth instar, the pupal commitment peak is correlated with sharply increased EcR-B1 *and* EcR-A expression, an altered USP-1 response. These altered patterns of expression apparently encode a change in downstream gene expression, leading to a pupal phenotype as well as imaginal disc differentiation during this stage.

During the pupal stage, ecdysteroid levels rise in the complete absence of JH (Fig. 3). This signals commitment to the adult phenotype and accelerated development of imaginal discs. It is remarkable that αE begins to rise on day 1 of the pupal stage, well before elevation of 20HE, and this is correlated with increases in both EcR-B1 and EcR-A expression. This suggests that αE may have a hormonal role itself in programming the adult stage. Elevation of 20HE occurs in two phases, one beginning on day 3 and a second, steeper rise on day 7. The slow gradual rise coincides with the adult commitment phase, whereas the steeper rise beginning on day 7 is associated with differentiation of new tissues. This latter phase coincides with a rapid rise of EcR-B1 and USP-1 expression.

The hormonal signaling mechanisms governing metamorphosis are complex and include a diversity of hormones, receptors, and varying temporal patterns of hormone release and receptor expression. Ecdysteroid signaling in the presence or absence of JH can set the stage for changes in the programming of target tissues, such as epidermal cells that secrete cuticle. Depending on the responses of EcR and USP subtypes, qualitatively different cellular programs are initiated.

See Also the Following Articles
Ecdysteroids • Embryogenesis • Imaginal Discs • Juvenile Hormone • Metamorphosis • Mating Behaviors • Temperature, Effects on Development and Growth

Further Reading
Baker, F. C., Tsai, L. W., Reuter, C. C., and Schooley, D. A. (1987). *In-vivo* fluctuation of JH, JH acid and ecdysteroid titer and JH esterase activity during development of fifth stadium *Manduca sexta. Insect Biochem.* **17**, 989–996.

Ewer, J., Gammie, S. C., and Truman, J. W. (1997). Control of insect ecdysis by a positive-feedback endocrine system: Roles of eclosion hormone and ecdysis triggering hormone. *J. Exp. Biol.* **200**, 869–881.

Gammie, S. C., and Truman, J. W. (1997). Neuropeptide hierarchies and the activation of sequential motor behaviors in the hawkmoth, *Manduca sexta. J. Neurosci.* **17**, 4389–4397.

Gilbert, L. I., Tata, J. R., and Atkinson, B. G. (1996). "Metamorphosis: Postembryonic Reprogramming of Gene Expression in Amphibian and Insect Cells." Academic Press, San Diego.

Lanzrein, B., Gentinetta, V., Abegglen, H., Baker, F. C., Miller, C. A., and Schooley, D. A. (1985). Titers of ecdysone, 20-hydroxyecdysone and juvenile hormone III throughout the life cycle of a hemimetabolous insect, the ovoviviparous cockroach *Nauphoeta cinerea. Experientia (Basel)* **41**, 913–917.

Lageaux, M., Hetru, C., Goltzene, F., Kappler, C., and Hoffmann, J. A. (1979). Ecdysone titer and metabolism in relation to cuticulogenesis in embryos of *Locusta migratoria. J. Insect Physiol.* **25**, 709–723.

Nijhout, H. F. (1994). "Insect Hormones." Princeton University Press, Princeton, NJ.

Riddiford, L. M., Cherbas, P., and Truman, J. W. (2001). Ecdysone receptors and their biological actions. *Vitam. Horm.* **60**, 1–73.

Temin, G., Zander, M., and Roussel, J.-P. (1986). Physico-chemical (GC–MS) measurements of juvenile hormone III titres during embryogenesis of *Locusta migratoriasta. Int. J. Invertebr. Rep. Dev.* **9**, 105–112.

Truman, J. W., and Riddiford, L. M. (1999). The origins of insect metamorphosis. *Nature* **401**, 447–452.

Truman, J. W., and Riddiford, L. M. (2002). Endocrine insights into the evolution of metamorphosis in insects. *Annu. Rev. Entomol.* **47**, 467–500.

Zitnan, D., Kingan, T. G., Hermesman, J., and Adams, M. E. (1996). Identification of ecdysis-triggering hormone from an epitracheal endocrine system. *Science* **271**, 88–91.

Zitnan, D., Ross, L. S., Zitnanova, I., Hermesman, J. L., Gill, S. S., and Adams, M. E. (1999). Steroid induction of a peptide hormone gene leads to orchestration of a defined behavioral sequence. *Neuron* **23**, 523–535.

Diapause

David L. Denlinger
Ohio State University

Diapause is a form of developmental arrest in insects that is much like hibernation in higher animals. It enables insects and related arthropods to circumvent adverse seasons. Winter is most commonly avoided in temperate zones, but diapause is also used to avoid hot, dry summers and periods of food shortage in the tropics. Unlike quiescence, which represents a halt in development elicited immediately at any stage by an adverse condition, diapause is a developmental response that is expressed only during a specific developmental stage, which depends on the species of insect. For example, the commercial silkworm *(Bombyx mori)* always diapauses as an early embryo, the European corn borer *(Ostrinia nubilalis)* as a fifth instar, the cecropia moth *(Hyalophora cecropia)* as a pupa, and the Colorado potato beetle *(Leptinotarsa decemlineata)* as an adult. A few species are capable of entering diapause several times, but this usually occurs only in species living at high latitudes for which several years may be required for the completion of development. If the diapause occurs in response to environmental cues it is referred to as "facultative diapause," but if it occurs during each generation regardless of the environmental cues it receives, it is considered an "obligatory diapause." Facultative diapause is by far the more common, but several important species such as the gypsy moth *(Lymantria dispar)* have an obligatory diapause.

Embryonic diapauses are common in many of the Lepidoptera, in many Hemiptera, and in some Diptera such as mosquitoes. The arrest can occur at any stage of embryonic development, from shortly after fertilization (e.g., commercial silkworm) until after the first instar has already been fully formed (e.g., gypsy moth). Larval diapauses, especially common in the Lepidoptera, are most frequent in the final instar but they sometimes occur in earlier instars as well, e.g., the second instar of the spruce budworm *(Choristoneura fumiferana)*. Pupal diapause is well known for the Lepidoptera and Diptera. Usually the arrest of pupal development occurs in the true pupal stage, but there are a few examples of diapause occurring in pharate adults (after completion of adult differentiation but before adult eclosion). Adult diapause is common in the Coleoptera, Hemiptera and Homoptera, Hymenoptera, Orthoptera, and Neuroptera, as well as some Diptera and Lepidoptera. Adult diapause, sometime referred to as a reproductive diapause, represents a halt in reproduction. Ovaries of females remain small, and the oocytes within the ovarioles contain little or no yolk. In males of some species, the testes remain small during diapause, but in others the testes are well developed and contain sperm. Male accessory glands, the organs that produce spermatophores and factors responsible for sperm activation, usually remain small and inactive during diapause. Mating behavior is strongly suppressed during diapause. In wasps, mating takes place in the autumn; males die soon thereafter and only the females overwinter in diapause. In many other insects, both sexes overwinter and mating takes place in the spring, after diapause has been terminated. Some species, such as lacewings and weevils, mate both before and after diapause.

In preparation for diapause, the insect usually sequesters additional energy reserves and moves to a site that is somewhat protected from the full onslaught of the inclement environmental conditions. Such sites may be underground, beneath debris on the soil, within galls and other plant tissues, or inside cocoons or other structures constructed by the insect. A migratory flight may be a preparatory step for diapause. This may include a short flight to a fence row or a local wooded area, but in the extreme it may be a long-distance flight, as made by the monarch butterfly *(Danaus plexippus)* when it leaves its summer habitat in Canada and the northern regions of the United States and flies to the highlands of Mexico or California to spend the winter in an adult diapause.

Upon entering diapause, development (or reproduction if it is an adult diapause) is halted and metabolic activity is suppressed. Usually, feeding ceases during diapause; thus, the insect is forced to survive on the energy reserves it has garnered prior to the onset of diapause. It is not unusual for an insect destined for diapause to sequester twice as much lipid reserves as its counterpart that is not programmed to enter diapause. The economic utilization of these reserves is enhanced by the suppression of metabolism, and for poikilotherms such as insects, the low temperatures prevailing during winter further serve to conserve energy reserves. Another challenge faced by diapausing insects is the lack of access to free water. Although some insects may drink during diapause, certain stages such as embryos and pupae do not have this option. This lack of water poses special constraints for an organism as small as an insect. Their large surface-to-volume ratios make insects particularly vulnerable to water loss across the surface of their integument. Two features appear to be common adaptations for maintaining water balance during the long months of diapause. The cuticles of many diapausing insects are coated with extra thick layers of wax that are effective in retarding water loss. In addition, a number of diapausing insects are capable of absorbing atmospheric water vapor directly through their cuticle using a mechanism that is not yet clearly understood.

Color changes are sometimes noted for diapausing individuals. For example, diapausing larvae of the southwestern corn borer, *Diatraea grandiosella,* are white, whereas their nondiapausing counterparts are brown. Reproductively active adults of a lacewing, *Chrysopa carnea,* are green but turn brown when they enter diapause in the autumn. In the spring, when the lacewings become reproductively active, they again turn green. Such changes presumably serve to camouflage the insect and help it blend with the dominant colors of the seasonal environment.

Flight muscles in many beetles and bugs degenerate when the adults enter diapause. Flight muscles are particularly expensive to maintain, thus their degeneration presumably saves energy that would otherwise be expended for maintenance of this tissue.

Several species that diapause as adults, especially beetles, bugs, and butterflies, are found in aggregations. For species that are distasteful, aggregations are likely to provide protection from predators. Such aggregations, however, may also provide another important function by providing a more stable microenvironment. In diapausing aggregations of a tropical fungus beetle, *Stenotarsus rotundus,* the beetle's metabolic rate is inversely related to group size and relative humidity. By forming an aggregation the beetles create a stable, high humidity in their environment, a feature that serves to reduce metabolic rate.

Being in diapause does not, by itself, ensure winter survival. The small size of insects implies that they quickly assume a body temperature close to that of the environment, and their body water is thus vulnerable to freezing. Diapausing insects that live in temperate and polar regions have a host of behavioral, physiological, and biochemical adaptations that enable them to survive at low temperature. A few insects such as the goldenrod gall fly, *Eurosta solidaginis,* are freeze tolerant, which implies that they can actually survive body freezing. But, the majority of insects cannot tolerate body freezing. Such freeze-intolerant or freeze-avoiding insects prevent body freezing by several mechanisms. For example, selection of a thermally buffered microhabitat is a first line of defense. Ice nucleators such as food particles or microbes are usually eliminated from the digestive tract to reduce sites for ice formation. Glycerol, sorbitol, or other polyols serve as classic antifreezes that are synthesized and released into the body to suppress the supercooling point. Several proteins, including thermal hysteresis proteins, ice nucleator proteins, and heat-shock proteins, also contribute to cold hardiness. In some insects, such as flesh flies *(Sarcophaga),* cold hardiness is directly linked to diapause, indicating that the same genetic program that dictates diapause also results in cold hardiness. In other insects, for example the European corn borer, the two programs are regulated independently: the European corn borer enters diapause without initially being cold hardy, but it becomes cold hardy later in the season in response to prevailing low temperatures.

Diapause thus represents a syndrome of developmental, physiological, biochemical, and behavioral attributes that together serve to enhance survival during seasons of environmental adversity.

ENVIRONMENTAL REGULATION

Obligatory diapause is not elicited by environmental cues. It simply occurs in each generation when the insect reaches a certain developmental stage. In the example of the gypsy moth, diapause occurs when the embryo has completed its development and the first instar is nearly ready to hatch. With the exception of a few aberrant individuals, the gypsy moth always halts development at this time, regardless of the environmental cues they receive. In this example, environmental conditions, mainly temperature, determine when diapause should be terminated but play no role in programming the insect to enter diapause.

This is in contrast to the majority of insects, those with a facultative diapause, which use environmental cues to decide whether to enter diapause. If a certain environmental cue is received during a sensitive period the insect will enter diapause, but if this cue is not received or not received at the correct time, development will proceed without interruption. This design feature enables an insect to track seasonal changes and regulate its development accordingly. Many insects can produce multiple generations each year, and insects with a facultative diapause frequently produce spring and summer generations without diapause and then produce a generation in late summer or autumn that enters an overwintering diapause. The environmental cue used most widely to signal diapause induction is photoperiod, but temperature, food quality, and other factors may contribute to the decision.

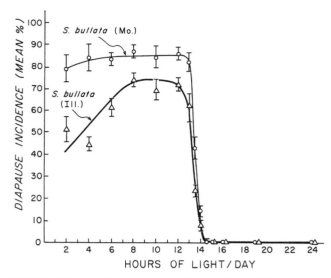

FIGURE 1 Photoperiodic response curves for pupal diapause induction in two populations of the flesh fly *S. bullata* from Illinois and Missouri. Fly cultures were maintained at 25°C under the range of daylengths indicated, and the incidence of pupal diapause was recorded. The critical daylength in this case is 13.5 h of light/day. (Reproduced, with permission, from Denlinger, 1972.)

Photoperiod

Seasonal change in daylength has all the design features that are desirable in a reliable indicator that can be used for predicting upcoming periods of inclemency. It is mathematically accurate and can be used to effectively foretell the advent of winter or other seasons that are to be avoided. The developmental period that is sensitive to photoperiod usually occurs far in advance of the actual diapause stage. Thus, diapause is not usually an immediate reaction to photoperiod but occurs in response to signals received at an earlier stage. Such early programming offers the insect a period to prepare for diapause by sequestering food reserves and making other preparatory adjustments prior to the actual onset of the developmental arrest.

For many of the insects that overwinter in diapause in the temperate regions, short daylengths dictate the expression of diapause. In the example shown in Fig. 1, flesh flies reared at long daylengths, those longer than 13.5 h, develop without interruption, but at daylengths shorter that 13.5 h, the majority enter diapause as pupae. The daylength marking the transition, 13.5 h in this example, is referred to as the critical daylength. The shape of the photoperiod response curve shown in Fig. 1 is common for temperate species that overwinter in diapause, but the curves may have different forms. For some species, especially those that undergo a summer estivation and reproduce in the autumn, long rather than short daylengths may be used to program diapause. Other species may respond to only a narrow range of daylengths for diapause induction, whereas daylengths both shorter and longer avert diapause.

Near the equator, seasonal changes in daylength are progressively less pronounced; yet insects living as close as 5°

north or south of the equator are still capable of using photoperiodic cues to regulate diapause. Diapause still exists in insects living in equatorial regions, but cues derived from temperature, rainfall, and food quality take precedence over photoperiod.

The photoperiodic response controlling diapause varies among geographic populations. Populations living at lower latitudes characteristically respond to shorter critical daylengths. An increase in latitude of 5° results in an increase in critical daylength of approximately 30 min. This pattern of variation is closely related to the latitudinal temperature gradient and is well documented in species of *Drosophila* that inhabit the Japanese archipelago. The species that occur in the subtropical zone exhibit only a weak diapause or no diapause at all. As one moves northward in the archipeligo, the diapause response becomes more pronounced and the flies use longer critical daylengths for diapause induction.

The period sensitive to photoperiod usually does not encompass the entire prediapause period, but instead a shorter interval, usually well in advance of the actual diapause stage. For example, the pupal diapause in the flesh fly *S. crassipalpis* is programmed during a photosensitive stage that includes the final 2 days of embryonic development and the first 2 days of larval life. In the tobacco hornworm, *Manduca sexta,* a species that also has a pupal diapause the photosensitive stage is much longer; it begins during embryonic development and continues through the feeding phase of the fifth instar. In the silkworm, *B. mori,* embryonic diapause is programmed during the mother's period of embryonic development. This timing of the photosensitive stage thus facilitates the channeling of development toward diapause at an early stage and allows sufficient time for the preparative phase of diapause.

The duration of diapause, often called diapause depth, also may depend on photoperiod. For example, in the lacewing, *C. carnea,* diapause depth is controlled by photoperiod in such a way that the adult diapause is deeper when it is induced earlier in the autumn, thus preventing an untimely termination of diapause before the onset of winter. And, in the tobacco hornworm, *M. sexta,* the duration of pupal diapause is a function of the number of short days the embryo and larva have received. Exposure to a few short days, such as would occur in mid- to late summer, results in a long diapause, while exposure exclusively to short days, an event that could occur only in early autumn, results in a short diapause. Such qualitative responses to photoperiod allow the insect to fine tune its development to fit the changing season.

Photoperiodic information is perceived through a receptor in the brain, integrated and stored in the brain, and then translated into the endocrine events that control the induction and maintenance of diapause. The location of the photoreceptor responsible for measurement of daylength has been studied in relatively few insects, but in most of them the compound eyes and ocelli are not the conduit for this information. Surgical destruction of these visual centers or coating the eyes with an opaque paint usually does not inter-

fere with the photoreception involved in the programming of diapause. The photoperiodic signal appears to impinge directly on the brain, but the exact location of these extraretinal photoreceptors has not been elucidated. As in many other plants and animals, the photoperiodic response in insects is primarily a blue-light response. Cryptochromes, proteins involved in photoperiodic responses in a diverse array of organisms, are present in insects and are likely to be implicated in this response. Several important clock genes have been identified in insects, but thus far their involvement in photoperiodism has not been well established.

The role for photoperiod in the environmental regulation of diapause is mainly in the inductive phase of diapause. There are a few species that use daylength as a direct environmental cue for diapause termination. More commonly, photoperiod may influence the rate of diapause development, which in turn does impact the duration of diapause, but frequently diapause development proceeds at a rate determined by temperature rather than photoperiod.

Temperature

Temperature provides another important seasonal cue for diapause induction, but the daily fluctuations in temperature mean that it is less reliable than photoperiod in this regard. Frequently, a short-day response is enhanced by low temperature. For example, the maximum diapause response observed for flesh flies shown in Fig. 1 is approximately 80%. But, this was for flies reared at 25°C, and if the temperature is lowered to 18°C, the diapause incidence is elevated to nearly 100%. In these flies the critical photoperiod is not influenced by temperature, but in some insects the critical photoperiod may shift as well.

Near the equator, where seasonal changes in daylength are too subtle to be used as environmental cues, temperature may replace photoperiod as the primary environmental regulator of diapause, as it does for flesh flies living in East Africa: daylength has no influence on the expression of diapause, but instead low daytime temperatures experienced in July and August are used to program the flies for pupal diapause.

A period of chilling may be essential for diapause termination. Diapausing insects often cannot resume development or reproduction immediately upon transfer to favorable conditions but require a period of chilling. Although some insects do not absolutely require chilling before initiating development many will terminate diapause more quickly if they have first been chilled for a few months.

HORMONAL REGULATION

The juvenile hormones (JHs) and ecdysteroids, two of the major families of insect hormones that direct insect development, metamorphosis, and reproduction, are intimately involved in regulating diapause. The JHs, which are isoprenoids secreted by the corpus allatum (CA), maintain the juvenile characters during the premetamorphic molts, while the steroid hormones from the prothoracic gland (PG), ecdysone and related compounds, dictate the decision to molt. In turn, the CA and PG are regulated by both neural and humoral factors from the brain. Brain neuropeptides governing the CA can exert either a stimulatory (allatotropins) or an inhibitory action (allatoinhibins) on the CA. The dominant regulator of the PG is the brain neuropeptide prothoracicotropic hormone (PTTH). These hormones, together with diapause hormone, a unique neuropeptide that regulates the embryonic diapause of the commercial silkworm, are the key hormonal regulators of insect diapause. In certain situations, the presence of one or more of these hormones promotes diapause, while in others it is the absence of a certain hormone that causes diapause.

Embryonic Diapause

The best understood hormonal mechanism regulating embryonic diapause is based on the silkworm. In this species diapause intercedes early during embryogenesis, just before segmentation. The developmental fate of the embryo is determined by the presence or absence of diapause hormone (DH), a neuropeptide secreted by the mother's subesophageal ganglion. In the presence of DH, the ovariole produces eggs that enter diapause, and when the hormone is not present the eggs develop without the interruption of diapause. Whether the mother releases DH is dependent upon the photoperiod she was exposed to as an embryo. Thus, the mother's photoperiodic history dictates whether she will release the DH needed to influence the diapause fate of her progeny.

The structure of DH has been defined, as well as the sequence of the cDNA that encodes the peptide. DH appears to exert its effect on diapause by influencing carbohydrate metabolism. In the presence of DH, the developing oocytes incorporate glycogen stores, which in turn are converted to sorbitol. Sorbitol was originally thought to function simply as a cryoprotectant, but recent work suggests that sorbitol may actually be involved in shutting down development in the embryo. The addition of sorbitol to an embryo that is programmed to develop without diapause elicits a developmental arrest; in contrast, the removal of sorbitol from diapause-programmed embryos enables the embryos to develop without diapause.

No other diapauses appear to rely on DH for diapause regulation. It appears to be a hormonal regulator unique to the silkworm. In the gypsy moth diapause occurs at the end of embryogenesis, just before hatching of the first instar. The diapause of this species appears to be regulated by maintenance of a high ecdysteroid titer. As long as the ecdysteroid titer remains high, the pharate first instar remains locked in diapause. Only when the ecdysteroid titer drops in the spring is the gypsy moth free to terminate its diapause and hatch. Yet another mechanism seems to operate in the giant silkmoth, *Antheraea yamamai.* In this insect, an unidentified

repressive factor from the mesothorax inhibits the action of a maturation factor from the abdomen. The fact that all three species that have been examined display different endocrine control mechanisms suggests a wealth of mechanisms operating in the regulation of these early stage diapauses.

Larval Diapause

Larval diapause frequently intercedes at the end of larval life, just before the onset of pupation and metamorphosis, but it is not at all uncommon in earlier instars as well. Common to most examples of larval diapause is a shutdown in the brain–prothoracic gland axis. In the absence of ecdysteroids from the PG, the larva fails to initiate the next molt. The failure of the brain to release ecdysteroids can usually be directly attributed to the brain's failure to release PTTH. In a number of species, JH may also play a role. For example, in the southwestern corn borer, *D. grandiosella,* the JH titer remains elevated throughout diapause, and the diapause can be terminated only when the JH titer drops. In some other species such as the European corn borer, *O. nubilalis,* the JH titer is high in early diapause but then declines and remains low throughout the remainder of diapause. No role for JH is apparent in several other insects: the larval diapause of both the parasitic wasp, *Nasonia vitripennis,* and the blow fly, *Calliphora vicina,* can be explained strictly as an ecdysteroid deficiency.

Pupal Diapause

Pupal diapause is the consequence of a shutdown in the brain–prothoracic gland axis. Thus, in the absence of ecdysteroids from the PG the progression of adult differentiation is halted. At the termination of diapause ecdysteroids are again released, triggering adult development. In *H. cecropia* a period of chilling is required before the brain can stimulate the PG to release ecdysteroids. Pupal diapauses can usually be quickly terminated with an injection of 20-hydroxyecdysone. Usually the absence of ecdysteroids can be attributed directly to a failure of the brain to release the neuropeptide PTTH needed to stimulate the PG to synthesize ecdysteroids, but in some insects, e.g., *Heliothis zea,* PTTH is released shortly after pupation, but pupa fail to develop until the PG has been chilled adequately.

Unlike larval diapause there is no evidence suggesting that JH regulates pupal diapause induction or termination, yet JH is indeed present during pupal diapause in some species. In flesh flies, cycles of JH activity apparently drive infradian cycles (4-day periodicity at 25°C) of metabolic activity that persist throughout diapause.

Adult Diapause

A shutdown in JH synthesis is a key feature in the regulation of adult diapause. The corpora allata, the endocrine glands that synthesize and release JH, are characteristically small during diapause. Application of exogenous JH or implantation of active corpora allata into a diapausing individual usually prompts the termination of diapause. Conversely, the surgical extirpation of the corpora allata from a nondiapausing adult causes the adult to enter a diapause-like state. Measurement of the JH titer also supports the idea that adult diapause is the consequence of a shutdown of the corpora allata: the titer of JH typically drops as the insect enters diapause and increases again when diapause is terminated.

It is the brain that regulates the corpora allata, and both nervous and humoral pathways are involved in its regulation. In the Colorado potato beetle, the brain exerts its control over the corpora allata by a humoral mechanism, but in the linden bug, *Pyrrhocoris apterus,* nervous control is also involved. Ecdysteroids may also be involved in some species. The ecdysteroid titer is nearly twice as high in Colorado potato beetles destined for diapause than in those that are not destined to enter diapause, and an injection of ecdysteroids can terminate adult diapause in *Drosophila melanogaster.*

MOLECULAR MECHANISMS

The environmental cues that regulate diapause have been well defined, and there is also a fairly good understanding of the downstream hormonal signals that serve to coordinate diapause. But, the molecular underpinning of diapause remains poorly understood. Is diapause simply a shutdown in gene expression or does it represent the expression of a unique set of genes? An examination of the synthesis of brain proteins in flesh flies suggests that far fewer proteins are synthesized in the brain during diapause but, in addition, the brains of diapausing flies synthesize a set of proteins not observed in brains of nondiapausing flies. This suggests that diapause represents both a shutdown in gene expression and the expression of a unique set of genes.

One of the most conspicuous groups of genes that are diapause up-regulated is that of the heat-shock proteins. Both heat-shock protein 70 (Hsp70) and one of the small heat-shock proteins (Hsp23) are up-regulated in flesh flies during diapause. The Hsps are up-regulated upon entry into diapause, remain elevated throughout diapause, and then drop sharply at diapause termination. But, not all heat-shock proteins are up-regulated during diapause. Hsp90, by contrast, is actually down-regulated. The Hsps may offer protection from environmental stresses during diapause and possibly contribute to the cell cycle arrest that characterizes diapause.

Genes that are diapause down-regulated are potentially of equal interest. Among the genes in this category is the gene that encodes proliferating cell nuclear antigen, a cell cycle regulator. The down-regulation of this gene during diapause may be important in bringing about the cell cycle arrest. As more genes are examined, it is evident that certain genes are expressed throughout diapause, others are turned off during diapause, while still others are expressed only during early or late diapause or may be expressed intermittently during diapause.

It is still too early to know if common sets of genes are expressed during diapause in different species and different life stages, but preliminary data suggest that the expression patterns of at least some of the genes, those that encode Hsp70, may be shared across species and life stages.

See Also the Following Articles
Aestivation • Bombyx mori • Cold/Heat Protection • Dormancy • Juvenile Hormone

Further Reading
Danks, H. V. (1987). "Insect Dormancy: An Ecological Perspective." Biological Survey of Canada, Ottawa.
Denlinger, D. L. (1972). Induction and termination of pupal diapause in *Sarcophaga* flesh flies. *Biol. Bull.* **142,** 11–24.
Denlinger, D. L. (1985). Hormonal control of diapause. *In* "Comprehensive Insect Physiology, Biochemistry, and Pharmacology" (G. A. Kerkut and L. I. Gilbert, eds.), Vol. 8, pp. 353–412. Pergamon Press, Oxford.
Denlinger, D. L. (2002). Regulation of diapause. *Annu. Rev. Entomol.* **47,** 93–122.
Denlinger, D. L., Giebultowicz, J. M., and Saunders, D. S. (eds.) (2001). "Insect Timing: Circadian Rhythmicity to Seasonality." Elsevier, Amsterdam.
Lee, R. E., Jr., and Denlinger, D. L. (eds.) (1991). "Insects at Low Temperature." Chapman & Hall, New York.
Saunders, D. S. (2002). "Insect Clocks," 3rd edition. Elsevier, Amsterdam.
Tauber, M. J., Tauber, C. A., and Masaki, S. (1986). "Seasonal Adaptations of Insects." Oxford University Press, Oxford.

Digestion

Walter R. Terra
University of São Paulo, Brazil

Digestion is the process by which food molecules are broken down into smaller molecules that are able to be absorbed by the gut tissue. Most food molecules requiring digestion are polymers such as proteins and starch, and are sequentially digested through three phases (Fig. 1). Primary digestion is the dispersion and reduction in molecular size of the polymers and results in oligomers. During intermediate digestion, these undergo a further reduction in molecular size to dimers, which in final digestion form monomers. Digestion usually occurs under the action of digestive enzymes from the midgut, with minor or no participation of salivary enzymes. In most insects, midgut pH is either mildly acidic or neutral. Lepidopteran and trichopteran larvae, scarabaeid beetles, and nematoceran flies have alkaline midguts, whereas cyclorrhaphous flies have a very acidic section in the middle of the midgut. The midgut is, as a rule, an oxidizing site, although in some wool-digesting insects it is a reducing site, a condition necessary to break disulfide bonds in keratin, thus facilitating enzymatic hydrolysis.

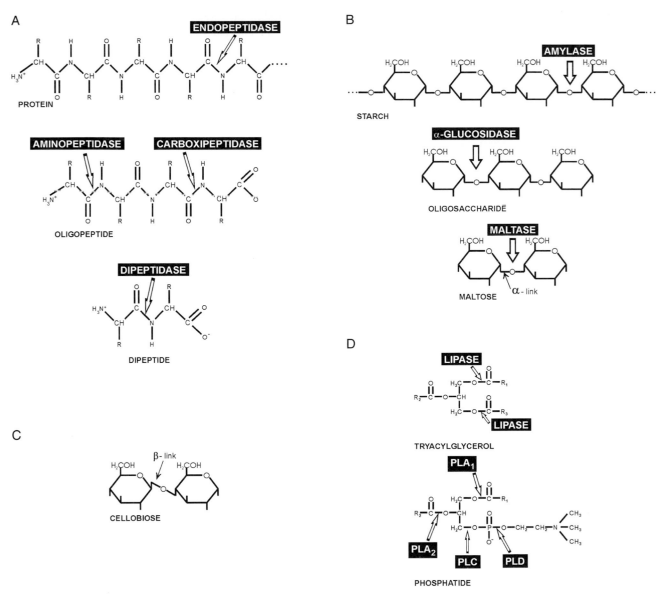

FIGURE 1 Digestion of important nutrient classes. Arrows point to bonds cleaved by enzymes. (A) Protein digestion; R, different amino acid moieties. (B) Starch digestion. (C) β-linked glucoside. (D) Lipid digestion; PL, phospholipase; R, fatty acyl moieties.

DIGESTION OF PROTEINS

Initial digestion of proteins is carried out by proteinases (endopeptidases), which are enzymes able to cleave the internal peptide bonds of proteins (Fig. 1A). Different endopeptidases are necessary to do this because the amino acid residues vary along the peptide chain (R is a variable group in Fig. 1A). Proteinases may differ in specificity toward the reactant protein (substrate) and are grouped according to their reaction mechanism into the subclasses: serine, cysteine, and aspartic proteinases. Trypsin, chymotrypsin, and elastase are serine proteinases that are widely distributed in insects and have molecular masses in the range 20 to 35 kDa and alkaline pH optima. Trypsin preferentially hydrolyzes (its primary specificity) peptide bonds in the carboxyl end of amino acids with basic R groups (Arg, Lys); chymotrypsin is preferential toward large hydrophobic R groups (e.g., Phe, Tyr) and elastase, toward small hydrophobic R groups (e.g., Ala). The activity of the enzymes also depends on the amino acid residues neighboring the bond to be cleaved. This may explain the differences in susceptibilities of insects to strains of *Bacillus thuringiensis*, because the deleterious effects depend on the previous proteolysis of the bacterial endotoxin. Related to this is the growing evidence that insects fed on trypsin inhibitor-containing food express new trypsin molecules insensitive to the inhibitors. These inhibitors are proteins and their binding to the enzyme has molecular requirements similar to those of the substrate.

Cysteine and aspartic proteinases are the only midgut proteinases in hemipterans and they occur in addition to serine proteinases in cucujiformia beetles. Their occurrence in Hemiptera is interpreted as a consequence of the loss of the usual digestive serine proteinases associated with the adaptation of hemipteran ancestors to a diet lacking proteins (plant sap), followed by the use of lysosome-like enzymes in adapting to a new predatory habit. The presence of cysteine and aspartic proteinases in cucujiformia beetles is likely an ancestral adaptation to circumvent proteolytic inhibition caused by trypsin inhibitors in ingested seeds. Cysteine and aspartic proteinases have pH optima of 5.5 to 6.0 and 3.2 to 3.5 and molecular masses of 20 to 40 kDa and 60 to 80 kDa, respectively. Because of their pH optima, aspartic proteinases are not very active in the mildly acidic midguts of Hemiptera and cucujiformia beetles, but are very important in the middle midguts (pH 3.5) of cyclorrhaphous flies.

Intermediate digestion of proteins is accomplished by exopeptidases, enzymes that remove amino acids from the N-terminal (aminopeptidases) or C-terminal (carboxypeptidases) ends of oligopeptides (fragments of proteins) (Fig. 1A). Insect aminopeptidases have molecular masses in the range 90 to 130 kDa, have pH optima of 7.2 to 9.0, have no marked specificity toward the N-terminal amino acid, and are usually associated with the microvillar membranes of midgut cells. Therefore, the action of aminopeptidase is restricted to the surface of midgut cells. Because aminopeptidases are frequently active on dipeptides, they are also involved in protein-terminal digestion together with dipeptidases. Aminopeptidases may account for as much as 55% of the midgut microvillar proteins in larvae of the yellow mealworm, *Tenebrio molitor*. Probably because of this, in many insects aminopeptidases are the preferred targets of *B. thuringiensis* endotoxins. These toxins, after binding to aminopeptidase (or other receptors), form channels through which cell contents leak, leading to insect death. The most important insect carboxypeptidases have alkaline pH optima, have molecular masses in the range 20 to 50 kDa, and require a divalent metal for activity. They are classified as carboxypeptidase A or B depending on their activity upon neutral/acid or basic C-terminal amino acids, respectively.

DIGESTION OF CARBOHYDRATES

Initial and intermediate digestion of starch (or glycogen) is accomplished by α-amylase. This enzyme cleaves internal bonds of the polysaccharide until it is reduced to small oligosaccharides or disaccharides (Fig. 1B). The amylases are not very active in digesting intact starch granules, making mastication prior to ingestion important. Insect amylases depend on calcium ions for activity or stability, they are activated by chloride ions (amylases in Lepidoptera are exceptions), their molecular masses are found in the range 48 to 68 kDa, and their pH optima vary widely (4.8–9.8) depending on the insect taxon. As described for trypsin,

insects feeding on amylase inhibitor-containing food express new amylase molecules insensitive to the inhibitors.

The final digestion of starch chains occurs under α-glucosidases, enzymes that sequentially remove glucosyl residues from the nonreducing ends of short oligomaltosaccharides. If the saccharide is a disaccharide, it is named maltose (Fig. 1B). Because of that, α-glucosidase is also called maltase. As a rule, sucrose (glucose α1,β2-fructose) is hydrolyzed by α-glucosidase. If an enzyme is able to hydrolyze sucrose, but not maltose, it is likely a β-fructosidase, an enzyme attacking sucrose by the fructosyl residue. Sucrose is found in large amounts in nectar and phloem sap and in lesser amounts in some fruits and leaves.

The important insect hemolymph and fungal sugar trehalose (glucose α1,α1-glucose) is hydrolyzed only by the specific enzyme trehalase. This digestive enzyme occurs in luminal contents or immobilized at the surface of midgut cells and also as an enzyme at the midgut basal cell membrane, making available glucose from hemolymph trehalose.

Although cellulose is abundant in plants, most plant-feeding insects, such as caterpillars and grasshoppers, do not use it. Cellulose is a nonramified chain of glucose units linked by β-1,4 bonds (Fig. 1C) arranged in a crystalline structure that is difficult to disrupt. Thus, cellulose digestion is unlikely to be advantageous to an insect that can meet its dietary requirements using more easily digested food constituents. The cellulase activity found in some plant feeders facilitates the access of digestive enzymes to the plant cells ingested by the insects. True cellulose digestion is restricted to insects that have, as a rule, nutritionally poor diets, as exemplified by termites, woodroaches, and cerambycid and scarabaeid beetles. There is growing evidence that insects secrete enzymes able to hydrolyze crystalline cellulose, challenging the long-standing belief that microbial symbionts are necessary for cellulose digestion. The end products of cellulase action are glucose and cellobiose (Fig. 1C); the latter is hydrolyzed by a β-glucosidase.

Hemicellulose is a mixture of polysaccharides associated with cellulose in plant cell walls. They are β-1,4- and/or β-1,3-linked glycan chains made up mainly of glucose (glucans), xylose (xylans), and other monosaccharides. The polysaccharides are hydrolyzed by a variety of enzymes from which xylanases, laminarinases, and lichenases are the best known. The end products of the actions of these enzymes are monosaccharides and β-linked oligosaccharides. The final digestion of those chains occurs under the actions of β-glycosidases that sequentially remove glycosyl (usually glucosyl, galactosyl, or xylosyl) residues from the nonreducing end of the β-linked oligosaccharides. As these may be cellobiose, β-glycosidase is frequently also named cellobiase. Thus, β-glycosidases complete the digestion of cellulose and hemicelluloses.

A special β-glycosidase (aryl β-glycosidase) acts on glycolipids and *in vivo* probably removes a galactose from monogalactosyldiacylglycerol that together with digalactosyl-diacylglycerol is a major lipid of photosynthetic tissues. Digalactosyldiacylglycerol is converted into monogalactosyl-diacylglycerol by the action of an α-galactosidase. The aryl β-glycosidase also acts on plant glycosides that are noxious after hydrolysis. Insects circumvent these problems by detoxifying the products of hydrolysis or by repressing the synthesis and secretion of this enzyme while maintaining constant the synthesis and secretion of the other β-glycosidases.

DIGESTION OF LIPIDS AND PHOSPHATES

Oils and fats are triacylglycerols and are hydrolyzed by a triacylglycerol lipase that preferentially removes the outer ester links of the substrate (Fig. 1D) and acts only on the water–lipid interface. This interface is increased by surfactants that, in contrast to the bile salts of vertebrates, are mainly lysophosphatides. The resulting 2-monoacylglycerol may be absorbed or further hydrolyzed before absorption.

Membrane lipids include glycolipids, such as galactosyl-diacylglycerol and phosphatides. After the removal of galactose residues from mono- and digalactosyldiacylglycerol, which leaves diacylglycerol, it is hydrolyzed as described for triacyl-glycerols. Phospholipase A removes one fatty acid from the phosphatide, resulting in a lysophosphatide (Fig. 1D) that forms micellar aggregates, causing the solubilization of cell membranes. Lysophosphatide seems to be absorbed intact by insects.

Nonspecific phosphatases remove phosphate moieties from phosphorylated compounds to make their absorption easier. Phosphatases are active in an alkaline or acid medium.

See Also the Following Articles
*Metabolism • Nutrition • Phytophagous Insects •
Salivary Glands*

Further Reading
Cristofoletti, P. T., and Terra, W. R. (1999). Specificity, anchoring and subsites in the active center of a microvillar aminopeptidase purified from *Tenebrio molitor* (Coleoptera) midgut cells. *Insect Biochem. Mol. Biol.* **29**, 807–819.
Kerkut, G. A., and Gilbert, L. I. (eds.) (1985). "Comprehensive Insect Physiology, Biochemistry and Pharmacology." Pergamon Press, Oxford. [See especially Vol. 4, Chaps. 5 and 7]
Lehane, M. J., and Billingsley, P. F. (eds.) (1996). "Biology of the Insect Midgut." Chapman & Hall, London. [See especially Chaps. 3, 6, 7, 11, and 14]
Silva, C. P., Terra, W. R., de Sa, M. F. G., Samuels, R. I., Isejima, E. M., Bifano, T. D., and Almeida, J. S. (2001). Induction of digestive alpha-amylases in larvae of *Zabrotes subfasciatus* (Coleoptera: Bruchidae) in response to ingestion of common bean alpha-amylase inhibitor 1. *J. Insect Physiol.* **47**, 1283–1290.
Terra, W. R., and Ferreira, C. (1994). Insect digestive enzymes: Properties, compartmentalization and function. *Comp. Biochem. Physiol.* **109B**, 1–62.
Vonk, H. J., and Western, J. R. H. (1984) "Comparative Biochemistry and Physiology of Enzymatic Digestion." Academic Press, London.
Watanabe, H., Noda, H., Tokuda, G., and Lo, N. (1998). A cellulase gene of termite origin. *Nature* **394**, 330–331.

Digestive System

Walter R. Terra and Clélia Ferreira
University of São Paulo, Brazil

The digestive system consists of the alimentary canal (gut) and salivary glands, and is responsible for all steps in food processing: digestion, absorption, and feces delivery and elimination. These steps occur along the gut. The anterior (foregut) and posterior (hindgut) parts of the gut have cells covered by a cuticle, whereas, in the midgut, cells are separated from the food by a filmlike anatomical structure referred to as the peritrophic membrane. Salivary glands are associated with the foregut and may be important in food intake but usually not in digestion. Remarkable adaptations are found in taxa with very specialized diets, such as cicadas (plant sap), dung beetles (feces), and termites (wood), and in insects with short life spans, as exemplified by flies and moths. Digestion is carried out by insect digestive enzymes, apparently without participation of symbiotic microorganisms.

GUT MORPHOLOGY AND FUNCTION

Figure 1 is a generalized diagram of the insect gut. The foregut begins at the mouth, includes the cibarium (preoral cavity formed by mouthparts), the pharynx, the esophagus, and the crop (a dilated portion, as in Fig. 2A, or a diverticulum, like Fig. 2K). The crop is a storage organ in many insects and also serves as a site for digestion in others. The foregut is lined by a cuticle that is nonpermeable to hydrophilic molecules and in some insects is reduced to a straight tube (Fig. 2F). The proventriculus is a triturating (grinding into fine particles) organ in some insects, and in most it provides a valve controlling the entry of food into the midgut, which is the main site of digestion and absorption of nutrients.

The midgut includes a simple tube (ventriculus) from which blind sacs (gastric or midgut ceca) may branch, usually from its anterior end (Fig. 2A). Midgut ceca may also occur along the midgut in rings (Fig. 2F) or not (Fig. 2H) or in the

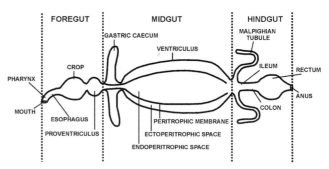

FIGURE 1 Generalized diagram of the insect gut.

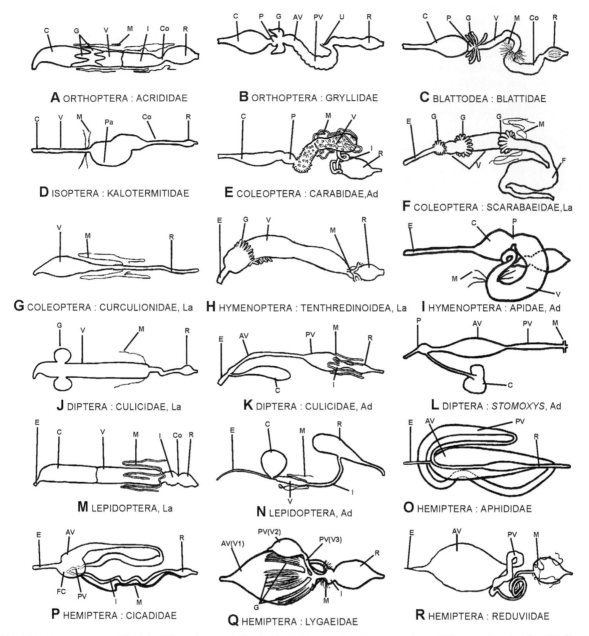

FIGURE 2 Major insect gut types: Ad, adult; AV, anterior ventriculus (midgut); C, crop; Co, colon; E, esophagus; F, fermentation chamber; FC, filter chamber; G, midgut (gastric) ceca; I, ileum; La, larva; M, Malpighian tubules; P, proventriculus; Pa, paunch; PV, posterior ventriculus (midgut); R, rectum; V, ventriculus. Not drawn to scale. [Based partly on Terra, W. R. (1988). Physiology and biochemistry of insect digestion: An evolutionary perspective. *Brazilian J. Med. Biol. Res.* **21,** 675–734.]

posterior midgut (Fig. 2Q). In most insects, the midgut is lined with a filmlike anatomical structure (peritrophic membrane) that separates the luminal contents into two compartments: the endoperitrophic space (inside the membrane) and the ectoperitrophic space (outside the membrane). Some insects have a stomach, which is an enlargement of the midgut to store food (Fig. 2R). In the region of the sphincter (pylorus) separating the midgut from the hindgut, Malpighian tubules branch off the gut. Malpighian tubules are excretory organs that individually empty in the gut and

may be joined to form a ureter (Fig. 2B); in some species, however, they are absent (Fig. 2O).

The hindgut includes the ileum, colon, and rectum (which is involved in the absorption of water and ions) and terminates with the anus. The hindgut is lined by a cuticle (usually impermeable); although in some insects it is reduced to a straight tube (Fig. 2G), in others it is modified in a fermentation chamber (Fig. 2F) or paunch (Fig. 2D), with both structures storing ingested food and harboring microorganisms that have a controversial role in assisting cellulose digestion.

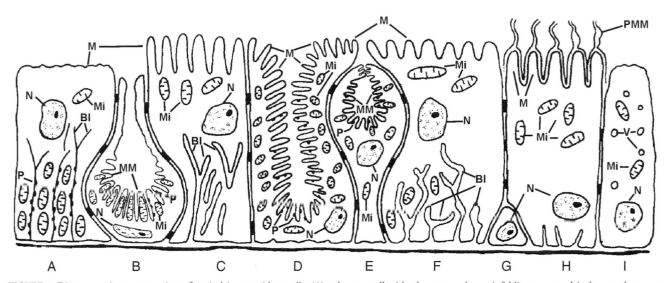

FIGURE 3 Diagrammatic representation of typical insect midgut cells: (A) columnar cell with plasma membrane infoldings arranged in long and narrow channels, usually occurring in fluid-absorbing tissues; (B) lepidopteran long-necked goblet cell; (C) columnar cell with highly-developed basal plasma membrane infoldings displaying few openings into the underlying space, usually occurring in fluid-absorbing tissue; (D) cyclorrhaphan dipteran oxyntic (cuprophilic) cell; (E) lepidopteran stalked goblet cell; (F) columnar cell with highly developed plasma membrane infoldings with numerous openings into the underlying space, frequently present in fluid-secreting tissue; (G) regenerative cell; (H) hemipteran midgut cell; (I) endocrine cell. Note particles (portasomes) studding the cytoplasmic side of the apical membranes in B, D, and E and of the basal plasma membranes in A. Abbreviations: Bl, basal plasma membrane infoldings; M, microvilli; Mi, mitochondria; MM, modified microvilli; N, nucleus; P, portasomes; PMM, perimicrovillar membranes; V, vesicles.

The gut epithelium is always simple and rests on a basal lamina that is surrounded by conspicuous circular and a few longitudinal muscles, the organization of which varies according to species. Wavelike contractions of the circular muscles cause peristalsis, propelling the food bolus along the gut. The gut is oxygenated by the tracheal system, and whereas the foregut and hindgut are well innervated, the same is not true for the midgut. The gut is also connected to the body wall through the extrinsic visceral muscles. These act as dilators of the gut, mainly at the foregut, where they form a pump highly developed in fluid feeders (cibarium pump), exemplified by sap (Hemiptera) and blood (Hemiptera and Diptera) feeders. However it is also present in chewing insects (pharyngeal pump), which are thus enabled to drink water and to pump air into the gut during the molts. The gut sensory system includes the chemoreceptors in the cibarium and stretch receptors associated with muscles of the foregut and hindgut.

Salivary glands are labial or mandibular glands opening in the cibarium. They are usually absent in Coleoptera. The saliva lubricates the mouthparts, may contain an array of compounds associated with blood intake, or may be used as a fixative of the stylets of sap-sucking bugs. Saliva usually contains only amylase and maltase or no enzymes at all, although in a few hemipteran predators it may have the whole complement of proteolytic enzymes.

The epithelium of the midgut is composed of a major type of cell usually named columnar cell, although it may have other forms (Fig. 3A, C, F); it also contains regenerative cells (Fig. 3G) that are often collected together in nests at the base of the epithelium, cells (Fig. 3I) whose purpose is not understood but are generally believed to have an endocrine function, and also specialized cells (goblet cells, Fig. 3B, E; oxyntic cells, Fig. 3D; hemipteran midgut cell, Fig. 3H).

The peritrophic membrane is made up of a matrix of proteins (peritrophins) and chitin to which other components (e.g., enzymes, food molecules) may associate. This anatomical structure is sometimes called the peritrophic matrix, but this term is better avoided because it does not convey the idea of a film and suggests that it is the fundamental substance of some structure. The argument that "membrane" means a lipid bilayer does not hold here because the peritrophic membrane is an anatomical structure, not a cell part. Peritrophins have domains similar to mucins (gastrointestinal mucus proteins) and other domains able to bind chitin. This suggests that the peritrophic membrane may have derived from an ancestral mucus. According to this hypothesis, the peritrophins evolved from mucins by acquiring chitin-binding domains. The parallel evolution of chitin secretion by midgut cells led to the formation of the chitin–protein network characteristic of the peritrophic membrane. The details of peritrophic membrane formation are not known, although there is evidence that peritrophins are released by exocytosis (Fig. 4A) in Diptera or by microaprocrine secretion (Fig. 4D) in Lepidoptera and somehow interlocked with chitin fibers that are synthesized at the luminal surface of midgut cells.

The formation of the peritrophic membrane may occur in part of the midgut or in the entire organ (type I), or only at the entrance of the midgut (cardia) (type II). The two types

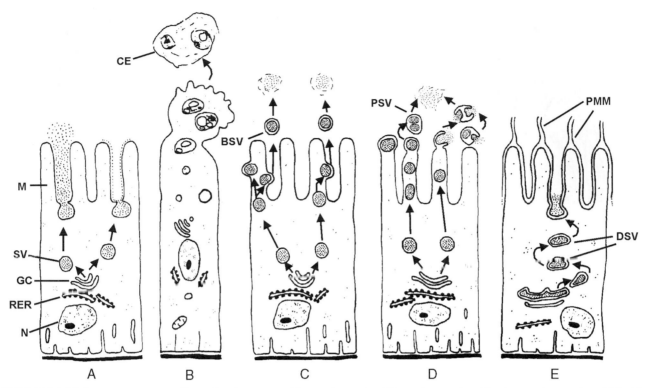

FIGURE 4 Models for secretory processes of insect digestive enzymes; (A) exocytic secretion, (B) apocrine secretion, (C) microapocrine secretion with budding vesicles, (D) microapocrine secretion with pinched-off vesicles, and (E) modified exocytic secretion in hemipteran midgut cell. Abbreviations: BSV, budding secretory vesicle; CE, cellular extrusion; DSV, double-membrane secretory vesicle; GC, Golgi complex; M, microvilli; N, nucleus, PMM, perimicrovillar membrane; PSV, pinched-off secretory vesicle; RER, rough endoplasmic reticulum; SV, secretory vesicle.

of membrane differ in their constituent peritrophins and in their supramolecular organization. Type I peritrophic membrane occurs in most insects, whereas type II is restricted to larval and adult (except hematophagous) mosquitoes and flies (Diptera) and a few adult Lepidoptera. Although a peritrophic membrane is found in most insects, it does not occur in Hemiptera and Thysanoptera, which have perimicrovillar membranes in their cells (Fig. 3H). The other insects that do not seem to have a peritrophic membrane are adult Lepidoptera, Phthiraptera, Psocoptera, Zoraptera, Strepsiptera, Raphidioptera, Megaloptera, and Siphonaptera as well as bruchid beetles and some adult ants (Hymenoptera). Most of the pores of the peritrophic membrane are in the range of 7 to 9 nm, although some may be as large as 36 nm. Thus, the peritrophic membrane hinders the free movement of molecules, dividing the midgut lumen into two compartments (Fig. 1) with different molecules. The functions of this structure include those of the ancestral mucus (protection against food abrasion and microorganism invasion) and several roles associated with the compartmentalization of the midgut. These roles result in improvements in digestive efficiency and assist in decreasing digestive enzyme excretion, and in restricting the production of the final products of digestion close to their transporters, thus facilitating absorption.

DIGESTIVE PHYSIOLOGY

Overview

The study of digestive physiology involves the spatial organization of digestive events in the insect gut. Digestive enzymes that participate in primary digestion (cleavage of polymers like protein and starch), secondary digestion (action on oligomers exemplified by polypeptides and dextrans), and final digestion (hydrolysis of dimers as dipeptides and disaccharides) are assayed in different gut compartments. Samples of the ectoperitrophic space contents (Fig. 1) are collected by puncturing the midgut ceca with a capillary or by washing the luminal face of midgut tissue. Midgut tissue enzymes are intracellular, glycocalyx-associated or microvillar membrane-bound. In addition to the distribution of digestive enzymes, the spatial organization of digestion depends on midgut fluxes. Gut fluid fluxes are inferred with the use of dyes. Secretory regions transport injected dye into the gut lumen, whereas absorbing regions accumulate orally fed dyes. Upon studying the spatial organization of the digestive events in insects of different taxa and diets, it was realized that the insects may be grouped relative to their digestive physiology, assuming they have common ancestors. Those putative ancestors correspond to basic gut plans from which groups of insects may have evolved by adapting to different diets.

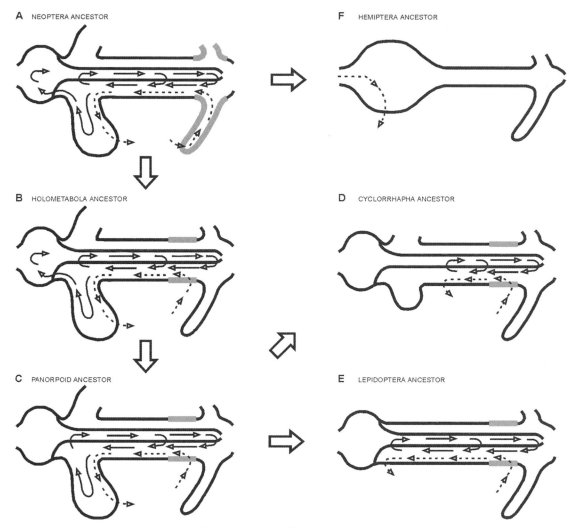

FIGURE 5 Diagrammatic representation of water fluxes (dashed arrows) and of the circulation of digestive enzymes (solid arrows) in putative insect ancestors that correspond to the major basic gut plans. In Neoptera ancestors (A), midgut digestive enzymes pass into the crop. Countercurrent fluxes depend on the secretion of fluid by the Malpighian tubules and its absorption by the ceca. Enzymes involved in initial, intermediate, and final digestion circulate freely among gut compartments. Holometabola ancestors (B) are similar except that secretion of fluid occurs in posterior ventriculus. Panorpoid (Lepidoptera and Diptera assemblage) ancestors (C) display countercurrent fluxes like Holometabola ancestors, midgut enzymes are not found in crop, and only the enzymes involved in initial digestion pass through the peritrophic membrane. Enzymes involved in intermediate digestion are restricted to the ectoperitrophic space and those responsible for terminal digestion are immobilized at the surface of midgut cells. Cyclorrhapha ancestors (D) have a reduction in ceca, absorption of fluid in middle midgut, and anterior midgut playing a storage role. Lepidoptera ancestors (E) are similar to panorpoid ancestors, except that anterior midgut replaced the ceca in fluid absorption. Hemiptera ancestors (F) lost crop, ceca, and fluid-secreting regions. Fluid is absorbed in anterior midgut.

Neopteran insects evolved along three lines: the Polyneoptera (which include Blattodea, Isoptera, and Orthoptera), the Paraneoptera (which include Hemiptera), and the Holometabola (which include Coleoptera, Hymenoptera, Diptera, and Lepidoptera). Polyneoptera and Paraneoptera evolved as external feeders occupying the ground surface, on vegetation, or in litter, and developed distinct feeding habits. Some of these habits are very specialized (e.g., feeding wood and sucking plant sap), implying adaptative changes of the digestive system. Major trends in the evolution of Holometabola were the divergence in food habits between larvae and adults and the exploitation of new food sources, exemplified by endoparasitism and by boring or mining

living or dead wood, foliage, fruits, or seeds. This biological variation was accompanied by modifications in the digestive system. Among the panorpoid Holometabola (an assemblage that includes Diptera and Lepidoptera), new selective pressures resulted from the occupation of more exposed or ephemeral ecological niches. Following this trend, those pressures led to shortening life spans, so that the insects may have more generations per year, thus ensuring species survival even if large mortality occurs at each generation. Associated with this trend, the digestive system evolved to become more efficient to support faster life cycles.

The basic plan of digestive physiology for most winged insects (Neoptera ancestors) is summarized in Fig. 5A. In these ancestors, the major part of digestion is carried out in the crop by digestive enzymes propelled by antiperistalsis forward from the midgut. Saliva plays a minor role or no role at all in digestion. After a while, following ingestion, the crop contracts, transferring digestive enzymes and partly digested food into the ventriculus. The anterior ventriculus is acid and has high carbohydrase activity, whereas the posterior ventriculus is alkaline and has high proteinase activity. This differentiation along the midgut may be an adaptation to the instability of ancestral carbohydrases in the presence of proteinases. The food bolus moves backward in the midgut of the insect by peristalsis. As soon as the polymeric food molecules have been digested to become small enough to pass through the peritrophic membrane, they diffuse with the digestive enzymes into the ectoperitrophic space (Fig. 1). The enzymes and nutrients are then displaced toward the ceca with a countercurrent flux caused by secretion of fluid at the Malpighian tubules and its absorption back by cells (similar to Figs. 3A, C) at the ceca (Fig. 5A), where final digestion is completed and nutrient absorption occurs. When the insect starts a new meal, the ceca contents are moved into the crop. As a consequence of the countercurrent flux, digestive enzymes occur as a decreasing gradient in the midgut, and lower amounts are excreted.

The Neoptera basic plan is the source of that of the Polyneoptera orders and evolved to the basic plans of Paraneoptera and Holometabola. Lack of data limits the proposition of a basic plan to a single Paraneoptera order, Hemiptera. Symbiont microorganisms may occur in large numbers in insect gut. For example, the bacteria *Nocardia rhodnii* may represent up to 5% of the *Rhodnius prolixus* midgut dry weight. The symbionts are believed to provide nutrient factors (such as B vitamins or fermentation end products) to the host. Microorganism symbionts have rarely been associated with digestion, and the few that are known are implicated with cellulose digestion only.

Polyneoptera

BLATTODEA AND MANTODEA Cockroaches, which are among the first neopteran insects to appear in the fossil record, are extremely generalized in most morphological features. They are usually omnivorous. In spite of the lack of detailed data on midgut fluxes and enzyme distribution, it is thought that digestion in cockroaches occurs as described for the Neoptera ancestor (Fig. 5A), except that part of the final digestion of proteins occurs on the surface of midgut cells. Another difference observed is the enlargement of hindgut structures (Fig. 2C), noted mainly in wood-feeding cockroaches. These hindgut structures harbor bacteria producing acetate and butyrate from ingested wood or other cellulose-containing materials. Acetate and butyrate are absorbed by the hindgut of all cockroaches, but this activity is more remarkable with wood roaches. Cellulose digestion

may be accomplished by bacteria, but there is evidence that wood roaches have their own cellulases. Mantids have a capacious crop, and a short midgut and hindgut. It is probable that the major part of digestion takes place in their crops.

ISOPTERA Termites are derived from and are more adapted than wood roaches in dealing with refractory materials such as wood and humus. Associated with this specialization, they lost the crop and midgut ceca and enlarged their hindgut structures (Fig. 2D). Termites digest cellulose with their own cellulase, and the products pass from the midgut into the hindgut, where they are converted into acetate and butyrate by hindgut bacteria as in wood roaches. Symbiotic bacteria are also responsible for nitrogen fixation in hindgut, resulting in bacterial protein. This is incorporated into the termite body mass after being expelled in feces by one individual and being ingested and digested by another. This explains the ability of termites to develop successfully in diets very poor in protein.

ORTHOPTERA Grasshoppers feed mainly on grasses, and their digestive physiology clearly evolved from the neopteran ancestor. Carbohydrate digestion occurs mainly in the crop, under the action of midgut enzymes, whereas protein digestion and final carbohydrate digestion take place at the anterior midgut ceca. The abundant saliva (devoid of significant enzymes) produced by grasshoppers saturate the absorbing sites in the midgut ceca, thus hindering the countercurrent flux of fluid. This probably avoids excessive accumulation of noxious wastes in the ceca, and makes possible the high relative food consumption observed among locusts in their migratory phases. Starving grasshoppers present midgut countercurrent fluxes. Cellulase found in some grasshoppers is believed to facilitate the access of digestive enzymes to the plant cells ingested by the insects by degrading the cellulose framework of cell walls. Crickets are omnivorous or predatory insects with most starch and protein digestion occurring in their capacious crop (Fig. 2B).

Paraneoptera

HEMIPTERA The characteristics of the Paraneoptera ancestors cannot be inferred because midgut function data are available only for Hemiptera. The Hemiptera comprise insects of several suborders (e.g., cicadas, leafhoppers, aphids, and fulgorids) that feed almost exclusively on plant sap, and insects of the taxon Heteroptera (e.g., assassin bugs, plant bugs, stinkbugs, and lygaeid bugs) that are adapted to different diets. The ancestor of the entire order is supposed to be a sapsucker similar to present-day cicadas and fulgorids.

The hemipteran ancestor (Fig. 5F) differs remarkably from the neopteran ancestor, as a consequence of adaptations to feeding on plant sap. These differences consist of the lack of crop and anterior midgut ceca, loss of the enzymes involved in initial and intermediate digestion and loss of the peritrophic membrane associated with the lack of luminal digestion, and, finally, the presence of hemipteran midgut cells (Fig. 3H), which have their microvilli ensheathed by an outer (perimicrovillar) membrane. The perimicrovillar membrane maintains a constant distance from the microvillar membrane, extends toward the luminal compartment with a dead end, and limits a closed compartment, the perimicrovillar space (Fig. 3H). Ongoing research suggests that aphids have modified perimicrovillar membranes.

Sap-sucking Hemiptera may suck phloem or xylem sap. Phloem sap is rich in sucrose (0.15–0.73 M) and relatively poor in free amino acids (15–65 mM) and minerals, whereas xylem fluid is poor in amino acids (3–10 mM) and contains monosaccharides (about 1.5 mM), organic acids, potassium ions (about 6 mM), and other minerals. Thus, except for dimer (sucrose) hydrolysis, no food digestion is necessary in sapsuckers. The major problem facing a sap-sucking insect is to absorb nutrients, such as essential amino acids, that are present in very low concentrations in sap. Amino acids may be absorbed according to a hypothesized mechanism that depends on perimicrovillar membranes. In phloem feeders such as aphids, this process may have an assimilation efficiency of 55% for amino acids and only 5% for sugars, whereas in xylem feeders such as leafhoppers, about 99% of dietary amino acids and carbohydrates are absorbed.

Organic compounds in xylem sap need to be concentrated before they can be absorbed by the perimicrovillar system. This occurs in the filter chamber (Fig. 2P) of Cicadoidea and Cercopoidea, which concentrates xylem sap 10-fold, or in the filter chamber of Cicadelloidea (phloem feeders), which is able to concentrate dilute phloem about 2.5-fold. The filter chamber consists of a thin-walled, dilated anterior midgut in close contact with the posterior midgut and the proximal ends of the Malpighian tubules. This arrangement enables water to pass directly from the anterior midgut to the Malpighian tubules, concentrating food in midgut.

The evolution of Heteroptera was associated with regaining the ability to digest polymers. Because the appropriate digestive enzymes were lost, these insects instead used enzymes derived from lysosomes. Lysosomes are cell organelles involved in intracellular digestion carried out by special proteinases referred to as cathepsins. Compartmentalization of digestion was maintained by the perimicrovillar membranes as a substitute for the lacking peritrophic membrane. Digestion in the two major Heteroptera taxa—Cimicomorpha, exemplified by the blood feeder *R. prolixus,* and Pentatomorpha, exemplified by the seed sucker *Dysdercus peruvianus*—is similar. The dilated anterior midgut stores food and absorbs water and, at least in *D. peruvianus,* also absorbs glucose. Digestion of proteins and absorption of amino acids occur in the posterior ventriculus. Most protein digestion occurs in lumen with the aid of a cysteine proteinase and ends in the perimicrovillar space under the action of aminopeptidases and dipeptidases. Many Heteroptera feed on parenchymal tissues of plants. In some of these insects, excess water passes from the expanded anterior midgut to the closely associated midgut ceca, which protrude from the posterior midgut (Fig. 2Q). These ceca may also contain symbiont bacteria.

Holometabola

The basic gut plan of the Holometabola (Fig. 5B) is similar to that of Neoptera except that fluid secretion occurs in the posterior ventriculus by cells similar to Fig. 3F, instead of by the Malpighian tubules. Because the posterior midgut fluid, unlike Malpighian tubular fluid, does not contain wastes, the accumulation of wastes in ceca is decreased. There is an evolutionary trend leading to the loss of anterior midgut ceca in holometabolous insects and an increase in the use of anterior ventricular cells for water absorption. Ceca loss probably further decreases the accumulation of noxious substances in the midgut, which would be more serious in insects that have high relative food consumption rates, such as is common among Holometabola. Digestive systems may change remarkably between larvae and adults of holometabolous insects. Despite these changes, adult digestive systems probably evolved in parallel to larval systems because, except for minor differences, the compartmentalization of digestion in larvae and adults seems to be similar.

The basic plan of Coleoptera and Hymenoptera did not evolve dramatically from the Holometabola ancestor, whereas the basic plan of Diptera and Lepidoptera (panorpoid ancestor, Fig. 5C) presents important differences. Thus, panorpoid ancestors have countercurrent fluxes like Holometabola ancestors but differ from these in the lack of crop digestion, in midgut differentiation in luminal pH, and in which compartment is responsible for each phase of digestion. In Holometabola ancestors, all phases of digestion occur in the endoperitrophic space (Fig. 1), whereas in panorpoid ancestors only initial digestion occurs in that region. In the latter ancestors, intermediate digestion is carried out by free enzymes in the ectoperitrophic space and final digestion occurs at the midgut cell surface by immobilized enzymes. The free digestive enzymes do not pass through the peritrophic membrane because they are larger than the peritrophic membrane pores. Immobilized enzymes may be either soluble enzymes entrapped in the cell glycocalyx or membrane-bound enzymes, which are those embedded in the lipid bilayer forming the microvillar membranes (intrinsic proteins). As a consequence of the compartmentalization of digestive events in panorpoid insects, there is an increase in the efficiency of digestion of polymeric food by allowing the removal of the oligomeric molecules from the endoperitrophic space, which in turn is powered by the recycling mechanism associated with the midgut fluxes. Because oligomers may be substrates or inhibitors for some polymer hydrolases, their presence should decrease the rate of polymer degradation. A fast polymer degradation ensures that polymers are not excreted, hence increases their digestibility. Another consequence of compartmentalization is an increase in the efficiency of oligomeric food hydrolysis by allowing the transference of oligomeric molecules to the ectoperitrophic space and by restricting oligomer hydrolases to this compartment. In these conditions, oligomer hydrolysis occurs in the absence of probable partial inhibition (because of nonproductive binding) by polymer food and presumed nonspecific binding by nondispersed undigested food. This process leads to the production of food monomers only in the neighborhood of the midgut cell surface, causing an increase in the concentration of the final products of digestion close to their transporters, thus facilitating absorption.

COLEOPTERA   Larvae and adults of Coleoptera usually display the same feeding habit; that is, both are plant feeders (although adults may feed on the aerial parts, whereas the larvae may feed on the roots of the same plant) or both are predatory. Coleoptera ancestors are like Holometabola ancestors except for the anterior midgut ceca, which were lost and replaced in function by the anterior midgut. Nevertheless, there are evolutionary trends leading to a great reduction or loss of the crop and, similar to panorpoid orders, occurrence of final digestion at the surface of midgut cells. Thus, in predatory Carabidae most of the digestive phases occur in the crop by means of midgut enzymes, whereas in predatory larvae of Elateridae initial digestion occurs extraorally by the action of enzymes regurgitated onto their prey. The preliquefied material is then ingested by the larvae, and its digestion is finished at the surface of midgut cells. The entire digestive process occurs in the larval endoperitrophic space of Dermestidae. In Tenebrionidae, the final digestion of proteins takes place at midgut cell surface; in Curculionidae and Cerambycidae, the final digestion of all nutrients is carried out at midgut cell surface. It has been proposed that Cerambycidae larvae acquire the capacity to digest cellulose by ingesting fungal cellulases while feeding on fungus-infested wood. In contrast, Coccinellidae adults use their own cellulase to digest cellulose. The distribution of enzymes in gut regions of adult Tenebrionidae is similar to that of their larvae. This suggests that the overall pattern of digestion in larvae and adults of Coleoptera is similar even though (in contrast to adults) beetle larvae usually lack a crop. Insects of the series Cucujiformia (which includes Tenebrionidae, Chrysomelidae, Bruchidae, and Curculionidae) have cysteine proteinases in addition to (or in place of) serine proteinases as digestive enzymes, suggesting that the ancestors of the whole taxon were insects adapted to feed on seeds rich in serine proteinase inhibitors.

Scarabaeidae and several related families are relatively isolated in the series Elateriformia and evolved considerably from the Coleoptera ancestor. Scarabid larvae, exemplified by dung beetles, usually feed on cellulose materials undergoing degradation by a fungus-rich flora. Digestion occurs in the midgut, which has three rows of ceca (Fig. 2F), with a ventral groove between the middle and posterior row. The alkalinity of gut contents increase to almost pH 12 along the midgut ventral groove. This high pH probably enhances cellulose digestion, which occurs mainly in the hindgut fermentation chamber (Fig. 2F), likely through the action of bacterial cell-bound enzymes. The final product of cellulose degradation is mainly acetic acid, which is absorbed through the hindgut wall. Whether scarabid larvae ingest feces to obtain nitrogen compounds, as described above for termites, is a matter of controversy.

HYMENOPTERA   Hymenoptera comprise several primitive suborders (including sawflies and horntails) and Apocrita. Apocrita are divided into Parasitica, which are parasites of other insects, and Aculeata, in which the piercing ovipositor of Parasitica evolved into a stinging organ. The first Apocrita were probably close to the ichneumon flies, whose larvae develop on the surface or inside the body of the host insect. Probably because of that, the larvae of Apocrita present a midgut that is closed at its rear end, and remains unconnected with the hindgut until the time of pupation. Hymenoptera ancestors differ from the Holometabola ancestor in the lack of anterior midgut ceca, which are replaced by the anterior midgut in the function of fluid absorption, and in the absence of midgut enzymes in the crop. Wood wasp larvae of the genus *Sirex* are believed to be able to digest and assimilate wood constituents by acquiring cellulase, xylanase, and possibly other enzymes from fungi present in wood on which they feed. In larval bees, most digestion occurs in the endoperitrophic space. Countercurrent fluxes seem to occur, but the midgut luminal pH gradient hypothetically present in the Hymenoptera ancestor was lost.

Adult bees ingest nectar and pollen. Sucrose from nectar is hydrolyzed in the crop (Fig. 2I) by the action of a sucrase from the hypopharyngeal glands. After ingestion, pollen grains extrude their protoplasm into the ventriculus, where digestion occurs. Worker ants feed on nectar, honeydew, plant sap, or partly digested food regurgitated by their larvae. Thus, they seem to display only intermediate and (or) final digestion.

DIPTERA   The Diptera evolved along two major lines: an assemblage of suborders corresponding to the mosquitoes, including the basal Diptera, and the suborder Brachycera, which includes the most evolved flies (Cyclorrhapha). The Diptera ancestor is similar to the panorpoid ancestor (Fig. 5C) in having the enzymes involved in intermediate digestion free in the ectoperitrophic fluid (mainly in the large ceca), whereas the enzymes of terminal digestion are membrane bound at the midgut cell microvilli. Although these characteristics are observed in most nonbrachyceran larvae, the more evolved of these larvae may show reduction in size of midgut ceca (e.g., Culicidae, Fig. 2k). Nonhematophagous adults store liquid food (nectar or decay products) in their crops. Digestion occurs in their midgut as in larvae. Nectar ingested by mosquitoes (males and females) is stored in the crop, and digested and absorbed at the anterior midgut. Blood, which is sucked only by females, passes to the posterior midgut, where it is digested and absorbed.

The Cyclorrhapha ancestor (Fig. 5D) evolved dramatically from the panorpoid ancestor (Fig. 5C), apparently as a result of adaptations to a diet consisting mainly of bacteria. Digestive events in Cyclorrhapha larvae are exemplified by larvae of the house fly *Musca domestica.* These larvae ingest food rich in bacteria. In the anterior midgut there is a decrease in the starch content of the food bolus, facilitating bacteria death. The bolus now passes into the middle midgut where bacteria are killed by the combined action of low pH, a special lysozyme, and an aspartic proteinase. Finally, the material released by bacteria is digested in the posterior midgut, as is observed in the whole midgut of insects of other taxa. Countercurrent fluxes occur in the posterior midgut powered by secretion of fluid in the distal part of the posterior midgut and its absorption back into the middle midgut. The middle midgut has specialized cells for buffering the luminal contents in the acidic zone (Fig. 3D), in addition to those functioning in fluid absorption (Fig. 3A). Except for a few bloodsuckers, Cyclorrhaphan adults feed mainly on liquids associated with decaying material (rich in bacteria) in a way similar to house fly adults. That is, they salivate (or regurgitate their crop contents) onto their food. After the dispersed material has been ingested, starch digestion is accomplished primarily in the crop by the action of salivary amylase. Digestion is followed in the midgut, essentially as described for larvae. The stable fly, *Stomoxys calcitrans,* stores and concentrates the blood meal in the anterior midgut and gradually passes it to the posterior midgut, where digestion takes place, resembling what occurs in larvae. These adults lack the characteristic cyclorrhaphan middle midgut and the associated low luminal pH. Stable flies occasionally take nectar.

LEPIDOPTERA Lepidopteran ancestors (Fig. 5E) differ from panorpoid ancestors because they lack midgut ceca, have all their digestive enzymes (except those of initial digestion) immobilized at the midgut cell surface, and present long-necked goblet cells (Fig. 3B) and stalked goblet cells (Fig. 3E) in the anterior and posterior larval midgut regions, respectively. Goblet cells excrete K^+ ions, which are absorbed from leaves ingested by larvae. Goblet cells also seem to assist anterior columnar cells in water absorption and posterior columnar cells in water secretion. Although most lepidopteran larvae have a common pattern of digestion, species that feed on unique diets generally display some adaptations. *Tineola bisselliella* (Tineidae) larvae feed on wool and display a highly reducing midgut for cleaving the disulfide bonds in keratin to facilitate proteolytic hydrolysis of this otherwise insoluble protein. Wax moths *(Galleria mellonella)* infest beehives and digest and absorb wax. The

participation of symbiotic bacteria in this process is controversial. Another adaptation has apparently occurred in lepidopteran adults that feed solely on nectar. Digestion of nectar requires only the action of an α-glucosidase (or a β-fructosidase) to hydrolyze sucrose, the major component present. Nectar-feeding lepidopteran adults have amylase in salivary glands and several glycosidases and peptidases in the midgut. The occurrence of the whole complement of digestive enzymes in nectar-feeding moths may explain, at least on enzymological grounds, the adaptation of some adult Lepidoptera to new feeding habits such as blood and pollen.

ABSORPTION OF WATER AND NUTRIENTS

Overview and Absorption of Lipids

Absorption is the passage of molecules and ions from the gut lumen into the gut cells, thus traversing the cuticle (if present) and the cell plasma membrane. Absorption depends on the permeability of those barriers and on the concentration ratio of a compound in gut lumen and inside gut cells. The permeability of cuticles is variable, whereas that of the plasma membrane is greater for water (a fact not well understood) and for hydrophobic compounds, reflecting the ease with which they solubilize in the lipid bilayers characteristic of cell membranes. Thus, absorption of hydrophilic compounds requires special devices (transporters) to help the molecules find their way through the cell membrane. These transporters are transmembrane proteins that bind the molecule to be transported in a membrane face and, after suitable conformational changes, deliver the molecule from the other face. A uniporter is a transporter that carries a single solute, whereas symporters and antiporters are transporters that carry two solutes into the same and opposite directions, respectively. To transport molecules against a concentration gradient, the process must be energized by coupling with ATP hydrolysis or with cotransport of another molecule down its concentration gradient. Transporter-mediated absorption may be inhibited by molecules resembling those of the transported solute, and its velocity attains a maximum (transporter becomes saturated) at a high solute concentration. This behavior is not observed in the case of simple diffusion, exemplified by lipid absorption.

The study of gut absorption in insects is difficult because of the small size of these animals, which frequently hinders the use of methods developed for studying vertebrates. Absorption sites in insect guts are identified by feeding groups of insects with known dye solutions and then dissecting insects at different periods of time. If the insect is large enough, absorption studies can go further, using gut sections mounted as a sac and measuring the rates at which compounds traverse it under different conditions.

Tracer studies showed that lipid is absorbed more heavily in the anterior than in the posterior midgut of insects, thus following the tissue distribution of fatty acid binding proteins. These proteins are thought to facilitate fatty acid uptake by cells, by decreasing their diffusion back from cells to the gut lumen and by targeting them to specific metabolic pathways. The fatty acids acetate and butyrate are absorbed in significant amounts by the hindgut of insects utilizing cellulose.

Water

Water absorption in the midgut occurs associated with midgut fluid fluxes, but in large amounts it is characteristic of insects feeding on dilute diets, of blood feeders, and also of insects in which salivation is important (e.g., grasshoppers, seed-sucker bugs). Water absorption in the hindgut is part of the water conservation mechanism that is important in all terrestrial insects. Frequently, special cell aggregates (rectal pads) are involved in this process. Water uptake is thought to depend on salt being pumped into spaces enclosed by the basolateral infoldings (Fig. 3A, C) of the absorbing cell. This creates an osmotic pressure that moves water into these restricted spaces. The resulting hydrostatic pressure drives water into the hemolymph, with salts being absorbed from the water on its way out. In insects feeding on dry diets in dry habitats (e.g., *T. molitor, D. maculatus*), hindgut water absorption is improved by the cryptonephridial system, which consists of an association of Malpighian tubules and rectal pads. Less sophisticated forms of this system are found in other insects.

Ions, Amino Acids, and Sugars

In insects, as is usual for all animals, most nutrient absorption occurs in the midgut through symporters, with ions being cotransported down the concentration gradient. Favorable ion gradients are maintained by ion pumps. The most ubiquitous of these pumps is the ATP-driven Na^+, K^+-antiporter (Na^+-K^+-ATPase) localized in the midgut cell basal membrane. Another important ion pump is the H^+, K^+-ATPase found in the goblet cell (Fig. 3B, E) microvillar membranes of lepidopteran larvae. Those pumps maintain cell Na^+ and cell K^+ low in insects with Na^+-rich and K^+-rich diets, respectively. The midgut of lepidopteran larvae contains K^+-dependent symporters for amino acids, consistent with the K^+-rich plant diets of these larvae. Such symporters are relatively insensitive to amino acid sizes and shapes, but have narrow specificities towards charge. Thus, there are usually symporters for neutral, acid, and basic amino acids. Insects with high-Na^+ diets seem to have midgut amino acid symporters dependent on Na^+, as shown in cockroaches. The absorption of glucose was shown in several insects to depend on a difference of concentration between midgut lumen and cells. This transport increases as the concentration of luminal glucose increases (no saturation) and is not inhibited by molecules similar to glucose, at least in the range of concentrations tested. This finding led to the speculation that, in insects, glucose is absorbed by simple diffusion. Nevertheless, this is probably false, since a hydrophylic molecule is not expected to pass through membranes without the help of a transporter.

The primary urine produced in Malpighian tubules contains salts and amino acids, and passes into the hindgut together with food remains. Salts are absorbed in the hindgut by means of special pumps, like those for chloride and calcium. Amino acids, at least in locusts, are absorbed in the hindgut through a Na^+-dependent amino acid symporter.

MIDGUT SECRETORY MECHANISMS

Insects are continuous (e.g., Lepidoptera and Diptera larvae) or discontinuous (e.g., predators and hematophagous insects) feeders. Synthesis and secretion of digestive enzymes in continuous feeders seem to be constitutive; that is, these functions occur continuously, whereas in discontinuous feeders they are regulated. It is widely believed (without clear evidence) that putative endocrine cells (Fig. 3I) occurring in the midgut could, like similar cells in vertebrates, play a role in regulating midgut events. The presence of food in the midgut is necessary to stimulate synthesis and secretion of digestive enzyme. This was clearly shown in mosquitoes.

Mosquitoes express constitutively small amounts of a trypsin called early trypsin. After a blood meal, early trypsins generate free amino acids and small peptides from blood proteins. These compounds are the initial signals that induce the synthesis and secretion of large amounts of late trypsins, which complete protein digestion.

Like all animal proteins, digestive enzymes are synthesized in the rough endoplasmic reticulum, processed in the Golgi complex, and packed into secretory vesicles (Fig. 4). There are several mechanisms by which the contents of the secretory vesicles are freed in the midgut lumen. During exocytic secretion, secretory vesicles fuse with the midgut cell apical membrane, emptying their contents without any loss of cytoplasm (Fig. 4A). In contrast, apocrine secretion involves the loss of at least 10% of the apical cytoplasm following the release of secretory vesicles (Fig. 4B). These have previously undergone fusions originating larger vesicles that after release eventually free their contents by solubilization (Fig. 4B). When the loss of cytoplasm is very small, the secretory mechanism is called microapocrine. Microapocrine secretion consists of releasing budding double-membrane vesicles (Fig. 4C) or, at least in insect midguts, pinched-off vesicles that may contain a single or several secretory vesicles (Fig. 4D). In both apocrine and microapocrine secretion, the secretory vesicle contents are released by membrane fusion and/or by membrane solubilization due to high pH contents or to the presence of detergents.

Secretion by hemipteran midgut cells displays special features because the cells have perimicrovillar membranes, in addition to microvillar ones (Fig. 3H): double-membrane vesicles bud from modified (double-membrane) Golgi structures (Fig. 4E). The double-membrane vesicles move to the cell apex, their outer membranes fuse with the microvillar membrane, and their inner membranes fuse with the perimicrovillar membranes, emptying their contents (Fig. 4E). Because apocrine and microapocrine mechanisms waste membrane and cytoplasm material, these mechanisms are preferred only when they present advantages over the exocytic mechanism. This occurs when a burst of digestive enzymes is needed, as in hematophagous flies after a blood meal, and when secretion occurs in a midgut region responsible for water absorption, a common situation in the anterior midgut of most insects. An exocytic mechanism in a water-absorptive region is not efficient, because the movement of fluid toward the cells would prevent uniform diffusion of the material secreted. Fluid movement has little effect on apocrine and microapocrine secretion because the enzymes are released from budded or pinched-off secretory vesicles far from cells. Since posterior midgut cells usually secrete fluid, no problem arises in the dispersion of material released by exocytosis by these cells. Microapocrine mechanisms seem to be an improvement relative to apocrine mechanisms, because they waste less material. This is consistent with the observation that apocrine mechanisms were found in less evolved grasshoppers and beetles, whereas microaprocine mechanisms were described in the more evolved moths.

See Also the Following Articles
Blood Sucking • Excretion • Feeding Behavior • Symbionts

Further Reading
Chapman, R. F. (1998). "The Insects: Structure and Function." 4th ed. Cambridge University Press, Cambridge, U.K. (See especially Chaps. 2–4).

Cristofoletti, P. T., Ribeiro, A. F., and Terra, W. R. (2001). Apocrine secretion of amylase and exocytosis of trypsin along the midgut of *Tenebrio molitor* larvae. *J. Insect Physiol.* **47**, 143–155.

Daly, H. V., Doyen, J. T., and Purcell III, A. H. (1998). "Introduction to Insect Biology and Diversity." 2nd ed. Oxford University Press, Oxford, U.K. (See especially Chap. 15.)

Dow, J. A. T. (1986). Insect midgut function. *Adv. Insect Physiol.* **19**, 187–328.

Kerkut, G. A., and Gilbert, L. I. (eds.). (1985). "Comprehensive Insect Physiology, Biochemistry and Pharmacology," 13 vols. Pergamon Press, Oxford, U.K. (See especially Vol. 4, Chaps. 4–6.)

Lehane, M. J., and Billingsley, P. F. (1996). "Biology of the Insect Midgut." Chapman & Hall, London.

Silva, C. P., Ribeiro, A. F., Gulbenkian, S., Terra, W. R. (1995). Organization, origin and function of the outer microvillar (perimicrovillar) membranes of *Dysdercus peruvianus* (Hemiptera) midgut cells. *J. Insect Physiol.* **41**, 1093–1103.

Terra, W. R. (1990). Evolution of digestive systems of insects. *Annu. Rev. Entomol.* **35**, 181–200.

Terra, W. R., and Ferreira, C. (1994). Insect digestive enzymes: Properties, compartmentalization and function. *Comp. Biochem. Physiol.* **109B**, 1–62.

Terra, W. R. (2001). The origin and functions of the insect peritrophic membrane and peritrophic gel. *Arch. Insect Biochem. Physiol.* **47**, 47–61.

Vonk, H. J., and Western, J. R. H. (1984). "Comparative Biochemistry and Physiology of Enzymatic Digestion." Academic Press, London.

Wolfersberger, M. G. (2000). Amino acid transport in insects. *Annu. Rev. Entomol.* **45**, 111–120.

Diplura

Robert T. Allen
Author and Consultant, Paris, Arkansas

The Diplura are a group of primitive Arthropods usually included in the class Insecta. Typically they live in the soil, under the bark of decaying trees, under rocks, and under decaying leaf debris. The majority of the species are small, less than 5 mm, but some members of the superfamily Japygoidea, *Atlasjapyx atlas*, may be 60 mm in length. The Diplura are worldwide in distribution, with about 1000 described species assigned to eight families. The name Diplura is derived from the presence of paired caudal appendages.

CLASSIFICATION

All Diplura possess the following defining characters: (1) elongate body, (2) 10th abdominal segment with a pair of caudal cerci or one segmented forcep-like pincers, (3) absence of eyes, (4) entognathous mouthparts, and (5) two pair of spiracles on the thorax (Fig. 1).

There is no doubt that these organisms are primitive arthropods standing near the base of the evolutionary lineage that led to the class Insecta. Whether they should be included in the Insecta or merit a class unto themselves is debatable. Kristensen assigned them to their own class (class and order Diplura) in 1991. This suggestion, although it may be correct, has not been followed in general textbooks of entomology. The classification within the order Diplura has developed gradually over the past 100 years as our knowledge of the group increased. Table I lists the major taxa and their general distribution.

The 1000 or so species that have thus far been described may represent only 50% or less of the actual world fauna. Distribution records of the known taxa are poor, with many species known from a single locality. A great deal of work is left to be accomplished before an accurate idea of the diversity and distribution of the Diplura is known.

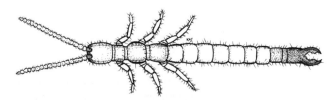

FIGURE 1 Dipluran (*Parajapyx* sp.). (Illustration by K. A. Justus.)

TABLE I   Classification of the Diplura

Class or order	Diplura
Suborder	Rhabdura
Superfamily	Projapygoidea
Family	Anajapygidae (1 genus *Anajapyx*, 4+ spp.; CA, MX)
Family	Projapygidae (2 genera *Projapyx*, 7 spp.; *Symphylurinus*, 19+ spp.; West AF, Brazil, MX, AU, CH, IN)
Superfamily	Campodoidea
Family	Procampodeidae (1 genus, *Procampodea*, 2 spp.; CA)
Family	Campodeidae
Subfamily	Campodeinae (30+ genera, 160–200+ spp.; cosmopolitan)
Subfamily	Lepidocampinae (1 genus; tropical, cosmopolitan)
Subfamily	Hemicampinae (2 genera, 4 spp.; US, MX, SA)
Subfamily	Plusiocampinae (5 genera, 40+ spp.; cosmopolitan)
Subfamily	Syncampinae (1 genus, 1 sp.; China)
Family	Octostigmatidae (1 genus, 1 sp.; South Pacific)
Suborder	Dicellurata
Superfamily	Japygoidea
Family	Parajapygidae (cosmopolitan)
Family	Dinjapygidae (Peru, Bolivia)
Family	Japygidae
Subfamily	Heterojapyginae (AU, NZ, Mad., Pamit, Tibet)
Subfamily	Japyginae (cosmopolitan)
Subfamily	Evalljapyginae (NA, CA)
Subfamily	Provalljapyginae (NA, Brazil)

Note. CA, Central America; MX, Mexico; AF, Africa; AU, Australia; CH, China; IN, India; SA, South America; NZ, New Zealand; Mad., Madagascar; NA, North America; US, United States.

COLLECTING AND SPECIMEN PREPARATION

Because of their small size and obscure living habitats Diplura are not well represented in most collections. They are, however, easily collected in most ecosystems. The collector needs small vials of ethyl alcohol, a size 00 or 000 camel hair brush, a small tool to turn rocks or pry away bark, and a keen eye. When a rock is turned over, the specimens may be on the underside of the rock or on the soil. They usually begin to move rapidly once exposed and it is necessary to act quickly. The brush is wetted in the vial of alcohol and then touched to the specimen. Usually the specimen will adhere to the brush and can then be transferred to the alcohol vial. Some of the larger Japygidae may be collected using the fingers or a pair of forceps. Once specimens are collected they must be mounted on microscope slides for study and identification.

There is no single source that allows ready identification of Diplura. The sources listed under Further Reading will assist in keying specimens to families and genera and provide a guide to the numerous papers necessary for species identification.

BIOLOGY

Relatively few studies have been done on the Diplura, and thus we know very little about their habits. However, the studies that have been published have recorded the basic biological characteristics of the group. The males deposit sperm bundles in the soil and females pick up these sperm bundles and become fertilized. Eggs may be deposited randomly and in clusters. Some japygids suspend eggs at the end of a filimentous stalk. The prelarvae hatch in 7 to 16 days depending on the species. The prelarva does not feed and moves very little. The prelarva molts in about 2 days. The newly molted immature is fully mobile and feeds readily on whatever food source is available. After the second molt the immature form possesses the major setae and other anatomical characters used for identification. During the fourth or fifth molt the individual becomes sexually mature as evidenced by the appearance of the sex organs along the posterior margin of sternum VIII. Diplura continue to molt throughout their lives, adding clothing setae on the various sclerites and regenerating damaged body appendages.

Both major groups of Diplura appear to be omnivores. Many species are predators as well as scavengers. Foods that have been recorded include other Diplura, mites, Collembola, Symphyla, Isopoda, fly and beetle larvae, small arthropods of any class, enchytraeid worms, fungal spores, and mycelia. Some species have been observed feeding on the roots of living plants, including peanuts, sugarcane, and melons.

See Also the Following Articles
Arthropoda and Related Groups • Protura

Further Reading
Allen, R. T. (1995). Key to the species of Campodea (Campodea) from eastern North America and description of a new species from Virginia (Diplura: Campodeidae). *Ann. Entomol. Soc. Am.* **88**, 255–262.

Ferguson, L. M. (1990). Insecta: Diplura. *In* "Soil Biology Guide" (D. L. Dindall, ed.), pp. 951–963. Wiley, New York.

Kristensen, N. P. (1991). Phylogeny of extant hexapods. *In* "The Insects of Australia" (CSIRO, ed.), 2nd ed., Vol. I, pp. 125–140. Melbourne University Press, Carlton.

Paclt, J. (1957). Diplura. *In* "Genera Insectorum" (P. Wytsman, ed.), pp. 1–123. Crainhem, Belgium.

Diptera
(Flies, Mosquitoes, Midges, Gnats)

Richard W. Merritt
Michigan State University

Gregory W. Courtney
Iowa State University

Joe B. Keiper
The Cleveland Museum of Natural History

The Diptera, commonly called true flies or two-winged flies, are a group of familiar insects that includes mosquitoes, black flies, midges, fruit flies, and house flies. The Diptera are among the most diverse insect orders, with approximately 124,000 described species. These insects are diverse not only in species richness but also in their structural variety, ecological habits, and economic importance. The group is ubiquitous and cosmopolitan, having successfully colonized nearly every habitat and all continents, including Antarctica. Although brachyptery (wings reduced) or aptery (wings absent) are known in some Diptera (e.g., some Mycetophilidae, Tipulidae, Phoridae, and Hippoboscidae), adults usually are winged and active fliers. Depending on the group, adults can be nonfeeding or feeding, with the latter including diets of blood, nectar, and other liquefied organic materials.

Larval Diptera are legless and found in a variety of terrestrial and aquatic habitats. Most larvae are free-living and crawl or swim actively in water (e.g., Simuliidae, Culicidae, Chironomidae, Ptychopteridae, Blephariceridae), sediments (e.g., Tipulidae, Psychodidae, Ceratopongonidae, Tabanidae), wood (e.g., Tipulidae, Mycetophlidae), fruit (e.g., Drosophilidae, Tephritidae), or decaying organic material (e.g., Muscidae, Ephydridae, Sphaeroceridae, Sarcophagidae). Other larvae inhabit the tissues of living organisms (e.g., Oestridae, Tachinidae).

As expected for a ubiquitous group with diverse habits and habitats, the Diptera are of considerable economic importance. Pestiferous groups can have significant impacts in agriculture (e.g., Agromyzidae, Tephritidae), forestry (e.g., Cecidomyiidae), animal health (e.g., Oestridae), and human health (e.g., Culicidae, Simuliidae, Psychodidae). Other groups can be a general nuisance if present in high numbers (e.g., Muscidae, Ceratopogonidae) or because of allergic reactions to detached body hairs (e.g., Chironomidae). Despite these negative impacts, flies can play a valuable role as scavengers (e.g., Mycetophilidae, Muscidae, Calliphoridae), parasitoids and predators of other insects (e.g., Tachinidae, Empididae, Asilidae), pollinators (e.g., Syrphidae, Stratiomyiidae, Bombyliidae), food for vertebrates (e.g., Chironomidae, Tipulidae), bioindicators of water quality (e.g., Chironomidae, Blephariceridae), and tools for scientific research (e.g., Drosophilidae).

MORPHOLOGY

Because of the structural variety in Diptera, especially among larvae, it is difficult to generalize about morphology. Despite this variety, flies share a number of features. Except for certain forms (e.g., cave-dwelling species), adult flies usually possess large compound eyes. In some species, eyes meet or almost meet dorsally (holoptic); in other groups, eyes are widely separated (dichoptic). Further modifications include eyes that are divided into distinct dorsal and ventral components, a feature found in many Simuliidae, Blephariceridae, and other groups. These modifications are among many that might be related to swarming behavior. The

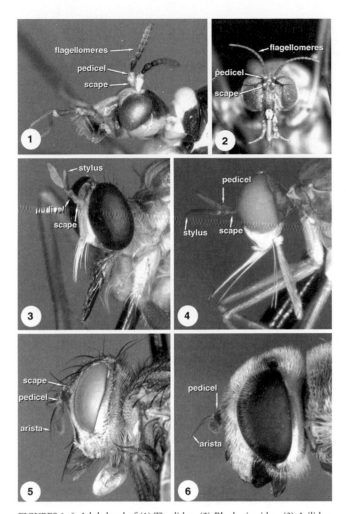

FIGURES 1–6 Adult head of (1) Tipulidae, (2) Blephariceridae, (3) Asilidae, (4) Empididae, (5) Tachinidae, (6) Syrphidae. (Photographs by G. Courtney.)

regions of a fly head include the vertex, a dorsomedial area above and posterior to the eyes; the frons, an area extending from the vertex to the antennal insertions; and the face, which extends from the antennal insertions to the clypeus, a region intimately associated with the mouthparts. All of these areas can bear a variety of setae, the number and position of which often are useful in identification.

Nearly all flies have well-developed antennae, with the flagellum being the most varied component. In nematocerous families, the antennae are usually composed of many segments and are filiform, plumose, or pectinate (Figs. 1–2), whereas brachycerous flies typically have the first flagellomere enlarged and the remaining flagellomeres stylate or aristate (Figs. 3–6). The mouthparts of adult flies also vary between groups, ranging from vestigial forms (e.g., Deuterophlebidae, Oestridae) to those that are well developed. The latter include two general types: (1) piercing and sucking, as seen in simuliids, culicids, and asilids, and (2) lapping and sucking, as seen in tipulids and most brachycerous groups. Typically, the proboscis comprises the unpaired labrum–epipharynx, labium, and hypopharynx and the paired mandibles and maxillae. In most groups, the base of each maxilla bears a distinct palpus and the apex of the labium is modified into a labellum, which consists of membranous lobes derived from the labial palpi.

Perhaps the most distinct feature of the adult fly is the single pair of wings (hence, the ordinal name, Diptera, meaning "two wings"). A related characteristic is the highly modified thorax, with a reduced prothorax and metathorax, and a greatly enlarged mesothorax. The latter includes several prominent dorsal and lateral sclerites and, internally, houses much of the wing musculature. Wing venation varies greatly throughout the Diptera and can be extremely important for identification. The metathoracic wings are modified into distinct club-shaped halteres, which are thought to play an important role as balancing organs. Interestingly, halteres are distinct in some groups that are otherwise wingless (e.g., Hippoboscidae). The legs of an adult fly are typical of most insects, each with a coxa, trochanter, femur, tibia, and, in nearly all groups, a tarsus comprising five tarsomeres. Beyond this basic arrangement, there is considerable diversity of leg structure in Diptera, with this diversity often providing useful taxonomic information.

The adult abdomen also shows considerable variety. In basic structure, the abdomen consists of 11 segments, the last 2 or 3 of which are highly modified for reproduction. Most abdominal segments consist of a dorsal and ventral sclerite, connected laterally by a pleural membrane of varying width. There is a general trend toward a shortening of the abdomen in Diptera (cf. Tipulidae and Muscidae). The terminalia of Diptera are complex, highly variable, and of considerable use in taxonomic and phylogenetic studies. Details of terminalic structure are beyond the scope of this article; however, the structural variety of Diptera terminalia and the controversy about interpreting their homologies can be found in some of the general references listed at the end.

The dipteran pupa also varies considerably in form. Some fly pupae look like a cross between the worm-like larva and the adult, whereas others are relatively featureless and seed-like in appearance. The former are typical of the Nematocera and are described as obtect, or having the appendages fused to the body (Figs. 7–10). For instance, a crane fly (Tipulidae) pupa has identifiable head, thoracic, and abdominal segments, but the antennal sheaths, legs, and wing pads adhere to the pupal body (Fig. 9). Nematocerous pupae are frequently leathery to the touch. The exterior of the nematoceran pupa may be adorned with spines, gill-like respiratory devices, or locomotory paddles (Figs. 7–10). The Brachycera and Cyclorrhapha form the pupal stage in a different, more concealed manner. Families of the so-called higher Diptera form pupae that are described as coarctate, which literally means "compacted" or "contracted" (Figs. 11–15). These taxa (e.g., Syrphidae, Drosophilidae, Muscidae) form a puparium that is composed of the hardened skin of the last larval instar (Fig. 14). This relatively tough, desiccation-resistant structure houses and protects the pupa; the adult

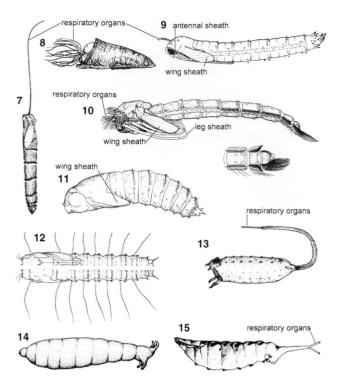

FIGURES 7–15 Pupa of (7) Ptychopteridae, (8) Simuliidae, (9) Tipulidae, (10) Chironomidae, showing anal division below, (11) Tabanidae, (12) Empididae, (13) Syrphidae, (14) Muscidae, (15) Ephydridae. (All illustrations modified, with permission, from Merritt and Cummins, 1996.)

also forms within the puparium. The enclosed adult must break through the puparial skin and does so by extruding a balloon-like structure from the frons called the ptilinum. The ptilinum is used to break the cephalic cap, a lid-like structure positioned anteriorly on the puparium, thus liberating the teneral (or newly emerged) adult. Very few external features are noticeable on the puparium, although careful examination will reveal the spiracles through which atmospheric air is obtained by the pupa.

Diptera larvae can be distinguished from the larvae of most other insects by the lack of jointed thoracic legs. In other features, larval dipterans show tremendous structural variety. This variation is exemplified by cranial structure. Larvae of most nematocerous flies are eucephalic, i.e., characterized by a complete, fully exposed, and heavily sclerotized head capsule (Figs. 17–19 and 24). Larval tipulids are special among nematocerous flies, as the head capsule often is fully retracted into the thorax (Fig. 16) and the posterior cranial margin may possess small to extensive longitudinal incisions (Fig. 23). In contrast to the condition in nematoceran larvae, the cranial sclerites of brachyceran larvae are greatly reduced or absent. The hemicephalic head capsule of many orthorrhaphous Brachycera consists of slender arms and rods that are partly retracted into the thorax (Figs. 25–26). The culmination of cranial reduction is in the acephalic head of larval Cyclorrhapha, in which the external portions of the head are membranous, and much of the head is retracted into the

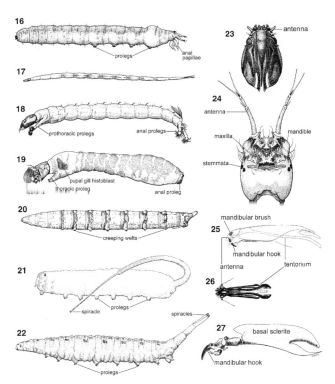

FIGURES 16–27 Larva of (16) Tipulidae, (17) Ceratopogonidae, (18) Chironomidae, (19) Simuliidae, (20) Tabanidae, (21) Syrphidae, (22) Ephydridae. Larval head capsule of (23) Tipulidae, (24) Chironomidae. Cranial sclerites and mouth parts of (25) Tabanidae, (26) Dolichopodidae. (27) Cephalopharyngeal skeleton of Sciomyzidae. (All illustrations modified, with permission, from Merritt and Cummins, 1996.)

thorax (Fig. 27). The internal portion, or cephalopharyngeal skeleton, is thought to comprise the remnants of internal cranial sclerites (tentorium) and various mouthparts. Although referred to as "acephalic," the primary difference between the head of a cyclorrhaphan larva and that of a nematoceran larva is that most of the constituent segments are withdrawn into the thorax and thus externally hidden (Fig. 22). Cranial modifications are accompanied by general changes in the shape and rotation of mandibles and other mouthparts. The mandible of larval nematocerans typically consists of a stout, toothed structure that moves in a horizontal or oblique plane and operates as a biting and chewing organ. The brachyceran larval mandible usually is more claw-like, has fewer teeth along the inner surface, moves in a vertical plane, and operates as a piercing or slashing organ.

In most Diptera larvae, the thorax and abdomen are soft, flexible, and only occasionally provided with sclerotized plates. The thorax usually consists of three distinct segments and the abdomen usually eight or nine segments (Figs. 17–19). Body form varies almost as much as does cranial diversity and ecological habits. In many nematoceran groups (e.g., most Chironomidae, Tipulidae, and Simuliidae), the body is subcylindrical (Figs. 16, 18, and 19). Other groups are predominantly fusiform (e.g., Cecidomyiidae) or elongated and serpentine (e.g., Ceratopogonidae) (Fig. 17). The latter body form is common in groups inhabiting soil and interstitial

aquatic habitats. The larvae of some groups (e.g., Culicidae) are unusual in that the thoracic segments are indistinctly differentiated and form a single large segment that is wider than the rest of the body (Fig. 48). The typical body shape of a cyclorrhaphan larva is that of a maggot (i.e., pointed at the anterior end, with the thoracic segments approaching the maximum body diameter). The variation in body form is particularly impressive in families whose larvae feed on a variety of substrates (e.g., Syrphidae). Cyclorrhaphan larvae can be dorsoventrally flattened, a feature often associated with the presence of segmental or branched body protuberances. The syrphid genus *Microdon* has one of the most unusual larvae, being ventrally flattened, dorsally dome-shaped, and sluglike in overall appearance. Larvae with parasitoid and parasitic life styles (e.g., Pipunculidae, Oestridae) are often extremely stout or pear-shaped, their body form being closely adapted to that of the host.

Despite the absence of jointed thoracic legs, locomotion is highly diverse in fly larvae, reflecting the group's diversity in habitat and habits. Locomotory appendages operate through a combination of turgor pressure and muscle action and include creeping welts, prolegs, and other specialized structures (e.g., suctorial discs). Creeping welts are transverse, swollen areas (ridges) that bear one to several modified setae or spines; creeping welts are characteristic of several groups, including many crane flies, dance flies, and deer and horse flies (Fig. 20). Among orthorrhaphous groups, ventral creeping welts are common in the larvae of Rhagionidae and Empididae. Cyclorrhaphan larvae typically use creeping welts as anchoring devices, with welts usually comprising bands of small spines on abdominal segments. The distribution and morphology of creeping welts vary considerably between families, species, instars, and segments. Prolegs usually are paired, round, elongate, fleshy, retractile processes that bear apical spines or crochets; prolegs come in a diversity of shapes, sizes, and positions and are typical of Chironomidae, Deuterophlebiidae, Simuliidae, Rhagionidae, and various members of other groups (Figs. 16, 18, 19, 21, and 22). Other specialized structures used for locomotion or attachment include friction pads and suctorial discs. Several genera of Psychodidae possess friction pads, which are areas of modified cuticle on the ventral surface of the thorax or abdomen. Functionally similar structures may occur in certain Ephydridae, particularly in groups inhabiting waterfalls and thin films of flowing water. Suctorial disks are true suction devices on the ventral body surface of larval net-winged midges and are an obvious adaptation to life in torrential streams.

Larval Diptera show a variety of respiratory adaptations, many a reflection of life in fluid or semifluid habitats. The basic respiratory system comprises an internal system of tracheae and the external spiracles. Respiration may be directly from the atmosphere, from plant tissues, or from oxygenated fluids. The presence of hemoglobin in the blood of some midges can assist the absorption of oxygen. Many aquatic larvae, particularly those from well-oxygenated streams, are apneustic (lack spiracles) and absorb oxygen directly through the skin. Some families (e.g., Psychodidae) possess spiracles on the prothorax and last abdominal segment, whereas others (e.g., Culicidae and most cyclorrhaphans) have spiracles on only the last segment. In several groups (e.g., many Ephydridae and Syrphidae), the spiracles are at the end of a retractile respiratory siphon (Figs. 21 and 22).

PHYLOGENY AND CLASSIFICATION

Traditionally the Diptera have been divided into two or three suborders: Nematocera ("lower" Diptera) and Brachycera ("higher" Diptera), with the latter sometimes divided further into the Orthorrhapha and Cyclorrhapha. Although there is general agreement that the Diptera, Brachycera, Cyclorrhapha, and a few other subordinate groups are monophyletic, there is comparably general agreement that the Nematocera is a paraphyletic or grade-level grouping. No synapomorphies (shared, derived characters) unite the Nematocera, and the Brachycera are thought to have originated from some subgroup within the Nematocera. Despite this, it is useful to mention some of the primitive features shared by most nematocerans. The name itself ("Nematocera") refers to the fact that adults of these flies typically have long, multisegmented antennae. Furthermore, adult nematocerans generally are slender, delicate, long-legged flies (e.g., Tipulidae and Culicidae); however, the group also includes some rather stout-bodied flies (e.g., Simuliidae and Ceratopogonidae). Larval nematocerans typically have a well-developed, sclerotized head capsule, and their mandibles usually rotate at a horizontal or oblique angle. Brachycera are characterized by the short, three-segmented antennae, the last segment of which is usually either stylate or aristate. Brachyceran larvae usually have a hemicephalic or acephalic head capsule, consisting mostly of slender, sclerotized rods that are partly or largely retracted into the thorax. Within the Brachycera, there are additional differences between orthorrhaphous and cyclorrhaphous groups. The former group, which includes Rhagionidae, Tabanidae, Stratiomyiidae, and a few other families, is similar to nematocerous Diptera in that it is considered a paraphyletic group. Finally, within the Cyclorrhapha are two major subgroups, the presumed paraphyletic Aschiza (includes Phoridae and Syrphidae) and the monophyletic Schizophora (includes the majority of Brachycera, such as Tephritidae, Drosophilidae, Ephydridae, Agromyzidae, Muscidae, and Tachinidae).

ECOLOGY

Life History

As a holometabolous insect, or one that undergoes complete metamorphosis, the dipteran life cycle includes a series of distinct stages or instars. A typical life cycle consists of a brief egg stage (usually a few days or weeks, but sometimes much longer), three or four instars (typically three in Brachycera, four in nematocerous flies, and more in simuliids, tabanids, and a few others), a pupal stage of varying length, and an adult stage that lasts from less than 2 h (Deuterophlebidae) to several weeks or even months (some female Culicidae). The eggs of aquatic flies are usually laid singly, in small clusters, or in loose or compact masses in or near the water and attached to rocks or vegetation. In Deuterophlebiidae and certain members of some other groups, the female crawls beneath the water to select oviposition sites, a behavior that ensures eggs are placed in a suitable larval habitat. The latter also is typical of many terrestrial flies, such as calliphorids, which will lay their eggs near the body openings (eyes, nose, mouth, anus) of carcasses. Some tephritid fruit flies use a rigid ovipositor to pierce plant tissue. Oviposition in parasitic flies can be complex and may involve placement of eggs in or on the host or in areas frequented by the host. Some parasitoids (e.g., some tachinid flies) produce eggs that are ingested by a feeding host, then larvae hatch inside the host and penetrate the gut wall. Furthermore, some parasitic groups will oviposit on a blood-feeding arthropod (e.g., tick or another fly), with the heat of the next host stimulating hatching.

All instars occur in the same habitat in most taxa. Exceptions include flies that demonstrate hypermetamorphosis, which is characterized by an active, slender first instar (planidium) and grublike, endoparasitic later instars. Acroceridae, Nemestrinidae, and Bombyliidae are among the better known groups with hypermetamorphic representatives. In general, the duration of the first larval stage is shortest, whereas that of the last instar is much greater, often several weeks or even months.

Habitat

The diversity of Diptera habitats is partly a reflection of the different ecological roles of larvae and adults, with larvae generally adapted for feeding and growth and adults for reproduction and dispersal. Whereas fly larvae occur in both terrestrial and aquatic habitats, virtually all adults are terrestrial and capable of flight. Wingless and, therefore, flightless groups include certain tipulids, marine chironomids, and phorids, as well as ectoparasitic adults of Hippoboscidae and Nycterobiidae. Adult flies are arguably one of the most aerial of organisms. Swarms of flies, which usually consist primarily of males, are a common sight in many areas. These aggregations, often for the purpose of enhancing male visibility to prospective female mates, may be seen along roadsides, over certain trees or bushes, above sunlit pools along streams, at the summits of hills, in sunny gaps of forest canopies, or at any number of other swarm markers. Swarming is probably a primitive feature of Diptera, which might explain the prevalence of this behavior in nematocerous groups. These Diptera and other flies share a number of structural features that might be adapted for swarming, including enlarged compound eyes and wings with well-developed anal lobes. These features and others are thought to assist flies in both maneuvering in flight and perceiving conspecific individuals in swarms. Swarming and related behaviors are especially developed in Bibionidae and Empididae. Males of the latter group are known for their predaceous habits and the elaborate behaviors and "nuptial gifts" for prospective female mates. Other groups (e.g., Bombyliidae, Syrphidae) are among the most agile flying insects, being particularly adept at hovering.

Diptera larvae have colonized a variety of terrestrial and aquatic habitats, including water (e.g., Simuliidae, Culicidae, Chironomidae), soil and damp sediments (e.g., Tipulidae, Ceratopogonidae, Tabanidae), rotting wood (e.g., Tipulidae, Mycetophilidae,), fruit (e.g., Tephritidae), decaying organic material (e.g., Muscidae, Sarcophagidae), and the tissues of living organisms (e.g., Sciomyzidae, Oestridae, Tachinidae). Despite this diversity of habitats, most larvae are in a broad sense aquatic. Even "terrestrial" groups from decomposing vegetation, carcasses, leaf litter, rotting wood, or soil often live in a rather aqueous environment. This requirement for a damp environment partly reflects that the larval cuticle is usually thin, soft, and susceptible to drying. Truly aquatic larvae occur in coastal marine, saline, and estuarine waters, shallow and deep lakes, ponds, cold and hot springs, plant cavities (phytotelmata), artificial containers, slow to torrential streams, groundwater zones, and even natural seeps of crude petroleum! Aquatic habits are most prevalent in larvae of nematocerous flies, including all or most Culicidae, Simuliidae, and Chironomidae. Among brachycerous flies, aquatic habits are most common in ephydrids, sciomyzids, and tabanids. In some groups, such as muscoid flies, only a few species are aquatic.

Trophic Relationships

Their trophic diversity and numerical abundance make the Diptera an important component in many ecosystems, both as primary consumers and as a food resource for other organisms. Trophic diversity is reflected in the wide range of larval feeding habits, which encompass nearly every category. In some groups (e.g., asilids, most empidids), larvae and adults belong to the same trophic category; in other groups (e.g., simuliids, tachinids) these life stages usually adopt different feeding strategies; in still others, feeding can be restricted to only the larvae or adults (e.g., chironomids, hippoboscids, and nycteribiids). The latter comprise primarily the so-called Pupipara, in which the females are hematophagous and do not lay eggs and instead give birth to fully formed larvae (i.e.,viviparous development). In addition to the above-mentioned variety of feeding habits, some groups may feed on multiple food resources during the same life stage (e.g., larvae that can be both saprophagous and predaceous and adults that are both nectarivorous and hematophagous). Larval sciomyzids may feed on dead or living mollusks, and some ephydrid larvae may consume algal, bacterial, or detrital resources during the same instar.

Saprophagous habits are among the most prevalent in Diptera, especially in brachycerous groups. Many fly larvae feed on decaying organic material or organic detritus, in which the resident bacteria and other microorganisms are the primary source of nutrition. Among the more common sources of these materials are animal carcasses, which are frequently colonized by callphorids, muscids, phorids, sphaerocerids, and others. The sequence of colonization is often quite predictable, which contributes to the use of Diptera in forensic studies. Decaying fruit and vegetable material also is colonized by many groups, including especially otitid, sphaerocerid, and muscid flies. Decomposing plant fragments can be an important food resource in aquatic habitats, where it is consumed by the larvae of tipulids, ephydrids, otitids, and other groups. These groups and others (e.g., Psychodidae, Syrphidae, Stratiomyiidae) also contain many species that feed on decaying, fine organic matter and associated microorganisms. Most Culicidae and Simuliidae consume fine particulate organic matter of varying size and quality, but use modified mouth-brushes or labral fans to extract particles from water. In most other saprophagous groups, including aquatic species, a sieve-like pharyngeal filter is used to concentrate microorganisms and other organic particles, whereas those feeding on carrion have well-developed mouthhooks for shredding and macerating raw meat.

Phytophagous groups, which consume live plants (including algae and fungi), are well represented by the larvae of bibionids, cecidomyiids, mycetophilids, tipulids, phorids, tephritids, and agromyzids. Many of these flies can be serious agricultural pests. Aquatic habitats contain numerous flies that consume the thin films of algae and organic matter that occur on rocks and other substrata. Among the more obvious of these aquatic grazers are blepharicerids and certain species of psychodids, simuliids, and ephydrids.

Most predaceous Diptera attack other invertebrates as their primary food. Many families (e.g., Chironomidae, Culicidae, Tipulidae, and Ephydridae) contain a few predaceous species, whereas other groups (e.g., Ceratopogonidae and nearly all noncyclorrhaphan Brachycera) feed primarily or exclusively on invertebrates. Vertebrate prey (frogs and salamanders) can be part of the diet of larval Tabanidae. Whereas predaceous larvae typically kill multiple hosts, parasitic and parasitoid larvae generally attack only one host. Parasitoids typically will kill that host, often after a long association with it. Twenty-two families of Diptera include parasitoid members, with tachinid flies perhaps the best known of these. Dipterans are parasitoids of other invertebrates, mostly other arthropods. Because other insects (some pests) often are attacked, parasitoids often are useful for biological control. The Diptera also includes several true parasites, which attack but do not kill the host, such as oestrids and various other groups that often exhibit distinct and complex migrations in vertebrate hosts.

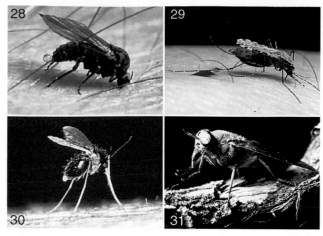

FIGURES 28–31 (28) Female black fly adult (Simuliidae) taking a blood meal. (29) Female mosquito adult (Culicidae: *Anopheles*) taking a blood meal. (Photographs by R. W. Merritt.) (30) Female sand fly adult (Psychodidae) taking a blood meal. (Photograph by B. Chaniotis.) (31) Female horse fly adult (Tabanidae). (Photograph by R. W. Merritt.)

ECONOMIC IMPORTANCE

Injurious Families

Several families of Diptera are of major economic importance and involved in the transmission of more disease pathogens to humans and other animals than any other group of arthropods. Biting flies cause annoyance that impacts tourism, recreation, land development, and industrial and agricultural production, whereas their effects on livestock can cause reduced milk, egg, and meat production.

The adults have mouthparts that have very effective piercing stylets, enabling these flies to "bite" and suck blood. Some major families with this characteristic include members of Simuliidae (Fig. 28), Culicidae (Fig. 29), Psychodidae (Fig. 30), Ceratopogonidae, Tabanidae (Fig. 31), and the blood-sucking Muscidae (Figs. 32 and 33). The bites from these groups can often cause severe allergic reactions, resulting in intense itching, rashes, and local swelling or, in some instances, hospitalization as a consequence of toxemia or anaphylactic shock.

Some of the major human and other animal diseases resulting from the transmission of causative organisms by Diptera include human onchocerciasis (river blindness) by Simuliidae; leishmaniasis (sand fly fever) by phlebotomine sand flies belonging to the family Psychodidae; several protozoan and viral diseases of domestic and wild animals, poultry, and waterfowl by Simuliidae and Ceratopogonidae; malaria, yellow fever, filariasis, dengue, dog heartworm, the encephalitides, and related viral diseases by Culicidae; and tularemia and animal trypanosomiases by Tabanidae. Several other species belong to the blood-sucking muscoid flies and include the tsetse fly of Africa, responsible for transmitting the pathogen causing human sleeping sickness, and the stable fly (Muscidae) (Fig. 32), whose vicious bites can annoy humans in recreational areas, and bother domestic animals such as

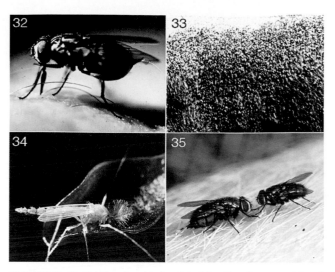

FIGURES 32–35 (32) Female stable fly adult (Muscidae: *Stomoxys calcitrans*) taking a blood meal. (Photograph by E. Hansens.) (33) Horn flies (Muscidae: *Haematobia irritans*) resting and feeding on the back of a bull. (Photograph by R. W. Merritt.) (34) Adult male midge (Chironomidae). (Photograph by R. F. Harwood.) (35) Adult blow flies (Calliphoridae: *Phaenicia sericata*) on a pig. (Photograph by M. J. Higgins.)

horses, cattle, and sheep. The horn fly (Muscidae) (Fig. 33) is a well-established biting cattle pest throughout many tropical and temperate areas of the northern hemisphere, whereas its close muscoid relative, the buffalo fly, is particularly important to cattle and dairy industries of Australia.

In addition to the biting habits and disease agent transmission of the above groups, flies can cause annoyance and interference with human comfort. Members of the genus *Hippelates* in the family Chloropidae are referred to as "eye gnats" because they frequently are attracted to the eyes of the victim, feed on secretions, and may assist in the entrance for pathogenic organisms. A muscoid fly having similar habits, known as the "face fly," has been associated with the transmission of "pink eye" to cattle. Several other species of muscoid flies (e.g., house fly, bush fly, latrine fly) generally breed in excrement and at times can be economically important pests of humans and/or domestic animals. Two families of Diptera that can cause annoyance and constitute a nuisance by their sheer numbers emerging from ponds and lakes are the Chironomidae (nonbiting midges) (Fig. 34) and the Chaoboridae (chaoborid gnats). These are commonly mistaken for mosquitoes (Culicidae), but do not bite. When one encounters swarms of these midges or gnats, it is difficult to keep them out of one's eyes or avoid inhaling them.

The dipteran families Calliphoridae (blow flies) (Fig. 35) and Sarcophagidae (flesh flies) (Fig. 36) are the major producers of myiasis, i.e., the infestation of organs and tissues of humans or other animals by fly maggots. The larvae of these groups feed on necrotic tissue and may accidently be ingested or invade wounds of humans and domestic animals, causing severe discomfort and subsequent secondary infections. The primary and secondary screwworm flies (Calliphoridae)

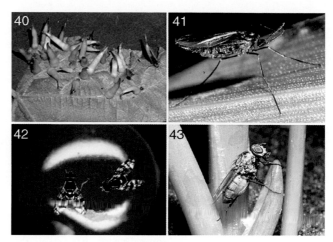

FIGURES 36–39 (36) Adult flesh fly (Sarcophagidae). (Photograph by R. W. Merritt.) (37) Secondary screwworms (Calliphoridae: *Cochlimyia macelleria*) on a pig. (Photograph by M. J. Higgins.) (38) Horse bot fly larvae (Oestridae: *Gasterophilus intestinalis*) attached to the stomach of a horse. (Photograph by R. W. Merritt.) (39) Human bot fly larvae (Oestridae: *Dermatobia hominis*) under the hide of an ox in Costa Rica. (Photograph by L. Green.)

FIGURES 40–43 (40) Cecidomyiid gall on grape leaves. (Photograph by R. Isaacs.) (41) Hessian fly (Cecidomyiidae: *Mayetiola destructor*). (42) Cherry fruit fly adults (Tephritidae: *Rhagoletis cingulata*) on cherry. (Photographs by Department of Entomology, Michigan State University.) (43) Onion maggot adult (Anthomyiidae: *Delia antiqua*). (Photograph by J. Spencer.)

(Fig. 37) are attracted to the wounds and sores of animals, and the former was one of the most serious pests of livestock in the United States until it was eradicated through the sterile male release program. In recent times, the identification and aging of the larvae of some species of blow (Calliphoridae) and flesh flies (Sarcophagidae) have proved useful in establishing the time of death in forensic investigations.

One other family, the Oestridae (cattle, sheep, horse, human, and rodent bot flies), is involved in enteric myiasis of animals and sometimes humans. Damage caused by horse bots (*Gasterophilus* spp.) (Fig. 38) varies from violent reactions by horses due to the flies ovipositing, to irritation by larvae when burrowing into the oral tissue and susequent interference with digestion. The larvae of cattle grubs (*Hypoderma* spp.) migrate through the host's body and eventually reach the upper back where they cut a small opening in the hide and remain there for some time. Economic losses in cattle result from reduction in milk production, weight loss, and damage to hides. Another species of bot fly, the human bot or torsalo *(Dermatobia hominis),* is common in parts of Mexico and Central and South America. It parasitizes a wide range of hosts, including humans, but is a more serious pest of cattle and oxen in these areas (Fig. 39).

Several families of Diptera are economically important to agriculture. The Cecidomyiidae or gall gnats "sting" the plant and make it grow a "gall home" for them (Fig. 40), within which they find not only shelter but also adequate and abundant food. Examples are the goldenrod ball gall and the pine cone gall. Some very destructive species in this family, such as the Hessian fly (Fig. 41), chrysanthemum gall midge, and wheat, pear, and cloverseed midge, feed on cultivated crops and do not always form galls. The Tephritidae, or fruit flies, contain some species whose larvae bore into the stems of plants; some produce galls, others are leaf miners, and most important of all are those that bore into the flesh of fruits and vegetables. The latter include some of the most important of all economic insects, specifically the apple maggot, cherry fruit flies (Fig. 42), walnut husk fly, and Mexican, Mediterranean, oriental, olive, and melon fruit flies. The Anthomyiidae, or root maggot flies, have larvae that feed on decaying vegetable matter from which a number have adopted the habit of attacking the roots of vegetables. These include the cabbage maggot, onion maggot (Fig. 43), seed corn maggot, and spinach leafminer. Larvae of the family Agromyzidae are known as leafminers and feed between the leaf surfaces, leaving light-colored, narrow, winding mines or large blotches that decrease photosynthesis and make produce unsalable. The leaves are weakened and the mines promote disease and decay.

Beneficial Families

The Diptera contain several families that can be considered beneficial to humans and their environment. First, and most important, is the role of all Diptera in food chains in nature. Groups such as Culicidae, Chironomidae, and Simuliidae occur in large numbers as larvae and adults and provide a major prey base for many other invertebrates as well as vertebrates such as fish, birds, bats, and amphibians. In turn, several families contain predators and parasitoids as larvae and adults, including the Asilidae, Empididae, Dolichopodidae, Syrphidae, and Tachinidae. Many families are important decomposers and recyclers of decaying organic matter of different types. Examples include the Psychodidae, Tipulidae, Stratiomyiidae, Mycetophilidae, Sciaridae, Sepsidae, Coleopidae, Muscidae, Calliphoridae, Sarcophagidae, Phoridae, Syrphidae, and Sphaeroceridae. Some Diptera are important pollinators of flowers and include some species of Syrphidae, Bombyliidae, and even adult male Culicidae who visit flowers to imbibe nectar.

Some families of aquatic Diptera have been important in water quality and bioassessment studies to classify the degree of pollution in a water body. For example, larvae belonging to the midge genus *Chironomus* in the family Chironomidae have been referred to as blood worms because of the hemoglobin in their blood. These and another group known as the "rat-tailed maggots" (Syrphidae: *Eristalis*) are often used as indicators of polluted water or water low in oxygen. The presence of Simulidae in a stream generally indicates clean well-aerated water. The Culicidae and Chironomidae have members that are associated with both polluted and clean water habitats. Finally, some Diptera have been the subject of study for scientists throughout the world. For example, chironomid midges are used in acute and chronic laboratory toxicity studies to compare toxicants and the factors affecting toxicity and to ultimately predict the environmental effects of the toxicant. The small fruit fly, *Drosophila* (Drosophilidae) (Fig. 44), has been the organism of choice in most genetic studies for years and has contributed significantly to studies ranging from neurobiology to evolutionary theory. Overall, the Diptera represent an order containing a variety of species that are economically very beneficial and equally injurious to humans.

BIOLOGY OF SELECTED FAMILIES

Suborder Nematocera

TIPULIDAE   Crane flies (Fig. 45) are a diverse group of 14,000 species that inhabit a variety of freshwater and terrestrial habitats. Larvae are significant shredders *(Tipula, Pedicia)* of leaves that enter streams and are predators *(Hexatoma, Dicranota)* in aquatic habitats. The moist transition zone between aquatic and terrestrial areas supports a distinct assemblage of species (e.g., *Erioptera, Ormosia*). Terrestrial habitats are home to species that feed on coniferous *(Limonia)* and deciduous *(Epiphragma)* rotting logs or decaying organic material *(Tipula)* and that may even be pestiferous consumers of sod *(Tipula)*. A few species can tolerate high salinity and inhabit the rocky intertidal zones of marine habitats. The adults generally do not feed, although they are frequently mistaken for "giant mosquitoes." A few taxa possess a long proboscis that presumably allows nectar feeding. Large and gangly, adult crane flies are easily taken by vertebrate predators such as birds.

PSYCHODIDAE   Sand flies (Fig. 30), drain flies, and moth flies are typical representatives of this family and contain 2500 species. Adult sand flies *(Phlebotomus)* are tropical hematophagous (blood-feeding) flies that can transmit leishmaniasis, a disease caused by parasitic protozoa spread by sand-fly bites. However, most psychodids do not bite and are harmless to humans and livestock. Drain flies *(Psychoda* and *Telmatoscopus)* and moth flies *(Psychoda)* resemble tiny moths (about 2–4 mm in length) with hairy, pointed wings. The former have larvae that develop on the rich organic material that builds up in domestic pipes and drains and can be abundant in households and public restrooms. Moth flies have aquatic to semiaquatic larvae that breathe atmospheric oxygen by maintaining contact with the atmosphere using hydrofuge hairs on their posterior spiracles. Eutrophic lakes, marshes, and wastewater treatment plants may produce large numbers of adults. As detritivores, the larvae of moth flies probably are significant nutrient recyclers in lentic ecosystems.

BLEPHARICERIDAE   The net-winged midges (300 species) have peculiar larvae (Fig. 46) that use ventral suckers (suctorial disks) to maintain their positions on rocky substrates in torrential streams. A hydraulic, piston-like apparatus gives the larvae the ability to generate suction that allows their suckers to work—even waterfall habitats are occupied by blepharicerid larvae. The mouthparts are positioned ventrally on the head capsule and are specialized for scraping thin algal films off of rocks within fast-flowing environments. Diatoms and other unicellular algae are most often consumed, but fungi and bacteria may also be included in the larval diet. Pupae are also firmly attached to rocks within the flow with permanent suction pads. The adult (Fig. 47) will emerge and maintain a brief grip on the attached pupal skin as the exoskeleton hardens prior to flight. Little deviation from these habits has been documented within the Blephariceridae. Adults are known as net-winged midges because of the finely divided venation of the wings.

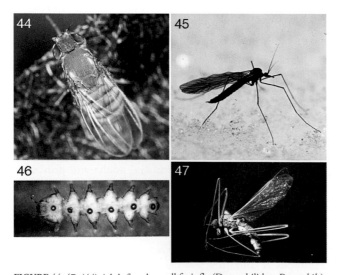

FIGURE 44–47 (44) Adult female small fruit fly (Drosophilidae: *Drosophila*). (Photograph by R. D. Akre.) (45) Adult crane fly (Tipulidae). (Photograph by Department of Entomology, Michigan State University.) (46) Ventral view of larva of Blephariceridae showing suctorial discs. (47) Adult net-winged midge (Blephariceridae). (Photographs by G. W. Courtney.)

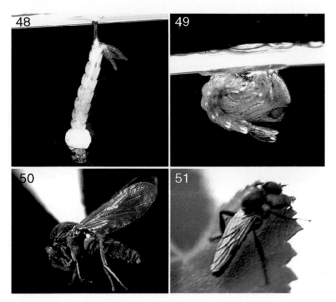

FIGURES 48–51 (48) Larva of mosquito (Culicidae: *Aedes aegypti*). (49) Pupa of mosquito (Culicidae: *Anopheles quadrimaculatus*). (Photographs by R. W. Merritt.) (50) Adult Ceratopogonidae. (Photograph by G. W. Courtney.) (51) Adult march fly (Bibionidae). (Photograph by Department of Entomology, Michigan State University.)

CULICIDAE Mosquitoes (3000 species) (Fig. 29) are well-recognized for their roles in disease agent transmission and as pests to humans, livestock, birds, and a variety of other vertebrate hosts. However, adults may emerge in high numbers and provide ample food for avian, bat, and certain predatory invertebrate populations. Mosquitoes exhibit the ability to colonize new aquatic habitats quickly and can survive in confined container habitats. In terms of mosquito control, the ecological importance of the larvae, pupae, and adults is rarely considered. The larvae are mostly filter feeders, but some scrape organic material and algae from solid substrates in standing water habitats. The clearance rate of particles from standing water is impressive and may alter the characteristics, such as turbidity, of the water the insects inhabit. Larval populations are a major component of the neuston, or water-surface inhabitants, and maintain contact with the atmosphere with their spiracles. Larvae (Fig. 48) are known as "wrigglers" because of their frantic swimming action that allows them to dive when threatened; lessening of light intensity by a mere shadow will initiate the wriggling action in some *Culex,* making them difficult to collect. Some taxa, such as *Culex, Culiseta,* and *Aedes,* have their spiracles positioned apically on respiratory siphons; others, such as *Anopheles,* lack this breathing-tube apparatus. *Mansonia* and a few other genera possess siphons that are specialized for piercing the roots of wetland plants such as cattails to obtain oxygen and therefore do not need to come to the water surface to breath. The mosquito pupa (Fig. 49) is free-swimming with respiratory trumpets that allow individuals to obtain atmospheric oxygen; pupae are known as "tumblers"

because of their tumbling action that propels them below the surface when disturbed. Emergence occurs quickly at the water surface as the pupal skin breaks to liberate the adult. Although females feed on sugar sources and may take a blood meal for the purpose of egg production, males feed only on nectar and lack bloodsucking proclivities.

CERATOPOGONIDAE This family is known as biting midges, punkies, and no-see-ums and contains 5500 species. The adults (Fig. 50) are minute bloodsuckers that swarm around mammalian hosts, including humans (e.g., *Culicoides, Lasiohelea*). Certain taxa also feed on other invertebrates as ectoparasites, including crane flies, dragonflies, and mantids. The tiny black to gray adults frequently have darkly patterned wings and relatively long antennae. Larvae are encountered in a variety of standing water habitats, including saturated mud and sand, tree holes (*Dasyhelea*), rain pools, marshes, lakes, and even hot spring algal mats (*Bezzia*). The genus *Leptoconops* can be pestiferous and biting adults are encountered at ocean-side beaches. The larval feeding habits of biting midges consist mostly of scavenging and predatory behavior.

SIMULIIDAE Although the general public is often aware of the pest nature of mosquitoes, knowledge of blood feeding by black flies (1500 species) is often restricted to anglers and those who recreate within or near aquatic systems. Like mosquitoes, the larvae play an important role as filter feeders; however, black flies are restricted to flowing water systems. Larval simuliids spin a patch of silken webbing on the surface of riffle rocks and maintain a hold on the webbing with hooks positioned on the posterior abdominal segment. The mouthparts are modified in many species (e.g., *Simulium, Prosimulium*) and resemble head fans that allow the larvae to capture organic particles, including materials as small as bacteria. Rocky substrates below dam spillways where organic-rich water flows may support tens of thousands of larvae per square meter. Other species are more mobile and scrape or collect food materials from benthic substrates (*Gymnopais* and *Twinnia*). The larvae are apneustic (i.e., lack spiracles) and therefore require moving water for cutaneous respiration. Pupae are firmly attached to areas of rocks exposed to current where thoracic pupal respiratory organs (gill-like structures) dangle in a downstream direction, supplementing spiracular respiration (Fig. 8). Most of the pupa is enclosed within a sheath-like cocoon. As in mosquitoes, males have weak mouthparts and may feed on nectar, whereas the females of most species have short probosci and cutting mouthparts for obtaining a blood meal from their vertebrate hosts. The adults are small and grayish black, lack distinct patterns on the wings, and have short antennae (Fig. 28). Some species can swarm in large numbers and are capable of causing shock in domestic animals (e.g., cattle) due to blood loss.

BIBIONIDAE   March flies (700 species) are named for their early spring appearance in temperate habitats. The stout, dark-colored adults (Fig. 51) feed on flowers; in contrast, the worm-like larvae are general detritivores and can be found in organic soils and compost heaps in abundance. The common genus *Bibio* overwinters as larvae prior to forming pupae after being exposed to cold temperatures. One predaceous species, *Plecia nearctica,* was introduced to the southeastern United States to control mosquitoes. Although its impact on mosquitoes is questionable, its impact on human residents is very real. The adults appear for brief periods (about 2 weeks) in such large numbers as to smear automobile windshields and clog radiators. The smashed bodies may even damage a car's paint if not washed off quickly. The adults are known as "love bugs" because males and females are frequently seen flying *in copula.*

SCIARIDAE   These are known as dark-winged fungus gnats (1000 species) because of their small, gnat-like size and smoky grayish-black wings. *Sciara* is the genus most frequently encountered by people, as the pale, slender larvae develop in a variety of materials, including potting soil used in greenhouses and household planters. In nature, larvae consume the fungus-rich detritus formed under the bark of rotting trees, as well as within the logs themselves. Organic-rich compost heaps and mushrooms are also inhabited by sciarid immatures. Although a few taxa are pests (e.g., *Pnyxia* attacks mushrooms), most species of this common family are harmless.

CECIDOMYIIDAE   Gall midges and gall gnats are minute flies that are abundant, species-rich (4500 species), and cosmopolitan. More than 1000 species occur in North America alone, and many undescribed species await taxonomic attention. Most species form distinctive galls within which the maggots develop (Fig. 40). Indeed, it is frequently easier to determine what species is attacking a plant based on gall morphology rather than adult or larval morphology. Many people are quite familiar with the plants that are affected, such as the damage from maple leaf spot *(Rhabdophaga),* or the attack of the Hessian fly *(Mayetiola destructor),* which can be a serious pest of wheat. However, the Cecidomyiidae as a family shows impressive breadth in the plant species it attacks. A few species (e.g., *Miastor*) exhibit paedogenesis, whereby the larvae reproduce. The "mother larva" produces a number of larvae within her body, which eventually consume the mother and then escape.

Suborder Brachycera

TABANIDAE   The horse flies (e.g., *Tabanus, Hybomitra*) (Fig. 52) and deer flies *(Chrysops, Silvius)* (Fig. 52) contain 3000 species and are a familiar insect group to people who frequent rural outdoor areas. The adults are rapid fliers; one species was estimated to fly over 150 km per hour! Eggs are normally laid in masses, frequently on vegetation overhanging water or saturated soils. The cryptic larvae are restricted to aquatic and semiaquatic habitats where most species are predators of other invertebrates. The life cycle generally takes about 1 year to complete, whereas some of the larger horse flies require up to 3 years. Although most species inhabit stagnant habitats, some are found at the margins of streams. Females are blood-feeders and may inflict a painful bite. Rather than puncturing a host's skin and sucking blood like mosquitoes, tabanids create a laceration on the host's skin and quickly lap up the pooling blood before retreating. The attack on livestock can be so severe as to reduce milk yields in dairy cattle. Like many other families of biting flies, females use visual cues to locate hosts and also sense plumes of carbon dioxide produced during vertebrate respiration. Horse flies tend to be large (about 10–25 mm) with nearly colorless or smoky wings, whereas deer flies are smaller (around 8 mm) and have yellow or black bodies that support darkly patterned wings. Human disease transmission by Tabanidae (i.e., tularemia, anthrax) is possible, but not significant in North America. However, transmission can be significant in other areas of the world (e.g., Africa).

RHAGIONIDAE   The snipe flies (500 species) superficially resemble some deer flies, but have a more slender body. Most common in woodlands, snipe flies are often dull yellow to brown (e.g., *Rhagio*), but the gold-backed snipe fly *(Chrysopilus ornatus)* of eastern North America has brilliant gold hairs adorning the thorax and abdomen. Most adults are nectar feeders, whereas a few taxa are predators of flying insects. Larval rhagionids tend to be predators of small invertebrates within masses of rotting wood, organic-rich soil, or compost. One genus *(Symphoromyia)* of western North America has blood-feeding adults that will bite humans in woodland areas.

FIGURES 52–55 (52) Adult deer fly (Tabanidae). (Photograph by R. W. Merritt.) (53) Adult robber fly (Asilidae). (Photograph by R. W. Sites.) (54) Adult bee fly (Bombyliidae). (Photograph by R. D. Akre.) (55) Adult long-legged fly (Dolichopodidae). (Photograph by Department of Entomology, Michigan State University.)

MYDIDAE The largest adult Diptera are the mydas flies, with 400 species. Some tropical species are over 50 mm. The adults are dark and have red to yellowish coloration on some abdominal segments. Little biological information is available on this family, although the larvae are predators found in decaying logs in woodlands. Pupae occur a few centimeters below the soil, and are adorned with heavy spikes for digging to the surface just prior to adult emergence. The adults are also thought to be predators that specialize in capturing other flying insects, but a fair number of species have vestigial mouthparts. The females of the latter may simply live on the accumulated fatty tissue in the abdomen.

ASILIDAE The robber flies (5500 species) occur in a vast number of terrestrial habitats; most adult activity occurs in areas that are sunny or at least partially sun lit. Adults (Fig. 53) may reach approximately 30 mm in length (e.g., *Proctacanthus*), whereas others are less than 10 mm in length (e.g., *Holocephala*). There is great morphological variation in this family among adults, but all species share in common a conspicuously sunken vertex. Adults are predators that are able to take larger prey such as dragonflies, but the selected prey size varies among species. The type of prey, whether stationary, crawling, or flying, is also species-specific among robber flies. The mouthparts contain a stout proboscis that the adult uses to exsanguinate prey species. Some species (e.g., *Laphria*) mimic bumble bees, which reduces predatory attempts on the adult. Most larvae live in soil or rotting wood where they hunt other insect larvae and nymphs; however, some species are ectoparasitic on Diptera, Coleoptera, Hymenoptera, and Orthoptera immatures. Very few life history studies have been done on the Asilidae.

BOMBYLIIDAE Bee flies (5000 species) are stout, hairy-bodied flies and, as the name implies, adults are frequently mistaken for hymenopterans because of their bee-like appearance (Fig. 54). Furthermore, the adult behavior often involves hovering at flowers, beelike, and extending a long proboscis to obtain nectar while in flight! Some taxa have bold patterns on the wings (e.g., *Anthrax, Exoprosopa*) or have the anterior margin of the wing darkened (e.g., *Bombylius*). Although a widespread family, most species occur in arid areas. The biology of the immature stages remains unknown for most species, but it appears that all species for which the larval feeding habits are known are parasitic on Diptera, Lepidoptera, Hymenoptera, Coleoptera, and Neuroptera larvae or pupae. Many larvae have relatively large, tong-shaped mandibles, presumably suited to aid the parasitic life history. A few bombyliid species consume grasshopper eggs.

DOLICHOPODIDAE The long-legged flies (5000 species) are small to minute flies that are often brilliant green, blue, or copper colored (Fig. 55). Males have genitalia that are nearly as long as the other abdominal segments combined (e.g., *Dolichopus*). Adults participate in complex courtship rituals, and males of some species have legs adorned with flattened hair-like scales used as flags to communicate with females during courtship. The family is impressively diverse in its habitat use, as adults can be abundant in freshwater marshes and lake edges, stream margins, woodlands, open fields, and coastal marine areas. Adults appear to be exclusively predaceous. Larvae have been taken from water, damp soil, grass stems, under bark, and other places. Most taxa are predaceous, but a few (e.g., *Thrypticus*) are phytophagous. One genus (*Medetera*) has predaceous larvae that feed on bark beetles.

Suborder Cyclorrhapha

PHORIDAE This family, also known as humpbacked flies and scuttle flies (3000 species), is another group of flies that exploits a wide range of habitats and exhibits diverse feeding habits. The humpbacked appearance and reduced venation make the adults easy to identify. Many species are consumers of decaying organic matter and can infest household garbage cans on occasion; the females are strongly attracted to the odor of decay. Other species are more unusual, specializing on the consumption of slug eggs (*Megaselia*) or parasitic on spiders, millipedes, and at least nine insect orders. Some species are currently targeted as potential biocontrol agents of fire ants, a serious pest in the southern United States. One species is known as the coffin fly (*Conicera tibialis*) because it was reported to maintain many generations on a single human body in the confines of a buried casket.

SYRPHIDAE Like the bee flies, flower fly adults (6000 species) resemble Hymenoptera and can mimic bees, bumble bees, hornets, and others (Fig. 56). Syrphids have the ability to hover (thus, they are also known as hover flies), and

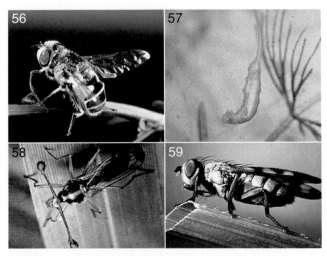

FIGURES 56–59 (56) Adult flower fly (Syrphidae: *Eristalis* sp.). (Photograph by R. F. Harwood.) (57) Rat-tailed maggot (Syrphidae). (Photograph by R. W. Merritt.) (58) Adult stalk-eyed fly (Diopsidae). (Photograph by R. D. Akre.) (59) Adult picture-winged fly (Otitidae: *Melieria similies*). (Photograph by J. A. Novak.)

individuals are frequently found flying near flower heads where they obtain nectar. Adults are common near wetlands and lakes (e.g., *Eristalis, Allograpta*) but are also abundant in terrestrial areas (e.g., *Merodon, Syrphus*) where appropriate flowering vegetation grows. In aquatic habitats the larvae, or rat-tailed maggots, as they are called, are collector–gatherers and may use retractable siphons for respiration (Fig. 57). Emergent vegetation, or vegetation at the aquatic–terrestrial interface, may be infested with aphids and other homopterous herbivores that certain flower fly larvae devour upon discovery. Some aphids enter plant stems compromised by boring larvae of other orders (e.g., Lepidoptera) to feed on decaying plant juices; it is not uncommon to find flower fly larvae that have also entered the damaged areas of plants to obtain prey. Predatory habits are also seen in terrestrial habitats, where larvae of certain species will inhabit dung, rotting logs, and decaying vegetation, as well as the exterior of plants. Overall, larval habitats are diverse, including dung, rotting cactus, peat, and hymenopteran nests. Some species have also been implicated in intestinal myasis.

DIOPSIDAE   The stalk-eyed flies (Fig. 58) are one of the most aptly named and morphologically unusual dipteran families (150 species). Each eye and its antenna are positioned at the end of individual stalks that protrude laterally from the head; the distance from eye to eye may be approximately equal to the entire body length! North America's one species (*Sphaerocephala brevicornis*) has very short eye stalks and breeds in decaying organic matter. The adults of this species exhibit no particular courtship displays, whereas highly adorned males of Afrotropical species (*Diopsis*) battle with one another using their stalks as levers during aggressive "wrestling matches." The larvae of *Diopsis* are herbivorous, and some species develop within the stems of rice plants.

OTITIDAE   These flies are also known as picture-winged flies because their boldly patterned wings are used in courtship and species recognition (Fig. 59). Adults are commonly found walking along vegetation flashing their wings. Otitids are abundant in both aquatic and terrestrial habitats (800 species). In marshes and vegetated lake margins, picture-winged flies are herbivores (*Eumetopiella*), secondary invaders of damaged plants (*Chaetopsis*), and general scavengers (*Seioptera*). Herbivory also occurs in terrestrial species (*Tetanops, Tritoxa*), but scavenging of decaying organic material appears to be more common (*Delphinia, Euxesta, Notogramma*). Some species also attack fungi (*Pseudotephritis*).

PYRGOTIDAE   Although some species of certain fly families (e.g., Tipulidae) are attracted to collecting lights at night, pyrgotid flies are unusual in that they are exclusively nocturnal. These flies (200 species) are relatively large and usually have strongly patterned wings (Fig. 60). Adults (e.g., *Pyrgota, Sphecomyiella*) seek scarab beetles, most notably June beetles, and apparently attack flying beetles by laying a single

FIGURES 60–63 (60) Adult of Pyrgotidae. (61) Adult marsh fly (Scromyzidae: *Limnia*). (62) Adult shore fly (Ephydridae: *Ochthera mantis*). (Photographs by The Cleveland Museum of Natural History.) (63) House fly adult (Muscidae: *Musca domestica*). (Photograph by R. W. Merritt.)

egg on the dorsum of the thorax or abdomen that is exposed when a beetle's elytra and wings are spread. The larva hatches from the egg and burrows into the body, acting as a parasitoid. The feeding larva eventually kills the host and consumes the remaining tissue. Larvae pupate within the hollowed host, and the adult exits the beetle exoskeleton to continue the life cycle.

TEPHRITIDAE   These true fruit flies (4000 species) are essentially entirely terrestrial in their habitat selection, although the host plants exploited by the family sometimes grow at the margins of lakes and marshes. Adults (Fig. 42) oviposit on the flower heads of the plant family Compositae or on fleshy fruits. Like the Otitidae, the wings of most adults are distinctly patterned, and adults flash their wings during courtship; this behavior has earned the Tephritidae a second common name, "peacock flies." Species tend to be fairly specific in their host plant preferences or at least attack a narrow spectrum of plant taxa. Fruit fly species are also specific in the area of a plant that they infest. Some species are frugivorous (*Ceratitis, Rhagoletis*), seed-head predators (*Euaresta, Trupanea, Tephritis*), gallmakers (*Eurosta*), or leafminers (*Euleia*). Frugivorous larvae damage the host fruit, causing it to rot quickly; seed predators of select young, developing seeds. Galls may be formed on a variety of plant areas, including stems, leaves, and flower heads.

DRYOMYZIDAE   These are relatively uncommon flies (300 species), with the biology of only 2 of the 8 North American species known. *Dryomyza anilis* is a scavenger and breeds in decaying mammalian carcasses; it can be reared on raw ground beef. A contrasting life history is found in *Oedoparena glauca*, which preys on barnacles in the intertidal zone of western North American shorelines. This character makes *O. glauca* one of the truly marine insects, as it is tied intimately to an ocean-inhabiting invertebrate. Adults lay their eggs into the barnacle's operculum when dropping tide levels expose them. Larvae consume the soft tissue, and mature larvae frequently move to new barnacles to continue feeding. Pupariation occurs within the final host.

SEPSIDAE   The black scavenger flies are fairly abundant in both aquatic (*Enicomira, Themira*) and terrestrial (*Sepsis*) environments (250 species), where the larvae are scavengers of decaying organic matter. Dung of a variety of mammalian animals, carcasses, rotting snails, and washed-up seaweed have been exploited. The adults of many species are easily recognized by their rounded heads and the presence of a black dot at the apex of each wing.

SCIOMYZIDAE   Called marsh flies and snail-killing flies, neither name encompasses all the habits of this well-studied family (500 species). Some species (e.g., *Dictya, Limnia*) are found in marshes, but some species of certain genera (e.g., *Sciomyza, Pherbellia*) are fully terrestrial. Many species are larval parasitoids or predators of snails, and some attack slugs (e.g., certain *Tetanocera, Euthycera*) or fingernail clams (*Renocera*); the larvae of one genus (*Antichaeta*) prey on the eggs of aquatic and semiaquatic snails. The adults range in color from yellowish brown to brownish black, have antennae that may be long or short, and vary in size from a few millimeters to nearly 1 cm (Fig. 61). However, the trophic niche of exploiting freshwater or terrestrial Mollusca (i.e., snails, slugs, and clams) ties all Sciomyzidae together evolutionarily. Ovipositional habits vary from certain species that lay their eggs directly on the host (e.g., *Sciomyza*) to species that lay eggs on plants, thus requiring larvae to search for their hosts (e.g., *Tetanocera*). Only one species (*Sepedonella nana*) from Africa seems to deviate from the trophic tie to mollusks, as laboratory-reared larvae have fed and survived on aquatic oligochaetes in the laboratory.

CHAMAEMYIIDAE   Aphid flies are predators of aphids, mealy bugs, and other homopterous herbivores (250 species). Adults lay their eggs on plant surfaces, in galls, or in the egg sacs of scale insects, and the maggots can reduce homopteran populations (*Leucopis*). One report from Mexico showed that the adults of an aphid fly fed on the secretions of vertebrate animals and that the larvae may have developed in bird nests (*Paraleucopis*). Some taxa are found within emergent and shoreline vegetation of aquatic habitats, whereas others are encountered in woodland or open fields.

PIOPHILIDAE   These flies (70 species) are most commonly represented by the cheese-skipper fly (*Piophila casei*), a cosmopolitan consumer of proteinaceous materials. Larvae frequently infest cheese and exhibit the rather peculiar escape strategy of grabbing the posterior body segment with their mouth hooks to form a U shape and then releasing their grip, which causes the larvae to propel, or skip, away from their original location. These behaviors give them the name cheese-skipper, even though the larvae are also known to consume the drying tissues of aging mammalian carcasses and dung. While most species appear to be scavengers of decaying materials and carcasses (*Protopiophila*) or mushrooms (*Amphipogon*), the larvae of some species are parasitic on avians (*Neottiophilum*).

SPHAEROCERIDAE   The small dung flies represent a speciose family (2500 species) that have predominantly scavenging larval feeding habits. One genus (*Leptocera*) consistently appears to be scavengers; however, the spatial niches inhabited by the larvae are highly diverse. Larvae have been found in decaying vegetable matter, sewage, dung, dung beetle broods, stranded masses of seaweed, fungi, slime molds, carrion, and the organic matter accumulated within cups of bromeliads. Muddy, organic-rich margins of aquatic habitats, such as marshes and ponds, will support virtual clouds of adults. Some species are quite habitat-specific, such as those that inhabit bogs. The adults of common species have long, stiff bristles dorsally and are black to gray in color, and the arista is several times longer than the other segments of the antenna.

EPHYDRIDAE   These flies are also known as shore flies, and most taxa are associated with aquatic habitats. This is one of the most genera-rich families of Diptera (1300 species) and one of the most diverse in feeding habits. Larvae are consumers of decaying organic matter (e.g., *Discocerina*), secondary stem borers of damaged plants (e.g., *Typopsilopa*), primary herbivores (e.g., *Hydrellia*), generalist feeders of algae (e.g., *Scatella*), specialist consumers of algae and cyanophytes (e.g., *Hyadina*), diatom specialists (e.g., *Parydra*), predators (e.g., *Ochthera*), and consumers of spider eggs (*Trimerina*). Virtually all aquatic habitats, from flowing water to stagnant environments, temporary to permanent, fresh water to hypersaline, and cold water to hot springs, are occupied by ephydrids. Terrestrial environments are less likely to support ephydrid populations, but these flies are found in moist woodlands and even in sod from suburban areas. Shore flies can be abundant in human-made habitats, including constructed wetlands and sewage treatment plants. It is difficult to make generalizations about the overall morphology of this family, except that the adults tend to be small; the smallest adults (*Lemnaphila*), only a couple of millimeters across, mine the thalli of duckweed plants. Adult body color ranges from silvery gray to jet black, and the wings are completely colorless to highly patterned with various shades of gray and brown (Fig. 62). The species, habitat, and feeding diversity have led the dipterist Harold Oldroyd to state that the shore flies are currently "in the flower of their evolution."

DROSOPHILIDAE   Pomace flies, vinegar flies, and small fruit flies are another highly species-rich family (3000 species). The latter name has led to some confusion as the Tephritidae are also known as fruit flies. The adults (Fig. 44) are small (generally only a few millimeters in length), but can disperse about 10 km in 1 day. Some drosophilids are frugivorous, but a vast array of food sources are used. For example, many feed on fungi (e.g., *Amiota, Mycodrosophila, Stegana, Scaptomyza*), living flowers (e.g., *Apenthecia, Styloptera*) or are predaceous on other invertebrates (e.g., *Rhinoleucophenga, Cacoxenus, Acletoxenus*). Indeed, the genus *Drosophila*, mostly known for the experimental studies of *D. melanogaster*, in the wild exhibits a vast trophic ecology and includes species that develop in rotting vegetation, rotting fruit, tree sap, fungi, living flowers, and plant stems and that prey on other invertebrates. Two *Drosophila* species are commensal with crabs: larvae live attached to the crab exoskeleton and consume semiliquid excretions from the crab or develop in the crab's branchial chamber and consume its microflora.

CHLOROPIDAE Flies is this family are called chloropid flies or frit flies (2000 species). Larvae are generally scavengers of decaying organic matter, secondary invaders of damaged plants, or primary herbivores. Adults are common and abundant among vegetation in terrestrial and aquatic situations. Species have been reared from dozens of different substrates around the world. Grasslands commonly support populations of certain genera (e.g., *Meromyza, Parectecephala*), and fungi are the sole food of others (e.g., *Fiebrigella, Apotropina*). However, a great diversity occurs in vegetated areas of aquatic habitats (e.g., *Chlorops, Epichlorops, Eribolus, Diplotoxa*) where most taxa are detritivores in decaying masses of vegetation, secondary stem borers, and primary herbivores of aquatic or semiaquatic plants. A few taxa are predatory on Homoptera (e.g., *Thaumatomyia*). Other, less common, larval food sources include dung (*Cadrema*), decaying wood, and bird nest debris (*Gaurax*), and one Australian genus (*Batrachomyia*) is subcutaneously parasitic on frogs and toads. The adults tend to have rounded flagellomeres and range from dull colored to bright yellow or green, and many species have a distinctly shiny triangle positioned at the vertex of the head.

AGROMYZIDAE These are known as the leaf-miner flies because of their highly herbivorous nature. Like the Chloropidae, agromyzids are well represented in both aquatic and terrestrial environments (2000 species). Herbaceous and woody plants are both attacked, but larvae tend to feed on a single host plant, or a narrow spectrum as host plants as leaf miners, stem borers, or seed head predators. Wetland taxa can form large populations in which both monocot and dicot flora are used as host plants (e.g., *Agromyza, Cerodontha, Liriomyza, Phytomyza*). One species (*Melanagromyza dianthereae*) is a specialist stem borer of water willow, a flowering plant found at the edge of streams. The females of this fly lay eggs on the exterior of the plant, and upon hatching the larvae burrow into the stem to initiate feeding. Leaf mines are frequently seen as dead or brown areas on a leaf surface, and mine morphology is sometimes distinctive enough to determine which agromyzid species is responsible for plant damage.

ANTHOMYZIDAE These common flies are an example about which little is known of their biology (50 species). One

genus (*Anthomyza*) has small yellowish adults that may feed on the culms of wetland sedges, but it is unclear if they are herbivorous or act as secondary stem borers after plants have been attacked by other herbivorous insects.

MUSCIDAE This large family (4000 species) includes anthropophilic species such as the house fly (*Musca domestica*) (Fig. 63) and the stable fly (*Stomoxys calcitrans*) (Fig. 32). The house fly is well known for its "filthy habits," and the stable fly bites both humans and livestock. The reproductive rate of the house fly is noteworthy, as one female can eventually give rise to 2 billion other female flies after several summer generations are produced (assuming all flies live, which is never the case). A short life cyle (12–14 days required for development from egg to adult in summer temperatures) is at least partially responsible for the success of this species and is necessary for developing in such ephemeral, human-made habitats such as dung heaps, garbage cans, and mammalian road kill. However, most muscid species are not directly associated with human populations. The larval feeding habits found among the Muscidae include herbivory (*Atherigona, Dichaetomyia*), scavenging (*Graphomyia*), and predatory behaviors (*Coenosia, Lispe, Spilogona*). A few taxa cause myiasis in birds (*Muscina*) or are avian blood feeders (*Philornis*). Some Muscidae form a cocoon prior to pupariation (formation of puparia), which is uncommon among Diptera. Adult muscid flies may be predaceous on other insects, but most are generalized scavengers or feed on pollen.

OESTRIDAE These are commonly known as bot or warble flies (40 species). The larvae of all species are endoparasites. Species that attack livestock burrow into the host skin to feed on living tissue and either form their pupae under the skin, forming warbles (*Hypoderma*), or drop off the host and pupariate in soil (*Oestris*). Four species of the horse bot fly (*Gasterophilus*) infest the alimentary tract of horses (Fig. 38), donkeys, and mules. One genus (*Cuterebra*) (Fig. 64) infests lagomorphs and rodents and is among the biggest bot flies (about 2.5 cm). The human bot fly (*Dermatobia*) lays eggs on mosquitoes and other biting flies. When a larva hatches, it hangs onto the bloodsucker's leg until it lands on a human to take obtain a blood meal. The maggot then drops onto the host and burrows into the skin. Human bot flies are restricted to the Neotropical areas of the world and use a variety of mammalian hosts in addition to humans. North American vacationers and

FIGURE 64–65 (64) Adult rodent bot fly (Oestridae: *Cuterebra jellisoni*). (Photograph by C. Baird and R. D. Akre.) (65) Adult of Tachinidae. (Photograph by E. A. Elsner.)

workers visiting the fly's home range frequently return home with a painful welt, under which lays a feeding maggot that respires through a small hole in the person's skin. The experience is painful, and most infected travelers have the larva removed surgically prior to pupariation or adult emergence.

NYCTERIBIIDAE Bat flies (250 species) are specialized ectoparasites of bats. The spider-like adults lack wings, which probably reduces the host's chances of removing the fly. Females bear mature living young (i.e., they are pupiparous), which is uncommon among Diptera. Larvae receive nutrients produced by glands within the female abdomen. The female bat flies deposit the larva on the walls of bat roosts, and females of some species will sit on top of the larva and briefly press it to the wall to ensure good adhesion. Pupariation occurs quickly without larval feeding. The adult emerges, and then seeks a host to continue the bloodsucking habit. Therefore, larval feeding does not occur outside of the adult female fly.

CALLIPHORIDAE The blow flies are thought to have been given their name from Homer's classic book, *The Iliad*, in which he wrote about the "blows of flies" infesting the wounds of injured and dead soldiers. Most of the 1000 species of this cosmopolitan family are attracted to rotting flesh and can sense the chemical scent of decay within minutes of death. In nature, the adults accelerate the decomposition of all types of vertebrate carcasses, and most blow flies (e.g., *Calliphora, Cochliomyia, Lucillia, Phaenicia*) are specialist scavengers. Because of the ability to find dead bodies rapidly, forensic scientists use the stage of larval development (i.e., age of a larva) found on corpses of people who died from suspicious causes as a way to determine the time between death and corpse discovery. It is well documented that by using this method, the time of death often can be estimated with a fair amount of accuracy. Other taxa of blow fly, however, exhibit other feeding habits, such as parasitism of land snails *(Helicobosca)*, earthworms *(Pollenia)*, and amphibians *(Bufolucilia)*. The adults are also known as blue bottle and green bottle flies because some taxa have metallic brightly colored bodies (Figs. 35 and 37).

SARCOPHAGIDAE Flesh flies have been given a name that often contrasts their biology. Only a few of the 2500 species invade or consume carrion *(Sarcophaga)* or living tissue *(Wohlfahrtia)*. Dung (e.g., *Ravinia*) is more commonly used. Many taxa are parasitic on other invertebrates (snails, earthworms, insects, and others), whereas some specialize in consuming the decaying bodies of insects found in the bottoms of pitcher plants of wetland habitats. Females are viviparous, young hatch within the female's abdomen, and she deposits them as first instars on the desired substrate. This may give flesh flies an advantage over potential competitors for food because mortality of eggs by predation or parasitism is avoided, and larvae can feed immediately rather than waiting to hatch for some days prior to feeding.

The adults of most genera are easily recognizable by the gray thorax possessing longitudinal black stripes (Fig. 36).

RHINOPHORIDAE This is an unusual fly family in that nearly all of the 100 species for which biological details are known are specialist endoparasitoids of terrestrial isopods (also known as sow bugs, pill bugs, and potato bugs). This family represents the only dipterans that attack isopods. The larval life is tenuous, because larvae hatch from eggs laid in moist soil and must wait for a passing isopod. Perhaps the proleg-like apparatus present on the first instar is an adaptation to securing itself to a host. Both species of rhinophorids found in North America were probably introduced from Europe.

TACHINIDAE These flies (9500 species) (Fig. 65) are important parasites of a variety of other insects and are used in biological control programs against pestiferous Lepidoptera. Eggs are deposited on hosts or in areas where hosts are common. Some species retain their eggs so that they will hatch almost immediately after being laid; this strategy prevents the loss of the egg if the host happens to molt shortly after oviposition. Insertion of the egg through the epidermis of the host has evolved in a few species *(Phorocera)*. An alternative strategy used by some taxa is to lay many eggs on partially consumed plants; when a potential host returns to continue feeding, the eggs are consumed along with the plant material. A few eggs survive maceration by the mandibles, and the larvae hatch within the host's foregut. Other genera broadcast their eggs, and the larvae burrow selectively into soil or rotting wood where they actively seek a host insect. Most often, tachinid flies attack only one species or a narrow spectrum of hosts; however, a small number *(Compsilura)* have been reared from some 200 different animal host species.

See Also the Following Articles
Drosophila melanogaster • *House Fly* • *Mosquitoes* • *Tsetse Fly*

Further Reading
Brown, B. V. (2001). Flies, gnats, and mosquitoes. *In* "Encyclopedia of Biodiversity" (S. A. Levin, ed.), pp. 815–826. Academic Press, London.
Cole, F. R. (with collaboration of E. I. Schlinger) (1969). "The Flies of Western North America." University of California Press, Berkeley.
Courtney, G. W., Teskey, H. J., Merritt, R. W., and Foote, B. A. (1996). Aquatic Diptera, Part One, Larvae of aquatic Diptera. *In* "An Introduction to the Aquatic Insects of North America." (R. W. Merritt and K. W. Cummins, eds.), pp. 484–514. Kendall–Hunt, Dubuque, IA.
Dethier, V. G. (1963). "To Know a Fly." McGraw-Hill, Columbus, OH.
Feener, D. H., Jr., and Brown, B. V. (1997). Diptera as parasitoids. *Annu. Rev. Entomol.* **42,** 73–97.
Hennig, W. (1973). Diptera (Zweiflügler). *In* "Arthropoda," Hälfte 2, "Insecta," No. 2, "Spezielles, 31," (Helmecke *et al.,* eds.), Vol. IV of "Handbuch der Zoologie: Eine Naturgeschichte der Stämme des Tierreiches." Berlin/New York.
James, M. T. (1948). "The Flies That Cause Myiasis in Man." U.S. Department of Agriculture, Washington, DC. [Publication No. 631]
McAlpine, J. F., Peterson, B. V., Shewell, G. E., Teskey, H. J., Vockeroth, J. R., and Wood, D. M. (coordinators) (1981). "Manual of Nearctic Diptera," Vol. 1. Research Branch, Agricultural Canada. [Monograph 27]

McAlpine, J. F., Peterson, B. V., Shewell, G. E., Teskey, H. J., Vockeroth, J. R., and Wood, D. M. (coordinators) (1987). "Manual of Nearctic Diptera," Vol. 2. Research Branch, Agricultural Canada. [Monograph 28]

McAlpine, J. F., and Wood, D. M. (coordinators). "Manual of Nearctic Diptera," Vol. 3. Research Branch, Agricultural Canada. [Monograph 32]

Merritt, R. W., and Cummins, K. W. (1996). "An Introduction to the Aquatic Insects of North America." Kendall–Hunt, Dubuque, IA.

Merritt, R. W., Webb, D. W., and Schlinger, E. I. (1996). Aquatic Diptera, Part Two, Pupae and adults of aquatic Diptera. *In* "An Introduction to the Aquatic Insects of North America" (R. W. Merritt, and K. W. Cummins, eds.), pp. 515–548. Kendall–Hunt, Dubuque, IA.

Oldroyd, H. (1964). "The Natural History of Flies." Norton, New York.

Oosterbroek, P., and Courtney, G. W. (1995). Phylogeny of the nematocerous families of Diptera (Insecta). *Zool. J. Linn. Soc.* **115,** 267–311.

Papp, L., and Darvas, B. (eds.) (2000). "Manual of Palaearctic Diptera," Vol. 1. Science Herald, Budapest.

Yeates, D. K., and Wiegmann, B. M. (1999). Congruence and controversy: Toward a higher-level phylogeny of Diptera. *Annu. Rev. Entomol.* **44,** 397–428.

Diversity

see *Biodiversity*

Division of Labor in Insect Societies

Gene E. Robinson

University of Illinois, Urbana-Champaign

Division of labor is fundamental to the organization of the insect societies and is thought to be one of the principal factors in their ecological success. Different activities are performed simultaneously by specialized individuals in social insect colonies, which is more efficient than if tasks are performed sequentially by unspecialized individuals.

Division of labor is one of the defining characteristics of the most extreme form of sociality in the animal kingdom, "eusociality." Eusociality is defined by three traits: (1) cooperative care of young by members of the same colony, (2) an overlap of at least two generations of adults in the same colony, and (3) division of labor for reproduction, with (more or less) sterile individuals working on behalf of fecund colony members. It is now recognized by many biologists that eusociality extends to taxa beyond the ants, bees, wasps (Hymenoptera), and termites (Isoptera). This article focuses on the societies of the classic social insects, particularly the Hymenoptera, because they have the most elaborate and well-studied systems of division of labor.

DIVISION OF LABOR FOR REPRODUCTION

Females dominate the functioning of insect societies, even in termite societies, in which males play more diverse roles than in hymenopteran societies. There are two types of females in an insect society, queens and workers. Queens specialize in reproduction and may lay up to several thousand worker eggs per day. Workers are either completely or partially sterile, engage in little if any personal reproduction, and perform all tasks related to colony growth and maintenance. Worker sterility occurs because the ovaries do not develop or because critical steps in oogenesis do not occur. Worker sterility occurs either during preadult stages or during adulthood.

In many species of social insects, queens and workers are distinguished by striking morphological differences. A queen can have huge ovaries and a sperm storage organ that maintains viable sperm for years. The most striking morphological differences between queens and workers occur as a result of caste determination, which occurs during preadult stages. Caste determination has an endocrine basis. Research on the honey bee, *Apis mellifera,* and the bumble bee, *Bombus terrestris,* has shown that a high hemolymph titer of juvenile hormone (JH) during a critical period of larval development induces queen development. JH and presumably other hormones trigger a variety of processes that ultimately result in the production of either a worker or a queen. For example, caste-specific apoptosis (cell death) occurs in the ovaries of worker-destined honey bees and is associated with low titers of JH and ecdysteroid. Molecular analyses of endocrine-mediated caste determination have just begun. Some of the first findings involve caste-specific differences in the expression of genes that are associated with metabolism and protein synthesis, reflecting the fact that developing queens are metabolically more active than developing workers.

Little is known about how extrinsic factors act on endocrine-mediated developmental processes to influence caste determination. There is a strong circumstantial link between diet and JH in honey bee larvae, but how nutritional information acts to elevate JH levels is still largely unknown. In other species, extrinsic factors that influence caste determination include temperature and social factors such as behavioral interactions and pheromones released by adult colony members. These might affect the larvae directly or might influence the treatment accorded them by a colony's workers.

In societies in which queens and workers have strong morphological differences, the major mechanisms for queen domination of reproduction appear to be primer pheromones produced by queens. However, only one queen primer pheromone has been well characterized, that being the mandibular pheromone of the queen honey bee. Workers exposed to queen pheromones show little or no ovary development or egg-laying behavior. In other species of social insects, the physical differences between queens and workers can be very slight. Division of labor for reproduction in these "primitively eusocial" species is achieved by a dominance hierarchy that is established and maintained by direct behavioral mechanisms, including pushing, biting, and physical prevention of egg laying. Behavioral domination is an ongoing process because some workers are physiologically capable of producing offspring and do, under some circumstances.

Queen behavior and pheromones affect adult worker neuroendocrine systems to reduce reproductive potential. JH has been implicated in the regulation of division of labor for reproduction in some, but not all, species studied to date, especially *B. terrestris;* the paper wasp, *Polistes gallicus;* and the fire ant, *Solenopsis invicta.* This is consistent with the function of JH as a hormone promoting reproductive development. JH does not appear to play this traditional role in adult *A. mellifera.* Ecdysteroids and biogenic amines also are suspected of being involved in the regulation of division of labor for reproduction among adult queens and workers, but a clear picture has not yet emerged.

DIVISION OF LABOR AMONG WORKERS

In most insect societies there also is a division of labor among the workers for tasks related to colony growth and maintenance. The evolution of a highly structured worker force is generally seen as an evolutionary consequence of the developmental divergence between queens and workers. Once workers were limited to serving largely as helpers, their characteristics could be shaped further by natural selection acting at the level of the colony to increase colony fitness. This perspective is consistent with the observation that the most intricate systems of division of labor among workers are found in species with the strongest division of labor for reproduction.

Age-related division of labor is the most common form of worker organization. Workers typically work inside the nest when they are young and shift to defending the nest and foraging outside when they are older. In the more elaborate forms of age-related division of labor, such as in honey bee colonies, workers perform a sequence of jobs in the nest before they mature into foragers. Physiological changes accompany this behavioral development to increase the efficiency with which particular tasks are performed. Among these are changes in metabolism, diet, and glandular secretions.

A less common but more extreme form of division of labor among workers is based on differences in worker morphology. This is seen in a minority of ant species and nearly all termites. Morphological differences among workers result from processes similar to worker–queen caste determination and morphologically distinct worker castes are recognized. For example, small ant workers (minors) typically labor in the nest, whereas bigger individuals (majors) defend and forage. Sometimes this form of division of labor also involves dramatic morphological adaptations in some worker castes, such as soldiers with huge and powerful mandibles and the ability to release a variety of potent defensive compounds.

A third form of division of labor among workers involves individual variability independent of age or morphology that results in an even finer grained social system. There are differences in the rate at which individual workers grow; some show precocious behavioral development, while others mature more slowly. There also are differences between individuals in the degree of task specialization. For example, foragers may specialize in the collection of a particular resource, such as some honey bees that collect only nectar or only pollen. It also has been found that some workers simply work harder than others.

The prevailing behavioral explanation for these three forms of division of labor among workers involves the application of the stimulus–response concept. Workers are thought to differ in behavior because of differences in exposure to, perception of, or response thresholds to stimuli that evoke the performance of a specific task. These differences can result from differences in worker genotype, age, experience, or morphological caste. There is some behavioral evidence for differences among workers in stimulus perception and response thresholds; challenges for the future are to more precisely define the nature of the stimuli and extend these analyses to the neural levels.

Some endocrine and neural mechanisms regulating age-related division of labor have been discovered, primarily in honey bees. Changes in hemolymph titers of JH act to influence the rate and timing of behavioral development, but JH is not required for a worker to mature into a forager. Evidence for a similar role for JH has been found in the advanced eusocial tropical wasp *Polybia occidentalis.* JH also affects the activity of exocrine glands that produce brood food and alarm pheromones in honey bees, apparently acting to ensure that physiological changes are coordinated with behavioral development. As JH receptors have not yet been identified in any insect, it is not known whether JH exerts its effects on division of labor directly in the brain, on other target tissues, or at a variety of sites. Octopamine acts as a neuromodulator in honey bees. Higher levels of octopamine, particularly in the antennal lobes of the brain, increase the likelihood of foraging. Changes in brain structure also occur as a worker bee matures into a forager, particularly in the antennal lobes and mushroom bodies, but the functional significance of these changes is unknown. As with caste determination, molecular analyses of behavioral development have only recently been initiated. Differences in the expression of several genes have been detected in the brains of younger and older honey bee workers. The orchestration of the neural and behavioral plasticity that underlies age-related division of labor is undoubtedly based on changes in the expression of many genes in the brain and other tissues as well.

Mechanisms underlying morphologically based systems of worker division of labor also have been studied. Morphological differences among adult workers have their origin in pathways of development that diverge during the larval stage. Information on worker caste differentiation, drawn largely from studies of *Pheidole* ants, suggests mechanisms similar to those involved in queen–worker caste determination. Both larval nutrition and JH have been implicated in the differentiation of *Pheidole* minors and soldiers.

Genetics is one factor influencing the third form of division of labor among workers, individual variability among workers that is independent of age or morphology. Genotypic variation within colonies arises as a consequence of multiple mating by queens or multiple queens in a colony. This genotypic variation is strongly associated with behavioral differences between individuals within a colony. Genotypic variation in honey bee colonies is known to influence how specialized a worker becomes on a particular task or the age at which it shifts from nest work to foraging. For example, quantitative trait loci have been found that are associated with variation in the tendency of honey bees to collect either nectar or pollen. These findings can lead to the identification of differences in specific genes that contribute to individual differences in behavioral specialization. Genotypic effects on division of labor also have been documented in several ant and wasp species.

PLASTICITY IN COLONY DIVISION OF LABOR

Colony division of labor, though highly structured, also shows great plasticity. Colonies respond to changing needs by adjusting the ratios of individual workers engaged in different tasks. This is a consequence of the flexibility of the individual workers. For example, there is plasticity in age-related division of labor, with workers able to respond to changes in colony age demography with accelerated, retarded, or reversed behavioral development. Workers also can shift to emphasizing a different task that is part of their age-specific repertoire or they can simply work harder. Morphologically specialized workers can be induced to shift their behavior; majors, normally specialized in foraging or defense, can care for the brood in the absence of minor workers. This plasticity in division of labor contributes to the reproductive success of a colony by enabling it to continue to grow, develop, and ultimately produce a new generation of reproductive males and females during changing colony conditions.

Plasticity in division of labor in advanced eusocial species is achieved by a variety of mechanisms of behavioral integration. These mechanisms enable workers to respond to fragmentary information with actions that are appropriate to the state of the whole colony. This makes sense because it is unlikely that any individual workers have the cognitive abilities to monitor the state of their whole colony and then perform the tasks that are needed most or direct others to do so.

Mechanisms of worker behavioral integration often involve social interactions. For example, in many species, including *Polybia* wasps and honey bees, nest workers routinely relieve the foragers of their newly acquired loads, whether nest material or food. Foragers that are unloaded immediately upon their return to the nest are likely to continue foraging for the same resource, apparently because the quick unloading signals to them that they have brought something of high value back to the colony. In contrast, foragers that experience a significant time delay before being unloaded respond by changing their behavior, perhaps shifting to the collection of another resource. The nutritional status of a fire ant colony strongly influences the behavior of its foragers, with the relevant information transferred during social feeding. In colonies of the desert-dwelling red harvester ant, *Pogonomyrmex barbatus,* workers obtain information on the needs of the colony by changes in their encounter patterns with members of various task specialist groups. For example, red harvester ant foragers are more likely to leave the nest to forage when they encounter greater numbers of successful returning foragers.

Social inhibition is a potent mechanism of integration in insect colonies. In colonies of honey bees, social inhibition acts to keep the division of labor synchronized with changes in colony age demography. Older workers inhibit the rate of maturation of younger workers. Some young workers in a colony deficient in older workers, for example, exposed to lower levels of social inhibition, respond by becoming precocious foragers. The specific honey bee worker factor that causes this inhibition has not yet been identified, but other sources of social inhibition have, emanating from the queen and the brood. The regulation of the size of the soldier force in *Pheidole* colonies also is based on a process of social inhibition. In this case, the presence of adult soldiers inhibits the production of new soldiers. Involvement of a pheromone is suspected, but no specific soldier inhibition pheromone has been identified yet.

The integration of activity in primitively social insect societies appears to be more centralized than in advanced eusocial societies. Primitively eusocial colonies often contain only a few dozen individuals, making centralized control more feasible. Queens act as central pacemakers and modulate worker activity via behavioral interactions in sweat bees and polistine wasps. Queens do not appear to be able to get workers to shift to different tasks, but they do cause them to work harder at the tasks they are already doing.

We are far from understanding how the behavior of individual workers is integrated into a well-functioning colony. Studies of behavioral integration are aided by various kinds of theoretical models. In some models, an insect colony is likened to a developing organism, i.e., the "superorganism" metaphor. In other models, an insect colony is analyzed with perspectives from neural network theory, with individual workers serving as analogs of individual neurons. Still other models view an insect colony as a self-organizing entity and use complex systems theory to develop ideas on colony function.

See Also the Following Articles

Ants • Caste • Colonies • Hymenoptera • Isoptera • Recruitment Communication • Sociality

Further Reading

Beshers, S. N., Robinson, G. E., and Mittenthal, J. (1999). The response threshold concept and division of labor. *In* "Information Processing in Social Insects" (C. Detrain *et al.*, eds.), pp. 115–141. Birkhauser, Basel.

Bloch, G., Wheeler, D. L., and Robinson, G. E. (2002). Endocrine influences on the organization of insect societies. *In* "Hormones, Brain and Behavior" (D. Pfaff *et al.*, eds.), pp. 195–237. Academic Press, San Diego.

Detrain, C., Deneubourg, J. L., and Pasteels, J. M. (eds.) "Information Processing in Social Insects." Birkhauser, Basel.

Page, R. E., and Robinson, G. E. (1991). The genetics of division of labour in honey bee colonies. *Adv. Insect Physiol.* **23**, 117–171.

Robinson, G. E. (1992). The regulation of division of labor in insect societies. *Annu. Rev. Entomol.* **37**, 637–665.

Robinson, G. E. (1998). From society to genes with the honey bee. *Am. Sci.* **86**, 456–462.

Wilson, E. O. (1971). "The Insect Societies." Harvard University Press, Cambridge, MA.

Dobsonfly

see *Megaloptera*

Dog Heartworm

Edward G. Platzer
University of California, Riverside

Dog heartworm, *Dirofilaria immitis,* is an important filarial nematode infection of dogs and canids primarily, but can occur in other mammals and occasionally humans. It usually occupies the pulmonary arteries and the right ventricle of the heart. Dog heartworm is transmitted by mosquitoes.

BIOLOGY

D. immitis is a filarial nematode in the superfamily Filarioidea (order Spirurida, class Secernentea). The males are 12 to 20 cm in length and 0.7 to 0.9 mm in diameter, with a spirally coiled posterior end. The females are 25 to 31 cm in length and 1.0 to 1.3 mm in diameter. *D. immitis* was first found in the United States, but it occurs globally, with a tendency for increased prevalence in humid warm regions conducive to abundant mosquito populations. In the United States, the prevalence can be as great as 45% in dogs within 150 miles of the Gulf of Mexico coast, Atlantic coast north to New Jersey, and Mississippi River and its tributaries. The prevalence elsewhere in the United States is generally less than 5%. Although dogs are the primary host, *D. immitis* has been found in coyotes, wolves, dingoes, foxes, sea lions, harbor seals, wolverines, ferrets, and cats. The number of nematodes per dog is variable, ranging from single nematodes to as many

as 250. Cats are less tolerant, with maximum parasite loads of 1 to 3 nematodes. *D. immitis* can persist for 5 to 7 years in dogs and 2 to 3 years in cats.

Mature female nematodes reproduce ovoviparously and the microfilaria are released from the uterus of the worm via the vaginal opening into the host blood. Microfilaria are 220 to 330 μm in length and 5 to 7 μm in width. Under laboratory conditions, over 60 species of mosquitoes are competent hosts for *D. immitis.* When mosquitoes take a blood meal, microfilaria are ingested and reside in the mosquito midgut briefly and then migrate to the Malpighian tubules, where they enter the cells and shorten. After a short developmental period, the juvenile nematodes leave the cells day 6 to 7 postinfection and enter the lumen of the tubules where the first molt (day 10) takes place, with the formation of the second-state juveniles (J2). After further growth and differentiation, the J2 molts (days 10–14 postinfection) to the infective stage (J3), which reaches a length of 1.3 mm. The J3 migrate through the hemocoel to the proboscis sheath of the mosquito. Development of the juvenile stages ceases if ambient temperatures decline below 15°C, which constitutes a constraint on the distribution of *D. immitis.* When the infected mosquito takes a blood meal, the J3 escape from the proboscis sheath, dropping onto the host in a droplet of hemolymph; they enter the host through the wound made by the piercing mouthparts of the mosquito. The J3 enter the subcutaneous tissue where they undergo the third molt to the fourth-stage juveniles, which reside in subcutaneous tissues or muscle of the abdomen or thorax for about 60 days, at which time the last molt to the adult stage occurs. The nematodes are now 12 to 15 mm long and enter the pulmonary arteries and attain lengths of 3.2 to 11 cm by 85 to 120 days postinfection. Fertilized females can be found 120 days postinfection and microfilaria enter the blood 6 to 9 months postinfection. Microfilaria can survive in the blood for $2^1/_2$ years. Host treatment with tetracycline leads to the loss of the endosymbionts and a concomitant reduction in survival and reproduction of *D. immitis.*

HOST PATHOLOGY

Usually no signs of infection are present in dogs until 8 to 9 months postinfection. Symptoms are generally related to the intensity of infection. The presence of 25 worms may be tolerated by the dog with no signs of disease. Symptoms such as reduced exercise tolerance and coughing increase significantly when the nematode burden doubles and serious symptoms, such as dyspnea (difficulty in breathing), hepatomegaly (enlargement of the liver), syncope (temporary loss of consciousness), and ascites (fluid accumulation in the abdomen) appear. At this level, death may occur. Pathogenesis is related primarily to inflammation of pulmonary arteries and lungs induced by the adult stage of *D. immitis.* This chronic inflammatory process in combination with the physical obstruction by the nematodes of blood flow leads to pulmonary hypertension and heart failure. Diagnosis depends on an accurate history, recognition of symptoms, and diagnostic procedures such as microfilarial detection, serology, clinical laboratory tests, radiology, ultrasonography, and angiography. Microfilaria are detected through concentration of the microfilaria from a small quantity of blood by the Knott's test or filtration. In some dog heartworm infections (occult heartworm infection), microfilaria are not detectable and diagnosis relies on serology and other diagnostic tools.

MANAGEMENT

Treatment is effective. The goal is elimination of the primary disease agent, the adults. Two arsenical drugs are currently approved for use in dogs, melarsomine hydrochloride (Immiticide; Merial) and thiacetarsamide sodium (Caparsolate; Merial). Treatment requires concomitant restriction of exercise and use of anti-inflammatory support to reduce the possibility of pulmonary thromboembolism from the dead heartworms. Microfilaria are eliminated by secondary treatment with ivermectin or milbemycin. Prevention of heartworm infection is safer and more economical than treatment and is accomplished readily by routine administration of diethylcarbamazine (daily in the diet) or one of the macrolide anthelminthics (monthly treatment; ivermectin, milbemycin, moxidectin, selamectin) during the transmission season.

See Also the Following Articles

Blood Sucking • Mosquitoes • River Blindness • Veterinary Entomology • Wolbachia • Zoonoses, Arthropod-Borne

Further Reading

American Heartworm Association (2001). http://www.heartwormsociety.org/.
Anderson, R. C. (2000). "Nematode Parasites of Vertebrates: Their Development and Transmission," 2nd ed. CAB Int., Wallingford, UK.
Bowman, D. D. (1999). "Georgis' Parasitology for Veterinarians," 7th ed. Saunders, Philadelphia.
Taylor, M. J., and Hoerauf, A. (1999). *Wolbachia* bacteria of filarial nematodes. *Parasitol. Today* **14**, 437–442.

Dormancy

Richard E. Lee, Jr.
Miami University, Oxford, Ohio

D ormancy is an inactive state associated with metabolic depression and arrested development that promotes the survival of insects during periods of harsh environmental conditions, including high or low temperatures or moisture conditions and reduced food quality or availability. Diapause may occur in any life stage (egg, larva, pupa, or adult); however, for a given species it generally occurs in only one stage.

Dormancy may be manifested in a variety of forms that differ widely in their intensity and duration. Consequently, it has proved difficult to classify these forms into discrete categories, and general agreement on terminology is lacking. Nonetheless, dormant states range from a moderate depression of short duration (quiescence) to a profound and extended period of metabolic suppression and developmental arrest (diapause).

Quiescence commonly refers to short periods of dormancy that are directly induced by adverse environmental conditions, principally low or high temperature. It also has the advantage of being quickly reversible upon the return of favorable conditions; this rapid response may be especially important in extreme environments, such as alpine regions or deserts, where access to food and favorable conditions are intermittent and unpredictable.

In contrast, diapause is not directly induced, but is triggered by genetically programmed responses to environmental cues that occur in advance of adverse conditions. Anticipatory induction allows time for substantial physiological changes prior to the arrival of adverse conditions. These changes may include accumulation of lipid and glycogen reserves, deposition of cuticular lipids that enhance desiccation resistance, suppression of gametogenesis, decreased metabolic rate, increased tolerance of anoxia, and low temperature.

Although moisture conditions, temperature, and host-plant quality may serve as cues for the induction of diapause, photoperiod is the factor that has been identified most commonly in this regard and is the one that has been the subject of the most investigation. Furthermore, these environmental conditions may interact to promote or inhibit the induction of diapause; an unusually cool autumn may induce diapause sooner than would be expected based solely on photoperiod. The cue for diapause induction need not be received by the life stage that enters diapause. For example, in many species adults that experience short daylengths produce diapause eggs, while those exposed to longer photoperiods do not.

See Also the Following Articles

Aestivation • Cold/Heat Protection • Diapause

Further Reading

Danks, H. V. (1987). "Insect Dormancy: An Ecological Perspective." Biological Survey of Canada (Terrestrial Arthropods), Ottawa.
Denlinger, D. L. (1986). Dormancy in tropical insects. *Annu. Rev. Entomol.* **31**, 239–264.
Leather, S. E., Walters, K. F. A., and Bale, J. S. (1993). "The Ecology of Insect Overwintering." Cambridge University Press, New York.
Lee, R. E., and Denlinger, D. L. (eds.) (1991). "Insects at Low Temperature." Chapman & Hall, New York. [See especially Chaps. 8, 9, 10, and 14]
Tauber, M. J., Tauber, C. A., and Masaki, S. (1986). "Seasonal Adaptations of Insects." Oxford University Press, New York.

Dragonfly

see *Odonata*

Drosophila melanogaster

Patrick M. O'Grady

American Museum of Natural History

When biologists refer to *"Drosophila"* they usually mean *Drosophila melanogaster*. This small, inconspicuous species has become one of the premiere model systems in modern biology. Research on *D. melanogaster* over the past century has led to better understanding of virtually every discipline of biology, especially genetics and developmental and evolutionary biology. This work has applications not merely to the biology of flies and other insects, but also into the causes of a variety of human diseases. The most powerful aspect of *Drosophila* as a model system is the ease with which its genome can be manipulated through a variety of genetic techniques, including germline transformation with transposons. The genomics revolution promises to expand the utility of *D. melanogaster* and make this species not only a pivotal tool in understanding the evolution and working of the genome but also an important model for bioinformatics and genome annotation.

HISTORY OF *DROSOPHILA* RESEARCH

D. melanogaster was first described by Meigen in 1830. Subsequent taxonomists described this species under at least five different names from 1830 to 1941. The profusion of names was likely due to the quick spread of this species throughout the world as a result of the fruit trade.

Drosophila research began in the early 1900s when a number of scientists, most notably T. H. Morgan, began to use *D. melanogaster* as a model organism for studies of genetics. W. E. Castle was the first to bring this species into the lab and develop many of the culture techniques still used today. It was Morgan's group at Columbia University, however, that fully took advantage of this species as a research model. Morgan, up to that time, had been experimenting with marine invertebrates in an effort to understand a number of developmental processes. He was looking for a small, rapidly developing species that produced large numbers of progeny and was both easy and inexpensive to maintain and manipulate in the laboratory. Early in these studies, it became clear that *D. melanogaster* was just such a model system. In 1912 Morgan's group had isolated roughly two dozen mutants. Morgan and his colleagues began to use these mutants to provide experimental evidence for the chromosome theory of inheritance, and they devised methods for gene mapping that are still used today.

Drosophila was an important model organism throughout the 20th century. Ed Lewis began working on homeotic mutants in the 1950s. His work focused on the bithorax gene complex. Most Diptera have only a single set of wings on the mesothoracic segment, but these mutant flies had two pairs

FIGURE 1 The ultrabithorax mutant. [From Lawrence P. A. (1992). "The Making of a Fly: The Genetics of Animal Design." Blackwell Scientific Publications, Oxford, U.K., with permission of the publisher.]

of wings, one each on the meso- and metathoracic segments (Fig. 1). This set of genes has since proved to be the major control switch for body axis development and is conserved in many organisms, including humans. The Nüsslein–Volhard and Wieschaus screens of the early 1980s further advanced the use of *D. melanogaster* as a model system to study the development of more complex organisms. The future Nobel laureates elegantly showed the genetic control of development, mapping many of the genes involved in forming the major body axes in nearly all metazoans.

D. melanogaster continues to be an important model system in biological research, and the *Drosophila* Genome Project has completed the entire genome sequence of *Drosophila melanogaster*. This work, described by Adams and colleagues in 2000, has provided researchers with an immense amount of data that can be used to understand the mechanisms of development and the evolution of the genome. As of late 2002, a reference search of Flybase (http://flybase.bio.indiana.edu/) recovers about 20,000 papers with the query terms *"Drosophila melanogaster."* Furthermore, GenBank (http://www.ncbi.nlm.nih.gov/) currently contains over 322,000 nucleotide entries for this species. Several stock centers around the world are dedicated to maintaining live cultures of *D. melanogaster* and its relatives for research. For example, the Bloomington Stock Center (http://flystocks.bio.indiana.edu/) currently has about 8700 different lines, mostly mutants of *D. melanogaster*, and the Tucson Stock Center (http://stockcenter.arl.arizona.edu/) maintains about 1300 cultures from nearly 300 species in the family Drosophilidae.

ECOLOGY AND LIFE CYCLE

Drosophila melanogaster originated in tropical west Africa and has spread around the world, primarily through its commensal associations with humans. This species is a generalist and breeds in a variety of rotting fruits in its natural environment. It was first recorded on the east coast of North America in the 1870s following the end of the American Civil War and the expansion of the fruit trade.

Like all other members of the family Drosophilidae, *D. melanogaster* is holometabolous and undergoes a complete metamorphosis. Development times vary, depending on temperature. Typical *Drosophila* laboratories maintain flies between 18 and 25°C. Stocks or infrequently used strains are usually kept at lower temperatures to slow development and reduce the amount of stock changing required. Complete development takes about 3 weeks at 18°C. At 25°C, embryonic development is completed roughly 1 day after the egg is laid. The fly then goes through three larval stages prior to pupation. Larvae are motile and work their way through the food media feeding on yeast and bacteria. After 4 days, the larvae enter a stationary pupal stage. Pupation takes approximately 4 days, after which time adults emerge from the pupal case. After they eclose, females require about 2 to 3 days to develop mature eggs. Therefore, at 25°C about 10 or 11 days is required to complete a cycle from egg to egg. At higher temperatures (29–30°C), pupal lethality and female sterility begin to have an effect on culture viability.

After the adult ecloses, it takes between 6 and 12 h for both males and females to begin mating. Genetic crosses require known paternity. Females are collected prior to reaching sexual maturity and isolated from males, so controlled crosses can be made. Mean adult life span is 40 to 50 days, although some individuals may live up to 80 days. A single female can lay as many as 75 eggs in a day, for a total of perhaps 500 eggs in a 10-day period.

MORPHOLOGY AND PHYLOGENY

The family Drosophilidae is divided into a number of genera, subgenera, species groups, and species subgroups; this system gives each species a "taxonomic address" that loosely defines relationships within the family. For example, *D. melanogaster* is placed in the genus *Drosophila*, subgenus *Sophophora*, and *melanogaster* species group and subgroup (Fig. 2). *D. melanogaster* is a typical drosophilid and possesses a number of the characteristics, such as red eyes and plumose arista, that delineate this family. Along with the other taxa in the *melanogaster* and *obscura* species groups, *D. melanogaster* bears a single sex comb on its first tarsal segment. These are 7 to 12 thickened setae (hairs), which are closely set in a row, or comb. The number and position of the setae diagnose *D. melanogaster* from all but the most closely related species.

Within the Afrotropical *melanogaster* species subgroup, *D. melanogaster* is most closely related to the triad of species containing *D. simulans*, *D. sechellia*, and *D. mauritiana,* the common ancestor of which is thought to have diverged between 2 and 3 mya. *D. simulans,* a closely related species that is also cosmopolitan, can be differentiated only by examining the characters of the male genitalia, namely the number of prensisetae and the shape of the epandrial lobes.

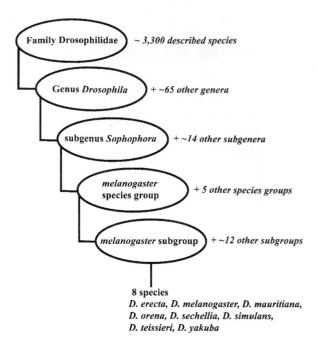

FIGURE 2 Placement of *D. melanogaster* within the family Drosophilidae. [Modified after Powell, J. R. (1997). "Progress and Prospects in Evolutionary Biology. The *Drosophila* Model." Oxford University Press, New York.]

DROSOPHILA MELANOGASTER AS A GENETIC MODEL

Over the past 100 years, geneticists have built a large "toolbox" of specialized methods that allow them to manipulate the genome of *D. melanogaster* with more deftness than is possible with any other organism. These methods have largely taken advantage of some of *Drosophila*'s inherent characteristics, such as the lack of recombination in males. Some widely used techniques include polytene chromosome visualization and *in situ* hybridization, using balancers and other cytological aberrations for genetic crosses, and germline transformation using *P* elements and other transposons to examine gene expression and to tag genes for cloning.

The chromosomes found in the larval salivary glands are highly duplicated, allowing a characteristic banding pattern to be visualized with a compound light microscope (Fig. 3). Polytene chromosomes, which allow researchers to observe and study large-scale genetic rearrangements such as inversions, duplications, translocations, and deletions, have been used by geneticists to answer a variety of questions. Early work focused on understanding chromosome mechanics and using deletions to map the location of specific genes. Molecular geneticists have used the polytene chromosome in conjunction with *in situ* hybridization to more specifically localize the chromosomal site of specific cloned genes or gene fragments. For example, small fragments of DNA can be amplified by using the polymerase chain reaction (PCR), incorporating radioactive or bioluminescent probes as labels and with hybridization to the polytene chromosomes. In addition, evolutionary and population geneticists have used inversion patterns to reconstruct the history of species and populations.

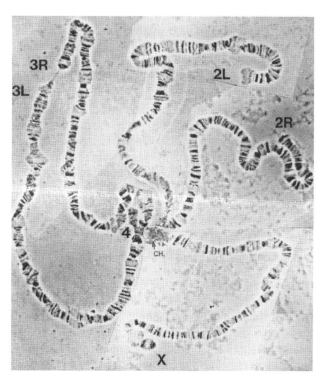

FIGURE 3 Polytene chromosome of *D. melanogaster:* X, X chromosome; 2R, right arm of second chromosome; 2L, left arm of second chromosome; 3R, right arm of third chromosome; 3L, left arm of third chromosome; 4, fourth chromosome; CH, chromocenter. [From Krimbas C., and Powell, J. R. (eds.) (1992). "*Drosophila* Inversion Polymorphism," Fig. 2B, p. 344. CRC Press, Boca Raton, FL, with permission.]

Balancers are multiply inverted chromosomes that repress recombination and are useful for making controlled genetic crosses as well as keeping homozygous lethal mutant genes in culture. In addition to being marked with a visible phenotype, such as curly wings, balancer chromosomes are often homozygous lethal, making crosses and the establishment of multiple mutant stocks much simpler.

Transposable elements (TEs) are native components of the genomes of nearly all organisms. TEs typically encode a protein, called transposase (some also move via a method that is mediated by reverse transcriptase), which can catalyze the movement of the element throughout the genome. Transposons have been used to mutagenize and clone genes, as well as to study spatial and temporal patterns of gene expression. The *P* transposable element was isolated after several researchers noticed an aberrant syndrome of hybrid sterility when certain geographic strains were crossed. This sterility was caused by the introduction of *P* elements into a genetic background lacking these transposons. This transposon has become the most versatile and widely used tool in modern *Drosophila* genetics.

Since their discovery, *P* elements have been heavily modified and are now used extensively to manipulate *Drosophila* germline DNA. The transposase coding regions have been removed and replaced with a wild-type marker gene, resulting in an inactive transposon with a dominant marker. It is possible to introduce such a *P* element into the germline of mutant embryos by injecting the embryos with cloned *P* element DNA and a buffer containing the active transposase. The offspring carrying the *P* element construct will have a wild-type phenotype because the marker will rescue the mutant phenotype of the recipient strain. Such transformed lines can be used in mutagenesis screens by crossing the inactive *P* element construct line to a stock engineered to contain active transposase. Because *P* elements insert at random into the genome, they are very effective mutagenic agents and can insert into a gene, thereby disrupting its function. Once a phenotype has been observed, the transposase can be "crossed out," leaving a stable *P* element insertion into a gene of interest. The mutagenized gene can be easily cloned by means of a variety of techniques (e.g., inverse PCR) because the sequence of the *P* element is known and a "transposon tag" is present in the gene of interest. Other powerful techniques that exploit transposons are enhancer trapping and the flipase recombination system.

THE GENUS *DROSOPHILA* AS A MODEL SYSTEM

In addition to referring to the single species *D. melanogaster,* "*Drosophila*" can also refer to the entire genus *Drosophila,* a spectacular radiation of roughly 1500 described species. This genus can be found throughout the world in every conceivable habitat, from tropical rain forests to subarctic regions. Generally, these species are saprophytic, feeding and ovipositing in rotting plant and, sometimes, animal material. Members of this genus have been used as a model system for understanding evolutionary biology. A number of *Drosophila* groups, such as the *obscura, repleta,* and *virilis* species groups, have become prominent model systems in evolutionary biology. Such studies include chromosome and molecular evolution, the mechanisms of species formation, phylogeny, ecology, and behavior.

See Also the Following Articles
Chromosomes • Diptera • Genetic Engineering • Research Tools, Insects as

Further Reading
Adams, M. D., Celniker, S. E., Holt, R. A., Evans, C. A., Gocayne, J. D., Amanatides, P. G., Scherer, S. E., Li, P. W., Hoskins, R. A., Galle, R. F., *et al.* (2000). The genome sequence of *Drosophila melanogaster. Science* **287,** 2185–2195.

Ashburner, M. (1989). "*Drosophila*: A Laboratory Handbook." Cold Spring Harbor Laboratory Press, Cold Spring Harbor, New York.

Ashburner, M., Carson, H. L., and Thompson, J. N., Jr. (eds.). (1981–1986). "The Genetics and Biology of *Drosophila*." Academic Press, New York. (See especially Wheeler, M. R., pp. 1–84, and Lemeunier *et al.,* pp. 147–256.)

Barker, J. F. S., Starmer, W. T., and MacIntyre, R. J. (eds.). (1990). "Ecological and Evolutionary Genetics of *Drosophila*." Plenum Press, New York.

Kohler, R. E. (1994). "Lords of the Fly: *Drosophila* Genetics and the Experimental Life." University of Chicago Press, Chicago.

Krimbas, C., and Powell, J. R. (eds.). (1992). "*Drosophila* Inversion Polymorphism." CRC Press, Boca Raton, FL.

Lawrence P. A. (1992). "The Making of a Fly: The Genetics of Animal Design." Blackwell Scientific, Oxford, U.K.

Patterson, J. T., and Stone, W. S. (1952). "Evolution in the Genus *Drosophila.*" Macmillan, New York.

Powell, J. R. (1997). "Progress and Prospects in Evolutionary Biology: The *Drosophila* Model." Oxford University Press, New York.

Sullivan, W., Ashburner, M, and Hawley, R. S. (eds.). (2001). "*Drosophila* Protocols." Cold Spring Harbor Laboratory Press, Cold Spring Harbor, NY. (See especially, Pardue, M.-L., pp. 119–129.)

Throckmorton, L. H. (1975). The phylogeny, ecology, and geography of *Drosophila. In* "Handbook of Genetics," Vol. 3, "Invertebrates of Genetic Interest." (R. C. King, ed.), pp. 421–469. Plenum Press, New York.

Dung Beetles

James Ridsdill-Smith
Commonwealth Scientific and Industrial Research Organisation, Australia

FIGURE 1 Adult *S. sacer.* Unworn tibiae indicate a newly emerged beetle. (Photograph courtesy of CSIRO Entomology.)

Dung beetles are specialized to feed and breed on an ephemeral and discrete food resource, namely, the piles of dung produced by herbivorous warm-blooded animals. Adults are strong flyers and can search for some distance to find fresh dung. Most species make tunnels in the soil and remove dung from the pat, which is packed into the tunnel to form a brood mass in which a single egg is laid. Ecosystems contain many coexisting dung beetle species, particularly in tropical grasslands. Intraspecific and interspecific competition for dung is high, and beetles show diverse behavior to reduce its effects. Fecundity of dung beetles is extremely low, the eggs are relatively large, and adult investment in nesting behavior is high, as is illustrated in male/female adult reproductive behavior and in brood care by female beetles.

Dung beetles have a place in history, in which the ball-rolling species, *Scarabeus sacer* (Fig. 1), was sacred to the early Egyptians (Fig. 2). The ability of large ball-rolling beetles to create a perfect sphere, which is then rolled along the soil surface, was taken to be representing Khepri, a great scarab beetle, rolling the globe of the rising sun. The scarabeus, symbol of the sun, is often depicted hovering with outstretched wings. The new beetle emerges from the inactive pupa, representing rebirth or reincarnation. Scarab amulets often were placed over the heart of the dead to simulate rebirth or worn widely by the living to bring good luck.

BIOLOGY

Dung as a Resource

When scarabaeine dung beetles are abundant, numbers can be observed swarming upwind in the odor plume from fresh dung. Volatile compounds produced from the fresh dung attract adult dung beetles, and most arrive within the first few hours after the dung is deposited. Up to 16,000 beetles have been recorded at a single elephant dropping, to which

4000 beetles were attracted in 15 min. Beetles leave when most of the dung is buried or when feeding activity has removed most of the moisture from the dung (referred to as shredding). Adult beetles feed on the liquid "soup" in the dung. The incisor lobe of the adults is flattened and fringed for handling soft food, and the particulate components of the food are filtered out before being ingested. The galae and laciniae of the adults have special brushes for collecting food. Dung produced by herbivorous animals is highly variable in size and consistency and may range from mounds weighing over 1 kg from an elephant to pellets of about 1 g from a rabbit. The water content of dung is high at deposition (90% water), but dung dries out quickly, the rate depending on both temperature and the size of the dung pat. Typically, dung is used as a resource by dung beetles for 1 to 4 weeks, although small sheep pellets may dry out in 3 h in summer and are relatively little used by dung beetles after this time.

FIGURE 2 Egyptian sacred scarabs. Top center is a winged scarab pectoral in blue faience with holes for attachment. Bottom left is a heart scarab in blue faience. Bottom right is an inscribed heart scarab in felspar. (Photograph courtesy of Ashmolean Museum, University of Oxford.)

Dung Burial for Brood Masses

The behavior for which the dung beetle is best known is the removal of dung from the pat and burial in the ground as provisioning for their offspring. Three groups are distinguished based on their behavior in creating a brood mass, the compacted dung in which a single egg is laid. In teleocoprids, a sphere is made from pieces of dung at the dung pat or from pellets of dung (Fig. 3A). Beetles roll the ball away from the dung pat, usually with the hind legs. It is buried in the soil, and a single egg is laid in a small cavity. In paracoprids, the beetle digs a tunnel in the soil under a dung pat, cuts off pieces of dung using its front legs, head, and body, and carries them down the tunnel where they are packed into the end to form a compacted brood mass. As each brood mass is completed, a single egg is laid. Soil is then placed over the brood mass and another brood mass is made. Branching tunnels may be made containing many brood masses with eggs (Fig. 3B). The size and shape of the brood mass, and the depth of the brood mass in the soil, are characteristic for each species. These will be affected by soil moisture and soil hardness. Endocoprid species construct brood balls in cavities within the dung pat.

Life Cycle

When a female beetle is ready to lay her eggs, she constructs a chamber in the top of the brood mass and lays an egg on a small pedestal, which prevents it from coming into contact with the surrounding dung (Fig. 3C). The larva hatches after a week or two and feeds on the dung of its brood mass (Fig. 3D). Larvae have biting mouthparts, unlike the adults, and can use the fiber content of the dung. They typically complete three instars over about 12 weeks and then undergo a pupal stage (Fig. 3E) 1 to 4 weeks before turning into an adult (Fig. 3F). The adults emerge from the brood shells, dig their way to the soil surface, and then fly off and find fresh dung on which to feed. Depending on the biology of the individual species, there may be periods of diapause or quiescence by mature larvae, pupae, or adults during development. Such adaptations are usually related to enhancing survival over, for example, a dry summer or a cold winter and can delay the completion of the life cycle by several months. Dung beetle species are univoltine, completing one generation a year, or multivoltine, completing several.

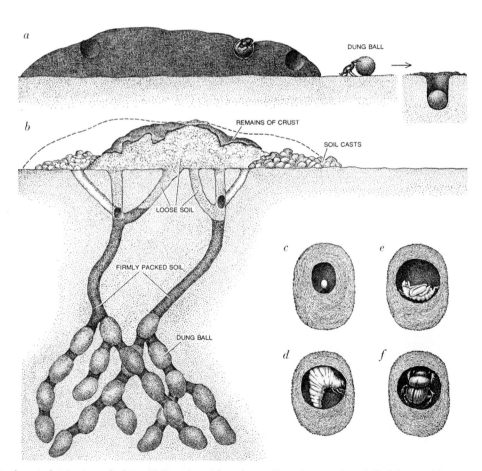

FIGURE 3 Dung beetle reproduction in cattle dung. (a) Dung pat with a teleocoprid species removing a ball of dung and burying it. (b) Dung pat with a paracoprid species producing brood masses in tunnels beneath the pat. Brood mass containing: (c) egg, (d) larva, (e) pupa, and (f) young adult. (Illustration by Tom Prentis from Waterhouse, 1974; reproduced with permission.)

Dung Quality

Beetle egg laying is very sensitive to seasonal changes in dung quality. This quality is influenced by several factors, including rainfall, which affects the growth of plants on which grazing animals are feeding and hence the quality of dung they produce, and the plant species grazed upon. Under favorable laboratory conditions, the rate of egg production of *Onthophagus binodis* on dung collected from cattle grazing on dry summer annual pasture in winter rainfall regions of Australia is 7% of that on dung from cattle grazing on green spring annual pasture. *Onitis alexis,* a larger species, is somewhat less affected by the same seasonal changes in dung quality. The rate of egg production of *Euoniticellus intermedius* on dung collected from cattle grazing on dry winter pasture in summer rainfall regions of Australia is 30% of that on dung from green summer pastures.

Egg Laying

The adult female reproductive system of Scarabaeinae has only a single ovary, consisting of a single ovariole. Newly emerged beetles have no differentiated oocytes in their single ovary. The eggs develop sequentially during a period of maturation feeding. The terminal oocyte is the only one ready to be laid at any one time. If conditions are unsuitable for oviposition, the oocyte is extruded from the ovariole into the hemocoele and nutrients are resorbed.

Fecundity of scarabaeine dung beetles is very low, but they produce relatively large eggs. The length of an average egg is about 33% of the adult female body, and the volume of the egg is about 2.5% of that of the female. Most species probably produce as few as 20 eggs/female/year in the field, because weather conditions or dung quality are rarely ideal for adult reproduction, and some species produce 5 or fewer. Competition for dung at times beetles are ovipositing is high. Adult investment in nesting behavior is high to enhance the success of the offspring that are produced. This is illustrated here in terms of adult male/female reproductive behavior and in female brood care.

ADULT REPRODUCTIVE BEHAVIOR

Recognition of beetles of the same species as mates is important, as beetles are frequently present in large numbers in fresh dung. Male beetles court females by tapping them with their head and forelegs prior to successful copulation. The males of many species produce pheromones that are probably involved in close-range species recognition and in sexual attraction. Pheromones are released via pumping movements from forelegs, or from abdominal sternites, depending on the species.

Male-to-male intraspecific competition occurs in tunnels in the soil for the possession of a female making a brood mass where she is to lay her egg. Of the two males, the larger is usually successful in pushing the smaller beetle away from the

FIGURE 4 A pair of *K. nigroaeneus* rolling a brood ball. The female sits on the ball while the male rolls it backward (beetles approx 2 cm in length). (Reproduced, with permission, from Edwards and Aschenborn, 1988.)

tunnel and the female. This behavior is widely reported in species in both the larger genera such as *Scarabaeus, Kheper,* and *Typhoeus,* and the smaller genera such as *Onthophagus.* The size of male beetles and of horns arising from their head, thorax, and clypeus can be very variable. In many of the smaller species, beetles exhibit dimorphic male morphology. Both large horned major and small hornless minor morphs (forms) coexist in the field. Females produce more brood masses, and the brood masses are larger, in the presence of horned males than in the presence of hornless males. Offspring size is determined by the size of the brood mass used to provision the larva. The horned males assist the females in dung provisioning, providing each egg with more dung. They also guard the tunnels where the females are producing brood masses from other males. The hornless males show alternative mating strategies, by sneak mating with unguarded females, but they do not assist the females in providing dung for brood masses after mating.

BROOD CARE AND SUBSOCIAL BEHAVIOR

Investment in nesting behavior that enhances the survival of offspring is evident in brood care by adult female beetles. The female remains underground with the brood ball, providing care of her offspring during its development. In *Copris,* a male and a female make a chamber in the soil and carry dung down to construct a large dung cake of about 100 g. The male then leaves and the burrow is sealed. The female cuts up the cake into about four brood balls and lays one egg in each. The mother then remains with the brood balls during the development of her larvae, caring for them, repairing damage to the balls by the larvae, and removing fungal growth from the outside. Such behavior doubles the survival rate of the offspring. The mother emerges from the soil after her offspring have emerged from their brood balls.

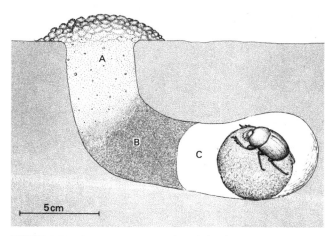

FIGURE 5 Female *K. nigroaeneus* in a brood chamber with a single brood ball. Male has left and packed the chamber entrance closed. (A) Loose soil, (B) hard packed soil, and (C) chamber. (Reproduced, with permission, from Edwards and Aschenborn, 1988.)

An extreme case of low fecundity occurs with *Kheper nigroaeneus,* a large ball-rolling species, which produces only a single offspring on each nesting occasion. A pair of beetles forms a single large dung ball over 4 cm in diameter. The male rolls the ball away from the dung pat while the female clings to the top of it (Fig. 4). The ball is buried via a diagonal tunnel to a chamber about 14 cm deep. The male then leaves the nest while the female beetle stays in the chamber and lays a single egg. She remains underground with the ball for the 12 weeks taken for egg and larval development (Fig. 5). As with *Copris* spp., the presence of the female greatly enhances offspring survival. After the new adults have emerged, the mother leaves the nest.

ECOSYSTEM-LEVEL PATTERNS

Species Richness and Competition

Worldwide there are about 4000 species of scarabaeine dung beetles. Local species richness is generally related to the species richness of large mammalian grazing animals, although cattle dung does support a rich dung beetle fauna around the world. The most competitive assemblages of dung beetles occur in tropical grasslands, where up to 120 species can be present in local areas; here, competition for the dung resource is high.

Intraspecific competition for dung occurs in a range of beetle species, expressed as a reduction in the number of eggs laid/female at high densities. There is also evidence for interspecific competition in which the presence of beetles of one species reduces egg production of a second species. In these instances the competition is frequently asymmetric, and the larger species has a greater effect on the smaller species than the other way round, particularly at high beetle densities. The competitive advantage of larger beetles is associated with preemptive dung burial, whereby they bury a greater proportion of dung in the first day.

Interspecific competition is avoided where there are differences in patterns of dung use and reproductive strategies. Species can be allocated into functional groups on the way they use dung and where it is buried under the pats. Competition is reduced also by the occurrence of aggregated spatial distributions between the discrete dung pats as a result of timing of optimal flight activity and of dung or soil preferences. Of the species present in a region, only some will have a preference for the niche represented by each dung pat, including the time and place at which it is dropped.

Australian Dung Beetle Program

The arrival of farmers from Europe in the 19th century led to a major change in the Australian landscape, resulting from planting pastures and introducing domesticated grazing animals such as sheep and cattle. Native beetles, active only for restricted periods of the year, and occupying mainly heath and other undisturbed habitats, were unable to use effectively the large quantities of cattle dung in the newly created pasture habitats. Dung fauna in these pastures thus consisted of a high abundance of dung-breeding pest flies, but few dung beetles or predatory beetles. George Bornemissza of the Commonwealth Scientific and Industrial Research Organisation (CSIRO) suggested introducing to Australia exotic scarabaeine dung beetles adapted to open pasture habitats and active mainly at the time of year native beetles were not, to correct this imbalance in pastures, to improve nutrient cycling, and to control nuisance flies breeding in the dung. A total of 46 species of scarabaeine dung beetles were introduced into Australia between 1967 and 1995, of which 26 species are established. The project has greatly increased the rate of dung recycling in Australia and reduced the population of at least one important dung-breeding nuisance pest, the Australian bushfly, *Musca vetustissima.*

See also the Following Articles

Coleoptera • Cultural Entomology • Mating Behaviors • Parental Care

Further Reading

Bornemissza, G. F. (1976). The Australian dung beetle project—1965–1975. Australian Meat Research Committee Review No. 30, 1–30.

Doube, B. M. (1990). A functional classification for analysis of the structure of dung beetle assemblages. *Ecol. Entomol.* **15,** 371–383

Edwards, P. B., and Aschenborn, H. H. (1988). Male reproductive behaviour of the African ball-rolling dung beetle, *Kheper nigroaeneus* (Coleoptera: Scarabaeidae). *Coleopterists Bull.* **42,** 17–27.

Halffter, G., and Matthews, E. G. (1999). "The Natural History of Dung Beetles of the Subfamily Scarabaeinae." Reprint Medical Books, Palermo.

Hanski, I., and Cambefort, Y. (eds.) (1991). "Dung Beetle Ecology." Princeton University Press, Princeton, NJ.

Ridsdill-Smith, T. J. (1991). Competition in dung insects. *In* "Insect Reproductive Behaviour" (W. J. Bailey and T. J. Ridsdill-Smith, eds.). Chapman & Hall, London.

Waterhouse, D. F. (1974). The biological control of dung. *Sci. Am.* **230,** 100–109.

Earwig

see *Dermaptera*

Ecdysis

see *Molting*

Ecdysteroids

Michael E. Adams
University of California, Riverside

The ecdysteroids are steroid hormones that, in combination with juvenile hormones, program gene expression appropriate for each stage of insect development. Beginning at embryogenesis, ecdysteroid levels rise transiently during each stage to initiate molting. Depending on the level of juvenile hormone, elevated ecdysteroids mobilize nuclear receptors of several types that bind directly to DNA to either promote or suppress gene expression. During the adult stage of some insects, ecdysteroids play important roles in reproduction, principally in the development of gametes.

DISCOVERY

Early in the 20th century, scientists began to demonstrate roles for circulating hormones in the development and maturation of insects. Stefan Kopec, working in Poland,

observed that removal of the brain in gypsy moths (*Lymantria dispar*), caused developmental arrest. Amazingly, reimplantation of the brain allowed development to resume. In other experiments, Kopec used an experimental tool called a ligature to demonstrate the timing of hormone release from the brain. A ligature is applied with a loop of string that is pulled tight, like a tourniquet, interrupting blood flow between front and back blood compartments. If a ligature is tied around a last instar before the "critical period" of 5 to 7 days, the front part of the animal will develop pupal cuticle, whereas development in the back part will be arrested (Fig. 1). However if the ligature is applied after the critical period, both sides will develop. These observations implied that something was released from the front end of the animal during the interval between days 5 and 7 of development. Consistent with Kopec's earlier work, this signal was later shown to come from the brain. These ingenious experiments illustrated the surprising fact that the insect brain, in addition to its well-known function in electrical signaling, is a secretory organ, controlling developmental processes through release of a hormone into the bloodstream.

The experiments of Kopec were published in two classic papers in the early 1920s, but little happened for about a

FIGURE 1 Ligature tied prior to the critical period (arrow) leads to arrested development in the posterior part of the animal, where green larval cuticle is retained. The thorax synthesizes a new, dark pupal cuticle in response to ecdysteroid release by the prothoracic glands. [Modified from Farb, P. (1962). "The Insects." Time, Incorporated, New York.]

decade. The notion that hormones program insect development was not readily accepted because thinking was dominated by mid-19th century studies of reproductive hormones in birds. These experiments showed that implantation of the avian male testes causes masculinizing effects, whereas removal of the testes produced loss of male characteristics. Such experiments performed in insects had no apparent effect, and as a consequence it became accepted that insects did not engage in hormonal signaling.

Nevertheless, Vincent Wigglesworth, S. Fukuda, and later Carroll Williams extended Kopec's work, by providing evidence for a second signal located in the thorax of the insect that is released in response to the brain hormone. It became evident that implantation of the brain, as Kopec had done, worked only if the brain was placed in the thoracic area. When the brain was implanted into an isolated insect abdomen, developmental arrest persisted. The source of this second factor was the prothoracic gland. Williams showed that, if both the brain and the prothoracic glands were implanted into an isolated abdomen, development resumed. The prothoracic glands alone could accomplish this, provided they had prior exposure to the brain. It therefore became evident that the brain provides a hormonal signal that induces release from the prothoracic glands of a "molting hormone" that is critical to promoting growth.

CHEMICAL NATURE OF ECDYSTEROIDS

Attempts to isolate the molting hormone, or "ecdysone" as it came to be known, began in the 1940s and continued for the next 10 years, until the efforts of two German chemists, Butenandt and Karlson, yielded microgram quantities of pure ecdysone. Because of its water solubility, ecdysone at first was not recognized as a steroid. Soon X-ray studies of ecdysone crystals revealed it as a unique steroid with five hydroxyl substituents, accounting for its ability to dissolve in water (Fig. 2).

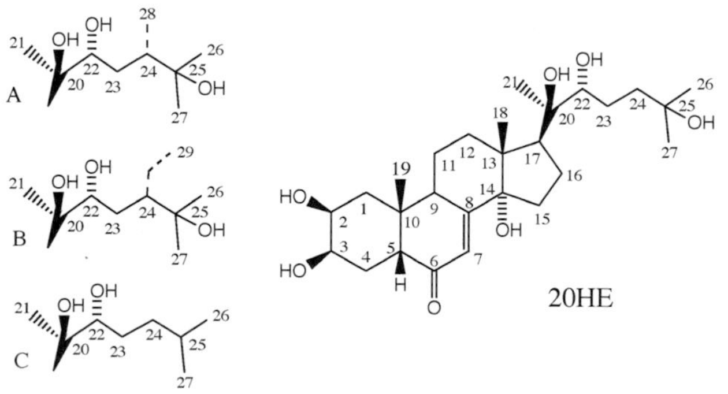

FIGURE 2 The structure of 20-hydroxyecdysone (20HE) and related ecdysteroids. The hydroxyl substituent of 20HE at position 20 confers biological activity to ecdysone. Many insects obtain phytoecdysones from plants and convert them to biologically active forms. These molecules differ from 20HE only in the number of carbons in the alkyl side chain, shown for (A) makisterone A (28 carbons), (B) makisterone C (29 carbons), and (C) ponasterone A, which lacks the hydroxyl group at position 25.

The first ecdysone structure soon was recognized as a precursor to the biologically active material. Upon release from the prothoracic glands, ecdysone, or its 3-dehydro form in some insects, is converted to the active form upon arriving at target tissues. The tissues capable of responding to the hormone produce the enzymes needed to attach a single hydroxyl group to the carbon at position 20, making it 20-hydroxyecdysone (20HE). This substance had already been identified in crustaceans as crustecdysone. Identical molecules function as the biologically active hormone in insects and in crustaceans.

Not long after the identification of ecdysone, phytoecdysones were discovered in plants. The first, ponasterone A, was discovered by Koji Nakanichi, and an identical molecule was later found in some crustaceans. Ponasterone A, which differs from ecdysone in lacking a single hydroxyl at position 25, eventually proved useful in radiolabeled form for the characterization of ecdysteroid receptors. Hundreds of phytoecdysones have been identified, including 20HE itself. Reasons for the presence of phytoecdysones in plants are unclear, but they may serve defensive roles by disrupting the growth of herbivorous insects.

As more insects were examined, additional configurations of the basic ecdysone structure were found. This group of molecules now is collectively known as ecdysteroids. Insects require cholesterol in the diet to synthesize ecdysteroids. Often phytosterols such as campesterol and β-stigmosterol are converted to produce hormonally active makisterones A and C, respectively (Fig. 2).

SOURCES AND FUNCTIONS FOR ECDYSTEROIDS

During the immature stages of development, the chief source of ecdysteroids is the prothoracic gland, a diffuse organ located in the thorax. In higher insects such as flies (Diptera) and bees (Hymenoptera), the prothoracic gland has become part of a composite structure called the ring gland. The precursor ecdysone or 3-dehydroecdysone is synthesized and immediately released into the blood. Then 20-hydroxymonoxgenase is converted to 20HE in target tissues such as epidermal cells, salivary glands, fat body, nervous system, gut, and imaginal discs.

During each stage of development, feeding and growth are followed by a sudden elevation of ecdysteroids, which induces animals to stop feeding and to engage in a new round of gene expression appropriate for the next stage. The epidermal cells begin to secrete a new layer of cuticle and to take back most of the old cuticle, recycling the chitin and protein recovered into the new layer. If the next stage is to be larval, ecdysteroids circulate with high levels of juvenile hormones, and a new set of larval characters is expressed. However, when larval development is complete, metamorphosis is signaled by short ecdysteroid pulses in the absence of juvenile hormones. Completely new structures are created as the insect undergoes the process of changing from a caterpillar into a reproductive, winged adult.

The actions of ecdysteroids can be observed almost immediately after their release as "puffs" on the large polytene chromosomes in the salivary glands of some flies. Even before ecdysteroids were chemically identified, Clever and Karlson noticed puffs within hours of injecting the natural hormone and suggested that 20HE acted to regulate gene expression. This was an excellent insight, for it is now known that ecdysteroids have direct actions to either activate or suppress the expression of many genes via nuclear receptors and that the puffs correspond to gene loci at which transcription takes place.

A model explaining relationships between temporally distinct puffs was proposed by Michael Ashburner in the early 1970s. He found that just after ecdysteroid release, a set of early puffs could be observed, followed by early-late and late puffs. With the identification of ecdysteroid receptors, his model now views early and early-late puffs as evidence of transcriptional activity induced by ecdysteroid receptors (EcRs) and other nuclear receptors. The gene products resulting from these events then give rise to late puffs and also repress further transcriptional activity at early puffs. In this way, a complex but coordinated series of gene expression events occurs, initiated by EcRs.

Having been present throughout larval development, the prothoracic glands degenerate during metamorphosis. Nevertheless, ecdysteroids persist in the adult stage, where they play important reproductive functions. The source of ecdysteroids in adults remained a mystery for many years, although anecdotal accounts implicated the mobile oenocytes as a possibility. In the 1970s, Henry Hagedorn provided a breakthrough, showing that mosquito ovaries produce large quantities of 20HE and that the hormone is required for vitellogenesis, or yolk deposition in developing oocytes. The precise source of ecdysteroids is the follicle cell layer surrounding the oocyte. Since Hagedorn's initial finding, ecdysteroids have been identified also in the testes, where they are involved in sperm maturation.

MOLECULAR BASIS OF ECDYSTEROID SIGNALING

Ecdysteroids belong to a large class of steroid chemical signaling molecules. Because of their lipophilic character, they pass through the cell membrane easily. Whether they affect the cell upon entry depends on the presence or absence of specialized proteins belonging to a large class of soluble, diffusible nuclear receptors. Nuclear receptors get their name from the part of the cell in which they conduct their business, which is the regulation of gene expression. Early in the 1990s, David Hogness and colleagues discovered in fruit flies a class of nuclear receptors they called EcRs. Upon their activation by ecdysone binding, EcRs bind directly to DNA at "ecdysone response elements" (EcRs) to turn genes on or off. Further work showed that to bind with high affinity to DNA, the EcR first finds a partner protein to form a doublet, or "dimer" complex. It is this protein dimer that, together with coactivator proteins, binds to EcREs, resulting in regulation of gene expression.

It is well known that EcRs affect cells in many different ways, causing some to differentiate into muscles and some into glands, and others to form particular kinds of cuticle and cuticular structures appropriate for a larva, a pupa, or an adult. The process by which ecdysteroid receptors encode this diversity of effects is very complicated, and many questions are under current investigation. But it is known already that several different types of ecdysteroid receptor occur in insects, including EcRA, EcrB1, and EcRB2. These receptor "subtypes" occur at different stages of development and can be specific to particular tissues. Thus at least some of the stage-specific effects seem to depend on this diversity of receptors. Another point is that the partners with which they bind to DNA probably vary substantially, providing a further level of combinatorial diversity.

ECDYSONE-BASED INSECT CONTROL

One way of managing pest insect populations is to target unique aspects of their physiology. Because molting is a particularly unique aspect of insect biology, scientists have attempted to learn more about hormonal control systems to be able to design "magic bullets" targeting only insects. The first complete synthesis of ecdysone was accomplished by John Siddall in the late 1960s, but development of such a complicated molecule has not proven to be commercially viable. As insecticides go, the ecdysone structure itself is rather complex and would be expensive to produce on a large scale; moreover, it is too unstable in the environment to be useful in field applications.

Nevertheless, chemicals with unexpected biological activities are produced every year by the chemical industry, and it has become routine to test these compounds, using all available biological assays. This has led to unexpected successes on many occasions, including the serendipitous discovery of the first ecdysone agonists, or "ecdysanoids," by Rohm & Haas in the 1980s. Keith Wing and colleagues at Rohm & Haas found that a series of bisacylhydrazines had astounding ecdysone-like activity, even though their structures were not recognizable as steroidlike. Application of these compounds caused premature insect molting, differentiation of cells, and death of insects due to improper

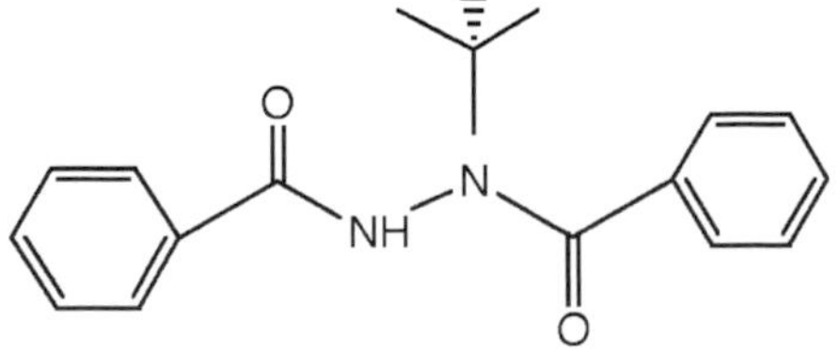

FIGURE 3 The structure of RH5849, one of the first ecdysone agonists developed at Rohm & Haas.

programming of development. The ecdysanoids have been commercially developed for agricultural pest control and represent an encouraging example of how targeting basic insect physiological processes can lead to safer, more environmentally sound insect control agents.

See Also the Following Articles

Development, Hormonal Control of • Juvenile Hormone • Metamorphosis • Molting

Further Reading

Ashburner, M., Chihara, C., Meltzer, P., and Richards, G. (1973). Temporal control of puffing activity in polytene chromosomes. *Cold Spring Harbor Symp. Quant. Biol.* **38,** 655–662.

Clever, U., and Karlson, P. (1960). Induktion von Puff-Veranderungen in den Speicheldrusenchromosomen von *Chironomus tentans* durch Ecdyson. *Exp. Cell. Res.* **20,** 623–626.

Farb, P. (1962). "The Insects." New York, Time Incorporated.

Hagedorn, H. H., O'Connor, J. D., Fuchs, M. S., Sage, B., Schlaeger, D. A. and Bohm, M. K. (1975). The ovary as a source of alpha-ecdysone in an adult mosquito. *Proc. Natl. Acad. Sci. U.S.A.* **72(8),** 3255–3259.

Koelle, M. R., Talbot, W. S., Segraves, W. A., Bender, M. T., Cherbas, P., and Hogness, D. S. (1991). The *Drosophila* EcR gene encodes an ecdysone receptor, a new member of the steroid receptor superfamily. *Cell* **67(1),** 59–77.

Nakanishi, K. (1992). Past and present studies with ponasterones, the first insect molting hormones from plants. *Steroids* **57(12),** 649–657.

Riddiford, L. M., Cherbas, P., and Truman, J. W. (2000). Ecdysone receptors and their biological actions. *Vitam. Horm.* **60,** 1–73.

Wigglesworth, V. B. (1983). Historical perspectives. *In* "Comprehensive Insect Physiology, Biochemistry, and Pharmacology" (G. A. Kerkut and L. I. Gilbert, eds.), Vol. 7, pp. 1–24. Pergamon Press, Oxford, U.K.

Wing, K. D., Slawecki, R. A., and Carlson, G. R. (1988). RH 5849, a nonsteroidal ecdysone agonist: Effects on larval Lepidoptera. *Science* **241,** 470–472.

Eggs

Diana E. Wheeler
University of Arizona

Most insects use fertilized, nutrient-rich eggs to reproduce. In some insects, however, eggs can develop into embryos without fertilization or egg nutrients. And in a few insects, embryos develop directly inside the female insect's body. Nutrient-rich eggs are a resource to parasitoids and predators, as well as to embryos, and so insects use a variety of methods to protect their eggs.

TYPICAL INSECT EGGS

Typical insect eggs contain nutrients to support embryogenesis and produce newly emerged first instars. Most eggs contain large amounts of lipid, for use as building material and energy, and yolk proteins, for the amino acids needed to build a larval insect body. Eggs also contain a cytoplasmic "starter kit" for development that includes cellular machinery such as ribosomes. In species with symbiotic bacteria or protists, eggs are inoculated with a small population of the mutualistic microbes.

Insects typically have internal fertilization, and fertilized eggs contain one set of chromosomes from each parent. Eggs are laid in protected places in environments where young are likely to find food. For example, many butterflies lay their eggs on larval food plants, mosquitoes lay their eggs in water in which larval food grows, and parasitoids lay eggs in, on, or near a host insect. Because mature eggs are usually covered by a thin shell, there must be a way for sperm to penetrate the shell before it is laid, and a way to accommodate water balance and respiratory needs afterward. Sperm enter through an opening called the micropyle. Water and air can pass through specialized regions of the eggshell and embryonic membranes. Finally, some insect groups, remarkably, produce offspring without sperm, egg nutrients, or both.

INSECTS FROM EGGS WITHOUT FERTILIZATION

The sex of hymenopteran insects normally is determined by the number of sets of chromosomes. Unfertilized, haploid eggs have only their mother's set and develop as males. Fertilized, diploid eggs have chromosome sets from both their parents and develop as females. Mated females have control over when sperm is released from the spermatheca to fertilize eggs. Therefore, they can adjust their offsprings' sex ratio in response to a variety of cues. Hymenoptera are particularly susceptible to manipulation of sex determination by parasitic microbes. *Wolbachia,* for example, can alter sex determination so that haploid, and therefore unfertilized, eggs develop as females.

Aphids have complex life cycles that often include female forms that reproduce parthenogenetically, producing female clones of themselves. In such aphids, diploid oocytes form in the germarium and begin development without fertilization.

INSECTS FROM EGGS LACKING NUTRIENTS

Stored nutrients are a major feature in typical insect eggs, but diverse insects have reduced amounts of yolk or lack it entirely. Obviously, if nutrients for embryonic development are not provided in eggs, they must come from another source. Two major alternate sources are the mother and other insects, which can serve as hosts for both embryonic and larval development.

Females that provide nutrients to their embryos, in addition to or instead of egg materials, are termed viviparous. The parthenogenetic aphids described earlier are viviparous: they produce first instars rather than eggs. Young aphid embryos shed their covering of follicle cells and break the strands of tissue that connect them to the germarium. Then, they absorb the necessary nutrients directly from the mother's body.

The viviparous Pacific beetle cockroach, *Diploptera punctata,* produces eggs with insufficient yolk to support complete embryonic development. During pregnancy, females make a supplementary protein, termed roach milk, that is taken up by embryos. Females in a few groups, such as the order Strepsiptera, are neotenous. Neotenic insects do not go through a full metamorphosis, and they reproduce during the larval stage. Strepsipteran females lack oviducts, so that eggs produced in the ovaries are released into the blood. The eggs can contain some, but insufficient amounts of yolk. To fertilize these eggs, sperm must move from the genital canal into the blood. When embryogenesis is complete, larvae use the genital canals to leave the mother's body.

Some parasitoids also develop inside other insects, but here the insect is the parasitized host. Female parasitoids that oviposit directly inside other insect eggs, larvae, or adults are likely to produce small eggs with little or no nutrients. The parasitoid embryos, lacking their own supply of nutrients, then use the host's body to supply materials for their own development.

ECOLOGY OF EGGS

Eggs, the first life stage of insects, can be important ecologically. For example, eggs are the diapausing stage in many insects, with embryogenesis stopping at a species-specific point. Eggs of silkworms (*Bombyx mori*) have an obligatory diapause that coincides with winter under natural conditions. Embryonic diapause has been studied extensively in silkworms because delayed development can be a nuisance from an industrial perspective. Gypsy moth eggs also diapause during winter, but they arrest development at a later stage. Embryos complete embryogenesis and overwinter as unhatched larvae.

Daylength is the most common cue for inducing diapause, but moisture, temperature, and food quality can also be important. The environmental cues that cause eggs to stop developing can be detected by females and then passed on to signal the eggs. Alternatively, the cues can be detected directly by the eggs and embryos. Later, eggs must break diapause in response to another environmental combination of daylength, temperature, and moisture.

Eggs are rich sources of nutrients and therefore pose a great "temptation" to parasitoids, parasites, and predators. Insects protect their eggs in a variety of ways. For example, eggshells can be thick and protective, or cryptic (difficult to detect). Eggs can be laid in protected places. Primarily females, but sometimes males, can contribute chemical repellents or toxins to eggs to deter attacks. A variety of insects stay with their egg masses and actively protect them.

See Also the Following Articles

Diapause • Ovarioles • Parthenogenesis • Spermatheca • Symbionts • Vitellogenesis

Further Reading

Chapman, R. F. (1998). "The Insects: Structure and Function," pp. 298–312. Cambridge University Press, Cambridge, U.K.

Hinton, H. E. (1981). "Biology of Insect Eggs," Vols. 1 and 2. Pergamon Press, Oxford, U.K.

Egg Coverings

Diana E. Wheeler
University of Arizona

The protective coverings females provide their eggs reflect the full range of environments exploited by insects. Egg coverings fall into two major categories: those produced by follicle cells and those produced by accessory glands of the reproductive tract. Follicle cells secrete the chorion, or insect eggshell. The design of the chorion is important in fertilization, egg respiration, and water balance. Coverings produced by accessory glands provide additional protection from the elements, predators, and parasites.

THE EGGSHELL

Layers of the Eggshell

As the oocyte develops, follicle cells secrete structural proteins, along with other substances, as layers. The making of insect eggshell has been best documented in the fruit fly *Drosophila melanogaster.* The vitelline envelope, sometimes considered to be the first layer of the chorion, is the first and innermost layer produced. Next, a layer of wax is secreted, giving the egg greater resistance to desiccation. Then, several more chorionic layers are produced, commonly with sheetlike inner and outer layers separated by a pillared region enclosing air spaces. In insects such as Odonata and Ephemeroptera, which lack accessory glands for producing additional secretions, follicle cells may also produce gelatinous or adhesive coatings. Figure 1 shows the chorionic structures of *Antherea polyphemus,* the polyphemus moth.

Functions of the Eggshell

The eggshell is a layer of armor protecting the egg and developing embryo from the elements, predators, and parasites. Eggshell shape, texture, and color can also provide protection through camouflage and warning coloration.

The protection must be breached, however, to allow for vital functions. First, sperm must be able to enter and fertilize mature eggs. Micropyles, which are openings allowing sperm access to the egg interior, often have distinct architecture, such as the conical protrusion in *D. melanogaster.* Some insect eggs have multiple micropyles.

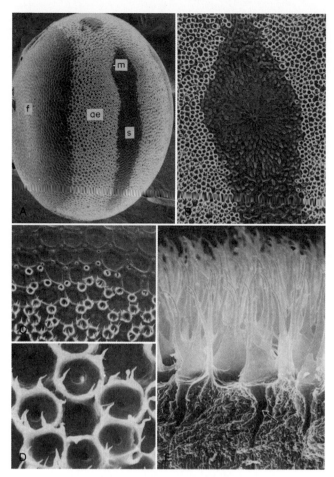

FIGURE 1 Chorionic structures of the polyphemus moth *Antherea polyphemus.* (A) Regional differences in chorion structures. Note micropyle (m) and aeropyle (ae) regions. (B) Micropyle region magnified. (C–E) Aeropyle region. [From Margaratis, L. H. (1985). Structure and physiology of eggshell. *In* "Comprehensive Insect Physiology, Biochemistry, and Pharmacology," Vol. 1 (Kerkut, G. A., and Gilbert, L. I., eds.), Figs. 35 and 72A–C. Pergamon Press, Oxford, U.K.]

Second, the developing embryo is very active metabolically and must respire. Various architectural features of the chorion allow the embryo to exchange gases in both air and water. The interior chorion layer, with its pillared structure, harbors a thin layer of air that can connect directly to the atmosphere by openings termed aeropyles. In moist environments, when air trapped by the inner chorion cannot access atmospheric air directly, the trapped air layer functions as a plastron. Plastrons allow diffusion of gases between water and the air space based on differential partial pressures. The proportion of the egg surface that functions as plastron or aeropyle reflects moisture levels generally found in a particular habitat. To enhance respiratory capacity, eggshells may have extensions, called respiratory horns, that increase the surface area for gas exchange.

Third, some insect eggs are able to absorb environmental water to replace that lost by evaporation. Hydropyles, which are regions specialized for water uptake, may include

chorionic layers in addition to the oocyte membranes interior to the vitelline envelope.

Finally, the embryo itself must be able to emerge from the eggshell when development is complete. Lines of weakness built into the eggshell, based on the interruption of some of the chorionic layers, create a "door," or operculum, through which the first-stage larva can emerge.

OTHER EGG COVERINGS

The colleterial glands (based on *colle* from Greek for glue) are accessory glands of the female's reproductive system that produce egg coatings. Most simply glue is produced to attach eggs to a substrate. In addition, some coatings deter predators or parasites chemically.

Insects in the orthopteroid orders Blattodea, Mantodea, and Orthoptera, as well as some beetle taxa, secrete an egg case or pod surrounding the eggs, to give additional protection from desiccation and predation. As with eggshells, these egg casings must be constructed to allow respiration and hatching. In cockroach eggs, the air space surrounding each embryo opens into a ventilated air duct in the keel. In grasshoppers, the colleterial gland secretions are churned into a froth in which the eggs are suspended. The entire oviposition hole is filled with the frothy material, which then hardens to form a plug. Mantids produce a similar substance from their colleterial glands, which they mold into an egg case that is attached to a flat surface or suitable vegetation.

Beetles are also known to produce egg cases, with the most complex occurring in the cassidine Chrysomelidae. Less complex ootheca have been reported in a variety of other beetle groups as well. Colleterial glands in hydrophilid beetles produce silk that is used to form a cocoon for the egg mass.

Various insect taxa use additional material to enhance protection for eggs. In some beetles, fecal material and/or secretions from anal glands apparently provide chemical defense. Some Lepidoptera use urticating (or irritating) hairs from the larval skin to protect eggs. As adult females emerge from the cocoons, they pick up discarded larval setae (hairs) with their anal tufts and later deposit them on eggs. Nonurticating adult scales, most commonly from the anal tuft, are also used by some species to create an effective physical barrier.

See Also the Following Articles
Accessory Glands • Ovarioles • Reproduction, Female

Further Reading
Hinton, H. E. (1981). "Biology of Insect Eggs," Vols. 1 and 2. Pergamon Press, Oxford, U.K.
Margaratis, L. H. (1985). Structure and physiology of eggshell. *In* "Comprehensive Insect Physiology, Biochemistry, and Pharmacology," Vol. 1 (Kerkut, G. A., and Gilbert, L. I., eds.), pp. 153–230. Pergamon Press, Oxford, U.K.

Embiidina
(Embioptera, Webspinners)

Edward S. Ross
California Academy of Sciences

Embiidina (Embioptera, "webspinners," "foot spinners," "embiids") are warm-climate-adapted insects. Only about 300 species are named, but the order perhaps comprises about 2000 species. Embiids are perhaps most closely related to stoneflies and stick insects (Phasmida) but, probably since Carboniferous or Permian times, have followed their own specialized evolutionary line. The most peculiar feature of all embiids, regardless of developmental stage or sex, is an ability to spin hundreds of strands of silk with each stroke of the greatly enlarged, gland-packed basal segment of the foretarsi.

The silk is formed into narrow galleries serving as protective runways in or on the food supply—weathered bark, lichens, moss, or leaf litter. In arid regions the galleries extend deep into soil and there serve as refuges from heat and desiccation. The primordial habitat is tropical forest, where predation-reducing galleries of most species radiate on edible surfaces of tree trunks. When disturbed, an embiid quickly darts backward into the depths of the labyrinth or into a crevice beneath such cover.

HABITS AND SPECIALIZATION

Except for short, hazardous dispersal of adults, embiids almost never leave the shelter of their self-created microenvironment, and most of the order's anatomical and behavioral characteristics foster very smooth, rapid, reverse movement in narrow galleries. Such specializations include the following:

1. A linear, short-legged, supple body with the head projected forward (Figs. 1 and 2).

FIGURE 2 Typical adult male, wings in repose: *Antipaluria caribbeana* (family Clothodidae) of northern Venezuela; body length, 20.00 mm. Males of species inhabiting arid regions often are apterous, or subapterous.

2. Rapid reverse movement aided by great enlargement of depressor muscle of the hind tibiae.

3. Highly sensitive cerci serving as tactile guides during backward movement.

4. Complete apterism, and thereby elimination, of projecting structures in all females due to endocrinal arresting of development of adult anatomy (neoteny or pedomorphosis) at an early nymphal stage. Males of many species, especially in arid regions, also are apterous or subapterous.

5. Flexibility and forward folding of wings of adult males (Fig. 3)—an advantage in reverse movement as a means of reducing the barb effect against gallery walls and thus a slowing of reverse movement and thereby increased predation.

6. A compensating ability temporarily to stiffen wings for flight by increasing the blood pressure in the full length of the anterior radius (RA), the cubitus, and the anal vein (Fig. 4). This unique wing specialization must have early evolved in both sexes but has been supplanted in females by complete apterism through neoteny.

FIGURE 1 Typical adult female: *"Aposthonia"* n. sp. (family Oligotomidae) of Thailand; body length, 18.00 mm. Females of all species are apterous, and neotenic.

FIGURE 3 Forward wing-flip during defensive, reverse movement of a male: *"Aposthonia"* n. sp. (family Oligotomidae) of Thailand. In repose, wings are flexible; when extended for flight they are temporarily stiffened by blood pressure, particularly in the full length of the anterior radius vein.

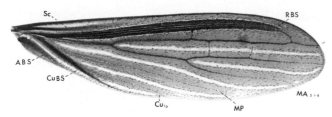

FIGURE 4 Typical embiid forewing: *Pararhagadochir trinitatis* (family Embiidae), Venezuela; wing length 10.0 mm. Most important are the blood sinus veins, especially the anterior radius (RBS, radial blood sinus); less important are the cubital blood sinus (CuBS) and the anal blood sinus (ABS). Hyaline stripes between veins also characterize embiid wings.

In spite of the great antiquity of the order, one that must predate the fragmentation of Pangaea, embiids constitute a single adaptive type of organism, one with a single "ground plan." As in earthworms, no great diversity of body form has occurred because of the physical uniformity of the galleries. As a result, it is difficult to sight-recognize various higher taxa. However, ages of evolutionary diversification are reflected in complex external genitalia of adult males, their head structures, and other characters useful in classification. Females and nymphs are difficult to identify without associated adult males. Adult females offer some anatomical characters in their paragenital sternites and hind tarsi; but size and coloration are most useful for species recognition of females.

RANGE

Embiidina are endemic to all continental landmasses presently in tropical latitudes. The principal evolutionary centers, in order of importance are Africa, the Americas, tropical Asia, and Australia. There also are natural extensions into adjacent temperate regions, such as southern United States and Europe. Several species, particularly of the Asian family Oligotomidae, have widely spread in both ancient and modern commerce and are the most frequently collected species of the order (males are attracted to lights).

Recent extensive, worldwide collecting by Ross has greatly increased the number of known taxa, but most await descriptions now in progress.

The most generalized species (family Clothodidae) occur in tropical South America. The large, diverse family Embiidae comprises a number of subfamilies found in South America, Africa and adjacent Palearctic regions, and Asia as far east as Myanmar. Southeastern Asia and Australia have the peculiar families Embonychidae and Notoligotomidae, and others soon to be described in the literature.

Anisembiidae are confined to the New World tropics and adjacent warm regions. The peculiar family Australembiidae is restricted to the eastern portion of Australia. The Oligotomidae occur almost entirely in tropical Asia and Australia. The large family Teratembiidae is mostly Neotropical and Afrotropical; only a few of its species occur in tropical Asia.

See Also the Following Articles
Phasmida • Silk Production

Further Reading
Ross, E. S. (1970). Biosystematics of the Embioptera. *Annu. Rev. Entomol.* **15**, 157–171.
Ross, E. S. (2000). EMBIA. Contributions to the biosystematics of the insect order Embiidina. 1: Origin, relationships and integumental anatomy of the insect order Embiidina. *Occas. Pap. Calif. Acad. Sci.* **149**, 1–53.
Ross, E. S. (2000). EMBIA. Contributions to the biosystematics of the insect order Embiidina. 2: A review of the biology of the Embiidina. *Occas. Pap. Calif. Acad. Sci.* **149**, 1–36.

Embryogenesis

Lisa Nagy
University of Arizona

Miodrag Grbić
University of Western Ontario, Canada

Embryogenesis is the process by which a larva or a juvenile is built from a single egg. The fertilized egg divides to produce hundreds of cells that grow, move, and differentiate into all the organs and tissues required to form a larva or juvenile. Embryogenesis is extremely diverse in different insect species. For each generalization that can be made about some aspect of insect development, numerous variations and multiple exceptions exist. In some species a single egg gives rise to several thousand larvae, in others, embryos devour their mothers prior to hatching. The most extreme variations are found among insects that parasitize other insects. This article presents a generalized view of some of the more regular features of insect development.

EGG MEMBRANES

Most insects lay eggs in terrestrial environments. For the most part, an insect egg forms a self-reliant developmental system that is generally impervious to the external environment, although sensitive to temperature, which serves as an important cue for many developmental events. Insect eggs are typically quite large, both in absolute dimensions and relative to maternal body size, and well-provisioned with yolk. Eggs vary from about 0.02 to 20 mm in length. To prevent desiccation, they are covered by some of the most resistant and impenetrable egg coverings found in the animal kingdom. Egg contents are protected by a vitelline membrane and covered by an external hard shell, the chorion. The chorion, vitelline membrane, and egg membrane itself surround the internal contents of the fertilized egg: the zygote nucleus and two types of macromolecule. Nutritive material such as yolk proteins, lipids, and carbohydrates are used for nourishment and growth of the embryo. Patterning molecules (such as specialized proteins and mRNAs) direct major events in embryogenesis including establishment of embryonic polarity, segmentation, and gastrulation.

EGG CLEAVAGE

Development in nearly all animals involves a period in which the egg is subdivided into increasingly smaller cells. Compared with other animals, insect eggs undergo an unusual type of cleavage. In most animals, cleavage involves subdivision of both cytoplasm and nuclear material, to form individual cells called blastomeres. In contrast, the early cleavages of most insects involve only nuclear subdivisions (karyokinesis) and are not accompanied by cleavage of the cytoplasm (cytokinesis). This type of cleavage is called syncytial cleavage and results in the formation of a common compartment (syncytium), where up to several thousand nuclei reside (Fig. 1).

It is unclear how syncytial cleavage evolved in insects. The sister group to insects, the entognathans, has both total egg cleavage (Collembola) and syncytial cleavage (Diplura). Basal arthropods such as chelicerates, which include horseshoe crabs, spiders, and scorpions, have both syncytial and total cleavages. Even Onychophora, which are believed to be a sister group to Arthropoda, exhibit both syncytial cleavage (in oviparous species with yolky eggs) and total cleavage (in placental, viviparous species). However, it is reasonable to believe that syncytial cleavage in arthropods and insects evolved from total cleavage in the ancestor to the arthropods.

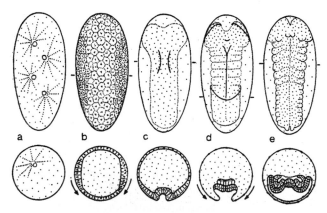

FIGURE 1 Diagram of the basic pattern of early insect embryogenesis: ventral views of eggs, anterior poles at top, are shown above cross sections at the levels indicated by bars in top row. (a) Syncytial cleavage. (b) Formation of the cellular blastoderm: arrows show that the lateral cells are coalescing toward the ventral surface to form the germ anlage. (c) Gastrulation. The prospective mesoderm begins invagination along the midline of the germ anlage. (d) Germ band after gastrulation, with segment borders (dotted) and amniotic folds forming: arrows indicate the movement of the serosal cells to enclose and cover the developing germ band. (e) Advanced germ band stage, with appendage buds, and transient coelomic sacs formed by the mesoderm. [Adapted from Sander, K., *et al.* (1985).]

BLASTODERM FORMATION

In most insects, the early syncytial cleavages proceed rapidly to form a syncytium with up 6000 nuclei. Syncytial nuclei are surrounded by islands of cytoplasm that separate the nuclei from one another. In general, these early cleavages are synchronized and, as the nuclei divide, they separate, such that there is regular spacing between them. Nuclei and associated cytoplasm are referred to as energids. Upon reaching a critical density, the energids migrate to the periphery of the egg. The arrival of the energids at the surface is sometimes visually apparent as bumps along the surface. At the periphery, the energids continue to undergo several rounds of division. In most insects, some of the nuclei remain in the yolk mass. These nuclei subsequently cellularize and become vitellophages, which serve to break down yolk to be used for embryo nutrition. Following the arrival of energids at the egg periphery, the egg membrane invaginates from the egg surface to surround each of the individual syncytial nuclei, marking the end of the syncytial stage of development. The single sheet of cells thus formed at the periphery of the egg is the cellular blastoderm (Fig. 1).

FORMATION OF GERM CELLS

In some species, distinctive granular inclusions can be found in the posterior cytoplasm of the egg. The cells that inherit these granules become the germ cells and eventually migrate into the ovaries or testes to become sperm and eggs. When the germ cells are ablated, the germline is missing and the individual is sterile, as noted in 1911 by Hegner. The nuclear energids (in some species, such as *Drosophila*) arrive at the posterior before reaching any other egg region, and the cells that will include this specialized cytoplasm cellularize earlier than any of the other cells. Germ cells rarely grow or divide during embryogenesis. The early segregation of these cells is thought to protect them from potential errors incurred during division and differentiation that might damage the genetic material necessary to build the next generation. In other species (e.g., most Lepidoptera), segregation of the germline occurs in the middle of the blastoderm; in other species no apparent germ cells can be detected at the blastoderm stage.

SEROSAL FORMATION

Only blastoderm cells destined to form the embryo coalesce to form the germ anlage, which later develops into the germ band. The cells that do not contribute to the germ anlage form an extraembryonic membrane called the serosa. In most species, the boundary between the future serosa and the future embryo ruptures, and the serosal cells migrate over and envelope the embryonic primordium and yolk cells (Fig. 1). However, there is variation in how the serosa is formed. In extreme cases like dipterans, the serosa cells do not migrate over the germ anlage but remain as a cluster of cells on one side of the egg. In addition to the serosa, a second protective membrane, the amnion, forms later from the cells immediately adjacent to the germ anlage. These cells proliferate, flatten, and elongate. As they extend over the germ band, they resemble a sleeping bag being pulled up from the posterior and down from the presumptive head lobes of the germ band. Ultimately, the amnion cells meet in the middle of the embryo and form a single cell layer that lies between the embryo and the now separate serosa. In some derived, holometabolous species (e.g., *Drosophila*), this membrane has become vestigial, and the cells never migrate over the germ band.

GERM ANLAGE FORMATION

The size of the germ anlage varies relative to the length of the egg. In nearly all species, the nuclei arrive at the periphery to form a blastoderm that encompasses the whole surface of the egg. In metamorphic species, such as fruit flies and honey bees, the germ anlage forms from nearly the entire blastoderm surface. However, in direct developing species (such as the grasshopper and cricket), after the formation of a uniform synctyial blastoderm, nuclei migrate and aggregate near the posterior pole, where the germ anlage forms. The germ anlage thus forms from a relatively small proportion of the blastoderm. In the former case, called long-germ-type embryos, the complete body pattern (head, gnathal, thoracic, and abdominal segments) is patterned at the blastoderm stage and all segments appear nearly simultaneously in development. In contrast, in short-germ-type embryos, the head lobes, the most anterior trunk segments, and the posterior terminus are patterned first. Additional segments are added progressively, through proliferative growth. Some insects develop with germ types intermediate between these two extremes. The pleisiomorphic condition for insects is believed to be an intermediate-sized germ anlage. Short/intermediate germ embryogenesis is predominant in direct-developing hemimetabolous insects; more derived, metamorphic insects exhibit long-germ development. However, this division is not clear-cut. In some insect families, closely related species can exhibit both short- and long-germ types of development.

GASTRULATION

Formation of the cellular blastoderm is followed by gastrulation, the process of cellular invagination that results in the formation of a layered embryo comprising two germ layers. Cells that remain at the blastoderm periphery will form the ectoderm, and cells that invaginate below the ectoderm will form the mesoderm. The presumptive mesoderm in most species consists of a strip of cells along the ventral midline (Fig. 1). Gastrulation can happen in any number of ways: by the mesodermal cells invaginating simultaneously, as in *Drosophila,* or by cells invaginating sequentially, beginning at the anterior, while the gastrulation furrow progresses toward the posterior. Most often the presumptive mesoderm lies along the ventral midline, but in the apterygote thysanuran *Thermobia domestica,* cells migrate inward from every part of the germ band. Regardless of the mechanism of gastrulation, the end result is a bilayered embryo, with mesodermal precursors underlying the ectoderm.

SEGMENTATION

Segmentation refers to the process by which repeated units of similar groups of cells, the metameres, are created. Segmentation proceeds nearly simultaneously with gastrulation. Current understanding of the process of segmentation comes from the genetic dissection of development in *Drosophila* by Ed Lewis, Christiane Nüsslein-Volhard, and Eric Wieschaus. These researchers used a large-scale mutant screen to uncover developmental defects in *Drosophila*. They found a complex genetic regulatory cascade that specifies the insect body plan (Fig. 2), which has since been shown to have many commonalties with molecular patterning in vertebrate embryos. The three were awarded the Nobel Prize for their efforts. However, even before the genetic dissection of segmentation, the elegant work of Klaus Sander had defined the basic mechanisms of embryo patterning that helped in the interpretation of the new genetic data. By analyzing the outcomes of embryonic manipulations of leafhopper embryos *(Eucelis),* Sander concluded that two morphogenetic gradients specify the pattern elements along the anteroposterior axis of

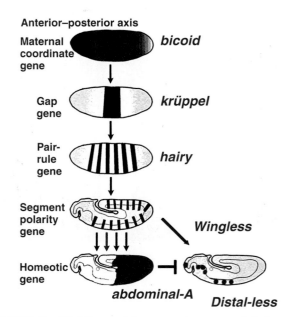

FIGURE 2 Simplified diagram of the segmentation gene cascade in *Drosophila melanogaster* and its relation to limb development. Diagrammatic representation of some of regulatory interactions between genes in the *Drosophila* segmentation cascade: a maternal coordinate gene, a representative gap gene, a pair-rule gene, the segment polarity gene, a homeotic gene, and the limb-patterning gene *Distal-less.* [Modified from Nagy, L. (1998). *Am. Zool.* **38**(6).]

the germ band: one gradient with a high point in the posterior pole and another with a high point at the anterior pole (Fig. 3). In 1976 Sander reviewed these experiments, and a large body of other experimental work on insect embryos, setting the stage for modern understanding of the molecular basis of insect development.

MOLECULAR CIRCUITS THAT REGULATE SEGMENTATION

In the cascade of gene activity that generates the segmental pattern of the embryo (Fig. 2), segmentation proceeds by a progressive refinement of positional information that will eventually specify groups of cells that form metameric units. Refinement is initiated with maternally provided proteins that form gradients from the anterior to the posterior of the egg and early cleavage stages. These gradients of maternal proteins provide the coordinates that position the front and

back of the embryo; hence the genes that encode these proteins are called the maternal coordinate genes. The function of these maternal gradients is to activate the gap genes, which are so named because when their function is lacking, the segmental pattern of the embryo has large gaps in it (e.g., the three thoracic segments may be missing, or the first few abdominal segments.

The proteins encoded by the gap genes in turn activate the pair-rule genes. The pair-rules genes are expressed in every other segment and represent the first apparent metameric pattern. It was somewhat surprising that the first metameres produced by the pair-rule genes during embryogenesis do not correspond to the adult segments, but rather consist of a unit with double-segment periodicity. When pair-rule genes are absent, the larva has only half the normal number of segments. The pair-rule proteins then activate the segment polarity genes, a set of genes expressed in a segmentally reiterated manner. Finally, the homeotic genes are activated in a region-specific manner. The homeotic genes are a well-studied group of genes that are responsible for conferring segment character. They provide information on whether an individual segment will be a specific mouthpart, thoracic, or abdominal segment.

Much of what has been learned about the molecular process of segmentation is from *Drosophila;* how much is representative of a general process for all insects is not yet known. The segment polarity and homeotic genes, as well as their presumed functions, seem to be conserved in all insects examined so far; however, the activity of the maternal coordinate, gap, and pair-rule genes is more variable (Fig. 4). Because of the variation in the formation of germ anlage, it is not surprising that the earliest stages of the segmentation gene cascade established in *Drosophila* do not function in more ancestral insects. Exactly how short-germ-type embryos establish their segmental pattern remains to be discovered.

THE GERM BAND AND DORSAL CLOSURE

The germ band is a two-layered structure, comprising both ectoderm and mesoderm, that represents the outline of the final body plan along both axes. As the embryo grows, the germ band transforms from this essentially two-dimensional, two-layered sheet into a three-dimensional larva. From anterior to posterior, all the segments are represented. Individual segments first become visible near the anterior end, where the ectoderm differentiates into the brain and compound eyes. Protrusions develop anterior to the mouth opening that will eventually grow to form the labrum (front lip of mouthparts) and the antennae. The next segment, the intercalary segment, develops a transient limb bud, which is later retracted. This bud may be a remnant of a second pair of antennae found in this position along the anterior–posterior axis in crustaceans. Each of the first three segments behind the mouth form paired appendages that become the mouthparts: mandibles, maxillae, and labium. The next three segments

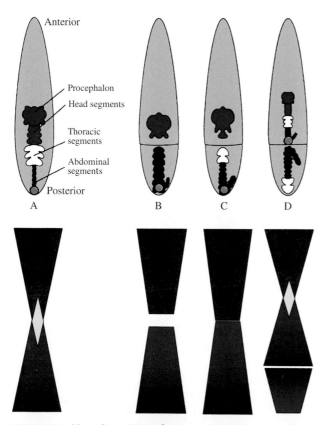

FIGURE 3 Double gradients. (A) Leafhopper embryo consists of head (red), thoracic (white), and abdominal segments (black). Posterior pole of the egg is marked by bacterial symbiont (green). (B) After early egg ligation, anterior and posterior fragments form fewer segments than in the normal embryo. (C) Late ligations result in more segments formed in anterior and posterior fragments. (D) Finally, when posterior material has been displaced anteriorly and the egg ligated just below the symbiont marker a mirror-image duplication was formed. Schematic of corresponding anterior (blue) and posterior (red) gradients and their overlap (yellow) below images illustrates possible anterior and posterior gradients and their interactions in each experimental intervention.

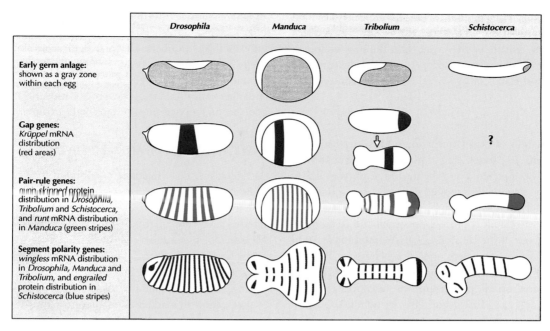

FIGURE 4 Comparative germ band and expression patterns of segmentation genes. [Modified from Nagy, L. (1994). *Curr. Biol.* **4**, 811–814.]

develop into the thorax and form appendages that become walking legs. During the remainder of embryogenesis, as organs develop and differentiate, the flanks of the germ band, both ectoderm and mesoderm, grow laterally and extend around the yolk. The two edges of the germ band meet and fuse along the dorsal midline, such that the mesodermal and ectodermal layers now enclose the yolk.

ORGANOGENESIS

When the germ band is fully segmented and gastrulation is complete, the remainder of embryogenesis involves the differentiation of the ectoderm and mesoderm into the organ systems of the larva or juvenile. The ectoderm gives rise to the bulk of the larval or adult form. Most obviously the ectoderm forms the "skin" of the larvae, marked by numerous bristles and hairs. In addition, the nervous system develops from the ventral ectoderm, and the tracheal system develops from invaginations of the lateral ectoderm. Ocelli, salivary glands, a prothoracic gland, corpora allata, molting glands, oenocytes, and silk glands also develop as ectodermal invaginations. Finally, two additional invaginations of the ectoderm occur:

1. The stomadeum occurs in a central position near the anterior of the germ band, and once invaginated, these cells proliferate in a posterior direction to form the foregut.

2. The proctodeal invagination occurs in the terminal segment, and these cells grow anteriorly to form the hindgut.

Malpighian tubules, the insect excretory organ, develop from outpocketings of the proctodeum. The invaginated mesoderm initially forms a pair of transient coelomic sacs in each segment (Fig. 1E). From these, the dorsal vessel, or

heart, the internal reproductive organs, muscles, fat body, subesophageal gland, and hemocytes will form. The midgut arises from a third germ layer, the endoderm, that develops at the edge of the fore- and hindgut invaginations and eventually fuses with them to complete the gut. During the remainder of development both the mesodermal and ectodermal organ primordia all undergo differentiation into tissue-specific cell types and cell rearrangements required to form the final organ structures.

APPENDAGE DEVELOPMENT

In direct-developing hemimetabolous insects, leg and wings develop as direct outpocketings from the lateral embryonic ectoderm. Leg buds appear early, just after the completion of gastrulation, whereas wing buds appear later in development, after the lateral ectoderm has grown dorsally. In many metamorphic insects, rather than outpocketing, a cluster of cells that will form the adult leg and wing invaginates below the ectoderm. These cells become the leg and wing imaginal discs and do not undergo any further differentiation until later larval stages.

The molecular basis of positioning the limb primordia within the embryo is also well established in *Drosophila* and seems to be similar in many respects throughout both hemi- and holometabolous insects. The same information required to pattern the body axis (Fig. 2) is used to pattern the limb primordia. Every segment has the capacity to form a limb, and limbs appear at a discrete boundary formed at the intersection of the segment polarity genes, and the graded signals that are used to pattern the dorsal ventral axis of the embryo. Limb primordia are marked by the expression of the

Distal-less gene. The absence of *Distal-less* gene function results in the loss of distal limb structures up to the proximal limb segment, which is the coxa. The limbless abdomen characteristic of insects is created by a subsequent repression of *Distal-less* gene activity by several of the homeotic genes.

HATCHING

Upon completion of organ formation and histogenesis, the embryo begins to stretch and contract its newly formed muscles, and gas is secreted into the trachea. Embryogenesis is over when the maternally supplied yolk has been consumed. Hatching is achieved by any number of means, but typically, hatching is a mechanical process, in which the larva either chews its way out of the chorion, grows by imbibing air until the chorion cracks, or uses a special egg burster. There is sometimes an enzymatic digestion of the eggshell, but complicated hydrostatic mechanisms are used, as well. The hatchling emerges as a first instar (larva or nymph).

See Also the Following Articles

Eggs • Imaginal Discs • Segmentation • Vitellogenesis

Further Reading

Cohen, S., and Juergens, G. (1989). Proximal–distal pattern formation in *Drosophila:* Graded requirement for *Distal-less* gene activity during limb development. *Roux's Arch. Dev. Biol.* **198,** 157–169.
Counce, S. J., and Waddington, C. H. (1972). "Developmental Systems: Insects," Vol. 2. Academic Press, London.
Hegner, R. W. (1911). Experiments with Chrysomelid beetles. III: The effects of killing parts of the eggs of *Leptinotarsa decemlineata. Biol. Bull.* **20,** 237–251 (1911).
Kume, M., and Dan, K. (1968). "Invertebrate Embryology." Garland Publishing, New York.
Nüsslein-Volhard, C., and Wieschaus, E. (1980). Mutations affecting segment number and polarity in *Drosophila. Nature* **287,** 795–801.
Lewis, E. B. (1978). A gene complex controlling segmentation in *Drosophila. Nature* **276,** 565–570.
Sander, K. (1976). Specification of the basic body pattern in insect embryogenesis. *Adv. Insect Physiol.* **12,** 125–238.
Sander, K., Gutzeit, H. O., and Jaeckle, H. (1985). Insect embryogenesis: Morphology, physiology, genetical, and molecular aspects. *In* "Comprehensive Insect Physiology, Biochemistry, and Pharmacology." (G. A. Kerkut and L. I. Gilbert, eds.). Pergamon Press, Oxford, U.K.

Endangered Insects

Scott Hoffman Black and Mace Vaughan
The Xerces Society

The Xerces blue butterfly, Antioch katydid, Tobias' caddisfly, Roberts's alloperlan stonefly, Colorado burrowing mayfly, and Rocky Mountain grasshopper all were driven extinct by humans, and all foreshadow the fate of the world's endangered insects. With almost 1 million described species, insects eclipse all other forms of animal life on Earth, not only in sheer numbers, diversity, and biomass, but also in their importance to functioning ecosystems. However, human-induced changes to the natural environment endanger vast numbers of these organisms, threatening them and the vital services they provide with extinction.

INSECT DIVERSITY AND IMPORTANCE

As biologist J. B. S. Haldane noted more than 60 years ago, "The creator must have an inordinate fondness for beetles." The more than 300,000 species of beetle to which Haldane referred are representative of the great diversity of insects. Measured by the number of formally described species, insects are by far the most diverse group of organisms on Earth. More than 950,000 species of insects have been described, comprising 72% of the total identified animal species on Earth.

Even more remarkable are the estimates of how many insects we have not cataloged. Most insect species that have been classified and named to date are from temperate zones, but tropical habitats harbor far more. Smithsonian Institution entomologist Terry Erwin has suggested that as many as 30 million insect species may exist based on extrapolations from the number of beetles found in particular tropical tree species. The most conservative estimates suggest that 5 to 8 million insect species have not been discovered. This number contrasts sharply with the 5,000 to 10,000 species of vertebrates that may await discovery and description around the world.

The sheer number and mass of insects reflect their enormous ecological impact. The world's ecosystems depend upon insects for pollination, decomposition, soil aeration, and nutrient and energy cycling. As Harvard biologist E. O. Wilson wrote, "So important are insects and other land dwelling arthropods, that if all were to disappear, humanity probably could not last more than a few months."

INSECT ENDANGERMENT

A report by the World Commission on Environment and Development noted, "there is a growing consensus that species are disappearing at rates never before witnessed on the planet" but that "we have no accurate figures on current rates of extinctions, as most of the species vanishing are the least documented, such as insects in tropical forests." Scientists and conservationists agree that insect species are going extinct. But how many have been lost and how many more are at risk remains unclear.

Extinct Insects

The International Union for Conservation of Nature and Natural Resources (IUCN) lists 72 insects as extinct worldwide. In the United States, the Natural Heritage Program lists 160 insect species either as presumed extinct or as missing and possibly extinct. Many scientists believe that these numbers drastically underestimate actual insect extinction and that many hundreds, or perhaps thousands, of species have gone extinct unnoticed in North America and Europe in the past 2 centuries. The loss in tropical areas has probably been much greater.

For example, the Antioch katydid, *Neduba extincta,* from California was described in 1977 from preserved specimens collected 40 years earlier. Searches of its sand dune habitat, now largely destroyed, have proved fruitless. The Tobias' caddisfly, *Hydropsyche tobiasi,* was described in 1977 from specimens collected on the Rhine River in the 1920s. None have been seen since.

In some instances, insects that at one time were very common have disappeared. During the mid-1800s, immense swarms of the Rocky Mountain grasshopper, *Melanoplus spretus,* periodically migrated from the northern Rocky Mountains and destroyed crops throughout the western and central portions of the United States and Canada. However, in the late 1880s this species began a precipitous decline. Some believe that a natural population crash combined with habitat destruction and introduced species led the Rocky Mountain grasshopper to extinction. If a widespread species can vanish because of human activity, the fate of many endemic tropical species must hang in the balance as their only habitat is destroyed.

Endangered Insects

Based on available information we can deduce that a very large number of insects are endangered. The majority of animals on the planet are insects and, if the factors that endanger other animals also affect insects, the number of endangered insects must be very large.

According to the 2000 IUCN *Red List of Threatened Species,* 163 insects are listed as critically endangered or endangered worldwide. In 1987, West Germany classified 34% of its 10,290 insect and other invertebrate species as threatened or endangered and, in Austria, this figure was 22% of 9694 invertebrate species. More recent figures from 2000 for Great Britain show that 10.8% of its 14,634 described insect species are rare, vulnerable, or endangered. In the United States, both the U.S. Fish and Wildlife Service (USFWS) and the Natural Heritage Program track endangered species, including insects. The USFWS lists 44 insects as either endangered or threatened, whereas the Natural Heritage Program lists 165 insects as either critically imperiled or imperiled.

Are these figures on endangered insect species realistic? Because we lack an enormous amount of information on the taxonomy, life history, and distribution of insects and because endangered species documentation is biased in favor of vertebrates, we certainly are underestimating the number of at-risk insect species. To illustrate, only 7 and 4% of the endangered animal species listed by the IUCN and USFWS,

respectively, are insects, yet insects make up more than 72% of global animal diversity. Of all the vertebrates described in the United States, 17.9% are listed as threatened or endangered. If we assume that insects and vertebrates face similar destructive forces at similar levels of intensity, then one should expect to find on the order of 29,000 at-risk insects in the United States alone. Although this assumption oversimplifies the situation, it shows that the 44 insects listed as endangered and threatened by USFWS are a significant underestimate. The Natural Heritage Program may be closer to the mark for select groups of insects for which we have more information. It estimates that 43% of stoneflies, 19% of tiger beetles and butterflies, and 17% of dragonflies and damselflies are critically imperiled or imperiled in the United States. In addition, according to the IUCN *Red Book of Swallowtails,* 10% of swallowtail butterflies are considered threatened. Swallowtails are the only group of insects to have been assessed worldwide.

IMPORTANCE OF ENDANGERED INSECTS

A rare and endangered species of insect is unlikely to determine the fate of a large ecological system, but as a group they may have a large effect. Ecosystem functions, such as the recycling of nutrients, often are done by specialists like the American burying beetle rather than generalists. There are innumerable specialized insects that feed on particular kinds of wood, dung, or carrion. For instance, the plates that cover the shells of tortoises are made of keratin, a protein few scavengers can digest. However, in Florida there is a moth, *Ceratophaga vicinella,* whose caterpillar appears to have specialized on a diet of dead gopher tortoise shells.

Endangered species also can play a linchpin role in small, specialized systems, such as caves, oceanic islands, or some pollinator–plant relationships. For example, many plant species rely on one or a few pollinators. Decreased abundance or loss of any of these pollinators can have dramatic consequences, especially if a plant depends on a single, obligate pollinator.

Some endangered species might provide useful products, such as new defenses against diseases and tools for studying various ecosystem or organismal processes, as well as direct material benefits. For instance, the conservation of several species of butterflies is helped by the market value of aesthetically pleasing specimens or of live specimens for butterfly houses that charge admission.

In addition to these material reasons for conserving endangered insects, we also have the responsibility of caring for the rich biological heritage we leave to future generations. At this time, we cannot begin to grasp the full value of biodiversity and, thus, it is in our best interest to be conservative.

CAUSES OF ENDANGERMENT

Insects become endangered because of the same destructive forces faced by many other animals. According to the IUCN, the leading causes of animal endangerment are habitat destruction, displacement by introduced species, alteration of habitat by chemical pollutants (such as pesticides), hybridization with other species, and overharvesting. Many at-risk insects are threatened by more than one of these causes. For example, according to the Natural Heritage Program there are six tiger beetles and 33 butterflies that are imperiled or federally listed under the U.S. Endangered Species Act. The major threat to all six tiger beetles is habitat degradation and loss. Two of these beetles also are threatened by overcollecting. For the 33 butterflies, 97% are threatened by habitat loss, 36% by alien species, 24% by pollution, and 30% by overcollecting.

Insects as a group are not at risk because many species are generalists or widely distributed. A significant proportion of the total diversity of insects, however, is composed of species that are highly specialized or are restricted to one or a few small patches of habitat. The giant flightless darkling beetle, *Polposipus herculeanus,* for instance, lives only on dead trees on tiny Frigate Island in the Seychelles. The stonefly *Capnia lacustra* exists only in Lake Tahoe and is also the only stonefly in the world known to be fully aquatic in the adult stage. Another unusual stonefly, *Cosumnoperla hypocrema,* is known from only one intermittent spring in the Cosumnes River Basin in California.

Habitat Destruction

Agriculture, commercial development, outdoor recreation (including off-road vehicles), pollution, and water development rank as the most frequent causes of habitat degradation affecting federally listed endangered and threatened insect species in the United States (Fig. 1). Commercial and residential developments often are situated on sites that have naturally high diversity, such as along rivers or near bays and estuaries. Urban development in the southeastern United States and California has had particularly strong impacts on native insects because of the high rates of insect endemism

where these cities were built. The best known case is that of San Francisco, California, which now almost entirely covers what was once one of the major coastal dune ecosystems in western North America. Three dune butterflies, which were endemic to this region, are now extinct: *Cercyonis sthenele sthenele, Glaucopsyche xerces* (Fig. 2), and *Plebeius icarioides pheres.* Three other butterflies, *Speyeria callippe callippe, Callophrys mossi bayensis,* and *Plebeius icarioides missionensis,* are now limited to the San Bruno Mountains just south of San Francisco, the last remnant of the San Francisco hills ecosystem.

Conversion of natural habitats for agriculture, particularly for planted food and fiber crops (e.g., cotton), is one of the most extensive land uses and, according to Robert Pyle (a noted lepidopterist and author), has resulted in the greatest loss of native insect populations. The most serious losses of endemic insects to agricultural conversion have taken place in the tropics, but because of the lack of knowledge of insects in these regions, it is impossible to know the extent of this destruction.

Dams and other water development are implicated in the decline of 21% of federally listed insect species. Impoundments destroy habitat for native aquatic organisms, such as stoneflies, as well as some terrestrial insects. For example, the damming of the Columbia River in Oregon and Washington resulted in the destruction of much of the sand bar habitat of the tiger beetle, *Cicindela columbica.*

Although we have no numbers, insects most likely are lost to large-scale timber management. Studies have shown that there is higher invertebrate diversity, as well as endemism, in late successional forests than in younger stands, and less than 10% of U.S. native forests remain intact. Widespread use of off-road vehicles also threatens some species. For example, vehicles have crushed the larval burrows of the tiger beetle, *Cicindela dorsalis,* along beaches to such an extent that this

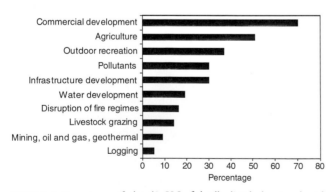

FIGURE 1 Percentages of the 43 U.S. federally listed threatened and endangered insect species affected by different causes of habitat destruction or degradation, as of December 2001. It is important to note that the habitats of most listed species are being degraded by more than one cause. (Bar graph format modified from B. A. Stein *et al.,* 2000. Data modified from D. S. Wilcove *et al.,* 1998, Quantifying threats to imperiled species in the United States. *Bioscience* **48,** 607–615.)

FIGURE 2 The Xerces blue butterfly *(G. xerces),* was one of the first butterflies in North America known to become extinct as a result of human interference. It was driven to extinction as San Francisco expanded over the butterfly's habitat. (Photograph courtesy of C. B. Barr and the Essig Museum of Entomology, University of California, Berkeley.)

once widespread, abundant species has been eliminated throughout most of its range. Wetland draining also has taken its toll. The draining of fens in England caused the extirpation of the butterfly *Lycaena dispar* in 1851 and possibly other insects as well. Capping of springs led to the loss of the fritillary butterfly, *Speyeria nokomis coerulescens,* in the U.S. portion of its range.

The biggest unknown is, of course, the loss of tropical rainforest. Tropical rain forests may hold the majority of terrestrial insect diversity and are being converted to agriculture and other uses at an alarming rate. As rainforests around the world are clear-cut, insects are bound to go with them.

Alien Species

The introduction of various exotic organisms (whether intentional or not) has affected native insects, both directly and indirectly. For example, introduced plants may out compete native plants and, thus, lead to the loss of insect host plants or habitat. Introduced plant diseases also can wreak havoc on insect populations. A classic example involves the American chestnut. Mature examples of the tree disappeared throughout its range following the accidental introduction of chestnut blight. At least five microlepidopterans, including the chestnut borer, *Synanthedon castaneae,* are believed to have gone extinct because of the loss of their host plant. Some aquatic insect species are restricted to small mountain lakes in the United States and have been impacted by introductions of nonnative fish. On the Island of Oahu, a species of *Megalagrion* damselfly is uniformly absent in stream reaches where nonnative mosquitofish in the family Poeciliidae have been introduced.

Intentional introductions of insects also many harm native insects. Over the past 50 years, nonnative insects often have been released to control nonnative pest insects. Although the damage to nontarget, native insects from these biological controls is rarely documented, some evidence is surfacing that it may be significant. For example, a parasitoid fly, *Compsilura concinnata,* that was released repeatedly in North America from 1906 to 1986 as a biological control against several pests, including the introduced gypsy moth, is implicated in the declines of four species of giant silk moths (Lepidoptera: Saturniidae) in New England. Another study in Hawaii found that 83% of parasitoids reared from native moths were former biological control agents.

Overcollecting

Although overcollecting has not been shown to harm healthy populations of insects, it may be an important threat to insect species with very small populations and is included in the list of threats to many of the federally protected insect species in the United States. The Endangered Species Act expressly forbids the collection of endangered or threatened species, and most insect conservationists feel that collecting from small populations should be done only for well-designed, hypothesis-driven, scientific studies. It is not too much to ask that scientists rise to this standard when studying populations that are at risk.

Other Potential Threats

Pesticides and other pollutants are implicated in the decline of many native bees and some aquatic insects, although the degree of impact is not conclusive. Lights along streets and highways also have been implicated in losses of nocturnal insects, particularly large moths. Finally, even though we cannot specify the exact effects of climate change at this time, it could lead to endangerment of endemic insects with specific, narrow habitat requirements. A changing climate may be especially detrimental to species that cannot disperse, like the Uncompahgre fritillary butterfly *(Boloria improba acrocnema),* which is restricted to high mountain slopes in southern Colorado.

PROTECTING AT-RISK INSECT SPECIES

Conservationists have concluded that the current, widespread destruction of the earth's biodiversity must be matched by a conservation response an order of magnitude greater than currently exists.

Protecting Habitat

Ultimately, to protect any species one must protect its habitat. Some insects need only small areas to thrive, and even backyard gardens may help some pollinator insects. Large swaths of land set aside as reserves, wilderness, national parks, and conservation easements ultimately may benefit insects and other invertebrates. Recent evidence, however, shows that some reserves, with management plans tailored to vertebrates, do little to protect insects such as butterflies.

One important caveat for setting aside land for insects is that species often have subtle habitat requirements and can be lost even from reserves because of apparently minor habitat changes. For example, larvae of the large blue butterfly *(Maculinea arion)* are obligate parasites of red ant colonies *(Myrmica sabuleti).* In 1979, this butterfly went extinct in England because habitat was not managed for these red ants. The large blue subsequently has been reintroduced successfully to appropriately managed sites in England using a subspecies from Sweden.

Federal Laws and Legislative Efforts

Federal legislation is vital to the protection of endangered insects. In the United States, the formal listing of species as threatened or endangered under federal or state endangered species legislation has been an extremely effective habitat protection tool because (1) these species are protected by law and (2) money is allocated for recovery efforts. In addition to this protection, a listing as "sensitive" or "indicator species" under U.S. Forest Service National Forest Management Act regulations, or even a formal listing from nongovernmental organizations such as IUCN and the Natural Heritage Program, raises visibility and an awareness of these species. This increased attention may lead to the stricter legal protection of a federal listing under the U.S. Endangered Species Act.

Other countries also have legislative efforts to protect insects and other invertebrates. In 1986, the Committee of Ministers of the Council of Europe adopted a charter favoring the protection of invertebrates. This charter has raised awareness to the plight of endangered invertebrates and, in some cases, led to habitat protection. For most developing countries in the world, protective legislation for insects is either lacking or only sporadically applied. One exception is Papua New Guinea, where there is legislation, as well as a management program, that protects the rarest birdwing butterflies, allows only citizens to sell native insects, and protects some insect habitat.

Research

Before we can work to protect insects and other invertebrates we need to know, at least, what species are present, if populations are stable or declining, and the habitat needs of these populations. In the long run, more emphasis needs to be placed on invertebrate survey, systematics, taxonomy, and population ecology so that these species can be identified and cataloged and their life histories understood. Research needs to go hand in hand with conservation, for a catalog of extinct species is of little use.

Insects as Commodities

Conservation-based ranching of butterflies and other charismatic insects, like scarabs, can protect and conserve critical habitat for threatened species where the appropriate tropical forests remain intact and where live insect export is legal. The tropical forests of Central and Latin America, the Philippines, Madagascar, Kenya, Malaysian Borneo, Jamaica, and Indonesian Irian Jaya meet these criteria. These ranches not only offer protection to these charismatic insects and their habitat, but also serve as a sustainable means of economic development.

We differentiate between butterfly farming and ranching. According to the Convention on International Trade in Endangered Species (CITES) "farming" operations are essentially closed systems, no longer dependent upon regular infusions of wild stock to produce successive generations in captivity. Ranching operations, on the other hand, are open-ended and depend upon a recurrent infusion of wild stock (such as by harvesting early instar larvae in the wild and then growing them out in controlled environments). Using the CITES terminology, butterfly ranching is preferable to farming because the viability of ranching efforts depends upon the continued availability of wild habitat from which to take the needed stock. This assumes, of course, that any harvest from the wild is sufficiently controlled so as not to be excessive.

Education

To conserve insects successfully, the general public, scientists, land managers, and conservationists need to understand the extraordinary value that these organisms provide. It is unlikely that very many people will develop an affinity for these animals, but it is plausible that a more compelling depiction of the contributions insects make to human welfare and survival will improve the public's attitude toward these organisms. An ambitious public education program would enhance recognition of the positive values of invertebrates and, indeed, all biological diversity.

THE TIME IS NOW

The number of endangered insects is large and growing. The rate of destruction and degradation of natural habitats currently is so great that there are not nearly enough biologists to even catalog, much less study, the species that are suddenly on the edge of extinction. In Indonesia, approximately 1.3 million hectares of tropical forest were cut in 2001. In Argentina, 7964 metric tons of insecticides were used in 1998. In the United States, imported red fire ants have infested over 260 million acres in the southeast. These examples of threats to endangered insects continue to mount across the world. The time is now for agencies, scientists, conservationists, and land managers to promote the conservation of imperiled insects.

See Also the Following Articles
Biodiversity • Conservation • Greenhouse Gases, Global Warming, and Insects • Pollution, Insect Response to

Further Reading
Bean, J. M. (1993). Invertebrates and the Endangered Species Act. *Wings,* Summer.
Buchmann, S. L., and Nabhan, G. P. (1996). "The Forgotten Pollinators." Island Press, Washington, DC.
Collins, N. M., and Thomas, J. A. (eds.) (1991). "The Conservation of Insects and Their Habitat." Academic Press, London.
Deyrup, M. (2001). Endangered terrestrial invertebrates. *In* "Encyclopedia of Biodiversity" (S. A. Levin, ed.), Vol. 2. Academic Press, New York.
Deyrup, M., and Eisner, T. (2001). Interviews at the edge of a cliff. *Wings,* Fall.
Hoffman Black, S. C., Shepherd M., and Mackey Allen, M. (2001). Endangered invertebrates: The case for greater attention to invertebrate conservation. *Endangered Species Update* **18,** 41–49.
Kellert, S. R. (1993). Values and perceptions of invertebrates. *Conservat. Biol.* **7,** 845–855.

Opler, P. (1995). Conservation and management of butterfly diversity in North America. *In* "Ecology and Conservation of Butterflies" (A. S. Pullin, ed.). Chapman & Hall, London.

Pyle, R. M., Bentzien, M., and Opler, P. (1981). Insect conservation. *Annu. Rev. Entomol.* **26**, 233–258.

Samways, M. J. (1994). "Insect Conservation Biology." Chapman & Hall, London.

Stein, B. A., Kutner, L. S., and Adams, J. S. (eds.) (2000). "Precious Heritage: The Status of Biodiversity in the United States." Oxford University Press, Oxford.

Wilson, E. O. (1992). "The Diversity of Life." Norton, New York.

Endopterygota

Endopterygota is a division of the class Insecta in the phylum Arthropoda. The orders of insects in this division have wing rudiments that develop internally, and external wings are not evident until pupal metamorphosis occurs. There are three superorders in this division: the Neuropteroidea (which includes the orders Coleoptera, Megaloptera, Neuroptera, Raphidioptera, and Strepsiptera), the Mecopteroidea (Diptera, Lepidoptera, Mecoptera, Siphonaptera, and Trichoptera), and the Hymenopteroidea (Hymenoptera). There is no morphological resemblance between larvae and adults, and habitats and biology of adults differ greatly from larvae.

Entomological Societies

Alan I. Kaplan

East Bay Regional Park District, Berkeley, California

Entomological societies, as a category of learned societies, grew out of small, localized groups whose members lived less than a day's carriage ride from each other. In an age when specimens could not be entrusted to an irregular (or nonexistent) postal service, visiting the "cabinet" (collection) of a fellow member to see an actual specimen was necessary. Today, international societies have hundreds to thousands of members; activities range from having only a journal subscription in common to annual meetings with thousands of participants. From a largely amateur base in the 19th century, entomological societies have grown increasingly professionalized, a pattern similar to the societies covering ornithology and botany, which also had their beginnings as organized sciences with broad, nonprofessional participation.

In response to increased professionalization of entomology in the 20th century, a large number of regional and international specialized societies have arisen, to serve both pro-fessional and amateur entomologists having particular systematic or disciplinary interests. From a time when all entomologists were amateurs (mid-19th century), through a period of increased professionalization resulting in marginalization of amateurs (early to late 20th century), amateur entomologists continue their contributions to the field today.

Herbert Osborn, in his *Brief History of Entomology* published in 1952, wrote, "the origin of entomological societies is to me still a mystery." We now know why: the first entomological society in the world was founded in London, sometime between 1720 and 1742. The exact date is uncertain because the collection, books, and regalia (and presumably the minutes) of this group, the (first) Aurelian Society, were destroyed in the Great Cornhill Fire of March 25, 1748. Its meeting place, Swan Tavern on Exchange Street, was burnt to the ground; the members, then in session, barely escaped with their lives.

GENERAL FEATURES OF SOCIETIES

Entomological societies share some common features: Membership requires payment of dues to maintain the organization, there is often a "pro forma" election to membership held at a meeting of the society, and prospective members are rarely refused. For example, the only person ever turned down for membership in the New York Entomological Society was "the author of a new version of the theory of spontaneous generation!" Honorary membership (usually limited to a small number) is offered to accomplished and distinguished entomologists in the home country of the society or from other countries. Distinguished Regular members may be elevated to Fellow status. There is often a category of nonlocal membership, usually referred to as "corresponding." Regular meetings are held, at least annually, often more frequently, with guest speakers and the opportunities for members to provide a greater number of shorter presentations. There are constitutions and by-laws, with officers who preside over business meetings. Field trips ("field days") to collect insects were a major feature of 19th and early 20th century society meetings, and annual meeting circulars and programs will suggest collecting opportunities near to meeting sites. Some societies have a tradition of insect protection: As early as 1896, the Royal Entomological Society (London) (RES) had a committee to look into protecting British insects from extinction. In 1988, the RES became the first entomological society to join the International Union of Conservation of Nature. But as far back as the second International Congress of Entomology, held at Oxford in 1912, N. C. Rothschild spoke on steps taken to protect insects in Great Britain. The British Entomology and Natural History Society formed several Conservation Working Groups in 1994 to bring the expertise of its members to bear on matters relating to conservation of the invertebrate fauna of the United Kingdom and to express the field naturalists' views of which species deserve special attention.

The publications of these societies—as proceedings, journals, memoirs, annals, bulletins, and newsletters—have been the main vehicle for dissemination of scientific information and more personal information about the work and lives of entomologists since the founding of the early societies. For example, many societies begin a publishing program the same year or soon after their founding (the French began *Annales* in 1832, the year the Société Entomologique de France was founded; the Royal Entomological Society (London) began its *Transactions* in 1834, a year after its founding). Publications have served as a medium of exchange with other societies in order to build up another feature, that of society library. The American Entomological Society library, with over 15,000 volumes, has been incorporated since 1947 into the Academy of Natural Sciences (Philadelphia, PA) library; the Pacific Coast Entomological Society does not maintain a separate library, but journals received in exchange for its publication, the *Pan-Pacific Entomologist,* and books received for review therein are deposited in the library of its host institution, the California Academy of Sciences.

Several societies (for example, the Amateur Entomologists' Society, Orthopterists' Society, Entomological Society of America) have produced a series of handbooks and guides for identifying insects and have a regular publishing program outside of the usual journal- and memoir-type series. The Brooklyn Entomological Society (BES) took as a goal the publication of alphabetical lists of scientific terms used in technical descriptions in entomology. The first entomological vocabulary published in North America was 800 terms and definitions, in Vol. 6 of the *Bulletin* of the BES in 1886. This was followed in 1906 by the BES-sponsored *Glossary, an Explanation of Terms Used in Entomology* by J. B. Smith (4000 entries). The BES published J. A. Torre-Bueno's *A Glossary of Entomology* in 1937 (10,000 terms, 12,000 definitions), and a supplement to it in 1960 by G. S. Tulloch added 500 new terms and revised 160 others. The Entomological Society of America publishes the only society-sponsored list of insect common names; in other countries this is usually a function of the department or ministry of agriculture.

Insect collections of societies and their members have become important components of the holdings of large institutions: For example, the Academy of Natural Sciences in Philadelphia has the collection of the American Entomological Society; the New York Entomological Society collection has been incorporated into those of the American Museum of Natural History. Members of the Pacific Coast Entomological Society often deposit type specimens of species described in the Society's journal (*Pan-Pacific Entomologist*) with the collection of the California Academy of Sciences. Society collections have at times been controversial: The (third) Aurelian Society in England, founded in 1801, dissolved 5 years later because of the odious requirement that members donate their best specimens to a central society collection. Disagreements over the deposition and loan of specimens of the Entomological Society of Philadelphia collection led to its

expulsion in 1862 of its first president, the eminent coleopterist John L. LeConte.

Another feature of some societies is a youth program: Membership is offered to young people at a discounted rate, special publications are aimed at them, occasional exhibits are developed to tour schools or be displayed at annual meetings, and field days featuring insect collecting trips are planned. The New York Entomological Society formed a Junior Division in 1958. The Entomological Society of America has had a Youth Membership category since 1989. The Young Entomologists' Society (U.S.A.) traces its origin to the Teen International Entomology Group, founded in 1965 by a teenager as a worldwide correspondence club to exchange letters and specimens with like-minded teens around the world.

These youth programs may take time to develop. For example, the Royal Entomological Society (London) Youth Development Scheme of 1990 had hopes of local and regional participation by its Fellows, which did not materialize, and the program failed a year later. But from it came the Bug Club, now a national organization in Great Britain.

ORIGINS

The earliest scientific societies were founded in Europe in the mid-16th and early 17th centuries, but the first entomological societies came about in England in the mid-18th century for the purpose of sharing knowledge of the Lepidoptera. These were the Society of Aurelians (also called the [first] Aurelian Society, formed sometime between 1720 and 1742) and its successor, the (second) Aurelian Society, formed in 1762. "Aurelia" is a classical name for the chrysalis of a butterfly; an aurelian is a butterfly collector. The (first) Aurelian Society was finished by the Great Cornhill Fire of 1748; the second ceased in 1767 because of personality clashes among members; a third Aurelian Society, founded in 1801, disappeared by 1806.

The oldest entomological society still in existence—the Entomological Club of London, founded in 1826—has had only eight members at a time since its inception and meets one evening each month to dine at members' homes or other places. It also hosts the annual Verrall Supper for entomologists, a tradition since 1887. The oldest existing national entomological societies are Société Entomologique de France (1832), the Royal Entomological Society (London) (1833), and the Nederlandsche Entomologische Vereeniging (1845). See Table I for a list of societies that are 100 years old or older and are still in existence as of 2001.

NORTH AMERICAN SOCIETIES

The first North American entomological society was the (first) Entomological Society of Pennsylvania which, from 1842 to 1853, took as its only mission the taxonomic description of American insect species. It did not try to spread knowledge of insects to the general public, nor to encourage the study of economic (applied) entomology, nor to investigate insect natural history. Although there were between 40 and 70 agricultural societies in the United States at this time (chiefly with the aim of county fair exhibitions), the first U.S. society's members feared that an association with this applied sphere of entomology would weaken the efforts of American entomologists to gain the respect of European entomologists. The Entomological Society of Pennsylvania had as its main project the publication of a catalog (a list of species with nomenclatural data, such as author names, dates of publication, synonyms, and taxonomic references) of American Coleoptera. It was defunct by 1853, when the Smithsonian Institution published the founding member and only president F. E. Melsheimer's *Catalogue of the Described Coleoptera of the United States.*

The oldest North American entomological society still in existence is the American Entomological Society, founded in 1859 as the Entomological Society of Philadelphia and renamed in 1867. For several years (1865–1867), this society published the first journal to be devoted to economic entomology, *The Practical Entomologist.* Its *Transactions* have been published since 1867, and it has published *Entomological News* since 1890.

The entomological societies of Canada began with the (first) Entomological Society of Canada, formed in Toronto in 1863 (an organizing meeting had been held the year before). Its journal, *The Canadian Entomologist,* has been published continuously since 1868. When support was obtained for this publication from the Council of Agriculture and Arts Association of Ontario, the society's name was changed to Entomological Society of Ontario in 1871 to reflect this support, but it still served as something of a national organization. Autonomous regional societies grew up across Canada (for example, in British Columbia in 1902, in Nova Scotia in 1914, and in Manitoba in 1945). A (second) Entomological Society of Canada emerged in 1950 to link the provincial societies into a truly national organization.

The current Entomological Society of America (ESA) was formed in 1953 from the union of the American Association of Economic Entomologists (AAEE) and the (first) Entomological Society of America. The AAEE was formed in 1889 (originally for state and federal entomologists, its first title was Association of Official Economic Entomologists; "Official" was soon dropped and "American" added in 1909). It grew out of the Entomological Club, a subsection of the Natural History Section of the American Association for the Advancement of Science (AAAS), which itself originated at an AAAS meeting in Hartford, Connecticut, in 1874, the year when entomologist John L. LeConte was AAAS president.

Dissatisfaction with the applied emphasis of AAEE and need for a societal home for academic and noneconomic entomologists led to the formation of the (first) Entomological Society of America in 1906, with Professor J. H. Comstock of Cornell University as its president. In 1908, both societies began publishing journals that are still printed today: the *Annals of the Entomological Society of America* and the *Journal of Economic Entomology.* In addition, the merged ESA produces other journals, including (since 1972) *Environmental Entomology* and, since 1986, has owned and published the *Journal of Medical Entomology* (originally published by the Bishop Museum of Hawaii).

Throughout the period 1906 to 1953, the two U.S. national entomological societies often held joint annual meetings, so members of both (and there was considerable overlap in membership) could participate in each meeting. By the time of the 1953 merger, membership in the AAEE was triple that of the ESA, reflecting the expansion of the applied entomology field in the age of modern insecticides. The strong regional (branch) divisions of the reorganized ESA closely follow the premerger AAEE structure, as do most of the subject sections within the organization (for example, sections for Regulatory and Extension Entomology and Crop Protection Entomology). Today's ESA sponsors a unique program of board certification for professional entomologists. In the past, the ESA had been the institutional home of the American Registry of Professional Entomologists (about 15% of ESA members in 1989 were registered with ARPE). A code of ethics and education and experience requirements, plus testing and continuing education, are elements in the process of "professionalization" that have emerged in only a few scientific fields (medicine being the best example), but these are hallmarks of a technology-oriented profession (civil, structural, and geologic engineering, for example). Entomology's unique status as a technology *and* a science emerged in the late 20th century, resulting in some confusion over status and prestige, which a process of professional certification clarified for some ESA members.

SOCIETAL GROWTH

In 1956, Curtis Sabrosky published a near-exhaustive list of entomological societies that had existed and/or were still in existence. By 1956, at least 70 entomological societies had begun, blossomed (or not), and then faded away. That year, there were 96 active regional or national entomological societies (not counting those devoted to applied aspects such as apiculture or pest control). Only 10 specialty societies, devoted to a taxonomic group or some other special, non-applied aspect, were listed: 3 devoted to Coleoptera (in Austria, the United States, and Japan), 4 for Lepidoptera (3 in Japan and 1 in the United States), and 1 international society for the study of social insects. The post-1956 period has seen a huge development of specialty societies. Today, there are at least 92 specialty societies (including 1 for conservation of invertebrates in general and 1 devoted to young entomologists). There has been increased worldwide interest in Lepidoptera; at least 22 new societies formed in the period from 1970 to 1999. Although there are specialty societies devoted to at least nine orders of insects, the majority are devoted to just three: Coleoptera (12), Lepidoptera (39), and Odonata (12). In addition, regular international symposia are held on other, smaller orders (for example, Trichoptera and Ephemeroptera). Just as improved optics (binoculars, spotting scopes, cameras) and better field guides stimulated the mid-20th century interest in bird-watching, similar improvements in entomological materials (for example, a recent series of books on identifying butterflies and dragonflies with binoculars) have made Lepidoptera and, more recently, Odonata popular subjects of observation and interest to larger numbers of people, who in turn have formed many new entomological societies devoted to their insect passions.

ROLE OF AMATEUR ENTOMOLOGISTS

Professionalization and an improved image as scientists has been an issue for entomologists since the 19th century. From their origin in amateur lepidopterist clubs and local societies of collectors interested in the taxonomic position of their specimens and little else, entomological societies grew into associations of applied scientists who recognized the contribution of a client base of agriculturalists but did not make a place for them or for hobbyist entomologists (amateurs) in their national organizations. Once applied entomologists in the United States organized into the AAEE, they neither encouraged nor discouraged nonprofessional participation, but instead created two classes of membership. One was a professional category, which required educational qualifications and vocational activity; the other was an "associate" category, which was a second-class membership for amateurs and others with inadequate qualifications.

Yet, amateur entomologists, who had been the founders of the field and its earliest supporters as collectors and bene-factors, continued to make contributions. Early societies at times depended on wealthy amateurs for support. The best example is the support given to the American Entomological Society (AES) by one of its founders, Thomas B. Wilson, an executive of the Pennsylvania Railroad. He paid for the AES's building and was patron of its library and collection. Wilson provided a sinecure for E. T. Cresson, Sr. (one of North America's greatest hymenopterists), as his private secretary, which in reality supported Cresson as curator of the AES collection and its corresponding secretary for many years. When Wilson died, Cresson worked for an insurance company for the next 40 years (1869–1910). "The Wilson Fund" was still supporting AES publications in 1984, almost 120 years after their provider's death.

The first national Canadian entomological society was formed in 1863 by a 25-year-old divinity student, Charles J. S. Bethune, and a 28-year-old pharmacist, William Saunders. They founded its journal, *The Canadian Entomologist,* in 1868 and were the sole contributors to its first two numbers. Each went on to distinguished careers in Canadian entomology (Bethune as Professor of Entomology and Zoology at Ontario Agricultural College, Saunders as the first Director of Experimental Farms [agricultural experiment stations] for the Dominion of Canada) but they both had begun the Entomological Society of Canada as amateurs.

Amateur entomologists were always welcome to publish in the *Journal of the New York Entomological Society.* Annie T. Slosson was a NYES founder (in 1892) and the largest financial supporter of its journal—her donated specimens of Lepidoptera raised the most money at the Society's auctions. She was a well-known collector and contributed many journal articles, though she did not publish new taxonomic names for the species she gathered; she preferred to send them to specialists to describe.

The Cambridge Entomological Club was saved from extinction by an infusion of amateur members. The Club was founded at Harvard in 1874 with 12 members. It reached 48 members by the time of incorporation in 1877, but had declined in 1902 to just 7 (of whom only 3 or 4 attended meetings at any one time). The Club combined with a local amateur society, The Harris Club, with its 38 members, in 1903, and active amateur members have been an important component of the Cambridge Entomological Club ever since. The participation by amateurs was encouraged in these early days by ant specialist, Professor William M. Wheeler, who supported amateur naturalists and said, "We have all known amateurs who could make an enthusiastic naturalist out of an indifferent lad in the course of an afternoon's rambling, and, alas, professors who could destroy a dozen budding naturalists in the course of an hour's lecture."

In the first volume (1908) of *Annals of the Entomological Society of America,* the Canadian entomologist H. H. Lyman, himself an amateur, urged the society "to secure the support and cooperation of the great body of amateur entomologists." This has not been accomplished. Almost 80 years later, in 1986, a survey of the ESA's 9111 members found only 31 (0.5% of the 5505 respondents) who described themselves as amateurs. But amateur entomologists have found welcome and a home in regional and local societies devoted to taxonomic specialties (Coleoptera and Lepidoptera predominately, but Odonata have become popular) or geographically restricted. Since 1939, the Amateur Entomologists' Society (United Kingdom) has been a flagship of the great amateur enterprise, publishing its bulletin and a large number of identification guides and handbooks.

A survey of adult amateur entomologists in 1987 by Janice Matthews found that they often suffer from being stereotyped by professional entomologists as less qualified or educated and get a cool (or even hostile) reception from professionals. Amateurs actually produce the great bulk of educational outreach on entomological topics (for example, programs for school children, other amateur naturalists, and the public at large). Adult amateur entomologists' professional lives align very closely with those of professional entomologists: Amateur entomologists are doing science and math in their daily work; they are in education; they are in service occupations (by comparison, pest control work is also a service occupation). The science background of amateur entomologists can be as strong as that of professionals, but is often in a related field. None of the respondents to Matthews' study reported that a professional entomologist influenced their childhood interest in insects; the failure of the Youth Development Scheme of the Royal Entomological Society (London) in 1990 was attributed to just this kind of lack of interest on the part of its members toward young entomologists.

That amateur entomologists have made, and continue to make, great contributions to entomology is unquestioned. From the great coleopterist P. F. M. A. DeJean (Napoleon's general and aide-de-camp at Waterloo), to 19th century lepidopterist William H. Edwards (a lawyer and coal company president), to civil engineer Richard H. Stretch (who first warned of the economic dangers of cottony cushion scale in California in 1872), through a long list of physician–entomologists (for example, H. Bernard Kettlewell, who was a general practitioner while pursuing his studies of melanism in Lepidoptera), to the great student of leaf-mining flies, Kenneth Spencer (an electronics executive, he published 74 papers before retiring in 1969 and then published 45 more papers in the next 20 years), to the Parisian taxi driver Pierre Morvan (honored with the Rolex Enterprise Award in 1987 for his biogeographic study of Asian ground beetles, he is a self-taught entomologist and author of over 50 scientific publications), entomology advances through the efforts of its many amateur practitioners.

See Also the Following Article
History of Entomology

Further Reading
Allen, D. E. (1976). "The Naturalist in Britain: A Social History." Princeton University Press, Princeton, N.J.
Connor, J. T. H. (1982). Of butterfly nets and beetle bottles: The Entomological Society of Canada, 1863–1960. *HSTC Bull.* **6**, 151–171.
Mallis, A. (1971). "American Entomologists." Rutgers University Press, New Brunswick, NJ.
Matthews, J. R. (1988). Adult amateur experiences in entomology: Breaking the stereotypes. *Bull. Entomol. Soc. Am.* **34(4)**, 157–161.
Osborn, H. (1952). "Brief History of Entomology." Spahr and Glenn, Columbus, OH.
Sabrosky, C. W. (1956). Entomological societies. *Bull. Entomol. Soc. Am.* **2(1)**, 1–22.
Salmon, M. A. (2000). "The Aurelian Legacy: British Butterflies and Their Collectors." University of California Press, Berkeley.
Scientific Reference Resources (2001). http://www.sciref.org.
Smith, E. H. (1989). The Entomological Society of America: The first hundred years, 1889–1989. *Bull. Entomol. Soc. Am.* **35(3)**, 10–32.
Sorensen, W. C. (1995). "Brethren of the Net: American Entomology, 1840–1880." University of Alabama Press, Tuscaloosa.

Ephemeroptera
(Mayflies)

John E. Brittain
Natural History Museums and Botanical Garden, University of Oslo

Michel Sartori
Museum of Zoology, Lausanne

Mayflies (order Ephemeroptera) date from Carboniferous and Permian times and represent the oldest order of the existing winged insects. They are unique among the insects in having two winged adult stages, the subimago and imago (Fig. 1). Adult mayflies do not feed; instead, they rely on reserves built up during their nymphal life. As adults they generally live from 1 to 2 h to a few days, and mayflies spend most of their life in the aquatic environment, either as eggs or nymphs. The nymphal life span in mayflies varies from 3 to 4 weeks to more than 2 years. The length of egg development varies from ovoviviparity (i.e., the release of live offspring) to a period of up to 10 to 11 months in some arctic/alpine species.

Because of their winged adult stage and a propensity for drift (i.e., downstream movements) as nymphs, mayflies are often among the first macroinvertebrates to colonize virgin habitats. However, over longer distances their dispersal capacity is limited, owing to their fragility and short adult life. Mayflies are found in almost all types of freshwater habitat throughout the world, although in the Arctic and in mountain areas above the tree line there are few species. Mayfly faunas on oceanic islands and isolated mountain areas have few species, and they are usually restricted to the Baetidae and/or Caenidae. Their greatest diversity is in lotic habitats in temperate and tropical regions, where they are an important link in the food chain, from primary production by algae and plants to secondary consumers such as fish. Mayflies are used extensively as indicators of pollution and environmental change.

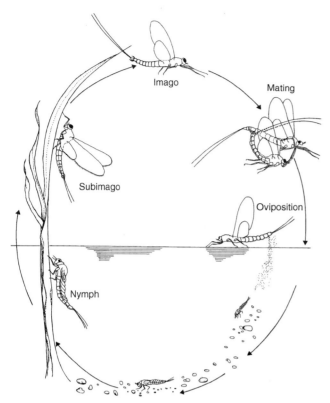

FIGURE 1 Mayfly life cycle showing the alternation between the aquatic and terrestrial environments. Mayflies are unique in having two winged stages, the subimago and imago. The adult life is very short and most of the time is spent in the aquatic environment.

ORIGINS AND EVOLUTION

Ephemeroptera are among the oldest known winged insects still extant. Carboniferous fossils have been ascribed to mayfly precursors or even mayflies. Permian data confirm that the order was already present at the end of the Paleozoic. Ephemeroptera reached their highest diversity during the Mesozoic, mainly in the Jurassic and Cretaceous. All these species belong to extinct families. The Tertiary fauna, as documented by for instance Baltic amber, is undeniably modern, with both the extinct and living genera of modern families.

The relationship of Ephemeroptera with other modern winged insects is still a subject of debate. Together with the Odonata, mayflies were traditionally placed in the Paleoptera, which was considered the sister group of all other extant primarily winged orders. More recently, it was suggested that Ephemeroptera per se are the sister group of Odonata + Neoptera. This idea is based on a number of features unique to mayflies, such as the presence of a subimaginal stage, the nonfunctionality of the adult mouthparts, and the presence of only one axillary plate in the wing articulation. This hypothesis is also supported by anatomical data: female mayflies exhibit telotrophic meroistic ovaries instead of panoistic ones as found in Odonata.

CLASSIFICATION AND PHYLOGENY

The Ephemeroptera are numerically a small order of insects, with about 3000 described species within more than 375 genera and 37 families (Table I). About 350 species occur in Europe, and 670 in North America. During the 1990s, partly as a result of the discovery of new taxa, especially in tropical areas, where the mayfly fauna is still poorly known, 10 new families, 75 genera, and more than 500 species were added. The expansion of the order is also the result of several phylogenetic analyses that led to a narrower concept of supraspecific taxa. As a consequence of these important changes, there is no real consensus about the higher classification of Ephemeroptera (superfamilies, suborders, or infraorders). Based on the structure of the nymphal wing pads, mayflies were traditionally divided into two suborders, Pannota (with fused wing pads) and Schistonota (with free wing pads). That the latter suborder is paraphyletic is now well documented, but there is no agreement about the composition and even the names of these higher taxa.

ADULTS

The adult mayfly has two main functions, mating and oviposition, which produce a general uniformity in structure. The prominent turbinate eyes of males, especially well-developed in the Baetidae and some Leptophlebiidae, provide both high acuity and good sensitivity. This enables them to detect and capture single females in a swarm at low light intensities.

The forelegs of most mayflies also show sexual differences; those of the male are unusually long for grasping and holding the female during mating. In the Polymitarcyidae, the middle and hind legs of the male and all the legs of the female are reduced, and in *Dolania* (Behningiidae) all the legs of both sexes are reduced. In *Dolania* and several members of the Polymitarcyidae and Palingeniidae, the females remain in the subimaginal stage. The reason for two winged stages has provoked much discussion. It has been suggested that this primitive trait is maintained because there has not been the selective pressure on the short-lived stages to produce just a single molt. Another explanation is that two molts are necessary to complete the elongation of the caudal filaments and forelegs of the adults. Most mayflies have two pairs of wings, but in the Caenidae, Tricorythidae, Baetidae, and some Leptophlebiidae, the hind wings are reduced or even absent.

Fecundity

Spermatogenesis and oogenesis are generally completed in the final nymphal instar, and the eggs and sperm are physiologically mature in the subimago. Most species produce 500 to 3000 eggs, but values range from less than 100 in *Dolania* to 12,000 in *Palingenia,* and the fecundity values recorded for the females of the larger species of mayfly are higher than in most other insect groups except the social

TABLE I   Overview of the Mayfly Families and the Approximate Number of Genera and Species

Family	Genera	Species	Biogeography
Acanthametropodidae	2	5	Asia and North America
Ameletidae	2	45	Asia, Europe, and North America
Ameletopsidae	4	10	Australia, New Zealand, and South America
Ametropodidae	1	5	Asia, Europe, and North America
Arthropleidae	1	5	Europe and North America
Baetidae	95	700	Worldwide
Baetiscidae	1	12	North America
Behningiidae	3	5	Asia, Europe, and North America
Caenidae	13	100	Worldwide
Coloburiscidae	3	5	Australia, New Zealand, and South America
Coryphoridae	1	1	South America
Dipteromimidae	1	1	Japan
Ephemerellidae [a]	16	300	Worldwide except Australia and New Zealand
Ephemeridae [b]	8	30	Worldwide except Australia
Ephemerythidae	1	5	Africa
Euthyplociidae	5	15	Asia, Madagascar, and South America
Heptageniidae	28	500	Africa, Asia, Europe, and North America
Isonychiidae	1	20	Asia, Europe, North and South America
Leptohyphidae	7	120	North and South America
Leptophlebiidae	120	900	Worldwide
Metretopodidae	2	10	Asia, Europe, and North America
Neoephemeridae	3	10	Asia, Europe, and North America
Nesameletidae	3	5	Australia, New Zealand, and South America
Oligoneuriidae	11	45	Worldwide except Australia and New Zealand
Oniscigastridae	3	10	Australia, New Zealand, and South America
Palingeniidae	7	30	Asia, Europe, and Madagascar
Polymitarcyidae [c]	7	65	Worldwide except Australia and New Zealand
Potamanthidae	3	25	Africa, Asia, Europe, and North America
Prosopistomatidae	1	15	Africa, Asia, Australia, Europe, and Madagascar
Pseudironidae	1	1	North America
Rallidentidae	1	1	New Zealand
Siphlaenigmatidae	1	1	New Zealand
Siphlonuridae	4	30	Asia, Europe, and North America
Teloganellidae	1	1	Asia
Teloganodidae	7	15	Africa, Asia, and Madagascar
Tricorythidae [d]	6	30	Africa, Asia, and Madagascar
Vietnamellidae [e]	2	5	Asia and Australia
Total	**376**	**3083**	

[a]Including Melanamerellinae. [b]Including Pentageniinae and Ichthybotinae. [c]Including Exeuthyplociinae. [d]Including Dicercomyzinae and Machadorythinae. [e]Austremerellidae.

Compiled with the assistance of Jean-Luc Gattolliat (Lausanne) and Jan Peters and Michael D. Hubbard (Tallahassee).

Hymenoptera. In species with a long emergence period or with a bivoltine life cycle (having two summer emergence periods), early emerging females are larger and therefore more fecund than those emerging later.

Mating and Swarming

Swarming in adults is a male activity, apart from the Caenidae and Tricorythidae, where both males and females may participate. The females fly into these swarms, and mating occurs almost immediately and usually in flight. Swarming may take place over the water itself, over the shore area, or even away from the water. Most swarms are positioned according to terrain markers such as areas of vegetation, the shoreline, and trees. The time of swarming varies considerably, although dusk is the most common time of day in temperate regions.

Parthenogenesis has been reported in about 50 mayfly species, although it is not obligatory as a rule.

Oviposition

The majority of mayflies, including most Ephemeridae, Heptageniidae, and Leptophlebiidae, oviposit by descending to the water and releasing a few eggs at a time by dipping their abdomen into the water. Species of *Ephemerella, Siphlonurus,* and *Centroptilum,* however, release all their eggs in a single batch that separates immediately on contact with water. In *Habroleptoides* and some Heptageniidae the female rests on a stone above the water, and dips her abdomen into the water to lay the eggs. This is taken a stage further in several species of *Baetis* in which the female actually goes underwater and lays her eggs on suitable substrate, often under stones.

EGGS

Mayfly eggs have a variety of attachment structures that enable them to adhere to submerged objects or to the substrate. Differences in egg morphology have enabled the construction of identification keys, purely on the basis of eggs. This has provided a useful complement, not only to studies of phylogeny, but also to taxonomy, since identification of female adults by means of external characters is often difficult.

Development

Most nymphs hatch at temperatures in the range of 3 to 21°C. However, in the North American *Hexagenia rigida,* the nymphs hatch successfully between 12 and 32°C and even at 36°C if incubation is started at lower temperatures. In *Tricorythodes minutus,* nymphs hatch between 7.5 and 23°C, but mortality is least at 23°C. Hatching success is variable, ranging from over 90% in several *Baetis* and *Hexagenia* species to less than 50% in the Heptageniidae studied. Excluding the few ovoviviparous species, the total length of the egg development period varies from a week in *H. rigida* to almost a year in *Parameletus columbiae.* Temperature is the major factor determining the length of the period of egg development in mayflies. There is no indication that photoperiod influences egg development time. Ovoviviparity is rare in the mayflies and is restricted to the Baetidae. In North America, a number of species in the genus *Callibaetis* are ovoviviparous.

NYMPHS

In contrast to the adults, mayfly nymphs show considerable diversity in habit and appearance. Differences do not always follow taxonomic lines, and convergent and parallel evolution seems to be common (Fig. 2).

Growth and Development

Mayflies have a large number of postembryonic molts. Estimates of the number of instars vary between 10 and 50; most are in the range 15 to 25. The number of instars for a particular species does not seem to be constant, but probably varies within certain limits. Environmental conditions, such as food quality and temperature, may affect instar number. Because of its simplicity, by far the most common measure of development and growth in mayflies has been body length, although head width and other body dimensions also have been used. However, growth of the various body parts is not always isometric. Many authors have also used body weight, and the length–weight relationship is usually well expressed by a power function.

Nymphal growth rates are influenced by several environmental factors, although the major growth regulator is mean temperature, the scale of diurnal fluctuations, or total degrees-days. Other factors, such as food and current velocity, may exert a modifying influence on growth rates. No true diapausing nymphal stage has been reported in the Ephemeroptera, although growth rates often are very low during the winter.

Respiration

The gills of mayflies are very diverse in form, ranging from a single plate in *Ameletus* to fibrillar tufts in *Hexagenia.* Respiratory tufts are sometimes developed on other parts of the body besides the abdomen, such as those at the base of the coxa in *Isonychia* and *Dactylobaetis.* In several families the second abdominal gill has developed into an operculate (lidlike) gill cover for the remaining gills, and in certain Heptageniidae the gills are markedly expanded so that they together form an adhesion disc. In many of the Siphlonuridae, the gills are used as swimming paddles, which has been put forward as their original function. In respiring, the gills may function either as respiratory organs or as ventilatory organs for other respiratory exchange surfaces.

High rates of oxygen consumption are often reported in association with emergence and gonad maturation. High water temperatures at that time may mean that low oxygen concentrations can be critical. Many burrowing Ephemeridae and pond-dwelling Baetidae are able to survive moderately low oxygen concentrations, especially for short periods. However, so far only one species, the European *Cloeon dipterum,* has been shown to survive long-term anoxia.

Population Movements

During the final stages of nymphal life there is a movement to and a concentration in the shallower areas of lakes and rivers. In running waters, springtime mass movements of mayfly nymphs along the banks of the main river and into slower flowing tributary streams or into areas flooded by spring snowmelt have been observed. In running water, mayfly nymphs may move down into the substratum in response to spates or as part of a daily rhythm. Generally, however, mayflies do not extend far down into the substratum (i.e., the hyporheic zone).

Mayflies, especially Baetidae, are a major component of invertebrate drift in running waters. Their drift shows a strong diel periodicity, with a peak during the hours of darkness. Drift rates are not constant for a particular species, and the larger size classes are usually more in evidence. Other factors that have been shown to influence mayfly drift include changes in current velocity and discharge, increased sediment loading, temperature changes, oxygen conditions, density, food availability, and predators.

EMERGENCE

Emergence, the transition from the aquatic nymph to the terrestrial subimago, is a critical period for mayflies. Their movement up to the water surface makes them especially vulnerable to aquatic and aerial predators. Shedding of the nymphal skin usually occurs at the water surface on some object, such as a stone or macrophyte stem, or in midwater.

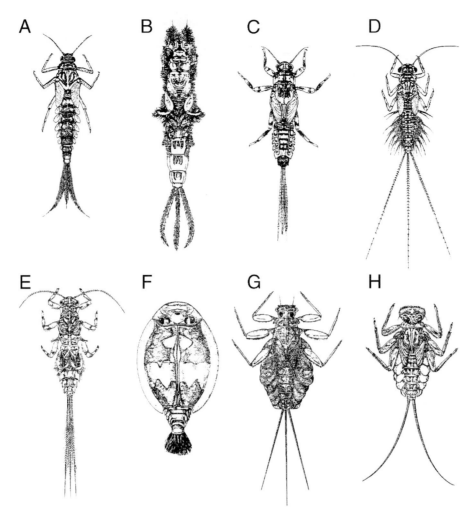

FIGURE 2 Mayfly nymphs: (A) *Baetis subalpinus* (family Baetidae), (B) *Ephemera danica* (family Ephemeridae), (C) *Ephemerella mucronata* (family Ephemerellidae) (D) *Leptophlebia vespertina* (family Leptophlebiidae), (E) *Caenis robusta* (family Caenidae) (F) *Prosopistoma boreus* (family Prospistomatidae), (G) *Lepeorus thierryi* (family Leptophlebiidae), and (H) *Epeorus alpicola* (family Heptageniidae). Illustrations show some of the large range in morphology, often related to habitat and food habits and not necessarily to family relationships. For example, *L. thierryi* and *E. alpicola* are morphologically similar and adapted to fast-running waters but belong to different families.

The latter location is more typical of the burrowing species that inhabit deeper waters and of a number of river species. Genera such as *Siphlonurus, Isonychia,* and *Baetisca* crawl completely out of the water before they molt.

Diel Patterns

In temperate regions, the crepuscular emergence of mayflies is well known. However, dusk is not the only time of day that mayflies emerge, although most species exhibit clear diel patterns of emergence that are, with few exceptions, characteristic for a given species, genus, or even a whole family. For example, the emergence of the short-lived Caenidae invariably takes place either at dawn or dusk and seems to be controlled by light intensity. Several baetid and leptophlebiid genera emerge around midday. In temperate areas, the higher daytime air temperatures are less restrictive for flight activity, although the adults are probably more susceptible to predation.

In the tropics and warm temperate regions, night air temperatures are less restrictive, and to escape from daytime predators it seems that most longer-lived forms emerge during the first two hours of darkness. The shorter-lived genera, such as *Caenis,* are subject to fewer restraints on their emergence, and there are few constant differences between tropical and temperate species.

The daily emergence of males and females is usually synchronous, especially in the short-lived forms, although there may be an excess of males at the start of the day's emergence. In species in which the females oviposit as subimagos, the males, which molt to imago, emerge well before the females.

Seasonal Patterns

Mayflies have distinct and finite emergence periods, especially in temperate and arctic areas. In the tropics, emergence is often nonseasonal, although some species have clear emergence patterns. The lunar rhythm of emergence from a number of lakes of the African species *Povilla adusta,* is well known. The burrowing mayflies of the Ephemeridae, Polymitarcyidae, and Oligoneuriidae are noted for their sporadic mass emergence. The mass emergence of *Hexagenia* from the Mississippi River has been well documented. There are latitudinal and altitudinal gradients in the timing of emergence. For example, in both North American and European *Leptophlebia,* emergence occurs progressively later as one moves northward. In a similar way, the onset of emergence is delayed with increasing altitude. In habitats with several mayfly species, peak emergence of the major species may be separated in time, especially in congeneric species.

It has been suggested that emergence falls into two main categories: synchronized and dispersed, and represents two approaches for reducing adult mortality. Synchronous emergence attempts to saturate a potential predator, and dispersed emergence seeks to lower the possibility of predator–prey encounters. However, emergence pattern can vary with abundance and locality, and from year to year within the same species.

Water temperature thresholds, often in conjunction with rising temperatures, are important for both seasonal and daily emergence of many mayflies. Photoperiod has also been suggested as a potential factor regulating seasonal emergence in mayflies; few concrete data are available, however, and successful emergence occurred when nymphs were reared in complete darkness. Other abiotic factors may also affect daily emergence totals.

LIFE CYCLES

There is an extensive literature on mayfly life cycles, although mostly from temperate areas in Europe and North America. However, care should be taken in the interpretation of mayfly life cycles, especially when only field observations are available. Particular care is necessary in interpreting the length of time for egg development from field data.

Several authors have classified mayfly life cycles; most have used a combination of voltinism, duration of egg development, and nymphal growth rates as criteria. Multivoltine species usually have two or three generations in temperate regions, often a slowly growing winter generation and one or two rapidly growing summer generations. Limited data from the tropics, where many species are nonseasonal, indicate that some species go through about four and possibly up to six generations during the course of a year.

In temperate areas, the univoltine life cycle is the most widespread type. Several authors have distinguished two main types of univoltine cycle: when overwintering occurs during the nymphal stage after a relatively short egg developmental period, and when hatching occurs in the spring after a long period of egg development. Semivoltinism, with generation times up to 3 years, is relatively uncommon in mayflies.

Mayfly life cycles show a distinct trend from the tropics to the Arctic. In the tropics, nonseasonal multivoltine cycles predominate, with seasonality becoming more distinct in mountainous and continental areas. As one approaches the Arctic, univoltine cycles dominate.

Many mayflies exhibit flexibility in life cycle, whereas some mayflies (e.g., the widespread species *Leptophlebia cupida*) have a univoltine winter cycle over a wide range of latitudes and climates. However, a number of common and widespread species display a considerable degree of life cycle flexibility throughout their distributional range. This is perhaps best exemplified by many Baetidae, which may switch from multivoltine to univoltine depending on climate. The North American *Hexagenia* show a similar flexibility.

ABIOTIC AND BIOTIC RELATIONSHIPS

Nutrition

The majority of mayfly nymphs are herbivores, feeding on detritus and periphyton (algal communities on stones and plants). This explains their relative uniformity in mouthparts. The modifications that are present are a result of different food-gathering mechanisms rather than differences in diet. The herbivorous mayflies fall into two main categories, collectors and scrapers. Among the collectors, several genera are filter feeders, with setae on the mouthparts or forelegs acting as filters. Oligoneuriidae, Leptophlebiidae, Siphlonuridae, and the Heptageniidae have several genera that are probably filter feeders. By using their gills to produce a current of water through their burrows, several of the Ephemeridae and Polymitarcyidae may, at least for part of their food supply, be regarded as filter feeders. To supplement their diet, *Povilla* nymphs, especially the larger ones, leave their burrows at night and graze on periphyton. Most mayflies, however, are fineparticle detritivores. These include many Siphlonuridae, Baetidae, Leptophlebiidae, Metretopodidae, Ephemerellidae, Caenidae, and Baetiscidae, as well as some Heptageniidae. Members of the other major feeding group within the mayflies, scrapers, feed on the periphyton present on mineral and organic surfaces. These include representatives of several mayfly families, notably the Baetidae, Heptageniidae, Leptophlebiidae, and Caenidae. Shredders are probably also represented among mayflies.

True omnivory is of limited occurrence in the mayflies and is restricted to some species in genera such as *Isonychia, Siphlonurus, Stenonema,* and *Ephemera.* The predatory habit is also relatively uncommon in the mayflies. In North America, *Dolania, Analetris,* and the heptageniid, *Pseudiron, Spinadis,* and *Anepeorus* feed largely on chironomids. The baetid genera *Centroptiloides* and *Raptobaetopus* have carnivorous nymphs. Within the Prosopistomatidae there are also carnivorous species. Several species, such as *Siphlonurus occidentalis* and *Stenonema fuscum,* may change from a predominantly detrital diet in the early instars to one containing a significant proportion or even a dominance of animal material in the mature nymphs.

The time for food to pass through the gut is often short, and in *Baetis, Cloeon,* and *Tricorythodes* it has been shown to be only about 30 mins. *Hexagenia* nymphs feed continuously during the day and night, and at most temperatures they ingest over 100% of their dry body weight per day. In contrast, values for the surface-dwelling collector *Stenonema* are much lower and vary between 2 and 22% of dry body weight per day. The carnivorous *Dolania*, feeding more intermittently but on a higher energy diet, has consumption indices similar to those of *Stenonema*. Studies have shown little or no cellulase activity in mayflies, whereas the proteolytic activity of trypsin- and pepsinlike enzymes is very high.

Predation

Mayfly nymphs are eaten by a wide range of aquatic invertebrate predators, including stoneflies, caddisflies, alderflies, dragonflies, water beetles, leeches, triclads, and crayfish. Mayflies are also important food organisms for fish. Birds and winged insects, such as Odonata, also prey on mayfly adults. Birds may take both the aquatic nymphs and the aerial adults. Several other animal groups, including spiders, amphibians, marsupials, and insectivorous mammals such as bats and shrews, have been reported to take mayflies. Many parasites also utilize these food chain links.

Symbiosis, Phoresy, and Parasitism

There is a wide range of organisms that live on or in mayflies. They include the normal spectrum of protozoan, nematode, and trematode parasites, and phoretic and commensal relationships with other organisms occur, as well. Chironomids in the genus *Symbiocladius* are ectoparasites and may cause sterility, although ectoparasites in the genus *Epoicocladius* do not seem to be detrimental to their host. In fact the cleaning effect, especially of the gills, may facilitate oxygen uptake in the mayfly. Mayflies can also be commensal, and two baetid genera, *Symbiocloeon* from Thailand and *Mutelocloeon* from West Africa, live between the gills of freshwater mussels.

DISTRIBUTION AND ABUNDANCE

Because of their fragility and short adult life, mayflies are generally rather limited in their dispersal powers. Together with their ancient origin and the strict association of larvae with freshwaters habitats, Ephemeroptera represent an interesting group for biogeographical analyses. The Siphlonuridae and allied families, typically cool-adapted mayflies, are mainly distributed in the temperate Northern Hemisphere, except for the Oniscigastridae, Nesameletidae, Rallidentidae, and Ameletopsidae, which are confined to New Zealand, Australia, and southern South America. We can hypothesize that this lineage was already present on the Pangaea, and radiated later on in Laurasia (Northern Hemisphere continent). Gondwanian representatives (Southern Hemisphere continent) expanded over the transantarctic land bridge and were confined to cool habitats.

The weak dispersal power of mayflies also results in a high percentage of endemism. Many species colonizing cool running waters in the European Alps are found nowhere else, but have related species in the Pyrenees or the Carpathians. On some islands, such as Madagascar and New Caledonia, endemism in mayflies reaches 100%. In contrast, many species that are effective dispersers may have very wide distributions.

Worldwide, two families, the Leptophlebiidae and the Baetidae, are especially important both in terms of abundance and diversity, representing half of the known species. In contrast, the Siphlaenigmatidae (New Zealand) and Dipteromimidae (Japan) encompass only one species apiece.

The distribution and abundance of mayflies has received considerable attention. Within the basic zoogeographical limitations, abiotic factors, notably temperature, substratum, water quality, and, in running water, current speed, seem to be the most important. Other factors, such as ice, floods, drought, food, and competition, may also influence abundance and distribution. Generally, the number of mayfly species decreases with increasing altitude.

Many lotic mayflies are either dorsoventrally flattened or streamlined as an adaptation to life in swift current. The physical substratum also traps different amounts of detritus and silt, and this is a major factor influencing microdistribution. The richest mayfly community is often found in association with aquatic vegetation, which, as well as providing shelter, functions as a detrital trap and as a substratum for periphyton. For burrowing mayflies, the presence of the correct substratum is obviously a major determinant of both macro- and microdistribution. In lakes, the highest mayfly diversity occurs in the shallow littoral areas. At deeper levels, the mayfly fauna, although often reaching high densities, is usually poor in species. Mayflies are generally absent from the profundal (the deep waters where light does not penetrate) of lakes. Many mayflies can tolerate a wide range of salinities, and a few species within the Baetidae, Caenidae, and Leptophlebiidae occur in brackish water.

Mayflies constitute a major part of the macroinvertebrate biomass and production in freshwater habitats. Seasonal variation in density, biomass, and annual production are strongly influenced by life cycle parameters, indicating the importance of correct life cycle information in production studies. Most mayfly production values, expressed in terms of dry weight per square meter per year, are in the range of 0.1 to 10.0 g.

HUMAN INFLUENCE

Humans increasingly affect the distribution and abundance of mayflies and, by virtue of their widespread occurrence and importance in aquatic food webs and particularly in fish production, mayflies have been widely used as indicators of water quality. Mayflies often occur in habitats of a particular trophic status, and increased eutrophication due to human activities can lead to the reduction or even extinction of certain species. *Baetis* species are often among the most tolerant of mayflies to pollution. In North America, the use of mayflies as indicators of water quality has not escaped attention. The mass emergence of burrowing mayflies from Lake Erie and the Mississippi River has provided a useful barometer of water quality. Organic and nutrient enrichment of Lake Erie in the 1940s and 1950s led to an increase in the intensity and frequency of mass emergence of *Hexagenia* until 1953, when prolonged periods of oxygen depletion in the hypolimnion (the lower layer of cold water in lakes that stratify) caused the population to crash to virtual extinction. However, improvement of water quality has now led to a resurgence of emerging swarms. Mayflies, particularly *Hexagenia,* have been used in numerous bioassays for various pollutants. Pesticides also affect nontarget organisms such as mayflies, and Canadian studies in connection with blackfly control have demonstrated catastrophic drift and reduced biomass in mayfly populations over long distances in rivers treated with methoxychlor. Although most mayflies are adversely affected by petroleum products, a few species may show small increases owing to the extensive algal growth that often occurs on oiled substrates.

Acidification of fresh waters is a major threat to mayfly communities. Many mayflies are affected adversely by low pH, and emergence is a particularly critical period. The genus *Baetis* seems to be particularly sensitive and is often replaced by less sensitive *Leptophlebia* and *Siphlonurus.*

River and lake regulation (e.g., by impoundment in reservoirs) for water supply and power can have profound effects on the mayfly community, especially when there is a hypolimnion drain. For example, an increase in winter temperatures and a fall in summer temperatures may remove obligatory life cycle thresholds, leading to extinction. Fecundity may also be influenced by changes in water temperature. In reservoirs themselves, lentic (still water) conditions and increased water level fluctuations usually produce a reduced mayfly fauna, although there may be an increase in the abundance of burrowing and silt-dwelling species. The flooding of new areas can also create new habitats for mayflies, and in many of the large African reservoirs the mayfly *Povilla adusta* has developed large populations, which burrow into the submerged trees and play an important role in tree breakdown. It has recently been demonstrated that ovipositing mayflies are deceived by asphalt roads because the strongly polarized light reflected from the surface mimics a water surface, thus representing a threat to successful reproduction. Climate change scenarios involve changes in water temperatures, which in turn will affect many of the facets of mayfly biology and lead to changes in mayfly communities.

See Also the Following Articles

Aquatic Habitats • Pollution • Respiratory System

Further Reading

Alba-Tercedor, J., and Sanchez-Ortega, A. (eds.). (1991). "Overview and Strategies of Ephemeroptera and Plecoptera." Sandhill Crane Press, Gainesville, FL.

Brittain, J. E. (1982). Biology of mayflies. *Annu. Rev. Entomol.* **27,** 119–147.

Campbell, I. C. (ed.). (1990). "Mayflies and Stoneflies: Life Histories and Biology." Kluwer, Dordrecht, the Netherlands.

Ciborowski, J. H., and Corkum, L. D. (eds.). (1995). "Current Directions in Research on Ephemeroptera." Canadian Scholars' Press, Toronto.

Florida Agricultral & Mechanical University, http://www.famu.edu/mayfly/

Hubbard, M. D. (1990). "Mayflies of the World—A Catalogue of the Family and Genus Group Taxa (Insecta: Ephemeroptera)." Sandhill Crane Press, Gainesville FL.

Landolt, P., and Sartori, M. (eds.). (1997). "Ephemeroptera & Plecoptera: Biology–Ecology–Systematics." MTL Fribourg, Switzerland.

McCafferty, P. (1981). "Aquatic Entomology." Science Books International, Boston.

Purdue University, Department of Entomology. http://www.entm.purdue.edu/Entomology/research/mayfly/mayfly.html

Estivation

see *Aestivation*

Evolution

see *Phylogeny*

Excretion

Timothy J. Bradley
University of California, Irvine

Excretion is the elimination from the body of excess ions, water, and metabolic wastes. As in all organisms, excretion in insects serves to promote the appropriate regulation (homeostasis) of the intracellular environment as the key to organismal well-being and survival. Regulation of the intracellular environment depends in turn on appropriate regulation of the extracellular fluids. In insects, these extracellular fluids are contained in the hemolymph and in the interstitial fluids surrounding the cells. Excretion therefore refers to the processes by which wastes, as well as excess ions and water, are eliminated from the hemolymph. Wastes produced in the cells and transported into the hemolymph are concentrated and excreted by the excretory organs.

STRUCTURE OF THE EXCRETORY ORGANS

The processes of excretion in insects are carried out largely by the organs of the insect gut. These include the midgut, the

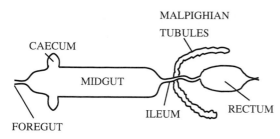

FIGURE 1 A diagrammatic representation of the insect gut.

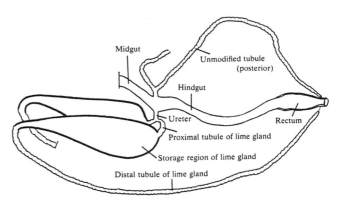

FIGURE 2 A schematic diagram of the Malpighian tubules of *E. hians*. The Malpighian tubules are differentiated on the left and right side. The hindgut is composed, from anterior to posterior, of the ileum, colon, and rectum. [From Herbst, D. B., and Bradley, T. J. (1989). A Malpighian tubule lime gland in an insect inhabiting alkaline salt lakes. *J. Exp. Biol.* **145**, 63–78.]

Malpighian tubules, and the structures in the hindgut, namely the ileum and rectum (Fig. 1).

Midgut

The midgut is a tubular epithelium. Upon ingestion, food and fluids move through the esophagus and pass directly into the midgut. Because the cells in the midgut epithelium are derived from embryonic endoderm, the midgut is not lined with cuticle. Most of the cells in the midgut are involved in the secretion of digestive fluids and the absorption of nutrients from the midgut lumen. These secretory and absorptive cells have apical microvilli that greatly increase the surface area available for inward and outward transport. In insects that feed periodically (such as adult mosquitoes), the microvilli shorten during nonfeeding periods and lengthen following ingestion. Many insects have additional cell types termed goblet cells that are thought to be involved in the secretion of fluids that modify the acidity and alkalinity (pH) of the luminal fluid. These goblet cells have been intensively investigated in Lepidoptera, where they serve to produce a markedly alkaline pH.

The basal surface of the midgut cells possesses a network of longitudinal and circular muscles that upon contraction can produce peristaltic waves. These contractions serve to move the food along the gut and stir the midgut contents during digestion. Many insects possess globular outpocketings in the anterior region of the midgut, termed ceca. The cells types in the ceca are generally differentiated from those in the midgut proper.

Malpighian Tubules

The Malpighian tubules are the site of urine formation in all insects except the Collembola, Thysanura, and aphids. The Malpighian tubules are tubular epithelia that are diverticulae (outpocketed extensions) of the gut itself. The tubules open into the gut near the midgut–hindgut junction, and the lumina of these two tubular epithelia are continuous. The contents of the tubules flow into the gut lumen; the ends of the tubules distal to the gut are closed. Fluid is produced in the Malpighian tubules by secretion; and because the tubules are closed at the distal end, hydrostatic pressure builds up and fluid flows through the tubules into the gut.

The number of Malpighian tubules is quite variable depending on the insect species. Bloodsucking Hemiptera

(e.g., *Rhodnius prolixus*) and higher Diptera (e.g., *Drosophila melanogaster*) have as few as four tubules, whereas the desert locust *(Schistocerca gregaria)* has hundreds. Attached to the Malpighian tubules of many insects are longitudinal muscles. When these muscles contract, the tubules are waved about in the hemolymph, presumably for the purpose of stirring the fluid adjacent to the tubules and promoting fluid and solute transport. These tubules may also serve the more general function of promoting hemolymph circulation throughout the abdomen.

The Malpighian tubules of all species examined to date contain more than one cell type. In some cases, a single epithelial region contains two or more cell types (regions with heterologous cell types) reflecting, presumably, separate physiological roles for each cell type. In other species, the tubules are divided into distinct regions, each consisting of a single cell type (regions with homologous cell types). In these insects, each tubule region has a distinct function in transport. Finally, in many insects, the tubules show regional specialization as well as multiple cell types within a region. It is presumed that each cell type in these tubules performs a distinct function.

As an example of cell type heterogeneity, consider the Malpighian tubules of the larvae of the brine fly, *Ephydra hians* (Fig. 2). The tubules in this insect are differentiated on each side of the body, as well as along their length. On one side of the body is a pair of tubules called the lime gland tubules. The distal ends of these are secretory and contain two regions that can be differentiated on the basis of cell color: one white, one yellow. More proximal to the gut are expanded regions of the tubules that serve to store concentric concretions in the tubule lumen. Finally, two of these cells combine in a common ureter that empties into the gut. On the opposite side, the tubules have only the yellow and white regions of the tubules, with no storage section. This example illustrates the variety of cell types that can exist in a single tubule. The details of transport function in these and other highly complex tubules have not been fully elucidated.

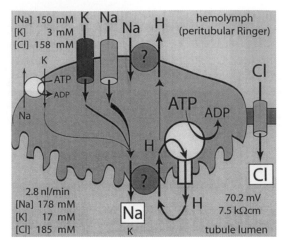

FIGURE 3 Model of the transport processes occurring in the Malpighian tubules of adult mosquitoes, based on the work of Klaus Beyenbach using the species *A. aegypti*. Upon stimulation with mosquito natriuretic peptide (MNP), rates of transepithelial fluid secretion increase from an unstimulated rate of 0.4 nl min^{-1} to 2.8 nl min^{-1}. In parallel, Na$^+$ concentrations in the secreted fluid rise, and K$^+$ concentrations fall. Electrophysiological studies reveal that MNP, working via cyclic AMP, induces an increase in basolateral membrane Na$^+$ conductance, presumably through the actions of Na$^+$ channels in the basolateral membrane of principal cells. The hyperpolarization of the transepithelial voltage and the decrease in transepithelial resistance are consistent with the activation of Na$^+$ channels in the basolateral membrane of principal cells. (Figure and legend provided by Klaus Beyenbach.)

In most Malpighian tubules, formation of the primary urine occurs in a cell type often referred to as the primary cell (Fig. 3). This cell type has extensive apical microvilli, often containing a central core of microfilaments. Frequently, these microvilli contain fingerlike extensions of the mitochondria and even of the endoplasmic reticulum. The basal surface of the cells exhibits deep infolds, often again closely associated with mitochondria by means of structures termed scalariform junctions. The intercellular spaces are occupied apically by septate or continuous junctions. More basally, the intercellular space contains gap junctions or undifferentiated basolateral membranes indistinguishable from the basal membrane infolds.

Numerous other cell types occur in the Malpighian tubules. It is presumed that each histologically distinguishable cell type performs a unique function within the tubules. In addition, distinct functions have been found in some cell types in the absence of histological or ultrastructural differentiation. A common cell type in Malpighian tubules is the stellate or secondary cell. This cell type possesses smaller microvilli than the primary cells, and these microvilli contain no mitochondria. As described later in the section on function, the secondary cells may be involved in modification of the primary urine produced by the primary cells.

Ileum

Posterior to the midgut, most insects possess a segment of gut referred to as the ileum. Because this region is part of the hindgut, it is covered on the apical surface by cuticle. The cells often show deep apical and basal membrane infoldings reflecting the role of these cells in fluid and solute transport. The basal surface of the cells is covered by layers of longitudinal and circular muscle that serve to generate peristaltic movements of the gut. These muscular contractions move the gut contents through the gut and may also serve an important role in reducing unstirred layers adjacent to both the apical and basal membranes of the epithelium.

Historically, the Malpighian tubules and rectum have been assumed to carry out most of the fluid transport in the posterior regions of the gut. The ileal epithelium is smaller in diameter, with less highly developed apical and basal membrane infolds, and a lower mitochondrial density than is observed in rectum. Nonetheless, when the transport properties of the ileum are investigated, this region of the gut is always found to be carrying out important transport functions.

Rectum

All insects possess an enlarged chamber called the rectum near the posterior end of the gut. The structure of the rectum can vary substantially from species to species. The rectal lumen is covered by a thick cuticle. Posterior to the rectum, insects possess an anal canal through which the feces and urine are eliminated. The strong bands of muscle surrounding the rectum contract during defecation, expelling the feces and urine into the external environment through the anus.

The recta of terrestrial insects are large and very active organs. Regions within the rectum are highly differentiated and contain cells with deep membrane folds. If (as in many orthopterans and lepidopterans) these cells are contained in thickened ridges that extend into the rectal lumen, they are referred to as rectal pads. If (as in many adult Diptera) the cells extend into the lumen as fingerlike structures, often on a thin stalk, they are referred to as rectal papillae. The rectal pads and the papillae are the major sites of fluid resorption and urine concentration. The cells in the rectal pads have a complicated array of intercellular junctions associated with the active sites of ion secretion and resorption. The evolution of a rectal structure permitting the formation of a urine hyperosmotic to the hemolymph was, along with the evolution of a waxy cuticle, a major adaptive event permitting insects to invade drier terrestrial habitats.

Both rectal pads and rectal papillae are covered with thick cuticle to protect the underlying epithelial cells from abrasion by the fecal material. Between the rectal pads, the rectum of terrestrial insects possesses a thin, cuticle-lined epithelium that, upon unfolding and stretching, permits rectal swelling during feces and fluid accumulation. In *E. hians,* the rectum is merely a thin, distensible epithelium that expands upon filling with feces prior to defecation. In this insect, the task of modifying the urine is carried out by the colon, which lies just anterior to the rectum. The presence of a rectum without transport capabilities is very unusual. In most insects the rectum is the major organ responsible for osmotic regulation of the urine.

Excretion outside the Gut

Storage excretion of concentric concretions containing calcium salts of urate and carbonate occurs in the fat body cells of most insects. Some insects contain nephrocytes, cells in the head that also store wastes and toxic elements. Finally, in collembolans and thysanurans, cephalic nephridial glands, sometimes termed labial glands, are responsible for excretory function, including the excretion of nitrogenous waste.

FUNCTION OF THE EXCRETORY ORGANS

Midgut

TRANSPORT   The primary function of the midgut in insects is the digestion and uptake of ingested nutritive materials. There are two processes that occur in the midgut and contribute to excretion: pH regulation and storage excretion. The midguts of many insects secrete fluids that contribute to extreme alkalinity or acidity in the gut. In Lepidoptera, for example, the midgut epithelium consists of goblet and columnar epithelial cells. The goblet cells are responsible for transporting ions and creating a highly alkaline solution in the midgut lumen. The columnar cells contribute to digestion of the food and uptake of ions.

The apical membrane of the goblet cell contains hydrogen pumps (H^+-ATPases) that use adenosine triphosphate (ATP) as their source of energy. These pumps transport hydrogen ions into the cuplike apical crypt of the goblet cell. The resulting electrical and pH gradient is used in an exchange process (antiporter) to exchange potassium ions for the hydrogen ion. Potassium ions associated with hydroxyl ions remain in the lumen following the hydrogen/potassium exchange process. These ions diffuse from the goblet cavity into the midgut lumen, with the result that the potassium hydroxide (KOH) causes the lumen to be very alkaline. The high potassium concentration in the lumen is, in turn, used as an energy source for the uptake of amino acids from the digested foodstuffs. This uptake occurs by means of transporters (symporters) in the apical cell membranes of the columnar cells that use the potassium gradient to cotransport potassium ions and amino acids from the lumen into the cell interior.

Various functions have been proposed for the highly alkaline pH in the midgut of Lepidoptera. Clearly, such an extreme pH would serve to kill pathogens and to saponify lipids in the ingested food material. It has also been proposed that this pH serves to reduce the solubility and activity of tannic acids in the food of Lepidoptera, protecting herbivorous larvae from the high concentrations of these toxic compounds found in the leaves of many plants.

In other insects, for example, mosquito larvae, low-pH (acid) conditions are observed in the midgut. The precise mechanisms by which this pH is produced and maintained remain to be elucidated, but there is evidence in mosquitoes that the midgut epithelium is also energized by the H^+-ATPase.

STORAGE EXCRETION IN THE MIDGUT   The columnar midgut cells of insects often contain concentric mineralized concretions. The major cation in these concretions has been shown to be calcium. The anion can be quite variable and has been found to be largely urate or carbonate, depending on the cell type in which the concretions are located. In the midgut, the concretions have also been shown to play a role in the storage excretion of toxic cations such as copper, zinc, iron, and selenium. Each of these elements is toxic in high concentrations but plays a critical role in metabolism in trace amounts. The concretions may therefore play both a protective and a homeostatic role in insects by regulating the free concentration of these ions and metals in the body. They reduce toxic concentrations and serve as a reservoir for these elements, which can be used for physiological purposes when required. Storage excretion in the form of concentric concretions also occurs in the Malpighian tubules.

Malpighian Tubules

TRANSPORT   The production of urine in insects occurs by the active transport of ions across the epithelium from hemolymph to tubule lumen. This process generates an osmotic concentration that drives the movement of water across the epithelium as well. Generally, Malpighian tubules have a high permeability coefficient for water (a low osmotic resistance), and as a result water moves rapidly across the epithelium in response to relatively low osmotic concentration gradients. These concentration differences across the epithelium are indeed so low that they have been difficult to measure. Recent experimental results have led to a general consensus, however, that an osmotic gradient of a few milliosmoles is sufficient to account for the observed rates of water movement across Malpighian tubules.

Let us consider first the active transport of ions across the epithelium, and then the passive movement of water that follows. In insects, potassium is the predominant cation transported across the epithelium of the Malpighian tubules. Insects evolved as a distinct clade on land, feeding on plants and detritus. It may be that their dependence on potassium as the major cation used for fluid transport evolved at this time as well. Certainly, animals of marine origin, such as the vertebrates and crustaceans, rely on sodium as the principal cation for driving fluid movements.

In the cell interior, the Malpighian tubule cells have a negative electrical charge relative to the hemolymph. This electrical potential facilitates the entry of potassium into the cells. Thermodynamically speaking, therefore, the most energetically costly transmembrane movement for potassium in the epithelium occurs as this cation crosses the apical membrane. The process by which this occurs has been very difficult to unravel, but in recent years it has been demonstrated that a very active hydrogen ATPase, related to vacuolar H^+-ATPase found widely in eukaryotic cells, is located on the apical membrane of Malpighian tubule cells. This transporter moves hydrogen ions from the cell interior into the tubule lumen, thereby setting up a large electrochemical gradient for hydrogen ions. Depending on the circumstances, this electrochemical gradient can be predominantly expressed as a pH gradient or as an electrical gradient. In either circumstance, this electrochemical gradient is thought to serve as an energy reservoir that can be subsequently used for a variety of coupled transport processes.

For example, if the energy contained in the hydrogen ion gradient is used in an antiporter (a transporter that couples ion movement in one direction to ion movement in the opposite direction) that exchanges hydrogen for potassium at the apical membrane, the result of hydrogen transport with subsequent hydrogen exchange for potassium is a net active transport of potassium. Potassium is moved against its electrochemical gradient by the antiporter, using the energy contained in the transmembrane hydrogen ion gradient. In some Malpighian tubules, similar mechanisms may exist for the coupled transport of other cations, (e.g., sodium, calcium, or magnesium). In bloodsucking insects such as adult mosquitoes and the reduviid *R. prolixus,* the plasma portion of the blood meal also provides the insect with a substantial intake of sodium. Any amount of sodium that exceeds the physiological needs of the insect must be excreted. In these insects, a substantial active transport of sodium occurs accords the Malpighian tubules. This process is thought to be driven across the apical membrane by a hydrogen–sodium exchange mechanism.

Cation transport across the epithelium also requires a process for entry of the ions into the cytoplasm from the hemolymph. In most insects, the basal membrane contains an enzyme (Na^+-K^+ ATPase) that uses the energy in ATP to transport sodium actively out of the cell and potassium actively in. As a result, the cytoplasm is greatly enriched in potassium. The passive outward diffusion of this ion through barium-sensitive channels produces an electrical potential across the basal membrane, the inside of which is negatively charged. This potential can be used for a variety of transport functions; one that seems to be almost universally present in Malpighian tubules is the bumetanide-sensitive $Na^+/K^+/2Cl^-$, cotransporter. This transporter uses the energy contained in the sodium gradient to move one sodium, one potassium, and two chloride ions simultaneously from the hemolymph into the cytoplasm. This process serves to provide chloride to the cell interior, as well as sodium in the cell types to which sodium is moved transepithelially.

The movement of anions across the epithelium involves one or more of three distinct transport pathways, depending on the characteristics of the tubules. In the first, chloride is thought to move across the same cells as those in which the cations are transported. As mentioned earlier, the interior of the cells is electrically negative relative to the cell exterior. The movement of chloride into the cell is therefore a thermodynamically active process. As already discussed, it may be driven by the sodium electrochemical gradient in a cotransport process. In other insect species in which the Malpighian tubule cells in a single region of the tubule are differentiated into two or more cell types, chloride ions have been shown to enter the lumen via a cell type distinct from that involved in active cation transport. For example, in *D. melanogaster,* the fluid-transporting segment of the tubules contains both primary and stellate cells. The former cells are the sites of active potassium transport, the latter the site of chloride flux from the hemolymph to the tubule lumen. Although this movement of chloride into the tubule lumen is thermodynamically downhill, the precise mechanism of chloride transport is presently unknown.

A third process has been described in the Malpighian tubules of adult mosquitoes. Although these insects also possess primary and stellate cells in the Malpighian tubules, it has been proposed that chloride moves into the lumen of the tubules via the intercellular junctions. In fact, this process has been shown to be under hormonal control. The movements of chloride into the Malpighian tubules may therefore be quite variable depending on cell types. The movements of anions are much less well characterized at this time, with regard to the molecules that drive the process, than are the movements of cations. A model of ion transport at both the apical and basal membrane of the Malpighian tubules of adult mosquitoes is shown in Fig. 3.

Regardless of the mechanisms by which cations and anions enter the Malpighian tubules, it is clear that the types of ions transported can vary greatly with the species of insect and will depend on an individual's physiological needs and demands. Thus since blood-sucking insects ingest a large amount of sodium compared with other insects, the Malpighian tubules of bloodsuckers contain specific mechanisms designed to reduce the large sodium load. Species of mosquitoes whose larvae can survive in salt water must ingest the medium and eliminate the ions as means of obtaining water. Those species that have been investigated can excrete magnesium and sulfate via the Malpighian tubules. The larvae of brine flies (ephydrids) generate crystals in the lumina of the tubules that are rich in calcium and carbonate. Both these ions must be transported across the epithelium, although the combination forms insoluble crystals that reduce the activity of these ions in the lumen.

The Malpighian tubules of insects are also the site of excretion of the waste products of energy and nitrogen metabolism. Acid by-products of energy metabolism have been shown to be actively transported into the lumen from the hemolymph. The precise molecular mechanisms of the process remain unclear, but the process is of paramount importance for the insects in maintaining acid/base balance and energy homeostasis. The by-products of nitrogen metabolism are also excreted by the Malpighian tubules.

In aquatic insects, ammonia may be excreted, but in most insects and certainly in terrestrial forms, urea and uric acid predominate. Both these compounds are actively removed from the hemolymph by the Malpighian tubules. The transport of uric acid has been investigated in *R. prolixus,* in which the blood meal provides a very protein-rich meal requiring intense capacity for the elimination of nitrogenous waste. In *Rhodnius,* the primary urine is produced in the most upstream portions of the Malpighian tubules, the upper tubule. This urine is modified in the downstream section (the lower tubule) through the resorption of potassium and chloride. This process serves to return potassium to the hemolymph, and to remove waste from the hemolymph, through the retrieval of an isosmotic fluid. Thus hemolymph volume is retained and the sodium in the urine is concentrated. In addition, uric acid is transported in the lower tubule from hemolymph to urine. Potassium urate is fairly insoluble, particularly at neutral to acid pH. As a result, crystals of uric acid form in the urine. This process further removes osmotically active compounds from the urine, allowing the additional movement of water from the urine to the hemolymph by osmosis.

CONCENTRIC CONCRETIONS IN THE MALPIGHIAN TUBULES Concentric concretions occur in the midgut, where they are thought to contribute to excretion by storage in an insoluble form of salts containing calcium, magnesium, manganese, copper, cadmium, and zinc. Identical concretions are found intracellularly in the Malpighian tubules. These concretions are thought to perform an identical function, namely storage of ions in an insoluble form either for subsequent use or as a means of removing the ions from the body. In the Malpighian tubules, however, these concretions also appear in the tubule lumen, a location from which they can move into the gut and be eliminated with the excrement. It has been suggested by many authors that the intracellular concretions in the cells of the Malpighian tubules can be transported by exocytosis into the lumen of the tubules. Although there are occasionally physiological conditions in which the concretions disappear from the cells and appear in the lumen, it has not been unambiguously demonstrated that the crystals move from one location to the other intact. Instead, it is likely that the crystals are dissolved within the cells of the Malpighian tubules, that the soluble ions are transported into the lumen, and that the crystals are formed anew in the tubule lumen. Crystals are formed in some tubule segments (e.g., in the lower tubule of *R. prolixus*) where no crystals exist in the cells.

The crystals in the midgut, fat body, and Malpighian tubules are concentric and perfectly round. This is in marked contrast to the natural structure of the crystals formed by the same salts in solution. Uric acid crystals, for example, have sharp corners and sometimes take a needlelike form. It is thought that the concentric concretions avoid acicularity through the activity of organic compounds that are known to be a substantial component of the concretions. The compounds are thought to nucleate and direct crystal formation, leading to the formation of round concretions. This spherical shape is less damaging to the cells of the tubules and can be excreted from the tubules and gut with little tissue damage. Ultrastructurally identical concretions are observed in the urine of birds, which is rich in uric acid. It has been proposed that organic compounds are excreted into the tubule lumen, where they nucleate and guide the formation of the concretions. The ions contained in the concretions can vary greatly, ranging from potassium urate in some tissues to calcium carbonate in others. Even though such crystals should be quite distinct in shape, the concretions produced by the insects all have the same distinct concentric, spherical shape. This set of properties argues that the organic compounds have a profound effect on crystal form and formation.

Ileum

The principal function of the ileum is to act as a tubular epithelium that serves to transport to the rectum the undigested remains of the food from the midgut and fluid from the Malpighian tubules. This transport occurs by peristaltic movements of the circular and longitudinal muscles surrounding the ileum. The ileum also engages in important transepithelial transport functions. This has been investigated in considerable detail in the locust *Schistocerca gregaria.* In this species, potassium and chloride ions are transported from the lumen of the ileum into the hemolymph. This transport is iso-osmotic. It therefore does not contribute directly to osmotic regulation but serves instead to reduce the volume of the urine and to retain valuable ions and water in the hemolymph. This transport is under hormonal control, presumably to allow the insect to modulate the return of water to the hemolymph depending on whether osmotic condition of the animal dictates a diuretic or an antidiuretic response.

In some insects, an additional segment of the hindgut exists, which is termed the colon. Although this segment is hard to distinguish with the unaided eye, it is functionally and histologically distinct from the ileum. In larvae of *E. hians,* for example, an ileal segment occurs near the midgut, while a colonic segment of the hindgut lies between the ileum and the rectum. It has been shown that active ion transport occurs in the colon. The colon has a relatively low osmotic permeability, thus allowing the secretion in this segment of a fluid that is strongly hyperosmotic to the hemolymph. Production of hyperosmotic excreta is crucial for the osmotic regulation in this species because the insects live in the waters of a saline lake, the osmotic concentration of which is six times more concentrated than the hemolymph. The ions transported in the colon include sodium, chloride, and sulfate. Sulfate ions are large in comparison to other transported ions; therefore the transport of sulfate through an epithelium capable of maintaining a substantial osmotic gradient is unusual. The larvae of the blowfly, *Sarcophaga bullata,* have also been shown to engage in active transport in the colon. In this species, the colon is a major site for the excretion of nitrogenous waste in the form of ammonium ion. Because these larvae feed in rotting flesh, the active transport of ammonium is a critical adaptive feature in the physiology of the species.

Rectum

In most insects, the rectum is the most active ion-transporting organ on a per-gram basis. All fluids and solids deriving from the midgut and Malpighian tubules pass through the ileum and enter the rectum before being excreted. The rectum is therefore the last location in the gut in which the ionic and osmotic concentration of the excreta can be modified to meet the regulatory needs of the insect.

In terrestrial animals, the requirements for osmotic homeostasis vacillate between the production of a dilute excreta (diuresis) and the production of a concentrated excreta (antidiuresis). Control of the rectum is therefore a critical element in the maintenance of osmotic homeostasis in the hemolymph. None of the other elements of the excretory system discussed thus far are capable of producing a fluid differing in osmotic concentration from that of the hemolymph.

The role of the rectum in terrestrial insects has been most intensively studied in *S. gregaria*. In this insect, the cells in the rectal pads serve to transport a hypo-osmotic fluid from the lumen into the hemolymph. This serves to produce excreta with a very high osmotic concentration and, in the process, conserve water in the hemolymph.

The process by which the locust transports a hypo-osmotic fluid is complicated and, unlike a functionally analogous process in the kidney of mammals, it requires cells of only a single type, the cuticle-covered rectal pad cells. In the parts of the rectum differentiated into rectal pads, the cells underlying the cuticle have deep apical infolds associated with numerous mitochondria. The rectal epithelium in the regions of the rectal pads is thick, meaning that the rectal pad cells comprise a tall, columnar epithelium. The intercellular junctions in these cells are highly convoluted and contain open spaces or intercellular swellings in the clefts between the cells.

The process of fluid resorption from the lumen begins with the active transport of ions across the apical membrane (i.e., from the lumen to the intracellular compartment). Once in the cytoplasm, the ions are transported across the intercellular membrane into the enlarged spaces in the intercellular clefts. The compounds transported are principally potassium and chloride, although other compounds including acetate and proline are actively transported out of the lumen as well. These transported compounds produce a fluid with high osmotic concentration. It is thought that water is drawn from the lumen into the intercellular clefts, probably through the apical septate junctions. As a result, fluid accumulates in the intercellular clefts and in the open spaces in the intercellular regions. From here, the fluid flows extracellularly between the cells in a basal direction toward the hemolymph. It is thought that as this fluid flows, transporters within the lateral cell membranes remove ions. If these membranes have a low osmotic permeability, ions can move across with little water following. As a result, ions are removed faster than water can follow, resulting in a fluid that is hypo-osmotic not only to the lumen but also to the hemolymph.

Under conditions in which the insect is well hydrated (e.g., after eating lush vegetation), the rectum removes ions from the rectal lumen but little water follows, presumably because either the site or the rate of transport in the more lateral and basal membranes has been modified. This produces a dilute urine, the excretion of which serves the osmotic needs of the insect.

Aquatic insects are similarly dependent on the rectum for the final modification of the urine prior to excretion. In freshwater insects, the fluid derived from the midgut and Malpighian tubules is iso-osmotic to the hemolymph. Excretion of this fluid would lead to rapid loss of ions and the death of the animal. The rectum serves to transport ions from this primary urine back into the hemolymph.

Transport of potassium and chloride has been documented for number of freshwater insects. These transport mechanisms are relatively easy to demonstrate because the fluid entering the rectum from the Malpighian tubules is enriched in these two ions, and the excreted urine leaving the rectum much depleted.

Rectal function has also been investigated in aquatic insects residing in hyperosmotic media, for example, in saline-tolerant dipteran larvae inhabiting coastal and desert saline waters. In species of *Aedes* inhabiting these waters, the rectum is differentiated into two segments. The anterior rectal segment is identical in function to the rectum of freshwater species and serves to remove ions from the urine under conditions in which the larvae find themselves in hypo-osmotic media (i.e., fresh water). When the larvae hatch in saltwater, or when the medium becomes concentrated because of evaporation, the posterior rectal segment becomes active. This segment has a single cell type, which is characterized by deep apical and basal infolds associated with numerous mitochondria. The cells actively transport ions from the hemolymph into the rectal lumen. Because the epithelium has a low osmotic permeability, ions are transported faster than water can follow. As a result, a concentrated urine is produced by secretion in this segment, which has been called the salt gland.

The ions transported in the posterior rectal segment vary with the environment in which the larvae occur. In seawater, sodium, magnesium, and chloride predominate. In bicarbonate-rich waters, a concentrated fluid is secreted, and the urine is rich in sodium and bicarbonate. The precise molecular mechanisms of ion transport in the rectum, as well as their neuronal or hormonal control, are poorly known for aquatic insects.

See Also the Following Articles

Digestion • Fat Body • Hemolymph • Water and Ion Balance

Further Reading

Beyenbach, K. W. (1995). Mechanisms and regulation of epithelial transport across Malpighian tubules. *J. Insect Physiol.* **41,** 197–207.
Bradley, T. J. (1985). The excretory system: Structure and physiology. *In* "Comprehensive Insect Physiology, Biochemistry and Pharmacology" (G. A. Kerkut and L. I. Gilbert, eds.), pp. 421–465. Pergamon Press, Oxford, U.K.

Bradley, T. J. (1998). Malpighian tubules. *In* "Microscopic Anatomy of the Invertebrates," Vol. XI, "Insecta" (M. Locke and F. W. Harrison, eds). pp. 809–829. Liss, New York.

Chapman, R. F. (1998). "The Insects: Structure and Function." Cambridge University Press, Cambridge, U.K.

Herbst, D. B., and Bradley, T. J. (1989). A Malpighian tubule lime gland in an insect inhabiting alkaline salt lakes. *J. Exp. Biol.* **145**, 63–78.

Phillips, J. E., and Audsley, N. (1995). Neuropeptide control of ion and fluid transport across locust hindgut. *Am. Zool.* **35**, 503–514.

Wigglesworth, V. B. (1965). "The Principles of Insect Physiology," 6th ed. Methuen., London.

Exopterygota

Exopterygota is a division of the class Insecta in the phylum Arthropoda. The orders of insects in this division have wings that develop externally during the maturation of the larva (which is variously referred to as a larva, nymph, or naiad). There are two superorders in this division: the Orthopteroidea (which includes the orders Blattodea, Dermaptera, Embiidina, Grylloblattodea, Isoptera, Mantodea, Mantophasmatodea, Orthoptera, Phasmatodea, and Plecoptera) and the Hemipteroidea (Hemiptera, Phthiraptera, Psocoptera, Thysanoptera, and Zoraptera). Except for the developed wings and genitalia, there is a strong morphological resemblance between larvae and adults (although habitats and biology may differ greatly).

Exoskeleton

Svend O. Andersen

Copenhagen University

The exoskeleton is noncellular material that is located on top of the epidermal cell layer and constitutes the outermost part of the integument. The local properties and appearance of the exoskeleton are highly variable, and nearly all visible features of an insect result from the exoskeleton. The exoskeleton serves as a barrier between the interior of the insect and the environment, preventing desiccation and the penetration of microorganisms. Muscles governing the insect's movements are attached to the exoskeleton.

Although the exoskeleton is a continuous structure, its mechanical properties differ from region to region. Sometimes the transition between regions is gradual, but often it is quite abrupt; pliant and elastic regions can thus border on hard and heavily sclerotized regions. Most exoskeletal regions of soft-bodied larvae, such as larvae of moths and flies, are soft

and pliant, and only restricted regions of their exoskeletons are hard and stiff, such as legs, head capsule, and mandibles. Most of the body surface of adult, winged insects is covered by a stiff exocuticle, which can be somewhat flexible and bendable but also serves as a hard protective armor. The exoskeleton covering the dorsal abdomen of many beetle species is thin and easily flexed, whereas the ventral abdominal exoskeleton of the same animals is hard and resistant. The mechanical properties of all exoskeletal regions are precisely adapted to be optimal for the lifestyle of the insect.

FORMATION OF THE EXOSKELETON

The exoskeleton is produced and modified by the epidermal cell layer, and each cell in the epidermis must have the necessary information for producing and depositing the right amount of the right cuticular components at the right time; some of them will later have to modify the secreted products to give a mature material. The timing of the various events is often hormonally controlled, but the quantitative information on how much to produce must be inherent in individual epidermal cells.

A new exocuticle is produced at each molt. A thin, lipid-rich epicuticle is initially secreted from the epidermal cells and deposited beneath the old cuticle, followed by secretion of a thicker procuticle, consisting of chitin and proteins. To allow growth, the total surface area of the new cuticle is larger than that of the old one, and expansion and stretching of the new cuticle take place during and after emergence from the old cuticle (exuvium). Some exoskeletal regions, such as the head capsule, mouthparts, and spines, may be sclerotized before ecdysis; this will aid emergence from the old cuticle. These regions cannot be further expanded but will keep their pre-ecdysial size and shape. Other exoskeletal regions are soft and pliant at ecdysis and are sclerotized soon after emergence when cuticular expansion is complete; as soon as the sclerotization process has started, these regions are irreversibly locked in their new shape.

Sclerotization not only makes the exoskeleton harder and stiffer, it also makes the proteins inextractable and more resistant to enzymatic digestion. Before sclerotization, the exoskeletal proteins are bound to each other and to chitin by various noncovalent links, such as electrostatic interactions, hydrogen bonds, and hydrophobic interactions. Such links can be weakened by changes in pH and ionic strength, making the cuticle more pliant, because displacements of the cuticular components will be easier. During the sclerotization process the proteins are linked firmly to each other, polymerized sclerotizing material fills the voids between proteins and chitin molecules, the cuticle is dehydrated, and deformations of the material will be more difficult.

The sclerotization precursors, *N*-acetyldopamine (NADA) and *N*-β-alanyldopamine (NBAD), are synthesized from tyrosine in the epidermal cells. The tyrosine molecules are transformed by decarboxylation and hydroxylation to dopamine, which is acylated to NADA and NBAD. These precursors are secreted from the epidermal cells into the cuticular matrix, where they encounter enzymes (phenoloxidases), which oxidize them to the corresponding orthoquinones. Oxidases of different types (tyrosinases, laccases, peroxidases) have been reported and characterized from cuticle. The quinones produced are highly reactive; they will react spontaneously with histidine and lysine residues in the matrix proteins, resulting in cross-links between neighboring proteins, and they will also react with each other, resulting in complex phenolic polymer mixtures. Depending on the precise reaction conditions, the exoskeleton may remain colorless, or a lighter or darker brown coloration may appear during sclerotization.

The water content of the exoskeleton decreases during incorporation of the sclerotizing precursors into the matrix, probably from a decrease in the number of positively charged amino acid residues in the cuticular proteins, which makes the matrix proteins less hydrophilic. Exclusion of water from the intracuticular voids from accumulation of polymerized material also presumably contributes to dehydration of the exoskeletal material. Often only the exocuticular layer of the sclerites is sclerotized, but in some insects the sclerotization process continues for extended periods after ecdysis, resulting in sclerotization of parts of the endocuticle, although to a lesser extent than the exocuticle.

Both the loss in cuticular water content and the formation of cross-links between proteins contribute to a stabilization of the exoskeletal material. The amounts of sclerotizing material incorporated into the various exoskeletal regions varies from less than 1% to more than 10% of cuticular dry weight. These differences are assumed to be responsible for most of the variation in hardness and stiffness of the various exoskeletal regions. Exocuticle tends to be harder and more difficult to deform than endocuticle, presumably because of more extensive sclerotization. The endocuticular layer will tend to be compressed when a piece of exoskeleton is bent, whereas the stiffer exocuticle will be little deformed, although it will be in tension.

MUSCLE ATTACHMENTS

The muscles that act on the exoskeleton are connected to the basal surface of the epidermal cells by means of desmosomes. The muscular forces are transferred through the cells by a rich array of microtubules, running in parallel from the basal to the apical surface of the cells, where they attach to tonofilaments stretching into the cuticular material. The muscles are often attached to infoldings of the exoskeleton, the apodemes, which can stretch deep into the body of the insect, allowing larger muscles to act on the same skeletal region.

ELASTIC EXOSKELETONS

Some small exoskeletal regions are characterized by a rubberlike elasticity; they can undergo considerable deformation when exposed to mechanical stresses and return to their original shape when unstressed. The amount of energy used for deformation is almost completely recovered during relaxation. Its elasticity is the result of the matrix protein resilin. Resilin-containing ligaments are used for energy storage when a fast release of mechanical energy is needed: for example, in the flight system of insects and in the jumping systems of fleas and click beetles. Most resilin-containing ligaments contain chitin microfibrils, making them inextensible, but readily flexible, but there are some ligaments that consist of nearly pure resilin and are devoid of chitin. Such ligaments can be reversibly stretched to three to four times their unstrained length before breaking. The protein chains in resilin are cross-linked by a mechanism different from that used for the solid cuticle; the chains are linked together by covalent bonds formed between side chains of tyrosine residues during the secretion of soluble resilin from the epidermal cells. The elastic properties of the cross-linked material are due to the flexibility and random coiling of the chain segments between cross-links.

PLASTICIZATION

Sometimes the mechanical properties of the exoskeleton can be changed rapidly and reversibly. In bloodsucking bugs (e.g., nymphs of *Rhodnius prolixus*), the abdominal cuticle is stiff and inextensible before a blood meal. When a meal is initiated, the abdominal cuticle is plasticized, enabling the animal to gorge itself with a volume of blood 10 to 12 times larger than the total volume of the animal before the meal. To do this, stretch receptors send nerve impulses via the central nervous system to axons terminating in the abdominal epidermis. A neurohormone is released from these nerve endings, and the epidermal cells respond by effecting a slight decrease in intracuticular pH. The water content of the abdominal cuticle increases simultaneously, probably owing to the pH change, and the interactions between cuticular proteins decrease, resulting in increased plasticity of the cuticular material.

To facilitate emergence from the old cuticle during ecdysis, the stretchability of the new, pharate cuticle may be temporarily increased to make it easier for the animal to escape from the rather stiff exuvium and facilitate expansion of the new cuticle after emergence. In the tobacco horn worm *Manduca sexta,* and probably in many other insects, the plasticization of the pharate adult cuticle is triggered by release of eclosion hormone into the hemolymph. As in *Rhodnius* nymphal abdominal cuticle, the plasticization of *Manduca* pharate cuticle at emergence is probably due to an intracuticular pH decrease in combination with increased hydration.

Newly emerged blowflies, which must dig free of the soil before they can expand to their proper size, have a relatively stiff cuticle until they have reached the surface and can begin to swallow air for expansion. For a brief period, their cuticle is plasticized, from release of the neurohormone bursicon. This hormone also plays a role in initiating sclerotization and deposition of endocuticle in the blowflies and probably in other insects.

VISCOELASTICITY

Most types of cuticle are more or less viscoelastic; when exposed to a deforming force for extended periods, they will suffer a slight, time-dependent elongation, and recovery after release of the force may not be complete. A special type of highly stretchable, viscoelastic cuticle is found in the abdominal intersegmental membranes of sexually mature female locusts. This stretchability allows elongation of the abdomen necessary for depositing eggs in the soil at a sufficient depth. The membranes in both male and female locusts are soft and pliable, but not very stretchable, as long as the animals are sexually immature. When sexual maturation is initiated in the females by resumed production of juvenile hormone, the organization of the chitin microfibrils in the intersegmental membranes changes from a helicoidal arrangement to one that is perpendicular to the long axis of the animal; at the same time, special hydrophilic proteins are deposited in the membranes. The fully mature intersegmental membranes stretch when loaded, but recover only partly when the load is released. When reloaded with the same load as before, they elongate significantly more than during the first load, and by repeated application of even small loading forces the females can elongate the membranes to about 10 to 15 times their relaxed length, corresponding to a threefold elongation of the total abdomen. Such stretching enables the female locust to deposit eggs in the soil to a depth of 10 to 12 cm.

METAL REINFORCEMENT

The mandibles of plant-eating insects are often extremely hard and abrasion resistant because of incorporation of metals, such as zinc and manganese, in the cuticular matrix of the cutting edge of the mandibles. Up to 5% zinc has been registered in some mandibles.

PROTECTIVE BARRIER

The exoskeleton serves also as a water-impermeable barrier, protecting the insect against desiccation. The main part of the barrier is located in the wax-covered epicuticle.

An important function for the exoskeleton is to act as a barrier preventing microorganisms from access to interior of the animal. Soft, pliant cuticles are more easily damaged and penetrated by microorganisms than the sclerotized regions, but they contain a defense system of inactive precursors of phenoloxidases. When the cuticle is damaged, these precursors are activated by limited proteolysis to active phenoloxidases, which will oxidize tyrosine and other phenols to highly reactive quinones. The reaction products are toxic for microorganisms, and they will close minor wounds in the cuticular surface.

COLORATION

Often the result of various pigments present in granules in the epidermal cells, the colors of insects can also be due to colored material in the cuticle, diffraction or interference of light caused by special cuticular structures, or the Tyndall effect.

A brown coloration in the cuticle develops often during sclerotization of the exocuticle, especially when NBAD is used as precursor for the sclerotization agents, whereas uncolored and transparent cuticles results when NADA is the sole sclerotization precursor. The intensity of the color varies from very light brown over tan to a very dark brown, which can be difficult to discern from the genuinely black cuticles that contain melanins. Melanins are formed when free tyrosine or dopamine is oxidized to orthoquinones, which readily polymerize to complex, black, intractable materials. Melanins either are diffusely distributed in the cuticle or occur in discrete, membrane-bounded granules.

Structural colors of the cuticle from interference of light can be caused by regularly spaced layers in the cuticle in, for example, the cornea of the compound eyes in many flies. Light reflected from the individual layers will interfere to give colors varying with the angle of reflection. Structural colors may also be produced by diffraction of light by regularly spaced microscopic structures on the cuticular surface. The brilliant colors of many beetle species are due to such surface diffraction.

Light scattered by sufficiently small particles ($< 0.7\ \mu m$ in diameter) looks blue because of the Tyndall effect, as in the blue colors of many dragonflies. The light-scattering particles may be located in the epidermal cells underlying a transparent cuticle, or the light may be scattered by a very fine bloom of wax filaments deposited on the cuticular surface after emergence.

SENSE ORGANS

Several exoskeletal structures are involved in sense perception. Various types of mechanoreceptor are involved in registering the exact position of, and deformation in, the various exoskeletal regions and body parts, movements of surrounding objects, currents of air or water, vibrations in the substrate, and sound oscillations. Chemoreceptors are involved in registering and discerning the presence of various chemical substances; these receptors can be contact chemoreceptors (taste) or olfactory chemoreceptors (smell). Many of the sense organs take the form of setae (bristles, hairs, etc.), which are sensilla consisting of an elongated cuticular structure in connection with the sensory cell(s). A trichogen cell in the epidermis produces a more or less elongated structure, which can be variously shaped, often as a flexible hair, a rigid spine, or an arched dome. The hairs are usually connected to the surrounding cuticle by a joint, flexible membrane, and the sensory cell responds to deformations of the cutaneous membrane. The campaniform sensilla are rigidly connected to the surrounding cuticle, and they respond to tensions in the dome shaped cuticle.

The cuticle covering the elongated sensilla of olfactory chemoreceptors contains numerous narrow pores, allowing access for the airborne stimulatory molecules into the interior of the sensilla, where they come in contact with and stimulate the dendritic membrane of the sensory cell. The contact chemoreceptors are constructed according to the same principle, but they often contain a single larger pore through which molecules can get access to the sensory cell.

A characteristic feature of the visual system in insects is that both the compound eyes and the single eyes (ocelli) are covered by a transparent cuticle, the lens or cornea, through which light reaches the light-sensitive cells. Both the corneal cuticles and the cuticles used for construction of the other sense organs are constructed according to the common cuticular plan.

See Also the Following Articles

Chemoreception • Coloration • Cuticle • Mechanoreception • Molting

Further Reading

Andersen, S. O. (1985). Sclerotization and tanning of the cuticle. *In* "Comprehensive Insect Physiology, Biochemistry, and Pharmacology" (G. A. Kerkut, and L. I. Gilbert, eds.), Vol. 3, Chap. 2. Pergamon Press, Oxford, U.K.

Bereiter-Hahn, J., Matoltsy, A. G., and Richards, K. S. (eds.) (1984). "Biology of the Integument," Vol. 1, "Invertebrates." Springer-Verlag, Berlin. (See especially Chaps. 27–35.)

Blomquist, G. J., and Dillwith, J. W. (1985). Cuticular lipids. *In* "Comprehensive Insect Physiology, Biochemistry, and Pharmacology," (G. A. Kerkut, and L. I. Gilbert, eds.), Vol. 3, Chap. 4. Pergamon Press, Oxford, U.K.

Hepburn, H. R. (ed.) (1976). "The Insect Integument." Elsevier, Amsterdam.

Hepburn, H. R. (1985). Structure of the integument. *In* "Comprehensive Insect Physiology, Biochemistry, and Pharmacology," (G. A. Kerkut, and L. I. Gilbert, eds.), Vol. 3, Chap. 1. Pergamon Press, Oxford, U.K.

Kramer, K. J., Dziadik-Turner, C., and Koga, D. (1985). Chitin metabolism in insects. *In* "Comprehensive Insect Physiology, Biochemistry, and Pharmacology," (G. A. Kerkut, and L. I. Gilbert, eds.), Vol. 3, Chap. 3. Pergamon Press, Oxford, U.K.

Neville, A. C. (1975). "Biology of the Arthropod Cuticle." Springer-Verlag, Berlin.

Wigglesworth, V. B. (1972). "The Principles of Insect Physiology." 7th ed. Chapman & Hall, London.

Extension Entomology

Nancy C. Hinkle and Beverly Sparks
University of Georgia

Linda J. Mason
Purdue University

Karen M. Vail
University of Tennessee, Knoxville

Land-grant institutions have teaching, research, and outreach (service) as their missions. Cooperative extension is the university's face to the state's citizenry, just as teaching faculty are the university's face to students and research faculty are the component visible to their academic peers around the world. In linking the university to the public, extension entomologists translate research results into practical applications and convey them to end users, while simultaneously apprising university researchers of real-world needs.

HISTORY OF COOPERATIVE EXTENSION

The Smith–Lever Act created the Cooperative Extension Service in 1914. However, several key legislative acts preceded Smith–Lever and these acts were critical in leading to the formation of the Cooperative Extension Service. The Morrill Acts of 1862 and 1890 (also known as the Land-Grant Acts) authorized that each state be granted 30,000 acres (12,141 ha) of public land for each senator and representative of the states in Congress at that time. Revenue generated from these lands was to be used for endowment, support, and maintenance of at least one college to teach fields of study related to agriculture and mechanical arts "to promote the liberal and practical education of the industrial classes in the several pursuits and professions in life."

The second Morrill Act provided funding to establish the 1890 land-grant institutions. Under the conditions of legal racial separation in the South during the late 1800s, black students were not permitted to attend the original land-grant institutions. Passage of the second Morrill Act expanded the 1862 system of land-grant universities to include historically black institutions.

The Hatch Act is often likened to a sturdy bridge between the Morrill Acts and the Smith–Lever Act. Signed on March 2, 1887, the Hatch Act gave this nation its network of agricultural experiment stations. The Hatch Act states that experiment stations should "conduct original and other research, investigations and experiments bearing directly on and contributing to the establishment and maintenance of a permanent and effective agricultural industry." These experiment stations were charged with conducting research for effective and efficient production of food and fiber. Research findings from systems across the country revised farming methods to fit America's diverse geography, making farmers more productive.

The federal–state research partnerships funded through the Hatch Act supported research that addressed "hunger and poverty and the drudgery of subsistence agriculture production." From its inception, research stations created by the Hatch Act were designed to meet the needs of agriculture in the areas in which the experiment stations were located, but the research generated often has far-reaching applications. In fact, research supported by Hatch Act funding benefits every person in the United States and much of the world.

The Smith–Lever Act of 1914 created the Cooperative Extension Service. Senator Hoke Smith (Georgia) and Representative Frank Lever (South Carolina) introduced this act "to aid in diffusing among the people of the United States useful and practical information on subjects relating to agriculture and home economics, and to encourage application of the same." This legislation created a partnership between the U.S. Department of Agriculture, the land-grant universities, and the 1890 institutions that was charged to provide outreach education to the citizens of each state. In practical terms this legislation created the ability for representatives of land-grant universities and 1890 institutions to work with farm families on their farms to introduce research-based advances in agriculture, home economics, and other fields.

Today, this educational system includes professionals in each of America's land-grant universities (in the 50 U.S. states, Puerto Rico, the Virgin Islands, Guam, Northern Marianas, American Samoa, Micronesia, and the District of Columbia) and in 16 1890 historically black, land-grant universities plus Tuskegee University.

The Cooperative Extension Service is a partnership between the U.S. Department of Agriculture, the land-grant institutions, and the 1890 institutions. Legislation in various states has also enabled local governments in the nation's counties to become a fourth legal partner in this educational endeavor. Organization of the Cooperative Extension Service at national, international, state, regional, and county levels is discussed below.

ORGANIZATION AT THE NATIONAL LEVEL

At the national level, the Cooperative Extension Service is an integral part of the Cooperative State Research, Education, and Extension Service (CSREES). The CSREES is a national research and education network that links education programs of the U.S. Department of Agriculture with land-grant institutions, with 1890 institutions, with agricultural experiment stations, with Cooperative Extension Services, with schools of forestry, and with colleges of agriculture, colleges of veterinary medicine, and colleges of human sciences. CSREES, in cooperation with all these partners, develops and supports research and extension programs in the food and agricultural sciences and related environmental and human sciences. Examples of some program areas in which CSREES and its partners are currently working include improving agricultural productivity; protecting animal and plant health; promoting human nutrition and health; strengthening children, youth, and families; and revitalizing rural American communities.

CSREES serves as a critical connection between research and extension. CSREES works with extension educators on identifying and communicating agricultural, environmental, and community problems (Table I). These problems are then relayed to researchers at the land-grant institutions and agricultural experiment stations. Working together, these partners initiate and stimulate new research that provides solutions to real-world problems.

TABLE I Extension Educational Programs within CSREES

1. Provide model education programs on food safety; sustainable agriculture; water quality; children, youth, and families; health; environmental stewardship; and community economic development in all 50 states, all U.S. territories, and the District of Columbia.
2. Represent over 9600 local extension agents working in 3150 counties.
3. Engage 5.6 million youth in 4-H programs for personal development and community service.
4. Involve 3 million trained volunteers who work with outreach education programs nationwide.
5. Provide farm safety education programs in all 50 states and Puerto Rico.
6. Provide pesticide applicator programs that train over half a million people each year in safe and environmentally sound pesticide use.
7. Participate in international education programs taught by over 200 extension professionals in 17 countries.

INTERNATIONAL EXTENSION ORGANIZATION

Although some industrialized countries have attempted to reduce costs by delegating extension responsibilities to the private sector, with varying degrees of success, most developing countries have modeled their extension systems on the U.S. paradigm. Frequently, extension outreach in third-world nations is funded by such agencies as the U.S. Agency for International Development, the World Bank, and the United Nations' Food and Agriculture Organization. These technology transfer programs typically are most effective when closely linked with university research programs, permitting rapid transmittal and adoption of research results. Alternatively, the outreach may be handled by such governmental entities as the ministry of agriculture.

STATE, REGIONAL, AND COUNTY ORGANIZATION

Organizational structure of the Cooperative Extension Services varies greatly in size from state to state. In general, leadership of Cooperative Extension Services within each state is the responsibility of the dean and/or director of the agricultural college of the land-grant university and/or 1890 institution within each state. These directors provide leadership to an administrative staff that often includes associate and/or assistant deans of extension, directors of county operations, department heads and/or extension program leaders within various scientific disciplines, and directors of units that support programming. In states with numerous counties, the organizational structure often includes regional administrators that serve under the Cooperative Extension director.

The extension director, along with administrators within each scientific discipline, also oversees a faculty of extension specialists. These specialists serve as educational resources to county agents and their clientele in various subject matter/disciplines. Extension specialists are most often administratively based within their academic department and may be located on the main campus of the land-grant university, at experiment stations or, occasionally, within county extension offices.

The organization of an extension office at the county level also varies greatly from state to state and county to county. In areas of the United States that have very low populations there are county offices with only one county extension agent, responsible for the administration and delivery of programming in all subject areas. In more populated areas, county extension offices often house several agents with program areas divided among agents.

Cooperative Extension at the County Level

The interface between extension entomologists on the state and county staff tends to follow a similar model in the majority of states. At the state level the positions are usually tied to a university academic unit and filled with a Ph.D.-level entomologist. These persons would either be full-time extension or have a partial extension appointment combined with other duties, including teaching and/or research. At the county level, job responsibilities and qualifications may vary; however, some common models are evident. County-level extension entomologists usually are termed agents, advisors, or educators, terms used synonymously within this article. Additionally, entomology positions typically fall under either Agriculture and Natural Resources (ANR) or 4-H and Youth programmatic areas.

County agents initially were itinerant teachers hired for their practical farm and home experiences. Today extension educators are highly trained, often specialized, professionals. Generally extension educator positions require a master's degree or a bachelor's degree with significant related experience. At least one degree in a discipline related to the specialty area is usually required. Specialty areas may include entomology but could be any related field such as botany, plant pathology, agronomy, horticulture, general agriculture, soil science, or animal science. Forty-one percent of educators have one-half or more of their job assignments in agriculture. County agents with agricultural backgrounds are expanding their roles to serve urban/suburban clientele as programs such as Master Gardeners become more successful.

County-based extension entomologists, whether ANR or 4-H and youth based, need to be highly skilled, technically based professionals with excellent people, writing, and presentation skills; multitasking abilities; and willingness to work flexible hours.

Extension Specialists

Although extension agents are located in the counties and are expected to have broad expertise, extension specialists typically are housed on university campuses and specialize in discipline areas. The position of Cooperative Extension Specialist is one of statewide leadership toward university colleagues, agricultural industries, consumers, youth, policymakers, and governmental and other agencies. The specialist keeps campus and county colleagues and clientele apprised of emerging issues and research findings and directions, works with them to develop applications of research knowledge to specific problems, and provides educational leadership and technical information support for county staff/clientele.

A Cooperative Extension Specialist is a primary liaison with university research units, providing leadership, facilitating teamwork, developing collaborative relationships with colleagues, and ensuring appropriate external input into research and educational program planning by the Agricultural Experiment Station (AES) and Cooperative Extension. Ideally, the AES–extension relationship is a seamless continuum, with extension identifying timely research opportunities to AES colleagues and conveying research results to clientele. The specialist also defines and considers needs of relevant clientele groups in planning, development, and execution of applied research and education programs.

EXTENSION TEACHING Specialists provide leadership for nonformal education of end users, intermediate users, and the public. In addition to directing planning and coordination of statewide extension education and information transfer programs related to areas of responsibility, specialists facilitate coordination of work group activities with appropriate internal and external organizations. Specialists serve as scientific and technical resources on work groups, providing disciplinary input and perspective.

Specialists' education efforts are directed toward four main clientele groups—county agents, producer/professional groups, public/private agencies, and the general public. They educate and serve as teaching resources in areas of responsibility for extension county/area personnel via individual consultations, conferences, and workshops. In addition to formal teaching at training sessions, specialists provide one-on-one consultation in person, electronically, and by telephone.

Specialists prepare and evaluate educational materials, such as publications, newsletters, slide sets, videotapes, computer software, and other learning aids, to extend subject matter information to county staff and the public sector. Because county agents are the main public interface, specialists focus on "training the trainers," developing county skills to serve clientele. In addition, specialists assist agents in customizing materials for their clientele and disseminate industry-appropriate articles through relevant channels.

Although term-length, resident classroom instruction is not the norm for full-time extension specialists, they may participate in teaching programs (via lectures and seminars) of relevant campus-based courses. Doing so permits specialists to serve as models for students developing careers in extension while fostering interactions with undergraduate and graduate students, providing these groups a vision of the third function of a university. In addition, specialists train graduate students, serve on advisory committees, and participate in other graduate education activities.

APPLIED RESEARCH AND OTHER CREATIVE WORK
Like their AES counterparts, specialists are expected to plan, conduct, and publish results of applied research/creative activity directed toward resolution of important issues or problems, independently or, more commonly, in collaboration with other research and extension personnel (including county agents). In addition, specialists provide leadership for planning and coordination of applied research activities related to areas of responsibility with departmental and other researchers, encouraging interdisciplinary collaboration and work-group participation.

Research and creative activity include synthesis and interpretation of extant knowledge, an integral aspect of the Smith–Lever mission. Extension fulfills its role by assisting in formulating policy and establishing regulatory standards and mechanisms, providing science-based information upon which policy decisions are made, and serving as the university's liaison with nongovernmental organizations and historically underserved groups.

PROFESSIONAL ACTIVITY
Specialists participate in appropriate professional societies and educational organizations and serve on state, regional, national, and international committees; review panels; and editorial boards. Enhanced professional stature accrues to the reputations of specialists' home institutions in addition to reflecting positively on CSREES.

UNIVERSITY AND PUBLIC SERVICE
As good university citizens, specialists participate in activities of committees within the department, college, campus, and other university entities. Serving as liaisons, specialists respond to regulatory and state and federal agencies, external groups, industry organizations, and the media on issues related to areas of expertise, as well as representing the university to producer groups and other organizations.

The value of Cooperative Extension is its ability to design, develop, and deliver educational programs that meet the unique needs of people as they adjust to change. The Smith–Lever Act specifies that the main function of Cooperative Extension is synthesis of existing knowledge, ancillary to creation of new knowledge. The complementarity of AES and Cooperative Extension is demonstrated not only in that extension takes AES's discoveries to the people but also in extension's conveying the needs of the citizenry to AES researchers, ensuring that these issues are addressed.

See Also the Following Articles
Agricultural Entomology • Regulatory Entomology

Further Reading
Bartholomew, H. M., and Smith, K. L. (1990). Stresses of multicounty agent positions. *J. Extension* **28(4)**, 1–6.
Bushaw, D. W. (1996). The scholarship of extension. *J. Extension* **34(4)**, 5–8.
Cooper, A. W., and Graham, D. L. (2001). Competencies needed to be successful county agents and county supervisors. *J. Extension* **39(1)**, 1–11.
Gray, M. E., and Steffey, K. L. (1998). Status of extension entomology programs: A national assessment. *Am. Entomol.* **44**, 9–13.
Jones, M. P. (1944). Extension entomology activities in wartime. *J. Econ. Entomol.* **37(3)**, 354–356.
Jones, M. P. (1950). Extension entomology. *J. Econ. Entomol.* **43(5)**, 736–739.
Lincoln, C., and Blair, B. D. (1977). Extension entomology: A critique. *Annu. Rev. Entomol.* **22**, 139–155.
Palm, C. E. (1954). The growing responsibility of entomology to human welfare. *J. Econ. Entomol.* **47(1)**, 1–6.
Patton, M. Q. (1986). To educate a people. *J. Extension* **24**, 21–22.
Steffey, K. L., and Gray, M. E. (1992). Extension–research synergism: Enhancing the continuum from discovery to delivery. *Am. Entomol.* **38**, 204–205.

Eyes and Vision

Michael F. Land
University of Sussex, Brighton

Insect eyes are of two basic types: compound (or multifaceted) and simple (or single chambered). In adults, the principal organs of sight are nearly always compound eyes, although simple eyes—often quite good ones—are frequently present in immatures. Despite the major differences in their form and construction, compound and simple eyes perform essentially the same job of splitting up the incoming light according to its direction of origin (Fig. 1). Compound eyes are of two distinct and optically different kinds: apposition eyes, in which each receptor cluster has its own lens, and superposition eyes, in which the image at any point on the retina is the product of many lenses.

APPOSITION EYES

History of Insect Optics

The facets of compound eyes of insects are too small to be resolved with the naked eye, and it required the invention of the microscope in the 17th century before they could be properly depicted. The process of working out how compound eyes functioned took more than 2 centuries from Robert Hooke's first drawing of "The Grey Drone Fly" (probably a male horse fly) in his *Micrographia* of 1665 to

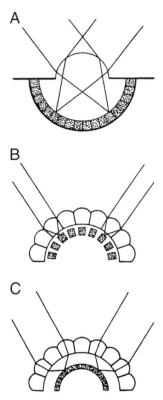

FIGURE 1 The three types of eye found in insects. (A) Simple, or single-chambered, (B) apposition compound, (C) superposition compound. The receptors are shown stippled. (Reproduced, with permission, from Land and Nilsson, 2002.)

the essentially modern account by Sigmund Exner in 1891. The first person to look through the optical array of an insect eye was Antoni van Leeuwenhoek, and his observations caused a controversy that was not fully resolved until the 1960s. The following quotation comes from a letter from Leeuwenhoek to the Royal Society of London, which was published in 1695.

Last summer I looked at an insect's cornea through my microscope. The cornea was mounted at some larger distance from the objective as it was usually done when observing small objects. Then I moved the burning flame of a candle up and down at such a distance from the cornea that the candle shed its light through it. What I observed by looking into the microscope were the inverted images of the burning flame: not one image, but some hundred images. As small as they were, I could see them all moving.

Evidently, each facet of the eye (at least in apposition eyes) does produce an inverted image, even though the geometry of the eye as a whole dictates that the overall image is erect (Fig. 1). What, then, does the insect see? Do the receptors (typically eight) beneath each lens resolve the inverted images, or do they just indicate the average intensity across the field of view of the ommatidium? (An ommatidium is the "unit" of a compound eye, consisting of the lens, receptors, and associated structures. See Fig. 2A).

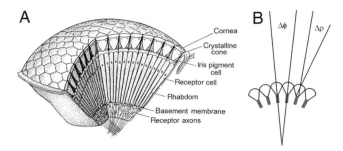

FIGURE 2 (A) Basic structure of an apposition eye, showing its construction from ommatidial elements. (B) Definitions of the interommatidial angle, $\Delta\varphi$, and rhabdom acceptance angle, $\Delta\rho$. (Reproduced, with permission, from Land and Nilsson, 2002.)

Remarkably, the answer depends on the animal. By the 1870s histological studies had shown that in most apposition eyes the eight receptor cells in each ommatidium contribute to a single radial structure, known as a rhabdom (Greek for rod; Figs. 2 and 3). Much later, in the 1950s, this material was found to be made up of photoreceptive membrane covering large numbers of long narrow microvilli, but even by the time that Exner wrote his monograph in 1891 it was

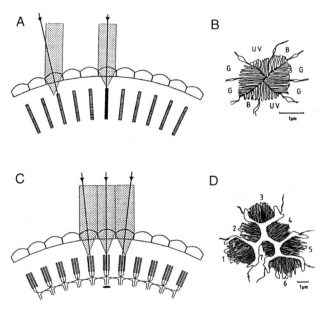

FIGURE 3 Optical comparison of an apposition eye (A,B) and a neural superposition eye (C,D). In an apposition eye each rhabdom (hatched) views light from a slightly different direction (arrows), and the rhabdoms (B), although made up from eight receptors, have a fused structure that acts as a single light guide. UV, B, and G indicate the receptor elements that respond to ultraviolet, blue, and green in an ommatidium from the eye of a worker bee. In neural superposition eyes, light from a single direction is imaged onto different rhabdomeres in adjacent ommatidia (C). The axons from all receptors imaging the same point collect together in the first synaptic layer (the lamina, Fig. 5) so that here the image has the same structure as in an ordinary apposition eye. The section (D) shows the arrangement of the separated rhabdomeres in an ommatidium from a fly. The six outer rhabdomeres (1–6) all send axons to different adjacent laminar "cartridges" (as in C). The central pair (7 overlying 8) bypass the lamina and go straight to the next ganglion, the medulla. (Reproduced, with permission, from Land and Nilsson, 2002.)

clear that the rhabdom was the structure sensitive to light. Optically, each ommatidium works as follows. The inverted image that Leeuwenhoek saw is focused onto the distal tip of the rhabdom. Having a slightly higher refractive index than its surroundings, the rhabdom behaves as a light guide, so that the light that enters its distal tip travels down the structure, trapped by total internal reflection. Any spatial information in the image that enters the rhabdom tip is lost, scrambled by the multiple reflections within the light guide, so that the rhabdom itself acts as a photocell that averages all the light that enters it. Its field of view is defined, in geometric terms, by the angle that the tip subtends at the nodal point of the corneal lens ($\Delta\rho$; Fig. 2B), and in a typical apposition eye this acceptance angle is approximately the same as the angle between the ommatidial axes (the interommatidial angle, $\Delta\varphi$ Fig. 2B). Thus the field of view of one rhabdom abuts (or "apposes," hence the name) the field of its neighbor, producing an overall erect image made up of a mosaic of adjacent fields of view.

Although the eight receptors that contribute to the rhabdom share the same visual field, it does not mean that they supply the same information. The labels UV, B, and G on the cross section of a bee rhabdom in Fig. 3B indicate the regions of the spectrum that the cells respond to best. Most

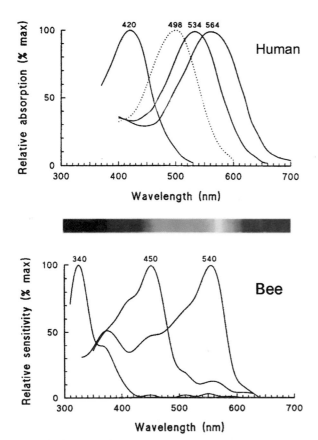

FIGURE 4 The spectral sensitivity curves for the three human cone mechanisms (and rods, dotted) and the corresponding three curves for a bee. The spectrum shows the colors as they appear to human eyes. (Reproduced, with permission, from Land and Nilsson, 2002.)

insects have trichromatic color vision, just as humans do, although their visible spectrum is shifted toward shorter wavelengths compared with ours (Fig. 4). Some butterflies and dragonflies have four-color vision.

The second feature of the bee rhabdom (Fig. 3B) is that the microvilli making up the structure are arranged in orthogonal sets. It has been known since the work of Karl von Frisch in the 1940s that bees can navigate using the pattern of polarized light in the sky. This capacity arises from the way the photoreceptor molecules are arranged on the microvilli. A geometric consequence of the cylindrical shape of the microvilli is that there will be twice as many light-sensitive chromophore groups of the rhodopsin molecules aligned parallel to the long axis of each microvillus than at right angles to it. This, in turn, means that the receptors respond best to light polarized parallel to this axis. In fact bees use a special dorsal region of the eye (the POL area) to analyze sky polarization; in the rest of the eye the receptors are twisted to abolish polarization sensitivity, so that it does not interfere with color vision. Polarization vision is also used by some insects, such as the water bug *Notonecta,* to detect water surfaces, which polarize light strongly.

The description of apposition optics given above holds for most diurnal insects (e.g., bees, grasshoppers, and dragonflies), but it does not apply to the true (two-winged) flies, the Diptera. Since 1879, when Grenacher observed that the receptors in fly ommatidia have separate photoreceptive structures (rhabdomeres) that do not contribute to a common rhabdom, there had been suspicions that flies might actually be resolving the Leeuwenhoek images. In the focal plane of the lens of a fly ommatidium, the distal tips of the rhabdomeres are separated from each other and form a characteristic pattern (Fig. 3D) that resolves the image into seven parts (there are eight receptors, but the central pair lie one above the other). This raises the obvious question: how are these seven-pixel inverted images welded together to form the overall erect image, if indeed that is what occurs? Kuno Kirschfeld finally solved this conundrum in 1967. It turns out that the angle between the fields of view of adjacent rhabdomeres within an ommatidium (about 1.5° in a blow fly) is identical to the angle between neighboring ommatidial axes. Furthermore, the fields of each of the six peripheral rhabdomeres in one fly ommatidium are aligned, in the space around the fly, with the field of the central rhabdomere of one of the neighboring ommatidia (Fig 3C). Thus, each point in space is viewed by seven rhabdomeres in seven adjacent ommatidia. What does this complicated and seemingly redundant arrangement achieve? To answer this it is necessary to know what happens to the signals from the seven receptors that view the same point, and that turns out to be the most astonishing part of the story. Beneath each ommatidium, the emerging receptor axon bundle undergoes a 180° twist before the individual neurons disperse to nearby regions of the first optic ganglion (the lamina) that correspond to the adjacent ommatidia. The net result of this impressive

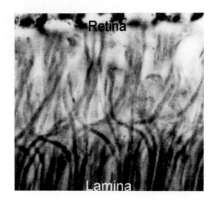

FIGURE 5 The interchange of axons that occurs between retina and lamina of a blow fly *(Calliphora)*, which makes possible the neural superposition mechanism of Fig. 3C.

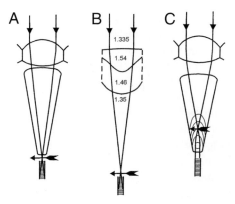

FIGURE 6 Four mechanisms of image formation in apposition eyes. (A) Corneal lens (bee, fly). (B) Multisurface lens (water bugs). (C) Lens/lens-cylinder afocal combination (butterflies). Details in text. (Reproduced, with permission, from Land and Nilsson, 2002.)

feat of neural knitting (Fig. 5) is that all the axons that "look at" the same point in space finish up making connections with the same cells in the lamina. Thus, as far as the lamina is concerned, the image is exactly the same as it would be in a conventional apposition eye, except that the signal, in terms of photon captures, is seven times stronger. One advantage of the extra signal is that it provides flies with a short period at dawn and dusk when they can see well, but when the eyesight of their predators and competitors is less sensitive and so less effective at detecting small objects.

Kirschfeld called this arrangement "neural superposition," because, as in optical superposition (see later), the contributions of a number of ommatidia are superimposed in the final image. One might ask: could the signal not have been made stronger simply by increasing the diameter of the rhabdom in a conventional apposition eye? Indeed it could, but that would mean increasing the rhabdom acceptance angle ($\Delta\rho$; Fig. 2B) at the same time, which in turn would mean a loss of resolution for the eye as a whole. The beauty of the fly solution, and undoubtedly the reason why it evolved, is that it involves no increase in acceptance angle, provided the rhabdomeres are properly aligned. There are strong hints that something like neural superposition occurs in other insect groups (some beetles, earwigs, water bugs, and crane flies) but it is only in the advanced flies that the perfect nearest-neighbors arrangement is known to be achieved.

Imaging Mechanisms

The structures that form the images in the ommatidia of apposition eyes are quite varied (Fig. 6). In terrestrial insects, as in terrestrial vertebrates, the simplest way to produce an image is to make the cornea curved (Fig. 6A). Ordinary spherical-surface optics then apply, and an image is formed about four radii of curvature behind the front face. In aquatic insects such as the water bug *Notonecta,* the external surface of the cornea has little power because of the reduction in refractive index difference (Fig. 6B). It is augmented by two other surfaces, the rear of the lens and an unusually curved

interface in the center of the lens whose function may be to correct one of the defects of spherical surfaces—spherical aberration.

The eyes of butterflies, which resemble ordinary apposition eyes in nearly all respects, have an optical system that is subtly different from the arrangement in Fig. 6A. Instead of forming an image at the rhabdom tip, as in the eye of a bee or locust, the image lies within the crystalline cone. The proximal part of the cone contains a very powerful lens cylinder that makes the focused light parallel again, so that it reaches the rhabdom as a beam that just fits the rhabdom (Figs. 6C, and 17). This arrangement, known as afocal apposition because there is no external focus, has much in common with the superposition optical system of moths, to which butterflies are closely related, and will be considered later.

Resolution

For any eye, the resolution of the image seen by the brain is determined by the fineness with which the ommatidial mosaic samples the environment, represented by the interommatidial angle, $\Delta\varphi$ (Fig. 2B), and by the quality of the image received by each rhabdom, represented by the rhabdom acceptance angle $\Delta\rho$ (Fig. 2B). (Although the eight receptors that contribute to each rhabdom usually have different spectral and polarization responses, they all share a common field of view.) In asymmetric eyes (which most are) $\Delta\varphi$ may be different along different axes of the facet array, but for present purposes $\Delta\varphi$ is taken to be the average of the angle measured along each of the three axes of the array. In the central region of a bee eye, $\Delta\varphi$ is about 1.7°. An extensive table of values can be found in a recent review by Land in 1997.

One would expect that apposition eyes would show a rough match between the interommatidial angle and the acceptance angle ($\Delta\rho$) of a single rhabdom, the argument being that no individual rhabdom can resolve detail finer than $\Delta\rho$, so there is no point spacing the directions of view of ommatidia closer than this angle. The acceptance angle $\Delta\rho$

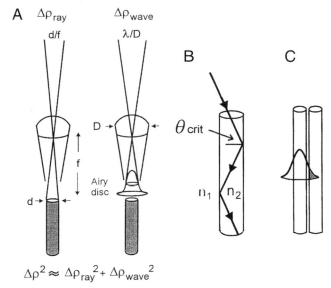

$$\Delta\rho^2 \approx \Delta\rho_{ray}^2 + \Delta\rho_{wave}^2$$

FIGURE 7 (A) The acceptance angle ($\Delta\rho$) of an ommatidium results from a combination of the Airy diffraction pattern (point-spread function), given by λ/D (right), and the geometrical angular width of the rhabdom (*d/f*) at the nodal point of the lens (left). (B) Light is trapped in a rhabdom by total internal reflection, which occurs when the angle the light makes with a normal to the wall is greater than the critical angle, given by $\sin \theta_{crit} = n_1/n_2$, the ratio of the refractive indices outside and inside the rhabdom. A typical rhabdom can trap a cone of light about 22° wide. (C) In narrow light-guiding structures some of the light is actually outside the fiber, and can potentially be caught by adjacent fibers and so spoil resolution.

is actually a combination of the contributions of ray and wave optics (Fig. 7A). Geometrically, $\Delta\rho_{ray}$ is the angle subtended by the rhabdom tip at the nodal point of the facet lens, i.e., the rhabdom diameter divided by the focal length (*d/f* radians). Typical values (for a bee) are 2 μm for *d* and 60 μm for *f*, which makes $\Delta\rho_{ray}$ 0.033 radians, or 1.9°. In wave optics, the limit to image quality is set by diffraction, specifically by the angle subtended by the Airy disk (the diffraction image of a point source), and this is given by λ/D radians. If the wavelength (λ) is 0.5 μm and the facet diameter (*D*) is 25 μm, then $\Delta\rho_{wave}$ is 0.02 radians, or 1.1°. To obtain the final value for $\Delta\rho$, $\Delta\rho_{ray}$ and $\Delta\rho_{wave}$ have to be combined, and unfortunately the proper way of doing this (convolution, taking the wave-guide properties of the rhabdom into account) is very complicated. A simple approximation is given by $\Delta\rho^2 = \Delta\rho_{ray}^2 + \Delta\rho_{wave}^2$. This is adequate for most purposes but tends to overestimate $\Delta\rho$ slightly. Using this approximation, $\Delta\rho$ for the bee data is 2.2°, somewhat larger than $\Delta\varphi$. Typically in light-adapted diurnal insects the ratio of $\Delta\rho$ to $\Delta\varphi$ is about 1:1.

The neural superposition eyes of dipterans have an additional constraint, namely that the separation of the tips of the rhabdomeres must match the interommatidial angle. In a house fly, $\Delta\varphi$ is about 2°, and with an ommatidial focal length of 70 μm, this means that the tip separation must be 2.4 μm, which does not leave a great deal of room (Fig. 3D). Because narrow light guides, such as rhabdomeres, tend to be

"leaky," with a substantial fraction of the light energy outside the guide itself (Fig. 7C), there needs to be an adequate gap between one rhabdomere and the next to prevent cross talk. In flies there is a 1-μm gap between adjacent rhabdomeres (Fig. 3D), which means that the rhabdomeres themselves must be very narrow. They have a distal tip diameter that is also about 1 μm, making them among the narrowest photoreceptors in any animal. In most other respects, however, neural superposition eyes are optically similar to other apposition eyes.

Diffraction and Eye Size

In a short and remarkable article titled "Insect sight and the defining power of compound eyes," published in 1894, Henry Mallock, an optical instrument maker, described insect vision in these terms: "The best of the eyes...would give a picture about as good as if executed in rather coarse wool-work and viewed at a distance of a foot."

Why is insect vision so poor? The problem, as Mallock recognized for the first time, is diffraction. Compound eyes have very small lenses compared with the lenses of single-chambered eyes, and because the size of the diffraction blur circle (the Airy disk) is inversely proportional to aperture diameter, the blur circles are large and the resolution correspondingly poor (Fig. 7A). A 25-μm diameter facet of a bee produces an Airy disc that is just over 1° wide in angular terms. One degree is about the size of a thumbnail at arm's length, so one can imagine a bee's world made up of pixels of about that size. In terms of the acuity of our own eyes ($\Delta\varphi$ about 0.01°), this is not very good at all.

Mallock's article goes on to discuss what a compound eye with human resolution would look like, and he came to the astonishing conclusion that it would need to be more than 20 m in diameter, or bigger than a house. The reason for this is clear: the human eye achieves high resolution by having a daylight pupil diameter of 2 mm, 80 times the diameter of a bee lens. For a bee to have the same resolution, diffraction requires that all its lenses would need to have this diameter, and to exploit all the detail in the scene they would need to be spaced at 0.5 arcmin angular intervals, the same as the receptors in our fovea. In a spherical eye, the interommatidial angle ($\Delta\varphi$) is the angle subtended by one lens diameter at the center of the eye (*D/r* radians, where *r* is the eye radius), which gives $r = D/\Delta\varphi$. With $\Delta\varphi$ = 0.5 arcmin of arc (0.000145 radians; 1 radian = 57.3° and 1° = 60 arcmin), and *D* = 2 mm, the radius of curvature will be 13.8 m and the diameter twice this. (Kirschfeld has pointed out that this calculation is a little unfair because resolution in the human eye falls off dramatically away from the fovea, to a tenth of its maximum value at 20° from the fovea, and even less farther out. Taking this into account the "human" compound eye can be shrunk in size considerably, to an irreducible 1 m diameter, which still looks very clumsy). Dragonflies seem to approach the limit of what it is possible with an apposition eye. Their eyes are 8 mm or more in diameter, have up to 30,000 facets each, and resolve about 0.25° in their most acute region. This is still poor compared with what is achievable by any camera-type eye of the same diameter.

The outcome of this discussion is that it is very hard for an apposition eye to improve its resolution; it simply gets too big. Space is thus at a premium; a little extra resolution here must be bought by a bit less there, and for this reason the different visual priorities of arthropods with different lifestyles show up in the distribution of interommatidial angles, and often facet sizes, across the eye.

Sensitivity

The sensitivity of an eye is the ratio of the amount of light received by a single photoreceptor to the amount emitted by the surface that eye is imaging. It can be used to work out the numbers of photons that individual receptors receive, and this determines the way in which the eye will perform under dim light conditions. Sensitivity can be calculated from the formula $S = 0.62\ D^2\ \Delta\rho^2$, where D is the lens diameter and $\Delta\rho$ the rhabdom acceptance angle (Figs. 2B and 7A) (we ignore the effect of receptor length here). Although D is roughly 100 times greater in a human eye than in a bee ommatidium, $\Delta\rho$ is about 100 times smaller (approximately 0.015° compared with 1.5°), so that the value of S is very similar in the bee and the human. Thus, the range of illumination conditions over which an insect with an apposition eye can operate is similar to that of a mammal using its cone system. Mammals can also see at much lower intensities, by pooling the responses of rods over quite large retinal areas (effectively increasing $\Delta\rho$). It is unlikely that pooling occurs to any great extent in insect eyes.

When discussing sensitivity, "adaptation" can have two meanings. Different eyes may be adapted in the evolutionary sense to work permanently under conditions of high or low illumination, e.g., night or day, deep sea or surface. Alternatively, the same eye can be said to be light- or dark-adapted via reversible and temporary changes in its optical anatomy. In both cases, the above equation is the key to interpreting changes and differences.

Light and Dark Adaptation

Temporary light and dark adaptation mechanisms take a number of forms in apposition eyes. Some are illustrated in Fig. 8 and include the following: (A) an iris mechanism just above the distal tip of the rhabdom that restricts the effective value of $\Delta\rho$. In the case of crane flies (Tipulidae), which have an arrangement of six outer and two central rhabdomeres, the iris cuts off the outer six in the light, leaving only the central pair. (B) A "longitudinal pupil" consisting of large numbers of very small pigment granules that move into the region immediately around the rhabdom in the light and withdraw in the dark is a second form. The main effect of

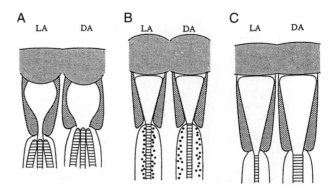

FIGURE 8 Three mechanisms of dark adaptation in apposition eyes of insects (see text). (Reproduced, with permission, from Land and Nilsson, 2002.)

this is to absorb the wave-guided light that travels just outside the rhabdom. This is replaced with light within the rhabdom, and this is absorbed in turn, so that light is progressively "bled" out of the rhabdom. This mechanism is particularly important in higher Diptera (house flies, etc.) and in butterflies, and it can work in a matter of seconds. (C) The rhabdom dimensions may themselves change, usually over a period of hours. This mechanism may involve the resynthesis of photoreceptive membrane in the dark and its sequestration in the light. In addition to these changes there are electrical and enzymatic changes in the receptors themselves that alter the gain of transduction and increase response time in the dark.

Ecological Variations in Apposition Design

As we have seen, the optical design of apposition eyes means that there is no spare room on the head surface, and what there is needs to be used as efficiently as possible. A survey of the apposition eyes of insects and crustaceans leads to the conclusion that there are three main patterns of acuity distribution that one can identify fairly easily. These are identified in Fig. 9, which illustrates the ecological reasons for these patterns (Figs. 9A–9C) and examples of the distributions themselves (Figs. 9D–9F). Figure 9D shows the pattern related to the motion across the eye encountered in forward locomotion, especially flight. Figure 9E has an "acute zone" associated with predation or sex, these zones sometimes developing into separate components of a double eye. In Fig. 9F the narrow horizontal strip of high resolution is associated with environments such as water surfaces and sand flats, where almost all important activity takes place around the horizon.

THE FORWARD FLIGHT PATTERN   When an animal is moving through the world, the objects in the world appear to move backward across the eye. Objects to the sides move faster than those in front, and there is a point in the direction of the animal's travel (the "focus of expansion") where there is no image motion. Objects farther away move more slowly

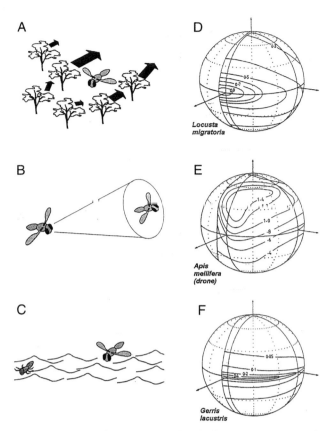

FIGURE 9 (A–C) Three situations that lead to asymmetries in the distribution of resolution in apposition compound eyes. (A) Flight through vegetation. (B) Chasing mates or prey. (C) Flight close to flat surfaces. (D–F) Plots of the density of ommatidial axes around the eyes of three insects, corresponding to the three situations in A–C. (D) Locust (forward flight pattern). (E) Drone bee (chasing females). (F) Water strider (hunting on water surface). Contours show the numbers of ommatidial axes per square degree of space around the animal. (Reproduced, with permission, from Land and Nilsson, 2002.)

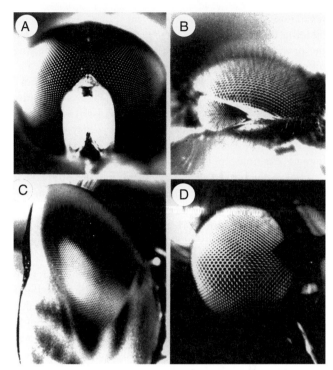

FIGURE 10 Eyes in which facet size reflects local resolution. Paradoxically, large facets produce high resolution. (A) *Syritta* (syrphid male), (B) *Dilophus* (bibionid male), (C) *Aeschna* (dragonfly), (D) *Hilara* (empid fly). See text. (Reproduced, with permission, from Land and Nilsson, 2002.)

than near objects. Clearly, near objects to the side are likely to move so fast across the retina as to cause blurring, and if this is the case it would be economical to use fewer receptors there, as high resolution is not usable. For a bee or butterfly flying half a meter from foliage, the blur streak can be estimated to be about 2.3° long. It follows that there is little point in having lateral-pointing receptors closer together than 2 or 3°, however good the resolution at the front of the eye may be. This seems to be borne out in practice. In the butterfly *Heteronympha merope*, for example, the horizontal interommatidial angle decreases from 1.4° in front to 2.6° at the side. Bees, butterflies, and acridid grasshoppers are flying insects, and their eyes all show decreasing horizontal interommatidial angles from front to rear, consistent with these ideas. Nonflying insects, e.g., many tettigonid grasshoppers, have more or less spherical eyes, without this gradient. In all the flying groups there is another, separate gradient of vertical interommatidial angles; they are smallest around the eye's equator and increase toward both dorsal and ventral poles. This results in a band around the equator with enhanced vertical acuity. The most likely reason for this vertical gradient is that the region around the eye's equator contains the highest density of information important to the animal, especially if it is an insect that feeds on flowers.

The combined effects of these two gradients on the overall density of ommatidial axes are shown for a locust in Fig. 9D, in which the contours represent the number of ommatidial axes per square degree on the sphere surrounding the animal. Worker bees and female blow flies *(Calliphora)* show a similar pattern, although in male flies and drone honey bees, this pattern is distorted to give a more pronounced acute zone concerned with mate capture (also Fig. 9E).

ACUTE ZONES CONCERNED WITH PREY CAPTURE AND MATING Many insects have a forward- or upward-pointing region of high acuity, related either to the capture of other insect prey or to the pursuit in flight of females by males (Fig. 9E). When both sexes have the specialization (mantids, dragonflies, robber flies), predation is the reason, but more commonly it is only the male that has the acute zone (simuliid black flies, hover flies, mayflies, drone bees), indicating a role in sexual pursuit. The acute zones vary considerably. In male house flies and blow flies, they may involve little more than a local increase in the acuity of the "forward flight" acute zone common to both sexes (see earlier). However, in other insects the acute zone may be in a separate eye, as is the case with the dorsal eyes of male bibionid flies (Fig.10B). In these more extreme double eyes, the upward-pointing part is often specialized for detecting other small animals against the sky.

Good examples of forward-directed acute zones are found in the praying mantids, predators in which both sexes ambush prey. The eyes have large, binocularly overlapping acute zones that are used to center potential prey before it is struck with the spiked forelegs. Mantids provide the only known example in insects in which prey distance is determined by binocular triangulation. The interommatidial angle ($\Delta\varphi$) in *Tenodera australasiae* varies from 0.6° in the acute zone center to 2.5° laterally. Facet diameters decrease from 50 μm in the acute zone to 35 μm peripherally, but this is less of a decrease than would be expected from diffraction considerations alone.

In many male dipterans an acute zone associated with sexual pursuit is typically situated 20 to 30° above the flight direction. In *Calliphora* flies it is characterized by a low value for $\Delta\varphi$ of 1.07° compared with 1.28° in the female. In house flies and probably in other flies there are also anatomical differences at the receptor level that suggest that this region (it has been called the "love spot") is specifically adapted for improved sensitivity. This is no doubt caused by the very fast response times required for high-speed chasing. Male flies also have a number of "male-specific" interneurons in the optic ganglia, which are undoubtedly involved in the organization of pursuit behavior.

In the small hover fly *Syritta pipiens* the sex difference is particularly striking. In the male's acute zone, $\Delta\varphi$ is about 0.6°, nearly three times smaller than elsewhere in the eye or anywhere in the female eye (Fig. 10A). Drone bees have a similar anterodorsal acute zone, where the density of ommatidial axes is three to four times greater than anywhere in the female eye (Fig. 9E). They use this region when they chase the queen and can be induced to chase a dummy queen on a string subtending only 0.32°, much smaller than the ommatidial acceptance angle of 1.2°. This implies that the trigger for pursuit is a brief decrease of about 6% in the intensity received by single rhabdoms.

Most of the animals just discussed have to detect their prey or mates against a background of foliage, a far from easy task. However, many insects have simplified the problem by using the sky as a background, against which any non-luminous object becomes a dark spot. Thus, one finds not only upward-pointing acute zones but also double eyes with one component directed skyward (Figs. 10B and 10C). For example, dragonflies hunt other insects on the wing and have acute zones with a variety of configurations. Many in fact have two acute zones, one forward pointing, and presumably concerned with forward flight as discussed above, and another directed dorsally and used to detect prey. The migratory, fast-flying aeschnids have the largest eyes and most impressive acute zones. Exactly 28,672 ommatidia have been counted in one eye of *Anax junius,* which has the smallest interommatidial angles of any insect (0.24° in the dorsal acute zone) and facets of corresponding size (62 μm). The dorsal acute zone takes the form of a narrow band of high resolution extending across the upper eye along a great circle, 50 to 60° up from the forward direction. The axis density (five per square degree) is twice that in the forward acute zone and five times higher than in a male blow fly. The dorsal acute zone is easily visible as a wedge of enlarged facets (Fig. 10C). Presumably the great high-acuity stripe in *Anax* is used to trawl through the air, picking out insects against the sky much as the scan line on a radar set picks up aircraft.

Simuliid flies have divided eyes and use the upper part to detect potential mates against the sky. They can do this at a distance of 0.5 m, when a female subtends an angle of only 0.2°. As in drone bees, this is a small fraction of an acceptance angle. The eyes of male bibionid flies are similarly divided (Fig. 10B), with larger facets and smaller interommatidial angles in the dorsal eye (1.6° compared with 3.7°, in *Bibio marci*). The upper eyes are used exclusively for the detection of females; movement of stripes around the lower eye evokes a strong optomotor turning response (the almost universal visual behavior used by insects to prevent involuntary rotation) but the dorsal eye is quite unresponsive to this kind of stimulus.

HORIZONTAL ACUTE ZONES   As we have seen, many flying insects have a zone of increased vertical acuity around the horizon, no doubt reflecting the visual importance of this part of the surroundings. The visual field of the locust in Fig. 9A shows this clearly. There are environments where this region is even more important.

Insects that fly over water have a similarly narrow equatorial field of interest. Empid flies hunt close to the surfaces of ponds, again looking for stranded insects, and they have a horizontal acute zone that can be recognized by a linear region of enlarged facets around the eye (Fig.10D). In *Rhamphomyia tephraea,* vertical interommatidial angles are only 0.5° in this 15°-high region, rising to 2° above and below it.

Water surfaces themselves provide a similarly constrained field of view, and water striders *(Gerris)* that hunt prey stranded in the surface film have a narrow acute band imaging this region, as shown in Fig. 9F. This has a height of only about 10°, centered on the horizon, and within this the vertical interommatidial angle in the frontal region is only 0.55°, which is close to the diffraction limit and impressive in an eye with only 920 ommatidia.

SUPERPOSITION EYES

The Nature of Superposition Imagery

From the outside, apposition and superposition eyes are almost indistinguishable. Both are convex structures with

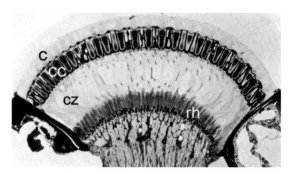

FIGURE 11 Section through the superposition eye of a dung beetle (*Onitis westermanni*). c, cornea; cc, crystalline cones; cz, clear zone; rh, rhabdoms. (Photograph by Dr. S. Caveney. Reproduced, with permission, from Land and Nilsson, 2002.)

facets of similar dimensions and are clearly variants of the same general design. But there the resemblance ends. Internally, there are several crucial anatomical differences: the retina is a single sheet, not broken up into discrete ommatidial units as in apposition eyes, and it lies deep in the eye, typically about halfway between the center of curvature and the cornea. Between the retina and the optical structures beneath the cornea there is a zone with very little in it, the clear zone, across which rays are focused—the equivalent of the vitreous space in a camera-type eye (Fig. 11). The optical devices themselves are complex—in insects they are nearly always tiny refracting telescopes—although to a cursory examination most do not look very different from the lens structures of apposition eyes.

The real surprise is optical. All superposition eyes produce a single deep-lying erect image in the vicinity of the retina. This distinguishes them not only from apposition eyes, which have multiple inverted images, but also from camera-type eyes in which the image is inverted. Clearly, we are dealing here with something quite out of the ordinary. Around the turn of the 20th century there were a number of successful attempts to photograph these images. A recent attempt by the author to re-create this photographic feat, in a firefly eye, is shown in Fig. 12 (right), in which the single erect image should be contrasted with the multiple inverted

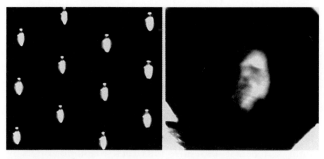

FIGURE 12 Left: Apposition-type inverted images photographed behind the cleaned cornea of a robber fly (Asilidae). Right: Photograph of an influential 19th century naturalist, taken through the superposition optics of the cleaned cornea of a firefly (*Photuris* sp.) (Reproduced, with permission, from Land and Nilsson, 2002.)

images of an eye of the apposition type (Fig. 12, left). It turns out that it is important to use a beetle (such as a firefly) for this. Other insects, in particular moths, have superposition eyes but there the optical structures that create the image are not joined to the cornea, and they are swept away when the eye is cleaned to make a lens for photography. In beetles, however, the optical elements are continuous with the cornea and so survive the removal of the eye's internal structures.

The credit for the discovery and elucidation of this remarkable piece of optics is due to Sigmund Exner, who worked on the problem throughout the 1880s and published his complete findings in 1891. Exner showed that the only way an erect image could be formed was for the optical elements to behave in a rather strange way, as shown in Fig. 13A. Basically what each has to do is not form an image from a parallel beam as in a conventional lens, but redirect light back across the element's axis, to form another parallel beam on the same side of the axis (Fig. 13B). Exner realized that although a single lens would not do the job, a two-lens telescope would, and he went on to demonstrate (as well as he could

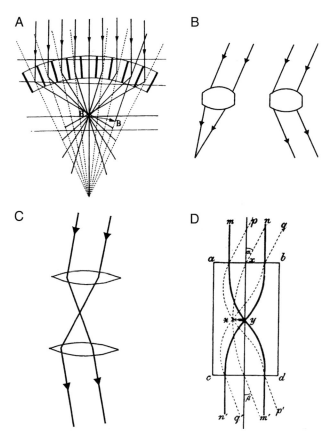

FIGURE 13 (A) Sigmund Exner's diagram of ray paths in a superposition eye. Note that the rays are bent in a "dog-leg" path by the optical elements. (B) An ordinary lens (left) will not produce the ray bending at the right, as required in (A). (C) A two-lens telescope is needed to redirect a light beam back to the same side of the axis, as in (A). (D) Exner proposed a lens cylinder equivalent to the telescope, in which rays are bent within the structure by a parabolic gradient of refractive index, highest in the center. (Reproduced, with permission, from Land and Nilsson, 2002.)

with the technology of the time) that such structures were indeed present in the superposition eyes of insects.

Telescopes and Lens Cylinders

In a lens-based superposition eye, the optical elements need to act as simple inverting telescopes that redirect the entering beam of light back across the axis, as shown in Fig. 13C. The most straightforward way to do this is to have two lenses separated by the sum of their individual focal lengths, with an image plane between them (Fig. 13C). Exner realized that, given plausible refractive indices and the curvatures of the structures revealed by histology, there was not enough ray-bending power in each element of a beetle eye to make this possible. He came up with the idea that structures must have an internal refractive index gradient similar to that in the *Limulus* horseshoe crab eye. The result would be that most of the ray bending would occur within the tissue, rather than at its external surfaces. The pure form of this structure, a flat-ended cylinder with a radial parabolic refractive index gradient, Exner called a lens cylinder. He showed that, depending on its length, it could act as a single lens or as a pair of lenses making up an inverting telescope of the kind required for superposition optics (Fig. 13D). Although Exner did not have the means in his time of establishing whether beetles and moths had optical elements with the required refractive index gradient, numerous studies since the advent of interference microscopy have shown that his brilliant conjecture was correct.

Resolution and Sensitivity

The geometry of a superposition eye is shown in Fig. 14. The peculiarities of this type of image formation mean that the nodal point of the eye (the point through which rays pass undeviated) is at the center of curvature, and the focal length is the distance out from the center to the image. The interrhabdom angle ($\Delta\varphi$) is s/f, where s is the rhabdom separation, just as in a camera-type eye. As in apposition eyes,

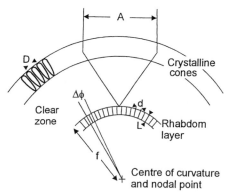

FIGURE 14 Optical definitions that apply to superposition eyes. D, facet diameter; A, superposition aperture; $\Delta\varphi$, interreceptor angle (compare Fig. 2B); d, rhabdom diameter; L, rhabdom length; f, focal length. (Reproduced, with permission, from Land and Nilsson, 2002.)

the rhabdom acceptance angle is a combination of the geometrical subtense of a rhabdom *(d/f)* and the width of the blur circle provided by the optics (Fig. 7A).

In the past, there has been a belief that superposition eyes suffer from poor resolution, mainly because of the difficulty of conceiving how the large numbers of ray bundles contributing to a single point on the image could be directed there with sufficient accuracy. However, this reputation seems not to be justified, except perhaps in extreme cases. A careful study by Peter McIntyre and Stan Caveney on the eyes of dung beetles that fly at different times of the day and night found that in the day-flying *Onitis belial* about 50 optical elements (the effective superposition aperture) contributed to the image at any one point, and in the nocturnal *O. aygulus* the number was close to 300. *O. belial* had a calculated rhabdom acceptance angle ($\Delta\rho$) of 2.2°, which is comparable with values from many apposition eyes, and in *O. aygulus* $\Delta\rho$ was somewhat larger, 3.0°, which is still quite impressive for an eye with such a huge aperture. These modeling studies have since been confirmed by electrophysiological recordings from single receptors. In the Australian day-flying moth *Phalanoides tristifica*, the image quality has been measured directly with an ophthalmoscopic method that uses the eye's own optics to view the retina and images on it. The result was that $\Delta\rho$, the acceptance angle of a rhabdom when viewing a point in space, was 1.58°, of which optical blur contributed only 1.28°. This is itself only slightly larger than the half-width of the Airy diffraction image from a single facet. Thus, a superposition eye in which 140 elements contribute to a point image has optics that are almost as good as is physically possible. (Although the superposition pupil is many times wider than an individual facet, it does not behave for diffraction purposes as a single large lens, and the Airy disk diameter depends on the diameter of single facets, just as in apposition eyes.)

Size for size, superposition eyes are more sensitive than apposition eyes, which is why they are most commonly encountered in animals such as moths and fireflies that are active at night. For an apposition eye and a superposition eye of the same size and the same resolution, the sensitivity of the superposition eye (with an aperture 10 facets wide) is 100 times that of the apposition eye, meaning that it will work just as well at light levels 100 times lower.

Eye Glow and the Superposition Pupil

Most moths have a reflecting layer (tapetum) behind the rhabdoms. Its function is the same as the tapetum in the eye of a cat: to double the light path through the photoreceptors and so to improve their photon catch. In some diurnal moths, a reflector also surrounds each rhabdom, optically isolating it from its neighbors. In dark-adapted eyes, the tapetum causes the eye to glow when viewed from the same direction as the illuminating beam (Fig. 15). In some diurnal moths, such as the sphingid *Macroglossum*, the glow is always

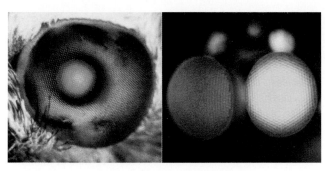

FIGURE 15 Left: Blue light reflected from the tapertum of the day-flying hummingbird hawk moth *(Macroglossum)*. The bright area corresponds to the superposition pupil. Right: Superposition dorsal eyes of a male mayfly *(Centroptilum)*. The yellow color is not from a tapetum, but results from the scattering of long wavelengths by screening pigment. (Photographs by Dr. D.-E. Nilsson. Reproduced, with permission, from Land and Nilsson, 2002.)

visible. The mechanism is similar to that in a cat's eye. The optical system forms a point image of the light source on the tapetum, or close to it, and this point acts as an emitter of light which, on passing through the optics again, emerges as a roughly parallel beam.

If the optics are good, that is to say they really do bring a parallel beam to a point in the image, then the patch of glow seen at the surface of the eye will have the same diameter as the beam that entered the eye. This is the superposition pupil (i.e., the amount of eye surface from which rays contribute to each point on the image (Fig. 15). Eye glow can also provide a useful test of image quality. If the glow can be seen only over a narrow angle (a few degrees) from the direction of the illuminating beam, then the retinal image must itself be very small. On the other hand, if the glow can be seen over a wide angle, this indicates either that there is a large blur circle on the retina or that the tapetum is situated a long way from the focus.

Light and Dark Adaptation

The high sensitivity of most superposition eyes means that they must protect their visual pigment in daylight and so need adaptation mechanisms that can reduce image

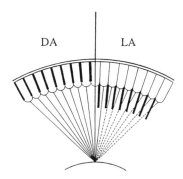

FIGURE 16 Mechanism of dark and light adaptation (DA, LA) in superposition eyes. Screening pigment migrates inward, cutting off the outer rays in the image-forming bundle. (Reproduced, with permission, from Land and Nilsson, 2002.)

brightness by several orders of magnitude. The main mechanism of light adaptation in superposition eyes consists of pigment movements that result in the progressive interception of rays from the outer zones of the superposition pupil (Fig. 16). This reduction may ultimately result in light from only a single facet reaching a single point in the image, which is essentially the apposition condition.

The eye glow (Fig. 15) provides a means of monitoring the process of light and dark adaptation. As oblique rays across the clear zone are cut off during light adaptation (Fig. 16), the brilliance of the glow and the size of the patch are reduced, often disappearing completely. In the dark, these slowly return. In insects with refracting superposition eyes, the main pigment movement is a longitudinal inward migration of granules in both the primary and the secondary pigment cells. In the dark, the granules are bunched up between the crystalline cones, and with the onset of light they extend inward, over a matter of minutes, to occupy much of the clear zone.

Interestingly, the trigger for pigment migration in some moths is not provided by photoreception in the rhabdoms themselves. In the crepuscular sphingid moth *Deilephila,* a region immediately beneath each crystalline cone initiates pigment migration, when illuminated with ultraviolet light, and the much deeper lying rhabdoms are not involved. However, in the owl fly *Ascalaphus,* a day-flying neuropteran with double superposition eyes, the pigment movements can be triggered from both the region below the cones and the rhabdoms themselves.

Single and Double Eyes

In superposition eyes, major departures from spherical symmetry are rare because the geometry of the eye is constrained by the shared optics (the hummingbird hawk moth *Macroglossum* is an exception in this respect, with a visibly asymmetric eye, but excellent resolution everywhere). One way around this problem is the use of double eyes, in which each part is essentially separate from the other and has its own radius of curvature. Although common among crustacean groups such as mysids and euphausiids, double superposition eyes are uncommon among insects. As mentioned earlier, owl flies *(Ascalaphus)* have double superposition eyes. Male mayflies have a pair of dorsal superposition eyes, which they use for sighting females against the sky, in a way similar to that of bibionid flies (Fig. 10B). However, the lower eyes, present in both sexes and responsible for other visual activities, are of the apposition type. The field of view of the dorsal eye is small, and it is adjusted to the environmental circumstances of the species; those species swarming in woods with small gaps in the canopy have the narrowest fields.

Afocal Apposition: The Eyes of Butterflies

Butterflies and moths are classified together in the Lepidoptera and are undoubtedly very closely related. Most butterflies [skippers (Hesperidae) are the exception] have eyes that behave in most respects as apposition eyes. They have long narrow rhabdoms abutting the bases of the crystalline cones, no clear zone, and complex pseudopupils. Many moths, on the other hand, have refracting superposition eyes with wide, deep-lying rhabdoms, clear zones, and eye glow. Transitions between the eye types must have occurred a number of times within the moths, as well as between moths and butterflies. A very similar picture emerges in the beetles, most of which have apposition eyes, but a substantial number of nocturnal and crepuscular groups, including the dung beetles and the fireflies, have superposition optics.

It is not very easy to see how it is possible to get from one type of eye to the other, without going through an intermediate that does not work. Apposition eyes use simple lenses and superposition eyes two-lens telescopes (or the equivalent lens cylinder devices), and there does not seem to be much room for compromise. In the case of butterflies we do know the answer: in 1984 Dan-Eric Nilsson and his colleagues discovered that their apposition eyes actually have an extreme form of superposition optics in the ommatidia, in which the proximal lens in each telescopic pair has become not weaker, as one might have guessed, but extremely powerful (see Fig. 6C).

The way this works is shown in Fig. 17A. As in a normal superposition eye, a combination of the curved cornea and a weak lens cylinder in the distal region of each crystalline cone results in the formation of an image within the crystalline cone, about 10 μm in front of its proximal tip. This focused

light then encounters a lens with an extraordinarily short focal length, about 5 μm. The discovery of this lens involved taking thin frozen sections from the tiny region at the base of the crystalline cone and examining their image-forming properties. The last 10 μm of the cone produced excellent images. The effect of this second lens is to bring the light focused by the first (distal) lens back into a parallel beam, just as in a superposition eye. The essential difference is that, whereas in a superposition eye the magnification of the telescopic pair of lenses rarely exceeds –2, here it is much greater. The large difference in the focal length of the distal and proximal lenses gives an overall magnification of –6.4 in the nymphalid butterfly *Heteronympha merope*.

This high magnification has two important consequences, illustrated in Fig. 17B. The first is that the beam that emerges from the proximal tip makes an angle with the axis that is 6.4 times greater than the beam that entered the facet from outside. A ray making an angle of 1° with the facet axis emerges at 6.4°, and similarly a beam 3° wide at the cornea emerges into the rhabdom as a 19.2°-wide beam. The significance of this is that a rhabdom with a refractive index of 1.36 will just contain (by total internal reflection; Fig. 7B) a beam 22° wide, which in turn means that the acceptance angle of the ommatidium will be limited to just over 3°: light making higher angles with the rhabdom wall will escape and be absorbed by the surrounding pigment. Thus, in this kind of eye, the ommatidial acceptance angle is limited principally by the refractive index of the rhabdom, not (as in a conventional apposition eye) by its diameter (Fig. 7A). The second effect of the magnification is to reduce the diameter of the beam leaving the base of the crystalline cone by a factor of about 9 (angular magnification × refractive index), compared with that entering the facet. The entering beam is limited by the facet diameter, typically about 20 μm. The beam leaving the crystalline cone and entering the rhabdom is squashed down to a diameter of 2.1 μm, which is indeed close to the diameter of a butterfly rhabdom. Thus rhabdom diameter and facet diameter are related and between them determine the effective aperture of the ommatidium and hence its sensitivity. Bright-light butterflies tend to have smaller facets (20 μm) and narrow rhabdoms (1.5–2 μm), whereas the crepuscular Australian butterfly, *Melanitis leda*, has 35-μm facets and 5-μm rhabdoms. A further consequence of this optical system is that the rhabdom tip is imaged onto the cornea (Fig. 17C), which means that one can sometimes see magnified versions in the cornea of the wave-guide mode phenomena that occur in the rhabdom.

What we have seen is that butterfly eyes behave as apposition eyes, because light entering a single facet is received by a single rhabdom. They are called afocal because light is not focused on the rhabdom tip as in most apposition eyes, but enters the rhabdom as a parallel beam. In their fundamental optical design, however, these ommatidia remain of the superposition type, constructed from two-lens telescopes. This makes it easy to understand how different

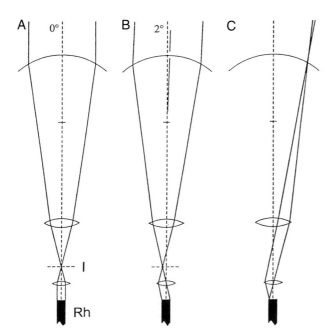

FIGURE 17 "Afocal" apposition in butterfly eyes. (A and B) Although each ommatidium acts independently, like an apposition eye, the optical elements function as telescopes with an internal image, as in superposition eyes (Fig. 13). The wide beam of light reaching the cornea is reduced to "fit" the rhabdom (see text). (C) A consequence of this arrangement is that the rhabdom tip is imaged onto the cornea. I, image plane; Rh, rhabdom. (Reproduced, with permission, from Land and Nilsson, 2002.)

lepidopteran groups managed to switch readily from the diurnal (apposition) version of the afocal eye to the nocturnal (superposition) version. To become nocturnal, the powers of the distal and proximal lenses must become more equal, the receptor layer moves to a deeper location, and gradually more and more facets contribute to the image. There are no blind intermediaries.

SIMPLE CORNEAL EYES IN INSECTS

Insect simple eyes, or ocelli, fall into two main groups: the larval eyes of holometabolous insects and the dorsal ocelli present in most winged adult insects. In both, the curved air/tissue cornea interface is the main refracting surface, although as in vertebrate eyes, a lens of some kind often augments the optical power of the system and aids in the formation of the image.

Larval Ocelli

In insects with a distinct larval stage, the ocelli are the only eyes the larvae possess. They vary greatly in size and complexity. The larvae of flies have no more than a small group of light-sensitive cells on either side of the head. Lepidopteran caterpillars, however, have ocelli with lenses and a structure resembling that of a single ommatidium from a compound

eye. In each ocellus in the *Isia,* seven receptors contribute to a two-tiered rhabdom containing the photopigment (Fig. 18A). There seems little possibility of spatial resolution within each ocellus, but as it appears that the fields of view of the 12 ocelli do not overlap, they are capable of providing a 12-"pixel" sampling mosaic of the surroundings. These ocelli do, however, resolve color; three spectral types of receptor have been found in butterfly larval ocelli.

The ant lion *Euroleon* (Neuroptera) also has six ocelli on each side of the head, borne on a small turret (Fig. 18B). Unlike caterpillars, however, each has an extended retina of 40 to 50 receptors, giving interreceptor angles ($\Delta\varphi$) of 5 to 10°. Although this resolution is not impressive, it is presumably enough to allow the animals to detect their prey, e.g., moving ants, at a distance of about 1 cm. Sawflies (Hymenoptera) have larvae with a single pair of ocelli, each with an in-focus retina covering a hemisphere (Fig. 18C). The rhabdoms in *Perga* are made up of the contributions from eight receptors (much as in an ordinary compound eye) and are spaced 20 μm apart, giving an interreceptor angle of 4 to 6°. These larvae are vegetarian, and it seems that the main function of the ocelli is to direct the larvae to their host plants. However, *Perga* larvae will also track moving objects with their head and defend themselves by spitting regurgitated sap.

The most impressive of all larval ocelli are found in tiger beetles *(Cicindela).* These have a lifestyle similar to that of ant lions, ambushing insect prey as they pass their burrows (Figs. 18D–18F). There are again six ocelli on each side of the head, but two are much larger than the others. The largest has a diameter of 0.2 mm and a retina containing 6350 receptors. The interreceptor angle is about 1.8°, comparable with or better than the resolution of the compound eyes of most adult insects. This raises the interesting question as to why the insects did not retain eyes like this into adult life.

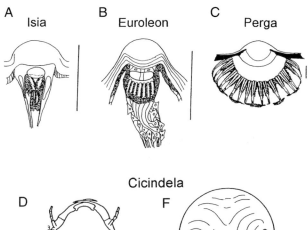
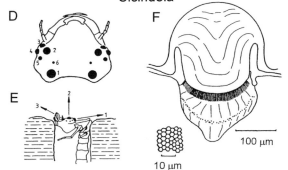

FIGURE 18 Simple eyes (ocelli) of increasing complexity in larval insects. (A) Lepidopteran, (B) Neuropteran, (C) Hymenopteran. Scale bars, 0.1 mm. (D–F) Large simple eyes of tiger beetle larvae *(Cicindela).* They are used to spot prey (usually ants), which they ambush from the burrow. (D) Head with six pairs of eyes. (E) Larva in ambush position. (F) Largest ocellus showing corneal lens and retina. Inset: Tangential section of retina. (Reproduced, with permission, from Land and Nilsson, 2002.)

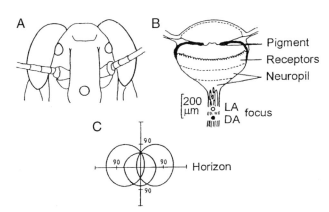

FIGURE 19 Dorsal ocelli of adult locust. (A) Positions of frontal and lateral ocelli on head. (B) Section of an ocellus, showing the different layers and the positions of the focus in light- and dark-adapted states. The focus is a long distance behind the receptor layers. (C) Fields of view of the three ocelli straddling the horizon. (Reproduced, with permission, from Land and Nilsson, 2002.)

Dorsal Ocelli of Adults

Adult insects that fly typically have three simple eyes on the top of their heads. These dorsal ocelli resemble larval ocelli in possessing a lens and (like sawfly larvae) an extended retina (Fig. 19), but they are not embryologically related to the larval eyes. Some dorsal ocelli have tapeta, and some a mobile iris. They each have a wide field of view of 150° or more and may have as many as 10,000 receptors. So far all this suggests that these are "good" eyes, like those of hunting spiders. However, they are profoundly out of focus, with the retina much too close to the lens. For example, in the blow fly *Calliphora* the receptors extend from 40 to 100 μm behind the lens, but the focus is at 120 μm.

What then are they for? Recent studies mainly support the idea that the ocelli are horizon detectors, involved in enabling an insect to make fast corrections for pitch and roll. The defocus then makes sense; high spatial frequency clutter such as leaves and branches will be removed, allowing the receptors to respond to changes in the overall distribution of light in the sky. The idea that these ocelli contribute to flight equilibrium is supported by the fact that the receptors converge massively onto a relatively few second-order neurons that project directly into the optomotor system.

See Also the Following Articles

Brain and Optic Lobes • Ocelli and Stemmata

Further Reading

Exner, S. (1989). "The Physiology of the Compound Eyes of Insects and Crustaceans." Springer-Verlag, Berlin. [Translated from the 1891 German ed. by R. C. Hardie.]

Land, M. F. (1997). Visual acuity in insects. *Annu. Rev. Entomol.* **42,** 147–177.

Land, M. F., and Nilsson, D.-E. (2002). "Animal Eyes." Oxford University Press, London.

Nilsson, D.-E. (1989). Optics and evolution of the compound eye. *In* "Facets of Vision" (D. G. Stavenga and R. C. Hardie, eds.), pp. 30–73. Springer-Verlag, Berlin.

Wehner, R. (1981). Spatial vision in arthropods. *In* "Handbook of Sensory Physiology VII/6C" (H. Autrum, ed.), pp. 287–616. Springer-Verlag, Berlin.

Fat Body

Ephraim Cohen
The Hebrew University of Jerusalem

The insect fat body is a mesodermal tissue composed of a meshwork of loose lobes suspended in the hemocoel and bathed in the insect hemolymph. The tissue is composed primarily of vacuolated rounded or polyhedral cells called adipocytes or trophocytes, which commonly harbor stored inclusions of proteins, lipids, and glycogen. In certain insect species, mycetocytes (cells containing symbiontic microorganisms) and urocytes (cells containing nitrogenous waste product in the form of uric acid) are present. The fat body is also associated with connective tissue and various blood cell types. Being a major biosynthetic and storage organ in insects, the insect fat body is equivalent to the vertebrate liver. It is the prime location of intermediary metabolism and detoxification processes, as well as storage and excretion of glycogen, lipids, and proteins. Storage of reserves is characteristic of the larval fat body cells. Such reserves are subsequently used for metamorphosis in holometabolous insects and for flight and reproduction in adults.

STRUCTURAL ORGANIZATION

Although the insect fat body is widely distributed throughout the hemocoel, two major regions can be distinguished. Near the integument and musculature is the peripheral (subcuticular) fat body, which largely functions for storage. The second layer, the perivisceral (gut) fat body, which surrounds the alimentary canal, is more metabolically active than the previous layer. The fat body tissue surrounds other insect organs such as brain and nervous tissues, gonads, and muscles. It is noteworthy that the fat body is intimately associated with nearly all vital tissues and organs in the insect body, including the tracheal system, the

musculature, the Malpighian tubules, and the hemolymph. This spatial organization is well adapted to the physiology and the open circulatory diffusion system of insects, thereby facilitating absorption and release of metabolites and nutrients.

FAT BODY CELLS

Adipocytes (trophocytes) are the predominant cell type associated with metabolic and storage functions. In young cells, a few inclusions can be detected and the nuclei are round. As the cells mature and accumulate nutritional reserves, they become vacuolated and the nuclei are compressed. The colors of adipocytes, which depend on the insect species and change with maturation, range from white, yellow, tan, and brown to blue.

Urocytes are special cells common in cockroaches, which sequester uric acid (the main end product of nitrogen metabolism in terrestrial insects) for excretion and storage. They are degenerate cells, which unlike adipocytes, lack organelles such as mitochondria, ribosomes, or the endoplasmic reticulum.

Mycetocytes are cells that harbor symbiontic microorganisms and may serve for nutritional purposes. Mycetocytes are in proximity to urocytes, a spatial organization that implies some sort of physiological–biochemical interaction.

The adipocytes are arranged in two or three layers in the periphery of the fat body lobe, and the more metabolically active cells face the circulatory system. The mycetocytes are located in the center of the lobe surrounded by urocytes.

Other cell types associated with the fat body, including various blood cells, can be found adhered to fat body cells. Oenocytes, which are large ectodermally derived cells, have also been observed to be attached to adipocytes. Their exact physiological role is unresolved.

METABOLIC FUNCTIONS

The fat body participates in myriad metabolic activities and functions. Absorption from hemolymph and buildup of intracellular storage nutrients in the form of lipid droplets, carbohydrate (glycogen) deposits, and protein granules during the immature stages are aimed at accumulating reserves for later stages, and primarily to serve adult activities. Fat body cells, having homeostatic functions related to metabolism, respond to nutritional and hormonal cues that regulate and modulate blood sugars, lipids, and proteins at larval and mature stages.

As in vertebrates, the oxidative metabolism is mediated via the tricarboxylic acid (TCA) cycle and the electron transport enzyme systems. The fat body contains enzymes mediating the gluconeogenesis process as well as enzymatic systems with a detoxification role to manage harmful endogenous metabolites and toxic xenobiotic compounds. Detoxifying enzyme systems include microsomal mixed function oxidases, in which the cytochrome P450 is predominant, various hydrolytic enzymes (esterases, phosphoesterases), and conjugating systems.

The cells synthesize the various blood proteins (lipoproteins, glycolipoproteins), which include juvenile hormone (JH) carrier proteins (protecting JH from degradation), diglyceride carrier proteins, diapause proteins, and, particularly at the adult stage, production of vitellogenins (yolk proteins) that are absorbed by the maturing oocytes. Fat body cells also synthesize JH esterase, which regulates levels of JH in the insect blood, and enzymes involved in purine metabolism. Generally, proteins released into the insect blood during larval development are sequestered by the adipocytes, forming large intracellular granules until their use during metamorphosis. Triglycerides, which are the major form of stored lipids, are mobilized when needed and released into the hemolymph in the form of diglycerides accompanied by the production of specific carrier proteins. Trehalose, produced by the fat body, constitutes the major disaccharide in the insect blood. Glycogen, which is the principal form of stored carbohydrates, is mainly present in the peripheral fat body adipocytes. Glycogen is synthesized (by glycogen synthase) and hydrolyzed (by glycogen phosphorylase), by these enzymes active in the fat body cells. The hydrolytic products are mobilized at molting and metamorphosis to serve as precursors required for chitin synthesis and formation of the new cuticle.

ENDOCRINE CONTROL OF FAT BODY METABOLISM

Neuroendocrine secretions from brain and ganglia, ecdysteroids (molting hormones), JHs, and the myriad corpora cardiaca neurosecretions affect the metabolic state of the adipocytes. These endocrine secretions are strongly influenced by stimuli from internal and external environments, and they function to coordinate and integrate crucial metabolic activities involved in molting, growth, metamorphosis, and reproduction. The fat body is a target tissue for endocrine regulation as is illustrated shortly. Stored glycogen and proteins are mobilized during the molting process to form the newly synthesized cuticular chitin–protein complex. The blood level

of trehalose is regulated by a corpora cardiaca neurohormone. The adipokinetic hormone from the corpora cardiaca stimulates the adipocytes to release diglycerides and the accompanied lipoprotein carrier, and enhances lipid oxidation to fuel flight in favor of carbohydrate oxidation. Synthesis and release of vitellogenins by the female fat body cells usually are under the control of JH, although in certain insect species also the molting hormone is involved.

FAT BODY DURING DEVELOPMENT AND METAMORPHOSIS

During the period of metamorphosis the fat body tissue undergoes extensive morphological, histological, biochemical, and organizational changes. These processes are triggered by the molting hormone on the background presence of extremely low levels of the JH. Such alterations have been thoroughly studied in dipterans and lepidopterans. Two major strategies for transforming the larval fat body into an adult tissue exist: (1) the histolytic pathway, in which the larval fat body adipocytes in dipteran species are completely histolyzed and the adult new tissue is formed from undifferentiated stem cells, and (2) the remodeling pathway, in which adipocytes in the larval stage of lepidopteran insects dissociate at metamorphosis into individual cells before being reassociated into the adult new tissue. In certain holometabolous insect species, a combination of the two processes takes place.

Dynamic exchanges of nutrients between fat body cells and the hemolymph compartments are evident throughout the life cycle of holometabolous insects (Fig. 1). Buildup of reserves and their partial use at the molting periods are characteristic of the larval stages. During the prepupal period, mass quantities of reserve material are accumulated in the fat body cells. Lysis of fat body cells in higher dipteran species at metamorphosis results in the discharge of stored reserves into the hemolymph. However, as the new adult fat body cells are

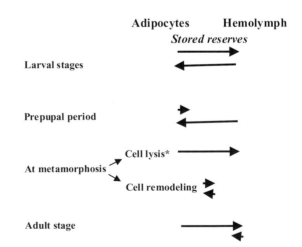

FIGURE 1 Exchange of stored reserves between fat body cells and hemolymph during the life cycle of holometabolous insects. Asterisk indicates that later, as stem cells are differentiated into adult fat body cells, a buildup of reserves occurs.

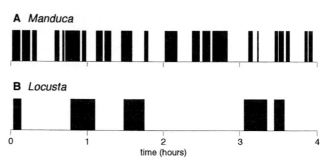

FIGURE 1 Patterns of meals taken by a caterpillar and a locust. Periods of feeding are shown by black bars. Notice that short pauses sometimes occur within meals, but the intervals between meals are markedly longer. (A) A final stage larva of *M. sexta* feeding on tobacco leaves. [Reproduced, with permission, from Reynolds, S. E., Yeomans, M. R., and Timmins, W. A. (1986). The feeding behaviour of caterpillars on tobacco and artificial diet. *Physiol. Entomol.* **11**, 39–51.] (B) A final stage nymph of *L. migratoria* feeding on wheat. [Reproduced, with permission, from Blaney, W. M., Chapman, R. F., and Wilson, A. (1973). The pattern of feeding of *Locusta migratoria. Acrida* **2**, 119–137.]

formed, nutrients are reabsorbed. In contrast, in lepidopteran species in which cell remodeling occurs, a status quo prevails in that fat body cells in adults are depleted of reserve materials, which are used for locomotion and reproduction.

See Also the Following Articles

Excretion • Juvenile Hormone • Metabolism • Vitellogenesis

Further Reading

Dean, R. L., Locke, M., and Collins, J. (1985). Structure of the fat body. *In* "Comprehensive Insect Physiology, Biochemistry, and Pharmacology," Vol. 3 (G. A. Kerkut and L. I. Gilbert, eds.), pp. 155–210. Pergamon Press, Oxford, U.K.

Haunerland, N. H., and Shirk, P. D. (1995). Regional and functional differentiation in the insect fat body. *Annu. Rev. Entomol.* **40**, 121–145.

Keeley, L. L. (1985). Physiology and biochemistry of the fat body. *In* "Comprehensive Insect Physiology, Biochemistry, and Pharmacology" Vol. 3 (G. A. Kerkut and L. I, Gilbert, eds.), pp. 211–248. Pergamon Press, Oxford, U.K.

Law, J. H., and Wells, M. A. (1989). Insects as biochemical models. *J. Biol. Chem.* **264**, 16355–16638.

Feeding Behavior

R. F. Chapman
University of Arizona, Tucson

The most extensive studies of insect feeding behavior and its regulation have focused on two insects with completely different feeding habits: the adult black blow fly *(Phormia regina)*, which is a fluid feeder, and the final nymphal stage of the migratory locust *(Locusta migratoria)*, a grass-feeding insect. Although their feeding habits are different, there are many common features in their feeding behavior patterns and the mechanisms that control their feeding behavior. It is likely, though unproven, that similar principles are involved in other insects, and this article focuses on these two insects, making reference to other species and other feeding habits where data are available. V. G. Dethier initiated and guided the earlier work on the blow fly, summarized later by Stoffolano. The studies on the locust were initiated by R. F. Chapman and developed by E. A. Bernays and S. J. Simpson.

WHY INSECTS HAVE DISCRETE MEALS

Although casual observation may give the impression that insects feed nonstop, many insects eat in discrete meals separated by periods of nonfeeding (Figs. 1A and 1B). This is most extreme in species that feed on vertebrate blood: larval *Rhodnius prolixus* take a single meal in each developmental stage, and adult female mosquitoes usually have a single blood meal associated with each vitellogenic cycle. Nectar-feeding insects and phytophagous insects also feed in discrete

meals, but the degree to which this is true in other insects has not been investigated. For predatory insects, a single prey item is commonly not sufficient to produce satiation, and it is likely that a "meal" would involve several prey, just as a "meal" for a nectar-feeding insect involves foraging from a number of flowers because no single flower contains sufficient nectar to produce satiation.

The underlying causes of this behavior are probably both physiological and ecological. Energy is expended in acquiring food and initiating digestion so that, when food is first ingested, there is a net loss of energy. Subsequently, as food is digested and absorbed, there is a gain in resources and energy, but as the process continues the rate of gain declines and the net gain plateaus as digestion and absorption are completed (Fig. 2). Consequently, there is an optimal period for which an insect should retain food in its gut before replacing it with

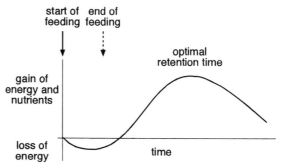

FIGURE 2 Nutrient and energy returns associated with eating a discrete meal. At first, the insect expends energy in obtaining its food. As the food is digested, nutrients are absorbed increasingly rapidly but as the nutrients are removed from the food the nutrient return decreases. To optimize the rate of nutrient return, most insects have discrete meals, with intervals between meals that approximate to peak rates of return. [Reproduced, with permission, from Sibly, R. M., and Calow, P. (1986). "Physiological Ecology of Animals." Blackwell Science, Oxford.]

newer, undigested food. Thus, it is advantageous for the insect to eat a discrete amount of food (a meal) and process it before taking more food.

The risk of predation has almost certainly also had a major role in shaping feeding behavior. Predation risks are much higher during feeding presumably because of the movements made by the insect that can be detected visually or mechanically by potential predators. For example, genista caterpillars, *Uresiphita reversalis,* feed for only about 3% of the day, yet 80% of predation by anthocorid bugs occurs during this period. Similarly, although tobacco hornworm caterpillars *(Manduca sexta)* on tobacco plants in a greenhouse fed for only about 7% of the time, 20% of predation occurred during this period.

Grasshoppers typically move away from a feeding site following a meal, sometimes backing down into the mass of vegetation and remaining unmoving and hidden until the time to feed again approaches. The caterpillars of *U. reversalis* move into silken shelters between meals. Most blood-sucking insects leave the host as soon as they are replete, usually moving to shaded places where they are inconspicuous, and it is probably true that most insects move away from the immediate area of feeding where food-related cues might reveal their presence to predators.

As a consequence of feeding in discrete, relatively short meals, the time spent feeding by most insects is only a small proportion of the available time; for most of the remainder they remain inactive and presumably minimize predation risks. Blood-sucking insects, which commonly ingest more than their own weight of food in a single meal, feed for less than 1% of the time; nectar-feeding butterflies and flies (feeding on unlimited supplies of nectar in the laboratory) feed for up to 14% of the time and this is true also for grasshoppers, both in the laboratory and in the field. All these insects have part of the gut modified for temporary food storage. Final-stage caterpillars of the tobacco hornworm spend about 35% of the time feeding in the field. In grasshoppers, the reduction in activity after feeding is controlled, at least partly, by a hormone released from the corpora cardiaca at the end of a meal. Hormonal release is induced by distension of the crop at this time.

Phloem-feeding homopterans appear to differ from most other insects. Planthoppers and aphids do not have discrete meals and ingest food more or less continuously. The phloem provides a continuous supply of sugars and free amino acids, requiring little or no digestion, so the availability of nutrients for absorption remains virtually unchanged over time. Under these circumstances the physiological necessity of eating discrete meals is eliminated. In these insects, the act of feeding is not associated with obvious body movements because once the feeding tube is plugged into a phloem sieve tube, the insect remains in one place for hours. This probably applies to xylem-feeding insects, which also need to process very large amounts of fluid because of the low concentrations of nutrients in xylem. Filter-feeding aquatic insects, such as some mosquito larvae, also probably feed continuously.

THE START OF FEEDING

As the time from the previous meal (the intermeal interval) gets longer, the likelihood that the insect will respond to food stimuli increases. A locust starts to move again and so the likelihood of encountering food is increased. Other factors, not related to the food, may also further increase the probability of feeding. In a locust, a sudden increase in light intensity or the act of defecation may have such effects. Conversely, an encounter with a highly unpalatable food source may delay the start of feeding and careless movements by an observer may have a similar effect. Simpson demonstrated that, in the migratory locust, there was in addition a tendency for meals to begin with some pattern of regularity which, in his observations, had a period varying from 12 to 16.5 min in different individuals. This does not mean that feeding or some other activity occurred every 15 min, but when it did so it was usually at some multiple of 15 min from a set time, which he determined to be lights-on in his experiments (Fig. 3). There is now evidence for similar rhythms in the caterpillars of an arctiid moth, *Grammia geneura,* and the sphingid *M. sexta.* The evidence for the latter is based on field observations, and the rhythm had a period of 3 to 4 min. The discovery of these rhythms was dependent on detailed, long-term observations on individual insects. Such sets of observa-

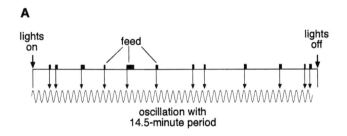

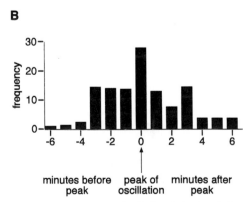

FIGURE 3 Oscillation underlying the feeding behavior of the migratory locust, *L. migratoria.* (A) Feeding record of an individual during a 12-h light phase. Notice that each meal begins close to the peak of a 14.5-min oscillation. (B) Times at which feeding started relative to the peak of the oscillation for eight insects on one 12-h day. [Reproduced, with permission, from Simpson, S. J. (1981). An oscillation underlying feeding and a number of other behaviours in fifth-instar *Locusta migratoria* nymphs. *Physiol. Entomol.* **6,** 315–324.]

tions are rare, and the extent to which similar rhythms occur in other insects is not known because observations are lacking.

The effects of these varying factors on feeding can be accounted for by an, as yet hypothetical, excitatory state in the central nervous system first postulated for the blow fly and subsequently elaborated for the migratory locust (Fig. 4). Only when the central excitatory state exceeds a certain threshold can feeding occur, but feeding is not an automatic consequence of reaching the threshold; it is a probabilistic event. At the end of a meal, the central excitatory state is assumed to be depressed below threshold. As time since the previous meal increases, so does the level of the central excitatory state so that it approaches and ultimately exceeds threshold. Rhythmic changes in the central excitatory state are presumed to account for the basic rhythmicity of feeding, and other events, such as defecation, may temporarily elevate it, while others (disturbance) may depress it.

THE SIZE OF A MEAL

The size of a meal, assuming the food supply to be unlimited, depends on the net phagostimulatory effects of the food and the nutrient requirements of the insect. Net phagostimulatory effect is the balance between nutrient components of the food that stimulate taste receptors leading to feeding and any other factors, such as the presence of toxic compounds or undue hardness, that tend to inhibit feeding. For many insects, sugars are major phagostimulants and higher concentrations in the food result in larger meals. Amino acids

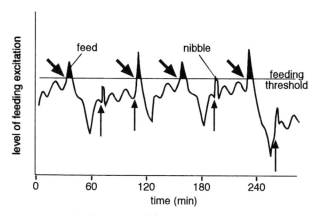

FIGURE 4 Model of the control of feeding in a locust eating wheat. Similar principles are believed to apply to other insects. The irregular line shows the level of feeding excitation (the central excitatory state). When this exceeds a threshold, the insect feeds. Notice that after a meal, the excitatory state declines sharply. Subsequently it rises slowly and the level oscillates with a period of about 15 min. Defecation (upwardly pointing arrows) causes a sudden rise in excitation. If this causes excitation to exceed the threshold, the insect may feed. Biting the food (oblique arrows) releases juices from the food and phagostimulants cause a sharp rise in the central excitatory state. (Reproduced, with permission, from Simpson, 1995.)

may also influence the phagostimulatory input. Plant secondary compounds, such as alkaloids, are often feeding deterrents even for insects, such as the blow fly, whose food does not normally contain them. Inorganic salts at higher concentrations are also deterrent although at low concentrations they may stimulate feeding. Phagostimulatory compounds do not just switch on feeding that then continues until the insect is replete; their continued input is necessary for feeding to continue and the behavior of some insects appears to reflect this. Chemosensory receptors usually adapt within a few seconds if continually stimulated, but the palpation behavior of grasshoppers and caterpillars appears to permit a continual flow of information by bringing the receptors into contact with the food for frequent very brief periods. The sensilla on the palp tips of a locust make about 10 contacts per second and each contact may be only 10 to 20 ms. As a result they remain largely unadapted.

Although continual stimulation is important to maintain feeding, meal size seems to be determined by the level of phagostimulation when the insect first bites into its food and releases the internal fluids containing a mixture of stimulating chemicals. This was demonstrated by an experiment in which the mouthpart chemoreceptors of locusts were stimulated with a highly phagostimulatory solution that they were not allowed to ingest. These insects subsequently ate larger meals than others stimulated with water alone, despite the fact that during the meal the receptors of both sets of insects were equally stimulated. Events before feeding started determined how much was eaten. Comparable experiments have shown that distasteful compounds can reduce meal size. Such experiments are interpreted as reflecting changes in the central excitatory state. A high concentration of phagostimulant is believed to elevate the central excitatory state well above threshold and feeding continues, provided some level of input is maintained, until the excitatory state declines to threshold. Thus, the higher the initial level, the longer it takes the excitatory state to reach threshold and the larger is the meal. It is supposed that a high level of deterrent compounds would inhibit feeding by depressing the level of the central excitatory state below threshold.

The elevated level of the central excitatory state is also believed to account for the "dances" of flies and the palpation behavior of locusts and grasshoppers following loss of contact with food early in a meal. When a fly loses contact with a drop of sugar it moves in an irregular path with frequent turns as if "searching" for the food. The more concentrated the solution, the more frequent the turns (Fig. 5A). If a locust loses contact with its food it palpates vigorously and such behavior lasts longer if loss of contact occurs earlier in a meal. Toward the end of a meal, loss of contact with the food results in only a limited period of palpation (Fig. 5B). The so-called searching behavior of other insects, such as that described for coccinellid larvae when feeding on aphids and temporarily losing contact with the prey, probably has a similar basis.

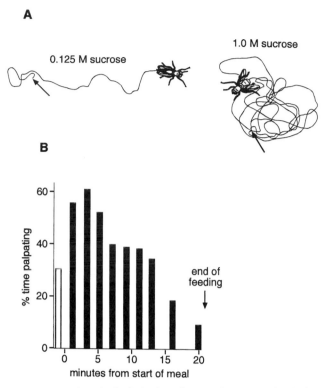

A

0.125 M sucrose

1.0 M sucrose

B

FIGURE 5 Searching for food after loss of contact during a meal. (A) Blow flies dance when they lose contact with a drop of sugar. After experiencing more concentrated sugar solutions the high rate of turning is much more sustained. [Reprinted, with permission, from Dethier, V. G. (1957). Communication by insects: Physiology of dancing. *Science* **125**, 331–336. Copyright 1957 American Association for the Advancement of Science.] (B) The migratory locust palpates when it loses contact with a blade of grass. Each bar represents the percentage of time palpating after losing contact with the food at a different stage of the meal. Soon after the start of a meal it palpates for most of the time, but toward the end of a meal it is less persistent. The open bar represents the time palpating just before starting to feed. [Reprinted, with permission, from Bernays, E. A., and Chapman, R. F. (1974). The regulation of food intake by acridids. *In* "Experimental Analysis of Insect Behaviour" (L. Barton Browne, ed.). Springer–Verlag, Berlin. Copyright Springer–Verlag.]

THE END OF A MEAL

The end of a meal is ultimately determined by the degree of distension of the part of the gut in which the food is temporarily stored. In grass-feeding grasshoppers and nectar-feeding flies, this temporary store is the crop (part of the foregut). In *R. prolixus,* feeding on vertebrate blood, the food is stored in the anterior midgut. In each case, the degree of distension is monitored by some form of stretch receptor. In locusts, these receptors are multipolar cells on the wall of the foregut, and receptors on the most anterior part of the foregut, which is the last part to fill; these receptors are responsible for inhibiting further feeding. *R. prolixus* has chordotonal organs in the body wall. The input from these stretch receptors has an inhibitory effect on feeding, presumably by leading to a decline in the level of the central excitatory state to below threshold. If food quality and the nutritional and feeding status of the insect are constant,

stretch receptor input determines that an insect ingests a similar amount of food at each meal.

CHANGES IN FEEDING BEHAVIOR

The pattern of feeding changes with the age of the insect, its previous experience, and its nutritional needs. Phytophagous insects in general tend to eat greater amounts in the middle of a developmental stage and more in the light than in the dark. The average meal size taken by the final-stage nymph of the migratory locust, for example, increases from about 50 mg on the day of molting to almost 100 mg 4 days later, whereas the average interval between meals declined from 82 to 71 min. At night, the insects take fewer meals even though the temperature may be constant.

Some phytophagous insects become less selective if they experience a long period without food and this has given rise to some confusion in the literature. For experimental purposes, it is often convenient to use insects that feed readily when presented with food. This is achieved by depriving them of food, often for 24-h periods. However, because such insects are less selective than insects with continual access to food, grasshoppers, for example, were generally considered to be unselective in their choice of foods. More critical observations, however, show that this is not accurate. With increasing periods of food deprivation, several grasshoppers have been shown to accept a wider range of food plants. It is probable that this acceptance of previously unacceptable plants reflects a need for water rather than for other specific nutrients, although this hypothesis has not been thoroughly investigated. It is, however, clear that a well-hydrated locust actively moves away from wet filter paper, whereas a dehydrated one attempts to eat it. Similarly, dehydrated flies drink water, whereas hydrated ones do not.

The tendency of grasshoppers and caterpillars, and probably other insects, that are deprived to sample food that would otherwise be rejected can play a major part in the subsequent acceptance of food. This becomes possible because taste receptors that initially signaled rejection because of some distasteful component of the food become habituated and are no longer stimulated by the distasteful compound. At the same time, detoxifying enzymes are probably mobilized within the insect, providing it with the capacity to minimize any harmful effects that the compound might have.

The nutritional requirements of insects vary through life and this is reflected by changes in their feeding behavior. During larval development, the amount of food consumed is usually maximal in the middle of each developmental stage, falling to zero for a period before each molt. Changes also occur in adults in relation to somatic development and, in females of many species, in relation to egg development. This variation is illustrated for adult red locusts (*Nomadacris septemfasciata*) in Fig. 6. When the insect first becomes an adult the cuticle is soft and the flight muscles are poorly developed. During this teneral period, both sexes feed actively

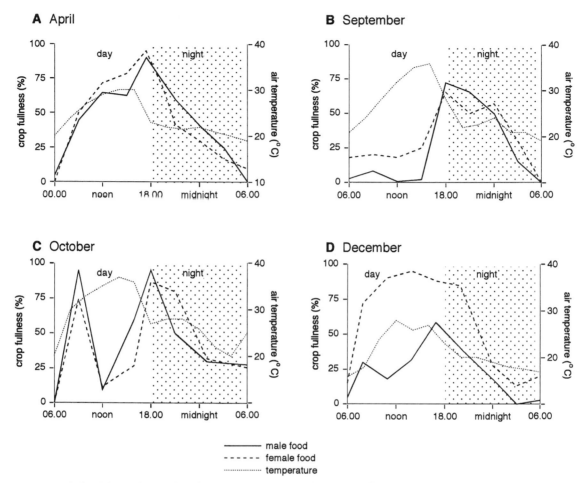

FIGURE 6 Variation in feeding behavior during adult life. The red locust, *N. septemfasciata,* in the field. (A) April. Soon after becoming adult both sexes feed for most of the day. This is a period of somatic growth when flight muscles and cuticle become fully developed. (B) September. Despite moderately high temperatures during the day, very little feeding occurs until late afternoon. The insects are in reproductive diapause. (C) October. Little feeding occurs in the middle of the day, perhaps because of the high temperature. The insects are beginning to become sexually mature. (D) December. Females eat much more than males during the period of egg development. All these samples were taken from the same generation and population of insects, which live for about 9 months as adults. Each graph shows the amount of food in the foreguts of a sample of insects taken at each time point over a 24-h period; 100% would indicate that all the locusts were full, 0% that they were all empty. When the temperature is 30°C or above, the foregut becomes more than half empty within an hour, so that crop fullness above 50% during the day indicates recent feeding. At 25°C and below, the food takes several hours to leave the foregut so that night time values largely reflect feeding before dark. [Reproduced, with permission, from Chapman, R. F. (1957). Observations on the feeding of adults of the red locust. *Br. J. Anim. Behav.* **5**, 60–75.]

during the day (Fig. 6A). Subsequently, the insects enter reproductive diapause and feeding is reduced to a single meal each day (Fig. 6B). During the reproductive period, females eat much more than males (Fig. 6D).

Among some adult flies and grasshoppers, there is good evidence that mature females change their feeding behavior to acquire protein for the synthesis of vitellogenin. This is most obvious in blood-sucking flies, such as mosquitoes and tabanids, females of which use nectar as a flight fuel, but vertebrate blood as their primary protein source. Males of these same species feed only on nectar. This is also true of blow flies. Mature female grasshoppers, given the opportunity, tend to select food with a higher protein level than do males or immature females. Thus, they tend to eat the seed heads of developing grain rather than foliage.

Under laboratory conditions, when fed on artificial diets,

locusts and caterpillars are able to select from the foods with different amounts of proteins and carbohydrates to maintain an appropriate balance of the two classes of compound. Locusts can make the adjustment from one meal to the next, with an interval of less than an hour between meals. The extent to which insects can fine tune their nutritional balance when feeding on natural food with much smaller deficiencies of protein or carbohydrate has yet to be demonstrated.

FEEDING BEHAVIOR UNDER NATURAL CONDITIONS

In the field, feeding behavior is determined to a large extent by environmental factors, although relatively few extensive studies have been carried out. Temperature has a major effect on feeding behavior, as it does on other insect activities, with

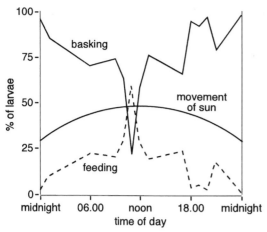

FIGURE 7 Feeding is limited by temperature. Most caterpillars of *G. groenlandica,* living within the Arctic Circle, feed during a 2-h window when the sun is at its zenith. For most of the time, the insects bask to raise their body temperatures, enabling them to feed efficiently; during feeding their temperature falls rapidly. [Reproduced, with permission, from Kukal, O., Heinrich, B., and Duman, J. G. (1988). Behavioural thermoregulation in the freeze-tolerant Arctic caterpillar, *Gynaephora groenlandica. J. Exp. Biol.* **138,** 181–193. Copyright Company of Biologists.]

little feeding occurring at low or at very high temperatures (Figs. 6A and 6C). The effects of temperature are most obvious in insects living under extreme conditions of low or high temperature. For *Gynaephora groenlandica* caterpillars living within the Arctic Circle, feeding is possible only when the insect has raised its body temperature by basking. As a result, most feeding occurs in a relatively narrow window of time around noon each day, when the sun is highest in the sky (Fig. 7).

Darkness also tends to reduce feeding. For many visually foraging insects, finding food at night is impossible, although night-flying moths obtain nectar only during darkness and some blood-sucking insects, such as mosquitoes, feed most actively at night or in the crepuscular periods. These insects locate their host primarily by odor, although night-blooming flowers often also present conspicuous targets because of their size and whiteness.

Biotic factors may also have a profound effect. For example, a caterpillar of *M. sexta* that has defended itself from the attack of a tachinid fly does not feed for some time after it has successfully repelled the attacker.

See Also the Following Articles

Blood Sucking • *Mouthparts* • *Salivary Glands*

Further Reading

Bernays, E. A. (1997). Feeding by lepidopteran larvae is dangerous. *Physiol. Entomol.* **22,** 121–123.

Bernays, E. A., and Simpson, S. J. (1995). Control of food intake. *Adv. Insect Physiol.* **16,** 59–118.

Chapman, R. F., and de Boer, G. (eds.) (1995). "Regulatory Mechanisms of Insect Feeding." Chapman & Hall, New York.

Dethier, V. G. (1976). "The Hungry Fly." Harvard University Press, Cambridge, MA.

Simpson, S. J. (1995). Regulation of a meal: Chewing insects. *In* "Regulatory Mechanisms of Insect Feeding." (R. F. Chapman and G. de Boer, eds.), pp. 137–156. Chapman & Hall, New York.

Simpson, S. J., Raubenheimer, D., and Chambers, P. G. (1995). The mechanisms of nutritional homeostasis. *In* "Regulatory Mechanisms of Insect Feeding" (R. F. Chapman and G. de Boer, eds.), pp. 251–278. Chapman & Hall, New York.

Stoffolano, J. G. (1995). Regulation of a carbohydrate meal in the adult Diptera, Lepidoptera, and Hymenoptera. *In* "Regulatory Mechanisms of Insect Feeding" (R. F. Chapman and G. de Boer, eds.), pp. 210–247. Chapman & Hall, New York.

Fire Ants

Les Greenberg

University of California, Riverside

Most residents of the southeastern United States are very familiar with fire ants. These reddish brown ants are well known for their aggressiveness and stings that produce a burning sensation. The term "fire ant" actually applies to a group of New World ant species in the genus *Solenopsis.* Many people refer to them as "red ants," although this term is also used to refer to the larger red harvester ants found in desert climates. In Spanish, the fire ant is sometimes called *hormiga colorada,* and in Portuguese it is *formiga de fogo.* In North America there are four native fire ant species, two introduced species, and two hybrid forms. The two imported species in the United States are the red imported fire ant, *Solenopsis invicta,* and the black imported fire ant, *S. richteri.* The former has spread throughout the southeastern part of the country, whereas the latter is restricted to northeastern Mississippi, northern Alabama, and northwestern Georgia. Seventeen fire ant species are currently described from South America.

ORIGIN AND SPREAD

There seems little doubt that the most important fire ant pest, *S. invicta,* traveled from South America to Mobile, Alabama, in ship ballast between 1930 and 1940. It spread in all directions from there, limited only by cold winters or desert drought conditions. *S. richteri* may have arrived earlier (perhaps in 1919), only to be largely displaced by *S. invicta.* The latter is currently established in 11 states (Florida, Georgia, Alabama, Mississippi, Louisiana, Texas, North Carolina, South Carolina, Arkansas, Tennessee, Oklahoma) and Puerto Rico. A new infestation was discovered in California in 1998, and eradication efforts are under way.

It was predicted that imported fire ants could not survive a winter when the minimum temperature was below −12°C. However, infestations now occurring in southern Oklahoma and Tennessee have led to a revised estimate of −18°C. Another limiting factor is rainfall. The deserts of west Texas have proven a barrier to the fire ant's progression westward. It is likely that annual rainfall of less than 25 cm precludes *S. invicta* from becoming established unless there is also irrigation.

IDENTIFICATION AND BEHAVIOR

All fire ants have two segments in their narrow waist and antennae with 10 segments. The workers range in size from small to large (for *S. invicta,* about 2–5 mm in length; queens are about 7 mm long). To be able to sting, fire ants must first gain leverage with their mandibles by biting; they then curve around the abdomen to insert the stinger. The fire ant injects venom consisting mainly of piperidine alkaloids that produce a burning sensation. Shortly thereafter, a red spot is usually visible. The burning sensation is short-lived, followed by itching. In most people, a white pustule will develop at the site within a few hours. These pustules are sometimes called "sterile pustules" because they are not produced by infectious bacteria. The pustules can last from days to weeks and can become infected if they are scratched. The venom also contains a small amount of protein (about 1%) that can cause anaphylactic shock in susceptible individuals. Fire ants can sting repeatedly; therefore, stinging ants should be brushed off rapidly.

Mounds and Foraging Behavior

When undisturbed, the typically dome-shaped mounds of *S. invicta* can reach heights of 30 cm or more above the ground. These mounds allow the workers to respond to local conditions by moving up and down with their brood and queens according to temperature and humidity. Exit holes are usually not apparent on the mounds themselves, but foraging trails extend outward from the mound just below the surface. During floods, fire ants move to the upper parts of the mound. If the water gets any higher, the ants grasp each other to form floating rafts that carry the brood and queens downstream. During droughts, fire ants can extend their tunnels down 6 m or more in search of moisture.

Fire ant workers can feed only on liquids: they have filters in their digestive tract that prevent the ingestion of solids. Only the fourth instar can digest solids directly, and it is the only path for processing of solid food particles in the colony. Workers deposit insect parts and other solids on the larva's "food basket." After feeding on these solid foods, the larva secretes liquids that are licked up by workers and distributed around the colony.

Mating Behavior

Like most other ants, fire ants have mating flights. In the southeastern United States, flights are most frequent in the spring following rain and subsequent sunshine. After the rain, workers fill the queens with food to prepare them for the flight. Dissection of queens at this time shows a large drop of yellow oil in their crops. In midafternoon the workers open large exit holes in the mounds to allow quick exit of the males and females. Workers become very agitated and start chasing the reproductives, which then climb vertical objects nearby from which they fly. Mating occurs in the sky. The males drop to the ground and die shortly thereafter. The queens also land, quickly shed their wings, and search for a place to dig a tunnel. The queen will close the tunnel and start to lay eggs, producing her first workers in about one month. Queens typically live 6 or more years. Because they mate only once, they must store live sperm for the rest of their lives. For this purpose, they have a transparent sac in their abdomens called the spermatheca that is filled with over a million sperm after mating. When the queen lays an egg, she can open a valve on the spermatheca, allowing the escape of sperm to fertilize her eggs. These diploid eggs give rise to females, either workers or new queens. If she does not release sperm, the egg she lays is haploid and becomes a male, as is typical for all haplodiploid social Hymenoptera.

Number of Queens

There are two forms of *S. invicta* in the United States. Originally this species was thought to be monogyne, having one queen per colony. Polygyne, or multiple-queen fire ant colonies, were first described from Mississippi in the early 1970s. In these colonies there can be dozens of fertilized queens. The queens are not aggressive toward each other and are frequently together in one part of the nest. One or more of these queens may be dominant, laying more eggs and receiving more food than the others. Polygyne queens in a colony are not closely related, suggesting that they are adopted from outside sources. They are also smaller on the average than monogyne queens and lay fewer eggs. However, the total number of eggs laid by all the polygyne queens in a colony exceeds that produced by a single monogyne queen. On the average, polygyne colonies also have smaller workers: there is a negative correlation between the number of queens and the average worker size. Furthermore, the polygyne form is not aggressive toward conspecifics, whereas the monogyne form will fight with nearby conspecifics. The polygyne form can bud off new colonies of queens and workers and is thereby able to quickly populate an area with fire ants. The polygyne form is now predominant in Texas and has also been found in Florida, Georgia, and even South America.

IMPACT ON PEOPLE AND THE ENVIRONMENT

Imported fire ants arrived in the United States without their native parasites and predators. In addition, few native ant species are able to compete with them. When fire ants encounter other ant species they vibrate their gasters (abdomens) and protruded stingers, spraying their venom and chasing away most other ants. They can displace native

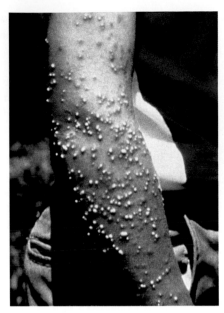

FIGURE 1 A child's arm showing many pustules after numerous fire ant stings. (Photograph courtesy of the U.S. Department of Agriculture.)

ant species and quickly become the dominant ant species and significant pests. Fire ants will attack newly hatched birds both on the ground and in trees. Fawns and calves can be stung in their eyes and blinded if they are dropped on fire ant mounds. Soft plant tissue, such as okra, can be destroyed. Ants in mounds at the bases of trees can eventually girdle and kill trees. The sick and elderly, pets and children in backyards, campers, and picnickers, all can become victims of fire ant stings (Fig. 1). One oddity about fire ants is their evident attraction to electrical fields. They frequently enter electrical boxes such as outside air conditioners, traffic boxes, and lights, where they chew wires and short out the circuits.

POSSIBLE REMEDIES

There are many insecticides that control fire ants. Baits are advantageous because it is not necessary to find the mounds; ants carry the bait back to their nests. Most fire ant baits consist of corncob grits coated with soybean oil as an attractant. Typically, a toxicant or insect growth regulator is dissolved in the oil. However, none of these solutions are permanent. Efforts are now under way to bring into the United States some of the fire ant parasites and predators from South America. Decapitating flies (Phoridae) are one promising predator now being released in Florida and elsewhere. These flies lay their eggs on fire ants. The grub that hatches invades the ant's head, where it consumes its brain. Ultimately the ant's head falls off and a new fly emerges. Although these flies parasitize only a small percentage of ants, they do interfere with the ant's foraging behavior and may make fire ants less competitive with other ants. Other potential biocontrol agents include protozoal

parasites, soil fungus, and even parasitic ants from South America. Ultimately a broad approach using chemicals and biological agents will best manage this invasive species.

See Also the Following Articles
Ants • *Introduced Insects* • *Regulatory Entomology* • *Venom*

Further Reading
Greenberg, L., Fletcher, D. J. C., and Vinson, S. B. (1985) Differences in worker size and mound distribution in monogynous and polygynous colonies of the fire ant *Solenopsis invicta* Buren. *J. Kans. Entomol. Soc.* **58**, 9–18.
Porter, S. D. (2000). Host specificity and risk assessment of releasing the decapitating fly *Pseudacteon curvatus* as a classical biocontrol agent for imported fire ants. *Biol. Control* **19**, 35–47.
Trager, J. C. (1991). A revision of the fire ants, *Solenopsis geminata* group (Hymenoptera: Formicidae: Myrmicinae). *J. N.Y. Entomol. Soc.* **99**, 141–198.
Tschinkel, W. R. (1998). The reproductive biology of fire ant societies. *Bioscience* **48**, 593–605.
Vander Meer, R. K. (1996). Potential role of pheromones in fire ant control. *In* "Pest Management in the Subtropics: Integrated Pest Management: A Florida Perspective" (D. Rosen, F. D. Bennett, and J. L. Capinera, eds.), pp. 223–232. Intercept Ltd., Andover, U.K.
Vargo, E. L., and Hulsey, C. D. (2000). Multiple glandular origins of queen pheromones in the fire ant *Solenopsis invicta*. *J. Insect Physiol.* **46**, 1151–1159.
Vinson, S. B. (1997). Invasion of the red imported fire ant (Hymenoptera: Formicidae): Spread, biology, and impact. *Am. Entomol.* **43**, 23–39.

Flea

see *Siphonaptera*

Flight

Michael Dickinson
California Institute of Technology

Robert Dudley
University of California, Berkeley

From their first appearance in the late Paleozoic, winged insects have emerged as critical components of nearly all terrestrial ecosystems. Many important behavioral features of insects, including evasion of predators, dispersal, and reproductive strategies, rely in some way on flight behavior. Wings themselves, as cuticular structures, have no intrinsic musculature and are moved instead via thoracic deformations and by muscles that insert directly at the wing base. The back and forth motion of the wing through the wingbeat, as well as the rotation about its longitudinal axis at the beginning and end of each stroke, creates unsteady aerodynamic forces of continuously changing direction and magnitude. Flight control relies on multiple sensory modalities to maintain stable trajectories and to maneuver via bilaterally asymmetric motions of the wings and body. Flight is energetically costly, and the delivery of oxygen to flight musculature is limited by diffusion in the fine branches of the tracheal system. Heat simultaneously produced by contraction of inherently inefficient muscles may be co-opted in the regulation of body temperature during flight to further enhance performance. Forces of both natural and sexual selection have contributed synergistically to the evolution of insect flight performance and maneuverability. Contemporary insect diversity largely comprises extensive radiations of miniaturized species. Flight biomechanics of these small insects is complicated by the viscous nature of airflows and depends in part on the high wingbeat frequencies enabled by a specialized muscle type termed asynchronous muscle. Asynchronous flight muscle has evolved independently more than eight times among the winged insects and enables muscles to generate increased mechanical power by trading sarcoplasmic reticulum for more contractile fibrils and mitochondria. The majority of insect species-level diversity appears to derive indirectly from this flight-related innovation that facilitates miniaturization.

EVOLUTION OF FLIGHT

Although many features have contributed to the radiation of insects in terrestrial ecosystems, the evolution of actively powered flight is almost certainly the key innovation responsible for their remarkable success. The relative abundance of extant winged (ptergygote) insects to wingless (apterygote) insects (a ratio of at least 500,000:1 in species richness) manifests the potent advantages of flight. Although the selective advantages of flight are obvious, the means by which ancestral hexapods evolved wings and associated flight behavior are not. Because flight is such a specialized form of behavior and is associated with morphological and physiological traits that represent extreme forms of the basic arthropod body plan, reconstructing the series of functional intermediates between flightless ancestors and flying insects continues to pose a challenging problem. The evolution of flight involves two distinct, but overlapping, questions. First, what is the morphological structure from which wings arose? And second, what suite of selective forces drove the evolution of wings as aerodynamic structures?

Morphological Origin of Wings

The morphological origin of the wing in pterosaurs, birds, and bats is unequivocal; in all these animals it arose from a modification of the forelimb. Insect wings are novel structures, at least in the sense that they are not homologous with the legs. Biologists have long debated which structure served as the anatomical precursor of insect wings. Of the various theories that have been proposed, the dominant view until recently was that wings arose from rigid lateral extensions of the notum. Such a scenario seems at first plausible, given the structural plasticity of the thoracic exoskeleton in extant insects. When considering function of flight morphology as a whole, however, the most complicated feature of the wing is not the flat distal blade that serves as the aerodynamic surface, but rather the complicated hinge with its associated muscle attachments that enables the wing to flap and rotate during the stroke. Over the past 15 years an alternative hypothesis, that the wings evolved from basal branches of the leg, has emerged from work in a number of disciplines. This theory owes much to the work of the paleontologist Jarmila Kukalova-Peck, who challenged the widely held view that insects, as distinct from other arthropods, possess unbranched limbs. According to her alternative view, the ancestors of winged insects possessed biramous appendages and used a developmental-genetic program for limb development that they shared with crustaceans and other arthropods. The structure that gave rise to the wing may have been a dorsal branch, or exite, of a precoxal segment of the leg called the epicoxa. Whereas the epicoxa has been lost or incorporated into the pleurum of the thorax, its exite has been retained as a wing. However, rather than classifying the wings of extant insects as direct morphological homologues of epicoxal podites in ancestral apterygotes, it may be more precise to view both as arising from homologous morphogenetic programs. The leg podite theory of wing origin solves an enigmatic step in the evolution of functional wings, the formation of the wing hinge and its complex arrangement of muscles. As a leg branch, the protowing would have been endowed with joints and muscles long before it ever took on an aerodynamic role. Further, because legs are replete with various mechanosensory structures, the protowing would have inherited the campaniform sensilla, stretch receptors, or chordotonal organs that may have mediated the reflexes and motor patterns that presumably served as the foundation of flight control circuitry. Although several lines of evidence support a leg podite origin for insect wings, this intriguing issue is far from resolved and the consensus may change with additional fossil evidence and further comparative studies of arthropod development.

Functional Origin of Flight Behavior

Hindered by the inherent difficulty of extracting behavior and physiology from fossil evidence, the functional origin of flight remains enigmatic. The fact that wings arose from small structures poses the same problem that Darwin first recognized for all organs of great complexity—it is difficult to reconstruct a series of functional intermediates between a tiny leg podite and an aerodynamic surface capable of sustaining active flight. The aerodynamic performance of a wing increases with length and surface area. Thus, a small wing is incapable of generating enough force to sustain active flight. Without a selective pressure driving the wing to larger sizes, how did the structure initially attain the size required to support active flight? It is unlikely that any single selective pressure was responsible for the hypertrophy of the wings. For example, if the direct ancestors of pterygotes possessed an aquatic nymph stage, protowings might have served as gill covers or respiratory paddles. Given the high density and buoyancy offered by an aquatic medium, it is even possible that wings may have functioned as hydrodynamic structures for underwater propulsion. However, no matter what role they may have played in the aquatic stage of life history, the use of wings in air would necessitate a substantial increase in size.

Hypotheses attempting to explain the early selective engine for true aerial flight segregate into two basic types. One set of hypotheses suggests that early selective pressures for an increase in wing size had nothing to do with aerodynamics per se, but rather with some other size-dependent selective force. For example, the use of wings as reflectors and conduits in basking butterflies has led to the proposal that wings first served a thermoregulatory role. Other possibilities include the use of wings in sexual displays or copulatory offerings by males. The second set of hypotheses asserts that protowings functioned aerodynamically before they were large enough to support active flight. For example, small wings might serve to increase glide angle or offer added stability during controlled descents. The utility of small protowings in gliding behavior might have been enhanced by their serial repetition, and fossil evidence indicates that protowings were present on the prothorax and abdominal segments in some groups of early insects. Vegetation and surface topography would have served as the most convenient launching points for gliding or parachuting insects. Another possibility is that protowings may have prolonged jumps, thereby serving as an important anti-predator behavior in response to the coevolutionary radiation of terrestrial predators at the time. Recently, James Marden suggested that protowings may have served as aerodynamic structures used to either sail or flutter insects across the surfaces of streams and ponds. This intriguing hypothesis is based on the behavior of extant stoneflies that skim across streams in this manner when the temperature is too low for their flight muscles to generate sufficient mechanical power to sustain flight. The atmospheric composition at the time, in which both oxygen levels and air density were elevated by today's standards, might also have aided the transition to active flight (see later). Whatever selective pressures led to the evolution of flight, analyses of insect phylogeny strongly suggest that flight evolved only once within the clade. However, no behavior that has been proposed as a model for ancestral pterygotes, such as sun basking or surface skimming, maps into the current phylogeny in a way that is entirely consistent with it being an ancestral trait. With no definitive means of excluding any of the proposed scenarios, the functional origins of insect flight are likely to remain alluring, controversial, and unresolved for years to come.

AERODYNAMICS

Conventional Aerodynamics

The scientific study of insect flight is haunted by the widely told story of an engineer who proved that a bumblebee could not fly. Although the flight of insects is indeed more complicated than that of airplanes, the underlying physics is nevertheless fully explicable within the rubric of modern fluid mechanics. To understand how insects fly by flapping their wings, it is useful to first consider the means by which fixed-wing aircraft create aerodynamic forces. The design of conventional airplanes is based on the steady-state principle that the flow of air around the wings and the resulting forces generated by that flow are stable over time. As the wing of a plane moves through the air, it meets the oncoming flow at a small inclination, termed the angle of attack. As the flow of air approaches the leading edge of the wing, it divides into two streams on the undersurface of the wing. Because of the viscous behavior of air (a general property of all "fluids," including liquids and gases), the two streams meet again smoothly at the sharp trailing edge. For the flow to separate under the wing, but meet again at the trailing edge, the upper stream must travel faster than the lower because it covers a greater distance. By Bernoulli's principle, this higher velocity generates lower pressure, which sucks the wing upward producing lift.

Although the explanation of flight based on Bernoulli's principle is sufficient for simple situations, engineers and physicists often use a mathematical transformation to quantify the velocity difference above and below the wing and analyze more complex situations. Subtracting the background flow caused by the speed of the airplane from the local flow near the wing uncovers a net circular movement of air around the wing called vorticity. Cohesive filaments or loops of vorticity are called vortices, a term that also applies to more familiar flow structures such as tornadoes, whirlpools, and smoke rings. Although the net circular flow of air around a wing is a mathematical abstraction, wings are, in effect, vortex generators. At a low angle of attack a wing creates a bound vortex, so named because the center of vorticity is located within the wing. The Kutta–Joukowski theorem, perhaps the most essential equation in aerodynamics, states that the lift generated by each section of a wing is proportional to the strength of the vorticity it creates, a quantity termed circulation. The simplest way of increasing the amount of circulation, and thus the lift, is to increase the angle of attack. At angles of attack above about 10°, however, the flow over the top surface separates as it rounds the leading edge, resulting in a catastrophic loss of lift known as stall. For a wing operating according to conventional aerodynamics, the stall angle places an upper limit on the amount of stable circulation, and thus lift, that a wing can continuously generate. Early analyses of insect flight aerodynamics applied conventional steady-state theory unto the complex motion of flapping wings. This approach, termed quasi-steady theory, is equivalent to "freezing" the wing at one position within the stroke cycle and then testing it at that particular velocity and angle of attack in a wind tunnel under steady flow. If conventional theory were sufficient, then a series of such measurements repeated for each point in the stroke cycle should sum up to the animal's body weight. In most cases such simple quasi-steady approaches cannot account for the forces required to sustain flight, indicating that unsteady aerodynamic mechanisms play an important role in insect flight.

Scaling Parameters

Before discussing such mechanisms in detail, it is useful to introduce two important parameters that help organize the great diversity of flight patterns in insects. The first term, the Reynolds number, quantifies how changes in body size, wingbeat frequency, and atmospheric conditions affect aerodynamic mechanisms. The wing or body of an insect encounters two forces as it moves through the air, a shear force caused by fluid viscosity and an inertial force from the fluid momentum. The dimensionless Reynolds number is simply the ratio between these two forces and, for insects, is equal to the product of wing velocity, wing length, and air density divided by air viscosity. Reynolds numbers vary among insects from about 10 for the tiniest to 10,000 for the largest insect. At high Reynolds numbers, the inertial behavior of the air dominates and wings generate pressure forces acting perpendicular to their surface. At a Reynolds number less than 1, a viscous shear force dominates, acting parallel and opposite to the direction of motion. Recent measurements of force production by flapping wings indicate that aerodynamic performance is remarkably constant across a range of Reynolds numbers spanning from about 100 to 5000—encompassing the operating range of most insects. Nevertheless, miniaturization is a common theme in insect evolution, and many species are so small that viscous forces, if not dominant, are large enough to greatly influence force production. The functional peculiarities of lower Reynolds numbers are manifest in the unique wing morphology of the smallest insects, including the brush-like wings of thrips and the whip-like wings of some miniaturized beetles. Although the kinematics used by these tiny insects is as yet unknown, it is possible that they flap their "wings" in such a way as to generate an excess of viscous drag during the downstroke, akin to the power strokes of aquatic plankton. Reynolds numbers are also used to construct large mechanical models of flapping insects for the purpose of directly measuring aerodynamic forces and visualizing flow. This technique, termed dynamic scaling, is based on the principle that the fluid-based forces acting on two geometrically similar but different-sized objects are the same as long as the Reynolds numbers are identical.

Another important dimensionless parameter, called the advance ratio, is useful in coarsely assessing whether conventional steady-state aerodynamics is sufficient to explain force production. The advance ratio is simply the animal's airspeed divided by the flapping velocity of its wings. At one extreme, an infinitely high advance ratio indicates that an animal is gliding, and all the air flowing past the wings derives from the motion of the body as a whole, which is a condition amenable to conventional steady-state aerodynamics. Even if the wings flap up and down, steady-state approximations may be valid as long as the forward speed is substantially greater than the velocity of the wings. The situation is much more complicated for hovering or near-hovering conditions, in which the insect is essentially stationary and most of the airflow encountered by the wings is generated by their back-and-forth motion. Under these conditions, the flow of air around the wing changes substantially throughout the stroke, and the analysis of aerodynamic forces is more complex. Low advance ratio flight is typical of many insects, particularly those of small body size, and is characterized by a motion in which the wings flap back and forth in a roughly horizontal plane. During the two strokes (somewhat inappropriately named the downstroke and upstroke), the wings translate through the air at high angles of attack creating elevated vorticity. At the end of each stroke, the wings rapidly flip over such that the dorsal surface of the wing faces upward during the downstroke, and the ventral surfaces faces upward during the upstroke. As it flips, the wing sheds the vorticity it created in the previous stroke, thereby adding to a complex vortex wake that forms underneath the stroke plane akin to the downwash beneath a hovering helicopter. Thus, at the start of each stroke the wing travels not through still air, but through its own wake. These three peculiarities of wing motion during flapping flight, (1) the high angle of attack during translation, (2) the rapid rotation between strokes, and (3) the influence of the wake on subsequent flow of air around the wings, all profoundly influence the manner by which insects create and modify aerodynamic forces.

Aerodynamic Mechanisms

The total force created throughout a stroke by a flapping wing may be conveniently separated into four main components: translational force, rotational force, wake capture, and inertial force (Fig. 1). Inertial force results from the acceleration of the wing back and forth during each stroke. Although the mass of the wings is small, the acceleration is great and the resulting inertial forces are substantial. Peak values during stroke reversal may be many times greater than the aerodynamic forces. However, because the flapping motion is largely sinusoidal, wing inertia averages close to zero over each stroke and thus does not contribute to the average forces acting on the body. Another component of inertial force derives from the acceleration of the air displaced by the wing as it accelerates, termed virtual mass. Although the precise volume of air disturbed by an accelerating or rotating wing is difficult to calculate, conservative estimates indicate that added mass inertia is relatively small compared with the wing mass inertia and other aerodynamic components. Thus, although wing and virtual mass inertia may complicate the

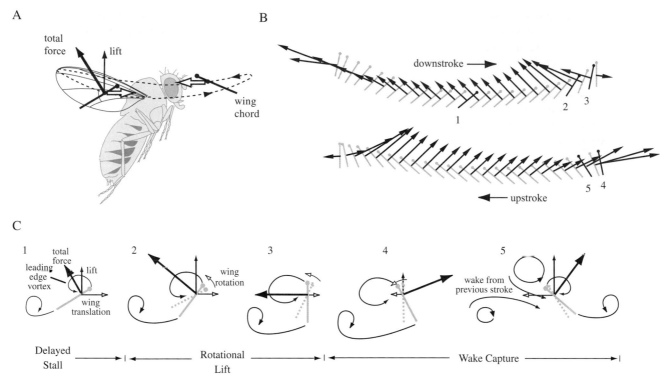

FIGURE 1 Summary of aerodynamic mechanisms used by insects. (A) Under hovering or near-hovering conditions, insects flap their wings back and forth at a high angle of attack during each stroke. The wing path is shown by the dotted line; white arrows indicate wing motion. Between strokes, the wing rapidly rotates so that the dorsal surface faces up during the downstroke, while the ventral portion faces up during the upstroke. The total aerodynamic force (thick arrow) acts perpendicular to the surface of the wing and may be decomposed into orthogonal lift and drag components (thin arrows). (B) Diagram of wing motion indicating magnitude and orientation of the total aerodynamic force vector (black arrows) generated throughout the stroke. Gray lines indicate instantaneous position of the wing at temporally equidistant points during each stroke. Small circles indicate the leading edge of the wing. Time moves left to right during downstroke and right to left during upstroke. (C) Drawings indicate air flow around wing and resulting forces at points within the wing stroke (as indicated in B). Delayed stall (1) results from the formation of a leading edge vortex on the wing. Rotational lift (2 and 3) results from the rapid rotation of the wing at the end of the stroke. Wake capture (4 and 5) results from the collision of the wing with the wake shed during the previous stroke.

precise time course of force production, they are typically ignored in biomechanical analyses of insect flight.

The salient feature that distinguishes the translational forces of insect wings from those generated by airplanes, helicopters, and most birds is that the wings flap back and forth at relatively high angles of attack (30–45°). At such high angles, the stream of air separates from the leading edge of the wing, forming a characteristic flow structure called a leading edge vortex. The lift resulting from the leading edge vortex is much greater than that produced by the bound circulation generated at lower angles of attack. This transient increase in lift at the start of motion at high angles of attack, termed delayed stall, was first recognized by aeronautics engineers in England in the early 1930s, but is too brief to be of use to most aircraft. After only a few moments, the vortex structure grows too large and is shed into the wake, resulting in a precipitous drop in lift. Insects, however, can make use of the initial stages of stall because their wings strokes are so brief. At the end of each stroke the wing sheds the vortex, rotates, and develops a new leading edge vortex swirling in the opposite direction. In addition, complex three-dimensional aspects of the flow, such as a base-to-tip helical flow of air

through the center of the vortex, may in some cases remove energy from the structure, enabling it to remain attached to the wing throughout the stroke.

Rotating objects are themselves sources of circulation and concomitant lift production, which is why a tennis ball hit with backspin rises as it moves through the air. The physics of circular balls and flat wings is somewhat different because the wing has sharp leading and trailing edges, but in both examples the act of rotation serves as a source of circulation, creating a faster flow of air over the top surface resulting in an elevated pressure force. The direction of this rotational force is such that the wing generates positive lift if it flips over before stroke reversal and negative lift if the wing flip is delayed until after stroke reversal, which are kinematic conditions roughly analogous to backspin and topspin on a tennis ball. Thus, unlike the translational component of the total aerodynamic force, the rotational component is strongly dependent on the precise timing of rotation between strokes. For this reason, rotational forces may be particularly important for steering and flight control.

After reversing direction, the wing does not move through undisturbed air, but rather collides with the wake generated during the previous stroke. Because the leading edge vortex moves downward after it is shed, its influence on the flow around the wing is maximal at the start of wing translation, but then rapidly diminishes. Nevertheless, the instantaneous air velocity experienced by the wing at the start of each stroke can be substantially greater than that caused by its own flapping speed. Under certain conditions, this increased flow can result in additional force by a mechanism called wake capture. Because a vortex wake represents the energy lost to the fluid by a moving object, wake capture is an aerodynamic mechanism that enables an insect to recover some of the energy otherwise lost to the air. As with rotational forces, wake capture may play a particularly important role in flight control and maneuverability. By changing the timing and speed of wing rotation, insects can manipulate the magnitude and direction of forces during stroke reversal, thereby manipulating force moments around the body's center of mass.

The wake generated by the wings influences aerodynamic forces in other ways. Vortices shed from the wings drive a column of air downward from the plane of wing motion, which is a change in fluid momentum that is equivalent to the average upward force on the wings. This downwash alters the flow around the wings, but reduces the effective aerodynamic angle of attack and thus attenuates the production of translational forces. In addition, flow interactions may occur among the wings on the same insect. For example, in some insects the close apposition and subsequent rotation of the wings at the beginning of the downstroke, termed the "clap" and "fling," augment force production at the start of the stroke by enhancing the development of the leading edge vortex. In four-winged insects such as dragonflies, the wake of the forewing might under certain conditions increase the forces created by the hind wing.

Although certain general aerodynamic principles apply to all insects, the precise details of flight aerodynamics likely vary in concert with the extreme morphological and behavioral diversity found among the species. The force-generating mechanisms described above, as well as additional mechanisms yet to be discovered, are best viewed as a palette from which the flight behavior of any given species is constructed. The long-term goal for the study of insect flight aerodynamics is not only to uncover the mechanism by which any particular species stays in the air but also to show how it manipulates various mechanisms to maneuver and accomplish the aerial behaviors that are necessary for its survival and reproduction. Recent work in elucidating specific aerodynamic mechanisms must be viewed as only a starting point toward a more comprehensive understanding of flight mechanics and behavior.

NEURAL CONTROL

Sensory Systems

The extreme morphological adaptations associated with flight behavior in insects are paralleled by equally impressive special-

izations within the nervous system. Perhaps most extreme among these alterations relative to the basic neural organization of wingless insects is the hypertrophy of the compound eyes and associated visual ganglia. Large eyes capable of rapid response and broad adaptation to ambient light level are characteristic of diurnal insects such as butterflies, dragonflies, bees, wasps, and true flies. The visual system provides essential sensory feedback for flight control in most diurnal species and is used for a variety of tasks, including velocity and altitude control, obstacle avoidance, landing responses, target recognition, and spatial memory. Features of the anatomy and physiology of the visual system of individual species correlate well with flight behavior and habitat. The elevated translational and rotational speeds characteristic of flight, particularly compared with those of walking and running, place a premium on rapid response time of the visual system. The enhanced visual processing speeds of insects is exemplified by the flicker fusion rate of house flies, which at roughly 300 Hz is the highest found among all animals.

In addition to the eyes, several other sensory modalities on the head provide critical feedback during flight. Although incapable of extracting detailed spatial information, output from the three ocelli helps to stabilize pitch and roll. Because the associated neural computations are relatively simple, the ocellar system can detect and process changes in body orientation more rapidly than can the visual system. Hair cells on the head and mechanoreceptors at the base of the antennae are capable of measuring the magnitude and direction of airflow during flight. In conjunction with visual measurements of ground speed, the input from these wind-sensitive cells is crucial for calculating ambient wind direction, an important capability for flying upwind or tracking odor sources, which are detected in part by chemosensory sensilla on the antennae.

Although sensory structures on the head provide relatively slow tonic cues used for modulating wing motion or body posture over many wingbeats, sensory input from mechanosensory cells on the thorax provides fast phasic input that can alter wing movements on a cycle-by-cycle basis. These mechanosensory structures include the tegula, an organ below the wing that is stimulated during the downstroke, and stretch receptors embedded in the wing hinge that fire during the upstroke. Wing veins contain arrays of tiny campaniform sensilla that encode deformations of the wing surface throughout the stroke. In flies, these arrays are greatly elaborated at the base of tiny drumstick-shaped hind wings called halteres, which function as equilibrium organs. Associated sensory fields detect the Coriolis forces that deflect the beating haltere when the animal's body rotates during flight. Remarkably, a similar specialization is found among stresipterans, but in these insects it is the forewing that has been transformed into an equilibrium organ, whereas the hind wings retain the aerodynamic function. Although the precise role of the thoracic mechanosensory organs varies from species to species, their general function is to tune the output

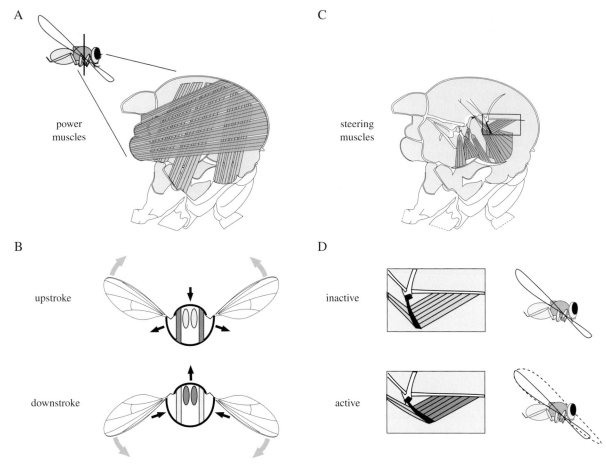

FIGURE 2 In insects using an asynchronous flight motor, the wing muscles are segregated into two anatomically, physiologically, and functionally distinct groups. (A) The large indirect power muscles, which fill the thorax, are arranged in two antagonistic groups. (B) A cross section through the thorax (as indicated by line in A) showing the action of the power muscles. The laterally placed dorsoventral muscles drive the upstroke, whereas the more medial dorsolongitudinal muscles drive the downstroke. The contraction of each muscle set stretches the antagonist group thereby activating the next phase of oscillation. The motion of the thorax is indicated by black arrows, motion of wings is shown by gray arrows. (C) The arrangement of direct steering muscles. (D) Illustration of how activity of a steering muscle changes wing motion (enlargement of rectangular region in C); as muscle becomes active (dark gray), wing trajectory changes.

of interneurons and motor neurons that pattern the activity of the flight muscles. The phasic, phase-locked nature of this feedback is important because the mechanical properties of steering muscles are extremely sensitive to the precise time at which they are activated within the wingbeat cycle.

Muscles

As with sensory systems, motor systems of insects exhibit many specializations related to flight behavior. Unlike the wings of birds, bats, and pterosaurs, insect wings contain no intrinsic muscles. The wing is attached to the thorax by a complicated hinge structure that amplifies the tiny strains of the flight musculature into the large sweeping motions of the wing. The hinge is composed of a connected set of hard sclerotized elements (the wing sclerites or pteralia) embedded within a matrix of more compliant cuticle. Flight muscles may be segregated into two morphological groups according to how they transmit force to the wing. Direct flight muscles insert upon apodemes connected directly to the wing sclerites. In

contrast, indirect flight muscles insert within the thorax some distance from the base of the wing. Odonates are distinct in possessing only direct flight muscles, whereas most insects possess some combination of direct and indirect muscles. In many of the most species-rich orders, including the Coleoptera, Hymenoptera, and Diptera, direct and indirect muscles differ physiologically and serve distinct functions (Fig. 2).

Large indirect "power" muscles provide the mechanical energy to drive the gross up-and-down motion of the wings, whereas a set of small direct "steering" muscles controls the fine changes in wing kinematics during flight. Each contraction in a steering muscle is activated one for one, by action potentials in presynaptic motor neurons, but contractions in the power muscles are asynchronous with motor input. By a molecular mechanism not yet fully understood, rapid stretch activates the crossbridges in asynchronous muscles, causing them to shorten after a brief delay. The low-frequency drive of motor neurons is sufficient to elevate calcium concentration within the sarcoplasm of asynchronous muscle to a level that maintains crossbridges in a stretch-activated state. During flight, contractions within sets of antagonist downstroke and upstroke muscles provide the requisite mechanical stretch to activate each other. Stretch activation frees muscles from the requirement of an extensive sarcoplasmic reticulum (SR), which is necessary in synchronous muscle for the release and subsequent uptake of calcium during twitches.

Asynchronous muscles are capable of generating elevated levels of mechanical power because their internal volume is filled almost exclusively with contractile fibrils and mitochondria. The advantage of stretch activation is especially strong at high frequencies for which typical twitch muscles would require an enormous surface area of SR, severely compromising their ability to generate power. Thus, asynchronous fliers can attain much higher wingbeat frequencies, and thus smaller body size, than can insects using synchronous flight muscles. The mechanical efficiency of asynchronous muscles should also be high because the normal costs associated with cyclic release and uptake of calcium through the SR are not incurred. Because their contraction is only partially controlled by the nervous system, indirect asynchronous muscles are ill-suited to mediating rapid changes in wing motion. The nervous system exerts its control of flight behavior primarily through the action of the direct synchronous steering muscles.

Pattern-Generating Circuits

The motor neurons that innervate insect flight muscles are driven by complex rhythm-generating circuits within the nervous system. Seminal studies by Don Wilson on locust flight led to the discovery of central pattern generators (CPGs), circuits consisting of interneurons and motor neurons capable of generating rhythmic patterns in the complete absence of phasic sensory feedback. Cells within CPGs excite and inhibit the motor neurons of upstroke and downstroke muscles so that they fire antiphasically during the stroke cycle. Even stretch-activated muscles are driven by CPGs, although the firing rate is roughly 10 times lower than wingbeat frequency. Although there is no doubt that insect nervous systems contain CPGs, research pioneered by Kier Pearson and colleagues has demonstrated that sensory feedback from thoracic mechanosensory structures plays an essential role in patterning motor output during flight. For example, electrical stimulation of wing stretch receptor cells can reset the timing of the flight rhythm in locusts—thus fulfilling a strict criterion that is used to test whether a neuron is a member of a CPG. The circuitry underlying flight behavior is best described as a distributed pattern-generating network, consisting of both central and peripheral neurons.

ENERGETICS

Fuel and Oxygen Delivery

Metabolic rates during flight exceed resting values by a factor of 50 to 200, and the thoracic muscles of flying insects exhibit the highest mass-specific rates of oxygen consumption known for any locomotor muscle. Mitochondrial densities within flight muscle fibers are correspondingly high, ranging in some insects to values as high as 45% of the total muscle volume. Energy during flight is derived almost entirely from the oxidation of chemical fuels; anaerobic pathways are absent from flight muscles. Metabolic fuels diffuse from the hemolymph surrounding muscle fibers to the point of oxidation within mitochondria, whereas bulk movement of hemolymph within the body cavity transports fuels from the abdominal fat body to the thoracic musculature. The type and composition of the fuel used in flight (i.e., lipids, carbohydrates, or amino acids) vary with phylogenetic association and may even change with time during a single flight duration in some species. Oxygen influx and carbon dioxide efflux during flight occur primarily via diffusion within tracheal pathways, but may be augmented by convective motion. For example, contraction of flight muscles and the associated deformations of the thorax can compress and expand internal air sacs and even first- and second-order tracheal branches. Although most higher order branches within the tracheolar network are unlikely to experience convective pumping, muscular contraction may augment diffusion by deforming tracheoles that invaginate muscle fibers.

One important issue relating to flight energetics concerns the limits of insect body size. In dragonflies, studies of tracheal geometry suggest an upper limit to thoracic radius of about 0.5 cm if diffusion alone supplies oxygen during flight. The thoraces of many extant insects are well above this limit, however, and the relative contribution of convection to oxygen supply has yet to be determined for any insect. The existence of some flight-related constraint on maximum body size is supported by the observation that many large insects (e.g., the giant stick insects of Southeast Asia) are secondarily flightless. In a modern species of dragonfly, flight metabolic rates vary in direct proportion to ambient oxygen concentration, a result that is consistent with diffusion-limited oxygen transport. The existence of widespread gigantism in late Paleozoic insects (and among other arthropods) during periods of elevated atmospheric oxygen concentration provides further evidence for diffusive limits to flight metabolism, and thus body size, of flying insects.

Energy Requirements for Flight

Although selection has presumably acted to minimize mechanical power expenditure, most of the energy consumed during flight is lost as heat in the flight musculature. Estimates for the mechanical efficiency of insect flight muscle range from only 4 to 30%, depending on taxon and assumptions as to the amount of elastic energy storage within the thorax. Thus, a comparatively small fraction of the fuel an insect consumes is available as mechanical power to drive the wings. This mechanical energy must support three requirements: parasite, inertial, and aerodynamic power. Inertial power is the power required to accelerate the wings back and forth during the stroke. Unless inertial power is substantially greater than aerodynamic power, even moderate elastic storage within the thorax renders inertial costs small. Parasite power is the work required to overcome the drag on the animal's body as it moves through the air. Thus, parasite power is negligible at low advance ratios, but increases with the cube of flight speed. The aerodynamic power is the rate of the work the wings perform on the air, which may be further subdivided into induced power, the cost of generating lift, and profile power, the cost of overcoming drag on the wings. Because the lift-to-drag ratio for most wing kinematic patterns capable of generating sufficient lift is quite low, profile power requirements may substantially exceed the induced power, especially in smaller insects. Also, recent measurements of drag on dynamically scaled model insect wings indicate that values of profile power may be two to three times higher than previously thought. Underestimates of aerodynamic power resulting from unrealistically low values for wing drag may explain the low estimates of mechanical efficiency for asynchronous flight muscle.

The variation in power requirements with forward airspeed is of ecological and evolutionary interest because of its implications for optimal foraging and dispersal strategies. Both direct measurements and aerodynamic modeling of bumble bees in forward flight suggest that mechanical power requirements are approximately constant over an airspeed range of 0 to 4.5 m/s. In contrast, calculations for various lepidopteran and odonatan species show substantial increases in mechanical power expenditure, with forward airspeed due to the rise in parasite power. In situations in which parasite power is large relative to aerodynamic power, the choice of airspeed during flight has significant energetic implications. One study with dragonflies suggests, in fact, that maximum flight speeds are determined predominantly by the dramatic increase in body drag and associated power requirements at extreme airspeeds.

Temperature Effects

As with many features of flight muscle physiology, power production is strongly temperature-dependent, an effect that has several implications for overall flight performance. Measurements on isolated muscles show that mechanical power output typically increases with temperature and is maximal near muscle temperatures characteristic of the free-flying insect. However, the temperature dependence of power output differs greatly among taxa, and although some insects can instantly take off from the surface of glaciers, others must warm their thoraces to 40°C before their muscle generates sufficient power to sustain flight. In insects for which the flight muscles require elevated temperatures to attain adequate performance, the heat generated during flight that results from low muscle contractile efficiency is available as a source with which to regulate thoracic temperature. In small insects, most metabolic heat generated during flight is lost via convective cooling, and body temperature is close if not equal to ambient

air temperature. In larger insects, however, metabolic heat gain is high relative to convective loss and body temperatures are correspondingly elevated. Many large insects regulate internal heat distribution via control of hemolymph circulation between the thorax and the abdomen, using the latter to radiate excess heat. The dramatic amounts of heat produced by muscular contraction are illustrated by the capacity of bumble bees and of some moths to maintain thoracic temperatures exceeding 30°C when ambient air temperature is only 2 to 3°C. Evolution of such thermoregulatory capacity in many insects is consistent with strong historical selection on muscle performance to meet the exacting energetic demands of flight. Further evidence supporting the link between thermoregulation and flight is the phenomenon known as preflight warm-up. In larger insect taxa, pronounced contraction of the thoracic muscles and low-amplitude wing vibrations precede flight. These actions elevate thoracic temperature to values at which the muscles yield sufficient mechanical power for takeoff. Ontogenetic variation in the temperature dependence of muscle power output can also be substantial. In some dragonflies, for example, thermal sensitivity of force production by flight muscle is correlated with changes in the expression of myosin isoforms through development. This finding suggests that physiological features of flight performance are matched to particular environmental conditions and selective demands.

ECOLOGY AND DIVERSITY

Wing Arrangement

The origin of wings was followed by an explosive diversification of insect orders. Many Carboniferous insects possessed wings of approximately equivalent size, shape, and aerodynamic function that were probably limited to low-amplitude flapping. Equivalently sized fore- and hind wings persist to this day in at least seven orders. However, major differences in the sizes of meso- and metathoracic wings are evident in both contemporary fauna and fossils from the Paleozoic. With the exception of the Coleoptera and Strepsiptera, enlarged hind wings are for the most part confined to extant exopterygote orders. Many endopterygote orders (Hymenoptera, Lepidoptera, Diptera), by contrast, reduce the aerodynamic role of the hind wings. In many insects in which the hind wings provide aerodynamic force, the forewings have been modified for supplemental function. Far from isolated events, the evolutionary transformation of the forewing into either a tegmen or an elytron has occurred at least three times at the ordinal level. Elytra of the Coleoptera have much reduced aerodynamic roles relative to the hind wings and provide for greater mechanical resistance to crushing in conjunction with increased sclerotization of the body as a whole. A similar functional role may be hypothesized for tegminized forewings (e.g., Blattodea and Orthoptera) and for the hemelytra of Hemiptera. Insect wings may also serve a variety of behavioral functions unrelated to flight, including sound production and visual communication. None of these functions are mutually exclusive, although the role of aerodynamic force production remains paramount for at least one wing pair.

Flightlessness

The behavioral and ecological advantages of flight notwithstanding, flightlessness has evolved independently many times in insects. Approximately 5% of the extant insect fauna may be classified as flightless, if all forms of variable wing expression and of reduced flight musculature are included. One common feature of the otherwise diverse manifestations of flightlessness is a reduced need for locomotor mobility. Selection for maintaining flight may be weak if this capability is not required for dispersal, reproductive behavior, or predator avoidance. Even in flying species, the costly development of wings and associated musculature may not occur under all ecological conditions.

Flight Diversity and Body Size

Changes in body size represent major trends in the evolution of winged insects. Although direct paleontological evidence is not available, body lengths of the first flying insects were probably in the range of 2 to 4 cm. Substantial increases in body length appear to have occurred by the mid-Carboniferous, and gigantism relative to today's forms was typical of many late Paleozoic insects as well as of other arthropods. The most parsimonious explanation for Paleozoic gigantism is a contemporaneous increase in atmospheric oxygen concentrations, possibly to values as high as 35% relative to today's 21%. Such high oxygen concentrations, together with higher diffusion constants due to an increase in total atmospheric pressure, would have relaxed diffusional constraints on flight metabolism and thus would have permitted the evolution of giant flying forms. Increased atmospheric density would also have yielded increased augmented lift production during flight, both effects possibly being advantageous during the initial periods of wing evolution. Furthermore, geophysical evidence suggests a decline in atmospheric oxygen concentration through the mid- to end-Permian. As would be consistent with asphyxiation on a geological time scale, all giant terrestrial arthropod taxa of the late Paleozoic went extinct by the end of the era.

In sharp contrast to the late Paleozoic giants, the contemporary insect fauna is characterized by a diversity of miniaturized forms. For example, mean adult beetle body length lies between 4 and 5 mm. Much of the wealth of dipteran and hymenopteran diversity is similarly associated with small body sizes, particularly among the parasitoid and hyperparasitoid taxa. Wingbeat frequencies vary inversely proportional to body size, and today's small insects typically fly with wingbeat frequencies in excess of 100 Hz, rates achievable only with the use of asynchronous muscle. Thus,

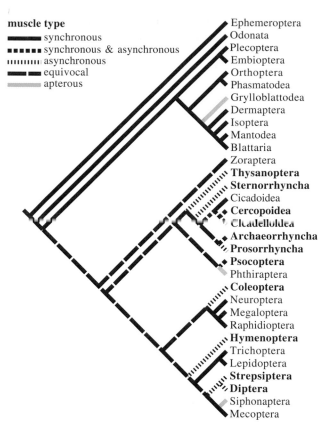

FIGURE 3 Phylogenetic distribution of asynchronous flight muscle. The paraphyletic assemblage Homoptera is here represented at lower taxonomic levels of suborders and superfamilies. Equivocal branch designations indicate either an unknown (e.g., Zoraptera) or an unresolved character state.

the acquisition of asynchronous flight muscle has played a major role in morphological diversification among different insect orders. Asynchronous flight muscle is phylogenetically derived relative to synchronous precursors and has evolved repeatedly among pterygote lineages (Fig. 3). Because flight at small body sizes mandates elevated wingbeat frequencies, this repeated evolutionary acquisition of asynchronous muscle may have facilitated taxonomic radiations of small insects. For example, systematic comparison of sister insect lineages that differ in muscle type statistically demonstrates a decrease in mean body size and an increase in species number if asynchronous flight muscle is present. Three-quarters of all described insect species and three of the four largest orders (i.e., Coleoptera, Diptera, and Hymenoptera) are characterized by asynchronous muscle. Because higher wingbeat frequencies yield increased aerodynamic force, asynchronous muscle may also permit a reduced wing area relative to body mass. This effect may help to explain how one wing pair in many insect groups evolved nonaerodynamic roles.

Flight Behavior and Ecology

Flight plays a central role in the life history patterns of most pterygote insects. A partial list of important insect behaviors mediated by flight includes pollination, phytophagy, hematophagy, escape from predators, mate acquisition, and migration. Forces of both natural and sexual selection have demanded ever-increasing flight performance from insects through evolutionary time, whereas different selective agents are often mutually reinforcing. For example, intra- and intersexual selection often acts synergistically on maneuverability, as does escape from predation attempts by bats, birds, and other insects. Coevolutionary defensive responses among insects, including increased maneuverability and erratic flight styles, parallel the diverse radiations of insectivorous vertebrates worldwide. The morphological and behavioral mimicry among certain chemically defended insects provides wonderful testimony to the strength of such natural selection.

Another major coevolutionary theme in the terrestrial biosphere concerns relationships between flying insects and plants. Phytophagy and pollination by insects are particularly influenced by three-dimensional aerial mobility, the capacity for which dramatically increases access to nutritional resources and suitable oviposition sites. The antiquity of such interactions is well demonstrated by fossil evidence for feeding on plants in the Upper Carboniferous, whereas high rates of herbivory imposed by insects characterize most present-day floras. The evolutionary presence of flying insects has similarly influenced the reproductive biology of many plants. Contemporary angiosperms are pollinated primarily by a broad diversity of insect taxa, most of which are miniaturized forms that can hover at flowers either before or during pollination. Small body size facilitates both incidental and intentional dispersal by wind, and as a consequence tiny insects can act as long-distance pollen vectors.

Continuous aerial entrainment by winds interacts with the large individual numbers of insects worldwide to result in a transient but substantial population of insects moving at heights up to 10 km from the earth's surface. Remarkably, insects from continental faunas have been captured in the mid-Pacific far from any land mass or island. The ability to decouple the flight trajectory from ambient winds depends on the relative magnitude of insect airspeeds, which but rarely exceed typical wind speeds. Thus, directed movement is likely only a few meters from the ground or within canopies of vegetation. Dispersal, on the other hand, is readily attained simply by flying upward into moving air masses. Even migratory flights of larger, more powerful insects (such as locusts and butterflies) are influenced by the directionality of prevailing winds.

See Also the Following Articles

Anatomy • Migration • Muscle System • Odonata • Swimming • Walking and Jumping • Wings

Further Reading

Brodsky, A. K. (1994). "The Evolution of Insect Flight." Oxford University Press, Oxford.
Dalton, S. (1975). "Borne on the Wind: The Extraordinary World of Insects in Flight." Reader's Digest Press, New York.
Dickinson, M. H. (2001). Solving the mystery of insect flight. *Sci. Am.* **284**, 34–41.
Dickinson, M., Lehmann, F.-O., and Sane, S. (2001). Wing rotation and the aerodynamic basis of insect flight. *Science* **284**, 1881–2044.
Dudley, R. (2000). "The Biomechanics of Insect Flight: Form, Function, Evolution." Princeton University Press, Princeton, NJ.
Ellington, C. P. (1999). The novel aerodynamics of insect flight: Applications to micro-air vehicles. *J. Exp. Biol.* **202**, 3439–3448.
Ellington, C. P., Van den Berg, C., Willmot, A. P., and Thomas, A. L. R. (1996). Leading edge vortices in insect flight. *Nature* **384**, 626–630.
Harrison, J. F., and Roberts, S. P. (2000). Flight respiration and energetics. *Annu. Rev. Physiol.* **62**, 179–205.
Heinrich, B. (1993). "The Hot-Blooded Insects: Strategies and Mechanisms of Thermoregulation." Harvard University Press, Cambridge, MA.
Josephson, R., Malamud, J. G., and Stokes, D. R. (2000). Asynchronous muscle: A primer. *J. Exp. Biol.* **203**, 2713–2722.
Marden, J. H. (2000). Variability in the size, composition, and function of insect flight muscles. *Annu. Rev. Physiol.* **62**, 157–178.
Nachtigall, W. (1974). "Insects in Flight: A Glimpse behind the Scenes in Biophysical Research." McGraw–Hill, New York.
Wootton, R. J. (1992). Functional morphology of insect wings. *Annu. Rev. Entomol.* **37**, 113–140.
Young, D., and Simmons, P. (1999). "Nerve Cells and Animal Behavior," 2nd edition. Cambridge University Press, Cambridge, UK.

Fly

see *Diptera*

Folk Beliefs and Superstitions

James N. Hogue
California State University, Northridge

An integral part of any society's cultural heritage is the collection of stories and traditions passed from generation to generation through the ages. Folklore serves to define a people's identity by mirroring its beliefs, concerns, and fantasies. The passing on of traditional tales thus creates a lasting, tangible bond between the living and their ancestors and provides meaning to a people's existence in the present.

As with other aspects of human culture, the characters, lessons, and motivation of human folklore are greatly influenced by a people's surroundings. It is therefore not surprising that ubiquitous insects are common elements in the variety of traditional stories told by people from both past and present societies. Among the variety of loosely categorized stories, traditions, and beliefs typically passed orally through time in human societies are folk tales and superstitious beliefs.

Although such a medium for passing on information often results in inconsistent transmission across time, it provides an avenue for the creative embellishment of stories that reflects the ideals of the teller and the contemporary state of the particular culture. Other aspects of a people's culture passed on in this way, such as religion or mythology, can be thought of as fundamentally different and thus treated separately.

FOLKLORE AND MYTHOLOGY

Although one certainly grades into the other and it is impossible to generalize across every situation, there are some marked differences between folk tales and stories classified as mythology. Myths are typically more infused with expressions of the unconscious and have more symbolic or religious significance than folk tales. Entire societies are grounded in myths, not folklore. Myths also accompany rituals and ceremonies much more so than folk tales. For example, myths are a common medium for reenactments of the past, such as the creation of the world or other significant events in a culture's history. This is true sometimes for stories treated as folklore, but they are much less engrained with ritual or any extended meaning much beyond the particular focus of the tale. In general, folk tales tend to be more for entertainment, whereas myths tend to be more for spiritual instruction.

Folk tales and superstitions serve a variety of purposes, some of which reveal possible reasons for their origin. Some beliefs and tales deal with societal problems and incompatibilities between culture and nature. Some are used as vehicles for wish fulfillment or as a means to speculate on explanations for phenomena observed in nature. They also may serve as mere tales for the enjoyment and entertainment of both the teller and the listener. Other tales serve to instill moral lessons or provide instructions for living one's life in a particular way. Such tales, because they are inherently interesting and entertaining, are particularly suited to providing historical or moral instruction to young minds that might otherwise be unreceptive to such teachings.

TALES THAT EXPLAIN OBSERVED PHENOMENA

Many folk tales dealing with insects are based on fanciful explanations of natural phenomena. The ancient tale of the bugonia apparently originated from such confusion. Bugonia comes from a Greek word that means ox progeny, and is based on the notion that a swarm of honey bees could be spontaneously generated from the rotting carcass of an ox (Fig. 1). This was not merely a description of something perceived to occur in nature, but was a means whereby people could generate many new individuals of these beneficial insects. For this to be successful, precise instructions had to be followed regarding the proper methods and timing of the slaughter and preparation of the ox carcass. This European tale is also found in Chinese and Japanese folklore, and similar beliefs existed for the generation of other bee-like insects from the

FIGURE 1 A 16th century depiction of spontaneous generation of honey bees from a dead ox. [Illustration modified from Bodenheimer (1928).]

carcasses of other animals, namely wasps from horses and hornets from mules.

The bugonia tale originated in ancient Egypt, in a place and time at which the ox and the bee were revered as gods. A Biblical reference to this phenomenon (Joshua 14:8) attests to the antiquity of this belief. This tale of ancient times persisted well into the 1600s, when more careful observations of insect biology led to other explanations. In 1883, the eminent dipterist C. R. Osten Sacken proposed an explanation for the origin of the bugonia story that led to it being discredited. The supposed bees and wasps occurring in the carcasses of dead animals were in fact the drone fly, *Eristalis tenax* (Syrphidae). This fly, a Batesian mimic of honey bees, breeds in putrefying organic matter and could easily be mistaken for a bee by the untrained eye.

The presence of particular forest clearings in western Amazonia is ascribed to the activities of forest gnomes known as *chulla chaqui*. These mischievous creatures live near the clearings and eat the fruits of the only shrub that grows in such places, *Duroia* spp. (Rubiaceae). Like many other figures in zoological folklore, these gnomes can take on the appearance of other forest creatures. They take particular delight in transforming into a brilliant blue morpho butterfly, whereby they attract the attention of human visitors and lure them into the forest, only to disappear and leave the disoriented humans lost. In reality, these clearings are formed and maintained by ants that live in a symbiotic relationship with *Duroia*. The ants clear potential competitor seedlings from areas around their myrmecophytic host plants in exchange for a place to live. This folk explanation is similar to that for small clearings in temperate forests or fields that serve as

FIGURE 2 A typical insect-winged fairy. [Illustration by Ellen Edmonson from "Honey Bees and Fairy Dust," by Mary Geisler Phillips (1926).]

places where woodland fairies commonly gather. Fairies, those furtive, entomologically inspired imaginary beings of diminutive human form that typically bear insect-like wings (Fig. 2), are common figures in European folklore. In fields, the clearings known as fairy rings, supposedly caused by dancing fairies, are in reality caused by underground fungi that make their appearance in the form of an ever-expanding ring of mushrooms that encircles a bare patch among the surrounding vegetation.

Another example of transformation surrounds the explanation for the name of a famous insect used by people as an object of adornment. In Mexico, the jewel-adorned ironclad beetles that serve as living brooches are known as *Ma'Kech* after a legend about a Mayan Prince of Yucatan who is said to have escaped his lover's guards by transforming, with the help of the Moon Goddess, into this beetle. His lover was so impressed by his resolve that she uttered *"Ma'Kech."* This phrase not only means "you are a man" but also means "does not eat," and refers to this insect's and the Prince's ability to go without food for long periods of time.

The phenomenon of crypsis (imitating the background in form, color, pattern, or behavior by an organism to avoid detection) is explained by some indigenous peoples using yet

another example of transformation. It is said that leaves can transform into insects such as katydids and mantids. This is a reasonable explanation given the striking leaf-like appearance of these insects. The transformation of plants into insects was also implicated by some early European naturalists in their explanation of the issuance of insects from galls, nuts, and fruit. This fanciful theory supposed that the various insects brought forth from these sources were generated by the "vegetative and sensitive soul" of the plant. The origin of another organism associated with insects is similarly explained. The elongate fruiting bodies of certain fungi (*Cordyceps* spp.) that commonly attack insects in the American tropics are thought to be the first stage in the development of particular jungle vines that are used for binding poles.

Explanations for insect behavior often take the form of folk tales. Several stories and rhymes tell the tale of the origin of the katydid's song. These short tales typically center on a girl or young woman named Katy who is accused of committing some bad deed such as deceiving or killing another person. The shameful act is immortalized by the singing insects in the trees that continue to debate whether "Katy did" or "Katy didn't." Some insects spend the greater part of their lives boring through and feeding upon living or dead wood. According to a story from the Tahltan of British Columbia, these insects were tricked into searching for their food in this manner by another insect. Long ago a beetle larva and a mosquito lived together. Every day, the larva watched his friend the mosquito come home engorged with blood. Upon being asked by the beetle larva where he was able to regularly find food, the mosquito, not wanting to give up his secret, replied that he sucked his meal out of trees. The next day the larva began boring into wood looking for food, an activity that continues to this day.

In addition to biological phenomena, stories about the origins of some geophysical entities similarly incorporate insects. The origin of fire has been attributed to the actions of fireflies that were responsible for starting the mythological first campfire. According to the Yagua Indians of the upper Amazon, the origin of the river is a result of the misguided actions of insects. Before the existence of the river, the water used by people came from the "tree of water" that, when cut, would release some of this precious liquid. In an effort to liberate more water, wood-boring insects were deceptively used by some children in a plan that damaged the tree such that it released all its waters at once. This resulted in the formation in the mighty Amazon River.

BENEFICIARY TALES AND ENTOMOPHOBIC LEGENDS

Another force behind some folklore is a means of obtaining some diffuse or ancillary benefit for the originator or propagator of the tale. The tale of the Machaca among some inhabitants of Amazonia is a good example. The purportedly deadly consequences of the bite from the Machaca, which in actuality is the harmless but menacing looking fulgorid *Fulgora laternaria,* can be thwarted by having sexual relations within 24 h. These insects instill fear and should be avoided, but should the unfortunate happen, a cure is available. Such "sex antidotes" are fairly widespread among folk cures. The potential benefits to those disseminating such tales are obvious.

Other superstitious beliefs benefit particular insects by protecting them from undue harm from people. The Cornish believed that fairies were the souls of ancient heathen people that were too good for Hell but too bad for Heaven. These beings had gradually shrunk from their natural size to that of ants. It was therefore unlucky to kill ants. Similar tales of bad luck when people willingly or inadvertently step on or otherwise harm particular insects are found throughout the world. This is particularly true for insects perceived as beautiful or beneficial to human endeavors such as butterflies and ladybird beetles.

Some insect folklore stems from a general dislike of insects by people and serves to pass this feeling on to others and propagate fright and ill will toward insects. In some stories, insects may be stigmatized with imagined, dangerous qualities. This is most common for insects that have a frightening appearance and gives reason for them to be despised and avoided. Dragonflies and damselflies, for example, are the bearers of nearly 100 English folk names related to their appearance or supposed behaviors. One of their names is "the devil's darning needle," referring to their ability to sew closed the mouth, nostrils, and eyelids of someone unfortunate enough to be the focus of their displeasure. Other examples focus on fanciful abilities of certain pestiferous species to invade nearly any aspect of human life. One fictitious tale describes the plight of an unlucky woman who kept her hair pinned up for such a long time that it became infested with cockroaches.

A little known legend surrounds the comings and goings of body lice, *Pediculus humanus humanus,* an ectoparasite long associated with humans. There was a belief during the 16th century that during trans-Atlantic voyages, lice on the heads and bodies of mariners would miraculously disappear from the westward traveler at a line of longitude roughly 100 leagues west of the Azores. Furthermore, these parasites would return to the eastbound sailors at the same meridian. The basis of this sailor's tale is unclear, but it may be loosely related to the effects that the increase in ambient temperature and the associated shedding of clothing had on the number of observed lice as ships approached more tropical climes.

INSECTS AS OMENS AND SOOTHSAYERS

Insects that are most commonly featured in human folklore are those that most closely associate with humans or impact human affairs. It is not surprising then that insects such as cockroaches, mosquitoes, and bees are some of the most common subjects in stories and superstitions in which an insect's presence or activity is related to significant events in people's lives.

Because humans have practiced honey hunting and beekeeping for thousands of years, it is not surprising that there is much folklore surrounding these activities. The discovery and collection of honey is reason for merriment and joy in many hunter–gatherer societies and much significance has been attributed to the presence of bees and their role as makers of honey. The activities of foraging honey bees are used to predict the weather. When bees forage far from the hive, good weather is expected, but when they forage nearby, poor weather is sure to come. In ancient Rome, swarms of bees foretold impending misfortune. The significance of the timing of bee swarms is exemplified by the following rhyme:

A swarm of bees in May, is worth a load of hay.
A swarm of bees in June, is worth a silver spoon.
A swarm of bees in July, is not worth a fly.

This saying is relevant to the beekeeper whose summer swarms of bees heading off into the distance mean lost assets.

In addition to bees, the presence and behavior of other insects are used to predict the weather. The most widely known insect-mediated weather forecaster is the larvae of some tiger moths (Arctiidae), known as woollybear caterpillars. These caterpillars, in particular those of the banded woollybear, *Pyrrharctia isabella,* are thickly covered with erect black hairs and have a band of reddish brown hairs encircling the middle of their body. The width of the central band supposedly predicts the weather conditions of the coming winter. Narrow bands indicate a long, cold winter, whereas wide bands indicate a short, relatively warm winter. Other insects associated with weather forecasts are butterflies, flies, wasps, and ants. The Zuni of the American Southwest say to expect rain when the white butterfly flies from the southwest. American folklore tells us that when the gnats swarm, rain and warmer weather are believed to be coming, and when hornets build nests near the ground a harsh winter is expected. Rain is expected when ants withdraw into their nests or if someone steps on an ant. The European stag beetle, *Lucanus cervus,* is supposed to be able to attract thunderbolts. This association is perhaps explained because these beetles were commonly found in old oak trees that were often struck by lightning. Because of this belief, these beetles were sacred to Thor, the Germanic god of thunder.

The association of particular insects with common events in distant parts of the world sometimes depends on the characteristics of a particular taxon. Praying mantids are considered pious prophets or soothsayers in various parts of the world. There is also a considerable body of folklore associated with ladybird beetles. Named after the Virgin Mary (Our Lady), these beetles are widely equated with good luck and are often associated with the ability to portend happy events. These beetles are reputed to have been sacred to Freyja, the ancient Norse goddess of love. To harm one of these insects would certainly bring bad luck. That the most common European species of ladybird has seven spots is the basis for one explanation why this beetle is venerated in this part of the world. The number 7 has long been considered a mystical, powerful, and "perfect" number. In southeastern France, a young girl can predict the year when she will marry by placing a ladybird beetle on her finger and counting the years aloud until the beetle flies away. In other instances, an insect's significance depends on characteristics or behaviors shared between quite different taxa.

In general, the insects found in the folklore of a particular place are drawn from the local fauna. Consequently, significant events common to people worldwide are associated with different species of insect. In British folklore, the presence of deathwatch beetles (Anobiidae) is correlated with the demise of someone in the household. These beetles that live in wood, such as that framing an old English house, send telegraphic messages to each other by tapping their heads on the tunnel walls. This tapping sound is audible to people when all else is quiet, such as in a silent room during a bedside deathwatch. In parts of the Neotropics, the activities of termites fulfill this role as a harbinger of death in a similar manner. Other insects associated with impending death include the appearance of lice in one's dreams, cockroaches flying in one's room, and the sighting of a death's head hawk moth *(Acherontia atropos)*. The scales on the dorsum of the thorax of this moth form the readily recognizable image of a whitish human skull against a dark background. The association of this moth with death in the minds of humans was inevitable.

Often the appearance of a given insect conveys a different meaning in different places or at different times. For example, in some parts of the world, a cricket in the house means good luck, but in other places the presence of this insect means ill fortune. According to one superstition in Brazil, careless contact with fireflies can cause blindness, but in the hands of a curandeiro (folk healer or medicine man), fireflies can be used to cure blindness.

INSTRUCTIONAL TALES

In addition to being entertaining, some folk tales serve as a useful means of instruction. Many tales are told to convey a moral message or pass on useful information in an interesting, amusing, and hence more easily remembered format. An example is that of Aesop's fable of the ant and the grasshopper. While the ant concerned himself all summer with gathering provisions for the upcoming winter, the grasshopper spent his time in leisure and song. The grasshopper even derided the ant for spending so much of his time at work instead of play. When winter came, the grasshopper was not prepared and suffered the consequences of his folly. The ant on the other hand, lived comfortably through the winter on the stores he gathered all summer. The activities of these insects in this story are used to show the importance of preparation for future times of necessity. In addition to ants, the behavior of other social insects such as termites, honey bees, and wasps is commonly used to exemplify the benefits of cooperation, diligence, and hard work.

FOLK MEDICINE

Folk remedies for the treatment of the innumerable ailments that befall humans and their animals are found worldwide. Although less important than herbal remedies, insects play a role in the folklore of healing and drug use. One of the most well-known insect-derived folk medicines is cantharidin. This powerful vesicant is derived from dried blister beetles, particularly *Lytta vesicatoria*. Although cantharidin can be extremely toxic to humans, as recently as the early 1900s cantharidin was used to treat a variety of ailments such as asthma, epilepsy, warts, sterility, and bedwetting. In Europe, where the drug as well as the beetle is known as "Spanish fly," powdered cantharidin was taken orally for its purported qualities as an aphrodisiac. Cossinus, a close friend of the Roman Emperor Nero, reportedly died when an Egyptian doctor gave him "cantharis" to drink for treatment of a skin disease.

Many other insects and insect-derived products have been, and sometimes continue to be, used to improve health and treat disease. One product of insects that is widely used today in the context of what might be called folk medicine is bee pollen. The consumption of bee pollen is said to improve general health and increase stamina. Tonics and teas derived from nearly every insect order, from bedbugs to beetles and cicadas, have found their way into the human apothecary. In China, exuviae left behind by newly emerged adults are used to prepare a tonic to treat eye disease and ailments of the lungs and liver and to soothe crying children. Another particularly interesting use of insect-derived pharmaceuticals in China has recently received much publicity. A tonic made from the fruiting body of the entomophagous fungus *Cordyceps sinensis* is considered a general-health and stress-relieving tonic. The fungus is collected in the wild from the dead caterpillar hosts of the hepialid moth *Hepilus fabricius*. In addition to the variety of ailments purportedly treated with this tonic, caterpillar fungus is also used to improve stamina and endurance. The tremendous performances of Chinese female distance runners in the early 1990s were attributed in part to the use of this caterpillar fungus tonic as part of their training regimen.

Although generally based on some empirical observation some time in the distant past, the validity of insect-based folk medicines should not be assumed, even on the grounds of widespread and long-term use. This is particularly true of aphrodisiacs. The symbolic, religious, and ceremonial associations common to the historical use of many drugs tend to obscure evidence on actual potency. On the other hand, the medicinal use of insects in folk remedies should not be dismissed outright as untrue. Each insect species possesses a unique biochemistry that has the potential to perform any number of medicinal tasks. Some insect-based folk remedies, such as the use of bee venom to treat arthritis and rheumatism, may eventually find a place in modern medicine or may at least serve as the basis for the derivation of modern treatments.

In addition to folk remedies that use insects to cure ailments, another body of insect folklore deals with ways to rid ourselves of pestiferous insects. Pliny the Elder wrote that one sure way to rid one's fields of pests, particularly plagues of cantharid beetles, is to have a menstruating woman walk through the field. This treatment was said to cause the "caterpillars, worms, beetles, and other vermin to fall to the ground." One widespread remedy for an infestation of cockroaches is to seal a few roaches and three coins in an envelope and leave it outside. Whoever picks up the envelope would not only be a little richer, but would also be the new owner of your roaches. A simpler remedy was to slip some roaches to some unsuspecting acquaintance to take home with them with the assurance that your roaches would soon follow. Problem ants can be dealt with in much the same way. By rolling several of the bothersome ants in a leaf and leaving it at a neighbor's house, you could be sure that the ants in your house would soon depart to take up residence with the neighbor. Similarly, some folklore deals with how other animals rid themselves of pestiferous insects. Scottish foxes infested with fleas were said to hold a lock of wool in their mouth and then slowly submerse themselves in water until only the nose and the wool were above water. In trying to escape the water, the fleas end up on the fox's nose and the wool. To finish the job, the fox puts its nose under water and releases the wool along with its passengers of fleas.

CONCLUSION

The acculturation of indigenous peoples worldwide, the disconnection of people from the natural world, and the spread of scientifically based knowledge facilitated by modern means of communication all work in concert to diminish the generation and proliferation of folklore in modern societies. With the ongoing loss of folk traditions in the modern world, so goes the place of insects in traditional folk beliefs and superstitions. Some continue to live on in various forms among contemporary societies, especially among indigenous people in places where the impact of the modern world has yet to take hold. The tidbits of factual observations and the incorrect information that came from antiquity through folklore formed the basis on which the first truly scientific naturalists started their work in the 16th century and led to modern scientific investigation. What was once a way of dealing with problems of unknown nature in our world is now being supplanted by stories of mere curiosity, which although largely untrue, provide interesting insight into the origin and development of human societies, cultures, and religions and are aesthetically pleasing to study.

See Also the Following Articles

Beekeeping • Cultural Entomology • Food, Insects as • Ladybugs • Medicine, Insects in • Teaching Resources

Further Reading

Bodenheimer, F. S. (1928). "Materialien zur Geschichteder Entomologie bis Linné," Vol. I. Junk, Berlin.
Clausen, L. W. (1954). "Insect Fact and Folklore." Macmillan Co., New York.
Hogue, C. L. (1981). Commentaries in cultural entomology. 2. The myth of the louse line. *Entomol. News* **92**, 53–55.
Hogue, C. L. (1985). Amazonian insect myths. *Terra* **23**, 10–15.
Hogue, C. L. (1987). Cultural entomology. *Annu. Rev. Entomol.* **32**, 181–199.
Kirk, G. S. (1970). "Myth: Its Meaning and Functions in Ancient and Other Cultures." Cambridge University Press, London.
Kritzky, G., and Cherry, R. (2000). "Insect Mythology." Writers Club Press, San Jose, CA.

Food, Insects as

Gene R. DeFoliart
University of Wisconsin, Madison

Insects in certain taxonomic groups have played an important role in the history of human nutrition. Although their use as food has long been taboo in almost all Western cultures, their traditional use in tropical and subtropical countries continues to be widespread and to provide significant benefits—nutritional, economic, and ecological—especially for rural communities. The potential benefits of continued or wider use are obvious enough that there seems to be a lessening of the negative attitude in the West.

The type of metamorphosis undergone by an edible species determines which life stage(s) is likely to be consumed. In the insect orders with simple or incomplete metamorphosis (i.e., the Hemimetabola), the life stages usually eaten are the nymphs and/or adults. These orders include the Orthoptera (grasshoppers, locusts, katydids, crickets), Isoptera (termites), Heteroptera (true bugs), and Homoptera (cicadas). Legs, wings, head, and any other hard parts are usually removed before cooking. Orders having complex or complete metamorphosis (i.e., the Hemimetabola) include the Lepidoptera (moths, butterflies), Coleoptera (beetles, weevils), and Hymenoptera (bees, ants, wasps). The life stage usually eaten is the larva, but sometimes it is the pupa or, rarely, the adult.

The insects used as food are, for the most part, clean-living in their choice of food and habitat. Most feed on leaves or other parts of plants. Some of the coleopterous and lepidopterous larvae are wood borers in either dead or living trees and bushes; some, such as cicada nymphs, feed on plant roots. Some hemipterans and coleopterans are aquatic, and some of these and other edible insects are predaceous. Some hymenopterans such as wasps provision their nests with insect prey upon which the young feed. Some edible species have other aesthetic qualities. Some African termites are architects, erecting earthen cathedral-like termitaria that may rise to heights of 3 or 4 m or more. Cicadas and crickets are songsters.

To collect wild insects for use as food, one should be knowledgeable about which local species are edible, particularly in Western cultures in which insects are not among traditional foods that are widely recognized. Some insects secrete toxins or sequester toxic chemicals from food plants or serve as a source of injectant, ingestant, contactant, or inhalant allergens. Bright colors, especially red, or showy behavior such as slow, deliberate flight may suggest that an insect contains toxins, or is unpalatable, and should be avoided.

There are many environmental and ecological ramifications relevant to the use of insects as food. Because of the large number of insect species and the consequently wide variety of plants used as hosts, in general, insects are potentially capable of converting a much wider range of vegetation and waste substances into animal biomass than are the animals currently considered acceptable as food by Western cultures. Many plants that either are not used efficiently or are not used at all in food production serve as hosts for edible insects. In Mexico, it has been suggested that some plants that are widespread and characteristic of arid regions, but of limited food value, such as mesquite, madrono, and some cacti, could be used for cultivation of their associated insects, the weevil *Metamasius spinolae* and the larva of the skipper butterfly, *Aegiale hesperiarus*. The protein and fat content of these insects is many times higher than that of their plant hosts. In general, insects also are higher in their food conversion efficiency than are other food animals when both are fed diets of high quality (see the house cricket, *Acheta domesticus*).

ORTHOPTERA (GRASSHOPPERS, LOCUSTS, KATYDIDS, CRICKETS)

Family Acrididae (Shorthorned Grasshoppers)

Grasshoppers and locusts are included among the foods of almost every culture having any history of using insects as food. In southern Africa, before there were crops to protect, the arrival of a locust swarm, some of which were dense enough to block out the sun, was hailed with rejoicing as a time of harvest. Villagers collected them in the evenings after the swarms had alighted and were benumbed by the cool of the night. The locusts were roasted or boiled or, when plentiful, dried and crushed in mortars to make a much appreciated flour. Sometimes the flour or porridge was mixed with honey to make a sort of cake. Early reports noted that indigenous populations with access either to these vast locust swarms or to winged termites soon grew "visibly fatter and in better condition than before." Grasshoppers were also an important food of Indian tribes in western North America. Various methods of harvest were used, but, most frequently, the grasshoppers were encircled by a number of people and driven into a pit previously dug or onto a bed of coals. Thus, slightly roasted, they could either be eaten or dried and kept for winter food.

In more modern times, within the past 20 years, grasshopper harvest has at one time or another replaced insecticide spraying in parts of Mexico, Thailand, and the Philippines.

Sphenarium is the grasshopper genus of greatest commercial food importance in Mexico. The rice grasshopper, *Oxya velox*, was formerly widely eaten in Japan and Korea. Following reduced use of pesticides on rice in both countries, it is again increasing in numbers. Known as *inago* in Japan it is now found in supermarkets as a luxury item; known as *metdugi* in Korea, it is considered a health food.

Family Gryllidae (Crickets)

Several species of crickets are important as food. In Southeast Asia, *Brachytrupes portentosus* lives in tunnels that are about 30 cm deep, usually one cricket per hole, and comes out only at night. They feed on young plants and are an agricultural pest. They are collected by digging, by filling the holes with water, or as they fly around lights at night. After the wings are removed they are eviscerated, then fried, grilled, or put into curry as a substitute for meat. They are sold by villagers in the markets. In the market at Chiang Mai in Thailand, the shopkeeper takes the crickets live from a plastic bag and spits them longitudinally from head to abdomen on a bamboo stick, three or four crickets per stick. They are then fried in oil in front of shoppers.

Another species of *Brachytrupes*, the sand cricket *(B. membranaceus)*, occurs widely in eastern Africa. Like its cousin in Asia, its presence is indicated by a small heap of soil pushed out from its burrow. It is usually collected by the women and children, and as many as 100 can be collected in a day. It has been said of the sand cricket, "When well prepared it is considered a delicacy, for it turns an ordinary meal into a dinner." In Zimbabwe and likely elsewhere, *B. membranaceus* is one of the species that has increased in numbers in recent years because it is particularly suited to the new kinds of agroecosystems. It is now a significant pest in sand-soil fields, and it is sold in urban markets.

The cricket most readily available to Western insect gourmets is the cosmopolitan house cricket, *A. domesticus* (Fig. 1), which is widely reared commercially as food for pets and other small animals. Studies in the United States led to estimates that this cricket, when kept at temperatures of

FIGURE 1 Mass-reared edible house crickets, *A. domesticus.*

30°C or higher and fed diets equal in quality to those used in bringing conventional livestock to market condition, shows a food conversion efficiency about twice as high as those of broiler chicks and pigs, four times higher than sheep, and nearly six times higher than steers when losses caused by dressing percentage and carcass trim are taken into account. In addition, female crickets have much higher fecundity than beef animals; each cricket lays 1200 to 1500 eggs over a period of 3 to 4 weeks. In beef production, by contrast, four animals exist in the breeding herd for each market animal produced, thus giving crickets a true food conversion efficiency close to 20 times better than that of beef.

ISOPTERA (TERMITES)

Termites are a highly regarded food throughout sub-Saharan Africa. They are eaten raw, fried, or roasted and are found widely in village markets. The fungus-growing termites of the genus *Macrotermes* (family Termitidae) are the most widely used as food.

The large winged adults (sexual forms) are collected as they emerge from the nests on their mating flights at the beginning of the rainy season. They are strongly attracted to light and this behavior is utilized in harvesting them. The high termitaria of some species of *Macrotermes* are a spectacular feature of the African landscape; they are even considered private property in some areas. In southern Congo (Kinshaza), the termitaria average three to five per hectare and may cover 4.3 to 7.8% of the *miombo* woodland. The flora of the mounds is characteristic and quite different from that of the surrounding *miombo*, thus increasing habitat diversity. This diversity is in danger of being lost, however, because in suburban regions and towns, the characteristic flora and fauna are being destroyed and the mounds converted to other uses.

Winged termites are a rich source of fat; analysis of *M. subhyalinus* in Angola yielded an energy value of 613 kcal/100 g and *M. falciger* in Zimbabwe a value of 761 kcal/100 g on an ash-free basis. Protein and fat content of the latter were shown to be 41.8 and 44.3%, respectively, on a dry weight basis. In addition to the winged adults, soldier termites are also eaten.

The high regard in which winged termites are held in Zambia has been documented by Silow in 1983 and appears typical for other countries in Africa: "The Mbunda, Nikangala, Lucazi, Luvale, Cokwe, and Yauma generally agree that the meat of *Macrotermes* spp. [winged sexuals] is better than meat of animals, birds, [or] fish. Perhaps one or another of the edible caterpillars is comparable with them, but most of my informants are of the opinion that [*Macrotermes*] or honey is the best existing food." Silow notes that there are a few people who simply do not like termites and that some missionaries have condemned termite eating as a heathen custom. But he further states that Bemba, Namwanga, Nyanja, and Nsenga whom he has met unanimously declare that *Macrotermes* winged adults "are more delicious than anything else, or at least among the most delicious dishes."

HETEROPTERA (TRUE BUGS)

Family Belostomatidae (Giant Water Bugs)

A giant water bug, *Lethocerus indicus,* is widely eaten in Southeast Asia and is especially popular in Thailand, where it is known as *ma-lang-da-na.* The bugs are 5 to 8 cm in length and fierce predators on frogs, large insects, and small fish. They are caught using nets or at lights, to which they are attracted. There are many methods of preparation, including roasting, frying, steaming, and grilling. After cooking they may be pounded and used for flavoring sauces and curries. The males secrete a fragrant liquid from two abdominal glands and are made into a much-prized sauce to accompany meat and fish. In the markets, males sell for three or four times the price for females. Artificial water bug flavoring is now produced, but people still prefer to eat the real bugs.

Imported bugs (known as *mangda*) from Thailand and extracts of the bug (known as "mangdana essence") can be found in Southeast Asian food shops in California. They are popular with Thai and Laotian customers who use them to make a bug paste called *nam prik mangda* prepared by mashing a whole bug with salt, sugar, garlic, shallots, fish sauce, lime juice, and hot Thai capsicum peppers. The *nam prik mangda* is commonly used as a vegetable dip and as a topping for cooked rice. The extracts known as mangdana essence can be used as a substitute for a whole bug in the preparation of *nam prik mangda,* but they are considered inferior in taste to that prepared from a whole bug.

Families Corixidae (Water Boatmen) and Notonectidae (Backswimmers)

The famous Mexican "caviar," also known as *ahuahutle,* is composed of the eggs of several species in these families. These insects formerly bred in tremendous numbers in the alkaline lakes of central Mexico and were the basis of aquatic farming for centuries. Lake water pollution has now reduced their numbers. The eggs are harvested by what amounts to setting oviposition trap lines. Bundles of shore grass are tied together and weighted with a stone and then distributed by canoe. They are left in place for about 3 weeks during which the adult bugs swim up and lay their eggs on the submerged grass. The bundles are then collected, brought to shore, and dried in the sun. When dry, they are shaken and the eggs fall off. The "caviar" is a true delicacy that appears on the menus of the finest restaurants in Mexico.

HOMOPTERA (CICADAS AND OTHERS)

When there is an emergence of one of the species of periodical cicadas (family Cicadidae), many Americans, for whatever reasons, seem to regard them as legitimate fun food. During a recent (1990) emergence in Chicago and northern Illinois, for example, the *Chicago Sun-Times* carried several articles, the second of which began: "Millions of tasty, entrees-if-you-dare will be available for the gathering during the next month in northern Illinois, and some Chicagoans will want to know how cicada fanciers prepare them." Several recipes were provided. Articles described cicada biology and how to prevent damage caused by egg laying on very young plants and urged Chicagoans to forego the use of insecticidal sprays. There were many radio reports, a cicada hotline, and even *Time* magazine published a recipe.

There are six species of periodical cicadas *(Magicicada)* in North America, three with a 13-year cycle and three with a 17-year cycle. The nymph remains in the soil, feeding on the roots of various plants until ready for the final molt. It then digs itself out of the ground, climbs the nearest tree or shrub, and attaches itself firmly. The adult lives for a month or longer. The so-called dog-day cicadas, such as those of the genus *Tibicen,* have shorter life cycles, but even they require at least 4 years. Cicadas are eaten in many countries, but probably most widely in the countries of southeastern Asia.

LEPIDOPTERA (BUTTERFLIES, MOTHS)

Lepidopterans reach their maximum food importance in Africa where, in many countries, more than 20 species are consumed. In the southern part of Congo (Kinshasa), for example, caterpillars of at least 35 species are consumed.

Family Saturniidae (Giant Silkworms)

In 1980, Malaisse and Parent analyzed 23 species (17 of which were Saturniidae) eaten by humans, using samples that were prepared in a manner identical to that preceding their culinary preparation and then dehydrated. Crude protein content averaged 63.5%, kcal per 100 g dry weight averaged 457, and most species proved to be an excellent source of iron, 100 g averaging (in 21 species analyzed for iron) 335% of the daily requirement. In Angola, the saturniid caterpillar *Usta terpsichore* was found to be a rich source of zinc, iron, thiamine, and riboflavin.

Probably the best known of the edible caterpillars is *Gonimbrasia belina,* the so-called "mopanie worm," which is widely eaten in southern Africa. The South African Bureau of Standards has estimated annual sales of mopanie through agricultural cooperative markets at about 40,000 bags, each containing 40 kg of traditionally prepared, dried caterpillars. This total represents only those entering reported channels of commerce and does not include those privately collected and consumed or sold. The caterpillars, up to 10 cm in length, grip the host plant tightly and cannot be shaken off; they must be picked by hand. A good picker in an average infestation can collect 18 kg per hour. In areas where they are abundant and bulk-dried, they are first eviscerated and then roasted for 15 min before being spread out to dry. About 2 days are required for the product to become dry enough for storage.

The mopanie caterpillar is an important food item and is actively traded not only in South Africa but also in Botswana and Zimbabwe and is exported by the ton to Zambia. From extensive studies in South Africa, Dreyer and Wehmeyer concluded in 1982 that "the consumption of mopanie caterpillars can to a substantial degree supplement the predominantly cereal diet with many of the protective nutrients." The amino acid composition of the dried caterpillars is relatively complete, with high proportions of lysine and tryptophan (which are limiting in maize protein) and of methionine (limiting in legume seed proteins). There is increasing concern in South Africa that the mopanie caterpillar might be collected to the point of extinction.

In Malawi, *G. belina* and another saturniid, *Gynanisa maia,* still occur abundantly in Kasungu National Park; the larvae are in season from mid-October to December, a time of year when food stocks of families living near the Park are running low. The caterpillars are nonexistent outside the Park because of the absence of host trees, which have been displaced by extensive agriculture. A study has shown that opening Kasungu National Park to controlled sustainable use, such as caterpillar harvest, by local people can reduce the problems of poaching in parks and other protected areas. Similarly, of ecological benefit, observations in Zambia have shown that there are very few late bushfires in areas where the caterpillars of *Gy. maia* are found. Fires late in the dry season cause considerable damage by killing trees, reducing regrowth, and increasing erosion. The caterpillars are a highly prized food, and in areas where they are abundant they provide the incentive for people to burn early, thereby protecting the caterpillars and enhancing woodland regeneration. There are other examples in Africa where protection of caterpillars as a food resource enhances biodiversity.

Family Bombycidae (Silkworm Moths)

A by-product of the silk industry, pupae of *Bombyx mori* remain after the silk is reeled from the cocoons. These pupae are widely sold, often canned, in markets in Eastern Asia. In China, the pupae, along with waste materials from the reeling factories and from the silkworm rearing, are also used as fish food in pond-fish culture. Canned pupae are exported, especially from Korea, and can be found in Asian food shops in the United States.

Family Cossidae (Carpenterworms, Leopard Moths)

Many insects were important foods for the Aborigines of Australia and among the most prized were the *witchety* or *witjuti* grubs, several species of root-boring cossid larvae belonging to the genus *Xyleutes.* Tindale conducted in 1953 the definitive study on these insects and stated, "Aborigines with access to *witjuti* grubs usually are healthy and properly nourished. …Women and children spend much time digging for them and a healthy baby seems often to have one dangling from its mouth in much the same way that one of our children would be satisfied with a baby comforter." Over a period of several months spent observing nomadic Pitjandjaras in the Mann and Musgrave Ranges, it was noted that part of nearly every day's diet consisted of these larvae. Tindale states elsewhere that the taste of *witchety* grub, "when lightly cooked in hot ashes, would delight a gourmet."

Recently in Australia there has been an explosion of interest in native, or "bush tucker," foods, including *witchety* grubs and other insects such as *bardi* grubs (Cerambycidae) and honey ants. Bush food is increasingly found in restaurants frequented by tourists, and book stores are well-stocked with books on bush tucker. *Witchety* grubs are on the menu of the posh restaurant Rountrees on Sydney's North Shore; the chef says of them, "They have a nice, nutty flavor when roasted."

Family Megathymidae (Giant Skippers)

The larva of the giant skipper butterfly, *Aegiale hesperiaris,* known as *gusano blanco de maguey,* or the white agave or maguey worm, is in demand by people of all social classes in Mexico. Whereas campesinos with access to maguey plants can collect their own larvae to eat or to sell, restaurants in the larger towns and cities charge as much as U.S. $25 per plate. The *gusanos* are served fried or roasted in butter, chili, or garlic sauce. They are also exported as gourmet food. Two other edible insects are associated with the maguey. The pink worm of the maguey, *Xyleutes redtenbachi* (family Cossidae), also called the red agave worm or *gusano rojo de maguey,* is the larva used in bottles of tequila. They are sold in the markets and are also used to season sauces or they may be roasted or fried with salt and eaten in tacos. Along the maguey's roots are often colonies of ants, which serve as a source of the prized *escamoles,* or so-called "ant eggs," which actually are ant pupae.

Family Pyralidae (Wax Moths, Grass Moths)

Taylor and Carter wrote in 1976 as follows: "Larvae of the greater wax moth (*Galleria mellonella*) are tasty and, fortunately, easily reared, hardy and odorless. If only they were commercially available, we would probably have centered most of our recipes around them. They are our favorite insect. They are thin-skinned, tender, and succulent. They would appear to lend themselves to commercial exploitation as snack items." The authors note that the larvae, when dropped into hot vegetable oil, immediately swell, elongate and burst, looking then not like an insect, but like popcorn, and having the flavor of potato chips, corn puffs, or the like. These larvae, known as wax moths, are now available from various dealers in North America.

COLEOPTERA (BEETLES, WEEVILS)

Family Curculionidae (Snout Beetles, Weevils)

The larvae of palm weevils, several species of *Rhynchophorus,* also called palm worms, are widely eaten and greatly esteemed.

A modern cookbook on Cameroon cuisine includes a recipe describing "coconut larvae" as "a favorite dish offered only to good friends." The major species are *Rhynchophorus palmarum* in the Western Hemisphere, *R. phoenicis* in Africa, and *R. ferrugineus* and *R. bilineatus* in southeastern Asia, Indonesia, and the western Pacific. All of these species have long been semicultivated or "farmed" by indigenous peoples and are excellent examples of how harvests of edible insects from natural populations can be increased by intentional creation of additional breeding sites. Cultivation consists basically of cutting down palms and leaving the logs in the forest with the expectation that larvae will be ready to harvest from the decaying pith 1 to 3 months later. The flavor of the sago grub (*R. ferrugineus papuanus*) in Papua New Guinea has been described as "tender and sweet with a slightly nutty flavor." The insect not only sells regularly in local food markets and is bought by foreigners as well as Papua New Guineans, it also is the focus of annual "grub festivals."

Palm weevils are also destructive pests of palms and, in the Western Hemisphere, are vectors of the nematode *Bursaphelenchus cocophilus,* the causal agent of red-ring disease. Although insecticides have been used in attempts to control the weevils, emphasis is on cultural methods. With the palm worms considered such a delicacy, it has been suggested it might be possible to combine increased production with more efficient recycling of dead and diseased palms and as part of reduced-pesticide integrated pest management (IPM) programs and disease control on coconut and other palms.

Family Scarabaeidae (June Beetles, Dung Beetles, Rhinoceros Beetles)

Of the several edible groups within this family, the most interesting is probably the subfamily Dynastinae or giant rhinoceros beetles, particularly the genus *Oryctes.* Three species, including two that breed mainly in dead standing palms, are eaten in Africa, whereas *Oryctes rhinoceros* is a major pest of palm in Asia and the western Pacific. Main hosts of the adult beetles are coconut, oil, and date palms, whereas the larvae live in a variety of dead but not yet decomposed plant material, including dead standing coconut palms, stumps and logs on the ground, and other types of decaying wood, as well as compost, dung heaps, rotting straw, rotting coconut husks, coffee and cacao pulp waste, refuse from sugar cane factories, ricemills, and sawmills, and other wastes from agricultural processing. Control of rhinoceros beetles is based on sanitation and cultural practices similar to those recommended for *Rhynchophorus* weevils, suggesting that *Oryctes* might also be incorporated into palm IPM programs, recycling an endless variety of tropical wastes into animal protein and fat.

Family Cerambycidae (Longhorned Beetles)

In this family, it is the larvae, primarily, that are used as food. They are wood borers in both living and dead trees and in logs and stumps. They have long life cycles, a year or more, so would not be good candidates for mass-rearing under controlled conditions. A major genus, with edible species, is *Batocera* in Asia.

Family Tenebrionidae (Darkling Beetles)

Tenebrionids have a bad reputation as pests of meal, flour, and other stored and packaged cereal foods, but, despite this, the yellow mealworm, *Tenebrio molitor,* has been reared by zoos, aquaria, and commercial dealers as food for birds, fish, and a variety of small animals since at least the 18th century. Their easy availability makes them one of the insects most commonly recommended for inclusion in recipes in the West. There is a problem of quinone contamination in some tenebrionid-infested food products, but this appears to be much less a problem in *T. molitor* than in species of the genus *Tribolium.*

HYMENOPTERA (ANTS, BEES, WASPS)

Family Apidae (Honey Bees)

Honey is prized by many indigenous cultures, and bee pupae/mature larvae, sometimes called "grubs" or "brood," are often as highly prized as the honey. In southeast Asia, three species of wild bees, *Apis dorsata, A. florea,* and *A. indica,* are important sources of honey, wax, and brood. *A. dorsata* is the largest species and its nests, in the higher branches of large trees, may be up to 2 m in diameter. Its honey is also the most expensive, but honey from *A. florea* is most commonly found in the markets. People often eat the grubs uncooked, but they are also fried or put into soup. In Latin America, the grubs of *A. mellifera* and of species in several genera of stingless bees (subfamily Meliponinae) are used as food, and some of the bees, in Brazil and Mexico, for example, are semidomesticated. Bees, including stingless species, are also important in Africa. In some places, such as the Congo (Kinshasa), honey and brood are still harvested by cutting down the tree although the practice has been much criticized. Apiculture in the United States is based on the introduced honey bee, *A. mellifera,* and it has been suggested that, because of its good public image, this species might be a valuable tool in helping to reshape attitudes toward insects as food in the United States.

Family Formicidae (Ants)

Many kinds of ants serve as food in different parts of the world and they are generally considered delicacies. In Colombia, for example, toasted leafcutter ants (genus *Atta*) are said to constitute the highest attainment of Colombian cookery. A campesino, by collecting and selling *Atta* ants, can earn during the 3-month season the equivalent of a year's wages for the average rural worker. The genus is restricted to the Western Hemisphere. Only the alates are eaten, the large

females being especially prized. They are collected as they swarm from the nest by the thousands on their mating flights during the early part of the rainy season. Two species, *A. cephalotes* and *A. sexdens,* are the most widely consumed, being relished across northern South America, with the former extending up into Mexico.

Fungus gardens grown on chewed leaf fragments are tended in the underground chambers of the large nests of the leafcutter ants. The fungus converts cellulose into carbohydrates that can be metabolized by the ants, thus allowing them to tap the virtually inexhaustible supply of cellulose in their forest environment. Forest trees are able to survive the grazing pressure of the ants, but the ants are serious pests of many cultivated trees and other crops when nests are located at the edges of forests adjacent to cultivated areas.

Escamoles are eaten by all social classes in Mexico and the ants have been described as the most enjoyable and expensive edible insect in the markets. Although called "ant eggs," *escamoles* are mainly mature larvae/pupae of two species of the genus *Liometopum, L. apiculatum* and *L. occidentale* var. *luctuosum.* Digging out the underground nest where the *escamoles* are found is very labor intensive. After harvest of ants from the nest (two or three times per year between February and June), the nest is covered with nopal, dried grass, and fresh weeds to maintain an environment suitable for survival and regrowth of the colony. People who collect *escamoles,* known as *escamoleros,* sometimes make more money during the harvest season than other rural people make during the entire year. The *Liometopum* ant is considered such a special treat in Mexico that it is the subject of songs, dances, and festivities.

Honey ants are a source of sweet treats in Mexico (species of *Myrmecocystus*) and Australia (several species in the genera *Camponotus* and *Melophorus*). Specialized worker ants, called repletes, store the honey in the abdomen, which may become the size of a small marble. They are eaten by grasping the head of the ant and sucking the honey from the abdomen. In Australia, *Camponotus inflatus* develops the largest repletes and they are considered a great luxury by the indigenous population. The repletes are found in galleries in the underground nests, where they are immobile and must be fed by the workers. Some aborigines expend much time and effort digging for the repletes but they only partially dig up the nests so as not to destroy the colonies and thus to preserve this valuable resource.

See Also the Following Articles
Commercial Products from Insects • Cultural Entomology

Further Reading
Bukkens, S. G. F. (1997). The nutritional value of edible insects. *Ecol. Food Nutr.* **36,** 287–319.
de Conconi, J. R. E. (1982). "Los Insectos como Fuente de Proteinas en el Futuro." Editorial Limusa, Mexico City.
DeFoliart, G. R. (1995). Edible insects as minilivestock. *Biodiversity Conserv.* **4,** 306–321.

DeFoliart, G. R. (1997). An overview of the role of edible insects in preserving biodiversity. *Ecol. Food Nutr.* **36,** 109–132.

DeFoliart, G. R. (1999). Insects as food: Why the Western attitude is important. *Annu. Rev. Entomol.* **44,** 21–50.

Dreyer, J. J., and Wehmeyer, A. S. (1982). On the nutritive value of mopanie worms. *S. Afr. J. Sci.* **78,** 33–35.

Finke, M. D., DeFoliart, G. R., and Benevenga, N. J. (1989). Use of a four-parameter logistic model to evaluate the quality of the protein from three insect species when fed to rats. *J. Nutr.* **119,** 864–871.

Malaisse, F., and Parent, G. (1980). Les chenilles comestibles du Shaba meridional (Zaire). *Nat. Belges* **61(1),** 2–24.

Menzel, P., and D'Aluisio, F. (1998). "Man Eating Bugs: The Art and Science of Eating Insects." Ten Speed Press, Berkeley, CA.

Oliveira, J. F. S., Passos de Carvalho, S. J., Bruno de Sousa, R. F. X., and Magdalena Sinao, M. (1976). The nutritional value of four species of insects consumed in Angola. *Ecol. Food Nutr.* **5,** 91–97.

Paoletti, M. G., and Bukkens, S. G. F. (eds.) (1997). Minilivestock. *Ecol. Food Nutr.* **36,** 95–346. [Special Issue]

Ramos-Elorduy, J. (1998). "Creepy Crawly Cuisine: The Gourmet Guide to Edible Insects." Park Street Press, Rochester, VT.

Silow, C. A. (1983). Notes on Ngangala and Nkoya ethnozoology: Ants and termites. *Ethnol. Stud.* **36,** 1–177.

Taylor, R. L., and Carter, B. J. (1976). "Entertaining with Insects, or: The Original Guide to Insect Cookery." Woodbridge Press, Santa Barbara, CA.

Tindale, N. B. (1953). On some Australian Cossidae including the moth of the witjuti (witchety) grub. *Trans. R. Soc. S. Aust.* **76,** 56–65.

Forensic Entomology

M. Lee Goff
Chaminade University of Honolulu

In its broadest sense, forensic entomology includes any situation in which insects or their actions become evidence within the legal system. Medicocriminal entomology involves insects as evidence in a criminal case, most frequently homicide, and this is the area that has been most closely associated with the term "forensic entomology" by the general public and, in fact, most entomologists. The use of insects and other arthropods as evidence in criminal investigations dates from 12th century China. Other records appear sporadically in both the forensic and the entomological literature from various parts of the world until a resurgence of interest in the field in the mid-1980s. Prior to this period, the primary application of entomological evidence was to determine the postmortem interval of decomposed bodies. Although this remains the primary application of forensic entomology in criminal investigations, it is now recognized that insects and other arthropods can provide insights into movement of a corpse following death, assessment of wounds (antemortem versus postmortem), characteristics of a crime scene, and abuse and neglect of children and the elderly, as well as serving as alternate specimens for toxicological analyses and sources of human DNA.

INSECTS AS INDICATORS OF THE POSTMORTEM INTERVAL

Decomposing remains provide a temporary microhabitat offering a progressively changing food source to a variety of organisms, ranging from bacteria and fungi to vertebrate scavengers. The arthropods constitute a major element of this fauna, with the insects as the predominant taxa in terrestrial environments worldwide, in terms of both numbers of individuals and species diversity. In North Carolina, for example, 522 species in three phyla were recovered from decomposing pig carcasses, and of these, 84% were insects. In the Hawaiian Islands, 133 different kinds of arthropods were collected from pig and cat carcasses and of these, 83% were insects. There have been numerous decomposition studies conducted worldwide using different animal models, ranging from lizards and toads to elephants. There has been considerable variation in the numbers of different taxa recovered. These differences may be related to both geographic variation and differences in the animal models used.

RELATIONSHIPS OF INSECTS TO THE REMAINS

The use of insects to estimate the postmortem interval requires an understanding of the insect's life cycle, the relationship of the insect to the remains, and the relationship of the remains to the habitat in which they are discovered. Insects pass through a number of distinct stages during their life cycle. Using a blow fly in the family Calliphoridae as an example, the female fly arrives at the body and deposits eggs in body openings associated with the head, anus, and genitals, or in wounds. After hatching, larvae or maggots feed on the decomposing tissues. There are three larval stages, with a molt in between each stage. Once the maggot is fully developed, it ceases to feed and moves away from the remains before pupariation. The puparium is an inactive stage during which the larval tissues are reorganized to produce the adult fly.

The insects encountered on a corpse in any given habitat consist of species unique to that particular habitat and those having a wider distribution. The unique components may be restricted to a particular geographic area or a particular habitat type within a given geographic area. Those taxa having wider distributions are frequently encountered in several different habitat types and are typically highly mobile species. Many of those taxa closely tied to carrion show this wider pattern of distribution. In estimating the postmortem interval, taxa from both components may, under given circumstances, provide essential information on the history of the corpse.

Of those insects having a direct relationship to the corpse, there are four basic relationships, as described below.

Necrophagous Species

Those taxa feeding on the corpse compose this group. This includes many of the Diptera [Calliphoridae (blow flies) and

Sarcophagidae (flesh flies)] and Coleoptera (Silphidae and Dermestidae). These species may be the most significant taxa for use in postmortem interval estimates during the earlier stages of decomposition, defined here as days 1 to 14.

Parasites and Predators of Necrophagous Species

This is the second most significant group of carrion-frequenting taxa and includes Coleoptera (Silphidae, Staphylinidae, and Histeridae), Diptera (Calliphoridae and Stratiomyidae), and hymenopteran parasitoids of larvae and puparia of Diptera. In some instances, dipteran larvae that are necrophages during the early portions of their development turn into predators.

Omnivorous Species

Ants, wasps, and some beetles, which feed on both the corpse and associated arthropods, compose this group. Large populations of these may severely retard the rate of carcass removal by depleting populations of necrophagous species.

Adventive Species

This category includes those taxa that use the corpse as an extension of their own natural habitat, as in the case of the Collembola, spiders, and centipedes. Acari in the families Acaridae, Lardoglyphidae, and Winterschmidtiidae that feed on molds and fungi growing on the corpse may be included in this category. Of less certain association are the various Gamasida and Actinedida, including the Macrochelidae, Parasitidae, Parholaspidae, Cheyletidae, and Raphignathidae, that feed on other acarine groups and nematodes.

DECOMPOSITION

Although there have been many decomposition studies conducted in different parts of the world and under different environmental conditions, most studies have been conducted in temperate areas and fewer in tropical and subtropical habitats. Common to the majority of these studies has been

an attempt to divide the decompositional process into a series of discrete stages. Decomposition is, in nature, a continuous process and so discrete combinations of physical parameters and arthropod assemblages do not occur. There is a value to these stages, however, in providing reference points when faced with the problem of explaining the events associated with decomposition to a jury.

Regardless of locality, there are certain common patterns. The faunas involved tend to be regional, except for some widely distributed species of Diptera and Coleoptera, but the families involved are somewhat stable. The division of decomposition into five stages can be applied to most studies.

Fresh Stage

This stage begins at the moment of death and ends when bloating is first evident. The first insects to arrive at the corpse are flies in the families Calliphoridae and Sarcophagidae. Adult females investigate the corpse, frequently feed, and then, depending on the species of fly, deposit either eggs or larvae around the natural body openings associated with the head (eyes, nose, mouth, and ears) and anogenital regions. Wounds are secondary sites of attraction to tropical species but also may be of major significance in temperate environs.

Bloated Stage

Putrefaction, the principal component of decomposition, begins. Gases produced by the metabolic activities of anaerobic bacteria first cause a slight inflation of the abdomen and, later, the corpse appears balloon-like and fully inflated. Internal temperatures rise during this stage as the result of bacterial decay and metabolic activities of feeding dipteran larvae. Calliphoridae are strongly attracted to the corpse during this stage. As the corpse inflates, fluids are forced from natural body openings and seep into the soil. These fluids combined with the by-products (ammonia, etc.) produced by the metabolic activities of the dipteran larvae cause the soil beneath the corpse to become alkaline and the normal soil fauna departs.

FIGURE 1 Pig carcass during decay stage of decomposition. (A) Day 8: active maggot mass consists primarily of third instar *Chrysomya rufifacies*. (B) Day 13 (end of decay stage): maggots have completed development and migrated away from carcass for pupariation.

Decay Stage

This is the only stage in the decomposition process which has a distinct starting point. The decay stage begins when the skin is broken, allowing gases to escape and the corpse to deflate. Dipteran larvae form large feeding masses and are the predominant taxa present (Fig. 1A). Although some predatory forms, such as beetles, wasps, and ants, are present during the bloated stage, both necrophagous and predatory taxa are observed in large numbers during the later portions of the decay stage. By the end of this stage, most Calliphoridae and Sarcophagidae have completed their development and departed the corpse for pupariation (Fig. 1B). Dipteran larvae will have removed most of the soft tissue from the corpse by the end of the decay stage.

Postdecay Stage

As the remains are reduced to skin, cartilage, and bone, Diptera cease to be predominant. In xerophytic and mesophytic habitats, various Coleoptera predominate throughout this stage, and the diversity of these taxa increases. Associated with this increase is an increase in the numbers of parasites and predators of beetles. In wet habitats (swamps, rain forests, etc.), however, other taxa, primarily Diptera, and their predator/parasite complexes predominate.

Skeletal Stage

This stage is reached when only bones and hair remain. No obvious carrion-frequenting taxa generally are present and there is a gradual return of the normal soil fauna to the area under the corpse. An examination of the soil during the early portions of this stage will reveal various acarine groups that may be of use in estimating the postmortem interval. There is no definitive end to this stage and changes in the soil fauna may be detectable months or even years following the death.

BIOLOGY OF DIPTERA

One of the major problems facing forensic entomologists is the accurate identification of the larvae collected from the remains. Too frequently, the entomologist must work with dead specimens collected by crime scene investigators and submitted in marginal states of preservation. Even when the local faunas are well known, identification of these specimens is difficult, especially for early instars. Work by Erzinclioglu in England and Liu and Greenberg in the United States has provided identification keys to larvae of forensically important taxa. Given the wide distributions of many of the sarcosaprophagous taxa, these keys have wider application than their regional nature implies.

Seasonal variation in the populations of Calliphoridae have been documented in North America and Europe. There are also successional patterns in Calliphoridae. In northern Mississippi, for example, *Phaenicia caeruleiviridis* was the first species to arrive during spring, whereas *Cynomyopsis cadaverina* was the first during the fall and winter months.

Considerable emphasis has been placed on how temperature influences the duration of the various stadia for different species of flies. Accumulated degree hours (ADH) or accumulated degree days (ADD) can be used to estimate the postmortem interval. Ambient temperature data from weather stations in the vicinity of the corpses are used as an indicator of the temperatures at which larvae developed. These temperatures, however, may not reflect the temperatures at which the larvae actually develop. Based on a Hawaiian study, internal temperatures associated with the maggot masses can be as much as 22°C above ambient. Similar observations have been made for human corpses in the former Soviet Union.

In instances of heavy maggot infestations, it is obvious that there is little, if any, direct relationship between ambient air temperature and the temperatures at which the maggots are developing. Although heat generated by maggot masses influences the rate of larval development, this heat generation may not occur immediately, but requires a period of several days to develop. Such a delay may be because of a lack of an organized maggot mass during the early instars. For corpses found during cool weather when colonizing fly populations are low, ADH or ADD calculations generally are more accurate than in higher temperatures with high fly populations. In addition to the rate of development, temperatures may also serve to limit the species that can use the corpse for development. Only some species of Calliphoridae can tolerate high temperatures inside a maggot mass during development. Thus, maggot-generated heat can influence the rate of maggot development, the nature of the corpse arthropod community, the character of the subcorpse community, and the validity of ADH or ADD calculation-based postmortem interval estimates.

FACTORS DELAYING INVASION OF THE CORPSE

The initial invasion of the corpse by insects and other arthropods starts the clock that is ultimately interpreted to give the estimate of the time since death. The basic assumption underlying this is that the invasion occurs soon after death. In decomposition studies, fly activity begins as soon as 10 min after death, but there are factors that may delay this invasion. Invasion can be delayed by wrapping of the corpse or submersion in water. Adverse climatic factors such as cloud cover, temperature, and rainfall may inhibit adult fly activity. Darkness has long been believed to inhibit calliphorid activity, but nocturnal oviposition has been observed in Calliphoridae commonly associated with decomposing human remains in North America and Hawaii. In these instances, temperatures during oviposition were above 20°C. Although nocturnal activity and oviposition may be expected for some species in tropical habitats, it appears to be the exception rather than the rule in temperate regions. Clearly, care must be taken in forensic interpretations dealing with nocturnal activity of Calliphoridae.

APPLICATIONS TO DEATH INVESTIGATIONS

In the preceding sections, the biological basis for using insects as evidence in estimating postmortem interval (time since death), patterns of decomposition, and several of the problems have been presented. An understanding of insect biology and ecology is necessary as a template for determining postmortem interval, but it does not always provide an accurate estimate. In processing insect data, a forensic entomologist must bear in mind that an estimate based on insect activity only determines the period of insect activity. This time frame may not correspond to the entire period of time since death. Each case is unique and must be analyzed on its own merits with careful attention to all data available.

As noted earlier during discussions of decomposition, there is great geographic and seasonal variation in insect populations associated with a corpse. Even in areas where the species populations are similar, there may be seasonal changes in developmental patterns which may complicate the analyses.

In addition to gathering and processing of entomological data, it is necessary to develop protocols for cooperation between the forensic entomologist and various law enforcement agencies, medical examiners, coroners, and the courts. As these individuals better understand the functioning of the forensic entomologist, the end results of the investigations will improve.

COLLECTION PROTOCOLS

Several workers have independently formulated their own protocols for collecting and processing entomological evidence. Among those published, Lord and Burger provide the best general procedures.

ESTIMATION OF THE POSTMORTEM INTERVAL

In the actual process of estimating the postmortem interval from entomological data, each case will be unique. Regardless, the process tends to follow the same general sequence of events:

1. The stage or physical state of decomposition based on physical parameters for the corpse should be determined, and any indications of disturbance or dismemberment of the corpse that may have occurred following death are noted. If the community collected under the corpse does not conform to the observed stage during later analysis, the possibility of movement of the corpse following death must be considered.

2. Specimens collected from the corpse and crime scene must be identified as completely as possible. Immature stages must frequently be reared to the adult stage for final species-level identifications. Representative samples of immatures (species and stages) must be preserved properly to provide a record of what was present on the corpse at the time of examination.

3. For outdoor scenes, appropriate climatic data (temperatures, rainfall, cloud cover, etc.) from weather stations and on-scene observations must be obtained. Aspects of the scene that may serve to influence the effects of these climatic factors on arthropod invasion of the corpse should be considered (cover by vegetation, shading by trees, slope of the ground, burial of the corpse). For indoor scenes, the temperatures (automatic heating, thermostat settings, air conditioner) possible for the time period in question should be noted; positioning of the corpse relative to windows and doors may be significant in terms of both heat and solar radiation.

4. From the autopsy report or one's own observations, the sites of infestations by arthropods should be noted.

5. The postmortem interval is estimated. In the earlier stages of decomposition, this estimate may be based on the developmental cycles of dipteran larvae. In simplest form, the time required to reach the most mature stage of development of the earliest arriving species on the corpse under prevailing conditions would correspond to the minimum postmortem interval. Further consideration must be given to factors that could delay the onset of insect activity (climatic factors, wrapping of the corpse, seasonal variation). When these factors are considered, the final postmortem interval estimate may be greater than the estimated faunal ages. During this comparison, both presence and absence of taxa and developmental stage must be considered. For this reason, it is essential that collections from the corpse and the surrounding area be as complete as possible. In general, the parameters for the postmortem interval estimate will become wider as the time since death increases. During the earlier stages of decomposition, the estimate may be expressed conveniently in terms of hours, whereas later it may be in days, months, or even seasons of the year.

MOVEMENT OF THE BODY FOLLOWING DEATH

Although insects and other arthropods are among the most widely distributed organisms on earth, they are often quite specific in their distributions. As noted earlier, the fauna of the decomposing corpse is composed of species having a wide distribution and species specific to the particular area in which the body is discovered. Presence of species not typically associated with the habitat in which the body is discovered is an indication that the victim died in one location and the body was exposed to insect activity for a period of time. Following this, the body was transported to another location where a second colonization by insects took place.

ASSESSMENT OF TRAUMA

The initial sites of fly egg-laying activity on a decomposing body will normally be the natural body openings associated with the head, anus, and genitals. For many species, the presence of blood associated with wounds is also attractive and egg-laying occurs at those sites. A wound that occurs prior to death (antemortem), while the heart is still beating, produces blood, which is attractive to the flies. If the wound occurs following death (postmortem), similar quantities of blood are not associated with the wound and it will be less attractive to flies for oviposition. If a body is encountered during the earlier stages of decomposition with significant infestations of maggots in areas other than the natural body openings, the possibility of antemortem wounds must be considered. In like manner, if there are wounds present on the body, but infestations are primarily restricted to the natural body openings, these wounds may well be postmortem artifacts.

ABUSE AND NEGLECT

Although most of the applications of entomological evidence involve the dead, in some instances entomological evidence may involve living victims. In these instances, the maggots are encountered feeding on live tissues. This is a condition known as myiasis and is an obligate condition in the life cycles of many flies, primarily in the families Calliphoridae and Sarcophagidae. In other species of flies, myiasis may be a facultative situation. Stages of development recovered from wounds or sores on children, the elderly, or those otherwise unable to care for themselves can be used to document the period of abuse or neglect.

An additional potential problem is in situations in which myiasis occurs prior to death. If the sites of infestations are not noted, the period of development of the maggots on the living individual may be added to the postmortem development, thus increasing the estimated time since death. Any departure from the normal pattern of arthropod invasion of a corpse should be a cause for care in interpretation of entomological evidence.

ENTOMOTOXICOLOGY

Over the past 2 decades, drug-related deaths have increased in the United States and other countries. In many instances, these deaths are not immediately reported and the remains may be undiscovered for several days. Because of decompositional processes, estimations of the time of death or postmortem interval are based on analyses of insects and other arthropods infesting the remains. The data most frequently employed are those associated with insect development rates and successional patterns. Recently, the accuracy of these estimations has been questioned in deaths involving narcotic intoxication. Relatively few studies are currently available detailing the effects of drugs, such as cocaine and heroin, in decomposing tissues on the rates and patterns of development of carrion-feeding arthropods. Additionally, there are few data dealing with effects of other tissue contaminants, such as toxins and environmental pollutants, in decomposing tissues on rates and/or developmental patterns of arthropods

using such tissues as food. Interest also has focused on the potential use of arthropods as alternate specimens for toxicological analyses, in situations in which more normal specimens of blood, tissue, or body fluids are not available. These two areas now comprise entomotoxicology.

Detection of Drugs and Toxins

It is not unusual for remains to be discovered in a highly decomposed or skeletal stage, when there is insufficient tissue for toxicological analyses. There frequently are, however, arthropods or their cast larval or puparial skins still associated with the remains. Various toxic and controlled substances can be detected by analyses of these arthropods and their residues. Generally, arthropod materials have been homogenized and then processed in the same manner as other tissues or fluids of toxicological interest. Analytic procedures include radioimmunoassay, gas chromatography, thin layer chromatography, and high-performance liquid chromatography–mass spectrometry.

Effects of Drugs on Development of Insects

Although many of the studies mentioned documented the potential for use of maggots and puparia as alternate specimens for toxicological analyses, few were concerned with the potential effects of these drugs on the development of the insects ingesting them. In providing an estimate of the postmortem interval, particularly within the first 2 to 4 weeks of decomposition, it has been assumed that the insects will develop at predictable rates for given environmental conditions. That this might not always be true was established by studies on the effects of cocaine on development of the sarcophagid *Boettcherisca peregrina*. In this example, maggots were reared on tissues from rabbits that had received known dosages of cocaine, corresponding to 0.5, 1.0, and 2.0 times the median lethal dosage by weight. Two patterns of development were noted. Control and sublethal-dosage colonies developed at approximately the same rate, as indicated by total body length. In contrast, the colonies fed on tissues from the lethal and twice-lethal dosages developed more rapidly. This difference continued until maximum size was attained and the postfeeding portion of the third instar was reached. Due to the increased rate of development during the feeding stages, pupariation occurred first in the lethal and twice-lethal colonies, but the actual duration of the puparial period was the same for all colonies and there were no detectable differences in puparial mortality.

The potential significance of these alterations in the rates of larval and puparial development is illustrated in trying to establish the postmortem interval of a Caucasian woman, approximately 20 years of age, discovered in a pine woods area northeast of Spokane, Washington. The body was in the early bloated stage of decomposition and had extensive populations of maggots on the face and upper torso. Maggots were submitted to the entomologist after being refrigerated for 5 days and reared to the adult stage. Two species were identified from the adults: *C. cadaverina* and *Phaenicia sericata*. Typically, *P. sericata* oviposits within 24 h following death, whereas *C. cadaverina* oviposits 1 to 2 days following death. Three size classes of maggots were present on the corpse. The first consisted of maggots measuring 6 to 9 mm in length that were consistent of a period of development of approximately 7 days. The second consisted of smaller maggots, consistent with continued oviposition by adult flies. The third consisted of a single maggot measuring 17.7 mm in length and indicative of a developmental period, under prevailing conditions at the scene, of approximately 3 weeks. Given the other data associated with the case, this period did not seem possible. The possibility that this maggot had migrated from another nearby source was eliminated, as no carrion could be located nearby and the probability of only a single maggot migrating was low. The alternate explanation was that the maggot's growth rate had been accelerated in some manner. It was learned that the victim had a history of cocaine abuse and that she had snorted cocaine shortly before her death. This maggot had most probably developed in a particular pocket in the nasal region containing a significant amount of cocaine.

SOURCES OF HUMAN DNA

Recent advances in technology have provided means to identify individual hosts from analyses of hematophagous arthropod blood meals. Using polymerase chain reaction and amplified fragment length polymorphism DNA characterization procedures, two human DNA genetic markers, D1S80 and HUMT01, have been typed from human DNA material derived from excreta from the human crab louse, *Pthirus pubis*, fed on human volunteers. Although these results are preliminary, they demonstrate the potential for this technology to provide individual characterizations for cases of rape or homicide in which hematophagous arthropods are encountered by investigators at the scene. More recently, human mitochondrial DNA (mtDNA) has been isolated, amplified, and sequenced from crab lice fed on human volunteers. This study demonstrates the potential for mtDNA analyses to characterize individual hosts even from desiccated and frozen arthropod blood meals, and it is yet another example of the stability of mtDNA.

CONCLUSIONS

Applications of entomological evidence have increased significantly since the 1980s. Although the major application is estimation of the postmortem interval, entomological evidence also can be applied to toxicology, DNA analyses, and aspects of crime scene assessment. Although there have been many advances, there is still a need for much basic research. Questions remain with regard to the life cycles of many of the necrophagous arthropods, particularly

Coleoptera. One problem in application of entomological evidence is the diversity of the insect fauna which changes from one geographic area to the next. Databases are still needed for many parts of the world where the sarcosaprophagous fauna is poorly known. Even in areas relatively well investigated, there may be significant microgeographic variation in the insect fauna. Additional investigations are needed in the areas of toxicological analyses and applications of DNA technology. A current underlying problem is the relative lack of trained forensic entomologists. Most individuals now working in the field are employed in academic institutions where only a small part of their research effort is assigned to forensic concerns.

See Also the Following Article
Medical Entomology

Further Reading
Catts, E. P., and Goff, M. L. (1992). Forensic entomology in criminal investigations. *Annu. Rev. Entomol.* **37,** 253–272.
Early, M., and Goff, M. L. (1986). Arthropod succession patterns in exposed carrion on the island of O'ahu, Hawaiian Islands, USA. *J. Med. Entomol.* **23,** 520–531.
Goff, M. L. (1993). Estimation of postmortem interval using arthropod development and successional patterns. *Forens. Sci. Rev.* **5,** 81–94.
Goff, M. L. (2000). "A Fly for the Prosecution." Harvard University Press, Cambridge, MA.
Goff, M. L., Charbonneau, S., and Sullivan, W. (1991). Presence of fecal material in diapers as a potential source of error in estimations of postmortem interval using arthropod development rates. *J. Forens. Sci.* **36,** 1603–1606.
Goff, M. L., Omori, A. I., and Goodbrod, J. R. (1989). Effect of cocaine in tissues on the rate of development of *Boettcherisca peregrina* (Diptera: Sarcophagidae). *J. Med. Entomol.* **26,** 91–93.
Greenberg, B. (1991). Flies as forensic indicators. *J. Med. Entomol.* **28,** 565–577.
Liu, D., and Greenberg, B. (1989). Immature stages of some flies of forensic importance. *Ann. Entomol. Soc. Am.* **82,** 80–93.
Lord, W. D., and Burger, J. F. (1983). Collection and preservation of forensically important entomological materials. *J. Forens. Sci.* **28,** 936–944.
Payne, J. A. (1965). A summer carrion study of the baby pig *Sus scrofa* Linnaeus. *Ecology* **46,** 592–602.
Schoenly, K., Goff, M. L., and Early, M. (1992). A BASIC algorithm for calculating the postmortem interval from arthropod successional data. *J. Forens. Sci.* **37,** 808–823.
Smith, K. V. G. (1986). "A Manual of Forensic Entomology." British Museum (Natural History), London.

Forest Habitats

David L. Wood
University of California, Berkeley

Andrew J. Storer
Michigan Technological University, Houghton

Forest habitats exhibit extraordinary diversity. Within them are complex assemblages of species, and myriad interactions occur among these species. Forest habitats comprise not only forested landscapes but also associated lakes, streams, and meadows. All orders of insects can be found in forested habitats. Even insects that occur in salt water can be found in some forest habitats, such as in tropical mangrove forests. All feeding habits are thus represented, that is, scavengers (especially in soils and beneath the bark of dead trees) and phytophagous, entomophagous, and parasitic species. In California, over 5000 species of insect inhabit oak forests alone. In this article we give greatest emphasis to those orders that feed on trees and the predators and parasitoids that consume these phytophagous species. Insects found primarily in soils, lakes, and streams; those that feed on other vegetation found in forests; and those that parasitize other animals are considered elsewhere in this encyclopedia.

THE DIVERSITY OF FOREST HABITATS

Forest habitat diversity is a reflection of geographic location of the forest, as well as the influences of humans on the forest. The broadest definition of a forest habitat is that of a habitat in which trees are a significant component. Natural forests also vary in complexity based on species composition resulting from a combination of environmental factors. A number of classification systems for natural forests have been proposed, but perhaps the most useful is that developed by the Food and Agriculture Organization of the United Nations in the 1970s. This system divides the natural forests of the world into five broad categories:

1. The cool coniferous forests are a circumpolar belt of boreal forests across northern latitudes. They occur between tundra to the north and temperate mixed forests to the south. Conifers are the dominant species in these forests, but they also include a few species of broadleaf trees.

2. The temperate mixed forests are found south of the cool coniferous forests and also are found in parts of the southern hemisphere. Pines and deciduous and evergreen broadleaf trees tend to dominate these forests.

3. Tropical moist evergreen forests are rainforests characterized by high annual precipitation (>2000 mm) evenly distributed during the year. Amazonia, western equatorial Africa, and the Indo-Malayan region are the three main regions for this forest type.

4. Tropical moist deciduous forests are found in tropical regions with 1000–2000 mm annual rainfall and a dry season for 1 or more months. These forests include monsoon forests in Asia where a dry period of 2 to 6 months is followed by heavy rains.

5. Dry forests are found in both temperate and tropical zones where annual precipitation is less than 1000 mm and are low and simply structured wooded areas.

These forest types may be further subdivided according to dominant species types and climatic differences among other factors. Additional forest types can also be recognized, including those heavily influenced by humans, such as urban forests and plantations.

The species of host plants naturally influence the species of phytophagous insects present in a forest. Additionally, human influences affect the species and density of insects in a forest. Forest management in natural forests may involve fire suppression that may favor insects that can exploit shade-tolerant tree species. For example, the fir engraver, *Scolytus ventralis,* has increased in density in mixed conifer forests of the Sierra Nevada in California because fire suppression has led to a dense understory of white fir, *Abies concolar.* Regeneration of shade-intolerant pines is suppressed and this in turn leads to an overly dense forest dominated by white fir, *Abies concolar.* Periodic logging or thinning may result in a build up of insect populations in cut material or associated waste. For example, neglect of forest hygiene following thinning operations in European pine forests may result in an increase in populations of the European pine shoot beetle, *Tomicus piniperda,* which subsequently may damage the shoots of standing trees during its adult feeding stage. Clear-cutting and replanting or natural regeneration of forests favors insects that feed in young trees. For example, the pine weevil, *Hylobius abietis,* is perhaps the most significant forest insect pest in Europe. Adults feed on the bark at the base of conifer seedlings, resulting in seedling mortality. In many locations, any efforts to establish a plantation or naturally regenerated forest must have a management protocol to reduce the impact of this weevil.

In human-made forests, the species composition may be very simple, as in monocultures, or more complex, depending on the management goals of the plantation. Age and size classes of trees are frequently very limited. These plantations may comprise native or exotic tree species and may be on previously forested land or on previously unforested land. Perhaps the simplest forest habitat is a plantation monoculture of even-aged trees (such as in Monterey pine, *Pinus radiata,* plantations in New Zealand or loblolly pine, *P. taeda,* plantations in areas of the southeastern United States). This habitat is nevertheless very diverse as a result of the array of different ecological niches present on the dominant tree species. Usually, plantation monocultures involve areas of forest that are even-aged and that therefore encounter pressure from phytophagous insects that are adapted to, or more pestiferous during, certain stages of forest growth. For example, the gouty pitch midge, *Cecidomyia piniinopsis,* is a pest of young ponderosa pines, *Pinus ponderosa,* in the Sierra Nevada but is less significant on mature trees. Plantation forestry may involve tree species that are not native to the location where a forest is planted. Any insects that occur in these forests, but which were not present prior to the existence of that forest, are therefore introduced. Often these insect populations are not under natural regulatory pressures that limit their numbers. For example, as spruce forests have been planted in areas of Europe, using spruce as exotic species, the European spruce beetle, *Dendroctonus micans,* has moved westward from Eurasia during the 20th century. Outbreaks of this bark beetle have been most severe at the edge of its expanding range before natural enemies spread or are introduced into the areas now exploited by *D. micans.* In the creation of new forest habitats, the interactions among the species in the forest are unpredictable, because an essentially novel habitat has been created.

DIVERSITY OF INSECTS IN FORESTS

Among all plants, trees present the most diverse habitats for insects to occupy. Insects feed on all parts of the tree, i.e., vegetative structures such as leaves, stems, and roots and reproductive structures such as flowers, fruits, and seeds. Some insects are specialized to feed on phloem and/or xylem tissues, dead sapwood, and heartwood. Insects that feed on these structures and tissues vary in size from 1–2 mm (scale insects) to 6 cm (longhorned beetles). Life cycles (from egg to adult) can be completed in a few days or weeks (aphids) or be prolonged for 50 years (metallic wood borers).

Although all orders of insects are found in forest habitats, only a small number feed on the trees or are the predators and parasitoids of these taxa. Species from the following orders are generally referred to as "forest insects": Hemiptera, Isoptera, Orthoptera, and Thysanoptera (in the Exopterygota, which undergo incomplete metamorphosis) and Coleoptera, Diptera, Hymenoptera, Lepidoptera, and Raphidioptera (in the Endopterygota, which undergo complete metamorphosis).

Habitat diversity is greatly favored by the large size of trees, both in mass and in height. Thus, both the abiotic and the biotic environment can vary considerably from the roots to the upper canopy. For example, a bark beetle, *Hylastes nigrinus,* spends its life cycle feeding beneath the bark on the roots of Douglas-fir, *Pseudotsuga menziesii,* growing in California, whereas a cone midge, *Contarina oregonensis,* feeds in a gall in the seed coat within a cone. Old-growth Douglas-firs with a height over 100 m and a diameter of 3 m are not uncommon. Over 240 insect species from the above-mentioned orders are listed as feeding on Douglas-firs. However, only a few of these would be found at the same time on a large, living tree. Similar diversity of insects feeding on trees can be found in broadleaved deciduous forests and broadleaved evergreen forests. Deciduous oaks in England are fed upon by over 280 insect species, and some evergreen oaks in coastal California provide food for about 300 insect species.

This complexity is magnified further when the guilds of insects in the canopy are considered. The phytophagous insects are represented by chewing, mining, gall-making, and sap-sucking species. Chewing insects are found in the Coleoptera, Lepidoptera, Hymenoptera, and Orthoptera; mining species in the Hymenoptera, Lepidoptera, and Diptera; gall-making species in the Diptera and Hymenoptera; and sap-sucking species in the Hemiptera (Homoptera and Heteroptera) and Thysanoptera. Predators are found in the Coleoptera, Diptera, Hymenoptera, and Raphidioptera and parasites in the Diptera and Hymenoptera. Insects that feed on epiphytes such as lichens, algae, and mosses are found in the Psocoptera

(bark lice), Collembola (springtails), Dermaptera (earwigs), and Plecoptera (stoneflies). Mosquitos (Diptera) breed in water contained in tree holes, which are decayed cavities in the wood.

Many detritivores are found in the canopy as well as on other parts of the tree. They feed on many food sources such as protozoa, bacteria, fungi, nematodes, and small particles of plant and animal tissues. These species are found for example in the Psocoptera, Collembola, and Blattoidea.

Finally, many insects that do not feed on the trunk or foliage use them for a resting or hiding place, mating, pupation and/or hibernation, or estivation. These temporary residents have been classified as "tourists." Many of these insects feed on the surrounding vegetation. Their predators and parasitoids would be included as tourists also. For example, ants and spiders, which are very important predators of tree-inhabiting insects, use tree trunks as roadways from the forest floor to the canopy in search of prey. Another example, a sawfly, *Strongylogaster distans,* that feeds on bracken fern in the Sierra Nevada of California, pupates in the bark crevices of nearby ponderosa pines.

In summary, about half of the insect orders are directly or indirectly associated with trees. As with humans, insects use trees for food, shelter, support, and travel.

ROLE OF INSECTS IN FOREST SUCCESSION

Although trees can live for thousands of years (e.g., giant sequoia, *Sequoiadendron gigantea*), natural forests are dynamic plant communities that can change very slowly over thousands of years or very quickly over a few days to a few years. Fires, volcanic eruptions, strong winds, and snow and ice often have dramatic effects on forest succession. Insects can also have this effect during outbreaks (referred to as epidemics or gradations). For example, bark beetles, together with their associated blue-stain fungi, are known to kill millions of trees over thousands of hectares. The forest shown in Fig. 1 may change from an old

FIGURE 1 Mountain pine beetle, *Dendroctonus ponderosae* (Coleoptera: Scolytidae), killed lodgepole pine, *Pinus contorta* var. *latifolia,* in 1982 in Glacier National Park, Montana. [Photograph by Mark D. McGregor, www.forestryimages.org.]

growth (ca. 80 years), even-aged, predominantly lodgepole pine forest to a younger forest of lodgepole pine if a fire occurs subsequent to the mountain pine beetle infestation. Lodgepole pine cones often remain closed for many years and thus can store a large seed crop on the tree. Following a fire, which opens the cones, an almost pure forest of lodgepole pine is established over very large areas. Thus, the stage is set for another mountain pine beetle outbreak in 80 to 100 years.

Insects and fungi are the most important biotic agents that affect forest succession. In general, these organisms influence the number and growth rate of trees through space and time. Insects that feed on seeds influence reproduction of trees, shoot-feeding and defoliating insects influence growth rates of trees, and insects that infest the main stem of trees accelerate the mortality of trees in a forest. All of these effects occur through interactions with other biotic and abiotic agents as part of the complex forest cycle that involves gaps, building, and mature and degenerate phases. Insects therefore impact forest succession by influencing the forest cycle. The changes in these parameters affect the distribution and abundance of trees at a given moment in time and can be projected into the future using mathematical modeling techniques.

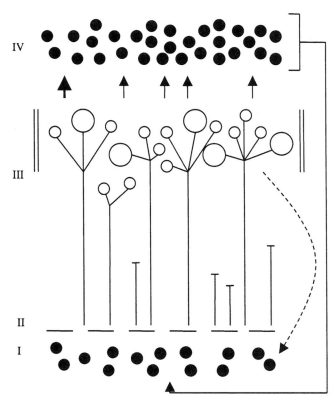

FIGURE 2 Diagrammatic representation of the plant life cycle, illustrating a tree as a series of modular units (the shoots). There are four basic components in the life cycle: (I) the seed bank, (II) the recruitment and establishment of individuals of the population from the seed bank, (III) the growth of individuals, and, (IV) reproduction and dispersal. (Reproduced from Coulson and Witter, "Forest Entomology: Ecology and Management. Copyright 1984. This material is used by permission from John Wiley & Sons, Inc.)

Insects affect the growth rate and mortality of trees by feeding on various parts of the tree. Thus they affect the life cycle of trees (Fig. 2) by influencing the size of the seed bank, the amount of recruitment, the growth of individuals in height and volume, and the reproduction and dispersal of the tree species. Insects consume seeds within fruiting structures or on the ground, thus reducing the size of the seed bank. They consume young seedlings, thus decreasing the recruitment of new trees into the population. They kill tips and shoots, suck plant fluids from the phloem and xylem, and consume the foliage, thus reducing the photosynthetic capacity of the tree and, as a result, reduce the growth of individuals. Insects also kill trees by reducing their growth rate so they cannot compete with other individuals of the same or different species for light, space, water, and nutrients. Trees are killed quickly, often in a few weeks, following severe defoliation by moths or sawflies and by bark beetles that introduce pathogenic fungi into the phloem and xylem. Direct and indirect interactions occur between insect species and between insects and forest pathogens. Root disease pathogens may indirectly affect bark beetle populations by weakening trees and thus predisposing them to attack by bark beetles. Direct interactions with pathogens may occur when, for example, insects transmit the pathogen, as with the pitch canker pathogen, *Fusarium circinatum,* in California and the Dutch elm disease pathogen, *Ophiostoma novo-ulmi,* in Europe and North America.

The examples drawn upon in this article emphasize those insect taxa that affect the survival of trees and in turn influence successional patterns of forests. This treatment thus includes phytophagous species and their natural enemies. Some authors refer to this interaction between insects and plants as a predaceous or parasitic relationship. The predator kills the host but the parasite does not kill the host directly. Insects may also form commensalistic and mutualistic associations with trees. In a commensalistic interaction, the tree is not affected, for example, when it is used as a resting place, whereas the tree benefits in a mutualistic interaction, such as with a pollinator. Our objective then is to emphasize the negative interactions between insects and their host trees, in which only the insect benefits and the tree is debilitated or killed. The tree does, however, benefit in the long term at the population level, at which succession allows nutrient turnover and reproduction allows adaptation of the tree species to environmental changes.

FEEDING GROUPS

Insects can be assigned to feeding groups based on the part of the tree they attack and the method of feeding they use. In forest habitats, these feeding groups are:

A. Insects that feed on cones and seeds
B. Insects that feed on shoots and tips
C. Insects that feed on foliage
D. Insects that feed on the trunk and large branches
E. Insects that feed on roots

In the following discussion, species have been selected from each feeding group to demonstrate the great diversity of insect adaptations to their tree hosts. An example is given from each of the major orders of insects associated with each feeding group. The examples are taken mostly from the Old (Palearctic) and New (Nearctic) World fauna that have received the greatest scientific attention to date. We select examples based on insect orders, knowledge base, and a diversity of biological attributes, all within each feeding group. This approach should give the greatest insights into the mechanisms of insect interactions with their host trees. Thus, the reciprocal responses of the host and the insect are considered at the level of the individual tree and how this affects forest successional patterns.

Insects That Feed on Cones and Seeds

Important species of cone and seed insects are found in the Coleoptera, Diptera, Heteroptera, Hymenoptera, and Lepidoptera. They damage the reproductive structures of trees, resulting in reduced seed production. These insects may feed on the seed itself or on structures associated with the seed. The number of insects feeding in conifer cones is very large when the small size of the feeding substrate is considered. For example, in western Europe, cone- and seed-feeding insects number some 59 species from 30 genera, 13 families, and 4 orders.

Conophthorus **spp. (Cone Beetles); Coleoptera: Scolytidae (Bark and Ambrosia Beetles)** Species in this genus colonize only pine species in North America. Morphological divergence is greatly reduced and therefore, species separation is difficult. Thus, many species were described based on their host associations. Recent studies of pheromone components and DNA are helping to establish the status of these sibling (closely related and difficult to distinguish) species.

Except for the time spent between emergence from the brood cone and colonization of a new cone, the entire life cycle is spent within the cone. A single female tunnels into a new cone in the spring through the cone scales near the supporting stem (as in the "hard" pines, e.g., ponderosa pine) or into the supporting stem of "soft pines" (e.g., eastern white pine, *Pinus strobus*). She excavates a gallery along the cone axis and deposits eggs in the gallery wall. The gallery is packed with frass, which is composed of uneaten fragments of cone tissues and excrement. The male mates with the female inside the cone prior to gallery elongation. One female can infest more than one cone. In general, infested cones from the soft pines fall prematurely from the tree, while infested hard pine cones remain attached to the branch until they drop in the fall. Beetle attacks cause the death of the cone before and after the cone reaches full size. New brood adults can overwinter in the cone or emerge and mine twigs on living trees (*C. ponderosae* on sugar pine, *Pinus lambertiana*) or mine first-year cones (*C. radiatae* on Monterey pine) and spend the winter as nonreproducing adults.

Other insect species enter the galleries of the cone beetle. The dry twig and cone beetle, *Ernobius punctulatus* (Anobiidae), feeds in the drying cone tissues that follow cone beetle feeding activity. A parasitic wasp (Bethylidae: *Cephalonomia* species) enters the cone beetle tunnel to lay eggs. The wasp larvae then feed on the cone beetle larvae.

Contarina oregonensis **(Douglas-Fir Cone Midge); Diptera: Cecidomyiidae (Gall Midges)** This midge infests cones of Douglas-fir in California, Oregon, Washington, and British Columbia. Eggs are laid in young female cones as they open for pollination in the spring. Larval feeding tunnels stimulate gall formation on the seed coat, which destroys the seed. Several galls may form on the seed coat. In the fall, the mature larvae leave the cone and fall into the litter beneath the tree where they spend the winter. There is one generation each year; however, some individuals remain in a resting stage in the litter for more than 1 year. Arrested development, termed diapause, is common in insects that infest cones and seeds. This mechanism permits the insects to survive through periods when few cones are produced. Many other species of insects can inhabit these cones with this midge, including moths, beetles, and wasps.

Megastigmus spermatrophus **(Douglas-Fir Seed Chalcid); Hymenoptera: Torymidae (Torymids)** Most species in this family are parasites of gall-forming insects, which are usually other wasps or flies. However, in the Douglas-fir seed chalcid, the female inserts her long ovipositor through the cone scales and into the seed where she deposits an egg. The larva develops entirely within the seed coat. The seed coat continues to develop normally. In the spring of the following year the adult cuts a round hole in the mature seed coat and emerges to begin a new generation. Adult emergence may be delayed 2 to 3 years. Cones are harvested in the fall to extract seed for future plantings. Infested seeds cannot be distinguished from uninfested seeds by external appearance. This species has been introduced into Europe and has become a serious pest in plantations of Douglas-fir. Parasites reared from infested cones in western North America have been introduced into France to help reduce populations of this seed chalcid. *M. speculatris* was also introduced from North America and now infests cones of Siberian fir, *Abies sibirica* and Nordmanns fir, *A. nordmanniana* in Europe.

Leptoglossus occidentalis **(Western Conifer Seed Bug); Heteroptera: Coreidae (Coreid Bugs)** This insect is named the "leaf-footed bug" because the tibiae of the hind legs are broad and flat while the tibiae of the first two pairs of legs are tubular in shape. This insect feeds on the seed of many conifer species, including Douglas-fir, incense cedar, *Libocedrus decurrens*, and ponderosa pine. This bug sucks plant juices, which results in damage to the seed. The eggs are barrel-shaped and are deposited in rows on the needles. The adults overwinter in protected areas including inside buildings. The broad host range of this species favors survival during years with low cone production in one or more host species.

Cydia strobilella (**Spruce Seed Moth**); **Lepidoptera: Tortricidae (Tortrix Moths)** The female spruce seed moth oviposits between the cone scales of spruce flowers. After hatching, the larvae feed on the central stalk of the cone before moving to the developing seeds. Each larva consumes several seeds to complete its development. Fourth instars return to the cone stalk to overwinter and may remain in extended diapause for 2 or more winters. Higher than average summer temperatures result in extensive cone production in the following year and also stimulate the larvae of spruce seed moth to break diapause. However, even in good years, not all larvae within a cone break diapause. In some European seed orchards, over 50% of cones may be damaged by this species.

Insects That Feed on Shoots and Tips

Insects feeding in or on shoots and tips are found in the orders Coleoptera, Diptera, Hemiptera, and Lepidoptera. These insects may feed externally or internally on these structures. In addition, they may induce the tree to produce a gall as a result of their feeding activities. This gall may have nutritional as well as protective benefits to the insect. Some insects that feed externally gain protection by, for example, producing a spittle mass, as in spittlebugs, or wax, as in woolly aphids. Feeding may be on foliage in addition to the shoot and/or tip.

T. piniperda (**Pine Shoot Beetle**); **Coleoptera: Scolytidae (Bark and Ambrosia Beetles)** In spring, adults of the pine shoot beetle emerge from overwintering sites in the "duff," the needle and bark litter on the ground. They also overwinter in short galleries below the litter layer. They attack trees weakened by root disease (e.g., annosus root disease caused by *Heterobasidion annosum*) and broken and cut trees, but are also capable of killing apparently healthy trees. There can be two or more early emergence periods that result in "sister" broods. This bark beetle infests several pine species, but Scots pine, *Pinus sylvestris,* is its principal host in Europe. The female excavates an egg gallery in the phloem, which is parallel with the grain of the wood. The male mates with the female in the egg gallery. She cuts niches in the gallery wall and deposits one egg in each niche. Larvae then tunnel laterally from the egg gallery. The larval gallery gradually becomes wider as the larvae develop. Pupation occurs in the early summer and emergence occurs in midsummer. These adults fly to living trees and tunnel into the axils of young shoots, which causes the shoots to die and break off. The feeding that occurs in the shoots is necessary for maturation of the gonads. The beetles emerge from the dying shoots and attack host material as described above. This habit has been termed "maturation feeding." Death of the terminal shoot causes flattened canopies. The pine shoot beetle was discovered in Ohio in 1992 and has since spread to adjoining states and Ontario, Canada. It infests eastern white pine and Scots pine Christmas tree plantations.

C. piniinopsis (**Gouty Pitch Midge**); **Diptera: Cecidomyiidae (Gall Midges)** The gouty pitch midge feeds on a large number of pine species in North America. Only the hard pines are infested. It is usually found feeding on young, open-grown pines in both natural stands and plantations. Damaged shoots are first observed in early summer, when new shoots droop and turn yellow and then red. Repeated infestations in northern California slow tree growth and eventually kill the tree. Heaviest infestations occur on ponderosa pines exhibiting sticky twigs while lightest infestations occur on dry, powdery stems, suggesting that some genotypes are resistant to the feeding-induced damage caused by this midge. The bright red larvae overwinter in small resinous pits beneath the bark. In spring larvae migrate to the needles where they pupate. There is one generation per year. Dead shoots resemble frost-damaged tips.

Adelges cooleyi (**Cooley Spruce Gall Adelgid**); **Homoptera: Adelgidae (Adelgids)** The Cooley spruce gall adelgid is distributed from coast to coast in North America where it infests many species of spruce, including Sitka, *Picea sitchensis*, Engelmann, *P. engelmannii*, white, *P. glauca*, blue, *P. pungens*, and Brewer, *P. brewerana*. In the west, Douglas-fir is the alternate host. It was introduced into Europe along with Douglas-fir. In Great Britain, it is named the Douglas-fir woolly aphid. Infestations are also found on Sitka spruce, which has also been introduced into Europe as a plantation species. This aphid has a complex life cycle. When both hosts are present, there can be six life stages in addition to eggs and crawlers. The life cycle, which includes all life stages, is completed in 2 years. Immature females overwinter under bark scales near the tips of spruce twigs. In the spring, they develop into mature females, termed stem-mothers. They each lay up to several hundred eggs under a white cottony, waxy mass. The newly hatched nymphs feed at the base of needles. Light green to purple elongate galls are formed that enclose the nymphs. A chamber at the base of the needle may contain 3 to 30 wingless adelgids, which are covered with a white wax. A few winged adelgids may also be produced. The galls vary between 12 and 75 mm in length. After the nymphs leave the gall, it turns brown and hardens, often remaining on the tree for many years. Infestations kill branch tips, which stunts and deforms the trees. Nymphs transform to adults on needles and fly to Douglas-fir and lay eggs on newly developing needles, shoots, and cones. Infested needles are twisted and chlorotic and can drop from trees in large numbers. No galls are formed on Douglas-fir. Winged adults produced on this host fly back to spruce to oviposit. Where each host occurs independent of the other, continuous generations are produced on one host. In areas of western North America where spruce is rare or absent, two parthenogenetic generations are produced on the needles of Douglas-fir in one season. No galls are produced on spruce where Douglas-fir is absent.

Rhyacionia buoliana (**European Pine Shoot Moth**); **Lepidoptera: Olethreutidae (Olethreutid Moths)** This moth occurs throughout Europe where it feeds on almost all pine species. It was first discovered in America in New York State in 1914 and later was found in British Columbia, Washington, and Oregon. It has not spread into other western North American states. This moth is a serious pest of pine plantations in Europe and North and South America. Recently, it has spread rapidly throughout the extensive Monterey pine plantations in Chile. Lodgepole pine planted in France are often heavily damaged, as are red pine, *Pinus resinosa,* plantations in the states and provinces that border the Great Lakes in North America. In summer, the female lays yellowish, disk-shaped eggs in rows on needles as well as under and on bud scales and on small twigs. The larvae hatch from the eggs and feed on the needles where they are attached to the twig. The larvae then spin a silken web between the needle sheath and the twig. The web is usually coated with resin. The larvae tunnel into the bud, which produces a resinous crust. These larvae overwinter under this crust or within the bud. In the spring, they mine another bud and then feed on the base of needles. They pupate within mined shoots, but the pupal skin protrudes from the shoot so that adult emergence is not impeded by the shoot or resin. The mined shoots are killed and buds formed later produce new shoots that grow around the dead shoot. Repeated infestations produce a bush-shaped tree. A crook is formed on the main stem when the terminal bud is killed.

Insects That Feed on Foliage

The foliage is the predominant photosynthetic part of the plant. Loss of foliage results in loss of photosynthetic area with resulting reduction in the production of carbohydrate. Defoliation may be tolerated in some species of trees, but less so in others. Successive cycles of defoliation, however, are rarely tolerated and often result in tree death, usually in association with other insect groups. Insects that feed on the foliage of trees are predominantly found in the orders Coleoptera, Diptera, Hemiptera, Lepidoptera, and Orthoptera. Most of these feed externally on the foliage, though some mine into the foliage for part or all of their feeding stage. Various larvae in the Coleoptera, Diptera, and Lepidoptera mine the leaves of broadleafed plants, including trees. This activity protects the insect from desiccation and from predators. Many Hymenoptera induce galls to form on leaves as a result of oviposition into the leaf. These galls can take on many shapes and all protect the developing larva from desiccation and predation. Additional protection from abiotic and biotic effects may be provided by webbing to hold needles together, leaf rolling, or producing a silken nest in which insects congregate for protection from predators, as in the case of tent caterpillars.

Pyrrhalta luteola (Elm Leaf Beetle); Coleoptera: Chrysomelidae (Leaf Beetles)   The elm leaf beetle is a native European chrysomelid beetle. It infests all species of elm. This beetle often causes severe growth loss, which weakens trees and, in turn, makes them susceptible to invasion by bark beetles and the Dutch elm disease fungus, which they carry into the tree.

The elm leaf beetle overwinters as an adult in sheltered, dry habitats, such as inside houses and barns and under loose bark on trees. They can become a nuisance in fall and spring as they enter and leave hibernation sites inside residences. They can be active in houses during the winter. Adults emerge from overwintering quarters and fly to elms at the time leaves are beginning to emerge from buds in spring. They lay eggs (up to 800 per female) in groups or irregular rows, along the major leaf veins on the underside of leaves and on nearby twigs. Larvae emerge from eggs in late spring and feed on the underside of leaves, causing skeletonization, where small veins are visible between the major lateral veins that emanate from the midrib. The later instars consume some of the area between these lateral veins. The leaves soon become desiccated and turn brown. At high densities, the entire tree turns brown and appears to be dead. However, in late summer some trees refoliate with smaller leaves. Repeated infestations can kill trees. Two or more generations are produced each season. This insect was introduced into the United States in the late 1890s and has since spread to the west coast. It feeds on all species of elm; however, American elm, *Ulmus americana,* and the introduced Siberian, *U. pumila,* and European elms are severely damaged by this beetle. Introduction of parasites into the United States has met with mixed control results.

Elatobium abietinum (Green Spruce Aphid); Homoptera: Aphididae (Aphids)   The green spruce aphid feeds on native and introduced spruce species in Europe and frequently occurs in forest nurseries. Introduced North American species such as Sitka spruce are most susceptible, whereas most European species, with the notable exception of Norway spruce, *Picea abies,* are less suitable as hosts. In parts of Europe a sexual as well as an asexual stage is found, and overwintering occurs as eggs. Elsewhere, such as in Great Britain, the sexual stage is absent, and parthenogenetically reproducing females remain on the needles during the winter. In North America, where this aphid is presumed to be an introduction, only the parthenogenetic form is known. Most feeding occurs when amino acid levels in the leaves are high in the winter and early spring. In late spring/early summer, longer day length coincides with a drop in amino acid levels, and winged individuals (alatae) are produced, which disperse to new hosts. Most outbreaks coincide with mild winters. The nymphs and adults feed on the lower side of older leaves where they use piercing mouthparts to access the phloem sap. They may completely defoliate trees and attack current-year needles later in the year.

Pristophora erichsonii (Larch Sawfly); Hymenoptera: Tenthredinidae (Sawflies)   The larch sawfly is a native European species. It was first recorded in North America in 1880, in Alberta, Canada, in 1930, in Oregon in 1964, and in Alaska in 1965. Some strains are now believed to be native to North America. The prepupal stage overwinters in a cocoon in the ground or duff under defoliated trees. Adults emerge in spring and as late as late summer depending on location. They lay eggs in rows under the bark of newly elongating shoots. This oviposition behavior causes the shoots to droop. The common name of this family is derived from the saw-like ovipositor that females use to cut slits into the bark where they lay eggs. This species reproduces parthenogenetically. Larvae emerge from eggs and feed on the needle margins and then move to the older needle clusters. They feed gregariously and eat most of the needles on one shoot before moving to another. Between summer and fall, mature larvae drop to the ground and spin a paper-like cocoon in the forest litter. Here, they enter a diapause where they spend the winter. However, some individuals overwinter 2 or 3 years.

Successive defoliation results in tree mortality, especially eastern, *Larix laricina,* and western larch, *L. occidentalis,* in North America, and European larch *L. decidua.* It has been most destructive in the states around the Great Lakes. Where black spruce, *Picea mariana,* and eastern larch co-occur, black spruce is favored in succession because the larch sawfly kills both larch seedlings and mature trees. In the Alps, when aphid densities are high, larch sawfly densities are low, due in part to predation by ants that are tending aphids seeking their honeydew. Biological control efforts in the western United States have been successful in lowering the overall average densities of the larch sawfly. This program is considered one of the most successful biological control programs aimed at a widely distributed forest insect. During studies of the parasitoids of this sawfly, encysted larvae were discovered in certain populations. A cellular layer surrounding the larva is produced by the hemolymph (phagocytosis), a rare phenomenon in insects.

Lymantria dispar (Gypsy Moth); Lepidoptera: Lymantriidae (Tussock Moths)   The gypsy moth is an important defoliator of hardwood trees in Europe and North America. It was introduced into Massachusetts in 1869. Since then it has slowly spread southward to North Carolina and westward into Wisconsin. Small populations have been eradicated repeatedly in California, Oregon, Washington, and British Columbia. In North America, massive defoliation has occurred over thousands of hectares. This species feeds on over 50 tree species, including mostly hardwoods, such as apple, beech, basswood, *Tilia americana,* elm, hornbeam, oak, poplar, and willow. During outbreaks, larch and pine are consumed. In southern Europe and northern Africa, outbreaks occur at irregular intervals. In North America, considerable resources are directed toward control and slowing the spread to uninfested areas. The Asian form of this species has been found in the western United States and Canada, but apparently has not become established.

In spring, larvae emerge from overwintering egg masses, at about the time oak leaves emerge from buds. The young caterpillars move to the tops of trees where they feed on the bases of young leaves or they chew small holes in the leaf surface. They also drop from trees on silken threads and are windblown. Older larvae feed mostly on the leaf margins. The larger veins and midribs of leaves are usually not consumed. At high densities (tens of thousands per hectare), the larvae literally "eat themselves out of house and home." They move to new areas in search of food or pupate, often giving rise to small adults. The larvae are very colorful, exhibiting six longitudinal rows of tubercles. On the thorax and first two abdominal segments the tubercles are blue and on the rest of the abdomen they are red. Larvae crawl down the tree and aggregate in sheltered places. At night they crawl up the tree to feed. In June, larvae pupate on many substrates, including rocks, limbs and trunks of trees, picnic tables, automobiles, and forest debris. In North America, females are flightless, and after emergence from the pupae they crawl a short distance and emit a sex pheromone that attracts male moths. After mating, the females lay eggs in a mass and cover them with larval hairs and a frothy substance. These egg masses may be transported on vehicles or on lawn furniture and rocks that are moved inside vehicles. This means of dispersal gave rise to the common name, "gypsy" moth. The eggs overwinter in a diapause state. The Asian gypsy moth is of major concern because the females are excellent fliers. It has frequently been intercepted on the west coast of the United States, and flying females enhance the dispersal of this species if it becomes established. Many natural enemies of this moth, including parasitoids, predators, and protozoa, have been collected in Europe and introduced in North America. These biological control studies are among the earliest attempts to use indigenous natural enemies to lower densities of an introduced pest.

Coleotechnites milleri (Lodgepole Needleminer); Lepidoptera: Gelechiidae (Gelechiid Moths)   The lodgepole needleminer has a unique life history in the Sierra Nevada of California. The adults are synchronized to emerge in midsummer in odd-numbered years. Adults are not known from even-numbered years. However, another population of this species (or an undescribed species) that occurs in a nearby mountain range in western Nevada also has a 2-year life cycle, but adults emerge in even-numbered years. Another undescribed species feeds on lodgepole pine in central Oregon, but this species has a 1-year life cycle.

After mating, females lay eggs in mined needles or on branches near the needle fascicles. First instars enter near the tip of a single needle and remain in the mine through the winter. During the even-numbered season, fourth instars mine several needles and overwinter in a single, mined needle. In spring of the odd-numbered years, fifth instars feed on many needles and finally pupate in a mined needle.

The most severe outbreaks occur in mature lodgepole pine stands and occur over large areas. In the high-elevation forests (4000 m) near Yosemite National Park, many outbreaks have been recorded, some lasting 16 to 18 years, before returning to low densities. There does not appear to be a regular pattern to these epidemics. Severe defoliation weakens trees, which predisposes them to infestation by the mountain pine beetle, *Dendroctonus ponderosae*. Thus these two insect species convert these extensive mature to overmature lodgepole pine forests to a young forest of lodgepole pine. The Sierra Nevada form of lodgepole pine, *Pinus contorta* var. *murrayana*, produces largely nonserotinous cones that are opened by sunlight. Furthermore, lodgepole pine is moderately shade tolerant and can thus regenerate in the understory of larger trees. When needle miner populations increase to epidemic levels, young trees are released to grow and seeds germinate to produce new seedlings.

The decline of outbreaks has been attributed to rain, and often snow, during the mating period, abundance of parasites and predators, and reduction in amount of host foliage for colonization. The biology of host-specific and generalist parasites has been investigated in an attempt to determine the basis for the 2-year synchronized life cycle.

Diapheromera femorata (Walkingstick); Phasmatodea: Heteronemiidae (Walkingsticks)   Walkingsticks are very unusual insects because, as their name signifies, they are long, slender, and oval and thus resemble a stick or defoliated branch. Most species are wingless. Their body color varies from gray to green to brown to red, which is similar to the leaves and branches on which they are feeding. Through this camouflage they may escape predation from birds. In late summer and fall they deposit their hard, seed-like eggs directly on the ground or by dropping them from trees where they are feeding. At high population levels, the large number of eggs falling on the forest floor literally sounds like rain. Most young hatch during the following spring throughout the southern United States, whereas in the northern portion of its range, the young hatch in the second spring. Occasionally, severe outbreaks occur in the northern United States that last several years, causing extensive tree mortality. Two defoliations can occur during the same season. They feed on a variety of hardwood species, including cherry, basswood, birch, aspen, elm, hickory, oak, locusts, and dogwood.

Insects That Feed on the Trunk and Large Branches

Insects that feed on the trunk and larger branches of trees are found in the Coleoptera, Hemiptera, Hymenoptera, Isoptera, and Lepidoptera. Insects that tunnel into the trunk and large branches of trees may exploit two resources: they may derive most of their nutrition from the phloem, or they may tunnel into the wood and derive their nutrition from the xylem. Insects that do not tunnel into the host, such as scale insects, are sucking insects and use nutrients in the xylem or phloem depending on the species. Many insects that infest the wood of trees, such as termites, serve as primary decomposers of the woody material.

D. ponderosae (Mountain Pine Beetle); Coleoptera: Scolytidae (Bark and Ambrosia Beetles)   The mountain pine beetle is one of the most important forest insects in North America. It is widely distributed in many pine species throughout western North America, from southeastern Alaska to northern Baja California, and eastward through the Yukon territory in Canada and the Rocky Mountains of both the United States and Canada to the Black Hills of South Dakota. The eight-toothed spruce bark beetle, *Ips typographus,* shows a similarly wide distribution in Europe and Asia but it has a more limited host range than *D. ponderosae.* During outbreaks, this bark beetle kills millions of trees over large areas. It infests some of the most widely distributed and important timber-producing species, including ponderosa pine, western white pine, *P. monticola,* and sugar pine. Extensive mortality of ponderosa pine at the turn of the 19th century in the Black Hills of South Dakota attracted public attention that led to the establishment of the Federal Forest Insect Research Program in the United States. The effects of mountain pine beetle are greatest in climax lodgepole pine forests where lodgepole pine is self-perpetuating or in even-aged stands where shade-tolerant species are not abundant enough to replace lodgepole pine. These even-aged forests are usually created by a stand-replacement fire. Mountain pine beetle infestations kill the largest trees and leave behind small-diameter, low-vigor, and mistletoe-infested trees. Young seedlings that have survived in the understory because of their shade tolerance are then released. New seedlings appear in new openings where trees have died. The resultant forest following mountain pine beetle outbreak is uneven aged and multistoried.

The life histories of tree-killing *Dendroctonus* (meaning "tree killers") species are generally similar. The female emerges from overwintering sites beneath the bark of trees killed in the previous year. She penetrates the outer bark and begins feeding in the phloem. She releases a pheromone component, *trans*-verbenol, and a host monoterpene hydrocarbon, myrcene (in lodgepole pine) or α-pinene (in western white pine). The male joins the female gallery and releases *exo*-brevicomin, another pheromone component. This mixture of compounds is highly attractive to other mountain pine beetles and a massive aggregation of beetles occurs on the tree. The attracted beetles introduce a pathogenic bluestain fungus, *Ophiostoma clavigerum,* which, together with tunneling females, causes the death of the tree. The growth of this bluestain fungus in the sapwood interrupts water conduction (a vascular wilt) to the crown. As the tree begins to die, females excavate egg galleries that are positioned vertically in the trunk, 30 to 90 cm in length. The egg gallery etches the sapwood superficially and may be quite sinuous, as with the western pine beetle, *D. brevicomis.* Eggs are laid in niches cut in the phloem on both sides of the gallery. Larvae hatch from the eggs and they excavate lateral galleries that increase in size as the larvae grow. A pupal cell is excavated at the end of the gallery, and the larvae transform to pupae and then to adults in these cells. Both larvae and adults are known to overwinter. Depending on the latitude and altitude one generation may take 2 years to complete or there may be as many as two generations and a partial third in 1 year.

Cryptococcus fagisuga (Beech Scale); Hemiptera: Eriococcidae (Eriococcids)   The beech scale is a native European species that was introduced into Halifax, Nova Scotia, about 1890. Since then it has slowly spread eastward to Toronto, Ontario, and Ohio and south to West Virginia and western Virginia. Infestations are found in North Carolina, Tennessee, and Michigan. The beech scale in combination with the fungus *Nectria coccinea* var. *faginata* causes beech bark disease, which has killed a large number of native and ornamental beeches in Europe and the northeastern United States. European beech, *Fagus sylvatica,* and American beech, *F. grandifolia,* are severely impacted by this scale and its associated fungus, and it also infests other native as well as ornamental beeches from the Orient and Europe.

In late spring and early summer, pale-yellow females deposit up to 50 yellow-colored eggs in groups of 5 to 8 that are coated with a white-colored wax-like substance. Only wingless females are known for this species. The newly hatched nymphs or "crawlers" search out a location on the bark to settle and insert their tubular mouthparts. During this crawler stage they can be dispersed by wind, often over 100 meters. This sessile stage overwinters. After egg-laying the female dies. There is one generation produced each year. At high densities, the bark on the trunk and lower portion of the branches is completely white with scales. Under these conditions, the bark is killed and forms pits and ultimately ruptures, allowing entry of *N. coccinea* var. *faginata*. This fungus, along with the native species, *N. galligena* and *N. ochroleuca,* kills the cambium and sapwood in these areas, which results in interruption of water and food transport. The trunk often turns red with the fruit bodies produced by this fungus. Tree mortality often ensues after a few years. Extended periods of drought increase the rate of mortality caused by the interaction of this scale and fungus with their host. Cold temperatures (–38°C) often kill the scales above the snow pack. In Europe, the ambrosia beetle, *Trypodendron domesticum,* attacks beeches weakened by this scale and its associated fungus. Several decay-causing fungi are also associated with dying trees.

Sirex noctilio (Steely-Blue Wood Wasp); Hymenoptera: Siricidae (Horntails/Wood Wasps)   *S. noctilio* is indigenous to Europe where it infests pines weakened by fire, insects, and diseases. In its native habitat it seldom causes tree mortality. However, this species was introduced into New Zealand in the late 19th century, where it became a serious agent of mortality in planted forests of Monterey pine. It has since been introduced into Monterey pine plantations in Australia and South Africa. Females have a long ovipositor that extends straight back from the anus and is often mistaken for a stinger. The female drills a hole about 12 mm deep through the bark and into the sapwood. As many as four side holes are drilled from the primary entrance hole. One hole is filled with mucus from the female's accessory glands, as well as spores of the symbiotic fungus, *Amylosterium areolatus,* that reside in a specialized gland (mycangium) at the base of the

FIGURE 3 Female horntail, *Urocerus gigas* (Hymenoptera: Siricidae), ovipositing in the end grain of a European larch log. [Photograph by Andrew J. Storer.]

ovipositor. Eggs are deposited in the other side tunnels. Adults tend to aggregate on weakened trees for oviposition. The mucus secretion has toxic properties that cause interruption of water conduction in the sapwood (i.e., a vascular wilt). Needles are subsequently killed and they fall from the tree. Young larvae feed on the fungus growing in the egg chamber. Older larvae tunnel deeply into the sapwood and their finely divided frass (mostly woodchips) is tightly packed into their galleries. They pupate near the surface and excavate a round hole in the bark through which they emerge.

Horntails occur in both hardwood and softwood species and they are found in forests throughout the northern hemisphere (Fig. 3). The life cycle can be completed in 1 year, but in some cases development is delayed for several years.

A major biological control program for *Sirex noctilio* was undertaken and many species of hymenopteran parasitoids were collected from Europe and North America and introduced into New Zealand and Australia. Two species of nematodes, *Deladenus siricidicola* and *D. wilsoni,* were introduced into Australia where they have caused a significant reduction in tree mortality. These nematodes infect the reproductive system and prevent ovarian development.

Reticulitermes flavipes (Eastern Subterranean Termite); Isoptera: Rhinotermitidae   Subterranean termites are social insects that live in colonies in the soil. They feed on a variety of cellulosic materials as their principal energy source. Termites and fungi are the most important organisms that can digest plant cell walls and thus reduce cellulose from a complex polysaccharide to simple sugars. Symbiotic fungi and protozoa live in the hindguts of termites and possess the enzymes needed to digest cellulose. Thus termites play an important role in recycling nutrients in a forest. Subterranean termites coexist in soils with many other microorganisms, especially fungi. Some of these fungi have been shown to produce chemicals that are the same as the trail-following pheromones produced by the termites, as well as chemicals that act as feeding stimulants and deterrents to termites.

The eastern subterranean termite is distributed throughout the eastern and mid-western part of the United States where it feeds on most species of wood found in forested habitats. Related species of subterranean termites are found in western North America and in Europe and Asia. Subterranean termite colonies contain seven castes: larvae or immatures, workers, soldiers, nymphs, winged (alate) primary reproductives, wingless (dealate) primary reproductives, and supplementary reproductives. Workers feed all the castes. Supplementary reproductives can occur in colonies with primary reproductives. They take over the oviposition role if the primary "queen" reproductive dies. Soldiers have large hardened (sclerotized) heads and mandibles, which they use to defend the colony from invaders, especially ants.

With the onset of warm rains in spring or fall (western United States), large numbers of winged females and males (termed swarmers) emerge from earthen tubes extending out of the ground or from the surface of logs, stumps, and wooden structures. These adults are weak fliers and soon drop to the ground where they lose their wings. Copulating adults paired end-to-end can be seen running about. The mated female seeks a damp piece of wood and excavates a chamber where she lays her first eggs. The young hatch into a worker caste and when large enough, they begin foraging for food to feed the queen and other castes as they are produced.

Subterranean termites are a critical component of forest habitats because of their important role in recycling wood. However, they are also the most destructive pests of human habitations made from wood. They construct earthen tubes from the ground into wooden structures and thus can consume wood in dry habitats long distances from the colony in the ground. They often escape notice because they travel in hollowed out timbers and in tubes constructed in wall cavities. Their galleries parallel the grain of the timbers and are excavated first in the early wood (springwood) and later in the late wood (summerwood). This damage to structures can be extensive before it is discovered. A large pest control industry has developed to protect wooden structures from termites and other wood-destroying insects and fungi.

Synanthedon sequoiae **(Sequoia Pitch Moth); Lepidoptera: Sesiidae (Clearwing Moths)**   The sequoia pitch moth is attracted to the resin that flows from wounded pine trees. Larvae hatch from eggs deposited individually on the bark of limbs and the trunk and tunnel into the phloem–cambial layer beneath the outer bark. As they feed on the phloem, they excavate a chamber that fills with resin and frass. This material is deposited on the outer bark surface over the chamber and forms a white resinous mass that is interspersed with reddish brown fecal material. As this resin mass ages, the surface turns reddish brown. Resin streaming is occasionally observed from these masses. Larvae pupate beneath the resinous mass, but with the pupal integument extending through the mass. This enables the moth to emerge from the mass without contacting the resin. These resinous masses often exhibit two or three pupal cases, indicating that these areas are often recolonized by later generations of this species. Oviposition occurs in summer. The life cycle is completed in 1 or 2 years.

Infestations of this species rarely cause tree mortality directly. However, heavy infestations may result in attraction of tree-killing bark beetles. This moth infests many native species of pines as well as European species such as Scots pine, Austrian pine, *Pinus nigra*, maritime pine, *P. pinaster*, and Italian stone pine, *P. pinea*. Pines growing in urban environments can be heavily infested around wounds caused by pruning and by vehicles. The sequoia pitch moth is distributed throughout western North America, including California, Oregon, Washington, Idaho, Montana, and British Columbia. Coast redwood, *Sequoia sempervirens*, is not a known host for this moth. Similar pitch masses are produced by a related moth, *S. pini*, which infests many pine and spruce species in eastern North America and in the midwestern states.

Insects That Feed on Roots

Insects that feed on the roots of trees may be sap feeders or may enter the host tissue and exploit the phloem, cambium and/or xylem. Representatives are found in the Coleoptera, Hemiptera, and Hymenoptera. The underground portion of the tree represents a protected environment for root-feeding insects.

H. abietis **(Pine Weevil); Coleoptera: Curculionidae (Weevils)**   The pine weevil is native to the forests of northern Europe and Asia where it colonizes roots of weakened and recently dead conifers. Larvae emerge from eggs laid in the bark and tunnel through the phloem and score the sapwood surface. They pupate in cells excavated in the outer sapwood. Its abundance has been increased dramatically by thinning and clear-cutting of large forested areas and replanting with pines. Under these conditions, the weevils breed in the stumps and emerge in the next season to feed on seedlings planted after logging. This species has the reputation of being one of only a few species of which one individual can kill one or more healthy trees, although they are small trees. Grasshoppers (Acrididae) and scarabs (Scarabaeidae) also have this capability. Adults feed on the tender young bark of most coniferous species. They are present from early spring through early fall. Control of this species is necessary in new conifer plantations. One generation is produced in 1 to 3 years, depending on latitude. The pales weevil, *H. pales*, has a comparable biology in the eastern United States. These insects are considered to be among the most destructive pests of conifer plantations in both Europe and North America.

Pachypappa termulae **(Spruce Root Aphid); Homoptera: Pemphigidae (Pemphigids)**   The spruce root aphid is cream colored and feeds in colonies on the fine roots of standing trees. These colonies are covered with waxy wool. Spruce trees are most commonly infested, though other conifers may also be affected. As with all aphids, this is a sap-feeding species that has sucking mouthparts. Aboveground symptoms are usually absent except in nursery stock, where an overall decline of plant health may be observed. Often these insects are noticed only during transplantation of nursery stock. This species can be found year round in parts of Europe and North America and is thought to be associated with dry conditions.

Camponotus modoc **(Carpenter ant); Hymenoptera: Formicidae (Ants)**   *C. modoc* excavates galleries in fire-scarred and rotted standing and fallen trees of many coniferous species growing in western North America, including pines, true firs, Douglas-fir, western redcedar, *Thuja plicata*, and giant sequoia. In a study of uprooting and breakage of over mature giant sequoia in the Sierra Nevada of California, carpenter ants and decay fungi were often associated with tree failure. Most of the roots of recently uprooted trees showed evidence of advanced decay. Carpenter ant galleries were observed in the few functional roots that were present. These roots showed evidence of early-to-moderate stage decay and carpenter ants were observed tunneling in these areas. This excavation activity is probably associated with the establishment of subsidiary colonies. These ants are found tending aphids, e.g., *Cinara* spp., in colonies on understory white fir and sugar pine trees. *C. modoc* was also observed excavating cavities in the bark of the root collar of young white firs. These cavities were later colonized by *Cinara* spp. that were tended by these ants. These shade-tolerant conifers have greatly increased in abundance because of fire exclusion in the old-growth giant sequoia groves. Thus carpenter ant abundance may have also increased as a result of increased aphid colonies on these understory trees.

Many insect families are found almost exclusively in forest habitats and have not been mentioned above because of space limitations. These include, for example, the wood-boring beetles in the families Buprestidae, Cerambycidae, Platypodidae, Micromalthidae, Anobiidae, Bostrichidae, and Lyctidae. In the Lepidoptera, the family Cossidae is almost entirely found infesting wood. Similarly, a very large number of species of moths, wasps, aphids, and leafhoppers feed on the foliage of trees.

FOREST PEST MANAGEMENT

Insect populations may reach levels in both natural and human-made forests where they are considered pests. High populations of cone and seed insects in a conifer seed orchard, mortality of seedlings resulting from feeding by weevils, defoliation of mature trees, and bark beetle outbreaks are all examples for which a management activity aimed at reducing the density of the insect population may be appropriate. These management activities all have a financial cost associated with them, which needs to be weighed against the anticipated benefit from the activity. With a few notable exceptions, the use of traditional insecticides to control forest insects is rarely practical, because of cost, or desirable because of impacts on nontarget organisms. Exceptions may be high-value urban trees and

trees in seed orchards. In these situations, the area being targeted with the pesticide is limited. Biological pesticides are, however, very useful in limiting the impacts of defoliating Lepidoptera. Extracts from the bacterium *Bacillus thuringiensis* (Bt) are used in aerial applications where defoliators threaten to kill large tracts of trees. Extensive areas in North America are treated with Bt to reduce the damage from the gypsy moth and to slow its spread. Insect outbreaks may be a result of fire suppression that creates an overmature forest that is susceptible to these insects. Long-term management strategies should therefore address this issue rather than use short-term solutions such as insecticides.

In populations of native and introduced forest insects, especially defoliators, it has been possible to identify viruses that are specific to their host insect. Rearing and release of these viruses offer promise for pest control in some situations, especially for defoliating Lepidoptera and Hymenoptera. The nuclear polyhedrosis virus (NPV) of the European pine sawfly, *Neodiprion sertifer,* has been applied to thousands of hectares of forests in Europe over the past 35 years. An NPV that infects gypsy moth has also been used in efforts to slow the spread of this species in North America. The accidental introduction of an NPV in the 1930s, along with a parasite of the European spruce sawfly, *Diprion hercyniae,* into eastern Canada has reduced this species to very low densities.

Biological control has achieved considerable success in managing forest insect populations. It is especially useful when addressing problems caused by introduced insects. For example, as the European spruce beetle spread westward across Europe during the 20th century, outbreaks were most severe on the edge of its range, in part because of the lack of natural enemies in newly invaded areas. Mass releases of the predatory beetle *Rhizophagus grandis* had some success in limiting the impact of this species in many European countries. Similarly, in urban forests where many tree species are introduced, management of exotic pests is often achieved by introduction of predatory or parasitic insects. In California, eucalyptus are widely planted in urban landscapes, and in the latter part of the 20th century, many new pests were introduced. Prompt identification and monitoring of pest populations coupled with introduction of appropriate biological control agents have successfully reduced the adverse impacts of many of these introduced insects, especially the eucalyptus longhorned beetle, *Phorocantha semipunctata.*

Much scientific research has focused on the pheromone systems of forest insects, especially defoliators and bark beetles. Pheromones have been incorporated into management plans aimed at reducing the impacts of the target pest species. For example, mountain pine beetle pheromones are used to bait living lodgepole pines that are to be logged following infestation. This treatment attracts beetles into plots that will be harvested, and the beetles are removed with the logs. The treatment is aimed at reducing mortality in the unharvested areas. It prevents trees from dying at more scattered locations in these unharvested areas where most trees would not be harvested because of cost and accessibility. Other techniques for reducing the impacts of target pest species can be achieved in a number of ways. Pheromones can be used to monitor insect populations as a guide for when to apply pesticides or embark on an alternative management strategy. They are also used to detect invasive pest species in new areas. Throughout the western states of North America, pheromone traps are placed to detect gypsy moth so that early detection can be met with rapid response eradication programs. Similar use of pheromone traps and traps baited with other behaviorally active compounds occur at ports to detect the arrival of exotic organisms.

Attempts have been made to use pheromones to mass trap insects and thereby reduce their density to nonpestiferous levels. Because insect populations naturally decline at some point following an outbreak due to a range of factors, it is often difficult to determine whether such mass trapping efforts were in part responsible for a return of the insect population to endemic levels. In Scandinavia, mass trapping of the eight-toothed spruce bark beetle occurred during 1979 and 1980, and billions of insects were caught. The subsequent drop in population size of this insect may have been in part a result of this intensive mass trapping effort. Mass trapping may be useful locally where a small-scale disturbance may have resulted in elevated activity of a particular insect. This situation may occur when management activities have produced a large amount of host material. Pheromone traps may be deployed in this situation to reduce populations. Trap logs that have become infested with bark beetles are removed from the forest in an attempt to lower population levels. Pheromones may be used to bait the logs or trees to enhance the effectiveness of this technique.

Pheromones have also been used in larger doses to confuse or disrupt their natural function. Male moths following pheromone plumes to locate females for mating are not able to find the female if there is a large concentration of the pheromone present, such that the plume is effectively hidden. Western pine shoot borer, *Eucosma sonomana,* pheromone release of 10 to 20 g per hectare has been shown to reduce damage by over 65% in ponderosa pine plantations in the western United States.

Many forest management practices are available for use to reduce the impacts of insect populations. Removal of susceptible trees, thinning of the stand to increase tree vigor, prescribed burning to eliminate susceptible host material, and enhancement of natural enemy populations are all tools available for the management of forest insect populations. The thinning of lodgepole pine stands reduces tree mortality from mountain pine beetle. Prescribed fire following harvesting operations in ponderosa pine forests reduces populations of the California five-spined ips, *Ips paraconfusus,* that breeds in logging debris. This practice lowers the probability of these populations emerging and killing living trees in the area.

THE FUTURE

Insects perform essential roles in forests as part of these complex ecosystems. However, when transplanted from one forest habitat to another, or when the natural forest habitat is disturbed, their role in the ecology of the forest may change. For example, the eastern five-spined ips, *I. grandicollis,* generally infests stressed or diseased trees in its native habitat in the southeastern United States and is not considered a primary agent of tree mortality. Its introduction into Australia in the mid-20th century has resulted in significant tree mortality. This beetle infests many pine species throughout the eastern United States but is now infesting a Californian pine, Monterey pine, in Australia. Efforts have been made to find biological control agents in North America that are effective in Australia. Increased global transport of forest products undoubtedly will result in introductions of both pestiferous and potentially pestiferous insects into different parts of the world. Ongoing efforts to reduce this occurrence are essential if we are to maintain forest habitats that meet environmental and economic goals. Furthermore, research that addresses environmentally sound methods for the integrated management of forest pests is essential if humans are to minimize their indirect impacts on forest habitats resulting from the introduction of new insect species. Some insect outbreaks are important natural processes in forests, whereas others are the product of human activities. Recognizing the causes of outbreaks, detecting signs of imminent outbreaks, and having appropriate management options available help to ensure the viability of forest habitats and the environmental and economic benefits they provide.

See Also the Following Articles

Biological Control • Gypsy Moth • Phytophagous Insects • Soil Habitats

Further Reading

Bevan, D. (1987). "Forest Insects: A Guide to Insects Feeding on Trees in Britain." Forestry Commission, London. [Handbook 1]

Coulson, R. N., and Witter, J. A. (1984). "Forest Entomology, Ecology and Management." Wiley–Interscience, New York.

Dajaz, R. (2000). "Insects and Forests." Lavoisier, Paris.

Drooz, A. T. (ed.) (1985). "Insects of Eastern Forests." U.S. Department of Agriculture, Forest Service, Miscellaneous Publication No. 1426.

Furniss, R. L., and Carolin, V. M. (1977). "Western Forest Insects." U.S. Department of Agriculture, Forest Service, Miscellaneous Publication No. 1339.

Johnson, W. T., and Lyon, H. H. (1988). "Insects That Feed on Trees and Shrubs," 2nd edition. Cornell University Press, Ithaca, NY.

McCullough, D. G., Werner, R. A., and Neumann, D. (1998). Fire and insects in northern and boreal forest ecosystems of North America. *Annu. Rev. Entomol.* **43,** 107–127.

Paine, T. D., Raffa, K. F., and Harrington, T. C. (1997) Interactions among scolytid bark beetles, their associated fungi, and live host conifers. *Annu. Rev. Entomol.* **42,** 179–206.

Speight, M. R., and Wainhouse, D. (1989). "Ecology and Management of Forest Insects." Clarendon Press, Oxford.

Fossil Record

David Grimaldi

American Museum of Natural History, New York

Insects are the most evolutionarily successful group of organisms in the 4-billion-year history of life on earth, with perhaps 5 million species alive today and untold millions of extinct species. Although fossils of insects are not as abundant as has been found for some other types of organisms, the insect fossil record extends back for 400 million years, making them among the oldest terrestrial animals known, and the fossils contribute unique insight into the evolutionary history of insects. Particularly significant periods in the evolution of insects are the Paleozoic, Triassic, and Cretaceous. Key features that gave rise to their spectacular success, notably flight and complete metamorphosis, originated at least 300 and 250 million years ago (mya), respectively.

FOSSILS

Fossils are generally, but not necessarily, extinct species whose remains have been preserved for thousands to millions of years. The remains are most commonly mineralized replacements of original tissues (Figs. 1 and 2); in rare situations portions of the original organism are preserved (Figs. 2b–2e, 2g, and 5d–5f). Remains of apparently existing (or extant) species that are thousands to several million years old are sometimes called *subfossils*. Earth's fossils are dominated by organisms from marine continental shelves, where deep sediments efficiently preserved durable calcified parts such as shells and skeletons. Terrestrial life is less well known in the fossil record and largely is preserved as bones, leaves, and pollen in freshwater sediments. Well-preserved insects, by comparison, are rare; their occurrence depends on conditions under which

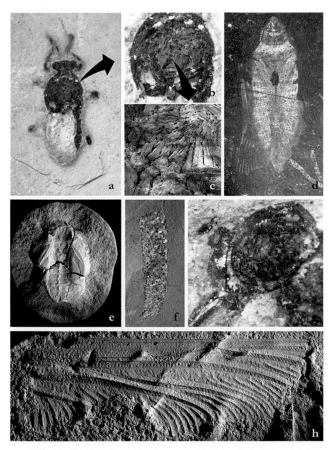

FIGURE 1 Various kinds of fossil insects, modes of fossilization, and degree of preservation. Different scales. (a–c) Iron hydroxide concretion of a heteropteran in Cretaceous limestone from Brazil, showing preservation of thoracic muscles [b, light micrograph; c, scanning electron micrograph (SEM)]. (d) Silvery carbon film of a belostomatid on fine-grained, Triassic shale (Virginia). (e) Nymph of †*Herdina* (in this article, † signifies an extinct group) in ironstone concretion from the Carboniferous of Mazon Creek, Illinois. (f) Trichopteran case of sand pebbles in volcanic shale from Florissant, Colorado (late Eocene/early Oligocene). (g) Part of the head of tabanid fly from Florissant, showing the eye facets. (h) Portion of the wing of †*Typus* ("Protodonata"), from the Permian of Elmo, Kansas, with the wing fluting preserved.

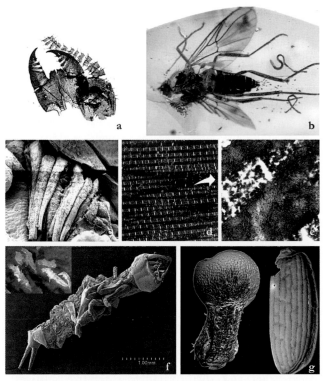

FIGURE 2 Exceptional preservation of fossil terrestrial arthropods. (a) Centipede from Upper Devonian (New York). (Courtesy of William Shear, Hampton–Sydney College.) (b) *Mycetobia* woodgnat in Miocene amber (Dominican Republic), with parasitic nematodes bursting from abdomen. (c–e) Flight muscles of meliponine bee in Dominican amber, showing ultrastructural preservation of myofibrils (d) and even the fingerprint-like mitochondria (e). (f) Silicified replicas of early instar dytiscoid beetle, from Miocene of California (inset, photomicrograph; SEM is larger). (g) Cuticular remains of extant beetle species from the Wisconsin stage (ca. 10,000–80,000 years ago) of Alaska (left, weevil head; right, carabid elytron). (Courtesy of Scott Elias, University of Colorado.)

the sediments were fine grained, anoxic (i.e., lacking oxygen), and deposited rapidly but without significant disturbance. Even amber is usually deposited in lacustrine or swampy sediments; it would otherwise completely decompose from oxidation and other processes. Insects preserved in such sediments were aquatic or semiaquatic and died *in situ* (autochthonous), or their bodies were transported via winds or water from surrounding habitats (allochthonous).

SIGNIFICANCE OF FOSSILS

There are at least five reasons why fossils are uniquely significant for understanding the evolutionary history of organisms:

1. Fossils provide the only direct record of extinct lineages, such as giant dragonfly-like forms from the Carboniferous and Permian (e.g., †Meganeuridae) (in this article, † signifies an extinct group).

2. Fossils reveal patterns and timing of extinctions and radiations. The mass extinction at the end of the Permian, for example, was the most cataclysmic event in the history of life and may have caused the extirpation of the †Paleodictyopteroidea (known almost exclusively from the Carboniferous and Permian). The extinctions at the Cretaceous/Tertiary boundary that extinguished the remaining dinosaurs, ammonites, and other groups, appear to have had little impact on families of insects. Although insects have been affected by some mass extinction events, major lineages of insects appear particularly resistant to extinction.

3. Fossils provide the only direct information on the ages of lineages. Because there is never assurance that a fossil is the earliest, original occurrence of a taxon, the age of the earliest fossil is the minimum age of a taxon. This information, in conjunction with the phylogenetic positions of all fossils in a group, can be used to estimate actual ages and significant gaps in the fossil record. Such information is further useful for calibrating and then estimating rates of change, such as rates of genetic change among living species and dates of divergence.

4. Fossils may assist in reconstructing the phylogeny of a group. While fossils are widely acknowledged to possess combinations of characters unique from those of living species, their significance in reconstructing phylogenetic relationships is controversial. Nonetheless, fossils also provide the only direct evidence for the evolutionary sequence of character change. For example, in the fossil record of the Blattaria (cockroaches and their primitive roach-like relatives), tegminous forewings and a large, discoid pronotum appear well before the loss of an external ovipositor—a sequence not revealed by the study of living species alone.

5. Fossils can provide evidence that a taxon is old enough for its distribution to have been affected, for example, by Cretaceous continental drift. Indeed, many families of insects extend to the Cretaceous or even earlier, whereas some large groups are apparently too young to have been affected by continental drift, such as the ditrysoan Lepidoptera and the schizophoran Diptera. Often, too, a fossil is found outside the present-day range of its group, indicating formerly widespead distribution. A famous example is the occurrence of Glossinidae (tsetse flies) in the Cenozoic of North America and Europe.

PRESERVATION OF FOSSIL INSECTS

The small size and external cuticle of insects are largely responsible for the many modes of fossilization, which are much more varied than for vertebrates and plants. Insect fossils are most commonly encountered as impressions or compressions in sediments (Figs. 1d, 1g, 1h, 5a and 5b), generally as disarticulated cuticle and particularly as wings because these are especially resistant to decay. Because wing venation has many systematically significant features, isolated wings often can be identified at least to family level. Generally, remains in sediments are highly compressed, but can still reveal microscopic structures such as flagellomeres, tarsomeres, microtrichia, wing scales, and even color patterns. Some are preserved as concretions, which are three-dimensional permineralized replicas of the original animal (Figs. 1a–c, 1e and 2f).

The finest preservation of insects is in amber (Figs. 2b–2e and 5d–5f). These formed when the resin was originally viscous and sticky, and small organisms became mired and then engulfed by the flows; they were embalmed so thoroughly as to preserve parasites, soft internal organs and tissues, and even organelles of cells (Figs. 2b–2e). The putatively most ancient DNA in the geological record is reported from insects preserved in amber, but authenticity of the DNA is disputed by those who unsuccessfully attempted to replicate these results.

Exceptional preservation is also seen in some insects preserved free in sediments. Terrestrial arthropods in several Devonian deposits of eastern North America are preserved as original cuticle, with even microscopic sensilla and setae preserved (Fig. 2a). In several Miocene deposits from California, insects are preserved in nodules as perfect three-dimensional silicified replicas (Fig. 2f). Similar relief and microscopic fidelity are found in carbonized remains in Cretaceous clays, rendered by ancient forest fires that charcoalified small organisms buried in leaf litter. Traces of insects have also been preserved as tracks, burrows, nests, galleries, feeding damage, and larval cases (e.g., Fig. 1f). Lepidoptera, for example, are very rarely preserved in rocks, probably because they are so soft-bodied, but larval mines characteristic of various microlepidopterans occur in some fossil leaves.

The various modes of fossilization each have their biases. Entrapment in amber is biased against larger insects that could extract themselves from the resin and against insects that live in open, nonforested habitats. Also, the earliest insects preserved in amber are from only the Lower Cretaceous, some 275 million years after the earliest known hexapods appeared.

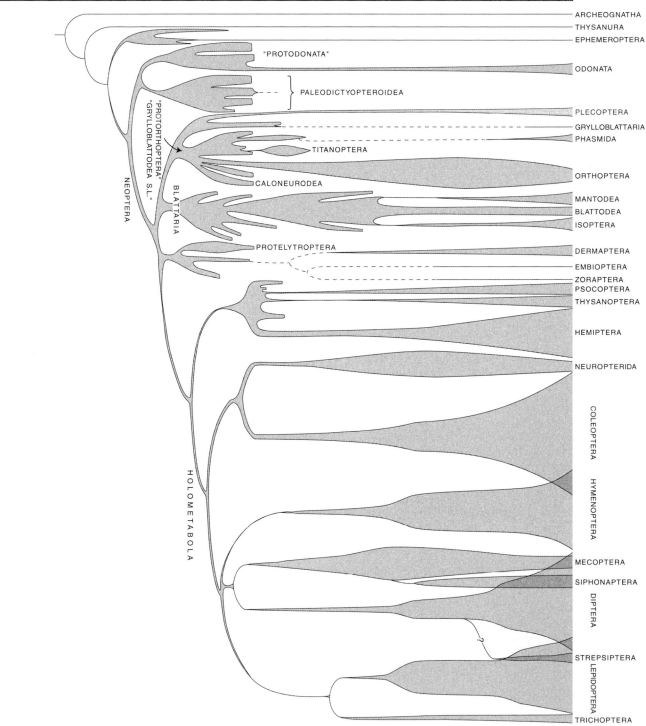

FIGURE 3 Living and extinct orders of insects, their possible relationships, and chronology. Width of lineages is a rough approximation of diversity. Some groups with a meager or nonexistent fossil record (i.e., Phthiraptera, related to Psocoptera) are not included.

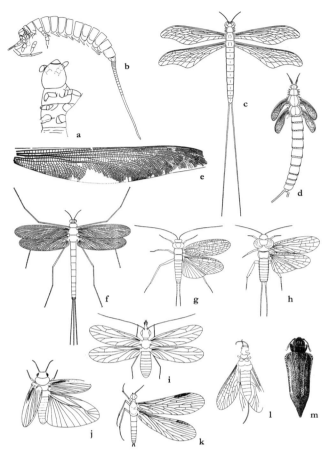

FIGURE 4 Representative Paleozoic hexapods. (a) Devonian. (c and d) Carboniferous. All others Permian. (a) †*Rhyniella* (Collembola). (b) †*Dasyleptis* (Archaeognatha). (c, d) †*Mischoptera* adult (c) and nymph (d) (†Megasecoptera). (e) †*Meganeuropsis* ("Protodonata"). (f) †*Protereisma* (near Ephemeroptera). (g) †*Liomopterum* (Paraplecoptera). (h) †*Lemmatophora* (near Plecoptera). (i) †*Dichentomum* (near Psocoptera). (j) †*Protelytron* (†Protelytroptera). (k) †*Permopanorpa* (Mecoptera). (l) †*Sojanoraphidia* (Raphidioptera). (m) †*Sylvacoleus* (Coleoptera). Not to the same scale. (Reproduced, with permission of the publisher, the Geological Society of America, Boulder, Colorado, from Carpenter, 1992.)

Preservation in rock is biased against smaller insects, and microscopic features are usually not visible against the grain of the matrix. Collectively, though, the fossil record of insects is actually much better than most paleontologists realize.

SUMMARY OF THE INSECT FOSSIL RECORD

Figure 3 summarizes the chronology, approximate diversity, and possible relationships of major groups of living and extinct insects.

Paleozoic (570–245 Mya)

The earliest remains of terrestrial animals are arthropods from the Silurian, including primitive scorpions, millipedes, and †Trigonotarbida (extinct, primitive arachnids). Two major hypotheses on the origins of the hexapods (including insects) are that they are most closely related to either the myriapods (centipedes and/or millipedes) (all comprising the Tracheata, or Atelocerata) or the Crustacea. Crustacea may actually have the oldest fossil record of all animals (formerly held by the trilobites) because some Precambrian fossils have recently been reinterpreted as crustaceans. If hexapods are closely related to crustaceans, it is most likely to be a group within Crustacea, and the earliest evidence of this stem group will probably be found in the Silurian.

THE DEVONIAN (408–362 MYA) The Devonian was a period when the continents were largely inundated and joined into the large supercontinent Pangaea. Lycopods, lycopsids, and horsetails (Equisitoidea) were common terrestrial plants; primitive groups of fishes radiated, and amphibians first appeared. The earliest record of hexapods is from the Rhynie chert of Scotland, ca. 400 mya, and is the collembolan †*Rhyniella praecursor* (Fig. 4a), although a pair of mandibles in this chert is suggestive of dicondylic insect mandibles. The Rhynie chert also contains mites, trigonotarbids, and a primitive spider.

Approximately 10 million years younger is an undescribed archaeognathan (=Microcoryphia) from the Gaspé Peninsula in eastern Canada. Apterygotes, each with a single, long caudal filament ("Monura" = Archaeognatha), occur in the Carboniferous and Devonian (Fig. 4b). Unidentified insect remains from 378 mya are known from Gilboa, New York, along with centipedes (Chilopoda) (Fig. 2a), true spiders (Araneae), trigonotarbids, oribatid mites, and pseudoscorpions. The Gaspé and Gilboa remains are original cuticle.

THE CARBONIFEROUS (362–290 MYA) The Carboniferous period is famous for the wet, warm climates and extensive swamps of mosses, ferns, seed ferns, horsetails, and calamites. Remains of insects are scattered throughout Carboniferous coal deposits (particularly blattarian wings); two particularly important deposits are Mazon Creek, Illinois, and Commentry, France. The earliest pterygotes appear in the Carboniferous, including the Blattaria, †Caloneurodea, primitive stem-group ephemeropterans (Fig. 4f), Orthoptera, †Paleodictyopteroidea (Figs. 4c and 4d), †"Protodonata" (Fig. 4e), and †"Protorthoptera"; the latter two are paraphyletic assemblages of primitive pterygotes.

Hypotheses on the evolution of insect wings include their use originally as gills or gill covers, or for mating displays, but early outgrowths of the insect pleuron most plausibly served in gliding. Feeding damage on plants is also recorded first in the Carboniferous, in the form of punctures and deep holes probably made by the long, beaked mouthparts of paleodictyopteroid insects. Thus, insects have been evolving in close association with plants for at least 350 million years, which is longer than any other group of terrestrial animals. Arborescent plants appear in the Upper Devonian, and as Carboniferous insects increasingly dwelled in them to feed, gliding probably became so adaptive for escape and dispersal that flapping wings and powered flight evolved rather suddenly.

Putative Holometabola are recorded from the Carboniferous. One is a larva from Mazon Creek, *Srokalarva berthei,* many features of which are inconsistent with extant holometabolan larvae, including segmented abdominal legs, ocelli, and possible compound eyes. Legs and body segments of *Srokalarva* are undifferentiated, as in myriapods. Some tree fern galls (ca. 300 mya) are attributed to the Holometabola on the basis of size of frass pellets in the galls. Some Paleozoic arthropods were considerably larger than living relatives, and Carboniferous gall-making mites are also known, and so it is possible that large mites caused these ancient galls. The earliest definitive Holometabola occur in the Permian.

Blattaria (Blattoptera)   This group consists of the Dictyoptera (mantises, termites, and ovipositorless roaches) plus the Paleozoic and Mesozoic "roachoids" that possessed an ovipositor. Very early, Carboniferous blattarians possessed a large, discoid pronotum and coriaceous forewings with a distinctive CuP vein. From the Carboniferous to the Cretaceous there was a gradual diminishment in the length of the ovipositor. The first true, ovipositorless roaches as well as probable oothecae appeared in the Triassic. Thus, the common view that modern roaches are exceptionally ancient is inaccurate.

†Caloneurodea and †Miomoptera   Both orders are known only from the Upper Carboniferous to Permian and, with Orthoptera and Blattaria, were among the earliest known Neoptera. These insects had homonomous wings with small anal lobes. In Caloneurodea the wings had many crossveins that formed numerous square cells. Wing structure and unsegmented cerci (the latter known from only few genera) suggest relationships with the orthopteroids. The wings of Miomoptera were shorter and broader and had very reduced venation. A close relationship of Miomoptera is plausibly with the Psocoptera, but probably with the Hemiptera. With a wing length of 3 to 5 mm, these are the smallest Paleozoic pterygotes known.

Orthoptera   This is an ancient Recent order extending from the Carboniferous, from which time even the distinctive synapomorphy (advanced defining feature) of saltatorial hind legs is preserved (†Oedischiidae). By the Triassic, stridulatory organs on wings evolved, and some extant families had first appeared (Haglidae, Gryllidae). †Elcanidae and related families (Permian to Cretaceous) are distinctive for the reduced forewing venation with numerous parallel M and Cu veins; they were probably the most diverse and abundant orthopterans in the Late Paleozoic and Early Mesozoic. Many extant families do not appear until the Cretaceous (e.g., Eumastacidae, Gryllotalpidae, Tridactylidae, Tetrigidae) or even the Cenozoic (Acrididae, Tettigoniidae, Gryllacrididae).

†Paleodictyopteroidea   This group includes the orders †Diaphanopterodea, †Megasecoptera, and †Paleodictyoptera, which had very long cerci, an ovipositor, and wings with little or no anal lobe (all primitive features). It is the only definitively monophyletic group from the Paleozoic, defined by a long, rigid beak with five stylets, as well as (primitively) a well-developed pair of possible maxillary palps. Nymphs were terrestrial, with large, free wing pads bearing rudimentary venation (Fig. 4d) and mouthparts similar to those of the adults. The group is among the most primitive pterygotes, and their distinctive parapronotal lobes (often with a rudimentary venation) probably are part of the ground plan of pterygote insects. The Paleodictyoptera are probably paraphyletic with respect to the other two orders, because they possess complete wing venation with an archedictyon. Diaphanopterodea folded wings over the abdomen at rest, an ability convergent with Neoptera. Megasecoptera had substantially reduced venation, often with a graded series of crossveins (Fig. 4c). Some had striking color patterns on the wings (e.g., Spilapteridae), and some were impressively large (e.g., *Homioptera,* approximately 40-cm wing span).

†"Protodonata"   This group is a paraphyletic assemblage of primitive pterygotes similar to true Odonata but lacking the derived features of modern, true Odonata, such as the nodus, pterostigma, and arculus. Some (i.e., Permian Meganeuridae) were the largest insects ever, with a 70-cm wing span (Fig. 4e). They existed from the Upper Carboniferous to the Permian and were clearly the dominant aerial predators for some 100 million years until flying reptiles appeared in the Triassic. Fossils with the venation of true Odonata did not appear until the Triassic.

†"Protorthoptera" (=†"Paraplecoptera") (Fig. 4g)   This is another paraphyletic assemblage of primitive pterygotes, with affinities not necessarily suggested by their name. Some refer to this group as the Grylloblattida *(sensu lato),* but there are no derived features that indicate a relationship of these extinct insects with this relict, extant order. Some had forewings that were coriaceous (i.e., sclerotized and leathery), but usually with an archedictyon. Unlike Orthoptera, most did not have hind wings with expanded anal lobes, and none are known to have folded their forewings roof-like over the abdomen (instead, folded flat and over each other), none had saltatorial hind legs nor a tarsomere number reduced to 3. Some had raptorial forelegs; †Geraridae had large pronotal spines.

PERMIAN (290–245 MYA)   The Permian was a relatively short but very important period, when extensive mountain ranges were formed, such as the Appalachians; this caused the interior climates of continents to become cooler and drier. Extensive glaciation of the Southern Hemisphere also occurred. Voltziales and glossopterid plants radiated. Insect orders from the Carboniferous extended into the Permian, and many even extensively diversified. Numerous new orders appeared in the Permian: the †Protelytroptera (Fig. 4j) and primitive relatives of the Plecoptera ("Paraplecoptera") (Fig. 4h), Psocoptera (Fig. 4i), Mecoptera (Fig. 4k), Raphidioptera (Fig. 4l), Neuroptera, and Coleoptera (Fig. 4m), the last four orders being the earliest definitive records of the Holometabola. The most significant and diverse Permian insect deposit is from Elmo, Kansas (ca. 260 mya); others are from New South Wales, Australia (ca. 240 mya) and central Eurasia (ca. 250 mya).

The Permian is biologically most notable for the mass extinction that marks the brief interval between the end of this period and the beginning of the Mesozoic. It is estimated that as much as 95% of all Permian life forms became extinct by the early Triassic, although this is based largely on the marine fossil record. It is difficult to estimate the impact of the Permo-Triassic extinction on insects because many of the Paleozoic orders are paraphyletic, and it is likely that particular lineages within each survived into the Mesozoic (Fig. 3). One definitive monophyletic group, the Paleodictyopterodea, may have become extinct at the end of the Permian, but a possible survivor of this group *(Thuringopteryx)* occurred in the Triassic. Otherwise, major groups of insects show few effects of the most cataclysmic extinction known.

Hemiptera   The earliest Permian records are the auchenorrhynchan-like wings of †Dunstaniidae, †Palaeontinidae, and †Prosbolidae, which also extend into the Mesozoic. True auchenorrhynchans from the Triassic are Cercopoidea, Cicadoidea, Cicadellidae, Cixiiidae, and Membracoidea. The earliest Sternorrhyncha are †Archescytinidae, †Pincombeidae, †Boreoscytidae, and †Protopsylidiidae (all from the Permian into the Mesozoic). The first diverse records of true aphids and coccoids occur in Cretaceous ambers; psylloids are older. Heteroptera are slightly younger than homopterans. Presumed Permian heteropterans are †*Actinoscytina* (†Progonocimidae) and †*Paraknightia* (†Paraknightiidae). Venation of the former is barely different from Auchenorrhyncha; the latter had expanded parapronotal lobes, a large ovipositor, and forewings with unusual venation (perhaps an aberrant roach). The earliest true Heteroptera are various predatory aquatic bugs (Nepomorpha) from the Triassic of Virginia (Fig. 1d). By the Jurassic, phytophagous pentatomorphs and cimicomorphs appeared, and modern families were widespread in the Cretaceous.

Neuropterida   This group includes the orders Raphidioptera and Neuroptera. Systematic position of the one Permian family of putative raphidiopteran (†Sojanoraphidiidae) (Fig. 4l) has been considered doubtful, although it had a long ovipositor distinctive to this order and a series of short costal crossveins (distinctive to Neuropterida); the wing venation, however, is quite primitive. Jurassic and Cretaceous raphidiopterans were diverse, belonging to the †Alloraphidiidae, †Baissopteridae, and †Mesoraphidiidae. Raphidioptera is relict today, with three genera in two families having a disjunct distribution in the Northern Hemisphere. The only Southern Hemisphere raphidiopterans are from the Lower Cretaceous of Brazil, indicating that the group was formerly more widespread.

Definitive Neuroptera also appear in the Permian (†Archeosmylidae, †Palaemerobiidae, †Permithonidae, and †Sialidopsidae). Triassic diversity is very poorly known (e.g., Psychopsidae), and by the Jurassic there appeared forms that are primitive relatives or members of the Chrysopidae, Coniopterygidae, Nymphidae, Osmylidae, and Polystoechotidae. An impressive Mesozoic family was the †Kalligrammatidae, with broad, patterned wings and a long, rigid proboscis. In the Cretaceous are the earliest records of the Ascalaphidae, Berothidae, Mantispidae, and Myrmeleontidae. Berothidae were particularly abundant and diverse in Cretaceous ambers. An unusual, dipterous mantispid (†*Mantispidiptera*) is known in Cretaceous amber from New Jersey.

Coleoptera   The earliest Permian forms (i.e., †Tshekardocoleidae) had long, pointed, coriaceous forewings with definitive venation and arrays of small cells and pits (Fig. 4m). Beetles similar to modern cupedoids were diverse in the Upper Permian and the Mesozoic. Cupedoidea today are a small, relict group. Fossil forms were probably all wood borers as larvae and adults, like modern relatives. By the Triassic, Adephaga appeared (Carabidae, Trachypachidae) as did some basal Polyphaga (definitive Staphylinidae). By the Jurassic the first Hydradephaga (†Coptoclavidae, Dytiscidae, Gyrinidae) appeared, as well as other living families (Elateridae, Hydrophilidae, Silphidae). The earliest records of many living families are from the Cretaceous and the lower Tertiary (especially Baltic amber). Interpretation of compression fossil beetles is greatly compromised by the lack of wing venation characters and restriction to the use of simple features such as elytral structure.

Plecoptera   Permian †Palaeoperlidae, †Perlopseidae, Eustheniidae, and Taeniopterygidae had venation consistent with this order. Several extant families are recorded from the Jurassic and Cretaceous. †Lemmatophoridae and †Liomopteridae (Figs. 4g and 4h) had expanded parapronotal lobes (sometimes with distinctive venation like the wings) and have been placed in "Protorthoptera." These families are probably related to the Plecoptera because they had nymphs with abdominal tracheal gills and so were probably aquatic.

†Protelytroptera   Found only in the Permian, this group is characterized by narrow, elytrous/tegminous forewings and with hind wings with an expansive anal lobe and radiating venation (the latter one reason they have been allied with Dermaptera) (Fig. 4j). The forewings of some species have very few veins; others have intricate reticulation. The †Umenocoleidae from the Cretaceous, formerly placed in this order, are actually blattarians.

Psocoptera   The earliest putative members of this order are the Permian †Psocidiidae, but these have cerci and five tarsomeres (vs lack of cerci and generally three tarsomeres in living species) (Fig. 4i). †Psocidiidae do possess the areola postica (short, terminal branch of vein CuA), although this feature also occurs in the Zoraptera and Embioptera. Psocoptera are poorly known in the Triassic, but are diverse and abundant in Cretaceous ambers, which include many living families (e.g., Prionoglariidae, Lepidopscocidae). †Lophioneuridae (Permian–Cretaceous) have traditionally been placed in this order, but the reduced wing venation (especially in Permian †*Zoropsocus*) and (where preserved) the

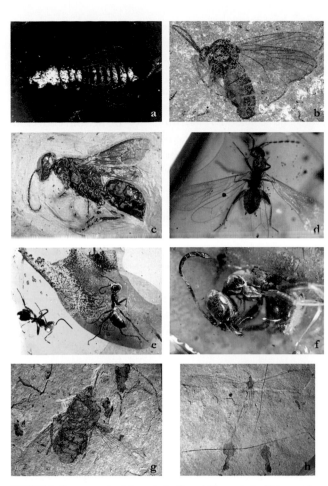

FIGURE 5 Representative Mesozoic (a–f) and Cenozoic (g, h) insects (not to the same scale). (a) Primitive, oldest known thysanopteran (Triassic, Virginia). (b) †Protorhyphidae (Diptera) (Jurassic, Kazakhstan). (c) Sphecidae (Cretaceous, Brazil). (d) Oldest known zorapteran (Cretaceous amber, Burma). (e and f) Oldest definitive ants [Formicidae; *Sphecomyrma* (e), *Kyromyrma* (f)] (Cretaceous amber, New Jersey). (g) Large, extinct tsetse fly (Diptera: Glossinidae) (U. Eocene/L.Oligocene, Colorado). (h) Spoon-winged lacewing (Neuroptera: Nemopteridae) (Colorado).

insects are from the Upper Triassic (especially Carnian, ca. 230 mya), including South Africa; Virginia; Queensland, Australia; Tadjikistan; and France. The first evidence of a diverse freshwater insect fauna appears during the Triassic, as did the oldest living families: Cercopidae, Cicadellidae, Cixiidae, and Membracidae (Auchenorrhyncha); Belostomatidae (Fig. 1d); Naucoridae (Heteroptera); Carabidae; Staphylinidae; Trachypachidae (Coleoptera); Anisopodidae, Chironomidae, and Tipulidae s.l. (Diptera); and Xyelidae (Hymenoptera). The first true Odonata, Heteroptera, Thysanoptera (Fig. 5a), Diptera, and Hymenoptera appeared during this time. The Triassic Hymenoptera consisted entirely of primitive symphytans; the Diptera consisted of diverse nematocerans. †Titanoptera were restricted to the Triassic and the size of some species fit their name (e.g., *Gigatitan,* ca. 33-cm wing span). Others were no larger than typical orthopterans, to which they are probably most closely related. They may have been Early Mesozoic analogues of the predatory mantises because their forelegs, when preserved, were spiny and apparently raptorial.

JURASSIC (208–145 MYA) One of the more significant events during this period was the origin of birds in the Upper Jurassic. Birds and (much later) bats are the only other skilled fliers and are perhaps the most important predators of modern insects. Numerous Jurassic deposits of insects occur in Europe and Asia: Grimmen and Solnhofen, Germany (the latter famous for the earliest birds, *Archaeopteryx*); Dorset, England; Issyk-Kul, Kirghizstan; and, the most diverse and productive site of all, Karatau, Kazakhstan. Jurassic deposits of insects are barely known in North America and in the Southern Hemisphere. The earliest aculeate (stinging) Hymenoptera appeared in the Upper Jurassic and were of the family †Bethylonymidae. Diptera in the Jurassic were diverse nematocerans (i.e., Fig. 5b), and the oldest definitive brachycerans evolved, dominated by Rhagionidae.

CRETACEOUS (145–65 MYA) In many respects, the insect fauna of the earlier part of the Cretaceous has more similarity to the Jurassic than to the later part of the Cretaceous. No doubt this is the result of the radiations of angiosperms 120–100 mya. Today, insects pollinate some 85% of the angiosperms, and so pollinating Cretaceous insects must have helped spawn the diversification of angiosperms; also, the diversification of phytophagous insects (and probably their parasitoids) was promoted by angiosperms. Radiations of some very speciose insect groups began during the Cretaceous, including the Scarabaeoidea, Cerambycidae, Chrysomelidae, and Curculionoidea (Coleoptera) and the Lepidoptera (the largest lineage of phytophagous organisms). The Yixian Formation of China (Lower Cretaceous, ca. 130 mya) has yielded important vertebrate fossils, as well as early angiosperms and nemestrinid flies with long proboscides; these are the earliest records of specialized insect pollination. Early sphecoid wasps (Fig. 5c), empidid and other flies, and

narrowed, projected mouthparts indicate relationship to the thrips (Thysanoptera). †Lophioneuridae lack extensive vein reduction, the marginal fringe, and the flagellum and tarsal structure distinctive to true thrips.

Mesozoic

The Mesozoic was an era when terrestrial ecosystems became modern, with the rise of cycads and ginkgos and the diversification of conifers in the Triassic and Jurassic. In the Lower Cretaceous there occurred what was probably the most profound evolutionary event: the explosive radiation of the angiosperms.

TRIASSIC (245–208 MYA) This was a period when arid and semiarid savannas developed and when the first mammals, dinosaurs, and pterosaurs also appeared. Most deposits with some beetles were probably very important, generalized pollinators of early angiosperms. Other important Cretaceous deposits are from Koonwarra, Victoria, Australia; Orapa, Botswana; Ceará, Brazil; Purbeck, Dorset, and Weald, England; Baissa, Russia; and Llérida, Spain. Isoptera appeared first in the Cretaceous, represented entirely by the primitive families Hodotermitidae, Termopsidae, Mastotermitidae, and Kalotermitidae. True, eusocial termites and ants existed for at least 50 million years before they became abundant in the Cenozoic. Eusocial wasps and bees did not first appear until the Upper Cretaceous.

Cretaceous continental drift fragmented Gondwanaland and Laurasia into the continents seen today. One result was fragmentation of ancestral ranges, the areas most famous for this being disjunct temperate regions of the southernmost regions of Africa, South America, Australia, and New Zealand (the "Austral Region"). This region harbors many primitive, relict insect groups.

Conifers were still diverse in the Cretaceous, which produced virtually all of the amber found during the Cretaceous. The oldest amber with insect inclusions is from the Lower Cretaceous of Lebanon, Japan, and England. Highly fossiliferous, somewhat younger amber deposits occur in northern Burma, western Canada, New Jersey, northern Spain, and the Taimyr Peninsula in northern Siberia. These have been exceptionally important in the study of smaller insects, having revealed the earliest bee (Apoidea), ants (Formicidae) (Figs. 5e and 5f), Zoraptera (Fig. 5d), and Stepsipteran, as well as the oldest definitive Embioptera and Lepidoptera and many families of insects. All Cretaceous Lepidoptera belong to phylogenetically basal families. Cretaceous ambers have also revealed a great diversity of primitive parasitoid Hymenoptera, sternorrhynchan hemipterans (especially Coccoidea), and empidoid flies. Cyclorrhaphan Diptera were rare and primitive in the Cretaceous.

Many of the oldest records of blood-sucking insects are from the Cretaceous: Ceratopogonidae, Culicidae, phlebotomine Psychodidae, Tabanidae, and a possible stem group to the Siphonaptera (†*Tarwinia,* from Koonwarra). Unlike modern fleas, †*Tarwinia* did not have jumping hind legs and had long antennae, and so it is at best a very primitive relative of fleas. Apterous mecopteroids from the Cretaceous of Baissa and Transbaikalia, Russia, are also believed to have been vertebrate ectoparasites, perhaps of pterosaurs or feathered dinosaurs. Simuliidae appeared first in the Upper Jurassic but were still scarce and primitive in the Cretaceous. Vertebrate ectoparasitism probably first appeared in the Jurassic, but were fully developed in the Cretaceous.

Cenozoic (65 Mya–Recent)

The sudden radiation of orders of modern mammals occurred in the Paleocene (65–56 mya). This period is very poorly known from the insect fossil record, the most diverse deposit being the Fur Formation from Denmark, which preserved giant ants, noctuid moths, and others. The first butterflies (Papilionoidea) are known from the upper Paleocene.

The Eocene (56–35 mya) is far better known for insects, with extensive compression deposits from British Columbia, Canada (Horsefly, Merritt, Princeton); Green River, Wyoming; Gurnet Bay and Bembridge Marls, Isle of Wight, England; and Florissant, Colorado. Florissant (e.g., Figs. 5g and 5h) was an ancient lake inundated with volcanic ash and is probably the most diverse compression fossil insect deposit from the Cenozoic. The greatest diversity of fossil insects is preserved in Baltic amber, huge deposits of which occur throughout northern Europe (from Eocene to Oligocene). The Baltic amber has preserved the first diverse faunas of schizophoran Diptera, ditrysian Lepidoptera, advanced termites, bees, and ants, all of which are relatively young, very speciose groups whose greatest radiations occurred in the Cenozoic. At least 30 species of bees have been found in Baltic amber; paradoxically, the great majority of them are corbiculate bees belonging to extinct genera. The diversity of advanced eusocial bees was much greater than today. The oldest definitive muscoid flies and fleas occur in Baltic amber. Other important Eocene amber deposits are from Fushun, China, and Arkansas.

Deposits from the Oligocene (35–23 mya) and the Miocene (23–5 mya) indicate that the insect fauna was essentially modern, with virtually all species (extinct) belonging to modern genera. Particularly rich Oligocene deposits are from the Ruby River Basin of Montana (compression) and Chiapas, Mexico (amber). Rich Miocene deposits occur in Oeningen, Switzerland (compression), and the Dominican Republic (amber). The age of the Dominican amber has been cited as Eocene but is definitively Miocene. The Dominican amber forest was very similar to contemporary neotropical forests, but there were some groups now extinct from the Caribbean (meliponine and euglossine bees, certain genera of ants, and others) or even from the Western Hemisphere (*Mastotermes* termites and some genera of ants, acrocerid and phorid flies, and others). Dramatic extirpation of ranges has also been found in other Cenozoic deposits, such as tsetse (now entirely African) (Fig. 5g) and nemopterid lacewings (Fig. 5h, now Southern Hemisphere) from Florissant and many currently austral taxa preserved in Baltic and Cretaceous ambers of the Northern Hemisphere. Although some specimens of insects preserved in Cenozoic ambers are very similar to modern species, it is unclear (especially genetically) if these are conspecific. The average duration of insect species is difficult to estimate, but the upper limit probably extends to 10 mya.

Study of Pliocene (5–1.5 mya), Pleistocene (1.5–0.01 mya), and Holocene (10,000 years ago–Recent) deposits has been extremely useful along with that of fossil pollen in reconstructing paleoclimates and ecological succession. These remains are preserved in existing or ancient lakebeds, bogs, and tarpits. Paleoclimatic use of these remains depends on the availability of indicator taxa or readily identified extant species whose distributions are well documented. Particularly persistent and abundant in lakebeds, and therefore commonly used, are fragments of beetles (Fig. 2g) and the larval head capsules of chironomid midges.

See Also the Following Articles

Amber • Coevolution • Endangered Insects • Nomenclature and Classification • Phylogeny • Wings

Further Reading
Carpenter, F. M. (1992). "Superclass Hexapoda," Vols. 3 and 4 of "Treatise on Invertebrate Paleontology," Part R, "Arthropoda 4." University Press of Kansas, Lawrence, and Geological Society of America, Boulder, CO.

Elias, S. (1994). "Quaternary Insects and Their Environments." Random House (Smithsonian Inst. Press), Washington, DC.

Evenhuis, N. L. (1994). "Catalogue of the Fossil Flies of the World (Insecta: Diptera)." Backhuys, Leiden.

Grimaldi, D. A. (1996). "Amber: Window to the Past." Abrams/Am. Mus. Nat. History, New York.

Grimaldi, D. A., and Cumming, J. (1999). Brachyceran Diptera in Cretaceous ambers and Mesozoic diversification of the Eremoneura. *Bull. Am. Mus. Nat. Hist.* **239.**

Hennig, W. (1981). "Insect Phylogeny." Wiley, New York.

Kukalova-Peck, J. (1991). Fossil history and the evolution of hexapod structures. *In* "The Insects of Australia," Vol. 1. Cornell University Press, Ithaca, NY.

Labandeira, C. C. (1998). Early history of arthropod and vascular plant associations. *Annu. Rev. Earth Planet. Sci.* **26,** 329–377.

Larsson, S. G. (1978). "Baltic Amber: A Paleobiological Study." Scandinavian Sci. Press, Copenhagen.

Rasnitsyn, A. P., and Quicke, D. L. J. (eds.) (2002). "History of Insects." Kluwer Acad, Publ., Dordrecht.

Rohdendorf, B. B. (1962). "Fundamentals of Paleontology," Vol. 9. "Arthropoda, Tracheata, Chelicerata." 1991 English translation of original Russian published by Amerind Pub., New Delhi.

Freshwater Habitats

see *Aquatic Habitats*

Fruit Fly

see *Drosophila melanogaster*

Gallmaking and Insects

Katherine N. Schick and Donald L. Dahlsten
University of California, Berkeley

Plant galls, which are abnormal growths of plant tissue that often resemble plant organs, can be induced by a wide variety of different insect species. The gallmaking insect stimulates the host plant through a complex chemical interaction so that the resulting gall is much more than a simple response to wounding or feeding damage. The precise gall form and position of the gall on the host plant is consistent and characteristic for each species of gallmaking insect.

Although cecidology, the formal study of plant galls, was initiated in 1679 with Malpighi's study of gallmaking insects, humans have admired and utilized galls for thousands of years. For example, Gallic acid (3,4,5-trihydroxybenzoic acid) was first derived from an oak gall induced by the cynipid wasp *Andricus gallaetinctoriae.* These galls have been commercially traded from source trees in the Middle East and the gallic acid derived from them has been used historically as a dye as well as an antiseptic astringent skin medication. Derivatives are used as photographic developers and the ink base made from these galls has been used to make permanent inks for such purposes as printing the U.S. dollar bill.

In 16th century England, the cynipid gall, *Biorhiza pallida,* was used for personal ornamentation. On May 29, the English parliament's official "Oak Apple Day," sprigs of oak leaves and gilded galls were worn to commemorate restoration of the English monarchy.

CECIDIA: PLANT GALLS

Cecidia, or plant galls, are abnormal growths of plant tissue under the influence of a parasitic organism. Within the growing cecidium, plant cells proliferate (hyperplasy) and enlarge (hyptertrophy) into a characteristic structure specific to that particular gallmaking organism. The organism inducing cecidogenesis (gall formation) receives nourishment and shelter while the host plant seldom benefits. Plant galls are induced by a variety of organisms including bacteria, fungi, nematodes, and arthropods.

Insect gallmakers span seven orders (see Table I) within which two entire insect families (Cecidomyiidae, the gall midges, and Cynipidae, the gall wasps) are found only within

TABLE I   Insect Gallmakers

Insect order	Gallmaking insects: galls
Thysanoptera (thrips)	Thrips: roll and fold galls on leaves and buds, mostly tropical.
Heteroptera (true bugs)	Tingidae (lace bugs): galls on flowers.
Homoptera (aphids, hoppers, and scales)	Adelgidae, Aphididae, Asterolecaniidae, Cercopidae, Cicadellidae, Coccidae, Diaspididae, Eriococcidae, Eriosomatidae, Kermidae, Phylloxeridae, and Psyllidae (leafhoppers, aphids, and scales): pouch and roll galls on leaf, stem, and root; mostly of woody plants.
Coleoptera (beetles)	Curculionidae (weevils): galls on Brassicaceae.
Lepidoptera (moths)	Aegeriidae, Coleophoridae, Cosmopterygidae, Gelechiidae, Heliozelidae, Lycaenidae, Orneodidae, Pterophoridae, and Tortricidae (mostly small moths): mostly fusiform galls on stems or petioles.
Diptera (flies)	Agromyzidae, Anthomyzidae, Cecidomyiidae, Chloropidae, Platypezidae, and Tephritidae (gall midges, fruit flies, and leafminers): a variety of galls on woody and herbaceous dicots and monocots.
Hymenoptera (wasps)	Agaonidae, Cynipidae, Eurytomidae, Pteromalidae, and Tenthredinidae (sawflies, fig wasps, seed chalcids, and gall wasps): galls on all plant parts of mostly woody plants and a few herbaceous plants.

cecidia. While individual insect species make a characteristic gall on only one part of a single plant species, the thousands of gallmaking insects induce cecidia on nearly all plant parts of a wide variety of plant species worldwide.

CECIDOGENESIS: MECHANISMS OF GALL INDUCTION

The process of cecidogenesis (gall induction) involves increased levels of plant growth regulators (auxins, cytokinins, gibberellins, abscisic acid, etc.), the stimuli for which differ among taxa of gallmaking insect. For example, galls induced by tenthredinid sawflys (Hymenoptera) form in response to chemicals produced in the female accessory gland and placed on the plant at the time of oviposition. However, the chemical stimulus for galls induced by cynipid gall wasps (Hymenoptera) is released with larval feeding, and gall formation ceases if the larva dies. The exact mechanism by which insects induce gall structures characteristic to that insect species and markedly different from those of other gallmaking insects is still poorly understood.

The plant tissue stimulated to form a gall is always unspecialized parenchyma. As these plant cells undergo hyperplasy and hypertrophy, some cells may specialize to form the characteristic structures associated with that gall. However, some of the gall cells always remain unspecialized. These parenchyma cells sequester macronutrients (such as amino acids) and micronutrient minerals (such as calcium, iron, and magnesium) so that the galls act as physiological sinks in the host plant.

ECOLOGY OF GALLMAKING

Biology of Gallmakers

Formation of the gall has an adaptive advantage to the gallmaking insect, for nutritive gall tissues feed the growing larva and the gall structure hides it from natural enemies. These insect benefits of gallmaking are produced at a cost of photosynthate and energy to the host plant. The majority of insect gallmakers are plant parasites with a notable exception: fig wasps in the family Agaonidae (Hymenoptera) form mutualistic associations with their host plant (in the genus *Ficus*) as pollinators. Agaonid wasps from male flowers are introduced into fig orchards in the ancient agricultural process of caprification to allow pollination of the female flowers necessary to produce the fruit.

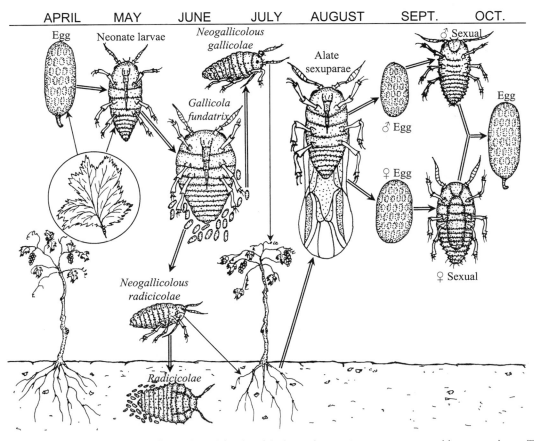

FIGURE 1 Heterogeny in grape phylloxera. In October males and females of the bisexual generation emerge, mate, and lay eggs on leaves. Their offspring form blister galls on grape leaves. In April the all-female leaf gallers lay eggs on leaves or drop to roots to lay eggs. In June all-female root gallers lay eggs on roots. Their offspring form nodular galls on grape roots. In late August a winged generation of females crawls out of the soil to lay their eggs (of the bisexual generation) under bark.

A number of gallmaking insect species exhibit heterogeny, alternating generations that include both sexes with generations including only females. Alternate generations frequently make very different galls on different parts of a plant, as does the homopteran grape phylloxera, *Daktulosphaira vitifoliae,* of which one generation induces galls on the leaves and stems while another generation induces galls on the roots of grapes. (Fig. 1). Some alternating generations of gallmaking insects form galls on different host plants, as is found in the cynipid gall wasp, *Andricus quercuscalicis,* with the unisexual generation forming galls on acorns of *Quercus robur* (English oak) and the bisexual generation forming galls on staminate flowers of *Q. cerris* (Turkey oak).

Other Insects in the Gall Community

Although a plant gall would appear to offer a place for an insect to escape predators and parasites, it is also a sedentary structure where natural enemies can predictably locate insect larvae. Predators of hidden insect larvae, such as woodpeckers, regularly prey upon galls. Many parasitoids, insects which receive their nutrients from a single host insect, eventually killing the host, have adapted to the predictability of plant galls.

In addition to gallmakers and parasitoids, the gall community also contains inquilines, which are insects which live in the gall and consume the plant gall tissue. Some inquilines deliberately kill the gallmaker, probably to prevent lignification of the gall tissues. However, many galls have both gallmaker and inquiline emerging as adults unless one or both have succumbed to parasitoids. The presence of inquilines, and the chemical stimulants they secrete, can sometimes alter the shape of the final gall, especially among cynipid gall wasps.

ECONOMIC IMPACT OF GALLS

The majority of plant galls harm the host plant only by diverting plant resources and thus have little economic impact. The economic impacts of gallmaking insects include the benefits of fig pollination as well as some negative economic effects in the form of crop losses.

Three examples of gallmaking insects with negative economic impacts are the Hessian fly (Diptera: Cecidomyiidae), the grape phylloxera (Homoptera: Phylloxeridae), and the oriental chestnut gall wasp (Hymenoptera: Cynipidae). The Hessian fly, *Mayetiola destructor,* is well known for the damage it causes to wheat crops. This fly does not cause galls on wheat, but rather exhibits a kind of facultative cecidogenesis by inducing galls only when the eggs are laid on barley, where the galls are much more damaging to the host plant than the necroses it forms on wheat. Grape phylloxera, *D. vitifoliae,* nearly destroyed the wine industries of Europe in the late 19th century. Although the leaves of European vines are resistant and are not infested with the leaf- and stem-galling forms, generations of wingless females gall the

roots, eventually killing the vines. The only successful treatment has been to graft grapes onto certain resistant root strains. The economic impact of the oriental chestnut gall wasp, *Dryocosmus kuriphilus,* results from the loss of nut crop as galls form on the buds and flowers of chestnut trees (*Castanea* spp.), although the trees remain healthy.

See Also the Following Articles

Agricultural Entomology • Commercial Products from Insects • Host Seeking, for Plants • Phytotoxemia

Further Reading

Ananthakrishnan, T. N. (1984). "The Biology of Gall Insects." Oxford & IBH, New Delhi.

Csóka, G. (1997). "Plant Galls." Forest Research Inst., Budapest.

Csóka, G., Mattson, W. J., Stone, G., and Price, P. (1998). "The Biology of Gall-Inducing Arthropods." Forest Service, USDA, St. Paul.

Darlington, A. (1975). "The Pocket Encyclopaedia of Plant Galls in Colour." Blandford Press, Poole.

Meyer, J. (1987). "Plant Galls and Gall Inducers." Gebrüder, Borntraeger, Stuttgart.

Redfern, M., and Askew, R. R. (1992). "Naturalists' Handbooks." 17. "Plant Galls." Richmond Pub., Slough, England.

Shorthouse, J. D., and Rohfritsch, O. (1992). "Biology of Insect-Induced Galls." Oxford University Press, New York.

Williams, M. A. J. (1994). "Plant Galls: Organisms, Interactions, Populations," Special Vol. 49 of the Systematics Association. Clarendon Press, Oxford.

Genetically Modified Plants

D. A. Andow

University of Minnesota, St. Paul

Although there is no one absolute definition of a genetically modified (GM) plant, to some it is any plant that has had its genes deliberately altered by humans, by whatever means. This definition includes all plants produced by conventional plant breeding. Even though specific genes cannot be altered deliberately using conventional plant breeding, these conventional processes modify many genes simultaneously in statistically predictable ways. Hence conventionally produced plants can be considered GM plants, broadly speaking.

To most others, a GM plant is more narrowly defined as a plant that has been produced using transgenic methods. These plants are also called transgenic or genetically engineered plants. Transgenic methods are molecular methods that enable the transfer of a gene or potentially a group of genes from an individual of one species to an individual or individuals of a different species. Currently, there are two common methods by which purified genes are introduced into plant cells: one uses the Ti plasmid of *Agrobacterium tumefaciens* to transfer the gene as a part of the plasmid; the other uses a metal particle or fiber or an electric pulse to pierce the cell wall and carry the gene into the nucleus (also called gene gun or electroporation). Transgenic methods enable humans to alter specific genes deliberately. The term transgenic is sometimes restricted only to genetic transfers across the species boundary, but usually includes molecular gene transfers within species as well.

The European Union uses GM plants in the narrow sense in discussing the regulation of biotechnology. Within the Cartagena Protocol on Biosafety under the Convention on Biodiversity, the term living modified organism is defined as a GM plant (narrow sense) that is intended to be grown, which excludes grain shipments and most other trade from consideration. The U.S. Office of Science and Technology Policy, which authorized the Coordinated Framework for the regulation of biotechnology in the United States uses GM plants in the narrow sense, except that it is broadened slightly to include a couple of methods that would normally be considered conventional methods. All of these organizations have chosen a definition of GM plants to exclude conventional plant breeding, in part because they do not want to regulate conventional plant breeding.

Although it has become less common, several others have used GM plants (broad sense) to blur the distinction between transgenic plants and conventionally produced plants. Because commercialized conventionally produced plants and their food products have generally been assumed to be safe for the environment and human consumption, blurring the distinction has often been a device to suggest that there are few legitimate concerns about transgenic plants. More recently, this argument has been reversed. Some conventionally produced plants are in fact potentially harmful to the environment or human health, and it may become useful to evaluate the potential dangers associated with some of these plants.

WHY THERE ARE SO MANY *BACILLUS THURINGIENSIS* PLANTS

The first commercial transgenic crops were planted in China during the early 1990s. These were primarily virus-resistant tobacco and tomato. In the United States, the first commercialized crop was Calgene's FLVR SAVR tomato in 1994. This product was not a commercial success in part because it did not pack well for shipping. Initially, a variety of transgenic crops were planted (Table Ia); by 1999, however, four crops dominated: soybean, corn, cotton, and canola. The primary traits of these GM plants are herbicide tolerance and insect resistance (Table Ib). In 1999, herbicide-tolerant soybeans, Bt *(Bacillus thuringiensis)* corn, herbicide-tolerant corn, Bt cotton, herbicide-tolerant cotton, and herbicide-tolerant canola accounted for over 99% of the commercial transgenic crops grown worldwide. All of the insecticidal-transgenic crops currently available are based on *cry* toxin genes from *B. thuringiensis,* and a few now under development are based on other toxin-coding genes from *B. thuringiensis.* The *cry* genes code for crystalline proteins that are toxic to some insects.

Bt genes have been incorporated into broccoli, cabbage, canola, cotton, corn, eggplant, poplar, potato, soybean, tobacco, and tomato, and the commercially available crops during 2001 in the United States are Bt corn and Bt cotton. Since their introduction during 1995, the cropping area of all of these transgenic crops has grown substantially (Table II). By 1999, Bt corn was grown on 9.6 million ha. Bt cotton lagged behind substantially in total area because about five times more corn than cotton is grown in the United States. Clearly, Bt corn in the United States is one of the dominant transgenic crops in the world today. Interestingly, Bt corn area has decreased during 2000, probably in response to market uncertainty. In contrast,

TABLE I   Area of Transgenic Crops in the World from 1996 to 2000 (in Millions of Hectares)

	1996	1997	1998	1999	2000
(a) Crop					
Soybean	0.5	5.1	14.5	21.6	25.8
Maize	0.3	3.2	8.3	11.1	10.3
Cotton	0.8	1.4	2.5	3.7	5.3
Canola	0.1	1.3	2.5	3.4	2.8
Tobacco	1.0	1.7	+	+	+
Tomato	0.1	0.1	+	+	+
Potato	+	+	+	+	+
(b) Trait					
Herbicide tolerance	0.7	6.9	20.1	31.0	35.8
Insect resistance	1.0	4.7	8.0	11.8	11.5
Virus resistance	1.1	1.8	+	+	+
Quality traits	+	+	+	+	+
(c) Total	2.8	12.8	27.8	39.9	44.2

Note. +, <100,000 ha were grown. The first commercial crops were planted in China during the early 1990s. The first commercial production in the United States was tomatoes during 1994. Several crops were first commercialized during 1995, including Bt corn. (a) Area by crop. Several minor crops are not listed. (b) Area by transgenic trait. Values do not always sum to the worldwide total because some crops have more than one transgenic trait. (c) Total area worldwide.

TABLE II   Area of Bt Corn, Bt Cotton, and Bt Potato, 1995–1997 and 2000 for the United States Only; 1998–1999 for the World, Although Nearly All Was Planted in the United States

	1995	1996	1997	1998	1999	2000	2001
Bt maize	+	0.3	2.8	6.7	9.6	6.3	5.3
Bt cotton	+	0.7	0.8	na	2.2	2.2	2.3
Bt potato	+	3650	10,000	20,000	23,000	na	+

Note. Corn and cotton data are in million hectares. Potato data are in hectares. na, data not available; +, <100,000 ha for maize and cotton and <1000 ha for potato.

Bt cotton area increased, perhaps tied to increased demand for herbicide-tolerant cotton.

Several factors probably account for the predominance of Bt genes in transgenic crops. First and foremost, transgenic technology is relatively new, and consequently the products are those that are technically feasible, are readily accomplished, and have a clear path to commercialization. The *cry* genes present in Bt have been technically easier to use than genes from plant or animal sources. The *cry* gene structure is simple, and Cry toxins require no posttranscriptional or posttranslational processing to be functional. Initially, *cry* genes did not express very well in plants, in part because bacterial DNA is A-T rich, whereas plant DNA is G-C rich. By directed mutagenesis, the "wobble" codon and other triplet redundancies were converted from A or T to G or C, and the resulting *cry* genes were expressed more consistently in plants. The commercialization path for Bt crops was believed to be relatively clear. Several Cry toxins have been used in commercial formulations of insecticides and these have been considered much safer for the environment and much less toxic to humans than nearly all other synthetic chemical insecticides. More recently, however, the differences between the Cry toxins in the insecticides and the Cry toxins in the transgenic plants have received increasing scientific and regulatory attention. Another reason for the predominance of *cry* genes is that as each particular Bt crop variety was approved for commercial use by U.S. regulatory agencies, the regulatory requirements became increasingly clear, making it increasingly easy to prepare for regulatory evaluation. Thus, the transformation technology, the regulatory environment, and the toxicological and insecticidal characteristics of Cry toxins have contributed to the present abundance of Bt crops.

THE POTENTIAL OF THIS TECHNOLOGY

There are many different kinds of transgenic crops under development, including some that are expected to be commercialized in the next few years, others that may reach commercial status on a midterm horizon sometime during the next decade, and finally others that are mere ideas that require significant research breakthroughs before they can be realized. Most of these aim to address one of four broad needs: improved agricultural characteristics, improved postharvest processing, improved food quality and other novel products for human use, and improved mitigation of environmental pollution.

Among the transgenic traits near to commercial release are new Bt genes that provide protection against additional types of insect pests. One of these traits is a gene that protects corn against rootworm damage. Some traits are based on Cry toxins, but others are based on novel Bt toxins (a binary toxin requiring two components that are toxic to insects). Other agriculturally useful traits include tolerance to abiotic stresses, such as drought, salt, and cold. These traits, however, probably require significant research breakthroughs before they can become realities.

Transgenic technology is also being applied to several commercially important tree species, including poplar, eucalyptus, aspen, sweet gum, white spruce, walnut, and apple. The first traits being genetically engineered into trees are herbicide tolerance and insect resistance, which may be useful for establishing and maintaining young trees and protecting valuable fruits. Several traits are under development to better adapt trees to postharvest processing. For example, the lignin content of certain tree species is being engineered to improve pulping, the process by which wood fibers are separated to make paper. Reduced lignin may improve the efficiency of paper production and reduce pollution from the paper-production process.

Transgenic technologies are being applied to alter macronutrients, starch, protein, oil, and micronutrients in several food crops, such as maize and soybean. For example, soy protein is deficient in the essential amino acids that contain sulfur. Increased production of these essential amino acids has been approached through genetic engineering by altering the activity of the enzymes that synthesize them, by overproducing a protein that contains them, or by blocking production of major proteins that lack them, which thereby increases the percentage of these amino acids in the remaining proteins. There are 17 minerals and 13 vitamins required at minimum levels to prevent nutritional disorders, and all of these have attracted biotechnology research. For example, increased levels of tocopherol, the lipid-soluble antioxidant vitamin E, has been engineered in the model plant *Arabidopsis thaliana* by overexpressing the gene responsible for the last step in vitamin E synthesis. Using a similar approach, current efforts are creating transgenic soybean and canola plants with enhanced levels of vitamin E.

In the future, transgenic plants may be grown to mitigate pollution. For example, it has been proposed that transgenic plants could contribute to removing or detoxifying heavy metal pollutants in contaminated soils ("phytoremediation"). A particular problem in some locations is mercury. Plants have already been created that can accumulate mercury, and genetic engineering may be able to enhance this ability so that growing such plants could become a potential solution for cleaning up mercury pollution at despoiled sites.

Few of these future potential applications are directly related to insects or insect control. However, each of them is likely to create significant needs for entomological investigations. Traits such as abiotic stress tolerance, altered lignin content, altered macronutrient or micronutrient content, and metal accumulation are all likely to have profound effects on plant physiology and growth. Because insect herbivores and their natural enemies are very sensitive to changes in plant physiology, growth rates, and morphological structure, these new transgenic crops are likely to create new kinds of pest control problems that will require entomological solutions. Indeed, it may become necessary to devise variety-specific pest management systems for some of these novel plants.

THE LIMITS OF THIS TECHNOLOGY

The rate at which new transgenic traits can be expected to appear in the near future depends largely on the number of genes encoding them. Traits controlled by single genes, or traits that can be altered or eliminated by the loss of expression through silencing of a single gene or group of related genes, have been the first developed and commercialized. Genetically complex traits probably will require additional years of research to understand them, let alone to express and regulate them in a genetically engineered crop species. Nevertheless, many complex traits, including those controlling adaptation to abiotic stresses such as drought and salinity, flowering and reproduction, and hybrid vigor, are being investigated, and it is possible that some of these traits could appear in transgenic crop plants during the next decade.

The social acceptability of products from transgenic plants has affected and may continue to affect their adoption. Social acceptability has many components, including environmental and human health risks, food choices, the ownership of agricultural inputs and production process, the future structure of agriculture, and so on. This issue is extremely complex and volatile, and it may take several more years before it has stabilized sufficiently so that it will be possible to anticipate how various societies around the world might accept or reject transgenic plants.

Although these and other factors are involved currently in limiting the application of transgenic plants, the central trade-off that may limit it ultimately may involve a classic gene–environment trade-off in crop production. A broadly adapted trait may not be able to be used optimally without a corresponding management system. For example, the short-statured rice varieties of the Green Revolution required an intensive management system oriented around high fertilizer and pesticide use to attain their high yields across vast areas

of Asia. These varieties, whether transgenic or conventional, are limited by the applicability of the attendant management system, which creates the environment in which they can flourish. Conversely, a plant with a trait adapted to specific environments associated with a single field or valley would not be broadly applicable to vast growing regions because the environment would not be equally suitable throughout. The scale of use of such locally adapted varieties depends on a system of seed production and distribution that reliably delivers the appropriate seed at the proper time. The system of self-propagated land races admirably meets these needs, but it remains to be seen whether a commercial seed distribution system is able to deliver varieties that are so locally adapted. Finally, as noted above, transgenic plant varieties are being developed for extremely specialized uses on areas of only hundreds of hectares. The scale of use of these specialty varieties will probably depend on how many different such uses are commercially successful as well as the intensivity of management needed to produce them. In conclusion, if technical factors and social acceptance do not limit the adoption of transgenic crops, the costliness of management associated with the specialty transgenic crops, the applicability of management of the broadly adapted transgenic crops, and the delivery of seed of locally adapted transgenic crops likely will limit their applicability, much like conventional varieties are limited today.

POTENTIAL ADVERSE EFFECTS

The risks associated with transgenic plants stem from, but are not directly caused by, the nature of the transformation process. First, transgenic methods enable traits to be expressed that have never before been expressed in a plant. This widened range of traits creates potential risks that should be evaluated. Second, the present transgenic methods cannot incorporate foreign DNA into precise locations in the plant genome. Because expression of genes can depend on where the gene occurs in the genome, and because the incorporation can be complex, the expression of the transgene cannot be predicted completely. The scope of potential traits and the uncertainty associated with trait expression create the circumstances requiring the evaluation of risks to human health or the environment.

Because transgenes code for proteins, human health risks associated with these proteins and products produced by these proteins are possible. These potential risks include creation of novel toxicants, possible shifts in the nutritional content of food, and the possible creation of novel allergens. Most of the scientific attention has focused on allergens, because they are difficult to assess and there has been an increase in the incidence of food allergies. Novel proteins and their products can be altered after synthesis by alterations in amino acid sequence and by reactions with other chemicals, such as glycosylation. Assessing each of these possibilities will be challenging.

Environmental risks stem from several types of potential effects: (1) effects associated with the movement of the transgene itself and its subsequent expression in a different organism or species, (2) effects associated directly or indirectly with the transgenic plant as a whole, (3) nontarget effects associated with the transgene product outside of the plant, (4) resistance evolution in the targeted pest populations, and (5) indirect effects on human health that are mediated by the environment. The European Union (EU) recognizes affects on genetic diversity as a separate category of environmental effect in the modified 90/220 directive. The United States government has not recognized this as an environmental effect because it believes that it is the effects of altered genetic diversity, such as increased extinction rate, a compromised genetic resource, inbreeding depression, or increased vulnerability to environmental stresses, that are the actual environmental hazards. The EU recognizes this category as a precautionary measure, because the effects of movement of the transgenes are uncertain and are at present incompletely characterized. By recognizing the more easily measured, intermediate effects on genetic diversity as a potential effect, the EU risk analysis will address all of the effects caused by movement of transgenes without having to assess them specifically.

Hazards Associated with Movement of the Genes

Horizontal transfer is the nonsexual transfer of genetic material from one organism into the genome of another. Although there are no cases of transgenes moving horizontally from plants to any other organism at rates higher than normal, new discoveries could change the assessment and significance of this risk. Pollen dispersal provides an opportunity for the sexual transfer of transgenes to relatives of the crop, including other varieties of that crop, related crops, and wild relatives. Potential effects include the evolution of increased weediness (i.e., more vigorous agricultural weeds, more invasive plants) or increased risk of extinction of native species by hybridization.

Hazards Associated with the Whole Plant

The transgenic plant itself may become an environmental hazard if the traits it receives improve its fitness and ecological performance. Although many crop plants may pose little hazard, insofar as they are unable to survive without human assistance, most crops have weedy and/or wild populations in some part of their global distribution. In these areas, transgenes that improve fitness could increase weediness of the crop. In addition, because transformation includes forage grasses, poplars, alfalfa, sunflowers, wild rice, and many horticultural species, the risk of invasiveness may increase.

Nontarget Hazards

Nontarget organisms are any species that are not the direct target of the transgenic crop, and consequently, the list of potential nontarget species is very long. These organisms can be grouped conveniently into five categories: (1) beneficial species, including natural enemies of pests (lacewings, ladybird beetles, parasitic wasps, and microbial parasites), and pollinators (bees, flies, beetles, butterflies and moths, birds, and bats); (2) nontarget pests; (3) soil organisms, which usually are difficult to study and identify to species; (4) species of conservation concern, including endangered species and popular, charismatic species (monarch butterfly); and (5) biodiversity, which is the entire group of species in an area.

Hazards of Resistance Evolution

Resistance evolution can occur in pests that are targeted for control by or associated with the transgenic crop. If the pest becomes resistant, then alternative, more environmentally damaging controls may be used. Insects, weeds, and microbial pathogens all have the potential to overcome most control tactics used against them. Insect resistance to Bt crops is considered inevitable, and efforts are being made to manage resistance evolution to these transgenic crops.

Indirect Hazards

Transgenic crops can have indirect environmental impacts, especially when scaled up for commercial production. Many of these effects are associated with changes in production practices or cropping systems. For example, transgenic maize resistant to corn rootworms may lead to an expansion of continuous corn (corn planted after corn) and its attendant environmental risks, such as soil erosion. In addition, it is possible that crops transformed to produce pharmaceutical or other industrial compounds might mate with plants grown for human consumption with the unanticipated result of novel chemicals in the human food supply.

See Also the Following Articles

Biotechnology • Genetic Engineering • Insecticides • Pathogens of Insects

Further Reading

Andow, D. A. (2001). Resisting resistance to *Bt* corn. *In* "Genetically Engineered Organisms: Assessing Environmental and Human Health Effects" (D. K. Letourneau and B. E. Burrows, eds.), pp. 99–124. CRC Press, Boca Raton, FL.

DellaPenna, D. (1999). Nutritional genomics: Manipulating plant micronutrients to improve human health. *Science* **285**, 375–379.

Ellstrand, N. C., and Elam, D. R. (1993). Population genetic consequences of small population size: Implications for plant conservation. *Annu. Rev. Ecol. Sys.* **24**, 217–242.

Georghiou, G. P. (1986). The magnitude of the resistance problem. *In* "Pesticide Resistance: Strategies and Tactics for Management," pp. 14–39. National Academy Press, Washington, DC.

Green, M. B., LeBaron, H. M., and Moberg, W. K. (eds.) (1990). "Managing Resistance to Agrochemicals: From Fundamental Research to Practical Strategies." American Chemical Society, Washington, DC.

Hansen, L. C., and Obrycki, J. J. (2000). Field deposition of Bt transgenic corn pollen: Lethal effects on the monarch butterfly. *Oecologia* **125**, 241–248.

Hilbeck, A., Moar, W. J., Pusztai-Carey, M., Filippini, A., and Bigler, F. (1998). Toxicity of *Bacillus thuringiensis* Cry1Ab toxin to the predator *Chrysoperla carnea* (Neuroptera: Chrysopidae). *Environ. Entomol.* **27,** 1255–1263.

Losey, J. E., Rayor, L. S., and Carter, M. E. (1999). Transgenic pollen harms monarch larvae. *Nature* **399,** 214.

Munkvold, G. P., Hellmich, R. L., and Rice, L. G. (1999). Comparison of fumonisin concentrations in kernels of transgenic Bt maize hybrids and nontransgenic hybrids. *Plant Dis.* **93,** 130–138.

National Research Council (2002). "Environmental effects of transgenic plants: The scope and adequacy of regulation." National Academy Press, Washington, DC.

Saxena, D., Flores, S., and Stotzky, G. (1999). Insecticidal toxin from *Bacillus thuringiensis* in root exudates of transgenic corn. *Nature* **402,** 480.

Shintani, D., and DellaPenna, D. (1998). Elevating the vitamin E content of plants through metabolic engineering. *Science* **282,** 2098–2100.

Snow, A. A., and Moran-Palma, P. (1997). Commercialization of transgenic plants: Potential ecological risks. *BioScience* **47,** 86–96.

Genetic Engineering

Peter W. Atkinson
University of California, Riverside

David A. O'Brochta
University of Maryland Biotechnology Institute

Tremendous progress has been made in the development of genetic engineering technologies in insects. This article emphasizes studies with the vinegar fly, *Drosophila melanogaster,* because, as a result of its successful use as a genetic model for eukaryotic genetic systems, developments in genetic engineering with this species establish a benchmark for what can, or could, be done in other insect species. The article also discusses the more recent use of transposable-element-based genetic transformation procedures in nondrosophilid insects and concludes that many of the tools required for genetic manipulations of nondrosophilid insects are now available.

The term "genetic engineering" is typically taken to refer to the direct manipulation of genes. It has become synonymous with a more general term, "DNA technology," which has come to encompass all contemporary molecular-based techniques. However, many insect geneticists were using "DNA technology" before the development of recombinant DNA technology in the 1970s and 1980s, Genetic control approaches applied to such insect pests as the Mediterranean fruit fly *(Ceratitis capitata)* (medfly), the mosquito *(Culex tarsalis),* and the Australian sheep blowfly *(Lucilia cuprina)* used genetics to develop new strains that could be used in insect control and/or eradication programs. The tools of these pioneers were not DNA modification and restriction enzymes, thermocyclers, or DNA sequencers, but rather radiation sources, microscopes, and the knowledge that mutations and chromosomal rearrangements could be created and selected for. These tools have now been surpassed, but one aim remains the same: the generation and application of new genetic strains of insects that can be used to control pest insect species. The development of sophisticated genetic tools, in conjunction with the rapid progress being made in genomics, will provide insect scientists with the ability to characterize and manipulate, in hitherto unimaginable ways, insect genes.

GENETIC ENGINEERING IN *DROSOPHILA MELANOGASTER*

Genetic Technologies Are More Advanced in *Drosophila* Than in Other Insect Species

One cannot discuss the genetic manipulation of insects without describing the molecular genetic tools that are available in *D. melanogaster.* Traditionally, a gulf has existed between entomologists who view the harmless vinegar fly as being distant to the problems of insect control and *Drosophila* geneticists who utilize the many biological attributes of *Drosophila* to understand the basis of gene action. This gulf will close as comparative genomics reveals similarities and differences in the conservation of many genes and molecular pathways between *Drosophila* and other insect species. The power of this comparative approach to modern biology will offer insect scientists and traditional entomologists exciting opportunities to bring the power of genetics and molecular biology to the control of insects. The development and application of these tools is what insect scientists seek to achieve in pestiferous and beneficial insects.

Genetic engineering in *D. melanogaster* is an extremely mature technology. It is founded on several independent phenomena:

1. The presence of a transposable element, called the *P* element, which is an efficient genetic transformation vector. This vector has been available and exploited since the early 1980s.

2. The ability to create and maintain genetic mutants by traditional techniques such as chemical- or radiation-induced mutagenesis or by transposon insertion mutagenesis, and the construction and availability of balancer chromosomes to maintain many of these mutants.

3. The presence of strains that lack the *P* element, thus providing recipient strains suitable for *P* element transformation.

4. The completion of the *Drosophila* genome project and the public availability of the data generated.

These planks of achievement are a consequence of the intense and sustained research that has been invested into *Drosophila* over the course of the last 90 years. The picture in all other insect species is, by comparison, sparse. For example, transposable elements capable of transforming nondrosophilid species have been available only since 1996.

Also, traditional mutagenesis approaches have been used to generate mutations in a handful of insect species. Many of these have been lost because of problems arising from the rearing of these species (it should be noted that what attracted T. H. Morgan to *Drosophila* was the ease with which it could be reared and mated in the laboratory) and, often, because it had been necessary to depend on a handful of dedicated workers to maintain these strains. (In *Drosophila,* by contrast, there are central repositories for strain maintenance as well as hundreds, if not thousands, of researchers who maintain even the most problematic genetic stocks.) Except for medfly, balancer chromosomes have not been constructed in nondrososphild insects.

Two other factors are important. The interactions, if any, of the transposable elements so far known to transform non-drosophilids with components of the insect genome remain unknown, as do the molecular mechanisms by which these elements move both within and between insect genomes. Second, to date, no insect species other than *Drosophila* has had its entire genome sequenced. Some mosquito genomes are the target for current and future genomic projects.

Transformation Technologies in *D. melanogaster*

P ELEMENT TRANSFORMATION   Population geneticists in the 1970s had observed that when males from certain strains of flies recently established from wild populations were mated to females from long-established laboratory populations, a number of abnormal traits were observed in the progeny. These traits included high rates of mutation, sterility, and recombination in males, traits that are not usually seen in this species. Collectively the traits that arose only when specific hybrid insects were created were thought to be manifestations of a single syndrome that became known as "hybrid dysgenesis." Because the factors responsible for this syndrome were transmitted by males, they were referred to as paternal or "*P*" factors. Those working on *P*-factor-mediated hybrid dysgenesis quickly realized that there were multiple factors that mapped to many different locations within the genome. Some of the mutations that were induced during hybrid dysgenesis were very unstable and were themselves capable of mutating further to result in more extreme phenotypes or to revert to wild type. This instability as well as other genetic observations suggested that the *P* factors were mobile genetic elements or transposons. As long ago as 1989, Engels gave a comprehensive description of the *P* element, *P* factors, and their use in genetic transformation.

Concurrent with the efforts of population geneticists to understand the phenomenon of hybrid dysgenesis were efforts of molecular geneticists to clone genes from *D. melanogaster.* The eye color gene known as *white* was one of the first genes to be cloned from this species, largely owing to the great amount of genetic analysis that had been done on this locus. Having the *white* gene cloned provided a unique opportunity to isolate *P* factors. Because *P* factors were responsible for

causing mutations, a genetic screen was performed to seek mutations induced by hybrid dysgenesis of the *white* gene. The reasoning behind this experiment was that once a *P*-factor-induced mutation of the *white* gene was obtained, it should, in theory, be readily cloned by conventional genomic DNA library screening using the wild-type allele of the *white* gene as a probe. By comparing the mutant allele with the wild-type allele, the nature of the *P* factor might be deduced. As expected, mutations of the *white* gene induced by hybrid dysgenesis contained insertions, and the insertion sequences had all the characteristics of a transposable element. In fact, *P* factors were transposable elements and became known as *P* elements. Complete *P* elements were about 3 kb in length and contained four open reading frames encoding for a protein essential for *P* element movement. The terminal sequences of *P* elements consisted of inverted repeat sequences of 31 bp. In structure, the *P* elements generally resembled other transposable elements that had been isolated from bacteria and were referred to as short, inverted repeat–type transposable elements.

The physical isolation of an active transposable element from *D. melanogaster* provided researchers with a unique opportunity to integrate foreign DNA into the chromosomes of this species. Efforts to integrate exogenous DNA into the chromosomes of insects can be traced back to the late 1960s. While there was an interest in genetically transforming insects and a few reports of minor successes, there was no reliable method for creating transgenic *D. melanogaster*. The *P* element solved that problem. It was reasoned that if the terminal, noncoding sequences of the element, which serve as signal sequences directing the cutting and pasting of the element, were attached to any piece of DNA, that piece of DNA would acquire the mobility properties of a *P* element. Furthermore, if this altered transposable element could be introduced into a cell that was going to form gametes and the element jumped (i.e., transposed) onto one of the chromosomes of this cell, then the gametes arising from this cell would be transgenic and would give rise to transgenic progeny. This reasoning proved to be precisely correct. Under the appropriate conditions, genes to which the terminal noncoding sequences of a *P* element have been attached can readily integrate into the chromosomes of presumptive germ cells and lead to the efficient creation of transgenic insects (Fig. 1).

This relatively simple technology helped fuel a revolution in the study of this model organism. Today, this transposable element forms the basis for a suite of technologies that allow researchers to identify and analyze genes in a variety of ways. The *P* element gene transformation system has also served as a paradigm for the development of similar technologies applicable to other species of insects.

GENE TAGGING WITH TRANSPOSABLE ELEMENTS
The key to the identification and isolation of *P* elements was the availability of the cloned *white* gene. The *white* locus was used as a trap; once the *P* element had been identified and cloned, it could be used as a way of identifying and isolating

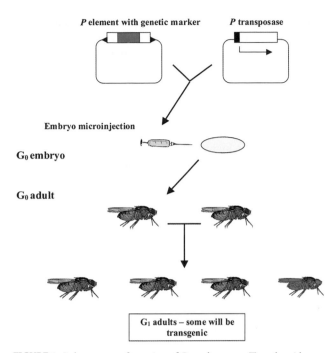

FIGURE 1   *P* element transformation of *D. melanogaster*. Two plasmids, one containing a *P* element into which has been cloned a genetic marker (green) and a helper plasmid containing the *P* element transposase (white) placed under the control of an inducible promoter (blue) are coinjected into embryos. The *P* element inverted terminal repeats are shown as black arrowheads. G_0 adults arising from injected embryos are not transgenic but some will contain a percentage of gametes containing the *P* transposable element. These adults are outcrossed and G_1 progeny are examined for the presence of transgenic individuals (green fly).

genes. As already discussed, one of the prominent features of *P* element movement (as revealed by the phenomenon of hybrid dysgenesis) was the creation of mutations. These mutations are caused by the insertion of the *P* element into an essential region of a gene, thereby altering its level or pattern of expression.

Mutations and their associated phenotypes define genetic loci. The existence of a mutant insect with an altered eye color defines a locus that plays some role in eye pigmentation. Although the existence of a mutant reveals the presence of a gene and its location, it does not provide researchers with a means of readily isolating the DNA containing the gene. If, however, the mutation is caused by the insertion of a sequence, such as a *P* element, and we know the sequence of the insertion sequence, we can use this information to isolate the DNA of the gene that was mutated. By making a genomic DNA library from the mutant insect, one can use conventional DNA hybridization techniques to identify sequences in the library that contain the *P* element. Because the mutation was caused by the insertion of the *P* element into a gene, the DNA adjacent to the *P* element is likely to be the gene responsible for the mutant phenotype. This methodology of transposon tagging is very powerful and has been used not only in *D. melanogaster* but in a number of other organisms as well. The

requirements for an effective transposon-tagging system that ensures unambiguous gene identification are an active transposon that has little integration site specificity and insect strains that contain few or only one transposon-tagging transposable element. Roberts has described the use of the *P* element for gene tagging and enhancer trapping.

ENHANCER TRAPPING   Transposable element-based mutagenesis or transposon tagging is a powerful technology with one limitation: it can identify only genes that have a recognizable mutant phenotype following element integration. Many of the genes that one mutates either do not result in a visible phenotype or cause the death of the organism. Such genes will never be recovered from a screen based on transposon tagging.

A complementary methodology that does not rely on mutagenesis for gene identification is called enhancer trapping. Enhancers are gene expression regulatory elements, and they function to fine-tune the control of gene expression, temporally and spatially. They are quite distinct from gene promoters in that enhancers are not sites of RNA polymerase binding but are instead sites for protein binding that influence when and how often RNA polymerase will associate with a promoter. A remarkable and useful feature of enhancers is their ability to act over long distances by mechanisms that are not entirely clear. That is, an enhancer may be located hundreds or even thousands of bases away from its target promoter. If a new promoter is inserted near the enhancer, it too will become regulated by that enhancer. This phenomenon provides a clever, nonmutagenic method for gene identification based on patterns of gene expression called enhancer trapping.

Like transposon tagging, enhancer trapping relies on the movement of a transposable element. The element in this case has been engineered to contain a gene whose expression is readily detected. Today the green fluorescent protein from the jellyfish is a common choice. The reporter gene has been engineered to contain a minimal basal promoter, meaning that it contains an RNA polymerase binding site but no associated enhancers. Consequently, this enhancerless gene construct does not result in reporter gene expression unless the transposon in which it is contained integrates near an active enhancer. The presence of enhancers can be detected by moving the transposon around the genome and looking for expression of the reporter gene. By identifying enhancers with particular properties, one then has indirectly identified the genes controlled by these enhancers. Often, the genes regulated by enhancers identified by using this method are located in the proximity of the enhancer. The significant difference between this method of gene identification and transposon tagging is that enhancer detection does not require mutating the enhancer or its associated gene. Consequently, genes that may not have been detected by a transposon tagging screen might be detected using an enhancer trap (Fig. 2).

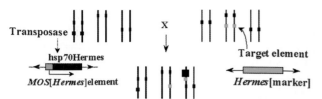

Target element inserts adjacent to an enhancer

FIGURE 2 Example of enhancer trapping in insects: three pairs of chromosome with their centromeres (purple). One strain contains a *MOS* element into which the *Hermes* transposase has been cloned (orange). The *MOS* inverted terminal repeats are shown as pink arrowheads. A second strain contains a *Hermes* element containing a genetic marker (blue) placed under the control of a weak promoter. The two strains are crossed, whereupon the *Hermes* transposase causes the *Hermes* elements to move to new regions of the insect genome. Should a *Hermes* element insert near an enhancer element (black box), the genetic marker in the *Hermes* element would show the same tissue- and stage-specific expression of the gene controlled by the enhancer. The gene and the enhancer can then be cloned by standard gene tagging techniques.

Transposon tagging and enhancer trapping are rather intense genetic methods for gene identification. Such methods require the ability to efficiently perform genetic crosses, to recognize mutants or desirable reporter gene expression patterns, and then to maintain large numbers of distinct genetic lines of insects. Although *Drosophila* is readily amenable to such manipulations, other insects may be less so. Nevertheless these methods will be of great value to those entomologists working on a variety of insect species.

HOMOLOGOUS RECOMBINATION Transposon tagging and enhancer trapping are methods of identifying genes based on a phenotype: a mutant phenotype in transposon tagging, an expression phenotype in enhancer trapping. The availability of essentially the entire DNA sequence of the genome of *D. melanogaster* has permitted the identification of genes based entirely on DNA sequence patterns. Often the role of these genes is completely unknown because flies with mutations in these genes have not been identified. Without the ability to examine the phenotypes of flies with mutant alleles of the gene, gene function must be deduced entirely by other means, such as patterns of expression or analysis of the protein gene product. Today, however, it is possible for researchers who know the DNA sequence of a specific gene to create *D. melanogaster* with mutations in that gene. This method of targeted mutagenesis relies on the process of homologous recombination.

Homologous recombination, the process of gene exchange that typically occurs during meiosis, depends on the association of DNA sequences that are identical or nearly identical. Breaks in one of the strands of a DNA duplex can result in this strand becoming associated with its homologue on another chromosome, leading to gene exchange. It is now possible to exchange a gene located on a chromosome of a fly with a nearly identical gene created in the laboratory. This somewhat involved process relies on the use of a site-specific recombinase and a site-specific endonuclease, but it is potentially a method that will be generally applicable to any insect. Rong and Golic have described this technology in *D. melanogaster.*

The strategy behind using homologous recombination takes advantage of the high recombinogenicity of linear molecules of DNA. Such molecules will preferentially recombine with sequences homologous to the sequence at the end of the linear molecule. Gene targeting by homologous recombination in *D. melanogaster* is based on a clever method for generating the highly recombinogenic targeting molecule *in vivo.* The process begins by creating a transgenic insect using, for example, a *P* element gene vector that contains the targeting sequences flanked by site-specific recombination sites such as the FRT sites of the FLP recombinase system. When FLP recombinase is expressed (from a previously integrated transgene) in the insect, the FRT sites will recombine causing the targeting gene to be excised from the integrated gene vector. This recombination event results in the creation of extrachromosomal circular molecules in the nuclei of the insect. These extrachromosomal circles are then linearized by expressing a site-specific endonuclease (from a previously integrated transgene) that recognizes a DNA sequence that has been placed in the targeting gene in such a way that digestion results in the target gene sequences being located at the ends of the linearized circle. This highly recombinogenic molecule will then recombine with the chromosomal homologue, resulting in gene disruption.

Homologous gene replacement has been achieved for two *Drosophila* genes, the *yellow* gene and the *pugilist* gene, and most likely will be applicable to a large number of *D. melanogaster* genes. In particular it should enable gene function to be assigned to the thousands of new genes identified in the *Drosophila* genome project through replacing the wild-type forms with nonfunctional mutations that have been created *in vitro.*

A prerequisite for targeted gene replacement is a set of transgenic insects that can express the appropriate restriction enzyme and the FLP recombinase. This is readily achieved in *D. melanogaster* and now can also be accomplished, in principle, in other insect species because transposable elements exist that can be used to genetically transform them. The FLP recombinase system has been shown to function correctly in the yellow fever mosquito, *Aedes aegypti,* and most likely will function in all insects into which it is placed. Similarly, the ability of a yeast restriction enzyme to function in *Drosophila* suggests that it should also function correctly in a range of insect species into which it is placed.

GENETIC ENGINEERING IN NONDROSOPHILID INSECTS

Genetic Transformation of Nondrosophilid Insects

The *P* element paradigm is successful in nondrosophilid insects. Despite many attempts, the *P* element was found to be unusable as a gene vector in nondrosophilid insect species.

The reason for the narrow host range of *P* is unknown; however, it has been proposed that *P* is dependent for its mobility, in part, on the presence of host-encoded factors. These are thought to be absent, or at least sufficiently diverged, to prevent the mobility of *P* in these species. The *P* element is, however, not required for insect transformation because of the discovery and performance of four transposable elements, each from a separate family of transposable elements. Each of these is endowed with a broad host range, and each can transform *D. melanogaster* as well as a number of nondrosophilid species. They are briefly described below.

What is conserved between drosophilid and nondrosophilid transformation has been described as the *P* element paradigm. This refers to the mode of transformation. The *P* element and the four elements described shortly are class II transposable elements. They all transposase by a "DNA-only" type of mechanism—no production of an RNA intermediate is needed. These elements have an overall structure that is shared between them. They are short (< 4 kb), have inverted terminal repeated sequences, and encode a transposase enzyme that catalyzes the movement of the transposable element from one genomic location to the next. The same methodology is used to introduce these transposable elements regardless of species. Typically two plasmids are coinjected into preblastoderm embryos. One plasmid contains the transposable element, into which has been placed a genetic marker and an effector gene—a gene meant to alter the phenotype of the insect in a desired way. The placement of the marker gene and the effector gene interrupts and inactivates the transposase gene within the element, necessitating the use of a second plasmid containing the corresponding transposase, which is typically placed under the control of an inducible promoter such as the hsp70 promoter of *D. melanogaster*. This transposase mediates the transposition of the transposable element from the donor plasmid to the genome of the developing germline cells. As for *D. melanogaster* transformation, the individual arising from the injected embryo is not transformed; rather, it contains genetically transformed gametes. Individuals are mated, and transgenic insects are screened for in the next generation.

TRANSPOSABLE ELEMENTS USED FOR NONDROSOPHILID INSECT TRANSFORMATION Four
transposable elements can be used to genetically transform nondrosophilid insects: *piggyBac*, *Hermes*, *Mariner*, and *Minos*.

piggyBac The 2.5-kb *piggyBac* element has 13-bp inverted terminal repeats and 4-bp direct repeats located proximally to these. It contains a 2.1-kb open reading frame that encodes a transposase enzyme. *piggyBac* was discovered through its ability to transpose from the chromosomes of the Cabbage looper *Trichoplusia ni* into the genome of a baculovirus that had infected this TN368 cell line. Transposition of *piggyBac* into the baculovirus genome led to a mutation that resulted in few polyhedra being generated, in turn causing a clear change in cell morphology. *piggyBac* inserts only at TTAA sites and generates duplications of this sequence at the target site. Excision of *piggyBac* is precise—unlike other class II insect transposable elements, no deletions or additions of DNA remain at the empty excision site. *piggyBac* has found wide use as a gene vector in insects and has been used to genetically transform the flies *C. capitata*, *Bactrocera dorsalis*, *Anastrepha suspensa*, *Musca domestica*, *L. cuprina*, and *D. melanogaster*; the mosquitoes *Anopheles albimanus*, *An. stephensi*, *An. gambiae*, and *Ae. aegypti*; the moths *Bombyx mori* and *Pectinphora gossypiella*; and the beetle *Tribolium castaneum*. Little is known about the distribution of *piggyBac* throughout insects, although highly similar elements have recently been found in three strains of *B. dorsalis*. Over the 1.5 kb of nucleic acid sequence examined, these *B. dorsalis* elements are 95 to 98% identical to the element originally isolated from *T. ni* cells. Two of these *B. dorsalis piggyBac*-like sequences contain small deletions that interrupt the open reading frame, whereas the third has an intact open reading frame over the region examined. Conceptual translation of this region yields a sequence identity of 92% compared with the corresponding region of the *T. ni piggyBac* transposase. The basis of the distribution of *piggyBac*-like elements combined with the possible effect that incumbent *piggyBac*-like sequences may have on introduced elements in transgenic lines is a fertile field for investigation.

Hermes *Hermes* elements are members of the *hAT* family of transposable elements that are widely dispersed in animals and plants. Some members of this family, such as the *Ac* element of maize and the *Tam3* element of snapdragon, have a broad host range, and this attribute is shared with the *Hermes* element. *Hermes* was isolated from the house fly, *M. domestica*, and was first recognized by its ability to cross-mobilize the related *hobo* element when this was introduced into house fly embryos by microinjection. The 2.7-kb *Hermes* elements contain 17-bp inverted terminal repeats and a 1.8-kb open reading frame that encodes a transposase of 70 kDa. *Hermes* elements exhibit a preference for inserting at 5'-GTnnnAC 3' sites and create 8-bp duplications of these sites upon insertion. They have been used to genetically transform *D. melanogaster*, *C. capitata*, *Stomoxys calcitrans*, *Ae. aegypti*, *Culex quinquefasciatus*, and *T. casteneum*. Plasmid-based transposition assays have shown that *Hermes* can transposase in several other insect species as well. *Hermes* transposes by a cut-and-paste mode of transposition in higher Diptera but seems to integrate by another, transposase-dependent mode in mosquito germlines. The molecular basis of this remains unknown. *Hermes* elements can interact with the related *hobo* element (and vice versa) when both are present in the genome of *D. melanogaster*.

Mariner *Mariner* elements are widespread among arthropods. They are approximately 1.3 kb with inverted terminal repeats typically around 30 bp long. *Mariner* elements can be present in an extremely high copy number in some species; however, it seems likely that only a handful (if any) of these may contain a single open reading frame that encodes an active transposase of approximately 33 kDa. Based on DNA sequence comparisons, five different subfamilies of *Mariner* elements exist in arthropods. The distribution of members of these subfamilies is inconsistent with the established evolutionary histories of their host species and it is now accepted that *Mariner* elements have been horizontally transferred throughout evolutionary time. At present only one naturally occurring, active *Mariner* element has been discovered. This is the *MOS* element from *Drosophila mauritiana* and has been used to genetically transform *D. melanogaster*, *Ae. aegypti*, and *M. domestica*. Indeed *MOS* displays a broad host range and has been used to genetically transform *Leishmania*, chickens, and zebrafish. The mobility characteristics of *MOS* are preserved in these species; it transposes by a cut-and-paste mechanism and inserts at, and duplicates, TA nucleotides. A second active element, *Himar*, was constructed based on a consensus of *Mariner* sequences obtained from the horn fly, *Haemotobia irritans*. *Himar* is active in *Escherichia coli* but so far is inactive in insects.

Minos   The *Minos* element is a member of the *Tc*1 family of transposable elements. The *Tc*1 family of elements is related in sequence and mobility properties to the *Mariner* family of elements, and both are grouped into a single superfamily of elements. *Minos* elements are approximately 1.8 kb and possess long, 254-bp inverted terminal repeats. *Minos* contains two long open reading frames that are interrupted by an intron. Conceptual translation of the *Minos* transposase gene reveals a greater than 40% identity with the *Tc*1 transposase of *Caenorhabditis elegans*. *Minos* has been used to genetically transform *C. capitata*, *D. melanogaster*, and *An. stephensi*.

Transposable Elements in New Hosts

These four transposable elements just discussed provide the means by which genes can be introduced into pest insect species. Although these elements represent four different transposable element families, the transformation frequencies achieved, with some exceptions, are in the range of 1 to 10%. It seems likely that all will enjoy use as gene vectors in a range of insect species, and all may well be subject to interactions with endogenous transposable elements or other host factors present in these species. This is an important point that is not encountered by geneticists working on *Drosophila*. The recipient strains used for *P* element transformation are devoid of *P* elements (and any other related elements) and are deliberately chosen for this reason. This is not possible in other insect species in which the composition of the target genome with respect to transposable elements is unknown.

Whether interactions with endogenous transposable elements and/or host factors occur at levels that detrimentally affect transgenic stability is an issue that must be addressed.

Central to this is development of a complete understanding of how these transposable elements are regulated both in their original host species and in species into which they have been introduced.

Genetic Markers

The development of universal genetic marker genes, together with the identification of promoters to drive their expression in heterologous species, has played a major role in the extension of genetic engineering into nondrosophilid insects. Natural and modified forms of the green fluorescent protein (GFP) gene of the jellyfish, *Aequeria victoria*, have enabled transgenic insects in several species to be easily identified from nontransgenic siblings at most stages of development. These include *D. melanogaster*, *C. capitata*, *B. dorsalis*, *Ae. aegypti*, *An. stephensi*, *Cx. quinquefasciatus*, *P. gossypiella*, *T. casteneum*, and *S. calcitrans*. In these species, the GFP gene has been placed under the control of a promoter that enables either organelle-specific or tissue-specific expression of the marker gene to occur. Examples of the former are the actin5C and polyubiquitin promoters of *D. melanogaster*. Examples of the latter are the Pax6 and actin88 promoters. The actin88 promoter is from *D. melanogaster* and is specifically expressed in the indirect flight muscles of the pharate adult and adults. Pax6 is a member of the Pax family of transcription factors and is specifically involved in the development of the eye and central nervous system.

The enhanced GFP (EGFP), cyan fluorescent protein (CFP), yellow fluorescent protein (YFP), and Ds Red forms of the fluorescent protein genes can also function as genetic markers in insects.

OTHER APPROACHES TO GENETIC ENGINEERING IN INSECTS

FLP/FRT Recombinase in Nondrosophilid Insects

The FLP/FRT recombinase system of the yeast *Saccharomyces cerevisiae* can also function correctly in at least one nondrosophilid species. Plasmid-based excision and integration assays showed that the FLP recombinase enzymes could recognize and recombine FRT sites in the soma of developing *Ae. aegypti* embryos. Excision at the FRT sites was high—60% of plasmids examined had undergone an excision event mediated by FLP recombinase. Integration, as measured by the formation of heterodimeric plasmids arising from the recombination between two plasmids each containing an FRT site, occurred at a low, but statistically significant, frequency. The ability of the FLP/FRT recombinase system to function correctly in *Drosophila* and *Aedes* suggests that it should function across a range of insect species. It cannot, however, be used to directly genetically transform an insect species because to achieve this, FRT sites must first be introduced into the target genome by some other means, such as transposable elements. When combined with transposable element technology, this yeast recombination system should allow investigators to undertake precise manipulations of both introduced and host DNA. This ability will be of particular importance if DNA sequences necessary for the movement of transposable elements need to inactivated, (e.g., for regulatory reasons) following initial integration of the element into the target genome.

RNA-Mediated Interference (RNAi) in Insects

RNA-mediated interference (RNAi) refers to the targeted disruption of gene expression arising from the introduction of double-stranded RNA (dsRNA) into the cell. This disruption is targeted only to RNA molecules homologous to the invading dsRNA. It was initially characterized in plants and in the nematode *C. elegans* but is now thought to be a general phenomenon of eukaryotic cells that enables them to overcome invasions of RNA viruses. The mechanism by which RNAi works is unknown. It does not work through a simple titration of nascent or messenger RNA as would occur for an antisense RNA-based mechanism because the RNAi response can be elicited by far fewer dsRNA molecules per cell than, target RNA molecules. A catalytic mechanism in which the presence of dsRNA induces the destruction of homologous cellular RNAs has been recently proposed. RNAi technology has been harnessed to allow the targeted inactivation of specific genes and will prove to be a valuable component of genomics projects in those species in which nucleic acids can be introduced into cells. In its original experimental design, the effects of RNAi were not inherited. RNAi technology has recently been combined with *P* transposable element technology in *D. melanogaster* to produce heritable RNAi-mediated gene inactivation. Thus it is possible to examine the function of genes expressed in later stages of development of this insect and also the generation of genetically stable mutant lines in which production of the dsRNA can be induced or terminated based on the promoter used to drive expression of the extended hairpin loop RNA. RNAi technology should be extendable into other insect species in which transformation systems exist, and its exploitation in insects such as mosquitoes will enable the effects of the selective inactivation of specific genes to be quickly determined. This will represent a significant advance over traditional methods of creating and isolating mutants in these insect species that have not traditionally been amenable to genetic analyses.

EXAMPLES OF INSECT GENETIC ENGINEERING FOR INSECT POPULATION CONTROL

Transgenic technology in nondrosophilid insects has already been applied to examine promoter function and gene expression in transgenic lines of *Ae. aegypti* and *C. capitata*. In addition, recent work performed in *D. melanogaster* illustrates how transgenic approaches may be applied to pest insect

control in the foreseeable future. This approach involves using transgenic technology to develop new genetic sexing strains. Although these experiments have been performed in *D. melanogaster,* the concepts are applicable to any insect species in which transgenic technology has been developed, and the ability to generate and test novel genetic strains in pest insect species should result from such additional experiments.

Both systems were centered on exploiting the tetracycline-controlled transactivator (rTA) gene, which is inactivated in the presence of tetracycline. As a dietary component, tetracycline can readily be administered to *Drosophila* larvae in measured doses. Both systems consist of two components, which are combined in a single strain when transgenic lines containing each component are crossed. The rTA gene was placed under the control of the enhancer from the yolk protein 1 (*yp1*) gene of *D. melanogaster.* This enhancer results in fat-body- and female-specific expression of the *yp1* gene. The second component of their system was a proapoptosis gene *(head involution defective—hid),* the expression of which leads to apoptosis and the death of the organism. The *hid* gene was placed under the control of the tetracycline operator (tetO), which contains the binding site for the rTA protein. Thus, in females the yp1-rTA gene is induced and, in the absence of tetracycline in the diet, the rTA protein binds to the tetO sequence and so induces the expression of the *hid* gene. All transgenic females that were raised in the absence of tetracycline and possessed both components of this lethal genetic system died. When tetracycline was added to the diet, the rTA protein was inactivated, and there was no female lethality. Males containing both components were unaffected on either diet.

These experiments clearly demonstrate that transgenic technology can be used to construct efficient genetic sexing strains in at least one species of insect—*D. melanogaster.* The genes, promoters, and enhancers chosen to do so are predicted to be of generic use in insects. The tetracycline-controlled transactivator system is from bacteria and, given that it functions correctly in *Drosophila,* will most likely be applicable to all insects in which tetracycline, or its analogues, can be delivered in measured doses. Female-specific enhancers would be expected to exist in nondrosophilids, should the *D. melanogaster* enhancers not function correctly in these species. Similarly, should conditional lethal alleles of *Drosophila* genes not function in other species, it should be possible to generate analogous mutants either by established procedures or by employing an RNAi-based approach. The extension of these strategies into nondrosophilid insects requires, in the end, genetic transformation procedures and, as already discussed, several of these now exist for nondrosophilid insect species.

CONCLUDING REMARKS

For many years the absence of genetic transformation techniques for nondrosophilid insect species was seen as bottleneck for the full extension into these important pest species of strategies based on molecular genetics. The development of successful transposable-element-based transformation technologies enables the potential of these strategies to be tested at last. Insect geneticists have at their disposal gene vectors, universal genetic markers, promoters that can be utilized in heterologous insect species, and many target genes to test and manipulate. In addition, as outlined here, there is reason to be confident that generic techniques such as gene tagging, enhancer trapping, homologous recombination, FRT/FLP recombination, and RNAi-based gene silencing can now also be applied to insects other than *D. melanogaster.* Reports of sex-specific lethal genetic systems working in *Drosophila* have been published, and there is every expectation that similar systems will soon be established and tested in pest insects. All these technologies are precise—targeting only the genes that investigators seek to change—and the effects on a laboratory population can be predicted and are unambiguous. How successfully these technologies can be extended into pest insects, both in the laboratory and in the field, will be a matter of some interest in the years ahead.

See Also the Following Articles

Drosophila melanogaster • *Sterile Insect Technique*

Further Reading

Atkinson, P. W., and James, A. A. (2002). Germ-line transformants spreading out to many insect species. *In* "Advances in Genetics" (J. C. Hall, ed.), Vol. 47, pp. 49–86.

Atkinson, P. W., Pinkerton, A. C., and O'Brochta, D. A. (2001). Genetic transformation systems in insects. *Ann. Rev. Entomol.* **46,** 317–346.

Berghammer, A. J., Klingler, M., and Wimmer, E. A. (1999). A universal marker for transgenic insects. *Nature* **402,** 370–371.

Collins, F. H., and James, A. A. (1996). Genetic modification of mosquitoes. *Sci. Med.* **3,** 52–61.

Engels, W. R. (1989). *P* elements in *Drosophila. In* "Mobile DNA" (D. Berg and M. Howe, eds.), pp. 437–484. American Society for Microbiology, Washington, DC.

Handler, A. M., and James, A. A. (2000). "Insect Transgenesis—Methods and Applications." CRC Press, Boca Raton, FL.

Heinrich, J. C., and Scott, M. J. (2000). A repressible female-specific lethal genetic system for making transgenic insect strains suitable for a sterile-release strain. *Proc. Natl. Acad. Sci. USA* **97,** 8229–8232.

Horn, C., and Wimmer, E. A. (2000). A versatile vector set for animal transgenesis. *Dev. Genes Evol.* **201,** 630–637.

Kennerdell, J. R., and Carthew, R. W. (2000). Heritable gene silencing in *Drosophila* using double-stranded RNA. *Nat. Biotechnol.* **18,** 896–898.

Montgomery, M. K., Xu, S., and Fire A. (1998). RNA as target of double-stranded RNA-mediated genetic interference in *Caenorhabditis elegans. Proc. Natl. Acad. Sci. USA* **95,** 15502–15507.

Morris, A. C., Schaub, T. L., and James, A. A. (1991). FLP-mediated recombination in the vector mosquito, *Aedes aegypti. Nucleic Acids Res.* **19,** 5895–5900.

O'Brochta, D. A., and Atkinson, P. W. (1998). Building the better bug. *Sci. Am.* **279,** 90–95.

Roberts, D. B. (1998). "*Drosophila*—A Practical Approach." 2nd ed. IRL Press at Oxford University Press, Oxford.

Rong, Y. S., and Golic, K. G. (2000). Gene targeting by homologous recombination in *Drosophila. Science* **288,** 2013–2018.

Rong, Y. S., and Golic, K. G. (2001). A targeted gene knockout in *Drosophila. Genetics* **157,** 1307–1312.

Genetic Variation

George K. Roderick
University of California, Berkeley

The genome, the entire collection of an organism's genetic material, provides the blueprint containing information that dictates all biological forms and functions. Without change, this blueprint would be passed identically to future generations, preserving past genetic structures that have proven to be successful, but also constraining potential future adaptation to new situations. Only when genetic variability is present can processes such as selection, genetic drift, and migration act to change the frequencies of genetic variants and in so doing allow evolution that may lead to adaptation and ultimately even speciation. Thus, understanding genetic variation—its origins, maintenance, and pattern of change—is critical to understanding the diversity of life.

GENETIC VARIABILITY: TYPES AND ORIGINS

How much genetic variation exists in natural populations? Before easy access to DNA sequences themselves, genotypic variants or polymorphisms were examined at the level of chromosome banding, particularly in the salivary glands of *Drosophila,* and through genetically based variation in enzymes as revealed through allozyme electrophoresis. Studies of enzyme variation in the mid-1960s by Lewontin and Hubby working with *Drosophila pseudoobscura* showed that an unexpectedly high number of loci were polymorphic (two or more alleles were found in 30% of all loci examined) and that over all loci nearly every individual was genetically unique. This work prompted the question of what was responsible for all this genetic variation, and in particular, did natural selection maintain this polymorphism or was the variation selectively neutral, being influenced only by processes of random genetic drift? That many loci are in fact polymorphic has been confirmed more recently with information directly from DNA sequences of both protein-coding and noncoding regions of DNA.

How is this variation then expressed in the observable phenotype? The link between the genotype and the phenotype is often relatively straightforward. Many phenotypic traits are determined by only one or a few genetic loci. However, other traits are influenced not by one or two loci, but by many loci, each with a relatively small effect; here, the link between genotype and phenotype is described by quantitative genetics. Usually such traits have measurable genetic and environmental components and frequently an interaction between the two. For such traits, the amount of variation that is "genetic" is described as the heritability, usually denoted h^2, which can be estimated through breeding studies or by examining relatives of known genetic relatedness. Crosses of inbred lines that differ in traits of interest can be used to determine the inheritance, approximate number, and relative importance of loci responsible for variation in those traits. In such analyses, loci responsible for variation in quantitative traits are termed quantitative trait loci.

The origins of genetic variability rest in the processes of mutation, which are typically classified by the type of change caused by the mutational event. Mutations can arise through substitution (one nucleotide is replaced by another), recombination (crossing over and gene conversion), deletion (one or more nucleotides are removed), insertion (one or more nucleotides are added), or inversion (180° rotation of a double-stranded DNA segment). In protein-coding regions, some nucleotide substitutions do not change the amino acid for which they code and such substitutions are termed synonymous or silent. Substitutions that change the amino acid are termed nonsynonymous. Rates of mutation vary widely and can be influenced by the internal genetic environment as well as by the external environment. Aspects of the genetic environment that have been shown to affect rates of mutation include the functional role of the region (whether coding or not) and, for a given base, its position within a gene (e.g., stems or loops of ribosomal DNA) or within a codon (e.g., third positions change much more frequently than first or second positions due to redundancy in the genetic code). Rates of mutation are also affected by genome size and type, i.e., whether organellar or nuclear. For example, in insects, rates of mutation for protein-coding regions are typically higher for haploid mitochondrial DNA than for diploid nuclear DNA, a fact usually attributed to a lack of an efficient mechanism for DNA repair in insect mitochondrial DNA. Recent findings have shown that parts of the genomes of organisms may originate from other sources. For example, recent research on Orthoptera and Diptera has revealed nuclear DNA inserts of what were previously mitochondrial genes, and transposable elements, highly mobile pieces of DNA, are likely widespread in insect nuclear genomes. Of course, rates of mutation are also affected by external environmental variables, such as temperature and radiation.

MAINTENANCE OF GENETIC VARIABILITY

At equilibrium, the gene frequencies in a population will not change from one generation to the next. This important concept has been formalized in the Hardy–Weinberg Principle (H–W), which forms the foundation for the general understanding of population genetics of sexually reproducing organisms and, by extension, the understanding of the genetic theory of evolution. H–W is best thought of as a population genetic "null" model. The basic idea is that a single generation of random mating results in genotype frequencies that are directly predictable by the frequency of alleles in the population, no matter what history gave rise to the current mixture. This concept results in the familiar equation for a system of two alleles at one locus,

$$p^2 + 2pq + q^2 = 1, \tag{1}$$

where p and q are the frequencies of the two alleles in the population and p^2, q^2, and $2pq$ are the frequencies of the three possible genotypes (two homozygotes and heterozygote, respectively). A number of assumptions are explicit in H–W, including random mating, infinite population size (i.e., no random genetic drift), no gene flow or migration, no mutation, and individuals all having equal probabilities of survival and reproduction (i.e., no natural selection). It is the study of deviations from these assumptions that makes the principle so useful, particularly for examination of nonrandom mating, genetic drift (chance events), gene flow and migration, and selection.

ADAPTATION

Genetic variation is fundamental to Darwin's theory of evolution through natural selection, although when the idea was initially developed the mechanisms of inheritance were not yet known. Selection acts to favor some phenotypes over others, resulting in differences in relative fitness. The extent to which these phenotypes have a genetic basis determines whether those phenotypes that survive will pass on their attributes to their offspring. This process of genetic change through natural selection is termed adaptation. It can most readily be studied in recognizably polymorphic species, i.e., those in which genetic variability can be monitored over space and time. Accordingly, perhaps the best known study of natural selection in the wild is that of industrial melanism in the peppered moth, *Biston betularia,* in Britain. The typical form *(typica)* of this moth is light colored with black, pepper-like spots and is well camouflaged on lichen-covered birch trees. Following industrialization of certain areas of England, a darker form *(carbonaria)* started to appear in great abundance. This prompted the question, were the typical forms more subject to bird predation where the lichen on trees was soot covered and darker, leaving the darker forms in greater numbers? This issue was addressed by Kettlewell, in the 1950s, who manipulated the relative frequency of morphs and monitored their success. His findings were consistent with the hypothesis of predation—*typica* are eaten from dark backgrounds much more often than from light backgrounds, and vice versa for *carbonaria.* Further, a more recent reanalysis has demonstrated that, associated with the postindustrial reduction in pollution, there has been a shift to a lower frequency of the darker *carbonaria* forms. Still, a number of issues remain unresolved, demonstrating the difficulty involved in studying such systems in nature. For example, the *carbonaria* form was been found to persist even in areas relatively unaffected by pollution and two melanic forms have been shown exist in nature.

Adaptation has been studied using a number of other genetically determined polymorphisms in nature as indicators of variability. These studies include not only the now familiar color polymorphisms, such as are found in butterfly wing patterns, ladybird beetles, walking sticks, and happy face spiders, but also behavioral polymorphisms such as caste structure in social insects. Because of the relative ease by which polymorphisms can be measured phenotypically, and the often relatively simple genetic basis that underlies the variation, studies of polymorphisms have contributed much to the general understanding of natural selection in the field, particularly in demonstrating that selection can be a very powerful force and that balancing selection can maintain genetic variability.

SPECIATION

How much genetic divergence is necessary for two populations to maintain themselves as separate species? In part, the answer to this question depends upon how the species themselves were formed. In a study of allozyme polymorphism in allopatric populations (those found in different localities) within the *Drosophila willistoni* group in Venezuela, Ayala and colleagues showed that isolated populations of the same species were very similar genetically, with a mean genetic identity (the proportion of loci identical in two samples) of $I = 0.970$. However, the genetic identity was lower in populations that were practically reproductively isolated (termed "semispecies"; $I \approx 0.8$) and even lower in well-recognized, reproductively isolated sibling species ($I = 0.517$) and closely related, but nonsibling, species ($I = 0.352$).

Species formed in sympatry (i.e., in the same locality) might potentially be more similar genetically than those formed in allopatry, particularly if only a small number of loci are important for initial divergence. Sympatric speciation is most commonly associated with true fruit flies (Tephritidae) in the genus *Rhagoletis*. Bush in the 1960s suggested that speciation in these flies may have occurred following shifts in host use within the same habitat. He offered as an example of the process the host shift of the apple maggot, *Rhagoletis pomonella*, from its native hawthorn host to introduced apples in New York State in the 1860s. There was considerable resistance initially to the concept of a nonallopatric mode of speciation. However, more recent work by Bush and colleagues has provided a convincing case for sympatric speciation in *Rhagoletis*, with differences between species being maintained through genetic control of emergence times on the different hosts. There also exists at least partial premating reproductive isolation associated with host-plant fidelity. Berlocher in 1976 measured the mean genetic identity between *R. pomonella* and two other different species of *Rhagoletis* that were likely formed in sympatry through host shifts and found the species to be nearly identical genetically ($I = 0.980$ and $I = 0.989$) and significantly more similar than were *Rhagoletis* species formed in allopatry and the *Drosophila* species noted above.

The conclusion from these studies is that the average genetic divergence itself is not critical for speciation, nor for the maintenance of reproductively isolated species, but rather that the effects of a small set of nonrandom loci may be important in species formation. Recent advances by Feder, facilitated through better understanding of the *Rhagoletis* genome, and by others studying *Heliconius* butterflies and pea aphids suggest that changes in only relatively few functional loci can lead to rapid speciation.

THE IMPORTANCE OF GENETIC VARIABILITY

How important is genetic diversity itself to the persistence of insect populations? This question is difficult to answer. Clearly, the presence of resistant alleles in pest populations has led to the development of resistance to numerous insecticides or to the development of virulence on resistant plant varieties. For example, it is not uncommon to find alleles for resistance in mass screening that takes place through field applications of insecticides or in the field use of transgenic plants expressing insecticidal proteins. By contrast, it is more difficult to detect such alleles in laboratory studies in which only a relatively small handful of individuals are sampled. Yet, for many species, genetic diversity seems to matter little. Many parasitoids, for example, are notoriously lacking in genetic polymorphisms, perhaps as a result of years of brother–sister mating. Invasive species such as the Mediterranean fruit fly, *Ceratitis capitata*, have reduced genetic variability, likely as a result of successive population bottlenecks, yet survive well while expanding their range. Indeed, the lack of variability may explain the success of some invasive species. Invasive populations of the Argentine ant, *Linepithema humile*, for example, have lost the genetic ability to distinguish one colony from another, thus escaping from population control imposed by intercolony conflict.

In these examples, high levels of genetic variability do not seem to be important for survival. However, it should be noted that the pests and invasive species that are the largest economic problems are only a very small sample of the diversity of insect species and that these species have experienced selection over many generations to be successful as pests or invasives. During this period such species may have overcome potential genetic obstacles associated with low levels of genetic variability. In natural populations of insects, however, the importance of genetic diversity still remains poorly understood.

APPLICATIONS

The genetic variability held within individuals and populations can provide critical insights into the structure and dynamics of populations that would otherwise be difficult or impossible to study. For example, variation in allele frequencies among populations can be used to assess the genetic structure of populations, summarized by F_{ST} (calculated as the variance in allele frequencies standardized by the mean) or related measures. From these measures, with appropriate assumptions, it is often possible to infer additional information on the biological dynamics in those populations, including attributes such as effective population size (N_e), degree of inbreeding, and rates of gene flow or migration. Genetic variability also provides the opportunity for tracing the history of populations, species, and their ancestors through methods that recognize the genealogical nature of genetic material.

A second application that uses information on genetic variation is in the area of insect pest management. For example, an understanding of the genetic basis of resistance to chemical insecticides (whether administered externally or through genetic modification of plants) has been critical in the development of strategies to delay the evolution of resistance in herbivorous insects. Genetic variability is also important in the ability of insects imported for biological control to establish themselves, as well as in their potential to attack nontarget hosts.

Finally, genetic information has been of great value in the area of conservation biology and biodiversity, as for example in efforts to determine which insect species or populations are most worthy of protection. Specific applications include genetic estimation of population sizes and spread and assessment of genetic or phylogenetic uniqueness for assignment of conservation priorities. The full value of such measures based on genetic variability has yet to be realized, and future developments will almost certainly provide further insights into both past histories and future trajectories.

See Also the Following Articles

Chromosomes • *Conservation* • *Industrial Melanism* • *Insecticide and Acaricide Resistance* • *Ladybugs*

Further Reading

Berlocher, S. H., and Feder, J. L. (2002). Sympatric speciation in phytophagous insects: Moving beyond controversy? *Annu. Rev. Entomol.* **29**, 403–433.
Falconer, D. S., and MacKay, T. F. C. (1996). "Introduction to Quantitative Genetics," 3rd ed. Longman, Harlow.
Ford, E. B. (1971). "Ecological Genetics," 3rd ed. Chapman & Hall, London.
Futuyma, D. J. (1997). "Evolutionary Biology," 3rd ed. Sinauer, Sunderland, MA.
Graur, D., and Li, W.-H. (1999). "Fundamentals of Molecular Evolution," 2nd ed. Sinauer, Sunderland, MA.
Hartl, D. L., and Clark, A. G. (1997). "Principles of Population Genetics," 3rd ed. Sinauer, Sunderland, MA.
Howard, D. J., and Berlocher, S. H. (eds.) (1998). "Endless Forms: Species and Speciation." Oxford University Press, Oxford.
Kettlewell, H. B. D. (1973). "The Evolution of Melanism." Clarendon Press, Oxford.
Price, P. W. (1996). "Biological Evolution." Saunders College, Fort Worth.

Grasshopper

see **Orthoptera**

Grassland Habitats

Teja Tscharntke, Ingolf Steffan-Dewenter,
Andreas Kruess, and Carsten Thies
University of Göttingen

Grasslands are plant communities that are based on grasses and herbs, and in which shrubs are rare and trees are absent. Perennial grasses represent the dominant species of grasslands, and make up the largest portion of their biomass, but not necessarily of their species richness. Grasses are often followed by legumes in abundance and herbs from many other plant families. Grassland is the natural vegetation in areas of low or strongly seasonal rainfall (250–1000 mm), but naturally occurring mammalian herbivory (e.g., by elephants) may also effectively suppress establishment of trees. Grasslands naturally encompass a wide range of habitat and vegetation types and span a large latitudinal gradient, from tropical grassland (savannas) to temperate grassland (the prairie in North America and the steppe in Eurasia) to the arctic tundra, totaling about 25% of the earth's land surface. Herbivory in temperate grasslands is dominated by insects, whereas large ungulate herbivores dominate in tropical grasslands. The temperate meadows and pastures are seminatural grasslands growing in essentially deforested areas with a forest climate, and their succession to forests is inhibited by mowing, burning, and human-controlled grazing.

INSECT COMMUNITIES ON GRASSLANDS

Grasslands are habitats for many insects and may harbor an extraordinarily species-rich community. One temperate old-field grassland may be habitat for more than 1500 insect species, whereas cereal fields, which are monocultures of annual grasses, may contain 900 species. Compared to forests, the structural complexity of the grassland vegetation is obviously simpler, so that the insect diversity is reduced. Similarly, the litter layer of forests is larger and more heterogeneous, with a correspondingly richer decomposer community. Further features of grassland-specific insect communities include the dominance of species adapted to feed on grasses.

Plant and insect communities of grasslands greatly differ depending on climate, soil type, and management practices. Some marked differences are apparent between the plant–insect communities of temperate and tropical habitats. Plant species richness, which determines much of the insect diversity, may be only 10 to 15 species in intensively managed and highly fertilized grasslands, but 50 to 70 in extensively managed and low-input temperate grasslands. In contrast, tropical grasslands may contain over 200 plant species. Chalk-rich temperate grasslands with abundant earthworm populations tend to support the highest faunal biomass (often >100 g fresh mass per square meter), whereas arid and semiarid steppe and desert soils, dominated by microfauna such as protozoans and nematodes, may have a biomass of only 1 g/m². Tropical grasslands and tundra tend to be somewhere in-between. Termites and ants are dominant groups in tropical and subtropical grasslands, some surface-dwelling predatory arachnids such as scorpions and solifugids are restricted to warm, arid soils, and cold tolerance limits the range of many species in arctic and antarctic conditions. The ways in which insect communities of grasslands are influenced will be the subject of the remainder of this article.

Which insect species attack grasses, and what are the typical plants of grasslands? Ectophages, which feed externally on leaf tissue by chewing, scraping, or sucking, are distinct from endophytic feeders, which include leafminers, gallers, and borers. Grass foliage-chewing insects belong primarily to the Orthoptera, Lepidoptera, Coleoptera (mainly Chrysomelidae and Curculionidae), Hymenoptera (Tenthredinidae), and Phasmida. Of the specialized grass chewers in Great Britain <2% are Coleoptera, 6% Lepidoptera, 6% Hymenoptera, and 41% Orthoptera (the grasshoppers). Specialization on grasses appears to be particularly important in grasshoppers, and their abundance in grasslands is high. Sap-feeders on grasses are Homoptera (Auchenorrhyncha, Sternorrhyncha, Pseudococcidae), Heteroptera (mainly Miridae), and Thysanoptera. The endophagous, mostly stem-boring herbivores belong primarily to the the Diptera (mainly Cecidomyiidae, Chloropidae, Agromyzidae), Hymenoptera (Cephidae, Eurytomidae), Lepidoptera (mainly Pyralidae, Noctuidae), Coleoptera (mainly Cerambycidae, Mordellidae, Chrysomelidae), and mites (Acari).

The number of endophagous insect species associated with grass species can be predicted by (1) the annual–perennial dichotomy (annuals, in contrast to perennials, support almost no endophagous insects); (2) the mean shoot length; and (3) the abundance of the host plant. Annuals have impoverished communities and both shoot length and abundance are positively correlated with insect diversity. Plant height is usually a surrogate for the complexity of the plant architecture and, generally, a well-known predictor of plant–insect ratios. Furthermore, the more widely distributed and abundant a plant is, the more insects it should encounter in its evolutionary history. Annuals, which typically dominate in early-successional habitats, are characterized by a faster relative growth rate than perennials and, therefore, a short exposure time, so that they are a spatiotemporally unpredictable resource for insects.

Grasses (which are monocots) are hosts of many specialized endophagous insects and a multitude of ectophagous insects, a pattern that shows no principal difference from that of dicots. Host-plant preferences of oligophagous grass feeders often differ between grass species. Even more, variability among the many commercially available strains of the perennial ryegrass *Lolium perenne* to frit fly (the stem-boring chloropid fly *Oscinella frit*) attack is

greater than the variability between many pasture species. Attack of many species, such as frit fly, are negatively correlated with silica content, which presumably influences the females' choice of oviposition site and larval performance. Further, wild biotypes are often better resources than grasses grown from commercially available seeds and support richer insect communities, which may be of importance for sowings with a nature-conservation background.

GRASSES AS FOOD RESOURCE

Grasses make up the largest portion of the grassland biomass; consequently, the insect communities of grasslands are determined more by the monocotyledonous Poaceae than by the dicotyledonous herb families. Grasses differ from the Dicotyledonae in that their architecture is simple, and the intercalary meristems, which substitute for growth from terminal buds, are protected by hard leaf sheaths. Most grasses lack the variety of secondary compounds that deter herbivory in most dicotyledons. For example, cyanogens and toxic terpenoids are rare, and alkaloids are present in <0.2% of grass species but in 20% of all vascular plants. Grass-feeding insects such as the oligophagous grasshoppers select their pooid-grass host plants in that they simply reject plant tissues enriched with secondary compounds (deterrents), while no phagostimulants characterizing grasses as a group have been found. Grasses are not toxic, but this does not mean that they are little protected from herbivory; just the contrary is true (see below).

Endophytic fungi have been considered acquired chemical defenses in grasses, and the main mechanism is the production of mycotoxins, notably alkaloids. The presence of these seed-borne *Neotyphodium* endophyte fungi may cause dramatic toxicosis to grazing livestock, best known from *L. perenne* and *Festuca arundinacea*. In addition to deterring vertebrate herbivory, these endophytes are also well known for increasing resistance to insect pests, microorganisms, and drought. Endophytes may also alter attack of natural enemies in that they enhance larval development time of the herbivore (the slow growth–high mortality hypothesis) or directly affect immature enemies, e.g., parasitoids feeding on the toxic tissues of their hosts.

Within and among grass species, a considerable chemical and morphological variability may be found. Nutrient availability of grass shoots is greatly determined by the shoots' age. Fresh internodes have high concentrations of the major nutrients (water, protein, minerals) and reduced concentrations of plant-resistance factors (raw fiber, silicate). High levels of plant nitrogen are generally associated with a high assimilation efficiency and density of phytophagous insects.

HERBIVORE–PLANT INTERACTIONS

Long-term experiments with chemical control to eliminate insect herbivores indicated an average annual yield loss of

15%, and nematode control increased biomass by 12 to 28%. The biomass losses appeared to be mainly the result of frit fly and other stem-boring Diptera, which kill the central grass shoots, root-feeding wireworms (*Agriotes* spp., Elateridae), root-feeding scarabeid grubs, the range caterpillar *Hemileuca oliviae,* armyworms (*Spodoptera* spp.), grass worms (*Crambus* spp.), the Mormon cricket (*Anabrus simplex*), and leatherjackets (*Tipula* spp., in wetter soils). Planthoppers (Auchenorrhyncha), grass bugs (genera *Labops, Irbisia, Leptopterna*) and grasshoppers (Acrididae), and plant-feeding nematodes may also be important pests. In Sweden, the grass-feeding antler moth *Cerapteryx graminis* may reach densities of 100 to 1500 individuals per square meter; their corresponding effects on grass biomass consequently enhance herb populations. In the years following *C. graminis* outbreaks, shifts from herb dominance to renewed grass dominance show effects of competitive release and the return to competitive exclusion.

In temperate grasslands, the below-ground standing crop of insects is 2 to 10 times greater than the aboveground insect mass, although the effects of below-ground insects remain largely unseen, unless scarabeid beetle larvae or nematodes cause heavy decreases in shoot growth or even kill grass over large areas. In a latitudinal gradient across North American grasslands, root-to-shoot ratios vary from 2:1 to 13:1, with high values in cooler climates; tropical grasslands have even lower ratios (0.2:1 to 2.6:1). As can be expected from these data, the soil fauna is less abundant in tropical savannas and forests compared to temperate ecosystems. Earthworms usually dominate the soil biomass, but in the tropics, termites and ants are particularly important. These below-ground species can be a key in nutrient dynamics determining plant growth and aboveground plant–insect interactions.

Grasses are well adapted to herbivory and, in general, tolerate grazing better than herb species; therefore enhanced grazing pressure increases the fraction of grasses in pastures. The high resistance, tolerance, and compensatory ability of grasses are the result of (1) the generally high silicate content, lignification of vascular bundles, and additional sclerenchyma in mature leaves that make foliage hard to chew and digest; (2) the rapid induction of dormant buds that develop into lateral shoots following defoliation or destruction of apical meristems, which is based on the below-ground nutrient reserves; (3) the location of meristematic zones that are in many instances near the ground and not at the top of the plant, where they would be better accessible to grazers; and (4) the compensatory photosynthesis and growth stimulation by bovine saliva, which may also play a role. However, the concept of a herbivore-optimization curve or even grass-grazer mutualisms overestimates the compensatory abilities of grasses and grasslands.

Grazing causes much sprouting from dormant buds and converts tall canopies into shorter and denser grazing lawns. Heavily grazed pooid populations are smaller and have higher silicate concentrations, exhibiting ecotypic variation as a result of different grazing histories. The mitigation of predation by the highly silicified grasses is presumably not confined to mammals, because the mandibles of many grass-chewing insects are adapted to biting and grinding and are analogous to the teeth of grazing mammals. Although the evolution of siliceous grass leaves appears to be driven by many stress factors (including drought and fungal attack), both mammal and insect herbivory may have been important factors.

CONSERVATION OF SPECIES-RICH GRASSLANDS

Insect diversity in grassland ecosystems can be best predicted by floral diversity or related characteristics of vegetation structure, especially biomass and structural heterogeneity of the plant community. Species richness of butterflies, wild bees, phytophagous beetles, true bugs, etc., was found to be positively related to the species richness of plants. However, age of the habitat as well as fragment size is known to disproportionally enhance the number of species in higher trophic levels. The fraction of specialized predators and parasitoids increases greatly with area and age of grasslands, although the plant species richness may respond little.

Intermediate levels of vegetation disturbance, caused by ants, rodents, foxes, rabbits, sheep, and other mammals, significantly increase species richness of vegetation with consequent effects on the insect community. For example, gaps reduce the likelihood of competitive exclusion in a plant community when space is monopolized by a few dominant species. The openings are rapidly exploited by seedlings. Rotational management also may enhance grassland heterogeneity, creating a mosaic of old and young, tall and short, early and late successional patches.

Mineral or organic fertilization of meadows or rangeland increases biomass and may also enhance palatability of the nitrogen-rich foliage, resulting in higher insect densities. But the main result of continued grassland fertilization is a steady reduction in plant species richness with a corresponding loss of insect species.

Grasslands established by sowing are colonized in the beginning by relatively few insects. As these grasslands age, communities become more species rich in both plants and insects, and the biotic interactions, such as between predators and their prey or parasitoids and their hosts, increase. Ants and subterranean insects are absent on newly created fields because establishment of nests and populations needs time, and their highest densities occur in mature grasslands. Percentage of macroptery (i.e., those with full wings) in dimorphic insects such as grass-feeding planthoppers is high in early successional habitats, whereas brachypterous (short-winged) species dominate in persistent habitats.

The destruction and fragmentation of habitats has become one of the major threats to biodiversity. Not all insect species are equally affected by habitat fragmentation: species of higher trophic levels, rare species, species with specific habitat requirements, species with greatly fluctuating populations, and species with poor dispersal abilities are expected to be more prone to extinction. For example, butterfly communities on calcareous grasslands show positive species–area relationships, and the most specialized and endangered butterflies profit most from large grassland fragments. For a few butterflies, morphological characters associated with flight ability have been shown to change with isolation of limestone habitat fragments. This indicates that habitat fragmentation in simple, human-dominated landscapes may also have evolutionary consequences for the life-history traits within populations. As a result of changes in community structure, interspecific interactions such as plant–pollinator interactions may be disrupted. Grasses are wind-pollinated, but most herbs, such as the many legume species that typically play a major role in nutrient-poor or extensively managed grasslands, depend on insect pollination. Populations of pollinating bees need nectar and pollen resources as well as suitable nesting sites. Both may be limiting in small grassland fragments, and so very small plant patches usually receive fewer pollinator visits. Plant ecologists have found clear evidence that pollination efficiency, gene flow by pollen dispersal, and seed set are reduced in small calcareous grasslands. Habitat fragmentation is known to also affect specialized populations of higher trophic levels, for example, in a plant–herbivore–parasitoid food chain. Communities of monophagous butterflies show a steeper increase with the area of grasslands than communities of plants. Theoretical models and empirical evidence show that specialized parasitoids (and predators) suffer even more, so that food chain length tends to be shortened and herbivores tend to become released from possible control of their natural enemies.

Habitat quality of species-rich grasslands such as the calcareous grasslands mainly depends on the opposing forces of management (see below) and succession. Speed of succession may be related to fragment size because late-successional shrubs and trees often invade from the edge. Abandoned grasslands will often increase in species richness of both plants and insects, but they will certainly decrease on late-successional grasslands (for example after 10–20 years of abandonment), when shrubs and trees become dominant. Many specialized butterflies mainly occur on regularly mown or grazed calcareous grasslands as they appear to rely on warm microclimates and host plants associated with only sparse vegetation (Fig. 1). The rare British butterfly *Hesperia comma* prefers small plants of the grass *Festuca ovina* surrounded by sunny bare ground and nectar resources as oviposition sites. Death of rabbits from the myxomatosis virus appeared to enhance population declines of this butterfly, because the reduced rabbit populations caused less grazing. Ground-nesting species such as solitary bees are also more abundant on regularly mown or grazed grasslands, because the sparse vegetation and open soil provide nesting sites and thereby greatly enhance populations. In contrast, aboveground-nesting solitary bees are enhanced by dense, high, and woody vegetation that offers the necessary plant

FIGURE 1 Calcareous grasslands belong to the most species-rich habitat types in Central Europe and depend on annual cutting or grazing (near Göttingen, Germany, photograph by Jochen Krauss).

material for nest construction. Altogether, species will profit from early-, mid-, or late-successional stages depending on their life-history traits, and highest overall diversity should be conserved with a mosaic of different successional stages.

MANAGEMENT OF GRASSLAND

Cutting, grazing, and burning are typical methods of grassland management. As management alters plant growth and vegetation structure profoundly, the community of associated insects will also change. The insects' responses greatly differ between functional and taxonomic groups, and consequently it is often difficult to decide which management strategy is best in the conservation of overall diversity. Closely cut (or grazed) grasslands typically have an impoverished insect fauna. This is partly the result of pronounced vertical stratification of species using different parts of the sward canopy during the growing season (Fig. 2) and has been shown for planthoppers and leafhoppers (Auchenorrhyncha) as well as phytophagous beetles (Coleoptera). In particular, cutting affects flower visitors, pollen feeders, and grass-seed feeders among the gall midges (Cecidomyiidae) and other groups (Miridae, Chloropidae, Thripidae). In general, vegetation height and, therefore, the structural complexity of grasses decrease with intensity of grazing or mowing, and the complexity of plant architecture is a good predictor of insect species richness. The positive correlation between aboveground plant biomass and insect species richness is well established, whereas root feeders and other soil invertebrates are often more abundant at intermediate levels of grazing or mowing (of temperate grassland) than on unaffected patches. When grassland has been left unmanaged for a few years, the hemipterous and coleopterous fauna quickly recovers. With increasing number of mowings per year, which may be best observed in the sometimes extremely often mown urban turf-grass areas, species richness of both plants and insects (e.g., planthoppers

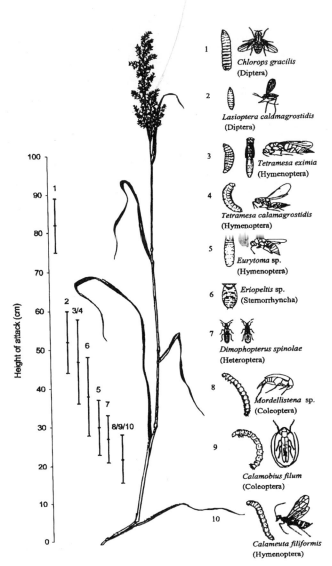

1 *Chlorops gracilis*
(Diptera)

2 *Lasioptera caldmagrostidis*
(Diptera)

3 *Tetramesa eximia*
(Hymenoptera)

4 *Tetramesa calamagrostidis*
(Hymenoptera)

5 *Eurytoma* sp.
(Hymenoptera)

6 *Eriopeltis* sp.
(Sternorrhyncha)

7 *Dimophopterus spinolae*
(Heteroptera)

8 *Mordellistena* sp.
(Coleoptera)

9 *Calamobius filum*
(Coleoptera)

10 *Calameuta filiformis*
(Hymenoptera)

FIGURE 2 The stem-boring insects feeding on pure stands of the grass *Calamagrostis epigeios*. Height of attack (arithmetic means and 95% confidence limits) is given for each of the 10 species. (Reproduced, with permission, from M. Dubbert, T. Tscharntke, and S. Vidal, 1998, *Ecol. Entomol.* **23**, 271–280.)

and true bugs) becomes extremely poor. Moderate cutting or grazing may promote grasshopper populations, possibly through tillering rejuvenation or through changes in the proportions of nutritious grasses. Because the quality of grass shoots as a food resource declines with age, the induction of tillers and side shoots by cutting make nutritious food available later in the season (for example, for the populations of many grasshoppers and enhanced infestations of stem-boring Diptera). In conclusion, the effects of cutting or grazing on insect communities can be divided into short-term effects (simplification of plant architecture, regrowth of young and nutrient-rich plants) and long-term effects caused by changes in the structure of plant communities.

Grazing adds to the effects of cutting in that grazers feed selectively on the more palatable plants, compress or loosen the ground by trampling, and fertilize grassland patches by

urination and the deposition of dung. Accordingly, the changes in the plant community following grazing affect insect community structure in a complex way and make the habitat more heterogeneous than a homogeneous cutting regime. Further, grazing is a gradual form of vegetation removal, except at high stocking densities, and thereby differs from the large-scale disturbance of cutting (or burning). Cattle feed on taller vegetation than sheep and may open up tall vegetation. The cattle's trampling effects are usually high compared to those of sheep and enhance vegetation heterogeneity with disturbed and bare areas that improve habitat quality of many invertebrates. High cattle densities, however, lead to short and uniform swards and create problems due to vegetation damage, especially on wetter soils and on slopes. Most of the nutrients removed by grazing are returned through the deposition of urine and dung. Cow dung harbors a unique and speciose insect community. The breakdown of ungulate dung in temperate environments is enhanced by fly maggots, such as *Scatophaga* sp. and dung-burying beetles such as the scarabeid *Geotrupes* sp. Deposition of bovine dung poses no problems where bovines have an evolutionarily associated fauna that exploits the fecal resources. However, in Australia, native detritivores could not process cow dung because cows were brought over by the first English colonists only at the end of the 18th century. The loss of pasture under dung has imposed a huge economic problem to agriculture in Australia, and only the decision in 1963 to establish African dung beetles there led to a solution.

Burning grassland is less common in Europe than in America or Australia, where burning is a widespread natural phenomenon. Burning, like cutting and grazing, tends to produce a greater floristic uniformity, and it is considered to be very detrimental to grassland invertebrates. Controlled burning has been suggested as an alternative to chemical or biological control of pest arthropods. Direct effects are diverse, depend on the intensity of the burn, and include the escape of many flying insects as well as few changes in many soil insects. Indirect effects are the xeric conditions after burning and the mineral-rich regrowth after burning, which for many animals is a superior resource quality.

See Also the Following Articles
Biodiversity • Plant–Insect Interactions • Soil Habitats

Further Reading
Curry, J. P. (1994). "Grassland Invertebrates." Chapman & Hall, London.
Siemann, E. (1998). Experimental tests of effects of plant productivity and diversity on grassland arthropod diversity. *Ecology* **79**, 2057–2070.
Stanton, N. L. (1988). The underground in grasslands. *Annu. Rev. Ecol. Syst.* **19**, 573–589.
Steffan-Dewenter, I., and Tscharntke, T. (2002). Insect communities and biotic interactions on fragmented calcareous grasslands—A minireview. *Biol. Conserv.,* **104**, 275–284.
Tscharntke, T., and Greiler, H.-J. (1995). Insect communities, grasses, and grasslands. *Annu. Rev. of Entomol.* **40**, 535–558.
Watts, J. G., Huddleston, E. W., and Owens, J. C. (1982). Rangeland entomology. *Annu. Rev. Entomol.* **27**, 283–311.

Greenhouse Gases, Global Warming, and Insects

Peter Stiling
University of South Florida

Greenhouse gases, the gases involved in determining the Earth's average temperature and climate, are accumulating at a rapid rate within the atmosphere. Such gases include carbon dioxide, methane, nitrous oxide, ozone, and chlorofluorocarbons. By far the most important of these is carbon dioxide, CO_2, whose contribution to the total greenhouse gas warming effect is at least 50%. For this reason, nearly all research pertaining to insects and greenhouse gases has focused on the response of insects to elevated levels of CO_2. Global atmospheric carbon dioxide levels are increasing at an astonishing rate, mainly because of the burning of fossil fuels. The atmospheric concentration of CO_2 has increased from a preindustrial level of about 270 ppm to a current level of about 365 ppm, an increase of nearly 100 ppm or 35%. According to some reports, the atmospheric concentration of CO_2 will likely stabilize at four times the preindustrial levels. Most studies indicate that CO_2 levels will at least double from preindustrial levels over the next five to ten decades. This increase represents one of the most large-scale and wide-reaching perturbations to the environment.

Many of the changes in insect populations likely to result from elevated CO_2 will be brought about by changes in plant chemistry. The chemical changes in plants result from increases in plant carbon, decreases in nitrogen, and increases in levels of defensive compounds such as phenolics. In addition, global warming, which will result from elevated levels of greenhouse gases, may increase the reproductive capabilities of some insects and change their distributional ranges. This could change the abundance of some pest species and disease vectors. Some of the myriad effects of elevated CO_2 on insects are summarized in Fig. 1.

STUDYING THE EFFECTS OF GREENHOUSE GASES ON INSECTS

Most published studies on the effect of CO_2 on insects tell of experiments in which plants and insects are confined to CO_2 levels of 700 to 710 ppm, or about double the current level. Such experiments are typically conducted in the laboratory, where well-watered potted plants are grown in nutrient-rich soil and maintained under elevated CO_2 for several months. Insects are introduced onto these experimental plants, and their feeding rates and performance are measured.

Studying the effects of elevated CO_2 on altered temperature and rainfall patterns, and the effects on insects of these modifications of the environment is much more problematic. It is not easy to warm whole communities in the field, except through the use of greenhouses—which tend to change many other features such as precipitation patterns. Thus the effect of temperature is often studied on laboratory populations. In addition, mathematical models are used to determine the likely range alterations of plants and insects in the face of increased global temperatures and changes in precipitation patterns.

CHANGES IN PLANT CHEMISTRY

Plants commonly respond to elevated CO_2 by increasing their rates of photosynthesis. Higher rates of photosynthesis usually result in higher accumulations of carbon-rich carbohydrates. Furthermore, the increased atmospheric CO_2 levels mean that stomatal conductance is reduced because plants can get sufficient atmospheric CO_2 into their leaves even when their stomates are closed more often. A reduction in stomatal conductance results in greater efficiency of water use by plants, because less water is lost through transpiration. Both these factors have important effects on plant chemistry.

First, increased carbon uptake by plants results in higher plant growth rates, with leaf area index, woody biomass, and below-ground biomass sometimes increased by as much as 25 to 50%. Despite the increase in plant growth, there is usually no increase in the availability of soil nutrients, particularly nitrogen, and these nutrients must be spread further among the available plant biomass. The usual result is a decrease in total plant nitrogen because nitrogen is diluted over the entire plant. Herbivore growth is most often limited by nitrogen rather than by carbon, so that plants grown in atmospheres of elevated CO_2 become poorer quality forage. Plant–water content usually affects digestibility, so that the poorer quality diet is partly offset by an increased ease of digestion.

The second major change in plants grown under conditions of elevated CO_2, namely, is a change in the ratio of carbon to nitrogen (C:N), as described in the preceding section. This has major implications for the concentration of defensive compounds in the leaves, the so-called secondary chemicals. Carbon-based secondary chemicals often increase and deter insect feeding. The overall effect of increased CO_2 on insect herbivores is to decrease plant palatability because of decreases in nitrogen levels and increases in secondary chemicals.

For secondary chemicals, the increases seem to be greatest for soluble phenolic compounds, especially condensed tannins, which are found in a variety of trees, especially oaks. These compounds are known to negatively affect many herbivorous insect species. Yet for other defensive compounds, such as linear furanocoumarins, found in celery, and monoterpenes and sesquiterpenes, found in peppermint, little increase has been noted.

CHANGES TO INSECT HERBIVORE FEEDING

There have been over 40 studies of the performance of insect herbivores of various types under conditions of elevated CO_2.

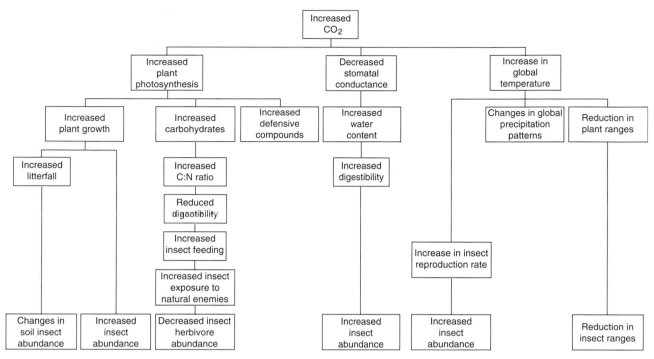

FIGURE 1 Schematic representation of the effects of elevated CO_2 on insects.

The majority of these, over 80%, have been conducted with leaf-chewing insects, especially lepidopteran caterpillars. The most commonly reported change is that food consumption increases as the insects struggle to obtain sufficient nitrogen in their diet. The efficiency of food conversion to insect biomass (conversion efficiency) decreases, probably because of the increased concentration of secondary chemicals, such as tannins, which bind digestive enzymes and render them less effective. Thus, it takes insects much longer to develop, and their final weight is often reduced. Early instars seem to be more susceptible than late instars. Of course such changes in diet could, in theory, be partly offset by the increase in digestibility due to the increased water content. However, at least in the studies done so far, the net outcome of elevated CO_2 on herbivorous insect digestibility has been negative.

The responses just outlined may vary somewhat according to the feeding guild of insects involved. Thus, chewing insects, which often digest the whole leaf and encounter both reduced nitrogen levels and increased defensive compounds, are particularly susceptible to changes in nitrogen and phenolics. Insects that feed in a different way may be less susceptible. Phloem and xylem feeders in particular may be less affected by CO_2 because they feed on plant sap, which is low in defensive compounds. Seed feeders also may be less affected by increased CO_2 because these plants try to maintain high levels of nitrogen in their reproductive parts. In cotton, for example, the C:N ratio of cotton balls is unaffected by elevated CO_2 and lepidopterans feeding there are unaffected. The concern is that pest insects could be stimulated to feed on these reproductive parts when the quality of the remainder of the plant decreases, which in turn would increase the pest status of some insects.

Of course CO_2 has also the direct effect of increasing temperature via the greenhouse effect, which may stimulate feeding activity because of increased metabolic rate in higher temperatures. Studies on the green peach aphid, *Myzus persicae,* a pest of many crops, suggest that elevated temperature increases aphid population growth rate and thus the likelihood that aphids will become more important pests in the future. In this case, both elevated CO_2 and elevated temperature increased aphid densities in experiments. Since, however, very few experiments have examined both CO_2 concentration and temperature in factorial experiments, the generality of the aphid results is unknown. It is also possible that the effects of elevated CO_2 and elevated temperature could cancel each other out for other insect species, especially leaf chewers.

EFFECTS OF CHANGED PLANT CHEMISTRY ON INSECT DENSITIES AND MORTALITIES

There has been relatively little research into how CO_2-mediated changes in plant chemistry affect insect densities and mortalities. This is because most plant–insect work has been done in laboratory conditions, where insects are fed foliage grown in elevated or ambient CO_2 and insect weight gains, losses, and digestibility coefficients are measured. To predict the effects of elevated CO_2 on insect densities, a population must be established on CO_2-treated foliage. However, in the few cases where insects have been reared from first instars through to pupae and adults, a significant decrease has been found in resultant population sizes in over 30% of the cases. This is usually because nutritionally inadequate foliage kills the immature insects. In some species (e.g., leafminers), we can get a good estimate of host-plant-induced mortality. Here, larvae that die from nutritional inadequacy are entombed with the leaf and can be counted, permitting an accurate assessment of deaths induced by the host plant.

To study the effects of elevated CO_2 on the interactions of insect herbivores with their natural enemies, such as predators and parasites, whole communities containing insect herbivores and their predators, parasites, and diseases are exposed to elevated CO_2. Such community-wide exposure has proved to be very difficult to achieve in the laboratory. Only where whole communities of plants and insects are exposed to elevated CO_2 in the field is it possible to fully address the effects of CO_2 on natural enemies. Experiments like this are very costly to do because of the huge quantities of CO_2 needed to arrive at a large enough increase in CO_2 under field conditions. However, it is widely thought that the net result of increased plant consumption and slower growth by herbivorous insects in elevated CO_2 is likely to result in increased exposure to natural enemies. For example, consumption of additional foliage increases the probability of ingestion of viruses or pathogenic bacteria, such as *Bacillus thuringiensis,* which can cause death. Once again, leafmining insects are a valuable study organism with which to examine the effects of elevated CO_2 on attack rate by natural enemies. This is because the leafmines themselves leave a permanent record of the fate of the insect inside. Parasite larvae can often be found within a mine, or the emerging adult parasitoids leave characteristic small shotgun-like holes in the upper mine surface. A recent study by the author, at Kennedy Space Center, was able to examine attack rates of leafminers by parasitic Hymenoptera in field chambers under conditions of ambient and elevated CO_2. The open-topped chambers contained the full complement of herbivores and their natural enemies on naturally occurring oak vegetation. Leafminer density was reduced inside the chambers, and leaf nitrogen content was reduced. The leafminers died more frequently inside the mines in elevated CO_2 and the mine area was bigger, indicating that larval leafminers had to eat more. Attack rate by natural enemies, particularly parasitoids, was significantly increased inside the chambers in which CO_2 was elevated. Perhaps the leafminers had created bigger, more obvious mines. Alternatively, their developmental time might have been slower in elevated CO_2, exposing them to natural enemies for a longer time, or they might have been physiologically less well able to resist attack.

In other systems, aphids known to produce alarm pheromones show a reduced capacity to do so under elevated CO_2. Once again, the result is an increased susceptibility to natural enemy attack. Finally, increased global temperatures are also likely to increase parasite and predator abundance as a result of increased population growth rates. This in turn could also lead to higher insect herbivore mortalities.

EFFECTS OF TEMPERATURE CHANGES ON INSECT DISTRIBUTION PATTERNS

Greenhouse gases are likely to change insect distribution patterns both directly, via increases in temperature and rainfall, and indirectly, via changes in the distribution of host plants. Recent research on a sample of 35 nonmigratory European butterflies showed that 63% had ranges that shifted to the north by 35 to 240 km during the 20th century, while only 3% shifted to the south. Thus for many insects, global warming has already changed range boundaries. The data appear to be robust because for most of these species, northward shifts have been shown in more than one country. Furthermore, the data appear to be robust across families, with many members of the Lycaenidae, Nymphalinae, Satyrinae, and Hesperiidae showing such range shifts. The northward shifts of the butterflies are of the same magnitude as the shift in climatic isotherms, which have moved about 120 km north as Europe has warmed by about 0.8°C.

Changes in rainfall, likely to have at least as big an impact as rising temperatures, have not been much studied. Global rainfall patterns clearly will change as a result of changes in global temperature, with many coastal areas becoming wetter and many interior continental areas becoming drier. This set of changes will affect the distribution of host plants and the insects that live on them. In addition, rainfall changes can directly affect the hatching of immatures from eggs laid in the soil, including eggs of many species of locust. Increased soil moisture increases the likelihood of locust outbreaks because it increases hatching and stimulates growth of host plants on which the locusts feed. The threat of locust plagues in new areas of the globe is therefore very real.

Margaret Davis, a paleobotanist from the University of Minnesota, showed that in the event of a CO_2 doubling, beech trees, presently distributed throughout the eastern United States and southeastern Canada, would die back in all areas except northern Maine, northern New Brunswick, and southern Quebec. Of course favorable new locations would develop in central Quebec, but the trees would take a long time to colonize such areas. Presumably the animals that feed on beech trees, including insect herbivores, would suffer a severe range contraction too, though this has not yet been studied.

EFFECT OF GREENHOUSE GASES ON SOIL INSECTS

It is doubtful that soil-inhabiting insects will respond directly to increased levels of CO_2 because of existing high concentrations in the soil. However, there are many likely indirect effects of CO_2 on soil insects. Light interception by a larger canopy may lower soil temperature and moisture. The most important change, though, is likely to be increased litterfall. Soil organic matter is likely to accumulate, rendering grasslands and forests net sinks of carbon under conditions of elevated CO_2. However, before senescence, most leaf nitrogen is reabsorbed by the plant, so that whereas living leaves in elevated CO_2 generally have a lower nitrogen content than leaves in ambient CO_2 conditions, litter quality remains unchanged. However, the increased volume of litter is likely to increase the number of litter-decomposing insects there.

In addition, increased root production may benefit root-feeding insects. Total numbers of Collembola per kilogram of soil have been shown to be significantly higher in experimental laboratory-based mesocosms where CO_2 levels were 60% above ambient. Species composition of Collembola also changed. Part of these increases in Collembola may be due to changes in abundance of mycorrhizal and nonmycorrhizal fungi on which they feed.

AQUATIC COMMUNITIES

Aquatic insect communities are unlikely to be directly influenced by increased CO_2 as much as terrestrial systems are. However, the increased litterfall associated with forest productivity is likely to increase allochthonous (i.e., leaf fall from riparian zones) litter input into forest streams and lakes. Such increased litter input is likely to increase stream insect populations. Litter quality itself, because it does not generally differ between ambient and elevated CO_2 treatments, is unlikely to affect aquatic decomposer communities. This has been verified by adding litter from ambient and elevated CO_2 to laboratory microcosms (simulated treeholes) and examining effects on eastern treehole mosquitoes, *Aedes triseriatus*. No differences in mosquito development time or survival were found. However, the elevated water temperatures and precipitation may increase the abundance of disease vectors such as mosquitoes. On the other hand, some cold water species may be reduced in abundance.

ELEVATED CO_2 AND DISEASE VECTORS

One of the main concerns voiced about global warming is that the delicate balance between diseases, their vectors, and humans might be upset as tropical climates that are so hospitable to spawning and spreading diseases move poleward. The spread of infectious diseases is controlled by the range of their vectors—mosquitoes and other insects. Increases in temperatures mean increases in the activity and ranges of these vectors.

Data on recent trends support this observation. An increase of one degree Celsius in the average temperature in Rwanda in 1987 was accompanied by a 337% rise in the incidence of malaria that year as mosquitoes moved into mountainous areas they had not previously inhabited. Also, *Aedes aegypti*,

a mosquito that carries dengue and yellow fever, has extended its range high into the mountain areas of such diverse areas as Colombia, India, and Kenya. Although global warming is expected to deliver its most deadly punch in the tropical areas of the world, where over 500-million people are affected (and 2.7 million die), the United States is not immune. A computer model by a Dutch public health team proposed that an average global temperature increase of 3°C in the next century could result in 50 to 80 million new cases of malaria each year. In the United States, public health facilities are likely to keep new incidences of disease in humans to a minimum, because of vaccinations. But disease outbreak in wildlife, which is not vaccinated, could be more severe.

See Also the Following Articles
Aquatic Habitats • Growth, Individual • Malaria • Pollution • Temperature, Effects on Development and Growth

Further Reading
Bezemer, T. M. and Jones, T. H. (1998). Plant–insect herbivore interactions in elevated and atmospheric CO_2: Quantitative analysis and guild effects. *Oikos* **82**, 212–222.
Coviella, C. E., and Trumble, J. T. (1999). Effects of elevated atmospheric carbon dioxide on insect–plant interactions. *Conserv. Biol.* **13**, 700–712.
Drake, B., Gonzalez-Meler, M., and Long, S. P. (1997). More efficient plants: A consequence of rising atmospheric CO_2. *Annu. Rev. Plant Physiol. Plant Mol. Biol.* **48**, 607–637.
Houghton, J. T., Meira Filko, L. G., Callander, B. A., Harris, M., Kattenburg, A., and Maskell, K. (1995). "Climate Change 1995. Science of Climate Change." Cambridge University Press, New York.
Keeling, C. D., and Whorf, T. P. (2000). Atmospheric CO_2 records from sites in the SIO via sampling network. *In* "Trends: A Compendium of Data on Global Change." Carbon Dioxide Information Analysis Center, Oak Ridge National Laboratory, U.S. Department of Energy, Oak Ridge, TN.
Parmessan, C., Ryrholm, N., Stefanesu, C., Hill, J. K., Thomas, C. P., Descimon, H., Huntley, B., Kaila, L., Kullberg, J., Tammaru, T., Tennett, W. J., Thomas, J. A., and Warren, M. (1999). Poleward shifts in geographical ranges of butterfly species associated with regional warming. *Nature* **399**, 579–583.
Stiling, P. (2002). "Ecology: Theories and Applications." 4th ed. Prentice Hall, Upper Saddle River, NJ.
Stiling, P., Rossi, A. M., Hungate, B., Dijkstra, P. D., Hinkle, C. R., Knott, W. M., and Drake, B. (1999). Decreased leaf-miner abundance in elevated CO_2: Reduced leaf quality and increased parasitoid attack. *Ecol. Appl.* **9**, 240–244.

Growth, Individual

Martin B. Berg
Loyola University Chicago

Richard W. Merritt
Michigan State University

The growth of individual insects proceeds in a progressive manner throughout the immature period of development, although the rate of growth can vary depending on a variety of factors such as molting frequency, temperature, and nutrition. Growth can be measured as an increase in biomass or body size, although biomass can be more variable than body size owing to differences in food and water intake. Growth usually is distinguished from development in that the latter refers to the various morphological and physiological changes that occur throughout the life span of an insect as it progresses toward maturation.

HOW INSECTS GROW

The limited ability of the rigid exoskeleton of insects to expand imposes a considerable constraint on individual growth resulting in the necessity to shed (i.e., to molt) the old exoskeleton, a process termed ecdysis, for growth to continue. Some insects (e.g., collembolans, diplurans, and thysanurans) exhibit indeterminate growth and continue to molt even after reaching the adult stage, although little if any increase in biomass occurs. In contrast, the majority of insects exhibit determinate growth in which both growth and molting cease upon reaching the reproductively mature last instar (i.e., the adult).

The pattern of individual growth differs depending on whether growth is measured as an increase in biomass or as an increase in body size. When measured as an increase in biomass, individual growth occurs between molts and is more or less continuous in most insects, although decreases in biomass often occur at the time of molting. When measured as an increase in body size, however, individual growth is largely dependent on the amount of sclerotization of the insect or of a particular body part. Membranous body parts, and those insects that are not highly sclerotized, can continuously increase in size between molts as folds in the cuticle expand, whereas insects or body parts that are more highly sclerotized increase in size immediately following each molt and exhibit a more discontinuous pattern.

Different body parts of insects may exhibit either isometric or allometric growth compared with the body as a whole. Isometric growth occurs when body parts grow at the same rate as the body as a whole, i.e., body length. Allometric growth occurs when body parts grow at rates different from that of the body and can be expressed as a power function of the form $x = ky^a$, where x is the dimension of the whole, y is the dimension of the part, a is the growth coefficient, and k is a constant. Thus, a straight line results from a log–log plot of body part size vs body length (Fig. 1).

INCREASES IN BIOMASS

Biomass increase of insects during development can be appreciable, with immature final instars weighing as much as 1000 or even 10,000 times greater than first instars. Changes in biomass can be expressed as either an absolute (weight/time) or a relative (weight/weight/time) increase in

biomass. Absolute growth usually is greatest in later instars and for the tobacco hornworm, *Manduca sexta,* larval weight increases as much as 90% (10-fold) in the final two instars. Similarly, 90% of the growth of larval *Paratendipes albimanus* (Diptera: Chironomidae) occurs in the final 10% of the life cycle. In contrast, relative growth rates normally decrease in later stages of development as the insect increases in size.

INCREASES IN BODY SIZE AND DYAR'S LAW

In 1890, H. G. Dyar noted that the head capsule widths of lepidopteran larvae followed a geometric progression in growth. During development of an immature insect, increases in body size occur in discrete steps, with highly sclerotized body parts exhibiting predictable and regular increases by a relatively constant factor, subsequently known as Dyar's Law (or Rule). Although initially based on observations of lepidopteran larvae, Dyar's Law has been applied to immature insects in general and refers to the geometric progression in the size of sclerotized structures that is constant throughout development. Dyar's Law has been widely used in entomological studies to discern instars of immature insects and also has been used to predict the size of instars missing from samples. The ability to distinguish instars is crucial to accurately describe insect life histories and growth patterns and is widely used in secondary production studies.

Because membranous portions of the less sclerotized cuticle (e.g., intersegmental membranes of the abdomen) allow the body to grow more or less continuously, overall body size is not considered a good indicator of instar. In addition, the increase in body size at each molt varies with different species and the growth of various body parts of many insects may differ from the growth rate of the body as a whole (i.e., allometric growth). In contrast, the rigid exoskeleton of

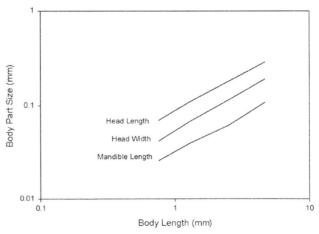

FIGURE 1 Allometric relationships between body part length and body length for *P. albimanus* (Diptera: Chironomidae) (Data from Ward and Cummins, 1978.)

immature insects prevents expansion and results in discontinuous growth of highly sclerotized cuticular parts (e.g., head capsule, legs). Thus, these structures increase in size incrementally in a stepwise manner following a molt. It is these distinct increases in the size of sclerotized structures that allows for the distinguishing of different instars with little overlap occurring between size classes.

The publication of Dyar's observations was in response to two previous papers that presented contradictory data concerning the number of molts in other species. Dyar studied the number of molts in 39 individuals of 28 species of caterpillars and chose the head capsule as the structure to measure for ease of measurement and because it was not subjected to growth during each stadium as was body length. The taxa chosen by Dyar ranged from 4 to 10 instars. Dyar calculated head-width ratios of successive instars and found that the progression was often nearly constant for a given species (mean = 1.5; range 1.3–1.7), what one would expect from a geometric progression. Dyar then calculated expected head capsule widths of each instar by multiplying the width of the final instar by this ratio and then back-calculating to the first instar. To test the applicability of the ratio, Dyar compared calculated head widths to those observed from reared specimens. Using this approach, it was possible to detect whether some instars had been missed or mismeasured. The most common method for detecting these problems is to plot the logarithm of the head capsule width measurement (or a measurement of another highly sclerotized structure) against the appropriate instar (Fig. 2). Conformity to Dyar's Law results in a straight line the slope of which is constant for a given species. According to Dyar's Law, deviations from a straight line indicate potentially missed instars or errors in measurement.

Although Dyar's Law has been widely used in entomological studies, the progression in the size of sclerotized body parts is not always constant and can be influenced by abiotic and biotic factors such as temperature and food. In addition, apparent contradictions to Dyar's Law occur when two requisite conditions are not met: (1) the number of instars is constant and (2) head capsule growth occurs only at ecdysis. Despite these constraints, approximately 80% of the entomological studies published from 1980 to 2000 that have examined the validity of Dyar's observations provided support for his law.

There has been some disagreement as to whether Dyar should be credited with the findings of geometric progression in growth. In a paper published 4 years prior to Dyar's article, Brooks reported that total larval length of a species of crustacean stomatopod (Stomatopoda: Squillidae) sequentially increased in size by a factor of 1.25 at each molt. Brooks also noted, as did Dyar, that this relationship could be used to determine whether larval stages were missing from the series. Thus, Dyar's Law may occasionally also be referred to as Brooks' Law (or Rule) in the literature. It is likely that entomologists were unaware of Brooks' observations because they were documented in a specialized publication on stomatopods.

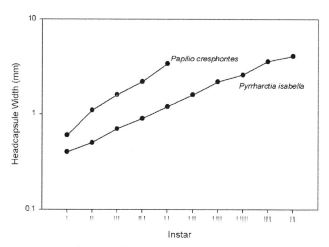

FIGURE 2 Conformity of head capsule width to Dyar's Law for giant swallowtail (*Pa. cresphontes,* Lepidoptera: Papilionidae) and banded woollybear (*Py. isabella,* Lepidoptera: Arctiidae) (Data from Dyar, 1890.)

EFFECTS OF TEMPERATURE ON INDIVIDUAL GROWTH

Because insects are ectotherms, temperature can have a profound impact on individual growth. In general, insect growth is correlated with environmental temperature; however, the strength of this relationship may be species- and habitat-specific. For example, many terrestrial insects exhibit little to no growth at low temperatures (0–4°C) because of either reduced feeding rates or overall low metabolic activity. Aquatic insects, in contrast, particularly those whose evolutionary ancestral habitat was in cold streams or lakes, can exhibit high growth rates during winter, assuming the presence of an adequate food supply. Although feeding rates of these insects may be low at reduced temperatures, basal metabolic needs also are low and result in the ability of the insect to direct more of the energy derived from ingestion to growth. The interactions between food quantity and quality and temperature are complex. In some streams, water temperature can influence the growth of microbial populations attached to detrital particles that are ingested by filter-feeding aquatic insects, and in turn, this can enhance their growth rate. This type of indirect control on aquatic insect growth and survivorship is difficult to separate from the direct effects of temperature on insect metabolism.

Elevated environmental temperatures also can result in either high or low larval growth rates because of the influence of temperature on larval ingestion, digestion, and development time. In general, larval ingestion and digestion are positively correlated with temperature and should result in increased growth rates at higher temperatures. The effect of temperature on growth rate, however, is confounded by an inverse relationship between ingestion rates and assimilation rates. In contrast, high temperatures can accelerate developmental time, resulting in the inability of larvae to maximize absolute growth rates.

EFFECTS OF NUTRITION ON INDIVIDUAL GROWTH

Individual growth of immature insects is strongly influenced by food availability, feeding selectivity, and food quality. Insect growth is often directly related to food availability, in that larval growth rates are highest in the presence of an abundant food supply. Because availability of food resources can vary temporally, growth responses also can be expected to vary throughout the year. For example, several species of stream chironomids (Diptera: Chironomidae) that feed primarily, but not exclusively, on attached algae, exhibit periods of maximal larval growth that coincide with times of the year when instream algal production is highest.

Feeding selectivity and food quality, however, also influence the relationship between growth and food availability. Even in the presence of an apparently abundant food supply, larval growth rates may be reduced if that food resource is not preferred or if it is of low quality or lacking essential nutrients. For example, the leaf-shredding crane fly, *Tipula abdominalis,* shows a strong preference for hickory, maple, and American chestnut leaves and a low preference for American beech, white oak, and red oak leaves. Larval growth rates are highest on the more preferred leaf types and lowest on the less preferred leaves. The higher growth rates on more preferred leaves are not due to higher food conversion efficiencies, but rather result from increased consumption rates due to a more palatable food source.

Although food availability may not be a limiting factor for growth in some insects, food quality may impose a substantial constraint to larval growth. Differences in food quality also can affect larval growth rates and the ability to complete development and reproduce. Absolute and relative growth rates of the caddisfly *Clistoronia magnifica* (Trichoptera: Limnephilidae) reared on diets differing in quality were significantly higher on a diet rich in triglycerides (conditioned alder leaves plus whole wheat grains) compared with diets of conditioned alder leaves, conditioned alder leaves with a fatty acid mixture, conditioned alder leaves plus hyphomycete fungi, or hyphomycete fungi alone. In addition, only larvae fed the high-triglyceride diet successfully completed development and reproduced. Some insects, however, even in the absence of higher quality food, can maintain a relatively uniform growth rate throughout development by increasing ingestion rates of lower quality food.

See Also the Following Articles

Body Size • Development, Hormonal Control of • Exoskeleton • Feeding Behavior • Nutrition • Temperature, Effects on Development and Growth

Further Reading

Berg, M. B., and Hellenthal, R. A. (1992). Life histories and growth of lotic chironomids (Diptera: Chironomidae). *Ann. Entomol. Soc. Am.* **85,** 578–589.

Chapman, R. F. (1998). "The Insects: Structure and Function." Cambridge University Press, Cambridge, U.K.

Crosby, T. K. (1972). Dyar's rule predated by Brook's rule. *N. Z. Entomol.* **5,** 175–176.

Dyar, H. G. (1890). The number of molts of lepidopterous larvae. *Psyche* **5,** 420–422.

Epstein, M. E., and Henson, P. M. (1992). Digging for Dyar. *Am. Entomol.* **34,** 148–169.

Gillott, C. (1980). "Entomology." Plenum Press, New York.

Richards, O. W., and Davies, R. G. (1979). "Structure, Physiology, and Development," Vol. I of "Imm's General Textbook of Entomology." Chapman & Hall, London.

Sweeney, B. W. (1984). Factors influencing life-history patterns of aquatic insects. *In* "The Ecology of Aquatic Insects" (V. H. Resh and D. M. Rosenberg, eds.), pp. 56–100. Praeger, New York.

Ward, G. M., and Cummins, K. W. (1978). Life history and growth pattern of *Paratendipes albimanus* in a Michigan headwater stream. *Ann. Entomol. Soc. Am.* **71,** 272–284.

Wigglesworth, V. B. (1965). "The Principles of Insect Physiology." Methuen, London.

Grylloblattodea
(Rock Crawlers, Ice Crawlers)

D. C. F. Rentz
California Academy of Sciences

The Grylloblattodea, or ice crawlers, are a small group of soft-bodied, apterous, terrestrial and termitelike insects confined to the Northern Hemisphere. They occur under rocks in forest leaf litter or above the treeline in the high mountains. In parts of the United States, Canada, and China they are found under rocks in soil at or below freezing temperature. In Japan and Korea, they are active in midsummer in the deep leaf litter of mixed conifer and deciduous forests, where daily ambient temperatures approach 30°C. In Japan and Korea some have been found in caves; in the western United States, they are known from subterranean lava tubes. Grylloblattids are slender, depressed insects covered with fine hairs and having reduced eyes. Adults range from 2 to 3.5 cm in length. Ice crawlers appear to be primarily nocturnal. They are considered to be some of the most primitive of orthopteroid insects and have been thought to be related to cockroaches (Blattodea) and stick insects (Phasmida).

CLASSIFICATION AND FEATURES OF THE ORDER

The Grylloblattodea comprise a single family, Grylloblattidae, which includes only four genera. *Grylloblatta* occurs in North America and Canada, where at least 11 species are known. *Grylloblattina* is known from a few species from Siberia. *Grylloblattella* has species known from Korea, Japan, and Siberia. *Galloisiana* contains at least 10 species from Japan. The group, however, is undoubtedly larger, since new species are discovered every few years.

FIGURE 1 Copulating pair of grylloblattids, *G. nipponensis:* the female is the lower individual (note ovipositor); the male has the eversible sac of the left phallomere exposed. [From Nagashima *et al. In* Ando, H. (ed.). (1982). "Biology of the Notoptera," p. 48, Kashiyo-Insatu, Nagano, Japan.]

Grylloblattids are cryptic, ground-dwelling insects that prefer wet habitats and cool temperatures. They shun light and occur under stones or in dense leaf litter. Species of *Grylloblatta* that live under rocks under or near snowbanks emerge after dark and feed as scavengers or predators on dead or dying insects that have been blown onto the snow from lower elevations. During winter, the species probably occupy the airspace between the ground and overlying snowpack, where they remain active at temperatures of 0°C. Massive fat bodies build up prior to winter; during winter, the insects may feed on decaying plant material. Korean species live under debris on the floor of caves at only 200 m altitude and apparently never venture forth from the cave habitat. The most widespread species in Japan, *Galloisiana nipponensis,* is found at elevations ranging from 300 to 3000 m, where the insects live under stones and in the leaf litter of thick, mixed coniferous and hardwood forests. These species consume both insects and plant material day and night.

The resemblance of grylloblattids to Dermaptera has been shown to be superficial and is associated only with the fact that both have a projecting head. The head, however, is typical of orthopteroid insects. The antennae are elongate and thin. The mouthparts are structured like those of a predator. Ocelli are absent, and the eye comprises fewer ommatidia in young instars than in adults. The legs are simple, slender, and not suited for jumping. The abdomen comprises 11 segments, with the cerci long and flexible and the male genitalia asymmetrical. The ovipositor comprises three pairs of slender, tapering, partly free valves.

Although grylloblattids are normally considered to be cool-adapted insects, they cannot withstand temperatures much below 0°C. At −5.5 to 8.0°C they stress. Contrary to the popular belief that they can withstand very low temperatures, they can be killed by ice formation within the body as a result of their low levels of glycerol, sorbitol, or erythrol.

Copulation has been observed in a few species. In *G. nipponensis,* the female is chased and seized by the male. The resulting copulation can last from 30 min to 4 h. Males always assume a position on the right side of the female as a response to the male's asymmetrical genitalia. Oviposition occurs 10 to 50 days after copulation. Females lay eggs with the elongate ovipositor in wood or under stones and decaying plant material. None have been found in moss. Oviposition for each egg takes about 3 min, and females lay 5 or 6 eggs per day to a total of about 30 eggs. A captive female laid 145 eggs in her lifetime. The large, black eggs, develop over periods of from 5 months to 3 years.

See Also the Following Articles

Blattodea • Orthoptera • Phasmida

Further Reading

Ando, H. (ed.). (1982). "Biology of the Notoptera." Kashiyo-Insatsu, Nagano, Japan.
Storozhenko, S. (1989). A review of the family Gryllop(b)lattidae (Insecta). *Articulata* **13,** 167–181.

Gypsy Moth

Joseph S. Elkinton
University of Massachusetts

The gypsy moth, *Lymantria dispar,* is one of the world's most damaging defoliators of hardwood forest trees. It is native to Europe and Asia. It was introduced from Europe to North America near Boston, Massachusetts, in 1869 and has been spreading slowly south and west ever since. A large body of research has focused on the biology, management, and population dynamics of this species.

GEOGRAPHIC RANGE AND SPREAD

Gypsy moth occurs throughout much of the Northern Hemisphere. Its native range stretches from Japan, China, and Siberia across Russia to western Europe and as far south as the Atlas Mountains of North Africa. In North America, gypsy moth has spread over much of the eastern United States and Canada. Currently, the leading edge of the infestation stretches from North Carolina to Wisconsin and adjacent regions of Ontario. Because female gypsy moths from Europe have wings but do not fly, the rate of spread of this insect has been extremely slow. The spread occurs when newly hatched larvae spin down on silken threads and are blown in the wind. Most of this dispersal is less than 50 m, although some larvae are carried by wind currents for greater distances. Natural spread has been augmented by inadvertent human transport of egg masses laid on vehicles or other backyard objects, often to locations well outside the region infested by gypsy moth in the northeastern United States. Such new infestations have been eradicated at many sites

FIGURE 1 Life stages of gypsy moth. (1,2) Adult female; (3,4) adult male, (5) pupae; (6,7) larvae; (8) egg mass; (9,10) individual eggs. (Reproduced from J. Bridgham, 1896, 43rd Annual Report, Massachusetts State Board of Agriculture.)

throughout North America. In recent years, attention has shifted to new introductions of gypsy moth from Asia, particularly the Russian Far East. Adult female gypsy moths from these regions are able to fly. Consequently, the rate of spread would be much faster, if they became established. These introductions have occurred at several locations in North America, mainly in the Pacific Northwest, but thus far, none have become established.

LIFE HISTORY AND HOST TREES

The gypsy moth female lays a single egg mass (Fig. 1), usually on the stems of trees, and she covers the eggs with her body hairs. The egg mass typically contains from 100 to 600 eggs. The eggs are laid in midsummer, but overwinter in this stage. Larvae developing within the egg enter diapause and hatch

the following spring at the time of host-tree budbreak. Emerging larvae climb to the tops of trees, where many of them spin down on silken threads and are borne away by the wind. If larvae land on an acceptable host tree, they begin feeding. They develop through five instars for males or, frequently, six for females throughout May and June. Beginning in the fourth instar, larvae seek resting locations during daylight hours, either in the forest litter at the base of trees or under bark flaps on tree stems. They usually pupate in these same locations. The adults emerge after about 12 days in the pupal stage. Soon after eclosion, the female releases a sex pheromone from a gland on the tip of her abdomen. Males locate females by flying upwind when they detect the pheromone. After mating, the female lays her egg mass, often just a few centimeters from where she eclosed, and dies soon thereafter. There is one generation per year.

Gypsy moth larvae feed on a wide range of tree species. Favored tree species include oaks (*Quercus* spp.), aspen (*Populus* spp.), and, in Japan, Japanese larch, *Larix leptolepis*. Gypsy moth outbreaks occur in forests that are dominated by these species. Gypsy moth will feed on many other tree species, such as maple (*Acer* spp.) and many conifers, but significant damage to these trees usually occurs only in gypsy moth outbreaks, when more favored hosts have already been defoliated. If defoliation is complete, most deciduous hardwood trees will put out a new set of leaves. Most trees will survive one defoliation, but if outbreaks persist for several years in a row, a significant proportion of the trees may die. Trees that survive defoliation suffer growth loss in subsequent years.

DISEASES OF GYPSY MOTH

As with most insects, gypsy moths are host to a suite of natural enemies and these play a pivotal role in the dynamics of the gypsy moth populations. There are two major diseases: a nuclear polyhedrosis virus and a fungal pathogen. The virus, *LdMNPV,* causes epizootics that are largely responsible for the collapse of gypsy moth outbreaks. Similar viruses terminate the outbreaks of many defoliating Lepidoptera. High mortality from these viral diseases occurs only in dense populations, because transmission of the virus takes place when larvae feed on leaves contaminated by cadavers of larvae that have previously died from the disease. Encounters with cadavers are only likely in dense populations. Transmission of *LdMNPV* from one generation of gypsy moths to the next occurs primarily by way of external contamination of the egg mass; larvae become infected as they emerge from the mass in the spring. It is not entirely clear how the virus persists at low density, but it does survive in the forest litter for several decades.

The fungal pathogen *Entomophaga maimaiga* was, until recently, known only in the Far East, especially Japan. In 1989, a dramatic epizootic of *E. maimaiga* occurred throughout the northeastern United States from Pennsylvania to Maine. In subsequent years, the fungus spread across the mid-Atlantic states and was introduced intentionally by researchers to Virginia and Michigan. It is now established throughout the region infested by gypsy moth in North America. *E. maimaiga* produces two kinds of spores: conidia and resting spores. The conidia are released from cadavers and are carried by wind currents to uninfected larvae, which they infect by penetrating the cuticle; these conidia are responsible for the rapid spread of *E. maimaiga* in North America. Late instars produce resting spores that overwinter in the forest litter, where they persist for up to 10 years before germinating to infect new gypsy moths. A. Hajek and colleagues analyzed the DNA of *E. maimaiga* and showed that the pathogen in North America is identical to *E. maimaiga* in Japan. How *E. maimaiga* was introduced into North America is unknown. Since 1989, it has continued to cause high levels of mortality in gypsy moth populations, particularly in years with high rainfall in May and June. A key difference from *LdMNPV* is that *E. maimaiga* causes substantial mortality in low-density as well as in high-density populations of gypsy moth. This means that *E. maimaiga* can prevent outbreaks from occurring, whereas *LdMNPV* can only cause the collapse of outbreak populations.

PARASITOIDS

As with most insects, various parasitoid species attack the different life stages of gypsy moth. In North America, efforts to introduce parasitoids of gypsy moth from Europe and Asia began around 1905, and 10 species have been established. The egg parasitoid *Ooencyrtus kuvanae* (Encyrtidae) from Japan is frequently observed on gypsy moth egg masses in late summer and may cause as much as 30% mortality of the eggs. Larval parasitoids include *Cotesia melanoscela* (Braconidae) and the tachinids *Blepharipa pratensis, Compsilura concinnata,* and *Parasetigena silvestris.* The most common pupal parasitoid is *Brachymeria intermedia* (Chalcididae). The impact of these parasitoids on gypsy moth populations remains equivocal. Total mortality caused by parasitoids in North America is typically below 50% and is not consistently density dependent, so that their ability to regulate gypsy moth populations is in doubt. In Europe, on the other hand, parasitism of gypsy moth is often much higher than that observed in North America. European gypsy moths are attacked by several parasitoid species that were never established successfully in North America. It seems likely that parasitoids are responsible for preventing gypsy moth outbreaks, which are rare in western Europe, but more common in central and southern Europe.

PREDATORS

Compared with parasitoids, an even larger community of vertebrate and invertebrate predators feeds on gypsy moth. Very little is known about the impact of most predators, because predation is extremely difficult to measure. Many bird species worldwide feed on gypsy moth, but it is generally believed that most birds dislike the hairy cuticle of gypsy moths and avoid them.

Research groups led by H. Bess in the 1940s and R. Campbell in the 1970s concluded that predation by small mammals, particularly the white-footed mouse, *Peromyscus leucopus,* has a major impact on low-density gypsy moth populations. The mice feed on late instars and pupae, particularly on the forest floor. Both research groups demonstrated an increase in gypsy moth survival in forest plots from which small mammals had been removed or excluded. More recently, J. Elkinton and colleagues showed that predation on gypsy moth pupae was strongly correlated with density of mice and that gypsy moth densities increased when mouse densities declined. The density of mice, in turn, was correlated with the abundance of acorns, which, in oak-dominated forests, are their principal overwintering food. Indeed, there are many studies that link forest-dwelling mice to abundance of acorn crops. Poor acorn crops, which occur on a regional scale and are caused by a variety of weather events, could thus be the ultimate trigger of gypsy moth outbreaks. C. Jones and colleagues provided further experimental proof of these ideas. They removed mice from experimental plots and observed an increase in gypsy moth and they augmented food in other plots and observed an increase in mice.

There are a number of invertebrate predators of gypsy moth. One of these is the introduced ground beetle, *Calosoma sycophanta* (Carabidae). R. Weseloh has shown that this insect becomes quite abundant in outbreak populations of gypsy moth and may cause substantial mortality. Several researchers have documented predation by ants as a significant source of mortality in low-density populations, but it is usually much less than predation by mice.

POPULATION DYNAMICS

In the 1970s R. Campbell developed the first comprehensive theory of gypsy moth dynamics, wherein populations alternate between low-density and high-density phases, each maintained by different factors and sources of mortality. At low density, predators, particularly mice, maintain gypsy moth populations indefinitely at a low-density equilibrium. The concept of an equilibrium implies that predation is density dependent, which means that it increases as gypsy moth density increases until total mortality balances fecundity of gypsy moth and the population density stops growing. The equilibrium is an unstable one, however, because the density-dependent response of most natural enemies is constrained by a variety of factors. For example, most predators or parasitoids have their own natural enemies. These constraints produce a threshold density above which gypsy moth population growth outpaces the mortality caused by natural enemies, and as a result densities of gypsy moth increase rapidly into an outbreak phase. At the much higher outbreak densities, a different set of natural enemies becomes predominant. These natural enemies, coupled with competition among gypsy moths for available foliage, limit further increases in gypsy moth density. The outbreak population either persists for several generations at high density or collapses back to the low-density phase. The model is particularly appropriate when the principal mortality factors maintaining the low-density equilibrium are generalist predators, such as mice. Unlike specialist natural enemies, whose densities often track those of their hosts, the densities of mice are not determined by gypsy moths. Instead their densities are determined by their overwintering food supply, mainly acorns. Whether this conceptual model is an accurate description of the gypsy moth system remains to be demonstrated.

Gypsy moth outbreaks are spatially synchronized (Fig. 2). Outbreaks tend to occur simultaneously over a large region. A cause of this pattern may be acorn crops, which are heavy or meager over a region in response to regional weather patterns. Indeed, as indicated by P. Moran, populations of many species are synchronized by a variety of weather-related influences. Weather may synchronize gypsy moth populations by exerting a common influence across populations on many of the factors that affect gypsy moth growth and survival. Many gypsy moth researchers believe that there is a 10-year cycle of abundance of gypsy moth embedded in the erratic temporal pattern evident in Fig. 2. Analyses of these data by D. Williams and A. Liebhold provide some support for this view. If these cycles exist, the factors that cause them remain unknown. Analyses by P. Turchin of decade-long records of

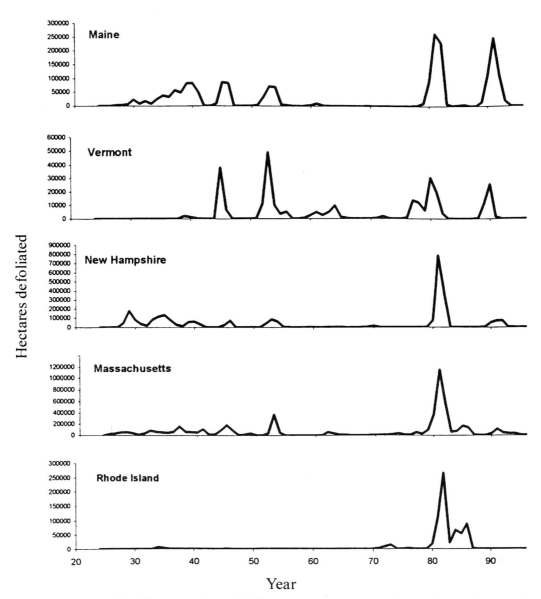

FIGURE 2　Historical record of area defoliated by gypsy moth in each of five states in northeastern United States from 1924 to 1996. (Reproduced from Liebhold *et al.*, 2000, *Pop. Ecol.* **42**, 257–266, © Springer-Verlag, with permission.)

gypsy moth egg-mass density in Yugoslavia provide more convincing support for regular cycles in that region, evidently caused by fluctuations in parasitism.

Adding to the complexity of gypsy moth dynamics in North America is the recent appearance of the fungal pathogen *E. maimaiga.* In the northeastern United States, epizootics of this fungus have occurred nearly every year since 1989, except in years that were extremely dry. We still have much to learn about this agent, but it causes substantial mortality in both low- and high-density populations and appears to have prevented several incipient outbreaks. It appears that incidence of *E. maimaiga,* which is largely determined by rainfall in May and June, is now a prime determinant of whether outbreaks occur.

MANAGEMENT OF GYPSY MOTH

Management of gypsy moth in North America and elsewhere has evolved over time as different tools became available and as public attitudes toward pesticide use have changed. In the 1960s, large areas were sprayed by air with DDT. DDT was banned in the late 1960s and was supplanted by other chemical pesticides, such as carbaryl. In the late 1980s, *Bacillus thuringiensis* (Bt), a bacterial pesticide, became a viable alternative to chemical pesticides and became the material of choice in many regions. The advantage of Bt is that it is more selective than most chemical pesticides; it affects only larval Lepidoptera that feed on Bt-contaminated foliage. In addition, government agencies in the generally infested region in northeastern United States concluded that large-scale application of

pesticides against gypsy moth was neither ecologically acceptable nor worth the considerable expense. Most forest trees survive gypsy moth outbreaks, and the outbreak populations soon collapse on their own. Control activities in these regions generally aim at foliage protection on high-value trees, rather than suppression of gypsy moth populations.

See Also the Following Articles
Biological Control • Forest Habitats • Pathogens of Insects • Population Ecology

Further Reading
Campbell, R. W., and Sloane, R. J. (1977). Forest stand responses to defoliation by gypsy moth. *Forest Science Monogr.* No. 19.
Doane, C. C., and McManus, M. L. (eds.) (1981). The gypsy moth: Research toward integrated pest management. USDA Forest Service Tech. Bull. 1584.
Elkinton, J. S., and Liebhold, A. M. (1990). Population dynamics of gypsy moth in North America. *Annu. Rev. Entomol.* **35,** 571–596.
Hajek, A. E. (1999). Pathology and epizootiology of *Entomophaga maimaiga* infections in forest Lepidoptera. *Microbiol. Mol. Biol. Rev.* **63,** 814–835.
Liebhold, A. M., and McManus, M. L. (1999). The evolving use of insecticides in gypsy moth management. *J. Forestry* **97,** 20–23.
Liebhold, A. M., Elkinton, J. S., Williams, D., and Muzika, R.-M. (2000). What causes outbreaks of gypsy moth in North America? *Populat. Ecol.* **42,** 257–266.
Wallner, W. E. (ed.) (1989). Lymantriidae: Comparisons of features of new and old world tussock moths. USDA Forest Service Gen. Tech. Rep. NE-123.
Williams, D. W., Fuester, R. W., Metterhouse, W. W., Balaam, R. J., Bullock, R. H., Chianese, R. J., and Reardon, R. C. (1992). Incidence and ecological relationships of parasitism in larval populations of *Lymantria dispar* (Lepidoptera: Lymantriidae). *Biol. Control* **2,** 35–43.

Hearing

Jayne Yack
Carleton University

Ron Hoy
Cornell University

Among all terrestrial animals, only vertebrates and insects are richly endowed with a sense of hearing. By "hearing," we usually mean the ability to detect minute, time-varying changes in air pressure that we familiarly experience as "sound." Under this restricted definition, we can say that audition has evolved in at least seven orders of insects, including all of the major orders except the Hymenoptera (wasps, ants, bees). However, if we were to include under "hearing" the ability to detect sound waves in water and solids, or the displacement of molecules in a sound's near field, then the number of "auditive" insects would expand enormously to include not only the Hymenoptera, but even small orders such as Plecoptera (stone flies) and Isoptera (termites). Initially, we focus on the form and function of tympanal ears, which are organs that are sensitive to sound signals that are propagated through the air or water as fluctuations in pressure and which come to mind when we (humans) use the term "hearing with ears." Following this, we provide some examples of nontympanal hearing organs.

Using sound, vertebrates and insects are often capable of sensing, identifying, and locating their predators, prey, conspecific rivals, and mates by hearing their intentional or unintentional acoustic signals. As might be expected, natural selection has shaped the form and function of hearing organs ("ears") in insects over evolutionary time. In this respect, the ears of insects show much greater diversity than those of vertebrates, for reasons that will be apparent in our discussion. However, it must be emphasized that despite the scope of morphological diversity among insect ears, there is a morphological "bauplan" (structural design) that underlies their great range in behavioral and physiological function.

MORPHOLOGICAL REQUIREMENTS

There is tremendous morphological diversity of insect ears (Fig. 1). The multitude of different ear designs and locations reflects the unique physical and behavioral challenges faced by each insect. Yet despite their many differences, most ears follow a similar morphological plan. Each typically consists of three identifiable substructures: a tympanal membrane, a tracheal air chamber, and a chordotonal sensory organ.

Tympanal Membrane

The tympanal membrane (eardrum) is a thinned region of exoskeleton, typically supported by a chitinous ring and stretched across an enlarged air-filled cavity. Sound impinges upon the membrane, setting it and its associated nerve cells into motion. The thickness of the membrane can vary from 1 to over 100 μm. The ultra-sound-sensitive ears of many nocturnal Lepidoptera (butterflies and moths), for example, are so thin that they are transparent. Such fragile membranes are typically protected within body cavities or by external flaps of cuticle. In contrast, the thicker, opaque tympanal membranes of some diurnal butterflies or grasshoppers are conspicuously positioned on the outer surface of the body.

Tracheal Air Chamber

The internal face of the eardrum backs onto an enlarged air-filled chamber, which forms part of the tracheal respiratory system. In some ears, the air chambers are connected directly to other sound input sources (spiracles or contralateral ears) or resonating chambers via the tracheal system. In a few rare cases [e.g., green lacewings (Chrysopidae) and some water bugs (Corixidae)] the tympanic chambers are largely fluid filled.

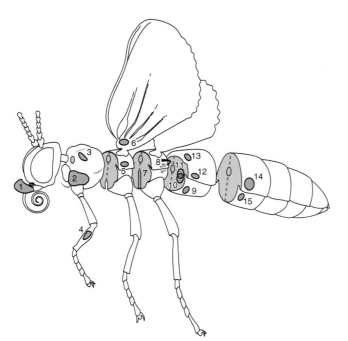

FIGURE 1 A schematic drawing of a generalized insect showing 15 body locations where tympanal ears have been identified. Each number represents a position on the body where an ear has evolved independently in one or more taxa, although all species within a taxonomic division do not necessarily possess ears. (1) Lepidoptera: Sphingidae (Choerocampini, Acherontini). Location: palp–pilifer region. (2) Diptera: Sarcophagidae, Tachinidae. Location: ventral inflation of prosternum, between coxa. (3) Coleoptera: Scarabidae, Dynastinae. Location: dorsolateral region of prosternum. (4) Orthoptera: Ensifera: Gryllidae, Tettigoniidae. Location: tibia of foreleg. (5) Heteroptera: Hydrocorisae (water boatmen). Location: lateral mesothorax, ventral to wing base. (6) Lepidoptera: Papilionoidea, Hedyloidea; Neuroptera: Chrysopidae. Location: base of ventral forewing. (7) Dictyoptera: Mantodea. Location: within a deep groove between the metathoracic legs. (8) Lepidoptera: Noctuoidea. Location: within a cavity on the posterior metathorax. (9) Lepidoptera: Pyraloidea. Location: within a cavity on ventral surface of first abdominal segment. (10) Lepidoptera: Geometridae. Location: within a cavity on anterior side of first abdominal segment. (11) Lepidoptera: Drepanidae. Location: internalized tympanal membrane located between two air-filled chambers on first abdominal segment. (12) Orthoptera: Acrididae. Location: lateral surface of first abdominal segment. (13) Coleoptera: Cicindelidae. Location: dorsal surface of first abdominal segment, beneath the elytra. (14) Homoptera: Cicadidae. Location: within cavity on lateral second abdominal segment. (15) Lepidoptera: Uraniidae. Location: within cavity at the anterior (females) or posterior (males) end of the second abdominal segment. (Illustration by M. Nelson.)

Chordotonal Organ

Associated with the inner surface of the tympanal membrane is one to several chordotonal organs. Chordotonal organs are specialized mechanoreceptors unique to insects and crustaceans, but not unique to ears. Each chordotonal organ comprises one or more individual sensory units called scolopidia, and each scolopidium consists of three cells arranged in a linear array: a sensory cell, a scolopale cell, and an attachment cell (Fig. 2A). The total number of scolopidia in an ear ranges from one in some moths (Notodontidae) to almost 2000 in the bladder grasshopper *(Bullacris membracioides)*. A chordotonal organ may attach directly to the inner surface of the tympanic

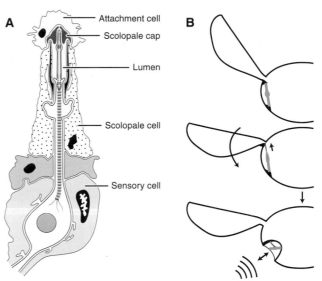

FIGURE 2 Insect tympanal sensory receptors. (A) A typical tympanal scolopidial organ, consisting of three cell types. The dendrite of a bipolar sensory neuron projects into a fluid-filled space (lumen) formed by the walls of a enveloping scolopale cell. The distal tip of the dendrite inserts into the scolopale cap, an extracellular secretion of the scolopale cell. The attachment cell connects the sensory neuron and scolopale cell to the tympanal membrane, either directly or indirectly via a tracheal air sac. A chordotonal organ may have from one to several thousand scolopidia. (B) A schematic diagram depicting the hypothetical transition from a wing-hinge proprioceptive chordotonal organ to a tympanal hearing chordotonal organ. The top two images show a chordotonal organ functioning as a proprioceptor monitoring wing movements. At the bottom the chordotonal organ has been mechanically isolated within a rigid tympanal cavity and attaches to a thinned region of cuticle (the tympanic membrane) that detects sounds. (A was modified, with permission, from E. G. Gray, 1960, The fine structure of the insect ear, *Philos. Trans. R. Soc. B* **243,** 75–94. Illustrations by M. Nelson.)

membrane (e.g., Fig. 4E) or to tracheal air sacs indirectly associated with the tympanum. The axons of the sensory neurons collectively make up the auditory nerve that forms the neural link between the mechanosensory stimulation of the eardrum and its "perception" by the central nervous system. Vibrations of the eardrum and/or air chamber cause bioelectric currents to flow in the sensory cell, initiating action potentials in the auditory nerves and signaling neurons in the auditory pathways of the nervous system.

MECHANISMS

Sound waves are fluctuations in pressure traveling through a medium away from a source of mechanical disturbance. Tympanal ears are designed to detect these minute pressure changes traveling through air or water. Depending on the structural design of the ears, they may convey information about the location, frequency, and intensity of an acoustic stimulus.

There are two distinct types of insect tympanal ears: pressure receivers and pressure difference receivers (or pressure gradient receiver). In a pressure receiver, the tympanal membrane forms one side of an otherwise enclosed air chamber. The

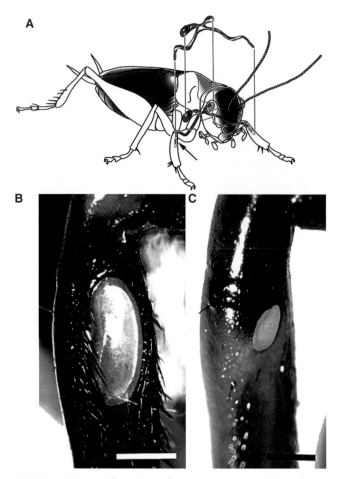

FIGURE 3 Schematic illustrations of structures associated with hearing in crickets (Ensifera). (A) The locations of the tibial ears and the H-shaped tracheal system connecting the contralateral tympanal membranes and spiracles are shown. (Illustration by M. Nelson.) (B and C) Light micrographs of the larger posterior and smaller anterior tympanal membranes located on the tibia of the forelegs in *Gryllus bimaculatus*. (Courtesy of A. C. Mason.)

acute sense of directionality such that they can detect and localize their calling hosts/prey from a distance of tens of meters, even from high in the air. Such ears appear to be unique to these parasitoid flies.

The frequencies detectable by insect tympanal ears range from a few kilohertz (e.g., the water boatman *Corixa*) to over 100 kHz (e.g., some tettigoniid species). This broad bandwidth of frequency sensitivity is used by some insects for mate calling and in others it enables them to detect their predators (e.g., echolocating bats). Some insects have the ability to discriminate between different sound frequencies, which may be achieved at the level of the tympanal membrane or the chordotonal organ. To date, most insect ears studied appear to be tone deaf (i.e., different tones are indistinguishable).

LOCATION

Ears can be located just about anywhere on insects (Fig. 1). There is one striking difference between insect ears and vertebrate ears: those of the latter, because of developmental constraints, are always on opposite sides of the head and always on the head, whereas insect ears have been found on virtually every part of the body. Of course, in any given species of insect, ears are always found on the same part of the body, but from group to group, ears may be found on the head, legs, wings, thorax, or abdomen.

Although the ears of most insects are clearly recognized by the presence of a conspicuous tympanal membrane either on the body's external surface or within an ear cavity, the ears of other insects are morphologically cryptic. For example, some hawk moths (superfamily Sphingoidea) possess hearing organs in their mouthparts, and by inflating their palps while they are feeding on flowers at dusk, they create functional ears that alert them to the echolocation calls of bats. The "cyclopean" ear of the praying mantis is located within a deep groove between the hind legs, and the tympanal membranes are not morphologically distinct from surrounding body parts, even to the trained eye. Even more cryptic are the ears of the Madagascar hissing cockroach, *Gromphadorhina portentosa,* in which there is no obvious eardrum or tympanal membrane overlying the internal chordotonal organs contained within the tibiae of its forelegs.

Although most insects possess a single pair of hearing organs, there are at least two reports of insects with multiple ears. One lineage of praying mantids has two sets of ears: one located between the mesothoracic legs and tuned to ultrasonic frequencies (25 to 40 kHz) and another between the metathoracic legs and tuned to lower frequencies (2–4 kHz). The multieared bladder grasshopper *(B. membracioides)* of South Africa has six pairs of serially repeated abdominal ears that function in detecting mating calls at distances of ~2 km.

DEVELOPMENTAL ORIGINS

In vertebrates, ears are always located on the head, behind the eyes, and above the jaw, but in insects they are virtually

membrane vibrates in response to sound waves arriving only to the external surface of the tympanal membrane. In insects whose ears are sensitive to high-frequency sounds (like many bat-detecting moth ears), the sound shadow of the body is sufficient to cause a difference in sound amplitudes arriving to each ear, which provides useful directional cues to the animal. Many insects using lower frequency sounds are faced with the difficulty of being able to localize sounds that arrive almost simultaneously to both ears. They overcome this problem using pressure-difference receivers in which the sound impinges on both the outer and the inner surfaces of the membrane, and the membrane vibrates in response to the difference in pressure between the two sides (e.g., Fig. 3).

A third and very unusual way to localize sounds has been "invented" by certain parasitic tachinid and sarcophagid flies. These flies localize their hosts (field crickets, katydids, and cicadas) by hearing and then homing in on the mating calls. The ears possessed by these flies are conventional chordotonal organs, containing 70 or more scolopidia. However, the fly's eardrums are mechanically connected, which confers an anywhere on the body. Why are they not confined to one place like those of vertebrates? Evolutionary and developmental origins of the vertebrate ear, drawing from embryological anlagen (organs or structures in their earliest state) such as the gill arches, neural crest, and the otic capsule, have constrained the ears to their canonical position in the head. Insect ears require but two structural modifications, cuticular and spiracular, combined with innervation by a chordotonal organ. In the insect body, chordotonal organs do not function only as hearing organs, but are actually widely distributed throughout the body, where they act as detectors of self-induced body movements (proprioceptors) or substrate vibrations. Insect bodies are made up of a series of hard cuticular plates joined by flexible membranes, and chordotonal organs are frequently suspended between moving joints. We now know through developmental studies and comparative anatomy that it is not so difficult to "construct" an ear by making a few peripheral modifications to an existing proprioceptor and its surrounding cuticular and tracheal structures. By simply thinning the cuticle, enlarging the surrounding tracheal air sacs to allow membrane vibration, and mechanically isolating the sensory organ from body movements, a proprioceptor can be converted into a sound-pressure receiver (Fig. 2B). For example, the bat-detecting ears on the thorax of noctuoid moths are thought to have evolved from proprioceptors monitoring wing movements. Given the jointed, segmental body plan of any insect, the ubiquitous branching of the respiratory tracheae lining the inner face of the cuticle, and the widespread occurrence of chordotonal organs that span different segments in the body and appendages, the precursors of an insect ear can be found virtually at every joint in the body and appendages. There seem to be few developmental constraints in positioning an ear, should an adaptive need arise.

Given so many possibilities for developing an ear, how does it come about that a particular insect possesses an ear in one place, say its forelegs, and another species in another, say its abdomen? One can imagine a host of anatomical, biophysical, ecological, and evolutionary constraints and advantages that might play important roles in the selection process. Such factors as the distances between the ears, the degree of protection offered by surrounding structures, the availability of tracheal air sacs, or the preexisting connections to the central nervous system may all be important. For example, if the function of hearing in a flying nocturnal insect like a moth is to detect and avoid echolocating bats, then a proprioceptor with preestablished neural connections to wing flight musculature would be a better ear candidate than, say, a leg proprioceptor. Or, possibly, the reason the thorax and abdomen are so "busy" with ears (Fig. 1) could be because these locations offer maximum interaural distance and a high degree of protection.

EVOLUTION

Tympanal ears have evolved independently at least 20 times within the class Insecta, and this number surely underestimates the number of ears that do exist. Still, hearing appears to be the exception, not the rule, for insects, with only 7 of the 25 recognized extant neopteran orders having tympanate species. The Orthoptera (crickets, grasshoppers, and katydids) and the Lepidoptera (moths and butterflies) boast the largest number of eared species. In two of the most speciose orders, the Diptera (flies and mosquitos) and the Coleoptera (beetles), tympanal ears are rare, and surprisingly, the Hymenoptera (wasps, ants, and bees) are completely atympanate as far as we know. Following is a brief introduction to the major taxonomic groups for which tympanal ears have been identified to date.

Orthoptera

Orthopteran ears come in two different forms and are divided nicely by the taxonomy of the order. The Acridoidea (locusts, grasshoppers) have ears on either side of the first abdominal segment. The tympanal membranes are nearly circular, opaque, and clearly visible upon inspection with the naked eye. Tracheal sacs connect both ears, allowing the locust to determine the direction of a sound source. The acridid ear is one of the few insect ears known to have the capability of pitch discrimination. About 60 to 80 scolopidia form four separate groups that attach to different regions of the eardrum, which, in turn, resonate to different sound frequencies. The ears of extant acridids function primarily in conspecific communication, but comparative evidence suggests that the primitive function was for predator detection.

In the suborder Ensifera [crickets (Gryllidae), katydids (Tettigoniidae)] the ears occur just below the "knee" region, on the tibia of the forelegs. Each leg has two eardrums—one on each side of the leg (Fig. 3). The tympanal membranes are connected to other sound input sources (the spiracles, contralateral ear) via a system of tracheal tubes and air chambers, which play important roles in directional hearing. The ensiferan auditory chordotonal organ (crista acoustica) has typically between 60 and 80 sense cells arranged in a linear array down the leg, connecting indirectly to the tympanal membrane by tracheal air sacs. Like the acridid ear, the ensiferan ear is capable of pitch discrimination. Presumably, the function of hearing in primitive Ensifera was conspecific communication, which remains the primary function in extant species. Some species are sensitive to ultrasound and use their ears to detect bats in addition to communicating with conspecifics. Fossil records demonstrate that the ancestors of modern Ensifera, which predate the appearance of bats by at least 100 million years, had ears on their legs and therefore, ultrasonic hearing for defense against bats seems to have evolved secondarily.

Lepidoptera

Tympanal ears have evolved more times within the Lepidoptera than any other insect order. To date, at least seven ears of independent origins have been described. Ears are located in the mouthparts (Sphingoidea), at the base of the forewings (butterflies: Hedyloidea, Nymphalidae), in the thorax (Noctuoidea), or on anterior abdominal segments (Tineoidea, Pyraloidea, Drepanoidea, Geometroidea, Uraniidae). Moth ears are among the simplest of all insect ears in that they have very few auditory sensory cells (one, two, or four). K. D. Roeder first demonstrated that the primary function of hearing in moths is to detect the ultrasonic echolocation cries of insectivorous bats. Despite their simplicity, moth ears are capable of determining the distance and direction of an approaching bat. Flying moths evade bats by either turning away from a distant bat or engaging in evasive and erratic flight maneuvers to avoid a sudden and unexpected attack. W. E. Conner has shown that some species, including many day-flying tiger moths (Arctiidae), use their hearing in social interactions. In most Arctiidae, ultrasound production functions to either jam a bat's echolocation calls or warn the bat of a distasteful meal (aposomatic display), but in those species that use sounds for social interactions, both hearing and sound production have secondarily taken on different roles.

FIGURE 4 Nocturnal butterflies of the superfamily Hedyloidea possess ultrasound-sensitive ears on their wings that mediate evasive flight maneuvers to avoid bats. (A) A male *Macrosoma heliconiaria* (Hedylidae). (B) Lateral view of *M. heliconiaria,* showing the general location of the right ear. An arrow points down the canal toward the tympanic cavity where the tympanic membrane resides. Scale bar, 3 mm. (C) Scanning electron micrograph of the right tympanal ear. The hind wing and a dense fringe of scales have been removed to reveal the tympanic cavity. Ca, canal; Cj, conjuctiva. Scale bar, 110 μm. (D) Consecutive video images (30 frames/s) of a free-flying *M. heliconiaria* responding to a short (~250 ms), high-frequency (25 kHz), high intensity (>100 dB) sound. The direction of flight is marked with an arrow and the stimulus onset with an arrowhead. (E) Scanning electron micrograph of the two most proximal chordotonal organs (arrows), viewed from inside the tympanic chamber. The largest organ attaches to the proximal border of the tympanal frame (black arrow). TM, tympanic membrane. Scale bar, 50 μm. (A and E courtesy of J. Yack. B–D were modified, with permission, from J. E. Yack and J. H. Fullard, 2000, Ultrasonic hearing in nocturnal butterflies. *Nature* **403**, 265–266.)

Hearing in butterflies has only begun to be explored. The hedylid butterflies of the neotropics, unusual because of their nocturnal habits, have ultra-sound-sensitive ears on their wings that function as bat detectors (Fig. 4). Some diurnal butterflies of the family Nymphalidae possess ear-like structures on their wings, and there is evidence that the "cracker" butterfly *(Hamadryas feronia)* uses its ears for conspecific communication. No doubt there will be further examples of butterfly hearing in the near future, since tympanal-like structures have been described anatomically in many species.

Homoptera and Heteroptera

Most people are familiar with the loud "buzzing" sounds of cicadas (Cicadidae) during hot weather; these are typically males calling to females. Cicada ears are located within cavities on the ventral side of the second abdominal segment. They are among the largest of all insect ears, with over 1000 scolopidia in each ear. It has been suggested that the large number of sense cells enhances the ear's sensitivity for long distance communication, but cicadas do not appear to have more sensitive hearing organs than other insects with far fewer scolopidia.

Water bugs (subfamily Hydrocorisae) are reported to have ears on various body parts, including the mesothorax, metathorax, or first abdominal segment. The best known is the ear of the water boatman, *Corixa* (Corixidae), which occurs on the lateral mesothorax between the wing and the leg. The ears, like those of moths, are simple, with only two auditory cells, both tuned to low-frequency sounds (1–2 kHz), within the range of conspecific calls. The insect carries a bubble of water with it to allow the membrane to vibrate under water. Unlike for most other insect ears, the tympanal membrane is backed by fluid, not air. Hearing in corixids appears to function primarily for mate attraction.

Diptera

Although it had been well documented that certain parasitic flies were attracted to the songs of crickets, katydids, and cicadas, until recently, it was not known how these flies were eavesdropping on their hosts. Two families of parasitoid flies (Sarcophagidae and Tachinidae) have independently evolved a pair of peculiar ears on their prosternum, just below the head, in the "neck" region (Fig. 5). The gravid females use their ears to locate singing insect hosts upon which to lay their predaceous larvae. The ears of the two parasitoid groups are described earlier in this article. The design features of these dipteran ears show remarkable convergence in anatomy and function, despite the fact that tachinids and sarcophagids are only distantly related. This suggests that evolutionary and developmental constraints are at work here in ways that we do not yet understand.

Mantodea

Until recently, praying mantids were thought to be deaf; now we know that 65% of all mantid species can hear. The ears

FIGURE 5 The prothoracic tympanal ear of a parasitoid tachinid fly, *Ormia ochracea* (Tachinidae). (A) A female fly resting on its host (*Gryllus integer*). (B) A light micrograph of the prosternal ear in a female. The outline of the tympanal membrane is indicated with arrows. (Photographs provided courtesy of D. Robert, R. Hoy, and G. Haldimann.)

occur in the most bizarre location: the two tympanal membranes face one another inside a narrow groove between the metathoracic legs. The mantid ear is functionally an "auditory cyclops" because the close proximity of the ear drums (less than 150 μm) provides no directional information to the animal. The ears function as bat detectors and are most sensitive between 25 and 50 kHz.

Coleoptera

Tympanal ears have been described in two beetle families to date. Several species of the genus *Cicindela* (Cicindelidae) have ears on the dorsal surface of the first abdominal segment, beneath the wings. Some scarabs (two tribes in the subfamily Dynastinae) have ears located just beneath the neck membranes (pronotal shield). The ears of both families are tuned to ultrasonic frequencies, and there is strong evidence that they function as bat detectors. It is a little bit surprising that there are not more examples of hearing in beetles. Given the large number of species, the wide diversity of niches worldwide, and numerous reports of sound production, we expect that more examples will be uncovered.

Neuroptera

Green lacewings (Chrysopidae) have an ear near the base of each forewing in a location similar to that of the ears of some butterflies. The ear consists of a swelling of the radial vein, with a region of very thin cuticle on the ventral side that functions as a tympanal membrane. Like the corixid ear, the tympanal chamber is predominantly fluid filled. The ears respond to sounds between 40 and 60 kHz and are sufficiently sensitive to detect echolocating bats at close distances.

NONTYMPANAL HEARING ORGANS

Until now we have focused on tympanal ears, which are sensitive to traveling waves of changing pressure in air and water, known as the acoustic far field. In the broadest sense of the word, however, hearing encompasses the detection of near-field sounds, as well as vibrations traveling through solid substrates. By and large, the near field can be thought of as a short distance, a few body lengths, from the sound source. Substrate vibrational signals have also been described as "seismic communication." In this larger sense, then, we could argue that most insects can hear. We highlight a few recent developments in the study of these alternative, but no doubt widespread, forms of hearing.

Detecting Near-Field Sounds

When a sound is produced, air particles are being pushed back and forth near the source of disturbance. These particle movements, or near-field sounds, are generally of low frequency (typically below 1 kHz) and, unlike pressure waves, do not travel far from the sound source, in many instances, just a few body lengths in distance. A small, light, and ponderable object occurring within the near-field sound will move in response to the vibrating air molecules. In insects, loosely attached setae or antennae are commonly used for detecting particle velocity. Depending on the structure of the receptor organ, and its position relative to the sound source, near-field receptors can offer information about the direction and the intensity of a sound source.

Some caterpillars can detect the near-field sounds produced by the beating wings of a flying wasp up to distances of 70 cm. Specialized hairs on the dorsal thorax of the caterpillar are displaced in the sound's near field, eliciting an evasive response, such as freezing or dropping from a leaf. Similar particle-displacement-sensitive setae on the cerci of crickets and cockroaches function in predator avoidance and possibly for close-range conspecific communication.

The antennae of many insects also function as near-field sound detectors. In many Diptera (mosquitos, chironomids, and fruit flies), for example, the males are attracted to the near-field "buzzing sounds" of females (Fig. 6). The long flagella of the antennae in male mosquitoes resonate to the tune of flying females, and these antennal movements in turn stimulate many thousands of scolopidia in the Johnston's organ, a chordotonal organ located at the base of the

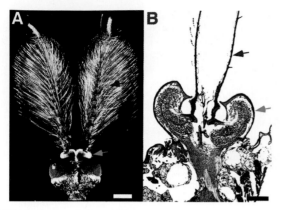

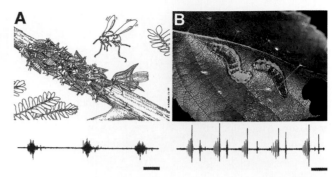

FIGURE 6 The near-field sound hearing organs of a mosquito. (A) Micrograph of the head of a male *Toxorhynchites brevipalpis,* showing the plumous flagella of the antennae (blue arrows), which resonate in response to the near-field sounds of flying females. (B) A cross section through the antennal base, showing the location of the Johnston's organ (green arrows), where the auditory scolopidia are located. (A is courtesy of D. Huber. B was adapted, with permission, from M. C. Göpfert and D. Robert, 2001, Active auditory mechanics in mosquitoes, *Proc. R. Soc. London B* **268**, 333–339.)

FIGURE 7 Acoustic communication through substrate-borne vibrations. (A) A group of treehopper nymphs (*Umbonia crassicornis*) use coordinated substrate-borne vibrations to warn their mother of an approaching predatory wasp. The mother detects the signals and rushes to the defense of her offspring. The waveform represents three group signals from an aggregation of nymphs. Scale bar, 560 ms. (Reproduced, with permission, from R. B. Cocroft, 2002, Antipredator defense as a limited resource: Unequal predation risk in broods of an insect with maternal care, *Behav. Ecol.* **13(1),** 125–133. Waveform courtesy of R. B. Cocroft.) (B) Two masked birch caterpillars (*Drepana arcuata*) engaged in an acoustic "duel" over a silken nest on a birch leaf. The waveform depicts the three signal types (green, anal scrapes; blue, mandible drumming; orange, mandible scraping) used by the caterpillars. (Reproduced, with permission, from J. E. Yack *et al.,* 2001, Caterpillar talk: Acoustically mediated territoriality in larval Lepidoptera, *Proc. Natl. Acad. Sci. USA* **98(20),** 11371–11375.)

antennae. The Johnston's organ may also be involved in the detection of near-field sounds produced by honey bee waggle dances. It is also the sensory organ by which the well-known *Drosophila melanogaster* detects the patterned wingbeats that make up the mating songs of these species.

Detecting Substrate-Borne Vibrations

The detection of vibrations traveling through solid substrates may be one of the most ubiquitous but least appreciated forms of acoustic communication in insects. With recent technological advances in the detection of substrate-borne signals (e.g., laser vibrometry and piezoelectric sensors), we are learning that a large number of insects can detect vibrations produced by both intentional and unintentional senders. In fact, most insect orders probably include species capable of detecting vibrations. At present, however, there are few complete studies on this subject. We provide two recent examples out of many possibilities.

Membracid treehoppers (Homoptera: Membracidae) transmit alarm calls through tree stems to communicate with conspecifics. Nymphs living in colonies of up to 100 individuals on the stem of a host plant produce coordinated waves of vibrations when threatened by a predator. The mother treehopper detects these alarm calls and rushes to defend her offspring, by kicking her hind legs and fanning her wings at the intruder (Fig. 7A). For many species of caterpillars that live in social or crowded conditions, vibrational signaling may be a principal means of communication. The common North American masked-birch caterpillar (*Drepana arcuata*) engages in acoustic "battles" with invading conspecifics. The nest owner drums and scrapes its mandibles and scrapes modified "oars" against the leaf in ritualized acoustic displays. Acoustic "duels" between residents and intruders can last

from a few minutes to a few hours (Fig. 7B). At present, we know little about how insects detect substrate vibrations. The subgenual organ (a chordotonal just "below the knee" in many insects) functions as a vibration receptor in some groups (like some crickets and termites), but for most insects, the receptor organs are yet to be identified. Clearly further research is required before we gain a full appreciation of this important form of communication in insects.

FUNCTION

Because of their physical properties, acoustic signals are highly adaptive for certain kinds of behavioral interactions: sound waves can travel at any time of the day or night, through thick vegetation or muddy water; they convey information instantaneously and can be transmitted over long distances; and sounds are easy to localize, do not leave lingering traces, and can transmit large amounts of information per unit time. For the majority of insects, acoustic communication functions primarily in reproductive behavior and predator avoidance, but may also be used for detecting prey or host species (parasitic flies, wasps; predatory water striders, ant lions) or calling to conspecifics to form aggregations (sawfly larvae) or warn of danger (termites, treehoppers).

For humans, the most conspicuous sounds commonly heard from insects are the loud chirps and trills of field crickets, the long raspy choruses of katydids by night, and the intense, shrill-like buzzes and rattles of cicadas by day. These are the mating calls emitted by males in order to attract conspecific females. Sounds used in reproductive interactions function in species recognition, courting, pair maintenance, female mate choice, and male–male competition. The hearing organs of crickets, katydids, grasshoppers, mosquitos, and cicadas are used primarily for these purposes and are sharply tuned to the calls of conspecifics. The features of these mating calls have surely been shaped by sexual selection.

Many insects have ears for the sole function of detecting predators. Many nocturnally active insects (most Lepidoptera, some mantids, beetles, and lacewings) have ears tuned to the ultrasonic vocalizations of insectivorous bats that use biosonar to detect and home in on their prey. Unlike ears specifically designed for conspecific communication, the ears of predator detectors are usually more broadly tuned and more simple in their design, sometimes having only a few auditory cells per ear.

CONCLUSION

Since antiquity we have known that many insects produce sounds, but only during the past 150 years have scientists realized that some insects can hear. Detailed descriptions of ear anatomy, and the behaviors associated with hearing, began in the early 1800s, providing the basis for current developments in the field of insect bioacoustics. Over the past 40 years there have been significant advances in the field: many new ears have been discovered, and previous claims to tympanal hearing (based on morphological studies) have been validated. With the development of new instruments for detecting acoustic signals outside the realm of human perception (e.g., ultrasound, solid-substrate-borne vibrations), and for recording neurophysiological responses to sound, we are now beginning to better appreciate the immense diversity of insect sound receptor organs.

There is still much to learn about insect hearing. We know little, for example, about the chain of physical and bioelectrical events leading to sound reception at the level of the auditory cells or how acoustic sensory responses are integrated at the level of the central nervous system to promote adaptive behaviors. New tympanal ears will no doubt turn up in the years ahead, but perhaps most significantly, future explorations into substrate-vibrational and near-field sound communication are sure to yield exciting insights into how insects communicate acoustically.

See Also the Following Articles

Mating Behaviors • Mechanoreception • Orientation • Vibrational Communication

Further Reading

Bailey, W. J. (1991). "Acoustic Behavior of Insects." Chapman & Hall, London/New York.
Ewing, A. W. (1989). "Arthropod Bioacoustics." Cornell University Press, Ithaca, NY.
Fullard, J. H., and Yack, J. E. (1993). The evolutionary biology of insect hearing. *Trends Ecol. Evol.* **8**(7), 248–252.
Hoy, R. R., and Robert, D. (1996). Tympanal hearing in insects. *Annu. Rev. Entomol.* **41**, 433–450.
Hoy, R. R., Popper, A. N., and Fay, R. R. (eds.) (1998). "Comparative hearing: Insects. Springer Handbook of Auditory Research." Springer-Verlag, New York.
Michelsen, A. (1979). Insect ears as mechanical systems. *Am. Sci.* **67**, 696–706.
Michelsen, A., and Larsen, O. N. (1985). Hearing and sound. *In* "Comprehensive Insect Physiology, Biochemistry and Pharmacology" (G. A. Kerkut and L. I. Gilbert, eds.). Pergamon, Oxford.
Robert, D., and Hoy, R. R. (1998). The evolutionary innovation of tympanal hearing in Diptera. *In* "Comparative Hearing: Insects" (R. R. Hoy, A. N. Popper, and R. R. Fay, eds.), Chap. 6. Springer-Verlag, New York.
Roeder, K. D. (1967). "Nerve Cells and Insect Behavior." Harvard University Press, Cambridge, MA.
Scoble, M. J. (1995). Hearing, sound and scent. *In* "The Lepidoptera. Form, Function and Diversity." Oxford University Press, Oxford.
Yack, J. E., and Fullard, J. H. (1993). What is an insect ear? *Ann. Entomol. Soc. Am.* **86**(6), 677–682.
Yager, D. D. (1999). Structure, development, and evolution of insect auditory systems. *Microsc. Res. Tech.* **47**, 380–400.

Hemiptera

see *Auchenorrhyncha; Prosorrhyncha; Sternorrhyncha*

Hemolymph

Michael R. Kanost
Kansas State University

Hemolymph is the circulating fluid or "blood" of insects. It moves through the open circulatory system, directly bathing the organs and tissues. Insect hemolymph differs substantially from vertebrate blood, with the absence of erythrocytes and a high concentration of free amino acids being two of the common distinguishing features. The main component of hemolymph is water, which functions as a solvent for a variety of molecules. Water in hemolymph makes up 20 to 50% of the total water in insect bodies, with larval stages generally having a larger relative hemolymph volume than adults. Hemolymph serves as a water storage pool for use by tissues during desiccation and as a storage depot for other types of chemicals. It also contains circulating cells called hemocytes. Hemolymph can function as a hydraulic fluid, for example, in the expansion of a newly molted butterfly's wings. Hemolymph serves important roles in the immune system and in transport of hormones, nutrients, and metabolites.

INORGANIC COMPONENTS

The composition of inorganic ions in hemolymph varies widely among different insect groups. The pH of the hemolymph of most insects is in the range of 6.4 to 6.8. Apterygotes contain high levels of sodium and chloride, similar to mammalian blood. In hemolymph of exopterygotes, sodium and chloride are also high but magnesium makes up a large portion of the total inorganic cations. In endopterygotes, particularly Lepidoptera, Coleoptera, and Hymenoptera, concentrations of sodium and chloride tend to be much lower and are replaced with high levels of potassium, magnesium, and organic anions. This difference has been attributed to the coevolution of these insect groups with flowering plants and the consequent dietary importance of leaves (which contain high concentrations of magnesium and potassium). However, the concentration of inorganic ions is not a function of only the diet, because insects are able to regulate the ion composition of hemolymph to some degree.

LOW-MOLECULAR-WEIGHT ORGANIC COMPONENTS

Citric acid and other organic acids and organic phosphates (such as glycerol 1-phosphate and sorbitol 6-phosphate) account for much of the anion content in hemolymph from many insect species. The most abundant carbohydrate in hemolymph of most insects is the disaccharide trehalose. Transport of trehalose as an energy source for tissues is an important function of the hemolymph. Trehalose levels are hormonally regulated and can be increased through synthesis from glucose phosphate derived from glycogen stored in the fat body. Glucose may also be present in hemolymph, although generally at a lower concentration than trehalose. In some insects, diapause or exposure to low temperatures can stimulate synthesis of glycerol and sorbitol (from glycogen stored in fat body). The resulting high concentration of these compounds in hemolymph depresses the freezing point and protects the insects from damage that would occur if ice crystals were to form in hemolymph.

Hydrophobic lipoidal compounds present in hemolymph are carried by specific transport proteins. Diacylglycerol is the major transported form of lipid in most insects, but triacylglycerol, fatty acids, phospholipids, and cholesterol are also present. Pigments such as β-carotene, riboflavin, and biliverdin, which give hemolymph of many insects a characteristic yellow or green color, are also carried by specific proteins.

Free amino acids are present at high concentration (up to 200 mM) in hemolymph and make a major contribution to hemolymph osmolarity. All 20 of the amino acids found in proteins exist as free amino acids in hemolymph. Although the relative concentrations of the amino acids vary in different species, glutamine and proline are typically abundant. Proline is known to serve as an energy source for flight muscles in some species. Hemolymph may also contain some amino acids that are not found in proteins, such as β-alanine and taurine. Tyrosine, which is metabolized for use in cuticle sclerotization, often occurs in hemolymph as a conjugate with glucose, phosphate, or β-alanine, which increases its solubility. The

phosphate and glucose substituents are removed from tyrosine by specific enzymes when tyrosine is needed for sclerotization. Catecholamines derived from tyrosine, which are used in cuticle sclerotization and pigmentation, are also present in hemolymph as conjugated forms.

PLASMA PROTEINS

Proteins are a major component of the hemolymph plasma. Typical protein concentrations in plasma range from 10 to 100 mg/ml. In most species, the concentration of proteins in plasma increases during each instar and decreases at each molt. The fat body is responsible for the synthesis of the majority of plasma proteins, but there is also a contribution of some specific proteins from epidermis and hemocytes. Plasma from each species contains a few very abundant proteins and more than a hundred other proteins at much lower concentrations. Although the identities and functions of the major proteins are understood, many of the minor hemolymph proteins have not yet been thoroughly investigated.

Storage Proteins

The most abundant proteins in larval hemolymph belong to a class known as storage proteins or hexamerins (because they are assembled from six ~80-kDa polypeptide subunits). The storage proteins are synthesized by the fat body and reach extremely high concentrations in the last instar. At the end of this stage, most of the storage proteins are taken back into the fat body, through interaction with specific receptors, and stored in protein granules. During metamorphosis the storage proteins are broken down into free amino acids, which are used for synthesis of other proteins required in the adult stage. In some exopterygotes, hexamerins are again synthesized by the adult, although their function at this developmental stage is unclear. The hexamerins can be classified according to their amino acid compositions. Those rich in the aromatic amino acids (phenylalanine, tyrosine, and tryptophan) are called arylphorins, whereas another group of hexamerins are known as methionine-rich storage proteins. In addition, some other proteins that function as storage proteins but are not similar in sequence to the hexamerins have been identified in lepidopterans.

Transport Proteins

Several hemolymph proteins function to transport small molecules that have low solubility in water. Insect plasma contains two proteins that specifically bind iron; ferritin appears to sequester dietary iron, whereas transferrin acts as a shuttle to transport iron between tissues.

The most abundant transport protein in hemolymph is lipophorin, which transports lipids between tissues. Like lipoproteins in mammalian plasma, lipophorin is composed of proteins that complex with lipids in such a way that the lipids are protected from contact with the surrounding water.

Lipophorin docks with specific receptors on the surface of tissues to either accept or unload diacylglycerol. Lipophorin contains two polypeptide subunits, apolipophorin-I and apolipophorin-II, which are produced by proteolytic cleavage of a larger protein precursor. In insects that use lipids as a fuel for flight muscles, diacylglycerol is released from the fat body into the hemolymph under control of a peptide hormone known as adipokinetic hormone. As lipophorin accepts large amounts of diacylglycerol, its volume increases and its density decreases as it is converted from high-density lipophorin to low-density lipophorin. Low-density lipophorin contains a third type of protein subunit, apolipophorin-III, which binds to the surface to stabilize the expanding lipid–water interface.

Juvenile hormone (JH), a sesquiterpenoid lipid, has low solubility in water and is transported through hemolymph bound to a specific carrier protein. JH binding proteins of ~30 kDa have been well characterized from plasma of lepidopterans, whereas in other insect orders lipophorin or a specific hexamerin takes on the role of JH transport. In addition to keeping JH in solution, these proteins also protect the hormone from degradative enzymes that help to regulate JH concentration in plasma. JH binding proteins may also aid in delivery of the hormone to target tissues.

Egg Yolk Proteins

In adult female insects, certain proteins synthesized by the fat body and secreted into the hemolymph are delivered to the ovary, where they are taken up by developing oocytes. The most abundant of these is called vitellogenin. Once vitellogenin becomes a part of the egg yolk, it is called vitellin. Vitellogenins are typically large, phosphorylated lipoglycoproteins that are expressed specifically in adult females. Lipophorin is also taken up from hemolymph into eggs and provides additional lipids for use by the developing embryo. Vitellogenin and lipophorin are related in their amino acid sequences, indicating that they have a common ancestral gene. Vitellogenin, lipophorin, and a few other plasma proteins are taken up into oocytes by receptor-mediated endocytosis.

Proteins and Peptides Involved in Immune Responses

A group of plasma proteins functions in defense against microbial infection. Hemolymph of many insects contains lysozyme, an enzyme that degrades bacterial cell walls. In addition, low-molecular-weight antimicrobial peptides are synthesized in response to bacterial or fungal infection. Many of these peptides act by disrupting the integrity of bacterial cell membranes. Phenoloxidase, an enzyme present in plasma of some species and stored in hemocytes of others, is synthesized as an inactive precursor, prophenoloxidase. In response to infection or injury, prophenoloxidase is activated and catalyzes the production of quinones that polymerize to form the pigment melanin, which helps to trap and kill invading organisms. The tendency of hemolymph to darken has been known for more than 100 years, but this melanization has only recently become understood at a molecular level. Plasma contains proteins that bind to carbohydrates on the surface of microorganisms. This causes activation of a cascade of proteases that results in the proteolytic activation of prophenoloxidase. To regulate this immune response, plasma contains several types of proteins that function as protease inhibitors.

HEMOCYTES

The circulating cells in hemolymph are called hemocytes. Insects lack erythrocytes, and hemocytes cannot be directly equated with vertebrate leukocytes. Some fraction of hemocytes remains sessile and attached to the surfaces of tissues, and in some species (mosquitoes, for example) such cells may account for a majority of the hemocytes. Several different morphological types of hemocytes can be identified in each insect species. Some commonly observed hemocyte types are illustrated in Fig. 1. Prohemocytes are small, round cells that may be precursors from which some other cell types develop. Granular hemocytes contain conspicuous cytoplasmic granules that can be discharged as part of a defensive response

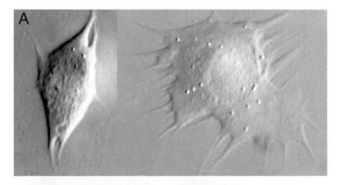

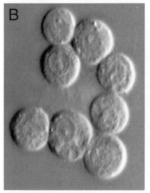

20µm

FIGURE 1 Examples of hemocyte types from a lepidopteran, *Manduca sexta.* (A) Plasmatocytes. The plasmatocyte shown on the left has just begun to spread, whereas the one on the right has spread extensively. (B) Granulocytes. (C) Oenocytoids. (D) Spherulocytes.

to invading parasites. Plasmatocytes usually contain few granules and are characterized by their ability to change from round or spindle-shaped cells in suspension to extensively flattened, ameboid cells after attaching to a substrate. Spherule cells contain very large cytoplasmic granules, which may contain mucopolysaccharides. Oenocytoids are large cells that synthesize prophenoloxidase.

Plasmatocytes and granular hemocytes are usually the two most abundant hemocyte types, although their proportions can vary between species and within a species at different developmental stages. These two hemocyte types participate in immune responses, including: (1) phagocytosis of small organisms such as bacteria; (2) nodule formation, in which multiple hemocytes aggregate to trap microorganisms; and (3) encapsulation, in which hemocytes attach to the surface of a larger parasite and form a multilayered hemocyte capsule, in which the parasite is killed. Nodules and capsules often become melanized through the action of phenoloxidase. Hemocytes, especially plasmatocytes, also aggregate in a type of coagulation response, sealing wounds to prevent hemolymph loss. Another function of hemocytes is in synthesis of the extracellular matrix that covers tissues exposed to the hemolymph. Granular hemocytes appear to be the primary cell type involved in this aspect of hemocyte function.

See Also the Following Articles

Circulatory System • Fat Body • Immunology • Vitellogenesis • Water and Ion Balance, Hormonal Control of

Further Reading

Gillespie, J. P., Kanost, M. R., and Trenczek, T. (1997). Biological mediators of insect immunity. *Annu. Rev. Entomol.* **42**, 611–643.
Gupta, A. P. (1985). Cellular elements in the hemolymph. *In* "Comprehensive Insect Physiology, Biochemistry, and Pharmacology" (G. A. Kerkut and L. I. Gilbert, eds.), Vol. 3. Pergamon, New York.
Haunerland, N. H. (1996). Insect storage proteins: Gene families and receptors. *Insect Biochem. Mol. Biol.* **26**, 755–765.
Kanost, M. R., Kawooya, J. K., Law, J. H., Ryan, R. O., Van Heusden, M. C., and Ziegler, R. (1990). Insect haemolymph proteins. *Adv. Insect Physiol.* **22**, 299–396.
Lackie, A. M. (1988). Haemocyte behaviour. *Adv. Insect Physiol.* **21**, 85–178.
Mullins, D. E. (1985). Chemistry and physiology of the hemolymph. *In* "Comprehensive Insect Physiology, Biochemistry, and Pharmacology" (G. A. Kerkut and L. I. Gilbert, eds.), Vol. 3. Pergamon, New York.
Ryan, R. O., and van der Horst, D. J. (2000). Lipid transport biochemistry and its role in energy production. *Annu. Rev. Entomol.* **45**, 233–260.
Telfer, W. H., and Kunkel, J. G. (1991). The function and evolution of insect storage hexamers. *Annu. Rev. Entomol.* **36**, 205–228.
Wyatt, G. R. (1961). The biochemistry of insect hemolymph. *Annu. Rev. Entomol.* **6**, 75–102.

Heteroptera

see *Prosorrhyncha*

Hibernation

Richard E. Lee, Jr.
Miami University, Oxford, Ohio

Hibernation refers to the state in which animals pass the winter. In discussion of insects overwintering is often used as a synonym for hibernation; usually hibernation is associated with entry into a dormant state. Estivation is a term used for animals that become dormant in the summer. Some insects enter an extended period of dormancy, referred to as estivohibernation, which begins in summer and continues through the winter. This term also is used commonly in connection with those mammals that lower their body temperature slightly (e.g., carnivorean lethargy in bears and skunks) or extensively (e.g., hibernation *sensu stricto* in ground squirrels) during the winter.

It is critical for insects to synchronize their periods of feeding, growth, and reproductive activities with those times of the year when food is available and environmental conditions are suitable. Hibernation generally includes entry into diapause, a dormant state that promotes survival by depressing metabolism and energy utilization when host plants and other food sources are unavailable. Typically, this also includes reduced morphogenesis in immature stages; hibernating adults typically hibernate before reproducing.

Specific behavioral changes are often associated with movement to overwintering sites, termed hibernacula. One of the most extreme examples is the monarch butterfly *(Danaus plexippus),* which may migrate more than 5000 km from southern Canada and New England to mountain sites in central Mexico. Other insects migrate locally as they move to hibernacula within the soil where they may burrow to avoid exposure to winter cold. Still others seek sites beneath rocks, logs, bark, and leaf litter.

In temperate regions, insects typically hibernate in a specific life stage. The egg, larva, and pupa are more common stages for overwintering than is the adult. However, alpine and polar insects or others living in extreme environments, in which growing seasons are short or unpredictable, and having life cycles that may be extended over several years may hibernate multiple times in one or more life stages.

In temperate regions, most hibernating insects enhance their resistance to environmental extremes, particularly cold and desiccation. During autumn, many species markedly enhance their tolerance to low temperature, termed cold-hardening. Most insects are freezing intolerant and are unable to survive freezing within their body fluids. These species typically enhance their capacity to supercool (i.e., remain unfrozen at temperatures below the melting point of their hemolymph) by synthesizing glycerol, sorbitol, trehalose, or other cryoprotective compounds, often at high concentrations of 1 M or more. Production of antifreeze proteins, avoidance of inoculative freezing by external ice, and ridding the body of ice nucleators that may catalyze ice formation are other mechanisms used to avoid lethal freezing.

In contrast, a few insects are freeze tolerant and can survive the freezing of 70% or more of their body water. Cryoprotectants are also commonly synthesized by freeze-tolerant species, as are ice-nucleating proteins that induce ice formation at high subzero temperatures. Overwintering insects also may acquire exceptionally high levels of desiccation resistance, comparable to those of desert species.

Behavioral and physiological changes associated with hibernation, diapause, and cold-hardening are commonly triggered by environmental cues, including photoperiod, temperature, moisture conditions, and changes in host plant quality. These cues ensure that adaptive responses occur before severe winter conditions arrive.

See Also the Following Articles

Aestivation • Cold/Heat Protection • Diapause • Dormancy • Monarchs

Further Reading

Danks, H. V. (1987). "Insect Dormancy: An Ecological Perspective." Biological Survey of Canada (Terrestrial Arthropods), Ottawa.

Denlinger, D. L., Giebultowicz, J. M., and Saunders, D. S. (eds.) (2001). "Insect Timing: Circadian Rhythmicity to Seasonality." Elsevier, Amsterdam.

Hallman, G. J., and Denlinger, D. L. (eds.) (1998). "Temperature Sensitivity in Insects and Application in Integrated Pest Management." Westview, Boulder, CO.

Leather, S. E., Walters, K. F. A., and Bale, J. S. (1993). "The Ecology of Insect Overwintering." Cambridge University Press, New York.

Lee, R. E., and Denlinger, D. L. (eds.) (1991). "Insects at Low Temperature." Chapman & Hall, New York.

Tauber, M. J., Tauber, C. A., and Masaki, S. (1986). "Seasonal Adaptations of Insects." Oxford University Press, New York.

History of Entomology

Edward H. Smith
Cornell University (Emeritus)
Ashville, North Carolina

Janet R. Smith
Asheville, North Carolina

This brief history traces the interactions of humans and insects dating from the adoption of agriculture and its inherent ecological disruptions. Humankind's early preoccupation with survival focused on insects as relentless pests, competitors for food and fiber, threats to health and comfort. The high hopes following World War II for relief from the bondage of insects through the use of chemical insecticides such as DDT proved unrealistic. The reassessment that followed led to a concept based on ecological principles which is referred to as integrated pest management (IPM). In this system, multiple control technologies are used, with the additive effect being to hold insect injury at acceptable levels while avoiding excessive environmental insult. The age-old struggle continues; entomologists are now armed with the lessons of the past; advances in insecticidal chemistry, biological control, and cultural methods; and visionary new technologies based on genetic modification of plants and animals. Simultaneously, the rise of the environmental movement and ecological awareness has placed insects in a new context, highlighting their essential role in biodiversity on which the viability of the Earth depends. The vision for the 21st century calls for compatibility between insect control and conservation; both are prerequisites to human well-being. Stewardship of the Earth is the greatest challenge ahead and one that places awesome responsibility on the shoulders of entomologists.

IN THE BEGINNING

The history of life on earth reaches back some 4 billion years. From this beginning the long evolutionary trail unwound. Along the way, 99% of the forms that appeared met with extinction.

The great exterminations that have occurred since the appearance of insects in the Devonian period, 400 mya, revealed insects' remarkable survival qualities. Insects witnessed the last of the trilobites that preceded them by 175 million years. By the time the dinosaurs appeared in the Triassic period, 210 mya, the major orders of insects existing today were already well established. Dinosaurs became extinct 66 mya, leaving a niche occupied in time by mammals. The mammals, in turn, provided a niche for insects, offering furry cover and warm meals. The disappearance of the dinosaurs coincided with a great radiation of insects based on insects' symbiosis with flowering plants. For the past 150 million years, the flowering plants and insects have honed their intricate coevolution, which accounts for their immense biodiversity on which human habitability of the earth depends.

Insects have withstood trial by ice and fire, meteorite strikes, volcanic eruptions, global dust veils, acid rain, and continental upheavals. This evolutionary experience is encoded in their DNA and attests to the advantage of their small size, external skeleton, flight, metamorphosis, and specialized systems of reproduction. These are significant credentials in insects' rivalry with *Homo sapiens,* a species that draws on an evolutionary history of a scant 7 million years.

COEXISTENCE, HUMANS AND INSECTS

Class Insecta has plagued and fascinated humans for all of their history. The most striking features of the Insecta are diversity and numerical superiority. Of the 5 to 30 million species estimated to compose the global flora and fauna, approximately 1.7 million have been named, more than half of them being insects. It is estimated that insects make up 75% of the known animal kingdom.

Because insects occupy almost every conceivable terrestrial niche, they interact with humans in countless ways that accord them status as "pests." This same diversity bestows on insects essential roles in the functioning of the biosphere as a sustainable biological system. Considering the countless interactions between humans and insects, it is not surprising that insects have become fixed in the fabric of human culture. They have become important components of our art, language, literature, music, philosophy, and religion. In addition, insects are remarkable sources of knowledge, ideal models for the study of biological processes including genetics, physiology, and molecular biology.

Professional entomologists find challenge in our universities where they engage in teaching, in conducting research to advance knowledge, and in extension, applying knowledge to the solution of applied entomological problems. In addition to professional entomologists, amateur naturalists are drawn to the study of insects because of their form, color, and behavior. As the expanding human population continues its modification of the natural habitat, the interface between humans and insects will become more problematic.

ENTOMOLOGICAL ROOTS

A mere 10,000 years ago, primitive hunter–gatherers made a great leap forward; they entered into partnership with plants, and agriculture was born. Thereafter, humans would seek to alter ecosystems to their own advantage. They would intervene to favor some plants and animals over others, thereby altering the evolutionary process that had shaped the biological world for the preceding 4 billion years.

Insects in their coevolution with plants and animals had been a powerful force in shaping the biosphere. They posed a primary threat to humans' alteration of ecosystems to provide food, fiber, shelter, and comfort for themselves and their domestic animals. The struggle that followed moved through stages of ignorance, myth, religion, and then enlightenment through science and technology.

Our brief historical sweep will jump the seemingly long sleep of ancient civilizations and go to the Greek civilization in the time of Aristotle (384–322 B.C.). The orderly study of biology began with his speculations. He relied on his own observations, defined the field, posed the questions, and accumulated evidence to answer them. Aristotle's vision of rationality lay dormant for centuries until the Renaissance. In the meantime, Judaism and Christianity imposed a new concept, one focused on God and creation as depicted in the book of Genesis. Accordingly, God created the world, directing Man to "be fruitful and multiply…" and "with dominion over every living thing that moves upon the earth" (Genesis 1:28). Man was not part of nature. Nature was subservient to Man.

The Scientific Revolution of the 16th and 17th centuries marked the beginning of modern science and included mathematics, mechanics, and astronomy but had little impact on biology. While the revolution rejected superstition, magic, and the dogma of medieval theologians, it did not reject the ideological bias of the Judeo–Christian religion. The hand of God was still directing the course of the natural world.

Not until the 17th and 18th centuries was entomology advanced as a field of study within zoology. Anton van Leeuwenhoek (1632–1723) of Holland used the microscope to extend the power of the human eye. He was obsessed with the study of detail, including the morphology and specialized organs of insects. His revelations established insects as proper subjects for scientific study. Francesco Redi (1626–1697) of Italy demonstrated in 1668 that insects arose, not from spontaneous generation, but from eggs laid by fertilized females. Jan Swammerdam (1637–1680) of Holland did superb anatomical work on insects, including the honey bee.

The excitement of these discoveries was further enhanced by the flow of exotic plants and animals brought back from voyages to other continents. Charles Darwin's voyage of *HMS Beagle* in 1831–1836 followed this tradition. The wealth of material acquired made students aware of the need to classify the organisms collected and to assemble specimens in orderly collections. Other investigators focused on the activity of insects in the field and their role as pollinators and as agricultural pests.

A prerequisite to advancing the study of insects was the development of a classification system that would bring order out of chaos. The Swedish naturalist Carl Linnaeus (1707–1778) met this need. Although trained in medicine, he studied botany extensively and turned to the classification of plants, animals, and minerals. His *Systema Naturae* (10th ed., 1758) is still regarded as the foundation stone of zoological nomenclature. He greatly simplified insect classification by using insect wings (hence the suffix *-ptere,* meaning wing, for most order names) as the basis for classification. The other great feature of his system was the designation of genus and species each by a single word, thus providing a binomial system to replace the unwieldy descriptive names employed earlier. Linnaeus's "artificial" system of insect systematics based only on wings was in time modified by adding other characters to construct a "natural" system.

Another great naturalist, René Antoine Ferchault de Réaumur of France (1683–1757), infused a new perspective into the emerging study of insects. He deplored the confusion that existed regarding metamorphosis, distribution, and "industries" of insects. He championed the study of insects out of sheer curiosity, claiming that useful discoveries would be made in the process. His six volumes of *Memoires pour Servir a l'Histoire des Insectes* (1734–1742) with their exacting attention to morphology and function, complete with accurate drawings, established a new standard of excellence.

The work of Linnaeus and Réaumur provided the templates for orderly classification and elucidation of fundamental and applied aspects of entomology. Their works were extended and refined by French naturalists Pierre André Latreille (1762–1833), Georges Cuvier (1769–1832), and Jean Lamarck (1744–1829). By the 19th century, entomology was firmly established in European zoological science. The taxonomic treaties established in this process were to provide the guides to the classification of American insects. These sources were augmented by two sources in Great Britain: Gilbert White's (1720–1793) *The Natural History and Antiquities of Selborne* (1789) and William Kirby (1759–1850) and William Spencer's (1783–1860) *Introduction to Entomology* (1816–1826). The writings of these field naturalists on the biological characteristics of insects made insects, at one point, the most popular component of natural history in Victorian England. In addition, they contributed to the development of the biological species concept. This concept, essential to the understanding of biological communities, recognizes a species as a reproductively (genetically) isolated group of inbreeding populations.

The next step in the unfolding of the biological sciences was a giant one: the publication, in 1859, of Charles Darwin's (1809–1882) theory, *On the Origin of Species.* This event placed conceptual biology in a new light. In a single stroke, Darwin's work challenged the natural theology that had dominated biological thought for 3 centuries. Natural theology had been elaborated in John Ray's (1627–1705) *The Wisdom of God Manifested in the Works of the Creation* published in 1691. The concept provided a truce between science and religion. It contended that God created the world and the evidence of His omnipotence was to be found in the study of His creatures.

There was no middle ground between Darwin and Ray. Darwin provided a new way of viewing biology. The living world had evolved; it could be explained on the basis of descent with change. It was noteworthy to entomologists that much of Darwin's supporting evidence was derived from his study of insects dating from his days at Cambridge University (1828), where he was an avid insect collector.

Intense debate followed the publication of Darwin's theory. Nowhere was the debate more intense than at Harvard University, where Asa Gray (1810–1888), a botanist and staunch Darwinian, challenged Louis Agassiz (1807–1873), the foremost naturalist of the world and unrelenting defender of the creationist view.

An unlikely pair of outspoken individuals led the pro-Darwinian entomologists of North America. Benjamin D. Walsh (1801–1869), who had collected beetles with Darwin at Cambridge University, emerged from obscurity on the Illinois frontier to assert his long dormant entomological interest and declare his support for Darwinism. He was joined by the youthful Charles V. Riley (1843–1895), a fellow Englishman and self-taught entomologist who was then writing on entomology for the *Prairie Farmer,* the leading farm journal of the Midwest (Fig. 1). Their early collaboration on the issue of Darwinism bode well for the future. In time, evolution was accepted by biologists as a

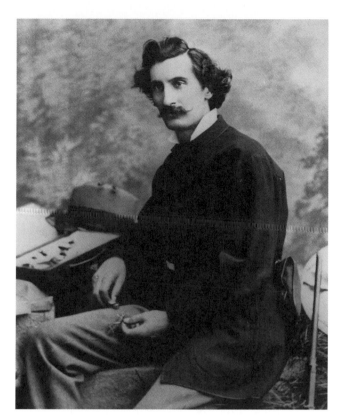

FIGURE 1 Charles V. Riley's nine annual reports as the state entomologist of Missouri (1868–1876) figured prominently in the knowledge base on which the philosophy of insect control in North America was based.

fundamental basis of the discipline, although segments of the public still oppose it as a challenge to the Creation story as reported in the Old Testament book of Genesis. The distinguished geneticist, Theodosius Dobzhansky (1900–1975), summarized the views in the major fields of biology in his essay entitled, "Nothing in Biology Makes Sense Except in the Light of Evolution" (1973).

ENTOMOLOGY IN THE NEW WORLD

Nurturing Environment and Supporting Institutions

EMPIRE AND INSECTS Early entomological developments in the United States occurred in the climate of Thomas Jefferson's America. Jefferson took office as the nation's third president in March 1801 and proceeded to sell his vision to his countrymen, then numbering just over 5 million. The United States had only recently gained its independence from England. Louisiana, the land stretching from the Mississippi to the Pacific Ocean, although claimed by Spain, was available and being considered by Russia, France, and England. Of these three, France, energized by Napoleon, was feared the most. Despite the tenuousness of the situation, Jefferson clung to his vision of this vast land mass, stretching from sea to sea, being united under a stable government of the United States. Napoleon's agreement to sell Louisiana came as a surprise; its purchase, although not applauded by some leading politicians, was a masterful stroke. It doubled the land mass of the United States and resolved the feuding over control of the Mississippi River. But what had been purchased? To answer that question, Jefferson dispatched an expedition, led by Captain Meriweather Lewis and Lieutenant William Clark (1804–1805). Commissioned to explore the new acquisition, their report, on their return, removed some of the mystique of the Pacific Northwest but provided only tantalizing glimpses of the area's natural history. Unfortunately, the expedition included no trained naturalists.

The follow-up to Lewis and Clark's report came in 1819–1820. Major Stephen H. Long, under authorization of President Monroe, led an expedition of "Gentlemen of Science" to study this vast unexplored territory, its natural history, and the American Indians of the Rocky Mountain region. Entomology was well represented by Thomas Say, whose affiliation with the Academy of Natural Sciences of Philadelphia had won him the reputation as "perhaps the most brilliant zoologist in the country." Say epitomized the confidence and vision of the young nation's leaders in natural history; he was tall, handsome, and resplendent in the uniform of the Long Expedition, with its high-collared, gold-buttoned, green jacket and black military trousers.

Although Say's primary interest was insects, he covered the entire field of botany and zoology and conducted studies on the American Indians. His entomological studies provided the foundation for his *American Entomology*, published in three volumes (1824, 1825, and 1828). These were the first books on North American insects. They were beautifully illustrated by the artistic talents of his wife, Lucy, and Titian Peale of the distinguished family of Philadelphia artists. They provided a stimulus for American entomologists and signaled their emancipation from the European centers to which the study of American insects had previously been consigned, named by Europeans, and retained in their collections. With the spirit of the Revolution still vibrant, consigning American insects to European collections offended national pride.

The Long Expedition imparted an American quality to the study of the nation's natural history. The Gentlemen of Science who manned the expedition were trained in the nation's centers of learning. They were not closet naturalists; they were adventurers. Their spirit was described by Say himself: "If our utmost exertions can perform only a part of a projected task, they may, at the same time, claim the praise due to the adventurous pioneer, for removing the difficulties in favor of our successors" *(American Entomology)*.

THE CULTURAL CENTERS   Natural history found strong support in the cultural centers of Philadelphia, Boston, and New York. Philadelphia led the way, with the American Philosophical Society and the influence of the distinguished Benjamin Franklin (1706–1790). The Academy of Natural Sciences, founded in 1812, nurtured the founding of the

Entomological Society of Philadelphia in 1859, which in turn launched the *Practical Entomologist* in 1865.

Boston looked to Harvard College and the Massachusetts Society for Promoting Agriculture. William D. Peck (1763–1822) was appointed Professor of Natural History at Harvard in 1805 and offered the first lectures in entomology in North America. Thaddeus William Harris (1795–1853), a physician turned Harvard librarian, found time to become entomological author, teacher, collector, and correspondent. His report on *Insects Injurious to Vegetation* (1841) summarized the knowledge of insect control in Europe and North America, earning him the title, "Father of Economic Entomology."

New York asserted its interest by appointing Amos Eaton and John E. LeConte to the Lyceum of Natural History of New York. Other distinguished leaders included John Abbot (1751–1840), Thomas Say (1787–1834), and Frederick V. Melsheimer (1749–1814). A striking feature of these men and their institutions was their support of both classical and applied entomology. The individuals were well trained by the standards of the day, often completing training in medicine or theology, because there was no specific training in entomology.

Systematics was the primary entomological interest, followed by aid to agriculture, which was beset with countless insect pests. These leaders experienced the frustration of gaining access to the European literature, founding periodicals for the publications of their own findings, and establishing reference collections.

In the 1840s, American entomologists turned to the task of establishing the institutions that would sever their European dependence. The institutional framework took shape rapidly, led by the American Association for the Advancement of Science (AAAS), founded in 1847. It marked a transition from amateur to professional status, provided a national scientific forum, and nurtured the founding of professional societies. Within approximately 2 decades, 1859 to 1881, five additional societies were established in North America: the Entomological Society of Philadelphia (1859), the Entomological Society of Canada (1862), the Cambridge Entomological Society (1874), the Brooklyn Entomological Society (1872), and the Entomological Club of AAAS (1872). The institutional framework was now in place to expand the scientific and technical dimensions of entomology.

STATE AND FEDERAL ACTION IN THE UNITED STATES
Agriculture held the key to moving the nation from an agrarian to an industrial society. The farmer was viewed as the noblest and most independent man in society. Unlike in Europe, the availability of fertile soil was seemingly unlimited. While great physical obstacles lay in the path of progress, the greatest was the limits of the human intellect. With these elements in the national outlook, it followed that state and federal action would augment the private efforts in support of agriculture and entomology. In 1854, two landmark appointments were made: Townend Glover was appointed to

the Federal Patent Office for work in the newly established Bureau of Entomology, and New York State, responding to pressure from the New York State Agricultural Society, appointed Asa Fitch as its first state entomologist. Illinois and Missouri followed suit in 1868 with the appointments of Benjamin D. Walsh and Charles V. Riley, respectively.

These appointments represented historic landmarks, because state and federal funds were appropriated in support of agriculture with entomology in the vanguard. These men were able individuals whose evangelical zeal and sound professional grounding were attuned to national goals. Their publications, with Charles V. Riley's nine Missouri reports forming the core, laid the foundation for applied entomology in North America.

National goals for agriculture led to enabling federal legislation in three steps. First, the Morrill Land Grant Act of 1862 provided grants to each state, the proceeds from which funded a college, "to teach such branches of learning as are related to agriculture and the mechanic arts...." A research dimension was added in 1887 by the provisions of the Hatch Act, a state experiment station being added to each college and coordinated by a central office in the Department of Agriculture in Washington, DC. The events cited in the foregoing, occurring within approximately 3 decades, provided an impetus for applied entomology that was unprecedented in the world.

Cooperative Extension, the outreach arm of the Land Grant University, which had been active from the start, was formally recognized and funded by the Smith-Lever Act of 1914. This institutional framework with its catalytic feedback from teaching, research, and extension has been recognized as one of the greatest educational innovations of all time.

With economic entomology rapidly expanding under the stimulus of the experiment stations, Charles V. Riley, now Chief of the Bureau of Entomology, perceived the need for a national organization to advance the goals of economic entomology. His organizational abilities and partnership with his Canadian counterpart, James Fletcher, led to the establishment of the American Association of Economic Entomologists in 1889. At Riley's insistence, the association focused on economic entomology, leaving unmet the needs of the broader dimensions of biology, taxonomy, morphology, and faunistic studies of insects. In 1906, the Entomological Society of America was organized to meet these needs, with John Henry Comstock of Cornell University serving as president. With the various forces that shaped these professional and governmental institutions in mind, we can examine how the institutions responded to the challenge posed by insect pests.

Insect Pests

That the world is not awash in insects, despite their remarkable potential for reproduction, attests to the "balance of nature." But nature's balance, while avoiding extremes, does not preclude insect activity that is annoying to humans. Insects that take humans' crops or blood, and invade their dwellings, are termed "pests." The term has no biological significance; it only expresses a human perception.

Let us examine a few insect outbreaks that occurred in agricultural, medical, and veterinary entomology in the late 1800s. They were to test the mettle of the institutions crafted to address such problems. They taught us much about insects, ourselves, and our vast land mass with its unique biomes and punctuated with its geographical features: the Rockies in the west, the Appalachian range in the east, the Great Plains, the Great Lakes, and the Mississippi. It was this abundance of land that appealed to the early settlers from land-poor Europe. It was from the vastness of the land with its rich flora and fauna coupled with the democratic spirit of its people that the American dream was fashioned. However, the dream's social fabric was not matched by its concept of stewardship of the land. New England's forest primeval had to be breached to make way for agriculture. Conquering nature was viewed as a prelude to progress, and the American Indian and the bison fared poorly under this credo.

AGRICULTURAL ENTOMOLOGY The Colorado potato beetle, *Leptinotarsa decemlineata*, existed in the foothills of the Rockies on the buffalo bur, *Solanum rostratum*. As pioneer settlers pushed westward with their crops, the beetle colonized the cultivated potato, *Solanum tuberosum*, and began its eastern migration along the "potato trail." It was observed as a potato pest in Nebraska in 1859, reached the Atlantic coast by 1874, and traveled thence to Europe in 1876, where it remains an important pest.

The early search for control measures established the arsenical, Paris green, an industrial pigment, as an effective poison. It soon became the standard treatment and was the first widely used poison to kill by ingestion.

The early marketing of insecticides invited fraud through adulteration and false claims. It was not until 1910 that federal legislation was passed requiring labeling to reveal efficacy and ingredients in the two most widely used insecticides, Paris green and lead arsenate.

The boll weevil, *Anthonomous grandis grandis*, crossed the Rio Grande to Texas in 1894 and began its eastward trek, occupying the entire 1,500,000-km^2 cotton belt by 1925. Efforts to impede its progress by establishing no-cotton barriers failed for lack of community compliance. Countless control measures were tried but insecticides eventually won as the first line of defense. Calcium arsenate was adopted for control in about 1920, and its use soon reached 20,000 tons per year. This marked a new scale of area-wide pesticide treatment with its attendant environmental and human safety problems.

The social and economic impact of the weevil was incalculable. The prosperity of the south evolved around a single crop, cotton. With its loss, the economic infrastructure collapsed, and panic ensued. Black laborers left, mortgages were foreclosed, and banks failed. The potential for economic disaster in the wake of insect outbreak was seared in the memory of the people of the cotton-producing states. Only the Civil War had greater impact on the economic and social life of the southern states than did the boll weevil.

The Rocky Mountain grasshopper, *Melanoplus spretus,* appeared in an epidemic eastward migration, borne on wind currents from the foothills of the Rockies to the Mississippi valley in 1874–1876. Presumably, this migration was in response to the agricultural disruption of the ecosystem that had been dominated by the American bison. The ravages of these hordes of airborne insects created a crisis for the affected states, whose governors appealed to Washington, DC, for federal intervention. In response, the U.S. Entomological Commission was created with the colorful Charles V. Riley as its chairman. This was not a staid Washington bureaucracy; it was a mobile force that reached out to the crisis whenever it arose. Riley scoffed at Sundays devoted to prayer for divine intervention to restrain the pest. Rather, he urged the people to adopt control measures based on intricate knowledge of the life history of the pest. His insights led to bold predictions of the pest's demise from natural causes. With some sound observations and a modicum of luck, his predictions held. The Commission, despite its denials, was credited with solving the problem, bringing new credibility to entomologists and federal aid to the states.

The gypsy moth, *Lymantria dispar,* was introduced from France, not by accident but by design, by Leopold Trouvelot (1827–1895), a Harvard astronomer and amateur entomologist, who was interested in silk-producing moths. Larvae, emerging from egg masses he had imported, escaped from his Medford, Massachusetts, residence in 1869. After a period of 20 years, the moth reappeared in an epidemic outbreak, it having been mistakenly overlooked as a native species. With this head start, the scorched earth practice of cutting and burning infested trees, augmented by arsenical sprays, failed. The effort did stimulate advances in the technology of spray machines. Today, this introduced pest has spread westward and southward, occupying a swatch from the Great Lakes to the Carolinas.

Biological control was enthusiastically touted following the spectacular success achieved by Charles V. Riley's innovative introduction of the Vedalia, *Rodolia cardinalis* (Coleoptera), into California to destroy the cottony cushion scale, *Icerya purchasi.* The pest had been introduced from Australia about 1868 and soon threatened destruction of the state's citrus industry. Two years after the introduction of the predator, the pest was miraculously under control.

The whole array of control measures, cultural, mechanical, chemical, plant resistance, and biological, was employed in seeking to cope with these problems. The entomologists were influenced by expectations and perspectives of their farmer clientele. The farmer's time frame was established by harvest date and sale of the crop; his risk tolerance was low. Insecticides provided immediate and predictable results and they became the backbone of control programs.

Meanwhile, several factors intensified the pressure for insect control, including monoculture, susceptible crops, exacting market standards, and introduced pests; all required greater intervention and modification of the agroecosystem.

MEDICAL AND VETERINARY ENTOMOLOGY The foundation for modern medical and veterinary entomology was laid by Louis Pasteur, a French microbiologist, who formulated the theory of microbial causation of disease based on his work with the silkworm *Bombyx mori* in 1887. Without benefit of the germ theory, Josiah Nott, a Mobile, Alabama, physician, proposed (1848) that the causative agents of malaria and yellow fever were transmitted by mosquitoes. In 1881, Carlos Finlay, a Cuban physician, postulated that mosquitoes transmitted the yellow fever agent, setting the stage for Major Walter Reed and associates to verify his claim. In 1897, Ronald Ross demonstrated the occurrence of the malaria parasite in mosquitoes that fed on a human patient whose blood contained the parasite, thus leading to the elucidation of the epidemiology of malaria.

In 1889, Theobold Smith discovered the causative agent of Texas cattle fever and, working with F. L. Kilbowen, showed in 1893 that the cattle tick, *Boophilus annulatus,* was the vector. Their work paved the way for tick prevention and development of the cattle industry in the southern United States.

The experiences cited above in control of insects of agricultural, medical, and veterinary importance were unprecedented in the American experience and left no region untouched. They revealed the social, political, biological, economic, and environmental dimensions of insect problems. A nation leaning so heavily on agriculture was sensitive to the impact of these problems on the nation's well-being.

The fundamental principles gleaned from these experiences were to shape the philosophy of insect control for the future. They included the following: (1) taxonomic knowledge of the vast insect fauna is a prerequisite for detection and development of control programs; (2) advances in international commerce breach the ancient oceanic barriers to the dispersal of insects and increase the likelihood of introducing exotic species; (3) introduced species, uninhibited by their natural controls, often become major pests in their new habitat; (4) the economic well-being of vast regions of the nation is vulnerable to insect attack; (5) intervention at the federal level is required for insect problems beyond the scope of individual states; (6) alterations of ecosystems trigger changes in patterns of insect behavior; (7) the use of insecticides requires federal regulations to protect the user, the public, and the environment; and (8) an informed public will underwrite sound programs of insect control.

PARTNERSHIP IN PEST CONTROL, 1880 TO WORLD WAR I

The institutional framework for colleges of agriculture was well established by 1880. The first department of entomology was founded at Cornell University in 1874 under the leadership of John Henry Comstock; others followed shortly.

The primary objective of the department was to train students who wished to become farmers, to identify and classify the insect fauna, to study life cycles, to devise control measures, and to train farmers in their use.

As insecticides became a more important component of production technology, an alliance of increasing importance developed among the agricultural constituency, the agricultural colleges, and the chemical industry. As this partnership developed, agribusiness expanded to provide the goods and services required for agricultural production and grew even more mechanized, technical, and capital intensive. The three partners shared a common objective, pooling their resources to increase the efficiency of agricultural production that accrued ultimately to the benefit of the consumer.

The arrangement involved the agricultural constituency, lending political support to the agricultural colleges in return for their services. The colleges then aided the chemical industry by testing their products and giving their stamp of approval, which enhanced their marketability. A grateful chemical industry provided grants to the entomology departments, which were always short of operational funds. The deans at the agricultural colleges had the difficult task of being broker between the college faculty, with its leaning toward basic research, and the farm constituency seeking low-risk pest control programs. The arrangement was an American innovation that seemed to please everyone. Furthermore, the chemical industry was greatly stimulated by the economic and political activities of World War I. Food and fiber production was given high priority and new discoveries advanced the pesticide industry.

The period from 1880 to 1940 witnessed the maturing of the Agricultural Experiment Stations as a national, highly coordinated network. The extension entomologists became the connecting link between the agricultural college and the agricultural producer. Because insecticides had become the first line of defense, the growing chemical industry added strength to this already solid partnership with the colleges and farmers in the aftermath of World War II. Furthermore, this was the threshold of an era of discovery of new molecules that would affect biological processes of plants and animals.

Along the way some ominous straws in the wind signaled trouble ahead. In 1913, The San Jose scale, *Quadraspidiotus perniciosus,* developed resistance to lime sulfur but the phenomenon was not recognized as an expression of Darwinian selection. In 1928, the codling moth, *Cydia pomonella,* was shown to be resistant to lead arsenate. The apple industry was dealt a severe blow in the mid-1930s when British markets rejected fruit from the United States because of high arsenical residues. Simultaneously, fruit trees were showing loss of vigor because of insecticidal toxicity to foliage and accumulation of residues in the soil.

Although these disquieting revelations were not widely publicized, they were of concern to entomologists, as they and the pesticide industry were marshaled to meet the greater demands for food and fiber required for World War II. Even then, the sustainability of insecticidal control was clearly in doubt.

ENTOMOLOGY, POST-WORLD WAR II

Technology's Triumph

The explosion of the atom bomb over Nagasaki in 1945 brought a dramatic end to World War II and in so doing, highlighted the role of science and technology in the victory. Following this, the age-old ritual of "beating swords into plowshares" turned scientific and technical advances to peaceful ends.

No field emerged with more exciting prospects than did the field of entomology. DDT, with its wartime secrecy removed, was hailed as the answer to insect control. Its employment in arresting an epidemic of typhus in Naples in 1943–1944 dramatically neutralized the lethal companion of armed conflict, vector-borne disease.

Overnight the entomological community documented DDT's remarkable effectiveness in controlling insect pests of agricultural, medical, and veterinary importance. The race was on, and an old alliance assumed new vigor. The Land Grant Universities joined with industry and agriculture to exploit the new possibilities of chemical pest control.

Although industrial grants to the Agricultural Experiment Stations to fund trials of mutual interest dated from the early 1930s, they assumed a greater role in Experiment Station research as the partnership geared up for a new era in the synthesis of pesticides. The chlorinated hydrocarbons, with DDT their prototype, yielded related compounds followed by the development of organophosphates, methyl carbamates, and pyrethroids, all neuroactive chemicals. By the 1950s, post-WWII insecticides had become the mainstay of insect control, with the prewar calls for biological and cultural controls in eclipse.

Professional Societies

Professional societies perform important functions. They establish the ethical and intellectual standards of the discipline, provide liaisons with the scientific community and the public, and provide continuity and the written record. In North America, three major professional societies have served the entomologists of Canada, Mexico, and the United States: the Entomological Society of Canada, La Sociedad Mexicana de Entomologia, and the Entomological Society of America (ESA). Today, the membership in these three societies stands at about 8000. The new challenge to entomological societies will be to hold to their traditional goals while embracing the environmental and biodiversity crusades, advancing IPM, and increasing interdisciplinary collaboration. Because the ESA is the largest of the three societies and enjoys substantial joint membership from the other two, we will focus on its recent history.

As the scope of applied and basic entomology grew under the stimulus of post-World War II goals, the two primary entomological societies in the United States, the American Association of Economic Entomologists (AAEE) and the ESA, recognized two common needs. They needed professional management to better process publications and provide other services for their members. In addition, professional entomologists needed to address the big policy issues posed by the growing importance of the biological sciences in human affairs. Big science was emerging as a national goal, and the landscape was changing physically. For example, the stately elms were vanishing, casualties to an insect-vectored disease. The gypsy moth and the imported fire ant were on the move. The Green Revolution had captured the imagination of the great humanitarian organizations. Therefore, entomology was being summoned to new levels of leadership.

A merger of the two organizations seemed to offer advantages in both management and unity. This view was not unanimous, particularly among ESA members, who were outnumbered three to one by AAEE. After protracted debate and two ballots, the two societies merged in 1953 as the Entomological Society of America.

It soon became evident that although the new ESA gained administrative efficiency, it was not unified in its philosophy and the more numerous and vocal members oriented to applied entomology predominated. Stress arising from this dichotomy within the membership has characterized the ESA over the nearly 5 decades since the merger.

Enter Rachel Carson

Early warnings of the danger of insecticide mania were sounded within the entomological community, but these warnings were largely ignored. It was the publication of Rachel Carson's *Silent Spring* in 1962 that triggered the avalanche of public concern (Fig. 2). She lamented that "so primitive a science has armed itself with the most modern and terrible weapons, and that in turning them against the insect, it has turned them against the earth."

Overnight, her exhortation changed the public's perception of entomologists. Their traditional obscurity was swept away. They were in the public eye, viewed as allies with the corporate giants, poisoners of robins and the earth, all under the pious veil of aiding the consumer by aiding the farmer.

Response to Rachel Carson's charges came largely from industry, whose strategy was to discount the witness. This proved ineffective as the accumulating evidence reinforced her concerns. Practices that endangered birds, especially the national symbol, the American eagle, were certain to stir emotions.

The public debate on the pesticide issue did not reach the agenda of the Entomological Society of America. The Society formed by merger in 1953 had not become a forum for debate on issues concerning the field. Instead, plenary sessions at the Society's annual meetings were largely ceremonial with substantive debate occurring in the subject matter sections representing the various areas of specialization among their members. In short, there was no philosophical common denominator that united the Society.

Carson's *Silent Spring* became a corner stone of the environmental movement. Her thesis focused on the "web of

FIGURE 2 Rachel Carson's *Silent Spring* (1962) focused public attention on the pesticide issue. Her crusade catalyzed the environmental movement. (Harbrace photograph.)

life," which placed humanity's relationship to all forms of life in an ecological context. The concern for the environment and the natural world triggered by *Silent Spring* melded with other concerns for humans—women's rights, the war in Vietnam, and Native American rights—to give rise to the broadly based environmental movement embracing the rights of humans and nature, animate and inanimate. This great philosophical debate extending over the past 4 decades proceeded without the active involvement of the entomological community.

One of the immediate effects of *Silent Spring* was to make pesticide policy a matter of public debate. While focusing on DDT, the issue became broader and embraced the central tenet of the environmental movement: that human intervention had become the dominant environmental influence on the planet. The fact that, despite the restricted use of DDT, its residues could be found in Antarctica implicated entomologists in the global insult.

In 1967, a group of concerned individuals formed the Environmental Defense Fund, its object being to use litigation in defense of citizens' right to a clean environment. In protracted public hearings, entomologists were called to testify that their practices were not infringing on citizens' rights to a clean environment. It was an uncomfortable defensive position in which these dedicated "defenders of agriculture" were placed. In 1972, a decade after *Silent Spring,* the Environmental Protection Agency banned DDT. Its meteoric rise and fall, from discovery to banning, had spanned only 3 decades.

Economic entomologists in general viewed *Silent Spring* as an attack on their professional competence and integrity. Since the late 19th century, they had cultivated a self-image as dedicated public servants, bringing science "to the distressed husbandman" whose labors were closely aligned with the national interest. This explains in part their emotional response and sense of hurt that has lingered among entomologists of the DDT era.

Ecology's Promise

Economic entomologists of the 1960s faced two daunting challenges, the loss of public confidence in the aftermath of *Silent Spring* and the failure of their programs of insect control. These were powerful incentives for reassessment.

In the early 1950s, well before *Silent Spring,* the concerns regarding the insecticidal treadmill led a group of entomologists at the University of California at Berkeley and at Riverside to reassess control practices. Drawing on the biological control heritage pioneered by Harry Scott Smith (1883–1957) they sought to "integrate" features of biological and chemical control. This concept with further refinement led to the adoption by the late 1960s of IPM. In practice, IPM seeks to integrate multiple control measures into a cohesive package, the additive impact of which would hold pests within acceptable levels with minimum adverse environmental impact.

The abbreviation "IPM" was soon adopted worldwide to identify a holistic approach to pest control. Its enthusiastic reception reflected the optimism accorded a new paradigm, one that placed pest control on an ecological foundation.

The most impressive feature of the movement has been its evolving nature. The underlying theory and the fundamental question that has plagued population ecologists, "What factors determine the number and distribution of animals?" remain under debate. Views on the role of pesticides in IPM are likewise evolving. Progress has been made in tailoring pest-specific insecticides with reduced environmental disruption. Although impressive gains have been made in specific programs, the IPM era has not resulted in a major decline in the total quantities of pesticides used.

The euphoria induced by the IPM concept has run its course, and its promise after 3 decades is a subject of lively debate. One of the problems affecting acceptance and support of IPM is the difficulty of assessing its effectiveness. It is a complex system with many obstacles and a restricted database for evaluating a variety of constraints: technical, financial, educational, organizational, and social. Whatever the outcome, there is no turning back. The human intellect has been unable to construct a more promising strategy for keeping humans' age-old competitors in check. For millions of people threatened with disease and hunger, IPM constitutes their safety net for tomorrow.

In seeking to understand the strategy employed by applied entomologists, it is helpful to note historical perspective. Applied entomologists were late to embrace ecology despite the entreaties of their distinguished president of the

Entomological Society of America in 1912, Stephen A. Forbes. He insisted that "the economic entomologist is an ecologist pure and simple whether he considers himself so or not." In retrospect, it appears that entomologists opted for the certainty of insecticides favored by their farmer clientele over the uncertain promise of ecology. Their adherence to the conventional wisdom of insecticidal control in the DDT era tarnished their image as environmentalists.

In medical and veterinary entomology, the post-World War II experience with the miracle insecticides paralleled the experience with agricultural pests. First, there was euphoria following the miraculous effectiveness of the insecticides. So promising were the prospects that in 1955 the World Health Organization (WHO) proposed global eradication of malaria. However, the development of resistant strains of vectors and parasites as well as economic and political factors doomed the eradication program. In 1976, the WHO abandoned eradication in favor of more modest programs of control. Research languished under the demoralizing effect of this decision. With antimalarial drugs and insecticides losing their effectiveness, the battle against malaria was being lost. Alarmed at these developments, the WHO, in 1993, called for a renewed global effort. The initially slow response has gained support, with unprecedented funding available in 2000 for new initiatives.

An ambitious objective is the development of a vaccine against malaria. Although the scientific obstacles are enormous, researchers are now predicting a successful vaccine by about 2010. The most ambitious and futuristic of all approaches to combating malaria is creating a strain of *Anopheles gambiae* mosquito unable to transmit the parasite. To displace the native vectors involves three steps: find genes that interrupt the parasite's life cycle, develop techniques to transfer those genes into the mosquito, and finally, develop ways of replacing existing mosquito populations with the genetically engineered model. One additional hurdle remains. With such a mosquito in hand, there may be strong resistance to releasing such transgenic forms into nature. In the meantime, the disease continues to cast its shadow over the malaria-endemic areas of the world, which are home to 40% of the world's population.

Advancing the Science

The past 5 decades have witnessed remarkable advances in both applied and basic entomology. The collapse of chemical control of insects forced reassessment, which gave rise to IPM. While the pesticide issue dominated public interest, basic science was forging ahead.

The primary stimulus was the discovery in 1953 that the compound deoxyribonucleic acid (DNA) encodes genetic information that provides the blueprint for synthesis and cellular differentiation; this discovery elucidated the great mystery of life, the cell's ability to self-replicate. Many aspects of biology were catalyzed by the discovery. It dramatically reaffirmed Darwin's hypothesis of common descent and revealed evolutionary pathways. The studies of molecular systematics that followed have resulted in accumulation of much DNA sequence data from most insect groups. These data complement and enhance the morphological and ecological data of classical systematics, thereby making substantial contributions to evolutionary biology.

The DNA breakthrough also paved the way for biotechnology, the introduction of genes from various species into plant and animal species. With biotechnology, plants can be engineered to produce their own pesticides. Corn can be altered to contain a pesticide produced by the bacterium, *Bacillus thuringiensis,* with such corn being designated "Bt" corn.

This technology can place the cornucopia of biodiversity in the service of agriculture and medicine. But this novel technology comes with complex ethical and scientific issues in environmental stewardship and human health. The scientific community as well as the general public is in strong disagreement over the introduction of genetically modified organisms (GMOs) into the ecosystem. Caution should prevail until some basic questions are answered. For instance, what impact will the thousands of acres of Bt corn have on the complex of beneficial and injurious insects on the modified plants?

GMOs conferring drought tolerance to food crops of developing countries would be welcomed additions as the food supply grows more tenuous with rising populations and civil disruption. In weighing these options, it needs to be recognized that an infrastructure to monitor such crops is not in place at present. The role of GMOs in IPM is likewise uncertain in developed countries, and resistance to GMOs is particularly acute in European countries.

Industry has been quick to recognize biotechnology's commercial potential, and substantial segments of the seed market have been given over to GMOs. Questions of a scientific nature are joined by social and economic questions. For instance, should corporate interest determine and control the genetic profile of the three crops, corn, rice, and wheat, that provide sustenance for most of the peoples of the world?

The economic issues surrounding GMOs seem to overshadow the more basic environmental issues they pose. For instance, the biodiversity program seeks to conserve natural forms, whereas biotechnology seeks to replace natural forms with modified ones exempted from evolutionary testing. Over time, how will this practice affect the gene pool, the timeless and priceless biological resource?

Although elucidation of the structure and function of DNA is clearly the most important discovery of the 20th century, other discoveries have greatly advanced our understanding of insects. This progress is due in large measure to technical advances in fields such as insect olfaction, acoustics, flight, and communication (e.g., by pheromones).

Such advances have in turn altered the way scientists communicate in person and in professional literature. They became more informal and democratic. The excitement was often centered in youth, in graduate students, with women strongly represented.

The excitement of discovery and exuberant professional exchange produced masses of data leading to new specialized journals. The worldwide computer network has catalyzed the processing and exchange of data among colleagues on a global scale. The predominant use of English in scientific journals has reduced language barriers. Thus, in both applied and basic entomology, the latter half of the 20th century has represented a new order, new methodologies, new discoveries, and new organizational arrangements. The paths of progress in the multifaceted phases of entomology are well documented in the *Annual Review of Entomology,* published since 1956.

Historical Perspective

The preceding 2 centuries of entomological enterprise in North America have been directed primarily to two activities: (1) protecting humans' food, fiber, and health and (2) basic research to advance knowledge of insects. These were and continue to be appropriate objectives in the national interest.

FIGURE 3 Edward O. Wilson's prolific writings advanced the science of entomology, established the field of sociobiology, and led the 21st century movement to preserve biodiversity.

In the past half century the environmental movement and the emerging science of ecology have highlighted two salient points: (1) insects play a vital role in the sustainability of the global biosphere and (2) the biodiversity essential to sustainability is threatened by human intervention. The factors of habitat destruction, pollution, and introduction of exotic species are believed to account for extinction rates much higher than before the coming of humans. In addition, the ecological impact of global warming looms on the horizon. The movement to preserve biodiversity has been led by E. O. Wilson (Fig. 3) following publication of his *The Diversity of Life* (1992). The concept has been generally accepted and is now part of the American culture.

The extinction dilemma poses new challenges to the field of entomology and calls for modification of the image entomologists hold of themselves and of the institutions established in the past to deal with entomological matters. The new order calls for entomological statesmanship that looks beyond entomology's traditional agricultural constituency to the global environmental issues. Thus, the age-old challenge of insect control will be joined with the challenge of insect conservation.

See Also the Following Articles

Agricultural Entomology • Biological Control • Entomological Societies • Extension Entomology • Integrated Pest Management • Regulatory Entomology

Further Reading

"Annual Review of Entomology" (1956–2003). Vols. 1–48. Annual Reviews, Palo Alto, CA.
Berenbaum, M. R. (1995). "Bugs in the System, Insects and Their Impact on Human Affairs." Addison–Wesley, Reading, MA.
Dunlap, T. R. (1981). "DDT: Science, Citizen and Public Policy." Princeton University Press, Princeton, NJ.
Howard, L. O. (1930). "A History of Applied Entomology," Vol. 84. Smithsonian Inst., Washington, DC.
Mallis, A. (1971). "American Entomologists." Rutgers University Press, New Brunswick, NJ.
Mayr, E. (1982). "The Growth of Biological Thought. Diversity, Evolution, and Inheritance." Harvard University Press, Cambridge, MA.
Paladino, P. (1996). "Entomology, Ecology and Agriculture: The Making of Scientific Careers in North America, 1885–1985." Harwood Academic, Amsterdam.
Perkins, J. A. (1966). "The University in Transition," p. 43. Princeton University Press, Princeton, NJ.
Perkins, J. H. (1982). "Insects, Experts, and the Insecticide Crisis." Plenum Press, New York.
Smith, R. F., Mitter, T. E., and Smith, C. H. (eds.) (1973). "History of Entomology." Annual Reviews, Palo Alto, CA.
Sorensen, W. C. (1995). "Brethren of the Net: American Entomology, 1840–1880." University of Alabama Press, Tuscaloosa.
Stein, B. A., Kutner, L. S., and Adams, J. S. (eds.) (2000). "Precious Heritage: The Status of Biodiversity in the United States." Oxford University Press, Oxford.
Whorton, J. (1974). "Before Silent Spring: Pesticides and Public Health in Pre-DDT America." Princeton University Press. Princeton, NJ.
Wilson, E. O. (1992). "The Diversity of Life." Harvard University Press, Cambridge, MA.
Wilson, E. O. (1994). "Biophilia." Harvard University Press, Cambridge, MA.
Worster, D. (1988). "Nature's Economy: A History of Ecological Ideas." Cambridge University Press, New York.

Homeostasis, Behavioral

P. Kirk Visscher
University of California, Riverside

Behavioral homeostasis refers to mechanisms of behavior that allow an insect or group of insects to maintain conditions within a certain range of values. These conditions may be the temperature of the body or the environment, internal water balance or environmental humidity, nutritional state or food stores, the balance between different activities of the individual or of the group, or the number and composition of individuals in a group. Behavioral mechanisms of homeostasis are important to individual insects, whether solitary individuals or part of a group, and include such nearly universal behaviors as feeding and drinking, as well as behavior concerned with thermoregulation and habitat choice. This article, though, is mostly concerned with homeostasis in groups of insects, such as the colonies of bees, wasps, ants, and termites. Individual behavioral homeostasis in physiological regulation, thermoregulation, and habitat choice are covered elsewhere in this encyclopedia.

ENVIRONMENTAL REGULATION BY GROUPS OF INSECTS

Insects are relatively small animals, with high surface-to-volume ratios. Because of this, they readily lose body heat or water to the environment (or gain heat if the ambient temperature is high). However, a few species of insects form large groups that are able to exert some control over these processes. The most striking examples of this come from the social insects (the wasps, ants, bees, and termites), but some other insects also form groups that enhance homeostasis (Fig. 1).

The control of groups of insects over heat exchange may take two forms. First, they may form a cluster that effectively makes them more similar collectively to larger organisms. If the surface-to-volume ratio is of a cluster of insects rather than an individual, it has a smaller value, and heat exchange is slower. Second, most social insects construct nests, and the architecture of these nests can result in the interior environment being substantially different from the ambient environment outside the nest.

Honey Bees

Honey bees exhibit both of the above strategies. Honey bees (*Apis* spp.) arose in the tropics, but *A. mellifera* and *A. cerana*

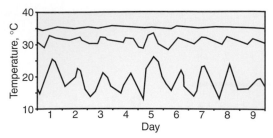

FIGURE 1 Honey bees (*Apis mellifera*, blue line) and yellowjacket wasps (*Vespula vulgaris*, red line) both maintain their nests at temperatures that fluctuate less than outside air temperatures (black line). At cool outside temperatures, as here, the nests are kept warmer than ambient. Note that the honey bee colony, with tens of thousands of workers, achieves more precise temperature homeostasis than the wasp colony, with only hundreds of workers. (Data from H. Kemper and E. Döhring, 1967, *Die Sozialen Faltenwespen Mitteleuropas*, Parey, Berlin.)

have colonized much of the temperate zone as well. These honey bees are unique among temperate insects in maintaining a high temperature in their nests throughout the winter, even when environmental temperatures are dramatically lower. For example, an *A. mellifera* colony can maintain a temperature in the center of its winter cluster inside the nest of 35°C, even when the temperature outside the nest is −40°C. The bees accomplish this by clustering together tightly so that the bees themselves, as well as the nest structure (often a hollow tree or wooden beehive) serve as insulation. The bees consume honey as metabolic fuel and contract their large flight muscles to create heat. Bees on the outside do get chilled, but they trade places with bees in the warm interior from time to time. Even when a bee colony is not in a nest, as when they are moving as a swarm to a new homesite, they maintain warm temperatures inside the cluster of thousands of bees.

The environment is not always cold, so temperature homeostasis for a bee colony sometimes involves cooling the nest. Honey bees fan their wings to move outside air through a colony to remove excess metabolic heat (and carbon dioxide). When this does not cool the colony enough, the bees begin to collect water and evaporate it within the nest to provide cooling. Also, when the nest becomes too warm, many bees leave the cavity and cluster outside the nest, reducing the heat input from their metabolism.

Termites

Many species of termites, like honey bees, live in large groups. Indeed, the largest colonies of social insects occur among the termites, some species of which may have several million individuals in a nest. Unlike honey bees, termite workers do not have wings, and so they cannot move air by fanning. Instead, some species of termites rely on the structure of the nest to regulate temperature and humidity. *Macrotermes subhyalinus* colonies, for example, construct tall "chimneys"

on their nests. These chimneys are thought to increase airflow in two ways. As the metabolic heat of the termite colony (and the fungus gardens that they cultivate in the nest) warms the air in the chimneys, it rises and is replaced by cooler air from passages near the ground. Also, when wind blows across the open tops of the chimneys, the Bernoulli effect causes lower air pressure at the chimney top and draws air upward.

The climate-control nest structure of another species of *Macrotermes*, *M. bellicosus*, was described by Martin Lüscher. These termites build nests with a closed-circuit air circulation system. Air warmed by metabolic heat rises in central galleries of the nest but then enters channels on the outer ribs of the nest. Here, it loses heat to the outside through the nest material, and the denser, cooled air settles to chambers at the base of the nest, from which it is drawn to replace the rising air in the central nest, over and over again. As the air passes through the thin outer channels, carbon dioxide diffuses out and oxygen diffuses inward. This system allows gas exchange and cooling, while limiting water loss.

In Australia, *Amitermes meridionalis* nests are constructed as flat towers, always oriented with their long axis north and south. The result is that they are warmed by sun as it rises in the east early in the morning and strikes their broad side, but they receive relatively little sunshine at midday when the sun is in the north and strikes the nest edge on. These termites are known to sense the earth's magnetic field and use it to coordinate the nest-building activity of the colony's many workers to achieve this striking geographic orientation of the nest.

Tent Caterpillars

Although less organized in their social behavior than most social insects, tent caterpillars use some of the same thermal strategies to get a jump on the warm season. The larvae of tent caterpillars cluster together and form tents from silk that they produce. A group of caterpillars clusters together inside the tent during the night, where both the tent and the presence of many clustered insects reduce heat losses. The higher temperatures that the caterpillars experience allow them to develop more quickly than they would if they were isolated and exposed to the low temperatures that are common, especially at night, in their environment. Tent caterpillar behavior is adapted to keeping with the group. They find their way back to the tent by trails of odors and silk that are laid down as the caterpillars move from the tent to the foliage on which they feed during the day.

COMMUNICATION AND GROUP ACTIVITIES

An individual organism must allocate its time and resources between food collection, reproduction, habitat selection, and other activities. In the social insects, one sees similar behavioral adaptations. There is added complexity, though, because in social insects they occur at both the level of the individual and the level of the group. Group-level adaptations include the regulation of numbers of individuals in the colony, the allocation of reproduction between workers and sexual forms, the division of labor among individuals (e.g., caste), and the social organization of food collection (e.g., recruitment). All of these homeostatic activities by colonies of insects require mechanisms of communication to coordinate the activities of multiple individuals. It is for this reason that the social insects provide so many of the examples of communication among insects, because in nonsocial insects, communication is largely restricted to behavior associated with mating or defense. As in other insect groups, much of the communication in social insects is carried on chemically, by means of pheromones.

Homeostasis is fundamental to the survival of organisms, because the processes of life occur in a well-regulated manner only within a certain range of conditions. The same could be said about the processes conferring advantages of group living on those insects that live in groups. If a colony is too large, or fails to coordinate its activities in foraging, reproduction, or defense, it may perish. It is the function of behavioral mechanisms of homestasis to regulate both the group environment and the properties of the group itself in a manner that preserves its efficient functioning.

See Also the Following Articles

Dance Language • Magnetic Sense • Nest Building • Recruitment Communication • Thermoregulation

Further Reading

Abe, T., Bignell, D. E., and Higashi, M. (2000). "Termites: Evolution, Sociality, Symbioses, Ecology." Kluwer Academic, Dordrecht.
Heinrich, B. (1993). "The Hot-Blooded Insects: Strategies and Mechanisms of Thermoregulation." Harvard University Press, Cambridge, MA.
Seeley, T. D. (1994). "The Wisdom of the Hive: Social Physiology of Honey Bee Colonies." Harvard University Press, Cambridge, MA.
von Frisch, K., and von Frisch, O. (1974). "Animal Architecture." Harcourt Brace Jovanovich, New York.
Wilson, E. O. (1971). "The Insect Societies." Harvard University Press, Cambridge, MA.

Homoptera

see *Auchenorrhyncha; Sternorrhyncha*

Honey

Eva Crane
International Bee Research Association

Honey is a sweet substance produced by social bees and some other social insects. They collect nectar or honeydew from living plants and transform it into honey, which they store in their combs. Most nectar collected by bees contains from 15 to 50% sugars; these are mainly sucrose in some plants, glucose and fructose in others, and all three in the rest. There are also many minor constituents.

THE ORIGIN OF HONEY

The Bees That Produce Honey and How They Do So

Certain social insects produce and store honey as a nonperishable food for use in dearth periods. The insects include all honey bees (*Apis* spp.) and stingless bees (*Melipona* and *Trigona* spp.) and also certain species of social wasps in South America (*Nectarina*) and honey ants, e.g., *Melophorus inflatus* in Australia. Honey-producing species whose colonies die out at the end of the active season, which are most social wasps and bumble bees (*Bombus* spp.), store comparatively little honey, and it is not economically important.

Production of honey by the honey bee, *A. mellifera,* has been studied most. Foraging workers collect nectar from plants and, when they return to their colony, "house bees" (young workers) take it from them and deposit it in cells of the comb. Bees evaporate water from it by manipulations that increase its surface area, while other bees fan to maintain a current of warm air through the hive. During this process (and even during its transport to the hive), secretions from the bees' hypopharyngeal glands are added to it. These contain the enzyme invertase, which inverts sucrose into fructose and glucose. At hive temperatures the solubility of glucose in a solution of fructose is unusually high, and the final honey has a very high sugar content, around or even above 80%. The relative amounts of the two sugars depend on the nectar sources. Fructose is more soluble in water than glucose, and high-fructose honeys (e.g., from *Robinia pseudoacacia*) rarely if ever granulate (crystallize), whereas high-glucose honeys (e.g., from dandelion, *Taraxacum officinale*) do so very quickly. Gentle warming of granulated honey redissolves the crystals.

Plant Sources of Honey

Most nectar is produced by flowers, although a few plants have extrafloral nectaries, including cotton (*Gossypium barbadense* and *G. hirsutum*) and rubber (*Hevea brasiliensis*). A number of world nectar plants have been classified according to the weight of honey that may be produced from a hectare of the plant in bloom, and the following are among those reported to be in the highest class (over 500 kg honey/ha or pounds/acre): *Epilobium angustifolium* (fireweed, rosebay willowherb), *Melilotus alba* (white melilot), *Phacelia tanacetifolia* (phacelia), *R. pseudoacacia* (false acacia, black locust), *Thymus vulgaris* (common thyme), and *Trifolium pratense* (red clover).

Where honeydew is available, bees collect it as well as nectar; it is sap from the host plants of certain plant-sucking insects in the order Hemiptera (Stenorhynchota) that excrete part of the sap they ingest. Honeydew honey, which contains certain sugars not in floral honey, lacks any floral fragrance. It is much favored in some regions where it is produced—such as the Black Forest in Germany—but elsewhere the more delicate flavor of honey from nectar is much preferred.

COMPOSITION AND PROPERTIES OF HONEY

The composition of honey varies according to its plant origin and the weather conditions when the honey was produced. An analysis of 490 U.S. honeys gave an average of 69.5% fructose + glucose, 8.8% other sugars, 17.2% water, and small amounts of free acids, lactones, ash, and nitrogen.

The color of fresh honey varies with plant source, and honeys become darker during storage, especially at high temperatures. The color of liquid honey is important in marketing, and a number of countries have established systems for color grading.

Crystallization of honey (often referred to as granulation) is of great importance. It is a reversible process that changes liquid (run) honey into solid (set) honey, and it consists of the spontaneous crystallization of glucose (dextrose) monohydrate from a supersaturated solution. In Europe, liquid honey has been preferred, and honey from *R. pseudoacacia*—which rarely granulates—is favored; on the other hand, in Canada granulated honey is the norm, possibly because a common source is alfalfa *(Medicago sativa),* and its honey granulates rapidly.

The aroma and flavor of a sample of honey depend on its plant source, and beekeepers learn to recognize the plant origins of the honeys their bees produce. Honey processed for sale on the mass market is usually blended to maintain a constant product.

HONEY PROCESSING

In a beekeeper's hives the bees store honey in the combs of an upper honey box that is removed when it is full. Bees may be cleared from combs in the honey box by various methods: brushing and shaking bees off combs, using a bee-escape board through which bees can leave the honey box but not return, using a bee repellent, or blowing the bees out of the boxes with a stream of air.

Most honey is separated from the wax comb and processed for sale in containers, without any comb or wax. Processing the honey is likely to consist of the following stages: (1) clearing bees from the combs to be harvested, which are then taken to the honey house; (2) warming the combs to 32 to 35°C; (3) uncapping the combs and dealing with the cappings; (4) extracting the honey from the combs in a centrifuge; (5) clarifying the honey by passing it through a strainer and/or baffle tank; (6) flash heating and pressure filtering (in large processing plants in some countries); (7) if desired, initiating controlled granulation, on a large or small scale.

Honey is hygroscopic, and it should not be exposed to air with a relative humidity above 60%, or it may absorb water. (Some operators reduce the water content of honey slightly during stage 2 of processing or between stages 5 and 7.)

The processing of honey for sale either liquid or granulated is obviated if honey combs themselves are sold. Traditionally, "sections" were miniature wooden frames fitted with a very thin wax comb foundation on which the bees built cells, filled them with honey and—the beekeeper hoped—completely sealed them; the weight of honey in each section sold was usually 0.5 kg or 1 lb. However, perfectly sealed sections are difficult to produce, and in the 1900s several easier ways were devised to prepare honey in the comb for sale.

One alternative is cut-comb honey. To produce it the beekeeper inserts large frames fitted with extra-thin unwired wax foundation in the hive, harvesting them when full of honey (unlike sections, they need not be entirely capped). Each frame is placed on a flat surface and the comb cut out of its frame with a heated knife. Fully capped areas of it are cut into portions for sale, and honey is allowed to drain from the cut edges. Each piece is packaged in a heat-sealed box or a sheet of transparent plastic.

An easier alternative is to sell a jar containing a piece of honey comb and filled up with liquid honey; this is referred to as chunk honey.

HONEY AS A PRODUCT

Present World Production and Consumption

According to figures available, 1.1 million tonnes were produced in 1999. Honey yields per hive are usually highest in countries with an extensive belt between latitudes 23 and 30° (N or S), including China, Argentina, Mexico, and Australia.

Honey as Food

Honey from bees' nests was probably eaten by some mammals, including bears, before humans did so, and chimpanzees have been observed using tools to get access to honey in bees' nests in a tree. In Africa, India, and Spain, rock art from Mesolithic times and later shows human honey hunters harvesting from nests in trees or rocks.

Within the historical period, the use of honey is recorded from around 3000 B.C. onward. In India it was used by the famous surgeon Susruta around 1400 B.C. and much praised in the Vedas, sacred Hindu books collected together about 1500 B.C. In Rome, Columella judged honeys by their plant source, that from thyme being the best. Honey from a few plants, including *Rhododendron,* is toxic; in 399 B.C. when Xenophon's army retreated from Persia across Pontus in Asia Minor, the soldiers ate honey near the Black Sea coast that probably came from *R. ponticum.* It made them very ill, but they recovered by the third day. Records of baking with honey survive from 1200 B.C. onward in ancient Egypt.

Honey in Medicine

Honey has been regarded as a health-giving substance since ancient times, and Pythagoras (ca. 530 B.C.) was said to have attributed his long life to his constant use of it. Honey is a common ingredient of cough mixtures and lozenges and is often recommended as a symptomatic treatment for dyspepsia and peptic ulcers; the organism *Helibacter pylori*, which is a common cause of peptic ulceration, is inhibited by honey. Some sufferers from hay fever may be helped by eating honey that contains pollen.

A beneficial effect on wound healing has been known since early times, and a mechanism for this was established in the 1960s. The hypopharyngeal glands of *A. mellifera* workers secrete the enzyme glucose oxidase; this enters the honey, and in the presence of water a small amount of hydrogen peroxide is produced, which is bactericidal. Also, honey is hygroscopic, so it extracts exudates from infected lesions. For these reasons, honey is currently used in a number of hospitals, especially on wounds that are difficult to dress.

Honey in Alcoholic Drinks

From ancient times onward, drinks have been made by fermenting fruit to make wine, or cereals to make ale or beer. In many regions where bees were kept in hives, an important use of honey was its fermentation to produce an alcoholic drink, often referred to as mead. Where vines were grown, wine had a higher social status than honey-based drinks and tended to displace them, but honey-based drinks remained important north of the warmer vine-growing areas. In tropical Africa "honey beer" was made by fermenting honey for a short period.

See Also the Following Articles

Beekeeping • Commercial Products from Insects • Medicine, Insects in

Further Reading

Crane, E. (1975). "Honey: A Comprehensive Survey." Heinemann, London.
Crane, E. (1999). "The World History of Beekeeping and Honey Hunting." Duckworth, London.
Crane, E., Walker, P., and Day, R. (1984). "Directory of Important World Honey Sources." International Bee Research Association, London.
Graham, J. G. (ed.) (1992). "The Hive and the Honey Bee," revised edition, Chap. 21. Dadant, Hamilton, IL.
Killion, E. E. (1981). "Honey in the Comb," revised edition. Dadant, Hamilton, IL.
Riches, H. R. C. (2000). "Medical Aspects of Beekeeping." HR Books, Northwood, U.K.

Honey Bee

see *Apis Species*

Honeydew

Thomas E. Mittler
University of California, Berkeley

Angela E. Douglas
University of York, United Kingdom

Honeydew is a sugar-rich liquid released from the anus of some phloem sap-feeding insects of the order Homoptera. It consists principally of the residue of ingested phloem sap after digestion and assimilation in the insect gut, but it also contains waste products of insect metabolism eliminated via the gut. Honeydew deposited onto plant or other surfaces is an important source of energy-rich food for other animals, including some flies, parasitoids, ants, and microorganisms, and is used as a foraging cue by insect predators and parasitoids of some homopterans. Ants also collect the honeydew directly as it is being released from the producing insect, a behavior known as ant attendance.

THE NATURE OF HONEYDEW

Honeydew has historically been a source of wonderment. For example, the honeydew produced by the coccid *Trabutina mannipara* on tamarisk trees may have been the "manna from heaven" on which the Israelites fed during their escape from Egypt; honeydew has also been described as "the milk of Paradise" by Samuel Taylor Coleridge. However, the biological nature of honeydew is more mundane. The phloem sap of plants contains very high concentrations of sugars, usually the disaccharide sucrose or oligosaccharides of the raffinose family. Phloem-feeding insects ingest very large amounts of sugars relative to other essential nutrients, and up to 90% of the ingested sugar may be egested via the anus, and this sugar-rich material is honeydew.

Honeydew is produced by phloem-feeding insects, not by xylem feeders. Most phloem-feeding insects are members of the homopteran suborder Sternorrhyncha, which includes aphids, whitefly, mealybugs, and psyllids, or the homopteran suborder Auchenorrhyncha, which includes planthoppers and leafhoppers. Honeydew production is not dependent on either gut anastamoses (filter chambers) or Malpighian tubules absent in many and all aphids, respectively. It is released exclusively from the anus. The cornicles of aphids (capable of discharging defensive secretions and pheromones) are not involved.

PRODUCTION

The amount of honeydew produced by insects can be substantial. For example, first instars of the willow aphid *Tuberolachnus salignus* release more honeydew than their own body weight on an hourly basis. However, the rate of honeydew production by other aphids is generally considerably lower.

The fluid that comprises honeydew accumulates in the rectum and is then ejected as a single droplet via the anus at fairly regular intervals, usually once every 15 to 40 min. Both the volume of each droplet and the frequency of production are influenced by many factors, including aphid age, size, species, and host plant.

Honeydew can be hazardous for the insect producing it because droplets may smother the insect and promote microbial growth on or near the insect. Various mechanisms reduce these risks in insects both on exposed plant surfaces and in confined spaces, including galls. On exposed surfaces, aphids generally project the honeydew droplet from the anus to a distance of up to several centimeters by either a kicking action of one of the hind legs or by using the cauda (a small appendage dorsal to the anus) to catapult the droplet ventrally. The legs and cauda bear hydrophobic cuticular hairs that prevent the sticky honeydew from adhering to the insect surface. Honeydew production by some species has been reported to be interrupted in windy conditions, presumably to avoid being smothered by the honeydew.

Insects in galls or other confined spaces (e.g., subterranean forms) are not wetted by their honeydew because they are coated with hydrophobic wax secreted from cuticular glands. In addition, each honeydew droplet produced by these insects tends not to be projected away from the insect body but remains poised at the anus until it is coated by wax. In some insects, notably certain nymphal psyllids, the honeydew and waxes coalesce to form a gelatinous or crystalline substance (called lerp), which acts as a solid protective covering for the growing insects.

Honeydew production may serve purposes additional to the elimination of waste compounds, with implications for the pattern of honeydew production. This is illustrated by two phenomena: "honeydew-panting" and ant-tending.

The former behavior is displayed by certain aphids, notably *T. salignus,* at elevated temperatures. The aphids raise their abdomen almost at right angles to the plant surface with their mid- and hind legs extended, and small honeydew droplets are alternately protruded and retracted from the anus. This behavior may cool the aphids as a result of evaporational water loss from the droplets.

The ejection of honeydew from some insects is modified by the attendance of ants or other insects. When solicited by an ant, the insect releases the honeydew droplet slowly and holds it at the anus while the ant imbibes, and, if the droplet is not removed by an ant, the insect may repeatedly extrude and retract the droplet, as if advertising the availability of honeydew. However, certain obligately ant-tended species apparently excrete honeydew droplets only in response to solicitations by their attending ants. The cauda and rectal musculature of some ant-tended aphids is much reduced, presumably reflecting their dependence on ants to remove their honeydew.

COMPOSITION

From a physiological perspective, honeydew is dominated by first, egesta, the components of ingested phloem sap that have not been assimilated by the insect (some phloem sap compounds may be enzymatically altered by gut enzymes prior to egestion), and second, excreta, waste products of insect metabolism that are eliminated via the gut after transfer from the body tissues to the gut lumen.

In quantitative terms, honeydew is dominated by sugars. In the best studied group, the aphids, the principal sugars in honeydew are typically different from the sugar ingested by the insect and usually of higher molecular weight. This reflects the vital osmoregulatory function of the gut in phloem-feeding insects. The osmotic pressure of phloem sap is generally considerably higher than the osmotic pressure of the insect body fluids, creating a tendency for the insect to lose tissue water to the gut.

Amino acids are the principal nitrogenous compound in phloem sap. Not all of the ingested amino acids are absorbed across the insect gut and assimilated (estimates of assimilation efficiency vary from 60 to 99%), and amino acids are routinely recovered from insect honeydew.

Honeydew also contains nitrogenous excretory compounds but generally at very low concentrations. This is because the symbiotic microorganisms in homopterans act as an internal sink for waste nitrogen compounds. For example, uric acid, the principal nitrogen waste compound of the planthopper *Niloparvata lugens,* is not voided in the honeydew, but retained within the insect body and metabolized by the insect's symbiotic yeasts. Similarly, ammonia, the dominant waste nitrogen compound of aphids, is in low concentration in their honeydew because their symbiotic bacteria *Buchnera* consume much of the ammonia synthesized by these insects.

Honeydew may contain microorganisms and viruses derived either from the ingested phloem sap (and passed directly through the gut) or from the resident insect microbiota. For example, aphids feeding on plants infected with the luteovirus barley yellow dwarf virus will ingest viral particles from the phloem sap; those particles that are not transported into the insect hemocoel are expelled in the aphid honeydew. Plant viruses multiplying in the insects may also pass into the gut and occur in honeydew.

HONEYDEW AS FOOD

Many insects in several orders, including Diptera, Hymenoptera, Lepidoptera, Coleoptera, and Neuroptera, feed on honeydew that has fallen onto plant or other surfaces. Among these insects are herbivores (e.g., tephritid flies, butterflies, and moths) and many entomophagous taxa, such as chrysopids, coccinellids, syrphids, tachinid flies, and hymenopteran parasitoids. A number of nectivorous birds in Mexico and Australia forage on honeydew and lerp; lerp is also consumed by flying foxes in Australia. Other small mammals and reptiles also feed on honeydew, and dipterous vectors of human diseases (e.g., mosquitoes and phlebotomine sand flies) may rely on honeydew for an energy source.

Because it is usually freely accessible on leaf surfaces, it can readily be imbibed by insects that lack the specialized mouthparts needed to exploit floral nectar. However, several features of honeydew reduce its availability and suitability as food. First, there is the tendency of honeydew sugars to crystallize. Second, the performance of various predators and parasitoids is generally lower on honeydew than on nectar. Third, plant-derived secondary compounds in certain honeydews are toxic to other insects. It has been suggested that insects have selection pressure to produce honeydew of little nutritional value; hence, potential competitors and natural enemies may be one factor shaping honeydew composition.

When animals consume honeydew as it is voided they are described as "tending." By far the most widespread group of tenders are the ants, including most species of the subfamilies Myrmicinae, Dolichoderinae, and Formicinae. Other insects reported to tend homopterans include polybiine wasps (e.g., *Brachygastra* and *Parachartergus* spp. associated with membracids and planthoppers, respectively) and silvanid beetles (e.g., *Coccidotrophus* spp. with the mealybug *Pseudococcus breviceps*). It is believed widely that only insects tend honeydew-producing homopterans, but it has been demonstrated recently that several Madagascan gekkoes stimulate planthoppers of the family Flatidae to release honeydew droplets on which they feed.

In the interactions involving ants, both the ants and their tended homopterans generally benefit from the association, which is therefore described as mutualistic. Access to honeydew has been shown to enhance the rate of increase of ant colonies, but the magnitude of the nutritional benefit varies widely with ant species and environmental circumstance. Predominantly predaceous species feed on honeydew only very occasionally; some ants switch between tending and preying on homopterans, depending on the nutritional quality of the honeydew (as shaped by plant physiology) and the nutritional needs of the ant colony, and honeydew accounts for more than half of the diet of many temperate wood ants of the genus *Formica* and is the dominant, even sole, food of certain subterranean ants, such as *Acropyga* spp., and of *Solenopsis* (fire ants).

An indication that honeydew is an important food source for many tending ants is that the ants protect the tended homopterans from predators such as lacewings, syrphids, and coccinellids. In addition, certain ant species enhance their supply and quality of honeydew by transporting their tended homopterans to suitable parts of the host plant where the phloem nitrogen content is high and concentration of toxic plant chemicals is low (e.g., *Lasius* and aphids of genus *Stomaphis*), and some members of the genus *Acropyga* that tend coccids bear live coccids in their mandibles during the nuptial flight. The homopteran partner benefits from the protection from natural enemies and ant-mediated removal of honeydew, as frequently indicated by elevated rates of population increase in field conditions.

Honeydew has been used as a source of food by people. Encrustations of honeydew produced by scale insects have been eaten since biblical times in the Middle East. Certain groups of Australian Aborigines and American Indians also used lerp from psyllids and honeydew from scale insects as a source of sugar. In Central Europe, large amounts of honeydew are consumed indirectly, because the honeydew of aphids on conifers is the principal, and sometimes sole, source of food for some honey bees. The honey produced from this source, often referred to as Wald Honig (forest honey), is considered of inferior in quality to floral honey but is, nevertheless, consumed extensively.

Honeydew may also be an important source of carbon and nitrogen for microorganisms. For 2 decades, this topic has been influenced by an as yet experimentally unsupported hypothesis that the use of nutrients in insect honeydew by soil microorganisms would mobilize soil nutrients and enhance nitrogen fixation and thus promote plant nutrition.

In agricultural contexts, the growth of molds on deposited honeydew can depress plant photosynthesis and crop yield and contaminate fruits, vegetables, and flowers, making them unmarketable. For example, sooty molds arising from untreated infestations of greenhouse whitefly can halve the yield of glasshouse tomato crops, and cotton growers in the United States have suffered financially as a result of the "gray cotton" caused by sooty mold growing on cotton lint contaminated with whitefly and aphid honeydew. Other detrimental effects of honeydew are sticky sidewalks, glazed windshields, and gummed up harvesting machines.

HONEYDEW AS A KAIROMONE

The smell or taste of honeydew on the plant surface is used as cues by various predators and parasitoids of homopterans to locate their hemipteran prey or, for reproductive females, as a stimulus for oviposition. This has led to the use of "artificial honeydew" sugar sprays onto crops to increase the numbers and effectiveness of natural enemies.

See Also the Following Articles
Aphids • Auchenorrhyncha • Food, Insects as • Sternorryncha

Further Reading
Budenberg, W. J. (1990). Honeydew as a contact kairomone for aphid parasitoids. *Entomol. Exp. Appl.* **55,** 139–148.
Douglas, A. E. (1998). Nutritional interactions in insect–microbial symbioses: Aphids and their symbiotic bacteria *Buchnera. Annu. Rev. Entomol.* **43,** 17–37.
Kiss, A. (1981). Melizitose, aphids and ants. *Oikos* **37,** 382.
Mittler, T. E. (1958). The excretion of honeydew by *Tuberolachnus salignus* (Gmelin). *Proc. R. Entomol. Soc. (London) A* **33,** 49–55.
Mittler, T. E. (1958). Studies on the feeding and nutrition of *Tuberolachnus salignus* (Gmelin). II. The nitrogen and sugar composition of ingested phloem sap and excreted honeydew. *J. Exp. Biol.* **35,** 74–84.

Molyneux, R. J., Campbell, B. C., and Dreyer, D. L. (1990). Honeydew analysis for detecting phloem transport of plant natural products. *J. Chem. Ecol.* **16**, 1899–1909.

Owen, D. F., and Wiegert, R. G. (1976). Do consumers maximize plant fitness? *Oikos* **27**, 488–492.

Stadler, B., Michalzik, B., and Muller, T. (1998). Linking aphid ecology with nutrient fluxes in a coniferous forest. *Ecology* **79**, 1514–1525.

Wackers, F. L. (2000). Do oligosaccharides reduce the suitability of honeydew for predators and parasitoids? A further facet to the function of insect-synthesized honeydew sugars. *Oikos* **90**, 197–201.

Way, M. J. (1963). Mutualism between ants and honeydew-producing Homoptera. *Annu. Rev. Entomol.* **8**, 307–344.

Hornet

see *Wasps*

Host Seeking, by Parasitoids

Ronald M. Weseloh
Connecticut Agricultural Experiment Station

Parasitoids are holometabolous insects that are free living as adults; their larvae are parasites within the bodies of other insects, which they invariably kill as they develop. Most parasitoids are small-to-large wasplike insects in the hymenopteran superfamilies Ichneumonoidea, Chacidoidea, Serphoidea, and Cynipoidea, or are flies in the dipteran family Tachinidae. Adult females, which are well suited for this task, almost always carry out host seeking in parasitoids. Most have wings and are active fliers, making it possible for them to explore large areas, relative to their body size. Most also have well-developed legs that facilitate the exploration of complicated surfaces. Typically, they possess tactile and chemosensory receptors on the antennae, feet, mouthparts, or ovipositor, and they have good visual acuity. Their ability to find specific host species may be important because many larval parasitoids exist inside other insects that often are capable of mounting immune responses unless the parasitoid is well adapted. Since, in addition, many parasitoids are small, it is easy for them to occupy restricted niches. Thus, the behaviors involved in host seeking in parasitoids are diverse and well developed. In fact, host seeking can be conveniently broken down into the overlapping, hierarchical categories of host habitat finding, host finding, and host acceptance.

HOST HABITAT FINDING

A female parasitoid may find herself far from potential hosts. This could occur if the host stage she emerges from is different from the one attacked. Also, many parasitoid females have a preoviposition period before eggs are ready to be laid. During this interval of a few days to several weeks, the parasitoid may leave the vicinity of the host to mate and obtain nourishment. For example, the ichneumonid wasp *Pimpla ruficollis* is a parasitoid of the European pine shoot moth, *Rhyaciona buoliana*. Yet for the first few weeks of her adult life, she is repelled by the odor of pine, and thus avoids the forest where the host is located. As a consequence, the initial stage in host seeking in many parasitoids is to search for locations where the host is likely to occur. Parasitoids often respond to general stimuli such as light, humidity, or vegetation form, leading them to meadows, forests, swamps, ponds, soil, or different vegetation strata. These behaviors considerably narrow the areas that must be actually searched for hosts. Many parasitoids are also attracted to volatile chemicals from plants. For example, after initially being repelled by pine, *P. ruficollis* females that are ready to oviposit are attracted by pine odors, and thus are drawn back to the forest. The ichneumonid *Itoplectis conquisitor* is attracted to the odor of Scots pine but not red pine, and does not attack lepidopterous hosts on the latter. In olfactometer tests, the aphid parasitoid *Diaeretiella rapae* is attracted to collard leaves. Wind tunnel experiments have shown that when such a parasitoid perceives a plant volatile, she reacts by walking or flying upwind (aenemotaxis), thus often leading to the plants where her preferred host feeds. In fact, in some cases the plant attraction is so important that the parasitoid host range encompasses the often diverse herbivores that feed on that plant rather than hosts that are taxonomically closely related. Plants are not the only habitat characteristics that can produce such attractive volatiles. Parasitoids of carrion-feeding flies are attracted to fresh or decaying meat, and parasitoids of *Drosophila* fruit flies respond to odors from yeast in decaying fruits where their hosts are likely to be present.

HOST FINDING

Unless random search is important, parasitoids usually find their hosts as a result of cues derived directly or indirectly from the host itself, often after they have entered the host habitat. Often host cues are perceived at close range. For example, the braconid *Cardiochiles nigriceps,* increases its searching when it contacts secretions produced by the mandibular glands of its host, *Heliothis* spp. caterpillars, as the latter feed on plants. The braconid *Microplitis croceipes* searches areas contaminated by chemicals contained in the feces of *Heliothis* caterpillars, whereas the braconid *Cotesia melanoscela* intensely searches leaf areas where host gypsy moth *(Lymantria dispar)* caterpillars have deposited silk strands. Once these materials have been perceived, stereotyped searching behaviors occur that typically consist of intense examination of the area with the antennae or tarsi. Parasitoids often also decrease their walking speed (orthokinesis) and/or increase turning rates (klinokinesis).

These behaviors, which serve to keep the parasitoid in the area having the host products, often lead to host discovery.

Other parasitoids are attracted from longer distances directly to hosts. Some parasitoid females respond to pheromones produced by their host. Parasitoids of the European elm bark beetle, *Scolytus multistriatus,* are attracted to "multilure," the aggregation pheromone of adult beetles. *Aphytis* spp. (Hymenoptera: Aphelinidae) are attracted to the sex pheromone produced by their host, California red scale *(Aonidiella aurantii).* A number of true bugs produce sex pheromones that are attractive to a variety of fly (Tachinidae) and to hymenopteran parasitoids. Some parasitoids are drawn from a distance to chemicals produced by plants in response to damage caused by host herbivores. For example, the braconid larval parasitoid *Cotesia marginiventris* responds to volatile terpenoids released from corn seedlings as a result of eating damage caused by host *Spodoptera* caterpillars. These chemicals may be components of the induced resistance that plants have developed against pathogens and herbivores. Even leaves not directly damaged by a herbivore may produce such materials. The induced plant chemicals may, along with materials directly produced by herbivores, serve to attract parasitoids. For example, parasitoids of bark beetles are attracted to a combination of plant chemicals produced by trees as a reaction to the mass attack of the beetles as well as to the aggregation pheromone produced by these beetles. Simultaneous responses of parasitoids to long-range host and plant cues illustrate that the division between the categories of habitat and host finding is often arbitrary.

HOST ACCEPTANCE

Although perhaps not technically part of host-seeking behavior, whether a host can be recognized as such after it has been contacted by a parasitoid is very important for the parasitoid. If a female cannot recognize a host as suitable for her progeny, habitat and host-finding activities would be wasted. Parasitoids have evolved behaviors that enable them to accurately choose suitable hosts. Many detect chemicals in the host cuticle or egg chorion that enable them to differentiate one potential host from another. These they usually detect with their antennae (Hymenoptera) or front tarsi (tachinid flies). Some parasitoids are also able to distinguish between hosts after insertion of the ovipositor by use of sense organs on the egg-laying organ itself. Acceptance of hosts via other sensory modalities, such as touch, sound, or sight, have also been documented. For instance, a *Trichogramma* wasp female examines a host egg with her antennae to determine its size. The ichneumonid parasitoid *Campoletis sonorensis* is influenced by host caterpillar shape. A cylindrical shape that approximated the shape of the *Heliothis virescens* host was more effective in stimulating oviposition than round or flat shapes. The egg–larval parasitoid *Chelonus texanus* accepts host lepidopteran eggs that have a rough or sculptured surface rather than a smooth one. Hairs from the body of gypsy moth larvae are enough to cause examination behavior in the parasitoid *C. melanoscela.* Movements perceived by sight or through vibrations of the substrate are important cues for a number of different parasitoids.

HOST DISCRIMINATION

A behavior related to host seeking involves the discrimination between hosts that have already been parasitized and those that have not. Some parasitoid females, after they have successfully parasitized a host, deposit a chemical marker on the surface that serves to tell other parasitoids of the same or other species that the host is already parasitized. Such marking pheromones have most often been found in parasitoids that attack sedentary hosts, such as eggs or pupae. Often the oldest parasitoid in a host is the one that survives; consequently, host marking saves other parasitoids from wasting time and eggs on a host in which their offspring are likely to perish. Also parasitoids whose ovipositor has been inserted into a host are often able to distinguish parasitized from unparasitized hosts. Whether the cues perceived are the result of marking materials specifically injected by the first parasitoid or of chemical changes in the host resulting from parasitoid development is generally not known.

VARIATIONS IN HOST-SEEKING BEHAVIOR

Although the host-seeking process in parasitoids may be very efficient, such that eggs are deposited only in host species suitable for their development, other strategies are used. Some parasitoids lay eggs in an area likely to be inhabited by their host. The larvae hatching from these eggs then must find their own way to the host. Members of the hymenopteran family Eucharidae are parasitic on ant larvae. Adult females lay eggs on or in plants. Each hatching larva is a planidium and so is free living and waits until it can attach to a passing adult ant, whereby it is taken into the nest and transfers to ant larvae. Many immature blister beetles (family Meloidae) are also parasitoids. Adult females lay eggs in the soil or sometimes on plants, and the emerging larvae are called triungulins. These active larvae find their own way to the eggs of locusts or nests of solitary bees, where they devour the eggs and/or provisions of the hosts. (Strictly speaking, meloids should probably be called "egg predators"; but their impact is much like that of true parasitoids.) Some tachinid flies lay large numbers of very small eggs on foliage that potential hosts may eat. In some cases, the females are attracted to damaged leaves, and this increases the chances of success for their larvae, but suitable hosts may never ingest many eggs. Nevertheless, these parasitoids may be as host specific as those that actively search for hosts.

FUNCTIONAL TERMINOLOGY FOR BEHAVIORAL CHEMICALS

A terminology has been developed for chemicals that function as signals between organisms, paralleling the activity of the chemical cues involved in host-seeking behavior of parasitoids. All chemical attractants, arrestants, and so on that are important as modifiers of behavior between different organisms are grouped under the general term "semiochemicals." Pheromones are semiochemicals that serve to communicate between organisms of the same species; sex pheromones are an obvious example. Allelochemicals have effects between species and are further divided into those depending on whether the producing or receiving organism is helped or hurt by the signal. If the species producing the material is helped and the receiving one is harmed, the chemical is called an allomone. Examples of allomones include repellents that a stinkbug may produce to ward off predators such as ants or birds. A substance that harms the producing species but helps the receiving one is called a kairomone. The chemicals produced by insect hosts that serve as cues to parasitoids are kairomones because the parasitoid exploits them to the host's detriment. Usage is important here. A sex pheromone may attract a male to a female moth; but if a parasitoid cues on this chemical, the substance also functions as a kairomone. There are some materials that benefit both sender and receiver species. These are called synomones, and in the present context, the plant volatiles that attract parasitoids to host plants are synomones because they make it easier for the parasitoid to find herbivores damaging the plant.

IDENTITY OF HOST-SEEKING CHEMICALS

Some progress has been made in identifying chemicals important in host-seeking behavior of parasitoids. Volatiles involved in host habitat finding include ethanol and especially acetaldehyde produced in rotting peaches that attract the braconid, *Biosteres longicaudatus,* a parasitoid of tephritid fruit flies. Allyl isothiocyanate produced by crucifers is attractive to the braconid parasitoid of aphids *D. rapae.* The straight-chain hydrocarbons docosane, tricosane, tetracosane, and pentacosane from the scales of adult *Helicoverpa zea* moths are cues used by *Trichogramma evanescens* to locate host eggs. In the frass of *H. zea,* 13-methylhentriacontane is an examination-stimulating cue for the braconid parasitoid *M. croceipes,* as is heptanoic acid in the frass of the potato tuberworm *(Phthorimaea operculella)* for the braconid parasitoid, *Orgilus lepidus.* Members of a series of methyl branched hen-, do-, and tritriacontaines from the mandibular glands of *Heliothis virescens* serve to intensify searching of the braconid, *C. nigriceps* on areas of leaves damaged by host feeding. Long-chain hydrocarbons (heptacosane, nonacosane, and several dimethyl compounds) in the cuticle of gypsy moth pupae are important in the host acceptance behavior of the chalcid, *Brachymeria intermedia.* The ichnuemonid parasitoid *I. conquisitor* oviposits into a wax-covered cylinder of water mixed with the amino acids arginine, isoleucine, methionine, lysine, leucine, and serine, as well as magnesium chloride. The tachinid fly *Cyzenis albincans* lays very small, microtype eggs on oak foliage that caterpillars of the winter moth *(Operophtera brumata)* eat. The parasitoid is stimulated to lay eggs in the presence of sugars exuded by damaged oak leaves, thus increasing the chance that its host will be nearby.

LEARNING AND HOST SEEKING

Insect behavior is sometimes perceived as a rigid, instinctive, inherited phenomenon not subject to change. However, there is ample evidence that many insects vary their behavior depending on circumstances, and that often learning is involved. This is also true for parasitoids. Many female parasitoids respond to host stimuli more strongly after they have parasitized a host. The heightened response, which may take the form of faster host finding and/or more intensive searching, may be considered to be a form of reward conditioning in that the female responds more avidly to host stimuli once she has been "rewarded" by being able to oviposit. Indeed, the response to host stimuli may wane if the parasitoid is prevented from oviposition. This occurs in the ichneumonid *Campolitis sonorensis* if it is not allowed to oviposit after contacting host frass or damaged plant material. Also, when the eucoilid parasitoid *Leptopilina heterotoma* is not able to oviposit, it becomes unresponsive to host cues (*Drosophila* larvae), but it searches more avidly if placed in a novel environment. Another ability that some parasitoids demonstrate is associative learning. In this type of learning, the parasitoid becomes able to associate a nonhost stimulus with the presence of hosts. For example, the ichneumonid parasitoid *I. conquisitor* can learn to distinguish between different shapes, sizes, and colors of artificial tubes holding host lepidopterous pupae, depending on which ones it has been allowed to oviposit in. Another ichneumonid, *Venturia canescens,* which attacks lepidopterous larvae in cereals, can learn to associate the presence of hosts with the odor of a nonhost chemical such as geraniol. Also, the braconid *Bracon mellitor* learned to associate with its host an antibiotic incorporated into the artificial diet of that host, the boll weevil. Thus, the sensory modalities of vision, olfaction, and contact chemoreception may be involved in the process of associative learning.

The advantage of such flexibility is probably greatest for parasitoids that are not strictly host specific. Suitably malleable behavior would help these parasitoids take advantage of changes in host and habitat composition. Learning in parasitoid searching behavior has recently generated much research interest, so it is likely that many more examples will be forthcoming.

TRITROPHIC INTERACTIONS

The involvement of plant volatile chemicals in the host-seeking behavior of parasitoids has an ecological and an evolutionary aspect. By facilitating the parasitization of herbivores feeding on a plant these synomones aid both the plant and the parasitoid. As such, plant and parasitoid would be expected to coevolve, resulting in some finely developed systems of signal and response. Examples of tritrophic interactions in which such coevolution may have occurred include situations in which a parasitoid is attracted to volatile chemicals produced by a plant only when that plant has been damaged by herbivores. This makes the signal more meaningful to the parasitoid than a signal produced by all plants at all times. The plant also presumably benefits by not wasting resources to produce a signal that is not needed. However, another possible explanation is that substances produced as a result of injury are part of an induced resistance response of the plant to herbivory; thus the primary purpose of the material would be to decrease foliage palatability or otherwise directly harm the herbivore. A parasitoid might evolve to use these materials, but still have little or no impact on the evolution of the plant responses. Indeed, most work on coevolution in insects and plants has emphasized the plant–herbivore interactions, yet there is little solid information about tritrophic interactions. However, coevolution between parasitoid and plants is still a theoretical possibility, and researchers are beginning to study this interaction.

IMPLICATIONS FOR BIOLOGICAL CONTROL

In biological control, high searching capacity is considered to be a very desirable trait of natural enemies. Host-seeking behavior influences searching capacity greatly. The more readily a parasitoid can find a host, the better it will realize its full reproductive potential. Also, for a parasitoid to control a host at low densities, it is necessary to find that host under conditions of scarcity. By using specific cues, especially volatile chemicals produced directly or indirectly by hosts, many parasitoids are able to find these hosts very well. Research on host-seeking behavior has progressed far enough to permit the development of some general concepts that should aid biological control workers as they evaluate the effectiveness of parasitoids. For example, chemicals are very prominent as host habitat-finding and host-finding cues. Chemicals involved in host habitat finding are usually perceived from relatively long distances, and they orient parasitoids to travel upwind. Chemicals involved in host finding may be perceived from long distances, but often these are plant materials induced by feeding damage by herbivore hosts. Many chemical cues emanating directly from the host are short range or can be perceived only upon contact. Thus, plants strongly influence initial stages of the host-seeking process. There has been much concern about whether exotic natural enemies imported for biological control can have the detrimental effect of attacking nonpest species, especially endangered species. Because host-seeking behavior effectively determines the host ranges of many parasitoids, the general concepts developed from studies of host-seeking behavior should aid in efforts to delineate host ranges of parasitoids. For example, candidate parasitoids are now often screened for their ability to attack nontarget hosts. These screening tests usually occur in a laboratory, often using insect hosts that are removed from their usual plant hosts. Thus, only host-seeking behaviors

associated with cues directly derived from hosts are assessed. This ignores the often strong winnowing effect that attractions to plants exert on potential host ranges, and the laboratory-derived host range may be substantially wider than the natural range. Also important is learning, because the degree of flexibility in host seeking could quantify the likelihood of host switching in parasitoids.

The above-mentioned considerations illustrate the most important implications that host-seeking behavior in parasitoids has on practical biological control, but there is another dimension. As already mentioned, *Trichogramma* egg parasitoids intensively search areas in which scales from female moth hosts have been deposited. The main attractive material in these scales is tricosane. When tricosane is artificially deposited on foliage containing eggs of the host moth, the resulting parasitism by *Trichogramma* is higher than in areas not having tricosane. Thus, the direct use of such kairomones could improve pest control by manipulating the behavior of natural enemies. Although such schemes have so far not been economically viable, similar manipulations may prove to be workable in situations not yet tested. Continued research may well lead to some useful control methods.

See Also the Following Articles
Coevolution • Hypermetamorphosis • Learning • Oviposition Behavior • Parasitoids

Further Reading
Nordlund, D. A., Jones, R. L., and Lewis., W. J. (eds.). (1981). "Semiochemicals: Their Role in Pest Control." Wiley, New York.
Vet, L. E. M., Lewis, W. J., and Cardé, R. T. (1995). Parasitoid foraging and learning. *In* "Chemical Ecology of Insects" (R. T. Cardé and W. J. Bell, eds.), Vol. 2. Chapman & Hall, New York.
Vinson, S. B. (1984). Parasitoid–host relationships. *In* "Chemical Ecology of Insects" (W. J. Bell and R. T. Cardé, eds.). Sinauer, Sunderland, MA.
Vinson, S. B. (1998). The general host selection behavior of parasitoid Hymenoptera and a comparison of initial strategies utilized by larvaphagous and oophagous species. *Biol. Control* **11**, 79–96.
Vinson, S. B. (1999). Parasitoid manipulation as a plant defense strategy. *Ann. Entomol. Soc. Am.* **92**, 812–828.

Host Seeking, for Plants

Elizabeth A. Bernays
University of Arizona

Plant-feeding insects may find their hosts by seeking appropriate habitats, by increases in activity that maximize the chances of encountering a plant, by completely random activity in combination with strong arrestant properties of the host, or by attraction to a plant from a distance by smell or vision or both. Often generalized plant odors are attractive, but commonly host-specific odors can be distinguished by specialist insects, and recently it has become known that most such insects are highly sensitive to one or a few host odors that are particularly attractive. Host color and shape can be important in plants with characteristic visual features and in insects that are day flying, although visually mediated responses are usually relatively unspecific.

FINDING HOSTS INDIRECTLY

Habitat location appears to be the first step for a number of species, although it is difficult to prove in practice. For example, grass-feeding grasshoppers are attracted to open habitats where grasses are generally abundant, but there are likely to be other reasons for this behavior. The pierid butterfly *Euchloe belemia,* however, is attracted to patches of thorn plants where, typically, its small host plants grow most densely.

Some small insects that find their hosts within fairly short distances may simply rely on increases in activity and turning behavior when they detect the appropriate odor, so that they are more likely to encounter their hosts. Among chrysomelid beetles, some engage in random movements that show little change with level of host odor; crucifer flea beetles in the genus *Phyllotreta,* for example, move randomly within and between host patches. It has been shown mathematically that random activity is important in overall efficiency of search strategies because, under many circumstances and especially when host signals are weak and plant suitability variable, exploratory search enhances the likelihood that an individual will contact the better plants.

USING ODORS TO FIND HOSTS

There are many examples of insects being attracted to the odors of their host plants, both by flying and by walking or crawling. Generalists such as the moths *Trichoplusia ni* and *Heliothis virescens* and the desert locust *Schistocerca gregaria* fly or walk upwind in wind tunnels toward general green plant odors, and there are examples among all orders of specialist herbivores being attracted to chemicals arising specifically from their host plants (Table I).

Because of air turbulence, concentration gradients that an insect might follow do not generally exist, except within centimeters of the plant. Instead, there are pockets of odor-carrying air that are carried downwind in a rough plume, and an insect encounters and perceives an irregular series of these pockets. An insect usually responds in two stages. First, there is "arousal," preparing the insect to respond to some further stimulus. Then orientation occurs, either on the substrate or in the air. Usually the orientation response is a response to the wind, with the insect turning upwind, and this is termed an odor-induced upwind (positive) anemotaxis. Among flying moths, this has been demonstrated clearly in wind tunnels, and the same kind of response may be seen when males fly upwind toward the source of female pheromone.

TABLE I Example of Host Plant Volatiles Attracting Specific Phytophagous Insects

Insect	Chemical(s) or host odors
Cavariella aegopodii (carrot aphid)	Carvone (one of the host volatiles)
Brevicoryne brassicae (cabbage aphid)	Isothiocyanates (host volatiles)
Aphis gossypii (cotton aphid)	Host plant odor
Leptinotarsa decemlineata (Colorado potato beetle)	Host plant odor
Ceutorhynchus assimilis (Cabbage seedpod weevil)	Isothiocyanates (host volatiles)
Psila rosae (carrot fly)	Mixture of five host volatiles
Delia antiqua (onion maggot)	Disulfides (host volatiles)
Acrolepiopsis assectella (leek moth)	Thiosufinates (host volatiles)
Plutella xylostella (diamondback moth)	Host plant odor
Manduca sexta (tobacco hornworm)	Host plant odor
Heliothis subflexa (groundcherry moth)	Host plant odor

Once airborne, the insect needs to monitor its ground speed, so that it can increase its airspeed if the wind is strong. To do this it uses visual information (i.e., image movement across the eyes from front to back). If the wind is too strong and the insect is unable to keep the images flowing, it turns and flies downwind or lands. The use of visual images by a flying or swimming insect to maintain orientation to a current flow is called an optomotor reaction. It enables the insect to maintain an orientation at any angle to the wind, not just directly up- or downwind. If the insect is unable to see the pattern of objects on the ground, it cannot orient. As well as generally flying upwind in response to a particular odor, many moths and beetles follow zigzag flight paths. This behavior, which evidently is programmed in the insect central nervous system, has the possible function of increasing the chances of encountering a pocket of odor.

Walking insects also show odor-induced anemotaxis. This has been demonstrated in locust nymphs, certain beetles, and aphids, for example, where individuals walk upwind in response to host odors.

In a number of smaller insects such as phytophagous flies, the odor-induced anemotaxis is slightly different; this is well studied are the onion maggot, *Delia antiqua,* and the cabbage maggot, *D. radicum.* In these species, after perception of the host odor, an individual fly turns into the wind and makes short flights. After landing, and again detecting the odor, it reorients into the wind and takes off. This tactic is particularly effective for host finding in vegetation, where the path to the food plant may be rather devious and the odor plume very broken.

A different response to host odor after the initial arousal is to move toward or land on a relevant visual target. This odor-induced visual orientation is believed to occur, for example, in the cabbage seed weevil, which uses odor-conditioned anemotaxis from a distance and then odor-conditioned landing responses on yellow targets close to the source. A number of insect species may be readily trapped by means of a yellow water trap combined with a host odor source, and it is probably generally true that landing responses induced by the host odor are responsible.

Insects living in soil use odors alone to find hosts. Since, the air moves little in soil, steep gradients of volatile chemicals can be achieved and maintained. Carbon dioxide is commonly used by such larvae, but for specialists, host-specific compounds are also used. Root-feeding larvae, such as that of the carrot fly, *Psila rosae,* and the corn rootworm, respond by moving directly up a concentration gradient. Larvae of the carrot fly respond to a mixture of five compounds found in carrot odor.

For insects that fly or walk, the distances from which olfactory cues elicit responses vary from less than a meter as in the Colorado potato beetle, *Leptinotarsa decemlineata,* to about 30 m in some bark beetles and 100 m in some flies such as the onion maggot. Those that crawl in soil respond from just a few centimeters.

USING VISION TO FIND HOSTS

Visual attraction can result from responding to the color or form of the host plant. Because these vary so greatly within a species, and because there is relatively little specificity of shape among plant species, visual responses often occur only in the presence of an appropriate olfactory signal.

In a few examples, visual responses to host features have been demonstrated without the presence of odors. Walking insects of several species are attracted to narrow vertical targets in a plain arena, but the precise significance of this attraction is unknown. Perhaps it is a response to potential vegetation or shelter. Several species of butterflies, however, have been shown to land preferentially on leaves of particular shapes, with further discrimination occurring only after landing. Shape may interact with color as in the apple maggot, *Rhagoletis pomonella.* Host odors play a role here, but when colored rectangles are offered, the only color to attract flies is yellow, perhaps representing vegetation. If colored spheres are presented, the red and black shapes attract flies, perhaps representing the host fruit.

With respect to color, both wavelength and intensity are important. *D. radicum* lands preferentially on leaves with a leaf reflectance pattern characteristic of its host, whereas the western flower thrips, *Frankliniella occidentalis,* land most on yellows and whites, and more at highest intensities of reflected light. Patterns can also matter. For example, females of *Heliconius* butterflies lay their eggs on *Passiflora* leaves but tend not to oviposit on leaves that already have eggs on them. This is known to be a visual response to the yellow eggs, because if the eggs are painted green to match the leaf, butterflies do not discriminate against them.

A response to color is often coupled with a chemical cue. *Pieris rapae* require the presence of glucosinolates to oviposit but still responds to these chemicals only if they are on blue, yellow, green, or white substrates. Females reject red or black substrates.

Visual cues are usually important only at close range, though occasionally they attract specific herbivores from 10 m or so. This is true for the apple maggot, which has a very clear signal in the bright red fruits of its host substrate.

LEARNING IN HOST SEEKING

Although the studies are few, it is clear that many insects take advantage of experience in their foraging activities and thus improve efficiency of host finding. For example, butterflies learn to land on leaf shapes that resemble their hosts' leaf shapes, making many fewer mistakes with experience, and they learn many visual cues, especially color, when these are coupled with nectar rewards. Grasshoppers have been shown to learn that certain colored backgrounds are associated with the presence of high-quality food, and the time taken to find the food inside colored boxes in laboratory training experiments with *Melanoplus sanguinipes* was reduced from about 40 min for naive individuals to less than 10 min after a single experience.

Less is known about olfactory learning, but the work so far suggests that it may be more important than visual learning. Grasshoppers in experiments have been trained with different food odors associated with high-protein, low-carbohydrate diets and low-protein, high-carbohydrate diets. They were then fed untreated diets of one or the other type of imbalance until they were relatively deprived of one or the other major nutrient. Then, given a choice, grasshoppers tended to select against the odor that had originally been paired with the unbalanced food. Thus, if they were overfed protein and underfed carbohydrate, they were more likely to avoid the odor that had originally been paired with high-protein food and instead be attracted to the odor that had originally been paired with high-carbohydrate food.

Food aversion learning has been demonstrated in grasshoppers and caterpillars, whereby individuals having a deleterious postingestive experience after eating a certain food thereafter reject it or eat little of it. However, the role of odor and the importance of the associated cues in behaviors prior to contact have not yet been investigated.

ECOLOGICAL INTERACTIONS

The abiotic environment and the presence of other organisms influence host-seeking behavior in nature. Among abiotic factors, temperature constraints and needs are probably the most important. For example, thermoregulating grasshoppers choose sites off the ground for cooling, and warm sunny substrates for basking. This can dictate the plants that are immediately available for feeding upon, so selection of thermoregulatory sites influences food selection. For example, the black lubber grasshopper, *Taeniopoda eques,* is highly polyphagous; when temperatures become very high in its desert environment in the middle of the day, however, it roosts as high off the ground as possible on mesquite or acacia bushes, and thus, any feeding is on these plants. During the cooler mornings and evenings it feeds only on plants at ground level. Many temperate butterflies seek out sunny or warm patches, and thus plants in those patches. For example, the meadow brown butterfly, *Pararge aegeria,* oviposits on various grasses but the actual choice depends on the temperature of the leaves, which in turn is influenced by whether the leaves are in sun or shade.

Wind is important for most insects. Wind speed and constancy influence odor plumes used by orienting insects. The wind speed also limits flight, with larger, stronger flying species remaining airborne at higher speeds. Very small insects are often carried by wind, and depending on the terrain, are deposited preferentially in certain places, such as the lee side of trees and hedges.

The presence of certain nonhost plants and the relative abundance or clumpiness of the host plant can alter the detailed behaviors involved in host seeking. For example, butterflies ovipositing in a habitat where two or more host plant species occur commonly tend to choose the species they laid eggs on previously, so that they land more often on the common host. In other insects, the host being selected for oviposition is dependent on factors such as the need for additional resources. In one example, the celery fly, *Phylophylla heraclei,* requires trees near to the celery host because this is where mating occurs and the adult food of aphid honeydew is available.

Insects that show odor-induced anemotaxis to their host plants presented alone in a wind tunnel in the laboratory do not always show the same behavior in field situations. For example, the Colorado potato beetle is attracted, at least from short distances, to its preferred host, potato. However, if nonhosts are also present, the response may be reduced or absent, and the host odor is said to be masked. Such interactions reduce the distance over which some host odors can be detected by phytophagous insects, and the phenomenon may be one of the mechanisms involved reduction of pest numbers in certain crop mixtures.

In addition, some insects are influenced by olfactory or visual evidence of prior occupation of a plant, competitors of the same or different species, and of the presence of natural enemies.

PHYSIOLOGY OF THE HERBIVORE

Host-seeking behavior is restricted to times when the ovipositing or feeding insect is in a suitable physiological state. For example, insects about to molt do not feed and are generally not responsive to host odors; in adult females, a load of eggs ready for laying alters motivation so that searching for a host takes priority over other behaviors. Similarly, an insect that has been deprived of food seeks hosts more readily than one that is replete. In nymphs of the desert locust, for example, positive anemotactic responses to the odor of grass in a wind tunnel were not seen in well-fed individuals but were dramatic in nymphs that had been deprived of food for 4 h.

In the bean aphid, *Aphis fabae,* winged individuals that fly distances from one host to another are attracted, when they take off, to the short wavelengths of the blue sky. After flying certain distances, they are preferentially attracted to the longer wavelengths of yellow, so that they then tend to land on plants in the vicinity. A number of aphid species bias their landings toward the yellower greens that often are associated with plants in an appropriate physiological state rather than toward plants of a particular species.

See Also the Following Articles
Eyes and Vision • Learning • Migration • Orientation • Phytophagous Insects

Further Reading
Barton Browne, L. (1993). Physiologically induced changes in resource oriented behavior. *Annu. Rev. Entomol.* **38,** 1–25.
Bell, W. J. (1991). "Searching Behaviour: The Behavioural Ecology of Finding Resources." Chapman & Hall, London.
Bell, W. J., Kipp, L. R., and Collins, R. D. (1995). The role of chemo-orientation in search behavior. *In* "Chemical Ecology of Insects" (R. Cardé and W. J. Bell, eds.), Vol. 2, pp. 105–152. Chapman & Hall, New York.
Bernays, E. A., and Chapman, R. F. (1994). "Host-Plant Selection by Phytophagous Insects." Chapman & Hall, New York.
Jones, R. E. (1991). Host location and oviposition on plants. *In* "Reproductive Behaviour of Insects" (W. J. Bailey and J. Ridsill-Smith, eds), pp. 108–138. Chapman & Hall, London.
Morris, W. F., and Kareiva P. M. (1991). How insect herbivores find suitable host plants: The interplay between random and nonrandom movement. *In* "Insect–Plant Interactions" (E. A. Bernays, ed.), Vol. III, pp. 175–208. CRC Press, Boca Raton, FL.
Murlis, J., Elkinton, J. S., and Cardé, R. (1992). Odor plumes and how insects use them. *Annu. Rev. Entomol.* **37,** 505–532.
Papaj, D. R., and Prokopy, R. J. (1989). Ecological and evolutionary aspects of learning in phytophagous insects. *Annu. Rev. Entomol.* **34,** 315–350.
Prokopy, R. J. (1986). Visual and olfactory stimulus interaction in resource finding by insects. *In* "Mechanisms in Insect Olfaction" (T. L. Payne, M. C. Birch, and C. E. J. Kennedy, eds.), pp. 81–90. Clarendon Press, Oxford, U.K.
Schoonhoven, L. M., Jermy, T., and van Loon, J. J. A. (1998). "Insect–Plant Biology: From Physiology to Evolution." Chapman & Hall, London.
Visser, J. H. (1988). Host plant finding by insects: Orientation, sensory input and search patterns. *J. Insect Physiol.* **34,** 259–268.

House Fly
(*Musca domestica*)

Gregory A. Dahlem
Northern Kentucky University

The house fly, *Musca domestica* (Fig. 1), is one of the best known and most widely distributed insects known to humans. It is a classic example of a synanthropic animal, one that lives in association with humans and their domesticated

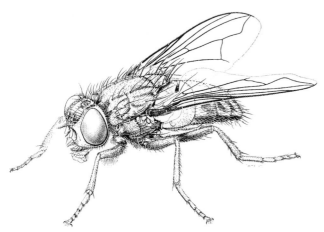

FIGURE 1 *M. domestica.* [After Huckett, H. C., and Vockeroth, J. R. (1987). Muscidae. *In* "Manual of Nearctic Diptera" (J. F. McAlpine, B. V. Peterson, G. E. Shewell, H. J. Teskey, J. R. Vockeroth, and D. M. Wood, eds.), Vol. 2. Biosystematic Research Institute Research Monograph **28.** For the Department of Agriculture and Agri-Food, Government of Canada. © Minister of Public Works and Government Services, Canada, 1987. Reproduced with the permission of the Minister of Public Works and Government Services, 2001.]

animals. House flies occur and thrive wherever humans are found but are very rare in natural or wild areas throughout the world. "Insects will survive long after humans disappear" is a common expression, but it is not true of the common house fly. House flies would likely not be able to survive in the absence of humans because their relationships are so closely linked.

The available literature on the house fly is vast. A computer search of only one database (BIOSIS) using *"Musca domestica"* as the key words, yielded well over 1000 references for the period between 1990 and 2000. A complete synopsis of the known information on this one species is not possible in this short article, but the following information should give the reader a better appreciation for this common species and the readings listed at the end will enable the curious to pursue this topic further. House flies are included in the group known as the calyptrate Diptera, which includes the Muscidae, Anthomyiidae, Calliphoridae, Sarcophagidae, Tachinidae, and several smaller families.

LIFE CYCLE AND BEHAVIOR

Mating Behavior

Courtship and copulatory behaviors are the most important and the most complex behaviors exhibited by the house fly. Visual, chemical, tactile, and auditory cues are all used, to various degrees, in courtship and copulation. The elimination of a male's production of, or a female's reception of, any one stimulus may not greatly affect mating success. However, if combinations of stimuli are simultaneously eliminated, mating can be significantly affected. In general, males mate as often as they can, whereas females mate just once.

The courtship behaviors of *M. domestica,* and many other calyptrate Diptera, are initiated when the male first sights a prospective femalelike object. Males are not very discriminatory in their initial choice of partners and strike other males, other species of flies, and small inanimate objects moving through their visual field.

The discovery of cuticular hydrocarbons that serve as sex pheromones in the house fly triggered a burst of investigations into the role of such pheromones in the mating behavior of the calyptrate Diptera. No evidence has been found that these pheromones are olfactory stimulants; rather, they appear exclusively to be contact excitants. In house flies, chemosensilla involved with contact chemoreception (or "taste") are located on both the mouthparts and the tarsi. Thus, house flies can "taste" with their feet. When a male touches a female with his tarsi, as he grasps her upon initial contact, he can use the female-produced sex pheromone to determine whether a potential mate is of the correct species, sex, and even mating status. This pheromone is a very important stimulus for the male, and a male repeatedly attempts to copulate with an object that is of appropriate size and "tastes right."

Behaviors that involve the touching or bodily movement of males and females beginning after the initial contact may be elicited by tactile cues. Tactile cues may be given by either sex during courtship, but it appears that the male's role is much more complex than that of the female. High-speed photography has shown that the house fly's courtship is extremely brief and complex. The highly ritualized sequence of movements that the male performs immediately after contact with the female seems to be very important to the female in her choice of potential mating partners. Males often strike females in midflight and perform the courtship ritual during their plunge to the ground. If the courtship is performed to the female's satisfaction, she allows the male to mate with her. If not, she can dislodge the male and stop mating from occuring by performing one of several different rejection maneuvers.

Mating pairs of *M. domestica* are normally quiescent during copulation. However, if the pair is disturbed they move, and the female flies, short distances carrying the male on her back.

Development

The number of eggs that mature in a fly's ovaries at one time is about 120. The female requires both sugar and protein meals for egg production. After copulation, egg laying takes place in 4 to 8 days. The female requires nearly a day to deposit the eggs, which may be deposited in a single mass or distributed in a number of locations. Each female is capable of developing several batches of eggs during her lifetime. Animal manure is the preferred ovipositional substrate, although a variety of decaying organic material can be used if fecal material is not available.

Developmental time is highly temperature dependent. Hatching usually takes place within one day after oviposition. Larval development occurs rapidly, with the larva (maggot) normally passing through all three instars in 5 to 9 days. After full larval development, the third instar turns into a dark, cylindrical puparium, which is composed of the sclerotized skin of the last (third) larval stage. The process of pupation normally lasts about 5-days.

Most adults live for 2 to 3 weeks at normal summertime temperatures in the temperate regions. During the summer in temperate regions of the world, the entire life cycle (from egg to egg-laying adult) can be accomplished in 10 days to 2 weeks.

Flight and Dispersal

House flies are comparatively slow fliers, with a normal flight speed of about 2 m s^{-1} (or 7.2 km/h^{-1}). They have an innate tendency to disperse from their rearing site, even when conditions are favorable. Capture–release studies with marked house flies indicate that 85 to 95% of the flies stay within a 3-km radius of their release point after 4 days, although a few individuals may travel over as much as 20 km.

House flies are one of very few species of fly that purposely enter human structures, such as houses and barns. This propensity for entering dark openings has implications on the dispersal of this species, because house flies readily enter cargo or passenger areas of trucks, trains, ships, and airplanes. By this means, gene flow between geographically distant populations is easily, although accidentally, promoted.

Ability to Land and Walk on Ceilings and Vertical Surfaces

A common question about house flies is, "How do the flies land on ceilings, and how do they walk up smooth vertical surfaces, such as glass windows?" In landing on ceilings, a house fly normally performs a "half-roll" and reaches its legs out to the ceiling. Contact of the tarsi with the ceiling inhibits flight and the fly comes to rest, generally facing the direction that it was flying. A close look at the structures found on the tips of the tarsi help to explain the fly's ability to cling and walk on ceilings or smooth glass windows. The apical tarsal segment bears a pair of curved claws that are used to cling to rough surfaces. At the base of each claw is a padlike structure, called the pulvillus, which bears a large number of glandular setae. These setae are coated with secretions that make them sticky, allowing the fly to walk on vertical, or even inverted, smooth surfaces.

THE HOUSE FLY AS A VECTOR OF HUMAN AND ANIMAL DISEASE

House flies, particularly in large numbers, are a nuisance to humans when they enter houses, land and feed on human food, and spot windows with their feces. Of greater importance to humans, however, is their ability to spread human and veterinary disease agents. House flies have been associated with over 100 pathogens that can cause disease in humans and animals. Unlike the pathogens responsible for many other insect-borne diseases, the pathogens spread by the house fly do not usually multiply within the fly, nor do they require association with the fly for part of their life cycle. Instead, the usual association between house flies and pathogenic organisms is one of physical transmission of pathogens the flies pick up on their bodies at one feeding site (e.g., a garbage can or manure pile) and transfer to human and/or animal food when they land and feed. House flies have been associated with the transfer of a variety of viral and bacterial diseases, such as typhoid fever, cholera, dysentery, and infantile diarrhea, as well as a variety of parasitic worms.

See Also the Following Articles

Chemoreception • Medical Entomology • Urban Habitats

Further Reading

Colwell, A. E., and Shorey, H. H. (1977). Female-produced stimuli influencing courtship of male house flies *(Musca domestica). Ann. Entomol. Soc. Am.* **70,** 303–308.

Crosskey, R. W., and Lane, R. P. (1995). House flies, blowflies and other allies (calyptrate Diptera). *In* "Medical Insects and Arachnids" (R. P. Lane and R. W. Crosskey, eds.). Chapman & Hall, London.

Greenberg, B. (1971). "Flies and Disease," Vol. I of "Ecology, Classification and Biotic Associations." Princeton University Press. Princeton, NJ.

Greenberg, B. (1973). "Flies and Disease." Vol. II of "Biology and Disease Transmission." Princeton University Press. Princeton, NJ.

Huckett, H. C., and Vockeroth, J. R. (1987). Muscidae. *In* "Manual of Nearctic Diptera" (J. F. McAlpine, B.V. Peterson, G. E. Shewell, H. J. Teskey, J. F. Vockeroth, and D. M. Wood, eds.), Vol. 2, pp. 1115–1131. Biosystematic Research Institute Research Monograph **28,** Ottawa, Ontario, Canada.

Tobin, E. N., and Stoffolano, J. G., Jr. (1973). The courtship of *Musca* species found in North America. 1. The house fly, *Musca domestica. Ann. Entomol. Soc. Am.* **66,** 1249–1257.

West, L. S. (1951). "The Housefly. Its Natural History, Medical Importance, and Control." Comstock, Ithaca, NY.

Hymenoptera
(Ants, Bees, Wasps)

Donald L. J. Quicke
Imperial College, University of London

The Hymenoptera are a major order of holometabolous insects. That is, they undergo complete metamorphosis with distinct egg, larval, pupal, and adult stages. They are one of the five megadiverse insect orders along with Coleoptera, Diptera, Lepidoptera, and Heteroptera, and perhaps even the most species rich of any insect order—certainly this is true at temperate latitudes. They are generally cosmopolitan and, except for some specialized groups, they are most speciose in the tropics.

Although after working with Hymenoptera for a while it becomes easy to recognize members of this order, there are almost no conspicuous defining characters, because the majority of hymenopteran attributes are plesiomorphic; that is, they are shared with the common ancestors of various other orders. It is not surprising therefore that although almost everyone on earth, save perhaps those living at extreme northern latitudes, is familiar with ants, bees, and social wasps and has vernacular names for these particular taxa, there is not a single vernacular name in any language that refers to them in toto. Hymenoptera are also diverse in terms of their life histories: they include phytophagous, parasitoid, and predatory taxa, both solitary and highly social species, and they range in size from the rather large and intimidating spider-hunting pompilid wasps that can reach 12 cm wingspan down to the tiniest parasitic wasps that are approximately 0.1 mm in length (males of the wingless mymarid chalcidoid, *Dicopomorpha echmepterygis*). It is hard to overstate their ecological importance because they are collectively involved in so many types of interaction, and it is likely that many are effectively keystone species in their own habitats.

The Hymenoptera get their name from the Greek words *humen* and *pteron,* meaning membrane and wing, respectively, and this gives the first clue to identifying them. Excluding the numerous exceptions of apterous and brachypterous species that are widely distributed through the order, hymenopterans possess two pairs of membranous wings that are devoid of scales. The forewings are larger than the hind wings, and the two are interlocked during flight by a row of special hooks called hamules (or hamuli) that are on the anterior margin of the hind wing; these hamuli engage (or interlock) with a fold on the posterior edge of the forewings. This system makes the Hymenoptera functionally dipterous (two-winged) during flight, since the wing surfaces on either side of the body acts as a single aerofoil. Hamules are unique to this order of insects.

GENERAL BIOLOGY

Since Hymenoptera is a very large order, it is not surprising that a considerable number of biologies and life history strategies are exhibited by its various taxa. Broadly speaking, the basal lineages are phytophagous as larvae, feeding both ecto- and endophytically on a large range of herbs, shrubs, and trees; few tropical pergid sawflies, even feed on slime molds! The great majority of the remaining species are either parasitoids of other insects or predators of insects (e.g., the yellow-jackets or social wasps, which are members of the Vespidae) or spiders. However, among the higher taxa, there have also been several reversals to phytophagy, especially through the formation of galls on plants (cecidogenesis). The bees (Apidae) and one other, small tropical group, the Masarinae within the Vespidae, have evolved to make use of pollen and nectar as a larval food source.

Development

Like other holometabolous groups, hymenopterans primitively pass through generally five instars, though the number of instars is typically smaller in endoparasitic taxa, and in one such genus there seems to be just a single instar. The final instars of hymenopterans are rather morphologically conservative, with most sawflies having rather caterpillar-like larvae with well-developed true legs and variously developed prolegs on several of the abdominal segments. Adoption of an endophytic way of life by cephoid sawflies and wood wasps was accompanied by a reduction in the prolegs and more generally by reduction of sensory structures. Final instar apocritan wasp larvae are all quite similar and are termed hymenopteriform. They are superficially rather maggotlike in that they lack legs and other processes and often have a rather reduced head. However, many endoparasitoids have highly bizarre, first instars characteristic of their particular families, and for which a variety of specific terms have been coined.

The pupal stage of hymenopterans is exarate; that is, the antennae, legs, and wings are free from the body (in contrast to the Lepidoptera, e.g., in which these components are fused with the body). The pupae tend to be rather delicate and are easily damaged. All sawflies and most members of the Ichneumonoidea + Aculeata clade produce a silken cocoon to protect the pupa. Most of the other parasitic taxa do not, however, probably because they pupate within the host remains or, if they pupate externally, do so in a location where the pupa is likely to be protected by the surroundings, such as within a leaf mine, gall, or wood boring.

Key Features in Hymenoptera Evolution

Given the huge size of the order, it is interesting to consider what features have enabled hymenopterans to be so successful in terms of both individuals and total number of species. Most attention has focused on a small number of features such as selection of oviposition site, modification of that site, the use of venoms, and the evolution of the thin wasp waist, all of which are discussed in this article. In addition, the unusual form of sex determination mechanism, haplodiploidy, may have been particularly important in the evolution of sociality. It is likely, however that few of these traits have operated in isolation, and it is the interactions of these and other factors that have been important. Thus, for example, evolution of sociality may have been facilitated by the sex determination mechanism but also requires the abilities to remember where the nest is, to recognize nestmates, and to be able to defend the nest.

Several studies have emphasized that the success of the Hymenoptera has probably been a consequence of the general tendency of these insects to provide their offspring with particularly nutritious food sources, and when necessary (and that has been often) to modify poorer foods to better ones. Although this may be most familiar in terms of the provisioning of larvae in the nests by the social wasps and bees, such

TABLE I   Most Recent Classification of the Sawflies and Wood Wasps

Superfamily	Family	Described extant species	Notes
Xyeloidea	Xyelidae	50	Most ancient family, with Holarctic distribution.
Pamphilioidea	Pamphiliidae	250	Sometimes called webspinning sawflies after the habit of early instars. Some are pest species.
	Megalodontesidae	40	Rare group with little known about biology; some feed on Apiaceae and on Rutaceae.
Tenthredinoidea	Argidae	800	A common group.
	Blasticotommidae	~10	Usually uncommon, larvae live in a ball of foam of ferns.
	Cimbicidae	130	Occasionally common, often rather large and beelike sawflies.
	Pergidae	500	Principally southern group, especially in Australia and South America.
	Diprionidae	90	Pine sawflies.
	Tenthredinidae	4000	Very common and speciose in temperate areas, uncommon but moderately diverse in the tropics. Most are exophytic with caterpillar-like larvae. Some are gall formers
Cephoidea	Cephidae	80	Stem sawflies, elongate, associated with grasses and rosaceous shrubs.
Anaxyeloidea	Anaxyelidae	1	A single species from western United States associated with fire-damaged trees.
Siricoidea	Siricidae	95	Horntail wood wasps.
Xiphydroidea	Xiphydriidae	100	Horntail wood wasps.
Orussoidea	Orussidae	75	Parasitic sawflies.

behaviors and physiological adaptations are to be seen all through the order and are manifested in many different ways.

First, there is egg placement and the larval food resource. The morphology of the ovipositor has been crucial in this respect. The hymenopteran ovipositor is used not only for laying eggs, it is also used to pass venom and/or other secretions to the place of oviposition. In the parasitoid taxa, these venoms either cause paralysis of the host or are important in overcoming the host's immune response against the parasitoid. The ovipositor is typically well supplied with sensilla, and the insects receive and interpret the resulting sensory information and use it in deciding whether they have located a site or host suitable for egg laying. This organ has been especially well studied in parasitoid taxa, and such observations have been used to test many evolutionary concepts.

In the majority of the aculeates (stinging wasps, bees, and ants) the egg-laying role has been lost, but the same structures are still present and are used for envenomation of prey or enemies. The venoms of most of these act on the nervous systems or nerve–muscle junctions of their prey insects, permanently paralyzing them. In this sense, the venoms are rendering their larval food sources manipulable and safe by preventing the prey insect from wriggling or moving to damage the wasp's developing young.

"Venoms" were important even before the evolution of parasitoidism. For example, at least some and possibly most wood wasps inject chemicals into their host trees along with their eggs and symbiotic fungi fragments, and these toxins probably either kill the living cambium cells or in some other way help the symbiotic fungi to overcome the trees' defenses so that the wood wasp larvae can feed on the developing nutritious fungal hyphae. As often happens, these conclusions are based on relatively few data, and observations of other species are very much needed.

Evolution of the thin wasp waist, which defines a large group of families called the Apocrita, was another absolutely key feature in that it greatly increased the mobility of the posterior abdomen relative to the thorax. This in turn allowed greater control of the ovipositor and greater variety in its use; later, it allowed the sting, which is in fact just a derived ovipositor, to be much more effective as a weapon of defense and offense. It is interesting that vertebrates can learn that a bee or wasp can deliver a sting and that part of the recognition of this ability involves the very conspicuous abdominal movements of the insect as it probes for a vulnerable spot with its sting. Because male Hymenoptera never possess stings (because the males do not have an ovipositor-derived apparatus!), they are harmless in this respect. Often, however, males very effectively mimic female wasp stinging movements such that people, and probably many experienced predators, do not take the risk and quickly release them—this behavior has been termed *aide-mémoire* mimicry.

The wasp waist, contrary to many people's initial expectations, is actually not located between the thorax and abdomen, but is a constriction between the first and second abdominal segments (Fig. 1). In the ants, posterior abdominal mobility is increased even more by second and sometimes third constrictions between the second and third, and third and fourth, abdominal segments, which give rise to the distinct node or nodes between the middle and posteromost body regions. There is a very good reason for the wasp waist to be positioned after the first abdominal segment. Higher hymenopterans are typically strong fliers, and their longitudinal flight muscles are consequently large. Because these muscles are attached internally on the anterior of thorax (actually the mesonotum) and posteriorly on a large internalized chitinous phragma that slants posteriorly, if there were a constriction immediately behind the last (third)

TABLE II   Generally Accepted Classification of the Apocrita[a]

Superfamily	Family	Described extant species	Notes
Stephanoidea	Stephanidae	200	No common name. Idiobiont parasitoids of wood-boring beetle larvae.
Megalyroidea	Megalyridae	50	No common name. Usually rare, presumed idiobiont parasitoids of wood-boring beetle and aculeate wasp larvae.
Trigonaloidea	Trigonalidae	100	No common name. Complex hyperparasitic (rarely primary parasitic) life cycle.
Evanioidea	Evaniidae	500	Ensign wasps. Often common in the tropics, endoparasitic or egg-predatory in cockroach oothecae.
	Gasteruptiidae	420	Moderately common, cosmopolitan, kleptoparasites of solitary bees. Their larvae kill the host bee's egg and consume the latter's pollen food store.
	Aulacidae	200	Koinobiont endoparasitoid of wood wasps and of wood-boring Coleoptera larvae.
Ceraphronoidea	Ceraphronidae	350	Small to very small, very common wasps with a wide range of parasitic biologies.
	Megaspilidae	450	Small to very small, common wasps with a wide range of parasitic biologies.
Platygastroidea	Platygastridae	1,100	Extremely common, usually small, cosmopolitan, mainly koinobiont endoparasitoids of Diptera larvae, but other biologies and hosts known.
	Scelionidae	3,000	Extremely common, usually small, consmopolitan, idiobiont egg parasitoids of many groups of insects and spiders.
Proctotrupoidea	Austroniidae	3	Extremely rare, Australian, biology unknown.
	Diapriidae	2,300	Very common, idiobiont and koinobiont endoparasitoids, mainly of Diptera larvae/pupae.
	Heloridae	7	Usually uncommon, koinobiont endoparasitoids of Neuroptera (Chrysopidae) larvae.
	Maamingidae	2	Most recently described family, known only from New Zealand, biology unknown.
	Monomachiidae	20	Parasitoids of Diptera (Stratiomyidae) in Australia and South America.
	Pelecinidae	3	Moderately common, large, entirely New World, koinobiont endoparasitoids of subterranean Coleoptera larvae.
	Peradeniidae	2	Very uncommon, Australian, biology unknown.
	Proctotrupidae	310	Common, mainly northern, koinobiont endoparasitoids, mainly of Coleoptera larvae, mostly in soil or litter layer.
	Renyxidae	2	Holarctic, extremely rare, biology unknown.
	Roproniidae	18	Usually uncommon, Holarctic and Oriental, parasitoids of sawflies.
	Vanhorniidae	5	Generally uncommon, New World parasitoids of eucnemid beetle larvae.
Mymarommatoidea	Mymarommatidae	14	Very uncommon, minute, biology unknown but guessed to be egg parasitoids.
Chalcidoidea	~20 families	19,000	Chalcids. Approximately 20 families are recognized, most with diverse biologies.
Cynipoidea	Ibaliidae	50	Egg-larval, koinobiont endoparasitoids of wood wasps.
	Liopteridae	50	Endoparasitoids (probably koinobiont) of wood-boring Coleoptera larvae in the tropics.
	Cynipidae	1,000	True gall wasps and also inquilines in other cynipid galls.
	Figitidae	1,500	Very common endoparasitoids of Diptera, of Neuroptera, and of hymenopterous parasitoids of aphids.
	Austrocynipidae	1	Extremely rare, parasitic on Lepidoptera larvae in *Araucaria* cones in Australia.
Ichneumonoidea	Ichneumonidae	22,000	Very common, biologically diverse though not including any egg parasitoids.
	Braconidae	20,000	Very common, biologically diverse though not including any egg parasitoids.
Chrysidoidea	Bethylidae	2,000	Common and widespread, small ectoparasitoids of small beetle and moth larvae in semicryptic locations; some show parental care.
	Chrysididae	3,000	Common, cosmopolitan. Chrysidines are mainly larval parasitoids of solitary vespid wasps and bees; cleptines attack sawfly prepupae; others are idiobiont egg parasitoids of stick insects.
	Dryinidae	950	Common, koinobiont parasitoids of larger Auchenorrhyncha (e.g., Cicadelloidea), some developing partially externally.
	Embolemidae	16	Uncommon, cosmopolitan; one species parasitic on Heteroptera nymphs.
	Plumariidae	20	Very rare tropical wasps, biology unknown.
	Sclerogibbidae	10	Very rare, koinobiont ectoparasitoids of webspinners (Embioptera).
	Scolebythidae	3	Very rare, tropical ectoparasitoids (probably idiobiont) of wood-boring beetle larvae.
Vespoidea	Bradynobaenodae	200	Very rare, cosmopolitan, may be koinobiont ectoparasitoids of sun-spiders (Solipugida). Great sexual dimorphism.
	Formicidae	10,000	Ants. Extremely common and cosmopolitan. Most are eusocial but also includes social parasites of other ants and slave makers.
	Mutilidae	5,000	Velvet ants. Not true ants, commonest in arid tropics, these are idiobiont ectoparasitoids of aculate larvae and pupae in their cells. Great sexual dimorphism.

(continues)

TABLE II   *Continued*

Superfamily	Family	Described extant species	Notes
Vespoidea (cont.)	Pompilidae	4,000	Spider wasps. Sometimes large. Some are ectoparasitoids of spiders *in situ;* most relocate spider prey to a new site.
	Rhopalosommatidae	35	Moderately common, ectoparasitoids of crickets (Gryllidae).
	Sapygidae	80	Kleptoparasitoids of solitary bees. Usually uncommon but occasionally a pest.
	Scoliidae	300	Often large, primarily tropical or warm temperate, idiobiont ectoparasitoids of subterranean beetle larvae.
	Sierolomorphidae	10	Very rare, Americas and Oriental region, biology unknown.
	Tiphiidae	1,500	Often common, mainly tropical, idiobiont ectoparasitoids mainly of subterranean beetle larvae. Great sexual dimorphism.
	Vespidae	4,000	Very common, includes the familiar social wasps or yellowjackets, mason or potter wasps. Females progressively provision their larvae with chewed insect/spider tissue, or in Masarinae, pollen.
Apoidea	Apidae	30,000	Bees, from solitary to highly social. Progressively provision larvae with pollen.
	Sphecidae	8,000	Almost entirely solitary, nest-building predators of insects and spiders.

*ª*The Chrysidoidea, Vespoidea, and Apoidea comprise the aculeate Hymenoptera. The true number of species in the Ceraphronoidea, Platygastroidea, Diapriidae, Chalcidoidea, and Ichneumonoidea, in particular, are likely to greatly exceed the figures given here.

thoracic segment, the size of the flight muscles would be greatly restricted. By having the first abdominal segment fused to the thorax, larger flight muscles can be accommodated. Thus, the middle body part of an apocritan hymenopteran is comprised of the pro-, meso-, and metathorax, plus the first abdominal segment, the latter being termed the propodeum. Of course, this nomenclature has often led to confusion among those who lack detailed familiarity with wasp physiology. Nowadays, to avoid ambiguity, it is becoming increasingly common to refer to the middle body region as the mesosoma and the part behind it as the metasoma. Further nomenclatural confusion can arise when, as in the ants and some parasitic wasps, the first metasomal segment (i.e., the second abdominal segment) is greatly reduced. The most conspicuous part of the metasoma is then referred to as the gaster.

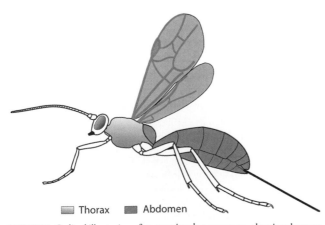

□ Thorax   ■ Abdomen

FIGURE 1 Stylized illustration of an apocritan hymenopteran showing the wasp waist between the first and second abdominal segments. The first abdominal segment, called the propodeum, is broadly attached to the thorax (colored blue), and the combined structure is referred to as the mesosoma. The abdomen (darker shading) posterior to the mesosoma is often called the metasoma.

Venoms

Most if not all Hymenoptera, even the sawflies (which are phytophagous), have a "venom" gland associated with the ovipositor or sting. In fact, most have at least two distinct glands, the venom gland proper, also referred to as the acid gland on account of its typical histological staining properties, and a second gland, referred to as the alkaline gland or Dufour's gland in the Aculeata (see later). Details of the function of the secretions of these glands are known for only a few species; for the sawflies, such knowledge is almost completely lacking. It may reasonably be assumed that the initial function was production of lubricants that assisted passage of the egg down the ovipositor; but natural selection would have acted quickly to favor organisms that showed an ability to benefit from modification of the substrate.

In the parasitic wasps, the venom gland products are thought to be primarily associated with two roles: to help overcome the host's immune defense against the parasitoid's egg or larva, and sometimes, especially among the ectoparasitic idiobionts, to paralyze the host. In some taxa, perhaps in many, there may also be secretory products from parts of the female reproductive tract itself, and these may play important roles in overcoming host immunity. Of great interest among these are the "polydnavirus soups" produced by the calyx gland in a few groups of Ichneumonoidea; this gland is a modified part of the lateral oviduct.

The venom glands in the aculeate wasps are the source of the well-known, pain-inducing toxins that many social and some solitary Hymenoptera use to such good effect in self-defense. One solitary aculeate, a mutillid wasp (or velvet ant), is commonly called the camel-killer because its venom is reputedly strong enough to have that effect, and there are anecdotal reports of soldiers who have been incapacitated by the pain caused by encounters with this substance. Typically, however, the venoms are rather less fearsome. These pain-causing venoms are very specialized and contain a variety of neurotoxins, including, in some taxa, small ringlike peptides that insert themselves in cell membranes and cause depolarization of nerve cells, and consequently pain. It is easy to envisage how these peptides could have evolved from toxins that were originally selected to cause paralysis of the arthropod prey of their ancestors, although probably most are so highly modified that any initial similarity has been lost. A few of the larger parasitic wasps have also developed pain-causing venoms for defense, but their stings are quite mild and the effects short-lived compared with those of many aculeates.

The main function of Dufour's gland seems to be the production of substances that are involved in intraspecific communication. In the parasitic wasps, the gland probably serves primarily as a source of marking pheromones that indicate where an egg has been laid, likely to minimize self- and intraspecific superparasitism. Among the social aculeates, the functions of this gland have been greatly elaborated, and it is the source of many other pheromones that are involved in colony organization.

The Ovipositor: A Key Organ

Since Hymenoptera in general are known to take great care in the placement of their eggs, it is not surprising that the ovipositor is an important organ, and one that has shown many specializations for particular modes of life.

The hymenopteran ovipositor is derived from abdominal appendages and comprises three independently movable parts, called valves, that together form the egg canal. The dorsal valve is a fused structure, but the ventral ones are separate. There is no intrinsic ovipositor musculature; rather, the movements of the valves depend on muscles within the abdomen that pull on the internal apodemes of the three valves. Nevertheless, several parasitic taxa have evolved mechanisms that enable them to steer their ovipositors and thus increase their chances of successfully attacking a mobile host that might otherwise be able to wriggle away from its attacker. Although the penetration of the substrate by wasp ovipositors is usually referred to as "drilling," it is important to realize that there is no circular motion: penetration is achieved by the to-and-fro motion of the three valves relative to one another. In the simplest mode of operation, one valve has a projection or nodus that interlocks with the substrate, and this acts as a support for the others to be pushed forward.

The sawflies get their name from the laterally compressed, strongly serrated, ovipositors with which they insert their eggs under plant cuticle. These ovipositors are unsuited, however, for penetration of wood, and the wood wasps' ovipositors are longer and rounder in cross section, with serrations used for rasping wood fibers, located just at the tip (see Fig. 2). Most of the parasitic Hymenoptera have a similar ovipositor except that in many of those with exposed hosts it is much shorter and has reduced serrations because there is no substrate to "drill" through. Some ovipositors are very

FIGURE 2 A large ichneumonid wasp, *Megarhyssa* sp., using its ovipositor to "drill" through a tree trunk to reach its host, a siricid wood wasp larva. The ovipositor is very thin and pointing between the fore legs; the large black structures are the protective ovipositor sheaths. (Photograph by Nathan Schiff.)

long (up to 12 times longer than the wasp's body), and various mechanisms and behaviors have evolved to enable the wasp to manipulate them.

The Road to Parasitoidism

The evolution of parasitoidism has long been of interest, and several possible scenarios have been discussed at various times. As our understanding of the phylogenetic relationships of the basal parasitic wasps has firmed up over recent years, it now seems most probable that the transition to a parasitoid way of life occurred first among some ancestral wood wasp, because the closest extant sister group of the parasitic Hymenoptera is almost certainly the Orussidae; which collectively are derived from wood wasp ancestors. The most widely discussed and generally accepted proposal for the evolution of parasitism in the Hymenoptera envisages an ancestral wood wasp gaining an advantage by producing a larva that could encounter, kill, and eat another wood-boring insect—possibly the larva of another wood wasp or of a beetle—because such a food item would have a greater nutritional value than the plant diet. An advantage therefore would have been gained if the female ancestral parasitoid were to seek out for oviposition sites branches where such food bonuses occurred. This could occur only if the prey item, such as a batch of eggs or perhaps a minimally mobile prepupa or pupa, did not pose a danger to the ancestral parasitoid. The next step would be from a facultative utilization of vulnerable prey insects to supplement a plant-based diet with an obligate one, eventually eliminating the need to consume plant material. So by evolution of the wood wasps, prey species become hosts. At first, as in extant orussids (see later) the larva might have done the final prey location, but the protoparasitoid in this scenario always evolves to use its ovipositor to injure or kill the prey, perhaps by stabbing. Subsequently, venoms evolved greater sophistication until they were able to induce permanent host paralysis, and even to help keep the host fresher for longer by preventing it from becoming infected with fungi and bacteria.

Gall Formers

As with several other insect orders, some groups of Hymenoptera have evolved ways of making plants produce especially safe and nutritious places (i.e., plant galls) in which to shelter their young. Because hymenopterans produce venoms that contain a wide range of pharmacologically active compounds that affect host insect physiology, it is not very surprising that some of these might affect plants—and the typical plant response to damage is cell proliferation, forming a callus. Thus, venom usage may have preadapted parasitic wasps for the evolution of gall forming (cecidogenesis). We cannot know whether this has always been true, and at least in the extant Cynipidae, the gall wasp family, the phytotoxins that stimulate gall production are produced by the wasp's larva and are not components of the female venom. Gall-forming sawflies, however, do seem to have undergone this preadaptation. True gall wasps (cynipids) have colonized and diversified upon only a few plant groups, notably Rosaceae and Fagaceae, although the other gall-forming Hymenoptera collectively attack a huge range of plants.

One family of chalcidoid wasps, which are also gall formers in a sense, are of particular interest. These are the pollinating fig wasps belonging to the Agaonidae. These insects develop within the ovaries of fig flowers, consuming as larvae the galled tissue. Many people eat figs, but few know that originally at least, the cultivated fig and all its relatives in the genus *Ficus* (Moraceae), relied absolutely on the pollinating activities of these specialized agaonid wasps. Figs are actually not fruit in the strict sense but are syconia—that is, a hollow flasklike structure containing many flowers. The female fig wasps collect pollen from the flowers within the fig in which they have developed, and upon entering another developing fig, actively pollinate the flowers. Figs and their pollinator wasps have evolved in a tight association, and each of the 400 or so species of fig (all members of the genus *Ficus*) has one or, rarely, a couple of pollinating wasp species associated with it that pollinate no other fig species. In these wasps, the males are especially highly modified and do not leave the interior of the fig as adults. Instead they compete with one another for mates and frequently kill competitors with their large mandibles. Males are also responsible for chewing an exit hole through the wall of the fig that enables the female pollinator fig wasps to escape and go in search of new figs.

The Road to Sociality

Perhaps the greatest claim to fame of Hymenoptera is that the order includes several highly social groups of insects,

various bees, yellowjacket (vespid) wasps, and ants. The only other insect order with such a large number of highly social species are the termites (Isoptera). Within the Hymenoptera, sociality has evolved on quite a few separate occasions, and much consideration has been given to the reasons for this circumstance. Probably a variety of factors have contributed.

Provision of a good food source for the larvae first involved selection of a suitable host plant species, but with the evolution of parasitoidism, the degree of selectivity increased. Further selectivity is apparent with the very narrow host ranges of many of the parasitic taxa and, also, their ability to assess the suitability of individual hosts of the right species. This may be viewed as a progression in the degree of individual attention provided to each individual offspring. In several lineages, all in the Aculeata, additional behaviors have evolved to make hosts, in a sense, more suitable—at first these changes consisted of moving a host to a slightly preferable location before oviposition. This is seen in several members of the Bethylidae. Some bethylids also show a degree of parental care in that the female remains with her single brood through their development to guard them against predators and to guard the host against other parasitoids including conspecific females. In the bethylid wasps that is about the limit of brood care, but in a number of groups, notably among the pompilids and sphecids, the female wasp prepares a hideaway in which to cache the insect or spider that will provide her offspring with food. Probably at first, once a host had been identified, the female would locate, dig, or modify a burrow; a further evolutionary step was likely the postponement of burrow construction until the search for a host had begun. This stage required the behavioral sophistication of being able to remember the location of the burrow and to relocate it once a prey had been found. This step was crucial for the evolution of sociality because nest members must be able to locate their nests after foraging expeditions. Another major development allowed by this evolutionary advance was the use of hosts, now usually called prey, that are smaller than what would be necessary for the development of the wasp, because it was now possible to bring back multiple individual hosts for each of the wasp's larvae. This is the stage exhibited by many sphecid wasps. In these, the paralyzed prey are first accumulated until there are sufficient, then an egg is deposited on the cache, and this set is sealed into a cell, after which the female starts collecting more prey for her next egg. Bees and vespid wasps have independently dispensed with sealing their larvae in individual closed cells with all the food that they need; instead, they provide food continuously upon demand—preprepared food, that is, rather than whole prey individuals.

Reproduction and Sex Determination

As far as is known, all hymenopterans have a haplodiploid sex determination system, which means that haploid individuals (having only one set of chromosomes and resulting from unfertilized eggs) are males, whereas diploid individuals (having two copies of each chromosome and resulting from fertilized eggs) are females. This sex determination system is found in a few other groups of organisms, notably in thrips and some rotifers. Development of unfertilized eggs into males is a form of parthenogenesis and is termed arrhenotoky.

There is certainly more than one sex determination mechanism even within this haplodiploid system, and this has important consequences in matters such as biological control. The best understood (or surmised) mechanism, called complementary sex determination, is characterized by the occasional occurrence of diploid (but infertile) males and by the tendency of the proportion of these to increase with inbreeding. Culturing species with this sex determination mechanism is difficult because small population sizes result in the gradual loss of sex alleles and so an increase in frequency of diploid males and loss of colony vigor. In the rearing of insects for biological control programs, the appearance of an abnormally high number of diploid males can be a very serious setback, because when colonies become inbred, they go extinct.

COMPLEMENTARY SEX DETERMINATION (CSD)

Although the exact molecular details are unresolved, there is a reasonable hypothesis that CSD may involve polymeric proteins, the most simple of which are dimers. A heterodimers (i.e., an association of two different protein chains) has a different shape from a homodimer (two identical protein chains), and this shape difference determines the sex of the offspring. Because haploid individuals can make only one form of the protein (they only have one gene locus), they must make the homodimeric form, and this means that they will develop as males. It seems that in natural populations there are typically quite a few sex alleles (roughly between 6 and 50), with the result that the proportion of fertilized eggs that contain two copies of the same allele is rather small. Thus the proportion of eggs that either fail to develop or produce infertile diploid males would be expected to be small, as well.

NONCOMPLEMENTARY SEX DETERMINATION

Not all Hymenoptera can possess CSD, as is evidenced by the routine inbreeding that occurs, for example, in some parasitic wasps. In these species, a female lays on or in a single host an often large brood of eggs consisting mostly of daughters (from fertilized eggs) and a single haploid male (or at least a very low number of sons), which fertilizes all his sisters when they emerge. This goes on for many generations and undoubtedly must lead to increased homozygosity, but these wasps show no progressive change in sex ratio or fecundity as CSD would necessarily cause. However, what sex determination system is involved in these insects is not known, and while a gene dosage mechanism is widely postulated and seems highly likely, it is not proven.

RELATEDNESS AND MATING

A particular consequence of haplodiploidy that has been invoked as a major reason for the multiple independent evolutions of sociality within the Hymenoptera is that it leads to a change in the normal degrees of relatedness between a mother and her sons and daughters and between siblings, and this difference is most pronounced if the female has mated only once. The reasons are as follows. A male offspring gets all his genes from his mother (because he is haploid and so has no male parent), but he gets only half of the mother's chromosomes; in the Hymenoptera, therefore, a son is 50% related to the mother just as in mammals. However, if the female has mated only once, her daughters, which come from fertilized eggs, contain half her set of chromosomes plus 100% of those from the male parent (because he is haploid and all his sperm are identical). Overall, therefore, each daughter is 75% related to each other daughter, whereas daughters are related to their mother only 50%. The argument regarding sociality is that because sisters in Hymenoptera are more closely related to each other than they are to any potential offspring they could have, their fitness will be better enhanced by helping their mother to produce more sisters than by reproducing themselves.

The vast majority of species of social ants, bees, and wasps only mate once. Thus the foregoing arguments may generally hold. However, there are exceptions, and honey bee queens typically mate about a dozen times. Thus for these insects the disparity in relatedness between offspring and sisters is much closer to the 50:50 ratio of normal diploid taxa.

PHYLOGENY, CLASSIFICATION, AND WHAT PARTICULAR GROUPS DO

The Hymenoptera are member of the monophyletic group of insects known familiarly as the Holometabola, and it is among these therefore that their relationships must be sought. The search for such relationships, however, has thus far proved to be rather difficult, and despite a growing body of molecular data, a consensus about the exact relationships of the order has yet to emerge. Weak evidence has been put forward to suggest a relationship with the Mecoptera, but the characters involved are liable to homoplasy. One possibility is that the Hymenoptera form a sister group to the whole of the rest of the Holometabola (or at least, of the extant holometabolan orders).

Classification

For a long time the Hymenoptera have been broadly divided into three groups: the sawflies or "Symphyta," the aculeates, or stinging and social wasps, bees, and ants and their close allies, and the remainder, which are typically referred to as the parasitic wasps or "Parasitica," even though many of them are not parasitic at all. It has long been realized that this is an unnatural arrangement, and slowly an attitude more consistent with our understanding of hymenopteran phylogeny has been filtering into the literature. Thus, although it is still useful to be able to refer to sawflies as a group, the use of a formal classificatory term for them to the exclusion of all other Hymenoptera (i.e., Symphyta) is rather unsatisfactory. This is because it has long been known that the "Symphyta" are a basal grade that leads to the wasp-waisted Hymenoptera; "Symphyta" are therefore not a monophyletic group (i.e., all the descendants of its single basal species are not included); rather, they are a paraphyletic one, and so from a cladistic standpoint, should be recognized as such. The situation, though, is complicated because most languages, lack a vernacular term for all the sawflies exclusive of the wasp-waisted taxa. Thus, use of informal terms, typically indicated as such by the use of quotation marks, is seen more and more. Terms like "Symphyta" indicate that the name is being used as a handle of convenience and should not be taken as a reference to a monophyletic group.

The wasp-waisted Hymenoptera have also been traditionally classified into two groups, the "Aculeata" and the "Parasitica," the former group being largely (though not entirely) distinguished from the latter by the loss of use of the ovipositor for egg laying, being used instead for stinging prey and/or potential enemies. As with the "Symphyta," the "Parasitica" are no longer recognized as a formal group because there is a reasonable consensus that the aculeates are derived from within them, so rendering them paraphyletic. The Aculeata are strongly supported as monophyletic, at least on morphological grounds.

Sawflies and Wood Wasps

The basal representatives of the order from within which the parasitic and social families of wasps have evolved are the sawflies and wood wasps. There is little doubt that the most ancient extant family of the Hymenoptera is the Xyelidae, represented today by two subfamilies, one whose larvae feed on gymnosperms (either pollen of male cones or on buds or young shoots), and the other having rather caterpillar-like larvae that feed exposed on the leaves of elms and walnut species. Xyelid sawflies possess distinctive antennae (Fig. 3) with the first section of the flagellum very enlarged (possibly as a result of the fusion of multiple segments). Fossil xyelids date from the middle or late Triassic, some 200 mya, and are the earliest known fossil Hymenoptera.

Three other superfamilies of typical sawflies are recognized: the Pamphiloidea, which are relatively uncommon (though they include a few pest species), the Cephoidea or stem sawflies, which are rather slender and feed endophytically in grasses and a few woody plant stems, and the very large and speciose Tenthredinoidea. All these are predominantly Holarctic in distribution, although it is likely that at least the tenthredinoids have moderate species diversity in the tropics, even though they are typically quite uncommon there. Some tenthredinoids form galls (e.g., the genera *Euura* and *Pontania*), and it appears that secretions from the female sawfly contain the cecidogenic compounds. A few are interesting because they show a degree of parental care (Fig. 4), and in some Australian pergids, the caterpillar-like larvae form

FIGURE 3 *Angaridyela vitimica*, a typical xyelid sawfly from the Lower Cretaceous of Baissa in Siberia (body 10.5 mm long). Note the enlarged third antennal segment. (Photograph by Alexandr P. Rasnitsyn, Paleontological Institute, Russian Academy of Sciences.)

resting aggregations during the day but disperse over the food plant at night to feed on the leaves, apparently communicating by means of vibrations.

The wood wasps were recognized as a group of three superfamilies, the Siricoidea, Xiphydroidea, and Anaxyeloidea (by Vilhelmsen in 2001), although the greater part of the literature on wasps refers to them simply as Siricoidea. This division is intended to reflect better the phylogeny of the group (Fig. 5), which forms a grade rather than a monophyletic group. Their endoxylous larvae are typically associated with fungus-infected wood, and the adults are responsible for transporting these fungi to their host trees in a special mycangial pouch that is associated with the reproductive system and lies near the base of the ovipositor. In addition to

FIGURE 4 A female pergid sawfly guarding her brood of first instars. (Photograph by Nathan Schiff.)

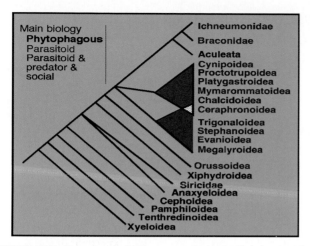

Main biology
Phytophagous
Parasitoid
Parasitoid &
predator &
social

Ichneumonidae
Braconidae
Aculeata
Cynipoidea
Proctotrupoidea
Platygastroidea
Mymarommatoidea
Chalcidoidea
Ceraphronoidea
Trigonaloidea
Stephanoidea
Evanioidea
Megalyroidea
Orussoidea
Xiphydroidea
Siricidae
Anaxyeloidea
Cephoidea
Pamphiloidea
Tenthredinoidea
Xyeloidea

FIGURE 5 Working scheme of Hymenoptera higher level phylogeny. The relationships among the basal, principally phytophagous sawflies and wood wasps are becoming well established, but only a few broad groupings are widely accepted among the Apocrita.

these fungi, at least in some siricid species, components of the secretions injected into the tree at the time of oviposition are actually phytotoxic and may be important in helping their fungi overcome the tree's defenses. Some siricids are important pests of conifer plantations.

There is a very large body of evidence that the Orussidae form the sister group of all the wasp-waisted or apocritan Hymenoptera and, therefore, its biology is of such interest that the family warrants some separate discussion, small though it is. There have been few detailed biological studies on the group, but it is clear that some, probably all, are parasitic on wood-boring insect larvae, although in at least one species the final instar continues to feed on and in its host when it is fairly well decayed. The orussids cannot be considered to represent the ancestral condition of the parasitic wasps as a whole because they show many very derived features not found in any other Hymenoptera. It is tempting, however, to consider at least that their biology is closely similar to and perhaps not modified much from the earliest parasitic wasps. Recent studies have shown that they use vibrational sounding (i.e., echolocation through a solid substrate) to detect their host boring. The females' antennal apices are massive and solid and are used to tap the substrate, and their foretibiae contain massive subgenual organs that are used to detect the vibrations that are transmitted through the wood. This form of host location has evolved on several independent occasions within the order, but the Orussidae are particularly interesting because the egg may be laid into a boring some distance from the host (in the studied species a beetle pupa), and it is the first instar orussid that seeks out the host.

Apocrita or Wasp-Waisted Hymenoptera

In terms of numbers of species, the Apocrita is dominated by parasitoids, but the very great biological developments

leading to the evolution of sociality in bees, yellowjacket wasps, and ants has had the effect of polarizing study and discussion, and, in the past, also classification. It has long been appreciated that the aculeate Hymenoptera, which include the social taxa, have evolved from within the Apocrita (in fact, they seem to be most closely related to one particular superfamily, the Ichneumonoidea).

THE "PARASITICA" The "Parasitica" is a paraphyletic group (with respect to the aculeates) that comprises some 11 or 12 superfamilies, the exact number being a little unstable because rapid advances in phylogenetic studies are tending to suggest that some previously recognized taxa are paraphyletic or polyphyletic. Consequently, the number is likely to increase by a few when new evidence leads to robust and better resolved phylogenetic trees. A rather conservative phylogenetic hypothesis is shown in Fig. 5, but even this is problematic. Areas of particular uncertainty are the monophyly of the Proctotrupoidea (even after the Platygastroidea, once included therein, have been removed to a separate group), and of the evaniomorph superfamilies (viz., Megalyroidea, Evanioidea, Trigonaloidea, Ceraphronoidea, and Stephanoidea). Even within this grouping it is not yet totally certain that the three families constituting the Evanioidea form a monophyletic group; they have very different biologies, and no unique synapomorphies have been found to unite the component families.

The vast majority of known and undescribed species are parasitoids of other insects. The term "parasitoid" is used almost universally nowadays to distinguish the interactions of these insects from those of parasites, although purely for reasons of euphony, we still often refer to them as parasitic wasps rather than parasitoid wasps. Whereas true parasites live off the living bodies of their hosts, they seldom kill them; indeed, it would usually be maladaptive for them to do so because the longer the host lives, generally the greater will be the reproductive opportunity of the parasite. Parasitoids, however, always kill their host and treat it as a single meal that will provide all the food necessary for their own development. Once the parasitoid wasp has eaten all of the host that it needs, the host is no longer of use, and so there is no need to leave it to recover. In this respect, parasitoids are rather akin to predators but, unlike predators, they require only a single host (i.e., prey) individual to provide all their needs. (Predators, on the other hand, eat multiple, often very many, prey during their life span.) Even so, a few groups of "parasitic" wasps actually behave more like predators. These include species that attack egg masses of, for example, spiders, and species whose larvae will eat not one egg but many or all in the spider's batch.

IDIOBIONTS AND KOINOBIONTS Almost all parasitoids can be classified into one of two classes defined by whether their hosts continue developing after parasitization (the koinobiont strategy) or whether further host development is curtailed at that time (the idiobiont strategy). These strategies explain many other aspects of the wasp's biology, such as longevity, egg development, egg size, and fecundity, and host range, as discussed in greater detail elsewhere in this volume and in works by Godfray and Quicke. There has been general agreement that the first parasitic wasps were probably idiobionts. However, it is far from proven that this means that within any given parasitic wasp family or superfamily, the idiobiont taxa are necessarily ancestral. Indeed, the most parsimonious explanation for the distribution of parasitoid life history strategies on several of the most widely cited phylogenies suggests that endoparasitoids have evolved into ectoparasitoids on many occasions within the Chalcidoidea, though probably not within the Ichneumonoidea.

Aculeate Hymenoptera

The very familiar, often social, ants (Formicidae), bees (Apidae or Apiinae depending on classification system), and yellowjacket wasps (Vespidae) have rather dominated the traditional classifications of Hymenoptera. These taxa are united with a number of other, rather less familiar ones to form a monophyletic group called the Aculeata. Most members of this clade have derived biologies, though some families are still functionally parasitoids. In these, however, the ovipositor proper is not used for reaching hosts, and only in a few (possibly derived) cases is it actually used for passing the egg. Even among the essentially parasitic species there is a strong tendency toward physically manipulating hosts. Many spider wasps (Pompilidae), for example, drag paralyzed hosts to a hiding place before ovipositing on them, sometimes biting off the spiders' legs to facilitate handling and carrying. The least-derived biology is to use a preexisting cavity as a hiding place, and subsequent evolution has led to many levels of modification or *de novo* nest construction.

SOCIAL KLEPTOPARASITES AND SLAVE MAKING   The great resources afforded by social insect nests have attracted the evolution of numerous thieves, some of which are other hymenopterans, often closely related to the species being robbed. The most interesting of these interactions are exhibited by social kleptoparasites when a female of a nonsocial species replaces the queen of a social one and makes use of the workers in the host colony to rear her own brood, rather than relying on more of their own siblings. This ability obviously depends on sophisticated mimicry, not just visual, but more particularly chemical and tactile, or otherwise the usurped workers would detect the intruder. Social kleptoparasitic taxa have evolved independently on numerous occasions and are known among bees (Apidae), vespid wasps, and ants. "Slave making," another form of social parasitism, is known only among the ants, but within this family it has been evolved by several different lineages. The workers of one species are abducted by the slave-making species and forced to forage for the latter.

RELATIONSHIPS TO MAN

Beneficial Species

The number of beneficial hymenopterans greatly outweighs the number of harmful ones, though some taxa fall into both categories depending on circumstance. Even the vespid wasps (yellowjackets), which are well-known nuisances at picnics and barbecues, and occasionally have serious medical consequences, are responsible for eating a very large number of other insects (their larvae are fed almost entirely on chewed-up insect muscle), and in agricultural settings undoubtedly devour many pest insects. Humans have made use of the voracious insect-eating capacities of some ants for many years. For example, Chinese citrus growers have traditionally transferred tree ants, *Oecophylla* spp., into their orange groves to consume potential pests, and central European foresters have had considerable success in controlling pest outbreaks by transporting into forests the nests of wood ants, *Formica rufa*.

Biological control programs have made a great deal of use of parasitic Hymenoptera to control host pest populations, sometimes with spectacular results, and when such controls work, the cost–benefit ratio is very favorable. It is becoming increasingly possible to use commercially produced parasitic wasps to control pests in private gardens and greenhouses as well as on large commercial enterprises, and this has obvious desirable features such as reducing the need to apply pesticides.

Bees and Honey

Probably best known of the beneficial affects of Hymenoptera is the production of honey, which people have valued as a sweetener since prehistory. Originally various honey-producing wild bee nests would have been harvested, and indeed are still harvested by indigenous peoples on several continents. However, one species, the European hive bee or honey bee, *Apis mellifera,* was found to be both productive and manageable, and it was effectively domesticated by getting it to nest in artificial hives several thousand years ago. The first records of beekeeping come from ancient Egyptian wall paintings, some 2500 years B.C. Bees also produce wax from which their larval cells are formed, and this is also widely used for a variety of purposes, from candles to cosmetics and from pharmaceuticals to polishes.

Although most people think of *A. mellifera* when they talk of bees, there are in fact more than 25,000 bee species in the world, the majority of which are solitary rather than social. Collectively, bees are of immense economic importance because of their major role in plant pollination, which includes the pollination of a large number of crop species, and this service is far more valuable than the production of honey and beeswax. In addition to the economic importance of bees, most ecosystems rely on them for pollination, largely because bees have evolved, diversified, and specialized together with the angiosperm plants they pollinate, and it cannot be doubted that loss of many bee species would have a long-term dramatic impact on the floral composition of many habitats. Interest has been increasing of late in non-*Apis* bees because of the realization that they may be excellent pollinators of many crops, sometimes better than honey bees, and that changes in habitat use have led to a decline in many of these useful species. Additionally, *Varoa,* a parasitic mite that has spread from its natural eastern honey bee host, *A. cerana,* onto the common honey bee, has resulted in serious losses of the latter and thus in pollination rates.

All bees feed their larvae on pollen that is harvested and processed by the adult females (worker bees among the highly eusocial taxa). Typically the pollen is mixed with nectar, although plant oils are sometimes used. The honey made, for example, by *Apis* is not a food for the bee larvae at all, but rather a high-energy food source that is produced and stored by the social bees for times when their normal source of energy food, nectar, is in short supply or simply unavailable—as in temperate winters.

Pests and Medical Importance

Considering the great abundance and species diversity of the Hymenoptera, it could easily be argued that the order includes very few pests. There is general awareness that many bees and most social wasps are capable of delivering a quite painful sting, but for most people these stings are relatively minor, if unpleasant, quickly forgotten occurrences. In the United States approximately 50 people die each year as a result of hymenopteran stings. To put this statistic into perspective, about 25 people are killed each year by lightning in the United States.

The major cause of death from hymenopteran sting is respiratory block, sometimes as a result of a sting in the inside the throat as a consequence of inhaling a wasp or bee, but also often because of more systemic reactions. Approximately 2% of the population are hypersensitive to Hymenoptera stings, but although such sensitization puts the victims at a greater risk, only about one death per year in the United States is attributable to a hyperallergenic response.

Some sawflies and gall wasps are economically important pests of crop and ornamental plants, but many of these have potential roles in the biological control of weeds as well. Notable pest sawflies include both external foliage feeders such as the pine sawflies (Diprionidae), which can be major defoliators of coniferous forests, and concealed feeders such as the wheat stem sawfly, *Cephus cinctus* (Cephidae) and the siricid wood wasp, *Sirex noctilio,* with the latter having caused considerable harm to pine *(Pinus radiata)* plantations in Australia and New Zealand.

Probably the largest number of pest taxa are to be found among the ants. There are a number of introduced taxa in many parts of the world that cause considerable damage to crops, are harmful to livestock, and have nuisance value to humans. Three of many possible examples will serve to illustrate the range of harmful interactions that can occur.

The tiny pharaoh ant, *Monomorium pharaonis,* a tropical species widely introduced into buildings in the Northern Hemisphere, often gets into hospitals, where it is a potential vector of bacteria—it has the habit of getting into almost anything in its foraging, including under bandages. The Argentine ant, *Linepithema humile,* is a highly invasive species that fights and outcompetes other ants in introduced regions, and its huge colonies cause damage in agricultural situations because they protect aphids and other honeydew-forming pest insects. Third, the red imported fire ant, *Solenopsis invicta,* which has spread alarmingly in the southern United States in recent years, is very aggressive, especially when its mounds are disturbed. Fire ants will sting people, pets, and livestock many times, with resulting pain, blisters, and even systemic reactions, and small pets are sometimes killed by them. Interestingly, many of these pests are far less harmful in their native regions.

IDENTIFICATION

As might be expected with such a large insect order, identification of hymenopteran species, but also genera and even families, is not without its difficulties. Even in parts of the world where the insect fauna is quite well known, such as Europe and North America, there will be many groups for which there are no satisfactory identification keys. In less well known parts of the world, almost any reasonably sized sample will contain numerous undescribed species and within some families even genera. The presence of so many unclassified hymenopterans reflects a combination of innate taxonomic difficulties such as the insects' small size, the large number of similar species, and often a great deal of superficial convergence. In addition, because the order, with few exceptions, has not attracted a great deal of amateur attention, relatively little work has been done on it. Recent years have, however, seen vast improvements to the situation. Several well-illustrated keys to all families have been published, as well as several major works on the more popular aculeates. Particularly useful works are by Gauld and Bolton and Goulet and Huber.

See Also the Following Articles

Ants • Apis Species • Division of Labor • Sex Determination • Sociality • Venom • Wasps

Further Reading

Askew, R. R. (1971). "Parasitic Insects." Heinemann Educational, London.
Askew, R. R. (1984). The biology of gall wasps. *In* "Biology of Gall Insects" (T. N. Ananthakrishnan, ed.), pp. 223–271. Edward Arnold, London.
Cook, J. M. (1993). Sex determination in the Hymenoptera—A review of models and evidence. *Heredity* **71**, 421–435.
Gauld, I. D., and Bolton, B. (1988). "The Hymenoptera." Oxford University Press, Oxford, U.K.
Godfray, H. C. J. (1993) "Parasitoids: Behavioral and Evolutionary Ecology." Princeton University Press, Princeton, NJ.
Goulet, H., and Huber, J. T. (eds.). (1993). "Hymenoptera of the World: An Identification Guide to Families." Agriculture Canada.

Greathead, D. J. (1986). Parasitoids in classical biological control. *In* "Insect Parasitoids." (J. Waage, and D. Greathead, eds.). Academic Press, London.

Hölldobler, B., and Wilson, E. O. (1990). "The Ants." Belknap Press of Harvard University Press, Cambridge, MA.

O'Toole, C., and Raw, A. (1991). "Bees of the World." Blandford, London.

Quicke, D. L. J. (1997). "Parasitic Wasps." Chapman & Hall, London.

Rasnitsyn, A. P., and Quicke, D. L. J. (eds.) (2002). "The History of Insects." Kluwer, Dordrecht, the Netherlands.

Ronquist, F., Rasnitsyn, A. P., Roy, A., Eriksson, K., and Lindgren, M. (1999). Phylogeny of the Hymenoptera: A cladistic reanalysis of Rasnitsyn's (1988) data. *Zool. Scripta* **28**, 13–50.

Thompson, G. H. (1960). "The Alder Woodwasp and Its Insect Enemies." Film rereleased as video by World Educational Films, http://www.worldeducationalfilms.com.

Vilhelmsen, L. (2001). Phylogeny and classification of the extant basal lineages of the Hymenoptera (Insecta). *Zool. J. Linnaean Soc. (Lond.)* **131**, 393–442.

Wilson, E. O. (1971) "The Insect Societies." Belknap Press of Harvard University Press, Cambridge, MA.

Hypermetamorphosis

John D. Pinto

University of California, Riverside

Hypermetamorphosis is a form of complete insect metamorphosis or holometaboly in which at least one of the instars in the life cycle differs considerably from the others. The term heteromorphosis, preferred by some entomologists, carries a degree of ambiguity in that it also refers to the relatively minor differences characterizing consecutive instars in virtually all insects, as well as to the phenomenon of organ replacement following mutilation. Hypermetamorphosis is most common in parasitoids, where it usually is the first instar that deviates structurally and behaviorally from the others. In some groups one or more of the subsequent instars also are distinctive. In the same way that holometabolous development allows a division of function between the larva and adult insect, hypermetamorphosis can be viewed as an exaggerated form of holometaboly characterized by additional division of function within the larval stage.

GENERAL CHARACTERISTICS AND TERMINOLOGY

Two broad categories of hypermetamorphosis can be recognized in insects. In the most widespread form, there is a decoupling of oviposition site and the larval food; in the other, the oviposition and larval feeding sites are identical. For convenience, these can be referred to as type I and type II hypermetamorphosis, respectively.

Type I adult females do not oviposit directly at the larval feeding site; instead, the first instars must find the food source. Such larvae are active, slender, and well sclerotized

(Fig. 1a); they are further characterized by their ability to exist for a considerable time without nourishment and without becoming desiccated. Depending on the group, the first instar may attain the feeding site by direct searching, or indirectly by phoresy, in which it is carried to its food usually by the host itself. Once on the larval food, the first instar begins feeding, and subsequently molts into a grublike and less mobile larva (Fig. 1b). Phoretic larvae are commonly equipped with elongate caudal cerci and a terminal suction process (or pygopod), which allows them to stand erect, thus facilitating contact with passing hosts. Although comparative studies are few, hypermetamorphic taxa seem to be more fecund than nonhypermetamorphic relatives. This apparent difference is attributed to the reduced likelihood of larvae finding suitable hosts. Two general characteristics of adult

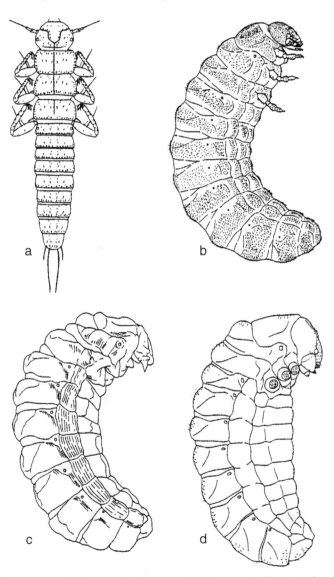

FIGURE 1 Larvae of *Meloe dianellus* (Coleoptera: Meloidae) illustrating the four types of meloid larvae: (a) first instar (planidium with legs), (b) first grub (fifth instar), (c) coarctate (sixth instar), and (d) second grub (seventh instar). Natural lengths: a, 2 mm; b–d, 7–10 mm. [From Pinto, J., and Selander, R. (1970). Illinois Biological Monographs, No. 42.]

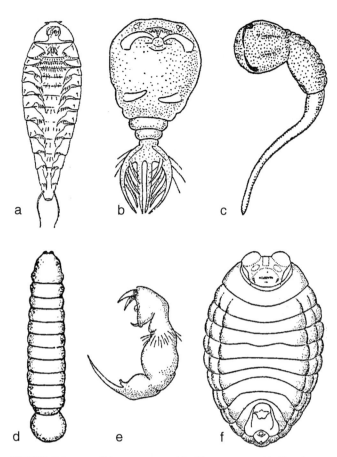

FIGURE 2 Larvae of hypermetamorphic Hymenoptera. (a) first instar *Perilampus hyalinus* (planidium without legs) [from Smith, H. (1912) *U. S. Bur. Ent. Tech. Ser.,* Bull. 19]; (b) first instar *Platygaster* sp. (cyclopiform) [from Kulagin, N. (1898), *Zeitschr. Wiss. Zool.* **63**, 195–235]; (c) first instar *Aridelus* sp. (mandibulate and caudate) [from Kirkpatrick, T. (1937), *Trans. Roy. Ent. Soc. London* **86**, 247–343]; (d) second instar *Apanteles* sp. (vesiculate) [from Allen, W. (1958) *Hilgardia* **27**, 1–42]; (e) first instar *Hadronotus ajax* (teleaform); and (f) third instar, *H. ajax* [from Schell, S. (1943) *Ann. Ent. Soc. Amer.* **36**, 625–635].

females with this form of development are the absence of a well-developed ovipositor and the laying of eggs in masses rather than one at a time. Type I hypermetamorphosis is found in the Strepsiptera (all groups), Neuroptera (Mantispidae), Coleoptera (several families), Diptera (Nemestrinidae, Acroceridae, most Bombyliidae, some Tachinidae), Hymenoptera (Perilampidae, Eucharitidae, some Ichneumonidae), and Lepidoptera (Epipyropidae).

Although there is some inconsistency in usage, the active first instar in type I hypermetamorphic taxa has generally been referred to as a *triungulin* or triungulinid if it has legs (Neuroptera, Coleoptera, Strepsiptera, Lepidoptera; Fig. 1a), or as a *planidium* if it is legless (Diptera, Hymenoptera; Fig. 2a). However, triungulin is an inappropriate term for most of the groups it is applied to. Usage stems from the fact that the first instar of *Meloe* (Coleoptera: Meloidae), the first hypermetamorphic group studied, has trident-shaped claws. Such claw structure does not occur in other hypermetamorphic families

and is uncommon in the Meloidae itself. Thus, planidium, signifying "little wanderer," although usually restricted to the larvae of Hymenoptera and Diptera, is perhaps more appropriately applied to all type I first instars. This usage is followed here and was by adopted R. E. Snodgrass in his 1954 review of insect metamorphosis.

In type II hypermetamorphosis the oviposition and larval feeding sites do not differ. In these insects the first instar (sometimes the second as well) has a distinctive morphology (Fig. 2b–e) compared with the relatively simple and often amorphous subsequent instars (Fig. 2f), but these modifictions are associated with functions other than host finding. The adaptive significance attributed to the often bizarre larvae of type II hypermetamorphic groups includes locomotion, protection and combat, ingestion, and respiration. This type of development occurs in several families of parasitic Hymenoptera and in the dipteran family Cryptochetidae *(Cryptochetum)*. Correlated with the greater possibilities of larval function, the morphology of type II first instars is considerably more variable than that found in type I taxa.

Reviews by Clausen and Hagen identified 12 structural forms of first instars in the parasitic Hymenoptera with type II hypermetamorphosis. The generalized type in nonhypermetamorphic species (and in hypermetamorphic species as well after the first instar) is referred to as hymenopteriform. It is ovoid or fusiform and relatively featureless (Fig. 2f). Examples of deviations in hypermetabolic groups include the following. The caudate type has a taillike prolongation of the terminal segment and is found in many Ichneumonoidea and Chalcidoidea (Fig. 2c). It is believed to be an adaptation for locomotion within the host and/or food absorption. The mandibulate and cyclopiform types have enlarged falcate mandibles and are found in several groups (e.g., Ichneumonoidea, Diapriidae, Platygasteridae) (Fig. 2b,c). They are associated with feeding as well as combat in cases of multiple or superparasitism. The vesiculate larva (several Ichneumonoidea) is similar to the hymenopteriform type, except its hindgut is evaginated to form an external vesicle to aid in respiration within the host (Fig. 2d). Several of these larval types are characteristic of taxonomic groups. For example, the cyclopiform larva (Fig. 2b) characterizes the Platygasteridae, and the teleaform larva (Fig. 2e) is typical of Scelionidae.

Hypermetamorphic development not falling into types I or II occurs in the Hydroptilidae in the order Trichoptera (caddisflies), a group with aquatic larvae. There are modifications in shape, setation, sclerotization, and leg development in the fifth instar not found in the four preceding instars. The first four instars of hydroptilids are free living, whereas the fifth lives within a case constructed of varying materials.

The term hypermetamorphosis generally is applied only to the Holometabola. However, the validity of the practice of excluding hemimetabolous groups belonging to the Sterrnorhyncha (Homoptera), namely, the Aleyrodidae (whiteflies) and Coccoidea (scale insects), is questionable.

Although these groups are phytophagous they also have instars with considerable differences in structure and function. The first instar, or crawler, is responsible for locating the feeding site and for dispersal. The following instars are sessile and modified for a sedentary existence.

Type I hypermetamorphosis with its planidium larva is widespread in insects and is an excellent example of convergence in evolution. It is convenient to summarize the occurrence of this form of development by surveying the groups where it is known to occur.

SURVEY OF INSECT TAXA DISPLAYING TYPE I HYPERMETAMORPHOSIS

Neuroptera

Only the Mantispidae have true hypermetamorphosis. Most of the species feed on spider eggs associated with a single egg sac. The first instar either enters a previously constructed egg sac or attaches onto a female spider and enters the sac as she constructs it. Other species feed on the larvae of various aculeate Hymenoptera. Several of these are phoretic and reach the food source by attaching to the adult bee or wasp.

Lepidoptera

The only lepidopteran family with hypermetamorphosis is the Epipyropidae. This group is parasitic on various Homoptera, a unique association for the generally phytophagous Lepidoptera. The active first instar seeks out a host and the grublike instars that follow occur on the body of a single homopteran.

Coleoptera

Hypermetamorphosis occurs in several families of beetles. Most are parasitoids of other insects. The larvae of Bothrideridae, and a few Carabidae and Staphylinidae, attack the larvae or pupae of other coleopterans or dipterans. The genus *Sandalus* (Rhipiceridae) feeds on cicada nymphs. All are hypermetamorphic to some degree, with active and relatively long-legged planidia followed by short-legged, grublike feeding instars.

Hypermetamorphosis perhaps is best known in the Meloidae and Rhipiphoridae. Most meloids parasitize grasshopper eggs or, more commonly, the larvae and provisions of soil- or wood-nesting bees. The planidia of several groups are phoretic on adults of their hymenopteran host. There are four distinctive larval types in the typical meloid life cycle: planidium, first grub, coarctate, and second grub (Fig. 1). The planidium (Fig. 1a) encounters the food source. The first grub (Fig. 1b), consisting of four instars, is the primary feeding stage. The coarctate (Fig. 1c), a quiescent instar, is adapted for overwintering or aestivation. The second grub (Fig. 1d) is a nonfeeding instar that precedes pupation. Each has a distinct phenotype, the most unusual

being the immobile coarctate with its thick, highly sclerotized cuticle, aborted appendages, closed mouth and anus, and vestigial musculature.

The Rhipiphoridae include parasitoids of immature Hymenoptera (Rhipiphorinae) and Blattodea (Rhipidiinae). The planidia of Rhipiphorinae are phoretic on adults of their host. After being carried to the host nesting cells, they burrow into the body of the host larva, eventually molting into a grub that emerges to feed externally. The planidia of Rhipidiinae attach directly to their cockroach host and eventually molt into a legless and amorphous larva, which enters the host to feed. The last instar regains poorly developed legs, exits the host, and moves away for pupation.

Among nonparasitoid groups of Coleoptera, hypermetamorphosis is known to occur in one genus of Eucnemidae (*Rhacopus*) and in the Micromalthidae. The larvae of both groups feed in decaying wood and have a planidial-like first instar. The single species of Micromalthidae, *Micromalthus debilis,* has perhaps the most complicated life cycle known in insects, with several distinct morphological and reproductive types of larvae. Hypermetamorphosis has also been reported in the Megalopodidae.

Strepsiptera

The Strepsiptera parasitize Thysanura, Orthoptera, Hemiptera, Diptera, and Hymenoptera. The free-living planidium encounters the host. If parasitic on a hemimetabolous host, the larva may immediately penetrate its body and develop. Those parasitizing Hymenoptera are phoretic and are carried to the nesting site, where the immature stages are attacked. The planidium molts into an endoparasitic legless grublike larva, which may have several instars. This secondary larva lacks mouthparts and feeds by diffusion through the cuticle. Male Strepsiptera are free living; females are neotenic and most remain on their host.

Diptera

Hypermetamorphosis is known in the Acroceridae, Nemestrinidae, most Bombyliidae, and some Tachinidae. All have legless, well-sclerotized, and active planida that search for the host, followed by soft-bodied maggotlike larvae. The Acroceridae are internal parasitoids of spiders. The Nemestrinidae are endoparasites of grasshoppers and beetles. The Bombyliidae are parasitoids of immature Lepidoptera, Hymenoptera, Coleoptera, Neuroptera, and Diptera; some prey on grasshopper eggs. They are either endo- or ectoparasitic.

Hymenoptera

A legless planidial larva (Fig. 2a) is known in three families of Hymenoptera: Perilampidae, Eucharitidae, and Ichneumonidae (*Euceros*). It is followed by a soft-bodied, relatively immobile larva. The Eucharitidae are endoparasitic on mature larvae or pupae of ants and are phoretic. Eggs are laid on vegetation. The planidia of most groups attach to worker ants and are carried to the nest, where they transfer to larvae for feeding. The Perilampidae is a closely related family, similar bionomically to the Eucharitidae but known to parasitize several orders of insects. In the Ichneumonidae, the planidium of *E. frigidus* attaches to the integument of sawfly larvae and eventually transfers to feed on the larva of other ichneumonid species that are primary parasites of the sawfly.

See Also the Following Articles

Host Seeking, by Parasitoids • Larva • Metamorphosis • Phoresy

Further Reading

Clausen, C. P. (1940). "Entomophagous Insects." McGraw-Hill, New York.

Clausen, C. P. (1976). Phoresy among entomophagous insects. *Annu. Rev. Entomol.* **21**, 343–368.

Crowson, R. A. (1981). "The Biology of Coleoptera." Academic Press, London.

Hagen, K. S. (1964). Developmental stages of parasites. *In* "Biological Control of Insect Pests and Weeds" (P. H. DeBach and E. Schlinger, eds.), pp. 168–246. Chapman & Hall, London.

Snodgrass, R. E. (1954). "Insect Metamorphosis." Smithsonian Miscellaneous Collections, 122 (9). Smithsonian Institution, Washington, DC.

Stehr, F. W. (ed.) (1991). "Immature Insects," Vol. 2 (see Chaps. 33–35, 37). Kendall Hunt, Dubuque, IA.

Hyperparasitism

Daniel J. Sullivan

Fordham University

Hyperparasitism is a highly evolved behavior in the Hymenoptera and in a few species of Diptera and Coleoptera, in which an adult hyperparasitoid (or secondary parasitoid) oviposits on or in a primary parasitoid host that has attacked another (usually herbivorous) insect species. The larval offspring of the hyperparasitoid cause the death of the primary parasitoid. Ecologists emphasize this interaction as a food-web "community." This article focuses on the hymenopteran microwasps in which hyperparasitism occurs.

There are a variety of behaviors by hyperparasitoids depending on the species of secondary and primary microwasp parasitoids which in turn are influenced by the species of phytophagous host, often an insect pest. In addition, there is an economic interest in hyperparasitism because if primary parasitoids are considered to be beneficial insects when used in biological control programs, it would seem that hyperparasitoids that attack primary parasitoids would be detrimental. However, hyperparasitoids may play a positive role by preventing extreme oscillations of the primary parasitoids that might reduce the numbers of the phytophagous host enough to cause the local elimination of both the insect pest and the beneficial primary parasitoid.

EVOLUTION

Hyperparasitism has evolved in only three insect orders: in Hymenoptera (in 17 families) and in a few species of Diptera and Coleoptera. Its evolution was preceded by that of primary parasitism that evolved in the Hymenoptera during the Jurassic, about 135 mya. In the primary parasitoids, ectophagous feeding probably evolved before endophagous, with the parasitoid egg deposited near or on the host rather than in it. Hence, ectophagous parasitoids usually attacked concealed hosts, often within galleries in wood or plant galls. The use of venom by primary parasitoids apparently developed very early and produced physiological changes in the host. Although the venom of the more ancestral ectophagous parasitoids resulted in idiobiosis (permanent paralysis or death), the venom of the specialized endophagous species tended toward koinobiosis (temporary or nonlethal paralysis).

Facultative hyperparasitism probably evolved from primary ectophagous parasitoids because few special adaptations are needed to oviposit and feed externally on a primary parasitoid as well as on the primary's phytophagous host. Obligate hyperparasitism has a wider taxonomic distribution and may have evolved via facultative hyperparasitism as an opportunistic behavior to specialize only in attacking readily available primary parasitoid hosts—especially if they share similar physiological and/or ecological attributes. Hence, it is not surprising that hyperparasitoid species can be either ecto- or endophagous, whereas some are idiobionts and others are koinobionts.

The host spectrum of hyperparasitoids is broader at the species level than that of primary parasitoids, but hyperparasitism is usually restricted to immature stages of hymenopteran hosts (larvae and/or pupae) that are natural enemies mainly of phytophagous insects in the Hemiptera (mainly suborder Sternorrhyncha), Lepidoptera, and the hymenopteran suborder Symphyta. Hyperparasitoids rarely attack the egg and adult stages of primary parasitoids. Also interesting is that some families of Hymenoptera that are well known for their species of primary parasitoids (Braconidae, Trichogrammatidae, Aphidiidae, Mymaridae, and almost the entire superfamily Proctotrupoidea) do not seem to have evolved any hyperparasitoids. Similarly, in the order Diptera, hyperparasitoids are absent in some important parasitic groups such as the family Tachinidae.

MULTITROPHIC ECOLOGY

There are two complementary ways that ecologists look at the interacting food-web community involving hyperparasitoids. One aspect is the "bottom-up" effect beginning with the first trophic level that shows both inter- and intraspecific plant variation influencing the ecology and behavior of the second trophic level of phytophagous (herbivorous) insects. This in turn is one of the fundamental determinants of the third trophic level of entomophagous (carnivorous) insects such as the primary parasitoid microwasps. Finally, insect hyperparasitism is the highly evolved fourth trophic level, wherein a secondary parasitoid microwasp oviposits on or in the primary parasitoid and kills it. The other complementary aspect is the "top-down" view by which the hyperparasitoid at the fourth trophic level exerts selective pressure on the primary parasitoid at the third trophic level. The next interaction is with the phytophagous insect at the second trophic level that in turn is feeding on the plant at the first trophic level. This food-web community can have economic importance if the plant is an agricultural food crop or even a forest used for commercial lumbering or as a park.

APHID COMPLEX

Over many decades, studies of hyperparasitism have been conducted on the primary parasitoid microwasps in the Hymenoptera that attack the Hemiptera in the suborder Sternorrhyncha, and in particular the superfamily Aphidoidea, with special emphasis on the family Aphididae. The aphid–primary parasitoid–hyperparasitoid food web has been used as a model system in community ecology partly because of the economic importance of aphids as worldwide pests on a variety of agricultural crops and forests, but also because of the relative ease of rearing aphids, their primary parasitoids, and hyperparasitoids in the laboratory and/or greenhouse for precise behavioral and ecological studies. Primary aphid parasitoids are found in only two families of Hymenoptera: all genera of the Aphidiidae (Braconidae: Aphidiinae), and the Aphelinidae (*Aphelinus* and related genera). These primary parasitoids of aphids are in turn attacked by many genera of hyperparasitoids in three hymenopteran superfamilies such as Chalcidoidea (Pteromalidae: *Asaphes*; Encyrtidae: *Syrphophagus*; Eulophidae: *Tetrastichus*), Cynipoidea (Alloxystidae: *Alloxysta*), and Ceraphronoidea (Megaspilidae: *Dendrocerus*).

Aphid hyperparasitoids can be divided into two categories based on adult ovipositional and larval feeding behaviors:

1. The female wasp of endophagous species such as *Alloxysta* (*Charips*) *victrix* (Fig. 1A) deposits her egg inside the primary parasitoid larva while it is still feeding on and developing inside the live aphid, but before the aphid has become mummified (a mummy is the hardened exoskeleton of the dead aphid that remains attached to the leaf). Being a koinobiont hyperparasitoid, the larva usually does not hatch until after the mummy has been formed by the primary parasitoid larva. Then the hyperparasitic larva feeds internally on the primary parasitoid larval host.

2. The female wasp of ectophagous species such as *Asaphes lucens* (Fig. 1B) and *Dendrocerus* (*Lygocerus*) *carpenteri* (Fig. 1C) deposits her egg on the surface of the primary parasitoid larva

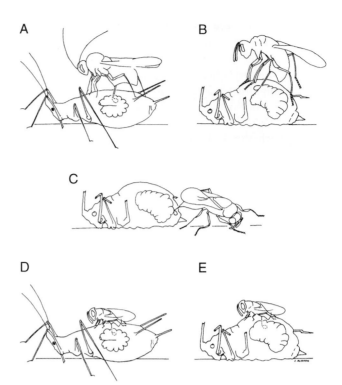

FIGURE 1 Female ovipositional behavior of four genera of aphid hyperparasitoids. (A) Endophagous koinobiont *Alloxysta victrix* jumps on a live parasitized aphid and deposits her egg internally inside the primary parasitoid microwasp larva while the aphid is still alive, but before mummy formation. (B) Ectophagous idiobiont *Asaphes lucens* stands on top of a dead aphid mummy, drills a hole, and deposits her egg externally on the surface of the primary parasitoid larva developing inside the mummy. (C) Ectophagous koinobiont *D. carpenteri* stands on the leaf, backs into the dead aphid mummy, drills a hole, and deposits her egg externally on the surface of the primary parasitoid larva developing inside the mummy. (D) and (E) "Dual ovipositional" behavior of endophagous koinobiont *S. aphidivorus*. (D) *Syrphophagus* stands on top of a live parasitized aphid and deposits her egg internally inside the primary parasitoid larva while the aphid is still alive, but before mummy formation. (E) *Syrphophagus* stands on top of a dead aphid mummy, drills a hole, and deposits her egg internally inside the primary parasitoid larva developing inside the mummy. [Reprinted from Sullivan, D. J. (1988). Aphid hyperparasites. *In* "Aphids, Their Biology, Natural Enemies and Control" (A. K. Minks, and P. Harrewijn, eds.), Vol. 2B, 192, with permission from Elsevier Science.]

after the aphid has been killed and mummified. To do this, the female must first drill a hole in the mummy in which to deposit her egg. Then the hyperparasitic larva feeds externally on the primary parasitoid larva while both are still inside the mummy. The venom of species varies such that *Asaphes* is an idiobiont, whereas *Dendrocerus* is a koinobiont. An unusual species is *Syrphophagus* (*Aphidencyrtus*) *aphidivorus,* whose females display a "dual ovipositional" behavior by attacking primary parasitoid larvae within both living aphids (Fig. 1D), and also inside dead aphid mummies (Fig. 1E). Either way, *S. aphidivorus* develops as an endophagous koinobiont hyperparasitoid.

However, in all species of aphid hyperparasitoids, further development is similar to that of primary parasitoids, such that pupation is also inside the dead aphid mummy. Then the single adult hyperparasitoid emerges by cutting a hole in the dorsum of the mummy. After pulling itself out, the new adult male or female is ready for mating. From the time of the female's attack on the primary parasitoid larva/pupa, the period of hyperparasitoid development from egg deposition to adult emergence varies with different hyperparasitoid species from as short as 16 days to 25 or more.

NONAPHID COMPLEX

Hyperparasitism exists not only with aphids, but also with almost all insect taxa. There is a complex of primary and secondary parasitoids associated with other members of the suborder Sternorrhyncha (scale insects, whiteflies, mealybugs, psyllids), as well as in various orders such as the Lepidoptera (gypsy moth, leafminers, budworms, stem borers), Diptera (gall makers, leafminers), Hymenoptera (cynipid gall makers, leafcutter bees, sawflies), and Coleoptera (lady beetles, weevils).

BIOLOGICAL CONTROL

Biological control uses predators, parasitoids, and pathogens to reduce an insect pest's population to an acceptable level. Because primary parasitoids are considered beneficial, hyperparasitoids may interfere with such programs. It has been the policy in biological control to exclude "exotic" (nonindigenous) *obligate* hyperparasitoids by enforcing quarantine procedures during importation. It is less easy to decide what to do with "exotic" *facultative* hyperparasitoids when no exclusively primary parasitoids are available, and so perhaps admitting these could be beneficial. Such dilemmas are evaluated with caution, and the solution depends on the seriousness of the insect pest problem. On the other hand, "indigenous" hyperparasitoids (whether obligate or facultative) already exist in the ecosystem and may or may not interfere with the beneficial "exotic" primary parasitoid.

CONCLUSIONS

Hyperparasitism intrigues entomologists because of its multidisciplinary relationship to evolution, ecology, behavior, biological control, taxonomy, and mathematical models. More field studies are needed to determine whether hyperparasitoids are always detrimental to biological control programs. Perhaps, instead, they could have a beneficial influence by regulating the extreme/detrimental population oscillations of the beneficial primary parasitoids.

See Also the Following Articles
Biological Control • Parasitoids

Further Reading

Godfray, H. C. J. (1994). "Parasitoids: Behavioral and Evolutionary Ecology." Princeton University Press, Princeton, NJ.

Gordh, G. (1981). The phenomenon of insect hyperparasitism and its taxonomic occurrence in the Insecta. *In* "The Role of Hyperparasitism in Biological Control: A Symposium" (D. Rosen, ed.), Publication 4103, pp. 10–18. University of California Press, Berkeley.

Holt, R. D., and Hochberg, M. E. (1998). The coexistence of competing parasites. II: Hyperparasitism and food chain dynamics. *J. Theor. Biol.* **193,** 485–495.

Mackauer, M., and Völkl, W. (1993). Regulation of aphid populations by aphidiid wasps: Does parasitoid foraging behaviour or hyperparasitism limit impact? *Oecologia* **94,** 339–350.

Price, P. W., Bouton, C. E., Gross, P., McPheron, B. A., Thompson, J. N., and Weis, A. E. (1980). Interactions among three trophic levels: Influence of plants on interactions between herbivores and natural enemies. *Annu. Rev. Ecol. Syst.* **11,** 41–65.

Stiling, P., and Rossi, A. M. (1994). The window of parasitoid vulnerability to hyperparasitism: Template for parasitoid complex structure. *In* "Parasitoid Community Ecology" (B. Hawkins, and W. Sheehan, eds.), pp. 228–244. Oxford University Press, Oxford, U.K.

Sullivan, D. J. (1987). Insect hyperparasitism. *Annu. Rev. Entomol.* **32,** 49–70.

Sullivan, D. J. (1988). Aphid hyperparasites. *In* "Aphids, Their Biology, Natural Enemies and Control" (A. K. Minks, and P. Harrewijn, eds.), Vol. 2B, pp. 189–203. Elsevier, Amsterdam.

Sullivan, D. J., and Völkl, W. (1999). Hyperparasitism: Multitrophic ecology and behavior. *Annu. Rev. Entomol.* **44,** 291–315.

Whitfield, J. B. (1998). Phylogeny and evolution of the host–parasitoid relationship in the Hymenoptera. *Annu. Rev. Entomol.* **43,** 129–151.

Imaginal Discs

Seth S. Blair
University of Wisconsin-Madison

The term "imaginal disc" is used to describe structures found in the larvae of the Holometabola. Holometabolous insects can be defined as those in which the final instar metamorphoses into a radically different adult during a quiescent pupal stage; they are thought to be a monophyletic group, distinct from the Hemimetabola. During the metamorphosis of Holometabola, the epidermis must form novel structures that were lacking in the larva and, in some insects larval tissues that were lost must be replaced. The cells that give rise to the new epidermal tissues of the adult (imago) are often referred to as histoblasts. When histoblasts are organized into morphologically distinct clusters, these structures are commonly referred to as imaginal discs or imaginal buds. This article discusses this definition and briefly reviews some of the experimental studies examining the biology and, especially, the development of imaginal discs.

WHAT IS AN IMAGINAL DISC?

The terms "imaginal disc" and "histoblast" are more pragmatic than precisely defined, and their usage varies from author to author because imaginal discs vary considerably between taxa in number, morphology, and development. For example, some authors have defined imaginal discs and histoblast cells based on their undifferentiated state and early appearance in development, since in some taxa imaginal discs are formed in the embryo. These early-developing discs may sometimes secrete a thin cuticle-like substance, but they do not contribute appreciably to larval life and thus can be considered to be specialized, relatively undifferentiated structures set aside for

adult development. However, in other taxa the imaginal discs cannot be detected until the final instar and are apparently derived from cuticle-secreting, differentiated larval cells. Several authors have argued that late-developing discs reflect the ancestral condition within the Holometabola.

The morphology of imaginal discs also varies. The typical imaginal disc is a pocket or sac of cells that has invaginated from the larval epidermis and is destined to form part or all of an adult appendage, compound eye, or genitalia. One portion of the sac has thickened to form the disc epithelium, whereas the rest forms a thinner peripodial membrane; the space within the sac is termed the peripodial cavity (Fig. 1). This sac evaginates ("everts") during metamorphosis and contributes to the adult cuticle (Fig. 1). However, the positions within the disc of the disc epithelial cells and peripodial cells vary, as does the disc's degree of invagination, and some authors have defined several categories of discs or quasidiscs. For example, some discs have invaginated a large distance from the larval epidermis and remain connected to it by only a thin peripodial stalk. At the other extreme, as in the wings of some Coleoptera, the regions of disc epidermis that form adult structures never invaginate from the larval epidermis. Some of these discs can still be recognized as thickenings of the larval epidermis. However, in other examples, such as the leg of the lepidopteran *Pieris brassicae,* the histoblasts cannot be easily recognized, and the novel portions of a single adult appendage develop from several zones of dividing cells within the larval appendage.

Thus, sometimes it is only its eventual contribution to a novel adult structure that makes a cell a histoblast, distinct from neighboring cells in the larval epidermis. Moreover, the organization of histoblasts into a morphologically identifiable imaginal disc is simply one of several ways these cells can be arranged. And though the term "imaginal disc" is used only for holometabolous insects, it is only the invaginated morphology of most discs that distinguishes them from structures such as the external wing pads of hemimetabolous

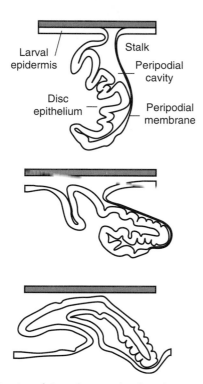

FIGURE 1 Eversion of *D. melanogaster* leg disc, shown in cross section. [After Blair, S. S. (1999). *Drosophila* imaginal disc development: Patterning the adult fly. *In* "Development-Genetics, Epigenetics, and Environmental Regulation" (V. E. A. Russo, D. Cove, L. Edgar, R. Jaenisch, and F. Salamini, eds.), p. 348. © Springer-Verlag GmbH & Co. KG, Heidelberg.]

pterogotes, which also are morphologically distinct and form adult structures. Thus, the evolution of invaginated wing imaginal discs from the evaginated wing pad may have required only a morphological change, rather than the evolution of a radically different cell type or organ. In fact, a similar evolution has apparently occurred in some Hemimetabola, such as thrips, which have invaginated wing precursors instead of external wing pads.

EXPERIMENTAL STUDIES ON IMAGINAL DISCS

The biology of imaginal discs has been described in a number of species but has been experimentally analyzed in only a few. As with other metamorphosing organs, there have been studies on the endocrine control of imaginal disc eversion, differentiation, and cuticle secretion. Other types of experiment, such as ablation or transplantation of portions of imaginal discs, have been used to study aspects of developmental patterning, such as the development of pigmentation patterns in the wings of butterflies and moths. And in a sense, all studies of the evolution of adult Holometabola reflect on the histoblast cells and discs from which portions of those adults are formed.

However, the most extensively studied imaginal discs are those of the fruit fly, *Drosophila melanogaster*. This species, which has been the subject of intense genetic studies for a

century, has become a model for the study of a variety of biological problems. Gaining an understanding of the imaginal discs in this species has been particularly critical because the entire adult epidermis is derived from imaginal discs or disclike structures. Thus, any mutation that alters the external form of the adult *D. melanogaster* does so by altering the development of these imaginal structures. Thus a large literature has appeared on the genetics, development, and cell biology of these structures and, via comparisons with other drosophilids, a number of evolutionary studies have been published, as well.

The imaginal discs of *D. melanogaster* are at one extreme in the spectrum of imaginal tissues, and it is not clear to what extent the mechanisms underlying this apparently derived state are shared by other Holometabola. As in other cyclorrhaphan Diptera, the metamorphosis of *D. melanogaster* is unusual in several respects. First, the cuticle of the last instar is not shed during the formation of the pupa (pupariation), as occurs in most Holometabola. Rather, the larval cuticle is retained and converted into an outer covering, the pupal case. Second, in *D. melanogaster* the nondisc cells of the larval epidermis are polyploid (containing more than the diploid number of chromosomes), and these die during the early stages of metamorphosis. In many other Holometabola, large portions of the larval epidermis and appendages are retained in the adult. Finally, in *D. melanogaster* the imaginal disc primordia are formed during embryonic development, rather than during the last instar as they are in some other Holometabola.

In *D. melanogaster* embryos each imaginal disc primordium contains 10 to 40 cells, which divide during the three instars to form as many as 50,000 cells by late third instar. The arrangement of imaginal discs in the late third instar is shown in Fig. 2. The adult head is derived from a pair of fused eye–antennal discs, as well as pairs of proboscis (labial) and labral (cibarial or clypeolabral) discs. The adult thorax is derived from the three pairs of ventral leg discs and, dorsally, pairs of prothoracic (humeral), wing, and haltere discs. Each of these contributes to the body wall, and also forms the appendage for which it is named. Each segment of the adult abdomen is formed from four pairs of small "histoblast nests" and the genitals from the genital disc.

At the late third instar each imaginal disc consists of a simple epithelial sac, invaginated from the larval epithelium (Fig. 1). One surface of each sac forms the thickened and folded disc epithelium, from which most of the adult structures are derived. The other surface of the sac is the thinner peripodial membrane, and each disc remains connected to the larval epithelium by a long, narrow stalk. During the first few hours of pupariation each disc everts through the stalk, expands, and eventually sutures together with adjacent discs. The peripodial membrane is lost during this process, and the polyploid cells of the larval epidermis are histolysed and replaced. The portions of each disc epithelium that are fated to form the appendages unfold and lengthen: the prospective legs and antennae form long tubes, whereas the prospective

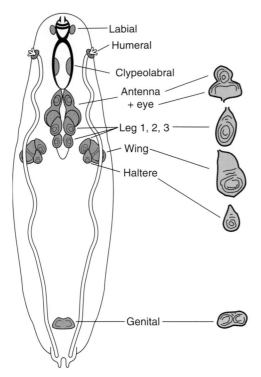

FIGURE 2 Imaginal discs in late third instar *D. melanogaster.* [From Blair, S. S. (1999). *Drosophila* imaginal disc development: Patterning the adult fly. *In* "Development-Genetics, Epigenetics, and Environmental Regulation" (V. E. A. Russo, D. Cove, L. Edgar, R. Jaenisch, and F. Salamini, eds.), p. 348. © Springer-Verlag GmbH & Co. KG, Heidelberg.]

dorsal and ventral surfaces of the wing lengthen and flatten together. The disc cells then secrete the pupal cuticle, which is separated from the epithelial surface shortly thereafter. After a delay during which there is further morphogenesis and differentiation, the adult cuticle is secreted.

Many aspects of the biology of imaginal discs of *D. melanogaster* have been examined over the years and continue to be the subjects of intense research. These include the hormonal control of disc development and cuticle secretion, the molecular genetics of cell division and cell growth, and various problems in cell biology. Perhaps the most striking advances, however, have been made in developmental patterning by examining for the most part the large wing, leg, and eye–antennal discs. Space constraints prevent the discussion of many fascinating and important research areas, such as the patterning of the compound eye, the formation of sensory organs, and the planar polarity of the disc epithelium. Instead, the following discussion briefly reviews some early developmental events that help establish the identities and axes of the discs.

PATTERNING IMAGINAL DISCS IN *D. MELANOGASTER*

The identity of each disc primordium is specified during embryonic development. This specification relies on the disc-

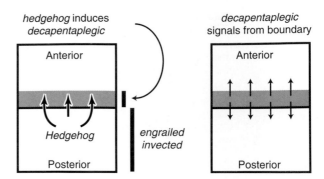

FIGURE 3 Signaling and anterior boundary cells in wing disc of *D. melanogaster.*

specific expression of a small number of genes, encoding largely transcription factors or their cofactors. The loss or misexpression of these genes causes dramatic homeotic transformations of disc identities. The localized expression of these transcription factors is to a large extent inherited from or signaled by the embryonic epidermis, which was subdivided into anterior–posterior and dorsal–ventral domains prior to the formation of the disc primordia.

How are different tissues formed within each disc? Even at late third instar most imaginal disc cells appear morphologically similar; only a few (mostly sensory) elements have begun to differentiate terminally. However, the premetamorphic imaginal disc is in no sense a *tabula rasa,* since it has already been subdivided into number of regions. In fact, most disc primordia are subdivided into anterior and posterior lineage "compartments" from the earliest stages of their development. This subdivision is controlled by the posterior-specific expression of the *engrailed* and *invected* transcription factors in each disc, which act as a binary switch, controlling the choice between posterior and anterior identities and preventing cells from crossing between compartments. The wing and haltere discs are further subdivided into prospective dorsal and ventral lineage compartments, in this case by the dorsal-specific expression of the *apterous* transcription factor.

Why have lineage compartments? Cells in different lineage compartments have different signaling capabilities, and this has important developmental consequences. Posterior cells secrete the signaling molecule Hedgehog, but only anterior cells are capable of responding to that signal (Fig. 3). Because Hedgehog diffuses only a short distance into the anterior compartment, it induces the formation of a specialized group of cells just anterior to the compartment boundary. The formation of this group of boundary cells is critical for appendage development because boundary cells in turn secrete several important signals (Fig. 3), including Decapentaplegic (a member of the BMP family of morphogens, generated in the wing, dorsal leg, and dorsal antenna), and Wingless (a member of the Wnt family of morphogens, generated in ventral leg and ventral antenna). Cells outside the boundary region judge their approximate position in the disc by the

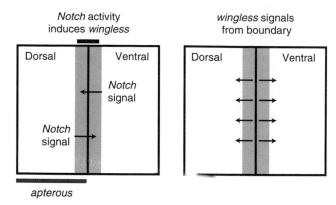

FIGURE 4 Signaling and dorsal–ventral boundary cells in the wing disc of *D. melanogaster.*

levels of boundary signal they detect. Thus, these boundaries define important axes used to pattern the entire disc.

The subdivision of the wing into dorsal and ventral compartments by the dorsal expression of *apterous* plays a very similar role. This subdivision establishes reciprocal signaling between dorsal and ventral cells, this time via the *Notch* pathway, which results in the specification of cells on either side of the dorsal–ventral boundary (Fig. 4). These cells secrete Wingless, which helps pattern the wing blade along the proximodistal axis.

Do the other Holometabola share these axis-defining patterning mechanisms? There have been to date only a few descriptions of gene expression in the imaginal discs of other taxa, and no experimental tests of the type that verified these patterning mechanisms in *Drosophila*. However, it seems likely that at least some of the fundamental features are shared. All insect appendages so far examined (both larval and adult) are subdivided into apparent anterior and posterior compartments by the posterior expression of *engrailed*-like transcription factors, including the wing discs of the lepidopteran *Precis coenia*. Similarly, the wing blade of *P. coenia* is subdivided into dorsal and ventral domains by the dorsal expression of an *apterous*-like molecule. As in *D. melanogaster,* a *wingless*-like molecule is expressed along the dorsal–ventral compartment boundary in *P. coenia.*

See Also the Following Articles

Development, Hormonal Control of • Drosophila melanogaster • Embryogenesis

Further Reading

Blair, S. S. (1999). *Drosophila* imaginal disc development: Patterning the adult fly. *In* "Development-Genetics, Epigenetics and Environmental Regulation" (V. E. A. Russo, D. Cove, L. Edgar, R. Jaenisch, and F. Salamini, eds.), pp. 347–370. Springer-Verlag, Heidelberg.

Carroll, S. B., Gates, J., Keys, D. N., Paddock, S. W., Panganiban, G. E. F., Selegue, J. E., and Williams, J. A. (1994). Pattern formation and eyespot determination in butterfly wings. *Science* **265**, 109–114.

Fristrom, D. K., and Fristrom, J. W. (1993). The metamorphic development of the adult epidermis. *In* "The Development of *Drosophila melanogaster*"

(M. Bate and A. Martinez Arias, eds.), pp. 843–897. Cold Spring Harbor Laboratory Press, Plainview, NY.

Kim, C.-W. (1959). The differentiation centre inducing development from larval to adult leg in *Pieris brassicae* (Lepidoptera). *J. Embryol. Exp. Morphol.* **7**, 572–582.

Švácha, P. (1992). What are and what are not imaginal discs: Reevaluation of some basic concepts (Insecta, Holometabola). *Dev. Biol.* **154**, 101–117.

Tower, W. L. (1903). The origin and development of the wings of Coleoptera. *Zool. Jahrb. Anat. Ontog. Tiere* **17**, 517–572.

Truman, J. W., and Riddiford, L. M. (1999). The origins of insect metamorphosis. *Nature* **401**, 447–452.

Immunology

Nancy E. Beckage
University of California, Riverside

Insects lack immunoglobulins but nevertheless mount a variety of effective immune responses to parasites and pathogens. In permissive or susceptible hosts, either no response is mobilized or the responses induced fail to counter the invader, so that the host is successfully infected. In resistant (also called refractory) hosts, the parasite or pathogen is thwarted, and infection is prevented. Regardless of the outcome, the result is a dynamic interplay of host and parasite genes, with the products of host resistance/susceptibility genes being counterbalanced by the virulence/avirulence characteristics of the invader (Fig. 1).

In insects, both cellular and humoral immune responses figure prominently in host defense, with many parasitic and pathogenic infections resulting in the deployment of defenses of both types. Insect hemocytes mobilize the cellular defenses, which include phagocytosis, nodulation, and encapsulation. In the main classes of humoral defenses, the following substances are produced: antibacterial, antifungal, and (presumably) antiviral molecules, the melanizing enzyme phenoloxidase, and agglutinins or clotting factors.

Although much recent progress has been made in deciphering the cellular and humoral aspects of insect immune responses, much less is known about the recognition mechanisms responsible for the initial discimination of "nonself" material at the host–invader interface that sets the stage for the immune response. This is especially true for the recognition of multicellular parasites, which seem to use a combination of active and passive strategies to avoid being detected as foreign.

CELLULAR IMMUNE RESPONSES

Insect Hemocytes

Initially, hemocytes were classified on the basis of morphological criteria alone, resulting in the publication of numerous

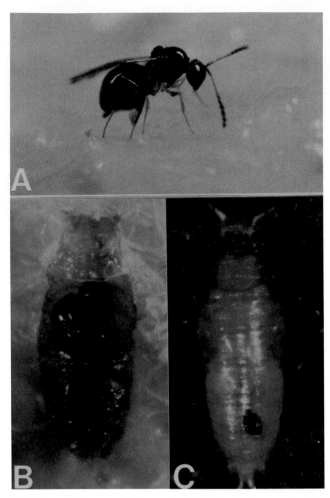

FIGURE 1 (A) The parasitoid wasp, *L. boulardi,* preparing to oviposit in *D. melanogaster* host larvae. If the host strain is of the susceptible genotype, and the wasp is of the virulent strain of this parasitoid species, the wasp will develop normally in the host (B) and successfully emerge while the host dies. However, if the host expresses resistance genes, the parasitoid will be encapsulated and melanized (C), and the host will survive to adulthood. (Photographs courtesy of Dr. Yves Carton, CNRS, Gif-sur-Yvette, France.)

conflicting classification schemes, even for the same species. The different hemocyte types display marked inter-, and even intra-, species variability in appearance and behavior, further complicating classification. Additionally, hemocytes that are examined before they have attached to a substrate are difficult to identify in the unspread state, in contrast to spreading cells (e.g., plasmatocytes). Today, tools such as monoclonal antibodies, which bind cell-type-specific epitopes, are used in combination with other biochemical markers to facilitate identification of the different hemocyte classes. In addition, techniques such as density gradient centrifugation have been employed to purify homogeneous populations of the individual hemocyte morphotypes, facilitating *in vitro* studies of their biochemistry and behavior.

In the Lepidoptera, the two most abundant hemocyte classes are plasmatocytes and granulocytes, which are the primary phagocytic and encapsulative cells. In the higher Diptera including *Drosophila melanogaster,* the multifunctional lamellocytes play these roles, and crystal cells carry the melanizing enzyme phenoloxidase. In contrast, in lepidopterans including the tobacco hornworm, *Manduca sexta,* the bulk of this enzyme is carried in the plasma. Taxonomic variations are seen in the total hemocyte count, with the Lepidoptera having abundant hemocytes in contrast to dipterans (e.g., mosquitoes), which have many fewer cells per microliter of hemolymph.

The fat body is also an immunoresponsive organ and synthesizes a variety of antimicrobial peptides, enzymes (e.g., lysozyme), and other immunoreactive molecules. Indeed, the fat body represents the primary source of hemolymph-borne macromolecules and is the most metabolically active tissue in the insect. The gut, which is constantly assaulted with pathogens ingested by the insect, also produces a battery of antibacterial and antifungal agents to counter these invaders. During molting, when the newly synthesized cuticle is most fragile and the cuticular linings of the fore- and hindguts are shed, bacteria are released into the lumen of the gut, resulting in enhanced production of antimicrobial peptides by the gut when the animal is most vulnerable to infection.

Hemocyte-Mediated Immune Responses

Phagocytosis is the process by which pathogens such as bacteria and small particles (< 1 μm in diameter) are engulfed by host hemocytes, culminating in death of the invader. The membrane of the cell invaginates, and the pathogen is engulfed in a membrane-bound vesicle into which lytic enzymes are released, causing the pathogen's demise. This process appears to be mediated by prostaglandins (eicosanoids) produced by the hemocytes. The phagocytic cells include plasmatocytes and granulocytes, although other cell types may also participate in this response to a lesser degree. Large numbers of bacteria-laden hemocytes may clump together to form nodules, which attach to the host's internal tissues and are removed from circulation. Frequently these nodules are melanized and deposited on lobes of fat body, the Malpighian tubules, or gut tissues.

To counter eukaryotic invaders (i.e., parasites) that are too large to be phagocytosed by a single cell, multiple classes of hemocytes cooperate in the mobilization of the multiphasic encapsulation reaction. Usually the capsule is formed from several hundred to several thousand cells that form a dense capsule enclosing the parasite. In the initial phases of encapsulation of parasitoid eggs/larvae, or encapsulation of abiotic (nonliving) implants such as Sephadex beads, granulocytes are the first cells to make contact with the target. These cells, which contain large numbers of refractile granules in their cytoplasm, attach to the target surface and then release their contents in a degranulation reaction, forming a sticky matrix on the surface of the target. The degranulation event triggers the attachment of multiple layers of plasmatocytes, which are fibroblast-like cells that spread, flatten out, and encase the parasite in a multilayered sheath of cells. Sometimes multiple

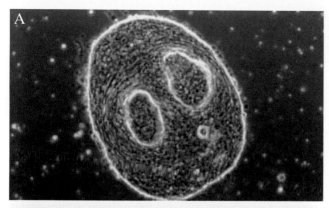

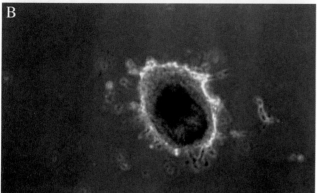

FIGURE 2 Light micrograph series showing fully encapsulated *C. congregata* eggs recovered from the nonpermissive host *P. occidentalis* 72 h postparasitization of the host larva. Multiple eggs are sometimes engulfed in a single capsule (A), which frequently shows signs of partial or complete melanization (B). [Reprinted from Harwood, S. H., *et al.* (1998). Production of early expressed parasitism-specific proteins in alternate sphingid hosts of the braconid wasp *Cotesia congregata. J. Inv. Pathol.* **71**, 271–279, with permission from Academic Press.]

targets are enclosed in a single capsule, as occurs with eggs of the parasitoid *Cotesia congregata,* in the nonpermissive host *Pachyshpinx occidentalis* (Fig. 2A). The encapsulation reaction is terminated by the adherence of additional granular cells to form a thin envelope around the completed capsule. Like nodules, fully formed capsules frequently adhere to the host's internal tissues (including fat bodies, Malpighian tubules, gut, or salivary glands) and are thereby removed from circulation. In rare instances, the encapsulated parasite may actually be extruded from the hemocoel and pass through the epidermis, to be shed with the host's exuvial cuticle in a molt during "cuticular encystment," which occurs when a parasitoid develops in a nonpermissive host.

During the encapsulation process or upon its completion, the innermost layer of cells deposits melanin or its toxic quionone precursors over the surface of the invader, regardless of whether the foreign material is an abiotic or biotic target (Figs. 2B, 3A,B). However, the occurrence of melanization is variable, and sometimes capsules persist for the remaining life span of the host, showing no signs of melanization. Some species of hosts with encapsulated parasitoids undergo metamorphosis and live to the adult stage (e.g., *D. melanogaster*

with encapsulated *Leptipolina boulardi* parasitoid egg; Fig. 1), whereas other hosts die prematurely, frequently showing symptoms of endocrine disruption. When the host dies prematurely, the parasitoid fails to survive. When the parasitoid *C. congregata* is encapsulated in host larvae of the tobacco hornworm, which is normally permissive for this parastoid, the "hosts" with encapsulated parasitoids molt to supernumerary instars and then to nonviable larval–pupal intermediates. These developmental symptoms suggest that the host's juvenile hormone titer is elevated to a level high enough to interfere with normal pupation. In these instances, endocrine disruption reems to be mediated by the factors (i.e., polydisperse DNA viruses: polydnavliusn) injected by the wasp into the host during parasitization and thus even in the absence of developing wasps, host development is arrested prematurely.

During parasitization, endoparasitoids belonging to the hymenopteran families Braconidae and Ichneumonidae inject into their lepidopteran hosts polydnaviruses that subsequently play a critical role in suppressing the host immune response. These viruses, which are integrated in the genomic DNA of the wasp and undergo replication only in the female's ovary, rapidly enter host hemocytes following parasitization, and viral genes are expressed. Depending on the host–parasitoid combination, the host's hemocytes either alter their behavior and fail to spread (thereby inhibiting the encapsulation response) or, alternatively, undergo fragmentation and programmed cell death. In *M. sexta,* larvae parasitized by *C. congregata,* massive numbers of dead and dying hemocytes undergo clumping and then are removed from circulation soon after parasitization, resulting in a dramatic drop in the host's total hemocyte count. By 8 days postparasitization, new cells have differentiated from prohemocytes, and the host regains its ability to encapsulate Sephadex beads. However, the living parasitoid larvae remain unencapsulated (Fig. 3A), suggesting that they escape being detected as foreign by another mechanism. However, if second-instar parasitoids are dissected from a host, killed, then implanted into a surrogate "host" caterpillar, the parasitoids are avidly encapsulated and melanized (Fig. 3B), suggesting that something unique about the living parasitoid surface suppresses a host immune response. These simple observations suggest that the living parasitoid larvae either escape being detected as foreign by mechanisms that may involve host antigen mimicry (or masking) or by the presence of specific as yet unidentified surface molecules that prevent their recognition as "nonself" by hemocytes.

The final biochemical events that culminate in death of the parasite remain unclear for the most part. Although melanin and its precursors are toxic, much recent evidence points to the potential role of other toxic molecules such as reactive intermediates of nitrogen (nitric oxide) or oxygen (superoxide), released by cells localized in the innermost layers of the capsule, in causing lethality. Although it was formerly presumed that death was induced by asphyxiation of the parasite or parasitoid inside the capsule, biochemically

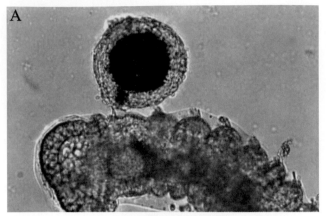

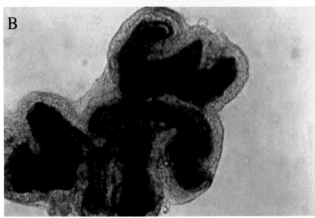

FIGURE 3 (A) Evidence that living *C. congregata* parasitoids are not encapsulated larvae of the host, *M. sexta,* which are permissive for this parasitoid, even though the host may mobilize hemocytes to encapsulate abiotic targets such as Sephadex beads. The bead shows hemocytic encapsulation and associated melanization, whereas the parasitoid larva has avoided this response, suggesting that the living parasitoids actively evade the host's immune response by antigen mimicry, secretion of immunosuppressive molecules, or other mechanisms. (B) Encapsulation of first-instar *C. congregata* that had been dissected from a fifth-instar host tobacco hornworm, killed by immersion in ethanol, then implanted into the hemocoel of a "surrogate host." Thick, agglutinated, and melanized capsules surround the larvae 24 h after implantation. [Photographs from Lavine, M., and Beckage, N. (1995). Polydnaviruses: Potent mediators of host insect immune dysfunction. *Parasitol. Today* **11,** 368–378, with permission from Elsevier Science.]

mediated parasite-killing strategies now seem to be important in causing the ultimate death of the invader. Phagocytosis is followed by the generation of these cytotoxic intermediates in the phagolysozome, indicating that this pathway is also important to the function of phagocytic cells.

The antiviral defenses of insects are just now beginning to be deciphered, although recent observations made in parasitized insects have yielded some insights. Larvae of *M. sexta* that are parasitized by *C. congregata* are dramatically more susceptible to the *Autographa californica* nucleopolyhedrovirus than nonparasitized larvae of the same age, which are normally semipermissive hosts of this pathogenic baculovirus. Parasitized larvae die faster and at higher rates than nonparasitized larvae infected with the same dose of the occluded form of the

baculovirus. This enhanced susceptibility appears because of the inactivation of the cellular immune response of the host induced by the wasp's polydnavirus. Thus, there is likely to be a cellular immune response to the baculovirus, which is suppressed in parasitized larvae. In nonparasitized larvae, cellular plaques composed of hemocytes clump on cells localized in the host's tracheal epithelium that harbor the virus. These plaques do not form in parasitized larvae, and the virus is rapidly disseminated throughout the body cavity via the tracheolar epithelium, uninhibited by the hemocytic response. Although the polydnavirus genes that render the host more susceptible to the baculovirus have yet to be isolated, they offer promise for formulation of baculovirus biopesticides with enhanced potency and a broader host range for control of lepiopteran insect pests.

HUMORAL IMMUNE RESPONSES

Plasma-Borne Factors

In addition to hemocyte-mediated immune reactions, insects possess a variety of potent plasma-borne defense molecules that are toxic to parasites and pathogens. Usually these are synthesized by the fat body or the hemocytes and secreted into the plasma, where they act either on the invader directly or via the hemocytes in altering their behavior to enhance the immune response. A battery of antibacterial proteins are produced by many insects, including defensins, drosocin, cecropins, attacins, and diptericins, depending on the species, and, in addition, the ubiquitous lysozyme family of antibacterial proteins. These proteins differ in their specificity for gram-positive versus gram-negative bacteria, with some acting on both types of bacteria with varying degrees of potency. Often the proteins disrupt the bacterial cell membrane function by inducing pore formation, causing lysis of the cell. Antifungal molecules are also produced by insects, providing a first line of defense against fungal invaders that often infect the insect via its cuticle, which is penetrated by the fungus. Although insects are not known to produce antiviral interferon-like molecules, the mobilization of biochemical defenses against viruses seems to be a likely component contributing to the evolution of viral resistance in insect populations treated with viral biopesticides. Although antiviral resistance has been characterized at the insect population level, the cellular and molecular mechanisms contributing to resistance remain relatively ill-defined.

Fortuitously, several antimicrobial peptides such as defensins have also been shown to have antiparasite activity, killing malaria parasites and filarial nematodes in insects that are injected with these molecules. Hence, molecular geneticists are now exploiting defensin genes in the production of transgenic mosquitoes that show up-regulation of defensin gene expression under regulation of tissue-specific promoters either in the gut (malaria) or flight muscle (filaria) where the parasites develop.

In *Drosophila,* mosquitoes, and other insects, the activation of transmembrane *toll* receptors mediates the physiological response to microbial ligands or septic injury, leading to activation of killing mechanisms such as production of nitric oxide, resulting in death of the invader. The *toll* receptors are conserved across a wide range of animal phyla including Mammalia, indicating that this pathway, which insects share with a variety of species, is likely of ancient evolutionary origin.

The phenoloxidase pathway is activated by the synthesis of DOPA from tyrosine via the action of the monophenoloxidase enzyme (also called tyrosinase). Then DOPA is converted to DOPA quinone by diphenoloxidase, and thence to melanin via a series of toxic intermediates. Phenoloxidase activity may be associated with hemocytes, as occurs in mosquitoes, or as in lepidopterans (e.g., *M. sexta*), it may be secreted into the plasma. The first step is the activation of prophenoloxidase by a serine protease, which cleaves a peptide from the proenzyme, generating the active phenoloxidase molecule.

In many species of parasitized lepidopterans, including *M. sexta* larvae parasitized by *C. congregata,* levels of hemolymph phenoloxidase activity have been found to be suppressed following parasitization, which benefits the parasitoid by inhibiting this immunoreaction. This effect seems to be expression of polydnavirus genes that inhibit translation of the phenoloxidase mRNA, thereby suppressing levels of this enzyme in the blood.

In refractory strains of mosquitoes, melanization of the malaria ookinete occurs in the midgut wall, apparently without the intervention of phenoloxidase derived from hemocytes. Disease transmission stops because melanized parasites die trapped in the gut without ever moving to the hemocoel and salivary gland. Hence, there is widespread interest in using phenoloxidase genes to bioengineer refractory transgenic mosquitoes to halt malaria transmission. One approach has been to link this gene to the vitellogenin promoter, which is activated when the mosquito takes a blood meal in preparation for production of eggs.

Cross-Talk between the Cellular and Humoral Immune Response Networks

Blood cells in vertebrates produce many cytokines, which act at close range on other immunocompetent cells. In insects, the characterization of cytokines is less well documented, but recent evidence indicates that factors such as plasmatocyte-spreading peptide (which was first isolated in the moth *Pseudoplusia includens*) are produced by plasmatocytes and act to stimulate spreading of the hemocyte over the surface of the parasite. Other cytokines, which likely play a role in the cell-to-cell communication events that accompany encapsulation, have yet to be characterized. In parasitized lepidopterans, these cytokines may include plasmatocyte and granulocyte depletion factors. Without these cell types, parasite encapsulation cannot occur, and thus the number of viable circulating cells available to mount the encapsulation response is drastically reduced.

UNRESOLVED QUESTIONS

The role of chemical mediators in the cell-to-cell communication pathways required for successful immune responses as complicated as encapsulation have only just begun to be explored. Also, the initial signaling events that discriminate "nonself" at the most basic level remain remarkably obscure. This is particularly true for the recognition of eukaryotic parasites, which seem to use a complex mix of strategies such as host immunosuppression, antigen masking, and antigen mimicry to avoid being detected as foreign. Progress in identifying resistance and susceptibility genes in insect hosts, and the complementary virulence/avirulence genes in parasites has begun only recently, although efforts are intensifying to clarify these genetic components, particularly in *Drosophila.* Although insects lack "memory cells" in their immune system, some long-lived insects such as cockroaches exhibit an enhanced response to a secondary challenge, raising the possibility that insects can show a sensitization response, albeit via as yet unknown mechanisms. Although the antibacterial defenses of insects have been well characterized in the past two decades, even at the molecular level, the defenses mobilized by insects against invasion by viral pathogens are only now being delineated, despite intensive efforts to clarify mechanisms of antiviral resistence. In most species, the roles of hormones such as juvenile hormone and ecdysteroid in regulating immunity have yet to be deciphered. The prospect of developing transgenic insects expressing refractoriness-related traits offers promise of the ability to control disease transmission by insect vectors. Likewise, the prospects of genetic manipulation of agricultural pests to enhance their susceptibility to biopesticides, and to increase the virulence of the pathogen, offer great promise to agriculture. Clearly, many challenging questions need to be addressed to decipher the tactics of parasite/pathogen offense and host defense.

See Also the Following Articles
Genetic Engineering • Hyperparasitism • Pathogens of Insects

Further Reading
Beckage, N. E. (1997). The parasitic wasp's secret weapon. *Sci. Am.* **277,** 50–55.
Beckage, N. E. (1998). Modulation of immune responses to parasitoids by polydnaviruses. *Parasitology* **116,** S57–S64.
Beckage, N. E., Thompson, S. N., and Federici, B. A. (eds.). (1993). "Parasites and Pathogens of Insects": Vol. 1, "Parasites," Vol. 2, "Pathogens." Academic Press, San Diego.
Carton, Y., and Nappi, A. J. (2001). Immunogenetic aspects of the cellular immune response of *Drosophila* against parasitoids. *Immunogenetics* **52,** 157–164.
Dimopoulos, G., Muller, H. M., Levashina, E. A., and Kafatos, F. C. (2001). Innate immune defense against malaria infection in the mosquito. *Curr. Opin. Immunol.* **13,** 79–88.
Gillespie, J. P., Kanost, M. R., and Trenczek, T. (1997). Biological mediators of insect immunity. *Annu. Rev. Entomol.* **42,** 611–643.
Hoffmann, J. A., and Reichhart, J. M. (1997). *Drosophila* immunity. *Trends Cell Biol.* **7,** 309–316.
Hoffmann, J. A., Kafatos, F. C., Janeway, C. A., and Ezekowiltz, R. A. (1999). Phylogenetic perspectives in innate immunity. *Science* **284,** 1313–1318.

Gupta, A. P. (ed.). (1991). "Immunology of Insects and Other Arthropods." CRC Press, Boca Raton, FL.

Imler, J. L., and Hoffmann, J. A. (2001). *toll* Receptors in innate immunity. *Trends Cell Biol.* **11**, 304–311.

Paskewitz, S. M., and Gorman, M. J. (1999). Mosquito immunity and malaria parasites. *Am. Entomol.* **45**, 80–94.

Pathak, J. P. N. (ed.). (1993). "Insect Immunity." Kluwer Academic Publishers, Dordrecht, Netherlands.

Schmidt, O., Theopold, U., and Strand, M. (2001). Innate immunity and its evasion and suppression by hymenopteran parastioids. *Bioessays* **2**, 344–351.

Vass, E. and Nappi, A. J. (2001). Fruit fly immunity. *BioScience* **51**, 529–535.

Wiesner, A., Dunphy, G. B., Marmaras, V. J., Morishima, I., Sugumaran, M., and Yamakawa, M. (eds.). (1998). "Techniques in Insect Immunology." SOS Publications, Fair Haven, NJ.

Industrial Melanism

Michael E. N. Majerus
University of Cambridge

Industrial melanism may be defined as a proportional increase of dark, or melanin, pigments in individuals of a population, caused by changes in the environment resulting from industrial pollution. Both increases in the frequencies of distinct melanic forms and the general darkening of some or all forms within a population may be involved.

The increase in dark forms of some species of moth in industrial regions of western Europe, and latterly elsewhere, has provided some of the best known, most easily understood, and most often quoted examples of evolution in action. Increases in pollution following the industrial revolution led to changes in the environment. In particular, sulfur dioxide denuded trees and other substrates of lichens, while particulate air pollution blackened the resulting surfaces. In response to these changes, many species of moth and some other invertebrates that rely on camouflage for defense against some predators have changed their coloration, becoming darker, in line with the darkening of the substrates that they rest upon by day. These changes have occurred largely in the past 150 years and are cited as examples that illustrate the central mechanism of Charles Darwin's theory of evolution: natural selection.

TYPES OF INDUSTRIAL MELANISM

Three categories of industrial melanism have been recognized:

A. Full industrial melanic polymorphism involves distinct melanic forms that have arisen since the industrial revolution and have increased as a consequence of the effects of industrialization on the environment.

B. Partial industrial melanic polymorphism involves polymorphic species that had melanic forms prior to the industrial revolution. These forms have increased in frequency following and as a consequence of the effects of industrialization.

C. Polygenic industrial melanism involves species in which the average ground color of some or all members of a population has darkened gradually as a consequence of the effects of industrialization.

It should be noted that melanism is a common phenomenon throughout the animal kingdom, with many factors unrelated to industrialization or pollution influencing the success of melanic forms in some species.

FULL INDUSTRIAL MELANIC POLYMORPHISM

The Peppered Moth

The peppered moth, *Biston betularia,* has dominated the literature on industrial melanism. In Britain, the ancestral form of this species (form *typica*) is white, liberally speckled with dark brown or black scales (Fig. 1). In 1848, a predominantly black form of *B. betularia,* form *carbonaria* (Fig. 2) was recorded in Manchester, England. Within 50 years, 98% of Mancunian peppered moths were black. From this original location, *carbonaria* spread to many other parts of Britain. The renowned Victorian lepidopterist J. W. Tutt was the first to suggest that camouflage and bird predation could be involved in the spread of *carbonaria.* In 1896, arguing that the typical form was camouflaged well on surfaces covered by foliose lichens, he noted that the nature of many natural surfaces had changed as a consequence of pollutants resulting from heavy industry. In particular, the combined effects of sulfur dioxide, which killed foliose lichens, and soot fallout, which blackened the denuded surfaces, had led to darker and more uniform substrates. He stated that on these surfaces the *carbonaria* form would be better camouflaged than *typica* and so gain protection from bird predation. Natural selection, through the medium of differential bird predation, augmented by "hereditary tendency," had led to an increase in the frequency of the black form.

FIGURE 1 The typical form of the peppered moth, *B. betularia.*

FIGURE 2 The *carbonaria* form of the peppered moth.

FIGURE 4 A peppered moth in its natural resting position, beneath a lateral tree branch.

Tutt's hypothesis was largely rejected, both at the time and for a considerable period thereafter, because most entomologists and ornithologists concurred in the view that birds are not major predators of cryptic, day-resting moths. A variety of other explanations of the increase in melanic forms of some moths were thus put forward during the first half of the 20th century (pollutants acting as mutagenic agents, Lamarckian evolution, heterozygote advantage).

One important advance during this period was the calculation by Haldane in 1924 that *carbonaria* would have to have been one and a half times as fit as *typica* to account for the rapidity of the rise in *carbonaria* frequency in Manchester. This fitness difference was much higher than most evolutionary biologists of the time thought feasible.

Not until the 1950s was Tutt's bird predation explanation of the rise of *carbonaria* in polluted regions tested by scientific experimentation. Dr. Bernard Kettlewell, using direct observation of the predation of live moths released onto tree trunks, and mark–release–recapture techniques, in two populations, one in a polluted and the other in a nonpolluted oak woodland, obtained strong evidence to support Tutt's

differential predation hypothesis. Both experiments showed that the *typica* form of the moth had lower fitness than *carbonaria* in the polluted woodland, but a higher fitness in the nonpolluted wood. It was the fact that Kettlewell obtained reciprocal results in the two environments that made his conclusions so convincing. Kettlewell also mapped the frequency of *carbonaria* against sulfur dioxide and soot fallout, finding a significant correlation between the frequency of *carbonaria* and both pollutants, that with sulfur dioxide being strongest. This correlation between high melanic frequencies and high levels of pollutants has been reinforced by the finding that *carbonaria* frequencies have declined following decreases in pollution levels as a result of anti-pollution and smoke control legislation.

The elements of the basic story of the peppered moth that are usually related are therefore:

1. The peppered moth has two distinct forms.
2. These forms are genetically controlled.
3. Peppered moths rest by day on tree trunks.
4. Birds find peppered moths on tree trunks and eat them.
5. The likelihood of a moth being found by a bird depends on its degree of camouflage.
6. Nonmelanic peppered moths are better camouflaged than melanics on lichen-covered tree trunks in rural areas. Melanic peppered moths are better camouflaged than nonmelanics in industrial areas where tree trunks have been denuded of lichens and blackened by soot fallout.
7. The frequencies of melanic and nonmelanic moths in a particular area depend on the level of bird predation of each form and the rate of migration of moths into the area from adjacent districts in which the form frequencies are different.

Since Kettlewell's research, other studies on the peppered moth, which have included work on the intermediate form, *insularia* (Fig. 3), and the natural resting behavior of the moth (Fig. 4), have refined some of the details of the case. However, Kettlewell's basic qualitative deductions remain valid.

FIGURE 3 The peppered moth has a third intermediate form, *insularia.*

TABLE I The Decline in the Frequency of the *carbonaria* Form of the Peppered Moth Since Anti-pollution Legislation at Two Sites in England

Year	Caldy Common, West Kirby, northwest England	West Cambridge, England
1960/1961	94.2	94.8
1965	90.2	—
1970	90.8	75.0
1975	86.6	64.7
1980/1981	76.9	45.9
1985	53.5	39.5
1990	33.1	22.2
1995	17.6	19.2
1998/2000	11.5	15.1

Note. Similar differences have been recorded in the United States (see text and Grant *et al.*, 1995, 1996).

Industrial Melanism in Reverse

In the 1950s, anti-pollution legislation was introduced in industrial countries on both sides of the Atlantic. This legislation led to declines in both sulfur dioxide and particulate soot emissions. Subsequently, the frequencies of the melanic forms have declined considerably in industrial regions in Britain (Table I). Current rates of decline are broadly in line with theoretical predictions using computer simulations. If the decline in *carbonaria* continues at its present rate, this form will be reduced in Britain to the status of a rare mutation by 2020. A similar decline in the frequency of the melanic form (f. *swettaria*) of the American subspecies of the peppered moth, *B. betularia cognataria,* has occurred in some parts of the United States.

Data on the declines of *carbonaria* in Britain and *swettaria* in America are important for three reasons. First, they show that evolution is not a one-way process. Evolutionary changes can be reversed if the selective factors that lead to them are reversed. Second, the data sets from different populations in Europe and America are, in effect, replicate natural experiments. The consistency in the patterns of increase and decrease in the frequencies of melanic forms correlated to pollution levels adds weight to the selective explanation of the evolutionary changes observed. Third, the accord between predicted decreases in melanic frequency and the observed frequency currently being obtained argues that the factors incorporated into the models are broadly correct.

Other Examples of Full Industrial Melanic Polymorphism

The case of the peppered moth is not unique. A small number of other examples of full industrial melanic polymorphism are known. The melanic forms in most of these cases are controlled by dominant alleles of single genes.

FIGURE 5 Nonmelanic and melanic forms of the brindled beauty, *L. hirtaria.*

An exception is that of the brindled beauty, *Lycia hirtaria,* in which the melanic form *nigra* (Fig. 5) is controlled by a recessive allele. That most recent melanic forms are genetically dominant is not surprising because a dominant mutation will be fully expressed as soon as it arises and will quickly be favored by selection if advantageous. Recessive mutations would not be exposed to selection until they occurred in homozygotes, in which their effects would be expressed phenotypically.

In some species showing full industrial melanic polymorphism, such as the lobster moth, *Stauropus fagi* (Fig. 6), melanism developed at roughly the same time as in the peppered moth. In others, industrial melanism has developed much more recently, as in the cases of the sprawler, *Brachionycha sphinx,* and the early grey, *Xylocampa areola,* in which industrial melanism developed only in the second half of the 20th century. The reason that industrial melanism did not evolve earlier in these species is probably serendipitous: a melanic mutation simply did not occur previously in an appropriate population.

The different timings of the initial occurrence of industrial melanics of different species help emphasize that natural selection cannot cause change unless phenotypic variation

FIGURE 6 Nonmelanic and melanic forms of the lobster moth, *S. fagi.*

exists. This is manifest in the oak beauty moth, *Biston strataria,* the closest British relative of the peppered moth. The oak beauty has a melanic form, *melanaria,* which is a common industrial melanic in Holland, but has never been recorded, except as a rare mutation, in Britain. In terms of its ecology, behavior, and distribution, the oak beauty is similar to the peppered moth. However, the *melanaria* mutation seems never to have arisen in Britain in favorable circumstances nor has this form reached Britain from continental Europe as a migrant. Melanism in the oak beauty in Europe can be contrasted with that of another moth, the figure of eighty, *Tethea ocularis.* The melanic form, *fusca,* of this species was known in Belgium and Holland in the early part of the 20th century, but was absent from Britain. This form arrived in southern England, by migration, in the mid-1940s. Following its arrival, f. *fusca* spread to many industrial parts of Britain and increased in frequency rapidly, although its frequency is now declining again in response to reductions in pollution.

The current declines in melanism seen in the peppered moth, the figure of eighty, and several other species, following anti-pollution legislation suggest that future studies of industrial melanism may have to shift to countries in which industrialization is still increasing and anti-pollution measures are as yet limited.

PARTIAL INDUSTRIAL MELANIC POLYMORPHISM

Melanic forms of many species of moth are independent of industrialization. The factors that can favor melanism are numerous and varied. These have been discussed in detail by Kettlewell and Majerus. Their relevance to industrial melanism is that in some moths, the presence of melanic forms prior to, and independent of, industrialization provided a repository of melanic variants that were favored as pollution levels increased.

Indeed, it is likely that the majority of moths that exhibit melanic polymorphism, with melanic frequency correlated to pollution levels, had melanic forms occurring at relatively low equilibrium frequencies prior to the industrial revolution. Changes in the environment resulting from increased pollution favored these dark forms and their frequencies increased. The willow beauty, *Peribatodes rhomboidaria,* illustrates the idea well. In Britain, this species has long had a nonindustrial melanic form, *perfumaria.* The *perfumaria* form greatly increased in frequency in industrial regions in the late 19th century. In the 20th century, *perfumaria,* which still occurs at low frequency in some rural areas, particularly in Scotland, was replaced in industrial areas, but not elsewhere, by an even darker form, f. *rebeli.* Here then, f. *perfumaria* should be regarded as a partial industrial melanic, while f. *rebeli* is a full industrial melanic.

Many probable instances of partial industrial melanic polymorphism could be cited, but rather few of the species that fall into this class of melanism have been investigated in any depth. Exceptions are the pale brindled beauty, *Phigalia*

pilosaria; the mottled beauty, *Alcis repandata;* and the green brindled crescent, *Allophyes oxyacanthae.* All are trunk-resters, the increase in melanics in industrial regions being attributed to increased crypsis.

Some of these species show morph-specific habitat preferences. Morph-specific habitat preferences in Lepidoptera showing melanic polymorphism were first suggested to explain abrupt differences in melanic frequencies of the mottled beauty and the tawny-barred angle, *Semiothisa liturata,* either side of sharp habitat boundaries. Such differences have subsequently been recorded in 14 species, in all cases melanics having higher frequencies in woodland with dense canopies than in adjacent more open habitats.

Many species that now have industrial melanics first evolved melanism in specific ecological circumstances prior to industrialization. It is known, for example, that a number of species now show melanic polymorphism in unpolluted ancient coniferous forests, such as Rannoch Black Wood in Scotland. Similar habitats were more widespread in the past and are likely to have supported melanic forms. These melanic forms would have been at a selective disadvantage if they moved from areas with the specific ecological circumstances to which they were adapted. Consequently, the melanics evolved behaviors that restricted them to such habitats. Recent changes in forestry and land usage and increases in pollution have provided new habitats (e.g., conifer plantations, polluted woodlands) with ecological conditions that favor melanics. The melanics have consequently spread and risen in frequency, producing examples of partial industrial melanic polymorphism in which morph-specific habitat preferences are retained to some extent.

POLYGENIC INDUSTRIAL MELANISM

Of all categories of melanism, polygenic industrial melanism has been the least considered and is the most difficult to address. Examination of specimens collected over the past century and a half suggests that many species have experienced a gradual darkening of the colors and loss of patterning in industrial regions, irrespective of morph. Although some of this change may be attributed to the gradual fading that occurs in museum specimens with time, it is difficult to ascribe all of the differences to this phenomenon. Comparison of series of specimens of six species, from rural and industrial regions, collected between 1880 and 1914 with those collected between 1992 and 1996 showed that the ground color had darkened more in industrial regions than in the rural areas.

This gradual darkening is probably the result of selection acting on polygenic variation. Small variations in the color patterns of many species are known to be controlled by many genes, each having a small effect. The selective predation of lighter and thus less cryptic forms in regions affected by particulate air pollution will result in those alleles which produce darker morphs increasing in frequency. It is difficult to see how this hypothesis can be tested. However, if it is correct, the recent decrease in pollution should lead to a reversal of this trend, with ground colors lightening and patterns becoming more clearly defined again. Novel, digital methods of measuring the spectral reflectance of surfaces and storing data should allow measurement without reliance on museum specimens or photographs, both of which may fade with time.

MELANISM AND THE STUDY OF EVOLUTION

The significance of industrial melanism in the Lepidoptera to evolutionary biology has been considerable. It has provided one of the best observed examples of evolutionary change caused by natural selection and has shown that Darwinian selection can be a strong force. In the peppered moth, differential bird predation, together with migration, has been primarily responsible for the rise and fall of the melanic form *carbonaria*.

Although the story of the peppered moth is undoubtedly more complex than usually related, data accumulated during the past 40 years have done nothing to undermine Tutt's initial hypothesis of the role of differential bird predation or Kettlewell's experimental demonstrations of this role.

Within the Lepidoptera, the factors responsible for melanism and the forms of melanism that result are very variable. Because a great variety of factors may promote melanism, it may be misleading to extrapolate from one population to another, let alone from one species to another. Even within one class of melanism, the relative influence of different aspects of a species' biology will vary between species. Each species that has evolved melanic forms will have done so in the presence of a variety of different intrinsic and extrinsic circumstances. The differences in the factors affecting melanism in even the few well-studied cases suggest that there is still enormous scope for original research into this phenomenon. However, in species in which melanism is strongly correlated with pollution levels, such as the peppered moth, we are rapidly running out of time to pursue research into this phenomenon as melanics decline.

See Also the Following Articles
Coloration • Crypsis • Genetic Variation • Lepidoptera • Pollution • Thermoregulation

Further Reading

Cook, L. M. (2000). Changing view on melanic moths. *Biol. J. Linn. Soc.* **69,** 431–441.
Grant, B. S., Owen, D. F., and Clarke, C. A. (1995). Decline of melanic moths. *Nature* **373,** 565.
Grant, B. S., Owen, D. F., and Clarke, C. A. (1996). Parallel rise and fall of melanic peppered moths in America and Britain. *J. Hered.* **87,** 351–357.
Haldane, J. B. S. (1924). A mathematical theory of natural and artificial selection. *Trans. Cambridge Philos. Soc.* **23,** 19–41.
Howlett, R. J., and Majerus, M. E. N. (1987). The understanding of industrial melanism in the peppered moth (*Biston betularia*) (Lepidoptera: Geometridae). *Biol. J. Linn. Soc.* **30,** 31–44.
Kettlewell, H. B. D. (1955). Selection experiments on industrial melanism in the Lepidoptera. *Heredity* **9,** 323–342.
Kettlewell, H. B. D. (1956). Further selection experiments on industrial melanism in the Lepidoptera. *Heredity* **10,** 287–301.
Kettlewell, H. B. D. (1973). "The Evolution of Melanism." Clarendon Press, Oxford.
Majerus, M. E. N. (1998). "Melanism: Evolution in Action." Oxford University Press, Oxford.
Majerus, M. E. N., Brunton, C. F. A., and Stalker, J. (2000). A bird's eye view of the peppered moth. *J. Evol. Biol.* **13,** 155–159.
Owen, D. F. (1961). Industrial melanism in North American moths. *Am. Nat.* **95,** 227–233.
Tutt, J. W. (1896). "British Moths." George Routledge, London.

Insecta, Overview

Vincent H. Resh
University of California, Berkeley

Ring T. Cardé
University of California, Riverside

The species-rich superclass Hexapoda includes all insects and their near relatives that share the characteristic arrangement of having, as adults, three major body regions and six legs. The number of described insect species has increased greatly from the time of the early catalogers of life. Those 18th-century pioneers in biodiversity, such as Carl Linné, would not have conceived that there would be nearly a million species described by the 21st century. Most estimates today suggest that this number represents only 10 to 30% of the actual number of insect species thought to exist. The richness of living things is essentially the result of insect richness; animal biodiversity is therefore, in reality, mainly insect biodiversity.

Within the class Insecta, major forms of insects are grouped in orders. Ordinal-level groups represent divergent lineages that are nearly always recognizable by a set of distinctive characteristics. Almost always, an adult insect can be readily determined to order at a glance. The number of recognized orders has fluctuated slightly as entomologists' understanding of the included taxa and methods for classifying have developed. Classification schemes are both organizational systems and true scientific hypotheses. In this way they are dynamic, changing as new information becomes available. There are several important ways in which a classification may evolve. One is the subjective change in taxonomic rank. For example, in the 1950s all Ephemeroptera (the order containing mayflies) in North America were assigned to three families, and today they are in 21. Mostly, this is the result of raising subfamilies to family status.

If substantial evidence is found that a group previously recognized as an order is paraphyletic (i.e., does not contain all descendants of that group), then new monophyletic arrangements will be proposed. A good example is the order

Hemiptera. The taxa included in the order now were traditionally divided into two groups (often given ordinal status): Heteroptera (true bugs) and Homoptera. Recent analyses suggest a more complicated pattern of relationships. Three groups within the order Hemiptera are treated separately as suborders in this encyclopedia: Auchenorrhyncha (cicadas, spittlebugs, leafhoppers, treehoppers); Prosorrhyncha (Heteroptera and Coleorrhyncha); and Sternorrhyncha (aphids, psyllids, scale insects, whiteflies).

Perhaps the most exciting way that classifications may change is by the discovery of something genuinely novel. However, in insects, finding a truly new order (i.e., a group of taxa that have a combination of characteristics unique at that

level) is an astounding event. At the time that this encyclopedia was ready for printing, a new order (Mantophasmatodea) was discovered in southern Africa. It is the first order of living insects to be described in over 80 years! This discovery reinforces the point that there is much left to learn about our earth's biodiversity.

In many classification systems, orders are grouped into superorders, but what comprises a superorder is far from fixed. For example, zorapterans are viewed as being in the superorder Orthopteroidea by some and in the superorder Hemipteroidea by others. Although a close link between Trichoptera (caddisflies) and Lepidoptera (moths and butterflies) is supported by a wealth of concordant evidence

TABLE I   The Orders of Insects and Other Members of the Arthropod Superclass Hexapoda

Hierarchical category	Taxon	Families	Species
Class	Parainsecta		
Order	Protura	8	600
	Collembola (springtails)	27	9,000
Class	Entognatha		
Order	Diplura	8	1,000
Class	Insecta		
Subclass	Apterygota		
Order	Archaeognatha (bristletails)	2	500
	Zygentoma (thysanurans, silverfish)	5	400
Subclass	Pterygota		
Infraclass	Paleoptera		
Order	Ephemeroptera (mayflies)	37	3,000
	Odonata (dragonflies, damselflies)	31	5,500
Infraclass	Neoptera		
Division	Endopterygota		
Order	Blattodea (cockroaches)	5	4,000
	Mantodea (mantids)	8	1,800
	Isoptera (termites, white ants)	7	2,500
	Grylloblattodea (rock crawlers)	1	25
	Dermaptera (earwigs)	7	2,000
	Plecoptera (stoneflies)	16	2,000
	Embiidina (webspinners)	8	300
	Orthoptera (grasshoppers, katydids)	23	20,000
	Phasmida (walkingsticks)	2	3,000
	Mantophasmatodea	?	?
	Zoraptera	1	32
Superorder	Hemipteroidea		
Order	Psocoptera (booklice, barklice)	17	4,400
	Phthiraptera (biting lice, sucking lice)	24	4,900
	Hemiptera (true bugs)	104	55,000
	Thysanoptera (thrips)	9	5,000
Division	Endopterygota		
Order	Megaloptera (alderflies, dobsonflies)	2	300
	Raphidioptera (snakeflies)	2	260
	Neuroptera (lacewings, ant lions)	17	6,000
	Coleoptera (beetles)	135	350,000
	Strepsiptera	8	550
	Mecoptera (scorpion flies)	7	550
	Diptera (flies)	117	125,000
	Siphonaptera (fleas)	15	2,600
	Lepidoptera (moths, butterflies)	120	160,000
	Trichoptera (caddisflies)	45	11,000
	Hymenoptera (ants, bees, wasps)	73	150,000

and consequently is undisputed, relationships among many endopterygote (e.g., Coleoptera and Hymenoptera) orders are unclear because different data sets present conflicting evidence. Some orders [e.g., Collembola (springtails) and Protura] are considered to be noninsects (placed in the class Parainsecta), but evidence clearly places them in the superclass Hexapoda with insects.

The classification presented here is a snapshot of the current hypotheses of insect relationships. Because the field of systematics that underlies this classification scheme is ever evolving, future arrangements will undoubtedly shift. The extant (or existing) orders of insects, their common names, and estimates of their worldwide species and family richness are listed in Table I.

See Also the Following Articles

Arthropoda and Related Groups • Phylogeny of Insects

Insecticides

Fumio Matsumura
University of California, Davis

Insecticides is the term coined to describe chemicals used to control pest insects and related invertebrate pest species. Insects are by far the most important species against which these chemicals are targeted. Other major groups of pest organisms include mites, ticks, and nematodes. Acaricides (for the control of mites and ticks) and nematocides (for the control of nematodes) are chemicals specifically used to control these pests, but they are still considered subgroups of the broadly defined "insecticides" group.

Not all insecticides are designed to kill pest insects, despite the use of the suffix "-cides" which gives the connotation of biocidal agents. Insecticides have been defined to include any chemical that can be used to reduce damage caused by insects. Thus, nonlethal chemicals such as pheromones, repellents, hormone mimics, growth regulators, feeding inhibitors, anorectic agents (which cause loss of appetite), behavioral disrupters, food attractants (used in traps and as bait), and anesthetics, as well as those causing physical problems such as surfactants, sticky substances, desiccants, and barriers (such as oil film on the surface of water for mosquito larval control) are considered to be insecticides.

BRIEF HISTORY

Insecticides used prior to the 1940s were mostly inorganic compounds such as arsenicals. After World War II, DDT and other chlorinated pesticides came on the market. There is no question about the spectacular insect-controlling effects

achieved on many crops, and populations of some pests that affect both public and veterinary health were greatly diminished. The shortcomings of these compounds, particularly their lack of selectivity and harmful environmental effects, were eventually realized, however, leading to the termination of their use by the late 1970s. Meanwhile, organophosphorus and carbamate insecticides gained in popularity and have established themselves as two of the major classes of insecticides. Many of them offer at least some degree of selectivity (malathion is particularly outstanding in this regard) and are less persistent in the environment. In more recent years, functional synthetic analogues of naturally occurring toxic chemicals were developed. Pyrethroids, for example, are essentially synthetic mimics of naturally occurring pyrethrins found in the flowers of species of chrysanthemum. The synthetic neonicotinoids mimic naturally occurring nicotine from tobacco plants. Useful microbial products were also developed in the 1980s and 1990s; examples are *Bacillus thuringiensis* (Bt) toxins, avermectins, and spinosyns. Modern insecticides used today are generally very selective, mostly affecting only the targeted pest insect. They are potent, requiring only small quantities to achieve their effects, and they are much less persistent in the environment.

CLASSIFICATION OF INSECTICIDES

Synthetic organic insecticides may be divided into several major classes: (1) chlorinated hydrocarbons, (2) organophosphorus compounds (often referred to as organophosphates), (3) carbamates, (4) pyrethroids, (5) nicotinoids, (6) fumigants, (7) GABA receptor antagonists, (8) chitin synthesis inhibitors (benzoylureas), (9) mitochondrial poisons, and (10) insect hormone mimics. These classifications are based on either group-specific chemical characteristics (classes 1–6) or their action mechanisms (classes 7–10).

Other insecticides belonging to minor classes (i.e., fewer compounds per class or less frequent use) are (11) botanically derived naturally occurring insecticides (other than pyrethroids and nicotinoids), (12) microbially produced insecticides, (13) synergists, (14) semiochemicals such as attractants, including pheromones, (15) insect repellents or feeding deterrents, and (16) behavior-modifying agents for use on insects.

USE PATTERNS

Insecticides as a class of pesticides constitute about one-quarter of total pesticides (approximately a billion pounds per year) used in the United States. By far the largest volume of pesticides used is herbicides (620 million pounds), followed by insecticides (247 million pounds) and fungicides (131 million pounds) (all 1993). Approximately 75% of all pesticides used in 1993 were for the control of agricultural pests. Other uses are for pests found in the home (including gardens), industry, commerce, and public and veterinary health. The top 17 insecticides (used in 1993) were (1) chlorpyrifos, (2) terbufos, (3) methyl parathion, (4) carbofuran, (5) carbaryl, (6) phorate, (7) cryolite, (8) aldicarb, (9) propargite, (10) acephate, (11) malathion, (12) fenofos, (13) methomyl, (14) dimethoate, (15) azinphos-methyl, (16) ethyl parathion, and (17) profenfos. Most of these are organophosphates (1–3, 6, 10–12, 14–17) or carbamates (4, 5, 8, 13), but propargite is a sulfite ascaricide and cryolite (sodium fluoroaluminate, Na_3AlF_6) is a naturally occurring inorganic fluoride compound. Of these, the use of methyl parathion (3) and ethyl parathion (16) has been phased out. Among organochlorine insecticides, most of which have been eliminated, the only ones remaining are endosulfan (19th) and dicofol (22nd). The most popular pyrethroid is permethrin (25th, approximately 1,000,000 pounds) followed by cypermethrin (225,000 pounds) and fenvalerate (66,000 pounds). Pyrethroids are used in much lower quantities than organophosphates and carbamates mainly because the former compounds are much more powerful than the latter, and therefore only small amounts of pyrethroids per hectare are needed to control insect pests.

MECHANISMS OF ACTION OF INSECTICIDES

The great majority of insecticides used today are nerve poisons. This is because insects have highly developed nervous systems and, furthermore, many of their sensory receptors are exposed to the atmosphere outside the insect body. The insect nervous system relies on several key functions that have been exploited as the targets of insecticides: the sodium channel, acetylcholinesterase, the γ-aminobutyric acid (GABA) receptor, and the acetylcholine receptor.

The sodium channel, which is the insecticidal target of DDT, pyrethroids, pyrethrins, and other minor classes of insecticides, lines the outer surface of the neurons and functions as the voltage-dependent sodium ion pore (i.e., the pore opens or closes depending on the change in voltage). Upon the arrival of stimuli, this pore allows the selective entry of sodium ions into the neuron for a brief moment and then abruptly shuts down the flow (this phenomenon is called "inactivation"). Thereafter, the sodium channel goes through an internal rearrangement to recover its original state. Such an action causes a brief local equalization of sodium ions between the outside and the inside of the neuron (depolarization), and this change is sensed as a local signal for excitation by the affected neuron. These insecticides delay the shutdown process and furthermore delay the recovery process, resulting in a prolongation of the period of excitation. Insects thus affected continue in a state of hyperexcitation, leading to exhaustion and, at high doses of the insecticide, death.

The next important insecticidal target is acetylcholinesterase, which is attacked by organophosphorus and carbamate insecticides. This enzyme, by inducing hydrolysis, inactivates the interneuron nerve transmitter acetylcholine.

This excitatory transmitter is released upon the arrival of a signal from the distal end of one neuron, travels across the intercellular gap, arrives at the frontal end of the second neuron, and reacts with its specific acetylcholine receptor on the surface that sends the signal of excitation to the second neuron. It is important to stress here that such a successful signal transmission must be followed with an abrupt termination of the action of the transmitter; this allows for the second neuron to recover quickly enough and thereby stay ready for the next message, maintaining the normal function of the message-transmitting neuron. This termination action is mainly carried out by acetylcholinesterase, which eliminates acetylcholine from the vicinity of the acetylcholine receptor of the second neuron. All organophosphorus and carbamate insecticides, or their active metabolites, show potent inhibitory actions on acetylcholinesterase of insects as well as other animals. The insects affected by these chemicals show overt signs of excitation, exhaustion and, at sufficient doses, death.

The acetylcholine receptor also can be deactivated to cause the same type of hyperexcitation. Indeed, nicotinoids (which include naturally occurring nicotine analogues and their modern derivatives, sometimes called "neonicotinoids"), such as imidacloprid, are known to directly activate the acetylcholine receptor, just like acetylcholine. Nicotine's excitatory action is well known. Neonicotinoid derivatives readily penetrate the insect's body and nerve sheath, arriving at critical sites of neurons, and persisting there long enough to exert a powerful excitant effect.

The GABA receptor, in contrast, acts as the receiver for the inhibitory transmitter, GABA. That is, unlike acetylcholine, it is not an excitatory transmitter. The signal generated by this GABA–GABA receptor interaction is converted to the opening of chloride channels, which upon the arrival of the signal permit Cl^- ions to come into the signal receiving cells (either neurons or muscle cells), to make them nonresponsive to excitation stimuli. Those insecticides—chlorinated hydrocarbon insecticides, cyclodienes (such as γ-HCH, dieldrin, endosulfan, toxaphene), and more modern insecticides (such as fipronil)—render the chloride channel inoperative so that chloride ions cannot come into the cells. Cells thus affected fail to receive the inhibitory signal of GABA and therefore cannot counterbalance any excitatory forces. One group of insecticides, avermectin analogues, keep the chloride channel stuck in the open position, an action opposite from that of the excitation-inducing insecticides. These compounds induce long-lasting inhibition of excitation in insects. Insects thus affected by avermectin analogues show diminished activities, nonresponse to stimuli, and slow death through paralysis.

Certainly there are other mechanisms by which normal functions of insects may be affected. The main ones are as follows:

1. Mitochondrial poisons, such as rotenone, which causes respiratory failure.

2. Inhibitors of cuticle formation, via the action of dimilin, including the rest of the diflubenzyron derivatives, which cause difficulty with molting and maintaining protective shields.

3. Insect hormone mimics such as juvenile hormone analogues that keep affected insects as immature forms (this method is effective against insects that cause damage only as adults, e.g., mosquitoes). Another group is ecdysone analogues, which affect insect development, including molting.

4. *B. thuringiensis* toxins, which mainly affect the potassium channel in insect digestive systems.

5. Formamidine analogues, such as chlordimeform, which mimics octopamine, a naturally occurring transmitter/hormone, by acting on its receptor. Octopamine is used by insects and mites to control their behavior (among many of its actions), and therefore chlordimeform analogues are known to modify many behavioral patterns of insects and mites, and thereby protect crops from those pests.

INSECTICIDE RESISTANCE

In 1958 A. W. A. Brown's landmark publication, *Insecticide Resistance in Arthropods,* established the principle that insects as well as other related invertebrates are capable of developing resistance to insecticides through natural selection. The probability of the development of resistance largely depends on (1) the frequency of the resistance-conferring gene in the given population, (2) the level of selection pressure, (3) the degree to which resistant gene density is diluted by susceptible genes through influx of individuals from untreated areas, and (4) the stability of the resistance gene in the given population. In some cases, once established, resistance genes may persist in the same locality for many years. A good example may be the pyrethroid resistance of the moth *Helicoverpa armigera,* in Australia.

How insects develop resistance to insecticides is a topic that has fascinated many entomologists. Basically, there are two major ways through which insect pests acquire resistance: increased detoxification capabilities and alteration of the insecticide target sites (target sensitivity). The first type of resistance occurs more frequently than the second type, as well as all others. Detoxification of toxic insecticidal chemicals is carried out by specialized enzymes designed to handle all chemicals toxic to insects, not just insecticides. Insects, particularly those feeding on plants that produce naturally large amounts of toxic chemicals, have well-developed detoxification enzymes. There are three major types of detoxification enzymes: (1) broad-spectrum oxidases such as mixed function oxidases catalyzed by cytochrome P450, (2) hydrolases that break up esters, ethers, and epoxides, and (3) conjugation systems such as glutathione *S*-transferase, which are mediated to cover up the reactive part of the toxic chemical and further facilitate its removal. Every type of detoxification enzyme has been documented to play a role in the development of some form of resistance against various classes of insecticides.

In determining which type of detoxification enzymes will become the key player in the development of resistance, the most important factor for consideration is the chemical properties of the insecticide. For instance, carbamates and pyrethrins are readily detoxified by mixed function oxidases; therefore, if resistance is reported against these insecticides, one must first look for increased activities of mixed function oxidases in the resistant insects. If higher activity levels are found, the resistance spectrum (i.e., cross-resistance of carbamate-resistant insects to other types of insecticide) is usually wide because mixed function oxidases are capable of detoxifying chemicals of many different types. In contrast, organophosphorus and pyrethroid insecticides are mainly degraded by hydrolases. Thus, the involvement of an increased hydrolytic enzyme activity may be suspected when insects develop resistance against these chemicals.

A good example of this is malathion resistance. Malathion molecules contain two extra carboxylic acid ethyl ester parts. Malathion-resistant insects always show increased carboxylesterase activity. Esterases of these types are not broad-spectrum enzymes, and therefore malathion resistance is usually specific (i.e., usually the insects resistant to malathion are not resistant to other insecticides). Insecticides with labile halogens, epoxides, methoxy unsaturation, and some aliphatic unsaturation may be degraded through these glutathione-mediated detoxification systems, and hence their elevated presence could be suspected to cause resistance. This scheme is, however, merely a rough guess about the possible mechanism of development of metabolic resistance. Indeed, unexpected and unique resistance mechanisms have been reported to occur in some combinations of insecticides and insects (e.g., DDT resistance in *Drosophila*). The recommended method of identification of the metabolic cause is to co-treat insects with the insecticide and specific inhibitors for each type of metabolic detoxification system, such as piperonyl butoxide for mixed function oxidase and DEF for esterases.

In studies of mechanisms for target insensitivity resistance, mutations occurring in the sodium channel, the GABA receptor, and acetylcholinesterase have been found in insects resistant to DDT/pyrethroids, cyclodiene insecticides, and organophosphorus and carbamate insecticides, respectively. Those resistances are characterized by their specificity (low degrees of cross-resistance) and the general stability of resistance among insect populations in given localities.

REGULATIONS OF INSECTICIDE USES

Insecticides, like all other types of pesticide, are highly regulated by governments in all countries. In the United States, the main law governing the use of insecticides is the Federal Insecticide, Fungicide, and Rodenticide Act (FIFRA), which mandates registration with the U.S. Environmental Protection Agency of all insecticides used in the country. The initial data requirements for successful registration (so-called Tier 1) depend on the extensiveness of the projected use, the levels of acute toxicity of the agent contemplated, its effectiveness as an insect control agent (called "efficacy"), the intended modes of usage, and the availability of background knowledge, among other requirements. Occasionally, experimental use permits are given after this Tier 1 examination/process (e.g., for insect pheromones, which are already known to be almost nontoxic and are to be used only for a specific pest in small areas). Usually, however, registrants are required to go through a much more extensive and rigorous process of registration, data procurement, and evaluation. For example, extensive tests are required for acute, chronic (such as carcinogenicity tests), genetic, pathogenic, reproductive, hormonal, and immune toxicities along with the environmental behavior of chemicals and limited wildlife toxicities. Such registration processes, which must be completed before a new chemical pesticide can be sold in the United States, typically require 7 to 10 years and roughly $100 million.

Despite the thoroughness of the registration processes, occasionally problems come to the attention of the scientific community or the regulatory agencies. Sometimes, for example, old pesticides are registered despite the availability of extensive records of their actual use. This is partly the result of the relative ease of the registration process in the past and partly from the absence of the main registrant, who is not economically motivated to reregister the compound because the patents for those chemicals (and thereby the exclusive marketing right) have expired. The second type of problem is due to the failure of the regulatory agency/scientific community to address the special vulnerability of certain groups of human populations or ecosystems. Examples include the lack of toxicological data on infants and embryos, women, and the aged, and science's incomplete understanding of the hormonal effects of pesticides on humans and wildlife. The third type of problem is caused mostly by unforeseen scientific or technological developments, or unfortunate circumstances that are difficult to predict. The question of the safety of genetically modified crops and the assessment of strategies to study the recently discovered skin-hypersensitizing action of some pesticides serve as examples.

A recent trend is to look at this issue from the consumer's side. A good example is the enactment of the Food Quality Protection Act (FQPA), which addresses the presence of pesticide residues and other toxic chemicals in food and drinking water. A key part of this regulation is the consideration of children's health. Here, an extra safety factor of 10× is demanded to accommodate the postulated extra vulnerability of embryos, infants, and developing children. This requirement is enforced unless registrants can provide actual safety data to demonstrate that the susceptibility of these groups to the hazardous effects of the compound is equal to or less than that of adults.

In the end, the toxicological methods of evaluation, including overall risk assessment approaches, address the majority of health concerns. Future improvements are needed, however, to deal with unresolved environmental and human health risks.

See Also the Following Articles
DDT • Integrated Pest Management • Regulatory Entomology

Further Reading
Brown, A. W. A. (1951). "Insect Control by Chemicals," pp. 117–207. Wiley, New York.
Crouse, G. D., and Sparks, T. C. (1998). Naturally derived materials as products and leads for insect control: The spinosyns. *Rev. Toxicol.* **2**, 133–146.
Elliott, M., Farnham, A. W., Janes, N. F., Needham, P. H., and Pulman, D. A. (1974). Synthetic insecticide with a new order of activity. *Nature* **248**, 710–720.
Forrester, N., Cahill, M., Bird, L. J., and Layland, J. K. (1993). Management of pyrethroid and endosulfan resistance in *Helicoverpa armigera* (Lepidoptera: Noctuidae) in Australia. *Bull. Entomol. Res. Suppl. Ser.* **O(1)**, I–VIII, 1–132.
Kasuda, Y. (1999). Development of and future prospects for pyrethroid chemistry. *Pestic. Sci.* **55**, 775–782.
Marrone, P. G. (1999). Microbial pesticides and natural products as alternative. *Outlook Agric.* **28(3)**, 149–154.
Plunkett, L. M. (1999). Do current FIFRA testing guidelines protect infants and children? Lead as a case study. *Regul. Toxicol. Pharmacol.* **29(1)**, 80–87.
Yamamoto, I., Yabuta, G., Tomizawa, M., Saito, T., Miyamoto, T., and Kagabu, S. (1995). Molecular mechanism for selectivity toxicity of nicotinoids and neonicotinoids. *J. Pestic. Sci.* **20**, 33–40.

Insecticide and Acaricide Resistance

Gregor J. Devine and Ian Denholm
Rothamsted Research, U.K.

Insecticide resistance is an example of a dynamic evolutionary process in which chance mutations conferring protection against insecticides are selected for in treated populations. This article reviews the origins and mechanisms of resistance, the factors that influence its severity, and the current options for combating its detrimental impact on agricultural productivity and human health.

INTRODUCTION

The genetic variation inherent in all populations is the consequence of random mutations within individuals, their recombination through meiosis, and the dispersal of genes between populations (gene flow). This variation is then shaped by the chance events of genetic drift and by the deterministic process of natural selection. The latter phenomenon eliminates alleles (gene variants) that reduce the fitness of an organism and preserve those that are neutral or that increase fitness. In eukaryotes, the phenotypic changes (adaptations)

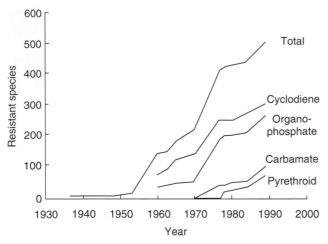

FIGURE 1 Increase in the number of arthropod species reported to resist insecticides over time, in total, and in response to the four most widely used classes of insecticide [Adapted with permission from Georghiou, G. P. (1990). Overview of insecticide resistance. *In* "Managing Resistance to Agrochemicals" (M. D. Green, H. M. Le Baron, and W. K. Moberg, eds.), pp. 18–14. ACS Symposium Series 421. Copyright (1990) American Chemical Society, Washington, DC.]

that result from this process are seldom visible over a human lifetime. The development of pesticide resistance by arthropods, however, is a spectacular exception to the rule.

Since the 1940s, synthetic insecticides have been used on an increasing scale to control the insects and mites that cause immense crop losses and pose major threats to public and animal health. However, because many of the target species have evolved resistance, some of these chemical control programs are failing. At the current time, more than 500 arthropod species have evolved resistance to at least one pesticide, and a few populations of some of those species are now resistant to all, or almost all, of the available products (Fig. 1).

The evolution of insecticide resistance has undoubtedly contributed to overall increases in the application of chemicals to crops. About 500,000 metric tons of insecticide is now applied each year in the United States alone, with obvious implications for both human health and the environment. Yet resistant insects continue to affect our agricultural productivity and our ability to combat vectors of disease. As a result, insecticide resistance imposes a huge economic burden on much of the world (in the United States alone, annual losses in crop and forest productivity have been estimated at $1.4 billion). Moreover, it is proving impossible to combat resistance by embarking on a chemical arms race. The development of a new insecticide takes 8 to 10 years at a cost of $20 to 40 million, and the rate of discovery of new insecticidal molecules, unaffected by current resistance mechanisms, seems to be on the wane.

Within just a few years of the registration of some of these new molecules, resistant insect populations have evolved.

DIAGNOSIS OF RESISTANCE

Although a large number of laboratory bioassay methods have been developed for detecting and characterizing resistance, most of these are limited to defining phenotypes and provide little or no information on the underlying genes or mechanisms. Thus, although bioassays remain the indispensable mainstay of most large-scale resistance monitoring programs, much attention is being paid to developing more incisive techniques that not only offer greater precision and turnover rates, but also diagnose the type of mechanism(s) present and, whenever possible, the genotypes of resistant insects. A variety of approaches are being adopted for this purpose, including electrophoretic or immunological detection of resistance-causing enzymes, kinetic and end-point assays for quantifying the activity of enzymes or their inhibition by insecticides, and DNA-based diagnostics for mutant resistance alleles.

The sensitivity of these techniques is exemplified by work on the green peach aphid, *Myzus persicae*. In northern Europe, this insect possesses three coexisting resistance mechanisms: an overproduced carboxylesterase conferring resistance to organophosphates, an altered acetylcholinesterase conferring resistance to certain carbamates, and target-site resistance (i.e., knockdown resistance, kdr) to pyrethroids. These mechanisms collectively confer strong resistance in this species to virtually all available aphicides. Fortunately, it is now possible to diagnose all three mechanisms in individual aphids by using an immunoassay for the overproduced esterase, a kinetic microplate assay for the mutant AChE, and a molecular diagnostic for the *kdr* allele. The combined use of these techniques against field populations provides up-to-date information on the incidence of the mechanisms and serves to inform growers of potential control problems and in the development of optimal strategies for the management of *M. persicae*.

EXTENT OF RESISTANCE

In some insects, resistance extends only to a few closely related compounds in a single chemical class. It may be very weak or restricted to a small part of the insects' geographical range. At the other extreme, some widespread pests, such as anopheline mosquitoes (e.g., *Anopheles gambiae*), the diamondback moth (*Plutella xylostella*), the Colorado potato beetle (*Leptinotarsa decemlineata*), and the sweet potato whitefly (*Bemisia tabaci*) now resist most or all of the insecticides available for their control. The most extensively used insecticide classes—organochlorines, organophosphates, carbamates, and pyrethroids—have generally been the most seriously compromised by resistance, and many principles relating to the origin and evolution of resistance can be demonstrated solely by reference to these fast-acting neurotoxins. In recent years, however, there has also been a worrying increase in resistance to more novel insecticides. These include compounds attacking the developmental pathways of arthropods (e.g., benzoylphenylureas), their respiratory processes [e.g., mitochondrial electron transport inhibiting (METI) acaricides], their digestive systems [e.g., *Bacillus thuringiensis* (Bt) endotoxins], and pathways associated with the regulation of their nervous processes (e.g., neonicotinoids).

ORIGINS AND BREADTH OF RESISTANCE

Insecticides are not considered to be mutagenic at their field application rates and are, therefore, not the causative agents of insecticide resistance. Rather they act to select favorable mutations inherent in the population to which they are applied. Some attempts to estimate the rates at which resistant mutations occur have been made. The treatment of blow flies (*Lucilia cuprina*) with a chemical mutagen resulted in the production of dieldrin-resistant target-site mutations in less than one per million individuals. Other studies, however, have found the incidence of resistant mutations to be worryingly high. A recessive allele conferring resistance to Bt toxins in unselected populations of the tobacco budworm, *Heliothis virescens*, was estimated to be present in about one in every thousand individuals in some areas of North America. Sixteen in every hundred insects were found to carry a Bt-resistant allele in unselected populations of the pink bollworm, *Pectinophora gossypiella*, in Arizonan cotton fields. Despite this, Bt cotton remains effective in the control of these species, suggesting that such estimates need to be interpreted carefully. Less empirical measures of mutation rates are extremely variable (10^{-3} to 10^{-16}), but they will undoubtedly be dependent on the resistance mechanism involved.

Resistant mutations seldom confer protection to just a single toxin. Most commonly, they exhibit differing levels of resistance to a range of related and unrelated insecticides. In its strictest sense, the term *cross-resistance* refers to the ability of a single mechanism to confer resistance to several insecticides simultaneously. A more complex situation is that of *multiple resistance*, reflecting the coexistence of two or more resistance mechanisms, each with its own specific cross-resistance characteristics. Disentangling cross-resistance from multiple resistance, even at the phenotypic level, is one of the most challenging aspects of resistance research.

Cross-resistance patterns are inherently difficult to predict in advance, because mechanisms based on both increased detoxification and altered target sites can differ substantially in their specificity. The most commonly encountered patterns of cross-resistance tend to be limited to compounds in the same chemical class (equivalent to the term "side-resistance" as used by parasitologists). However, even these patterns can be very idiosyncratic. For example, organophosphate resistance based on increased detoxification or target-site alteration can be broad ranging across this group or highly specific to a few chemicals with particular structural similarities. The breadth of target-site resistance to pyrethroids in houseflies is also dependent on the resistance allele present. The *kdr* allele itself affects almost all compounds in this class to a similar extent (~ 10-fold resistance), whereas resistance due to the more

potent *super-kdr* allele is highly dependent on the alcohol moiety of pyrethroid molecules, ranging from about 10-fold to virtual immunity. Cross-resistance between insecticide classes is even harder to anticipate, especially for broad-spectrum detoxification systems whose specificity depends not on insecticides having the same mode of action, but on the occurrence of common structural features that bind with detoxifying enzymes.

Empirical approaches for distinguishing between cross-resistance and multiple resistance include repeated back-crossing of resistant populations to fully susceptible ones, to establish whether resistance to two chemicals cosegregates consistently, and reciprocal selection experiments, whereby populations selected for resistance to one chemical are examined for a correlated change in response to another. If available, biochemical or molecular diagnostics for specific resistance genes can assist considerably with tracking the outcome of genetic crosses or with assigning cross-resistance patterns to particular mechanisms.

MECHANISMS OF RESISTANCE AND THEIR HOMOLOGY

Depending on the mechanism involved, resistance has been shown to arise through structural alterations of genes encoding target-site proteins or detoxifying enzymes, or through processes affecting gene expression (e.g., amplification or altered transcription). Examples of the former include the following.

- Enhanced metabolism of insecticides by cytochrome P_{450} monoxygenases can potentially confer resistance to most chemical classes. Much of the evidence for this mechanism is indirect, based on the ability of monoxygenase inhibitors to reduce the magnitude of resistance when used in combination with insecticides in bioassays.
- Enhanced activity of glutathione *S*-transferases (GSTs) is considered to be potentially important in resistance to some classes of insecticide, including organophosphates. Like monoxygenases, GSTs, exist in numerous molecular forms with distinct properties, making correlations of enzyme activity with resistance very challenging and often ambiguous.
- Enhanced hydrolysis or sequestration by esterases (e.g., carboxylesterases) capable of binding to and cleaving carboxylester and phosphotriester bonds undoubtedly plays an important role in resistance to organophosphates and pyrethroids. Biochemically, this is the best-characterized detoxification mechanism. Sometimes (e.g., for mosquitoes, blowflies, and *M. persicae*) the esterases have been identified and sequenced at the molecular level. Resistance caused by increased esterase activity can arise through a qualitative change in an enzyme, improving its hydrolytic capacity, or (as in mosquitoes and aphids) a quantitative change in the titer of a particular enzyme that already exists in susceptible insects.

The following examples appear to show that although some adaptations to the environment are unpredictable (e.g., the modifications of the forelimbs for flight are very different in birds, bats, and pterodactyls), the opportunities for insects to modify or reduce binding of insecticides, hence to develop target-site-based resistance mechanisms, are very limited indeed. It is conceivable that most of the mutations that confer such resistance do not allow the organism to retain normal functioning of the nervous system.

• Pyrethroids act primarily by binding to and blocking the voltage-gated sodium channel of nerve membranes. Knockdown resistance, or insensitivity of this target site, is now unequivocally attributed to structural modifications in a sodium channel protein. The same amino acid substitution (leucine 1014 to phenylalanine) in a sodium channel protein confers a "basal" kdr phenotype in a range of species including house flies, cockroaches, the green peach aphid, the diamond-back moth, and a mosquito *(A. gambiae)*. This phenotype may subsequently be enhanced (to "super-kdr" resistance) by further mutations that also recur between species.

• GABA receptors are targets for several insecticide classes including cyclodienes (a subclass of the organochlorines), avermectins, and fipronils. The primary mechanism of resistance to cyclodienes and fipronils involves modification of a particular GABA receptor subunit, resulting in substantial target-site insensitivity to these insecticides. The target-site mechanism of cyclodiene resistance has been attributed to the same amino acid substitution (alanine 302 to serine) in the GABA receptors of several species of diverse taxonomic origin including *Drosophila*, several beetles, a mosquito *(Aedes aegypti)*, a whitefly *(B. tabaci)*, and a cockroach *(Blatella germanica)*. When susceptible individuals of the sheep blowfly *(L. cuprina)* were exposed to the mutagen ethyl methanesulfonate (EMS), and their progeny screened for resistance to dieldrin (a cyclodiene), surviving insects exhibited an alanine-to-serine amino acid substitution in the GABA receptor identical to that found in nature.

• Organophosphates and carbamates exert their toxicity by inhibiting the enzyme acetylcholinesterase (AChE), thereby impairing the transmission of nerve impulses across cholinergic synapses. Mutant forms of AChE showing reduced inhibition by these insecticides have been demonstrated in several insect and mite species. Biochemical and molecular analyses of insecticide-insensitive AChE have shown that pests may possess several different mutant forms of this enzyme with contrasting insensitivity profiles, thereby conferring distinct patterns of resistance to these two insecticide classes.

Some of these resistance mechanisms are illustrated schematically in Fig. 2.

SPREAD OF RESISTANCE GENES

The recurrence of specific resistance mutations within and between taxa begs another question: Have such mutations arisen repeatedly within the same species, or have they appeared on a limited number of occasions and subsequently

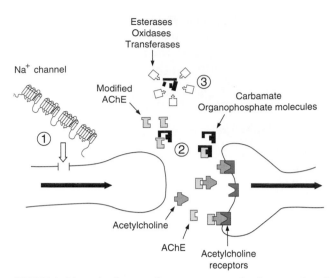

FIGURE 2 Schematic diagram of a nerve synapse showing examples of insecticide resistance mechanisms: (1) changes in the structure of the sodium channel confer kdr or super-kdr target-site resistance to pyrethroids; (2) modified AChE is no longer bound by organophosphates and remains available to break down acetylcholine molecules after neurotransmission across the synapse; (3) detoxifying enzymes degrade or sequester insecticides before they reach their targets in the nervous system.

spread through migration and/or human agency? Although there is molecular evidence for some resistance genes having several independent origins in the same species (e.g., for target-site resistance to cyclodienes in the red flour beetle, *Tribolium castaneum*), other examples suggest that some mechanisms have arisen only once.

Organophosphate resistance in the mosquito *Culex pipiens* is primarily conferred by allozymes at two closely linked loci (esterases A and B), coding for insecticide-detoxifying carboxylesterases. Overproduced allozymes (resulting from amplification of A or B genes) tend to recur in geographically disjunct areas. This situation could be explained by recurrent mutations generating each amplification event *de novo*, or by a nonrecurrent mutation that has spread within and between populations. Restriction mapping of DNA around the esterase genes suggest the latter, with large-scale gene flow attributable to passive migration of mosquitoes on ships and/or airplanes. A new resistance allele in southern France is known to have originated in the vicinity of the international airport and seaport at Marseilles.

Resistance to organophosphates in the aphid *M. persicae*, is also attributable to the amplification of a gene encoding an insecticide-detoxifying carboxylesterase. Despite the often widespread dispersion of these amplified genes in the aphid genome, restriction analyses have indicated that all copies are in the same immediate genetic background. This suggests that amplification occurred only once, whereupon the amplified DNA was moved intact around the genome through chromosomal rearrangements.

FACTORS AFFECTING THE EVOLUTION OF RESISTANCE

As an evolutionary trait, insecticide resistance is unusual in that we can identify the main selection pressure with ease, but the rate at which resistance develops is governed by numerous biotic and abiotic factors. These include the genetics and ecology of the pests and their resistance mechanisms, and the operational factors that relate to the chemical itself and to its application. To manage resistance effectively, an assessment of genetic, ecological, and operational risk is required. Although this can be done empirically on a species-by-species basis, one of the great challenges of the future is to understand why some species seem to have a greater tendency to become resistant than others.

Genetic Influences

To predict how quickly resistance will become established, it is necessary to understand how resistant alleles affect the survival of phenotypes in the field. For example, the dominance of resistance genes exerts a major influence on selection rates. In laboratory bioassays evaluating the relative survival of susceptible homozygotes (SS), heterozygotes (RS), and resistance homozygotes (RR) over several insecticide concentrations, RS individuals usually respond in an intermediate manner. In the field, however, dominance is dependent on the concentration of insecticide applied and its uniformity over space and time. Even when the initial concentration is sufficient to kill RS individuals (rendering resistance effectively recessive), upon weathering or decay of residues, this genotype may later show increased survival, with resistance becoming functionally dominant in expression. When resistance genes are still rare, hence mainly present in heterozygous condition, this sequence can have a profound effect in accelerating the selection of resistance genes to economically damaging frequencies.

The diverse mating systems of insects also influence the rate at which resistance evolves. Although most research has focused on outcrossing diploid species (typified by members of the Lepidoptera, Coleoptera, and Diptera), systems based on haplodiploidy and parthenogenesis also occur among key agricultural pests. In haplodiploid systems, males are usually produced uniparentally from unfertilized, haploid eggs, and females are produced biparentally from fertilized, diploid eggs. The primary consequence of this arrangement (exemplified by whiteflies, spider mites, and phytophagous thrips) is that resistance genes are exposed to selection from the outset in the hemizygous males, irrespective of intrinsic dominance or recessiveness. Whether a resistance gene is dominant, semidominant, or recessive, resistance can develop at a similar rate.

Most species of aphid undergo periods of parthenogenesis (in which eggs develop and give rise to live offspring in the absence of a paternal genetic contribution) promoting the selection of clones with the highest levels of resistance and/or the most damaging combination of resistance mechanisms. In fully anholocyclic (asexual) populations, such as those of *M. persicae* in northern Europe, the influence of parthenogenesis has led to strong and persistent associations between resistance mechanisms within clonal lineages.

Ecological Influences

Fecundity and generation times have a huge bearing on the evolution of resistance in a population. The greater the number of individuals, and the faster they reproduce and attain maturity, the higher the likelihood that a favorable mutation will occur, and be maintained in the population. Faster growth and higher population numbers will also have an effect on the size of a pest population, and therefore the need for insecticide treatment.

The dispersal capabilities of pests can also act as primary determinants of resistance development. Movement of pests between untreated and treated parts of their range may delay the evolution of resistance because of the diluting effect of susceptible immigrants. Conversely, large-scale movement can also accelerate the spread of resistance by transferring resistance alleles between localities. A good example relates to the two major bollworm species (Lepidoptera: Noctuidae) attacking cotton in Australia. Only the cotton bollworm *Helicoverpa armigera*, has developed strong resistance. *H. punctigera*, despite being an equally important cotton pest, has remained susceptible to all insecticide classes. The most likely explanation is that *H. punctigera* occurs in greater abundance on a larger range of unsprayed hosts than *H. armigera*, thereby maintaining a large pool of unselected, susceptible individuals, which dilute resistant mutations arising on treated crops.

Operational Influences

Operational factors are at human discretion and can be manipulated to influence selection rates. Factors exerting a major influence in this respect include the rate, method, and frequency of applications, their biological persistence, and whether insecticides are used singly or as mixtures of active ingredients.

Equating operational factors with selection is often difficult, since without detailed knowledge of the mechanisms present it is impossible to test many of the assumptions on which genetic models of resistance are based. If resistance alleles are present, the only entirely nonselecting insecticide doses will be ones sufficiently high to overpower all individuals, regardless of their genetic composition, or ones so low that they kill no insects at all. The latter is obviously a trivial option. Prospects of achieving the former depend critically on the potency and dominance of resistance genes present. A pragmatic solution to this dilemma is to set application doses as far above the tolerance range of homozygous, susceptible individuals as economic and environmental constraints permit, in the hope that any heterozygotes that do arise will be effectively controlled. However, this approach will obviously be ineffective if resistance turns out to be more common than suspected (resulting in the presence of homozygous resistant individuals) or if resistance alleles exhibit an unexpectedly high degree of dominance (and heterozygotes are therefore phenotypically resistant). Unless a high proportion of insects escape exposure altogether, the consequence could then be very rapid and effective selection for homozygous resistant populations.

In practice, concerns about optimizing dose rates to avoid resistance are secondary to those related to the application process itself. Delivery systems and/or habitats promoting uneven or inadequate coverage will generally be more prone to select for resistance, because, under these circumstances, pests are likely to encounter suboptimal doses of toxins that will permit survival of heterozygous individuals.

The timing of insecticide applications relative to the life cycle of a pest can also be an important determinant of resistance. A good example of this is found in the selection of pyrethroid resistance in *H. armigera* in Australia. On cotton foliage freshly treated with the recommended field dose, pyrethroids killed larvae up to 3 to 4 days old irrespective of whether they were resistant by laboratory criteria. Since the sensitivity to pyrethroids of larvae of all genotypes was found to decline with increasing larval size, the greatest discrimination between susceptible and resistant phenotypes occurred when larvae achieved a threshold age. Targeting of insecticides against newly hatched larvae, as is generally advocated for bollworm control, not only increases the likelihood of contacting larvae at the most exposed stage in their development but also offers the greatest prospect of retarding resistance by overpowering its expression. It may also have the effect of reducing genetic variation and therefore the potential number of resistant mutations. Indeed, it is also possible to impose genetic "bottlenecks" by applying pesticides when populations are already low (e.g, when they are overwintering). Although such a tactic might be beneficial where populations are fully susceptible, if resistant mutations are already present, it might act to increase their frequency.

In theory, the application of two or more unrelated chemicals as insecticide mixtures offers substantial benefits for delaying the selection of resistance. The underlying principle is one of "redundant killing," whereby any individuals already resistant to one insecticide are killed by simultaneous exposure to another, and vice versa. However, achieving this objective requires that each type of resistance be rare and that both ingredients persist throughout the effective life of an application. Otherwise, one compound will exert greater selection pressure than the other, and the advantage of applying a mixture will be lost.

Fitness of Resistant Individuals

In the absence of insecticidal selection pressure, resistance genes can impose fitness costs on their carriers. Sometimes these costs are quite subtle and difficult to determine. In *M. persicae,* resistant individuals are less inclined to move from senescing to younger leaves and are therefore more vulnerable to isolation and starvation after leaf abscission. These costs appear to contribute to a decline in the frequency of resistant insects between cropping seasons.

COMBATING INSECTICIDE RESISTANCE

Insecticide resistance management (IRM) aims to intervene in the evolutionary process and either overcome resistance or prevent its appearance in the first place. There are several practical, economic, and political constraints on the choice of possible IRM tactics and the precision with which they can be applied:

- The properties of any resistance genes present are often unknown, and knowledge of pest ecology may still be rudimentary.
- It is often necessary to contend with a whole pest complex rather then just a single pest species.
- There may be a very limited number of insecticides available for use in management strategies.
- For highly mobile pests, at least, countermeasures may need to be standardized and synchronized over large areas, sometimes whole countries.
- Resistance is a dynamic phenomenon; that is, any mechanisms already known to exist may change over time.
- To promote compliance with management strategies, the tactics adopted should be as unambiguous, rational, and simple as possible.

A strategy first implemented on Australian cotton in 1983 against *H. armigera* illustrated many features of large-scale attempts at resistance management. Introduced in response to unexpected, but still localized, outbreaks of pyrethroid resistance in *H. armigera,* the strategy was based primarily on the concept of insecticide rotation. The threat of pyrethroid resistance was countered by restricting these chemicals to a maximum of three sprays within a prescribed time period coincident with peak bollworm damage. To diversify the selection pressures being applied, farmers were required to use alternative insecticide classes at other stages of the cropping season.

Initially, this strategy had the desired effect of preventing a systematic increase in the frequency of pyrethroid-resistant phenotypes. Additional recommendations, including the targeting of insecticides against newly hatched larvae (the most vulnerable life stage) and the plowing in of cotton stubble to destroy resistant pupae overwintering in the soil, undoubtedly contributed to this success. Unfortunately, the restrictions placed on pyrethroid use were insufficient to combat resistance in the long-term, and it has been necessary to revise the strategy to place greater emphasis on the strategic use of nonpyrethroids against this pest.

Another strategy incorporating a wide range of chemical and nonchemical countermeasures was introduced on Israeli cotton in 1987. The primary objective was conservation of the effectiveness of insecticides against *B. tabaci*. Under recommendations coordinated by the Israeli Cotton Board, important new whitefly insecticides are restricted to a single application per season within an alternation strategy optimized to contend with the entire cotton pest complex and to exploit biological control agents to the greatest extent possible. One major achievement of this strategy has been a dramatic reduction in the number of insecticide applications against the whole range of cotton pests, but especially against *B. tabaci*. Sprays against whiteflies now average fewer than two per growing season compared with over 14 per season in 1986. Most importantly, the strategy has generated an ideal environment for releasing additional new insecticides onto cotton and for managing them effectively from the outset.

An integral part of delaying or preventing the evolution of resistance is the preservation of the innate "susceptibility" of a pest species. This is arguably as valuable a genetic resource as those of the rice, wheat or apple "gene banks" that are so carefully tended in institutes around the globe. The most effective way to conserve susceptibility, based both on evolutionary models and on empirical evidence, is to ensure the presence of pesticide-free "refugia" in which susceptible genotypes may survive and reproduce. The inclusion of refugia as essential components of IRM strategies is a recent phenomenon, signaling that pest management is no longer simply about eradication, but is now at least partially focused on conservation.

TRANSGENIC PLANTS

A recent development in crop protection has been the release of crop plants genetically engineered to express genes for insecticidal toxins derived from the microbe *B. thuringiensis*. In 2001 the total area worldwide planted to Bt plants was estimated to exceed 12 million ha. Existing toxin genes in Bt cotton and corn are active specifically against certain key lepidopteran pests (especially bollworms and corn borers); another engineered into potatoes provides protection against the Colorado potato beetle.

Aside from their commercial prospects, insect-tolerant transgenic crops offer numerous potential benefits to agriculture. By affording constitutive expression of toxins in plant tissues throughout a growing season, the incorporation of Bt genes into crops could reduce dramatically the use of conventional broad-spectrum insecticides against insect pests, as well as remove the dependence of pest control on extrinsic factors such as climate and on the efficiency of traditional application methods. However, this high and persistent level of expression also introduces a considerable

risk of pests adapting rapidly to resist genetically engineered toxins. To date, there are no substantiated reports of resistance selected directly by exposure to commercial transgenic crops, but resistance to conventional Bt sprays (selected in either the laboratory or the field) has been reported in more than a dozen insect species. Research into the causes and inheritance of such resistance is providing valuable insights into the threats facing Bt plants and the efficacy of possible countermeasures.

Tactics proposed for sustaining the effectiveness of Bt plants have many parallels with those considered for managing resistance to conventional insecticides. However, they are more limited in scope because of the long persistence and constitutive expression of engineered toxins, and because of the limited diversity of transgenes currently available. Indeed, for existing "single-gene" plants, the only prudent and readily implementable tactic is to ensure that substantial numbers of pests survive in nontransgenic refugia. These can be incorporated into the crop itself, or or they may comprise alternative host plants. The success of this strategy is dependent on some key assumptions: (1) that resistant mutations are recessive or at least only partially dominant, so that their heterozygous forms can be controlled by the toxins expressed; (2) that refugia will produce enough susceptible insects to ensure that insects carrying resistant alleles do not meet and mate; and (3) that resistant alleles will carry a fitness cost, rendering insects less fit when the selection pressure is removed (e.g, outside the growing season when the insect is dependent on other crops).

In the longer term, there are potentially more durable options for resistance management: stacking (or pyramiding) of two or more genes in the same cultivar, or possibly rotations of cultivars expressing different single toxins. Whatever measures are adopted, it is essential that plants expressing transgenes be exploited as components of multitactic strategies rather than as a panacea for resistance problems with conventional insecticides.

RESISTANCE IN NONPEST SPECIES

The ability of insect predators and parasitoids to develop pesticide resistance would be of enormous benefit to pest management strategies that are chemically dependent. Although pyrethroid and organophosphate resistance has been documented in predatory mites (e.g., *Typhlodromus pyri* in orchards and *Amblyseius womersleyi* in tea fields) and hymenopterous parasitoids (e.g., *Aphytis holoxanthus* in orchards and *Anisopteromalus calandrae* in grain stores), reports of insecticide-resistant beneficial species from the field are far rarer than they are for pest species. Reporting bias aside, the most likely reasons for this are the difficulty in host location when both natural enemy and host are under selection pressure and, in comparison with herbivorous species, the possibility that the enzyme systems of predators and parasites are less well adapted to detoxify xenobiotics.

Resistance may, therefore, be more likely to develop if the hosts or prey are themselves resistant, thereby making their location easier. For example, a parasitic wasp *(A. calandrae)* of a stored-grain beetle *(Sitophilus oryzae)* is resistant to insecticides, and it is thought that this adaptation has been encouraged because the host organism is sheltered from insecticides by the grain kernels it inhabits.

Many attempts have been made to select resistance in beneficial species in the laboratory, but limitations on the size of the populations (and hence their genetic variability) that can be maintained under these conditions means that resistance tends to arise through the development of polygenic traits. Once released into natural populations, these are more likely to fragment and dissipate than rarer, but generally more robust, single mutations.

In general, when resistance does occur in nonpest species, its mechanisms are similar to those exhibited by pest species. Organophosphate resistance in strains of *A. calandrae* has been linked to the presence of carboxylesterase-like enzymes similar to those conferring organophosphate resistance to the *M. persicae*. The expression level of the carboxylesterase-like enzyme in this wasp is approximately 30-fold higher in the resistant strain relative to that in the susceptible strain, and the mechanism seems to have its basis in a single nucleotide replacement. Organophosphate resistance in strains of the warehouse pirate bug *(Xylocoris flavipes)* has also been linked to the presence of a carboxylesterase. Resistance to this chemical group in the lacewing, *Chrysopa scelestes,* has been attributed to increased AChE activity.

CONCLUDING REMARKS

Research on the topic of insecticide resistance has provided invaluable insights into the origin and nature of adaptations, and these are proving to have broad significance for understanding genetic responses to change in the environment. In many respects the continuing battle against resistance is as good an example of coevolution as any and is a clear illustration of how such processes generate biological diversity. In this instance, however, the diversity being created is undesirable from a human standpoint and, because of the threat posed to susceptible genotypes, probably temporary.

It is important to note that the pest management problems posed by the evolution of resistance are not unique to control strategies that use conventional insecticides. The utilization of host plant resistance is a case in point. Resistance to insects in crop plants is selected by screening for genes that provide resistance in the laboratory or in field plots, then crossing those genes into crop strains with other desirable characteristics. At least six major genes for resistance to the Hessian fly *(Mayetiola destructor)* have been successively bred into wheat over the past two decades. In each instance, the introductions of new resistant mutations in the plant were rendered useless by the evolution of

corresponding protective adaptations in the fly. Another example of such coevolution comes from the use of semiochemical tools for pest control. In many parts of Asia, a synthetic pheromone is used to disrupt mating in the tea tortrix moth *(Adoxophyes honmai),* the larvae of which can cause severe damage in tea plantations. Researchers in Japan have recently reported the evolution of a new biotype of this species that exhibits reduced sensitivity to the pheromone. Such events make it clear that regardless of whether the major strategies for pest management continue to use conventional chemicals, the "arms race" between insect evolution and human ingenuity will continue to present major challenges.

See Also the Following Articles

Agricultural Entomology • Biotechnology and Insects • Genetically Modified Plants • Genetic Variation

Further Reading

Denholm, I., and Rowland, M. W. (1992). Tactics for managing pesticide resistance in arthropods—Theory and practice. *Annu. Rev. Entomol.* **37,** 91–112.

Devonshire, A. L., Field, L. M., Foster, S. P., Moores, G. D., Williamson, M. S., and Blackman, R. L. (1998). The evolution of insecticide resistance in the peach-potato aphid, *Myzus persicae. Philos. Trans. R. Soc. Lond. Ser. B, Biol. Sci.* **353,** 1677–1684.

ffrench-Constant, R. H., Pittendrigh, B., Vaughan, A., and Anthony, N. (1998). Why are there so few resistance-associated mutations in insecticide target genes? *Philos. Trans. R. Soc. Lond. Ser. B, Biol. Sci.* **353,** 1685–1693.

Forrester, N. W., Cahill, M., Bird, L. J., and Layland, J. K. (1993). Management of pyrethroid and endosulfan resistance in *Helicoverpa armigera* (Lepidoptera, Noctuidae) in Australia. *Bull. Entomol. Res.* suppl. 1.

Horowitz, A. R., Forer, G., and Ishaaya, I. (1994). Managing resistance in *Bemisia tabaci* in Israel with emphasis on cotton. *Pestic. Sci.* **42,** 113–122.

Martinez Torres, D., Devonshire, A. L., and Williamson, M. S. (1997). Molecular studies of knockdown resistance to pyrethroids: Cloning of domain II sodium channel gene sequences from insects. *Pestic. Sci.* **51,** 265–270.

McKenzie, J. A. (1996). "Ecological and Evolutionary Aspects of Insecticide Resistance." R. G. Landes, Austin, TX.

Pimental, D., Acquay, H., Biltonen, M., Rice, P., Silva, M., Nelson, J., Lipner, V., Giordano, S,. Horowitz, A., and Damore, M. (1992) Environmental and economic costs of pesticide use. *Bioscience* **42,** 750–760.

Raymond, M., Chevillon, C., Guillemaud, T., Lenormand, T., and Pasteur, N. (1998). An overview of the evolution of overproduced esterases in the mosquito *Culex pipiens. Philos. Trans. R. Soc. Lond. Ser. B, Biol. Sci.* **353,** 1707–1711.

Roush, R. T. (1989). Designing resistance management programs—how can you choose? *Pestic. Sci.* **26,** 423–441.

Roush, R. T. (1997). Bt-transgenic crops: Just another pretty insecticide or a chance for a new start in resistance management? *Pestic. Sci.* **51,** 328–334.

Soderlund, D. M., and Bloomquist, J. R. (1990). Molecular mechanisms of insecticide resistance. *In* "Pesticide Resistance in Arthropods" (R. T. Roush and B. E. Tabashnik, eds.), pp. 58–96. Chapman & Hall, London.

Tabashnik, B. E., Liu, Y. B., Malvar, T., Heckel, D. G., Masson, L., and Ferre, J. (1998). Insect resistance to *Bacillus thuringiensis:* Uniform or diverse? *Philos. Trans. R. Soc. Lond. Ser. B, Biol. Sci.* **353,** 1751–1756.

Insectivorous Plants

Lewis J. Feldman
University of California, Berkeley

The term "insectivorous" was used by Charles Darwin to characterize a group of plants that seemed to trap and feed on insects. Since Darwin's time, observations have revealed that these plants capture and interact with a greater variety of animals, which can include spiders, lizards, sow bugs, tadpoles, and frogs, and judging from some reports, even mammals such as rats and rabbits. Hence, because of this varied diet, many workers now prefer to describe such plants as carnivorous, rather than solely insectivorous. Yet the interactions between carnivorous plants and animals go beyond the presence of certain creatures as items on a plant's menu. While it is true that the most spectacular and usually the most obvious activities of carnivorous plants seem to be in the often elaborate mechanisms for capture and digestion of prey, many other (often more subtle) associations, occur between these plants and animals. Researchers are just at the beginnings of learning about these other fascinating interactions.

THE CARNIVOROUS HABIT

Plant carnivory is a rarity, occurring in only about 550 out of approximately 250,000 plant species. The carnivorous habit is not obligate, and carnivorous plants can grow without an insect meal, depending instead on photosynthesis and minerals supplied from the soil. In general, carnivorous plants grow in sunny areas, and in mineral-deficient, sometimes sandy soils. Often these soils have standing or gently moving water, with any dissolved minerals from the soil being easily carried away by the flowing water. The carnivorous plant habitat is typically low in nitrogen and phosphorus and, some reports suggest, in potassium as well. In this sort of habitat, plants that have alternative strategies for obtaining essential minerals are at a competitive advantage. The capture of insects and other animals thus provides carnivorous plants with a supplemental source of essential nutrients.

The carnivorous habit depends on an ability to trap prey. In the vast majority of carnivorous plants, the trap represents a modification of the entire leaf or of structures borne on the leaf. Given this rather straightforward requirement of a trap, it should be easy enough to characterize a plant as carnivorous, or not. However, the picture is not so simple: many plants can trap insects yet are not considered to be carnivorous. What truly distinguishes a plant as carnivorous is not only a trapping ability, but also a mechanism to digest prey and to absorb the prey's nutrients.

Digestion implies an ability of the plants to break down the trapped prey into its component chemicals, to be able to absorb them as nutrients. It is the specifics of this digestion that have caused some controversy. Some workers consider a plant to be carnivorous only if it has an inherent ability to digest prey—that is, if the plant itself produces enzymes to break down the insect. Other plants, sometimes considered to be semicarnivorous, are able to trap prey but depend on the assistance of other organisms, usually, but not always, microbes, to digest the captured insects. However, for this short article, a plant is considered to be carnivorous if it traps and has a means, of its own making or not, for digesting prey.

The trap of a carnivorous plant is a true marvel, designed to attract, capture, digest, and then absorb nutrients from the prey. Traps can be grouped by whether they are "passive," with no or slowly moving "parts," often relying on gravity to aid in capture of the insect, or "active," exhibiting some sort of usually rapid movement. Perhaps the most familiar examples of passive traps are the sundews *(Drosera)* and the pitcher plants (*Sarracenia* and *Darlingtonia* in temperate climes, and *Nepenthes* in the tropics).

PASSIVE TRAPS

The sundews capture their prey by producing from stalked glands an adhesive, or glue (the drop of "dew"), which captures and holds fast the insect. As the prey struggles, it is covered with the sticky mucilage, and as a consequence, suffocates. The stalked glands then bend in toward the prey; in some species, the entire leaf enfolds the prey. A second type of gland on the leaf secretes digestive enzymes and acids, initiating the breakdown and subsequent absorption of nutrients. Darwin was so enthralled with the sundews that about two-thirds on his book *Insectivorous Plants* is devoted to this group. He notes his surprise at "finding how large a number of insects were caught by the leaves of the common sundew," and speculates that "as this plant is extremely common in some districts, the number of insects thus annually slaughtered must be prodigious."

A fascinating variation on the carnivorous plant passive trap theme is shown by plants that comprise the genus *Roridula*. These plants, considered by some workers not to be truly carnivorous, are native to South Africa and may be near extinction. Individuals in this group have leaves covered with stalked mucilage-secreting glands, which as in the sundews, capture and hold fast insects. This is where the carnivorous story would end, were it not for another player, an assassin bug (Heteroptera: Reduviidae). Large numbers of these capsid bugs may inhabit *Roridula* and are able to traverse the leaves without themselves being ensnared by the glue. When other insects are captured by the plant, the assassin bugs move to the trapped prey, suck out their liquid contents, and, some time afterward, secrete a nutritious substance that is absorbed by the leaf and nourishes the plant.

The second major group of plants having passive traps are the pitcher plants. In this group the leaf becomes variously modified, often into a tube, and develops at the base of the

FIGURE 1 Habit view of the California pitcher plant *(Darlingtonia californica)* growing in a bog in northern California. The leaf is divided into a hood region and tube region. Digestion occurs in a well of water at the base of the tube.

tube a well that must fill with water for the pitcher to function as a trap. The temperate species of pitcher plants *(Darlingtonia* in the western United States and *Sarracenia* in eastern North America) (Figs. 1 and 2) are usually terrestrial. In these plants the leaf lures flying insects by producing nectar that sometimes covers colorful appendages. Crawling

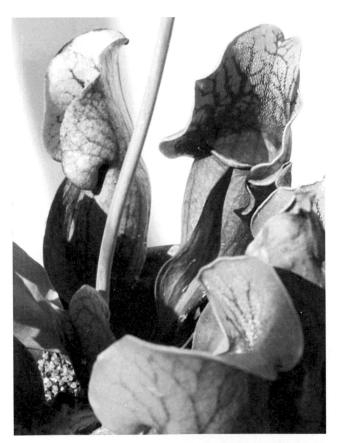

FIGURE 2 Eastern pitcher plant *(Sarracenia purpurea)* from Ohio. In this species a hood is absent.

insects follow nectar trails running along the outside of the leaf. The nectar trails lead to the mouth of the tube, where the surface is smooth and slippery and from which the insect can easily lose its foothold, thus falling into the watery well. Escape from the well is almost impossible, since the inside wall of the tube-leaf is lined with downward-pointing hairs. One would think that flying insects could fly out if they started to fall. To counter this possibility, pitcher plants such as *Darlingtonia* have developed a hooded leaf, transparent and sealed at its top. When an insect tries to leave the leaf, it flies toward light coming through the transparent upper portion of the hood. Since, however, the exit is sealed, eventually the insect becomes so exhausted that it falls into the well. There is some suggestion that pitcher plants may produce a "drug" to confuse the flying insect, and fluids in the well may contain substances that stun and quiet the fallen insect. In addition to nectar serving as an attractant, the possible development of ultraviolet signaling, as employed by flowers to attract pollinators, may also serve to lure insects to the trap. Many temperate pitcher plants secrete hydrolytic enzymes into the liquid in the well, thereby digesting the insect, whereas other pitcher plants (e.g., *Darlingtonia*) produce none of their own digestive enzymes but instead rely on bacteria to decay the insect. In either model, the digestive enzymes can be quite powerful, with only the hardest parts of insects, such as legs or shells, remaining undigested.

The tropical pitcher plants belong to the genus *Nepenthes,* so named by Linnaeus after the drug "nepenthe," which Helen of Troy was said to have dispensed in drink to soldiers to "relieve their sorrow and grief." In giving this name, Linnaeus noted, "What botanist would not be filled with admiration if, after a long journey, he should find this wonderful plant. In his astonishment, past ills would be forgotten when beholding the admirable work of the creator!" In *Nepenthes,* the pitcher develops at the end of a leaflike petiole. Indeed, the complexity and variety of pitchers in *Nepenthes* strains one's credulity, for it is hard to believe that what one is looking at is a leaf. Like their temperate cousins, *Nepenthes* spp. produce nectar to lure prey, which subsequently become intoxicated, lose their foothold, and fall into the trap. *Nepenthes* spp. generally produce climbing stems, thus elevating the pitchers, and perhaps thereby making them more accessible to potential prey.

Species of *Nepenthes,* and likely all carnivorous plants, do not seem to be designed to trap one particular species. An inventory of the traps shows that their diets are ever-changing and can be quite varied. For example, from 10 *Nepenthes* pitchers over a season, Erber found arthropods of 150 identified species, belonging principally to the orders Diptera, Hymenoptera, and Collembola, and the families Fromicidae, Aphididae, and Acarina. Similar tallies in *Sarracinea* reveal victims of 115 families belonging to 14 orders of insects, including several species of Mollusca. This strategy no doubt ensures some prey, if, for example, a particular insect species is not present one year, or becomes extinct, and also allows the plant to trap a variety of insects over a very long season. Additionally, such variety may be important in supplying a diversity of nutrients.

ACTIVE TRAPS

Plants with rapid movement usually first come to mind when carnivorous plants are mentioned. These so-called active traps imprison their prey by a quick movement of all or part of the leaf. Into this group are placed several genera; one well known, the Venus flytrap *(Dionaea)* (Fig. 3) and the others less familiar (e.g., the bladderworts, *Utricularia*) (Fig. 4). It is believed that the rapid movement comes about when the prey makes contact with a triggering mechanism, resulting in the generation of a small electric current and the activation or closure of the trap. As with the passive traps, the lure for the insect is usually some sweet nectar. In the case of the Venus flytrap leaf, in which the two halves of the blade are joined along one side, as in an open book, nectar is produced on the inner surface of the leaf (Fig. 3). As an insect wanders along this inner surface collecting nectar, it may contact the trigger hairs, of which there are about two to four on the inner surface of each half-leaf. For the trap to close, either an individual trigger hair "must be touched twice, or two different trigger hairs must be touched sequentially, within a time period that is neither too short (< 0.75 s) nor too long (> 20 s)." When the hair or hairs are touched within the right time interval, the trap literally snaps shut, though at first, not completely. Initially, small openings remain between the two halves, presumably to allow smaller insects to escape from the trap. When unsuitable prey gain their release, the trap reopens and awaits the main course. But if the insect is unable to escape through the small openings and continues to struggle, the trap closes more fully. Subsequently, enzymes are secreted by special gland cells, and the insect is digested and its nutrients absorbed by the leaf.

Less well known as active trappers but possessing traps more complex than the Venus flytraps are the bladderworts, which grow in wet or periodically wet areas. Bladderworts, the largest genus of carnivorous plants, grow worldwide, on every continent. They develop diminutive, often microscopic traps that cover the leaves (Fig. 4). The size of the trap determines what creatures will enter: paramecia, rotifers, water fleas, worms, and mosquito larvae, for example. As in the Venus flytrap, contact with trigger hairs initiates the trapping mechanism, which involves the opening of a "door" leading to a chamber maintained under a vacuum, a sucking in of the prey, and a resealing of the trap; all this occurs within 10 to 15 thousandths of a second! With the secretion of enzymes, the prey is digested, usually within hours. There is some speculation that the trap can also lure prey.

FIGURE 3 Habit view of the Venus flytrap (*Dionaea* sp.). In this genus the blade is divided into two halves, which are attached along one side. On the inner surfaces of the blade a lure is produced, and here too are located the trigger hairs.

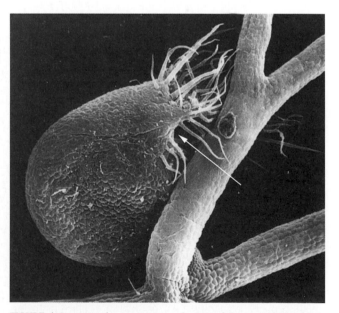

FIGURE 4 Scanning electron micrograph view of the trap of *Utricularia neglecta*. The large hairs ("antennae") may act as guides luring the prey to the trap mouth (arrow). [After Juniper, B. E., Robins R. J., and Joel, D. M. (1989). "The Carnivorous Plants." Academic Press, London. Reprinted with permission.]

OTHER INTERACTIONS BETWEEN CARNIVOROUS PLANTS AND INSECTS

Thus far, plant–insect interactions have been presented in the context of insects serving as prey. But the associations can be much more varied and complex, as seen in the instance of the assassin bug and *Roridula*. Entomologists are now only beginning to appreciate the many other ways in which carnivorous plants and insects/animals interact. Some of the more fascinating examples are found in the pitcher plants, where other animals turn the traps to their own advantage. Spiders often can be found prowling about the mouth of pitchers, then lowering themselves on silken strands to retrieve prey from the pitcher well. In *Nepenthes,* spiders also use the pitcher for protection from predators. If a predator is detected, the spider will lower itself on a silk thread to the pool and, if necessary, will even hide under the water until danger passes.

Other insects spend part or most of their lives in the wells of pitchers. For example, the pitcherplant mosquito *(Wyeomyia smithii)* lays its eggs on the moist inner surface of the leaf, or more often, in the pool of liquid. The larvae hatch and feed on detritus from trapped insects, bacteria, and protozoans. As winter approaches, the larvae go into a dormant state and overwinter in the pitcher, exiting the pitcher in the spring as adult mosquitoes. In climates where water freezes, the larvae spend winter frozen in the ice of the pitcher. Interestingly, the liquid that digests the trapped insects seems to have no detrimental effects on the larvae. Fish fly larvae also live in pitchers and, like the mosquito larvae, are not injured by the pitcher's digestive enzymes because, it is speculated, their bodies produce a protective substance.

One of the most fascinating examples of an association between plant and insect, benefiting the insect, is seen in species of *Exyra* moths (Noctuidae) that exploit the pitcher leaf to shelter their young. The cycle begins with the female moth entering an open leaf and laying its eggs on the inner wall of the pitcher leaf. When the larvae hatch, they move about on silken strands, feeding on the inner wall. As they grow, hence becoming more visible to predators, the larvae move to the top of the pitcher, severing vascular strands carrying water to the upper regions of the pitcher leaf, which causes the top of the pitcher leaf to dry, collapse, and fold over the opening. The developing larvae are now shielded from predators. Just before a larva prepares to pupate, it chews a hole in the wall of the leaf. Through this hole, the moth exits the leaf.

Yet another example of an insect exploiting its association with carnivorous plants is the solitary sarracenia wasp *(Chlorion harrisi).* This insect uses the pitcher as an incubator for its eggs. In preparation for the laying of eggs, the wasp packs into the bottom of the pitcher tube a layer of grass, which is then overlayered with freshly killed grasshoppers or crickets. This process may be repeated several times, resulting in alternating layers of insects and plant materials. Eggs are then laid among the dead insects and the whole construction covered by another layer of grass. The eggs can now develop protected, and when the young hatch, they have a food supply.

The last example represents what may come closest to a commensal, or symbiotic, association between plant and insect. Certain species of *Nepenthes* (e.g., *N. bicalarate*) have enlarged, hollow petioles in which ants take up residence. In return for this "home" (domatia), the ants, it is suggested, protect the plant from predators.

See Also the Following Article
Plant–Insect Interactions

Further Reading

D'amato, P. (1998). "The Savage Garden." Ten Speed Press, Berkeley, CA.
Darwin, C. (1875). "Insectivorous Plants." Appleton, New York.
Juniper, B. E., Robins, R. J. and Joel, D. M. (1989). "The Carnivorous Plants." Academic Press, London.
Lerner, C. (1983). "Pitcher Plants: The Elegant Insect Traps." Morrow, New York.
Lloyd, F. E. (1942). "The Carnivorous Plants." Chronica Botanica, Waltham, MA.

Insectivorous Vertebrates

Insects are a part of the diet of a vast array of animals. Indeed, there are almost no animal groups that do not include some representatives that consume insects, incidentally or intentionally, actively or accidentally. Even obligate herbivores consume insects, although never in any large number. Several groups of animals are insectivorous as juveniles and shift dietary focus as adults, whereas other groups become insectivorous only as adults.

DIVERSITY OF INSECTIVORY

A simple list of the insectivorous terrestrial animals would be lengthy and include many freshwater fishes, most frogs and salamanders, lizards and snakes, and birds and mammals. Because marine ecosystems do not include many insects, the incidence of insectivory among marine invertebrates and vertebrates is much less common.

In freshwater fishes, insects are ubiquitous and widely consumed. In streams, for example, 70 to 90% of the macroinvertebrates are insects, comprising as much as 99% of the numbers of individual organisms and 99% of the biomass. Fishes take advantage of these abundant resources and consume insects from all of the 13 orders of insects with aquatic life stages.

Terrestrial salamanders and almost all frogs, which are often filter feeders or herbivores as larvae, become predominantly insectivorous as adults. Frogs and salamanders that remain aquatic, or become so secondarily, possess a wide array of dietary choices that can include other aquatic life, including (but not limited to) insect adults or larvae.

Lizards, more so than snakes, include insects in the diet. Consumption of insects also occurs in other reptiles such as turtles and juvenile crocodilians. Insect prey selected by lizards is somewhat size dependent; smaller lizards consume more small insects, whereas larger lizards can also consume larger insects. Some lizards that are insectivores as juveniles become more herbivorous as adults. In snakes, smaller insectivores become more carnivorous as they get larger, focusing on other vertebrate prey, especially mammals, frogs, and other snakes. Among some groups of lizards and snakes, specialization for insectivory is a familiar pattern; in these instances, ants and termites are most frequently consumed.

There are many birds that consume insects as a dominant part of the diet. Some of these insectivorous lineages include pipits and wagtails (Motacillidae), bulbuls and allies (Campephagidae, Pycnonotidae, Chloropseidae), waxwings and allies (Ptilogonatidae, Bombycillidae, Dulidae), dippers (Cinclidae), warblers and gnatcatchers (Sylviidae, Parulidae), flycatchers (Muscicapidae), and titmice, nuthatches, and treecreepers (Paridae, Sittidae, Certhiidae). Many birds capture insect prey in flight, whereas other birds forage in shrubs and trees or on the ground. Some birds specialize by obtaining insect food grooming large mammals or following behind large mammalian herbivores and foraging on the insects disturbed by large mammal movements.

Mammals include many insectivorous groups, some generalists and others obligate specialists. Most of those that specialize in eating insects eat either ants or termites. Generalized insectivores will eat insects along with other arthropods such as centipedes, millipedes, spiders, and scorpions. Marsupials,

bats, primates, rodents, carnivores, and other groups of mammals include insectivorous lineages. Specialized feeding on particular insects, especially ants and termites, occurs among a few frogs, many lizards, and some snakes and has occurred in several different lineages of mammals (see Table I).

ANATOMICAL SPECIALIZATIONS FOR INSECTIVORY

Although eating insects is a dominant part of many diets, anatomical and behavioral specializations for insectivory are not as widespread. Many lizards, for example, feed on whatever suitable prey item might be available. Nevertheless, there are several anatomical specializations that seem to assist in the capture of insect prey. Perhaps the most obvious and remarkable trait is a highly projectile tongue. This sort of tongue evolved many times, in many different lineages, and in many different ways. Most frogs have a projectile tongue whose intrinsic muscles attach to the lingual edge of the symphysis of the jaw. The tongue is flipped out of the mouth, in the same way that a catapult works, so that the back of the tongue after it is extruded contacts the prey first. Salamanders have evolved several different types of projectile tongues. One group of lungless salamanders, the plethodontids, uses the hyobranchial skeleton and associated muscles, once used to ventilate lungs, in a tongue protrusion mechanism that is quite spectacular. Contraction of these muscles results in protrusion of the tongue as well as large parts of hyobranchial skeleton, resulting in an extruded tongue that can reach 80% of body length.

Chameleon lizards have an unmatched ability to accurately aim, project, and hit arboreal insect prey. The muscles that chameleons use to accomplish this ballistic feat contract faster than any other vertebrate muscle. Furthermore, the tongue

TABLE I Examples of Frog, Lizard, and Mammal Lineages in Which Termites and/or Ants Are a Significant Part of the Diet

Frogs	Mammals
Dendrobatidae—poison frogs	Canidae—dogs and foxes
Microhylidae—narrowmouth frogs and toads	Cercopithecidae—Old World monkeys
Pelobatidae—spadefoot toads	Cricetidae—New World mice, hamsters, etc.
Rhinophrynidae—Mesoamerican burrowing toads	Dasypodidae—armadillos
Lizards	Didelphidae—opposums
Agamidae—angelheads, calotes, dragon lizards, and allies	Herpestidae—mongooses
Amphisbaenia—wormlizards	Lorisidae—lorises
Gekkonidae—geckos and pygopods	Manidae—pangolins
Iguaninae	Muridae—Old World rats
Lacertidae—wall lizards, rock lizards, and allies	Myrmecobiidae—marsupial anteaters
Phrynosomatinae	Myrmecophagidae—anteaters
Scincidae—skinks	Orycteropodidae—aardvarks
Teiidae—whiptail lizards, tegus, and allies	Sciuridae—squirrels
Tropidurinae	Tachyglossidae—echidnas
Snakes	Talpidae—moles
Anomalepididae—early blindsnakes	Thylacomidae
Leptotyphlopidae—threadsnakes and wormsnakes	Vespertilionidae—bats
Typhlopidae—blindsnakes	

Note. Not all the taxa that are a part of these lineages eat only ants and/or termites. There are many other lineages not listed that eat insects.

can be accurately projected up to 200% of body length. The tongue has a large tip covered with viscous mucous. A muscle in the large fleshy tip contracts just after the tongue tip strikes a prey item, creating a slight vacuum that assists in prey capture.

Mammals eat insects and the most distant ancestors of mammals may have been insectivores. Generalized features derived from the primitive amniote condition are associated with mammalian feeding, including a longer secondary pallate, heterodont dentition, higher metabolic rates and more active foraging behaviors. Several of these features are reversed in obligate ant- and termite-eating mammals. For example, anteaters, pangolins, and the echidna, numbat, and aardvark possess highly simplified teeth few in number or lack teeth entirely. Some ant and termite specialists have lower metabolic rates, but it is not clear if these rates are retained from a primitive mammalian ancestor or if they are a response to prey that may present chemical challenges to typical mammalian digestive systems. A long, sticky, and protrusible tongue is a common feature among ant- and termite-feeding mammals. Details of tongue anatomy confirm that many of these lineages evolved these specializations independently.

See Also the Following Articles
Food, Insects as • Predation

Further Reading

Cushing, C. E., and Allan, J. D. (2001). "Streams: Their Ecology and Life." Academic Press, San Diego.
Gerking, S. D. (1994). "Feeding Ecology of Fish." Academic Press, San Diego.
Schwenk, K. (ed.) (2000). "Feeding: Form, Function, and Evolution in Tetrapod Vertebrates." Academic Press, San Diego.
Thorp, J. H., and Covich, A. P. (eds.) (2001). "Ecology and Classification of North American Freshwater Invertebrates," 2nd ed. Academic Press, San Diego.
Zug, G. R., Vitt, L. J., and Caldwell, J. P. (2001). "Herpetology: An Introductory Biology of Amphibians and Reptiles," 2nd ed. Academic Press, San Diego.

Insect Zoos

Leslie Saul-Gershenz
Center for Ecosystem Survival, San Francisco, California

Insect zoos, exhibits that display live insects and arthropods to the general public, have gained recognition for their educational value and broad appeal. They have demonstrated that insects are interesting to millions of people, not just a select few. Insects represent the majority of terrestrial species on earth and display a dazzling diversity of lifestyles, behaviors, and adaptations. The great naturalists Pliny, Fabre, Wallace, Darwin, Belt, and Wheeler possessed a boundless desire to observe living insects in the natural world. The observation of living insects in the field and lab still excites the imagination of naturalists today, and the general public has been increasingly infected with this enthusiasm through exposure to the lives of insects through superb nature films and photography. Insect zoos provide opportunities for positive firsthand observations of live insects to millions of people. These exhibits are intended to inspire a curiosity to learn more, and to balance the myths and misrepresentations promoted by the cinema and other commercial media. Insectariums have proven to be enormously popular with the public wherever they appear, whether modest or grand, whether showcasing native or exotic species. Today there are over 100 insect zoo exhibits throughout the world.

INSECT ZOOS DEFINED

The term "insect zoo" has been applied to facilities of many different types. Defined broadly, an insect zoo or insectarium is an exhibit facility dedicated to the display of live insects housed in a separate room, building, or distinct exhibit hall and maintained primarily for public visitation. Insect zoos typically are permanent, year-round facilities that house live insects and related groups of arthropods (arachnids, centipedes, millipedes, and crustaceans) and occasionally representatives of other invertebrate groups. Insect zoos have been built in zoological parks, natural history museums, botanical gardens, county parks, horticultural centers, amusement parks, nature reserves, and universities, and on privately owned land. Interest in their development has increased (excluding the period between the first and second World Wars) with growing public interest in biodiversity and the documented success of insect exhibits (Table I).

Most facilities contain a series of terrariums, where species are displayed in naturalistic mini-environments. A major insectarium is a comprehensive coverage of the class Insecta, representing many different orders such as Coleoptera, Hymenoptera, Orthoptera, and Mantodea. Observation honey bee hives, ant and termite colonies, walkingsticks, katydids, lubber grasshoppers, and assassin bugs are a few examples of typical displays. These facilities can be distinguished from collections of a few species of arthropod housed in a reptile house or aquarium or included in an exhibit that focuses on the interpretation of a particular ecosystem. Many tropical rain forest exhibits today include a few displays of insects, often a leafcutter ant exhibit and a few other invertebrates as nominal representatives of the vast diversity of invertebrates. However, overall these exhibits emphasize the vertebrate fauna of rain forests and present a relatively minor treatment of the subject of invertebrates. Major insectariums typically have between 30 to 100 live displays, exhibiting up to 100 species of arthropods.

A butterfly house or lepidoptery is a type of live insect facility that primarily displays members of the order

TABLE I   List of Selected Insect Exhibits Built Worldwide from 1797 through 2001

Exhibit	Date of establishment
Europe	
Small early acclimatization collections	
Jardin des Plantes, Muséum d'Histoire Naturelle et Ménagerie, Paris, France	1797
Jardin d'Acclimatisation, Paris, France	1860
Permanent exhibits	
Insect House, London Zoo, London, England (renovation 1913, Web of Life, 1999)	1881
Artis Zoo, Amsterdam, Netherlands	1898
Frankfurt Zoo, Frankfurt, Germany (renovation 1957)	1904
Zoologischer Garten Köln, Cologne, Germany (new Insektarium 1971)	1905–1929
Budapest Zoo, Budapest, Hungary (renovations: Vivarium, 1970; Butterflies, 2000)	1907
Zoo-Aquarium Berlin, Berlin, Germany (renovated 1978–1983)	1913
Zoologischer Garten, Leipzig, Germany	1913
Gruga Park, Essen, Germany	1956
Sherbourne Butterfly House, Dorset, England	1960
Löbbecke Museum and Aquazoo, Düsseldorf, Germany (renovation in 1987)	1970
Guernsey Butterfly Farm, England	1977
Insektarium, Zoological–Botanic Garden Wilhelma, Stuttgart, Germany	1980
London Butterfly House, Syon Park, England	1981
Stratford-upon-Avon Butterfly Farm, Stratford-upon-Avon, England	1985
Papiliorama-Nocturama, Marin Center, Neuchâtel, Switzerland	1988
Dortmund Zoo, Dortmund, Germany	1991
Bug World, Bristol Zoo, Bristol, England (renovated in 1996)	1992
Butterflies Center, Girona, Spain	1995
Mariposario del Drago, Tenerife, Spain	1997
Idea Schmetterlings-Paradies Neuenmark, Germany	1998
Krefeld Zoo, Krefeld, Germany	1998
Micropolis, Saint-Léon-en Lévezou France	2000
The Butterfly Arc, Montegrotto Terme, Italy (seasonal, 1988)	2001
Projeckledare Aquademin, Sweden "Science Centre," Göteborg, Sweden	2001
Asia	
Takarazuka Insectarium, Takarazuka Zoological and Botanical Garden, Japan	1954–1967
Toshima-en Insectarium Toshima-en Amusement Park, Tokyo, Japan	1957
Insectarium, Tama Zoo, Japan (renovations in 1966, 1975, 1988)	1961
Penang Butterfly House, Penang, Malaysia	1986
Fragile Forest, Singapore Zoo, Singapore	1998
North America	
United States	
Early short-lived or seasonal exhibits	
Bronx Zoo, New York Zoological Society, New York	1910, 1940, 1945
Goddard State Park, Providence, Rhode Island	1934–1937
Chicago (Brookfield) Zoo, Brookfield, Illinois	1938, 1947–1950
Flushing Meadow Zoo, New York	1969–1970
Insect Zoo, Smithsonian National Museum of Natural History, Washington, DC	1971
Permanent year-round exhibits	
Arizona–Sonoran Desert Museum, Tucson, Arizona	1952
Otto Orkin Insect Zoo, Smithsonian Museum of Natural History, Washington, DC	1976
World of the Insect, Cincinnati Zoo and Botanic Garden, Cincinnati, Ohio	1978
Insect Zoo, San Francisco Zoological Gardens, San Francisco, California	1979
Invertebrate Exhibit, National Zoological Park, Washington DC	1987
Butterfly World, Coconut Creek, Florida	1988
Cecil B. Day Butterfly Center, Callaway Gardens, Pine Mountain, Georgia	1988
Ralph K. Parsons Insect Zoo, Natural History Museum of Los Angeles County, Los Angeles, California	1992
Cypress Gardens, Winter Haven, Florida	1992
Terminix Insect City, Fort Worth Zoo, Fort Worth, Texas	1992
Butterfly Encounter, San Diego Wild Animal Park, Escondido, California	1993
Moody Gardens, Galveston, Texas	1993
Cockerell Butterfly House, Houston Natural Science Museum, Houston, Texas	1994
Butterfly Pavilion and Insect Center, Westminster, Colorado	1995
Detroit Zoo, Royal Oak, Michigan	1995
Bug World, Woodland Park Zoo, Seattle, Washington	1996
Sophia M. Sachs Butterfly House, Chesterfield, Missouri	1998

(continues)

TABLE I   *(Continued)*

Exhibit	Date of establishment
Chicago Academy of Sciences, Chicago, Illinois	1999
Tropical Butterfly House and Insect Village, Pacific Science Center, Seattle, Washington	1999
Mosanto Insectarium, St. Louis Zoo, St. Louis, Missouri	2000
Puelicher Butterfly Wing Milwaukee Public Museum, Milwaukee, Wisconsin	2000
Magic Wings, South Deerfield, Massachusetts	2000
Butterfly House & Insectarium, North Carolina Museum of Life & Science, Durham, North Carolina	2000
Canada	
L'Insectarium de Montréal, Montréal, Quebec	1990
The Niagara Parks Butterfly Conservatory, Niagara Falls, Ontario	1996
Newfoundland Insectarium, Deer Lake, Newfoundland	1997
Victoria Bug Zoo, Victoria, British Columbia	1997
F. Jean MacLeod Butterfly Gallery, Science North, Sudbury, Ontario	2000
L'Arche des Papillons c.m. Inc., Quebec	2000
Australia	
Butterfly House, Royal Melbourne Zoological Park, Parkville, Victoria	1985
Australian Butterfly Sanctuary, Kuranda, Queensland	1987
Insectarium of Victoria, Woodend	1993
New Zealand	
Butterfly and Orchid Garden, Thames	1999
Latin America	
The Butterfly Farm, La Guacima de Alajuela, Costa Rica	1990
The Butterfly Garden at Shipstern Nature Reserve, Belize	1990
Spirogyra Butterfly Garden, San Jose, Costa Rica	1992
La Selva Butterfly Farm—Primary Forest, Ecuador	1992
Green Hills Butterfly Ranch and Botanical Collection, Cayo District, Belize	1997
Tropical Wings Nature Center, Cayo District, Belize	1998
Africa	
Insectarium de la Réunion, Réunion	1992
Butterfly World, Klapmuts, South Africa	1996
French West Indies	
La Ferme des Papillons, St. Martin, French West Indies	1994

Lepidoptera, typically in a large walk-through immersion exhibit enclosed in glass or screening, such as the Cockerell Butterfly House in Houston, Texas. Some facilities are combinations of both insect zoos and butterfly houses. Year-round walk-through butterfly houses vary in size. The Sophia M. Sachs Butterfly House in Chesterfield, Missouri, covers 511 square meters, whereas the Penang Butterfly House in Penang, Malaysia, is much larger (1474 m^2). Seasonal exhibits are generally smaller. A few facilities, such as the insectarium at the Tama Zoo in Japan, display grasshoppers and aquatic species in walk-through immersion exhibits more typical of butterfly houses.

PURPOSE AND VALUE OF INSECT ZOOS

Live insects are one of the best teaching tools for children and adults alike. Despite an enormous range and intensity of educational, research, and conservation activities, most insectariums and butterfly displays offer some level of educational programming that promotes the appreciation and understanding of insect life. Aside from the actual live exhibits and graphics, these programs take the form of informal presentations, hands-on opportunities to touch live animals, formal classes and lectures from kindergarten to university level,

teacher training, field trips, outreach programs, printed educational materials, multimedia materials, Web-based resources, and special events such as insect fairs and film festivals. In addition to these more lofty goals, insects have proven to be enormously popular with the public and thus are used to generate increased visitation and revenue for various types of nonprofit organizations and commercial enterprises.

Basic and applied research is carried out by some facilities, resulting in presentations at conferences and publications in conference proceedings and scientific journals. Several insectariums participate in captive breeding programs for threatened or endangered species such as the American burying beetle, *Nicrophorus americanus,* the Italian ground beetle, *Chrysocarabus olympiae,* the giant wetas, *Deinacrida* spp., and the Polynesian tree snails, *Partula* spp. For example, London Zoo currently manages six invertebrate conservation programs, involving 38 species. *In situ* conservation programs are also supported by a number of insectariums and their parent organizations.

Some of the early insectariums began with the immediate goal of raising food for insectivorous zoo animals such as the Tama Zoo's Insectarium, which opened in its first form in 1961. It was felt that decreases in wild populations of grasshoppers due to widespread use of insecticides required the development of stable food sources. An earlier insectarium at Tokyo's Toshima-en Amusement Park was developed with the idea of using its live insect residents for making nature films.

DIVERSITY OF EXHIBIT TECHNIQUES

Facilities that house live insects are diverse in construction as well as in the type of sponsoring institution. To face the challenges of exhibiting small, short-lived, seasonally limited, diapausing animals with radically different life stages, major facilities rely heavily on the in-house maintenance of breeding colonies so that specimens are available year-round. Founder stock is collected from the wild or obtained through exchange or purchase from other insectaries to establish and maintain genetically healthy colonies. In contrast, many butterfly houses rely on independently run breeding facilities, often in the species's country of origin. Pupae are generally received weekly from various butterfly ranches or farms. Some butterfly houses also have supplemental in-house breeding colonies of selected species of butterflies. Many for-profit butterfly houses maintain breeding populations both for their own display and to sell to other butterfly houses.

Exhibit techniques have not changed much during the 120 years since the opening of the first insect exhibit at the London Zoo in 1881 (Fig. 1). Even in 1881, informative labels and preserved specimens of insects accompanied the live insect displays. The major change has been the inclusion of interactive techniques and graphics to enhance the interpretive experience of the display. Today, insectariums are really zoo–museum hybrids. Interactive computer modules, microscopes, audio tracks, video loops, models, robotics, and cultural artifacts are now used to enrich and enliven the educational messages. Special displays on topics related to

FIGURE 1 The London Zoo opened the first major insect exhibit in the world in 1881. This contemporary photograph was made by collector William Hornaday. [From "Zoological Gardens Illustrated" (photograph album), Vol. 1, "London Zoological Gardens." © Wildlife Conservation Society, headquartered at the Bronx Zoo.]

cultural entomology have been included in some facilities such as the Insect Zoo at the San Francisco Zoo, which has produced special exhibits on ancient cricket cages of China, insects as human food, and aquatic insects in terms of fly fishing and fly tying in North America. These techniques combine to address different visitor learning styles, ages, and interests. Some facilities have educational outdoor garden displays to focus attention on native insects and plant interactions, and a few facilities adjoin or are sister organizations to native wildlife reserves.

Most often, insect exhibit facilities are associated with larger institutions. Zoological gardens such as the London Zoo, Cologne Zoo, Berlin Zoo, Tama Zoo, Cincinnati Zoo, San Francisco Zoo, and St. Louis Zoo all contain major insectariums. General histories of zoos and aquariums, while focused on vertebrates, can be found in the encyclopedic memory and archives of Marvin Jones (San Diego Zoo) and in edited volumes by Kisling, and Hoage and Deiss. Natural history museums such as the Smithsonian National Museum of Natural History, the Natural History Museum of Los Angeles County, and the Houston Museum of Natural Science have insect zoos or butterfly houses; the Cecil B. Day Butterfly Center is located within Callaway Gardens in Pine Mountain, Georgia; universities such as Kansas State University, Michigan State University, the University of Joensuu in Finland, and the University of Alberta in Canada all have seasonal butterfly gardens or insect zoos. The Insectarium de Montréal (Canada), the Insectarium of Victoria (Australia), the Butterfly Pavilion and Insect Center (Colorado), and the Sophia M. Sachs Butterfly House (Missouri) are examples of independent stand-alone facilities. Insectariums vary from tax-supported municipal institutions to nonprofit organizations to for-profit corporations. The for-profit businesses such as the Penang Butterfly House in Malaysia, Stratford-upon-Avon Butterfly Farm in England, Butterfly World in Florida, and the Australian Butterfly Sanctuary in Australia serve as both public exhibits and commercial suppliers. While many are permanent, year-round facilities, the popularity of live insect exhibits (and perhaps the short life span and easy transportability of insects) has allowed for the explosion of temporary or seasonal exhibits, particularly butterfly displays.

INSECTARIUMS AROUND THE WORLD

During the 19th century, expanding empires, increased trade, and improved transportation and communication stimulated interest in exotic wildlife. European powers sent expeditions to bring back specimens for potential domestication and commercial use. Illustrated publications on biological subjects appeared, and books recounting the adventures of naturalist–explorers allowed the public the vicarious thrill of discovery. With the emergence of modern systematics, the number of described genera rose. This was a fertile time of new discoveries and new theories. Darwin,

Wallace, and others pondered the origin of species. From 1828 through 1914, each year an average of two zoological institutions opened throughout the world; the total was 168, with 86 in Europe alone. Themes of metamorphosis and evolution permeated art and literature. The Industrial Revolution transformed commercial production and altered attitudes toward nature as reflected in the diverse philosophies encompassed by the Art Nouveau movement (~1870–1914). Menageries toured, zoological gardens sprang up, and exotic nature became a source of fashionable urban pleasure and scientific study.

Europe

FRANCE Loisel's 1912 history of zoos documents the early development of zoos in general and includes some interesting details about a few insect exhibits. In France, the Jardin des Plantes was created in 1793 adjacent to the Muséum d'Histoire Naturelle and the Ménagerie d'Observation Zoologique. A chair of Insects and Worms was created, and its first occupant was Jean-Baptiste Lamarck. The Jardin's mission was exploration to find species of plants and animals for utility or ornament, with instruction of the public as a minor goal. In 1797 the visitors could view silkworms and stroll through the grounds past honey bees, which were housed in a large glass hexagonal structure. In the middle of the 19th century a disease killed almost all the silkworms in France. In 1860 the Jardin Zoologique d'Acclimatisation was created on the western fringe of Paris to focus on the study of animals of economic utility. New species of silkworm were imported from China and India and kept at the Jardin d'Acclimatisation. Two of these species were acclimatized, and a fertile hybrid was produced. French scientist Louis Pasteur, in 1870, rescued the silk industry by discovering that the then epidemic "pebrine disease" of silkworms, now known to be borne by *Nosema bombycis,* could be prevented through microscopic examination of adult moths and isolation of uninfected stock. These advances set the trend for a more scientific approach to silk production.

In France today, a handful of butterfly houses have appeared as independent facilities. In 2000, a new theme park called Micropolis opened in the town Saint-Léon-en Lévezou, which was home to the famous French entomologist Jean-Henri Fabre.

ENGLAND AND THE NETHERLANDS In 1828 the Zoological Society of London (London Zoo) opened, and visiting it became a popular recreational pastime. Commercial butterfly farms appeared as early as 1865 in Colchester, England. In 1881, a mere 53 years after its debut, the London Zoo opened the first major public insect house. In what had been a refreshment room, a series of terrariums on tables along the wall and in the center of the room displayed live specimens of silkworms, aquatic insects, and other invertebrates. Display tanks were well labeled, and mounted specimens enhanced the exhibit.

Seventeen years after the opening of London's insect zoo, in 1898, the Artis Zoo in Amsterdam opened the Insectarium founded by schoolteacher Rudolph A. Polak, who feared that the urban population was becoming alienated from nature. Polak was part of the "Biologisch Reveil" (reveil means wake-up call) movement—to remind people of their connection to the natural world. His goal was to use insects to bring children into close contact with nature, and this tradition is still carried on today at Artis Zoo. Amazingly, Polak continued to work as a teacher and ran the Insectarium as a hobby.

In the early 1900s, other insect-viewing opportunities existed as well. In 1900 a businessman operated a butterfly farm in Kent, where he bred various species to sell to collectors, museums, and universities in England and America. In 1908 Londoners could pay 6 pence (equivalent to perhaps $10 today) to visit the storefront menagerie display of an enterprising optician, filled with display cases of live bees and ants. The London Zoo, where the formal display of live insects really began, celebrated the new millennium by creating the Web of Life facility in 1999, boasting 65 live displays, 156 invertebrate species, a room-sized enclosure for desert locusts, *Schistocerca gregaria* (including a half-buried jeep for atmosphere), and a giant anteater.

Long involvement with the silk industry and an active amateur naturalist community seemed to make England fertile ground for the birth of modern-day butterfly houses. Businessman David Lowe, who created the Guernsey Butterfly Farm in 1977, went on to create seven other facilities in England by 1984. The London Butterfly House in Syon Park opened in 1981, and Stratford-upon-Avon Butterfly Farm, which followed in 1985, includes an extensive insect zoo exhibit adjoining the free-flight butterfly display. In 1986 there were approximately 40 butterfly houses in England. They number closer to 20 today.

GERMANY AND EASTERN EUROPE The Frankfurt Zoo opened its first insect house in 1904 as a seasonal exhibit in the summer (in the winter it was used for storks). In 1957 Frankfurt Zoo built a new large insectarium on top floor above the renovated aquarium. The Zoologischer Garten Köln (Cologne Zoo) opened an insect house in 1905 and maintained it through 1929. A new insectarium was built in 1971, with a butterfly room at its entrance on the first floor of the new aquarium-terrarium, with approximately 60 species maintained in this facility. The Budapest Zoo in Hungary opened its first insect exhibit in 1907 and renovated the vivarium in 1970, exhibiting 68 species. In 2000, a butterfly exhibit was added to the zoo. In 1913 the Berlin Zoo opened a large insect exhibit on the third floor of the new aquarium, as did the Zoologischer Garten Leipzig. The Berlin Zoo's building, destroyed during World War II, was rebuilt after 1945. The exhibit was renovated again in 1978 through 1983 and maintains over 35 species. An insect section was opened in Gruga Park in Essen in 1956, and in 1970 the Löbbecke Museum in Düsseldorf built a large insect exhibit associated with its aquarium. In 1987 the new Löbbecke Museum and Aquazoo opened an insectarium as the centerpiece of the whole new building. In 1980 the Zoological–Botanic Garden Wilhelma in Stuttgart built an insectarium exhibiting 35 species. The Dortmund Zoo and the Krefeld Zoo built butterfly exhibits in 1991 and 1998, respectively. The Noorder Zoo in the Netherlands, the Zoologicka Zaharada Praha in Czechoslovakia, and the Tiergarten Schönbrunn in Austria also have insect zoo exhibits.

Asia

Although no records have been found to document the early exhibition of live insects in China, the Chinese have long had a complex appreciation of the insect world. The development of silkworm culture and silk production, the development of bee culture, the early and extensive use of insects in traditional medicine, and the use of crickets as pets in the Táng dynasty (618–906) suggests an advanced appreciation of the utility and aesthetics of insect life.

Yajima describes the development of insectariums in Japan. The Insectarium at the Takarazuka Zoological and Botanical Garden opened in 1954 and was expanded in 1967. The Toshima-en Insectarium, at the Toshima-en Amusement Park in Tokyo, opened in the late 1950s. In 1961, Yajima went on to design the first incarnation of the Insectarium at the Tama Zoo. In 1966 Tama Zoo's Insectarium opened a walk-through grasshopper greenhouse filled with 6000 grasshoppers in addition to a walk-through butterfly house. A firefly building was added in 1975. In 1988 a new insectarium was built at Tama Zoo, which mixed traditional terrarium-type exhibits with walk-through exhibits without guardrails of Orthoptera, fireflies, butterflies, beetles, ants, and aquatic insects. Insect Ecological Land had a construction cost of over $5 million and increased attendance to the zoo by 20%. At its opening, 94 species were maintained by a staff of 12. As of 1995, there were 30 live insect exhibits in Japan alone.

A few other insectariums and butterfly houses can be found throughout Asia. The Penang Butterfly House in Malaysia was founded in 1986. Attached to the butterfly display is an exhibit of Malaysian insects and arthropods. The Fragile Forest exhibit, which opened at the Singapore Zoo in 1998, displays butterflies and other insects.

North America

Insectariums developed more slowly in the United States. Though the early entomologist Thomas Say contributed to bringing respect to the science of entomology in the early part of the 19th century, the zoological gardens in the United States remained focused on the display of vertebrates and the natural history museums focused on nonliving exhibits. Interest in

representing invertebrate biodiversity either taxonomically or zoogeographically in these institutions lagged behind that shown by their European counterparts. Enduring insectariums did not take hold in America until the late 1970s.

However, a number of early successful experiments at both zoos and museums proved extremely popular with the public. The New York Zoological Society Bulletin chronicles several early attempts. Raymond L. Ditmars, Curator of Reptiles at the Bronx Zoo and world-renowned herpetologist, actually began his career in science as an entomologist working as an assistant curator in the Entomology Department at the American Museum of Natural History. Throughout his tenure, he maintained a keen interest in insects. In 1910, under his direction, a live arthropod exhibit consisting of 56 cages and containing silk moths in various life stages, lubber grasshoppers, Hercules beetles, and tarantulas was set up as an experiment. A portion of the moths that emerged from the silk moth collections were mounted and sold as souvenirs by the Bureau of Information in the Lion House. The exhibit was so popular with visitors that plans were made to make it a permanent feature at the zoo. The permanent facility was never built, but Ditmars persevered and in 1940, a Department of Insects was created with Ditmars as curator. Ditmars died in 1942, however, without having brought the plan to fruition. Brayton Eddy, an entomologist for the state of Rhode Island, was hired as the new curator of insects in 1945. Eddy's experience included the development of a seasonal live insect zoo at Goddard State Park in Providence, housed in the first floor of the stately mansion on the property from 1934 to 1937. Then Eddy died unexpectedly in 1950, and again a planned permanent insect exhibit never got started. The new exhibit was to have featured local insects and tropical imports sent by Dr. William Beebe from his research station in Venezuela.

During the 1930s and 1940s, the Chicago Zoo (later renamed the Brookfield Zoo) was planning and experimenting with its own invertebrate exhibit. Construction on the Insect House (also called the Invertebrate House or the Special Exhibit and Demonstration Building) was completed between 1934 and 1938. Grace Olive Riley, acting as curator of reptiles and invertebrates, was succeeded by Robert Snediger as reptile curator. Snediger organized the "Animals Without Backbones" exhibit, which ran from 1947 into the early 1950s. Bees, cockroaches, aquatic insects, spiders, scorpions, centipedes, leeches, and other invertebrates were displayed, along with amoebas. Though planned as a permanent feature by Brookfield Zoo's early designers, the Insect Building was later converted into the zoo's library. The exhibits developed in the late 19th and early 20th centuries were not very different from the exhibits of today. Glass terrariums with screened lids displayed on tables containing local and exotic insect life were the state of the art in 1881, 1910, the 1940s, 1970s and 1990s.

Worth mentioning is the arthropod exhibit that was included as a permanent feature in the Arizona–Sonoran

Desert Museum from its opening date in 1952. Although not large, either in species number or in size of tanks, this collection was innovative in that it included invertebrates in the interpretation of an ecosystem. Then 20 years passed during which the early invertebrate exhibits seem to have vanished from the consciousness of zoo administrators.

Another period of experimentation with short-lived insect exhibitions emerged. An insect zoo was created at the Flushing Meadows Zoo from 1969 to 1970, and another was assembled at the Smithsonian National Museum of Natural History in 1971. The immense popularity of the exhibit at the Smithsonian resulted in the opening of the museum's permanent insect zoo in 1976, in time to greet the International Congress of Entomology, which held its meeting that year in Washington, D.C. A natural history museum had now stepped permanently into the world of live insect display, an endeavor formerly restricted to zoological gardens or commercial rearing facilities staffed by persons used to dealing with the challenges and demands of live animals. From the start, the Smithsonian staff interacted with visitors daily with informal hands-on demonstrations, and the species inventory included both local and exotic specimens collected by Smithsonian entomologists in the field.

Cincinnati Zoo followed the Smithsonian in 1978 and opened a new building containing 68 live displays and housing 70 to 100 species. This facility, the World of the Insect, set a new standard for insectariums in the United States. It also set a new standard for monetary commitment to insect exhibits, with a price tag of one million dollars, including the cost of the new stand-alone building, custom-made terrariums, colorful graphics, and educational interactive exhibits borrowed from museum methodology.

The Insect Zoo at the San Francisco Zoo opened in 1979 as a temporary summer exhibit in an unused auditorium building in the children's zoo section. Zoo attendance instantly increased by 50%, and the collection became a permanent facility, the second largest in the United States for two decades, containing 35 displays and maintaining 70 species.

In the United States, these three permanent insectariums created in the 1970s (at the Smithsonian, the Cincinnati Zoo, and the San Francisco Zoo) were really "arthropod zoos," with an occasional mollusk, leech, or annelid thrown in. These exhibits became models for many subsequent projects that followed throughout the country. Emphasis was placed on the development of year-round breeding colonies and use of native and exotic species.

The only other invertebrate exhibit to open in the following decade excluding the butterfly houses, was the Invertebrate Exhibit, which opened in 1987 on the lower floor of the Reptile House of the National Zoo, showing tropical insects alongside marine invertebrates such as cephalopods. A pollinarium exhibit was added in 1996.

The only nonbutterfly insect exhibit to emerge in the 1990s was the Ralph K. Parsons Insect Zoo at the Natural History Museum of Los Angeles County. Modest at its opening in 1992, it has grown to 40 live displays.

New walk-through butterfly exhibits dominated the late 1980s and early 1990s modeled on free-flight greenhouse exhibits in England. In 1988 the $3 million Cecil B. Day Butterfly Center at Callaway Gardens opened in Georgia. Two for-profit Florida endeavors, Butterfly World and the butterfly house at Cypress Gardens, opened in 1988 and 1992, respectively. Back in the zoo world, the San Diego Wild Animal Park's Butterfly Encounter debuted in 1993. In 1994, the Houston Museum of Natural Science opened the Cockerell Butterfly Center, a three-story immersion butterfly glasshouse attached to the museum, attracting 700,000 visitors during its first year of operation. A live arthropod exhibit room was later added to this facility. In 1995, the Butterfly Pavilion and Insect Center, became the first stand-alone nonprofit insect facility in the United States.

In the 1990s at least 20 seasonal butterfly houses emerged, including several supported by universities such as Michigan State University and Kansas State University under the auspices of their respective Departments of Entomology. The insectarium at the St. Louis Zoo opened in 2000. Costing $4 million and maintaining 80 to 100 species in addition to a butterfly display, it became the most significant facility to open in over a decade.

Parallel developments have occurred in Canada, but the most notable was the development of the Insectarium de Montréal founded by Georges Brossard in 1990. This $8 million exhibit mixed classic European museum design with multimedia interactive displays and challenged the live animal exhibit community to set a higher standard for the construction of new insectariums to educate the public about insects, which represent 80 to 95% of the animal species on terrestrial earth.

Australia

In Australia, the Melbourne Zoo was the first to get into the invertebrate business by opening a butterfly house in 1985. The Insectarium of Victoria and the Victorian Institute of Invertebrate Sciences opened at Heathcote in 1993, moving in 1998 to a permanent home in Woodend. On display is *Astacopsis gouldi,* or the giant yabby or crayfish, one of the largest freshwater invertebrates on earth.

Latin America and Africa

Butterfly exhibit houses in Latin America are most commonly derived from commercial butterfly ranching and farming operations. The Butterfly Farm in La Guacima de Alajuela, Costa Rica, founded in 1990, and Spirogyra Butterfly Garden in San Jose, Costa Rica, established in 1992, typify these facilities. The Green Hills Butterfly Ranch in Belmopan, Belize, is a display as well as commercial supplier, as is the La Selva Butterfly Farm in Ecuador. Butterfly World in Klapmuts, South Africa, opened in 1996.

BIODIVERSITY CONSERVATION AND INSECT EXHIBITS

Insectariums can play a strong role in conservation education. They can also play a role in the conservation of biodiversity in nature by providing economic opportunities for local populations that are alternatives to destructive resource extraction such as logging, mining, or conversion of forest land to agricultural enterprises such as cocoa, coffee, or oil palm plantations. It is difficult if not impossible to conserve natural resources where alternative economic opportunities are limited or absent. In response, as general awareness and concern for loss of tropical forest ecosystems and biodiversity emerged, the government of Papua New Guinea deemed insects a national resource and candidate for economic development. This policy resulted in the establishment of the Insect Farming and Trading Agency (IFTA) in 1978, to create income-producing opportunities for villages through nondestructive extraction of forest resources while at the same time creating incentive for preserving forest habitat. In 1983 a report published by an advisory committee of the National Research Council, in cooperation with the IFTA, promoted the idea of butterfly ranching and farming to supply research scientists, butterfly collectors, and other commercial uses. The report estimated that the current trade was between $10 and 20 million annually. Gram for gram, butterflies became more valuable than cattle. In 1981 it was estimated that an industrious butterfly farmer could earn from $100 to $3000 a year versus the mean per capita of $50. The IFTA sells $400,000 worth of stock annually and provides income for 1500 villagers in Papua New Guinea.

The worldwide explosion of public walk-through butterfly houses in the 1980s and 1990s created a new market for butterfly ranching and farming projects. Nairobi University scientists, working with the East African Natural History Society and the National Museums of Kenya, began the Kipepeo Project, (*kipepeo* is Swahili for butterfly) in 1993 to protect the Arabuko–Sokoke Forest in Kenya. A United Nations grant was awarded to develop sustainable utilization of butterfly biodiversity for the benefit of surrounding rural communities. In addition, a butterfly house was established as an ecotourism attraction to diversify the coastal tourism industry and to promote conservation of the Arabuko–Sokoke Forest. Several butterfly ranching projects have been created in Costa Rica to promote the conservation of remaining remnants of forest habitat in that country. These projects rely on the purchase of pupae by live butterfly exhibits. Organizations such as the U.S. Agency for International Development, The Nature Conservancy, Conservation International, and the World Wildlife Fund have now participated in the development of similar projects in Central America, and in Irian Jaya and Sulawesi in Indonesia.

Papiliorama-Nocturama Tropical Gardens in Neuchâtel, Switzerland, an exhibit and nonprofit organization that opened in 1988, invests income into a sister foundation, the International Tropical Conservation Foundation, which runs the 8.9 ha Shipstern Nature Reserve in Belize. Shipstern also has its own public butterfly house on the reserve grounds.

Approaches to conservation have shifted in the past decade from focusing on single-species protection to ecosystem preservation. In 1987, for example, the Center for Ecosystem Survival, a nonprofit consortium serving zoos, aquariums, museums, and botanical gardens, was created in San Francisco to raise funds for biodiversity conservation through habitat protection, conserving invertebrates and plants as well as vertebrates and other forms of life. In 2001, a total of 112 informal science institutions participated in this program and raised over $2 million for habitat purchase and protection, fueled in part by increased public awareness of the magnitude and importance of insect biodiversity.

See Also the Following Articles

Conservation • Museums and Display Collections • Photography • Rearing of Insects • Teaching Resources

Further Reading

Chicago Zoological Society. (1947). "Guide Book, Chicago Zoological Park." Brookfield, IL.
Hoage, R., and Deiss, W. A. (eds.), (1996). "New Worlds, New Animals." Johns Hopkins University Press, Baltimore.
Kisling, V. N. (ed.). (2001). "Zoo and Aquarium History—Ancient Animal Collections to Zoological Gardens." CRC Press, Boca Raton, FL.
Loisel, G. (1912). "Epoque contemporaine," Vol III in "Histoire des ménageries de l'antiquité à nos jours." Henri Laurens/Octave Doin et fils, Paris.
New York Zoological Society. (1911). "Zoological Park Notes," Zoological Society Bulletin 47, pp. 788–789.
New York Zoological Society. (1940). "Notes from the Park and Aquarium," Zoological Society Bulletin 43(6), pp. 198-203.
Parsons, M. (1992). Butterfly farming and conservation in the Indo-Australian region. *Trop. Lepid.* **3**, 1–28.
Saul-Gershenz, L. S., Arnold, R. J., and Scriber, J. M. (1995). Design of captive environments for endangered invertebrates. *In* "Conservation of Endangered Species in Captivity: An Interdisciplinary Approach," State University of New York Press (E. F. Gibbons, B. S. Durant, and J. Demarest, eds.). Albany.
Yajima, M. (1996). The development of insectariums and their future. *Int. Zoo News* **43**, 484–491.

Integrated Pest Management

Ronald Prokopy
University of Massachusetts

Marcos Kogan
Oregon State University

Across millennia, humans have used a variety of approaches in attempts to maintain pest insects at tolerable levels. The character of these approaches has evolved over time. Since the late 1960s, an approach termed integrated pest management (IPM) has ascended to dominance internationally. In essence, IPM is a decision-based process involving coordinated use of multiple tactics for optimizing the control of all classes of pests (arthropods, microbial pathogens, vertebrates, weeds) in an ecologically and economically sound manner.

HOW INSECT PESTS ORIGINATE

The manner by which an insect has become a pest influences the strategy used for managing it within an IPM framework. Some species have reached the status of pest because human thresholds for tolerating them have decreased as economic well-being has increased. Other species have become pests as a consequence of the availability of abundant resources for their survival and reproduction; resource concentration in time and space is conducive to rapid buildup of insect populations. Still other species have become pests in regions of their origin when they expanded their host range from comparatively unimportant native hosts to economically important introduced hosts. Finally, some species have become pests almost instantly upon being transported by humans to new locales having favorable resources but devoid of effective natural enemies.

CLASSIFICATION OF INSECT PESTS

For purposes of devising an appropriate IPM strategy, insect pests may be classified as key, secondary, or induced. Key pests are those whose populations, if unmanaged, repeatedly exceed tolerable levels; these are the principal focus of IPM endeavors. Secondary pests are those whose populations occasionally reach intolerable levels; their potential threat is recognized by the very act of devising IPM strategies. Induced pests are those whose populations rarely exceed tolerable levels under natural conditions, but if they become resistant to pesticides or other single-tactic control measures that harm their natural enemies, they can reach outbreak proportions. Usually induced pests return to nonpest status under a true IPM approach.

EVOLUTION OF CONCEPT OF IPM

Through trial and error across centuries, humans gradually came to use tactics such as cultural control, host resistance, and biological control in efforts to protect themselves, livestock, crops, and forests against pests. By the latter half of the 19th century, some pest control strategies that blended these tactics could rightfully be considered to be precursors of modern IPM strategies. During the 20th century, efforts to maintain pests at tolerable levels became more formalized as they became more intensive; these efforts can be considered as having progressed along the following four pathways.

"Pest control" was the terminology used during the first half of the 20th century to describe the set of actions taken to avoid, attenuate, or delay the impact of pests. Early in the century, inorganic and botanical insecticides gained increasing prominence as a control tactic against pest insects. By mid-century, use of organosynthetic insecticides supplanted virtually all other tactics in becoming the dominant approach to insect pest control, especially in developed countries.

"Integrated control" surfaced about 1950 in California as a concept of pest control aimed at combining and integrating biological and chemical control. The emergence and eventual popularity of this concept had roots in problems associated with overuse of organosynthetic insecticides, particularly a surge of key, secondary, and induced insecticide-resistant pests and environmental harm caused by insecticides. The publication of *Silent Spring* by Rachael Carson in 1962 was especially important in spurring widespread interest in the principles of integrated control.

"Pest management" represents a shortened version of "protective population management," a concept coined in 1964 by Australian ecologists to emphasize direct human interference in maintaining pests at tolerable levels, as opposed to reliance on unmanaged natural abiotic and biotic factors acting on pest populations. The concept of pest management embraced a broader range of pest control tactics than did the concept of integrated control. Until the late 1960s, both these concepts flourished simultaneously in accenting the need to move beyond use of insecticides as a sole pest control tactic.

"Integrated pest management" originated in 1968 as a contraction of "integrated pest population management," an expression used first in 1967 by R. F. Smith and R. van den Bosch. Soon afterward, the abbreviation "IPM" came into use worldwide for signifying a desirable and holistic approach to controlling pests. Numerous definitions have been put forward for IPM, but one that captures the broad and essential elements of many others is the following: IPM is a decision support system for the selection and use of pest control tactics, singly or harmoniously coordinated into a management strategy, based on cost–benefit analyses that take into account the interests of and impacts on producers, society, and the environment. No single definition, however, is likely to encompass all facets of IPM for all time. IPM has been and likely will continue to be an evolving concept. For example, a compendium of IPM definitions, available on the World Wide Web (http://www.ippc.orst.edu/IPMdefinitions/define.html?), listed 67 definitions in mid 2002.

ECOLOGICAL FOUNDATION OF IPM

Ecology is the study of relationships among organisms and their environment. Consideration of these relationships usually begins with focus on individuals or populations of a single species and subsequently broadens to include communities of organisms and, eventually, ecosystems.

Conceptually, the foundation of IPM is ecological. Ideal IPM programs are those that fully embrace ecosystem structure and processes in time and space. In reality, ecological complexity increases dramatically with each step from population to community to ecosystem. Such complexity challenges the realization of ideal IPM programs.

One approach that has been taken toward recognition of the ecological foundation of IPM, especially in developing countries, accentuates the pest-suppressing properties inherent in natural ecosystems as primary building blocks for the construct of human-designed ecosystems. After system construction based on ecological principles, the intent is to minimize human intervention to the greatest extent possible while still maintaining pest populations within tolerable levels.

Another approach, common in developed countries, takes as its starting point an existing ecosystem constructed by humans and aims at reducing negative impact in a succession of steps used in the management of pests. Such an approach usually commences with a focus on the ecology of a single-species population and may expand to consideration of structures and processes associated with communities and ecosystems.

ADVANCING LEVELS OF INTEGRATION IN IPM IMPLEMENTATION

A hierarchical structure results from the ecological foundation of IPM that lends itself to viewing IPM as progressing in scope through a series of ecologically rooted steps. These steps are characterized by ascending levels of complexity and spatial scale: from focus on a single-species population in a restricted locale to focus on a community of pest and other organisms in a larger area to focus on a whole ecosystem. Coincident with this ecocentric hierarchy is another stepwise hierarchy conceived of as a vehicle for measuring progress in achieving the goals of IPM. This hierarchy comprises a succession of levels from single-tactic (almost invariably based on pesticide use) to multitactic measures of pest and habitat management. The steps further involve ascending from focus on a single pest species in a single class of pests (e.g., insects) to multiple pest species across all classes of pests (insects, microbial pathogens, vertebrates, and weeds). These distinctive hierarchies can be blended in the form of a continuum of advancing levels of integration in IPM implementation. Three more or less distinctive levels along the continuum are described here (Fig. 1). Degree of success in integration at each level is shaped not only by ecological processes but also by government policy, regulatory legislation, social relations, economic forces, and cultural background that may enhance or constrain progress.

FIRST-LEVEL IPM

In the most basic and also the most widely practiced form of IPM, emphasis is on monitoring development and/or abundance of a single pest species at a single locale (e.g., a household, cow barn, greenhouse, cropped field, or woodlot) and using thresholds for deciding whether to take action.

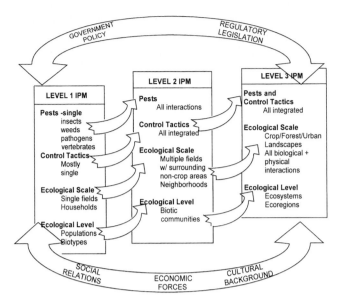

FIGURE 1 Levels of IPM integration: main targets, ecological scales, and levels of ecological complexity.

Application of a pesticide is by far the most common form of action taken under first-level IPM. Integration occurs when abundance of natural enemies of the pest in question also is considered in the decision-making process and when selection among candidate pesticides involves explicit attention to minimizing harm to these and other beneficial organisms. This form of IPM has been characterized by some as "integrated pesticide management."

Monitoring Pest Development

Because the developmental rate of an arthropod is regulated largely by temperature, the monitoring of developmental rate for pest management purposes usually takes the form of measuring accumulation of heat units above a threshold temperature at which development begins. At temperatures above these fostering the maximal developmental rate, development may decrease. Such decrease has not been investigated for most pest arthropods and has not yet played a significant role in making pest management decisions.

The simplest and most prevalent approach to measuring accumulation of heat units above developmental threshold temperature involves use of degree-days (DD). For a specific date, the number of accumulated DD equals the average temperature of that date minus the developmental threshold temperature of the arthropod. Several procedures have been devised to estimate average daily temperature. The most common one, albeit somewhat crude, consists simply of averaging the maximum and the minimum ambient temperature of the day. To illustrate the DD approach, if the high and low temperature for a given day were 30 and 20°C, respectively, with a developmental threshold temperature of 10°C, then 15 DD would have accumulated on that day.

Pest development as monitored by DD accumulation may benefit decision making under first-level IPM in several ways, particularly for optimal timing of management activities. For example, ability to predict when a majority of pest adults is about to emerge from pupae is useful for optimal timing of deployment of traps for monitoring adults. Knowledge about when oviposition is likely to begin and peak can facilitate optimal timing of pesticide application against newly hatched larvae, which often is the stage most vulnerable to pesticide treatment. Sometimes this determination is made in conjunction with date of first capture of adults by traps, known as a "biofix" point for initiation of DD accumulation. The ability to forecast when a majority of larvae or nymphs is at a particular growth stage can aid in optimal timing of sampling their abundance and the abundance of their natural enemies.

Monitoring Pest Abundance

Ideally, an IPM practitioner would have available a precise count of the number of individuals of an insect pest species present in an area of concern; realistically, obtaining information on absolute densities of pests is prohibitively costly. Therefore, most practitioners rely on imprecise estimates of pest population density obtained by using one or more population sampling techniques. The intent is to capture a more or less consistent, if unknown, proportion of the pest population. Choice of appropriate sampling technique varies considerably according to pest species and developmental stage.

For sampling comparatively mobile individuals such as adults, traps using odor and/or visual stimuli are common tools. Odor stimuli usually consist of synthetic equivalents of either attractive sex odors (sex or aggregating pheromones) or attractive food or host odors. Visual stimuli normally rely on synthetic mimics of visually attractive sites where feeding, mating, or egg laying occurs.

For sampling less mobile individuals such as larvae, common techniques include visual searching of the target area accompanied by direct counts of detected pests, use of a sweep net (especially effective for sampling individuals on foliage of nonwoody plants), and use of a loose or framed cloth placed beneath vegetation that is shaken or tapped to dislodge pests. Sampling immobile individuals such as eggs or pupae usually is done by visual inspection.

To obtain an acceptably accurate and cost-effective estimate of the size of a pest population by means of one of these techniques, careful attention must be given to the program under which sampling is conducted. Effective sampling programs take into account the daily activity pattern of the target species as well as its characteristic spatial distribution (uniform, random, or clumped). Historically, most programs have incorporated sampling at several or numerous sites in a target area to acquire sufficient representation of the size of a pest population; then

researchers have counted the sampled individuals of the target pest. New programs developed for some pests simplify these procedures. Sequential sampling is an approach that optimizes the number of sampling sites needed for classifying a pest population as below or above a density requiring action. Binomial sampling is an approach that classifies an individual species as either present or absent at a sampling site, thereby precluding the need to count all members of that species taken in a sample. Both these simplifying approaches require substantial species-specific background information for their development and use.

The emerging technologies of global positioning systems (GPS) and geographical information systems (GIS) offer unsurpassed capability of aiding in the mapping of site-specific variation in characteristics of areas under consideration for sampling. A GPS uses triangulation of signals from a constellation of satellites to identify the precise location (within a meter) of an area on the earth's surface. A GIS is a computer program for the mapping and spatial analysis of georeferenced information. GIS capabilities include assemblage, storage, manipulation, retrieval, and graphic display of information about attributes of precise locations identified through GPS. Such information can be exceptionally useful in forming associations between characteristics of a specific locale (e.g., terrain, soil, extent of vegetative growth, microclimate) and density of a population (Fig. 2). For pests, sampling can be directed toward specific sites in which densities are suspected to be highest.

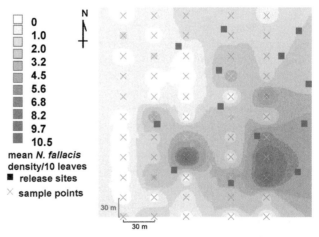

FIGURE 2 Dispersal of *Neoseiulus fallacis* for biological control of spider mites in a strawberry field, 8 to 15 weeks following release of 100 adult females at each of 15 sites: squares, release sites; crosses, sample points. This distribution is due to ambulatory foliar movement and aerial dispersal (dominant winds from south and southwest). Data represented using GIS (GRASS v. 4.1). [From Coop, L. B., and Croft, B. A. (1995). *Neoseiulus fallacis*: Dispersal and biological conrol of *Tetranychus urticae* following minimal inoculation into a strawberry field. *Exp. Appl. Acarol.* **19,** 31–43. Reproduced by permission from the authors and Chapman & Hall (now Kluwer Academic Publishers).]

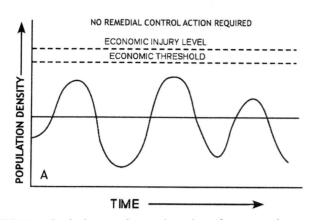
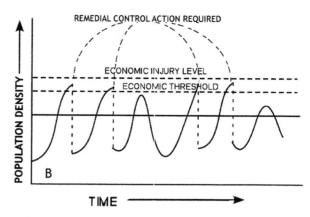

FIGURE 3 Graphs depicting theoretical population fluctuations of two insect pest species.

Deciding Whether to Take Action

Several approaches have been developed for deciding whether an insect pest population has or has not reached a level requiring intervention, such as an insecticide application. For agricultural purposes, the approach used most often centers on the concept of "economic injury level" (EIL), formalized in 1959 by V. H. Stern, R. F. Smith, R. van den Bosch, and K. S. Hagen and defined by them as the "lowest pest population density that will cause economic damage." These entomologists also proposed a related concept, which they termed the "economic threshold" (ET), defined as the "pest density at which control measures should be applied to prevent an increasing pest population from reaching the economic injury level" (Fig. 3).

The decision-making concepts of EIL and ET have been fundamental to the development and implementation of first-level IPM, particularly for insect management in agriculture (but less pronouncedly for disease, vertebrate, and weed management). They have been especially useful when insect pest populations are expected to increase over time within a crop, can be sampled reliably, can be related in a predictable way to reduction in crop yield or quality, and can be controlled readily by taking immediate action (e.g., application of insecticide) to prevent further damage. They are less valuable when human comfort or aesthetics, rather than economic damage, is paramount. Even for agriculture, the concepts of EIL and ET cannot be applied rigidly because of inherent unpredictability of such factors as future weather (which can markedly affect rate of pest population growth and degree of crop susceptibility to a pest) and future value of the crop in the marketplace. Also, a type of action that may require considerable time before reducing pest density, such as application of a biocontrol technique, is likely to be less appropriate than an insecticide application within an EIL/ET framework.

A refinement of the concept of EIL, put forward by L. P. Pedigo and L. G. Higley, introduces the element of environmental quality into the decision-making process. Negative effects of insecticides on natural enemies of pests and on other organisms in the environment are treated as costs in addition to monetary costs associated with insecticide application. Quantifying environmental costs has proven to be challenging and subject to much debate, but progress under this refined concept of EIL nonetheless has been made.

In 1984 W. L. Sterling advanced the concept of "inaction level," which is the density of natural enemies sufficient to maintain a pest below the EIL. M. P. Hoffmann and collaborators developed a sampling program for eggs of tomato fruitworm, *Helicoverpa zea,* that permits ready identification and quantification of parasitized eggs in addition to healthy eggs. If the level of egg parasitism is determined to be too low to prevent larval numbers from exceeding the EIL, the lowest adequate rate of a "soft" pesticide (one having least impact on parasitoids and other nontarget organisms) is recommended. This approach is an example of application of the inaction level concept and represents a high degree of first-level IPM implementation.

Area-wide IPM is an expansion of first-level IPM that may represent a significant transitional step toward second and third levels of IPM. Under area-wide IPM, the key pest of a crop is targeted for management by means of the most effective noninsecticidal approach. For example, the codling moth, *Cydia pomonella,* is managed in apple and pear orchards by tactics such as the pheromone mating disruption technique or the sterile insect release method, that impair normal reproduction. Such tactics are implemented over areas large enough to preclude recolonization by fertile females from adjacent areas. By reducing the impact of broad-spectrum insecticides, natural enemies are preserved and are usually capable of regulating most secondary pests in the crop. As area-wide IPM programs expand to incorporate multiple pest interactions, they become natural springboards to higher level integration in IPM systems.

SECOND-LEVEL IPM

Second level IPM is intermediate between basic and advanced. It is receiving increased research attention, but inherent complexities have greatly limited its effectiveness.

Emphasis is on management of key pests of all classes and their associated natural enemies comprising a community (e.g., a village of dwellings, an entire farm, a wooded area surrounding a village). Emphasis also is on substituting a variety of comparatively environmentally benign management tactics (e.g., cultural management, host resistance, biological control, behavioral control), to the greatest extent possible, for the therapeutic practice of pesticide application. Decision makers must determine how best to integrate these tactics to achieve long-term suppression of pests within a cost–benefit framework.

Cultural management is purposeful manipulation of the environment to reduce pest abundance. It is most effective when directed at the most vulnerable life stage of a pest. Four forms of habitat or environmental manipulation aimed at controlling pests have been practiced for centuries in agriculture: crop rotation, timing of planting or harvest to minimize pest damage, sanitation or elimination of noneconomic resources available for pest reproduction, and polyculture, or the interplanting of different crops to diffuse resource concentration. More recent practices include the planting or encouragement of selected types of noneconomic vegetation in the vicinity of crops or as cover crops to serve as harborage for natural enemies of pest insects. Analogues of these practices have been developed for managing insect pests in nonagricultural situations.

Host resistance is any inherited characteristic of a host that lessens effects of an attacking pest. For centuries, humans may have unknowingly or intentionally selected for cultivars of plants or breeds of animals that are best able to withstand pests. Modern breeding programs, however, often have placed more emphasis on increasing yields than on protection against pests. Entomologically, resistant traits are preadaptive characteristics of a host that reduce its detectability, acceptance, or nutritional value, or enhance its toxicity to a pest insect. Molecular genetics techniques that facilitate introduction of specific pest resistance genes into cultivars or breeds possessing desirable commercial or aesthetic traits are beginning to replace traditional resistance breeding approaches.

In biological control, parasitoids, predators, or pathogens are deployed as natural enemies in the reduction of pest populations. Of the myriad insect species that could become pests, most do not because they are suppressed effectively by naturally occurring populations of biological control agents. Natural levels of biocontrol, however, often are insufficient for IPM purposes. Biological control then takes the form of importing absent natural enemies from other locales (termed importation or classical biocontrol), augmenting existing natural enemies by rearing and then releasing substantial numbers into the target community (termed augmentation), or tailoring management tactics to reduce negative effects on existing natural enemy populations (termed conservation). The latter is the most widely practiced form of biocontrol.

Behavioral control is manipulation of the behavior of pest individuals to prevent them from causing harm or unpleas-antness. Because of the expense and technological challenges associated with its use, behavioral control usually is directed only at key pests. Behavioral control may involve use of natural or synthetic chemical or physical stimuli to lure pests to sites where they are killed, or use of such stimuli to disrupt the ability of pests to find or use a potential resource. An ideal form of behavioral control might involve joint use of disruptive and attractive stimuli to achieve maximum effect, but this form is not yet widespread in practice.

Vineyards in parts of Europe and North America represent one of the few areas in which pest management is practiced effectively under the second-level IPM concept. Besides using essential elements of first-level IPM for insect pests, certain practitioners of vineyard IPM in these locations blend host plant resistance with cultural, biological, and behavioral controls for suppression of key pest insects and also use a suite of cultural controls for managing key disease and weed pests. This approach has resulted in marked reduction in pesticide use and greater stability of relationships among organisms comprising vineyard communities.

THIRD-LEVEL IPM

Although many IPM practitioners aspire to implement third-level IPM, the most advanced form remains largely in an embryonic state of development. Emphasis is on using principles and practices of second-level IPM in harmony with all other elements that affect long-term productivity or well-being of an ecosystem. Such elements include sound horticultural or husbandry practices (for agriculture), sound forest management (for silviculture), and sound community health practices (for villages or subunits of cities). Third-level IPM features attention to environmental and societal costs and benefits in the making of pest management decisions. The focal ecosystem may be an entity no larger than a community, as considered under second-level IPM, or it could be an entity as extensive as an ecological region. For crops, third-level IPM is roughly synonymous with the concept of integrated crop management. It is not, however, synonymous with organic agriculture, which disallows some materials and practices acceptable under third-level IPM.

Spearheaded by P. E. Kenmore, an approach has been developed for growing rice in developing parts of Asia that reflects many of the tenets of third-level IPM. This approach acccentuates societal contribution to the IPM decision-making process. It involves weekly gatherings of small groups of rice farmers, accompanied by experienced pest management personnel, who jointly conduct agroecosystem observations, engage in data analysis, and consider local ecosystem structure, environmental health, and a range of immediate and long-term tactics before making pest management decisions. This process has resulted in dramatic increases in awareness by entire communities of an advanced form of IPM and often dramatic decreases in use of pesticides on rice. It stands in contrast to modes of decision making and levels of popular awareness characteristic of less advanced forms of IPM implementation in many developed countries, where it is commonplace for farmers to make IPM decisions either acting alone or at most interacting with a private consultant, government extension representative, or employee of a pesticide distributor.

IPM AND SUSTAINABLE DEVELOPMENT

In a broad sense, sustainable development is development that meets the needs of the present without compromising the ability of future generations to meet their own needs. Sustainable development is rooted in the concept of ecosystem integrity and permeates all facets of human endeavor, whether economic, social, or cultural.

The scope of concerns and practices of third-level IPM as described here corresponds closely to that of sustainable development. Each emphasizes preservation of processes associated with natural ecosystems, long-term well-being of humans as members of communities, economic viability, and deployment of exogenous resources only after careful consideration. For agriculture, concepts underlying third-level IPM can be equated with concepts underlying sustainable agriculture. For both, one can expect concepts to evolve further over time.

See Also the Following Articles

Agricultural Entomology • Biological Control • Extension Entomology • Insecticides • Physical Control • Population Ecology • Sterile Insect Technique

Further Reading

Altieri, M. A., and Nicholls, C. I. (1999). Biodiversity, ecosystem function, and insect pest management in agricultural systems. *In* "Biodiversity in Agroecosystems" (W. W. Collins and C. O. Qualset, eds.), pp. 69–83. CRC Press, Boca Raton, FL.

Benbrook, C. M. (1996). "Pest Management at the Crossroads." Consumers Union, Yonkers, NY.

Coop, L. B., and Croft, B. A. (1995). *Neoseiulus fallacis:* Dispersal and biological control of *Tetranychus urticae* following minimal inoculations into a strawberry field. *Exp. Appl. Acarol.* **19**, 31–43.

Dent, D. (1995). "Integrated Pest Management." Chapman & Hall, New York.

Higley, L. G., and Pedigo, L. P. (1993). Economic injury level concepts and their use in sustaining environmental quality. *Agric. Ecosys. Environ.* **46**, 233–243.

Kenmore, P. E. (1996). Integrated rice pest management. *In* "Biotechnology and Integrated Pest Management" (G. J. Persley, ed.), pp. 76–97. CAB International, Wallingford, Oxon, U.K.

Kennedy, G. G., and Sutton, T. B. (2000). "Emerging Technologies for Integrated Pest Management." APS Press, St. Paul, MN.

Kogan, M. (1998). Integrated pest management: historical perspectives and contemporary developments. *Annu. Rev. Entomol.* **43**, 243–270.

Lewis, W. J., van Lenteren, J. C., Phatak, S. C., and Tumlinson, J. H. (1997). A total system approach to sustainable pest management. *Proc. Natl. Acad. Sci. USA* **94**, 12243–12248.

Metcalf, R. L., and Luckman, W. H. (1994). "Introduction to Insect Pest Management." Wiley, New York.

Pedigo, L. P. (1996). "Entomology and Pest Management." Prentice-Hall, Upper Saddle River, NJ.

Prokopy, R. J. (1993). Stepwise progress toward IPM and sustainable agriculture. *IPM Pract.* **15(3)**, 1–4.

Sterling, W. (1984). Action and inaction levels in pest management. Texas Agricultural Experimental Station, Bulletin 1480. Texas A&M University, College Station.

Tan, K. H. (2000). "Area-Wide Control of Fruit Flies and Other Insect Pests." Pulau Pinang, Malaysia.

Zalom, F. G. (2000). Moving along the IPM continuum. *Proc. Am. Soc. Enol. Vitic.* 50th Meeting, Seattle, WA 2000, pp. 356–359.

Integument

Svend O. Andersen
Copenhagen University

The integument is the external layer of tissue that covers the outer surface of insects and the surfaces of the foregut and hindgut. It is composed of the epidermis, which is a continuous single-layered epithelium, and an underlying thin basal lamina plus the extracellular cuticle that lies on top of the epidermis.

BASAL LAMINA

The basal lamina separates the epidermal cells from the hemolymph in the body cavity; it varies in thickness from 0.15 to 0.5 μm and is composed of structural proteins including collagens, glycoproteins, and glycosaminoglycans. It is negatively charged and can act as a filter between the hemolymph and the epidermal cells, regulating which molecules gain access to the cells.

EPIDERMAL CELLS

The epidermal cells are attached to the basal lamina by hemidesmosomes, which anchor the cell membrane to collagen fibers in the basal lamina. Near their base, the cells are attached to each other by desmosomes; near their apical end, they are attached to each other by a narrow, impermeable zone (the adhering zonule), effectively separating the cuticular compartment from the lateral space between cells. Below the zonule are bands of septate desmosomes, which may be adhesive, and gap junctions through which the cells can communicate chemically with each other by interchange of low molecular weight compounds.

The cuticular materials (chitin and proteins) are secreted from the apical surface of the epidermal cells into the subcuticular space, or deposition zone, where they are assembled into an intact cuticle. The apical surface is folded into shorter or longer microvilli, depending on the stage of the molting cycle and the secretory activity of the cells.

THE MOLTING PROCESS

The cuticle is a rather inextensible structure, and to grow, insects need to shed their old cuticle at intervals after having produced a new one with a larger surface area. The whole process, from breaking the connections between the epidermal cells and the cuticle (apolysis) to emerging from the remnants of the old cuticle (ecdysis), is called molting. A molt is initiated by a brief increase in concentration of the molting hormone, ecdysterone. Whether the new cuticle will become a larval, a pupal, or an adult cuticle depends on the concentration of juvenile hormone during the ecdysterone surge.

The first microscopically visible sign of molting is apolysis, which is the formation of a narrow space between the apical membrane of the epidermal cells and the inner surface of the cuticle. The cell membrane is folded into rather low microvilli, which carry small, dense plaques at the tip. Patches of a new epicuticle form at the top of the plaques, and gradually these patches merge to form a thin continuous layer, called the outer epicuticle or the cuticulin layer, that effectively separates the old cuticle from the space in which new cuticle is deposited.

The epidermal cells undergo mitotic divisions at the onset of molting, resulting in an increase of cell number and total epidermal surface. To allow the animal to increase in size at molting, the epicuticle deposited above the apical cell surface has a larger area than the former epicuticle, and the epidermis together with the epicuticle is folded to be accommodated in the space available inside the old cuticle.

The epidermal cells secrete a mixture of hydrolytic enzymes (the molting fluid) into the space below the old cuticle. The enzymes (proteases, peptidases, chitinase, and glucosidases) degrade the endocuticular chitin and proteins in the old cuticle to free N-acetylglucosamine and amino acids to be resorbed by the insect and reused for building the new cuticle.

While the old cuticle is being digested, the new inner epicuticle is deposited beneath the outer epicuticle and the new procuticle starts forming. Chitin is synthesized by an enzyme complex (chitin–synthetase) located at the tip of the microvilli, and chitin microfibrils grow from here into the subcuticular space, the deposition zone. The cuticular proteins are synthesized intracellularly and transported via secretory vesicles from the Golgi complex to the apical plasma membrane, where by exocytosis they are secreted into the subcuticular lumen. The chitin and protein molecules are in some way organized into a macromolecular complex in the deposition zone between cells and cuticle, possibly by a process of self-assembly.

The procuticle grows steadily in thickness until ecdysis, when the partially digested old cuticle ruptures and the insect emerges. Free of the old cuticle (exuvium), the insect expands the new cuticle to a predetermined size, often dependent on the area of the epicuticle. Cuticular deposition of chitin and proteins is resumed and continues for several days after ecdysis, and extended regions of the new cuticle are hardened (sclerotized) by oxidative incorporation of phenolic compounds into the cuticular matrix. In many insects, sclerotization and deposition of endocuticle after ecdysis are governed by the neurohormone bursicon.

Secretion of material from the epidermal cells occurs not only at the apical surface. Some of the proteins synthesized by the cells are exported to the hemocoel via the basolateral membrane system, and others, such as arylphorins, are synthesized in the fat body and secreted into the hemolymph to be taken up by the epidermal cells and incorporated into the cuticle.

PORE CANALS

Most cuticles contain pore canals, minute ducts that traverse the cuticle from the apical surface of the epidermal cells to or close to the cuticular surface. When viewed with an electron microscope they may seem empty, but they often contain one or more cuticular filaments composed of wax and lipids. Cytoplasmic processes may extend into the ends of the pore canals during cuticle deposition. The pore canals are generally assumed to be a transport route for lipids and possibly also for sclerotizing agents and proteins to the epicuticle and outer exocuticle.

SENSE ORGANS

The integument of insects contains a large number of sensory cell types, involved in transferring information from the environment to the insect. The sensory cells in the epidermis are often connected to specific cuticular structures, forming sense organs of various types, such as contact chemoreceptors (taste), olfactory chemoreceptors (smell), and mechanoreceptors, which register any small distortion of the cuticle caused either by the movements of the animal or by influences from the environment.

OENOCYTES

Oenocytes, a special cell type, often are present between the basal region of the epidermal cells and the basal lamina, or they may adhere to the hemolymphal surface of the basal lamina. Electron microscope studies show that they have a highly developed smooth endoplasmic reticulum, characteristic of cells engaged in hydrocarbon synthesis. Oenocytes synthesize hydrocarbon waxes, which are transferred to the wax layer of the epicuticle, presumably via the epidermal cells and the procuticular pore canals. It has been suggested that the oenocytes also provide lipids to the outer and inner epicuticle as well as to the sclerotized regions of the exocuticle.

INTEGUMENTAL GLANDS

Epidermal glands are present in many types of integument; they often consist of a single cell surrounding a cavity, which functions as a product reservoir and is connected to the cuticular surface by a small duct. Glands of this type are assumed to be responsible for forming and maintaining the cement layer, which after ecdysis is spread as a protective layer on top of the epicuticular wax layer. Other integumental glands of this type produce various chemical defense secretions, which may be forcibly ejected when the insect is disturbed.

Integumental glands also may consist of epidermal cells without a cavity and duct, but with direct contact to the inner surface of the cuticle, through which the secretion passes. Often the secretions of such glands function as pheromones, playing a part in the communication between individuals.

COLORATION

The colors of most insects are the result of pigments located in cuticle and epidermis, or they may be physical colors caused by surface structures in the cuticle. Colored material in hemolymph or internal organs can also contribute to the insect's color if the integument is transparent. Ommochromes, pteridines, carotenoids, bile pigments, melanins, and urates are the most widespread and important of the epidermal pigments. The colored light reflected from such epidermal pigments passes the overlying cuticle before it reaches the eye of the observer, and the observed color is influenced by the amount of colored material present in the cuticle.

See Also the Following Articles

Chemical Defense • Coloration • Cuticle • Exoskeleton • Molting

Further Reading

Bereiter-Hahn, J., Matoltsy, A. G., and Richards, K. S. (eds.). (1984). "Biology of the Integument," Vol. 1 of "Invertebrates." Springer-Verlag, Berlin. (See especially Chaps. 27–35.)
Binnington, K., and Retnakaran, A. (eds.). (1991). "Physiology of the Insect Epidermis." CSIRO Publications, Melbourne, Australia.
Hepburn, H. R. (ed.). (1976). "The Insect Integument." Elsevier, Amsterdam.
Kerkut, G. A., and Gilbert, L. I. (eds.). (1985). "Comprehensive Insect Physiology, Biochemistry and Pharmacology," 13 vols. Pergamon Press, Oxford, U.K. (See especially Vol. 3, Chaps. 1–4.)
Ohnishi, E., and Ishizaki, H. (eds.). (1990). "Molting and Metamorphosis." Japan Scientific Societies Press Tokyo, Japan.

Introduced Insects

Daniel Simberloff
University of Tennessee

An introduced species is a species that did not achieve its current taxonomic status in some location by natural evolutionary processes. Such species are often said to be nonindigenous or adventive to the location of introduction. Among nonindigenous species, those carried to the location by humans, either deliberately or inadvertently, are called "introduced species," whereas those that arrived of their own volition are termed "immigrant." However, all nonindigenous species are often colloquially termed "introduced." "Exotic" and "alien" are sometimes used for all nonindigenous species.

Introduced insects generate major ecological and economic impacts by a variety of means. Many species are introduced in various regions, and they span most insect orders. Their means of introduction are numerous, but most arrive with human help. Only a minority become problematic, and the reasons why a newly arrived species survives or fails to establish a population, and, if established, has or does not have a major effect, are often mysterious. Some problematic introduced species can be eradicated, and several management procedures can adequately control undesirable species when eradication attempts are unsuccessful.

NUMBERS

Introduced species often comprise a substantial fraction of a regional entomofauna. For example, in Great Britain there are about 21,000 native insects and about 200 established, introduced ones (comprising ~1% of the total). For the contiguous United States, approximately 84,000 native species have been identified, as have 1862 established, introduced species (~2% of the total). Florida has relatively more introduced species: 11,509 native and 993 introduced (8%). For oceanic islands, the introduced proportion can be much greater. Thus, for the Hawaiian Islands, there are about 5400 native insects and 2600 introduced ones (32% introduced), whereas the mid-Atlantic island Tristan da Cunha has 84 native insects and 32 introduced ones (28% introduced). The greater proportions of introduced species on islands probably result more from the mathematics of smaller numbers of native species and relatively larger numbers of attempted introductions than from some inherent invasibility of island communities or stronger "biotic resistance" of continental ones.

SYSTEMATICS

Established introduced insects are not a systematically random group. An order may be over- or underrepresented, reflecting primarily the reasons for its introduction and/or the means of its arrival. This pattern is exemplified in Fig. 1, which depicts the proportional fractions, by order, of insects of the world compared with introduced insects of the contiguous United States. Thus, far more species of homopterans, hemipterans, and thysanopterans are introduced than would be expected based on their numbers in the world, because these orders are associated with introduced agricultural and ornamental plants. Hymenopterans are overrepresented

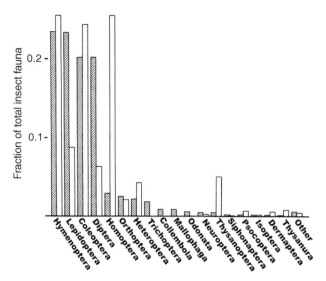

FIGURE 1 Fractional representation of different insect orders of the world (cross-hatched) and established introduced insects of the United States (clear). [From Simberloff, D. (1986). Introduced insects: a biogeographic and systematic perspective. *In* "Ecology of Biological Invasions of North America and Hawaii (H. A. Mooney and J. A. Drake, eds.), Fig. 1.1, p. 6. © Springer-Verlag GmbH & Co. K G, Heidelberg. Data from Arnett, R. H. (1983). Status of the taxonomy of the insects of America north of Mexico: a preliminary report prepared for the subcommittee for the insect fauna of North America project. (Privately printed) and Sailer, R. I. (1983). History of insect introductions. *In* "Exotic Plant Pests and North American Agriculture" (C. Graham and C. Wilson, eds.), pp. 15–38. Academic Press, New York.]

perhaps partly because of their wide use in biological control and partly because their haplodiploidy allows a single female, which can produce both male and female offspring, to found a population. Haplodiploidy may also lower inbreeding depression in founding populations. Haplodiploidy is also found among homopterans and thysanopterans, both overrepresented groups. At the family level, 26 families are overrepresented, and other patterns become clear. In the contiguous United States, coccinellid beetles and anthocorid bugs are both overrepresented; species in both groups were introduced as biological control predators. Tephritid flies were introduced deliberately for control of knapweeds and inadvertently on cultivated plants. Ten overrepresented families are in the Hymenoptera, and seven of these were introduced for biological control. Two thrips families in this list are associated with cultivated plants, as are the adelgids (pine and spruce aphids). Oestrid flies are overrepresented, and these are internal parasites of mammals, including livestock; they probably arrived with their hosts. Dermestid and anobiid beetles are overrepresented, and these stored product pests probably arrived with their food sources.

WHY, WHEN, AND HOW INSECTS ARE INTRODUCED

Some introduced insects arrive in new locations on their own and are true immigrants. Of course, any simple range

extension could bring a new introduced species in this sense, but as a rule this status is restricted to species arriving at a distant location by a discontinuous dispersal, rather than gradually diffusing from a neighboring site. Thus, the monarch butterfly *(Danaus plexippus)* in Australia might not be considered to be introduced. It arrived around 1870 and established a population, having spread through the Pacific during the 19th century largely unaided by humans. The monarch migrates long distances in its native North America, is an occasional straggler in Europe, and was recorded successively in Hawaii, the Caroline Islands, Tonga, and New Zealand before it reached Australia.

Some insects are introduced deliberately by humans. A few insects arrive as pets or pet food; recent pet price lists include mantids, walkingsticks, spider wasps, velvet ants, and dung and blister beetles. In Florida in 1989, giant Madagascan hissing cockroaches *(Gromphadorhina portentosa)* became highly popular as pets; at least some were released to the wild, where they survive. Similarly, caterpillars of a Chilean moth, *Chilecomadia morrei,* are sold as reptile food in the United States and Europe. The European honey bee *(Apis mellifera)* has been widely introduced for both the production of honey and crop pollination. The Asian silkworm *(Bombyx mori)* has been widely introduced along with its host plant, mulberry, for silk production. Even introductions that fail to establish a commercial industry can nonetheless establish a population. The Asian ailanthus moth *(Samia walkeri)* was brought to the United States in an attempt to found a silk industry on the ailanthus tree. Though the industry foundered, the moth remains. The gypsy moth *(Lymantria dispar)* was brought to North America to establish a silk industry; its predictable escape established one of North America's major pests.

Most deliberate insect introductions are for biological control. Although weeds and insect pests of agriculture are the usual targets, there are others. Thus, *Paratrechina fulva,* a Brazilian ant, was introduced to Colombia to control poisonous snakes, whereas over 45 dung beetle species were introduced to Australia to break down droppings of introduced livestock (~33% of established populations). Insects introduced to attack insects are either predators (primarily coccinellid beetles, but including other beetles, hemipterans, and neuropterans) or parasitoids (mostly hymenopterans, but including some dipterans). Insects introduced to control plant pests include primarily flies, beetles, and moths, although others such as bugs and thrips have been used. The number of insect species introduced for biological control purposes is substantial. For example, of approximately 2600 introduced insect species established in the Hawaiian islands, roughly 400 were introduced for biological control. About twice as many species introduced there for this purpose perished.

Far more insects are introduced inadvertently than deliberately by humans. Pathways are myriad. Soil ballast was an early predominant mode of entry to North America— many of the first introduced insects were soil beetles from southwestern England. This pathway is less common now, but insects are still carried in rootballs around cultivated plants and in soil on heavy equipment. Phytophages, particularly homopterans, dominated introductions to North America in the 19th century with the advent of fast steamships and a proliferation of imported nursery stock, and imported plants are still a major means of introducing insects worldwide. The grape phylloxera (*Daktulosphaira vitifoliae*) that devastated French vineyards in the 19th century arrived on saplings or cuttings of American vines. Insects can also be carried in water. The yellow fever mosquito *(Aedes aegypti)* probably arrived in colonial America in drinking water casks, while the Asian tiger mosquito *(A. albopicta)* reached North America in the 1980s in scrap tires from Japan. This is probably the route taken by *A. japonicus,* which arrived in the United States in 1998 and transmits West Nile virus. These latter two mosquitoes have recently been detected in used tires in New Zealand. Wooden packing material brought the Asian longhorned beetle *(Anoplophora glabripennis)* to the United States, whereas the European elm bark beetle *(Scolytus multistriatus)* that transmits Dutch elm disease arrived in North America on unpeeled veneer logs of European elm. The growth of international tourism can enhance the rate of insect introduction; in 1992 an Australian tourist returned from South America with a wound containing maggots of the New World screwworm *Cochliomyia hominivorax.*

ESTABLISHMENT AND SPREAD

The majority of introduced insects, like most introduced species, do not survive, although only biological control introductions generate substantial data on failed introductions. For parasitoid species introduced to control insect pests, only about 30% establish populations, whereas for all insects introduced for plant control, the comparable figure is about 60%. Because biocontrol candidates are chosen and often tested for survival in the target environment, one might expect failure rates for inadvertently introduced introductions to be even higher. For most taxa, invasion biologists believe that 5 to 20% of introduced species establish populations, although many of these may remain for years or in perpetuity near the point of introduction, and a large fraction are restricted to anthropogenous habitats such as human habitations or agricultural fields.

An arriving propagule must be large enough to survive the initial threat of demographic stochasticity—that is, random elimination of so many individuals during the first few generations that the population fails. Data from the biological control literature show that probability of establishment increases with propagule size and number of attempts, but many very small propagules have established large, widespread populations. For example, a single fertilized female of the cochineal insect *Dactylopius opuntiae* from Sri Lanka initiated a large, ongoing population on Mauritius. In Puerto Rico, two

females of a unisexual race of the encyrtid wasp *Hambletonia pseudococcina* were used to rear 7000 individuals, which were released, established, and quickly spread. For a parthenogenetic species, at least the difficulty of finding a mate is obviated, but demographic stochasticity has other components.

Even assuming an adequate propagule size, the environment, both physical and biotic, must be suitable for a species to survive and spread. Predators, parasites, competitors, and pathogens can eliminate an introduced species or restrict its ambit. For example, the Asian aphelinid wasp parasitoid *Aphytis fisheri,* introduced to California to control California red scale *(Aonidiella aurantii),* failed to establish because of competition from previously introduced *A. melinus* and *A. lingnanensis.* Conversely, the absence of natural enemies from its native range is often posited as the reason for the success of some invaders, such as the cynipid *Andricus quercuscalicis,* introduced into Great Britain. By contrast, the presence of another species, such as a food plant or a symbiont, might be necessary for an invader to survive. The monarch butterfly would not have survived in Australia but for the prior introduction of its host milkweeds. The physical environment is probably an even more frequent reason for introduced species to perish. A temperate climate does not augur well for a newly arrived tropical insect, but even subtler physical differences can be crucial to a species's survival. For instance, synanthropic species are unlikely to survive if they arrive in pristine natural habitats.

IMPACTS

Although quantifying the impact of a new species is an unsolved challenge, it is safe to say that most introduced species do not generate major impacts. However, some are enormously damaging, whereas others are highly beneficial. The variety of impacts is staggering.

Many introduced insects prey on natives. This activity can be useful, as in biocontrol introductions such as that of the Australian vedalia *(Rodolia cardinalis)* to attack the cottony cushion scale *(Icerya purchasi).* Predation can also be extremely damaging; on Christmas Island, the introduced yellow crazy ant *Anoplolepis gracilipes* has locally devastated populations of the dominant red crab *Gecarcoidea natalis.* Because the crab controls seedling recruitment and litter breakdown, the entire community is affected. Parasites can also be beneficial or harmful. The wasp *A. melinus* has effectively controlled California red scale in parts of California. Alternatively, sheep blowfly *(Lucilia cuprina),* introduced to Australia from Africa, caused staggering losses. Herbivory can similarly be beneficial or detrimental. The South American flea beetle, *Agasicles hygrophila,* effectively controls alligatorweed in Florida. However, phytophagous insect crop pests impose staggering costs. The alfalfa weevil *(Hypera postica)* caused $500 million in losses in the United States in 1990 alone.

Resource competition is subtler than predation, parasitism, and herbivory, but many introduced insects outcompete natives. The European sevenspotted lady beetle *(Coccinella septempunctata),* introduced to the United States for control of the Russian wheat aphid *(Diuraphis noxia),* has locally outcompeted several native lady beetles. European honey bees outcompete the native bee *Osmia pumila* for pollen in New York State. Introduced insects can even outcompete vertebrates. The introduced wasps *Vespula germanica* and *V. vulgaris* in New Zealand outcompete an endemic parrot for honeydew produced by a scale insect *(Ultracoelostoma assimile)* and have locally lowered parrot populations.

Introduced insects can transmit or be reservoirs of diseases of humans, domestic animals, cultivated plants, and wild animals and plants. The spread of yellow fever and dengue as the vector mosquito *A. aegypti* dispersed throughout the tropics, and of malaria to Brazil with the introduction of its vector *Anopheles gambiae,* are notable examples of human disease organisms transmitted by introduced insects. Cat fleas *(Ctenocephalides felis)* and dog fleas *(C. canis),* introduced to Australia with their hosts, are intermediate hosts for the dog tapeworm *(Dipylidium caninum).* The southern house mosquito, *Culex quinquefasciatus,* was accidentally introduced to the Hawaiian islands in 1826. Subsequently it transmitted the disease organism causing avian malaria, introduced with resistant Eurasian songbirds, to susceptible native birds and helped to exclude them from low elevations. The mosquito *A. japonicus* transmits West Nile virus to both birds and humans in the northeastern United States. Animal disease vectors may be useful biocontrol introductions. For example, the rabbit flea *(Spilopsyllus cuniculi),* the main vector of the organism causing myxomatosis in Europe, has been introduced (so far unsuccessfully) in Australia to attempt to boost disease transmission. In the lab, it also transmits calicivirus.

Among plants, Dutch elm disease was dispersed to and through North America with the European elm bark beetle. Beechbark disease spread throughout northeastern North America after the causal fungus was introduced from Europe around 1890 with its vector, the beech scale *(Cryptococcus fagisuga).* The wine industry in California is threatened by the recent introduction of the glassy-winged sharpshooter *(Homalodisca coagulata)* from the southeastern United States. The sharpshooter spreads an incurable bacterial disease of grape vines, a malady long present but rarely a problem until this vector arrived.

Introduced species often exacerbate one another's impacts, a process termed "invasional meltdown." Sometimes this interaction occurs when coevolved mutualists invade a region separately. In south Florida, over 60 species of ornamental figs were not invasive because their obligatory pollinating wasps were absent. Since the 1970s, three such wasps *(Parapristina* spp.) have arrived, and three formerly innocuous fig species have begun spreading in natural areas. However, invasional meltdown need not involve coevolved species. In California citrus orchards, the Argentine ant *(Linepithema humile)* tends and protects the Asian California red scale, thereby exacer-

bating its impact. Similarly, in Hawaii, the African bigheaded ant *(Pheidole megacephala)* protects the tropical American gray pineapple mealybug *(Dysmicoccus neobrevipes)* from coccinellids introduced for biological control.

Relative to the numbers of species introduced, insects rarely cause enormous ecological (as opposed to economic) damage. Introduced species whose impacts ripple through entire communities usually do so by changing the habitat dramatically, and such change agents are mostly plants (which become structural dominants or modify fire regimes) or pathogens, which attack dominant plants. Occasionally mammals can generate an enormous ecosystem impact by trampling or grazing. A recent list of the world's 100 worst introduced species included 15 insects, but at most one would qualify as having a huge ecosystem-wide impact: the yellow crazy (or long-legged) ant, which removes the keystone red crab species on Christmas Island. Of the 15 insects, five are ants. In addition to the yellow crazy ant, the Argentine ant, the bigheaded ant, the little fire ant *(Wasmannia auropunctata),* and the red imported fire ant *(Solenopsis invicta)* all affect other ants greatly, and sometimes other insects, but to date none has had the dramatic impact of certain plants and mammals. Some species among the 15 transmit organisms that cause human diseases *(Aedes albopicta* and *Anopheles quadrimaculatus)* and others are agricultural pests, such as the sweetpotato whitefly *(Bemisia tabaci)* and several of the ants. The Formosan termite *(Coptotermes formosanus shiraki)* has caused enormous damage to housing in New Orleans. With respect to ecosystem-wide damage to natural areas, however, the only members of the list that might qualify, aside from the yellow crazy ant, are the gypsy moth in North America, by virtue of its devastating impact on dominant trees, and the Argentine ant, because it has greatly lowered densities of native seed-carrying ants in the fynbos of South Africa. Other insects may have ecosystemic impacts by removing dominant plants. The beetle transmitting the Dutch elm disease organism has already been noted. The Asian balsam woolly adelgid *(Adelges piceae)* has eliminated the dominant Fraser fir throughout the high southern Appalachians, whereas the hemlock woolly adelgid *(A. tsugae)* has locally killed large fractions of hemlocks in much of eastern North America.

Impacts of an introduced species can occur after a substantial lag period during which the species can seem to be innocuous. For example, the beetle *Chrysolina quadrigemina,* introduced to Australia in 1939 to control St. John's wort, seemed to die out but resurfaced and spread in 1942. Such lags are mysterious; they are often attributed to favorable changes in the environment or to evolution of the invader, but evidence for these phenomena is generally lacking. Some introduced insects achieve great numbers and appear to have a major impact, but the population suddenly crashes, again for reasons poorly understood. The European browntail moth *(Euproctis chrysorrhoea)* followed this trajectory in New England and eastern Canada.

EVOLUTION

The conditions under which a species is introduced to a new region (isolation from parent population, small propagule size [usually], and different physical and biotic environment) should be conducive to rapid evolution. There has been little study of this phenomenon, but some striking examples have emerged. *Drosophila subobscura,* introduced to the Americas from the Old World around 1980, spread widely and by 2000 had evolved a cline of increasing total wing length with latitude in North America phenotypically similar to that in its native range. The Ichneumonid *Bathyplectes curculionis,* introduced to the western United States for biological control of introduced alfalfa weevils (*Hypera* spp.), evolved in less than 10 years to become less susceptible to the encapsulation reaction of its host.

A phenomenon widely reported among introduced vertebrates and plants, particularly in North America and Eurasia, is hybridization with native species, sometimes to the point of a sort of genetic extinction of the latter. Although hybridization is known to have played an important role in insect evolution, hybridization between native and recently introduced insects is rarely if ever reported. This absence of data may reflect biological differences or simply less genetic study of insects. There are instances of introduced populations hybridizing with one another, most notably the Italian and African strains of the honey bee. The red imported fire ant hybridizes extensively with the previously introduced black imported fire ant (*Solenopsis richteri*) in Tennessee.

ERADICATION AND MANAGEMENT

Eradication is one possible response to an introduced species, particularly one that has not dispersed widely. Many introduced insects have been eradicated, some from substantial areas. Perhaps the most impressive is the chemical eradication of the African malaria mosquito *A. gambiae* from 31,000 km^2 of northeastern Brazil in 1939–1940. The Mediterranean fruit fly (*Ceratitis capitata*) was eradicated over 18 months from a 20-county region of Florida by a strict quarantine, destruction of produce and plants, trapping, and insecticide sprays. More recent attempts to eradicate *C. capitata* in both Florida and California may not have been as successful. Although victory has been declared repeatedly, reappearances are frequent and may constitute either new invasions or simply recovery by uneradicated remnant populations.

The development of the sterile-male technique in the United States against the New World screwworm gave tremendous impetus to the eradication approach. Release of massive numbers of sterile males so reduced the probability of fruitful mating by females that this species disappeared totally from the island of Curaçao in 1954–1955, and this method greatly aided eradication of this fly from the southeastern United States in 1958–1959. The melon fly (*Bactrocera cucurbitae*) was eliminated from Rota Island by this method. The male annihilation method, in which males are attracted and destroyed,

has also succeeded in eradicating introduced fruitfly populations from islands, including the Oriental fruit fly (*Dacus dorsalis*) from Rota and Guam and the melon fly from Nauru. Male annihilation followed by release of sterile males eradicated the melon fly from the entire Ryukyu Archipelago. The white-spotted tussock moth (*Orgyia thyellina*) was eradicated from greater Auckland, New Zealand, by pheromone lures plus spraying of *Bacillus thuringiensis*.

There have also been disastrous failures of expensive eradication campaigns, such as the $200 million attempt to eradicate chemically the red imported fire ant from the southeastern United States, an effort that imposed greater mortality on native insects than on the invader. Quick detection, rapid response, sufficient resources to finish the project, and adequate regulatory power to enforce cooperation have proven most conducive to successful eradication.

If eradication fails or is not attempted, chemical and biological control are the two methods most commonly attempted to manage introduced insects. There are successes and failures for both methods. The nontarget and human health impacts of early generation insecticides such as DDT are legendary. Though more recent chemicals minimize or eliminate this problem, chemical control frequently is problematic for two main, related reasons. First, insects evolve resistance to chemicals; second, expense can be far too great, especially when ever larger amounts must be used because of resistance. For large natural areas, expense of continued chemical applications can be particularly prohibitive.

Biological control is attractive because the expense of development and deployment is lower, and because, although a host may evolve resistance, the biological control agent itself can evolve countermeasures (as witness *B. curculionis,* discussed earlier). However, the success rate of biological control is rather low. For instance, for parasitoids introduced for insect control, only 10% have been effective. Furthermore, biocontrol agents can affect nontargets, as the sevenspotted lady beetle has done, and these nontarget impacts can be generated by established biocontrol agents that are not even effective against their targets (about three times as many biocontrol parasitoids establish populations as actually control the target pest). The tachinid fly *Compsilura concinnata* has failed to control gypsy moths in North America, but it is believed to be responsible for the decline of several large native moths.

See Also the Following Articles
Conservation • Fire Ants • Gypsy Moth • Island Biogeography and Evolution • Neotropical African Bees • Regulatory Entomology

Further Reading
New, T. R. (1994). "Exotic Insects in Australia." Gleneagles, Adelaide.
Sailer, R. I. (1983). History of insect introductions. *In* "Exotic Pest Plants and North American Agriculture" (C. Graham and C. Wilson, eds.), pp. 15–38. Academic Press, New York.

Simberloff, D. (1986). Introduced insects: A biogeographic and systematic perspective. *In* "Ecology of Invasions of North America and Hawaii" (H. A. Mooney and J. A. Drake, eds.). pp. 3–26. Springer-Verlag, New York.

Simberloff, D. (1989). Which insect introductions succeed and which fail? *In* "Biological Invasions. A Global Perspective" (J. A. Drake, H. A. Mooney, F. diCastri, R. H. Groves, F. J. Kruger, M. Rejmánek, and M. Williamson, eds.), pp. 61–75. Wiley, Chichester.

Williams, D. F. (ed.). (1994). "Exotic Ants. Biology, Impact, and Control of Introduced Species." Westview, Boulder, CO.

Williamson, M. (1996). "Biological Invasions." Chapman & Hall, London.

Island Biogeography and Evolution

George K. Roderick and Rosemary G. Gillespie
University of California, Berkeley

Studies of insects have played a major role in the general understanding of the biota of islands, touching on all areas of biogeography, ecology, evolution, and conservation. The notable writings of Darwin and Wallace were influenced heavily by the biological diversity that each witnessed on islands and by the processes inferred to underlie that diversity. Rather more recently, studies exploiting the discrete nature of islands have given rise to pervasive organizing theories of community ecology, in particular MacArthur and Wilson's equilibrium theory of island biogeography (ETIB). With the advent of accessible molecular genetic tools, research on islands has allowed unique insights into the processes that generate biotic diversity, especially the mechanisms of speciation. Unfortunately, islands are also prime targets for biological invasions, mediated largely by anthropogenic disturbance. The severity of such impacts on island biotas may result from their evolution in isolation, but it is certainly compounded by their characteristically small population sizes. Yet, for many islands extinction among arthropods is largely unknown, although this circumstance may be attributable more to lack of knowledge than to any innate security that arthropods might possess. Indeed, it is likely that many island arthropods will go extinct before they have been collected and described.

THEORY OF ISLAND BIOGEOGRAPHY

Larger islands contain more species. This idea was formalized by MacArthur and Wilson in the 1960s with the development of the ETIB. This theory relates species and area by the formula $S = cA^z$, where S is species number, A is area, c is a constant measuring overall species richness, and z measures the extent to which increases in area have diminishing returns in terms of species number. Values of z tend to vary between 0.18 and 0.35; that is, doubling the species number requires increasing the area by a factor lying between 7 and 100. The premise of the theory is that the rate of immigration decreases with increasing distance from the source, whereas the rate of extinction decreases with increasing island size. The balance of these processes results in an equilibrium number of species on any one island. As the number of resident species on an island increases, the chance of an unrepresented species arriving on that island decreases and the likelihood of extinction of any one resident species increases. The predictions of the model are as follows: (1) the number of species on an island should change little once the equilibrium has been reached; (2) there should be continual turnover of species, with some becoming extinct and others immigrating; (3) small islands should support fewer species than large islands; and (4) species richness should decline with remoteness of the island, since islands farther from the source will have lower rates of immigration.

Rigorous tests of the ETIB have been surprisingly few, and they have supported some aspects of the theory but not others. For example, Simberloff used insecticides to defaunate mangrove islands and found that species of insects and spiders accumulated to an equilibrium number. However, contrary to expectation, turnover of species was not randomly distributed among species—species of particular types were likely to colonize or go extinct. Species numbers have been found to be affected unpredictably by both area and isolation; yet other work has shown that an equilibrium does not exist, or that parameters other than area per se may dictate species richness. Such factors include habitat diversity, climatic conditions, island age, and even the status of knowledge concerning the presence of resident species. However, the theory has proven to be remarkably useful and, although it was developed for islands, it has had relevance for the study of ecological communities of many kinds.

ADAPTIVE RADIATION

The ETIB assumes that islands are within the geographic distance into which a species is likely to disperse, thus maintaining genetic populations between source and island populations. On islands that are beyond the range within which populations can maintain genetic contact with source populations, one might predict (based on the theory) that few species should be present. But this tends not to happen. Isolated islands that are formed initially without life are often found to have large numbers of closely related species. When single colonists, isolated genetically from their source population, give rise to a series of species that have diversified ecologically, the phenomenon is termed adaptive radiation. Usually it occurs beyond the so-called radiation zone, or normal range of dispersal of a given organism. Species that form through adaptive radiation are typically neoendemics, formed *in situ* and found nowhere else. Among arthropods, the Hawaiian Islands hold the record in having the largest number of neoendemics, an extraordinary 98% of the fauna. Founder effects, behavioral isolation, ecological isolation, and host-associated isolation have all been implicated in the process of adaptive radiation. For insects, particularly noted examples include *Drosophila* flies, which are well known for their diversity of mating behaviors, as well as lineages of crickets that have diversified in song repertoire, sap-feeding planthoppers that have proliferated by switching between plant hosts, and beetles that have formed new species on different substrates. Diversification may follow a predictable pattern, at least in some groups; for example, among *Tetragnatha* spiders, similar ecological sets of species have evolved over and over again on each of the different Hawaiian Islands.

Compared with their hypothetical colonizing ancestors, species on remote oceanic islands are often characterized by a reduction in dispersal ability. Indeed, they are often found to have very narrow ranges of dispersal. Moreover, the individual species that colonize remote islands are a small sample of the continental source. They are therefore said to be "disharmonic" and not representative of the biotic diversity on continents, a phenomenon accentuated by the frequent proliferation of successful colonists.

NEOENDEMIC AND PALEOENDEMIC ISLANDS

Neoendemics typically form on isolated islands that have been created *de novo* and have abundant empty ecological space into which those few colonists can diversify. Besides Hawaii, other volcanic archipelagoes, including the Marquesas, Societies, and Galapagos in the Pacific and the Canaries in the Atlantic, have provided ideal conditions for the formation of neoendemics. However, species can also form on fragment islands, formed as a mass of land has broken away from a larger continental region. Examples of such islands include some of the Caribbean islands, and the islands of New Zealand and Madagascar. As these islands, formed upon losing connection with a continental source, become more isolated, gene flow between island and continental populations may become insufficient to overcome genetic divergence. Unlike volcanic islands that form in isolation, starting without any species and accumulating species through time, fragment islands are usually ecologically saturated at the time of separation and tend to lose species through ecological time, a process termed relaxation. Over evolutionary time, the species on these islands may change through relictualization, with the formation of paleoendemics, usually without adaptive radiation.

HABITAT ISLANDS

Many of the ideas originally developed for islands in the sea have been extended also to so-called habitat islands of a particular habitat type in a matrix of unlike terrain. Most such islands are fragments of habitats that were historically connected, such as remnant trees and forest patches. For habitat islands, as for islands in the sea, ecological and evolutionary processes are governed largely by isolation, time, and the nature of the matrix relative to the dispersal abilities of the organisms in question. Habitat islands, because they are discrete and easy to manipulate compared with islands in the sea, have been exploited in the development of many ecological principles, including those related to metapopulation dynamics and the physical design of nature reserves.

CONSERVATION

The biota of islands is often unique—for example, the islands of the Pacific have been designated a biodiversity hot spot. Assessing this diversity, particularly for arthropods, is problematic. The major impediment is a lack of taxonomic understanding of arthropods on many islands, particularly those that are more remote. New species are being collected at a remarkable rate in areas such as French Polynesia, Madagascar, and even the relatively well-studied Canary Islands, New Zealand, Hawaii, and the Galapagos, yet the training of arthropod systematists has lagged behind.

Anthropogenic disturbance has also had its impact not only in present times, but historically, as witnessed through the colonization of the Pacific by Polynesians several thousands of years ago. A number of characteristics of arthropods populations on islands, including high local endemism, limited dispersal abilities, and small population sizes, make them particularly vulnerable to both demographic accidents and environmental change. In addition, islands have been impacted heavily by invasive species, many of which are also arthropods. The impact has been both direct, such as through the extirpation of species by invasive predatory ants, and indirect, such as through diseases of vertebrates having mosquitoes as vectors.

Although islands have long served as extraordinary laboratories for studying processes associated with the generation of diversity, they are now contributing to understanding of processes leading to the loss of diversity. For example, studies of invasive species on islands have shown the importance of environmental factors as well as species-specific attributes that facilitate biological invasions and its negative effects. New tools are urgently needed: rapid biodiversity assessment techniques that bypass traditional taxonomic identification will be important in recognizing areas of high conservation priority, as will genetic or ecological approaches that can distinguish native species from those introduced in more recent history.

See Also the Following Articles
Biodiversity • Biogeographical Patterns • Cave Insects • Introduced Insects

Further Reading
Darwin, C. (1859). "The Origin of Species by Means of Natural Selection." John Murray, London.
Gillespie, R. G., and Roderick, G. K. (2002). Arthropods on islands: Colonization, speciation, and conservation. *Annu. Rev. Entomol.* **47,** 595–632.
Howarth, F. G., and Mull, W. P. (1992). "Hawaiian Insects and Their Kin." University of Hawaii Press, Honolulu.

MacArthur, R. H., and Wilson, E. O. (1967). "The Theory of Island Biogeography." Princeton University Press, Princeton, NJ.

Wagner, W. L., and Funk, V. A. (eds.). (1995). "Hawaiian Biogeography, Evolution on a Hot Spot Archipelago." Smithsonian Institution Press, Washington, DC.

Wallace, A. R. (1902). "Island Life." 3rd ed., Macmillan, London.

Whittaker, R. J. (1998). "Island Biogeography: Ecology, Evolution, and Conservation." Oxford University Press, Oxford, U.K.

Williamson, M. H. (1981). "Island Populations." Oxford University Press, Oxford, U.K.

Isoptera
(Termites)

Vernard R. Lewis
University of California, Berkeley

The ordinal name Isoptera refers to the two pairs of straight and very similar wings that termites have as reproductive adults. The common name, of Latin origin, translates as "woodworm." Termites are small and white to tan or sometimes black. They are sometimes called "white ants" and can be confused with true ants (Hymenoptera). However, a closer look reveals two easily observed, distinguishing features: termites have straight antennae and a broad waist between the thorax and the abdomen, whereas ants have elbowed antennae and a narrow waist. For reproductive forms, termites have four equally sized wings, whereas ants have two pairs of dissimilarly sized wings.

IMPORTANT FAMILIES OF TERMITES

The earliest known fossil termites date to the Cretaceous, about 130 mya. There are >2600 species of termites worldwide. Undoubtedly, more will be recognized with improved methods of discerning cryptic species and after intensive collecting of tropical and remote regions. Termites are most closely related to cockroaches and mantids. The greatest continental termite diversity is in Africa, where there are over 1000 species. Polar continents have none, and North America with 50 species and Europe with 10 species are intermediate in termite diversity.

Development is incomplete metamorphosis containing castes that include nymph, worker, pseudergate, soldier, and several types of reproductives (Figs. 1, 2, and 3). Nymphs hatch from eggs and molt at least three times before becoming functional workers. Workers are wingless, do not lay eggs, and, except for the family Hodotermitidae, are blind. Worker and pseudergate castes are the most numerous in a colony and conduct all major foraging and nest-building activities. Soldiers defend colonies with fearsome mandibles and/or chemical squirts from a nasus, a frontal projection

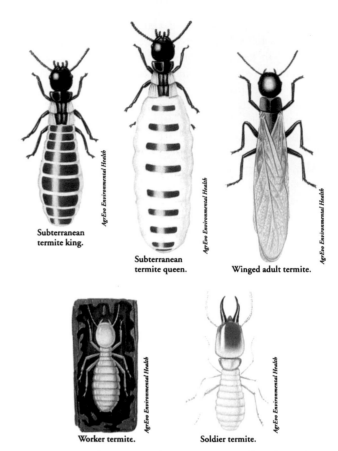

FIGURE 1 Castes for Isoptera. A lower termite group, *Reticultermes,* is represented. A large queen is depicted in the center. A king is to the left of the queen. A worker and soldier are below. (Adapted, with permission from Aventis Environmental Science, from *The Mallis Handbook of Pest Control,* 1997.)

from their heads. Soldiers, including nasutes, cannot feed themselves. Reproductives consist of a royal pair, the original colony founders, but supplementary and replacement reproductives (neotenics) can be generated from workers, nymphs, or other immatures dependent on pheromonal cues from the queen and environmental factors.

Termite families traditionally were categorized as lower or higher. However, this categorization may change soon as newer classification systems are adopted. Lower termites (families Mastotermitidae, Kalotermitidae, Termopsidae, Hodotermitidae, Rhinotermitidae, and Serritermitidae) have symbiotic intestinal protozoa and bacteria. Higher termites (Termitidae) have intestinal bacteria.

Termite identification at the family and genus level is determined using reproductive adults or soldiers or, in some groups, workers. All living termites can be divided into seven families as follows.

Mastotermitidae

This family contains the most primitive living termite, *Mastotermes darwiniensis* (Fig. 4), now limited to Northern

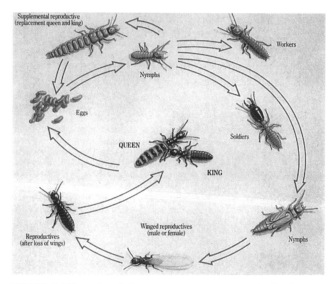

FIGURE 2 Life cycle of the termite. Lower termite family depicted. (Adapted, with permission from FMC Corp., from *The Mallis Handbook of Pest Control*, 1997.)

FIGURE 4 *Mastotermes darwiniensis,* the most primitive termite from Darwin, Australia. Tertiary-era fossils contain species from this family. Reproductive adults are in the center. Soldiers have large heads and mandibles. Smaller termites are workers. (Photograph courtesy of Dr. Barbara Thorne, University of Maryland.)

Australia. In appearance, these termites are light brown, robust, and about 8–10 mm in length. This family is recognized by the presence of an anal lobe in the hind wing of the reproductive adults and five-segmented tarsi. The hind wings are very similar to those of some cockroaches, considered a sister group to termites. Like cockroaches, reproductive females also lay egg cases containing up to 24 eggs arranged in two regular rows. Although egg masses contain few eggs, *Mastotermes* has many neotenic reproductives (no primary queen has ever been found in the field), and colonies can reach a population size of millions. Soldiers have powerful mandibles and excrete a toxic brown substance from their buccal cavity that repels intruders.

FIGURE 3 Castes of the Formosan subterranean termite, *Coptotermes formosanus* (Isoptera: Rhinotermitidae). In the center is a queen with large physogastric abdomen containing eggs. A king with a physogastric abdomen lies next to the queen. Soldiers have brown tear-shaped heads with sickle mandibles. A worker is also shown. (Photograph courtesy of Dr. Minoru Tamashiro, University of Hawaii.)

Kalotermitidae

Members of this family are commonly called "dry-wood termites" for their habit of nesting in wood above the soil level, although exceptions occur. Some dry-wood termites have subterranean habits, whereas others prefer rotten and damp wood. Dry-wood termites are brownish and are considered medium-sized termites, 10–13 mm in length. This family is recognizable by the presence of ocelli and two-segmented cerci in the alate form. There are more than 400 species worldwide. Dry-wood termites are common on most continents. Colonies are moderate in size and contain several thousand individuals, most of which function as workers. The queen lays about 1 dozen or so eggs per day.

Termopsidae

The damp-wood termites nest in wet and rotting wood, especially fallen logs and stumps in forests. Damp-wood termites were formerly grouped within harvester termites (Hodotermitidae), but now are considered a separate family. Damp-wood termites are among the largest termites, some reaching almost 25 mm in length. Most individuals retain marked developmental plasticity. There are about 20 species and they are limited to forests in the Americas, Eurasia, Africa, and Australia. Egg production per queen is relatively low (≤30 per day) and colony size is moderate, up to approximately 10,000.

Hodotermitidae

Members of this grass-harvesting family of 15 species are similar in appearance to damp-wood termites and are quite large (>15 mm in length). Reproductive adults lack ocelli and their cerci have three to eight segments. Modern species are savanna grass feeders and occur in Africa, the Middle East, and India. Queen egg production and colony size are similar to those of damp-wood termites.

Rhinotermitidae

Commonly called subterranean termites, this family typically requires its nest to contact the soil. However, exceptions occur (genera *Coptotermes* and *Recticulitermes*). Most species in this family are recognizable by their reproductive adults and a flat pronotum behind the head of soldiers. For some species, a fontanelle gland is present on the head of soldiers that produces a defensive fluid. Workers and soldiers are small (<5 mm) and are very pale white. More than 300 species are recognized. They occur on most continents except in polar and near-polar regions and are serious pests of structures. Queens of some species can produce more than 100 eggs per day, and colonies can number from the tens of thousands into the millions. Some mound-builders and aerial-nesters are found in this family. Aerial-nesting species still maintain contact to the soil for water by runways constructed from soil and saliva.

Serritermitidae

This family is very similar in appearance and closely related to subterranean termites (Rhinotermitidae). It also requires its nest to contact the soil. A single species occurs in South America. Soldiers have serrated teeth along the entire inner margin of the mandibles.

Termitidae

This family contains builders of the great mounds (up to 8 m high) that occur in the tropics, mainly in Asia, Africa, Australia, and South America. There are a few species in North America and none occur in Europe. More than 1800 species have been described, many from Africa. Termitids are distinguished by two prominent teeth on the left mandible of reproductive adults and a saddle-shaped pronotum. The Termitidae have a true worker caste. Workers are very small (<5 mm) and pale or dark in color. Many species have nasute soldiers. Members of this family are some of the most prolific producers of eggs in the animal kingdom. A queen can produce more than 10 million eggs in a single year.

TERMITE BIOLOGY AND ECOLOGY

Termites live in colonies that are social and can be long-lived. Colonies are composed of castes that conduct all tasks for survival (Figs. 1, 2, and 3). Some termite queens are larger than the length of a human thumb and can lay more than a thousand eggs per day. The king is also long-lived and mates intermittently to provide sperm to the queen. Some of the longest living insects are termites: some termite mounds and their queens are thought to be more than 70 years old and

Aborigine folklore claims some mounds in Australia are over 200 years old. There are no methods to age a queen.

Termites are herbivores, fungivores (i.e., plant or fungus feeders), and humivores (soil feeders). They feed on cellulose, directly from plants, dead or alive, or indirectly from fungus arising from decaying plant material within mounds. Plants are made of cellulose, a polysaccharide that is composed of glucose units. The traditional view is that termites rely on intestinal gut microorganisms for cellulose digestion. However, there is growing evidence that termites also use their own enzymes for cellulose digestion.

Before mating and starting new colonies, new kings and queens, called alates or swarmers, depart the colony and fly (Fig. 2). They mate after landing on the ground. Swarming behavior varies considerably among termite families and species, but occurs most frequently during the rainy season. However, dry-wood termites can swarm during hot days, or sometimes evenings, of summer. A mated king and queen lose their wings and find a suitable nesting site near or in wood where they construct a small chamber that they enter and seal. The queen soon begins laying eggs, and both the king and the queen feed the young predigested food until they are capable of feeding themselves. Once workers and nymphs are produced, the king and queen are fed by the workers and cease feeding on wood. The exchange of food among colony members is called trophallaxis. Social insects exchange food in two ways, stomodaeal and proctodaeal trophallaxis. Termites use the latter method for food and symbiont exchange, mouth to anus. Symbionts are protozoa and bacteria that occur in the hindgut of termites. These microbes help digest cellulose, the major food source for termites.

The reproductive adults have functional eyes, needed for flight and initial finding of nest sites. The blind workers and soldiers live deep in nests, soil, or mounds and do not require or need vision. They already are in contact with or close to their food source.

Termites can also communicate through chemical, acoustical, and tactile signals. Two termite trail pheromones, *(Z,Z,E)*-(3,6,8)-dodecatrienol and *(E)*-6-cembrene, have been identified. These messages are produced in a sternal gland on the underside of the termite's abdomen. However, other chemical signals, such as those used for alarm and colony recognition, are produced from other glands located throughout their body. Many termite behaviors (e.g., trail following, alarm, and sexual communication) are mediated by pheromones. Soldiers also produce chemicals that are important for colony defense. Colony recognition and colony spacing are thought to be regulated by cuticular hydrocarbons. These waxy compounds are produced over the exterior cuticle of termites and spread throughout the colony. Termites can also communicate danger by "head-banging" of soldiers, in which they tap their heads in galleries to alert their nestmates.

Termites play a major role in recycling wood and plant material, but their tunneling effort also ensures that soils are porous, aerated, and enriched in minerals and nutrients, all of which improve plant growth. For example, termite activity in the desert areas of west and north Africa helps to reclaim soils damaged by overgrazing. Termites are an important food source for many other animals, including reptiles, birds, and mammals. Termite mounds and trees hollowed out by termites provide shelter and breeding sites for birds, mammals, and other insects.

Termites also contribute to atmospheric gases. The most abundant gases produced are carbon dioxide and methane. Both are greenhouse gases, but they are not produced in sufficient quantities to have negative effects on the atmosphere.

TERMITES AS PESTS

Some termites are destructive feeders and consume homes, other wooden structures, and agricultural crops. In some regions of the world, tunneling by termites damages dams, which then results in flooding. Worldwide, several billion dollars is spent annually for the control and repair of damage caused by termites. In the United States alone, over $1 billion is spent annually for termite control and damage repairs. Globally, subterranean termites (Rhinotermitidae: genera *Reticulitermes, Coptotermes, Heterotermes,* and *Psammotermes*) are the most responsible for the control and damage costs. Dry-wood termites (Kalotermitidae: genera *Incisitermes* and *Cryptotermes*) have lesser importance as structural pests and are more prevalent in coastal, arid, or semiarid regions. Termites as agricultural pests are confined primarily to the Asia, Africa, South America, and Australia. The major pest species belong to the genera *Microtermes, Macrotermes,* and *Odontotermes* (Termitidae) in Africa and Indo-Malaysia. *Mastotermes* (Mastotermitidae) is an important pest in Australia, whereas *Cornitermes* and *Procornitermes* (Termitidae) are important pests in South America. Damage varies from superficial to killing the plant. Healthy plants can tolerate some termite damage with reduced yields. In general, exotic plants and stressed plants are most prone to termite attack.

TERMITE CONTROL/MANAGEMENT

Before termites in structures can be treated, the extent of the infestation must be assessed. Visual searching and probing of wood are the dominant means of inspection. However, the efficacy of visual searches is questionable, because structures have inaccessible areas. Several nonvisual detection methods are used, including electronic stethoscopes, dogs, methane gas detectors, and microwave and acoustic emission devices, but each of these technologies has some limitations. For subterranean termites, wood-baited monitoring stations can identify the presence and delimit the extent of colonies. Some species of subterranean termite have colonies as large as several million individuals, and these forage over an area of more than 10,000 m². Other termite species have much smaller colonies and forage within areas of only a few square meters. There is considerable debate about the methods and accuracy in reporting termite numbers and foraging behavior.

Termite control is most regulated in North America, Europe, and Australia. However, in many countries controlling termites is achieved by the hand removal of queens and nests, flooding nests, or drenching them with used motor oil. Soil drenches with liquid termiticides injected into the soil beneath structures to protect foundations and structural wood is the dominant control tactic for subterranean termites for several continents. Chlorinated hydrocarbon insecticides, such as chlordane, have been used extensively for subterranean termite control because of their long persistence, >30 years in the soil. Because of persistence and suspicions of health-related problems, chlordane has been removed from many markets. Chloronicotinyls and phenyl pyrazoles are new compounds marketed for termite control. The use of toxic baits (e.g., containing chitin and metabolic inhibitors) and physical barriers (sand and stainless steel mesh) for controlling subterranean termites are also gaining acceptance. Techniques to prevent infestations of subterranean termites include using wood pressure-treated with oil and water-soluble chemicals.

Surveys of pest control firms in the United States reveal that poor building practices, particularly wood in contact with soil and cracks in concrete foundations, lead to many of the subterranean termite infestations. Experimental efforts have been made to control soil-dwelling termites using biological control agents, such as argentine ants and nematodes. However, these methods have not yet been proven effective.

Dry-wood termite colonies are usually above soil level in structures, small, and difficult to detect. Treatments include whole-structure applications of fumigants (such as sulfuryl fluoride and methyl bromide) and heat. Chemicals, heat, freezing, microwaves, and electricity are used for localized or spot treatments of dry-wood termites.

See Also the Following Articles
Blattodea • Caste • Sociality • Urban Habitats

Further Reading
Abe, T., Gignell, D. E., and Higashi, M. (eds.) (2000). "Termites: Evolution, Sociality, Symbioses, Ecology." Kluwer Academic, Dordrecht.
Forschler, B. T. (1998). Subterranean termite biology in relation to prevention and removal of structural infestation. *In* "NPCA Research Report on Termites." NPCA, Dann Loring, VA.
Kofoid, C. A. (ed.) (1934). "Termites and Termite Control." University of California Press, Berkeley.
Krishna, K., and Weesner, F. M. (1969). "Biology of Termites," Vols. 1 and 2. Academic Press, New York.
Lewis, V. R., and Haverty, M. I. (1996). Evaluation of six techniques for control of the western drywood termite (Isoptera: Kalotermitidae) in structures. *J. Econ. Entomol.* **89,** 922–934.
Pearce, M. J. (1997). "Termites: Biology and Pest Management." CAB Int., Oxon, UK.
Potter, M. F. (1997). Termites. *In* "Mallis Handbook of Pest Control" (A. Mallis, ed.), 8th ed., pp. 232–333. Franzak & Foster, Cleveland.
Su, N.-Y., and Scheffrahn, R. H. (1990). Economically important termites in the United States and their control. *Sociobiology* **17,** 77–94.

Su, N.-Y., and Scheffrahn, S. H. (2000). Termites as pests of buildings. *In* "Termites: Evolution, Sociality, Symbioses, Ecology (T. Abe, D. E. Bignell, and M. Higashi, eds.), pp. 437–453. Kluwer Academic, Boston.

Thorne, B. L. (1998). Biology of subterranean termites of the genus *Reticulitermes*. *In* "NPCA Research Report on Subterranean Termites." NPCA, Dunn Loring, VA.

Thorne, B. L., Russek-Cohen, E., Forschler, B. T., Breisch, N. L., and

Traniello, J. F. A. (1996). Evaluation of mark–release–recapture methods for estimating forager population size of subterranean termite (Isoptera: Rhinotermitidae) colonies. *Environ. Entomol.* **25,** 938–951.

Watanabe, H., Noda, H., Tokuda, G., and Lo, N. (1998). A cellulase gene of termite origin. *Nature* **394,** 330–331.

Wood, T. G., and Pearce, M. J. (1991). Termites in Africa: The environmental impact of control measures and damage to crops, trees, rangeland and rural buildings. *Sociobiology* **19,** 221–234.

Japanese Beetle

David W. Held and Daniel A. Potter
University of Kentucky

The Japanese beetle, *Popillia japonica*, is among the most polyphagous of plant-feeding insects. The adults skeletonize the foliage, or feed on the flowers or fruits, of nearly 300 species of wild or cultivated plants. Favored hosts include many woody and herbaceous landscape plants, garden plants, fruits, and field crops. The larvae, or grubs, develop in the soil where they feed on roots of turf and pasture grasses, vegetables, nursery seedlings, and field crops. Hundreds of millions of dollars are spent annually for controlling the adults and grubs, and in state and federal regulatory efforts aimed at limiting the beetle's rate of spread in the United States and elsewhere.

The Japanese beetle was first discovered in the United States in 1916, near Riverton, New Jersey. How it was inadvertently transported from its native Japan is not known; however, the grubs may have arrived in soil around the roots of nursery plants. The species is not a major pest in Japan, where suitable grassland habitat is limited and natural enemies keep this beetle in check. The eastern United States, however, provided a favorable climate, with abundant moist turf as habitat for the eggs and larvae, numerous adult food plants and, at that time, no host-specific natural enemies. Populations increased and spread rapidly. By 2000, the beetle was established in all states east of the Mississippi River except for Florida, and in parts of Wisconsin, Minnesota, Iowa, and Nebraska. It also has spread north into southern Ontario and Quebec, Canada.

DESCRIPTION

Japanese beetles belong to the family Scarabaeidae, subfamily Rutelinae. Adults are broadly oval, 8 to 11 mm in length, metallic green, with coppery brown elytra that do not quite cover the end of the abdomen (Fig. 1A). The abdomen bears five patches of white hairs on either side, and another pair near its tip. Females, which tend to be slightly larger than males, have an elongate, spatula-shaped spur on the foretibia, used for digging. This spur is shorter and pointed in males.

Larvae are typical scarabaeiform grubs: C-shaped, grayish to cream colored, with three pairs of jointed legs, a distinct yellow-brown head capsule, and chewing mouthparts (Fig. 1B). Neonate grubs are about 1.5 mm in length, whereas the length of full-sized third instars is about 32 mm. The underside of the last abdominal segment, just anterior to the anal slit, bears two short rows of hairs forming a tiny, truncated V. This pattern distinguishes Japanese beetle larvae from the larvae of other common scarabs. The end of the abdomen appears dark because of ingested soil and food.

LIFE CYCLE

Japanese beetles have a one-year life cycle in most parts of their range. Adults occur from June to August. Upon emergence

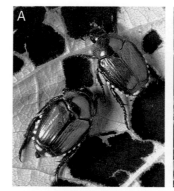

FIGURE 1 (A) Japanese beetles with characteristic feeding damage. (B) Japanese beetle grub.

from the soil, virgin females emit a volatile sex pheromone that attracts clusters of males. Subsequent matings occur on food plants. The beetles typically feed from the upper surface of leaves, chewing out the tissue between the veins and leaving a lacelike skeleton. Adults also feed on petals of flowers such as roses, and on developing fruits or berries. Food plants growing in sunny locations are preferred. Usually the beetles begin to feed on foliage near the top of a plant, regardless of its height. They often aggregate on particular shoots or plants. This phenomenon results from both sexes being attracted to blends of aromatic volatile compounds released from beetle-damaged leaves. Despite the beetles' broad host range, some plant species are rarely or never fed upon. Closely related cultivars within species may also differ in susceptibility. Resistance probably results from presence of feeding deterrents (e.g., certain phenolics) or other secondary plant compounds. Some plants (e.g., geranium, *Pelargonium hortorum*) are palatable to the beetles but cause paralysis or other toxic effects.

After feeding, gravid females fly to moist turf, pasture, or agricultural fields, where they burrow down to lay small clutches of eggs in the upper 8 cm of soil. Females alternate between feeding and egg laying; each female may emerge from the soil, fly to host plants, feed, mate again, and return to the soil 15 or more times, laying 40 to 60 eggs in her lifetime. The pearly white eggs, oval when first laid, swell with soil moisture to a diameter of about 1.5 mm. Larvae hatch in 2 to 3 weeks, usually by early to mid-August. Larvae feed just below the soil surface, consuming plant roots and organic matter. When grubs are numerous, the root system of turf grasses may be completely severed, such that the turf wilts and dies, and can be pulled from the soil like a loose carpet. Most grubs are third instars by September. About the time of first frost, the grubs move deeper (about 15–30 cm) to overwinter. In early spring, as soil temperatures warm to about 10°C, the grubs move back to the upper 2.5 to 5.0 cm of soil and resume feeding for about 4 to 6 weeks, after which they again go deeper and form an earthen cell in which to pupate. The first adults begin emerging a few weeks later.

PREDATORS AND PARASITIZERS

Vertebrate predators such as skunks, raccoons, birds, and moles may dig in infested areas to feed on the grubs. Indigenous predatory insects, including ants and ground beetles, feed on the eggs and young grubs. Birds, fish, and other insectivores eat the adults. From 1920 to 1933, entomologists searched for, and imported, numerous natural enemies from Asia and released them for biological control in areas infested with Japanese beetles. Only a few of these became established. The most widely distributed are two species of tiphiid wasps, *Tiphia vernalis* and *T. popilliavora*, whose larvae are ectoparasitoids of the grubs, and *Istocheta aldrichi*, a tachinid fly that parasitizes the adults. The grubs are susceptible to parasitic nematodes, as well as several lethal microbial pathogens, especially the milky disease bacterium, *Paenibacillus popilliae*.

See Also the Following Articles
Introduced Insects • Regulatory Entomology • Soil Habitats

Further Reading
Fleming, W. E. (1968). Biological control of the Japanese beetle. U.S. Department of Agriculture Technical Bulletin 1383.
Fleming, W. E. (1972). Biology of the Japanese beetle. U.S. Department of Agriculture Technical Bulletin 1449.
Loughrin, J. H., Potter, D. A., Hamilton-Kemp, T. R., and Byers, M. E. (1996). Role of feeding-induced plant odors in aggregative behavior of the Japanese beetle. *Environ. Entomol.* **25**, 1188–1191.
Potter, D. A., and Held, D. W. (1999). Absence of food-aversion learning by a polyphagous scarab, *Popillia japonica*, following intoxication by geranium, *Pelargonium hortorum*. *Entomol. Exp. Appl.* **91**, 83–88.
Potter, D. A., and Held, D. W. (2002). Biology and management of the Japanese beetle. *Annu. Rev. Entomol.* **47**, 175–205.
Vittum, P. J., Villani, M. G., and Tashiro, H. (1999). "Turfgrass Insects of the United States and Canada." 2nd ed. Cornell University Press, Ithaca, NY.

June Beetles

Daniel A. Potter and David W. Held
University of Kentucky

June beetles, sometimes called May beetles or june bugs, are heavy-bodied, brownish, plant-feeding scarab beetles (Fig. 1A). Almost all species are nocturnal in their habits. The adults are voracious feeders on leaves of many deciduous trees, shrubs, and some herbaceous plants. Their larvae, called white grubs, develop in the soil, where they feed on plant roots and can be pests of turf and pasture grasses, young nursery stock, corn, small grains, potatoes, strawberries, and other agricultural crops.

June beetles belong to the genus *Phyllophaga* (formerly *Lactosterna*) in the family Scarabaeidae, subfamily Melolonthinae. They occur in both the New and Old Worlds. In North America north of Mexico about 200 species are known, with many found in the north-central and eastern United States. They also have been reported from South and Central America, the West Indies, eastern and southern Asia, and the islands of the Pacific and Indian Oceans.

DESCRIPTION

June beetles average from 12 to over 25 mm in length, with a cylindrical or oblong body shape, dense hair on the metasternum, and lamellate antennae that end in a three-segmented club that is longer in males than in females. Each tarsal claw bears a small tooth near the middle. Coloration ranges from tan to mahogany to dark chocolate brown. The elytra of some species are hairy, whereas in others they are nearly smooth.

Larvae are typical scarab grubs: cream colored, C-shaped when feeding or at rest, with a brown head capsule, with

FIGURE 1 (A) A typical June beetle, *Phyllophaga* sp. (B) June beetle larva, or white grub.

chewing mouthparts and three pairs of jointed legs (Fig. 1B). The hind part of the abdomen usually appears dark because of ingested food and soil. The ventral surface of the last abdominal segment bears two parallel rows of short spines in a pattern that resembles a zipper. There are three instars. Full-sized grubs of most species are 25 to 38 mm in length.

LIFE CYCLE

Most June beetles have 2- or 3-year life cycles, although a few species have cycles lasting 1 or 4 years. Adults typically are active from April to June. The beetles emerge after sundown and fly to the tops of trees to feed and mate, returning to the soil before dawn. They are clumsy fliers and often are attracted to outdoor lights. Mated females fly to turf, pasture, or agricultural fields and burrow down 5 to 15 cm to lay eggs in the soil. Each female lays 20 to 50 eggs in her lifetime. Eggs are pearly white, about 2.5 mm long, and elliptical at first, becoming more spherical as the embryo develops. Hatching occurs in about 3 weeks, and the young grubs begin feeding on fine roots and organic matter.

Larvae of species with 2-year cycles typically overwinter as second instars. They resume feeding in early spring, molting again in April or May. Third instars attain their full size by summer's end. Pupation occurs underground, in an earthen cell. Most species transform to adults by late autumn, but the beetles remain underground until the following spring. Grubs of species with 3-year cycles feed throughout the first two summers, hibernating twice and pupating midway through the third summer. Their adults are usually fully formed by autumn but do not emerge from the soil until the following spring. Because of overlapping generations and presence of more than one species, several sizes of June beetle grubs may be found together at a given site.

ELIMINATING JUNE BEETLE GRUBS

June beetle grubs are susceptible to various microbial pathogens, including specific strains of the milky disease bacterium *Paenibacillus popilliae*. Wasps and flies of several kinds parasitize the larvae or beetles, and predatory insects (e.g., ants, carabids) feed on the eggs and young larvae. The grubs also attract vertebrate predators, including insectivorous birds, skunks, raccoons, moles, and armadillos. In the past, farmers were advised to clean June beetles out of heavily infested fields by pasturing the land with hogs, which would root out and eat the grubs. Today this objective is more typically accomplished through crop rotation, or with soil insecticides.

See Also the Following Articles

Coleoptera • Hibernation • Soil Habitats

Further Reading

King, A. B. S. (1984). Biology and identification of white grubs of economic importance in Central America. *Trop. Pest Manag.* **30**, 36–50.

Luginbill, P., and Painter, H. R. (1953). May beetles of the United States and Canada. U.S. Department of Agriculture Technical Bulletin 1060.

Pike, K. S., Rivers, R. L., Oseto, C. Y., and Mayo, Z. B. (1976). A world bibliography of the genus *Phyllophaga*. University of Nebraska Miscellaneous Publication 31.

Vittum, P. J., Villani, M. G., and Tashiro, H. (1999). "Turfgrass Insects of the United States and Canada." 2nd ed. Cornell University Press, Ithaca, NY.

Ritcher, P. O. (1966). "White Grubs and Their Allies." Oregon State University Press, Corvallis.

Juvenile Hormones

Michael E. Adams
University of California, Riverside

Juvenile hormones (JHs), acting in concert with ecdysteroids, orchestrate the expression of larval-specific genes each time the insect molts to a new stage. These morphogenetic effects include determination of an immature body form and internal organs, hardness and color of the cuticle, and accompanying physiology and behaviors. At the conclusion of larval development, juvenile hormone levels drop at critical times, allowing ecdysteroids to program expression of pupal and adult characteristics. Juvenile hormones return in the adult stage, in which they have gonadotropic functions in connection with reproduction. In addition to their morphogenetic and gonadotropic actions, juvenile hormones are involved in dormancy and various types of polyphenisms, including caste determination in social insects.

DISCOVERY

The discovery of juvenile hormones began in the 1930s, with a series of ingenious experiments conducted by Sir Vincent Wigglesworth aimed at elucidating the hormonal control of molting. Trained as a medical doctor, Wigglesworth dedicated his life to basic studies of insect physiology, believing that knowledge gained would hold the keys to controlling insect vector-borne disease and agricultural pests. As a model experimental insect, Wigglesworth chose the Chagas' disease vector *Rhodnius prolixus,* otherwise known as the "kissing bug" because of its habit of sucking blood from the lips of sleeping humans. The choice of *Rhodnius* was inspired, because its development is closely timed to its blood meals. This allowed Wigglesworth to precisely determine the physiological stage of the insects to coincide with his experimental manipulations. He found that 3 days after a blood meal, hormones are released into the *Rhodnius* system, stimulating the molt to the next stage. By performing a number of surgical procedures on the bug he demonstrated the source and timing of hormone release. Part of the advantage of working with insects as experimental animals is that they can survive for long periods without such seemingly vital organs as the brain, a fact that Wigglesworth took advantage of. He found that decapitation of animals prior to a 3-day critical period led to an arrest in development, even though the animal would remain alive for many months. If the brain was reimplanted, development resumed. He also found that the blood of a normally developing animal could reactivate development in the headless animal. This was achieved via a technique called parabiosis, in which the developmentally arrested animal was joined to the normal one by means of a tube, which allowed blood from the two animals to mix. With these experiments, Wigglesworth demonstrated that hormones released from the brain trigger molting. This discovery actually had been made more than a decade earlier by Stefan Kopec, working with gypsy moth, but Wigglesworth's experiments revealed a new type of hormone, one that influenced the form taken by the animal after each molt.

Wigglesworth fundamentally changed the thinking about insect development, specifically the distinction between regulation of growth and regulation of form by separate hormones. *Rhodnius* passes through five nymphal stages before molting to the adult form. It is easy to tell the adults from the nymphs, because of differences in pattern and color of the cuticle, as well as the fact that only adults have wings. Wigglesworth found that parabiosis of a fifth (last)-stage nymph with a young nymph prevented the former's metamorphosis to the adult stage. Instead, the animal molted to a sixth-stage nymph, an extra immature stage that never occurs normally. A chemical in the blood of the young insect promoted continued expression of larval characters, and this factor came to be known as the "juvenile hormone." The source of the juvenile hormone was traced to a pair of small glands behind the brain called the corpora allata (Fig. 1). Surgical removal of the gland did not interfere with molting, but drastically altered the form taken after the molt, causing animals to become precocious adults. Reimplantation of the glands led to the return of larval characters.

While the corpora allata proved to be the sole source of juvenile hormone, only very small amounts were available from the gland for chemical studies. The short supply of JH greatly constrained experimentation, slowing the process of discovery considerably. A breakthrough came with the discovery of

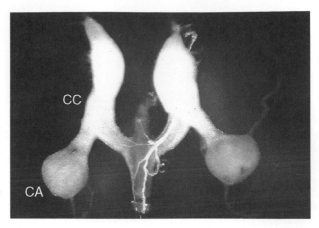

FIGURE 1 Photomicrograph of the corpora allata (CA), paired, spherical glands that are the sole source of the juvenile hormones in insects. Also shown are the elongated, white corpora cardiaca (CC). The CC and CA are positioned behind the brain, where they release hormones into the blood. Structures shown were dissected from the cockroach, *Periplaneta americana.* (Photograph courtesy of Dr. S. J. Kramer.)

large amounts of JH in abdomens of adult male silk moths by Carroll Williams. Ether extracts produced a dark orange material he called the golden oil. Such an abundance of juvenile hormone in a male adult at first was surprising, but already Wigglesworth had noted the essential gonadotropic role of JH associated with reproduction, that is, stimulation of egg and sperm development. The reason for enormous quantities of JH in male adult abdomens is likely caused by its inclusion in spermatophores, which contain sperm together with nutritive and hormonal stores and are provided to the female during mating for fertilization and nutrition of developing eggs. Williams is credited with stimulating the modern era of JH research, by making available enough of the natural hormone to conduct biological experiments on its modes of action in many types of insects. This work also provided quantities of starting material sufficient for the eventual isolation and chemical identification of the hormone, which occurred in the late 1960s and early 1970s.

CHEMISTRY OF JUVENILE HORMONES

The juvenile hormones are lipophilic sesquiterpenoid derivatives of farnesoic acid. Their chemical nature came into focus in the early 1960s, beginning with observations that farnesoid components of beetle excreta had JH-like activity in bioassays. Although far less potent than the native hormone, these substances, including farnesol and its oxidized form farnesal, were suspected to be JH precursors. Subsequent synthesis of methyl farnesoate with an epoxide at position 10–11 by William Bowers in 1965 gave a highly potent compound. It is indeed ironic that this compound was discovered 8 years later to be the most ubiquitous of the natural juvenile hormones, JH-III (Fig. 2).

Another interesting prologue to the discovery of the natural juvenile hormones was the discovery of a curious

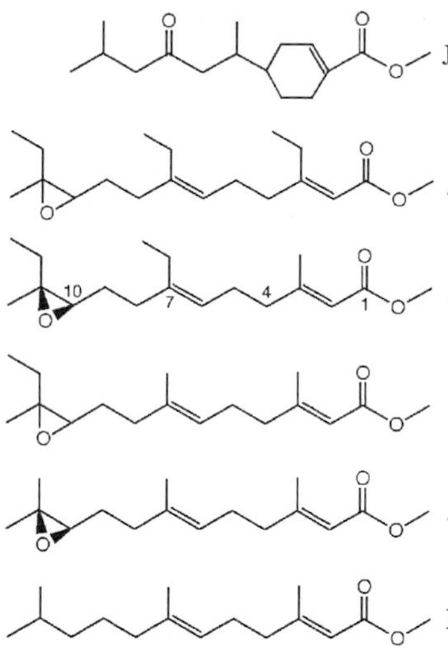

FIGURE 2 Structures of natural juvenile hormones and related compounds. Juvabione is the paper factor from North American balsam fir discovered by Williams, Slama, and Bowers. JH-I was the first natural JH discovered by Röller, and JH-II and JH-III soon followed. JH-III is present in most groups of insects, including the Lepidoptera. JH-0 is found in moth eggs, but its biological function is unclear. JH-0, JH-I, and JH-II are generally confined to the Lepidoptera. Methyl farnesoate, a JH precursor, has been isolated from some insects and crustaceans, in which it may serve as the active JH.

"paper factor" by Karel Slama and Carroll Williams in the mid-1960s. Slama had carried insects from his laboratory in Czechoslovakia to the laboratory of Williams at Harvard University for a series of joint experiments. Some weeks after arriving the insects began to develop extra stages and many died. Nothing of the sort was noticed in Czechoslovakia. It was eventually determined that the paper towels used to line the containers holding the insects contained compounds with JH-like biological activity. The substances were absorbed through the insect cuticle upon contact with the paper towels. The paper factor turned out to be a mixture of terpenoids in the wood pulp from which the paper towels were manufactured. These compounds were found only in American and Canadian balsam fir and not in European trees. One of these substances was chemically identified by Bowers and colleagues as "juvabione" (Fig. 2), which had a structure reminiscent of farnesol.

These studies provided new information on two critical issues of the time. First was the question of juvenile hormone structure. Observations that farnesol and methyl farnesoate, along with juvabione, possessed JH-like biological activity made it likely that terpenoid chemistry was involved. This guided further attempts to chemically define the natural material(s). The second issue was whether juvenile hormones or analogs such as juvabione could be used to produce, as Carroll Williams suggested, a new class of "third-generation" pesticides. Juvabione constituted a relatively stable terpenoid with potent insecticidal activity and hence stimulated further interest in this novel concept for new insecticides that would be both highly insect-selective and safe for warm-blooded animals.

With this as background, Röller and colleagues isolated sufficient quantities of juvenile hormone from silk moth abdomens for identification of the first natural juvenile hormone in 1967. The carbon skeleton was identified as a 15-carbon sesquiterpene substituted at positions 7 and 11 with ethyl groups. Further key structural features were the presence of a methyl ester and an epoxide at carbons 10–11. This landmark achievement was followed rapidly by publication of a second JH, also from moths. These molecules were named JH-I and JH-II, respectively, differing only at carbon 7, which was ethylated in JH-I and methylated in JH-II (Fig. 2). These two juvenile hormones are largely restricted to the Lepidoptera. Within a few years, a third hormone was identified by a completely different approach. In this instance, corpora allata from the moth *Manduca sexta* were removed and placed in organ culture containing a radiolabeled methyl donor, ^{14}C-labeled methionine. The corpora allata incorporated the ^{14}C-methyl group into the ester moiety of juvenile hormone. This isotope-labeled synthetic product was isolated and identified as JH-III. It is the most cosmopolitan of juvenile hormones, occurring in most insect groups, including the Lepidoptera. It has methyl groups at positions 7 and 11 (Fig. 2). Three additional JH structures have been identified: JH-0 and 4-methyl JH from moth eggs and JH-bisepoxide from the fruit fly *Drosophila*.

The juvenile hormones are derived from acetyl CoA and/or propionyl CoA via mevalonic acid and homomevalonic acid in the sterol biosynthetic pathway. The final steps of JH-III biosynthesis go by way of farnesol → farnesoic acid → methyl farnesoate, to which an epoxide is formed at carbons 10–11. Owing to their low aqueous solubility, the juvenile hormones are transported through the blood via binding proteins upon their release from the corpora allata. These binding proteins also protect JH from degradative enzymes.

In some insects, the corpora allata synthesize a precursor of the biologically active form of JH, which is converted to the active form in target tissues. For example, silk moth adults produce JH acid in the corpora allata and convert it to JH-I in the accessory glands of the abdomen. It is also known that the ovaries of certain species of mosquito can synthesize JH from precursors under culture conditions. Whether this occurs under natural conditions *in vivo* has not been demonstrated.

The levels of JH in the blood are regulated through a combination of synthesis and degradation. Synthesis by the corpora allata is promoted by neuropeptides called allatotropins. So far, only one allatotropin has been identified from the tobacco hawkmoth *M. sexta*. Surprisingly, the peptide is active only in the adult stage. More compelling evidence has been provided for the existence of allatostatins. These are neuropeptides synthesized in brain neurons that project to the corpora allata. Their release from nerve endings in the gland inhibits the synthesis of juvenile hormone.

The removal of juvenile hormones already in the blood, a necessary condition for metamorphosis, occurs through two enzymatic degradation pathways. One is through cleavage of the ester bond by JH esterase, the other through epoxide destruction by epoxide hydrolases.

BIOLOGICAL ACTIONS

Morphogenetic Effects

The presence of juvenile hormones in the blood promotes expression of juvenile characters, chief among these being an immature body form or morphology. For insects such as grasshoppers, which undergo incomplete metamorphosis, the effects are not so visible outwardly. The early stages look like miniature adults except for the absence of wings, but they also lack functional reproductive organs. However, for those insects such as flies, moths, and bees, which undergo complete metamorphosis, the effects are extreme. The immature stages are wormlike with no wings or legs.

Determination of holometabolous immature or larval body plan by juvenile hormone represents its morphogenetic action. The decision to develop larval characters during development is made near the end of each larval stage, when ecdysteroid levels increase to initiate the molt. Elevation of ecdysteroids causes a cessation of feeding and new round of gene expression appropriate for the next stage of development. If juvenile hormones are present at this time, genes for larval characters are expressed, whereas genes appropriate for pupal or adult characters are repressed. A primary larval character is the type of cuticle secreted by the epidermal cells. Larval cuticle is lighter and more flexible than pupal or adult cuticle, which are characteristics resulting from expression of larval cuticle protein genes that predominate under the influence of juvenile hormones. The flexibility of larval cuticle has to do with the absence of cross-linking between proteins and between proteins and chitin, the latter constituting the polysaccharide component of the cuticle. In contrast, pupal and adult cuticles are hard and dark, indicating a high degree of sclerotization and melanization.

Specification of pupal or adult features by JH, or lack thereof, is associated with transient, hormone-sensitive periods during development. It is important to note that the actions of JH depend not only on its presence in the blood, but also on the ability of cellular targets to respond. This latter condition presumably reflects the presence of suitable receptors required to mediate the action of the hormone. It has been observed in many studies that JH responsiveness occurs after priming, which could be associated with expression of receptor genes or other molecules necessary to complete the signaling pathway. For example, in moths, specification of a wandering period in preparation for pupation occurs upon the appearance of ecdysteroid peaks in the complete absence of JH. These "commitment" peaks of ecdysteroid prime the system to respond later in the same stage to elevated JH levels, which specify pupal features.

In many insects, considerable development of the gonads takes place during the larval stages. This is especially true for insects such as the silkworm *Bombyx mori,* which does not feed during the adult stage. Within hours of emergence, these animals mate and lay eggs. The ability to mate and produce viable eggs so soon after emergence means that gonadal development is well along during the larval and pupal stages. Juvenile hormones promote the development of gonads and gametes during the immature stages, but must disappear in order for final developmental steps to be completed. This drop in JH levels just prior to the pupal stage therefore serves both morphogenetic and gonadotropic functions.

Effects of JH in the Adult Stage

The decrease in JH levels just prior to metamorphosis is only a temporary condition. The corpora allata are retained in the adult stage, and JH eventually reappears to regulate adult reproductive functions. JH promotes sperm and egg development and hence is said to have "gonadotropic" functions. In the female, JH directly promotes the synthesis of lipo- and glycoproteins in the fat body and their uptake into the developing oocyte. This process, called vitellogenesis, is essentially yolk deposition. In many insects, JH levels rise and fall in a cyclic fashion as discrete batches of oocytes go through the vitellogenic process.

In other instances, as in some mosquitos, JH exposure leads to "competence" of the fat body to synthesize vitellogenic proteins upon later exposure to ecdysteroids. Likewise, JH exposure is required to induce competence of the ovaries to respond later to peptide hormones from the nervous system, thus stimulating uptake of vitellogenic proteins. In these instances, the gonadotropic actions of JH appear to be priming steps in preparation for ecdysteroid action.

Gonadotropic functions of JH in the male have to do with growth of the sperm. Sperm growth requires JH in many insects. However, maturation from spermatocytes to motile spermatids requires a drop in JH. As observed for oocyte development, JH exerts both positive and negative influences in sperm development.

Polyphenism and Caste Determination

Many insects have the remarkable ability to develop into alternate forms as they become adults. These alternate forms together with accompanying physiology and behavior, referred to as polyphenism, do not reflect differences in the genetic makeup of individuals. Rather, they result from a particular pattern of gene expression under hormonal control. Most polyphenisms are controlled by juvenile hormones acting at certain sensitive periods during immature development.

Some of the most common instances are caste polyphenisms observed in social insects such as bees, ants, and termites. In these insect societies, larvae can develop into workers, soldiers, or queens, depending on the diet they are raised on and the hormonal levels that result. If bee larvae are reared in a special cell in the hive and consistently fed a nutritious "royal jelly" beginning during the third instar, they develop into queens.

Treatment with JH will mimic this effect. If this feeding is delayed, larvae develop into workers instead. In certain ants, development of queens is regulated during embryonic development by JH levels. During postembryonic development, larvae fed a high-protein diet produce large amounts of JH, bringing blood levels to a threshold necessary for specification of soldier phenotype. If larvae are fed a diet lower in protein, JH levels are correspondingly lower, and development to worker is specified. The number of soldiers in the colony also is determined by a soldier-inhibiting pheromone, which elevates the JH threshold for soldier specification. Alternative body forms and behaviors in insect colonies provide for cooperative functions between members of the society to serve the greater whole.

Many types of phase polyphenisms occur in nonsocial insects. For example, locusts occur either in solitary or in migratory phases, depending on population density. Differences in both behavior and physiology are characteristic of these phases. Solitary locusts are sedentary, pale green, yellow, or brown, and have short wings and large ovaries. Crowding causes the switch to the gregarious phase, in which individuals are brightly colored, have longer wings and smaller ovaries, and are easily induced to engage in long flights. Both JH and peptide neurosecretory hormones from the brain are involved in the determination of these two phases.

Aphids exhibit at least two different types of phase polyphenism as a response to seasonal conditions: food quality and crowding. In one type, adults switch between winged or apterous (no wings) forms. The other type has to do with the mode of reproduction, either sexual or parthenogenetic. During the longer days of spring and summer, apterous, parthenogenetic females predominate, and juvenile hormone is involved in specification of these forms. As winter approaches, winged forms are produced, allowing for dispersal. Later, in autumn, males and females mate and lay eggs, which overwinter and hatch in the spring. In this context, body forms and accompanying dispersal or migratory behaviors maximize survival as the season changes.

In summary, JH and other neurosecretory hormones are important determinants of polyphenisms, which result in different body forms, reproductive physiologies, and behaviors in the adult stage. It is emphasized that such variability of form and function is not the result of genetic differences between individuals, which would be classified as polymorphisms. Rather, insects have the enormous potential to change form in response to environmental conditions through hormonal control mechanisms. Depending on the needs of a social colony, or changes of season and in food availability, the complex endocrinology of insects enables them to assume various alter egos to enhance success and survival.

Behavioral Effects of JH

The presence or absence of juvenile hormones has profound effects on behavior, some of which have been mentioned above. Throughout the stages of immature development, JHs program gene expression in the nervous system for the expression of behaviors appropriate for juvenile life, including, for example, locomotion, host or prey seeking, feeding, and silk spinning. From the point of view of behavior, the larva is an animal completely different from its later adult form.

In moths, the disappearance of JH at the end of the last instar allows ecdysteroids to program new behaviors appropriate for metamorphosis. Insects stop feeding, void their guts, and engage in wandering behaviors to locate a suitable pupation site. This accomplished, a series of behaviors leads to silk spinning for cocoon construction.

Upon becoming adults, female mosquitos initiate the search for a blood meal and become sexually receptive to males only after release of JH into the blood. In milkweed bugs, JH levels are influenced by daylength, temperature, and food quality. Under short daylengths, JH levels drop, and insects engage in migratory behavior immediately after molting to the adult stage. However, long days and warm conditions lead to high JH levels, whereby flight is inhibited and reproduction ensues.

Grasshopper females that have had corpora allata removed rebuff male sexual advances until JH is reintroduced by injection. In crickets, the male sings a species-specific calling song to attract the female for mating. The female responsiveness to this song is enhanced by elevated JH levels. These examples serve to illustrate the dramatic effects that JH has on the behavior of insects, effects that are specific and appropriate for each particular life stage.

Dormancy–Diapause

Insects are able to enter prolonged states of dormancy referred to as diapause, allowing them to resist freezing and low food supplies during the winter. Diapause can occur at any stage (egg, larva, pupa, or adult) and is triggered by decreasing daylength, low temperatures, decreased food or food quality, or a combination of these factors. The insect response to these environmental factors is mediated by a variety of hormones, depending on the stage and species.

Adult diapause is largely synonymous with reproductive diapause. Beetles, butterflies, and flies enter a reproductive diapause when the brain inhibits synthesis of JHs by the corpora allata. The lack of JHs leads to both physiological and behavioral changes, including cessation of vitellogenesis, loss of flight muscle, increasing stores of lipid in the fat body, burrowing, and construction of hibernacula (overwintering chambers). Implantation of corpora allata or injection of JHs reverses reproductive diapause.

JH involvement in larval diapause also has been documented. The southwestern corn borer *Diatraea grandiosella* enters diapause during the last instar when JH levels are depressed but are still high enough to inhibit development to the pupal stage. The animal spins a hibernaculum, exhibits a light pigmentation, and actually undergoes several "stationary" molts. Diapause in this stage lasts as long as JH levels remain elevated.

MOLECULAR BASIS OF ACTION

It is presumed that juvenile hormones exert their effects through receptor activation. However, it is a curious and surprising fact that, almost 70 years after its discovery by Wigglesworth, no definitive receptors have been identified. Even though no clear example of a JH receptor has been defined, a number of tantalizing possibilities have been suggested. One of these involves a nuclear receptor called ultraspiracle, or USP, a protein that is well known to regulate gene expression by forming a dimer complex with ecdysteroid receptors. The complex then binds to regulatory sequences on genes to turn them on or off. Grace Jones and Alan Sharp have demonstrated that JHs bind specifically to USP, although the affinity for this binding is lower than is generally expected for hormone–receptor interactions. It is proposed that JH binding to USP may influence how it interacts with ecdysteroid receptors (EcR) to regulate gene expression. The influence of JH and USP on EcR actions would seem to be a very plausible scenario for joint actions of JH and ecdysteroids, but further work is needed before USP is confirmed as a JH receptor.

Although early accounts of juvenile hormones focused on their uniqueness with respect to insect biology, the elusiveness of JH receptors has prompted a closer look at possible similarities between signaling mechanisms common to insects and mammals. Indeed, the chemical structure of JHs resembles those of retinoids and farnesoids, both of which function in mammalian nuclear signaling by activating retinoic acid receptors, retinoid X receptors, and the farnesoid X receptor. JH and farnesoids are capable of activating some of these receptors, and some retinoids are known to have JH-like activity. It also has been observed that vertebrate thyroid hormones mimic some of the actions of JHs. Efforts are under way to identify receptors homologous to their mammalian counterparts as possible JH receptors.

Recently, workers taking a genetic approach to the problem identified a strain of fruit flies resistant to methoprene, an insecticidal juvenile hormone analog (see next section for details). The resistant flies have a defect in a gene that encodes MET, a protein related to the vertebrate aryl hydrocarbon receptor, which upon binding a diverse range of hydrocarbons activates a battery of genes involved in their metabolism. If the MET protein has similar properties, this might help explain why many synthetic chemicals such as fenoxycarb and pyriproxyfen have very potent JH-like effects, but bear little obvious structural similarity to natural JHs.

The failure after so many years to define a receptor for JH may indicate that, for this particular hormone, signaling does not conform to conventional modes of action. Perhaps JH binds to certain proteins, which then act as coeffectors or adaptor proteins to amplify or modify transduction of signals initiated by other hormones at conventional receptors. The obvious example is modification of ecdysteroid receptor action. It turns out that MET also is related to steroid receptor coactivators, which could bind to EcR and/or USP

to modify their effects on gene expression. Although little is known about the specific actions of MET at the present time, the MET resistance gene may hold the key to understanding the elusive molecular action of JHs.

JUVENILE HORMONES AND INSECT CONTROL

The discovery of juvenile hormones in the late 1960s by Röller and others stimulated a period of great excitement regarding the concept of third-generation pesticides foreseen by Carroll Williams in the early 1960s. It was known that juvenile hormones and related substances such as juvabione could disrupt insect development with lethal effects. Likewise, surgical removal of the corpora allata led to precocious metamorphosis, also with lethal effects. It therefore seemed that insect hormones or their analogs could be synthesized and used to accomplish a form of "birth control" for insects.

This idea occurred to Carl Djerassi and Alejandro Zafferoni, two former colleagues at the Syntex Corporation, who were involved in synthesis of the first human contraceptives that led to the birth control pill. They formed a new company called Zoecon, a name chosen to denote "animal control" through the use of hormones and related chemical analogs. Their principal objective was to develop insect hormones for use as birth control agents specific for this group of animals. Chemists at Zoecon soon produced analogs of JH called "juvenoids" that were much more stable and could penetrate the cuticle. One of the first of these analogs to be granted a registration from the Environmental Protection Agency was methoprene (Fig. 3), a compound with outstanding biological activity against mosquitos, fleas, and biting flies. By mimicking JH, methoprene prevents treated insects from completing metamorphosis, and insects die during the pupal stage. Other juvenoids such as hydroprene (Fig. 3) are more effective against insects with incomplete metamorphosis, such as cockroaches. Treated cockroaches actually reach the

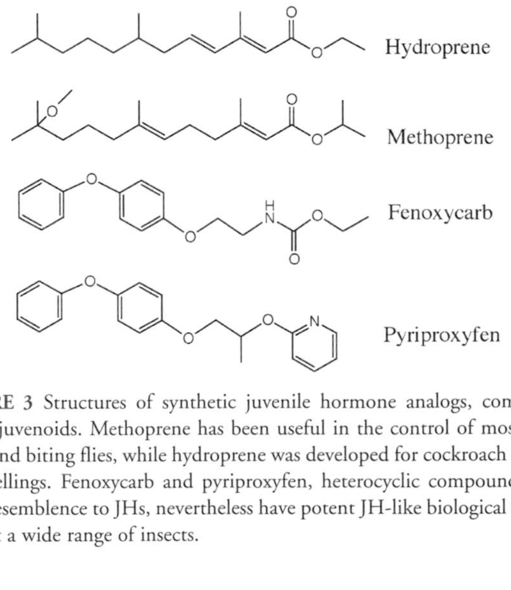

FIGURE 3 Structures of synthetic juvenile hormone analogs, commonly called juvenoids. Methoprene has been useful in the control of mosquitos, fleas, and biting flies, while hydroprene was developed for cockroach control in dwellings. Fenoxycarb and pyriproxyfen, heterocyclic compounds with little resemblence to JHs, nevertheless have potent JH-like biological activity against a wide range of insects.

adult stage, but the presence of juvenoid during the transition to the adult results in only a partial adult phenotype in which many adult features are abnormal. For instance the gonads are not fully developed, leading to sterility, and crinkled wings are an obvious morphological defect.

Subsequent efforts by several agrochemical companies have generated a variety of juvenoids, many with potent JH-like biological activity but with aryl rings substituted for isoprenoid units and without obvious similarities to natural JHs. These include fenoxycarb and pyriproxyfen (Fig. 3). Juvenoids have proved to be commercially successful for insects that are pests in the adult stage. However, because they do not control insects in the immature stages, they have not proved useful for large-scale agricultural pest control. For this purpose, juvenile hormone antagonists are needed for induction of precocious metamorphosis. So far, that goal remains as elusive as the search for the JH receptor.

See Also the Following Articles
Development, Hormonal Control of • Dormancy • Ecdysteroids • Hibernation • Mating Behaviors • Migration • Molting • Vitellogenesis

Further Reading
Ashok, M., Turner, C., and Wilson, T. G. (1998). Insect juvenile hormone resistance gene homology with the bHLH-PAS family of transcriptional regulators. *Proc. Natl. Acad. Sci. USA* **95**, 2761–2766.
Feyereisen, R. (1998). Juvenile hormone resistance: [No PASaran! *Proc. Natl. Acad. Sci. USA* **95**, 2725–2726.
Gilbert, L. I., Granger, N. A. and Roe, R. M. (2000). The juvenile hormones: Historical facts and speculations on future research directions. *Insect. Biochem. Mol. Biol.* **30**, 617–644.
Henrick, C. (1995). Juvenoids. *In* "Agrochemicals from Natural Products" (C. R. A. Godfrey, ed.), pp. 147–213. Dekker, New York.
Jones, G., and Sharp, P. A. (1997). Ultraspiracle: An invertebrate nuclear receptor for juvenile hormones. *Proc. Natl. Acad. Sci. USA* **94**, 13499–13503.
Nijhout, H. F. (1994). "Insect Hormones." Princeton University Press, Princeton, NJ.
Riddiford, L. M. (1994). Cellular and molecular actions of juvenile hormone. I. General considerations and premetamorphic actions. *Adv. Insect Physiol.* **24**, 213–274.
Truman, J. W., and Riddiford, L. M. (2002) Endocrine insights into the evolution of metamorphosis in insects. *Annu. Rev. Entomol.* **47**, 467–500.
Wigglesworth, V. B. (1983). Historical perspectives. *In* "Comprehensive Insect Physiology, Biochemistry, and Pharmacology" (G. A. Kerkut and L. I. Gilbert, eds.), Vol. 7, pp. 1–24. Pergamon Press, New York.
Williams, C. M. (1967). Third-generation pesticides. *Sci. Am.* **217**, 13.
Wyatt, G. R., and Davey, K. G. (1996) Cellular and molecular actions of juvenile hormone. II. Roles of juvenile hormones in adult insects. *Adv. Insect Physiol.* **26**, 1–155.

Katydid
see *Orthoptera*

Killer Bees
see *Neotropical African Bees*

Lacewing

see *Neuroptera*

Ladybugs

Michael E. N. Majerus
University of Cambridge

Ladybugs are one of the most familiar groups of insects. These beetles have received attention in both pure and applied areas of biological research. In some senses they are typical insects, having regular life cycles comprising egg, larval, pupal, and adult stages. However, close scrutiny of the behavior and habits of ladybugs has revealed a variety of fascinating evolutionary and ecological features, including color pattern polymorphism, extreme promiscuity, cannibalism, sexually transmitted diseases, and biased sex ratios, some of which seem to be contrary to theoretical expectation. Here the basic biology of ladybugs and some of these conundrums are considered.

As a group, the ladybugs are the most popular of beetles. The bright colors of many species and their reputation of being beneficial, because many species eat plant pests, are at the root of this popularity. In many parts of the world ladybugs are named after religious figures and are revered, often being considered harbingers of good fortune. Indeed, the common English-language name for this family of beetles derives from the Virgin Mary. Ladybirds are "Our Lady's birds."

DESCRIPTION

Ladybugs are beetles of the family Coccinellidae. This family consists of about 5200 known species of small to medium-sized, oval, oblong oval, or hemispherical beetles. The dorsal surface is convex and the ventral surface is flat. The forewings, or elytra, are strong and are often brightly colored, sporting two or more strongly contrasting colors in a bold pattern. Not all species are red with black spots. Almost every color of the rainbow is found as the predominant color of some species of ladybug. These ground colors are usually allied to a second color, which differs starkly from the first, particularly with respect to tone. Thus, ladybugs may be red and black, or yellow and black, or black and white, or dark blue and orange, and so on. Sometimes the spots are replaced with stripes or a checkered pattern. The elytra cover the membranous flight wings, which are folded away when the beetle is not in flight.

THE LADYBUG LIFE CYCLE

Ladybugs through the Year

The life cycle of ladybugs has four stages: egg, larva, pupa, and adult. The length and timing of the different stages varies greatly with geographic region. Mating usually occurs when food is available, and eggs are laid in the vicinity of larval food. In contrast to many other insects, the two feeding stages (larvae and adults) usually have the same diet. In regions with winter and summer seasons, reproduction usually occurs in late spring and early summer. In some climates, reproduction can continue throughout the summer, with several generations being produced. However, in places with hot summers, some ladybugs have a dormant period (or aestivation) in the hottest months, sometimes having a second period of reproduction in the fall. The winter is generally unfavorable for ladybug reproduction, and ladybugs usually pass the winter as dormant adults. In wet/dry seasonal climates, particularly in the tropics, many ladybugs are dormant through the dry season, beginning to reproduce at the start of the wet season when food becomes more readily available.

The rate of development of ladybugs, like that of other insects, depends largely on ambient temperature. In a species such as *Adalia bipunctata* in a temperate climate, the egg stage lasts from about 4 to 8 days; larvae feed for about 3 weeks. When they stop feeding, they form a humped prepupa and shed the final larval skin about 24 h later to produce a pupa that is attached to the substrate at its posterior. The pupal stage lasts 7 to 10 days. When the adult emerges, the elytra are pale yellow and unpatterned. Hemolymph is pumped into the elytra and flight wings to expand them, and the color patterns develop over the next day or two. Adult ladybugs live for up to a year.

The eggs of most species of ladybugs are bright yellow and are laid upright in batches (Fig. 1) in the vicinity of food. Newly hatched larvae habitually eat any remaining eggs in their clutch and then disperse to find food. For many species, this food is in the form of small sapsucking insects such as aphids or coccids. However, some species feed on fungi, while others are true vegetarians, eating the foliage of plants. The larvae (Fig. 2) are usually elongate, and the ratio of leg length to body length is variable, being correlated to diet. The pupa (Fig. 3) is usually formed on the host plant. Both larvae and pupae may be brightly colored and patterned.

Generalist and Specialist Ladybugs

Broadly, different species of ladybugs can be split into generalists and specialists on the basis of their dietary array and the range of habitats that they live in. Most of the commonest species feed on a variety of aphid species and move from one host plant to another as aphid colonies wax and wane. However, some species have a specialized diet and so are confined

FIGURE 2 Larva of *A. bipunctata*.

to specific habitats where their food occurs. This is true of some of the aphid feeders, as well as for many of the species that feed on other diets, such as coccids, mildews, the leaves of plants or the pollen, and nectar of flowers, as their principal food. Many of these species have evolved precise adaptations to their diets and habitats.

LADYBUG COLOR PATTERNS

Warning Colors and Chemical Defense

The bright, eye-catching color patterns of most ladybugs are their first line of defense against many predators. The bold markings of one bright color set on a background of another contrasting color provide a memorable image that warns potential predators that ladybugs have hidden defenses, being foul smelling and evil tasting.

The chemical defenses of ladybugs involve a range of chemicals: alkaloids, histamines, cardiac glucosides, quinolenes, and pyrazines, some of which are synthesized by the beetles while others are sequestered from food.

Anyone who has picked up a ladybug a little roughly will have noticed that they often secrete a yellow fluid. This

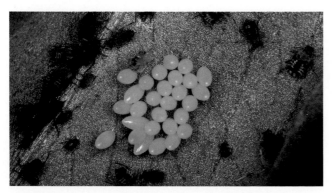

FIGURE 1 Egg clutch of *Harmonia axyridis* laid among aphids.

FIGURE 3 Pupa of *H. axyridis*.

FIGURE 4 *Anatis ocellata* reflex bleeding.

FIGURE 5 Melanic and nonmelanic forms of *A. bipunctata.*

behavior, called reflex-bleeding, is part of their defense. The fluid is filtered hemolymph and is exuded through pores in the leg joint (Fig. 4), whence it runs along grooves to form small droplets at the edge of the pronotum and elytra. This "reflex blood" contains a cocktail of volatile chemicals that have a strong and acrid scent to deter naïve predators.

Many species of ladybug share the same basic color combinations, red with black spots or yellow with black spots being the most common. From an evolutionary perspective, the similarities between many species are beneficial to all. The reasoning is simply that the more chemically defended species share the same color pattern, the smaller the number of individuals of each species likely to be harmed by naïve predators as the latter learn to associate a particular color pattern combination with unpalatability. This type of resemblance, involving a complex of species that resemble one another and are all unpalatable, is known as Müllerian mimicry.

Most of the generalist ladybugs have fairly simple patterns of just two strikingly different colors. However, some of the habitat specialists have more complex coloration. For example, some of the reed-bed specialists, such as *Anisosticta novemdecimpunctata,* have the ability to change color during their adult life. Through the fall and winter these ladybugs are beige with black spots and are well camouflaged between the old browning reed leaves where they overwinter. In spring, when the ladybugs move to new green reeds to feed and reproduce, their elytra become flushed with red pigment, thus giving the ladybugs a warning pattern.

Polymorphism

Many ladybug species have variable color patterns, and, in many cases, distinct color forms occur together as genetic polymorphisms. Thus, for example, in many parts of Asia, several different forms of *Harmonia axyridis,* can be found. The different forms are controlled genetically, with the inheritance of most depending on differences in just one gene. The existence of these genetic polymorphisms is surprising because the theory of warning coloration leads to the expectation that all members of the species will look the same. Considerable research time has been expended on this evolutionary conundrum, particularly in *A. bipunctata,* which has some forms that are mainly red with black spots and other forms that are black with red spots (Fig. 5). The factors implicated in the evolution and maintenance of the forms of this species include different levels of activity (black surfaces warm up more rapidly than red ones), sexual selection by female choice (some females have a genetic preference to mate with black males), and different levels of unpalatability to different predators. A fully convincing explanation for these polymorphisms has yet to be found.

REPRODUCTIVE BIOLOGY

Promiscuity

The reproductive biology of ladybugs raises several evolutionary problems. Both male and female ladybugs are highly promiscuous. Theoretically, females that produce large and energetically expensive germ cells should mate only often enough to ensure high fertilization rates. This theoretical limitation reflects the energetic and temporal costs of copulation and the possibility of contracting sexually transmitted diseases. Yet females of *A. bipunctata* mate about 10 times as often as they need to to fertilize all their eggs. Such promiscuity may represent a bet-hedging strategy, or it may afford a means of providing conditions for sperm competition. A hedging strategy addresses the unpredictability of the environments that will face a female's progeny: by mating with a wide variety of different males, she is assured of producing genetically diverse offspring, at least some of which may have genes appropriate to the unknown future habitat. With respect to sperm competition, female ladybugs store the sperm they receive from males in a storage organ called a spermatheca. Thus, after mating with many males, the spermatheca will contain contributions from a number of different males, and these sperm will have to compete for the opportunity to fertilize the eggs.

Female Mate Choice

The study of the reproductive biology of ladybugs has an important place in evolutionary biology, for one of Darwin's mechanisms of evolution was first demonstrated on a ladybug. In addition to natural selection, Darwin argued that some characteristics of some organisms were the result of sexual selection through either male competition or female choice of mates. That females may have a genetically controlled preference to mate with males of a particular genetic type was first demonstrated in *A. bipunctata*, over a hundred years after the theory was first proposed. In brief, it was shown that some females carry a single gene that is expressed as a preference to mate with melanic rather than nonmelanic males, irrespective of their own color. Subsequently, mating preferences were shown to exist in other species of ladybug.

Apparent Waste of Sperm

Male ladybugs also present some interesting problems. In a single copulation, for example, a male *A. bipunctata* can transfer to a female up to three sperm packages, or spermatophores. The spermatheca of a female can store about 18,000 sperm. An average spermatophore contains about 14,000 sperm. Therefore, a male that transfers three spermatophores passes more than twice the number of sperm a female can contain. This apparent waste is difficult to comprehend. Possibly by transferring an excess of sperm, the male is indulging in a coarse type of sperm competition, in which sperm in the female's spermatheca from previous matings are flushed out.

Consequences of Promiscuity: Sexually Transmitted Diseases

Not only do both sexes of many species mate many times, but the duration of each copulation is considerable, lasting several hours in many species. This promiscuity has had one obvious consequence: some species of ladybug are infected by sexually transmitted diseases. Such diseases are generally rare in invertebrates, yet both sexually transmitted mites and fungi infect ladybugs. The mite *Coccipolipus hippodamiae*, which appears to specialize on ladybugs, lives under the elytra, with its mouthparts embedded into the elytra, from which it sucks hemolymph. Mite larvae emerging from eggs produced by the adult females travel to the posterior end of their host before moving onto a new host when the ladybug copulates. A sexually transmitted fungus (in the *Laboulbeniales*) also occurs.

CANNIBALISM

Ladybugs indulge in cannibalism. Both adults and larvae will resort to eating conspecifics and sometimes other species, particularly when other food is scarce. The most vulnerable individuals are those that either are immobile (eggs, ecdyzing larvae, prepupae, pupae) or have a soft exoskeleton (recently ecdyzed larvae, newly formed pupae, newly emerged adults). Aphidophagous species tend to be more prone to cannibalism than those with other diets, largely because of the ephemeral nature of their prey and because they are more prone to large fluctuations in population density.

MALE-KILLING BACTERIA AND LADYBUG SEX RATIOS

The population sex ratio of the majority of sexually reproducing organisms is close to 1:1; selection will normally promote the production of the rarer sex, so that the stable strategy is for sex ratio equality. Female-biased sex ratios were first recorded in the ladybug *A. bipunctata* from Russia in the 1940s. Some females were found to produce only female offspring. The trait was inherited maternally. Subsequent research has shown that male embryos die while in the egg as a result of the action of bacteria such as *Wolbachia*. These male-killing bacteria live in the cytoplasm of cells and are transmitted from infected mothers to their eggs. Although the bacteria in male eggs die when they kill their host, they benefit clonally identical copies of themselves in their host's siblings, which consume the dead male eggs. The additional resources gained by these neonate female larvae increase their fitness and hence that of the bacteria that they carry.

PEST CONTROL

The benefits of allowing ladybugs to eat plant pests have long been recognized. Their importance in controlling aphids on hops in England was noted as early as 1815. For over a hundred years, many attempts have been made to use ladybugs as biological control agents of plant pests such as aphids and coccids. The first reported attempt, and still one of the most successful, was the introduction into California of an Australian ladybug, *Rodalia cardinalis,* to control the cottony cushion scale, *Icerya purchasi,* in 1888/1889. This project, costing just $1500, saw an almost immediate return because the orange crop in California increased threefold in 1890. It was the startling economic success of this project that began the biological control "explosion" that occurred through the first half of the 20th century, until the development of cheap and effective synthetic insecticides.

Not all attempts to use ladybugs in biological control have been as successful as that involving *R. cardinalis,* and in general, the successes reported have involved ladybugs that have been used to control scale insects (Coccidae) and mealybugs (*Pseudococcus* spp.). Ladybugs introduced to control aphids on a large scale have been less efficient, largely because aphid populations increase much more rapidly than do ladybug populations. This means that once aphid populations on a crop have reached sufficient density to attract ladybugs in numbers, the aphid population is already causing damage.

Despite this shortcoming, ladybugs are widely used on a smaller scale to reduce aphid populations. One species, the convergent lady beetle, *Hippodamia convergens,* is of particular note. This species aggregates in vast numbers in high mountain valleys, to pass the winter. Huge numbers are collected annually from these aggregations. The ladybugs are then packaged and stored under precisely controlled, cold conditions until the spring, when they are sold through garden centers or by mail order to ecologically minded gardeners and organic farmers.

See Also the Following Articles

Aposematic Coloration • Biological Control • Mating Behaviors • Mimicry • Predation • Sexual Selection

Further Reading

Gordon, R. D. (1985). The Coccinellidae (Coleoptera) of America north of Mexico. *J. N. Y. Entomol. Soc.* **93,** 1–912.

Hemptinne, J.-L. and Dixon, A. F. G. (1991). Why have ladybirds generally been so ineffective in biological control? *In* "Behaviour and Impact of Aphidophaga" (L. Polgar, R. J. Chambers, A. F. G. Dixon, and I. Hodek, eds.), pp. 149–157. SPB Academic Publishing, The Hague.

Hodek, I., and Honek, A. (1996). "Ecology of Coccinellidae." Kluwer, Dordrecht, Netherlands.

Iablokoff-Khnzorian, S. M. (1982). "Les Coccinelles Coléopteres-Coccinellidae Tribu Coccinellini des regions Paléarctique et Orientale." Société Nouvelle des Editions Boubée, Paris.

Majerus, M. E. N. (1994). "Ladybirds," Vol. 81 in New Naturalist Series. HarperCollins, London.

Majerus, M. E. N., and Hurst, G. D. D. (1997). Ladybirds as a model system for the study of male-killing symbionts. *Entomophaga,* **42,** 13–20.

Majerus, M. E. N., and Kearns, P. W. E. (1989). "Ladybirds," Vol. 10 in "Naturalists' Handbooks." Richmond, Slough, U.K.

Larva

Frederick W. Stehr

Michigan State University

The term "larva" is currently used for all immatures that are not eggs, pupae, or adults. When larva is used in this comprehensive sense, the subcategories include exopterygote larva (for Hemimetabola, which have the wingpads developing externally) and endopterygote larva (for Holometabola, which have the wings developing internally as histoblasts in the larva, becoming external wingpads in the pupal stage).

Larvae occur in a great diversity of sizes, shapes, and colors. Colorful ones almost always live in exposed habitats where their colors and shapes offer cryptic concealment or where their bright colors and spines warn potential predators that they are not to be eaten. Larvae that live in concealed habitats are nearly always combinations of white, gray, black, or brown.

Instar has been conventionally defined as the stage the larva is in between molts. Stadium is defined as the interval of time

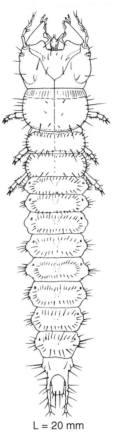

L = 20 mm

FIGURE 1 A campodeiform ground beetle larva, *Harpalus* (Carabidae). (Reproduced from A. Peterson, 1951, *Larvae of Insects,* Vol. 2, with permission of Jon A. Peterson.)

between molts. Others contend that the instar is properly defined as the stage the larva is in between apolysis (separation of the old cuticle) and the molt to the next stage, and a pharate (next-stage) larva would be present before the next molt.

There are some general terms used for types of holometabolous larvae that have broad usage. Campodeiform larvae (Fig. 1) are somewhat flattened and have an elongate body, thoracic legs that are well developed, a head that is directed forward, no abdominal prolegs, and antennae and cerci that are usually conspicuous. This larval type is common in the Coleoptera (beetles), Megaloptera (dobsonflies and fishflies), Neuroptera (lacewings and antlions), and Raphidioptera (snakeflies).

Elateriform larvae (Fig. 2A) are somewhat similar to campodeiform larvae, but their body is more elongate, subcylindrical, and more heavily sclerotized. This type is common in the Elateridae (click beetles) and other Coleoptera.

Scarabaeiform larvae (Fig. 3) have a C-shaped, whitish body, a dark head, and well-developed thoracic legs. White grubs (Coleoptera: Scarabaeidae) are the best example.

Eruciform larvae are caterpillar-like and have a cylindrical body and well-developed thoracic legs, and prolegs are present. This type is common in the Lepidoptera (butterflies and moths), Mecoptera (scorpionflies), and Hymenoptera (sawflies only).

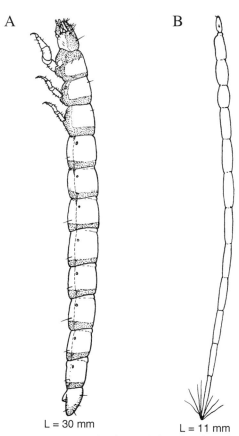

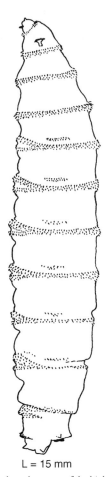

L = 30 mm

L = 11 mm

FIGURE 2 (A) A heavily sclerotized elateriform larva (lateral). (Reproduced from A. Peterson, 1951, *Larvae of Insects,* Vol. 2, with permission of Jon A. Peterson.) (B) An elongate, legless, vermiform "wormlike" larva. (Reproduced from A. Peterson, 1951, *Larvae of Insects,* Vol. 2, with permission of Jon A. Peterson.)

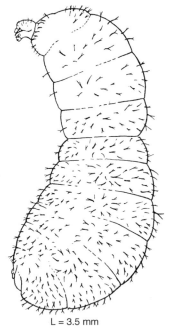

L = 15 mm

FIGURE 4 A legless, peg-shaped maggot of the higher flies. (Reproduced from A. Peterson, 1951, *Larvae of Insects,* Vol. 2, with permission of Jon A. Peterson.)

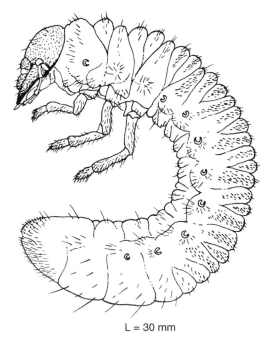

L = 30 mm

FIGURE 3 A C-shaped scarabaeiform larva, *Phyllophaga* (lateral). (Reproduced from A. Peterson, 1951, *Larvae of Insects,* Vol. 2, with permission of Jon A. Peterson.)

L = 3.5 mm

FIGURE 5 A baglike ant larva. It and other similar larvae are commonly called "grubs." (Reproduced from A. Peterson, 1951 *Larvae of Insects,* Vol. 2, with permission of Jon A. Peterson.)

Vermiform larvae (Fig. 2B) are "wormlike." This is an ill-defined term, but it is generally applied to an elongate, legless larva with or without a conspicuous head. Maggots (Fig. 4) are the larvae of higher Diptera. Their shape is peg-like and tapering toward the anterior end. They are legless, have a greatly reduced head (no head capsule), and have conspicuous mouthhook(s). The posterior end bears a pair of conspicuous spiracles.

Grub (Fig. 5) is an imprecise term that is often applied to "comma-shaped" larvae with or without legs or having greatly reduced legs. This term is commonly applied to weevil larvae and other Coleoptera larvae and to many larvae of the higher Hymenoptera (ants, bees, and wasps) that often have reduced or inconspicuous heads and appendages.

See Also the Following Articles

Caterpillars • Cocoon • Eggs • Metamorphosis • Pupa and Puparium

Further Reading

Stehr, F. W. (ed.) (1987) "Immature Insects," Vol. 1. Kendall/Hunt, Dubuque, IA.
Stehr, F. W. (ed.) (1992). "Immature Insects," Vol. 2. Kendall/Hunt, Dubuque, IA.

Learning

Daniel R. Papaj
University of Arizona

A butterfly learns to search for the shape of its preferred host plant's leaves and, contemporaneously, for the color of preferred nectar sources. A parasitoid wasp learns color, pattern, and odor components of its insect host's microhabitat. A grasshopper avoids feeding on a plant associated with a recent digestive malaise. An emerging adult paper wasp imprints on odor cues in its nest, using the odors to distinguish nestmates from nonnestmates. A male damselfy learns to recognize andromorphs (male-mimicking females) as females. A fly improves its depth perception with experience. A bee memorizes a sequence of visual landmarks between its nest and a patch of flowers, as well as the distance between landmarks. All of these are examples of learning, a phenomenon that is ubiquitous throughout the animal kingdom and, as these examples illustrate, well represented within insects. In fact, learning has been documented in all major insect orders. While best studied in the context of foraging for food or oviposition sites, evidence of learning has also been obtained in relation to water consumption, mate finding and choice, territoriality, predator avoidance, dispersal, migration, kin recognition, and thermoregulation.

LEARNING CHARACTERIZED

Characteristics

Learning eludes an easy, satisfying definition, but the following characteristics constitute a useful guide. Learning involves an enduring change in behavior with experience, the change usually progressing gradually with continued experience to some asymptote. Learned behavior is often modified by novel experiences, and effects of experience eventually wane if not reinforced.

Associative vs Nonassociative Learning

Learning can be categorized as nonassociative or associative. Nonassociative learning includes habituation and sensitization. Habituation involves the waning of a response to a stimulus upon repeated presentation of that stimulus. Alternately, repeated presentation of a stimulus sometimes enhances a response to that stimulus and often to related stimuli, a process termed sensitization. Associative learning involves pairing a stimulus with another stimulus, or with a motor pattern, such that the response to the first stimulus is altered as a consequence of the pairing. Associative learning is typically evaluated in two kinds of paradigms: classical (Pavlovian) conditioning and instrumental conditioning.

Classical and Instrumental Conditioning

In classical conditioning, an unconditioned stimulus (US) that elicits an unconditioned response is paired in time and space with a novel stimulus, the conditioned stimulus (CS). As a consequence of the pairing, the CS subsequently elicits a conditioned response. Both appetitive and aversive forms of classical conditioning have been documented in insects. Most of what we know about classical conditioning in insects has involved classical conditioning of the proboscis extension reflex (PER), principally in honey bees.

A case for associative learning is strengthened by evidence of discrimination learning. Discrimination learning (sometimes called differential conditioning) controls for effects of sensitization to a CS by training to two CSs, one which is reinforced with a reward (CS+) and one which is not (CS–). If learning is associative, response to the CS+ only is heightened, relative to controls. Discrimination learning is well documented in bees, hymenopterous parasitoids, moths, butterflies, cockroaches, and fruit flies. A case for associative learning is similarly supported if learning is restricted to forward pairing. In forward pairing, the CS is presented shortly before the US, whereas, in backward pairing, the CS is presented shortly after the US. Insects, like vertebrates, show strong learning in forward-pairing regimes but little or no learning in backward-pairing or random-pairing regimes.

In instrumental conditioning (roughly equivalent to operant conditioning), presentation of a reinforcing stimulus is contingent upon the insect's own motor actions. For example, an entirely novel motor pattern can be generated through a process of trial and error, as when a bee or butterfly learns how to extract nectar from a flower. A standard operant conditioning paradigm in the laboratory requires a tethered orthopteran to move its leg in response to an electric shock, heat, or access to food. A headless roach learns such a task, demonstrating that conditioning can occur at the level of ganglia.

Learning a given task in nature probably involves a combination of stimulus–stimulus and stimulus–response associations. When an insect pollinator learns nectar-extraction routines for different flowers, for example, it simultaneously learns identifying features of each type of flower, allowing the appropriate motor routine to be expressed on the appropriate flower.

Miscellaneous Types

Various forms of associative learning beyond the basic types have special meaning to students of learning. Food aversion learning, strongly implicated in work on vespid wasps, grasshoppers, mantids, and caterpillars, involves avoidance of food stimuli associated with a digestive malaise. Food aversion learning is noteworthy because an aversion can form even when a long period of time (hours) passes between ingestion of a food and the resulting illness.

Spatial learning is an important component of insect navigation. Commonly traversed routes are learned during homing by ants, bees, and wasps, and traplining is learned by bees and butterflies. Honey bees may additionally possess a topographically organized landscape memory that allows them to navigate along a novel route. Spatial learning is useful in contexts other than movement of the whole organism; for instance, bees learn to discriminate textures with their antennae and use such learning to evaluate the microtexture of flower petals.

One form of learning of significance in vertebrates which has not been documented to date in insects is observational learning, in which a subject imitates the motor actions of a demonstrator. Nevertheless, social interactions do influence what insects learn. Honey bees and bumble bees, for example, evaluate floral scents borne by returning foragers and forage selectively for those scents.

LEARNING PROCESSES

Many associative learning processes that have been described for vertebrates have also been shown in insects. The following list of selected processes is derived mainly from work on honey bees, unless otherwise noted.

Generalization refers to an animal's tendency to respond to stimuli that were not reinforced but that are related to a reinforced stimulus (A+) along some perceptual dimension. Moths and honey bees have been shown to generalize odors according to similarities in functional groups and carbon-chain length.

Blocking occurs when an animal that first learns to respond to a stimulus (A+), and is then reinforced on A and a novel stimulus, B, presented together ([AB]+), subsequently fails to show a heightened response to B alone, relative to controls. Learning of stimulus B has been blocked by coupling with the previously learned stimulus A. Blocking illustrates that temporal pairing between a CS and a US is not sufficient for associative learning to take place; rather, a new CS must convey new information in order to be learned. Whereas blocking is a robust phenomenon in vertebrates, studies of blocking in bees and fruit flies have yielded mixed results. Where blocking has been demonstrated, it seems to be restricted to intramodal stimuli (e.g., odor blocking in honey bees).

Overshadowing occurs when an animal reinforced on a compound of stimuli A and B ([AB]+) shows little response to B alone, relative to when reinforced on B alone (B+). As with blocking, overshadowing illustrates that temporal pairing between a CS and a US is not sufficient for associative learning to take place.

Sensory preconditioning occurs when an insect presented simultaneously with two stimuli in the absence of reinforcement ([AB]–), then reinforced on one stimulus (A+), subsequently shows a heightened response not only to A but also to B. During exposure to [AB], the insect learns that A and B belong together. Observed in *Drosophila* fruit flies and honey bees, sensory preconditioning illustrates that a stimulus does not have to be paired directly with a US in order for an association between the stimulus and the US to form.

Second-order conditioning refers to the capacity for a stimulus, once conditioned, to serve as a US in the conditioning of another stimulus. Second-order conditioning may play a major role in learning complex mixtures of stimuli, such as odor blends.

Patterning is evaluated by reinforcing two stimuli in turn (A+, then B+) and then explicitly not reinforcing a compound of those stimuli ([AB]–). Under this protocol, PER odor conditioning in bees shows "negative patterning," responses being greater to individual odors than to the compound. This result can be explained only if the insect treats the compound [AB] as a unit and relates it to the absence of reinforcement. Such learning is referred to as configural learning.

Rule extraction has been demonstrated with the use of delayed matching-to-sample tasks in which honey bees are required to respond to a stimulus that matches a sample stimulus recently experienced. Bees not only solve the task but also transfer the matching to stimuli not previously reinforced. For example, bees trained to match a color can subsequently match patterns of lines and, remarkably, bees trained to match an odor can subsequently match colors too. Such results have been interpreted to mean that insects can form a concept of "sameness." When trained in a delayed non-matching-to-sample task in which they must choose the stimulus that does not match the sample, bees again perform well and make similar transfers, showing a grasp of a "difference" relation. Bees also learn to extract bilateral symmetry from a series of rewarded patterns and subsequently transfer that extraction to evaluation of novel patterns.

MEMORY

Associative memory in insects, as in vertebrates and other animals, is time-dependent and phasic. Recent work on fruit flies and honey bees suggests as many as five memory phases: (1) an early and (2) a late form of short-term memory [eSTM and lSTM], (3) a midterm memory [MTM], and two forms of long-term memory (in honey bees, characterized as (4) an early form [eLTM] and (5) a late form [lLTM]; in *Drosophila*, characterized as (4) an anesthetic-resistant form and (5) a parallel, susceptible form). STM forms immediately upon association, is short-lived (seconds to minutes), and is relatively easily erased by conflicting information or treatment by cooling or shock. eSTM is characterized by a relatively nonspecific appetitive arousal and is highly suseptible to interference by new, conflicting information or by cooling. lSTM is more stable, is more specific, and takes longer to form than eSTM. The transition from STM to MTM after a single learning trial requires several minutes. MTM is more resistant to interference than STM, requiring hours to decay.

LTM takes longer still to form than either STM or MTM, involves longer lasting changes (hours to weeks), and is relatively resistant to interference. In bees, formation of LTM requires multiple learning trials. LTM is highly context-specific; landmarks learned by bees around their feeder, for example, may be entirely ignored when presented at a novel location. eLTM and lLTM have been distinguished in terms of the effects of inhibiting protein synthesis: synthesis inhibition after 24 h degrades memory, whereas inhibition after 3 days does not. Effects of inhibition depend on the time between learning trials, with closely spaced trials (termed "massed trials") resulting in memory that is independent of protein synthesis.

The underlying processes involved in memory formation are beginning to be revealed. In honey bees, a "value" neuron, the VUM_{mx1} of the subesophageal ganglion, which fires in response to sucrose stimulation, is proposed to be part of the US pathway. In PER odor conditioning studies, artificial depolarization of the VUM_{mx1} neuron just following presentation of an odor generates a conditioned response to the odor. The VUM_{mx1} neuron, which uses octopamine as a neurotransmitter, converges on two brain neuropils, the antennal lobe and the mushroom bodies. Consistent with these observations, olfactory memories can be established by odor-coupled injection of octopamine into either the antennal lobe or the mushroom bodies. The pattern of octopamine effects suggests that antennal lobe processes may relate more to eSTM, whereas mushroom body processes may relate more to lSTM and LTM.

To what degree these findings pertain only to honey bees or only to odor learning is uncertain. Analysis of *Drosophila* mutants suggests that the mushroom bodies are important for odor learning but dispensable for visual or tactile learning. Studies of locusts have indicated effects of feeding experience on diet choice that resemble discrimination learning, but are based on an entirely novel mechanism. This taste-feedback mechanism involves adjustments in the level of sensitivity to nutrients in the hemolymph.

FUNCTION OF LEARNING

In a sense, the function of associative learning is obvious. Animals learn by association to orient toward stimuli predicting positively rewarding resources (such as sugar, pollen, food plant, hosts) and away from stimuli predicting negatively rewarding events (shock, heat, toxins, predators). Likewise, habituation is a means for reducing energy-wasteful, time-consuming responses to meaningless stimuli. In either case, however, learning is needed only if the appropriate responses cannot be predicted without benefit of experience, else an insect could respond (or not respond) innately. Even in an unpredictable environment, whether learning yields higher fitness than innate behavior depends on the relative costs of learning. A robust assessment of costs and benefits of learning has proved elusive, perhaps in part because individual fitness in nature is especially difficult to measure in *Drosophila* and honey bees, the systems in which learning processes and mechanisms have been best studied.

ADAPTATION, CONSTRAINT, AND LEARNING

Limits to Learning and Memory

Of interest to behavioral ecologists is the degree to which learned behavior reflects adaptation by natural selection versus constraints on selection. Generalization (see above), for example, may seem at first to reflect a constraint on learning, but conceivably represents an adaptive mechanism of imprecision. A pollinator, for example, that responded only to the precise odor blend emitted by the first rewarding flower encountered might never visit another flower, owing to among-flower variation in the blend.

A classic case study of limits on learning and memory in nature that interested Darwin himself concerns the tendency for bees, butterflies, and other pollinators to show greater fidelity to one or a few floral species than expected based on the profitabilities of those species. According to one point of view, this so-called floral constancy is dictated by limits on the acquisition, retention, and/or retrieval of stored information about the floral resource.

That foraging success in insects is limited in terms of acquisition and retention seems unlikely at the level of LTM. LTM in insects, as mentioned above, is extraordinarily durable and the amount of information that can be maintained in LTM, as currently understood, is extremely impressive. Butterflies can learn visual cues in two foraging modes (nectar collection and oviposition) simultaneously, showing meaningful responses in each instance in just a single trial. Bees can be trained to distinguish multiple rewarded stimuli from multiple unrewarded ones and to link features of eight or more different flower species to the time of day at which nectar is available. These features include flower color, odor, pattern, and microtexture. In addition, a bee learns the location, profitability, and visual landscape associated with a rewarding patch of flowers, as well as the route between hive and patch and, in conjunction with the sun compass used to navigate, even the pattern of movement of the sun through the sky.

Retention at the level of LTM is similarly impressive. Bees have been shown to retain LTM without reinforcement for several weeks, a period of time comparable to average worker life expectancy. In *Tribolium* beetles and *Drosophila,* there is evidence that memory formed in the larval stage persists through metamorphosis.

If pollinators are limited at all in memory, it may be at the level of STM. As noted above, STM is particularly vulnerable to conflicting information; this fact may make it difficult for a bee once fixed on a flower type to switch to a novel one. Alternatively, the key to floral constancy may lie in the retrieval of stored information, specifically a constraint on the minimum time required to activate information stored in LTM and a limited capacity to activate multiple memories at once (together, limits on what for vertebrates has been referred to as working memory).

Learning and Memory as Products of Adaptation

An alternative, albeit not mutually exclusive, view holds that natural selection generates an adaptive balance between activation and suppression of memory, tuning that balance finely to the specific ecological requirements of a given species. For example, floral constancy might conceivably permit workers in a colony to partition floral resources efficiently, in which case the properties of learning and memory that contribute to constancy would be viewed as adaptive. It has even been proposed that memory dynamics in bees are tightly matched to foraging activity rhythms as well as the spatial patterning of the floral resource.

Abundant propositions as to adaptive specialization in learning have been made, especially from a comparative standpoint: "Insects of a given species should be prepared to learn particularly well those stimuli relevant to that species' needs." "Social insects should learn better than solitary ones (owing to the demands of a complex and unpredictable social environment)." "Generalist insects should learn better than specialists." For none of these propositions is there compelling evidence, nor will there be until better descriptions are made of learning in an ecological context, learning protocols are brought closer in rigor to those employed in comparative psychology, and more insect species are evaluated.

For now, the primary comparison to be made is a comparison between learning in insects and in vertebrates. Here, the pattern is one of shared features. Despite significant phylogenetic distance between insects and vertebrates, and despite substantial differences in their underlying physiology, there is a remarkable congruence in the diversity and form of learning processes in these taxa (see above). The similarities may reflect shared ancestry, evolutionary convergence, or both. A finding of evolutionary convergence would imply that certain universal, yet to be clearly defined functional principles govern the evolution of learning and memory processes.

See Also the Following Articles
Brain • Dance Language • Neuropeptides

Further Reading
Abramson, C. I., Yuan, A. I., and Goff, T. (1990). "Invertebrate Learning: A Source Book." Am. Psychol. Assoc., Washington, DC.
Bitterman, M. E. (1996). Comparative analysis of learning in honeybees. *Anim. Learn. Behav.* **24,** 123–141.
Chittka, L., Thomson, J. D., and Waser, N. M. (1999). Flower constancy, insect psychology, and plant evolution. *Naturwissenschaften* **86,** 361–367.
Dukas, R. (ed.) (1998). "Cognitive Ecology." University of Chicago Press, Chicago.
Menzel, R. (2001). Searching for the memory trace in a mini-brain, the honeybee. *Learn. Memory* **8,** 53–62.
Papaj, D. R., and Lewis, A. C. (eds.) (1993). "Insect Learning: An Ecological and Evolutionary Perspective." Chapman & Hall, New York.
Shettleworth, S. (1998). "Cognition, Evolution and Behavior." Oxford University Press, Oxford.
Smith, B. H. (1996). The role of attention in learning about odorants. *Biol. Bull.* **191,** 76–83.
Tully, T. (1997). Regulation of gene expression and its role in long-term memory and synaptic plasticity. *Proc. Natl. Acad. Sci. USA* **94,** 4239–4241.

Legs

Peter H. Adler
Clemson University

One of the most generally known and oft-repeated facts about insects is that they possess three pairs of legs, one pair each on the prothorax, mesothorax, and metathorax. Indeed, this condition is in the fundamental ground plan of insects and is amply represented in the fossil record. The condition inspired Latreille's taxon Hexapoda (Greek *hexa,* six, and *poda,* foot). Exceptions to the hexapodous condition are found in the apodous, or legless, insects that have secondarily lost their legs, typically as a result of selection for an obligatory parasitic or sedentary existence.

The six-legged condition is derived from an ancestral arrangement in which legs occurred on the majority of body segments. Over evolutionary time, the serially uniform legs became modified in the insectan lineage into the characteristic mouthparts, thoracic legs, and various abdominal appendages, such as cerci and genitalia, while typically becoming lost on other abdominal segments. Further evolution of the basic six-legged condition in the insectan lineage has resulted in an enormous diversity of structure and function. This structural and functional diversity of legs, along with the acquisition of wings without the loss of legs, which is a condition unique to insects, undoubtedly has been a key factor in the numerical success of insects and their representation in nearly every habitat on the planet. The exquisite diversity in leg structure plays an important role in the taxonomy and classification of insects.

STRUCTURE

In the classic textbook interpretation, the insectan leg has six well-sclerotized segments, arranged proximal to distal: the coxa, trochanter, femur, tibia, tarsus, and pretarsus. A more fundamental and complete segmentation scheme, which facilitates the recognition of leg and leg-derived homologies among all arthropods, involves 10 or 11 segments. These segments include the epicoxa (debatably present as the wing articulation and a fused portion of the tergum), subcoxa (absorbed into the pleuron), coxa, trochanter, prefemur (typically fused with the trochanter), femur, patella (fused with the tibia), tibia, basitarsus, eutarsus (often subdivided), and pretarsus. A more modern interpretation of the free leg of extant insects, therefore, depicts it as consisting of seven to eight distinct segments, which are the classical six plus a basitarsus and a prefemur, in some insects.

Each segment in the insectan leg, unless secondarily lost or fused, is independently movable by muscles inserted on its base. Thus, subdivisions of the eutarsus, marked by flexible cuticle but without corresponding internal muscles, are not true segments; these subdivisions are referred to as tarsomeres. The areas of flexion between segments are joints, and the well-sclerotized contact points in the joints are the condyles. The various joints contribute to the mechanical efficiency of the leg. The articulation between the coxa and the body, for example, allows the leg to move forward and rearward, whereas that between the coxa and the trochanter allows the leg to be lifted at the end of the backstroke and depressed at the beginning of the backstroke.

Leg joints are of two types. Monocondylic joints have a single point of articulation, somewhat like a ball-and-socket joint, and usually are situated dorsally. They allow considerable freedom of movement and are characteristic of the legs of larval insects. Dicondylic joints consist of an anterior and a posterior condyle, or a dorsal and ventral condyle in the case of the trochanterofemoral joint. They typically limit movement to that of a hinge. Adult legs usually have dicondylic joints, although the tibiotarsal joint is often monocondylic.

The coxa (plural coxae) is typically short and rather stout, although it varies in shape among taxa. It is set in a coxal cavity and articulates with the thorax at the coxal process of the pleural sulcus (groove). Quite often, it also articulates with the thoracic trochantin and sternum, somewhat restricting its movement. To withstand the forces of movement, the coxa is strengthened by a ringlike basicostal sulcus that sets off a basal sclerite, the basicoxite. Internally, the basicostal sulcus is expressed as a ridge, the basicosta, that provides for muscle attachment. Posterior to the point of articulation, the basicoxite is called the meron and in insects such as adult Neuroptera and Lepidoptera, it can be quite large. In higher Diptera, the meron is detached from the coxa and forms a plate in the mesothoracic pleuron. In some insects, an additional external groove, the coxal sulcus, divides the coxa lengthwise.

The trochanter is small and freely movable in a vertical direction on the coxa, but it is often rather fixed to the base of the femur. In the larvae and adults of numerous fossil insects and a few extant taxa, such as Odonata, two trochanteral segments are present, the distal one being the prefemur.

The femur (plural femora) is usually the largest and strongest segment of the leg. Its size is related to the mass of the tibial extensor muscles within it, varying from a small, thick segment in larval insects to the enormous segment in the hind leg of jumping Orthoptera. The femur often is equipped with spines and other cuticular modifications, especially in predatory insects.

The tibia (plural tibiae) typically is long and slender in adult insects. Proximally, it is bent slightly toward the femur, allowing the shaft of the tibia to be flexed close against the femur for more locomotory power in insects such as grasshoppers. It often bears spines for grooming or for engaging the substrate to aid in locomotion. Many insects also have apical or subapical movable spurs on their tibiae.

The tarsus (plural tarsi) is a simple, undivided segment in holometabolous larvae and basal hexapods such as Protura and some Collembola. In collembolans, the tarsus and tibia are fused into a single tibiotarsus. In most insects, a separate segment, the basitarsus, is present and the eutarsus is subdivided into two to four sections or tarsomeres. The ventral surface of the basitarsus and eutarsus often bears pads called tarsal pulvilli that aid movement on smooth surfaces and are especially well developed in some Orthoptera. The basitarsus and eutarsus generally are well endowed with sensory hairs and chemoreceptors. The ventral surface often has a secretory epithelium that produces a wax, possibly for waterproofing, inhibiting the uptake of undesirable water-soluble compounds, or preventing entrapment in surface films.

The pretarsus, also called the acropod or posttarsus, arises from the distal end of the eutarsus. In the Protura, Collembola, and larvae of many holometabolous insects, the pretarsus is a simple, clawlike segment. Typically, however, the pretarsus consists of a membranous base, a pair of hollow claws (ungues), and various sclerites and lobes. A sclerotized unguitractor plate articulates with the eutarsus into which the plate is partly invaginated. The muscles that flex the claws are inserted on a process of this plate. The claws articulate with the unguifer, a median process at the distal end of the eutarsus. A saclike, hollow lobe, the arolium, arises between the claws. In adult Diptera, other than crane flies, an arolium is absent. Instead, a padlike lobe called the pulvillus (plural pulvilli) arises from an auxiliary plate (auxilia) beneath the base of each claw, while an unpaired, lobelike or bristlelike process, the empodium, stems medially from the unguitractor plate. The pretarsal pads and lobes are covered with adhesive setae (tenent hairs) that allow the insect to climb and hold onto smooth surfaces.

Variation in structure, and hence function, can be found among the three pairs of legs within an individual, as well as between larvae and adults, between males and females, and among taxa. The thoracic legs of many larval insects are serially uniform or, sometimes, lacking. The legs of adults often vary in structure among the three pairs, although even adults of some insects (e.g., some female Coccidae, Psychidae, and Strepsiptera) are devoid of legs. The variation among the three pairs of legs in adults often is associated with acquisition of food, courtship, and mating. Developmental variation occurs in holometabolous insects, which have simple, rather generalized legs in the larvae and more specialized legs in the adults. Among the hemimetabolous groups, many Hemiptera–Heteroptera gain a tarsomere in the final molt.

Sexual dimorphism in leg structure is particularly prevalent. The reduced forelegs of nymphalid butterflies have short tarsomeres in females but lack all segments beyond the tibia in males. The forelegs of male Ephemeroptera are typically elongated to grasp the female. The slender, elongate legs of crane flies are even longer in males than in females for species in which a guarding male stands over the female during oviposition; in some crane flies the distalmost tarsomere is prehensile in males and used for holding females. The hind femora of males of many Coreidae are enlarged for intrasexual fighting. The pulvilli beneath the claws of some flies, such as Tachinidae, are considerably larger in males than in females.

FUNCTION

The majority of insectan legs are either elongate, slender, and designed for walking and climbing or cursorial, i.e., adapted for running, as in the cockroach (Fig. 1A). During walking, the legs form alternating triangles of support, with the fore and hind legs of one side and the middle leg of the opposite side contacting the substrate as the other three legs move forward. Various modifications allow the legs to be used in other forms of locomotion. Enlarged hind legs of many Orthoptera, fleas, and other insects are saltatorial, meaning they are designed for jumping. The jump of insects such as fleas is aided by a rubberlike protein called resilin in the cuticle that stores and subsequently releases energy for the jump. Powerful, spadelike forelegs of mole crickets, scarab beetles, burrowing mayflies, and other insects are fossorial, or adapted for digging and rapid burrowing (Fig. 1B). Flattened, fringed legs of aquatic insects such as dytiscid and gyrinid beetles and notonectid backswimmers serve as oars for paddling or swimming (natatorial legs), while long legs with hydrophobic tarsal hairs and anteapical claws, as seen in water striders (Gerridae), are for skating on the surface of water. The legs of some insects, although well developed, have lost their associated locomotory function. The spiny legs of Odonata, for example, are designed for perching or seizing and holding prey captured while the dragonfly is in flight; the legs are ineffectual for walking.

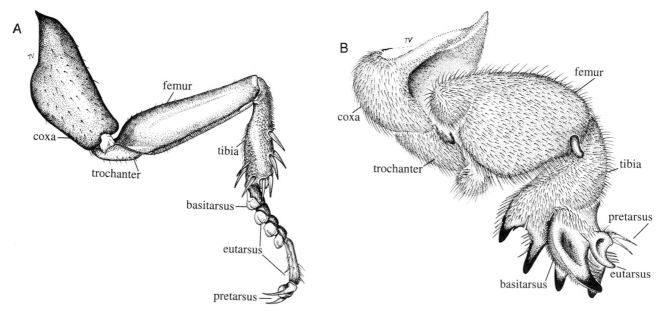

FIGURE 1 (A) Cursorial foreleg of the Madeira cockroach *(Leucophaea maderae)*. Illustration by T. S. Vshivkova. (B) Fossorial foreleg of the northern mole cricket *(Neocurtilla hexadactyla)*. Illustration by T. S. Vshivkova.

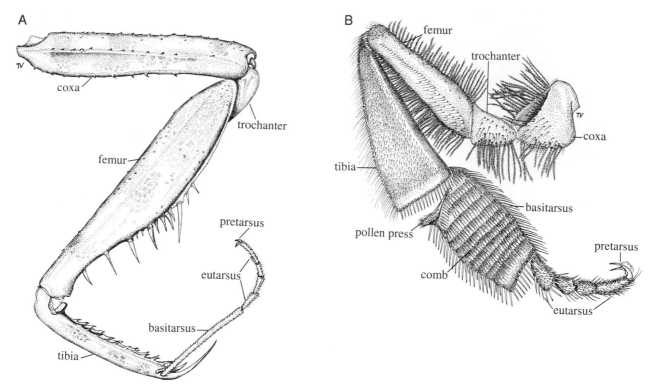

FIGURE 2 (A) Raptorial foreleg of the Carolina mantid *(Stagmomantis carolina).* Illustration by T. S. Vshivkova. (B) Inner surface of the hind leg of the honey bee *(Apis mellifera).* Illustration by T. S. Vshivkova.

Although typically regarded as agents of locomotion, the legs have assumed a wide range of additional, or altogether different, functions. Often, function can be inferred from structure; for example, the thickened, spinose legs of many insects signify a predacious mode of life. Raptorial legs (Fig. 2A), i.e., those designed to seize prey, have arisen independently in many insectan lineages. Either of the three sets of thoracic legs can be raptorial, but the trait is probably most often expressed in the forelegs (e.g., in Mantodea and Reduviidae) and less frequently in the middle legs (e.g., in some Empididae) and hind legs (e.g., in some Mecoptera).

Legs also play an important defensive role, not only in permitting escape by running, jumping, burrowing, and swimming, but also in ways such as kicking and slashing. The spines on the legs of many insects, when used in defense, effectively deter predators and competitors and can inflict considerable damage. Insects such as stink bugs and treehoppers deliver powerful kicks at parasitoids and predators that attempt to attack their young. Autotomy, or the loss of legs at predetermined points of weakness, often at the level of the trochanter, occurs in insects such as crane flies, leaving a predator with only a leg in its clutches as the insect escapes. Legs, whether lost through autotomy or accident, often can be regenerated to various degrees if one or more molts follow the amputation.

All legs are equipped with an extensive arrangement of sensory structures that allow the insect to feel, hear, and taste, providing the insect with its initial assessment of the environment. Chemoreceptors, which are especially prevalent on the basitarsus and eutarsus, provide sensory input on environmental substances and can be used to determine the acceptability of food, ovipositional substrates, and perhaps mates. Mechanoreceptors, most often in the form of hair organs, but also campaniform, chordotonal, and plate organs, provide sensory information on position, movement, and vibrations borne by air and substrate.

In many insects, the legs are used in sound production. Familiar examples include the shorthorned grasshoppers, which have a stridulatory mechanism on the hind femur, involving a series of pegs—the scraper—that is rubbed across a ridged wing vein. Some larval hydropsychid caddisflies have a scraper on the prothoracic femur that is rubbed against a file on the venter of the head. Legs also can be used to produce sound for intraspecific communication by drumming them against a substrate, as in some Orthoptera.

To maintain hygiene, insects spend considerable time preening and grooming their body and appendages. Grooming typically is effected by various leg structures, which can be in the form of cuticular combs (ctenidia), setal brushes, grooves, and notches. The cleaning setae on the foretibia of certain heteropterans are mirror images of the arrangement of antennal setae. The hind leg of honey bees is specially modified to groom pollen from the plumose hairs of the body (Fig. 2B). Combs on the inner surface of the hind basitarsus remove the pollen from the body hairs and pass it to the pollen press between the tibia and the basitarsus. Closure of the press forces the pollen into the pollen basket (corbiculum) on the outer surface of the tibia where the pollen bolus is held in place by rows of hairs. Once in the hive, the honey bee removes the pollen, with the aid of an apical spur on the middle tibia.

Legs often are used to hold onto objects, and they bear the relevant modifications, including enlarged segments to house increased musculature, various spines and setae, and adhesive organs. The grasping function is seen, for example, in the pincerlike, spiny raptorial forelegs of many predacious insects. It also is expressed dramatically in certain sucking lice in which the claw folds against a thumblike, spinose process of the enlarged tibia. Flies that feed on the blood of birds typically have a thumblike lobe at the base of each of their talonlike claws that helps them grasp feather barbules. Grasping is common during mating, and especially the males of many insects have legs designed to secure and hold their mates. Adhesion to objects such as mates and prey can be achieved with suction discs on the legs. Male dytiscid beetles have a flattened, disklike arrangement on each foreleg that is formed of the basitarsus and the succeeding two tarsomeres; all three structures bear minute suction cups ventrally that can be applied to the elytra of the female.

The colors and patterns of legs vary from subtle to stark, although their function is often poorly understood. Long-legged insects such as phantom crane flies (Ptychopteridae) and some mosquitoes often have banded legs that might render the insect less conspicuous through disruptive coloration. Other configurations of pattern and color play a role in camouflage, mimicry, and courtship. In flies such as some Syrphidae and Micropezidae, the forelegs resemble the antennae of aculeate Hymenoptera, reinforcing the remarkable overall resemblance of fly to wasp.

Other functions ascribed to the legs are often highly specialized. In the Embiidina, the basitarus of each foreleg houses multiple silk glands, and each gland is connected to a seta with an apical pore through which the silk is extruded. The inflated basitarsus of each leg in phantom crane flies contains a tracheal sac, perhaps aiding buoyancy during the driftlike flight. Some flies have specialized areas on their legs, particularly on the tibia, that possibly produce pheromones. Insects such as Chironomidae seem to use the legs much as a second set of antennae. In Protura, which lack antennae, the forelegs probably have assumed an antennal (i.e., sensory) function. Various ornamentations on insectan legs can serve a courtship or intrasexual combative role, as in some coreid bugs.

See Also the Following Articles
Anatomy • Segmentation • Swimming • Walking and Jumping

Further Reading
Adler, P. H., and Adler, C. R. L. (1991). Mating behavior and the evolutionary significance of mate guarding in three species of crane flies (Diptera: Tipulidae). *J. Insect Behav.* **4,** 619–632.
Chapman, R. F. (1998). "The Insects: Structure and Function," 4th ed. Cambridge University Press, Cambridge, UK.
Hlavac, T. F. (1975). Grooming systems of insects: Structure, mechanics. *Ann. Entomol. Soc. Am.* **68,** 823–826.
Kukalova-Peck, J. (1992). The "Uniramia" do not exist: The ground plan of the Pterygota as revealed by Permian Diaphanopterodea from Russia (Insecta: Paleodictyopteroidea). *Can. J. Zool.* **70,** 236–255.
McAlpine, J. F. (1981). Morphology and terminology—Adults. *In* "Manual of Nearctic Diptera" (J. F. McAlpine, B. V. Peterson, G. E. Shewell, H. J. Teskey, J. R. Vockeroth, and D. M. Wood, eds.), Vol. 1, Monogr. 27, pp. 9–63. Research Branch, Agriculture Canada, Ottawa.
Mitchell, P. L. (1980). Combat and territorial defense of *Acanthocephala femorata* (Hemiptera: Corediae). *Ann. Entomol. Soc. Am.* **73,** 404–408.
Snodgrass, R. E. (1935). "Principles of Insect Morphology." McGraw–Hill, New York.

Lepidoptera
(Moths, Butterflies)

Jerry A. Powell
University of California, Berkeley

Moths and butterflies make up the order Lepidoptera, and they are among the most familiar and easily recognized insects. The Lepidoptera is defined as a monophyletic lineage by a suite of more than 20 derived features, the most obvious of which are the scales and proboscis. The scales are modified, flattened hairs that cover the body and wings, shingle-like, and are the source of the extraordinary variety of color patterns typical of these insects. In all but the most primitive forms, feeding by adults is accomplished by pumping in liquid via a tubular proboscis (haustellum), which usually is elongate and coiled under the head. The sister group of Lepidoptera, the Trichoptera (caddisflies), lack this development of mouthparts and the covering of scales and possess caudal cerci on the abdomen, which are not present in Lepidoptera.

Like other holometabolous insects, lepidopterans pass through egg, larval, pupal, and adult stages. Mating and egg deposition are carried out by the adult moths and butterflies. Within the eggs, embryos develop to fully formed larvae. The larvae, commonly called caterpillars, feed and grow, which is accomplished by a series of stages (instars). At maturity they transform to pupae, usually within silken cocoons spun by the larvae, although many species pupate without a cocoon. Metamorphosis to the adult occurs during the pupal stage, and the fully developed adult breaks the pupal shell to emerge. Adults of most species feed, but they do not grow. Diapause, an arrested state of development, may occur in any of these stages, prolonging life and enabling the insect to bypass seasons that are unsuitable for growth and reproduction.

The Lepidoptera is one of the two or three largest orders of insects, with an estimated 160,000 named species. Based on specimens in collections and extrapolating from recent studies of Central American moths, we believe that fewer than one-half of the known species have been named by

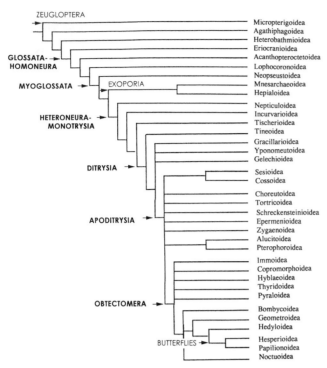

FIGURE 1 Hypothesis of phylogenetic relationships of extant lepidopteran superfamilies. Successively more derived clades representing major morphological changes are indicated in boldface to the left (modified from Kristensen and Skalski, 1999).

taxonomists; even in North America, an estimated one-third of the fauna is undescribed. Thus, a realistic projection of the total world Lepidoptera species number is not possible, but certainly it exceeds 350,000 and may be much larger. Much of this diversity can be attributed to the radiation of species in association with flowering plants. Lepidoptera represent the single most diverse lineage of organisms to have evolved primarily dependent upon angiosperm plants, and their numbers exceed those of the other major plant-feeding insects, Heteroptera, Homoptera, and Coleoptera (Chrysomeloidea and Curculionoidea). Figure 1 depicts the hypothesized evolutionary lineages and lists currently recognized superfamilies of Lepidoptera.

MORPHOLOGY

Adult

The body framework (Fig 2) consists of a hardened (sclerotized) exoskeleton made up of a head capsule with appendages; three fused thoracic segments, each with legs, and two pairs of wings, on the middle (mesothoracic) and third (metathoracic) segments; and an abdomen, which has 10 segments, is less sclerotized than the thorax, and is movable by intersegmental membranes. Complex genital structures of external origin arise from abdominal segments A8–10, and often there are accessory structures (pouches, glands, hair brushes) associated with sound reception, courtship, or other functions.

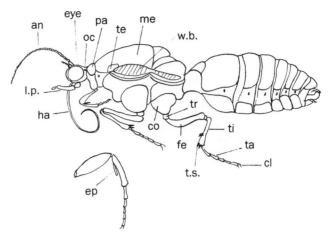

FIGURE 2 Schematic representation of the exoskeletal anatomy of a ditrysian moth, with prothoracic leg enlarged below. Head: an, antenna; eye, compound eye; oc, ocellus; l.p., labial palpus; ha, haustellum (proboscis); Thorax: pa, patagium; te, tegula; me, mesoscutum; w.b., wing base; co, coxa; tr, trochanter; fe, femur; ti, tibia; t.s., tibial spurs; ta, tarsomeres; cl, tarsal claws; ep, epiphysis. Abdomen: tergites and sternites 1–7 and spiracles shown.

HEAD   Structures include paired simple eyes (ocelli) and scaleless, raised spots (chaetosema), which are unique to Lepidoptera, although one or both are lost in many taxa (Figs. 2–4). There is enormous variation in the form of the antennae, often between the sexes of a species, being filiform or with the flagellar segments variously enlarged or branched. Antennae of butterflies are enlarged distally, forming apical clubs, while those of moths are not, although some moths have distally enlarged antennae that are tapered or hooked to the tip. The mouthparts of the most primitive moth families retain functional mandibles as in their mecopteroid ancestors, but in the majority of moths the mandibles are lost, and the maxillary galeae are elongate and joined to form a tubular

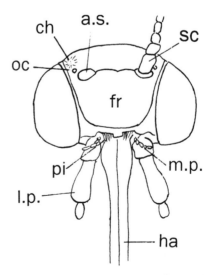

FIGURE 3 Descaled lepidopteran head, frontal aspect. ch, chaetosema; oc, ocellus; a.s., antennal socket; sc, scape; fr, frons; pi, pilifer; m.p., maxillary palpus; l.p., labial palpus; ha, haustellum, consisting of fused galeae.

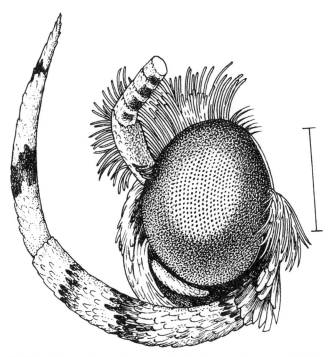

FIGURE 4 Head of ethmiid moth, showing the strongly upcurved labial palpus that is characteristic of most Gelechioidea. Scale bar = 1.0 mm.

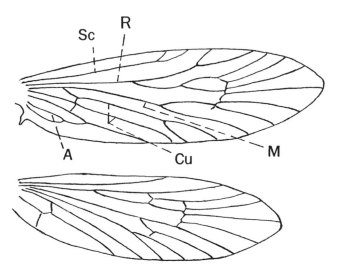

FIGURE 5 Wing venation of a homoneurous moth (Eriocraniidae). Vein systems: Sc, subcostal; R, radial; M, medial; Cu, cubital; A, anal.

proboscis (haustellum) with musculature that enables it to be coiled under the head when not being used to suck nectar from flowers or other fluids into the digestive tract by a pumping action. The maxillary palpi consist of one to five segments and in primitive moths are conspicuous, often folded. The labial palpi are more prominent in most Lepidoptera and vary in curvature and length, but they are not folded.

THORAX The pro-, meso-, and metathorax are fused, each consisting of a series of nonmovable sclerites (Fig. 2). In primitive groups the meso- and metathorax and their wings are similar in size, but in derived families the mesothorax is larger and has more powerful musculature, and the forewing has more rigid vein structure, especially on the leading edge. In the largest superfamily, Noctuoidea, the metathorax is modified posteriorly into a pair of tympanal organs. The tibia of the foreleg has an articulated epiphysis on the inner surface, a uniquely derived feature in Lepidoptera, usually with a comb of stout setae, that is used to clean the antennae and proboscis by drawing them through the gap between the comb and the tibia. The wings are tiny and soft at eclosion from the pupa, then rapidly expand by circulation of blood pumped into the flaccid veins, causing them to extend, stretching the wing membranes to full size, after which they rapidly harden, with the membranes pressed closely together, and the system of tubular veins provides structure. Homologies of the six vein systems are discernible across all families of Lepidoptera, and the configuration of veins has been used extensively in classi-fication. In the most primitive moths the fore wing (FW) and

hind wing (HW) are similar in shape and wing venation (homoneurous) (Fig. 5), while the more derived groups have lost parts of the vein systems and have fewer remaining in the HW than in the FW (heteroneurous) (Fig. 6). There are various wing-coupling mechanisms by which the FW and HW are linked to facilitate flight. Primitive homoneurous moths have an enlarged lobe at the base of the FW (jugum) that folds under the HW when the insect is at rest but extends over the HW in flight, which does not couple the wings efficiently. Most moths have the HW frenulum that hooks under the FW retinaculum, the development of which varies among taxa and between the sexes of many species.

In a few groups (e.g., Psychidae, Lymantriidae) females of many species are flightless, having very reduced wings (brachypterous), or are apterous and may not even shed the

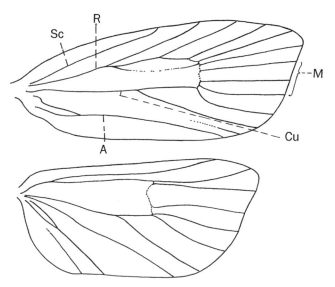

FIGURE 6 Wing venation of a heteroneurous moth (Tortricidae). Abbreviations as in Fig. 5.

pupal skin. Brachyptery has evolved many times independently, such as in high montane and winter-active species of various families in Europe, North America, and Australia. Both sexes are flightless in species of several families on remote southern oceanic islands and in one species of Scythrididae that occurs only on windswept coastal sand dunes in California.

ABDOMEN The abdomen has segments A7–10 or A8–10 modified to form external parts of the genitalia; the sternum of A1 in homoneurous families is small and is lost in other Lepidoptera. Articulation of the thorax and abdomen in derived families is accomplished by musculature attached to sclerotized struts (apodemes) that project from abdominal sternite 2. There are paired tympanal organs at the base of the abdomen in Pyraloidea and Geometroidea. Various male glandular organs associated with courtship occur on the abdomen in several families. Usually these are developed as expandable hair brushes or tufts, or as thin-walled, eversible sacs (coremata), from the intersegmental membrane at the base of the genitalia or on other segments.

The genitalia of Lepidoptera are highly complex and provide the basis for taxonomic species discrimination in most families and often generic or family-defining characteristics. In the male (Fig. 7) the valvae, which are thought to provide clasping stability during mating, usually are large, more or less covering the other structures in respose, and usually are densely setate on the inner surface, scaled exteriorly, and the most

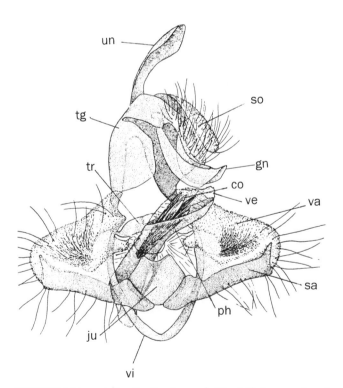

FIGURE 7 Male genitalia of a ditrysian moth (Tortricidae), venterolateral aspect with valvae reflexed. un, uncus; tg, tegumen; so, socii; gn, gnathos; tr, transtilla; ju, juxta; va, valva; sa, sacculus; vi, vinculum; ph, phallus (aedeagus); ve, vesica; co, cornuti.

visible part of the genitalia externally. The phallus, which is separately articulated and passes through the diaphragma, is sclerotized and contains the membranous vesica, the intromittent organ. The vesica often is armed with cornuti, which sometimes are deciduous and deposited in the female. Sperm are produced in paired testes and pass through a duct leading to the vesica and are deposited in a spermatophore produced by the male accessory glands during mating. The precise functions of most of the external, sclerotized parts of the genitalia are unknown, and they vary independently in form, being uniform in some taxa, variable in others, and thus of differing taxonomic value from one taxon to another.

In the female there are three fundamental types of genitalia. Primitive moths possess a single genital aperture near the posterior end of the abdomen, through which both copulation and oviposition occur (monotrysian). Other Lepidoptera have separate apertures for copulation and oviposition; Hepialidae and related families are exoporian (i.e., the spermatozoa are conveyed from the gonopore, or ostium bursae, to the ovipore via an external groove). All remaining families are ditrysian (i.e., having internal ducts that carry the sperm from the copulatory tract to oviduct) (Fig. 8). This feature defines the Ditrysia, comprising most of the superfamilies and more than 98% of the species. The papillae anales typically are soft and covered with sensory setae but in many taxa are modified for various kinds of oviposition, such as piercing. Both the ductus and the corpus bursae are variously modified in different taxa, the corpus often with one or more thorn-like sclerotized signa that may aid in retaining the spermatophore. Sperm are transported from the corpus bursae through the ductus seminalis to the bulla seminalis and ultimately to the oviduct. The musculature that controls the ovipositor and papillae anales, often involving extension and telescoping the abdomen, as well as the copulatory aperture, is inserted on the posterior and anterior apophyses.

INTERNAL ANATOMY Lepidoptera possess the same fundamental internal systems for breathing, blood circulation, digestion, excretion, central nerves, and endocrine functions as do other holometabolous insects (see relevant articles).

Egg

With few exceptions, female Lepidoptera produce eggs that are deposited externally after fertilization in the oviduct (Figs. 9 and 10). Moth and butterfly eggs vary enormously in size, shape, surface sculpture, and arrangement during oviposition. Within lineages such as families, larger species produce larger eggs, but depending upon the family, the sizes and numbers differ greatly. For example, females of hepialids, including some of the largest moths in the world, produce vast numbers of tiny eggs (20,000–30,000 or more by a single female) that are broadcast in the habitat. Conversely some small moths and butterflies produce few, relatively large eggs.

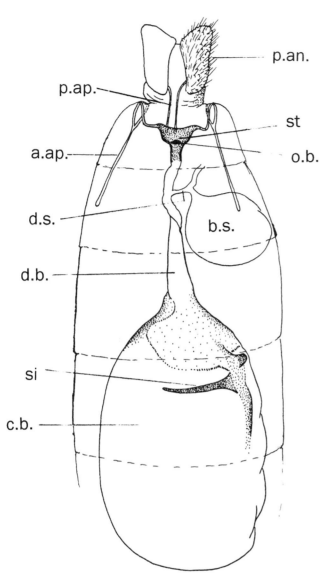

FIGURE 8 Female genitalia of a ditrysian moth (Tortricidae), ventral aspect; broken lines represent segments of abdominal pelt. p.an., papilla anale; p.ap., posterior apophysis; a.ap, anterior apophysis; st, sterigma; o.b., ostium bursae; d.b., ductus bursae; c.b., corpus bursae; si, signum; d.s., ductus seminalis; b.s., bulla seminalis.

FIGURE 9 Shells of the flat type ditrysian moth eggs (*Amorbia,* Tortricidae), which in this instance are deposited overlapping, in regularly arranged imbricate masses (photograph by A. Blaker).

deposited shingle-like, with the micropylar ends protruding partway over the preceding row (Fig. 9), while upright eggs are arranged side by side, like rows of miniature barrels (Fig. 10). Usually the eggs are glued to the substrate by a secretion of the female accessory (colleterial) glands, applied within the oviduct, sometimes forming a thick, paint-like covering to egg masses. Eggs may be covered with debris collected by the female or hairs or scales from her abdomen or wings or may be surrounded by fences of upright scales, but lepidopteran eggs are not tended or guarded by the adults.

Embryonic development is related to temperature, proceeding more rapidly under warmer conditions, but the rate is physiologically and hormonally controlled in many instances. It requires 7 to 14 days in most Lepidoptera but may be

The shell (chorion) is soft during development and quickly hardens after oviposition, assuming a regular form consistent for the species and often characteristic for genera or families. The chorion may be smooth or strengthened by raised longitudinal ribs or transverse ridges or both. At one end there is a tiny pore (micropyle), through which the sperm enters, surrounded by a rosette of radiating lines or ridges. Two types of egg form are defined, those laid horizontally, with the micropyle at one end, which are usually more or less flat, and those that are upright, with the micropyle at the top. Flat eggs are prevalent in the more ancestral lineages, microlepidoptera, while most derived groups, larger moths and butterflies, have upright eggs with more rigid and ornamented chorion. Eggs of either type are laid singly or in groups; flat eggs are sometimes

FIGURE 10 Eggs of the upright type of a ditrysian moth (Arctiidae) (photograph by R. Coville).

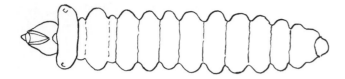

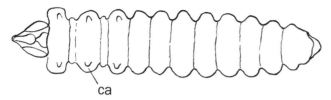

FIGURE 11 Flattened body form of a leaf mining larva (Tischeriidae), dorsal aspect above, ventral below. ca, ambulatory calli that represent vestigial remnants of the thoracic legs.

delayed for many weeks or months in species that overwinter in the egg stage.

Larva

The head (Figs. 12 and 14) is sclerotized, usually rounded (flattened in leaf-mining species, Fig. 11), with large lateral lobes, each bearing an ellipse of usually six simple eyes (stemmata) ventrolaterally and systematically arranged primary setae and are joined by a median suture, which is flanked by two narrow adfrontal sclerites. The mouthparts may be directed downward (hypognathous) or forward (prognathous). The labium is weak but carries a spinneret behind the mouthparts ventrally, which distributes the silk produced by modified salivary glands. The thorax has spiracles on the meso- and metathoracic segments, except in some aquatic pyraloids that have external gills. The abdomen usually has spiracles on segments 1 to 8, restricted to segments 1 to 3 or absent in some aquatic pyraloids. There are paired, ventral, fleshy, and nonsegmented leglike organs on all segments in the most primitive moths, while on others they are restricted to segments 3 to 6 (ventral prolegs) and 10 (anal prolegs), equipped with circles or bands of tiny hooks (crotchets) that aid in grasping and walking. The prolegs are fewer in Geometridae

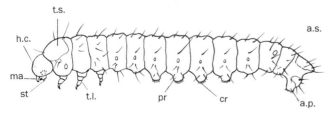

FIGURE 12 Typical form of a ditrysian caterpillar (Cossidae), lateral aspect. h.c., head capsule; ma, mandible; st, spinneret; t.s., thoracic shield; t.l., thoracic leg; sp, spiracle; pr, abdominal proleg; a.s., anal shield; a.pr., anal proleg; cr, crotchets.

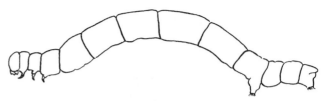

FIGURE 13 Body form of Geometridae larva (inchworm), lateral aspect, lacking prolegs on abdominal segments 1–5.

(Fig. 13) and some other groups and are lost in some borers (e.g., Prodoxidae), leaf miners (e.g., Eriocraniidae, Nepticulidae), and sand-dwelling larvae (a few Noctuidae). In some groups, A10 has a musculated anal fork used to flip frass away from the larval shelter.

There are sensory setae on the head and body integument, and the homology of their primary arrangements (chaetotaxy) (Fig. 15) can be compared in all but the few most primitive families. Their patterns have been valuable to understanding evolutionary trends and to identification of larvae, although the primary arrangement is lost or replaced by numerous secondary setae in many taxa, at least in later instars. The adfrontal sutures, arrangement of stemmata, and crotchet-bearing abdominal prolegs distinguish Lepidoptera from other insect larvae.

Pupa

The head, thorax, and abdomen of the pupa resemble those of the adult and can be recognized externally (Fig. 16). The mandibles of the most primitive families are functional and

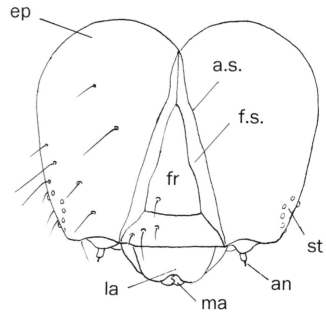

FIGURE 14 Schematic representation of the head capsule of a larval ditrysian moth, frontal aspect. ep, epicranial lobe; st, stemmata; a.s., adfrontal suture; f.s., frontal suture; fr, frons; la, labrum; ma, mandible; an, antenna.

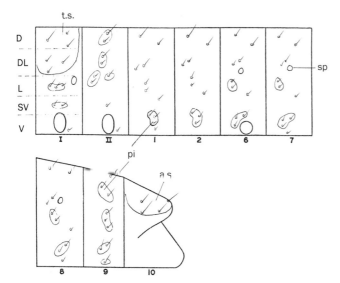

FIGURE 15 Chaetotaxy (setal map) of a larval ditrysian moth (Tortricidae); each rectangle represents one body segment from mid dorsum (upper border) to mid venter (lower border). I, II, pro- and mesothoracic segments; 1, 2, etc., abdominal segments. Setal groups: D, dorsal; DL, dorsolateral; L, lateral; SV, subventral; V, ventral; t.s., thoracic shield; a.s., anal shield; sp, spiracle; pi, pinacula, which are raised and often pigmented.

used to cut open the cocoon preceding eclosion of the adult. In other moths the head is sometimes provided with a beak or other armature that assists in the eclosion process. The appendages of the head and thorax are each encased in cuticle and in most Lepidoptera are fused to the venter of the body, with the wing cases wrapped around, adjacent to the antennae and mouthparts. Abdominal segments 7 to 10 are fused. In the more ancestral families some of the other

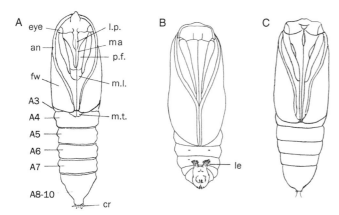

FIGURE 16 Pupae of ditrysian moths, ventral aspect. (A) Tortricidae, with abdominal segments 4–7 movable, enabling pupal movement forward at emergence. (B) Ethmiidae, with pupal movement restricted to flexible segments 5–6, and the pupa remains in place at emergence, a characteristic of Gelechioidea. (C) Noctuidae (Obtectomera) with all segments immobile. l.p., labial palpus; ma, maxilla including galeae (haustellum); p.f., prothoracic femur; m.l., mesothoracic leg; m.t., metathoracic tarsus; an, antenna; fw, forewing; A3–10, abdominal segments 3–10; cr, cremaster; le, leglike extensions of the 9th abdominal segment bearing hooked setae that anchor the pupa in lieu of a cremaster (A, C redrawn from Mosher, 1916).

segments are movable (Fig. 16A), usually provided with backwardly directed spines or spurs, and the pupa wriggles forward to protrude from the cocoon or burrow just before moth eclosion. Gelechioidea and derived moths (Obtectomera, Fig. 1) and butterflies are obtect, with fused abdominal segments (Figs. 16B and 16C). They remain in place, and adult eclosion occurs along a silken track or other means prepared by the larva or directly from the pupa, in butterflies and some moth groups that do not spin cocoons. Many species have a cremaster, hooked setae at the tip of the abdomen that anchor the pupa inside the cocoon or at the terminus of a silk emergence track, enabling pressure from the emerging adult to break the pupal shell. Others lack the cremaster but are held within a tight cocoon, in an earthen cell, or by a silk girdle. The integument is soft, smooth, and green or whitish when first formed but soon hardens and turns brown in most Lepidoptera. Those that pupate exposed, including butterflies, Pterophoridae, and some Gelechioidea, are mottled green or brownish and often have prominent spines or ridges that aid in camouflage.

BIOLOGY

Success of Lepidoptera populations is dependent upon several factors in the climatic and biotic environment, interrelated with the insects' behavior. First, larval foods, and for most species adult nourishment, must be available. Climatic conditions suitable for mating and oviposition, larval feeding, and pupation are necessary. Females must find appropriate places for deposition of eggs. Larvae must sense proper foods, eat, molt, grow, and pupate. Pupae need to avoid desiccation and other factors that might prevent successful adult eclosion. Finally, egg, larval, and pupal parasites and predators have to combine to take all but two of the offspring of each female (whose eggs may number 200–600 or more) that survive physical dangers, but on average they cannot exceed that, in order to maintain stable population levels.

Adult Behavior

Males usually begin emergence and peak in numbers a few days ahead of females. Both are sexually mature upon eclosion, and males of nearly all moths are attracted by chemical signals (pheromones) emitted by "calling" females. Hence, in most Lepidoptera mating takes place soon after female eclosion, and she has mature eggs ready to be fertilized and deposited within the first 24 h. Mate-seeking involves primarily visual cues in most butterflies, although there may be short-range pheromones produced by one or both sexes that mitigate courtship. Males, and females too in most species, mate more than once. It is assumed that sperm precedence prevails, wherein the most recent male's sperm is effective.

Adults of both sexes of most Lepidoptera feed and in confinement die quickly if water is not available. Feeding on honey-enriched fluids extends the life of some moths and increases fecundity. Most macromoths and butterflies feed at flowers, imbibing nectar, whereas most micromoths do not and apparently gain nourishment from extrafloral nectaries, sap flows, and honeydew secreted by aphids or other Homoptera. Exceptions occur in diurnal microlepidoptera (e.g., Adelidae, Sesiidae, Heliodinidae, Scythrididae, Plutellidae, and Tortricidae, but not nocturnal species of the latter three families), which visit flowers, often other than the larval hosts. The mouthparts are nonfunctional in a few families (e.g., Lasiocampidae, Lymantriidae) and in specialized species such as winter-active Geometridae and Ethmiidae, and females possess mature eggs upon eclosion.

Host-plant selection is made primarily by the female, which seeks by chemical and tactile cues the proper substrate or habitat for oviposition. This choice is made by instinct, inherited genetically, and the newly hatched larvae also require specific stimuli, detected by chemoreceptors on the antennae and mouthparts; in host-specific species, they starve if the proper plant is not available, ignoring plants or synthetic diets that are quite acceptable and sufficient for nourishment of generalist species.

Most butterflies and moths live only a few days, until mating and egg laying are accomplished, but some are active for several weeks, or they may overwinter as adults and become active on warm days. Some adult microlepidoptera enter a prereproductive state lasting through summer and winter, followed by mating and oviposition in early spring.

Larval Development

The newly formed larva, or caterpillar, first bites its way out of the eggshell, leaving a crescentic slit or ragged hole at the micropylar end. Some species then eat the reminder of the eggshell. All growth takes place during the larval stages, so caterpillars consume enough nutrients to carry through cocoon formation, pupation, and metamorphosis to the adult. It must be sufficient for the moth or butterfly to move to its first feeding or, in species with nonfeeding adults, enough to provide for complete egg development of the next generation. To accommodate growth, the larva molts its skin (cuticle) several times, through successively larger stages (instars). Most Lepidoptera undergo five or six instars, but many larvae that feed on detritus or dry plant material undergo indeterminate numbers of instars.

Silk is produced by paired labial glands. It is composed of two proteins secreted in a viscous fluid in two strands, which consolidate as they leave the spinneret and contact the air. Its functions are many: first instars of many species are dispersed by air currents on silk strands; many or most species lay down a silk line as they move, enabling them to cling to substrates; silk is used by most external-feeding micromoths to form shelters in foliage or other food sources, and some construct portable cases from which they feed; others line tunnels with silk in fruits, stems, roots, or soil from which they forage to feed. Finally, silk is used in cocoon formation

FIGURES 17–22 Leaf mines. (17) *Stigmella variella* (Nepticulidae) on *Quercus agrifolia;* (18) mature larvae of *Coptodisca arbutiella* (Heliozelidae, Incurvarioidea) and their abandoned mines, on *Arctostaphylos;* (19) *Cameraria gaultheriella* (Gracillariidae) on *Gaultheria shallon;* (20) *Marmara arbutiella* (Gracillariidae) on *Arbutus menziesii;* (21) *Phyllocnistis populiella* (Phyllocnistidae, Gracillarioidea) on *Populus tremuloides;* (22) *Epinotia nigralbana* (Tortricidae) on *Arctostaphylos* (photographs by J. Powell, all California, except Fig. 21, Alaska).

preceding pupation, within the larval shelter or gallery or separately, sometimes as a characteristically shaped structure.

Larval habits vary widely and often are quite specific for a family, genus, or species. These include leaf mining, in which a larva spends it entire life within a leaf, and the depth and form of the mines are consistent such that the moth family or genus often is recognizable from the mine (Figs. 17–22). Other types of internal feeding include stem mining; boring in seeds, stems, and roots (Figs. 25 and 26); or feeding in galls developed by plants, stimulated by the larvae (Figs. 27 and 28). Many external-feeding caterpillars avoid adverse conditions by seeking shelter in leaf litter at the base of the plant or in tunnels during the day and emerge at night to feed, when temperatures are cooler, humidity is higher, and diurnal predators are not active. Many macromoth and butterfly larvae remain exposed, motionless, protected by cryptic coloration, body form, and behavior (Figs. 29–32), or even camouflaged by a coat of flower bits or debris that collect on hooked body setae. Larvae of a few genera live gregariously in silken tents that shield them from climatic extremes (Fig. 33). Many others are protected from vertebrate predators by toxic chemicals they sequester, and advertize their presence by bright colors (aposematic) (Fig. 34).

The duration of larval development varies greatly with the feeding and life cycle types, even within families and genera. The time required to reach maturity also is dependent upon temperature within species, such as between seasonal generations. Most Lepidoptera grow slowly in early instars,

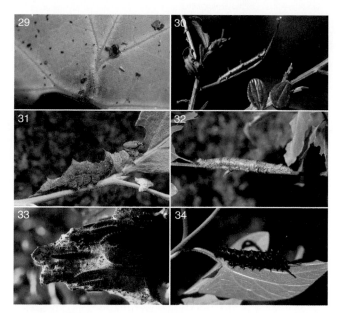

FIGURES 23–28 Case-bearers, borers, and gall inducers. (23) *Thyridopteryx meadii* (Psychidae, Tineoidea), case on *Larrea tridentata;* (24) *Coleophora* species (Coleophoridae, Gelechioidea) on *Malus;* (25) larva of *Synanthedon sequoiae* (Sesiidae, Sesioidea) under bark of a conifer; (26) larva of *Grapholita edwardsiana* (Tortricidae) in stem of *Lupinus arboreus;* (27) stem galls induced by *Gnorimoschema baccharisella* (Gelechiidae) on *Baccharis pilularis;* (28) stem galls caused by *Epiblema rudei* (Tortricidae), with newly emerged moth and its pupal shell on *Gutierrezia* (photographs by J. Powell, except 24, 25 by R. Coville, all California).

FIGURES 29–34 Cryptic and aposematic caterpillars. (29) *Oidaematophorus* species (Pterophroidae) on *Petasites palmatus;* (30) stick-like larva of *Sicya macularia* (Geometridae) on *Ceanothus thyrsiflorus;* (31) *Schizura unicornis* (Notodontidae, Noctuoidea) on unidentified tree; (32) *Catocala* species (Noctuidae) on *Quercus kelloggii;* (33) tent caterpillars, *Malacosoma californicum* (Lasiocampidae, Bombycoidea), on *Quercus agrifolia;* (34) *Battus philenor* (Papilionidae) on *Aristolochia californica* (photographs by J. Powell, except 31 by R. Coville, 32 by D. Wagner, all California except 31, British Columbia).

increasing in size much more rapidly in later instars, particularly the last. Growth after eclosion from the egg to maturity usually takes 30 to 50 days, but sometimes is more rapid, as few as 18 or 19 days. Larval life can extend much longer, particularly in species that enter quiescent phases at lower temperatures, intermittently feeding when warmer, or in detritus-feeders, which can simply wait long periods when food is not suitable. Such species may live 100 to 140 days before pupation, and those that enter obligate diapause, usually first or last instar, typically spend 9 or 10 months as inactive larvae in addition to their feeding and growth period.

Larval Foods

The nutritional requirements of many caterpillars are generally similar. Synthetic diets that contain the same basic elements, casein, sucrose, salt, cellulose, wheat germ, amino acids, and vitamins, incorporated in an agar base, are successfully used for rearing many kinds of Lepidoptera. However, sometimes species that are specific to particular plants do not accept a synthetic diet. Hence, nutritional value alone may not be sufficient to elicit feeding, and natural plant chemicals act either as cues for feeding or as deterrents, often the same chemical in both roles with different larval species.

The majority of Lepidoptera caterpillars are phytophagous, consuming living plants, almost exclusively flowering plants,

and primarily angiosperms. All parts of plants are eaten, each kind of caterpillar specializing on its particular niche, leaves, flowers, fruit, stems, or roots. Some species feed internally (endophagous) as leafminers and seed or root borers, others externally (exophagous), either concealed in shelters constructed with silk or exposed. Larvae of the most primitive family, Micropterigidae, consume liverworts and mosses or are general feeders on green plants, fern sporangia, or fungal spores in moist habitats. Some other groups of moths do not feed on flowering plants (e.g., Tineidae), but specialize on wood-rot fungi (Polyporaceae) or are detritivores on the ground, under bark of dead tree limbs, or in abandoned insect and spider nests or feed on animal products in mammal burrows, bird nests, or scats, and a few can digest wool. Many species feed on fallen leaves, notably Oecophoridae and Tortricidae on *Eucalyptus* (Myrtaceae) in Australia, and several groups of Noctuidae in wet forest habitats. Some Lepidoptera specialize on lichens (lithosiine Arctiidae, some Psychidae and Xylorictidae), mosses (some Crambidae), or ferns (unrelated species, mainly on oceanic islands). A few Lepidoptera are predaceous on scale insects or other Homoptera or in ant nests. A Hawaiian geometrid moth *(Eupithecia)* is predaceous on adult flies, which it catches by seizing the fly with elongate prolegs. Other members of the worldwide genus *Eupithecia* are plant feeders.

Virtually every kind of flowering plant is eaten by one or more species of caterpillar. Food preferences vary enormously among families; they are summarized in the accounts of the major families that follow. Nearly all internal feeders, such as leafminers, stem and root borers, and gall inducers, and most other microlepidoptera are specialists on one or a few related plants, whereas perhaps half or more of external-feeding macro-moth species are generalists within habitats, such as ground-dwelling cutworms feeding on low-growing herbaceous plants or shrub- and tree-feeding species. Most butterfly species are specialists.

Pupal Development

The duration of pupation during which metamorphosis to the adult occurs varies with temperature, usually requiring about 10 to 12 days, but many species require several weeks or hibernate as pupae, often for 10 months or more.

Pupal movement is an important adaptation in primitive moths and basal Ditrysia. The pupa moves forward just preceding adult eclosion and either anchors by the cremaster to silk or wedges in the emergence aperture, which is prepared by the larva to be slightly narrower than the pupal abdomen. This movement is aided by rows of dorsal, backwardly projecting spines. Gelechioidea and the Obtectomera (Figs. 1, 16B, 16C) have independently derived fusion of abdominal segments that restricts movement, enabling turning within the cocoon but not forward movement, and the adult emerges directly from the pupation site. Pupae respond to tactile stimuli, including potential predators and probing by a parasitoid wasp ovipositor, by turning or wriggling. Some moth pupae have special structures on the abdomen that produce clicking or rattling sounds when the wriggling abdomen strikes the walls of the pupal cells or parchment-like cocoon, or sounds are produced by rubbing fine pegs or rasp-like surfaces on adjacent segments. Such sounds may aid in pupal defense.

Life Cycle

Most Lepidoptera in temperate climates undergo a single annual generation (univoltine), although many have two discrete seasonal broods (bivoltine), and some produce continuous generations as long as favorable temperature conditions prevail (multivoltine). Diapause, a state of arrested development regulated by hormones, controls the life-cycle pattern and enables populations to survive during unfavorable times (winter, dry season, etc.) when necessary resources are not available. Diapause may be the single most important adaptation leading to species radiation of Lepidoptera in northern climates and high mountains, in the world's deserts and tropical dry season habitats, and in other places where insects could not grow and reproduce continuously. In Lepidoptera, diapause occurs primarily in eggs, in first or last instars, in pupae, or as a reproductive delay in adults, depending on the species. In Mediterranean climates, larval feeding typically occurs in spring when foliation peaks, and diapause lasts through the dry season in summer and hibernation in winter. Some species aestivate in diapause as prepupal larvae or pupae, fly

in autumn, and then hibernate as adults or eggs. Multivoltine species enter diapause at the end of the growing season, often triggered by decreasing day length, or the larvae simply wait in a quiescent state, feeding slowly on warm days through winter, and metamorphose, and adults eclose with warmer temperature in spring.

Most tropical Lepidoptera are too poorly documented to estimate the proportion of multivoltine to other life-cycle patterns. Some species migrate from wet regions to dry forest habitats at the beginning of the rainy season to take advantage of the newly available resources, but others undergo diapause through the dry season.

Many Lepidoptera are capable of maintaining the diapause to a second or later season if appropriate climatic conditions do not occur. This happens as a regular phenomenon in species adapted to seed feeding on plants with biennial crops such as conifers or sporadically in species that depend upon resources that are limited to a specific season but are erratic in abundance, such as flowering and fruiting by desert plants. Numerous prepupal larvae of yucca moths (Prodoxidae) have metamorphosed synchronously after 8 to 30 years in diapause under experimental conditions.

SIGNIFICANCE IN NATURAL AND HUMAN COMMUNITIES

The major role of Lepidoptera in natural communities is primary consumer of plants. Moths and butterflies make up the largest single evolutionary lineage adapted to depend upon living plants, in terms of species numbers and, in many communities, in biomass as well. Females of most species produce 200 to 600 eggs within a few days, vastly more in some species (1000–30,000), releasing a potentially enormous load of caterpillars onto particular plant species or plant groups such as herbs or woody shrubs and trees. Therefore, an important food resource is available for specialized parasitoid wasps and flies, general invertebrate predators such as spiders, mites, ants, and social wasps, and vertebrate predators, especially birds. There have been estimates of 80,000 caterpillars of several species feeding on a single oak tree and many times that number during outbreaks of single species that defoliate forest trees. Thus caterpillars comprise a major component of biological communities, affecting foraging by birds, buildup of yellow-jacket colonies, and insect disease epidemics. A secondary role as decomposers also is filled by Lepidoptera. Tineidae, several groups of Gelechioidea (particularly Oecophoridae in Australia), and some Noctuidae and other moths are detritivores and assist in reducing fallen leaves and fruit, fungi, and animal products (hair, feathers, predator scats) to humus. Finally, a few species are secondary consumers, predaceous on scale insects or other Homoptera in natural communities.

Lepidoptera larvae damage plants grown for human use (food, lumber, cotton, garden ornamentals) and our stored products (grain, flour, nuts, woolen clothes and carpets). Most agricultural damage occurs because monoculture crops are grown in places distant from the natural enemies of the pest species, which themselves usually have been introduced by human activities to a new region. Wide-scale insecticide suppression of pest species has further increased problems because local parasites and invertebrate predators are eliminated, and the pest species become resistant to the insecticides by selection for survivors of repeated treatments. Similarly, pests of stored food and wool products have been transported worldwide by human activities. Lepidoptera probably are the most important insect group as plant defoliators (e.g., spruce budworm, the economically most important insect in Canada; larch budworm in Europe) and they cause huge losses by damage to fruits (e.g., codling moth, the "worm" in apples), corn (corn earworm, European corn borer), potatoes (potato tuberworm), cotton (pink bollworm), and many other crops and garden plants. They are a major problem in stored meal, grain, and nuts (Angoumois grain moth, Indian meal moth, Mediterranean flour moth) and woolen products (casemaking clothes moth, webbing clothes moth, tapestry moth, and others). Still others infest bee nests, eating the combs (greater and lesser wax moths).

Conversely, some moths are believed to play significant roles in pollination in natural communities, especially Sphingidae and Noctuidae, and they may aid in crop pollination in some instances. Several Lepidoptera have been purposefully introduced to act as biological control agents against noxious plants. Notable examples include a pyralid, the cactus moth, from Argentina used to successfully suppress millions of acres of introduced prickly pear cactus in Australia; an arctiid, the cinnabar moth from Europe, on tansy ragwort in the Pacific states of North America; and several Mexican species against lantana in Hawaii.

FOSSIL RECORD AND EVOLUTION

A widely accepted phylogenetic hypothesis of relationships among lepidopteran evolutionary lineages, based on morphological characteristics in living forms, primarily of the adults, is shown in Fig. 1. The problem in such analysis is that we do not know what kinds of species might have preceded and interceded with the primitive extant lineages, each of which is now represented by one or a few relict genera that have divergent larval features not shared with other Lepidoptera. Moreover, the fossil record is of little use in revealing clues to "missing links," and the preservation usually fails to provide information on critical characteristics, particularly those of the larvae and pupae.

Fossil Record

There are fossils of Triassic age assigned to Trichoptera (caddisflies), the presumed sister group of Lepidoptera, and so branching of the two lineages could have occurred in the early Mesozoic (Fig. 35). The earliest fossil recognized as lepidopteran is a small scaled wing from the Lower Jurassic of

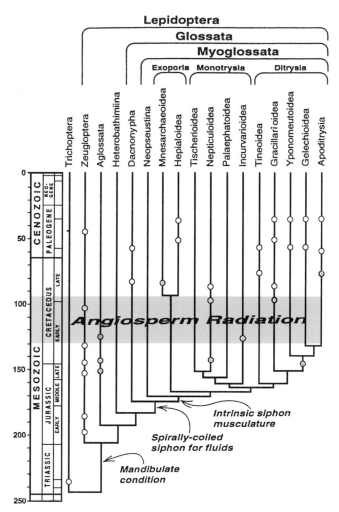

FIGURE 35 Phylogenetic hypothesis of major lepidopteran lineages superimposed on the geologic time scale, with fossil occurrences indicated. Open dots, reliable identifications; shaded dots, questionable assignments. Angiosperm radiation spans 130 to 95 mya from the earliest recognized occurrence of pollen to the time when angiosperms became the dominant vegetation (modified from Labandeira *et al.,* 1994).

Dorset, England. It was placed in a separate family, Archaeolepidae, suggested as a sister group to the Micropterigidae, but without characters known that might establish its relationships. Four genera were described from Upper Jurassic tuffites from Russia. Among these, two were assigned to Micropterigidae and two to Glossata and Ditrysia, but only one of them, *Protolepis,* possesses visible mouthpart structures. They were interpreted as a siphon formed of maxillary galeae, which would imply existence of Glossata, 20 to 30 mya, prior to the radiation of angiosperm plants during the early Cretaceous. That interpretation has been questioned, the structures possibly being maxillary palpi, and therefore the fossil may represent an extinct lineage of Aglossata. By the early Cretaceous there are well preserved Micropterigidae and an incurvariid (Heteroneura) in amber, and by the late Cretaceous several kinds of leaf mines representing modern families and host plant associations, both heteroneuran (Nepticulidae) and ditrysian (Phyllocnistidae, Gracillariidae), as well as a ditrysian larval head capsule of a free-living form such as Tineidae. That is, the fundamental clades of Lepidoptera are all represented before the beginning of the Tertiary. Hence, although Lepidoptera is the most recently evolved major insect order, its radiation was relatively rapid, paralleling that of the angiosperms, the major lineages having evolved between ca. 140 and 90 mya.

Morphological Evolution

Major changes in morphological adaptation in adult feeding, oviposition mode, wing structure, and larval locomotion are indicated by Figs. 1 and 35. The relict moths of ancient lineages (Micropterigidae, Agathiphagidae, Heterobathmiidae) share features of ancestral mecopteroids, functional mandibles in adults and pupae, similar fore- and hind wings with complete venation, and a single female genital aperture. However, larvae of their extant species differ greatly from one another, each adapted for a particular life-style. Micropterigid larvae are free-living ground dwellers in moist environments, with well-developed thoracic legs, no crotchet-bearing abdominal prolegs, and fluid-filled chambers in the cuticle. Agathiphagids are legless borers in primitive gymnosperm seeds with reduced head sclerotization and sutures and few stemmata. Heterobathmiids are flattened leafminers of southern beech, having a prognathous head with prominent adfrontal ridges, as well as seven stemmata laterally and thoracic legs with large, subdivided trochanters (unique in Lepidoptera), but no abdominal prolegs.

Adult Glossata (Eriocraniidae and all subsequent lineages) lack functional mandibles and feed by a proboscis formed of the maxillary galeae. Basal glossatan lineages have a piercing ovipositor and retain functional mandibles in the pupa, used to cut the cocoon at eclosion. The larvae have a spinneret. Several derived features occur beginning with the Exoporia (Mnesarchaeidae and Hepialidae): The ovipore and gonopore are separate, connected by an external groove for sperm transfer; the larvae have differentiated prolegs on abdominal segments 3 to 6 and 10, with circles of crotchets; and silk is used for various activities, not just cocoon formation, the ancestral condition in Lepidoptera. Functional pupal mandibles are lost and there is no piercing ovipositor. Differentiated size, shape, and venation between fore- and hind wings appear in the Heteroneura. The thoracic legs, crotchet-bearing larval prolegs, and silk webbing are lost by larvae of Nepticuloidea, which are severely modified for leaf mining. An independently derived piercing ovipositor occurs in Incurvarioidea, some of which have secondarily legless larvae.

The last fundamental change, leading to the Ditrysia, is the internal system for storage and transfer of sperm from the gonopore to oviduct. Evidently this had evolved by the mid-Cretaceous, when larval mines of Gracillarioidea appear in the fossil record. The most successful lineages, in terms of extant diversity, Pyraloidea, Geometroidea, and Noctuoidea, which are defined by independently derived tympanal organs, presumably originated coincident with radiation of the bats during the late Paleocene and early Eocene. The earliest butterfly fossils also date from late Paleocene–Eocene times.

Ecological Scenario

Questions remain concerning the origins of angiosperm feeding in basal lepidopteran lineages that led to major radiations of Lepidoptera. The ground-dwelling larvae of Micropterigidae are generalists, either detritivores or fungivores in leaf litter or feeding on low-growing green plants in moist habitats, including bryophytes and soft angiosperm leaves. Similar habits occur in Exoporia (Mnesarchaeidae and Hepialidae, except that many hepialids feed on roots or burrow into stems of woody angiosperms) and in basal Ditrysia (Tineidae, except that none feeds on green plants). By contrast, extant larvae of the other lower Lepidoptera are endophagous feeders that specialize on particular flowering plants (larvae of Lophocoronidae and Neopseustidae are unknown, but their ovipositor types indicate that at least early instars are internal feeders). We assume ground-dwelling, generalist habits are similar to those of mecopteroid ancestors of the Trichoptera–Lepidoptera clade, but we do not know if that mode of life persisted in basal members of all lineages through to the Ditrysia. If so, adaptation to endophagy and to specialist angiosperm feeding might have occurred at least four times, in heterobathmiids, in an eriocraniid + acanthopteroctetid + lophocoronid + neopseustid lineage, in nepticuloids, and, probably independently, in incurvarioids, when a piercing ovipositor reappears, and finally in a palaephatid + tischeriid lineage. If an unknown angiosperm-feeding lineage was the common ancestor, at least two reversals to ground-dwelling, external-feeding, generalist caterpillars characterized by multiple morphological reversals must be postulated for exoporians and again for Tineidae. In either scenario, there were independent origins of a piercing ovipositor (at least twice) and endophagous larval feeding accompanied by numerous derived morphological specializations in larvae (several times). Repeated shifts to angiosperm feeding (Fig. 36) may have been facultative, as it is in extant micropterigids, and multiple adaptations to endophagy imply parallel evolutionary trends, a more parsimonious scenario than multiple reversals to an ancestral morphological and behavioral ground plan.

CLASSIFICATION

Historically the Lepidoptera have been classified in four or five suborders, all but one of which are primitive moths that retain ancestral characteristics as relict, morphologically dissimilar groups. All the more derived moths and butterflies, more than 98% of the described species, comprise one evolutionary lineage, or clade, the Ditrysia. In recent decades, much progress has been made in detailed analyses of the relationships of the

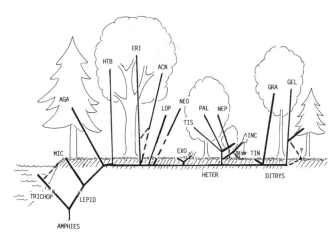

FIGURE 36 Cartoon representing a theoretical scenario of the origins of angiosperm feeding that led to the radiation of Lepidopteran during the Cretaceous. Ground-dwelling mecopteroid-like ancestor gave rise to the Trichoptera-Lepidoptera split, then successively to the ancestor of extant Micropterigidae (MIC) and several specialized, radically differing, angiosperm-feeding lineages. The ancestral ground dwelling caterpillar form is presumed to have been retained in Exoporia (EXO, Mnesarchaeoidea, Hepialoidea) and basal Ditrysia (Tineoidea). AGA, Agathiphagoidea; HTB, Heterobathmioidea; ERI, Eriocranioidea; ACN, Acanthopteroctetoidea; LOP, Lophocoronoidea; NEO, Neopseustoidea; HETER, Heteroneura; TIS, Tischerioidea; PAL, Palaephatoidea; NEP, Nepticuloidea; INC, Incurvarioidea; DITRYS, Ditrysia; TIN, Tineoidea; GRA, Gracillarioidea; GEL, Gelechioidea.

primitive groups, aided by discoveries of new taxa and previously unknown larvae and pupae. Phylogenetic analyses have shown the primitive lineages to be paraphyletic with respect to the rest of the Lepidoptera (Fig. 1), and consequently, the use of suborders and other ranks between order and superfamily has been abandoned by lepidopterists. On the other hand, we continue to recognize the obligate categories (family, genus, species) for purposes of names and communication across related lineages. Historically, the family has been the common denominator level for communication among entomologists, including for Lepidoptera, but in recent decades there has been a proliferation of both family and superfamily divisions such that the superfamily has become a commonly used and understood rank for lepidopterists. Recent authors have treated more than 120 families of Lepidoptera, and there is considerable discrepancy between analyses within some of the 45 to 48 superfamilies. Morphological and biological traits of the larger, worldwide superfamilies and families are summarized in the text that follows.

Primitive Lineages

ZEUGLOPTERA—MICROPTERIGOIDEA Micropterigidae are the most primitive lepidopterans, living fossils. There are micropterigids recognizable as modern genera preserved in amber dating back to dinosaur times in the early Cretaceous (125 mya). Adults (Fig. 37) are small (FW length 3–6 mm), often colorful, with metallic sheens of bronze or purple and yellow forewing markings, usually active in the daytime. They

are characterized by numerous ancestral traits not shared by other moths, most notably retention of functional mandibles, which are used to feed on pollen of various trees in Europe, more primitive plants, sedges, Winteraceae, and fern spores in New Caledonia and Madagascar. A more complete, Mecoptera-like wing venation led to proposal of this group as a separate order, the Zeugloptera, but overall evidence indicates the combined Zeugloptera + other Lepidoptera as a sister group to the caddisflies (Trichoptera). The larvae are wholly unlike caterpillars of other Lepidoptera; they are plump, somewhat hexagonal in cross section, with long antennae and short thoracic legs, and they lack the abdominal prolegs with crotchets typical of most Lepidoptera. The larvae live in moist leaf litter among mosses or in rotting wood, habitats with high moisture conditions; the cuticle has specializations unique among arthropods, with exo- and endocuticle separated by a fluid-filled space leading via pores to chambers in the exocuticle, overlaid by sticky pellicle to which particles of debris adhere. The pattern of primary setae on the body is unlike that of other moth larvae. Larvae of some species feed on liverworts, but most micropterigids are generalists, feeding on detritus, fungal hyphae, or angiosperm leaves. About 120 species are known worldwide, in a disjunct, relictual distribution pattern. More than half the named species are in the genus *Micropteryx* in the Palaearctic region, while only 2 are known in North America (*Epimartyria*); there is a greater diversity of genera in the Orient and southwest Pacific, particularly New Zealand, eastern Australia, and New Caledonia, which has about 50 species.

There are two other families of Aglossata: the Agathiphagidae (two species), caddisfly-like moths whose larvae are legless borers in the seed of primitive gymnoserms (*Agathis*) in Australia and south Pacific Islands, and Heterobathmiidae (nine species), which are similar moths to micropterigids but their larvae are leafminers in southern beech (*Nothofagus*) in Chile and Argentina.

GLOSSATA—HOMONEURA The majority of Lepidoptera comprise the Glossata, the monophyly of which is well supported by a suite of derived characters. The most obvious traits that distinguish glosssatans are the adult mouthparts: the mandibles are nonfunctional and maxillary galeae elongated, forming a proboscis that is coiled in repose, accompanied by reduction of the head capsule and its cuticular thickening associated with mandibular musculature. The basal lineages retain ancestral features of the wings: similarly shaped fore- and hind wings with relatively complete venation (homoneurous) and the jugal lobe at the base of the forewing. Females in these families have a flattened, sclerotized abdominal apex with serrate edges, forming a "saw," which is everted to cut into host-plant leaves to deposit the eggs.

Eriocranioidea Eriocraniidae form a Holarctic counterpart to the South American Heterobathmiidae, resembling them superficially as adults and larvae, mining primarily in birch and oak (Fagales) in early spring. Adults are small moths (FW length 4–6.5 mm) with relatively narrow wings covered by

FIGURES 37–64 Adults and larvae of microlepidoptera. Micropterigoidea: (37) *Epimartyria pardella* (Micropterigidae) (California). Incurvarioidea: (38) *Coptodisca arbutiella* (Heliozelidae) ovipositing into leaf of *Arbutus menziesii* (California); (39) *Adela septentrionella* (Adelidae) ovipositing into buds of *Holodiscus discolor* (California); (40) *Greya reticulata* (Prodoxidae), ovipositing into bud of *Sanicula* (California); (41) *Tegeticula maculata* (Prodoxidae) ovipositing into ovary of *Yucca whipplei* (California). Tineoidea: (42) *Tinea pellionella* (Tineidae) (Texas); (43) Larval cases of *Tinea pellionella* on wool fabric (Texas). Gracillarioidea: (44) *Caloptilia reticulata* (Gracillariidae) (California). Yponomeutoidea: (45) *Atteva punctella* (Yponomeutidae) nectaring (Illinois); (46) *Ypsolopha maculatella* (Plutellidae) nectaring at flower of Asteraceae, whereas the larval host is *Ephedra* (California). Gelechioidea: (47) *Antaeotricha* species (Stenomatidae) (Illinois); (48) *Ethmia arctostaphylella* (Ethmiidae), bird dropping-like resting posture on *Eriodictyon*, the larval host (California); (49) Larva of *Ethmia delliella* (Ethmiidae), which feeds on *Cordia* (Costa Rica); (50) *Arotrura longissima* (Scythrididae) nectaring at flowers of *Senecio*, whereas the larval host is *Lycium* (California); (51) *Esperia sulphurella* (Oecophoridae, Oecophorinae) (California); (52) *Callimima lophoptera* (Oecophorinae) (Australia); (53) *Coleophora* species (Coleophoridae) (California); (54) *Holcocera* species (Blastobasidae) (California); (55) *Telphusa latifasciella* (Gelechiidae) (Illinois). Choreutoidea: (56) *Tebenna gemmalis* (Choreutidae) nectaring at flowers of *Achillea*, whereas the larval host is *Wyethia* (California). Sesioidea: (57) *Synanthedon sequoiae* (Sesiidae) (California); (58) *Castnia* species (Castniidae) (French Guiana). Cossoidea: (59) *Acossus* species (Cossidae) (California). Tortricoidea: (60) *Argyrotaenia citrana* (Tortricidae, Tortricinae) (California); (61) *Synnoma lynosyrana* (Tortricidae, Tortricinae), flightless female in calling posture on *Chrysothamnus*, the larval host plant (California); (62) *Pseudatteria leopardana* (Tortricidae, Chlidanotinae), a diurnal and presumed distasteful species (Costa Rica). Alucitoidea: (63) *Alucita* species (Alucitidae) (Colorado). Pterophoroidea: (64) Platyptilinae species (Pterophoridae) (Costa Rica). (Photographs by: I. Common, 52; C. Covell, 58; R. Coville, 45, 47, 51, 53, 54, 55, 56, 57, 60, 64; H. Daly, 50; J. Hafernik, 42, 43, 59; P. Opler, 62, 63; J. Powell, 38, 39, 40, 41, 44, 46, 48, 49, 61; D. Wagner, 37).

iridescent, simple scales and hairs, often golden with purplish markings. Most are diurnal and fly in early spring just as the host trees are beginning to leaf out. The larvae are legless miners in newly expanded leaves, forming "baggy" full-depth mines. They mature quickly and enter the soil for pupation, and the mines dry and deteriorate after the leaf hardens. The pupae are mandibulate, and the emerging pharate adult uses the mandibles to cut through the cocoon and reach the soil surface the following spring. Larval foods are birch (Betulaceae), oak (Fagaceae), and other Fagales, or Rosaceae (1 species). There are about 20 species assigned to five genera, with about half the species in Europe and Asia and half in North America

A related family, Acanthopteroctetidae, with two species in the western United States, formerly was included in the Eriocranioidea, has been given superfamily status based on its more derived type of scales and first thoracic spiracle. The larvae of one species are miners in Rhamnaceae.

EXOPORIA Within the homoneurous Glossata, two superfamilies comprise the Exoporia, the Mnesarchaeoidea (15 species), a relict group of small, eriocraniid-like moths in New Zealand, and the worldwide Hepialoidea. Monophyly of the two is established by the unique configuration of the female genital system, which is shared by and interpreted as homologous in these otherwise quite dissimilar moths. The copulatory orifice is separate from the ovipore, but there is no internal connection between the two. Sperm is transferred via a groove in the body wall below the ovipore.

Hepialoidea The Hepialoidea is the most successful of the Homoneura and more primitive lineages in terms of extant diversity. The superfamily is characterized by having reduced mouthparts, with the proboscis absent or short and evidently nonfunctional. Hepialidae are large moths, even enormous in some genera, well represented on all nonpolar continents. Four other hepialoid families, Anomosetidae, Neotheoridae, Palaeosetidae, and Prototheoridae, are Southern Hemisphere relicts represented by one to a few species and are smaller moths.

Adults of the Hepialidae are large to very large (FW 10–120 mm), including some of the largest Lepidoptera in the world, *Trichophassus* in South America and *Zelotypia* in Australia, with a 10-in. wing span, and often beautifully colored in greens and pinks. The females carry enormous numbers of eggs—one female of *Trictena* in Australia laid 29,000 eggs and had another 15,000 in her ovaries when dissected—and therefore are bulky, heavy-bodied creatures, surpassing in weight the largest sphingid and saturniid moths. Hepialid males form groups, or leks, that fly together at dusk, as a ritual of courtship behavior; especially suggestive of the common name "ghost moths" is one European species that has white forewings and forms ghost-like clouds. The larvae are elongate, cylindrical, with fully developed thoracic legs and abdominal prolegs that bear rings of crotchets. They have primary setae distributed in patterns that are homologous with those of ditrysian larvae, and they lack secondary setae. Hepialid larvae are concealed feeders, living in silken galleries in leaf litter and

grasslands, in tunnels in roots or trunks, feeding indiscriminately on pteridophyte, gymnosperm, or angiosperm plants. Early instars of some species feed on decaying wood and fungi associated with it, and then bore into tree trunks in later instars. There are about 550 named species in 50+ genera worldwide, best developed in Australia and Africa.

HETERONEURA—MONOTRYSIA All the remaining Lepidoptera have different fore- and hindwing shapes and venation, with reduced radial system in the hind wing, and the hind wing usually is smaller. They possess a frenulum–retinaculum wing-coupling mechanism, and they have lost the first abdominal sternite. The five most basal superfamilies of the Heteroneura retain the ancestral monotrysian female reproductive system, but they share no derived characteristics that would unite them as monophyletic. The three most diverse of these are Nepticuloidea, Incurvarioidea, and Tischerioidea.

Nepticuloidea These are tiny moths whose larvae are leaf and stem miners. Specialization on diverse flowering plants has led this group to become the most speciose of the primitive Lepidoptera.

Nepticulidae: Adult nepticulids include the smallest Lepidoptera (FW length 1.5 to 4.5 mm), characterized by having the basal antennal segment (scape) usually greatly enlarged, forming a cap over the upper half of the relatively large eye. The head is rough-scaled, and the mouthparts are primitive, with long, folded maxillary palpi, and rudimentary proboscis with galeae not joined, used to lap up moisture and honeydew secreted by aphids. The FW is relatively broad with long scale fringes. The larvae are legless, obligate leafminers, typically forming a serpentine track beginning just below the egg cemented to the leaf surface, gradually enlarging to a full-depth tube or irregular blotch (Fig. 17). At maturity the larva cuts a crescentic slit in the upper cuticle and drops to the ground to form a tough silken cocoon. Larval foods usually are mature leaves of woody plants, although a few larvae mine stems or cause petiole galls. Individual species are specialists, using more than 40 families of angiosperms, primarily Fagaceae and Rosaceae in the Holarctic; some species groups are specialists on one plant family, such as Anacardiaceae, Polygonaceae, or Fabaceae. There are nearly 800 described species, placed in 11 genera, occurring in all nonpolar regions. No accurate estimate is available, but the total named includes fewer than 10% of the species in tropical regions and probably less than half the North American species.

A related family, Opostegidae (100+ species), are slightly larger moths (FW 1.8 to 8.3 mm), with enormous eye caps, completely obscuring the eyes from a frontal view. The FW is relatively broad, white with sparse black markings, and the apices often are strongly bent upward. The larvae are leafminers in Rutaceae and cambium miners in stems and fruit of Betulaceae, Ranunculaceae, Polygonaceae, Saxifragaceae in the Holarctic, and Fagaceae in Chile.

Tischerioidea Tischeriidae adults are very small (FW 2.7 to 5 mm), with lanceolate wings of white, gray, or yellow. They are nocturnal with large eyes and when at rest they perch with the head appressed to the substrate and tail end lifted at a 45° angle. The larvae are slightly flattened leafminers with thoracic legs reduced to two vestigial segments or ambulatory calli, abdominal prolegs rudimentary with crotchets (Fig. 11). The linear or blotch mines are characterized by a heavily silk-lined nest within which the larva retreats when not feeding; the mines have been recorded from nine angiosperm families, most commonly Fagaceae, Rosaceae, and Asteraceae, more diverse on the last than is true of other lepidopterous miners. This group is primarily Holarctic, with about 80 described species in one genus, a few of which are in the Neotropical, Ethiopian, and Indo-Malayan regions, none in Australia and Oceania.

Incurvarioidea   These are tiny to small moths having diverse biologies, but females all have a piercing ovipositor specialized for inserting the eggs into plant tissue, often the ovules or young seed. There are six families, Cecidosidae (seven species, gall inducers in South America and Africa), Crinopterigidae (one Mediterranean species, a larval case bearer with habits similar to some coleophorids), and four that are diverse and widespread: Heliozelidae, Incurvariidae, Adelidae, and Prodoxidae.

Heliozelidae: Species of this family occur worldwide but because of their minute size and diurnal habits they are rarely seen and many more species are known from the characteristic abandoned larval mines than there are named. Adults (Fig. 38) are tiny to small (FW 1.7 to 7.0 mm), typically with iridescent, metallic-appearing scaling. The eyes are small, characteristic of diurnal microlepidoptera. The larvae are flattened, usually legless leaf miners, having a thorax with paired ventral and dorsal movable calli; abdominal prolegs are absent. Early instars form a short, serpentine mine, and then enlarge it to a full-depth blotch (Fig. 18). The last instar constructs a portable case by cutting lenticular disks from the upper and lower epidermis and joining them with silk, giving rise to the common name "shield-bearers"; the abandoned mines, with their distinctive "shot holes" are highly characteristic of heliozelids (Fig. 18). The larva crawls off and descends by a silken thread to attach to a lower leaf or bark, where pupation occurs in the portable case. Heliozelids are host specific, using at least 17 families of usually woody angiosperms, with preponderance in Myrtaceae in Australia and Cornaceae and Vitaceae in the Holarctic, the only Lepidoptera to specialize on the latter. There are more than 100 described species in about 12 genera, distributed in all major faunal realms except New Zealand but poorly known in tropical regions.

Incurvariidae: Adults are small (FW 3.5 to 9 mm), with rough head scaling, relatively small eyes, and proboscis short, half the palpi length; maxillary palpi are elongate and folded. They are somber moths with dark, monochromatic wings, sometimes iridescent brown, bronze, or bluish. The larvae are moderately flattened, with well-developed thoracic legs and reduced abdominal prolegs. Early instars form blotch mines; later the larvae cut through the upper and lower epidermis to remove oval sections, which they sew together to form a portable case. Larvae of a few genera remain in the mines throughout feeding, and then cut out a case in which they pupate. Oviposition is host specific; the ancestral, southern-continent genera use Myrtaceae or Proteaceae, while Holarctic incurvariids use about 10 unrelated angiosperm families. There are about 100 described species in 11 genera, mostly Australian and Palaearctic; they are poorly represented in Africa and the Western Hemisphere.

Adelidae (Fig. 39) are best known for their enormously long antennae, often 2.5 or 3 times the forewing length. Usually they are much longer in the male, which in many species possesses greatly enlarged eyes, but the eyes are small in some species, irrespective of antennal length. Holarctic and Neotropical species (Adelinae) are small (FW 4.5–9 mm), diurnal moths, often brightly colored, iridescent green, blue, or purplish, with white antennae, while the primarily African Nematopogoninae are crepuscular or nocturnal and dull colored. Both sexes have a well-developed proboscis and seek nectar from various flowers other than the larval food plant. Males of the large-eyed species form small, dancing groups, reacting to one another during mate seeking. Females insert the eggs into the base of the ovaries of unopened flowers. First instars of the few species studied in detail feed in the developing ovules; after molting they drop to the ground and construct flat, portable cases from silk and debris and feed on fallen leaves or the lower leaves of the host plant, which often are short-lived annuals. Pupation the following spring occurs in the figure 8-shaped case, with the long antennae free and coiled several times around the abdomen. Oviposition is restricted to one or a few closely related plants, which include members of at least 18 angiosperm families. Biologies of the Nematopogoninae of the Southern Hemisphere are poorly known. There are more than 300 species in five genera, occurring in all faunal regions except New Zealand.

Prodoxidae are famous for the close symbiotic relationship between species of *Tegeticula* and yucca plants (Agavaceae). Females possess enormous "tentacles," appendages of the maxillary palpi, which are unique among all insects, used to gather pollen that is purposefully transferred to the stigmas while visiting other flowers for oviposition, thus ensuring cross-pollination. Other kinds of insects are not attracted to yucca flowers to collect pollen. Females are believed to leave a pheromone signal at the oviposition sites that deters later visiting females so that only a few larvae feed in any given seed pod and many unaffected seeds are produced. Adults (Figs. 40 and 41) are small (FW 4 to 16 mm), generally dull colored, white or gray, although a few *Greya* and *Prodoxus* species have patterned or iridescent bronze-colored forewings. The maxillary and labial palpi are relatively prominent but usually shorter than the proboscis. The Agavaceae-feeding prodoxines apparently do not seek nectar, although individuals of *Tegeticula maculata* have been recorded living up to 9 days in the field. Early instars of the more ancestral genera (*Lampronia, Greya*) feed in young ovules of the host plant and then leave to spin overwintering shelters on the ground.

In early spring they feed in flower beds or foliage shoots of the newly foliating host plants. These caterpillars and those of the pollen-carrying genera are stout, highly mobile, with well-developed thoracic legs, lacking abdominal prolegs, while those of the bogus-yucca moths *(Prodoxus)* are completely legless and apparently blind, living their entire life within the gallery and pupating there. Prepupal larvae of yucca moths are capable of maintaining the diapause for several years if optimal winter conditions are not experienced, up to 30 years followed by successful, seasonally synchronized development, in experimental trials. Holarctic Lamproniinae and species of the basal prodoxine genus *Greya* specialize on Rosaceae, Ericaceae, or Saxifragaceae, while the more derived prodoxines are Agavaceae specialists. Prodoxidae are predominately Holarctic, with Lamproniinae mainly Palaearctic and Prodoxinae largely Nearctic, with a few species ranging into southern Mexico. About 75 species in 10 genera are known.

Ditrysia

The Ditrysia includes 98% or more of the described species, most of the superfamilies and families, almost all of the external plant-feeding caterpillars, and most of the special adaptations for prey avoidance. All members possess reproductive systems based on separate female copulatory and oviposition orifices with internal ducts for transfer of the sperm.

TINEOIDEA   The tineoids are generally recognized as the most ancestral living group of the Ditrysia. Most tineoids have erect, roughened head scaling and elongate, five-segmented maxillary palpi that are folded, usually longer than the labial palpi, which have lateral bristles, while the haustellum has short, unconnected galeae, used to lap up surface moisture from detritus or fungi. Females of most species possess elongate apophyses of segments A9 and A10 that anchor musculature, enabling the ovipositor to be telescoped outward to inject the eggs into crevices or other niches in the habitat. Five families are regarded as comprising the superfamily, two of which are worldwide and more species rich, Tineidae and Psychidae. The others are smaller families of restricted distribution, Eriocottidae (70+ species) in the Mediterranean region and southern Africa to Australia and Taiwan, Acrolophidae (280 species) in the Neotropical and Nearctic regions, and Arrhenophanidae (30 species), Neotropical.

Tineidae (Fig. 42) are slender, small to moderately large moths (FW length 2.5–25 mm), usually shining brown, tan, or whitish, with FW patterns of black on pale or yellow on dark. Tineids are most easily recognized by the rough head vesture and the short (or absent) proboscis. They lack bipectination of the male antennae, characteristic of other tineoid families. The larvae are slender with integument usually lacking color pattern, often living within silken tubes or portable cases (Fig. 43). All instars have well-developed thoracic legs and abdominal prolegs with a single circle of crotchets. Larval foods—Tineids do not feed on flowering plants; they mostly are generalist detritivores or fungivores, and members of some

subfamilies tend to be specialists on animal products such as fur or feathers (e.g., Tineinae). Some are capable of digesting wool, including several cosmopolitan species that feed on woolen clothes and other manmade products. Others are primarily fungus-feeders (e.g., Scardiinae, Nemapogoninae), especially in sporophores of wood-rot fungi (Polyporaceae) or wood permeated by the hyphae, sometimes quite specialized in host preference. Fungivory or detrivory presumably was the ground plan for the family and therefore for the Ditrysia. Larvipary, wherein eggs mature and first instars emerge within an enlarged oviduct in the female, is known in numerous Andean and Indo-Australian Tineinae. Many fewer eggs are produced than by tineids with conventional reproductive systems. There are more than 3000 described species worldwide, probably less than half the number known in collections, especially in tropical regions. These are assigned to more than 300 genera in 15 subfamilies.

Psychidae: The common name "bagworms" applies to psychids because the larvae live in portable cases constructed from silk, plastered with debris or symmetrically arranged pieces of host plants (Fig. 23). Adults are small and slender to rather large and heavy bodied (FW length 4–28 mm). Males are fully winged, while females of some species may be fully winged, short-winged, wingless, or even larviform and never leave the larval case. Some species are female only (parthenogenetic) or bisexual only in some populations. The head vesture is roughened, with long, slender scales directed forward, and the antennae often are strongly bipectinate in males, particularly in species with flightless females, but are filiform in both sexes of species having winged females. Nearly all psychids are gray or brown without color patterns. Psychid larvae are stout compared to tineids, with the head and thorax larger and more heavily sclerotized than the posteriorly tapered abdomen and variously pigmented. The thoracic legs are well developed and are used to pull the cases along on the host plant, while the abdominal prolegs are reduced. Bagworms feed on lichens, grasses, conifer foliage, or leaves of angiosperm trees and shrubs, sometimes as specialists but often as generalists. At maturity the larva attaches the case to a substrate and then inverts itself and pupates in the case with the head toward the distal (older) end, whence the moth emerges. There are nearly 1000 described species from all faunal regions, about 85% in the Old World. Psychids are generally better studied than most microlepidopterans, owing to their fascinating behavior, biologies, and genetic complexity associated with the larviform females and parthenogenesis in five unrelated genera.

GRACILLARIOIDEA   This is the major clade of Lepidoptera adapted for larval mining in leaves (Figs. 19–21). Gracillarioids primarily mine woody trees, shrubs, and vines of angiosperms and conifers. The larvae are obligate leaf-, stem-, or fruit-miners in early instars; in many genera larvae leave the mines to feed exposed or in webs externally. The adults lack the tineoid lateral bristles of the labial palpi, have a smoothly scaled frons, and usually have a well-developed, elongate, coiled proboscis.

Gracillariidae: Adults (Fig. 44) are nocturnal, often are brightly colored, with the FW patterned in metallic orange, bronze, purple, or yellow. They are small, slender moths (FW 2–10 mm) with a head with smooth scaling directed forward over the front and tufts of erect scales on the crown in Lithocolletinae; the antennae are 0.8 times to much longer than the FW and filiform. The HW is lanceolate with scale fringe broader than the wing. The larvae characteristically are hypermetamorphic with more than one form in successive instars; early instars are modified for mining, flattened with legs reduced or lacking, transforming in the third or later instar to cylindrical caterpillars, with chewing mouthparts and fully developed legs. They use silk to buckle the mine into a tent-like shelter or feed externally, often folding a leaf into a tightly closed shelter in which they graze. In some genera a variously modified, nonfeeding instar spins the cocoon. Pupation occurs outside the mine in most genera. Nearly all gracillariids are specialists on one or a few closely related plants, typically woody angiosperms, including more than 80 plant families. More than 2000 species have been described from all major faunal regions, assigned to about 75 genera, and certainly a much greater number remain to be defined, especially in tropical forests.

Phyllocnistidae: Adults are tiny, slender moths (FW 2–3 mm) with long antennae, often with shining white or silvery FW, delicately banded with gray and rust distally; HW are lanceolate with a much broader fringe. The larvae are flat with legs reduced to stubs, mouthparts highly modified for sap feeding; they create extremely long, meandering or regularly zig-zag, subcutaneous mines (Fig. 21), often in new, still-soft leaves, causing them to curl conspicuously. More than 20 angiosperm families have been recorded as larval hosts, but many others in Central America are used, judging from the ubiquitous mines, probably all made by undescribed species. Phyllocnistid mines are described from mid-Cretaceous (97 mya) Magnoliidae, the earliest known ditrysian leaf mining. Fewer than 100 species are described, a small fragment of the fauna; mines are found on more kinds of plants at one lowland forest locality in Costa Rica than there are named New World species.

Bucculatricidae: Adults are tiny to small moths (FW 2.5–7 mm), most easily recognized by their elongate frons and large, erect tuft of scales on the vertex. The appendages are short, antennae 0.6 to 0.9 times the FW length. The wings are lanceolate, FW often with tufts of upraised scales. Larvae are hypermetamorphic; they are legless leaf miners in the first two instars and later have well-developed legs, feeding externally as exposed grazers. Bucculatricid species are host-plant specialists, with more than 20 angiosperm plant families recorded, Asteraceae and Fagaceae dominant in the Holarctic; many species use Cupressaceae. There are about 250 described species, distributed on all continents except New Zealand, most numerous in the Holarctic.

YPONOMEUTOIDEA   This superfamily includes a heterogeneous conglomeration of dissimilar microlepidopterans that are grouped by default, i.e., the nonapoditrysian Ditrysia that have nonmotile pupae and lack the scaled proboscis typical of Gelechioidea.

Yponomeutidae: Adults are slender moths with elongate FW, ranging from tiny (FW 3.2–6.8 mm), metallic golden, purple, or gray and white nocturnal *Argyresthia* to larger (FW 9–15 mm) and brightly colored, diurnal moths in *Atteva* (Fig. 45) and the white ermine moths in *Yponomeuta*. Larvae of typical yponomeutines live communally in extensive webs on trees and shrubs, sometimes causing economic damage to fruit trees; those of some *Zelleria* damage growing tips of conifers. Larvae of *Argyresthia* are miners in angiosperm buds or conifer foliage. Most species are specific to one or a few plants; Argyresthiinae feed on at least 13 gymnosperm and dicot families, with more than 40% of recorded species on conifers, including 25% on Cupressaceae, a greater degree of adaptation to conifers than by other Lepidoptera. There are about 600 described species, occurring in all biotic regions, with both the phylogeny and the taxonomy in tropical and south temperate faunas yet to be resolved.

Plutellidae and Ypsolophidae traditionally were treated as one family. The adults (Fig. 46) are small moths (FW 6–13 mm) with distinctive labial palpi, with the second segment broadly scaled and the third slender, smooth scaled and upcurved from the second preapically. FW are narrow to lanceolate, with a flared terminal fringe. Large pleural lobes enclose the male genitalia in ypsolophids but are small and narrow in plutellids. These are typically nocturnal moths with yellow, brown, or gray FW, often with linear markings. The diamondback moth *(Plutella xylostella)*, a ubiquitous pest of cabbage, cauliflower, and other plants of the mustard family, is the best known plutellid. The larvae are slender, tapered toward both ends, often with elongate abdominal prolegs, and pale green with unpatterned integument, and they live in slight webs as external feeders. Pupation occurs in large-meshed, open cocoons (*Plutella* group) or dense, envelope-like cocoons (*Ypsolopha* group). Most *Plutella* feed on Brassicaceae, the only moth lineage adapted to do so, while members of this family group as a whole use 50+ families of angiosperms, rarely monocots, and a few gymnosperms, including Ephedraceae and Cupressaceae. Nearly 300 described species are assigned to one or the other of these families, but the systematics relationships and descriptive inventory are incompletely known, especially in Southern Hemisphere faunas.

Glyphipterigidae are small (FW usually 3.2–10 mm), diurnal moths with a smooth-scaled head and porrect or decumbent, slightly upcurved labial palpi, often metallic gray; the FW have metallic markings and parallel, white chevron marks from costa and dorsal margin. The last abdominal tergum is greatly enlarged, forming a hood over the genitalia. The larvae are borers in seed, flowering stems, terminal buds, or leaves, primarily in sedges and rushes (Cyperaceae, Juncaceae), less commonly in grasses, and rarely in dicots, including Crassulaceae. Nearly 400 species are described in 20+ genera. The family is cosmopolitan and well represented in temperate, Palaearctic, Nearctic, and Australian regions, including New Zealand.

Heliodinidae: Adults are tiny, diurnal moths (FW mostly 3.2–5 mm; rarely to 8 mm), resplendent in shining metallic body and wing scaling, often red or orange with raised silver or lead-colored spots. Adults of many species hold the hind legs aloft when perched, which has been regarded as characteristic of heliodinids, but not all of them do so. Larvae are unpigmented, grub-like, and host specific as leafminers or stem or seed borers, or a few feed externally in flowers and fruit. This is the only lepidopteran family to specialize on Caryophyllales (90% of the known hosts), especially Nyctaginaceae. About 75 species are described worldwide, but the majority occur in the southwestern United States and Mexico.

GELECHIOIDEA This is the largest superfamily of micromoths by far, and because vast numbers of species remain undescribed, Gelechioidea may surpass Noctuoidea as the most diverse group of Lepidoptera. For example, even in North America only about one-third of the species known in collections are named, and a total of 4400 species is projected, 1000 more than Noctuoidea. In tropical regions the inventory is imponderably incomplete—more than 1100 species of Gelechioidea have been counted at one rainforest reserve in Costa Rica, comprising 20% of all Lepidoptera believed to occur there, the majority unnamed. In Australia, gelechioids are estimated to make up about 40% of the Lepidoptera species.

Gelechioids all have overlapping scales on the dorsal surface of the haustellum, up to half its length, and most have a smooth-scaled head, four-segmented maxillary palpi, and upward curved labial palpi with the third segment long and acute (Figs. 4 and 55). The pupa is nonmotile (obtect), remaining in the cocoon until emergence of the moth. There is general agreement on the phylogenetic unity of this superfamily, but there have been wide differences of opinion on the number and relationships of the included families in recent analyses. About 25 groups have been treated as families, of which only the larger, worldwide ones are mentioned here.

Elachistid Assemblage This group of taxa is defined by having modified abdominal articulation in the pupa, lateral condyles, on A5–A6 and A6–A7, that prevent lateral movement (Fig. 16B), although those of Elachistidae s. str. are polymorphic and questionably homologous. The first three groups have been treated as families or subfamilies of Oecophoridae or Elachistidae. They are broad-winged moths with strongly curved labial palpi, often exceeding the top of the head (e.g., Fig. 4).

Stenomatidae: Adults (Fig. 47) are small to moderate sized (FW 5–25 mm), with rectangular to nearly oval FW. The valvae of male genitalia have setae with prominent, multilobed apices. The male antennae usually have long cilia. Larvae of only a small proportion of described species are known; they are relatively stout caterpillars, often with heavily pigmented integument. They are external feeders in concealed shelters on diverse angiosperms (16+ families), predominately Myrtaceae in the Southern Hemisphere and Fagales in the Nearctic. More than 30 genera and 1200 species are named, 90% from the Neotropical region, where many are not yet described. This group and the Ethmiidae tend to be mutually exclusive on a broad geographical scale. Stenomatids are species rich in the southeastern United States and lowland wet forests of central America and northern South America, while ethmiids are speciose in arid parts of western North America and thorn forest regions of Mexico, Central America, and the Antilles.

Ethmiidae: Adults (Fig. 48) are small moths (FW mostly 4–16 mm, rarely 24 mm), having elongate, narrow FW, often dark with a sinuate pale band along the dorsal edge that renders a bird-dropping appearance when the moth rests, or white with black spotting, superficially resembling *Yponomeuta,* and some tropical species are colorful. Most are nocturnal, but some high montane species are diurnal, as are a group of species in the southwestern Nearctic that fly in early spring, adapted to use annual plants. The family is characterized by a strongly recurved basal part of the phallus, secondary SV setae of the larva, and two separately derived pupal-anchoring mechanisms, either development of "anal legs," ventral, setiferous, forward-directed extensions of the ninth segment (Fig. 16B) or grasping of the exuvial head capsule or cocoon silk between abdominal segments 6 and 7. Diapause occurs in the pupal stage, and development can be delayed several years, an adaptation to unpredictable, arid habitats. About 80% of the species for which the often colorful larvae (Fig. 49) are known feed on Boraginales (Boraginaceae, Ehretiaceae, Hydrophyllaceae). About 300 species are named worldwide, with the greatest richness in areas of seasonal drought, especially microphyllous thorn forests of the northern Neotropical Region. This group is better studied than most micromoths, excepting the African fauna, and most species in collections are described.

Depressariidae: Adults are small (FW 7–16 mm) and resemble the two preceding groups, with FW usually rectangular and labial palpi slender and strongly curved upward, lacking the multilobed setae of the valvae, the strongly recurved phallobase, and special pupal anchoring mechanisms that characterize stenomatids and ethmiids. The larvae mostly are leaf tiers but some bore into stems or seed, using at least 17 families of dicots, with strong specialization on Apiaceae and Asteraceae in the Holarctic. More than 600 species are described, assigned to 80+ genera, occurring in all major faunal realms and best represented in north temperate and tropical regions.

Elachistidae: Adults are tiny (FW 2.5–6.5 mm) with narrow wings; the HW fringe is much wider, and FW are usually white or black with white markings. The labial palpi are slender, strongly curved upward. The larvae are flattened, with head prognathous and recessed into the first body segment and legs short, adapted for mining. *Elachista* typically mine monocots (Poaceae, Cyperaceae, Juncaceae), while other species are miners in a few dicot families. There are about 250 described species worldwide, but they are mainly Holarctic.

Xyloryctid Assemblage The remaining families of Gelechioidea lack modified lateral articulation on the abdominal segments of the pupa but otherwise are not defined as a lineage by a derived feature.

Xyloryctidae: Adults are small to moderately large (FW 10–33 mm), nocturnal moths, often brightly colored, shining white or yellow patterned with black or brown, and relatively heavy bodied, having spiniform setae on the posterior part of abdominal terga 2 to 6. The labial palpi usually are long, strongly curved, and slender. Larvae are robust caterpillars with a bordered submental plate and secondary SV setae on abdominal segments 3 to 6. They form silken tubes or shelters in lichens, on bark, or among foliage. Some species feed in bark or tunnel in bark or stems and drag leaves back to the gallery at night. Larval foods are lichens, living angiosperm plants (20+ families), about half the known species on Myrtaceae and Proteaceae, or dead eucalypt leaves. More than 500 described species are assigned to 60+ genera, in Africa, the Indo-Australian region, and Polynesia, with the greatest numbers in Australia.

Scythrididae are tiny to small (FW 3–12 mm) stiletto-shaped moths with narrow wings that wrap around the body, rendering a tapered appearance from thorax to wing tips (Fig. 50). Most are somber colored, gray or brown, with darker or white FW markings, the nocturnal species tending to be white or pale gray with larger eyes, while diurnal scyth-rids are mostly dark brown with small eyes and visit flowers for nectar. The male genitalia display an astonishing array of forms, from a relatively unmodified gelechioid plan in most nocturnal species to an extremely reduced and modified by fusion, often asymmetrical form that defies interpretation of homologies of the structures, mostly in diurnal forms. A remarkable species on coastal sand dunes in California, *Areniscythris brachypteris,* is flightless in both sexes, has greatly enlarged hind legs that enable it to leap 20 times or more its body length, and buries itself at night. The larvae are slender, with a small head, tapered toward both ends, usually without integumental markings but with sclerotized rings around setae SD1 on the abdomen. The head capsule has a submental pit in most genera, at least in early instars, which with the setal rings indicates relationship to Blastobasidae. Most species live in frail webs and feed on growing tips of herbaceous plants; some are leafminers in early instars. At least 20 families of angiosperms are hosts, mostly dicots, with Cistaceae and Asteraceae prevalent in the Palaearctic; larvae of a few species eat lichens, mosses, grasses, or cacti. There are 700+ described species in 26 genera, with vast numbers awaiting study and naming (e.g., 90% of the North American fauna). Worldwide but most numerous in arid and seasonal drought areas such as the southwestern Nearctic, these moths are rare or lacking in wet tropical habitats.

Oecophoridae (s. str.): As now defined this is a worldwide assemblage of dissimilar moths, with a tremendous radiation of forms in Australia. Adults (Figs. 51 and 52) are small (FW 4–23 mm) with narrow to broad wings, mostly dull colored in Holarctic genera but wildly variable and colorful in Australia, where the FW are patterned in yellow, rose, rust, and browns. The abdominal terga are usually without spiniform setal bands. The larvae are cylindrical, with head often darkly sclerotized, sometimes with reduced numbers of stemmata and integument usually not pigmented; the thoracic legs and abdominal prolegs are well developed. Most feed on dead plant material, leaf litter, and other vegetative refuse, and the rich fauna in Australia depends mainly on *Eucalyptus* (Myrtaceae), with about 60% feeding on fallen leaves and 25% on living foliage. There are more than 3000 described species in 500+ genera worldwide; this is the dominant group in Australia, with 2200+ named species in 340 genera and a projected 35–40% of the Lepidoptera fauna. Several species are cosmopolitan household moths whose larvae feed in stored meal, potted plant humus, etc.

Coleophoridae are typically tiny to small (FW 3.5–13 mm), very slender moths with lanceolate wings (Fig. 53); the HW fringe is much wider than the wing. Usually these moths have rather long, nearly straight labial palpi that project forward, often slightly drooping. Paired patches of spiniform setae on the abdominal terga define their relationship with the Momphinae, which are tiny, stout moths with thick, diverging, upward-turned palpi. Mostly dull colored, yellowish, white, gray, or brownish, the FW often have linear, pale, or dark streaks. The larvae are slender, with very reduced abdominal prolegs in Coleophorinae, which are leafminers in the first instar and then live in a portable case constructed of silk covered with sand grains or pieces of plant material (Fig. 24); they feed by mining outward from the affixed case, forming characteristic, round mines with a central hole, as they move from spot to spot. Larvae of Momphinae are more stout, are grub-like, and feed within growing tips, stems, or galls they cause. Coleophorinae feed on more than 30 plant families, including conifers (rare); monocots, especially Juncaceae; and diverse dicots. Momphinae specialize on Onagraceae (70% of host records), the only Lepidopteran group to do so. More than 1100 species are described, with an estimated 500 unnamed species known in the Nearctic; they exist worldwide but are mainly Holarctic and are absent from Neotropical rainforests.

Blastobasidae: Adults (Fig. 54) are small (FW 4–15 mm), nocturnal moths with narrow wings and a short abdomen that bears a conspicuous, transverse row of stout, rust-colored or black, spiniform setae on each segment dorsally. The labial palpi are usually short, strongly curved upward, and appressed to the head. Blastobasids are uniformly dull colored, usually gray, having FW with whitish or black steaks, sometimes tan or yellowish. The larvae are slender, cylindrical, often with heavily pigmented integument, a labium with a submental pit, and SD1 setae usually with sclerotized rings. Larval foods—

The larvae are mainly scavengers, living in a wide variety of situations such as abandoned nests of insects, galleries of stem and root borers, and detritus associated with aphid and scale insect colonies, occasionally eating living insects, and a few species feed on living plant material. Owing to their consistently drab appearance and uniform genitalia structures, this family has been neglected in systematics studies. Worldwide, but more diverse in the Nearctic and Neotropical regions, there are 500+ described species and possibly 5 to 10 times that many awaiting study.

Cosmopterigidae: This is a diverse group not defined by any uniquely derived characteristic. The adults are tiny to small (FW 3–13 mm), slender with lanceolate or somewhat broad wings, lacking a gnathos, and having a strongly hooked aedeagus. The larvae are morphologically most similar to Gelechiidae. Habits are diverse, most of these insects are internal feeders, leafminers or bud, stem, bark, or root borers, sometimes causing gall formation by the host plant. Hence they tend to be stout with short legs, without secondary setae and little integumental pigmentation. More than 25 families of angiosperms are hosts. Typical cosmopterigines are leafminers, *Cosmopterix* often in monocots; others are seed feeders, and many are scavengers, in and under old bark, dead stems, etc., and in ferns and palms, especially on oceanic islands. Larvae of *Euclemensia* are predaceous on scale insects. Worldwide, there are 1650 described species in 100+ genera. The genus *Hyposmocoma* in Hawaii is the most famous example, with an estimated 450 mostly unnamed species that occupy diverse larval niches, including living plants, in dead wood or stems, on lichens, some feeding from a portable case, and in freshwater and littoral habitats, analogous to the Galapagos finches, but with a vastly more species-rich insular radiation.

Gelechiidae: One of the major families of micromoths, especially in temperate latitudes, the adults are most easily recognized by the hind wing shape, with the terminal margin indented below the acute apex. Adults (Fig. 55) are tiny to small (FW 3–12 mm, a few tropical species to 18 mm). The great majority are nocturnal, somber colored, brown, gray, or black, but some are colorfully patterned. The larvae usually form concealed shelters in new growing tips of trees and shrubs, but many are leafminers at least in early instars, or stem and root borers, and a few live in plant galls they cause (Fig. 27). Some feed in seeds or dead plant materials, while a broad diversity of living gymnosperms and angiosperms (80+ families) are used. A few feed on ferns or mosses, especially on oceanic islands. There are more than 4500 named species placed in 500+ genera and unknown numbers of undescribed species (e.g., estimated 60% of the North American species). These moths are most diverse in temperate zone areas, including deserts and other seasonally arid habitats. Several are important agricultural insects, including the pink bollworm (*Pectinophora gossypiella*), a threat to cotton growers worldwide; Angoumois grain moth (*Sitotroga cerealella*), which feeds in stored grains; potato tuber moth (*Phthorimaea operculella*); and many conifer needle miners.

APODITRYSIA All the more derived Lepidoptera are grouped in the Apoditrysia by possession of shortened apodemes on the second abdominal sternum that have enlarged bases, contrasted with the more ancestral state, continuations of longitudinal costae (venulae) of the sternal plate. Several superfamilies comprising the non-obtectomeran Apoditrysia retain the ancestral movable and spined abdominal segments in the pupa, which moves forward to protrude from the shelter, facilitating the moth's eclosion. The more derived superfamilies of this lineage, the Obtectomera (the pyraloids and macromoths), have nonmotile pupae.

Choreutoidea   The family Choreutidae is a small group of phenotypically similar moths that have a scaled proboscis, an independent development from the Gelechiidae and Pyraloidea, and deposit upright eggs. The adults (Fig. 56) are small (FW 2–9 mm), diurnal, with broad wings, which they twitch in a characteristic fashion as they strut jerkily about on host plant leaves. The male antennae usually have long ventral setae. Choreutids are mostly dark colored, with black or brown wings marked by metallic gray, white, or silver-white; some tropical species are orange, with harlequin patterns. The larvae are slender with elongate abdominal prolegs, living externally in slight webs, from which they graze on leaf surfaces. Larval foods include diverse dicot angiosperms (17+ families), concentrating on Moraceae in tropical regions, Fabaceae, Urticaceae, and Asteraceae in the Holarctic. About 400 species are described worldwide, with many undescribed tropical species.

Sesioidea   This superfamily consists of three families, Brachodidae, Sesiidae, and Castniidae, the adults of which are markedly dissimilar in appearance and behavior; their proposed relationship is based on subtle features: the eye is more strongly pigmented anteriorly, they have large patagia, and the larvae have an unusual crotchet arrangement, two transverse, uniordinal rows.

Sesiidae: These moths are wasp-like (Fig. 57), with FW basally narrow and relatively short HW, usually lacking scales except along the veins and distal margin. The wings are tightly coupled, with the posterior margin of the FW bent down, engaging with the upcurved costal margin of the HW, and both have rows of stiff scales that interlock, in addition to the normal frenulum and retinaculum. Sesiids are small to moderately large and heavy bodied (FW length 5–28 mm), diurnal or crepuscular, and almost all species resemble wasps or bees, often startlingly so. This involves not only clear wings and a colorful, banded abdomen, but the legs are modified with tufts, even to the extent of having yellow-tipped scales resembling pollen carried by bees. *Alcathoe* are black with bright orange wings, and the males have a long, slender, scaled process from the tip of the abdomen, which in flight resembles the trailing leg posture of tarantula hawks (Pompilidae, *Pepsis*). Sesiids often visit flowers with quick, darting flights. The larvae (Fig. 25) are borers in stems, bark, and roots; they are stout, with heavily sclerotized head and mandibles and unpigmented integument. Larval foods include 40+ families of flowering plants, including conifers but not monocots. There are more than 1100 described species in 120 genera, speciose in both temperate and tropical regions. Sesiidae may be the most completely known of any microlepidoptera. Many species are agricultural pests, borers in stems of berry and squash vines, fruit trees, and conifer trunks.

Castniidae are primarily tropical, large (FW 24–190 mm), strong-flying, diurnal moths that have broad, often colorful wings, and the antennae are swollen distally, so they resemble butterflies (Fig. 58). The antennal tip is abruptly narrowed and bears a tuft of long hairs, features shared with Sesiidae. Some species are crepuscular and occasionally come to lights. Larvae are stout, cylindrical, with short legs and a head retractable into the thorax. They feed in plant stems or form tunnels in soil and feed on subterranean plant parts. The larval life requires 4.5 months to 2 years. All confirmed feeding records are for monocots, and several species are pests of sugarcane, banana, or oil palm. There are about 180 species, placed in 33 genera. The distribution suggests a Gondwanan origin, with the subfamily Taschininae in Southeast Asia, the sister group of Castniinae in Central and South America and Australia. Castniids occur in tropical, subtropical, and warm temperate regions and are lacking from most of the Holarctic, southern South America, Tasmania, and New Zealand.

Cossoidea Cossidae: Goat moths are small to large, hawk-moth-like, robust moths (Fig. 59) (FW 4–70 mm, rarely to 125 mm), having short, usually bipectinate antennae in the male; the proboscis is not functional, reduced to a small triangular lobe; labial palpi are three-segmented, short, and upturned. The wing venation is more primitive than in related superfamilies, with the median vein complete and branched within the cell. Cossids typically are nocturnal, drab, mostly gray with black striae; a few have brown or orange patches. The eggs usually are laid in groups in crevices or under bark and may be produced in vast numbers, 18,000 counted for one species in Australia. First instars disperse and bore into branches or trunks of living shrubs or trees, sometimes living gregariously; later instars are stout, cylindrical, with heavily sclerotized head and mandibles and unpigmented integument. Larvae require 1 to 4 years to mature. Recorded hosts include at least 17 families of angiosperms, including 1 monocot, with woody legumes (Fabaceae) accounting for 25% of the records. More than 20% of the species are polyphagous. Cossids occur worldwide, with greatest numbers in tropical regions (40% Neotropical), including more than 670 described species placed in 83 genera.

Tortricoidea Tortricidae: This is a large and relatively homogeneous family, with three subfamilies, Chlidanotinae, Tortricinae, and Olethreutinae, each of which, along with several subordinate taxa, at times have been treated as families. Females possess modified papillae anales, which have been rotated 90% from the ancestral lateral position to form flat, expanded pads facing ventrally, usually with the outline of shoe soles. Adults (Figs. 60–62) are small to moderately large (FW mostly 3–25 mm, to 28 mm in the Asian Ceracini),

generally with rectangular FW and broader, plicate HW. The antennae are about 0.6 times the FW length, filiform, with sensory setae in males typically short, but long in some groups; the labial palpi are porrect or bent upward but not curved as in Gelechioidea. Most species are nocturnal, with the FW cryptically colored in gray, brown, rust, or tan, but some species have colorful markings. A few are spectacularly polymorphic, notably species of *Acleris* (Tortricini)—2 species in England have more than 100 named color forms. Because many tortricids are economically important as agricultural and forest pests, there is a vast literature on the biology, ecology, host-plant selection, oviposition behavior, pheromone chemistry, etc.—for example, more than 6000 references on the spruce budworms (*Choristoneura fumiferana* species complex) in North America. In all but the most ancestral tribes the eggs are flat, scale-like, and deposited singly in Chlidanotinae and Olethreutinae and the more ancestral tribes of Tortricinae, but derived Tortricinae deposit small to large, symmetrically shingled masses (100–150 eggs) (Fig. 9). Females of the Neotropical tribe Atteriini have thick mats of specialized (corethrogyne) scales of two types on sterna A6 and A7, which they deposit on and as upright fences around the egg masses. Two Australian Archipini and Epitymbiini fence the egg mass with scales from costal tufts on the HW. The larvae are cylindrical without secondary setae, with setal pattern and crotchet arrangements similar to those of Cossidae, usually with little or no integumental pigmentation other than the setal pinacula. Larvae of most Chlidanotinae and Olethreutinae feed as borers in stems (Fig. 26), roots, buds, or seeds, and most are specialists in host-plant selection. Larvae of a few species are miners in leaves (Fig. 22) or conifer needles or cause plant galls (Fig. 28). By contrast, nearly all Tortricinae are external feeders, often polyphagous, that form leaf rolls or other shelters in foliage, but species of Cochylini bore into buds and stems. Pupation usually occurs in the larval shelter or gallery, although some drop to the ground to pupate, especially those that diapause over winter as pupae or prepupal larvae in the Holarctic. External feeders of all three subfamilies possess an "anal fork," used to flip frass away from the larval shelter. An enormous array of plants serve as hosts. A few small tribes (few genera) are specialists, e.g., Bactrini on monocots. About 8500 described species are placed in 720+ genera, and incalculable numbers are unnamed in tropical regions—e.g., 70 to 80% of recently studied species of Neotropical tortricines have been previously unnamed. Rich faunas occur in all biogeographic regions. In addition to the spruce budworms, important economic Tortricinae include the light brown apple moth (*Epiphyas postvittana*) in Australia and fruittree leafroller (*Archips argyrospilus*) in North America, while Olethreutinae include the codling moth (*Cydia pomonella*), pea moth (*C. nigricana*), larch and spruce budworms (*Zieraphera*), several seed and cone borers (*Cydia*), and pine tip borers (*Rhyacionia*) in the Holarctic.

Zygaenoidea This is a conglomeration of families of very different appearing moths and larvae. All share two features, larval head retractile into thorax and second abdominal spiracle of the pupa covered by wings. The mouthparts are vestigial in all the families treated here except Zygaenidae. The larvae are stout with the ventral surface slug-like, having short prolegs or suckers, often resembling caterpillars of lycaenid butterflies. Feeding habits vary, including predators of Homoptera or ants. Defensive secretions containing cyanoglucides are produced by some zygaenoids. Some families have been intensively studied because of their interesting larval habits, reproductive behavior, or chemical ecology. Taxonomists define 12 families, most of them small, with fewer than 50 species and limited to one geographical region.

Epipyropidae: Adults are small (FW 4–10 mm), FW triangular, HW round, blackish or gray, rarely with white or orange HW. The antennae are bipectinate in both sexes, more broadly in the male. The moths are crepuscular or nocturnal but rarely come to lights. Females produce large numbers of eggs, up to 3000 in one African species, that are deposited on foliage of a plant frequented by the planthopper hosts. The larvae are hypermetamorphic, with first instar triungulin-like, slender, tapering posteriorly with long thoracic legs and ventral ambulatory setae. The first instar seeks the host, attaches by means of silk, and transforms to a grub-like larva, which may feed on secretions produced by the homopteran, and the host remains active. The larva is covered by wax secreted by glands in its integument. Pupation occurs away from the host in a dense cocoon impregnated with wax. Epipyropids feed on body fluids and secretions of fulgoroid leafhoppers of several families. There are about 40 described species assigned to nine genera, occurring in pantropical and warm temperate regions.

The Australian family Cyclotornidae (12 species) are similar to the Epipyropidae with an even more bizarre life cycle. Females deposit large numbers of eggs (1400 counted for one female) on vegetation infested with cicadellid leaf hoppers or Psyllidae (Homoptera). The active first instars follow *Iridomyrmex* ant trails leading to the homopterans. After feeding, the moth larva leaves and molts into a brightly colored, flat, broadened, scale-like larva, which curls its abdomen upward to expose the anus where a secretion is produced that is much sought after by the ants. The cyclotornid is seized by an ant and carried back to the nest, where it feeds on ant larvae and pupae.

Limacodidae: Adults (Fig. 65) are small to medium-sized (FW 6–35 mm), mostly nocturnal moths with a stout body, relatively short, broad wings, and densely scaled head and body. The antennae are bipectinate in male, often to about half the length of the body. Mostly brightly colored, in yellow, tan, browns, these moths assume a characteristic resting posture, with the body held at an acute angle to the substrate. Larvae are hypermetamorphic, with first instar, which often is nonfeeding, oval, flat, and bearing rows of large spines that are not homologous with basic setal patterns of other Lepidoptera. Later instars (Fig. 66) are lycaenid-like in form, lack abdominal prolegs, and often have ventral suckers on segments A1 to A7 that adhere to the foliage, aided by a fluid secreted over the cuticle, and they move slug-like by peristaltic waves passing along the sole-like venter. Different species exhibit a great variety of body form, smooth, sometimes with gelatinous warts, with protuberances, spines, and hairs, some of which are stinging (nettle caterpillars), or densely hairy. Later instars often are green, but in stinging species they are brightly colored. The fecal pellets are characteristically cup-shaped. The cocoon, which usually is hard, incorporating calcium oxalate that is secreted by the prepupal larva, has a circular, dehiscent lid. A variety of dicot trees and shrubs serve as larval hosts, with many species polyphagous, and some feed on monocots, including coconut, banana, rice, sugarcane, and other economically important plants. There are more than 1000 described species, occurring in all zoogeographical regions, with greatest numbers in the tropics.

The sister family to the Limacodidae, the Dalceridae (40 species), is limited to the Neotropical region. Adults (Fig. 67) are small moths (FW 6.5–24 mm), white, yellow, or orange, with broad, rounded wings and hairy bodies, similar to epipyropids and limacodids in having vestigial proboscis and small labial palpi; the antennae are bipectinate, more broadly in males. The larva (Fig. 68) has a brush-like spinneret, dorsum and sides with gelatinous humps secreted from glandular setae, venter with translucent cuticle, thoracic legs short, prolegs not developed, and crotchets absent in early instars, appearing in the final two.

Megalopygidae: Adults (Fig. 69) are also similar to Limacodidae but generally larger (FW 11–28 mm) and relatively heavy bodied, with FW markings often brightly colored and elaborately patterned and scales hair-like, cleft, or tripartite. The larvae (Fig. 70) are superficially similar to some limacodids, but with three rows of spined protuberances (scoli) bearing variable development of typically urticating setae, sometimes beneath tufts of long, silky hairs like a fur coat (puss caterpillars), with the unusual complement of seven pairs of prolegs, those of A2 and A7 lacking crotchets. Pupation occurs in a tough, tapered cocoon blended into a branch. There are about 260 species, primarily Neotropical, with several genera ranging into the United States. The small subfamilies Somabrachyinae, in the Mediterranean region and Africa, and the Neotropical Aidinae are sometimes separated as families.

Zygaenidae comprise a worldwide group of dissimilar moths exhibiting a large number of remarkable specializations. Nearly all are diurnal and many are slow-flying and brightly colored (aposematic) (Fig. 71) and are avoided by birds and other predators because both adults and larvae biosynthesize cyanoglucides. They are able to release hydrocyanic acid by enzymatic breakdown of the cyanoglucides. This has enabled many to take part in complex mimicry relationships. Moreover, adults and larvae are unusually resistant to cyanide, and naive collectors are startled to see a moth survive for half an hour in a potent vial that kills other moths in seconds. Adults are small to large (FW 5–50+ mm), with head and labial palpi smooth scaled, and they commonly visit flowers for nectar.

FIGURES 65–92 Adults and larvae of zygaenoid, pyraloid, and macro moths. Zygaenoidea: (65) *Parasa indeterminata* (Limacodidae) (New Jersey); (66) Larva of *Isa textula* (Limacodidae) (Maryland); (67) *Dalcerides ingenita* (Dalceridae) (Arizona); (68) larva of *D. ingenita* (Dalceridae) (Arizona); (69) *Trosia revocans* (Megalopygidae) (Amazonas); (70) larva of *Monoleuca semifascia* (Megalopygidae) (New Jersey); (71) *Zygaena ephialtes* (Zygaenidae), pair in copulo (France). Pyraloidea: (72) *Petrophila confusalis* (Crambidae, Nymphulinae) (California); (73) Pyraustinae species (Crambidae) (Costa Rica); *Crambus* species (Crambidae) nectaring at composite flower, whereas the larval host is a grass (Arizona). Geometroidea: (75) *Urania fulgens* (Uraniidae) (Ecuador); (76) *Dichorda illustraria* (Geometridae) (California); (77) *Neoterpes edwardsata* (Geometridae) (California); (78) flightless female of *Tescalsia giulianiata* (Geometridae), a winter moth (California). Bombycoidea: (79) *Phyllodesma* species (Lasiocampidae) (California); (80) *Bombyx mori* (Bombycidae), pair in copulo of the commercial silk moth; (81) *Eacles* species (Saturniidae, Citheroniinae) (Costa Rica); (82) *Hemileuca eglanterina* (Saturniidae, Hemileucinae), a diurnal species, female ovipositing (California); (83) *Argema maenas* (Saturniidae, Saturniinae) (Malaysia); (84) *Smerinthus cerisyi* (Sphingidae) in predator avoidance posture (Utah); (85) *Hemaris senta* (Sphingidae), a diurnal bumble bee mimic (California). Noctuoidea: (86) *Clostera apicalis* (Notodontidae) (California); (87) *Phryganidea californica* (Notodontidae, Dioptinae); (88) larva of *D. californica*, the California oak moth; (89) *Lymantria dispar* (Lymantriidae), mating pair of the notorious gypsy moth (Russia); (90) *Orgyia vetusta* (Lymantriidae), wingless female tussock moth (California); (91) larvae of *O. vetusta;* (92) *Horama panthelon* (Arctiidae) (Texas). (Photographs by: E. Buckner, 71; R. Cardé, 89; C. Covell, 75, 83, R. Coville, 72, 73, 76, 77, 79, 80, 81, 86, 87, 88, 90; J. Hafernik, 74; C. Hanson, 67; L. Penland, 65, 68; J. Powell, 78, 84, 91; D. Rubinoff, 82, 85, 92; J. Ruffin, 66, 70; K. Sandved, 69).

The antennae are often thick and either clubbed in both sexes or bipectinate in males and narrowly bipectinate or filiform in females. Glands located between the eyes and the base of the proboscis produce a whitish or yellow liquid or foam when the moth is disturbed. They are often metallic colored, particularly vivid in European *Zygaena,* in bright red and metallic blues. The eggs are deposited in rows, clusters, or overlapping patches and sometimes covered with scales from a special abdominal hair tuft, which in some Australian genera are urticating. The larvae are stout and broad, with the head usually retractile under the extended prothorax; the body is roughened and covered with dense secondary setae. Larval foods include numerous plant families, although the species are mostly host specific. European zygaenines specialize on Celastraceae, cyanogenic Fabaceae, and noncyanogenic Apiaceae; some Australian Procridinae rely on Dilleniaceae, Myrtaceae, or Vitaceae, and the last is used by some Nearctic species. About 1000 species are described, and the family is relatively well studied owing to the colorful forms, diurnal habits, and mimicry associated with the chemical ecology.

Alucitoidea   Alucitidae: The many-plume moths are so called because the wings are deeply cleft, so each wing has six fringed, plume-like segments (Fig. 63). The family was classified with the true plume moths (Pterophoridae) from which the alucitids differ by having discrete bands of spines on some or all of abdominal terga 2 to 7 and a relatively unspecialized pupa formed in a cocoon. Adults are small (FW 3–13 mm), slender moths that are unmistakable by their wing structure, but in two tropical genera with the largest members of the family, the wings are divided only a short distance. The antennae are filiform, proboscis is well developed, and labial palpi are usually fairly long, porrect or upcurved. Most are gray or brown, delicately banded with tan or white. They are nocturnal and collapse the wing plumes when at rest, holding them out from the body so as to resemble narrow-winged pyralids, but when active they strut about with the wings fully expanded, like miniature peacocks. The larvae are borers in flower buds, shoots, or in galls; the body is stout with short legs with short setae on inconspicuous bases. Pupation occurs in the larval shelter or in leaf litter in a cocoon. Larval hosts include at least eight dicot families, especially Caprifoliaceae in the Holarctic and Bignoniaceae and Rubiaceae in Australia and tropical regions; one species is a coffee pest in Africa. There are about 130 species described and likely there are many more in tropical regions.

Pterophoroidea   Pterophoridae: The plume moths are recognizable by their deeply cleft wings in all but the most ancestral genera. They lack proboscis scaling and abdominal tympana, the hind tibia is 2 or more times the length of the femur, and abdominal terga 2 and 3 are elongated. Adults (Fig. 64) are small (FW 4–18 mm), with long and slender bodies, legs, and wings; FW is cleft for about 0.25 to 0.33 its length and the HW deeply twice-cleft in most species. When at rest, the plumes are overlaid and rolled under the leading edge of the FW, resembling sticks held out from the body.

They usually are nocturnal and dull colored, tan, brown, or gray with paler and darker markings; although a few are colorful members of tropical mimicry complexes. The larvae (Fig. 29) typically are elongate and cylindrical with long prolegs. Most are external feeders on foliage and usually have dense setae that may be forked, clubbed, or glandular, secreting a sticky fluid. Some species are borers in stems and have short setae, even a strongly sclerotized anal shield with two stout horn-like processes resembling the urogomphi of coleopteran wood borers. The pupae are strongly spined or setose and either are formed in the galleries or are affixed to host-plant stems or debris by the anal cremaster to a silk pad, fully exposed, and they can bend and curl over by their movable abdomen. More than 20 dicot families are recorded as larval hosts, principally Asteraceae, usually herbs, but not monocots. The larvae of *Buckleria* in Europe are remarkable for feeding on sundews (Droseraceae). About 1000 species are described worldwide, with many unnamed tropical species.

APODITRYSIA: OBTECTOMERA
Pyraloidea   This is one of the largest superfamilies of Lepidoptera, with more than 17,500 species described and probably at least as many more from tropical regions awaiting study. The fundamental features that define the pyraloids are a basally scaled proboscis, well-developed maxillary palpi, and tympana consisting of paired chambers on the venter of abdominal segment 2. In recent decades specialists have agreed that differences in the morphology of the tympana and other adult and larval characters warrant treating the former Pyralidae as two families, the Crambidae and Pyralidae. While it is difficult to recognize crambids and pyralids as distinct groups on the basis of superficial appearance owing to the enormous variability within each, subfamilies of each family are distinctive. Pyraloids occur worldwide, other than Antarctica, and range from high alpine to low desert and tropical habitats but are most prevalent at low and middle elevations in the tropics. They are highly successful at dispersal and colonization and are especially well represented on oceanic islands.

Pyralidae: These are small to relatively large moths (FW 5–75 mm, mostly under 30 mm) that have the tympanal organs almost completely closed, with their conjunctiva and tympanum in the same plane; they have vein R5 of the FW stalked or fused with R3+R4 and lateral "arms" at the base of the uncus in the male genitalia; the larvae almost always have a sclerotized ring around the base of seta SD1 on abdominal segment 8 and often around SD1 of the metathorax. The larvae usually are stout and cylindrical, with relatively short legs and setae; the body typically is unpigmented, although some species are well patterned, even brightly colored. Almost all are concealed feeders, most often borers in seed, fruit, or stems or live in tunnels they construct in the soil beneath plants. Many others construct shelters among tied leaves, often of quite tough silk. A variety of flowering plants are hosts, as well as wood-rot fungi (Xylariaceae), dry vegetable matter including seeds, and the papery structure of social Hymenoptera nests (*Galleria, Achroia, Aphomia,* Galleriinae). Many are household and granary pests that have been transported worldwide by human activities (*Corcyra,* Galleriinae; *Pyralis,* Pyralinae; *Plodia, Ephestia, Anagasta, Ectomyelois,* Phycitinae). A few species are predaceous on scale insects (*Laetilia,* Phycitinae) or live in ant nests (some Chrysauginae); three genera of Chrysauginae feed in the dung of sloths, and the adults live in their fur. Several species of Phycitinae have been used for biological control of cactus. The majority of species feed on flowering plants, including conifers (*Dioryctria,* Phycitinae); and monocots, including pests of coconut and other palms (*Tirathaba,* Galleriinae) and corn (Epipaschiinae), and a wide range of forest trees, ornamental shrubs, and crops are damaged, especially in tropical regions. There are more than 6000 described species, about two-thirds of which are Phycitinae, and the tropical faunas are not thoroughly studied.

Crambidae: Crambids are more diverse and variable in morphology and biology than pyralids, but all share "open" tympanal organs (i.e., with a wide anterior aperture, and the conjectiva and tympanum meet at a distinct angle); vein R5 of the FW usually is not stalked with R3+4, male genitalia are without basal uncus "arms," and A8 of the larva lacks sclerotized rings around seta SD1. About 14 subfamilies are defined, and usually members of each can be recognized on superficial bases. Adults (Figs. 72–74) vary from small and slender to large and stout bodied (FW 3.5–47 mm), with FW narrow and HW plicate and folded under it at rest (e.g., Crambinae, Fig. 74), to broadly triangular with HW similarly colored and held flat (e.g., Pyraustinae, Fig. 73). The labial palpus is prominent, obliquely ascending or porrect (typically very long in Crambinae, snout moths), maxillary palpi are typically small, often with broadened scaling that connects a profile of frons with the labial palpus. Proboscis is usually well developed. Legs are long, and those of males often have structural modifications and/or androconial scale tufts. Larval food habits and morphology vary among subfamilies: Crambinae (lawn moths) live either as ground dwellers feeding primarily on grasses or as stem borers in various monocots; Schoenobiinae are borers in marsh grasses; Cybalomiinae and Evergestiinae specialize on Brassicaceae and Capparidaceae; Midliinae are borers in Araceae; Musotiminae larvae feed on ferns, a rare niche for Lepidoptera; many Odontiinae are leafminers or flower, bud, and stem borers, on diverse dicots; and Pyraustinae, which make up 65% of the known species of Crambidae, are mostly webworms but some borers, in an enormous variety of monocots and dicots, pest species of many crops, including corn, bananas, palms, pasture grasses, cucurbits, tomatoes and other solanaceous fruits, coffee and other tropical trees, garden mints, and conifers. Scopariinae are specialists on mosses and lycopods, tunneling in the roots and stems, and on ferns, or on seed-bearing vascular plants. Nymphulinae larvae are aquatic, living either in ponds on vascular plants,

often in cases, or in rapid streams, usually under webs on rocks and feeding on algae. Larvae either breathe through open spiracles, living in air-filled cases or stems, or absorb dissolved oxygen through tracheal gills. Pupae are formed in cases in chambers within the plants or in gas-permeable cocoons and breathe through the spiracles. In the Palaearctic *Acentria* (Schoenobiinae), females are wingless, enter the water, and are parthenogenetic until a bisexual generation of winged adults late in the season. More than 11,500 species are described, nearly 90% of which are members of Nymphulinae, Crambinae, or Pyraustinae.

Geometroidea   This group includes five families, Drepanidae, Epicopeiidae, Sematuridae, Uraniidae, and Geometridae, although they have been separated as two or three superfamilies by some authors, based primarily on the structure of the larval mandibles and tympanal organs. Geometroids typically are broad-winged, with slender bodies, small to large moths (FW 5–78 mm); they have abdominal tympana of structures different from those of pyraloids and lack scaling of the proboscis. Larvae of Uraniidae have well-developed abdominal prolegs, while most Drepanidae and Geometridae have some of the prolegs vestigial or absent.

Drepanidae: Adults have internal abdominal tympana unlike any other Lepidoptera, associated with the dorsal–ventral sclerites that connect tergum 1 with sternum 2, opening dorsally. Adults are medium sized to large (FW 8–31 mm), broad winged and geometrid-like in typical Drepaninae, often with the FW apex produced or curved; Thyatirinae are stout bodied and noctuoid-like. The antennae usually are short, lamellate or bipectinate to the tip for most of the length, sometimes filiform. The larvae have few secondary setae or rarely numerous but very short setae, sometimes with an eversible vesicle just above the prothoracic coxa in Drepaninae, often with notodontid-like protuberances in Thyatirinae; anal prolegs are usually vestigial but those of A3 to A6 are well developed, and the anal shield is conspicuously elongated. At least 20 families of diverse dicots and one monocot (Zingiberaceae) are hosts; some Holarctic species are generalists. More than 650 species are described in 120+ genera, mostly in the Holarctic and Oriental regions; Drepaninae are absent in the Neotropics. A few species are pests of coffee.

Uraniidae: This family is defined by the sexual dimorphism of the tympanal organs, which are on the lateral, posterior part of tergum A2 in males and on the lateral part of sternum A2 in females. Adults (Fig. 75) are small to large (FW 7–78 mm), broad winged, usually with a relatively slender body; wings are often resplendent in brilliant, iridescent colors; HW veins are sometimes produced into one or several swallowtail butterfly-like tails. Most are nocturnal, but 3 tropical genera are day flying, including *Urania,* some species of which are famous for their massive migratory flights, involving thousands of the spectacularly colored moths. Epipleminae are smaller, nocturnal, and dull colored; they rest with the FW extended and rolled, HW appressed to the body. The antennae are filiform, lamellate, or pectinate, sometimes thickened preapically. The larvae are more or less bare, with few secondary setae, occasionally with spatulate setae, and prolegs are well developed. Larval foods are recorded in about a dozen dicot families, including specializations on Oleaceae, Asclepiadaceae, and Euphorbiaceae (all known larvae of Uraniinae feed on euphorbs), unusual for Lepidoptera and likely sources of distasteful qualities, and many species appear to be aposematic. Worldwide, these moths are primarily pantropical; around 700 described species are assigned to 90 genera.

Sematuridae are tropical, similar superficially to Uraniinae, often brightly colored with HW tails, but adults lack abdominal tympana, and the antennae usually are distally thickened, like those of skipper butterflies.

Geometridae: This is one of the three most speciose families of Lepidoptera. Adults (Figs. 76–78) are small to large (FW 5–55 mm), typically with broad wings and slender body and abdomen with basal tympanal organs in deep, ventrolateral cavities. The great majority of geometrids are nocturnal and rest by day with the wings outspread, with cryptic resemblance to tree bark, lichens, green or fallen, brown leaves, often ornate with lines simulating leaf veins or spots resembling necrotic or eaten areas of foliage. Some genera hold the wings upright, butterfly-style, and their undersides are cryptically colored. Some geometrids are day flying, including early spring species in the Holarctic or mimetic species in tropical regions. A few species that are winter moths or occur at high elevations have flightless females (Fig. 78), and these have the mouthparts and tympana reduced or vestigial. The larvae usually have prolegs of segments A3 to A5 reduced or absent (Fig. 13), so the caterpillar walks by advancing in measured, looping steps, from which both the family name and the common name (inchworms) are derived. They are bare but with a variety of colors, protuberances, ornate body forms, and behavior; most are exposed feeders that depend upon cryptic resemblance to flowers, leaves, twigs, etc., to avoid predators (Fig. 30). Most species are general feeders on trees or shrubs. Owing to their worldwide species richness, an enormous variety of gymnosperms and angiosperms are eaten. At least 21,000 species and 1500 genera have been described, and no doubt many remain unnamed, especially in tropical regions. Several species are defoliators of hardwoods or conifers (e.g., spring and fall cankerworms, *Paleacrita vernata* and *Alsophila pometaria,* and the hemlock looper, *Lambdina fiscellaria,* in North America), but geometrids are not a major pest group.

Bombycoidea These are macromoths that have no thoracic or abdominal tympana. The group is distinguished by deep clefts between the prescutum and the mesoscutum of the mesothorax. There are 12 families, including 4 worldwide groups summarized here that are sometimes treated as superfamilies: Lasiocampidae, Bombycidae, Saturniidae, and Sphingidae. Other bombycoids are Mimallonidae, a mostly Neotropical family with 200 species; Anthelidae, a small Indo-Australian group; Eupterotidae, worldwide with 300 species; Endromidae, Mirinidae (Palaearctic), and Carthaeidae (Australian), each with 1 or 2 species; and 2 small Eurasian and African families related to Sphingidae, Lemoniidae and Brahmaeidae.

Lasiocampidae: Adults (Fig. 79) are small to large (FW 9–80 mm) with moderately broad wings and stout bodies. The eyes usually have fine hairs between the facets; mouthparts are absent or vestigial; antennae are bipectinate to the tip in both sexes; labial palpi are porrect, the first segment with a scaleless patch bearing sensory setae, unique in Lepidoptera. The wings have interlocking tiny setae but no retinaculum and frenulum. Lasiocampids are nocturnal, but males of a few species are primarily diurnal, mostly somber-colored moths, browns and tan. The female often is much larger than the male, with a stout abdomen having a fully developed complement of eggs upon emergence, sometimes deposited in a single large mass. The larvae do not have the fore coxae fused as in other bombycoids. The body is covered with dense, often long, unbranched secondary setae and frequently is brightly colored. Larvae of some genera are gregarious (e.g., Holarctic *Malacosoma,* tent caterpillars, Fig. 33), and some have urticating hairs that produce a skin rash in humans. Pupation occurs in a dense, parchment-like cocoon. Larval foods are diverse, mainly trees and woody shrubs, Betulaceae, Fagaceae, and Salicaceae in the Holarctic; low-growing Asteraceae, Brassicaceae, and grasses in Africa; and predominately Myrtaceae in Australia but also mistletoes (Santalaceae). Some species are polyphagous. There are about 1500 named species, in 150 genera, occurring worldwide, mostly in the tropics; they are absent from New Zealand.

Bombycidae: Including Apatelodidae of recent authors, this family is defined by a few subtle skeletal features and by short pupal galeae that fail to reach the foreleg apices. Adults are small to medium sized (FW 9–28 mm), nocturnal with broad wings, which usually are held outward from the body when at rest. The mouthparts are vestigial. The antennae are bipectinate nearly to the tip in both sexes, and the thorax and legs are clothed in long, hairlike scales. Larvae are similar to lasiocampids, but the secondary setae are either short and minute or long in Apatelodinae. The fore coxae are separate (Apatelodinae) or fused. Segment A8 has a middorsal scolus except in Apatelodinae. Larval foods are largely Bignoniaceae, Symplocaceae, Moraceae, and Theaceae—some species are reported to defoliate commercial tea. There are about 350 species referred to 40 genera, worldwide except in Europe, best represented in the Oriental and Neotropical regions. The silk moth (*Bombyx mori,* Fig. 80) is the most well known species; it has been domesticated for centuries and is not known in the wild, its adults no longer capable of flight.

Saturniidae: The emperor or giant silk moths include many of the world's most spectacular moths. They are medium sized to very large (FW 14–130+ mm), broad winged with highly variable color patterns (Figs. 81–83), often with eyespots with concentric rings (hence the family name), which presumably act as a defense mechanism against predators of the sluggish moths, many of which must warm themselves by pumping the abdomen before they can fly. Most genera are nocturnal, some mating only during early morning hours, while species of a few genera are diurnal (e.g., *Hemileuca* in North America, buck moths, Fig. 82). The proboscis and maxillary palpi are rudimentary or absent. The male antennae are bi- or quadripectinate, those of the female filiform to quadripectinate. The male FW usually is produced apically, while the hind margin of the HW sometimes has short to very long tails. The larvae are stout with a prominent, smooth head and body protuberances that often are branched; secondary setae are numerous but small, mostly on the ventral half of body and prolegs. The scoli of some genera bear poisonous spines that cause nettle-like stings in humans, and females of some species coat the eggs with urticating hairs that cause dermatitis. Abdominal segment 8 usually has a middorsal horn or scolus. Pupation occurs in a strong, sometimes double-walled silken cocoon that may be covered with plant fragments or in the soil without a cocoon. Many saturniids are polyphagous, and an enormous array of plant hosts are recorded—90 genera in 48 plant families recorded for one species of *Attacus*—but some are specialized feeders. There are about 1500 described species in 165 genera, occurring worldwide except at the highest latitudes, most abundant in moderate to high-elevation habitats, richest in the Neotropical region, particularly the South American Andes.

Sphingidae: Sphinx or hawk moths are among the largest, most easily recognized, and best known Lepidoptera. Adults (Figs. 84 and 85) are medium sized to very large (FW 16–90 mm), having a stout body with the abdomen typically tapering posteriorly. The FW is narrow and HW relatively short, its hind margin produced, angulate at the tip of veins 1A + 2A, emarginate beyond. The antennae are distinctive, usually lamellate ventrally or bi- to quadripectinate, tapering toward the apex, which is upturned or hooked; males with two rows of long sensillae that meet dorsally; shorter and filiform in females. The proboscis usually is well developed, sometimes much longer than the body, and used to imbibe nectar while hovering in flight, hummingbird-style, and in some habitats sphingids have significant roles in pollination. Most are nocturnal, extremely strong fliers, among the fastest insects, and several are well-known long-distance migrants. Some genera are diurnal, a few resembling bumble bees, with mostly transparent wings (e.g., *Hemaris*, Fig. 85); such species have fully scaled wings upon eclosion, but after drying, the scales are shed, all but along the margin and veins. Wing coupling is usually by frenulum–retinaculum, a long bristle in males, multiple setae in females. The larvae have a prominent, triangular or globose head; the body is covered densely with minute secondary setae and usually no other setae or protuberances except a middorsal horn or button on A8. The lateral markings are distinctive, each abdominal segment with an oblique stripe ascending posteriorly, those of A7 reaching the base of horn on A8. Color is often polymorphic; they usually feed completely exposed and rely upon cryptic coloration for protection. A distinctive characteristic is their resting pose, with the thorax raised and head turned down, resembling the pose of an Egyptian sphinx.

Pupation usually occurs without a cocoon, in soil or ground litter, but rarely in a silken cocoon. Sphingids feed on a very broad range of gymnosperms and angiosperms, often specialists on plants with chemical defenses that repel most insects, including Apocynaceae, Cleaceae, Solanaceae, Rubiaceae, and Violaceae. There are more than 1200 described species in about 200 genera, distributed worldwide, best represented in the tropics. It is probably better inventoried and cataloged than any other moth family.

Noctuoidea   This is the largest superfamily by far, with more than 7200 genera proposed for nearly 60,000 species, about 40% of all described Lepidoptera. As such, there is a tremendous variety of form, size, color, morphology, and behavior in both larvae and adults. Four major, cosmopolitan families are recognized, Notodontidae, Lymantriidae, Arctiidae, and Noctuidae, from which several regional lineages have been split as families by some authors. Monophyly of Noctuoidea is unequivocally based on complex metathoracic tympanal organs and associated abdominal structures. The tympana are assumed to have evolved in response to bat echolocation, and many species have been observed to engage in bat avoidance behavior upon receipt of their sounds. Tympana may also receive mating signals, especially in Arctiidae.

Notodontidae: With Thaumetopoeidae and Dioptidae included as subfamilies, this is a large and diverse family. Adults (Figs. 86 and 87) are medium sized to large (FW 16–50+ mm), typically with relatively long FW and stout body that extends 2 or more times the width of the HW. The head often has scale tufts or crests; antennae are usually bipectinate to the tip in the male, filiform or sometimes bipectinate in the female. Proboscis is usually well developed and coiled; labial palpi are often quite short. The abdomen is densely covered with long, slender scales and sometimes dorsal scale tufts at the base. The tips of the tibial spurs are serrated. These are mostly dull-colored, tan, brown, or gray moths, but many tropical dioptines are diurnal and brightly colored, involved in mimicry complexes. The larval body is stout, nearly bare, sometimes with long secondary setae, often possessing one or more protuberances, a modified body form (Fig. 31), a median knob or horn on A9, or anal prolegs modified into slender, single or double caudal processes (stenopods). All larvae except Dioptinae have two MD setae above the spiracle on abdominal segments, whereas other noctuoids have only one. Late instars have a smooth mandibular cutting edge, derived from the serrate ancestral state in other noctuoids. When disturbed, some species emit formic acid or ketones from a cervical gland (adenosoma). Many have various cryptic colors correlated with modified body forms, but others are brightly colored and aposematic, especially Dioptinae (Fig. 88). Larval foods include a wide diversity of dicot angiosperms, mainly woody shrubs and trees, and a few feed on grasses. Many specialize on plants containing toxic substances, including Anacardiaceae, Apocynaceae, Aristolochiaceae, Fabaceae, Passifloraceae, and Violaceae. There are more than 2800 described species, distributed worldwide except in New Zealand and the Pacific Islands and particularly rich in the Neotropics. A few are occasional defoliators of orchard or forest trees.

Lymantriidae: The tussock moths are so called because larvae of many have thick tufts of erect secondary setae on the dorsum, like those of a toothbrush (Fig. 91). Adults are small to moderately large (FW 7–45 mm), nocturnal, usually dull-colored moths, mostly brown or yellow; they are usually broad winged and densely hairy; the FW is triangulate, HW rounded, hidden under the FW at rest, with the wings appressed to the substrate and the densely hairy forelegs extended in front of the head. Females of some genera are flightless, their wings vestigial (Fig. 90). Antennae of males and usually females are bipectinate to tip. The proboscis is vestigial or absent. Females have a large, abdominal tuft of deciduous scales used to cover the egg masses. Males of many genera possess a tymbal organ, a pair of finely corrugated pockets on A3. The larval integument often is brightly colored, with hair tufts from verrucae, but the body form is not strongly modified as in notodontids. Lymantriid larvae all possess a mid-dorsal, eversible gland, often yellow or red, on segment A6 and usually another on A7. The larval hairs of *Euproctis* in Australia and some other genera are hollow, barbed, and urticating and cause a severe skin rash in humans. The body hairs are woven into the cocoons and often retrieved by the emerging female and redeposited on the egg masses and then used by first instars, which feed in protected aggregations. Many lymantriids are polyphagous, frequently on arborescent shrubs or trees, and a broad array of plants is eaten. Appreciable generic radiations feed on flowers and fruit of low herbs and grasses, and many Asian and African species feed on algae, fungi, and detritus. There are more than 2500 species placed in 360 genera, distributed in all geographic regions, reaching their greatest development in the Old World tropics. Many species in several genera are forest defoliators in Europe, North America, and Indo-Australian tropics, the most notorious being the Palaearctic gypsy moth (*Lymantria dispar,* Fig. 89), which was introduced into North America in the late 1800s.

Arctiidae: The tiger moth family includes Ctenuchidae and Pericopidae, formerly treated as families, which now are interpreted as artificial groupings. There are three subfamilies, Lithosiinae, Syntominae, and Arctiinae, all characterized by a pair of dorsal, eversible, single or branched pheromone glands from the terminal abdominal segment in the females. In addition, many arctiines and lithosiines and some syntomines have metathoracic tymbal organs in both sexes. Sound is produced by contraction of muscles that deform ridges on the tymbals rapidly to produce bursts of ultrasonic clicks, stimulated by tactile cues or in response to hunting signals of bats. Many species of all three subfamilies have prothoracic glands from which a liquid is extruded containing acetylcholines and histamines and probably pyrozines, the odor of which is believed to signal distasteful or toxic properties to predators. Adults (Figs. 92 and 93) are small to moderately large (FW 5–50 mm), usually brightly colored moths with a myriad of patterns, pre-dominately orange, red, and black, often aposematic, advertising their toxic qualities and involved in mimicry complexes, or they resemble wasps, with mostly scaleless wings. Females of some species have a large tuft of abdominal scales used to mix with or cover egg masses. Most arctiids are nocturnal and come to lights, even many tropical species with elaborate Hymenoptera resemblances. Some genera are strictly diurnal, such as North American *Ctenucha, Gnophaela,* and *Lycomorpha* that accompany aggregations of lycid beetles they mimic. Larvae of many arctiids have dense secondary setae over the body (woolly bears, Fig. 94), but setae are sparse in some arctiids, especially Lithosiinae. The body form is typically cylindrical with a full complement of prolegs, most are rapid crawlers and can move great distances in search of food or pupation sites. Lithosiinae possess an enlarged basal molar area of the mandibles, used to macerate algae and lichens, which are the principal foods (rarely liverworts and mosses). Some Syntominae also feed on algae and lichens, many are scavengers or fungivores, and some feed on flowers, especially Asteraceae, or grasses. Arctiinae are polyphagous plant feeders or specialize on one of a variety of angiosperms, some genera on latex-producing plants (Apocynaceae, Euphorbiaceae, Moraceae) or those with toxins (Asteraceae, Boraginaceae). More than 6000 species are described in 750+ genera, occurring worldwide, particularly rich in tropical regions. Tropical lithosiines are poorly studied, and there are numerous undescribed species.

Noctuidae: This is the largest family of Lepidoptera, with more than 35,000 species grouped into about 30 subfamilies, many of which have been regarded as families, including Pantheinae and Nolinae, by recent authors. The composition and classification of most of these subfamilies is debatable, with critical larval characteristics unknown for most genera. Members of this vast lineage do not share a distinguishing derived feature; they posses the wing venation and tympanal form of advanced noctuoids, retain a well-developed proboscis, and lack the eversible abdominal gland present in larval Lymantriidae and the eversible pheromone glands of arctiid females. Adults (Figs. 95–97) are small to very large (FW 4–140 mm), including *Thysania agrippina* in Central America, with the world's largest insect wing expanse, exceeding 10 in. The FW is triangular to narrow and the HW broad, usually folding fan-like under the FW when at rest. The body is relatively heavy, the thorax having powerful musculature in most noctuids, and many of these moths migrate long distances. The head has long scales, sometimes forming erect tufts or a conical projection in front; the proboscis is long and coiled, its tip armed with thorns for piercing fruit in some genera. Wing coupling is accomplished by frenulum–retinaculum, usually a single bristle in males, two or three finer bristles in females. The abdomen often has eversible coremata or tufts or pouches of specialized scales. Most noctuids are nocturnal and somber colored, with the FW cryptic against bark or leaf litter when the moths are at rest during the daytime. Some have brightly colored HW that are flashed when the insect is disturbed, presumably

FIGURES 93–116 Adults and larvae of Noctuoidea, Hedyloidea, and butterflies. Noctuoidea: (93) *Apantesis (Grammia) virgo* (Arctiidae) (E. U.S.); (94) larva of *Lophocampa maculata* (Arctiidae) on *Salix* (California); (95) *Megalographa biloba* (Noctuidae) (California); (96) *Catocala* species (Noctuidae) (California); (97) *Xanthopastis timais* (Noctuidae) (Florida). Hedyloidea: (98) *Macrosoma* species (Hedylidae) (Ecuador). Hesperioidea: (99) *Phocides* species (Hesperiidae, Pyrrhopyginae) (Ecuador); (100) *Autochton cellus* (Hesperiidae, Pyrginae) (Texas); (101) *Poanes melane* (Hesperiidae, Hesperiinae) (California); (102) larva of Hesperiidae (California). Papilionoidea: (103) *Papilio rutulus* (Papilionidae) (California); (104) larva of *Papilio polyxenes* (Papilionidae) (Costa Rica); (105) *Anthocaris stella* (Pieridae) (California); (106) *Lycaena (Tharsalea) arota* (Lycaenidae, Lycaeninae) (California); (107) *Callophrys (Incisalia) eryphon* (Lycaenidae, Theclinae) (California); (108) *Plebeius acmon* (Lycaenidae, Polyommatinae) (California); (109) larva of *Plebeius acmon* on *Eriogonum*, tended by ants (California); (110) *Apodemia mormo* (Lycaenidae, Riodininae) nectaring at *Eriogonum*, the larval host (California); (111) *Vanessa tameamea* (Nymphalidae, Nymphalinae) (Hawaii); (112) *Cercyonis oetus* (Nymphalidae, Satyrinae) feeding on ripe fruit (Nevada); (113) *Agraulis vanillae* (Nymphalidae, Heliconiinae) (California), (114) larva of *A. vanillae* on *Passiflora* (California); (115) Ithomiini species (Nymphalidae, Heliconiinae) (Costa Rica); (116) *Morpho peleides* (Nymphalidae, Morphinae) (Costa Rica). (Photographs by: C. Covell, 98, 99; R. Coville, 95, 96, 97; J. Hafernik, 100, 105, 106, 109; W. Hartgreaves, 111; W. Middlekauff, 102; P. Opler, 115; J. Powell, 94, 101, 103, 104, 107, 108, 110, 112, 113, 114, 116; unknown, 93.)

having a startle effect on would-be predators (e.g., *Catocala,* underwings, Fig. 96). Other Catocalinae hold the wings out flat at rest, with the HW cryptically colored, matching the FW. Some noctuids are diurnal and brightly colored, like flowers they visit, while many high-latitude and montane species are diurnal and dark colored. The larvae typically are cylindrical robust caterpillars, bare with only primary setae (cutworms). Those of some Catocalinae have lateral fringes that appress to the substrate, eliminating shadows (Fig. 32). Acronictinae, Pantheinae, and Nolinae have various secondary setae, sometimes in rows or as tufts on verrucae similar to the arctiids. The prolegs are reduced in some subfamilies, with those of segments A3 to A5 nonfunctional (semiloopers), especially in sedentary species that feed on herbs. Noctuids feed on all kinds of plants, probably nearly every gymnosperm and angiosperm family. Many are polyphagous, foraging on low-growing plants at night, while others are specialists on one or a few plants, including those with toxic chemicals or latex, such as Anacardiaceae, Apocynaceae, Asclepiadaceae, Euphorbiaceae, Moraceae, Urticaceae, and Vitaceae, as well as grasses and sedges, Liliaceae and Amaryllidaceae. Several groups feed on fallen leaves or in plant detritus (e.g., Hypenodinae) or on algae, lichens, and fungal-ridden plant matter. Several species are predaceous on scale insects or feed on detritus in spiderwebs or mammal nests. Many are important economically worldwide, especially polyphagous cutworms and armyworms (e.g., *Agrotis, Autographa, Trichoplusia, Heliothis, Pseudaletia*), eating soybean, sugarcane, cereal, legume, rice, and other field crops. There are more than 35,000 named species in 4200+ genera, worldwide, occurring from high elevations above timberline to low deserts and especially numerous in tropical regions.

Hedyloidea Hedylids are peculiar moths that superficially resemble geometrids, with which they were classified until recently proposed as the ancestral butterfly lineage. The hypothesis is based on 10 characteristics: mesothoracic aorta configuration, six features of the adult skeletal structure, an upright egg, larval anal comb, and pupal girdle. None, however, is unique or universal for all members of either Hedyloidea or Papilionoidea to affirm their monophyly. Probably these resemblances evolved independently, as they are not present consistently in ancestral members of the respective butterfly lineages. Adults (Fig. 98) are nocturnal, medium sized (FW 16–32 mm), with broad, semitranslucent wings, weakly scaled in patterns of gray or brownish and white. Wings coupled by a retinaculum and frenulum with a single bristle in the male, weak bristles in the female, typical of most ditrysian moths. The resting posture is characteristic, with the thorax titled so the HW nearly touch the substrate, and the slender abdomen is raised above them. The head is small, eyes are large, proboscis is well developed, labial palpi are ascending. The antennae are usually filiform, bipectinate in a few species, lacking the apical club of butterflies. There are small, tympana-like structures at the base of the FW, similar to those of some Nymphalidae, and there are no abdominal tympana that would link hedylids with Geometridae. The forelegs of males are reduced and not used for walking, like nymphalid butterflies. The egg is pierid-like, upright, spindle-shaped, and ribbed. The larval head is bizarre with elongate, trifid, barbed horns, similar to some nymphalids; the body is smooth and slender, and the last

abdominal segment is bifid, somewhat like some notodontid moths and satyrine butterflies. The pupa is exposed, anchored by a silken girdle, similar to Pieridae and some Papilionidae. Recorded larval foods are Euphorbiaceae, Malpighiaceae, Malvaceae, and Sterculiaceae. There are about 40 species, all in the genus *Macrosoma,* restricted to the Central and South American tropics, Cuba, and Trinidad.

Hesperioidea Hesperiidae includes six subfamilies, and the giant skippers, formerly accorded family status (Megathymidae), have been relegated to a subset of the subfamily Hesperiinae by recent authors. Adults (Figs. 99–101) are small to moderately large (FW 8–35 mm), stout bodied with powerful thoracic musculature; they are called skippers because of their quick, darting flights. The third axillary sclerite at the FW base is unusually wide and forms an irregular Y-shaped structure for muscle attachment, a defining character for Hesperiidae. The HW has an area of very small, specialized scales at the base of R + Sc, also not found in other Lepidoptera. Skippers typically perch with the wings outspread (Pyrginae, Fig. 100) or hold the HW out horizontally and the FW upright, slightly cocked open (Hesperiinae, Fig. 101), or they close both wings above the body like other butterflies, especially when taking nectar. The head is broader than the thorax, with the antennae widely separated and enlarged distally into a club with its tip (apiculus) attenuated and curved. The flagellum is strongly bent at the club in Pyrrhopyginae. The proboscis is well developed, and nearly all species feed at flowers, bird droppings, or other nutrient sources. Most skippers are rather drab, predominantly tan, brown, gray, or black, but many tropical species are colorful. The larval head is prominent (Fig. 102), frequently with protruding lobes, separated from the body by a constricted "neck" in all but the *Megathymus* group, a unique condition in Lepidoptera. In megathymids the head is narrower than the thorax, and the pupa moves in the larval tunnel, protruding at adult eclosion, a unique reversal among Obtectomera. The terminal body segment has an anal comb, analogous to that of some moths, used to flip frass from the larval shelter, sometimes remarkable distances. Trapetzinae in Australia feed on monocots, mainly Xanthorraceae, Poaceae, and Cyperaceae, as do Hesperiinae, mostly on grasses. Coeliadinae of the Old World tropics, Pyrrhopyginae in the New World, and Pyrginae specialize on dicots, with more than 50 plant families recorded, and a few feed on monocots. About 3500 species are described in 500+ genera, and many species complexes in tropical regions are not thoroughly studied. Hesperiidae are distributed worldwide except in New Zealand, with greatest richness in the Neotropical region.

Papilionoidea To most people butterflies are among the most conspicuous and recognizable insects. Their diurnal behavior, aesthetic beauty, and limited species numbers render them favorite subjects for beginning naturalists and amateur collectors and teaching insect metamorphosis to primary school children. Moreover, butterflies have been of special significance to biologists, in studies of geographical distribution patterns, chemical defenses and mimicry, migration, genetics and population biology, and host-plant relationships. In recent decades they have become poster children in conservation efforts. Butterflies are vastly better studied than most moths, but they are negligible in insect biodiversity, making up fewer than 0.1% of all insect species and 9% of described Lepidoptera (likely less than 4%, were moths equally well cataloged). Several adult skeletal features define members of this superfamily as monophyletic, and the antennae have apical clubs, without an apiculus like that of Hesperiidae and not subapically broadened as in sphingid, castniid, and sesiid moths. Systematists recognize four families of butterflies: Papilionidae, Pieridae, Lycaenidae (including riodinids), and Nymphalidae (including libytheids, satyrids, and danaids). There are an estimated 14,000 species, a total that is greatly inflated relative to moths and other insects, because they are more thoroughly studied, with a propensity by the industrious specialists to accord species status to geographically disjunct populations that differ in color patterns and size but not morphologically or by molecular analysis.

Papilionidae: The swallowtail family is one of the most easily recognizable of Lepidoptera, one whose phylogeny is well supported, based on several wing venation and skeletal features and the eversible gland of the larval prothorax (osmeterium). Adults (Fig. 103) are medium sized to very large (FW 14–105 mm), including the largest butterflies (birdwings of the Indo-Australian region); wings are broad, with the FW triangulate and the HW rounded, and often with one or more veins extended into "tails." Most are brightly colored, often aposematic, warning of their distasteful properties, and many swallowtails are models in mimicry complexes; some are polymorphic, with several forms each with different corresponding mimics, often nymphalids. Swallowtails are strong fliers and have been recorded dispersing several miles from a point of origin. Some tropical species engage in mass migrations. The larvae (Figs. 34 and 104) are plump, often ornate with filaments or protuberances. They appear bare but usually have numerous tiny secondary setae. They live exposed on foliage, inactive by day, and depend upon cryptic coloration for protection, often resembling bird droppings in early instars and then graduating to foliage or flower colors, as they grow, or they are aposematic, brightly colored (Fig. 34). The osmeterium is horn-like, forked, usually bright pink or orange, and everted to emit a foul aroma intended to ward off predators. The caterpillars feed on a wide variety of dicot angiosperms, including several groups on Aristolochiaceae and others on Magnoliaceae, Apiaceae, Rutaceae, Lauraceae, and other plants not used by most Lepidoptera. There are about 600 species in 26 genera, with virtually all the world's species described, distributed worldwide, with Parnassiinae in high latitudes and elevations of the Holarctic and Papilioninae mostly subtropical and tropical and with their greatest richness in the Old World tropics.

Pieridae: The whites and sulphurs make up a well-established, monophyletic family, based on several characters: the presence of pterin pigments in the wing scales; the foretarsi with inner claw subequal in length to the outer, whereas the inner is much shorter in other butterflies; and wing venation and thoracic skeletal features. Adults (Fig. 105) are small to medium sized (FW 11–48 mm), broad winged, mostly white, yellow, or orange, with some tropical species brightly colored, containing flavone pigments, mimicking other butterflies (e.g., South American Dismorphiinae). Most pierids display sexual dimorphism in color patterns, sometimes to the extreme, and many have marked seasonal variation. Remarkable mass migrations by some tropical pierids occur, often moving from seasonally dry to wet habitats. Larvae are slender caterpillars, relatively uniform in structure, without protuberances, and covered with short secondary setae, and each segment is divided superficially into six annulets. They are mainly green, including the head, or spotted with yellow and blue in species that feed in flowers. Some species possess an anal comb. Tropical Dismorphiinae feed on legumes, as do some Coliadinae, while most Pierinae specialize on Brassicaceae, Capparidaceae, Loranthaceae, or Santalaceae. The American genus *Neophasia* feeds on pines. Several species have achieved important pest status, particularly the cabbage white *(Pieris rapae),* which was introduced from Europe into North America in the 1880s, feeding on cabbage and other crucifer crops, and species of *Colias,* feeding on alfalfa. More than 1000 species in 75 genera have been described, including probably nearly all the world's species. This group is cosmopolitan except in New Zealand and the Pacific Islands, with greatest development in the tropics. Species range to the extreme limits of Lepidoptera habitats, *Colias* to 83° N latitude and *Baltia* to 5000 m (16,350 ft) in the Himalayas and several genera to similar elevations in the Andes.

Lycaenidae: The coppers (Fig. 106), hairstreaks (Fig. 107), blues (Fig. 108), and metalmarks (Fig. 110) together form a diverse butterfly family, with a remarkable array of larval biologies. Inclusion of the metalmarks (Riodininae) is debatable because they have several uniquely derived traits and because they have foreleg morphology and function that resemble those of Nymphalidae. However, exclusion of the riodinids leaves the remainder of the Lycaenidae an incomplete lineage (paraphyletic). Adults (Figs. 106–108, 110) are mostly small (FW 6–25 mm; Neotropical *Eumaeus* and African *Liphyra* reach 35 mm) and the upper surface of their wings is usually brightly colored, entirely or patterned, in blue, orange, or red, often brilliantly metallic, especially in the males, while the undersides, which are exposed when the butterfly is inactive and the wings are held together above the body, tend to be more cryptic, gray, brown, or green. The wings are relatively broad, the FW usually triangular and the HW rounded; most hairstreaks and a few blues have one or more slender filaments arising from the hind margin, often preceded by a colorful eyespot on the underside. During perching the wings are moved alternately, giving an impression of antennal movements, a behavior thought to deflect predator attack to the HW rather than to the head and thorax. Metalmarks exhibit a bewildering array of wing forms and color patterns, especially in tropical species, resembling diverse kinds of butterflies and moths. Lycaenid antennal bases are adjacent to and usually indenting the eyes. In Riodininae the antennae usually are long, more than half the FW length, and the forelegs are atrophied in males. The antennae are shorter and the male forelegs functional in other lycaenids. Lycaenid larvae are peculiar caterpillars, shaped like a sowbug, with the body segments broadened laterally and the small head retractable and hidden under the thorax; they are usually covered with short secondary setae, giving a velvety appearance (Fig. 109). Species that live in association with ants are bare, and Riodininae usually have long secondary setae. All lack eversible prothoracic glands characteristic of other butterfly caterpillars. Many Lycaenidae have evolved glands on the last abdominal segment that produce a sweet, honeydew-like fluid that is much sought after by ants, which display various behaviors. Some tend and "milk" the larvae on their food plants (thereby presumably warding off parasites and invertebrate predators) (Fig. 109), others transport the young caterpillars to their nests, where they are fed by the ants or eat the ant brood. A wide variety of flowering plants serve as hosts, including a few conifers and monocots. Most species are specialists, but some are polyphagous. Larvae of the African Poretiinae feed on algae and lichens. Those of Miletinae (African and 1 species in North America, *Feneseca*) feed exclusively on Homoptera or their secretions or in ant nests. The association with ants has developed in many Riodininae and unrelated genera of other lycaenids. Feeding on legumes has led to minor pest status for a few species, including the bean lycaenid *(Strymon melinus)* in North America, a polyphagous species also called the cotton square-borer. There are more than 6000 described species in 640+ genera, and many tropical taxa are not thoroughly studied. Lycaenids occur worldwide, with endemic species even in New Zealand and the Pacific Islands, but the majority occur in the Neotropics and Africa.

Nymphalidae: This is a large and diverse family that includes the typical Nymphalinae (brush-footed butterflies, admirals, checkerspots, Fig. 111), Libytheinae (snout butterflies), Satyrinae (wood nymphs, ringlets, Fig. 112), Heliconiinae (long wings, fritillaries, Fig. 113), Morphinae (morpho and owl butterflies, Fig. 116), and Danainae (milkweed and glasswing butterflies, Fig. 115). All possess three longitudinal ridges (carinae) on the ventral surface of the antennae that are unique in Lepidoptera, and the forelegs of males are reduced or modified (less so in Libytheinae), usually lacking claws and nonfunctional for walking. Adults are small to very large (FW usually 10–50 mm, ranging to 75 mm in tropical *Morpho* and *Caligo*), mostly broad winged except in Heliconinae, and usually brightly colored, often with orange, black, and white dominating, but mostly brown and tan in Satyrinae. Many tropical nymphalids are involved in mimicry complexes, either as models (Danainae, Heliconiinae) or as mimics of them or other distasteful butterflies and moths and/or they benefit in both roles. Glasswing butterflies (Ithomiini) live primarily in

deep shade of tropical forests and have sparsely scaled areas or transparent wings, with subtle color patterns, while owl butterflies fly at dusk. They, morphos, and satyrines have conspicuous eye-like spots near the margins of the wing undersides, presumably confusing would-be predators or diverting their attacks from the body. The larvae are cylindrical caterpillars with full complement of abdominal prolegs (Fig. 114), but there are diverse modifications, e.g., densely spinose or with dorsal projections (verrucae) that are spinose (Nymphalinae), smooth with filaments (Danainae), smooth with bifid caudal segment (Satyrinae), pubescent with hair tufts and usually bifid caudally (Morphinae). The pupa hangs head downward, attached by a cremaster, without a silken girdle. The larvae feed on a diverse array of flowering plants, with considerable specialization within subfamilies: Morphinae and Satyrinae almost exclusively on monocots, including Arecaceae, Bromeliadaceae, Heliconiacae, and Musaceae (a few species are pests on bananas) in the tropics, mostly Poaceae and Cyperaceae in the Holarctic, with 2 genera on Selaginellaceae; other nymphalids eat mostly dicot angiosperms, often specializing on plants with toxic chemicals (e.g., Heliconiinae on Flacourtiaceae, Passifloraceae, Urticaceae, Violaceae) or latex-producing plants (Danainae on Apocynaceae, Asclepiadaceae, Moraceae). About 6500 described species are placed in 630+ genera, occurring worldwide, ranging from Arctic–Alpine *Boloria* in the Holarctic to extremely rich tropical faunas in most subfamilies, several of which are not represented in New Zealand.

See Also the Following Articles
Caterpillars • Collection and Preservation • Scales and Setae • Trichoptera • Wings

Further Reading
Common, I. F. B. (1990). "Moths of Australia." Melbourne University Press and Brill, Leiden/New York.
Hering, E. M. (1951). "Biology of the Leaf Miners." Junk, The Hague.
Kristensen, N. P. (1984). Studies on the morphology and systematics of primitive Lepidoptera (Insecta). *Steenstrupia* **10**, 141–191.
Kristensen, N. P. (ed.) (1999). Lepidoptera, moths and butterflies, Vol. 1, Evolution, systematics, and biogeography. *In* "Arthropoda: Insecta" (M. Fischer, ed.), Part 35 of Vol. IV of "Handbook of Zoology." W. deGruyter, Berlin/New York.
Kristensen, N. P., and Skalski, A. W. (1999). Phylogeny and paleontology. *In* "Arthropoda: Insects." (N. P. Kristensen, ed.), Part 35 of Vol. IV of "Handbook of Zoology," Ch. 2, pp. 7–25. W. deGuyter, Berlin/New York.
Labandeira, C. C., Dilcher, D. L., Davis, D. R., and Wagner, D. L. (1994). Ninety-seven million years of angiosperm-insect association: Paleobiological insights into the meaning of coevolution. *Proc. Natl. Acad. Sci. USA* **91**, 12278–12282.
Medvedev, G. S. (ed.) (1987). "Lepidoptera," Part 1. Vol. IV of "Keys to the Insects of the European Part of the USSR." Oxonia Press, New Delhi.
Medvedev, G. S. (ed.) (1990). "Lepidoptera," Part 2. Vol. IV of "Keys to the Insects of the European Part of the USSR." Brill, Leiden/New York.
Mosher, E. (1916). A classification of the Lepidoptera based on characters of the Pupa. *Bull. Illinois State Lab. Nat. Hist.* **12**, 14–159.
Sbordoni, V., and Forestiero, S. (1984). "Butterflies of the World." Times Books, Random House, New York.
Scoble, M. J. (1992). "The Lepidoptera: Form, Function, and Diversity." Oxford University Press, Oxford.

Snodgrass, R. E. (1961). The caterpillar and the butterfly. *Smithsonian Misc. Collect.* **143**, 1–51.

Stehr, F. W. (ed.) (1987). Lepidoptera. *In* "Immature Insects," Chap. 26, pp. 288–596. Kendall/Hunt, Dubuque, IA.

Zimmerman, E. C. (1978). "Microlepidoptera," Parts 1 and 2. Vol. 9 of "Insects of Hawaii." University of Hawaii Press, Honolulu.

Lice

see *Phthiraptera*

Lice, Human
(*Pediculus* and *Pthirus*)

Terri L. Meinking

University of Miami School of Medicine

Lice are wingless, bloodsucking insects that belong to the order Phthiraptera. Although there are currently 4000 species of lice recognized, only 560 species suck blood and feed on mammals. Lice are very host specific; therefore human lice cannot be transmitted to or from other mammals. There are only three species of lice that infest humans: *Pediculus humanus humanus,* the body or "clothing" louse; *Pediculus h. capitis,* the head louse (Fig. 1) and *Pthirus* (or *Phthirus*) *pubis,* the crab or "pubic" louse (Fig. 3).

FIGURE 1 Head louse nit, adult female (left) and male (right). Body lice and head lice are identical in appearance except for size. The body louse is about 25% larger than the head louse. The female has an invaginated V-shape (center). (Photography by Bruce Hard.)

BIOLOGY

Lice, like all other insects, have six legs. They are hemimetabolous in development, meaning that they do not go through a complete metamorphosis like mosquitoes or fleas. There are three nymphal or "instar" stages that all look like a miniature adult. Within 7 to 12 days after eggs (nits) are laid, nymphs hatch, which will molt three times before becoming adults.

Within a few hours of hatching, the nymph must find a human blood meal or it will die of starvation and dehydration. At all stages, lice have a very tough, leathery cuticle capable of considerable expansion after feeding, usually taking in blood meals up to one-third of their own body weight every few hours. Regular feedings occur every 4 to 6 h with head and crab lice, although body lice can survive for days without a blood meal. Several feedings occur between the shed of each chitinous exoskeleton or cuticle. The first, second, and third nymphal stages last 3 to 4 days each. It is not until the final molt that the sex can be determined. Females are usually 20% larger than males of the same species, as well as longer, wider, and rounder, with the posterior portion of the female terminating in an invaginated V shape (Fig. 1). Gender identification is more difficult with *Pthirus* than with *Pediculus*.

Within 2 days of molting the female will feed several times, copulate, and begin laying an average of 3 to 6 eggs per day, with body lice laying more and crab lice laying fewer eggs. The female louse attaches the nits to hairs or fibers (in the case of body lice) by secreting a glue for which there is no solvent (Fig. 2). The life span of a louse from hatching through adult is 30 to 42 days.

HABITAT/EPIDEMIOLOGY

Pediculus h. humanus

The body louse lives in clothing or bedding and travels to the host only to feed. The nits are laid on the clothing fibers, especially in the seams and collars. This is truly an infestation of individuals who are unable to wash themselves or their clothing.

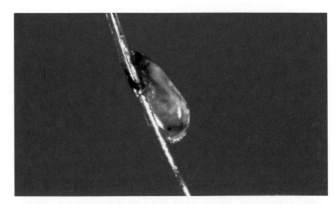

FIGURE 2 Head lice eggs are 0.8 mm in length. Note the glue that attaches the nit to the hair shaft. (Photography by Bruce Hard.)

Infestations with the body louse occur more commonly in individuals crowded together because of war, natural disaster, refugee status, or, in cooler climates, homelessness. They give all lice a "bad name" because people associate all lice with poor hygiene; however, this is the only human louse associated with lack of cleanliness.

Pediculus h. capitis

The head louse likes a clean healthy head and is found commonly on children 3 to 11 years of age. The nits are laid on the hair, usually close to the scalp for warmth. Head lice are primarily transmitted by contact, which is more frequent in younger children; however, older siblings, parents, childcare workers, and teachers may also be infested. Sharing of brushes, combs, hats, helmets, and other headgear and hair accessories also can result in transmission.

Pthirus pubis

The crab louse (Fig. 3) prefers the pubic and perianal areas, but contrary to older literature, can also be found on the beard, mustache, scalp, axillae, eyebrows and eyelashes, or any hairy part of the body. Unlike head lice, which tend to be somewhat particular, crab lice are found in all levels of society, and are generally transmitted by sexual contact. HIV-positive and immunocompromised individuals are more difficult to treat. Adult crab lice can remain alive off the host for at least 36 h; therefore, transmission by infested towels and bedding is more common than originally thought.

DISEASE TRANSMISSION

Pediculus h. humanus

Disease transmission primarily occurs as a result of contact with infested lice fecal pellets. The body louse is capable of transmitting the following diseases: epidemic typhus (caused by *Rickettsia prowazekii*), murine typhus (caused by *R. typhii*, trench fever (caused by *Bartonella quintana*), and relapsing fever (caused by *Borrelia recurrentis*).

Pediculus h. capitis and *Pthirus pubis*

Head lice and crab lice are capable of transmitting group A *Streptococcus pyogenes* and *Staphylococcus aureus*. Neither head nor crab lice have been studied as transmitters of blood-borne diseases, but this is an area of current interest.

TREATMENT

Pediculus h. humanus

Clothing and bedding should be disinfested by washing in hot water and then drying in a hot cycle (65°C, 149°F). Permethrin or malathion dusting powders are effective in treatment for mass eradication of body lice infestations. For individual cases, people should be treated with prescription permethrin 5% topical cream or oral ivermectin.

Pediculus h. capitis

Over-the-counter lice treatments that contain 1% permethrin or natural pyrethrin products can be used, but resistance has been reported in the United States and other countries. Prescription malathion lotion and oral ivermectin are the most effective treatments. Lindane products should not be used because of toxicity and resistance.

Pthirus pubis

Although head lice products may be used on crab lice, the most effective treatment is 5% permethrin cream or oral ivermectin. All hairy areas of the body should be treated, including the scalp.

Phthiriasis palpebrarum

Since topical treatments should not be used around the eyes, crab lice of the eyelashes should be treated with petroleum jelly, which suffocates lice and eggs and lubricates the lashes, making nit removal easier.

See Also the Following Articles

Blood Sucking • Medical Entomology • Phthiraptera

Further Reading

Alexander, J. O. (1984). "Arthropods and Human Skin." Springer-Verlag, Berlin.
Andrews, M. (1976). "The Life That Lives on Man." Taplinger, New York.
Meinking, T. L., and Taplin, D. (1995). Infestations. *In* "Pediatric Dermatology," 2nd ed. (L. A. Schachner and R. C. Hansen, eds.), pp. 1347–1392. Churchill Livingstone, New York.

FIGURE 3 Crab louse. *Pthirus pubis* is broader than *Pediculus* spp., and resembles a crab.

Meinking, T. L. (1999). Infestations. *Curr. Problems Dermatol.* **11(3)**, 75–118.

Meinking, T. L., and Taplin, D. (1995). Infestations: Pediculosis. *In* "Sexually Transmitted Diseases: Advances in Diagnosis and Treatment" (P. Elsner and A. Eichmann, eds.), pp. 157–163. Karger, Basel.

Meinking, T. L., Burkhart, C. G., and Burkhart, C. N. (1999). Ectoparasitic diseases in dermatology: Reassessment of scabies and pediculosis. *Adv. Dermatol.* **15**, 67–108.

Locusts

R. F. Chapman
University of Arizona

Locusts are medium-sized to large grasshopper-like insects that form swarms of hundreds of millions of individuals with the potential to migrate long distances in the tropics and subtropics. Locusts differ from grasshoppers in their responses to crowding. Locusts behave as solitary insects immediately after hatching or when maintained in isolation, but if they are forcibly crowded for as little as 6 h they subsequently tend to group together, or exhibit gregarious behavior. In contrast, if grasshoppers are kept in a crowd they usually remain as solitary insects and show no tendency to come together. A few grasshopper species, however, do have some tendency to gregarize when forcibly grouped, although this does not occur naturally. Thus, there is no absolute distinction between a "grasshopper" and a "locust."

Locusts do not comprise a single taxonomic group. Rather, they occur in three subfamilies of Acrididae, the Cyrtacanthacridinae, Oedipodinae, and Gomphocerinae. Even within a genus, some species exhibit the swarming habits of locusts, while others lack the habit and never swarm. This is most obvious in the genus *Schistocerca*, where a majority of the American species are nonswarming, and *Dociostaurus*, with many nonswarming species in Asia. *Locusta migratoria*, which extends from Australia and eastern Asia to Europe and West Africa, has a number of subspecies, which differ in their propensity to swarm. The species usually regarded as locusts and their distributions are given in Table I.

In addition to the behavioral change, locusts exhibit a marked color change when crowded. In isolation they are often green or exhibit a more or less uniform color matching that of the background; crowded locusts, however, exhibit a striking black and yellow or orange coloration in the nymphal stages. These changes, however, are not peculiar to locusts; similar changes are shown by some grasshoppers. The proportions of different parts of the body also differ between locusts reared in isolation and those reared in crowds. The most striking difference occurs in the migratory locust, where insects reared in isolation have a strongly crested pronotum (upper surface of the first thoracic segment), but in crowded insects the upper surface of the pronotum is saddle shaped.

PHASE THEORY OF LOCUSTS

Locusts do not swarm continuously; periods of swarming may last for several years but are separated by times when no swarms are reported (i.e., recession periods). The mystery of their apparent disappearance during recession periods was solved for *L. migratoria* by B. P. Uvarov in a paper published in 1921. He proposed the phase theory of locusts, suggesting that during recession periods the insects exist in a form that differs phenotypically and behaviorally from swarming locusts. He referred to these two forms as the "solitary" (later called "solitarious") and "gregarious" phases. *Locusta* in the solitarious phase had, until that time, been placed in a different genus, *Pachytylus*. Subsequent work showed that similar phases occur in the other locust species. Solitarious locusts are typically cryptically colored, relatively inactive as nymphs (but not necessarily as adults; see later), and live in isolation. Gregarious locusts are conspicuous, with contrasting colors; they form bands (as larvae) or swarms (as adults), and are usually highly mobile. The change from one form to another does not occur in a regular manner, but is dependent on environmental conditions. For example, a period of grouping may increase the tendency of nymphs to group (i.e., to become gregarious), but if they then become isolated again they will tend to lose these characteristics. Consequently, there is no regularity in the timing of outbreaks when swarms occur.

PHASE CHANGE IN THE FIELD

During recessions between outbreaks, solitarious locusts may be very widely distributed, but sometimes these insects are very uncommon. The transformation from solitarious to gregarious involves several discrete phenomena that were first recognized by J. S. Kennedy. Outbreaks are initiated by conditions that favor successful breeding, leading to an increase in the population size. Then the population becomes concentrated in particular areas as other parts become uninhabitable. This leads to aggregation in which the previously isolated individuals are forced into intimate contact with one another. Finally, in the process of gregarization, the behavior and physiology of the insects is changed, and they now tend to aggregate spontaneously. Concentration, aggregation, and gregarization are generally dependent on a drying out of the habitat following good breeding conditions.

Population increases may occur in any part of the distribution area of the insects, but swarming of most species seems to originate only in what are called outbreak areas. The peculiarity of these areas is that only here do the conditions for population increase, concentration, and aggregation coexist. The migratory locust has only a single outbreak area in Africa, even though the species is widespread and often common in many parts of the continent. This outbreak area is the delta of the middle River Niger in Mali. The area is unique because, in addition to local rain, it receives

TABLE I   The Species of Locusts

Subfamily/Species	Common name	Distribution
Cyrtacanthacridinae		
Anacridium melanorhodon	Sahelian tree locust	Sahel, eastern Africa, southwestern Arabian peninsula
Anacridium wernerellum	Sudanese tree locust	Sahel, eastern Africa
Nomadacris septemfasciata	Red locust	Southern Africa
Nomadacris succincta	Bombay locust	Southwest Asia
Schistocerca cancellata	South American locust	South America
Schistocerca gregaria	Desert locust	Northern Africa, Arabia, Indian subcontinent
Schistocerca piceifrons	Central American locust	Central America
Oedipodinae		
Chortoicetes terminifera	Australian plague locust	Australia
Locusta migratoria	Migratory locust	Southern Europe, Africa south of the Sahara, Malagasy Republic, southern Russia, China, Japan, Philippines, Australia
Locustana pardalina	Brown locust	Republic of South Africa, Mozambique
Gomphocerinae		
Dociostaurus maroccanus	Moroccan locust	Mediterranean countries, Middle Eastern countries east to Kyrgyzstan

considerable moisture from precipitation on the mountains of Senegal, the source of the Niger. From the mountains, the river runs inland to a low-lying area in Mali, where it branches to form a delta before flowing southwest through Nigeria to the Atlantic Ocean. The combined effects of rain and river flooding in the delta region produce an extended growing period for vegetation and so enable the locusts to have as many as four generations within a year, whereas elsewhere in Africa the species usually has only two. As a result, huge population increases can occur. However, as the floodplains dry out, suitable areas of vegetation become increasingly restricted, and the insects are first concentrated and then aggregated into smaller areas, where gregarization occurs. Biogeographical analysis of the occurrence of swarms during the last great outbreak of the migratory locust, which lasted from 1930 until 1940, shows clearly that the plague originated from the single outbreak area and spread progressively over Africa south of the Sahara.

The red locust, in southern Africa, unlike the migratory locust, has only a single annual generation. It has several outbreak areas in Tanzania and Zambia that, like the middle Niger, are floodplains. The red locust outbreak areas, however, have either no or very limited outflow of water. As a result, the water that accumulates and sometimes forms a lake has become salty over time. Nymphal development coincides with the rainy season, when extensive flooding produces lush, tall grasslands in which the locusts feed. As the vegetation dies, its distribution becomes more limited, perhaps as a consequence of increasingly saline conditions in the slightly lower parts of the floodplain and as the area of vegetation becomes more restricted, the locusts become concentrated, with the potential to give rise to swarms.

The desert locust differs from these species in that swarms arise in different places depending on the success of breeding and vegetation changes; there is no evidence of any single outbreak area from which the plagues of the 20th century originated.

MIGRATION

Locust swarms fly during the day and, if they are flying close to the ground, often tend to stream in one direction. This is still true at any one position within a higher flying swarm; in the swarm as a whole, however, the orientation of these streams is random. This would rapidly cause the swarm to disperse except that upon reaching the edge of the swarm, individuals turn back into it. It is not known what stimuli produce this behavior, but vision, sound, or even smell may be involved. Because the locusts within the swarm are, effectively, randomly oriented, the swarm itself has no directional movement and is carried downwind. The rate of displacement of swarms flying close to the ground is less than the airspeed because the insects tend to land at intervals, taking off again as the rest of the swarm passes. In high-flying swarms, however, this is not possible. The locusts may be carried on thermals as high as 3000 m above the ground, and then the swarms are displaced downwind at about the speed of the wind. If the winds are light and variable, swarm displacement is negligible. With sustained winds, however, displacements over hundreds or even thousands of kilometers can occur. This behavior is one of the factors enabling the desert locust to survive in some of the most arid regions on earth, the Sahara and Arabian deserts. Downwind displacement takes the insects to areas of wind convergence, where rain is most likely to occur, if it occurs at all, so that the chances of the insects breeding and producing viable offspring are greatly increased. Because wind patterns are not completely reliable, however, this strategy is not always effective. As a result, swarms of desert locusts in West Africa are sometimes carried out into the Atlantic or north to western Europe. The most spectacular recorded flight occurred in October 1988, when huge swarms were carried right across the Atlantic, with large numbers reaching the Caribbean and the northern coasts of South America, a distance of about 6000 km from the insects' source in West Africa.

The bulk of the work leading to our current understanding of swarm behavior was carried out by R. C. Rainey and Z. Waloff, working on the desert locust in the 1950s and 1960s. The assumption is that swarms of other locusts behave in the same way, although the other species do not generally form such massive swarms and are much less well studied.

It is also known that adults of the solitarious phases of at least some locust species migrate, but they do so at night. Evidence for night migration by solitarious individuals exists for *Anacridium* spp., *L. migratoria migratorioides* (the African subspecies), *Locustana pardalina,* and *S. gregaria. Nomadacris septemfasciata,* on the other hand, appears to be sedentary in the solitarious phase. In this respect, locusts are similar to tropical grasshoppers, many of which are sedentary, whereas a few are known to make extensive night migrations. These solitary migrants, unlike day-flying swarms, deliberately climb to relatively high altitudes (200–500 m above the ground) and then may maintain flight for some hours, although probably a majority of flights are relatively short. In the case of *L. m. migratorioides,* regular flights occur within the floodplains of the outbreak area and from them to the surrounding semiarid country, where breeding may occur. Return migrations to the floodplains also occur, and this strategy moves populations with the Inter-Tropical convergence, along which rain is likely. These flights are sometimes downwind, but there is also some evidence from radar observations that the insects can maintain a particular heading despite shifts in wind direction. These seasonal movements make an important contribution to the survival of the insects. Night flights by the Australian plague locust, *Chorthoicetes terminifera,* are also well documented, but these generally are of shorter range.

CONTROL OF PHASE

The physiology of phase change is not yet fully understood. It has been known for some time that grouping can be induced in isolated nymphs by touching individuals with fine wires dangling from a rotating circle, indicating that it is primarily physical contact with other locusts that initiates gregarization. More recent work has shown that touching the hind femora is more effective than touching other parts of the body. Presumably, the effects are registered by mechanoreceptors on the femora, leading to a change in the nervous system that alters the insect's behavior toward gregariousness. It is very likely, though not yet proved, that this sequence involves neuromodulators. A peptide hormone that enters the hemolymph via the corpora cardiaca induces the dark coloration of gregarious nymphs.

Pheromones play a part in the maintenance of gregarization. A number of experiments indicate that gregarious locusts of both sexes produce a gregarization pheromone. In adult desert locusts, benzaldehyde, veratrole, guaiacol, phenol, and phenylacetonitrile are its major components. This pheromone enhances the tendency to group as well as having some effect on color change. Solitarious locusts do not produce the full suite of compounds in comparable concentrations. It has been shown that the chemicals are produced from plant material ingested by the locusts and that bacteria are responsible for their production. Locusts reared on axenic (microbe-free) diets do not produce the pheromone. Mature males in the gregarious phase of both the desert and migratory locusts produce from epidermal glands a pheromone that accelerates maturation of insects of either sex. The major component of this pheromone in the desert locust is phenylacetonitrile. Its effect under natural conditions is, presumably, to tend to synchronize oviposition by the individuals in a swarm, which increases the likelihood that the first-stage nymphs, when they hatch, will be present in large numbers and so will be likely to interact with each other and gregarize. A chemical produced in the accessory glands of gregarious females of the desert locust promotes gregarious behavior and coloration in the nymphs hatching from the eggs; solitarious females do not produce the chemical. The chemical is contained in the frothy material that forms a plug above the egg mass and that is interpolated in spaces between the eggs. There is thus a marked intergenerational effect of phase with gregarious females producing offspring that already have some characteristics of gregarious individuals.

EVOLUTION OF SWARMING BEHAVIOR

It was once thought that the contrasting coloration of gregarious nymphs was likely to have a function in promoting gregarious behavior, but experimental evidence does not support this. Recent studies with the desert locust show that when the locusts feed on plants containing deterrent chemicals, such as the alkaloid hyoscyamine, predaceous lizards rapidly learn to avoid individuals with gregarious coloration but do not avoid solitariously colored nymphs even when they have eaten the same food. Other plants in the desert areas that are the habitat of *S. gregaria* also contain potentially noxious compounds, and it may be that the gregarious coloration results from selection for warning coloration. This, in turn, may have led to gregarious behavior, since aposematic insects commonly group together. Whether similar arguments can be applied to other locust species is not known.

The tendency to migrate is clearly an adaptation to living in arid habitats, enabling the insects to colonize new areas before the initial food supply is totally depleted. This is most clearly seen in the desert locust. Because some grasshoppers in these same habitats exhibit annual migrations, and some solitary locusts are also known to migrate, it must be supposed that swarm migrations arose from these individual movements. This, however, involved a switch from nighttime migration within the insects' boundary layer, where flight can be directed by the insect, to daytime flight that is often outside the boundary layer and displacement is largely determined by the wind.

See Also the Following Articles

Migration • Orthoptera

Further Reading

McCaffery, A. R., Simpson, S. J., Islam, M. S., and Roessingh, P. (1998). A gregarizing factor present in the egg pod foam of the desert locust *Schistocerca gregaria. J. Exp. Biol.* **201,** 347–363.

Pener, M. P., and Yerushalmi, Y. (1998). The physiology of locust phase polymorphism: An update. *J. Insect Physiol.* **44,** 365–377.

Sword, G. A., and Simpson, S. J. (1999). Is there an intraspecific role for density-dependent colour change in the desert locust? *Anim. Behav.* **59,** 861–870.

Sword, G. A., Simpson, S. J., el Hadi, O. T. M., and Wilps, H. (2000). Density-dependent aposematism in the desert locust. *Proc. R. Soc. Lond. A* **267,** 63–68.

Tawfik, A. I., Tanaka, S., de Loof, A., Schoofs, L., Baggerman, G., Waelkens, E., Derua, R., Milner, Y., Yerushalmi, Y, and Pener, M. P. (1999). Identification of the gregarization-associated dark-pigmentotropin in locusts through an albino mutant. *Proc. Natl. Acad. Sci. U.S.A.* **96,** 7083–7087.

Uvarov, B. P. (1966). "Grasshoppers and Locusts," Vol. 1. Cambridge University Press, Cambridge, U.K.

Uvarov, B. P. (1977). "Grasshoppers and Locusts," Vol. 2. Cambridge University Press, Cambridge, U.K.

Magnetic Sense

John Klotz
University of California, Riverside

Rudolf Jander
University of Kansas, Lawrence

Experimental evidence for magnetic field sensitivity has been reported in insects belonging to various orders, including Isoptera (termites), Diptera (flies), Coleoptera (beetles), Hymenoptera (ants and bees), and Lepidoptera (moths and butterflies). There is evidence that a few insect species obtain directional information from geomagnetic fields for compass orientation. Two alternative properties of the local geomagnetic vector could serve this purpose. Like a number of birds, animals either make use of the direction in which the dip angle points ("inclination compass") regardless of the field's polarity or sense the local declination and polarity ("polarity compass"). Which of these alternatives pertains to insects has been investigated in only one species, the yellow mealworm *(Tenebrio molitor)* (Coleoptera), which makes use of the polarity compass. The sensory system that mediates magnetoreception in insects has not been identified definitively, though one favored hypothesis is based on the detection of magnetic fields using particles of magnetite.

Magnetic compass orientation can be useful for insects in the context of home range (topographic) orientation and during long-distance migration, especially in the absence of visual compass cues. Both honey bees building combs in darkness and blind termites building oriented mounds appear to use magnetoreception for aligning their structures. On the other hand, it is difficult to imagine how an insect could make adaptive use of sensing the absolute strength of the local geomagnetic field.

MAGNETIC COMPASS ORIENTATION

Multiple directional orientation or compass orientation to artificially induced magnetic fields has been shown in several species. The insects always responded to changes in the magnetic field's declination, which implies sensing of magnetic polarity. In contrast, a geomagnetic inclination compass, as used by some migrating birds, has not been demonstrated for any insect.

Home Range Orientation in Social Insects

The magnetic sense of insects and its adaptive importance have been most thoroughly investigated in social insects such as ants, bees, and termites that require highly developed orientation skills to find and then communicate to their nestmates the location of resources within their home ranges. Much is known about how ants and bees use visual cues such as the sun, polarized light, the moon, and landmarks for spatial orientation. For most navigating insects, the primacy of visual cues must be taken into account in experiments designed to investigate magnetic field orientation. Indeed, an insect's competence in magnetic field orientation may be hidden if more salient cues such as light are present. However termite workers and soldiers, which have poor vision at best, rely more on nonvisual cues.

THE GEOMAGNETIC FIELD AS A BACKUP CUE FOR ANTS
Experiments with naturally foraging weaver ants, *Oecophylla smargdina,* revealed that their sense of direction is stronger and more accurate under clear than under overcast skies. In addition, when ants were tested for orientation indoors after displacement from their outdoor foraging trail, those exposed to overcast conditions maintained the correct trail heading but others exposed to clear skies did not. The difference in response indicated the ants' use of a nonvisual cue that is overridden by celestial cues if they are present. Support for

670

this hypothesis came with further experiments showing that ants trailing in dim, diffuse light reversed their heading when exposed to an artificially induced magnetic field with polarity opposite that of the geomagnetic field. Wood ants, *Formica rufa,* have also been shown to use magnetic field orientation when directional light cues are unavailable. These experiments with two species of ants suggest a hierarchically organized orientation system designed so that the primary light compass is more efficient than the magnetic compass, which serves as a backup when directional light cues are absent.

MAGNETIC DIRECTION AS A REFERENCE FOR LAND-MARK LEARNING IN HONEY BEES   In flight, the honey bee, *Apis mellifera,* also uses a magnetic compass in home range orientation. Foraging bees approaching the vicinity of their "target" learn the precise location of resources with respect to surrounding landmarks so they can return to the same place in the future. The most popular hypothesis assumes fast "snapshot"-like recall of near-target constellations of landmarks. The returning bee finds the target location by matching the current perception of landmarks with the "snapshot memories" of them. While learning the spatial relations of landmarks, bees face in a preferred compass direction, using directional light and the geomagnetic field. Honey bees trained in an artificial field with polarity reversed to the geomagnetic field face landmarks in the opposite direction. Hence, their magnetic compass may provide directional information as a frame of reference for the memorized landmarks.

A MAGNETIC CUE FOR HOMING TERMITES   All termites are social insects that have evolved a different set of adaptations for home range orientation. Termites are specialized for foraging underground and in enclosed spaces. The eyesight of workers and soldiers either has regressed or has been lost completely. All foraging termites depend heavily on pheromone trails for finding their way back home. However, as in ants, such trails do not provide any cue that helps to discriminate between the outward and homeward direction. The geomagnetic field could provide such a cue. This has indeed been demonstrated in the blind African grass-harvesting *Trinervitermes geminatus* (Termitidae: Nasutitermitinae) which, unlike the majority of termite species, is an open-air forager. Homing orientation in returning workers is substantially disturbed by distortions of the geomagnetic field due to weak bar magnets. Whether geomagnetic field orientation is widespread among termites is still an open question.

Migration in Moths and Butterflies

Long-distance compass migration has evolved in relatively few species of insects as an adaptation for dispersal and for coping with seasonal climatic changes. Examples are found among dragonflies (Anisoptera), true bugs (Heteroptera: e.g., the large milkweed bug, *Oncopeltus fasciatus*), and moths and butterflies (Lepidoptera). The implied geographic orientation mechanism

could, plausibly, make use of magnetic compass orientation, especially during nocturnal migration and migration under dense overcast. Some evidence supports this possibility.

Two nocturnal cross-country migrants, the large yellow underwing moth, *Noctua pronuba,* and the heart-and-dart moth, *Agrotis exclamationis,* reversed their direction of orientation in four-armed bioassay arenas when they were exposed to reversals of an artificial magnetic field.

The most spectacular example of geographic orientation in insects is the massive annual fall migration of the monarch butterfly, *Danaus plexippus.* Eastern North American populations of monarchs migrate over 3000 km to winter in the mountains of Mexico. Experimental evidence substantiates their use of a sun compass for geographic orientation, and some experiments suggest their use of a magnetic compass as well.

For years entomologists have speculated about magnetic compass orientation in migrating monarch butterflies. The first supportive evidence came in field experiments: migratory butterflies were exposed to a brief pulse of an induced magnetic field 15,000 times the intensity of the geomagnetic field, whereupon the treated butterflies were released and tracked to determine their direction of flight. Two control groups of butterflies were also tested. One of the control groups received the same treatment as the experimental group except for the magnetic pulse. The other group received no treatment and was composed of naturally occurring butterflies migrating through the test area. Both control groups of butterflies kept their normal migratory flight direction to the southwest, but directional headings of the magnetically treated group were randomly distributed, indicating disorientation. Because these experiments were conducted on clear days, however, the sun was also available as a cue. Thus conflicting information from the butterflies' sun and magnetic compasses may have caused the insects' disorientation.

OTHER EFFECTS OF MAGNETIC FIELDS ON THE ORIENTATION BEHAVIOR OF INSECTS

Some of the earliest and most detailed studies of magnetic field sensitivity were also conducted with social insects. Once again the honey bee was the focus of intense research, but this time the investigators studied its communication behavior. To recruit and guide nestmates to a newly discovered resource, a honey bee performs a dance indicating to her followers the direction and distance of the resource from the hive. The dance is usually performed in darkness on a comb's vertical surface. The flight direction to the resource in reference to the sun is transposed by the bee to the direction of her dance with respect to gravity on the comb. If the resource is in the direction of the sun, the dance is directed upward; if away from the sun, downward; and if in other locations, at various angles to the vertical. Small systematic errors in the directional component of this dance are correlated with daily fluctuations in intensity of the geomagnetic field. These errors disappear when the bees dance in an artificial magnetic field that compensates for the earth's field.

When honey bees are forced to dance on a horizontal surface in the dark, their dances become aligned with the cardinal and intercardinal axes of the geomagnetic field. This response intensifies when the magnetic field is artificially enhanced and disappears when the field is canceled.

Evidence also suggests that honey bees use magnetic fields in nest construction. Bees that are transferred to a new hive construct combs that are oriented in approximately the same magnetic direction as those in their old hive. In one study, bees built abnormal combs when they were exposed to magnets during construction.

Among the most spectacular and magnificent termite mounds are those of *Amitermes meridionalis* (Termitidae: Amitermitinae) in tropical Australia near the town of Darwin. These massive tombstonelike black structures reach up to 4 m in height, and their long horizontal axes align near perfectly north–south. Similar but less perfectly oriented and shaped mounds are constructed by *A. laurensis* on the Cape York Peninsula of Australia. It is more than tempting to refer to these mound builders as "magnetic termites."

Indeed some good evidence supports this label. If a strong, permanent magnet is buried underground where a new colony starts to build, the resulting structure is misshapen and lacks clear orientation.

In addition to nest alignment, numerous studies have identified insects that align the body axis to magnetic fields. Resting termites, flies, and honey bees adopt positions aligned with the cardinal axes of a magnetic field.

Finally, orientation transfer sometimes occurs from light orientation to magnetic compass orientation. When a yellow mealworm moves away from a light source, it remembers its current magnetic compass bearing. If the directional light is turned off, the course direction is maintained with the help of the remembered magnetic compass bearing.

POSSIBLE SENSORY MECHANISMS

Several hypotheses have been proposed to explain how animals sense magnetic fields. There is circumstantial but no definitive evidence in insects for two such sensory mechanisms. One type of mechanism could be based on the magnetic sensitivity of some chemical or photochemical reactions. If such reactions are linked to light reception in the eye, then changing the wavelength of ambient visible light could alter the directional orientation to the geomagnetic field. Such effects have been obtained in male *Drosophila melanogaster* (Diptera) as well as in some birds. A second mechanism could be based on the interaction between the geomagnetic field and intracellular, submicroscopic magnetite particles that have been found in some insects, including ants, honey bees and monarch butterflies.

See Also the Following Articles

Dance Language • Learning • Monarchs • Orientation

Further Reading

Jander, R. (1977). Orientation ecology. *In* "Encyclopedia of Ethology" (B. Grzimek, ed.), pp. 145–163. Van Nostrand Reinhold, New York.

Jander, R., and Jander, U. (1998). The light and magnetic compass of the weaver ant, *Oecophylla smaragdina*, (Hymenoptera: Formicidae). *Ethology* **104**, 743-758.

Perez, S. M., Taylor, O., and Jander, R. (1999). The effect of a strong magnetic field on monarch butterfly *(Danaus plexippus)* migratory behavior. *Naturwissen schaften* **86**, 140–143.

Walker, M. M. (1997). Magnetic orientation and the magnetic sense in arthropods. *In* "Orientation and Communication in Arthropods" (M. Lehrer, ed.), pp. 187–213. Birkhäuser Verlag, Basel.

Weaver, J. C., Vaughan, T. E., and Astumian, R. D. (2000). Biological sensing of small field differences by magnetically sensitive chemical reactions. *Nature* **405**, 707–709.

Wiltschko, R., and Wiltschko, W. (1995). "Magnetic Orientation in Animals." Springer-Verlag, New York.

Malaria

William K. Reisen
University of California, Davis

Malaria is a pyrogenic (fever-producing) disease caused by infection with one of four species of parasitic protozoa in the genus *Plasmodium* and is the most important arthropod-transmitted pathogen in the world today, in terms of numbers of cases, deaths, and economic burden. Acquired from the bite of an infective *Anopheles* mosquito or from infected blood products, malarial parasites continue to suppress development in Africa and parts of Asia and are emerging as a critical health issue in tropical Central and South America. Expanding and rapid global commerce and travel provide an effective conduit for malaria parasites to be reintroduced into currently malaria-free areas.

HISTORY AND DISCOVERY

There is little doubt that there has been a long evolutionary association between humans and malaria. The ascent of the human species and its dispersal from the African center of origin into Europe and Asia most likely were accompanied by host-specific and coevolved species of plasmodia. *Vivax* malaria possibly accompanied early Asian voyagers to the New World across the Pacific Ocean, whereas *falciparum* malaria probably was introduced into the New World from Africa with the post-Columbus slave trade.

Malarial disease has impacted human health throughout recorded history. References to seasonal intermittent fevers abound in the earliest Assyrian, Chinese, and Indian religious and medical writings; however, it was not until the 5th century B.C. that Hippocrates related the distribution of cases to specific seasons and residence near marshes. Malaria has altered the course of human history by afflicting political, scientific, and religious leaders as well as decimating invading armies. The Romans associated marshes with intermittent fevers and attempted to reduce their occurrence through swamp drainage. The term malaria was derived from the Italian *mal' aria* (bad air), drawing from the association between foul-smelling marsh gases and the occurrence of this disease. In the 1600s, powders from the bark of the Peruvian "quina-quina" tree (now known as quinine) were discovered in South America and shown to be therapeutic against certain seasonal fevers. Shortages of these natural medicinal powders and the resulting impact of malaria on military campaigns during World War I stimulated research to develop antimalarial drugs and resulted in the formulation of atebrin in 1930 and chloroquine in 1934. Although marshes, mosquitoes, poor living conditions, and malaria were associated throughout history, it was not until 1880 that Laveran first observed parasites in the blood of fever patients and 1897 that Ross found malarial parasites in an *Anopheles* mosquito that previously had fed on a malaria patient. The following year, Ross worked out the complex life cycle of the malaria parasite using a *Culex* mosquito–sparrow malaria model. Shortly afterward Grassi and colleagues elucidated the life cycle of the human parasite and with Manson demonstrated that protection from mosquito bites provided protection from infection.

The now-confirmed relationship between malarial infection and mosquitoes led to expanded control efforts by chemically treating or reducing surface water where larval mosquitoes occurred. In 1936, the insecticidal properties of DDT were discovered by Muller and Weisman. DDT spraying was used in the successful eradication of introduced *Anopheles gambiae* mosquitoes from Brazil in 1939–1940 and Egypt in 1942–1945. These successes and the eradication of malaria from the United States by spraying the inside walls of houses with DDT set the stage for the 14th World Health Assembly to adopt a global malaria eradication strategy that was implemented by the World Health Organization from 1957 through 1969.

However, failure to sustain the effort and funding after initial success, disregard for the magnitude of the malaria problem in Africa, and the onset of insecticide resistance in several key vector species resulted in a global collapse of this effort that was followed by a general resurgence of malaria throughout tropical regions of the world. In 1992, the World Health Organization again selected malaria as the target disease for a global initiative to improve human health and in 1998 launched the new "Roll Back Malaria" campaign to reduce malaria by 50% by the year 2010. Only time will determine if the human host finally will rid itself of its malaria burden.

PARASITES AND LIFE CYCLE

Four species of human *Plasmodium* may be identified, in part, by clinical symptoms such as the pattern of fever and chills (Table I), morphology and staining characteristics of the parasite within red blood cells, antigenic properties determined by serology, or genetic sequence. The *Plasmodium* life cycle is complex (Fig. 1). The female *Anopheles* becomes infected when gametocytes are ingested during blood feeding. Sexual union of gametocytes occurs in the mosquito midgut, after which the resulting ookinete penetrates the midgut wall and forms an oocyst. After asexual reproduction, the oocyst ruptures and the motile sporozoites make their way to the salivary glands. Humans become infected during blood feeding by the infective mosquito host when sporozoites are expectorated with mosquito saliva into the wound created by the mosquito bite. After entry into the circulatory system of the human host, sporozoites rapidly enter the liver where asexual reproduction occurs. Liberation from the liver may occur rapidly or be delayed, depending upon the species and strain of parasite (Table I). Once in the bloodstream, parasites rapidly enter red blood cells where they multiply asexually. The synchronous liberation of parasites from the red blood cells results in

TABLE I   Characteristics of Human Infection with Four Species of *Plasmodium*

Characteristic	*Plasmodium* species			
	vivax	*ovale*	*malariae*	*falciparum*
Incubation period (days)	13 (12–17)[a]	17 (16–18)	28 (18–40)	12 (9–14)
Exoerythrocytic cycle[b]	Present	Present	??	Absent
Merozoites/tissue schizont	>10,000	15,000	2,000	40,000
Erythrocytic cycle (h)[c]	48	49–50	72	ca. 48
Parasitemia (avg. per ml)	20,000	9,000	6,000	20,000–500,000
Attack severity	Mild–severe	Mild	Mild	Severe
Paroxysm duration (h)	8–12	8–12	8–10	16–36 or longer
Relapses	++	++	+++	None
Period of recurrence	Long	Long	Very long	Short
Duration of untreated infection (years)	1.5–3	1.5–3	3–50	1–2

Note. Modified from Bruce-Chwatt (1980).
[a]Strain dependent, may be up to 9 months.
[b]Continued production of merozoites within the liver.
[c]Time between red blood cell infection and rupture indicated by the pattern of paroxysms.

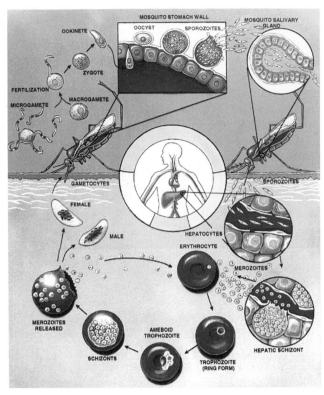

FIGURE 1 Generalized life cycle of the four human-infecting *Plasmodium* species. (From the U.S. Centers for Disease Control and Prevention.)

rhythmic paroxysms characteristic of malarial disease. As the infection progresses, gametocytes are formed in the peripheral bloodstream from where they are ingested by blood-feeding mosquitoes, thereby completing the life cycle.

MOSQUITO VECTORS

Only female mosquitoes in the genus *Anopheles* serve as definitive hosts for the four species of human malarial parasites. Of these, species in the subgenus *Cellia* account for most of the current global transmission and include members of the *A. gambiae* complex (*gambiae, arabiensis*) and *A. funestus* in subSaharan Africa and the *A. culicifacies* complex, *A. fluviatilis* complex, *A. stephensi*, and *A. minimus* in Asia. Historically, the *A. maculipennis* complex was important in the Mediterranean and Europe, whereas species in other subgenera such as *A. darlingi* and *A. albitaris* have been responsible for the resurgence of malaria in South America.

EPIDEMIOLOGY AND DISEASE

Malaria remains a critical health problem of global proportions, causing an estimated 500 million clinical cases and 2.7 million deaths annually. It is a health problem of crisis proportions and a severe economic burden in 90 countries inhabited by 2.4 billion people (roughly 40% of the world population). The temporal concordance between crop growing and malaria transmission seasons frequently results in a serious loss of agricultural productivity. The distribution of malaria in time and space and the efficiency of transmission are limited by temperature requirements for the development of the *Plasmodium* parasites within their poikilothermic *Anopheles* hosts and the abundance, bionomics, and behavior of the different *Anopheles* vectors. *P. vivax* can develop at temperatures as low as 14.5°C and is found at colder latitudes and higher elevations than *P. falciparum*, which requires temperatures above 16°C (Fig. 2). In addition to ambient temperature, transmission efficiency depends almost entirely on *Anopheles* bionomics expressed as vectorial capacity; species that are long lived, rapidly develop parasites, and feed frequently on humans are the most efficient transmitters of malaria parasites.

The incubation period between infection and clinical illness varies among malarial species and strains, being shortest for *P. falciparum* and as long as 9 months for some northern strains of *P. vivax* (Table I). Illness is characterized by the malarial paroxysm and, if untreated, increases in severity as the number of parasites multiplies logarithmically. Typical

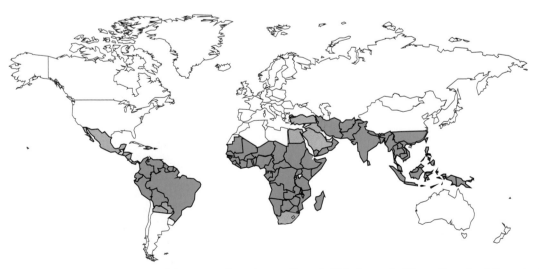

FIGURE 2 Global distribution of malaria. Extended *vivax* area shown in gray. (From the U.S. Centers for Disease Control and Prevention.)

complications include anemia and splenomegaly. In *P. falciparum* infection, changes in the structure of infected red blood cells creates congestion and blockages within the circulatory system, causing coma (brain hemorrhages), jaundice, and "blackwater fever" with the passing of black urine (liver failure, nephritis), and severe dysentery (dehydration, renal failure). Infection during pregnancy frequently leads to abortion, stillbirth, and neonatal mortality. Some liver stages of *P. vivax* and *P. ovale* remain dormant and, if untreated, may relapse for years after the initial infection.

Infection imparts transient immunity that is maintained in endemic areas by almost constant reinfection. In hyperendemic areas, morbidity and mortality are highest among the nonimmune, including travelers, infants, and pregnant women. Adults in these areas tolerate chronic infections and present a constant source of gametocytes for mosquito infection. Malaria mortality has selected for resistant and semiresistant phenotypes from the genome of affected human populations, leading to the evolution and persistence of traits such as sickle cell anemia and Duffy blood group antigen that alter the structure and surface of red blood cells, making them resistant to parasite infection.

TREATMENT AND CONTROL

Treatment has emphasized the use of chemical derivatives of the quinoline ring, originally found in quinine and present in chloroquine and primaquine. Primaquine has the important feature of destroying the liver stages of *vivax* and *ovale,* thereby eliminating relapses. Resistance has led to the development of alternative drugs, including proguanil, mefloquin, pyrimethamine, and sulfonamide; however, in some areas of Southeast Asia treatment of patients infected with resistant strains must revert to quinine with tetracycline. Extracts from plants of the genus *Artemisia* represent a new class of drugs from Asia that are undergoing clinical trials for the treatment of drug-resistant malaria.

Public health control efforts targeting eradication combined active-case detection and treatment with adult mosquito abatement. Active-case detection emphasized complete village-level surveys, the presumptive treatment of fever cases with chloroquine, and verification of malaria infection by slide examination. Residual house spraying with DDT and later malathion targeted indoor resting *Anopheles* females in an attempt to interrupt the transmission cycle. This combined approach resulted in remarkable successes in areas such as Sri Lanka, Pakistan, and India, where the primary vector, *A. culicifacies,* rests almost exclusively within houses and cattle sheds. However, interest and funding to sustain successful programs waned and eventually collapsed. Recently eradication has changed to control, is limited to passive case detection and treatment, and has been incorporated into general village-level health programs.

In addition to research to improve and expand the number of drugs for patient therapy, three control approaches currently are being investigated: (1) personal protection by sleeping under pyrethroid impregnated bed nets (the simplest and least expensive technology that is currently available), (2) vaccination (protection that probably will be of short duration and is expected within 7–10 years), and (3) genetic manipulation of vector competence in *Anopheles* (species-specific, costly, and untried). Success of malaria intervention in developing countries most likely will continue to be hindered by inadequate delivery systems, political unrest, and the low socioeconomic level of most rural populations. In the modern era, successes in malaria control typically have accompanied advances in education, economic well-being, and medical delivery systems.

See Also the Following Articles
DDT • Medical Entomology • Mosquitoes

Further Reading
Bruce-Chwatt, L. J. (1980). "Essential Malariology." Heinemann, London.
Garnham, P. C. C. (1966). "Malaria Parasites and Other Haemosporidia." Blackwell Sci., Oxford.
Macdonald, G. (1957). "The Epidemiology and Control of Malaria." Oxford University Press, London.
Malaria Foundation International. (1998). Worldwide impact of malaria. http://malaria.org/bginfo.html.
Oaks, S. C., Mitchell, V. S., Pearson, G. W., and Carpenter, S. C. J. (eds.) (1991). "Malaria: Obstacles and Opportunities. Report of the Committee for the Study of Malaria Prevention and Control: Status Review and Alternative Strategies." Division of International Health, Institute of Medicine, National Academy Press, Washington, DC.
Russell, P. F., West, L. S., Manwell, R. D., and Macdonald, G. (1963). "Practical Malariology." Oxford University Press, London.
World Health Organization. (1998). Malaria. Fact Sheet No. 94. www.who.int/inf-fs/en/fact094.html.

Mallophaga

see *Phthiraptera*

Mantodea
(Praying Mantids)

Lawrence E. Hurd
Washington and Lee University

The praying mantis (from the Greek for "prophet"), or mantid (from the name of one family in this group), has been mentioned in literature at least since the Egyptian *Book of the Dead*. However, surprisingly little formal scientific research has been done on mantids compared with many other orthopteroid insects, e.g., grasshoppers. No doubt at least some of the reason for this lies in the greater relative economic importance of insects that compete with humans for food. In any case, to date there has been but a single book summarizing research with mantids, *The Praying Mantids*. There are more than 1800 species worldwide, most of which inhabit the tropics. Among the 20 or so species that occur in the continental United States, the best known, most abundant mantids are 3 species that were introduced from Europe and Asia over the past century.

CLASSIFICATION OF MANTIDS

There is some disagreement as to the proper classification of mantids within the class Insecta. Although they clearly are related to cockroaches, grasshoppers, crickets, stick insects, and termites, most modern systematists recognize that mantids have sufficiently distinctive morphological characteristics to warrant taxonomic separation from these groups. The prevailing view is that praying mantids comprise several families (including Mantidae), either within the order Dictyoptera, suborder Mantodea, or else within a separate order, the Mantida. The most obvious morphological features that characterize this group are a highly mobile head, elongated prothorax (most anterior midbody segment), and especially the raptorial front legs attached to the prothorax. The forelegs are folded when the animal is at rest, giving it an attitude of being in prayer (hence, the common name for the group). The combination of these features is unique among insects.

Much of the uncertainty over classification of praying mantids is because a coherent phylogeny of this group is lacking. The fossil record for this group is both scanty, and recent: fossil mantids date no earlier than the Cenozoic. Therefore, scientists have had to infer phylogenetic relationships from a variety of features such as body shape, presence of auditory organs, and genetics of sex determination. These features have led to different conclusions as to whether, for example, modern mantids have monophyletic or polyphyletic origins. The recent discovery of the new order, Mantophasmatodea, hints at a close relationship between mantids and stick insects (Phasmida). However, based upon morphological and molecular evidence to date, the closest relatives of mantids appear to be Blattodea (cockroaches) and Isoptera (termites), both of which have much earlier origins in the fossil record.

Most biologists agree that there are many more species of insects on earth than have been identified so far, and this is almost certainly true with mantids. There are currently 8 families and 28 subfamilies in this suborder. The most important family (about 80% of all named species) is Mantidae, consisting of 21 subfamilies and 263 genera. This family contains some of the most abundant and widely distributed species on earth including the Chinese *(Tenodera aridifolia sinensis)* and European *(Mantis religiosa)* mantids that occur widely in the temperate zones of Europe, Asia, and North America.

Three families in Mantodea are well represented in the Old World in addition to Mantidae: Amorphoscelidae (two subfamilies found widely distributed in Africa and Australia), Empusidae (eight genera in Africa and Asia), and Eremiaphilidae (two genera of ground-dwelling desert species in Africa and Asia). Another tropical family, Hymenopodidae, contains three subfamilies that include some relatively rare and spectacular flower mimics such as *Hymenopus coronatus* (Asia) and *Pseudocreobotra ocellata* (Africa).

The remaining three families of Mantodea are less diverse tropical groups: Mantoididae, (with a single neotropical genus, *Mantoida*), Chaeteessidae (with only one neotropical genus, *Chaeteessa*), and the most primitive family in the order, Metallyticidae (with a single Malaysian genus, *Metallyticus*, named for their characteristic metallic coloration).

NEUROPHYSIOLOGY AND BEHAVIOR

Perhaps the majority of scientific studies of praying mantids, particularly during the past few decades, have involved the interaction of neurophysiology and behavior. These interactions include the role of binocular vision in estimating distance, hearing of ultrasound and its possible use for avoidance of bat predation, and behaviors associated with defense, sex, and prey capture. Mantids are models of behavioral complexity beyond the imaginations of earlier researchers. They are capable of integrating much detailed information from their environment and have exhibited an astonishingly sophisticated array of responses to stimuli.

Sexual behavior and cannibalism in mantids are particularly noteworthy, partly because they have received much anecdotal mention in the literature. Sexual behavior varies among species, but in general females attract males through a combination of airborne pheromones and visual cues. In many species females either cannot or do not fly, and so males find females by flying upwind along the pheromone plume. This places males at greater risk of predation than females because they are more apparent to birds while in flight. When a male finds a female he incurs even greater jeopardy from his intended mate, depending on her hunger level. Sometimes males are simply captured and eaten before they have a chance to mate, but unless the female is very hungry, he usually is able to mount her. However, a hungry female may decapitate and partially consume a male during copulation without interrupting the transfer of sperm.

The noted French naturalist of the late 19th and early 20th centuries, J.-H. Fabre, described in lurid detail the cannibalistic mating habits of female mantids in his laboratory. This behavior was once interpreted as "adaptive suicide" by the male, to invest both his sperm and his nutrients in the next generation. This requires the assumption that a cannibalized male can be sure he is the father of a female's brood and not simply a food item for a female that has already been fertilized by a previous suitor. The simpler, modern explanation for this behavior is that a female attracts males both for sperm and for nutrition at the end of the growing season when alternate prey are scarce and she has to gain significant mass to produce viable eggs. Males have no choice in the matter, because they cannot discern which pheromone-emitting females have been mated (Fig. 1).

FIGURE 1 A mating pair of Chinese mantids, *Tenodera aridifolia senensis*. The female's abdomen is already swollen with eggs. This is the most widespread and abundant species in the eastern United States and may have the widest global distribution of any mantid species.

ECOLOGY

Studies of the feeding in praying mantids link behavior and ecology. Praying mantids are bitrophic, feeding both on herbivorous arthropods and on other carnivores (e.g., spiders), including cannibalizing each other. The fact that all of these processes may be occurring simultaneously in the same ecosystem can complicate definition of the ecological role of these predators in ecosystem structure and dynamics.

Experimental studies show that bitrophic mantids have both direct (prey reduction) and indirect (prey enhancement) effects, because competition with, or predation on, other predators may reduce predation on some prey species. Mantids tend to eat many arthropods that are beneficial to plants, including pollinators such as bees and butterflies and predators such as wasps and spiders. Whether it is a good idea to add these predators to one's garden as an agent of biological pest control is not always clear. Much more evidence is required before generalizations can be made with confidence, but experiments have demonstrated that under natural conditions mantids can instigate a trophic cascade (top-down effect) whereby plant productivity is enhanced when mantids feed on herbivorous insects.

See Also the Following Articles
Hearing • Orthoptera • Predation

Further Reading
Helfer, J. (1963). "How to Know the Grasshoppers, Cockroaches and Their Allies." Brown, Dubuque, IA.
Hurd, L. E., and Eisenberg, R. M. (1990). Arthropod community responses to manipulation of a bitrophic predator guild. *Ecology* **76**, 2107–2114.
Prete, F. R., Wells, H., Wells, P. H., and L. E. Hurd (eds.) (1999). "The Praying Mantids." Johns Hopkins Press, Baltimore.

Mantophasmatodea

Klaus-Dieter Klass
Museum für Tierkunde, Dresden

Mantophasmatodea are hemimetabolous, wingless pterygote insects, 11 to 25 mm in length, found in Africa. Their body structure is fairly generalized, but a dorsal process on the tarsi, an unusual course of the subgenal sulcus on the head, and a medioventral projection on the male subgenital plate are unique features for this order. Discovered in 2001, the Mantophasmatodea are the most recently described order of insects. Knowledge about them is expected to increase rapidly.

SYSTEMATICS AND DISTRIBUTION

The description of the order and its first two species was based on the first extant specimens that were recognized: two museum specimens described as *Mantophasma zephyra* (collected in 1909 in Namibia; Fig. 1) and *M. subsolana* (collected in 1950 in Tanzania). Members of *Mantophasma* have fairly small eyes and ventral rows of spines on the fore- and midleg femora and tibiae. Two other extant species, yet undescribed, also occur in Namibia. One of them (the "gladiator") has larger eyes than *Mantophasma* and stout spines on the thoracic terga. Specimens recorded from the Western and Northern Cape Provinces of South Africa represent several additional undescribed species; some were collected 100 years ago. *Raptophasma* from Baltic Eocene amber (about 45 mya), with two known species, resembles *Mantophasma*, but has large eyes and stouter, spineless legs. These Tertiary fossils show that the order once also occurred in northern Europe, at a time when its climate was warm and humid.

FIGURE 1 *M. zephyra*, female. (Reprinted from Klass *et al.* 2002, *Science* **296**, 1456–1459. Copyright 2002 American Association for the Advancement of Science.)

ANATOMY

The hypognathous head lacks ocelli and bears orthopteroid mouthparts and long, multisegmented antennae. Wings are lacking. The pleuron of the prothorax is fully exposed. The legs have elongate coxae. In the five-segmented tarsi (Fig. 2A) the three basal tarsomeres are fused. The dorsal membrane beyond the third tarsomere (tm3) bears a characteristic triangular process (dpt in Fig. 2A), and the arolium (arl) of the pretarsus is conspicuously large.

In the abdomen, both tergum I and coxosternum I are free from the metathorax. The small spiracles lie in the pleural membrane and have a muscular closing device. In the male (Fig. 2B, C), coxosternum IX forms a subgenital lobe with a median spatulate process (spp in Fig. 2B) but without styli. The phallomeres (male genitalia) are reduced to membranous lobes around the gonopore. Behind them a transverse, asymmetrically produced sclerite articulates upon tergum X, which is similar to the vomer in Phasmatodea. The female (Fig. 2D) has a short subgenital lobe formed by parts of coxosternum VIII. The genital opening lies above it on segment VIII. The ovipositor comprises clawlike gonoplacs (third valves), blunt gonapophyses VIII (first valves), and gonapophyses IX (second valves) fused with the gonoplacs and interlocked with the gonapophyses VIII. The one-segmented cerci are long claspers in the male but short in the female (Fig. 2B–D).

The foregut has a large proventriculus (gizzard) armed with weak sclerites that terminate posteriorly in three successive whorls of lobes. Midgut ceca are a pair of short and wide lateral pouches. The abdomen has a ventral diaphragm. In the nerve cord, abdominal ganglion VII is free from the terminal ganglion including neuromeres VIII and the following. The egg lacks a defined operculum but has a circumferential ridge; the chorion displays a hexagonal pattern of grooves that are traversed by delicate bars.

PHYLOGENETIC RELATIONSHIPS

Mantophasmatodeans superficially resemble insects of the other "orthopteroid" orders. However, they lack the apomorphies (i.e., derived characters) of these, such as prognathous head, prothoracic repellent glands, and elongated female subgenital plate of Phasmatodea; the perforated tentorium, female subgenital lobe from coxosternum VII, and fused abdominal ganglia VII and VIIIff of Dictyoptera (mantises, cockroaches, and termites); the prognathous head, membranous sac on abdominal segment I, and loss of muscled closing devices of abdominal spiracles found in Grylloblattodea; and the pronotum overfolding the prothoracic pleura, the thickened hind femora, and the anterior intervalvula in the ovipositor found in Orthoptera.

Mantophasmatodea are assigned to the Pterygota by their lack of a noncuticular trunk endoskeleton and of the lateral parts of abdominal tergum XI, and to the Pterygota–Neoptera by the valvelike gonoplacs lacking styli and the presence of

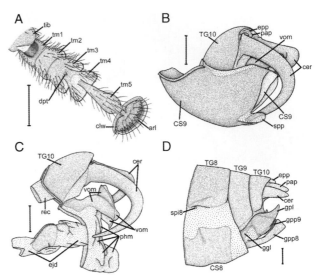

FIGURE 2 (A) *M. zephyra*, dorsal view of tarsus of female. Lateral views of male *M. subsolana* postabdomen with (B) and without (C) coxosternum IX, genitalia exposed. (D) Lateral view of female *M. zephyra* postabdomen. Scale: 0.5 mm. Abbreviations: arl, arolium; cer, cerci; clw, claw; CS8,9, abdominal coxosterna VIII and IX (subgenital plates); dpt, process beyond third tarsomere; ejd, ejaculatory duct; epp, epiproct; ggl, gonangulum; gpl, gonoplac; gpp8,9, gonapophyses VIII and IX; pap, paraproct; phm, phallomeres; rec, rectum; spi8, spiracle VIII; spp, spatulate process of male subgenital lobe; TG8,9,10, abdominal terga VIII, IX, and X; tib, tibia; tm1–5, tarsomeres 1–5; vom, vomerlike element. (Part A reprinted from the Lehrbuch der Speziellen Zoologie, Vol. 1, Part 5: Insecta. Copyright Spektrum Akademischer Verlag, Heidelberg, Berlin. Parts B–D reprinted with permission from Klass *et al.* 2002, *Science* **296**, 1456–1459. Copyright 2002 American Association for the Advancement of Science.)

apodemes on the abdominal spiracles. Otherwise, their phylogenetic position is unclear. Phasmida and Grylloblattodea are the most promising candidates as the sister group of Mantophasmatodea.

HABITATS AND LIFE HISTORY FEATURES

Mantophasmatodea prey on other insects, which they catch by means of their strong and usually spinose fore- and midlegs. They are found in relatively dry and stony habitats, where occasional heavy rain leads to formation of temporary pools of water. The structure of the egg chorion suggests that the egg can overcome temporary flooding by plastron respiration.

See Also the Following Articles
Grylloblattodea • Insecta, Overview • Mantodea • Phasmida

Further Reading

Klass, K.-D., Zompro, O., Kristensen, N. P., and Adis, J. (2002). Mantophasmatodea: A new insect order with extant members in the Afrotropics. *Science* **296**, 1456–1459.

Zompro, O. (2001). The Phasmatodea and *Raptophasma* n. gen., Orthoptera incertae sedis, in Baltic amber (Insecta: Orthoptera). *Mitt. Geol. Palaeontol. Inst. Univ. Hamburg* **85**, 229–261.

Zompro, O., Adis, J. and Weitschat, W. (2002). A review of the order Mantophasmatodea. *Zool. Anzeiger.* **241**.

Marine Insects

Lanna Cheng

Scripps Institution of Oceanography

Although insects are undoubtedly the most common animals on land, very few species appear to live in the sea. However, they are actually rather well represented in diverse coastal marine or saline habitats. A marine insect is any insect that spends at least part of its life cycle in the marine environment, which includes any habitat from the upper intertidal to the open ocean. Among the 30 or so recognized insect orders, marine members occur in more than one-half of them (Table I). The most important species are found in Collembola, Heteroptera, Homoptera, Coleoptera, and Diptera. There are also many species of Mallophaga (biting lice) and Anoplura (sucking lice) whose hosts live in or on the sea (mammals or seabirds). A Web page on marine insects, www.unk.edu/marineinsects, is available.

HABITATS

Marine habitats can be divided either by salinity or by their position relative to the tidal level. Three types of saline habitats are generally accepted, based on their salt content (in parts per thousand): brackish (0.5–32), sea (34–37), and inland saline (0.5–250). Marine biologists, on the other hand, have traditionally divided coastal habitats into various zones according to their coverage by seawater or exposure to the sun. Three major zones are recognized: supralittoral (covered only during highest spring tides), littoral or intertidal (covered regularly between high and low tides), and sublittoral (never exposed even during the lowest low tides). An additional important habitat for marine insects is the pelagic zone, which comprises the open ocean far from the shore.

The majority of marine insects occur in the intertidal zones, which can be further categorized by the types of vegetation associated with them, e.g., seagrasses and rushes *(Spartina, Juncus)*, seaweeds (green, blue-green, brown, or red), mangroves *(Rhizophora, Avicennia, Bruguiera, Sonneratia)*, or other higher plants *(Xylocarpa, Acanthus)*. The salinity of water in the various intertidal habitats tends to be variable or brackish. Larvae of several marine chironomids (Diptera) live among submerged vegetation in the sublittoral zone, which may include various green plants *(Enhalus, Halophila, Halodule)* and algae *(Halimeda, Corallina)*.

The occurrences of various insect orders in different marine habitats are given in Table I. Five habitat categories are used in this table: pelagic, coastal, intertidal, mangrove, and saltmarsh. Brackish water habitats are commonly associated with mangroves in the tropics but with saltmarshes in temperate regions.

TABLE I Occurrences of Insects in Marine Environments by Taxonomic Grouping and Habitats

Taxonomic group	Common name	Habitat				
		P	C	I	M	S
Subclass Apterygota						
Protura	proturans	–	–	–	–	–
Collembola	springtails	–	+	+	+	+
Diplura	diplurans	–	–	–	–	–
Microcoryphia	jumping bristletails	–	–	+	+	+
Thysanura	bristletails	–	–	+	+	+
Subclass Pterygota						
Ephemeroptera	mayflies	–	–	–	–	+
Odonata	dragonflies and damselflies	–	–	+	+	+
Blattodea	cockroaches	–	–	–	–	–
Isoptera	termites	–	–	–	l	+
Mantodea	mantids	–	–	–	–	–
Grylloblattodea	icebugs	–	–	–	–	–
Phasmatodea	stick insects (walkingsticks)	–	–	–	+	–
Orthoptera	grasshoppers, crickets, etc.	–	–	+	+	+
Dermaptera	earwigs	–	–	+	–	–
Embiidina	webspinners	–	–	–	–	–
Zoraptera	zorapterans	–	–	–	–	–
Plecoptera	stoneflies	–	–	–	–	–
Psocoptera	psocids, booklice	–	–	–	–	+
Mallophaga*	chewing lice	+	+	+	+	+
Anoplura*	sucking lice	+	+	+	+	+
Thysanoptera	thrips	–	–	–	–	+
Heteroptera	true bugs	+	+	+	+	+
Homoptera	cicadas, aphids, etc.	–	–	+	+	+
Megaloptera	alderflies, etc.	–	–	–	–	–
Raphidioptera	snakeflies	–	–	–	–	–
Neuroptera	lacewings, etc.	–	–	–	–	+
Coleoptera	beetles	–	–	+	+	+
Strepsiptera*	strepsipterans	–	–	–	–	+
Mecoptera	scorpionflies	–	–	–	–	–
Trichoptera	caddisflies	–	–	+	+	+
Lepidoptera	butterflies and moths	–	–	–	+	+
Diptera	flies	–	+	+	+	+
Siphonaptera	fleas	–	–	–	–	–
Hymenoptera	bees, wasps, and ants	–	–	+	+	+

Note. P, pelagic; C, coastal; I, intertidal; M, mangrove; S, saltmarsh; +, present; –, absent or no data; *, only habitats of hosts marine. From Cheng and Frank (1993).

TAXONOMIC GROUPS

Apterygota

Among the five known orders, marine members are found only in Collembola, Microcoryphia, and Thysanura. There are few marine thysanurans. The genus *Petrobius* has several marine species living on rocky shores in Europe, whereas species of *Neomachilis* can be found living under rock or in crevices in the upper intertidal in California and probably elsewhere in North America. Some species in at least six families of Collembola (Onychiuridae, Hypogastruridae, Neanuridae, Isotomidae, Entomobryidae, and Acraletidae) live in various intertidal habitats. The best studied and most widely distributed is *Anurida maritima* (Neanuridae), commonly found in rocky upper intertidal zones or saltmarshes. It lives in crevices among rocks and comes out to feed at low tide when the habitat is exposed. It is able to withstand submergence under seawater during high tide for periods up to 4 h by surrounding itself with an air bubble that acts as a compressible gas gill. Its orientation rhythm is endogenous and synchronized with the tides. These animals are usually negatively phototactic (i.e., going away from light), but between 2 and 7 h after low tide a large proportion of the population becomes positively phototactic (i.e., going toward light). Brightness of the horizon appears to be the main cue for the animals to move toward higher ground, where they seek shelter after foraging at low tide. When population densities become too high for any given crevice, these collembolans emerge and allow themselves to be dispersed at high tide by currents. They may then become stranded on the beach. Not infrequently huge aggregations consisting of millions of collembolans can be seen on beaches in various parts of the world.

Heteroptera

This is one of the most important orders in the marine environment, with nine families represented. Four families are semiaquatic and live at the water surface, five are shore dwellers, and only one, the Corixidae, is truly aquatic.

FAMILY GERRIDAE This is by far the most conspicuous and diverse family in the marine environment, with three subfamilies (Halobatinae, Trepobatinae, and Rhagadotarsinae), five genera (*Asclepios, Halobates, Stenobates, Rheumatometroides,* and *Rheumatobates*), and over 60 species. They can be found in various habitats ranging from near-shore mangrove streams, intertidal reefs, coastal lagoons, bays and estuaries, to the open ocean. The majority of the species are found in the Indo-Pacific region. The best studied genus in terms of taxonomy, distribution, ecology, phylogeny, and evolution is *Halobates* (Fig. 1). This genus is almost exclusively marine and contains 45 described species, including 5 that live a wholly pelagic life on the ocean surface. Although no living *Halobates* are known from the Mediterranean, a fossil species was described from an Eocene deposit (45 mya) in Verona, Italy.

FAMILY VELIIDAE This is also represented in the marine environment by three subfamilies (Rhagoveliinae, Microveliinae, and Haloveliinae), five genera, and more than 50 species. The genera *Trochopus* (5 species) and *Husseyella* (3 species) are confined to coastal bays, mangroves, and estuaries of the Americas, whereas *Xenobates, Halovelia,* and *Haloveloides* are know only from the Indo-Pacific. The latter three genera consist of small bugs, with adults measuring not more than 3 mm in body length. *Xenobates* spp., common among mangrove plants, are often overlooked, but there are at least 16 species. *Halovelia* is the largest genus of the Veliidae, with more than 35 species. These tiny bugs live in crevices among intertidal rocks and corals on tropical seashores and

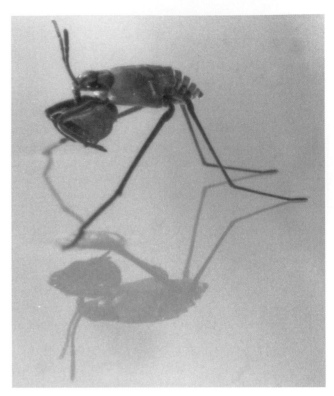

FIGURE 1 *H. sericeus* female (Heteroptera: Gerridae) feeding on *Drosophila* at the sea surface.

emerge to feed only at low tide. Although extant *Halovelia* species are confined to the Indo-Pacific, a fossil species was recently described from an Oligo-Miocene Dominican amber (15–45 mya).

FAMILY MESOVELIIDAE Only two genera are represented in the marine environment. While most species of *Mesovelia* are found in freshwater habitats, at least one is halophilous and regularly found in tidal water. However, both known species of *Speovelia* are marine.

FAMILY HERMATOBATIDAE This is a monotypic, exclusively marine family with eight recognized species, all except one being found in the Indo-Pacific. These unique bugs are associated exclusively with intertidal rocks or coral rubble, where they hide in crevices during high tide and come out to feed at low tide. However, some must remain active at the sea surface because individuals have been caught in net tows offshore or with light during the night at high tide.

FAMILY CORIXIDAE This is the only truly aquatic heteropteran family to be found in saline environments. Although about 60 species belonging to 12 genera have been reported from saline waters, most of them normally occur in freshwater habitats but are able to tolerate saline conditions. Many species are common in inland saline lakes, but only a few are found there predominantly. *Trichocorixa verticalis,* the most widely distributed species, is commonly found in saltmarsh pools, but specimens have occasionally been captured in plankton tows near shore. It is most tolerant of salinity changes and has been found to live and even to breed in ponds with a salinity approaching saturation (>300 ppt). Unlike most marine gerrids, which are wingless and flightless, adult corixids are winged and often disperse by flight.

SALDIDAE AND OTHER SHORE BUGS Five families of shore bugs (Saldidae, Gelastocoridae, Ochteridae, Omaniidae, and Aphelocheiridae) have representatives in marine environments. The most important family is Saldidae, which has at least 15 genera with more than 50 marine species worldwide. Most live in salt marshes. Winged forms occur in most shore bugs, and some species are rather strong fliers.

Homoptera

The Homoptera are rather poorly represented in the marine environment. In the Aphididae the best studied are *Pemphigus* spp. which feed on the roots of the saltmarsh *Aster* in Britain and probably elsewhere in Europe. Several species of herbivorous homopterans (Delphacidae, Issidae, Cicadellidae, and Cicadidae) feed and breed in seagrass beds in tidal saltmarshes. *Prokelisia marginata* (Delphacidae) is confined to beds of the seagrass *Spartina alterniflora* along the Atlantic coast of North America.

Coleoptera

Representatives of more than 20 families of beetles are found in marine environments, but none are truly aquatic. Most of the species occur in the intertidal zones among sand, rocks, algae, or wrack. Some are found in brackish waters, saltmarshes, or sand dunes. The most important marine families are Staphylinidae, Carabidae, Curculionidae, and Tenebrionidae.

FAMILY STAPHYLINIDAE This is the most important family, with more than 300 marine species. The genus *Cafius* is exclusively marine, with about 50 species. They are generally found on beaches associated with wrack (piles of stranded seaweeds), where they feed (as predators) and breed. The genus *Bledius,* with well over 400 species, has about 10% occupying marine habitats. Unlike most other staphylinids, they are not predatory. Females lay eggs in burrows and guard them from fungal attack and predation (a presocial behavior). The narrow openings of burrows of *B. spectabilis* have been shown to prevent flooding by tides. Adults and larvae of some other *Bledius* species dwell in inland salt flats. *Bryothinusa,* a genus of at least 24 species of small beetles, is exclusively marine.

OTHER FAMILIES Tiger beetles (Carabidae, subfamily Cicindelinae) include about 2300 species that occupy mainly terrestrial habitats. Adults of a few *Cicindela* species are common on sandy seashores and invade the intertidal areas during low tide. These predatory beetles run very fast and are difficult to catch. Females lay eggs in burrows where the larvae, after hatching, may remain for 2 or more years in the larval stage. Females of certain *Cicendela* species have been found to choose soils with specific salinity or shade conditions for laying eggs. Some genera of Curculionidae are known exclusively from driftwood or stranded seaweed on beaches, where they breed and the larvae develop. Beetles in several other families (Chrysomelidae, Cerambycidae, Curculionidae) are pests of tropical mangrove trees, feeding on flower buds, leaves, or bark and in some instances causing considerable damage.

Trichoptera

Caddisflies are predominantly freshwater insects, but several families breed in brackish water. The Chathamidae are exclusively marine and are known only from New Zealand and Australia. The four species belong to two genera, *Chathamia* and *Philanisus.* The adults are winged and can be found flying among intertidal rock pools. Eggs of the most widely distributed species, *P. plebeius,* are sometimes laid in the coelomic cavity of intertidal starfish or among coralline algal turf. The larvae use bits of coralline algae to construct their tubes (Fig. 2). They feed on various intertidal algae that may remain submerged at high tide.

Diptera

In addition to saltmarsh mosquitoes (Culicidae), biting midges (Ceratopogonidae), horse flies, and deer flies (Tabanidae), some of which are of great medical and/or economic importance, many other dipterans are found in various saline habitats. Almost all adult flies are winged, but the larvae of many species are truly aquatic and may remain submerged throughout their entire larval lives. The most commonly encountered nonbiting beach insects are probably seaweed flies belonging to at least five families (Coelopidae, Dryomyzidae, Muscidae, Borboridae, and Anthomyiidae). They are all associated with wrack or cast seaweeds where the adults feed and breed, and their life cycles tend to be synchronized with the tidal rhythm. Members of the

FIGURE 2 *P. plebeius* larvae in tubes constructed with coralline alga (Trichoptera. Chathamidae).

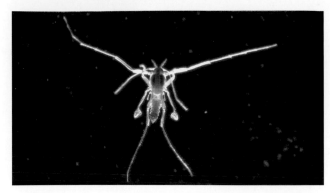

FIGURE 3 *Po. cottoni* male (Diptera: Chironomidae).

predatory family Canaceidae are exclusively marine and occur in the intertidal zone. Some of the most common invertebrates of inland salt lakes are brine flies in the family Ephydridae. Millions of these flies can sometimes be seen clustering along the lake shores, where they provide abundant food for shore birds.

Nonbiting flies in the families Chironomidae, Dolichopodidae, and Tipulidae are often associated with intertidal algal turf or submerged marine plants, where the larvae live and feed. The best studied is undoubtedly the chironomid *Clunio marinus,* whose life cycle is controlled by lunar as well as circadian rhythms. The most curious chironomid is probably *Pontomyia* (Fig. 3), an exclusively marine genus with four species. *Pontomyia* has an extremely short adult life span (30 min to 3 h), dying shortly after mating or egg laying. The eggs, embedded in a gelatinous matrix, are laid in a coil that sinks to the sea bottom. The larvae feed on submerged marine algae, and the pupae float to the sea surface shortly before emergence. The timing of emergence may be controlled by light, lunar, and tidal cycles.

Other Orders

Caterpillars of many species of moths feed on the leaves of mangrove trees, but their biology is poorly known. Other insect orders are represented in the marine environments by only a few species, e.g., the earwig *Anisolabis littorea* (Dermaptera) in New Zealand and the larvae of two sisyrids (Neuroptera) that live in brackish-water sponges. In addition, certain species of grasshopper (Orthoptera), ant (Hymenoptera), and even termite (Isoptera) can be common in some saltmarsh or mangrove areas.

See Also the Following Articles
Aquatic Habitats

Further Reading
Cheng, L. (ed.) (1976). "Marine Insects." North-Holland, Amsterdam.
Cheng, L. (1989). Factors limiting the distribution of *Halobates* species. *In* "Reproduction Genetics and Distributions of Marine Organisms" (J. S. Ryland and P. A. Tyler, eds.), pp. 357–362. Olsen & Olsen, Fredenborg, Denmark.
Cheng, L., and Frank, J. H. (1993). Marine insects and their reproduction. *Oceanogr. Mar. Biol. Annu. Rev.* **31,** 479–506.
Christiansen, K., and Bellinger, P. (1988). Marine littoral Collembola of north and central America. *Bull. Mar. Sci.* **42,** 215–245.
Denno, R. F., and Peterson, M. A. (2000). Caught between the devil and the deep blue sea, mobile planthoppers elude natural enemies and deteriorating host plants. *Am. Entomol.* **46,** 95–109.
Evans, W. G. (1980). Insecta, Chilopoda, and Arachnida: Insects and allies. *In* "Intertidal Invertebrates of California" (R. H. Morris, D. P. Abbott, and E. C. Haderlie, eds.), pp. 641–658. Stanford University Press, Stanford, CA.
Foster, W. A., and Benton, T. G. (1992). Sex ratio, local mate competition and mating behaviour in the aphid *Pemphigus-Spyrothecae. Behav. Ecol. Sociobiol.* **30,** 297–230.
Hoback, W. W., Golick, D. A., Svatos, T. M., Spomer, S. M., and Higley, L. G. (2000). Salinity and shade preferences result in ovipositional differences between sympatric tiger beetle species. *Ecol. Entomol.* **25,** 180–87.
Hogarth, P. J. (1999). "The Biology of Mangroves." Oxford University Press, Oxford.
Soong, K., Chen, G. F., and Cao, J. R. (1999). Life history studies of the flightless marine midges *Pontomyia* spp. (Diptera: Chironomidae). *Zool. Stud.* **38,** 466–473.
Zinkler, D., Rüssbeck, R., Biefang, M., and Baumgärtl, H. (1999). Intertidal respiration of *Anurida maritima* (Collembola: Neanuridae). *Eur. J. Entomol.* **96,** 205–209.

Mating Behaviors

Darryl T. Gwynne
University of Toronto

Mating behavior is typically viewed as comprising all events from pair formation through courtship to the final breakup of the mating pair. In most pterygote insects, sperm transfer is achieved through copulation. In contrast, in the few studied apterygotes, including both Insecta (Archaeognatha and Thysanura) and Ellipura (Collembola, Protura, and Diplura), sperm transfer is indirect; the spermatophore is placed on the substrate and is picked up by the female either following a period of courtship or with the pair making no contact at all. This article focuses on events occurring after the male and female have made physical contact; pair formation in insects is covered elsewhere. The main theme here is the function and adaptive significance of mating behaviors.

There is a vast amount of published information on the mating behaviors of insects. These behaviors have traditionally been viewed as relatively invariant within species. However, it is now evident that insect mating can show a great deal of adaptive variation and flexibility. As an introduction to this variation, consider insects that use carrion, a resource that can attract both males and females and thus serves as a location for mating. The complexity and plasticity of mating behavior observed in carrion insects easily rival those of other animals, including vertebrates (even Shakespeare's Romeo felt that "more courtship lives in carrion-flies than Romeo").

One courting carrion fly is attracted to the dry hide and bones of large old carrion sources, the main larval food for the

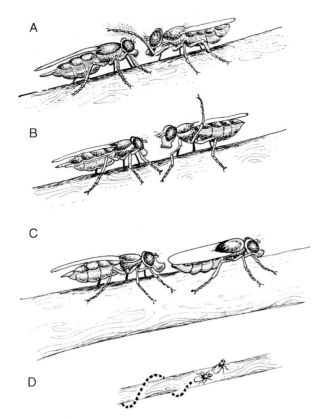

FIGURE 1 A fly similar to piophilids in exhibiting complex male courtship is the otitid, *Physiphora demandata*. The male first taps the female with a foreleg (A), then raises a middle leg (B), and turns and presents his abdomen to the female, who extends her proboscis to touch his abdomen (C). This can be followed by the female backing up in a spiral path, appearing to pull the male backwards (D). (Reproduced, with permission, from Alcock and Pyle (1979) *Z. Tierpsychol.* **49**, 354.)

species. The fly *Prochyliza xanthostoma* (Piophilidae) shows a remarkably complex courtship that can last for over 15 min (see Fig. 1 for another example). A courting male approaches the female while stepping rapidly from side to side and striking his abdomen downward. Males can repeat these vigorous movements and vary in the degree to which they display. If the female stops moving, the male stops courting, orients, and then slowly creeps toward her, occasionally repeating earlier parts of his routine. Individual males vary greatly in the vigor and length of the courtship display and this may represent variation in a signal of male quality used by females to select the best mates (Table I, Nos. 5 and 6).

Courtship by males of another temperate-zone carrion fly also appears to mediate female discrimination among males. Females of the fly *Dryomyza anilis* (Dryomyzidae) lay eggs on small carrion items such as dead fish. For this fly, courtship on the carrion occurs *after* copulation. A single courtship sequence consists of the male's genital claspers tapping vigorously on the female's external genitalia and then lifting and releasing her abdomen. Males vary in the number of genital tapping sequences performed and the number of sequences correlates with greater fertilization success. Bouts of tapping are followed by oviposition during which the male guards his mate from rivals. A male's success in fertilization is apparently achieved by the female biasing the distribution of sperm within her sperm storage organs (see Box 1). A similar influence on the success of courting male red flour beetles (*Tribolium castaneum*: Tenebrionidae) comes from a display in which the male rubs the female's elytra.

In a beetle that buries carrion, considerable variation occurs in reproductive behavior after the sexes have paired up and mated.

TABLE I   The Functions and Context of Mating Behaviors

	Before copulating	While inseminating (copulating)	After insemination and copulation
(1) To communicate information about sex (gender), possibly to suppress aggressive (in males) or cannibalistic tendencies (in predatory species).	X		
(2) To synchronize mating behavior, such as when physiological mechanisms synchronize the behavior of the sexes.	X	X	
(3) To perform movements associated with positioning of genitalia, transferring ejaculates, and uncoupling.	X	X	X
(4) To communicate species information to prevent costly interactions (e.g., mate-finding movements or inviable offspring) with the wrong species.	X	X	
(5) To communicate information about direct benefits (for mates or offspring) supplied during or after mating such as:			
(a) fecundity or number of ejaculated sperm (fertility) and the ability to supply nutrients (nuptial meals) and	X	X	
(b) territory quality, or level of parental care.	X	X	X
(6) To communicate information about indirect benefits (i.e., for offspring) such as compatibility of genotypes or genetic quality.	X	X	X
(7) To communicate competitive ability to rivals.	X	X	X
(8) To resolve struggles between the sexes that reflect conflict over whether to mate at all, when to terminate copulation, or whether the partner mates with another individual.	X	X	X

Box 1. A Broader View of Courtship: The Concept of "Cryptic" Sexual Selection

If males behave in order to maximize the number of surviving offspring, then male success is best estimated as fertilization success rather than success in mating many females. Thus, both direct competition between males and female discrimination should not end at copulation. The full development of this insight has coincided with the advent of a number of molecular-genetics methods to assign paternity. In insects the potential for paternity competition is high because females typically mate with more than one male and store their long-lived sperm in specialized organs. Indeed, males have been found to possess adaptations that incapacitate, physically displace, or remove rival ejaculates, examples being copulatory movements in dragonflies, such as *Calopteryx maculata,* in which penis brushes remove virtually all rival sperm from the female sperm storage organs. In fact, male insects can enhance fertilization sucess even after sperm transfer has occurred. For example, males are known to transfer chastity-enforcing chemicals to females or substances that cause the females to increase the rate of laying eggs. Ultimately, however, mechanisms that bias fertilization success are under female control and so are probably best viewed as female discrimination, a phenomenon that can be revealed by experimentally removing "male-control" effects on the paternity of offspring. Female adaptations include fertilization biases caused by moving favored ejaculates. Also, in species with "last ejaculate stored is the first to be used" mechanisms, females mating high-quality males (representing genetic quality in species with no paternal care, Table I, No. 6) can simply increase the rate of egg laying (e.g., *Oecanthus* tree crickets and *Hylobittacus* scorpionflies) or differentially allocate more resources to these eggs, thereby increasing offspring fitness (no examples from insects, but this is known in birds). Finally, the consequences of these "cryptic" sexual selection mechanisms are: (i) a male's courtship that signals his quality, e.g., genital copulatory displays (Table I, Nos. 5 and 6), can occur at any point during mating until the female oviposits (thus making courtship synchronous with mating) and (ii) the Darwinian division between primary (e.g., penes and testes) and secondary (e.g., the peacock's tail) sexual structures is blurred: male structures such as dragonfly penis brushes and large testes that are adapted to deliver large numbers of gametes into the sperm competition lottery are probably sexually selected devices. The vast diversity and complexity of insect genitalia may result from these processes: species with multiple-mating females are known to have more complex male genitalia than species in which females mate only once.

A male and female of the beetle *Nicrophorus defodiens* (Silphidae) cooperate both to defend a mouse-sized carcass from intrusions by other *Nicrophorus* and to bury the carrion in an underground chamber where it becomes food for the pair's offspring. However, the behavior of the sexes is quite different after interment of a rat-sized carcass, one large enough to support more offspring than can be produced by the initial pair. Here, conflict between the pair becomes evident when the male produces a pheromone to attract additional females. The male's signal causes his resident mate to try and thwart this signaling by mounting and biting the male (Table I, No. 8). Conflict stems from a potential sexual difference in success on larger carrion. On this food resource, the male stands to gain substantially from the increased number of larvae he fathers when mating several females, whereas his first mate can expect only decreased fitness owing to increased larval competition.

Sexual conflict and flexibility in mating behavior is also apparent in the postmating interactions of a neotropical rove beetle, *Leistotrophus versicolor* (Staphylinidae). Male and female *L. versicolor* are attracted to carrion (and occasionally dung) not as an oviposition site but as a place to prey on flies. After a pair copulates, a male can be observed to attack and bite his mate, often running after her for up to half a meter. However, this behavior occurs only when there are a number of rival males present. Male aggression appears to serve in driving the female away from the carrion, thus preventing her from mating with other males (Table I, No. 8). An alternative possibility

is that aggession toward females is a form of postcopulatory courtship (see Box 1 and Table I, No. 6).

SEXUAL CONFLICT DURING MATING

These episodes of insect mating reveal how Darwinian selection theory can be used to understand variation in behavior. The basic underlying assumption of this theory is that individuals behave in such a way as to yield the greatest number of surviving progeny. This theoretical insight suggests that courtship and copulation should rarely be a cooperative venture. Cooperation in courtship was the prevailing view among biologists at one time, in part because courtship was thought to synchronize mating events (Table I, No. 2). In contrast to this view, much research indicates that the sexes are often in conflict. Sexual conflict is expected to be common because the reproductive interests of male and female are so often at odds (Table I, No. 8). Conversely, cooperation is expected in the few cases in which male and female interests are similar. For example, interactions between a pair of burying beetles are mainly cooperative after they have interred a mouse-sized carcass. In contrast, when a larger carcass is buried, conflict is evident because, unlike the situation with small carcasses, the male has an opportunity to increase his reproductive success by attracting additional mates, whereas any added larvae from such matings probably decrease the initial female's success. The latter situation exemplifies a type of sexual

Box 2. *Risk of Predation and Mating Behavior*

Insects engaged in mating activities are known to assess the risks of predation and to adaptively change their behavior. For example, the typical song preference shown by female crickets, *Gryllus integer* (Gryllidae), can be overcome if the female can safely approach a less preferred song. In water striders, *Aquarius remigius* (Hemiptera: Gerridae), high predation risk appears to reduce male activity, thus decreasing their tendency to harrass females. This in turn allows large males to achieve high mating success (possibly because females can be more selective or can more easily avoid mating with smaller males). The threat of predation from fish, insect, and spider predators of this species and other gerrids can cause a decrease in mating frequency as well as in the duration of copulation.

FIGURE 2 Sexual conflict behavior in insects can be seen in mating struggles between a male and a female. Here a pair of scorpionflies, *Hylobittacus similis*, are using their hind tarsi in a tug-of-war over a nuptial prey (a blow fly) item captured by the male.

conflict expected in the reproductive activities of animals because of a sexual difference in reproductive strategy: males typically maximize the number of females mated so as to maximize fertilization success, whereas females maximize fecundity and offspring quality. This sexual difference also causes conflict when already-mated females are harrassed by promiscuously mating males. Examples come from the precopulatory struggles often observed in insects. A well-studied case involves water striders (Heteroptera: Gerridae). When a male uses forelegs and genitalia to secure a female for copulation, a vigorous struggle ensues during which the female attempts to dislodge him. Superfluous matings can be costly to females in terms of increased predation risk (Box 2) and energetic cost. To reduce such costs female *Gerris incognitus* have evolved upcurved abdominal spines that appear to function in thwarting male mating attempts. Another possible purpose for precopulatory struggles is that they test male quality (Table I, No. 6; see also the example of seaweed flies considered under Genetic Quality and Mate Choice).

Precopulatory struggles appear to be a result of sexual conflict in species in which males feed their mates, because females pay a cost if the size of their meal is reduced in any way. Thus a newly paired male and female scorpionfly (Mecoptera: Bittacidae) can both be seen to pull on the prey offering (Fig. 2). Conflict comes from males holding back the prey in order to conserve food for copulations with other mates and the female attempting to begin her meal as soon as possible. Conflict in some mate-feeding insects is particularly evident in the struggle between the sexes when a male attempts to force a copulation without providing the beneficial meal to his mate. To overcome female resistance, males of both panorpid scorpionflies and haglid orthopterans have specialized abdominal organs that function in holding onto females during forced matings.

A widespread form of sexual conflict arises from multiple mating by a female, which increases her success while compromising her mate's confidence of paternity. This conflict is evident in rove beetle males that drive their mates away from rivals. Striking examples also occur in male adaptations that not only enforce chastity in the female but also reduce her survival. Examples include toxic chemicals in the seminal fluid of fruit flies, *Drosophila melanogaster* (Drosophilidae), and damaging spines on the penis of lowpea weevils, *Callosobruchus maculatus* (Bruchidae), both of which decrease female life span. In the beetle, females appear to reduce injury to the reproductive tract by vigorously kicking males in order to terminate copulation (Table I, No. 8).

SEXUAL DIFFERENCES IN MATING BEHAVIOR

Advantages that females might obtain from choosing to mate with more than one male include acquiring goods and services—such as nuptial meals—or enhancing offspring quality by remating when a high-quality male is encountered (Box 1). This point highlights the basic sexual difference in mating behavior: typically females are choosy when it comes to the males that father their offspring, whereas males compete and display as a way to obtain multiple matings.

The factors controlling these typical sexual differences in behavior stem from the basic difference in the way males and females maximize reproductive success. Females usually invest more in individual offspring than males by providing materials for egg production and, in some species, caring for progeny. These maternal activities mean that fewer females than males are available for mating, thus causing males to compete for the limiting sex. Therefore sexual selection is greater on males than on females. This theory predicts that in species in which males invest more in offspring than females, sexual selection on the sexes will be reversed, causing a reversal in the mating roles, i.e., competitive females and choosy males. This prediction has been upheld in experiments with several katydid species (Orthoptera: Tettigoniidae). These species are useful experimental organisms because mating roles are flexible; when food in the environment becomes scarce females compete for mates and males are choosy. Hungry females fight to obtain matings because each copulation comes with a nuptial meal, a nutritious spermatophore (in contrast to sexual selection on males—to increase fertilizations—the sexual selection on meal-seeking females is to increase number of matings). In support of the theory, food scarcity causes an increase in relative investment in individual offspring (eggs) because there is an increase in the material in eggs derived from males—their spermatophore nutrients. The degree of choosiness shown by a sex should also be influenced by variation in the quality of the sex being chosen. Members of a sex are expected to be choosy if variation in the quality of potential mating partners is high.

SIGNALS TO MATES DURING MATING

Material Benefits and Mate Choice

Rejection of a mate (usually of males by females) is only one of the explanations for the failure of a pair to mate successfully. Other causes of a breakup of pairs are certain changes in the physical environment and a threat of harm from predators (see Box 2) or rival males. A number of studies have ruled out these alternatives and have thus shown that certain mating behaviors function in choosing mates, for example, individuals noted to move between signaling or swarming members of the opposite sex before mating with one of them and individuals pulling away from their mates after the mating sequence has begun.

As predicted by theory, female choice of mates is more widespread than male choice. Some of the clearest examples of mate choice come from species in which females obtain material benefits from males (Table I, Nos. 5a and 5b). One case involves scorpionflies (family Bittacidae), of which females attracted to males presenting food gifts of prey will manipulate the offering (Fig. 2) and reject males presenting small prey. Female scorpionflies also discriminate against such males by breaking off copulation prematurely. Only a male that transfers a large prey will complete copulation by supplying both a full complement of sperm and chastity-inducing substances.

Virtually all cases of male choice in insects appear to involve the acquisition of material benefits from large females, specifically the large number or size of eggs possessed by heavyweights. This has been noted in a number of insects, including tettigoniid orthopterans, cerambycid and brentid beetles, and empidid flies. Selection on males to mate with the most fecund females should mean that preferred females would not display their genetic quality (Table I, No. 6) because investment in costly displays probably indicates reduced fecundity. In fact, fecundity selection on females predicts that they will usually not evolve in displays of quality in the first place. In what may be

an exception to this prediction, female empidid flies, *Rhamphomyia longicauda,* display inflated abdomens and fringed legs to choosy males while flying in all-female swarms. The size of the inflated female, as perceived by a male entering the swarm, seems to be a poor predictor of her fecundity and may instead advertise her genetic quality.

As noted earlier, male choice is expected when the mating roles are reversed or when there is a high degree of variance in the quality of females. Examples of the former include male Mormon crickets (*Anabrus simplex:* Tettigoniidae) pulling away from mounted lightweight females—apparently after weighing them—and males of the empidid flies *Rhamphomyia sociabilis* and *Empis borealis* choosing large, fecund individuals from within all-female swarms. In contrast to male mate choice in role-reversed systems, male choice that evolves in response to a high variance in female quality typically is often found with a high degree of male–male competition, i.e., sexual selection on males. Indeed, male choice in this situation can be caused by sexual selection among males to mate with the highest quality females. An example of male choice when females vary in quality includes winter moths, *Operophtera brumata* (Lepidoptera, Geometridae), and red flour beetles. Finally, local population variation in the primary sex ratio can affect the likelihood of male choice; in red milkweed beetles, *Tetraopes tetraopthalmus* (Cerambycidae), a scarcity of males is associated with a higher degree of male choice.

Genetic Quality and Mate Choice

In theory, females are expected to show choice to obtain indirect benefits, i.e., benefits that enhance the genetic quality of offspring. Female *Dryomyza* flies appear to do this by biasing fertilization after evaluating male copulatory courtship. But what sorts of indirect benefits do choosing females obtain? In yellow dung flies, *Scathophaga stercoraria* (Scathophagidae), females can favor the stored sperm from males with genotypes likely to enhance offspring growth. For genotypes common in the population this involves mating with a male of a similar genotype when the environment is constant but choosing a male of a different genotype when the environment is unpredictable. In other insects, the cue to a male's genetic quality is consistent between females. For example, following a courtship consisting of wing-flicking and pheromone displays, older females of *Colias* butterflies (Pieridae) show mating preferences for genotypes that fly well and are long-lived. Male and female calopterygid dragonflies also court using wing displays and the size of wing spots. The latter is a sexually dimorphic trait in European *Calopteryx splendens* and is correlated with several aspects of male quality, including the level of immunocompetence, developmental stability, and resistance to gut parasites. In yet another dipteran, the seaweed fly, *Coelopa frigida* (Coelopidae), mating interactions involve prolonged premating struggles in which a mounted male can be dislodged as a result of kicking and shaking by the female. Such struggles favor matings with large males and this female bias enhances the genetic fitness of her offspring: progeny of large males tend to be heterozygous for a chromosomal inversion that increases offspring viability. However, female choice in this system may be maintained by more than good-genes sexual selection. Females carrying the inversion genotype show a strong preference for large males. Genes for preference thus appear to be linked with genes for the male display trait, suggesting a form of female-choice sexual selection, termed "runaway" or "Fisherian" sexual selection (after the originator of this idea, R. L. Fisher), in which mothers gain by producing "sexy sons," those that are highly attractive to females.

Mating Preferences for the Correct Species

One result of the expected rapid evolutionary change from runaway sexual selection may be speciation through behavioral isolation. Speciation results when there is sufficient between-population divergence in the female preference and the linked male display that a side effect of intraspecific mating preferences is discrimination against males from other populations (see also Box 1). This "effect" hypothesis for species discrimination differs from the hypothesis that certain female mating preferences have evolved to *function* in avoiding costly interactions with other species. An example of the latter involves the fruit flies *Drosophila pseudoobscura* and *D. persimilis,* in which hybrid matings result in decreased reproductive success because sons are sterile. Female *Drosophila* assess the wing-vibration displays of males, and *D. pseudoobscura* females collected from areas where the two species co-occur (sympatry) reject courting *D. persimilis* males more frequently than females collected from areas with no species overlap (allopatry). This result was not the result of differences in courtship by the males with the two types of females. These findings indicate that female discrimination against heterospecific male courtship has been reinforced in areas where maladaptive hybridization is likely to occur.

A high degree of discrimination against courtship by heterospecific individuals in sympatry has also been noted in *Calopteryx* dragonflies in which both sexes display patterned wings in precopulatory courtship. Compared with areas of allopatry, male *C. maculata* in sympatry discriminate more against the wing patterns of *C. aequabilis* females, and mate preferences during courtship appear to have reinforced wing-pattern differences between the species. For example, in a north–south transect in eastern North America, the proportion of pigmented wing area of both sexes is greater in areas of sympatry than in areas of allopatry.

SIGNALS TO RIVALS DURING MATING

In another North American calopterygid, *Hetaerina americana,* variation in wing-spot displays reflects an evolutionary history of competition between males; males with larger wing spots are more successful in defending mating territories than males with smaller spots. As males with experimentally enhanced spots suffer a cost (increased mortality), these signal patterns appear to have evolved as honest indicators of fighting ability. These signal indicators may convey information about male ability to rivals during wing-waving displays directed at females. Other behaviors taking place during mating appear to function in a competitive context (Table I, No. 7). For example, males of the carrion beetle mentioned previously *(L. versicolor)* will occasionally mimic female behavior during interactions with potential mates (Fig. 3). Pseudofemale behavior reflects a remarkable plasticity in the way the male beetles obtain matings. The most profitable way to gain access to females is to fight to defend the carcass that attracts them. However, males can obtain some matings subversively by mimicking female behavior and thus avoiding costly fighting. This form of behavior is conditional on the relative size difference between opponents: a male will engage a smaller male in a fight but will switch to pseudofemale behavior if his rival is larger. There are other mating behaviors that appear to reflect male–male competitive interactions. For example, after attracting females, males of some singing insects switch from song to a more reclusive signal such as substrate vibration, apparently as a way of avoiding courtship behavior that attracts rivals.

HOMOSEXUAL BEHAVIOR AND MATING MISTAKES

A male *L. versicolor* beetle can be duped into courting a small female-mimicking rival (Fig. 3) and there are a few other examples, such as in some butterflies and dragonflies, of homosexual mating mistakes when certain males adaptively resemble females. Homosexual mounting can also occur among insects of which the males have not evolved to mimic females. This male behavior is widespread in animals and appears to be simply an effect of poor sex recognition; strong selection on males to mate frequently causes them to mount any object that resembles a female. Examples of mating mistakes can even include inanimate objects, such as in the case of *Julidomorpha bakewelli,* an Australian buprestid beetle, the males of which attempt to copulate with beer bottles with a coloration and reflection pattern resembling the female's elytra (Fig. 4).

Poor sex recognition appears to be the explanation of why males of another beetle, *Diaprepes abbreviatus,* mount conspecific males. A big difference between this species and others, however, is that females also perform homosexual mountings. In this case, however, mounting appears to be an adaptive reproductive strategy rather than a mating mistake. Laboratory experiments with this species reveal that a mounted pair of females attracts males. In fact, large males attempt to mate more often with paired females than with single large or small females. As both the mounting and the mounted females had similar probabilities of copulating with the attracted male, it appears that the mounted pair mimics a heterosexual pair in order to incite the attraction of large, competitive (i.e., high-quality) males.

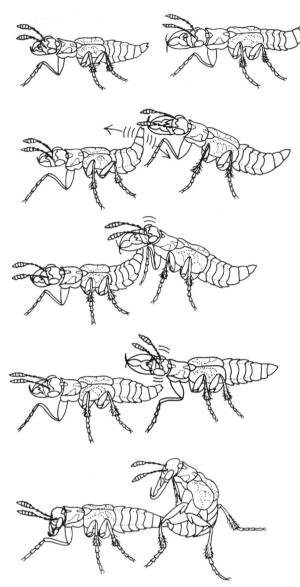

FIGURE 3 A male staphylinid beetle, *L. versicolor,* can avoid being chased by a rival from the carrion source by mimicking female behavior. The mimic male turns and presents his abdomen to the approaching rival, which antennates the abdominal tip and taps it with his head. Copulation (bottom) is the only stage of a heterosexual encounter that is not represented in these homosexual encounters (because the mimic male breaks up the encounter by walking away). (Reproduced, with permission, from Forsyth and Alcock (1990) *Behav. Ecol. Sociobiol.* **26,** 326.)

CONCLUSIONS

Observations of insect mating behaviors reveal a great diversity, some of which is a result of plasticity within species. The examples discussed here show how an understanding of the function of both inter- and intraspecific variation in mating behaviors can be gained by examining the consequences of behavior for the reproductive success of the mating male and female. Functions of insect mating that were proposed before the widespread use of the "selectionist" approach (Table I, Nos. 1 and 2) can be subsumed into this framework. For

FIGURE 4 A mating error by a male: a *J. bakewelli* male mounts a beer bottle. Note how the aedeagus (penis) is extended.

example, movements involved in delivering sperm are undoubtedly subject to sexual selection. And, any observations of synchronized courtship in a species inevitably lead to the question of how such synchrony enhances the reproductive success of the male and female. There is a wealth of behavioral diversity for future research, including apterygote insects, a virtually unstudied group and one of great interest because they lack copulation.

See Also the Following Articles
Hearing • Pheromones • Reproduction • Sexual Selection

Further Reading
Alcock, J. A., and Gwynne, D. T. (1991). Evolution of insect mating systems: The impact of individual selectionist thinking. *In* "Reproductive Behaviour in Insects: Individuals and Populations" (W. J. Bailey and J. Ridsdill Smith, eds.). Chapman & Hall, London.

Arnqvist, G. (1998). Comparative evidence for the evolution of genitalia by sexual selection. *Nature* **393,** 784–786.

Cade, W. H. (1985). Insect mating and courtship behaviour. *In* "Comprehensive Insect Physiology, Biochemistry and Pharmacology" (G. A. Kerkut and L. I. Gilbert, eds.). Pergamon Press, Oxford.

Choe, J. C., and Crespi, B. (eds.) (1997). "The Evolution of Mating Systems in Insects and Arachnids." Cambridge University Press, Cambridge, UK.

Eberhard, W. G. (1996). "Female Control: Sexual Selection by Cryptic Female Choice," Princeton University Press, Princeton, NJ.

Eggert, A.-K., and Sakaluk, S. K. (1995). Female-coerced monogamy in burying beetles. *Behav. Ecol. Sociobiol.* **37,** 147–153.

Lima, S. L. (1998). Stress and decision making under the risk of predation: Recent developments from behavioral, reproductive, and ecological perspectives. *Adv. Stud. Behav.* **27,** 215–290.

Lloyd, J. E. (1979). Mating behavior and natural selection. *Fla. Entomol.* **62,** 17–34.

Rowe, L., Arnqvist, G., Sih, A., and Krupa, J. J. (1994). Sexual conflict and the evolutionary ecology of mating patterns—Water striders as a model system. *Trends Ecol. Evol.* **9,** 289–293.

Simmons, L. W. (2001). "Sperm Competition and Its Evolutionary Consequences in Insects." Princeton University Press, Princeton, NJ.

Thornhill, R., and Alcock, J. (1983). "The Evolution of Insect Mating Systems." Harvard University Press, Cambridge, MA.

Mayfly

see *Ephemeroptera*

Mechanoreception

Andrew S. French and Päivi H. Torkkeli
Dalhousie University, Halifax, Nova Scotia

Mechanoreception is the sense that allows insects to detect their external and internal mechanical environments, including physical orientation, acceleration, vibration, sound, and displacement. The integument and internal organs contain a wide variety of mechanoreceptors. Prominent receptors, such as surface hairs that mediate touch or auditory organs, have been studied extensively, but many other physiological functions also depend on mechanosensory signals.

Arthropod mechanoreceptors are divided into two morphological groups: Type I, or cuticular, and Type II, or multipolar. Type I are ciliated receptors, associated with the cuticle, and have their nerve cell bodies in the periphery, close to the sensory endings. They can be subdivided into three major groups (Fig. 1). Hairlike receptors are found on the outer surface in a variety of shapes and sizes, from long, thin hairs to short pegs and scales. A sensory neuron is closely apposed to the base of the hair and its dendrite contains microtubules ending in a structure called the tubular body. It is assumed that movement of the hair compresses the ending, with the tubular body perhaps providing a rigid structure against which the compression can work. Hair receptors can contain additional sensory neurons, such as chemoreceptor neurons in taste hairs. Campaniform (bell-shaped) sensilla are also found on the outer surface, particularly in compact groups near the joints, where they detect stress in the cuticle. Stress moves the bell inward, compressing the dendritic tip containing the tubular body. Chordotonal receptors are generally found farther beneath the integument, although they can be connected to the integument by attachment structures. They serve several functions, including hearing and joint movement detection. They generally lack tubular bodies but have dense scolopales surrounding the dendrites and often have multiple mechanosensory neurons.

Type II mechanoreceptors are nonciliated neurons, whose central cell bodies have many fine dendritic endings, each of which is apparently mechanosensitive, but lacks the detailed structures seen in Type I receptors. Type II receptors are found in many internal structures, predominantly associated with mesodermal tissues, including the musculature, where they detect muscle tension.

Studies of mechanoreceptor morphology have used many techniques, including light microscopy, scanning and transmission electron microscopy, and immunohistochemistry. Receptor electrophysiology has been studied by three basic methods: (1) Extracellular recordings observe the receptor currents flowing along the axon. (2) Epithelial recordings measure the current flowing through the relatively low resistance of the thin socket tissue or through a cut hair. (3) Intracellular recordings give direct measurements of membrane potentials and currents.

Mechanosensation is commonly viewed as a three-stage process in which a mechanical event is first coupled to the receptor cell membrane by mechanical structures, then transduced into a receptor current at the cell membrane, and finally encoded into action potentials for transmission of information to the central nervous system.

DEVELOPMENT OF MECHANORECEPTORS

Type I sensory neurons are surrounded by specialized sheath cells of varying numbers and names, although the terms trichogen (hair-forming) and tormogen (sheath-forming) are commonly used for the innermost two layers of sheath cells. Development of these cells has been well characterized in several species, but especially in *Drosophila* external bristles, for which many of the genes involved have been identified. A single sensory organ precursor cell divides to give two different secondary precursors, IIA and IIB. IIA divides to form one trichogen and one tormogen cell. IIB gives rise to the neuron, another sheath cell, and sometimes an additional glial cell. The neuron then forms an axon that grows into the central nervous system. A variety of other noncellular structures, including sheaths, are also found, particularly in dendritic regions. The development of Type II receptors is less well understood.

MECHANICAL COMPONENTS

Extracellular tissues, often with elaborate structures, surround the sensory cells. These structures modify the spatial and temporal sensitivities of the receptors, and they are often designed to interact with the outside environment or other parts of the animal, such as cercal hairs detecting air movements or hair plates detecting joint rotation. External structures usually allow detection of mechanical events at some distance from

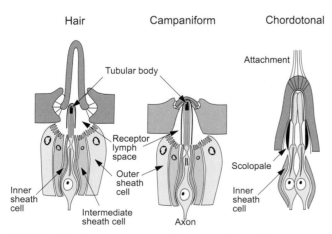

FIGURE 1 The three major groups of insect cuticular mechanoreceptors. The receptor lymph space surrounding the sensory ending is formed by a layer of sheath cells and epithelial cells connected by tight junctions. The numbers of sheath cells and their nomenclatures are both variable (see text).

the sensory cell but make the displacement at the receptor cell membrane smaller than the original movement. Estimates of this attenuation suggest that threshold movements at the cell membrane leading to sensation are in the range 1 to 5 nm.

TRANSDUCTION AND ENCODING

Mechanically activated ion channels, probably located in the tips of the sensory dendrites, transduce the mechanical stimulus into a receptor current. The channels are permeable to potassium ions and the receptor lymph spaces have high concentrations of potassium. The channels may have time-dependent properties that contribute to receptor behavior, but separating these from the time-dependent properties of any external mechanical structures is difficult. Current flowing through the channels causes a receptor potential that is encoded into action potentials. Mechanotransduction currents are more sensitive to temperature than most other membrane currents, with activation energy values of 12 to 22 kcal/mol, which are similar to those required to break chemical bonds and significantly higher than the energy barriers associated with ionic diffusion or conductance through ionic channels. Some crucial stage in the link between membrane tension and ion channel opening may lead to this high energetic barrier.

The receptor potential is encoded into action potentials using several different sodium and potassium currents. Action potentials propagate into the central nervous system along axons in nerve roots of the segmental ganglia. Afferent axons have a size range of 1 to 20 μm and conduction velocities are typically 1 to 5 m/s. Information is transmitted from mechanoreceptor axons into the central nervous system via cholinergic synapses.

CENTRAL, PERIPHERAL, AND HUMORAL MODULATION

Many mechanoreceptors receive GABAergic inhibitory efferent innervation close to the output synapses of their axon terminals. This presynaptic innervation modulates afferent mechanoreceptor information. Some mechanosensory neurons are also modulated by efferent innervation in the periphery and by circulating chemicals such as biogenic amines. Of these, octopamine has been most thoroughly studied, but without clear conclusions, because octopamine can increase or decrease firing frequency, even in the same neuron. Studies in locusts and cockroaches suggest that octopamine receptors are located on the peripheral regions of mechanoreceptors.

The extent and functions of peripheral modulation remain to be seen. It is the latest in a series of surprises about the complexity of insect mechanotransduction, but probably not the last.

See Also the Following Articles
Orientation • Vibrational Communication

Further Reading
Abdelilah-Seyfried, S., Chan, Y. M., Zeng, C., Justice, N. J., Younger-Shepherd, S., Sharp, L. E., Barbel, S., Meadows, S. A., Jan, L. Y., and Jan, Y. N. (2000). A gain-of-function screen for genes that affect the development of the *Drosophila* adult external sensory organ. *Genetics* **155**, 733–752.
Bräunig, P., and Eder, M. (1998). Locust dorsal unpaired median (DUM) neurones directly innervate and modulate hindleg proprioceptors. *J. Exp. Biol.* **201**, 3333–3338.
Burrows, M. (1996). "The Neurobiology of an Insect Brain." Oxford University Press, Oxford/New York/Tokyo.
Chapman, K. M., and Pankhurst, J. H. (1967). Conduction velocities and their temperature coefficients in sensory nerve fibres of cockroach legs. *J. Exp. Biol.* **46**, 63–84.
Field, L. H., and Matheson, T. (1998). Chordotonal organs of insects. *Adv. Insect Physiol.* **27**, 1–228.
French, A. S. (1988). Transduction mechanisms of mechanosensilla. *Annu. Rev. Entomol.* **33**, 39–58.
French, A. S. (1992). Mechanotransduction. *Annu. Rev. Physiol.* **54**, 135–152.
Höger, U., and French, A. S. (1999). Temperature sensitivity of transduction and action potential conduction in a spider mechanoreceptor. *Pflugers Arch.* **438**, 837–842.
McIver, S. B. (1985). Mechanoreception. In "Comprehensive Insect Physiology, Biochemistry, and Pharmacology" (G. A. Kerkut and L. I. Gilbert, eds.). Pergamon Press, Oxford.
Ramirez, J. M., and Orchard, I. (1990). Octopaminergic modulation of the forewing stretch receptor in the locust, *Locusta migratoria. J. Exp. Biol.* **149**, 255–279.
Spinola, S. M., and Chapman, K. M. (1975). Proprioceptive indentation of the campaniform sensilla of cockroach legs. *J. Comp. Physiol.* **96**, 257.
Torkkeli, P. H., and French, A. S. (1995). Slowly inactivating outward currents in a cuticular mechanoreceptor neuron of the cockroach *(Periplaneta americana). J. Neurophysiol.* **74**, 1200–1211.
Zacharuk, R. Y. (1985). Antennae and sensilla. In "Comprehensive Insect Physiology, Biochemistry, and Pharmacology" (G. A. Kerkut and L. I. Gilbert, eds.). Pergamon Press, Oxford.
Zhang, B. G., Torkkeli, P. H., and French, A. S. (1992). Octopamine selectively modifies the slow component of sensory adaptation in an insect mechanoreceptor. *Brain Res.* **591**, 351–355.
Zill, S. N., Ridgel, A. L., Dicaprio, R. A., and Frazier, S. F. (1999). Load signalling by cockroach trochanteral campaniform sensilla. *Brain Res.* **822**, 271–275.

Mecoptera
(Scorpionflies, Hangingflies)

George W. Byers
University of Kansas

Mecoptera (scorpionflies, hangingflies, and others) are holometabolous insects in which the head is characterized by a downward projecting rostrum or beak, at the end of which are chewing mouthparts. They are usually slender bodied and have four long, narrow wings that are membranous and often marked with dark bands, spots, or darkening along the numerous crossveins. There are exceptions, however. A few species are brachypterous or wingless; species in family Boreidae have hardened, highly modified wings, which are nearly oval and scalelike in females, slender and somewhat curved apically in males. The rostrum is not unusually prolonged in species of Nannochoristidae; and the wings of species of Meropeidae are only about 2.5 times longer than their greatest width.

Although Mecoptera are one of the minor orders of insects, with only about 550 living species so far made known, they are of much interest to entomologists. This is largely because fossil Mecoptera are among the oldest remains of insects with complete metamorphosis (of Permian age) and have been regarded by some as ancestral to the more recently evolved and vastly larger orders Diptera and Lepidoptera. Fossil evidence suggests that the Mecoptera were once one of the larger orders of holometabolous insects. Past diversity, based mainly on wing venation, has led to placement of some 350 fossil species in 87 genera in 34 families. Thus, modern Mecoptera are survivors of millions of years of evolutionary development.

FAMILIES AND GENERA

Following are the nine families of extant Mecoptera, with the number of genera included in each and the approximate geographical distribution: Panorpidae—3 genera, in Europe, Asia, and North America; Panorpodidae—2 genera, in easternmost Asia, and North America; Bittacidae—17 genera, in North and South America, Africa, Asia, Australia, and Europe; Boreidae—3 genera, in North America, Asia, and Europe; Choristidae—3 genera, in Australia; Nannochoristidae—2 genera, in Australia, New Zealand, and southern South America; Apteropanorpidae—1 genus, in Australia (Tasmania); Meropeidae—2 genera, in North America and southwestern Australia; Eomeropidae (formerly Notiothaumidae)—1 genus, in South America.

Most Mecoptera belong to the families Panorpidae (scorpionflies) and Bittacidae (hangingflies). Scorpionflies are so called because abdominal segments 7 and 8 of the male are slender, and segment 9 is abruptly enlarged and often held above the back, recalling the sting of a scorpion (Fig. 1). The female abdomen tapers to a slender tip. Both male and female hangingflies are slender bodied and, having a single, large (raptorial) claw at the end of each tarsus, are unable to stand on the upper surfaces of leaves but suspend themselves from twigs or edges of leaves (Fig. 2).

ZOOGEOGRAPHY

Mecoptera present a variety of zoogeographical patterns, from highly localized (as in *Notiothauma,* which is endemic in central Chile, or *Apteropanorpa,* found only in Tasmania) to virtually cosmopolitan (as the Bittacidae, which occur in the temperate and tropical parts of six continents). Only Bittacidae have been found in Africa. Some very disjunct occurrences suggest relics of ancient, more widespread ranges (as *Merope* in eastern North America and the very similar *Austromerope* in

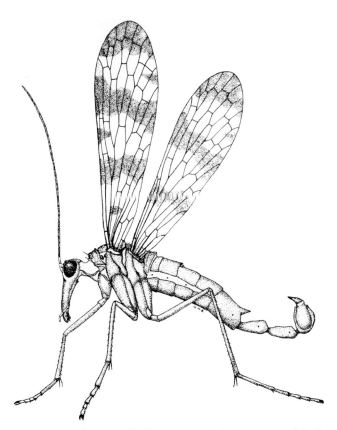

FIGURE 1 A male scorpionfly, showing anatomical characteristics of family Panorpidae. (Illustration courtesy of Holt, Rinehart and Winston, Inc.)

southwestern Australia; or Nannochoristidae in southeastern Australia, New Zealand, and southern South America).

HABITAT

Broad-leaved, herbaceous plants shaded by trees are the usual habitat of panorpids, bittacids, and most species in the smaller families. Some species, however, are most commonly found in vegetation near forest borders, whereas others occur in somewhat similar but more uniformly shaded habitats well within a forest. Elevation clearly affects the distribution of Mecoptera. Boreidae, for example, reach the adult stage in the cold part of the year (and being darkly colored are most often seen on the surface of snow) and at high elevations and high latitudes have a longer seasonal occurrence than boreids at lower elevations. In Mexico, Bittacidae are usually found at elevations below 1500 m, while Panorpidae occur above that level and up to over 3000 m.

DIET

Both adults and larvae of *Panorpa* are scavengers, feeding usually on dead insects, and less often on other dead organisms, including some small vertebrates. Adults occasionally eat pollen and associated parts of flowers; they may even invade the webs of spiders to feed on entangled

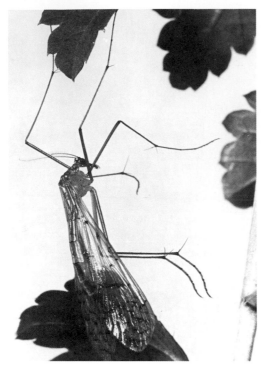

FIGURE 2 A hangingfly, *Bittacus* sp., in characteristic resting posture. (Photograph courtesy of D. W. Webb and Illinois Natural History Survey.)

FIGURE 3 Mating pair of Japanese bittacids feeding on nuptial prey (an opilionid) initially offered to the female by the male. (Photograph courtesy of Yasushi Iwasaki.)

insects (and are sometimes trapped themselves). Adult Bittacidae (Fig. 2) are predaceous, capturing a variety of insects and occasionally other arthropods by means of the raptorial hind tarsi. Larval bittacids, in contrast, are primarily saprophagous on dead insects but may occasionally feed on plant matter. Both adult and larval Boreidae feed on mosses. Adults of *Brachypanorpa* (Panorpodidae) have been observed scraping the upper epidermis from leaves of herbaceous plants; their larval diet is not known. There is some evidence that *Notiothauma* (Eomeropidae) is also a plant feeder. Larvae of aquatic Nannochoristidae feed on small larvae of chironomid flies, but nothing is known of the adult diet.

BEHAVIOR

Feeding by adults of some Mecoptera is often associated with mating behavior (Fig. 3). Some male bittacids, for example, capture insects as nuptial gifts, then extrude pheromone-dispersing vesicles between abdominal terga 6–7 and 7–8 to advertise their presence to females in the vicinity. Females seem to make an evaluation of the gift prey, leaving the male if his offering is too small or unpalatable. They may then respond to the pheromone of another male. Males of some Panorpidae, when they have found a good source of food, make known their presence by means of pheromonal vesicles within the enlarged genital segment. When a female is attracted closely enough, the male clamps the costal edge of her forewing in a structure formed by a peg on the anterior fourth abdominal tergum and an overlapping portion of the third (the notal

organ); mating then ensues as the female feeds. In other species of *Panorpa,* the male may build a small column of brownish saliva, gelatinous as it air-dries, which serves as a nuptial gift in place of a dead insect.

EGGS

Eggs of most Mecoptera are ovoid but approximately equally rounded at the ends. Those of *Bittacus* species are roughly cuboidal with the flattened surfaces shallowly impressed. The chorion of eggs of some species of *Panorpa* is smooth, while in others it is coated with a fine network of polygonal cells. In *Notiothauma,* the chorion is yellowish and granular; in *Austromerope,* the chorion is smooth. Eggs are ordinarily inserted into preexisting cavities in the soil by females of *Panorpa,* or in soil or rotten wood by *Brachypanorpa.* Females of these genera have been observed probing with the extended abdomen for some time before oviposition occurred. Female Bittacidae typically hang from the edge of a leaf and drop their eggs among dead leaves and other plant debris on the ground. Boreids deposit their eggs among the rhizoids of the mosses that will later be a source of food. As the embryonic larva develops and nears the time of hatching, the egg swells noticeably. In *Panorpa,* the increase is as great as 38%; larval eyes and mandibles become visible through the chorion. Eggs of *Bittacus,* roughly cuboidal when laid, become subspherical; those of *Harpobittacus* also become rounded and may double in size.

The egg stage is brief for some species (5–10 days for some American species of *Panorpa;* 14–16 days for panorpids living above 2250 m in Taiwan; 12–15 days in *Chorista*). In contrast, there may be an egg diapause, which together with larval development within the egg may last 216 to 256 days in one North American *Bittacus,* and up to 290 days in a Japanese *Bittacus.* Larvae from diapausing eggs of *Boreus* laid in early winter hatch the following spring.

LARVAE

Larvae of Panorpidae, Bittacidae, and Choristidae are eruciform; that is, they have somewhat the appearance of caterpillars. The head is well sclerotized, as is the dorsum of the prothorax; the legs are nearly conical, their skin largely membranous; and there are subconical prolegs on abdominal segments 1 to 8. Larval bittacids have paired, elongate, three-branched, fleshy structures on the back of the mesothorax, metathorax, and nine abdominal segments. Panorpids and choristids have setae in most of the corresponding positions. The larvae of Panorpodidae and Boreidae, described as scarabaeiform, lack abdominal prolegs and conspicuous dorsal setae. Larvae of Nannochoristidae are unlike those of the other families, so far as known. They are aquatic, with slender, elongate, almost eel-like bodies and no prolegs. Their mouthparts are directed forward, not downward as in eruciform larvae.

A striking characteristic of many larval Mecoptera is presence of compound eyes, whereas larvae of most holometabolous insects have only one or a few simple eyes at each side. The larvae of Panorpidae and Choristidae have 30 or more ommatidia in each compound eye, while those of the Bittacidae have seven, Boreidae usually only three, and larvae of Panorpodidae lack eyes altogether. Larvae of Nannochoristidae have eyes with indistinctly formed ommatidia. Because their larvae are so different from the larvae of other families in the order (and because of some characteristics of the adults, such as a short rostrum), the Nannochoristidae have sometimes been considered as a group distinct from other Mecoptera.

Duration of larval life varies with the species involved, with temperature and season, availability of food, and length of prepupal diapause. Under favorable conditions, a larva of *Panorpa* may pass through three growth stages and become a fourth instar in about a month. During the final larval stage, feeding and growth continue for several days. But then the larva prepares a cavity in the soil, or other material in its habitat, stops feeding and becomes quiescent as it begins the period of prepupal diapause (inactivity) in this cavity. In species of *Panorpa* that have spring and late-summer generations, prepupal diapause in summer lasts only about 5 weeks, while in overwintering larvae it requires about 6.5 to 7 months. In nearctic species of *Bittacus* it lasts about 7 months. The pupal stage is much shorter: 10 to 21 days in *Panorpa,* 15 to 24 days in *Bittacus,* and 37 to 40 days in *Boreus.*

PUPAE

Described as exarate, the pupa has the legs rather loosely drawn up against its ventral surface and the wings tightly folded within their sheaths, which are not closely adhered to the body. The pupal rostrum is shorter than that of the adult and the

mandibles larger. Otherwise, the pupal body has generally the form of the adult developing within.

See Also the Following Articles
Fossil Record • Mating Behaviors

Further Reading
Byers, G. W., and Thornhill, R. (1983). Biology of the Mecoptera. *Annu. Rev. Entomol.* **28,** 203–228.
Kaltenbach, A. (1978). Mecoptera (Schnabelhafte, Schnabelfliegen). *In* "Handbuch der Zoologie." Vol. 4, Part 2, Section 2, pp. 1–111. deGruyter, Berlin.
Setty, L. R. (1940). Biology and morphology of some North American Bittacidae (order Mecoptera). *Am. Midl. Nat.* **23,** 257–353.
Thornhill, R. (1980). Competition and coexistence among *Panorpa* scorpionflies (Mecoptera: Panorpidae). *Ecol. Monogr.* **50,** 179–197.
Thornhill, R. (1980). Mate choice in *Hylobittacus apicalis* (Insecta: Mecoptera) and its relation to some models of female choice. *Evolution* **34,** 519–538.

Medical Entomology

John D. Edman
University of California, Davis

Medical entomology is concerned with the impact of insects and related arthropods on the mental and physical health of humans, domestic animals, and wildlife. It is often subdivided into public health entomology and veterinary entomology. These divisions are tenuous since many of the same arthropods cause similar injuries and diseases in both humans and other animals. The history of medical entomology dates from the end of the 19th century, when arthropods were first shown to transmit important human diseases such as filariasis and malaria. Most arthropod-borne diseases are zoonotic infections that occur naturally in nonhuman hosts. Malaria, dengue fever, and most forms of filariasis are important exceptions. Household pests such as cockroaches and filth flies are sometimes included within medical entomology. When synanthropic flies and cockroaches mechanically contaminate food or other media with infectious organisms, they are clearly of medical importance. Nonetheless, these insects are generally treated in greater depth within the scope of urban entomology.

MEDICAL IMPORTANCE OF ARTHROPODS

Arthropods influence animal health in multiple ways. The most significant impact involves their role as primary vectors and alternate hosts of many devastating infectious disease agents. Parasitic agents transmitted by hematophagous arthropods include filariae, protozoa, bacteria, rickettsiae, and viruses. Arthropods also affect the health of vertebrates directly by triggering altered mental states (delusional parasitosis and entomophobia/arachnophobia), contact allergies, feeding annoyance and blood loss, envenomization, and myiasis.

Each year these arthropod relationships collectively cause the death of millions of humans and bring illness to hundreds of millions more. Their impact is greatest in poor tropical countries where they are one of the major factors limiting animal production, agricultural productivity, economic development, and well-being. Several vector-borne diseases (e.g., plague, malaria, leishmaniasis, yellow fever, and dengue) apparently were transported from the Old World to the Western Hemisphere via the slave trade or ship crews. Related species of insects in the New World were able to successfully maintain and transmit these introduced parasites. In recent years, several native but previously unrecognized arthropod-borne infections, including tick-borne Lyme disease and ehrlichiosis, have been discovered in the United States and elsewhere. Well-known diseases such as dengue and malaria have been reemerging or expanding in many regions of the world. Other diseases such as West Nile fever recently have been introduced into new regions with devastating effects. A combination of population growth, rapid movement of people and other animals, and environmental disruption has contributed to the growing threat posed by vector-borne diseases. Control of these diseases is complicated by a lack of investment in public health and development of drug-resistant parasites and insecticide-resistant vectors.

ORDERS AND FAMILIES OF MEDICAL CONCERN

Phylum Arthropoda contains several classes of invertebrates that have direct medical importance. In addition to insects, these include arachnids (spiders, mites, ticks, and scorpions), millipedes, and centipedes. Some crustaceans (sowbugs, copepods), molluscs (snails), arachnids (oribatid mites), and insects (mainly beetle larvae) are intermediate hosts for parasitic worms.

Orders and families of blood-feeding species are of major medical significance (Table I). Most but not all of these are involved in the transmission of microparasites. Nonetheless, the simple act of feeding by arthropod ectoparasites can result in blood loss, anemia, stress, discomfort, allergic reactions, and reduced productivity. A few insects are specialized ectoparasites on humans. Bed bugs and kissing bugs are nest parasites and some have permanently invaded human dwellings and feed at

TABLE I Orders and Families with Important Blood-Feeding Insects, Ticks, and Mites

Taxon	Common name	Blood feeding	Types of hosts
Insecta (=Hexopododa)			
Order Anoplura	Sucking lice	♂, ♀, nymph	Mammals
Order Mallophaga[a]	Chewing lice	♂, ♀, nymph	Birds/mammals
Order Heteroptera			
Family Cimicidae	Bed/bird/bat bugs	♂, ♀, nymph	Mammals/birds
Family Triatomidae	Kissing bugs	♂, ♀, nymph	Mammals/birds
Order Siphonaptera	Fleas	♂, ♀	Mammals/birds
Order Diptera			
Family Culicidae	Mosquitoes	♀	All vertebrates
Family Simuliidae	Black flies	♀	Mammals/birds
Family Ceratopogonidae	Biting midges	♀	All vertebrates
Family Psychodidae	Sand flies	♀	Mammals/reptiles
Family Tabanidae	Horse/deer flies	♂, ♀	Mammals
Family Rhagionidae	Snipe flies	♂, ♀	Mammals
Family Muscidae			
Stomoxys	Stable flies	♂, ♀	Mammals
Haematobia	Horn/bush flies	♂, ♀	Mammals
Musca	Cattle flies	♂, ♀	Mammals
Family Glossinidae	Tsetse flies	♂, ♀	Mammals/reptiles
Family Hippoboscidae	Louse flies, keds	♂, ♀	Mammals/birds
Family Nycteribiidae	Bat flies	♂, ♀	Bats
Family Streblidae	Bat flies	♂, ♀	Bats
Arachnida (subclass Acari)			
Family Ixodidae	Hard ticks	♂, ♀, L, N[b]	Birds/mammals/reptiles
Family Argasidae	Soft ticks	♂, ♀, L, N	Birds/mammals
Family Trombiculidae	Chigger mites	Larvae	Birds/mammals
Family Dermanyssidae	Mesostig mites	♂, ♀, L, N	Birds/mammals
Family Macronyssidae	Mesostig mites	♂, ♀, L, N	Birds/mammals
Family Laelapidae	Mesostig mites	♂, ♀, L, N	Birds/mammals
Family Demodicidae[a]	Follicle mites	♂, ♀, L, N	Mammals
Family Psoroptidae[a]	Mange mites	♂, ♀, L, N	Mammals
Family Sarcoptidae[a]	Scab mites	♂, ♀, L, N	Mammals

[a]Ectoparasites that feed mainly on skin/feather tissues rather than on blood.
[b]L, larvae; N, nymph.

night on sleeping people. Humans are also parasitized by three species of sucking lice (i.e., pubic louse, head louse, and body louse), each of which specializes in a different region of the body. Biting flies such as mosquitoes, black flies, deer flies, and biting midges can reach annoyance levels that make outdoor activities nearly impossible, but these hordes normally feed on a wide variety of domestic and wild animals. Only a few biting flies such as the yellow-fever mosquito *(Aedes aegypti),* the tropical house mosquito *(Culex pipiens quinquefasciatus),* and some vector species of *Anopheles, Simulium, Phlebotomus,* and *Lutzomyia* feed preferentially on humans.

Both insects and arachnids can cause harm because of the venoms they contain. Venomous insects are found mainly in the order Hymenoptera within the families Formicidae (ants), Vespidae (yellowjackets, hornets), Mutillidae (velvet ants), and Apidae (honey bees, bumble bees). Some Coleoptera (e.g., Meloidae, Staphlinidae, Chrysomlelidae, Dermestidae) and Lepidoptera (e.g., Noctuidae, Saturniidae, Sphingidae, Nymphalidae) also produce toxic defensive secretions through specialized glands and urticating hairs or have hemolymph that is toxic to vertebrates if the insect is crushed. These secretions are especially toxic when exposed to mucus or lachrymal glands. Spiders in the genera *Loxosceles* (recluses), *Latrodectus* (widows), *Atrax* (Australian funnel-webs), *Harpactirella* (South Africa), *Lycosa* (Central and South America), and *Phoneutria* (Brazil) include some of the most highly venomous species. All scorpions have venomous stings and those within the family Buthidae can be fatal to humans. Desert regions of the Americas, the Mediterranean, and northern Africa are home to most of the highly poisonous scorpions. Centipedes also have venomous bites and millipedes have toxic defensive secretions, but these normally are not severe or life threatening.

Arthropod allergens that cause acute asthma in humans are mostly associated with house dust mites *(Dermatophagoides* spp.) and cockroaches *(Periplaneta, Blattella);* however, the airborne wing scales or fine setae associated with large populations of other insects such as mayflies, caddisflies, and gypsy moths may invoke allergic reactions in sensitized individuals. Three families of dipterous insects (Calliphoridae, Sarcophagidae, Oestridae) include species whose larvae are obligate parasites living within the flesh of vertebrates, a condition referred to as myiasis. The unusual parasitic mites that live within the feathers, nasal passages, and lungs of birds are relatively benign.

DIRECT INJURIES CAUSED BY ARTHROPODS

The various direct effects of arthropods on humans are summarized in Table II. The most common of these is the asthma suffered by millions of people, especially children, who are allergic to the fine airborne particulates generated by insects and mites. These allergens are most often associated with feces or decomposing body parts. The advent of air conditioning and wall-to-wall carpeting seems to have exacerbated this problem especially among the middle and upper classes. Asthma is one of the fastest growing medical problems, particularly among children. Respiratory failure is not an uncommon outcome. Concern over life-threatening multiple stings and allergic reactions (anaphylaxis) to the venom of insects has increased as a result of the introduction and spread of the hybrid of the African honey bee *(Apis mellifera adansonii)* and the imported fire ant *(Solenopsis invicta)* in the Western Hemisphere.

Myiasis is a serious problem in animal production, especially in the neotropics where millions of dollars are lost annually due to these tissue-invading flies. These flies are often found within families that include species that normally feed on the decaying tissues of dead or wounded animals, i.e., flesh flies in the family Sarcophagidae and blow flies in the family Calliphoridae. Some species of Calliphoridae (and related families) sometimes will facultatively invade living tissues while others are so restricted to dead tissues that they are used in maggot therapy to clean deep wounds. All members of the four subfamilies of bot flies (Oestrinae, Gasterophilinae, Hypoderminae, and Cuterebrinae of the family Oestridae) are obligate parasites. The torsalo *(Dermatobia hominis)* is a neotropical dipteran whose eggs are glued to the abdomen of biting flies, and its larvae emerge during blood feeding by the host fly.

ARTHROPOD TRANSMISSION OF MICROPARASITES

Vector-borne diseases can be either biologically or mechanically transmitted. In mechanical transmission, vector mouthparts

TABLE II   **Direct Effects of Arthropods on Humans and Other Animals**

Condition	Health effects	Arthropods involved
Delusional parasitosis	Irrational or destructive acts	Imagined skin parasites
Entomophobia	Stress and mental fatigue	Spiders, wasps
Airborne allergies	Inflamation and respiratory distress	Usually cockroaches and house dust mites
Irritation and blood loss	Allergic skin reactions; pain, itching, inflammation; stress, anemia, and death	Blood-feeding and skin-invading ectoparasites
Envenomization	Arthus and anaphylactic reactions; neurological and cytolytic damage; pain, inflammation, and death	Those with toxic stings, bites, setae, or fluids
Myiasis	Tissue damage, prolonged pain; weight loss, stress, secondary infection, and death	Dipteran maggots

serve as contaminated hypodermic needles since there is no replication or development of the microparasite in the vector; the vector does not serve as an alternate host as in the case of biologically transmitted diseases. For infectious organisms to be mechanically transmitted by arthropods efficiently, they must be abundant in circulating blood or cutaneous tissues and able to survive external exposure. Diseases that are mechanically transmitted by arthropods generally have other transmission mechanisms as well.

Biological transmission takes one of three forms: *propagative* transmission, which involves the replication of eukaryotic parasites and dissemination to the salivary glands prior to transmission; *cyclodevelopmental* transmission, which occurs among filarial parasites, in which development of the parasite to the infective stage is required prior to transmission but there is no increase in the number of parasites; or *cyclopropagative* transmission, which involves both development and multiplication by the parasite as occurs with protozoan parasites such as *Plasmodium, Leishmania,* and *Trypanosoma.*

After the appropriate extrinsic incubation period for replication and/or development of the parasite, the arthropod vector is said to be infective, i.e., able to transmit. Infectious agents are passed to the vertebrate host through a number of different routes: (1) transovarially from the female to her offspring or transtadially from one stage to the next, (2) venereally from infected males to uninfected females, (3) through cofeeding when infected and uninfected vectors group feed, or (4) horizontally through (a) infective saliva injected during feeding, (b) regurgitation of parasites blocking the food canal, (c) defecation of infective feces on the skin, or (d) active escape from the mouthparts and invasion of the skin. In some cases, hosts must assist the transmission process by crushing the infective insect, scratching the contaminated area, and rubbing the eyes. Transmission may be promoted by the physiological or behavioral affect of parasites on the vector. Invasion of host cells or tissues by the parasite can be modulated by host immune responses to the salivary secretions of the vector.

The efficiency of transmission is determined primarily by the "competence" of the arthropod species to support development/replication of the parasite and by the ecology and behavior of the arthropod species. The latter determines the temporal and spatial connection between hosts and potential vectors. Thus, the density, feeding frequency, and host preferences of the vector play critical roles in establishing the *vectorial capacity* (number of infective bites received daily by a single host) of any given species. Environmental conditions also play an important role.

DISEASES TRANSMITTED BY ARTHROPODS

The major diseases transmitted by arthropods are listed in Table III. Protozoan parasites dominate this list in terms of worldwide importance. Malaria is the single most significant vector-borne disease, with an estimated 300 million people infected annually and over 1 million deaths among young children in Africa alone. It is endemic in most tropical and subtropical regions of the world, where it has been resurging since eradication attempts ended some 30 years ago. It is transmitted by *Anopheles* mosquitoes and can be successfully treated with drugs if promptly available. Drug resistance is a growing problem. Tsetse-transmitted African trypanosomiasis remains a serious human disease in parts of tropical Africa but its impact on domestic cattle production is even more severe. Wild bovines are the reservoir of the acute Rhodesian form, and humans and porcines are the reservoir of the chronic Gambian form. American trypanosomiasis is restricted to the mountain regions of tropical America and is often a silent disease that leads to early death. Great efforts are under way to control this zoonotic disease by spraying residual pesticides and by constructing houses with materials that prevent invasion by domesticated kissing bugs. Visceral and cutaneous forms of leishmaniasis affect millions of people and cause significant mortality in Africa, Asia, and South America. The reservoirs for these parasites are dogs, rodents, and a variety of other wild mammals. The cutaneous form also exists in southern Europe and Central America, but disfigurement rather than mortality is usually associated with this form.

Lymphatic filariasis (elephantiasis), transmitted by *Culex quinquefasciatus* and other human-biting mosquitoes, infects many millions of residents throughout the tropics. The long-lived nematode parasites cause debilitation, especially of the lower limbs, but seldom result in death. Onchocerciasis (river blindness), also a filarial infection, is limited to sub-Saharan Africa and some coffee-growing regions in Central and South America. Both of these nematode infections can be prevented by treatment with a new drug, ivermectin. River blindness has been effectively controlled in much of West Africa in recent years through pesticidal elimination of the black fly vectors breeding in streams and the use of anti-helminthics to treat the human population. There are a large number of bacterial and rickettsial infections transmitted by arthropods (mainly ticks, fleas, lice, and mites) annually but none compare to the impact of plague and typhus epidemics in earlier times. Some of these, such as tick-borne Lyme disease and ehrlichiosis, have just been recognized within the past 20 years. Lyme disease is the most prevalent vector-borne disease in the United States. Protective vaccines exist for few bacterial and rickettsial diseases, but all respond to timely antibiotic therapy.

Arthropod-borne viral diseases (arboviruses) are transmitted mainly by mosquitoes and biting midge, but Russian spring–summer encephalitis and several other zoonotic infections (especially of livestock) are transmitted by ticks. Historically, yellow fever was the most important human arboviral disease but today it has been replaced by dengue fever. Dengue viruses infect over a million people annually and can produce a fatal hemorrhagic disease, especially among children. This disease has reinvaded the Western Hemisphere in recent decades and now causes hundreds of thousands of cases annually. Vaccines exist for a

TABLE III Some Important Diseases Transmitted by Arthropod Vectors

Microparasite	Disease	Arthropod vector	Distribution
Transmitted biologically			
Nematodes			
Dirofilaria	Canine heartworm	Mosquitoes	Worldwide
Brugia, Wucheraria	Lymphatic filariasis	Mosquitoes	Tropics
Onchocerca	Riverblindness	Black flies	Africa, Central and South America
Protozoa			
Leishmania	Visceral and cutaneous leishmaniasis	Sand flies	Tropics and warm temperate areas
Trypanosoma spp.	Sleeping sickness and Nagana of cattle	Tsetse flies	Sub-Saharan Africa
Trypanosoma cruzi	Chagas disease	Kissing bugs	Neotropics
Plasmodium	Malaria	Mosquitoes	Mostly tropical
Theileria	Theileriosis	Hard ticks	Africa, Southern Europe
Babesia	Babesiosis	Hard ticks	Widespread
Bacteria			
Bartonella	Carrions disease	Sand flies	South America
	Trench fever	Body lice	Worldwide
	Cat-scratch fever	Fleas	Widespread
Borrelia	Lyme disease	Hard ticks	North America, Eurasia
	Relapsing fever	Soft ticks, lice	Worldwide
Yersinia	Plague	Fleas	Worldwide
Francisella	Tularemia	Hard ticks	Worldwide
Rickettsia and other obligate intracellular bacteria			
Rickettsia	Epidemic typhus	Body lice	Africa, Americas
	Murine typhus	Fleas	Widespread
	Spotted fevers	Hard ticks	Widespread
Orientia	Scrub typhus	Chigger mites	Southeast Asia
Cowdria	Heartwater	Hard ticks	Sub-Saharan Africa
Anaplasma	Anaplasmosis	Hard ticks	Worldwide
Ehrlichia	Ehrlichiosis	Hard ticks	Widespread
Arthropod-borne viruses (arboviruses)[a]			
Flaviviruses	Yellow fever, dengue, WN, SLE, JE, MVE, ROC, WSL	Mosquitoes	Tropics
	RSSE, OMSK, KFD, LI, POW	Hard ticks	Widespread
Bunyaviruses	CE, LAC, RVF	Mosquitoes	Africa, North America
	ORO	Biting midges	South America
	CCHF, NSD	Hard ticks	Africa, Eurasia
	SFF	Sand flies	South America, Africa, and Eurasia
Togaviruses	EEE, WEE, VEE, RR	Mosquitoes	Widespread
Rhabdoviruses	VSV, BEF	Biting flies	Widespread
Reoviruses	Bluetongue, AHS, EHD	Biting midges	Widespread
	Colorado tick fever	Hard ticks	Western North America
Unnamed	African swine fever	Soft ticks	Africa
Transmitted mechanically			
Protozoa			
Trypanosoma	Trypanosomiasis	Biting flies	Widespread
Bacteria			
Treponema	Yaws/pinta	Eyes gnats	Tropics
Bacillus	Anthrax	Biting flies	Widespread
Anaplasma	Anaplasmosis	Biting flies	Widespread
Various Anaerobes	Summer mastitis	Head flies	Widespread
Viruses			
Poxvirus	Myxomatosis, fowlpox	Biting flies	Worldwide
Retrovirus	Equine infectious anemia	Tabanids	Worldwide

[a] WN, West Nile; SLE, St. Louis encephalitis; JE, Japanese encephalitis; MVE, Murray Valley encephalitis; WSL, Wesselsbron; RSSE, Russian spring summer encephalitis; OMSK, Omsk hemorrhagic fewer; KFD, Kyasanur Forest disease; LI, Louping ill; POW, Powassan; CE, California encephalitis; LAC, La Crosse encephalitis; RVF, Rift Valley fever; ORO, Racio; CCHF, Crimean-Congo hemorrhagic fever; NSD, Nairobi sheep disease; SFF, sand fly fever; EEE, Eastern equine encephalitis; WEE, Western equine encephalitis; VEE, Venezuelan equine encephalitis; RR, Ross River; VSV, vesicular stomatitis virus; BEF, bovine ephemeral fever; AHS, African horse sickness; EHD, epizootic hemorrhagic disease.

few arboviral diseases (yellow fever and Japanese encephalitis) but vector control is often the only preventative measure.

DISEASE AND VECTOR MANAGEMENT

Efforts to develop vaccines for a wide range of vector-borne diseases, including malaria, have been vigorously supported but success has been slow. Antigenic variation in parasites has been a major deterrent. New drugs for treatment of parasitic diseases also are under development and several important new drugs for treatment of helminth and protozoan parasites have been marketed in recent years. Nonetheless, vector control is often the first line of defense against the transmission of these diseases and, during active epidemics, this is the only option outside of public education. Control of vertebrate reservoir animals has occasionally been practiced for diseases that are maintained by rodents. An early example was control of wild bovine reservoirs in parts of Africa to control trypanosomiasis in humans (sleeping sickness) and cattle (Nagana). Vector control programs are generally based on surveillance systems that monitor and report cases of disease or vector population levels. A variety of tools are available to manage vector populations, including chemical pesticides, biological controls, habitat alteration, and personal protection strategies (e.g., screens, bed nets, and repellents). Currently, genetic control strategies are receiving much attention. Optimal vector control programs utilize integrated vector management strategies and strive to maintain vector populations below the threshold densities required for transmission. Targeting the immature stages of vectors that blood feed and transmit disease only as adults normally is more efficient and cost effective. Lack of public health funds is the major limitation on surveillance and vector control programs, especially in developing countries where these diseases have the greatest impact.

See Also the Following Articles

Bubonic Plague • Chiggers and Other Disease-Causing Mites • Lice, Human • Malaria • River Blindness • Ticks • Tsetse Fly • Veterinary Entomology • Yellow Fever • Zoonoses, Arthropod-Borne

Further Reading

Busvine, J. R. (1976). "Insects, Hygiene and History." University of London Press, London.

Eldridge, B. F., and Edman, J. D. (2000). "Medical Entomology: A Textbook on Public Health and Veterinary Problems Caused by Arthropods." Kluwer Academic, Dordrecht/Norwell, MA.

Harwood, R. F., and James, M. T. (1979). "Entomology in Human and Animal Health," 7th ed. Macmillan, New York.

Kettle, D. S. (1994). "Medical and Veterinary Entomology," 2nd ed. Cambridge University Press, Cambridge, U.K.

Lane, R. P., and Crosskey, R. W. (1993). "Medical Insects and Arachnids." Chapman & Hall, London/New York.

Lederberg, J., Shope, R. E., and Oaks, S. C., Jr. (1992). "Emerging Infections." National Academy Press, New York.

Service, M. W. (1978). A brief history of medical entomology. *J. Med. Entomol.* **14,** 603–626.

Service, M. W. (2000). "Medical Entomology for Students," 2nd ed. Kluwer Academic, Dordrecht/Norwell, MA.

Medicine, Insects in

Ronald A. Sherman
University of California, Irvine

Throughout history, humans have used insects and their products therapeutically. Ingested, injected, or topically applied, insects have been used to treat an assortment of respiratory, gastrointestinal, cardiac, neuromuscular, and infectious diseases. To this day, therapeutic insects are prescribed worldwide. Medicinal maggots and honey bee venom therapy are two examples of therapeutic roles played by insects.

INTRODUCTION AND HISTORICAL OVERVIEW

Many insects are ingested for their medicinal value as well as their nutritional value. Roasted, boiled, or powdered cockroaches (Blattodea: Blattidae) have been ingested by people of many cultures to treat respiratory diseases. Various beetles (Coleoptera) are noted to be useful in the treatment of intestinal diseases. The blister beetle, *Lytta vesicatoria* (Spanishfly), is the source of cantharidin, a vesicant that was ingested in Europe as an aphrodisiac. Stinkbugs (Heteroptera: Pentatomidae) in China and termites (Isoptera) in India were used for the same function.

Up until a few decades ago, patients with neurosyphilis often were cured with an inoculation of malaria (*Plasmodium* spp.). The syphilis pathogen *(Treponema pallidum)* was killed by the recurrent fevers and/or other still unknown interactions with the malaria parasites. The malaria was then eradicated with quinine or another species-specific antimalarial agent. Initially, malaria inoculation was achieved by transferring the blood of a malaria-infected neurosyphilitic to a nonparasitized neurosyphilitic. This practice was soon replaced by mosquito-transmitted inoculations.

"Caterpillar fungus" *(dong chong xia cao)* is a Chinese moth larva (Hepialidae: *Hepialus oblifurcus*) infected with an entomopathogenic (insect-killing) fungus, *Cordyceps sinensis* (Clavicipatales: Ascomycotina). Ingestion of the caterpillar fungus reportedly strengthens and rejuvenates the body. The substance achieved international notoriety in 1993, when fungus-drinking Chinese athletes set new world track records. At a cost of approximately $1000 per kilogram, caterpillar fungus is often prepared as a broth; both the broth and the caterpillar are eaten.

Arthropods have played a role in wound care for centuries. The use of large ant or beetle mandibles for holding together wound edges has been documented in many countries. Honey and spiderwebs both have been used to dress wounds and prevent infection. Maggot therapy—the topical application of blowfly larvae (*Phaenicia, Lucilia,* and *Phormia*) to treat infected wounds—has been practiced around the world for at least 70 years.

MAGGOT THERAPY

The practice of maggot therapy is based on observations that wounds naturally infested with maggots (wound myiasis) often are free of infection and debris. For centuries, European military surgeons described how the maggot-laden wounds of soldiers not promptly removed from the battlefield often appeared clean, once the larvae were wiped away. Soldiers' maggot-infested wounds seemed to heal better than wounds that had not been infested. After his own observations of wound myiasis on the battlefields of World War I, William Baer intentionally placed blowfly larvae into the chronic wounds of his patients at Johns Hopkins and Children's Hospital in Baltimore. Baer first presented his results in 1929; by 1935, thousands of physicians and surgeons had embraced this practice. Many hospitals maintained their own therapeutic fly colonies; other practitioners obtained maggots from pharmaceutical companies.

Maggot therapy all but disappeared during the 1940s. The reasons are purely speculative but probably include the development of antibiotics and the refinement of surgical techniques that came about during World War II. The relatively high cost of maggots ($5 for a bottle of 1000 larvae) may have been another factor. Over the next several decades, therapeutic myiasis was performed only rarely, and only as a last resort, in patients who failed to respond to aggressive surgical and antibiotic treatments. The 1980s brought about the realization that surgery and antibiotics could not cure all wounds. Many infections were now resistant to the once omnipotent antimicrobials. The 1990s saw the reemergence of maggot therapy to treat many of these nonhealing wounds. Today, live fly larvae are once again used in over a thousand centers worldwide for treating chronic wounds.

Maggots effectively treat open wounds by removing dead and infected tissue (debridement), killing bacteria (disinfection), and stimulating the wound to heal. To understand the procedure of maggot therapy, it is necessary to review the natural history of the fly. Many species of blowflies (Calliphoridae) naturally "blow" or lay their eggs on carrion, feces, or the dead (necrotic) tissue of a living host. Upon hatching, the larvae ingest this tissue as it is liquefied by the maggots' digestive secretions. Within 3 to 7 days, the maggots leave what remains of their meal and pupate underground or in some other protected site. One to three weeks later, adult flies emerge. Therapeutic maggots—blowfly larvae that have been disinfected ("sterile" maggots)—are placed on wounds at a density of about 5 to 10 cm^{-2}. Larvae are retained on the wound for about 48 h, in cagelike dressings. After one or more such cycles of treatment, the wound is often free of necrotic tissue and able to accept a skin graft or heal spontaneously.

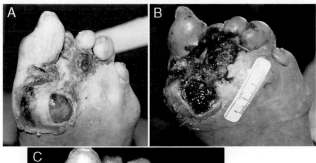

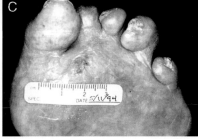

FIGURE 1 The right foot of a 73-year-old man who was treated for 3 years by orthopedic and podiatric surgery for his chronic foot ulcers. (A) before, (B) during, and (C) 1 year after maggot debridement therapy.

Treatments can be administered in the hospital, clinic, or nursing home, or at home.

Maggot therapy has been used to treat pressure ulcers, venous stasis ulcers, diabetic foot ulcers, burns, traumatic wounds, and nonhealing postsurgical wounds (Fig. 1). Compared with conventional wound therapy, medicinal maggots are credited with more rapid debridement and wound healing. Maggot therapy has reportedly saved numerous limbs from amputation and other surgical procedures. Other advantages of maggot therapy include its simplicity and safety, and (by current standards) the relatively low cost of treatment.

APITHERAPY

Honey bee (*Apis mellifera*) venom, propolis, royal jelly, beeswax, and honey all are used therapeutically. Medicinal use of any of these products can be considered to be "apitherapy," but many authorities use the term specifically to denote the clinical use of honey bee venom. Bee venom contains a multitude of polypeptides, enzymes (phospholipase A$_2$, hyaluronidase), and biologically active amines (histamine, dopamine, noradrenaline). The mechanisms by which venom exerts its beneficial actions are unknown, but might include an anti-inflammatory effect resulting from alterations seen in pituitary and adrenal gland function, local effects on the nerves and blood vessels, and stimulation of acupuncture-like pathways. A combination of these and other mechanisms may explain the diversity of benefits attributed to venom therapy.

Apitherapists have successfully treated rheumatological disorders (rheumatoid and psoriatic arthritis, gout, fibromyalgia), neurological diseases (multiple sclerosis, chronic pain syndromes), immunological diseases (scleroderma, systemic lupus erythematosis), and other chronic illnesses. Some therapists administer the treatments in the form of an increasing number of bee stings; other practitioners inject a partially purified extract of the bee venom, in gradually increasing doses. Serious toxic reactions to the venom are uncommon because patients are educated about the signs and symptoms of venom reactions; they are also supervised closely following treatment and are given ready access to medical care after leaving the therapist's office or apiary.

See Also the Following Articles
Bee Products • Beeswax • History of Entomology • Venom

Further Reading

Baer, W. S. (1929). Sacro-iliac joint—arthritis deformans—viable antiseptic in chronic osteomyelitis. *Proc. Int. Assembly Inter-state Postgrad. Med. Assoc. North Am.* **371,** 365–372.

Chernin, E. (1984). The malaria therapy of neurosyphilis. *J. Parasitol.* **70,** 611–617.

Gudger, E. W. (1925). Stitching wounds with the mandibles of ants and beetles; a minor contribution to the history of surgery. *J. Am. Med. Assoc.* **84,** 1861–1865.

Kim, C. M.-H. (1997). Apitherapy (bee venom therapy). *In* "Potentiating Health and the Crisis of the Immune System: Integrative Approaches to the Prevention and Treatment of Modern Diseases" (A. Mizrahi, S Fulder, and N. Sheinman, eds.), pp. 243–270. Plenum Press, New York.

Leclercq, M. (1969). "Entomological Parasitology; The Relations between Entomology and the Medical Sciences" (G. Lapage, transl.). Pergamon Press, Oxford, U.K.

Mizrahi, A., and Lensky, Y. (eds.) (1996). "Bee Products: Properties, Applications, and Apitherapy." Plenum Press, New York.

Mumcuoglu, K. Y., Ingber, A., Gilead, L., Stessman, J., Friedmann, R., Schulman, H., Bichucher, H., Ioffe-Uspensky, I., Miller, J., Galun, R., and Raz, I. (1999). Maggot therapy for the treatment of diabetic foot ulcers. *Diabetes Care* **21,** 2030–2031.

Sherman, R. A. (1998). Maggot therapy in modern medicine. *Infect. Med.* **15,** 651–656.

Sherman, R. A., Hall, M. J. R., and Thomas, S. (2000). Medicinal maggots: An ancient remedy for some contemporary afflictions. *Annu. Rev. Entomol.* **45,** 55–81.

Steinkraus, D. C., and Whitfield, J. B. (1994). Chinese caterpillar fungus and world record runners. *Am. Entomol.* **40,** 235–239.

Taylor, R. L. (1975). "Butterflies in My Stomach." Woodbridge Press, Santa Barbara, CA.

Thomas, S., Jones, M., Shutler, S., and Jones, S. (1996). Using larvae in modern wound management. *J. Wound Care* **5,** 60–69.

Megaloptera
(Alderflies, Dobsonflies)

N. H. Anderson
Oregon State University

The Megaloptera, which include the alderflies, dobsonflies, fishflies, and hellgrammites, are a small order of neuropterous insects with 250 to 300 species. They are often considered to be the most primitive group of insects with complete metamorphosis. In the fossil record, Megaloptera first occur about 250 mya in the late Permian. Their sister group is the Raphidioptera (snakeflies), and both these orders are closely related to the Neuroptera (lacewings and antlions). Among the major orders, they are most closely related to the Coleoptera (beetles). Larvae of all Megaloptera are aquatic predators. Their association with cool, well-oxygenated waters probably accounts for their greater species diversity in temperate regions than in the tropics. Although they are found throughout the world, the distributions are discontinuous, which is characteristic of a relict fauna.

DIAGNOSIS AND CLASSIFICATION

Megaloptera can be distinguished from other orders of insects by the following combination of characters: holometaboly, terrestrial adults and eggs, predaceous aquatic larvae, and exarate terrestrial pupae.

The adults are large, short-lived, and soft-bodied; the head is broad and flattened, with chewing mouthparts, large bulging compound eyes, and long many-segmented antennae; there are two pairs of similar membranous wings, held rooflike over the body, with all major veins present and many cross-veins; the five-segmented tarsi have paired apical

A   **B**

FIGURE 1 Mature larvae of *Sialis:* (A) *S. rotunda* and (B) *S. californica.* [From Azam, K. M., and Anderson, N. H. (1969). Life history and habits of *Sialis rotunda* and *S. californica* in Western Oregon. *Ann. Entomol. Soc. Am.* **62,** 549–558.]

FIGURE 2 *Sialis* adult.

FIGURE 3 *S. californica* egg masses. [From Azam, K. M., and Anderson, N. H. (1969). Life history and habits of *Sialis rotunda* and *S. californica* in Western Oregon. *Ann. Entomol. Soc. Am.* **62**, 549–558.]

claws; and the abdomen has 10 segments, eight pairs of spiracles, and lacks cerci.

The larvae have a well-sclerotized head, with long toothed mandibles, four- or five-segmented antennae, and six lateral stemmata (eyespots); the thorax is sclerotized dorsally, with a quadrate pronotum, meso- and metathoracic spiracles, and elongate five-segmented legs with two tarsal claws; and the abdomen is soft, with seven or eight pairs of lateral filaments, and spiracles on segments 1 to 8.

The order contains two families, Sialidae and Corydalidae. Sialids, or alderflies, are slow, awkwardly flying insects that range from 10 to 15 mm in length. Their bodies are black, brown, or yellowish orange. The wings are held tentlike over the abdomen so that they bear some resemblance to caddisflies (Trichoptera). The Corydalidae are much larger, 40 to 75 mm in length. Many species have pale, smoky wings mottled with brown, whereas others are nearly black, with white markings. This family is divided into two distinctive subfamilies: the Corydalinae or dobsonflies (larvae are hellgrammites), and the Chauliodinae or fishflies.

FAMILY SIALIDAE

Of the eight genera in the family Sialidae (Figs. 1–3), only the widespread Holarctic genus *Sialis* has received much attention. The life history of *S. lutaria* has been studied extensively in Europe, and the larvae have been used in behavioral and physiological experiments. *Sialis* larvae are found in many habitats, ranging from small springs to large rivers and from ponds to large lakes. They usually occur where the substrate

is soft and where dead leaves and other detritus have accumulated. Larvae may dig into the substrates to a depth of several centimeters. Sympatric species may be ecologically segregated based on habitat. For example, in Oregon *S. rotunda* occurs in ponds with a muddy bottom, whereas *S. californica* is usually found in pools or glides of streams. However, the two species can be found together in backwater reaches.

The life cycle takes 1 or 2 years depending on physical and biological conditions. The colder and less productive habitats such as trout streams and mountain lakes lengthen the period of larval life, whereas rapid development occurs in warm, productive habitats such as lowland ponds, warm lakes, and muddy rivers.

Larvae feed nonselectively on small invertebrates such as insect larvae, annelid worms, crustaceans, and mollusks. The prey is seized by the elongate mandibles and forelegs and worked into the mouth with the aid of labrum, maxillae, and labium. Sometimes only the softer abdominal parts of prey are eaten. Cannibalism occurs, especially in high-density situations.

During spring or early summer the final instars move to shallow areas near the shore. They then leave the water and prepare to pupate. The transformation also involves a switch in respiration from using aqueous dissolved oxygen via the gills to intake of atmospheric oxygen through the abdominal and thoracic spiracles. Pupation occurs in an unlined chamber dug 1 to 10 cm into soil or litter.

The adults emerge after a pupal instar of about 2 weeks. Adults are most active during midday. In mating, the male crawls beneath the female's abdomen from the rear and raises his abdomen upward and forward to couple the genitalia. A spermatophore is passed to the female within a few minutes, copulation is terminated, and oviposition occurs within a day.

Adult *Sialis* have biting mouthparts, and some observations suggest that they visit flowers to feed.

The eggs are deposited on a variety of substrates projecting over the water. The underside of overhanging leaves is the most common oviposition site. Egg masses contain 300 to 900 eggs. There are two distinctive types of egg mass; in one, the long axis of the egg is almost parallel to the substrate, and in the other it is almost upright. The incubation period is about 10 days to 2 weeks. Eggs are sometimes parasitized by tiny *Trichogramma* wasps.

In hatching, the larva pushes its head against the chorion below the micropylar projection. The toothed, V-shaped egg burster then ruptures the chorion, initiating a jagged tear through which the larva emerges. When the larva leaves the egg, a postembryonic molt occurs, and the egg burster and embryonic membrane are left attached to the eggshell. Then the appendages and abdominal filaments are expanded and the larva becomes active. It drops to the water surface, where it is quickly wetted to pass through the surface film, and then swims to the bottom. There are about 10 instars before pupation.

CORYDALIDAE

Subfamily Corydalinae

The subfamily Corydalinae includes nine genera and is found mainly in the tropics or subtropics. Only four species of *Corydalus* occur in America north of Mexico, whereas there are three genera and almost 50 species of Corydalinae described from the Neotropics. Corydalinae contain the largest Megaloptera, with some adults having a wingspan greater than 150 mm.

C. cornutus, the dobsonfly, ranges over most of North America east of the Continental Divide, from Canada to Mexico. The larvae, called hellgrammites, are large (to 65 mm), dominant predators in stream riffles that feed opportunistically on invertebrate prey. Final instars leave the stream to pupate in the soil under rocks, or in rotting logs.

C. cornutus is unusual in that the male has elongate, hornlike mandibles that are half the length of the body. The mandibles of the female are also large, but only as long as the head (Fig. 4). Mating occurs shortly after emergence. The male has scent glands between abdominal segments 8 and 9, which evidently produce a sex "stimulant." The male places his enlarged mandibles over the wings of the female for a short time before mating.

Females apparently feed on fruit juices or other liquid food, since the eggs are undeveloped at emergence and require considerable yolk deposition before oviposition. *Corydalus*

FIGURE 4 *Corydalus* (dobsonfly) female. (Photograph by B. M. Drees.)

females are reported to tear apart flowers to feed on nectar. Male dobsonflies do not feed, but they imbibe some water.

Females oviposit on objects overhanging the water. Eggs are encased in a white protective material in oval masses of one to five layers and contain over 1000 eggs. A female may deposit two or three egg masses. Adults live for about a week and females die shortly after ovipositing. The incubation period is 2 to 3 weeks.

The life cycle is temperature dependent. In north central Texas, dobsonflies are univoltine. However, northern populations may have a life cycle of 2 to 3 years and are larger as adults.

Subfamily Chauliodinae

There are 16 genera in the subfamily Chauliodinae, and the 18 species account for most of the diversity of megalopterans in America north of Mexico. Five genera and 10 species are known from the Neotropics.

The Chauliodinae have a wide geographical distribution and occur in a wider range of habitats than do sialids or corydalids. Several genera (e.g., *Neohermes, Dysmicohermes, Protochauliodes*) inhabit intermittent streams, whereas others (e.g., *Orohermes*) are found in cold, permanent streams and rivers as well as in the adjacent intermittent tributaries. *Chauliodes* occurs in slow waters, in swamps and ponds, and sometimes in intermittent habitats. These larvae can exploit low-oxygen habitats because the terminal spiracles are on contractile tubes.

The life cycle is also quite variable, with growth and development being temperature and habitat dependent. For example, the life cycle of *Neohermes* varies from 2 to 5 years depending on the duration of flow in temporary streams. Estimates of the number of instars range from 9 to 12. Females may have one more larval molt than males to account for their larger size.

Fishfly larvae are generalist predators, but in intermittent habitats they also may be scavengers, feeding on corpses of individuals stranded by receding waters.

Mating occurs in a tail-to-tail position with no display or special mating behavior (but pheromone attraction may be involved in some genera). The male *Nigronia* backs to the female, the genitalia are clasped, and they remain in copulation for several hours. Spermatophore transfer has not been observed for *Nigronia* or *Corydalus,* although it is reported for *Sialis.*

The final instar of *N. serricornis* leaves the water to pupate in a shallow cell that is dug into the soil. The exarate pupa is light colored but darkens in the last 2 days before emergence. The pupa is active; it turns in the cell, which helps in maintaining cell structure. Pupae do not feed but can use the sizable mandibles and will bite an object that is placed in contact. The pupal period ranges from 2 to 4 weeks depending on temperature.

Like most megalopterans, *Nigronia* adults are awkward and weak in flight. They fly throughout the day, with oviposition occurring in the afternoon. *Nigronia* adults readily take fluids, especially sugar solution and mashed fruit. Adult feeding is needed because some individuals live for 2 weeks.

Reproductive organs in *Nigronia* are mature at emergence, as is the case with *Sialis* but unlike *Corydalus*. A bimodal occurrence in egg deposition demonstrated that after their first oviposition females waited a week to lay a second mass. Egg masses are composed of one to five layers. Maturation of *Nigronia* eggs requires 2 to 3 weeks. Hatching begins with individuals in the uppermost layer and continues to the lower level.

Larvae of fishflies and corydalids often have colonies of epizooics attached to the exoskeleton. These growths, which are readily apparent at low magnification, include stalked protozoans, rotifers, and filamentous algae. Phoretic Chironomidae, for example, *Plecopteracoluthus downesi,* are often found strapped in cases attached to thoracic sternites.

See Also the Following Articles
Aquatic Habitats • *Neuroptera* • *Raphidioptera*

Further Reading
Azam, K. M., and Anderson, N. H. (1969). Life history and habits of *Sialis rotunda* and *S. californica* in western Oregon. *Ann. Entomol. Soc. Am.* **62,** 549–558.

Brigham, W. U. (1982). Megaloptera. *In* "Aquatic Insects and Oligochaetes of North and South Carolina" (A. R. Brigham, W. U. Brigham, and A. Gnilka, eds.). Midwest Aquatic Enterprises, Mohamet, IL.

Brown, A. V., and Fitzpatrick, L. C. (1978). Life history and population energetics of the dobsonfly, *Corydalus cornutus. Ecology* **59,** 1091–1108.

Contreras-Ramos, A., and Harris, S. C. (1998). The immature stages of *Platyneuromus* (Corydalidae), with a key to the genera of larval Megaloptera of Mexico. *J. North Am. Benthol. Soc.* **17,** 489–517.

Evans, E. D. (1972). A study of the Megaloptera of the Pacific coastal region of the United States. Ph.D. thesis, Oregon State University.

Evans, E. D., and Neunzig, H. H. (1996). Megaloptera and aquatic Neuroptera. *In* "Aquatic Insects of North America" (R. W. Merritt and K. W. Cummins, eds.). 3rd ed., Chap. 16. Kendall/Hunt, Dubuque, IA.

Kukalova-Peck, J. (1991). Fossil history and the evolution of hexapod structures. *In* "The Insects of Australia" (I. D. Naumann *et al.,* eds.), 2nd ed., Chap. 6. Melbourne University Press, Melbourne.

McCafferty, W. P. (1981). Fishflies, dobsonflies, and alderflies (order Megaloptera). *In* "Aquatic Entomology," Chap. 11. Science Books International, Boston.

Oswald, J. D., and Penny, N. D. (1991). Genus-group names of the Neuroptera, Megaloptera, and Raphidioptera of the world. Occasional Paper No. 147 of the California Academy of Sciences.

Petersen, R. C. (1974). Life history and bionomics of *Nigronia serricornis* (Say) (Megaloptera: Corydalidae). Ph.D. dissertation, Michigan State University.

Theischinger, G. (1991). Megaloptera. *In* "The Insects of Australia" (I. D. Naumann *et al.,* eds.). 2nd ed., Chap. 32. Melbourne University Press, Melbourne.

Metabolism

S. N. Thompson
University of California, Riverside

R. K. Suarez
University of California, Santa Barbara

Metabolism refers to the thousands of chemical reactions that occur in the cell. These reactions link together in defined series to form pathways. Metabolic pathways are interdependent and exquisitely regulated for the efficient extraction of energy from fuels, catabolism, and the synthesis of biological macromolecules, anabolism. Metabolism is a subject area of biochemistry, which also includes the structural chemistry of biological molecules and the chemistry of molecular genetics, the chemical processes involved in the storage and inheritance of biological information.

The biological diversity of insects has enabled focused study on the metabolic bases for many physiological capabilities that are unique to insects or their arthropod and close relatives. The literature on insect biochemistry is extensive. Early studies of insect biochemistry focused on chemical content, individual chemical reactions, metabolic rate, and respiration. Much of this was discussed in the seven editions of Sir V. B. Wigglesworth's *The Principles of Insect Physiology,* which first appeared in 1939. As advances were made other comprehensive treatments appeared, including the 1964 edition of M. Rockstein's *Physiology of Insecta* series and D. Gilmour's *The Biochemistry of Insects,* which first appeared in 1961. *The Biochemistry of Insects,* edited by M. Rockstein, appeared in 1978. More recently, insect metabolism was described in several volumes of the treatise *Comparative Insect Physiology, Biochemistry and Pharmacology,* edited by G. A. Kerkut and L. I. Gilbert (1985).

INTERMEDIARY METABOLISM

Insects share with other invertebrates most of the common pathways of carbohydrate, lipid, and amino acid metabolism. Although much has been presumed based on overt similarities to more extensive studies of mammals and other higher taxonomic groups, many aspects of intermediary metabolism have been examined in a number of insects and different insect tissues. Much of intermediary metabolism, including synthesis and storage of carbohydrate and fat, takes place in the fat body.

The metabolism and utilization of the glucose disaccharide trehalose as the principal hemolymph or blood sugar is unique to insects and some other invertebrates. Unlike glucose, trehalose is a nonreducing sugar, a sugar not readily oxidized by common oxidizing agents. First described from an insect by G. R. Wyatt and his associates in pupae of the polyphemus moth, *Antheraea polyphemus,* trehalose occurs in many insects at variable but high levels. In lepidopteran insects, trehalose levels are commonly as high as 100 mM. The concentration of trehalose in insects often greatly exceeds levels of glucose in the blood of mammals. Blood glucose in humans typically is about 5 mM, a value that would be considered very low for trehalose in hemolymph. With few exceptions, glucose occurs in insect hemolymph at levels less than 5 mM and often at less than 1 mM. Trehalose serves multiple functions, as a storage carbohydrate that serves as a fuel for flight and as a cryoprotectant, protecting insects from damage during overwintering in cold climes. The hemolymph level of trehalose plays an important role in regulating carbohydrate intake and maintaining nutritional homeostasis. Levels of trehalose in the hemolymph are maintained by a complex interaction of nutrient intake and metabolism.

Trehalose is synthesized in the fat body from two metabolic intermediates of glycolysis, glucose 1-phosphate and glucose 6-phosphate. The reactions synthesizing trehalose are catalyzed by trehalose-6-phosphate synthase and trehalose-6-phosphate phosphatase. The sources of glucose for trehalose synthesis include dietary sucrose, glycogen, and gluconeogenesis, dietary sugar being the sole source of glucose under fed conditions. Trehalose formation from glycogen has been described in several insects, including the American cockroach *Periplaneta americana* and tobacco hornworm *Manduca sexta* during starvation. The breakdown of glycogen to glucose is due to activation of the enzyme glycogen phosphorylase, first demonstrated by J. E. Steele and his associates to be under endocrine control by a neurohormone released from the corpora cardiaca in the brain. The induction of a "hypertrehalosemic" hormone RNA transcript in the central nervous system of the cockroach *Blaberus discoidalis* in response to starvation was recently demonstrated. Glucose synthesis, followed by trehalose formation, via gluconeogenesis has been reported only in *M. sexta* and was induced when nymphs were maintained on low-carbohydrate diets. Starvation did not induce gluconeogenesis.

Most insects obtain energy principally from aerobic respiration, but many species have some capacity for anaerobic energy metabolism when exposed to hypoxic or anoxic conditions. This is best known in aquatic insects such as midge larvae, in which the fermentation products may include lactate, ethanol, and acetate. For example, the midge *Chaoborus crystallinis* accumulates succinate, suggesting that this species is capable of anaerobic respiration, possibly involving fumarate reductase for ATP production. Polyol formation during diapause of some insects is another example of anaerobic metabolism.

CHEMISTRY AND METABOLISM OF SCLEROTIZATION

The hardness of cured exocuticle is the result of cross-linking or polymerization between molecules of the protein sclerotin and/or cross-linking between sclerotin and chitin. The chemical composition of sclerotin and the chemistry of the polymerization process were first understood from organochemical analyses of the products resulting from chemical degradation of cuticle. Current knowledge of cuticular structure came about through nondestructive nuclear magnetic resonance (NMR) analysis of intact cuticle. Although sclerotization is principally the result of the reaction of cuticular proteins and chitin with quinones derived from *N*-acetyldopamine, the metabolism of sclerotization is complex and sclerotized cuticles vary greatly in their physical and chemical properties because of metabolic variation.

The metabolism involves principally quinone methide intermediates first described by M. Sugumaran and his collegues. Reactions of quinone methides produce so-called β-sclerotin, with cross-linkages involving the β carbon of the catecholamine side chain, as well as dehydrodopamine intermediates producing linkages to the benzoyl ring. The metabolism of the quinone and dehydrodopamine intermediates is integrated, such that both the side chain and the ring structure may be involved in an individual cross-linked structure. Phenoloxidases, quinone isomerases, and quinone tautomerases are the principal enzymes involved in the metabolism of these various metabolic intermediates.

The structure of the cross-linkage between sclerotin and chitin in the pupal cuticle of *M. sexta* was recently established by K. Kramer, J. Schaefer, and associates through the use of solid-state NMR spectroscopy. As shown in Fig. 1, the structure demonstrates cross-linkage of phenolic and quinone intermediates with the imidizole ring of histidine residues of the protein and the β-hydroxyl group on Carbon 4 of *N*-acetylglucosamine units of chitin.

CHEMISTRY AND METABOLISM OF PIGMENTS

The diversity of insect coloration is in large measure because of an abundance of pigments. Combinations of pigments together with effects of light diffraction, refraction, and interference involving various anatomical structures produce the array of exotic colors familiar to insect observers. Many insects synthesize melanins, ommochromes, porphyrins, pteridines, and/or quinones. Other pigments such as flavonoids and carotenoids, although not synthesized, are often sequestered by insects from plants and contribute to coloration. Two pigment groups are notable: ommochromes, first reported from the eyes of insects, and papilochromes, a unique group

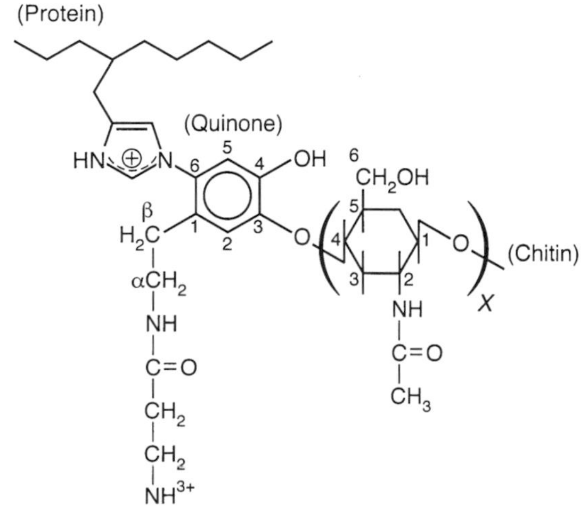

FIGURE 1 Structure of the protein–chitin cross-linkage in the pupal cuticle of *M. sexta*. Protein may be linked through the 1 or the 3 nitrogen of the imidizole ring to the 2, 5, or 6 ring carbon of the quinone derivative, and carbon 4, or other carbons of chitin may be linked to phenoxy carbon 3 or 4 of the quinone. (Adapted from Schaefer *et al.*, 1987.)

that occurs in the bodies and wings of the butterfly family Papilionidae.

Ommochromes are polymers of heterocyclic phenoxines, distributed among a variety of different insect tissues, producing yellow, red, and brown coloration. They are synthesized from tryptophan in a metabolic pathway involving kyneurenine derivatives. In the compound eye, ommochromes form the principal masking pigments that surround and isolate the individual ommatidia and thus, the origin of the name. Several eye-color mutants, described in several insect species, result from the absence of enzymatic function at specific steps in the synthetic pathway. Identification of these steps in *Drosophila* was one of the early comfirmations of A. Garrod's one gene–one enzyme hypothesis. The ommochrome biosynthetic pathway in the coloration of *M. sexta* larvae is hormonally regulated.

Papiliochromes are novel white, yellow, and red pigments whose synthesis intersects the well-known metabolic pathways the melanins and ommochromes. For butterflies of the genus *Papilio,* the precursors are β-alanine, tyrosine, and tryptophan. Papiliochromes accumulate in the wing scales and their distribution varies with the butterfly species. Recent studies on papiliochrome synthesis demonstrated that, as in the case of sclerotization, quinone methides derived from tyrosine are intermediates. The synthesis involves the nonenzymatic condensation of *N*-β-alanyldopamine quinone methide with L-kynurenine to produce a mixture of two diastereoisomers of papilochrome II, a white pigment. Papiliochrome II is a peptide in which the two aromatic rings are linked by a bridge between the aromatic amino group of kynurenine and the catecholamine side chain of norepinephrine derived from the quinone.

ENERGY METABOLISM DURING FLIGHT

Insect flight muscles are obligately aerobic, deriving energy from O_2-dependent substrate oxidation to CO_2 and H_2O. Small insects in flight achieve the highest known mass-specific rates of aerobic metabolism among animals. Of the estimated one-half million insect species capable of flight, the metabolism of only a few has been subjected to detailed examination. Insect species differ in the extent to which carbohydrates (principally trehalose), fats (mainly diacylglycerol), and proline (an amino acid) are used to fuel flight. A scheme summarizing the relevant pathways is shown in Fig. 2.

In some species of locusts and moths, flight is initially fueled by carbohydrate. Prolonged flight follows activation of

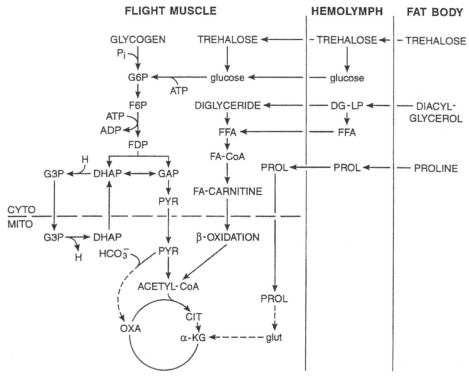

FIGURE 2 Metabolic scheme showing pathways of carbohydrate, fat, and proline oxidation in insect flight muscles. Included are the anaplerotic roles of pyruvate carboxylation and proline oxidation in some species, as well as the β-glycerophosphate shuttle (involving G3P and DHAP) for transferring reducing equivalents from cytoplasm to mitochondria. The contribution of each pathway varies according to species. In some species, substrate use may change with time in flight. Acetyl-CoA, acetyl coenzyme A; ADP, adenosine diphosphate; ATP, adenosine triphosphate; CIT, citrate; CYTO, cytosol; DG-LP, lipoprotein-bound diacylglycerol; DHAP, dehydroxyacetone phosphate; F6P, fructose 6-phosphate; FDP, fructose bisphosphate; FA-carnitine, fatty acid–carnitine; FA-CoA, fatty acid–coenzyme A; FFA, free fatty acids; G3P, glycerol 3-phosphate; G6P glucose 6-phosphate; GAP, glyceraldehyde 3-phosphate; glut, glutamate; HCO_3^-, bicarbonate ion; H, nicotinamide adenine dinucleotide; α-KG, ketoglutarate; MITO, mitochondrion; OXA, oxaloacetate, PROL, proline; PYR, pyruvate. (Adapted from Storey, 1985.)

fatty acid oxidation and inhibition of carbohydrate oxidation, which are processes triggered by adipokinetic hormones. Bees appear unable to make use of fats to fuel flight, but instead oxidize carbohydrate, while using proline as a carbon source for augmenting the concentration of Krebs cycle intermediates. Pyruvate carboxylation serves a similar, anaplerotic role in these species. Carbohydrate and proline are oxidized simultaneously to support steady-state flight by flies and some beetles, but there is considerable interspecific variation as to which fuel is used as the primary carbon source. The tsetse fly may be unique in using proline as the sole fuel for flight.

A huge increase in the rate of ATP utilization in muscle occurs during the transition from rest to flight. In species possessing synchronous flight muscles (e.g., locusts, butterflies, and moths), this is the result of increased activities of actomyosin-ATPase, Ca^{2+}-ATPase, and Na^+/K^+-ATPase. In insects with asynchronous muscles, actomyosin-ATPase accounts for most of the ATP hydrolyzed, and the energetic cost of excitation–contraction coupling (involving Ca^{2+}-ATPase and Na^+/K^+-ATPase) is lower in comparison with synchronous muscles. Contraction frequencies in asynchronous muscles are not limited by maximum rates of Ca^{2+} cycling, which may explain why asynchronous muscles, which are unique to insects, are found in 75% of insect species, including flies, beetles, and many species of wasps and bees.

ATPase activities increase upon the initiation of flight, and metabolic control mechanisms activate pathways of substrate catabolism, mitochondrial respiration, and oxidative phosphorylation to maintain ATP concentration and muscle function. Cellular rates of ATP hydrolysis and synthesis in insect flight muscle are so exquisitely matched that ATP concentrations are maintained within narrow limits, despite up to 100-fold increases in the rate of ATP turnover during the transition from rest to flight. The regulatory mechanisms that make possible these large flux changes between rest and flight remain poorly understood, and until recently, how such phenomenal steady-state rates of aerobic metabolism are achieved was largely unknown. Recent work has revealed that these are made possible by high concentrations of catabolic enzymes in the flight muscles, the operation of some of these enzymes at high fractional velocities, high mitochondrial content, large inner membrane surface areas per unit mitochondrial volume, and unusually high rates of electron flow between respiratory chain enzymes. Evolution has produced no locomotory muscles capable of higher rates of aerobic metabolism.

METABOLISM OF COLD HARDINESS

Many insects that exhibit arrested development during periods of cold, or routinely live in cold environments, display a resistance or tolerance to freezing. The phenomenon of supercooling was described for various insects over 100 years ago. Although supercooling temperatures are generally −10 to −15°C, supercooling temperature may approach −50°C.

In most cases, low-molecular-weight cryoprotectants, including some amino acids, mannitol, trehalose, sorbitol, and particularly glycerol, accumulate in the hemolymph and act as antifreeze. Cryoprotectants typically reach levels of 20 to 30% or more of fresh body weight as temperature falls and the level of cryoprotectant often is inversely related to the supercooling point. In freeze-tolerant species ice-nucleating proteins accumulate and the supercooling point increases.

Among the first metabolic studies on cryoprotectant synthesis were those of H. Chino on glycerol and sorbitol formation in diapausing *Bombyx mori* silkworm eggs. The metabolism is hormonally regulated and is summarized in Fig. 3.

Although most insects synthesize polyols from glycogen in a similar manner, their synthesis in *Bombyx* eggs not only provides cryoprotection, but also maintains redox balance in response to a reduced level of oxidative respiration. In this case, the formation of polyols is an example of anaerobic fermentation.

Regulation of polyol formation is poorly understood, but cold-induced activation of glycogen phosphorylase is involved. Studies with *Eurosta solidaginis* fly larvae demonstrate independent control over glycerol and sorbitol formation, and hormonal cues may be important. In insects that accumulate trehalose for cryoprotection, inhibition of phosphofructokinase has been suggested as the mechanism for shifting metabolism in the direction of sugar formation. A temperature-dependent change in the balance between glucose oxidation by glycolysis and the pentose phosphate pathway may also affect the balance between trehalose, sorbitol, and glycerol.

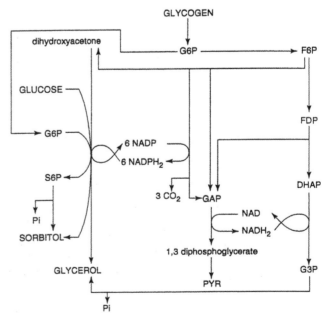

FIGURE 3 Metabolic scheme showing pathways of carbohydrate oxidation and polyol synthesis in *B. mori* silkworm eggs. CO_2, carbon dioxide; NAD, oxidized nicotinamide adenine dinucleotide; $NADH_2$, reduced NAD; Pi, inorganic phosphate; S6P, sorbitol 6-phosphate. Other abbreviations as in Fig. 2. (Adapted from Gilmour, 1965, "The Metabolism of Insects." W. H. Freeman, San Francisco.)

In conclusion, insects exhibit many unique metabolic characteristics. The brevity of this article allows discussion of only a few select examples. Much current molecular work focuses on silk fibroin synthesis and the molecular action of hormones. Relatively few investigations have been conducted on molecular-genetic aspects of metabolism and metabolic regulation. The available techniques in cell and molecular biology, including differential display and microarray analysis, offer marvelous opportunities to examine the "transcriptional physiology" of hormonal and nutritional regulation of intermediary metabolism. The metabolic diversity of the Insecta also awaits more detailed exploration within the context of their ecology and behavior.

See Also the Following Articles

Cuticle • Digestion • Fat Body • Flight • Muscle System • Temperature, Effects on Development and Growth • Thermoregulation

Further Reading

Bursell, E. (1981). The role of proline in energy metabolism. *In* "Energy Metabolism in Insects" (R. G. H. Downer, ed.), pp. 135–154. Plenum, New York.

Candy, D. J. (1985). Intermediary metabolism. *In* "Comprehensive Insect Biochemistry, Physiology and Pharmacology" (G. A. Kerkut and L. I. Gilbert, eds.), Vol. 10, pp. 1–41. Pergamon Press, New York.

Harshman, L. G., and James, A. A. (1998). Differential gene expression in insects: Transcriptional control. *Annu. Rev. Entomol.* **43**, 671–700.

Hoffman, K. H. (1984). Color and color changes. *In* "Environmental Physiology and Biochemistry of Insects" (K. H. Hoffman, ed.), pp. 206–224. Springer-Verlag, Heidelberg.

Josephson, R. K., and Young, D. (1987). Fiber ultrastructure and contraction kinetics in insect fast muscles. *Am. Zool.* **27**, 991–1000.

Josephson, R. K., Malamud, J. G., and Stokes, D. R. (2000). Asynchronous muscle: A primer. *J. Exp. Biol.* **203**, 2713–2722.

Kayser, H. (1985). Pigments. *In* "Comprehensive Insect Biochemistry, Physiology and Pharmacology" (G. A. Kerkut and L. I. Gilbert, eds.), Vol. 10, pp. 368–415. Pergamon Press, New York.

Lee, R. E., and Denlinger, D. L. (1991). "Insects at Low Temperature." Chapman & Hall, New York.

Lewis, D. K., Bradfield, J. Y., and Keeley, L. L. (1998). Feeding effects on gene expression of the hypertrehalosemic hormone in the cockroach, *Blaberus discoidalis. J. Insect Physiol.* **44**, 967–972.

Linzen, B. (1974). The tryptophan/ommochrome pathway in insects. *Adv. Insect Physiol.* **10**, 117–246.

Moore, M. V., and Lee, R. E. (1991). Surviving the big chill: Overwintering strategies of aquatic and terrestrial insects. *Am. Entomol.* Summer, 111–118.

Schaefer, J., Kramer, K. J., Garrow, J. R., Jacob, G. S., Stejskal, E. O., Hopkins, T. L., and Speirs, R. D. (1987). Aromatic cross-links in insect cuticle: Detection by solid-state ^{13}C and ^{15}N NMR. *Science* **235**, 1200–1204.

Simpson, S. J., and Raubenheimer, D. (1993). The central role of the haemolymph in the regulation of nutrient intake in insects. *Physiol. Entomol.* **18**, 395–403.

Storey, K. B. (1985). Metabolic biochemistry of insect flight. *In* "Circulation, Respiration and Metabolism" (R. Gilles, ed.), pp. 193–207. Springer-Verlag, Berlin/Heidelberg.

Storey, K. B., and Storey, J. M. (1991). Biochemistry of cryoprotectants. *In* "Insects at Low Temperature" (R. E. Lee and D. L. Denlinger, eds.), pp. 64–93. Chapman & Hall, New York.

Suarez, R. K. (2000). Energy metabolism during insect flight: Biochemical design and physiological performance. *Physiol. Biochem. Zool.* **73**, 765–771.

Sugumaran, M. (1998). Unified mechanism for sclerotization of insect cuticle. *Adv. Insect Physiol.* **27**, 229–334.

Metamorphosis

Frederick W. Stehr
Michigan State University

Metamorphosis means change in form. Most organisms undergo a change in form as they grow from an embryo to an adult. Some changes are radical and the immatures bear little resemblance to the adults; others are more gradual, with the immatures looking very much like the adults.

The term "larva" has very broad usage in invertebrate zoology, being applied to an assortment of forms (often the dispersive state) in virtually all invertebrate phyla. In Arthropoda other than insects, larvae is most often used for first stages, as it is in the mites and ticks, and for the first-stage hexapod larvae of millipedes, with "nymph" being used for second-stage mites and ticks. However, in continental Europe (especially France) "nymphe" refers to a pupa; in English-speaking countries "pupa" is universally used for the stage between the last instar and the adult of insects with complete metamorphosis (Holometabola).

In the insects, larva has been used in different ways, including such diverse forms as the immatures of the most primitive order Protura and of the most advanced order Hymenoptera (sawflies, ants, wasps, and bees). The termites present an interesting problem: some authors use nymph for all juvenile termites, whereas others use larva for those lacking wingpads, and nymph for those having wingpads. This is further complicated because the supplementary reproductives may be wingless or bear wingpads, even though the two forms are functionally equivalent.

The kind of metamorphosis insects undergo is closely related to which of the subclasses a taxon belongs to. In the Apterygota metamorphosis is either anamorphic in the Protura, in which three abdominal segments are added as the individual develops to an adult, or it is ametabolous, in which the number of molts is indefinite and molting may continue throughout life after sexual maturity (Collembola, Diplura, Archeognatha, and Zygotrema).

In the Pterygota there are two fundamental kinds of metamorphosis: the hemimetabola develop through the egg, larva, and adult stages, and the holometabola develop through the egg, larva, pupa, and adult stages. There are also other terms (defined below) that have been used to describe variations in metamorphosis.

LARVAE VS NYMPHS AND NAIADS

Defining a larva is also necessary because its use has been highly variable in the Insecta. In 1918, Comstock proposed restricting the term larva to juveniles of the holometabolous orders, nymph to the juveniles of his paurometabolous (nonholometabolous) terrestrial orders, and naiad to the juveniles of his hemimetabolous nonholometabolous aquatic orders (Ephemeroptera, Odonata, and Plecoptera). Because these three aquatic orders have a much greater change in form from the last instar to the adult than the terrestrial hemimetabolous orders, there was some basis for Comstock's proposal to call them naiads. However, the Ephemeroptera and Odonata are Paleoptera, which cannot fold their wings, whereas the Plecoptera are Neoptera, which can fold their wings over their back, so they are not closely related.

Currently there is a tendency to use larva for all immature insects that are not eggs, pupae, or adults and the term "immature insect" for all life stages except adults, no matter how many specialized names are applied to the various developmental forms in the different orders.

There is little difficulty in defining an egg or an adult, but naming and defining the instars or stages that may occur between egg and adult can be problematic. Some insects are larviparous, never depositing eggs; some multiply from a single egg by polyembryony, and some are sexually mature as immatures (paedogenesis or neoteny). Nevertheless, all of them undergo a series of molts as they grow. When larva is used in the comprehensive sense, the subcategories "exopterygote larva" (Hemimetabola that have the wingpads developing externally) and "endopterygote larva" (Holometabola that have the wingpads appearing externally in the pupal stage but having developed from internal larval histoblasts) are useful for pterygote immatures. A useful term roughly equivalent to larva in the comprehensive sense is "juvenile," which can be used as a general term for nonadult larvae of all orders.

KINDS OF METAMORPHOSIS

Below are terms that are currently widely used for different types of metamorphosis. Most species are either holometabolous or hemimetabolous, with more than 85% of them holometabolous and most of the rest hemimetabolous.

Anamorphosis

This term means development with fewer body segments at hatching than when mature, which is found in the Protura, in which three abdominal segments are added anterior to the tail as the individual develops to an adult. Because of this some workers believe proturans are not true insects.

Ametabolous

Ametabolous means development with the major change being an increase in size until sexually mature. The number of molts is indefinite, and molting may continue throughout life; it is found in Apterygota (excluding the Protura).

Simple Metamorphosis

This is a broad term covering all types of metamorphosis except holometabolous.

Hemimetabolous (Gradual, Incomplete, Direct, Paurometabolous)

Development through egg, larva, and adult is covered by this term, which includes everything except ametabolous, anamorphosis, and holometabolous. Among the hemimetabolous insects, most species are found in three orders, the Orthoptera, the Heteroptera, and the Homoptera.

Holometabolous (Complete, Indirect)

This means development through egg, larva, pupa, and adult.

All insects do not fit neatly into Hemimetabola or Holometabola. Some Hemimetabola are intermediate in having one or more nonfeeding stages before the adult instar and in having a last instar that forms into a pupa. For example, in the whiteflies (Homoptera: Aleyrodidae) the first instar is active, but subsequent instars are stationary, resembling and feeding like scale insects, and the last instar stops feeding and becomes a pupa, with the wings developing internally. This could be termed holometabolous, but the homopterans as a whole are certainly hemimetabolous. In the Thysanoptera (thrips), there are two feeding instars followed by two or three nonfeeding instars, the propupa and pupa, which may be contained within a cocoon formed by the last feeding instar. This is certainly closer to Holometabola than to Hemimetabola.

See Also the Following Articles

Development, Hormonal Control of • Hypermetamorphosis • Molting

Further Reading

Balls, M., and Bownes, M. (eds.) (1985). "Metamorphosis." Clarendon Press, Oxford.

Comstock, J. A. (1918). Nymphs, naiads and larvae. *Ann. Entomol. Soc. Am.* **2**, 222–224.

Etkin, W., and Gilbert, L. I. (eds.) (1968). "Metamorphosis." Appleton–Century–Crofts, New York.

Hall, B. K., and Wake, M. H. (eds.) (1999). "The Origin and Evolution of Larval Forms." Academic Press, New York.

Stehr, F. W. (ed.) (1987). "Immature Insects," Vol. 1. Kendall/Hunt, Dubuque, IA.

Stehr, F. W. (ed.) (1992). "Immature Insects," Vol. 2. Kendall/Hunt, Dubuque, IA.

Migration

Hugh Dingle
University of California, Davis

Migration, the major movement behavior of insects, allows them to escape deteriorating habitats, to colonize new areas, or to seek temporary shelter such as overwintering sites. It involves a complex of traits that include development, physiology, morphology, and behavior, and it is a major component of the life histories of many species. These trait complexes or syndromes are adjusted by natural selection in complex ways that increase the fitness and therefore the success of migrants.

MIGRATION AND OTHER MOVEMENTS

The movements characteristic of organisms can be roughly divided into two broad categories, those that are immediately responsive to resources and those that are not (Box 1). Within the category of immediate responses to resources are two further broad types. The first type consists of "station-keeping" responses that serve to keep the organism on its territory or within the home range in which it carries out most of its life functions and spends most of its time. Included within station-keeping movements are resource-sensitive behaviors crucial for survival. Examples are foraging, territorial behavior, and commuting, which is a periodic, often daily, round trip for resources. Foraging may be for any resource, including food, shelter, or mates; and commuting, which can also be considered to be a form of extended foraging, may involve travel over considerable distances. The commuting trails of leafcutter ants, for example, may extend for hundreds of meters both horizontally along the forest floor and vertically into the canopy. Foraging, commuting, and territorial behaviors are all readily responsive to resources: thus a female butterfly stops searching (foraging) upon encountering a host plant on which to oviposit, and a territorial forest drosophilid fly is bounded by the borders of its leaf display ground.

Ranging movements take organisms on exploratory journeys beyond the current home range and serve to locate a new home range or territory. Like station-keeping movements, ranging movement is a facultative response to resources, and like foraging or commuting, it ceases when a new resource (here in the form of previously unoccupied living space) is encountered. Ranging movements may also extend considerable distances but, like station keeping, still belong in the category of activities that are proximately resource sensitive.

Movement that is not immediately responsive to resources constitutes the distinct sort of behavior that is migration. Taking an organism beyond both its current home range and beyond neighboring potential home ranges, migration is physiologically distinct from all other movements. It is so distinct because sensory inputs from resources that would ordinarily cause movement to cease do not stop migration. Thus, a characteristic of migration is that the organism undertaking it is undistracted by and fails to respond to food or mates, otherwise so necessary a part of life functions. Furthermore, migration is usually triggered by environmental cues, such as photoperiod, which forecast habitat change rather than being directly responsive to the change itself (usually a deterioration in the quality or availability of resources). Other characteristics of migration include distinct initiating and terminating behaviors. Many insects climb to the top of a bush or tree branch to take off on migratory flights, behavior they show at no other time. Sensory responses may also change, as in aphids that are sensitive to blue light from the sky during the takeoff phase of migratory flight but become increasingly sensitive to yellow light, the characteristic wavelength of young host plants, as migration proceeds. Thus, migration is not defined by the distance traveled or by whether it is a "round trip." Rather, it is defined in terms of the physiological and behavioral responses to resources; this behavior is true of all organisms, not just insects.

The movement behavior of individuals also has an outcome for the population of which those individuals are a part. This outcome involves displacement for a greater or lesser distance, but at either distance it involves removal from the home range. It can also result in the scattering or dispersal of individuals within the population; thus "dispersal" is a population phenomenon, not an individual movement. Movement can also result in congregation by mutual attraction or aggregation in a habitat. Both tendencies result in a decrease in the mean distance among individuals and contrast with dispersal, which increases mean distances. Note that all the movements just described can contribute to aggregation, congregation, or dispersal, depending on species and ecological circumstance.

Three examples of the sorts of population outcome attributable to migratory behavior occur in locusts, aphids,

---

Box 1. *Types of Insect Movement*

Migration is a type of movement displayed by insects, but it differs from all other types because migratory insects (and other migrants as well) are unresponsive to suitable resources. There are two broad categories of movements:

I. Movements that are directed by resources and/or home range.
   A. Station keeping: examples are foraging, commuting (periodic, usually daily, movements), and territorial behavior.
   B. Ranging: movement to explore an area, often for a new home range or territory.
II. Movement not directly responsive to a resource or home range: here migration is undistracted movement with cessation primed (thresholds lowered) by the movement itself. Responses to resources are suppressed or suspended.

and moths of the genus *Heliothis* (and the very similar *Helicoverpa*). Locusts are a group of grasshopper species that under crowded conditions undergo remarkable behavioral and morphological changes known as phase transformation. In the desert locust, *Schistocerca gregaria,* perhaps the most extreme example of the phenomenon, crowded nymphs (hoppers) are strikingly black and yellow, whereas isolated individuals are a pale brown or green. Crowded adults are larger and display differences in body proportions that readily distinguish them from their isolated counterparts. It is behavior, however, that most distinguishes the two forms. Isolated individuals display no mutual attraction, forage independently, and migrate at night. Crowded individuals show a high degree of mutual attraction and form large swarms that can number in the millions. When a swarm is feeding, locusts at the rear are constantly running out of food, overflying the body of the swarm, and landing at the leading edge. The result is a "rolling" movement across country in extended foraging. If a swarm enters an area with no forage, it may rise in unison and be carried for some distance by the wind. If this happens for a long period or repeatedly, the individuals in the swarm may switch to migratory behavior and cover considerable distances to descend again in regions of fresh plant growth. These aggregated swarms are major pests over much of Africa, occasionally reaching adjacent areas of the Middle East and in several notable instances the New World. The arrival of a swarm can mean that "not any green thing" (Exodus 10:15) is left for human consumption. It is the aggregation and migration that make the desert locust such a notorious pest. It would be much less a pest if its characteristic behavior led to dispersal rather than swarming.

Various species of aphid are also capable of spreading far and wide by migration. In Europe, an extensive monitoring network coordinated by English and French entomologists has mapped the seasonal spread of bean aphid, *Aphis fabae.* Large concentrations appear first in central France in early to midsummer. The species then spreads westward and northward over succeeding generations so that by late summer the aphid has reached high densities as far north as Scotland. In North America, monitoring of the corn leaf aphid, *Rhopalosiphum maidis,* has indicated the arrival of large numbers in the cornfields of Illinois. Analysis of weather systems suggests these aphids have come from as far away as Texas and were transported on wind streams. Studies of a number of other insects indicate that the Mississippi Valley is a major spring migration route for wind-transported insects to the agricultural regions of the upper Midwest.

Heliothis moths breed following rainfall in tropical and subtropical arid regions. If productivity is high on the new vegetation on which they feed, large populations of migrating moths are produced and are carried by winds to agricultural areas. In Australia, moths are transported in spring to wheat- and cotton-growing regions in New South Wales from breeding areas in interior regions of New South Wales and Queensland. The location of rainfall in the interior of Australia is unpredictable from year to year, and considerable effort has gone into locating areas in which rain has fallen, determining whether this precipitation is sufficient to produce large moth populations, and forecasting the arrival of migrating moths in conjunction with weather systems so that necessary control measures can be undertaken and unnecessary ones avoided. In the spring in North America, there is similar breeding of *Heliothis* moths in northern Mexico and southern Texas and migration northward on winds.

HISTORICAL BACKGROUND

In the 1930s C. B. Williams collected and summarized the available information on insect migration. The two books that resulted were largely responsible for bringing to the attention of entomologists and other biologists the fact that the phenomenon was a common one. Williams focused on large insects such as butterflies and dragonflies, and he adopted the prevailing notion, derived largely from birds, that only round-trip movement could be called "migratory." The way entomologists now think about insect migration is primarily the result of the work of four Britons: C. G. Johnson, J. S. Kennedy, T. R. E. Southwood, and L. R. Taylor, beginning around 1960. Johnson and Kennedy stressed that insect movements vary with respect to physiology and function, and their ideas revamped notions concerning the behavioral and life history aspects of migration. Southwood showed that the type of habitat determines the likelihood of migration among insects, and Taylor noted the importance of movement to the dynamics of populations in both time and space. Combined, the work of all four made explicit that migration is a distinct behavior with consequences for populations.

The distinct nature of migratory behavior was precisely outlined by Kennedy in his studies of the flight of the summer parthenogenetic females of *A. fabae.* He used a flight chamber that allowed him to analyze the responses of free-flying aphids (Box 2). Key aspects of migratory flight that distinguished it from other types of flight were revealed by the flight chamber experiments. The aphids tested were the winged or alate form produced under crowded conditions. The uncrowded wingless females larviposit (bear live young) as soon as they make contact with a suitable host leaf. In contrast, the winged migrants do not larviposit until they have completed at least some flight. Furthermore, landing responses are primed by migratory flight: the longer the flight, the lower the threshold for landing. Finally, there is reciprocal interaction between flight and settling, since settling responses (i.e., probing a leaf to test its suitability and subsequent moving to the underside of the leaf to larviposit) can prime flight if they fail to be completed by attaching via the mouthparts and producing offspring. This flight after incomplete settling may actually be stronger than that occurring at the beginning of migration.

Box 2. The Kennedy Flight Chamber

J. S. Kennedy used this device in studies of insect migration. His experiments analyzed the performance of free-flying aphids and their landing and foraging responses. The lever arm can be twisted to shake the aphid off the platform, forcing it to fly, and it can be rotated out of the light and presented to the flying aphid again at will. Host plant leaves of different ages and leaves of different species of plant can be presented to permit investigators to observe variation in landing responses. Free flight is maintained by wind from the top of the chamber, and the wind speed is varied with the butterfly valve, whose setting thus indicates the strength of flight as measured by rate of climb which is balanced by the downward windspeed. (Figure reproduced from Dingle, H. (1996). "Migration: The Biology of Life on The Move." Used by permission of Oxford University Press, Inc.).

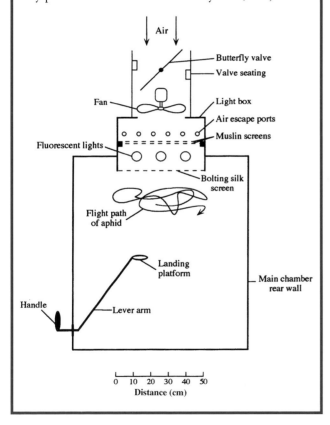

Migration thus is qualitatively different from other movement because station-keeping responses such as landing and probing (foraging) are inhibited by flight, but flight also primes them and promotes their later recurrence. Based on the behavior of migrating aphids, Kennedy provided a complete predictive definition of migration, as follows: migratory behavior is persistent and straightened-out movement effected by the animal's own locomotory exertions or by its active embarkation upon a vehicle; it depends on some temporary inhibition of station-keeping responses, but promotes their eventual disinhibition and recurrence.

By explicitly focusing on the interaction between growth and reproductive behaviors (station keeping) and migratory behavior, Kennedy put migration firmly in the context of life histories. This context was also emphasized by Johnson, who identified the "oogenesis-flight syndrome" as characteristic of insect migration. Johnson noted that in a high proportion of migratory insects, especially in females, flight is limited to individuals with immature reproductive systems. It thus seemed that migration was based on an interaction between flight and the maturation of reproduction. Implicit was the assumption that migration and reproduction were alternative physiological states, with trade-offs in the mobilization of energy and materials. Johnson further postulated that this life history syndrome would be mediated by juvenile hormone, a postulate now amply demonstrated (see later: Migratory Syndromes).

The population dynamical aspects of insect migration were assessed by Southwood and Taylor. Southwood placed migration into an ecological and evolutionary context by summarizing evidence that migration is characteristic of insects living in temporary habitats, such as seasonal pools or early successional fields. This condition is in contrast to that of insects with more permanent habitats such as forests or large lakes; such insects are nonmigratory and often even wingless. This pattern of migration as a response to transitory environments was later formalized by Southwood in 1977 in the ratio H/τ, with H the duration of the habitat and τ the generation time of the insect. The frequency of migration in populations or species increases as the ratio shrinks toward unity, as later nicely demonstrated, for example, in leafhoppers (Homoptera) by Denno and colleagues. Taylor stressed the role of migratory behavior in the spatial dynamics of insect migration. He noted that migrants can disperse or coalesce, depending on whether individuals attract or repel one another and on atmospheric dynamics. This behavior can create a mosaic of insect densities over the landscape. He also initiated the Rothamsted Insect Survey, an array of traps to sample insects in the air, from which data on numbers were taken and analyzed at the Rothamsted Experiment Station near London. This network allowed tracking and forecasting of insect pests such as aphids with major practical implications for insect control.

THE CAUSES OF MIGRATION

Since Southwood's original statement of the relationship, it has become apparent that the impermanence of habitats is indeed the primary selective force driving insect migration. Much of this impermanence is a function of season, and as with other well-known migrants such as many fish or birds, seasonality is a common factor in insect migrations. Most seasonal migrations are on a relatively small scale, with distances traveled only a few hundreds or thousands of meters, but others cover much greater distances. Examples of short-distance migrations to overwintering diapause sites include the Colorado potato beetle, *Leptinotarsa decemlineata,* several species of common seed-feeding hemipterans, including *Lygaeus kalmii,* a milkweed bug of Europe (in Sweden it often flies to lighthouses to diapause), and several species of lady beetles (Coccinelidae). In some lady beetles migration extends to several kilometers. In California, *Hippodamia convergens,* the convergent lady beetle, overwinters at high altitudes in the Sierra Nevada and migrates to agricultural areas in the Central Valley in March. Beginning in June, offspring of the early spring migrants fly back to intermediate altitudes and form aggregations. There is then a later movement to higher altitudes to overwinter so that the migration to overwintering sites is a two-step process.

Other insect migrants make spectacularly long journeys. The best known of these is that made by the eastern North American population of the monarch butterfly, *Danaus plexippus,* studied extensively by Lincoln Brower and F. A. Urquhart. This butterfly cannot overwinter in the temperate zone, so it must migrate to southern overwintering sites. The short days of autumn cause adult butterflies to enter reproductive diapause, and they undertake a southward journey of 3000 km or more. The majority of the eastern population overwinters en masse in a very few high-altitude protected sites in the Transvolcanic Range of central Mexico, where they arrive in the late autumn. Beginning in February, the aggregations break up, mating occurs, and the butterflies begin to move northward. Identification of the chemical cardenolide "fingerprints" of the milkweeds eaten by monarchs when they are caterpillars and stored in the adults has revealed that the overwintering generation stops and breeds on the spring flush of milkweeds along the coastal plain of the Gulf of Mexico. It is the offspring of these individuals that invade regions farther north beginning in late May and early June. Thus, as with fall California convergent lady beetles, the spring migration is at least a two-step process. A very similar migration pattern occurs in the same region in the large milkweed bug, *Oncopeltus fasciatus,* and it, too, occurs in two stages in the spring.

The migration of western populations of the monarch is more complicated. These populations overwinter in aggregations along the coast of southern California, where winter climatic conditions are similar to the aggregation sites in Mexico. When the aggregations begin to break up, as early as late January, the butterflies move to early sprouting milkweeds in the Coast Ranges and breed there. The next generation moves both to more coastal milkweeds and to milkweeds that grow farther inland as far east as the Rocky Mountains and as far north as the Canadian border, so that as in the eastern populations, the spring migration occurs in two stages over two generations. A very similar pattern occurs in the monarch population introduced into eastern Australia, with overwintering near the coast, a migration inland in the spring, and a return to the coast in the autumn. In more northern parts of the Australian range, there may be year-round breeding in coastal and subcoastal regions.

One way to assess the influence of ephemeral habitats on

the evolution of insect migration is to survey across species and populations occurring in different kinds of habitats. In Europe, a number of species and populations of water striders (Hemiptera: Gerridae) occur over an array of habitats, from small, temporary ponds to large lakes and permanent streams to isolated permanent bogs. Species in the more temporary bodies of water have wings and undertake regular migrations to locate their aquatic habitats as they appear and disappear in the landscape. At the opposite extreme in permanent lakes and bogs, there are species that are wingless. Across habitats with varying degrees of permanence are populations and species of water striders with varying proportions of winged and wingless individuals determined primarily environmentally (polyphenisms) where habitat change is predictable, and primarily genetically (polymorphisms), where change is increasingly random with respect to the life cycle.

A second example of the influence of habitat duration on the occurrence of migrants within a fauna occurs in Australian butterflies. Often, latitude predicts the amount of migration that will occur, especially where there is adequate rainfall and seasonal change is largely a function of temperature. In eastern North America, for example, 98% of the variance in the proportion of migratory birds is explained by latitude, with a higher proportion of migrants at northern latitudes. Similarly in eastern Australia, where the climate is warmer overall but still temperate with adequate rainfall, 72% of the variance in the proportion of butterfly migrants is explained by latitude. The situation is quite different in the dry regions of Australia west of the Great Dividing Range. Here latitude accounts for less than 1% of the variance in proportion of butterfly migrants, and climate variables that indicate rainfall patterns, such as soil moisture, which accounts for about 50% of the variance, are much better indicators of migration. The amount of rainfall is not correlated with latitude, and so latitude does not predict migration. In this dry climate it is the availability of erratic rainfall that counts, and only migrants that can take advantage of the ephemeral flushes of vegetation that follow such rainfall. Thus, as with *Heliothis* moths, migration allows some butterflies to exploit a dry and ephemeral habitat.

MIGRATORY SYNDROMES

Accompanying migratory behavior is a syndrome of traits that act in coordination to increase fitness. These traits vary from enzymes to life history characters and, being influenced by subsets of the same genes, are genetically correlated. At the physiological level, insects (like most other migrant organisms) use fat as fuel, primarily for two reasons. First, fat metabolizes to produce about twice as much energy as carbohydrate or protein; and second, it requires no water for storage (in contrast, storage of 1 g of carbohydrate requires 3 g of water). Insects such as the monarch butterfly and the large milkweed bug shift lipids from yolk formation to fat storage under the influence of the shorter photoperiods of autumn and just prior to migration. The flight muscles of migrants are also adapted to the energetic demands of lengthy flight. Enzymes active in oxidative metabolism, such as citrate synthase, and in fatty acid oxidation, such as β-hydroxyl coenzyme A dehydrogenase, or HOAD, tend to show higher levels of activity in the flight muscle of migrants compared with that shown in nonmigrants. This difference is most apparent where there are wing polymorphisms and migrants have longer wings.

The most important hormone involved with insect migration is juvenile hormone (JH). It has influence not only on the coordination of the various relationships of the oogenesis-flight syndrome but also has direct effects on migratory flight. In many insect migrants such as the monarch butterfly, short photoperiods result in reduced outputs of JH from the corpus allatum. This reduction in JH output in turn leads to a reduction in ovarian and egg development, which is then accompanied by migratory flight. In several species of migrant insects, prolonging of the prereproductive period by reduced JH titers results in the triggering and maintenance of migratory flight. At the same time it has been demonstrated in several migratory species (such as the large milkweed bug, the convergent lady beetle, and the monarch butterfly) that JH directly stimulates migration. Implants of corpora allata, the source of JH, or topical application of JH or some of its chemical analogues, are effective in increasing flight in migrants. In some insects such as the monarch, adipokinetic hormone (AKH—involved in promoting fat metabolism) also stimulates additional flight. The effects of JH and AKH in the monarch are illustrated in Fig. 1.

Because migratory flight occurs when reproduction is delayed by reduced JH concentrations, it is logical to inquire what level of JH determines migration. This question was answered for the large milkweed bug by M. A. Rankin. She selected for delayed onset of flight, which also resulted in delayed reproduction. Rankin measured JH titers in the blood during the prereproductive period and showed that JH titers were low when there was no flight or reproduction; inter-

mediate titers stimulated flight, and high titers stimulated oogenesis. Thus, if titers rose only to intermediate levels, as might occur under short days, for example, flight but not reproduction would be triggered. These JH titers may also be regulated by JH esterase, the enzyme that breaks down JH. In wing-polymorphic crickets, high concentrations of JH result in short-winged individuals. Artificial selection experiments that increased the frequency of long wings also resulted in increased amounts of JH esterase in the blood and so reduced amounts of JH. Selection also demonstrated that it was possible to change both mean and threshold JH esterase activity. The possible role of JH esterase in fully winged migratory insects remains to be studied.

An additional behavioral aspect of the migratory flights of many insects is the ability to maintain a more or less constant direction during migration. Mostly, this directionality has been studied in butterflies, although some other large insects such as dragonflies and larger Hymenoptera also seem to maintain a constant direction when migrating. The monarch butterfly in eastern North America flies in a steady southward or southwestward direction in the autumn, flight directions that lead to the overwintering sites in central Mexico. In the spring, the migratory flight is to the north. Compilations of observations of several species of Australian butterflies, including the monarch, reveal that the insects fly south or southwest in the spring and north or northeast in the autumn. The apparently migratory flights of a few species occur in the same direction no matter what the season, a phenomenon that has yet to be explained. In Europe migratory flights of the butterfly *Pieris brassicae* are consistent with both season and geography. Autumn migrants from northern Germany fly south or south by southeast, whereas migrants in the south of France fly to the southwest, which takes them to Spain rather than over the Mediterranean. Further experiments have demonstrated that butterflies that have diapaused, as they do during the winter, fly north when they migrate, but those emerging from nondiapause (summer) pupae fly toward the south. Seasonal winds also frequently carry migrating insects in the "correct" direction. Monarch butterflies in eastern North America frequently soar and are carried southward by northerly winds, and simulations of the migration of the large milkweed bug from the same region indicate that a portion of the population reaches southern overwintering areas regardless of whether they orient.

Where the mechanism of orientation has been studied, the evidence suggests that it is a time-compensated sun compass. To use the sun effectively for orientation, organisms must be able to compensate for its daily passage across the sky by reference to a "biological clock." To demonstrate that an organism is using a time-compensated sun compass, it is necessary to clock-shift it by maintaining it in a daily light cycle that is out of phase with the ambient cycle and to then show that its orientation is displaced by an amount consistent with the clock shift. A displacement of 6 h, for example, should lead to a directional change of 90°; whether

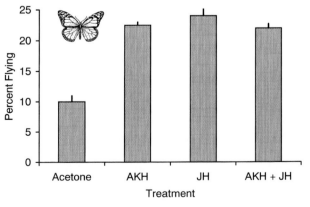

FIGURE 1 Influence of JH and AKH on tethered flight in the monarch butterfly. Flight (%) longer than 30 min is the index of migratory flight. The butterflies received topical application of hormones or the acetone control, and subsequent flight duration was determined. Both hormones, singly or together, increased flight over controls.

the change is plus or minus depends on the direction of the clock shift. Experiments with southward-moving monarch butterflies suggest that this pattern is indeed followed. In tropical Panama, two species of migrating pierid butterflies, *Aphrissa statira* and *Phoebis argante,* regularly maintain a directional flight across Gatun Lake in the Panama Canal. Clock-shift experiments resulted in changes in the direction of orientation that were consistent with a sun compass, even though there was also a component imposed by wind drift. The sophisticated orientation mechanisms of honey bees, ants, and certain other insects incorporating a sun compass, imply that orientation is probably widespread. The future will undoubtedly reveal the presence of a sun compass in other migrants, as well as the presence of other mechanisms, especially in nocturnal migrants. Radar observations indicate that the passage of many species of nocturnal migrants is specific to winds of a certain direction, but the means by which this preference is enforced are unknown.

In addition to behavioral and physiological characters, migration syndromes often include life history traits such as the age at first reproduction and fecundity, particularly in many wing-polymorphic insects. Typically, in these species the short-winged or wingless forms reproduce earlier and display higher fecundities than their long-winged counterparts. This dichotomy is at least in part because of trade-offs between flight and reproduction. The metabolically active flight muscles that accompany long wings and migration are costly to maintain, requiring considerably more maintenance energy than the thoracic musculature of wingless or short-winged individuals. In contrast, the later reproducing individuals, with lower egg production, are often longer lived.

Migration syndromes that include life history traits are the result of underlying genetic mechanisms, as revealed in artificial experiments using the large milkweed bug. Like all flying insects, this migrant can be induced to fly by removing substrate contact. Bugs that are glued at the prothorax to a tether will fly if contact with the tarsi is removed. An individual in the migratory state can fly on the tether for several hours, and the duration of flight can be used as an index of migration. Artificial selection can be used to increase the proportion of individuals making long (or short) flights, with the duration of flights also affected. Selection was used to both increase and decrease the proportion of bugs undertaking long flights. In addition to flight, wing length and fecundity responded to this selective regime. The bugs of the line with a higher proportion of long flights also had longer wings on average, and the females of this line produced more eggs during the first 5 days of reproductive life. This means that the genes influencing flight also influenced wing length and fecundity, most likely via pleiotropic effects. Longer term selection experiments on wing length, which also resulted in higher fecundities and increased flight as wing length increased, suggested that linkage disequilibruim is unlikely. Parallel selection experiments on a population that did not migrate failed to reveal genetic correlations among wing length, flight, and fecundity, indicating

that the genetically based syndrome of correlations among these traits is unique to migratory populations. The selection experiments reveal that natural selection has produced an adaptive migratory syndrome that includes wing length and fecundity. Interestingly, the age at first reproduction is unaffected by selection.

The conclusion from the brief survey of insect migration is that this behavior involves more than simply extended movement to escape to a new habitat. Rather, migration is a trait of considerable complexity, requiring knowledge of behavior, development, ecology, physiology, and genetics to provide a full understanding of its evolution and function.

See Also the Following Articles

Aphids • *Juvenile Hormone* • *Locusts* • *Magnetic Sense* • *Monarchs* • *Orientation*

Further Reading

Brower, L. P., and Malcolm, S. B. (1991). Animal migrations: Endangered phenomena. *Am. Zool.* **31,** 232–242.

Dingle, H. (1996). "Migration: The Biology of Life on the Move." Oxford University Press, New York.

Dingle, H. (2001). The evolution of migratory syndromes in insects. *In* "Insect Movement: Mechanisms and Consequences" (I. Woiwood and D. R. Reynolds, eds.), pp. 159–181. CAB International, London.

Johnson, C. G. (1969). "Migration and Dispersal of Insects by Flight." Methuen, London.

Kennedy, J. S. (1985). Migration: Behavioral and ecological. *In* "Migration: Mechanisms and Adaptive Significance" (M. A. Rankin, ed.). *Contr. Mar. Sci.* **27** (suppl.), 5–26.

Oliveira, E. G., Srygley, R. B., and Dudley, R. (1998). Do Neotropical migrant butterflies navigate using a solar compass? *J. Exp. Biol.* **201,** 3317–3331.

Rankin, M. A. (1991). Endocrine effects on migration. *Am. Zool.* **31,** 217–230.

Rankin, M. A., and Burchsted, J. C. A. (1992). The cost of migration in insects. *Annu. Rev. Entomol.* **37,** 533–559.

Zera, A. J., and Denno, R. F. (1997). Physiology and ecology of dispersal polymorphisms in insects. *Annu. Rev. Entomol.* **42,** 207–231.

Mimicry

Mathieu Joron

Leiden University, The Netherlands

Mimicry is the adaptive resemblance in signal between several species in a locality. The most spectacular and intriguing cases are those of accurate resemblance between distantly related animals, such as spiders mimicking ants. Closely related species can also benefit from mutual resemblance, in which case mimicry results from selection against signal divergence.

The vast majority of the hundreds of thousands of insect species are described and identifiable on the basis of morphological characters. This bewildering diversity is, however, ordered because species share characters with their relatives—and one of the taxonomist's tasks is indeed to recognize, among the shared and divergent characters, a sign of the relatedness of the taxa. Nevertheless, some distantly related species may share a common morphology. Such resemblance may be the result of evolutionary convergence, i.e., parallel lifestyles leading to the selection of similar morphological structures; in this case, resemblance per se is not under selection. On the contrary, when a character is taken as a signal between individuals, one species may benefit from bearing the same signal as one already used by another species; then selection acts directly to favor increased resemblance.

AN INTERACTION BETWEEN THREE PROTAGONISTS

The Discovery of Mimicry and the Development of Evolutionary Hypotheses

Mimicry in insects has been a puzzle for entomologists long before the Darwinian concept of natural selection, but the explanations for mimicry are tightly linked to the development of evolutionary thinking. While he was traveling in the Amazon with Alfred Russel Wallace in 1842, British entomologist Henry Walter Bates noted that distantly related butterfly species bore the same wing color pattern. Moreover, these communities of species changed their shared pattern in concert across localities. Among these species were the very abundant Ithomiinae (called "Danaoid Heliconiidae" then, now a subfamily in the Nymphalidae) and rarer Dismorphiinae (called Leptalidae then, now a subfamily in the Pieridae). Bates, as a pioneer evolutionist (but after Darwin published his *On the Origin of Species*), developed an adaptive explanation for the resemblance. Hypothesizing that ithomiines were inedible to most predators, he proposed that the edible pierids would benefit from being mistaken for their defended counterparts and would thus be selected to resemble them. Edward B. Poulton later named this kind of mimicry after him as Batesian mimicry, when an edible species mimics a distasteful one.

Bates also realized that some apparently inedible ithomiine species in the genus *Napeogenes* seemed to mimic other inedible Ithomiinae. He proposed that, in fact, rare species, whatever their palatability, should benefit from resembling defended common species. It was, however, more difficult to understand the resemblance of abundant and distasteful *Melinaea, Mechanitis* (Ithomiinae), *Lycorea* (Danainae), and some *Heliconius* (Heliconiinae) from Peru and Colombia, so he assumed the resemblance was the result of some inorganic or environmental factors. In 1879, German naturalist Fritz Müller was the first to develop a mathematical demonstration that two unpalatable prey could benefit from mutual resemblance. He understood that, if the community of predators had to kill a certain (fixed) number of prey to learn to avoid them, two indistinguishable distasteful species would together suffer this mortality and both reduce their death rate per unit time. Müller actually showed that this benefit was

biased in favor of the rarer species, to a factor equal to the square of the ratio of the species' abundance. Therefore, unequal population sizes translate into even more unequal, although still mutual, benefits: Müllerian mimicry, thus defined, could be beneficial for both species, and perhaps also for the predators, in contrast to parasitic Batesian mimicry.

Mimicry: An Interaction between Senders and Receivers

Mimicry typically involves at least three protagonists, two senders and one receiver, with the receiver judging the resemblance of the signals from the two senders (Fig. 1). Obviously, both the senders and the receiver should be found in the same locality for the mimicry to be possible, although time lags or geographic separation between senders may be plausible if receivers have a long-term associative memory and/or migrate. In a habitat, many senders will converge on the same signal, thereby forming what is called a mimicry complex, or mimicry ring. Signals may involve different sensory modalities, depending on the receiver's sensory ecology: static visual signals (e.g., warning color patterns in butterflies, recognizable body shapes in ants), motion (flight behaviors), acoustic signals (hissing and clicking in many Arctiidae moths), olfactory/chemical signals (pheromones or the so-called "cuticular hydrocarbon profiles" by which social Hymenoptera recognize one another), or tactile signals (used by brood parasites of ants to be allowed to enter their nests). Signaling is indeed often multimodal.

Apparent complications may arise when, for example, one of the senders is also the receiver. For example, a predator may mimic the appearance of its prey when approaching it (aggressive mimicry in some spiders or chemical/tactile mimicry for brood parasites); the prey is thus fooled by the predator via its own conspecific signal. The two senders can also be the

FIGURE 1 Conditioned predators and signaling prey. Predators are known to generalize their knowledge of distasteful prey to other resembling prey. Therefore, once predators recognize one prey as distasteful (prey A), other prey may gain from mimicry, whatever their palatability (prey B and C). If the prey is palatable (prey C), its mimetic gain becomes limited by its abundance in the locality. Finally, a conspicuous prey with a (nonmimetic) pattern new to the predator should suffer higher mortality, making the evolution of diversity in warning color and mimicry a puzzle.

same species. This is sometimes the case in chemical-sequestering phytophagous insects when unpalatability varies among individuals in the same population (e.g., *Danaus gillipus* in Florida), leading to so-called "automimicry" of palatable toward unpalatable individuals in the same species. Similarly, male Hymenoptera do not have the defenses that females have.

However, the present article is not organized around these numerous classificatory distinctions, which are based on subtle differences in the identities of senders and receivers or ecological situations. Instead, it highlights the important evolutionary dynamics that arise from whether receivers are expected to try to discriminate or generalize on the senders' signals or, in other words, from senders sending honest compared to dishonest signals. This should bring into perspective some of the main and still unresolved puzzles in mimicry theory, such as the rise and maintenance of diversity in mimicry signals. Most examples are chosen from the butterfly genera that represent today's best known mimetic organisms, such as *Papilio* and *Heliconius;* indeed, our knowledge of the ecology and genetics of mimicry in these genera is unequaled by any other group of insects.

FREQUENCY-DEPENDENT POPULATION PROCESSES

Batesian Mimicry and Negative Frequency Dependence

THEORY AND CONSEQUENCES In Batesian mimicry, one of the sender species, the mimic, sends a dishonest signal to deceive the receiver—e.g., a predator. It is thought that deception is possible only if the receiver has previously inherited or acquired knowledge about this signal. There is ample evidence that (1) vertebrate predators (birds, lizards) can learn to recognize distasteful prey, (2) that they can be deceived by mimicry, and (3) that mimics gain from the resemblance. The most famous Batesian mimic is probably the viceroy butterfly, *Limenitis archippus,* which mimics the monarch *Danaus plexippus,* although this relationship is now questioned (because viceroys can be unpalatable). Hoverflies (Diptera: Syrphidae), diurnal moths (Sesiidae, Sphingidae), striped beetles (Cerambycidae), or crane flies (Tipulidae) are well-known Batesian mimics of wasps and bees (Fig. 2).

Clearly, the efficiency of the deception is directly linked to the probability that predators have knowledge of the prey. It thus depends on the ratio of models and mimics in the population of prey (Fig. 1). As in host–parasite systems, the fitness of Batesian ("parasitic") mimics therefore depends negatively on their proportion in the prey community. Negative frequency dependence, the selective advantage to rare forms, is thought to be a strong force favoring and maintaining diversity in many ecological situations in nature. In Batesian mimics, any new (or rare) mutant resembling another protected model will be favored, leading to a balanced polymorphism between the two mimetic forms. Negative frequency dependence also

FIGURE 2 Batesian mimicry. The day-flying moth *Synanthedon tipuliformis* (Sesiidae) (top) is a Batesian mimic of stinging wasps in Europe. The resemblance is very accurate, and the moth is very rare compared to its wasp models, so that it is not often observed. Similarly, but in a totally different group, the beetle *Clytus arietis* (bottom) mimics wasps and is sometimes seen on blossoms. These two examples illustrate how the same general appearance can be achieved by morphological changes of totally different nature in different groups of insects. (Photographs copyright 1998–2002 Hania Arentsen and Hans Arentsen, reproduced, with permission, from The Garden Safari, http://www.gardensafari.net.)

predicts that the local number of Batesian species should be dependent on the abundance of the model(s).

Many, but by no means all, Batesian mimics are indeed polymorphic. Among the most famous is the African swallowtail *Papilio dardanus,* which may have three co-occurring forms that mimic different species of the Danaine genus *Amauris. Hypolimnas misippus* (Nymphalinae) is another African butterfly that has four forms mimetic of *Danaus chrysippus.* In South America, the swallowtail *Eurytides lisithous* has up to three forms that mimic the co-occurring *Parides* species (Papilionidae), whereas in Southeast Asia the famous *Papilio memnon* also mimics three or more different papilionid models. In the Diptera, the Old World hoverflies *Volucella bombylans* and *Merodon equestris* are examples of polymorphic species mimicking bumble bees.

EVIDENCE FOR NEGATIVE FREQUENCY DEPENDENCE Although experimental demonstration that Batesian polymorphisms stem from negative frequency dependence is still lacking, there is a lot of evidence for negative frequency dependence itself. A first line of evidence comes from the observation of patterns of abundance of models and mimics in nature. For example, the North American butterfly *Battus philenor* is known to be unpalatable to most birds and is believed to act as model for a number of edible mimics in the "black" mimicry ring. In one of them, *Papilio glaucus,* females are found as a mimetic and a nonmimetic (male-like) form, and the proportion of the mimetic form tends to be higher where its model *B. philenor* is more abundant. Similarly, the resemblance of the mimic *Papilio troilus* to *B. philenor* is higher where the latter is abundant. These give an overall pattern of mimics' occurrence consistent with negative frequency dependence. Moreover, field experiments directly showed a strong selective advantage to mimetic vs nonmimetic *Callosamia promethea* day-flying moths, another Batesian mimic of *B. philenor.*

Experimental approaches give more insight into the mechanisms involved in frequency dependence. In experiments, captive or wild predators can be tested with a variety of artificial or real prey, and the mimic/model proportions can be experimentally changed to explore how it affects the preys' survival. Traditional experiments were carried out in the 1970s with mealworms or pastry baits colored with food dyes, and/or dipped in quinine to make them distasteful, and exposed to garden birds in suburban Britain. Such experiments do suggest that a rare mimic has an advantage over a common one if the "model" is slightly distasteful, which demonstrates frequency-dependent selection. However, if the "models" were made very distasteful, the advantage of being rare decreased and eventually vanished. Laboratory experiments can also be used to search for evidence of frequency dependence, while avoiding potential confounding effects of field experiments. Experiments with captive great tits as predators showed that the mortality of both mimics and models depended on the frequency of the model and that both models and mimics survived better when mimics were fewer.

These experiments tell us that the intensity of frequency-dependent selection in mimics is highly dependent on the palatability of the models. To see its selective advantage decrease, the palatable mimic must become very common, or the model must be not very distasteful. This suggests there is some kind of effective "equivalence" between relative numbers of prey encountered and their relative levels of toxicity.

Positive Frequency Dependence in Müllerian Mimicry

THEORY: THE DISADVANTAGE OF RARE FORMS
Warning signals, or aposematism, evolve because prey bearing signals that predators associate better with unprofitability (e.g., harmful prey) survive better. The evolution of warning signals brings some apparent paradoxes that are not treated in that entry. However, there is plenty of evidence that aposematic prey are easily learned and subsequently avoided by vertebrate predators. Both the warning prey and the learning predator benefit from a correct interpretation of the signal. Under such an "honest signaling" framework, rare or new variants within a prey population should not be recognized as distasteful and should suffer higher predation (Fig. 1). This selection against rare forms translates into positive frequency-dependent selection: rare mutants are removed, leading to monomorphism in all populations.

Because predators select only on prey appearance, the selective pressure does not stop at the species boundary; several protected prey species may be selected to use the same warning signal, i.e., become Müllerian mimics. Although the phenomenon is not necessarily symmetrical, two or several defended species should all benefit from sharing a warning signal, which reduces their per capita predation rate. As more and more individuals join in the mimicry ring, the protection given by the signal becomes stronger. Therefore, the direct, and naïve, prediction is that all unpalatable prey of a similar size in a habitat should converge into a mimicry ring.

EVIDENCE FOR THE FREQUENCY-DEPENDENT BENEFITS OF MÜLLERIAN MIMICRY
Although comparative and/or biogeographical studies give strong support to the theory, the first convincing experimental evidence came from pastry-bait experiments with garden birds that tend to attack rare distasteful baits more often than common ones. Recently, laboratory experiments also showed strong selection against new rare warningly colored prey items. However, field evidence with free-living prey is crucial for a validation of these results. In one experiment, J. Mallet reciprocally transplanted *Heliconius erato* individuals between populations in which *H. erato* have different wing patterns, thus effectively releasing rare "mutant" and "control" butterflies into the host populations. A strong selective advantage of about 50% was calculated for the commoner form. More recently, to avoid the potential pathology of color patterns being adaptations to local habitat conditions in addition to mimicry, D. D. Kapan used a similar reciprocal release–recapture technique but used polymorphic populations of the butterfly *H. cydno*. In this species, two morphs coexist but participate in two different mimicry rings that differ in relative abundance in different locations in Ecuador. Life expectancy was 12 days for the locally common forms and only 2 days for the locally uncommon forms. These field data give unequivocal evidence for strong selection against rare forms in these Müllerian species.

CONSEQUENCES AND CHALLENGES
Strong purifying selection now seems well supported by theoretical, comparative, and experimental evidence. To evolve a new pattern, a toxic prey would have to pass an apparently impassable initial disadvantage, survive a transient polymorphism, and win the aposematic competition with alternative warning signals. It is therefore no surprise that most distasteful Müllerian mimics are indeed monomorphic in local populations (Fig. 3) and

FIGURE 3 Six butterfly mimicry rings from eastern Peru. The mimicry rings (groups of mimetic species) presented here are dominated by butterflies in the Ithomiinae and occur in the forests around the city of Tarapoto. Following G. W. Beccaloni's nomenclature, these mimicry rings are Tiger (1–16), Melanic tiger (17–21), Large transparent (22–24), Small transparent (25 and 26), Small yellow (27–31), and Orange-tip (32–34) mimicry rings. At least 5 other mimicry rings can be recognized involving Heliconiinae and/or Ithomiinae in this area, which brings the total to at least 11 mimicry rings for these two butterfly subfamilies. Many more species, not featured here, belong to these mimicry rings, particularly Ithomiines and especially in the Small transparent group. The Tiger mimicry ring involves a lot of species and the size distribution is almost continuous from small to very big. This may be because as more and more Müllerian mimics join the mimicry ring, predators might generalize more, and the selection for close resemblance could be somewhat relaxed. Note that some day-flying moths (6, 17, 22, 27) participate in these mimicry rings, probably as Müllerian mimics (they reflex-bleed bitter hemolymph when handled). Butterflies 13–16 and 31 are supposed to be Batesian mimics since they belong to palatable groups within their families. See more species belonging to these mimicry rings in Figs. 4 and 5. All butterflies are Nymphalidae: Ithomiinae, except 1–3 (Nymphalidae: Heliconiinae), 14 (Nymphalidae: Melitaeinae), 16 (Nymphalidae: Charaxinae), 15 (Papilionidae), 13 and 31 (Pieridae: Dismorphiinae), 34 (Riodinidae), and 6, 17, and 22 (Arctiidae: Pericopinae). Scientific names: 1, *Eueides isabella;* 2, *Heliconius pardalinus;* 3, *H. hecale;* 4, *Melinaea menophilus;* 5, *Tithorea harmonia;* 6, *Chetone histriona* sp.; 7, *Napeogenes larina;* 8, *Mechanitis lysimnia;* 9, *Mec. polymnia;* 10, *Mec. mazaeus plagifera* ssp.; 11, *Ceratinia tutia;* 12, *Hypothyris cantobrica;* 13, *Dismorphia amphiona;* 14, *Eresia* sp.; 15, *Pterouros zagreus;* 16, *Consul fabius;* 17, *Chetone histriona;* 18, *Mel. marsaeus;* 19, *Hyposcada anchiala;* 20, *Hypot. mansuetus;* 21, *Mec. mazaeus deceptus;* 22, *Notophyson heliconides;* 23, *Methona confusa;* 24, *Godyris zavaleta;* 25, *Greta andromica;* 26, *Pseudoscada florula;* 27, *Notodontid* moth; 28, *Aeria eurimedia;* 29, *Ithomia salapia;* 30, *Scada* sp.; 31, *Moschoneura* sp.; 32, *Hypos. illinissa;* 33, *Hypoleria sarepta;* 34, *Stalachtis euterpe.* Scale bar, 2 cm.

that polymorphisms are usually restricted to narrow hybrid zones between color-pattern races. In *H. erato,* in which two color races abut, frequency-dependent selection maintains a sharp boundary, alternative forms being positively reinforced on either side of a steep cline. Many species join Müllerian mimicry rings, which itself represents interspecific evidence for strong frequency-dependent selection.

However, in contrast with such extremely conservative forces, diversity is present at all levels in mimicry (Fig. 3). At a *macroevolutionary level,* aposematic and mimetic groups typically undergo rapid mimetic radiations into numerous species and races differing in color pattern, like heliconiine butterflies or pyrrhocorid red bugs. At the *community level,* many radically different mimicry rings coexist in the same habitat (e.g., five or six coexisting rings just within the *Heliconius* of Costa Rica, at least seven or eight rings just within the Ithomiinae of the Peruvian Amazon—Fig. 3.). At the *biogeographical level,* many aposematic species show a bewildering diversification in more or less sharply defined mimetic races. Finally, at the *population level,* several chemically defended species show mimetic polymorphism. For instance, the bumble bee *Bombus rufocintus* has two mimetic forms in North America, the burnet moth *Zygaena ephialtes* has two sympatric forms in Italy, the African monarch *D. chrysippus* has four main color forms coexisting in a large areas in East Africa, and the Amazonian *Heliconius numata* shows the most astounding polymorphic mimicry with up to 7 to 10 forms in the Andean foothills. In each of these cases, the different forms closely match the different local mimicry rings (Fig. 4).

This rampant diversity does not question the existence of frequency dependence itself, but the details of how purifying selection may or may not prevent the evolution of diversity. It may also question the validity of the two classical categories of protective mimicry (Batesian and Müllerian) and the existence of a sharp divide between them along the spectrum of prey palatability. Explaining these unexpected cases is therefore central to our understanding of signal evolution in distasteful insects.

The Palatability Spectrum and Predator Psychology

MODELS OF MIMICRY EVOLUTION   Case studies and experiments on mimicry are practically difficult, are time consuming, and inform us only on potential processes in particular cases. They are thus not always very informative as to which processes are generally important in the evolution of mimetic diversity. For these reasons, mathematical models simulating mimicry evolution have been widely used. Models of mimicry evolution have been traditionally of two different types: "evolutionary dynamics" models have concentrated on trait evolution in the prey populations, underestimating the effects of the details of predator behavior; "receiver psychology" models have concentrated on the effect of predator cognitive abilities in driving the costs and benefits to mimetic prey, but largely ignored evolutionary processes in the prey populations, particularly frequency or density dependence. The second category of models are those that "traditionally" pose a threat to the validity of the Batesian/Müllerian distinction, and M. P. Speed even coined the new term "quasi-Batesian mimicry" for the strange, though purportedly common, intermediate dynamics that his model highlighted.

The main discrepancies lie in the way predators are thought to respond to prey palatability and density. Speed's models assumed that predators attack a fixed fraction of a prey in a population, irrespective of their total number (linear frequency dependence), and that this fraction depends on the palatability of the species. In a mixture of prey of differing palatability, the resulting fraction killed would be intermediate between the fractions lost in each prey in the absence of mimicry, leading to one prey species benefiting and the other suffering from mimicry. This view, however, leads to the strange prediction that as more mildly unpalatable prey are present, the attacked fraction (per unit time) can increase. In contrast, J. Mallet and the author argued that predators are unlikely to be sensitive to frequency

FIGURE 4 Polymorphic Müllerian mimicry. The Amazonian butterfly *H. numata* (Nymphalidae: Heliconiinae—right column) is a Müllerian mimic in a variety of tiger-pattern mimicry rings. Each population (here around the city of Tarapoto in Eastern Peru) is polymorphic and up to seven forms may coexist, each being an exceptionally accurate mimic of species in the genus *Melinaea* (Nymphalidae: Ithomiinae—left column). Spatial variation in selection pressure is probably what maintains the polymorphism, by a balance between local selection for mimicry of the commonest *Melinaea* species and movement of individuals (gene flow) between neighboring localities selected for different wing patterns. From top to bottom (left column): *Melinaea ludovica ludovica, Mel. satevis cydon, Mel. marsaeus mothone, Mel. marsaeus phasiana, Mel. menophilus* ssp. nov., *Mel. menophilus hicetas,* and *Mel. marsaeus mothone.* (Right column) *H. numata* forms *silvana, elegans, aurora, arcuella, tarapotensis, timaeus,* and *bicoloratus.* Scale bar, 2 cm.

per se and should instead need only to attack a fixed number of prey before learning, making the "attacked fraction" a decreasing function of the total number of prey bearing the pattern. This should lead to a strongly nonlinear, effectively hyperbolic frequency dependence. The attacked fraction (per unit time) should always decrease when the total number of unpalatable prey increases, whatever their relative unpalatability.

The debate is still very much active, and decisive data are surprisingly scarce. In an experiment with pastry baits and wild passerines, Speed showed that the attack fraction of a mimetic pair was indeed intermediate between that of either "species" alone. Furthermore, birds seemed to learn only to a certain extent; that is, they never completely stopped attacking the unpalatable items. Despite some problems in the experimental design (no predator monitoring during the study, artificially high prey density, prey predictability, zero cost of experimenting for birds), these data remain a puzzle and may hint at more complex learning processes than a pure number-dependent dose response. More decisive evidence came from L. Lindström's study, in which novel toxic prey were introduced into a great tit's foraging arena at varying frequencies (=densities in this setting). Although the total number of attacked toxic prey increased with their initial frequency, the attack fraction decreased. Her data support the validity of nonlinear frequency dependence, although the idea of a strictly fixed number of prey killed could be an oversimplification. Absolute numbers of prey attacked may increase with warning signal density, but proportion will inevitably decrease, which should lead to a traditional Batesian–Müllerian distinction.

THE STRENGTH OF THE SELECTION Müller's number-dependent model also leads to a prediction that has hitherto been largely overlooked. At low densities, selection should act strongly against any transient polymorphism, but at higher densities, selection quickly becomes weak at intermediate form frequencies. This leads to effective neutrality of polymorphism once it is established in abundant species. Kapan's field experiments, in which *H. cydno* were released at varying density, showed precisely this trend. Polymorphism could therefore be nonadaptive but very weakly selected against by predators.

Numerical Mimicry and Density-Dependent Processes

The studies of J. Allen and his collaborators, and others, show that prey selection by predators can be frequency dependent in palatable, cryptic prey, i.e., even in the absence of mimicry of unprofitable prey. This is probably caused, in part, from predators using search images when foraging. For instance, at low densities of a particular kind of (palatable) prey, predators usually prey on the more common form, which corresponds to their search image, imposing a *negative*

frequency dependence. Cryptic prey may be globally numerous in a habitat, but because they are camouflaged, their apparent density to predators is bound to be low. This leads to the diversification of cryptic patterns, and perhaps the selection of plastic (partly environmentally induced) color-pattern genetic control, in prey. In contrast, at high density, predators usually prey on the odd phenotypes preferentially, even among perfectly palatable prey, effectively leading to a *positive* frequency-dependent selection on morphology.

Gregarious palatable prey that are at locally high density and that presumably rely on predator satiation to escape predation, might then be selected for mutual resemblance. Such a prey might be called "warningly colored," whereas the appearance itself is not protective. This idea led to the supposition that several prey species that co-occur at unusually high densities, like mud-puddling butterflies or schooling fishes, might evolve "numerical" or "arithmetic" mimicry by simple frequency-dependent predation unrelated to unprofitability. Prey traits like color, shape, and especially locomotor behavior are therefore thought to be under purifying selection in mixed-species aggregations. This attractive idea remains largely untested in insects, although R. B. Srygley proposed the pair of bright orange butterflies *Dryas julia* (Heliconiinae) and *Marpesia petreus* (Nymphalinae) as a potential candidate.

Female-Limited Mimicry

Some of the most spectacular and best studied cases of Batesian polymorphism are found in swallowtails, and in some species only the female is mimetic (see an example in Fig. 5). This peculiar tendency to sex-specific polymorphism seems to be restricted to butterflies (Papilionidae and Pieridae), and virtually no other case of sex-limited mimicry seems to be reported for other insects (except for male-limited mimicry in some moths). Female-limited mimicry was often viewed as a result of negative frequency dependence: if mimicry is restricted to one sex, the effective mimetic population size is only about half that of a nondimorphic species, reducing deleterious effects of parasitism onto the warning signal. But this group-selection argument cannot in itself explain why females tend to become mimetic more often than males and why mechanisms arise that restrict the mimicry to one sex. However, more proximal, individual-selection arguments are not lacking. First, mimicry may be more beneficial to one sex than to the other. For instance, female butterflies have a less agile flight because of egg load and a more "predictable" flight when searching oviposition sites, and they suffer higher rates of attacks by visual predators. Second, male wing patterns can be constrained by sexual selection, via either female choice or male–male interactions: males could not evolve Batesian mimicry without losing mating opportunities. In experiments with North American swallowtails (of which only females mimic *B. philenor*), male *P. glaucus* painted with the mimetic pattern

FIGURE 5 Female-limited mimicry in *Perrhybris pyrrha* (Pieridae), Eastern Peru. The female (top) is a Batesian mimic of the tiger-patterned Ithomiines and Helicomiines (see Fig.3), while the male (bottom) has retained a typical pierid white coloration. Scale bar, 2 cm.

had a lower mating success than normal yellow males; similarly, painted *P. polyxenes* males had a lower success in male–male fights and therefore held lower quality territories around hilltops. In these insects, the wing coloration appears to bear signals directed either to conspecific males or to predators, which creates a potential conflict leading to sex-limited polymorphism. It is interesting to note that *Papilio* and *Eurytides* species that mimic *Parides* (Papilionidae) in South America do not exhibit female-limited mimicry; different modes of sexual selection (e.g., absence of territoriality) may operate in the forest understory habitat. In a different ecological setting, diurnal males of the North American silkmoth *Callosamia promethea* are exposed to visual predators, and mimicry of *B. philenor* is limited to males; female *Callosamia* fly at night and benefit more by crypsis during the day.

MIMICRY AND THE EVOLUTION OF SIGNAL FORM

Resemblance and Homology

Mimicry can arise as soon as the signal is effectively copied, i.e., as soon as superficial resemblance is attained. Therefore, mimics usually bear characters similar to those of their models, but these are often clearly nonhomologous in terms of genes and mechanisms of development. For instance, red spots near the base of the wing in *P. memnon* mimic the spots on the bodies of their models. The translucency and iridescence of

distasteful Ithomiinae clearwing butterflies is mimicked by white raylets in dioptine and pyralid day-flying moths and provide the same impression in motion. Similarly, the black-wing patterning of some flies seems to mimic the superposition of wings over the abdomen in their wasp models. Therefore, mimics from distant phylogenetic groups are certainly under very different functional and developmental constraints to create a mimetic impression. Selection will retain the first characters that suddenly increase overall similarity. Which initial step is made will therefore strongly influence which route is selected to achieve mimicry.

The Genetics of Mimicry: Polymorphisms and Supergene Evolution

THE DEBATE   The genetical study of the evolution of mimicry was first dominated by a debate between gradualists (Fisher) and mutationists (Goldschmidt). Goldschmidt proposed that "systemic" mutations could affect the whole wing pattern of butterflies in one step and that models and mimics, although not using the same genes, were using at least the same developmental pathways. Because this view could not account for the obvious nonhomologies, like those pointed out above, Fisher and others claimed that mimicry was achieved by slow microevolutionary steps and the gradual accumulation of resemblance alleles.

Decisive steps toward a resolution of the debate came principally from the study, by C. Clarke and P. Sheppard in the 1960s, of Batesian butterfly mimics in which color pattern is easy to define and analyze and gene effects are straightforward to identify. Polymorphic mimics, particularly *Papilio* species, of which different forms could be crossed by breeding experiments (including hand pairing), were particularly useful. It appeared that color pattern is mainly inherited at one or few major loci, affecting the whole pattern. From rare recombinants, it could be shown that these loci were in fact supergenes, that is, arrays of tightly linked small-effect genes. Several additional unlinked "modifier" loci were also shown to increase resemblance via interaction and epistasis with the supergene. Goldschmidt's ideas seemed refuted.

However, although supergenes seem to be a necessary condition for the evolution of polymorphism (otherwise numerous nonmimetic, unfit recombinants would be produced), how they evolve is another issue. Theoretical models suggested that supergenes could not be achieved by simple gradual reduction in recombination. In the absence of spatial variation in selection pressures, tighter linkage cannot evolve by small steps via Fisherian gradual evolution, because good combinations of alleles are immediately broken up by recombination. Instead, gene clusters should preexist the evolution of polymorphism.

THE TWO-STEP HYPOTHESIS   These results led to a unifying, now widely accepted two-step mechanism of mimicry evolution: (1) mutations at genes of major effect first allow a phenotypic leap achieving an approximate resemblance to a particular model. Once these mutations have increased in the population, (2) resemblance can be enhanced through the gradual selection of epistatic modifiers. This two-step mechanism is supported by three lines of evidence. First, empirical evidence from butterflies suggests the existence of a small number of major-effect genes and numerous small-effect modifiers. In fact some of these genes of major effect could even include a series of regulatory upstream elements and transcription factors, now known to be involved in the development of butterfly color patterns. Pigment pathway genes and scale maturation regulators can also have very dramatic effects on the color patterns. Second, population genetics and dynamics models support the prediction that a major phenotypic jump is necessary to cross the deep fitness valleys in a rugged fitness landscape, after which gradual, Fisherian evolution may proceed to enhance resemblance. Finally, experiments show that birds associate cryptic patterns with edibility and generalize those in such a way that only profoundly deviant prey are treated as separate cases by the birds and memorized as warning patterns when appropriate. These experiments also indicate that increased resemblance is still significantly advantageous in imperfect mimics, supporting the second step of the two-step scenario.

LARGESSE OF THE GENOME Another, but not exclusive, route to supergenes for mimicry is called largesse of the genome, put forward by J. R. G. Turner. Under this scenario, it is believed that the modification of a trait can be achieved by so many different genes that some of them will inevitably happen to be linked. Among the many possible combinations of loci, selection could simply sieve out the ones that involve linked genes. This hypothesis is particularly likely for loss-of-function phenotypes that can be achieved by mutating any step in the development, like the loss of tail in the African swallowtail *P. dardanus*. Similarly, that different mimetic species use nonhomologous supergenes can be viewed as indirect evidence for the validity of the largesse of the genome hypothesis in the broad sense.

SUPERGENES IN MÜLLERIAN MIMICS: A PUZZLE Müllerian mimics being usually monomorphic locally, supergenes are not expected to control wing patterns, and multilocus control was hypothesized to be the norm. This basic prediction has, however, constantly been challenged by *Heliconius* color-pattern genetics, which show that a limited set of genes of large effect and supergenes control most of the racial color-pattern variation. In the polymorphic *H. numata*, one single gene seems to control the entire wing pattern, with as many as seven alleles, each allele bringing resemblance to a specific mimetic pattern (Fig. 4). Tight gene clusters are also found, to a lesser extent, in polymorphic *H. cydno*, in *H. melpomene*, and in *H. erato*. The existence of these supergenes seems puzzling. It is possible that butterfly color patterns in general are under the control of relatively few conserved

genes, at least in some lineages, such as developmental regulatory genes involved in eyespot formation.

In toxic prey, strong selection against any new form and the impossibility of gradual color-pattern changes have been theoretically and empirically demonstrated. It follows that, like Batesian mimics, Müllerian mimics seem to need an initial phenotypic leap, perhaps involving multimodal signal modifications, to jump either to an already protected pattern or away from predators' generalization of cryptic prey. Therefore, it is perhaps no surprise that most exaggerated signal forms studied are under the control of relatively few genes, following the same two-step scenario as in Batesian mimicry. Moreover, switches from one mimetic pattern to another are likely selected only if the new mutant's mimetic characters are not randomly recombined in its descendants. This imposes another constraint (or "sieve") on the genetic architecture for new mimetic patterns to be selected out of a transiently polymorphic population. It is therefore remarkable to note that although Batesian and Müllerian evolutionary dynamics are radically different, and are even perhaps engaged in an evolutionary arms race, the evolution of their signals might require a similar (though nonhomologous) genetic predisposition.

Myrmecomorphy

Ants represent the most abundant group of organisms in most biota and have powerful multimodal defenses such as acid taste, aggressive biting, painful sting, and social defense. For these reasons, foraging ants are generally little subject to predation and act as ideal models in mimicry rings. Many insects and spiders indeed have an altered morphology and resemble ants, a phenomenon called myrmecomorphy. For instance, several salticid spider genera such *Myrmarachne* or *Synmosyna* are bewilderingly good ant mimics. It is also common to spot ant-like myrid nymphs (Heteroptera) running among leafcutting *Atta* ants or *Ecitomorpha* staphilinid beetles among *Eciton* army ant columns. The adaptive significance of ant-like morphology has been the subject of considerable debate. For instance, several ant-like spiders are believed to mimic ants as a trick to approach and prey on their ant models ("aggressive mimicry"); some ant-like bugs use the same trick to approach and prey on ant-tended aphids. However, most ant-like insects are phytophagous, do not prey on foraging ants, and usually mimic the locally abundant ant species. They are therefore good Batesian mimicry candidates. The interesting aspect of ant mimicry is that, although small birds, lizards, or amphibians may be important predators on ant-sized insects, there are grounds to think that arthropod predators with developed visual skills could be the prime receivers selecting for ant mimicry. For instance, wasps in the Pompilidae are known as important predators of jumping spiders, but ignore ants, thus potentially selecting for ant-like morphology and behavior. Jumping spiders themselves are visual predators hunting insects and also tend to avoid stinging ants as prey. Although the cognitive abilities of arthropods are not well researched, several studies using mantids, assassin bugs *Sinea* sp., or crab spiders show that they are capable of associative learning and discriminate against ant-like prey. Despite the difference in visual acuity and cognitive abilities between vertebrates and arthropods, it is interesting to note that arthropod predators are likely responsible for visual mimicry that is very accurate to our eyes.

The Importance of Behavior and Motion

Myrmecomorphy highlights a crucial aspect of mimicry: the importance of behavior. Predators integrate many aspects of prey appearance when making a decision of whether to attack, and behavior is an important part of multimodal signals. Ants are characterized by jerky (e.g., *Pseudomyrmex* spp.) or zigzag (e.g., *Crematogaster* spp.) movements that their mimics adopt. Constant waving of antennae seems to be a common feature of ants, which mimics, such as ant-mimicking spiders (Salticidae) or spider-wasp-mimicking leaf-footed bugs (Coreidae), copy by waving their front legs. Because motion considerably enhances visibility, it is hardly surprising that details of the behavior make important identification cues for the predators. For instance, although slow flight in aposematic butterflies may save energy, slowness itself is certainly recognized as such by predators that can select on extremely minute details of flight unnoticeable to the human eye. R. B. Srygley's work on locomotor mimicry has shown that the two butterflies *H. erato* and *H. sapho* differ in the asymmetry of the upward and downward wing strokes, which their respective (Müllerian) mimics *H. melpomene* and *H. cydno* copy accurately in Panama. Batesian mimics usually retain escape behaviors characteristic of their groups: the lazily flying Neotropical butterfly *Consul fabius* (Nymphalidae: Charaxinae) (see Fig. 3) can start rapid escape flight when detected; ant-mimicking salticid spiders are also usually reluctant to jump unless attacked.

The tendency for predators to generalize the characteristics of palatable prey, on which they actually feed, probably selects aposematic signals away from these morphologies, and behavioral signals are no exception. Rapid jerky flight is usually characteristic of a tasty prey, a profit that predators have to weigh against the time and energy costs associated with catching the prey. Unconventional behaviors like the flight of *Heliconius* butterflies or the looping of honey bees make them highly noticeable to predators. This imposes an additional visibility cost on incipient mimetic prey; for the resemblance to be selected, such cost has to be offset by a significant reduction in predation. These considerations suggest that mimetic behavioral change probably evolves in much the same way as morphological characters do, i.e., a two-step process.

Escape Mimicry

Unpalatability is not the only way to be unprofitable to predators. Fast, efficient escape is another way for preys to teach predators that pursuit is useless and will bring no reward: predators unable to consume the desired prey may associate this frustration with the prey appearance and reduce their attacks on this prey altogether. Even if the prey can be seized, predators probably trade off the energy spent and the (often low) nutritional reward. In several experiments birds were shown to be able to decrease their attack rates when the presented prey would quickly disappear ("escape") during their attacks, and conspicuousness of the prey tended to enhance the response. Therefore, evasive prey could advertise their escaping abilities by color patterns, which other prey may mimic. At least three kinds of characters may enhance the difficulty of catching an evasive prey: erratic flight (like that of Pierids), fast and maneuverable flight (like that of charaxine butterflies), or high reactivity (like that of syrphid hoverflies). Typically, these escape specialists are all palatable to predators. Some species of the Neotropical butterfly genera *Adelpha* (subfamily Nymphalinae) and *Doxocopa* (subfamily Charaxinae) show convergent appearance and exhibit extremely quick escape when slightly disturbed, followed by very fast flight. Their resemblance is hypothesized by R. B. Srygley to be a case of escape mimicry. The poor resemblance of some hoverflies to their purported hymenopteran models has also led to the hypothesis that groups of syrphid species could represent an escape mimicry ring on their own.

Poor Mimicry

At least to our eyes, the model's color pattern is not always copied very accurately. Many syrphid flies, for instance, are difficult to assign to particular mimicry rings, although they seem to mimic the general appearance of Hymenoptera. The heterogeneity in mimetic accuracy has led biologists to propose adaptive and nonadaptive hypotheses, none of which seems very strongly supported at present. (1) The null hypothesis is that poor mimics are no mimics: many mimicry associations have been claimed on the general appearance of an insect, whereas careful examination of the geographic covariation of purported models and mimics may reveal evidence against them. In the case of inaccurate mimics, this method is not very powerful because the mimetic association itself is hard to define, so such covariation is difficult if not impossible to judge. (2) Another nonadaptive scenario is that accurate mimicry may not always be possible, either because of functional constraints/trade-offs on the modified organs or because of genetic or developmental constraints on the variation available in populations. Mimicry may then asymptotically reach a maximum level of resemblance, contingent on the route followed in the initial stages of the mimetic change. Again, this is theoretically plausible, but difficult to test. (3) Among the adaptive explanations for inaccurate mimicry is the hypothesis that these species are in the initial stages of their mimetic change and that our instantaneous view of evolution doesn't show us the complete picture. (4) Another adaptive scenario is that predators have biases and perceptions different from those of humans and are likely to generalize more in some directions than in others, leading to the possibility that mimics that look very inaccurate to us are in fact very good mimics for a predator. Generalization is also dependent on the strength of the harmfulness of the models, perhaps allowing lower levels of accuracy. This may be the case for poor mimicry in some hoverflies. The ultimate adaptationist hypothesis is that inaccuracy itself may be beneficial. It could either (5) allow the mimic to benefit from the protection of several different models, perhaps in a heterogeneous environmental context, or (6)—a related hypothesis—create conflict in the predators' recognition, which may give the mimic more time and chances to escape.

MIMICRY, COMMUNITY ECOLOGY, AND MACROEVOLUTIONARY PATTERNS

Habitat Heterogeneity, Spatial Dynamics, and the Coexistence of Mimicry Rings

The efficiency of a warning pattern depends on the abundance of that pattern in the habitat. Therefore, as new species join a particular mimicry ring, the protection given by the pattern increases, and more species should converge on this best protected pattern. Ultimately, all species should converge on a single mimicry ring. But nature seems to behave in a totally different way. In any one habitat, particularly in tropical environments, aposematic insects of similar size and shape usually cluster into a number of distinct mimicry complexes or mimicry rings.

MULTIPLE MIMICRY RINGS IN THE COMMUNITY
One possibility is that different mimicry rings are found in different microhabitats. If predators do not move between microhabitats, or retain microhabitat-specific information, insect species in different microhabitats could converge on different adaptive peaks. Flight height has been invoked as a possible explanation, following the rainforest stratification paradigm, but evidence from butterflies is rather equivocal. However, host-plant stratification and different nocturnal roosting heights in Neotropical butterflies have received empirical support. Forest maturity and succession stage influence the host-plant composition and may allow the maintenance of multiple mimicry rings in a mosaic habitat.

MULTIPLE MIMICRY RINGS WITHIN A SPECIES If some species are patchily distributed because of their microhabitat requirements, each "subpopulation" may be particularly sensitive to genetic drift and allow the local predators to learn and select a different color pattern in different patches. Once locally stabilized, the new pattern may be hard to remove. Indeed, local positive frequency dependence is both very efficient at stabilizing patterns around fitness peaks and slow at removing already established suboptimal patterns. Any slight difference in microhabitat

quality or patchiness of the species involved will increase the local apparent abundance of particular patterns to particular predators, further decreasing the power of selection to achieve ultimate convergence.

This "mosaic mimetic environment" theory can help explain some problematic cases of Müllerian polymorphism. For instance, *Laparus doris* is a Heliconiine butterfly (Nymphalidae) that has up to four coexisting forms in some populations, some of which are probably mimetic and others are not. The maintenance of polymorphism in this species could be attributed to its high larval and pupal gregariousness (several hundreds of individuals), which results in a patchy distribution of the adults. When hundreds of butterflies suddenly emerge from one single vine, they make up their own local mimetic environment, and the mimetic environment prior to the mass emergence might be effectively neutral to *L. doris*.

If the species composition and the resulting mimetic environment are spatially variable, polymorphism can evolve in microhabitat generalists, with gene flow across these microhabitats. For example, the Amazonian polymorphic species *H. numata* is selected toward different mimetic patterns in different localities that may represent different microhabitats for their more specialized models in the genus *Melinaea* (subfamily Ithomiinae) (Fig. 4). The balance between local selection and gene flow in a mosaic habitat (and perhaps weak selection against polymorphism as suggested earlier) can therefore maintain a nonadaptive, although widespread, polymorphism in *H. numata*.

Coevolution in Mimicry

EVOLUTIONARY RATES AND THE COEVOLUTIONARY CHASE Despite many potential sieves constraining mimicry, several to many edible species can end up mimicking a particular warning pattern in a parasitic way. In such cases, is it possible that a "Batesian-overload" threshold is reached, beyond which the efficiency of the signal is severely lowered? Batesian mimics are indeed parasites of the honest signals of their models, and so the models should escape their mimics by evolving a new warning pattern. However, this escape would be transient because the new pattern would soon attract new Batesian mimics, resulting in an evolutionary arms race, or coevolutionary chase, between the model and its mimics. Some authors suggested that this chase could be a cause of the mimetic diversity in both models and mimics and that cyclical interactions could arise in some cases. However, first, theory has shown that mimics always evolve faster than their models, because they gain a lot more from mimicry than models lose from being mimicked. Any gradual move of the model should be quickly matched by a similar evolution in the mimic. Second, the models, which are the prime educators of local predators, are under strong purifying selection against any new warning pattern. This strong intraspecific conservative force should in the vast majority of cases be stronger than the deleterious effects of being mimicked and preclude pattern change in the models. Coevolutionary changes between Batesian mimics and their models should therefore be stopped in their early stages by a stronger selection for the status quo, and both the models and their mimics should be trapped in the same warning pattern. Only by a phenotypic leap toward an already established warning pattern (Müllerian mimicry) or by crossing a fitness valley thanks to local genetic drift could the model ever escape its mimics.

MUTUALISM AND COEVOLUTION IN MÜLLERIAN MIMICRY   In contrast with the *unilateral* Batesian evolution in which mimics outrun their models, Müllerian mimicry was traditionally thought to involve *mutual* resemblance of the species involved, as if all had moved toward some halfway phenotype. Of course, Müller himself and others were quick to point out that the mutual benefits were not even, but lopsided, i.e., typically the rarer or the less distasteful species would benefit more than the more common or better defended one (respectively). However mutualistic the relation is, coevolution has often been assumed in Müllerian associations, and the protagonists are usually called "comimics" just because it is difficult to know if one species is driving the association. Coevolution also predicts that geographic divergence and pattern changes should be parallel in both species of comimics, like in the mimetic pair *H. erato* and *H. melpomene* in tropical America, presumably leading to parallel phylogenies. However, DNA sequences from mitochondrial and nuclear genes show distinct phylogenetic topologies in these two species and distinctly nonparallel evolution.

In fact, there are a number of grounds on which to believe that the asymmetrical relationship leads to one-sided signal evolution even in Müllerian mimicry, one species being a mimic and the other a model. First, because of number dependence, mimetic change of a rarer species toward a commoner species will be retained, but the reverse is not true: by mimicry of a less common species, the commoner species would lose the protection of its own ancestral pattern, and a change toward a rarer pattern would be initially disadvantageous. The commoner species is therefore effectively locked in its pattern, and initial changes are only likely in the rarer species. Second, given the selection against nonmimetic intermediates, the mutants in the rarer species will have to be roughly mimetic of their new model to be selected, thus bringing the ultimate shared signal closer to that of the common species. Once this initial step is made by the mimic, there could be gradual "coevolution" to refine the resemblance, but the resulting change in color pattern will inevitably be more pronounced in the mimic, the model remaining more or less unchanged. Because Müllerian pairs are of a mimic–model nature, even with mutual benefits, the prediction for parallel evolution is therefore not likely to be valid. Indeed, in the mimetic pair *H. erato/H. melpomene,* the phylogeography suggests that *H. melpomene* has radiated onto preexisting *H. erato* color-pattern races, thus colonizing all color-pattern niches protected by *H. erato* in South America.

Mimicry, Speciation, and Radiations

Racial boundaries in mimetic butterflies are usually very permeable to genetic exchange, since selection acts primarily on color-pattern genes. However, because clines moving geographically are likely stopped at ecological boundaries, the resulting racial boundaries are likely to rest on ecological gradients. Racial boundaries between mimetic color patterns could therefore be reinforced by adaptation to local ecology on either side of the cline, leading to speciation. Color-pattern diversification could then accelerate speciation by allowing both postmating reproductive isolation, because of a higher mortality of nonmimetic hybrid offspring, and premating isolation if color pattern itself is used as a mating cue by the insect. For these reasons, mimicry has the potential to accelerate speciation. The pattern of mimetic associations in *Heliconius* butterflies seems indeed to indicate that speciation and mimetic switches are usually coincident: sister species usually differ in their mimetic color pattern. Direct evidence of the role of color pattern in mate choice has been gathered for the sister species pairs *H. erato/H. himera* and *H. melpomene/H. cydno.* The first two species are geographically separated across an ecological gradient in the Andes. The second pair is sympatric, although the species also differ in ecological requirements in a patchy distribution. In both pairs, therefore, color-pattern and mimetic switches probably accelerated speciation initiated by ecological adaptation. It is unknown how general this mimicry-based speciation is in mimetic insects but it could be an important consequence of the rampant and apparently easy diversification of mimetic patterns at the intraspecific level. The genetic predisposition of mimetic species to evolve polymorphism—the first stage toward speciation—might explain why mimetic lineages are usually very speciose and undergo rapid radiations, both geographically and phylogenetically.

See Also the Following Articles
Aposematic Coloration • Coevolution • Defensive Behavior • Industrial Melanism • Monarchs • Predation

Further Reading
Bates, H. W. (1862). Contributions to an insect fauna of the Amazon valley: Lepidoptera: Heliconidae. *Trans. Linn. Soc. London* **23,** 495–566.
Bates, H. W. (1863). "A Naturalist on the River Amazon." Murray, London.
Beccaloni, G. W. (1997). Ecology, natural history and behaviour of ithomiine butterflies and their mimics in Ecuador (Lepidoptera: Nymphalidae: Ithomiinae). *Trop. Lepidoptera* **8,** 103–124.
Brower, L. P. (ed.) (1988). "Mimicry and the Evolutionary Process." University of Chicago Press, Chicago.
Edmunds, M. (1974). "Defence in Animals. A Survey of Anti-predator Defences." Longman, New York.
Jiggins, C. D., Naisbit, R. E., Coe, R. L., and Mallet, J. (2001). Reproductive isolation caused by colour pattern mimicry. *Nature* **411,** 302–305.
Kapan, D. D. (2001). Three-butterfly system provides a field test of Müllerian mimicry. *Nature* **409,** 338–340.
Mallet, J. (1993). Speciation, raciation, and color pattern evolution in *Heliconius* butterflies: Evidence from hybrid zones. *In* "Hybrid Zones and the Evolutionary Process" (R. G. Harrison, ed.). Oxford University Press, New York.

Mallet, J., and Joron, M. (1999). Evolution of diversity in warning color and mimicry: Polymorphisms, shifting balance and speciation. *Annu. Rev. Ecol. Syst.* **30**, 201–233.

Müller, F. (1879). *Ituna* and *Thyridia*: A remarkable case of mimicry in butterflies. *Trans. Entomol. Soc. London* **1879.**

Pasteur, G. (1982). A classificatory review of mimicry systems. *Annu. Rev. Ecol. Syst.* **13**, 169–199.

Rowe, C. (ed.) (2001). "Warning Signals and Mimicry." Kluwer Academic, Dordrecht. [Special issue of *Evolutionary Ecology,* 1999, **13(7/8)**]

Turner, J. R. G. (1977). Butterfly mimicry: The genetical evolution of an adaptation. *Evol. Biol.* **10**, 163–206.

Wallace, A. R. (1879). The protective colours of animals. *In* "Science for All" (R. Brown, ed.). Cassell, Petter, Galpin, London.

Wickler, W. (1968). "Mimicry in Plants and Animals." McGraw–Hill, New York.

Mites

Barry M. OConnor
University of Michigan

Mites comprise the Acari, which are the largest group within the arthropod class Arachnida, with over 48,000 described species. This number is misleading because it is estimated that only between 5 and 10% of all mite species have been formally described. In contrast with other arachnid groups such as spiders and scorpions, mites are distinctive in both their small size (adult body length ranging from 0.1 to 30 mm) and their ecological diversity. Some mites are predators, like almost all other arachnids, but mites may also feed on plants, fungi, or microorganisms or as parasites on or in the bodies of other animals. Mites are among the oldest known groups of arthropods, with a fossil record beginning in the Devonian period.

BODY STRUCTURE

Unlike insects, with bodies divided into head, thorax, and abdomen, the arachnid body is ancestrally divided into two functional units, the prosoma (the first six body segments) and the opisthosoma (the remaining segments). The body of a mite is further modified in that these original units are fused. A secondary subdivision separates the first two body segments into a structure termed the gnathosoma, specialized for feeding, and the remainder of the body, termed the idiosoma, containing organs of locomotion, digestion, and reproduction. Most mites show no evidence of external body segmentation, other than the serial appendages. The gnathosoma bears the first two pairs of appendages, the chelicerae, which may retain the ancestral chelate, or pincer-like form, or may be highly modified as stylets for piercing and sucking; the pedipalps, which may be almost leg-like, are strongly modified for grasping prey or attaching to a host or

highly reduced. The anterior idiosoma typically bears four pairs of walking legs, the first pair of which may be modified as antenna-like, sensory structures. Legs may also be modified for attaching to a host. Occasionally legs of males are modified for grasping a female during mating or for intraspecific combat.

The mite's body cuticle may be entirely soft, divided into a number of hard, sclerotized plates, or almost entirely sclerotized. In a few mites, crystalline, mineral salts also strengthen the cuticle. Such modifications balance the needs for flexibility in movement and protection from predators. The body surface bears setae, typically hair-like sensory organs, arranged in characteristic patterns in different subgroups of mites. Setae are primarily hair-like, but may take on an incredible variety of shapes, from thick spines, to flat plates, to highly branched, feather-like forms. The pedipalps and legs also bear tactile setae as well as chemosensory structures termed solenidia, which are organs of smell and taste, and other specialized sensilla that are sensitive to infrared radiation. Simple eyes, or ocelli, may be present on the anterior idiosoma, and specialized sensory organs, the trichobothria, on the anterior idiosoma or legs may detect vibrations or electric fields.

Like other arthropods, the inside of a mite's body is a hollow cavity, the hemocoel, in which the internal organs are surrounded by fluid, the hemolymph. Hemolymph distributes food materials and waste products and contains hemocytes, which are the cells that serve as the mite's immune system, but it does not contain oxygen-binding proteins as are found in the blood of vertebrates and some other arthropods. The mite's digestive system is divided into the three parts typical of arthropods: foregut, midgut, and hindgut. The midgut may be divided into diverticulae for food storage, particularly in parasitic mites. Some mites lack a connection between the midgut and the hindgut; these mites feed only on fluids and do not defecate. The hindgut in these mites is transformed into an excretory organ for elimination of nitrogenous wastes. Other mites, with entire guts, may have Malpighian tubules, like insects, extending from the junction of the midgut and hindgut as excretory organs. The internal reproductive system typically consists of a single ovary (paired in the Astigmata) in the female and paired testes in the male. Females typically possess a spermatheca for sperm storage after insemination, and both sexes have various accessory glands and ducts to the exterior as part of the system. Tracheal systems for respiration have evolved independently a number of times in the Acari. These open at spiracles, or stigmata, on various parts of the body in different groups. Other mites lack any respiratory system, and gas exchange occurs through the cuticle in these groups.

CLASSIFICATION OF MITES

The classifications of mites used by various authors vary considerably in the number of higher categories recognized and the hierarchical ranking of the various groups. The simplest system, used by Walter and Proctor (1999), recognizes three orders within Acari: Opilioacariformes, Parasitiformes, and Acariformes. The Opilioacariformes, comprising a single family with about 20 species, is considered the most primitive. These mites are relatively large (2–3 mm) and resemble small opilionids in their general form, having a leathery cuticle that retains traces of external segmentation. These mites resemble the Parasitiformes in having a tracheal system opening laterally on the body, but they have four pairs of stigmatal openings in contrast to the single opening of the Parasitiformes. Opilioacarids resemble some Acariformes in feeding on solid food particles and bearing a pair of rutella, which are sclerotized food-processing structures located near the ventral apex of the gnathosoma.

The order Parasitiformes is a diverse group comprising 76 families divided among three suborders: Gamasida (or Mesostigmata), Ixodida, and Holothyrida. Compared with the Acariformes, this order is morphologically relatively conservative, with most species retaining the same basic body plan. The Holothyrida includes 3 families and around 30 species of heavily sclerotized, predatory or scavenging mites of tropical regions. The Ixodida, or ticks, includes 3 families and around 850 species exclusively parasitic on vertebrate hosts. The vast majority of parasitiform mites are included in the Gamasida, with 70 families. Most gamasid mites retain the ancestral predatory life-style, but the group includes a number of parasites of vertebrates and other arthropods, a few mites which feed on pollen or fungi, and one small group of detritivores capable of feeding on solid food particles.

The order Acariformes is the largest and most diverse group of mites, in terms of both its morphology and its ecological diversity. Hundreds of families are recognized, and over 30,000 species are included. Acariform mites are characterized by the internalization of the basal leg segment, the coxa, leaving the next segment, the trochanter, as the first functional leg segment. Most acariform mites also possess structures termed "genital papillae." While associated with the genital region in the postlarval instars, these structures are actually osmoregulatory organs.

The order Acariformes is conveniently divided into two suborders, Trombidiformes (largely equivalent to the Prostigmata of some authors) and Sarcoptiformes (including the Oribatida and Astigmata of some authors). Most trombidiform mites have tracheal systems opening on or near the gnathosoma. Many have strongly modified chelicerae adapted for piercing animal prey, plant tissue, or the skin of a host animal. Sarcoptiform mites ancestrally feed on solid food and have gnathosomal rutella, like the Opilioacariformes. Tracheal systems opening at the leg bases or anterior dorsal idiosoma have evolved independently several times in this group. Sarcoptiform mites are most diverse in soil habitats, but many have adapted to patchy habitats and have developed commensal or parasitic associations with vertebrates and other arthropods (Fig. 1).

LIFE CYCLES, DEVELOPMENT, AND REPRODUCTION

Most mites exhibit a fixed developmental pattern, passing through the same number of instars regardless of how much food is available. The most complete pattern consists of egg, prelarva, larva, protonymph, deutonymph, tritonymph, and adult. The prelarva and larva are distinguished by having only three pairs of legs; the fourth pair is added at the protonymphal molt. Other immature stages are distinguished from each other by a characteristic pattern of additions of leg and body setae. The prelarva is typically a short-lived stage, either passed completely within the egg or, if it actually hatches, having a highly regressive morphology. The few active prelarvae known do not feed and typically begin the molt to the larval stage within hours after hatching from the egg. This life cycle is found in the Opilioacariformes and ancestrally in the Acariformes. Reductions from this number of instars appear in other groups of mites. Within the Parasitiformes, the prelarva is not observed, and the tritonymph is retained only in some Holothyrida. In one family of Ixodida, the Argasidae, the number of nymphal instars is not fixed. Molts take place after each blood meal in these ticks, but the adult morphology develops only when the mite reaches a minimum body size. In many trombidiform mites, the last nymphal instar is suppressed, and in some extreme cases, all immature stages are suppressed and passed within the body of the female mite. After an extreme form of engorgement termed "physogastry" on fungal food or host-insect hemolymph, these females give birth to fully developed adults. Another developmental pattern observed in the large trombidiform subgroup, the Parasitengona, involves alternation of active and inactive instars. Active stages in this life cycle include the larva, deutonymph, and adult, while the prelarva, protonymph, and tritonymph are morphologically regressive, inactive stages.

Mites exhibit a variety of reproductive strategies and modes of sperm transfer. Ancestrally, mites appear to practice indirect sperm transfer, with males producing and depositing a package of sperm, termed a spermatophore, on the substrate. Females then take the spermatophores into their reproductive tract. This type of reproduction is found in most acariform subgroups, and individuals of the two sexes may or may not be in close contact at the moment of insemination. In known parasitiform groups, males typically use their chelicerae to assist in directly inserting a spermatophore into the female's primary genital opening (as in Ixodida and primitive Gamasida), or the male chelicerae bear an organ termed the spermatodactyl which is used to transfer sperm from the male's genital opening into secondarily developed sperm induction pores near the bases of the female's legs. These paired openings lead to a median spermatheca which is connected directly to the ovary, where fertilization takes place. Direct insemination involving the development of an intromittent organ, the aedeagus in the male, has appeared independently in several groups of acariform mites. Secondary sexual dimorphism typically accompanies

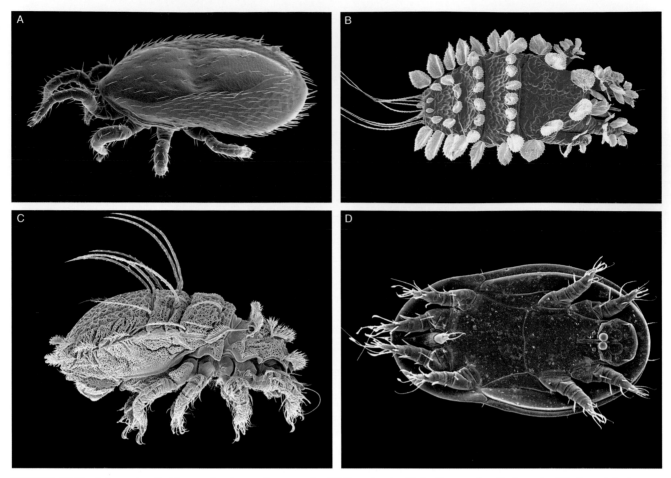

FIGURE 1 (A) Tropical rat mite, *Ornithonyssus bacoti,* a vertebrate parasite. (B) Peacock mite, *Tuckerella* sp., a plant-feeding mite. (C) *Cosmochthonius* sp., a soil mite. (Photographs by D. E. Walter and C. Meacham.) (D) Heteromorphic deutonymphs, phoretic on a predatory mite. (Photograph by D. E. Walter.)

direct mating, with males often having modified appendages for holding the female during mating. Many such males also practice precopulatory guarding of immature females, either merely waiting near a juvenile female about to molt or actively attaching to her. This last behavior is taken to an extreme in the sarcoptiform family Chirodiscidae, species of which live on the hairs of mammals, in which immature females are legless and unable to move. They must be found upon hatching by an adult male who uses an elaborate clasping organ to attach to, and carry about, the juvenile female until her legs appear at adult eclosion.

Sex determination mechanisms and reproductive modes also vary widely throughout the Acari. Some mites are diploid in both sexes, with males having either a Y chromosome or no sex chromosome. Other mites are arrhenotokus, in that females are diploid and males haploid. Such males develop from unfertilized eggs. An unusual reproductive mode, termed parahaploidy, is found in some Gamasida. In these mites, fertilization is necessary for egg development, but in males, the paternal genome is inactivated shortly after the first embryonic cell divisions, and adult males are functionally haploid. Finally, thelytoky, or all-female parthenogenesis, is found in many groups of mites. Such mites reproduce clonally, with diploid eggs developing directly into females without fertilization.

ECOLOGY

Mites exhibit a breadth of ecological interactions unmatched in any other arthropod group. Mites may be found in all geographic provinces, from tropical rainforests to arctic tundra and rocky outcrops in Antarctica and from desert habitats to the deep ocean trenches. They dominate the microarthropod fauna of the soil where they may be found several meters deep or even in groundwater. They occur in all types of aquatic habitats, including freshwater lakes, streams, seepage areas, and even hot springs. Unlike insects, mites are also quite diverse in marine habitats, ranging from the intertidal zone to the deep trenches.

Soil Mites

A single square meter of temperate forest soil may contain upward of 250,000 mites, belonging to a hundred different families. In the litter and upper layers of organic soil, mites play many roles in food webs based on decaying plant materials.

Gamasid and some trombidiform mites are the dominant predators in such systems, feeding on nematodes, small annelids, collembolans, other mites, and the eggs of insects. These predatory mites have developed several strategies for prey detection and capture, from active, foraging species in the gamasid families Laelapidae and Parasitidae, and the trombidiform families Raphignathidae and Cunaxidae, to more sedentary species in the trombidiform families Caeculidae and Cheyletidae, with palps or forelegs modified as traps for unwary prey. Most soil-inhabiting mites, however, are detritivores or fungivores, feeding directly on decaying organic materials or on fungi or microorganisms growing upon them. The greatest diversity of detritivores belongs to the sarcoptiform subgroups collectively known as oribatid mites. These mites are typically slow moving and may take up to 3 years to complete the life cycle. Adults tend to be well sclerotized as a defense against predators, while soft-bodied juveniles may burrow into substrates to avoid predation. Oribatid mites are primarily detritivores, feeding directly on particulate organic material. Others preferentially scrape decaying leaves for their microbial or fungal floras. Despite their numbers, compared with earthworms or other larger soil invertebrates, mites actually process a relatively smaller amount of organic material and are thus of less importance in converting biomass to nutrients again available to plants. However, in terms of the cycling of particular nutrients, notably calcium, mites play an essential role. Mites are also extremely important in the dispersal of bacterial and fungal agents of organic decomposition. Mites feeding on such substrates ingest bacteria or spores that can often pass undigested through the mites' guts. The movement of the mites through the soil, with the associated deposition of fecal pellets containing decomposer propagules, provides a much more efficient dispersal of these organisms than simple physical processes.

Specialized soil types have specialized mite faunas. Dry, sandy, and nutrient-poor soils typically harbor a fauna of primitive acariform mites that show little morphological change from their Devonian fossil ancestors. This entire community may consist of such living fossils, with this type of nutrient-poor soil likely similar to the original terrestrial environment at the time of the first land-colonizing animals. Another highly specialized fauna of mites inhabits the deeper layers of mineral soils. Because there is little organic material that filters down to these layers, many of the mites feed directly on the sparse microbial flora or are predators on nematodes that are able to extract nutrients from the limited resources. Deep soil mites tend to be quite small and soft-bodied and may be elongated to allow for movement through very tight spaces between mineral soil particles. Most are effectively aquatic because the deep soils are often saturated, with the interstitial spaces filled with water.

Mites in Patchy Habitats

Mites living in large, continuous habitats such as the soil and litter layers generally have limited dispersal capability. Being small and lacking wings, mites would seem to be limited to such habitats. However, mites also form major components of the microarthropod communities associated with patchy habitats, which are those separated by distances greater than mites' ability to walk. Such mite communities occur in habitats such as decaying logs, dung and manure, carrion, fungal fruiting bodies, nests of insects and vertebrates, and other concentrations of organic matter such as treeholes, sap flows, and other specialized habitats associated with plants.

Common in patchy habitats are specialized gamasid mites in the families Parasitidae, Macrochelidae, Laelapidae, Digamasellidae, and Uropodidae. These species are typically predators on insect eggs and larvae, particularly those of the Diptera that also frequent patchy habitats, and nematodes. Among the Trombidiformes, species in the subgroup called Heterostigmata are largely associated with patchy habitats, feeding primarily on fungi. The most diverse group of mites in patchy habitats is the sarcoptiform subgroup Astigmata, which appears to have had its origin in such associations.

All of these groups, and some others as well, are able to exploit these habitats, which are generally unavailable to most mites, through a specialized dispersal mode termed phoresy. Phoresy involves one organism utilizing another, larger organism to facilitate its dispersal. In all of the mentioned groups of mites, one life stage is typically specialized for phoretic dispersal on an insect, myriapod, crustacean, or mammal host. Gamasid mites disperse either as inseminated females or as deutonymphs, which is the final juvenile stage in this group. Female laelapid and macrochelid mites typically attach to insect carriers by grasping host setae or other structures with their chelicerae. Parasitid and digamasellid mites disperse as deutonymphs, often in the space under the elytra of beetles, and may roam freely over the insect's body. In the Uropodidae, the deutonymph is often specialized for dispersal and may attach to the host by secreting a sticky substance from posterior ventral glands. This material is drawn into a stalk that hardens in air and connects the mite to its host.

Heterostigmatid mites disperse as adult females, with many species exhibiting a polymorphism in this stage. Nondispersing females have normally developed anterior legs and are not attracted to insect or mammal hosts. Dispersing females, or phoretomorphs, have very enlarged forelegs with a grasping claw that allows attachment onto insect setae or mammalian hair. In the Astigmata, the deutonymph is highly specialized for dispersal. These deutonymphs look nothing like the preceding or following instars, having no mouth or mouthparts, but bearing suckers or claspers at the posterior end of the body for attaching to a host. They are typically heavily sclerotized and able to withstand major fluctuations in environmental conditions. Many astigmatid mites inhabit naturally occurring patches of organic matter such as decaying wood or mushrooms and disperse on any insects that frequent the habitat. Others have developed closer associations with particular insects, notably nest-building bees, wasps, ants, and termites, and depend on these insects not only for dispersal, but also for creation of the habitat in which the mites live. Still other astigmatid mites have adapted to the nests of vertebrates, with many species inhabiting mammal nests specialized for phoretic dispersal on the mammalian host itself. Species living in birds' nests still disperse on nest-inhabiting insects such as beetles and fleas.

A few of these phoretic associations between mites and insects have become mutualistic, with mites providing either "cleaning services" or "pest control" for their hosts. Some species of Old World carpenter bees may carry several species of mites on their bodies. The astigmatid mites in these communities are kleptoparasites, feeding on the provisions intended for the bee's offspring. A female bee may also carry trombidiform mites in the genus *Cheletophyes* (family Cheyletidae) in specialized pouches, termed acarinaria, on the thorax. These mites are obligate predators of the astigmatid mites. The same female bee may also have a large acarinarium in the anterior part of the abdomen, carrying large (2–3 mm) gamasid mites in the genus *Dinogamasus* (family Laelapidae). These mites have modified chelicerae that scrape the cuticle of the bee larva and remove potentially pathogenic microorganisms and fungal spores as well as cuticular exudates. Some astigmatid mites in the family Histiostomatidae are mutualists in the nests of sweat bees (family Halictidae). Feeding stages of the mites have highly modified chelicerae for filter feeding. These mites wander over the nectar and pollen provisions, straining potentially harmful microorganisms. Deutonymphal mites ride in a rudimentary acarinarium on the propodeum or anterior gaster of the female bees.

Aquatic Mites

A number of different groups of mites have successfully colonized and diversified in aquatic habitats. The most diverse of these, with over 40 families, and 5000 species, is a group within the trombidiform subgroup Parasitengona that is termed Hydracarina or, more simply, water mites. This lineage is characterized by the enlargement or multiplication of the genital papillae (often termed acetabula in this group), acariform organs of osmoregulation that allow these mites to maintain ionic balance in hypoosmotic freshwater environments. The parasitengone life cycle is unusual, with its alternation of active and inactive stages, and in most terrestrial and aquatic species, the larva is parasitic, typically on an adult, flying insect. This parasitic larva not only acquires nutrients by feeding on its host, but also is able to disperse over some distance while on the host. The deutonymph and adult stages are typically active predators on other arthropods or their eggs. Water mites are primarily inhabitants of freshwater habitats including temporary ponds, permanent ponds and lakes, streams and rivers, and interstitial waters. One family of water mites, the Pontarachnidae, has invaded marine, intertidal waters and has lost the genital papillae, while another, the Thermacaridae, is restricted to hot springs and capable of surviving temperatures close to 50°C. Mites in standing waters

may crawl about on the substrate or aquatic vegetation, but many species have morphological adaptations for active swimming. These include anterior displacement of the leg bases and long setae, termed swimming hairs, on the legs. These mites actively seek and capture aquatic crustaceans and small insect larvae. Mites inhabiting running waters are typically smaller, with flattened bodies, robust legs, and often sclerotized plates on the body. These mites crawl on and in the substrate, feeding on the eggs of aquatic insects and other microinvertebrates. Some of these mites are specialized predators of the eggs of the same insect species used as hosts by their larvae. Some stream-inhabiting species live deep in the interstitial waters, often having quite elongate bodies for squeezing through the spaces between rock and sand particles. Many water mites are brightly colored, either retaining the red color common among the ancestral, terrestrial Parasitengona or becoming a cryptic blue or green. Some water mites have modified the ancestral parasitengone life cycle by producing fewer, larger eggs. Larvae hatching from these eggs transform to deutonymphs without feeding or dispersing on a host.

Another relatively large group of aquatic mites forms a separate trombidiform lineage, the family Halacaridae. These mites are most diverse in marine habitats, with most species found in intertidal waters. Some halacarids, however, have been collected in abyssal depths up to 7000 m. Feeding ecology of halacarids varies, with some species retaining the ancestral predatory behavior, while others feed on algae or as parasites on crustaceans, echinoderms, or cnidarians. Some halacarids have reinvaded freshwater habitats, presumably via groundwater connections. Such mites are often collected from well water, and a number of species are restricted to freshwater habitats.

Other groups of mites contain aquatic taxa, but none has diversified to the extent seen in the water mites and Halacaridae. Mites in the oribatid family Hydrozetidae are often collected on aquatic vegetation, while those in the family Trimalaconothridae occur in the substrates of ponds and streams. The sarcoptiform group Astigmata includes the family Hyadesiidae, all species of which live in marine, intertidal habitats. These mites are unusual among the Astigmata in living in more or less continuous habitats, and they have lost the dispersing deutonymph from the life cycle. Some species in the family Algophagidae live in brackish waters, and one is known from a fast-flowing river. Other Astigmata live in temporary aquatic habitats, such as water-filled treeholes and other phytotelmata, or water-filled plant cavities, such as pitcher plants, bromeliads, the leaf axils of aroids, and the flower bracts of heliconias and related plants. These species still retain the phoretic deutonymph that disperses on an insect host. Relatively few gamasid mites have become aquatic, but some species in the family Ascidae live in phytotelm habitats or regularly flooded swamp or flood-plain soils. Some of these have a modified cuticle around their respiratory openings that functions as a plastron, holding a bubble of air against the spiracle when the mite is submerged.

Mites on Plants

Unlike all other arachnid groups, several groups of mites have evolved the ability to feed on living plant tissue. Most species belong to one of several lineages of Trombidiformes, each of which has independently evolved this capability, but all share the modification of the chelicerae into piercing stylets. One lineage, the superfamily Tetranychoidea, contains the spider mites and their relatives. Spider mites (family Tetranychidae) are so named because some species utilize silk in constructing webbing on leaves or pads for oviposition and also for dispersal via ballooning much in the manner of some spiders. Silk production is not unique to this group, however, because it is also found in related trombidiform groups not associated with plants. Tetranychoid mites have elongate cheliceral stylets that pierce leaf or root tissue and feed on cell contents or on interstitial fluids. Most species are relatively host specific and do little damage, but some, such as the twospotted spider mite, *Tetranychus urticae,* are polyphagous and are serious pests of agricultural crops, particularly herbaceous annuals, such as beans, and fruit trees. Another tetranychoid group, the false spider mites, or flat mites (family Tenuipalpidae), also includes serious agricultural pests.

A second lineage of plant-feeding mites, the Eriophyoidea, contains extremely tiny species with a highly modified body form. These elongate, worm-like mites have only two pairs of legs at the anterior end of the body, but possess a sucker at the posterior end and move inchworm fashion over plant surfaces. All are obligate plant feeders, using their short stylets to pierce individual cells. Most species are highly host specific, and a single plant species may harbor many species in this group, most of which simply wander over the leaf surfaces. Large populations of such mites may cause loss of color in leaves, leading to the common name rust mites. Another common name, gall mite, refers to the ability of some species to induce characteristic galls on leaves, buds, stems, flowers, or fruits of their host plants. Salivary chemicals mimic certain plant growth hormones and induce the formation of galls in which the mites live. Simple erineum galls form when epidermal cells produce elongate hairlike growths upon which the mites feed. Pouch galls are like erinea but actually form into elongate cavities within which the mites live. Mite-induced proliferation of woody tissue causes "witches' brooms" on trees. Although rusting and gall formation are often unsightly and may affect fruit set in orchard crops, the most important effects of eriophyoid mites on agricultural systems are as vectors of viral pathogens such as wheat streak mosaic virus. On the other hand, other, highly host-specific, eriophyoid mites have been used as virus vectors in the biological control of weeds.

Other plant-feeding mites occur in the families Pentheleidae, including the redlegged earth mite, *Halotydeus destructor,* a serious pest of grasses and herbaceous plants in the Southern Hemisphere, and Tarsonemidae. This last family includes such serious agricultural pests as the broad mite, *Polyphagotarsonemus latus,* which, true to its scientific name, is a polyphagous pest of many agricultural crops.

Other mites living on plants are beneficial as predators of phytophagous mites. Chief among these are species in the gamasid family Phytoseiidae. These mites range from generalist to specialist predators, often attacking economically important species of spider mites, flat mites, and eriophyoids. Several species of Phytoseiidae are commercially marketed for biological control of these pests.

Parasitic Mites

A great many lineages of mites contain parasites of vertebrate and invertebrate animals, some of which are of importance in human and veterinary medicine. Most important of these is the Ixodida, the ticks, but the Gamasida contains a diversity of parasites of reptiles, birds, mammals, insects, myriapods, and crustaceans, most of which belong to the mite superfamily Dermanyssoidea. Among the vertebrate parasites, several different types of parasitism occur. The simplest of these is facultative parasitism, in which typically nest-inhabiting predators may feed opportunistically from a wound on a bird or small mammal host. Other nest-inhabiting mites are obligate parasites, but get on the host only to feed. Notable among these are the northern fowl mite, *Ornithonyssus sylviarum* (family Macronyssidae), and the chicken mite, *Dermanyssus gallinae* (family Dermanyssidae), both of which parasitize a variety of wild birds and domestic poultry and will bite people. Many of these mites have chelicerae modified for piercing and sucking blood or tissue fluid. Finally, some gamasids have become permanent parasites, spending all their time on the host's body. These may have enlarged claws or spurs on the body for holding onto the host. Several different groups of dermanyssoid mites have become endoparasites, living in the respiratory tract of snakes, birds, and some mammals, notably dogs and seals. Some species in the family Rhinonyssidae can cause respiratory distress in cage birds. Other endoparasites inhabit the ear canals of ungulates such as cattle and goats. Some parasitic gamasids act as vectors of bacterial, viral, and protozoan pathogens to their normal hosts, but only one, the dermanyssid *Liponyssoides sanguineus,* acts as a vector for a bacterial pathogen from mice to humans, causing the disease known as rickettsialpox.

Other dermanyssoid mites are parasitic on arthropods, with the most important being the honey bee parasite, *Varroa destructor* (family Laelapidae). This mite is responsible for the worldwide decline in populations of the European honey bee, *Apis mellifera.* The mites feed on hemolymph of bee larvae, causing the adult bee that develops to have aborted wings that prevent the bee from foraging. Buildup of mites in a bee colony causes its destruction over time. This mite became a pest after colonizing *A. mellifera* from its ancestral host, the Asian honey bee, *Apis cerana.* In the normal host, this mite is not pathogenic to the colony because populations do not reach damaging levels.

Several groups of trombidiform mites have become parasitic, the largest of which is the Parasitengona. This group includes the water mites discussed above, but also a number of terrestrial groups. Larvae of most species parasitize insects, in which they may reduce fecundity or longevity. Larvae of one family, the Trombiculidae, or chiggers, parasitize vertebrate hosts. All groups of terrestrial vertebrates may serve as hosts for this very large group of parasites. Most of the over 5000 described species are known only from their parasitic larval stage. Chiggers feed on tissue fluid and lysed host tissue. On most hosts, they appear not to affect the host negatively, but species biting humans induce an immune reaction that not only causes the death of the chigger, but causes a relatively long-lasting, itchy lesion. Different species of chiggers have achieved "pest" status in various parts of the world. Most are merely irritants to human hosts, but some species in the genus *Leptotrombidium,* ranging from Japan and Korea, west to Pakistan, and south to northern Australia, act as vectors of a serious bacterial pathogen from rats to humans. The disease, termed scrub typhus or tsutsugamushi disease, can be fatal if untreated. A chigger is able to vector the pathogen, the bacterium, *Orientia tsutsugamushi,* despite parasitizing only a single host in its lifetime, because the pathogen remains in the mite's body and enters the eggs of the female mite. Thus, larval chiggers are capable of transmitting the pathogen at hatching.

Another diverse group of trombidiform parasites is included in the superfamily Cheyletoidea. Different families in this group parasitize reptiles, birds, and mammals, with the family Demodicidae containing two species specifically parasitic on humans. *Demodex folliculorum,* an elongate, worm-like mite, inhabits the hair follicles of the face and occasionally other body regions, whereas *Demodex brevis* lives in the sebaceous, or sweat, glands in the skin. Although heavy infestations have been linked to acne rosacea, most people harbor these mites with no discernable effect. Other demodicids can be more pathogenic in their normal hosts, such as *Demodex canis* in dogs and *Demodex bovis* in cattle. The former can cause a mange condition, with hair loss and irritated skin, especially in puppies, while the latter causes large nodules full of mites to form in the skin. Species in other cheyletoid families parasitize birds, living on the skin, in feather follicles, or inside feather quills. One interesting group in the family Cheyletidae lives within feather quills, but feeds as predators on other quill-inhabiting mites.

The trombidiform lineage Heterostigmata includes many parasites of insects. The honey bee tracheal mite, *Acarapis woodi* (family Tarsonemidae), is of considerable economic importance as a pest in the respiratory system of honey bees. Other parasitic Tarsonemidae inhabit the defensive glands of coreid Hemiptera, one of the most unusual habitats known, even among mites!

Among the Sarcoptiformes, the Astigmata includes a great diversity of parasitic species, the hosts of which include mammals, birds, and insects. Certain nest-inhabiting astigmatid mites that ancestrally dispersed via phoretic deutonymphs have modified the nature of the association. Instead of merely attaching to the hair or skin of the host and simply dispersing, deutonymphs in several groups in the superfamily Glycyphagoidea associated with small mammal hosts, and species in the family Hypoderatidae with bird hosts, enter either the hair follicles or the subcutaneous tissue of their host. Despite lacking a mouth and functional gut, these deutonymphs engorge, with some Hypoderatidae increasing their body volume up to 1000-fold. The mode of nutrient acquisition in these parasites is unknown, but some are able to complete the remaining, free-living part of the life cycle in the host's nest without additional food.

Other groups of astigmatid mites have become permanent parasites of birds or mammals, eliminating the deutonymph from the life cycle. Among mammal hosts, these mites are most diverse on marsupials, rodents, insectivores, primates, and bats, with relatively few occurring on carnivores or ungulates. Most are relatively nonpathogenic, feeding primarily on sebaceous materials on the hair shafts. Others, however, can cause problems for their hosts.

Species in the family Psoroptidae live on the host's skin or in the ears and feed by abrading the skin with their chelicerae and imbibing tissue fluids. These mites irritate the skin and cause itching. Several species of psoroptid mites occur on domestic animals, notably the carnivore ear mite, *Otodectes cynotis,* common in cats and dogs, and the scab mites in the genera *Psoroptes* and *Chorioptes* in horses, cattle, sheep, and others. The sheep scab mite, *Psoroptes ovis,* particularly causes economic damage by causing loss of wool.

Probably most important among parasitic astigmatid mites are species in the family Sarcoptidae. Commonly known as mange mites, species in several genera can parasitize humans and domestic animals. Naturally most diverse on marsupials, bats, primates, and rodents, several species have been able to colonize new hosts. *Sarcoptes scabiei* is ancestrally a parasite of humans, causing the skin disease scabies. Like other sarcoptids, these mites burrow into the superficial layers of the skin. In healthy humans, this disease is an itchy annoyance, but in immune-compromised individuals, a serious condition known as crusted scabies can develop in which the patient may harbor millions of mites in large, crusty lesions all over the body. *S. scabiei* has also been able to colonize many domestic animals, notably dogs, pigs, cattle, camels, and others, in which the disease known as sarcoptic mange can be fatal due to the large mite populations and secondary bacterial infections.

Many other astigmatid mites parasitize birds, in which, again, most do not cause harm. Feather mites may be very diverse on an individual bird, with one parrot species known to harbor almost 40 species. Like their fur mite counterparts on mammals, these mites feed on skin oils and do not harm the host. Others, however, may parasitize the feather follicles, skin, or respiratory tract. Skin-inhabiting species in the family Knemidokoptidae can be quite pathogenic in domestic poultry, cage birds, and wild species. Endoparasitic species in the family Cytoditidae live in the air sacs and can cause respiratory distress in poultry.

IMPORTANCE OF MITES

As indicated above, there are a number of instances in which mites are important to humans. Many species are serious pests of agricultural crops, either through direct damage or indirectly as vectors of plant pathogens. Other species are parasitic on domestic animals and cause losses in meat, egg, and fiber production. Others, such as the human scabies mite, are direct agents of human disease or, as in the case of chiggers and ticks, vectors of pathogens. Other mites may affect humans by infesting stored food products. Many species of Astigmata are known as stored-product mites because they have moved from their ancestral rodent nest habitats into human food stores. Such mites may also cause damage in animal feed by causing allergic reactions in livestock and are also known to cause skin irritation in humans handling infested materials. A related group of astigmatid mites, also ancestrally nest inhabiting, is the family Pyroglyphidae. These mites have colonized human habitations from bird nests and are the primary source for allergens in house dust. Commonly known as "house dust mites," species particularly in the genus *Dermatophagoides* produce many proteins that induce allergic responses in sensitive individuals. House dust allergy may take the form of respiratory distress or skin irritation. Mites typically inhabit beds, chairs, and carpets in houses, and their shed skins and feces provide the bulk of the allergens in house dust extracts.

On the other hand, as indicated above, some mites are beneficial to humans in their role as biological control agents against agricultural pests. Also, the natural role of mites in providing "ecosystem services" in the form of nutrient cycling cannot be overlooked.

See Also the Following Articles
Arthropoda and Related Groups • Chiggers and Other Disease-Causing Mites • Medical Entomology • Neosomy • Plant Diseases and Insects • Predation • Ticks

Further Reading
Evans, G. O. (1992). "Principles of Acarology." CABI, Wallingford, UK.
Krantz, G. W. (1978). "A Manual of Acarology," 2nd ed. Oregon State University Press, Corvallis.
Walter, D. E., and Proctor, H. C. (1999). "Mites: Ecology, Evolution and Behaviour." CABI, Wallingford, UK.
Woolley, T. A. (1988). "Acarology: Mites and Human Welfare." Wiley, New York.

Molting

Lynn M. Riddiford
University of Washington

Molting is the process of producing a new cuticle and the subsequent shedding (or ecdysis) of the old cuticle. This molt is orchestrated by a series of hormones so that it can be triggered by both internal and external cues.

THE MOLTING PROCESS
Cuticle Production

The cuticle is the outer covering of the insect and is its exoskeleton to which the muscles are attached (Fig. 1). The outermost layer is called the epicuticle; under this is the exocuticle followed by the endocuticle. In some systems, the exo- and endocuticle are classed together as the procuticle. In some insects, only the epi- and exocuticle are deposited before ecdysis, with the endocuticle following ecdysis, whereas in others, some endocuticle may be deposited before ecdysis. The epicuticle is composed of only protein, whereas the exo- and endocuticle contain both chitin and protein in varying proportions depending on the type of cuticle, i.e., whether rigid or flexible. Chitin is a polymer of *N*-acetylglucosamine and can be cross-linked to the protein components of the cuticle in a process called sclerotization or hardening, which usually occurs in the exocuticle just after the shedding of the old cuticle and expansion of the new cuticle. After sclerotization the insect is able to move, feed, fly, etc. The rigid parts of the cuticle are then set and cannot be expanded, whereas flexible cuticle may expand either by a simple unfolding of the new epicuticle or in response to hormonal signals. When the epicuticle has completely unfolded, further expansion is impossible and the larva must molt in order to grow further. Molting is also necessary at the end of larval life for metamorphosis.

The epidermis is a single cell layer that produces the cuticle that lies above it (Fig. 1). During the intermolt period, the epidermis actively deposits lamellate endocuticle, especially in those regions where the cuticle is extensible. The chitin and protein are secreted as plaques at the tips of the

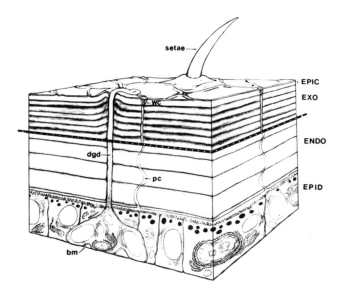

FIGURE 1 Diagram of the relationship of the epidermis (EPID) to the overlying cuticle that it produces. EPIC, epicuticle; EXO, exocuticle; ENDO, endocuticle; bm, basement membrane; dgd, dermal gland (Verson's gland) duct; pc, pore canal; setae, cuticular hair; wc, wax canal. [From Hadley, N. (1982). *J. Exp. Zool.* **222**, 239–248, Copyright © 1982. Reprinted by permission of Wiley-Liss, Inc., a subsidiary of John Wiley & Sons, Inc.]

microvilli at the apical surface of the epidermal cells. Above the plaques in the extracellular space, the cuticle arises by self-assembly of the chitin microfibrils and the secreted proteins. As the larva grows, the epidermal cells underneath the extensible cuticle also grow. During this cell growth, there is DNA synthesis and the epidermal cells may become polyploid (i.e., >2*n*). In the case of a soft-bodied insect such as the larva of the tobacco hornworm, *Manduca sexta,* the epicuticle is deposited in folds to allow for growth during the succeeding intermolt period during which time the epicuticle then unfolds. In this instance, the underlying endocuticle expands via a series of apical expansion points created by the deposition by the epidermal cell of vertical chitin microfilaments during the first day after ecdysis. To accommodate these two processes that must occur at the same time, an unstructured layer is deposited between the epicuticle and the lamellate procuticle.

At the onset of the molt, the epidermal cells detach from the overlying cuticle and go through a burst of RNA synthesis including that of ribosomal RNA. At this time, the cells (including those that are polyploid) may divide. Endocuticle synthesis ceases and is followed by secretion of inactive proteolytic and chitinolytic enzymes, and other proteins, to form the "molting gel" that fills the space between the old cuticle and the apical border of the epidermis. Later activation of these enzymes at the end of the molt leads to digestion of the old endocuticle. Other cellular events preparatory to the deposition of the new cuticle, such as cellular shape changes that prefigure the surface conformation, also occur. Then, cuticulin is deposited first at the tips of the plasma membrane plaques followed by deposition between the plaques to form a complete layer. Under this layer the epicuticle precursors (lipid, protein, and polyphenols) are secreted and self-assemble on the inner face of the cuticulin layer. The whole structure is then stabilized by the action of phenoloxidases that cross-link the polyphenols and the proteins. Subsequently, the apical membrane of the epidermal cells withdraws from the patterned surface and begins to form the procuticle, beginning with the exocuticle.

Digestion and Ecdysis of the Old Cuticle

During most of the formation of the new cuticle, the old cuticle remains intact and the muscles remain attached to this cuticle to allow the insect to move. At the end of the molt shortly before ecdysis, specific proteases are secreted into the molting gel to clip off a part of the inactive chitinases and proteases to render these enzymes active. These enzymes work together to digest both the protein and the chitin in the old endocuticle down to its component amino acids and *N*-acetylglucosamine sugars. This molting fluid then is resorbed into the hemolymph for its components to be recycled for production of the next cuticle or for other uses. Resorption is thought to occur in one of two ways, either back through the new cuticle and the epidermis or through the gut via swallowing and uptake in the hindgut.

Near the end of molting fluid resorption, the insect begins the process of shedding the old cuticle or ecdysis. This shedding occurs in a stereotyped sequence of behaviors. The preecdysis behavior is characterized by a series of coordinated movements that serve to loosen the muscle attachments to the old cuticle. This phase is followed by ecdysis behavior itself, which often is a series of peristaltic waves that travel from posterior to anterior and cause the animal to rupture the old cuticle anteriorly and to escape headfirst. The cuticle opens at ecdysial sutures that are areas of the old cuticle lacking exocuticle so that all but the epicuticle has been digested. In insects with rigid head capsules such as lepidopteran caterpillars, the head capsule has slipped down over the forming mandibles early in the molt to allow the formation of a larger head capsule. At the time of ecdysis, the old head capsule separates from the remainder of the old cuticle and falls off as the new larva walks out of its old cuticle.

At the time of ecdysis, a waterproofing cement layer is deposited on top of the epicuticle by the secretion of dermal glands known as Verson's glands (Fig. 1). This layer is spread over the surface by the movements of the animal under the old cuticle as it sheds its old cuticle. In some cases, a waxy layer is secreted on top of this layer in the first few days after ecdysis for further prevention of desiccation. This secretion occurs through the pore canals that traverse the cuticle from the epidermal cell to the surface of the cuticle (Fig. 1).

Postecdysial Expansion and Sclerotization

After ecdysis the animal fills its tracheae with air and also swallows air in order to expand the new larger cuticle. When it attains its final size, the new cuticle hardens and may also darken (tan) to varying degrees depending on whether the cuticle is to be flexible or rigid. In many insects there is preecdysial tanning and hardening of certain key structures such as the mandibles or the crochets on the abdominal prolegs of caterpillars used for grasping.

Sclerotization is the process of hardening the exocuticle by cross-linking the proteins together and the proteins with chitin to form a stabilized structure suitable for an exoskeleton that anchors the muscles to allow movement. The primary cross-linking agents are *N*-acetyldopamine and *N*-β-alanyldopamine. The latter is found in tan cuticles such as those of many lepidopteran pupae. Both compounds are derived from the amino acid tyrosine through a series of enzymatic steps of which the key enzymes are phenoloxidase for conversion of tyrosine to dopa and dopa decarboxylase for conversion of dopa to dopamine.

HORMONAL CONTROL OF MOLTING

Prothoracicotropic Hormone and Ecdysone

The molt is initiated by the release of prothoracicotropic hormone (PTTH), a neuropeptide, from the brain (Fig. 2). In

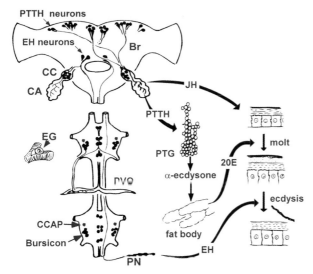

FIGURE 2 Diagram of the insect neuroendocrine system that is involved in molting. Br, brain; CA, corpora allata; CC, corpora cardiaca; CCAP, crustacean cardioactive hormone; EG, epitracheal glands; EH, eclosion hormone; JH, juvenile hormone; PN, proctodeal nerve; PTG, prothoracic gland; PTTH, prothoracicotropic hormone; PVO, perivisceral organ; 20E, 20-hydroxyecdysone.

Lepidoptera there are only two pairs of neurosecretory cells that make and release PTTH (Fig. 2). These cells send their axons out to the corpora cardiaca (or corpora allata in Lepidoptera) where they terminate and store the PTTH in vesicles until either internal or external environmental signals stimulate the cells to fire action potentials that cause release of their vesicular contents. In the blood-sucking bug *Rhodnius prolixus*, Vincent B. Wigglesworth in his classic studies of the insect endocrine system showed that the blood meal is the signal for PTTH release. Distention of the abdomen by the blood is relayed to the brain by stretch receptors. In many larvae such as the large milkweed bug, *Oncopeltus fasciatus*, and *M. sexta*, size determines the time of PTTH release although the sensory pathway utilized has not been determined. In *Manduca*, the time of day at which the larva attains critical size is also important because PTTH can be released only during a particular time period during the night called a "gate." Thus, if the larva attains critical size after the gate closes, PTTH is not released until the gate opens the following night. During the intervening time the larva continues to feed and grow and so will be larger than one that attained critical size during the open gate period and immediately released PTTH. Thus, the brain can integrate the sensory input from both the internal and the external environments and direct the time of molting.

PTTH acts on the prothoracic glands to cause the synthesis and secretion of the steroid hormone α-ecdysone into the hemolymph. α-Ecdysone (E) is converted to 20-hydroxyecdysone (20E) (Fig. 2) by the fat body, the Malpighian tubules, and sometimes other tissues. During the rise of the ecdysteroids for molting, one observes that E appears first in the hemolymph (blood) followed by 20E and later 20,26-dihydroxyecdysone (20,26E). 20E is the main component at the peak of the titer, and 20,26E is the primary ecdysteroid at the end of the molt.

Both E and 20E play a role in molting, with E being important in early cellular changes in the initiation of the molt, such as proliferation, and 20E necessary for the differentiative changes. Experiments with tissues in culture show that low levels of 20E can mimic the early predifferentiative changes caused by E, whereas high levels of E are unable to cause the differentiative events caused by 20E. Although high 20E initiates the deposition of the cuticulin and epicuticle, procuticle consisting of chitin and protein can be deposited only when the 20E levels in the hemolymph have declined. Experiments both with imaginal discs (precursors of adult structures found in larvae of moths and flies) and with *Manduca* abdominal epidermis in culture have shown that in the continuous presence of 20E, neither chitin nor the exo- and endocuticular proteins are deposited, although the epicuticle can be. To form a new cuticle *in vitro*, tissues taken during the intermolt period must first be exposed to 20E for a time commensurate with their exposure during the molt, and then they must be transferred to hormone-free medium. Although 20,26E is present for a long time during the decline of ecdysteroids at the end of the molt, so far no role has been found for it either in the timing of events that occur at the end of the molt, such as the production of new endocuticle or of dopamine, or in any aspect of ecdysis behavior. Therefore, at the present time, 20,26E is still considered to be an inactive metabolite of 20E.

Juvenile Hormone

Juvenile hormone (JH) is a sesquiterpenoid produced by the corpora allata (Fig. 2) and is present throughout nymphal and larval life in all insects. Its primary action is to prevent metamorphosis in response to ecdysone at the time of the molt as first demonstrated by V. B. Wigglesworth with *Rhodnius* and Carroll Williams in the wild silk moth *Hyalophora cecropia*. Consequently, Williams called it the "status quo" hormone.

At the beginning of or during the final nymphal or larval stage, the corpora allata cease production of JH, and JH in the hemolymph declines to undetectable levels. Then when ecdysone next rises, it causes metamorphosis. In most insects metamorphosis of the epidermis consists of switches in developmental programs from nymph to adult or from larva to pupa to adult. This switch is best understood in the epidermis of *Manduca* in which a combination of *in vivo* and *in vitro* experiments by Lynn Riddiford and her colleagues has shown that 20E acts directly on the epidermal cells to cause them to become pupally committed and that JH prevents this switch. The pupally committed epidermis no longer can form a larval cuticle, but can form a pupal cuticle only during a larval molt (as assayed by implantation into a penultimate stage larva). A similar critical ecdysone-induced switch to adult commitment is seen at the onset of the adult molt of the nymph or of the pupa.

In *Drosophila* and other higher flies, there is massive cell death of larval tissues at metamorphosis except for the nervous system and the Malpighian tubules. The imaginal discs that have only proliferated during larval life then take over and form the pupal and adult structures. The larval abdominal epidermis, however, makes the pupal cuticle while the abdominal histoblasts proliferate during the formation of the pupa. These new cells then spread over the abdomen, displacing the larval cells (which die) during the adult molt and producing the adult abdominal cuticle. JH is present during larval life in these flies and presumably suppresses precocious metamorphosis, but cannot prevent the onset of differentiation of the discs in response to ecdysone at the end of the final larval instar. By contrast, JH given at this time can prevent the metamorphosis of the abdominal histoblasts so that a fly is formed with a normal head and thorax but a pupal-like abdomen.

Molecular Basis of the Action of Ecdysone and Juvenile Hormone in Molting

The molecular basis of the action of ecdysone was first studied using the giant polytene chromosomes of the salivary glands of flies, namely the midge *Chironomus tentans* and the fruit fly *Drosophila melanogaster*. During larval life the gland cells enlarge and the chromosomes replicate but the DNA strands do not separate. Consequently, by the final larval stage the chromosomes are readily visible in nuclear squashes and show bulges at specific locations known as "puffs." These puffs appear and disappear in a dynamic fashion and are sites of messenger RNA (mRNA) transcription from specific genes. When these glands are exposed to ecdysone, a few new puffs appear within 15 to 30 min ("early" puffs) followed by their regression and a second series of puffs in 3 to 4 h ("late" puffs). The early puffs appear in response to ecdysone even when protein synthesis is prevented, whereas the late puffs do not. In the early 1960s these findings led Peter Karlson to suggest that ecdysone acted directly on genes to regulate their activity, a hypothesis that has since been proven true for all steroid hormones. Based on the precise timing of the effects on puffing seen after the addition of 20E to the salivary glands, Michael Ashburner suggested in 1974 that ecdysone acted by directly activating the early genes that produced mRNA to make proteins that in turn activated the late genes and inhibited the early genes. This "Ashburner cascade" was essentially confirmed in the 1990s when the genes involved were isolated and their products identified. The following is the modern version of ecdysone action.

Ecdysone enters the cell and goes to the nucleus where it combines with the ecdysone receptor (EcR), a protein in the nuclear receptor superfamily that has the characteristic structure seen in Fig. 3 (top). The DNA binding domain (C in Fig. 3) consists of 66 amino acids and contains two cysteine–cysteine "zinc fingers" by which the zinc is held coordinately by the four cysteines. The first zinc finger is

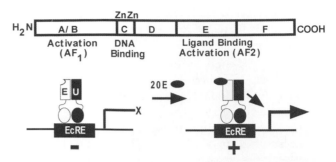

FIGURE 3 Diagram of the modular nature of the nuclear receptor molecule based on the structure of the ecdysone receptor (top) and of the heterodimeric complex of the ecdysone receptor (E) and ultraspiracle (U) on an ecdysone response element (EcRE) in the promoter of an ecdysone-activated gene in the absence (–) and the presence (+) of the active hormone 20-hydroxyecdysone (20E) (bottom). See text for details. AF1 and AF2, activation domains that interact with other proteins to cause activation of transcription when the hormone ligand is bound. (Modified, with permission, from Truman and Riddiford, 2002.)

involved in binding to the DNA, whereas the second is necessary for stabilizing that binding by protein–protein interactions. The ligand-binding domain (E in Fig. 3) has a pocket that binds the ecdysone and on so doing the molecule changes its shape and may interact with coactivator proteins. The N-terminal A/B domain (Fig. 3) is also important for interactions with other proteins that are necessary to activate a gene. Normally the ecdysone receptor forms a heterodimer with ultraspiracle (USP), another member of the nuclear receptor family. This heterodimer binds to the DNA at a particular sequence called the ecdysone-response element in the gene "promoter" (a region usually but not always upstream of the gene that is necessary to turn the gene on and off) (Fig. 3, bottom). In the absence of hormone, the EcR/USP heterodimer keeps the gene suppressed; then when ecdysone appears and binds to EcR, the gene may be activated.

Most of the early genes activated by ecdysone encode transcription factors, proteins that bind to DNA in promoter regions and may activate or inactivate transcription of those genes. In the ecdysone cascade, these transcription factors activate the late genes, which are either general or tissue-specific genes that are involved in the molting process for a particular tissue. They also inactivate the early genes and the intermolt genes such as those for endocuticle synthesis in the epidermis. In addition, these early transcription factors are important in the regulation of the so-called "delayed early" genes that encode transcription factors and are activated by ecdysone but require protein synthesis for their full activation. All of these early and delayed early factors require continuous exposure to ecdysone for their continued transcription, but may be present for differing periods of time depending on the dynamics of their inhibition by the various ecdysone-induced factors. The few early genes that do not encode transcription factors instead encode proteins that are likely critical to the molting process, such as a calcium-binding protein, an ATP-binding cassette membrane transporter, and a protein in imaginal discs important for the cell shape changes that occur at metamorphosis.

The coordination of events such as endocuticle synthesis and later dopamine production that occur during the decline of the ecdysteroid titer also is dependent on ecdysone-regulated transcription factors. These factors appear only after exposure to 20E followed by its removal. One can show that the timing of this appearance varies among the factors in a sequence similar to that seen in the animal and is presumably dependent on the levels of both 20E and the various inhibitory transcription factors that 20E has induced.

JH has been found to inhibit the appearance of a set of ecdysone-induced transcription factors named the broad complex (BR-C) in the epidermis during larval molts. In both *Manduca* and *Drosophila* the BR-C factors appear in this tissue at the time of pupal commitment and are also present during pupal cuticle synthesis. In *Drosophila* mutants that lack the *broad* gene, larvae develop to the final larval stage but cannot metamorphose because the BR-C factors are active in various tissues at this time. The BR-C factors are members of a family of chromatin-associated transcription factors. In *Drosophila* salivary glands at the onset of metamorphosis, they are associated with both the switching off of a larval-specific gene and the switching on of some of the glue protein genes. Because of the close correlation between the appearance of BR-C and the pupal commitment of *Manduca* epidermis in response to 20E in the absence of JH, the inhibition of BR-C transcription by JH may be one of the key roles of JH in preventing epidermal metamorphosis. The molecular mechanism of JH action in preventing the 20E induction of BR-C or any of the other switching actions of 20E is not yet known.

Hormonal Control of Ecdysis: ETH, EH, and CCAP

The molt culminates in the shedding of the old cuticle during ecdysis, which is followed by the expansion of the new cuticle and then its hardening and often darkening or tanning. A cascade of small peptide hormones that are released after the new cuticle is formed and the ecdysone titer has declined below a threshold level initiates ecdysis. This cascade has been studied in detail by James Truman and co-workers and by Dushan Zitnan. The precise nature of the signal that initiates this cascade is not yet understood, but at a certain time near or at the end of molting fluid resorption, the epitracheal glands just below the spiracles (the openings of the tracheae to the outside air) release a small peptide, ecdysis-triggering hormone (ETH) (Figs. 2 and 4). ETH enters the central nervous system (CNS) and initiates a sequence of behavior called the preecdysis behavior that serves to loosen the muscle attachments from the old cuticle. ETH also acts on a set of neurosecretory cells in the ventromedial region of the brain to cause the release of eclosion hormone (EH) both into the ventral nervous system and into the hemolymph from their endings in the proctodeal nerve along the hindgut (Fig. 4). In the ventral

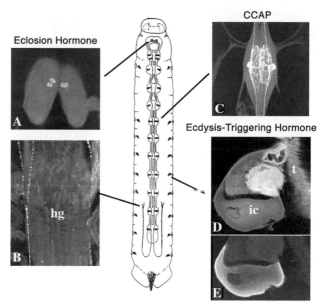

FIGURE 4 Diagram of the neuroendocrine system controlling ecdysis in the tobacco hornworm larva, *M. sexta,* with pictures of the different cells producing the peptides involved. (A) Eclosion hormone (EH) neuron cell bodies in the brain. (B) EH stores in the proctodeal nerve along the hindgut (hg). (C) Neurons in the ventral nerve cord that contain crustacean cardioactive peptide (CCAP). (D) The epitracheal gland showing the Inka cell (ic) that contains ecdysis-triggering hormone (ETH) and the autofluorescent secretory duct complex. t, trachea. (E) ETH in the Inka cell. [Modified, with permission, from Ewer *et al.* (1997). *J. Exp. Biol.* **200,** 869–881, © 1997 by Company of Biologists Ltd.]

nervous system, EH activates a network of neurons that contain crustacean cardioactive peptide (CCAP) (so named because it was first isolated from the crab in which it increases the beating of the heart) (Fig. 4). In response to EH release within the CNS, CCAP is released both into the CNS and into the hemolymph via the neurons' endings in the perivisceral organ (PVO) (Fig. 2). The action of CCAP in the CNS is to trigger the ecdysis behavior that allows the animal to shed its old cuticle. Some of these neurons also apparently contain and release bursicon, the tanning hormone (see below), into the hemolymph at the same time as they release CCAP.

In moths and flies, the sequences of the preecdysis and ecdysis behaviors are found to be stereotyped programs that once initiated by the hormone in question run their course. This behavior can even be seen in the isolated nervous system in the absence of sensory feedback, indicating a preprogrammed network that has only to be triggered by an external signal. The absence of ecdysone is necessary for this system to function since ecdysone suppresses the release of ETH from the epitracheal glands, thereby preventing this cascade of peptides. Ecdysone also decreases the excitability of the EH cells.

EH release also is controlled by photoperiod in many species so that its release, like that of PTTH, can occur only during a certain gate during the day. A good example of this control was shown by the classical studies of Truman on the

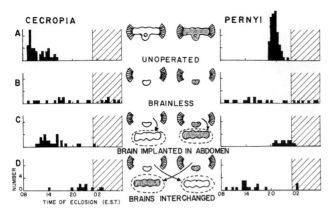

FIGURE 5 Role of the brain in causing timed eclosion of the two wild silk moths, *A. pernyi* (pernyi) and *H. cecropia* (cecropia). See text for details. [Reproduced from Truman (1972), with permission of PUDOC Press, Wageningen, The Netherlands.]

eclosion (adult ecdysis) of two species of wild silk moth, *H. cecropia* and *Antheraea pernyi* (Fig. 5). *H. cecropia* normally emerges in the morning and *A. pernyi* emerges in the late afternoon. When the brain was removed, both species emerged randomly throughout the day and night and showed very uncoordinated ecdysis behavior. When the brains were switched between the species, *H. cecropia* containing implanted *A. pernyi* brains emerged late in the day and *A. pernyi* with *H. cecropia* brains emerged in the morning in a coordinated manner. Thus, EH release from the implanted brain occurs at the time of day dictated by the donor brain, indicating that both detection of the photoperiod signals and the clock that determines the timing of EH release are located in the brain. Also, these studies indicated that EH is necessary for the coordination of ecdysis behavior. Specific destruction of the EH-releasing cells in *Drosophila* brains in the embryo and early larva resulted in about half the animals being unable to complete the larval molts, whereas the remaining flies emerged in an uncoordinated manner, again indicating that EH is necessary for behavioral coordination.

In the hemolymph EH acts back on the epitracheal glands in a positive feedback to cause further release of ETH so that within a few minutes, both the ETH and the EH cells are depleted of their stored peptides. EH also causes the filling of the new tracheae with air and the Verson's glands to release their waterproofing products over the surface of the animal as the insect is shedding its old cuticle.

Hormonal Control of Cuticular Expansion and Hardening: Bursicon, CCAP

After ecdysis the insect expands the new cuticle, then the cuticle hardens. Both CCAP and bursicon released from the PVOs into the hemolymph are involved. CCAP stimulates an increase in heart rate that is associated with the expansion of the new cuticle.

Bursicon is thought to initiate both of the above processes. For example, *Rickets* mutants in *Drosophila* lack the putative bursicon receptor ecdyse but show no wing expansion or other expansion behaviors and do not harden their cuticle. Bursicon is a large protein that is synthesized by various neurosecretory cells within the CNS (Fig. 2) but has not yet been chemically characterized. It is thought to act via cyclic AMP in the epidermal cell to trigger the cross-linking activities of *N*-acetyldopamine and *N*-β-alanyldopamine that cause sclerotization and tanning. Its plasticizing role in cuticular expansion has not been well studied.

See Also the Following Articles

Chitin • Cuticle • Development, Hormonal Control of • Ecdysteroids • Exoskeleton • Hemolymph • Juvenile Hormone • Metamorphosis

Further Reading

Bayer, C., von Kalm, L., and Fristrom, J. W. (1996). Gene regulation in imaginal disc and salivary gland development during *Drosophila* metamorphosis. *In* "Metamorphosis: Postembryonic Reprogramming of Gene Expression in Amphibian and Insect Cells" (L. I. Gilbert, J. R. Tata, and B. G. Atkinson, eds.), pp. 321–361. Academic Press, San Diego.

Ewer, J., Gammie, S. C., and Truman, J. W. (1997). Control of insect ecdysis by a positive-feedback endocrine system: Roles of eclosion hormone and ecdysis triggering hormone. *J. Exp. Biol.* **200,** 869–881.

Gammie, S., and Truman, J. W. (1999). Eclosion hormone provides a link between ecdysis-triggering hormone and crustacean cardioactive peptide in the neuroendocrine cascade that controls ecdysis behavior. *J. Exp. Biol.* **202,** 343–352.

Hadley, N. F. (1982). Cuticle ultrastructure with respect to the lipid waterproofing barrier. *J. Exp. Zool.* **222,** 239–248.

Henrich, V. C., Rybczynski, R., and Gilbert, L. I. (1999). Peptide hormones, steroid hormones, and puffs: Mechanisms and models in insect development. *Vit. Horm.* **55,** 73–125.

Hopkins, T. L., and Kramer, K. J. (1992). Insect cuticle sclerotization. *Annu. Rev. Entomol.* **37,** 273–302.

Locke, M. (1998). Epidermis. *In* "Insecta," Vol. 11A of "Microscopic Anatomy of Invertebrates," pp. 75–138. Wiley–Liss, New York.

Nijhout, H. F. (1994). "Insect Hormones." Princeton University Press, Princeton, NJ.

Reynolds, S. E., and Samuels, R. I. (1996). Physiology and biochemistry of insect moulting fluid. *Adv. Insect Physiol.* **26,** 157–232.

Riddiford, L. M. (1994). Cellular and molecular actions of juvenile hormone. I. General considerations and premetamorphic actions. *Adv. Insect Physiol.* **24,** 213–274.

Riddiford, L. M., Hiruma, K., Lan, Q., and Zhou, B. (1999). Regulation and role of nuclear hormone receptors during larval molting and metamorphosis of Lepidoptera. *Am. Zool.* **39,** 736–746.

Truman, J. W. (1972). Circadian rhythms and physiology with special reference to neuroendocrine processes in insects. *In* "Proceedings: International Symposium on Circadian Rhythmicity," pp. 111–135. PUDOC Press, Wageningen.

Truman, J. W., and Riddiford, L. M. (2002). Insect developmental hormones and their mechanism of action. *In* "Hormones, Brain and Behavior" (D. Pfaff, A. Arnold, A. Etgen, S. Fahrbach, and R. Rubin, eds.), Vol. 2, pp. 841–873. Academic Press, San Diego.

Zhou, B., and Riddiford, L. M. (2001). Hormonal regulation and patterning of the broad complex in the epidermis and wing discs of the tobacco hornworm, *Manduca sexta. Dev. Biol.* **231,** 125–137.

Zitnan, D., Kingan, T. G., Hermesman, J. L., and Adams, M. E. (1996). Identification of ecdysis triggering hormone from an epitracheal endocrine system. *Science* **271,** 88–91.

Monarchs

Lincoln P. Brower

Sweet Briar College

The monarch butterfly (*Danaus plexippus,* Nymphalidae) belongs to the tropical subfamily Danainae, the members of which are called milkweed butterflies because their larval host plants occur mainly in the milkweed family, Asclepiadaceae. With the exception of the monarch, most of the 157 known Danainae species are limited to tropical regions in Malaysia, Africa, South America, and the Greater Antilles. The adults of several species exhibit both short-distance migrations during the dry and wet seasons and social clustering behavior. This suggests that the long-distance migration and overwintering-aggregation behavior of the monarch butterfly in North America was evolutionarily elaborated from an ancient (plesiomorphic) character of the taxon.

During the late Cenozoic, the milkweed genus *Asclepias* underwent an adaptive radiation that produced 108 species in temperate North America, ranging from Mexico to the boreal forests of Canada and from the Atlantic to the Pacific coasts. As the climate alternated between hot and cold periods during the Pleistocene, the North American flora periodically advanced and retreated. The author hypothesizes that the monarch tracked the geographic expansion and retraction of its milkweed hosts and in the process refined its inherited ancestral ability to move between habitats. Because the monarch cannot tolerate temperate zone winters, natural selection would have favored those individuals that moved southward as summer waned. As time passed, the migration syndrome gradually evolved to become increasingly sophisticated, ending in the present round trip migration, one of the most complex in the animal kingdom (Figs. 1A and 1B).

UNPALATABILITY AND MIMICRY

C. V. Riley, the most famous 19th century entomologist in North America, proposed in 1871 that the monarch was distasteful and advertised its unpalatability with its conspicuous behavior and its spotted pattern of bright orange, black, and white. Time has proven him correct and a remarkable coevolutionary interaction of the monarch with the North American *Asclepias* species resulted in great refinements of the

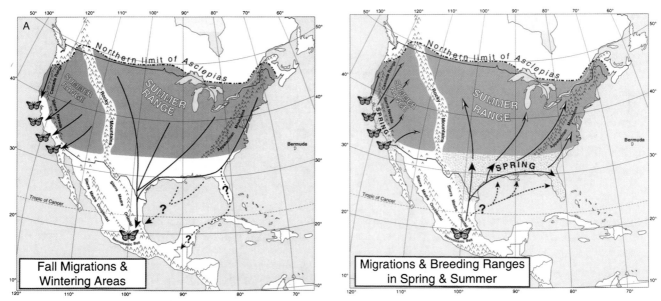

FIGURE 1 (A) Two migratory populations of the monarch butterfly occur in North America. The western population breeds west of the Rocky Mountains during the spring and summer and migrates to numerous overwintering sites, mainly along the California coast. The second, much larger eastern population breeds over several generations east of the Rocky Mountains and in the autumn migrates southward to overwintering sites in the high peaks of the Transverse Neovolcanic Belt, south of the Tropic of Cancer in central Mexico. Migration across the Gulf of Mexico and through Florida and to Guatemala remains hypothetical. (Reproduced, from Brower, 1995, with permission of the Lepidopterists' Society). (B) Spring remigrations of the monarch butterfly in North America. Western monarchs leave the coastal overwintering areas in early spring and reestablish their summer breeding range as shown. Monarchs that overwintered in Mexico remigrate at the end of March to the Gulf Coast states, where they oviposit on southern milkweeds *(Asclepias)* and produce the first new spring generation of adults by the end of April to early May. These butterflies migrate northeasterly across the midwestern states to southern Canada, laying eggs along the way and establishing a large second generation in the western and central Great Lakes region. The midwestern component of the second-generation monarchs is produced in June and they appear to continue the migration eastward over the Appalachians. One or two more summer generations (depending on temperature) follow in the Midwest and east of the Appalachians, with the last generation entering reproductive diapause and migrating southward in the autumn. Spring remigrations over the Gulf of Mexico and through Cuba and Florida remain hypothetical. (Reproduced from Brower, 1995, with permission of the Lepidopterists' Society.)

FIGURE 2 (a) A blue jay eats a monarch butterfly that contains the emetic heart poisons that its larva had sequestered from a milkweed plant. (b) About 15 min later, the jay sickens and vomits. One unpleasant experience is sufficient for most jays to avoid all further monarchs on sight. (Photographs by L. P. Brower.)

monarch's chemical defense. Milkweeds synthesize differing arrays and amounts of vertebrate heart poisons, known as cardenolides. These are bitter-tasting chemicals that cause severe vomiting when ingested. Monarch larvae are insensitive to these molecules and, as they feed on the milkweeds, they sequester and store them in their bodies. The poisons are passed on into the chrysalids and then to the adults in sufficient amounts to sicken vertebrate predators, especially birds and mice. Laboratory experiments with blue jays (*Cyanocitta cristata*) have shown that some monarchs are so toxic that once a bird has eaten one, the noxious experience is so intense that the bird not only refuses monarchs in future

encounters, but may actually retch at the sight of another (Figs. 2A and 2B).

Riley also proposed that the unrelated viceroy butterfly (*Limenitis archippus*) had evolved through natural selection to mimic the color pattern of the monarch. The viceroy was originally considered palatable and a so-called Batesian mimic, but recent studies suggest that it is also unpalatable. Therefore the monarch and viceroy have most likely converged on a common warning color pattern and are Müllerian mimics. However, the situation is more complex because some milkweeds lack cardenolides. As a result, monarch larvae that feed on nontoxic milkweeds produce palatable butterflies, whereas those that feed on toxic species produce unpalatables. This discovery gave rise to the concept of automimicry, in which palatable members of a species are identical in appearance to, but are protected by their exact resemblance to, the unpalatable ones. Thus in the wild, monarchs exhibit a "palatability spectrum," with the result that mixed populations of monarchs and viceroys may simultaneously exhibit Batesian mimicry, Müllerian mimicry, and automimicry.

THE EASTERN AND WESTERN MIGRATORY POPULATIONS

Monarchs that breed west of the Rocky Mountains have been known since the mid-19th century to migrate during the fall from their breeding areas to numerous overwintering sites along the coast of California, the most famous of which is in Pacific Grove on the Monterey Peninsula. Almost certainly attracted to each other by visual and pheromone signals, the

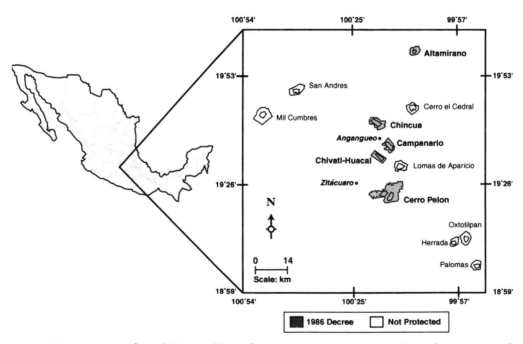

FIGURE 3 Locations of 12 mountain massifs on which monarch butterfly overwintering occurs in the oyamel–pine forest ecosystem of central Mexico, about 120 km west of Mexico City. The total area shown is approximately 11,700 km². The five shaded polygons are overwintering areas protected by presidential decree in 1986 (16,100 ha). The unshaded polygons are seven additional overwintering areas that currently are not protected. (From Brower *et al.*, 2002, with permission of Odyssey Press.)

butterflies aggregate from mid-October through February in spectacularly dense clusters on the live branches of Monterey pines, cypress, and eucalyptus. As spring advances, the surviving monarchs mate and migrate inland, where the females seek out the spring flush of milkweeds on which they lay singly up to 400 eggs before they die.

It was long suspected that the much larger eastern population of monarchs—those that breed during the summer in a 2.6 million-km^2 area east of the Rocky Mountains—migrates to Mexico. Fred and Norah Urquhart at the University of Toronto developed a wing-tagging program involving hundreds of collaborating amateurs that finally led one of their associates, Kenneth Brugger, to discover the first overwintering site in Mexico on January 2, 1975. Subsequent searching by Lincoln Brower, William Calvert, and their Mexican colleagues located overwintering sites on 12 separate mountain ranges within a 30 by 60-km area of central Mexico (Fig. 3). Unlike the coastal overwintering sites in California, the overwintering areas in Mexico all occur above 3200 m altitude in a coniferous oyamel fir–pine ecosystem. This is a very limited ecosystem in the Transverse Neovolcanic Belt of mountains that run across Mexico, just south of the Tropic of Cancer.

THE MEXICO OVERWINTERING PHENOMENON

The numbers of butterflies in the Mexican overwintering colonies are astoundingly large (Fig. 4). Early research, based on mark, release, and recapture studies in California, suggested that the butterfly densities in Mexico are 10 times greater and led to an estimate of 10 million monarchs per hectare (2.47 acres) of forest. However, a catastrophic winter storm in January 2002 killed so many butterflies (one sample had more than 50,000 dead butterflies in a single square meter!) that the revised density estimate is now at least 5 times this. In other words, the fir and pine trees are festooned with 50 million monarchs per hectare. So far the maximum combined area occupied by all known colonies is about 20 hectares, a total of at least 1 billion monarchs. The extraordinary beauty and mystery of the densely aggregated cluster of monarchs in these colonies is now well known. They represent one of the greatest biological wonders on this planet.

The extreme unpalatability of monarchs was probably one factor that allowed the species to elaborate the extremely dense winter aggregation behavior. Were they not chemically protected, the aggregations—an enormous potential food supply—would be exploited by vertebrate predators. In fact, field studies have shown that most, but not all species of birds and mice in the overwintering areas in both California and Mexico avoid eating monarchs. However, in Mexico, orioles and grosbeaks have broken through the chemical protection and killed an average of 15,000 monarchs per day in one colony, i.e., more than a million butterflies during the overwintering season. Again, the old adage holds that no protection, no matter how sophisticated, can ever be perfect.

FIGURE 4 Monarch butterflies festooning oyamel trees in the Sierra Chincua on a clear day, 1991. (Photograph by L. P. Brower.)

THE MULTIGENERATIONAL MIGRATION SYNDROME

Because various milkweed species fed upon by monarch larvae synthesize arrays of chemically distinct cardenolides, it is possible to extract the molecules from the butterflies and by thin-layer chromatography to obtain a chemical fingerprint that indicates which species of milkweed each monarch ate when it was a caterpillar. Using this technique, Brower and his colleagues determined that individual monarchs that have survived the winter in Mexico remigrate in the spring to the Gulf coastal states where they lay their eggs and then die. The ensuing new spring generation then continues the migration northward into the Great Lakes region and establishes the first summer generation.

By July, the first summer generation of monarchs disperses east to the Atlantic coast and west to the Rocky Mountains and produces at least one more generation of adults. By mid-August, shortening daylength and colder nights reduce juvenile hormone production in the final summer generation. This prevents gonadal maturation and the butterflies become gregarious and begin their fall migration to Mexico. As shown by Gibo, these migrants are adept at using thermal lift and tail winds. As dusk approaches, the butterflies drift down and aggregate on trees where they spend the night. When the winds blow from the south, the butterflies interrupt their migration and seek out fields of flowers and assiduously drink nectar. They convert the sugar in the nectar into lipid and store it in their abdominal fat bodies. By the time the migrants reach central Texas, they have increased their lipid content by 500%. This reserve is crucial both to sustain them over their 5 months' overwintering period and to fuel the subsequent 1 month's spring remigration.

UNRESOLVED QUESTIONS

The individual monarchs of the last summer generation are genetically programmed to perform this migration to central Mexico in the fall, to survive the winter, and then to remigrate to the southern United States in the spring. Thus adult monarchs born in the Toronto area in August and returning to central Texas the following April traverse a distance of more than 5000 km. Their ability to find their way through deserts and mountain passes, to compensate for wind drift, and finally to locate the very small areas in Mexico that they have never before encountered remains a mystery. It will probably be solved when satellites can monitor electronic tags placed on individual butterflies.

The migratory orientation of individual monarchs shifts from south in the fall to north in the spring. How they maintain a particular course is poorly understood. While recent evidence casts doubt on the possibility that they may use magnetic orientation, there is strong evidence that sun compass orientation is involved. A complementary hypothesis is that monarch individuals have an internal clock that ticks away in all life stages and shifts the potential angle of the migration direction 1° per day throughout the year. Thus, at the spring equinox, the monarchs head out of their Mexican overwintering areas on a due north course (0°). The new spring generation, about 45 days later, heads northeastward toward the Great Lakes (45°), and the next generation that is produced about 90 days after the spring equinox heads due east (90°). By the fall equinox (September 21), their heading would be 180°, i.e., due south, changing to southwesterly as they migrate southward and finally reach the overwintering areas in November and December (Fig. 5).

Another unresolved question is the degree to which the eastern and western North American migratory populations are geographically isolated from each other. Molecular evidence suggests little differentiation, and it is possible that the western population is derived from and ultimately dependent on monarchs that get displaced westward by occasional strong northwesterly winds during the spring remigration from Mexico. Until the natural interchange between the eastern and the western populations is better understood, it seems prudent to avoid experimental and frivolous commercial transfers between them.

MIGRATION AND OVERWINTERING: ENDANGERED BIOLOGICAL PHENOMENA

The overwintering monarchs in Mexico are highly adapted to the oyamel fir–pine forest ecosystem. The forest provides a microclimatic envelope that protects the butterflies during

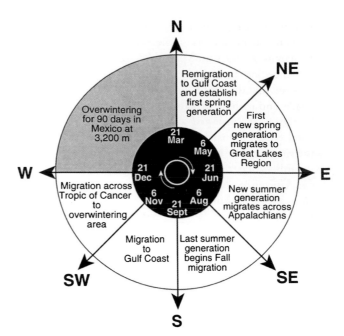

FIGURE 5 The rotational orientation hypothesis holds that all generations of monarch butterflies are migratory and that their orientation shifts clockwise at a rate of 1° per day throughout the year. The number of generations produced in each annual cycle varies from 3 to 5 and is temperature dependent. The spring equinox apparently triggers the northward remigration from Mexico. As time proceeds, the hypothesis holds that the orientation of subsequent generations shifts as shown, with some degree of scattering in each successive geographic displacement. Once the last generation has reached the overwintering sites, their migratory activity is repressed, but their orientation clock is assumed to continue running. By the time the cycle is complete at the new spring equinox, the overwintered butterflies are primed to migrate due north. (Reproduced from Brower, 1996, with permission of the Journal of Experimental Biology.)

the 5 winter months and allows them to remain in reproductive dormancy until the northern milkweed flora resurges in the spring. The high altitude and low latitude selected by the butterflies combine to provide a microclimate beneath the forest canopy that is generally wet enough to prevent the butterflies from desiccating, warm enough to prevent their freezing, and cool enough to preserve their lipid reserves. However, their overwintering can be precarious.

Anderson and Brower determined that adult monarchs can tolerate temperatures to about −8°C if they are dry, but when wetted by rain and exposed to the night sky they lose most of their natural freezing resistance. They concluded that an intact forest serves both as a blanket and as an umbrella for the overwintering monarchs and that removal of even a single large tree exposes and jeopardizes thousands of monarchs during winter storms. Unfortunately this warning was realized in January 2002 when a severe northern cold front penetrated the overwintering region. The ensuing storm soaked and then froze more than a quarter of a billion butterflies in the Chincua and Campanario colonies. The severity of this kill was exacerbated by the fact that the surrounding forests have been thinned and severely fragmented, and despite three presidential decrees supposedly protecting the overwintering forests, extensive legal and illegal logging is accelerating throughout the region.

Monarchs are also becoming rapidly imperiled in their eastern breeding range by industrialized agriculture that is eliminating milkweeds and nectar sources. An assay based on hydrogen and carbon isotope ratios that vary systematically with respect to the geographic origin of the monarch's *Asclepias syriaca* food plants has determined that the major area of summer breeding in the United States coincides with the corn belt that has replaced the former grassland ecosystem in the midwestern United States. This has ominous implications because of the genetic engineering of crops to be resistant to herbicides. Massive and increasingly sophisticated herbicide spraying is killing the principal milkweed food plant of the monarch *(A. syriaca)* as well as most of the native flora over tens of thousands of hectares. Thus monarchs are losing both their larval food resources and their access to the diversity of flowers that provide critical nectar resources.

Because of these combined pressures on the breeding, migratory, and overwintering habitats, the migration of the monarch butterfly in North America has become an endangered biological phenomenon. The remarkable syndrome manifested by the monarch butterfly is too great a cultural and scientific treasure to allow these rampantly destructive processes to continue. Time is rapidly running out.

See Also the Following Articles

Chemical Defense • Lepidoptera • Migration • Mimicry

Further Reading

Ackery, P. R., and Vane-Wright, R. I. (1984). "Milkweed Butterflies: Their Cladistics and Biology." Cornell University Press, Ithaca, N Y.

Alonso-Mejia, A., Rendon-Salinas, E., Montesinos-Patino, E., and Brower, L. P. (1997). Use of lipid reserves by monarch butterflies overwintering in Mexico: Implications for conservation. *Ecol. Appl.* **7**, 934–947.

Brower, L. P. (1995). Understanding and misunderstanding the migration of the monarch butterfly (Nymphalidae) in North America: 1857–1995. *J. Lepidopterists' Soc.* **49**, 304–385.

Brower, L. P. (1996). Monarch butterfly orientation: Missing pieces of a magnificent puzzle. *J. Exp. Biol.* **199**, 93–103.

Brower, L. P. (1999). Biological necessities for monarch butterfly overwintering in relation to the oyamel forest ecosystem in Mexico. *In* "Paper Presentations: 1997 North American Conference on the Monarch Butterfly (Morelia, Mexico)" (J. Hoth, L. Merino, K. Oberhauser, I. Pisanty, S. Price, and T. Wilkinson, eds.), pp. 11–28. Commission for Environmental Cooperation, Montreal, Canada.

Brower, L. P., and Malcolm, S. B. (1991). Animal migrations: Endangered phenomena. *Am. Zool.* **31**, 265–276.

Brower, L. P., Fink, L. S., Brower, A. V. Z., Leong, K., Oberhauser, K., Altizer, S., Taylor, O., Vickerman, D., Calvert, W. H., Van Hook, T., Alonso-M., A., Malcolm, S. B., Owen, D. F., and Zalucki, M. P. (1995). On the dangers of interpopulational transfers of monarch butterflies. *Bioscience* **45**, 540–544.

Brower, L. P., Castilleja, G., Peralta, A., Lopez-Garcia, J., Bojorquez-Tapia, L., Diaz, S., Melgarejo, D., and Missrie, M. (2002). Quantitative changes in forest quality in a principal overwintering area of the monarch butterfly in Mexico: 1971 to 1999. *Conserv. Biol.* **16**, 346–359.

Gibo, D. L. (1986). Flight strategies of migrating monarch butterflies *(Danaus plexippus L.)* in southern Ontario. *In* "Insect Flight: Dispersal and Migration" (W. Danthanarayana, ed.), pp. 172–184. Springer-Verlag, Berlin.

Herman, W. S. (1993). Endocrinology of the monarch butterfly. *In* "Biology and Conservation of the Monarch Butterfly" (S. B. Malcolm and M. P. Zalucki, eds.), pp. 143–146. Los Angeles County Museum, Los Angeles.

Malcolm, S. B., Cockrell, B. J., and Brower, L. P. (1993). Spring recolonization of eastern North America by the monarch butterfly: Successive brood or single sweep migration? *In* "Biology and Conservation of the Monarch Butterfly" (S. B. Malcolm and M. P. Zalucki, eds.), pp. 253–267. Natural History Museum of Los Angeles County, Los Angeles.

Mouritsen, H., and Frost, B. J. (2002). Virtual migration in tethered flying monarch butterflies reveals their orientation mechanisms. *Proc. Natl. Acad. Sci. USA* **99**, 10162–10166.

Riley, C. V. (1871). Two of our common butterflies. Their natural history; with some general remarks on transformation and protective imitation as illustrated by them. *In* "Third Annual Report on the Noxious, Beneficial and Other Insects, of the State of Missouri" (C. V. Riley, ed.), pp. 142–175. Missouri State Board of Agriculture, Jefferson City.

Urquhart, F. A. (1987). "The Monarch Butterfly: International Traveler." Nelson–Hall, Chicago.

Wassenaar, L. I., and Hobson, K. A. (1998). Natal origins of migratory monarch butterflies at wintering colonies in Mexico: New isotopic evidence. *Proc. Natl. Acad. Sci. USA* **95**, 15436–15439.

Mosquitoes

Bruce F. Eldridge

University of California, Davis

Mosquitoes are small flying insects and are related to other members of the order Diptera, the "two-winged flies." The immature stages of larvae are aquatic and live in stagnant water sources in every biogeographic region of the world. Adult female mosquitoes of most species feed on blood of vertebrates, including humans, and this habit has resulted in great economic and public health significance for this group of insects.

There are well over 3000 species and subspecies of mosquitoes in the world. They occur in a variety of habitats, ranging from deserts at or below sea level to high mountain meadows at elevations of 3000 meters or more. Adult mosquitoes are terrestrial flying insects; immature stages are aquatic. Larvae and pupae of the various species can be found in ponds, ditches, puddles, swamps, marshes, water-filled rot holes of trees, rock pools, axils of plants, pools of melted snow, discarded tires, tin cans, and many other types of standing water. Some of the species are most active in the warmest part of the year, whereas others are adapted to cool temperatures. Many species of mosquitoes are rarely encountered and seldom pose a threat to the health or well-being of humans and domestic animals. However, other species are abundant, frequently encountered, and readily attack people, their pets, and their livestock. Some of these species are capable of transmitting microbial organisms that cause malaria and encephalitis and other severe diseases of humans and other vertebrates.

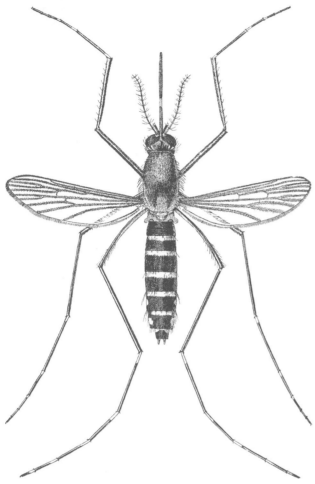

FIGURE 1 Adult female *Ochlerotatus taeniorhynchus,* a saltmarsh mosquito. (From King, Bradley, and McNeel, 1939, "The Mosquitoes of the Southeastern States," USDA Miscellaneous Publication No. 336.)

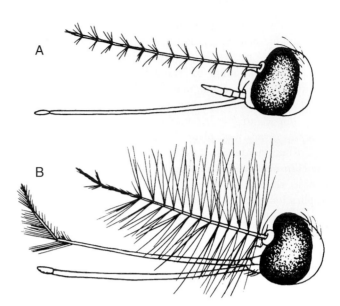

FIGURE 2 (A) Female and (B) male heads of adult *Aedes* mosquitoes. (From Gjullin and Eddy, 1972, "The Mosquitoes of the Northwestern States," USDA Technical Bulletin No. 1447.)

CLASSIFICATION AND IDENTIFICATION

Mosquitoes are classified into three subfamilies, with different characteristics in all of their life cycle stages. The species of importance from the standpoint of public health are contained in the subfamilies Anophelinae (referred to as anophelines) and Culicinae (referred to as culicines). Females of species in a third subfamily, Toxorhynchitinae, lack mouthparts adapted for sucking blood from vertebrates. The larvae of this subfamily are predaceous on other aquatic organisms and have been proposed as biological control agents of mosquito larvae.

Common genera of the Culicinae include *Culex, Aedes, Ochlerotatus* (formerly included in *Aedes*), *Psorophora, Mansonia, Haemagogus, Sabethes, Coquilletidia,* and *Culiseta.* Most species in the Anophelinae are contained in the genus *Anopheles.* The subfamily Toxorhynchitinae contains only the genus *Toxorhynchites.*

Mosquito adults are small flying midge-like insects. Most female mosquitoes can be differentiated from similar insects by the presence of a long slender proboscis, which is adapted

for piercing skin and sucking blood, and long slender wings that are covered with small scales (Figs. 1 and 2A). Male mosquitoes have scale-covered wings, but their probosces are adapted for sucking plant juices and other sources of sugars (Fig. 2B). Most male mosquitoes can also be differentiated from females of the same species by their generally smaller size and by the presence of much longer and hairier maxillary palps. The immature stages of mosquitoes, the larvae (Fig. 3) and pupae (Fig. 4), vary in color from yellowish tan to black. Most mosquito larvae have a distinctive siphon, or air tube, at the rear of their bodies (Fig. 3B), but some species lack this tube (Fig. 3A).

Culicine larvae have an air tube extending from the posterior section of their body and in most species hang at rest from water surfaces at an angle of approximately 45°. Larvae of *Coquilletidia* and *Mansonia* have air tubes adapted for piercing submerged plants to obtain air for breathing. They are rarely found at water surfaces. Anopheline larvae lack an air tube and consequently rest parallel to water surfaces.

Culicine adult females have proboscis developed for piercing the skin of vertebrates and sucking their blood. While feeding, their bodies are usually arranged somewhat parallel to the skin surface of their hosts. Anopheline adult females also have proboscis adapted for piercing vertebrate skin, but they orient themselves at about a 45° angle while blood feeding.

The eggs of mosquitoes also vary (Fig. 5). Females of culicine species deposit single eggs (*Aedes, Psorophora*), boat-shaped rafts of 100 or more eggs (*Culex, Culiseta*), or clusters of eggs attached to floating plants (*Mansonia, Coquilletidia*). Anopheline eggs are also laid singly, but have elaborate floats extending to the sides of the eggs. Anopheline eggs are often found in clusters on water surfaces, forming interesting

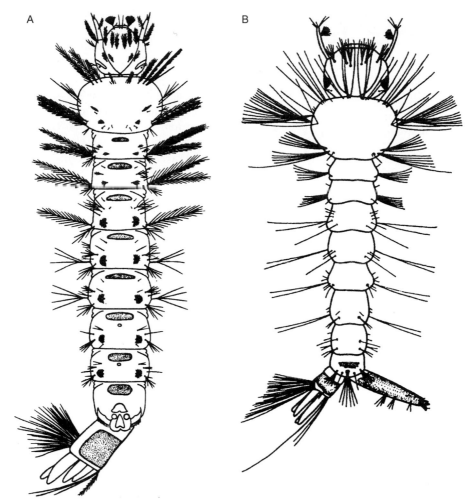

FIGURE 3 Larvae of mosquitoes. (A) An anopheline larva *(Anopheles quadrimaculatus)*. (B) A culicine larva *(Culex quinquefasciatus)*. (Illustrations by Harry D. Pratt, courtesy of U.S. Centers for Disease Control and Prevention.)

geometric patterns. *Toxorhynchites* eggs are also laid singly, usually on water surfaces.

LIFE CYCLE

Egg Stage

The egg-laying habits of female mosquitoes vary widely from species to species. Some female mosquitoes lay eggs on water surfaces (e.g., *Anopheles*), others lay single eggs on moist soil where later flooding is likely (e.g., *Aedes*). From eggs deposited on water surfaces, larvae usually hatch within a day or so, but from eggs laid on soil surfaces, larvae do not hatch until eggs are flooded, which may occur months, or even years, later. The environmental cues female mosquitoes use to find suitable sites for oviposition remain only partially known. Color, moisture, and volatile chemical stimulants appear to play a role in certain species. Efforts to explain the occurrence of various mosquito species in different aquatic habitats based strictly on oviposition cues have been unsuccessful.

Larval Stages

Small larvae that are nearly invisible to the naked eye hatch from eggs. Larvae molt three times to become fourth-stage larvae. Several days later, this larval form molts again to become a pupa. The time required for development of the larval stages depends on several factors, the most important of which is water temperature. Availability of food and larval density are also factors. Water temperature and food are inversely related to time of development; larval density is directly related.

The majority of mosquito species have larvae that are restricted to fresh water. However, larvae of a few species can develop under other conditions, e.g., brackish or salt water or water polluted with organic solids. Species with larvae adapted to salt water can maintain osmotic pressure within their bodies by drinking substantial amounts of water and by removing ions from their hemolymph through their Malpighian tubules and rectum. Generally, saline species can also develop in fresh water, but do not compete well with freshwater species. The

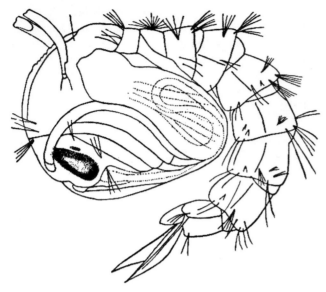

FIGURE 4 A mosquito pupa. (Original illustration from Howard, Dyar, and Knab, 1912, "The Mosquitoes of North and Central America and the West Indies." Reprinted from USDA Handbook No. 336, 1939.)

inverse is not true and consequently, various kinds of water usually have a characteristic mosquito fauna.

Adults

Adult mosquitoes emerge 1–2 days after the appearance of pupae, with males emerging first. In the summer, the entire life cycle, from egg to adult, may be completed in 10 days or less. Females feed on vertebrate blood for the development of eggs. This behavior by females is the single most important

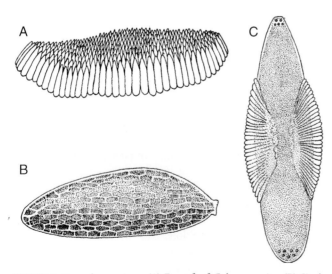

FIGURE 5 Eggs of mosquitoes. (A) Egg raft of *Culex* mosquito. (B) Single egg of *Aedes* mosquito. (C) Single egg of anopheline mosquito. (Original illustration from Howard, Dyar, and Knab, 1912, "The Mosquitoes of North and Central America and the West Indies." Reprinted from USDA Handbook No. 336, 1939.)

characteristic of mosquitoes from the human standpoint. Blood feeding in insects is believed to have evolved several times independently from ancestral forms adapted for sucking plant juices or for preying on other insects. The means by which female mosquitoes locate suitable hosts for blood feeding has been studied for many years, but there are still many unknown features to this behavior. The best explanation is that females are attracted by warmth, moisture, and carbon dioxide from hosts, but other factors are involved. There have been studies that have suggested that substances such as lactic acid, a component of human sweat, may serve as an attractant.

Ordinarily, a female mosquito cannot develop a batch of eggs unless she has taken a blood meal to obtain nourishment for ovarian development. However, some strains or individuals of several species can develop eggs without a blood meal, which is called autogeny. The nourishment for egg development is carried over from the larval stages, and consequently, only the first batch of eggs can develop in this way. The usual situation, in which a blood meal is required for the development of all batches of eggs in an individual female, is called anautogeny.

Some mosquitoes take blood only from certain groups of vertebrate animals. For example, *Culex pipiens,* the northern house mosquito, feeds almost entirely on birds. *Ochlerotatus sierrensis,* the western treehole mosquito, feeds only on mammals. *Culex tarsalis,* the encephalitis mosquito, feeds on birds and mammals (this dual host preference is one characteristic of an effective vector of disease pathogens). In the past, this relative host specificity has been called host preference. However, this term is not appropriate, because it ignores availability of hosts, host defensive behavior, and other factors unrelated to the mosquitoes themselves. The blood-feeding drive is controlled by neurohormones and can be induced artificially by treatment with juvenile hormone or one its analogs. This hormonal influence is why mosquitoes that have recently had a blood meal and are developing a batch of eggs do not usually seek another blood meal. However, multiple blood meals (more than one blood meal in a single gonotropic cycle) do occur at times in nature in some species.

Blood feeding by mosquitoes is a complex process. It is facilitated by the infection of saliva into the feeding wound of the vertebrate host. Saliva comes from organs in the thorax of mosquitoes called salivary glands. Saliva may contain a variety of substances, including chemicals that reduce clotting of vertebrate blood. Digestion of a blood meal usually takes 2–3 days, depending on the ambient temperature. The uptake of blood is accomplished by the action of muscular pumps in the head of female mosquitoes. Blood travels through the digestive tract of the mosquito into a structure called the midgut. After blood reaches the midgut it is soon surrounded by a thin sheath, the peritrophic membrane, that is secreted by cells at the front of the midgut. Digestion of the blood takes place within this structure.

Seasonal Development

Some species of mosquito have but a single generation per year (univoltine), whereas others have many (multivoltine), depending upon the length of the season favoring the activity of the adult stages. To avoid seasons of the year not favorable to adult activity (usually the winter), mosquitoes may have some kind of diapause mechanism. In *Aedes* and related genera, the diapause mechanism usually involves the egg stage. In temperate and subarctic zone *Aedes,* populations may survive winters as desiccation-resistant eggs, sometimes under the surface of snow or along river flood plains. The larvae then hatch in the spring after the eggs are flooded from melted snow or after flooding of the riverbanks.

Culex and *Anopheles* females usually survive unfavorable periods as diapausing or quiescent adult females. Male mosquitoes usually do not survive unfavorable periods, so it is necessary for insemination to occur before the onset of diapause.

Some mosquito species survive unfavorable periods as diapausing larvae (e.g., species of *Aedes, Anopheles, Culiseta*). Diapause can be variable in some species, depending upon the latitude at which they occur, with diapause occurring in the larval stage at warmer latitudes and in the egg stage at cooler ones.

There is considerable variation in the environmental and physiological control of diapause. In nearly all diapausing mosquitoes studied, diapause is triggered by exposure of one or more of the life cycle stages to daylength. In *Culex* species, and other mosquitoes that overwinter as adults, exposure of late-stage larvae and of pupae to short daily photophases occurring in autumn results in diapause in adult females. This diapause is manifested by lowered activity levels, inhibition of blood-feeding drive, and arrestment of follicle development in ovaries. In some *Aedes* species, the short autumn days experienced by females result in deposition of eggs that are in the diapause state. The larvae in these eggs do not hatch until a period of exposure to near-freezing temperatures lasting several months. In other species of *Aedes,* diapause results from exposure of the eggs themselves to short daylengths. Still other *Aedes* species have larvae that enter diapause triggered by their exposure to short daylengths.

As with other aspects of reproduction and development, diapause is controlled directly by neurohormones. Diapause can be induced in most diapausing species by exposure to juvenile hormone or one of its analogs.

Many tropical and subtropical species, such as *Aedes aegypti,* the yellow-fever mosquito, do not have a diapause mechanism. Still other tropical species have mechanisms for avoidance of hot, dry seasons, but these mechanisms have been little studied.

PUBLIC HEALTH AND VETERINARY IMPORTANCE

As discussed earlier, female mosquitoes of nearly all species require blood from vertebrate animals to develop their eggs, and many species bite people, their pets, and livestock for this purpose. The most important result of this behavior is the transmission of microorganisms that cause diseases such as malaria, filariasis, yellow fever, and dengue. These and other mosquitoborne diseases can have serious and sometimes fatal consequences in people. These diseases can also have an impact on livestock, pets, and wildlife. Even when no infectious disease pathogens are transmitted by mosquitoes, they can be a serious health problem to people and livestock. Biting of people by mosquitoes can result in secondary infections, allergic reactions, pain, irritation, redness, and itching. Mosquito biting of beef cattle can result in reductions in weight gains and in dairy cows, reduction in milk production.

The interactions between mosquito hosts and the pathogens they transmit are highly variable. Three basic types of transmission mechanisms are involved: (1) propagative transmission, in which the pathogen multiplies within the mosquito but does not undergo any changes in developmental form; (2) developmental transmission, in which the pathogen undergoes developmental changes, but does not multiply; and (3) propagative-developmental transmission (also called cyclodevelopmental transmission), in which the pathogen multiplies and undergoes changes in developmental forms. Transmission of the yellow-fever virus by the yellow-fever mosquito is an example of propagative transmission. The virus is taken up by a female mosquito from a viremic host during blood feeding, multiplies many times, and eventually infects the salivary glands of the host. When the female mosquito takes another blood meal, she may infect a new host by injection of saliva.

Some pathogens are transmitted to the offspring of female mosquitoes via infected eggs. This type of transmission is known as transovarial transmission.

Filarial worms, the cause of the disease filariasis (a type of which is called elephantiasis) in humans and other vertebrates, are transmitted by developmental transmission. In this example, very small immature forms of the worms, called microfilariae, occur in the blood of infected vertebrate hosts and are taken up by female mosquitoes in a blood meal. Within the mosquito, the filariae molt several times until they eventually become infectious larvae. These larvae migrate down the proboscis of the mosquito and enter the feeding wound caused by the mosquito during a subsequent blood feeding. Within the vertebrate host, these larvae may eventually develop into adult male and female worms that mate and produce microfilariae. It is the presence of large numbers of adult worms that results in the symptoms of filariasis.

Malarial parasites have a very complex life cycle, involving both multiplication of parasites and development of life cycle stages. Anopheline mosquitoes are the vectors of human malaria, and because the sexual stages and fertilization occur within mosquitoes, they are the definitive hosts. Parasite forms called microgametocytes (male sex cells) and macrogametocytes (female sex cells) occur in the peripheral blood of humans and are taken up by mosquitoes. Fertilization of the female cells by the male cells occurs within the gut of the vector mosquito. After several life cycle changes, and multiplication of forms within cysts on the gut wall, forms of the parasite called sporozoites enter salivary glands of the mosquito and infect new hosts during blood feeding.

There are hundreds of types of microorganisms that are transmitted by mosquitoes to vertebrates that result in diseases. A few are extremely important worldwide because of their high incidence and the severity of their symptoms in humans.

Malaria is one of the most important diseases in the world. Several hundred million people are infected with malarial parasites, resulting in over 2 million fatalities annually, mostly in tropical countries in Africa and Asia. Malaria is especially serious in pregnant women and young children. Typically, more than a million children die each year from this disease. The economic development of a number of tropical countries is badly hindered by malaria because of the burden of chronic malaria infections in working-age men and women.

The virus disease known as dengue, transmitted mostly by the yellow-fever mosquito, is a rapidly expanding problem in the world and now is considered second in importance only to malaria among mosquitoborne diseases. The increase in global human travel resulting from expanded rapid air transportation has been paralleled by the increase in the number of viral strains causing dengue and the increase in the number of cases of a particularly serious form of the disease called dengue hemorrhagic fever. This form of the disease is most serious in children and is a significant cause of mortality.

Filariasis is a general term applied to infection of vertebrate animals by many different species of parasitic worms belonging to the superfamily Filaroidea. A form of mosquitoborne filariasis is called lymphatic filariasis because infection can cause impairment of the lymphatic system. Lymphatic filariasis is a chronic disease that can lead to the well-known disfigurement of humans called elephantiasis. Another type of filariasis called dog heartworm occurs in dogs, other canids (e.g., wolves and coyotes), and felids (e.g., cats). Heavy infections can result in large buildups of adult worms in the cardiopulmonary system and can be fatal.

Yellow fever, a virus disease, has virtually disappeared from the United States because of the availability of an extremely effective vaccine. This vaccine may provide lifelong immunity from a single inoculation. Unfortunately, the availability of the vaccine is limited on a worldwide basis and there are many unvaccinated people living in areas where the mosquito vector, *A. aegypti,* is common. Yellow fever is an extremely serious disease. There is no available treatment, and infections in humans are frequently fatal. Periodic epidemics continue to occur in various tropical countries. *A. aegypti* is common in urban and suburban areas of the tropics and subtropics. The larvae of this species occur in water in various types of artificial containers such as shallow wells, water urns, discarded containers, and tires. It is very difficult to control.

There are many other mosquitoborne diseases, several of them caused by viruses. Some of these viral diseases, such as Japanese encephalitis, La Crosse encephalitis, West Nile fever, Ross River disease, and Rift Valley fever, affect large numbers of people in parts of the world where they occur.

CONTROL OF MOSQUITOES

In most industrialized nations, mosquito control is done by government-supported agencies that are either components of health agencies or separate agencies organized specifically for that purpose. In the United States, states that have the most serious mosquito problems (e.g., New Jersey, Florida, Texas, Louisiana, California) have many such agencies. Some are small and have responsibility for mosquito abatement in a few hundred square kilometers, whereas the activities of others may encompass one or more entire counties. However, even in states that have many mosquito abatement districts, many people live in areas with no organized mosquito control. In underdeveloped areas of the world, organized mosquito control is rare except for scattered programs aimed at specific diseases such as malaria.

Most organized mosquito control is accomplished by searching out mosquito larvae in standing water and then treating the water with some kind of material that kills the larvae. Modern materials are highly specific for mosquitoes and ordinarily have little or no effect on other organisms. One such material is a bacterial product called Bti *(Bacillus thuringiensis israelensis)* that produces a toxin that kills only larvae of mosquitoes, black flies, and certain midges. Mosquito abatement agencies may also apply chemical pesticides to kill adult mosquitoes, but ordinarily only when adult populations become so high that they cause extreme annoyance to many people or when the threat of disease transmission to people is high. Therefore, the most common method for this is known as ultralow volume, or ULV. This approach involves using special equipment to spray extremely small volumes of small particles of highly concentrated insecticides. When used properly, it is a safe and highly specific method of mosquito control.

Control of irrigation water in agricultural areas to avoid excess runoff is an important mosquito control method, but in recent years elimination of small bodies of water that can serve as wildlife habitat has ceased to be a mosquito control option.

In the first half of the past century, elimination of bodies of temporary and permanent water (swamps, marshes, vernal pools) was an accepted form of mosquito control. Recent years have seen the realization that such habitats are valuable and irreplaceable components of the environment and that a variety of activities have resulted in the permanent loss of many of these wetland habitats. This loss has resulted in the development of mosquito management strategies that are much more ecologically sound. Considerable research has been conducted on management strategies that enhance wetland habitats while minimizing problems from mosquito breeding.

Biological Control

A method that is a preferred alternative to chemical control is the use of live organisms to control mosquitoes, either by predation or by infection. Mosquitofish (the common guppy) have been used for many years for this purpose, often with effective results. However, because mosquitofish are generalist predators, they must be used with great care to avoid damage to other aquatic organisms. Many other forms of biological control for mosquitoes have been tried, including other types of fishes, fungi, bacteria, nematode worms, flat worms, protozoan parasites, and predaceous insects (including some mosquitoes). Some of these organisms have been effective under special circumstances, but few of them have been adopted widely. Microbial organisms such as Bti and *Bacillus sphaericus* may be considered biological control agents, and these are used to great advantage in a variety of aquatic habitats.

Insecticides

At one time there were dozens of insecticides available for killing both adults and immature stages of mosquitoes. However, because of economics, primarily the costs involved in developing, testing, and registering new materials, and the development of resistance to insecticides by mosquitoes, the number of available materials is now down to a handful. A class of insecticides known as insect growth regulators has been highly effective and specific for mosquitoes, but the development of resistance to even these materials has clouded the future of these so-called third-generation pesticides. The best hope for circumventing resistance to pesticides is the use of a combination of approaches referred to as pesticide resistance management. Frequent testing for susceptibility in mosquito populations, alternation of pesticides, and avoidance of methods that result in the persistence of low dosages of pesticides are examples of this approach. Insecticide resistance is under genetic control, and the goal of insecticide resistance management is preservation of genes in mosquitoes associated with susceptibility.

Protection from Mosquito Bites

People living in areas lacking organized mosquito control must protect themselves from bites of mosquitoes by using a variety of strategies. Probably the most effective method of personal protection from mosquito bites is to avoid places where mosquito densities are high and to avoid being out-of-doors at times of the day when mosquito activity is at its highest. In mountainous areas, most species of mosquitoes bite during the morning and afternoon and often not at all during periods of darkness. In low-elevation areas some mosquitoes tend to bite at night, whereas others bite during the day. The species of mosquito present in a given area varies from place to place, and it is necessary to learn the activity patterns of the mosquitoes present to develop avoidance strategies.

If exposure to biting mosquitoes cannot be avoided, there are several ways to minimize discomfort. The most important of these is to reduce exposed skin surfaces by wearing hats, long trousers, and long-sleeved shirts. Some mosquitoes bite through light clothing, but the number of bites received is definitely reduced if most areas are covered. When mosquito densities become very high, application of a mosquito repellent may be needed to avoid bites. Currently, the only really effective repellents on the market are those that contain the material DEET. Skin repellents have some drawbacks. After application, they are effective for only about 4 h at the maximum. Other factors, such as wind, high temperature, high humidity, and sweating, reduce this time even further. When applying DEET, the material must be thoroughly applied to all exposed skin, including behind the ears. In recent years, longer lasting formulations of DEET have been developed by the incorporation of various additives such as lotions and polymers.

Many people have tried gadgets such as ultrasonic emitters, electric grids, aromatic plants, and even vitamins for mosquito protection. Research has shown that most such methods are of little or no value in repelling mosquitoes, but such devices continue to appear on the market. In some areas of the world incense coils are sold for avoidance of mosquitoes. They may afford protection within a short distance of the burning coils.

Bednets can provide excellent protection from mosquito bites at night if used properly. The use of bednets treated with insecticides has been shown to afford excellent protection from attack by malaria mosquitoes. When they are available, vaccines may protect humans from mosquito-borne disease (e.g., yellow fever) and prophylactic drugs may be used to avoid some diseases (e.g., malaria).

See Also the Following Articles
Aquatic Habitats • Blood Sucking • Dengue • Dog Heartworm • Malaria • Medical Entomology • Salivary Glands • Yellow Fever

Further Reading
Bates, M. (1954). "The Natural History of Mosquitoes." Macmillan Co., New York.
Clements, A. N. (1963). "The Physiology of Mosquitoes." Macmillan Co., New York.
Clements, A. N. (1992). "The Biology of Mosquitoes," Vol. 1. Chapman & Hall, New York.
Clements, A. N. (1992). "The Biology of Mosquitoes," Vol. 2. CAB Int., Oxon, U.K.
Darsie, R. F., Jr., and Ward, R. A. (1981). "Identification and Geographical Distribution of the Mosquitoes of North America, North of Mexico." American Mosquito Control Association, Fresno, CA.
Eldridge, B. F., and Edman, J. D. (eds.) (2000). "Medical Entomology." Kluwer Academic, Dordrecht.
Knight, K. L., and Stone, A. (1977). "A Catalog of the Mosquitoes of the World." Thomas Say Foundation, Entomol. Soc. Am., College Park, MD.
Service, M. W. (1993). "Mosquito Ecology: Field Sampling Methods," 2nd ed. Elsevier, London/New York.

Moth

see *Lepidoptera*

Mouthparts

R. F. Chapman
University of Arizona

The mouthparts of insects are structures surrounding the mouth that are involved in the mechanics of feeding and processing and manipulating the food so that it can be ingested. Although functionally equivalent to the jaws of vertebrates, they lie outside the mouth, not within a buccal cavity. Good basic accounts of insect mouthpart structure are to be found in most textbooks of entomology. The aim of this article is to supplement these basic accounts by briefly considering some of the variation associated with different feeding habits and different types of food. It also gives some information on the functioning of the mouthparts.

BITING AND CHEWING INSECTS

Insect mouthparts are derived from the appendages of four of the segments forming the insect head. They surround the mouth and are external to it, unlike the condition in vertebrates in which the teeth are within the oral cavity. The basic segmental character of the mouthparts is most apparent in insects that bite off fragments of food and then chew it before ingesting it (Fig. 1). Insects that do this are said to be "mandibulate" because the mandibles are relatively unmodified compared with those of fluid-feeding insects (see below). These are also commonly called biting mouthparts, although there is some risk of confusion with blood-sucking insects, such as mosquitoes, which bite! In this article the latter are distinguished as "piercing." The mandibulate arrangement occurs in the primitively wingless insects (Apterygota), in the cockroaches and grasshoppers and their allies, in larval and adult beetles (Coleoptera) and most Hymenoptera, and in caterpillars (larval Lepidoptera), among the more advanced groups of insects.

Immediately in front of the mouth is the labrum formed from the fusion of the appendages on either side of the labral segment. It comprises a flat sclerotized plate of cuticle continuous with the cuticle of the front of the head (clypeus). Its inner side (toward the mouth) is known as the epipharynx, and it is formed from membranous cuticle-bearing tracts of noninnervated hairs, all pointing toward the mouth. In grasshoppers, and probably in other insects with similar mouthparts, the hairs are easily wetted, whereas the other parts of the cuticle are water repellent. The hydrophilic hairs may serve to

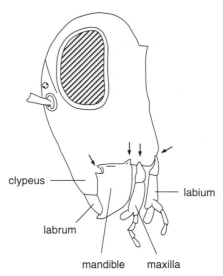

FIGURE 1 A lateral view of the head of a grasshopper showing the segmental arrangement of the mouthparts: labrum, mandible, maxilla, and labium. Arrows show the points of articulation (condyles) with the head capsule. The mandible has two condyles (dicondylic), the maxilla only one, and the labium one on each side (modified after Snodgrass, 1935, "Principles of Insect Morphology," McGraw-Hill).

direct the flow of fluid from the food toward the mouth and also to groups of contact chemoreceptors (taste receptors) that occur just outside the mouth. Contact chemoreceptors also often occur along the distal edge of the labrum. At rest, the labrum presses back on the mandibles, which are immediately behind it, being held in this position by a rubber-like protein, called resilin, in its connection with the clypeus.

The mandibles, one on each side, are hinged to the head capsule by one or two condyles. Archaeognatha have only one condyle (monocondylic), whereas Thysanura and all mandibulate pterygote insects have two (dicondylic). The change from one to two condyles represents a considerable evolutionary advance because it gives the mandibles a much firmer base and so facilitates feeding on hard materials. The mandibles of the two sides swing transversely to meet below or in front of the mouth, depending on the orientation of the head, and are opened and closed by a pair of muscles, one inserted on either side of the axis of mandibular attachment at the condyles. The opener muscle is called the abductor, whereas the closer is the adductor. The latter is the larger of the two because it provides the force necessary to bite into or through material. Both muscles arise on the cuticle at the top of the head and, in grasshoppers and caterpillars, the head capsule grows bigger to accommodate the increased size of the adductor muscle if the insect feeds on tough food.

The two mandibles are asymmetrical so that where they meet in the midline the cusps on the biting surface of the two sides fit between each other (Fig. 2). These cusps are extremely hard. In addition to being sclerotized like the hard cuticle elsewhere in the body, their cuticle contains zinc or manganese or, occasionally, iron, which are assumed to contribute to the hardness. The form of the biting cusps varies from species to

A

right
mandible

left
mandible

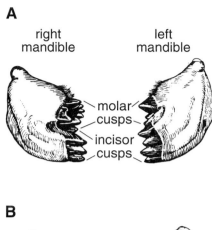

molar
cusps

incisor
cusps

B

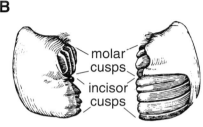

molar
cusps

incisor
cusps

FIGURE 2 Mandibulate mouthparts. Mandibles, seen from in front with the labrum removed, of grasshoppers with different feeding habits. Notice that the mandibles of the two sides are asymmetrical. (A) A grasshopper that feeds on soft, broad-leaved plants. (B) A grasshopper that feeds on grasses. (Reproduced, with permission, from F. B. Iseley, 1944, Correlation between mandibular morphology and food specificity in grasshoppers. *Ann. Entomol. Soc. Am.* **37**, 47–67.)

A

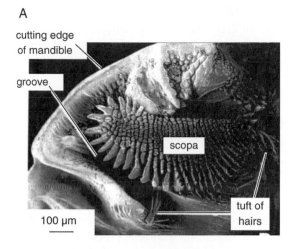

cutting edge
of mandible

groove

scopa

tuft of
hairs

100 µm

B

10 µm

FIGURE 3 Mandibulate mouthparts. Specialized mandible of a pergid sawfly larva. These insects feed on *Eucalyptus* leaves containing large quantities of essential oils. The oils are apparently squeezed from the leaf tissue by the action of the scopa mandibularis (scopa) and conducted along the groove to a pharyngeal diverticulum where they are stored. (A) Surface view of the biting face of the right mandible showing the rows of spines forming the scopa mandibularis. (B) Spines of the scopa mandibularis, which presumably abrade the leaf surface to release the oils. (Reproduced, with permission, from S. Schmidt, G. H. Walter and C. J. Moore, 2000, Host plant adaptations in myrtaceous-feeding Pergid sawflies: Essential oils and the morphology and behaviour of *Pergagrapta* larvae (Hymenoptera, Symphyta, Pergidae). *Biol. J. Linn. Soc.* **70**, 15–26.)

species in relation to feeding habits in a way that, superficially, is comparable with the adaptations in the jaws of mammals. Predaceous tettigoniids, for example, have a sharply pointed cusp distally and powerful blade-like cusps more proximally that have some resemblance to the canine and carnassial teeth of carnivorous mammals and presumably serve similar functions of grasping and tearing the prey. Among grasshoppers, species feeding on soft, broad-leaved plants have small, sharply pointed cusps that cut the food into very small fragments. Grass feeders, on the other hand, have very long, chisel-edged incisor cusps distally with short, flattened molar cusps proximally, which can superficially be compared with the chisel-shaped incisors and grinding molars of mammalian herbivores. Other insects also exhibit food-related modifications of the mandibles. The cusps become worn down with use, especially if the insect is feeding on hard foods, and there is some evidence that this wearing down reduces the rate of food intake. The cusps can be renewed only at a molt, when new cuticle is formed.

In insects eating food that requires special treatment during ingestion, the mandibles may become highly modified. An example occurs in the larvae of pergine sawflies (Fig. 3). These Australian insects feed on *Eucalyptus* and related trees, the leaves of which are rich in essential oils. The insects sequester the oils in a diverticulum of the foregut and use them for defense. The mandibles are apparently adapted for separating the oils from the leaf tissue. Sticking out from the center of the mandible is a structure called the scopa mandibularis. It is covered by

rows of pointed setae and these, perhaps by scraping and shredding the leaves, seem to be involved in extracting the oils.

In the midline, immediately behind the mouth and probably also derived from the mandibular segment, is the hypopharynx. This structure is a lobe of mostly membranous cuticle but with rods of sclerotized cuticle to which muscles are attached. Like the epipharynx, it bears tracts of hairs pointing toward the mouth and these hairs probably help to move food toward the mouth as the hypopharynx is moved by its muscles.

Behind the mandibles are the maxillae, one on each side of the head. Each maxilla articulates with the head capsule by a single condyle so that it is extremely mobile. This high degree of movement allows the maxillae to manipulate food between the mandibles and move it toward the mouth. The laciniae at the distal ends of the maxillae are especially important for this and they are usually curved and pointed with the tip hardened like the mandibles. The maxillary palps are leg-like structures often with three to five segments and they have an important sensory function. At the tip of each palp is an array of contact chemoreceptors; in a large grasshopper there may be as many as 400 chemosensilla on the tip of each palp. These receptors have an important role in food selection. Grasshoppers drum on a leaf surface with the palps before accepting or rejecting it as food, and they continue to drum at intervals during feeding. Chemoreceptors are also present on the galea, a distal lobe of the maxilla immediately lateral to the lacinea.

The labium is essentially similar in structure to the maxillae but with the appendages of the two sides fused together in the midline behind the hypopharynx. There is a single articulation with the head capsule on each side, which allows the labium to swing beneath the head in the vertical plane of the body. It provides a scoop that prevents food from spreading backward from between the mandibles. As with the maxilla there are two terminal extensions on each side, known as the glossa and paraglossa. There is also a leg-like labial palp with chemoreceptors at the tip.

The labium is uniquely developed in larvae of dragonflies and damselflies, forming their prey capture equipment. The form is basically similar to the labium of other insects except that the basal parts are lengthened and the palps are claw-like. It is sometimes called a labial mask because the distal parts cover the lower part of the face when the labium is folded beneath the head. The mask can be suddenly extended by hemolymph pressure, enabling the larva to capture prey a little distance in front of it without moving its body.

FLUID-FEEDING INSECTS

Many insects feed on liquid food and their mouthparts are modified to form a tube through which fluid can be drawn into the mouth and, often, another tube through which saliva is injected into the food so that it is digested to some extent before being ingested. In most fluid-feeding insects the basic segmental arrangement and appendicular form of the mouthparts are no longer obvious, but some predaceous larvae that feed on the body fluids of their hosts are mandibulate, with mandibles resembling those of insects feeding on solid food. These are larvae of lacewings and ant lions (Neuroptera), glowworms (Lampyridae), and dytiscid beetles. In all of them, the mandibles are sickle shaped with a groove along the inner edge. In the beetles, the two sides of the groove arch over to meet, or almost meet, so that a tube is formed. In ant lions, the lacinea of the maxilla is also sickle shaped and it fits in the mandibular groove to form an enclosed canal. These insects can pump the fluid contents of their prey into the foregut through the tubes.

In other fluid-feeding insects, the basic segmental arrangement of the mouthparts is not apparent and in many insects the mouthparts themselves are drawn out into long, slender structures called stylets. The food and salivary canals are formed in different ways in different insect groups. In Hemiptera, both canals are formed between the styliform maxillae, which interlock by a tongue-and-groove mechanism that permits them to slide lengthwise with respect to each other but prevents them from coming apart (Fig. 4). They are supported in a groove along the anterior margin of the elongate labium, which is often referred to as the rostrum. The food canal is formed between the maxillary galeae in Lepidoptera, but here the two sides are linked by a series of cuticular hooks and plates that hold the two sides together while allowing them to coil up beneath the head when not in use (Fig. 5). This device makes it possible for some lepidopterans to have an extremely long tongue, which would not be possible if the insect were unable to coil it. The longest examples are in the hawk moths, the Sphingidae. Many of these have a tongue 30 mm or more in length, but one species, *Cocytius cluentis,* from South America has a tongue 250 mm long! Lepidopterans have no salivary canal in the tongue because the nectar on which they feed does not require digestion before being ingested. Each galea contains an extension of the hemocoel and the proboscis is uncoiled by an increase in pressure generated at the base of each galea. A series of short muscles extends across the galea and these muscles are involved in coiling the proboscis beneath the head. There are contact chemoreceptors at the tips of the galeae and the axons from the sensory receptor cells combine to form a nerve, which also

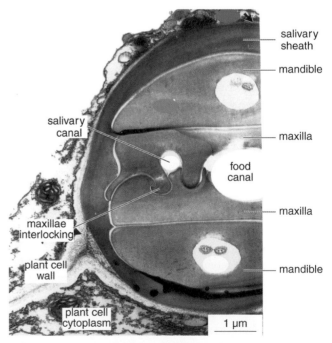

FIGURE 4 Piercing and sucking mouthparts of a hemipteran. Electron micrograph of a transverse section through part of the stylet bundle of an aphid within a leaf. Notice how the maxillary stylets interlock to form the food canal and the salivary canal. The stylets are surrounded by a sheath of solidified saliva that is produced as they penetrate the plant. The stylets are within the cell wall, which is seen in its normal form at lower left. (Reproduced with permission, from W. F. Tjallingii and T. H. Esch, 1993, Fine structure of aphid stylet routes in plant tissues in correlation with EPG signals. *Physiol. Entomol.* **18,** 317–328.)

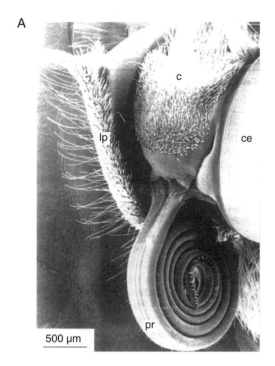

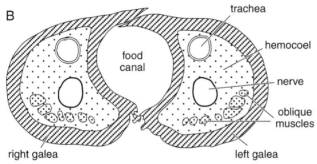

FIGURE 5 Sucking mouthparts of a butterfly. (A) Proboscis coiled beneath the head. The labial palp on the left (near) side has been removed. Abbreviations: c, clypeus; ce, compound eye; lp, labial palp; pr, proboscis. (Reprinted from *Int. J. Insect Morphol. Embryol.*, **27**, H. W. Krenn and C. M. Penz, Mouthparts of *Heliconius* butterflies (Lepidoptera: Nymphalidae): A search for anatomical adaptations to pollen-feeding behavior, 301–309, copyright 1998, with permission from Elsevier Science.) (B) Transverse section through the proboscis near the base. The galea of either side fit together to form the food canal. Each galea is blood filled and contains a nerve and trachea running the full length of the proboscis and short muscles (seen in oblique cross sections in the diagram) that run obliquely across the galea and are involved in coiling the proboscis. (Reproduced, with permission from Springer-Verlag, from H. W. Krenn, 1990, Functional morphology and movements of the proboscis of Lepidoptera. *Zoomorphology* **110**, 105–114. Copyright Springer-Verlag.)

contains motor axons to the muscles, running the length of each galea. Oxygen is supplied to these tissues via a longitudinal trachea.

Among the flies (Diptera), the feeding canal is a groove along the underside of the long labrum, closed behind by the other mouthparts, whereas the salivary canal is a narrow tube running through the styliform hypopharynx. The labium forms a sheath that encloses the stylets formed by the other

mouthparts and is called the haustellum in the higher Diptera. In addition, in cyclorrhaphous flies the distal part of the labium is extended to form a flattened membranous lobe called the labellum. It is conspicuous in house flies and blow flies (Fig. 6). The ventral surface of each labellum is invaginated to form a series of channels that collect together medially where they make contact with the food canal in the labrum. The walls of the channels are supported by a series of incomplete rings of sclerotized cuticle. These rings prevent the channels from collapsing when suction is exerted by the pumps in the head and give the channels an appearance that is superficially like that of tracheae and so they are called pseudotracheae. The labellum, with the pseudotracheae, enables the fly to draw fluids from a relatively large surface. The channels open to the exterior via a narrow groove that is closed off during feeding except for occasional openings through which fluid can pass freely. Fleas have the food (blood) channel between a highly developed epipharynx and the two maxillae. There are two salivary canals, one in each maxilla.

The mouthparts of bees are unusual among the fluid-feeding insects. They retain normal mandibles that are used for wax and pollen manipulation, but are not involved in nectar feeding. The other mouthparts retain some semblance of the appearance in biting and chewing insects, but are elongate. The two glossae (parts of the labium) are fused together to form an elongate tongue with an open gutter posteriorly. The glossal tongue is surrounded by the lengthened and flattened galeae (of the maxillae) and labial palps. The food canal is formed by the space between the glossal tongue and the other components.

Lepidoptera, bees, and some flies feed from fluids, often nectar, that is present on a surface, but other fluid-feeding insects obtain their food from within plants or other animals and so must pierce the host tissues before being able to feed. This is true of all Hemiptera, fleas, and some flies. In Hemiptera, the mandibular stylets are the main piercing structures. The relatively stout labium does not enter the wound, but folds up beneath the insect as the mandibles and maxillae penetrate deeper into the host tissues. The stylets of aphids and coccids are very flexible and usually follow intercellular pathways that may be quite tortuous. The watery saliva of aphids contains a pectinase that degrades the pectin of the cell walls and facilitates movement of the stylets.

Among blood-sucking flies, the maxillae are the primary piercing organs of female mosquitoes (Fig. 7). They have recurved teeth distally that probably anchor the stylets in position in the wound so that when the retractor muscles contract they pull the down toward the host skin, pushing the labrum into the wound at the same time. Male mosquitoes do not feed on blood and, in many species, the piercing stylets are greatly reduced. Horse flies and deer flies (Tabanidae) have a completely different mechanism. Their somewhat elongate mandibles are articulated to the head capsule so that they move transversely, like the mandibles of biting and chewing insects. This scissor-like motion cuts through the skin of the host and the labrum and maxillae are forced into the wound. The labium

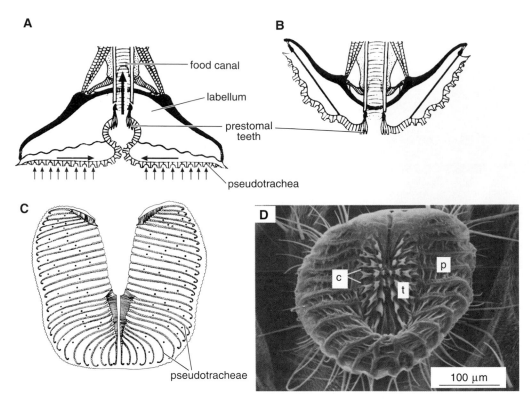

FIGURE 6 Sucking mouthparts of a blowfly. (A) Feeding by suction. With the labellar lobes spread flat on a surface, the openings of the pseudotracheae are brought into contact with fluid on the surface. Suction exerted by the pharyngeal pump draws liquid up the food canal and so through the pseudotracheal openings and along the pseudotracheae. Notice that the prestomal teeth are not exposed. Arrows show the direction of flow. (B) Feeding by rasping. By pulling back the labellar lobes, the prestomal teeth can be brought into contact with the substrate and are used to rasp at solid food. (A and B reproduced with permission from Cambridge University Press, from G. S. Graham-Smith, 1930, Further observations on the anatomy and function of the proboscis of the blowfly, *Calliphora erythrocephala* L. *Parasitology* **22**, 47–115.) (C) View of the ventral surface of the labellum showing the arrangement of pseudotracheae. The smaller branches join with major collecting trunks, which are functionally connected to the food canal. (Reproduced with permission from Wiley–Liss, Inc., a subsidiary of John Wiley & Sons, Inc., from M. Wilczek, 1967, The distribution and neuroanatomy of the labellar sense organs of the blowfly *Phormia regina* Meigen. *J. Morphol.* **122**, 175–201.) (D) Ventral view of the labellum with the prestomal teeth everted for rasping. Abbreviations: c, openings of collecting channels into which the pseudotracheae open; p, pseudotracheae; t, prestomal teeth. (Reproduced with permission from Elsevier Science, from Smith, 1985.)

does not enter the wound and blood is taken directly into the food canal on the inner side of the labrum. Tsetse flies *(Glossina)* and stable flies *(Stomoxys)* penetrate the host tissues by a rasping movement of prestomal cuticular teeth on the labellar lobes (Fig. 8). House flies have similar, but much smaller teeth that they may use to rasp the surface of solid food, but in their blood-sucking relatives the teeth are stronger and are accompanied by banks of cuticular spines that form rasps. When these flies feed, the teeth and rasps are rapidly rotated round the tip of the labellum in a series of rasping movements that enable them to tear through the host's skin. For example, compare Fig. 8B, which depicts the rasps and spines on the inside of the labellar lobe and pointing downward, with Fig. 8C, in which they are on the outside and pointing upward. Contact chemoreceptors are also exposed as the teeth are moved round the tip of the labellar lobes so that they are in a position to detect blood as the host's capillaries are damaged.

Once the insect starts to feed, the properties of the fluid and dimensions of the food canal affect the rate of uptake and so the rate of nutrient intake. The more viscous a fluid, the more slowly it flows, so that although nectar containing high concentrations

of sugars has more nutrients per unit volume, it also is taken up more slowly than a more dilute solution. The flow rate is also negatively correlated with the length of the proboscis, but positively correlated with the diameter of the food canal; the greater the diameter of the canal, the faster the fluid flows.

Movement of fluid into the gut is affected by three factors: the hydrostatic pressure of the fluid in the host organism, capillarity, and muscular activity. If the fluid is under high pressure, simply piercing the vessel containing it is sufficient for the fluid to be forced out, just as water gushes out from a burst water main. Phloem, the fluid carrying sugars and amino acids away from photosynthetic tissues to other parts of a plant, is under such positive pressure, up to 1 MPa, and, consequently, phloem-feeding insects, such as most aphids, have simply to penetrate a sieve tube and the phloem is forced through the food canal and into the gut. If the stylets are cut experimentally, phloem continues to ooze out, and this oozing provides a method for obtaining samples of phloem. Although vertebrate blood is under pressure, pressure in the blood capillaries is probably too low to play a major part in forcing blood into the insect's gut.

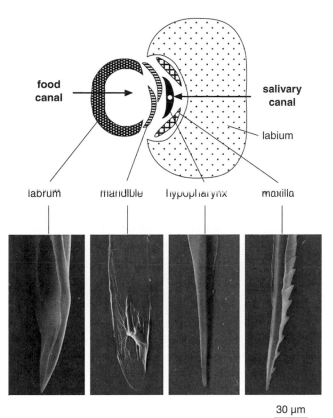

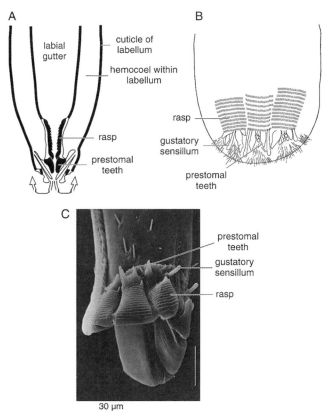

FIGURE 7 Piercing and sucking mouthparts of a female mosquito. Above is a transverse section through the proboscis. Below are electron micrographs of the tips of the stylets. (Reproduced, with permission from Elsevier Science, from Smith, 1985.)

FIGURE 8 Piercing and sucking mouthparts of a tsetse fly. (A) Longitudinal section through the tip of the labellum at rest. The rasps are internal, and prestomal teeth are concealed within the opening. Arrows indicate the directions of movement of the rasps when the labellar lobes are everted. (B) Inside of one labellar lobe in the rest position. Each rasp is made up of rows of downwardly pointing cuticular spines. (A and B reproduced, with permission from Cambridge University Press, from B. Jobling, 1933, A revision of the structure of the head, mouth-part and salivary glands of *Glossina palpalis* Rob-Desv. *Parasitology* **24,** 449–490.) (C) Side view of the labellum with rasps everted. The rasps and prestomal teeth are now on the outside of the labellar lobe. Notice that the teeth now point upward. In moving from the position shown in (A) and (B), the pointed ends have scraped the skin of the host. This movement, and the anatomical arrangement, are basically similar to that seen in the blow fly (Fig. 6). (Reproduced, with permission from Elsevier Science, from Smith, 1985.)

Xylem, in contrast to phloem, is under high negative pressure which may exceed −1 MPa. [Xylem is the fluid imbibed by the roots of plants and drawn upward through xylem vessels as a result of water loss (transpiration) from the leaves.] Consequently, insects that feed on xylem require a powerful pump to overcome this negative pressure and draw the fluid into the mouth. Cicadas have a highly developed cibarial pump made obvious externally by the inflated clypeus. The cibarium is the space between the mouthparts, outside the mouth. In fluid-feeding insects, this space forms a continuum between the mouthparts and the mouth. Blood-sucking insects such as mosquitoes also have a cibarial pump, but it is less well developed than the pharyngeal pump formed by the first part of the foregut.

The importance of capillarity in insect feeding is not well understood. Insect cuticle, in general, tends to be water repellent but, if the cuticle lining the food canal in the mouthparts is wettable, capillarity might be important. In honey bees, when the glossal tongue is dipped into nectar, the fluid adheres to it, being held in place by hairs that project from the surface of the tongue. The glossa is drawn in and out between the folds of the galeae and labial palps and it is probable that capillarity is important in drawing the fluid toward the mouth.

See Also the Following Articles
Feeding Behavior • Rostrum

Further Reading

Chapman, R. F. (1995). Mechanics of food handling by fluid-feeding insects. *In* "Regulatory Mechanisms of Insect Feeding" (R. F. Chapman and G. de Boer, eds.), pp. 3–31. Chapman & Hall, New York,

Kingsolver, J. G., and Daniel, T. L. (1995). Mechanics of food handling by chewing insects. *In* "Regulatory Mechanisms of Insect Feeding" (R. F. Chapman and G. de Boer, eds.), pp. 32–73. Chapman & Hall, New York.

Labandeira, C. C. (1997). Insect mouthparts: Ascertaining the paleobiology of insect feeding strategies. *Annu. Rev. Ecol. Syst.* **28,** 153–193.

Smith, J. J. B. (1985). Feeding mechanisms. *In* "Comprehensive Insect Physiology, Biochemistry and Pharmacology" (G. A. Kerkut and L. I. Gilbert, eds.), Vol. 4, pp. 33–85. Pergamon Press, Oxford.

Movies, Insects in

May R. Berenbaum and Richard J. Leskosky
University of Illinois

The ubiquity of insects and the frequency of their interactions with humans virtually ensure that they will feature prominently in cultural contexts. Throughout history, insects have conspicuously appeared in a range of visual media, including painting, sculpture, printing, and engraving. Thus, with the advent of film in the late-19th century, insects were depicted in some of the earliest efforts; since then, they have made appearances in virtually every form of this modern medium. As the 21st century begins, insect images are common in film and television, and their role in cinema is firmly established. In fact, their impact on culture has been so pronounced that references to insect movies even serve on occasion as punch lines in jokes and cartoons, and the expression "big bug film" is widely recognized.

THE TAXONOMY OF CINEMA

What constitutes an insect in cinema is not necessarily consistent with scientific standards. In the taxonomy of cinema, any jointed-legged, segmented organism with an exoskeleton is likely to be classified as an insect, irrespective of how many legs or how few antennae it possesses. For example, in *Sherlock Holmes and the Spider Woman,* Holmes identifies a spider (used by the Spider Woman to dispatch her victims) as *Lycosa carnivora* from the Obongo River in Africa, the "deadliest insect known to science"; in actuality, spiders are classified as arachnids and not as insects at all. Taxonomic categories are also ill defined in cinema; in the film *Tarantula,* for example, the artificial-nutrient-enhanced giant spider is identified as being "from a species called Arachnida—a tarantula to be exact." The taxon Arachnida is not a species; rather, it is a class, containing thousands of species. In *Mimic,* the eminent entomologist Dr. Gates, mentor of a young scientist engaged in genetic engineering experiments with cockroaches, makes reference to the "Phylum Insecta." Again, "Insecta" is the name of the taxonomic unit called a "class"; the phylum to which insects belong is Arthropoda.

Insect morphology in the movies reflects the relatively sketchy familiarity most filmmakers have with entomological reality. As is the case for real-life insects, most movie insects have six legs, whereas movie arachnids often have eight. In general, movie insects also have the three characteristic body regions—head, thorax, and abdomen—that differentiate them from other arthropods. Even at the ordinal level, many morphological features are depicted with some degree of accuracy. Movie mantids can have raptorial forelegs (e.g., *The Deadly Mantis*), and movie lepidopterans (e.g., *Mothra,* the giant moth that attacked Tokyo in a series of Japanese films

from the 1960s) possess scales. Large flat objects the size of pie plates scattered around the countryside (eventually identified as oversize scales) provide evidence of an enormous moth in *The Blood Beast Terror.*

Other aspects of insect anatomy, however, are not so accurately portrayed. Compound eyes are cause for some confusion; many films present an "insect-eye" view of a particular scene (usually a victim-to-be) through a Fresnel lens, to simulate what is imagined to be the image created by compound eyes (e.g., in *Empire of the Ants*). In reality, these images appear to insects to be more like mosaics than repeated images. In *Monster from Green Hell,* the compound eyes of the cosmic wasps roll in their sockets; real compound eyes are incapable of such motion. Antennae are also poorly understood anatomical features; on occasion, movie arachnids are equipped with a pair, even though antennae are lacking in real-life arachnids. Not surprisingly, mouthparts (whose intricacies in real insects are rarely visible to the naked eye) in movie arthropods often bear little resemblance to real arthropod mouthparts.

Insect physiology in movies often bears only a passing resemblance to the physiology of real arthropods. According to Dr. Elliot Jacobs, the entomologist in *Blue Monkey* who assists in attempting to control an outbreak of genetically engineered mutant cockroaches in a hospital, "Insects aren't like humans or animals. They're 80% water and muscle. They have very few internal organs." A recurring conceit in insect films is the violation of the constraint imposed by the ratio of surface area to volume—movie arthropods routinely grow to enormous size without suffering the limitations of tracheal respiration or ecdysis and sclerotization experienced by real-life arthropods. Nonetheless, there are physiological attributes of film arthropods that are reproduced with some degree of fidelity. Insect pheromones figure prominently in insect fear films (although they are not always identified as such; in *The Bees* they're called "pherones"). In *Empire of the Ants,* for example, giant ants use pheromones to enslave the local human population and to compel the humans to operate a sugar factory for them. The explanation provided for the response is that a pheromone "causes an obligatory response—did you hear that? Obligatory. It's a mind-bending substance that forces obedience...." Although they have long been documented to exist in a wide range of organisms (including humans), pheromones rarely appear in science fiction films outside an entomological context.

As is the case with insect physiology and morphology, insect ecology takes on different dimensions in the movies. Life cycles are unorthodox and generally dramatically abbreviated by entomological standards. In *Mosquito,* for example, mutated mosquitoes, the offspring of normal mosquitoes that had consumed the blood of aliens in a crash-landed UFO, have a life cycle consisting only of egg and adult stages. In *Ticks,* full-grown ticks eclose from what seems to be a cocoon. Population dynamics differ as well. A number of movie arthropods seem to have a population size of one (as evidenced by the titles—e.g., *Tarantula, The Deadly Mantis*), and reproduction does not seem to occur (at least over the 2-h span of the movie). At the other extreme, populations often build up to enormous sizes without depletion of any apparent food source. Bees blacken the sky in *The Bees* and *The Swarm* in a remarkably short period of time with no superabundance of nectar sources in evidence. It must be assumed that food utilization efficiencies of virtually all film arthropods are far higher than they are in real life because arthropods in films, giant or otherwise, rarely produce any frass (in *Beginning of the End,* giant grasshoppers that consume several tons of wheat in a 3-month period with little or no frass to show for it). In *Starship Troopers,* it is unclear what the giant arthropods living on a planet that is bereft of other life-forms eat to attain their large size. However, because they are alien insects, terrestrial biological standards may not necessarily be applicable.

Insect behavior in big bug films is often biologically mystifying. Screen insect predators and herbivores alike almost invariably announce their presence with an ear-piercing stridulating sound (e.g., *Them, The Deadly Mantis, Beginning of the End, Empire of the Ants*); in reality, such behavior would alert prey to danger and elicit escape or defensive behavior (which, on the part of humans in many films, involves machine guns and bazooka fire directed at the insect). For example, in *Beginning of the End,* a television newscaster updates viewers in Chicago on the Illinois National Guard's efforts against hordes of gigantic radiation-induced mutant grasshoppers descending on the city, reassuring them that "the one advantage our forces hold over the enemy is that they ALWAYS reveal their intention to attack. Before every attack the locusts send forth this warning in the form of a high-pitched screech. Now, this screech increases in intensity until it reaches ear-shattering proportions. And it's when this screech reaches its full intensity that the locust attacks." Such maladaptive behavior is unlikely to persist in nature.

INSECTS IN ANIMATED FILMS

Until the mid-20th century, insect representation in cinema was restricted largely to animated films. The small size of insects presented challenges to the standard equipment of the time that could not be met without either a disproportionate increase in cost or a decrease in visual quality. In animated films, however, technical limitations could be avoided; to create the illusion of a close-up, the animator can simply draw a larger image. In animated films, one or two frames are exposed at a time, and between exposures small changes are introduced; for example, one drawing may be substituted for another slightly different drawing or a puppet or clay model slightly repositioned. When the film is projected at normal speed, the image appears to move. An insect may even have inspired one film pioneer to become one of the first animators. Segundo de Chomon, a Spanish filmmaker of the late 19th and early 20th centuries, allegedly conceived of the animation process while shooting intertitles for a silent film and noticing that a fly, included on the footage exposed a frame at a time, appeared to move in a jerky fashion when the film was projected.

This is not to say that insects were not a challenge to animators. Because of their many moving body parts—six legs, two antennae, and from two to four moving wings—many animators simplified their drawing by reducing appendages. Thus, in animated films, insects may be depicted with four instead of six legs and spiders with six instead of eight. The first appearance of an insect in an animated cartoon was in a 1910 film by Winsor McCay titled *How a Mosquito Works*—the second American animated cartoon. Although McCay accurately portrayed his mosquitoes with six legs and two wings, in contrast with later animated films featuring insects with reduced appendages, even this early film contains many of the other conventions typically found in insect cartoons—an adversarial relationship between humans and insects, as well as the depiction of insect mouthparts as tools. McCay showed the film in vaudeville houses to large crowds and later returned to the use of insect characters with his 1921 film *Bug Vaudeville.*

Another very early example of puppet animation was provided by entomologist-turned-animator Wladislaw Starewicz. In attempting to film the mating behavior of stag beetles, Starewicz discovered that the hot lights used to illuminate his subjects caused them to stop moving altogether; accordingly, he killed and dismembered the beetles and wired their appendages back onto their carcasses, painstakingly repositioning them for sequential shots in the short film *The Fight of the Stag Beetles.* That film and its fictionalized sequel, *Beautiful Lucanida or the Bloody Fight of the Horned and the Whiskered,* proved to be quite popular with audiences. Starewicz expanded his efforts, eventually abandoning real insects and constructing puppets *de novo* for his later films with more complex plots (as in *Revenge of the Kinematograph Cameraman,* a story of love and betrayal among a variety of insect species).

Arguably the most well-known animated arthropod in animation was Jiminy Cricket, who initially appeared in a supporting role in the 1940 Walt Disney feature *Pinocchio.* Disney animators used a talking cricket, a minor character that appeared in the original Pinocchio story by Carlo Collodi, to unify disparate elements within the film. The character proved to be popular as a "voice of conscience" and appeared in several series of subsequent short subjects and educational films. Jiminy exemplifies the liberties taken with insect morphology by animators; although early sketches depicted the character with more insect-like features, the final film version, with its two arms and two legs, eyes with pupils, and morning coat and vest, resembles a dapper elf more than any arthropod.

Computer animation developed at a rapid pace during the 1980s and has proved particularly well suited to depicting insects. Modern methods of computer-generated imagery (CGI) have become particularly effective at creating shiny metallic surfaces and at joining slender rodlike structures to larger volumes—exactly what is needed to depict an insect's exoskeleton and multiple appendages. The first computer-animated insect was Wally-Bee, in the 1984 short film from

Pixar titled *The Adventures of Andre and Wally-Bee;* this film was the first computer-animated short film with a plot line. That CGI offers technological advantages over traditional animation is not to say that it has resulted in more realistic animated insects. *A Bug's Life* (1998) from Disney/Pixar continued to depict insects with anthropomorphized faces and four limbs to ensure audience empathy with the characters; the DreamWorks film *AntZ* (1998) gave its ants six legs but provided them with similarly humanized faces and raised the head and thorax into a vertical position, making them look like tiny centaurs. CGI is not limited to what is basically caricature, however; the otherwise live-action *Joe's Apartment* (1996) featured hundreds of computer-rendered cockroaches which were indistinguishable from the real thing, except for their ability to sing and dance.

INSECTS IN FEATURE FILMS

Big Bug Films

Frequent appearances by insects in live-action films are a relatively recent phenomenon in the history of film. For many years, the technical challenges of filming very small, largely untrainable, fast-moving creatures proved a disincentive for incorporating them into films. The pioneering efforts of special-effects genius Willis O'Brien, starting in the 1930s, and of his protégé Ray Harryhausen, as well as technical advances in the production of film stock and traveling matte techniques, gradually made the incorporation of insect images in film economically attractive, or at least reasonable. Moreover, competition for audiences, particularly with the rise of television, led the major film studios to increase investment in hitherto minor genres, such as science fiction. With bigger budgets, more elaborate effects became feasible.

The year 1954 was a watershed year; *Them!* was released by Warner Brothers Studios, featuring giant ants mutated by exposure to atomic testing in the Arizona desert (Fig. 1). The film, tapping into widespread fears of atomic power in the aftermath of World War II, was an enormous success, grossing more money for the studio that year than any other and winning an Academy Award for special effects. Its success is understandable in retrospect: its use of large mechanical models was innovative and dramatic, its screenplay was tight and well written, it featured several big-name actors of the era, and its subtext about invasion disrupting the fabric of American life played well to American fears of communist powers.

The "big bug films" inspired by the success of *Them!* were by and large lesser efforts. Many of these were the work of director/producer Bert I. Gordon, who made so many films with big animals that he was known as "Mr. Big" (a reference as well to his initials—B.I.G.); his big bug films included *Beginning of the End* (1957), featuring giant radiation-induced grasshoppers threatening to destroy Chicago, and *Empire of the Ants* (1977), about giant radioactive-waste-induced ants threatening a real estate development in

FIGURE 1 Lobby poster from the science fiction classic *Them!* (1954), noted for its dramatic special effects and suspenseful screenplay. The firm depicted an attack on the city of Los Angeles by ferocious giant ants, which have been enlarged to greater than human size by the mutating effects of radiation exposure. Made during the height of the 1950s Red Scare, *Them!* attacted large audiences with its ability to link the imaginary threat of gigantic, murderous insects with America's very real fears of nuclear fallout, foreign invasion, and scientific manipulation of the natural world. (*Them!* © 1954 Warner Bros. Pictures, Inc. All rights reserved.)

Florida. Other notable titles of the fifties in the "big bug" genre (Table I) include *Tarantula, The Deadly Mantis, The Black Scorpion, Monster from Green Hell,* and *Earth vs the Spider.* The Japanese film industry did not embrace this genre until the 1960s but made up for the slow start in volume; the first Japanese big bug film, *Mothra,* featuring a giant radiation-induced moth, was released in 1962 and was followed by four sequels, in which Mothra appeared with other "big" science fiction stars such as Godzilla and Rodan.

Transformation/Metamorphosis Films

Metamorphosis is a characteristic of a substantial proportion of movie arthropods, although the process differs on screen.

TABLE I   Live-Action Feature Films with Insects as Major Components

Year	Film
1938	*Yellow Jack*
1944	*Sherlock Holmes and the Spider Woman*
1944	*Once upon a Time*
1953	*Mesa of Lost Women*
1954	*Them!*
1954	*Naked Jungle*
1955	*Tarantula*
1955	*Panther Girl of the Kongo*
1957	*The Black Scorpion*
1957	*Beginning of the End*
1957	*Earth vs the Spider*
1957	*The Deadly Mantis*
1957	*Monster from Green Hell*
1958	*The Cosmic Monsters*
1958	*She Devil*
1958	*The Fly*
1959	*Return of the Fly*
1959	*The Brain Eaters*
1959	*Wasp Woman*
1962	*Mothra*
1964	*Godzilla vs The Thing*
1965	*Horrors of Spider Island*
1966	*The Deadly Bees*
1968	*Destroy All Monsters!*
1969	*The Blood Beast Terror* (aka *The Vampire Beast Craves Blood*)
1970	*Flesh Feast*
1971	*The Hellstrom Chronicle*
1971	*The Legend of Spider Forest*
1972	*Kiss of the Tarantula (Shudders)*
1973	*Invasion of the Bee Girls*
1974	*Phase IV*
1974	*Locusts*[a]
1974	*The Killer Bees*[a]
1975	*Bug*
1975	*The Giant Spider Invasion*
1975	*Food of the Gods*
1976	*The Savage Bees*[a]
1976	*Curse of the Black Widow*[a]
1977	*Empire of the Ants*
1977	*Exorcist II—The Heretic*
1977	*Ants: It Happened at Lakewood Manor*[a]
1977	*Kingdom of the Spiders*

continues

TABLE I   (*Continued*)

Year	Film
1977	*Terror out of the Sky*[a]
1978	*Tarantulas: The Deadly Cargo*[a]
1978	*The Bees*
1978	*The Swarm*
1978	*Curse of the Black Widow*[a]
1980	*Island Claws* (aka *Night of the Claw*)
1982	*Creepshow*
1982	*Legend of Spider Forest*
1985	*Flicks*
1985	*Creepers* (aka *Phenomena*)
1986	*The Fly*
1987	*Blue Monkey*
1987	*The Nest*
1987	*Deep Space*
1989	*The Fly II*
1990	*Arachnophobia*
1991	*Meet the Applegates*
1991	*The Age of Insects*
1993	*Cronos*
1993	*Ticks*
1994	*Skeeters*
1995	*Mosquito* (aka *Nightswarm*)
1996	*Angels and Insects*
1996	*Wax, or the Discovery of Television among the Bees*
1996	*Joe's Apartment*
1996	*Wasp Woman*[a]
1997	*Starship Troopers*
1997	*Men in Black*
1997	*Mimic*
1998	*X-Files: The Movie*
1999	*Deadly Invasion: The Killer Bee Nightmare*[a]
1999	*Atomic Space Bug*
2000	*They Nest*[a]
2000	*Bug Blaster*[a]
2000	*Spiders* (aka *Cobwebs*)
2001	*Evolution*
2001	*Bug*
2001	*Mimic II: Hardshell*
2001	*Spiders II*
2002	*Men in Black II*
2002	*Spiderman*

[a] Made-for-television movie.

The transformation most frequently depicted in films is insect to human or human to insect, generally involving some form of exchange of body fluids—"*Drosophila* serum" in the case of *She Devil* (which allows the patient to transform herself at will from brunette to blonde), "spider hormones" in *Mesa of Lost Women,* "royal jelly" in *Wasp Woman,* and "DNA" in the 1986 remake of *The Fly.* Insects most likely appear frequently in films involving metamorphosis because of the shock value—the transformation of a human into a life-form radically different in appearance. Generally, transformations of humans into other animal forms in films involve magic or reincarnation (*The Shaggy Dog, The Shaggy D.A., Oh, Heavenly Dog, Lucky Dog*) or genetic predisposition (*Teen Wolf* and its sequel, *The Howling,*

Cat People) rather than mediation by hormones (with the exception of the early films of Bela Lugosi, including *The Ape Man* and *Return of the Ape Man,* which involve serum exchanges between humans and apes).

In the 1980s, insect fear films acquired a new life with the release of David Cronenberg's *The Fly.* Although as scientifically as inaccurate as earlier efforts with respect to surface area/volume rules, it was generally regarded by critics as an artistic success, thematically depicting physical transformation leading to mental and emotional change. Although *The Fly II* (directed Chris Walas, special-effects artist on the earlier film) was not embraced as enthusiastically by critics, it nonetheless was perceived as more than just a horror film, with allegorical elements relating the physical and emotional changes of adolescence with the metamorphic transformation of the protagonist. Despite the presence of redeeming intellectual content in these films, these films attracted considerable attention for their graphic special effects, far surpassing earlier efforts.

In 1983, the first report of successful genetic transformation of an insect *(Drosophila melanogaster)* was published, and by 1987 genetically engineered insects (specifically, mutant killer cockroaches) made their first appearance in a science fiction film, in the otherwise unremarkable film *The Nest.* Genetic engineering techniques advanced more quickly on screen than in real life; by 1997, in *Mimic,* the young entomologist Susan Tyler (Mira Sorvino) is able to incorporate termite and mantid DNA into cockroaches, with the goal of creating a "Judas bug" to bring contagion to the cockroach vectors of a human illness but instead unleashing a plague of six-foot-tall people-eating cockroaches in the subway system of New York City. In *Spiders,* unspecified alien DNA is incorporated into the titular arthropods to wreak havoc in a secret government laboratory.

"Social" Insect Films: Small Size, Large Numbers

One of the largest orders of real-life arthropods, the Hymenoptera, is in fact the most frequently depicted in insect fear films. There may be several reasons for this proportional similarity. For example, bees are relatively easily manipulated for the camera in comparison with other insects, can be produced commercially, and can be reared in enormous numbers with comparative ease. Perhaps an even more important factor, however, is their familiarity to the audience. Encounters with bees, ants, and wasps are part of the normal course of life for most moviegoers. Such an explanation also can account for the proliferation of films involving cockroaches, although these, too, share the practical advantage of ease of rearing in enormous numbers and affordability.

Films using footage of real insects engaging in more or less normal insect behaviors rose to prominence in the 1970s and included such efforts as *Phase IV,* featuring documentary-quality footage of ants, and *Bug,* featuring Madagascar hissing cockroaches (albeit engaged in some unusual behaviors, such as spelling out death threats with their bodies on the wall of a house). The appearance of so-called killer bees on a container ship in San Francisco harbor in 1974 may have inspired filmmakers to capitalize on a real threat—the introduction of African honey bees with a reputation for defensive behavior often lethal to animals and sometimes to humans. The films proved popular with filmmakers in part because audience members enter the theater with at least passing familiarity with the film's antagonists (in contrast with giant arthropods). As well, bees can be controlled chemically—by pheromones—to cluster or land in a particular spot and so are more easily manipulated for special effects. Five films were made about killer bee invasions between 1974 and 1978 (Table I), although none of them was particularly successful at the box office (surprisingly for

the 1978 film *The Swarm,* with a screenplay by Arthur Herzog and a cast including such Academy Award-caliber actors as Henry Fonda and Michael Caine).

In many of the films featuring large numbers of small insects, ecological disruption is a recurring theme. Biomagnification, accumulation of toxins up a food chain, is the focus of several. In *Kingdom of the Spiders,* tarantulas take over a town and start consuming livestock because "DDT" destroyed the food chain and deprived them of their normal prey. Other films depicting altered food web dynamics as a result of pollution (radioactive and/or toxic waste) include *Skeeters* and *Empire of the Ants.* In *Ticks,* fertilizers and other chemicals used by illegal marijuana growers are encountered by ticks, which grow to enormous size and terrorize a group of inner-city teens in the woods on a wilderness survival trip.

Another ecological phenomenon of concern both in the movies and in real life is the accidental introduction of alien species (although in the movies these are more likely to be real aliens, from outer space, not just a foreign country). *Arachnophobia* depicts the fictional consequences of the accidental introduction of a South American spider species to the Pacific Northwest. The many killer bee movies pointedly make reference to the dangers of accidental importation of strains of bees into new habitats (although in *The Bees* their introduction is no accident; greedy cosmetics magnates import killer bees in the hope of producing large amounts of profitable royal jelly).

INSECTS IN DOCUMENTARY FILMS

Although educational shorts for school and extension markets often deal with entomological topics, documentary filmmaking, which combines information and art, has tended to ignore this area. For a long time, documentary filmmakers faced many of the same challenges faced by feature filmmakers with an interest in insects. Only until the latter half of the 20th century did developments in technology permit the capture of small moving objects (such as insects) on film in a compelling and effective manner. Yet another obstacle, particularly problematic for documentary filmmakers, was audience interest; whereas audiences could accept insects bent on destruction of the human race in science fiction or horror films, they generally showed considerably less interest in the accurate depiction of the lives of real-life insects. Animal documentary filmmakers have long had to accept the fact that audiences prefer drama to accuracy in depictions of nature. Walt Disney, with his groundbreaking *True Life Adventure* nature films made between 1948 and 1970, relied in many of his nature films on personification and anthropomorphism to make the animal subjects of studio films more appealing to audiences.

Arguably the first "documentary" films involving insects were the pioneering efforts of F. Percy Smith, who in 1912 created films aimed at illustrating the physical prowess of the common house fly. Smith enclosed a fly inside a dark box equipped with a thin glass door at one end; the door in turn had a small opening into which was fitted a toothed wheel that was free to rotate. The fly, orienting to the light entering through the glass door at one end of the box, would move toward the light; when it encountered the glass door obstructing its escape, it was struck on the head by a tooth in the wheel which rotated as a consequence of the fly's movements. Eventually, via conditioning, the fly simply walked up the wheel, which would rotate, creating a treadmill and providing the photographer an opportunity to film the fly walking in place. Smith modified his approach to film flies outside the box, tethered in place, and in this way was able to obtain footage of them seemingly juggling dumbbells, corks, bits of vegetables, other flies, and sundry other objects. When the film was released newspaper reports accredited the cinematographer with strange powers, and the capacity to train house flies as others do circus animals.

Audience reluctance to accept insects for their own sake is the explanation for the peculiar framing device used in the first big-budget feature-length documentary about insects, *The Hellstrom Chronicle*. This film was originally conceived as a straightforward documentary and featured what was at the time state-of-the-art macrophotography that provided startling and dramatic close-ups of its arthropod subjects. The extraordinary inventiveness of cinematographer Ken Middleham led to spectacular images of insects engaged in a wide range of behaviors. However, the studio heads were unconvinced that a documentary about insects could bring in an audience and insisted on adding to the film a fictional storyline, about an academic, Dr. Nils Hellstrom (Lawrence Pressman), denied tenure because of his insistence that insects were bent on human destruction. As a result, the hybrid film was a commercial success as well as an artistic success of sorts (earning a Grand Prix de Technique award at the 1971 Cannes Film Festival for its remarkable images), although it was panned by critics, in part because of its sensationalistic tone.

A general awakening of the American public to environmental issues in the 1970s did little to inspire interest in insect biology, and insect documentaries have been few and far between since *The Hellstrom Chronicle*. Insects figured peripherally in the documentary *Cane Toads, An Unnatural History,* directed and written by Mark Lewis, and first shown in 1988. The cane toad *Bufo marinus* was deliberately introduced into North Queensland, Australia in 1935 to control *Lepidoderma albohirtum* (a beetle larva) and its relatives. Although the cane toads were ineffectual biocontrol agents, they were exceptionally effective colonizers, which now populate much of Queensland, northern New South Wales, and eastern Northern Territory, wreaking ecological and environmental havoc. The history of this ill-conceived biocontrol effort and its consequences are the subject of the documentary.

Microcosmos (1996) is similar to *The Hellstrom Chronicle* in that its success was due largely to quantum improvements in capturing insect images and behavior on film. Filmmakers Claude Nuridsany and Marie Perennou spent 15 years researching, 2 years designing equipment (including inventing a remote-controlled helicopter for aerial shots), 3 years shooting, and 6 months editing a masterpiece of insect cinematography. Yet the concept underlying the film was quite novel. Instead of a "superdocumentary" of amazing insect feats, the filmmakers settled on the idea of telling the story of a single summer day (albeit in reality filmed over a much longer period of time) in a field in the countryside of Aveyron, France (where they lived and worked). Their goal was to depict insects and other small creatures not as "small bloodthirsty robots" but rather as individuals with unique abilities. Instead of narration, there was a simple introduction, 40 words, spoken by actress Kristin Scott Thomas. *Microcosmos* was well received by critics (although it failed to win a nomination for best documentary at the Academy Awards), and it performed respectably at the box office. Although some notable entomologists bemoaned the absence of voiceover and the loss of an opportunity to educate the public about the insect lives captured on film, the extraordinary images depicted on screen will likely set the standard for excellence in insect documentary filmmaking for years to come.

INSECT WRANGLERS AND SPECIAL EFFECTS

Because handling insects and other arthropods and eliciting appropriate behaviors from them on cue is beyond the experience and training of most directors, these responsibilities are frequently delegated to a specialized crew member known in the profession as an "insect wrangler" or "bug wrangler." Since the early 1960s, only a handful of individuals have engaged in this occupation in a conspicuous way. Some insect wranglers specialize in handling a narrow range of taxa. Norman Gary has been a bee wrangler for more than a quarter-century. Currently an emeritus professor at University of California at Davis, he served as a faculty member in bee biology from 1962 to 1994. His research interests have been in the area of bee behavior, and he has written or coauthored over 100 publications on bees. Since 1966, he has been a consultant for legal, industrial, film, and television productions about bees. His ingenuity in developing methods for manipulating bees and their behavior has led him to develop methods of narcotizing queens to facilitate instrumental insemination, as well as vacuum devices for handling, tagging, counting, confining, and otherwise handling bees. An abbreviated filmography for Gary includes *My Girl, Fried Green Tomatoes, Candyman, Beverly Hillbillies, Man of the House, X-Files, The Truth about Cats and Dogs, Leonard Part VI, A Walk in the Clouds,* and *Invasion of the Bee Girls*.

Another individual with an affinity for a particular taxon is Ray Mendez, who worked as an entomologist at the American Museum of Natural History in New York. Mendez, along with colleague David Brody, provided over 20,000 cockroaches for the film *Creepshow* in 1982; in 1996 Mendez wrangled 5000 live cockroaches and provided advice on animated and puppet cockroaches for the film *Joe's Apartment*.

Mendez is also an authority on naked mole rats and was featured in the documentary *Fast, Cheap and Out of Control.*

Steven R. Kutcher, a consulting entomologist in Arcadia, California, and part-time biology instructor at West Los Angeles College, is notable for the range of arthropods with which he has worked. Kutcher obtained a bachelor's degree in entomology at University of California at Davis and a master's degree in biology with an emphasis on insect behavior and ecology at California State University, Long Beach. Since 1976 he has been involved in arthropod wrangling for many movies and commercials. He has worked with a variety of arthropods, including spiders, yellowjackets, cockroaches, mealworms, grasshoppers, and several species of butterflies. Among his film credits are *Extremities, Exorcist II: The Heretic, Arachnophobia, Race the Sun, Jurassic Park,* and *Spiderman.* His unusual vocation has made him the focus of more than 100 print articles, and in 1990 his work was the subject of a short documentary by National Geographic.

INSECT FEAR FILM FESTIVALS

The idea of using insects in movies as a means of entomological outreach apparently dates back to the origins of the annual Insect Fear Film Festival at the University of Illinois at Urbana-Champaign. The first festival, brainchild of then assistant professor of entomology May Berenbaum, was held in March 1984. The goal of the festival has been to use insect fear films to draw in an audience and to use the films as a means for highlighting scientific misconceptions about insects. At each festival, two or three feature-length films are shown, interspersed with animated shorts. Before the festival begins, and between films, the audience is invited to see and handle a variety of live specimens as well as pinned specimens. Generally, the festivals are organized around themes, which have included female insects, noninsect arthropods, orthopteroids, social insects, cockroaches, flies, and mosquitoes.

Other events that have been held in conjunction with the festival included a thematically relevant blood drive, held in cooperation with Community Blood Services of Champaign, for the 1999 mosquito film festival. Attendance at these festivals can exceed 1000. Over the years, the festival has been featured in a wide range of media throughout the world.

Other insect fear film festivals per se are few in number; Iowa State University has conducted an Insect Horror Film Festival since 1985, and Washington State University has hosted its Insect Cinema Cult Classics festival since 1990. Insect films, however, have been elements of insect expo and public outreach efforts in many venues, including museums, science centers, and universities across the country.

There is one legitimate insect film festival in the traditional sense, in which films are submitted in competition and are judged and awarded prizes. FIFI, organized by l'Office pour les Insectes et leur Environnement du Languedoc-Roussillon (OPIE LR) and the the Regional Natural Park of Narbonne and the city of Narbonne, France, is a biannual international film festival dedicated to insects and other small animals. The FIFI, in its fourth year in 2001, is the result of a partnership with the Institute for Research and Development (IRD), the French National Center for Scientific Research (CNRS), the National Institute of Agronomic Research (INRA), the City of Sciences and Industry (Paris), the National Museum of Natural History (Paris), and the Agronomic University of Gembloux (Belgium). Its stated objectives are to increase the sensitivity of the media and the public to the ecological importance of continental invertebrates as well as to encourage and promote the making of films or videos dedicated to insects.

FUTURE OF INSECTS IN CINEMA

With the continuing development of CGI and the veritable explosion of such outlets for film as cable stations, satellite television, DVD and video markets, the future of insects and other arthropods in the movies looks assured. Arthropods will certainly continue to be objects of distaste and unease for audiences throughout the world and so will remain staples of horror films and certain types of science fiction adventure. Moreover, CGI and developments in macrophotography techniques ensure that insect images on screen will become increasingly sophisticated, although scriptwriting will almost assuredly remain as resolutely unrealistic as it has since the earliest days of insects in cinema.

See Also the Following Articles

Cultural Entomology • Folk Beliefs and Superstitions • Teaching Resources

Further Reading

Berenbaum, M. R. (1995). "Bugs in the System: Insects and Their Impact on Human Affairs." Addison Wesley, Reading, MA.
Berenbaum, M. R. (2000). See you in the movies. *Am. Entomol.* **46,** 210–212.
Berenbaum, M., and Leskosky, R. (1991). Mosquitoes in the movies. *Vector Control Bull. North Central States* **1(2),** 94–98.
Berenbaum, M. R., and Leskosky, R. J. (1992). Life history strategies and population biology in science fiction films. *Bull. Ecol. Soc. Am.* **73,** 236–240.
Brosnan, J. (1991). "The Primal Screen. A History of Science Fiction Films." Little Brown, Boston. (See Chap. 6, The Metaphor That Ate Tokyo: Monster Movies of the Fifties.)
Kottmeyer, M. S. (1997). Bugs invade! A cultural history of horror. *UFO Mag.* **12,** 20–25.
Leskosky, R. J. and Berenbaum, M. R. (1988). Insects in animated films; or, not all "bugs" are bunnies. *Bull. Entomol. Soc. Am.* **34,** 55–63.
Mertins, J. W. (1986). Arthropods on the screen. *Bull. Entomol. Soc. Am.,* **32,** 85–90.
Warren, W. (1982). "Keep Watching the Skies!" McFarland, Jefferson, NC.
Warren, W. (1986). "Keep Watching the Skies! II." McFarland, Jefferson, NC.

Musca domestica

see *House Fly*

Muscle System

Robert Josephson

University of California, Irvine

Muscle is the excitable, contractile tissue of animals that is responsible for movement and behavior. Although there is great variability in structure and performance among different insect muscles, many basic features of biochemical composition, ultrastructural organization, and contractile performance are common among insect muscles and indeed are similar between muscles of insects and those of vertebrates.

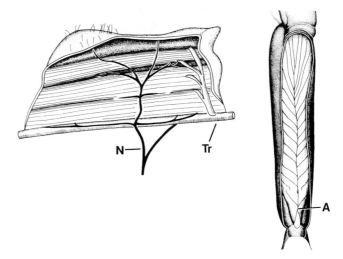

FIGURE 1 Muscle with parallel fibers (left) and one with pinnate fibers (right). The parallel-fibered muscle is the mesothoracic dorsal longitudinal flight muscle of the tettigoniid *Neoconocephalus robustus.* (Modified from Stokes, Josephson and Price, 1975, *J. Exp. Zool.* **194,** 379–407.) The dark structure coursing across the muscle surface is the motor nerve that innervates the muscle. The pinnately fibered muscle is the metathoracic extensor tibia of the cricket *Teleogryllus oceanicus.* (Modified from Donaldson and Josephson, 1981, *J. Comp. Neurol.* **196,** 735–742.) Abbreviations: N, motor nerve; Tr, trachea; A, apodeme.

MUSCLE STRUCTURE AND ULTRASTRUCTURE

Muscle Fibers and Fiber Bundles

The skeletal muscles of insects are bundles of elongate, multinucleate cells called muscle fibers. The fibers attach at each end to the exoskeleton. The muscles typically span joints of the exoskeleton and, when active, cause bending of the joint or stabilization of the joint against external forces. Skeletal muscles are the muscles of behavior, the muscles involved in posture and locomotion. In addition to skeletal muscles, insects contain visceral muscles that cause movement of the gut, Malpighian tubules, and parts of the reproductive system; there are also cardiac muscles that cause contraction of tissue sheets and vessels associated with the circulatory system. The visceral and cardiac muscle cells are typically small, spindle shaped, and with a single nucleus.

An individual insect contains many morphologically identifiable skeletal muscles. The large number of muscles is a consequence of the segmental organization of insects and the serial replication of parts associated with segmentation. Each of the wing-bearing segments of a cockroach contains about 50 separate muscles, an abdominal segment a somewhat smaller number. In a classic anatomical study, Lyonet, in 1762, noted that the larva of the goat moth, *Cossus,* contains three times the number of anatomically distinct skeletal muscles as does a human!

In most insect muscles the fibers lie parallel to one another, and when the muscle contracts it shortens along the long axis of the fiber bundle. Such muscles are spoken of as being parallel-fibered muscles (Fig. 1, left). In some muscles, in particular peripheral leg muscles, the fibers attach obliquely at one of their ends onto an internal, cuticular extension called an apodeme (Fig. 1, right). When these muscles are activated, the muscle as a whole shortens along the axis of the apodeme, oblique to the fiber axis. The oblique insertion of fibers onto the apodeme is remindful of the oblique junction between lateral filaments and the main shaft of a feather, hence muscles with an oblique fiber arrangement are called pinnate (L. *pinna* = feather). The force that a muscle can generate increases with increasing cross-sectional area. The pinnate arrangement of muscles increases the effective cross-sectional area and hence the force that the muscle can produce.

Filaments and Fibrils

Muscle shortening in insects as in other animals results from sliding movement between interdigitating thick and thin filaments contained within the muscle fibers. The force of contraction is a shearing force developed between these filaments. The thick filaments are made up largely of the protein myosin, the thin filaments of the protein actin. A single thick filament is composed of many individual myosin molecules and, similarly, a thin filament contains many actin molecules. Projections of the myosin molecules from the thick filaments toward the thin filaments, called cross-bridges, are the sites of interaction between the two and are the force generators for contraction.

The thick and thin filaments are grouped into longitudinal bundles called fibrils. The filaments within a fibril are grouped precisely, both longitudinally and transversely (Figs. 2 and 3). The thin filaments attach to and project from both sides of transverse structures called Z disks. The Z disks occur regularly along the length of the fibril. The interval from one Z disk to the next is called a sarcomere. The thick filaments lay side by side in the middle of the sarcomere. The sarcomere lengths in fibrils of fast muscles such as flight muscles are 2 to 4 μm; those in leg muscles, body wall muscles, and visceral muscles tend to be longer, up to 7 to 10 μm. The regular longitudinal arrangement of Z disks, thin filaments, and thick filaments creates a striped pattern along the length of a fibril

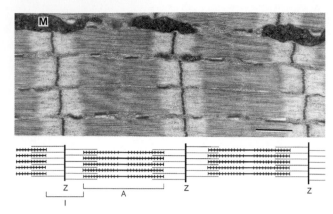

FIGURE 2 Origin of the transverse striations in skeletal muscle. The upper electron micrograph is a longitudinal section of a somewhat stretched fiber from the mesothoracic dorsal longitudinal muscle of the tettigoniid *Neoconocephalus ensiger*. The scale bar represents 1 μm. Abbreviations: M, mitochondrion; I, I band; Z, Z disk; A, A band.

(Figs. 2 and 3). The most obvious components of the striped pattern are (1) the Z disks; (2) the A bands, corresponding to that part of a sarcomere containing thick filaments; and (3)

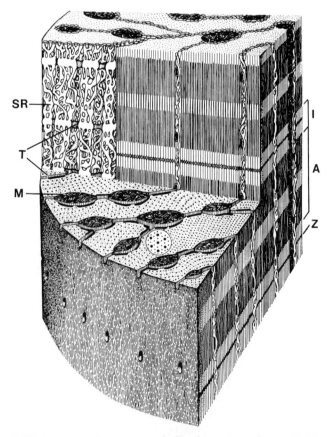

FIGURE 3 Structural organization of a fiber from an insect fast muscle. The drawing is based on electron micrographs from a tettigoniid singing muscle. The fibrils here are radial–lamellar. Abbreviations: A, A band; I, I band; M, mitochondrion; SR, sarcoplasmic reticulum; T, transverse tubule; Z, Z disk. (Modified from Josephson, 1975.)

the I bands, corresponding to that part of a sarcomere without thick filaments. When a muscle shortens, the thin filaments slide toward the center of the sarcomere and the I bands become shorter. Because of their transverse banding pattern, muscles in insects (and skeletal muscle in vertebrates, which have a similar organization) are described as being striated muscles. The visceral muscles of insects are similar in function to vertebrate smooth muscles and in many ways similar in physiology as well. But although vertebrate smooth muscles lack striations, the visceral muscles of insects, like the skeletal muscles, are striated.

The thick filaments of insect muscles, and of vertebrate striated muscles, occur in a regular, hexagonal array. In vertebrate muscles a thick filament is surrounded by 6 thin filaments, each of which lies at the midpoint between three adjacent thick filaments (Fig. 4), and the overall ratio of thin to thick filaments is 2:1. In fast muscles of insects, for example flight muscles, there are also 6 thin filaments surrounding each thick filament, but these occur at the midpoint between two thick filaments and the thin-to-thick ratio is 3:1. In slower insect muscles, such as body wall muscles, the usual pattern is for each thick filament to be surrounded by a circle of up to 12 thin filaments.

The fibrils of insect muscles occur in two basic patterns, cylindrical and radial–lamellar (Figs. 3 and 5). In muscles with cylindrical fibrils the bundles of filaments forming the fibrils occur as elongate cylinders that are often polygonal in cross section. In radial–lamellar fibers the fibrils are ribbon-shaped structures arranged radially about the center of the fiber.

Other Components

The cellular components of muscle fibers seen in electron micrographs fall into four functional groups. First are those structures directly involved in the generation of force and mechanical power. These structures are the thick and thin filaments that collectively form the fibrils. Second are those components involved in the control of contraction. The most obvious structures involved in the control of contraction are the transverse tubular system (T tubules) and the sarcoplasmic reticulum (SR). The T tubules are membrane-bound tubular structures oriented perpendicular to the fiber axis. The membrane of a T tubule is continuous with the surface membrane of the fiber, and the T tubule can be regarded as an inwardly directed

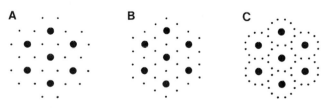

FIGURE 4 Organization of thick and thin filaments as seen in cross sections of fibers from (A) a vertebrate skeletal muscle, (B) a fast insect muscle, and (C) a slow insect muscle.

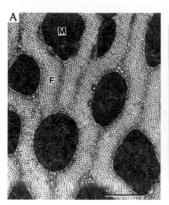

FIGURE 5 Transverse electron microscope sections through (A) a fiber with radial–lamellar fibrils (a flight muscle of the tettigoniid *Euconocephalus nasutus*) and (B) a fiber with columnar fibrils (from the tymbal muscle of the cicada *Abricta curvicoasta*). The scale bars indicate 1 μm. Abbreviations: M, mitochondrion; F, fibril; Tr, intracellular tracheole.

extension of the surface membrane. In most insect muscles there are two tubules per sarcomere, lying in the overlap areas between thick and thin filaments (Fig. 3), but in the fibers of some muscles there is a single, centrally located T tubule per sarcomere. Within the fiber the T tubules make specialized junctions with the SR, which is an internally closed, membrane-bound compartment within the fiber. The function of the T tubules and SR is considered further below. Other elements involved in the control of contraction are the surface membrane of the muscle fibers and membrane specializations at the sites at which nerve processes contact muscle fibers. Third are the structural elements of the metabolic power supply. These are the mitochondria, which provide ATP, and glycogen granules. ATP is the immediate energy source for contraction; glycogen is a stored fuel for cellular metabolism. It would be appropriate to include among the elements involved in metabolic power the tracheoles, the terminal portions of the gas-exchange system that ramify throughout muscle fibers, even though topologically tracheoles are external to and not really part of the muscle fibers. Fourth are the structures involved in long-term maintenance of muscle, specifically the many nuclei of the fibers.

The relative abundance of different cellular components in muscle is tightly correlated with the functional capacity of the muscle fibers. SR and T tubules are particularly abundant in muscles that can produce brief contractions, that is, in muscles in which the contractile apparatus can be rapidly activated and inactivated. Muscles capable of sustained activity at high power output are particularly well supplied with mitochondria and tracheolar endings. Mitochondria make up 30 to 40% of the muscle volume in wing muscle of active fliers and in sound-producing muscles that are active continuously and at high frequency. Such muscles are often pink, because of the cytochromes in the abundant mitochondria. It should be noted that hypertrophy of mitochondria, and of T-tubules and SR, is at the expense of myofibrillar volume, so fast and fatigue-resistant muscles are likely to be relatively weak.

Muscle Attachments

Skeletal muscles attach to the cuticle of the exoskeleton through specialized epidermal cells. The muscle fibers are joined to these cells by specialized junctions. called desmosomes. The terminal sarcomeres of the fibrils lack a final Z disk; instead, the thin filaments are attached to the muscle portion of the terminal desomsome through a band of what have been called junctional filaments. Visceral muscles are frequently joined to one another by desmosomes, and cardiac muscle fibers are joined by structures resembling the intercalary disks of vertebrate cardiac muscle.

INNERVATION AND ACTIVATION

There is an electrical potential across the surface membrane of a living, resting muscle fiber; the interior of the fiber is typically 30 to 70 mV electrically negative with respect to the extracellular solution. Nerve cells in the central nervous system send out long processes (motor axons) to the muscle fibers where they make specialized contacts termed synapses. A motor axon makes many synaptic contacts along the length of each muscle fiber that it innervates (multiterminal innervation), and a single muscle fiber may receive inputs from more than one motor axon (polyneuronal innervation). Impulses initiated in the central nervous system travel along the motor axons and cause the release of specific chemical signals (transmitters) from the motor axon terminals at the synapses. The transmitter released from the terminals of most motor axons leads to a reduction (depolarization) in the transmembrane potential of the muscle fiber in the vicinity of the nerve terminal. Muscle fiber depolarization initiates contraction of the fiber. Motor axons that depolarize muscle fibers and cause muscle contraction are called excitatory axons. Some axons, termed inhibitory axons, release transmitters that stabilize the transmembrane potential of the muscle fiber or even make it greater, thus antagonizing excitatory inputs. In addition to excitatory and inhibitory neural inputs, many muscles receive inputs from modulatory motor neurons, activity that releases chemicals that modify muscle performance, for example, increasing muscle force and work output or speeding relaxation.

Insects, like other arthropods, manage their muscles using relatively few motorneurons. Some major muscles, for example tymbal muscles of cicadas, are innervated by a single motor neuron. Many muscles receive 2 to 4 motor neurons. The largest number of motor neurons yet described for an insect muscle is 16, to the flexor muscle in the leg of a locust.

The processes linking membrane depolarization and contractile activation have been little studied in insect muscles, but the ultrastructure, biochemistry, and contractile performance of insect muscle are so similar to those of the far better studied frog, cat, and rodent muscles that one can predict with confidence that the basic principles worked out for vertebrate muscles apply to insects as well. The expected scheme is as follows. Membrane depolarization spreads inwardly into the fiber along the T tubules. Depolarization of the T tubules, which are coupled to the SR through specialized junctions, leads to release of calcium from the SR. Released calcium reversibly binds to regulatory sites in the fibrils, turning on the contractile machinery. Relaxation occurs as the SR takes up the released calcium and reduces the calcium concentration in the cytoplasm below that needed for contractile activity.

MUSCLE MECHANICS

Muscle Force and Muscle Length

The muscle contraction initiated by a single stimulus, or by a single impulse in an innervating motor neuron, is termed a twitch; that evoked by repetitive input at a frequency high enough to maintain full activation of the muscle is termed a tetanus. A response in which a stimulated muscle develops force while held at constant length is called an isometric contraction. The isometric force generated by a muscle stimulated to contract in a tetanus is maximal at about the normal muscle length in the insect body and declines at longer and shorter lengths. The decline in force with increasing muscle length beyond the optimum is thought to be caused by a reduction in the overlap between the thick and the thin filaments and therefore in the number of myosin cross-bridges that can interact with the actin filaments. The decrease in force at short muscle lengths is probably a consequence of the thick filaments running into and being impeded by the Z disks, of collision of thin filaments in the middle of the sarcomere, and, at still shorter lengths, of overlap of thin filaments with portions of thick filaments of inappropriate polarity on the far side of the center of the sarcomere.

Some muscles in insects and elsewhere can shorten to a small fraction of their resting length, a response termed supercontraction. The capacity for supercontraction appears to involve modifications in the structure of the Z disk such that there is not a collision between the Z disks and the thick filaments at short muscle lengths. In the supercontracting muscles that have been examined, the Z disk becomes perforate at short muscle lengths and the thick filaments slide through the spaces in the disks.

The posterior, intersegmental, abdominal muscles of female locusts are of particular interest for the wide range of lengths over which they can operate. During oviposition, appendages on the end of the abdomen dig and pull the posterior abdomen down into a relatively deep hole. The intersegmental muscles become stretched to about nine times their resting length. During this stretch, called superextension, the Z disks become broken up into discontinuous, nonaligned elements to which the thin filaments are attached. Muscle contractility is not lost, and contraction of intersegmental muscles returns the abdomen to its normal length following oviposition. The latter part of the recovery may be supercontraction, for in the resting state the posterior intersegmental muscles are normally supercontracted, with thick filaments protruding through gaps in the Z disks.

Force, Shortening Velocity, and Power

There is an inverse relationship between the force on a muscle and the velocity with which it can shorten, a relationship conveniently expressed in a force–velocity plot (Fig. 6). To facilitate comparison of muscles of differing size, force in a force–velocity plot is usually expressed as stress (force per unit cross-sectional area) and shortening velocity as strain rate (shortening velocity per unit muscle length). Two points on a force–velocity plot are frequently used to characterize a muscle's contractile properties: the maximum isometric stress of the muscle (F_{max}, the intercept of the curve with the 0 velocity axis) and the maximum shortening velocity (V_{max}, the intercept of the curve with the 0 force axis). Values for the maximum force in insect muscles are generally 5 to 35 N cm^{-2} (N, Newton; 1 N is approximately the downward force exerted by a mass of 100 g in the gravitational field at the earth's surface). The few available measurements of the maximum shortening velocity for insect muscles, all from fast muscles, are on the order of 3 to 15 lengths/s.

The product of force and shortening velocity has dimensions of force × distance per time (work per time) and is the rate of doing work, i.e., the mechanical power output. The product of stress (force per area) and strain rate (shortening distance per second per unit muscle length) is the mechanical power per unit volume of muscle, which is readily convertible to power output per unit muscle mass. Thus each point on a force–velocity plot (or a plot of stress against strain rate) represents a power output. The power predicted from a force–velocity curve is the instantaneous power output. The peak instantaneous power is substantially greater than the sustainable power from a muscle, for during maintained activity a muscle goes through repeated contraction–relaxation cycles and therefore shortens and produces power for only part of the total time. For fast muscle, including fast insect muscles, the peak power output is 100 to 500 W kg^{-1} (1 W = 1 Joule s^{-1} = 1 N m s^{-1}).

Muscles in insects may be divided into synchronous muscles and asynchronous muscles on the basis of the relationship between the patterns of neural activation and of contraction (see below). The sustainable power available during repetitive, cyclic contraction has been determined for several synchronous insect muscles using the work loop approach, in which the muscle is subjected to length changes simulating those during normal activity and stimulated phasically during the length cycles. A plot of muscle force against muscle length for a full cycle produces a loop, the area of which is the net work output of the muscle for that cycle. The product of work per cycle and cycle frequency is the power output. The mechanical power available from synchronous flight muscles of several locusts and katydids and of a moth measured in this way ranges from 50 to 120 W kg^{-1} at normal operating temperature.

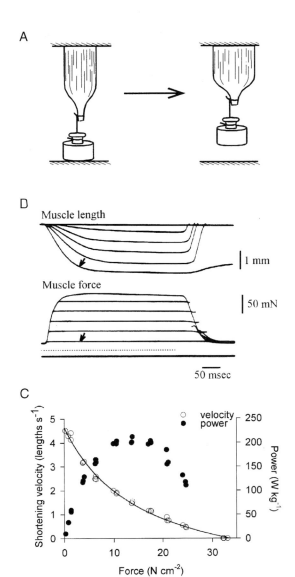

FIGURE 6 Relationships between muscle force, shortening velocity, and power output. (A) Elements of a method used to determine the relationship between force and shortening velocity. The muscle is attached to a load that is supported from below. When stimulated the muscle develops force without shortening until the force equals that of the load, following which the muscle shortens under constant load. (B) Results from an experiment examining force–velocity relations using a tettigoniid wing muscle. (Modified from Josephson, 1984, *J. Exp. Biol.* **108**, 77–96.) The lowest trace indicates the times at which the muscle was stimulated. The force trace marked by an arrow is the contraction with the smallest load of the series; the corresponding shortening trace, which has the shortest latency and the highest initial velocity, is similarly marked. (C) Force–velocity plot and a corresponding plot of power output for a wing muscle of the locust *Schistocerca americana*, 30°C (data provided by J. Malamud).

Asynchronous and Synchronous Muscles

Most insect muscles are like vertebrate skeletal muscles in that each contraction is initiated by depolarization of muscle fibers, and there is a 1:1 relationship between muscle electrical activity and muscle contraction. Such muscles may be termed synchronous muscles, reflecting the congruence between electrical and mechanical activity. The major flight muscles in

several insect groups are different in that there is no synchrony between electrical and mechanical events. These muscles are known as asynchronous muscles. Neural input and fiber depolarization are needed to activate an asynchronous muscle, but when it is activated an asynchronous muscle can contract in an oscillatory manner if it is attached to a mechanically resonant load. The resonant loads for the flight muscles are the wings, which may be regarded as small, somewhat dampened tuning forks. The frequency of the oscillatory contraction is the mechanically resonant frequency of the load, which is greater than the neural input frequency required to keep the muscle fully activated. The contraction frequency of asynchronous wing muscles during flight is typically 3 to 10 times higher than the neural input frequency in each of the motorneurons activating the muscle. The main singing muscles in some but not all cicadas are asynchronous muscles; the resonant load here is the cuticular tymbal to which the muscle is attached and whose inward movement produces the sound pulses.

The features of asynchronous muscle that allow oscillatory contraction are stretch activation and shortening deactivation. When allowed to shorten rapidly an asynchronous muscle becomes deactivated, and while deactivated it can be stretched out to its original length, developing less force than it did while shortening. Stretching the muscle, in turn, reactivates it. Because of shortening deactivation, less work is required to restretch an asynchronous muscle than is produced by the muscle during shortening, and there is net work output when the muscle undergoes a shortening–lengthening cycle. It is this net work that is available to drive the wings and power flight.

Asynchronous muscles occur in several of the most successful insect groups. They power flight in beetles, flies, bees, and wasps and many of the true bugs. The distribution of asynchronous muscles among insect taxa suggests that this mode of muscle control has evolved independently as many as 7 to 10 times. It is likely that asynchronous muscle has been favored by evolution because it is more powerful and more efficient than is synchronous muscle for operation at the high frequencies characteristic of insect flight. It is more powerful, in part, because asynchronous control does not demand rapidity in the rate at which muscle is turned on and off by neural input. High-frequency contraction is achieved without hypertrophy of the sarcoplasmic reticulum, leaving more room in muscle fibers for fibrils, which are the power-producing component. It is more efficient because a relatively low-frequency neural input is needed to maintain full activation, which reduces the amount of calcium that is released and re-bound during activity and the associated metabolic costs of calcium cycling.

Are Insect Muscles Unusual as Motors?

In the minds of many people, insects are extraordinary athletes. One sometimes hears it said that if a person were as strong as an insect, he or she could carry enormous weights or leap over tall buildings. Such assertions are largely based

TABLE I Contractile Properties of a Locust Flight Muscle (Metathoracic Second Tergocoxal Muscle of *S. americana*) and the Frog Sartorius Muscle

	Locust	Frog
Twitch time course, ms		
Rise time	20	21
Onset to 50% relaxation	39	52
Tension, N cm^{-2}		
Twitch	17	5
Tetanic	30	25
Maximum shortening velocity, length s^{-1}	4.1	6.4
Power, W kg^{-1}		
Peak instantaneous	150	250
Cyclic, sustained	48	50

Note. Values were collected at 25°C or adjusted to the expected value at 25°C from measurements made at 20 or 30°C using an assumed Q_{10} of 2.

Locust data are from Malamud, Mizisin, and Josephson (1988, *J. Comp. Physiol. A* **162**, 827–835) and Malamud, unpublished; frog data are from Renaud and Stevens (1981, *Am. J. Physiol.* **240**, R301–R309), Rome (1983, *Physiol. Zool.* **56**, 33–40), and Stevens (1988, *J. Muscle Res. Cell Motil.* **9**, 329–333).

on incorrect application of principles of scaling. Consider, for example, jumping ability. A 1-g locust can develop enough power to lift its 1 g of mass to a height of about 1 m. A 70-kg person can develop enough power in a jump to lift his or her 70 kg to a height of 1 m. The work done is 1 g m for the locust, 70 kg m for the person, and the power required per mass of animal is the same.

The most often studied and certainly the most completely analyzed muscle for any animal is the frog sartorius muscle. The most complete body of information on contractile properties for an insect muscle is probably for the wing muscles of locusts, both *Schistocerca gregaria* and *S. americana*. The frog sartorius muscle is not the strongest or the fastest vertebrate muscle known, but it is a good representative of a fast vertebrate muscle. Similarly locust flight muscles are neither the strongest nor the fastest insect muscles, but they are reasonable representatives of fast insect muscles. Some of the contractile properties of frog and locust muscle are tabulated in Table I. The vertebrate muscle and the insect muscle are surprisingly similar in many of their contractile properties. The muscles of insects share the same capacities and are subject to the same limitations as are muscles elsewhere throughout the animal kingdom.

See Also the Following Articles
Flight • Walking and Jumping

Further Reading
Dudley, R. (2000). "The Biomechanics of Insect Flight: Form, Function, Evolution." Princeton University Press, Princeton, NJ.

Hoyle, G. (1983). "Muscles and Their Neural Control." Wiley, New York.

Josephson, R. K. (1975). Extensive and intensive factors determining the performance of striated muscle. *J. Exp. Zool.* **194**, 135–154.

Josephson, R. K. (1993). Contraction dynamics and power output of skeletal muscle. *Annu. Rev. Physiol.* **55**, 527–546.

Josephson, R. K., Malamud, J. G. and Stokes, D. R. (2000). Asynchronous muscle: A primer. *J. Exp. Biol.* **203**, 2713–2722.

Smith, D. S. (1984). The structure of insect muscles. *In* "Insect Ultrastructure" (R. C. King and H. Akai, eds.), Vol. 2, pp. 111–150. Plenum Press, New York.

Museums and Display Collections

Gordon M. Nishida
University of California, Berkeley

Insect collections are often oversimplified as either a precious and essential, rapidly developing treasure or a musty, dusty, moribund assemblage of archaic specimens. Those who now prefer the latter characterization have not been paying attention. Entomological collections and their curators are experiencing a steady, if not spectacular, evolution in the way collections operate, and perhaps the best is yet to come.

The term "insect collections" is often loosely used, and these collections usually contain other arthropods, including arachnids (e.g., spiders, mites), myriopods (e.g., centipedes, millipedes), and terrestrial crustaceans (e.g., sowbugs and pillbugs, amphipods). So most entomological collections are actually arthropod collections.

Collections of insects and related arthropods constitute an enormous resource for biological information and are an irreplaceable tapestry documenting Earth's entomological natural history. The numbers of specimens in collections are staggering, as is the amount of work yet to be accomplished just describing and cataloging millions of new species. Despite the revolutionary changes energizing these collections, still daunting are the myriads of problems, particularly financial, that continue to debilitate many collections' operations.

SCOPE OF THE WORLD'S INSECT COLLECTIONS

Insects are the largest, most diverse group of organisms in the world. Over 900,000 species have been described, and current estimates on the number still without names range as high as 30 million. That theoretical number is based on work in the New World tropics. However, more recent work testing that hypothesis, including the Old World tropics and measuring host specificity, now suggests a more modest but still astounding 4 to 6 million unnamed species. Reflecting the number of species potentially involved, insect collections maintain a large number of specimens. Based on numbers and extrapolations primarily from Arnett, Samuelson, and Nishida and adjusted for the last 8 years using an annual growth rate of 2% (average over a 20-year period), there are conservatively 724 million specimens in entomological collections worldwide. This figure does not include the unknown number of specimens in private collections, nor does it include a significant number of unprocessed specimens. In 1991 Miller reported the numbers of specimens for entomological collections in the United States and Canada and also included figures on unprocessed specimens (i.e., backlog). In 1976 reports indicated that 26% of specimens were reported unprocessed, increasing to 28% in 1981 and 30% in 1986. If these percentages are extrapolated worldwide and added to the processed specimens, there may be nearly a billion arthropod specimens housed in entomological collections.

As might be expected, a great many collections house the rather large number of specimens that have been amassed. The insects and spider collections of one Web site (http://www.bishopmuseum.org/bishop/ento/codens-inst.html) lists 904 institutional or organizational insect collections. Despite the plethora of collections, the largest house a disproportionate number of the specimens. Table I lists the largest entomological collections, that is, those reporting over 5 million specimens in their holdings. The largest collections are found in Europe, North America, Australia, and New Zealand. The collections' locations are likely influenced by the historical origins of insect collections, centers for entomological research, cultural interests, and economic restrictions. These collections do not coincide geographically with the regions of highest insect biodiversity, which tend to be in the tropics.

The total number of specimens may appear excessive given the number of species described, but the total figure includes many undescribed species. Also, long series are required to study and account for the morphological, geographic, and seasonal variability within a species. The specimens are also widely spread throughout many collections, enabling multiple centers for study and providing insurance against loss of species representation in case of natural or man-made disaster.

TABLE I Largest Entomological Collections[a, b]

Collection	Country	Number of specimens
Muséum d'Histoire Naturelle	France (Paris)	30,000,000[b]
The Natural History Museum	United Kingdom (London)	30,000,000
Smithsonian Institution	United States (Washington, D.C.)	30,000,000
Zoologische Staatssammlung	Germany (Munich)	16,566,000
American Museum of Natural History	United States (New York)	16,204,000
Canadian National Collection	Canada (Ottawa)	15,000,000
Alexander Koenig Zoological Museum	Germany (Bonn)	14,000,000
Bernice P. Bishop Museum	United States (Hawaii, Honolulu)	13,250,000
Musée Royal de l'Afrique Centrale	Belgium (Tervuren)	10,510,000
Australian National Insect Collection	Australia (Canberra)	10,000,000
Museum für Naturlunde der Humboldt University	Germany (Berlin)	10,000,000
Zoologisch Museum, Universiteit van Amsterdam	Netherlands (Amsterdam)	9,685,000
Field Museum of Natural History	United States (Chicago)	9,000,000
Institut Royal des Sciences Naturelles	Belgium (Brussels)	8,000,000
Museum of Comparative Zoology	United States (Massachusetts, Cambridge)	7,601,000
California Academy of Sciences	United States (San Francisco)	7,000,000
Hungarian Natural History Museum	Hungary (Budapest)	6,700,000
New Zealand Arthropod Collection	New Zealand (Auckland)	6,560,000
Naturhistoriska Riksmuseet	Sweden (Stockholm)	6,500,000
Finnish Museum of Natural History	Finland (Helsinki)	6,500,000
Florida State Collection of Arthropods	United States (Florida, Gainesville)	6,500,000
Bohart Museum of Entomology	United States (California, Davis)	6,241,000
Naturhistorisches Museum Wien	Austria (Vienna)	6,000,000
National Museum of Natural History	Bulgaria (Sofia)	6,000,000
Zoological Museum, University of Copenhagen	Denmark (Copenhagen)	6,000,000
Illinois Natural History Survey	United States (Illinois, Champaign)	6,000,000
Carnegie Museum of Natural History	United States (Pennsylvania, Pittsburgh)	5,500,000
National Natuurhistorische Museum	Netherlands (Leiden)	5,200,000
South African National Collection of Insects	South Africa (Pretoria)	5,000,000
Staatliches Museum für Tierkunde	Germany (Dresden)	5,000,000
Los Angeles County Museum	United States (Los Angeles)	5,000,000

[a]From Arnett, R. A., Jr., Samuelson, G. A., and Nishida, G. M. (1993). "The Insect and Spider Collections of the World." 2nd ed. Sandhill Crane Press, Gainesville, FL and Miller, S. E. (1991). Entomological collections in the United States and Canada. *Am. Entomol.* Summer, 77–84.

[b]Collections reporting over 5 million specimens; does not include large Russian collections such as Leningrad whose holdings are not available.

In 1993 Arnett, Samuelson, and Nishida reported an estimate of 100 million. However, this is generally considered to be in error, hence the more conservative figure provided here.

BEGINNINGS OF INSECT COLLECTIONS

The beginningss of insect collections are lost in unrecorded history. The Chinese used silkworms as early as 4700 B.C., honey bees by the fifth century, and scale insects by the 13th century. A treatise describing insects and their pharmaceutical properties had been published in China by at least A.D. 200. Cuneiform texts found in Mesopotamia dating to earlier than 669–626 B.C. contain systematically arranged names of insects. Aristotle (384–322 B.C.) studied insects and taught entomology. In A.D. 77 Pliny produced an encyclopedia that included entries on insects. Insect collections were no doubt made during antiquity, but no record of them has been found.

The early periods saw interest in insects primarily for their practical use or as pests. European explorations in the 15th and 16th centuries opened up shipping lanes for commerce and trade. The explorers and adventurers brought back hordes of items including insect novelties, piquing the interest of many. The first entomological collections were included in the cabinets of "curiosities" assembled by wealthy Renaissance families to show to friends and associates. These cabinets were to eventually lead to modern natural history museums.

The first compound microscopes were made by Hans and Zacharias Jansen in Holland in 1590. This major technological advance, coupled with the improvements made by Antony van Leeuwenhoek in the late 1600s, permitted the observation of tiny insects and their minute parts, advancing their study.

As the number of curiosity cabinets continued to grow in the 16th and 17th centuries, a means to organize their contents became necessary. John Ray (1628–1705) attempted a classification of insects in 1705, which was published posthumously in 1710. Carolus Linnaeus (Carl von Linné) (1707–1778) published the first edition of his *Systema Naturae* in 1735, the 10th edition, published in 1758, became the basis for modern insect classification.

SOURCES OF EARLY SPECIMENS

Linnaeus surrounded himself with students not only to carry on his work but also to provide specimens for study. Linnaeus was particularly interested in obtaining species of practical use, and he arranged for his students to go on voyages and encouraged them to send material from their travels. From 1744 to 1796, students sent back to Linnaeus and his associates material not only from Europe but from the Middle East, Africa, India, Asia, South America, Africa, and the Pacific (Cook's first two voyages).

Some of these students of Linnaeus eventually produced works of their own: for example, Fredrik Hasselquist's *Iter Palaestinum eller Resa til Heliga Landet* in 1762 and Pehr Forsskål's *Descriptiones animalium* in 1775. Other Linnaean students, for example, Johann Christian Fabricius (1745–1808), were extremely productive in discovering and describing new species, continuing the Linnaean traditions and fostering this era of cataloging.

Wealthy scholars and others amassed large collections that eventually wound up in institutional collections. Catherine the Great started Peter Simon Pallas on collecting and exploring travels throughout the Russian Empire between 1767 and 1810. A. M. F. J. Palisot de Beauvois described insects from Africa and America collected on his own travels from 1781 to 1797. Guillaume Antoine Olivier, sent on expedition to Turkey, Asia Minor, Persia, Egypt and the Mediterranean islands (1792–1798), later became a patron for other naturalists. Pierre François Marie Auguste Dejean (1780–1845), a soldier of fortune, collected in Austria and by exchange or purchase amassed the greatest collection of beetles in the world at that time. Thomas Say (1787–1834), the father of American entomology was appointed naturalist for Long's expeditions to the Far West and visited the Rocky Mountains and the sources of the St. Peters River in 1823. Victor Ivanovich Motschulsky (1810–1871) was a Russian military officer who traveled through Europe, the Caucasus, Siberia, the Kirghiz steppes, Egypt, India, the United States, and Panama. Thomas de Gray, Lord Walsingham (1843–1919), was an English nobleman wealthy enough to travel extensively (United States, North Africa, Europe) and also purchase specimens.

The 19th century was a fertile time for voyages and expeditions. Alexander von Humboldt visited the Spanish colonies of the American tropics between 1799 and 1804. Otto von Kotzebue sailed around the world from 1815 to 1818. On board was J. Friedrich von Eschscholtz who collected in California, Hawaii, the Philippines, Brazil, Chile, and other places. On a later voyage, again with Kotzebue, Eschscholtz amassed a large collection from the tropics, California, and Sitka (Alaska). Other important voyages for entomological specimens included those of the *Astrolabe* (1826–1829), the *Astrolabe* and the *Zelee* (1837–1840), the Swedish frigate *Eugenie* (1851–1853), and the Austrian frigate *Novara* (1857–1859). The United States Exploring Expedition (1838–1842) visited Madeira, Brazil, Chile, California, Oregon, Pacific Islands (including Hawaii, Australia, the Philippines, and Singapore), South Africa, and St. Helena. Sources of entomological collections mirrored the spread of empire and the pursuit of national interests.

Rather than circumnavigating, some targeted specific locations. For example, Henry Walter Bates spent the years 1851 to 1870 in South America, mostly in the Amazon, collecting and sending material back to England. Giacomo Doria (1840–1913), who founded the Genoa Museum, funded expeditions to areas particularly rich in insect diversity (Africa, Southeast Asia, New Guinea, Indian Ocean islands). Others such as Lionel Walter Rothschild (1868–1937) focused on specific groups. Lord Rothschild concentrated on butterflies and moths and funded expeditions to the far corners of the world; in 50 years' time he amassed the greatest personal collection ever (including 2.25 million butterflies and moths).

The amount of material grew exponentially, and as the Linnaean system became entrenched, secure and centrally located places to store the reference collections and unprocessed materials were sought.

DEVELOPMENT OF FORMAL COLLECTIONS

The first natural history museum probably was that of Conrad Gessner a scholar of mid-16th-century Zurich. Very few present-day natural history museums were established before the mid-18th century. The Muséum National d'Histoire Naturelle in Paris was established in 1635 and was the first natural history museum established in the form we recognize today. Others were the Staatliches Museum für Tierkunde in Dresden established in 1650, the Zoologiceskii Instituti Zoologiceskii Muzei in Leningrad in 1727, the Zoologiska Museet in Lund in 1735, the Naturhistorisches Museum in Austria in 1748, and the British Museum in 1753. Scientific academies, beginning with the Accademia dei Lincei in Rome in 1603, fostered and often housed early collections. The Royal Society in London was founded in 1662, and the Académie Royale des Sciences of Paris in 1666. The Academy of Natural Sciences in Philadelphia, founded in 1812, is the oldest North American academy. The academies were later augmented by natural history societies that often performed similar functions. The First Aurelian Society (early insect collectors were known as "Aurelians") was founded in London in 1745. The Entomological Society of Philadelphia, established in 1859, was the first American entomological society.

The great explosion of natural history museums occurred in the latter part of 18th century and into the 19th century, with the continuation of exploration and collecting. Coincidentally, this proliferation of museum collections coincided with the earliest use of persistent poisons such as arsenic to protect biological specimens from damage by the pests that had destroyed many earlier collections (a Western discovery, but the Chinese had written about the use of arsenic and mercury for control of human parasites in A.D. 100–200). Perhaps the concentration of the entomological collections in temperate Europe and North America was a result not only of European influence but also of the climate, which likely was less favorable to potential pests of museum specimens than in the warmer regions of the world.

Growth continued through the 20th century with the exception of periods of global conflict. The 19th and 20th centuries saw increasing participation of institutions and government in organizing and funding expeditions. Expatriates and professionals on foreign assignment were also great sources of collections. The 20th century later saw a focus on regional faunas, opportunistic trips, and taxon-based initiatives. Institutions became less involved with collecting efforts and individuals increased their efforts, particularly to aid their research goals. Toward the end of the 1900s, cooperative efforts returned as the cost of fieldwork increased. Large national inventories such as the Instituto Nacional de Biodiversidad (INBio) established in 1989 in Costa Rica were undertaken, and ATBIs (all taxa biological inventories) were begun to inventory areas with a combination of high biodiversity and high threat of loss due to that biodiversity.

As the great expeditions filled the collections' coffers and the continuous additions provided more than enough work for taxonomists, collections also needed technological support to improve storage and retrieval capabilities.

TECHNOLOGICAL ADVANCES IN METHODOLOGIES USED BY COLLECTIONS

Technological advances, from the microscope to finding effective pesticides, had and still have a profound influence on the development of entomological collections. Although the basic methods of collections and specimen preparation have not changed radically, new innovations usually made the process quicker, more efficient, and more inclusive. Steady changes have taken place in the past three decades, and in retrospect, a major revolution has occurred.

Many innovations have been implemented, including the use of glass-topped drawers for specimen storage rather than just simple boxes. It is likely that insects were placed in containers to take advantage of the efficacy of pesticides by maintaining the specimens in a closed environment. Dried specimens are still prepared with pins, and paper labels are affixed to the pin beneath the specimen. However, the materials used have been much improved. Insect pins are of higher quality and are less likely to corrode.

Along with the advances in technology, the last few decades have seen a concerted effort to introduce materials conservation techniques into collections management. Recent advances in storage technology in insect collections include the development of the unit tray system. This is a system of topless boxes made of cardboard and with a material that functions as a pinning base at the bottom of the box. This innovation permits the rapid rearrangement of the collections as new studies modify the organization of a group. Placing like elements in a single unit tray (e.g., specimens of a species all collected in one area) also permits the use of a header card, or tray identifier, that permits rapid recognition and retrieval of information. The material first selected for tray bottoms was cork. Within several decades, however, cork was found to be unacceptable because its acidic nature affected the pins and labels and even perhaps the specimens. A new material was searched for, and most collections have settled on a cross-linked polyethylene product that is inert and pliable.

In the late 19th century, new methods helped streamline the papermaking process. Unfortunately, the new papers are slightly more acidic and tended to break down faster, with the ironic result that older 17th and 18th-century labels are more permanent. Labels have changed in both substance and content. Some early collections were not labeled at all to show the origin of specimens. As systematic work progressed and some species were found to be restricted in their distribution, labels were added. These labels were laboriously handwritten in ink and often would indicate only a country, an island, or a region. As even more species of insects were discovered, classifications became larger, and more specimens were collected, labels became more specific, adding ecological data, collecting information, and details of the collecting locality. With the development of GPS (global positioning systems), collecting location coordinates are routinely being added to label data.

Hand-printed labels gave way to mass-printed forms. Typesetting took time, and so photoreduction methods were used to speed the process. Since the 1980s, the desktop computer has become a mainstay in collections work. Initial efforts to use the computer to generate labels were restricted to providing printed copies since the early printers did not have a font size small enough to fit all the necessary information onto the tiny specimen labels. As laserwriters developed, collections personnel began experimenting with producing labels on them. The first efforts were encouraging until it was discovered that the toner (laser cartridge "ink") did not stick to the paper used for labels in fluid (ethanol) collections. At this time, inkjet printers seem to be a useful alternative.

The last two examples serve to show why collections often embrace technology slowly. Yet despite the pitfalls inherent in adopting new technologies, most collections today are in much better shape for long-term survival than they were before. Other advances in storage and maintenance include new or retrofitted buildings for many collections, use of climate control to reduce fluctuations in temperature and humidity, and installation of compactor storage systems to make better use of space and improve access to collections. Repellents used in collections that are potential human health risks, such as naphthalene (usage introduced in the late 19th century) and paradichlorobenzene, are being replaced with freezing procedures and integrated pest management monitoring techniques such as sticky traps to intercept possible problems sooner.

Collecting methods have improved and can easily overwhelm the preparation capabilities of most present-day collection staffs. Mass collecting techniques took a great leap forward with the invention of a flight intercept trap by René Malaise in 1937 (prototyped in 1933). The Malaise trap is a tentlike structure placed in a position to intercept flying insects and have them self-collect in strategically placed containers. Today's arsenal of passive traps includes yellow pan traps, pitfall traps, Berlese samplers, and innumerable other specialty traps and methods, including modifications of the original Malaise trap. For those unwilling to wait for their specimens to come to them, proprietary devices (e.g., D-Vac) allow vacuuming of vegetation. Even more efficient is the use of pyrethrum fog to assemble vast quantities of specimens from tree canopies.

As collections techniques evolved, methods of studying and interpreting species have influenced collections also.

CHANGING ROLES OF COLLECTIONS

Modern entomological collections are rooted in Linnaean classification. The century following Linnaeus was devoted to describing and cataloging the massive amount of material gathered, but there were too many species and not enough taxonomists. Other disciplines began studies of arthropods, and many researchers had to develop taxonomic expertise on their own to have names for the organisms they were studying.

Around the mid-19th century, large-scale agriculture and expanding horticultural efforts underscored the importance of insects as pests and opened up an entirely new area for collections, namely, the role of identifying insect pests and establishing the authenticity of such identification. The use of parasitoid and predatory insects for biological control began a new type of biological exploration, sending professional entomologists around the world to find control agents, which were subsequently reared and released. Governments established sections of entomology in agricultural divisions and often associated them with national collections such as the U.S. Department of Agriculture and the Smithsonian Institution or the Commonwealth Agricultural Bureaux and the British Museum of Natural History.

Between 1897 and 1900, Ronald Ross and Patrick Manson discovered and experimentally proved that mosquitoes act as vectors in transmitting malaria, launching yet another era in collections development: the collection and study of medically important arthropods. Identification and research units were often established within military units or centers, such as the Walter Reed Army Medical Center in the United States.

Evolutionary biology has had a profound influence on the development of insect collections. Both Charles Darwin and Alfred Russel Wallace had been influenced by insects when they proposed their theory of natural selection in 1858. Further evidence for evolution was sought within insects, enhancing collections in the process. In the past 50 years, areas such as genetics, population ecology, and even bioprospecting have used existing insect collections and developed additional collections as adjuncts to their research. Insect collections have provided support for unexpected areas such as medical forensics. Within the last two decades, molecular biology has been asserting its influence on entomological collections, similarly using available specimens and gathering more, making use of DNA evidence to establish better understanding of species relationships.

New approaches to taxonomic studies have appeared in the decades since 1980. Phenetics (sometimes referred to as numerical taxonomy) bases classifications on overall similarities. Cladistics places emphasis for classification on branching points of a phylogeny. Evolutionary taxonomy adds degree of similarity to evolutionary origin. The advent of computers has helped the establishment of these systems by facilitating the manipulation of data. Ernst Mayr perhaps summed up the current status in insect classification best: "Taxonomy has been more active, more in ferment, in the last 50 years, than ever before in its history." Today these words are even more applicable.

As these new sciences developed, positions in classical taxonomy dropped steadily, many workers transferred to the newer "cutting edge" sciences. An increase in efforts to stimulate appreciation of entomology was one of the responses to the perceived loss of positions and funding.

COLLECTIONS ON DISPLAY

Early collections were displayed in their entirety. In 1864 John Edward Gray of the British Museum proposed to store the collections away from public view, and Richard Owen eventually produced an "index collection" that became the model for display of collections specimens. This began the period of exhibit collections, with only a small portion of the material placed on view and the remaining material devoted to research. With little variation, this was the extent of entomological specimen exhibition for the next century.

Toward the latter half of the 20th century, new entomological exhibits with educational themes began appearing in an effort to interest the public. As operating costs rose and income stagnated, museums began adding an entertainment component to their exhibits in hopes of attracting more visitors, and to this end they allied themselves more closely with educational institutions. Exhibitors and collections staff quickly discovered that people respond positively to the display of live arthropods.

In 1976 the first insect zoos (e.g., Smithsonian Insect Zoo) appeared in North America. These living exhibits were extremely successful, spawning dozens of replicates. The first butterfly house debuted in 1976 on the British island of Guernsey; over a hundred soon followed throughout the world. Determining the exact number of living insect exhibits throughout the world is difficult because many zoological parks have an insect component that is not readily apparent. Both conventional and insect zoos have live specimens on display, but whereas the older style zoos maintain special enclosures and habitats for the animals, butterfly houses permit visitors to walk through the enclosure. These two innovations sparked a flurry of activity in arthropod husbandry. An indication of the popularity of living collections just in North America is the number (328) of organizations represented at the year 2000 Invertebrates in Captivity Conference sponsored by SASI (Sonoran Arthropod Studies Institute).

SHIFTING FUNDING

As museums began appearing in the late 1700s and 1800s, funding for collections shifted from the province of wealthy patrons, who were largely replaced by institutions supported by government or public funds. As governments centralized and became more active, particularly in the areas of medicine and agriculture, funding support shifted. Beginning in the late 1800s, agriculture became a major source of economic support, and indeed, modern land grant college collections owe their existence to agricultural funding. Today most large museums in the United States survive on a mix of revenue from endowment, gifts, income from visitors, grants from government or private sources, and direct support from government. In the rest of the world, funding for collections more likely comes from government sources, with less dependence on private sources. Traditional sources of funding such as agriculture or military research have waned recently, as has government spending in general. Though it is likely that funding sources such as biodiversity or conservation have not yet realized their full potential, many insect collections are now seeking supporting funds from private foundations as well as continued support from customary sources such as the National Science Foundation.

FUTURE OF ENTOMOLOGICAL COLLECTIONS

Inventory of the world's biodiversity is far from complete, and present work is being accomplished in the face of extraordinarily high rates of extinction. Despite the large holdings presently in collections, many more species probably have not been collected yet, or even discovered. Financing for large-scale inventory projects has been minimal at best. However, a revival of taxonomy is under way, with funding coming from more diverse sources than before and with entomological collections reasserting their roles. Interest remains in endangered species and interest in alien species and their impact is growing rapidly.

Collections have always been in the information business. Specimens are about data, and collections are in the business of brokering information and storing and retrieving data. Present information technologies are revolutionizing data-sharing capabilities, and many collections are furiously converting their older information retrieval systems (card files, specimen data, etc.) into electronic systems that can be used internally and shared internationally. Bioinformatics tools such as electronic catalogs [e.g., Biosystematic Database of World Diptera (http://www.sel.barc.usda.gov/diptera/biosys.htm) or Orthoptera Species File Online (http://viceroy.eeb.uconn.edu/Orthoptera)] are rapidly coming online. Large clearinghouses for biological information such as the National Biological Information Infrastructure (NBII) are available on the World Wide Web.

The global availability of this information will provide more opportunities for cooperative ventures and at the very least will make the process of scientific inquiry a lot less time-consuming and less costly. Perhaps even more intriguing is the potential to overlay information from other disciplines—for example, adding geographic capabilities to enhance understanding of species distributions, or plant data to further understand host associations—and provide a historical perspective to boot. Ambitious systems such as Species 2000 and the Global Biodiversity Information Facility will permit this type of knowledge synthesis and interoperability. Perhaps an opportunity to better understand human impact on the world's ecosystems from an arthropod perspective is finally at hand. This is an exciting time for insect museums.

See Also the Following Articles

Collection and Preservation • Entomological Societies • History of Entomology • Insect Zoos • Photography of Insects • Teaching Resources

Further Reading

American Association for the Advancement of Science. (2000). Bioinformatics for biodiversity: A Web registry. *Science* **289,** 2229–2440.

Arnett, R. A., Jr., Samuelson, G. A., and Nishida, G. M. (1993). "The Insect and Spider Collections of the World." 2nd ed. Sandhill Crane Press, Gainesville, FL.

Duckworth, W. D., Genoways, H. H., and Rose, C. L. (1993). "Preserving Natural Science Collections: Chronicle of Our Environmental Heritage."

National Institute for the Conservation of Cultural Property, Washington, DC.

Ghiselin, M. T., Leviton, A. E. (eds.) (2000). "Cultures and Institutions of Natural History." Memoir 25. California Academy of Sciences, San Francisco.

Gilbert, P., and Hamilton, C. J. (1983). "Entomology. A Guide to Information Sources." Mansell, London.

Mayr, E. (1988). Recent historical developments. *In* "Prospects in Systematics" (D. L. Hawksworth, ed.), pp. 31–43. Clarendon Press, Oxford, U.K.

Miller, S. E. (1991). Entomological collections in the United States and Canada. *Am. Entomol.* Summer, 77–84.

Smith, R. F., Mittler, T. E., and C.N. Smith, (C. N. eds.), (1973). "History of Entomology." Annual Reviews, Palo Alto, CA.

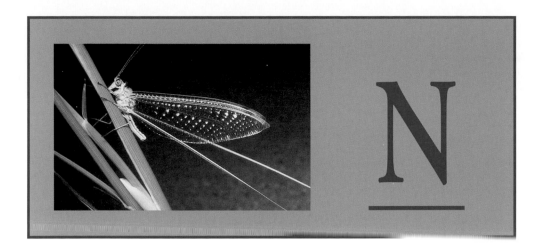

Neosomy

Frank J. Radovsky
Oregon State University

Neosomy in an arthropod is a remarkable enlargement or formation of new external structure, or both, resulting from the secretion of new cuticle unrelated to a molt, during an active instar or in the adult. It is distinct from the more modest addition of cuticle that contributes to intermolt growth of many immature insects (notably larvae of Endopterygota) and that can be detected in some adult insects. Neosomy usually is associated with symbiosis and often with tachygenesis. *Physogastry* has been used overlappingly, but this term etymologically indicates abdominal swelling, and it usually has been defined as distension and not in relation to cuticular growth.

EXAMPLES

Neosomy is widely present in the Acari, including some parasitic larval mites and all stages (larva, nymph, adult female) of hard ticks (Ixodidae). Female ixodid ticks generally increase in volume about 100× or more, after first doubling cuticular thickness during the principal time they are attached to the host. In the chigger genus *Vatacarus* taken from the lungs of sea snakes, larval volume increases 1500× or more and neosomules (the new external structures) form as papillae that aid worm-like movement. All feeding is larval, and the adult casts larval and nymphal exuviae together when it emerges. In the Crustacea, some ectoparasitic copepods are also neosomatic.

Neosomy is seen in adults, primarily females, of some holometabolous insects. It occurs in females of four families of fleas. For example, *Tunga monositus*, embedded in the skin of rodents, grows more than 1000× until it no longer has any surface resemblance to a flea; addition of cuticle is centered in the second abdominal segment. *Tunga penetrans* (Fig. 1A), in mammals, including humans, increases in size less than other *Tunga*. Flies of the genus *Ascodipteron* in bats develop much as *Tunga* do.

Some termite- and ant-associated beetles and flies are neosomatic. Social parasites among staphylinid beetles mimic termites in order to be accepted by the colony. The termite-mimicking cuticle that grows from the abdomen of an initially normal-looking adult beetle has paired "legs" and "antennae" (Fig. 1B). Neosomatic cuticle usually grows in soft areas between sclerites, but it can also involve sclerotized parts in these beetles. Many queen termites, sometimes >12.5 cm in length, and some queen ants are neosomatic.

FUNCTIONAL SIGNIFICANCE

Combination of neosomy with symbiosis results from the abundance of food provided by a host or host colony. Neosomy in a termite or ant queen suggests that, although the association is intraspecific and hence not symbiotic, the

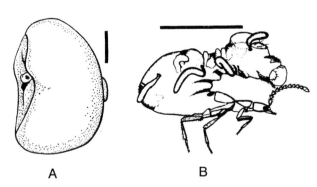

FIGURE 1 Female neosome, lateral view (scale bar, 1 mm). (A) *T. penetrans*, with part of head visible on left. (Reproduced, with permission, from Audy *et al.*, 1972.) (B) *Coatonachthodes ovambolandicus*. (Reproduced, with permission, from Kistner, 1979.)

queen's relationship to the colony as a whole parallels symbiosis functionally. A female ixodid tick or a queen termite, for example, has unusual access to food and produces many eggs compared with related groups.

Tachygenesis has a comparable effect in sheltering the species from risks. *T. penetrans* has two rather than the usual three instars of fleas. The more advanced *T. monositus* does not feed as a larva, and all stages except the female in the host are sheltered in a rodent burrow, nonfeeding, and quiescent (except the larva spinning its cocoon and the male mating). The mimicry seen in termite-associated beetles presumably could not be achieved other than by neosomy.

See Also the Following Article
Mites

Further Reading
Audy, J. R., Radovsky, F. J., and Vercammen-Grandjean, P. H. (1972). Neosomy: Radical intrastadial metamorphosis associated with arthropod symbioses. *J. Med. Entomol.* **9**, 487–494.
Kistner, D. H. (1979). Social and evolutionary significance of social insect symbionts. *In* "Social Insects" (H. R. Hermann, ed.), Vol. 1. Academic Press, New York.
Nijhout, H. F. (1994). "Insect Hormones." Princeton University Press, Princeton, NJ.
Rothschild, M. (1988). Giant polyploid cells in *Tunga monositus* (Siphonaptera: Tungidae). *Syst. Assoc. Spec.* **37**, 313–323.
Williams, C.M. (1980). Growth in insects. *In* "Insect Biology in the Future" (M. Locke and D. S. Smith, eds.). Academic Press, New York.

Neotropical African Bees

Orley R. Taylor
University of Kansas

Neotropical African bees are derived from an intentional introduction of bees from South Africa into Brazil. These bees are famous for their stinging behavior and they are known as "killer" or "Africanized" bees in the media.

The introduction from South Africa into Brazil of a subspecies of the common honey bee, *Apis mellifera*, resulted in a spectacular biological invasion that has had profound effects on agriculture, beekeeping, and human and animal health. Establishment of the African bees led to hybridization with bees of European subspecies maintained by beekeepers for honey production and pollination. The resulting apiary bees became extremely defensive, and honey production declined as beekeepers abandoned beekeeping. Stinging incidents that resulted in human and animal deaths soon led the press and the public to refer to these invaders as "killer bees." The perception that the feral African-derived bees were of hybrid origin led to the term "Africanized bees," which is a misnomer.

PREVIOUS INTRODUCTIONS

Honey bees *(A. mellifera)* are native to the Old World, and all honey bees present in the Americas are descendants of bees introduced from Europe, Africa, and the Middle East. Managed honey bees in the Americas are derived from at least five introduced European honey bee (EHB) subspecies belonging to two major lineages of *A. mellifera* (west European bees, *A. m. mellifera* and *iberica;* eastern European bees, *A. m. ligustica, carnica,* and *caucasica*). These temperate subspecies are notably successful in other temperate regions (e.g., Australia) and have been used with some success in tropical areas. However, they are not well adapted to tropical conditions and did not establish large self-sustaining feral populations in the tropics as they have done in subtropical and temperate regions. In many tropical areas, European bees would not persist without human assistance and, in most habitats (with the exception of regions with pronounced dry seasons) honey production was marginal. In 1956, to improve honey production in Brazil, Warwick Kerr intentionally introduced to Brazil *A. m. scutellata,* a subspecies belonging to the African lineage of *A. mellifera.* The introduced subspecies became established in southern Brazil in early 1957 and, because this bee was well adapted to tropical conditions, a large feral population soon developed and began to spread at rates of 100 to 300 miles per year. No other invading species has expanded so rapidly into new habitats.

DEFENSIVE BEHAVIOR

The Neotropical African honey bees (AHBs) are well known for their defensive behavior, and the deaths of hundreds, perhaps thousands, of people (and certainly thousands of domestic animals) have been attributed to these insects. These bees are not always defensive, but under certain conditions they will attack people and animals near their nests in massive numbers, inflicting hundreds and even thousands of stings. The venom (per bee) of AHBs is less toxic than that of EHBs; nevertheless, human victims of massive stinging require immediate medical attention to minimize lysis of blood cells, breakdown of muscle tissue, and kidney damage, which can result in acute kidney and multiple organ failure.

IMPACT ON BEEKEEPING

Beekeepers accustomed to dealing with relatively gentle and manageable EHBs were unable to adapt to the defensive behavior as AHBs advanced through the Americas, and many abandoned beekeeping. Honey production declined and many countries became honey importers rather than exporters. For example, in the Yucatan peninsula of Mexico, an area of intense beekeeping with both EHBs and native stingless bees (*Melipona* and *Trigona*), both types of beekeeping declined precipitously following the arrival of AHBs. In much of the Americas, as new beekeepers adapted to AHBs, beekeeping and honey production recovered, most notably in Brazil.

AHBs are now used for honey production in many areas of the Americas where EHBs were ineffective, and it appears that Kerr's goal of improving honey production in the New World tropics will be realized.

DIFFERENCES BETWEEN AFRICAN AND EUROPEAN HONEY BEES

Honey bees subspecies from tropical and temperate regions have evolved adaptations that are suitable for their respective environments. AHBs are smaller, and they have higher metabolic rates, more rapid development, reduced longevity, smaller nest sizes, greater brood production, and lower honey storage than EHBs. These traits combine to limit the ability of AHBs to overwinter in areas where the interval between first and last frost is longer than 3.5 months. In contrast, feral EHBs can exist in areas where this winter interval is 6 months. Adaptations to tropical conditions that give AHBs advantages include higher rates of swarming (reproduction) and the ability to abandon the nest (abscond) and move to new habitats under unfavorable conditions. Also affording AHBs a distinct competitive advantage in the tropics are these insects' abilities to find pollen and nectar, to increase their brood production under conditions in which EHBs are unable to do so, and to mount an intense nest defense that repels predators.

HYBRIDIZATION AND GENETIC DIFFERENCES

When two formerly isolated species or populations come into secondary contact there are four possible outcomes: coexistence with complete reproductive isolation, replacement of one population by the other, fusion of the two populations and complete mixing of the two gene pools (sometimes referred to as "dilution"), and establishment of a more or less permanent hybrid zone. The first scenario, coexistence with reproductive isolation between these biotypes, has not developed anywhere in the Americas. The second scenario seems to be the rule in nonmanaged populations: African-derived bees establish large feral populations and replace any resident European feral honey bees.

Gene flow between neotropical AHBs and EHBs seems to be strongly asymmetrical. AHBs have maintained their genetic integrity, in spite of hybridizing with EHBs, as they have expanded their distribution. Even after 45 years of interaction with EHBs, these bees are indistinguishable in behavior and so similar genetically to bees in the Transvaal of South Africa that it is more appropriate to refer to them as Neotropical African bees or as African-derived bees rather than Africanized bees. Low acquisition of EHB traits into the AHB population can be attributed to pre- and postzygotic isolating mechanisms (i.e., mate selection, queen developmental time, and hybrid dysfunction). For example, AHB queens mate predominantly with AHB drones even in the presence of large numbers of EHB drones. When AHB queens are inseminated with semen from drones of both types, the AHB queen progeny

develops faster than the hybrids, assuring that in most cases the next queen would be an AHB rather than hybrid.

Colonies from backcrosses of F1 hybrid queens to either parental genotype have unusual metabolic patterns, low honey storage, and high rates of mortality. This finding suggests possible incompatibility of the nuclear and mitochondrial genomes of these biotypes. EHBs, in contrast, rapidly become Africanized, and nearly all traces of the EHB nuclear and mitochondrial genome disappear from the feral bee populations following the arrival of AHBs. Disappearance of the European traits seems to result from a lack of prereproductive isolation, which results in extensive mating by EHB queens with AHB drones. This is followed by a pattern of queen development that favors hybrid rather than EHB queens. Matings by these F1 queens to AHB drones results in colonies with low fitness and the eventual loss of EHB mitochondrial DNA from the population. Displacement of EHBs therefore seems to result, in part, in a type of "genetic capture" in which one form, *A. m. scutellata,* eliminates the others by hybridizing with their females. The genetic and population consequences of the interactions between *A. m. scutellata* and *A. mellifera* subspecies from Europe suggest that *A. m. scutellata* deserves the status of a semispecies.

HYBRID ZONES

A hybrid zone formed at the southern limit of AHBs in northern Argentina in the late 1960s and early 1970s. Although many anticipated that a similar zone would form in the United States, this has not happened. Coincident with the arrival of AHBs in the United States, the mite, *Varroa destructor,* an introduced brood parasite from Asia that kills EHB colonies, spread rapidly throughout the country eliminating feral EHB colonies. At present, feral EHBs, which usually are escaped swarms from managed apiaries, are transitory and persist only for a short time; this precludes the formation of a persistent hybrid zone. In Argentina, it is likely that *Varroa,* which arrived after the formation of the hybrid zone, has changed the dynamic of the interaction of both types of bee as well.

FUTURE

In the future, the only significant feral bee populations in the United States will consist of AHBs in the southwestern states and possibly Florida, until or unless EHBs acquire sufficient tolerance of *Varroa* mites to once again establish feral populations (Fig. 1). AHBs will not become more European through hybridization or selection and move further north. Barriers to gene flow into the AHB genome, coupled with selection against any AHBs with EHB genes that make them susceptible to *Varroa* mites, assures the continuation of a nearly separate Neotropical African bee genome. AHBs have many useful attributes but hesitancy on the part of beekeepers to work with these bees, partly because of familiarity with EHBs but also because of concerns of legal issues should their bees be linked

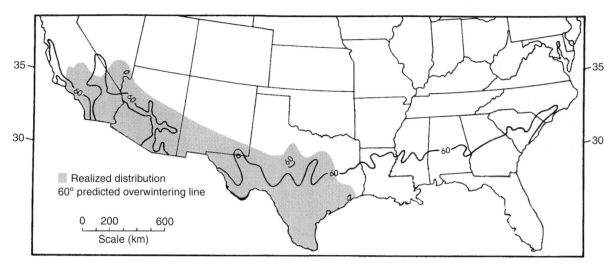

FIGURE 1 The known summer (shaded) distribution and predicted overwintering limit (60-degree line for January) of Neotropical African honey bees in the United States.

to stinging incidents, will keep American beekeepers from adapting this bee to apiculture.

See Also the Following Articles
Apis Spieces • Beekeeping • Introduced Insects • Venom

Further Reading
Hall, H. G., and Muralidharan, K. (1989). African honey bees spread as continuous maternal lineages. *Nature* **339,** 211–213.

Kerr, W. E., Del Rio, S., and De Barrionuevo, M. D. (1982). The southern limits of the distribution of the Africanized bee in South America. *Am. Bee J.* **122,** 196–198.

Michener, C. D. (1975). The Brazilian bee problem. *Annu. Rev. Entomol.* **20,** 399–416.

Taylor, O. R. (1985). African bees: Potential impact in the United States. *Bull. Entomol. Soc. Am.* **31,** 14–24.

Taylor, O. R., and Spivak, M. (1984). Climatic limits of tropical African honeybees in the Americas. *Bee World* **65,** 38–47.

Nervous System

see *Brain and Optic Lobes*

Nest Building

Robert L. Jeanne
University of Wisconsin

Insects manipulate their environment for a variety of purposes: to trap prey, attract mates, and provide shelter from the elements and protection from predators and parasites. Nests are a special category of environmental manipulation. A nest may be defined as any modification of the environment by adult insects that provides shelter for the rearing of their offspring. In most nest-building insects, the nests are simple excavations or small constructions that provide temporary protection for eggs or larvae, with or without the adult parent(s) in attendance to provide continuing parental care. In the numerous lineages of nest-building insects, an increase in parental care has generally been accompanied by the evolution of more elaborate nesting behavior. The trend has climaxed numerous times in the eusocial insects such as termites, ants, wasps, and bees, whose nests may be very large and architecturally complex and may house the colony for many years under homeostatically controlled physical conditions.

TAXONOMIC DISTRIBUTION OF NEST-BUILDING

Nest building has evolved in only a handful of mandibulate insect orders. In this brief survey, only a few selected examples of nest-building species are given for each.

Orthoptera

True nests have evolved in a few species of locusts and crickets. In the burrowing cricket, *Anurogryllus muticus* (Gryllidae), nesting behavior reaches the highest point found in the order. The female of this species excavates a brood chamber in the soil and then seals herself inside and lays her eggs. When the nymphs hatch, the mother feeds them with special trophic eggs and later with grass that she gathers outside and brings into the nest.

Coleoptera

A number of beetles manipulate the environment so as to provide shelter and/or food for their young. The female of the leaf roller, *Deporaus betulae* (Attelabidae), cuts across a leaf along a precise trajectory, then rolls the leaf into a tube, inside of which she lays her eggs. The larvae feed on the inner layers of the leaf roll, while being protected by the outer layers. The dung beetles (Scarabaeidae) excavate nests in the ground and provision them with balls of dung rolled to the site. Some of the carrion beetles (Silphidae) form the body of a dead mouse or other small mammal into a ball, drop it into an excavated chamber, then lay eggs on it. In some species, the female remains in the nest and feeds the young larvae by regurgitation until they are large enough to feed on the carrion directly.

Embiidina

Both sexes of webspinners, adults as well as nymphs, produce silk from the swollen metatarsal glands in the forelegs, which they use to spin a network of galleries on tree trunks or in leaf litter. Not only do the galleries serve as a center for brood-rearing, but they also provide a shelter within which the family of webspinners grazes on bark, dead leaves, moss, or lichens.

Isoptera

All the termites are eusocial (reproductive division of labor, cooperation in brood care, and overlap of at least two generations capable of contributing labor to the colony) and all live in nests. In the more primitive species, the colony nests in the wood source it feeds on. Such "single-site nesting" is exemplified by the small colonies formed by species of Termopsidae, most genera of Kalotermitidae, and the less derived members of Rhinotermitidae. The colony spends its entire life in its log, the nest consisting simply of the irregular galleries excavated by the feeding termites. The "higher" termites, belonging to the family Termitidae and others, have evolved the ability to nest independent of their food source, in the soil or arboreally. Dissolving the identity between food source and nest freed these species to evolve larger colony size and to exploit a wider range of cellulose sources, including wood fragments of all sizes, grass, seeds, leaf litter, and humus.

Hymenoptera

Nesting behavior in the Hymenoptera is limited to three super-families: Sphecoidea, Vespoidea, and Apoidea. The ancestors of nest-building aculeates were nonnesting parasitoids of other arthropods. Nesting behavior probably got its start when a female parasitoid dragged her paralyzed prey into a crudely excavated nest in the ground and laid an egg on it, much as some sphecoids do today. Many of the solitary sphecoids and vespoids (sand wasps, digger wasps, spider wasps) excavate a subterranean nest and stock it with one or more paralyzed prey, on which an egg is laid, whereas others (mud daubers, potter wasps) construct aerial nests of mud. A few sphecids nest in hollow stems or other natural cavities.

Except for the parasitic "cuckoo bees," all bees (Apoidea) make nests. Most are solitary, the female excavating a nest in the ground or using hollow stems or other natural cavities. Carpenter bees excavate burrows in solid wood. Some solitary bees construct nests of resin or a mixture of resin and pebbles, leaf pulp, or mud on rocks, stems, or leaves.

Eusocial behavior has arisen in all three superfamilies. Nesting behavior, a prerequisite, was already established well in advance of the numerous origins of eusociality in these taxa. In the ants (Formicidae) and bees (Apidae), the evolution of eusocial behavior occurred in subterranean nests, while in the wasps (Sphecidae, Vespidae) it took place in constructed, aerial nests of naked brood cells. In each group, as social life became more elaborate, nests increased in size and complexity and adapted to new nesting sites. Although many species of ants nest in the ground, many others, especially in the tropics, construct arboreal nests or nest intimately with plants. Many of the eusocial bees construct their nests in cavities, whereas others construct aerial nests, either with combs exposed or enclosed in a heavy involucrum. With larger colony size in the wasps came the evolution of protective nest envelopes and/or the move to cavities in the soil or in trees.

NEST-BUILDING BEHAVIOR

Materials and Tools

For the majority of social species that excavate nests in soil or wood, the nest consists merely of the cavity left after the removal of material. In contrast, constructed nests, which have evolved in all four eusocial groups, require a combination of exogenous structural material and adhesive to bind the particles of material together. A variety of materials are used: termites use soil or wood particles cemented together with saliva and/or fecal material. Ants use wood or other vegetable fiber or mud. *Lasius fuliginosus,* for example, fills its nest cavity with an irregular carton meshwork glued together with honeydew. The matrix is strengthened by the penetration of the hyphae of a symbiotic fungus. Social wasps (Vespidae) are known as "paper wasps" because familiar species use wood pulp as a structural material, although many tropical species use plant hairs and some even use mud. The fibers are chewed and mixed with a proteinaceous secretion of the labial gland that dries into a plastic-like matrix, giving the finished carton strength and a modicum of water repellency. Wasps that build exposed, pedicellate combs construct the pedicel primarily of this secretion, giving it toughness and a dark, shiny appearance (Fig. 1). Honey bees are unusual in using wax, secreted by wax glands on the abdomen, rather than collected material, to construct their brood and storage cells. Other social bees also use wax, but mix it with exogenous materials, including pollen, plant resins (propolis), vegetable material, mud, or even feces. *Microstigmus* wasps (Sphecidae) produce silk from glands at the tip of the abdomen and use this to glue together the leaf pubescence from which their delicate nests are sculpted. Weaver ants (*Oecophylla* spp.) sew living leaves together with

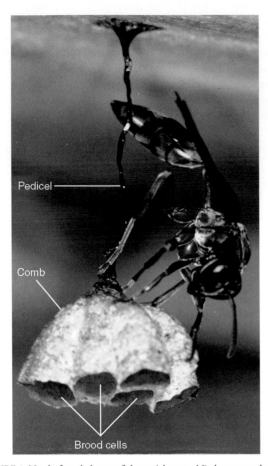

FIGURE 1 Newly founded nest of the social wasp, *Mischocyttarus drewseni,* from Brazil. The founding female is shown wiping an ant-repelling secretion (the gland opens at the base of the terminal abdominal sternite) onto the pedicel of the nest, where it reduces the likelihood that ants will discover the comb of brood cells while the queen is away on a foraging trip. (Reprinted, with permission, from R. L. Jeanne, Chemical defense of brood by a social wasp. *Science* **168,** 1465–1466. Copyright 1970 American Association for the Advancement of Science.)

strands of larval silk to create multiple arboreal nesting chambers in which the young are reared.

In all nest-building insects the mandibles are chisel and trowel, the primary tools used to excavate, collect, carry, and mix materials and shape them into the nest. Other tools are important in a few species: sand wasps use the legs to kick excavated sand out of the burrow, and paper wasps use the forelegs to help manipulate wads of nest material during chewing and mixing with oral secretion. Sensory feedback is critical for precision construction. Wasps use the antennae as calipers to control the size of brood cells in the comb. Honey bees measure brood cell diameter with the prothoracic tarsi, while they sense the thickness of wax in the cell walls via pressure receptors on the antennae.

Information Sources

The information required to construct the nest ultimately resides in the genome of individuals, not as a blueprint of the finished nest, but as a set of one or more kinds of construction acts combined with a set of decision rules. The decision rules determine the location and orientation of material added in relation to environmental cues that include gravity, the current structure of the nest, and the location and state of brood and food stores in the nest. It is the interaction of innate rules of behavior and feedback from external cues that results in the species-typical form of the nest.

The simplest nests of solitary species are constructed by following a linear (nonbranching) sequence of steps. A sand wasp digging her nest, for example, need only decide when to switch from extending the burrow to excavating a brood cell. In contrast, nonlinearity characterizes the construction behavior of all social insects. Rather than following a programmed linear sequence, workers make choices among several types of building behavior according to the current state of construction of the nest. Thus, a *Polistes* wasp can use her load of pulp to thicken the pedicel, lengthen a brood cell, initiate a new brood cell, or cover the silken cap of a cell containing a pupa. A social insect worker in a large colony may, in the course of her entire lifetime, perform only one or a small subset of the kinds of construction acts and decision rules in her species' repertory.

Social Organization of Building

In the eusocial insect colony, workers specialize on different elements of nest construction. Older *Polybia* wasp workers collect materials, some specializing in water, others in wood pulp. Back at the nest, these materials are turned over to younger workers, the builders, who keep the pulp moist with water as they add it to the appropriate places on the nest. The builders regulate the overall rate of activity, for it is they who have direct contact with the construction site and can determine the level of demand for materials. Foragers gain information about demand for their material as they seek builders to unload to.

Nest Architecture and Expansion

Termites and ants tend their brood in loose piles in nursery chambers. In contrast, eusocial bees and wasps rear their offspring individually in cylindrical cells (bumble bees rear several immatures per cell). In all but the simplest bee and wasp societies, brood cells are grouped into combs of various sizes, shapes, and orientations (Fig. 1). The most space-efficient way to close-pack cylindrical cells is to surround each cell with six others. Since adjoining cells share walls, this results in the familiar "honeycomb" pattern of hexagonal cells.

From the core of the nest outward, the typical arrangement is brood, then food storage areas, and finally the defensive structures. In the honey bee hive, for example, the central brood cells are surrounded by a concentric layer of pollen-storage cells and then an outer layer of honey-storage cells. The entrance to the nest cavity (or a hive box) is secured by guard bees against intrusion by predators and parasites. A similar arrangement is also seen in the nests of termites (Fig. 2) and wasps.

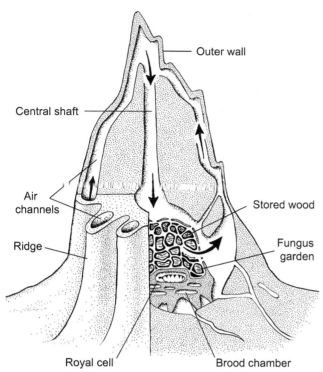

FIGURE 2 Simplified diagram of the nest of the termite, *Macrotermes bellicosus*, from the savanna of Ivory Coast. The front half of the nest is cut away, except for the lower left quarter, which shows the external surface. In the lower part of the mound, just at ground level, is the nest proper, a construction consisting of the central royal cell (containing the queen, king, and attending workers), surrounded by chambers containing brood, fungus gardens, and stored food. Surrounding this is the ridged outer nest, whose design enables it to function as a giant air conditioner. During the day in the dry season, ventilation within the nest is externally driven by the sun, which warms the air in the peripheral air channels, causing it to rise. This sets up a convective circulation within the mound (arrows). CO_2 produced in the central nest diffuses out through the walls of the ridges. Air temperatures are highest and CO_2 levels lowest in the upper portions of the peripheral air channels. Air temperature within the fungus gardens is kept within 29–31°C. (Illustration by Lee Clippard; based on Collins, 1979, and Korb and Linsenmair, 2000.)

Most social insects are able to expand their nests to accommodate colony growth. In some species, such as yellowjacket wasps, growth is continuous throughout the life the colony, whereas in others it occurs in bouts separated by periods of no growth. Honey bees expand the combs in the hive when there is a strong nectar flow coupled with a shortage of honey storage cells. In ants and termites, nest expansion may occur opportunistically after rains soften the soil.

FUNCTIONS OF NESTS

Social Functions

The nest is the information center of the colony. It is here that information is communicated about resource supply and demand and about the status of the queen. The nest itself is involved in the distribution of information about colony membership. Each colony has a unique mix of chemicals that labels every individual as belonging to that colony. This "colony odor" resides on the cuticle of each individual as well as in the nest material, and it has been shown for wasps and honey bees that newly emerging workers learn to recognize their colony odor from the nest.

Food Storage

Bumble bees construct specialized wax pots in which they store honey and pollen during periods of good foraging. Stingless bees and honey bees store pollen and enough honey to sustain the adult population through the unfavorable season. The "honey wasps" (*Brachygastra* spp.) of the Neotropics also store large amounts of honey in their brood cells for the same purpose. Many other species of social wasps store enough honey as droplets in empty brood cells to get the colony through several days of poor foraging. Desert seed harvester ants stockpile seeds in chambers in their nests, and honey ants (*Myrmecocystus* and others) store large amounts of honey in the crops of specialized workers called repletes. The fungus ants and higher termites in the subfamily Macrotermitinae (Termitidae) grow specialized fungus as food in chambers in their nests (Fig. 2).

Defense

Nests often incorporate or accommodate some means of defense against natural enemies. The broods of small, newly initiated colonies are especially vulnerable when the founding queen must leave the nest to forage. Wasps in the genera *Polistes* and *Mischocyttarus* suspend their uncovered combs from a narrow, tough pedicel, which they coat with an ant-repelling secretion produced by an exocrine gland at the base of the terminal sternite (Fig. 1). Among the swarm-founding wasps of the tropics are several species that surround the access to the nest with "ant traps" made of carton bristles several millimeters long, each tipped with a sticky droplet.

Bees and wasps that nest in cavities or construct protective outer covers reduce access by ants and parasitoids to a narrow entrance that can be guarded by a few defending workers. Some stingless bees cover the entrance tube with sticky propolis as a barrier against ants, whereas others pull the soft, waxy tube closed each night. The outer layer of the involucrum of arboreal nests of *Trigona corvina* and *T. spinipes* is thin and easily broken by an intruder, allowing defending bees to swarm out through passageways in the tough, inner layer and launch an attack.

Homeostasis

Social species that form small colonies can exert little control over temperature, humidity, or atmospheric gas concentration in the nest, but some compensate by placing their nests in favorable microhabitats. By building their nests where sun-warmed air collects, such as under eaves on the east and south sides of outbuildings in sunny locations, *Polistes* wasps at higher latitudes achieve shorter egg–adult development times than they would at ambient temperatures.

By virtue of a larger metabolizing biomass and lower nest surface/volume ratios, social species with larger colonies are better able to regulate nest conditions, and nest architecture is often adapted to enhancing homeostatic control. Large colonies produce considerable amounts of metabolic heat, raising nest core temperature well above ambient. Thick nest cavity walls or insulating envelope reduce the loss of this heat to the environment. By combining metabolic heating with evaporative cooling, honey bees can regulate the temperature in the core of the nest to within half a degree of 35°C, even if the outside temperature is many degrees lower or higher. The multiple layers of paper envelope of yellowjacket wasps *(Vespula)* enclose dead air spaces that insulate the nest against heat loss, enabling the colony to maintain steady nest temperatures well above ambient.

Subterranean nests of termites and ants have less control over the temperature in the chambers of their nests, but they can construct the nest to take advantage of solar heating. Some ground-nesting ants of temperate regions excavate chambers under flat rocks lying on sunny ground. As the rock warms in the sun during the day, heat is conducted downward to the ground below. By moving the brood up into these warm but moist chambers during the day, the ants accelerate the development of their immatures. At night, as the rock loses its heat, the brood is moved down to relatively warmer chambers deeper in the soil.

Some ants in higher latitudes build honeycombed mounds of soil or plant detritus. In some of the *Formica* species these can be over 2 m in height. The sun warms the mound to several degrees above ambient, and the colony incubates pupae by moving them up into chambers in the mound during the day.

The most spectacular examples of homeostatic control of nest conditions are the large epigeal (aboveground) mounds built by termites in the savannas of the tropics. Homeostatic mechanisms vary across species, habitat, season, and even time of day. One example is *Macrotermes bellicosus,* found on the savannas of western Africa, whose colonies can reach 2 million workers living inside large, cathedral-like towers that are 3 m or more in height. Air circulation within the mound during the day in the dry season is shown in Fig. 2.

See Also the Following Articles

Ants • Beekeeping • Homeostasis • Isoptera • Parental Care • Sociality • Wasps

Further Reading

Abe, T., Bignell, D. E., and Higashi, M. (2000). "Termites: Evolution, Sociality, Symbioses, Ecology." Kluwer, Boston.
Collins, N. M. (1979). The nests of *Macrotermes bellicosus* (Smeathman) from Mokwa, Nigeria. *Insect. Soc.* **26**, 240–246.
Edgerly, J. S. (1997). Life beneath silk walls: A review of the primitively social Embiidina. *In* "The Evolution of Social Behavior in Insects and Arachnids" (J. C. Choe and B. J. Crespi, eds.). Cambridge University Press, Cambridge, UK.
Evans, H. E. (1958). The evolution of social life in wasps. *In* "Proceedings of the 10th International Congress of Entomology," pp. 449–457.
Evans, H. E. (1966). "The Comparative Ethology and Evolution of the Sand Wasps." Harvard Univ. Press, Cambridge, MA.
Hansell, M. H. (1984). "Animal Architecture and Building Behaviour." Longman, New York.
Hölldobler, B., and Wilson, E. O. (1990). "The Ants." Harvard University Press, Cambridge, MA.
Korb, J., and Linsenmair, K. E. (2000). Ventilation of termite mounds: New results require a new model. *Behav. Ecol.* **11**, 486–494.
Michener, C. D. (2000). "The Bees of the World." Johns Hopkins University Press, Baltimore.
Ross, K. G., and Matthews, R. W. (1991). "The Social Biology of Wasps." Cornell University Press, Ithaca, NY.
Seeley, T. D. (1995). "The Wisdom of the Hive: The Social Physiology of Honey Bee Colonies." Harvard University Press, Cambridge, MA.
Shellman-Reeve, J. S. (1997). The spectrum of eusociality in termites. *In* "The Evolution of Social Behavior in Insects and Arachnids" (J. C. Choe and B. J. Crespi, eds.). Cambridge University Press, Cambridge, UK.
Wilson, E. O. (1971). "The Insect Societies." Harvard University Press, Cambridge, MA.

Neuropeptides

Miriam Altstein
Agricultural Research Organization, Israel

Neuropeptides (Nps) are extracellular chemical messengers, found throughout the animal kingdom, forming a most structurally and functionally diverse group of compounds. Nps have been very well conserved during the course of evolution, indicating their major role as regulators of physiological processes. Nps may act as neurotransmitters, neurohormones, or neuromodulators and, in the hierarchy of entities that regulate endogenous biochemical control functions, the Np messengers rank the highest. The original definition of Nps covered small molecules (< 50 amino acids) of a peptidic nature, synthesized in specialized nerve cells termed neurosecretory cells (NSC) and released from their axon terminals, either into the intracellular space of an adjacent cell (nerve, endocrine, or nonendocrine) or into the circulatory system. Nps released into intracellular spaces, affect proximal effector sites; those that enter the general circulation reach peripheral organs, where their activity can be manifested either directly by activation of a distal target organ or indirectly via signals to nonneuronal internal secretory glands. In recent years, during which many nonneuronal tissues have been found to produce the same peptides as neural tissues, the Np concept has been widened to include peptides that serve to integrate the brain and other tissues for the maintenance of normal physiology, homeostasis, and behavioral patterns.

In insects, Nps were found to regulate a long list of physiological and behavioral processes during development, reproduction, and senescence, and to maintain growth, homeostasis, osmoregulation, water balance, metabolism, and visceral activities. In the past two decades, a large number of insect Nps have been identified, some of which are similar in structure to vertebrate Nps. The study of insect Nps is diverse and multidisciplinary. It integrates cellular and molecular studies of the basic principles of Np action (e.g., biosynthesis, posttranslational processing, release, transport, activation of the target cell, and degradation), chemical approaches for their identification and characterization, immunochemical studies for anatomical localization, and physiological, behavioral, and pharmacological approaches to study their roles in the physiology of organisms. This article describes the distribution and localization of the insect neuroendocrine system, lists the various Np families, and considers the cellular and molecular basic steps of Np action, providing insights into the common properties of the large number of Nps presently identified and the approaches taken to study their regulatory functions.

NP RESEARCH IN HISTORICAL PERSPECTIVE

The concept of neuroendocrine control dates back to the beginning of the 20th century (1922) when Stephan Kopeč first suggested that metamorphosis in insects is regulated by brain factors that are released into the hemolymph. Further progress in the field came from the studies of Berta and Ernst Scharrer, who introduced the basic concepts of neurosecretion and NSC, and described the similarities between the retrocerebral complex in insects and the hypothalamic–hypophysial system in vertebrates. Although neurosecretion was first observed in insects, invertebrate neuropeptide research lagged behind the vertebrate studies, mainly because of low availability of biological material and the lack of sensitive techniques for isolation, sequencing, and synthesis of peptides. The development of chemical, biochemical, and genetic engineering technologies as well as the growing awareness of the major role Nps play in the physiology of organisms, stimulated active interest in insect Np studies, and indeed, in 1975 Starratt and Brown released their pioneering publication announcing the initial determination of a primary structure of the insect Np proctolin. Since then, nearly 150 insect Nps have been reported in the literature, most of which have been isolated from cockroaches (e.g., *Leucophaea maderae, Periplaneta americana, Diploptera punctata*), locusts (e.g., *Locusta migratoria, Schistocerca gregaria*), moths (e.g., *Manduca sexta, Bombyx mori,* and various Heliothinae species), and the fruit fly, *Drosophila melanogaster.*

ANATOMY OF THE NEUROENDOCRINE SYSTEM IN INSECTS

The cellular distribution of Nps in insects has been mapped by histochemical and immunocytochemical techniques as well as by *in situ* hybridization studies. On the basis of this work it was found that the main localization of NSC and their release sites are in the brain–subesophageal ganglion

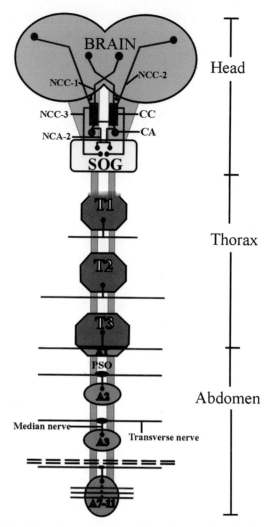

FIGURE 1 Schematic representation of the major neurohemal release sites of the central nervous system of insects. Most of the thoracic and abdominal release sites have not yet been fully characterized. CC, corpora cardiaca; CA, corpora allata; SOG, subesophageal ganglion; NCC, nervus corporis cardiaci; NCA, nervus corporis allati; PSO, perisympathetic organs; T1–T3, thoracic ganglia; A1–A11, abdominal ganglia. [From Predel, R., and Eckert, M. (2000). Nemosecretion: peptidergic systems in insects. *Naturwissenschaften,* **87,** 343–350. © Springer-Verlag GmbH & Co. KG, Heidelberg.]

(SOG)–corpora cardiaca–corpora allata complex (Fig. 1). This complex comprises six clusters of NSC: a pair of medial NSC that originates in the pars intercerebralis (PI), a pair of lateral NSC that originates in the protocerebral region of the brain, and a pair of NSC that originates in the SOG. These six clusters form axon bundles termed nervi corporis cardiaci (NCC): NCC1, NCC2, and NCC3, respectively. Each of the nerve bundles terminates in a pair of retrocerebral neurohemal glands termed corpora cardiaca (CC). The nerve terminals form the storage lobe of the CC through which Nps are released into the circulatory system. Additional neuroendocrine (intrinsic) cells are present in another lobe of the CC glands (the glandular lobe), which is the site of synthesis and release of other Nps. Two additional clusters of

NSC that originate in the SOG extend axons that form the nervi corporis allati 2 (NCA2), which terminate in another pair of endocrine glands (of non-nervous-tissue origin) termed corpora allata (CA) and form another neurohemal region. Another pair of dorsolateral NSC that originate in the PI region of the brain, extend axons that terminate in the CA. The CC and CA are adjacent glands, partially fused with the ventral wall of the aorta, which enables release into the circulatory system of the Nps that are synthesized in the glandular cells, as well as those from the brain/SOG nerve terminals.

Another much smaller neuroendocrine structure consists of the segmentally arranged perisympathetic organs (PSO), which serve as the storage and release site of Nps produced in the ventral nerve cord. Most of the Nps that have been detected in the PSO are not homologous with those found in the retrocerebral complex. A newly discovered endocrine structure is the epitracheal system, which consists of segmentally arranged nerve cells located at the trachea near the spiracles that form the epitracheal glands (EGs). The system produces two blood-borne Nps that trigger pre-ecdysis and ecdysis behavior: pre-ecdysis triggering hormone (PETH) and ecdysis triggering hormone (ETH). Other NSC are distributed in the insect body: in the central nervous system, in the sympathetic nervous system and the peripheral nervous system, on the aorta, and at the ampullae of the antennal heart. Most of these release sites are situated in well-circulated regions of the body. Np-containing cells have also been reported in visceral organs such as the gut, oviduct, accessory glands, and even hemocytes. Raabe has provided a detailed description of the insect neuroendocrine anatomical structure.

NP FAMILIES

Categorization of Nps is usually based on their main action or the one for which a given Np is best known. The major groups of Nps involved in development, reproduction, homeostasis, myotropic activity, and coloration are as follows.

Developmental Nps: The main Nps in this category are the allatotropins/allatostatins, which stimulate/inhibit synthesis of juvenile hormones by the CA; PETH, ETH, crustacean cardioactive peptide (CCAP), and eclosion hormone (EH), which are involved in controlling pre-ecdysis and ecdysis behavior; prothoracicotropic hormone (PTTH), which stimulates molting by initiating biosynthesis and release of ecdysone by the prothoracic gland and diapause hormone that arrests development.

Reproductive Nps: This family includes the ovary-maturating peptide (OMP) and egg development neurosecretory hormone (EDNH), which stimulate egg development; oostatic hormone (OH), which inhibits maturation of ovaries; trypsin modulatory oostatic factor (TMOF), which regulates egg development by modulating trypsin biosynthesis in the gut; neuroparsin, which affects gonad activity; PTTH, which

affects egg development; and pheromone biosynthesis activating neuropeptide (PBAN), which elicits sex pheromone biosynthesis in female moths.

Homeostatic Nps: The homeostatic group includes adipokinetic hormone (AKH), hypertrehalosaemic peptides, bombyxin, ion transport peptide (ITP), and other insulin-related peptides that control fat, carbohydrate, and protein metabolism. Additional members of the family are the diuretic and antidiuretic peptides, which are involved in ion and water balances.

Myotropic Nps: This family is one of the largest Np families in insects. It includes peptides such as proctolin and cardiostimulatory peptides, myokinins, sulfakinins, pyrokinins, myotropins, tachykinins, periviscerokinins, accessory gland and midgut myotropins, myoinhibitory peptides, and FMRFamide-related peptides.

Chromatotropic Nps: Members of this family include melanization and reddish coloration hormone (MRCH), pigment-dispersing hormone (PDH), and corazonin, a cardioactive peptide that has recently been found to exhibit dark pigmentation properties.

Recent immunocytochemical studies have indicated the presence of insect Nps that are comparable to vertebrate Nps. Their functions, however, have not been discovered. Most of the Np families in insects have not yet been detected in vertebrates.

Most of the above-mentioned insect Nps have been characterized, their amino acid sequences have been determined, and their cDNA and genes have been cloned from various insect species. The studies revealed that some Nps may occur in multiple forms (e.g., AKH; allatostatins; myotropic and FMRF-related peptides), a well-known phenomenon among invertebrate Nps. The multiple peptide forms are often encoded by the same gene and result from repeated internal gene duplication and subsequent diversification. Np diversity can also result from a duplication of the whole gene and subsequent mutations. Many Nps elicit more than one biological response in the same or different insect species, and several biological activities may be regulated, in the same insect species, by more than one peptide. A detailed, well documented, review on the structural, biochemical, and physiological characterization of insect Nps has been presented by Gäde.

NOMENCLATURE

The nomenclature of insect Nps is usually based on two primary characteristics: the Np source, which is indicated by the first two letters of the genus name (with the first letter capitalized) and the first letter of the species name, and the first-reported or the major biological function. For example, a peptide isolated from *Helicoverpa zea,* which was first reported to have a pheromonotropic activity, would be designated Hez-PBAN. A detailed explanation of insect peptide nomenclature was presented by Raina and Gäde.

CELLULAR AND MOLECULAR ASPECTS OF NP ACTIVITY

The cellular and molecular components of Np activity include biosynthesis, release, transport, activation of the target cell, and degradation of the Np to terminate its action. Nps are synthesized as large precursor polypeptide chains (termed pre-prohormones) that include a signal sequence that is removed by an endopeptidase during translation. The remaining peptide chain, the prohormone, is transported to the Golgi apparatus, where it is packed into secretory vesicles in which it is further processed proteolytically into smaller fragments by prohormone-converting enzymes to yield biologically active (and inactive) peptides. During the transit through the Golgi network, the precursors may be subjected to posttranslational modifications such as glycosylation, phosphorylation, sulfation, or hydroxylation. Gene expression is regulated by a complex series of factors controlled by other Nps and neurotransmitters. Nps are secreted by a regulatory secretory pathway in which peptides, stored in secretory vesicles, are released in response to secretagogues. Secretion of Nps is usually Ca^{2+} dependent and, unlike neurotransmitters, Nps are not recycled at the nerve terminals but are newly synthesized in the cell body. Upon secretion (either to the circulatory system via neurohemal organs or to the intracellular space of adjacent cells), Nps reach their target organ, where they activate the target cells by binding to cell surface proteins (termed receptors) and exciting second-messenger systems, thus initiating a variety of cellular responses. At the end of the activation, Nps dissociate from the receptor and are rapidly inactivated by peptidases present in the plasma, or the intercellular space or in the target cell membrane. All the above-mentioned events are common to all Nps, regardless of their origin or biological function. Strand has provided a detailed summary of these processes.

FUTURE PROSPECTS

Despite the enormous amount of structural information that has been accumulated on insect Nps, our understanding of their mode of action is very limited and rudimentary. It is still necessary to develop additional *in vivo* bioassays that involve whole organisms and to create novel tools and technologies to unravel the complex coordination of the many Nps involved in the regulation of the physiological processes and behavioral patterns. There is also a need to study the cellular and molecular factors that underlie their activity, and to discover their pathways of synthesis and release and their targets of action. These issues are being addressed with the help of highly advanced molecular biology and genetic engineering techniques (e.g., gene cloning and expression, *in situ* hybridization, gene transfer, gene knockouts, site-directed mutagenesis), immunocytochemical techniques, and advanced chemical (e.g., liquid chromatography in combination with mass spectrometry), immunochemical, and biochemical methods. Of great importance is the study of Np receptors and the

development of selective agonists and antagonists, which may serve not only as research tools but also as a basis for the design of novel, environmentally friendly insect control agents. Altstein has summarized a novel approach to the exploitation of this avenue.

See Also the Following Articles
Brain and Optic Lobes • Development, Hormonal Control of • Juvenile Hormone

Further Reading
Altstein, M. (2001). Insect neuropeptide antagonists. *Biopolymers (Pept. Sci.)* **60**, 460–473.
Gäde, G. (1997). The explosion of structural information on insect neuropeptides. *In* "Progress in the Chemistry of Organic Natural Products," (W. Herz *et al.,* eds.) Vol. 71, p. 128. Springer-Verlag, New York.
Kopeč, S. (1922). Studies on the necessity of the brain for the inception of insect metamorphosis. *Biol. Bull.* **42**, 323–342.
Raabe, M. (1989). "Recent Developments in Insect Neurohormones." Plenum Press, New York.
Raina, A. K., and Gäde, G. (1988). Insect peptide nomenclature. *Insect Biochem.* **18**, 785–787.
Scharrer, B., and Scharrer, E. (1944). Neurosecretion. IV: Comparison between the intercerebralis–cardiacum–allatum system of the insects and the hypothalamo-hypophysial system of the vertebrates. *Biol. Bull.* **87**, 242–251.
Starratt, A. N., and Brown, B. E. (1975). Structure of the pentapeptide proctolin, a proposed neurotransmitter in insects. *Life Sci.* **17**, 1253–1256.
Strand, F. L. (1999). "Neuropeptides. Regulators of Physiological Processes." MIT Press, Cambridge, MA.

Neuroptera
(Lacewings, Antlions)

Catherine A. Tauber and Maurice J. Tauber
Cornell University

Gilberto S. Albuquerque
Universidade Estadual do Norte Fluminense, Brazil

The Neuroptera, also known as Planipennia, is one of the oldest insect orders with complete metamorphosis; it includes the green and brown lacewings, antlions, owlflies, dustywings, mantidflies, and allies. Although it is a relatively small order, with about 6000 species distributed among 17 families, its members occur in a variety of habitats throughout the world and their habits are diverse and interesting. Largely because of their lacey and colorful wings, delicate bodies, and fascinating biology, neuropteran adults are attractive to both biologists and laypersons (see Fig. 1).

Neuropteran larvae, which are less noticeable than adults, have received much less attention. Unlike the adults, which may or may not be predaceous, almost all neuropteran larvae are predaceous; they feed on a variety of soft-bodied arthropods. Because of this predaceous habit, several families

FIGURE 1 Neuropteran adult in the family Crocidae. Note the colorful forewings, long slender hind wings, and delicate body. (Photograph by E. S. Ross, California Academy of Sciences.)

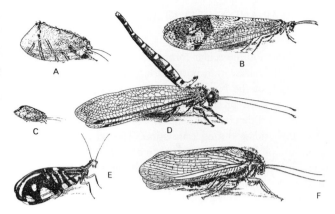

FIGURE 2 Adult Neuroptera. (A) Psychopsidae; (B) Myrmeleontidae; (C) Hemerobiidae; (D) Ascalaphidae; (E) Osmylidae; (F) Ithonidae. (Reproduced, with permission, from New, 1991, © CSIRO Australia.)

(primarily Chrysopidae, Hemerobiidae, and Coniopterygidae) are very useful in the natural, biological, and integrated control of many economically significant insect pests. But, despite their actual and potential importance, they have received less emphasis than other groups, such as the predaceous lady beetles.

The name "Neuroptera" is rooted in two Greek words: "neuron" meaning "sinew" and "pteron" meaning "wing." It refers to the netlike arrangement of the veins and crossveins of the wings and comes from an old usage of the word "nervation," meaning "strengthening by sinews." The name does not refer to nerves, as suggested by many authors; the order received its name long before nerves were recognized as tissue.

CHARACTERISTICS

Typically, neuropteran adults are soft-bodied and have four membranous wings that are similar in size, structure, and venation and that are held roof-like over the body while at rest (Fig. 2). Branches of the veins are generally bifurcated at the margins of the wings. The adults have chewing mouthparts, large lateral eyes, and multiarticulate antennae that are usually filiform (threadlike) or moniliform (with beadlike segments); except for one family (Osmylidae), neuropteran adults lack ocelli. The mesothorax and metathorax are similar in structure, and the abdomen is cylindrical and without cerci.

Unlike megalopteran and raphidiopteran larvae, neuropteran larvae differ distinctly from their adult stages. Larval mandibles and maxillae are usually elongate, slender, and modified for sucking, whereas maxillary palpi are absent. The larval thorax bears walking legs, and the one-segmented tarsus usually ends in two claws that function in locomotion. The terminal adhesive disks of the abdomen also aid locomotion. Like the adult abdomen, the larval abdomen does not have cerci.

Neuropteran pupae are enclosed within silken cocoons and they have characteristics that help distinguish them from the pupae of other insects. For example, they are exarate, i.e., the legs and wings are free from the body, the abdomen is moveable; as a result the pupa itself is capable of limited locomotion.

Also, they are decticous, i.e., they have strong mandibles that are used to open the cocoon during emergence.

Neuroptera possess some interesting cytological characteristics. Although some variation occurs, almost all neuropteran species studied thus far have XX/XY sex determination. However, the sex chromosomes of most neuropteran species have an unusual type of pairing, called distance pairing, in which the chromosomes do not align to form bivalents during meiosis; rather, they are pulled from within the spindle to stabilized positions at the poles. This characteristic form of meiosis is shared with Raphidioptera. In many neuropteran taxa, meiosis occurs early in development, e.g., during the last instar or the pupal stage. In some taxa, the adult males have completely degenerate testes; mature sperm bundles are stored in seminal vesicles.

FOSSIL RECORD AND GEOGRAPHIC DISTRIBUTION

Although the fossil record of the Neuroptera is small and fragmentary, ancient (extinct) neuropteran families have been traced back with certainty to the Lower Permian in Kansas and the Upper Permian in Australia and Russia. The affinities of these archaic forms to modern taxa are unknown.

Modern families (e.g., Nymphidae, Psychopsidae, Chrysopidae, Osmylidae) appear in mid-Mesozoic fossils; probably the earliest and most diverse of these fossils are allied to the Psychopsidae. The largest known neuropteran, a psychopsid-like lacewing with a wingspan of 24 cm and large conspicuous eyespots on the wings, existed in the Jurassic. Recognizable examples of berothids and coniopterygids were also present during the Lower Jurassic. Cretaceous amber includes specimens from an array of modern families, including Berothidae, Mantispidae, Sisyridae, Chrysopidae, and Hemerobiidae, and the diverse Neuroptera found within Baltic amber can be placed in modern families.

Today, the order Neuroptera is distributed worldwide, with the exclusion of Antarctica. Europe, North America, and Asia have rich neuropteran faunas as do southern Africa, South America, and Australia. Australia probably has the broadest diversity; it lacks representatives in only two families (Dilaridae and Polystoechotidae) and most of the presumed archaic families are represented there (e.g., Nevrorthidae, Ithonidae). In contrast to this abundance and diversity, New Zealand and the South Pacific islands have only meager neuropteran faunas. Remarkably, Hawaii seems unique in that it is the only island group where complexes of endemic species (Hemerobiidae and Chrysopidae) have evolved.

EVOLUTIONARY RELATIONSHIPS WITH OTHER ORDERS

Taken together, the three orders, Megaloptera (dobsonflies, alderflies), Raphidioptera (snakeflies), and Neuroptera (Neuroptera *sensu stricto,* Planipennia) (lacewings, antlions, dustywings, and allies) form the superorder Neuropterida.

Because of its ancient fossil record and the generalized body structure of its larvae and adults, this superorder is considered to be among the most primitive of the Holometabola (insects with complete metamorphosis). Significant morphological and molecular evidence indicates that this superorder has a sister relationship with the Coleoptera.

A number of synapomorphic (shared, relatively derived or specialized) characteristics distinguish the Neuroptera as a monophyletic order that is separate from Megaloptera and Raphidioptera; notably, almost all of these distinguishing features occur in the larvae. For example, megalopteran and raphidiopteran larvae have biting–chewing mouthparts and the mouth opens anteriorly. In contrast, the mouthparts of neuropteran larvae are suctorial and consist of elongate and pointed mandibles and maxillae whose adjacent grooved surfaces form a feeding tube. The mouth, instead of opening anteriorly, connects to the feeding tubes at the sides of the head.

Other larval characteristics distinguish the Neuroptera from the other two neuropteridan orders. Neuropteran larvae do not have contiguous intestinal tracts. Rather, the midgut and hindgut remain separate until pupation. As the larva feeds, feces accumulate in the midgut, and only after metamorphosis from the larval stage to the adult, during which the midgut and hindgut become connected, does the newly emerged adult expel the feces in the form of a meconial pellet. Neuropteran larvae use the hindgut and associated structures (Malphigian tubules) to produce silken cocoons within which pupation occurs. In contrast, megalopteran and raphidiopteran larvae have contiguous intestines. Moreover, they do not form cocoons; rather they pupate within earthen chambers or wooden cells.

Another synapomorphy of neuropteran larvae is an articulated, neck-like cervix. This contrasts with the ribbon-like cervical sclerite of megalopteran and raphidiopteran larvae.

Two recent cladistic analyses provide conflicting results concerning the relationships among the three neuropteridan orders. One study, based on morphological characteristics, indicates that contrary to traditional thought, the Megaloptera and Neuroptera are more closely related to each other than either is to the Raphidioptera. The subsequent cladistic analysis, which was based on both morphological and molecular data, yielded a trichotomous relationship among the three neuropteridan orders. The relationships remain unresolved; but, in this article we chose to use the first study; it yielded dichotomous relationships among the three neuropteridan orders and a fairly well supported sister relationship between Megaloptera and Neuroptera (Fig. 3).

EVOLUTIONARY RELATIONSHIPS WITHIN NEUROPTERA

The order Neuroptera encompasses 17 families that currently fall into three more or less well-supported suborders (see below). Several strong larval synapomorphies support the hypothesis of a sister relationship between Myrmeleontiformia and Hemerobiiformia. The Nevrorthiformia appears to be ancestral.

Order Neuroptera (=Neuroptera *sensu stricto,* Planipennia)
  Suborder Nevrorthiformia (=Neurorthiformia)
    Family Nevrorthidae (=Neurorthidae) (11 species)
  Suborder Myrmeleontiformia
    Family Psychopsidae (silky lacewings) (26 species)
    Family Nemopteridae (spoon-winged lacewings) (100 species)
    Family Crocidae (thread-winged lacewings) (50 species)
    Family Nymphidae (including Myiodactylidae of some authors) (split-footed lacewings) (35 species)
    Family Myrmeleontidae (including Stilbopterygidae of some authors) (antlions) (2100 species)
    Family Ascalaphidae (owlflies) (400 species)
  Suborder Hemerobiiformia
    Family Ithonidae (including Rapismatidae of some authors) (moth lacewings) (32 species)
    Family Polystoechotidae (giant lacewings) (4 species)
    Family Chrysopidae (green lacewings) (1200 species)
    Family Osmylidae (160 species)
    Family Hemerobiidae (brown lacewings) (550 species)
    Family Coniopterygidae (including Brucheiseridae of some authors) (dustywings) (450 species)
    Family Sisyridae (spongillaflies) (50 species)
    Family Dilaridae (pleasing lacewings) (50 species)
    Family Mantispidae (mantidflies) (400 species)
    Family Berothidae (including Rhachiberothidae of some authors) (beaded lacewings) (115 species)

SUBORDER NEVRORTHIFORMIA

The Nevrorthiformia (=Neurorthiformia, which is a misspelling in the literature) contains a single, very small family, Nevrorthidae (=Neurorthidae, a misspelling). This family shares several characteristics with the Megaloptera and is considered basal among the Neuroptera; e.g., its aquatic larvae have distinctive head structures and Malphigian tubules

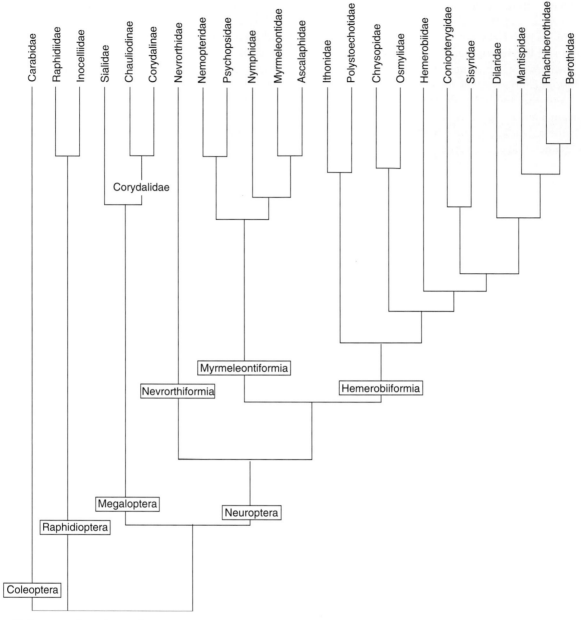

FIGURE 3 Phylogenetic relationships within Neuropterida (as proposed by Aspöck *et al.,* 2001). Note the sister relationship between Neuroptera and Megaloptera, the composition of the three neuropteran suborders, and the relationships of the families within the suborders.

that lie free distally (i.e., without cryptonephry). Thus, the Nevrorthiformia emerges as the basal neuropteran suborder.

Among the Neuropterida, a terrestrial lifestyle without cryptonephry is considered plesiomorphic (primitive); this lifestyle is typical of Raphidioptera. The evolution of an aquatic larva (without cryptonephry) is considered a synapomorphy that supports the sister-relationship between the megalopteran and the neuropteran lineages. Apparently, the aquatic lifestyle was retained in the Nevrorthiformia, whereas a reversal to the terrestrial lifestyle characterizes the remainder of the Neuroptera (the Myrmeleontiformia and Hemerobiiformia). Moreover, in these two groups (Myrmeleontiformia and Hemerobiiformia), cryptonephry arose (i.e., all of the Malphigian tubules except

two are fused to the hindgut). The only exception is found in Sisyridae, which falls within the Hemerobiiformia; here, the apparent partial loss of cryptonephry (all but one Malphigian tubule are free of the rectum) may be regarded as an evolutionary reversal associated with the secondary evolution of an aquatic larva. The single remaining fused Malphigian tubule may be evidence of ancestral cryptonephry.

Nevrorthidae

Eleven species are known from three genera of Nevrorthidae; these occur in the Mediterranean Region, eastern Asia, and Australia. Nevrorthid adults are small (forewing length 6–10

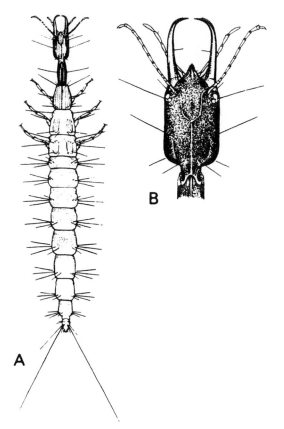

FIGURE 4 Nevrorthidae. (A) Larval body; (B) larval head. (Reproduced with permission, from New, 1991, © CSIRO Australia.)

mm) and delicate, and although they resemble sisyrids, they have unique, defining characteristics in the head structure and male and female genitalia.

Nevrorthid larvae, like sisyrid larvae, occur in aquatic habitats, but morphologically they are very distinct. For example, unlike sisyrids, they lack abdominal gills and their mouthparts are robust and curved inward (not needle-like) (Fig. 4).

Little is known about the biology or behavior of nevrorthid adults or larvae. Larvae of a southern European species have been collected on the stony bottoms of cold, swiftly moving mountain streams; those of an Australian species are believed to live in moist litter. Adults are usually found near streams or in wet, forested areas.

SUBORDER MYRMELEONTIFORMIA

Although its constituent members have changed over the years, the suborder Myrmeleontiformia has long been recognized as a valid grouping. Currently, it is composed of six families: Psychopsidae, Nemopteridae, Crocidae, Nymphidae, Myrmeleontidae, and Ascalaphidae.

Although myrmeleontiform adults are morphologically diverse and without clearly defined synapomorphies, the larvae of the six families share many morphological and biological characteristics. Primary among these are a head capsule with a highly sclerotized tentorium and a prementum that resembles

a segment of the labial palp. Generally, myrmeleontiform larvae ambush, rather than pursue, their prey, and the head capsules and jaws are modified into a "trap"-like mechanism that can close very quickly. Their long, robust, inwardly curved mandibles are frequently toothed and constricted rather than enlarged basally. The maxillae are lance-like, and the head capsule is robust, quadrate, or cordate (Fig. 5). The body form is short, broad, and powerful. The antennae are short and have only 10 to 12 segments, with a thick scape and a narrow distal portion. There are two tarsal claws and the empodium is absent (except from psychopsid larvae).

A sister relationship between Myrmeleontiformia and Hemerobiiformia appears to be well supported by a suite of presumed apomorphies in both male and female genitalia and in the larvae (e.g., terrestrial lifestyle, cryptonephry).

Within the Myrmeleontiformia, the Nymphidae, Myrmeleontidae, and Ascalaphidae form a reasonably well-defined group, within which the Myrmeleontidae and Ascalaphidae are very closely related and seem to have a sister relationship. In fact, there are probably no absolute criteria for separating owlfly and antlion larvae or adults because most of the distinguishing traits are shared by some members of the other family. Nevertheless, in the majority of myrmeleontid and ascalaphid species throughout the world, both adults and larvae exhibit the suite of characters that are typical of one family or the other, and so it is prudent to retain the two families until a thorough cladistic analysis is completed.

Currently, thoughts differ as to whether the Psychopsidae is the basal group of the Myrmeleontiformia or has a sister relationship with the Nemopteridae or Crocidae. Detailed studies of additional species from all three families will help resolve this issue.

Psychopsidae (Silky Lacewings)

The Psychopsidae forms a small family of approximately 26 species within five genera. They are large, attractive, and moth-like insects (forewing length 10–35 mm). The family is restricted to sub-Saharan Africa, southeast Asia, and Australia. However, fossils are known from North America, Europe, Asia, and Australia.

Very little is known about psychopsid biology. Eggs are unstalked and attached to the substrate. They are laid singly or in groups on the bark of trees, and they are covered with a secretion that is presumably derived from plant material or sand but whose origin is unknown. Larvae of Australian species are found under loose bark (Fig. 5A). There are three instars and the life cycle may take 2 years. The cocoon has two layers of silk. Adults have been collected in river valleys.

Nemopteridae (Spoon-Winged or Ribbon-Winged Lacewings)

Until recently, this small family of approximately 100 species included species now comprising the Crocidae. Both groups

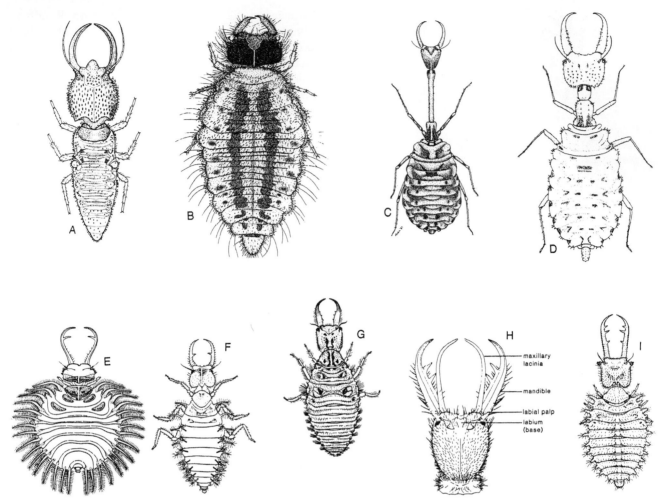

FIGURE 5 Myrmeleontiform larvae. (A) Psychopsidae; (B) Nemopteridae; (C) Crocidae; (D) Crocidae; (E) Nymphidae (Myiodactylinae); (F) Nymphidae (Nymphinae); (G) Myrmeleontidae; (H) Myrmeleontidae; (I) Ascalaphidae. (A, D–G, and I reproduced with permission, from New, 1991, © CSIRO Australia. B and C reproduced with permission, from Mansell, 1996. H modified from Tauber, 1991.)

are extraordinary looking lacewings that are easily distinguished by their remarkably elongate and narrow hind wings and by a very short metathorax. Nevertheless, there are morphological and ecological differences between the two groups that are consistent with family status. Externally, nemopterid adults are distinguished from crocid adults by their large size (forewing length 13–35 mm) and hind wings that are ribbon-like, with or without dilation, but with distinct dark areas. Nemopterid larvae are myrmeleontid-like (Fig. 5B). In contrast, adults of Crocidae have smaller bodies and their hind wings are thread-like, white, and rarely with dark shading. Crocid larvae are very unusual, especially in the elongation and sclerotization of the cervix (Figs. 5C and 5D).

The Nemopteridae occurs in the arid and desert zones of the southern borders of the West Palaearctic and West Oriental Regions, and in dry areas of the Neotropical, Afrotropical, and Australian Regions. The world's richest nemopterid fauna is concentrated in southern Africa (>60% of the species).

Although nemopterid larvae have not been collected in ant nests, myrmecophily is strongly suspected. Nemopterid

eggs are unstalked, small, spherical, or ovoid, and laid singly in the sand or soil. They have a micropyle and an oviruptor (an internal, toothed structure that tears open the chorion during hatching), but the surface of the chorion is smooth and seems to lack the aeropyles that are present on crocid eggs. The eggs may be very hard and lack adhesives, and it is believed that they are harvested and introduced into ant colonies by foraging ants. Granivorous ants apparently collect the eggs.

Larvae may also be harvested by ants and those of at least some nemopterid species live in or very near ant nests. Apparently, the larvae avoid detection by the ants through chemical mimicry and also through the covering of sand that adheres to their bodies. Young larvae may burrow with the front legs and push with the other legs and they may climb to the soil surface. Mature third instars become very rotund and movement is slow and awkward. Ants appear to be the main food and it is thought that toxic materials are injected into prey, because prey become immobilized immediately after attack. Larvae can withstand long periods (up to several months) without food, and they require few prey items to complete their development. Apparently they have very low rates of metabolism. Pupation occurs within a silken cocoon.

Adults are mostly crepuscular or nocturnal, and they feed on pollen. The modified hind wings appear to have a defensive function against aerial predators; they provide camouflage and crypsis or they give the illusion of greater size. Like myrmeleontid males, some nemopterid males have membranous sacs between tergites 5 and 6 that may emit pheromones. The elongated rostrum probably functions in collecting pollen and nectar. Adults occur during a short period of the year, perhaps in synchrony with the ephemeral burst of flowers in their inhospitable habitat.

Crocidae (Thread-Winged Lacewings)

Until very recently, this small family of approximately 50 species was included in the Nemopteridae. The two families share elongate hind wings, but crocids are distinguished by the thread-like hind wings (Fig. 1) and several larval characteristics. The distribution of Crocidae overlaps that of the Nemopteridae: arid and desert zones on the southern borders of the West Palaearctic and West Oriental Regions and dry areas from the Neotropical, Afrotropical, and Australian Regions. Crocid adults are medium to large sized (forewing length 7–15 mm).

The unstalked eggs of Crocidae differ markedly from nemopterid eggs: they have a sponge-like micropyle, the chorion has aeropyles, and no oviruptor is present. The larvae of some crocid species have elongated cervical regions (Fig. 5C); these species are frequently associated with caves or dwellings. Species that have a shorter cervix usually live in detritus, under rocks or in crevices (Fig. 5D). Crocid larvae are largely "sit-and-wait" predators and they occur beneath sand or soil surfaces; however, some may pursue prey. In both cases, they are able to survive for long periods without food, but little is known of the larval diet. Pupation occurs within a silken cocoon that incorporates sand or debris externally. Adults are either crepuscular or diurnal, and they may feed on pollen and/or nectar. The hind wings apparently play sensory and stabilizing functions, which enable the lacewings to detect vertical and horizontal surfaces and to fly in confined spaces, such as caves. They may also function in mate attraction and courtship.

Nymphidae (Split-Footed Lacewings)

This small neuropteran family is restricted to the Australian Region (Australia, New Guinea, and nearby islands); it contains about 35 species in seven genera. Currently the Nymphidae includes two well-defined lineages, the subfamilies Myiodactylinae and Nymphinae. Nymphid adults are large (forewing length 18 to >40 mm).

Nymphid eggs are laid on slender, filamentous stalks. In Myiodactylinae, the stalks are either pendant or looped, so that the egg contacts the substrate. In Nymphinae, the eggs are arranged in intricate patterns. In both subfamilies, the stalk may be coated with beads of liquid that may serve nutritional and/or defensive functions. Hatching occurs via an oviruptor. Myiodactyline larvae are very flat, and the margins of their bodies have long scoli (Fig. 5E); they are green and arboreal, and they rest on the surface of leaves with their jaws at an angle of ~180°. Nymphine larvae (Fig. 5F) occur in litter or on the bark of trees where they are camouflaged by the debris they carry or by the markings on the body. Cocoons probably are spun in sand. Adults of one species form large aggregations, but the function (e.g., mating, defense) of the aggregations is not known. Some species occur in association with acacias. Adults produce an odor from eversible abdominal glands and copulation involves enlarged and elaborate male terminalia that presumably have a grasping function.

Myrmeleontidae (Antlions, Doodlebugs)

With approximately 2100 species in 300 genera, the antlions constitute the largest neuropteran family. Members of this family have intrigued naturalists from the earliest times; imaginative accounts were summarized by W. M. Wheeler in his 1930 book, *Demons of the Dust*. Most people know antlions because the larvae of some species have pit-building habits. In truth, both the name "antlion" and the assumption that all antlions construct pits are misleading; myrmeleontid larvae do not feed exclusively on ants and most do not construct pits. Adults are slender bodied and medium sized to large (forewing length 10–70 mm).

Four subfamilies of Myrmeleontidae are generally recognized: Myrmeleoninae, Palparinae, Acanthaclisinae, and Stilbopteryginae (formerly a separate family, Stilbopterygidae). Larval morphological (Figs. 5G and 5H) and biological characteristics are crucial in the classification of the family, especially at the tribal level but also for many genera.

This family has a worldwide distribution, notably in the arid and semiarid areas of subtropical and tropical Africa, Australia, Asia, and the Americas. They inhabit open woodlands, scrub grasslands, and dry sandy areas. Efforts have been made in South Africa to assess the taxonomic richness of the large myrmeleontid fauna and to help conserve it.

Myrmeleontid eggs are unstalked and relatively large; they are laid singly in open areas or tree holes, under bushes, in caves, under rock overhangings, or in areas sheltered by buildings. The eggs are covered with a glandular secretion that may facilitate adhesion of sand or soil particles, and they lack an oviruptor.

The larvae of most species appear to be sit-and-wait predators. In most taxa they live beneath the soil surface, on trees, in tree holes, under stones, or in debris. Larvae of a very small number of genera construct pits in sand or soil where they capture insects that fall into the pits. For some of these species, pit architecture and the influence of prey availability on pit size, location, and relocation, as well as on larval growth itself, have been studied extensively.

Although the larvae of some myrmeleontid species can travel quickly across the surface of the sand, others have slow, creeping movements or fast backward movements under the sand that are aided by the forward-directed terminal segments (fused tibia–tarsus) of the hind legs. These patterns of behavior have led to the common name, "doodlebugs." Ingestion is accomplished by the injection of digestive enzymes from the midgut into the prey, followed, after several minutes, by sucking. The regurgitated gastric juice is not mixed with the contents of the crop or the midgut. Instead, it seems to be extruded from the space between the peritrophic membrane and the epithelium of the midgut; then contractions of the crop's muscular system transport the fluid forward through a fold in the wall of the crop.

Larvae pass through three instars. In some species, larval development can be protracted over several years depending upon the availability of prey; overwintering occurs in the larval stage. Univoltine or semivoltine life cycles and the synchrony of adult emergence are maintained by photoperiodic and thermal responses during early and late larval stages and by the physiological effect of pupal body size. Larvae generally spin a single-walled cocoon; a double-walled cocoon occurs in one unusual South American species.

Adults are largely nocturnal and presumed predaceous. Their flight resembles that of a damselfly. Sexual communication involves the extrusion of "hair pencils" and abdominal glands or sacs on the male (analogous to those found in lepidopteran males), as well as the release of volatile substances from thoracic glands in both males and females.

Ascalaphidae (Owlflies)

The owlflies constitute a medium-sized neuropteran family of about 400 species assigned to 65 genera. There are three subfamilies: one with bisected eyes (Ascalaphinae) and two with entire eyes: Haplogleniinae and Albardiinae. The subfamily Albardiinae consists of a single, very unusual, large-bodied Brazilian species.

The two major subfamilies (Ascalaphinae and Haplogleniinae) are widely distributed in warm regions of the world, but of the two only the Haplogleniinae occurs in Australia. Ascalaphids inhabit grasslands and warm dry woodlands. Most species are nocturnal or crepuscular, but some Eurasian species are diurnal and have pigmented wings that resemble those of butterflies. Adults are relatively large (forewing length 15–60 mm).

Clusters of 20 to 75 large, unstalked eggs are laid on twigs in spirals or rows. Individual eggs are reported to have two micropyles, but lack an oviruptor. Females of many species place small, modified eggs (repagula) on or around egg batches; these repagula reportedly have defensive and nutritional functions—they serve to repel predators and/or provide food for newly hatched larvae. The two major subfamilies in the New World possess this habit, but it is absent from Old World and Australian ascalaphids.

Newly hatched larvae often remain together near the egg cluster for a week or more before dispersing. Larvae are either terrestrial (in the soil or litter) or arboreal (on leaves or tree trunks), and most seem to be sit-and-wait predators (Fig. 5I). Characteristically, the larvae hold their jaws open at very wide angles; some New World species are able to open their jaws beyond 270°. When prey makes contact with the larva, the jaws can close very rapidly and larvae can take relatively large prey. Considerable evidence shows that the larvae paralyze their prey with toxins from the midgut, not from glands. As in other myrmeleontiform larvae, there are three instars.

Larvae of some species that live in the soil or sand camouflage themselves with sand grains or debris. Such behavior shares features with the "camouflaging" behavior of chrysopid larvae, but ascalaphid larvae use their flexible foretarsi, rather than the jaws, to place material on their dorsa. A thick mat of tangled threads anchors the debris to the dorsal surface.

Second and third instars resist starvation well, and development may be extended for 1 or 2 years. Because ascalaphid adults typically occur at very specific times of the year, diapause probably intervenes in some larval stages. Photoperiod or other factors may be involved in regulating the occurrence of diapause, but the responses and mechanisms have not been studied. Pupation occurs within silken cocoons that are spun either on the ground or on trees, sometimes incorporated with sand or debris.

Adults remain motionless in a characteristic head-downward position for most of the day; flight is restricted to a relatively short period around dusk and is preceded by ~10 min of muscle-warming via wing vibration. Compared to that of other neuropterans, the flight of ascalaphids is very strong and agile and similar to that of dragonflies. Adults attack and feed on large numbers of flying insects (e.g., caddisfly adults); prey capture and mating occur on the wing.

SUBORDER HEMEROBIIFORMIA

The Hemerobiiformia is currently considered to be a monophyletic but heterogeneous grouping having a sister relationship with the Myrmeleontiformia. It contains 10 families whose larvae are diverse in lifestyles and structure (Fig. 6), but which share a number of distinguishing features. For example, hemerobiiform larvae have a head capsule whose posteroventral region is composed primarily of the maxillae, cardines that are elongate, and a cervix that is cushion-like.

Within the Hemerobiiformia, two groupings emerge. First, Ithonidae + Polystoechotidae appear as a basal sister group. Among the Neuroptera, ithonid and polystoechotid larvae are unique in that they feed by sucking on plant tissues. Given the close relationship between *Rapisma,* which has been designated a separate family (Rapismatidae), and *Adamsiana,* a newly described genus of Ithonidae from Honduras, it is difficult to justify retaining the family Rapismatidae. Thus, we include both genera in the Ithonidae while awaiting discovery and study of their larvae and a thorough phylogenetic analysis of the Neuroptera.

The second hemerobiiform grouping harbors the remaining families, all of which are carnivorous in the larval stages. Within

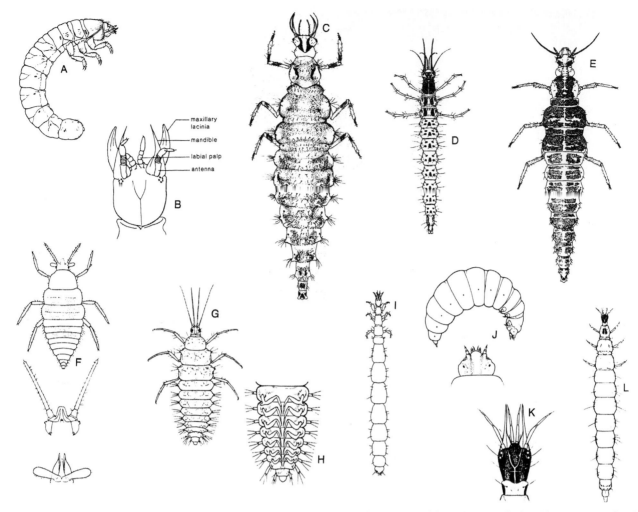

FIGURE 6 Hemerobiiform larvae. (A) Ithonidae; (B) Polystoechotidae; (C) Chrysopidae; (D) Osmylidae; (E) Hemerobiidae; (F) Coniopterygidae; (G) Sisyridae; (H) Sisyridae (venter of abdomen; note gills); (I) Dilaridae; (J) Mantispidae; (K) Berothidae; (L) Berothidae. (A, D, F–H, and J–L reproduced, with permission, from New, 1991, © CSIRO Australia. B and I modified from Tauber, 1991. C reproduced from Tauber, 1974. E reproduced, with permission, from Tauber and Krakauer, 1997.)

this grouping, the Chrysopidae and Osmylidae appear to form a sister group, although the evidence for this relationship is not strong. In turn, these two families emerge as a sister group to the poorly defined group of six families that remain. The reasonably well-defined dilarid lineage [Dilaridae, Mantispidae, and Berothidae (including the Rhachiberothinae)] apparently has a sister relationship with the Coniopterygidae–Sisyridae sister group, but the relationship is not strongly supported. Finally, the Hemerobiidae emerge as a sister-group to the dilarid + (coniopterygid + sisyrid) lineages, but again the relationship is tentative.

Ithonidae (Moth Lacewings)

This very small family includes approximately 32 species in three genera from Australia, southern Asia, southwestern United States, Mexico, and Central America. Adults are large and moth-like (forewing length 15–30 mm); they share a variety of plesiomorphic characters, but few apomorphic

characters are known. The larvae are subterranean and scarabaeiform (Fig. 6A). The abdomen is very large and swollen; the legs are short and fossorial. The short mandibles curve inward and slightly upward. The maxillae are broad and robust; the mandibles are narrow. Eyes are absent.

Little is known about ithonid biology. Unstalked eggs are laid singly in the soil, where their adhesive surface gathers soil and sand particles. The larvae of one species in Australia are associated with the roots of *Eucalyptus* trees and a North American species occurs near creosote bushes. The specific food source of these larvae (plant, mycorrhizae or other fungi, or associated herbivores) is not known. In one Australian species, five instars have been demonstrated. Larval development may take 2 or more years; apparently, mature larvae undergo diapause. Pupation occurs within silken cocoons in the soil and adults emerge synchronously in large numbers, usually following a period of rainfall. Males emerge first and form aggregations that attract females; adults live for only a few days.

FIGURE 7 Despite their common name, green lacewings, adults of many species of Chrysopidae, like this *Nothochrysa* adult, are darkly colored. (Photograph by E. S. Ross, California Academy of Sciences.)

Polystoechotidae (Giant Lacewings)

The Polystoechotidae constitutes an extremely small family (four species) of lacewings that are characterized by a very large body and long wings (forewing length 15–40 mm). The adults resemble ithonids, but they have larger wings, more complex wing venation, and genital differences. This family is restricted to the New World.

Eggs are like those of ithonids: large, unstalked, and covered with sticky material. Only the first instar of *Polystoechotes* is known. Unlike the scarabaeiform ithonid larvae, this larva is hemerobiid-like: it is elongate and has relatively long legs, a short multisegmented antenna, and robust jaws (Fig. 6B). Little is known about polystoechotid biology; phytophagy is suspected but not confirmed. It is noteworthy that populations of one species in North America have experienced large reductions in their numbers since the 1900s. Thus, lack of knowledge concerning this family is especially unfortunate.

Chrysopidae (Green Lacewings)

Chrysopidae, with its approximately 1200 recognized species, is one of the largest families of Neuroptera, second only to the Myrmeleontidae. The larvae of many chrysopid species feed on insect and mite pests of agricultural crops or horticultural plantings, and because of their importance in biological control, chrysopids are the most frequently studied of the Neuroptera.

Adults are medium-sized to large, delicate insects with four subequal wings (forewing length 6–35 mm) and relatively long filiform antennae. In most species the adults are green with large golden eyes, but some species have black, brown, or reddish adults (Fig. 7). Larvae vary in shape and habits; some are voracious, active, and free-living general predators (thus the name "aphis-lions") (Fig. 6C); others are slower moving, cryptic, trash-carrying predators that are associated with specific types of prey or habitats; still others live in ant nests where they feed on the inhabitants.

Currently, the Chrysopidae comprises three subfamilies (Nothochrysinae, Apochrysinae, and Chrysopinae), all of which are based strictly on adult morphological characteristics and none of which are well defined. Comparative studies of the larval morphology and biology are needed to clarify the systematics of the group and to facilitate the identification and use of these predators in biological control.

The Nothochrysinae includes only a small number (9) of genera; it is believed to be the basal chrysopid lineage. Defining characteristics may occur in the larval stages, but larvae from very few genera are known. Apochrysinae and Chrysopinae are probably not monophyletic. Apochrysinae has about 13 genera that are based largely on somewhat variable characters in the wing venation; this subfamily contains the largest and visually most spectacular green lacewings. The larvae of one apochrysine species have been described, but distinguishing subfamilial traits were not apparent. The large subfamily Chrysopinae encompasses over 97% of the known chrysopid species; it includes about 60 genera distributed among four poorly defined tribes. The tribe Chrysopini contains almost all of the lacewings of economic importance.

As a group, the Chrysopidae is cosmopolitan; similarly, all of the subfamilies are widely distributed. Nevertheless, many of the genera are restricted to small regions of the earth. For example, most genera of Nothochrysinae are endemic to very small geographic ranges; many species are known solely from a very few specimens. Among the Apochrysinae, one genus occurs only in South Australia, two only in Central and South America, and one only in the Oriental Region. Two other genera are widespread. The chrysopine genera range from cosmopolitan to narrowly endemic.

There are two basic larval forms: debris-carrying and naked. The debris carriers construct and carry large packets of exogenous material (e.g., plant parts, exuviae, waxy secretions, or remains of prey) on their dorsal surfaces. Usually, they have humped (gibbous) abdomens with rows of hooked setae and long thoracic tubercles bearing numerous, long setae. In contrast, naked larvae have more or less flat abdomens and short, straight setae. The thoracic tubercles are reduced in size and also bear short setae. In the most extreme cases, the lateral tubercles are absent. Given the wide range of morphological and behavioral variation among chrysopid larvae, it is clear that inclusion of all life stages is crucial for advancing the systematics of the family. However, except for the European and Japanese faunas for which larvae of approximately 80% of the species are described, the world's chrysopid larvae are poorly known.

Typically, chrysopid eggs are laid at the end of long stalks, singly, in groups, or in clusters, with the stalks loosely or tightly intertwined. The egg stalks can be naked or they may bear oily droplets; the droplets may contain nutrients or defensive substances that protect the egg or the newly hatched larva from natural enemies.

Chrysopid larvae feed on a variety of soft-bodied arthropods; they may be generalist predators or they may have very strong association with a particular type of prey. For example, prey specialization in *Chrysopa* is based on a number of intrinsic and extrinsic factors, including maternal oviposition behavior, larval size, morphology and behavior, prey influence on life-history traits, responses to natural enemies that are associated with specific prey, and phenology.

Adults of some chrysopid genera are predaceous. Those of other genera take honeydew and pollen; in this group the dorsal crop diverticulum has numerous tracheae and is filled with symbiotic yeast that aid in digestion. Chrysopid adults are not strong flyers; nevertheless, they are known to move considerable distances with the wind. Adults of some species emit foul-smelling defensive odors when they are disturbed.

Overwintering may occur in the larval, prepupal, or adult stages; the overwintering stage is a generic characteristic. Usually a photoperiodically induced diapause is involved.

Chrysopine lacewings have two types of hearing. The "ear" (tympanal organ) is located at the base of the radial vein in each forewing. It is the smallest tympanal organ known, and functions in the perception of ultrasound signals produced by bats that prey on small flying insects. When a lacewing perceives ultrasound signals that are emitted at low rates (1–50 pulses per second), flight ceases; this response causes the lacewing to begin falling. As the bat continues to approach, its signal increases in frequency; the high-frequency signal causes the lacewing suddenly to flip its wings open and fly, thus aiding its escape. The second type of hearing, the perception of low-frequency, substrate-borne sounds that are emitted during courtship, is accomplished through scolopidial organs in the legs. Such sounds are an integral part of courtship in a number of *Chrysoperla* species; both species-specific and geographic variations in the production of these sounds appear to be considerable.

The endemic complex of green lacewings on the Hawaiian Islands belonging to the genus *Anomalochrysa* has evolved several unique characteristics and exhibits an extraordinary range of variation in morphology and behavior. For example, unlike any other known chrysopids, *Anomalochrysa* females lay sessile (unstalked) eggs, either singly or in batches. Larvae range in body form from fusiform with greatly reduced lateral tubercles and few, short setae, to flattened with well-developed lateral tubercles and numerous, robust, long setae. In continental lineages, such broad variation is found only among genera. In some species, adults or larvae are very bright and colorful; in others they are cryptic and polymorphic. Males and females may produce conspicuously loud clicking sounds during courtship and mating; how these sounds are perceived is unknown.

A number of species in the genus *Chrysoperla* are mass reared and released for use in the biological control of agricultural and horticultural pests. Among those in North America are *Chrysoperla carnea* and *Chrysoperla rufilabris*. These species possess a number of characteristics that are excellent for mass rearing, including the ability of adults to use artificial diets for reproduction and to be stored for long periods without significant loss of reproductive potential and the ability of larvae to feed on artificial or factitious prey. Larvae of *Ceraeochrysa* species, which are trash carriers, share many of these traits and are also excellent candidates for mass produc-

tion and release. They have the added advantage of being camouflaged and thus protected from their own natural enemies, e.g., ants. The role and efficacy of lacewings in pest management are beginning to be evaluated quantitatively under field conditions, but additional studies in a variety of crops are needed.

Osmylidae

Although the evidence for a relationship is weak, this family of 160 species may be the sister taxon of Chrysopidae. Eight poorly defined subfamilies of osmylids are recognized, but their systematics needs considerable reassessment.

Osmylids are slender, moderate-sized lacewings (forewing length 15–30 mm), with broad pigmented wings. The family is distributed over much of the Old World. Five of the subfamilies occur in Australia and two in South America, but apparently osmylids are absent from North America.

Little is known of osmylid biology. Elongate, knobbed, unstalked eggs are laid with their sides attached to foliage. Larvae live under stones or at the water–land interface near streams or under the loose bark of trees. Osmylid larvae have long slender stylets like those found in sisyrids (and berothids) (Fig. 6D), but unlike sisyrids, they lack gills and breathe through thoracic and abdominal spiracles.

Hemerobiidae (Brown Lacewings)

The third largest neuropteran family, with approximately 550 species, Hemerobiidae constitutes a cosmopolitan clade that is relatively well known and easily recognized. Adults are generally small (forewing length 3–18 mm), brown, and inconspicuous. The approximately 27 extant genera of hemerobiids fall into 10 reasonably well-defined subfamilies. The Carobiinae and Psychobiellinae each consist of one genus that is restricted to the Australian Region. The Hemerobiinae, Sympherobiinae, Notiobiellinae, and Microminae each include 3 to 5 genera; all four subfamilies are cosmopolitan, although some of the small genera that they encompass have very restricted distributions. The Drepanacrinae and Drepanepteryginae each contain three genera with restricted distributions, and the Megalominae comprises one genus with broad distribution. The most recently described subfamily, Adelphohemerobiinae, consists of one genus that is known only from South America.

Despite the fact that hemerobiid larvae offer a rich suite of traits for phylogenetic analysis, the larvae of only nine genera (from 7 of the 10 subfamilies) have been described. There are three instars. In the first instar, body setation is sparse, and trumpet-shaped empodia are present between the tarsal claws. Second and third instars are similar to each other except in size; they may have numerous short setae, and their empodia are short (Fig. 6E).

Mainly because the systematics of the family was neglected until recently, the life cycles of relatively few hemerobiid genera are known, and the groups that have been studied occur largely in the Northern Hemisphere. In general, hemerobiid eggs are sessile (unstalked) and laid singly or in clusters. Hatching is accomplished by means of an oviruptor.

Larvae prey upon a variety of small, soft-bodied arthropods and eggs. Little is known about the range of larval diet or its specificity, but some species show strong association with a particular type of plant or habitat. Pupation occurs within thinly spun cocoons. Pupae have a peculiar set of hooks on the dorsum of the abdomen; their function is unknown. Most species seem to be predaceous in the adult stage, but there are records of extensive honeydew feeding by adults. Life cycles range from univoltine to multivoltine, but for most taxa the overwintering stage is unknown.

Flightlessness has evolved several times in the Hemerobiidae; it is largely confined to species that occur on islands or are restricted to isolated mountains. In flightless forms, the hind wings are greatly reduced or absent, or the forewings are hardened or fused. Modifications associated with flightlessness are probably most extremely exhibited in the endemic Hawaiian *Micromus*. Spectacular sculpturing of the wings also occurs in winged (and flighted) endemic Hawaiian *Micromus* species.

Many species of Hemerobiidae are believed to be important natural enemies of insect pests on agricultural and horticultural crops or in forests. Hemerobiids often are active at relatively low temperatures; thus they can be useful as biological control agents in temperate regions early in the season when other natural enemies remain inactive.

Coniopterygidae (Dustywings)

Because of their smallness and cryptic nature, coniopterygids are generally overlooked and thus considered rare. However, with approximately 450 species, the Coniopterygidae constitutes a relatively large family and is perhaps the best known systematically. Although they clearly belong within the Neuroptera, coniopterygids differ in a number of ways from other neuropteran families. Previously, they were considered the sole family of a separate primitive suborder (superfamily), the Coniopterygoidea. However, a recent cladistic analysis provides some evidence that the Coniopterygidae and Sisyridae may form a derived sister group within the Hemerobiiformia.

The Coniopterygidae is generally a very homogeneous family characterized by very small adults (forewing length 2–5 mm) with bodies covered by white waxy ("dusty") secretions. The secretions originate from hypodermal wax glands on the sternites and tergites of the abdomen and are spread over the body by the hindlegs. Other than in the coniopterygids, such glands are found only in the homopterans Aleurodina and Coccina. This similarity represents a remarkable example of convergent evolution especially because coniopterygids frequently are associated with these waxy homopterans.

Currently, the Coniopterygidae contains three well-defined, probably monophyletic subfamilies: Coniopteryginae, Aleuropteryginae, and Brucheiserinae. Both the Coniopteryginae and the Aleuropteryginae are large groups with cosmopolitan distributions. They are distinguished from the Brucheiserinae by their predominantly longitudinal wing venation, whereas the Brucheiserinae have highly unusual reticulate wing venation. Brucheiserinae is known only from the neotropics and its larvae are not described. Some authors have considered it to comprise a separate family (Brucheiseridae), but this distinction is probably not justified. In this regard, discovery of the larvae is likely to prove very valuable.

The life histories of very few coniopterygid species are known. Both larvae and adults occur on trees and bushes (sometimes on low vegetation). Many species appear to be associated with specific types of vegetation, and this habit may indicate a degree of prey specialization. Eggs are unstalked and laid near prey; there are three or four instars (Fig. 6F).

Adults and larvae are predaceous on small, soft-bodied arthropods (aphids, scales, mites); adults probably also feed on honeydew and perhaps pollen. Flat cocoons with double walls are spun on foliage or tree trunks. Adults are usually active at dusk or at night. Life cycles and overwintering stages vary (prepupae within cocoons, free-living second instars) and have not been well studied.

Many species of coniopterygids are considered important natural biological control agents, but unfortunately, their role has not been evaluated. Other species have great potential for use in classical biological control, as well as in commercial mass production and release. To date, their potential remains undeveloped.

Sisyridae (Spongillaflies)

The Sisyridae constitutes a small but cosmopolitan family that contains about 50 species in four genera: *Climacia,* which is restricted to the New World; *Sisyra,* which is cosmopolitan; and *Sisyrina* and *Sisyrella,* which are small Australian and Asian genera. This is the only neuropteran family with truly aquatic larvae. Sisyrid larvae are believed to feed exclusively on freshwater sponges, and they are unique among the Neuroptera in having segmented abdominal gills that function in breathing (Fig. 6H). Adults are dull colored and relatively small (forewing length 4–10 mm). They closely resemble brown lacewings, but the simple, open venation of the forewing and the branching pattern of the radial sector distinguish them.

The sessile eggs of sisyrids are laid singly or in groups on objects that overhang water. A flat layer of silk covers the eggs. Hatching is aided by an oviruptor, and subsequently the neonate larvae walk or drop to the water where they seek a sponge colony.

Sisyrid larvae probe sponges by means of the long, flexible mouthparts (Fig. 6G). Only one of the Malphigian tubules is attached distally to the rectum, a condition that is probably related to the aquatic lifestyle. After feeding and development, the mature larvae swim to the shore, attach to objects close to the water, and spin a double-layered cocoon within which they pupate. Adults forage on nectar, pollen, algae, fungi, aphids, and mites.

Dilaridae (Pleasing Lacewings)

This small family of approximately 50 species has well-defined affinities with the Berothidae and Mantispidae. It is divided into two subfamilies: Dilarinae, which is confined to the Old World, and Nallachiinae, which is restricted to the New World (with one species known from South Africa). It is one of the few neuropteran families that does not occur in the Australian region.

Adults resemble small, delicate hemerobiids (forewing length 3–16 mm in males and 5–22 mm in females). But, they are differentiated by ocelli-like tubercles on the head of both sexes (functional ocelli are absent), a long ovipositor in females, and pectinate antennae in males.

Dilarid eggs are elongate and unstalked. Those of *Nallachus* are laid in association with dead trees. The larvae of *Nallachus* inhabit the galleries of insects in decaying logs or the area beneath the tightly adhering bark of erect, recently dead trees (Fig. 6I). Larvae of *Dilar* have been found in the soil.

The larval diet is not known, but it does not appear to be highly specific. Development probably takes 1 year. Larvae may undergo supranumerary molts; i.e., if undernourished, they may continue to molt as many as 12 times. However, those that did so under laboratory conditions did not metamorphose successfully. Pupation occurs within cocoons.

Mantispidae (Mantidflies)

With approximately 400 species, Mantispidae is the largest family in the dilarid lineage (Dilaridae, Berothidae, Mantispidae). Adults are recognized by raptorial forelegs that resemble those of a mantid (Fig. 8); their simple, subequal wings are narrow and elongate and they have a distinct pterostigma and chrysopid-like venation. They are moderate-sized to large lacewings (forewing length 5–30 mm). Larvae are similar to those of the Berothidae in that they are hypermetamorphic; however, in the case of the mantispids the first instars are campodeiform and the second and third instars are grub-like (Fig. 6J). The mantispids differ from the Rhachiberothinae in their wing venation, terminalia, and larval characteristics, but they are of similar size and also have raptorial forelegs.

FIGURE 8 Mantispid adult in the genus *Plega*. Note the mantid-like raptorial forelegs and narrow forewings. (Photograph by E. S. Ross, California Academy of Sciences.)

The family contains four apparently monophyletic subfamilies: Symphrasinae, Drepanicinae, Calomantispinae, and Mantispinae. Symphrasinae encompasses a large, diverse assemblage of species that are distributed throughout South America and southern North America. Drepanicinae is a smaller subfamily that occurs in restricted areas within South America and mainland Australia. Calomantispinae, another small subfamily, has a disjunct distribution: eastern Australia (including Tasmania) and southern North America. Finally, the large subfamily Mantispinae ranges between 50°N and 45°S throughout much of the world. The biology and immatures of Drepanicinae and Calomantispinae remain largely unknown, whereas those of several genera in Symphrasinae and Mantispinae have been studied.

Larvae in the subfamily Mantispinae usually inhabit the egg sacs of spiders where they feed upon the contents, although some may be subterranean predators or possibly generalist predators. Numerous (200–2000) stalked eggs are laid randomly (and sometimes communally) on leaves and wooden structures.

The newly hatched campodeiform larvae find their hosts (spider eggs) via one of two methods. Either they actively seek a previously constructed spider egg sac that they enter through direct penetration or they climb onto a female spider and enter the egg sac as the female builds it. Those that board spiders can feed on the hemolymph of their host, but they do not molt until they enter an egg sac. If the campodeiform larva attaches to a male spider, it may transfer to a female during copulation. After an egg sac is found, the larva feeds on the contents (predation) and undergoes hypermetamorphosis.

The mature larva spins a cocoon, and pupation occurs within the larval skin, within the cocoon. Adults are predaceous and they are active during the day or night. Overwintering in some species occurs in the first instar, and there may be one to several generations per year.

Several species within the Symphrasinae have been reared from nests of aculeate Hymenoptera or reared in the laboratory on larvae or pupae of Lepidoptera, Coleoptera, and Diptera. First instars may find their hosts by attaching to an adult bee or wasp and moving into the cell when the egg is laid. Subsequently, they feed on a single host (parasitism), have the typical mantispid hypermetamorphosis, and adhere to their host with a sticky, yellow secretion.

Adult *Climaciella* can be highly polymorphic, with each of several morphs mimicking a different species of polistine wasp. The proportion of the different morphs may vary depending on the number and aggressiveness of the various wasp species at each locality.

The larval habits of the Drepanicinae are unknown; Calomantispine larvae may be generalist predators.

Berothidae (Beaded Lacewings)

The Berothidae is a small family of approximately 115 species in four more or less distinct subfamilies: Rhachiberothinae, Berothinae, Nosybinae, and Cyrenoberothinae. Adults typically are small to medium-sized (forewing length 6–15 mm) lacewings with brown wings and bodies. Frequently, the outer margin of the forewing is deeply incised. Larvae are associated with termite nests (Figs. 6K and 6L). The family is cosmopolitan and occurs predominantly in warm temperate, tropical, and subtropical areas; the greatest number of species is found in Africa.

One subfamily of berothids, the Rhachiberothinae, differs considerably from the others. Its adults have raptorial forelegs that are very similar to those of mantispids. As a result, some authors consider them to be a separate family. However, because of the close similarity between the few known larvae of Rhachiberothinae and the few known larvae of other Berothidae, we treat the Rhachiberothinae as a specialized subfamily of the Berothidae, while awaiting further study.

Little is known of the biology of the Berothidae except for the North American genus *Lomamyia* (Berothinae). Eggs of three subfamilies are laid in clusters attached to one or several long, silken stalks; those of the fourth subfamily, Rhachiberothinae, are sessile. *Lomamyia* eggs are laid on dead trees or logs that contain colonies of subterranean termites. First instars are mobile and after entering the termite colony they feed and molt. The second instar does not feed, but hangs immobile from the roof of the termite tunnel. Third instars resume feeding on termites. They subdue their prey with an allomone that is emitted from the tip of the abdomen and/or a neurotoxin that is injected through the mouthparts. Pupation occurs within silken cocoons that are spun in the termite nest. Adults are primarily nocturnal. As many as three generations can occur per year; the prepupal stage overwinters.

See Also the Following Articles

Hypermetamorphosis • Megaloptera • Raphidioptera

Further Reading

Aspöck, U., Plant, J. D., and Nemeschkal, H. L. (2001). Cladistic analysis of Neuroptera and their systematic position within Neuropterida (Insecta: Holometabola: Neuropterida: Neuroptera). *Syst. Entomol.* **26**, 73–86.

Brooks, S. J., and Barnard, P. C. (1990). The green lacewings of the world: A generic review (Neuroptera: Chrysopidae). *Bull. Br. Mus. Nat. Hist.* **59**, 117–286.

Brushwein, J. R. (1987). Bionomics of *Lomamyia hamata* (Neuroptera: Berothidae). *Ann. Entomol. Soc. Am.* **80**, 671–679.

Canard, M., Séméria, Y., and New, T. R. (1984). "Biology of Chrysopidae." Junk, The Hague.

Hagen, K. S., Mills, N. J., Gordh, G., and McMurty, J. A. (1999). Terrestrial arthropod predators of insect and mite pests. *In* "Handbook of Biological Control" (T. S. Bellows and T. W. Fisher, eds.), pp. 383–503. Academic Press, San Diego.

Mansell, M. W. (1996). Unique morphological and biological attributes: The keys to success in Nemopteridae (Insecta: Neuroptera). *In* "Pure and Applied Research in Neuropterology." Proceedings of the Fifth International Symposium on Neuropterology, Cairo (M. Canard, H. Aspöck, and M. W. Mansell, eds.), pp. 171–180. M. Canard, Toulouse.

McEwen, P., New, T. R., and Whittington, A. E. (eds.). (2001). "Lacewings in the Crop Environment." Cambridge University Press, Cambridge, UK.

Meinander, M. (1990). The Coniopterygidae (Neuroptera, Planipennia). A check-list of the species of the world, descriptions of new species and other new data. *Acta Zool. Fennica* **189**, 1–95.

Monserrat, V. J. (1996). Larval stages of European Nemopterinae, with systematic considerations on the family Nemopteridae (Insecta, Neuroptera). *Dtsch. Entomol. Z.* **43**, 99–121.

New, T. R. (1991). Neuroptera. *In* "The Insects of Australia," Vol. I, pp. 525–542. Melbourne University Press, Carlton, Victoria.

Oswald, J. D. (1993). Revision and cladistic analysis of the world genera of the family Hemerobiidae (Insecta: Neuroptera). *J. N. Y. Entomol. Soc.* **101**, 143–299.

Stange, L. A., and Miller, R. B. (1990). Classification of the Myrmeleontidae based on larvae (Insecta: Neuroptera). *In* "Advances in Neuropterology." Proceedings of the Third International Symposium on Neuropterology, Berg en Dal, Kruger National Park, Republic of South Africa (M. W. Mansell and H. Aspöck, eds.), pp. 151–169. Department of Agricultural Development, Directorate of Agricultural Information, Pretoria.

Tauber, C. A. (1974). Systematics of North American chrysopid larvae: *Chrysopa carnea* group (Neuroptera). *Can. Entomol.* **106**, 1133–1153.

Tauber, C. A. (1991). Neuroptera. *In* "Immature Insects" (F. W. Stehr, ed.), Vol. 2, pp. 126–143. Kendall–Hunt, Dubuque, IA.

Tauber, C. A., and Krakauer, A. H. (1997). Larval characteristics and generic placement of endemic Hawaiian hemerobiids (Neuroptera). *Pacific Sci.* **51**, 413–423.

Tauber, M. J., Tauber, C. A., Daane, K. M., and Hagen, K. S. (2000). Commercialization of predators: Recent lessons from green lacewings (Neuroptera: Chrysopidae: *Chrysoperla*). *Am. Entomol.* **46**, 26–38.

Nomenclature and Classification, Principles of

F. Christian Thompson
Systematic Entomology Laboratory, U.S. Department of Agriculture

Classification has two meanings in English: the process by which things are grouped into classes by shared characters and the arrangement of those classes. Identification is the process of observing characters and thereby classifying things. Biological classifications are arrangements of organisms. The ability to classify is common to all animals, for to survive animals must group other organisms into at least three classes: Those to be eaten, those to be avoided, and those to associate with, especially members of their own class. For scientists, classification is formalized into a nested or hierarchical set of hypotheses: hypotheses of characters, groups (taxa), and relationships among the groups. Individual specimens of organisms are observed and characteristics noted. So, for example, we may observe that some people are black, others yellow or white, and conclude as Linnaeus did that there are different groups of humans *(Homo sapiens)*. This is a hypothesis that skin color is a useful character. Further testing of this character hypothesis has shown that skin color among humans does not delimit natural groups; hence we reject skin color as a character for humans as well as those groups this character defines. Color, however, is a very useful character for classifying many other groups. Then there is the hypothesis of a group. Groups of biological organisms are called taxa (taxon, singular) and these taxa are hierarchically arranged in our classifications. Taxa in a classification have rank, with the basic (basal, bottom) rank being designated as species. Some of the higher ranks are genus, tribe, family, order, class, phylum, and kingdom.

Nomenclature is a system of names along with the procedures for creating and maintaining that system. Classification, in its second definition, is the structure for nomenclature, being the model on which names are arranged. Names form the essential language of biology and are the means that we use to communicate about our science. To avoid a Tower of Babel, a common system of nomenclature is required, especially an effective, efficient, system that has a minimal cost.

NAMES AND CLASSIFICATION

Names are tags. Tags are words, short sequences of symbols (letters) used in place of something complex, which would require many more words to describe. Hence, tags save time and space. Instead of a long description, we use a short tag. A scientific name differs from a common name in that the scientific name is a unique tag. In nonscientific languages, such as English, there may be multiple tags (common names) for the same organism. For example, imagine the various words in English that are used to describe *H. sapiens*. In computer (database) jargon, data elements that are used to index information are termed keys, and keys that are unique are called primary keys. Scientific names are primary keys. The word "key" has another meaning in English, which is "something that unlocks something." Scientific names are those critical keys that unlock biosystematic information, which is all that we know about living organisms. To repeat: scientific names are tags that replace descriptions of objects or, more precisely, concepts based on objects (specimens). Scientific names are unique within a classification, there being only one valid scientific name for a particular concept, and each concept has only one valid scientific name.

Scientific names are more than just primary keys to information. They represent hypotheses. To most systematists, this is a trivial characteristic that is usually forgotten and thereby becomes a source of confusion later. To most users, this is an unknown characteristic that prevents them from obtaining the full value from scientific names. If a scientific name were only a unique key used for storing and retrieving information, it would be just like a social security number. *H. sapiens* is another unique key used to store and retrieve information about humans, but that key also places that information into a hierarchical classification. Hierarchical classifications allow for the storage at each node of the hierarchy of the information common to the subordinate nodes. Hence, redundant data, which would be spread throughout a nonhierarchical system, are eliminated. Biological classifications, however, do more than just hierarchically store information. If one accepts a single common (unique) history for life (phylogeny) and agrees that our biological classifications reflect this common

history in their hierarchical arrangement, then biological classifications allow for prediction, namely that some information stored at a lower hierarchical node may belong to a higher node; that is, may be common to all members of the more inclusive group. Such predictions take the following form: if some members of a group share a characteristic that is unknown for other members of the same group, then that characteristic is likely to be common to all members of the group. So scientific names are tags, unique keys, hierarchical nodes, and phylogenetic hypotheses. Thus systematists pack a lot of information into their names and users can get a lot from them.

Scientific names are hypotheses, not proven facts. Systematists may and frequently do disagree about hypotheses. Hypotheses, which in systematics range from what is a character to what is the classification that best reflects the history of life, are always prone to falsification, hence to change. Disagreements about classification can arise from differences in paradigm and/or information. Systematists use different approaches to construct classifications, such as cladistic versus phylogenetic versus phenetic methods. Given the same set of data that underlies a given hierarchy, cladists will derive classifications different from those derived by pheneticists (Fig. 1). Even among cladists, there can be differences about the rank (genus, family, order) and thereby the hierarchical groups used. These are disagreements based on paradigm. There can be disagreement about the hypotheses that underlie the information used to construct the classifications, such as what are the characters. And disagreement can arise among systematists because individuals use different information. While disagreements will affect the ability to predict, they need not affect the ability to retrieve information.

The desirable attribute that must be preserved to ensure complete access to information across multiple classifications is uniqueness. Our scientific nomenclature must guarantee that any scientific name that is used in any classification is unique among all classifications. This can be assured by having two primary keys. Unfortunately, having two keys increases the overhead of our information systems. So most systematists and *all* users want to avoid this problem by mandating that there be only *one* classification. Although in theory there is only one correct classification, as there was only one history of life, in reality there have been multiple classifications in the past, there may be multiple classifications in use today, and there will be multiple classifications in the future. That is the price of scientific progress, of the increase in our knowledge of the world. If information is to be retrieved across time—that is, if we want to extract information stored under obsolete classifications, and if we want to avoid dictating "the correct" classification—then we need a nomenclatural system that supports two unique keys.

The two keys for our language of biodiversity are the valid name and the original name. The valid name is the correct name for a concept (taxon) within a classification; the original name is the valid name in the classification in which it was

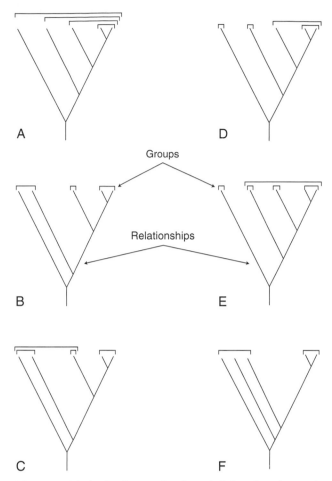

FIGURE 1 Multiple classifications for identical cladistic hypotheses. The brackets along the top of the trees indicate the groups recognized. Cladistic classifications are shown for trees A, D, and E and phenetic classifications for B, C, and D.

proposed. Valid names may be different among classifications, but the original name is invariant across all classifications (Table I). Valid names are the best names to use because they provide the full value of scientific names. These are the names that provide a basis for prediction. The original name is useful only for information retrieval across multiple classifications. Although valid and original names may be and frequently are the same, users must know the differences between them. Specifically, they need to know that a valid name is a powerful

TABLE I   Multiple Classifications (rows) and Primary Keys (columns) to Information

Year	Valid name	Original name	Authority
1776	*Musca balteata*	*Musca balteata*	De Geer
1822	*Syrphus balteatus*	*Musca balteata*	Meigen
1843	*Scaeva balteata*	*Musca balteata*	Zetterstedt
1917	*Episyrphus balteatus*	*Musca balteata*	Matsumura
1930	*Epistrophe balteata*	*Musca balteata*	Sack
1950	*Stenosyrphus balteatus*	*Musca balteata*	Fluke
Today	*Episyrphus balteatus*	*Musca balteata*	Vockeroth

inference tool, that a valid name provides for prediction about unknown attributes of the organism that bears the name. But they must understand that there may be multiple valid names in the literature and/or in use and that valid names represent hypotheses that may change as our knowledge is tested and improved. So most importantly, if there are multiple valid names in use, then there are conflicting scientific hypotheses being advocated, and users must select the name that best serves their purpose. If users do not want to decide, do not want to use classifications to organize and synthesize their information, then they may use the original name to index their information, being assured that it will always be a unique key.

There are other problems today with our classifications: synonymy, having two names for the same concept, and homonymy, having the same name for different concepts. These problems are, however, largely the result of ignorance. If we knew all names and their types and could agree on what are species, then by applying the rules of nomenclature we could immediately eliminate all synonymy and homonymy problems. Homonymy is eliminated by the rule of uniqueness. Synonymy is addressed by the rules of typification, which tie a physical instance of a concept to a name, and is resolved by logic of circumscription and the convention of priority (or usage). The name of a concept is the name affixed to one and only one of the types that falls within its circumscription (Fig. 1). The name used is determined by which name is the oldest (priority) or most widely used (usage). The specific rules for resolving homonymy and synonymy, as well as for the proper formation and documentation of names are our codes of nomenclature. These rules, however, do not address the problem of multiple classifications, nor can they establish order under conditions of ignorance of the universe of applicable names and their typification.

There is one final problem: the species problem. This is the problem of what is the basic unit of information and/or data. There is also the question of what species are and whether species are real or hypotheses. Species may be a category (rank) in classifications or a unit of information. The best current review on these questions is by Wheeler and Meier, but for nomenclature the species (or more precisely the species group, which includes the subspecies category) is considered to be a basic unit of information. The problem is that the species is not a data element. The species is not an indivisible unit, but consists of information, that is, data derived from specimens that have been identified as belonging to that species. Mistakes can be made during this identification, which is after all another hypothesis. Information is ultimately not derived from species, but from specimens. Biological information management really begins with specimen data management. The problems of specimen-based data management are not intractable but are readily addressed by the use of unique identifiers, such as bar codes, another form of unique keys.

The species problem is also one of circumscription, the definition of the limits of a taxon. A group with the same name and type may be more or less inclusive depending on the characters used to define its limits. Zoologists differ from botanists in not considering circumscription to be a problem, since minimally all identically named taxa have at least some characteristics in common. The problem of how much is held in common, therefore, is best resolved by enumeration of the included taxa or specimens. The history of circumscription can be tracked by use of an additional key that uniquely identifies the person who defined the limits and the date of that action. Sufficient for our purposes is to know that specimen-based data will always be summarized into species-based information units and that all species-based information should be specimen based.

PARADIGMS AND CLASSIFICATIONS

The information that is embedded in nomenclature comes from the classification used. As noted, classifications consist of hierarchically nested groups of taxa, with the basic unit being the taxa ranked as species. Paradigms are theories about scientific knowledge and its organization. The first classifications developed by Linnaeus and Fabricius were largely based on Aristotelian essentialism/typology. Things were grouped together because they shared the essences of the group, which is the type. Later, when evolution was articulated as a paradigm, classifications were based on phylogeny, which is the genealogical hypothesis of relationship. More recently, when computers began to appear, classifications were proposed on the basis of statistical measures of overall or phenetic similarity. Finally, different ways of deciphering phylogeny were developed, and so, different ways of translating phylogenetic information into a hierarchical classification were proposed (phylogenetic vs cladistic methods). Over the past half-century, much has been written about the relative merits of phenetics, evolutionary systematics, and cladistics, but the inescapable conclusion for predictive and, therefore, maximally informative classifications, is that the cladistic paradigm is mandatory. Schuh provides a good summary of the arguments for cladistic classifications.

Regardless of the paradigm followed, all approaches leave unsolved the problem of how to translate the result of taxonomic analysis, be it a tree or a branching diagram of overall similarities, into a hierarchical classification. There are only two approaches to the translation of an analysis into a classification: subordination or sequencing. For subordination, each clade/branch becomes a recognized (named) taxon and a rank indicator provides a key to the relative level of subordination. Subordination works best when the phylogeny/branching diagram is balanced, that is, when each branching point divides the remaining terminal taxa into equal sized groups. For example (Fig. 2, Table II), 8 species could be clustered into 4 genera and 2 subfamilies, whereas a fully pectinate analysis would yield 7 genera clustered into 5 named ranks (subfamily, infrafamily, supertribe, tribe, subtribe). For sequencing, only the terminal clades/branches are recognized, but their order is preserved and suitably indicated to encode their sequential level of subordination. This method is highly efficient for analyses that result in

pectinate trees. The pectinate example could be reduced to 7 sequenced genera. Sequencing does not work when the analysis is balanced. Given that most analyses are neither fully balanced nor fully pectinate, a mixture of subordination and sequencing should be used as long as the classification properly identifies which methodology was used for each portion. Wiley provides a full set of conventions to deal with these issues as well as others that involve the placement of fossil groups *(plesion)* or groups of uncertain or changeable position *(sedis mutabilis)* or unknown relationships *(incertae sedis)*.

Beyond the translation of a taxonomical analysis into a hierarchical classification, another challenge remains, that is, what groups to formally name and what ranks to assign to

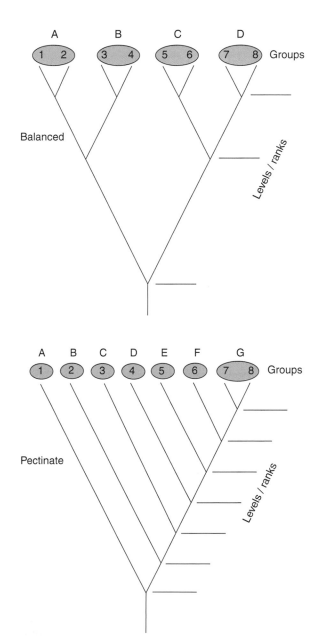

FIGURE 2 (Top) Balanced and (bottom) pectinate cladistic hypotheses. See Table II for the different classifications that result from these hypotheses.

TABLE II Classification: Sequencing and Subordination

Subordinated classification for Fig. 2A balanced analysis	Subordinated classification for Fig. 2B pectinate analysis	Sequenced classification for Fig. 2B pectinate analysis
Family A-idae	Family A-idae	Family A-idae
Subfamily A-inae	Subfamily A-inae	Genus A
Genus A	Genus A	Genus B
Genus B	Subfamily B-inae	Genus C
Subfamily C-inae	Infrafamily B-ites	Genus D
Genus C	Genus B	Genus E
Genus D	Infrafamily C-ites	Genus F
	Supertribe C-idi	Genus G
	Genus C	
	Supertribe D-idi	
	Tribe D-ini	
	Genus D	
	Tribe E-ini	
	Subtribe E-ina	
	Genus E	
	Subtribe F-ina	
	Genus F	
	Genus G	

those named groups. Obviously, when a group is fully resolved taxonomically, there could be as many named groups as there are terminal taxa. No school of taxonomy insists on naming all of them, but other than that there is no method nor any consensus among taxonomists on what taxa to name. This general problem is usually referred to by the names of the extreme views on either side, the "splitters" and the "lumpers," or those who would recognize many groups versus those who would recognize only a few. The merit of splitting is that the more taxa named, the more hierarchical information is embedded into those names and classification itself. Unfortunately that also leads to a loss of utility inasmuch as less information is summarized in each taxon.

Consider birds, the best-known group of organisms. Some 9700 species are clustered into 204 families and 2004 genera. Their scientific nomenclature is largely meaningless to many users, such as bird-watchers. For bird-watchers, common names, which more closely follow the original Linnaean classification, such as ducks *(Anas)* or hawks *(Falco)* or hummingbirds *(Trochilus),* are more meaningful groups than the oversplit genera. On the other hand, mosquitoes, some 3500 species, are clustered into only 34 genera. The important disease vectors, such as *Anopheles* for malaria and *Aedes* for yellow fever and dengue, remain large groups where the scientific name and common name are the same and are useful to doctors, public health workers, and other entomologists. The problem of the appropriate rank for groups recognized is similar. Naturally, splitters must have a greater series of rank indicators to express their fully named hierarchies. So, although there are relatively few species of birds, they are clustered into a large number of families (204), whereas flies (order Diptera) comprise 16 times as many species clustered into fewer families (142)!

The ranking issue also brings with it the question of equivalency. Obviously a family of birds is not an equivalent

unit of biodiversity or of anything else in comparison to a family of flies. Rank equivalence is an important issue because many biologists want to make comparisons across different groups of organisms. Biological comparison should never be made on the basis of taxonomical categories above the rank of species. For example, studies that base conclusions on the circumstance that one treatment or niche has more families, than another are totally meaningless because the units being compared are not equivalent. Biological comparison should be made only on the basis of cladistically defined sister-group relationships, since sister groups are of equal age.

The entomologist Willi Hennig proposed in 1966 an objective method for assigning ranks that also allowed for biological comparisons: rank should reflect the hypothesized age of origin of the taxon. His suggestion has been rejected by all on the ground that the approach would cause a major upheaval in the traditional ranks of groups. For example, humans, placed in a separate kingdom by some (Psychozoa by Huxley in 1957), would be clustered among the apes and lemurs as nothing more than a species group. For entomology, some of the larger groups, like Coleoptera, Diptera, and Lepidoptera, which go back to Aristotle, would change in rank if not content. So after more than 2000 years of using those concepts, no one wants to split up the groups or change their rank.

These issues of classifications are largely ignored by working taxonomists, most of whom focus on their specialty and do not concern themselves with global classifications. Entomologists generally do not care how birds are classified, nor do beetle workers even worry about how flies are classified. Entomologists also tend to take a pragmatic, utilitarian approach, such that conservative ranks and grouping are used among mosquitoes and other economically important insects. In summary, a few general guidelines should be followed:

1. Only monophyletic taxa should be recognized and named.

2. Subordination or sequencing should be used as is most appropriate given the analysis and always should be annotated.

3. "Empty" taxa should never be named (i.e., if a family contains only a single genus, there is no a need to name a subfamily or tribe simply because these ranks are used elsewhere in the classification).

4. The fewer taxa named, the more useful the classification generally will be to nonspecialists.

5. Traditional groups and ranks should be preserved where possible.

CODES OF NOMENCLATURE

Because names are critical for communication and information retrieval, nomenclature needs to be universal, precise, and accurate. Universality requires that the same methodology be used by all and that methodology ensure stable nomenclature over time. Precision requires that only one result be derived from an individual application of the rules of nomenclature. Accuracy requires that names be consistently and precisely tied to the hypotheses they denote. The International Code of Zoological Nomenclature (ICZN) ensures the implementation of these basic functions in our scientific names and classification. This is achieved through a series of rules organized into chapters and articles.

Stability of nomenclature should not be confused with stability of taxonomic hypotheses (taxa) and classifications. As knowledge improves and more characters are discovered and analyzed, resulting in improved understanding of relationships among organisms, taxa and classifications will change. So, as more is known about the history of life, old Aristotelian groups like reptiles will be replaced by better defined ones, and the name Reptilia will drop from our classifications. But in other well-characterized groups, such as spiders (order Araneae) or flies (order Diptera), which have proven to be natural, the names shall remain unchanged in our classification.

The current ICZN is the product of a long evolution that began with the system of binominal nomenclature introduced by Carolus Linnaeus, a Swedish professor of natural history. This system was the direct result of an earlier government biodiversity project. The Swedish crown had some far-flung possessions and wanted to know what use could be made of them. Linnaeus was sent to investigate, to survey what today is called biodiversity, and to write a report characterizing his findings with recommendations on how to use them. At the time, there was only a binary system of nomenclature: one word for the genus, with the species being described by a series of adjectives. Given the diversity Linnaeus found, he did not want to waste time repeating long strings of adjectives that were required to characterize the biodiversity. So, because the base characterizations were in his flora of Sweden, he used a combination of the genus name and single word (an epithet)

for each species to form a unique key to those descriptions (Stearn gives more details).

The system was an immediate success. Linnaeus codified the system, built and maintained a universal information database for all names (his *Systema Naturae,* 10th edition in 1758), and trained a cadre of students to carry on his work. The students dispersed and converted others. But since there could be only one master, Linnaeus, they divided nature up. There was to be no more *Systema Naturae.* For entomology, the student in charge was Johann Christian Fabricius. Fabricius defined his principles in his *Philosophia entomologica* and produced a series of *Systemae* for insects, the last comprehensive one being published in 1792 to 1794.

For the next 50 or so years, there was a significant increase in the number of animals discovered, described, and named, but little concern for nomenclature, which became muddled. This led a group of British zoologists, in 1843, to propose a formal set of rules, now known as the Strickland code, from which the modern ICZN evolved. After their effort, there was another half-century of new codes being proposed for various groups of animals (birds, fossils, insects) or nationalities (English, French, German). This proliferation led to zoologists joining forces and working toward an international code for all animals, resulting in the establishment of the International Commission on Zoological Nomenclature in 1895 and *Règles Internationales de la Nomenclature Zoologique* in 1905. Although a few entomologists (e.g., Banks and Caudell in 1912) continued to work on a specialized code for insect names, these development efforts were quickly abandoned.

For the next half-century, the *Règles* and the commission operated well, but clearly improvements were needed. So after the World War II, the task of revision began. After a series of international meetings, the American entomologist J. C. Bradley, then president of the commission, wrote out a draft that in 1962 became the second edition. The next edition, in 1985, and the current one, in 1999, were largely the work of Curtis W. Sabrosky, an American entomologist, David Ride, an Australian mammalogist, and Richard Melville, a British paleontologist.

The challenge in writing codes of nomenclature is making a set of rules that demand the best practices of taxonomists today, but also preserve the names created by past workers. Hence, to accommodate the work of the past, a code makes general provisions followed by a series of exceptions qualified by dates. Also, in zoology, there are two options for preserving history. A provision can be made in the code to solve a problem, or an appeal can be made to the commission to set aside the code to preserve an old name. Changing the code affects all occurrences of a problem, but a ruling of the commission applies only to a particular occurrence. In the past the commission frequently took many years to rule on cases. Hence, for Sabrosky and others, changing provisions of the code became the preferred method of addressing problems of old names. Hence, the current ICZN has many clauses that exist only to make old names available and to preserve their customary usage. Unfortunately, in adding to the ICZN provisions of these kinds, Sabrosky and others frequently created unforeseen problems. So the ICZN must be used carefully, since for almost every positive statement there is usually an exception. Other linguistic constructions may also be confusing, such as the frequent use of the phrase "as such." For example, this phrase, used in Article 1.3.2, requires subsequent workers to decipher the intent of the original author. If the original author knowingly was describing an aberrant specimen, then the scientific name does not enter into nomenclature, but if the author thought the specimen was typical of the taxon being described, then the name must be considered to be available for use in nomenclature. Finally, the ICZN uses a number of specialized terms or special definitions for words; these are all covered in its glossary.

INTERNATIONAL CODE OF ZOOLOGICAL NOMENCLATURE

The current 4th edition of the ICZN consists of a preamble, a series of 18 chapters comprising 90 articles, recommendations and examples, and a glossary. The book also contains a preface, an introduction, and three appendices (the first two are general recommendations and the last is the Constitution of the International Commission on Zoological Nomenclature). The text is in both English and French, and there is a single combined index. This book, bounded in green, is the "official" edition published by the commission, but other official versions have been approved and published in different languages.

The preamble sets the objectives and basic principles of the code: "to promote stability and universality" of names and to ensure that the name of each taxon is unique. The preamble also declares that taxonomy is independent of nomenclature. The articles are the definitive rules, with examples of how they are applied in specific cases as well as recommendations of appropriate practices. The glossary defines each term used so that the rules can be interpreted consistently. The following is a summary of the code by major topics (in parentheses after the topic are the articles covered).

Scope (Arts. 1–3)   The scope of zoological nomenclature is restricted to names for animals published starting in 1758, the date of the 10th edition of Linnaeus' *Systema Naturae*. Here, the ICZN uses the verb "deemed" to declare for nomenclature that *Systema Naturae* was published on 1 January and before Clerck's *Aranei Svecici* (spiders of Sweden), neither of which is true, since Clerck's work was actually published in 1757! Exclusions are also listed, such as hypothetical concepts. That simply means that if the Loch Ness monster is not real, then its name *Nessiteras rhombopteryx* Scott, is not a name covered by the ICZN.

Publication (Arts. 7–9), Dates (Arts. 21–22), and Authorship (Arts. 50–51)   Although zoological nomenclature is a language for communication, the names that are regulated are those "published." Since taxonomy is based on some 250 years of work, the definition of publication used by the ICZN is based on printed works. For names and nomenclatural acts to be within the coverage of the ICZN, they must have been first published in a printed work in numerous copies available to the public and for the permanent scientific record. This definition excludes some printed works, such as daily newspapers, which are not published for the permanent and scientific record. The ICZN rules exclude the evolving digital world, such as the Internet. This assures that all users read the same material in determining what are the appropriate names. The ICZN does accept new digital media such as CD-ROM or DVD disks that are "stamped out," not printed. The ICZN provides rules to determine who is the author of these works and their dates of publication.

Names (Arts. 4–6) and Their Formation [Arts. 11 (11.2–11.3, 11.7–11.9), 25–49]   Beyond falling within the scope of zoological nomenclature and having been published, names must be properly formed. They must be written with the Latin alphabet, and they must agree with various other requirements, many of which reflect the origin of the system at a time when all scholarly works were written in Latin. The ICZN groups scientific names into three kinds: family-group names, the names of taxa above the genus and species; genus-group names, the names for groups of species that form the first part of the binomen; and species-group names, the specific (epithet) names. The ICZN prescribes that names of higher taxa, such as superfamilies (-oidea), families (-idae), subfamilies (-inae), tribes (-ini), and subtribes (-ina), have specific suffixes; that generic and subgeneric names have gender and be nouns or be treated as such; and that specific names (epithets) be either nouns (and invariant) or adjectives (and whose ending agrees with the gender of the generic name with which it is combined).

Availability (Arts. 10, 12–20)   Given that all the foregoing conditions are fulfilled, names and nomenclatural acts must meet additional requirements if they are to be made available for consideration under the ICZN and if they are to be held to be valid. The distinction between available and valid is critical. A valid name is the correct name to be used for a taxon, that is, a hypothesis of a group; an available name is any name that meets the requirements of the ICZN.

The additional requirements that names must follow to be available are as follows: (1) they must be formed as part of the system of binominal nomenclature, and (2) they must represent taxa that are considered to be valid. How taxa are defined is further regulated in a series of articles that are applied according to time. Before 1930 the standards were simple, such as attaching a previously unpublished name to an illustration, but current requirements are much more rigorous. For example, one must explicitly declare that a new name is being proposed, fully document the hypothesis (a description that "purports" to differentiate the taxon from all others), and designate the type for the name (which for a species is usually a dead specimen); if extant, the specimen must be deposited in a named, bona fide collection, or plans for such deposition must be furnished. (see Arts. 13.1.1, 16.1, and 16.4).

Validity (Arts. 23–24) A key problem for nomenclature is the existence of two or more names for the same taxon, for only one name can be considered to be correct or valid. This reflects the more general problem in science of who is to be given the credit for new ideas when multiple people claim to have arrived at them first. The general principle involved is that of priority—credit should go to the first person who published the idea. This principle was set forth by Henry Oldenburg, the first secretary of the Royal Society and editor of its *Philosophical Transactions* (1664–1677). Unfortunately, sometimes priority fails because the first to publish may have been forgotten, and someone else has become recognized as the first. In scientific nomenclature, this means that one name may become so familiar to many people that to change it to a name that is less widely known would cause instability. Hence, the ICZN provides a saving clause to allow for a widely used and familiar name to be preserved as the valid name when an older, obscure name is rediscovered. So, the ICZN provides a statement of the principle of priority (Art. 23.1), how it is to be applied in various situations, and, finally, when the principle should not be used (reversal of precedence, Art. 23.9).

Homonymy (Arts. 52–60) Another key problem of nomenclature occurs when two or more names that are the same apply to different taxa. This is known as homonymy. Because some names consisting of different Latin letters may mean the same thing, the ICZN defines "same": for example, the epithet pairs *microdon* and *mikrodon,* and *litoralis* and *littoralis* are considered to be the same (Art. 58). Then the ICZN dictates how homonymy is to be resolved: the senior name is to be retained, and the junior name is to be replaced; but there are exceptions as explained in Articles 52 to 60.

Typification (Arts. 61–76) Since names are only tags from scientific hypotheses, the question of whether two or more names are synonyms involves both nomenclature and taxonomy. Taxonomy in this regard can be defined as the circumscription of character space (Fig. 3), that is, the definition of taxa. Nomenclature is then the name, the designation of types for the name, and the rules for selecting among multiple types (see earlier remarks under Validity). Thus, a nominal taxon is only a name and a type; a taxon, the hypothesis, includes at least one nominal taxon. For species-group names, the type is a specimen (holotype, neotype, or lectotype) or a group of specimens (syntypes), which is or are the ultimate source of character information. For genus-group names, the type is a species name (nominal taxon), and for family-group names the type is a genus-group name. The determination of the type/genus of a family-group name is self-evident because that genus is basis for the family-group name itself (the type of the family Muscidae is the genus *Musca*). For both genus-group and species-group names, the types are designated either by the original author of the name or by subsequent workers.

An author may declare that a particular species/specimen is the type (original designation/holotype), may include only one species/specimen in the taxon (monotypy/holotypy) or may

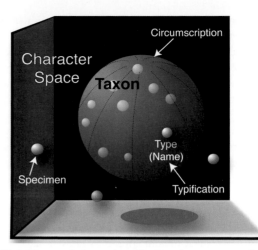

FIGURE 3 Circumscription and typification. Axes represent characters; small spheres represent specimens plotted against those axes; the large sphere represents the circumscription of a taxon; one specimen sphere is a type and its location determines the name for the taxon.

for genus-group names include a species-group name that is the same as the genus-group name (absolute or Linnaean tautonymy). If the type is not clearly fixed in the original publication, the ICZN provides rules for determining what species/specimens may be designated the type by subsequent workers.

For genus-group names, all species included in a newly defined genus are eligible to be designated as the type species. But this applies only to genus-group names proposed before 1931, for as noted earlier, after 1930 type designation became a requirement of availability. If no species were originally included, then those first subsequently included are to be considered. To subsequently designate a type species, a worker merely, but unambiguously, declares one of these originally included species as the type.

For species-group names, when no type is designated in the original publication, all specimens upon which the author based the name (including specimens not seen by the author but referred to by bibliographic citation) are eligible to be designated lectotype and are called syntypes until such a lectotype is designated.

Collectively all specimens that are either holo-, lecto-, neo-, or syntypes are termed primary types. Other specimens studied by the author may be termed paratypes (or one may be an allotype if of a sex different from the holotype), but these secondary types have no nomenclatural significance. When all primary types are no longer extant (lost or destroyed), a subsequent worker may designate any specimen as the neotype to objectively define the nominal species-group taxon. Naturally, there are recommendations and restrictions about which specimen may make a more appropriate neotype. Provisions are also made for types that have been misidentified; that is, the characters used by an author to define a taxon do not agree with those of the nominal type. When this happens, workers are free to select as type either one that agrees with the characters or the named type.

Exceptions (Arts. 78, 80–83) and Registration (Art. 79)
The stated objectives of the ICZN·are to promote stability and universality in the names of animals, and thus all its provisions must be subservient to these goals. Hence, the ICZN provides means by which any provision of the ICZN (except those that deal with its authority and exception handling) can be set aside in a particular situation. These articles outline how one appeals to the commission and how the commission arrives at its opinion, which is then published in its *Bulletin of Zoological Nomenclature*. These plenary powers and specific powers are restricted to specific cases, usually involving only a few names or works. The ICZN has provided a procedure to rule on a whole set of names at once and has created a *List of Available Names in Zoology*. This specific power allows international groups of zoologists to propose a set of names that may be approved by the commission, thereby fixing all the relevant details about those names (their spelling, authorship, place and date of publication, and typification) and giving those names precedence over other names. Names not on the list are thereafter excluded from zoological nomenclature.

Authority (Arts. 77, 84–90) The last section of the ICZN includes a series of rules explaining the derivation and perpetuation of its authority, as well as the various regulations governing the particular edition of the ICZN. These state that the ICZN is prepared by the International Commission on Zoological Nomenclature with the participation of the zoological community under the authority of a single organization (originally the International Congresses of Zoology, now the International Union of Biological Sciences). For the future, authority can be delegated to other international organizations as specified. The effective date of the fourth edition is given as January 1, 2000, and all previous editions no longer have any force.

OTHER CODES

Currently there are five different codes of nomenclature in use for organisms-one each for plants, cultivated plants, bacteria, viruses, and animals. All these codes address the same problem, the need for universal, stable, and precise sets of names for organisms, and all are similar in their methodology. However, differences are significant and can cause difficulties for those developing databases that cover all life. Hence, in the early 1990s, an effort was undertaken to develop a single code of nomenclature for all life. Meetings were held and a draft BioCode was published, but nothing further has happened.

All the codes in use today are based on and have their origin in the binominal system established by Linnaeus. Although the Linnaean system has evolved from its topological roots into one adapted to the Darwinian evolutionary model, some believe that the system cannot fully express human knowledge about the cladistic relationships among organisms. A PhyloCode has been proposed to address these perceived failures. Unfortunately, the PhyloCode adds more uncertainty, since names for clades can be based on three different methods for defining groups, and clade names have no rank, which means that virtually all information content is lost (as discussed by Benton, Forey, and Platnick).

INTERNATIONAL COMMISSION ON ZOOLOGICAL NOMENCLATURE

Zoologists realized early on that no code of nomenclature could be perfect, always able to resolve all situations in a manner that promotes stability and universality in scientific names. Hence, zoologists established an international group of specialists with the authority not only to develop and maintain the ICZN, but to make rulings on specific names and situations. What the commission is empowered to do has been outlined in this article and is covered in Articles 78 to 81. How the commission operates is set in Articles 77 and 83 and in its constitution, which is published as Appendix 3 of the ICZN. To appeal to the commission, a worker prepares a proposal and submits it to the commission. The proposal is then published in the *Bulletin of Zoological Nomenclature* for public comment. After 6 months, the commission may vote on the proposal, and the ruling will later be published in the *Bulletin*. Each proposal is treated separately on its own merits. The commission does not set precedents or follow case law.

CONCLUSIONS

No modern science places as much emphasis on priority as taxonomy and nomenclature. This emphasis requires specialists to be familiar with at least a century of published work and sometimes 300 years' worth. Some may question the value of such a long view, in as much as most sciences look back only a decade or so. Beyond the moral and ethical considerations, however, much can be gained by understanding the past. Recognition of taxa is an innate ability of humans. Ernst Mayr once noted that the primitive natives of New Guinea knew and had names for 137 of the 138 local species of birds that took western scientists years to formally describe. So, previous workers who failed to generate cladistic classifications and were not aware of the proper names for their taxa may well have recognized and characterized natural groups. So, one tries to understand how one's predecessors, who looked at the same organisms, decided how to organize their observations into taxa. Thus all who want to truly master nomenclature and classification are well advised to examine carefully what their precedessors did, appreciating what Newton once wrote: "If I have seen further, it is by standing upon the shoulders of Giants."

See Also the Following Articles
Biodiversity • History of Entomology • Museums and Display Collections • Phylogeny

Further Reading
Banks, N., and Caudall, A. N. (1912). The entomological code—a code of nomenclature for use in entomology. Washington, DC.

Benton, M. J. (2000). Stems, nodes, crown clades, and rank-free lists: is Linnaeus dead? *Biol. Rev.* **75**, 633–648.

Forey, P. L. (2001). The PhyloCode: description and commentary. *Bull. Zool. Nomencl.* **58**, 81–96.

Hennig. W. (1966). "Phylogenetic Systematics." University of Illinois Press, Urbana.

Huxley, J. S. (1957). The three types of evolutionary process. *Nature* **180**, 454–455.

International Commission on Zoological Nomenclature. (1999). "International Code of Zoological Nomenclature," 4th ed. International Trust for Zoological Nomenclature, London.

Mayr, E., Linsley, E. G., and Usinger, R. L. (1953). "Methods and Principles of Systematic Zoology." McGraw Hill, New York.

Melville, R. V. (1995). "Towards Stability in the Names of Animals. A History of the International Commission on Zoological Nomenclature 1895–1995. International Trust for Zoological Nomenclature, London.

Schuh, R. T. (2000). "Biological Systematics. Principles and Applications." Cornell University Press, Ithaca, NY.

Stearn, W. T. (1957). An introduction to the *Species Plantarum* and cognate botanical works of Carl Linnaeus. *In* "C. Linnaeus, *Species Plantarum.*" [A facsimile of the first edition of 1753]. The Ray Society, London.

Thompson, F. C. (1996). Names: the keys to biodiversity. *In* "BioDiversity II" (Reaka-Kudla, M. L., Wilson, D. E. and Wilson, E. O. eds.), pp. 199–212. Joseph Henry Press, Washington, DC.

Wheeler, Q. D., and Meier, R. (eds.). (2000). "Species Concepts, Phylogenetic Theory: A Debate." Coumbia University Press, New York.

Wiley, E. O. (1981). "Phylogenetics. The Theory and Practice of Phylogenetic Systematics." Wiley, New York.

Nutrition

S. N. Thompson
University of California, Riverside

S. J. Simpson
University of Oxford

The nutrition of animals reflects their heterotrophic character and thus focuses on the need for animals to obtain many preformed organic substances that they lack the ability to synthesize from simpler carbon sources. Nutrients then are environmental factors that connect and intersect an animal's physiology and ecology. In the broadest sense, nutrition refers to the taking in and processing of substances that fuel the organism's energy needs for growth, maintenance, and reproduction. A strict definition of nutrition, however, is not possible and any specific view of nutrition depends on perspective. Regarding nutritional requirements, insects share much in common with other animals and nutritional studies with insects have contributed significantly to our general understanding of nutrition.

HISTORICAL OVERVIEW

Early interest in insect nutrition aimed at understanding the dietary requirements of insects, an investigative focus often called dietetics, that is, identifying and characterizing the food that insects eat to satisfy their nutritional needs, as well as the feeding behaviors and sensory physiology associated with obtaining those foods. A principal purpose of this research was to rear insects in the laboratory on artificial foods, thereby avoiding the often costly effort necessary to maintain natural foods for this purpose. Among the first successes was E. Bogdanow's 1908 report on rearing of a dipteran, *Calliphora vomitoria,* on a diet containing meat extract, starch, and peptone. Similar accounts followed, describing the rearing of many other insect species from a variety of taxa. Numerous scientists are associated with this early work, including S. Beck, R. Craig, G. Fraenkel, W. Trager, and B. Uvarov. Subsequent research attempted to refine the diets by using more purified natural products or even pure chemicals and nutrients. The effort was greatly aided by new information gleaned from research on the basic nutritional requirements of insects. A major advance was the discovery by E. Hobson in 1935 that insects require cholesterol. This clearly distinguished the nutrition of insects from that of most mammals and other vertebrate animals, whose nutritional requirements were, at that time, far better known. Further, the discovery quickly led to significant advances in the study of insect dietetics and nutrition.

ARTIFICIAL SYNTHETIC DIETS

Despite the diversity of insect dietary habits, insect nutritionists have been remarkably successful at developing artificial diets and rearing programs employing artificial diets for insects. Various compilations of insect diets suggest that several hundred species can be reared partially, or through their entire life cycles, without their natural foodstuffs, which was based on a collective effort of over 1000 scientific contributors. A variety of literature is now available describing various applications of this nutritional technology. A recent catalogue of a well-known commercial concern lists for sale artificial diets for over 50 species of insects, and artificial diets have been employed for mass culture of several insects. A U.S. Department of Agriculture laboratory in Phoenix, Arizona, for example, is currently rearing, on a completely artificial diet, 6 to 7 million pink bollworms, *Pectinophora gossypiella,* daily, for use in an autocidal pest management control program in California.

QUALITATIVE NUTRITIONAL REQUIREMENTS

The nutritional requirements of insects were at first inferred from knowledge of the chemical compositions of natural foodstuffs. On this basis, insects were categorized as carnivorous, omnivorous, or phytophagous, with the appropriate inferences regarding the nutritional content of the animals, plants, and other foods eaten. Studies of food utilization, and the analysis and comparison of foodstuff eaten and excreted or unabsorbed, allowed further assessment of the relative importance of various nutrients for insects with different feeding habits. A more detailed and complete understanding of insect nutritional requirements finally emerged following the successful development of artificial diets, particularly chemically defined diets. The essentiality of individual nutrients could then be established by dietary deletion, the sequential elimination of individual chemicals, potentially nutritious, from a diet. This advancement involved many scientists now recognized for their contributions to the foundation of insect nutrition—J. S. Barlow, R. H. Dadd, S. Friedman, H. T. Gordon, H. L. House, J. G. Rodriquez, T. Ito, E. S. Vanderzant, G. P. Waldbauer, and others.

A nutrient is deemed *essential* if, when deleted, further growth, development, and/or reproduction is prevented. Almost all insects have a common set of essential nutritional requirements. Nutrients demonstrated to be necessary for optimizing growth, development, and/or reproduction are *required* nutrients. The value of specific nutrients, however, often depends on total dietary content. Dietary carbohydrate, for example, is often unnecessary, but may be required or even become essential for providing energy in the absence of dietary fat, or protein, in excess of the amount required for normal growth. The following summarizes the essential qualitative requirements of insects.

Nitrogen

Insects consume and utilize a wide variety of proteins to satisfy their nutritional requirements for amino acids. Of the commonly occurring 20 amino acids that comprise most proteins, 10 are not synthesized by insects, or most other animals, and are essential nutrients. These include arginine, histidine, isoleucine, leucine, lysine, methionine, phenylalanine/tyrosine, threonine, tryptophan, and valine. Beyond the bulk requirement for protein synthesis, these essential amino acids serve a variety of additional physiological functions of particular note in insects and other invertebrates. Arginine, for example is a precursor for the principal invertebrate muscle phosphagen phosphoargininine. Tyrosine is important for production of phenolic and quinone metabolites that are critical components for cross-linking of protein during sclerotization.

Although the other 10 commonly occurring proteinaceous amino acids are generally nonessential, they are nevertheless required to some degree for normal growth and development, because few insects will develop on diets containing only the 10 essentials. Moreover, many insects will do poorly on diets containing amino acids as the sole source of nitrogen and require protein or a mixture of protein and amino acid for normal growth and development.

Insects can generally synthesize their own nucleosides, nucleotides, and nucleic acids, although several species, particularly various dipterans, have been shown to benefit by inclusion of nucleic acid or some constituents. In a few cases, nucleic acid is considered essential for completion of development.

Vitamins

Vitamins, particularly water-soluble B vitamins, including biotin (vitamin H), folic acid (B11), niacin, pantothenic acid, pyridoxine (B6), riboflavin (B2), and thiamine (B1), or close chemical relatives of these, are essential nutrients for all insects. These serve principally as precursors for the nearly universally needed coenzymes of intermediary metabolism. Regarding the fat-soluble vitamins of other animals, only tocopherol (E) and retinol (A) have proven beneficial for reproduction and sight, respectively, by some insects. Tocopherol also plays an important role as a lipid antioxidant.

Carbohydrate

Carbohydrate nutrients, although often required as an energy source, are rarely essential. An exception may be the case of some adult lepidopterans that feed solely on plant nectars. Indeed, some of these insects are thought to be capable of digesting sucrose alone. The same may hold true for some adult dipterans and hymenopterans. In addition to their role as nutrients, sugars, particularly sucrose, are powerful phagostimulants, without which some insects feed poorly or not at all.

Sterols

Insects also have an essential requirement for a dietary sterol. Cholesterol appears to be widely acceptable, but a number of other sterols, particularly β-sitosterol and other plant sterols, can also serve. Among a few exceptions is the interesting example of *Drosophila pachea,* the senita cactus fly, which requires Δ7 dietary sterols, metabolic derivatives of schottenol, a Δ7 sterol found in its natural food, the senita cactus. In addition to the bulk requirement for sterol utilization in membrane formation, sterol is also important for the production of ecdysone and other molting hormones in insects.

Fatty Acids

Fatty acids are not essential for most insects, but several mosquitoes and some lepidopterans require a polyunsaturated fatty acid. This requirement is associated with one of the few nutritional disease syndromes—the "crumpled wings" syndrome—in which absence of a polyunsaturated fatty acid results in adult insects that fail to fully expand their wings and are thus unable to fly. The chemical nature of this requirement is poorly understood. In the case of mosquitoes, arachidonic acid or some closely related fatty acids with an ω-6 double bond is essential. In those Lepidoptera requiring a polyunsaturated fatty acid, however, some species utilize an ω-6 fatty acid, whereas others utilize ω-3 fatty acids such as α-linolenic acid. Moreover, in those Lepidoptera requiring an ω-6 fatty acid, α-linoleic is generally preferred, and arachidonic acid is unsuitable. Nothing is known of the physiological basis for this requirement, but in the mosquitoes it may be linked to the synthesis of prostaglandins, local regulators that target a wide variety of cellular functions in vertebrate animals.

Inorganics

The complex of mineral ions nutritionally essential for other animals is likewise essential for insects. Here the balance is often dramatically different from the well-known salt requirements established for mammals. Many insects, for example, require much greater proportions of potassium, magnesium, and phosphate relative to sodium, calcium, and chloride.

Ascorbic Acid and Other Water-Soluble Growth Factors

L-Ascorbic acid, vitamin C in vertebrate animals, is an essential growth factor for many phytophagous insects. In its absence, these insects generally fail to grow and/or develop. The pattern of this requirement among insects is thus similar to that of the higher animals of which species that have adapted to a diet of fruit and/or vegetables have apparently lost the ability to synthesize ascorbic acid. In contrast to the vitamin C requirement of vertebrates, ascorbic acid is required by insects in relatively large amounts, although this may in part reflect the necessity for a high antioxidant activity in synthetic artificial diets employed for testing. Moreover, unlike vertebrates requiring L-ascorbic acid, some insects utilize dehydroascorbate as effectively as L-ascorbate, as well as use the D geometric forms of closely related lactones, although these were generally not as effective. Although there is little understood about the role of ascorbic acid in insects, beyond its potential antioxidant action, it may play the same role as in vertebrates, that is, as a factor necessary for enzyme activities involving hydroxylation. Deficiency in insects is often associated with abnormalities of molting, possibly due to the absence of ascorbic acid effects on diphenyloxidases, perhaps in a manner analogous to its action on the synthesis of serotonin from tryptophan.

The lipogenic growth factors, choline and inositol, principal components of phospholipids, are required by many insects. The essentialness of these remains in question, although choline is likely an essential nutrient for most insects.

Other Unique Essential Nutrients

In addition to the above essential requirements, several nutrients may be uniquely essential for some insects. Several mosquitoes and one tachinid species, for example, appear unable to synthesize asparagine and this is an essential dietary amino acid. An essential requirement for proline by several taxonomically disparate insect species may be related to a limited activity of the urea cycle. For unknown reasons, glutamic acid or aspartic acids, normally nonessential, have been reported as essential for a few insects.

A unique and essential requirement for carnitine is known for several tenebrionid beetles, for which this usually nonessential nutrient is called vitamin B_T. Normally derived from choline, carnitine plays an important role in fatty acid transport.

Essential and nonessential nutrients are required in specific amounts, but as implied above, optimal levels of individual nutrients often depend on the dietary concentrations of other nutrients. Early studies with artificial diets purported to formulate optimal levels of nutrients, based on the relative concentrations found in natural foodstuffs. Diets employing these concentrations of nutrients often produced adequate results, but subsequent studies have demonstrated that the quantitative aspects of insect nutrition were far more complex than that approach suggested. Moreover, ecological, behavioral, and physiological factors are also important for optimal nutrition.

QUANTITATIVE NUTRITION

As we have seen, if an insect is to survive, grow, and reproduce, it must ingest several dozen different types of nutrient molecule. These molecules come packaged in varying amounts and ratios within foods, along with nonnutritive (and sometimes toxic) compounds. Foods in turn are distributed through time and space, and their finding, ingestion, and processing engender metabolic and ecological costs. Added to that, the nutritional needs of insects change with age, stage of development, reproductive status, etc. Matching the multidimensional and changing nutritional demands of the insect against the complex and changing composition of the nutritional environment has posed one of the greatest challenges to evolution—and to scientists who study insect nutrition. The problem is particularly difficult for herbivorous insects, for which the nutritional composition of host plants may be highly variable and there is the added complexity of secondary plant metabolites serving as antifeedants and toxins. There is a growing realization, however, that predators too may face considerable variation in food quality and may therefore have to regulate their intake and use of multiple nutrients rather than relying on more general food properties.

A powerful way of defining and exploring nutritional regulation that has arisen from work on insects has been to represent the animal, its intake and growth requirements, and the foods in its environment using multinutrient plots. This "geometric framework" has enabled the identification of the key elements in complex nutritional systems and the quantification of the interactions among them. These include interactions among the multiple constituents of the food as well as between behavioral and physiological regulatory mechanisms. The resulting descriptions provide a powerful means to study the mechanisms, ecology, and evolution of nutritional systems.

Quantifying Intake Requirements

Estimating the intake requirements (intake target) is a primary aim of any nutritional study. One way to do this is to allow the insect to demonstrate whether it is able to regulate its nutrient intake and if so, in which nutrient dimensions. A well-documented study system is the locust, which has been shown to regulate its intake of both protein and carbohydrate under several challenges:

1. Pairs of complementary foods. When locusts were provided with one of four complementary food pairings (28:14 or 14:7% protein:% digestible carbohydrate vs either 14:28 or 7:14) they adjusted the amount and ratio of the two foods eaten to maintain a constant point of protein and carbohydrate intake (Fig. 1, top).

2. Food dilution. When locusts were given one of five foods containing a near-optimal ratio of protein to digestible carbohydrate (1:1), but diluted up to fivefold with indigestible cellulose (35:35, 28:28, 21:21, 14:14 or 7:7% protein:% digestible carbohydrate), they adjusted their consumption across all dilutions to maintain a constant point of nutrient intake (Fig. 1, middle).

3. Food frequencies. When two complementary foods (31:11 and 7:35) were provided in relative abundances of 1:3, 2:2, or 3:1 dishes of one versus the other food type, locusts precisely selected a point of protein-to-carbohydrate intake by adjusting their distribution of consumption between dishes (Fig. 1, bottom).

These remarkable feats of homeostasis were found to extend to regulation of salt versus macronutrient intake. Moreover, other studies indicate that such capabilities are by no means restricted to locusts.

Changes in Intake Requirements with Time

The intake requirements of insects are not static. They change with recent nutritional experience and level of activity. For instance, locusts and caterpillars select a protein-rich food following a short experience (only one meal in the case of the locust) of a protein-depleted food and show a similar preference for carbohydrate-rich food after a 4-h period of carbohydrate deprivation. Nutrient requirements also vary with stage of development, reproduction, and diapause. Over a longer time scale, nutritional requirements evolve to track changing nutritional environments and life histories.

Mechanisms Regulating Intake

Regulating nutrient intake requires two sources of information, the first being the composition of the food and the second the nutritional state of the animal. The responsiveness of an animal to a food of given composition should reflect, through feedbacks, the animal's nutritional state. Insects are able to taste certain key nutrients, notably sugars, amino acids, salts, and water. Studies on locusts, caterpillars, and blow flies have shown that the responsiveness of taste receptors to sugars and amino acids varies with nutritional state, as represented by concentrations of these nutrients in the hemolymph. Such nutrient-specific feedbacks enable insects to make sophisticated behavioral decisions about

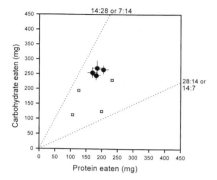

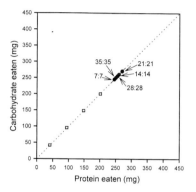

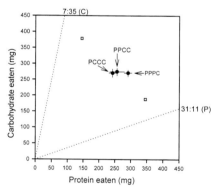

FIGURE 1 Three experiments indicating regulation of intake to a point ("intake target") in a carbohydrate–protein plane. The plots are of the form used in the geometric framework. Foods are shown as lines ("rails") radiating out from the origin. An insect feeding from a single food is constrained to move along that food rail in intake space, while two foods provide the opportunity to move anywhere in between by food mixing. In the top plot locusts were provided with one of four food choices (14:28 or 7:14 % protein:% carbohydrate vs 28:14 or 14:7) and altered relative amounts eaten from the two foods, thus reaching the same place in nutrient space. Open symbols indicate where insects would have arrived were they to have eaten indiscriminately between the two foods provided. The middle shows locusts that were given one of five foods containing a 1:1 ratio of protein to digestible carbohydrate, but diluted to various degrees with indigestible cellulose. Those on food 7:7 ingested five times more food than those on 35:35, thus achieving the same intake of both protein and digestible carbohydrate. Open symbols show points of intake were insects not to have compensated for dilution but eaten the same amount of food on each treatment. Locusts in the experiment shown at the bottom were given four food dishes, containing either 7:35 % protein:% carbohydrate (food C) or 31:11 (food P). The two food types were provided at different frequencies (all four dishes contained C, all four contained P, or two contained C and two P). Locusts adjusted the amounts eaten from each dish and regulated nutrient intake. Open symbols indicate points of intake were locusts to have distributed feeding equally among dishes and not shown frequency-dependent food selection.

what foods to eat to regulate nutrient intake. In addition, learning of various sorts, including aversion, learned specific appetites, and induced neophobic responses, also plays an important role in regulating food selection in insects. An impressive example is the ability of locusts to learn to associate the odor of a food with its protein content and to be attracted by odors previously paired with high-protein food, but only when in a state of protein deficit.

Nutrient Balancing on Suboptimal Foods

If an insect is restricted to suboptimal foods and is unable to reach its intake target, it must balance undereating some nutrients against overeating others, reaching some "point of compromise." Such a situation exposes whether and how the mechanisms regulating intake of different nutrient groups interact. A simple means of exploring the interactions between regulatory systems for different nutrients is to provide insects with one of an array of foods of varying composition and to measure intake. Collectively, the resulting points of nutrient intake across the array of foods form a pattern that describes the relationship between the mechanisms regulating intake of the nutrients concerned.

Various such relationships have been described to date in insects. The simplest outcome is that in which the insect abandons regulation of one nutrient when forced to balance it against regulation of another. For instance, locusts regulated macronutrient (protein and carbohydrate) intake and let salt intake vary passively when fed single foods containing suboptimal salt levels, even though they regulated both salt and the macronutrients when allowed to switch between complementary foods. In other cases, one nutrient does not overwhelm the other. When locusts were fed foods varying in protein and carbohydrate content they balanced over- and undereating the two nutrients, with the precise balancing rule depending on the species of locust. The grass-feeding specialist *Locusta migratoria* was less able to overeat unbalanced foods to gain more of the limiting nutrient than was the host-plant generalist *Schistocerca gregaria* (Fig. 2). It appears that the generalist species is better able to capitalize on excess ingested nutrients than is the specialist.

Regulation of Growth and Metabolism

Whereas an insect may be constrained from reaching its intake requirements by available foods, it may still be able to regulate postingestive processing and thus achieve its growth target (Fig. 3). Regulation of growth and body composition involves differentially using ingested nutrients: ridding nutrients in excess and conserving those in deficit. Locusts are extremely effective at regulating growth, excreting excess ingested nitrogen, and respiring excess ingested carbohydrate or converting it to lipid. Deamination of excess ingested protein, principally by oxidation or transamination to the corresponding keto acid, may serve to augment limiting carbohydrate for energetic purposes.

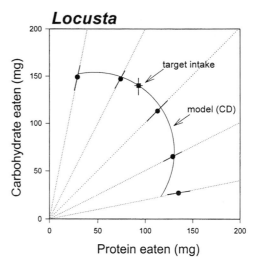

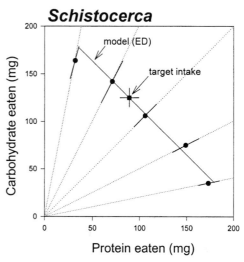

FIGURE 2 Arrays of nutrient intake shown by locusts when fed single foods and hence forced to balance their intake of two nutrient groups. The two plots show how the grass-specialist *L. migratoria* and the host-plant generalist *S. gregaria* have different balancing rules for protein and carbohydrate. *L. migratoria* minimized the distance from the intake target in nutrient space when unable to reach its intake target (the CD rule), while in comparison *S. gregaria* ate more of the nutritionally unbalanced foods (the ED rule).

NUTRIENT–ALLELOCHEMICAL INTERACTIONS

As well as primary nutrients, the foods of many animals contain harmful or unpalatable nonnutritive materials. This is especially the case for plants, which contain an abundance of defensive secondary metabolites (often termed allelochemicals). In some instances, insects make use of such compounds either as cues for host-plant recognition or as resources in their own right for defense or communication. In the main, however, allelochemicals are an impediment to nutrient regulation. A key point is that the effectiveness of allelochemicals as antinutritive compounds depends on the nutritional context in which they occur in the food. For example, locusts are not affected adversely by the presence of tannic acid in their food, even up

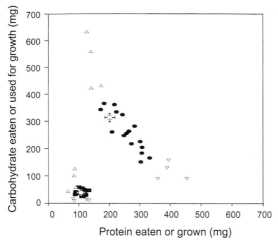

FIGURE 3 Patterns of intake and growth in locusts fed 1 of 25 diets varying in protein and digestible carbohydrate content. Intake data are across the entire fifth stadium. Note that protein- and carbohydrate-derived growth was regulated despite insects on the different diets having eaten widely different amounts of protein and carbohydrate. Locusts on all but the most extremely unbalanced diets regulated growth to a high degree in both nutrient dimensions. Regulation of growth involved differential utilization of ingested protein and carbohydrate. On diets with a very low protein-to-carbohydrate ratio (upright triangles), development was extended and body lipid content was high, whereas on foods with a very high protein-to-carbohydrate ratio (inverted triangles), there was no lipid deposition. Insects able to regulate their intake by selecting between complementary foods are indicated with the open square symbol.

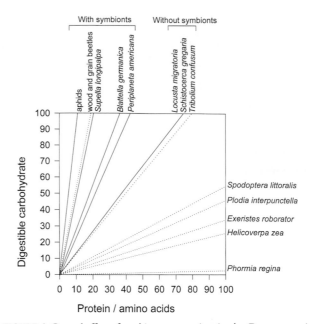

FIGURE 4 General effect of symbionts on nutrient intake. Data summarize the protein (or amino acid)-to-carbohydrate ratios in diets that support good larval development in a selection of insect species. Note how insects with microbial endosymbionts develop best on diets with a lower protein (or amino acid)-to-carbohydrate ratio (steep rails) than do other species. Dotted lines indicate endopterygote species while unbroken lines are for exopterygotes.

to 10% dry weight, if the protein and carbohydrate ratio and concentration are near optimal. When foods contain less than an optimal protein-to-carbohydrate ratio, tannic acid serves as a powerful feeding deterrent and thus causes high mortality and extended development. However, at higher than optimal protein-to-carbohydrate ratios tannic acid does not reduce intake but instead results in high mortality by disrupting protein utilization.

MICROBIAL ASSOCIATIONS

Symbionts, principally actinomycete fungi and bacteria, play a critical role in insect nutrition, enabling many species to develop normally on foods of limited nutritional value. Well-known examples of such foods include wood, blood, phloem, and plant litter. Many insects would quickly perish in the absence of these symbiotic relationships. Symbionts often provide nutrients directly as a result of synthetic capabilities that their insect hosts lack and/or allow, through the production of gut enzymes, insects to digest otherwise indigestible foodstuff. Alternatively, the symbiont itself may serve as food. Many symbiotic relationships are casual, involving ectosymbionts, usually comprising a rich gut microflora. Endosymbiosis is also common, and many insects have developed highly specialized anatomical and behavioral features for optimizing the benefits of the relationship and for efficiently transmitting their symbionts between generations.

Among the most significant endosymbioses are those involving mycetocytes and bacteriocytes, host cells specialized for housing symbiotic microorganisms—fungi or bacteria. Often these polyploid cells are associated with the midgut epithelium, although in the case of fungal-infected cells, aggregates called mycetomes may sometimes be found in the hemocoel. The genomes of the symbiont and the host cell are closely coordinated, forming a functional unit known as the symbiocosm. Such an endosymbiont is found in the rice weevil *Sitophilus oryzae,* in which it is referred to as the SOPE or *S. oryzae* principal endosymbiont. This bacterium (Family Enterobacteriaceae), whose expression is partly regulated by the host, occurs in the cytoplasm of the bacteriocyte (2×10^3 bacteria/host cell) and is known to be critically important in the insect's biology. It is, for example, the source of several vitamins, including riboflavin, pantothenic acid, and biotin. Moreover, the presence of the symbiont alters the balance of amino acid metabolism and mitochondrial phosphorylation, thereby affecting flight ability and performance.

The impact of symbionts on the quantitative nutrition of insects was recently made apparent by an analysis of the optimal dietary protein:carbohydrate ratio for 117 insect species. Insects housing symbionts displayed very steep intake target rails (Fig. 4), strongly suggesting that symbionts add considerably to the nitrogen nutrition of such species and, moreover, that this may generally be the case.

CONCLUSION

A thorough knowledge of insect nutrition is essential for understanding the biology of insects. The study of insect nutrition has recently undergone a metamorphosis, in that information gleaned from earlier investigations that focused principally on basic nutritional requirements and rearing technology is now being applied to understanding the feeding strategies, nutritional ecology, and evolution of insects. Nutritional physiology and biochemistry are also advancing. The neurological bases for food selection and the role of biogenic amines in regulating food choice are beginning to be understood. The chemical composition of the hemolymph is now recognized as a dynamic indicator of nutritional status, affecting food selection and nutrient intake. The metabolic responses of insects to altered nutritional status and the effects of fat-body metabolism on hemolymph composition are also being investigated. Future studies employing multidisciplinary approaches will continue to unravel the mysteries of insect nutrition and its consequences and significance to insect biology.

See Also the Following Articles

Digestion • Feeding Behavior • Metabolism • Plant–Insect Interactions • Rearing of Insects • Symbionts, Bacterial

Further Reading

Amakawa, T. (2001). Effects of age and blood sugar levels on the proboscis extension of the blow fly *Phormia regina. J. Insect Physiol.* **47,** 195–203.

Anderson, T. E., and Leppla, N. C. (1989). "Advances in Insect Rearing for Research and Pest Management." Westview Press, Boulder, CO.

Bernays, E. A., and Chapman, R. F. (1994). "Host Plant Selection in Phytophagous Insects." Chapman & Hall, New York.

Chapman, R. F. (1995). Chemosensory regulation of feeding. *In* "Chemical Ecology" (R. T. Cardé and W. J. Bell, eds.), pp. 101–136. Chapman & Hall, New York.

Dadd, R. H. (1985). Nutrition: Organisms. *In* "Comprehensive Insect Physiology, Biochemistry and Pharmacology" (G. A. Kerkut and L. I. Gilbert, eds.), Vol. 4, pp. 313–390. Pergamon, London.

Douglas, A. E. (1995). "Symbiotic Interactions." Oxford University Press, Oxford.

Heddi, A. Grenier, A.-M., Khatchadourian, C., Charles, H., and Nardon, P. (1999). Four intracellular genomes direct weevil biology: Nuclear, mitochondrial, principal endosymbiont, and *Wolbachia. Proc. Natl. Acad. Sci. USA* **96,** 6814–6819.

Raubenheimer, D., and Simpson, S. J. (1999). Integrating nutrition: A geometrical approach. *Entomol. Exp. Appl.* **91,** 67–82.

Reinecke, J. P. (1985). Nutrition: Artificial diets. *In* "Comprehensive Insect Physiology, Biochemistry and Pharmacology" (G. A. Kerkut and L. I. Gilbert, eds.), Vol. 4, pp. 391–419. Pergamon, London.

Simpson, S. J., and Raubenheimer, D. (2000). The hungry locust. *Adv. Stud. Behav.* **29,** 1–44.

Simpson, S. J., and Raubenheimer, D. (2001). The geometric analysis of nutrient–allelochemical interactions: A case study using locusts. *Ecology* **82,** 422–439.

Simpson, S. J., Raubenheimer, D., and Chambers, P. G. (1995). The mechanisms of nutritional homeostasis. *In* "Regulatory Mechanisms of Insect Feeding" (R. F. Chapman and De Boer, eds.), pp. 251–276. Chapman & Hall, New York.

Singh, P., and Moore, R. F. (1985). "Handbook of Insect Rearing," Vols. I and II. Elsevier, Amsterdam.

Slansky, F. (1992). Allelochemical–nutrient interactions in herbivore nutritional ecology. *In* "Herbivores: Their Interactions with Secondary Plant Metabolites" (G. A. Rosenthal and M. R. Berenbaum, eds.), Vol. 2, pp. 135–174. Academic Press, New York.

Thompson, S. N., and Redak, R. A. (2000). Interactions of dietary protein and carbohydrate determine blood sugar level and regulate nutrient selection in the insect *Manduca sexta* L. *Biochim. Biophys. Acta* **1523,** 91–102.

Ocelli and Stemmata

Frederick W. Stehr
Michigan State University

The number, size, and arrangement of ocelli, or simple eyes, are important diagnostic characters for many larvae. Until relatively recently, the simple eyes of both larvae and adults have been termed "ocelli" (ocellus), although it has been recognized that these are two different groups that are innervated from different parts of the brain.

OCELLI

One group of simple eyes, found in adult insects and larvae of nonholometabolans, is termed "dorsal ocelli" or simply "ocelli." These structures are innervated dorsally from the protocerebrum between the optic lobes. There are basically four dorsal ocelli, but one pair is fused to form the median ocellus; thus, there are typically three ocelli located near the midline of the head, but the number varies from zero to three (eight in Collembola). Their function is apparently visual, but knowledge is sparse about precisely what they see and how they interact with the compound eyes.

STEMMATA

The second group of simple eyes, formerly termed "lateral ocelli" but now termed "stemmata" (singular stemma), is found in the larvae of Holometabola. Stemmata are innervated laterally from the optic lobes, and typically there is a group on each side of the head. The number of stemmata is variable, ranging from zero to seven, and the number and arrangement can be diagnostic. They are most highly developed in externally feeding larvae such as caterpillars, sawfly larvae, and predaceous larvae, and are frequently less well developed, fewer in number, or absent in larvae found in concealed situations. Stemmata are used as horizon detectors, but the sharpness of their perception is no doubt limited.

See Also the Following Article
Eyes and Vision

Further Reading
Gilbert, C. (1994). Form and function of stemmata in larvae of holometabolous insects. *Annu. Rev. Entomol.* **39,** 323–349.
Paulus, H. F. (1881). Eyestructure and the monophyly of the Arthropoda. *In* "Arthropod Phylogeny" (A. P. Gupta, ed.), pp. 299–383. Van Nostrand Reinhold, New York.
Snodgrass, R. E. (1935). "Principles of Insect Morphology." McGraw-Hill, New York.
Stehr, Frederick W. (ed.). (1987). "Immature Insects," Vol. 1 Kendall/Hunt, Dubuque, IA.

Odonata
(Dragonflies, Damselflies)

K. J. Tennessen
Tennessee Valley Authority

Odonata (dragonflies) are paleopterous, exopterygote aquatic insects, related to the Ephemeroptera (mayflies). Dragonfly adults are predaceous, relatively long-lived insects. Their large compound eyes, strong chewing mouthparts, long legs, and unparalleled flight capabilities are ideal adaptations for catching and consuming insect prey. Although adult dragonflies have mastered the air, the larvae are aquatic and are usually much longer lived. Adaptation to an underwater existence has resulted

FIGURE 1 Examples of Zygoptera: (A) *Calopteryx dimidiata* and (B) *Argia* sp. (Photographs by R. S. Krotzer and K. J. Tennessen, respectively.) Examples of Anisoptera: (C) *Cordulegaster obliqua* and (D) *Libellula flavida*. (Photographs by R. S. Krotzer.)

in striking differences in form among larvae, whereas adults are much more uniform in shape.

Dragonflies are quite harmless insects; they do not sting and will try to bite humans only when held captive. However, they are hosts of trematodes (flukes in the family Lecithodendriidae) in Southeast Asia, and when eaten raw, they can be a source of infection in humans (by ingestion of Metacercariae). On the whole, dragonflies are considered to be beneficial insects for several reasons. In both larval and adult stages, they feed on many insects that are pests of humans and domestic animals, such as mosquitoes (Culicidae), deer-flies (Tabanidae), blackflies (Simuliidae), and other Diptera. They are important components of aquatic food webs and are used as indicators of ecological health of streams and lakes; in some areas, larvae are used as fish bait or as food. They make good subjects for behavioral and ecological studies, and poets and visual artists are often inspired by their beauty and behavior. Because some species are quite large and many are beautifully colored (Fig. 1), dragonflies have become fairly popular with the public. The relatively large size and distinctive color patterns allow the identification of many species of Odonata in the field, especially through binoculars. The recent appearance of field guides such as Dunkle's will facilitate natural history and behavioral studies also. As dragonflies become more popular, they may join butterflies as "ambassadors" for insect conservation and appreciation.

Dragonflies are a warm-water-adapted group that probably originated in a tropical environment, as evidenced by lower present-day species diversity in cooler climates (i.e., with increasing latitude and increasing altitude). At least 75% of the world fauna is tropically distributed, although a few genera (*Aeshna, Somatochlora,* and *Leucorrhinia*) have diversified in cooler climates, and their centers of distribution are located at

higher latitudes and altitudes. Most high-latitude species in these genera are centered in the warmer parts of their geographic ranges, but a few species (e.g., in genus *Somatochlora*) are distributed mainly north of 60°.

PALEONTOLOGY

Fossil wings of several types of predatory, dragonfly-like insects from the Carboniferous (about 320 mya) have been found. The Eugeropteridae of the mid-Carboniferous are the most archaic members of the superorder Odonatoidea known. They had prothoracic winglets, but they also had pleated wings and a basal wing complex that included a cell in the shape of a parallelogram that allowed changes in camber of the wings and therefore maneuverable flight. Apparently the early odonatoids radiated and flourished throughout the Permian, when the large landmass Pangaea was still intact. They include the broad-winged Protodonata and Protanisoptera, and the petiolate-winged Protozygoptera. These primitive species lacked one or more of the diagnostic wing characters of all extant Odonata, including the modern basal wing complex (arculus and triangular or quadrangular conformation of veins), nodus, and pterostigma (see later) that together provide for highly maneuverable, swift flight. Some of these extinct "dragonflies" were probably the largest insects ever to have existed; for example, *Meganeuropsis* had a wing span of about 75 cm. The archaic odonatoids persisted until the Permo-Triassic extinction, a span of roughly 70 million years.

Although the phylogenetic relationships of the Paleozoic representatives are still poorly understood, the Odonatoidea as a whole almost unquestionably form a monophyletic group. Modern Odonata probably did not stem directly from these early odonatoids. Instead, it is probable that the ancestor of modern Odonata was similar to some of the extinct Jurassic groups (e.g., the Tarsophlebiidae, with nonpetiolate wings) previously placed in the "Anisozygoptera," now known to be a nonmonophyletic grouping. Therefore, the broad wings of Anisoptera and the petiolate wings of Zygoptera are equally derived characters, which probably evolved in the Jurassic. Anisoptera first appear in the fossil record in the Jurassic (150 mya), whereas Zygoptera do not appear until the Cretaceous period (120 mya).

Another puzzling question concerning Odonata paleontology is the appearance of aquatic larvae. The earliest fossil evidence of an aquatic larval existence is from the Mesozoic, and it has been suggested that they were semiaquatic before that period. However, the sister group of the Odonata, the Ephemeroptera, show evidence of larvae with gills in the early Permian at least 270 mya; no fossil mayfly larvae have been regarded as terrestrial.

SYSTEMATICS

Although systematics of Odonata is relatively advanced compared with most other insect orders, classification at the

TABLE I   Odonata Families of the World[a]

Family	Distribution by continent	Genera	Species
Zygoptera			
Amphipterygidae	Africa, Asia, North and South America	4	9
Calopterygidae	Africa, Asia, Europe, North and South America	15	158
Chlorocyphidae	Africa, Asia	15	133
Chorismagrionidae	Australia	1	1
Coenagrionidae	Worldwide	82	1067
Dicteriadidae	South America	2	2
Diphlebiidae	Asia, Australia	2	9
Euphaeidae	Asia	8	63
Hemiphlebiidae	Australia	1	1
Isostictidae	Australia	12	45
Lestidae	Worldwide	9	155
Lestoideidae	Australia	1	4
Megapodagrionidae	Africa, Asia, Australia, North and South America	43	255
Perilestidae	Africa, North and South America	3	20
Platycnemididae	Africa, Asia, Australia, Europe	23	198
Platystictidae	Asia, North and South America	5	159
Polythoridae	North and South America	8	56
Protoneuridae	Africa, Asia, Australia, North and South America	24	240
Pseudostigmatidae	North and South America	5	18
Synlestidae	Africa, Asia, Australia, North America	7	33
Anisoptera			
Aeshnidae	Worldwide	49	411
Austropetaliidae	Australia, South America	7	15
Chlorogomphidae	Asia	2	40
Cordulegastridae	Asia, Europe, North America	4	51
Corduliidae	Worldwide	44	358
Epiophlebiidae	Asia	1	2
Gomphidae	Worldwide	91	952
Libellulidae	Worldwide	140	962
Neopetaliidae	South America	1	1
Petaluridae	Asia, Australia, North and South America	5	11
Synthemistidae	Australia	7	42
Total		621	5471

[a]North America includes Central America and the West Indies. Based in part on Davis, D. A. L., and Tobin, P. (1984). "The Dragonflies of the World," Vol. 1, "Zygoptera, Anisozygoptera." Societas Internationalis Odonatologica Rapid Communications (suppl.), No. 3, Utrecht; Davis, D.A.L., and Tobin, P. (1985). "The Dragonflies of the World," Vol. 2, "Anisoptera." Societas Internationalis Odonatiologica Rapid Communications (suppl.), No. 5, Utrecht; and Schorr, M., Lindeboom. M., and Paulson, D. (2000). List of Odonata of the world. http//www.ups.edu/biology/museum/worldodonates.html.

suborder and family level is still controversial. Until recently, three extant suborders were accepted, the Anisoptera, Zygoptera, and Anisozygoptera, with the latter group being represented by only two extant species of the Asian Epiophlebiidae. However, recent analyses indicate that "Anisozygoptera" is not a monophyletic group and that all taxa originally placed in that group are extinct. Therefore, the two living Epiophlebiidae are considered Anisoptera.

Classification at the family level also has been unstable. At present, 31 families are generally accepted (Table I), although there is some disagreement with this arrangement, especially within the Zygoptera. It is likely that DNA methods will create changes in classification and in the understanding of taxonomic relationships. Worldwide nearly 5500 species of Odonata have been described. The two suborders (Anisoptera and Zygoptera) have approximately equal numbers of known species. The rate of species description has remained fairly constant throughout

recent decades (average nearly 350 new species per decade in the 20th century), an indication that the order is far from being completely known.

Larvae are much less well known than adults, especially in the tropics. However, larvae of nearly all the 427 North American species and 120 European species have been discovered, and much knowledge exists on their habitat requirements and life histories. Larvae of approximately 25% of the more than 1200 South American species have been described.

CHARACTERIZATION AND MORPHOLOGY

Characterization of the Order

The order Odonata is characterized by a prognathous head with chewing mouthparts, large compound eyes, three ocelli, small bristlelike antennae, a small prothorax, the meso- and metathoracic segments fused into a large pterothorax, relatively long legs with three-segmented tarsi, two pairs of elongate wings, elongate abdomen, accessory male genitalia including the intromittent organ (not homologous with the penis of other insects) on the venter of the second abdominal segment, and one-segmented cerci. The odonate pterothorax is unusual in several features: (1) the bases of the legs are crowded forward, an arrangement conducive not to walking but to grasping; (2) the sternal sclerites make up most of the lateral and dorsal walls, with the mesepisterna meeting dorsally to form a middorsal carina; and (3) the wing bases are positioned posteriorly, and the tergal sclerites are extremely reduced. All these thoracic modifications facilitate strong flight and the pursuit and handling of prey. The huge flight muscles are connected directly to the bases of the wings. Therefore, the front and hind wings can be moved independently. The wing beat rate is relatively slow in comparison to neopterous insects (20 to 40 beats s^{-1} vs nearly 1000 beats s^{-1}), but dragonflies can fly almost as fast and as agilely as any other insect. Although most Zygoptera are relatively slow fliers, they can navigate precisely among stems and tangles of vegetation. The larvae are unique among Insecta in possessing an elongate, hinged labium that is folded under the head when not in use; it may be as long as one third of the larval body when extended to capture prey.

External Morphology

The suborders differ in two major characters. In Zygoptera (Fig. 2), the head is wider than the thorax and the hind wings are similar to the forewings in basal width and the orientation of the quadrangles. In Anisoptera (Fig. 3), the head is not wider than the thorax in dorsal view and the hind wings are wider at the base than are the forewings; also, the triangle of the forewing lies perpendicular to the long axis of the wing, whereas it usually lies parallel to the long axis in the hind wing. Larvae of the two suborders differ in that Zygoptera have three elongate, platelike or saclike anal gills and two comparatively short cerci (Fig. 4A,B), whereas Anisoptera have five pointed anal appendages (Fig. 4C,D) and an internal rectal gill chamber. Although the term "dragonfly" typically is used for the entire order, some authors restrict this term to the Anisoptera and use "damselfly" for the more slender Zygoptera.

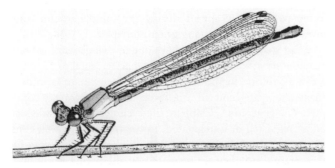

FIGURE 2 Zygoptera adult.

FIGURE 3 Anisoptera adult.

The wings of Odonata are richly veined to support the wing membrane. Wing venation is extremely important in systematics, especially at the family level. The beautiful colors of the body are produced by pigments under the cuticle and by diffraction of light by the cuticle. Females are often less strikingly colored than males of the same species, particularly in the large families Libellulidae and Coenagrionidae. Coloring may also be produced by pruinescence, a white or bluish white exudate of the hypodermis, which forms with sexual maturation, especially in males. Larvae are usually darkly colored, probably an adaptation to the colors of their microhabitat. In contrast to adults, larval body shape is highly variable, ranging from slender cylinders to nearly flat circles, and undoubtedly reflects specific habits of foraging and escaping predation. Larval antennae are more developed than in the adults. The most distinguishing characteristic of larvae is the prey-capturing labium, which is highly variable in its morphology and ranges from flat to cup- or spoon-shaped and from very elongate to short and wide. The palpal lobes and prementum are armed with a highly variable number of raptorial setae, though these may be absent.

BIOLOGY

Life Cycle

Odonata are hemimetabolous. The larvae have external wing sheaths, and although there is no pupal stage, the larva in its final stage differs greatly in form from the adult. Several developing adult structures, such as the labium, often can be

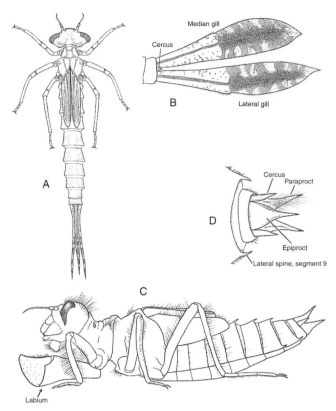

FIGURE 4 Odonata larvae: (A) Zygoptera larva in dorsal view, (B) anal gills of Zygoptera in lateral view, (C) Anisoptera larva in lateral view, and (D) anal pyramid of Anisoptera in dorsal view.

seen at this stage through the larval integument. Of the three life stages (egg, larva, adult), the larvae show the greatest diversity in functional morphology.

EGG STAGE   Eggs are laid in or above permanent or temporary water bodies. The Zygoptera and a few Anisoptera families (Aeshnidae, Petaluridae) lay their eggs in plant tissue (endophytic oviposition); most Anisoptera families lay their eggs in open water (exophytic oviposition), although some species attach their eggs to plant tissue. Eggs are spindle shaped in endophytic species and usually round to ellipsoidal in exophytic species; they range in size from about 0.23×0.48 mm to 0.60×1.40 mm. Eggs are fertilized as they pass through the female's vagina during oviposition, and embryogenesis begins immediately after the eggs are laid. Fertilized eggs change from creamy white or light gray to a light or reddish brown or dark gray within the first 24 h. However, eggs of some tropical species are brightly colored (e.g., blue, green, pink) throughout the egg stage. Embryos that undergo direct development hatch within 5 to 60 days, whereas those undergoing delayed development (diapause) hatch between 80 and 200 days after oviposition depending on temperature.

The hatching process begins several hours before actual eclosion. Peristaltic movements of the esophagus result in swallowing of amniotic fluid, which causes water to enter the egg

through the micropyles. In Zygoptera, increasing pressure within the egg causes the chorion to rupture, usually along curved lines of weakness. Continual swallowing and abdominal distension move the embryo forward, pushing the head forward into a chamber formed by the vitelline membrane. The embryo then swallows water, bursting the vitelline membrane, and the first instar slips most of the way out the chorion, but typically the tip of its abdomen remains inside the egg. In Anisoptera, a sclerotized frontal crest, called an "egg burster," is used to produce a longitudinal slit in the chorion, through which the larva exits the egg.

LARVAL STAGE   The first instar, known also as the prolarva, is extremely brief in duration. It does not feed, and the legs are seldom functional. It may last for a few seconds to a few hours, depending on whether the egg was deposited in or out of water. In most species, the prolarva molts to the second instar while the tip of the abdomen is still within the egg. In species that hatch above the waterline, the prolarva drops out of the egg and reaches the water primarily by jumping. Second instars retain some yolk in their midgut to provide nutrition for a day or so, allowing them to become adept at their predatory habits. They are usually fairly pale, becoming darker in succeeding instars, usually matching their microhabitat.

The number of instars is highly variable within the order, ranging from 9 to 17. Most species have 11 to 13 instars, but even siblings treated identically can undergo a different number of molts. Instar classification is especially difficult with field-collected larvae, except that the first three and last three instars usually can be determined.

Duration of each instar is also highly variable and is dependent on species, temperature, and food availability. The final instar is the longest in duration, lasting as little as 5 days in rapidly developing species to a year or more in others. Growth occurs immediately after each molt for about an hour, while the integument is still pliable. Growth ratios (proportional increases in linear dimension from one instar to the next) for Odonata range from 1.2 to 1.3 and are usually very close to the average for Hemimetabola (1.27). Certain body structures change with successive molts; for example, the number of antennal segments and the number of palpal and premental setae increase, dorsal protuberances on the head of early instars disappear, the compound eyes become larger, color patterns develop, and rudiments of the sexual gonapophyses appear. Wing pads usually appear during the middle instars, and grow more rapidly than any other body part.

Most species of Odonata have one or two generations per year, but many are semivoltine. A higher percentage of species are multivoltine in the tropics than in temperate latitudes; many temperate-centered Anisoptera take 4 to 6 years to complete one generation. Odonate life cycles can be classified as either regulated or unregulated. In the tropics, life cycles are regulated by alternating wet and dry seasons, whereas in temperate zones they are regulated by alternating warm and cold seasons. In regulated types of life cycles, the dry season is usually passed as prereproductive adults or as eggs, whereas the cold season is passed as larvae or as diapausing eggs. Species that occupy ephemeral habitats undergo rapid larval development, but they may or may not be multivoltine, depending on other environmental conditions (combinations of photoperiod and temperature). In continuously available habitats, life cycles are unregulated.

Although a few species of Odonata are known to have terrestrial larvae (e.g., *Megalagrion* in Hawaii), the vast majority require fresh water for functions such as respiration and feeding and to prevent desiccation. Likewise, a few species occupy brackish water habitats (the libellulid *Erythrodiplax berenice* can tolerate truly saline conditions and occupies coastal marshes but not the open ocean).

Respiration is largely through the integument, augmented by an internal rectal chamber in Anisoptera and by external anal and rarely lateral abdominal gills in Zygoptera. Anisoptera larvae can be readily observed breathing: as the abdomen enlarges, water is taken in through the anal opening; contraction of the abdominal muscles forces water out. The rectal epithelium of Anisoptera is developed into a specialized, richly tracheated branchial basket, into which oxygenated water is drawn by pumping action. Different families, and different genera within families, can differ greatly in tolerance of water low in dissolved oxygen. For example, many species of Libellulidae can thrive in low dissolved oxygen levels. In the family Gomphidae, larvae of *Aphylla* bury deep in the soft substrate of lake and pond habitats where oxygen levels are low, whereas species in the genus *Ophiogomphus* lie shallow in sand/gravel substrates typically in swift, highly aerated streams. The latter group appears to have narrow environmental requirements compared with a species such as *Dromogomphus spinosus*, which can occupy fast or slow streams and lakes; it even colonizes newly formed ponds.

Shortly after hatching and throughout their larval life, dragonflies must capture living prey and escape predation, often they are at or near the top of the aquatic food web. Prey includes many kinds of Diptera, but probably all other kinds of aquatic insect are consumed (e.g., mayflies, heteropterans, caddisflies) as well as many other invertebrates (protozoans, oligochetes, crustaceans, mollusks). Some species prey on small vertebrates such as larval fish and amphibians. Prey are detected usually by their movement, either tactually or visually. Motionless prey, such as snails, may be detected visually by recognition of their shape. Tactile detection is more important in earlier instars, vision becoming more keen after the first few molts. Larvae either stalk or ambush their prey. The prehensile, protractile labium (Fig. 4C), unique to the Odonata, strikes within milliseconds to capture prey. The labium grasps the prey and brings it to the mandibles, where it is chewed or engulfed whole. Because odonates are generalist predators, the potential for greatly affecting a particular prey population is low. Predators of dragonfly larvae include fish, frogs, a few reptiles and birds, other odonates, aquatic beetles, and heteropterans. Escape mechanisms include reduction of movement, even feigning death and seeking cover. When a predator grasps a zygopteran by a gill or leg, the appendage may be autotomized and later regenerated. Anisoptera are also capable of leg autotomy, or when grasped by the head or thorax, may make stabbing movements with the abdomen, using the sharp tips of the anal appendages to deter a predator.

In general, habitats range from streams of all sizes to seeps, bogs, ponds, lakes, swamps, and marshes. Some families, such as the Calopterygidae and Cordulegastridae, are restricted to flowing waters. Within the broad habitat categories, various species occupy different microhabitats. For example, elongate, cylindrical larvae, such as aeshnids and coenagrionids, usually cling to stems and sticks, whereas more flattened species, such as gomphids and libellulids, dwell near or in the bottom. A number of specialists occupy phytotelmata, which includes leaf axils (such as provided by bromeliads), rotten holes in tree trunks, bamboo internodes, and depressions at the bases of large trees, mostly in tropical areas. There is great diversity in larval form and behavior that allows dragonflies to occupy different types of habitat. Four categories of larvae (claspers, hiders, sprawlers, and burrowers) are based on how the microhabitat is occupied, although many species could be put in more than one of these categories, and larvae may move from one type of microhabitat to another, depending on age and season. Categorization is based mainly on leg shape and how the legs are used to situate the larva in its resting position.

Claspers, which cling to rocks, stems, or logs, have stout curved legs and cylindrical abdomens. Hiders are less elongate and conceal themselves among dead leaves or other debris by using strong legs; they usually have many stout setae to accumulate mud particles for better concealment. Sprawlers usually are flattened dorsoventrally and lie flat at the water–substrate interface. Burrowers, which may be semicylindrical or flattened, dig into the substrate to hide; the tip of the abdomen is often elongated to protrude above the substrate for respiration. Some burrowers propel themselves through the water by forcing water out of the anus, a form of jet propulsion (e.g., *Progomphus* can move several inches horizontally with one pump). A very few burrowing larvae make an actual burrow (e.g., some Petaluridae). The function of dorsal protuberances and lateral spines on the abdomen, common especially in Anisoptera, is not clear, although it is likely they serve multiple functions such as aiding in concealment and defense against attackers. Experimental evidence indicates that larvae grow longer abdominal spines in the presence of fish predators than larvae raised in the absence of such predators, suggesting that habitat shifts may be responsible for some of the differences observed within and between closely related taxa.

EMERGENCE   Late in the final instar, when the larva is completing development but with a week or more to go before metamorphosing to the adult stage, the wing pads begin to thicken. The final instar usually does not leave the water until the day it is to metamorphose. At this time, the pharate adult is encased in the exuviae of the final instar. The main requirement for the dragonfly at this time is to find proper support structure, as it must cling tightly for metamorphosis to proceed successfully. The emergence support can include any physical object, ranging from the horizontal, terrestrial substrate adjacent to the water body to upright (vertical or inclined) objects such as rocks, plants, or synthetic structures. Shortly after leaving the water and securing a support grip, the dragonfly splits the integument along the middorsal line of the thorax and begins to push the adult thorax through the narrow opening. A slit in the integument of the head then appears, and the head and thorax push out. Shortly afterward, the legs, wings, and anterior portion of the abdomen appear. The dragonfly then remains motionless with legs folded for 10 to 20 min, hanging downward if oriented vertically, or protruding upright if oriented horizontally. When this apparent rest period is over, the legs are extended to grasp the exuviae and the rest of the abdomen is quickly withdrawn. Usually the cloudy wings then expand, followed by lengthening of the abdomen, each process taking about 15 min. As the wings become clear, drops of water are emitted from the anus and the abdomen becomes more slender, slowly taking final adult shape. The wings are suddenly spread out horizontally, and begin to vibrate. The full emergence process lasts from about 30 min to 2 h, ending with the maiden flight.

Some species emerge at night, apparently to escape predation, although many emerge at dawn or in full daylight. Some species have relatively synchronized emergence, all individuals within a population emerging within a day to about 2 weeks of each other. Other species are unsynchronized, emerging throughout the warmer seasons. Periodic exuviae collections can reveal such trends in emergence curves and population size. In nearly all species studied, the ratio of males to females is close to 1:1, although usually a slightly higher percentage of males occurs in Zygoptera, contrasted by a slightly higher percentage of females in Anisoptera.

Adult Behavior

The two main phases of adult life are the prereproductive (or maturation) period and the reproductive period. The prereproductive period lasts from the completion of emergence to the onset of sexual maturity. A brief postreproductive period, after reproductive capacity has passed, is sometimes observable.

PREREPRODUCTIVE PERIOD   Upon reaching safety following the maiden flight, dragonflies remain in a teneral condition for about a day, during which time colors begin to develop and the integument begins to harden. The prereproductive phase then lasts from 2 days to several months, depending on species and environmental conditions; females usually take slightly longer to mature than males. During the prereproductive period, the gonads mature, the thoracic musculature becomes fully developed for agile flight (necessary for attaining a mate), weight increases, and mature colors are attained (for sex and maturity recognition in some species).

REPRODUCTIVE PERIOD   Males and females of nearly all species mate with more than one individual, although monogamy has been reported in the coenagrionid genus *Ischnura*.

Male and Female Encounters   Vision is the main sense used by Odonata for finding mates. The sexes usually meet at or near the aquatic oviposition site. Males that occupy a fixed territory can be categorized either as "fliers," patrolling continuously along the proper habitat, or as "perchers," making short defensive flights from a convenient perch. Territories are maintained by flying at or even clashing with intruders, and pursuit often results in both individuals leaving the habitat temporarily. Resident males usually return quickly to their territory. Mating usually occurs when a female arrives at the water, although males will attempt to acquire females several hundred meters from an oviposition site.

Recognition of Conspecifics   Males recognize females mainly by color, color pattern, body shape, and flight style. Males of most species directly pursue and attempt to grasp any female that comes within sight, and if successful in achieving tandem, will attempt immediately to initiate copulation. In such species, males sometimes take heterospecific females into tandem.

Courtship has been described in a few families (Calopterygidae, Chlorocyphidae, Euphaeidae, Hemiphlebiidae, Platycnemididae, Libellulidae). In some species within these families, some males (but not all) present color and/or posture displays to induce a female to copulate. For example, in *Perithemis tenera* (a small North American libellulid with sexual wing color dimorphism), males establish territories around oviposition sites that consist of some sort of vegetation protruding from the water surface. On detecting a female near his site, a male will fly toward her and follow, moving from side to side. He then turns and flies to the oviposition site. The female may or may not follow him, depending on suitability of the site. If acceptable, she will follow, whereupon the male hovers over the site, fluttering his wings. The female slows her wing beat frequency and may even perch on the site. At this signal, the male initiates tandem linkage and copulation follows. When females are unreceptive to a male's approach, they perform distinctive refusal behaviors. Female Anisoptera usually fly very rapidly or erratically to escape, although some simply hide from males. Female Zygoptera usually show refusal by remaining perched and spreading their wings, sometimes also raising the abdomen or curving down the posterior portion of the abdomen. Males respond to such displays often by leaving, although some still attempt to achieve tandem.

Tandem Linkage and Sperm Transfer   The postures adopted by Odonata in male–female tandem linkage and in copulation are unique. Tandem linkage, a necessary precursor of copulation, is initiated by the male. Males land on and seize females usually in flight (Fig. 5A), although they may grasp perched females also. The male, after landing on the female's pterothorax, brings his abdomen up and forward so that he can grasp the female's head or thorax with his anal

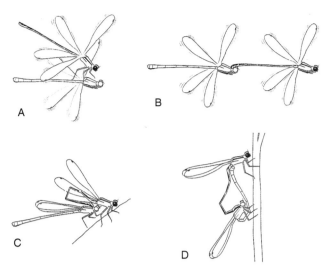

FIGURE 5 Mating sequence of Odonata (male with dark markings, female pale): (A) male grasping female, (B) tandem linkage, (C) intramale sperm translocation, and (D) copulatory wheel.

appendages; he then lets go with his legs and straightens his body, thus having achieved tandem linkage (Fig. 5B). In most Anisoptera, the male appendages fit over the dorsum and rear of the female head, whereas in Zygoptera they fit over the dorsum of the female prothorax. In the anisopteran family Aeshnidae, the male cerci fit tightly on the rear of the female head but also touch the anterior part of the prothorax; in some Zygoptera, the male cerci touch the anterior portion of the pterothorax. Shortly after achieving tandem, the male transfers sperm from the gonopore on abdominal segment 9 to the penis on segment 2 (Fig. 5C); this act is called intramale sperm translocation. Males sometimes translocate sperm before acquiring a female. The tandem pair either copulates in flight (most Anisoptera) or flies to a perch to copulate (Zygoptera and several Anisoptera families). The female swings her abdomen forward from underneath so that her genital aperture engages the venter of segment 2 of the male; the pair is now in the copulation wheel (Fig. 5D). Sperm are then transferred from the male's intromittent organ to the female's sperm storage organ. Copulation is usually extremely brief in flight (3–20 s) but can last from a few minutes to over an hour in perched pairs.

The male intromittent organ of Odonata is designed not only to inject sperm into the female, but also to remove or displace sperm of previous males. Jonathan Waage's discovery of this dual function led to evolutionary understanding of the whole suite of reproductive phenomena, from sexual selection to sperm competition and mate guarding. Sperm displacement may be achieved by removal, repositioning, or dilution. The structure and shape of the penis are vital to the mechanism employed: penes with backwardly directed barbs or hooks remove sperm from the bursa copulatrix, whereas those that are rounded pack sperm. Because most eggs are fertilized with

sperm from the most recent insemination as they pass the female's fertilization pore during oviposition, the last male to copulate with a female is most likely to leave progeny. This phenomenon, termed sperm precedence, explains why males guard females after mating with them. By such postcopulatory association, a male protects his genetic investment by preventing other males from overtaking the female and by inducing the female to deposit (and therefore fertilize with his sperm) most of the eggs she is carrying.

Oviposition When copulation is terminated, pairs may break tandem linkage or remain together. Oviposition usually takes place shortly after copulation regardless of whether the tandem linkage is maintained, although females do lay eggs at times when males are not at oviposition sites. Ovipositing late in the day is a fairly common way for females to avoid interference from males. Males of many species guard females with which they mate, either by maintaining tandem contact with or by remaining near them, or both. Attempts to guard females are not always successful, as intruding males sometimes grasp and copulate with guarded females. Typically, guarded females oviposit more rapidly than those unguarded; their fitness thereby is enhanced.

Contact guarding Males of many Zygoptera remain in tandem with their mated female while she oviposits, even when she submerges. Males of many Coenagrionidae project vertically in the air with legs and wings folded, their only support provided by the grip of their anal appendages on the female thorax. In many Libellulidae, male and female fly in tandem low over the oviposition site, the male lowering his abdomen to cause the end of the female's abdomen to dip into the water and release eggs.

Noncontact guarding When males do not maintain females in tandem, they guard females by flying or perching nearby and warding off any intruding males. Males display toward and chase intruding males; male-to-male body and wing clashes may ensue. Some males guard multiple females. When male density is high, intensity of guarding ovipositing females probably increases in most species, but it has been reported to decrease in at least one species. In many Zygoptera, the male takes the female in tandem to the oviposition site, then releases her and either does or does not guard her. In some Trameinae, a subfamily of Libellulidae, the male releases the female as they fly in tandem over the oviposition site; she drops down to the water surface and releases a few eggs, then she flies back upward and the male takes her back into tandem. These latter two behaviors illustrate combinations of contact and noncontact guarding.

FORAGING Odonata feed on living prey throughout their adult life. When foraging, dragonflies can be categorized as "perchers" or "fliers." Perchers spend much of their time stationary, making short flights from perches to capture prey and then perching to consume it. In contrast, fliers are on the wing for a large part of their feeding activity, capturing their prey in the air and swallowing small prey while in flight; they do, however, perch to consume larger prey. The "flier" mode requires much more energy, but fliers are more opportunistic feeders, able to forage later in the day. Perchers typically capture most of their prey during midday. This dichotomy of foraging styles results because fliers generate more body heat than perchers and can therefore remain active at lower air temperatures.

The major stimulus for detecting prey is movement. Odonata have very large eyes with many ommatidia, a specialization for detection of movement. However, a few species take stationary prey, apparently recognizing the prey by its shape. Prey such as small flying insects may be captured directly with the mouthparts, but the legs are also used for subduing certain types of prey. Odonata are typically generalists with few exceptions. Diptera, especially mosquitoes and midges, are a major component of the adult diet. One analysis found that Chironomidae constituted a significantly higher proportion of the gut contents than did Culicidae, probably reflecting differences in the flying and perching habits of midges and mosquitoes. Some species take mainly large prey, such as Lepidoptera and Odonata. For example, the large North American gomphid *Hagenius brevistylus* often has been observed feeding on other Anisoptera and has been dubbed the "dragonhunter." Members of the Neotropical family Pseudostigmatidae are specialist feeders. They glean small spiders in the rain forest by vertically searching trees, hovering near webs found on leaf tips, then flying directly up to the webs and snatching the spiders from their perches.

THERMOREGULATION   Although insects are basically ectothermic, large species are able to generate body heat or adopt body positions to absorb sunlight and are able to maintain this heat gain via certain behavioral mechanisms. Odonata are among those with ecto- and endothermic thermoregulatory capabilities. The two basic behavioral styles, fliers and perchers, use different strategies to prolong activity under less than optimal ambient temperatures. Under cool conditions, fliers warm the thorax by wing-whirring (endothermy), whereas perchers expose as much of the surface area as possible to solar radiation. In some species of perchers, hairs on the thorax serve as insulators, or the wings may be deflected downward to insulate the thorax. Such species are among the first to appear in the spring at higher latitudes. Under very warm or hot conditions, perchers remain stationary longer and posture their bodies to absorb less solar radiation. A common posture is the obelisk position, in which the abdomen is raised to expose the minimum surface area to the sun and the wings are lowered to reflect sunlight away from the thorax. Fliers generally become inactive during midday and hang up in the shade. However, some species of Libellulidae glide, and some species of Aeshnidae are able to continue flying by shunting warm blood from the thorax to the abdomen, where excess heat is dissipated. By prolonging activity at the breeding site, dragonflies increase their chances of obtaining mates, feeding, and escaping predation.

Dispersal

Most flight involves small-scale, intrahabitat movements for immediate needs (feeding, finding mates, and escaping predators) that directly affect individual survival and reproductive success. Such flights usually result in dispersal distances up to a few hundred meters. Large-scale flight resulting in interhabitat displacement is regarded as migratory flight. Corbet defined migration as "spatial displacement that typically entails part or all of a population leaving the habitat where emergence took place and moving to a new habitat in which reproduction ensues." These dispersal movements also have consequences for survival and reproductive success.

In examples of migration thus far elucidated for tropical species, migration is a means of overcoming drought in the area where the species developed. For example, in ephemeral lentic habitats in Africa, the aeshnid *Hemianax ephippiger* develops rapidly (within 60–90 days) and upon emergence flies with rain-developing systems several hundred kilometers to areas that will receive the rainfall, as far north as Europe. There they feed, mate, and lay eggs in newly filled water bodies. Temperate species migrate to circumvent cold temperatures. For example, the wide-ranging aeshnid *Anax junius* emerges early in the year in southern North America, and arrives in the northern United States and southern Canada during warm periods as early as March and April. These immigrants mate and lay eggs in shallow lentic habitats; their progeny complete development in late summer or fall. The second generation then flies south; large numbers of "green darners" have been observed flying overhead as late as mid-November. Migratory flights may be made up of several species, usually from the families Aeshnidae and Libellulidae. Very few Zygoptera are known to migrate, and all recorded so far are in the subfamily Ischnurinae (Coenagrionidae).

ODONATA SURVIVAL IN A CHANGING WORLD

Habitat creation, loss, and alteration are the major causes of odonate population changes. Some odonate species have increased their geographic ranges and population numbers in response to man-made changes in habitat. For example, in the United States many pond-dwelling Libellulidae that were historically centered in the eastern part of the country have moved far west of the Mississippi River with advent of irrigation. Furthermore, exotic species can immigrate when gravid females ride tropical storms or can be introduced as eggs and larvae with the aquarium trade, their ranges thereby increasing dramatically. Alternately, many riverine and wetland species have undoubtedly declined because of habitat degradation and drainage changes. For example, some riverine Gomphidae are extremely rare, but could be protected by conserving the remaining habitat (e.g., in the eastern United States, *Ophiogomphus edmundo* is known from three localities, and *Gomphus sandrius* is known from seven localities). Only one species in the United States has federal protection status as a threatened and endangered species, Hine's emerald *(Somatochlora hineana)*. Many species in this genus are locally distributed, inhabiting lakes and bogs in different stages of succession, and their populations depend greatly on the availability of the proper microhabitat.

There is awareness of the threat to dragonfly diversity and populations of sensitive species in most countries, especially in Europe and Japan, where protection efforts are designed to heal or prevent damaged ecosystems. In tropical areas, however, where diversity is highest and incompletely known, habitat destruction continues at alarming rates. There has been some effort toward habitat conservation, as national parks and preserves have been established in many tropical countries. For example, in Thailand, almost all remaining forest areas are protected by parks, wildlife sanctuaries, and a ban on logging; this effort amounts to nearly 15% of the total land area. In other countries, the situation is less promising, and even preserves afford no insurance against habitat alteration. Odonata have existed for many millions of years, undoubtedly surviving small and massive extinction episodes; however, the magnitude of present-day environmental change may be without parallel.

See Also the Following Articles
Aquatic Habitats • Fossil Record • Mating Behaviors • Sexual Selection • Thermoregulation

Further Reading

Alcock, J. (1994). Postinsemination associations between males and females in insects: The mateguarding hypothesis. *Annu. Rev. Entomol.* **39**, 1–21.

Arnquist, G., and Johansson, F. (1998). Ontogenetic reaction norms of predator-induced defensive morphology in dragonfly larvae. *Ecology* **79**, 1847–1858.

Corbet, P. S. (1980). Biology of Odonata. *Annu. Rev. Entomol.* **25**, 189–217.

Corbet, P. S. (1999). "Dragonflies: Behavior and Ecology of Odonata." Cornell University Press, Ithaca, NY.

Davies, D. A. L., and Tobin, P. (1984). "The Dragonflies of the World: A Systematic List of the Extant Species of Odonata," Vol. 1, "Zygoptera, Anisozygoptera." Societas Internationalis Odonatologica Rapid Communications (Suppl.) No. 3, Utrecht.

Davies, D. A. L., and Tobin, P. (1985). "The Dragonflies of the World: A Systematic List of Extant Species of Odonata." Vol. 2, "Anisoptera." Societos Internationalis Odonatologica Rapid Communications (suppl.) No. 5, Utrecht.

Dunkle, Sidney W. (2000). "Dragonflies Through Binoculars: A Field Guide to Dragonflies of North America." Oxford University Press, New York.

Fincke, O. M. (1987). Female monogamy in the damselfly *Ischnura verticalis* Say (Zygoptera: Coenagrionidae). *Odonatologica* **16**, 129–143.

Jacobs, M. E. (1955). Studies on territorialism and sexual selection in dragonflies. *Ecology* **36**, 566–586.

Schorr, M., Lindeboom, M., and Paulson, D. (2000). List of Odonata of the world. Available at http//www.ups.edu/biology/museum/worldodonates.html

Tennessen, K. J. (1997). Aquatic insect resource management. *In* "Aquatic Fauna in Peril: The Southeastern Perspective" (G. W. Benz and D. E. Collins, eds.). Special Publication 1, Southeast Aquatic Research Institute, Lenz Design & Communications, Decatur, GA.

Tillyard, R. J. (1917). "The Biology of Dragonflies (Odonata or Paraneuroptera)." Cambridge University Press, Oxford, U.K.

Waage, J. K. (1979). Dual function of the damselfly penis: Sperm removal and transfer. *Science* **203**, 916–918.

Wootton, R. J. (1988). The historical ecology of aquatic insects: An overview. *Palaeogeogr. Palaeoclimatol. Palaeoecol.* **62**, 477–492.

Orientation

Ring T. Cardé
University of California, Riverside

Orientation refers to the way in which organisms direct their course of movement—not to their body or positional orientation per se, although clearly, if an organism is to aim its course, it must align its body's long axis with its intended track. Although orientation maneuvers thus involve an organism's orientation of body position, this article is concerned mainly with movements in location that range in scale from millimeters to thousands of kilometers. For example, a minute parasitoid wasp may walk to a resource, such as a prospective host, only millimeters away. On the other hand, movements of insects also can cover considerable distances, such as the long-distance migration of several thousands of kilometers that is undertaken in autumn by monarch butterflies *(Danaus plexippus)* flying from northeastern North America to their overwintering site in central Mexico. The mechanisms that insects and other organisms use to accomplish such feats are enormously variable. Moment-to-moment steering typically relies on simultaneous inputs from multiple sensory modalities, such as chemical cues, light, and wind. Most organisms use some stored information about very recent encounters with these cues and the organism's past position. In many parasitic and social Hymenoptera, learned information, including spatial maps and landmarks, plays a crucial role in these insects "knowing" either where they have been or their destination.

CLASSIFICATION OF ORIENTATION MANEUVERS

The modern system of categorizing orientation mechanisms by their forms of locomotion and their presumed sensory inputs dates to Fraenkel and Gunn's *The Orientation of Animals,* first published in 1940. These authors' classification of maneuvers relies on two distinctive patterns of movement. The first kind of maneuver is termed a kinesis (pl. kineses); it is defined as an undirected response in which the body's long axis exhibits no consistent relationship to the direction of the stimulus and the direction of locomotion is random. If a gradient of stimulus intensity regulates either the frequency of turns or the amount of turning per unit of time, the reaction is termed a klinokinesis. If the gradient of stimulus intensity regulates either speed or the frequency of locomotion, then the reaction is termed an orthokinesis. Both kinds of kinesis require a minimum of one sensory detector to monitor stimulus intensity, although multiple detector systems (such as paired antennae) are the norm. Randomly directed movement would seem to be an ineffective means for moving toward or away from a stimulus. Klinokinetic maneuvers can facilitate movement either toward or away from a stimulus gradient as illustrated in Fig. 1 where simply turning more frequently in bright light

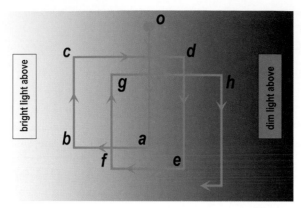

FIGURE 1 Track that could be expected from a hypothetical animal that always changes its direction by 90° to the right at a rate that is dependent on light intensity. The animal starts at point *O*. Turns occur more frequently when the animal is in bright light. Because of this tendency, the path *cd* is longer than *ab*. Although this is a simplistic representation of a klinokinetic reaction (in reality, turns will vary in angle and may be random to the right or left), it nonetheless demonstrates how klinokinesis can result in an animal orienting along a shallow stimulus gradient. [Modified from Fraenckel, F., and Gunn, L. L. (1940).]

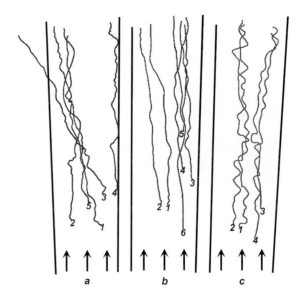

FIGURE 2 Courses (viewed from above) of blowfly *(Lucilia)* larvae crawling away from a light source (arrows depict direction of horizontal rays of light). The three individual larvae recorded are denoted "a" through "c" in different colors, and the tracks taken in repeated trials are indicated by numbers. Individuals in "a" and "b" seemed to deviate to the left in some trials and to the right in others. The track of larva "c" is represented in more detail than in the other tracks, showing alternating right and left movements of the head. Although J. Loeb stated that larvae "move as though they were impaled on a ray of light which passed through their medial plane," the larvae clearly have some variability in their paths. Based on the detailed head movements of the track of larva "c," larvae seem to orient by transverse klinokinesis. [Modified from Mast, S. O. (1911). "Light and the Behavior of Organisms." J. Wiley, New York.]

inevitably results in motion by the organism away from the bright light.

The second kind of maneuver is called a taxis (pl. taxes); it is defined as a directed reaction in which the organism's long body axis is aligned with the stimulus and movement is more or less directed toward or away from the stimulus. In klinotaxis the organism has available two strategies for sampling the intensity of the stimulus. In transverse klinotaxis the sampling occurs by moving the entire body or a part of it from side to side along the path. Alternatively, in longitudinal klinotaxis the organism samples intensity successively along its path. Both forms of klinotaxis require only a single detector capable of measuring stimulus intensity. A classic transverse klinotactic reaction is the movement of blow fly larvae *(Lucilia)* away from light (Fig. 2). A close examination of the movement of larvae along their path reveals that although their tracks are nearly straight, their heads move from side to side. A similar pattern is seen in ants following a pheromone trail (Fig. 3).

Tropotaxis, in contrast, relies on a paired detector system (such as the antennae); by balancing the stimulus intensity on the two sides of the organism, the heading can be aligned with a relatively steep stimulus gradient. Honey bee workers *(Apis)*, for example, can center their body's long axis between balanced inputs of odor to each antenna. Telotaxis is a "direct" form of orientation in which the stimulus intensity is processed by a receptor system that has an array of directionally sensitive receptors, so that setting of a course toward a stimulus involves the relatively simple navigational task of holding a certain part of the receptor array in alignment with the stimulus. Sometimes termed "goal orientation," telotaxis is known only for orientation along a beam of light.

In klino-, tropo-, and telotactic reactions, the organism's long body axis is aligned with the direction of the stimulus, such as a beam of light or a fairly steep gradient of odor. In menotaxis, orientation of the long body axis is at a fixed angle to the stimulus, and course setting is maintained by stimulation of the sensory apparatus in a manner similar to telotaxis. Menotaxis is commonly called "compass orientation," and some still unresolved form of it is used by monarch butterflies to head toward their wintering grounds, and by honey bees in their dance language. A menotactic reaction also seems to be responsible for the attraction to lights seen in many moths and other nocturnal insects. The assumption is that moths follow a straightened-out path at night by using celestial cues as a menotactic guide. When they encounter an artificial point source of light, they attempt to maintain the same angle with respect to the fixed point as in menotaxis, but in so doing they inevitably spiral toward the light source. Such a spiral path is indeed seen in the approach of some insects to a light.

CLASSIFICATION OF SENSORY INPUTS

In describing how organisms orient, it is common to create terms that combine the kinds of environmental cue used in orientation with the form of taxis or kinesis. Common prefixes used include anemo (wind), chemo (odor or taste), mechano (pressure), phono (sound), photo (light), rheo (water flow) and scoto (darkness). The *Lucilia* maggot moving away from light

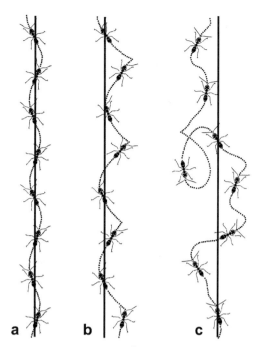

FIGURE 3 Trail following in worker ants *(Lasius fuliginonus)* in relatively still air. The straight red line marks the centerline of an odor trail. The dotted blue line denotes the path of the ant. The ant in "a" swings right and left across the trail, presumably bringing one antenna to an area of discernibly lower concentration before turning in the opposite direction. In "b," the ant's left antenna has been removed and she overcorrects her course to the right. In "c," the antennae were crossed and then glued, but the ant is able to orient along the trail, albeit with difficulty, aided partially by a light compass reaction. [Modified from Hangartner, W. (1967). *Z. Vgl. Physiol.* **57,** 103–136.]

ALTERNATIVE CLASSIFICATIONS

"Attraction" and "aggregation" are often used to categorize orientation, but these terms describe end points of orientation and tell us little about the preceding maneuvers. Attraction and aggregation nonetheless remain widely used to describe the consequences of taxes and kineses. Pheromones, for example, are often labeled as attraction pheromones or aggregation pheromones.

In another approach to classifying orientation mechanisms, Jander emphasized in 1975 the distinction between the two broad categories of information used in orientation: information that is based on immediate sensory processing (for extrinsic or exokinetic orientation) and information that is stored centrally (for intrinsic or endokinetic orientation). Information that is stored may be subdivided into memory and that which is genetically determined. Jander also stressed the importance of ecology in studying orientation, and so his other categories included positional orientation (either staying in place or exhibiting locomotion), object orientation (movements with respect to the spatial position of either resources or sources of stress), topographic or home-range orientation (learned spatial orientation), and geographic orientation (migration over considerable distances).

Bell disavowed the time-tested system of taxes and kineses in his 1991 synthesis of foraging behavior. In analyzing the vast literature on foraging movements, Bell advocated describing the kinds of locomotory paths that were observed, the kinds of information available to mediate the motor output, and the presumed guidance system. Bell eschewed terming any of these maneuvers "taxis" or "kinesis." Despite such attempts to devise a new terminology, however, the system of taxes and kineses is likely to remain prevalent for some time because no clear alternative has emerged.

can be said, for example, to be a navigating by a negative transverse photoklinokinesis. It is obvious that the seeming precision of this classification scheme makes for an unwieldy terminology. This deficiency was noted in 1984 by Bell and Cardé, who highlighted the need for "more practical, functionally related terms" that are not "teleological, poorly defined, non-probabilistic and difficult to spell." A related problem is that these terms tend to define the reactions so precisely that they can dictate the boundaries of experimental investigations, such that these may either neglect the integration of mechanisms or fail to consider mechanisms that do not fall within these constructs.

These classifications also neglect the role of internally stored information. The system of taxes and kineses assumes that an animal steers its path entirely in reference to the position of the external stimuli. However, the maneuvers can involve as well some self-steering that is governed by stored information about the animal's previous path. Such information can be classified as either idiothetic (i.e., information that is internally stored) or allothetic (i.e., information that is external, such as visual features of the environment). Kineses are, for example, clearly self-steered, whereas transverse and longitudinal klinokineses are partially self-steered, and menotaxis is not self-steered.

ODOR-INDUCED OPTOMOTOR ANEMOTAXIS

Among the best-studied orientation systems are those that enable organisms to locate upwind resources by flying along a plume of odor to the odor's source. Examples of such maneuvers include male moths flying over distances of hundreds and perhaps thousands of meters to a pheromone-releasing female, tsetse flies and mosquitoes flying over tens and perhaps hundreds of meters to a prospective vertebrate host, and parasitoid wasps flying over several meters to their intended host. All these reactions are mediated by odors that are released by the resource and form an odor plume as they are carried downwind. It is crucial to note that maneuvers cannot be governed by orienting to a gradient of odor. A gradient that would be sufficiently steep for such directional information exists only within a meter or less of the odor's source. Instead, insects and other organisms orient by advancing upwind when they encounter an above-threshold concentration of the odor linked to the resource.

The nonintuitive mechanism permitting in-flight anemotaxis is the optomotor response. An organism immersed in air

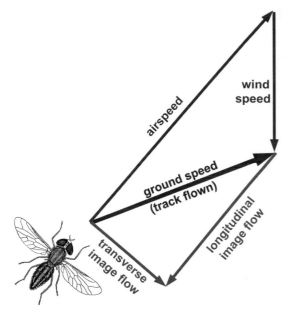

FIGURE 4 The relationship between a flying insect's body heading and the track taken when flying at an angle to the wind. The vectors depict the wind direction and velocity and the fly's direction and velocity along the track. The image flow the fly sees directly below has both longitudinal and transverse components, and therefore flows obliquely across the fly's eyes. When the fly heads directly upwind, the image flow is longitudinal, i.e., front to rear.

or water cannot discern the direction of the flow of these fluids by mechanosensory input, although it can use mechanosensory information to gauge its airspeed or water speed (i.e., its movement relative to the fluid flow). Instead, it detects its displacement relative to its ground position by literally seeing how the flow alters its path. For example, if an organism's long body axis is aligned directly with the fluid flow and the organism is making progress along the plume, then the image directly below or above the organism flows from front to rear. If the organism is moving at an angle to the fluid flow, then the image flows obliquely across the eyes (Fig. 4).

Optomotor anemotaxis was first demonstrated experimentally in 1939 by John Kennedy, working in England with the yellow fever mosquito, *Aedes aegypti*. Kennedy's ingenious wind tunnel used a movable pattern, projected onto the tunnel's floor, to manipulate the visual feedback a flying mosquito would experience. He was able to show that the upwind flight of females induced by carbon dioxide (the activating ingredient in human breath) was governed by the optomotor reaction. When the projected floor pattern was moved in the same direction as the airflow, mosquitoes decreased their airspeed, apparently perceiving by visual feedback from below that their airspeed had increased; conversely, when the image flow was reversed to the opposite direction, mosquitoes immediately increased their airspeed. Mosquitoes regulate their airspeed to maintain a relatively constant rate of longitudinal image flow. These simple manipulations verified that the mosquitoes' perception of movement relative to their visual surroundings dictates their airspeed, rather than some

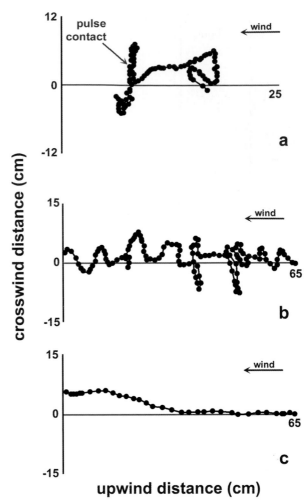

FIGURE 5 Top view of flight tracks of males of the moth *Cadra cautella* flying toward a source of a sex pheromone. The moth is traveling from left to right. The dots represent the moth's position every 1/30th of a second. The blue line shows the time-averaged centerline of the plume. Track "a" shows the path of a moth after intercepting a pulse of pheromone. [Redrawn from Mafra-Neto and Cardé (1994), *Nature* **369**, 142–144.] About 200 ms after intercepting a pulse, the male surges upwind. Track "b" shows a male flying along a very narrow ribbon plume of pheromone, sporadically encountering puffs of pheromone. Track "c" shows a male flying along a wide turbulent plume of pheromone, encountering many filaments of pheromone per second. [Redrawn from Mafra-Neto and Cardé (1995), *Physiological Entomology* **20**, 117–133.]

form of mechanosensory feedback.

The task an insect faces in finding an upwind source of odor, however, is much more complicated than simply flying upwind when an appropriate odor is encountered. Because turbulent forces cause the plume to meander and undulate, the direction of the plume's long axis is not always aligned with the upwind direction. Thus, an insect proceeding upwind often exits the plume. Thus many insects "lose" the plume and then "cast"; that is, they stop moving upwind and instead move back and forth lateral to the direction of wind flow. If they recontact the plume during such to-and-fro maneuvers, upwind flight may resume. A further difficulty is that turbulent forces fragment the plume's internal structure. Plumes therefore have patchy distributions of odor, so that even within the plume's overall boundaries, insects encounter filaments of odor interspersed with gaps of clean air. Filaments can be encountered many times a second; for moths, whenever the encounters with filaments of odor are frequent, say above 10 Hz, flight can be aimed rapidly upwind, but when the encounter rate falls below 5 Hz, the flight can exhibit a much more substantial crosswind component (Fig. 5).

Odor-induced, optomotor anemotaxis as used by flying insects exemplifies several points common to orientation mechanisms. Several kinds of input (here visual, mechanosensory, olfactory) and self-steering all contribute to course setting and motor output. It is the integration of information that allows organisms to set their course.

Several related situations illustrate the variations on this theme. A flying insect also can orient along a plume of odor for moderate distances by flying a course that is set upon takeoff. After detecting the odor, the insect gauges upwind direction by mechanoreceptors and takes off due upwind. This is called an "aim-and-shoot" reaction, and the straight-line course is maintained by using visual cues perceived ahead to set the course. Flight continues as long as the insect remains in the plume. Tsetse flies (*Glossina* spp.) are thought sometimes to use aim-and-shoot upon takeoff, but other observations support a conventional optomotor anemotaxis maneuver during flight. Tsetse flies may shuttle between these two orientation strategies. Walking insects use a nonoptomotor version of anemotaxis. Upon detection of odor, the insect simply heads upwind, using mechanosensory input to determine wind direction.

CONCLUSION

Taxes and kineses remain the principal organizing system for understanding and investigating how insects and other organisms "know where to go." Discovering how these maneuvers work—what information is extracted, how it is processed, and the nature of guidance systems—remains an active area of inquiry.

See Also the Following Articles

Dance Language • Host Seeking • Magnetic Sense • Migration • Monarchs • Mosquitoes • Pheromones • Tsetse Fly

Further Reading

Bell, W. J. (1991). "Searching Behaviour: The Behavioural Ecology of Finding Resources." Chapman & Hall, London.
Bell, W. J., and Tobin, T. R. (1982). Chemo-orientation. *Biol. Rev. Cambr. Philos. Soc.* **57**, 219–260.
Bell, W. J., and Cardé, R. T. (eds.). (1984). "Chemical Ecology of Insects." Chapman & Hall, London.
Cardé, R. T. (1996). Odour plumes and odour-mediated flight in insects. *In:* "Olfaction in Mosquito–Host Interactions." Ciba Foundation. Symposium. 200, pp. 54–70. Wiley, Chichester, U.K.
Fraenckel, G. S., and Gunn, D. L. (1961). "The Orientation of Animals. Kineses, Taxes and Compass Reactions," 2nd ed. Dover, New York.
Jander, R. (1975). Ecological aspects of spatial orientation. *Annu. Rev. Ecol. Syst.* **6**, 171–188.
Kennedy, J. S. (1986). Some current issues in orientation to odour sources. *In:* "Mechanisms in Insect Olfaction" (T. L. Payne, M. C. Birch, and C. E. J. Kennedy, eds.), pp. 00–00. Clarendon Press, Oxford, U.K.
Schöne, H. (1984). "Spatial Orientation. The Spatial Control of Behavior in Animals and Man." Princeton Univresity Press, Princeton, NJ.

Orthoptera
(Grasshoppers, Locusts, Katydids, Crickets)

D. C. F. Rentz
California Academy of Sciences

You Ning Su
Australian National University

Orthoptera is considered here in the restricted sense; that is, we are not including the cockroaches, mantids, and stick insects, and these orders are covered elsewhere. The Orthoptera include terrestrial insects commonly known as short-horned grasshoppers, katydids, bush crickets, crickets, and locusts. The adult size range is from a few millimeters to some of the largest living insects, with bodies over 11.5 cm in length and wing spans more than 22 cm. Orthopterans occur all over the world except in the coldest parts of the earth's surface. They are best developed in the tropics, especially the New World tropics. In terms of numbers they are among the most common insects and are an important component of the fauna in most parts of the world. The order is readily identified by the characteristic hind legs that are developed for jumping. Summer nights are often dominated by the songs of many species in several families. Locusts are among the world's most important economic insects, and many species cause devastation in many parts of the world. Orthopterans are mentioned in biblical writings and in the earliest of Chinese literature. In recent times they have been important elements in the development of several fields of biology. Biological lifestyles in the Orthoptera include phytophilous (leaf-living), geophilous (living on and in the ground), cavernicolous (living in caves), and myrmecophilous (living with ants). Species can be diurnal and nocturnal. More than 20,000 species are known, but it is estimated that this figure may double when a thorough census has been made of uncollected regions of the globe.

CLASSIFICATION

There are several disparate classifications of orthopteroid insects that are used simultaneously, depending on preference. There has been an overall escalation of rank of categories in recent years above the tribal level. One of the extremes of these views was expressed by Dirsh, who created 10 orders from what was

TABLE I   Some Characters Used to Separate the Two Suborders of the Order Orthoptera

Character	Ensifera	Caelifera
Antenna	More than 30 segments	Less than 30 segments
Auditory structure (when present)	On foretibia	On first abdominal tergite
Alary stridutatory structures (when present)	Forewings specialzed with file and scraper	Forewings modified laterally and ventrally
Ovipositor	Elongate, sword like or sickle-shaped	Short and stub-like
Molting	Skin usually eaten	Skin never eaten

traditionally considered one! Kevan provides a synthesis of the classifications to 1982. The advent of cladistics and molecular phylogeny has spawned additional changes to orthopteroid classification, and these are noted in the discussions of the respective groups. In general, except where noted, the conservative approach will be adopted here, since this is a period of flux and there is no consensus on which classification to adopt. The Orthoptera usually are divided into two suborders: the Ensifera (long-horned Orthoptera) (Table I) and the Caelifera (short-horned Orthoptera). The ensiferans are considered to be the more ancient group, with fossils dating from the

Carboniferous, whereas caeliferans are known only from as far back as the Triassic (Fig. 1A).

FEATURES OF THE ORDER

General Comments

Orthoptera have been popular subjects for the behaviorist. Much has been done of an interdisciplinary nature relating to natural and sexual selection, signaling behavior, acoustic and vibrational communication, and displays.

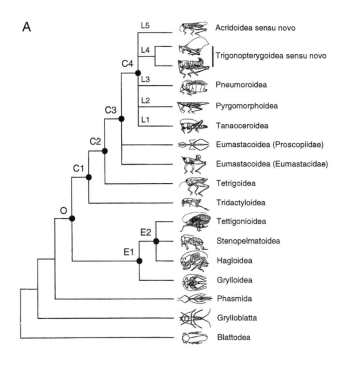

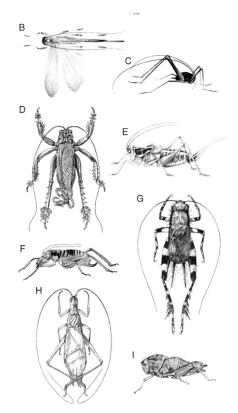

FIGURE 1 (A) Proposed phylogeny of the orthopteroid insects based on molecular studies. Note the arrangement of the Caelifera (C1). [From Flook, P. K., *et al.* (1999). *Syst. Biol.* **48,** Fig. 2, with permission of P. K. Flook.] (B) *Xanthogryllacris punctipennis,* an example of a gryllacridid with patterned wings. (C) *Hadenoecus puteanus,* a camel cricket illustrating the typical humpbacked appearance. [B and C modified from "Genera Insectorum." (1937). Vol. 206.] (D) *Schizodactylus monstrosus,* an unusual orthopteroid. [Modified from Karny, H. H. (1929). *Lignan Sci. J.*] (E) *Apote notabilis,* a large North American tettigoniine. [From "Genera Insectorum." (1908). Vol. 72.] (F) *Henicus prodigiosus,* a wingless South African henicine. [Modified from "Genera Insectorum." (1937). Vol. 206.] (G) *Dianemobius fascipes.* [From Chopard, L. (1969). "Fauna of India and the Adjacent Countries."] (H) *Oecanthus pellucens,* a typical oecanthine, male. [From Chopard, L. (1943). "Orthoptèroides de L'Afrique du Nord." Paris.] (I) *Stolliana sabulosa,* female wingless species. [From "Genera Insectorum." (1916). Vol. 170.]

Many orthopterans have excellent sight and hearing and are wary and difficult to approach. Others appear sluggish and rely on a number of cryptic strategies and distasteful properties for protection. For example, if seized, most species will kick out with their spiny hind legs and regurgitate acrid contents of the crop. Because orthopterans are fed upon by a vast number of vertebrates and many other insects, such behavior is not entirely successful. If a grasshopper, cricket, or katydid is seized by the hind leg, the leg will generally be lost at a point between the femur and trochanter. A number of unrelated species will squirt or discharge repugnatorial secretions from intersegmental glands.

The majority of orthopteran species are phytophagous, feeding on the foliage of higher plants. A number feed on roots and others on fungi. Many species are predaceous, while others are omnivorous. Relatively few species have been studied for the purpose of discerning their feeding activities, but some are very highly specialized, feeding only on seeds, pollen, or nectar or flowers of certain plant types. The fore and middle legs are used by some predators to form a "clap-trap" to catch unwary insects.

Orthopterans are renowned as examples of forms of crypsis. They span the range resembling leaves, twigs, bark, stones, or flowers. These appearances are meant to deceive vertebrate predators. To achieve this, various parts of the body are modified, and these combine with camouflaging colors and patterns, accompanied by the appropriate behavior. Mimicry of other insects abounds in the orthopteroids. Mimicry in the nymphal stages is often based on very different models from the adults. In a few examples, males have a different mimic model from females. The distasteful properties involved with aposematic coloration are exhibited in many orthopterans in a wide range of families and genera. The normally associated behavioral traits of sluggish behavior, conspicuous situations, and gregariousness are also seen. Flash or frightening colors are exhibited in a number of species. Several unrelated species in different parts of the world bury themselves in sand overnight or for short periods during the day when danger threatens. Many species overwinter in cracks or under leaves in leaf litter. A number of katydids and grasshoppers are semiaquatic. Some swim or skate over the surface film, and even nonaquatic species can swim freely in midwater. These species have subtle structural adaptations for aquatic life.

The majority of orthopterans are cryptozoic, lacking bright colors and patterns. They rely on pale, drab, and dull colors to conceal their presence. Cavernicolous species have pigmentation associated with their particular lifestyle in their habitat. The more highly specialized types lack pigmentation and have thin integument, as well as extraordinarily long antennae and long, delicate appendages. Extremes in structural modification are types that remain larviform as adults, having reduced eyes and antennae and nonjumping hind legs, but powerful burrowing adaptations. Some small crickets are dorsoventrally flattened and wingless and are myrmecophilous.

Reproduction

A diversity of attitudes is adopted in courtship and copulation. Sperm are transferred mostly in spermatophores, and much has been written regarding male investment, female selection of mates, and sperm competition. In the Ensifera, the spermatophore is transferred at copulation, and the vesicle usually remains attached externally to the vulva; it is eaten by the female and serves an important nutrient function in the development of the fertilized embryos. In the Acridoidea several small spermatophores may be inserted into the female tract, or the tubular part of the single spermatophore may penetrate the spermathecal duct while the vesicle remains in the phallus. Some other grasshoppers (the Australian endemic Morabinae of the Eumastacidae) produce no spermatophore, and their sperm is delivered directly to the bursa copulatrix, not the spermathecal duct.

Orthoptera eggs are most often laid in soil, but many other media are used. Many Ensifera insert them singly or in small numbers, into stems, leaves, or roots, or cement them to stems, twigs, or bark. One Australian species coats the exposed egg laid on a twig with soil particles, apparently to protect the egg from the elements and possibly from parasites. Acridoidea lay their eggs in oothecae or pods, in groups of more than 10 to 200, in pithy stems, in soft portions of dead timber, at the bases of grass tufts, or in animal dung; a number of species cement eggs to water plants below the surface. Burrowing forms deposit eggs in special chambers, lay them in the sides of chambers, or place the eggs around roots.

The ovipositor is often highly modified. In the Ensifera it is mostly a needlelike spear that is inserted into the substrate, usually the soil, wood, or other plant material. Many species lay specially hardened eggs that are glued to twigs or leaves. Many select a thick leaf, and with alternating penetrations of a highly modified, laterally compressed ovipositor, insert a disk-like egg into the leaf's edge. In Caelifera, the ovipositor is relatively similar among the wide range of species. It is short and "pronged," and it penetrates the substrate by opening, closing, and extending the abdomen. In the Acridoidea, the abdomen is frequently stretched to twice its normal length during oviposition.

Development

A newly hatched immature, or nymph, is enclosed in the embryonic cuticle and is called a pronymph or vermiform larva. At hatching the eggshell is fractured by pulsations of an extrusible structure known as the cervical ampulla, which is part of the dorsal membrane of the neck. This work is assisted by the cutting action of a ridge or row of teeth positioned at the front of the head. The ampulla is also involved in the emergence of the pronymph from the egg repository in the soil or plant tissue and in the intermediate molt by which the embryonic cuticle is cast when the insect is freed.

Nymphs differ from adults outwardly mainly in the reduced development of wings and reproductive organs, the less developed pronotum, and a reduced number of antennal segments. Nymphs undergo several molts, usually fewer in males than females. The intermediate stages of the developmental process are called instars. Molting occurs between instars and usually takes place at night under the protection of darkness or early morning, when the relative humidity is favorable for such processes. The wing buds and terminal reproductive structures increase with each molt. Antennal segments are added at molting. Many entomologists have great difficulty in attempting to determine whether a short-winged grasshopper is a nymph of a mature individual of a species. In actuality, it is a relatively easy determination but requires a little experience: grasshoppers have two pairs of wings; the first pair is the tegmina, and the second pair the wings. In immature grasshoppers, the wings are lateral with the costal margin positioned ventral, as in the adult, but in usually the penultimate molt, they rotate and assume a reversed, more dorsal position in which the costal margin is dorsal and the morphological ventral surface is external with the hind wing overlapping the tegmen.

REVIEW OF TAXA

Suborder Ensifera

Crickets and katydids (bush crickets, long-horned grasshoppers, and relatives) comprise the suborder Ensifera. This is the classical division of the order, but in the system using Grylloptera as a higher taxon, this group is treated as an order. Ensiferans share important characters with members of the other suborder, the caeliferans, such as the biting–chewing mouthparts, the modified pronotum, "leaping" hind legs, the similarities in wing venation and shape, and the sound-producing capacities (stridulation) of males.

More than 10,000 species are known in about 2000 genera with four infraorders, recognized here as superfamilies: Stenopelmatoidea (Gryllacridoidea), Tettigonioidea, and Grylloidea (including the Schizodactyloidea). The classification we follow is illustrated in Fig. 1A.

STENOPELMATOIDEA (GRYLLACRIDOIDEA, IN PART)

Variously known as raspy crickets, leaf-rolling crickets, wood crickets, Jerusalem and sand crickets, king crickets, cave and camel crickets, and other names, the Stenopelmatoidea form a distinctive, conspicuous group with many species quite large and presenting a ferocious appearance. Most, if not all, are nocturnal, with all needing to seek shelter during daylight because of the threat of desiccation due to the thin integument. When the Grylloptera is recognized as a distinct order, this taxon with the Tettigoniidea constitutes the suborder Tettigonioidea. Because of their nocturnal and subterranean habits, most species are light colored, with brown or grays dominating the color scheme. Cave-dwelling species are often pale or white: green forms are very rare.

Species occupy most habitats, including the driest deserts and the wettest rain forests. The majority of species seem to be concentrated in the Old World, especially in the Southern Hemisphere. A few are considered to be minor crop pests, many attract public attention because of their size, and several are considered to be endangered by reason of habitat deprivation or loss due to introduced organisms like rats. The group comprises about 1000 species, but there are many undescribed species known from many parts of the world, especially Australia.

Gryllacrididae   Recently considered as a subfamily of the Stenopelmatoidea, this taxon has been recognized as a family for more than 50 years. Gryllacridids are known by a variety of names. They are best known as raspy or leaf-rolling crickets but some are called king crickets or tree crickets. Most instars and adults spin threadlike "silk" from the mouthparts and use this material in reinforcing burrows and tying leaves and detritus together to form shelters. The integument of the body is soft and pliable (Fig. 1B), and the legs are adorned with many spines, some on the hind legs being modified for digging. They range in size from about 5 to 75 mm in body length. This family is best represented in the Old World. The majority of species are probably in Australia, with other regions of the Southern Hemisphere contributing considerable numbers.

Of the few species of gryllacridids that have been studied, all have been found to have peculiar life histories. The majority of these studies come from Australia, where the insects comprise a notable portion of the orthopteran fauna at a given locality. Species may construct burrows or tie leaves and detritus together, forming individual enclosures in which they reside during the day. Others live in the hollow branches or twigs and others assume a commensal lifestyle with termites.

Raspy crickets feed on a wide range of material. Some are specific seed eaters, others predaceous, and others specialize on flowers or fruits. This latter activity can be of economic concern when ripe fruit is chewed and when cut flowers or orchids are damaged by feeding activities. At times when gryllacridids inadvertently enter houses, curtains and draperies may be ruined by their chewing and clipping as bits of material are tied together with silk to make shelters.

Raspy crickets produce sounds in more than one way. All species possess a femoroabdominal stridulatory mechanism featuring a hind femur that is rough on the inside, with a pattern or shagreening of dorsal surface. This roughened area is rubbed against a pattern of pegs or modified hairs on the side of one or more abdominal tergites. These stridulatory mechanisms seem to show species-specific patterns and have been used to distinguish species. Vibrational sounds are produced, often by both sexes, during courtship activities. Depending on the species, sounds are generated by rhythmic drumming of the abdomen on the substrate or by "stomping" of the hind feet against the substrate. Some species use use combinations of drumming, stomping, and rasping during courtship. The fast-paced drumming is quite audible to the human ear for a short distance.

Surprisingly, none of these insects possess any obvious organs that would enable them to hear the sounds they produce. Perhaps, they detect the vibrations through sensory hairs on the pads of the tarsi.

Rhaphidophoridae Camel, cave, and sand-treader crickets are fairly similar in appearance (Fig. 1C). An extinct subfamily is known from amber inclusions. All members are apterous, but some can produce sounds by rubbing the inner faces of the hind femora against the opposing side of the abdomen and by rhythmic drumming of the abdomen against a substrate, be it the ground, a twig, or a branch. All have a humpbacked body structure with very long hind legs and antennae. Some of the sand-dwelling species have the hind legs modified into "sand baskets" for digging. Some groups are wholly confined to caves and others are obligate burrowers; the majority, however, live in leaf litter of dark crevices, where they spend the daylight hours and emerge only on humid nights to feed on detritus and leaf litter. Some feed on fungi, and there is at least one record of a cave-dwelling species that feeds on newly hatched birds. One camel cricket is cosmopolitan in its distribution, being moved in commerce, and is said to be a pest of greenhouses. About 350 species are known, with most species coming from the Indo-Australian area and Polynesia.

Schizodactylidae The splay-footed crickets are among the most bizarre of orthopteroids (Fig. 1D). There are fully winged as well as apterous species. They are broadly expanded by the possession of Lobelike or digitiform processes, which enable the crickets to run across dry sandy surfaces with efficiency. One interesting feature of the group is that members that have been studied have lower chromosome numbers than most typical orthopteroids. The splay-footed crickets are primarily predaceous, but one is considered to be a minor crop pest. They occur in parts of India, Myanmar, southwest Asia, and South Africa.

Anostostomatidae (Stenopelmatidae, in part of authors) The Anostostomatidae is a relatively recently proposed taxon accommodating a large range of genera formerly included in the Stenopelmatidae. Included here are the giant king crickets of Australia, New Zealand, and southern Africa. The odd genus *Cooloola* from Australia and a variety of peculiar genera from western North America have been placed here, along with an odd entity from Chile.

This family includes the Parktown prawn, *Libanasidus vittatus,* which lives in the Johannesburg area, where it is very well known. Females measure in excess of 6 cm, and most people fear them. They enter houses and beds of sleeping residents at night, and their scuttling and kicking is a common cause of concern. Their vile-smelling feces, exuded when disturbed, contribute to the unpleasantness. Myths have arisen regarding this species: some people feel it is an alien invader or the result of a freak mutation. The notoriety of this cricket has been a useful vehicle for educating the public about the biodiversity crisis. One of the most interesting features of this cricket is the tusklike mandibles of males. In this respect, they are similar to those of the New Zealand weta. The mandibles of *L. vittatus* serve in digging burrows as well as in battles with other males.

The New Zealand tree weta of the Deinacridinae are often large (8 cm or more in length), secretive, and aggressive species. The biology of *Hemideina* spp. is very well known. Some live in tree holes in galleries initially made by beetles or moth larvae. Others live in natural crevices and cavities. They prefer living trees to dead wood. These insects are herbivorous, feeding on leaves, flowers, and fruit. They may scavenge recently killed invertebrates. They remain arboreal except for a short period when females descend to the ground to oviposit. Some species have a unique size polymorphism related to social behavior. There is allometric growth of the head and jaws, much as the Australian king crickets. Certain males have extra instars to gain the larger head and mandibles. There is a dominance hierarchy, with the largest megacephalic males commanding the most desirable resources, which include galleries and mature females. Smaller males spend more time defending their galleries. These males live in galleries with smaller apertures and have been advantaged in dealing with alien predators such as rats. Tree wetas are smooth and shiny, with contrasting bands on the abdomen. Other New Zealand weta live on the ground. Several species raise their hind legs vertical to the substrate when alarmed. With a female weighing 50 g and measuring 7.5 cm in body length, this is one of the more formidable insects in defense.

All members of the Anostostomatidae are vulnerable to alien predators. The large size of the adults of many species may be effective in battles with rats, for example, but other stages from eggs to moderate-sized nymphs are extremely vulnerable. As a result, several species are threatened with extinction.

Stenopelmatidae This family is now considered by at least one authority to comprise five subfamilies, including the Gryllacridinae and Schizodactylinae. However, these groups are so different from typical stenopelmatids in morphology and biology that they are considered here as separate families. The remaining three subfamilies include the well-known Stenopelmatinae, comprising the Jerusalem crickets. This subfamily will be included here. The largest genus, *Stenopelmatus,* occurs in North and Central America. These insects are often known locally as "potato bugs" because they have been dug up in garden or potato fields. The derivation of the common name of the group, Jerusalem crickets, is shrouded in mystery. These insects have behavior patterns similar to those described for several anostostomatids.

There may be 100 species of the genus *Stenopelmatus,* but only a handful have been described. They are all very similar in overall appearance and differ only in their sexual drumming activities. Sympatric species have different drumming patterns. Both sexes and nymphs can produce rhythmic sounds by drumming their abdomens against the sides of their burrows or on the surface of the ground. Some of the sounds are audible from 20 m distance. Surprisingly, the nymphs produce species-specific drumming patterns in their later instars. The production of these sounds may serve to keep the species together and may be effective in areas where there are sympatric species.

Copulation is distinctive in this group. In at least one species, the male rolls on his side and, if receptive, so does the female. Mating will not occur unless both partners "roll." Then, facing in opposite directions, the male grabs one of the hind tibiae of the female, not damaging her, positions his hind tarsi near her coxae, and curls his abdomen between his and her hind legs toward her subgenital plate. After several minutes, the male's telescoping abdomen nears the female's subgenital plate, and he grasps her with his hooks. With this anchor he everts the phallic lobes and attacks a spermatophore with a larger spermatophylax. In many ensiferans the spermatophylax is eaten, providing a source of nutrition for the mother and her eggs, the Jerusalem cricket female, however, does not consume the organ. Sometimes, though, the female consumes the male after mating; males offer no resistance to cannibalistic females.

TETTIGONIIOIDEA   This is the largest superfamily, with more than 6000 species. Katydids (bush crickets or longhorned grasshoppers) can be of economic concern at times. Some flying species can swarm in the manner of locusts. With the "proper" environmental conditions, even flightless species can build in numbers that can affect crops and cause serious losses. The most notable example in the latter category is the Mormon cricket *(Anabrus simplex)* of western North America. It damages a variety of crops and rangeland plants. In Australia, Asia, and Africa, meadow katydids of the genus *Conocephalus* can build in numbers and move in large swarms. On a local level, many species cause damage to rice and crops from time to time.

On the other hand, the same species are often predaceous, feeding on eggs and larvae of other more important crop pests. In tropical regions rice is attacked by the copiphorine *Euconocephalus* sp. Phaneropterines seem to provide the majority of species that damage crops. *Scudderia* spp. in North America and *Caedicia* spp. in Australia both feed on the developing fruits and new leaves of citrus varieties. *Phaneroptera* species feed on a wide range of crops as well as citrus and coffee, kapok, and mimosas. Another phaneropterine, *Holochlora pygmaea,* feeds only on tea. *Ducetia* species feed on rice on two continents. *Elimaea chloris* feeds on a range of economic plants, the most notable of which are soybean, sugarcane, tobacco, and tea. The pseudophylline *Chloracris prasina* damages cacao, dadap, and rubber by its ovipositional habits. The widespread tropical mecopodine *Mecopoda elongata* feeds on beans, betel, cassava, castor, dadap, maize, potato, rice, sorghum, and tobacco. There are abundant literature references to "sexavae" or coconut treehoppers of two genera, *Sexava* and *Segestidia.* They are major pests of coconut and oil palm. They also feed on banana, karuka, manila hemp, and sago palm. With agriculture ever expanding to more remote areas, additional species not usually associated with economic damage can be expected to cause economic problems.

Katydids are widely distributed throughout the world, but the majority of species can be expected in tropical regions,

especially the New World tropics. Many species are arboreal or bush dwelling. Some live in reeds or grasses, and many live on the ground. A few species can be found in alpine regions far above treeline. Most species exhibit cryptic behavior, especially during the daytime when they are inactive. Other species are aposematic and display warning colors. These species are primarily diurnal in their habits. Immature stages (nymphs) of some species mimic wasps, ants, beetles, bugs, and spiders. Ancestors of katydids probably were predaceous, but this role is minor among the extant forms. The majority feed on foliage, flowers, and seeds or are omnivorous.

Haglidae (Prophalangopsidae of authors)   The ambidextrous crickets or hump-winged crickets comprise a single living family that is divided into two subfamilies. One subfamily, the Prophalangopsinae, consists solely of the species *Prophalangopsis obscura,* represented by a unique specimen, which is recorded from northern India. The specimen is fully winged and appears to resemble a mixture of a tettigoniid and a gryllacridid. The second subfamily, the Cyrtophyllitinae, is represented by two living genera. All species are ground dwelling and live in coniferous forests, where they ascend tree boles after dark to broadcast their loud stridulations.

Tettigoniidae   The katydids, bush crickets, and longhorned grasshoppers differ from the Haglidae by having a more advanced stridulatory specialization. Many subfamilies are recognized, with several having been escalated to family rank as a result of recent investigations. Following the pattern adopted at the beginning of this article, these controversial changes are not recognized here.

The "primitive" katydids have been considered to be those in which the antennal attachment to the head is low on the head, that is, below the halfway point of the eyes. But this placement is open to interpretation, and it is often difficult to determine just where the critical position is. As a result some "primitive" katydids may not be primitive at all. Some of the more notable subfamilies are highlighted here.

The Bradyporinae is represented by a single genus found in the eastern Mediterranean as far as Iran. The Ephipperinae are represented by eight genera, mostly of Mediterranean origin. The Hetrodinae contribute 13 genera in Africa that often occur in aggregations. In some species both sexes can stridulate. They are characterized by a spiny appearance. The Acridoxeninae or dead-leaf katydids, are remarkable insects. The group is represented by a single genus with two species in equatorial West Africa, where they are apparently found on spiny plants.

The Phaneropterinae, or leaf katydids, bush katydids, or lyre bush crickets, is the largest assemblage of genera in the family. It has been accorded family status by some investigators. About 2000 species are known, distributed throughout the world, but the majority of species are in tropical climes. Many species are involved in mimicry complexes. Most nymphs resemble foliage or plant parts, but many are mimics of ants, bugs, spiders or cicindelid ground beetles.

The Pseudophyllinae, true katydids or bark crickets, include giant fully winged, leaflike species (with wings spanning more than 20 cm) and smaller, micropterous ones. Most species are splendid examples of twig, foliage, and bark mimicry. With some species that resemble leaves, transparent holes and irregular, "bitten" pieces, and even colors resembling fungus or lichens are not uncommon in this group. About 1000 species are known in 250 genera with the majority in the New World tropics. The species that provided the name "katydid" is a member of this subfamily. All known species are phytophagous, and all known species use the often large falcate ovipositor to deposit eggs in plant tissue, either in bark, twigs, or dead wood. Some species have unusual lifestyles. A Mexican species of *Pterophylla* has been observed crossing a stream underwater. A species of the southeast Asian genus *Callimenellus* occurs in marine littoral rock crevices.

The Microtettigoniinae are represented by a single genus, *Microtettigonia* from southern Australia. These are minute (males as small as 5 mm) micropterous diurnal katydids that are extraordinarily fast moving. They live in grasses and lilies, and they seem to feed on floral parts.

The Conocephalinae is a large cosmopolitan group that comprises at least four tribes with more than 1000 species. The Gondwanan-distributed Coniungopterini is represented by three genera: two occur in Australia and New Guinea, and a third is found in Chile. The genus *Conocephalus* is represented by more than 50 species in Australia alone. They are small, agile katydids with a relatively broad, blunt fastigium. They are similar in appearance and habit and often occur in large numbers. Some species are diurnal, others nocturnal, and still others are active both day and night. *Conocephalus* species can be of economic concern at times. At least one species has been distributed through commerce.

Other genera have a much smaller number of species and are widely distributed. The Copiphorini are often associated with grasses, where their slender, bladelike appearance renders them almost invisible as they perch on the stems. The species associated with grasses and reeds feed largely on the floral parts of the host plants, preferring seeds. The mandibles are unusually strong, and this is an adaptation for seed predation. The buzzing calls are familiar sounds to most residents and visitors to appropriate habitats on all continents, but few people have ever seen the producers of the sounds they frequently hear because of the insects' secretive and cryptic habits. Thus many attribute their sounds to cicadas. The other subfamily, the Agraeciini is a disparate assemblage that most likely comprises a number of higher taxa. Members of this group have an extraordinary size range, with some of the world's most robust species represented.

The Mecopodinae or Kutsuwa bush crickets are usually large (some species have wingspans > 20 cm), and most species are fully winged and resemble either dead or living leaves to a remarkable degree. Others are short winged or wingless in at least one sex. Females sometimes stridulate. Most species are confined to the Old World. The subfamily is represented by two tribes, the Mecopodini and Moristini (Sexavini). Members of three genera are called coconut treehoppers because of the damage they cause to coconuts. Some species of this subfamily are kept in cages in Asia for the songs they produce.

The Phyllophorinae, or giant leaf katydids, are among the largest of all tettigoniids, with wingspans great than 25 cm. Some 10 genera and 60 species are known, mostly from the rain forests of Indo-Malaysia, New Guinea, and northern Australia. They are related to the mecopodines and the phaneropterines.

The Phasmodinae, or stick katydids, are wholly confined to southwestern Australia. They occur in winter and spring and are largely gone as the hot days of summer approach. They are very elongate and wingless in both sexes. They can cause economic concern when they feed on wildflowers in plantings in parks adjacent to natural areas. The remarkable resemblance to Phasmatodea is one of the most striking examples of convergence in the orthopteroid insects.

The Zaprochilinae, or pollen- and nectar-feeding katydids, are represented by four genera comprising 17 species from Australia. They are usually gray, setose, soft-bodied insects, and some species are fully winged, with the tegmina held at an angle and rolled; some species are micropterous in males and wingless in females. *Kawanaphila* species have been important study organisms in studies of sexual selection.

The Tettigoniinae, comprising shield-backed katydids, wartbiters, or great green grasshoppers, is one of the largest and most widespread of tettigoniid subfamilies, with the majority of representatives in temperate climes of both hemispheres. The common name, shield-backed katydids, is derived from the development of the pronotum, which is often extreme. While most are predaceous or specialized feeders, the Mormon cricket *(Anabrus simplex),* the Coulee cricket *(Peranabrus scabricollis)* of North America, and *Decticoides brevipennis* of Africa can occur in large numbers during favorable years, causing enormous damage to crops. The wart-biter *(Decticus verrucivorus)* was used during the Middle Ages to "cure" warts by allowing the aggressive insects to bite them off. Perhaps a substance in the insect's saliva contributed to the cure. The subfamily is diverse in form, ranging from among the smallest of tettigoniids to some of the largest. Fully winged and micropterous (Fig. 1E) species are known.

The Onconotinae are unusual micropterous katydids living in shrubbery. The Saginae is a small but characteristic subfamily of voracious predators or large insects. Four genera are represented, all from the Old World. *Saga* species can enter a cataleptic state characterized by complete immobility lasting some 20 min followed by a slow recovery. The function of this behavior is unknown.

The Austrosaginae comprises five genera, some of which use the powerful mandibles to crack seeds and feed on fruits. Listroscelidinae is a disparate assemblage of genera, many of which will probably be moved to other subfamilies when the group is fully revised. All species appear to be predaceous, some nocturnal and others diurnal. Some of the world's most spectacular katydids are members of this subfamily. Many are capable of delivering a painful bite when handled. The center of distribution for this group seems to be Australia.

The Meconematinae, or oak crickets or swayers, are small greenish-yellow katydids that are mostly diurnal; their distribution is tropical or subtropical. The notable exception is *Meconema thalassinum,* which occurs in Europe and Asia and has been introduced into the eastern United States. These delicate katydids have large eyes on the top of an unusual heart-shaped head. The few observations that have been made record the katydids catching their small prey on the "jump," in midair. This is not the manner used by phisidine katydids to secure their prey. The meconematines have been confused with the Phisidini of the Listroscelidinae, which, although somewhat similar in appearance, have an entirely different lifestyle. The phisidine species are nocturnal predators and have typical listroscelidine eggs. Eggs of meconematines are incredibly large for the size of the katydid and can be seen through the thin integument of the female. They are laid in decaying wood or other plant tissue, such as galls made by wasps, with the flattened, hard cap protruding from the substrate. The Tympanophorinae, or timbrel bush crickets, comprise a single genus confined to temperate Australia. The tegmina are unique in that some species, bear four kinds of stridulatory teeth on an expanded "rib." There is evidence that soft structures on the rib may provide openings for a lubricant or a substance released from a chamber or reservoir beneath it. This feature is probably associated with the unusual reproductive biology of species. In other tettigoniids, females are drawn to stridulating males, or answer them with their own calls, and the pairs are eventually brought together. The situation must be different in the tympanophorines, since the females are flightless and incapable of producing sounds. Male stridulatory behavior is unusual: they sing from perches for a short time and then fly 30 m or more to another perch and continue their song. How a flightless female can attract the attention of the transient male is unknown. All tympanophorine species are nocturnal and predaceous. They use the short forelegs to capture small-insect prey.

GRYLLOIDEA   The true crickets and mole crickets range in size from less than 1 mm to more than 6 cm. Most species possess "ground colors," and few green forms are known. This is an adaptation to living on or in the ground.

The Grylloidea can be divided into a few very unequal families. More than 350 genera are known, encompassing 3000 species that comprise the Gryllidae, only about six genera and some 70 species comprise the Gryllotalpidae, and only a few species are known in the Myrmecophilidae. Although the majority of species are tropical, large numbers occur in all the temperate parts of the world. All terrestrial habitats seem to be inhabited except the highest mountain peaks. A few species can "skate" on water surfaces, and several live in mangrove swamps, where they use the stems to submerge themselves in saltwater when danger threatens. Many species burrow deep into the ground and seldom emerge. Others are blind, lack pigmentation, and live deep in subterranean caves. A number of species are commensal, living with rodents, ants, termites, and even

mankind. Eggs are laid singly and deposited in the ground or in plant tissue such as decaying wood or grass stems.

Many crickets are crop pests, and population explosions sporadically occur with devastating results. Invasions of human habitations by crickets cause angst owing to the interminable chirping. Crickets around the world, have been known to ravage foods and furnishings.

Several radical classifications have raised "classical" subfamilies to familial status. However, the boundaries between these taxa are often unclear and a broader, more classical view is adopted here.

Gryllidae   The true crickets comprise the principal family of the superfamily. Body size can range from less than 5 mm to more than 50 mm. The family has been divided into a number of subfamilies, and various classifications elevate some of these taxa to family level. The Gryllinae contain the crickets known to almost everyone. There are more than 550 species in more than 75 genera known from all parts of the world. Field crickets and house crickets belong to this group. Almost all species live on the ground, and some construct elaborate burrows that are often modified to amplify sound production. The house cricket, *Acheta domesticus,* can be considered to be domesticated insect. It is used in commerce as a food for mammals, birds, and reptiles, and seems unable to exist for any length of time in nature.

Several cricket species are kept, mostly in Asian countries, for the songs they produce. They are the source of a rich folklore and are regaled in poetry and song. Fighting crickets are an important part of the social scene in many Asian countries. A number of the subfamilies are highlighted shortly. The short-tailed crickets, a group that is quite prominent in Asian culture, are notable for the extensive galleries they make. There are brood chambers where the young are looked after. Several species cause damage to crops by feeding or destroying the roots of the plants. Some authors have accorded them a subfamily of their own, Brachytrupinae, but the differences between them and the Gryllinae are small, and they are probably best regarded as a separate tribe, the Brachytrupini.

The Nemobiinae, or pygmy or dwarf crickets, comprise more than 200 species and, as the name suggests, are small. While most forms are winged, there are many apterous species only a few millimeters in body length (Fig. 1G) that can easily be mistaken for nymphs of other species. Several genera exist in marine habitats. A number of genera in several parts of the world are adapted for life deep in caves and lava tubes. The majority of species live on the ground, often in moist habitats. Their numbers can be incredible in certain situations. Both diurnal and nocturnal species are known.

The Trigonidiinae, or leaf-running or sword-tailed crickets, are small, ground-dwelling or bush-dwelling crickets that are usually diurnal. They are active and often brightly colored. Some have a metallic sheen. Many species mimic spiders, ants, or wasps to mask their edibility.

Eggs are deposited in stems and bark. More than 300 species are known in more than 25 genera. Many genera are cosmopolitan or at least occur on more than one continent.

Preferred habitats are in rank vegetation, including shrubs, grasses, and small trees, but several genera occur in leaf litter. One genus with long tibial spurs can skate over water.

The Eneopterinae or bush crickets are small to medium-sized slender crickets. There is a range of color in this group depending on the habitats occupied. About 100 genera with more than 500 species are known, with the majority of species in the tribe Podoscirtini. Most of the many endemic crickets of Hawaii are in this subfamily. Most examples are from the tropics, with only a few making it to the temperate climes. These crickets usually live in shrubs, trees, herbaceous vegetation, or grasses. A small number can be found in leaf litter. Virtually none are of economic significance.

The Phalangopsinae, or spider crickets, are long-legged, rather flimsy crickets that are often gregarious in habit. The subfamily comprises more than 60 genera and 300 species, with the majority in the Old World tropics. Some species occur on the ground or in tree holes or decaying logs. Large numbers live in caves, crevices, or in the cavities created by large animals. Similar habitats afforded by dams, bridges, and buildings attract these crickets. True cave crickets belong to this subfamily. The cheerful songs of oriental species are prized, and these crickets are often kept in cages for such reasons.

The Sclerogryllinae, or stiff-winged crickets, are represented by only a single genus and several species. These crickets occur in Africa and southern and eastern Asia, where they live in leaf litter. The Pteroplistinae, or feather-winged crickets, are represented by three genera and only seven species. The Cachoplistinae are sometimes called beetle crickets because members of at least one species strongly resemble small beetles.

The Mogoplistinae, or scale crickets, are a widespread and often common assemblage of small flattened species, all of which are covered with minute scales. About 15 genera are known, comprising nearly 200 species. They are very widespread, with the majority of species in the Old World tropics. Many species live on the leaf surfaces, and there is an abundance of species that live in leaf litter. A single European species is considered to be intertidal. One tribe is known to be associated with rodent burrows. Both nocturnal and diurnal species are known. They are mostly small, ranging from 1.0 to 20 mm. The beautiful tones and sequences of the male song are valued in the Orient, and several species are kept in cages and sold in market for their songs.

The Oecanthinae, or tree crickets, are often given familial status. This subfamily has a worldwide distribution, but only seven genera and many species are known. Their nocturnal calls are well known to many people, although the insects themselves are seldom seen by humans. Some males utilize holes in leaves to amplify and direct their songs. They are kept also in cages in Asian countries because of the melodious songs they produce. Tree crickets are both beneficial and important local crop pests. Nymphs are often predaceous, feeding on a variety of insects, including aphids and scales. At times females oviposit in developing fruit and have been involved in transmitting plant diseases. Tree crickets have a characteristic appearance (Fig. 1H)

and are usually pale greenish white, ranging from 1.0 to 1.5 cm in length.

Myrmecophilidae The Myrmecophilidae, or ant crickets, are minute, scale-covered, wingless crickets. There are five genera in this small family, with about 50 species found throughout the world, although most are known from tropical and subtropical regions. They are associated with ants, often living in the nests with them or following them along their trails. Some species have a wide range of hosts; others are known from only one species. Their commensal relations are unclear, but they seem to be unable to live independently for any period of time. A few species exist with termites. Some species seem to be parthenogenetic. Eggs are relatively large for the size of the female producing them. They are robust crickets that live in association with ants.

Gryllotalpidae The mole crickets comprise five genera with about 60 species worldwide. Most species are found in the cosmopolitan genus *Gryllotalpa*. Mole crickets use the extraordinarily developed forelegs for digging deep, permanent galleries and foraging for plant roots. Some mole crickets are known to collect seeds and store them in larders in circular chambers underground for future use. Some species brood eggs in chambers, and in many species both sexes stridulate. The calling songs are often of short duration and very loud. The horn-shaped entrance chamber of the burrow is used differently by different species to increase the male's acoustical output. Most species are herbivorous, but a few are carnivorous. Several cause major damage to crops by feeding on roots, on seedlings, or both.

Suborder Caelifera

The short-horned grasshoppers and locusts and their relatives comprise the large and well-known suborder Caelifera. Rarely, this group is treated as a separate order.

Most caeliferans are diurnal, but increasingly investigators are discovering that many species are active both day and night. Males of many species stridulate in bright sunshine; rarely, those of others sing on warm nights. Although primarily tropical, many species occur in all parts of the world. Some are semiaquatic, and only a few are true burrowers. Almost all species feed on plant material, but many feed on dead members of their own or other species. None are commensal. Aposematic coloration is a feature among many species.

Copulation in the Caelifera is rather uniform, with the male clinging to the back of the female for a considerable period of time with the spermatophore elongated and occupying the female's genital tract. There is no visible external spermatophylax. Eggs are laid in pods or oothecae that are enclosed in a reticulate membrane and covered by a foamy secretion that dries out eventually. Eggs of most species are deposited directly into the ground, but others lay in plant tissue or in cracks in bark.

Many species are disposed to gregarious behavior and swarming, with locusts offering the best example of this behavior. All true locusts are placed in the Caelifera. However, many other species not properly designated locusts can become extraordinarily numerous and cause great damage to crops and other vegetation. Many species are eaten in Asia and Africa, usually fried. Many others are dangerously poisonous, and deaths have been recorded from eating these insects in Africa. After fires, the predominant colors among nymphs are dark, often black. This characteristic, called fire melanism, is found in almost all grasshoppers in all areas. The time span for these changes is extremely short. However, some surviving adults that become very dark after a fire produce nymphs that show little or no melanism.

The more than 2000 genera and 11,000 species comprising the Caelifera are arranged in superfamilies, suborders, or infra-orders depending on the classification followed. Here they are regarded as superfamilies and are as follows (Fig. 1): Acridoidea, Eumastacoidea, Pyrgomorphoidea, Tanaoceroidea, Pneumo-roidea, Trigonopterygoidea, Tetrigoidea, and Tridactyloidea.

ACRIDOIDEA *(SENSU LATO)* The grasshoppers and locusts comprise the largest superfamily of the suborder Caelifera. More than 8000 species in more than 1500 genera are known worldwide. Adults range in size from less than 1 cm to more than 25 cm.

Many grasshoppers have a disparity in size between males and females, with the former being much smaller than the latter. In many species males ride "piggyback" on females, not always copulating but "protecting" their prize and defending their potential progeny from other males with similar intent.

Grasshoppers have a short, almost invisible ovipositor. The eggs are laid in a pod covered by a protective coating. Many species oviposit on grasses, wood, and other plant tissue. A North American species has been observed ovipositing in dry buffalo dung. A number of species, especially locusts, utilize "lekking sites" to oviposit. Large numbers of females lay in the same place year after year. Knowledge of such behavior can be valuable in planning the control of pest species. Many grasshoppers take advantage of walking tracks or nonpaved roads for such activity. In oviposition, the female uses muscular contractions to extend the abdomen into the ground, often telescoping many times its length. The digging process also is performed by contractions and is aided by the teeth at the end of the abdomen. In tropical species the eggs can hatch in 3 to 4 weeks. In temperate species the eggs enter a period of diapause and hatch later. In some species, only a portion of the eggs hatch the following year, with the remainder hatching over a period of years. This serves as a "safety valve," ensuring that the species survives in periods of unfavorable weather. Upon hatching, the vermiform larva wriggles to the surface of the soil and upon reaching the surface, molts into the miniature form that will eventually become the adult. Nymphs undergo a series of approximately six molts, each one resembling more and more the adult in color pattern and shape. Gregarious grasshoppers illustrate synchronized molting, with all nymphs shedding their skins within hours of one another.

Locusts are species that occasionally form dense migratory swarms. These are often so large that they cross oceans but most occur in inland regions. Worldwide, more than 20 species from several subfamilies of Acrididae form migratory locust swarms.

The number of families depends on the classification followed. The scheme followed here (Fig. 1A) reflects the latest molecular attempts to establish a phylogenetically based classification based on defensible evidence.

Pamphagidae The Pamphagidae comprises generally larger, sluggish grasshoppers often resembling stones and bark. They have an array of body shapes (Fig. 1I). The family has an Old World distribution with the majority of species occurring in Africa and a few in Europe and Asia. It is absent from Australasia.

Lentulidae The lentulid grasshoppers have been called "nymphlike" grasshoppers and range in size from 8 to 25 mm in body length. These (Fig. 2A) grasshoppers are wingless. They occur in southern, eastern, and central Africa. Most species live in bushes, and at least one species causes feeding damage to nursery stock. One species prevents a weedy shrub from becoming a pest.

Pauliniidae The Pauliniidae, or aquatic grasshoppers, are small to medium-sized species with a smooth body integument. This family is known only from South America. Although other grasshoppers are aquatic, these grasshoppers are more wholly aquatic than any others. They can skate on the surface or dive and swim beneath it. Eggs are laid on the submerged parts of water plants. Their terrestrial behavior seems to be mostly nocturnal. They feed on aquatic plants and one species, *Paulinia acuminata,* has been introduced into Africa for the control of *Salvinia.*

Tristiridae The Tristiridae, or Andean wingless grasshoppers, are small to moderate in size. The body shape is variable, but the integument is always wrinkled and the color brown or grayish. The family has three subfamilies found only in the mountains of Cordilleran and Patagonian South America. They are found at high altitudes (2800–3000 m). Their protective coloration renders them almost invisible on pebbles and gravel, but aside from that nothing is known of their biology.

Ommexechidae The Ommexechidae are called South American toad-hoppers. The family includes 12 genera with some 30 species, all from South America. They live on the ground, inhabiting "coarse" vegetation in dry, sandy, or stony areas. They are of no economic importance.

Romaleidae The Romaleidae, or lubber grasshoppers, are regarded as a subfamily of the Acrididae in most older works. Recent molecular investigations, however, show that full family status is more appropriate. These grasshoppers are moderately large to very large in size and are often very colorful. The family has about 200 species in three subfamilies with the majority of species found in the Americas. A small number of examples occur in the Old World, in India, Afghanistan, Iran, and eastern Africa. They are notably absent from Africa and Australia. They occur in many habitats from desert to tropical rain forest. Some live on the ground, closely resembling stones

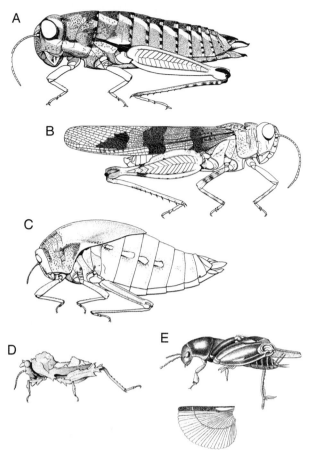

FIGURE 2 (A) A lentulid grasshopper, *Lentula callani* from Africa. [From Dirsh, V. M. (1975). "Classification of Acridomorphid Insects." Classey, Faringdon, U.K.] (B) *Oedipoda miniata*, now considered to be in the Acridinae, but many experts prefer to include it in a separate subfamily, Oedipodinae, with many other species. [From Dirsh, V. M. (1975). "Classification of Acridomorphid Insects." Classey, Faringdon, U.K.] (C) Flying gooseberry, *Bullacris unicolor*, male; tegmina and wings absent from females. [From Dirsh, V. M. (1975). "Classification of Acridomorphid Insects." Classey, Faringdon, U.K.] (D) *Cota saxosa*, a Peruvian species. Note flanges on legs and pronotum. [From "Genera Insectorum." (1906). Vol. 48.] (E) *Bruntridactylus tartarus*. Note fan-shaped hind wing. [From Saussure, H. (1874). "Voyage au Turkestan, Orthoptères."]

or "toads"; others are found in bushes, and still others high in trees. Many species show aposematic coloration, if not in the adult stage, then as nymphs. Some species can produce a hissing sound when disturbed, and bubbles can be produced from spiracles when the insects are greatly disturbed. In many aspects of both appearance and behavior, these grasshoppers resemble the pyrgomorphs. Few romaleines, however, are associated with grasses or sedges. The lubber grasshopper, *Brachystola magna*, of the southwestern United States, can be a serious road hazard when killed by automobiles. Their crushed bodies have caused cars to skid and accidents have occurred. The largest of grasshoppers occur in this family. The tropical American genera *Titanacris* and *Tropidacris* normally feed on vines, but at times they can become pests of plantations by defoliating plantation trees. They are called "giant locusts," but they are not really locusts at all because they do not form swarms.

Lathiceridae   The Lathiceridae have been called desert gravel-hoppers. They are another peculiar small group known only from Africa. There are only four genera and five species, and nothing is known of their biology except that they burrow in sand and spend considerable time underground. They strongly resemble pebbles.

Pamphagodidae   The Pamphagodidae (Charilaidae) are called the twin-keeled grasshoppers because of the parallel longitudinal carinae in the surface of the pronotum. There are four genera, comprising only five species, in this very small group from Africa.

Acrididae   The Acrididae, or true locusts or grasshoppers, comprise the largest family of the suborder, and a full range of most of the features already described can be found here. Species range in size from less than 5 mm to more than 12 cm. The body form reflects grasshopper's role in the habitat. Short, stout, stone- or toadlike species are known, as well as stem- or twiglike grasshoppers. The colors span the complete range for the order, with browns or earth colors predominating.

Grasshoppers occur everywhere that orthopteroids are found. Many of the higher groups have restricted distributions, but the majority of species are tropical. Until recently, around 17 subfamilies were recognized, with more than 1500 genera. The characteristic songs are produced by males to attract females. Since these grasshoppers are primarily diurnal, color and behavior play important roles in species recognition. Some species have more than a dozen movements necessary for mate recognition. Elaborate colors and patterns and modifications of body pairs are associated with this behavior. But not all courtship occurs during the day. Some groups mate under the cover of darkness and rely on chemical clues to find mates. With these species, color is mostly protective, and bright colors are associated with aposematic features. Grasshoppers have a worldwide distribution and extend into some of the very cold regions of north and south. In this respect they occupy more territory than the Ensifera. Only a few subfamilies can be highlighted here.

The Oxyinae include many species associated with water and grasslands. *Oxya japonica*, one of the most destructive grasshoppers in rice in Southeast Asia, does only minimal damage in Australia. *Bermiella acuta* from Australia occurs on sedges, rushes, and grasses near the water and has adaptations for aquatic life such as dense, water-excluding patches of hairs on the distal abdominal sterna and tegmina, and an "air chamber" formed by the doming of the costal area of the tegmen over the first abdominal spiracle.

The Gomphocerinae are an important component of the acridid grasshoppers. The songs of stridulating males are a summertime characteristic in the meadows of the Northern Hemisphere. Several species of the same genus may occupy similar habitats and have similar feeding habitats. The calling songs of the respective species are different and serve as isolating mechanisms. Females are attracted only to the males that perform the calls of their own species.

The Catantopinae, or spur-throated grasshoppers, comprise many species (84% of the fauna in Australia). Some are very colorful. Many catantopines produce no audible sound, but some perform drumming actions on their host plants with the hind legs, thereby announcing the presence of a mate without the need of acoustical amplification. The responding members of the opposite sex merely move to the point of the drummer to consummate the union. A few other species produce a soft sound by rubbing the mandibles together. This can serve two functions. Some do this only when grasped, a startling reaction that might cause a would-be predator to drop the grasshopper. Others produce a similar sound from perches in shrubbery, obviously as an attractant to potential mates.

The Cyrtacanthacridinae, or large spur-throated grasshoppers, comprise some of the world's most important locust pests. *Valanga irregularis* from Australasia is one of the world's largest grasshoppers. It feeds on the leaves of trees and shrubs and can cause damage to fruit, nut, and plantation trees. Grasshoppers and locusts of the genus *Schistocerca* are members of this subfamily.

The Acridinae is a large group (Fig. 2B) with a worldwide distribution. Several locusts are included. Many are colorful and have brightly colored hind wings and slanted heads. This group contains some of the most phylogenetically advanced grasshoppers. Australia's most important locusts pest, *Chortoicetes terminifera*, is a member of this subfamily. The plague locust, *Locusta migratoria*, is also a member of the Acridinae.

TANAOCEROIDEA   This superfamily includes a single family, the Tanaoceridae. Once considered to be related to both the eumastacids and xyronotids, the family Tanaoceridae comprises two genera with a few species from the deserts of western North America and Mexico. Members of *Mohavacris* have been found on sagebrush (*Artemisia tridentata*), where they closely resemble the bark of the thick stems. *Tanaocerus* species occur on the ground or on shrubbery. Both genera are nocturnal and are active on the cold nights of winter. Eggs are apparently laid in the ground and hatch in early autumn.

PYRGOMORPHOIDEA   This superfamily now includes only the Pyrgomorphidae, comprising a most diverse assemblage of genera. The size and shape of the grasshoppers is extremely variable, but the head has a characteristic fastigial furrow. The family comprises some 30 tribes, and two, at the most, subfamilies are recognized. The family is mostly tropical, with the greatest number of species being from Africa and Madagascar, but the greatest diversity of genera is from the Australasia. In the Americas, the family is represented only in Mexico.

Pyrgomorphs live on bushes, herbage, grasses, and sedges and on soil and sand. Although many are camouflaged and show adaptation in body shape with twigs, grasses, leaves, and so on, many are stunning examples of aposematic coloration. They perform impressive displays when annoyed or threatened. Several species eject irritant fluids or a foamy froth when irritated. Not only are some of these substances toxic to mammal, birds, and reptiles, they are poisonous to humans. The majority of species oviposit in the ground like other grasshoppers, but some have been observed ovipositing in rotting logs, epiphytes and, most likely, the soil in trees caused by the presence of epiphytes. A few species are facultatively aquatic. Many species are gregarious, especially in the juvenile stages, and move together in the manner of locusts. At times they cause damage to crops but, in general, they are not pests.

PNEUMOROIDEA   The bladder hoppers, or flying gooseberries, comprise a single, most peculiar family, the Pneumoridae. They are usually large with some species reaching 10 cm or more in length. The body shape of males is characteristic (Fig. 2C), that of females more resembling a normal grasshopper. The major part of the abdomen in males forms a huge, inflated, resonating chamber that is highly translucent. Females produce sounds but in a different way. Biological observations made more than 200 years ago indicate that males produce a loud noise after dark from shrubs in dry habitats. The sound, with its deep resonance, is often mistaken that of for a larger animal such as a bullfrog. There are two subfamilies with nine genera and some 20 species. They are all confined to southern and eastern Africa. They are of no economic importance.

EUMASTACOIDEA   The Eumastaciodea consists of two families, the Proscopiidae and the Eumastacidae. Both have short antennae and an angular head that often appears too large for the body. Most species are wingless, but there are winged species that often resemble damselflies. The proscopiids, or false stick insects, are readily distinguished by elongate and twiglike appearance. They have an exaggerated head that appears too large for the body. All species are wingless, and some can be quite large (2.5–16.5 cm). Sixteen genera are all confined to South America.

The Eumactacidae, or monkey grasshoppers, is a much larger group. Many subfamilies have been recognized with hundreds of species. Some of these taxa have been regarded as separate families in some publications. Most species are small, with the body seldom exceeding 4.5 cm. Many are wingless, but there are many Old World forms that are fully winged and readily fly. Most species have a characteristic way, shared with the proscopiids, of sitting, exposed, with the hind legs splayed akimbo. They are diurnal and readily fly in the sun, and they resemble damselflies in several respects. Eumastacids feed on a variety of plant types ranging from grasses and sedges to desert shrubs and ferns in Old World rain forests. Probably the majority of species are nocturnal.

The Australian endemic subfamily, Morabinae, comprises 41 genera with 243 species. They are elongate, matchsticklike plain brown or green grasshoppers that do not sit with legs akimbo. They live on a wide variety of plants and are often very localized in their distribution. A fossil species of the extant genus *Erucius*, of the Eruciinae, has been discovered in Oligocene deposits in western North America. Present-day species in this genus are found in Malaysia and the Philippines.

TRIGONOPTERYGOIDEA This superfamily comprises two families: Trigonopterygidae and the Xyronotidae. The Trigonopterygidae are now not believed to be related to the pyrgomorphs or pneumorids (Fig. 1A). They have been called broad-leaf bushhoppers. There are two subfamilies. The group is confined to Asia, and several of the species occur in Borneo, where they live on the ground in dead leaves they resemble.

The Xyronotidae, or razor-backed bushhoppers, comprise a single genus with only two species with no known relatives. These grasshoppers live on the ground in leaf litter and have continuous generations. They are found only in Mexico.

TETRIGOIDEA This suborder contains the grouse locusts or pygmy grasshoppers. These are small usually gray, black, or mottled grasshoppers seldom exceeding 20 mm in body length (Fig. 2D). Depending on the classification, there are two families, or one family and many subfamilies. About 1000 species are known in 185 genera, with the majority of species in the Tetriginae. Most species live on the ground, most often on moist ground or along streams and ponds, where they feed on algae and diatoms. Members of the tribe Scelimini have members that are fully aquatic and can swim effectively underwater. In tropical climes some species are arboreal and live among lichens and mosses in tree buttresses or even higher in the canopy. Eggs of the terrestrial species are laid in the soil and bear a peculiar terminal filament that is directed upward when the eggs are laid. They are mostly of little economic concern, although some species are said to feed on rice.

TRIDACTYLOIDEA This small group contains some of the most bizarre of the orthopteroids. The pygmy mole crickets, pygmy sand crickets and mud crickets, and sand gropers are peculiar in a many respects. They range in size from less than 4 mm to more than 80 mm. Three families are recognized. The Tridactylidae and Rhipipterygidae are more closely related to one another than to the Cylindrachetidae.

Tridactylidae and Rhipipterygidae The tridactylids and rhipipterygids are small, usually variegated black, yellowish, or reddish minute cricketlike insects (Fig. 2E). The two families comprise about 210 species in 11 genera.

Most of the tridactylids and rhipipterygids are associated with damp habitats. They seem to be gregarious, and those that live in these situations can construct "nests" out of mud and debris to spend the night or to "hibernate." Many are active swimmers. There is another group that is arboreal, living on leaf surfaces in tropical climes, where they feed on the rain of particulate matter from the canopy. Many of these species have wasplike color patterns and wasplike "jerky" movements. Others living in the same habitats are dark blue and have white tips to the antennae.

Cylindrachetidae The Cylindrachetidae or sand gropers have an elongate or wormlike appearance. There are nine known species in three genera. They reflect a Gondwanan distribution, with species in Australia, New Guinea, and Patagonia. At least one species causes considerable damage to wheat in Western Australia, where large populations build up in the soil and feed on the root of the plants.

See Also the Following Articles
Crickets • Locusts

Further Reading
Bailey, W. J., and Rentz, D. C. F. (eds.). (1990). "The Tettigoniidae: Biology, Systematics and Evolution." Crawford House Press, Bathurst, Australia.
Chapman, R. F., and Joern, A. (eds.). (1990). "Biology of Grasshoppers." Wiley, New York.
Dirsh, V. M. (1975). "Classification of Acridomorphid Insects." E. W. Classey, Faringdon, U.K.
Field, L. (2001). "The Biology of Wetas, King Crickets and Their Allies." CAB International, Wallingford, Oxon, U.K.
Flook, P. K., Klee, S., and Rowell, C. H. F. (1999). Combined molecular phylogenetic analysis of the Orthoptera (Arthropoda, Insecta) and implications for their higher systematics. *Syst. Biol.*, **48**, 233–253.
Gwynne, D. T. (2001). "Katydids and Bush-crickets: Reproductive Behavior and Evolution of the Tettigoniidae." Cornell University Press, Ithaca, NY.
Naumann, I. D. (ed.). (1991). "The Insects of Australia, a Textbook for Students and Research Workers," Vols. 1, 2. Melbourne University Press, Carlton, Australia.
Preston-Mafham, K. (1990). "Grasshoppers and Mantids of the World." Blandford, London.
Rentz, D. C. F. (1996). "Grasshopper Country: The Abundant Orthopteroid Insects of Australia." New South Wales University Press, Sydney, Australia.

Ovarioles

Diana E. Wheeler
University of Arizona

Ovarioles are egg-producing tubules that are the fundamental units of ovaries in female insects. The number of ovarioles in each ovary is typically 4–8, but varies widely depending on the particular insect and its ecology.

GENERAL ARRANGEMENT AND STRUCTURE

Each ovariole is a tube in which oocytes form at one end and complete development as they reach the other. The terminal filament and the germarium, which contains germ cells, are at the distal end. Ovarioles may have one of several topological arrangements within an ovary. In some species ovarioles join the end of an oviduct radially around a central point. In others, ovarioles arise in single file off the oviduct, like teeth on a comb.

NUMBER

The number of ovarioles per ovary varies with taxon, size, and life history. All Lepidopteran females have four ovarioles, but many groups tend to be more variable, both within and across species. Variability in ovariole number is particularly spectacular in social insects. Obligately sterile workers in ants can lack ovarioles entirely, and the most fecund queens in ants and termites have about 1200 ovarioles per ovary.

DEVELOPMENTAL ORIGIN

The period during which ovarioles form varies widely in insects, ranging from embryonic development in aphids to the pupal stage in flies. In some taxa, the number of ovarioles can be adjusted based on environmental factors. In *Drosophila,* for example, ovarioles form during the pupal period. This timing provides the opportunity for the number of ovarioles constructed to be adjusted based on previous diet and temperature. In honey bees, however, ovarioles form in early larval development. The number of ovarioles formed is at first the same in future queens and workers. In workers, however, most ovarioles undergo cell death, whereas those in developing queens persist.

SOMATIC TISSUE AROUND DEVELOPING OOCYTES

Each ovariole is made up of both somatic and germ cells. The somatic tissue includes a tubular sheath surrounding all the developing eggs as well as follicle cells around each oocyte. The sheath consists of inner and outer layers. The outer sheath is an open network of cells, sometimes containing muscle. The tissue is rich in lipids and glycogen and is metabolically active. Even so, there is no evidence of direct involvement in oocyte development. Tracheoles form part of the outer sheath but do not penetrate below it. The outer sheath can be important in sequestering bacterial symbionts that will be passed on to offspring. The inner sheath is a layer of extracellular matrix. In addition to physical support, the inner sheath can function as a sieve.

A layer of follicle cells surrounds each developing oocyte. Follicle cells are very active metabolically, contributing a variety of materials essential to developing eggs. During yolk uptake, follicle cells can separate slightly, allowing vitellogenin-laden hemolymph to contact the oocyte surface directly. The oocyte then can take up vitellogenin and other nutrients. As egg development nears completion, follicle cells secrete the eggshell, which consists of vitelline envelope and layers of chorion.

PATTERNS OF OOCYTE DEVELOPMENT

Ovarioles can be categorized based on how oocytes are produced from stem germ cells (Fig. 1). A stem germ cell produces two daughter cells when it divides; one remains a stem cell and the other becomes a cystoblast. Most commonly, cystoblasts undergo rounds of division but remain connected by intercellular bridges. This process is called cluster formation.

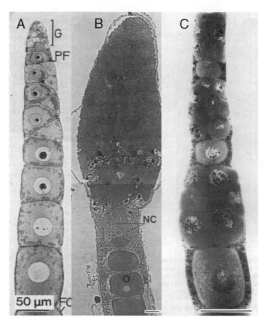

FIGURE 1 Patterns of oocyte development. (A) Panoistic ovary of an apterygote insect, *Lepisma saccharina.* G, germarium; FC, follicle cell. (Reprinted from R. C. King and J. Büning, 1985, The origin and functioning of insect oocytes and nurse cells. *In* "Comprehensive Insect Physiology, Biochemistry, and Pharmacology," Vol. 1, pp. 37–82, with permission from Elsevier Science.) (B) Teletrophic ovary of a heteropteran insect, *Dysdercus intermedius.* The nutritive cords (NC) extending from the nurse cells to the oocyte are clearly visible. (C) Polytrophic ovary of a carabid beetle, *Nebria brevicollis.* (B and C reproduced, with kind permission from Kluwer Academic Publishers, from J. Büning, 1994, Figs. 3.35 and 3.57.)

Generally, only one of the daughter cells in a cluster becomes an oocyte, while the remainder degenerate or become nurse cells. The type of oocyte development in which an oocyte is connected to sister cells that contribute to the contents of the egg is termed meroistic.

Stem cells in ovarioles of the most primitive insects do not undergo cluster formation. Instead, each cystoblast develops into an oocyte. Such ovarioles are called panoistic. Insect orders that have primitively panoistic ovaries are Archeognatha, Thysanura, Odonata, Embioptera, and most of the orthopteroid orders. Plecoptera shows an interesting intermediate condition, in which cluster formation occurs, but all the daughter cells become oocytes. Panoistic ovaries are also found in more recently derived insect groups, in which the design has evolved secondarily, and these taxa include Siphonaptera, Strepsiptera, and Thysanoptera.

Meroistic ovaries can be subdivided into two types based on the location of accessory germ cells relative to the oocyte. In polytrophic ovaries, accessory germ cells maintain short connections and accompany the oocyte as it moves down the ovariole. In telotrophic ovaries, accessory tissue remains at the distal end of the ovariole.

In polytrophic ovaries, follicle cells encapsulate both the nurse cells and the oocyte. Nurse cells generally are polyploid and express genes whose products are transported to the

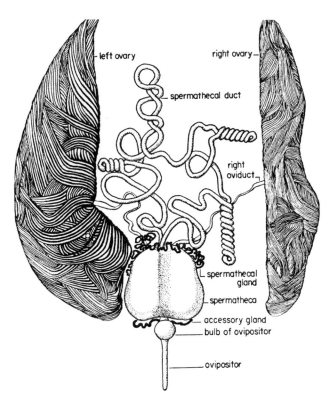

FIGURE 2 An army ant queen *(Eciton burchelli)* shows extreme development of ovarioles in both number and length. [Reproduced from H. R. Hagan, 1954, The reproductive system of the army-ant queen, *Eciton (Eciton)*. American Museum Novitates No. 1663. Courtesy of the American Museum of Natural History.]

growing egg. Trichoptera, Lepidoptera, Hymenoptera, Diptera, adephagan Coleoptera, Mecoptera, and Neuroptera all have polytrophic ovaries.

In telotrophic ovarioles, accessory nurse cells remain in the anterior part of the ovariole and so, as an oocyte moves away, cords of oocyte tissue lengthen to maintain the connection. Teletrophic ovaries have apparently evolved independently several times. In Heteroptera, Coleoptera (Polyphaga), Raphidioptera, and Megaloptera, they evolved from polytrophic ovaries. Telotrophic Ephemeroptera evolved directly from panoistic ancestors.

Ovariole architecture is related to both phylogeny and life history. Both panoistic and meroistic ovarioles can support very rapid egg production as illustrated by the more than 40,000 eggs/day produced by both the panoistic ovarioles of the most fecund termite queens and the polytrophic ovarioles of the most fecund ants (Fig. 2). An important consequence of panoistic vs meroistic eggs is duration of embryonic development. The design of meroistic ovarioles facilitates the production of eggs well-stocked with ribosomes, mRNA, tRNA, and other macromolecules. Such eggs generally develop rapidly, as exemplified by *Drosophila melanogaster*. In contrast, autonomous oocytes produced by panoistic ovaries require longer periods to complete embryogenesis.

See Also the Following Articles
Egg Coverings • Reproduction, Female • Vitellogenesis

Further Reading
Büning, J. (1994). "The Insect Ovary." Chapman & Hall. London.
Chapman, R. F. (1998). "The Insects: Structure and Function," pp. 298–312. Cambridge University Press, Cambridge, UK.
Engelmann, F. (1970). "The Physiology of Insect Reproduction," pp. 4–7 and 45–49. Pergamon, Oxford.
Snodgrass, R. E. (1935). "Principles of Insect Morphology," pp. 552–561. McGraw–Hill, New York.
Stys, P., and Bilinski, S. (1990). Ovariole types and the phylogeny of hexapods. *Biol. Rev.* **65,** 401–429.
Wigglesworth, V. B. (1939). "The Principles of Insect Physiology," pp. 376–377. Methuen, London.

Oviposition Behavior

Marc J. Klowden
University of Idaho

Oviposition behavior comprises one of the final steps in insect reproduction. It involves the deposition of the mature egg outside the body of the female and includes a series of behavioral and physiological events that begin with the movement of the egg through the oviduct and end with the placement of the egg on a substrate that will support the development of the larva. Specialized behaviors and structures on the female allow her to place the eggs within a protected environment during oviposition.

OVULATION

Insect eggs develop within the ovaries, the reproductive structures of the female that are composed of tapering units called ovarioles. The oocytes differentiate from germ cells within the germarium of the ovariole, and as they begin their downward movement in the ovariole they are first completely surrounded by a monolayer of follicle cells (Fig. 1). These follicle cells are involved in the transport from the hemolymph into the oocyte of substances that are stored for later use and nourish the embryo during its development. Nurse cells may also be present to provide the oocyte with other maternal contributions, such as messenger RNA and mitochondria. The nurse cells subsequently degenerate, and the bulk of the yolk proteins deposited in the oocyte cytoplasm must then cross the layer of follicle cells after their synthesis in the fat body. During a later stage of development, the follicle cells also synthesize the eggshell (or chorion), which provides protection and waterproofing once the egg has been laid. After deposition of the chorion, the follicle cells degenerate, leaving the chorion as the outermost egg layer. With the follicle cells no longer on the outside of the egg, the egg is free to move out of the ovariole

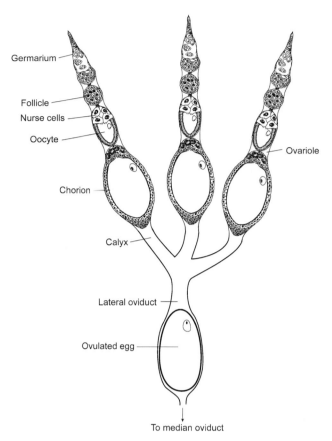

FIGURE 1 A cluster of three ovarioles that make up the larger ovary. Each has a calyx that connects to the lateral oviduct. Stem cells give rise to oocytes within the germarium, and as the oocyte descends within the ovariole, it becomes enclosed by follicle cells and deposits yolk into its cytoplasm. Upon degeneration of the follicle cells, the egg is free to move into the lateral oviduct in the process of ovulation. [Modified from Schwalm, F. E. (1988). "Insect Morphogenesis." Karger, Bosel.]

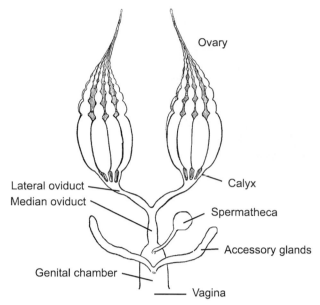

FIGURE 2 The female reproductive system, consisting of the ovaries that contain the ovarioles. As the egg moves down the median oviduct, it is fertilized by sperm released from the spermatheca. The egg is oviposited when it is deposited outside the body. [Modified from Snodgrass, R. E. (1935). "Principles of Insect Morphology." McGraw-Hill, New York.]

and into the oviduct, under the impetus of contractions of muscles in the oviduct walls. This movement of the egg to the outside of the ovariole is termed ovulation. Ovulation of the egg must occur before oviposition can take place.

The muscular contractions of the ovariole and oviduct that propel the egg through the reproductive tract are coordinated by hormones called myotropins. Myotropins are secreted by neurosecretory cells in the brain once the central nervous system has received the physiological confirmation that mating has occurred and that the eggs are mature. For example, in the bloodsucking bug *Rhodnius prolixus,* ovulation is initiated by a myotropin that is released only after the spermatheca has been filled with enough sperm and male accessory gland substances to produce a factor that induces the maturing eggs to begin producing 20-hydroxyecdysone. In the tsetse fly, *Glossina,* ovulation is initiated when a mature egg is present in the ovariole and the female has mated, but the stimulus from mating is not released until the female has received prolonged mechanical stimuli associated with copulation in addition to substances

from the male accessory glands. Even several short copulatory experiences that fail to transfer sperm can initiate ovulation if their total duration is as long as a single successful copulation.

FERTILIZATION

As the egg passes through the median oviduct, it is fertilized by sperm already stored within the female's capsulelike spermatheca (Fig. 2). Sensory receptors within the oviduct are mechanically stimulated by the distension the egg creates as it descends; motor neurons activate the muscular walls of the spermathecal duct that allow the sperm to exit the spermatheca.

In hymenopterans, where sex determination occurs by haplodiploidy, the process of sperm release is controlled more directly by the female. Eggs that are fertilized develop into female workers, but those that are not fertilized develop into male drones. The workers ultimately control whether the queen fertilizes the egg; they determine the size of the cells for rearing the larvae into which the queen deposits the egg. Before the egg is laid, the queen evaluates each cell with the sensory receptors on her ovipositor, opening her spermathecal duct only when placing an egg into the larger cell for female rearing.

The sperm enter the egg through the micropyle, a modification of the normally impermeable chorion. Just prior to fertilization, the oocyte has been arrested in metaphase of the first meiotic division, but shortly after the sperm has penetrated the egg and oviposition has occured, the oocyte completes its meiosis. The meiotic division results in a haploid oocyte nucleus and three polar nuclei that inhabit the periplasm at the periphery of the egg. The oocyte nucleus, surrounded by an island of cytoplasm, then moves to the interior of the egg, where it meets the sperm that has already entered. Syngamy, the union of sperm and egg, occurs at the interior. In some insects whose eggs develop by parthenogenesis (i.e., without the fertilization by sperm), a haploid polar nucleus combines with the haploid oocyte nucleus to restore the diploid number without requiring fertilization by male gametes.

OVIPOSITION

After ovulation and fertilization, the eggs are usually deposited outside the female's body during oviposition. The eggs move down the common oviduct by peristaltic waves of muscle contractions and out of the body through the ovipositor. The movement of the egg downward through the oviducts is facilitated by backwardly directed scales inside the oviduct, which act like a ratchet mechanism, allowing the egg to move in only one direction, down toward the genital opening.

Modified dermal glands known as female accessory glands may also be present on the common oviduct. These glands produce cement that allows the deposited eggs to be glued together or attached to the substrate. In some insects that retain their eggs after the young hatch, such as *Glossina* the accessory glands produce a nutritive secretion that nourishes the larvae during their entire larval period. In cockroaches and mantids, the female accessory glands produce the hardened egg case, or ootheca.

The spermathecae are used for the storage of sperm in the inseminated female. Also ectodermal in origin, the spermatheca generally opens into the common oviduct and releases sperm as the fully formed egg passes by. The spermathecal duct may contain glycogen deposits that can serve as an energy source for the sperm as they pass through to the egg. Nearest its opening to the outside, the common oviduct may be modified into a genital chamber that can be used to incubate eggs internally. The bursa copulatrix is an additional pouch within the chamber that is present in some insects into which sperm may first be deposited after mating. The sperm leave the bursa and then move into the spermatheca, where they are permanently stored. The bursa may have a series of toothlike structures that disrupt the spermatophore, the vessel in which the sperm are contained in more primitive insects, and facilitate their release. It may also secrete chemical signals into the hemolymph when it is filled with sperm to signal to the female that mating has occurred.

The control of oviposition in the cockroach *Spodromantis* is a good example of the integration of environmental and physiological information that must occur during egg laying. This insect normally lays its eggs at the beginning of the photophase. The brain integrates the information it receives about the photoperiod and the presence of mature oocytes and triggers the release of an oviposition-stimulating hormone that activates the ovipositor, ovariole, and oviduct muscles. As the insect probes the substrate with its ovipositor, tactile sensations from sensilla are sent to the terminal abdominal ganglion,

which controls further movements of the egg and secretions by the accessory glands.

In most species, ovulation and oviposition may be part of a continuum, but in others they may be separate events in which the eggs are retained for a variable period between ovulation and oviposition. Some species of ovoviviparous cockroaches retain in a brood sac eggs that have been ovulated until the female finds a suitable place to lay them. The nymphs hatch inside the female. The viviparous tsetse retain their larvae after they have hatched from the egg and provide them during the larval stage with food that is secreted from their accessory glands. "Larviposition" occurs toward the end of the larval stage, and after oviposition the larvae pupate in the soil.

Most insects lack specialized structures for the deposition of eggs, but in some species the terminal segments of the abdomen form a telescoping ovipositor that can be used to deposit eggs into crevices or even, when the tip is modified into sawlike blades, into hardened substrates such as wood. This ovipositor may be derived from the modified appendages just described or from the modified terminal abdominal segments themselves. Special muscles within the ovipositor allow it to engage in superextension, lengthening considerably to reach hidden sites into which the eggs can be placed. Both the tips of the abdomen and the ovipositor may contain sensory receptors that can provide the female with information about the nature of the oviposition substrate. Sensory cues that are evaluated by the receptors are important in locating an oviposition site and in initiating oviposition once the site has been found.

Prior to depositing her eggs, the female must engage in behaviors that will bring her to an environment that is suitable for larval development. This is especially necessary for the eggs of holometabolous insects, since the larvae are very different from the adults. Being relatively immobile, the larvae depend on the adult to locate a site on which they can develop. For example, adult mosquitoes are terrestrial but their larvae must develop in water. Although the adults are generally attracted to stimuli from vertebrate hosts for blood, when they carry mature eggs they become more sensitive to the stimuli from potential oviposition sites. The presence of mature eggs appears to trigger a physiological switch that changes the behavior of the female, attracting her to oviposition sites suitable for her eggs. In the absence of mature eggs, the female is more attracted to a host for a blood meal that would support another batch of eggs.

Lepidopterans undergo a sequence of behaviors, including searching, orientation, encounter, landing, evaluation of the surface, and acceptance. Stimuli received during each phase are integrated by the central nervous system, which processes the sensory information from sensory receptors on the antennae, tarsi, and ovipositor. Visual cues such as plant color and shape are important during searching behavior. Volatile chemicals produced by the plant initiate the orientation and encounter behaviors that attract the female from a distance of several meters. Visual cues are once again important in landing behavior, with the involvement of chemical cues that act as attractants, arrestants, or repellents. Once on the host plant, the female evaluates the physical and chemical cues from the plant surface and often begins drumming her front legs on the leaf surface, perhaps to provide more chemical information about the plant to tarsal sensory receptors. By using the total sensory input to the central nervous system, the female is able to recognize the suitability of the substrate for her eggs before they are laid.

The torsalo, *Dermatobia hominis,* has an unusual way of finding an oviposition site. The gravid female captures another bloodsucking fly and glues her eggs to the underside of the carrier's body. When the carrier seeks out a vertebrate host for a blood meal, the larva hatches and burrows through the skin of the host. The larva develops in the vertebrate host, exits prior to the pupal stage, and pupates, in the soil.

See Also the Following Articles

Accessory Glands • Egg Coverings • Reproduction, Female: Hormonal Control of

Further Reading

Austin, A. D. and Browning, T. O. (1981). A mechanism for movement of eggs along insect ovipositors. *Int. J. Insect Morphol. Embryol.* **10,** 93–108.

Curtin, T. J., and Jones, J. C. (1961). The mechanism of ovulation and oviposition in *Aedes aegypti. Ann. Entomol. Soc. Am.* **54,** 298–313.

Davey, K. G. (1985). The female reproductive tract. *In* "Comprehensive Insect Physiology Biochemistry, and Pharmacology (G. A. Kerkut and L. I. Gilbert, eds.), Vol. 1, pp. 1–14. Pergamon Press, Oxford, U.K.

Hinton, H. E. (1981). "The Biology of Insect Eggs," Vol. 1. Pergamon Press, Oxford, U.K.

Mesnier, M. (1984). Patterns of laying behaviour and control of oviposition in insects: Further experiments on *Sphodromantis lineola* (Dictyoptera). *Int. J. Inverteb. Rep. Dev.* **7,** 23–32.

Mesnier, M. (1985). Origin and release sites of a hormone stimulating oviposition in the stick insect *Clitumnus extradentatus. J. Insect Physiol.* **31,** 299–306.

Ramaswamy, S. B. (1988). Host finding by moths: Sensory modalities and behaviours. *J. Insect Physiol.* **34,** 235–249.

Renwick, J. A. A., and Chew, F. S. (1994). Oviposition behavior in Lepidoptera. *Annu. Rev. Entomol.* **39,** 377–400.

Schwalm, F. E. (1988). "Insect Morphogenesis." Karger, Basel.

Snodgrass, R. E. (1935). "Principles of Insect Morphology." McGraw-Hill, New York.

Sugawara, T., and Loher, W. (1986). Oviposition behaviour of the cricket *Teleogryllus commodus:* Observation of external and internal events. *J. Insect Physiol.* **32,** 179–188.

Thompson, J. N., and Pellmyr, O. (1991). Evolution of oviposition and host preference in Lepidoptera. *Annu. Rev. Entomol.* **36,** 65–89.

Parasitoids

Nick Mills
University of California, Berkeley

Parasitoids are insects with a parasitic larval stage that develops by feeding on the body of a single host insect or other arthropod. Feeding by the larval parasitoid invariably results in the death of its host, and the resulting adult parasitoid is a free-living insect. Thus parasitoids occupy an intermediate position between predators and true parasites; in contrast to parasites they kill their host like a predator, but in contrast to predators they require only a single host to complete their development, as is the case for parasites.

Nearly 10% of described insect species are parasitoids and, as they often belong to poorly known groups of insects, it has been suggested that they are more likely to represent 20–25% of all insect species. Since the first accounts of insect parasitism at the beginning of the 18th century, parasitoids have attracted considerable attention because of their potential to regulate the abundance of insect hosts in both natural and managed ecosystems. Numerous insect pests have been effectively controlled through releases of insect parasitoids in biological control programs around the world. Although not always applicable or successful, biological control continues to provide some dramatic examples of sustained long-term control of invasive insect pests. As a result of the research conducted through biological control programs, parasitoids have also become important model organisms for the study of ecology, behavior, and evolution.

ORIGIN AND DIVERSITY OF PARASITISM

About 74% of the known parasitoid species belong to the parasitic Hymenoptera, an arbitrary division of the suborder Apocrita of the order Hymenoptera. It is generally believed that parasitism evolved just once in the Hymenoptera and that the Apocrita together with the sawfly family Orrusidae form a holophyletic group that includes all of the known parasitic wasps. Although the larvae of some species may initially feed on microorganisms in the tunnels of wood-boring insects, the young larvae of others feed externally and subsequently internally on larvae of the borers themselves. Many horntails and wood wasps (sawfly superfamily Siricoidea) carry symbiotic fungi in pouches located at the base of the ovipositor (mycangia) that are inoculated at oviposition, allowing the larvae to feed on infected and partially digested wood. However, not all siricoids carry fungal symbionts and some may have evolved to kill those that did and subsequently feed on the more nutritious dead insect rather than the wood. This trait is seen among the present-day orrusids.

Within the parasitic Hymenoptera there are close to 64,000 described species of parasitoids in 10 superfamilies that are, with rare exception, exclusively parasitic (Table I). However, parasitoids are also known from five other insect orders. In contrast to the Hymenoptera, parasitism has evolved repeatedly within these orders, from a fungal-feeding (e.g., Rhipiphoridae), dead-organism-feeding (e.g., Phoridae, Sarcophagidae), predatory (e.g., Carabidae, dipteran families), or phytophagous (e.g., Lepidoptera) ancestor. For example, parasitism is estimated to have occurred 21 times in the Diptera, producing nearly 15,000 described parasitoid species, and 15 times in the Coleoptera, producing a further 3400 parasitoid species (Table I). Parasitism is unusual and rare among the remaining three orders.

PARASITOID LIFESTYLES

Parasitoids are frequently categorized as having larvae that are either ectoparasitic (external feeders) or endoparasitic (internal feeders) and either solitary (one per host) or gregarious (several per host) in their development. However, a more useful categorization of the lifestyles of parasitoids is the dichotomy between idiobiosis and koinobiosis. Idiobionts paralyze and/or

TABLE I   The Main Superfamilies of Hymenoptera and Families of Other Insect Orders That Contain Parasitoids, with an Indication of either the Total Described Species for Taxa That Are Exclusively Parasitic or the Number of Described Parasitoid Species for Taxa (Followed by *) That Include Many Nonparasitic Species

Order	Superfamily/family	World species
Hymenoptera	Orussoidea	75
	Trigonaloidea	70
	Evanioidea	1,050
	Cynipoidea	2,335*
	Chalcidoidea	17,500*
	Proctotrupoidea	6,135
	Ceraphronoidea	250
	Ichneumonoidea	25,000
	Chrysidoidea	5,850*
	Vespoidea	5,500*
Diptera	Cecidomyiidae	6*
	Acroceridae	475
	Bombyliidae	3,000
	Nemestrinidae	300
	Phoridae	300*
	Pipunculidae	600
	Conopidae	800
	Sarcophagidae	1,250*
	Tachinidae	8,200
Coleoptera	Carabidae	470*
	Staphylinidae	500*
	Rhipiphoridae	400
	Meloidae	2,000*
	Mengeidae	10*
Lepidoptera	Pyralidae	1*
	Epipyropidae	10*
Neuroptera	Mantispidae	50*
Trichoptera		1*

Note. Data from Godfray (1994).

arrest the development of a host at oviposition, providing their larvae with an immobilized static resource on which to feed. In contrast, koinobionts allow the host to continue to feed and/or develop after oviposition, such that their larvae feed on an active host that is killed only at a later stage. Although ecto- and endoparasitism are distinct modes of parasitism, the separation of idiobiont and koinobiont lifestyles is less clear. There should be no problem in correctly categorizing parasitoids that attack an active stage of the host life cycle (larva or adult) or those that allow a transition between life cycle stages of the parasitized host (egg–prepupal and pupal–adult parasitoids). However, the distinction is not always so obvious for parasitoids that attack and complete their development in an inactive stage of the life cycle (egg or pupa).

Idiobionts

Idiobiont parasitoids frequently attack hosts that are concealed in plant tissues or exposed hosts that provide some other form of physical protection, such as the scale covering of armored scale insects. Thus, many have long ovipositors to reach their concealed hosts and strong mandibles to escape from concealed locations. The majority of idiobionts that attack concealed hosts have larvae that are ectoparasitic and feed on hosts in the later stages of their development. An ovipositing adult first injects a venom into the host, to induce temporary or permanent paralysis, and then oviposits on or near to the immobilized host. In a few cases (some Bethylidae, Braconidae, and Eulophidae in the Hymenoptera), the parent female remains with the parasitized host either to defend her young offspring against competitors or hyperparasitoids or to ensure continued paralysis of the host. Ectoparasitic idiobionts are often long lived, feeding for somatic maintenance on honeydew, nectar, or other plant exudates. They are synovigenic, meaning that they continue to mature eggs throughout adult life and produce large yolk-rich eggs by using nutrients gained from host feeding (see below). Consequently, ectoparasitic idiobionts have a relatively low rate of host attack and a low fecundity. The hatching larvae are protected from desiccation by the host concealment, but develop continuously and rapidly to consume the host before it is attacked by scavengers or microbial decay.

Endoparasitic egg and pupal parasitoids are also considered to be idiobionts, although as noted above, this categorization is less clear. In this case the hosts are frequently exposed rather than concealed, and in place of paralysis, development of the host appears to be arrested by secretions either from the ovipositing adult (egg parasitoids) or from the young larva (pupal parasitoids). Endoparasitism of host eggs is facilitated by the lack of an immune defense, but it is not known how pupal idiobionts avoid such defenses in their hosts.

The absence of an intimate association between the juvenile stages of an ectoparasitic idiobiont and its host allows this group of parasitoids to have a broader host range, using a variety of hosts that share a common habitat or host plant. Some endoparasitic idiobionts are more restricted in their host range, however, perhaps because of a need for more specific cues in host recognition or for detoxification of the chemical defenses of exposed hosts.

Koinobionts

All dipteran parasitoids, the majority of coleopteran parasitoids, and many hymenopteran parasitoids adopt a koinobiont way of life. Koinobiont parasitoids attack both exposed and concealed hosts, and the majority of species have endoparasitic larvae. Endoparasitic koinobionts attack a broad range of developmental stages of their hosts. Access to concealed hosts is facilitated either by attacking the more accessible egg or young larval stage of their host (hymenopteran koinobionts only) or by production of a free-living first instar that can complete the location of a suitable host.

Koinobionts typically oviposit into the body of their hosts with minimal disruption to the normal activity of the host. The hemocytic immune response of the host can result in the encapsulation of parasitoid eggs and larvae and is a significant obstacle to endoparasitism. This response can be overcome by (1) avoidance, by placing eggs in host tissues such as the salivary gland or nerve ganglia; (2) evasion, by producing an egg with a fibrous coat or a coating of proteins that are not recognized as foreign to the host; (3) suppression, by injecting immuno-suppressive polydnaviruses or virus-like particles at oviposition; or (4) subversion, by allowing host hemocytes to form a sheath around the developing larva that is attached to the host tracheal system to avoid asphyxiation. Koinobionts are also mostly synovigenic, but differ from idiobionts in having a much greater fecundity, with the majority of their eggs produced early in adult life. Koinobiont eggs are relatively small and enrichment occurs after oviposition through absorption of nutrients from the host hemolymph. The greater emphasis on early reproduction is often associated with a shorter adult life, although longevity is greatly influenced by access to a sugar food source.

Many koinobionts exhibit protracted larval development, remaining as a first instar within the host until the latter has retreated to its pupation site. This both allows the host larva to complete its development without suffering debilitating effects from parasitism and permits the parasitoid to remain in its most competitive stage, because first instars bear strong defensive mandibles, until the host has gained maximum size as a resource for parasitoid larval development. Thus, koinobiont parasitoids are often more specialized in their host range than idiobionts, because of their more intimate relationship with an actively feeding or reproducing life stage of the host and the need to overcome its immune defenses.

BEHAVIOR AND INTERACTIONS

Host Feeding

Many idiobiont adult females acquire nutrients from the host by feeding on the body fluids that exude from wounds inflicted by the ovipositor, a process known as host feeding. Some small gregarious parasitoids are able to host feed on the same host individual used for oviposition, but in most cases host feeding is destructive and can be responsible for substantial levels of host mortality. Destructive host feeders select smaller host individuals for host feeding, because they are unsuitable for parasitism, and then use larger hosts for oviposition.

Interactions among Parasitoids

Because the majority of host insects are attacked by several different parasitoid species, a variety of trophic and competitive interactions occur among them. Hyperparasitism is a trophic interaction that occurs when a secondary parasitoid parasitizes a primary parasitoid. Competitive interactions include superparasitism and multiparasitism. Superparasitism occurs with more than one oviposition by one or more individuals of the same parasitoid species into the same host individual. The resulting intraspecific competition between parasitoid larvae leads to the death of all but one individual, in the case of solitary parasitoids, and to male bias in the sex ratio and reduced

adult size among the progeny of gregarious parasitoids. Multiparasitism is the corresponding interspecific competition that results from oviposition by two or more different parasitoid species in the same host individual. The outcome of multiparasitism is often indeterminate; it can favor the species that attacked first but it can also be fixed, with a strong competitor being the victor whatever the sequence of attack. A particularly interesting form of the latter is cleptoparasitism, in which the success of host location by a cleptoparasitoid is facilitated by its response to chemical markers used by an inferior competitor to avoid reattacking a previously parasitized host. The cleptoparasitoid is able to steal the host from its inferior competitor by having an aggressive first instar that is able to kill the original occupant of the host.

Clutch Size and Sex Ratio

One of the most important "decisions" for a gregarious parasitoid is how many eggs to lay on a particular host. Clutch size increases with the size and quality of a host, but decreases as the rate of host encounters increases and often decreases with the age of the parasitoid. Gregarious parasitoids adjust clutch size to match the quality and frequency of hosts encountered, thereby balancing the need to maximize reproductive output and to minimize intraspecific competition among the larval brood. Because hymenopteran parasitoids use haplodiploid reproduction (males develop from unfertilized haploid eggs, females from fertilized diploid eggs), parent females can choose the sex of their offspring. Solitary parasitoids tend to allocate male eggs to small or low-quality hosts and female eggs to large or high-quality hosts, but typically produce a balanced sex ratio. In contrast, female bias is frequent in the sex ratio of gregarious parasitoid broods in which the probability of sibling mating is high and sons compete within broods for mates. Thus, local mate competition within broods of gregarious hymenopteran parasitoids leads to the allocation of just enough sons to be able to mate effectively with all the daughters in the brood.

PARASITOID COMMUNITIES

Parasitoids can utilize hosts in a variety of ways to support the development of their progeny. For example, a lepidopteran host such as the gypsy moth *(Lymantria dispar)* is attacked at different stages through its life cycle by a series of 13 species in six parasitoid guilds (Fig. 1). Very few terrestrial insects escape the attention of parasitoids, exceptions being the few taxa that are either too small (e.g., Adelgidae) or too well defended (e.g., *Dactylopius* mealybugs). However, the parasitoid load, or number of parasitoid species supported by a host species, varies tremendously. For example, the average number of parasitoid species associated with an aphid is 1.7 in comparison with 12.4 for bivoltine Lepidoptera. In contrast, aquatic insects support relatively few parasitoids and are most vulnerable during the nonaquatic stages of their life cycle. It is notable, however, that a few mostly very small parasitoids

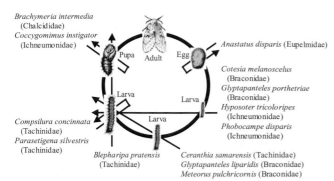

Brachymeria intermedia
(Chalcididae)
Coccygomimus instigator
(Ichneumonidae)

Anastatus disparis (Eupelmidae)

Cotesia melanoscelus
(Braconidae)
Glyptapanteles porthetriae
(Braconidae)
Hyposoter tricoloripes
(Ichneumonidae)
Phobocampe disparis
(Ichneumonidae)

Compsilura concinnata
(Tachinidae)
Parasetigena silvestris
(Tachinidae)

Blepharipa pratensis
(Tachinidae)

Ceranthia samarensis (Tachinidae)
Glyptapanteles liparidis (Braconidae)
Meteorus pulchricornis (Braconidae)

FIGURE 1 The parasitoid assemblage associated with the gypsy moth *(L. dispar)* as it passes through its life cycle in Eurasia. Arrows indicate the host stages attacked and killed by the six different guilds of parasitoid species.

are known to be truly aquatic, using their legs or wings to swim through the water to locate submerged host eggs.

Parasitoid load is determined by a combination of phylogeny, feeding niche, abundance, and chemical defense. The absence of a pupal stage greatly reduces the parasitoid load of a host, but even among the holometabolous insects, beetles consistently support far fewer parasitoids than moths, indicating the importance of host phylogeny. The feeding niche of a larval host also has an important influence on parasitoid load, being greatest among those hosts that have restricted mobility and poor protection (e.g., leafminers, casebearers) and reduced either by greater mobility (e.g., external feeders) or by greater protection (e.g., borers). The greater the abundance of a host, the greater its parasitoid load, because a number of less specialized parasitoids and even incidental species are able to make use of an abundant resource. Then, finally, the sequestration of defensive plant chemicals by externally feeding host larvae appears to offer a further line of defense that can lead to a reduction in parasitoid load.

It has frequently been suggested that parasitoid diversity declines from temperate zones to the tropics. Such a pattern is evident for the very species-rich superfamily the Ichneumonoidea (Ichneumonidae and Braconidae), but is not upheld among the Chalcidoidea. Nonetheless, the decline, or absence of an increase, in overall parasitoid diversity in the tropics is an interesting anomaly in comparison to the increase in diversity of their insect hosts in the tropics. An increased level of predation (notably by ants) in the tropics, a general reduction in the abundance of each host species due to fragmentation of resources, and a greater availability of plant-based chemical defenses for host insects to use for protection may all help to account for this anomaly.

See Also the Following Articles

Biological Control • Gypsy Moth • Host Seeking, by Parasitoids • Population Ecology

Further Reading

Eggleton, P., and Belshaw, R. (1992). Insect parasitoids: An evolutionary overview. *Philos. Trans. R. Soc. London B* **337**, 1–20.

Feener, D. H. (1997). Diptera as parasitoids. *Annu. Rev. Entomol.* **42**, 73–97.
Gauld, I. D., and Hanson, P. E. (1995). The biology of Hymenoptera. *In* "The Hymenoptera of Costa Rica" (P. E. Hanson and I. D. Gauld, eds.). Oxford University Press, Oxford, UK.
Godfray, H. C. J. (1994). "Parasitoids: Behavioral and Evolutionary Ecology." Princeton University Press, Princeton, NJ.
Gordh, G., Legner, E. F., and Caltagirone, L. E. (1999). Biology of parasitic Hymenoptera. *In* "Handbook of Biological Control" (T. S. Bellows and T. W. Fisher, eds.). Academic Press, San Diego.
Mills, N. J. (1994). Parasitoid guilds: A comparative analysis of the parasitoid communities of tortricids and weevils. *In* "Parasitoid Community Ecology" (B. A. Hawkins and W. Sheehan, eds.). Oxford University Press, Oxford, UK.
Quicke, D. L. J. (1998). "Parasitic Wasps." Chapman & Hall, London.
Strand, M. R. (1995). Immunological basis for compatibility in parasitoid–host relationships. *Annu. Rev. Entomol.* **40**, 31–56.
Whitfield, J. B. (1998). Phylogeny and evolution of host–parasitoid interactions in Hymenoptera. *Annu. Rev. Entomol.* **43**, 129–151.

Parental Care

Michelle Pellissier Scott
University of New Hampshire

Parental care in insects ranges from covering eggs with a protective coating to remaining to feed and protect young to forming eusocial societies with life-long associations of parents and offspring and alloparental care. Most commonly care is provided solely by females; males rarely care for eggs and young alone, but in some cases males and females form long associations to rear young to adulthood (Table I). Parental care has evolved independently many times, and examples do not follow along phylogenetic lines. This trait is widespread taxonomically and is most developed in Hemiptera (true bugs), Thysanoptera (thrips), Embioptera (webspinners), Coleoptera (beetles), Hymenoptera (ants, bees, and wasps), and Isoptera (termites).

FORMS OF PARENTAL CARE

Maternal Care

The most rudimentary form of maternal care is provided by females that incorporate toxins into their eggs, oviposit them in protected places, or cover their eggs with a hard shell or waxlike compound before abandoning them (Table I). For example, embiopteran webspinner females *(Antipalurai urichi)* cover their eggs with layers of macerated bark and other substrate materials and silk to protect them from hymenopteran parasites. Webspinner maternal care is more complex and extensive than egg protection, and these females construct silk galleries and remain with their eggs and nymphs.

Many species of insects guard their young against predators by using chemicals or defensive behaviors. Care

TABLE I Types of Parental Care by Insects

Behavior[b]	Number of orders[a]		
	Maternal care	Paternal care	Biparental care
EP	6	0	0
EC	12	2	3
YC	9	2	4
YP	6	0	3
OV	10		
PZ		9	

[a]Number of orders in which parents exhibit the behaviors listed.
[b]Key: EP, either or both parents cover eggs before and emerge; EC, either or both parents remain and guard eggs; YC, either or both parents remain with young and care for them; YP, either or both parents provision young or regurgitate food to them; OV, females extend development internally and are ovoviviparous, larviparous, or viviparous; PZ, males provide prezygotic investment.

may end when young hatch, or it may extend until larvae or nymphs are mature. For example eggplant lace bugs (*Gargaphia solani*) guard their eggs and gregarious nymphs until maturity; if a predator approaches, the female rushes at it, fanning her wings.

A second major function of maternal care is to facilitate feeding. A plant-feeding membracid bug, *Umbonia crassicornis,* cuts slits in the bark with her ovipositor and remains with the nymphs, actively maintaining feeding aggregations until the young reach adulthood. Parental care often comprises a suite of adaptations of multiple behaviors that serve multiple functions. A well-cited example of this behavior is the female salt-marsh beetle, *Bledius spectabalis,* which maintains a burrow shaped in a way that prevents flooding during high tide. She also provisions the young with algae, prevents mold, and protects the vulnerable first instars from attack by parasitic wasps.

In some insects, maternal investment takes the form of a period of internal development. Among insects, cockroaches carry this investment to the extreme and show the entire range of reproductive modes and maternal care. The oviparous German cockroach *Blatella germanica* carries her egg sac externally until nymphs hatch, and *B. vaga* produces maternal secretions that her neonates feed on briefly. In ovoviviparous species, eggs develop inside the body of the mother and have sufficient yolk to complete development. The viviparous cockroach *Diploptera punctata* displays a remarkable form of parental care. Females undergo a 60-day "pregnancy" during which a highly nutritious milk, secreted from the walls of the brood sac, is ingested orally by the developing young. At birth young are in an advanced state of development, and care is terminated shortly after birth.

Paternal Care

Exclusive paternal care of eggs or larvae is restricted to about 100 species of insects, almost all within the Hemiptera. For example, in a giant water bug, *Abedus herberti,* females adhere their eggs to the wingcovers of a male, which then stops feeding and instead spends his time aerating the eggs at the water surface, protecting them from predators until young emerge. Males of the subsocial spider-hunting wasp, *Trypoxylon superbum,* are an unusual example from the Hymenoptera. After the female has provisioned and sealed the cells, males remain to guard nests against parasitism and ant predation.

Indirect paternal contributions to offspring are widespread in a number of taxa. Males may invest in offspring with nutritional offerings to the female in the form of nuptial gifts of captured prey items or even their own bodies. They may transfer proteins or protective substances in a spermatophore. Male katydids are excellent examples because they provide a spermatophore during copulation that may be as much as 40% of their body mass; spermatophore nutrients have been shown to be important to the reproductive success of females. Male arctiid moths, *Utetheisa ornatrix,* provide a different sort of indirect paternal contribution when they transfer protective pyrrolizidine alkaloids to females during mating. These alkaloids are passed to the eggs, which are then unappealing to predators.

Biparental Care

Biparental care of offspring is restricted to beetles, termites, and cockroaches, and may include earwigs. It can be very elaborate and extensive. For example, the woodroach, *Cryptocercus punctulatus,* and all termites form life-long family associations. Male and female construct and guard an extensive tunnel system or a nest, and they protect and facilitate feeding of young until the offspring reach maturity. Woodroaches care for a single brood for 3 or more years, feeding them on hindgut secretions containing symbiotic fauna necessary to digest their wood diet. In many of the "higher termites," (e.g., Rhinotermitidae and the Termitidae), few or no workers or soldiers reproduce; rather, they remain as alloparents. Task specializations based on morphology and sex is strongly expressed. Primary reproductives may live 20 years or more, whereas workers and soldiers often live less than a year.

In another well-studied example, male and female burying beetles cooperate to bury and prepare small vertebrate carcasses to serve as the food source for their young. Both parents treat the carcass with preservative anal and oral secretions; both regurgitate semidigested protein to the begging larvae (Fig. 1). In the burying beetle, *Nicrophorus orbicollis,* males commonly remain in the nest until larvae are half-grown and the carcass is substantially consumed; females remain until larval development is complete and may even accompany larvae during the wandering stage. As with most species with biparental care, male and female burying beetles do not have exclusive, specialized tasks. When both parents are present, females feed larvae more often; but if the male becomes a single parent, he compensates for the loss of a mate with increased feeding rates. However, in another species with biparental

FIGURE 1 Female *N. orbicollis* regurgitates to larvae while the male, in the background, keeps the carcass free of fungi. (Photograph courtesy of Mark W. Moffett, University of California, Berkeley.)

care, the dung beetle, *Cephalodesmius armiger,* there are some task specializations. Male and female form a permanent pair-bond to rear one brood in a subterranean nest. Males forage outside for plant material that females process into brood balls into which they lay a single egg. Males continue to forage, and females enlarge the brood ball as the larvae grow.

EVOLUTION OF PARENTAL CARE

E. O. Wilson identified four ecological pressures that select for parental care in insects: (1) stable, structured environments such as wood, (2) stressful environments such as tidal salt marshes, (3) the need for an unpredictable but valuable resource such as carrion, and (4) high predation pressure. Parental care is often associated with territoriality and spatial fidelity. It is common when there is something that parents are able to do to increase the survival of young. Thus, wood is an abundant resource and supplies both food and protection, but it is difficult for immature insects to access and requires gut symbionts to use. Foliage- and sap-feeding insects may also be considered to live in a stable environment. In many of these species parents defend young and may facilitate feeding. At the other extreme, stressful environments may require parental care if young are to survive at all. However, the evolution of parental care involves a suite of changes, including greater dependence on the part of the young and the loss of adaptations for independence from parents. Many environments, both

stressful and stable, harbor parental and nonparental members of a family.

Valuable but unpredictable resources (carrion or dung) or resources that are hard to acquire (a tunnel system) are especially potent selective forces for parental care. Both involve the construction of elaborate nests that represent a substantial investment. These resources are associated with the production of fewer and smaller clutches of larger young. Burying beetles, dung beetles, and the woodroach produce on average only a single brood in a lifetime. Biparental care is associated with these species both because male assistance in guarding and provisioning can often greatly increase the survival of the young and because there are few additional mating/breeding opportunities. The improved survival of his young offsets the male parent's potential gain from leaving and searching for rare carrion or females when most individuals in the population are mated.

Exclusive paternal care is rare in insects because external fertilization is rare. Internal fertilization both reduces the certainty of paternity if females have multiple mates and disassociates the father from his offspring when they appear. These factors discourage paternal care and encourage maternal care in other taxa. Exclusive paternal care is associated with fidelity to a nest site, including a "nest" of his own wingcovers, and with the ability of the male to guard the clutch of multiple females. Under these conditions, a male does not forgo additional matings; in fact, the demonstration of paternal behavior may increase his attractiveness to females.

Thus, insects not only show a variety of forms of parental care but also many demonstrate behavioral plasticity that allows them to adjust the level or form of care to changing circumstances, such as the loss of a mate. As more examples come to light and familiar ones are better studied, it becomes clear that the functions of care are also varied and complex even within a single species. This variety is not surprising, since parental care has evolved independently many times in insects in response to different selective forces.

See Also the Following Articles

Defensive Behavior • Division of Labor in Social Insects • Dung Beetls • Egg Coverings • Mating Behaviors • Nest Building • Spermatophore

Further Reading

Choe, J. C., and Crespi, B. J. (1997). "Social Behavior in Insects and Arachnids." Cambridge University Press, Cambridge, U.K.

Clutton-Brock, T. H. (1991). "Parental Care." Princeton University Press, Princeton, NJ.

Scott, M. P. (1998). The ecology and behavior of burying beetles. *Annu. Rev. Entomol.* **43,** 595–618.

Tallamy, D. W., and Wood, T. K. (1986). Convergent patterns in subsocial insects. *Annu. Rev. Entomol.* **31,** 369–390.

Trumbo, S. T. (1996). Parental care in invertebrates. *In* "Parental Care: Evolution, Mechanisms, and Adaptive Significance" (J. S. Rosenblatt and C. T. Snowdon, eds.), pp. 3–51. Academic Press, San Diego.

Wilson, E. O. (1971). "Insect Societies." Belknap Press, Cambridge, MA.

Zeh, D. W., and Smith, R. L. (1985). Paternal investment by terrestrial arthropods. *Am. Zool.* **25,** 785–805.

Parthenogenesis

Lawrence R. Kirkendall
University of Bergen, Norway

Benjamin Normark
University of Massachusetts

Most insects, like most other animals and plants, reproduce sexually. Each gamete (egg or sperm) contains one complete set of chromosomes, and the fusion of a sperm and an egg results in a zygote, which then develops into a new individual. In some insects, however (as in many other groups), offspring can be produced in a way that circumvents mating. Parthenogenesis (*partheno,* virgin, + *genesis,* to give birth, from *gen,* to be produced) is the development of an egg cell into a new individual without fertilization.

Insects have always played a central role in our understanding of parthenogenesis; Charles Bonnet first demonstrated the occurrence of this phenomenon by careful experiments with isolated females of three species of aphids, in the 1740s. During the next hundred years, further experiments with aphids and with bagworm moths, drone bees, and silkworm moths verified the reality of virgin birth, though the term "parthenogenesis" did not come into common usage until the 1850s. Parthenogenesis has fascinated biologists since it was first described, in part because it involves doing away with mating and with males and in part because clonal reproduction challenges cherished assumptions about the necessity of genetic variability in nature.

Over the past century, we have come to realize that there are numerous reproductive systems in insects that deviate from outbreeding sexual reproduction. Modern discussions of parthenogenesis adhere less strictly to the criterion of fertilization (or lack thereof), focusing instead on whether genomes are passed on intact. Whereas sexually produced offspring may be highly variable and have unique combinations of genes, parthenogenetically produced offspring typically have a genotype identical to that of their mother or only slightly different from hers.

Parthenogenesis has dramatic consequences for individuals, populations, and species. It allows females to: (1) pass along their successful genotypes to all of their offspring; (2) produce only daughters, maximizing the rate of increase; and (3) eliminate the need for finding (or being found by) a mate. Sexual reproduction, in contrast, results in: (1) offspring being different from each other and from their mothers; (2) production of 50% males, which cannot themselves produce offspring; and (3) an inability to reproduce without males being locally present and without diverting a certain amount of time and energy to the mating process.

Specialized Terms

apomixis Parthenogenesis in which eggs are produced without meiosis.

automixis Parthenogenesis in which eggs undergo meiosis.

diploid Having two complete sets of chromosomes (like a typical adult animal).

haplodiploidy (arrhenotoky) A genetic system in which unfertilized, haploid eggs develop into males and fertilized, diploid eggs develop into females.

haploid Having one complete set of chromosomes (like a typical egg or sperm cell).

parthenogenesis Reproduction without fertilization (in the sense of fusion of sperm and egg nuclei).

polyploid Having more than two complete sets of chromosomes (e.g., triploid, having three sets, tetraploid, having four sets).

pseudogamy A form of sperm-dependent parthenogenesis in which eggs require activation by entry of sperm, but only maternal chromosomes are expressed and passed on.

thelytoky Parthenogenesis in which only female offspring are produced.

tychoparthenogenesis The rare or occasional production of eggs that start developing without having been fertilized.

FORMS OF PARTHENOGENETIC REPRODUCTION

In most species of insects, every individual develops from a diploid zygote formed by the union of haploid egg with haploid sperm. But there are many alternative life cycles in which some or all individuals develop from cells that are not zygotes in this sense, and most of these life cycles come under the heading of "parthenogenesis." Thus, insect parthenogenesis encompasses much of the diversity of insect life cycles.

Sex of Offspring

Eggs laid by a virgin female may develop into other females (thelytoky), into individuals of both sexes (deuterotoky), or into males (arrhenotoky). Arrhenotoky and deuterotoky are only partially parthenogenetic systems, because males still occur. Only thelytoky is potentially a completely parthenogenetic system, and the term "parthenogenesis" is often used to refer specifically to thelytoky.

Facultative or obligate thelytoky occurs sporadically but is found in over 80 families of the superclass Hexapoda and is also scattered throughout the mites (Acari; we use mites as a collective term for all acarines, including ticks). Generally, thelytoky occurs as scattered instances in hexapods, though there are several families of mites (in the suborder Oribatida) that are strictly thelytokous.

Thelytoky is found in most orders of hexapods but the orders with the highest frequency of strictly thelytokous species are Thysanoptera, Psocoptera, Hemiptera (*sensu lato*, with thelytoky concentrated in the "homopterous" suborder Sternorrhyncha), and Phasmatodea. The largest insect orders (Lepidoptera, Diptera, Coleoptera, Hymenoptera) have a low incidence of thelytoky overall, but very high rates of thelytoky in some families, such as weevils (Coleoptera: Curculionidae), bagworm moths (Lepidoptera: Psychidae), and chironomid midges (Diptera: Chironomidae) (Diptera). Thelytoky has never been reported from several species-poor orders (Protura, Diplura, Zoraptera, Grylloblattodea, Megaloptera, Raphidioptera, Mecoptera), from the strictly parasitic orders (Siphonaptera, Phthiraptera), or from Plecoptera, and there have been only dubious reports of thelytoky in Demaptera, Neuroptera, and Strepsiptera.

Deuterotoky differs from thelytoky only in that males as well as females are produced from unfertilized eggs. There are a few insect life cycles in which deuterotoky is a normal feature: cynipid wasps, micromalthid beetles, and a few cecidomyiid flies. Reports of deuterotoky in several groups of mites have been proposed, but none have yet been confirmed. However, the boundary between thelytoky and deuterotoky is sometimes unclear; intensive study of parasitic wasp species thought to reproduce by thelytoky often turns up occasional males, which may or may not be able to mate. The offspring of matings with such males are thelytokous females. It is unclear whether an otherwise thelytokous species with rare males (which may or not be sexually competent) should be classified as deuterotokous.

In arrhenotoky, fertilized eggs develop into females while unfertilized eggs become males. Males are haploid (the only known exception occurs in the scale insect *Lecanium putmani*) and the females diploid; this is why arrhenotoky is commonly referred to as "haplodiploidy." In two orders (Hymenoptera and Thysanoptera), haplodiploidy is the only known sexual system. Haplodiploidy also occurs in some scale insects (Hemiptera: Margarodidae), whiteflies (Homoptera: Aleurodidae), some bark beetles (Curculionidae), and the bizarre beetles of the monotypic family Micromalthidae.

Arrhenotokous males arise parthenogenetically; indeed, the production of drones by queen bees prevented from mating was one of the earliest experimental demonstrations of parthenogenesis (by Dzierzon, in the 1840s). However, a haplodiploid species as a whole reproduces sexually, and the genotype of a parthenogenetically produced male is a unique recombinant product of meiosis. Thus, although parthenogenesis occurs in a haplodiploid species, clonal reproduction does not. Because males must always mate to produce offspring, under arrhenotoky there are never two successive parthenogenetic generations. Genetically equivalent to haplodiploidy is the production of male offspring in lineages that engage in paternal genome elimination (also known as pseudoarrhenotoky), in which the paternal genome in males is genetically inert and is ultimately eliminated entirely, such that males pass on only their maternally inherited genes. Pseudoarrhenotoky occurs in some

mites (Acari), in most scale insects (Hemiptera: Coccoidea), in fungus gnats (Diptera: Sciaridae and Cecidomyiidae), and in one group of bark beetles (Curculionidae, Scolytinae, *Hypothenemus*). Because arrhenotoky and pseudoarrhenotoky are essentially obligately sexual systems of reproduction that have little in common with the departures from sexuality represented by thelytokous systems, this entry focuses primarily on thelytoky.

Occurrence of Meiosis

The most fundamental feature of the canonical eukaryote life cycle is the alternation between meiosis, in which the DNA content of a diploid nucleus is halved, and syngamy, in which two haploid nuclei (usually, from a sperm and an egg) fuse to form a new diploid nucleus. This alteration is commonly called "the sexual cycle," or simply "sex." But since the word sex has even more different meanings than the word parthenogenesis, it is less confusing to use the technical term mixis to refer to this cycle. In many insects, the ancestral mictic cycle has been replaced by apomixis, in which there is neither meiosis nor syngamy; instead, new individuals arise from mitotically produced cells that are genetically identical (except for new mutations) to the parent that produced them. "Vegetative" reproduction in insects is in the form of embryos undergoing fission, a process known as polyembryony. As many as 2500 individuals can develop in this way from a single egg. Polyembryony in insects occurs regularly in parasitoids in a few families of Hymenoptera and in Strepsiptera and has been reported to occur sporadically in grasshoppers. Polyembryonically produced insects will be genetically identical to each other, but not to their parents.

Apomixis is a form of parthenogenesis, but the term parthenogenesis encompasses mictic cycles as well, in which an alteration of gamete production or of the early stages of embryo development allows the production of offspring with the same ploidy as their parents; if syngamy occurs, it is a fusion of two maternally derived nuclei. Thus, in automixis, a normal meiosis occurs in the course of oogenesis but the resulting offspring have only one parent; the contrasting term is amphimixis, in which gametes from two different parents join in syngamy. Automixis itself includes a diverse group of systems, which vary as to how and when diploidy is restored; the exact mechanism of diploidization determines whether homozygosity is enforced in every generation or heterozygosity is maintained for at least some time (Fig. 1).

In premeiotic doubling, chromosomes are replicated prior to meiosis, so that the reduction–division process that is meiosis results in a diploid egg with the same chromosome complement as in the parent. *Warramaba virgo* (an acridid grasshopper) and the pseudogamous triploid ptinid beetle *Ptinus clavipes* are the only known insect examples. Maternal ploidy (in the following examples, diploidy) can also be restored by several postmeiotic means, all of which result in instant homozygosity at all loci.

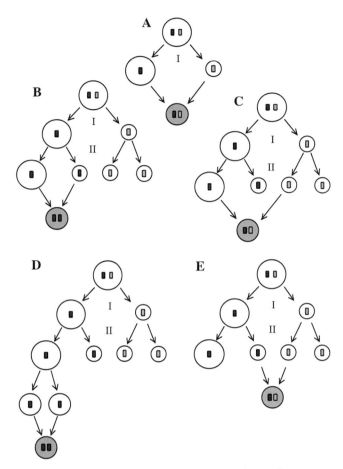

FIGURE 1 The meiotic process in five common forms of automictic parthenogenesis. One set of homologous chromosomes is followed up to the point of the restoration of maternal ploidy (the shaded circle), for each form; "I" indicates the first meiotic division, "II" the second. (A) The two products of the first meiotic division fuse. (B) The second polar nucleus fuses with the egg nucleus, an automictic process called "terminal fusion." (C) A derivative of the first polar body fuses with the egg nucleus. (D) The products of a division of the egg nucleus fuse, "gamete duplication." (E) The two central polar nuclei fuse, "central fusion." Note that in A, C, and E, the zygote has the same chromosome constitution as the mother; heterozygosity is maintained indefinitely in A, but C and E allow heterozygosity to gradually decay through crossing-over. In B and D, the zygote consists of duplicated chromosomes, producing instant homozygosity. (Modified from Suomalainen *et al.,* 1987.)

In a process sometimes referred to as gamete duplication, the haploid egg nucleus replicates but the two products fuse. Examples are found in the Phasmatodea, Hemiptera (one aleurodid, various coccids), Lepidoptera (occasional in silkworm moths), Diptera (three *Drosophila* species), and one Hymenoptera (a cynipid wasp). In terminal fusion, syngamy occurs between genetically identical cells, the egg nucleus, and the second polar nucleus. Examples are found in Orthoptera (various tychoparthenogenetic tettigid grasshoppers), Homoptera (four coccid species from three different genera), Thysanoptera (one *Heliothrips* species), Diptera (the same three *Drosophila* species as above, one or two *Lonchoptera* species), and Hymenoptera (two tenthredinids, one aphelinid).

Central fusion represents a postmeiotic diploidization process in which all maternal chromosomes (and hence heterozygosity) are retained. It is characterized by the fusion of the two central polar nuclei, and examples include Lepidoptera (*Solenobia triquetella,* a well-studied psychid moth), Diptera (the parthenogenetic *Drosophila* spp. and *Lonchoptera* spp.), and Hymenoptera (an ichneumonid wasp *Venturia canescens* and occasionally in the cape honey bee *Apis mellifera capensis*). A final postmeiotic process is the fusion of the egg nucleus with a product of the first polar nucleus, again restoring the original chromosome complement, a process that occurs in a collembolan and in Lepidoptera (four psychid moth species from three genera).

Although the above-mentioned forms of automixis are considered to be parthenogenetic in the traditional sense, an exception is self-fertilization by an individual that produces both eggs and sperm (a hermaphrodite). Hermaphroditism in insects has been found only in three species of *Icerya* scale insects, all of which are known to self-fertilize.

Obligate, Cyclical, and Facultative Parthenogenesis

Modes of reproduction can be thought of as arrayed along a continuum, in which the faithfulness of mother-to-egg reproduction of a chromosome or set of chromosomes varies from about 50% (outbreeding sexuality) to about 100% (obligate apomictic parthenogenesis). Many of the intermediate modes of reproduction fall under the (again heterogeneous) categories of cyclic or facultative parthenogenesis. Cyclical parthenogenesis involves an alternation of one generation of sexual reproduction with one or more generations of parthenogenetic reproduction. The term facultative parthenogenesis potentially covers a great many possible life cycles, but it implies that a given individual can reproduce either sexually or asexually. Within the category of facultative parthenogenesis, an important subcategory is tychoparthenogenesis, in which eggs are typically fertilized, but if a clutch of eggs is left unfertilized for sufficiently long, a small proportion of them will begin development. The proportion of eggs that develop successfully to adults is usually extremely low; for example, for caught-in-the-wild *Drosophila mercatorum,* the chance of a female developing from an unfertilized egg who is herself capable of parthenogenesis is 1/10,000. Rare overall in insects, mites, and ticks, this capability is rather common in some polyneopteran orders (in orthopterans, cockroaches, stick insects, and mantids) and also occurs in psocopterans, lepidopterans, and dipterans. Where the cytology is known, in all cases oogenesis is automictic. Tychoparthenogenesis provides evidence for genetic variability for parthenogenesis within normally amphimictic species. Artificial selection experiments with drosophilids and stick insects have been able to increase by up to a thousandfold the proportion of eggs that develop parthenogenetically.

The categories of cyclical and facultative parthenogenesis intergrade when the number of parthenogenetic generations intervening between sexual ones is variable, as in certain cecidomyiid midges, all aphids, and micromalthid beetles, all of which switch from parthenogenesis to sexuality in response to environmental cues. However, there is never more than one sexual generation before the switch back to parthenogenesis occurs. Only cynipid gall wasps are regularly cyclical in the sense that there are never two successive parthenogenetic generations, although in cynipids, as in every other group with cyclic or facultative parthenogenesis, obligately parthenogenetic lineages have arisen multiple times.

Occurrence of Mating

Thelytoky is characterized by absence of mating; indeed, the first indication that an insect species might be thelytokous is the observation that no males exist in collections. In pseudogamy (also called gynogenesis), mating with males of the same or a different species is necessary; egg development is activated by contact with or penetration by sperm. After entry into the egg, paternal chromosomes degenerate, and only maternal chromosomes survive in the offspring. This process can be considered pseudofertilization because offspring develop clonally, and all will be female, and is thus genetically a form of parthenogenesis. Pseudogamy is difficult to observe, since mating does occur; it is usually discovered by noting that mated females regularly produce all-female broods. In insects, pseudogamy is found only in Coleoptera (two origins in *Ips* bark beetles and one species of ptinid beetle), two species of Lepidoptera, and one species each of Collembola, Orthoptera, and Homoptera. Possible cases of pseudogamy in acarines have not been confirmed.

In hybridogenesis, the paternal genome is excluded during oogenesis such that ova have only maternally derived chromosomes. The intact maternal genome is paired in each new generation with sperm from males of a sexually reproducing host species. Such hemiclonal reproduction was first discovered in vertebrates and to date is known in only one group of insects, stick insects of the genus *Bacillus*. Although classed along with pseudogamy as "sperm-dependent parthenogenesis," it is actually a hybrid form of reproduction; it combines sexual reproduction, with respect to spermatogenesis, with nonrecombinant, clonal maternal genomes.

Finally, regular inbreeding between close relatives (such as self-fertilization or brother–sister mating) has genetic consequences virtually identical to those of some forms of automictic parthenogenesis: extreme inbreeding should result in homozygous clones. Because of this, some evolutionary biologists include extreme inbreeding under a broader definition of parthenogenesis, despite the involvement of mating. The similarity to automictic parthenogenesis is enhanced by the fact that species with regular sibling mating usually produce only one or a few males per brood, such that populations are nearly all-female. The taxonomic distribution of extreme inbreeding is not precisely known; it has evolved frequently in bark beetles (Scolytinae), occurs sporadically in Hymenoptera (in which it has been intensively studied in parasitic wasps,

but also is found in wasps, bees, and a few ants), eusocial Thysanoptera, and mites.

ORIGINS AND GENETICS OF PARTHENOGENESIS

Relatively little is known of the details of how parthenogens arise or of the genetic changes necessary to subvert the mictic cycle. Parthenogenesis seems often to have resulted from the genetic disturbances that accompany the intrusion of foreign chromosomes; all vertebrate parthenogens appear to have arisen from interspecific hybridization, as shown by studies of chromosomes, protein variation, and DNA sequences. A hybrid origin for thelytoky has also been shown for some of the best studied cases in insects, *Warramaba* grasshoppers and *Bacillus* stick insects (nonhybrid thelytokes also occur, in the latter). Experimental hybridization between presumed parental species has generated parthenogenetic forms indistinguishable from naturally occurring ones, in apomictic triploid pseudogamous *Muellerianella* delphacid leafhoppers. Interspecific hybridization in *Oncopeltus* milkweed bugs also has produced pseudogamy.

However, at least three lines of evidence argue that parthenogenesis can evolve without hybridization, in at least some groups. In species with facultative parthenogenesis (such as tychoparthenogenesis), parthenogenesis clearly arises without interspecific hybridization. Also, clonal lineages that are genetically and morphologically similar or nearly identical to known sexual species have presumably originated without interspecific matings (e.g., thelytokous "races" of aphids and cynipid wasps that occur in otherwise cyclically parthenogenetic species). Finally, some species with thelytokous races are taxonomically isolated; *Bromius obscurus* exists as diploid bisexuals in North America but as triploid apomicts in Europe and is the only species in its genus.

There is a third way in which parthenogens arise. Recently, the intracellular parasitic proteobacterium *Wolbachia pipientis* has been shown to induce automictic thelytoky in various wasps, certain thrips, and some mites, all of which are haplodiploid. *Wolbachia* causes gamete duplication in unfertilized eggs, leading them to develop as diploid females rather than haploid males. Since *Wolbachia* is not universally present in these haplodiploid groups, thelytoky has evolved both with and without the help of these microorganisms.

Intensive cytological and genetic investigations frequently uncover the presence of multiple mechanisms for parthenogenetic reproduction in a species or group of species. The wingless stick insect genus *Bacillus* provides an excellent example. Endemic to the Mediterranean region, the genus includes two sexual species, *B. rossius* (which also has thelytokous females) and *B. grandii* (strictly bisexual), and a thelytokous lineage known as *B. atticus*. Where two or three species occur in sympatry (as on the island of Sicily), hybridization occurs, which has resulted in the production of two hybridogens, the diploid automictic *B. whitei* and the trihybrid apomictic *B. lynceorum*. Each *Bacillus* hybrid uses a different egg maturation process; however, they share the common cytological feature of an intrameiotic extra doubling of DNA, resulting in four-stranded chromosomes, which enables the meiotic process to produce balanced chromosome complements in gametes.

Examples of multiple mechanisms within single species are provided by the extraordinary life cycles of *Micromalthus debilis* beetles and cecidomyiid midges *Heteropeza pygmaea*, *Miastor metralaos*, and *Mycophila speyeri* (all living in dead wood). In each of these species there are four different kinds of reproductive females: an adult female; a thelytokous paedogenetic larva (i.e., a reproductively mature larva), an arrhenotokous paedogenetic larva, and a deuterotokous paedogenetic larva.

A few studies have revealed something of the genetic basis for parthenogenesis, and it is likely that it will prove as varied as are the mechanisms of parthenogenesis. In *Rhopalosiphum padi* (the bird cherry-oat aphid), obligate parthenogenesis appears to be determined by a single locus and is recessive to cyclical parthenogenesis. In contrast, the predisposition for parthenogenesis in *D. mercatorum* was induced by genes at a number of independent loci. Some lineages of *R. padi* reproduce largely by parthenogenesis but do produce some males (although no sexual females). There is some evidence that in several cases genes carried by these males have "converted" cyclically parthenogenetic lineages of *R. padi* into obligate parthenogens.

CONSEQUENCES OF PARTHENOGENESIS: GENETIC VARIATION

Parthenogenetic insect lineages usually consist of a variety of genetically distinct clones, each of which was derived independently from the ancestral sexual populations. Clones may vary in ploidy and in their genetic composition, and as a result parthenogenetic populations can exhibit considerable diversity in morphology, behavior, and life history. Although much less than the variability within an outbreeding amphimictic population (in which each individual is genetically unique), the clonal polymorphism of thelytokous forms implies a potential for adaptation to a wider range of ecological conditions than would be possible for an invariant population.

PATTERNS IN PARTHENOGENESIS: BIOGEOGRAPHY AND ECOLOGY

Many parthenogenetic hexapods and mites, whether mictic or amictic, are common and abundant. In many cases, clonal forms are more widespread than their closest sexual relatives. Frequently, parthenogens have geographic distributions different from those of the sexual taxa to which they are most closely related, a phenomenon known as geographic parthenogenesis. Geographic parthenogenesis in hexapods (but not acarines) often takes the form of parthenogens being closer to the poles (a latitudinal gradient) and at higher altitudes. Thus, the bisexual forms of several European weevils (e.g., *Otiorhynchus scaber*, *Polydrosus mollis*) and *Solenobia* bagworm

moths are found only in a few alpine sites thought to have been glacial refugia, while thelytokous forms of the same species are widespread in central and northern Europe. Psocoptera species with both thelytokous and bisexual forms follow a similar pattern in North America.

A number of ecological patterns have been elucidated, to describe the distribution of parthenogenesis in hexapods and mites. For example, parthenogenesis seems to be associated with low dispersal capabilities (with winglessness in Phasmatodea, Orthoptera, and Lepidoptera) and disturbed or ephemeral habitats (parthenogens often being categorized as "weedy"). Among mites, clusters of thelytokous taxa (including entire families and genera of "Endeostigmatida," Mesostigmatida, Prostigmatida, and Oribatida) are strongly associated with soil dwelling (particularly with stable soil horizons), and thelytokous (vs nonthelytokous) oribatids are strongly overrepresented on oceanic islands. In Collembola, too, there is a strong association between soil dwelling and thelytoky, and the only cockroach with thelytokous races *(Pycnoscelis indicus)* is a burrower.

CONCLUSION

Parthenogenesis encompasses a variety of reproductive systems and is often considered synonymous with "clonal reproduction." Indeed, the central feature of thelytokous parthenogenetic reproduction is that maternal genomes are normally passed on intact through a series of all-female broods. It is important to emphasize, however, that there are forms of automictic parthenogenesis in which recombination is possible and that in pseudogamous parthenogenesis, mating is necessary even though reproduction is essentially clonal.

Parthenogenetic reproduction requires a mechanism to circumvent the normal halving of ploidy that results from gametogenesis. In insects, many mechanisms for the preservation or restoration of diploidy have evolved. Either meiosis is eliminated (apomixis) or diploidy is restored (automixis) during or after meiosis. Apomixis and some forms of automixis result in maintenance of heterozygosity, whereas other forms of automixis result in instant homozygosity. Far from being a reproductive curiosity, parthenogenesis has arisen in most insect groups, and many parthenogenetic species are both abundant and widespread. Thelytokous and pseudogamous taxa are so ecologically successful that we cannot simply view them as reproductive aberrations. However, because parthenogenesis leads instantly, or relatively quickly, to lineages of genetically identical individuals, parthenogens should be depauperate in genetic variation relative to comparable sexual forms. Given the success of clonal forms, we cannot always assume that genetic variability is vitally important in nature. We are encouraged, then, to search for patterns in the occurrence of parthenogenesis that might explain when and why it is successful.

The study of parthenogenesis can illuminate one of the central problems in biology, that of explaining the ubiquity of sex and recombination, by revealing when and where genetic variability is important in nature. Insects and mites, because of their short life cycles and often large population numbers, are ideal organisms for studying the evolutionary and ecological consequences of parthenogenesis.

See Also the Following Articles

Genetic Variation • Polyembryony • Sex Determination • Wolbachia

Further Reading

Bell, G. (1982). "The Masterpiece of Nature: The Evolution and Genetics of Sexuality." Croom Helm, London.

Beukeboom, L. W., and Vrijenhoek, R. C. (1998). Evolutionary genetics and ecology of sperm-dependent pathenogenesis. *J. Evol. Biol.* **11**, 755–782.

Normark, B. B. (2003). Evolution of alternative genetic systems in insects. *Annu. Rev. Entomol.* **48**, 397–423.

Parker, E. D., Jr. (2000). Geographic parthenogenesis in terrestrial invertebrates: Generalist or specialist clones? *In* "Progress in Asexual Propagation and Reproductive Strategies" (R. N. Hughes, ed.). Oxford–IBH, Oxford.

Suomalainen, E., Saura, A., and Lokki, J. (1987). "Cytology and Evolution in Parthenogenesis." CRC Press, Boca Raton, FL.

Wrensch, D. L., and Ebbert, M. A. (eds.) (1993). "Evolution and Diversity of Sex Ratio in Insects and Mites." Chapman & Hall, New York.

Pathogens of Insects

Brian A. Federici

University of California, Riverside

Pathogens are viruses or microorganisms that cause disease. Like all other organisms, insects are susceptible to a variety of diseases caused by pathogens. Many of these pathogens cause diseases that are acute and fatal and therefore are used as models to study processes of infection and pathogenesis as well as to control populations of insects that are pests or vectors of plant and animal diseases. Generally, insect pathogens have a relatively narrow host range and thus are considered to be more environmentally friendly than synthetic chemical insecticides. The pathogens that cause disease in insects fall into four main groups: viruses, bacteria, fungi, and protozoa. This article discusses the primary biological properties of each of these pathogen groups, with specific emphasis on how these pathogens have been used to benefit humans.

VIRUSES

Viruses are obligate intracellular parasites, meaning that they can reproduce only in living cells and are composed in the simplest form of a nucleic acid, either DNA or RNA, and a protein shell referred to as the capsid. More complex viruses also contain a lipoprotein envelope. Insect viruses can be cultured in living hosts (i.e., *in vivo*) or in cultured insect cells *(in vitro)*. In general, insect viruses are divided into two broad nontaxonomic categories, the occluded viruses and the nonoccluded viruses. Occluded viruses are so named because after formation in infected cells, the mature virus particles (virions) are occluded within a protein matrix, forming paracrystalline bodies that are generically referred to as either inclusion or occlusion bodies. In the nonoccluded viruses, the virions occur freely or occasionally form paracrystalline arrays of virions that are also known as inclusion bodies. These, however, have no occlusion body protein interspersed among the virions.

The five most commonly encountered types of insect viruses are iridoviruses, cytoplasmic polyhedrosis viruses, entomopoxviruses, ascoviruses, and baculoviruses.

Iridoviruses

Nonoccluded viruses with a linear double-stranded DNA genome, the iridoviruses (family Iridovirdae) produce large, enveloped, icosahedral virions (125–200 nm) that replicate in the cytoplasm of a wide range of tissues in infected hosts. Virions form paracrystalline arrays in infected tissues, imparting an iridescent hue to infected hosts, from which the name of this virus group is derived. Over 30 types are known, and these have been most commonly reported from larval stages of Diptera larvae, such as mosquito larvae, as well as from larvae of Coleoptera and Lepidoptera. Generally, the iridoviruses occur very broadly, and they are known from other invertebrates, such as isopods, as well as from certain vertebrates including frogs and fish. Observations of natural occurrence in host field populations suggest that one host range of each type is quite narrow, although in the laboratory iridoviruses are easy to transmit from one insect species to another by inoculation. Prevalence and mortality rates in natural populations of host insects are typically less than 1%.

Cytoplasmic Polyhedrosis Viruses

The cytoplasmic polyhedrosis viruses (family Reoviridae) are occluded double-stranded RNA viruses with a genome divided into 9 or 10 segments of RNA. These viruses, commonly referred to as CPVs, cause a chronic disease and reproduce only in the stomach of insects, where typically they form large (ca. 0.5–2 μm) polyhedral to spherical occlusion bodies in the cytoplasm of midgut epithelial cells. Infection in early instars retards growth and development, extending the larval phase by weeks. The disease is often fatal. In advanced stages of disease, the infected midgut is white rather than translucent brown because large numbers of polyhedra have accumulated there. This virus type is relatively common among lepidopterous insects and among dipterous insects of the suborder Nematocera (e.g., mosquitoes, blackflies, midges). CPVs are typically easy to transmit by feeding to species that belong to the same family of the host from which they were isolated, and thus the host range of this virus type quite broad.

Entomopoxviruses

The entomopoxviruses (family Poxviridae) are occluded double-stranded DNA viruses that produce large, enveloped virions (150 nm × 300 nm) that replicate in the cytoplasm of a wide range of tissues in most hosts, causing an acute, fatal disease. Occlusion bodies vary from being oval to spindle shaped and generally occlude 100 or more virions. These viruses have been most commonly reported from coleopterans, from which there are over 30 isolates, but they are also known from lepidopterous, dipterous (midges), and orthopterous (grasshoppers) insects. This virus type is easily transmitted by feeding, although where the experimental host range of individual isolates has been tested, it has been found to be relatively narrow, generally being restricted to closely related species. Insect poxviruses are related to vertebrate poxviruses, such as the variola virus, the etiological agent of smallpox, and they may be the evolutionary source of the vertebrate poxviruses.

Ascoviruses

The ascoviruses are a new family of DNA viruses (family Ascoviridae) at present known only from larvae of species in the lepidopteran family Noctuidae, where they have been reported from several common pest species such as the cabbage looper, cotton budworm, corn earworm, and fall armyworm. Ascoviruses cause a chronic, fatal disease of larvae. The virions of ascoviruses are large (130 nm × 400 nm), enveloped, and reniform to bacilliform in shape; they exhibit complex symmetry and contain a linear, double-stranded DNA genome. During the course of ascovirus disease, large numbers of virion-containing vesicles accumulate in the blood of infected caterpillars, changing its color from translucent green to milky white. These virion-containing vesicles are formed by a unique developmental sequence in which each infected host cell cleaves into a cluster of vesicles as virion assembly proceeds.

An interesting ascovirus feature is that transmission from host to host depends on vectoring by female endoparasitic wasps. Ascoviruses are extremely difficult to transmit by feeding, with typical infection rates averaging less than 15% even when larvae are fed thousands of vesicles in a single dose. In contrast, infection rates for caterpillars injected with as few as 10 virion-containing vesicles are typically greater than 90%, and experiments with parasitic wasps show that these insects can transmit ascoviruses.

Baculoviruses

Baculoviruses (family Baculoviridae) are large, enveloped, double-stranded, occluded DNA viruses. These viruses are divided into two main types, commonly known as the nuclear polyhedrosis viruses (NPVs) and the granulosis viruses (GVs). Both NPVs and GVs are highly infectious by feeding, and in some insect species periodically cause epizootics, or widespread outbreaks of disease, that result in significant (> 90%) declines in caterpillar populations.

NUCLEAR POLYHEDROSIS VIRUSES The NPVs (Fig. 1) are known from a wide range of insect orders but have been most commonly reported by far from lepidopterous insects, from which well over 500 isolates are known. Many of these are different viruses (i.e., viral species). NPVs replicate in the nuclei of cells, generally causing an acute fatal disease. The virions are large (80–200 nm × 280 nm) and consist of one

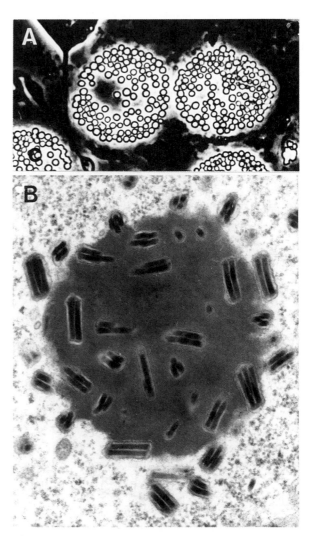

FIGURE 1 Nuclear polyhedrosis virus polyhedra. (A) Wet mount preparation viewed with phase microscopy showing refractile polyhedra in two infected nuclei. (B) Transmission electron micrograph through a single polyhedron showing the enveloped rod-shaped virions, characteristic of NPVs, occluded within the polyhedral matrix. Upon ingestion, this matrix dissolves in the insect midgut, and the virions invade the host through midgut microvilli.

or more rod-shaped nucleocapsids with a double-stranded circular DNA genome enclosed in an envelope. The occlusion bodies of NPVs are referred to commonly as polyhedra because typically their shape is polyhedral. Polyhedra are large (ca. 0.5–2 μm) and form in the nuclei, where each occludes as many as several hundred virions. The NPVs of lepidopterous insects infect a range of host tissues, but those of other orders are typically restricted to the midgut epithelium. Some NPVs have a very narrow host range and may replicate efficiently only in a single species, whereas others, such as the AcMNPV (i.e., the NPV of the alfalfa looper, *Autographa californica*), have a relatively broad host range and are capable of infecting species in other genera.

GRANULOSIS VIRUSES   The GVs, of which over 100 isolates are known, are closely related to the NPVs but differ from the latter in several important respects. The virions of GVs are similar to those of NPVs but contain only one nucleocapsid per envelope. GVs are known only from lepidopterous insects. Like NPVs, they initially replicate in the cell nucleus, but pathogenesis involves early lysis of the nucleus (as virions begin to assemble), which in the NPVs occurs only after most polyhedra have formed. After the nucleus has lysed, GV replication continues throughout the cell, which now consists of a mixture of cytoplasm and nucleoplasm. When completely assembled, the virions are occluded individually in small (200 nm × 600 nm) occlusion bodies referred as granules. Many GVs primarily infect the fat body, whereas others have a broader tissue tropism and replicate throughout the epidermis, tracheal matrix, and fat body. One, the GV of the grapeleaf skeletonizer *Harrisina brillians*, is unusual in that it replicates only in the midgut epithelium.

Use of Viruses as Insect Control Agents

The best example of the use of a virus as an insect control agent is the use of the NPV of the European spruce sawfly, *Gilpinia hercyniae*, as a classical biological control agent. The European spruce sawfly was introduced into eastern Canada from northern Europe around the turn of the century and had become a severe forest pest by the 1930s. Hymenopteran parasitoids were introduced from Europe in the mid-1930s as part of a biological control effort, and inadvertently along with these came the NPV, which was first detected in 1936. Natural epizootics caused by the virus began in 1938, by which time the sawfly had spread over 31,000 km^2. Most sawfly populations were reduced to below economic threshold levels by 1943 and remain under natural control today, the control being effected by a combination of the NPV, which accounts for more than 90% of the control, and the parasitoids.

Although viruses, particularly NPVs, are frequently associated with rapid declines in the populations of important lepidopterous and hymenopterous (sawfly) pests, *G. hercyniae* NPV is the only example of a virus that has proven effective as a classical biological control agent. Another putative baculovirus,

the "nonoccluded" baculovirus of the palm rhinoceros beetle, *Oryctes rhinoceros*, has been a quasi-classical biological control success in that once introduced into populations, can yield control for several years, but ultimately it dissipates and must be reapplied. Moreover, augmentative seasonal introductions have been effective only rarely and are not well documented. Thus, the control potential of most viruses is best evaluated by assessing their utility as microbial insecticides. From this perspective, the iridoviruses are essentially useless because of their poor infectivity by feeding. Cytoplasmic polyhedrosis viruses are not much better because, although highly infectious by feeding, the disease they cause is chronic. CPVs have, however, been useful in some situations, such as for suppression of the pine caterpillar, *Dendrolimus spectabilis*, in Japan. Ascoviruses and entomopoxviruses have not been developed as control agents for any insect owing to lack of efficacy.

For several reasons, the viruses most commonly used or considered as microbial insecticides in industrialized as well as less developed countries are the NPVs. First, NPVs are common in and easily isolated from pest populations. In addition, production in their hosts is cheap and easy, and the technology for formulation and application is simple and adaptable to standard pesticide application methods. Most NPVs, however, are narrow in their host range, infecting only a few closely related species. Furthermore, although several can be grown *in vitro* in small to moderate volumes (ca. 20- to 300-liter cell cultures), no fermentation technology currently exists for their mass production on a scale that would permit repeated applications to hundreds of thousands of acres, which is possible with *Bacillus thuringiensis* (Bt) chemical insecticides. These two key limitations have been major disincentives for the commercial development of NPVs, especially in industrialized countries.

Despite these drawbacks, several NPVs have been registered as microbial insecticides even though the market size for most is small. And registered or not, several are used in many less developed countries, particularly for control of lepidopteran pests of field and vegetable crops. Moreover, over the past decade there has been renewed interest in developing NPVs because recombinant DNA technology offers potential for improving the efficacy of these viruses.

Other Uses

In addition to the use of NPVs in insect control, one baculovirus, the AcMNPV noted earlier, has been developed as an expression vector for producing a large number of foreign proteins *in vitro*. This expression system takes advantage of the strong polyhedrin promoter system, which in the wild-type viruses produces large amounts of the polyhedria used to occlude virions. By substituting foreign genes for the polyhedrin gene, it is possible to synthesize in insect cell cultures large quantities of foreign proteins, such as the capsid proteins of viruses that attack the vertebrates used for vaccine development.

BACTERIA

Bacteria are relatively simple unicellular microorganisms that lack internal organelles such as a nucleus and mitochondria, and reproduce by binary fission. With a few exceptions, most of those that cause disease in insects grow readily on a wide variety of inexpensive substrates, a characteristic that greatly facilitates their mass production. A variety of bacteria are capable of causing diseases in insects, but those that have received the most study are spore-forming bacilli (family Bacillaceae), especially *B. thuringiensis.* Many subspecies of Bt are used as bacterial insecticides and as a source of genes for insecticidal proteins added to make transgenic plants resistant to insect attack, especially attack by caterpillars and beetles. The other bacterial insect pathogens that have received various degrees of study are *B. sphaericus, Paenibacillus popilliae,* and *P. larvae,* the latter being the etiological agent of foulbrood, an important disease of honey bee larvae, *Serratia entomophila* and *S. marcescens.* Several of these, in order of importance, are discussed here to represent the diversity of bacteria that cause disease insects.

Bacillus thuringiensis

B. thuringiensis is a complex of bacterial subspecies that occur commonly in such habitats as soil, leaf litter, on the surfaces of leaves, in insect feces, and as a part of the flora in the midguts of many insect species. Bts are characterized by the production of a parasporal body during sporulation that contains one or more protein endotoxins in a crystalline form (Fig. 2). Many of these are highly insecticidal to certain insect species. These endotoxins are actually protoxins activated by proteolytic cleavage in the insect midgut after ingestion. The activated toxins destroy midgut epithelial cells, killing sensitive insects within a day or two of ingestion. In insects species only moderately sensitive to the toxins, such as *Spodoptera* species (caterpillars commonly known as armyworms), the spore contributes to pathogenesis. Bt also produces other insecticidal compounds including β-exotoxin, zwittermicin A, and vegetative insecticidal proteins (Vips).

The most widely used Bt is the HD1 isolate of *B. thuringiensis* subsp. *kurstaki* (Btk), an isolate that produces four major endotoxin proteins packaged into the crystalline parasporal body (Fig. 2B). This isolate is the active ingredient in numerous commercially available bacterial insecticides used to control lepidopterous pests in field and vegetable crops, and in forests. Another successful Bt is the ONR60A isolate of *B. thuringiensis* subsp. *israelensis* (Bti), which is highly toxic to the larvae of many mosquito and blackfly species. This isolate also produces a parasporal body that contains four major endotoxins (Fig. 2B), but these are different from those that occur in Btk. Several commercial products based on Bti are available and are used to control both nuisance and vector mosquitoes and blackflies. A third isolate of Bt that has been developed commercially is the DSM2803 isolate of *B. thuringiensis* subsp. *morrisoni* (pathovar *tenebrionis*). This

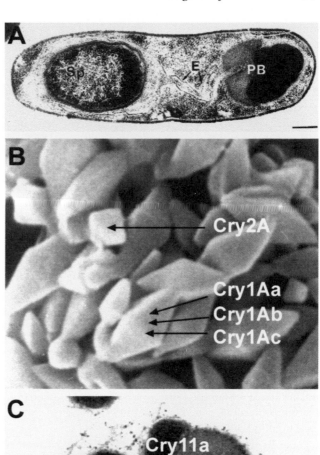

FIGURE 2 Sporulating cell of *Bacillus thuringiensis* and insecticidal parasporal bodies. (A) Transmission electron micrograph through a cell of *B. thuringiensis* subsp. *israelensis* illustrating a developing spore (Sp) and endotoxin-containing parasporal body (PB) outside the exoporium membrane (E). Bar, 250 nm. (B) Scanning electron micrograph of parasporal bodies (crystals) of *B. thuringiensis* subsp. *kurstaki,* a subspecies used widely to control caterpillar pests. The bipyramidal crystals contain three endotoxins (Cry1Aa, Cry1Ab, and Cry1Ac), whereas the smaller cuboidal crystal contains a single endotoxin (Cry2A). The bipyramidal crystals contain three endotoxin proteins (Cry1Aa, Cry1Ab, and Cry1Ac), and the cuboidal crystal has an additional toxin (Cry2A). This toxin complexity accounts for the broad spectrum of activity of many isolates of *B. thuringiensis* subsp. *kurstaki.* (C) Transmission electron micrograph of a parasporal body of *B. turingiensis* subsp. *israelensis* used widely to control the larvae of mosquitoes and blackflies. This parasporal body is also composed of four major endotoxins, a large semispherical inclusion containing Cyt1Aa, a dense spehrical body that apparently contains the Cry4Aa and Cry4Ba proteins, and a bar-shaped body that contains Cry11Aa. The endotoxin inclusions of this subspecies are held together by an envelope of unknown composition. This parasporal body has the highest specific toxicity of known Bt species, and this is due to synergistic interactions between the Cyt1Aa and Cry proteins as well as synergistic interactions among the Cry proteins. Bt endotoxins act by destroying the insect midgut epithelium (stomach).

isolate produces a cuboidal parasporal body toxic to many coleopterous insects and is used commercially to control several beetle pests.

All the above-mentioned isolates are essentially used as bacterial insecticides, applied as needed. A variety of formulations are available, including emulsifiable concentrates, wettable powders, and granules, for use against different pests in a variety of habitats. On a worldwide basis, millions of hectares are treated annually with products based on Bt. Recent estimates indicate the worldwide market is about $80 to $100 million and growing. Although most use is currently as a bacterial insecticide, plants have been engineered to produce Bt proteins for resistance to insects, and this use will probably surpass the use of Bt insecticides during the first decade of this century.

Bacillus sphaericus

Since the mid-1960s it has been known that many isolates of *B. sphaericus* (Bs) are toxic to certain mosquito species. Over the past three decades, three isolates have been evaluated for their mosquito control potential, 1593 from Indonesia, 2297 from Sri Lanka, and 2362 from Nigeria. The 1593 and 2297 isolates were obtained form soil and water samples at mosquito breeding sites, whereas 1593 was isolated from a dead adult blackfly.

Like Bt, Bs acquires its toxicity as the result of protein endotoxins that are produced during sporulation and assembled into parasporal bodies. Bs is unusual in that the main toxin is a binary toxin, (i.e., composed of two protein subunits). These are proteolytically activated in the mosquito midgut to release peptides having molecular masses of, respectively, 43 and 39 kDa, that associate to form the binary toxin, with the former protein constituting the binding domain, and the latter the toxin domain. The toxins bind to microvilli of the midgut epithelium, causing hypertrophy and lysis of cells, destroying the midgut and killing the mosquito larva.

Paenibacillus popilliae

P. popilliae is an highly fastidious bacterium that is the primary etiological agent of the so-called milky diseases of scarab larvae. These insects are the immature stages of beetles, such as the Japanese beetle, *Popillia japonica,* that are important grass and plant pests belonging to the coleopteran family Scarabaeidae. The term "milky disease" is derived from the opaque white color that characterizes diseased larvae and results from the accumulation of sporulating bacteria in larval hemolymph (blood). The disease is initiated when grubs feeding on the roots of grasses or other plants ingest the bacterial spores. The spores germinate in the midgut and vegetative cells invade the midgut epithelium, where they grow and reproduce, changing in form as they progress toward invasion of the hemocoel (body cavity). After passing through the basement membrane of the midgut, the bacteria colonize the blood over a period of

several weeks and sporulate, reaching populations of 100,000,000 cells per milliliter. For larvae that ingest a sufficient number of spores early in development, the disease is fatal. Dead larvae in essence become foci of spores that serve as a source of infection for up to 30 years.

Despite decades of research, suitable media for the growth and mass production of *P. popilliae in vitro* have not been developed. Thus, the technical material (i.e. spores) used in commercial formulations is produced in living, field-collected scarab larvae. Nevertheless, a small but steady market remains for *P. popilliae* in the United States because of serious problems due to scarab larvae, such as damage to turf grass by larvae of the Japanese beetle.

Serratia entomophila

A novel bacterium named. *S. entomophila* causes amber disease in the grass grub, *Costelytra zealandica,* an important pest of pastures in New Zealand, and has been developed as a biological control agent for this pest. This bacterium adheres to the chitinous intima of the foregut, where it grows extensively, eventually causing the larvae to develop an amber color; the result of infection is death. The bacterium is easily grown and mass-produced *in vitro* and can now be grown to densities as high as 4×10^{10} cells ml^{-1}. Successful mass production of *S. entomophila* led to its rapid commercialization. It is now used to treat infested pastures in New Zealand at a rate of one liter of product per hectare. Liquid formulations of this living, non-spore-forming bacterium are applied with subsurface application equipment. The rapid development and commercialization of the bacterium, even though the use is rather restricted, shows how microbials can be successful in niche markets, where there are few alternatives, and mass production methods, the most critical factor, are available.

FUNGI

The fungi constitute a large and diverse group of eukaryotic organisms distinguished from others by the presence of a cell wall, as in plants, but lacking chloroplasts and thus the ability to carry out photosynthesis. Fungi live either as saprophytes or as parasites of plants and animals, and require organic food for growth, obtained by absorption from the substrates on which they live. The vegetative phase, known as a thallus, can be either unicellular, as in yeasts, or multicellular and filamentous, forming a mycelium, the latter being characteristic for most of the fungi that attack insects. During vegetative growth, the mycelium consists primarily of hyphae, which may be septate or nonseptate, and these grow throughout the substrate to acquire nutrients. Reproduction can be sexual or asexual, and during this phase the mycelium produces specialized structures such as motile spores, sporangia, and conidia, typically the agents by which fungi infect insects. Fungi usually grow best under wet or moist conditions, and those that are saprophytic as well as many of the parasitic species are easily cultured on artificial media.

The fungi are divided into five major subdivisions, and these reflect the evolution of the biology of fungi from an aquatic to terrestrial habitats. For example, species of the genera *Coelomomyces* and *Lagenidium* (subdivision Mastigomycotina) are aquatic and produce motile zoospores during reproduction, whereas members of the genera *Metarhizium* and *Beauveria* (subdivision Deuteromycotina) are terrestrial and reproduce and disseminate via nonmotile conidia.

Unlike most other pathogens, fungi usually infect insects by active penetration through the cuticle. The typical life cycle begins when a spore, either a motile spore or a conidium, lands on the cuticle of an insect. Soon after, under suitable conditions, the spore germinates, producing a germ tube that grows and penetrates down through the cuticle into the hemocoel. Once in the hemolymph, the fungus colonizes the insect. Hyphal bodies bud off from the penetrant hyphae and either continue to grow and divide in a yeastlike manner or elongate, forming hyphae that grow throughout the insect body. Complete colonization of the body typically requires 7 to 10 days, after which the insect dies. Some fungi produce peptide toxins during vegetative growth, and in these strains death can occur within 48 h. Subsequently, if conditions are favorable, which generally means an ambient relative humidity of greater than 90% in the immediate vicinity of the dead insect, the mycelium will form reproductive structures and spores, thereby completing the life cycle. Depending on the type of fungus and species, these will be produced either internally or externally as motile spores, resistant spores, sporangia, or conidia.

Fungi are one of the most common types of pathogen observed to cause disease in insects in the field. Moreover, outbreaks of fungal diseases under favorable conditions often lead to spectacular epizootics that decimate populations of specific insects over areas as large as several hundred square kilometers. As a result, there has been interest in using fungi to control insects for well over a century; the first efforts, in Russia in the late 1880s, used *Metarhizium anisopliae* to control the wheat cockchafer *Anisoplia austriaca*. Though there have been numerous attempts since then to develop fungi as commercial microbial insecticides, very few of these efforts have met with success. Thus, at present barely a handful of commercially available fungal insecticides are available for use in industrialized countries, and true commercial success has remained elusive. On the other hand, in developing countries (e.g., Brazil and China), "cottage industry" technology like that used to produce viruses has been turned to the production fungi such as *M. anisopliae* and *Beauveria bassiana*. A quasi-commerical product Boverin, developed and used in Russia for control of the Colorado potato beetle, proved ineffective in the United States. Current efforts to find alternatives to chemical insecticides have intensified research on fungi, with the aim of identifying new isolates or improving existing strains through molecular genetic manipulation. Researchers hope to obtain products that will prove more successful as either classical biological control agents or mycoinsecticides. The subsections that follow summarize briefly the critical biological features of selected fungi to illustrate the advantages and disadvantages of these as control agents.

Aquatic Fungi

Aquatic fungi of two types that attack mosquito larvae have received considerable study: species of *Coelomomyces* (class Chytridiomycetes: order Blastocladiales) and *Lagenidium giganteum* (class Oomycetes: order Lagenidiales).

The genus *Coelomomyces* comprises over 80 species of obligately parasitic fungi that have a complex life cycle involving an alternation of sexual (gametophytic) and asexual (sporophytic) generations. The sexual phase parasitizes a microcrustacean host, typically a copepod, whereas the asexual generation develops, with rare exception, in mosquito larvae. In the life cycle, a biflagellate zygospore invades the hemocoel of a mosquito larva, where it produces a sporophyte that colonizes the body and forms resistant sporangia. The larva dies and subsequently the sporangia undergo meiosis, producing uniflagellate meiospores that invade the hemocoel of a copepod host, where a gametophyte develops. At maturation, the gametophyte cleaves, forming thousands of uniflagellate gametes. Cleavage results in death of the copepod and escape of the gametes, which complete the life cycle by fusing to biflagellate zygospores, which then seek out another mosquito host. The life cycles of these fungi are highly adapted to those of their hosts. Moreover, as obligate parasites these fungi are very fastidious in their nutritional requirements, and as a result no species of *Coelomomyces* has been cultured *in vitro*.

Coelomomyces, the largest genus of insect-parasitic fungi, has been reported worldwide from numerous mosquito species, many of which are vectors of important diseases such as malaria and filiariasis. In some of these species, *Anopheles gambiae* in Africa, for example, epizootics caused in some areas by *Coelomomyces* kill greater than 95% of the larval populations. Such epizootics led to efforts to develop several species as biological control agents. For several reasons, however, these efforts were discontinued. One important factor was the discovery that the life cycle requires a second host for completion. Also contributing were the inability to culture these fungi *in vitro* and the development of Bti as a bacterial larvicide for mosquitoes.

Although it is unlikely that *Coelomomyces* fungi will be developed as a biological control agents, interest remains in developing *L. giganteum.* This oomycete fungus is easily cultured on artificial media and does not require an alternate host. In the life cycle, a motile zoospore invades a mosquito larva through the cuticle. Once within the hemocoel, the fungus colonizes the body over a period of 2 to 3 days, producing an extensive mycelium consisting largely of nonseptate hyphae. Toward the end of growth, the hyphae become septate, and out of each segment an exit tube forms which grows back out through the cuticle and forms zoosporangia at the tip. Zoospores quickly differentiate in these, exiting through an apical pore to seek out a new substrate. In addition to this asexual cycle, thick-walled resistant sexual oospores can be formed in the mosquito cadaver.

Terrestrial Fungi

The fungi that have received the most attention for use in biological control are terrestrial fungi, with most emphasis placed on the development of selected species of hyphomycetes such as *M. anisopliae* and *B. bassiana* for use as microbial insecticides. In addition, the more specific and nutritionally fastidious entomophthoraceous fungi continue to receive attention, but for their potential use as classical biological control agents rather than as microbial insecticides. Representative examples of these terrestrial fungi are discussed in the subsections that follow.

ENTOMOPHTHORALES   These fungi comprise a large order of the class Zygomycetes that contains numerous genera, many species of which are commonly found parasitizing insects and other arthropods. The fungi routinely cause localized and sometimes widespread epizootics in populations of hemipterous and homopterous insects, particularly aphids and leafhoppers, but also in insects of other types such as grasshoppers, flies, beetle larvae, and caterpillars. In addition, a few species of the genus *Conidiobolus* are able to cause mycoses in some mammals, including humans. Apart from these few species, most of the entomophthoraceous fungi are highly specific, obligate parasites of insects and therefore their use for biological control poses no threat to nontarget organisms. As with *Coelomomyces,* however, the complex nutritional requirements, which have thus far prevented mass production *in vitro,* and high degree of host specificity, make these fungi poor candidates for development as microbial insecticides. Moreover, the conidia are very fragile, providing a challenge to formulation, and the resistant spores, like the oospores of *L. giganteum,* are difficult to germinate in a predictable manner. Nevertheless, there is evidence that if cultural practices in crop production are modified, these fungi can provide effective insect control where they occur naturally, and through the introduction of foreign strains and species (i.e., a classical biological control approach).

The most important genera found attacking insects in the field are *Conidiobolus* (aphids), *Erynia* (aphids), *Entomophthora* (aphids), *Zoophthora* (aphids, caterpillars, beetles), and *Entomophaga* (grasshoppers, caterpillars). Although many species of these genera cause epizootics and have received considerable study, none really seems to have much potential for development as a commercial microbial insecticide. On the other hand, cultural control, classical biological control, and environmental monitoring methods continue to show promise for using entomophthoraceous fungi for insect control. For example, the introduction of *Erynia radicans* from Israel into Australia to control the spotted alfalfa aphid, *Therioaphis maculata,* has proven a classical biological control success.

A recent example of apparent classical biological control can he found in the natural outbreaks of *Entomophaga miamiaga* in larval populations of the gypsy moth, *Lymantria dispar,* an important pest of deciduous forests throughout several states comprising the middle Atlantic and New England regions of the United States. Outbreaks of *E. miamiaga* have reduced larval populations to below economic thresholds, and the fungus is spreading westward naturally, and with human assistance, to gypsy moth populations established in other states. The source of this fungus is Japan, although it is not clear when the fungus causing present outbreaks of disease first appeared in the United States. The fungus was purposely introduced into the United States around the turn of the century but seems not to have become established at that time. Then in the late 1980s, outbreaks of *E. miamiaga* began to occur in Connecticut and New York, and later in Virginia. In areas where it has established, given sufficient rainfall, the fungus seems to be capable of keeping the gypsy moth population below defoliation levels. It will require another 10 years of evaluation to determine whether this is a valid instance of classical biological control by a fungus.

CLASS HYPHOMYCETES   The hyphomycete fungi belong to the fungal subdivision Deuteromycotina (imperfect fungi), a grouping erected to accommodate fungi for which the sexual phase (perfect state) has been lost or remains unknown. This group contains the fungal species that most workers consider to have the best potential for development as microbial insecticides, *B. bassiana* and *M. anisopliae,* the agents of, respectively, the white and green muscardine diseases of insects. Unlike the fungi already discussed, these two species have very broad host ranges and probably are capable of infecting insects of most orders.

With respect to the general life cycle of these fungi, the process of invasion, colonization of the insect body, and formation of conidiophores and conidia is similar to that described for the other fungi. During invasion and colonization, some fungal species produce peptide toxins that quicken host death. The infectious agent is the conidium (Fig. 3), and the taxonomy for the hyphomycetes is based primarily on the morphology of the reproductive structures, particularly the conidiophores and the conidia. Most of the hyphomycete fungi used or under development grow well on a variety of artificial media, and this attribute, along with their ability to infect insects via the cuticle, favors commercial development. In the "cottage industry" commercial operations in Brazil, China, and the former Soviet Union, solid or semisolid substrates are used for production, and the primary ingredients are grain or grain hulls.

In general, the development of *B. bassiana* and *M. anisopliae* is being targeted for control of insects that live in cooler and moist environments, such as beetle larvae in soil and planthoppers on rice, though the former species is also being evaluated against whiteflies in glasshouses, as well as grasshoppers, especially locusts, in field crops. In addition to these two species,

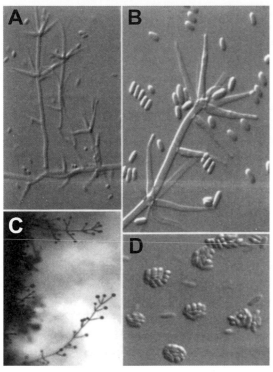

FIGURE 3 Typical reproductive structures of deuteromycete (imperfect) fungi. (A–D) Wet mount preparation of conidia-generating cells and conidia of *Verticillium lecanii,* which commonly attacks aphids and whiteflies. The conidia visible as free conidia and conidial clusters in (B) and (D) are the principal infective units. When these come in contact with an insect host, they germinate and penetrate the body, forming a mycelium that colonizes the insect over a period of several days. When conditions are appropriate, typically meaning high relative humidity, hyphae penetrate back out through the cuticle, producing conidiophores, the visible branched structures in these panels (A–C), which form reproductive conidia at their tips. (Photomicrographs courtesy of Richard A. Humber, U.S. Department of Agriculture).

several species with much narrower host ranges are considered to have potential for development, including *Paecilomyces fumoso-rosea* (for whiteflies), *Verticillium lecanii* (for aphids and whiteflies in glasshouses), *Hirutella thompsonii* (for mites), and *Nomurea rileyi* (for noctuid caterpillars).

With these apparent advantages, it is natural to ask why none of the hyphomycete fungi have been commercially successful as microbial insecticides in developed countries. There are several reasons related to their biological properties. First and foremost is that the production of conidia or mycelial fragments that are used as the active ingredient of formulations is not cost-effective because too much material is required to allow the achievement of an acceptable level of control. In addition to the problem of inefficient yields, the formulations are bulky, and preservation of fungal viability beyond a few months is low because the conidia are fragile. In mosquito and blackfly control, similar constraints apply. In addition, the discovery of cost-effective strains of *B. thuringiensis* and *B. sphaericus* has generally eliminated imperfect fungi, as well as many other microorganisms, for consideration as biological control agents for these important nuisance and vector insects.

In developing countries, *B. bassiana* and *M. anisopliae* have been used in some crops with considerable success. For example, in China *B. bassiana* has been used to control the European corn borer, *Ostrinia nubilalis,* in maize. The fungus is produced in large covered pits on maize stalks. In Brazil, a preparation of *M. anisopliae* known as Metaquino has been used for many years on sugarcane plantations and in pastures to control the spittlebug, *Mahanarva posticata.* Fungal conidia are produced in sealed plastic bags on rice. Figures indicate that as much as 50,000 ha is treated annually, and reductions in spittlebug populations are sufficient to keep populations below damaging levels. In the South Pacific, *M. anisopliae* has also been used to assist control of the rhinoceros beetle, *Orycetes rhinocerous,* a serious pests of coconut palms. Application of conidia at a rate of 50 g per square meter of soil yielded 80% larval mortality and improved cococnut yields by 25%. While these are examples of local successes, their applicability to agricultural production in developed countries is questionable.

PROTOZOA

Protozoa is a general term applied to a large and diverse group of eukaryotic unicellular motile microorganisms that belong to what is now known as the kingdom Protista. Members of this kingdom can be free living and saprophytic, commensal, symbiotic, or parasitic. The cell contains a variety of organelles, but no cell wall, and cells vary greatly in size and shape among different species. Feeding is by ingestion or more typically by adsorption, and vegetative reproduction is by binary or multiple fission. Sexual reproduction, often useful for taxonomy, can be very complex, but asexual reproduction occurs as well. Many protozoa produce a resistant spore stage that is also used in taxonomy. Divided into a series of phyla based primarily on mode of locomotion and structure of locomotory organelles, the kingdom includes the Sarcomastigophora (flagellates and amoebae), Apicomplexa (sporozoa), Microspora (microsporidia), Acetospora (haplosporidia, now thought to be a type of parasitic alga), and Ciliophora (ciliates). Protozoa of some types, such as the free-living amoebae and ciliates, are easily cultured *in vitro,* whereas many of the obligate intracellular parasites have not yet been grown outside cells.

As might be expected from such a large and diverse group of organisms, many species of protozoans are associated with insects, and the biology of these associations covers the gamut from being symbiotic to parasitic. Those that are parasitic have the general feature of causing diseases that are chronic. Many of the parasitic types, especially the microsporidia, build up slowly in insect populations, eventually causing epizootics that lead to rapid declines in populations of specific species. These epizootics attracted interest in the possibility of using protozoa to control pest insects, and over the past several decades numerous studies have been aimed at evaluating this potential. In general, these studies have shown that protozoa hold little potential for use as fast-acting microbial insecticides because of the chronic nature of the diseases they cause and because commercially suitable methods for mass production are lacking. However, as with the entomophthoraceous fungi, the possibility exists that protozoans, particularly microsporidia, may be useful as classical biological control agents. Clear examples of the effectiveness of such strategies remain to be demonstrated.

The life cycles and biologies that occur among the various protozoa that attack insects are too diverse in relation to their pest control potential for even a few to be covered here. Instead, the group with the most potential—the microsporidia—is described in terms of general biology and possible use in insect control.

General Biology of Microsporidia

The microsporidia (phylum Microspora) are the most common and best studied of the protozoans that cause important diseases of insects. Well over 800 species are known, and most of these have been described from insects. Microsporidia have been most commonly described from insects of the orders Coleoptera, Lepidoptera, Diptera, and Orthoptera, but they are also known from other orders and probably occur in all. The epizootics in insect populations caused by protozoa are usually due to microsporidia. All microsporidia are obligate intracellular parasites and are unusual in that they lack mitochondria. In addition, they produce spores that are distinguished from the spores of organisms of all other known types by the presence of a polar filament (Fig. 4), a long coiled tube inside the spore used to infect hosts with the sporoplasm.

The typical microsporidian life cycle begins with the ingestion of the spore by a susceptible insect. Once inside the midgut, the polar filament everts, rapidly injecting the sporoplasm into host tissue. The sporoplasm is unicellular but may be uni- or binucleate. Upon entry into the cytoplasm of a host cell (e.g., the fat body in many species of insects), the sporoplasm forms

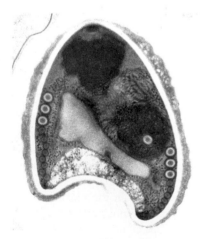

FIGURE 4 Representative microsporidian spore: transmission electron micrograph through a uninucleate spore of *Amblyospora abserrati* from a larva of the mosquito *Ochlerotatus abserratus*. The circular structures on each side of the spore are cross sections through the polar filament that is used to inject the contents of the spore into the mosquito body after ingestion and activation of the spore. (Photomicrograph courtesy of Dr. Theodore Andreadis, Connecticut Agricultural Experiment Station.)

a plasmodium (meront), which undergoes numerous cycles of vegetative growth (merogony). During these, the cells multiply extensively, dividing by binary or multiple fission and spreading to other cells, and, in many species, to other tissues of the host. After several mergonic cycles, the microsporidian undergoes sporulation. This consists of two major phases, sporogony—a terminal reproductive division committed to sporulation—and spore morphogenesis. In the sexual phase of reproduction, meiosis occurs early during sporogony. The spores, which in general measure several micrometers in diameter and length, have a thick wall and are highly, refractile when viewed by phase microscopy. The disease often lasts for several weeks during which billions of spores may accumulate in the tissues of a single infected host.

Microsporidian systematics is based on the size and structure of the spores, life cycles, and host associations. In addition to transmission by ingestion, many microsporidia are transmitted vertically from adult females to larvae via the egg (transovarially). With respect to host range, some species are species specific, whereas others occur in many species of the same family or order, and some can be transmitted to insects of different orders.

Microsporidia as Biological Control Agents

Naturally occurring epizootics caused by microsporidia are periodically very effective in significantly reducing insect pest populations. The problem is that these epizootics cannot be predicted with any degree of accuracy, nor can they be relied upon for adequate control, even though many of the conditions that facilitate their occurrence are known. The epizootics caused by *Nosema pyrausta* in populations of the European corn borer often a classic example of this unreliability. These epizootics are useful when they occur, but because this often happens too late to prevent economic damage, reliance on *N. pyrausta* alone is insufficient. Thus, efforts have been directed toward developing methods for amplifying spore loads in the field through inundative releases, in essence using microsporidia as microbial insecticides.

Because they are obligate intracellular parasites that lack mitochondria, microsporidia cannot be grown on artificial media. Several species have been grown, however, in established insect cell lines, although this is not practical for field use. For field application, whether for microbial insecticide trials or for introductions into populations, spores are grown in living hosts. With such methods the yield can be quite high (10^9–10^{10} spores per host). These yields in terms of the number of larvae that must be grown to treat a hectare and infect most of the target population are comparable to the requirements for nuclear polyhedrosis viruses. Thus, if the microsporidia could cause acute diseases, they would be on an equal footing with many of the NPVs. However, the diseases are chronic, and even if a high percentage of the target pest population is infected, there all too often is little, if any, crop protection. In fact, if advanced instars such as thirds and fourths are treated, the larvae may live longer and cause greater crop damage than if the fields were left untreated. Thus, microsporidia are not useful as microbial insecticides.

There is now a general realization that microsporidia and other protozoans have virtually no potential for use as microbial insecticides. They may, however, be useful as population management tools.

See Also the Following Articles

Biological Control • Cell Culture • Genetically Modified Plants • Integrated Pest Management

Further Reading

Adams, J. R., and Bonami, J. R. (eds.) (1991). "Atlas of Invertebrate Viruses." CRC Press, Boca Raton, FL.

Anthony, D. W., and Comps, M. (1991). Iridoviridae. *In* "Atlas of Invertebrate Viruses" (J. R. Adams and J. R. Bonamri, eds.), pp. 55–86. CRC Press, Boca Raton, FL.

Brooks, W. M. (1988). Entomogenous protozoa. *In* "Handbook of Natural Pesticides. Microbial Pesticides," Part A, "Entomogenous Protozoa and Fungi" (C. M. Ignoffo and N. B. Mandava, eds.), pp. 1–149. CRC Press, Boca Raton, FL.

Federici, B. A. (1998). *Bacillus thuringiensis* in biological control. *In* "Handbook of Biological Control" (T. S. Bellows, G. Gordh, and T. W. Fisher, eds.), pp. 519–529. Academic Press, San Diego, CA.

Fuxa, J. R., and Tanada, Y. (eds.) (1987) "Epizootiology of Insect Diseases." Wiley, New York.

Entwistle, P. F., Cory, J. S., Bailey, M. J., and Higgs, S. (eds.) (1993). "*Bacillus thuringiensis,* an Environmental Biopesticide: Theory and Practice." Wiley Chicester, U.K.

Hajek, A. E., Elkinton, J. S., and Witcosky, J. J. (1996). Introduction and spread of the fungal pathogen *Entomophaga miamiaga* (Zygomycetes: Entomophthorales) along the leading edge of gypsy moth (Lepidoptera: Lymantriidae) spread. *Environ. Entomol.* **25**, 1235–1247.

Henry, J. (1990) Control of insects by protozoa. *In* "New Directions in Biological Control" (R. Baker and P. Dunn, eds.), pp. 161–176. Liss, New York.

Hunter-Fujita, F. R., Entwistle, P. E. Evans, H. F., and Crook, N. E. (eds.) (1998). "Insect Viruses and Pest Management." Wiley, Chichester, U. K.

Lacey, L. A. (1997). "Manual of Techniques in Insect Pathology." Academic Press, San Diego, CA.

McCoy, C. W., Samson, R. A., and Boucias, D. G. (1988). Entomogenous fungi. *In* "Handbook of Natural Pesticides, Microbial Pesticides," Part A, "Entomogenous Protozoa and Fungi" (C. M. Ignoffo and N. B. Mandava, eds.), pp. 151–236. CRC Press, Boca Raton, FL.

Miller, L. K. (ed.). (1997). "The Baculoviruses." Plenum Press, New York.

Tanada, Y., and Kaya, H. K. (1993). "Insect Pathology." Academic Press, San Diego, CA.

Phasmida
(Stick and Leaf Insects)

Erich H. Tilgner
University of Georgia

Phasmida are nocturnal exopterygote insects. They exhibit a variety of unpredictable and bizarre shapes. Some look

FIGURE 1 A female of *Oxyartes spinosissimus,* a lichen mimic.

like twigs or tree bark (Figs. 1, 2) and may seem to be covered by lichens or moss. Others are indistinguishable from living or dead leaves (Fig. 3), mimicking even leaf veins and mildew spots to perfection. Phasmida are large, and a few are remarkable for their gigantic size. The longest insect species in the world is *Phobaeticus kirbyi* from Borneo, with one documented female measuring 55 cm in length. Phasmida inhabit tropical, subtropical, and temperate forests, savannas, grasslands, and chaparral; their diversity is highest in the tropics.

PHYLOGENY AND CLASSIFICATION

Over 3000 species of Phasmida have been described. The genus *Timema* from the western United States is considered to

FIGURE 2 A male and female of *Aplopus* sp. in the act of mating. The smaller male is hanging off the back of the female.

FIGURE 3 A female of the walking leaf *Phyllium bioculatum.*

be sister group to the remainder of the order, which is referred to as the Euphasmida. *Timema* are small, wingless, and cryptically colored. Euphasmida are larger, winged or wingless, usually possessing an elongated mesothorax, and are stereotyped as stick or leaf insects (Fig. 4). *Timema* have no fossil record. The oldest Euphasmida fossils date to the middle Eocene, 44 to 49 mya. Oligocene and Miocene fossils are known from Florissant shale, Baltic, and Dominican Republic amber.

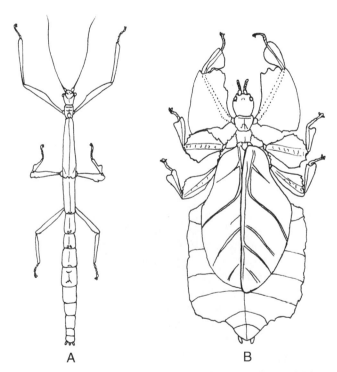

FIGURE 4 Female Euphasmida: (A) *Phenacephorus auriculatus* and (B) *P. bioculatum.*

The taxonomy of the order is problematic. No workable classification scheme exists, and those that are available are not based on phylogenetic relationships. Assignment to a category such as family, tribe, even suborder provides so little information that it is almost meaningless. This is in contrast other insect orders, such as Coleoptera, where a suborder, or family-level identification, say, provides a wealth of biological information about the specimen. In spite of the lack of an acceptable classification, the fauna of a few areas, (e.g., Europe, Malaysia, Borneo, Japan, United States, Canada, New Zealand) have been sufficiently studied to permit tentative identification of species by nonspecialists.

BIOLOGY

Sexual dimorphism is extreme in the Phasmida, and it is difficult to associate the sexes unless mating adults are found under natural conditions, or if males and females are obtained from the rearing of eggs in captivity. Reproduction is usually sexual, but many species are parthenogenetic. Eggs resemble plant seeds, are laid singly, and are either dropped, flicked, buried, glued to a surface, or riveted to a leaf. Some species rely on ants to disperse them. After successive molts, nymphs regenerate limbs lost by autotomy (the purposeful shedding of appendages). The entire life cycle takes from several months to several years depending on the species. Phasmida feed primarily on flowering plants, but a few eat either gymnosperms or ferns. The primary defense against predation is crypsis. Secondary defenses can include catalepsy (e.g., death feigning), startle displays, or the ejection of an irritating spray fired from a pair of prothoracic exocrine glands.

See Also the Following Articles
Crypsis • Parthenogenesis

Further Reading
Brock, P. D. (1999). The amazing world of stick and leaf-insects (R. Fry, ed.). Cravitz Printing, Essex, U.K.
Key, K. H. L. (1991). Phasmatodea (stick-insects). *In* "The Insects of Australia," Vol. I (CSIRO, ed.). Melbourne University Press, Carlton, Australia.
Tilgner, E. (2001). Fossil record of Phasmida (Insecta: Neoptera). *Insect System. Evol.* **31**, 473–480.

Pheromones

Ring T. Cardé and Jocelyn G. Millar
University of California, Riverside

Pheromones are chemical messages that induce a behavioral reaction or developmental process among individuals of the same species. The term is derived from the Greek for "carrier of excitation" and was coined in 1959 by the German biochemist Peter Karlson and the Swiss entomologist Martin Lüscher during their investigations of the chemicals that regulate caste development in termites. In 1963 E. O. Wilson and W. H. Bossert of Harvard University formally distinguished two classes of pheromones. Releaser pheromones are messages that induce an immediate behavioral reaction in the receiver. The kinds of behavioral response evoked in insects are incredibly diverse, and they include alarm, defense, aggregation, attraction, kin and colony recognition, marking of territories and egg deposition sites, mating behaviors, recruitment, trail following, and even thermoregulation. In contrast, primer pheromones cause a physiological change in the receiver, such as development of a particular caste or sexual maturation, which eventually modifies the organism's behavior.

All pheromones fall under the broader umbrella classification of semiochemicals—chemicals that are involved in communication. The two major classes of semiochemicals besides pheromones are allomones and kairomones. These are solely interspecific cues, in contrast to pheromones, which are always intraspecific cues. Allomones are chemicals that provide some advantage to the emitter (e.g., defensive secretions), whereas kairomones are signals that confer an advantage to the receiver (e.g., emanations used by a parasite to locate a host). This article only touches on the diversity and complexity of pheromone-mediated behaviors and developmental changes in insects. Communication among social insects, especially among ants, bees, wasps, and termites, involves a highly sophisticated pheromonal language, in which the interpretation of the individual chemical constituents or "words" depends on their particular combinations, ratios, concentrations, and even order of presentation. Context, that is, the recent experiences of the receiver and its physiological state, is all-important in response.

SEX ATTRACTANT PHEROMONES OF LEPIDOPTERA

The first definitive evidence of pheromone communication dates to experiments performed with the great peacock moth *Saturnia pyri* by the French naturalist Jean-Henri Fabre in the 1870s. Fabre sequestered a female moth in a screened cage following her morning emergence, to permit her wings to expand and harden. That evening more than 40 male moths arrived at Fabre's study, "eager to pay their respects to their marriageable bride born that morning." Further observations showed that cages that had housed virgin females also were attractive; this and other observations led Fabre to conclude that "effluvia of extreme subtlety" mediated attraction. Nearly 90 years would pass before microanalytical techniques would permit identification of the minute quantities of pheromone involved.

The first pheromone to be identified was the sex attractant pheromone of *Bombyx mori,* the commercial silkworm. This silkworm is an entirely domesticated species that is no longer capable of flight; its female-emitted

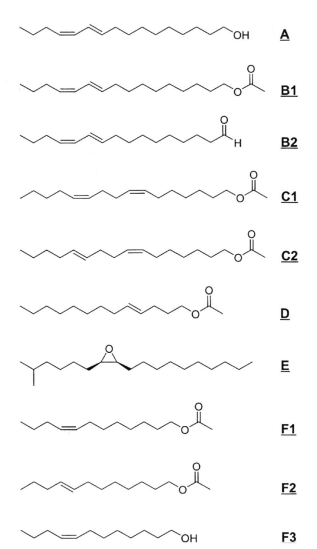

FIGURE 1 Representative structures of lepidopteran (moth) pheromones. (A) *Bombyx mori,* the commercial silkworm (Bombycidae), (B) *Hemileuca electra* (Saturniidae), (C) *Pectinophora gossypiella,* pink bollworm (Gelechiidae), (D) *Keiferia lycopersicella,* tomato pinworm, (Gelechiidae), (E) *Lymantria dispar,* gypsy moth (Lymantriidae), and (F) *Grapholita molesta,* oriental fruit moth (Tortricidae).

pheromone (Fig. 1) induces upwind walking and courtship behaviors in conspecific males. The German biochemist Adolf Butenandt (who received the Nobel Prize in 1939 for his work identifying the human sex hormones) spent more than two decades in this quest. In 1959 he identified (*E*10,*Z*12)-hexadecadienol as the single compound causing upwind walking and copulatory attempts and named it "bombykol." To provide some perspective on this remarkable achievement, Butenandt and his coworkers extracted a half million female moths, finally isolating a few milligrams of the pure pheromone. Today's modern methods of isolation and characterization (especially coupled gas chromatography and mass spectrometry) were not yet available.

The amount of pheromone that is secreted from or is present in a pheromone-producing gland varies enormously with species and, to some extent, behavioral function. Sex attractant pheromones may be present in microgram, nanogram, and even picogram (10^{-12} g) quantities per individual. Microanalytical techniques are now so advanced that identifications on occasion can be made from either gland extracts or airborne collections from a few individuals, and with just nanogram or even lower quantities of natural chemical. Even the always tedious behavioral bioassays, long used to monitor for behaviorally active components of gland extracts and airborne collections, have been largely supplanted by using a living insect antenna as a detector. The electroantennogram (EAG) was pioneered in the mid-1950s by Dietrich Schneider, working at the Max Planck Institute near Munich, Germany. In the 1970s, Wendell Roelofs of Cornell University adapted this assay to speed up identifications. A moth antenna was used to monitor which fractions separated by gas chromatography contained the active components. Later applications mounted the EAG apparatus at the outflow of a gas chromatograph column, and this living detector indicated the presence (and the retention times) of compounds that were likely to be behaviorally active by means of an electrical signal elicited by the interaction of pheromone and the antennal receptors. These advances allowed chemists to zero in quickly on the compounds present in crude extracts that were most likely to comprise the pheromone.

The sensitivity of a male silkworm to bombykol is legendary. It has been investigated by recording the electrical response of individual sensory hairs on their antennae (each antenna is equipped with 40,000 such hairs) and by monitoring a single male's change from quiescence to wing fanning and upwind walking. The estimates are astonishing: one bombykol molecule is sufficient to induce the firing of an individual receptor, and a behavioral response can be evoked with only 200 molecules (~10^{-19} g!).

Pheromone structures now have been described for several hundred species of moths. Nearly all these pheromones induce upwind flight by the male to the pheromone-releasing female. The majority of known structures for moth pheromones (examples in Fig. 1) are hydrocarbon chains, usually 10 to 18 carbons in length, with 1 to 3 double bonds and a terminal acetate, alcohol, or aldehyde. Less common structural motifs in moth pheromones include epoxides, ketones, and hydrocarbons with one or more double bonds or methyl branches; chain lengths known so far range from C_{10} to C_{23}. Many pheromones, such as those of the moths *Hemileuca electra* and *Grapholita molesta*, comprise blends of two, three, or even more components. Specificity of the chemical message is accomplished in many species by females emitting and males responding to precise ratios of their pheromone blend. For example, for males of *G. molesta*, the ratio of the *(Z)*-8- and *(E)*-8-dodecenyl acetate components must be very close to the 95:5 mix produced by the female for maximum attraction. The use of blends and in some species precise ratios allows many closely related moth species to have "exclusive" communication channels, even though they share some

FIGURE 2 Female of the day-active saturniid moth *Hemileuca electra* exposing her pheromone gland, located at the tip of her abdomen. Such pheromone-releasing behavior, termed "calling," and the male's mate-finding behaviors typically occur at set times of the day or night. In the Mojave Desert of California *H. electra* calls from midmorning to early afternoon; the closely related species *H. burnsi*, which shares pheromone components with *H. electra*, calls from midafternoon to dusk. Without exclusive times for mating activities, these species would cross-attract. (Photograph courtesy of Chris Conlan.)

components of their respective blends. Other strategies for partitioning of the communication channel include restricting sexual activity to specific times of the day or night (Fig. 2).

Pheromones of other types are produced by males of many moths and facilitate close-range recognition and acceptance by the female. In a few species the sexual roles are reversed, with male moths being the attractive sex and recruiting females. Many male butterflies also use pheromones in courtship, disseminating an "aphrodisiac" scent from scales on their wings or, in some butterflies, from specialized paired brushes located at the tip of the abdomen. However, butterflies do not attract mates with long-distance pheromones; instead they rely on visual cues for mate finding.

ATTRACTANT AND AGGREGATION PHEROMONES

Although long-distance communication by attractant pheromones is well established in nearly all moth lineages, pheromones are widely used by many insect groups in mate finding. Such messages are categorized as either sex attractant pheromones, if one sex attracts the other (as in moths), or aggregation pheromones, if both sexes are attracted. Feeding on a plant host and release of aggregation pheromones typically are linked, and mating often occurs in such aggregations. Therefore aggregation pheromones can play a multifunctional role. Representative structures (Fig. 3) of sex attractants of insects other than moths include those of the cockroach, aphid, scale insect, caddisfly, sawfly, beetle, and true fruit fly. The chemistries of these messages are diverse, as are the locations of the glands responsible for their production.

BARK BEETLE PHEROMONES

Bark and ambrosia beetles (Scolytidae) use pheromones to facilitate colonization of host trees (aggregation) and to attract

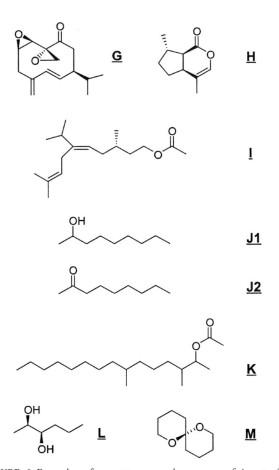

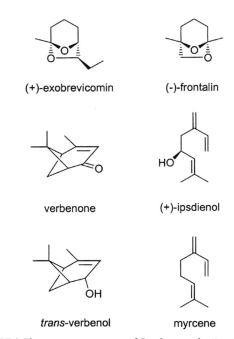

FIGURE 4 Pheromone components of *Dendroctonus brevicomis,* the western pine beetle. Myrcene is emitted by the beetle's principal host, ponderosa pine *(Pinus ponderosa).*

FIGURE 3 Examples of sex attractant pheromones of insects from nonlepidopteran orders: (G) periplanone B, from the American cockroach, *Periplaneta americana,* (H) nepetalactone, a pheromone component of a number of aphid species, (I) pheromone of California yellow scale, *Aonidiella citrina,* (J1 and J2) pheromone components of caddisflies, (K) pheromone component of diprionid sawflies, (L) (2*R*,3*R*)-2,3-hexanediol, a pheromone component of the cerambycid beetle *Hylotrupes bajulus,* and (M) *(R)*-1,7-dioxaspiro[5,5]decane, from the olive fruit fly *Dacus oleae.*

mates. Many scolytid species must attack a tree en masse if they are to overwhelm the tree's defense, which consists of exuding sap into the tunnel that each beetle bores. The first beetles to arrive may identify the host by means of chemicals emitted by the host tree itself; as they bore into the tree, they release pheromones and increase emission of tree chemicals that together attract both male and female beetles. David Wood of the University of California at Berkeley and Robert Silverstein, then at Stanford Research Institute, worked out these intricate interactions in *Dendroctonus brevicomis,* the western pine beetle. Infestations begin when females are attracted to their principal host, ponderosa pine, by myrcene, a monoterpene the tree emits as a defensive compound (allomone) when injured or stressed, and by the tree's silhouette. As the "pioneer" females bore into the tree's bark, they release their pheromone, (+)-*exo*-brevicomin (Fig. 4), which is augmented by increased release of myrcene from the host tree. Males are attracted and, after one enters the female's tunnel, he emits (–)-frontalin. The combination of

host-, female-, and male-released volatiles attracts many more males and females, ensuring that the tree's sticky sap defense will be insufficient. Once males and females have mated, they alter their chemical message: males and females emit verbenone and *trans*-verbenol and males release (+)-ipsdienol (Fig. 4). Together, these three chemicals interrupt further attraction of males and females, thereby helping to regulate the level of infestation and avoiding overexploitation of the tree.

PHEROMONES OF SOCIAL INSECTS

Pheromones mediate many activities of social insects, including defense of the colony, recruitment to food, recognition of individuals and nestmates, and regulation of caste development. The exocrine glands that produce the various pheromones are dispersed throughout the body, as exemplified by those of leafcutter ants (Fig. 5).

Alarm and Defense

Charles Butler was the first to describe the behavioral effects of an insect alarm signal. He recognized in 1609 that the stinger of a honey bee *(Apis)* worker impaled on human skin or clothing attracts more bees to sting that site. The multifunctional role of this signal is shown by the reactions to the same chemical signal of guard honey bees at the entrance to their hive. The presence of an intruder can cause a guard bee to release alarm pheromone from her sting chamber; she disseminates this message into the hive by wing fanning, thereby summoning many bees. These alerted bees seem to be "agitated," with rapid movements and mandibles

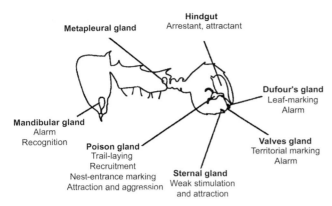

FIGURE 5 Location and function of pheromone-producing glands in leafcutter ants. [After Howse *et al.* (1998).]

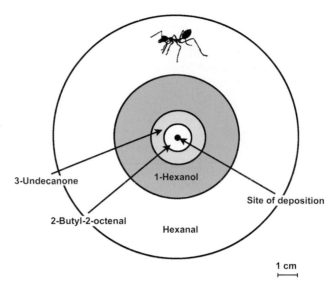

FIGURE 6 Active spaces in still air of the principal components from the mandibular gland of the weaver ant, *Oecophylla longinoda*. The pheromone has been deposited in the center and the boundaries of behavioral activity of each component 20 seconds later are represented. [After Howse *et al.* (1998).]

agape, poised to defend their colony. The sting chamber has more than 20 known pheromone components, most of which induce either alerting or stinging. The other behavioral reactions evoked by components of the sting chamber include lowering the number of foragers departing the hive and repelling foragers that have arrived at a food source. The main constituent of this mixture-evoked alerting and, to a lesser extent, stinging is isopentyl acetate, and the amount per worker changes dramatically with the behavioral task that bees are performing. In the first few days of an adult worker's life, when she is confined to housekeeping and brood-tending tasks inside the hive, essentially no isopentyl acetate is present; the amount rises to 4 to 5 μg at several weeks of age, when she either assumes guard duties or starts foraging outside the hive; for the remainder of her life as a forager, the amount falls to approximately 2 μg per stinger. Stinging itself seems to be released by many components of the sting chamber, including isopentyl acetate, and especially *n*-butyl acetate and 1-pentanol. As with many reactions of social Hymenoptera, the context in which the signal is released is crucial to the kinds of behavior evoked. Context must be taken into account in interpreting behavior and in devising diagnostic behavioral bioassays. For assays of honey bee stinging, for example, one standardized procedure is to provide a vibration of set intensity to the hive, followed by presentation in front of the hive entrance of a swinging target such as a cotton ball or piece of leather containing a candidate alarm pheromone. The number of bees attacking and stinging the target is used to score the level of response.

Defensive behaviors of social insects are quite varied and difficult to categorize into mutually exclusive behaviors. Detection of an alarm pheromone by an ant, for example, can cause it to splay its mandibles, raise its head, bite, and spray an odoriferous and irritating defensive secretion toward a perceived enemy. Unlike attractant and aggregation pheromones, which typically are carried downwind in a turbulent wind flow, alarm pheromones are often released either in relatively still air within the confines of the colony or at ground level, where wind is attenuated. In such situations, molecular diffusion largely or exclusively governs the distribution of pheromone, and the resulting concentration gradient of pheromone supplies potentially useful information about the direction toward the source of alarm. Thus, ants may run toward the source of pheromone (essentially up the odor gradient) or, at lower concentrations farther away from the odor source, movements may seem to be undirected with respect to the odor source. There are many kinds of defensive reactions and, in leafcutter ants, several glandular sources for alarm pheromones (Fig. 5).

An instructive example is provided by the defensive reactions evoked by some of the more than 30 mandibular gland components of the weaver ant, *Oecophylla longinoda*, worked out in considerable detail by John Bradshaw, Philip Howse, and Ray Baker at the University of Southampton. If a droplet of mandibular gland secretion is daubed onto a flat surface in still air, the volatile pheromone diffuses outward at a rate that is dependent on its molecular weight and its diffusion coefficient. A region in which the concentration of pheromone is above the minimum required to elicit a particular behavioral reaction is termed an "active space." The active spaces of each of the four principal components of the mandibular gland secretion 20 s after its deposition are shown in Fig. 6. The sequence of defensive activities seems to be ordered by proximity to the odor source. At the outer limit of the active space, worker ants encounter only hexanal above threshold levels. Ants show heightened levels of running with open mandibles, but their trajectory is not aimed toward the odor's source. Ants that enter the 1-hexanol region, however, move up the odor gradient toward the odor's source. Once they have reached the active space of 3-undecanone, this compound further facilitates orientation and lowers the threshold for biting, as does 2-butyl-2-octenal. Together these four compounds ensure that the ants are recruited to the site at which the alarm pheromone was released and that they attack an adversary that has been marked with this secretion. Mandibular gland components involved in defense, including other active constituents in addition to the four main constituents, vary within a colony among castes and even between major and minor workers. There also is substantial variation in a given caste among colonies, suggesting that different colonies may have unique defensive codes.

Trail Following and Recruitment

Social insects use trails of varying permanence to exploit food resources and sometimes for colony movement and relocation. E. O. Wilson's exhaustive study of what governs the persistence of the food trail of the fire ant, *Solenopsis invicta,* provides an example of how such systems function. A foraging worker that has encountered a suitable food source lays down a chemical trail by dragging its stinger sporadically along the ground as it returns to the nest. The trail pheromone is a mixture of farnesenes [mainly *(Z,E)*-α-farnesene] from their Dufour's gland, and at any given moment each ant contains only about a nanogram of trail pheromone. The trail from one individual does not persist for long—the active space falls below threshold in less than 2 min, and its effective length is not much more than a meter. *Solenopsis* can even adjust the amount of pheromone deposited on the trail by altering how firmly it drags its sting. The amount of pheromone on the trail is regulated by three factors: the number of ants returning, the proportion of ants laying a trail, and the amount that each ant contributes to the trail. When the food is gone or an ant cannot reach the food source because of other ants, any ant that cannot feed simply does not reinforce the trail.

The number of ants recruited to leave the nest for foraging is a direct function of the quantity of trail pheromone released by a returning forager: to induce nestmates to leave the nest and forage along the trail, the returning forager releases much more pheromone than is found along the trail itself. These simple rules followed by individuals allow mass recruitment, a sophisticated system whereby one group of ants transfers information about the quality of a food source some distance away to another group of ants. The seeming disadvantage of the impermanence of such trails is actually a useful feature of the system that permits ants to match the number of foragers to the quantity of the resource.

Trail communication also is used in relocation of *Solenopsis* colonies. Scouts that have discovered a suitable nest location lay down a trail that other workers follow to the same site. If the location is indeed deemed favorable by new workers inspecting the site, these workers add pheromone to the trail upon their return trip to the nest; this leads to an exponential increase in traffic. Eventually the brood is transferred to the new nest, and the queen follows. Trails close to the nest of some ant species can be relatively permanent, lasting days, and these are called trunk trails. The constituents of the trail pheromones are known for many ant groups, and they are produced from a variety of glandular sources, including the Dufour's, poison, and sternal glands, and the hindgut. Trail pheromones also are widely used by termites in foraging activities. Chemical trails are also important to regulating foraging activities of colonial tent caterpillars (Lasiocampidae). Caterpillars that have located food on distant branches add pheromone to the silken trail on their return trip to the silk nest at which they spend their nonfeeding time. These marked paths are then followed preferentially by future foragers.

Queen Pheromone of the Honey Bee

As emphasized for alarm pheromones, a pheromone can communicate many meanings depending on context. The queen pheromone of the honey bee *(Apis mellifera)* exemplifies this principle, which has been termed "pheromonal parsimony" by Murray Blum of the University of Georgia. The queen pheromone is produced by the queen's mandibular glands and its five known components are 9-oxo-*(E)*-2-decenoic acid, (+)- and (−)-9-hydrox-*(E)*-2-decenoic acid, methyl *p*-hydroxybenzoate, and 4-hydroxy-3-methoxyphenylethanol. The queen produces about 500 μg of this mixture daily, most of which is picked up by the continually changing retinue of a dozen or so workers that constantly groom the queen. Trophallaxis (interchange of food), antennation, and grooming among colony members in turn disperse queen pheromone throughout the colony. The releasing functions of the queen pheromone include the "retinue" behavior (attendance and grooming of the queen—these behaviors require all five components), suppression of construction of new queen cells, and a delay in swarming.

Outside the hive, swarming bees without a queen are attracted to a source of 9-oxo-*(E)*-2-decenoic acid, but they will not form a cluster without the addition of 9-hydrox-*(E)*-2-decenoic acid. Drones (males) are attracted to virgin queens flying 10 or so meters above ground level by her release of 9-oxo-*(E)*-2-decenoic acid. Perhaps the most dramatic effect of the queen pheromone is in its governance of colony productivity. Without queen or queen pheromone in the colony, many workers remain idle. Queen pheromone stimulates comb construction, brood rearing, foraging, and food storage. Artificial application of queen pheromone increases all these activities. The queen pheromone also has a clear primer effect inasmuch as it inhibits development of the workers' ovaries. In the absence of the queen (and the queen pheromone), egg production is triggered in up to one-quarter of the workers, and these "false queens" and other workers in turn produce some queen pheromone.

Pheromones of Honey Bee Workers

Honey bee workers produce diverse messages from a number of exocrine glands. The Nasonov gland (which queens and drones lack) is situated on the seventh abdominal tergite. A worker exposes this gland by flexing its abdomen, usually while wing fanning and elevating its abdomen. The secretion contains mainly geraniol, geranial and geranic acid, and these influence foraging, marking and, when coupled with queen pheromone, clustering. Nasonov pheromone also is important to swarming. After a swarm has departed the hive, the first workers to arrive at a clustering site expose their Nasonov gland, and the scent attracts other flying workers. Nasonov secretions also are used in "house hunting." A scout that has found a potentially suitable nest site returns to the swarm and communicates direction and distance by the dance language. Scouts release Nasonov pheromone at the site, thereby helping to attract more bees to evaluate its suitability. Nasnonov pheromone also is released by bees fanning at the hive's entry; this odor (probably mixed with odors from the hive) seems to aid disoriented foragers in finding the hive's entrance.

The mandibular gland produces mainly 2-heptanone, which is released by guard bees as an alerting pheromone, and possibly to mark an intruder. This compound, perhaps in combination with other components of the mandibular gland, also may be used to mark flowers that are no longer productive, thereby improving foraging efficiency. Other pheromones labeled "footprint" pheromones mark the nest entrance, and a thermoregulatory pheromone causes nurse bees that are incubating pupae to raise their body temperature by means of muscle contractions. Capping of brood cells is induced by a pheromone consisting of mixture of four fatty acid esters. The examples of pheromonal communication in *A. mellifera* considered here provide a glimpse into the pervasiveness of pheromonal communication in this insect, and the diversity of reactions that can be mediated by pheromonal messages.

Termite Pheromones

The development of castes in termites seems to be governed by complex interactions between juvenile hormone, pheromones, and environmental conditions such as food availability and time of year. For example, in a colony of *Kalotermes* (a "lower termite"), the absence of a king and queen in the colony causes development of replacement (supplementary) reproductives from pseudergate workers, but with the establishment of a reproductive pair (or more), they secrete pheromones that induce pseudergates to eat the excess of reproductives. A queen-produced pheromone inhibits female pseudergates from becoming reproductively competent, and a male-produced pheromone similarly inhibits male pseudergates from becoming reproductive. In the absence of a queen, the king secretes a pheromone that stimulates production of females. The proportion of soldiers in the colony is also regulated by similar interactions. The identity of the pheromones that modulate the proportion of castes in a colony remains unknown.

WHEN PHEROMONES BECOME KAIROMONES

It is also worth noting that parasitoids and predators have coevolved to exploit and manipulate the pheromones of their prey. For example, a group of clerid beetles uses the pheromonal signals of bark beetles to locate and invade the tunnels of their prey in the bark and cambium layers of conifers. Similarly, pentatomid bug species frequently suffer high levels of parasitism from parasitic flies from several families, or from specialist wasp predators. It has been unequivocally demonstrated that these parasitoids use the bugs' pheromones as kairomonal cues to locate hosts. The flies and wasps are attracted specifically to components of their host's pheromone blend. For both the predatory clerids attacking bark beetles and the fly and wasp species attacking pentatomid bugs, the attraction to the pheromones of their prey can be so strong that traps baited with the prey pheromones actually catch more of the parasitoids or predators than the target species.

However, illicit use of the pheromones of prey can go well beyond simply eavesdropping on pheromonal signals. In a fascinating example of coevolution, bolas spiders in the genus *Mastophora* (and several other genera) produce the pheromonal signals of their prey to lure the prey close enough to be caught by a swinging, sticky thread of silk. The prey are male noctuid moths responding to copies of the female pheromone. Even more extraordinary, there is evidence to suggest that within an hour or so, the spiders can change the pheromone lures they produce, to maximally exploit the different times of flight of prey species that respond to differing pheromone blends.

A variety of insects that live inside and sometimes parasitize social insects such as ants and termites also have developed the ability to aggressively mimic the pheromonal signals of their hosts. For example, several staphylinid beetle species live inside termite nests, where they receive all their food from their hosts, and are groomed and cared for by their hosts as though they were indeed termite brood. The chemical cues that both prevent the termites from recognizing the inquilines and induce the feeding and grooming behaviors closely mimic the true pheromones used by the termites for these functions. In an even more aggressive example, the larvae of some syrphid flies are obligate predators on the brood of their ant hosts. The fly larvae produce a blend of cuticular hydrocarbons that closely matches the hydrocarbon profile of the host's brood, effectively camouflaging the fly. The camouflage is so good that if the nest is attacked, the worker ants will carry the fly larvae to safety as though they were indeed ant brood.

APPLICATION OF PHEROMONES IN PEST MANAGEMENT

Insect pheromones have proven useful in pest control. Most of these applications use synthetic copies of pheromones that mediate either attraction or aggregation. Compounds are formulated in protective matrices or reservoirs that emit the pheromone over weeks or months. Pheromone-baited traps are used to detect exotic invaders, to decide whether pest levels are sufficient to warrant intervention, and to time the application of conventional insecticides or other control measures. For example, the spread of the gypsy moth (*Lymantria dispar*) in the United States is monitored with inexpensive pheromone-baited traps whose sticky internal surface ensnares males. Approximately 350,000 traps are deployed yearly to determine the extent of spread of the European strain of the gypsy moth from the eastern portions of the United States to the Midwest and South, or, especially along the west coast, to signal the occasional invasion of the Asian gypsy moth strain.

Pheromones also are used for direct population control. The tomato pinworm *(Keiferia lycopersicella)*, for example, is a devastating pest of tomatoes in Mexico, largely because this moth is highly resistant to insecticides. Application of "cocktails" (mixtures of two or more insecticides) as many as 40 times during a crop cycle may not prevent complete crop loss. Ideally, however, if a tomato field is blanketed with plastic dispensers that release micrograms per hour of synthetic pheromone, the emitted pheromone will interfere with normal mate-finding activities of males, even if the pinworm population is initially at high density. Just how mating disruption works is not fully established, but efficacy likely involves the additive effects of habituation of responsiveness (a presumed central nervous system phenomenon) and some competition for the male's attention between the natural emitters, the females, and the numerous sources of synthetic pheromone. The amount of synthetic pheromone needed to disrupt mate finding is quite small: typical application rates are several grams per hectare per week, and nearly all pheromones are nontoxic and nonpersistent. This technique is now commonly used to control several dozen moth species.

CONCLUSION

Pheromones are a dominant form of communication in most insects, and the messages conveyed serve myriad behavioral and physiological functions. Current microanalytical techniques permit identification of many of these messages, even though they occur in minuscule quantities. Studies expanding our understanding of the chemistry of these messages continue, although in many pheromone systems our ability to characterize the chemicals produced is in advance of our progress in understanding the evoked behaviors, particularly among the complex communication systems of social insects. Current frontiers of investigation in insect pheromones include establishing how genes control biosynthesis, how these chemicals are transduced into an electrical signal in the receptor cells of the responder, and how the signals are processed in the brain, leading to a behavioral output. Only a few structures of primer pheromones have been elucidated, and consequently much remains to be learned about their mode of action.

See Also the Following Articles
Antennae • Caste • Chemoreception • Defensive Behavior • Orientation • Recruitment Communication

Further Reading
Agnosta, W. C. (1992). "Chemical Communication: The Language of Pheromones." Scientific American Library, New York.
Cardé, R. T., and Bell, W. J. (eds.). (1995). "Chemical Ecology of Insects," Vol. 2. Chapman & Hall, London.
Cardé, R. T., and Minks, A. K. (1995). Control of moth pests by mating disruption: Successes and constraints. *Annu. Rev. Entomol.* **40**, 559–585.
Cardé, R. T., and Minks, A. K. (eds.). (1996). "Insect Pheromones. New Directions." Chapman & Hall, New York.
Eisner, T., and Meinwald, J. (eds.). (1996). Chemical Ecology. The Chemistry of Biotic Interaction." National Academy of Science Press, Washington, DC.
Francke, W., and Schulz, S. (1999). Pheromones. *In* "Comprehensive Natural Products Chemistry," Vol. 8 "Miscellaneous Natural Products Including Marine Natural Products, Pheromones, Plant Hormones, and Aspects of Ecology." (K. Mori, ed.), pp. 197–261. Pergamon Press 5, Amsterdam.
Free, J. B. (1987). "Pheromones of Social Bees." Chapman & Hall, London.
Hansson, B. S. (ed.). (1999). "Insect Olfaction." Springer-Verlag, Berlin.
Hardie, J., and Minks, A. K. (1999). "Pheromones of Non-Lepidopteran Insects Associated with Agricultural Plants." CABI Publishing, Wallingford, Oxon, U.K.
Holldöbler, B., and Wilson, E. O. (1990). "The Ants." Bellknap Press of Harvard University Press, Cambridge, MA.
Howse, P., Stevens I., and Jones, O. (1998). "Insect Pheromones and Their Use in Pest Management." Chapman & Hall, London.
Vander Meer, R. K., Breed, M. D., Winston M. L., and Espelie, K. E. (eds.). (1998). "Pheromone Communication in Social Insects." Westview Press, Boulder, CO.

Phoresy

Marilyn A. Houck
Texas Tech University

Phoresy is a special kind of commensal relationship in which one organism (the phoretic or phoront) attaches to another (the host) for a limited time period to enhance dispersal of the phoront from the natal (or birth) habitat, resulting in colonization of a new and potentially better habitat. In addition to transport, the phoretic host may incidentally provide substrate, shelter, and even some indirect defense or protection for the phoront, but the strict definition of phoresy excludes any direct physiological benefit during transit. For example, the host does not provide the phoretic with food while in transit nor does it contribute to the ontogeny (development) of the phoretic during transit. If feeding does occur, the more appropriate term to describe this relationship would be parasitism. Although phoresy is not a form of parasitism, phoresy can eventually extend into a parasitic association (see below). Alternatively, if the host receives a benefit from its passenger the relationship is again not phoresy, but a form of mutualism. Thus, this discrete definition of phoresy separates phoresy from all other forms of symbiotic interactions.

The term phoresy is uniformly applied throughout the plant and animal kingdoms and does not exclusively apply to interactions of, or with, insects. Seeds that hitchhike on fur and pants are phoretic. The remora fish (Echeniedea: *Remora remora*) has a dorsal fin modified into a sucker that allows it to attach to the sides of larger fish and turtles, using them for phoretic transport. However, if the remora also snatches pulverized leftovers created during feeding by the host, this act violates the strict definition of phoresy. It soon becomes clear that it can be difficult to distinguish phoresy from other forms of symbiosis. Detailed aspects of behavior, natural history, and physiology are essential to a firm understanding of interspecific associations.

PHORESY AMONG INSECTS AND ARACHNIDS

Phoresy among arthropods has been recognized since at least the mid-1700s. It was formalized and defined by Lesne in 1896 as "those cases in which the transport host serves its passenger as a vehicle." The etymology of the word unites the perspective of both the phoretic (*phor* (Gr.) = thief) and the host (*phoras* (Gr.) = bearing) within the interaction and has a counterpart to most forms of human transport: a "boat" in aquatic phoretic interactions, a "bus" in terrestrial settings, and an "aircraft" in aerial transport. Despite the medium being traversed, there are general principles that seem to be consistent among phoretic relationships, but these are not inclusive and exceptions occur: (1) the phoretic is usually much smaller than its host; (2) there are often several phoretic passengers on any individual host, and mass transit is not uncommon; (3) the phoretic does not have an effective means of independent dispersal (e.g., wings or oars), whereas the host usually is quite mobile; (4) phoresy has played a role in dispersal for a long time and evidence of phoretic relationships can be found in Baltic and Dominican amber from as far back as the early Tertiary (40 mya); (5) phoresy is a response to degradation of an ephemeral habitat (transient habitats available for relatively short periods of time, e.g., beach wrack, dung, etc.) or depletion of a limited resource; (6) phoretics dismount from the host when a suitable new habitat is encountered by the host, indicating that the phoretic has sensory recognition and interpretation of some element in that habitat; (7) usually only one member of a complex life cycle participates in the phoresy and the other members in the life cycle are not phoretic; in such cases, the phoretic may come from among any of the life stages (adult, nymph, or larva); (8) phoresy may be highly coevolved and stenoxenic when both the phoretic and the host are trophic specialists or very indiscriminant, in which case many "buses" may lead to alternate and appropriate habitats; (9) phoresy may be obligate (required) or facultative (occurring under some conditions); (10) phoretics may have little morphological adaptation specific to the attachment to the host (e.g., hold on with mouthparts or

clasp with legs) and some have extensive modifications specific to attachment (e.g., extensive sucker plates or highly modified grasping legs); (11) enhanced by wind currents, phoresy is effective across impressive distances and there is even evidence of transoceanic voyages; (12) phoresy may be continual or seasonal, period, or cyclical; (13) the phoretic and the host may come from very different branches of the "tree of life," as divergent as humans and plants, or from within related lineages (e.g., different insect orders); and (14) phoresy has originated independently several times within one lineage, in some instances (e.g., Meloidae or the blister beetles).

Most insect orders have members that participate in phoresy; however, the Diptera and Coleoptera form some of the most extensive phoretic associations with vertebrates, other insects, and mites. They can participate in phoresy as phoronts, as well as phoretic hosts. An interesting example of an insect as a phoretic host is the case of a common phoretic nematode, *Pelodera coarctata,* and its dung beetle host, *Aphodius.* As a dung pat deteriorates and dries, a special resistant phoretic nematode larva is produced that attaches to visiting dung beetles. The phoretic nematodes remain in a dormant state on the beetles until the beetles arrive at a fresh dung pat. Then the nematodes emerge, become active, and begin a new population of free-living nematodes.

Pseudoscorpions (arachnids that looks like small scorpions) are notorious phoronts, found on an impressively large array of insect hosts: Diptera, Hymenoptera, Coleoptera, Odonata, Orthoptera, Heteroptera, Lepidoptera, Trichoptera, harvestmen (Opiliones), spiders, birds, and even small mammals. Pseudoscorpions attach by the chela, or venomous pedipalps, and hold on tightly enough to prevent being brushed off or blown off the host during transit. In one interesting phoretic interaction, a neriid fly that began as a phoretic host for a species of pseudoscoprion then becomes a postdispersal meal at the end of the journey: a case of turning the bus into a lunch wagon at the point of destination.

However, by far the most impressive radiation of phoretic associations occurs among the mites (Fig. 1). Intense selection pressure results in phoresy when organisms are of such extremely small size and they do not possess wings for dispersal. Small body size allows mites to exploit limited ephemeral resources that would be too small to be useful to larger organisms (e.g., a dead snail or nectar within a single flower). Because these resources are small, they degrade quickly and disappear rapidly. And, there may be large distances between them.

To survive, mites must travel to a better resource. Thus, they spend their lives tracking transient habitats. Establishing phoretic relationships with other organisms traveling among the same kinds of habitats gives them more rapid and direct access to a potentially better future and enhances their chances of survival.

Dung beetles, for example, thrive on dung, but as the dung dries and turns to soil it is no longer useful to the dung beetles. As the beetles depart, mites using the dung patties climb on board and hitch a ride to the next site. The journey

FIGURE 1 Example of a mite–beetle phoretic relationship. The mites on the head and body of the *Nicrophorus* beetle are likely *Poecilochirus* (Mesostigmata: Parasitidae), which feed on nematodes in the beetles' nest chamber. (Photographed in Carrer County, Minnesota, by Raphael Carter.)

would be perilous if these soft-bodied mites had to depend on walking the distance to their next meal. As the phoretic association progresses, the mites becomes increasingly dependent on those organisms that provide the most direct route to the best habitats.

EVOLUTION OF PARASITISM FROM PHORESY

Phoresy may begin with unrelated organisms moving independently among shared habitats. A relationship between a phoretic and a potential host can be established when some of the members of the population incidentally and randomly climb on board and are then delivered fortuitously to a habitat suitable for population growth. If encounters are repeated and consistent, mites develop cues to the most productive of these associations. Successful relationships become even more specialized and eventually stenoxenic. In some instances, phoretic association can become an intermediate step that grades into parasitism when the phoretic finds a way to get a meal, as well as transport, from the host.

A well-documented case of a phoretic relationship becoming parasitic is that of the mite *Hemisarcoptes* and the coccinellid beetle *Chilocorus*. Both feed on diaspid scale insects and the association originated as a phoretic relationship. However, coccinellid beetles are reflex bleeders and *Hemisarcoptes* has become adapted to the reflexed alkaloid toxins in the hemolymph and now requires it to molt and complete its development. Because feeding and completion of ontogenesis are part of the contribution of the host, this relationship has graded into parasitism. Other related members of the same mite family (Hemisarcoptidae), which use other phoretic hosts, remain phoretic. This is good evidence that phoresy can be an end point and that it can also progress to other forms of symbiosis (e.g., parasitism). Phoresy thus benefits the individual, but it can also act to enhance the diversity and complexity of community interactions within and among habitats.

See Also the Following Articles
Hypermetamorphosis • Migration

Further Reading
Bologna, M. A., and Pinto, J. D. (2001). Phylogenetic studies of Meloidae (Coleoptera), with emphasis on the evolution of phoresy. *Syst. Entomol.* **26**, 33–72.

Holte, A. E., Houck, M. A., and Collie, N. L. (2001). Potential role of parasitism in the evolution of mutualism in astigmatid mites: *Hemisarcoptes* as a model. *Exp. Appl. Acarol.* **1720**, 1–11.

Houck, M. A., and O'Connor, B. M. (1991). Ecological and evolutionary significance of phoresy in the Astigmata. *Annu. Rev. Entomol.* **36**, 611–636.

Klepzig, K. D., Moser, J. C., Lombardero, F. J., Hofstetter, R. W., and Ayres M. P. (2001). Symbiosis and competition: Complex interactions among beetles, fungi, and mites. *Symbiosis* **30**, 83–96.

Poinar, G. O., Ćurčić, B. P. M., and Cokendolpher, J. (1998). Arthropod phoresy involving pseudoscorpions in the past and present. *Acta Arachnol.* **47**, 79–96.

Schwarz, H. H., and Huck, K. (1997). Phoretic mites use of flowers to transfer between foraging bumblebees. *Insectes Soc.* **44**, 303–310.

Zeh, D. W., and Zeh, J. A. (1992). On the function of harlequin beetle-riding in the pseudoscorpion, *Cordylochernes scorpioides* (Pseudoscorpionida: Chernietidae). *J. Arachnol.* **20**, 47–51.

Photography of Insects

Mark W. Moffett
University of California, Berkeley

Quality photography of small animals in the field depends on rapid response to opportunities; this requisite is often impeded by the need to constantly root around in overstocked camera bags for the "right" piece of equipment. This article describes the author's personal approach to insect photography, which is to improvise with as little gear as possible. If it doesn't fit comfortably in your backpack, why have it at all? This philosophy may not be relevant to the intended audience of most insect photography essays, namely, the museum curator, studio photographer, artist, or laboratory scientist, who have time to "set up" a picture. Specific brands are not considered here. In part this is because the market is in flux with the ongoing technological shift to autofocus and digital cameras and, in part, because the quality of lenses produced by any of the major camera lines today is sufficient to produce excellent photographs. Expensive gear is not a necessity for the best results.

For all but the largest insects or insect-produced structures such as nests, capturing insects on film typically requires macrophotography, which is photography at a magnification of 1:1 ("life size") or greater, which is the focus of this article. At 1:1, subjects are the same size on the film as they are in life. That is, the image of a 30-mm-long beetle measures 30 mm in length on the film negative or slide, which means it fills most of the frame in a photograph on 35-mm film. For a 3-mm beetle to appear just as large (i.e., so that it is likewise 30 mm in length on the film) requires 10 times life-size magnification, which is typically described as "×10," or 10:1. The measurements are taken from the original film exposed in the camera, so prints made from a negative or slide would enlarge the subject beyond this size. For example, a 35-cm-wide print of the 3-mm-long beetle at ×10 on 35-mm film shows the beetle 30 cm in length, or 100 times its natural size.

One of the wonders of a well-executed insect photograph is that the subject's original size is forgotten: the small beetle is just as imposing as the large one. This actually makes photography a wonderful medium for entomologists. Not realizing this, photographers often concentrate on insects that are impressive merely for their size. In fact, when not photographed on the photographer's hand, many large insects can look small on film. Similarly, elephants may be spectacular in life, but on film we can only make one appear *smaller* than it really is. I recall my mentor at National Geographic, editor Mary G. Smith, taking her first look at my first "macro" images in 1986 (Fig. 1), which were photographs of marauder ants killing prey. She exclaimed that they reminded her of the

movie "Terminator," even though the ants looming over doomed prey in my pictures were just a few millimeters long. A criterion for a good insect photograph is that viewers are as surprised as Mary when a subject's size is revealed.

LENSES

Although regular lenses can be used in combination with other equipment to produce images of life size or greater, for example, by adding extension tubes or magnifying filters, the results can be inferior to the photographs made with standard "macro lenses." Although the term "macro" has been watered down in recent years by its application to lenses that focus relatively closely to the subject, true macro lenses focus all the way to 1:2 or 1:1 (sometimes an extension tube may be required for 1:1). These macro lenses are the sharpest, most optically corrected lenses produced by most companies and thus can be a wise investment. Many of them (typically those with focal lengths of about 50, 100, or sometimes 200 mm) can also be used as regular lenses, in which case they completely replace any other lens of a similar focal length. It should be noted that some of the optical precision of macro lenses relates to their having a "flat field" of focus. This characteristic is seldom important when working with live insects, unless perhaps one is in the habit of photographing straight down on flat insects living on the kitchen floor. Thus, when used with care, some less expensive lenses may be just as good in the field as true macro lenses, even when used at magnifications as high as 1:1.

A constant problem in insect photography is the distance from the front of the lens to the subject. With a macro lens set to a high magnification this distance may be only a few millimeters, so it takes practice not to disturb the insect when preparing to take a picture. In this regard stalking an insect is little different from stalking a leopard or a deer. As one moves the camera until the quarry comes into focus, one can learn to recognize through the viewfinder the moment when an insect detects the photographer. Each insect can require a different stalking technique.

If a short working distance causes the subject to be knocked or startled, the photographer should shift to a longer focal-length lens. A 200-mm macro lens provides a much greater working distance than a 50-mm macro set to the same magnification. Yet the longer lens is likely not to be quite as sharp, is harder to hold steady enough to focus precisely, requires heftier (and more unwieldy) brackets to hold the flashes, and needs more extension tubing to achieve magnifications beyond 1:1. If one can handle its short working distances, a 50-mm lens is therefore the ideal macro, but the 100-mm lens comes a very close second.

To achieve magnifications beyond life size with standard macro lenses, extension tubes are placed between a lens and the camera mount. Bellows are a more flexible alternative to tubes, because they provide every conceivable magnification between some upper and lower limit, but they are unwieldy and fragile

FIGURE 1 An 8-mm-long major worker ("soldier") marauder ant, *Pheidologeton diversus,* killing a centipede that has been pinned down by numerous smaller minor worker ants in Malaysia.

FIGURE 2 Soldiers of the aphid *Pseudoregma* jamming their needle-like "horns" into the cuticle of a syrphid fly larva that was attacking their colony in Japan. Photographed at ×14 magnification.

in the field. The fixed lengths of the tubes are seldom a liability outdoors where being able to quickly frame and shoot insects is more important than precise cropping. In fact, tubes shorter than about 25 mm generally have little value with macro lenses because their effect on magnification is so slight.

In addition to standard macro lenses some manufacturers offer lenses designed exclusively for macrophotography (for example, the ×1 to ×5 zoom lens made by Canon). Most of these lenses are built for high magnifications. An article by the author on soldier aphids in 1989 (Fig. 2) contains images between ×5 and ×20 on the film taken with a hand-held camera in the field. Whereas a standard macro lens can be used with only two or three extension tubes before working distance becomes too small to be practical or image quality declines, with some of these specialized lenses it may be possible to use several tubes at once.

Rotating the lens barrel either manually or with autofocus accomplishes focusing in normal photography, but it changes the magnification of an image in macrophotography, making it difficult to achieve the desired results. Autofocus therefore should be disengaged. Instead, focusing must be done as a manual, multistep process. Begin by deciding what magnification is desired, based on a subject's size. Select the lens and extension tubes needed to achieve that magnification. Then aim the camera toward the subject, rocking slowly in and out until the subject appears in focus.

Photographs at the highest magnifications require a steady hand, a sharp eye, and a lot of practice. Each photographer needs to understand his or her own limits from experience. The problem is not only in seeing and composing the shot but also in the limited depth of field at higher magnifications. The depth of field is the depth (from front to back) that appears in focus within an image. All the significant parts of the image must fall within this area for a photograph to make sense. At times the author has used a half meter of extension tubing in the field, and in these situations the depth of field can be the length of a paramecium! To increase the chance of success, take a few images after focusing on the subject by rocking back and forth slightly to subtly change the plane of focus each time.

As in all high-magnification photography, automatic film advance comes in handy not so much to allow rapid-fire picture taking but to allow one to keep a subject correctly framed and in focus while these frame-to-frame shifts in the film plane are made. Another aid is to use a flashlight, even at midday (illuminate crevices and other small shadows or the shadow of the lens itself). Using Velcro, tape, or glue, attach a small penlight to either the lens or one of the flashes so that it can be aimed at the subject. Camping stores have many flashlights that can work. It is best to bring in a camera and try out specific designs.

FLASHES

It is possible, but seldom advisable, to photograph a large insect with natural light using a tripod. With sufficient photographic skill, any image of an insect can be improved by the addition of flashes. Flashes provide a consistent and high quality of light, though flashes with colder (bluer) light can be improved by leaving a "warming gel," like a Kodak CC10Y filter, taped over the flash window (one layer of frosted scotch tape can by itself sometimes do the trick). Flashes freeze any motion of the subject or camera—including most importantly any motion caused by the trembling of the photographer's hands—and do so far more effectively than a tripod (particularly at high magnification). Furthermore, in the time it takes to set up a tripod, most insect subjects will have moved on to greener pastures. Flashes allow one to move in quickly and to constantly adjust the angle of approach as the action unfolds. For these reasons, the author seldom carries a tripod and even then never for macro work. Nonetheless, views of large insects in their environment, often at a high aperature with a wide-angle lens focused on a close subject, can be taken using natural light and a tripod, though with ground- or bark-dwelling species the camera may be steadied sufficiently by pressing it firmly against the stable substrate.

Not only can flash light increase the quality of any insect photograph, but when used with skill, two flashes are always an improvement over one flash, and two regular flash units are always an improvement over any method using a ring flash. The second flash serves as a "fill light," that is, it is weaker (usually one-half the strength or one stop weaker) than the other, "main flash." A fill flash results in photographs with nicely defined, pleasant shadows and lots of three-dimensional information about the subject. The fill flash can be a weaker flash model, but often it is the same kind of flash placed at the same distance as the primary flash, but set at a lower power output or filtered to reduce the light (a sheet of tissue or artist's tracing paper over the fill-flash head can do the trick). The flashes should be aimed toward the subject at about 45° from the axis of the lens and positioned anywhere from 90 to 120° apart around the lens barrel. For insects on reflective surfaces such as flowers and bright leaves, the light bounces off the surfaces and "fills" shadows somewhat, reducing the necessity of a fill flash. For shooting under these conditions, a single flash method as described by Shaw may suffice.

Ring flashes present problems with insects. The lower part of the ring impedes one from shooting low to the ground, blocking the most dramatic views. Also, the ring projects forward beyond the front of the lens all around the circumference, reducing the working distance. Further, because their light circles the lens and is directed forward rather than angled at the insect, ring flashes reduce the image's three-dimensional content, making a subject look relatively shapeless and flat (ring lights were developed for flat objects such as stamps). This problem can be partially remedied by taping over parts of the ring. On the positive side, ring flashes provide good color saturation for photographers unsure of their photographic technique, especially when they are used in combination with a fill light.

The greatest stumbling block to great macrophotography is the scarcity of good two-flash systems. Some flash brackets have been marketed, but most are bulky (making it hard to photograph in the tight corners, where many insects dwell), are difficult to hold steady for long periods because of their weight, and are easily knocked out of position in the field. The author makes his own systems, from various flashes, power packs, brackets, cords, and slaves. Because flashes can be used in their nonautomatic (manual) settings, items from various camera brands can be mixed to achieve the lightest, most compact results with the flashes at optimal positions. Camera brands must be mixed with care, for example, by taping over flash shoe connections if flashes of one brand are used with a camera body of another brand. No design is perfect, but some are better than others, and all of them can be modified.

The position of the flashes is most critical. When a small flash head is even a few centimeters away from the insect, viewed from the subject's position it appears as a point of light—much as the sun (although it is in fact a huge disc) appears as a point in the sky because of its distance from us. This kind of "point source" of light causes the most intensely illuminated spots to be "burned out" (too bright) and casts deep shadows that no film can handle (as occurs with any sunlit object on a cloudless day). To avoid such problems flashes should be placed as close to the subject as comfortably possible. So positioned, the lights resemble the light boxes used in studio photography for portraits. What is desired, in fact, is to replicate such a studio in miniature. One can even add a third light traditional in portrait studios, the "hair light," that is aimed from behind to define the edges of the subject, but for most field photographers this light is often unnecessary.

SHARPNESS AND EXPOSURE

Achieving a sharp, correctly exposed macro image usually takes knowledge that comes from testing the system being used. Exposure tables and other textbook information are seldom accurate and are no substitute for judging for oneself what is pleasing. Tests that are done carefully the first time will never have to be done again—good results are guaranteed.

Light meters are not always effective with many macro scenes, which often include objects that can be either black or brightly lit depending on slight changes in camera position relative to the subject and its background. For this reason, manual exposure techniques provide greater accuracy, but if a camera meter is preferred, these problems in a scene must be recognized and exposure bracketed accordingly. The best flashes for manual exposure work have multiple manual settings, that is, full power, half-power, quarter-power, and so on.

To test both the flashes and the lens and to develop a technique, select a fine-grained slide film (many with an ISO of 100 or less work well). Slides allow one to accurately gauge the exposure and sharpness of the images. As a test subject, put a dead insect of a kind likely to be photographed on an 18% gray card (available at most major photography stores). Then take a series of test photographs, recording magnification, flash position, flash power, and f-stop for each frame of the film as follows:

1. Set the lens so that it gives a certain magnification (say, 1:1) or has a certain number of tubes that can be remembered (say, one 25-mm tube).

2. Select the power settings of the two flashes (perhaps put one on full power and the fill flash on half-power).

3. Take a series of photographs of the insect on the gray card at a standard angle (say about 45° from the horizontal), the first at the lens' minimum aperture (say, f32) and then at one-f-stop intervals below that, down to f8.

4. Select another set of power settings for the flashes, say making both of them one stop weaker (that is, one at half-power and the other at quarter-power).

5. Repeat steps 2 and 3 with weaker flash power settings.

The developed slides should be checked for image detail (such as the texture of the gray card and the sculpturing on the insect) and exposure (have the original gray card on hand to see if the brightness of the slide matches the gray of the card). Pick the combination of flash powers and f-stop that gives the most pleasing result (see below). That result might be, for example, f22 with the flashes set at half- and quarter-power (but if f22 looks a bit too dark and f16 looks too light, record the intermediate setting as correct, i.e., f18). Write the settings down and use them thereafter for that magnification (although, as in any photographic situation, one can bracket slightly, for example, by opening up to f16 when the subject or its background is very dark). If it turns out the test photographs are all dark, move the flashes closer or purchase stronger units. If they are all too bright (or if the correct exposure occurs at an f-stop that has too little depth of field, as is explained below), weaken both flashes and test again.

Results from the first magnification can be used as the starting point for testing other magnifications, because results tend not to differ radically from one magnification to the next. For example, try a series of magnifications such as ×2, ×3, and ×6 (or, if preferred, the same lens with 50, 100, and 200 mm of lens tubing). After working out the correct exposure for each magnification, write up an exposure table and tape it to the back of the flash heads for reference.

Most people choose images that are overexposed (i.e., too bright), which for macrophotography means that valuable flash battery power has been wasted in producing too much light. If in doubt, choose a slightly darker image over a slightly bright one. To correctly judge exposure and image quality use a color-corrected (5500K) light table and a photographer's loupe. For slide film the highlights (bright parts of the image) should not be entirely burnt out except perhaps for tiny areas, and the colors over most of the image should be saturated (richly hued and not faded by strong light). Meanwhile, the darkest parts of an image should be inky black, but not so much as to lose detail by rendering an image blotchy. If burnt areas or blotchiness occur, try repositioning the flashes.

The tests may show the classic trade-off in macrophotography between depth of field and image quality. Thus, even though most macro lenses tend to close down to 32 and this f-stop provides the most depth of field, to produce sharp images it is best to open the aperture at least one stop from this setting (in this example, f22). The best quality—highest resolution and contrast—may actually occur at a stop lower that that (e.g., at f16), but the difference may be marginal enough so that the best choice is to use f22 because of the greater depth of field it provides. If a lot of extension tubing is added to a lens, image quality for the same f-stop setting on the lens may drop further, perhaps to an aperture of f11. This adjustment is a problem because the depth of field also declines as extension tubes are added or magnification is otherwise increased. Therefore the highest magnification attainable by a lens depends on the accuracy of the photographer with focusing and the optical limits of that lens. At some point it becomes necessary to purchase a lens better designed for the magnification in question.

TECHNIQUE

The most difficult subjects require considerable patience and a lot of time—sometimes a hundred attempts for every usable image. Ways of improving one's chances can be found, such as having on hand a supply of food items in photographing predation, but such techniques must be used with

care. Usually, the best image results from capturing the animal in the act of normal behavior in its normal habitat. This image is "best" not necessarily because of its technical perfection (on the contrary, the gritty realism of a slightly imperfect image sometimes enhances its drama, as is often true in photojournalism), but because of its accuracy. For example, a knowledgeable person might detect that the prey provided to the subject for a photograph is not a species that it normally would find and catch. Even more egregious is the refrigeration of specimens to slow them down for a picture. Despite their stiff exoskeletons, insects express themselves by subtle postures and actions (see Fig. 1). To the expert, a staged picture of a chilled insect appears as unnatural as one of a frozen human being.

With time and experience, one may want to attempt a photoessay or lecture on a particular subject. To hold a viewer's interest, try to incorporate a variety of compositions and magnifications. A critical overview of nature photojournalism was provided by the author in a 1995 article.

See Also the Following Articles

Collection and Preservation • Cultural Entomology • Museums and Display Collections

Further Reading

Constant, A. R. (2000). "Close-Up Photography." Focal Press, Woburn, MA.
Davies, P. W. (1998). "The Complete Guide to Close-Up and Macro Photography." David & Charles, Devon.
Moffett, M. W. (1986). Marauders of the jungle floor. *Natl. Geographic* **170**, 272–286.
Moffett, M. W. (1989). Samurai aphids: Survival under siege. *Natl. Geographic* **176**, 406–422.
Moffett, M. W. (1995). Essay. *In* "Magnificent Moments: The World's Greatest Wildlife Photographs" (G. H. Harrison, ed.), pp. 106–111. Willow Creek Press, Minocqua, WI.
Shaw, J. (2000). "Nature Photography Field Guide." Amphoto, New York.
Sholik, S., and Eggers, R. (2000). "Macro and Close-Up Photography Handbook." Amherst Media, Buffalo, NY.

Phthiraptera
(Chewing and Sucking Lice)

Ronald A. Hellenthal
University of Notre Dame

Roger D. Price
University of Minnesota

Phthiraptera are obligatory, lifelong ectoparasites of birds and mammals. They are hemimetabolous and wingless, with dorsoventrally flattened bodies, three pairs of well-developed legs, and a single- or double-segmented tarsus usually with one or two claws, but occasionally with claws absent. Ocelli are absent, eyes are reduced or absent, and antennae are short with three to five segments. Body length of adults ranges from 0.3 to 12 mm.

EVOLUTION AND DIVERSITY

Over 4900 louse species are recognized. Lice formerly were classified in two orders, Mallophaga (chewing lice) and Anoplura (sucking lice). However, they now are combined into the order Phthiraptera, with the former Mallophaga comprising three suborders (Amblycera, Ischnocera, and Rhynchophthirina) and the Anoplura a fourth. Although the suborders show substantial differences, they are believed to form a hemipteroid monophyletic unit whose sister group is the Psocoptera (psocids or booklice). Chewing lice have mandibulate mouthparts and probably evolved on birds. They are thought to have fed initially on skin and feathers, with some groups ultimately expanding their diets to include tissue fluids and blood. Some chewing lice eventually made a transition from birds to mammals and some Ricinidae (Amblycera) have mouthparts adapted to pierce host skin.

Sucking lice are restricted to mammalian hosts and have unusual piercing–sucking mouthparts that consist of three stylets, probably derived from the fused maxillae, hypopharynx, and labium. The stylets, retracted into the head when not in use, are everted to pierce skin and feed on host blood. Salivary secretions injected into the host include an anticoagulant to prevent blood coagulation during feeding.

Lice are recorded from approximately 5000 bird and mammal species, encompassing fewer than 30% of mammal species and about 36% of bird species. As many as 50% of the extant louse species remain undescribed. Lice are found on all bird orders but are not known to be associated with the mammalian order Monotremata (monotremes), the marsupial orders Microbiotheria (the monito del monte) and Notoryctemorhpia (marsupial "moles"), Chiroptera (bats), Cetacea (whales, dolphins, and porpoises), Sirenia (dugong, sea cow, and manatees), or Pholidota (scaly anteaters).

LIFE HISTORY

Details of louse life history are known for only a relatively few species, especially those of economic importance. Eggs (also known as nits) typically are cemented to hairs or feathers of the host and all life stages are confined to a single host. Some notable exceptions to this egg-laying behavior include the human body louse, *Pediculus humanus humanus* (Anoplura: Pediculidae), which attaches its eggs to clothing fiber and spends most of its time between feedings on clothing rather than on the host itself, and several genera in the chewing louse family Menoponidae (Amblycera) that spend their entire life cycles and deposit eggs inside the quills of primary and secondary feathers.

Immatures emerge from eggs by exerting pressure on an operculum at the free end of the egg. The timing and duration

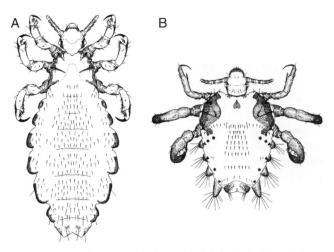

FIGURE 1 Lice of humans (dorsal view of adult females). (A) Head louse, *P. humanus capitis* (Anoplura: Pediculidae), adult body length 2.1–3.3 mm (body lice look similar to head lice, but tend to be slightly larger). (B) Crab louse, *P. pubis* (Anoplura: Pthiridae), average adult body length about 1.75 mm for females and 1.25 mm for males. (Illustrations adapted from Ferris, 1951, courtesy of the Pacific Coast Entomological Society and Stanford University Press.)

of the egg, three instars, and adult stages vary among species and may be affected by environmental temperature and humidity. Nymphs of the human head louse *(P. humanus capitis)* take from 6 to 9 days to hatch and reach the adult stage in about 1 week. Adults remaining on the host live about 1 month, with each female laying as many as eight eggs per day. Immatures and adults normally feed twice each day and cannot survive off of the host for more than a few days.

Dispersal of lice typically requires close contact between hosts, although immature lice have been found attached to flying insects that move between hosts, and phoresy may play a role in the dispersal of some species. The human body louse is dispersed by sharing infested clothing, the head louse (Fig. 1A) by personal contact or sharing combs and brushes, and the crab louse, *Pthirus pubis* (Anoplura: Pthiridae) (Fig. 1B), by sexual contact or, less commonly, through infested bed linens, towels, or clothing.

LOUSE–HOST SPECIFICITY AND COSPECIATION

Many lice show a high degree of host specificity, causing them to play an important role in studies of host–parasite cospeciation. For example, the 36 species of pocket gophers (Rodentia: Geomyidae) are parasitized by 122 species and subspecies of *Geomydoecus* and *Thomomydoecus* chewing lice (Ischnocera: Trichodectidae), many of which have distributions that conform closely to host–subspecies groups that are genetically similar. However, host specificity is not universal because many louse species occur on more than a single host species. *Anatoecus dentatus* and *A. icterodes* in the Philopteridae (Ischnocera), for example, each occur on more than 60 different species of ducks, geese, and swans (Anseriformes: Anatidae).

Many host species are parasitized by more than one louse species. For example, a single subspecies of brown tinamou, *Crypturellus obsoletus punensis* (Tinamiformes: Tinamidae), is infested by 11 species of lice, representing 10 genera in two families. When multiple louse species infest a single host individual, the different species may congregate in different areas of the body. In mammalian lice, habitat specialization often is associated with hair diameter. For example, the crab louse tends to be confined to the relatively coarse hairs associated with the genital areas, the face, and the underarms, whereas the head louse tends to be associated with the smaller diameter hairs of the scalp.

ECONOMIC IMPORTANCE

The economic importance of most louse species is unknown, although they are known to cause irritation, inflammation, and itching and serve as vectors of diseases and other parasites. High louse densities often are found on weakened or sick hosts, especially on birds with damaged bills or those unable to perform normal grooming activities. Lice infest a number of domestic animals including poultry, domestic dogs and cats, cattle, sheep, goats, horses, pigs, and rabbits. Animals in zoos and laboratory colonies of rats and mice also may become infested. Economic loss to the livestock and poultry industry in the United States caused by lice has been estimated at more than $550,000,000 per year, with about two-thirds of this resulting from weight and egg production losses in poultry caused by chewing louse infestations. More than $350,000,000 is spent annually on control of lice on humans.

The human body louse serves as the vector of epidemic typhus and epidemic relapsing fever and as an occasional vector of murine typhus. Epidemic louse-borne typhus, a rickettsial disease caused by *Rickettsia prowazekii,* is an important scourge of humans often associated with wars and disasters. It still is endemic in cold areas of Africa, Asia, and Central and South America. Zinsser gives a remarkable account of the impact of this disease on human history. For example, the disastrous failure of Napoleon's Grand Armée to conquer Russia during the campaign of 1812 is attributed largely to the effects of epidemic louse-borne typhus fever combined with malnutrition, dysentery, and exposure. The effectiveness of the pesticide DDT was demonstrated in 1943 during World War II in the control of lice responsible for an outbreak of typhus fever after the Allied bombing of Naples, Italy.

The name typhus is derived from Greek *typhos,* meaning stupor or fever. Infected humans experience high fever often accompanied by severe headaches, mental confusion, chills, coughing, and muscular pain. A rash generally appears on the 5th to 6th day and usually spreads to much of the body except the face, palms, and soles of the feet. The mortality rate usually is 1 to 20%, but can be much greater under epidemic conditions in populations that are weakened by malnutrition or other diseases. Transmission is by contamination of wounds with louse feces rather than

through the bite or feeding of the louse. Brill–Zinsser disease is a relatively mild form of typhus that can remain viable in humans for years and can serve as a reservoir of the disease and the source of future epidemics. Although not associated with disease transmission, outbreaks of the human head louse are epidemic throughout the world, with treatment complicated by the development of louse resistance to some commonly used chemical control agents.

CLASSIFICATION AND HOST ASSOCIATIONS

Amblycera is regarded as the most primitive louse suborder and the Rhynchophthirina and Anoplura are thought to be most advanced. The suborder Amblycera (Fig. 2A) includes 6 families: Boopiidae (mostly on marsupials, with one species infesting dogs and another the cassowary), Gyropidae (mostly on rodents, but with one genus occurring on New World monkeys and another on peccaries), Laemobothriidae (found on six bird orders), Menoponidae (widely distributed on birds), Ricinidae (on passerines and hummingbirds), and Trimenoponidae (on marsupials and rodents). The suborder Ischnocera (Fig. 2B) includes 2 families: Philopteridae (widely distributed on birds, with two species on primates, one on lemurs and the other on the indri) and Trichodectidae (on seven mammalian orders). The suborder Rhynchophthirina (Fig. 2C) contains a single family, Haematomyzidae, found on elephants, wart hogs, and the red river hog. The suborder Anoplura (Fig. 2D) includes 15 families, all restricted to mammalian hosts: Echinophthiriidae (on seals, sea lions, the walrus, and the North American river otter), Enderleinellidae (on squirrels), Haematopinidae (on horses and their relatives, pigs, cattle, and deer), Hamophthiriidae (on flying lemurs), Hoplopleuridae (on 8 families of rodents, a few shrews, and one species of pika), Hybophthiridae (on the aardvark), Linognathidae (on deer, cattle and their relatives, camels, and dogs and their relatives), Microthoraciidae (on the llama, alpaca, guanaco, and dromedary), Neolinognathidae (on elephant shrews), Pecaroecidae (on peccaries), Pedicinidae (on Old World monkeys), Pediculidae (on humans, chimpanzees, and New World monkeys), Polyplacidae (on 13 families of rodents, shrews, tree shrews, hares, and 5 families of primates), Pthiridae (on humans and gorillas), and Ratemiidae (on horses and their relatives).

There is disagreement as to whether the human head louse should be recognized as a distinct species *(Pediculus capitis)* or as a subspecies of *P. humanus (P. humanus capitis).* Those favoring its recognition as a separate species cite differences in behavior, size, coloration, and ability to transmit disease. Those favoring subspecies recognition note that populations of head and body lice are separable only statistically on the basis of minor differences in body size; that coloration in human lice is highly variable, with lice often taking on the color of their surroundings; and that hybridization between head and body lice has been demonstrated in the laboratory and is thought to occur in nature.

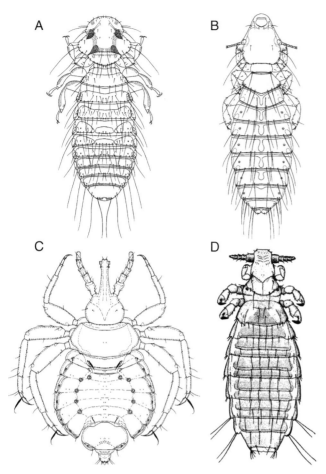

A B
C D

FIGURE 2 Louse suborders (dorsal view of adults). (A) Amblycera: *Colpocephalum fregili* (Menoponidae), male from the red-billed chough *(Pyrrhocorax pyrrhocorax)*. (Adapted from Price and Beer, 1965, *Proc. Entomol. Soc. Wash.* **67,** 7–14.) (B) Ischnocera: *Quadraceps crassipedalis* (Philopteridae), female from the least seed-snipe *(Thinocorus rumicivorus)*. (Adapted from Emerson and Price, 1985, *Proc. Entomol. Soc. Wash.* **87,** 395–401.) (C) Rhynchophthirina: *Haematomyzus porci* (Haematomyzidae), male from the red river hog *(Potamochoerus porcus)*. (Adapted from Emerson and Price, 1988, *Proc. Entomol. Soc. Wash.* **90,** 338–342.) (D) Anoplura: *Polyplax spinulosa* (Polyplacidae), female of the spiny rat louse from the Norway rat *(Rattus norvegicus)*. (Adapted from Kellogg and Ferris, 1915, Leland Stanford Junior University Publication, courtesy of Stanford University Press.)

See Also the Following Articles

Lice, Human • Medical Entomology • Psocoptera • Salivary Glands • Veterinary Entomology

Further Reading

Durden, L. A., and Musser, G. G. (1994). The sucking lice (Insecta, Anoplura) of the world: A taxonomic checklist with records of mammalian hosts and geographical distributions. *Bull. Am. Mus. Nat. Hist.* **218,** 1–90.
Ferris, G. F. (1951). The sucking lice. *Mem. Pacific Coast Entomol. Soc.* **1,** 1–320.
Hellenthal, R. A., and Price, R. D. (1991). Biosystematics of the chewing lice of pocket gophers. *Annu. Rev. Entomol.* **36,** 185–203.
Hopkins, G. H. E., and Clay, T. (1952). "A Check List of the Genera & Species of Mallophaga." Br. Mus. (Nat. Hist.), London.
Kim, K. C., Pratt, H. D., and Stojanovich, C. J. (1986). "The Sucking Lice of North America: An Illustrated Manual for Identification." Pennsylvania State University Press, College Park, PA.
Peterson, R. K. D. (1995). Insects, disease, and military history: The Napoleonic campaigns and historical perceptions. *Am. Entomol.* **41,** 147–160.
Price, M. A., and Graham, O. H. (1997). "Chewing and Sucking Lice as Parasites of Mammals and Birds." USDA Technical Bulletin No. 1849.
Zinsser, H. (1934). "Rats, Lice and History: A Study in Biography." Little, Brown, Boston.

Phylogeny of Insects

Peter S. Cranston and Penny J. Gullan
University of California, Davis

Ideas concerning the phylogenetic relationships among the major taxa of arthropods, and the included insects, are dynamic. Although there is a single evolutionary history, efforts to uncover this phylogeny vary between different researchers, techniques, and character systems studied. No technique or character system alone can guarantee that it reveals the true relationships of the studied taxa; in actuality, convergent (homoplastic) similarity that confuses relationships is common to all data. The evidence behind traditional systems, representing perhaps the thorough understanding of a single character system rather than an integration of all knowledge, sometimes cannot withstand detailed scrutiny. Molecular sequence data often appear to overturn previous ideas derived from morphological interpretation, but may be misleading because of undersampling, unrecognized sampling of alternative gene duplicates (paralogs), and/or inappropriate analyses. In this article, the different sources of evidence for the phylogenies that we have chosen to portray are assessed critically. Well-founded and less well founded traditional, even refuted, relationships are discussed, and where resolution appears to be lacking, this lack is identified.

RELATIONSHIPS OF THE HEXAPODA TO OTHER ARTHROPODA

Insects belong to arguably the most successful major lineage of the phylum Arthropoda, the joint-legged animals. This clade comprises myriapods (centipedes, millipedes, and their relatives), chelicerates (horseshoe crabs and arachnids), crustaceans (crabs, shrimps, and relatives), and hexapods (the 6-legged arthropods, Insecta and their relatives). Lobopods (onychophorans) sometimes have been included, but now almost universally are considered to lie among likely sister groups outside Arthropoda. Although traditionally each major arthropod group has been considered monophyletic, most have been suspected of nonmonophyly by at least a few investigators. Results of molecular analyses have provided frequent challenges, particularly in suggesting the possible paraphyly of myriapods and of crustaceans. Even if considered monophyletic, estimation of interrelationships has been a most

apomorphy (-ic) A feature of an organism in the derived state, contrasted with an alternative one in the ancestral (primitive) state—a **plesiomorphy** (-ic). For example, with the character of forewing development, the sclerotized elytron is an apomorphy for Coleoptera, and the alternative, a conventional flying forewing, is a plesiomorphy at this level of comparison.

cladogram Diagramatic illustration of the branching sequence of purported relationships of organisms, based on distribution of shared derived features (synapomorphies).

monophyletic Referring to a taxonomic group (called a clade) that contains all descendants derived from a single ancestor and recognized by the possession of a shared derived feature(s). For example, the clade Diptera is monophyletic, recognized by shared derived development of the hind wing as a haltere (balancing organ).

paraphyletic Referring to a taxonomic group (called a grade) derived from a single ancestor but not containing all descendants; grades share ancestral features (e.g., Mecoptera relative to Siphonaptera).

polyphyletic Referring to a taxonomic group derived from more than one ancestor and recognized by the possession of one or more features evolved convergently. For example, if the primitively wingless silverfish were united with secondarily wingless grasshoppers, beetles, and flies, the resulting group would be polyphyletic.

sister groups Species or monophyletic groups that arose from the stem species of a monophyletic group by a singular, identical splitting event. For example, the Lepidoptera and Trichoptera are sister groups; they shared a common ancestor that gave rise to no other lineage.

synapomorphy (-ic) A derived state shared among the members of a monophyletic group, in contrast to a **symplesiomorphy**—a shared ancestral (plesiomorphic) state from which phylogenetic relationships cannot be inferred.

taxon (pl. taxa) The general name for a taxonomic group at any rank.

taxonomic rank The classificatory level in the taxonomic hierarchy, e.g., species, genus, family, order. No rank is absolute, and comparisons between ranks of different organisms are inexact or even misleading; despite this, traditional ranks used for insects—notably orders and families—have useful didactic and synoptic value.

contentious issue in biology, with almost every possible higher level relationship finding some support. A once-influential "Mantonian" view proposed three groups of arthropods, each of which was derived from a different nonarthropod group, namely Uniramia (lobopods, myriapods, and insects, united by single-branched legs), Crustacea, and Chelicerata. More recent morphological and molecular studies reject this hypothesis, proposing instead monophyly of arthropodization, but postulated internal relationships are diverse. Part of Manton's Uniramia group—Atelocerata (also known as Tracheata), comprising myriapods and hexapods—finds support from morphological features including the occurrence of a tracheal system, Malpighian tubules, unbranched limbs, eversible coxal vesicles, postantennal organs, and anterior tentorial arms, lack of any evidence of the second antenna of crustaceans or a homologous structure, and the mandible comprising a complete limb rather than the limb base of the crustacean mandible. Proponents of this relationship saw Crustacea either grouping with chelicerates and extinct trilobites, separate from Atelocerata, or forming its sister group, in a clade called Mandibulata. Among all these schemes, the closest relatives of Hexapoda were proposed to be with, or possibly within, Myriapoda.

In contrast, novel and some rediscovered shared morphological features, including some from the nervous system (e.g., brain structure, neuroblast formation, and axon development), the visual system (e.g., fine structure of the ommatidia, optic nerves), and the process and control of development, especially segmentation, argued for a close relationship of Hexapoda to Crustacea, termed Pancrustacea, and exclusion of myriapods. Furthermore, all analyzed molecular sequence data with adequate signal to resolve relationships support Pancrustacea and not Atelocerata. As more nuclear, mitochondrial gene order and protein-encoding gene data have been examined for an ever-wider set of taxa, little or no support has been found for any of the possible groupings alternative to Pancrustacea. This does not imply that such analyses all identify Pancrustacea—sometimes certain "problematic" taxa have had to be removed, even from sparsely sampled data sets, and evidently certain genes do not retain strong phylogenetic signal from very old radiations.

If molecular-derived relationships are correct, features understood previously to infer monophyly of Atelocerata must be reconsidered. Postantennal organs occur in Hexapoda, but only in Collembola and Protura, and are suggested to be convergent with the organs in Myriapoda. Shared absence of features (such as lack of second antenna) cannot be taken as positive evidence of relationship. Malpighian tubules also are present (surely convergently) in arachnids, and evidence for homology between their structure in hexapods and myriapods remains inadequately studied. Coxal vesicles are not developed in all clades and may not be homologous in Myriapoda and those Hexapoda possessing these structures. Thus, morphological characters traditionally used to support Atelocerata include states that may be nonhomologous and convergently acquired through terrestriality, not distributed across all included taxa, or inadequately surveyed across the immense morphological diversity of the arthropods. A major finding from molecular embryology, that the developmental expression of a homeotic gene *(Dll—Distal-less)* in the mandible of studied insects was the same as in sampled crustaceans, refutes the independent derivation of hexapod mandibles from those of crustaceans. This developmental homology for mandibles substantiates an earlier morphological hypothesis and undermines Manton's argument for arthropod polyphyly. In summary, data derived from the neural, visual, and developmental systems, even though sampled across relatively few taxa, appears to reflect more accurately phylogeny than much of the earlier external morphological studies.

The question remains as to whether part or all of the Crustacea constitute the sister group to Hexapoda. Morphology generally supports a monophyletic Crustacea, but inferences from some molecular data imply paraphyly, including a suggestion that Malacostraca alone are sister to Hexapoda (see below). Combined morphological and molecular data support both Crustacea and Pancrustacea monophyly, and Crustacea monophyly is thus preferred.

THE EXTANT HEXAPODA

Hexapoda (ranked usually as a superclass) contains all 6-legged arthropods; diagnosis includes possession of a unique tagmosis, namely specialization of successive body segments that more or less unite to form three sections or tagmata: head, thorax, and abdomen. The head is composed of a pregnathal region (often considered to be 3 segments) and 3 gnathal segments bearing mandibles, maxillae, and labium, respectively; the eyes are variously developed and sometimes absent. The thorax comprises 3 segments each of which bears one pair of legs, and each thoracic leg has a maximum of 6 segments in extant forms, but was primitively 11-segmented with up to 5 exites (outer appendages of the leg), a coxal endite (an inner appendage of the leg), and 2 terminal claws. Primitively the abdomen has 11 segments plus a telson or homolog; abdominal limbs, if present, are smaller and weaker than those of the thorax and primitively occurred on all segments except the 10th.

Basal hexapods undoubtedly include taxa whose ancestors were wingless and terrestrial. This grouping is not monophyletic, being based entirely on evident symplesiomorphies or otherwise doubtfully derived characters. Included groups, treated as orders, are Protura, Collembola, Diplura, Archaeognatha, and Zygentoma (Thysanura). True Insecta are the Archaeognatha, the Zygentoma, and the huge radiation of Pterygota (primary winged hexapods). Because Insecta is treated as a class, the successively more distant sister groups Diplura and Collembola (with or without Protura) are of equal rank.

Relationships among the component taxa of Hexapoda are uncertain, although the cladogram shown in Fig. 1 and the classification presented in the sections that follow reflect a current synthetic view. Traditionally, Collembola, Protura,

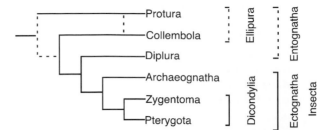

FIGURE 1 Cladogram depicting relationships among, and inferred classification of, higher ranked Hexapoda (six-legged Arthopoda). Dashed lines indicate uncertainty in cladogram or paraphyly in classification.

and Diplura were grouped as "Entognatha," based on the apparently similar morphology of the mouthparts. The mouthparts of Insecta (Archaeognatha + Zygentoma + Pterygota) are exposed (ectognathous), whereas those of Entognatha are enclosed in folds of the head. However, two different types of entognathy now are recognized, one shared by Collembola and Protura and the second found only in Diplura. Other morphological evidence indicates that Diplura may be closer to Insecta than to other entognathans and thus Entognatha may be paraphyletic (as indicated by the broken line in Fig. 1).

Protura (Proturans)

Proturans are small, delicate, elongate, mostly unpigmented hexapods, lacking eyes and antennae, with entognathous mouthparts consisting of slender mandibles and maxillae that slightly protrude from the mouth cavity. Maxillary and labial palps are present. The thorax is poorly differentiated from the 12-segmented abdomen. Legs are 5-segmented. A gonopore lies between segments 11 and 12, and the anus is terminal. Cerci are absent. Larval development is anamorphic, that is, with segments added posteriorly during development. Protura either is sister to Collembola, forming Ellipura in a weakly supported relationship based on similarity of the entognathous mouthparts and lack of cerci, or is sister to all remaining Hexapoda.

Collembola (Springtails)

Collembolans are minute to small and soft bodied, often with rudimentary eyes or ocelli. The antennae are 4- to 6-segmented. The mouthparts are entognathous, consisting predominantly of elongate maxillae and mandibles enclosed by lateral folds of head and lacking maxillary and labial palps. The legs are 4-segmented. The abdomen is 6-segmented with a sucker-like ventral tube, a retaining hook, and a furcula (forked jumping organ) on segments 1, 3, and 4, respectively. A gonopore is present on segment 5, the anus on segment 6. Cerci are absent. Larval development is, epimorphic, that is, with segment number constant through development. Collembola form either the sister group to Protura comprising Ellipura or a more strongly supported relationship as sister to Diplura + Insecta.

Diplura (Diplurans)

Diplurans are small to medium sized, mostly unpigmented, possessing long, moniliform antennae (like a string of beads), but lacking eyes. The mouthparts are entognathous, with tips of well-developed mandibles and maxillae protruding from the mouth cavity and maxillary and labial palps reduced. The thorax is poorly differentiated from the 10-segmented abdomen. The legs are 5-segmented and some abdominal segments have small styles and protrusible vesicles. A gonopore lies between segments 8 and 9, the anus is terminal. Cerci are filiform to forceps-like. The tracheal system is relatively well developed, whereas it is absent or poorly developed in other entognath groups. Larval development is epimorphic. Diplura forms the sister group to Insecta.

Class Insecta (True Insects)

Insects range from minute to large (0.2 to 300 mm in length) and are very variable in appearance. They typically have ocelli and compound eyes, at least in adults, and the mouthparts are exposed (ectognathous), with the maxillary and labial palps usually well developed. The thorax is variably developed in immature stages, but distinct in adults with degree of development dependent on the presence of wings. Thoracic legs have more than 5 segments. The abdomen is primitively 11-segmented with the gonopore nearly always on segment 8 in the female and segment 9 in the male. Cerci are primitively present. Gas exchange is predominantly tracheal with spiracles present on both the thorax and the abdomen, but variably reduced or absent (e.g., in many immature stages). Larval/nymphal development is epimorphic, that is, with the number of body segments constant during development.

The insects may be divided into two groups. Monocondylia is represented by just one small order, Archaeognatha, in which each mandible has a single posterior articulation with the head, whereas Dicondylia (Fig. 1), which contains the overwhelming majority of species, is characterized by mandibles with secondary anterior articulation in addition to the primary posterior one. The once traditional group Apterygota comprising the primarily wingless taxa Archaeognatha + Zygentoma is paraphyletic and rejected (Fig. 2).

ARCHAEOGNATHA (ARCHAEOGNATHANS, BRISTLE-TAILS) Archaeognathans are medium-sized, elongate cylindrical apterygotes, with some 500 species in two extant families. The head bears 3 ocelli and large compound eyes that are in contact medially. The antennae are multisegmented; the mouthparts project ventrally and can be partially retracted into the head and include elongate mandibles, one with 2 neighboring condyli each, and elongate 7-segmented maxillary palps. Often coxae II and III or III of legs bear a coxal stylet, tarsi 2- to 3-segmented. The abdomen, which continues in an even contour from the humped thorax, bears ventral muscle-containing styles (representing reduced limbs) on

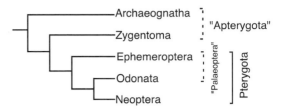

FIGURE 2 Cladogram depicting relationships among, and inferred classification of, higher ranked Insecta. Dashed lines indicate paraphyly in classification.

segments 2 to 9 and generally one or two pairs of eversible vesicles medial to the styles on segments 1 to 7. Cerci are multisegmented and shorter than the median caudal appendage. Sperm transfer is indirect via sperm droplets attached to a silken line or by stalked spermatophores. Development occurs without change in body form.

The two families of recent Archaeognatha, Machilidae and Meinertellidae, form an undoubted monophyletic group, whose position at the base of Ectognatha and as sister group to Dicondylia (Zygentoma + Pterygota) should be carefully investigated (Figs. 1 and 2).

ZYGENTOMA (THYSANURA, SILVERFISH) Zygentomans (thysanurans) are medium-sized, dorsoventrally flattened apterygotes with almost 400 species in four extant families. Eyes and ocelli are present, reduced, or absent and the antennae multisegmented. The mouthparts are ventrally to slightly forward projecting and include a special form of double-articulated (dicondylous) mandibles and 5-segmented maxillary palps. The abdomen continues the even contour of the thorax and includes ventral muscle-containing styles (representing reduced limbs) on at least segments 7 to 9, sometimes on 2 to 9, with eversible vesicles medial to the styles on some segments. Cerci are multisegmented and subequal to the length of the median caudal appendage. Sperm transfer is indirect via a spermatophore that the female picks up from the substrate. Development occurs without change in body form.

PTERYGOTA Pterygotes are winged or secondarily apterous insects, in which the thoracic segments of adults are usually large with the meso- and metathorax variably united to form a pterothorax. The lateral regions of the thorax are well developed. The 8 to 11 abdominal segments lack styles and vesicular appendages, and only most Ephemeroptera have a median terminal filament. The spiracles primarily have a muscular closing apparatus. Mating is by copulation. Metamorphosis is hemi- to holometabolous, with no adult ecdysis, except for the ephemeropteran subimago (subadult).

Informal Grouping "Palaeoptera" Palaeopteran wings are unable to be folded against the body at rest because articulation is via axillary plates that are fused with veins. Extant orders typically have triadic veins (paired main veins with intercalated longitudinal veins of convexity/concavity opposite to that of the adjacent main veins) and a network of crossveins. This wing venation and articulation, substantiated by paleontological studies on similar features, has suggested that Odonata and Ephemeroptera form a monophyletic group, Palaeoptera, which is a sister group to Neoptera, which contains all remaining extant and primarily winged orders. However, reassessment of morphology of extant basal lineages and much existing molecular evidence appear to reject a monophyletic Palaeoptera. Here Ephemeroptera is treated as sister to Odonata + Neoptera, with Odonata alone as the sister group to Neoptera, giving a higher classification of Pterygota into three Divisions.

Ephemeroptera (Mayflies)   Ephemeroptera has a fossil record dating back to the Carboniferous and is represented today by a few thousand species. In addition to their palaeopteran features, mayflies display a number of unique characteristics including the nonfunctional, strongly reduced adult mouthparts, the presence of just one axillary plate in the wing articulation, a hypertrophied costal brace, and male forelegs modified for grasping the female during copulatory flight. The retention of a subimago (subadult stage) is unique. Nymphs (larvae) are aquatic and the mandible articulation, which is intermediate between monocondyly and the dicondylous ball and socket joint of all higher Insecta, may be diagnostic. Historic contraction of ephemeropteran diversity and remnant high levels of homoplasy render phylogenetic reconstruction difficult. Ephemeroptera traditionally has been divided into two suborders: Schistonota (with nymphal forewing pads separate from each other for over half their length), containing the superfamilies Baetoidea, Heptagenioidea, Leptophlebioidea, and Ephemeroidea, and Pannota ("fused back"—with more extensively fused forewing pads), containing Ephemerelloidea and Caenoidea. Recent studies suggest that this concept of Schistonota is paraphyletic. Families Baetiscidae and Prosopistomatidae, whose nymphs have unusually developed thoracic shields, have been withdrawn from the Caenoidea and are placed now in the suborder Catapacea.

Odonata (Dragonflies and Damselflies)   Odonates have palaeopteran as well as many additional unique features, including the presence of two axillary plates (humeral and posterior axillary) in the wing articulation and many features associated with specialized copulatory behavior, including possession of secondary copulatory apparatus on ventral segments 2 and 3 of the male and the formation of a tandem wheel during copulation. The immature stages are aquatic and possess a highly modified prehensile labium for catching prey.

Traditonally odonatologists have recognized three groups of taxa, Zygoptera (damselflies), Anisozygoptera, and Anisoptera (dragonflies), generally ranked as suborders. Assessment of their monophyly or paraphyly relies very much on characters derived from the very complex wing venation, but homology of these features within the odonates and between other insects has been substantially prejudiced by prior phylogenetic ideas. Thus, the Comstock and Needham wing-vein naming system implies that the common ancestor of modern Odonata was anisopteran and the zygopteran venation arrived by reduction. In contrast, the Tillyard system implied that Zygoptera is a grade on the way to Anisozygoptera, which itself is a grade on the way to Anisoptera. A well-supported view, including information from the substantial fossil record, has Zygoptera probably paraphyletic, Anisozygoptera undoubtedly paraphyletic, and Anisoptera as the monophyletic sister group to some extinct anisozygopterans.

Zygoptera contains three broad suprafamilial groupings, the Coenagrionoidea, Lestoidea, and Calopterygoidea. The first is centered around the family Coenagrionidae, but the group is made paraphyletic by excision of Eurasian Platycnemididae (for their dilated tibiae), South American Pseudostigmatidae (for their huge size), Platystictidae (for a narrow wing), and Protoneuridae and Isostictidae (for wings narrowed in a different way). Those latter two families differ only in the degree of wing narrowing, and the narrower one, Australasian Isostictidae, often is treated as a subfamily of the less narrow one, Protoneuridae.

The Calopterygoidea usually is regarded as centered around Calopterygidae. This specious family is made paraphyletic by excision of Chlorocyphidae (whose frons has a unique shape), Dicteriadidae (with legs lacking setae), Euphaeidae (a Southeast Asian variant on Calopterygidae, with abdominal gills in the larvae), and South American Polythoridae, with brightly colored wings.

Lestoidea contains the families Lestidae and Synlestidae, which clearly are related to each other. Perilestidae may belong here, and the enigmatic Hemiphlebiidae may be sister to this grouping. The quite different-looking Megapodagrionidae are related in some way to Amphipterygidae and Lestoideidae (the latter being an unstable mix of small genera allocated to either Lestoidea or Calopterygoidea).

Among Anisoptera four major lineages can be recognized, but their relationships to each other are obscure. Three aeshnid families, Aeshnidae, Neopetaliidae (evidently a subset of Aeshnidae), and Cordulegastridae (aeshnids with a secondarily elongate ovipositor), form a clade. The small (10 species) but very distinct Petaluridae forms a distinctive group. Gomphidae forms a large family all on its own. The superfamily Libelluloidea traditionally is divided into two large families, Cordulidae and Libellulidae, but the limits of each division are unclear, and no single character separates them. Chlorogomphidae, Macromiidae, and Synthemistidae are small, local "families" often separated out as near the corduliids.

Sister to Anisoptera is the minor suborder Anisozygoptera containing one extant genus with two species.

Neoptera   Neopteran insects diagnostically have wings capable of being folded back against their abdomen when at rest, with wing articulation deriving from separate movable sclerites in the wing base and wing venation with fewer (or lacking completely) triadic veins and mostly lacking anastomosing (joining) crossveins.

The phylogeny (and hence classification) of the neopteran orders is still the subject of debate, mainly concerning (a) the placement of many extinct orders described only from fossils of variably adequate preservation, (b) the relationships among the Polyneoptera (orthopteroid and plecopteroid orders), and (c) the relationships of the highly derived Strepsiptera.

However, the summary that follows reflects one possibility among current interpretations, based on both morphology and molecules. No single or combined data set provides unambiguous resolution of insect order-level phylogeny and there are several areas of controversy (such as the position of the Strepsiptera) arising from both inadequate data (insufficient or inappropriate taxon sampling) and character conflict within existing data. In the absence of a robust phylogeny, ranking is somewhat subjective and "informal" ranks abound.

A group of 11 orders termed the orthopteroid–plecopteroid assemblage (if monophyly is uncertain) or Polyneoptera (if monophyletic) is considered to be sister to the remaining Neoptera. The remaining neopterans can be divided readily into two monophyletic groups, namely Paraneoptera and Endopterygota (Holometabola). These three clades may be given the rank of subdivision.

Polyneoptera (or Orthopteroid–Plecopteroid Assemblage of Basal Neopteran Orders) [Isoptera, Blattodea, Mantodea, Dermaptera, Grylloblattaria (Grylloblattodia), Plecoptera, Orthoptera, Phasmatodea, Embiidina (Embioptera), Zoraptera, Mantophasmatodea] The relationships of the basal neopteran orders are poorly resolved with several, often contradictory, relationships being suggested by morphology. The 11 included orders may be monophyletic, based on the shared presence of tarsal plantulae (lacking only in Zoraptera) and limited, but increasing, molecular information. Within Polyneoptera only the grouping comprising Blattodea (cockroaches), Isoptera (termites), and Mantodea (mantids)—the Dictyoptera (Fig. 3)—is robust. Although each of these three orders is distinctive, features of the head skeleton (perforated

tentorium), mouthparts (paraglossal musculature), digestive system (toothed proventriculus), and female genitalia (shortened ovipositor above a large subgenital plate) demonstrate monophyly of Dictyoptera, substantiated by nearly all molecular analyses. However, as seen below, views on the internal relationships are changing. Dermaptera (earwigs) is sister to Dictyoptera, and Grylloblattarea (rock crawlers; now apterous, but with winged fossils) may be sister to this grouping.

Some molecular data suggest that Orthoptera (crickets, katydids, grasshoppers, locusts, etc.), Phasmatodea (stick insects or phasmids), and Embiidina (webspinners) may be closely related, forming Orthopteroidea in the sense of Hennig. The relationships of Plecoptera (stoneflies), orthopteroids, Zoraptera (zorapterans), and the recently discovered Mantophasmatodea to one another and to the above groupings are less well understood.

Isoptera (Termites, White Ants). Isoptera forms a small order of eusocial hemimetabolous neopterans, with some 2600 described species, living socially with polymorphic caste systems of reproductives, workers, and soldiers. The mouthparts are typically blattoid, being mandibulate but varying between castes, with some soldiers having bizarre development of mandibles or a nasus (snout). The compound eyes are frequently reduced, the antennae are long and multisegmented, and the fore- and hind wings are generally similar and membranous and have restricted venation. *Mastotermes* (Mastotermitidae) has complex wing venation and a broad hind-wing anal lobe and is exceptional among termites in that the female has a reduced blattoid-type ovipositor. The male external genitalia are weakly developed and symmetrical, in contrast to the complex, symmetrical genitalia of Blattodea and Mantodea.

Isopteran relationships are somewhat controversial, although they have always been considered to belong in Dictyoptera close to Blattodea. Recent studies that include the structure of the proventriculus and molecular sequence data suggest that termites may have arisen from within the cockroaches, thereby rendering that group paraphyletic. Under this scenario, the (wingless) wood roaches of North America and eastern Asia (genus *Cryptocercus*) form the sister group to Isoptera. This contrasts with alternative suggestions that the semisociality (parental care and transfer of symbiotic gut flagellates between generations) of *Cryptocercus* was convergent with certain features of termite sociality and independently originated within the true cockroaches. These two contrasting views are shown in Fig. 4. The social system and general morphology of *Mastotermes* suggests a cockroach-like condition, and most phylogenies place this group as sister to remaining extant Isoptera. Of considerable interest is the wide distribution and species richness of Mastotermitidae in Cretaceous times, compared with the reduced diversity of the extant family, which comprises just one species in northern Australia.

Blattodea (Cockroaches). Blattodea contains over 3500 species in at least eight families worldwide. They are hemimetabolous, dorsoventrally flattened insects with filiform, multisegmented antennae, and mandibulate, ventrally projecting mouthparts.

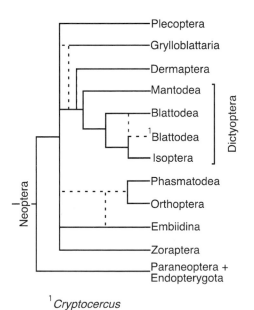

^{1}*Cryptocercus*

FIGURE 3 Cladogram depicting relationships among, and inferred classification of, orders of Neoptera: Polyneoptera. Dashed lines indicate uncertainty in cladogram. Mantophasmatodea not included.

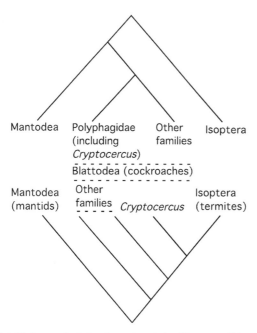

FIGURE 4 Cladogram depicting alternative relationships among Dictyoptera. Dashed lines indicate paraphyly in classification.

The prothorax has an enlarged, shield-like pronotum, often covering the head; the meso- and metathorax are rectangular and subequal. The forewings are sclerotized as tegmina, which protect the membranous hind wings folded fan-like at rest beneath the tegmina and characterized by many vein branches and a large anal lobe; wings are often reduced. Often the legs are spiny and the tarsi are 5-segmented. The abdomen has 10 visible segments, with a subgenital plate (sternum 9), bearing in the male well-developed asymmetrical genitalia, with one or two styles, and concealing the reduced 11th segment. Cerci have 1 or, usually, many segments; the female ovipositor valves are concealed beneath tergum 10.

Although long considered an order (and hence monophyletic) convincing evidence has been produced that the termites arose from within the cockroaches, and the "order" thus is rendered paraphyletic. The sister group of the Isoptera may be *Cryptocercus,* a social cockroach (Fig. 4). Blatellidae and Blaberidae form a derived sister grouping.

Mantodea (Mantids). Mantodea contains some 2000 species in eight families. They are hemimetabolous predators, with males generally smaller than females. The small, triangular head is mobile, with slender antennae, large, widely separated eyes, and mandibulate mouthparts. The prothorax is narrow and elongate, with the meso- and metathorax shorter. The forewings form leathery tegmina with a reduced anal area; the hind wings are broad and membranous, with long unbranched veins and many crossveins, but often are reduced or absent. The forelegs are raptorial, whereas the mid- and hind legs are elongate for walking. The abdomen has a visible 10th segment, bearing variably segmented cerci. The ovipositor is predominantly internal and the external male genitalia are asymmetrical.

Mantodea forms the sister group to Blattodea + Isoptera and shares many features with Blattodea, such as strong direct flight muscles and weak indirect (longitudinal) flight muscles, asymmetrical male genitalia, and multisegmented cerci. Derived features of Mantodea relative to Blattodea involve modifications associated with predation, including leg morphology, an elongate prothorax, and characteristics of visual predation, namely the mobile head with large, separated eyes. Internal relationships of Mantodea are uncertain and little studied.

Grylloblattaria (Notoptera, Grylloblattodea) (Grylloblattids or Rock Crawlers). Grylloblattaria contains one family (Grylloblattidae) with 20 species, restricted to western North America and central to eastern Asia and particularly adapted to cold and high elevations. Grylloblattids are moderate-sized, soft-bodied insects with anteriorly projecting mandibulate mouthparts and compound eyes that are either reduced or absent. The antennae are multisegmented and the mouthparts mandibulate. The quadrate prothorax is larger than the meso- or metathorax, and the wings are absent. The legs are adapted for running, with large coxae and 5-segmented tarsi. There are 10 visible abdominal segments with rudiments of segment 11, including 5- to 9-segmented cerci. The female has a short ovipositor, and the male genitalia are asymmetrical.

The phylogenetic placement of Grylloblattaria is controversial, as they are generally argued to be relicts that either "bridge the cockroaches and orthopterans" or are "primitive among orthopteroids." The antennal musculature resembles that of mantids and embiids, mandibular musculature resembles that of Dictyoptera, and the maxillary muscles resemble those of Dermaptera. Embryologically grylloblattids appear closest to the orthopteroids. The only molecular phylogenetic study that included a grylloblattid implied a sister group relationship to Dictyoptera, instead of one lying more basal in the Neoptera as is implied by the morphology. However, sampling in this analysis lacked some important plesiomorphic taxa, such as *Cryptocercus, Mastotermes,* and Embiidina. A tentative relationship of Grylloblattaria as sister to Dermaptera + Dictyoptera remains a favored hypothesis.

Dermaptera (Earwigs). Dermaptera is a worldwide order, modest in size, with some 10 families and about 1800 species. Adult earwigs are elongate and dorsoventrally flattened with mandibulate, forward-projecting mouthparts, compound eyes ranging from large to absent, no ocelli, and short annulate antennae. The tarsi are 3-segmented with a short second tarsomere. Many species are apterous or, if winged, the forewings are small, leathery, and smooth, forming unveined tegmina, and the hind wings are large, membranous, semicircular, and dominated by an anal fan of radiating vein branches connected by crossveins; when at rest, the hind wings are folded fan-like and then longitudinally, protruding slightly from beneath the tegmina.

The five species of suborder Arixeniina are commensals or ectoparasites of bats in Southeast Asia. A few species of semiparasites of African rodents have been placed in a suborder,

Hemimerina. These earwigs are blind, are apterous, have rod-like forceps, and exhibit pseudoplacental viviparity. Recent morphological study of Hemimerina suggests derivation from within Forficulina, rendering that suborder paraphyletic. The relationships of Arixeniina to more "typical" earwigs (Forficulina) are uninvestigated. Within Forficulina, only four (Karshiellidae, Apachyidae, Chelisochidae, and Forficulidae) of eight families proposed appear to be supported by synapomorphies. Other families may not be monophyletic, as much weight has been placed on plesiomorphies, especially of the penis specifically and the genitalia more generally, or homoplasies (convergences) in furcula form and wing reduction.

A sister group relationship to Dictyoptera is well supported on morphology, including many features of the wing venation.

Plecoptera (Stoneflies). Plecoptera is a minor order of some 16 families, predominantly living in temperate and cool areas. The adult is mandibulate with filiform antennae, bulging compound eyes, 2 or 3 ocelli, and subequal thoracic segments. The fore- and hind wings are membranous and similar except that the hind wings are broader; when folded, the wings partly wrap the abdomen and extend beyond the abdominal apex; aptery and brachyptery are frequent. The abdomen is soft and visibly 10-segmented, although remnants of segments 11 and 12 are present, including cerci. Nymphs have many (up to 33) aquatic instars, which have fully developed mandibulate mouthparts, and wing pads first become visible when the young are half-grown.

Monophyly of the order is supported by few morphological characters, including in the adult the looping and partial fusion of gonads and male seminal vesicles and the absence of an ovipositor. In nymphs the presence of strong, oblique, ventrolongitudinal muscles running intersegmentally and allowing lateral undulating swimming and the probably widespread "cercus heart," an accessory circulatory organ associated with posterior abdominal gills, support the monophyly of the order. Gills may be present in nymphal Plecoptera on almost any part of the body or may be absent, causing problems of homology of gills between families and between Plecoptera and other orders. Whether Plecoptera are derived from an aquatic or terrestrial ancestor is debatable.

The phylogenetic position of Plecoptera is certainly among "lower Neoptera," possibly as sister group to the remainder of Neoptera. However, some molecular and combined molecular plus morphological evidence tends to support a more derived position, including as sister to (i) Embiidina or, more likely, (ii) Dermaptera + Dictyoptera.

Internal relationships have been proposed as two predominantly disjunct suborders, the austral Antarctoperlaria and northern Arctoperlaria. The monophyly of Antarctoperlaria is argued based on the unique sternal depressor muscle of the fore trochanter, lack of the usual tergal depressor, and presence of floriform chloride cells, which may have a sensory function. Some of the included taxa are the large-sized Eustheniidae and Diamphipnoidae, the Gripopterygidae, and the Austroperlidae—all families with a Southern Hemisphere "Gondwanan"-type distribution. Recent molecular studies support this clade.

The sister group Arctoperlaria lacks defining morphology, but is united by a variety of mechanisms associated with drumming (sound production) used in mate-finding. The component families Scopuridae, Taeniopterygidae, Capniidae, Leuctridae, and Nemouridae (including Notonemouridae) are essentially of the Northern Hemisphere with a lesser radiation of Notonemouridae into the Southern Hemisphere. Molecular studies suggest the paraphyly of Arctoperlaria, with most elements of Notonemouridae forming the sister group to the remainder of the families. Relationships among extant Plecoptera are proving important in hypothesizing the origins of wings from "thoracic gills" and in tracing the possible development of aerial flight from surface flapping, with legs trailing on the water surface, and forms of gliding.

Zoraptera (Zorapterans). Zoraptera is one of the smallest and probably the least known pterygote order. Zorapterans are small, rather termite-like insects, found worldwide in tropical and warm temperate regions except Australia. Their morphology is simple, with biting, generalized mouthparts, including 5-segmented maxillary palps and 3-segmented labial palps. Sometimes both sexes are apterous, and in alate forms the hind wings are smaller than the forewings; the wings are shed as in ants and termites. Wing venation is highly specialized and reduced.

Traditionally the order contained only one family (Zorotypidae) and one genus (*Zorotypus*), but has been expanded to include seven genera delimited predominantly on wing venation. The phylogenetic position of Zoraptera based on morphology has been controversial, ranging through membership of the hemipteroid orders, sister to Isoptera, an orthopteroid, or a blattoid. Analysis of major wing structures and musculature imply that Zoraptera belong in the blattoid lineage. Although the wing shape and venation resemble those of narrow-winged Isoptera, cephalic and abdominal characters indicate an early divergence from the blattoid stock, prior to the divergence of Dermaptera and much before the origin of the Dictyoptera lineage.

Orthoptera (Grasshoppers, Locusts, Katydids, Crickets). Orthopterans belong to at least 30 families and more than 20,000 species, and most are medium-sized to large insects with hind legs enlarged for jumping (saltation). The compound eyes are well developed, the antennae are elongate and multisegmented, and the prothorax is large with a shield-like pronotum curving downward laterally. The forewings form narrow, leathery tegmina, and the hind wings are broad, with numerous longitudinal and crossveins, folded beneath the tegmina by pleating; aptery and brachyptery are frequent. The abdomen has 8 or 9 annular visible segments, with the 2 or 3 terminal segments reduced, and 1-segmented cerci. The female has a well-developed ovipositor formed from highly modified abdominal appendages.

Virtually all morphological evidence, and much of the newer molecular data suggest that the Orthoptera form the sister group to Phasmatodea. Some authors have united the orders, but the different wing-bud development, different egg morphology, and lack of auditory organs in phasmatids suggest separation. Molecular evidence indicates that Embiidina may be sister to the orthopteran–phasmatid clade, but the support for this relationship is weak.

The division of Orthoptera into two monophyletic suborders, Caelifera (grasshoppers and locusts—predominantly day-active, fast-moving, visually acute, terrestrial herbivores) and Ensifera (katydids and crickets—often night active, camouflaged or mimetic, predators, omnivores, or phytophages), is supported on morphological and molecular evidence. Grylloidea probably is the sister group (but highly divergent, with a long branch separation) of the remaining ensiferan taxa, Tettigonioidea, Hagloidea, and Stenopelmatoidea. On grounds of some molecular and morphological data Tettigoniidae and Haglidae form a monophyletic group, sister to Stenopelmatidae and relatives (mormon crickets, wetas, cooloola monsters, and the like), but alternative analyses suggest different relationships, and conservatively, an unresolved group is perhaps appropriate at this stage.

In Caelifera a well-supported recent proposal for four superfamilies, namely [Tridactyloidea (Tetragoidea (Eumastacoidea + "higher Caelifera"))] reconciles molecular evidence with certain earlier suggestions from morphology. The major grouping of acridoid grasshoppers (Acridoidea) lies in the unnamed clade "higher Caelifera," which also includes the superfamilies Tanaoceroidea, Pyrgomorphoidea, Pneumoroidea, and Trigonopterygoidea.

Phasmatodea (Phasmatids, Phasmids, Stick Insects or Walking Sticks). Phasmatodea are a worldwide, predominantly tropical order of more than 2500 species of hemimetabolous insects, conventionally classified in three families (although some workers raise many subfamilies to family rank). Body shapes are variations on elongate cylindrical, and stick-like or flattened, or often leaf-like. The mouthparts are mandibulate. The compound eyes are relatively small and placed anterolaterally, with ocelli only in winged species and often only in males. The antennae are short to long, with 8 to 100 segments. The prothorax is small, and mesothorax and metathorax are long in wingless species and shorter in apterous ones. The wings, when present, are functional in males, often reduced in females, but with many species apterous in both sexes; the forewings form short leathery tegmina, whereas the hind wings are broad with a network of numerous crossveins and with the anterior margin toughened to protect the folded wing. The legs are elongate, slender, and adapted for walking, with 5-segmented tarsi. The abdomen is 11-segmented, with segment 11 often forming a concealed supra-anal plate in males or a more obvious segment in females.

Phasmatodea are sister to Orthoptera in the orthopteroid assemblage. Novel support for this grouping comes from the dorsal position of the cell body of salivary neuron 1 in the subesophageal ganglion and presence of serotonin in salivary neuron 2. Phasmatodea are distinguished from the Orthoptera by their body shape, asymmetrical male genitalia, proventricular structure, and lack of rotation of nymphal wing pads during development.

Embiidina (Embioptera) (Embiids, Webspinners). Embiidina comprise some 200 described species (perhaps up to an order of magnitude more remain undescribed) in at least eight families. The body is elongate, cylindrical, and somewhat flattened in males. The head has kidney-shaped compound eyes that are larger in males than in females and lacks ocelli. The antennae are multisegmented and the mandibulate mouthparts project forward (prognathy). All females and some males are apterous, but if present, the wings are characteristically soft and flexible, with blood sinus veins stiffened for flight by blood pressure. The legs are short, with 3-segmented tarsi, and the basal segment of the fore tarsi is swollen because it contains silk glands. The hind femora are swollen by strong tibial muscles. The abdomen is 10-segmented with rudiments of segment 11 and with 2-segmented cerci. The female external genitalia are simple (no ovipositor), and those of the males are complex and asymmetrical.

Embiids are undoubtedly monophyletic based on, *inter alia,* the ability to produce silk from unicellular glands in the anterior basal tarsus. They have a general morphological resemblance to Plecoptera based on reduced phallomeres, a trochantin–episternal sulcus, separate coxopleuron, and premental lobes. However, molecular evidence suggests a closer relationship to Orthoptera and Phasmatodea; they also have some similarity to the Dermaptera, notably deriving from their prognathy. Internal relationships among the described higher taxa of Embiidina suggest that the prevailing classification includes many nonmonophyletic groups. Evidently much further study is needed to understand relationships within Embiidina and among it and other neopterans.

Mantophasmatodea. Mantophasmatodea has been recognized recently for a species in Baltic amber and museum specimens representing two species from southwest and east Africa and from freshly collected material from Namibia. The taxon cannot be placed within any of the existing orders and initial estimates of relationships are unclear. Some resemblances to Grylloblattarea and Phasmatodea are evident, but more study, including molecular sequencing, is required.

Paraneoptera (Acercaria or Hemipteroid Assemblage) This group contains Psocoptera + Phthiraptera, Thysanoptera, and Hemiptera and is defined by derived features of the mouthparts, including the slender, elongate maxillary lacinia separated from the stipes, and the swollen postclypeus containing an enlarged cibarium (sucking pump), and the reduction in tarsomere number to 3 or less.

Within Paraneoptera, the monophyletic superorder Psocodea contains Phthiraptera (parasitic lice) and Psocoptera (book lice). Although Phthiraptera is monophyletic, the clade arose from within Psocoptera, rendering that group paraphyletic. Although sperm morphology and some molecular sequence data imply the relationship Hemiptera (Psocodea + Thysanoptera), a grouping of Thysanoptera + Hemiptera

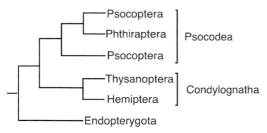

FIGURE 5 Cladogram depicting relationships among, and inferred classification of, Paraneoptera.

(superorder Condylognatha) is supported by head and mouthpart synapomorphies, including the stylet mouthparts, features of the wing base, and the sclerotized ring between antennal flagellomeres. Condylognatha thus forms the sister group to Psocodea (Fig. 5).

Psocoptera (Plant Lice). Psocoptera is a worldwide order of cryptic small insects, with a large, mobile head, bulbous postclypeus, and membranous wings held roof-like over the abdomen. Some 3000 species are described in 35 families. Evidently Psocoptera belong with Phthiraptera in a monophyletic clade, Psocodea. However, Psocoptera is rendered paraphyletic by a postulated relationship of Phthiraptera to the psocopteran family Liposcelidae. Internal relationships in Psocoptera are poorly known.

Phthiraptera (Lice). Phthirapterans are wingless obligate ectoparasites of birds and mammals, lacking any free-living stage, with some 3000 species in some 17 families. Monophyly of, and relationships among, the traditional suborders Anoplura, Amblycera, Ischnocera, and Rhyncophthirina are poorly understood and nearly all possible arrangements have been proposed. The latter three suborders have been treated as a monophyletic Mallophaga (biting and chewing lice) based on their feeding mode and morphology, in contrast to the piercing and blood-feeding Anoplura. Cladistic analysis of morphology has disputed mallophagan monophyly, suggesting the relationship Amblycera [Ischnocera (Anoplura + Rhyncophthirina)]. The only molecular data adduced thus far found unequivocal monophyly only for Amblycera, with a placement that rendered Ischnocera paraphyletic. The data neither supported nor refuted mallophagan monophyly. Resolution of these issues is important in estimation of degree of cospeciation between lice and their bird and mammal hosts.

Thysanoptera (Thrips). Thysanoptera is a worldwide order of just over 5000 species in nine families. The development of thrips is intermediate between hemi- and holometabolous. Their head is elongate and the mouthparts are unique in that the maxillary laciniae formed grooved stylets, the right mandible is atrophied, but the left mandible forms a stylet; all three stylets together form the feeding apparatus. The tarsi are 1- or 2-segmented, and the pretarsus has an apical protrusible adhesive ariolum (bladder or vesicle). Female thrips are diploid, whereas males (if present) are haploid, being produced from unfertilized eggs.

Molecular evidence supports one of the traditional morphological divisions of the Thysanoptera into two suborders, Terebrantia and Tubulifera, containing the sole, speciose, family Phlaeothripidae. Terebrantia includes three speciose families, Thripidae, Heterothripidae, and Aeolothripidae, and five smaller families. Relationships among families in Terebrantia are poorly resolved, although phylogenies are being generated at lower levels concerning aspects of the evolution of sociality, especially the origins of gall-inducing thrips and of "soldier" castes in Australian gall-inducing Thripidae.

Hemiptera (Bugs, Cicadas, Leafhoppers, Planthoppers, Spittlebugs, Aphids, Psylloids, Scale Insects, Whiteflies, Moss Bugs). Hemiptera is the largest of the non-endopterygote orders, with more than 50,000 species in approximately 100 families. Hemipteran mouthparts diagnostically have the mandibles and maxillae modified as needle-like stylets, lying in a beak-like, grooved labium, collectively forming a rostrum or proboscis within which the stylet bundle contains two canals, one delivering saliva and the other taking up fluid. Palps are lacking. The thorax usually has a large prothorax and mesothorax and a small metathorax. Both pairs of wings often have reduced venation, some species are apterous, and male scale insects have only one pair of wings. Legs often possess complex pretarsal adhesive structures. Cerci are lacking.

Hemiptera and Thysanoptera are sister groups within Paraneoptera. Hemiptera used to be divided into two groups, Heteroptera (true bugs) and "Homoptera" (cicadas, leafhoppers, planthoppers, spittlebugs, aphids, psylloids, scale insects, and whiteflies), treated variously as suborders or orders. All homopterans are terrestrial plant feeders and many share a common biology of producing honeydew and being ant attended. However, although possessing defining features, such as wings held roof-like over the abdomen, forewings in the form of a tegmina of uniform texture, and a rostrum arising ventrally close the anterior of the thorax, "Homoptera" represents a paraphyletic grade rather than a clade. This view is supported by reinterpreted morphological data and by cladistic analysis of nucleotide sequences from the nuclear small subunit ribosomal RNA gene (also called 18S rRNA). These data also suggest a much more complicated pattern of relationships among the higher groups of hemipterans (Fig. 6).

The ranking of the various clades is much disputed and thus the more stable superfamily and family names have been used here. Four suborders appear warranted on phylogenetic grounds: Archaeorrhyncha, Clypeorrhyncha, and Prosorrhyncha collectively form the Euhemiptera, which is the sister group to the fourth suborder, Sternorrhyncha. The latter contains the aphids, psyllids, scale insects, and whiteflies, which are characterized principally by their possession of a particular kind of gut filter chamber, a rostrum that appears to arise between the bases of their front legs, and, if winged, the absence of vannus and vannal fold in the hind wing. Some relationships among Euhemiptera are unsettled. A monophyletic Auchenorrhyncha, morphologically defined by their possession of a tymbal acoustic system, an aristate antennal

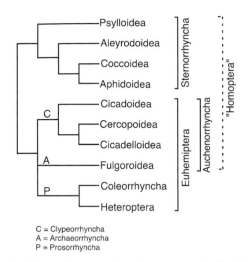

C = Clypeorrhyncha
A = Archaeorrhyncha
P = Prosorrhyncha

FIGURE 6 Cladogram depicting relationships among, and inferred classification of, Hemiptera. Dashed line indicates paraphyly in classification.

flagellum, and reduction of the proximal median plate in the wing base, contains two suborders, Archaeorrhyncha (planthoppers, often called Fulgoromorpha) and Clypeorrhyncha (cicadas, leafhoppers, and spittlebugs, often called Cicadomorpha). Molecular data tend to refute this monophyly, implying that Archaeorrhyncha is closer to Heteroptera, but relationships depend upon sampling and more traditional arrangements are minimally less parsimonious.

Heteroptera (true bugs, including assassin bugs, backswimmers, lace bugs, stink bugs, waterstriders, and others) and its sister group, variously called Coleorrhyncha, Peloridioidea, or Peloridiomorpha and containing only one family, Peloridiidae, or moss bugs, form the suborder Prosorrhyncha. Although small, cryptic, and rarely collected, moss bugs have generated considerable phylogenetic interest due to their combination of ancestral and derived hemipteran features and their exclusively "relictual" Gondwanan distribution. Heteropteran diversity is distributed among some 75 families, forming the largest hemipteran clade. Heteroptera is most easily diagnosed by the presence of metapleural scent glands, and monophyly is never disputed.

Endopterygota (Coleoptera, Neuroptera, Megaloptera, Raphidioptera, Hymenoptera, Trichoptera, Lepidoptera, Mecoptera, Siphonaptera, Diptera, Strepsiptera) Endopterygota comprise insects with immature (larval) instars that are very different from their respective adults. The adult wings and genitalia are internalized in their preadult expression, developing in imaginal discs that are evaginated at the penultimate molt. Larvae lack true ocelli. The "resting stage" or pupa is nonfeeding and precedes an often active pharate ("cloaked" in pupal cuticle) adult. Unique derived features are less evident in the adults than in immature stages, but the clade is consistently recovered from morphological, molecular, and combined analyses.

Two or three groups currently are proposed among the endopterygotes, of which one of the strongest is a sister group

relationship termed Amphiesmenoptera between the Trichoptera (caddisflies) and Lepidoptera (butterflies and moths). A plausible scenario of an ancestral amphiesmenopteran taxon has a larva living in damp soil among liverworts and mosses followed by radiation into water (Trichoptera) or into terrestriality and phytophagy (Lepidoptera).

A second (usually) strongly supported relationship is between three orders, Neuroptera, Megaloptera, and Rhaphidioptera—Neuropterida (sometimes treated as a group of ordinal rank)—showing a sister group relationship to Coleoptera.

A third, postulated relationship—Antliophora—unites Diptera (true flies), Siphonaptera (fleas), and Mecoptera (scorpionflies and hangingflies). Debate contines about the relationships of these taxa, particularly concerning the relationships of Siphonaptera. Fleas were considered sister to Diptera, but molecular and novel anatomical evidence increasingly points to a relationship with a curious-looking mecopteran, *Boreus*.

Strepsiptera is phylogenetically enigmatic, but resemblance of their first-instar larvae (called triungulins) to certain Coleoptera, notably parasitic Rhipiphoridae, and some wing base features have been cited as indicative of relationship. This placement is becoming less likely, as molecular evidence (and haltere development) suggests a link between Strepsiptera and Diptera. Strepsiptera is highly derived in both morphological and molecular evolution, and thus possesses few features shared with any other taxon. The long-isolated evolution of the genome can create a problem known as "long-branch attraction," in which nucleotide sequences may converge by chance events alone with those of an unrelated taxon with a similarly long evolution, for the strepsipteran notably with Diptera. The issue remains unresolved.

The positions of two major orders of endopterygotes, Coleoptera and Hymenoptera, remain to be considered. Several positions have been proposed for Coleoptera but current evidence derived from female genitalia and ambivalent evidence from eye structure supports a sister group relationship to Neuropterida. This grouping forms the sister to the remaining Endopterygota in most analyses. Hymenoptera may be the sister to Antliophora + Amphiesmenoptera; the many highly derived features of adults, and reductions in larvae, limit morphological justification for this position.

Within the limits of uncertainty, the relationships within Endopterygota are summarized in Fig. 7.

Coleoptera (Beetles). Coleoptera undoubtedly lie among the basal Endopterygota. The major synapomorphic feature of Coleoptera is the development of the forewings as sclerotized rigid elytra, which extend to cover some or many of the abdominal segments and beneath which the propulsive hind wings are elaborately folded when at rest. Some molecular studies show Coleoptera polyphyletic or paraphyletic with respect to some or all of the Neuropterida. However, this is impossible to reconcile with the morphological support for coleopteran monophyly, and a sister group relationship to Neuropterida is accepted.

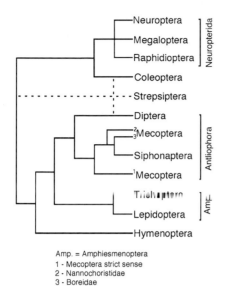

Amp. = Amphiesmenoptera
1 - Mecoptera strict sense
2 - Nannochoristidae
3 - Boreidae

FIGURE 7 Cladogram depicting relationships among, and inferred classification of, Endopterygota. Dashed lines indicate three possible alternative placements of Strepsiptera.

Within Coleoptera, four modern lineages (treated as suborders) are recognized: Archostemata, Adephaga, Polyphaga, and Myxophaga. Archostemata includes only the small families Ommatidae, Crowsoniellidae, Cupedidae, and Micromalthidae and forms the sister group to the remaining extant Coleoptera. The few known larvae are woodminers with a scerotized ligula and a large mola on each mandible. Adults have movable hind coxae with usually visible trochantins and 5 (not 6) ventral abdominal plates (ventrites) but share with Myxophaga and Adephaga wing-folding features (apex spirally rolled, major transverse fold crossing vein MP), absence of cervical sclerites, and the external prothoracic pleuron. In contrast to Myxophaga, the pretarsus and tarsus are unfused.

Adephaga is diverse, second in size only to Polyphaga, and includes ground beetles, tiger beetles, whirligigs, predaceous diving beetles, and wrinkled bark beetles, among others. Larval mouthparts are adapted for liquid-feeding, with a fused labrum and no mandibular mola. Adults have the notopleural sutures visible on the prothorax and have 6 visible abdominal sterna with the first three fused into a single ventrite, which is divided by the hind coxae. Pygidial defense glands are widespread in adults. The most speciose included family is Carabidae, or ground beetles, with a predominantly predaceous feeding habit, but Adephaga also includes the aquatic families, Haliplidae and Noteridae, which are algivorous, and the mycophagous Rhysodidae. Morphology has suggested that Adephaga is sister group to the combined Myxophaga and Polyphaga, but molecular data (18S rDNA) suggest Adephaga as sister to Polyphaga, with Myxophaga sister to them.

Myxophaga is a clade of small, primarily riparian aquatic beetles, comprising the families Lepiceridae, Torridincolidae, Hydroscaphidae, and Microsporidae, united by the synapo-

morphic fusion of the pretarsus and tarsus. The 3-segmented larval antenna, 5-segmented larval legs with a single pretarsal claw, fusion of trochantin with the pleuron, and ventrite structure support a sister group relationship of Myxophaga with the Polyphaga. This has been challenged by some workers, notably because wing venation and folding provide evidence for (Polyphaga (Archostemata (Myxophaga + Adephaga))).

Polyphaga contains the majority (>90% of species) of beetles, with about 300,000 described species. The suborder includes rove beetles (Staphylinoidea), scarabs and stag beetles (Scarabaeoidea), metallic wood-boring beetles (Buprestoidea), click beetles and fireflies (Elateroidea), as well as diverse Cucujiformia, including fungus beetles, grain beetles, ladybird beetles, darkling beetles, blister beetles, longhorned beetles, leaf beetles, and weevils. The prothoracic pleuron is not externally visible, but is fused with the trochantin and remnant internally as a "cryptopleuron." Thus one suture between the notum and the sternum is visible in the prothorax in polyphagans, whereas two sutures (the sternopleural and notopleural) often are visible externally in other suborders (unless secondary fusion between the sclerites obfuscates the sutures, as in *Micromalthus*). The transverse fold of the hind wing never crosses MP, cervical sclerites are present, and hind coxae are mobile and do not divide the first ventrite. Female polyphagan beetles have telotrophic ovarioles, which is a derived condition within beetles.

The internal classification of Polyphaga involves several superfamilies or series, whose constituents are relatively stable, although some smaller families (whose rank even is disputed) are allocated to different clades by different authors. Large superfamilies include Hydrophiloidea, Staphylinoidea, Scarabaeoidea, Buprestoidea, Byrrhoidea, Elateroidea, Bostrichoidea, and the grouping Cucujiformia. This latter includes the vast majority of phytophagous (plant-eating) beetles, united by cryptonephric Malpighian tubules of the normal type, a cone ommatidium with open rhabdom, and lack of functional spiracles on the eighth abdominal segment. Constituent superfamilies of Cucujiformia are Cleroidea, Cucujoidea, Tenebrionoidea, Chrysomeloidea, and Curculionoidea. Evidently adoption of a phytophagous lifestyle correlates with speciosity in beetles, with Cucujiformia, especially weevils (Curculionoidea), forming a major radiation.

Neuropterida or Neuropteroid Orders: Megaloptera (Alderflies, Dobsonflies, Fishflies), Raphidioptera (Snakeflies), and Neuroptera (Lacewings, Antlions, Owlflies). Neuropterida comprise three small orders with holometabolous development, with approximately 6000 species of Neuroptera in about 20 families, 300 species of Megaloptera in 2 widely recognized families, and 200 species of Raphidioptera in 2 families. Adults have multisegmented antennae, large, separated eyes, and mandibulate mouthparts. The prothorax may be larger than the meso- and metathorax, which are about equal in size. Legs sometimes are modified for predation. The fore- and hind wings are quite similar in shape and venation, with folded wings often extending beyond the abdomen. The abdomen lacks cerci.

Megalopterans are predatory only in the aquatic larval stage; although adults have strong mandibles, they are not used in feeding. Adults closely resemble neuropterans, except for the presence of an anal fold in the hind wing. Raphidiopterans are terrestrial predators as adults and larvae. The adult is mantid-like, with an elongate prothorax, and the head is mobile and used to strike, snake-like, at prey. The larval head is large and forwardly directed. Many adult neuropterans are predators and have wings typically characterized by numerous crossveins and "twigging" at the ends of veins. Neuropteran larvae usually are active predators with slender, elongate mandibles and maxillae combined to form piercing and sucking mouthparts.

Megaloptera, Raphidioptera, and Neuroptera may treated as separate orders or united in Neuropterida, or Raphidioptera may be included in Megaloptera. There is little doubt that Neuropterida is monophyletic, with new support from wing base morphology. This latter feature also supplements data supporting the long-held view that Neuropterida forms a sister group to Coleoptera. Each component appears monophyletic, although a doubt remains concerning megalopteran monophyly. There remains uncertainty about internal relationships, which traditionally have Megaloptera and Raphidioptera as sister groups. Recent reanalyses with some new character suites have postulated Megaloptera as sister to Neuroptera and proposed a novel scenario of the plesiomorphy of aquatic larvae (all Megaloptera and Sisyridae in Neuroptera) in Neuropterida.

Strepsiptera. Strepsiptera form an enigmatic order of nearly 400 species of highly modified endoparasites, most commonly of Hemiptera and Hymenoptera, and show extreme sexual dimorphism. The male has a large head with bulging eyes comprising few large facets and lacks ocelli; the antenna are flabellate or branched, with 4 to 7 segments; the forewings are stubby and lack veins, whereas the hind wings are broadly fan-shaped, with few radiating veins; the legs lack trochanters and often also claws. Females are either coccoid-like or larviform, wingless, and usually retained in a pharate (cloaked) state, protruding from the host. The first instar is a triungulin, without antennae and mandibles, but with three pairs of thoracic legs; subsequent instars are maggot-like, lacking mouthparts or appendages. The pupa, which has immovable mandibles but appendages free from its body, develops within a puparium formed from the last instar.

The phylogenetic position of Strepsiptera has been subject to much speculation because modifications associated with their endoparasitic lifestyle mean that few characteristics are shared with possible relatives. In having posteromotor flight (only metathoracic wings) they resemble Coleoptera, but other attributes traditionally argued to be synapomorphous with Coleoptera are suspect or mistaken. The forewing-derived halteres of strepsipterans are gyroscopic organs of equilibrium with the same functional role as the halteres of Diptera, although the latter are derived from the hind wing. Molecular sequence studies indicate that Strepsiptera possibly is a sister group to Diptera, and some tantalizing information from developmental biology suggests that wings and halteres might be "reverse-expressed" on meso- and metathoracic segments.

Mecoptera (Scorpionflies, Hangingflies). Mecopterans are holometabolous insects comprising about 500 known species in nine families, with common names associated with the two largest families—Bittacidae (hangingflies) and Panorpidae (scorpionflies). Adults have an elongate ventrally projecting rostrum, containing elongate, slender mandibles and maxillae, and an elongate labium. The eyes are large and separated, the antennae filiform and multisegmented. The fore- and hind wings are narrow, similar in size, shape, and venation, but often are reduced or absent. The legs may be modified for predation. Larvae have a heavily sclerotized head capsule, are mandibulate, and may have compound eyes comprising 3 to 30 ocelli (absent in Panorpidae, indistinct in Nannochoristidae). The thoracic segments are about equal and have short thoracic legs with fused tibia and tarsus and a single claw; prolegs usually are present on abdominal segments 1 to 8, and the terminal segment (10) has either paired hooks or a suction disk. The pupa is immobile, mandibulate, and with appendages free.

Although some adult Mecoptera resemble neuropterans, strong evidence supports a relationship to Diptera. Intriguing recent morphological studies, plus robust evidence from molecular sequences, suggest that Siphonaptera arise from within Mecoptera, as a sister group to the "snow fleas" (Boreidae). The phylogenetic position of Nannochoristidae, the southern hemisphere mecopteran taxon currently treated as being of subfamily rank, has a significant bearing on internal relationships within Antliophora. Molecular evidence suggests that it lies as sister to Boreidae + Siphonaptera and therefore is of rank equivalent to the boreids, the fleas, and the residue of Mecoptera—and logically each should be treated as an order or Siphonaptera should be reduced in rank within Mecoptera.

Siphonaptera (Fleas). Siphonaptera is a highly modified order of holometabolous insects, comprising some 2400 species, all of which are bilaterally compressed, apterous ectoparasites. The mouthparts are specialized for piercing and sucking, lack mandibles, but have an upaired labral stylet and two elongate serrate, lacinial stylets that together lie within a maxillary sheath. A salivary pump injects saliva into the wound, and cibarial and pharyngeal pumps suck up the blood meal. Fleas lack compound eyes and the antennae lie in deep lateral grooves. The body is armed with many posteriorly directed setae and spines, some of which form combs, especially on the head and anterior thorax. The metathorax houses very large muscles associated with the long and strong hind legs, which are used to power the prodigious leaps made by these insects.

After early suggestions that the fleas arose from a mecopteran, the weight of evidence suggested that they formed the sister group to Diptera. However, increasing molecular and novel morphological evidence now points to a sister group relationship to a subordinate component of Mecoptera, specifically Boreidae (snow fleas). Internal relationships of the fleas are under study, and preliminary results imply that monophyly of many families is uncertain.

Diptera (True Flies). Diptera is a major order of insects, with perhaps as many as a quarter of a million species in some 120 families. Dipterans are holometabolous and readily recognized by the development of hind (metathoracic) wings as balancers, or halteres (halters), and in the larval stages by lack of true legs and the often maggot-like appearance. Venation of the fore (mesothoracic) flying wings ranges from complex to extremely simple. Mouthparts range from biting-and-sucking (e.g., biting midges and mosquitoes) to "lapping"-type with a pseudotracheate labella functioning as a sponge (e.g., house flies). Dipteran larvae lack true legs, although various kinds of locomotory apparatus range from unsegmented pseudolegs to creeping welts on maggots. The larval head capsule ranges from complete, through partially undeveloped, to complete absence of a maggot head, retaining only the sclerotized mandibles ("mouth hooks") and supporting structures.

Traditionally Diptera had two suborders, Nematocera (crane flies, midges, mosquitos, and gnats), with a slender, multisegmented antennal flagellum, and the heavier built Brachycera ("higher flies," including hover flies, blow flies, and dung flies), with shorter, stouter, and fewer-segmented antenna. However, Brachycera is sister to only part of "Nematocera," and thus Nematocera is paraphyletic.

Internal relationships among Diptera are becoming better understood, although with some notable exceptions. Ideas concerning basal Diptera are inconsistent: traditionally Tipulidae (or Tipulomorpha if subordinate groups are given family rank, but nonetheless undoubtedly monophyletic) is a basal clade, particularly on evidence from the adult wing and other morphology. Such an arrangement is difficult to reconcile with the much more derived larva, in which the head capsule is variably reduced. Furthermore, some molecular evidence casts doubt on the basal position of the crane flies, but as yet does not produce a robust estimate for any alternative basal grouping. Alternative views based on morphology have suggested that the relictual family Tanyderidae, with complex ("primitive") wing venation, lies close to the base of the order. In this instance support comes both from the tanyderid larval morphology and from the putative placement in Psychodomorpha, considered a probable basal clade.

There is strong support for Culicomorpha, comprising mosquitoes (Culicidae) and their relatives (Corethrellidae, Chaoboridae, Dixidae) and their sister group the blackflies, midges, and relatives (Simuliidae, Thaumaleidae, Ceratopogonidae, Chironomidae), and for Bibionomorpha, comprising the fungus gnats (Mycetophilidae *sensu lato*), Bibionidae, Anisopodidae, and possibly Cecidomyiidae. However, in both groups internal relationships remain a matter of debate, which molecular evidence may help to resolve.

Monophyly of Brachycera, comprising higher flies, is established by features including, in the larva, the posterior elongation of the head into the prothorax, the divided mandible, and the loss of premandible and, in the adult, the 8 or fewer antennal flagellomeres, 2 or fewer palp segments, and separation of the male genitalia into 2 parts (epandrium

and hypandrium). Possible relationships of Brachycera include sister to Psychodomorpha or even to Culicomorpha (molecular data only) but strong support is provided for sister taxon to the Bibiomorpha or to subordinate Anisopodidae. Brachycera contains four equivalent groups with internally unresolved relationships: Tabanomorpha (with brush on larval mandible and larval head retractile), Stratiomyomorpha (with larval cuticle calcified and pupation in last instar exuviae), Xylophagomorpha (with distinctive elongate, conical, strongly sclerotized larval head capsule and abdomen posteriorly ending in sclerotized plate with terminal hooks), and Muscomorpha (adults with tibial spurs absent, flagellum with no more than 4 flagellomeres, and female cercus single segmented). This latter speciose group contains Asiloidea (robber flies, bee flies, and relatives) and Eremoneura (empidoids and Cyclorrhapha). Eremoneura is a strongly supported clade based on wing venation (loss or fusion of vein M_4 and closure of anal cell before margin), presence of ocellar setae, unitary palp, and several genital characters, plus larval features including maxillary reduction and presence of only three instars.

Cyclorrhaphans, united by their pupation within a puparium formed by the last instar skin, include a heterogeneous aschizan group comprising Phoridae and Syrphidae (hover flies) and the Schizophora, defined by the presence of a balloon-like ptilinum that everts from the frons to assist the adult escape from the puparium. Higher flies include the ecologically very diverse acalypterates and blow flies and relatives (Calypteratae), treated here as sister groups (Fig. 8), but with alternative views suggested.

Hymenoptera (Wasps, Bees, Ants, Sawflies, and Wood Wasps). Hymenoptera contains at least 100,000 described species of holometabolous neopterans, varying from minute (e.g., Trichogrammatidae) to large-sized (0.15–120 mm in length) and slender (e.g., many Ichneumonidae) to robust (e.g., certain bees). The hymenopteran head has mouthparts ventrally directed to forward projecting, ranging from

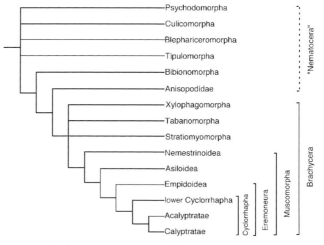

FIGURE 8 Cladogram depicting relationships among, and inferred classification of, Diptera. Dashed line indicates paraphyly in classification.

generalized mandibulate ("Symphyta"—woodwasps and sawflies) to sucking and chewing, with mandibles often used for killing and handling prey and, in Apocrita (ants, bees, and wasps), in defense and nest building. The compound eyes often are large; the antennae are long, multisegmented, and often prominently held forwardly or recurved dorsally. "Symphyta" has a conventional 3-segmented thorax, but in Apocrita the propodeum, abdominal segment 1, is included with the thorax into a mesosoma or, in ants, the alitrunk. The wing venation is relatively complete in large sawflies but is reduced in Apocrita in correlation with body size, such that very small species of 1 to 2 mm have only one divided vein or none. The hind wing has rows of hooks (hamuli) along the leading edge that couple with the hind margin of the forewing in flight. In Apocrita, the second abdominal segment (and sometimes also third) forms a constriction, or petiole. Hymenopteran female genitalia include an ovipositor, comprising 3 valves and 2 major basal sclerites, which may be long and highly mobile, allowing valves to be directed vertically between legs. The ovipositor of aculeate Hymenoptera is modified as a sting associated with a venom apparatus.

Symphytan larvae are eruciform (caterpillar-like) with 3 pairs of thoracic legs bearing apical claws and with some abdominal legs; most are phytophagous. Apocritan larvae are apodous, with the head capsule frequently reduced but with prominent strong mandibles; larvae may vary greatly in morphology during development (heteromorphosis). Apocritan larvae have diverse feeding habits and may be parasitic, gall-inducing, or fed with prey or nectar and pollen by their parent or, if a social species, other colony members. Adult hymenopterans mostly feed on nectar or honeydew; only a few consume other insects.

Hymenoptera is considered to form the sister group to Amphiesmenoptera (Trichoptera + Lepidoptera) + Antliophora (Diptera + Mecoptera/Siphonaptera), although a more basal position in the Holometabola has been advocated. Hymenoptera often are treated as containing two suborders, Symphyta (wood wasps and sawflies) and Apocrita (wasps, bees, and ants). However, Apocrita appears to be sister to one family of symphytan only, the Orussidae, and thus "Symphyta" form a basal paraphyletic group, whose basalmost clade is the Xyeloidea. This is sister to a monophyletic tenthrinoid (sawfly) clade, in turn sister to weakly supported Pamphiliodea [Cephoidea (possible grade Siricidae and relatives (Orussidae + Apocrita))] (Fig. 9).

Within Apocrita, aculeate (Aculeata) and parasitic (Parasitica or terebrant) wasp groups were considered each to be monophyletic, but Parasitica evidently is rendered paraphyletic by Aculeates originating from somewhere within Parasitica. Some traditional groupings also are nonmonophyletic, including Proctotrupoidea, but Proctotrupomorpha, comprising superfamilies Proctotrupoidea, Chalcidoidea, Platygasteroidea, and Cynipoidea, appears to be monophyletic. From morphology, Ichneumonoidea were argued to be sister to Aculeata, but molecular data and reanalysis refute this. The

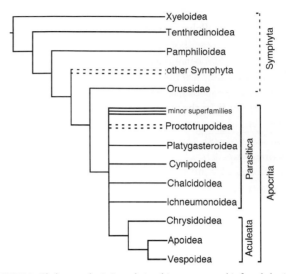

FIGURE 9 Cladogram depicting relationships among, and inferred classification of, Hymenoptera. Only major superfamilies are depicted. Dashed lines indicate uncertainty in cladogram or paraphyly in classification.

monophyletic aculeates, defined by their distinctive ovipositor construction and loss of female cerci, comprise Chrysidoidea (Vespoidea + Apoidea). Internal relationships of aculeates, including vespids (wasps), formicids (ants), and apids (bees), and the monophyly of subordinate groups are under scrutiny. Apidae evidently arose from within Sphecidae (digger wasps), but the precise relationships of another significant group of aculeates, Formicidae (ants), within Vespoidea is less well established.

Trichoptera (Caddisflies). Trichoptera contains about 45 extant families containing some 10,000 described species, with estimates of undescribed (mostly Southeast Asian) species diversity some four- to fivefold higher. Trichoptera are holometabolous; the moth-like adult has reduced mouthparts lacking any proboscis, but with 3- to 5-segmented maxillary palps and 3-segmented labial palps. The antennae are multisegmented and filiform and often as long as the wings. The compound eyes are large, and there are 2 or 3 ocelli. The wings are haired or less often scaled and differentiated from all but the most basal Lepidoptera by the looped anal veins in the forewing and absence of a discal cell. The larva is aquatic, has fully developed mouthparts, has 3 pairs of thoracic legs (each with at least 5 segments), and lacks the ventral prolegs characteristic of lepidopteran larvae. The abdomen terminates in hook-bearing prolegs. The tracheal system is closed and associated with tracheal gills on most abdominal segments. The pupa is also aquatic, enclosed in a retreat often made of silk, and possesses functional mandibles to aid in emergence from the sealed case.

Amphiesmenoptera (Trichoptera + Lepidoptera) is now unchallenged, despite earlier suggestions that Trichoptera arose from within Lepidoptera. Proposed internal relationships within the Trichoptera range from stable and well supported to unstable and anecdotal. Monophyly of the suborder

Annulipalpia in its strictest sense, comprising families Hydropsychidae, Polycentropodidae, Philopotamidae, and some close relatives, is well supported by larval and adult morphology—including the presence of an annulate apical segment of both adult maxillary and larval palp, absence of male phallic parameres, presence of papillae lateral to the female cerci, and, in the larva, presence of elongate anal hooks and reduced abdominal tergite 10.

The monophyly of the case-making Integripalpia (comprising families Phryganeidae, Limnephilidae, Leptoceridae, Sericostomatidae, and relatives) is supported *inter alia* by the absence of the *m* crossvein; hind wings broader than forewings, especially in the anal area; the female lacking both segment 11 and cerci; and larval character states including the usual complete sclerotization of the mesonotum, hind legs with lateral projection, lateral and middorsal humps on abdominal segment 1, and short and stout anal hooks.

Monophyly of a third putative suborder, Spicipalpia, is more contentious. Defined for the Glossosomatidae, Hydroptilidae, and Rhyacophilidae (and perhaps the Hydrobiosidae), the proposed uniting features are the spiculate apex of the adult maxillary and labial palps, ovoid second segment of the maxillary palp, and eversible oviscapt (egg-laying appendage). Morphological and molecular evidence fail to confirm Spicipalpia monophyly, unless at least Hydroptilidae is removed. All possible relationships between Annulipalpia, Integripalpia, and Spicipalpia have been proposed, often associated with scenarios concerning the evolution of case-making. An early idea that Annulipalpia are sister to a paraphyletic Spicipalpia + monophyletic Integripalpia finds support from some morphological and molecular data.

At lower phylogenetic levels, several interfamily relationships have been explored and congruent findings made, for example, concerning the families of Sericostomatoidea, including several from the landmasses formerly constituting Gondwana, including South Africa. In contrast, relationships of taxa associated with the family Leptoceridae vary dramatically between researchers. Forthcoming molecular evidence may be expected to assist in resolution of some issues and perhaps confuse otherwise robust relationships.

Lepidoptera (Moths and Butterflies). Lepidoptera, with some 140,000 described species in 70 families, is one of the major orders of Holometabola. Adults range from very small to large, with wings always covered in scales. The head bears a long, coiled proboscis formed from greatly elongated maxillary galeae; large labial palps usually are present, but other mouthparts are absent, except that mandibles are present primitively in some groups. The compound eyes are large, and ocelli frequently are present. The multisegmented antennae are often pectinate in moths and knobbed or clubbed in butterflies (Papilionoidea + Hedyloidea + Hesperioidea). The wings are completely covered with a double layer of scales (flattened modified macrotrichia), and the hind and forewings are linked by either frenulum, jugum, or simple overlap. Lepidopteran larvae have a sclerotized, head capsule with mandibulate

mouthparts, usually 6 lateral ocelli, and short 3-segmented antennae. The thoracic legs are 5-segmented with single claws, and the abdomen is 10-segmented with short prolegs on some segments. Silk-gland products are extruded from a characteristic spinneret at the median apex of the labial prementum. The pupa sometimes is contained within a silken cocoon.

The basal radiation of this large order is considered well enough resolved to serve as a test for the ability of particular molecules to recover phylogenetic signal. Although over 98% of the species of Lepidoptera belong in Ditrysia (Fig. 10), the morphological diversity is concentrated in a small nonditrysian group. Three of the four suborders are species-poor (Micropterigidae, Agathiphagidae, Heterobathmiidae), lie sequentially at the base of the Lepidoptera, and lack the synapomorphy of the megadiverse suborder Glossata, namely the characteristically developed coiled proboscis formed from the fused galea. The highly speciose Glossata contains a comb-like basal series of species-poor taxa and a subordinate clade (Neolepidoptera) defined by the possession in the larva (caterpillar) of abdominal prolegs with muscles and apical crochets (hooklets). Much of the glossatan diversity is found in the derived Ditrysia, defined by the unique two genital openings in the female, one the ostium bursae on sternite 8, the other the genitalia proper on sternites 9 and 10. Additionally the wing coupling is always frenulate or amplexiform and not jugate, and the wing venation tends to be heteroneuran (with venation dissimilar between fore- and hind wings). Trends in the evolution of Ditrysia include elaboration of the proboscis (haustellum) and the reduction or loss of maxillary palpi. Relationships between the smaller superfamilies (not shown in Fig. 10) and the few highly diverse ones (Tineoidea, Gelechioidea, Tortricoidea, Pyraloidea, Noctuoidea, and Geometroidea) are not well understood and susceptible to change. However, one of the best supported relationships in Ditrysia is the grouping of Hesperioidea (skippers) and Papilionoidea (butterflies), united by their clubbed, dilate antennae, lack of frenulum in the wing, and large humeral lobe

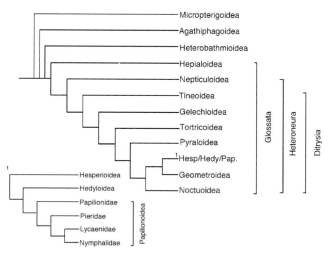

FIGURE 10 Cladogram depicting relationships among, and inferred classification of, Lepidoptera. Only selected superfamilies are depicted.

on the hind wing. To this has been added the neotropical Hedyloidea to form the monophyletic grouping we know as the butterflies.

See Also the Following Articles
*Arthropoda and Related Groups • Biodiversity •
Biogeographical Patterns • Fossil Record • Insecta, Overview*

Further Reading
Bitsch, C., and Bitsch, J. (2000). The phylogenetic interrelationships of the higher taxa of apterygote hexapods. *Zool. Scripta* **29**(2), 131–156.

Kristensen, N. P. (1999). Phylogeny of endopterygote insects, the most successful lineage of living organisms. *Eur. J. Entomol.* **96**(3), 237–253.

Lyal, C. H. C. (1985). Phylogeny and classification of the Psocodea, with particular reference to the lice (Psocodea: Phthiraptera). *Syst. Entomol.* **10**, 145–165.

Ouvrard, D., Campbell, B. C., Bourgoin, T., and Chan, K. L. (2000). 18S rRNA secondary structure and phylogenetic position of Peloridiidae (Insecta: Hemiptera). *Mol. Phylogenet. Evol.* **16**, 403–417.

Ronquist, F., *et al.* (1999). Phylogeny of the Hymenoptera: A cladistic reanalysis of Rasnitsyn's (1988) data. *Zool. Scripta* **28**, 13–50.

Schultz, J. W., and Regier, J. C. (2000). Phylogenetic analysis of arthropods using two nuclear protein-encoding genes supports a crustacean-hexapod clade. *Proc. R. Soc. Biol. Sci. B* **267**(1447), 1011–1019.

Sorensen, J. T., Campbell, B. C., Gill, R. J., and Steffen-Campbell, J. D. (1995). Non-monophyly of Auchenorrhyncha ("Homoptera"), based upon 18S rDNA phylogeny: Eco-evolutionary and cladistic implications within pre-Heteropterodea Hemiptera (s.l.) and a proposal for new monophyletic suborders. *Pan-Pacific Entomol.* **71**, 31–60.

Vilhelsen, L. (2000). Before the wasp-waist: Comparative anatomy and phylogenetic implications of the skeleto-musculature of the thoraco-abdominal boundary region in basal Hymenoptera (Insecta). *Zoomorphology* **119**, 185–221.

von Dohlen, C. D., and Moran, N. A. (1995). Molecular phylogeny of the Homoptera: A paraphyletic taxon. *J. Mol. Evol.* **41**, 211–223.

Physical Control of Insect Pests

Charles Vincent and Bernard Panneton
Agriculture and Agri-Food Canada, Quebec

Physical control is one of the four main approaches to crop protection against insects; the other three are the chemical, biological, and biopesticide approaches. From a theoretical and a technical point of view, all of these approaches have limits that make them more or less suitable against a given pest control target. In practice, the relative merits of each approach are also weighted against numerous factors before an actual decision is made regarding the most appropriate method to implement. A majority of agricultural commodities are protected using chemical control but ideally, all components and technologies should be blended optimally and harmoniously into an integrated pest management (IPM) program.

PHYSICAL CONTROL METHODS

Definition, Context, and Literature

Physical control methods in crop protection comprise all pest management techniques that rely on the use of physical processes to damage, kill, or induce behavioral changes in target organisms. The primary action may have a direct impact, for example, when insects are killed immediately by mechanical shocks. In other instances, the desired effect is attained through stress responses.

Various physical control methods have been used throughout the long history of plant protection (Table I). With the rapid advances that have occurred in the physical, chemical, and biological sciences since the late 19th century, agriculture has been transformed from a strictly empirical activity, largely based on tradition and aimed primarily at staying off famine, to a quantitative form of agriculture focused on producing a certain amount of food. During this transition, which has been sustained at an increasing rate over the past 50 years, physical control methods have been set aside because of the tremendous success of chemical control. It is only natural that some people should view the use of physical control methods as a step backward to those distant ancestral practices. Thanks to technological advances and greater precision in the implementation of such methods, physical control now has all the necessary attributes for incorporation into IPM strategies.

Use in Agricultural Production

The different methods of physical control used against crop pests have some common characteristics. One of the characteristics that differentiates physical tactics from the other control methods (Table I) is the absence of persistence. In almost every case, the effect of a treatment is limited to the period of application. When treatment stops, the stressor disappears immediately or dissipates quickly. From the standpoint of exercising control over the treatment and its secondary effects, the absence of a residual action is an advantage. However, this characteristic can also be regarded as a drawback, because the treatment may have to be repeated every few days to control crop pests that emerge and are active for a few days or a few weeks. In such cases, persistent chemicals constitute a much more convenient approach, although they are often undesirable from an environmental standpoint.

In addition to being restricted to the time of application, the impact of a physical control method is limited spatially. Mechanical, pneumatic, electrical, and thermal energies are dissipated locally over a distance of up to a few meters from the site of application. Electromagnetic radiation, which propagates over considerable distances and is subject to numerous restrictions (reserved frequency bands, maximum power, absence of interference), is an exception. Some pesticides have the unfortunate characteristic of dispersing over considerable distances. Similarly, many biological control agents can disperse or become dispersed beyond the treatment area.

TABLE I   **Comparison of Control Methods for Crop Protection**

Characteristic	Method		
	Chemical	Biological	Physical
Advent	20th century	20th century	With agriculture
Registration	Required	A few cases	Never
Supporting sciences	Analytical chemistry, chemical synthesis, biology	Biology, biotechnology, ecology	Engineering (mechanical, electrical, electronic), biology
Scientific references	Very abundant	Abundant	Few
Residual action (residues and persistence)	Yes (variable)	Yes, if reproduction occurs	Negligible
Possibility of combining with another method	Yes (sometimes difficult with biological methods)	Yes	Yes
Active or passive method	Active	Active	Active and passive
Application to field crops	High level	Low level	Low to moderate level
Application to crops with a high profit margin per hectare	High	Moderate to high	Moderate to high
Safety for crop	Moderate to high (phytotoxicity)	High	High (passive) Low (active)
Labor requirements	Low	High	Medium to high
Work rate (hectares treated per hour)	High	Variable	Low (active) High (passive)
Site of action	Photosynthetic system, nervous system (few genes involved)	Systems allowing adaptation to biotic stresses	Systems allowing adaptation to abiotic stresses
Environmental or toxicological requirements, safety	High and costly	Moderate (e.g., virus)	Low (exception: electromagnetic radiation)
Geographic impact	Drift, run-off, evaporation, food chain	Colonization of nontarget habitats by parasites or predators	Restricted to area treated (exception: electromagnetic radiation)
Energy requirements	High for production	Low	Low (passive) High (active)
Machinery required	Ground or aerial sprayer	Little or none	Many types of equipment, few machines are suited to more than one purpose

Note. Reproduced, with permission, from Vincent, C., Panneton, B., and Fleurat-Lessard, F. (eds.) (2001). "Physical Control Methods in Plant Protection." Springer/INRA. Copyright Springer-Verlag.

Compared with traditional chemical control, present methods of physical control are more labor intensive and often time consuming (Table I). This drawback is one of the main reasons physical control techniques have had little success in penetrating the field crop market. Given these circumstances, only crops with a high profit margin per hectare represent an obvious market for physical control methods. From the viewpoint of implementation, physical methods compare favorably with biological methods (other than biopesticides), which often entail labor-intensive field observations and are difficult to apply in a field-crop setting.

Modes of Action and Classification

There are two basic types of physical control, i.e., active and passive (Table II). Active methods use some form of energy to destroy, injure, or induce stress in crop pests or to remove them from the environment. This type of approach has an effect at the time of application, with virtually no residual action. Passive methods, in contrast, cause changes in the environment and

have a more lasting effect. Physical methods of control also can be classified according to the mode of energy use: thermal shock (heat), electromagnetic radiation (microwaves, infrared and radiofrequencies), mechanical shock, and pneumatic control (blowing or vacuuming tools) (Table II). Mechanical barriers to keep pests out, combined with physical suppression techniques, are the cornerstone of the approach adopted by industrialized countries to replace methyl bromide. Other techniques such as diatomaceous earth, hydrophilic particle films, sticky traps, and oils also are passive techniques.

Various applications that use thermal shock for crop protection in the field have been developed and research is in progress. This type of approach is based on the premise that the commodity or crop to be protected will be less sensitive than the target pest to an abrupt change in temperature. Research on thermal sensitivity thresholds and physiological reactions to short-duration thermal stresses is central to the development of control methods based on thermal shock.

Several avenues have been explored for applying the different forms of electromagnetic radiation as a tool for controlling

TABLE II   Classes, Families, and Examples of Physical Control Methods

Class, family	Method	Examples of target insects	Situation	Comments
Passive				
	Physical barriers	Several mosquito species of medical importance, Colorado potato beetle, chinch bugs	Pr, PH	Efficacy limited to area
	Trapping	Apple maggot	Pr	In apple orchards
	Mulching	Aphids	Pr	
	Spatial barrier	Screwworm		
	Inert dusts	Stored-product insects	Pr	Including diatomaceous earth, silica dust
	Thin films (kaolin)	Codling moth, leafroller, mites	Pr	
	Mineral and edible oils	Phytophagous mites, scales	Pr, PH	
	Adhesives	Ants, cockroaches, flies		Glues, petroleum jelly, creosote
	Slippery surfaces	Ants, cockroaches		Made of Fluon, Teflon, or dust
	Fences	Chinch bugs, Colorado potato, beetle, cutworm	Pr	
	Windbreak	Aphid vectors		
	Wrapping	Gypsy moth, forest tent caterpillar, insects of transformed products		
	Organic mulch	Melon worm, Colorado potato beetle		
	Inorganic mulch	Aphid vectors, potato tuberworm		
	Trench	Chinch bugs, Colorado potato beetle,		
	Water	Chinch bugs		
	Modified atmosphere		PH	Modulating CO_2
	Inert gases	Stored-product insects	PH	
Active				
Mechanical	Mechanical impacts			
	Pneumatic (vacuuming, blowing)	Colorado potato beetle, *Lygus* spp.	Pr	
	Dislodging	Plum curculio	Pr	In apple orchards
	Leaf shredding	Spotted tentiform leafminer	Pr	
	Disturbing	Stored-product insects	Pr	
	Forced air	House flies		Building entrances
Thermal	Flaming	Colorado potato beetle	Pr	
	Hot/cool air	Stored-product insects	PH	
	Hot water–steam		PH	
	Rapid freezing	Stored-product insects	PH	
	Postharvest chilling	Stored-product insects	PH	
	IR heating		PH	
Electromagnetic	Microwave		Pr, PH	Also in museums
	Radiofrequencies		PH	
	Ionizing radiation		PH	
	UV and visible light	Mosquitoes		In combination with electricity
Other	Flooding	Cranberry insects	Pr	

Note. Pr, preharvest control; PH, postharvest control.

insects. Nonionizing electromagnetic radiation kills insects by raising their internal temperature. The utilization of radio, microwave, and infrared frequencies is based on a principle similar to that of thermal shock methods except that, with applications involving electromagnetic radiation, the transfer of energy occurs without using a heat transfer fluid. Technologies that harness electromagnetic radiation are often too expensive for use in the field. Furthermore, existing regulations restrict the available frequency bands, either for reasons of user and environmental safety or because certain frequency bands have been set aside for specific applications that do not tolerate interference (e.g., microwave-based landing guidance systems for aircraft).

There is a wide variety of physical barriers used as physical control techniques. The technology associated with physical barriers can be applied in the field or in greenhouses. Barriers used in the field can take several forms (trenches, vertical nets, etc.) and can be deployed on a range of scales to protect a complete field, a crop row, or a group of plants. Passive methods should be used whenever possible, because they extend the length of the treatment. For example, plastic-lined trenches along field boundaries trap Colorado potato beetles during the whole migration period.

Pneumatic control consists in using an airstream to dislodge insect pests. Insects that are removed by vacuum pressure are killed when they pass through the moving parts of the blower (mechanical shock). After being dislodged by a blowing device, individuals of some insect species are injured and die because they are unable to climb back onto the host plant. Other machines are equipped with a device for collecting the dislodged insects, which are subsequently killed. Sound knowledge of the target insect's behavior is necessary in order to enhance the effectiveness of this type of approach as exemplified by the management of tarnished plant bug as a pest of strawberries.

Like any pest management approach, physical methods have strengths and weaknesses, and some of them are likely to have secondary effects on nontarget organisms, e.g., pollinators of strawberry. In an IPM context, the decision to use a physical control tactic must therefore be made on a case-by-case basis according to the same criteria as in decision-making regarding the appropriateness of pesticide applications: efficacy, cost-effectiveness, and undesirable effects. In addition, no physical control technique is sufficient on its own for all pest control treatments in a given crop.

Postharvest Physical Control

Postharverst situations offer ideal opportunities to research and implement physical control methods. In long-term storage of nonperishable agricultural commodities (seeds and grain, dried fruits, by-products, dried and dehydrated plants, spices, herbs, coffee, cocoa), the most serious losses are due to the action of insects and mites or the spoilage by certain microorganisms (e.g., fungi). Chemical control using persistent insecticides is the most commonly used approach for preventing damage to grain and seeds by insect pests. Some of the benefits of this strategy are low cost, ease of implementation, and protection that lasts for several months, until the quantity of active residues falls below the lethal threshold for the target species.

Physical control applications developed for postharvest treatments have focused on procedures for controlling physical conditions in stocks of stored grain (temperature and water content), thermal or mechanical shock, the establishment of extreme conditions for insect pests (anaerobiosis, pressure, and modified atmospheres), the use of abrasive or dehydrating minerals, and the erection of physical barriers to keep insects out. Postharvest control approaches are essentially based on passive methods, with the exception of thermal and mechanical shock treatments. In postharvest situations, most physical methods are suitable solely for prevention against pests and hence cannot be compared with classical chemical control. A thorough knowledge of integrated pest management strategies is required, since these techniques afford no protection following application, unlike persistent insecticides.

Practical use of physical control necessitates multiple verifications and supporting data, particularly in relation to secondary effects on the quality of treated products (for example, the germinating power of malting barley or the baking quality of bread-making wheat). Nonetheless, the prospect of registration reviews for several pesticides (e.g., methyl bromide) that are currently used in postharvest protection of raw foodstuffs or in processing and storage facilities for intermediate or finished products has revived interest in postharvest physical control. Furthermore, because the use of persistent contact insecticides on processed food products (e.g., wheat, semolina, dried fruit) is prohibited worldwide, the industry has to rely exclusively on fumigants or physical procedures to eradicate insects in such products. Fumigants are used only marginally in many countries such as France, the United Kingdom, and Canada, and their use is likely to decline further when methyl bromide production and use are phased out, sometime between 2002 and 2005. In view of this situation, physical control is the only strategy likely to ensure successful control of insect and mite pests in postharvest food stocks. Physical methods hold promise as a complement to chemical pesticides and a means of moving away from excessive reliance on chemical control. Furthermore, the most accessible physical methods for this pest management sector, such as dry heat treatments or airtight storage using inert gases, should help to diminish the secondary risk of spoilage of raw foodstuffs by storage molds.

GENERAL CONSIDERATIONS

Physical control deserves to be recognized as a well-defined area of expertise as is the case for biological control. This recognition is bound to come as the quest for alternatives to chemical pesticides intensifies. Although physical control went out of use with the advent of chemically based pest control methods in the middle of the 20th century, the limitations of pesticide use, coupled with the difficulties of implementing biological control, have created a crossroads for the renewed development of physical control.

Most physical methods of control can be used in a crop protection program incorporating both chemical and biological controls. A potential problem occurs when physical barriers are still in place during chemical or biological treatments. Chemical and biological methods are sometimes incompatible, particularly in production systems eschewing chemical pesticides. In the latter case, only biological and physical methods can be applied. Although not economically significant at present, organic crops represent a growing market segment. This is a niche that will definitely provide leverage for the development of physical control measures.

The regulatory framework for physical control differs markedly from that for agrochemical products (Table I). First, many physical techniques are subject to rules concerning their use (i.e., the registration process), which are designed to protect users and the general public. Sometimes, such as with the use of propane gas, specialized training is required. The use of electromagnetic radiation is constrained by telecommunications regulations, some of which stem from international agreements. In the case of microwave energy, for instance, only a handful of frequencies has been set aside for industrial, scientific, and medical applications. With regard to the regulatory framework for physical control technologies, it is completely defined *a priori*. In short, the equipment employed must meet the applicable standards (mostly related to user safety). With chemically or biologically based methods, the difficulty in anticipating secondary effects precludes the establishment of comprehensive specifications that would be known *a priori*. This explains the need for increasingly costly test protocols designed to evaluate pesticide safety from the standpoint of human health as well as ecotoxicology.

A number of factors tend to complicate the implementation of physical control methods, and physical tactics cannot be readily compared with crop protection systems based solely on the use of an agricultural sprayer to apply pesticides in liquid form. For agricultural operations, this system entails low variable costs and fixed costs. In contrast, the equipment used for physical control is often very specific: cultivators for weeds, vacuuming device for Colorado potato beetles, and so on (Table I). Very few physical control tools offer the operational versatility that would allow them to perform several types of crop protection operations. Integration efforts, such as research aiming to design burners for use in controlling Colorado potato beetles, killing weeds, and performing top-killing, are needed to enable physical control tools to penetrate the crop protection market.

See Also the Following Articles

Agricultural Entomology • Biological Control • Colorado Potato Beetle • Integrated Pest Management

Further Reading

Banks, H. J. (1976). Physical control of insects—Recent developments. *J. Aust. Entomol. Soc.* **15**, 89–100.

Bostanian, N. J., Vincent, C., Chouinard, G., and Racette, G. (1999). Managing apple maggot, *Rhagoletis pomonella* (Diptera: Tephritidae), by perimeter trapping. *Phytoprotection* **80**, 21–33.

Chiasson, H., Vincent, C., and de Oliveira, D. (1997). Effect of an insect vacuum device on strawberry pollinators. *Acta Horticult.* **437**, 373–377.

Davidson, N. A., Dibble, J. E., Flint, M. L., Marer, P. J., and Guye, A. (1991). "Managing Insects and Mites with Spray Oils." Publication No. 3347, IPM Education and Publications, University of California, Oakland.

Fields, P. G., and Muir, W. E. (1995). Physical control. *In* "Integrated Pest Management of Insects in Stored Products" (B. Subramanyam and D. W. Hagstrum, eds.), pp. 195–222. Dekker, New York.

Glenn, D. M., Puterka, G. M., VanderZwet, T., Byers, R. E., and Feldhake, C. (1999). Hydrophilic particle films: A new paradigm for supression of arthropod pests and plant diseases. *J. Econ. Entomol.* **92**, 759–771.

Hallman, G. J., and Denlinger, D. L. (eds.) (1998). "Temperature Sensitivity in Insects and Application in Integrated Pest Management." Westview Press, Boulder, CO.

Jayas, D., White, N. D. G., and Muir, W. E. (1995). "Stored-Grain Ecosystems." Dekker, New York.

Korunic, Z. (1998). Diatomaceous earths, a group of natural insecticides. *J. Stored Prod. Res.* **34**, 87–97.

Lurie, S. (1998). Postharvest heat treatments. *Post Harvest Biol. Technol.* **14**, 257–269.

Nelson, S. O. (1996). A review and assessment of microwave emergy for soil treatment to control pests. *Trans. ASAE* **39**, 281–289.

Oseto, C. Y. (2000). Physical control of insects. *In* "Insect Pest Management, Techniques for Environmental Protection" (J. E. Rechcigl and N. A. Rechcigl, eds.), pp. 25–100. Lewis, Boca Raton, FL.

Panneton, B., Vincent, C., and Fleurat-Lessard, F. (2001). Plant protection and physical control methods: The need to protect crop plants. *In* "Physical Control Methods in Plant Protection" (C. Vincent, B. Panneton, and F. Fleurat-Lessard, eds.), pp. 9–32. Springer-INRA, Heidelberg.

Vincent, C. (2002). Pneumatic control of agricultural insect pests. *In* "Encyclopedia of Pest Management" (D. Pimentel, ed.), pp. 639–641. Dekker, New York.

Vincent, C., and Chagnon, R. (2000). Vacuuming tarnished plant bug on strawberry: A bench study of operational parameters versus insect behavior. *Entomol. Exp. Appl.* **97**, 347–354.

Vincent, C., Panneton, B., and Fleurat-Lessard, F. (eds.) (2001). "Physical Control Methods in Plant Protection." Springer/INRA, Heildelberg. [Translated from Vincent, C., Panneton, B., and Fleurat-Lessard, F. (eds.) (2000). "La Lutte Physique en Phytoprotection." Editions INRA, Paris.]

Vincent, C., Hallman, G., Panneton, B., and Fluerat-Lessard, F. (2003). Management of agricultural insects with physical control methods. *Annu. Rev. Entomol.* **48**, 261–281.

Phytophagous Insects

Elizabeth A. Bernays
University of Arizona

Phytophagous insects are generally considered to be those that feed on green plants. They include species that attack roots, stems, leaves, flowers, and fruits, either as larvae or as adults or in both stages. Leaf feeders may be external (exophytic) or they may mine the tissues, sometimes even specializing on a particular cell type. Typically nectar and pollen feeders are not included. "Phytophagous" is often synonymous with "herbivorous," although the latter term is sometimes restricted to those species feeding on herbs (i.e., herbaceous plants). Commonly, species that use only one plant genus or species are called monophagous and species that use plants within a tribe or family are called oligophagous. The term stenophagous is less commonly used and includes both of these. Polyphagous species are those that use plants in several to many plant families.

DIVERSITY

Phytophagous insects are highly diverse and the total species number is at least 500,000. This represents about 25% of known multicellular animals. There are phytophagous insect species in the majority of insect orders, including Orthoptera, Lepidoptera, Coleoptera, Heteroptera, Hymenoptera, and Diptera but there are very large differences in numbers of species in the different groups (Table I). All green plants are eaten by one or more species of phytophagous insects.

Major differences occur among orders in the ways in which hosts are selected and the life forms that use plants. Thus, grasshoppers usually lay eggs in the soil so that the nymphal stages must select the plant resource. In the Orthoptera and Hemiptera, which are hemimetabolous, the nymphal and adult stages generally have similar feeding habits. By contrast, among Lepidoptera and Hymenoptera adults commonly feed on pollen or nectar while only the larvae are phytophagous, and among Coleoptera the larvae

TABLE I Approximate Proportions of Species of Phytophagous Insects from Different Insect Orders

Order	Common example	Proportion of species in order that are phytophagous	Proportion of all phytophagous insect species
Orthoptera	Grasshoppers	95	3
Heteroptera	Bugs	90	14
Hymenoptera	Sawflies	11	9
Diptera	Flies	29	15
Lepidoptera	Moths and butterflies	99	26
Coleoptera	Beetles	35	28

and adults of leaf beetles have similar habits, while larvae of wood borers often have adults with different feeding habits. Thus, there is a dichotomy (with some notable exceptions) between species in which the mother chooses for her offspring and species in which all stages are independent.

The great diversity of insects feeding on plants is matched by a remarkable diversity of lifestyles, mouthparts and gut morphological adaptations to the food eaten, cuticular morphology and coloration adapted for crypsis or aposematism, and behavioral adaptations for use of particular plants and escape from natural enemies. Many of the remarkable pictures of insects in popular journals involve surprisingly effective visual and behavioral crypsis.

Insects feeding on plants had their origin early in the history of life on land, with all the major orders that feed on plants today being present 300 million years ago (with the possible exception of Lepidoptera). This means that partitioning of food resources had occurred then, because there were spore feeders, sap suckers, and gall makers as well as miners and external feeders. This can be determined from information provided by fossils of insects and damaged plants and fossilized remains of insect feces. The diversity of insect mouthparts required for the different feeding guilds was established well before the appearance of angiosperms 200 mya, but diversification of families, genera, and species has continued unabated ever since. The diversification is now known to be clearly related to angiosperm diversity and believed to be related largely to the diversity of plant secondary metabolites. These chemicals provide the signals for acceptance or rejection of potential host plants.

Mouthparts and Feeding

Although each of the major phytophagous orders of insects has distinctive biting and chewing or sucking mouthparts, the structures are highly diversified to handle every type of physical problem. For example, caterpillar species feeding on different plants can often be recognized by their mandibular morphology, and species feeding on physically similar plant parts tend to have similar morphology, whether they have a common ancestry or not. Among grasshoppers (Acrididae), those that feed only on grasses have highly specialized mandibles with incisor regions for snipping through the tough parallel veins and molar regions for grinding the tissue.

These highly characteristic mandibles have evolved independently at least eight times during the evolution of grasshoppers. It is probable that the details of mouthpart structure evolve quite quickly to suit changing diet, because it has been found in certain seed-sucking bugs (e.g., *Jadera*) that beak length has evolved to suit different fruit sizes within the past 100 years. Such rapid evolution may reflect the need to process food efficiently to maximize growth as well as the need to ingest food very rapidly to minimize predation risk.

Unless they utilize such structures as seeds and pollen, most phytophagous insects deal with the low protein levels characteristic of much plant tissue by eating relatively large amounts, and the gut throughput rates are high, with food in some cases taking only a couple of hours to pass through the digestive system. Some supplement their diets with carnivory or use symbionts to upgrade the levels of essential amino acids. Aphids, for example, which feed on phloem, tend to be particularly short of certain essential amino acids such as tryptophan, and their symbiotic bacteria commonly have multiple copies of genes involved in making tryptophan, so that the aphid obtains its requirements with the help of the symbionts.

Unlike many vertebrate herbivores, insects that feed on plant tissues often do not make nutritional use of the cellulose, which makes up a large proportion of plant bulk. This may partly reflect the fact that for phytophagous insects, protein is more likely to be limiting than carbohydrate. As heterotherms (animals whose body temperature varies with that of the environment) they are not concerned with the use of diets that are high in calories for maintenance of body temperature. Those species that do use cellulose, such as beetles feeding on wood or termites feeding on dead and decaying plants, often have symbionts that break down the cellulose, releasing nonprotein amino acids as well as sugars. It is possible that the digestion of cellulose in these cases is a mechanism for releasing nonprotein amino acids that are bound to cellulose but that can be converted to useful amino acids for the insects by the resident symbionts.

Apart from the need to obtain sufficient quantities of major nutrients such as protein, insects (like other animals) often require nutrients in suitable ratios. For example, the proportions of protein and carbohydrate required for maximal growth vary among taxa; if a particular resource is limited in one respect the balance can be improved behaviorally. Thus, individual insects with a choice of high-protein/low-carbohydrate food and low-protein/high-carbohydrate food can eat a mixture of both. Studies indicate that species in different taxonomic groups are able to select among the foods available to optimize the balance ingested. This ability depends on a variety of mechanisms, including nutrient feedbacks that influence taste receptor sensitivity to sugars and amino acids, a tendency to move away from a food that has recently proved unsuitable, and a tendency to select foods with new and different flavors following experience on an unsuitable food. In addition, it has been demonstrated that insects can learn to avoid unsuitable food and to associate particular odors with high-quality foods.

THE IMPORTANCE OF PLANT SECONDARY METABOLITES

The great diversity of secondary metabolites in plants profoundly affects the behavior of phytophagous insects and, thus, the evolution of that behavior. These compounds may be repellent, or deterrent after contact. In addition, the deleterious postingestive effects of them enable insects to learn to reject a plant. Such food aversion learning has been demonstrated particularly in grasshoppers.

Many plant secondary metabolites serve as relatively nonspecific attractants or feeding stimulants for insects that feed on plants, although more commonly one or several particular compounds in a plant species serve as highly specific attractants or stimulants for feeding/oviposition by insects adapted to that plant. Thus, specialist phytophagous insects generally have a genetic predisposition to accept plants with a particular chemical or suite of chemicals present, the so-called sign stimuli; indeed, the sign stimuli may act as valuable signals in the sense of improving the speed of decision-making and discrimination by these insects. Some examples of sign stimuli are shown in Table II.

Apart from Orthoptera (specifically grasshoppers), the majority of phytophagous insect species tend to be specialists.

TABLE II   **Examples of Sign Stimuli That Are Particularly Important in Host Recognition by Phytophagous Insects**

Insect	Diet breadth	Chemicals
Junonia coenia, buckeye butterfly	Several families	Iridoid glycosides
Plutella xylostella, diamondback moth	Family Brassicaceae	Glucosinolates
Pieris rapae, imported cabbage worm	Family Brassicaceae	Glucosinolates
Uresiphita reversalis, genista caterpillar	Tribe Genisteae	Quinolizidine alkaloids
Delia antiqua, onion maggot	Genus *Allium*	Disulfides
Chrysolina brunsvicensis, beetle	Genus *Hypericum*	Hypericin (quinone)
Plagioderma versicolora, willow beetle	Genus *Salix*	Salicin (phenolic glycoside)

That is, they feed on just a few species, genera, or tribes of plants. It is common among Lepidoptera (moths and butterflies), for example, to find species that feed on plants in one family, one tribe, one genus, or one species of plant. It appears that the degree of specialization is related to the occurrence of one or a few chemicals that characterize that plant group. In addition, the narrower the host range, the more likely it is that non-host chemicals will repel or deter the insects at relatively low concentrations. Many of these chemicals that reduce feeding or oviposition behavior are not noxious if ingested, suggesting that their role in these cases is more as signals of non-hosts than as signals of toxicity in specifically evolved plant defenses. Nonetheless, there are situations in which plants probably evolved high concentrations of particular chemicals in response to the attack of insects. Insects, over time, would be likely to evolve countermeasures. Thus one can envisage, as many have done, that a chemical arms race between plants and phytophagous insects has occurred (and is occurring).

Diet breadth variation occurs in all phytophagous insect orders, and there is evidence from molecular and other studies that evolution of diet breadth can occur in either direction. What drives these changes has been a subject of much controversy. Included in the hypotheses are the following: arms race coevolution in which specialists are in some way more successful, sequential evolution of insects in which species benefit from the use of specific plant signals to improve behavioral efficiency, adoption of specific plant hosts from which specialists may sequester high levels of particular protective chemicals, and the selection pressure of parasites and predators that involve adoption of limited host plant species as a means of better avoiding attack (for example, by enhanced visual or chemical crypsis).

The study of host plant selection by phytophagous insects has been important in theories of resource use and whether it should be flexible in ecological or evolutionary time. For example, a change in host use may involve a change in specificity (how many different resources are acceptable) or a change in preference (which of a limited number of available resources is ranked highest). A change in specificity could result from a simple change in gustatory or olfactory sensitivity to many plant secondary metabolites or a change in the central nervous system affecting the relative importance of negative inputs from chemoreceptors. A change in preference, however, probably involves a change in receptor conformation or proportions of receptors with different conformations, at the level of the sensillum.

VARIATION IN HOST USE

Advances in how evolutionary changes occur or have occurred are being studied by examining genetic variation currently seen within populations of particular species, geographic variation that occurs in host specificity or preference, and historical changes in host use together with the physiological mechanisms underlying them. Experiential change can influence host preference, in turn altering the nature of selection pressure on those individuals and their offspring.

Host shifts or increases in host range have provided models of evolutionary change including the study of sympatric speciation, whereby populations of an insect species become associated with different hosts. Because many species mate on or near their hosts, gene flow between populations using different hosts drops and speciation becomes possible. An example of this kind of divergence is with the apple maggot *Rhagoletis pomonella* in North America. Some populations moved from the ancestral *Crataegus* to apple and, currently, populations are diverging and evolving additional differences.

The interaction of phytophagous insects and plants is greatly influenced by predators and parasites. This is because they are major mortality factors for the phytophagous insects and yet use the plants themselves for sources of nectar and places to shelter, as well as places where they find their prey and hosts. In many cases, they are attracted to the odors of plants that are being attacked by plant-feeding species. The phytophagous insects that can remain visually or chemically cryptic by being specialists, or sequester plant toxins and become warningly colored, will be selected for by these natural enemies.

AGRICULTURE

The study of phytophagous insects has been very important in agriculture. Probably from ancient times, humans have selected varieties of crop plants that are minimally attacked by insects, and in the last 100 years breeding programs have been important in specifically increasing plant resistance. For example, resistance in rice to the rice brown planthopper (*Nilaparvata lugens*) and resistance in wheat to the greenbug (*Schizaphis graminum*) have resulted from dedicated research effort. Today, genes that express resistance factors against particular insect pests are inserted into some crops. For example, a toxic protein from the insect disease bacterium *Bacillus thuringiensis* can be produced in plants genetically modified by introduction of the bacterial gene.

Research in all areas of the biology of phytophagous insects has found application in agriculture. For example, behavioral studies of attractants has led to the use of traps for specific pests, and the study of antifeedants has had use in development of such materials in crop protection. Plant resistance is sometimes indirect, with plants expressing characteristics that favor natural enemies of the phytophagous species. For example, some crop plant varieties have hairs distributed at such densities that the parasitoids of pest whiteflies are slowed down in their running on a plant to a speed that improves host recognition. In other crops, surfaces lacking a wax bloom enable small predators to run more easily and thus efficiently find small caterpillars.

See Also the Following Articles
Coevolution • Digestion • Feeding Behavior • Genetically Modified Plants • Host Seeking, for Plants • Mouthparts • Symbionts Aiding Digestion

Further Reading
Bernays, E. A. (ed.) (1989–1993). "Insect–Plant Interactions," Vols. I to V. CRC Press, Boca Raton, FL.
Bernays, E. A. (1998). Evolution of feeding behavior in insect herbivores. *BioScience* **48**, 35–44.
Bernays, E. A. (2001). Neural limitations in phytophagous insects: Implications for diet breadth and evolution of host affiliation. *Annu. Rev. Entomol.* **46**, 703–727.
Bernays, E. A., and Chapman, R. F. (1994). "Host-Plant Selection by Phytophagous Insects." Chapman & Hall, New York.
Farrell, B. D., Mitter, C., and Futuyma, D. J. (1992). Diversification at the insect–plant interface. *BioScience* **42**, 34–42.
Fritz, R. S., and Simms, E. L. (eds.) (1992). "Plant Resistance to Herbivores and Pathogens." Chicago University Press, Chicago.
Panda, N., and Khush, G. S. (1995). "Host Plant Resistance to Insects." CAB Int. Wallingford, UK.
Price, P. W., Lewinsohn, T. M., Wilson Fernandes, G., and Benson, W. W. (eds.) (1991). "Plant–Animal Interactions." Wiley–Interscience, New York.
Rosenthal, G. A., and Berenbaum, M. R. (eds.) (1991). "Herbivores: Their Interactions with Secondary Plant Metabolites," 2nd ed., Vols. I and II. Academic Press, New York.
Schoonhoven, L. M., Jermy, T., and van Loon, J. J. A. (1998). "Insect–Plant Biology: From Physiology to Evolution." Chapman & Hall, London.
Stamp, N. E., and Casey, T. M. (eds.) (1993). "Caterpillars: Ecological and Evolutionary Constraints on Foraging." Chapman & Hall, New York.
Strong, L., Lawton, J. H., and Southwood, R. (1984). "Insects on Plants: Community Patterns and Mechanisms." Harvard University Press, Cambridge, MA.

Phytotoxemia

Alexander H. Purcell
University of California, Berkeley

The feeding of insects and mites, especially those that suck fluids from plants, can cause symptoms of distress such as yellowing (chlorosis), silvering, bronzing or necrosis of foliage, wilting or discoloration of shoots, and malformation of leaves, stems, roots, fruits, and other plant organs or tissues. Phytotoxemia (or phytotoxicity) is the production of plant symptoms of distress caused by the reaction of plants to chemicals (toxins) produced by insect feeding.

DEFINITION AND IMPORTANCE

Insects damage plants in many ways, but the most common type of damage is the removal of plant tissues, as in the familiar examples of caterpillars, beetles, and grasshoppers, whose feeding creates noticeable holes or even the wholesale removal of leaves, fruits, or other plant parts. Minute arthropods such as mites and thrips remove plant parts on a smaller scale by evacuating the contents of individual plant cells by their feeding. In contrast to damage caused by the removal of plant tissue, a phytotoxemia or phytotoxic reaction is the reaction of a plant to a chemical toxin introduced by insect feeding. Sucking insects in the orders

Hemiptera (part of which has been called Homoptera in the past), such as plant bugs, aphids, leafhoppers, and psyllids, are the best known insects that cause phytotoxicity by their feeding. Examples of external phytotoxins include chemical air pollutants and pesticides. Abnormal plant growths or deformities called galls can be caused by certain insects.

The symptoms or appearance of insect phytotoxemias are often very similar to symptoms caused by infections with plant viruses or other infectious microorganisms that cause plant disease. Phytotoxemias can be distinguished from diseases caused by pathogens by the following criteria: (1) the capacity of the insect to produce symptoms is characteristic of the species, not restricted to certain individuals within the species; (2) the incidence and severity of symptoms are directly related to the number of insects (even where a single insect may cause symptoms) and the amount of time spent feeding; and (3) plants recover at least partly from further development of symptoms when the insects are removed. In contrast, plant viruses transmitted by insect vectors are not present in all individuals in a vector species; each individual must acquire or inherit the virus. Furthermore, plants develop symptoms of viral disease as a result of infection introduced by a single vector insect that might have transmitted the virus to the plant. Finally, plants will not recover from virus diseases when the vectors are removed from the plant.

The appearance of a phytotoxemia can be small, localized lesions of discolored or dead (necrotic) tissue; the formation of cork, scabs, or pits; premature fall of leaves or fruits; or curling or malformation of plant organs. These are often symptoms of plant diseases caused by pathogenic microorganisms and thus may be easily confused with symptoms caused by pathogens. Symptoms may be systemic, that is, far removed from the site of insect feeding. For example, leaves may be discolored or distorted by phytotoxic feeding in roots or stems.

EXAMPLES AND CAUSES OF INSECT PHYTOTOXICITY

The causes of phytoxemias are not well understood. Mechanical damage to plant tissues caused by sucking insects is not as easy to observe as for chewing insects because the damage occurs at the cellular level and is internal. For example, hopperburn is a condition caused in a variety of plants by small leafhoppers (family Cicadellidae) in the genus *Empoasca*. The most widely studied type of hopperburn is in alfalfa and is caused by the potato leafhopper *Empoasca fabae*. Laceration of the vascular tissues of the plant disrupts the flow of water and plant sap to leaves, but there is also evidence that chemicals within the leafhoppers' saliva induce yellowing and necrosis typical of hopperburn. Simulation of leafhopper feeding with a fine needle causes some but not all symptoms of hopperburn. However, introducing crushed salivary glands of the potato leafhopper more closely simulates hopperburn. Other species of *Empoasca* cause hopperburn in beans in South America *(E. kreameri)* and grapes in Europe and North Africa *(E. lybica)*. The phytotoxicity of the leafhopper *Sophonia rufofascia,* which

FIGURE 1 The myrtle bushes on the left are damaged by the feeding of the leafhopper *S. rufofascia* (inset). (Photograph used with permission of Vincent Jones, Washington State University. Insect photograph by Walter Nagamini, Hawaii Department of Agriculture.)

became established in Hawaii in the 1980s, caused extensive damage to many native plants in Hawaii, threatening the continued survival of some rare species. Its phytotoxicity is similar in many respects to hopperburn: halted growth and necrosis in some plants, yellowing of leaf tissues, and stunting of new growth in other plants (Fig. 1).

The grape phylloxera *(Daktulosphaira vitifoliae)* is a devastating pest of grapevines that ruined millions of hectares of European vineyards in the 19th century and continues today to be a serious pest of vines worldwide. The aphid-like phylloxera have a complex life cycle, with some forms (gallicoles) causing galls on leaves and other forms (radicoles) causing galls and necrosis in roots. Phylloxera damage to roots kills European vines unless they are grafted onto rootstocks of wild grape species native to North America or onto hybrids of tolerant wild species (Fig. 2). Resistant grape species may have fewer phylloxera on their roots than European grapes, but it is their tolerance of grape phylloxera infestation without damage that is the main basis of their usefulness in avoiding phylloxera phytotoxicity. Feeding causes small roots to develop deformities and also kills root cells just beneath the bark. Soil-inhabiting fungi that invade roots may promote the decline of phylloxera-damaged vines.

The feeding of the silverleaf whitefly *(Bemisia argentifolii)* causes silvering of the leaves of squash plants and reduces fruit size and color. Its feeding also causes tomatoes to ripen more slowly than normal. Unidentified chemical components of whitefly saliva seem to be involved, but the plant's reaction is genetically controlled. The potato psyllid *(Paratrioza cockerelli)* causes a condition of potato and tomato called psyllid yellows, which, in addition to causing foliage of affected plants to become yellowish, reduces the size and quality of the fruits. The immature stages of potato psyllid rather than adults cause the most damage. Because individual psyllids may vary markedly in their ability to cause damage and only a small number of psyllids are sufficient to cause symptoms, a virus was suspected

FIGURE 2 Dead grapevines attest to the severity of phylloxera feeding causing damage to roots. The healthy vines next to the declining vines have only the initial stages of phylloxera damage to roots. (Photograph by A. H. Purcell.)

Plant Diseases and Insects

Alexander H. Purcell
University of California, Berkeley

Insects and mites can cause plant diseases directly through feeding or can transport and inoculate viruses and microorganisms such as bacteria and fungi that cause plant disease. Insects are the most important vectors of plant viruses and the main or sole means of spread of many plant pathogens. Plant diseases spread by insect vectors can be crucially important to the profitable production of some crops. Insecticidal or biological control of vectors often does not control the diseases spread by some vectors; consequently, physiological and ecological relationships between vectors and the pathogens they transmit are important to understand.

The direct damage to plants caused by insect feeding (herbivory) generally is considered to be in a separate category from plant disease. However, some insect feeding causes plant responses (phytotoxic reactions or gall formations) that are very similar in appearance to plant diseases caused by microorganisms and may be difficult to distinguish from diseases caused by microbial pathogens such as viruses or fungi.

VECTOR TRANSMISSION OF PLANT VIRUSES

Arthropod transmission of plant viruses illustrates the complexity and variety of relationships between plant pathogens and the arthropods that transport and introduce viruses to the plants. Because plants and plant viruses cannot move by themselves, plant viruses usually have mobile vectors. Most plant virus vectors are insects, mites, or nematodes. Arthropods are not thought to be important in the spread of the smallest plant pathogens—viroids (infectious small ribonucleic acids or RNAs); these are transmitted by gall mites (family Eriophyidae). However, they are important in the transmission of viruses. Viruses consist of protein-coated nucleic acids [either RNA or deoxyribonucleic acid (DNA)] that provide the genetic information for host cells to generate new copies of the virus (replication). In some cases, the viral coat contains lipids or glycoproteins.

Vector transmission of a pathogen is usually characterized with respect to vector efficiency or competence. Efficiency is usually estimated on the basis of how likely transmission is to occur during each opportunity that a vector has for transmission. Usually, this is estimated by determining how many individuals of a particular species are able to transmit a pathogen to plants during a given time interval. An important aspect of transmission efficiency is that only a single insect species is known to transmit some viruses. Such viruses are said to be highly vector specific. Other viruses have less vector specificity; for example, many species of insects within a family or subfamily may be vectors of a particular virus. Such

to be involved in causing psyllid yellows. However, other criteria, such as the direct relationship between damage and the number of psyllids feeding on a plant and the disappearance of symptoms after removing the psyllids, provide evidence against involvement of a virus or other pathogen.

Other insects that cause phytotoxemias include species of aphids, mealybugs, planthoppers, treehoppers, spittlebugs, mirid bugs, coreid bugs, and stink bugs. Blister or gall mites (family Eriophyiidae) are also well known for inducing abnormal growth in many uncultivated plants as well as ornamentals and some crop plants.

See Also the Following Articles
Gallmaking • Plant Diseases and Insects • Salivary Glands

Further Reading
Carter, W. (1973). "Insects in Relation to Plant Disease," 2nd ed. Wiley, New York.
Cohen, S., Duffus, J. E., and Liu, H. Y. (1992). A new *Bemisia tabaci* biotype in the southwestern United States and its role in silverleaf of squash and transmission of lettuce infectious yellow virus. *Phytopathology* **82**, 86–90.
Dixon, A. F. G. (1994). Insect-induced phytotoxemias: Damage, tumors, and galls. *In* "Current Topics in Vector Research," (K. F. Harris, ed.). Springer-Verlag, Berlin.
Granett, J., Walker, M. A., Koesis, L., and Omer, A. D. (2001). Biology and management of grape phylloxera. *Annu. Rev. Entomol.* **46**, 387–412.
Jones, V. P., Anderson-Wong, P., Follett, P. A., Yang, P., Westcot, D. M., Hu, J. S., and Ullman, D. E. (2000). Feeding damage of the introduced leafhopper *Sophonia rufofascia* (Homoptera: Cicadellidae) to plants in forests and watersheds of the Hawaiian Islands. *Environ. Entomol.* **29**, 171–180.

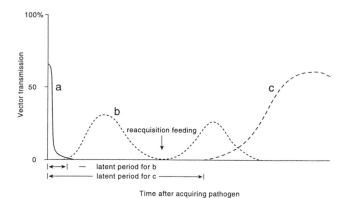

FIGURE 1 Vector transmission efficiency changes over time after acquisition. (a) Nonpersistent transmission; (b) persistent transmission; (c) persistent (over weeks), circulative, or propagative transmission. The latent periods for b and c are indicated. (Reproduced from Daly, Doyen, and Purcell, 1998, "Introduction to Insect Biology and Diversity," with permission of Oxford University Press, Inc., New York.)

viruses are classified as having group specificity or low vector specificity, respectively.

Transmission efficiency often changes dramatically over time. For example, vectors such as the green peach aphid *(Myzus persicae)* transmit viruses such as potato virus Y (PVY) most efficiently within seconds after acquiring the virus from a plant. In addition, after only a few minutes or at most hours of feeding on plants, the aphids (family Aphididae) no longer transmit the virus to plants. In fact, even the aphid's efficiency of acquiring PVY varies with time. Aphids that are fed only for a brief interval on the virus-infected source plant more frequently transmit virus than do aphids that are fed for longer intervals on the source plant. Aphid transmission of PVY is said to be nonpersistent, meaning that virus transmission rapidly declines after acquisition (Fig. 1a), although air-borne aphids lose transmission efficiency at much slower rates than aphids that are feeding on plants or even probing an inert surface such as glass. Nonpersistently transmitted viruses generally have low vector specificity; that is, many aphid species can transmit them. At the other extreme, the green peach aphid transmits potato leaf roll virus (PLRV) only after an interval of many hours or even days after it acquires PLRV from feeding on virus-infected plants; it then continues to transmit for many days (persistent transmission). Only a few aphid species can transmit PLRV. The time required between the vector's acquiring the virus and its successful introduction (inoculation) of the virus to a plant is called a latent period (LP).

Changes in vector transmission efficiency over time (Fig. 1) provide clues as to the nature of the relationship between the vector and the virus and help to explain how vectors transmit the virus. For example, the persistent type of transmission typified by aphid transmission of PLVR often results from the virus having to circulate within its aphid vector before it can be transmitted, explaining the delay or LP required for transmission. A plant virus, such as the aphid-transmitted lettuce necrotic yellows virus (LNYV), that must multiply within a

vector before it can be transmitted will typically have a median or average LP of days or even weeks. LNYV is also persistently transmitted by its aphid vectors after the LP is completed.

In contrast, nonpersistently transmitted viruses such as PVY seem to be carried in or on the needle-like mouthparts of its aphid vectors. Many such viruses produce a viral-encoded polypeptide or small protein "helper factor" that is thought to act as a bridge to aid the attachment of the virus to the aphid vector's mouthparts. The helper factor of one virus may also act as a helper factor for a different virus. Certain viruses may require the presence of another virus in the same plant to be transmitted to another plant. It is not known if the assisting virus or an extraviral helper factor provided by the assisting virus is what aids transmission of the dependent virus. For example, rice tungro disease is caused by the rice tungro bacilliform virus (RTBV), which can be transmitted only by its vector, the rice green leafhopper (*Nephotettix cincticeps,* family Cicadellidae), along with another virus, the associated rice tungro spherical virus (RTSV). By itself, RTBV can cause tungro disease but cannot be transmitted to other plants, and RTSV by itself does not cause a disease in rice.

Some viruses, such as maize chlorotic dwarf virus (MCDV), persist for only hours to days in the blackfaced leafhopper, *Graminella nigrifrons,* vector and are classified as semipersistently transmitted. The shedding of the vector's exoskeleton during molting from one growth stage to another halts the transmission of nonpersistently and semipersistently transmitted viruses. Because the lining of the foregut is shed during molting, PVY and MCDV are thought to be transmitted to plants from a location within the mouthparts or foregut of the vector. Electron microscopy and the labeling of viruses with fluorescently or colloidal gold-tagged antibodies that bind to specific viral proteins have been used to identify areas where viruses accumulate or attach within the foregut. The same approach can be used for viruses that circulate within the vector's body cavity.

The circulative viruses, such as luteoviruses, that are transmitted by aphids and gemini viruses that are transmitted by leafhoppers or whiteflies do not appear to multiply within their vectors; instead, their transmission is thought to entail efficient methods of viral uptake and translocation within the vector's hemocoel (body cavity). Luteoviruses appear to be taken up by a process of endocytosis, in which virus particles are engulfed in a portion of the external cell membrane of intestinal cells, transported into the cell, and expelled from the cell into the body cavity. By processes that are less well understood, viruses can penetrate the membranes surrounding the salivary glands of the vector. Virus particles then enter plants as a result of the vector's salivation while feeding on plants. Luteoviurses such as PLRV can bind to the major protein (symbionin) associated with bacteria (called symbionts) that that live in specialized tissues within aphids and provide required nutrients to their aphid hosts (Fig. 2). The attachment of virus particles to the symbionin molecules may aid in the efficient circulation and persistence of virus from its entry via the aphid's digestive tract

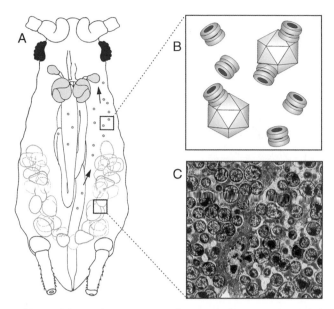

FIGURE 2 The symbionin protein of symbiotic bacteria within aphids binds luteoviruses in the hemolymph of aphids carrying the virus. [From Van den Heuvel *et al.,* (1999) *Trends in Microbiology* **7,** 71–76. Reproduced with permission from Elsevier Science.]

to its entering the accessory lobes of the aphid's salivary glands. Disruption of the symbionts with antibiotic chemicals greatly reduces aphid transmission efficiency of the luteovirus.

Experimental manipulations can result in a "heterologous" virus made up of the genetic information (DNA) of one virus encased in the protein coat of another virus. A heterologous virus consisting of the DNA of a whitefly-transmitted virus encased in the coat of a leafhopper-transmitted virus can be transmitted by leafhoppers from plants that contain the heterologous virus to a new plant. However, once the virus begins to replicate in the plant inoculated by the leafhopper, it constructs the proper coat protein directed by the DNA of the whitefly-transmitted virus. Leafhoppers no longer can transmit this virus from plant to plant; only whiteflies (family Aleyrodidae) transmit it. Experiments of this sort demonstrate that the viral protein coat, not the viral genome, determined the vector specificity of the virus. The protein coat probably does this by mediating passage of viral particles through the midgut and the salivary glands of the leafhopper, even though the viral DNA encoded whitefly-transmitted virus. Once the whitefly-transmitted virus replicated and produced its corresponding protein coat, only whiteflies could transmit it from plant to plant.

Some of the persistently transmitted plant viruses, such as the reoviruses, replicate within their insect vectors. It is remarkable that the same virus can subvert the genetic and protein processing systems of both plant and animal cells for viral replication. Although vectors normally acquire viruses by feeding on virus-infected plants, some plant viruses can invade the developing eggs or embryos within a female vector insect. The rice dwarf virus (RDV) is an example of a reovirus that passes from virus-infected female rice green leafhoppers to their

offspring by this route. Invariably these transovarially transmitted viruses, such as RDV, multiply within their vectors, yet most viruses that multiply in vectors are not transmitted transovarially. The tomato spotted wilt virus (TSWV) is unusual because it can be acquired only by immature stages of thrips. An adult thrip can transmit this virus only if it fed as a nymph on a plant with TSWV. Thus, the vectors of TSWV have vector specificity not only for particular vector species but also for the immature stages, at least in the acquisition phase.

Insects that feed on plants' vascular tissues (xylem and phloem) appear to be the most common vectors of plant viruses. The greatest number of insect vectors of plant viruses are sucking insects in the order Hemiptera. Within the Hemiptera, aphids (Aphididae) transmit the greatest number of different plant viruses, followed by whiteflies (Aleyrodidae), leafhoppers (Cicadellidae), and planthoppers (Fulgoroidea). Mealybugs (Pseudococcidae) and various other hemipteran, families have species that are virus vectors as well. Thrips (order Thysanoptera) transmit only a few viruses, but these can be of great economic importance worldwide. Mandibulate, or chewing insects, mostly beetles (order Coleoptera), transmit a relatively small number but varied types of plant viruses. Among the mites, the minute bud or gall mites (family Eriophyidae) are the most important virus vectors.

ARTHROPOD VECTORS OF BACTERIAL PATHOGENS

Unlike viruses, most bacterial (prokaryotic) diseases of plants do not require insects as vectors, relying instead on rain, wind, soil, seed dispersal, or other means of transport and entry to plants. However, insect vectors do contribute to the spread of some bacterial pathogens of plants. Fire blight is an important bacterial disease of pome fruits, such as pears and apples, in which flower-visiting insects may have an important role in disseminating the causal bacterium *(Erwinia amylovora)* among blossoms. Insects are not essential, however, for fire blight to spread within plants once the bacteria are established, and there is little vector specificity among flower-visiting insects. Bacteria that rot potatoes *(Erwinia caratovora)* may be transported from infested potato tubers to uninfested tubers by flies whose maggots feed on plant roots or seeds beneath the soil. There is much greater vector specificity in corn flea beetle transmission of the bacterium *(Erwinia stewartii)* that causes Stewart's wilt of corn and in cucumber beetle transmission of the bacterium *(Erwinia tracheiphila)* that causes cucurbit wilt, which is an important disease of melons, squash, and cucumbers. The bacteria enter feeding wounds made by the beetle vectors, but not much is known of how the beetles introduce the bacteria into plants. Overwintering adult beetles provide an important way for these bacteria to survive the winter season without host plants.

Some bacterial pathogens are specialized parasites of plant vascular systems and require insect vectors for plant-to-plant movement and to enter and infect plants. These bacterial pathogens are specialized for vector transmission and for living in plant vascular systems. Examples are the mollicutes (bacteria that lack a rigid cell wall) that live exclusively in the nutrient-rich phloem tissues. A few bacterial pathogens with rigid cell walls, such as the bacterium that causes citrus greening disease, also specialize in living within plant phloem sap. The citrus greening bacterium is transmitted by psyllids (superfamily Psylloidea). The mollicute plant pathogens include phytoplasmas and spiroplasmas. Most of the helical-shaped spiroplasma pathogens of plants, such as the spiroplasma that causes citrus stubborn disease *(Spiroplasma citri)* and the corn stunt spiroplasma *(Spiroplasma kunkelii),* can be cultured on artificial media. So far, none of the phytoplasma (formerly known as mycoplasma-like organism) plant pathogens have been cultured. Examples of economically important phytoplasmas are aster yellows phytoplasma in lettuce, carrot, celery, and other flower and vegetable crops and X-disease phytoplasma in stone fruits such as peach or cherry. Lethal yellowing disease of palms has been a major factor in killing coconut palms in Africa and the Caribbean. Both phytoplasmas and spiroplasmas are more specialized for parasitizing insects rather than plants because they can successfully colonize and, more importantly, can be transmitted by only a few species of insects. The most important vectors of mollicute plant pathogens are leafhoppers and planthoppers, but psyllids are an important third group of Hemiptera that are vectors. The pear psylla *(Cacopsylla pyicola)* transmits the pear decline phytoplasma, which causes the widespread pear decline disease. Typically, only one or a few species of insects within one of these families have been shown to transmit any particular mollicute. In contrast to their high degree of vector specificity, phytoplasmas and spiroplasmas can parasitize a typically wide range of plant species if the vectors can feed successfully on the plants.

Transmission appears to require that the mollicutes be taken up by vector feeding, penetrate the gut and multiply within the vector's body cavity, enter the salivary glands, and be expelled with saliva during vector feeding into functioning phloem tissues. Thus, it not surprising that vector transmission of various phytoplasmas or spiroplasmas requires a latent period ranging from 1 to over 4 weeks. The length of the latent period may be very sensitive to temperature, probably because the mollicutes must multiply within the vector for transmission to occur and multiplication is temperature sensitive.

Vector-borne bacterial species that parasitize the water-conducting part of the plant's vascular system (xylem) are less numerous but cause some important plant diseases. One such pathogen is *Xylella fastidiosa,* best known as the cause of Pierce's disease of grapes, but other strains of this bacterium cause important other diseases of citrus, coffee, peach, and other crop and forest plants. Sucking insects in several families that feed primarily on xylem sap are *Xylella* vectors. This includes sharpshooter leafhoppers in the subfamily Cicadellinae of the leafhopper family Cicadellidae and spittlebugs (family Cercopidae). Vectors appear to transmit the bacterium from their foregut without any required latent period, but continue to transmit for weeks or even months as adults. An immature vector (nymph) stops transmitting after molting its exoskeleton. Sumatra disease of cloves in Indonesia, caused by the xylem sap-inhabiting bacterium *Pseudomonas syzygii,* is spread by tube-building spittlebugs (family Machaerotidae), which are also xylem sap feeders.

Arthropods and Fungal Plant Diseases

The fungi are the most varied, common, and important plant pathogens, but the great majority of fungal pathogens do not require mobile vectors such as insects or mites. Instead fungal pathogens disperse to plants mainly in wind, rain, or soil. A large variety of fungi colonize plant wounds, including those made by arthropod feeding. However, some fungi are specialized for transmission by insect vectors.

Dutch elm disease is the best known example of a fungal disease of plants transmitted by an insect vector. The causal fungus, *Ophiostoma ulmi,* grows into a spore-bearing fungal mass (mycelium) under the bark and into the water-conducting woody tissues of elms. Adult bark beetles, such as the European elm bark beetle *(Scolytus multistriatus),* are especially attracted to distressed elms or freshly cut elm logs. The adult beetles excavate a tunnel by feeding beneath the bark and deposit eggs along the tunnel. Beetle larvae hatch from the eggs, tunnel farther under the bark, pupate, and then emerge as adults the following year. The Dutch elm disease fungus grows throughout brood chambers excavated by the beetles and produces sticky spores that attach to the body and mouthparts of the adult beetles that bore out of the bark to exit the tree. The beetles transmit the fungal spores to wounds they inflict while feeding on elm twigs. The fungus gradually spreads from the point of infection into the larger branches of the tree and then to the tree's trunk, where its action on the woody tissues eventually kills the tree. In very cold climates of North America, the native elm bark beetle *(Hylurgopinus rufipes)* is the main Dutch elm disease vector. Its transmission of fungal spores to elms leads to more rapid development of disease because it principally feeds on the trunk and large branches of the elm tree rather than small branches. Oak wilt disease, caused by the fungus *Ceratocystis fagacearum,* is spread by sap beetles (family Nitidulidae) that vector the spores from oozing cankers on diseased trees to fresh wounds on trees to which these beetles are attracted.

Some insects can create wounds, which fungi can then colonize without transport by the insects. Yet, even though insects in these cases are not vectors of the fungi, they can be important in determining how severe fungal infestation becomes. An example is a variety of fungi that can colonize the feeding wounds of caterpillars that feed on maize or peanuts. Some of these fungi (notably *Aspergillus* species) can produce powerful toxins, called aflatoxins, that sicken or even kill animals that are sensitive to the toxins. Preventing insect damage to grain in the field or in storage is an important step in preventing high levels of aflatoxins.

Insects as Vectors of Trypanosomes and Nematodes

Plant diseases caused by trypanosomes are not as well understood as trypanosome diseases of animals and humans, such as sleeping sickness. Trypanosomes are protozoans of variable body shape, which depends on their developmental stage and environment. Most insect-associated trypanosomes have stages that are elongated or leaf-like and are propelled by a centrally attached flagellum. Milkweed bugs (family Lygaeidae) transmit trypanosomes to milkweeds, in which they harmlessly occupy the interconnected latex system. A variety of other plant-parasitic trypanosomes inhabit the phloem systems of their plant hosts, causing severe disease. Phloem necrosis disease of coffee in northern South America and heartrot of palms are spread by sucking bugs in the family Pentatomidae and other related families.

Wood-boring beetles are vectors of pinewood nematodes (*Bursaphelenchus xylophilus*) that cause a severe disease of conifers in Asia and, more recently, North America. The juvenile nematodes enter the tracheae (breathing tubes) of adult long horned beetles (family Cerambycidae), as the beetles emerge from the dead trees in which they breed. As the beetles bore into new trees, the nematodes disperse from the beetles into the tree's woody tissues, causing blockage of the water-conducting system.

Control of Diseases Spread by Arthropods

The most obvious first step in controlling diseases caused by insect-borne pathogens might seem to be the elimination of vectors with insecticides. Although they are very effective in some situations, insecticides usually are not the best tools for control of most vector-borne pathogens of crops. The most effective approaches combine multiple methods and integrated approaches.

Sanitation to eliminate nearby sources of the pathogen (usually diseased plants) reduces the number of infective vectors near the crop to be protected. Physical isolation to prevent disease spread may be achieved in some cases by growing susceptible crops as little as 100 m or so from infected sources, but normally much greater isolation or separation is required. Therefore, area-wide cooperation may be necessary for sanitation to control some diseases.

The use of virus-free plants is probably the most widespread method of preventing virus spread. This is especially important for perennial plants such as fruit trees or plants propagated from vegetative cuttings, such as potatoes, strawberries, or sugarcane. Heat therapy or antiviral chemical treatments may be used to produce virus-free new plant growth that can be grafted or rooted to create virus-free plants for nursery propagation. As new infections of virus appear in trees, the trees can be removed; for some diseases, such as swollen shoot disease of cacao, trees within a specified radius of a newly diseased tree are also removed. Some viruses are transmitted via seeds from infected plants, and control may be based on planting virus-free seed. Lettuce mosaic virus (LMV), for example, is best controlled by using seed with less than 1 in 10,000 seedlings infected with LMV via seed. Above this threshold, aphid spread of the nonpersistently transmitted LMV will be economically damaging. The production of virus-free seed may require growing the seed crop in isolated areas that are otherwise free of the crop and the viruses of concern. Insuring virus-free foundation plants or seed to seed producers or nurseries is an important service usually provided by government or grower cooperatives.

Establishing a period of the year that is completely free of the targeted crop may reduce virus spread by breaking the transmission cycle. For example, this approach has proven effective for yellows viruses of sugar beet and celery mosaic virus, both of which are aphid transmitted. The growing of these crops for several months during the year is prohibited on an area-wide basis to prevent the carryover of virus in crop plants from one season to another. New fast-maturing varieties of rice that allowed multiple crops per year rather than a traditional single crop introduced new problems with long-established viruses because virus-infected crops coexisted next to newly planted fields. The solution was to have at least one period of the year free of all rice crops. Peak infective periods can be avoided for some virus diseases by planting after peak vector flight periods if the late planting still produces a profitable crop.

Removing diseased plants as soon as symptoms appear is an important step in preventing further spread of Dutch elm disease. The bark beetle vectors of the fungus that causes the disease are attracted to weakened trees, so removing diseased elms reduces populations of the beetles as well as reducing the percentage of beetles carrying fungal spores. Sanitation may also limit the spread of phytoplasma diseases of trees, such as X-disease of stone fruits like cherry.

The effects of the crop environment on vector flight behavior or plant choice may be effective in slowing virus spread, even for nonpersistently transmitted viruses. For example, reflective plastic sheeting used as a mulch (ground cover) repels aphids from landing on melon crops. The resulting delay in virus infection usually prevents the virus from reducing average fruit quality or yield. Sprays of 1% emulsions of paraffin oils on peppers or tomatoes reduces aphid transmission of nonpersistently transmitted viruses. Plants must be sprayed frequently with oil sprays to cover new growth because the oil directly affects the inoculation and acquisition of virus by aphids.

Insecticides generally are most effective in controlling disease spread where vector-borne pathogens are persistently transmitted or are spread mostly within the crop (termed secondary spread) rather than being carried into the crop from outside sources (primary spread) or where the most important vectors reproduce on the crop. Insecticides are usually not effective against nonpersistently transmitted viruses unless they quickly reduce or inhibit vector probing on treated plants. Some pyrethroid insecticides can rapidly intoxicate aphids after they land on plants, reducing even nonpersistent virus transmission. Examples of successful insecticidal control of vectors that achieved economic control of viruses are the persistently transmitted barley yellow dwarf luteovirus in grain crops, potato leaf roll virus, and leafhopper-transmitted beet curly top virus. These are all viruses that are acquired or inoculated into plants only during relatively long feeding probes by aphids. Insecticides also reduce the spread of the leafhopper-transmitted aster yellows phytoplasma if vector numbers are not too high.

Genetically based plant resistance to pathogens or tolerance of infection without loss of yield provides the basis for the most successful control programs for vector-borne plant pathogens. A drawback is that breeding resistant plant varieties that are commercially acceptable has proven to be difficult or impossible to achieve for some crops. Molecular methods of introducing novel genes for resistance to viruses directly into crop plants promise to provide resistance to virus diseases for which no genetic resistance has yet been discovered.

See Also the Following Articles
Borers • Gallmaking and Insects • Genetic Engineering • Phytotoxemia

Further Reading

Camargo, E. P., and Wallace, F. G. (1994). Vectors of plant parasites of the genus *Phytomonas* (Protozoa, Zoomastigophorea, Kinetoplastida). *Adv. Dis. Vector Res.* **10**, 333–359.

Carter, W. (1973). "Insects in Relation to Plant Disease," 2nd ed. Wiley, New York.

Gray, S., and Banerjee, N. (1999). Mechanisms of arthropod transmission of plant and animal viruses. *Microbiol. Mol. Biol. Rev.* **63**, 128–48.

Hadidi, A., Khetarpal, R. K., and Koganezawa, H. (eds.) (1998). "Plant Virus Disease Control." APS Press, St. Paul, MN.

Harris, K. F., and Maramorosch, K. (eds.) (1982). "Pathogens, Vectors, and Plant Diseases: Approaches to Control." Academic Press, New York.

Matthews, R. E. F. (1992). "Fundamentals of Plant Virology." Academic Press, New York.

Perring, T. M., Gruenhagen, N. M., and Farrar, C. A. (1999). Management of plant viral diseases through chemical control of insect vectors. *Annu. Rev. Entomol.* **44**, 4457–4481.

Pirone, T. P., and Blanc, S. (1996). Helper-dependent vector transmission of plant viruses. *Annu. Rev. Phytopathol.* **34**, 227–247.

Purcell, A. H. (1989). Homopteran transmission of xylem-inhabiting bacteria. *Adv. Dis. Vector Res.* **6**, 243–266

Purcell, A. H. (1998). Insects and microbes. *In* "Introduction to Insect Biology and Diversity" (H. V. Daly, J. T. Doyen, and A. H. Purcell, eds.), Chap. 13. Oxford University. Press, New York.

Van den Heuvel, J. F. J. M., Hogenhout, S. A., and van der Wilk, F. (1999). Recognition and receptors in virus transmission by arthropods. *Trends Microbiol.* **7**, 71–76.

Plant–Insect Interactions

J. Mark Scriber
Michigan State University

The interactions of plants and their herbivores center upon the primary (nutritional) and secondary (allelochemical) composition of plants. The fundamental limitations of insect use of plants as food involve nutrients (nitrogen/protein, waters, lipids, and various minerals) as well as various classes of secondary chemical defenses (including alkaloids, cyanogenic glycosides, glucosinolates, terpenoids, phenolics, phytoecdysteroids, and polyacetates). The location ("findability") and utilization (suitability) of plant parts as insect food may depend on phenotypic variations induced by previous herbivores and microbes, as well as a wide array of interactions involving environmental factors such as nutrient availability, light regime, water, temperature, carbon dioxide, and various pollutants. Seasonal changes in plant growth, reproduction, and chemical/physical defenses also are important. Natural declines during plant maturation in the concentration of many low-molecular-weight allelochemicals (often called qualitative defenses) are contrasted with other higher molecular weight chemicals (such as tannins, lignins, and fiber; sometimes called quantitative defenses). In leaves of plants, a general pattern of decline in the concentrations of total nitrogen, water, and many qualitative chemical defenses usually accompanies leaf maturation. The phytochemical suitability of leaves for insect herbivores (the leaf-chewing guild in particular) also has genetically based biochemical variation. Together these factors affect the physiological and ecological suitability of the plant for supporting herbivore feeding, growth, survival, and reproduction.

HOST PLANT RESISTANCE AND INSECT COUNTERADAPTATIONS

In natural terrestrial communities, approximately 10% of the annual plant production on average is consumed by herbivores, a percentage that is generally greater than the plant biomass allocated to reproduction. In addition to the well-known defensive structures of thorns, barbs, spines, trichomes, hairiness or fuzziness, and physical toughness, plants possess a large array of chemicals that defend against the herbivore and pathogen enemies. Artificial selection also has produced insect-resistant genotypes and cultivars of crop plants, which has helped reduce the reliance on broad-spectrum synthetic pesticides. Plant breeders have recently been able to use techniques of molecular biology to incorporate new arrays of biochemical or microbial "pesticides" for plant defenses that have not been previously evolved by the plants (but that may occur naturally in the plant environment or even on the leaf phylloplane).

Insects may respond to these secondary plant products physiologically (e.g., by sequestration or enhanced excretion rates), biochemically with resistance (e.g., via target site insensitivity or enzymatic detoxification), or behaviorally (e.g., by reducing exposure or consumption by changes in feeding behavior). Many of these phytochemicals may be used by adapted herbivores that sequester the bioactive compounds in their wings or other body parts where they may serve a protective function from herbivore enemies (such as in distasteful models with aposematic or warning coloration) and often in various insect mimicry complexes.

PLANT DEFENSE AND HERBIVORE OUTBREAK THEORY

An understanding of the full array of potential insect–plant interactions is probably beyond comprehension. The geographic, altitudinal, and seasonal variations in plant chemistry in even a single species and the associated responses of insect herbivores (each species with its own geographical and genetic variation) make the task truly daunting. However, out of this complexity, ecologists have attempted to identify general patterns and organizing principles.

It is especially useful to examine the development of a series of general explanatory hypotheses or proposed models of insect herbivory, population dynamics, and plant defenses. These hypotheses have been mostly generated since the 1970s and are not mutually exclusive and have often built upon the theories of their predecessors. Details of the historical development of plant defense theory were nicely summarized in 1997 by Price.

Climatic Release Hypothesis

Outbreaks of insect herbivores on plants following periods of atypically warm, dry weather are numerous and have been documented for nearly a century. Up to the 1950s, herbivorous insect life table analysis suggested bad weather, lack of food, or lack of their natural enemies (predators, parasites, or diseases) were the primary insect population regulators. The indirect effects of the climate and abiotic environment as mediated through changes in host plant nutritional quality will likely be of increased significance in the near future given increasing concentrations of certain atmospheric gases (e.g., carbon dioxide), acid rain, global warming, and increased pollution.

Plant Stress Hypothesis

The connection between warm, dry weather and insect population eruptions, combined with similar outbreaks on plants in poor, dry soils, led to the hypotheses that water stress in plants may affect the availability of soluble nitrogen (especially for those in the plant-sucking or sap-feeding guilds). Nitrogen generally limits insect herbivores and population growth rates.

Plant Apparency Hypothesis

Following study of insect herbivores on oak trees in England and those on the cabbage family (Cruciferae) in North America, P. P. Feeny noted the divergent patterns of chemical defense used by these two plant types. The tree leaves were composed of relatively high concentrations (2.5–5% dry weight) of compounds believed to be digestibility reducers (tannins, lignans, resins, cellulose, silica). In contrast, herbaceous crucifers (forbs) had low concentrations (usually lower than 1%) of biosynthetically "less expensive" toxins (such as mustard oil glycosides or other low-molecular-weight chemicals such as alkaloids, cyanogenic glycosides, and coumaric acids).

Herbivore food was prevalent and predictable (i.e., apparent) in trees and mature plant leaves containing these convergent digestibility reducers, or quantitative defenses or hurdles to herbivores, whereas the food resources for forb-feeding herbivores seemed less predictable, with divergent, qualitative (toxic) barriers in annual plants/herbs and very early immature (i.e., unapparent) plant parts. The "bound-to-be-found" trees were late successional, frequently in pure stands, large, and long-lived, with large amounts of general chemical defenses (effective against specialists as well as generalists) that acted in a dose-dependent (i.e., quantitative) manner. In contrast, the annual plants were short-lived, hidden from enemies in space and in time (unapparent), and defended by small quantities of qualitative allelochemicals such as mustard oils that repel nonspecialized insects (but are not effective against adapted herbivores) and that are effective at very low levels (basically not dose dependent).

The explanatory value of the apparency concept has been questioned because of the difficulty in its quantification. Most plants and plant parts have a dynamic continuum of both quantitative and qualitative chemical defenses, as well as phenological changes in the nutritional quality of leaves (as indexed, for example, by leaf water and total leaf nitrogen concentrations). Tannins were also shown not to be the general dose-dependent, digestibility-reducing chemicals they were originally believed to be. Instead, tannins evoke a wide variety of physiological effects such as increased mortality, decreased consumption rates, histopathological effects in the gut, and elevated metabolic costs for insect herbivores.

Induced-Defenses Hypothesis

The occurrence of phytochemical induction with leaf damage has been observed since the late 1970s. Plant-to-plant chemical communications and plant-to-insect parasite/predators have subsequently been included as multitrophic-level chemical synomones (plant volatiles that benefit both the sender and its receiver). Herbivore-damaged plants have been shown to provide carnivorous enemies of insect herbivores with important volatile chemical cues. These chemicals are detectable from a distance and aid natural enemies in locating suitable herbivore prey. However, plant volatiles from particular herbivores have not conclusively been documented to be consistently capable of providing critical species-specific herbivore information to the predator/parasites. Nevertheless, it is clear that multitrophic-level interactions are directly and indirectly influenced by damage-induced chemical responses of plants.

Expanded Hypotheses on Resource Availability and Plant Vigor for Defense against Herbivores

Although the value of the apparency hypotheses was weakened by difficulties in empirically assessing apparency, it nonetheless stimulated a great deal of research. Some of this resulted in promising alternatives relating to a causal relationship between "resource availability" and plant allocations to antiherbivore defense.

Reduced nitrogen availability for plants usually resulted in reduced nitrogen-based defenses (usually toxins), but not necessarily in reduced carbon-based defenses (digestibility reducers). Thus, on nutrient-poor or late-successional sites, it was suggested that the inherently slow growth rates of plants may select for more carbon-based herbivore defenses (e.g., resins or phenolics) that could be reduced with fertilization with nitrogen (with corresponding increases in the nutritional value for herbivores). Low carbon conditions (e.g., shade and reduced photosynthesis) may result in slow growth despite high nitrogen, which could then be used for N-based defenses (e.g., alkaloids, cyanogenic glycosides).

These hypotheses about the carbon-nutrient balance and resource availability for slow-growing and fast-growing plants received support from many researchers in the 1980s. In many instances, predictions of the apparency, resource availability, and growth–differentiation balance hypotheses are in agreement; for example, trends in the types of chemical defense of early successional plants and late successional plant communities are congruent. However, equally apparent plants in resource-rich and resource-poor environments suggest that the resource availability hypothesis may have greater explanatory power than the apparency concept because the fast growers in resource-rich environments (nutrients and light) seem to support high herbivory (and may be predisposed to rapidly recyclable defenses such as alkaloids or other toxins) that can contrast with slow growers in the tropics, temperate zone, and arctic communities. This preference of herbivores for fast-growing plants has led to the suggestion that interactions may relate most simply to "plant vigor" (as a general hypothesis) to explain not only persistent differences in plant defenses between species, but also quality differences within a single plant species, genotype, or even individual.

In summary, despite the different roles of an overwhelming diversity of secondary plant chemicals, the fundamental limitations on herbivore growth rates seem to relate to the nutritional suitability of the insect food. Different leaf water and leaf nitrogen contents for different plant tissues correlate well with insect growth rates and efficiencies for most guilds and hundreds of different herbivore species. These indices of resistance can be induced by herbivory itself, as well as being constitutive.

Voltinism-Suitability Hypothesis

In addition to nutritional value, plant tissue suitability to an insect herbivore also depends on the degree of physiological and behavioral adaptations to a variety of plant secondary chemicals as well to leaf water, nitrogen, lipids, vitamins, and minerals. The question even arises as to whether low concentrations of nutrients may sometimes actually serve as plant defenses. At certain latitudes or in geographically localized cold pockets, seasonal thermal unit constraints (degree days as a resource) can determine whether an additional insect herbivore generation is feasible in any given year, depending on its selection of the most nutritional host plant species, which varies with the timing of leaf bud break and phenological (seasonal) patterns, which differ at various locations. Thus, abiotic factors have been shown to affect host-plant choice (acceptability) and host-plant suitability for herbivores. High nitrogen and high water content generally reflect the most rapid leaf and cell growth and presumably plant vigor as well. Insect growth performance usually correlates well with these plant quality indices.

The range of host plants utilized at a given latitude (or local climatic zone) may be the result of natural selection in relation to these abiotic factors. Thermal constraints for the summer growing season (as in Alaska or in similar localized cold pockets in the continental United States) contrast ecologically with thermally relaxed zones (i.e., where choice of either excellent or poor host-plant species or leaves does not influence the possibility of an extra generation per season). The difficulty in the voltinism-suitability model is that a good host typically is more than just the plant species, more than its allelochemical acceptability, more than its nutritional suitability, and more than the abiotic thermal regime in which the herbivores are trying to optimize their growth and survival. The biotic community of natural enemies (e.g., enemy-free space as a resource) must also be considered as a critical determinant of the real ecological/evolutionary suitability. The relative roles in plant defense played by natural enemies, weather-induced stress, herbivore-induced stress, and various abiotic factors remain complex, with many unique and dynamic variations on the suitability hypothesis.

EVOLUTIONARY HISTORY AND PHYTOCHEMICAL FUTURE

Coevolutionary or reciprocal changes between plants and insects are the foundation of numerous phytochemical defense theories. However, there is surprisingly little direct evidence that insects select for plant phytochemical defenses. Most insect–plant interactions are diffuse without mutual counteradaptations. They will be, at best, a geographic mosaic with isolated and dynamic hot spots. Additional studies of different geographical populations (with and without herbivore selection pressure), plant and herbivore genetic analyses, phytochemical dynamics in relation to abiotic factors, and historical biogeography all seem to be warranted and critically needed. The ecologically enigmatic problem seems to be that our understanding of both plant resistance and insect counteradaptations ultimately depends on the identification of specific molecular pathways and an elucidation of the relative roles of genetic and environmentally induced variation between interactive populations.

INSECT HERBIVORES CAN BENEFIT PLANTS

It has been generally accepted that insect herbivory results in plant tissue damage that is detrimental for plant growth, survival, or reproduction. It has been argued that this is not true in all situations and that insect herbivory may sometimes be beneficial for plant productivity. Plant responses to herbivory are determined by many different habitat, plant, nutrient, and herbivore specifics that are biologically variable.

The frass fall and uneaten leaf pieces that reach the ground from insect herbivores release nutrients for plant growth throughout the season and delay plant leaf senescence. The carbohydrates in aphid honeydew drippings can stimulate soil microbes and enhance nitrogen fixation, which benefits the adjacent plants. Such enhancement of nutrient cycling by herbivore feeding can be important in regulation of ecosystem productivity, especially in grasslands and forests. Although many crops and other plants can sustain 30 to 40% defoliation with little obvious impact on production, such feeding, if repeated annually or if on flowers and/or seeds, can be much more damaging.

Insects provide direct nutritional benefit to some plants. Carnivorous plants that "digest" insects and use chemicals from their bodies for nutrients (especially nitrogen) are represented by many species, including pitcher plants, bladderworts, Venus flytraps, and sundews. These carnivorous plants are often found in soils that are low in available nitrogen, which may have been an important selection pressure for the evolution of these botanical life history traits.

SUMMARY

Insect–plant interactions involve a wide array of biotic and abiotic environmental influences as well as geographical and temporal variations built upon the diverse genetic foundations and inducible phenotypic plasticity of species, populations, and individuals. It therefore seems very appealing when theories arise that seem to have predictive power for these complex interactions. Such complexity is amplified when the variety of insect feeding guilds and variations in response of different plant parts and tissues are considered. The relationships between normal phenological changes in plant leaf (or part) composition throughout the growing season, carbon-nutrient stress, mineral nutrition, plant vigor, phytochemical induction of resistance in damaged/diseased leaves, increases in certain atmospheric gases, global warming, metabolism of different forms of carbon, and annual versus perennial growth forms need coordinated biocomplexity studies. With such knowledge, the suitability of such plant tissues for insect and other herbivores (or the resistance of plants to their enemies) may become much more predictable.

See Also the Following Articles

Honey • Host Seeking, for Plants • Nutrition

Further Reading

Ayers, M. P., Clausen, T. P., MacLean, S. F., Redman, A. M., and Reichardt, P. B. (1997). Diversity of structure and antiherbivore activity in condensed tannins. *Ecology* **78**, 1696–1712.

Bernays, E. A. (1989–1994). "Insect–Plant Interactions," Vols. 1–5. CRC Press, Boca Raton, FL.

Dicke, M. (1999). Are herbivore-induced plant volatiles reliable indicators of herbivore identity to foraging carnivorous arthropods? *Entomol. Exp. Appl.* **91**, 131–142.

Fritz, R. S., and Simms, E. (1992). "Plant Resistance to Herbivores and Pathogens: Ecology, Evolution, and Genetics." University of Chicago Press, Chicago.

Herms, D. A., and Mattson, W. J. (1992). The dilemma of plants: To grow or to defend. *Q. Rev. Biol.* **67**, 283–335.

Karban, R., and Baldwin, I. T. (1997). "Induced Responses to Herbivory." University of Chicago Press, Chicago.

Price, P. W. (1997). "Insect Ecology," 3rd ed. Wiley, New York.

Price, P. W., Lewinsohn, T. M., Fernandes, G. W., and Benson, W. W. (1991). "Plant Animal Interactions: Evolutionary Ecology in Tropical and Temperate Regions." Wiley, New York.

Rosenthal, G. A., and Berenbaum, M. R. (1991 and 1992). "Herbivores: Their Interactions with Secondary Plant Metabolites," Vols. 1–2. Academic Press, New York.

Schoonhoven, L. M., Jermy, T., and van Loon, J. J. A. (1998). "Insect–Plant Biology: From Physiology to Evolution." Chapman & Hall, New York.

Scriber, J. M. (1984). Insect–plant interactions: Host plant suitability. *In* "The Chemical Ecology of Insects" (W. J. Bell and R. T. Cardé, eds.), pp. 159–202. Chapman & Hall, London.

Scriber, J. M. (2002). Latitudinal and local geographic mosaics in host plant preferences as shaped by thermal units and voltinism in *Papilio* spp. (Lepidoptera). *Eur. J. Entomol.* **99**, 225–239.

Scriber, J. M., and Lederhouse, R. C. (1992). The thermal environment as a resource dictating geographic patters of feeding specialization of insect herbivores. *In* "Effects of Resource Distribution on Animal–Plant Interactions" (M. R. Hunter, T. Ohgushi, and P. W. Price, eds.), pp. 429–466. Academic Press, New York.

Scriber, J. M., and Slansky, F., Jr. (1981). The nutritional ecology of immature insects. *Annu. Rev. Entomol.* **26**, 183–211.

Slansky, F., and Rodriguez, J. G. (1987). "Nutritional Ecology of Insects, Mites, Spiders, and Related Invertebrates." Wiley, New York.

Slansky, F., and Scriber, J. M. (1985). Food consumption and utilization. "Comprehensive Insect Physiology, Biochemistry, and Pharmacology" (G. A. Kerkut and L. I. Gilbert, eds.), Vol. 4, pp. 87–163. Permagon, Oxford.

Plecoptera
(Stoneflies)

Kenneth W. Stewart
University of North Texas

Stoneflies comprise a hemimetabolous order of 16 families and more than 2000 species of aquatic insects distributed on all continents except Antarctica and most major islands except notably Cuba, Fiji, Hawaii, and New Caledonia. They are primarily associated with running water, where nymphs inhabit mineral or organic substrates of streambeds, and the winged adults rest throughout their seasonal lives in streamside microhabitats such as rocks, moss, debris, leaf packs, and riparian vegetation. A few species occur in waveswept substrates of cold alpine and boreal lakes or in intermittent streams.

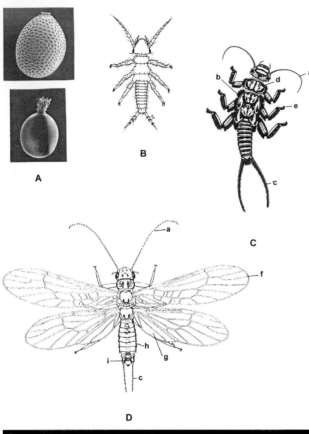

FIGURE 1 Life stages of stoneflies. (A) Eggs. (B) Hatchling (first instar). (C) Late instar. (D) Adult male. a, antennae; b, gills; c, cerci; d, prothorax; e, tarsal claws; f, forewing; g, hindwing anal area; h, abdomen; i, epiproct of male genitalia; (E) Adult of the stonefly *Calineuria californica* (Banks). (Photograph by B. P. Stark.)

Stonefly adults (Figs. 1D and 1E) are variable in size from about 5 to 50 mm and in color from black to green or yellow, often marked with distinctive light or dark patterns. The aquatic adult of one species known from the depths of Lake Tahoe *(Capnia lacustra)* and a few other species are apterous (wingless), but most adults are winged. The wings of males and females of some species, or particular populations of a species, are shortened (brachypterous) and they do not fly, but the typical condition is of two pairs of wings as long as or longer than the abdomen (macropterous). As the ordinal name (Plecoptera = folded wings) describes, the hind wings

typically have an expanded posterior (anal) lobe that folds longitudinally under the main wing (Fig. 1D).

Stoneflies are relatively slow, somewhat awkward fliers that typically fly short distances to disperse, to search for mates, or, for females, to deposit eggs.

Adults (Fig. 1D) have ten abdominal segments; the genitalia of males are distinctive at the generic and species levels and consist mainly of various external manifestations of the ninth and tenth segments, such as paired hooks, lobes (paraprocts), or sclerotized stylets and in some taxa a median probe (epiproct) of various shapes. During copulation, the copulatory organ (aedeagus), normally inside the abdominal cavity, is everted ventrally from the genital opening on the ninth sternum. The external female genitalia consists of a flap-like subgenital plate covering the genital opening on the eighth abdominal sternum, a structure that the male grasps or holds with his hooks or lobes during copulation. The wings, 3-segmented tarsus, genitalia, and a pair of usually multisegmented tails (cerci), arising from the 10th abdominal segment, generally characterize stonefly adults. Nymphs (or larvae) (Figs. 1B and 1C) may or may not generally resemble their adults. They are gill-less or have diagnostic simple or branched tracheal gills arising from different parts of the body such as near mouthparts, ventral head, thorax, coxae, or abdomen, and they always have multisegmented tails (cerci). Stubs or the basal remnants of gills are retained as vestigial structures in the adults of some taxa and aid in their identification to family and genus. The long, multisegmented nymphal cerci become reduced to fewer segments in some adults or to a single segment in adult males of the families Leuctridae, Nemouridae, and some Taeniopterygidae.

TAXONOMY AND GENERAL DISTRIBUTION

The Plecoptera is divided into the two suborders Arctoperlaria and Antarctoperlaria. The Arctoperlaria are distributed in the Northern Hemisphere, except the family Notonemouridae, which occurs only in southern South America, southern Africa, Madagascar, Australia, Tasmania, and New Zealand, and some genera of Perlidae such as *Anacroneuria* and *Neoperla* that have moved south across the equator in recent times, perhaps from 15 to 30 mya. The Arctoperlaria is further divided into the group Euholognatha (containing six families: Capniidae, Leuctridae, Nemouridae, Notonemouridae, Taeniopterygidae, and Scopuridae) and the group Systellognatha (containing six families: Chloroperlidae, Peltoperlidae, Perlidae, Pteronarcyidae, and Styloperlidae). The Euholognatha have mouthparts adapted for herbivory (scrapers, grazers, collector–gatherers, shredders, gougers, and detritivores), including molariform mandibles, and its species occur with few exceptions in streams of various sizes. The Systellognatha, except Peltoperlidae and Pteronarcyidae whose mouthparts are similar to those of Euholognatha because of convergent evolution to herbivorous food habit, have mouthparts mainly adapted for predation, including sharp-cusped mandibles and toothed lacinia for grasping and holding prey. The systellognathan families

Perlidae and Peltperlidae have very few species in arctic and subarctic streams.

The suborder Antarctoperlaria, as the name implies, is restricted in distribution to the Southern Hemisphere. In some areas, recently invading genera of Arctoperlaria, such as *Anacroneuria* (Perlidae) in South America and *Neoperla* (Perlidae) in Africa, have outcompeted them. The suborder contains four families: Austroperlidae, Diamphipnoidae, Eustheniidae, and Gripopterygidae. The Austroperlidae and Gripopterygidae live in a wide variety of habitats and the Diamphipnoidae and Eustheniidae are restricted to relict populations in the southern Neotropical and Australian regions. Each of the 16 families of Plecoptera has unique combinations of wing venational, genital, gill, and other characteristics.

ECOLOGICAL IMPORTANCE

Stoneflies are integral and important food web components of most stream ecosystems throughout the world and therefore are almost exclusively beneficial insects. The various taxa have radiated to use virtually every type of food and substrate habitat resource available to them. The nymphs are variously detritivores, herbivores, insectivores, or omnivores, and in some species the diets of nymphs shift from detritivory or herbivory, through omnivory, to strict insectivory as development proceeds. In turn, they become food for larger insectivores and fishes and are therefore important in the energy dynamics of stream food webs. Particular taxa are usually associated with particular stream microhabitats, such as the interspaces of loose gravel or cobble substrates, leaf packs, detritus, debris, or logs. The nymphs of numerous species have evolved to coexist in a relatively harmonious, noncompetitive way in given stream ecosystems by partitioning their food, space, and time resources. Most stonefly species require relatively undisturbed conditions of the streams they historically inhabit and therefore are important biological indicators of stream water quality. They constitute the "P" component of one of the major biomonitoring indexes of "clean water species," called the EPT Index ("E" for Ephemeroptera, "T" for Trichoptera), used for assessing water quality and degree of stream disturbance by humans.

LIFE HISTORY AND BEHAVIORS

Adults

Adult stoneflies (Figs. 1D and 1E) usually emerge during the night from nymphs that have crawled out of the water onto objects projecting from streams or on the stream bank, such as entrained leaf packs, logs or debris, rocks, or riparian vegetation. Some species of Euholognatha are black and emerge under ice or snow cover in winter. A light colored (teneral), clumped-winged, soft adult emerges from the last instar skin during the molt through a split in the head and dorsal thoracic segments. Males of many species emerge a day or more before females, so that they are present and searching when females appear. After some degree of hardening, both sexes become cryptic, hiding in crevices or vegetation during inactive periods, usually during the day, and becoming active in mate-finding and other activities typically at night. Adults of Systellognatha have reduced mouthparts and typically do not feed, but may ingest liquids or nectar. Adults of Euholognatha feed variously on algae, lichens, flower pollen and nectar, or soft fruits. For most species of stoneflies, the details of transformation, dispersal, feeding inactivity periods, and longevity are unknown. Generally, adults live only one to a few weeks and are mostly actively engaged during that time in the reproductive activities of mate-finding, copulation, and oviposition.

Communication, Mate-Finding, and Mating

The primary method of communication for locating mates in the Northern Hemisphere stonefly suborder Arctoperlaria is vibrational signaling through substrates. The vibrations are produced by tapping or rubbing the abdomen on the substrate or by body tremulations transferred to the substrate. The signals of most insects that use this method consist of simple volleys of evenly spaced vibrations, but stoneflies have evolved a much more diverse and complex system of vibrational communication than is known for other insect groups. The currently accepted paradigm of how this behavior, generally known as drumming, evolved suggests that the ancestral method of signal production was percussion and that signals were monophasic volleys of evenly spaced drumbeats. Natural selection favored increasingly complex signals, leading to greater specificity of communication and mate-finding among species and possibly increased capability for sexual selection to measure reproductive fitness. An increasing complexity of particularly male call signals may have evolved through the following three behaviors: (1) more sophisticated signaling methods, sometimes associated with specialized coevolving ventral abdominal structures, (2) rhythmic patterning of signals, and (3) possible use of selected natural substrates for signal transmission. The result has been that current species signal variously by percussion, scraping, or rubbing the abdomen on the substrate (abdominal–substrate stridulation) or tremulation (vibrations produced by push-ups or rocking motions of the body without abdominal contact with the substrate). Signal rhythms are species-specific and vary from evenly or unevenly spaced monophasic volleys to variously spaced diphasic or grouped signals.

The entire mating system of stoneflies involves communication as well as aggregation and movement behaviors of both sexes, the actual copulation, and postmating behaviors. The typical system in Arctoperlaria involves the following complex of behaviors: (1) initial aggregation of sexes at encounter sites near streams, (2) calling by males with species-specific signals during ranging search, (3) duet establishment by virgin females answering the male call if effective vibrational communication distance from her is achieved, (4) a localized search by the male in a "triangulation" or other pattern for the now stationary female while both continue dueting, and (5) almost immediate mating after the male locates and contacts the female. Males are polygamous and presumably continue calling and searching during their short reproductive lives.

Typically, mated and unguarded females reject subsequent male advances by raising and curving their abdomens. Southern Hemisphere stoneflies of the suborder Antarctoperlaria have never been documented to drum; therefore, their communication–search system for mate-finding is unknown, but they may have evolved a highly specific encounter site aggregation behavior that enables sufficient mate-locating ability without vibrational or other forms of intersexual communication.

Mating in stoneflies involves the male mounting the female, curving his abdomen around her left or right side, and engaging the subgenital plate, pulling it down with his external genitalia. This effectively matches her genital opening beneath the plate with a dorsal position between his cerci where the aedeagus will project. His aedeagus is everted from beneath the ninth sternum and expands backward and upward between his cerci into the female. Sperm are usually conveyed into the female by this intromittent aedeagus, but in some species sperm are conveyed through a hollow male epiproct or are externally deposited onto the female opening to be subsequently aspirated into the bursa (vagina) by telescoping movements of her abdomen.

Eggs and Oviposition

The eggs of stoneflies (Fig. 1A) vary considerably in size, shape, and details of chorionic (eggshell) ornamentation and sculpturing. Commonly, eggs are spindle-shaped, but they also may be spherical, flattened or three-sided. Frequently, an anterior collar is present and the shell surface may be smooth or ornamented either with ridges or the hexagonal pattern of impressions formed by the cells lining the ovarian chambers where the eggs are produced. Micropyles (sperm entrance holes) penetrate the chorion completely and may have associated surface grooves or ornate projections that serve as sperm guides. Actual penetration and fertilization by sperm of the egg is, as in most insects, delayed until the eggs are being stored or passed through the oviduct just prior to oviposition. The sperm are stored in the female spermathecum between copulation and fertilization. There also may be shallow pores leading to elaborate respiratory networks within the chorion. Eggs have sticky membranous or gelatinous surface coverings and sometimes filament-like projections with hooked tips that swell and help the eggs attach to substrates under water close to the selectively optimal sites where females deposit them.

Eggs are deposited by females in pellets or masses, each containing numerous eggs, that the females hold on the subgenital plate. They release the egg masses directly into the water by splashing into the surface during an oviposition flight, or by contacting shallow water while running near the shore, or by dropping eggs from the air while flying over water. Females of some Capniidae are also known to completely submerge themselves and crawl along the bottom and scrape the egg mass off onto the substrate.

In most species, embryonic development proceeds directly and is complete within 3 to 4 weeks. In other species, particularly those adapted to intermittent streams or streams subjected to extremes in temperature, embryonic development may be arrested for from 3 months to 1 or more years, and hatching is thus delayed until environmental conditions are favorable for nymphal survival.

Nymphs

Hatchlings (Fig. 1B) emerge from the egg by pushing on the chorion with the front of their head. The shell breaks into two halves or splits leaving a hinged cap and opening through which the first instar crawls out. The first instars are unpigmented with few body hairs, have fewer than 12 antennal and 6 cercal segments, and gills and wingpads are absent, reduced, or represented only by knobs or stubs. Little is known about the food, habitat, or behavior of hatchling and early instar stoneflies. The few species that have been studied feed mainly on detritus or the microflora–fauna on the surface of decomposing leaves.

Nymphs progressively develop and grow through about 10 to 24 sizes (instars). Full development of particular species requires from 4 months to 3 to 4 years. During this time there is a molt between each instar, addition of antennal and cercal segments, usually the addition of body hairs, a progressive increase in size of wingpads, and appearance and development of gills (if present in particular taxa) and characteristic pigment patterns. Growth of a particular species may be sustained at an even pace or seasonal, with fast and slow stages. In temperate climates growth is generally rapid during spring and fall and slowed or arrested (diapause) during extreme temperatures in summer or winter. But, interestingly, a number of euholognathan species, particularly in the families Capniidae and Taeniopterygidae, have adapted to experience their major growth in late fall and winter and emerge as adults during winter on ice or snow or in early spring during ice breakup. Completion of development and subsequent emergence as adults in temperate climates, therefore, may occur during any season, depending on altitude, latitude, and species.

Stoneflies have diversified their food habits such that different species fill about every conceivable major food niche in streams. Some species are herbivore–detritivores throughout their development, some are insectivores throughout development, and some experience an ontogenic (developmental) shift from herbivory–detritivory through omnivory and finally to carnivory. Characteristics of mandibles give a clue to food habits. The mandibles of herbivores have molariform surfaces or scraping ridges and those of carnivores sharp teeth for grasping and tearing. The food of carnivorous species consists primarily of the other aquatic insects of their communities such as midge larvae (Chironomidae), mayfly nymphs (Ephemeroptera), caddisfly larvae (Trichoptera), and occasionally the smaller nymphs of other stoneflies. Nymphs may be opportunists or in some cases are very selective for the size, behavior, and taxa of their prey. Prey are captured, grasped by the head with the lacinia and mandibles, and usually swallowed whole, headfirst.

Particular species of nymphs are found in certain types or sizes of streams at particular latitudes or elevations and often specific microhabitats. Rare and endemic species have very specific biological and physical requirements and therefore continue to exist only in pristine or little-disturbed sections of streams that in many instances are now found only in remote areas or in, or adjacent to, protected national parks or preserves. Many species have broader requirements and are more widespread or ubiquitous over large areas of continents in a wide variety of habitats. Only a few species are tolerant of the conditions of streams disturbed by siltation, alteration of natural temperature regimes, or chemical pollution. Stoneflies depend on substrates as a place in which to live. Slender species live in the interspaces of gravel, cobble, or vegetable debris such as leaf packs. Partitioning of microhabitats, and consequent microdistribution, is characteristic of most stonefly assemblages in a given stream. Most species live in the surface layers of a streambed, but a few live deep in loose mineral substrates such as glacial till and sometimes in the water-filled spaces of such substrates for considerable distances deep and lateral from the margins of the surface stream. Nymphs are sometimes found drifting in the water column of streams. This results from being dislodged by some physical disturbance or entering the water column as a behavioral means of dispersal, using the flow of water. Drifting enables nymphs to escape predators or move to a less populated habitat where food and/or space resources are more available or of higher quality.

Life Cycles

With few exceptions, a full generation of the egg, nymph, and adult stages of a stonefly species requires 1 to 4 years. For one-year (univoltine) cycles, the nymphal growth portion may be "fast," requiring only 4 to 7 months, or "slow," requiring nearly a full year. The fast type is characteristic of species that diapause for variable times, up to 8 months during warm or dry periods of streams; the slow type characterizes species whose nymphal stage requires about 11 months. Species requiring more than 1 year for a generation are termed semivoltine. Those that live in intermittent streams may diapause in the egg stage during drought periods for more than a year, then have a fast growing nymph for only 4 to 6 months. Semivoltine species living in cold streams may have a short egg stage, with nymphs requiring 2 or more years to develop, or a year-long diapausing egg stage, with nymphs requiring 1 to 3 years to develop. The life cycles and resource requirements of stoneflies are important considerations for developing stream management strategies.

See Also the Following Articles

Aquatic Habitats • Pollution • Swimming, Stream Insects • Vibrational Communication

Further Reading

Resh, V. H., and Rosenberg, D. M. (1984). "The Ecology of Aquatic Insects." Praeger, New York.
Stark, B. P., Szczytko, S. W., and Nelson, C. R. (1998). "American Stoneflies: A Photographic Guide to the Plecoptera." Caddis Press, Columbus, Oh.
Stewart, K. W. (1993). Theoretical considerations of mate-finding and other adult behaviors of Plecoptera. *Aquat. Insects* **16**, 95–104.
Stewart, K. W. (1997). Vibrational communication in insects: Epidomy in the language of stoneflies? *Am. Entomol.* **43**, 81–91.
Stewart, K. W., and Harper, P. P. (1996). Plecoptera. *In* "An Introduction to the Aquatic Insects of North America" (R. W. Merritt and K. W. Cummins, eds.). Kendall/Hunt, Dubuque, IA.
Stewart, K. W., and Stark, B. P. (1988). Nymphs of North American stonefly genera (Plecoptera). *Entomol. Soc. Am.* **12**, 1–460.
Zwick, P. (2000). Phylogenetic system and zoogeography of the Plecoptera. *Annu. Rev. Entomol.* **45**, 709–746.

Pollination and Pollinators

Gordon W. Frankie
University of California, Berkeley
Robbin W. Thorp
University of California, Davis

Pollination in its most basic sense is the transfer of pollen from the male sex organ (anther) to the receptive portion of the female sex organ (stigma) in flowering plants (Fig. 1).

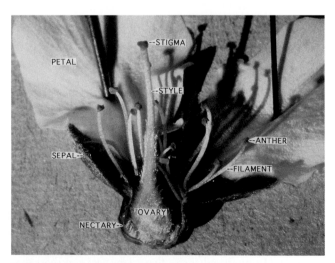

FIGURE 1 Floral parts of an almond blossom. The petals are color signals, the male stamen (anther and filament) and the female pistil (stigma, style, and ovary) are the reproductive parts. Nectar from the nectary and pollen from the anthers are food rewards to pollinators.

If the transfer is successful, it leads to fertilization, production of seed, and reproduction of the plant. This process often involves some sort of external vector such as wind, water, or animals. Some flowering plants may reproduce without the aid of pollen vectors, using mechanisms such as vegetative reproduction, apomixis, or automatic selfing. But our concern in this article is with animal vectors of pollen. Many kinds of animals may perform the ecological service known as pollination, including birds, bats, and some nonflying mammals. However, the dominant group of pollinators is the insects, especially bees.

ENTOMOPHILY

Many flowering plants are adapted to insects as pollinators and provide primary rewards that attract and keep the pollen vectors returning to flowers. Rewards include nectar, pollen, lipid secretions, food bodies, scents, resins, and material for nest building. In additon to primary rewards, most insect-pollinated flowers also produce a number of cues or signals that distinguish them from other species and promote the ease by which an insect can relocate a rewarding flower, thereby encouraging the insect to move pollen from flower to flower of the same species. These signals include odors, colors, shapes, textures, and tastes. These signals are often combined into patterns that have been recognized as syndromes related to the type of pollen vector. For example, a typical butterfly-pollinated flower that would be red, have little odor, possess a landing platform, have nectar hidden at the base of a deep tube that can be reached by the long coiled butterfly proboscis, have nectar that would be high in amino acids, and have flowers that would be open during the day. In contrast, a typical hawk moth-pollinated flower would be white, have a strong sweet odor, lack a landing platform, have long stamens with freely swinging anthers, and would bloom nocturnally. Nectar position and composition would be similar to those of a butterfly flower. All these characteristics are well suited to the hovering flight and extremely long proboscis of a hawk moth (Fig. 2).

Pollinator syndromes are perhaps most readily distinguished in the tropics. However, not all flowers are easily classified in a pollinator syndrome. Many flowers are visited by guilds of visitors that include diverse taxa of insects. For these reasons some pollination biologists see little value in the use of pollination syndromes even for teaching. However, even when using a pollination syndrome approach, it is important to distinguish which taxa among the visitor guilds are actually effective pollinators of a flowering plant.

Many insect taxa visit flowers and thus are potential pollinators; however, only a few taxa are of prime importance. Minor groups include those Orthoptera that feed on pollen and some Heteroptera that visit flowers for nectar or those that use flowers as sites that attract prey items (e.g., Phymatidae, Reduviidae). Thrips (Thysanoptera) are common on flowers and often feed on pollen and other flower tissues. They may

FIGURE 2 (A) Uncoiled proboscis (tongue) of a hawk moth, Costa Rica. (B) Tubular flower of *Lindenia rivalis* (Rubiaceae), pollinated at night by hawk moths, Costa Rica.

do more damage than good, but their positive contribution as pollen vectors is understudied. The major groups of insects that pollinate plants belong to the four largest orders of insects: Coleoptera (beetles), Diptera (flies), Lepidoptera (butterflies and moths), and Hymenoptera (bees, ants, and wasps).

Coleoptera

Beetles are often considered "mess and soil" pollinators in that while rumaging around in flowers feeding on pollen and other flower parts they pick up pollen on their bodies that is transferred to other flowers on subsequent visits. This type includes many that destroy some flowers by feeding on them (e.g., Scarabaeidae, Meloidae), but in the process others get pollinated. Some beetles are associated with pollination of some "primitive" flowers and have been considered responsible for pollination and diversification of early flowering plants (Angiospermae). Beetles of the family Nitidulidae feed on specialized food bodies on anther tips of the spice bush *Calycanthus* (Fig. 3).

FIGURE 3 White-tipped "food bodies" on anthers of *Calycanthus* (Calycanthaceae) that attract beetles (Nitidulidae) that pollinate the flower.

Diptera

Diverse flies, including male mosquitoes, various midges, carrion flies, pollen-feeding Syrphidae, and long-tongued nectar feeders (e.g., Bombyliidae, Acroceridae, Nemastrinidae), pollinate flowering plants. Pollination by flies is greatly understudied and underrated. Although flowers of pipevines are considered classical "trap" flowers that imprison flies with inward directed hairs until the flower has released pollen on them, flowers of the California pipevine, *Aristolochia californica,* exhibit a different mechanism involving a reward to retain flies until pollen is released. Midges of the family Mycetophilidae are the primary visitors and pollinators of *A. californica.* When flowers of *A. californica* first open, the stigma is receptive and a dark ring of glandular trichomes encircles the outer wall of the flower at the level of the stigma. Flies descend through the hooded entrance to the lower bowl and are attracted to the area of the stigma and trichomes by a light window. Flies feed on the trichome surface, contacting the stigma and depositing pollen from previous flower visits. The stigma closes, the trichomes wilt, and the anthers shed pollen into the bowl and onto the flies. Because there is no more food available, flies exit the flower and seek another, thereby pollinating the next flower.

Lepidoptera

Most adult butterflies and hawk moths are well-known flower visitors. Other day-flying moths (e.g., *Schinia* of the Noctuidae, *Adela,* Incurvariidae) and nocturnal moths pollinate while settling or perching on flowers during feeding or oviposition. A very specialized relationship exists between yucca moths (*Tegiticula,* Incurvariidae) and their host yucca flowers (*Yucca,* Liliaceae). Female *Tegiticula* enter *Yucca* flowers, collect pollen into a ball in specialized maxillary palps, move to the apex of the pistil where pollen is deposited on the stigma, and oviposit into the base of the pistil where seeds will develop from the pollination behavior. Larvae of the moth develop in the fruit pod, feeding on a portion of the seeds. Thus, both insect and plant benefit from this highly mutualistic association; the plant gets pollinated and produces seed, some of which goes to producing new moths.

Hymenoptera

Some sawflies and parasitoid wasps feed on pollen, especially on open shallow flowers. Classical mutualism occurs with the pollination association of fig wasps (Agaonidae) and their floral hosts, figs (*Ficus,* Moraceae). Similar to the situation with yucca moths and yucca, a portion of seeds in fig host flowers provides nourishment for development of the pollinating wasps. The story is often more complex and involves more than one generation of wasp per year and more than one host.

Other aculeate wasps, both social and solitary, augment insect prey diets with nectar. One genus of solitary wasps, *Pseudomasaris* (Masaridini, Vespidae), is completely dependent on flowers for nectar and pollen as food for their young. Ants may visit flowers but because their metathoracic glands secrete mold inhibitors they inhibit pollen germination and are unlikely pollen vectors. Bees (superfamily Apoidea) are the single most important taxon of pollinating insects with 20,000 to 30,000 species worldwide. Bees are derived from wasps and highly adapted to gathering pollen as brood food and nectar for flight fuel.

EXPLOITATION RELATIONSHIPS

Not all pollination relationships are mutualistic, i.e., beneficial for both partners. Some are based on deceit or robbery, in which only one partner benefits and the other may even be injured.

Thievery

Insect visitors to flowers may obtain the food items they seek without transferring pollen in the process. Some insects with mouthparts too short to reach nectar sequestered in the bottom of long tubes or spurs are able to penetrate the nectar-bearing structures with strong mandibles or maxillae. Such behavior is well documented for bumble bees, such as *Bombus occidentalis* in western North America or the related *Bombus terrestris* of Europe, when they encounter long-tubed flowers. This behavior is commonly exhibited by carpenter bees, especially in the tropics (Fig. 4). Insects that are mismatched in size with the flowers they visit may be effective gleaners of pollen from the anthers, but rarely if ever contact the stigmas in the flowers they visit. These thieves often scavenge pollen from flowers adapted to other types of pollinators. For example, the evening primrose of the southwestern deserts of North America are typically adapted for pollination by night-flying hawk moths, but they are visited early in the morning after they have opened by solitary ground-nesting bees of the genus *Andrena* for pollen. In fact, these bees have become so completely adapted to collecting this source of pollen, and their seasonal synchrony

FIGURE 4 Female carpenter bee (*Xylocopa tabaniformis orpifex,* Apidae) robbing nectar from base of a California fuchsia (*Epilobium canum,* Onagraceae) flower.

and the morphology of their pollen transport structures are so specialized, that they visit no other plants for pollen. So although the bees specialize on these flowers, they are not effective pollinators because they are small enough that they rarely contact the stigmas with the pollen they are collecting.

Floral Deceit

Many floral deceit mechanisms in flowers take advantage of basic behaviors and instincts in insects, especially feeding, mating, and oviposition. Some flowers are green or brownish rather than being colorful and have putrid or rotten meat aromas rather than sweet odors. These are highly attractive to carrion flies seeking source foods rich in amino acids and suitable as sites for oviposition and rearing of their young. Other flowers mimic females of bees or wasps in such fine details of form, color, odors, and texture that male insects actually attempt to mate with these models; in the process they pick up and distribute pollen from one flower to another. This process is called pseudocopulation because it relies on mating attempts by male wasps or bees. The flowers bloom during the brief time when male insects are on the wing before females of the species emerge. During this period, flowers are the only potential sources of "mates" for the male insects. Once females of the hymenopteran species emerge, the floral mimics are forsaken by males for the real female.

CROP POLLINATION

Economic Value

The most recognized benefit of pollinators to humanity is their value as pollinators of many of the crop plants that we use for food and fiber. The principal pollinator managed for crop pollination has been and currently is the honey bee, *Apis mellifera*. Although calculating the value of pollination by honey bees is far from exact, the most recent estimate of the annual value of increased production of crops contributed by honey bee colonies rented for pollination in the United States is over $14.5 billion, on average, over 1996 to 1998. The "free" contribution from native pollinators, especially other bee species, is even less measurable, but attempts are under way to estimate their value to sustainable agriculture farms in central California.

Exemplar Crops

Calfornia is perhaps the leading state in rentals of honey bee colonies for crop pollination. In large measure this is the result of the continuing increase in the acreage of almond, which has increased from about 36,000 hectares in the mid-1960s to over 200,000 hectares in 2001. At the recommended five to eight honey bee colonies per hectare, more than twice as many commercial colonies as exist in the state are required to accomodate the demand. Thus, there is a mass movement of colonies into

California each year from as far as the Dakotas and Texas and beyond to pollinate the crop.

Alfalfa is another crop traditionally pollinated by honey bees that was widespread in California in the mid-1960s. However, the honey bee is not an effective pollinator of alfalfa over much of the crop's range. When two more efficient alternative pollinators (the introduced alfalfa leafcutting bee, *Megachile rotundata,* and the native alkali bee, *Nomia melanderi*) came under management for pollination of the alfalfa seed crop, much of the production shifted to the Pacific Northwest. Only the more southern areas of California continued to produce alfalfa seed solely with honey bees. Currently even some of these areas are augmented with alfalfa leafcutting bees.

Crop Pollinators Other Than Honey Bees

Although honey bees are readily available, easily transportable in large quantities, and generalist pollinators, they are not universal pollinators. There are some crop flowers, such as figs, that require insect pollination (specialized wasps) and cannot be pollinated by honey bees. There are other crop flowers that can be pollinated by honey bees but for which honey bees are not the most effective pollinators; these include crops such as alfalfa, squash, and greenhouse tomatoes. Another concern about excess reliance on a single pollinator for a wide variety of crops has been the widespread decimation of feral honey bees by the "vampire" mite, *Varroa;* increased cost of treating colonies to maintain healthy pollinating units; and reduction in numbers of beekeepers and colonies available for pollination. Warnings about this overdependence on honey bees and the general decline of pollinators because of factors such as loss of habitat and pesticides were issued in 1996 in a landmark publication by Buchmann and Nabhan.

One of the first insects introduced into North America specifically to pollinate a crop was the fig wasp, *Blastophaga psenes,* for production of edible Smyrna figs in southern California in 1899. Attempts to produce edible figs in California in the late 1800s failed until it was recognized that wasps from the wild ancestral caprifig, *Ficus carica,* were required. Edible figs contain predominantly pistilate flowers; pollen from male flowers of the caprifig is vectored by fig wasps. The growing of caprifigs containing introduced fig wasps, harvesting the fruits with a new generation of fig wasps, and then hanging them in baskets in trees of edible figs became a common practice in southern California, a process called "caprification."

Other insects that have been used for commercial pollination on a small scale include various flies, especially for breeding hybrid seed crops in cages by seed companies. Most of these have been muscoid flies, the pupae of which are readily available from insectaries. Results of large-scale open-field trials using carrion or other baits to attract flies have been equivocal for crop pollination.

Various species of non-*Apis* bees have been and are being studied for their management potential for pollination of crops. In the late 1950s, studies were begun to manage two bee species (*M. rotundata* and *N. melanderi*) that are more effective than honey bees as pollinators of alfalfa for seed production.

More recently mason bees in the genus *Osmia* have been studied for pollination of crops in North America and Europe. *Osmia lignaria propinqua,* referred to as the "blue orchard bee," has been successfully managed to pollinate tree fruits in western North America. It is a cavity nester, like the alfalfa leafcutting bee, but uses mud partitions to create brood cells. Many of the management techniques were adapted from those used for *Megachile,* but modified to accomodate specific life history and behavioral attributes, including early spring activity and a single generation per year. Successes in managing other species of *Osmia* include *O. cornifrons* in Japan, *O. cornuta* in Spain, and *O. rufa* in Britain and France. In North America, *Osmia* are being studied for pollination of blueberries (*O. ribifloris*) and clovers (*O. sanrafaelae*).

In the late 1980s, major breakthroughs in year-round production of bumble bee colonies completely altered and expanded hothouse production of tomatoes. Tomato flowers require "buzz" pollination (i.e, vibration of flowers to release pollen from the apical pores of their specialized anthers) and bumble bees are much more effective at this than are honey bees or humans who hand pollinate with vibrating tools. This led to extremely large-scale movements of bumble bee colonies and queens, primarily *B. terrestris,* from central Europe, New Zealand, and Israel to many nations throughout the world. Some of this trafficking was unnecessary because closely related species were available for use at some locales. In Japan, environmental concerns have been expressed over the thousands of imported *B. terrestris* colonies and the subsequent establishment of this species outside the greenhouse environment. Males of this species will mate with queens of local species and produce viable offspring. Similar environmental concerns are being raised in other countries. In Canada and the United States, importation of *B. terrestris* was not sanctioned, but local bumble bees have been successfully reared and used in hothouse tomato production. East of the 100th meridian (a line that runs from central North Dakota to central Texas), *B. impatiens* was the species of choice and west of this line, *B. occidentalis*. These bees were used in their respective areas of distribution until 1998, when a disease outbreak was reported in the western species. Since then, the eastern *B. impatiens* has been imported into all western states, again causing concern in some areas that establishment outside its normal range may produce environmental damage.

Although pollinators are generally considered beneficial insects, importations to new areas should be done with considerable care to avoid environmental risks, such as introduction of disease organisms, nectar thieving, decreased pollination of nontarget native plants, enhanced pollination of introduced weeds, and genetic contamination of and competition for food and/or nest sites with native pollinators. The local fauna should be studied and searched for candidate species that could be suitably managed before any exotic

species are introduced. Rearing technology is already available for several species of cavity-nesting bees, for bumble bees, and for some soil-nesting species. These methods may be modified and applied to local species that show promise for solving difficult pollination problems.

If it is deemed necessary to introduce a new pollinator, these should be thoroughly screened for biotic enemies (e.g., parasites, disease organisms) before being introduced. They should be monitored after release in the new environment to determine their efficiency in pollinating the target crop and to detect any adverse enviromental effects.

POLLINATOR DECLINE

Pollination biologists have worked for years studying the behavioral, ecological, and evolutionary intricacies of animal–plant pollination relationships. With the environmental movement of the early 1970s came awareness that mistreatment of the environment could also negatively affect these unique relationships. However, very few biologists sounded the alarm at that time that trouble was brewing for pollination relationships and pollinators, especially at the ecosystem or landscape level.

In the early to mid-1990s, the issue of pollination/pollinator problems was again brought to the attention of the biological community and the informed public, through the Island Press publication of *The Forgotten Pollinators* in 1996 by Stephen Buchmann and Gary Nabhan. Although the book contains anecdotal accounts of pollinator problems, the message was clear—everyone concerned with the pollination of plants needed to pay serious attention to what appeared to be an emerging picture of global pollinator decline. A subsequently important publication by Allen-Wardell *et al.* in 1998 pointed out the potential threat of pollinator decline to the human food supply. Adding to the general concern for declining pollinators was the fact that European honey bees in the New World tropics, and several western and southwestern U.S. states, were being systematically replaced with Africanized honey bees, and that all honey bees were being attacked and significantly reduced in numbers by two species of parasitic mites. Continued careless use of pesticides and bacterial infections were also cited as factors in the decline of honey bees.

The above trends have led to many scientific conferences worldwide to address the issue of pollinator decline and its potential consequences to crop plants and to native wildland plants. The first important global conference was held in 1998 in São Paulo, Brazil, where numerous relevant issues on pollinator decline were discussed by more than 60 pollinator/pollination professionals representing several New and Old World countries. The São Paulo meetings and subsequent meetings in other parts of the world have put into motion new research directions in the field, which are listed below.

1. Documenting pollinator decline. One of the main recommendations emerging from São Paulo was to seek

quantification of pollinator declines through careful case history studies. Long-term monitoring of pollinators and comparative assessments at specific study areas were recommended approaches for gathering the needed data on decline. This has proved to be quite a challenge because of natural fluctuations in pollinator populations and the lack of long-term baseline data. Where decline has been detected, loss of habitat is suggested as the main cause.

2. Causes of decline and restoration of pollinators. Habitat loss is a convenient general explanation for the cause of decline, but detailed information is needed on the precise factors causing decline. Further, specific causes of decline are not always obvious, and this becomes an important issue when considering projects to restore or establish pollinators and their required resources (often diverse) to an area.

3. More research on non-honey-bee pollinators. Honey bees can be likened to a monoculture in agriculture; overdependence on a single organism in agriculture can lead to disastrous results when natural mortality factors get out of balance. Such was the case with blight on a large portion of the U.S. corn crop several years ago. Because honey bees are now showing great vulnerability to two species of parasitic mites, more research on these parasites and on other bees (especially native solitary species) is being conducted.

4. Conservation of pollinators. Much has been written on conserving pollinators and especially bees. Unfortunately, there is little evidence to suggest that conservation recommendations, which would result in measurable increases in pollinator numbers, have been put into action. Many possibilities exist for increasing pollinators through manipulations of preferred food plants, nonfloral plant products (e.g., resins), and planned efforts to increase nesting sites and other requisites such as alternate food plants for moths, beetles, wasps, flies, bats, etc.

5. Increasing awareness of pollinator services. The conservation of pollinators and calling attention to vital services provided by pollinators become issues of information transfer that biologists must address. They are the only professionals who know the needs and fragilities of small organisms such as bees, flies, beetles, and nocturnal organisms such as moths and bats. Pollinator/pollination professionals will need to form closer working relationships with policy-makers, land stewards, and a wide variety of government and nongovernmental organizations to realize future successes in the management of pollinators.

There are at least two courses of action that pollinator biologists could pursue now to assist declining pollinator populations. First, they can collaborate with other biologists who are also concerned about decline of their specific organisms (e.g., birds, mammals) and habitat. Building a coalition of concerned biologists with integrated management plans for habitat protection for several threatened species could be effective if land stewards, associated with the habitat, were receptive and willing to participate in some way as stakeholders in the project. Second, biologists could also work in a variety of ways toward conserving areas known to naturally harbor healthy populations of pollinators, preferably several types. Biologists are aware, through years of field experience, which areas have good diversity and abundance of, for example, bees and moths.

TROPICAL POLLINATION

The literature is filled with fascinating case histories of individual tropical plants and their pollinators. More recently, researchers have been investigating pollination systems involving groups of prominent pollinators (e.g., bees, bats, moths, hummingbirds) and their plants in major tropical life zones. A few larger, long-term studies from the New and Old World tropics have also provided the first community pollination patterns for a high percentage of the representative plant life forms. These latter studies are particularly helpful in elucidating diversity and frequency of pollination systems for conservation work, as well as for interesting comparisons with temperate environments.

In Table I, pollination systems of two lowland forest sites and one midelevation cloud forest (1200–1800 m) in Costa Rica are compared and contrasted with one lowland forest site in Malaysia. From about 40 to 70% of the surveyed plant species in each of the four sites were pollinated by bees. Although less frequent than bees, birds were important pollinators in Costa Rica's cloud forest and wet forest and in the Malaysian forest. Beetles were important in the Costa Rican wet forest and in Malaysia; moths were important in all three Neotropical forests. The four sites had numerous plant species that were visited (and pollinated) by a variety of general insects.

The importance of animals, especially insects, as pollen vectors of tropical plants was clearly demonstrated in a classic paper by Bawa in 1974, in which he reported that most tree species in a lowland dry forest of Costa Rica were obliged to outcross. Through controlled pollinations, he demonstrated that a high percentage of the tree species tested were self-incompatible (incapable of self-pollination) or dioecious (having separate male and female plants of a species). Until that time, many biologists believed that self-pollination was probably the rule in tropical forests. Flower-visiting animals, and especially insects, were not viewed as capable travelers between widely distributed tropical plants. Subsequent studies on interplant movements and foraging patterns of these animals have substantiated their capacities to move among flowering plant species at levels required to produce abundant fruit crops.

Some general, but limited, comparisons of tropical versus temperate pollination systems are possible with the information available in the literature. First, the flora of many low- to midelevation temperate habitats is mostly pollinated by bees. Estimates vary widely from 70 to more than 90%, depending on locality. Flies and Lepidoptera may also be important in

TABLE I Percentages of Pollination Systems Represented in Four Tropical Forests

	Costa Rica			Malaysia
	Dry forest[a]	Cloud forest[b]	Wet forest[c]	Dipterocarp forest[d]
Bee	69.5%	44.4%	38.4%	50.4%
Moth	7.3	6.2	8	1.1
Bat	2.4	3.1	3.6	1.5
Bird	2.2	9.6	14.9	7
Beetle	1.1	2.4	12.7	20.7
Wasp (large)	0.2	1.6	2.5	0
Butterfly	0.9	1.9	4.3	2.2
Fly	0	0.2	1.8	0
Fig wasp	0.4	0.5	?	0
Mammal (arboreal)	0	0.1	0	0.4
General insect	15.3[e]	20.4[e]	11.2[e]	13.7
Wind	0.9	5.2	2.5	0
Miscellaneous	0	4.5	0	3

[a] N = 465 plant species. Source: Frankie *et al.* (2003).
[b] N = 1100 plant species. Source: Frankie *et al.* (2003).
[c] N = 276 plant species. Source: Kress and Beach (1994).
[d] N = 270 plant species. Source: Momose *et al.* (1998).
[e] Several plant species listed in this category may prove to be primarily bee pollinated.

the pollination of temperate plants. Second, the diversity of pollination systems is comparatively lower in temperate environments because most lack, for example, the more specialized bird, bat, beetle, and fig wasp systems. Third, when lowland tropical and temperate forests are compared, the high diversity of tropical trees and their dependence on animal pollination become immediately apparent. Temperate forests have relatively low tree species diversity and most, such as conifers, oaks, willows, elms, and maples, are wind pollinated. In contrast, wind pollination is rare in tropical forests (Table I).

SPECIALIZED VS GENERALIZED POLLINATION

As mentioned previously, there is ongoing controversy about specialized versus generalized pollination systems and the associated concept of pollination syndromes that propose to characterize a plant as to a particular pollinator type. There can be little doubt that some plants have highly specialized systems that can be easily characterized by floral morphology and behavior alone, such as fig flowers and fig wasps, certain orchids and their specific bee relationships, and long, white-tubed fragrant flowers that open at night and are pollinated by long-tongued hawk moths (Fig. 2). There are, however, many instances in which flowers attract a wide variety of visitor types, making characterization of pollination syndromes difficult. For example, there are examples of "large-bee flowers" that are regularly visited by small bees, butterflies, and wasps; hawk moths that visit "bat flowers"; bees that visit both "hawk moth flowers" and "bat flowers" the morning after. Further, there are flower types that attract a wide diversity of insect visitors (Table I). Who are the pollinators and who are the

visitors to these flower types? Do many or most plants have the option of being pollinated by a variety of potential vectors?

The answers to these questions will be forthcoming through carefully planned experimental studies that include evaluations of all visitors and their capacity to transport pollen on their body parts, as well as between plants. Floral behavior must be studied simultaneously, especially with regard to breeding system and period of stigmatic receptivity. These case studies should take much of the speculation and guesswork out of pollinators, visitors, and pollination ecology.

CHEMICAL ECOLOGY AND POLLINATION

Chemical communication between flowering plants and their pollinators and between conspecific pollinators in relation to floral resources is commonplace in many natural communities. Flowers release a variety of odors that attract pollinators and other visitors. Some of these are sweet and highly fragrant, as with many "moth flowers" and some "bee flowers." Some are unpleasant to humans but highly attractive to flies and beetles. Other floral fragrances fall somewhere between fragrant and unpleasant such as musky odors associated with some flowers that attract a variety of visitor types.

In the New World tropics, chemicals are emitted from certain orchid species that attract only males of the Eugolssini tribe of bees, better known as orchid bees. These orchids and "their" bees have evolved a unique relationship in which chemical substances are scratched from special regions of the orchid flowers by male bees. The compounds are collected in special leg glands that have unique apertures on the hind legs. According to theory, the scratched compounds are metabolized and transformed into chemical messages or pheromones that male bees use for attracting females for mating. Some male orchid bees form elaborate leks or mating rituals, which they use to lure females. During the process of scratching chemicals, orchids belonging to genera such as *Catesetum, Cycnoches,* and *Stanhopea* have cleverly evolved elaborate mechanisms for attaching or actually gluing a pollen packet (or pollinia) on the bee for transport to the next orchid, thereby effecting cross pollination. Each orchid species glues its pollinia on a characteristic location of the bee. Where orchid diversity is rich, it is common to see some bees with pollinia on their heads, other bees with pollinia on their thoraces or abdomens, and still other bees with pollinia on more than one body location, indicating their visits to more than one orchid species.

There is also chemical communication among members of some social and solitary bee species. Some honey bee species scent mark flowers, and this informs others of the same species of a very recently visited flower. The mark also serves to alert a bee that it has just visited a particular flower that it marked, thereby conserving its energy and time. This behavior has also been observed frequently in large carpenter bees *(Xylocopa);* chemicals of the mandibular and mesosomal glands were found responsible for the scents. Stingless bees in the Neotropics regularly mark flowers and nearby vegetation in establishing a scent trail back to inform the nest where a good floral source can be located. Some bumble bees *(Bombus)* also scent mark flowers, making these flowers less attractive to other foraging bumble bees.

CONCLUDING REMARKS

Despite years of study and an enormous literature, we still have much to learn about pollination and pollinators. The field of study is particularly challenging today because of the many questions surrounding pollination/pollinator relationships in human-impacted environments. Many questions (and problems) will require new approaches and methods and will need to be better integrated with societal needs and structures.

Suggested studies for the future include: (a) more detailed work on chemical relationships between pollinators, visitors, and flowers; (b) more attention paid to actively conserving, protecting, and restoring pollinators at local, regional, national, and international levels; (c) improving methods for managing pollinators for production of human food crops; (d) development and transfer of information on pollination and pollinators to a wide variety of new audiences such as policy/decision-makers, government agencies, nongovernmental conservation organizations, and managers; and (e) developing monitoring methods to gauge effects of global warming on pollinators and the plants they pollinate.

See Also the Following Articles

Apis Species • Beekeeping • Conservation • Neotropical African Bees

Further Reading

Allen-Wardell, G., *et al.* (1998). The potential consequences of pollinator declines on the conservation of biodiversity and stability of food crop yields. *Conserv. Biol.* **12,** 8–17.

Bawa, K. S. (1974). Breeding systems of tree species of a lowland tropical community. *Evolution* **28,** 85–92.

Buchmann, S. L., and Nabhan, G. P. (1996). "The Forgotten Pollinators." Island Press, Covelo, CA.

Daily, G. C. (ed.) (1997). "Nature's Services: Societal Dependence on Natural Ecosystems." Island Press, Covelo, CA.

Delaplane, K. S., and Mayer, D. F. (2000). Crop pollination by bees. CAB Int., Tucson.

Frankie, G. W., *et al.* (2003). Flowering phenology and pollination systems diversity in the seasonal dry forest. *In* "Biodiversity Conservation in Costa Rica: Learning the Lessons in a Seasonal Dry Forest" (G. W. Frankie, A. Mata, and S. B. Vinson, eds.). University of California Press, Berkeley.

Free, J. B. (1993). "Insect Pollination of Crops," 2nd ed. Academic Press, San Diego.

Kress, W. J., and Beach, J. H. (1994). Flowering plant reproductive systems. *In* "La Selva, Ecology and Natural History of a Neotropical Rain Forest" (L. A. McDade *et al.,* eds.). University of Chicago Press, Chicago/London.

Matheson, A., Buchmann, S. L., O'Toole, C., Westrich, P., and Williams, I. H. (eds.) (1996). "The Conservation of Bees." Academic Press, San Diego. [Linnean Soc. Symp. Ser. No. 18].

Michener, C. D. (2000). "Bees of the World." Johns Hopkins University Press, Baltimore.

Momose, K., Yumoto, T., Nagamitsu, T., Kato, M., Nagamasu, H., Sakai, S., Harrison, R. D., Itioka, T., Hamid, A. A., and Inoue, T. (1998). Pollination biology in a lowland Dipterocarp forest in Sarawak, Malaysia. I. Characteristics of plant-pollinator community in a lowland Dipterocarp forest. *Am. J. Bot.* **85,** 1477–1501.

Morse, R. A., and Calderone, N. W. (2000). The value of honey bees as pollinators of U.S. crops in 2000. *Bee Cult.* March 2000, 1–15.

O'Toole, C., and Raw, A. (1991). "Bees of the World." Facts on File (http://www.factsonfile.com/).

Proctor, M., Yeo, P., and Lack, A. (1996). "The Natural History of Pollination." Timber Press, Portland, OR.

Roubik, D. W. (1989). "Ecology and Natural History of Tropical Bees." Cambridge University Press, Cambridge, UK. [Cambridge Tropical Biol. Ser.]

Roubik, D. W. (ed.) (1995). "Pollination of Cultivated Plants in the Tropics." FAO, Rome. [Agric. Services Bull. No. 118].

Pollution, Insect Response to

David M. Rosenberg
Freshwater Institute, Winnipeg, Canada

Vincent H. Resh
University of California, Berkeley

Pollution is essentially the wrong substance, in the wrong place, in the wrong concentration, at the wrong time. More formally, pollution can be defined as the introduction of human-made substances (or natural substances released by humans) and forms of energy into the environment that are likely to damage ecosystems or their constituents, amenities, or structures.

Insects, like other living organisms, are affected by pollution. However, insects are also used to assess the effects of pollution as surrogates or representatives of the larger assemblages of organisms in communities and ecosystems. We refer to insects in this latter role as biomonitoring agents.

Pollution can be caused by a variety of substances and activities: sewage and other organic enrichment, fertilizers (e.g., nutrients such as phosphorus), siltation (e.g., from erosion), pesticides (e.g., herbicides, fungicides, insecticides), metals (e.g., cadmium, mercury, selenium), organic compounds (polychlorinated biphenyls, polycyclic aromatic hydrocarbons, industrial atmospheric emissions (e.g., sulfur dioxide and NO_x, which are precursors to acid rain; greenhouse gases such as carbon dioxide and methane), radiation (as in the Chernobyl disaster), heat (e.g., thermal pollution from power plants), and habitat destruction (e.g., clear-cutting, stream channelization, reservoir creation). Disturbances that have consequences similar to those of human activities can also be caused by natural events such as volcanic eruptions or forest fires, but the products of these disturbances do not fit our definition because they are not deliberately introduced by humans.

POLLUTION EFFECTS ON INSECTS

The effects of pollution on insects occur at a variety of spatial and temporal scales. For example, effects can occur at the molecular level in fractions of seconds (i.e., biochemical effects), at the ecosystem level over several decades, and at various scales in between these extremes (Fig. 1). The most easily detected responses are at the level of the individual (as in bioassays), in which evaluation is often based on whether an insect lives or dies when exposed to a contaminant. However, population and community levels are generally used when effects are examined in nature (Fig. 1). Within a population, we can see shifts occurring in the frequency of organisms of different sizes. For example, early instars in a population seem to be more susceptible than later instars because early instars have higher surface-to-volume ratios, they are more active, and they have thinner cuticles. Eggs, pupae, and diapausing insects are usually more resistant stages.

Common community-level changes observed with the onset of pollution involve decreases in species richness (i.e., the total number of species), decreases in species evenness (i.e., the distribution of numbers of individuals among species), and alterations in species composition. Severe sewage pollution in fresh waters presents a good example of the kinds of changes that can occur. With sewage input, the total number of species decreases immediately downstream of a sewage source and species evenness decreases because the remaining tolerant organisms proliferate. Also, pollution-tolerant organisms such as dronefly maggots *(Eristalis tenax),*

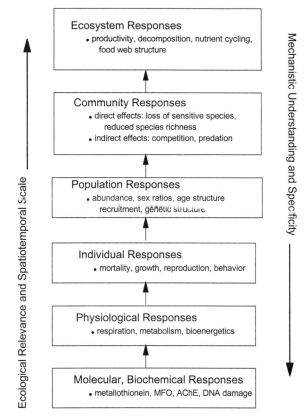

FIGURE 1 Relationship between ecological relevance, spatiotemporal scale, mechanistic understanding, and specificity across levels of biological organization. MFO, mixed-function oxidase; AChE, acetylcholinesterase; DNA, deoxyribonucleic acid. (Reproduced from W. H. Clements, 2000, Integrating effects of contaminants across levels of biological organization: An overview. *J. Aquat. Ecosyst. Stress Recovery* **7,** 113–116, with kind permission from Kluwer Academic Publishers.)

some species of midge (Chironomidae) larvae, and aquatic earthworms replace less tolerant ones, and these tolerant organisms can reach extremely high densities.

Alteration of feeding-group structure is another response to pollution. For example, the removal of a riparian zone surrounding a stream can lead to increased production of algae because of more open conditions and greater sunlight reaching the stream substrate, a decline in the leaves falling into the stream, and an increase in grazing insects (feeding on algae), which replace the insects that would normally feed on leaves.

More subtle community-level effects are also possible. For example, the suppression of parasites by air pollutants may enhance outbreaks of forest insect pests whose populations are normally held in check by the parasites. These indirect effects (Fig. 1) are often more difficult to detect than the direct effects discussed above.

Natural variability is a problem in evaluating the effects of pollution on insects. Only the accrual of long-term baseline data on insect species present and their abundances, and the use of an experimental approach to establish causation, can lead to understanding of the effects of pollutants on insects.

USE OF INSECTS IN BIOMONITORING

The responses of insects to pollution can be used in biomonitoring of air, land, and water quality. The most extensive biomonitoring has been developed in aquatic habitats, perhaps because aquatic insects are directly exposed to water pollution and their responses are easy to measure.

Insects present a number of advantages for biomonitoring: (1) they are ubiquitous, so they are exposed to pollution in many different habitats; (2) the large number of species offers a range of responses; (3) the sedentary nature of many insects allows spatial analysis of pollutant effects; and (4) their long life cycles allow temporal analysis of pollutant effects. Unlike relying on instantaneous measurement of physical and chemical variables, the use of living organisms, like insects, provides a temporal integration of pollutant effects over their life span.

Biomonitoring can be done at a variety of spatial and temporal scales (Fig. 1). The smallest scale is biochemical; for example, exposure of stonefly species to the pesticide fenitrothion can be measured by depression of acetylcholinesterase activity in the stonefly head. The largest scale is the ecosystem, in which measures of processes such as productivity and decomposition can indicate pollutant stress. However, biomonitoring is most commonly done at individual, population (or species assemblage), and community levels. Common examples of biomonitoring at the individual level include the use of insects as sentinel organisms and for measuring morphological deformities. The use of morphological deformities to measure pollutant effects is more common in freshwater than in terrestrial habitats. The head capsules of midge larvae are frequently used for this purpose. Bioassays often involve exposure of individuals to potential pollutants to detect sublethal responses (e.g., diminished growth or fecundity) or lethal responses (e.g., mortality).

Biotic indices are popular ways to summarize information and assess pollution at the population or species assemblage levels. For example, indices of the trophic status of lakes have been developed using midges, and an index of acid stress is available that uses a group of macroinvertebrate organisms, each of which has a different sensitivity to acidification.

However, biomonitoring is most frequently done at the community level, and measures vary from simple taxa richness to biotic indices that use the whole community (e.g., Family Biotic Index) to complex multivariate statistics that use the "reference condition" approach.

Bioassays are widely used in determining toxicity of industrial by-products before and after they are released into the environment and can be a valuable adjunct to field biomonitoring. They can be used to answer questions such as (1) At what concentration does a pollutant become toxic to a certain insect species? (2) What are the physical and chemical conditions under which the pollutant is most harmful? (3) Which stage of the life cycle is most susceptible to the pollutant? (4) What are the effects of acute versus chronic exposure? (5) Which species are most susceptible to the pollutant? Bioassays can be as simple or as complex as needed (e.g., single versus multiple species, static or flowthrough setup, laboratory beaker versus field mesocosm).

Biomonitoring using insects can involve volunteers as well as professionals. Work with insects is usually labor intensive, and volunteers can provide much of this labor. In fact, volunteer biomonitoring programs involving school classes and local community groups are currently very popular throughout the world.

In conclusion, although pollutants have affected living organisms, habitat destruction may be more important in determining whether organisms can recover. For example, it may be possible to clean air and water of their pollutants but recovery of ecosystems is impossible if the habitat remains impaired.

See Also the Following Articles

Aquatic Habitats • Conservation • Forest Habitats • Grassland Habitats • Greenhouse Gases, Global Warming, and Insects • Soil Habitats • Urban Habitats

Further Reading

Fjellheim, A., and Raddum, G. G. (1990). Acid precipitation: Biological monitoring of streams and lakes. *Sci. Total Environ.* **96,** 57–66.

Flannagan, J. F., Lockhart, W. L., Cobb, D. G., and Metner, D. (1978). Stonefly (Plecoptera) head cholinesterase as an indicator of exposure to fenitrothion. *Manit. Entomol.* **12,** 42–48.

Heliövaara, K., and Väisänen, R. (1993). "Insects and Pollution." CRC Press, Boca Raton, FL.

Hellawell, J. M. (1986). "Biological Indicators of Freshwater Pollution and Environmental Management." Elsevier, London.

Hilsenhoff, W. L. (1988). Rapid field assessment of organic pollution with a family-level biotic index. *J. North Am. Benthol. Soc.* **7,** 65–68.

Kerr, M., Ely, E., Lee, V., and Desbonnet, A. (1994). "National Directory of Volunteer Environmental Monitoring Programs," 4th ed. U.S. Environmental Protection Agency and Rhode Island Sea Grant, University of Rhode Island, Narragansett. [EPA 841-B-94-001]

Reynoldson, T. B., Norris, R. H., Resh, V. H., Day, K. E., and Rosenberg, D. M. (1997). The reference condition: A comparison of multimetric and multivariate approaches to assess water-quality impairment using benthic macroinvertebrates. *J. North Am. Benthol. Soc.* **16,** 833–852.

Rosenberg, D. M., and Resh, V. H. (eds.) (1993). "Freshwater Biomonitoring and Benthic Macroinvertebrates." Chapman & Hall, New York.

Saether, O. A. (1979). Chironomid communities as water quality indicators. *Holarctic Ecol.* **2,** 65–74.

Polyembryony

Michael R. Strand
University of Georgia

Polyembryony is a form of clonal development in which a single egg produces two or more genetically identical offspring. Many animals are sporadically polyembryonic, including humans, who occasionally give birth to identical twins. However, a few groups of parasites (some cestodes, treamatodes, and insects), colonial aquatic invertebrates (oligochaetes, bryzoans), and mammals (armadillos) are obligately polyembryonic. Among these obligately polyembryonic animals, insects produce the largest broods, with some species generating more than 3000 offspring per egg. There are hundreds and perhaps thousands of polyembryonic insects, but all known species occur in either the order Hymenoptera (bees, wasps, and ants) or the order Strepsiptera. Several species of polyembryonic wasps have been studied in detail, whereas little is known about polyembryonic strepsipterans.

In the Hymenoptera, polyembryony occurs in selected genera from four families: the Braconidae, Platygasteridae, Encyrtidae, and Dryinidae. Polyembryony in these groups clearly arose from ancestors that produced only a single offspring per egg (monoembryonic), because the most basal members of these large families are all monoembryonic. The phylogenetic distance between these families also indicates that polyembryony has evolved independently at least four times in the Hymenoptera. Despite multiple independent origins, all polyembryonic wasps share the common biology of being parasitoids that lay their eggs into the egg or larval (nymphal) stage of their insect hosts. After oviposition, the egg of a polyembryonic wasp develops into a single embryo. This embryo then proliferates into an assemblage of embryos called a polygerm or polymorula. The majority of embryos in the polymorula develop into larvae when the host molts to its final instar. These so-called reproductive larvae consume the host, pupate, and emerge as adult wasps that seek mates (if male) or new hosts (if female). In some species from the family Encyrtidae, however, a small proportion of the embryos in the polymorula develop into what are called precocious larvae. Precocious larvae are morphologically distinct from reproductive larvae and function as a sterile soldier caste that defends their reproductive-caste siblings from competitors. Embryonic development of polyembryonic wasps differs in several respects from other insects. The presence of a caste system in some polyembryonic wasps also offers insights into conditions favoring the evolution of extreme reproductive altruism.

LIFE HISTORY AND DEVELOPMENT OF POLYEMBRYONIC WASPS

Polyembryony in insects was first described by the Italian naturalist F. Caldani almost 200 years ago. The development of polyembryonic wasps has been most thoroughly studied in the encyrtid wasp *Copidosoma floridanum,* which parasitizes the eggs of plusiine moths such as the cabbage looper *Trichoplusia ni* (Fig. 1A). Unlike terrestrial insects, whose eggs contain large amounts of yolk and are enclosed in a rigid chorion, *C. floridanum* eggs contain no yolk and are surrounded by a thin chorion (Fig. 1B). After oviposition, the egg undergoes complete (holoblastic) cleavage to form a single embryo that ruptures out of its chorion to continue development unconstrained by the eggshell. This embryo is called the primary morula and consists of approximately 200 embryonic cells surrounded by an extraembryonic membrane of polar body origin (Figs. 1C and 1D). After the host emerges from its egg, the cells of the primary morula continue to divide inside the host larva to form additional embryonic masses that consist of embyronic cells surrounded by the extraembryonic membrane. These embryonic masses are called secondary morulae and the entire mass is called a polymorula (Fig. 1E). All morulae formed during this period of proliferation result from the invagination of the extraembryonic membrane and subsequent partitioning of embryonic cells (Figs. 1F–1H). The process of membrane invagination and cell partitioning repeats itself an indeterminate number of times during the first through the fourth instars of the host caterpillar. This results in an increasing number of secondary morulae per host, but each morula contains progressively fewer embryonic cells. For example, from 5 to 10 secondary morulae that each contain several hundred embryonic cells are present in a first-instar host, whereas an average total of 1200 morulae that each contain about 20 cells are present in a fourth-instar host.

A few morulae in the polymorula differentiate during the host's first through fourth instar and develop into precocious larvae (Fig. 1I). One or two precocious larvae develop when the host is a first-instar larva and from 2 to 10 precocious larvae develop during the host's second through fourth instar. As a result, the number of precocious larvae increases progressively as the host caterpillar grows. An average total of 40 precocious larvae develop, which equals approximately 4% of the total number of larvae produced per host. The remaining embryos always differentiate in the host's fifth instar and develop into reproductive larvae (Fig. 1J). Reproductive larvae rapidly consume the host and then pupate inside the remnant cuticle, forming a mummy (Fig. 1K). In contrast, precocious larvae never molt and die from desiccation after the host is consumed by their reproductive-caste siblings. Although no detailed embryological information is available outside *C. floridanum,* descriptions of polyembryonic braconids, platygasterids, and dryinids suggest that these wasps also lay small, yolkless eggs that undergo complete cleavage to form a single embryo. This embryo then proliferates into additional embryos. However, polyembryonic wasps in these families produce only embryos that develop into reproductive larvae (i.e., no precocious larvae or caste system exists).

REGULATION OF POLYEMBRYONIC DEVELOPMENT

Caste formation in social insects like ants, bees, or termites is usually mediated by environmental factors (photoperiod, crowding, pheromones, nutrition) acting on endocrine physiology. Environmentally induced alterations in hormonal

Host stages

Egg   **Larva**   **Mummy**

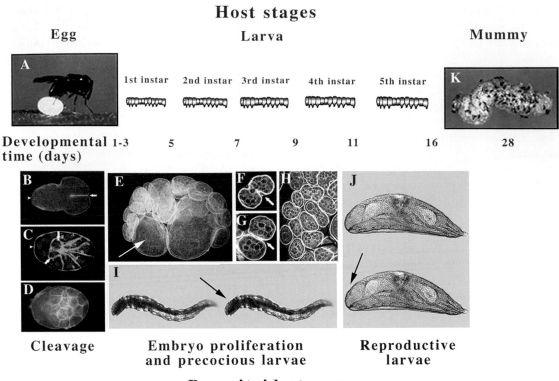

	1st instar	2nd instar	3rd instar	4th instar	5th instar	

Developmental 1-3   5   7   9   11   16   28
time (days)

Cleavage   **Embryo proliferation**   **Reproductive**
                 and precocious larvae   **larvae**

Parasitoid stages

FIGURE 1 Development of the polyembryonic wasp *C. floridanum*. The top shows the egg, larval, and mummy stages of the host, the middle shows developmental time from egg laying to emergence of adult wasps from the mummy, and the bottom shows key events during embryogenesis of the wasp. (A) The female wasp oviposits into the egg stage of the host, *T. ni*. (B) Soon after oviposition, the 50-μm-long wasp egg undergoes the first cleavage that results in two equal-sized blastomeres at the posterior (arrow) and a polar cell containing the polar nucleus at the anterior (arrowhead). (C, D) The syncytial extraembryonic membrane (arrowhead) derived from the polar cell begins to envelop the blastomeres that continue to divide (arrows). Soon after envelopment begins, the embryo ruptures out of the chorion, and by 24 h after oviposition the primary morula has formed. (E) At the end of the host's egg stage and throughout the first through fourth instar, embryo proliferation occurs to produce an increasing number of secondary morulae. Some embryos during this period undergo morphogenesis (arrow) and develop into precocious larvae. (F, G) Additional secondary morulae are formed during proliferation by ingrowth of the extraembryonic membrane (arrows) surrounding an embryo, which partitions the enclosed embryonic cells and results in formation of two embryos. (H) More than 1000 secondary morulae are formed in this manner by the host's fourth instar. (I) Precocious larvae formed during the host first through fourth instar are serepentine in shape and possess a well sclerotized head (arrow). (J) Reproductive larvae that emerge during the host fifth instar are rounded in shape and have a very weakly sclerotized head (arrow). These larvae consume the host at the end of the fifth instar and pupate inside the host cuticle to form a mummy. (K) One-millimeter-long adult wasps emerge from the mummy in approximately 28 days.

state in turn affect the development and/or behavior of individuals in the colony to produce distinct castes. In contrast, insect embryogenesis is primarily regulated endogenously by the coordinated expression of maternal and zygotic factors known as patterning genes. Signaling molecules and determinants associated with specific cell lineages also play a significant role in regulating embryogenesis of other animals. Within this framework, caste formation and embryogenesis in polyembryonic wasps could be regulated by both environmental signals from the host and endogenous factors in embryos.

Polyembryonic wasps differ from other caste-forming insects in that both castes develop from the same egg, coexist in the same environment (the host), and develop in proximity to one another in the polymorula. Nonetheless, the development of precocious larvae during the host first through fourth instar and reproductive larvae in the host fifth instar suggests that host environment may influence caste formation in *C. floridanum*. Molting and metamorphosis of the host, like in

most insects, are regulated by the steroid ecdysone and sesquiterpenoid juvenile hormone (JH). A rise in the ecdysteroid titer stimulates the host to molt, whereas the JH titer influences the type of molt that occurs. *T. ni* larvae parasitized by *C. floridanum* molt and develop almost identical to normal, unparasitized caterpillars. The obvious exception is that *C. floridanum* reproductive larvae consume the parasitized larva at the end of its fifth larval instar, whereas unparasitized *T. ni* pupate after the fifth instar. The fact that host JH titers are elevated in early larval instars and decline in the final instar suggests that embryos develop into precocious larvae when the host JH titer is elevated, whereas embryos develop into reproductive larvae under conditions of low host JH titer.

However, a number of different experiments collectively indicate that neither JH, ecdysone, nor other environmental factors are the primary factors that regulate morula proliferation, morphogenesis, or the proportion of embryos that develop into each caste. Instead, these events appear to be regulated by endogenous factors in individual embryos. One class of endogenous factors involved in regulation of morphogenesis contains the same genes that control pattern formation in all insects. Most insects (*Drosophila* is the best studied insect in this regard) lay eggs that are rich in yolk and that have a rigid chorion that protects the embryo from injury and desiccation. At the beginning of embryogenesis, *Drosophila* and most other insects undergo a syncytial phase of development whereby nuclear divisions occur without cell division to produce thousands of nuclei in a single-celled embryo. This syncytial phase is crucial for axial patterning as it allows RNAs of the maternal coordinate genes *bicoid* and *nanos* to be properly distributed within the egg. These maternal RNAs are then translated to form a gradient of regulatory proteins that control where the head, thorax, and abdomen of the embryo will develop. Localization of these factors in the posterior of the egg is also coupled to localization of another set of determinants that specify germ cells, which will give rise to sperm or eggs when the insect becomes an adult. In effect, the entire body plan and reproductive capacity of the future insect are determined immediately after the egg is laid. After syncytial cleavage, insect embryos cellularize to form a blastoderm. Subsequent specification of individual thoracic and abdominal segments is then regulated by a cascade of zygotic genes (*gap, pair-rule, segment polarity,* and *Hox* genes).

The embryos of polyembryonic wasps differ from those of *Drosophila* and most other insects in several ways. First, the eggs of polyembryonic wasps lack yolk and a rigid chorion, and embryogenesis proceeds after the embryo emerges from the eggshell. Second, the polarity of the egg is not directly linked to the axial polarity of future larvae, given that the primary morula proliferates to form thousands of embryos before morphogenesis. Since each embryo is randomly oriented within the polymorula, axial polarity must be reestablished after proliferation. Third, polyembryonic wasp eggs lack a syncytium and instead cellularize at the four-cell stage. Despite these differences, many of the genes that regulate pattern formation in other insects are also operative in polyembryonic wasps. However, all of these patterning genes are expressed after embryos begin morphogenesis and none of them are expressed during embryo proliferation, when the number and types (precocious or reproductive) of embryos increase. This suggests that patterning genes have not been co-opted to perform these novel functions in polyembryonic wasps. However, other factors that could play a role in regulating proliferation and caste fate are genes that control germ cell specification and certain cell signaling molecules that establish pattern in tissues such as imaginal discs.

THE EVOLUTION OF POLYEMBRYONY

The idea that polyembryony has evolved several times in the Hymenoptera but has not arisen in any other insects outside the Strepsiptera suggests that wasps possess unique preadaptations that have favored the evolution of this unusual form

of development. The Hymenoptera exhibit incredible diversity in life history (Fig. 2). Most hymenopterans are in the suborder Apocrita, which consists of free-living species (pollen-nectar feeders and predators) and parasitoids (Fig. 2). Typical free-living apocritans such as ants and bees are restricted to a single group known collectively as the Aculeata (Chrysidoidea, Vespoidea, and Apoidea), whereas the other major superfamilies comprise primarily parasitoids (Fig. 2).

Parasitoids exhibit two different developmental strategies. Some species develop as ectoparasitoids that lay their eggs externally on hosts and feed as larvae by rasping a hole through the host's cuticle. Other species develop as endoparasitoids that inject their eggs into the hemocoel of the host where the larvae feed on blood or tissues. Phylogenetic analyses indicate that all aprocritans likely evolved from an ectoparasitic ancestor related to contemporary orussoids (Fig. 2).

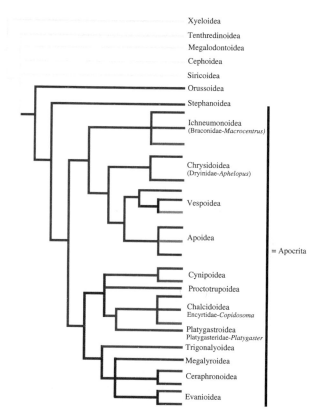

FIGURE 2 Consensus phylogeny and the evolution of different life histories in the order Hymenoptera. The listed life histories are herbivores (phytophagous) (yellow), ectoparasitoids (red), endoparasitoids (blue), gall formers (green), pollen-nectar feeders (magenta), and predators (gray). Unknown relationships are indicated in black. The specific families and genera in which polyembryony has evolved are indicated under the appropriate superfamily. The most primitive hymenopterans include several superfamilies of herbivores commonly known as sawflies. The advanced hymenopterans (Apocrita) form a monophyletic assemblage derived from an ectoparasitoid ancestor likely similar to orrusoids, which parasitize wood-infesting beetles. Endoparasitism and other life histories have thereafter arisen multiple times in different superfamilies from ectoparasitic ancestors. Among endoparasitic lineages, polyembryony is known to have evolved four times. (Figure developed from the consensus phylogenies and discussion of Whitfield, 1998).

Thereafter, endoparasitism evolved independently at least eight times from different ectoparasitic ancestors. The four hymenopteran families that contain polyembryonic species are in different superfamilies (Ichneumonoidea, Platygastroidea, Chalcidoidea, and Vespoidea), and the nearest relatives to each polyembryonic taxon are monoembryonic endoparasitoids. Thus, polyembryony itself has evolved at least four times and in each case this form of development arose from a monoembryonic, endoparasitic ancestor (Fig. 2).

The correlation between endoparasitism and the subsequent evolution of polyembryony is unlikely to be coincidental. Free-living insects such as *Drosophila,* ants, or honey bees develop in a terrestrial environment independent of the parent. Adaptations for survival include a thick chorion to protect the embryo and an abundant yolk source to supply nutrients for development. Ectoparasitoids encounter the same environmental conditions, but endoparasitoids develop in the nutrient-rich blood of another insect where protection from desiccation and prepackaging of a yolk source are no longer required. Unconstrained by the need for a prepackaged source of nutrition (yolk) and a chorion, the potential exists for embryos of endoparasitoids to shift away from syncytial cleavage and toward early cellularization and complete cleavage. Unconstrained by a rigid chorion, the embryos of endoparasitoids also have the potential to increase significantly in volume during embryogenesis. If endoparasitism is an important preadaptation for the ultimate evolution of polyembryony, the loss of yolk and alterations in early development associated with polyembryonic wasp eggs should also occur in at least some lineages of monoembryonic endoparasitoids. This in fact is seen in several different families of endoparasitic Hymenoptera, which supports the idea that endoparasitism has selectively favored alterations in early development essential for the evolution of polyembryony.

THE FUNCTION AND EVOLUTION OF A LARVAL CASTE SYSTEM

The evolution of specialized castes has occurred most often in species that live in groups comprising closely related individuals and that occupy resource-rich but defensible resources. These conditions exist among well-known eusocial insects such as ants and termites as well as among less studied caste-forming groups such as gall-making thrips and aphids. These conditions also exist among polyembryonic wasps that propagate clonally inside the nutrient-rich but defensible environment of the host. Reproductive larvae are so named because they are the only offspring that molt and develop into adult wasps. However, the insects parasitized by polyembryonic wasps are also commonly parasitized by other parasitoids, resulting in frequent opportunities for interspecific competition to occur. Cruz reported in 1981 that precocious larvae function as soldiers that kill interspecific competitors by piercing their cuticle with their mandibles. This altruistic behavior is advantageous, because precocious larvae increase

their own fitness by ensuring the survival of their reproductive siblings. Since selection acts at the level of both the individual and the brood, the ratio of investment in reproductive and precocious larvae would also be predicted to vary depending on how large or small the threat from competitors might be. Such adaptive phenotypic plasticity occurs in *C. floridanum,* in which the proportion of embryos developing into soldiers changes from 4% in hosts that are not attacked by another parasitoid to 24% in hosts that are attacked by a competitor.

A second, more complex function of precocious larvae is that in some species, they kill male siblings and significantly distort the sex ratio of adult wasps that emerge from the host. Like most Hymenoptera, polyembryonic wasps are haplodiploid, which means that male offspring develop from unfertilized eggs and female offspring develop from fertilized eggs. *C. floridanum* female wasps usually lay two eggs per host (one male and one female) and as a consequence most broods are composed of both sexes. However, precocious larvae develop almost exclusively from female eggs such that soldiers are fully related to their sisters but not to their brothers. Moreover, female precocious larvae attack their brothers while still embryos, but not their sisters, resulting in most of the adult wasps (>95%) that emerge from the host being females. These females then mate with the surviving male wasps before dispersing to find new hosts. The tendency to kill brothers but not sisters likely arose as a consequence of genetic conflict between siblings. In effect, the host is a finite resource and precocious larvae make available more resources for their reproductive sisters to consume by killing their brothers. However, a few males almost always survive and these individuals are able to mate with most of their sisters as well as disperse and mate with females from other hosts.

See Also the Following Articles

Development, Hormonal Control of • Embryogenesis • Parasitoids

Further Reading

Baehrecke, E. H., Aiken, J. M., Dover, J. M., and Strand, M. R. (1993). Ecdysteroid induction of embryonic morphogenesis in a parasitic wasp. *Dev. Biol.* **158,** 275–287.

Cruz, Y. P. (1981). A sterile defender morph in a polyembryonic hymenopterous parasite. *Nature* **294,** 446–447.

Gilbert, S. F. (1994). "Developmental Biology." Sinauer, Sunderland, MA.

Grbic', M. (2000). "Alien" wasps and the evolution of development. *BioEssays* **22,** 920–932.

Grbic', M., Ode, P. J., and Strand, M. R. (1992). Sibling rivalry and brood sex ratios in polyembryonic wasps. *Nature* **360,** 254–256.

Grbic', M., and Strand, M. R. (1998). Shifts in the life history of parasitic wasps correlate with pronounced alterations in early development. *Proc. Natl. Acad. Sci. USA* **95,** 1097–1101.

Harvey, J. A., Corley, L. S., and Strand, M. R. (2000). Competition induces adaptive shifts in caste ratios of a polyembryonic wasp. *Nature* **406,** 183–186.

Ivanova-Kasas, O. M. (1972). Polyembryony in insects. *In* "Developmental Systems, Insects" (S. J. Counce and C. H. Waddington, eds.), Vol. 1. Academic Press, New York.

Nijhout, F. (1994). "Insect Endocrinology." Princeton University Press, Princeton, NJ.

Riddiford, L. M. (1985). Hormone action at the cellular level. *In* "Comprehensive Insect Physiology, Biochemistry, and Pharmacology" (G. A. Kerkut and L. I. Gilbert, eds.). Pergamon, Oxford.

St Johnston, D., and Nüsslein-Volhard, C. (1992). The origin of pattern and polarity in the *Drosophila* embryo. *Cell* **68**, 201–219.

Strand, M. R. (2000). Developmental traits and life-history evolution in parasitoids. *In* "Parasitoid Population Biology" (M. E. Hochberg and A. R. Ives, eds.). Princeton University Press, Princeton, NJ.

Strand, M. R., and Grbic', M. (1997). The development and evolution of polyembryonic insects. *Curr. Top. Dev. Biol.* **35**, 121–158.

Whitfield, J. B. (1998). Phylogeny and evolution of host–parasitoid interactions in Hymenoptera. *Annu. Rev. Entomol.* **43**, 129–151.

Population Ecology

Joseph S. Elkinton
University of Massachusetts

Population ecology deals with questions related to the density or number of individuals of a species in a habitat or location. Insect population ecologists try to understand why population densities of some insects fluctuate dramatically, but others show little variation in density. They want to know why some species are rare, whereas others are common. They seek to understand the mixture of factors affecting mortality and fecundity that together determine the characteristic density of a species and whether the density will decline or increase. Applied population ecologists try to understand the conditions that allow densities of pest insects to reach damaging levels, and they seek to modify those conditions to keep the densities below those levels. To isolate the effects of particular variables from the multitude of factors influencing population growth and survival, population ecologists apply a range of tools to understand population systems, including gathering data on density, fecundity, and mortality from natural populations, construction of mathematical models of population interactions, and experimental manipulation of populations in the field.

ESTIMATING INSECT ABUNDANCE

Research on the population ecology of an insect typically begins with the development of sampling techniques to determine the abundance of the insect in the habitat under study. When the habitat has definable boundaries, we can estimate the number of individuals in the population. More frequently, we estimate density or number of individuals per unit of habitat. Sampling methods vary tremendously with habitat type. Specialized sampling techniques have been developed for insects that live in the soil, in the air, on foliage, and on vertebrate hosts. A large scientific literature exists on the methods developed for each of these habitats.

One can make several general distinctions about different methods of sampling insects. One of these is to distinguish between absolute and relative measures of density. Absolute measures quantify numbers per unit area or volume of habitat. Examples are numbers of grasshoppers per square meter in quadrat samples or numbers of Collembola per cubic centimeter of soil. Relative measures express numbers per sample unit, such as numbers of insects captured in a sweep-net sample or numbers of male moths captured in pheromone-baited traps. Relative measures may or may not reflect absolute density. They are frequently much easier to obtain than absolute measures, but they may be influenced by many factors (e.g., air temperature) that affect the activity of the insect and its likelihood of being captured. Consequently, much research is needed to be sure that relative measures give an accurate indication of differences in absolute population density. In addition, there are population indices, which are indirect measures of density, such as measures of defoliation.

Another general characteristic of sample methods is that some, such as sticky traps or pheromone-baited traps, obtain data continuously over an extended interval of time, whereas others obtain an essentially instantaneous "snapshot" of a population at a particular moment. Examples of the latter include sweep-net samples, quadrat samples, and insecticide knockdown samples. Such instantaneous samples are susceptible to the effects of time of day or weather conditions at the time of sampling, which in turn may influence the number captured. Continuous samples may also be influenced by weather conditions but, because capturing occurs over an extended interval, these effects are averaged over a range of such conditions.

Any investigator planning a sampling program must think carefully about how the samples will be selected. Consultations with statisticians with expertise in experimental design prior to collection of data is always wise, as is preliminary sampling or pilot testing to estimate the expected amount of sample variability. Most sampling schemes may be classified as random, systematic, or stratified random designs. In random sampling, the sample units are chosen or placed at randomly selected locations in the sample area. Random selection is typically done by choosing coordinate points from a table or list of random numbers. In a systematic design, sample units are placed at regular intervals across the sample universe (e.g., every 20 m or every 10th plant). Systematic samples are frequently much easier to conduct, and they ensure that the samples are distributed evenly across the sample universe. However, systematic samples violate the requirement that samples be selected independently and at random from the sample universe, imposed by most statistical analyses. Whether violation of this requirement leads to erroneous conclusions in any particular system is usually debatable. A reasonable compromise between the two approaches is a stratified random or randomized block design in which the sample universe is divided into regularly spaced subunits or blocks and samples are selected at random from each block. Differences in density between subunits caused by edge effects or density gradients across the field can be detected with appropriate statistical analysis.

Mark–recapture techniques are sometimes used to estimate insect abundance. Individuals are marked with a permanent mark that does not significantly alter the behavior of the insect or affect its survival. Marked individuals are counted and released into a population. Subsequently, a sample of individuals is collected from the population and the proportion *(m/n)* of marked individuals in the sample is recorded. Nearly all mark–recapture methods rest on the basic assumption that *m/n* of the sample is an estimate of the proportion of marked individuals in the larger population *M/N*. If one assumes that *M,* the number of marked individuals in the population, is equal to the number released, then the population size *N* can be calculated directly: *N* = *M(n/m)*. This calculation is known as the Lincoln index or Peterson estimate. As a rule, it cannot be assumed that *M* is equal to the number of marked individuals released because some of them will have died or emigrated in the interval between release and recapture. Several fairly elaborate techniques involving multiple rounds of mark and recapture have been developed to estimate *M* when it is unknown.

Applied ecologists frequently use sequential sampling to classify populations into density categories such as high (requiring treatment) or low (requiring no treatment). The basic idea is to minimize the cost of the sampling effort by collecting only the number of samples needed to make this decision with sufficient statistical reliability. Development of a sequential sampling scheme for a particular insect and crop or habitat requires research to define the population density that forms the dividing line between the high and low categories, a definition of the degree of allowable risk for making an incorrect classification, and an estimate of the degree of clumping or patchiness that typifies the species.

POPULATION GROWTH AND LIMITS TO GROWTH

A fundamental property of the population dynamics of all species is that the number or density of individuals will grow at an ever-increasing rate when conditions are favorable. The simplest example of such growth is illustrated by the replication of single-celled organisms. If a bacterium divides every hour, a colony that began with one individual would grow to 2, 4, 8, 16, ..., 2^t, where *t* is the number of hours or replications. With insects, the rate of replication with each generation is potentially much faster. The house fly, *Musca domestica,* for example, has generation time of approximately 2 weeks and lays more than 200 eggs. If half those eggs were females and all survived to maturity, the population would increase by more than a hundred fold each generation. By the end of one year the population would have increased to more than 10^{52} individuals (a mass of flies much larger than the earth). Even if mortality were much higher, (e.g., only 1 fly in 10 survived to maturity) the end result would be similar but would take longer to achieve. Mathematically, we refer to this process as geometric or exponential growth, and for

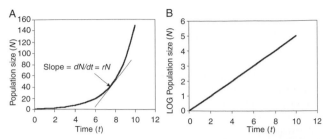

FIGURE 1 (A) Exponential growth of a population yields a straight line when density *(N)* is plotted versus time *(t),* as in (B), on a logarithmic scale.

insects that reproduce continuously, it is typically expressed with the following equation (Fig. 1):

$$\frac{dN}{dt} = rN \qquad \text{or} \qquad \frac{dN}{dt}\frac{1}{N} = r$$

Here *N* is the population size or density, *dN/dt* is the growth rate (the change in density per unit time), and the constant *r* is the instantaneous per capita rate of increase. The parameter *r* equals instantaneous per-capita birthrate minus the instantaneous per-capita death rate. When the rates of birth and death are equal, *r* = 0 and the population ceases to grow. When the death rate exceeds the birthrate, *r* is negative and the population declines.

The notion of exponential population growth was first expressed in 1798 by T. R. Malthus, who observed that human populations increased exponentially, whereas food production increased in a linear fashion. As a result, Malthus predicted that mankind would be doomed forever to a life close to the edge of starvation. Malthus did not foresee the large increase in agricultural production that accompanied the Industrial Revolution, but he may yet prove to be correct. The ideas of Malthus were fundamental to the development of Darwin's theory of evolution: all organisms must struggle to survive because all species produce more offspring than can survive.

It is obvious that no population can continue growing forever; sooner or later, it will reach a density above which individuals can no longer obtain the resources they need to survive. This density is known as the carrying capacity of the environment. For different species in different habitats, the carrying capacity will be determined by competition for particular resources. For desert plants, water is typically the limiting resource. For many animals, food supply determines the carrying capacity. As a population expands toward the carrying capacity, the rate of growth slows down. This process is typically represented by the following logistic equation (Fig. 2), which was first applied to population growth by P. F Verhulst in 1838 and independently by R. Pearl and L. J. Reed in 1920:

$$\frac{dN}{dt} = rN - \frac{rN^2}{K}$$

Here *K* is the carrying capacity and *r* and *N* retain their earlier definitions. The first term *(rN)* represents exponential

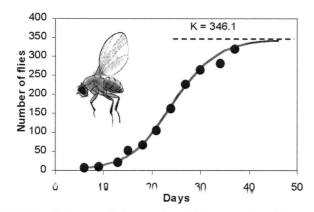

FIGURE 2 Logistic population growth model fit to data from laboratory colony of fruit flies. [Adapted from Krebs (2001) and Pearl, R. (1925). "Biology of Population Growth." Alfred A. Knopf, New York.]

growth. The effect of the second term (rN^2/K), often called environmental resistance, increases as N becomes large. As N approaches K, the rate of growth *(dN/dt)* approaches zero.

There are a number of assumptions inherent in the use of the logistic equation to represent population growth. The first is that unless disturbed, population density will approach the carrying capacity *(K)* and then remain stable about it. In actuality, most populations fluctuate in density, even populations kept in the laboratory under constant environmental conditions. Another assumption is that the shape of the curve is symmetrical above and below the midpoint. In fact, few population systems, even in the laboratory, follow the trajectory predicted by the logistic equation. Rather, the importance of the equation is its contribution to theoretical ecology. It captures the most basic processes of population dynamics: exponential growth and the effects of factors that limit growth. Variations of the logistic equation have been explored by many individuals; indeed, it is the foundation of a large body of work in theoretical population ecology. A. J. Lotka in 1925 and V. Volterra in 1926 extended the logistic to describe both competition between species and predator–prey interactions. More recent applications include models of food webs and interactions between many species in a community.

REGULATION OF POPULATION DENSITY

Although no one doubts that competition for resources confers an upper limit on the growth of all populations, it seems clear that many populations of insects persist at densities far below any obvious carrying capacity. Furthermore, the densities of most species fluctuate within a fairly narrow range of values. For a population to remain at constant density, the birthrate must equal death rate. Each individual must on average replace itself with one surviving offspring. Indeed, for any species to persist over evolutionary time, the average birthrate must equal the average death rate, although these quantities may vary considerably from year to year. Organisms that experience high

mortality compensate by producing lots of young. For this fundamental reason, most population ecologists believe that most populations are stabilized by factors that are called density dependent. Such factors influence birthrates or death rates in a way that varies systematically with density, such that populations converge to densities at which rates of birth and death are equal and the density is at equilibrium. Such factors act as a negative feedback system that is analogous to the regulation of room temperature by a thermostat. If densities rise above the equilibrium value, the death rate exceeds the birthrate and the population returns to equilibrium. If densities fall below the equilibrium value, the birthrate exceeds the death rate and the population increases.

Predators, pathogens, and parasitoids usually cause mortality to insects that is density dependent. The proportion or percentage of the population killed by these factors varies systematically with density. An increase in the proportion dying with increasing density is called positive density dependence; a decrease in the proportion dying with increasing density is termed negative or inverse density dependence. A mortality factor is density independent when the proportion killed varies in ways that are unrelated to population density. Many abiotic factors, such as mortality due to subfreezing temperatures, act in ways that are density independent. Although many insect population ecologists focus on sources of mortality, density-dependent changes in fecundity may either lead to changes in density or serve to stabilize densities. Certainly, competition for resources is a density-dependent process that will stabilize a population at the carrying capacity, if other factors do not intervene at lower densities.

For more than 50 years ecologists have debated whether population densities of most species are stabilized by such density-dependent factors. L. O. Howard and W. E. Fiske in 1911 were the first to articulate the idea that populations cannot long persist unless they contain at least one density-dependent factor that causes the average fecundity to balance the average mortality. Other early proponents of this idea in the 1950s were A. J. Nicholson and D. Lack. In contrast, H. G. Andrewartha and L. C. Birch argued in 1954 that most populations are not held at equilibrium density. Rather, densities merely fluctuate. In their view, most species avoid extinction because they comprise what we now call metapopulations, that is, a series of subpopulations that are linked by dispersal but whose densities fluctuate independently of one another. Extinction of subpopulations occurs quite frequently, but these are recolonized by individuals dispersing from other subpopulations, thereby allowing the species to persist indefinitely over the entire region. In recent years metapopulation dynamics has been explored by way of computer simulations, and these have revealed that in the absence of density dependence, such systems eventually go extinct.

The debate about the ubiquity of density-dependent processes has persisted despite the efforts of various ecologists to terminate the discussion either because it was bankrupt or because they deemed that prevalence of density dependence was too obvious to deny. Many ecologists have insisted that no conclusion can be made regarding the existence of density dependence in a population system unless its action can be demonstrated statistically in data collected from the populations. This has proved difficult to achieve. Until recently, adequate methods for detecting density dependence in population systems have been lacking, and earlier methods are now known to be statistically flawed. Several new methods were proposed in the decade following 1985, most of which involved a variety of computer-based resampling procedures, such as the bootstrap.

LIFE TABLES

Population ecologists typically summarize their collected mortality data in life tables, which express the proportion of the insects that survive to successive instars or life stages. The proportion dying in each instar or life stage can be partitioned into components caused by different agents of mortality. Techniques for quantifying mortality in this way were pioneered in the 1960s by R. F. Morris, who worked on spruce budworm, *Choristoneura fumiferana,* in Canada, and by G. C Varley and G. R. Gradwell in England, who studied the winter moth *Operophtera brumata,* a defoliator of oak trees in Europe. In 1960 Varley and Gradwell reported a 13-year study of winter moth survival to successive life stages at one site. They showed that predation on pupae was density dependent and was responsible for maintaining a low-density equilibrium. Their studies illustrated the benefits of collecting data on density, fecundity, and mortality from the same populations over many generations. They developed a graphical procedure known as key-factor analysis to identify the life stages or causes of mortality most responsible for observed fluctuations in density. In the winter moth system, the key factor was overwintering mortality. Subsequent work offered analytic procedures for key-factor analysis, but others, such as E. Kuno and T. Royama, noted statistical short-comings; as a result this procedure is rarely used.

Mortality from parasitoids and disease is typically measured by collecting a sample of hosts and dissecting them, or rearing them to determine the proportion that is parasitized or infected. This proportion, however, may or may not accurately reflect the total proportion of hosts attacked over a life stage or generation, since attack rates vary over time with the number of hosts available to attack and the number of adult parasitoids searching. In 1992 J. T. S. Bellows and colleagues reviewed the techniques that have been developed to convert proportion infected in samples to the summary values appropriate for life tables or for comparisons of parasitism between populations or treatments in a study.

PREDATORS, PATHOGENS, AND PARASITOIDS

Nearly all insect species are attacked by a suite of natural enemies and, as indicated earlier, these frequently act in a

density-dependent manner that serves to maintain populations at densities well below the carrying capacity determined by resource limitation. Different natural enemies will attack different life stages. Predators are mobile organisms that feed on many prey. Predators of insects include many vertebrate and invertebrate species. Parasitoids are insects whose immature stages develop on or within a single host individual, usually killing it in the process. The majority of parasitoids belong to the order Hymenoptera or Diptera. Pathogens are microorganisms including viruses, bacteria, microsporidia, or fungi that infect a host and cause disease. Biological control focuses on the use of natural enemies to maintain the densities of pest insects at levels below that at which they cause damage.

Density-dependent predation or parasitism may arise from two different sources: the numerical response and the functional response. The numerical response is an increase in the density or number of predators or parasitoids in response to increasing prey density. The numerical response can arise from increased reproduction or survival of predator or parasitoid offspring induced by increases in prey availability, or it can arise from an aggregative response whereby predators and parasitoids are attracted to sites with high densities of prey.

The functional response is an increase in the number of prey taken per predator or parasitoid at increasing prey density. Important contributions to the understanding of the functional response were made by C. S. Holling beginning in 1959. In laboratory experiments, Holling presented individual predators with different numbers of prey. He showed that the number of prey consumed over a specified time interval increased with number of prey available, but at a decreasing rate towards an upper maximum (Fig. 3A). This effect is caused by an upper limit in the predator's capacity for consumption and by the increasing proportion of time

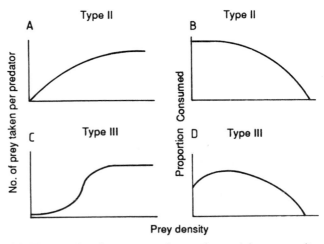

FIGURE 3 Number of prey consumed per predator and the corresponding proportion consumed for type II (A, B) and type III (C, D) functional responses. [Adapted from Holling, C. S. (1965). *Mem. Entomo. Soc. Can.* **45**, 3–60.]

devoted to handling the large number of prey at the expense of time spent searching for prey. Above this limit, further increases in prey density will not cause higher consumption. The proportion of prey consumed plotted versus prey density steadily declines (Fig. 3B), illustrating that the functional response is inherently inverse density dependent. Without a numerical response, predators and parasitoids are unlikely to stabilize a host population. Further work by Holling showed that under some important conditions, the functional response can lead to positive density-dependent predation. Whenever increases in prey density result in some change in the foraging behavior of the predator or parasitoid, such that foraging becomes more efficient or efforts are concentrated on the particular prey species, the number taken will accelerate with increasing host density (Fig. 3C), and the proportion taken will increase (Fig. 3D) over the lowest range of prey densities. Holling termed this a type III functional response in contrast to type II, which is the continuous decline in proportion taken evident whenever there is no change in foraging behavior in response to changes in prey density (Fig. 3A,B). Holling demonstrated a type III response for shrews foraging for sawfly pupae. He envisioned type III responses to be characteristic of vertebrate predators, which have a relatively high capacity for learning and behavioral change. However, the type III functional response has subsequently been demonstrated in many insect predators and parasitoids.

Specialist or monophagous natural enemies are those that attack a single host species. Oligophagous natural enemies restrict their attacks to a closely related group of species. Generalist or polyphagous natural enemies attack a wide range of host species. The distinction is important because generalists and specialists typically respond very differently to changes in host density. Specialists are most likely to exhibit a numerical response to changes in density of their prey because they depend on no other food sources and their seasonal development is closely linked with that of their prey. Generalists may exhibit little or no numerical response, because they depend on many types of prey and may shift from one prey to another, depending on which species are available. In fact, it is very common for many natural enemies, especially generalists, to exhibit inverse density dependence, wherein mortality declines as prey density increases. Such mortality cannot stabilize prey densities unless accompanied by a numerical response.

Many systems exhibit complex density dependence; for example, mortality from particular natural enemies may switch from positive to negative as host density increases. Thus, bird predation on forest-dwelling caterpillars that is density dependent at the lowest density may shift to inverse density dependence as the densities exceed the capacities of the predators to respond numerically and the functional response approaches the upper limit of prey consumption. Under such conditions, the prey densities may "escape" into an outbreak phase, which is characteristic of a few species. Outbreak populations are typically subject to a different suite of density-dependent factors, such as viral diseases and starvation, which become major sources of mortality only when densities are high. These factors can maintain populations at a high-density equilibrium, but more frequently they cause the collapse of populations back down to a low-density, endemic phase. In contrast, generalist predators, which might consume the majority of prey individuals at low density, are likely to consume a tiny fraction of the population at high density even though they are attacking the same or higher numbers of prey individuals at these high densities.

In 1976 T. R. E. Southwood and H. N. Comins proposed a "synoptic model" as a general feature of insects that occasionally go into an outbreak phase (Fig. 4). Earlier expressions of this idea can be found in the writings of R. F. Morris and R. M. Campbell. The model is depicted by plotting R_0, the net reproductive rate, against density. At intermediate densities in the "natural enemy ravine," density-dependent mortality caused by natural enemies maintains the population at equilibrium ($R_0=1$). The natural enemy ravine separates two "ridges," one at high and one at low density, where mortality is lower and population densities increase. At very high density, other mortality factors such as starvation and disease cause the populations to collapse. At the extremes of low density, an "Allé effect" comes into play, caused by the failure of individuals to find mates and reproduce. Populations in this range decline inexorably to extinction. Such low densities are infrequent in most natural populations.

POPULATION MODELS

Theoretical ecologists have developed numerous population models to study the effects of natural enemies on their prey. Much insight has been gained from simple mathematical expressions that relate prey density to that of changes in density of specialist predators or parasitoids. One approach, pioneered by A. J. Lotka and V. Volterra in the 1920s, entailed a simple modification to the logistic equation by

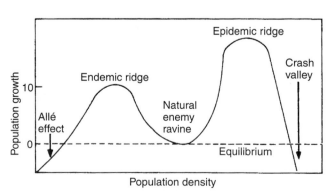

FIGURE 4 Synoptic model of complex density dependence. [Adapted from Southwood, T. R. E., and Comins, H. N. (1976). *J. Anim. Ecol.* **45**, 949–965, British Ecological Society.]

adding a term that represents prey consumption on densities of both host and predator:

$$\frac{dN}{dt} = r_1 N - k_1 PN$$

and

$$\frac{dP}{dt} = -r_2 P + k_2 PN$$

where N and P are the respective densities of host and predator, and the rates of population growth are given by dN/dt and dP/dt. In the first equation, the first term represents exponential growth of the host ($r_1 N$) in the absence of the predator, whereas in the second equation, the first term represents exponential decline in the predator ($-r_2 P$) in the absence of the host. The second term in each equation represents the effects of predation, which is determined by the encounter rate of host and predator and is proportional to PN. The constant k_1 is a measure of the ability of prey to escape predators, and k_2 is a measure of the ability of predators to capture the prey. The model predicts a predator–prey oscillation (Fig. 5A). The changes in density of the predator or parasitoid lag behind those of its host. The highest rates of attack on the host occur at peak predator density, which is observed after the host population density

has declined. Plotting percent mortality from the predator or parasitoid against host density would not reveal a consistent pattern of positive density dependence, even though the predator or parasitoid is clearly maintaining the host within a narrow range of density. This type of response is termed delayed density dependence. In the 1990s, P. Turchin and others developed new statistical methods based on time-series analysis for detecting delayed density dependence in annual census data. Laboratory studies of predator–prey interaction frequently show such predator–prey oscillations (Fig. 5B).

A different class of models appropriate for parasitoids and host populations with discrete generations was initiated by W. R. Thompson in 1924 and by A. J. Nicholson and V. A. Bailey in 1935. These models were difference equations, in contrast to the differential equations of the Lotka–Volterra type. The general form of the model expresses host or prey density N in generation $t + 1$ as follows:

$$N_{t+1} = \lambda N_t f(N_t P_t)$$

Here λ is the rate of increase per generation of the host in the absence of parasitism and $f(N_t P_t)$ is the proportion of hosts surviving parasitism in the preceding generation *(t)*. Similarly, the number of parasitoids in the next generation P_{t+1} is given by

$$P_{t+1} = cN_t[1 - f(N_t P_t)]$$

where $(1 - f(N_t P_t)$ is the proportion of hosts attacked by parasitoids in generation t and c is the number of surviving parasite progeny produced per parasitized host. The notation $f(N_t P_t)$ stands for any function of N_t and P_t. Variations in the model involve incorporating different factors into $f(N_t P_t)$. The simplest version for $f(N_t P_t)$, proposed by Nicholson and Bailey, assumes that all hosts are equally likely to be attacked and that parasitoids search at "random," such that the proportion of hosts that escape is given by the zero term of the Poisson distribution. This model predicts that hosts and parasitoids will experience density oscillations of ever-increasing amplitude until both go extinct. Obviously, this does not occur routinely in nature.

In the 1960s, M. P. Hassell, R. M. May, and colleagues began an exploration, extending over several decades, of the various factors that would stabilize host–parasitoid interactions in models of this type. These factors included mutual interference of parasitoids, patchiness of hosts or parasitoid attacks, and variation in host susceptibility. For example, Fig. 6 illustrates how an increase in the magnitude of mutual interference between parasitoids will reduce parasitoid efficiency at high host density and thus stabilize the system. When mutual interference is low (Fig. 6A,B), or absent altogether ($m = 0$, as in the original Nicholson–Bailey version), the model oscillates with increasing amplitude and both host and parasitoid go extinct. Other factors such as patchiness in parasite attacks have a similar effect.

The models described so far are relatively simple. They contain a small number of parameters or variables and leave out much of the known biology of the host and its natural

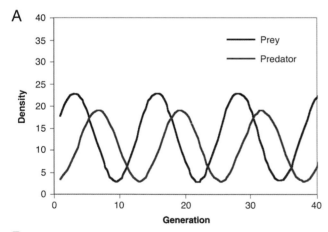

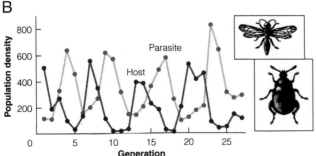

FIGURE 5 (A) Predator–prey oscillation as predicted by a Lotka–Volterra model and (B) Host–parasitoid oscillation of the Azoki bean weevil in laboratory culture. [From Krebs (2001), adapted from Utida, S. (1957). *Ecology* **38**, 442–449.]

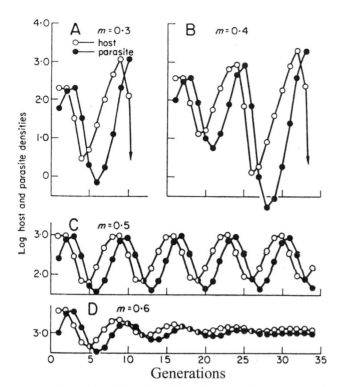

FIGURE 6 Stabilizing effects of mutual interference *(m)* between parasitoids on models of the Nicholson–Bailey type. [From Varley *et al.* (1973); adapted by permission from Hassell, M. P., and Varley, G. C. (1969). *Nature* **223**, 1133–1137, copyright 1969 Macmillan Publishers Ltd.]

enemies. Theoretical ecologists focus on such models because they can be analyzed by a variety of mathematical tools and can be used to address questions of general ecological significance. They hope that the models capture the essential features of the systems they represent. Applied ecologists, in contrast, are often drawn to more complex models, because they wish to understand the complex interplay of environmental and biotic variables that account for the density fluctuations of particular species of interest. With modern computers, there is virtually no limit to the complexity that can be built into such models, but this does not mean that the resulting simulations will necessarily be useful or revealing. Many highly complex models constructed in the 1970s, when high-speed computers first became widely available, were abandoned because they failed to accurately predict the behavior of the systems they represented and were too complicated to understand. Many ecologists have advocated models of intermediate complexity.

COMPETITION AND TROPHIC INTERACTIONS

When resources are limiting, intraspecific competition prevents growth of a population beyond the carrying capacity of the habitat. Interspecific competition occurs when two or more species compete for the same resource. Ecologists have long appreciated the importance of interspecific competition in shaping community structure and the interaction between species. Models of competition developed in the 1920s by

Lotka and Volterra, analogous to those for predation, predicted that when two species compete for the same resource, under most conditions one species will win out and the other will go extinct. This phenomenon, which became known as competitive exclusion, was readily demonstrated in laboratory populations by G. F. Gause in the 1930s with microorganisms, and by T. Park and associates in the 1950s with flour beetles (*Tribolium* spp.). Coexistence occurs only when species do not overlap completely in their use of resources. To put it another way, two species cannot coexist if they occupy the same niche. Clear examples of competitive exclusion of insects in the field are harder to come by. Most examples involve introductions of insects to regions outside their native range. For example, the Argentine ant *Linepithema humile* has been introduced to many regions of the world and has outcompeted and excluded many native ant species. In California around 1900, biological control introductions of *Aphytis* parasitoids of California red scale, *Aonidiella aurantii,* resulted in the establishment on citrus of *Aphytis chrysomphali.* After further introductions in the 1940s and 1950s, this species was displaced by *A. lingnanensis,* which in turn was displaced by *A. melinus* in the hot interior, but not the coastal, regions of southern California.

In many other systems, however, different insect species coexist, even though they appear to use the same resource in the same way. For example, in 1981 J. H. Lawton and D. R. Strong examined the various coexisting herbivores of bracken fern and found no evidence for competition. A famous experiment in the rocky intertidal habitat reported by R. T. Paine in 1974 suggested a general explanation. The intertidal community consists of various species of filter feeders (e.g., barnacles) and other invertebrates that coexisted, even though they competed intensely for space on the rock surfaces. Paine removed starfish, the top predator in this system, and as a result, one species (a mussel) outcompeted and excluded all the other invertebrates and took over the site. Paine termed the starfish a keystone predator, meaning that it had dominant effect on the number of species present in the entire community. The implication was that such predators in many communities prevent competitive exclusion among their prey by maintaining densities below the level at which competition would cause one species to predominate.

Population and community ecologists frequently portray the interaction of organisms in a community in terms of trophic levels or food webs. At the bottom trophic level are the primary producers, the green plants that use photosynthesis to sequester energy from the sun. The next trophic level is occupied by herbivores that feed on the plants. Carnivores that feed on the herbivores or on other carnivores occupy higher trophic levels. A typical food web involving insects on collard plants is illustrated in Fig. 7.

An influential paper by N. G. Hairston and colleagues in 1960 proclaimed that the importance of competition for resources, as opposed to regulation by natural enemies, varies with trophic level. These authors proposed that densities of primary producers are typically governed by competition for

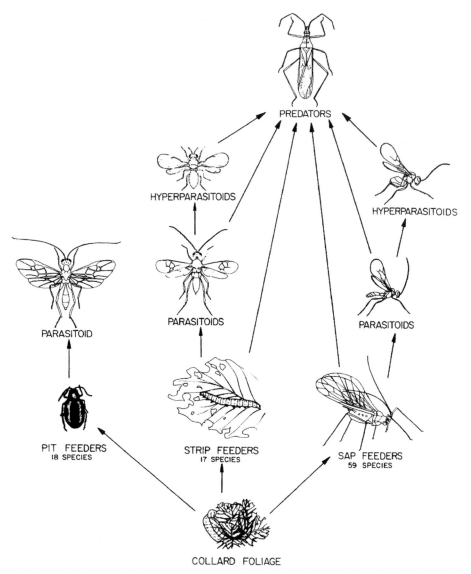

FIGURE 7 Schematic of arthropod food web in collards. [From Price, P. W. (1984). *In* "Ecological Entomology" (Huffaker, C.B., and Rabb, R. L., eds.), pp. 20–50. Wiley Interscience, NY; adapted from Root, R. B. (1973). *Ecological Monographs* **43**, 95–124.]

light, nutrients, or water. Herbivores, on the other hand, are typically regulated by their natural enemies at densities well below carrying capacity, so that competition for resources is unimportant. In contrast, carnivores typically limit the densities of their prey and thus are regulated by competition for resources. Regulation of population density by natural enemies has been termed "top-down" control. Regulation by way of competition for resources or by other interactions between an animal and its food supply has been termed "bottom-up" control.

The generalizations of Hairston and colleagues were widely criticized and many obvious exceptions exist, but variations on this idea have persisted. More recent versions have included effects of several levels of carnivores and other factors such as community productivity or environmental stress. Experimental manipulations, particularly in aquatic systems, have provided

support for the general idea in the form of trophic cascades. For example, in 1990 M. E. Power reported the experimental exclusion of top carnivores (large fish) from a food web in a California river. The results were a 10-fold increase in smaller fish and predatory insects that in turn caused an 80% decrease in herbivores (larval chironomid midges) and a 3- to 120-fold increase in the primary producers (algae). These results were the opposite of those predicted by Hairston and coworkers because there were four trophic levels, instead of three. Omnivores that feed simultaneously on several trophic levels complicate our understanding of most natural food webs. Research in several agricultural crops shows that intraguild predation can cause counterintuitive effects of predators on herbivorous insects. For example, in 2001 W. E. Snyder and A. R. Ives showed that carabid predators in alfalfa feed on pea aphids. Since, however, they feed even more heavily on the aphid "mummies" that contain parasitoids of pea aphids, their net impact is to increase aphid densities.

The relative importance of top-down versus bottom-up regulation of herbivores has been debated for several decades. A review of population studies by D. R. Strong and colleagues in 1984 concurred with Hairston and colleagues that competition among most insect herbivores was rare, because herbivore densities were typically kept far below carrying capacity by natural enemies. More recent reviews (e.g., by R. Denno *et al.* in 1995) have challenged this view. One issue is that many plants have sophisticated chemical defenses and many plant parts, such as foliage, are nutrient poor. Many herbivores consume specialized plant parts that are either nutrient rich (e.g., seeds) or are poorly defended (e.g., new leaves). Competition for these resources may be intense. Furthermore, plants influence both the survival and fecundity of herbivores in many subtle ways. For example, all herbivores disperse and must locate suitable host plants; failure to find host plants may be a dominant source of mortality and may vary from one habitat to another. Spring-feeding foliage feeders must synchronize their emergence with that of host foliage; failure to synchronize may cause mortality or may reduce fecundity because of consumption of inadequate or poor-quality foliage. Such factors are rarely documented in life table studies. For most insect herbivores, a mixture of top-down and bottom-up forces determines density.

TEMPORAL PATTERNS OF FLUCTUATION

Density fluctuations, whether they are large or small, may occur at erratic or at regular intervals. Simple population models of predators and prey or laboratory studies, as already discussed, often exhibit regular oscillations. Evidence, however, for regular oscillations in nature is rare. The density of most species fluctuates erratically (Fig 8). This is not surprising because the density of most species is influenced by weather conditions in a multitude of ways. Except for obvious regularity of seasonal variation, weather conditions vary from week to week and year to year in ways that are mostly erratic. Furthermore, most species are influenced by a large suite of natural enemies. Predator–prey oscillations, such as those that occur in simple theoretical models or simple laboratory systems, become erratic and irregular when many natural enemies are involved.

Although the density fluctuations of most species are erratic, there are a few that exhibit regular cycles of abundance, similar to what is observed in laboratory populations or in theoretical models. For many others it is not obvious whether the fluctuations are or are not regular. For example, which of the insects in Fig. 8 have a regular component in their fluctuations, and what is the cycle period? Time-series analysis provides the statistical tool to answer these questions. Perhaps the most famous example among insects of regular cycles is that of the larch budmoth, *Zeiraphera diniana*, an insect that periodically defoliates larch forests in the European Alps (Fig. 9). The defoliator reaches peak density at 8- to 9-year intervals and seems to increase or decrease exponentially in the intervals between (linear on a log scale, as in Fig 1B). The changes in budmoth density were associated with increases in both parasitism and mortality from a virus disease in what looked like a classic case of delayed density dependence. In 1981 R. M. Anderson and R. M. May proposed a host–pathogen model for this system that has served as a template for similar models of other insect host–pathogen systems. Anderson and May believed that the pathogen alone could account for the regular cycles. Other

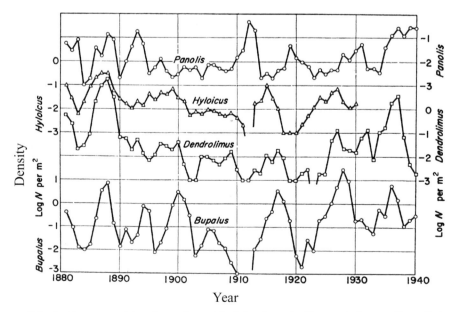

FIGURE 8 Sixty years of density fluctuations (plotted on a log scale) of four forest Lepidoptera in Europe. [From *Varley et al.* (1973); adapted from Schwerdtfeger, F. (1935). *Z Forst-u. Jagdw.* **67**, 15–38, 85–104, 449–482, 513–540 and Schwerdtfeger, F. (1941). *Z. Angew. Ent.* **28**, 254–303.]

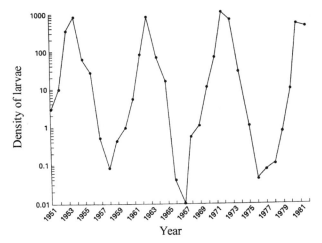

FIGURE 9 Regular cycles of larch budmoth density in the European Alps. [From Speight *et al.* (1999); adapted from Baltensweiler (1964). *Canad. Entomol.* **96**, 792–800.]

data suggested that effects of defoliation on host plant quality, as well as genetic variation in the budmoth itself, may also account for the density oscillations. This famous example illustrates the difficulty in distinguishing cause from effect in the confluence of bottom-up and top-down forces that account for cyclic behavior in some population systems.

Prior to the 1970s, ecologists assumed that erratic fluctuation evident in most populations was caused by erratic weather conditions or other sources of randomness. They assumed that the density fluctuations embodied in mathematical models or in simple laboratory populations would have simple dynamics. Densities of such systems would either remain at equilibrium or exhibit regular oscillations about the

equilibrium value, as in all the examples given earlier (Figs. 5, 6). The pioneering work on deterministic chaos by R. M. May in 1974 taught us otherwise. May studied the behavior of versions of the logistic equation appropriate for organisms with discrete or nonoverlapping generations (Fig. 10). There is a family of such models, of which the simplest is

$$X_{t+1} = aX_t - aX_t^2$$

where a is reproductive rate analogous to r of the continuous logistic, and X_t is density in generation t as a proportion of the carrying capacity K (i.e., $X_t = N_t/K$). When the reproductive rate was low (Fig. 10A), the population remains at equilibrium. At higher values of a, the system alternates between high and low values, a pattern known as a 2-point cycle (Fig. 10B). As a increases further, the system shifts to 4-point (Fig 10C), and then 8-point, ..., 2^n cycles. Above $a = 3.57$ however, an entirely new behavior, the chaotic regime, prevails, and densities fluctuate erratically without ever repeating themselves (Fig. 10D). These results had profound implications. They suggested that the erratic behavior characteristic of most natural populations (e.g., Fig. 8) might be due to the inherent mathematical properties of the interaction with natural enemies and other density-dependent effects, instead of resulting from random forces such as weather.

There ensued an effort to determine whether natural populations were indeed chaotic. Early studies concluded that most populations were not chaotic. These were based on attempts to fit natural populations to simple models and then to see whether the values of model parameters representing, for example, density dependence, time delays, or reproductive rate would elicit the expectation of chaotic behavior. The problem was that the conclusion depended on the particular

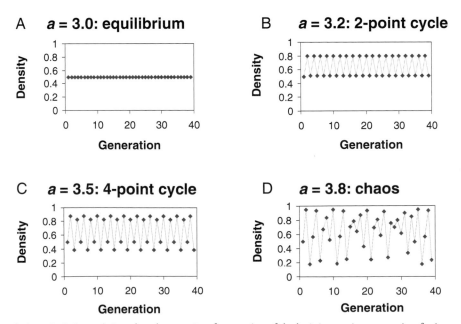

FIGURE 10 Density of a hypothetical population plotted versus time for a version of the logistic equation appropriate for insects with discrete generations under different values of the reproductive rate a illustrating (A) a steady state, (B) a 2-point cycle, (C) a 4-point cycle, and (D) chaos. [Adapted from May, R. M. (1976). *Nature* **261**, 459–467.]

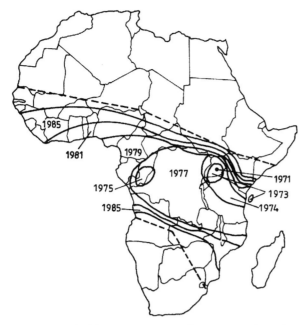

FIGURE 11 Spread of the cassava green mite from its point of introduction near Kampala in 1971. [Reproduced from Yaninek, J. S., de Moraes, G. J., and Markham, R. H. (1989). "Handbook on the Cassava Green Mite *(Mononychellus tanajoa)* in Africa: A Guide to Its Biology and Procedures for Implementing Classical Biological Control." International Institute of Tropical Agriculture, London.]

SPATIAL PROCESSES

Species that are introduced into new habitats will spread from the point of introduction to surrounding areas in a pattern that frequently resembles concentric rings (Fig. 11). Theoretical population biologists, beginning with J. G. Skellam in 1951, have developed models of population spread that couple processes representing population growth with that of dispersal. Dispersal is typically modeled as a random movement process that is analogous to molecular diffusion. These models typically predict that the leading edge of the infestation forms a traveling wave that moves across the landscape at a constant speed in any given direction. As illustrated in Fig. 11, speed of movement will in fact vary markedly in different directions with differences in prevailing winds or with the presence of barriers such as the Sahara Desert.

Density fluctuations of different populations of the same species frequently exhibit spatial synchrony, meaning that densities go up and down together over a large region, as illustrated by gypsy moth *Lymantria dispar,* in the northeastern United States (Fig. 12). Since there are several possible causes of spatial synchrony, including dispersal of animals between populations, increases in density in one population will produce emigrants that trigger increases in surrounding populations. By far the most common cause of spatial synchrony, however, is the Moran effect, named after the individual who studied the synchrony of lynxes and hares in boreal Canada. In 1953 P. A. P. Moran showed mathematically that the density fluctuations of two or more populations will synchronize, provided they are governed by the same factors, such as the same natural enemies and are also influenced by some common factor such as the weather, which varies from year to year in the same way on a regional

model used, which was always a simplistic abstraction of the inevitably complicated dynamics of real populations. Subsequent investigators offered techniques that were more general and free of assuming particular population models. Applications of these techniques have indicated that some population systems are chaotic, but most are not.

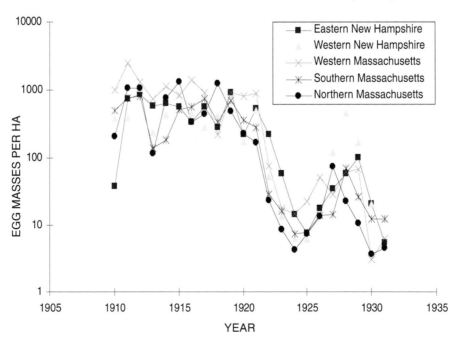

FIGURE 12 Spatial synchrony of gypsy moth populations in the northeastern United States. [Adapted from Campbell, R. W. (1981). *In* "The Gypsy Moth: Research Toward Integrated Pest Management" (Doane, C.C., and McManus, M.L., eds.), *USDA For. Serv. Tech. Bull. 1584,* pp. 65–86.]

spatial scale. The important point is that although variations in weather may cause spatial synchrony, we should not infer that weather is also responsible for fluctuations in density, which might typically be caused by interactions with predators, pathogens, and parasitoids.

CONCLUSION

Population ecology forms the theoretical foundations upon which management of insects is based. Much has been learned from the development of mathematical models of population systems, studies of laboratory populations, analyses of data from natural populations, and experimental manipulation of populations in the field. Despite decades of research, we are still far from understanding the population dynamics of most insects, even those such as gypsy moth or spruce budworm, which have been studied by several generations of researchers. Although various individuals have proposed theories to account for the dynamics of these species and have collected supporting data, rigorous demonstrations that these theories are correct elude us. For most insect species we have neither theories nor data to account for their population dynamics. The reason for this state of affairs is that it is extremely difficult to accurately measure the density of natural populations, especially when densities are low, and even more difficult to measure the impact of various factors causing mortality or variation in fecundity. Each insect and its natural enemies are embedded in a web of interactions whose intricacies must be disentangled. To understand how these intricacies vary over time, such information must be gathered for decades and coupled with experimental manipulations, that are often prohibitively expensive to conduct. The financial and human resources necessary to conduct such research are rarely assembled.

See Also the Following Articles

Biological Control • Insectivorous Vertebrates • Parasitoids • Pathogens of Insects • Predation

Further Reading

Begon, M. (1979). "Investigating Animal Abundance: Capture–Recapture for Biologists." University Park Press, Baltimore, MD.

Begon, M., Harper, J. L., and Townsend, C. R. (1996). "Ecology: Individuals, Populations and Communities," 3rd ed. Blackwell Science, Oxford, U.K.

Begon, M., Mortimer, M., and Thompson, D. J. (1996). "Population Ecology: A Unified Study of Animals and Plants," 3rd ed. Blackwell Science, Oxford, U.K.

Bellows, J. T. S., Van Driesche, R. G., and Elkinton, J. S. (1992). Life-table construction and analysis in the evaluation of natural enemies. *Annu. Rev. Entomol.* **37,** 587–614.

Denno, R., McClure, M. S., and Ott, J. R. (1995). Interspecific interactions in phytophagous insects: competition reexamined and resurrected. *Annu. Rev. Entomol.* **40,** 297–331.

Gilman, M., and Hails, R. (1996). "Ecological Modeling: Putting Theory into Practice." Blackwell Science, Oxford, U.K.

Hassell, M. P. (2001). "The Spatial and Temporal Dynamics of Host-Parasitoid Interactions." Oxford University Press, Oxford, U.K.

Krebs, C. J. (2001). "Ecology: The Experimental Analysis of Distribution and Abundance," 5th ed. Benjamin Cummings, San Francisco, CA.

Krebs, C. J. (2001). 'Ecological Methodology," 2nd ed. Benjamin/Cummings, Menlo Park, CA.

Price, P. W. (1997). "Insect Ecology," 3rd ed. Wiley, New York.

Royama, T. (1992). "Analytical Population Dynamics." Chapman & Hall, New York.

Seber, G. A. F. (1982). "The Estimation of Animal Abundance and Related Parameter," 2nd ed. Griffin, London.

Southwood, T. R. E., and Henderson, P. A. (2001). "Ecological Methods," 3rd ed. Blackwell Science, London.

Speight, M. R., Hunter, M. D., and Watt, A. D. (1999). "Ecology of Insects: Concepts and Applications." Blackwell Science, Oxford, U.K.

Strong, D. R., Lawton, J. H., and Southwood, T. R. E. (1984). "Insects on Plants. Community Patterns and Mechanisms." Harvard University Press, Cambridge, MA.

Turchin, P. (1995). Population regulation: old arguments and a new synthesis. *In* "Population Dynamics: New approaches and Synthesis" (N. Cappuccino and P. W. Price, eds.), pp. 19–41. Academic Press, San Diego, CA.

Varley, G. C., Gradwell, G. R., and Hassell, M. P. (1973). "Insect Population Ecology." Blackwell Science, Oxford, U.K.

Praying Mantid

see *Mantodea*

Predation/Predatory Insects

Ronald M. Weseloh
Connecticut Agricultural Experiment Station

P redators are animals that must consume more than one individual of another animal to complete their life cycle. Under this definition, the predatory habit occurs in a wide range of insect groups. Some orders, such as Odonata and Neuroptera, are wholly or predominately predaceous. Other orders contain large numbers of species that are predatory, such as Hemiptera, Coleoptera, Mecoptera, Diptera, and Hymenoptera. Even a few species in the orders Ephemeroptera, Orthoptera, Plecoptera, Thysanoptera, Trichoptera, and Lepidoptera are predaceous. This article emphasizes the diverse ways in which predaceous insects live.

MODES OF PREDATION

Hunting

Many insect predators carry out active hunting. The adults of tiger beetles (Cicindellidae), ground beetles (Carabidae), and many ants (Formicidae) actively run over the ground to capture prey. Many walking or crawling predators, including ladybird beetles (Coccinellidae), lacewing larvae (Chrysopidae), and syrphid fly larvae (Syrphidae), feed on sedentary insects such as aphids and scale insects. Other insect predators are agile flyers and actively snatch insects out of the air. Dragonflies and damselflies (Odonata) have large eyes and strong wings and feed mainly on mosquitoes and other small flying insects. Robber flies (Asilidae) also catch their prey on the wing, but will often tackle and subdue insects at least their own size, even bees and wasps. Their legs are very strong and they have piercing mouthparts to suck up prey juices. Other insects, such as many social wasps (Vespidae), pluck insects from the ground or vegetation while flying.

Many predaceous insects live in fresh water and pursue prey by swimming. The giant water bugs (Belostomatidae) are excellent swimmers and often capture and subdue small fish and tadpoles. Backswimmers (Notonectidae) hang ventral side up at the water surface and dart after aquatic organisms using powerful, oar-like hind legs. The diving water beetles (Dytiscidae) are also voracious predators that actively capture prey under water. Water striders (Gerridae) are true bugs that skim over the water surface and prey on small organisms that fall from above. Whirligig beetles (Gyrinidae) also hunt on the water film, but sometimes dive below the surface. Each eye is a double structure, with half adapted to seeing above the surface and the other half below.

Other active hunters are less visible. A large number of beetle and fly larvae, as well as immature insects from other orders, prey on small organisms by burrowing through soil, wet ground, rotting logs, or other vegetation. The immatures of many horse flies and deer flies (Tabanidae), for instance, are predaceous on organisms in moist soil and water. The larvae of some common midges (Chironomidae) feed on other organisms in the mud at the bottom of ponds or along its margins, as do the larvae of some crane flies (Tipulidae). Click beetle larvae (wireworms, Elateridae) are generally plant or detritus feeders, but some are predatory in soil or rotting logs.

Stalking and Ambush

Many predaceous insects subdue active prey, but not by hunting. These are the ambush predators or slow-moving stalkers. Many have special adaptations to efficiently capture prey. Praying mantids (Mantidae) have large, grasping forelegs, superbly fitted for grasping and holding prey, and remain motionless on vegetation until an insect comes near enough that they can strike out with their forelegs. A number of the true bugs also capture prey this way, including assassin bugs (Reduviidae), Nabidae, and ambush bugs (Phymatidae). Dragonfly and damselfly nymphs are successful ambushers that live in water. Their labium, or lower lip, is modified as a grasping tool that can be rapidly shot out to impale an insect, tadpole, or small fish. Tiger beetle adults are hunters, but their larvae dig vertical burrows in the ground and wait with their heads plugging the entrance. Such larvae grab small organisms walking near the burrow. They have back spines that help anchor them so they are not pulled out. After capture, the prey is dragged to the bottom of the burrow and devoured.

Trapping

Insects that trap their food are similar to ambush predators, but go a step further and use inanimate materials to help snare their prey. The ant lions (Myrmeleontidae) are the larvae of insects in the order Neuroptera that as adults look somewhat like damselflies. Larvae burrow into sand and form conical pits. Small, walking organisms, such as ants, fall into these pits and have difficulty escaping because of the loose material of the walls. In addition, the ant lion at the bottom actively tosses its head to remove sand dislodged by the victim. When the prey falls to the bottom, it is grabbed in the long mandibles of the predator, dragged under, and sucked dry. Some snipe fly larvae (Rhagionidae) build similar pits and capture prey the same way.

A few larvae of fungus gnats (Mycetophilidae) capture prey by secreting webs of mucilaginous materials in moist locations, such as beneath rocks, under bark, or in caves. In New Zealand caves, individuals of *Arachnocampa luminosa* not only spin webs that hang from cave ceilings, but also glow to attract adults of chironomid midges that breed in the water below. The colonies of the gnats are tourist attractions, and the ceilings of the caves are covered with pinpricks of light reminiscent of starry nights.

Provisioning

Solitary wasps in the families Vespidae, Sphecidae, and related families are often provisioners—the adult females capture prey that they paralyze with a sting and place the prey in a closed cell along with an egg. The larva that hatches feeds on the enclosed prey until fully developed. Different provisioner species specialize on different prey and construct different kinds of cells. For instance, the mud-dauber sphecid wasp, *Sceliphron caementarium,* constructs cells out of mud and provisions these with spiders. *Ammophila,* another sphecid, constructs cells by burrowing into the ground and stocks them with caterpillars. Others may burrow into wood or plant stems. Prey species vary from aphids to grasshoppers, true bugs, caterpillars, or spiders. Such provisioning wasps as the cicada killer, *Sphecius speciosus,* stretch the definition of predator. The female usually places one paralyzed cicada per underground cell, but sometimes she puts in two. Thus, the resulting progeny are sometimes parasitoids that can complete development on one host individual and sometimes predators that consume more than one prey.

Host Feeding

This is another behavior that blurs the line between parasitoids and predators. Host-feeders are female adult hymenopterous parasitoids, especially of the family Pteromalidae, that feed on host fluids oozing out of oviposition wounds. Such females host-feed to obtain protein for egg maturation. In many cases the host is parasitized as well. But some parasitoids have developed a habit of host-feeding without oviposition, and in some cases the host (which in this case becomes the prey) may die from excessive feeding. Some parasitoids even use fast-drying fluids secreted from the base of their ovipositors to construct "feeding tubes" from a concealed prey to the surface through which host fluids can be drawn up.

Mass Foraging

There are many social insects, primarily wasps and ants, that actively hunt prey for food. Most of these do so as individuals, capturing small prey and returning directly to the nest. But some ants are able to overcome vertebrate and large invertebrate prey. These are the mass foragers, the army ants and driver ants. These ants move over the ground in large swarms or columns and subdue prey, such as large scorpions and lizards, that other ants cannot. They also devour many small arthropods and deplete an area of prey so rapidly that they must frequently move from one nest site to another. The army ants of the New World tropics and the driver ants of Africa have been depicted in fiction as very dangerous predators that may even threaten humans. While it is possible that penned animals may be killed, the ants are not generally dangerous. The swarm front or hunting column moves forward only relatively slowly, and it is easy to step out of the way. In fact, a swarm of army ants removes vermin as they raid through a house and may be beneficial, so long as the house is unoccupied while the ants are there.

Deception and Predation

Some insect predators use deceptive practices. Firefly (Lampiridae) adults use bioluminescence to attract mates. Each species has a set of recognition flashes that are exchanged between flying males and the sedentary and often wingless females. If the sequence is correct, a male is guided to a receptive female. However, the females of some species mimic the recognition flashes of other species and so lure fooled males to their death.

Perhaps some of the most sophisticated deceptive predation practices occur among the insects that live in or near ant nests. These "ant guests" or "myrmecophiles" may survive by stealing food from ants by soliciting, but many are also predators. To be successful they must deceive the ants into accepting or at least tolerating them. Some of these predators live on the periphery of nests and prey on foragers that they encounter. The reduviid *Acanthaspis concinnula* fools ants by placing the bodies of its prey on its back. Another reduviid, *Ptilocerus ochraceus,* secretes "tranquilizing" attractive substances from trichomes on its abdomen that apparently subdue ants long enough for the bug to kill them.

Other insects, particularly beetles of the family Staphylinidae, live in the brood chambers of ants, where, among other things, they prey on ant larvae. Most of these predators apparently use chemical subterfuge to gain entry and stay in the brood chambers. Apparently, the chemicals mimic recognition substances produced by the ant brood. In fact, ants may pay more attention to the larvae of the beetles than they do to their own larvae. Some lycaenid butterfly larvae also have this habit. Early instars of these caterpillars live on plants and secrete attractive substances that ants collect. Larger caterpillars move to the brood chambers where they prey on ant larvae.

BEHAVIORAL ECOLOGY AND PREDATORS

Behavioral ecology is a field in which actions and movements of individual organisms are studied in order to understand interactions and impacts between organisms and their environment. Predaceous insects are often used as subjects in behavioral ecological studies because they are numerous, generally have a limited number of responses, and are small enough to be easily manipulated and yet large enough for easy observation. For instance, numerous studies have investigated how the number of prey eaten by a predator varies as prey numbers change. These functional responses usually show an increased tendency of predators to devour more prey as the number of prey individuals increases until the predators become satiated, so that the number of prey eaten increases rapidly at first as prey numbers increase and then becomes steady. However, backswimmers (Notonectidae), which feed on a variety of prey, may switch from one prey type to another as the latter become more abundant. In such a case, the functional response is S shaped, at first starting out slowly and then increasing rapidly until slowing again at high prey numbers. Such S-shaped, or sigmoid, functional responses are usually more typical of vertebrate predators than insects.

An important aspect of behavioral ecology has to do with searching by organisms in areas that contain various amounts and quality of food. Many predators, such as ladybird beetles (Coccinellidae), aggregate in places containing many prey. Finding the most profitable way in which a predator can divide its time between encountering prey within an area of prey abundance and searching for other concentrations of prey is an exercise in optimal foraging theory. Predators such as backswimmers (Notonectidae) have been found to do this rather well. The actual time they spend foraging in an area varies with prey abundance in a way that is close to optimal.

Ants have been used to investigate the kind of prey searching called center place foraging. In this situation, an organism departs and returns to a fixed location, often a nest, during a foraging session. Studies of ant foraging show that a center place forager often restricts its searching to only part of the available area, foraging in different areas at different times. Also, such foragers often do not search near the nest, but only at some distance from it. Both these characteristics may increase the efficiency of foraging by decreasing searching overlap with nest mates.

ECONOMIC IMPORTANCE OF PREDATORS

While most biological control work has involved parasitoids, some predators have been used as well. In fact, the first really successful classical biological control project was the control of the cottony cushion scale, *Icerya purchasi,* by the vedalia beetle (Coccinellidae), *Rodolia cardinalis.* Other coccinellids are important predators of aphids and mites. Ground beetles are important generalist predators, and at least one of these (*Calosoma sychophanta*) is an effective predator of the gypsy moth. Other predators used in biological control practice include the true bugs *Orius* (Anthocoridae) and *Geocoris* (Lygaeidae), clerid beetles, green lacewings, syrphid flies, and marsh flies (Sciomyzidae). On the other hand, some robber flies are voracious predators of bees and are considered to be economically harmful. Also, because of their propensity to attack nontarget prey, generalist predators are not considered good prospects for use in programs that introduce exotic natural enemies into new areas.

Insect predators may be more important to our well-being than is commonly thought. Because they often leave no identifiable remains after devouring prey, the extent of predation is hard to quantify. It is probably true that insect predators help keep the majority of potential pest organisms under control. Without them, the world likely would be quite different.

See Also the Following Articles

Biological Control • Feeding Behavior • Insectivorous Vertebrates

Further Reading

Bell, W. J. (1990). Searching behavior patterns in insects. *Annu. Rev. Entomol.* **35,** 447–467.

Caltagirone, L. E., and Doutt, R. L. (1989). The history of the vedalia beetle importation to California and its impact on the development of biological control. *Annu. Rev. Entomol.* **34,** 1–16.

Clausen, C. P. (1940). "Entomophagous Insects." Hafner, New York.

Cohen, A. C. (1995). Extra-oral digestion in predaceous terrestrial Arthropoda. *Annu. Rev. Entomol.* **40,** 85–103.

Dixon, A. F. G. (2000). "Insect Predator–Prey Dynamics: Ladybird Beetles and Biological Control." Cambridge University Press, Cambridge, U.K.

Hölldobler, B., and Wilson, E. O. (1990). "The Ants." Harvard University Press, Cambridge, MA.

Lövei, G. L., and Sunderland, K. D. (1996). Ecology and behavior of ground beetles (Coleoptera: Carabidae). *Annu. Rev. Entomol.* **41,** 231–256.

Obrycki, J. J., and Kring, T. J. (1998). Predaceous Coccinellidae in biological control. *Annu. Rev. Entomol.* **43,** 295–321.

Pearson, D. L. (1988). Biology of tiger beetles. *Annu. Rev. Entomol.* **33,** 123–147.

Prosorrhyncha
(Heteroptera and Coleorrhyncha)

Carl W. Schaefer
University of Connecticut

The Hemipteran suborder Prosorrhyncha comprises two groups, the Coleorrhyncha (family Peloridiidae) and the Heteroptera. Like all hemipterans, members of these two groups have elongated mouthparts for sucking fluids; unlike all other hemipterans, many heteropterans ("true bugs") suck the fluids of animals—mostly of arthropods but in a few instances of vertebrates. Heteroptera contains 8 infraorders, 79 families, and about 38,000 described species. The families range from the very small (one species) to the very large (> 10,000 species), and heteropterans as a group live in all habitable (and many apparently uninhabitable) parts of the world, from deserts to forest canopies, and on and below the sea. Heteropterans feed on all parts of plants (including on fungi and freshwater algae); some are sufficiently serious predators of pests to be biocontrol agents; and a few (bed bugs, triatomine bugs) feed on human blood. As a whole, Heteroptera is the most diverse of all hemimetabolous insect groups. Nevertheless, we know amazingly little about heteropteran biology and ecology, and the more we learn, the greater this diversity is shown to be.

CLASSIFICATION

The classification of the order Hemiptera is confused, sometimes defies the available evidence, and always arouses controversy. That there *is* an order Hemiptera, few doubt; moreover, that the group of true bugs, Heteroptera, is a phylogenetically valid taxon, no one doubts. Questions arise, however, concerning what had been considered the other suborder, Homoptera, a group that some (mostly homopterists) elevate to ordinal rank; indeed, some homopterists elevate groups within Homoptera to ordinal rank. However, recent work has shown that Homoptera and some of its subordinate groups are paraphyletic (and may not even be monophyletic). As a result, confusion reigns.

The order Hemiptera is characterized by being hemimetabolic (three life stages: egg, nymph, adult), having wings (with a few exceptions), and especially in possessing elongated mouthparts ("beak," or "rostrum") designed for the piercing of plants or animals and the sucking up of their juices. The Heteroptera differ from the majority of other hemipterans. In these other insects (mostly homopterans) the forewings are usually opaque (hence, "Homoptera"), but they are half opaque and half membranous in Heteroptera (again, hence the name). The two groups differ also in the apparent location of the mouthparts, which arise from the ventral surface of the head in both groups but, in Homoptera, arise from the back of the head (sometimes appear to arise from the thorax) and, in Heteroptera, arise from the front of the head. In addition, all homopterans, but not all heteropterans, are herbivorous. Nearly all nonheteropterans, except coleorrhynchans, have a filter chamber, which allows water ingested with the dilute plant juices to be "short-circuited" in the digestive system for rapid elimination. Heteropterans also differ from homopterans in that many feed on animal juices, a few indeed on the blood of vertebrates (including humans). It is likely that animal-feeding characterized the early heteropterans, although this view is controversial.

There is some confusion over the meaning and extent of "Hemiptera" and over just what a "bug" is. To some, "Hemiptera" refers only to Heteroptera, which is treated as an order the equivalent of Homoptera. To others, including most if not all students of Heteroptera, "Hemiptera" is an order with two suborders, Heteroptera and Homoptera. The features (i.e., synapomorphies) that the two groups share seem to be far more basic and significant than the two groups' differences, a circumstance that is best represented by treating all members of the two groups, Heteroptera and Homoptera, whatever this latter group may in fact contain, as belonging to an order, Hemiptera.

Because the group known as "Homoptera" (= non-heteropteran) is either paraphyletic or not monophyletic, the higher classification of Hemiptera is in flux. One consequence of the recent analysis of these relationships (based particularly on molecular evidence) has been the recognition of a small formerly homopteran group, Coleorrhyncha, as the sister group of Heteroptera. These two—Coleorrhyncha and Heteroptera —have been placed by some as a single hemipteran suborder, Prosorrhyncha. Although this idea was based on molecular evidence (and therefore likely to be discounted by some), morphological evidence also supports it. This article considers the hemipteran suborder Prosorrhyncha. Because the Coleorryncha is very small and of limited distribution, the focus is on Heteroptera.

Coleorrhyncha

The Coleorrhyncha is a group that contains the single family Peloridiidae, which had once been treated as a major homopteran group (Coleorrhyncha) and then as a group equal in taxonomic status to Heteroptera and Homoptera *(sensu antiquo)* and perhaps "bridging the evolutionary gap" between them.

The dozen or so species of Peloridiidae have a classic Gondwana distribution; they occur in southern South America, New Zealand, and Australia and are associated with southern beech trees *(Nothofagus)*, where they feed probably on moss (like idiostolids, mentioned later under Pentatomomorpha). They are small (2–5 mm long) and flattened, and many have expansions on the head and/or thorax which, semitransparent, are quite beautiful.

Heteroptera

All insects are not bugs. Heteropterans, however, are indeed bugs, and so are often called "true bugs" (i.e., "real" bugs, not "faithful" ones). The word *bug* derives from the Middle English word *bugge,* meaning a "spirit" or "ghost." If one awakens in the morning with red itching welts, one clearly has been visited in the night by a malevolent wraith or spirit—that is, by a bugge. This, of course, is correct: one has indeed been visited by a bug, by a bed bug *(Cimex lectularius),* which has fed and fled, leaving behind a red welt and a mystery. Bed bugs

are "bugs" and so, by extension, are all their relatives—all heteropterans. (The original meaning of *bugge* is retained in English in such words as *bugbear* and *bugaboo,* and in the name of the insect-repelling plant, bugbane. And the scientific name of the bed bug, *Cimex lectularius,* means "bed bug," *cimex* being the Latin for *bug,* and *lectularius* describing a small couch or bed. The word "cimex," by the way, was sometimes used by the Romans as a derogatory epithet.)

The suborder Heteroptera is perhaps the largest of the hemimetabolous groups (with about 38,000 species, and probably considerably more are undescribed), and also among the most diverse. The group's diversity is reflected in the variety of organisms fed upon, the variety of habitats lived in, and the variety of habits; these varieties are in turn reflected in the variety of shapes, sizes, and ornamentation found among the Heteroptera. Many heteropterans are aquatic or semiaquatic, and in this feature alone they differ from other hemimetabolans. Many are predaceous (upon other arthropods) and some are hematophagous (upon vertebrate blood), and in this feature too they differ.

Like all hemipterans, heteropterans have elongated mouthparts for piercing organisms and the drawing up of those organisms' juices; predaceous bugs reduce to juice the soft internal organs of their prey. The organisms fed upon include all terrestrial groups of plants (except perhaps algae; but Corixidae include freshwater algae in their diet), other arthropods, snails (fed upon by some giant water bugs), and of course vertebrate blood.

Heteroptera is further distinguished by having scent glands. These occur on the abdominal dorsum in immatures and on the metathoracic pleura in adults. Nymphs lack the thoracic glands, but the abdominal glands occur in adults usually as nonfunctional scars; however, considerable evidence has accumulated showing that in many heteropteran groups one or more of the abdominal glands are functional in adults as well as in nymphs. Another distinctive feature of heteropterans is the scutellum ("little shield"). This triangular structure (broad base anterior) is a modification of the mesonotum; it may be quite small relative to the body, or quite large. In a few families and in a few members of a very few other families, the scutellum is large enough to cover the wings and the entire dorsum of the abdomen.

Regarding the general account of heteroperan biology that follows, and also the more specific accounts of the different groups, it cannot be too heavily emphasized very little is known about most heteropterans: many of the generalities given here are based on small samples, and what we do not know far outweighs what we do know. The reader is urged to take all general statements, if not with a pillar, then with several grains of salt.

Predaceous bugs, and those feeding on vertebrate blood, are not particularly host specific, although some degree of specificity may be imposed upon them by their habitats: bat bugs, living in bat caves, are restricted to feeding on bats, although given the chance they may feed on other warm-blooded vertebrates; many triatomine bugs live in or with their vertebrate hosts and are perforce restricted to them. A few predaceous bugs are more clearly host specific: the bed bug (everybody's favorite heteropteran) lives only with humans and does poorly when fed other kinds of blood; certain minute flower bugs (Anthocoridae) appear to specialize on certain species of scale insects (although here the specificity may be habitat driven, not prey driven); and some predaceous mirids prefer certain species of lace bugs as prey. There are few other examples.

The host specificity of herbivorous heteropterans is in general greater than that of predaceous ones. Not all herbivorous bugs are host specific, and among those that are, the specificity is sometimes at the species–species, or species–genus level (bug–plant), but more commonly at the family group–family group level (e.g., Alydinae and Plataspidae on Leguminosae; Rhopalidae: Serinethinae on Sapindaceae). Many bugs, and family-level groups of bugs, are oligo- or polyphagous. Of course, the heteropterans that compete with humans for their food and are agricultural pests are those that are either host specific on crop species or, although polyphagous, can build up large populations quickly on crops.

Measuring host specificity can be impeded by the willingness of bugs to probe nonhost plants, either to test their suitability as food or merely to get water. Too many host plant records are of this sort; the best records are those of the feeding of immatures, since immatures, being wingless, cannot go elsewhere to better their lot.

Most heteropterans (perhaps two-thirds) are herbivorous. Many pierce plants to their circulatory systems and feed on the contents of phloem or (less often) xylem; a few very small heteropterans feed on plant cell contents. These heteropterans thus feed not unlike most homopterans. However, plant sap is low in nutrients, and many heteropterans feed on reproductive tissues such as flowers, ovules, and ripe and unripe seeds, which are richer in nitrogen; another group of heteropterans (Miridae and some Tingidae) feeds on somatic tissues that, to the delectation of the bugs, mobilize nitrogen when the host is wounded. A very few heteropterans feed on bryophytes, and a few Tingidae are the only heteropterans to induce plant galls. Many herbivorous bugs, especially in the infraorder Pentatomomorpha, contain bacterial symbionts in gastric ceca or mycetomes; these presumably supply necessary trace nutrients.

As mentioned earlier, several heteropterans may become pests on agricultural crops. In the Old World tropics, cotton stainers (Pyrrhocoridae) feed on developing cotton bolls and may allow entrance of destructive pathogens; the bugs' excreta also stain the cotton (hence the bugs' name). Sunn pests, species of the genera *Eurygaster* and *Aelia* (Scutelleridae and Pentatomidae), are among the most serious pests of small-grain crops (especially wheat) in a broad band of countries from eastern Europe through the Middle East into western Asia. Chinch bugs similarly afflict wheat in North America, and a close relative attacks lawn and golf course grasses in Florida. Lygus bugs (several species), a leaf-footed bug (Coreidae), and

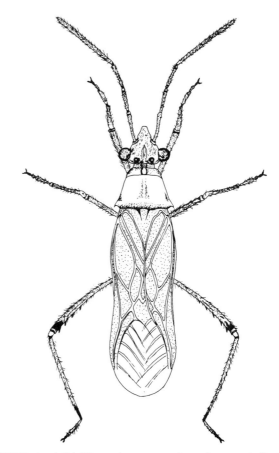

FIGURE 1 An alydid *(Neomegalotomus parvus),* a soybean pest in Brazil.

southern green stinkbugs (Pentatomidae) feed on a variety of vegetables and fruits and are pests worldwide (lygus especially in the north-temperate regions). Species of *Riptortus* and *Neomegalotomus* (Alydidae) may become pests of legume crops (Fig. 1). Several burrower bugs (Cydnidae) are pests of many crops and range grasses, upon whose roots they feed. The red pumpkin bug (Dinidoridae) may seriously damage cucurbit crops in Asia. There are many other major pests, and very many other minor or regionally restricted ones; nearly 1000 are discussed in a recent book by Schaefer and Panizzi. As the foregoing list suggests, most pests specialize on the plant family to which the crop belongs. A few (e.g., lygus, southern green stink bug, and the leaf-footed bug *Leptoglossus gonagra*) are exceptions, feeding on many unrelated plants and quickly achieving large populations.

Only a few heteropterans are vectors of plant diseases; one important group is Piesmatidae, one of whose members transmits a serious viral disease to sugar beets. Overall, however, heteropterans are far less important than homopterans as vectors of plant diseases, perhaps because bugs, unlike many homopterans, do not feed directly in plant cells.

The few bloodsucking bugs may cause damage by withdrawing excessive amounts of blood and thus weakening people who may, in addition, be weakened by disease, or frail in health because very old or very young. Bed bug populations may increase to the point where such damage may occur.

Of the two important groups of blood-sucking bugs, the bed and bat bugs (Cimicidae) do not transmit disease pathogens, although there is some slight evidence that the hepatitis B virus may be transmitted by bed bugs in southern Africa. On the other hand, triatomine bugs (Reduviidae, subfamily Triatominae) are vectors of *Trypanosoma cruzi. T. cruzi* causes Chagas' disease, a debilitating and often fatal disease of the New World tropics and subtropics (a handful of cases reported in the United States); major international efforts are now under way to eliminate this disease by eliminating the places where the bugs hide. These efforts have been successful in many areas of South America.

Although most heteropterans having an economic impact on humans are harmful, some are beneficial. Chief among the latter are the predaceous bugs that feed on insect pests. Because predators as a rule are not host specific, there are no one-to-one relationships of predator to prey in Heteroptera or in any other group considered for biological control. Nevertheless, many groups of heteropterans have been studied for, and some successfully used in, biological control programs: asopine pentatomids (Pentatomidae, subfamily Asopinae) against many pests, especially garden pests (indeed, an artificial aggregation pheromone has been developed to "call in" these predators); certain minute flower bugs (Anthocoridae) to combat stored-product pests; some predaceous mirids on vegetable and greenhouse crops; and an array of predaceous bugs in field crops. The one exception—remarkably, considering its wide distribution, its variety and numbers of species, and the size (and presumably appetite) of its members—is the Reduviidae. The biological control potential of this family has been surprisingly little explored. It should also be mentioned that the value of very small predators—Anthocoridae, small mirids, early instars of other predators—is unknown; these feed on arthropod eggs and small instars, whose absence goes unnoticed because their presence was overlooked in the first place.

In 1998 Cohen showed that predaceous bugs feed differently than had been thought: instead of merely sucking up the fluids inside a prey, these bugs inject salivary enzymes that digest all soft tissue; the digested material is then sucked up. This discovery has profound implications for biological control: a single predator will be satiated by a far smaller number of prey than had been believed, and will therefore be less effective in controlling pests. This realization is only slowly making its way into the biological control community.

Of less importance is the control of harmful plants, weeds, by plant-feeding heteropterans: Some tingids have been used in various parts of the world to control (not very successfully) the invasive plant lantana; cactus bugs (*Chelinidea,* Coreidae) have been tested—with small success—against invasive cactus in Australia; and several other bugs have been and are being tested for possible control of other weeds. So far, no major plant pest has been controlled by a heteropteran, but the exhaustive study of biology and ecology necessary for success in such an endeavor is only now being undertaken.

Detailed accounts of economically important heteropterans are provided in the book by Schaefer and Panizzi.

COURTSHIP, MATING, OVIPOSITION Unlike feeding, the sexual lives of heteropterans show little variety. There is little courtship, although males of several species may fight (usually briefly) for the attention of a female. The sexes are brought together initially probably by sex pheromones and, in a few species, by sex-neutral aggregation pheromones; however, only a tiny handful of species, all of them land bugs, has been shown to have pheromones of both types. Mating occurs end to end or with the male directly or diagonally across the female; the position seems to vary with the family group or even the infraorder of the bugs. Mating may be brief or long; if long, the larger female may move about and even feed, while the hapless male scrambles along behind, walking backward.

Mating by bed bugs and some relatives in Anthocoridae and Nabidae, is (to quote from another context) nasty, brutish, and short. The male mounts the female and, with a scimitar-shaped clasper, pierces the side of her abdomen and deposits sperm; in primitive species this traumatic insemination (as it is called) may occur in various parts of the abdomen, allowing sperm to enter the abdominal cavity; in more advanced species (including the bed bug itself), sperm is deposited in a special patch of tissue. Such insemination leaves small healed wounds in the female's cuticle; counting these, the number of times an individual has been mated can be determined. Sperm then progresses to a sperm storage organ at the base of the common oviduct.

The internal reproductive structures are very conservative—that is, they resemble those of very many other insects. There is a pair of ovaries or testes, each almost always consisting of seven ovarioles or testicular follicles. Paired ducts lead from them to a common duct, with which may also be associated various glands and (in females) organs for sperm storage. The external genitalia are more complex, consisting in the male of a genital capsule ("pygophore") containing a pair of claspers and the intromittent organ and, in the female, of a genital chamber into which sperm is deposited and a complex ovipositor designed in some bugs for the laying of eggs on surfaces ("platelike ovipositor") and, in others, for inserting eggs into crevices or actually into slits in vegetation ("laciniate ovipositor").

As is true of many other groups, the external genitalia of male heteropterans are usually more complex and varied than those of the female. In Heteroptera as in other groups, genitalia provide very useful characters for taxonomic and phylogenetic studies.

The egg is often the stage that survives inhospitable seasons (winter, dry season), and so eggs are laid where they may best be protected from the elements as well as from predators and parasites. Eggs are laid also where food will be available, because the immatures, wingless (of course), cannot seek food themselves. The latter requirement presents certain difficulties to predaceous bugs, which cannot be certain where prey will occur. Perhaps this accounts for the habitat specificity (resulting in a certain degree of prey specificity) of some predators, as mentioned earlier: predators' eggs are laid on plants on which certain prey species are host specific; the predator then in turns appears itself to be host specific, on the herbivore. Perhaps.

The egg itself varies considerably in shape from group to group; it may be round or oval, and in stinkbugs especially the eggs are barrel shaped. Each egg bears small openings through which it was fertilized, and many have other small openings for oxygen intake. Carbon dioxide is released through the shell (chorion). The few eggs that have been studied have a complex architecture of the chorion, presumably related both to gas exchange and to maintaining an appropriate humidity balance.

Eggs are laid singly or in small batches (rarely more than 100), and if in batches may be laid apparently at random or in tidy rows, sometimes more than a single egg deep. They may be deposited on surfaces (of leaves, stems or trunks, ground, stones, etc.) or placed into crevices; the crevices may in some groups be created by the female's ovipositor. The eggs of some mirids and tingids absorb water in the spring, and thus nymphs hatch just as the plant becomes suitable for feeding.

The females of some heteropteran groups (some Tingidae, many—all?—Acanthsomatidae; a few others) guard their eggs and the early immature stages. Guarded eggs have been shown to be less heavily parasitized, and guarded nymphs less heavily preyed upon, than unguarded eggs and nymphs. A single female will lay several batches of eggs. In a few groups [some giant water bugs (Belostomatidae), at least one coreid], females lay eggs on the backs of males. Belostomatid males then aerate the eggs, sweeping fresh water over them; without such aeration, the eggs might die of oxygen lack and/or of fungal infection. In some species each new batch requires a fresh mating; in bloodsucking species each new batch requires a blood meal.

Hatching occurs sometimes through areas of weakness in the egg's chorion; not all species' eggs have such areas. In some groups (Pentatomoidea, some Coreoidea, and others?) the new hatchling (first instars) do not feed, although they may take water, and pentatomoids may suck up symbionts from the emptied eggs. Nearly all heteropterans have five immature stages (instars) (those that do not, have four; very rarely are there six), and except for blood-feeding bugs, the stimulus to molt from one instar to the next is unknown. Blood-feeding bugs require the stimulation (manifested as a distension of the abdomen) of a blood-meal to molt.

As the instars grow larger at each successive molt, the wing pads develop; as these become more and more distinct, one can determine the instar by the degree of their development.

The instars (except sometimes the first one) feed for the most part on the same food as their adults. Exceptions include predaceous bugs, whose smaller instars may take smaller prey (although a group of small instars may attack a single large slow prey, like a caterpillar); and some herbivores, whose instars do not feed on the plants chosen by than their adults, probably

because of seasonal differences in the plants' availability, and/or because some plants may provide nutrition to the young and nutrients needed for reproduction to the adult.

Development (egg to egg) in temperate climes usually takes a year, the egg overwintering or sometimes a mated female (less often, other stages). In regions of less predictable or less severely contrasting seasons, there may be several—rarely, many—generations in a year. Even when there is only one annual generation, however, populations of some pest heteropterans can build up over a very few years.

A distinctive feature is the possession by some heteropterans of a pair of small (micro- or *m*-) chromosomes. The sex-determining mechanism is XY, although XO occurs in some groups and, in others (especially the bed bugs, Cimicidae), multiple X chromosomes may occur. The chromosomes are holocentric, and numerical doubling of the autosomal complement has occurred in several groups.

DISPERSAL As with all living things, dispersal is an important aspect of heteropterans' lives, necessary for the leaving of places no longer suitable and for the discovery of places as yet unexploited; it is necessary, too, for the finding of nonsibling mates. The wingless immatures depend upon the mother to place them where resources—biotic and abiotic—are suitable for development.

Dispersal for mate finding appears to be far more frequent among bugs than dispersal for the discovery of new habitats, although of course it may often serve both purposes. One important exception is the dispersal of cotton stainers (*Dysdercus* spp., Pyrrhocoridae) from areas where high populations have depleted food to new areas, which may include cotton fields. After the move, females resorb the flight muscles, converting the products into eggs; thus such dispersal occurs but once. (Evidence suggests that not all *Dysdercus* spp. do this.) Another exception may be the moonlit flights of some giant water bugs (Belostomatidae) in Africa; it is not clear that these flights correspond to drying of habitat. Like other insect groups, Heteroptera contains some groups that are drawn to light traps (white or black light) and some groups that are not; no one has listed these groups, much less sought a correlation between being attracted to light and having a propensity to disperse.

Of course, for many heteropterans resources are replenished: for predators, new prey is continually being produced; and many smaller herbivores feed on annual plants that appear and disappear too rapidly for their populations to be seriously reduced. In addition, many herbivorous bugs feed on several different species and turn easily from one host to another. Thus for these bugs, dispersal may at times be necessary, but it is not a constant need.

Far better documented for far more heteropteran species is movement in response to sex pheromones, often produced by both sexes. In addition, aggregation pheromones occur in some species. This type of dispersal is of course a "pull" dispersal, not a "push" dispersal: the insect is drawn toward something rather than being driven away. Nevertheless, like the messengers in *Through the Looking-Glass* (one to fetch and one to carry), movement toward a conspecific may draw one also to fresher resources. The pheromone of at least one beneficial predaceous bug has been synthesized, patented, and marketed to draw these bugs into users' gardens.

ENEMIES AND DEFENSE   Heteropterans have the same enemies as other organisms: predators that would eat them from the outside and parasites that would eat them from the inside. The major predators—again like other insects'—are insects; vertebrates seem to be less important than they are for some other insect groups. Bugs may be protected from vertebrate predators by the secretions of their metathoracic scent glands, which also (it has been suggested) protect bugs from ants; the experimental work demonstrating these possibilities is scant and mostly anecdotal.

Many bugs are also protected from visually orienting vertebrates by being bad tasting and warningly colored. Aposematic coloration occurs in nearly every group of land bug as an arresting contrast of black and red, yellow, orange, or white; the similarity of color and pattern has little to do with genetic or phylogenetic relatedness, and much to do with the strictures placed on color by limited metabolic pathways, as well as the limits placed on pattern and design by the strictures of small body size. Thus unrelated aposematic bugs may resemble one another merely because the possibilities for variety are few. However, as in other groups of animals, both Batesian and Müllerian mimicry occur in Heteroptera: the edible look like the inedible (Batesian) and the inedible look like other inedibles (Müllerian). Mimetic assemblages occur, although they have been little studied. One comprises some cotton stainer relatives (Pyrrhocoridae) and their predators, certain Reduviidae. It has been suggested that the reduviids can thus slip unnoticed upon their prey ("aggressive mimicry"). This seems outrageously unlikely: more probably, the group as a whole benefits from a collective aposematicism, from which the pyrrhocorids benefit more than they lose from being preyed upon.

Few creatures like to eat ants, and it is not surprising that ant mimicry has arisen often in the Heteroptera. The immatures of some Alydidae resemble specific species of local ants; and as the immatures become larger, they resemble increasingly larger species of ants. One assumes that looking like ants is good; no one has tested the idea.

None of these defenses (except perhaps scent) is especially effective against other insects. More effective (guessing here) are the thickened cuticles of many heteropterans, the coriaceous hemelytra, as well, perhaps, as the enlarged scutellum of several bugs.

Body shape may also play a part: the round sleek surfaces of Canopidae, Plataspidae, Megaridiidae, and a few other heteropterans offer the jaws of small predators (especially mandibulate insects) little purchase. And the many small spines of coreid nymphs (and others), and the single large spine of adult Cyrtocoridae and some Podopinae

(Pentatomidae), must also prove to be unpleasant surprises to vertebrate predators.

It is difficult to know whether heteropterans are more or less parasitized than other insects. The eggs of certain species are often well parasitized by wasps, and the immatures and occasionally the adults by tachinid flies. Two indirect bits of evidence suggest that parasitizing bugs is an old and successful way of life: certain family groups of parasites have evolved that parasitize only heteropterans (Diptera: Tachinidae: Phasiinae; several genera of Hymenoptera). And parental care has evolved several times in Heteroptera, to ward off potential parasites (and predators, but these seem to be a lesser evil).

It has been carefully shown that guarding by parents (usually but not always females) protects certain lace bugs (Tingidae), coreids, and many acanthosomatids from parasitism and predation; and the eggs placed upon the backs of some giant water bugs (Belostomatidae) are protected from fungal parasitism by the water currents generated by the male's rowing movements.

THE INFRAORDERS AND FAMILIES OF HETEROPTERA Cladistic work on the morphology of heteropterans has resulted in a new classification into eight infraorders: Enicocephalomorpha, Dipsocoromorpha, Gerromorpha, Nepomorpha, Leptopodomorpha, Cimicomorpha, Pentatomomorpha, and Aradomorpha. In a commonly accepted (but not thoroughly tested) analysis, each of these infraorders is the sister group of those succeeding. Aradomorpha, the most recently proposed, is not universally accepted.

To some extent, these groupings correspond to earlier ones: Preceding this arrangement, for example, Heteroptera was divided into Hydrocorisae (= Nepomorpha, with some adjustments), Amphibicorisae (= Gerromorpha, with some adjustments), and Geocorisae (the remainder, with some adjustments): "water," "amphibious," and "terrestrial bugs," respectively. The last was divided into Cimicomorpha and Pentatomomorpha, names retained in the new classification with the same family composition (with some adjustments in Cimicomorpha); Aradomorpha was included in the earlier Pentatomomorpha. Each -morpha name is based on a genus in the infraorder.

The following treatment of infraorders and families of Heteroptera, is presented, once again, with the admonition to recognize that little is known in this area and often details are generalized to an entire family from what is known about a tiny fraction of its members.

Enicocephalomorpha, Dipsocoromorpha These small infraorders are the most primitive in Heteroptera. The latter is usually assumed to have arisen from the former, but there is some suggestion that each may have arisen separately from one or more related preheteropteran ancestors. Each is small: the two families of Enicocephalomorpha, Aenictopecheidae and Enicocephalidae, contain about 20 and 400 described species, respectively; however, it is probable that several thousand species remain to be described.

The same is true of the Dipsocoromorpha, whose five families are Ceratocombidae (50 described species), Dipsocoridae (30), Schizopteridae (120), Hypsipteryigdae (3), and Stemocryptidae (1). As in Enicocephalomorpha, there remain to be described many hundreds, if not thousands, of dipsocoromorphan species.

Members of both groups are small (Enicocephalomorpha, 2–16 mm long; Dipsocoromorpha, perhaps the smallest of the heteropterans, 0.4–4 mm long). They are all predaceous (as far as is known) and live in or on the soil surface, in leaf litter, or in the interstices of mosses and similar low-growing plants. Their small size, and their life in a habitat of remarkably small interest to biologists or ecologists, render them apparently rare, but actually locally abundant yet nearly wholly unknown.

This is unfortunate, because not only are these bugs of great ecological interest (as predators—and perhaps scavengers—in an ancient and poorly known habitat), but because many occur worldwide and study would yield valuable biogeographic information. Moreover, since they are the most primitive of heteropterans, careful study should reveal much about the origins of this group, as well as its relationships to certain homopteran groups: some enicocephalomorphans share some genitalic characteristics with some auchenorrhynchous homopterans; and Aenictopecheidae is defined wholly by plesiomorphies.

Some enicocephalids swarm, either in single-sex or mixed-sex swarms, and have been confused with midges! This statement sums up our knowledge of enicocephalomorphan reproductive biology!

Gerromorpha Members of this infraorder are the semiaquatic bugs whose members live near or on water, but never in it. Several members of several gerromorphan families live near or on the surface of the sea and are among the only heteropterans that are, if imperfectly, marine; halobatines (a subfamily of the Gerridae) may occur many kilometers from land. These gerromorphans are *semi*aquatic bugs—they do not get wet, except occasionally; and indeed a few live not where it is wet, but where it is merely damp. Like other water-associated animals (e.g., Nepomorpha), gerromorphans are usually dark above and pale below.

All are predaceous, those living on the water surface feeding on land arthropods that fall onto and are caught by the water's surface film; there is some evidence that these bugs may also feed on aquatic crustaceans that are seized upon rising to the water's surface. Prey is located by many bugs by ripples caused by its struggling. The bugs probably secrete something unpleasant because they themselves are only occasionally preyed upon by fish.

Many gerromorphans are fully or partly winged, some are wingless, and populations of some are pterygopolymorphic (some members wingless, some winged), a phenomenon that may also vary seasonally. The hormonal and ecophysiological bases for pterygopolymorphism has been studied in a few gerrid species. This variety, as well as variety in diapause physiology, is doubtless related to the temporary nature of the habitat, which may dry up or freeze at various locations and times of the year.

In temperate regions, many gerromorphans overwinter in debris away from, but close to, their water habitat.

Because of their ubiquity, and especially because of their tight adaptations to an unusual way of life, gerromorphans (especially gerrids) are being used more and more to work out questions in communication and reproductive evolutionary ecology (including courtship). These efforts are helped of course by the wide interest in aquatic ecology, and by the excellent phylogenetic foundations provided by Andersen in 1982 and much recent work as well.

Being semiaquatic predators, a few gerromorophans have been studied for their ability to control rice pests and the aquatic larvae of biting flies; in each situation, success is at best limited, probably because rice pests do not fall into the water often enough and aquatic larvae do not spend much time at the water's surface.

There are eight families: Mesoveliidae (35–40 species), Hebridae (150), Paraphrynoveliidae (2), Macroveliidae (3), Hermatobatidae (8), Veliidae (720), Gerridae (620), and Hydrometridae (110–115). Several of these families—or components of them—were once contained within a more exclusive Gerridae or Veliidae, but the thorough phylogenetic studies just mentioned have yielded a firm and stable systematic classification.

The insects range in size from very small (1 mm long) to the aptly named *Gigantometra* (Gerridae), which is up to 36 mm long. Most gerromorphans are at the small end of the range.

Members of the family Gerridae occur worldwide and are among the most frequently seen and frequently admired inhabitants in—or rather on—bodies of fresh water. As they skate about on the surface, their shadows are patterned dark on submerged objects in the water body. This aesthetic appeal, and the bugs' ability to remain on the water, not in it, have earned them various names, such as "water bugs," "pond skaters," "wherrymen" (in Britain), and even "Jesus bugs."

Their shape is characteristic and easily recognized: the head extends somewhat forward of the eyes, the abdomen is often truncate and its segments compressed, and the middle and hind legs are greatly elongated, sometimes looking as if they had been designed for a larger bug. All legs are used in moving on the water's surface, and the front legs are short, often stout, and used too for prey capture. The claws of all legs are subapical (not apical, as is usual). Probably this is an adaptation for moving on water surface films—something they do with remarkable speed and agility.

Much work has been done on communication and courtship/reproduction behavior in a few gerrid species: communication is via wave patterns of surface waves and ripples created by the bugs; these ripples apparently serve several functions, including location and courting of potential mates, defining territories, and perhaps even recognizing conspecifics for the avoidance of cannibalism. The extent to which other gerromorphans use similar signals is unknown, although there is some evidence that a veliid does, and it seems reasonable that other surface skaters would also.

Veliidae (riffle bugs, broad-shouldered bugs) is the largest family; the genus *Rhagovelia* alone contains more than 200 species. Like gerrids' claws, veliids' claws are subapical; the legs are not relatively so long (not greatly surpassing the body), the abdomen is more often elongate, the pronotum is larger (covering the rest of the thoracic terga), and the middle legs' claws of some veliids are modified into a fanlike structure for pushing against the water's surface. In general, the veliids are smaller than gerrids: 1–10 mm vs 1.5–35 mm in length.

Like gerrids, veliids for the most part occur on the surface of ponds or of the quiet stretches of moving water; also like a few gerrids, some veliids live on ocean surfaces, and others are nearly terrestrial, occurring in damp situations or intertidally, or in mangrove swamps. Veliids too feed on smaller animals trapped in the surface film.

Again like gerrids, veliids can move very quickly, and they move even faster when they secrete from the mouthparts a fluid that lowers the surface tension and causes them to "scoot" away.

Veliids may form large assemblages of very many individuals, sometimes almost seeming to cover a stretch of water. What brings and keeps the bugs together is unknown, nor is the advantage of these assemblages clear.

Members of Hebridae are the velvet water bugs, so called because of their coating of fine (presumably hydrofuge) hairs. Quite small (1.2–3.5 mm long), they are not often collected and live in damp places such as the edges of water bodies, in moss, or in and among waterside vegetation. The family occurs worldwide. Hebrids look somewhat like small veliids, both being somewhat broader and thus relatively shorter than gerrids.

As the names suggest (in English and Latinized Greek), water treaders or marsh treaders (Hydrometridae) walk with measured tread on surfaces close to the water's edge, and sometimes even on the surface itself. They are classic sit-and-wait predators, motionless for long periods until some small prey comes near. Hydrometrids are elongate insects, small to moderate in size (2.5–20 mm or so long); a distinctive feature is the position of the eyes about halfway along the length of the head. The genus *Hydrometra,* with 80 to 85 species, is found worldwide, but the rest of the family is tropical.

Mesoveliidae may be the most primitive family in the infraorder, possessing as it does a number of primitive character states and sharing only a few advanced states with other gerromorphans; indeed, the group has at times been excluded from the infraorder, although its inclusion there is now firmly based. Like so many gerromorphans, mesoveliids are small and pterygopolymorphic (some species); *Mesovelia* occurs worldwide, but most of the remaining genera are localized. Mesoveliids vary in their habitats: some (many *Mesovelia*) are active on the water surface, but others live in damp habitats some distance from water; a few are cavernicolous, and a few are intertidal.

The other three families (Paraphrynoveliidae, Macroveliidae, and Hermatobatidae) are small in number of species (113 species in all) and in size (1.5–2.5, 2.5–5.5 and 2.5–4 mm long, respectively). The first has so far been found only in southern Africa; the second only in the New World; and members of the third live near coral reefs, where they remain in air bubbles during high tide and feed during low.

Nepomorpha   Most of these are the truly aquatic bugs—that is, bugs that actually occur under water and get wet; the group is the same as the earlier Hydrocorisae (water bugs). Most are predaceous, but corixids are, as a group, more omnivorous. Although nepomorphans live much of their lives in water, they also are active aerial dispersers. It is hard to imagine, therefore, why most of these bugs lack ocelli: even if unnecessary below the water's surface, ocelli ought to be useful in the air (e.g., the riparian families have ocelli). Water bugs range from small (1 mm long, some Helotrephidae) to huge, the largest of the Heteroptera (112 mm long, some Belostomatidae). Like gerromorphans, and for the same reason, most nepomorphans are much paler ventrally than dorsally. Unlike gerromorphans, most (not all) nepomorphans are streamlined, fusiform; those that are not (Gelastocoridae, Ochteridae) are riparian, not wholly aquatic. The antennae of nepomorphans are short and often concealed; hence an earlier name, "Cryptocerata" ("hidden horns").

Living as most nepomorphans do, under water, the problem of breathing arises and has been solved in several ways. Two families are nearly terrestrial: Gelastocoridae and Ochteridae live on the edges of streams and breathe like terrestrial bugs. Belostomatids and nepids have extrusible (belostomatids) or permanently exserted (nepids) "airstraps" or siphons, which can be thrust above the surface. Members of the other families rise to the surface periodically to trap air in hydrofuge hairs; oxygen is then taken into the body (much carbon dioxide here as in most animals is released through the body wall); in some groups and to some extent, plastron respiration occurs (i.e., as the oxygen tension in the air bubble drops, oxygen is drawn into the bubble directly from the water). Plastron respiration is so efficient in Aphelocheiridae that these bugs may remain permanently submerged.

Because of these bugs' importance as fish food (and sometimes as human food), because they occur in aquatic habitats, which have always held a fascination for ecologists, biologists, and folks in general, and because some are so large, nepomorphans have probably been better studied as a group than any other infraorder. The literature on the systematics, biology, and ecology of the group and its member is very large.

There are nine families: Corixidae (at least 600 species), Nepidae (225 species), Belostomatidae (150), Naucoridae (nearly 500), Notonectidae (350), Pleidae (40), Helotrephidae (44–120 species: authorities differ), Ochteridae (50–55), and Gelastocoridae (100). All but the last two live below the water's surface; Ochteridae and Gelastocoridae are riparian (ripicolous).

Corixids (family Corixidae) are the water boatmen, found commonly in ponds, lakes, and (less often) streams throughout the world. More than 600 species make this the largest nepomorphan family. Aspects of their morphology differ so greatly from those of other bugs that Corixidae has sometimes been placed in its own group, separate from other heteropterans. In particular, the labium of the mouthparts is broad and fused to the head, and the foretarsi are modified (enlarged, somewhat flattened, with an array of long setae) for food gathering. Like other nepomorphans, the hind legs are flattened and hairy, for swimming; in particular, corixids look superficially like notonectids, but corixids are flatter and swim rightside up. Under magnification, the dorsum of many corixids has many fine horizontal zigzag and anastomosing pale lines; this feature too distinguishes these bugs from notonectids and, indeed, from other bugs.

Water boatmen are 2.5 to 15 mm long and occur throughout the world; they live in aquatic habitats of all kinds, including the intertidal. Unlike other nepomorphans, corixids probably feed mostly on algae and (one suspects other small bits of living or dead organic matter). Some are carnivorous, however; indeed, probably many are (only a few have been studied) and are important predators in waters containing few others (e.g., acidified or saline inland waters). Excellent fliers over long distances, water boatmen are often among the first animal colonizers of new aquatic habitats and have even been captured landing on the shiny roof of an automobile in the desert far from the nearest water.

Corixids are of some small importance in feeding on aquatic larvae of noxious dipterans; they are of greater importance as food for commercially valuable fish (and in some cultures, for humans). In addition, they have been used as indicators of organic pollution.

The function of stridulation has been more thoroughly studied in Corixidae than in any other heteropteran group. (The so-called stridulitrum on the male's abdomen is not sound producing but probably aids in clasping the female.)

Superficially similar to corixids are the Notonectidae, backswimmers [wherrymen (cf. also Gerridae), boat flies] which, however, swim upside down; notonectids swim ventral-side down because that is where the buoyant air bubble is. Backswimmers, like water boatmen, are fusiform and also have flattened hirsute hind legs for swimming. However, they have typically heteropteran mouthparts, are wholly predaceous, lack modified foretarsi, and never have the pale irregular linear markings of many corixids. They range from 5 to 15 mm in length. Moreover, unlike corixids, notonectids bite viciously and painfully, as many a hydrobiologist has learned. The largest genus, *Notonecta*, occurs worldwide, and the other genera are more localized, although sometimes within entire continents.

Like corixids, notonectids are excellent fliers and are also early colonizers of new bodies of water. Some notonectids also may use sound in courtship and mating, but the evidence here is far more slight than that for corixids. Backswimmers inhabit nearly all types of water, although records from saline or near-saline waters are scarce.

These bugs are eager predators and they prey near the water's surface, whence they may drive their prey, a procedure aided by the bugs' buoyancy. They feed on other arthropods (and occasionally fish: see later), including small crustaceans and the larvae and pupae of blood-feeding flies. Mosquito larvae and pupae are favored by some notonectids, but over long periods they do not provide consistent biological control; however, female mosquitoes may recognize notonectid-infested ponds and refuse to lay eggs there.

Backswimmers are themselves food for several kinds of fish (good), but may also feed upon fish larvae (bad); however, neither being fed upon nor feeding is of great economic importance.

Pleidae (pygmy backswimmers) and Helotrephidae (no common name) resemble very small (both are about 1–4 mm long) notonectids; however, pleids lack the flattened hirsute hind legs of notonectids (and helotrephids). Both are oval to globular and very streamlined; helotrephids indeed have head and thorax fused. Like backswimmers, these bugs (both families) swim upside down (although they can at times swim rightside up), and they presumably get oxygen and prey in much the same way as do notonectids. Pleids occur almost exclusively in still waters, permanent or temporary; helotrephids live in almost any kind of water, still, or running, permanent or temporary, hot springs, or water seeps; one species can even wait in the desert for ephemeral waters. Members of both families may be of some small value in controlling mosquito larvae. Pleidae is wholly, and Helotrephidae mostly, tropical.

The giant water bugs comprise the Belostomatidae; here are the largest of Heteroptera, ranging from 9 to 112 mm in length. All are predaceous, and the largest may be important pests in fish hatcheries; one group of species feeds on snails in Africa and is of some value in suppressing populations of snails that carry bilharzia (schistosomiasis); the report of a giant water bug attacking baby ducklings may (or may not) be a rural legend. Although active swimmers, belostomatids usually wait on submerged materials to capture prey swimming by. The bugs are attracted to lights, which explains a common name, "electric light bugs."

The belostomatid body is broadly oval and somewhat flattened; the forelegs are modified for grasping, and one claw is reduced. The eighth abdominal tergum is modified laterally into a pair of extrusible airstraps, which can be thrust above the water's surface and through which air is moved to an air bubble below the wings. In several subfamilies several antennal segments bear fingerlike projections.

Giant water bugs are well known for the remarkably painful "bite" (actually, stab) they can inflict. But they are even better known because females of one subfamily (Belostomatinae) deposit their eggs on the backs of males. The males care for the eggs by swishing water over them; this aerates the eggs and (it has been shown) prevents fungal growth.

These bugs occur throughout the world and are the most diverse in the tropics. In some cultures giant water bugs are dried and eaten; in parts of Asia they are a delicacy.

Closely related to Belostomatidae is Nepidae, the water scorpions; they range in length from 15 to 50 mm. Nepines (subfamily Nepinae) look rather like small belostomatids except for the terminal air siphon, which may be more than twice the body's length. Ranatrinae also has a long air siphon, and its body is long and very narrow, nearly round in cross section. Like the antennal segments of belostomatids, those of nepids have projections; unlike belostomatids, nepids have one-segmented tarsi.

Water scorpions do not swim actively; they are, rather, "sit-and-wait" predators, often sitting patiently on vegetation in quiet ponds until appropriate prey wanders past. They grab the prey swiftly with the forelegs, rather like aquatic preying mantids.

Like Belostomatidae, Nepidae occurs worldwide but is particularly abundant in the tropics.

The toad bugs, Gelastocoridae (6–15 mm long), are riparian, not truly aquatic; they look (at least to the extremely untrained eye) like tiny toads, and they jump. The body is rough-textured and lumpy ("warty"); the eyes large, and the forefemora and foretibiae modified for prey capture. They occur throughout the world, and most species are tropical.

Ochterids (Ochteridae), or velvety shore bugs, are also riparian, living along the margins of water bodies. Unlike most other nepomorphans, ochterids have antennae that are visible (although small); the bugs are 4 to 9 mm long; and are mostly but not wholly tropical. With their visible antennae, suboval shape, and often mottled color patterns, ochterids look rather like saldids (see later, under Leptopodomorpha), a leptopodomorphan group that indeed is ecologically replaced by ochterids in the tropics.

Naucoridae, which contain the creeping water bugs, or saucer bugs, are 5 to 20 mm long and elongate-oval, with hidden antennae and enlarged forefemora. They look like small belostomatids; however, naucorids lack airstraps, their eyes are relatively larger than belostomatids', and their dorsal surface is sometimes mottled.

Two naucorid subfamilies, Aphelocheirinae and Potamocorinae, are often treated as separate families in the recent literature, partly because they have small but visible antennae and other nonnaucorid features. The former subfamily (60 species) ranges in length from 3.5 to 12 mm and contains the best plastron breathers in Heteroptera. So effective in trapping and keeping the air bubble are their hydrofuge hairs that these bugs can remain their entire lives below water, using the bubble as a physical gill. This ability allows aphelocheirines to live deep in the water, a niche unavailable to other aquatic bugs which must surface at least occasionally to replenish their air supply. Aphelocheirinae, like Nepinae, are primarily tropical, but with significant representation in the Holarctic (Nepinae) or Palearctic (Aphelocheirinae) region.

The eight potamocorine species are all Neotropical and look like very small (2.5–3 mm long) naucorids except for the visible if small antennae. Nothing is known of their biology.

Leptopodomorpha   With one exception, the families of this infraorder have few species and are poorly known. The family groups in Leptopodomorpha have been arranged variously by various authors. This article follows Schuh and Slater, who in 1995 listed four families: Saldidae (265+ species), Aepophilidae (1), Omaniidae (5), and Leptopodidae (about 40). Most leptopodomorphs live near water, in damp places, and members of the Omaniidae and Aepophilidae are intertidal; many leptopodids, however, live in quite dry habitats. These bugs are small (about 1–7.8 mm long), with large eyes (except Aepophilidae) and a broad head. All are predaceous (as far as is known).

The largest and best known family is Saldidae, whose members range in length from 2 to 7.8 mm. Known as shore bugs, saldids occur in damp areas near water; they are excellent jumpers, fly readily, and are more difficult to collect than they at first appear to be. The eyes are indented medially (kidney-shaped), and the dorsum of many species is mottled dark and light. The male has a grasping structure on the side of the abdomen. Shore bugs occur mostly in temperate regions (somewhat unusual for heteropterans) and may occur at fairly high elevations or quite far north (e.g., Alaska); a few are intertidal or found in salt marshes.

Aepophilidae and Omaniidae are wholly intertidal. Indeed, the single aepophilid species lives below the water, breathing (like Aphelocheiridae) via plastron respiration with an air bubble permanently trapped in a dense mat of hydrofuge hairs. Omaniids seek prey on exposed rocks at low tide; at high tide they wait in crevices, probably tapping air bubbles for oxygen. Members of both families are very small (2 and 1–2 mm long, respectively), and presumably feed on even smaller intertidal invertebrates. Aepophilids have quite small eyes, as do aphelocheirids, suggesting that a life permanently under erratically moving water is not sight based.

Aepophilus bonnairei occurs in Europe south to coastal North Africa. Omaniids are known from the Red Sea, south to Aldabra on the African coast, and east to Samoa; it seems likely that they occur on the Indian Ocean coast, and on coasts to the east.

Species of Leptopodidae are about the same size as saldids, ranging from about 2 to 7 mm in length. These bugs vary in shape, and males have a grasping organ similar to but different from that of saldids. Most species are tropical, and many occur near water, but others are in quite dry habitats. Some have been collected in ant lion pits, which suggests they may be scavengers.

Cimicomorpha   This is by far largest of the heteropteran infraorders, containing as it does Miridae and Reduviidae, the two largest families. Within the infraorder's 14 families is the widest variety of foods and of habitats: primitively predaceous on other arthropods, three major groups have moved to feeding on vertebrate blood, and others to feeding on plants. Habitats of various cimicomorphans range from human homes and caves to the webs of spiders, embiids, and psocids; one reduviid "fishes" for termites and another is "led" to its food by ants.

The group has long been recognized (before the current classification of Heteroptera) but is held together only loosely by shared advanced character states, not all of which are possessed by all members. The head *usually* bears some long-socketed hairs (trichobothria) and is *usually* prominent and extends directly forward; the forewing's membrane *usually* has several closed cells, the claws *usually* lack pads between them; and the eggs *often* have a characteristic type and arrangement of microstructures. However the sperm storage organ of other heteropterans (the spermatheca) is in cimicomorphs nonfunctional or either greatly reduced or absent. The cimicomorphan families are united by this last character. Moreover, since each family is more or less phylogenetically related to another, the first and last are joined less by their relationship to each other and more by their relationship one at time to those in between (rather like a group of people in a line holding hands).

With its 6700 species, Reduviidae is the second largest of the Heteropteran families; it is also perhaps the most diverse, as is suggested by the 25, or 22, or 21, or 29, or 23 subfamilies—and the confusion over the number! Suggested by the number is the systematic confusion at the higher taxonomic levels, a situation similar to that in the Pentatomidae (another large family) and—until recently—in the Lygaeidae (yet another).

Reduviids, also called assassin bugs, for the fierce way some attack prey, vary greatly in shape and in size, ranging in length from 3–4 mm to 40+ mm. All (or most) have a short, stubby, slightly curved beak, designed for being stabbed downward through the cuticle of arthropod prey; well-developed eyes (sight-orienting predators); forefemora often enlarged and with spines for prey capture; forewing membrane with two closed cells; glandular setae and a spongy pad on the foretibiae, presumably to help in grasping; paired first-abdominal glands (Brindley's glands) whose secretion apparently wards off enemies (but, then, what is the function of the metathoracic scent glands?); and a stridulitrum on the prosternum whose plectrum is on the beak, also presumably to frighten would-be vertebrate predators. Not all reduviids—even not all reduviid subfamilies and tribes—have all these characters, but most species are two to four times longer than wide, with a large head, bulging eyes, and a stout beak. Most assassin bugs are easily recognized.

All reduviids are predaceous, the great majority on other arthropods. There is some degree of specialization: Ectrichodiinae (645 species) on millipedes, Peiratinae (47 species) on hard-bodied prey like grasshoppers and beetles, Harpactocorinae (2060 species) on soft-bodied prey like grubs and caterpillars, Reduviinae (980 species) on social insects, Emesinae (920 species) on flies, and other subfamilies with usually lesser degrees of specialization (but again it must be emphasized: much generalization here is based on scant data).

One group, the subfamily Triatominae, feeds upon the blood of vertebrates (mostly warm-blooded vertebrates). These are the only bugs responsible for a human disease: Chagas' disease is a serious trypanosomiasis in the neotropics, costing many deaths every year. Triatomines that live near or in human dwellings are vectors of the protozoan parasite *Trypanosoma cruzi*. An intense cooperative effort by Latin American countries and several other national and international agencies is gradually reducing the range of this disease.

Triatomines are large (50–20 mm long, 5–8 mm wide) bugs, usually brown or patterned light and dark brown. They occur throughout the New World tropics and subtropics (including well into the United States) and the "wild" (sylvatic) ones feed on small mammals (often burrowing rodents) and (some species) on birds. Domestic triatomines live in the thatch or in litter on the floor of dwellings. One genus (*Linshcosteus*) occurs only in India; a species complex of *Triatoma* occurs in Malaysia, Indonesia, New Guinea, and surrounding areas; *T. rubrofasciata* is pantropical (and may be the ancestor of the species in the second group). The possibility that Triatominae is paraphyletic (or polyphyletic)—that its members are descended from several different reduviid groups and independently evolved blood feeding—is being investigated.

Certain reduviid species and smaller groups feed rather strangely: An Indian reduviine (*Acanthaspis siva*) hunts honey bees at their nests, and some neotropical apiomerines, which wait near resin sites for certain bees to arrive to harvest nest-constructing resin, may even release a kairomone to attract the bees, which are eaten; fossil evidence indicates that the latter relationship goes back at least 25 million years. The Indian harpactocorine *Lophocephala guerini* feeds on the juices emanating from cow feces (or perhaps on arthropods found within?) and is led to this food by ants, which in turn feed upon the bugs' feces. Many emesines live unentangled in spider or psocid webs, and they feed on their hosts or on prey trapped therein. *Salyavata variegata* (Salyavatinae, neotropical) waits by crevices in termite nests and disguises itself with bits of the nest itself; it eats termites that emerge to repair the crevice and even uses the remains of fed-upon termites as bait to lure more termites out.

Most reduviids actively seek prey and seize it. Some wait where prey is sure to be found (e.g., termite or bee nests); others wait more patiently for less specific prey. Once seized, prey may be fed upon at once or, often, dragged to a secluded spot. Nymphal reduviids may engage in communal feeding: several bugs feeding on a single prey. This occurs in other predaceous heteropterans too, and probably the pooling of saliva with its digestive enzymes is of benefit to all (except the prey). Several reduviids resemble other heteropterans (e.g., alydids, pyrrhocorids) and may be associated with them. Some have thought this a ruse to lure these others to their doom (see earlier comments on this "aggressive mimicry").

Reduviids are abundant in most habitats. They are large and presumably eat a lot. They are aggressive predators. It is therefore surprising that their possibilities as biological control agents have been so little studied (except—recently—in southern India, as described by Ambrose).

Phymatidae is a family often included in Reduviidae, and as often treated separately. Also known as ambush bugs, phymatids lie in wait for their prey, grabbing them with ferociously enlarged forelegs. In the United States, a common ambush bug is yellowish and lurks in late-summer yellow flowers, waiting for hapless bees and other insects seeking nectar where "rosae … sera moror" (to twist Horace a bit).

The small family Pachynomidae contains 15 described species of tropical predators. They look like certain nabids and share characters with several different cimicomorphan groups; recently they have been placed as the sibling group of Reduviidae. They and another family, Velocipedidae, are the only cimicomorphans with ventral abdominal trichobothria (long-socketed hairs, regularly arranged), a feature found more commonly in Pentatomomorpha. Pachynomids range from 3 to 15 mm in length. Nothing is known of their biology—not even the immatures (that they are predaceous is inferred from the thick curved beak).

The family Miridae [Capsidae (Britain), plant bugs], with its 10,000 species, contains nearly one-third of all heteropterans. So vast is the family and so diverse in habits and habitats, that it cannot be covered here. Luckily, Wheeler's excellent book on the family's biology is available.

All but one group of mirids lack ocelli (hence the German name *Blindwanzen,* "blind bugs"); all are relatively small (1.5–15 mm long), delicate in appearance, usually brown or greenish but often brightly and contrastingly colored, and range in shape from elongated to round to myrmecomorphic. Many (most?) have setae on their surfaces, a "break" in the forewing where leathery part (corium) and membranous part meet, one or two closed membranal cells, and asymmetrical genitalia in the male.

The primitive groups are predaceous, reflecting the predacity of their infraorder (Cimicomorpha). Many mirids are herbivorous, however, although even many of these have reverted to predacity ("secondarily predaceous"). Overall, the percentage of predaceous (and scavenging) mirids is quite high; and many others are omnivorous, supplementing a plant diet with arthropod prey, or vice versa. Moreover, many mirids are cannibalistic. In general, then, a great many mirids are opportunistic feeders and, therefore, may be crop pests under one set of circumstances and useful control agents (of weeds or pests) under other circumstances. However, mirids are small and feed on small prey (eggs, small arthropods). Thus they are rarely seen feeding, and their impact on natural ecosystems and on agroecosystems, although great, is much underestimated (this of course is true of all small predaceous arthropods).

Many mirids are somewhat host specific, at least at the bug-species to plant-genus/tribe level. The bugs feed, usually intracellularly, on the reproductive parts of plants and on new, still-growing somatic tissues. The bugs' enzymes cause the plant to mobilize nitrogen to these feeding wounds, which the bugs suck up. Many mirids feed on annual plants, and often on plants that appear only briefly (e.g., early successional plants); the brief presence of such plants, and their great diversity, may help explain the diversity of mirids: this type of association would seem to promote speciation.

Some of the predaceous mirids also appear to be host specific, but the specificity would seem to be more one of habitat than of prey: associated with a particular habitat (rhododendron plants, tree bark), the mirids perforce feed on prey found only there (cf. Anthocoridae). Members of one subfamily, Cylapinae, occur with fungi, and although they may feed on arthropods also found there, at least one species feeds directly on the fungus.

Because of their numbers and diversity, mirids occur throughout the world in all manner of habitats (except aquatic, but some are found in salt marshes). Probably a majority of plant species is fed upon by mirids, and their populations may become high locally; mirids are therefore often the most frequently collected of heteropterans.

As mentioned, the family is very large; one genus (*Phytocoris*) contains more than 600 species. Eight subfamilies are accepted (but not universally), but it is not surprising that there is less agreement about the number and validity of tribes. Much work is needed on cladistic analyses and higher group systematics of Miridae. And, because of the variety in food preferences, in habitats, in degrees of food and habitat specificity, and in other aspects of their biology, biological and ecological data should be used in these analyses.

Related to the mirids are the Tingidae, known as lace bugs because of the elaborate expansions of the thoraces (and sometimes heads and abdomens) of many of these bugs. Although small (1.5–10 mm long), these are among the loveliest of Heteroptera; some vaguely resemble large coleorrhynchans, but there is no phylogenetic relationship.

Because of their often bizarre shape and strange (and attractive) outgrowths, it is not easy to characterize lace bugs. The head is small or of moderate size, the body (when shorn of outgrowths) usually oval. In most tingids the expansions of the thorax conceal the scutellum and, in some, a pronotal "hood" may partly cover the head; some species are coleopteroid (forewings completely leathery or hard, and closely appressed; the bug looks beetlelike).

The family, with nearly 2000 species, is distributed worldwide; all tingids are herbivorous, even members of a small subfamily that live in ants' nests and (probably) feed on rootlets that penetrate therein. As a group, lace bugs prefer woody plants (but there are many exceptions) and are quite host specific, either at the plant-specific or -generic level. Species in one genus (*Acalypta*) are among the few heteropterans that feed on mosses; and members of two other genera (*Copium* and *Paracopium*) are the only heteropterans that induce gall formation in their host plants. Tingids' preference for woody (perennial) plants sets them apart from mirids (which seem to prefer annuals) and perhaps partly explains why there are so many more of the latter.

This feeding preference of lace bugs also explains why so many are pests of ornamental and fruit-bearing shrubs and small trees, where they feed mostly on somatic tissue. Large populations can arise quickly, the buildup aided in part by various defensive mechanisms of the nymphs. These range from maternal care through several defensive pheromones and kairomones, to an array of sharp or defensive-liquid-producing sctae on the body. Another strategy is the laying of small batches of eggs in different places, a form of bet hedging (this of course occurs in many other groups). These population buildups often lead to major problems localized in time and space.

Maternal care has been carefully worked out in the genus *Gargaphia,* especially *G. solani.* In this species several females may oviposit in a single cluster, which they then communally guard, as well as the resulting nymphs. This opens up the possibility of "cheating": a female that lays eggs in the cluster but does not help guard it can thus devote more time and resources to producing more eggs.

One other family of Cimicormorpha is herbivorous, the 19 species of Thaumastocoridae feed on a variety of plants, but many feed on palms, and the name palm bugs has been given to the family. These are small (2–5 mm long) insects, broadly to narrowly oval, and flattened. The external genitalia are much reduced: the ovipositor is gone in females, one or both parameres are gone in males; the genital capsule is highly asymmetrical. Like tingids, these bugs feed on the cells of plants' somatic tissues (usually leaves), leaving behind pale spots. Populations occasionally build upon ornamental plants to the point of visible damage. One such culprit is the royal palm bug, *Xylocoris luteolus,* which may at times cause unsightly damage to a decorative palm.

The approximately 400 species of Nabidae are about 8 to 12 mm in length and most are slight in appearance, which perhaps accounts for the derivation of their name, damsel bugs. All are predaceous, but some will probe plants, probably for water (none survive on plant food alone). The majority of species seem to prefer seeking prey on low vegetation or in fields, and for this reason nabids are an important component of agroecosystems. Nabids also fly readily and disperse well. As a result, several species are cosmopolitan. One genus (*Arachnocoris*) lives in spiderwebs, probably stealing prey.

Most nabids are brown or light brown, although a few are black and red. The males of most species have a specialized group of tibal setae which, upon being rubbed across part of the genital capsule, spread an attractant pheromone produced in rectal glands.

In many species copulation is normal, but in some a form of internal traumatic insemination (cf. Cimicidae) occurs, wherein the male aedeagus pierces the wall of the female's genital chamber and deposits sperm in her hemocoel.

Two small cimicomorph families, Medocostidae (one species) and Velocipedidae (discussed earlier), have been placed in Nabidae from time to time. The arguments for excluding them are slightly more persuasive than the arguments for including them (more persuasive for the first than the second); and the evidence for their phylogenetic relationships is equally insecure.

The single medocostid species is about 9 mm length, elongated oval, and presumably predaceous. It has an unusually long fourth rostral (beak) segment and lives under bark in western and central Africa.

Three other small families are Plokiophilidae (a half-dozen or so species), Microphysidae (25–30 species), and Joppeicidae (one species). The first appears to be related to Anthocoridae–Cimicidae, but the relationships of the other two are not clear, their affinities with other groups consisting mostly of character-state losses. All are small (1–3 mm long) and predaceous. Of these, the most interesting are the plokiophilid bugs, which live in webs, one species in those of embiids, where they feed on eggs or dead embiids. Other species live in spiderwebs (like some nabids), where they feed alongside the spider on trapped prey. Microphysids are found on tree trunks, and the joppeicid in dry situations.

Closely related to one another are Polyctenidae (bat bugs) and Cimicidae (bed bugs, although most—like all polyctenids—are ectoparasites of bats). Both groups are well adapted as ectoparasites of vertebrates: They are wingless and flattened, the latter an adaptation both for slipping through the hair or feathers of a host and for expansion of the body during engorgement. Cimicids are temporary ectoparasites, moving onto the host only to feed.

As noted earlier, it is cimicids, the bed bugs, that have given their name to heteropterans: true bugs. The 100 or so species are temporarily ectoparasitic on bats, birds (mostly those that roost in groups), and humans. Of the human bed bugs, *C. lectularius* (worldwide, especially in cooler regions) and *C. hemipterus* (tropics), the former is among the very few arthropods associated *only* with humans (like the house fly and the head/body louse). *C. hemipterus,* the tropical bed bug, may also be found on bats, but it seems to prefer humans.

These human bed bugs are small (5 mm long), round to broadly oval, flat (except when engorged with blood), and brown; they are well known, to the extent that although no one claims to have seen one, everyone knows someone who has. Because bed bugs are secretive and feed at night, they are not uncommon but are rarely seen, spending the day hidden in crevices in dwellings and in bedclothes. Despite their ubiquity, they usually do little harm. They spread no pathogen (although they have been implicated in transmission of hepatitis B virus in southern Africa, the evidence offered is at best equivocal), and for the most part they are annoying rather than harmful.

Cimicids that feed on birds can cause harm: some feed on poultry and may take enough blood to decrease egg and meat production; and swallow bugs (genus *Oecacious*) may cause serious damage to populations of these attractive and useful birds.

It is in the human bed bug that traumatic insemination has been studied in the most detail. The male punctures the venter of the female's abdomen with his scimitar-shaped left clasper. Sperm are deposited either directly into the hemocoel (more primitive cimicids, some noncimicids) or into a patch of tissue specialized for their reception. In either case, the sperm then travel (for part of their journey, from cell to cell) to the base of the ovarioles, where they are stored; this journey cannot occur if the sperm are not activated by an agent in the seminal fluid. Eggs are fertilized as they pass the storage depot. This process has not been worked out for the majority of traumatically inseminated species but is probably the same in broad outline.

Unlike bed bugs, polyctenids live permanently on their bat hosts and are so well adapted to this way of life that they closely resemble not cimicids but bat flies (Streblidae, Nycteribiidae). They are eyeless and have claws and stiff setae "designed" as in other ectoparasites to catch on fur and prevent them from being captured and dragged away by the host. Bat bugs have also carried further the traumatic insemination of their relatives: the female is mated before molting to adult, through the base of the right midcoxa into the hemocoel, whence the sperm move close to the oviducts. Here (apparently) fertilization takes place, and here too occurs not only embryogenesis but hatching. Nymphs are then "born," and because some of these births may occur before the female is adult, they represent a form of pedogenesis (i.e., reproduction by immature stages). The embryos are nourished in part directly from the oviduct wall, and the entire process has therefore been termed pseudo-placental viviparity, a phenomenon that occurs in other groups of insects (cf. the tsetse fly). (It is not clear that mating *must* occur before molting; presumably an unmated adult female can be mated.)

Bat bugs are 3 to 5 mm in length and occur in the tropics. As a group they show some host specificity, and it would be interesting to compare their host associations with those of streblids and nycteribiids.

Anthocoridae contains 500 to 600 species and here includes the subfamilies Lyctocorinae and Lasiochilinae, sometimes treated as separate families. Anthocorids are sometimes called minute pirate bugs, perhaps because they are predaceous and some of the more common ones are picturesquely black and white (like the Jolly Roger). Anthocorids' eyes are prominent, suggesting they orient to prey visually "up close"; some evidence indicates they are initially attracted by the odor of the leaves on which prey are feeding. The left paramere of the male is in many anthocorids modified for traumatic insemination of the female.

Although small (1.5–6 mm long), anthocorids are efficient predators on small arthropods and upon the eggs and early instars of larger arthropods. They are also abundant in many habitats, and for these reasons are quite useful in biological control. In Europe especially, anthocorids are used for control of pests in such confined growing areas as greenhouses. Some members of the Lyctocorinae live in dry areas and have proven useful in the control of stored-food pests.

Many anthocorid species seem to be habitat specific. One impressive example is *Elatophilus,* whose species are found on the trunks of north temperate pines, where they feed on *Matsucoccus* scale insects; the specificity appears to be one of habitat, not prey. Other anthocorids live on the trunks of other trees, feeding there on small homopterans and sometimes bark beetles.

Also known as flower bugs (again perhaps for their daintily contrasting colors), a few anthocorids—mostly of the large and widespread genus *Orius*—live on annual plants; but this is not the most important habitat of anthocorids. Most species (as far as is known: not very far) live on tree trunks, some in leaf and other ground litter and the bases of grass clumps, and some in natural accumulations of seed or berries (and in the nests of rodents that have carried these foods in—perhaps evolutionarily en route to living where grain is stored). Because of the considerable value of these bugs in biological control, the literature on their biology is vast. Unfortunately, the literature is concerned with relatively few species, and the biologies of most are wholly unknown. It is known that some species (perhaps many?) supplement their diet with pollen; and at least one may be entirely herbivorous.

Pentatomomorpha   This is the second largest of the heteropteran infraorders. Unlike the others, it (and Aradomorpha) is primitively herbivorous: that is, the ancestor, or the earliest members, of the infraorder, were themselves almost certainly feeders on plant juices. With few exceptions, members of Pentatomomorpha feed on plants. The exceptions are certain Rhyparochromidae that feed on vertebrate blood; members of Geocoridae, which prey on small arthropods as well as plants; a very few Pyrrhocoridae that may be either obligate or facultative predators (it is not always clear which); and Pentatomidae of the subfamily Asopinae, all of whose members are predaceous. For the most part, pentatomomorphs feed on the nitrogen-rich reproductive parts of plants: flowers, ovules, and ripe and unripe seeds. Many, however, feed on plant somatic tissues.

Pentatomomorphs range from the very small (some Lygaeoidea) to the largest of the land bugs. The beak is long, extending usually at least to the mesosternum and sometimes beyond the abdomen; the abdomen's venter bears a regular number of regularly arranged socketed setae. Associated with the metathoracic scent gland opening is one or more raised structures, and the shape is rounded to narrowly oval (rarely long and thin).

Pentatomomorpha contains five superfamilies—six if one includes Aradoidea. (The recent elevation of this group to infraorder level has not met with universal acceptance, and in this article "Aradoidea" is treated as the infraorder Aradomorpha).

The Idiostoloidea contains two families, Idiostolidae (four species) and Henicoridae (one species); the latter was until recently considered to be a subfamily of Lygaeidae (in the broad sense, as discussed shortly). In some features (e.g., abundance of trichobothria), Idiostoloidea appears to be the most primitive of the pentatomomorphan superfamilies; however, in Henry's analysis, this group is the sister group of the families formerly included in Lygaeidae. The biology of Idiostoloidea is not known, although specimens have been captured on the ground, not up on plants. Idiostolids are associated with the moss and litter of southern beech *(Nothofagus)* stands, and may indeed feed on mosses. The superfamily has a Gondwana distribution occurring in southern South America and in Australia (see also Coleorrhyncha, earlier).

Until recently most members of the superfamily Lygaeoidea were classified in the family Lygaeidae. However, it had long been recognized that the members of this family were not all closely related and that indeed Lygaeidae was paraphyletic and, probably, polyphyletic: that is, Lygaeidae did not include all descendents from the common ancestor, and indeed different members of Lygaeidae were descended from several different ancestors. In 1997 T. J. Henry applied cladistic analysis to Lygaeidae, as well as to related families, and presented a new classification of the members of the former Lygaeidae. This classification, which has been generally (but not universally) accepted, is used here.

The family Lygaeidae (*sensu* Henry) now consists of 15 families: Artheneidae, Blissidae, Cryptorhamphidae, Cymidae, Geocoridae, Heterogastridae, Lygaeidae (*sensu strictu*), Ninidae, Oxycarenidae, Pachygronthidae, Rhyparochromidae, Berytidae, Colobathristidae, Malcidae, and Piesmatidae. The last four had already been treated as separate families in the old Lygaeoidea. The superfamily contains roughly 4400 species, most of them in the former "Lygaeidae."

Lygaeoids are relatively small heteropterans (1–20 mm long); the largest are the milkweed bugs, several of which are common in North America. Milkweed bugs are Lygaeidae (*sensu strictu*), and many are warningly colored because they are distasteful. Of greater importance are the chinch bugs (Blissidae); all feed on grasses, and several are serious pests of graminaceous crops (wheat, sugarcane, rangeland and lawn grasses, etc.). Several other families are of some economic importance (Geocoridae, one of the few facultatively predaceous groups in Pentatomomorpha, in biological control; some Colobathristidae on sugarcane, Fig. 2).

Members of the largest lygaeoid family, Rhyparochromidae, with nearly half the superfamily's species, feed on fallen seeds; for this reason they are called seed bugs, a name that by extension is often applied to lygaeoids generally. Members of other lygaeoid families feed on seeds still attached to plants, and others feed directly on somatic tissues (mostly leaves, sometimes in vascular tissue, sometimes in plant cells).

Members of the rhyparochromine tribe Cleradini, and one species in the tribe Udeocorini, feed on vertebrate blood, and it has been suggested they might be able to vector *T. cruzi*, the pathogen of Chagas disease (see Reduviidae).

Of the five families in the superfamily Coreoidea (Coreidae, Alydidae, Rhopalidae, Stenocephalidae, and Hyocephalidae), the first is by far the largest in number of species (1800–1900) and in size; indeed, members of the neotropical coreid genus *Thasus* may be the largest land heteropterans known (Fig. 3); up to 40 mm in length, being exceeded in this respect only by some giant water bugs (Belostomatidae). The other families are smaller in numbers and size: Alydidae (about 250 species), Rhopalidae (somewhat more than 200), Stenocephalidae (about 30), and Hyocephalidae (3). All are phytophagous.

Members of Coreidae are sometimes called leaf-footed bugs because several species, occurring in several apparently unrelated tribes, have hind tibial expansions (and sometimes antennal expansions) as adults, as nymphs, or as both; their function is obscure, although they may serve to deflect the attacks of birds. Males of several groups of coreids have enlarged hind femora, which are used to defend territories from other males. The nymphs of several species are gregarious, and it would be interesting to learn whether this gregariousness occurs in species whose males are territorial. The coreid head is small relative to the body (cf. Alydidae), and the bugs' length ranges from about 6 to 40 mm.

Members of the family feed up on plants, some species on ripening seeds and other reproductive parts, and many species on the juices from vascular tissue. Varying degrees of host specialization occur in the family, from mono- through oligo- to polyphagy. Some species feed on cucurbits (hence another common name, squash bugs). Some unrelated coreid groups (Pseudophloeinae, and three coreine tribes) are among the few heteropteran family groups to have broached the defenses of the Leguminosae and to have radiated upon these plants. Several of these bugs are serious pests of legume crops (e.g., pigeon pea) in the Old World tropics.

FIGURE 2 A lygaeid, Wekiu bug, *Nysius wekiuicola,* feeding on Calliphoridae. (Photograph by Peter Oboyski.)

FIGURE 3 Mesquite bug (adult), *Thasus neocalifornicus.* (Photograph by John H. Acorn.)

FIGURE 4 Broad-headed bug (nymph), probably *Megalotomus quinque-spinosus*. Note ant mimicry. (Photograph by John H. Acorn.)

FIGURE 5 Box elder bugs (adult and nymphs), *Boisea trivittatus*. (Photograph by John H. Acorn.)

The males of one species, *Phyllomorpha laciniata,* are heavily spined and, somewhat like some giant water bug males, carry among the dorsal spines their females' eggs.

Alydidae—called broad-headed bugs because the head is relatively wide (it also resembles an ant's)—comprises three groups. The subfamily Alydinae is another group specializing on legumes, upon whose ripening seeds the bugs feed. Several species are pests of legume crops (e.g., soybean, pigeon pea) in the tropics. Immature alydines mimic ants, often successfully enough to fool heteropterists; some species mimic small species of ants as early instars and other larger ant species when older (Fig. 4).

Some members of the second alydid group (Micrelytrinae: Micrelytrini) mimic ants both as adults and as nymphs. The biology of this pantropical tribe is almost completely unknown, although one species may become abundant on range grasses in northern South America. It has been speculated that many micrelytrines are grass feeders (on grass seeds?), as are members of the other tribe (Leptocorisini) of Micrelytrinae.

All members of Leptocorisini feed on grasses, as far as is known. Some species (genera *Leptocorisa, Stenocoris*) are elongated and are often serious pests on rice in the Old World. Leptocorisines feed on the rice panicles and, when rice of the appropriate stage is not available, rice bugs feed on wild grasses found nearby. These grasses thus serve as a reservoir for the rice pests; some leptocorisines prefer these grasses to rice, however.

Members of Rhopalidae are called scentless plant bugs because all are herbivorous and the external opening of the metathoracic scent gland is small and placed more ventrally (hence harder to see) than are the openings of other bugs. Rhopalids are not scentless, however, and indeed the scents of a few species have been analyzed.

The subfamily Serinethinae (60–65 species, mostly tropical and subtropical) includes the box elder bug (Fig. 5), which in North America sometimes seeks warmth in houses in late fall. Serinethines feed on sapindalian plants and occasionally become minor pests on these ornamental plants. The 150 or so members of Rhopalinae are more drab than the red and black Serinethinae, and far more of them occur in the Holarctic region. They feed on a variety of plants and rarely become pests even locally. At least one species has helped biologically control a weed, velvetleaf, in the United States.

Stenocephalidae occurs in the Old World tropics, with a few members in the Palearctic region; the single species on the Galápagos Islands may have been introduced by ship from Africa. Stenocephalids feed on euphorbs, as far as is known. The family is of interest because it possesses certain features of Coreoidea (many) and of Lygaeoidea (ovipositor type, egg type); the phylogenetic significance is unclear.

The three species of Hyocephalidae are all Australian. The family is related to Stenocephalidae, and both families seem to be primitive in the Coreoidea. Little is known of hyocephalid biology except that some occur under stones, where they apparently feed on fallen seeds. Both stenocephalids and hyocephalids are relatively large (8–15 mm long) and slender.

Pentatomoidea is a well-defined superfamily, although there is some disagreement (not to say confusion) about the higher classification of the groups within: some heteropterists treat some groups as tribes, subfamilies, or even families, and others treat them differently. The taxonomic limits of Pentatomoidea are clear and agreed upon; and there is general agreement that the various tribes, and so on are indeed worthy of suprageneric rank: the level of that rank is sometimes argued. A forthcoming catalog of the largest family, Pentatomidae (by D. A. Rider), will help settle some of these questions.

In general, pentatomoids are larger than many other bugs. Most (but not all) have five-segmented antennae (hence the name); most other heteropterans have four. Most are also wider than many other heteropterans, and, ranging in length from about 4 to 30 mm, many pentatomoids appear somewhat "squat." Quite a few are very brightly colored. All except one subfamily (Pentatomidae: Asopinae) are herbivorous; the

FIGURE 6 Adult predaceous pentatomids *(Eocanthecona furcellata)* feeding collectively on a caterpillar *(Eupterote mollifera)* in southern India.

FIGURE 7 *Coleotichus blackburniae* (Scutelleridae). (Photograph by Peter Oboyski.)

superfamily contains general feeders and quite host-specific ones. Within both groups are some serious pests of crops, and asopines have shown some success in biological control.

The largest family, Pentatomidae, whose approximately 4500 species make it slightly larger than Lygaeidae (in the old sense, as discussed earlier), occurs abundantly throughout the world. One species, the southern green stinkbug, *Nezara viridula,* is cosmopolitan, feeds on just about anything green, is a major pest in some areas and a minor one in many others, and in Asia has several striking color varieties. This species has been so widespread for so long that its place of origin remained unknown until recently: it probably arose in or near Ethiopia.

Pentatomids are for the most part relatively large, ranging in length from 8 to 20 mm. They are also herbivorous, with the exception of one subfamily (Fig. 6). Most of the few pentatomids whose feeding habits are known are polyphagous; some feed more narrowly. Members of one group, related to *Aelia,* feed on grasses; and several species of *Aelia* itself are serious pests of small-grain crops (Sunn pests, as discussed shortly in connection with "Scutelleridae"). Pentatomids are among the most serious pests of soybean worldwide: when a new area opens up to soybean production (e.g., southern Brazil several years ago, central Brazil a few years ago), local legume-feeding or polyphagous pentatomid species rapidly become soybean pests. Because of the family's size and the number of crops its members feed on, Pentatomidae is one of the most important heteropteran families.

The subfamily Asopinae is predaceous, a fact easily recognized by its members' short stubby beak, well designed for stabbing into prey. Asopines are useful in biological control, and one *(Perillus bioculatus)* specializes on the Colorado potato beetle and related chrysomelids; the red of this bug is derived from ingested pigment of the beetle prey.

Many pentatomids are green or (more often) brown, perhaps as camouflage on plants and ground (some feed up on plants and hibernate or estivate on the ground). However, quite a few are brightly and contrastingly colored and are presumably aposematic. Many pentatomids live on tree trunks and indeed may be the most common group of herbivorous bugs to use this habitat; it is not clear what they feed on, perhaps mosses and other epiphytes and perhaps (occasionally) these bugs can penetrate through to the tree's cambium.

Most pentatomids are somewhat broad and flattened, but some (like the grass feeders) are more or less elongated. Some members of Podopinae are almost globose and may bear thornlike spines dorsally (see also Cyrtocoridae).

Various degrees of parental care have arisen in this family (or have been retained from parentally caring ancestors; the point is controversial) and in other pentatomoid families (and elsewhere, rarely). For the most part, this care protects eggs and early instars from parasitism and predation. In many species, however, this effort denies the female an opportunity to feed and, as a result, she produces fewer eggs or batches of eggs than her less caring sisters. The significance of this trade-off remains to be worked out for most species.

Scutelleridae contains 400 to 500 species in which the scutellum is greatly extended to cover the abdomen (hence the common name, shield bugs); these bugs range from 5 to 20 mm in length and are somewhat more globose than pentatomids (although some of the more spectacular ones are broadly elongate). Several species are brilliantly colored (one genus is *Chrysocoris,* or golden bug), either in solid sometimes iridescent colors (blues, violets, reds) (Fig. 7) or in bold patterns of spots and stripes. These are among the most beautiful of heteropterans and, in some cultures, considered to be among the most tasty. Most however are drab brown or tan, the color sometimes patterned.

Despite their relatively large size and often striking appearance, shield bugs remain poorly known biologically, and very poorly known ecologically. The females of a few care for their young, as do those of a few other pentatomoid families. All scutellerids are herbivorous, and a few are pests of several crops. One group is particularly important.

Several species of *Eurygaster* are very serious pests of small-grain crops in a broad swath from eastern Europe through the Middle East (where they are the most serious) into eastern Asia. In the spring, these Sunn pests move from upland wild plants down to the fields with the young grain;

there they feed first on the young shoots and then on seeds. In late summer or early fall, they move back into the hills and mountains, sometimes undertaking a trip of many miles. Crops may be completely destroyed over a very large expanse.

The family is represented throughout the world, but members of each of the four subfamilies are concentrated in different continents, with a few representatives in one or two other continents (the historical biogeography of the family is needed; the pretty ones are all tropical or subtropical).

Acanthosomatids (Acanthosomatidae; there is no common name) look much like pentatomids, but are somewhat larger (7–20 mm long) and more elongated. Their tarsi have only two segments, and the bugs have several other advanced features. The latest revision (1974) recognizes about 45 genera; there are about 200 species, although no one has listed or cataloged them. As far as is known, acanthosomatids feed on woody plants (trees and shrubs). Maternal care by the females of several species has been thoroughly studied; as in Pentatomidae, protection is provided eggs and early instars against predators (chiefly ants) and parasites; the body-jerking and wing-fanning movements are so strong that small predators may actually be hurled from the leaf.

The 250 species of Tessaratomidae are in general larger (15–26 mm long), and paleotropical (a few species in neotropics). The second abdominal spiracle is exposed, and a projection from the mesosternum extends anteriorly sometimes beyond the tip of the beak. One tessaratomid, the bronze orange bug *(Musgraveia sulciventris),* is a frequent pest on citrus in Australia.

Possibly related to this family is Dinidoridae, whose 85 to 90 members are also large (10–30 mm long) and robust, and whose distribution is roughly that of Tessaratomidae. Here too the second abdominal spiracle is exposed (although this may in both families be plesiomorphic); and often the pro- and mesosternum bear a midline groove. Dinidorids in general prefer cucurbits as food, and one species, the red pumpkin bug, *Aspongopus* (formerly *Coridius) janus,* is a pest in India on cucurbit crops.

Urostylinae and Saileriolinae, the two subfamilies of Urostylidae, share several features but look very different and differ considerably in general appearance and in other morphological features; most urostylids are Asian. Urostylines (80–90 species) are elongated, (4–15 mm long) and look more like coreids than pentatomoids. Saileriolines are small (2.5–3.8 mm long), more pentatomoid in shape, and with the anterior part of the forewing (the corium) less "leathery" than in most heteropterans. Some urostylines occasionally attack ornamental trees; in great numbers they may become a problem. One early (and successful!) control method was the use of gunpowder.

Cydnidae (burrower bugs) is the only heteropteran group most of whose members live below ground, sometimes several feet below the surface. Most cydnids are dark brown or black, 1.5 to 25 mm in length (most in the range of 5–12 mm), and have a wing stridulitrum. Their smooth bodies,

flattened heads, and strong forelegs are adapted to digging; so also probably is the coxal "comb" (not a "coxcomb"), an array of stout setae perhaps used for cleaning soil particles from the antennae. The foretibiae of one subfamily (Scaptocorinae) are greatly developed for digging. There are four subfamilies: Sehirinae (60 species), Thyreocorinae (5), Corimelaeninae (200), and Parastrachiinae (2). All live above ground. Females of the Sehirinae and Parastrachiinae care for their eggs and early instars. The Thyreocorinae and Corimelaeninae have sometimes been grouped as a single family; and the two *Parastrachia* species are large and brightly red and black and should certainly be elevated to family rank.

The remaining 350 species are subterranean, feeding on roots. They may at times become serious, although localized, pests of both crops and rangeland grasses.

Cydnidae may be phylogenetically close to the "base" of the Pentatomoidea and may also be related to the group of small families described next.

Within the Pentatomoidea is a number of small families, mostly restricted to the neotropics [Cyrtocoridae (11 species), Canopidae (8), Megarididae (16), Phloeidae (3)] or to Australia [Aphylidae (2 species), and Lestoniidae (2)]. Plataspidae (500 species, a very rough estimate) is primarily paleotropical, but a few species occur in the Palearctic region. Characteristic of many of these families are small size and a rounded globose body, with forewings and/or scutellum enlarged (and often ornamented) to cover the entire abdomen. The cyrtocorid scutellum bears a stout impressive spine. Some (all?) may be related to Cydnidae and to Scutelleridae and as, Schuh and Slater say, the phylogeny and biogeography of these groups need to be worked out. Of phylogenetic importance too is the primitive family Thaumastellidae, whose few members live in southern Africa.

Aradomorpha This infraorder contains two families, Aradidae (1800–2000 species, worldwide) and Termitaphididae (9 species, pantropical). These families were included in Pentatomomorpha, from which Aradomorpha probably evolved.

Aradomorphans are very flattened, and (most striking) their feeding stylets are very long, very narrow, and stored coiled within the head. Termitaphidids are small (2 to 3 mm long), wingless, eyeless, and ovipositorless; they live in termites' nests and probably feed, like aradids, on fungal mycelia.

Most flat bugs (Aradidae) are larger (3–11 mm long) and live under the bark of dead or dying trees, a habitat to which their flat body and shortened legs and antennae fit them; other aradids live in the litter of the forest floor; and a few live in termites' nests, or vertebrates' burrows. All these places are closely confined and thus both temperature and humidity vary little, conditions conducive to the growth of fungi, within whose long tubular mycelia the flat bugs feed. This way of life is unique to aradids, which have successfully exploited it (as witness the many aradid species).

One species is an exception: *Aradus cinnamomeus* (actually, a three-species complex) feeds on the circulatory tissues of living pines (and occasionally *Larix*), and in central and eastern Europe can become a serious pest. Some other aradids also may feed on tree fluids, but the frequency of such feeding, the number of species that so feed, and the phylogenetic significance of such feeding are all unexplored.

ENVOI

Heteroptera not only comprises the most species of any hemimetabolous group (except perhaps the paraphyletic "Homoptera") but contains the most biological, structural, and ecological diversity. It is a measure of this diversity, this evolutionary versatility, that so many ecological niches are filled by bugs. From the Alaskan cold to the webs of embiids to the bottoms of streams and the surface of the sea, and to the beds of people, heteropterans live where few other insect groups (and no other hemimetabolans) occur. Moreover, a vast number of heteropterans remain to be described, many from unusual habitats (tree canopies, leaf litter); and the biologies of most heteropterans remain to be worked out. The diversity we see, although great, is less than the diversity to be discovered. Why this group of insects should be so diverse cannot be answered here. But that it is so diverse explains the group's great and continuing fascination.

See Also the Following Articles
Aposematic Coloration • Aquatic Habitats • Auchenorrhyncha • Plant–Insect Interactions • Rostrum • Sternorrhyncha

Further Reading

Ambrose, D. P. (1999). "Assassin Bugs." Science Publishers, Enfield, NH.

Andersen, N. M. (1982). "The Semiaquatic Bugs (Hemiptera, Gerromorpha). Phylogeny, Adaptations, Biogeography and Classification." Entomonograph 3. Scandinavian Science Press, Klampenborg, Denmark.

Blatchley, W. S. (1926). "Heteroptera or True Bugs of Eastern North America." Nature Publishing, Indianapolis, IN.

Butler, E. A. (1923). "Biology of the British Hemiptera-Heteroptera." H.F. & G, Witherby, London.

Cohen, A. C. (1998). Solid-to-liquid-feeding: the inside(s) story on extra-oral digestion in predaceous Arthropoda. *Am. Entomol.* **44**, 103–116.

Dolling, W. R. (1991). "The Hemiptera." Oxford University Press, Oxford, U.K.

Henry, T. J. (1997). Phylogenetic analysis of family groups within the infraorder Pentatomomorpha (Hemiptera: Heteroptera), with emphasis on the Lygaeoidea. *Ann. Entomol. Soc. Am.* **90**, 275–301.

Polhemus, J. T. (1996). Aquatic and semi-aquatic Hemiptera. *In* "An Introduction to the Aquatic Insects of North America" (R. W. Merritt and K. W. Cummings, eds.), Chap. 15. Kendall/Hunt, Dubuque, IA.

Schaefer, C. W., and Panizzi, A. R. (2000). "Heteroptera of Economic Importance." CRC Press, Boca Raton, FL.

Schuh, R. T., and Slater, J. A. (1995). "True Bugs of the World (Hemiptera: Heteroptera)." Cornell University Press, Ithaca, NY.

Slater, J. A., and Baranowski, R. M. (1978). "How to Know the True Bugs." Brown, Dubuque, IA.

Weber, H. (1930). "Biologie der Hemipteren: eine Naturgeschichte der Schnabelkerfe." Springer-Verlag, Berlin.

Wheeler, A. G., Jr. (2001). "Biology of the Plant Bugs (Hemiptera: Miridae): Pests, Predators, Opportunists." Cornell University Press, Ithaca, NY.

Protura

Robert T. Allen
Paris, Arkansas

Members of the order Protura are small (usually < 1.0 mm), pale to white arthropods that live in the soil and ground litter debris. Because of these characteristics, the group was not discovered and recognized until 1907, long after almost all the other insect orders had been described and classified. Two notable Italian entomologists, F. Silvestri and A. Berlese, recognized the group at about the same time. In 1907 Silvestri described the order Protura and created the first family, Acerentomidae. This work was quickly followed by the description of a second family, Eosentomidae, and a number of new species by Berlese in 1908. In 1909 a comprehensive monograph on the order was published by Berlese. After almost 100 years of study by many workers, over 600 species have been described and placed in two suborders and eight families (Table I).

Since their discovery and characterization, the Protura have always been recognized as a very ancient group of arthropods that evolved early in the history of the phylum. Early workers placed the group with the Collembola and

TABLE I   The Order Protura

Suborder Acerentomoidea
Family Fujientomidae
Family Hesperentomidae
Subfamily Hesperentominae
Subfamily Huhentominae
Family Protentomidae
Subfamily Condellinae
Subfamily Proentominae
Family Acerentomidae
Subfamily Acerentulinae
Subfamily Tuxenentulinae
Subfamily Acerentominae
Family Berberentomidae
Subfamily Proacerelinae
Subfamily Berberentulinae
Subfamily Silvestridinae
Family Acerellidae
Subfamily Alaskaentominae
Subfamily Acerellinae
Subfamily Nipponentominae
Suborder Eosentomoidea
Family Eosentomidae
Subfamily Isoentominae
Subfamily Eosentominae
Subfamily Anisentominae
Subfamily Antelientominae
Family Sinentomidae

Source: After Yin, W. Y. (1984). A new idea on phylogeny of Protura with approach to its origin and systematic position. *Sci. Sin. (B)* **27**, 149–160.

FIGURE 1 Scanning electron microscopy of a proturan (family Acerentomidae) found in subtropical rain forest litter from Lamington National Park, Queensland, Australia. (Photomicrograph courtesy of D. E. Walter.)

Diplura and referred to the three orders as the "pouched head" or entognathous orders because the mouthparts were enclosed by the sides of the head capsule. Most recently, the Collembola and Protura have been removed from this association and placed in the suborder Ellipurata (some workers consider the group to be a class), standing between the class Symphyla and the true insects or hexapod orders.

CLASSIFICATION

The Protura may be readily recognized by the following characters: antennae and eyes absent, head conical, mouthparts enclosed by the sides of the head, first pair of legs with numerous sensory organs, and first three abdominal segments with paired styli (Fig. 1).

COLLECTING AND SPECIMEN PREPARATION

Protura are most readily collected by means of Berlese or Tullgren funnels. Flotation techniques have also been used to extract specimens from soil samples. Specimens may be collected in 70% ethanol and held indefinitely. For identification, specimens must be mounted on microscope slides. In properly mounted specimens, a compound microscope equipped with phase contrast may be used to study characters at 1000 ×, under oil immersion.

IDENTIFICATION

For North American forms, the keys published by Copeland and Imadate in 1990 are indispensable for generic identification. Once the genus has been correctly identified, it is necessary to refer to a number of articles describing individual species in that genus. General works published during the latter half of the 20th century by Tuxen, Nosek, Imadate, and Houston will also be helpful.

BIOLOGY

After insemination by males, the females lay their eggs in the soil or litter. Protura prelarvae hatch in 8 to 12 days under favorable conditions. The abdomen of prelarvae consists of nine segments,

and the mouthparts and thoracic and abdominal appendages are not fully developed. The prelarva molts in 2 to 3 days and changes to the larva I stage. During this stage the appendages and setae on the body develop more fully. After molting a second time, larva II adds an abdominal segment (10-segmented abdomen) and the chaetotaxi (specifically arranged hairs on the body) continue to develop. The next state, termed the maturus junior, is characterized by the addition of the last two abdominal segments (12-segmented abdomen), but the genitalia remain undeveloped. The preimago is the fifth-instar and quickly molts to become the imago or adult form. Adults are characterized by the 12-segmented abdomen and fully developed genitalia in both sexes.

It is now known that Protura can be among the most abundant arthropods inhabiting soil and litter. As many as 150 individuals representing several species have been collected in forest soil samples in eastern North America. They may occur in soil to a depth of 0.5 m or more.

The diet of the Protura is uncertain. Some species have been observed feeding on mycorrhizal fungi; other possible food sources are unknown. Other aspects concerning the biology and ecology of these animals have not been studied.

See Also the Following Article
Diplura

Further Reading
Bernard, E. C. (1990). New species, clarifications, and changes in status within *Eosentomon* Berlese (Hexapoda: Protura: Eosentomidae) from the United States. *Proc. Biol. Soc. Wash.* **103,** 861–890.
Copeland, T. P. (1962). A taxonomic treatment of *Eosentomon* Berlese (Protura) of east Tennessee. Ph. D. dissertation, University of Tennessee, Knoxville.
Copeland, T. P. and Imadate, G. (1990). Insecta: Protura. *In* "Soil Biology Guide" (D. I. Dindal, ed.), pp. 914–933. Wiley, New York.
Houston, W. W. K. (1994). Zoological Catalogue of Australia 22, Protura. CSIRO, Information Service, East Melbourne, Victoria 3002, Australia.
Imadate, G. (1974). "Fauna Japan, *Protura*." Kelgaku Publishing, Tokyo.
Nosek, J. (1973). "The European Protura." Muséum d'Histoire Naturelle, Geneva.
Tuxen, S. L. (1944). "The Protura." Hermann, Paris.
Yin, W. Y. (1984). A new idea on phylogeny of Protura with approach to its origin and systematic position. *Sci. Sin. (B)* **27,** 149–160.

Psocoptera
(Psocids, Booklice)

Edward L. Mockford
Illinois State University

Psocoptera (Corrodentia, Copeognatha) constitutes an order of neopterous, exopterygote insects commonly called psocids, barklice, or booklice. Their closest relatives are the

FIGURE 1 Psocid, *Graphopsocus cruciatus.* (Photograph courtesy of Ken Gray Collection, Entomology Department, Oregon State University.)

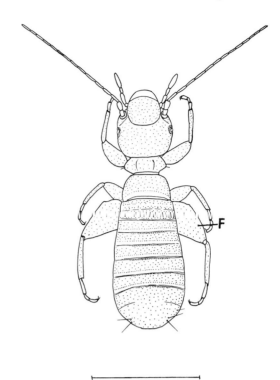

FIGURE 3 Habitus of a booklouse, *Liposcelis corrodens,* showing enlarged hind femur (F) and neotenic features: absence of wings and ocelli and reduced compound eyes. Scale, 0.5 mm.

Phthiraptera or true lice. Psocids are small and soft-bodied and therefore have received little attention from collectors. Only recently, when several dozens of species have been found in stored products, and the tropical forests have proved to harbor a highly diverse fauna, has greater interest been focused on this group.

EXTERNAL ANATOMY

Adult psocids range from about 1 to 10 mm in body length. Most adults are fully winged with the forewings longer and more complexly veined than their hind wings (Fig. 1). The forewings at rest generally exceed the tip of the abdomen. Antennae are long and slender, in the larger forms tending to be longer than the body. The head is rounded with compound eyes often large and bulging but sometimes greatly reduced. The postclypeus (Fig. 2) is usually swollen to accommodate the well-developed cibarial pump muscles. Often the postclypeus bears pigmented chevron marks (Fig. 2) indicating the attachment points of these muscles. Mouthparts are of the chewing type with large, well-developed mandibles. The maxilla contains an elongate, slender lacinia. The hypopharynx includes

structures unique to these insects that are important in maintaining water balance. In fully winged forms the prothorax is frequently reduced while the meso- and metanota are swollen to accommodate the flight muscles. Wing venation tends to be relatively simple. Legs tend to be slender, with the hind femur somewhat swollen in some forms, allowing short hops or initiation of flight by jumping. In the small booklice of the genus *Liposcelis* (Fig. 3), the greatly swollen hind femora are thought to facilitate the ability to change course rapidly and crawl backward. In many winged adult psocids, the hind coxae each bear a rasp and tympanum; the two structures together are called the coxal organ. This is thought to be a stridulatory organ, although as yet no sound has been detected from it.

The abdomen consists of 11 segments. The first 7 are usually membranous, although the terga of the first 2 may be sclerotized and fused together. The clunium, formed by fused segments 8 to 10, is the bearer of external genitalic structures. Distal to the clunium are three semimembranous flaps guarding the anus—the dorsal epiproct and the lateral paraprocts. These may be modified in various ways and probably are involved with copulation.

LIFE CYCLE

Most psocids are oviparous. In various taxa, eggs may be laid singly or in groups and either bare or with a covering of encrusting material, webbing, or both. Eclosion (emergence

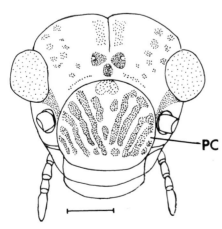

FIGURE 2 Face of a psocid, *Aaroniella badonneli,* showing prominent postclypeus (PC) with chevron marks. Scale, 0.2 mm.

of the pronymph from the egg) is facilitated by means of an oviruptor, a blade-like or saw-like structure on the head of the pronymph. By means of a slight rocking motion, imbibing of air, and hemolymphatic pressure on the head, the insect pushes its oviruptor against the thin egg cuticle, creating a slit through which it exits. Immediately it molts again, casting off the pronymphal exuviae (the cast cuticle), which remain partially caught in the exit slit of the egg.

Eggs of some species undergo a winter diapause and in the Northern Hemisphere this is brought about by the mother's perception of decreasing day length in late summer.

Nymphs undergo five to six instars, requiring 4 to 6 weeks to reach maturity. Adults remain teneral for a day or two prior to engaging in sexual activity. Females of some sexual species produce a sex pheromone that, at a short distance (1–2 cm), brings about immediate, rapid search behavior in mature males—involving them running about and vibrating their wings. This reaction may be elicited by a piece of substrate on which an "advertising" female has been standing or even by the tips of a pair of forceps that have just held such a piece of substrate. Behavior after contact is tremendously varied. Frequently a male "antennates" a female, i.e., touches his antennae to hers and then crawls over her back and forces himself under her body from in front. The copulatory structures then grasp together, after which the male may stay under the female or move—180° to face opposite the female (suborder Trogiomorpha) or back on top of the female (family Lachesillidae). Copulation lasts only a few seconds in many groups of the suborder Psocomorpha, although in the genus *Lachesilla* it generally requires some 35 min. In the suborder Trogiomorpha, it may last up to 2 h during which individual sperm pass through the slender spermathecal duct, then clump together, and become enclosed in "spermatophores" in the spermathecal sac. This is probably the sequence of events in any form in which the spermathecal duct is slender and spermatophores are seen in the spermathecal sac.

Oviposition may start within a day or 2 of copulation. Under favorable conditions, oviposition may continue over a period of 2 months.

EVOLUTION TOWARD NEOTENY

Although most adult psocids are fully winged, a variety of levels of wing reduction are seen throughout the order. Associated with wing reduction are reductions in several other structures: compound eyes, ocelli, paraproctal sensorium, and ctenidia (comb-based setae) of the basal hind tarsomere. The resulting adult appears nymph-like in these characters, and these changes can thus be regarded as neotenic. Several patterns of change are seen, which can be summarized as completely wingless, *Liposcelis* (Fig. 3); with very short scale-like or button-like wings, *Trogium, Cerobassi;* wing development variable in both sexes, *Rhyopsocus, Lachesilla* (some species); with sexual dimorphism in wing development and males wingless or nearly so, females variable in wing development, *Embidopsocus,*

FIGURE 4 Male (a) and female (b) of a psocid, *Lachesilla pallida,* showing extreme sexual dimorphism in wing development.

Psoquilla, Archipsocus; sexual dimorphism and males fully winged, females variable in wing development, *Peripsocus* (some species), *Valenzuela* (some species); and sexual dimophism with males fully winged, females wingless or nearly so, *Camelopsocus, Mesopsocus* (most species), *Reuterella, Lachesilla* (some species; Fig. 4). Except for the first two categories, these reductions have occurred repeatedly in the evolution of the order. Thus, they appear to offer advantages in energy conservation and keeping the organism closely associated with its habitat.

HABITATS AND FEEDING PREFERENCES

Psocids live in a great variety of habitats, including living leaves, especially of monocotyledonous plants and conifers; dead foliage of all plants, both hanging and in ground litter; trunks and branches of trees and shrubs (open surfaces and under bark); rock surfaces; and human dwellings. In general, psocids feed on algae, fungi, lichens, small eggs of insects, particles of organic debris, and dead bodies of insects. Some seem to be strict alga feeders (some Psocidae, Myopsocidae, Peripsocidae), and it is likely that they are somewhat specific in their choice of algae. There are also some strict lichen feeders, and some specificity has been observed among them. Some of the (micro)fungus feeders are not so specific. In culture, the corticolous lepidopsocid *Echmepteryx hageni* takes lichens, algae, yeasts, and pollen and occasionally nibbles on its own eggs when these are not well concealed. The species of leaf dwellers are primarily microfungus feeders, whereas the species living on open surfaces of tree trunks and branches and on rock outcrops are primarily lichen and alga feeders.

ECONOMIC IMPORTANCE

The booklice, genus *Liposcelis,* are frequently household pests and may be among the causative agents of asthmatic reactions. Of greater monetary importance is their tendency to enter and reproduce rapidly in food storage and food processing facilities, where they may render the products unfit for human consumption. Control methods emphasize sanitation and reduction of relative humidity.

See Also the Following Article
Phthiraptera

Further Reading
Eertmoed, G. (1978). Embryonic diapause in the psocid *Peripsocus quadrifasciatus:* Photoperiod, temperature, ontogeny and geographic variation. *Physiol. Entomol.* **3**, 197–206.
Glinyanaya, Y. I. (1975). The importance of day length in the regulation of seasonal cycles and diapause in some Psocoptera. *Entomol. Rev.* (transl. of *Entomologicheskoye Obsrenıye*) **54**, 10–13.
Mockford, E. L. (1991). Psocids (Psocoptera). *In* "Insect and Mite Pests in Food, an Illustrated Key" (J. R. Gorham, ed.). U.S. Department of Agriculture and U.S. Department of Health and Human Services, Washington, DC.
Mockford, E. L. (1993). "Flora and Fauna Handbook," No. 10, "North American Psocoptera (Insecta)." Sandhill Crane Press, Gainesville, FL/Leiden.
New, T. R. (1987). Biology of the Psocoptera. *Orient. Insects* **21**, 1–109.
Smithers, C. N. (1967). A catalog of the Psocoptera of the world. *Aust. Zool.* **14**, 1–145.
Smithers, C. N., and Lienhard, C. (1992). A revised bibliography of the Psocoptera (Arthropoda: Insecta). *Tech. Rep. Aust. Mus.* **6**, 1–86.

Pterygota

Pterygota is a subclass of the class Insecta of the phylum Arthropoda. It contains two divisions: the Exopterygota and the Endopterygota. Most extant orders of insects are pterygotes.

Puddling Behavior

Scott R. Smedley
Trinity College, Connecticut

Puddling is a behavior in which insects, chiefly adult lepidopterans, drink from mud puddles or moist soil. Although broadly distributed geographically, puddling is particularly spectacular in the tropics where numerous individual butterflies representing multiple species often gather to imbibe from damp ground along river banks. The behavior generally is strongly sex biased, with typically only males participating. Puddling is associated with both the insects' nutritional ecology and their reproductive biology. In certain moths, the behavior is rather herculean, as individuals imbibe several hundred times their body mass in puddle fluid in a single drinking bout!

GENERAL BACKGROUND

Puddling is widely distributed taxonomically among the Lepidoptera, occurring diurnally in butterflies (5 families reported) as well as nocturnally in moths (at least 12 families). Puddling takes place in tropical, temperate, and boreal ecological regions. Interestingly, in light of the discussion of sodium below, puddling may be less common in coastal areas. While studying butterflies puddling in Amazonia in the 1860s, H. W. Bates, the British naturalist best known for his work on insect mimicry, was the first to note that the behavior is strikingly male biased. Subsequently, numerous studies have confirmed that this is generally true, although there are certain species in which females often exhibit the behavior. In addition to lepidopterans, nocturnal aggregations of leafhoppers (Homoptera: Cicadellidae), also predominately male, drink from moist soil surrounding puddles.

NUTRITIONAL ECOLOGY

Insect herbivores experience limited access to certain nutrients because of the paucity of these materials in their diet. Sodium as well as proteins and amino acids are substances of physiological importance that are not readily available from the foliar diets of larval lepidopterans. Puddling is a means by which adults augment the larval intake of these scarce materials. Since the 1910s, beginning with the British entomologist E. B. Poulton, there was speculation that puddling might allow lepidopterans to obtain sodium. The demonstration in the 1970s that sodium was a necessary stimulus to evoke puddling in the tiger swallowtail *(Papilio glaucus)* supported the notion. This and subsequent studies with butterfly species showed that when presented an array of various ionic solutions, the insects preferred those with sodium as the cation, and among sodium solutions, those with higher concentrations were favored. In the 1990s it was established that puddling indeed leads to sodium uptake: while drinking from mud puddles, the notodontid moth *Gluphisia septentrionis* showed a gain of sodium, matched evenly (on a molar basis) by a loss of potassium. Lepidopterans typically have an abundance of potassium, because the mineral is plentiful in the larval diet and consequently in the adult body.

Puddling can further result in the acquisition of nitrogenous nutrients. Tiger swallowtails drinking from moist soil laced with tritiated glycine and leucine incorporated these labeled amino acids into body proteins. Studies of Malaysian butterfly communities show that representatives of certain families (Papilionidae and Pieridae) preferentially visit sodium sources, while members of other families (Lycaenidae, Nymphalidae, and Hesperiidae) typically prefer a protein source, suggesting that nutritional needs may vary among taxa. Other drinking behaviors of moths and butterflies are also likely specializations to procure sodium or nitrogenous substances and thus may be functionally related to puddling. These insects often imbibe fluids that arise from vertebrates, including urine, feces, perspiration, blood, and lachrymal secretions.

REPRODUCTIVE BIOLOGY

Puddling is linked to lepidopteran reproduction. The male *Gluphisia* moth sequesters about 17 μg of sodium through puddling. Approximately 10 μg of this puddle-derived material is conveyed nuptially to the female, apparently via the spermatophore. About 5 μg of the transferred sodium is incorporated into the eggs. A nuptial transfer of sodium is also known from the European skipper *(Thymelicus lineola),* another puddling species. In this butterfly, there is apparently no endowment of eggs with male-derived sodium, but access to the mineral does enhance male mating success. To date, no studies have directly examined whether male Lepidoptera contribute puddle-derived amino acids to their mates. However, such a bestowal is quite plausible, given that males of certain nonpuddling lepidopteran species transfer amino acids via the spermatophore. The observed relationship between paternal consumption of an amino acid solution and increased egg viability in the tiger swallowtail may reflect such a nuptial transfer.

EXTREME PUDDLING BEHAVIOR

When puddling, a number of butterfly and moth species pump fluid through the digestive tract, emitting droplets from the anus. This behavior is displayed in extreme form by the male *G. septentrionis,* which forcibly releases anal jets at approximately 3-s intervals while imbibing from mud puddles. In this insect's quest for sodium, drinking often persists for hours, resulting in the passage of immense volumes that can amount to over 600 times the moth's body mass. Coupled to this behavior is a sexual dimorphism of the ileum (anterior hindgut). The male's ileum is longer and wider than that of the nonpuddling female. Numerous villi are present in the male ileum, but are virtually absent in the female's. Thus the surface area of the male ileum is nearly 20 times that of the female, making this enteric region a likely site for the observed sodium absorption.

See Also the Following Articles

Digestion • Digestive System • Feeding Behavior • Lepidoptera • Nutrition

Further Reading

Adler, P. H. (1982). Soil- and puddle-visiting habits of moths. *J. Lepidopterists' Soc.* **36,** 161–173.
Adler, P. H., and Pearson, D. L. (1982). Why do male butterflies visit mud puddles? *Can. J. Zool.* **60,** 322–325.
Arms, K., Feeny, P., and Lederhouse, R. (1974). Sodium: Stimulus for puddling behavior by tiger swallowtail butterflies, *Papilio glaucus. Science* **185,** 372–374.
Beck, J., Mühlenberg, E., and Fiedler, K. (1999) Mud-puddling behavior in tropical butterflies: In search of proteins or minerals? *Oecologia* **119,** 140–148.
Boggs, C. L., and Jackson, L. A. (1991). Mud puddling by butterflies is not a simple matter. *Ecol. Entomol.* **16,** 123–127.
Lederhouse, R. C., Ayers, M. P., and Scriber, J. M. (1990). Adult nutrition affects male virility in *Papilio glaucus* L. *Funct. Ecol.* **4,** 743–751.

Pivnick, K. A., and McNeil, J. N. (1987). Puddling in butterflies: Sodium affects reproductive success in *Thymelicus lineola. Physiol. Entomol.* **12,** 461–472.
Smedley, S. R., and Eisner, T. (1995). Sodium uptake by puddling in a moth. *Science* **270,** 1816–1818.
Smedley, S. R., and Eisner, T. (1996). Sodium: A male moth's gift to its offspring. *Proc. Natl. Acad. Sci. USA* **93,** 809–813.

Pupa and Puparium

Frederick W. Stehr
Michigan State University

A pupa is the stage in the development of holometabolous insects between the mature larva and the adult wherein major morphological reorganization takes place. Pupation usually occurs in a protected location (in a cell or cocoon), but in some groups, such as many butterflies, the pupa (chrysalis) is suspended openly and is usually well camouflaged by its shape and color.

There are two basic kinds of pupae, exarate (Fig. 1) and obtect (Fig. 2). An exarate pupa has free appendages. An

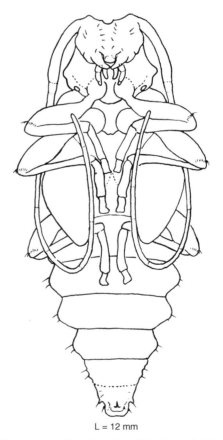

L = 12 mm

FIGURE 1 Exarate pupa of long-horned wood-boring beetle (Cerambycidae), showing free appendages (ventral). [From Peterson, A. (1948). "Larvae of Insects," Vol. 1. With permission of Jon A. Peterson.]

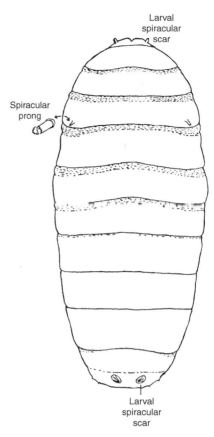

FIGURE 2 Obtect pupa of the monarch butterfly, *Danaus plexippus,* with appendages adhering to the body wall (lateral). The ready-to-emerge adult can be seen through the cuticle. (Photograph by Larry J. West, Mason, MI.)

FIGURE 3 Puparium of a higher fly, formed from the last larval skin, showing the larval spiracular scars and pupal spiracular prongs that anchor the exarate pupa within (dorsal). [From Peterson, A. (1948). "Larvae of Insects," Vol. 1. With permission of Jon A. Peterson.]

obtect pupa has the appendages adhering to the body wall. Most Lepidoptera, most lower Diptera, some chrysomelid and staphylinid beetles, and many chalcidoid Hymenoptera have obtect pupae; nearly all other pupae are exarate.

BEHAVIORS

Most pupae are inactive, their body movements often limited to the abdominal segments. However, pupae in some groups are capable of locomotion, and some have functional mandibles that enable them to cut their way out of the pupal cell, cocoon, or chamber. These active pupae are sometimes referred to as pharate adults because the adult is enclosed in the pupal cuticle.

DESCRIPTION

Pupae may have articulated mandibles (decticous) or nonarticulated mandibles (adecticous). Decticous pupae are capable of chewing their way out of cells or cocoons and may be active. They occur in the Mecoptera (scorpionflies), Megaloptera (dobsonflies and fishflies), Neuroptera (lacewings, ant lions, and relatives), Raphidioptera (snakeflies), Trichoptera (caddisflies), and primitive Lepidoptera (Micropterigidae).

In the higher Diptera (Muscomorpha: Schizophora), the exarate pupa (also called a coarctate pupa) is enclosed in a delicate membrane within the hardened and barrel-shaped last (third) larval skin termed the puparium (puparia) (Fig. 3), which surrounds and protects the pupa. The adult fly emerges by forcing the end of the puparium off with the ptilinum, a membranous eversible pouch between the eyes and above the antennae that is expanded outward by blood pressure from the abdomen and thorax. After emergence, the ptilinum is withdrawn into the head, resulting in the frontal suture that arches over the antennae. If the puparium is buried in the soil, the ptilinum may also be used to help the adult fly crawl or force its way to the surface.

A prepupa is the last instar that has completed feeding. It may wander in search of a pupation site, but generally it becomes nonmobile before pupation. It is easily observed in Lepidoptera, wherein the mature larva becomes shortened and the prolegs and crochets become progressively retracted before pupation.

See Also the Following Articles
Chrysalis • Cocoon

Further Reading

Stehr, F. W. (ed.). (1987). "Immature Insects," Vol. 1. Kendall/Hunt, Dubuque, IA.

Queen

see *Caste*

Raphidioptera
(Snakeflies)

Ulrike Aspöck
Naturhistorisches Museum Wien

Horst Aspöck
University of Vienna

The Raphidioptera is a small order of winged, holometabolous insects. Adults (Fig. 1) have an elongate pronotum and two pairs of subequal wings of about 5 to 20 mm in length. Females have an elongated ovipositor. Larvae (Fig. 2) are terrestrial, living under bark or in detritus, and have biting mouthparts. Pupae are decticous (with articulated mandibles).

Raphidioptera is presumed to be the sister group of Megaloptera + Neuroptera and comprises two families: the Raphidiidae with 185 described species and the Inocelliidae with 21.

The Raphidioptera is distributed throughout the Holarctic region, except for the northern and eastern parts of North America; the southernmost records are from Mexico, northwest Africa, northern India, Indochina, and Taiwan. They are restricted to woodland habitats and occur in almost all Holarctic types of forests and forest-like habitats. In southern parts of their distribution they live mainly at high altitudes, up to about 3000 m.

FIGURE 1 Female adult of *Turcoraphidia acerba* (Raphidioptera: Raphidiidae) from Anatolia. Length of forewing 8.5 mm.

HISTORY OF RESEARCH

Snakeflies first appear in the literature in 1735, when Linnaeus described an insect that he called Raphidia. By 1800 only 3 species had been described; in the 19th century snakeflies were known from southern Europe, Anatolia, and North America and, by the beginning of the 20th century, also from northern and central Asia and from northern Africa. By 1900, 31 species were known; in 1950, there were 60. At the beginning of the 1960s, a worldwide search for snakeflies and

FIGURE 2 Larva of *Indianoinocellia mayana* (Raphidioptera: Inocelliidae) from Mexico. Length of body 22.5 mm. (Reproduced, with permission, from U. Aspöck and H. Aspöck, 1996.)

revision of the order was started and about 140 species were discovered and described between 1960 and 2000. Many species were reared from the egg, so that preimaginal stages and biologies of the majority of species are documented.

CHARACTERIZATION

An extremely flexible head characterizes the adults. It is prognathous, elongate, flat, and strongly sclerotized and may be broad or tapering basally. The large compound eyes are situated laterally. Inocelliidae lack dorsal ocelli. The mouthparts are of the biting type. The prothorax is remarkably elongated, particularly in Raphidiidae, and very mobile (hence the common names snakeflies and camelneck flies). All three pairs of legs are similar and cursorial. Both pairs of the elongate subequal wings are membraneous, and the venation is simple, with few crossveins. The abdomen consists of 10 (visible) segments. Terminal sclerites constituting the external genitalia are extremely complex in males; in females they are equipped with a long ovipositor.

The larvae are elongate and flattened, with a prognathous head, biting mouthparts, and five to seven lateral stemmata. The head and the only moderately elongate prothorax are strongly sclerotized, and the 10-segmented abdomen is of soft consistency.

BIOLOGY

Adult Raphidiidae are day-active entomophagous insects, preying on soft-bodied arthropods and pollen; the natural food of Inocelliidae is virtually unknown, but in captivity they feed on artificial diets and pollen has very rarely been found in their gut. All stages of larvae of both families are entomophagous, feeding on a variety of soft-bodied arthropods; the spectrum of prey is, however, considerably different in bark-inhabiting larvae on one hand and in larvae living in the soil on the other.

There is a long courtship before mating, including a highly sophisticated cleaning behavior with legs and antennae. Two positions of copulation have been observed: a "dragging position" in Raphidiidae, in which the male hangs head first from the female and is carried by her, and a "tandem position" in Inocelliidae, in which the male crawls under the female attaching his head in fixed connection to the fifth sternite of the female. Copulation lasts from a few minutes to $1\frac{1}{2}$ hours in Raphidiidae, but up to 3 hours in Inocelliidae. Spermatophores have been observed and studied only in Raphidiidae.

The egg stage lasts a few days to 3 weeks. The number of instars varies around 10 to 11 and may reach 15 or more. The larval period lasts 1 year in a few species, in most species it is 2 or 3 years, and under experimental conditions it may be up to 6 years. The prepupal stage lasts a few days. In the majority of species, pupation takes place in spring and lasts a few days to 3 weeks. In some species pupation starts in summer or autumn and the pupal stage lasts several (up to 10) months; in a few others pupation starts in summer and adults hatch in late summer after a short pupal stage. Hibernation thus usually takes place in the last larval stage, the penultimate stage, or the pupa, but never as eggs, prepupae, or adults. The pharate adult (the active pupa) is very mobile.

Snakeflies need a period of low temperature (around 0°C) to induce pupation or hatching of the imago. Larvae that are continuously kept at room temperature will usually not pupate but become prothetelous, i.e., they develop pupal or imaginal characters, such as compound eyes, wingpads, and appendages on the abdomen, and may live for years.

Parasites, parasitoids, and hyperparasites are a frequent phenomenon among Raphidioptera. Hymenoptera are of considerable significance as parasitoids of larvae, and species of the genus *Nemeritis* (Ichneumonidae) comprise 90 to 95% of them; other ichneumonids, braconids, and chalcidids contribute about 1%.

Snakeflies are effective predators; all larval stages of all species of both families, and at least the adults of Raphidiidae, feed on (mainly soft-bodied) arthropods. Snakeflies are believed to be rare insects. This is true for many species and many regions, but a number of species often occur in large numbers. There have been several attempts to use snakeflies as biological control agents: an unidentified North American species was introduced in Australia and New Zealand 100 years ago, but did not become established. The use of snakeflies for future biological control efforts is, however, hampered by the long developmental period and narrow food preference of these insects.

DISTRIBUTION AND BIOGEOGRAPHY

Extant Raphidioptera are confined to the Northern Hemisphere and almost exclusively to the Holarctic region. In Central America the southernmost records are from high altitudes at the Mexican–Guatemalan border. In Africa snakeflies have been found only in arboreal regions north of the Sahara, and in Asia the southernmost records are from altitudes above 900 m in transition areas from the Palaearctic to the Oriental Region in northern India, Myanmar, and northern Thailand. The northern and eastern parts of North America lack snakeflies. Almost all species are restricted to very limited areas of a refugal nature. Only three species represent a Eurosiberian type of distribution occurring throughout northern Asia to central and northern Europe, and a few Nearctic species with distribution centers in the southwest have reached Canada.

SYSTEMATICS, TAXONOMY, AND FOSSILS

The order Raphidioptera is a relic group of "living fossils" that comprises two extant families and 206 described species.

Taxonomy of adults has been well established. Because shape and coloration of body structures and of wing venation are highly variable, the most powerful and reliable tool is the morphology of the genital sclerites, in particular of the males. Taxonomy of larvae, mainly based on patterns and coloration of the abdomen, still remains difficult because of the similarity of related species.

The fossils lead to the conclusion that there was an enormous biodiversity of Raphidioptera in the Mesozoic. The majority of species and genera (in several families) are known from Jurassic and Cretaceous deposits. The rich and diverse Raphidioptera fauna of the Mesozoic died out at the end of the Cretaceous, probably due to the K/T event (that is the worldwide catastrophe resulting from an asteroid of about 10 km diameter that hit our planet) 65 mya and its climatic consequences. In particular, all snakeflies of tropical climates disappeared and apparently only the few representatives adapted to a cold climate survived. Fossil snakeflies from Tertiary layers, as well as from Baltic amber, belong to the two extant families.

See Also the Following Articles
Megaloptera • Neuroptera

Further Reading
Aspöck, H. (1998). Distribution and biogeography of the order Raphidioptera: Updated facts and a new hypothesis. *Acta Zool. Fenn.* **209**, 33–44.
Aspöck, H., and Aspöck, U. (1991). Raphidioptera (snakeflies, camelneck-flies). *In* "The Insects of Australia. A Textbook for Students and Research Workers" (I. D. Naumann *et al.,* eds.), 2nd ed., Vol. I, pp. 521–524. Melbourne University Press, Melbourne.
Aspöck, H., Aspöck. U., and Rausch, H. (1991). "The Raphidioptera of the World. A Monograph of the Systematics, Taxonomy, Biology, Ecology and Chorology of the Extant Raphidioptera of the World, with a Summarizing Account of the Extinct Raphidioptera (Insecta: Neuropteroidea)." Goecke & Evers, Krefeld, Germany. [In German]
Aspöck, U., and Aspöck, H. (1996). Raphidioptera. *In* "Biodiversidad, Taxonomía Biogeografía de Artrópodos de México: Hacia una Síntesis de Su Conocimiento" (J. L. Bousquets, A. N. García Aldrete, and E. G. Soriano, eds.), pp. 277–286. Universidad Nacional Autónoma de México, México, D.F.
Aspöck, U., Aspöck, H., and Rausch, H. (1992). Extant southern boundaries of the distribution of the order Raphidioptera in America (Insecta: Neuropteroidea). *Entomol. Gen.* **17**, 169–184. [In German].
Aspöck, U., Aspöck, H., and Rausch, H. (1994). New species of the family Raphidiidae from Mexico and evidence of a spermatophore in the order Raphidioptera (Insecta: Neuropteroidea). *Entomol. Gen.* **18**, 145–163. [In German].
Aspöck, U., Plant, J. D., and Nemeschkal, H. L. (2001). Cladistic analysis of Neuroptera and their systematic position within Neuropterida (Insecta: Holometabola: Neuropterida: Neuroptera). *Syst. Entomol.* **26**, 73–86.
Kovarik, P. W., Burke, H. R., and Agnew, C. W. (1991). Development and behavior of a snakefly, Raphidia bicolor Albarda (Neuroptera: Raphidiidae). *Southwest. Entomol.* **16**, 353–364.
Tauber, C. A. (1991). Order Raphidioptera. *In* "Immature Insects" (F. W. Stehr, ed.), Vol. 2, pp. 123–125. Kendall–Hunt, Dubuque, IA.

Rearing of Insects

Norman C. Leppla
University of Florida

The goal of organized insect rearing is to provide reliable, affordable sources of high-quality insects for their many important uses. We are now able to rear literally thousands of insect species through multiple generations and many more for part of their life cycles. The greatest difficulty is to provide fresh host material or to develop a diet that is nutritionally complete and induces feeding, especially for parasites or parasitoids. Precautions must be taken to start colonies with an adequate number of uncontaminated specimens and maintain them with very limited levels of mortality. These measures help to limit genetic bottlenecks caused by inbreeding, competition with other species, and disease epizootics. As production levels and the number of species increase, rearing must be organized into systems with discrete operations or activities. These operations incorporate the rearing procedures necessary for each stage of the insect's life cycle plus maintenance of supplies, equipment, and facilities. Facilities must be designed and constructed to contain and maintain the insects under specified conditions. Problems encountered in established insect rearing systems invariably are caused by inattention to procedural details or lack of environmental control. For the foreseeable future, advances in insect rearing will be focused primarily on culturing new species, natural enemies, and pest species that have been genetically modified for pest management.

PURPOSES OF REARING INSECTS

Insects are reared for many reasons that may not be obvious to the nonentomologist. Certainly, insects are aesthetically pleasing and, therefore, reared by those who appreciate their many shapes, colors, features, and behavior patterns. This appreciation may be shared by means of personal collections, static displays, insect zoos, butterfly houses, and even household pets. Many of these and similar experiences would not be possible without insect rearing.

Insects are reared for a wide range of primarily agricultural and medical applications, an unusual example being the recently publicized biological warfare against plants that produce cocaine. Chemical insecticides are developed by using laboratory colonies of insects to mechanically screen massive numbers of candidates, more than 1 million insecticides per year in some instances. Similarly, plants are screened for resistance to insects or the disease organisms they transmit. Nontarget insects and plants are tested as possible hosts before nonindigenous natural enemies are released into nature. Insects have been reared, marked, and released to understand their orientation, dispersal, and migration. The cells of insects are used to study physiological processes such as reproduction, ontogeny, growth, aging, and cold tolerance. The fruit fly *Drosophila melanogaster* is the "white rat" of the geneticists and the nerves of large cockroaches are used by sensory physiologists. Students learn morphology by dissecting lubber grasshoppers, *Romalea* spp., and other insects reared in the laboratory. A variety of educational subjects involve insect colonies, including the process of insect identification, principles of insect taxonomy and systematics, and engineering aspects of insect flight. A somewhat gruesome but effective practice is the postoperative surgical use of the blow fly maggots *Lucilia sericata* and *Phormia regina* to maintain clean wounds.

The major use for laboratory-reared insects is in biologically based methods of pest management. One of these methods, biological control, is the rearing and release of parasitoids, predators, and pathogens to suppress pest insects. A global industry has developed to rear and sell these natural enemies (www.anbp.org). Autocidal control is accomplished using the sterile-insect technique or inherited sterility. Male insects are reproductively sterilized with γ-irradiation or chemicals and released by the thousands per hectare to mate with wild females. This ensures that most of the wild females mate with sterile males and do not produce offspring. Inherited sterility is a variation using partially sterile males to induce sterility in subsequent generations. Genetic control takes advantage of altered genes to disrupt the insect's physiology and behavior. In the near future, we may use mass-reared insects to widely distribute aberrant endosymbionts via paratransgenic strains.

Insects have been reared to enhance wild and domestic populations, particularly for improving their products and benefits. Silk and honey production are obvious examples but pollination and pest management are equally important. Elaborate strain development methods have been used for silkworms, honey bees, screwworms (autocidal control), parasitic wasps (biological control), and others. Field insectary populations of expensive, showy butterflies and beetles have been used to augment natural populations that may become depleted. Eventually, the use of reared insects in conservation may include protection of rare and endangered species and repopulation, as is currently practiced with birds and predators.

KINDS OF INSECTS THAT ARE REARED

Insect rearing may be partial (egg to larva or nymph and adult) or complete (egg to egg). Virtually any free-living insect can be collected as an egg, larva, or nymph and maintained through stages in metamorphoses until it becomes an adult. However, this often requires considerable knowledge about the species' habitat and natural food. Re-creation of soil or aquatic environments, symbiotic relationships, and specialized foods can make rearing difficult. Some insects undergo temperature- and photoperiod-dependent diapause or require host plant cues to terminate multiyear cycles. Trophallactic feeding may necessitate maintenance of an entire colony, as in termites, ants, and other social insects. Because of these and other peculiar life history characteristics, the easiest immature insects to rear are relatively small, multivoltine (more than one generation per year), plant-feeding, terrestrial species with wide host ranges and no unusual environmental requirements. Species that infest common crops, landscape plants, or stored products are particularly suited to artificial rearing.

Complete rearing of an insect for one or more generations is complicated by the mating and oviposition requirements of the adults. Species-specific temperature, humidity, amounts of space, light characteristics, photoperiod, population size, food, oviposition stimulants and substrates, and other environmental conditions all must be understood and provided. Insects may swarm and couple in flight, form mating aggregations on host plants, orient to each other by means of pheromones or auditory signals, transfer spermatophores (sperm packets), engage in postmating female guarding to protect their paternity, and perform other unimaginable actions to produce another generation. Fortunately, most of the species we rear for multiple generations have less complicated requirements; these are primarily butterflies and moths (Lepidoptera, 300+ species have been reared), beetles (Coleoptera, 200+ species), flies (Diptera, nearly 200 species), bugs (Heteroptera, <100 species), and bees and wasps (Hymenoptera, <100 species). Grasshoppers and katydids (Orthoptera), lacewings (Neuroptera), cockroaches (Blattodea), termites (Isoptera), and fleas (Siphanoptera) are also reared but in reduced numbers (roughly 10–20 species). Many more species undoubtedly could be reared using the techniques developed for their close relatives.

NATURAL AND ARTIFICIAL DIETS

Immature herbivorous (plant-feeding) insects can often be reared by feeding them clean, fresh-cut, or potted versions of the plant material on which they feed in nature. The roots, stems, leaves, flowers, or fruit must be readily available or grown in sufficient quantities. Examples are larval silkworms raised on mulberry leaves, boll weevils on cotton squares and bolls, tropical fruit flies on papaya, and monarch butterflies on milkweed leaves. Similarly, medical and veterinary insects, i.e., adult mosquitoes and biting flies, are fed on their hosts or suitable surrogates, such as rodents, sheep, goats, or pigs.

Substitute plants can also be used to maintain herbivorous insects that are adaptable, usually readily available human and animal food. Green beans can be used for plant bugs, lettuce for grasshoppers, dry dog food for cockroaches, and cow manure for house flies. Rearing natural enemies of plant- or animal-feeding insects is considerably more difficult because three trophic levels must be synchronized: the plant or animal, the insect host, and the natural enemy.

Artificial diets have been developed to simplify and improve the rearing of both plant- and animal-feeding insects. Henry Richardson's development in 1932 of a bran, alfalfa meal, yeast, and diamalt formula for rearing house flies eliminated the objectionable mess and odor of cow manure. A commercial diet made of wheat bran (33.3%), alfalfa meal (26.7%), and brewer's grain (40%) is now available for rearing house fly larvae (CSMA medium; Chemical Specialties Manufacturer's Association, Ralston-Purina, St. Louis, MO). Another historical advancement was M. H. Haydak's 1936 grain flower, milk powder, honey, and glycerine diet for stored-grain insects, such as mealworms and flower moths. Gelled diets were developed for rearing insect larvae that require large quantities of contained water in their diets. The first was Pearl's 1926 diet for *Drosophila* spp., followed by Botger's

1942 larval medium for the European corn borer, *Ostrinia nubilalis*. Interestingly, these and subsequent diets have been gelled with agar, a polysaccharide derived from seaweed that was previously developed for use in bacteriology by Robert Koch in the late 1800s. Agar remains the standard gelling agent for culturing both microorganisms and insects; however, its cost and requirement for heating have led to the development of alternative materials: polysaccharides (industrial gums, cellulose, pectin, plant starches), heteropolysaccharides (carrageenan, sodium and calcium alginate), scleroproteins (gelatins, animal albuminoids), starch polymers (polyacrylonitrile graft copolymers), crude fibrous plant products, and waxes. Ground plant fibers, such as sugarcane bagasse (pulp remaining after the sugar is extracted), corncobs, and carrot powder are used to rear tropical fruit fly larvae. The awful stench of using a mixture of dried blood, milk, and yeast in water for rearing screwworm larvae was virtually eliminated by a starch polymer-gelled diet developed primarily by David Taylor of the USDA, Agricultural Research Service (ARS), in 1988. A practical artificial diet (primarily of beef liver, ground beef, and sucrose) for predaceous insects has recently been perfected and patented by Allen Cohen, also with the ARS.

INITIATION AND MAINTENANCE OF COLONIES

Insect colonies are initiated from field-collected specimens or samples from previously established colonies. Any developmental stage can be used to start a colony, but surface-sterilized eggs are generally preferred because they are durable, easy to ship, and less likely than other stages to carry a pathogen. However, eggs may be difficult to find in nature and often larvae suffer high levels of mortality as first instars because they are not yet adapted to the laboratory. It is generally advisable to use late instars, hold them in individual containers for parasitoid and pathogen screening, combine the adults in mating cages with a suitable oviposition substrate, and collect and treat the eggs before colonization. From either source, field or insectary, the degree of success achieved over multiple generations will depend on the quality of the colonized insects and the skill with which they are reared. Many species that can be colonized and reared for multiple generations are much larger, healthier (free of malnutrition, pathogens, parasitoids, and predators), uniform in growth, and more active and fertile than those in nature. Special consideration must be given to rearing insects that are required to behave normally, particularly those destined for release to suppress wild populations. Fruit flies, screwworms, and other species that have been mass reared for 20 or more generations in isolation may no longer interact and mate with their target populations in nature. To avoid this so-called "strain deterioration," insectary populations must be recolonized or infused periodically using specimens collected from the targeted wild population. Yields will be relatively low for several generations in a new colony destined for mass production, unless a previously isolated strain has been adapted to the insectary in anticipation of its use for colonization. Infusion is accomplished best by holding field-collected larvae, obtaining adults, and combining their eggs with those of the mass-reared colony, as explained below. A large number of existing new relative insects must survive to ensure that the colony has been infused. Also, genetic bottlenecks (sources of selective mortality) must be avoided because these will hasten the selection of insects that no longer behave normally.

Colony starts can be obtained by contacting entomologists who publish on species of interest or are involved in large-scale, biologically based pest management programs. In the 1980s the Entomological Society of America published lists of *Arthropod Species in Culture* and specialized directories still exist for *Drosophila* strains and other species. *Arthropod Species in Culture* listed colonies of the following taxonomic orders (number of species, colonies): Acari (41, 77), Anoplura (1, 1), Coleoptera (78, 266), Diptera (168, 301), Heteroptera (90, 206), Hymenoptera (119, 169), Lepidoptera (101, 308), Mallophaga (3, 3), Neuroptera (2, 2), Orthoptera (64, 203), Siphonaptera (2, 7), and Thysanura (4, 11). *Suppliers of Beneficial Organisms in North America* is maintained by Charles Hunter of the California Environmental Protection Agency (http://www.cdpr.ca.gov). This publication lists more than 130 species of beneficial organisms available from 142 suppliers. The most popular species offered for sale were the green lacewing, *Chrysoperla carnea* (65 suppliers); brown lacewing, *C. rufilabris* (54); mealybug destroyer, *Cryptolaemus montrouzieri* (52); whitefly parasitoid, *Encarsia formosa* (54); convergent ladybeetle, *Hippodamia convergens* (56); predaceous mite, *Phytoseiulus persimilis* (54); and egg parasitoid, *Trichogramma pretiosum* (77). There are also many commercial sources of insects, particularly for classroom education, including Carolina Biological Supply, Entomos, and Combined Scientific Supplies. Pioneers in supplying diets for research are BioServe and Southland Products. It is easy to colonize expensive pet food insects, such as mealworms and crickets, that can be purchased at local stores. Insect strains are not patented like plant varieties, so their use is not restricted.

CATEGORIES OF INSECT REARING

There are three distinct approaches to rearing insects: single species, multiple species, and mass rearing. In single-species rearing, immature stages are usually fed on host plants or animals, although artificial diets may be substituted. Seminatural environments and oviposition substrates are duplicated from nature and all rearing operations can be performed by a single individual. Multiple-species rearing is usually accomplished in centralized facilities to support research. There is economy of scale in these rearing operations, i.e., diet preparation, egg treatment, larval rearing, harvesting of pupae, and adult colony maintenance can be combined for similar species. Multiple-species rearing is common in research laboratories, such as those used for insecticide or transgenic crop development, and is typically performed by a small staff. Mass rearing involves a single species produced for biologically based pest management that is reared in factory-like facilities with controlled environments, artificial diets and oviposition substrates, mechanized equipment, and operations performed by work units. The largest facilities are used to rear the screwworm and Mediterranean fruit fly. Although based on single-species rearing, these three approaches are distinct in design and implementation, not merely multiplications of scale.

BASIC INSECT REARING OPERATIONS

Regardless of the size of the colony, rearing procedures are organized into operations based primarily on the species' life history. At some level of complexity, insect rearing operations include inventory, acquisition, and storage of supplies; diet preparation and containerization; egg collection and treatment; larval or nymphal development; pupal or adult recovery and distribution; adult colony maintenance; quality control; and facility and equipment maintenance. Depending on the species, these operations are subdivided into procedures for which space is allocated and personnel are trained and assigned. "Traffic patterns" are established in the facility that flow from relatively clean areas to those that are potentially contaminated, i.e., diet preparation through larval development and diet disposal. Diet and eggs are usually handled more carefully until they are sealed in clean rearing containers. Isolating the larvae, pupae, and adults in containers protects them from contamination and prevents workers from being exposed to potentially harmful microorganisms and allergens. Rearing operations are performed in synchronized sequences, so they can be interfaced at the most appropriate times.

REQUIREMENTS FOR INSECT REARING FACILITIES

Insect rearing facilities have evolved from "field insectaries" with outdoor temperature and humidity to controlled labratories that contain insects under high security. Field insectaries provide adequate environments for rearing insects under seminatural conditions on host plants or animals. However, pathogens, parasitoids, and predators are not controlled, and workers are exposed to potentially dangerous pathogens and allergens. Conversely, temperature, humidity, air quality and quantity, light quality and photoperiod, sanitation, and security are closely maintained in laboratory insectaries. Insects and supplies are carefully screened for contaminants before they are admitted and human access and exposure are limited. All openings to a laboratory insectary are sealed or filtered to prevent insects from entering or leaving, particularly in quarantine facilities. Quarantine and containment facilities have strict construction requirements and operational protocols. Laboratory insectaries must be well insulated and have highly filtered, recirculated air to be efficient and cost effective.

PROBLEMS ENCOUNTERED IN MAINTAINING INSECT COLONIES

Virtually all severe insect rearing problems result from failure to perform standard operating procedures or defective environmental controls. Once established, rearing operations become routine and individual procedures are easy to forget. For example, a dietary ingredient may be omitted or destroyed by overheating, eggs may be accidentally desiccated following surface sterilization (exchorionation), closely related species may be mixed unintentionally, or larval densities may be more or less than required per container. Additionally, dietary ingredients can deteriorate after prolonged storage. These kinds of problems can be detected and corrected before a colony is lost. However, temporary loss of temperature control can destroy all of the insects. Pathogens can also be devastating, although diseased insects are usually confined to certain containers that can be eliminated before others are contaminated. Parasitoids and predators must be detected and similarly discarded. Genetic deterioration from inbreeding and genetic drift has been blamed for declines in insect colonies, but this is not a typical problem in large colonies. Insect rearing is generally safe for humans unless they are hypersensitive to insect proteins, react to the physical irritation of insects or their body parts, work with insects that sting or bite, or expose themselves to toxic substances used in the operations.

THE FUTURE OF INSECT REARING

Insect rearing is in transition, along with the science and technology it supports. Curiosity about the natural history of insects and their rearing is increasing steadily as people enjoy ecotourism, butterfly houses and gardens, insect zoos, and educational products based on insects. Insects are commonly used as baits for fishing and, in certain countries, have become popular pets. Rearing is becoming more important as field collection of insects is restricted to preserve habitat biodiversity and protect germplasm ownership. This is analogous to the collection of orchids and other showy organisms. Insects are no longer major sources of natural products, such as silk, wax, or dyes, and their use as human food is very limited. However, bird-watching and exotic pet ownership have become extremely popular in affluent countries, causing a significant increase in the use of insects as animal food. Rearing insects to produce bioactive substances remains a research support activity in the fields of biology and medicine. Living as well as dead insects are used increasingly in classroom education.

Agricultural uses for insects have expanded dramatically during the past 40 to 50 years in the development and support of new pest management technologies. Every major company that produces chemical insecticides or pest-resistant plants maintains a multispecies insect rearing facility. Although insecticides have provided effective insect control at individual farms and residences, overall losses to insects have not declined and we need new options for sustainable pest management. Large federal and state entomological research laboratories and university entomology departments also have laboratory insectaries, although the current trend is toward decentralization and outsourcing. The sterile insect technique, pioneered by E. F. Knipling of the United States Department of Agriculture, has proven too expensive, unless used on an area-wide basis with stringent regulatory controls, as in the screwworm, *Cochliomyia hominivorax,* and Mediterranean fruit fly, *Ceratitis capitata,* eradication programs. Advances in insect rearing have enabled the eradication of these species and others from vast geographical areas. As a result, area-wide approaches to pest management are increasing, along with new methods for using genetically modified organisms. Reliable, cost-effective rearing will be required for these technologies to be successful.

Major advances in insect rearing are currently being made in support of augmentative biological control. Natural enemies are reared and released to prevent rather than cure pest problems, and they rarely have unacceptable nontarget effects. Another major advantage in using natural enemies is that they do not induce the pest resistance that eventually makes insecticides ineffective. We now have efficient host-rearing systems for many parasitoids and new artificial diets for predators that have greatly increased shelf life. Advancements are also being made in the mechanization of rearing operations, large-scale release technology, and rearing of newly discovered natural enemies. A global industry has developed during the past 10 years to ensure the quality of natural enemies and expand their use. Markets are increasing in organic food production, ornamental and vegetable greenhouse crops, urban pest management, filth fly control, home gardening, and other specialized areas. Biological control will continue its expansion as insect resistance to chemical insecticides increases, worker protection and food safety regulations are enforced, people avoid real or perceived environmental contamination, and the efficacy of natural enemies improves.

Insect rearing will play a critical role in the future of entomology. Insects will always be appreciated for their intrinsic value, used as a source of useful materials, and produced as food for wildlife. However, they will become more important for pest prevention in natural areas, crops, and buildings and on human and animal wastes. Other uses will include the production of transgenic biological control agents, autocidal agents (sterile insect technique, paratransgenesis), and new species and strains for biological control. Insects will be reared to restore and supplement insect populations in nature; control pests over vast, low value-per-acre lands; and eliminate chemical insecticides in specialized cropping systems.

Insect rearing can be enhanced most by learning more about the natural history of insects, emphasizing their ecology, behavior, and systematics. This knowledge can be used to create artificial diets and environments that separate species from limiting factors, biotic (pathogens, parasitoids, and predators) and abiotic (temperature, humidity, and light). As information is gained, it must be published, communicated at meetings, and widely distributed to help advance the field. The Entomological Society of America and International Organization for Biological Control have been particularly helpful in publicizing insect rearing information. Unfortunately, however, insect rearing is often considered a support activity most appropriately described in the methods sections of articles on other subjects. This makes the information difficult to retrieve and has led to relatively obscure publications on insect rearing. Another major limitation has been a general lack of formal education and training in insect rearing. However, Frank Davis conducts a popular annual short course on the subject at Mississippi State University and David Dame covers the subject in his biannual FAO, International Atomic Energy Agency short course at the University of Florida. It is essential for us to continue discovering, documenting, and sharing insect rearing knowledge and preserving the rich history of this field.

See Also the Following Articles

Biological Control • Cell Culture • Genetic Engineering • Medicine, Insects in • Nutrition • Pathogens of Insects • Sterile Insect Technique

Further Reading

Anderson, T. E., and Leppla, N. C. (eds.) (1992). "Advances in Insect Rearing for Research and Pest Management." Westview Press, Boulder, CO.

Edwards, D. R., Leppla, N. C., and Dickerson, W. A. (1987). "Arthropod Species in Culture." Entomol. Soc. Am. College Park, MD.

Grenier, S., Greany, P. D., and Cohen, A. C. (1994). Potential for mass release of insect parasitoids and predators through development of artificial culture techniques. *In* "Pest Management in the Subtropics: Biological Control—A Florida Perspective" (D. Rosen, F. D. Bennett, and J. L. Capinera, eds.). Intercept, Andover, MA.

King, E. G., and Leppla, N. C. (eds.) (1984). "Advances and Challenges in Insect Rearing." USDA, ARS, U.S. Gov't Printing Office, Washington, DC.

Leppla, N. C., and Ashley, T. R. (eds.) (1978). "Facilities for Insect Research and Production." USDA Technical Bull. No. 1576.

Needham, J. G., Galtsoff, P. S., Lutz, F. E., and Welch, P. S. (eds.) (1937). "Culture Methods for Invertebrate Animals." Comstock, Ithaca, NY.

Peterson, A. (1964). "Entomological Techniques: How to Work with Insects." Entomological Reprint Specialists, Los Angeles.

Poinar, G. O., and Thomas, G. M. (1978). "Diagnostic Manual for the Identification of Insect Pathogens." Plenum, New York.

Ridgway, R. L., Hoffmann, M. P., Inscoe, M. N., and Glenister, C. S. (eds.) (1998). "Mass-Reared Natural Enemies: Application, Regulation, and Needs." Entomol. Soc. Am., Lanham, MD.

Singh, P. (1977). "Artificial Diets for Insects, Mites, and Spiders." Plenum, New York.

Singh, P., and Moore, R. R. (eds.) (1985). "Handbook of Insect Rearing," Vols. 1 and 2. Elsevier, Amsterdam.

Smith, C. N. (ed.) (1966). "Insect Colonization and Mass Production." Academic Press, New York.

Villiard, P. (1969). "Moths and How to Rear Them." Funk & Wagnalls, New York.

Recruitment Communication

James F. A. Traniello
Boston University

In the social insects, the daily demands of colonial life are often met cooperatively by directing workers to sites where work is required, such as an energetically rewarding food source, an alternate nest site, or the site of a territorial dispute. The behavior that mobilizes nestmates is termed recruitment, and it is a type of communication commonly used by social insects to accomplish work. The mechanisms that mediate recruitment communication can be understood by analyzing the behaviors of individuals and the signals they produce. Numerous studies have described how pheromones, sometimes in conjunction with nonchemical signals, underscore insect social integration. Although systems as "linguistically" elaborate as the honey bee waggle dance have evolved to communicate information about the location of food sources or nest sites, it is pheromones that serve the primary signaling role in wingless social species such as ants and termites.

The majority of research on recruitment behavior concerns communication during foraging and defense. Workers serve as scouts, searching for food sources and patrolling the territory of a colony; when food is located or a competitor from an alien colony is identified, nestmates are recruited. Scouts that have located food or a territorial intrusion return to the nest, laying a chemical trail and sometimes performing behavioral displays to alert nestmates, which then leave the nest and orient along the scout's trail to the target area. Recruited nestmates, in turn, may continue the process by reinforcing the trail. Regulatory mechanisms turn off recruitment when the food has been harvested or the threat no longer exists.

Ecology influences the evolution of recruitment communication, and the adaptiveness of recruitment behavior can be studied in reference to patterns of food distribution, predation, and competition. Foraging behavior and space-use patterns are the result of community-level interactions such as interference competition, and recruitment communication is one behavioral mechanism that mediates interactions between sympatric species. Recruitment signals, in turn, are generally trail pheromones that "excite" nestmates and orient them to a locus of activity. The physical characteristics of trail (e.g., how long it lasts as a signal) and the response it induces reflect the ecological function of the pheromone.

This article considers recruitment behavior in social insects, focusing primarily on ants and termites, groups that frequently employ these communication systems and for which the greatest level of ethological and ecological understanding has been achieved.

PHYSIOLOGY AND BEHAVIOR OF RECRUITMENT

Exocrine Gland Sources of Recruitment Pheromones; Trail-Laying and Trail-Following Behavior

Recruitment pheromones are discharged from exocrine glands, which are anatomical structures often specialized for their synthesis and secretion. Since the first identification of the source of the trail pheromone in the fire ant by E. O. Wilson more than four decades ago, ants have served as excellent models for the study of the organization of recruitment. Thanks to the research of Bert Hölldobler and Hiltrud Engel-Siegel, among others, the structure of ant exocrine glands and the function of their secretions have been described in detail for many species. In ants, the accessory glands to the sting (the Dufour's gland and the poison gland), the pygidial gland and sternal glands, the hindgut, the rectal gland, and the tibial and tarsal glands are known to produce substances that serve to recruit nestmates (Fig. 1). In ants in the subfamily Formicinae, the hindgut is the source of trail pheromone, which is emitted through the acidopore located at the tip of the gaster. In another diverse group of ants, the subfamily Myrmicinae, the poison gland and the Dufour's gland secrete recruitment chemicals. Other ants rely on a variety of glands to produce trail substances. In termites, only one source of trail pheromones, the sternal gland, has been described. The structure of the sternal gland, which is composed of modified epidermal cells, varies in different genera.

The secretions of the trail-substance-producing exocrine glands are deposited as a worker travels from a target area to the nest, or vice versa. In some ant species (e.g., myrmicine ants) the sting, serving as a conduit for the secretions of exocrine glands, is extruded and dragged over the substrate to release trail pheromones. Chemical trails may be deposited continuously or as a series of point sources between the nest and the target area, sometimes in conjunction with other secretions. The cuticle may be adapted as an applicator for the secretions of exocrine glands beneath. In termites, the sternal gland is pressed against the substrate, and sensory structures monitor contact between sternites and the substrate to regulate pheromone deposition. Some termites mark areas around the nest entrance or a food source by dotting the tip of the abdomen and laying directional trails using the sternal gland. There is significant convergence in the trail-laying behaviors of ant and termite species.

The spatial processing of the information encoded in an odor trail is by tropotaxis, which is a sampling of the trail pheromone by means of the paired antennae. Antennal chemoreceptors perceive variation in pheromone concentration along the trail's semiellipsoidal active space, the area in which the concentration of the pheromone is sufficient to elicit following behavior. Theoretically, diffusion yields a gradient of odor molecules; pheromone concentration is highest at the point of application (the centerline of the trail) and symmetrically decreases on either side, defining the boundaries of the active space. The odor gradient is sampled by the antennae as

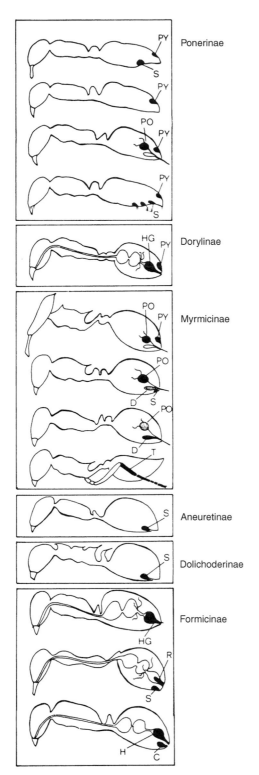

FIGURE 1 Exocrine gland sources of trail pheromones in the different subfamilies of ants: C, cloacal gland; D, Dufour's gland; HG, hindgut; PO, poison gland; PY, pygidial gland; R, rectal gland; S, sternal gland; T, tibial gland. [From Hölldobler, B., and Wilson, E. O. (1990). "The Ants," p. 268. Belknap Press of Harvard University, Cambridge, MA. Originally published in Hölldobler, B. (1984). Evolution of insect communication. *In* "Insect Communication" (Symposium of the Royal Entomological Society of London) (T. Lewis, ed.), pp. 349–377. Academic Press, London. Reprinted with permission of Harvard University Press and Prof. Hölldobler.]

the insect travels along the trail. When lower concentrations are sensed at the lateral edges of the active space, opposing movements are made so that position within the active space is maintained.

Determining the Source and Behavioral Effects of Trail Pheromones

Understanding the glandular sources and chemistry of trail substances is at the heart of the study of recruitment communication. The nature of the bioassay, the behavioral test used to determine the effectiveness of different substances as recruitment or trail pheromones, is critically important. The bioassays used in trail pheromone isolation and identification must distinguish among the range of behavioral responses involved in trail communication. It must be noted that trail pheromones can have both recruitment and orientation effects. A recruitment pheromone induces nestmates to leave the nest to travel to a work site. An orientation pheromone has no such stimulatory effect, but it can serve as a chemical "guide" for worker traffic. Trail substances, if they have a recruitment effect, will stimulate nestmates to leave the nest or otherwise alter their task performance in the context of a current need. In some ants, nestmates are recruited with a motor display delivered in the nest by a recruiting worker. The trail substance in this case does not have the ability to draw ants out from the nest; rather, it is used as an orientation cue by nestmates that have contacted a recruiting ant. Some trail pheromones can alone elicit both excitation and orientation in the absence of any other behavioral display or stimulus. If an artificial trail (one prepared from a solvent extract of the appropriate exocrine gland) is drawn out from the nest entrance and ants leave the nest to follow it, a recruitment effect has been demonstrated. If the artificial trail cannot induce inactive workers to leave the nest, yet the trail is able to orient workers alerted by either a motor display or some other trail chemical that has an alerting property, an orientation effect is occurring. Careful dissection of the kinds and sequences of behaviors in the recruitment process and detailed chemical analyses have revealed that several pheromone constituents may control a number of behaviors associated with recruitment and trail following.

TRAIL CHEMISTRY AND RECRUITMENT BEHAVIOR
Social insects may mix the secretions of different exocrine glands to induce recruitment and trail-following behaviors, or the chemical output of a single gland may be composed of more than one substance, each having a distinct role in releasing behavior. Trail communication can therefore be a multisource phenomenon or a process that involves a series of chemical homologues produced in the same exocrine gland. For example, the Dufour's gland of the fire ant is the source of a trail pheromone that induces both recruitment and orientation behaviors in workers. Dufour's gland chemistry is varied: the constituents of this gland's secretion regulate

different behaviors, which have been called "subcategories" of trail following. These chemicals include recruitment primers, synergists, and orientation inducers. Primer and inducer substances together release recruitment and orientation behaviors. *(Z,E)-a*-Farnesene is the principal trail orientation component isolated from the Dufour's gland. Another chemical fraction acts together with *(Z,E)-a*-farnesene to increase the effectiveness of the mixture in inducing trail communication. Homofarnesenes of presently unknown function and an orientation inducer present in the secretion also increase trail following. In the ant *Myrmica,* homofarnesenes in the Dufour's gland may be added to 3-ethyl-2,5-dimethylpyrazine (EDMP), which is the poison gland trail pheromone, as part of a multicomponent trail system.

Similar findings have been made in other ant species. Pure chemicals seem to induce lower responses than gland extracts, indicating the importance of the naturally occurring chemical mixtures in trail communication. Constituents present in different ratios sometimes show synergistic effects. Artificial trails prepared from extracts of the poison gland of the harvester ant *Pogonomyrmex badius* have a recruitment effect lasting approximately 20 mins, whereas artificial trails prepared from Dufour's gland secretions and aged for longer periods of time have elicited orientation responses. In the ant *Leptogenys diminuta* the poison gland and pygidial gland produce (3R,4S)-4-methyl-3-heptanol and isogeraniol, respectively, to regulate orientation and recruitment. Similarly, pheromone blends are known to mediate alarm communication, including alarm–recruitment systems involved in defense.

The sternal gland secretions of termites stimulate recruitment and may have highly durable orientation effects. Single chemicals such as *(E)*-6-neocembrene A and dodecatrienol have been isolated from whole-body extracts, but termite sternal gland secretions probably are more elaborate mixtures of pheromones that have different functions. Researchers have described "recruitment" and "basic" trails in termites; "basic" trails have only an orientation effect. The sternal gland secretion of *Nasutitermes costalis,* for example, can induce recruitment (drawing undisturbed soldiers and workers from the nest) and can orient searching and/or homing termites. Sternal gland material collected from trails aged for more than 20 years can orient, but not recruit, termites. Thus although the chemical that regulates recruitment dissipates in minutes, the orientation component of the secretion is a remarkably stable pheromone.

The persistent orientation components of a trail substance can "channel" foragers away from neighboring nests to minimize aggressive confrontations and can also serve as territorial recognition cues or as an initial guide for naive foragers. Some species of desert ants have trunk trail systems (a network of trails emanating from the nest entrance and arborizing at their distal portions) marked with Dufour's gland secretions composed of durable blends of hydrocarbons that are specific for species, populations, and colonies. In other ant species, different glands may produce trail chemicals with different behavioral effects. The ecological significance of trail structure in termites is not well understood, although apparently foraging galleries divide foraging space to increase the efficiency of harvesting food.

Trail Pheromone Specificity

Pheromone specificity is achieved through chemical mixtures and molecular structure. Early studies in ants suggested that trail substances were highly species specific, but results of more recent work do not support such a conclusion. One striking example of this lack of specificity is that ants in as many as six different genera in the subfamily Myrmicinae use the same trail pheromone, EDMP. And dodecatrienol, a trail pheromone in the termites *Reticulitermes virginicus, R. speratus,* and *Coptotermes formosanus,* provides a nonspecific orientation cue in these species, in other species of *Reticulitermes,* and in a cluster of geographically and phylogenetically diverse termites. *(E)*-6-Cembrene A, which has been isolated from whole-body extracts of the Australian *Nasutitermes exitiosus,* can induce orientation in African nasutitermitines, rhinotermitids, and to a lower degree in African macrotermitines (fungus-growing termites). At present, it is challenging to explain the significance of trail pheromone specificity in termites.

Metabolic end products, dietary differences, and genetics may produce variation in the chemistry of trail pheromones. In myrmicine ants, pyrazines and farnesenes are shared by different species and genera; these chemical constituents, present in small quantities and serving no function in one species, may have a prominent role in another. Biochemical variation may provide a substrate for evolution to act on in the selection of trail pheromones.

There have been many analyses of the level of specificity of trail substances, but the ecological significance of specificity is to a great extent unknown. Research on the behavioral ecology of trail pheromones indicates that chemical specificity may play a role in community structure. For example, competition may have selected for differences in the trail communication systems of sympatric species. Indeed, variation in recruitment communication systems in desert ants have been correlated with resource use and competition. Desert harvester ants forage as individuals on evenly distributed seed resources and recruit nestmates to cooperate in collecting seeds from dense patches. Different foraging strategies may be adapted to the exploitation of resources with different density distributions, serving as a mechanism of resource partitioning in granivorous ant communities. Foraging systems and their pheromonal regulatory mechanisms may also enhance food defense and retrieval, thus reducing interference competition.

The behavioral mechanisms of recruitment that are the basis for enhanced competitive ability may be associated with trail pheromone chemistry and response specificity. The harvester ants *Pogonomyrmex rugosus* and *P. barbatus,* which are very similar ecologically, have trunk trail systems that divert groups of foragers away from each other. Their trunk routes, which are composed of Dufour's gland and are colony specific, suggest chemical differentiation resulting from competition. The specificity of the trunk routes may also give an advantage in territorial defense if fighting ability is greater when ants are on their own territory. Other ant species are known to mark trails with persistent colony-specific pheromones.

Colony specificity in the trail pheromones of termites has rarely been examined. In *Trinervitermes bettonianus,* workers do not discriminate their own trails from those of neighboring colonies, although their trail pheromones and the trail substances of other sympatric termites appear to be species specific. Some termites distinguish between intra- and interspecific competitors. Workers deposit rectal fluid and sternal gland secretions on trails, a mixture that may encode colony identity.

Surprisingly, the chemical trails of some ants may also encode information about the identity of the individual that laid the trail. Individually specific trail markings are used during nest emigration by workers of the ant *Pachycondyla tesserinoda.* Individually specific trails, used for food recruitment, have also been described in the ant *Leptothorax affinis.* Individuality in a trail pheromone could provide a finely tuned mechanism of directional discrimination and maintain the path fidelity of individual foragers.

If ants can decipher the chemical code of another species' trail substance, they may be able to exploit information about the location of food sources. Interspecific trail-following is uncommon but is known in ants in parabiotic associations (i.e., different species of ants living together in a compound nest). In these species, foragers lay and follow their own chemical trails but are also able to interpret the trail pheromones of other species and to exploit their food discoveries.

Regulation of Recruitment and Foraging Activity

Trail pheromones communicate information about food quality to nestmates that have not had direct experience with a food source. Trail-laying behavior regulates pheromone concentration (i.e., the amount of pheromone deposited on a trail), which in turn controls a colony's response. After deposition on the substrate, a trail pheromone diffuses. The chemical properties of the pheromone both determine the spatial and temporal structure of its active space and regulate foraging activity at the colony level. The concentration of trail pheromone mediates communication between groups of individuals, those that have fed and any potential foragers within the nest. In fire ants, the continuity of the sting trail, measured by causing a scout ant to walk over a smoked glass slide (removing soot from the slide where the ant's sting is dragged) depends upon the concentration of a sugar solution offered as a food source. The more concentrated and thus rewarding the solution, the greater the extent to which the sting is extruded and dragged continuously over the substrate, and the greater the number of workers that will lay such a trail after they have contacted the food source (Fig. 2). The distance between the food source and the nest can also

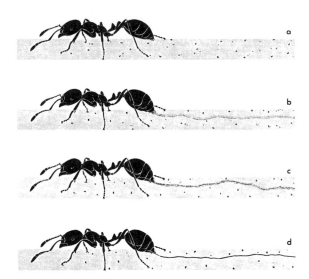

FIGURE 2 Fire ants deposit trail pheromones by discharging the contents of the Dufour's gland through the extruded sting. The continuity of the trail is made visible when a worker walks over a treated glass slide, removing soot where her body contacts the substrate. Marks made by the tarsi and hairs on the tip of the gaster are visible. Increase in the continuity of the sting trail can be seen. [From Hölldobler, B. and Wilson, E. O. (1990). "The Ants," p. 270. Belknap Press of Harvard University, Cambridge, MA. Originally published in Hölldobler, B. (1970). Chemische Verständigung im Insektenstaat am Beispiel der Hautflugler (Hymenoptera). *Unschau* **70(21),** 663–669. Reprinted with permission of Harvard University Press and Prof. Hölldobler.]

affect the rate at which food is retrieved, and thus the profitability of the colony's foraging.

The regulatory mechanism underlying foraging organization is called mass communication. Foraging is initiated when scouts locate new food, determine its quality, and deposit trail pheromone according to the food's energetic value and the colony's nutritional needs. This trail induces recruitment in nestmates, which repeat the cycle of food quality assessment and trail phermone deposition until the food source has been depleted or the colony satiated. The entire process is regulated by the concentration and evaporation of the trail pheromone. If the recruitment trail is not reinforced, the trail substance evaporates and foraging rate decays over time until food collection ends. Different ant and termite species show variations on the theme of mass communication that likely are associated with a the foraging ecology of individual species.

DEFENSIVE RECRUITMENT AND DIVISION OF LABOR BETWEEN CASTES

Social insects can be the most important competitors and predators of other social insects. Army ants, for example, exert significant predation pressure on social wasps and ground-dwelling ant species, favoring the evolution of adaptive systems of predator recognition and response. Some responses involve a division of labor among castes. Castes are groups of individuals specialized for a given set of tasks; they may have specific functions related to foraging or colony defense, and different castes may be recruited according to their specialization. Upon encountering a particularly important competitor or predator, ants that are patrolling the territory of a colony may prempt an attack through a caste-specific alarm– recruitment system. For example, the ant species *Pheidole dentata* and *Solenopsis geminata* are sympatric (i.e., occur together) in the southeastern United States and use similar nest sites. *S. geminata,* the native fire ant, as well as *S. invicta,* the imported fire ant, may attack colonies of *P. dentata* and destroy the colony (Fig. 3). Because fire ants are fierce predators of *P. dentata,* it is important to quickly respond to a potential threat. In *P. dentata,* minor workers usually care for brood, maintain the nest, and forage, whereas major workers, which have large heads and mandibles, have a significant role

in colony defense. When minor workers encounter as few as one or two fire ants, they return to the nest, laying pheromone trails to recruit minor and major workers. Major workers recognize the threat as emanating from fire ants through the odor of the predator carried on the messengers' bodies, and together with the excitatory behavior and trail substances deposited by minor workers, are recruited to the site at which the enemy has been located (Fig. 4). Because majors are recruited to respond defensively following contact with fire ants but not other ant species, this defensive recruitment system is said to have "enemy specification."

Species of termites that forage above the ground face a higher degree of predation than species that forage in the confines of subterranean gallery systems, and the social architecture of these ecologically different groups of species seems to have undergone adaptive modification. For example, species whose foragers harvest food above the ground have a high proportion of soldiers that use chemical defense in combating predators. Species that nest and seek food below ground have relatively low investment in soldiers, which rely on mechanical defense and are not involved in foraging. The recruitment communication systems that organize foraging in aboveground species appear to be adapted to reduce predation during the time period between the discovery of food and the construction of covered protective galleries, when termites are

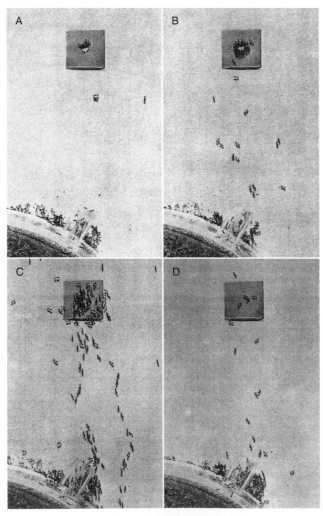

FIGURE 3 Recruitment and feeding behavior in a laboratory colony of fire ants. (A) Scout ants feeding at a drop of sucrose. (B) Ants initially recruited from the nest by trail-laying workers that had fed now feed and lay trails to the nest. (C) Recruitment increases as more ants arrive at the food and contribute pheromone to the trail. (D) The food is depleted, and new recruits arriving at the food do not lay trails on their return trip to the nest. Foraging now stops because the trail pheromone has evaporated. [From Wilson, E. O. (1963). Pheromones. *Sci. Am.* May, p. 110. Reprinted with the permission of Miriam F. Rothman for the estate of the photographer, Sol Mednick.]

FIGURE 4 Enemy specification in the recruitment behavior of the ant *P. dentata*. In response to contact with fire ants (light shading), major workers are recruited to defend the colony by attacking with their well-developed mandibles. [Reprinted with permission from Wilson, E. O. (1976). The organization of colony defense in the ant *Pheidole dentata* Mayr. *Behav. Ecol. Sociobiol.* **1,** 66. © Springer-Verlag GmbH & Co. KG.]

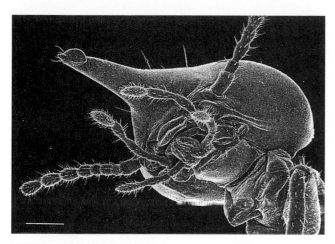

FIGURE 5 Head of a soldier of *Nasutitermes*. The mandibles are vestigial and the head is shaped to discharge defensive secretions. Scale bar = 0.25 mm. [Reprinted with permission from Eisner, T. *et al.* (1976). The organization of colony defense in the termite *N. exitiosus. J. Comp. Physiol.* **90**, Fig. 1. © Springer-Verlag GmbH & Co. KG.]

exposed and are vulnerable to predation by ants. In *Nasutitermes*, for example, the most diverse genus of the higher termites, there are worker and soldier castes: the ampule-headed soldiers are highly modified for chemical defense but also have an important role in organizing foraging (Fig. 5). Soldiers of the neotropical *N. costalis* scout in groups for food sources, and upon locating food communicate its location to nestmates. Groups of soldiers of *N. costalis* move in amoeboid fashion from the nest, ends of foraging galleries, and currently used food sources, recruiting other soldiers with sternal gland secretions as they slowly explore the environment. Parties of soldiers search as groups along trails in different areas, but all trails generally coalesce when a food source is located; subsequently additional soldiers and then workers are recruited.

There are three phases of foraging organization in *N. costalis,* each involving recruitment communication within and between castes. First, soldiers search for and discover new food sources and communicate information about their location to other soldiers. Next, workers are recruited in large numbers. Finally, the recruitment of workers increases further and soldiers flank both sides of the foraging trail to protect the more vulnerable workers traveling between the nest and the food. This pattern of soldier and worker recruitment shows that soldiers, which themselves do not feed directly, are nevertheless scouts, which assess food quality and recruit workers that will harvest food. Soldiers first recruit other soldiers to ensure an adequate defense at the food source and then communicate with workers to begin food collection. This intercaste communication is chemical; both soldiers and workers have a sternal gland, which produces a trail pheromone that induces the recruitment of soldiers and workers depending upon the quantity of pheromone deposited. The sternal gland secretion is not caste specific, but the volume of the sternal gland varies in soldiers and workers (large workers have a significantly larger sternal gland volume

than soldiers). Caste differences in pheromone perception and/or responsiveness to the sternal gland pheromone appear to regulate the prominent division of labor during foraging organization in nasute termites.

Nasute termites also have an alarm–recruitment response. When disturbed, soldiers discharge frontal gland terpenoid secretions; soldiers nearby then are recruited over short distances. Frontal gland and sternal gland secretions function in defensive recruitment and cause soldiers to remain in an area where a disturbance has been signaled. Similar foraging and defensive recruitment communication systems have been identified in other termites.

Defensive recruitment and the recognition of specific enemies occur in some termites. *N. costalis* responds defensively to the presence of a single soldier or worker of an alien conspecific colony by recruiting large third-instar workers, which attack the intruders with their mandibles. This response is not seen if other species of termites are encountered.

THE EVOLUTION OF RECRUITMENT COMMUNICATION

Ethological Models

Ethologists have long hypothesized that the origins of behavior can be traced from comparisons of similar actions in groups of closely related species. The history of recruitment behavior and chemical trail communication has been examined in ants with such an approach. The ancestral condition is thought to involve a behavior called tandem running. In this mode of recruitment, a single nestmate is led "in tandem" from the nest to a new nest or food source: a "leader" guides a "follower" to a target area. Tandem running involves motor displays that initiate pairing (Fig. 6) and surface pheromones and other exocrine gland secretions to maintain the communicative tie while the ants move pairwise, outbound from the nest. This type of recruitment communication is considered to be basal because it commonly occurs in ponerine species, which are themselves ancestral in the evolutionary history of ants.

In chemical mass communication, excitatory and orienting information is contained in the structure of the trail substance, and no other signals are required to control recruitment activity. In tandem running, behavioral displays alert recruitees to the need for their assistance; chemical signals on the body of the recruiter, as well as physical contact between the members of the tandem pair, provide directional guidance. In the most derived state of recruitment communication, all information required to regulate group action is encoded by pheromones. This type of trail communication is characteristic of many species in the majority of ant subfamilies and in the termites appears to be the only method of trail communication.

The origin of trail communication in termites has received relatively little attention. The secretions of the sternal gland are known to inhibit fungal growth; thus the ancestral function of this gland may have been the production and deposition

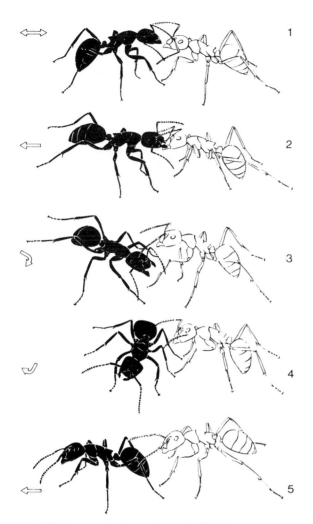

FIGURE 6 Tandem running in ants. A recruiter (black) contacts a nestmate (white) and shows a jerking behavior, pulling the nestmate by the mandibles to invite her to follow (1, 2). The recruiter then turns and offers the gaster (3, 4). When the recruited ant contacts the gaster and hind legs of the recruiter, the pair move in tandem to a new nest site (5). Arrows illustrate the direction of motion of the recruiter. [From Hölldobler, B. *et al.* (1974). Communication by tandem running in the ant *Camponotus sericeus. J. Comp. Physiol.* **90,** 105–127, © Springer-Verlag GmbH & Co. KG.]

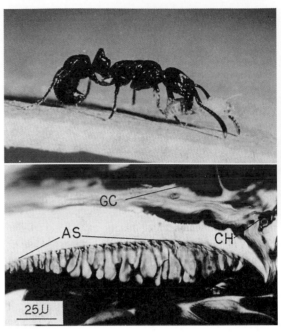

FIGURE 7 The pygidial gland and trail laying. (A) The ant *P. laevigata* applies the pygidial gland, located dorsally, to the substrate to lay a trail. (B) Structure of the gland: AS, cuticular applicator; GC, gland cells, CH, gland channels in the intersegmental membrane. Scale bar = 2.5 mm. [Modified from Hölldobler, B., and Traniello, J. (1980). The pygidial gland and chemical recruitment communication in *Pachycondyla* (= *Termitipone*) *laevigata. J. Chem. Ecol.* **6,** pp. 886, 887, with permission of Kluwer Academic/Plenum Publisher.]

of antibiotics to control microbial growth in the nest. Recruitment communication became more elaborate with the separation of nesting and feeding ecology and the evolution of a sterile worker caste.

Resource Distribution and the Evolution of Recruitment Communication

Some species of ants have no recruitment communication; these species offer important insights into the role of food distribution in the evolution of recruitment behavior. The relative significance of phylogeny and ecology can be separated in some of these species because the absence of recruitment communication has been noted in ants in derived subfamilies. The ant *Cataglyphis bicolor,* a formicine species that nests in

the ground and forages on the arid salt pans of North Africa, does not mobilize nestmates to food sources greater in size than what a single worker can transport. Recruitment does not occur naturally and cannot be induced experimentally. The diet of this ant is primarily composed of arthropods that have desiccated; the distribution of these prey is unpredictable in space and time, suggesting that the ecology of prey distribution has been the major selective force for individual retrieval without the possibility of recruitment.

Among the basal ponerine ants, the same point concerning food distribution and phylogeny can be made. Ponerine ants in the genus *Pachycondyla* generally show no food recruitment, employing tandem running only during nest emigration. These ants feed on randomly distributed individual prey. Yet chemical recruitment behavior occurs in species that utilize clumped food resources, such as *P. laevigata,* an obligate termite predator. *P. obscuricornis* huntresses collect prey as individuals but do not have food recruitment communication. Nest relocations involve tandem running; secretions from the pygidial gland, which is located dorsally beneath the seventh tergite on the gaster, hold the tandem pair leader and her follower together as they make their way to the new nest. In *P. laevigata,* an obligate termite predator, scout ants initiate the recruitment of nestmates after only a single termite has been found. Workers of this species apply the potent secretions of their pygidial gland to the substrate by curling the gaster forward ventrally and dragging the dorsal surface (Fig. 7). One trail-laying ant can induce the formation of a foraging column of several dozen workers. Shifts in diet in the genus therefore appear to have selected for the changes in trail-laying behavior and the chemistry of the pygidial gland secretion. Another basal ant that has well-developed recruitment communication is the ambyloponine *Onychomyrmex,* which is a specialist on large prey and has an army ant-like life cycle.

Resource distribution is associated with the use of chemical communication during foraging in desert seed-harvesting ants. *Pogonomyrmex rugosus* and *P. barbatus* occur sympatrically and feed on seed clumps. These ants have well-developed trail communication. On the other hand, *P. maricopa* collects scattered seeds, primarily through individual retrieval, and has a comparatively narrow diet breadth and a relatively weak recruitment response to seed patches.

RECRUITMENT, COMPETITION, AND COOPERATIVE RETRIEVAL

Recruitment communication allows the diet of a species to be expanded to include food items greater in mass than a single forager's load size limit. This is accomplished by recruiting nestmates to help transport prey. In this way, small-bodied ants can cooperate in prey retrieval, sometimes greatly increasing diet breadth. Recruitment communication is flexible; the number of ants in the cooperative retrieval group can be adjusted to prey size and thus allow a colony to efficiently collect resources of different sizes. There may be an energetic advantage to cooperative foraging, reflected in reduced search and retrieval costs and the caloric benefit of successfully acquiring large prey. Recruitment may also improve competitive success; workers recruited to a contested resource can enhance prey defense as well as assist in transport. Short-range recruitment, mediated by acoustical and/or chemical signals, may be used to attract nearby ants to a food find. Such is the case in the foraging organization of the desert ant *Aphaenogaster cockerelli.* Following short-range recruitment, a small group of recruits will cooperatively transport the food to the nest to minimize competition with sympatric ants.

The recruitment response of a colony is influenced by a number of ecological factors that can alter the way in which resources are partitioned between species. Recruitment varies with temperature and food item size, and each species in a community may have different thermal and prey size optima. This will cause competitive ability to vary among species and may result in some species "specializing" on prey of a certain size while foraging within a given temperature range. In open-field north temperate ant communities, the small-bodied *Monomorium minimum* is more tolerant of higher temperatures and can use a mass recruitment pheromone from the Dufour's gland to recruit rapidly to large prey, which workers defend with a chemical repellent originating in the poison gland. *M. minimum* foragers also dissect prey more rapidly than sympatric ants. Temperature preferences, recruitment, prey defense, and dissection together represent a foraging strategy that allows this ant to successfully retrieve large prey.

While the benefit of recruitment is diet expansion, its costs lie in the time required to assemble a cooperative retrieval group. During this time, competing ant species may discover the prey and interfere with its exploitation. Indeed, the probability of interference competition from sympatric ant species can increase from 0% to 100% when prey items exceed the upper size limit for individual worker carriage. The ability to move prey is an important factor in decreasing losses to competitors.

SELF-ORGANIZATION AND RECRUITMENT BEHAVIOR

Theories of foraging strategy note the significance of time, energy, and food profitability in the evolution of feeding behavior. Social insect colonies adjust their foraging (i.e., regulate the number of workers feeding at a given food source) according to food profitability, the risk of worker loss to predation, and competition. Recruitment communication, through the physical properties of trail pheromones, allows foraging adjustments to be made adaptively. It is at the colony level that such adjustments in worker feeding behavior, or decisions, are made. The concept of self-organization has been used to explain how collective action at the level of the colony can be the result of simple rules followed by individual workers. The complex foraging activities of a colony thus result from the interactions of simple workers, whose behavior is determined by trail pheromone concentration.

Self-organization has been used to explain the foraging behavior of mass-recruiting species such as fire ants, which have large colonies composed of workers with limited behavioral repertoires. In addition, their communication systems rely almost entirely on the deposition and diffusion of trail pheromones. Computer simulations and mathematical models have been used to describe the mechanisms involved in dividing foragers between two food sources, as well as the structure of raids of army ants and the rotation of foraging columns of desert seed-harvesting ants. Self-organization theory has also attempted to explain why mass-recruiting ant species have larger colony size: the greater complexity of larger colonies and their efficiency of colony operations may be dependent upon the individual simplicity of the workers that make up these social groups. Individual simplicity may allow greater flexibility in colony-level behavior than would be possible in a social group composed of complex individuals. This reinforces the notion that social insect colonies have a decentralized rather than hierarchical system of control. It has long been known that the queen does not "hand down" orders to workers to direct colony activites. The idea of self-organization can provide detailed descriptions of how efficiency is achieved through decentralization and the chemical regulation of worker behavior.

See Also the Following Articles
Ants • Caste • Dance Language • Division of Labor • Hymenoptera • Isoptera • Orientation • Pheromones • Sociality

Further Reading

Detrain, C., Pasteels, J. M., and Deneubourg, J.-L. (1999). "Information Processing in Social Insects." Birkhäuser-Verlag, Basel.

Hölldobler, B., and Wilson, E. O. (1990). "The Ants." Belknap Press of Harvard University, Cambridge, MA.

Traniello, J. F. A., and Leuthold, R. (2000). The behavioral ecology of foraging in termites. *In* "Termites: Evolution, Sociality, Symbioses, Ecology" (T. Abe, M. Higashi, and D. Bignell, eds.), pp. 141–168. Kluwer, The Hague.

Traniello, J. F. A., and Robson, S. (1995). Trail and territorial pheromones in the social insects. *In* "Chemical Ecology of Insects," Vol. 2 (R. T. Cardé and W. J. Bell, eds.), pp. 241–285. Chapman & Hall, New York.

Regulatory Entomology

Robert V. Dowell

California Department of Food and Agriculture

Regulatory entomology is concerned with preventing the unwanted movement of insects and related invertebrates from areas where they occur to areas where they do not occur. Nonnative species have caused untold amounts of damage to our crops, homes, environments, and persons. Officials in regional, state, or national governments achieve this goal with a five-component prevention program of exclusion, detection, eradication, identification, and public awareness. Exclusion is the primary prevention level. It uses phytosanitary regulations, including quarantines and other legal actions, to prevent the introduction of the quarantine pests. Detection is the secondary prevention level, which uses visual surveys or traps to find infestations of quarantine pests that have penetrated the exclusion barriers. Identification provides the name of the invading organism and information about its biology. It also provides intelligence to assist exclusion and eradication activities. Eradication is the tertiary prevention level. Eradication is the elimination of populations of quarantine pests found by the detection program. Public awareness strives to make the citizens aware of the regulatory laws concerning the importation of quarantine pests and items and enlists their cooperation in making the prevention efforts successful. The components work together much like a ladder with exclusion, detection, and eradication being the rungs, which are held together by identification and public awareness. Working together, these components can prevent the unwanted immigration of quarantine pests.

PEST EXCLUSION

The basic premise of pest exclusion is that it is better to live without pests than to live with them. History provides numerous examples of nonnative species that have caused extensive damage when brought into new areas. Examples of invasive species that have caused damage in nonnative areas include the Formosan termite *(Coptotermes formosanus)* in Hawaii and Louisiana, gypsy moth *(Lymantria dispar)* in the eastern United States, opuntia cactus *(Opuntia* spp.) and rabbits *(Oryctolagus cuniculus)* in Australia, brown tree snake *(Boiga irregularis)* in the Pacific region, American cockroach *(Periplaneta americana)* throughout most of the world, and Colorado potato beetle *(Leptinotarsa decemlinata)* in Europe. Pest exclusion uses phytosanitary regulations, including quarantines and other legal actions, to prevent the immigration of quarantine pests. Quarantines prohibit the movement of the quarantine pest or plants or other items known to be infested or liable to be infested from crossing the quarantine boundaries.

Quarantines may identify specific organisms (e.g., San Jose scale, *Quadraspiditotus perniciosus*) or groups of organisms (e.g., fruit flies in the family Tephritidae). Pest insects attack human food, fiber, shelter, or persons; carry diseases of people or other organisms, including domestic livestock or native biota; or are pests of human or natural environments (Table I).

Plants or other regulated articles are prohibited from crossing the quarantine boundaries unless they meet compliance procedures specified in the phytosanitary regulations that render them free of the quarantine pest. Compliance agreements and permits are authorizations to move the prohibited items under conditions prescribed by the quarantine issuer. Quarantine compliance may be performed at origin, the preferred option, or at destination, which is the less preferred option. The compliance requirements vary from those designed to keep the pest out of the regulated items, such as growing nursery stock in insect-proof screenhouses or applying pesticide sprays while the plants are growing, to those designed to kill any pests associated with the regulated items prior to or during shipment, including fumigation, hot/cold treatments, or irradiation (Table II).

Permits are also used to regulate the importation and holding of live insects for exhibit or research by insect zoos, butterfly houses, public or private research facilities and universities, and private citizens. The regulations that allow the

TABLE I   Examples of Quarantine Pests at Local, State, and National Levels

Scientific name	Common name	Reason
Aedes albopictus	Asian tiger mosquito	Human disease vector
Anoplophora malasiaca	White-spotted longicorn beetle	Wood-boring beetle
C. formosanus	Formosan termite	Dwelling pest
Dacus ciliatus	Lesser pumpkin fly	Fruit pest
Danaus plexippus	Monarch butterfly	Butterfly disease carrier
Diaphorina citri	Citrus psyllid	Citrus disease vector
Le. decemlineata	Colorado potato beetle	Vegetable defoliator
Ly. dispar	Gypsy moth	Forest defoliator
Q. perniciosus	San Jose scale	Fruit tree/ornamental pest
Trogoderma granarium	Khapra beetle	Grain pest

TABLE II Examples of Accepted Measures Used to Meet Phytosanitary Regulations

Pest	Origin	Measure used
Wood-boring beetles in dunnage	Asia	Fumigation
Fruit flies (Tephritidae)	Various	Vapor heat, fumigation, cold treatment, pesticide sprays
Thrips (Thysanoptera) on flowers	Hawaii	Irradiation
Caribbean fruit fly *(Anastrepha suspensa)*	Florida	Systems approach
Plum curculio *(Conotrachelus nenuphar)*	Eastern United States	Controlled atmosphere
Corn stalk borers	Eastern United States	Hot water dip
Japanese beetle *(Popillia japonica)* in nursery stock	Eastern United States	Pesticide drench of pots, plants grown in insect-proof screenhouses

insects to be held safely are specified in the permit issued to the person or entity requesting permission to import or hold live insects that are subject to regulation. Another form of pest exclusion uses the continual release of sterile insects throughout an area of high risk of invasion by that pest to preclude establishment. These sterile insects mate with wild adults as they emerge, producing infertile eggs. This preventative release approach stops the invading pest from developing infestations that may require eradication. This preventative release approach is being used against the Mexican fruit fly *(Anastrepha ludens)* along the Mexico–California border and against the Mediterranean fruit fly *(Ceratitis capitata)* in the greater Los Angeles area of California, along the Peru–Chile border, and along the Mexico–Guatemala border.

PEST DETECTION

Pest detection programs serve two purposes: to ensure that a political entity is free of the quarantine pest(s) regulated by their phytosanitary regulations and to find infestations of quarantine pests that have penetrated the exclusion barrier. There are two basic forms of pest detection: traps and visual surveys. The use of traps has many advantages and traps are the preferred method of detection of insect pests. However, not all pests respond to traps or lures. For these pests visual

TABLE III Advantages and Disadvantages of Traps and Visual Surveys Used to Detect Quarantine Pests

Advantages	Disadvantages
Traps	
Can be highly attractive, may detect pest at low densities	Not all pests can be trapped
Operate continuously when in field	Often limited to small number of species
Can be left in field for days to months	
One person can operate many traps	
Can cover large area quickly	
Visual surveys	
Can detect many pests at once	Labor intensive
Can detect any pest insect	Operate only when people are in field
	Cover limited area slowly

surveys, with their own advantages and disadvantages, are used (Table III).

Traps may be completely visual, such as yellow traps used to catch aphids and whiteflies, but most often they contain one or more chemicals attractive to the target species (Table IV). These chemical lures include sex pheromones, pheromone precursors, parapheromones (attractant not of botanical origin or a pheromone precursor that attracts males), food lures, chemicals of unknown action, and combinations of visual traps and lures.

Visual surveys can look for the pest itself, such as the white garden snail *(Theba pisana)*; the damage the pest can cause, such as that caused by the Asian longhorn beetle *(Anoplophora glabripennis)*; or the abode of the pest, such as the galls of the balsam gall midge *(Paradiplosis tumifex)*. Visual surveys may use food lures to attract and hold the pest as is done with the red imported fire ant *(Solenopsis invicta)*.

Detection efforts are concentrated in those areas where the target pest might enter and where the pest may become established. In general, large urban areas have a greater number of invading exotic species than rural or agricultural areas. This is the result of people directly bringing in exotic invertebrates in the fruit, flowers, and other items that they obtain while traveling and the importation of large volumes of goods (food, plants, etc.) that are needed to service large population centers (Table V).

If a pest is found during a detection program, a delimitation survey is used to determine whether an infestation exists

TABLE IV Examples of Lures Used in Pest Detection Programs

Lure	Type	Pest
Disparlure	Pheromone	Gypsy moth
Methyl eugenol	Pheromone precursor	*Bactrocera* fruit fly species
Trimedlure	Parapheromone	*Ceratitis* spp. fruit flies
Cuelure	Unknown	*Bacterocera* fruit fly species
Cracked grain	Food	Khapra beetle
Decaying protein in water	Food	*Anastrepha* fruit fly species
Ammonium carbonate and yellow traps	Food lure/foliage mimic	*Rhagoletis* fruit fly species

TABLE V   Examples of Locales in Which Detection Efforts Are Conducted

Locale	Pest
Urban areas	Many pests, including fruit flies
Campgrounds	Gypsy moth
Corn fields	European corn borer *(Ostrinia nubilalis)*
Slate yards	Snails and lygaeid bugs
Ports	Wood-boring beetles
Airports	Japanese beetle
Nurseries	Many pests, including scale insects
Almond orchards	Red imported fire ant *(S. invicta)* in palletized loads of honey bee colonies
Cotton fields	Boll weevil *(Anthonomus grandis grandis)*

TABLE VI   Examples of the Eradication Tactics Used against Invertebrate Pests

Pest	Tactic[a]
Japanese beetle	Cover sprays of pesticides
Olive fly *(Bactrocera oleae)*	Pesticide and bait sprays
Mediterranean fruit fly	Use of sterile males
Melon fly *(Bactrocera cucurbitae)*	Mass trapping using Cuelure and an insecticide
Oriental fruit fly *(Bactrocera dorsalis)*	Male annihilation using methyl eugenol and an insecticide
Asian longhorned beetle	Host removal
Gypsy moth	Cover sprays of microbial insecticides
Boll weevil/pink bollworm *(A. grandis grandis/Pectinophora gossypiella)*	Cultural controls such as specific plow-down dates

[a] One or more tactics are often used against a pest.

and if so, its physical boundaries. Unlike detection efforts that try to uniformly cover an area, delimitation efforts strongly target the area immediately around where the pest was found, with decreasing effort as one moves away from that point. For example, Mediterranean fruit fly detection in urban areas of California deploys a uniform 4 Jackson traps baited with Trimedlure per square kilometer. For delimitation, 100 Trimedlure-baited Jackson traps are deployed in the 2.56 km² area (core area) centered on the site where the first fly was found. An additional 50, 25, and 20 Trimedlure-baited Jackson traps per 2.56 km² are deployed respectively in three 1.6-km-wide rings around the core area.

Delimitation surveys are conducted until an infestation is confirmed and a decision about further actions is made or until it is determined that no infestation exists. The latter requires that no more target insects be found for a time period equal to several generations, typically two to three, of the pest. At the end of the delimitation effort, a trapping or visual survey program resumes at the detection level.

Delimitation efforts using visual surveys follow the same format, with a greater effort expended in the area immediately around the site where the pest was detected and diminishing effort as one moves away from that point.

PEST IDENTIFICATION

An important part of any detection program is the rapid and accurate identification of organisms that are found. Professional scientists provide these identifications and background information on the biology of the organism, which are used to help decide what, if any actions, may be taken against the pest. These professionals also provide information used to develop phytosanitary regulations and quarantine pest detection and delimitation programs.

PEST ERADICATION

Eradication of infestations of exotic pests is the most controversial aspect of regulatory entomology. Eradication is designed to eliminate a pest from a proscribed area, usually within a given time. Eradication programs are conducted at the point where the pest is found, not necessarily where it may do the most damage. Eradication programs using pesticide sprays in urban settings often generate considerable public outcry and opposition. Programs using nonpesticidal tactics or less intrusive methods of pesticidal applications engender little to no public concern. Thus the aerial application of pesticides or the extensive use of ground applications of pesticides in urban areas is strongly contested by the public, whereas mass trapping or male annihilation programs in the same areas are ignored. In general, successful eradication programs have the following components: the organism poses a clear-cut threat, an effective detection technique is available—usually not only visual surveys, the organism is limited in its distribution in the newly invaded area, continuous natural invasion of the organism cannot occur, and effective techniques exist to reduce the target population below the point at which reproduction can occur (Table VI).

Eradication procedures are applied to the target population for several life cycles of the organism beyond the last individual found. Posttreatment monitoring at delimitation levels is conducted for at least one additional life cycle of the pest. If no further individuals of the target organism are found within the treated area, eradication is considered to have been successful.

PUBLIC OUTREACH

Public outreach is conducted in a number of venues. Among the most frequently encountered are the forms that are required to be filled out before leaving a plane that has landed at a foreign airport or crossing a border. Information flyers handed out during eradication programs, on cruise ships, and at other sites are another venue through which the public is made aware of the exclusion efforts conducted on their behalf. Regular inspections of products moving into or out of an area keep professional importers and exporters aware of the quarantine regulations of their trading partners.

The five-component pest exclusion program described above, with modifications for local weather, topography, target species, and so on, is used by all countries that adhere to the International Standards for Phytosanitary Standards published by the Food and Agriculture Organization of the United Nations.

See Also the Following Articles

Agricultural Entomology • Extension Entomology • Fire Ants • Gypsy Moth • Sterile Insect Technique

Further Reading

Collard, S. B., III (1996). "Alien Invaders." Grolier Press, Danbury, CT.

Dowell, R. V., Siddiqui, I. A., Meyer, F., and Spaugy, E. L. (2000). Mediterranean fruit fly preventative release programme in southern California. *In* "Area-wide Control of Fruit Flies and Other Pests" (K.-H. Tan, ed.), pp. 369–376. Penerbit Universiti Sains Malaysia, Pulau, Pinang.

Hancock, D. L., Osborne, R., Boughton, S., and Gleeson, P. (2000). Eradication of *Bactrocera papayae* (Diptera: Tephritidae) by male annihilation and protein baiting in Queensland, Australia. *In* "Area-wide Control of Fruit Flies and Other Pests" (K.-H. Tan, ed.), pp. 381–388. Penerbit Universiti Sains Malaysia, Pulau, Pinang.

"Harmful Non-indigenous Species in the United States" (1993). Office of Technology Assessment, U.S. Govt. Printing Office, Washington, DC. [OTA-F-565].

"Insect Trapping Guide" (1998). California Department of Food and Agriculture (www.cdfa.ca.gov).

"International Standards for Phytosanitary Measures" (2000). Food and Agriculture Organization of the United Nations (www.fao.org).

Pegram, R. G., Gersabeck, E. F., Wilson, D. D., and Hansen, J. W. (2000). Progress in the eradication of *Amblyomma variegatum* Fabricius, 1794 (Ixodoidea, Ixodidae) from the Caribbean. *In* "Area-wide Control of Fruit Flies and Other Pests" (K.-H. Tan, ed.), pp. 123–130. Penerbit Universiti Sains Malaysia, Pulau, Pinang.

"Plant Quarantine Manual" (1998). California Department of Food and Agriculture (www.cdfa.ca.gov).

"Quarantine Pests for Europe" (1997). CAB Int. and European and Mediterranean Plant Protection Organization, University Press, Cambridge, UK.

Villasenor, A., Carrillo, J., Zavala, J., Stewart, J., Lira, C., and Reyes, J. (2000). Current progress in the medfly program Mexico–Guatemala. *In* "Area-wide Control of Fruit Flies and Other Pests" (K.-H. Tan, ed.), pp. 361–368. Penerbit Universiti Sains Malaysia, Pulau, Pinang.

Wyss, J. H. (2000). Screw-worm eradication in the Americas—Overview. *In* "Area-wide Control of Fruit Flies and Other Pests" (K.-H. Tan, ed.), pp. 79–86. Penerbit Universiti Sains Malaysia, Pulau, Pinang.

Reproduction, Female

Diana E. Wheeler

University of Arizona

In female insects, reproduction generally involves producing yolky eggs, mating, and then laying fertilized eggs. Across the diversity of insects, however, different ways of reproducing illustrate an astounding variation in this simple series of events as well as divergence from it. In the most extreme examples, females can reproduce without supplying eggs with yolk, without mating, and even without laying eggs.

Female reproduction has been one of the most intensively studied aspects of insect biology in the past 50 years for two reasons. First, frequent confrontations between humans and insects in the arenas of agriculture and health make understanding insect reproduction of great practical importance. Production of the next generation of insects has several steps that are centered on the female. To reproduce, females need to make eggs or provide their embryos with nutrition in other ways. Once made, females must find an appropriate spot to deposit their eggs. For entomologists concerned with problem insects, these steps offer opportunities to disrupt reproduction and reduce the number of insects in the next generation. Second, the diversity of ways that insects reproduce provides a rich source of material for discovering the underlying rules of biology. For example, the extraordinary effectiveness of female insects in converting resources into eggs led to their use as an intensively studied model system. The process by which yolk is taken up into insect eggs serves as a model for how cells take up large molecules from the surrounding environment.

STRUCTURE OF FEMALE REPRODUCTIVE SYSTEMS

Female insects can make eggs, receive sperm, store sperm, manipulate sperm from different males, and lay eggs. Their reproductive systems are made up of a pair of ovaries, accessory glands, one or more spermathecae, and ducts connecting these parts. Ovaries make eggs, and accessory glands produce substances to help package and lay the eggs. Spermathecae store sperm for varying periods of time and, along with portions of the oviducts, can control sperm use. The ducts and spermathecae are lined with cuticle.

The ovaries are made up of a number of egg tubes, called ovarioles. The number of ovarioles varies with the type of insect, its size, and its particular life history. Clearly, the number of ovarioles and the number of eggs that can be produced by each set an upper limit to the total number of eggs, or young, an insect can produce. The rate that eggs can develop is also influenced by ovariole design. In meroistic ovaries, the eggs-to-be divide repeatedly and most of the daughter cells become helper cells for a single oocyte in the cluster. In panoistic ovaries, each egg-to-be produced by stem germ cells develops into an oocyte; there are no helper cells from the germ line. Production of eggs by panoistic ovaries tends to be slower than that by meroistic ovaries.

Accessory glands or glandular parts of the oviducts produce a variety of substances for sperm maintenance, transport, and fertilization, as well as for protection of eggs. They can produce glue and protective substances for coating eggs or tough coverings for a batch of eggs called oothecae.

Spermathecae are tubes or sacs in which sperm can be stored between the time of mating and the time an egg is fertilized. Paternity testing of insects has revealed that some, and probably many, female insects use the spermatheca and various ducts to control or bias sperm used in favor of some males over others.

COMMON AND DIVERSE MODES OF REPRODUCTION

The most common mode of reproduction in insects is by yolked eggs, fertilized internally, that are laid outside the body. However, other modes of reproduction are not uncommon. Insects with unusual and even unique modes of reproduction are interesting as examples of extreme biological forms and processes. Modes of reproduction vary in three important aspects: whether eggs are fertilized, whether eggs are provisioned, and where embryonic development takes place.

First, reproduction without fertilization, or parthenogenesis, is common in insects as a normal means of reproduction in addition to or instead of sexual reproduction. For example, the system of sex determination in the Hymenoptera relies on parthenogenetic production of males. Virtually the entire order produces males from unfertilized haploid eggs and female from fertilized diploid eggs. This type of sex determination, called haplodiploidy, is also found in some Sternorryncha, Thysanoptera, and Coleoptera.

In another common type of parthenogenesis, germ stem cells in the ovary do not go through meiosis before they start development. As a result, offspring are clones of the mothers, having a full copy of her genes. Aphids are the premier example of this type of parthenogenesis. In a third type, meiosis occurs but the diploid number of chromosomes is restored by fusion of two nuclei. This type of parthenogenesis has been particularly well studied in stick insects (Phasmida). A fourth type requires mating, but fertilization is not completed. Sperm is necessary for development to begin, but the male's genes are discarded. Sperm-dependent parthenogenesis is found in some bark beetles.

A second important factor that defines the mode of development is the amount of yolk material provided to the egg. Amount of yolk can vary from none to more than enough to complete development. When eggs are laid without provisioning, nutrition must be obtained from elsewhere. Females of some insects, such as aphids, can keep developing embryos in their bodies and supply them directly with needed nutrients. In contrast, some parasitic and parasitoid wasps (e.g., Trigonalidae, Braconidae) and flies (Tachnidae) produce tiny eggs lacking yolk and lay them inside other insects that can provide for them.

A third important descriptor of developmental mode is the site of embryonic development. Most eggs are laid outside the mother's body but, in ovoviparous insects, eggs can be retained inside the body of the mother where the embryos develop. At hatching, they are released to the outside world. In true viviparity, embryos also develop in the mother's body but there is no intermediate egg stage. The site of development of parasitoids is similar to that in viviparous insects, in that it is inside another insect. Development of parasitoids, however, takes place in the bodies of host insects, rather than in the mother.

MANAGEMENT OF NUTRIENTS FOR EGGS

In the common mode of reproduction by yolked eggs, female insects accumulate large amounts of macronutrients, especially protein and fat. Lipids are usually derived from carbohydrates in the diet and are not generally in short supply. Amino acids, particularly essential amino acids, can be limiting. Therefore these are a particularly important part of the yolk.

Eggs may be provisioned with nutrients obtained in either the larval or the adult stage or both. Insects that do not feed as adults and have their lifetime's egg production completed when they become adults can draw only on larval nutrients. Even insects that feed after eclosion can use excess larval nutrients to provision eggs.

Mosquitoes and other blood feeding Diptera provide examples of the often interlocking roles of larval and adult feeding. Female mosquitoes feed on nectar and vertebrate blood to obtain nutrients for egg production. In some mosquitoes, however, food eaten during the larval stage supports the production of at least some eggs. The ability to produce eggs without blood feeding is called autogeny. Some autogenous species can mature their eggs only this way and have lost the ability to feed on hosts. Other species are more flexible and can use leftover reserves, if they are available, but can feed on blood immediately if they are not. In addition, when species occur over a broad geographical area, they can be locally adapted. The pitcherplant mosquito, *Wyeomyia smithii*, is completely autogenous in the northeastern United States, where larval densities are low and food more abundant, but must feed on blood in the southeast, where larval resources are more scarce.

Aspects of both larval and adult environments can favor autogeny. A larval environment that offers more consistent resources than the adult one will favor obligate autogeny, whereas more predictably abundant food in the adult environment will favor obligate blood feeding. In the Northern Hemisphere, autogeny becomes more common toward the arctic, where host vertebrates are less abundant. Short-term variability in nutritional resources, either in a patchy spatial environment or over time, can make physiological flexibility, termed facultative autogeny, a better strategy than obligate autogeny.

Males can be an important source of nutrients especially when the female's resources are limited. Nutritional contributions are especially conspicuous in Orthoptera, which transfer large spermatophores that can weigh over 20% of the male's body weight. In a variety of insects, proteins transferred to females during mating have been found in their eggs, ovaries, blood, and various body parts.

TRANSFER OF SPERM AND POSTMATING MANAGEMENT

Fertilization of eggs is the focus of sexual selection. Natural selection biases survival toward individuals that are the most successful in their environment. Sexual selection biases its rewards toward those individuals that enhance the success of their own genotype in the next generation through any aspect of mating and subsequent fertilization.

Classically, males compete with each other for access to females, and females can choose to mate with them or not. Because insects have internal fertilization, there are many possibilities for females to manipulate sperm after mating. When females mate multiple times, male competition can take place between sperm in the female reproductive tract. Overlooked for many years, it is now known that the female may also have control over which male's sperm will fertilize her eggs. Mechanisms that females use to bias paternity include active elimination of sperm, digestion of sperm, lack of sperm transport to the spermatheca, and decreased use of some sperm batches in fertilization. Postmating female choice is often called cryptic female choice, although it is no more cryptic than postmating competition between sperm.

Females are generally believed to choose their mates based on features that indicate their quality as parents. For example, a choice can be made based on behaviors such as gifts of food or some other indicator of resources, on the amount of sperm transferred, or on the concentration of a protective chemical given to the female. In addition to features that indicate success under the rules of natural and sexual selection, females may also choose sperm or mates based on genotypes that are most complementary with their own.

Highly social insects, especially those in perennial colonies, provide an interesting variation in female reproductive strategies. For example, to produce enough worker insects, queens must have a large supply of sperm. Only a small proportion of total sperm is used to produce new queens and males; most of it is used to make more workers. In social Hymenoptera, which includes ants, bees, and wasps, males can be produced without sperm because males develop from unfertilized eggs. As far as is known, new queens mate only before starting or joining a colony and must, therefore, at that time store as much sperm as they will need for the remainder of their lives. One important factor that affects colony longevity is the amount of sperm available for fertilizing eggs. Queens of large leafcutting ants *(Atta)* mate many times and store as many as 250 million sperm. In contrast, termites do not have to store large amounts of sperm because termite colonies are headed by both a queen and a king, and mating takes place throughout their lives.

Ants, the most speciose of the social insects, have a wide range of colony sizes and rates of worker production. Across the group, the number of sperm stored is correlated with the number of ovarioles, which suggests that there is a cost to long-term sperm storage and that sperm storage is matched to the lifetime needs of a successful colony queen.

OVIPOSITION

The culmination of insect reproduction is typically the oviposition of mature, fertilized eggs in an environment that will support their development. Eggs can be placed on surfaces, in crevices, in soil, and in animal or plant tissue, and accessory glands produce secretions used to protect eggs. Females often have a structure called an ovipositor, which is made up of modified appendages on the last abdominal segments and serves as a tool for penetrating substrates. For example, many grasshoppers use their ovipositors to dig into soil, where they lay eggs in a frothy matrix. Also, some parasitic wasps have long, needle-like ovipositors that can drill through wood to reach host insects.

Oviposition can be linked physiologically to prior events in egg maturation. For example, insects that mature eggs throughout the adult stage can often adjust their egg production based on available oviposition sites as well as on available nutrition. When oviposition sites are rare, unlaid eggs can inhibit the hormonal control network that guides egg maturation. Delayed oviposition can go further than slowing egg production and lead to resorption of eggs, which reallocates resources away from reproduction and toward survival.

Oviposition sites also provide sensory cues that can stimulate further egg maturation. For example, egg production in newly eclosed females in the diamondback moth, *Plutella xylostella,* is accelerated by the presence of single volatile components of host cabbage plants. Ovarian response to oviposition and related cues can reduce the opportunity costs females suffer when the supply of completed eggs does not match the availability of oviposition sites.

Protection of eggs, particularly those laid in clusters, can be extended by parental care. For example, egg guarding reduces the risk of eggs being used by parasitoids or eaten by predators.

See Also the Following Articles

Accessory Glands • Eggs • Ovarioles • Parthenogenesis • Sexual Selection • Spermatheca • Vibrational Communication • Vitellogenesis

Further Reading

Chapman, R. F. (1998). "The Insects: Structure and Function," pp. 298–312. Cambridge University Press, Cambridge, UK.
Eberhard, W. G. (1996). "Female Control: Sexual Selection by Cryptic Female Choice." Princeton University Press, Princeton, NJ.
Engelmann, F. (1999). Reproduction in insects. *In* "Ecological Entomology" (C. B. Huffaker and A. P. Guttierez, eds.), 2nd ed., pp. 123–158. Wiley, New York.
Leather, S. R., and Hardie, J. (eds.) (1995). "Insect Reproduction." CRC Press, Boca Raton, FL.
Nijhout, H. F. (1994). "Insect Hormones," pp. 147–156. Princeton University Press, Princeton, NJ.
Papaj, D. R. (2000). Ovarian dynamics and host use. *Annu. Rev. Entomol.* **45**, 423–448.
Thornhill, R., and Alcock, J. (2001). "The Evolution of Insect Mating Systems." iUniverse.com.
White, M. J. D. (1973). "Animal Cytology and Evolution." Cambridge University Press, London.

Reproduction, Female: Hormonal Control of

Diana E. Wheeler
University of Arizona

Female reproduction has been a major focus of entomological research for the past century, driven by the need to control populations of insect pests. The central process in female reproduction in insects, the production of eggs, is hormonally regulated. To reproduce successfully, females must coordinate egg production with other aspects of reproduction such as dispersal, the availability of resources, and selection of mates and oviposition sites. Environmental signals are effectively translated into physiological processes by networks of hormonal signals.

Egg development in insects has become a model experimental system studied to understand the general principles of stage-, sex-, and tissue-specific responses to hormones. In insects, juvenile hormone (JH) and ecdysone typically play important roles in orchestrating egg development. The development of improved analytical methods has led to the elucidation of the roles of other key hormones, particularly a variety of neurosecretory hormones. Common themes in the hormonal control of egg production are becoming clearer, as are important differences between insect groups.

PATTERNS OF FEMALE REPRODUCTION

To reproduce successfully, female insects must coordinate feeding, mating, and locating places to lay their eggs. Mating and egg laying almost always take place in the adult stage, but female insects vary widely in the manner in which they feed and accumulate nutrients for egg production. The accumulation of nutrients for eggs takes place during larval or nymphal development, as well as during adult life.

Eggs are generally filled with bulk nutrients to support embryonic development, and so it is not surprising that the timing of egg production is often tightly linked to the timing of feeding. The earlier nutrients required for egg production are eaten, the earlier egg production can begin. Starting egg production during the immature stage and completing it before eclosion allows female insects having very short adult lives to focus on mating and laying eggs. In fact, adults in many insects do not feed at all and even lack mouthparts. Mayflies are an example. At the other end of the spectrum, insects may take only sufficient nutrients during the larval stage to support their own larval or nymphal development. As adults, they must compensate for the lack of stored reserves by additional feeding. Female mosquitoes that emerge ready to find a blood meal are one example of this pattern. The relative advantages of these contrasting strategies depend on the opportunities

for acquiring nutrients, as well as other ecological factors that exist in both the immature and adult habitats. When egg development is telescoped into preadult stages, hormonal signals that control larval or nymphal development and metamorphosis must be coordinated with those of ovarian and reproductive development. Not surprisingly, setting the timing of egg development to different points relative to preadult development requires different hormonal controls.

Female insects that feed as adults differ in the size and number of their meals. At one extreme, females feed fairly continuously. At the other extreme, females take very large meals and have long periods of fasting between them. In blood-feeding insects particularly, a single blood meal can supply the bulk of nutrients necessary for a batch of eggs. Hormones signal the results of feeding to ensure that egg development is matched to an adequate supply of nutrients.

Egg development begins in the germarium, when stem germ cells produce daughter cells that become oocytes. Depending on the type of egg development, oocytes occur alone or in association with sister nurse cells, and follicle cells surround either the oocyte or the oocyte–nurse cells complex. Both nurse cells and follicle cells make and transfer materials important for future embryonic development to developing oocytes. In addition, vitellogenins, fats, and carbohydrates are all taken up into the egg, mostly from the blood. Most or all of the vitellogenin is made by the fat body, a complex organ analogous to the vertebrate liver. Fat body cells release yolk proteins into the blood, from which they are taken up by rapidly growing eggs. When uptake of nutrients is complete, the follicle cells secrete egg coverings in preparation for oviposition.

Hormonal control of egg production is well understood in only a few species. Nevertheless, enough is known about a wide variety of species to infer some general patterns. Four check points in ovarian development are commonly regulated by hormonal signals: the formation of new oocytes by stem cells in the germarium, initial growth of the oocyte, vitellogenin synthesis by the fat body, and vitellogenin uptake by the oocyte. The hormonal signals that break these checkpoints reflect various aspects of the external environment and internal conditions.

HORMONES

The two major hormones that control female reproduction, ecdysone and JH, also control preadult development and metamorphosis. In adult females, ecdysone is produced by ovaries, rather than by the prothoracic glands as in nymphs and larvae. JH is produced by a pair of glands called the corpora allata in both preadult and adult stages. The corpora allata are located near the brain and are connected to it by tracts of neurosecretory cells. Neurosecretory hormones made in the brain and in other parts of the central nervous system are also important controlling factors. Nerves are also part of control networks, especially in relaying sensory information to the brain cells, which affect the hormone-producing organs.

EXAMPLES OF HORMONAL CONTROL OF EGG PRODUCTION

Hemimetabolous Insects That Feed Continuously

Migratory locusts, *Locusta migratoria,* feed almost continuously if they can, and females produce and lay eggs in batches. The importance of JH in controlling egg production in this species has long been recognized. Shortly after eclosion, JH levels rise and stimulate synthesis of vitellogenin by the fat body. Mating and plant odors also affect JH level through neurosecretory cells in the brain. JH alone, however, is insufficient to complete a batch of eggs. An ovary-maturing parsin (OMP), which is a neurohormone, is required, in addition to JH, to stimulate sufficient vitellogenin synthesis and uptake.

In the viviparous Pacific beetle cockroach, *Diploptera punctata,* JH also regulates vitellogenin synthesis and uptake. Signals associated with mating and pregnancy result in increased and decreased levels of JH, respectively. Mating stimulates the release of JH through mechanical stretch of the reproductive tract. The signal is transmitted to the brain by nerves. After mating, females produce a set of eggs and retain them in a pouch off the oviduct called the brood chamber until embryonic development is complete. After the hatching and deposition of nymphs, the female can reproduce again. As in many insects, lack of sufficient nutrients causes JH levels to fall, which delays production of the next batch of eggs.

Hemimetabolous, Blood-Feeding Insects

The bloodsucking bug *Rhodnius prolixus* takes blood meals throughout its life. For nymphs, blood meals are required for growth and molting; for adults they are needed for egg production. In adult females, abdominal stretching associated with feeding stimulates the release of a peptide hormone from the thoracic ganglion. As a result, the corpora allata release JH, which causes vitellogenin synthesis by the fat body and vitellogenin uptake by growing oocytes. Only the terminal, largest oocyte in each ovariole develops, however. Nerves that stretch as the ovaries grow secrete a neurohormone, called an oostatic hormone, that prevents smaller oocytes from taking up vitellogenin. The result is a synchronously produced batch of eggs. As the eggs mature, the ovary secretes ecdysone, which triggers the release from the brain of a neurohormone that stimulates contractions for laying the eggs.

Lepidoptera: Holometabolous Insects That Vary in Timing of Egg Production

Lepidoptera are holometabolous insects, which means that preadults (larvae) and adults differ greatly in form, function, and ecology. Larvae are specialized for feeding, and adults are specialized for dispersal and reproduction. Despite the tidy appearance of such discrete stages, an extensive part of egg production can occur before the adult stage. Enough Lepidoptera species have been studied to allow some appreciation of the variation in egg production patterns across the group. Development and egg production are orchestrated by the same toolbox of hormones, so it is not surprising to find that Lepidoptera in which development and egg production occur simultaneously have control networks different from those in which these processes are sequential.

At one end of the spectrum, females complete egg production before eclosion. Adults of these species do not feed; they mate immediately after eclosion and are short-lived. Examples are silkworms (*Bombyx mori*) and gypsy moths (*Lymantria dispar*). JH does not stimulate egg production and can even inhibit some aspects of it. Pulses of ecdysone associated with pupal development apparently stimulate vitellogenin synthesis and uptake.

In some moths, only part of egg production is completed before eclosion. Examples include pyralid moths such as the Indian meal moth (*Plodia interpunctella*) and the southwestern corn borer (*Diatraea grandiosella*). Synthesis and uptake of vitellogenin take place before the molt to adult, and egg coverings are added afterward by the follicle cells. The portion of egg development that occurs after eclosion can be controlled by JH.

Finally, in many Lepidoptera, such as the monarch butterfly (*Danaus plexippus*), development and egg production are sequential and nonoverlapping, and hormones regulate egg development during the adult stage. In these insects, JH typically stimulates both vitellogenin synthesis and uptake.

Holometabolous, Blood-Feeding Insects

Hormonal control of egg production has been studied extensively in *Aedes aegypti,* the yellow fever mosquito, spurred by the ability of these and other mosquitoes to transmit diseases. Most mosquitoes must drink blood to obtain sufficient amino acids to make eggs. Females in a few species of mosquitoes have the ability to carry over reserves from the larval stage to make part or all of entire first batch without blood feeding.

Eclosion of female mosquitoes is usually followed closely by an increase in JH titer. The rise in JH induces target tissues to become responsive to hormonal signals that occur later. Specifically, the ovaries become sensitive to ovarian ecdysiotropic hormone (OEH), the primary oocyte grows slightly, and the fat body becomes responsive to ecdysone. The posteclosion peak of JH also causes behavioral changes: females search first for mates and then for hosts. Egg development is blocked until a blood meal can be taken. This checkpoint ensures that the intensive metabolic activity required for making eggs will not occur unless enough nutrients are available. When the mosquito has found a sufficiently large blood meal or series of meals, OEH is released into the blood by neurosecretory cells in the brain. OEH stimulates the ovary to produce ecdysone, which then stimulates the fat body to synthesize vitellogenin and the ovary to take it up. The ecdysone peak also causes new, secondary oocytes to separate from germaria, the first step in preparing for the next cycle of egg production. Growth of the primary oocyte ends at about 36 h after blood feeding. JH begins to rise again, and this stimulates the secondary oocytes to grow slightly and become ready for the next round of egg production. Finally, the eggshell is produced, the eggs are laid, and the mosquito is then ready to seek another blood meal.

Stimulation of vitellogenin synthesis by ecdysone has been studied at the molecular level. Analyses of the nucleotide sequence upstream of the vitellogenin gene show several broad regions of control. First, closest to the gene, is a binding site for ecdysone bound to its receptor complex. Binding to this site is necessary for expression. Second, more distally, are binding sites for two transcription factors, E74 and E75, that are known to be expressed quickly in response to an ecdysone signal. Therefore, the regulatory region of the gene has both direct and indirect interactions with ecdysone.

Holometabolous, Feeding Continuously

In contrast to mosquitoes, the fruit fly *Drosophila melanogaster* feeds continuously, and eggs develop serially as resources become available. The chain of oocytes in each ovariole includes a range of developmental stages. A balance between JH and ecdysone levels controls how many eggs produced.

In *D. melanogaster,* yolk proteins are made not only by the fat body but also by ovarian follicle cells. JH and ecdysone stimulate synthesis in the fat body, but only JH stimulates follicle cells.

JH appears to enhance egg maturation, whereas ecdysone inhibits it. The antagonism in the effects of the two hormones can be seen by manipulating the amount of food and mating. When food is withheld, JH level drops relative to ecdysone levels, and oocytes on the verge of taking up yolk proteins die. In addition, formation of new oocytes is reduced as more die than progress. Providing sufficient food again increases JH levels and reduces cell death. Mating enhances oocyte progression. Sex peptide, a hormonelike substance in seminal fluid, stimulates a rise in JH. Therefore, the increase in egg production seen after mating seems to act through the same hormonal mechanism used in nutrition. In this way, a sexually mature virgin female will not produce new eggs if she lacks either food or a mate. The two points of cell death, at separation of oocytes from the germarium and at the initiation of yolk protein uptake, correspond with similar checkpoints in mosquitoes.

HORMONAL LINKS BETWEEN EGG PRODUCTION, RECEPTIVITY, OVIPOSITION, AND PARENTAL CARE

Hormonal control of egg production in insects can be linked to other activities or states related to reproduction such as receptivity to mating, feeding, oviposition, and migration. In mosquitoes and *D. melanogaster,* receptivity and feeding are associated with an increase in JH level, which triggers early

events in the production of eggs. In other insects, however, JH can be inhibitory or have no effect.

Oviposition, too, can be under hormonal control. Oviposition by *L. migratoria* requires extreme extension of the abdomen. Muscles in the appropriate segments respond to JH by altering their contractile properties so that they can function when greatly extended.

Adult insects are often specialized for migration in addition to reproduction. These two functions can create a conflict between protein needs for flight muscles and eggs, and between lipid and carbohydrate needs for flight fuel and eggs. Hormonal coordination can minimize such physiological conflicts. Both the large milkweed bug, *Oncopeltus fasciatus,* and the boll weevil, *Anthonomus grandis,* use JH to stimulate egg production and long-duration flights. Stimulation of flight apparently requires low levels of hormone, whereas higher levels stimulate oogenesis but inhibit flight. High levels of JH in some insect species lead to both yolk uptake and breakdown of flight muscle.

Burying beetles *(Nicrophorus orbicollis)* use carcasses of small vertebrates for their own food, for an oviposition site, and as food for their offspring. Immediately after females have found a suitable dead body, their JH levels rise and egg production is stimulated. JH cannot, however, stimulate yolk uptake in the absence of a carcass, indicating that other factors are important in integrating the discovery of the food source and oviposition site with egg production. Juvenile hormone surges even higher in females when the larvae hatch. Egg production is not stimulated by this rise, which is necessary to stimulate and maintain parental care of developing larvae.

SUMMARY

Female reproduction in insects is controlled in diverse ways. The most fundamental process, egg production, is linked to other aspects of reproduction and reproductive behaviors by hormonal controls. Common themes are the central roles of juvenile hormone and ecdysone, with control points modulated by neurohormones. Hormonal control of egg production is relatively well understood in a few insects because of its use as a model system for studying hormone action. The integration of hormonal controls with other aspects of reproduction is more complex and is less well understood.

See Also the Following Articles
Ecdysteroids • Juvenile Hormone • Mating Behaviors • Migration • Vitellogenesis

Further Reading
Archives of Insect Biochemistry and Physiology, **35** (4) (1997). Special issue: Juvenile Hormone Revisited: Dynamics of Its Involvement in Reproductive Physiology and Behavior of Insects.

Chapman, R. F. (1998). "The Insects: Structure and Function," pp. 298–312. Chapman & Hall. London.

Engelmann, F. (1970). "The Physiology of Insect Reproduction," pp. 4–7, 45–49. Pergamon Press. Oxford, U.K.

Hagedorn, H. H. (1985). The role of ecdysteroids in reproduction. *In* "Comprehensive Insect Physiology, Biochemistry, and Pharmacology," Vol, 8 (G. A. Kerkut and L. I. Gilbert, eds.), pp. 205–262. Pergamon Press, Oxford, U.K.

Koeppe, J. K., Fuchs, M., Chen, T. T., Hunt, L. M., Kovalick, G. E., and Briers, T. (1985). The role of juvenile hormones in reproduction. *In* "Comprehensive Insect Physiology, Biochemistry, and Pharmacology," Vol, 8 (G. A. Kerkut and L. I. Gilbert, eds.), pp. 166–204. Pergamon Press, Oxford, U.K.

Nijhout, H. F. (1994). "Insect Hormones," pp. 147–156. Princeton University Press, Princeton NJ.

Raikhel, A. S. (2002). Vitellogenesis of disease vectors, from physiology to genes. *In* "Biology of Disease Vectors." 2nd ed. Harcourt Academic.

Soller, M., Bownes, M., and Kubli, F. (1999). Control of oocyte maturation in sexually mature *Drosophila* females. *Dev. Biol.* **208**, 337–351.

Wheeler, D. E. (1996). The role of nourishment in oogenesis. *Annu. Rev. Entomol.* **41**, 407–431.

Wyatt, G. R., and Davey, K. G. (1996). Cellular and molecular actions of juvenile hormone. II. Roles of juvenile hormones in adult insects. *Adv. Insect Physiol.* **26**, 1–155.

Reproduction, Male

Marc J. Klowden
University of Idaho

I nsect reproduction is accomplished with both enormous numbers and an elegant simplicity. In partnership with the female reproductive system, the male reproductive system of insects produces haploid gametes, the spermatozoa, that fertilize the haploid female gamete, the oocyte. Sperm develop in the testes and, when mixed with secretions from accessory glands, travel through the female genital tract when the insects mate, to be stored for later egg fertilization.

STRUCTURE OF THE MALE REPRODUCTIVE SYSTEM

The basic component of the male reproductive system is the testis (Fig. 1A), suspended in the body cavity by tracheae and fat body. The more primitive apterygote insects have a single testis, and in some lepidopterans the two maturing testes are secondarily fused into one structure during the later stages of larval development, although the ducts leading from them remain separate. However, most male insects have a pair of testes, inside of which are sperm tubes or follicles that are enclosed within a membranous sac. The follicles connect to the vas deferens by the vas efferens (Fig. 1B), and the two tubular vasa deferentia connect to a median ejaculatory duct that leads to the outside. A portion of the vas deferens is often enlarged to form the seminal vesicle, which stores the sperm before they are discharged into the female. The seminal vesicles have glandular linings that secrete nutrients for nourishment

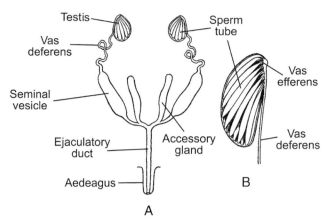

FIGURE 1 (A) The generalized structure of the male insect reproductive system. (B) A testis containing sperm tubes. (Modified from Snodgrass, 1935.)

and maintenance of the sperm. The ejaculatory duct is derived from an invagination of the epidermal cells during development and, as a result, has a cuticular lining. The terminal portion of the ejaculatory duct may be sclerotized to form the intromittent organ, the aedeagus. The remainder of the male reproductive system is derived from embryonic mesoderm, except for the germ cells, or spermatogonia, which descend from the primordial pole cells very early during embryogenesis.

PRODUCTION OF SPERMATOZOA

Within the testes are numerous tubular follicles that contain the developing male gametes. These follicles are in turn enclosed by a peritoneal sheath. The follicle number varies considerably; there is only 1 follicle in some coleopterans and over 100 in some orthopterans. At the anterior region of the follicle within the germarium are the spermatogonia, the undifferentiated germ cells, which are interspersed with somatic cells (Fig. 2). As the spermatogonia divide and move down the follicle, they become enclosed by the somatic tissue to form cysts. The development of the spermatozoa occurs within these cysts, which elongate as development proceeds. As more cysts are synthesized, they move the more mature ones posteriorly so that a range of cysts that contains gametes in progressive developmental stages is produced.

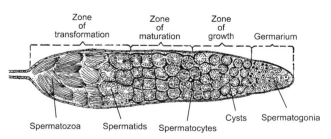

FIGURE 2 Longitudinal section of one sperm tube showing the development of spermatozoa from spermatogonia within cysts. (Modified from Snodgrass, 1935.)

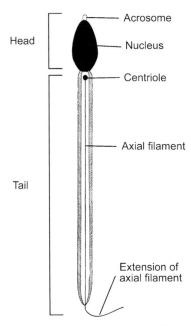

Head — Acrosome
— Nucleus
— Centriole
Tail — Axial filament
Extension of axial filament

FIGURE 3 A generalized insect spermatozoon. (Modified, with permission, from Breland *et al.*, 1968.)

In the zone of growth (Fig. 2), a spermatogonium within its cyst divides mitotically six to eight times, ultimately resulting in 64 to 256 diploid spermatocytes, each of which remains connected by cytoplasmic bridges, or ring canals. Farther down in the zone of maturation, the spermatocytes undergo meiotic division to become haploid spermatids. At the posterior end of the cyst, the spermatids complete a developmental transformation known as spermiogenesis to become the spermatozoa, with heads that contain the nuclear material and elongated flagella for motility. Finally, the spermatozoa break through the cyst and migrate to the seminal vesicles where they remain as a reservoir of sperm until the insect copulates with a female.

Insect spermatozoa have elongated heads and especially long tails (Fig. 3). Within the head are the nucleus and the acrosome, which contains the enzyme acrosin, which is associated with the penetration of the plasma membrane of the egg for fertilization. Insect sperm are unique in that they are the only cells of these animals that bear a flagellum. Within the flagellum is an axial filament that develops from the centriole and consists of an array of microtubules in a 9+9+2 arrangement, with nine outer accessory tubules, nine inner doublets, and a pair of central tubules. One member of each inner doublet also bears the protein dynein, which breaks down ATP and converts the energy released into flagellar movement. At the sides of the axoneme are elongated crystalline structures, the Nebenkern, that are derived from mitochondria.

Another feature of the reproductive tract of male insects is a pair of male accessory glands. These glands can open into either the vas deferens or the ejaculatory duct. Those arising from the vas deferens during development are mesodermal in origin, whereas in some insects in which they originate from the ejaculatory duct they have an ectodermal origin. The male

accessory glands serve a variety of functions, including producing the seminal fluid that serves as a transport, activation, and nourishing medium for sperm; acting as a vaginal mating plug that temporarily blocks sperm from another male from entering; and forming spermatophores, which are proteinaceous secretions of the glands that enclose the sperm and protect them while they are being transferred to the female.

The male accessory glands also produce substances that affect the behavior and physiology of the female. During insemination, peptides produced by the accessory glands that are contained in the semen are transported to various sites within the female. They are frequently involved in triggering her reproductive processes, including oogenesis and oviposition, or terminating subsequent female mating behavior so as to prevent her from acquiring the sperm of another male. Male accessory gland substances are frequently responsible for monogamy in females.

EXTERNAL GENITALIA AND MALE COURTSHIP

The external genitalia of males show an extreme morphological divergence. Indeed, their shapes often serve as the only reliable basis for species identification in taxonomic keys. In species of which the females mate often, there may be selective pressure for males to develop genitalia that compete for more effective ways to inseminate the females. There is a selective pressure for the evolution of male genitalia that offer males the greatest success; males with genitalia that more effectively transfer sperm are more likely to have their genetic contributions remain in the gene pool. Because copulation may last for a long period, males must have a way of avoiding detachment by other males that are also attempting to mate. Male insects have thus evolved a diverse array of grasping structures that allow the male to remain attached to the female; these devices include claspers, spines on the aedeagus, and even the use of mandibles to grasp a portion of the female's body. Structures have also evolved for the removal of existing sperm from a prior mating. For example, the aedeagus of the damselfly male scoops out the sperm present in the female before he adds his own. He thus eliminates any competition among sperm that may already be present in the female. Some male bedbugs inseminate the female by piercing her cuticle and depositing sperm directly into the body cavity.

Male reproductive behavior typically involves first searching for a female and, once she is found, precopulatory behaviors that increase her receptivity for copulation. This may include the presentation of a nuptial gift that occupies the female and increases her nutritional state. Once copulation is initiated, the actual transfer of sperm may require from a few minutes to a few hours. To protect their genetic investment, some males remain with the female for a variable interval after insemination has taken place. Male termites guard their mates for life. Some male scarab beetles gather dung for the female to lay her eggs on or help her to build a burrow and even remain inside the burrow with her for months.

Male paternal investment in offspring is most highly expressed in the giant water bugs of the family Belastomatidae.

Females lay their eggs on the backs of partially submerged males, interrupted only to mate repeatedly with the male. The males carry the eggs on their backs until the young hatch, actively maintaining their own position in the water in order to aerate the eggs.

See Also the Following Articles

Accessory Glands • Embryogenesis • Mating Behaviors • Spermatophore

Further Reading

Ben-Ari, E. T. (1999). Paternity battles. How males compete for fatherhood via sperm competition. *BioScience* **49**, 951–956.

Breland, O. P., Eddleman, C. D., and Biesele, J. J. (1968). Studies of insect spermatozoa. I. *Entomol. News* **79**, 197–216.

Davey, K. G. (1985). The male reproductive tract. *In* "Comprehensive Insect Physiology, Biochemistry and Pharmacology" (G. A. Kerkut and L. I. Gilbert, eds.), Vol. 1, pp. 1–14. Pergamon, Oxford.

Jamieson, B. G. M. (1987). "The Ultrastructure and Phylogeny of Insect Spermatozoa." Cambridge University Press, Cambridge, UK.

Phillips, D. M. (1970). Insect sperm: Their structure and morphogenesis. *J. Cell Biol.* **44**, 243–277.

Tallamy, D. W. (2001). Evolution of exclusive paternal care in arthropods. *Annu. Rev. Entomol.* **46**, 139–165.

Thornhill, R., and Alcock, J. (1983). "The Evolution of Insect Mating Systems." Harvard University Press, Cambridge, MA.

Reproduction, Male: Hormonal Control of

Marc J. Klowden
University of Idaho

The endocrine control of reproduction in female insects has been relatively well studied, perhaps because egg production is often cyclical and the experimental manipulation of hormones at key developmental periods has a direct effect on the number of offspring the females produce. This has not been the case for the males of most insect species, in which the production of sperm may begin early during the larval, nymphal, and pupal stages and often continues throughout adult life. Given the vastly different hormonal conditions that exist during the immature and adult periods, a unifying scheme for the control of male spermatogenesis has not been forthcoming. There are only a few examples of hormones controlling the reproductive processes of male insects.

PRODUCTION OF SPERM

The male gametes are produced in the paired testes from the spermatogonia. As the spermatogonia divide and move down the sperm tube, they become surrounded by tissue to form cysts. Their subsequent development to mature spermatozoa occurs within these cysts. First, the spermatogonia divide mitotically to produce diploid spermatocytes, and these spermatocytes undergo a meiotic division to become haploid spermatids. The spermatids then undergo a developmental transformation to become differentiated spermatozoa that ultimately break through the cyst and migrate to the seminal vesicles where they remain until the male copulates with a female.

The development of the testes during metamorphosis responds to the insect hormone 20-hydroxyecdysone (20HE). The synthesis of proteins by the testes of newly emerged adult *Tenebrio molitor* beetles is dependent on their exposure to this hormone.

Both 20HE and juvenile hormone (JH) have been implicated in the regulation of spermatozoa within the testes. High levels of 20HE increase the rate at which the spermatogonia undergo mitotic divisions to form spermatocytes, but this increase is abolished by high levels of JH. When spermatogenesis begins during the larval period, as in some Lepidoptera, the spermatocytes initiate their meiotic divisions but are arrested at prophase until the end of the larval period is reached. In many holometabolous insects, a peak of 20HE occurs before the end of the larval period that induces the "wandering" behavior that allows the larva to find a secluded spot in which to pupate. A postwandering peak of 20HE unblocks the meiotic division of the spermatocytes and allows the cells to proceed to metaphase. In some insects, JH has also been found to accelerate spermatogenesis.

When spermatogenesis begins during the larval stage, as it does in most Lepidoptera, it is interrupted if the insect diapauses as a larva or pupa. However, it resumes once diapause has been completed. The interruption is not caused by a pause in developmental activity but rather from the lysis of developing gametes before they become mature. The renewal of spermatogenesis when diapause has been terminated occurs when the titers of 20HE increase.

The developmental sequence in which spermatozoa develop from spermatids involves an elongation of both the nucleus and the flagellum. In Lepidoptera, the nuclear elongation is triggered by the declining concentrations of 20HE, but the elongation of the flagellum appears to be independent of any hormone.

Lepidopteran males produce two different types of sperm. Eupyrene sperm have a nucleus and fertilize the eggs, but apyrene sperm are anucleate. Both types of sperm are transferred to the female spermatheca after copulation. The function of the apyrene sperm is not well understood but they may provide nutrients required by the eupyrene sperm or aid in the migration of the eupyrene sperm within the female genital tract. The developmental pathway leading to the differentiation of apyrene sperm is regulated by a humoral factor that is present at the time of pupation.

ACCESSORY GLANDS

The accessory glands of the male reproductive system produce semen, accessory structures such as spermatophores, and various peptides that regulate female behavior and physiology. The interaction between JH and 20-hydroxyecdysone during postembryonic development regulates the development and differentiation of the glands. Juvenile hormone alone may also control the synthesis of those specific proteins that are transferred to the female. The accumulation of some secretory peptides in the glands that are enhanced by JH may be either enhanced or inhibited by the simultaneous presence of 20-hydroxyecdysone. In the German cockroach, *Blattella germanica,* the activity of the corpora allata, which is the source of JH, declines during the formation and transfer of the spermatophore and may thus initiate a new cycle of male accessory gland maturation.

See Also the Following Articles

Accessory Glands • Ecdysteroids • Juvenile Hormone • Spermatophore

Further Reading

Chapman, R. F. (1998). "The Insects: Structure and Function," 4th ed. Cambridge University Press, Cambridge, UK.

Dumser, J. B. (1980). The regulation of spermatogenesis in insects. *Annu. Rev. Entomol.* **25,** 341–369.

Friedländer, M. (1997). Control of eupyrene–apyrene sperm dimorphism in Lepidoptera. *J. Insect Physiol.* **43,** 1085–1092.

Gillott, C., and Gaines, S. (1992). Endocrine regulation of male accessory gland development and activity. *Can. Entomol.* **124,** 871–886.

Klowden, M. J. (2002). "Physiological Systems in Insects." Academic Press, San Diego.

Research Tools, Insects as

Kipling W. Will
University of California, Berkeley

In addition to serving as classic experimental laboratory animals (e.g., *Drosophila* flies in genetics, *Periplaneta* cockroaches in neurophysiology, *Manduca* moths and *Schistocerca* grasshoppers in physiology), insects have been essential to the formulation and testing of many general theorems in ecology and evolutionary biology. Scientists seek to develop general synthetic theories in biology, just as in physics and chemistry, that provide answers to questions of how and why things are as they are. Perhaps more importantly, these generalities make predictions that allow us to test what we think we know. The use of data to constantly reevaluate our theories divides empirical science from personal belief systems and popular metaphysics. The latter two are concerned with understanding the fundamental nature of all reality and often are based on abstract elements. In contrast, hypotheses in mainstream science are typically explanations of the processes that exist in nature and are tested against empirical observations.

To develop these testable hypotheses in science a constellation of data from model systems is needed, especially in biology.

Insects are used as many of these model systems. In part this is because they are so abundant and species rich that, in terms of diversity, they make up the bulk of terrestrial species and, in many habitats, the greatest number of individuals. Because of this dominance, comprehensive biological hypotheses must account for insects if the hypotheses are to be generally accepted. Insects provide the numerous observations that are essential for developing and supporting broadly applicable hypotheses explaining the diversity and distribution of life on earth.

Insects are a magnificent source of observations. The sheer number of units for study at all levels—species, populations, and individuals—provide the repeated patterns of variation that provoke questions and provide data for hypothesis testing. Also, insects usually have a relatively short life cycle, often more than one generation per year. This allows scientists to make multiple observations of all life stages of a species in a relatively short time period. Insect species can be widespread, but typically they are localized, making it easy to accumulate distributional data for at least the more conspicuous taxa. Certainly there is an important human factor as to why insects are so important in the study of biology. Insects delight us with their forms and behaviors and seem to embody all that fascinates humans about the natural world—beauty, diversity, mystery, and perhaps most of all discovery. Once a biologist, or any naturalist, is exposed to the wonders of insects they are usually hooked for life.

Biology has benefitted greatly from both reductionist and integrative research. The reductionist program attempts to minimize the number variables in the study system and identify causal mechanisms. Stunning success has been achieved using *Drosophila* as a laboratory animal to investigate developmental and genetic systems and to discover basic mechanisms from which inferences about general principles in biology are made. An integrative or synthetic approach is also essential in biology. Insects have been crucial model organisms, providing some of the most important advances in synthetic theory.

THEORETICAL WORK IN THE 19TH CENTURY

Evolution, or the theory of natural selection, is the most influential of all biological theories, and the two men who codified the basic mechanisms of descent with modification were dedicated observers and collectors of insects. Charles Darwin and Alfred Wallace shared what Wallace referred to as a "child-like" passion for beetles; Wallace even suggested this may have been a common thread that helped to lead both of them to arrive independently at similar conclusions about the evolutionary origin of species. Both present colorful stories of collecting insects. Wallace wrote wonderful passages on "one good day's work" collecting in Borneo, recalling species by species those collected and those that escaped one day, to be pursued the next. Darwin recalls with great passion his beetle collecting and the unfading thrill of discovering rare or new species. Wallace earned much of his livelihood collecting and providing specimens to museums and private collectors. He sold thousands of specimens at about 2 cents each to fund his tropical expeditions. For both Darwin and Wallace, attention to details necessary for separating species, subspecies, and varieties of insects was fundamental to developing their ideas. The diversity of forms and sheer reproductive output of insects provided examples that cultivated in their minds the theory of natural selection.

Wallace traveled with another entomologist and great naturalist, Henry Bates. In 1842 Wallace and Bates went to the Amazon to explore and to collect insects. These explorations and Bates' collections (Wallace's were unfortunately lost in a ship fire) became incredibly valuable in terms of insights into natural history and evolution. For 11 years, Bates collected insects, primarily butterflies and beetles, that were and remain a source of awe and study material for students of insects and users of European museum collections. Bates readily accepted Darwin's and Wallace's ideas of natural selection as the mechanism of evolution and went on to develop his theory of mimicry. Known as Batesian mimicry, this concept stemmed from his experience with tropical butterflies. This theory is widely applied throughout biology as an explanation for the similar and convergent appearance of some organisms.

THEORETICAL WORK IN THE 20TH CENTURY

Willi Hennig is best known for developing a coherent theory for phylogenetic systematics, a field of research that investigates and presents relationships among taxa. Hennig's theoretical works form the core of modern cladistic methods (use of shared derived characteristics to elucidate sister group relationships of taxa). Hennig was also the foremost authority on flies (Diptera) and produced many publications, including his series of publications on maggots (dipterous larvae), which became the standard work on the subject. Throughout his classic work *Phylogenetic Systematics* he relied on insect examples. Today our ideas about how to develop hypotheses of relationships for animals in an evolutionary scheme and how to classify them are largely based on theories and methods developed with insects as models.

Biogeography is a major field of biology that strives to understand the spatial relationships of organisms and looks at both historical and contemporary questions regarding biodiversity. Darwin, Wallace, and Hennig were all prominent contributors to this field, each drawing on ample observations from the insect world. Wallace was particularly influential in developing ideas that are still important in biogeographical studies. Biogeographical regions of the earth presented by Wallace, which were modified from Philip Sclater's previously published scheme, are still a standard part of describing the geographic distribution of animals. Most prominent is Wallace's observation of a distinct change in fauna between Bali and Lombok in the East Indies, known as Wallace's line. Wallace drew heavily on his knowledge of insect life histories, dispersal abilities, and distribution of some conspicuous insects (beetles and butterflies) to develop his biogeographical ideas.

A significant change from thinking about biogeography only in terms of evolutionary and historical scenarios to looking at ecological dynamics began in the 1960s. Robert MacArthur and E. O. Wilson published the equilibrium theory of island biogeography, a model based on land area and distance from source populations that explained how newly available islands could become populated with plants and animals, ultimately coming to a point of equilibrium in terms of species number. This became one of the most influential works in the field. Noted biologist and entomologist Ed Wilson is an ant systematist and he used these insects to support the development of this theory.

Wilson's contributions to biology are many, but some of the best known and most controversial are the ideas presented in sociobiology. In general this field seeks explanations of social behavior in animals based on common biological and evolutionary concepts. Largely this synthesis is based on his knowledge of ants, animals that are truly social. Many other insect groups, including beetles, butterflies, termites, dragonflies, and bees, just to name a few, are exemplar taxa used to illustrate concepts of sociobiology. Sociobiology is broad and interdisciplinary; insects are a vital part of the hypotheses that span many fields of inquiry.

LOOKING FORWARD IN THE 21ST CENTURY

Whether through reductionist approaches or the development of synthetic theories of biology, it is clear that the study of and passion for insects is an incredibly important part of our understanding of the world we live in and realizing our place in the natural system. Just as insects have proven to be essential in developing theories using the integrative approach, they will continue to be prominent in studies at all levels of organization from the molecular to the ecosystem.

See Also the Following Articles

Biodiversity • Drosophila melanogaster • Industrial Melanism • Mimicry

Further Reading

Bates, H. W. (1862). Contributions to an insect fauna of the Amazon valley: Lepidoptera: Heliconidae. *Trans. Linn. Soc. London* **23,** 495–566.
Darwin, C. (1859). "The Origin of Species by Means of Natural Selection: Or, the Preservation of Favoured Races in the Struggle for Life." Murray, London.
Darwin, C., and Wallace, A. R. (1858). On the tendency of varieties to depart indefinitely from the original type; and on the perpetuation of varieties and species by natural means of selection. *J. Proc. Linn. Soc. Zool.* **3(9),** 53–62.
Dethier, V. G. (1962). "To Know a Fly." Holden–Day, San Francisco.
Hennig, W. (1966). "Phylogenetic Systematics." University of Illinois Press, Urbana. [Translated by D. Dwight Davis and Rainer Zangerl]
MacArthur, R. H., and Wilson, E. O. (1967). "The Theory of Island Biogeography." Monographs in Population Biology, No. 1. Princeton University Press, Princeton, NJ.
Rogers, B. T., and Kaufman, T. C. (1997). Structure of the insect head in ontogeny and phylogeny: A view from *Drosophila. In* "International Review of Cytology" (K. W. Jeon, ed.). Academic Press, San Diego.

Sclater, P. L. (1858). On the general geographical distribution of the members of the class Aves. *J. Linn. Soc. Zool.* **2**, 130–145.

Statzner, B., Hildrew, A. G., and Resh, V. H. (2001). Species traits and environmental constraints: Entomological research and the history of Ecological Theory. *Annu. Rev. Entomol.* **46**, 291–316.

Wallace, A. R. (1855). Concerning collecting dated 8 April 1855, Si Munjon Coal Works, Borneo. *Zoologist* **13**, 4803–4807. [Letter]

Wallace, A. R. (1876). "The Geographical Distribution of Animals, with a Study of the Relations of Living and Extinct Faunas as Elucidating the Past Changes of the Earth's Surface." Macmillan & Co., London.

Wilson, E. O. (1975). "Sociobiology: The New Synthesis." Harvard University Press, Cambridge, MA.

Respiratory System

Jon F. Harrison
Arizona State University

The primary goals of the insect respiratory system are to deliver oxygen from the air to the tissues and to transport carbon dioxide from the tissues to air. Gases are transported through the tracheal system by both diffusion and convection, with the relative importance of these two mechanisms varying across and within species. Aquatic and endoparasitic insects exchange gases by a variety of mechanisms; most have tracheae and tracheoles within thin-walled appendages that function as gills. Within individual insects, the structure of the tracheal system can be altered dramatically during ontogeny and in response to rearing conditions, such as low oxygen. With a given tracheal structure, the ability of the tracheal system to transport gases can be modulated dramatically by varying spiracular opening, ventilation, and the fluid level in the tracheoles. Control of flexibility in tracheal gas exchange capacity depends strongly on neuroendocrine control of muscles that drive convection or control spiracular opening.

DEVELOPMENT

In most insects, the tracheal system first appears in the embryo. In general, the size of the tracheal system increases with age in order to support the increased gas-exchange needs of the larger insect. However, major changes in tracheal structure, including changes in spiracle number and tracheal system organization, can occur at each molt and during the pupal period for endopterygote insects. During the molts, the cuticular lining of the trachea is drawn out of the spiracle with the old integument.

Changes in tracheal system structure are not limited to molting periods, because the tracheoles can change structure within an instar. In the event of injury or oxygen deprivation, local tracheoles grow and increase in branching. If no undamaged tracheoles are nearby, damaged tissues produce cytoplasmic threads that extend toward and attach to healthy tracheoles. These threads then contract, dragging the tracheole and its respective trachea to the region of oxygen-deficient tissue.

Both tracheoles and trachea are fluid-filled in newly hatched insects, and fluid fills the space between the old and the new trachea at each molt. Usually, this fluid is replaced with gas shortly after hatching or molting. In some cases, the spiracles must be open to the air for gas-filling of the tracheae to occur, suggesting that gas-filling occurs as fluid is actively absorbed by the tracheal and tracheole epithelia, with air entering through the spiracle to replace the absorbed fluid. However, in aquatic insects that lack spiracles, the tracheae also become gas-filled, indicating that this gas can be generated by the tissues or hemolymph.

MECHANISMS OF GAS EXCHANGE

Insect gas exchange occurs in a series of steps. Oxygen molecules first enter the insect via the spiracle, then proceed down the branching tracheae to the tracheoles. The terminal tips of the tracheoles are sometimes fluid-filled, so at this point gas transport may occur in a liquid medium rather than air. Oxygen then must move across the tracheolar walls, through the hemolymph, across the plasma membranes of the cells, and finally through the cytoplasm to the mitochondria. Carbon dioxide generally follows a reverse path.

Diffusion

Diffusion is the passive movement of molecules down their concentration gradient, driven by random molecular motions. Because oxygen is transported to the tissues as a gas and the diffusion rate of oxygen is much more rapid in air than in water, the insect tracheal system is capable of high rates of gas exchange by diffusion. Consumption of oxygen by the tissues lowers internal oxygen levels, creating a concentration gradient from air to tissues that drives oxygen through the tracheae. The converse occurs for carbon dioxide. The final steps of oxygen delivery, from the tracheoles to the mitochondria, may occur by diffusion in all insects, because diffusion operates rapidly over micrometer distances. In the initial steps of oxygen delivery (across the spiracles, through the tracheae), the importance of diffusion is more variable. Simple diffusive gas exchange through the tracheae and spiracles likely occurs in some pupae, as washout rates of inert gases are similar to those predicted from their diffusion coefficients. Additionally, in a variety of insects, no ventilatory movements have been discerned, which may mean that these insects exchange gases by diffusion through their tracheae and spiracles.

Convection

Convection is the bulk movement of a fluid (gas or liquid) driven by pressure. Differential air pressures can drive gas movement through the tracheae and spiracles at much higher rates and over longer distances than diffusion. In many insects, well-coordinated actions of muscles and spiracles produce regulated convective air flow through the tracheae and spiracles.

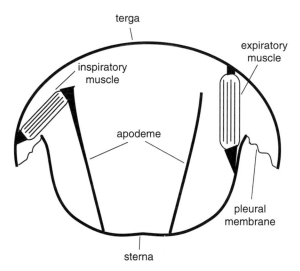

FIGURE 1 Respiratory muscles that drive the dorsoventral movements of the abdomen during abdominal pumping in a grasshopper. (Adapted from Hustert, 1975, and Chapman, 1998, with permission of Cambridge University Press.)

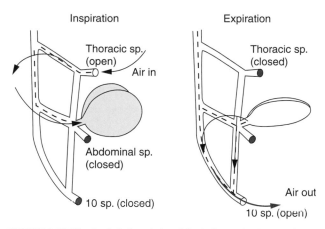

FIGURE 2 Unidirectional air flow during abdominal pumping in a grasshopper. During inspiration, air flows in through open thoracic spiracles (sp), along the longitudinal trachea, and into the air sacs. At low metabolic rates, air flows out only through the 10th abdominal spiracles; in more active animals, air flows out all abdominal spiracles.

Convection can be driven by a variety of mechanisms in insects. Most commonly, convection is driven by contractions of respiratory muscles attached to the body wall, which produce increases or decreases in body volume, causing compressible portions of the tracheal system to inflate or deflate.

One common mechanism by which insects accomplish convective air flow through the trachea is abdominal pumping. Expiratory muscles connect the ventral and dorsal cuticular plates and also span adjacent abdominal segments. When they contract, they pull the dorsal terga and ventral sterna together (Fig. 1) and the tip of the abdomen inward as the cuticular plates slide over the flexible pleural and intersegmental membranes. When the body volume decreases, the cuticle pushes on hemolymph and tissues, which in turn compress the collapsible air sacs. Inspiration may be passive and result from cuticular elasticity. Alternatively, contractions of inspiratory muscles attached to tall sternal apodemes and the lower edge of the terga can lift the terga relative to the sterna and expand abdominal volume and the air sacs (Fig. 1). Similar movements of the terga and sterna produce abdominal pumping in Orthoptera and in adult Odonata, Hymenoptera, and Diptera, whereas in adult Coleoptera and Heteroptera only the terga moves. In adult Lepidoptera, Orthoptera (Tettigoniidae), Neuroptera, and Trichoptera, movements of the sterna, terga, and lateral pleural membranes all contribute to abdominal pumping.

In many of the insects that have been examined, abdominal pumping is coordinated with spiracular opening in a manner that produces unidirectional air flow through the longitudinal tracheal trunks. During inspiration, abdominal spiracles are closed and air flows in through open thoracic spiracles (Fig. 2). During expiration, air flows out open abdominal spiracles, while the thoracic spiracles are closed. There is evidence for unidirectional air flow in adult cockroaches (Blattodea), grasshoppers (Orthoptera) and mantids

(Mantodea), honey bees (Hymenoptera), and dragonflies (Odonata). In some pupae, abdominal pumping is coordinated with the opening of one or a few "master spiracles" that exchange all gases.

Muscles that have the primary purpose of driving hemolymph circulation, such as the heart and ventral diaphragm, also play a role in creating convective ventilation. Pumping of the heart and accessory muscles at the base of the wings, antennae, and legs pushes hemolymph and air into these appendages. In a variety of adult Lepidoptera, Diptera, and Hymenoptera, the heart occasionally reverses pumping direction, shifting hemolymph from the thorax to the abdomen. As the hemolymph accumulates in one body compartment, it compresses air sacs in that compartment, causing convective air flow through tracheae and spiracles. Similar hemolymph transfers between body compartments coupled to ventilation may occur because of intermittent heart activity in bees (Hymenoptera).

DISCONTINUOUS GAS EXCHANGE

For many insects, especially those that are highly active, the spiracles close for only brief periods, if at all, and oxygen and carbon dioxide are exchanged relatively continuously. However, many insects exhibit discontinuous gas exchange. During discontinuous gas exchange, periods of spiracular closure (in which no or reduced gas exchange occurs) alternate with periods of spiracular opening (and greatly increased gas exchange; Fig. 3).

Discontinuous gas exchange has been most extensively studied in ants and diapausing lepidopteran pupae. In lepidopteran pupae, spiracular openings can be separated by hours. After the burst of gas exchange, the spiracles hermetically seal (closed phase). While the spiracles are closed, oxygen is consumed from within the tracheal system (Fig. 3). The oxygen removed from the tracheal air space is not completely replaced

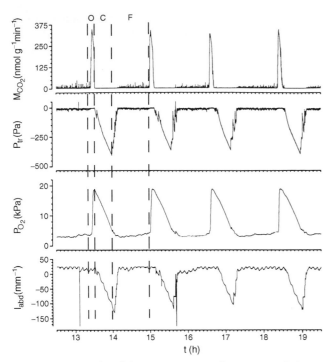

FIGURE 3 Four cycles of discontinuous gas exchange in pupal *Attacus* (Lepidoptera). From top to bottom: carbon dioxide emission (M_{CO_2}), tracheal pressure (P_{tr}), tracheal oxygen concentration (P_{O_2}), and abdominal length (I_{abd}). For the first cycle, the open (O), closed (C), and flutter (F) phases are indicated. (Reproduced with permission from Hetz, 1994.)

by carbon dioxide primarily because a large fraction of the carbon dioxide produced dissolves in the tissues and hemolymph (carbon dioxide, in contrast to oxygen, is highly soluble in biological fluids). Therefore, pressures fall below atmospheric within the tracheal system and abdominal length is reduced (Fig. 3). When internal oxygen tensions reach a low threshold, the spiracles begin to open slightly at high frequency (flutter phase), and tracheal pressures rise to near-atmospheric levels. During the flutter phase, high-frequency, but minutely subatmospheric, air pressures allow the animal to convectively take in oxygen with minimal emission of carbon dioxide. The spiracles remain sufficiently closed so that internal oxygen levels remain low (Fig. 3). Carbon dioxide accumulates throughout the closed and fluttering phases, eventually triggering the spiracular open phase, when tracheal oxygen levels are restored to near-atmospheric concentrations.

Discontinuous gas exchange patterns and mechanisms are quite variable among and within species. Many insects exhibit discontinuous gas exchange only when at rest or during diapause. However, some highly active insects, such as running ants, also exchange gases discontinuously. When metabolic rates increase (at higher temperatures, or during activity), the time between spiracular phases tends to decrease. In contrast to the pattern diagrammed in Fig. 3, significant carbon dioxide is lost during the flutter phase of many insects, suggesting a predominance of diffusive gas exchange. Some species use abdominal pumping to enhance gas exchange during the open phase. The ecological and evolutionary significance of

this variation is unclear. Although many researchers have hypothesized that discontinuous gas exchange functions to reduce respiratory water loss, most recent tests of this hypothesis have failed to support it.

AQUATIC AND ENDOPARASITIC INSECTS

Insects can obtain oxygen while living in fluid environments and are common in fresh waters, in brackish estuaries, and as endoparasites. Aquatic and endoparasitic insects generally retain an internal, air-filled tracheal system. Maddrell has suggested that the lack of insects in deep-water environments occurs because the air-filled, buoyant tracheal system precludes insects from being able to dive deeply and avoid predators.

Many aquatic and endoparasitic species have anatomical features that allow them to feed under water (or within the host) while maintaining contact with air. For example, in many aquatic dipteran larvae such as mosquitoes, the posterior spiracles are surrounded by water-repellent hairs and are kept in the air while the animals feed in a head-down position. In water scorpions (Heteroptera: Nepidae), spiracles are located on the end of a long tube which is extended up to the water surface. Similarly, endoparasitic insects such as chalcid (Hymenoptera) larvae and tachinid (Diptera) larvae connect to the air using posterior spiracles inserted through the host's integument or tracheal system.

Some beetles (Coleoptera) and true bugs (Heteroptera) use their hairs or wings to carry air bubbles adjacent to their spiracles when they dive. These air bubbles serve as oxygen stores and as temporary gas exchange structures. Oxygen is removed from the air bubble by the diving insect, causing oxygen concentrations in the air bubble to fall below that in the surrounding water and oxygen to move from the water into the air bubble by diffusion. The removal of oxygen from the bubble also causes the nitrogen concentration in the air bubble to rise above that in the surrounding water, so nitrogen leaves the bubble by diffusion. Eventually, when all the nitrogen leaves, the bubble disappears and the insect must return to the surface.

Many aquatic and endoparasitic insects never access air and must obtain oxygen directly from water or the blood of the host. Some aquatic insects that rarely visit the surface obtain oxygen from the water by having specialized structures called plastrons which hold a thin film of air on the outside of their body. These insects have a thick (as many as 2 million hairs per square millimeter) layer of short water-repellent hairs that resists wetting and does not collapse under pressure. Spiracles open directly into the air space of the plastron. Plastrons are thought to behave as gas exchange structures in a manner similar to that of air bubbles. However, because the dense hairs make the plastron incompressible, oxygen removal lowers both the total pressure and the oxygen concentration in the air of the plastron. Thus nitrogen levels are likely to remain near those in water, and the air and gas exchange function of the plastron can persist indefinitely.

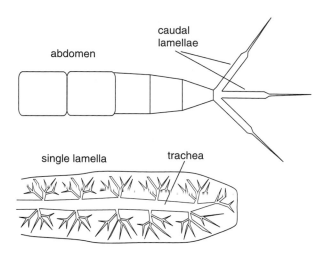

caudal
lamellae

abdomen

single lamella   trachea

FIGURE 4 Caudal lamellae, which function as tracheal gills. (Modified with the permission of Cambridge University Press, from Chapman, 1998).

The majority of aquatic or endoparasitic insects that do not access air lack spiracles, so that oxygen must be transported by diffusion from the water, across the cuticle, and then into the tracheae. A common feature of these insects is tracheal gills, leaf-like structures of thin cuticle containing many tracheae and tracheoles, which increase the available surface area for obtaining oxygen from water (Fig. 4). Gills may be on the abdominal tip (Diptera, Odonata: Zygoptera), laterally along the abdomen (Ephemeroptera, Trichoptera), or within the rectum (Diptera, Odonata: Anisoptera).

As oxygen is absorbed across a gill, oxygen level in the boundary layer of water over the gill can become low, slowing diffusive influx of oxygen. To overcome this problem, muscles drive rhythmic waving of abdominal gills, forcing fresh, oxygenated water to flow over the gills, thereby reducing boundary layer thickness. For similar reasons, rectal gills are ventilated by alternate contractions and relaxations of the rectum, which drive water in and out of the anus. Such ventilatory movements are enhanced in water containing low levels of oxygen.

Finally, some aquatic and endoparasitic insects have evolved the use of respiratory pigments to enhance oxygen storage and delivery. Some *Chironomus* larvae (Diptera) have hemoglobin in their hemolymph. Many of these species live in mud under stagnant water with low oxygen content, and the possession of hemoglobin with a very high affinity for oxygen allows these insects to obtain, store, and transfer oxygen at very low oxygen levels. Larval *Gasterophilus* (Diptera), an internal parasite in the hypoxic stomach lumen of horses, also possesses a hemoglobin with very high affinity for oxygen. In early instars, the hemoglobin of *Gasterophilus* is in the hemolymph, and so it may function in oxygen transport through the blood, whereas in later instars hemoglobin is concentrated in cells that are surrounded with tracheoles. It is thought that oxygen availability in the stomach varies with the presence of swallowed air bubbles and that possession of hemoglobin in cells allows *Gasterophilus* to store oxygen until the next air

bubble appears. The diving bug *Anisops* (Heteroptera) stores hemoglobin in hemoglobin cells surrounded with tracheae. *Anisops* obtains its oxygen from the air, carrying an air bubble under its wings when diving. The hemoglobin of *Anisops* has a lower affinity for oxygen than that of *Chironomus* or *Gasterophilus,* and it functions in a manner similar to that of hemoglobins in the blood of diving mammals, becoming oxygenated at the surface and releasing oxygen during the dive. Possession of hemoglobin allows *Anisops* to sustain a fivefold longer dive.

PLASTICITY IN TRACHEAL STRUCTURE AND FUNCTION

An insect's need for oxygen and production of carbon dioxide vary tremendously, as metabolic rate varies strongly with transitions from rest to flight, unfed to fed state, or diapause to growth. When metabolic rates increase, there is an increased need for oxygen uptake and carbon dioxide removal. Insects also require a higher capacity to exchange gases when they are exposed to low oxygen conditions such as hypoxic waters or high altitudes.

In the short term, increased gas exchange with a set tracheal system structure can be accomplished by at least four nonexclusive mechanisms. First, insects can simply tolerate lower internal oxygen and higher internal carbon dioxide concentrations. The increased gas concentration gradients will enhance gas exchange by either diffusion or convection. Second, insects can increase diffusive gas exchange by opening the spiracles to a greater degree or for longer time periods. Third, insects can increase convective ventilation through the spiracles and tracheal system (terrestrial insects) or over the gills (aquatic insects) by mechanisms such as abdominal pumping or gill waving. Fourth, insects can reduce fluid levels in the tracheoles, enhancing diffusive oxygen delivery within the tissues because of the faster diffusion of oxygen through air than through water. In the longer term, alterations in tracheal morphology (e.g., tracheal diameter, number of tracheoles) can strongly affect gas exchange capacity.

Locomotion

Locomotion requires muscular activity and increased ATP turnover, which increases the organism's need for oxygen. Running insects increase their gas exchange rates by 2- to 10-fold relative to resting conditions. Flight is usually associated with larger increases in gas exchange, especially in insects that maintain thoracic temperatures near 40°C, such as bees and dragonflies. These insect athletes have oxygen consumption rates among the highest known in the animal kingdom, 2 to 100 times resting values.

Relatively little is known of the mechanisms by which insects increase gas exchange during terrestrial locomotion. In a cockroach (Blattodea) and a grasshopper (Orthoptera) internal carbon dioxide levels rise while oxygen levels fall during running and hopping, respectively. In the two-striped grasshopper *Melanoplus bivittatus,* abdominal pumping rates are low during hopping, but do increase relative to resting rates afterward, increasing convection and restoring tracheal gases to normal, resting levels. Even though there is no evidence for increased abdominal pumping during hopping, increased convective ventilation may occur due to pressure fluctuations associated with cuticular deformations associated with jumping.

During flight, convective gas exchange can be increased by thoracic autoventilation and abdominal pumping. Thoracic autoventilation occurs when flight muscle contractions cause compression of the air sacs within the thorax, producing strong convective ventilation. Such increases in thoracic autoventilation have been shown for grasshoppers (Orthoptera); Cerambycidae, Elateridae, and Anthribidae (Coleoptera); moths (Lepidoptera); and dragonflies (Odonata). Another method for increasing gas exchange has been shown for the gigantic cerambycid beetle *(Petrognatha gigas),* in which wind pressure generated by forward flight drives convective air flow through the major tracheal trunks. Abdominal pumping is considered the major mechanism by which convective ventilation is increased during flight in large Hymenoptera, Diptera, and Coleaptera (Scarabaeidae and Buprestidae), and abdominal pumping supplements autoventilation during flight in dragonflies (Odonata) and grasshoppers (Orthoptera). At least in one insect studied (a moth), oxygen concentrations in active flight muscles are maintained at levels similar to those in resting animals, suggesting that increases in the capacity of the tracheal system to conduct gases match increased tissue needs for oxygen during flight.

During Hypoxia

Insects may encounter low environmental oxygen availability in a number of locations, including hypoxic waters, burrows, feeding sites within large dense structures such as granaries, or high altitudes. Terrestrial insects are generally quite good at coping with hypoxia and, usually, can maintain resting metabolic rates down to atmospheric oxygen levels of 1 to 5 kpa (normal oxygen concentration is 21 kpa). The large safety margin for oxygen delivery in resting terrestrial insects probably reflects the fact that tracheal systems must be designed to allow the much higher rates of gas exchange during activity. In support of this hypothesis, metabolic rates during flight are generally more sensitive to hypoxia, with flight metabolism being inhibited at 8 kpa in hovering honey bees and at 10 kpa in tethered flying flies and being stimulated by hyperoxia in a dragonfly. Growth rates are inhibited by relatively mild hypoxia (10 kpa), at least in larval mealworms (Coleoptera) and fruit flies (Diptera), suggesting that such moderate hypoxia limits some physiological processes even in nonlocomoting insects.

Insects respond to hypoxia by increasing tracheal gas exchange capacity. In a variety of insects, hypoxia induces spiracular opening. In ants and lepidopteran pupae exhibiting discontinuous ventilation, exposure to hypoxia increases the frequency and duration of spiracular opening. In adult grasshoppers, exposure to hypoxia causes a strong increase in convective ventilation, mostly accomplished by an increase in the frequency of abdominal pumping. Aquatic insects with gills generally increase convective water flow past the gills in response to hypoxia, by increasing the frequency of their gill beating (Ephemeroptera), body undulations (Plecoptera), or rectal pumping (Odonata). Aquatic insects exposed to hypoxia often move to faster flowing water, to the water surface, or even into the air.

Insects can exhibit changes in tracheal system structure in response to longer term exposure to hypoxia. For example, the transverse tracheae of mealworms (Coleoptera) increase in size when reared under conditions of low oxygen availability. Tracheole growth and branching are stimulated by hypoxia in the epidermis of *Rhodnius* (Heteroptera). Similarly, in some immature Ephemeroptera, gill area is inversely proportional to environmental oxygen levels.

CONTROL OF RESPIRATORY FUNCTION

The spiracular muscles are primarily controlled by nerves from the central nervous system. The motor neurons to the spiracular muscles arise from ganglia in the same segment or that immediately anterior. In dragonflies, *Periplaneta* cockroaches, and *Schistocerca* grasshoppers, the closer muscles are innervated by two motor nerves that branch from the median nerve, sending an axon to each side of the animal, so that both spiracles on a segment receive similar neural input. In the prothoracic spiracle of *Schistocerca,* two motor nerves innervate the opener muscle from the prothoracic ganglia, and one motor nerve arrives from the mesothoracic ganglia. Increasing action potential frequencies in these nerves stimulate increasing muscle activity, resulting in gradations in the magnitude and duration of spiracular closing.

Current information suggests that for spiracles with two muscles, effects of carbon dioxide or oxygen are mediated centrally. When high levels of carbon dioxide or low levels of oxygen are applied to the ventral ganglia or head, the frequency of action potentials to the closer muscles decreases, and the frequency of action potentials to the opener muscles increases, increasing spiracular opening. Temperature increases or flight also affect the frequency of action potentials to these spiracular muscles.

In contrast, data for spiracles with only a closer muscle suggest that carbon dioxide acts peripherally, while oxygen acts centrally. Central application of hypoxic gases stimulates a fall in the action potential frequencies to the closer muscle, whereas carbon dioxide has little effect. However, when carbon dioxide is applied directly on the spiracular muscle, there is a fall in the muscle membrane depolarization resulting from nerve stimulation, a decrease in closer muscle tension, and increased spiracular opening. Spiracle opening is stimulated by a rise in carbon dioxide but not a decrease in extracellular pH, suggesting that the carbon dioxide effect is not mediated by pH changes.

The rhythmic abdominal pumping movements that drive convective ventilation in many insects are initiated by central pattern generators in the metathoracic ganglia or the first abdominal ganglia. These rhythmic ventilatory bursts can be demonstrated in nerve cords isolated *in vitro*. The motor output from these central pattern generators passes down the ventral nerve cord via interneurons and stimulates rhythmical sequences of respiratory muscle activity.

In grasshoppers (Orthoptera) and probably many other insects, the rate of abdominal pumping can be altered by sensory output from stretch receptors in the abdomen, by chemosensors located in the thorax and head, and by feed-forward control. Increasing carbon dioxide or decreasing oxygen concentrations in the tracheae stimulate the frequency of abdominal pumping. Depression of carbon dioxide below normal levels or elevation of oxygen above normal levels inhibits abdominal pumping, indicating homeostatic regulation of internal gas levels. Feed-forward control of abdominal pumping occurs when neural centers that control flight muscles also stimulate the central pattern generators, increasing the rate of abdominal pumping.

In a series of classic experiments, V. B. Wigglesworth demonstrated that the terminal ends of the tracheoles can contain fluid that disappears in response to hypoxia or activity. Changes in the fluid levels in the tracheoles have the potential to strongly affect the ability of the tracheoles to conduct gases, which would provide a highly significant control mechanism for the tracheal system. Further experiments by Wigglesworth suggested that the withdrawal of fluid from the tracheoles could be due to elevations in hemolymph osmotic pressure.

In summary, the tracheal respiratory system of insects is a dynamic system, capable of a tremendous range of function and fine control. This light-weight, adaptable, high-capacity respiratory system is certainly one of the major traits that underlie the ecological and evolutionary success of insects.

See Also the Following Articles

Aquatic Habitats • Flight • Hemolymph • Muscle System • Tracheal System

Further Reading

Chapman, R. F. (1998). "The Insects: Structure and Function." Cambridge University Press, Cambridge, UK.
Harrison, J. F. (2001). Insect acid–base physiology. *Annu. Rev. Entomol.* **46**, 221–250.
Harrison, J. F., and Roberts, S. P. (2000). Flight respiration and energetics. *Annu. Rev. Physiol.* **62**, 179–206.
Hetz, S. (1994). "Untersuchungen zu Atmung, Kreislauf und Säure-Basen-Regulation an Puppen der Tropischen Schmetterlingsgattungen *Ornithoptera, Troides* und *Attacus.*" Friedrich-Alexander-Universität Erlangen–Nürnberg, Erlangen. [Ph.D. dissertation]
Hustert, R. (1975). Neuromuscular coordination and proprioceptive control of rhythmical abdominal ventilation in intact *Locusta migratoria migratorioides. J. Comp. Physiol.* **97**, 159–179.
Kestler, P. (1985). Respiration and respiratory water loss. *In* "Environmental Physiology and Biochemistry" (K. H. Hoffman, ed.). Springer-Verlag, Berlin.
Komai, Y. (1998). Augmented respiration in a flying insect. *J. Exp. Biol.* **201**, 2359–2366.
Lighton, J. R. B. (1996). Discontinuous gas exchange in insects. *Annu. Rev. Entomol.* **41**, 309–324.
Maddrell, S. H. P. (1998). Why are there no insects in the open sea? *J. Exp. Biol.* **201**, 2461–2464.
Manning, G., and Krasnow, M. A. (1993). Development of the *Drosophila* tracheal system. *In* "The Development of *Drosophila melanogaster*" (M. Bate and A. M. Arias, eds.). Cold Spring Harbor Laboratory Press, Cold Spring Harbor, NY.
Mill, P. J. (1985). Structure and physiology of the respiratory system. *In* "Comprehensive Insect Physiology, Biochemistry and Pharmacology," pp. 517–593. Pergamon Press, New York.
Slama, K. (1988). A new look at insect respiration. *Biol. Bull.* **175**, 289–300.
Wasserthal, L. T. (1996). Interaction of circulation and tracheal ventilation in holometabolous insects. *Adv. Insect Physiol.* **26**, 297–351.
Wigglesworth, V. D. (1983). The physiology of insect tracheoles. *Adv. Insect Physiol.* **17**, 85–149.

River Blindness

Vincent H. Resh
University of California, Berkeley

Onchocerciasis, or river blindness, is a nonfatal human disease that affects the skin and vision and can ultimately lead to blindness in infected persons. The name river blindness reflects both the place where the disease is most common and severe, and its ultimate outcome.

The disease is caused by a filarial (threadlike) worm in the phylum Nematoda, *Onchocerca volvulus*. The disease is found in 37 countries, 30 of which are in Africa, 6 in the Americas, and 1 in the Arabian Peninsula; however, Africa, and particularly sub-Saharan West Africa, is by far the most affected area in terms of clinical manifestations of the disease, number of affected persons, and widespread occurrence.

The worm is most commonly transmitted to humans in Africa by black flies in the *Simulium damnosum* species complex, which contains about 40 different forms, several of which are distinct species. Other species of *Simulium* are vectors in other parts of the world. Larval and pupal stages of these species occur attached to near-surface substrates in fast-flowing rivers. Because the adults are capable of long flights (>200 km), the aquatic stages are considered the most vulnerable for control.

In humans, the disease takes three progressive forms: first, skin lesions and intense, often violent, itching occur; second, painless nodules containing adult worms form where tissues are thin over bones, such as pelvis, ribs, and scalp, and these nodules are the sites of reproduction for the worms; and third, eye lesions form that can lead to blindness. The third phase can be devastating to communities; 10% of the people in vast areas can be blind, with levels of adults being blind in individual villages sometimes exceeding 30%. This disease is most prevalent among the rural villages that occur in some of the poorest areas of the poorest countries of the world. In these areas, subsistence (whether through agriculture or freshwater fishing) is difficult to achieve; with large portions of the adults being blind, the burden of the disease is devastating to these communities.

The life cycle involves both humans and black flies, and there are no known animal reservoirs of this disease. First, a biting, black fly female sucks blood from a human host; if that person is infected, she may also suck up some microfilaria (which are best thought of as worm embryos) of the parasite. Interestingly, evidence indicates that the worms migrate to the skin surface during the day, when black flies bite, which increases their chance of uptake by the biting females. Second, the microfilaria move from the fly's stomach to its thoracic muscles, where they pass through three larval stages (L_1, L_2, L_3), and the infective L_3 larva moves into the black fly vector's head and mouthparts. Third, in the course of biting, L_3 larvae are transferred from female black flies to humans. Fourth, the larval worm matures to an adult (which can be 40–45 cm in length) and adults mate (in the nodules that form), producing millions of microfilaria (each about 3 mm in length) that can be picked up from the skin of that human by a biting black fly. The death and disintegration of microfilaria result in an inflammatory reaction, which ultimately leads to itching, visual impairment, and eventually blindness.

The control of river blindness in West Africa has been one of the success stories of public health and economic development of the 20th century (Fig. 1). In 1974, the Onchocerciasis Control Programme in West Africa (OCP) began, with its objective being to eliminate onchocerciasis as a disease of public health importance and as an obstacle to socioeconomic development in this region. Eleven countries participated in this program, in a geographical region ranging from Senegal in the west to Benin in the east. Under the jurisdiction of the World Health Organization, and funded through the World Bank and 20 donor countries and organizations, the strategy was to interupt transmission of the blinding strain of the worm *O. volvulus* for a period of about 12 years, which is the life span of the adult worm. This was done by aerial application of seven different selective insecticides, weekly, to rivers infested with black flies. Because the female black fly lays batches of eggs at or near the water surface in fast-flowing water, these were the sites sprayed with insecticides. Much of the current understanding of the environmental impact of pesticides on aquatic ecosystems comes from techniques developed as part of this control effort. Beginning in the late 1980s, an ivermectin drug (a microfilaricide originally used as a veterinary product for the treatment of dog heartworm) was also distributed annually to people living in infected areas.

The success of the OCP is especially impressive when one considers the scope of the project: Onchocerciasis was eradicated in most parts of 11 different countries; at times, over 50,000 km of rivers had to be sprayed weekly for more than 10 years; no permanent environmental damage resulted from the spraying activities because of the increased reliance on insecticides that selectively targeted the black fly vectors;

FIGURE 1 Blind adults being led through villages were a common sight in West Africa before the control of river blindness. Photograph courtesy of The World Health Organization (WHO/TDR/W. Imber).

the drugs were distributed to almost 7 million people by over 22,000 community volunteers; millions of children have been born that will be free of this disease; the drugs were donated to the program without charge by the producer of them, Merck; the drug distribution program has recently been expanded to cover 19 more countries (and 157 million people); and plans include using this same system for dispensing drugs against lymphatic filariasis (elephantiasis) and pesticide-impregnated bed nets against malaria. The OCP ended in 2002; perhaps its most significant accomplishment is not only that the disease is under control, but also that the previously abandoned valleys, now free of river blindness, grow food for 17 million people in areas inhabited by the poorest of the poor, the people truly living at the "end of the road."

See Also the Following Article
Medical Entomology

Further Reading
Crosskey, R. W. (1990). "The Natural History of Blackflies." Wiley, Chichester, UK.

Crossky, R. W. (1993). Blackflies *(Simuliidae)*. *In* "Medical Insects and Arachnids" (R. P. Lane and R. W. Crossky, eds.). Chapman & Hall, London.

Davies, J. B. (1994). Sixty years of Ochocerciasis vector control: A chronological summary with comments on eradication, reinvasion, and insecticide resistance. *Annu. Rev. Entomol.* **39**, 23–45.

Hougard, J. M., Yaméogo, L., Sekétéli, A., Boaten, B., and Dadzie, K. Y. (1997). Twenty-two years of blackfly control in the Ochocerciasis Control Programme in West Africa. *Parasitol. Today* **13**, 425–431.

Rostrum

R. F. Chapman
University of Arizona

The term rostrum is derived from a Latin word meaning a beak or snout and it is somewhat loosely used in entomology to refer to development of the head or mouthparts, which have some resemblance to a beak.

In Heteroptera, the mouthparts are modified for piercing and sucking. The mandibles and maxillae are slender stylets that pierce the host tissues. They are supported by the labium (rostrum) and slide in a groove that runs along its anterior face, being concealed within it when the insect is not feeding. The rostrum has between one and five segments. It is three-segmented in Reduviidae and four-segmented in Coreidae and Pentatomidae. In psyllids, aleyrodids, aphids, and coccids it appears to arise from between the anterior coxae, giving rise to the name Sternorrhyncha for the suborder. The rostrum does not enter host tissue as the stylets penetrate and commonly telescopes within the lower part of the head or becomes folded beneath the head, with its tip still surrounding the stylets (Fig. 1). Most reduviids have a ridged file between the forelegs against which they can rub the tip of the rostrum to produce an irregular pattern of sound that probably functions to deter potential predators but may also serve as an alarm signal.

In various other insect groups, the anterior or ventral part of the head capsule is elongate to form a rostrum. Among the weevils (Curculionidae), the frons and vertex are lengthened dorsally, and the biting and chewing mouthparts are at the extreme tip of the forwardly extending rostrum. In many species, the female has a longer rostrum than the male and in some the terminal mouthparts permit the insect to bore a hole in which eggs are placed. A rostrum is also characteristic of most Mecoptera (scorpion flies), but here it points downward rather than forward and the elongate basal sclerites of the mouthparts contribute to it. The rostrum of Cyclorrhaphous Diptera is also a ventral extension of the head that bears the maxillary palps anteriorly and the rest of the mouthparts ventrally.

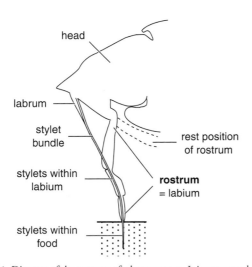

FIGURE 1 Diagram of the rostrum of a heteropteran. It is segmented and as the stylets enter the host tissues it folds beneath the head, still holding the stylets at its tip. When the insect is not feeding, the rostrum is held parallel with the body in the position shown by the dashed line. (Reproduced, with permission, from H. Weber, 1930, "Biologie der Hemipteren." Springer-Verlag, Berlin.)

Among termites, the rostrum of soldier termites in the subfamily Nasutitermitinae is different from these others. It is formed as a pointed projection of the frons and is independent of the mouthparts. Unlike the soldiers of most other termites, nasute termites have poorly developed mandibles but carry out their defensive function by ejecting a sticky spray from the opening at the tip of the rostrum.

See Also the Following Articles
Blood Sucking • Feeding Behavior • Mouthparts

Further Reading
Chapman, R. F. (1998) "The Insects: Structure and Function," 4th ed. Cambridge University Press, Cambridge, UK.

Royal Jelly

Eva Crane
International Bee Research Association

Royal jelly is secreted by the hypopharyngeal gland located below the pharynx in the head of young worker honey bees (brood food glands). The term royal jelly was probably first used by François Huber in Switzerland in 1792, as *gelée royale*. Young *Apis mellifera* worker bees secrete royal jelly into a queen cell, an extra large brood cell in which a female larva is fed on special food and develops into a queen (a sexually reproductive female). The workers then seal the larva inside, and it develops into a pupa and then an adult queen. A worker or drone larva receives food similar to royal jelly only for the first 3 days, and thereafter modified less rich food, which includes pollen and honey. A queen larva consumes about 25% more food than a worker larva, and its weight increases by 1300 times in 6 days.

COMPOSITION OF *A. MELLIFERA* ROYAL JELLY

Royal jelly, like beeswax, is a secretion of worker bees and as such has a more constant composition than honey or pollen, which are derived from plants. Of the components of royal jelly, the most interesting are perhaps 10-hydroxy-2-decenoic acid and vitamins in the B complex, which include thiamine, riboflavin, niacin, pyridoxine, pantothenic acid, biotin, and folic acid. Royal jelly contains very little vitamin C, and vitamins A and E are absent or nearly so; D and K are probably also absent.

PRODUCTION OF *A. MELLIFERA* ROYAL JELLY BY HUMANS

Well before the 20th century, the dramatic effect of royal jelly on female larvae that developed into queens created much interest. In the 1950s royal jelly was the first of the newer hive products to be exploited. One factor that stimulated beekeepers to move toward royal jelly production was that it could be accomplished in areas where plant sources of nectar and pollen were inadequate for profitable honey production. Provided colonies were fed sufficient sugar syrup and pollen, they could be managed so that they produced royal jelly for the beekeeper. However, to produce it, skilled operators must carry out a sequence of labor-intensive procedures tied to a rigorous timetable.

Each hive used may be organized as follows. The queen of each colony is confined in a lower brood box by a queen excluder placed over it. Above the excluder is a box of honey combs, and above that the box in which the royal jelly will be produced. This contains framed combs of honey and pollen and also unsealed brood to attract nurse bees. Some of the frames of comb are replaced by frames containing three cross-bars, each supporting on its underside about 15 "queen cups," synthetic shallow wax cups with an opening on the bottom, similar to the cells that bees build when they start to rear queens. An operator transfers a worker larva 18–24 h of age from a worker comb into each queen cup, using a "grafting tool" that is rather like an insect pin bent into the required shape and mounted on a handle. There are a hundred or more of the queen cups in the box.

On 3 successive days a frame of "grafted" cells is placed in the top brood box. On day 4 the cells in the first frame contain the maximum amount of royal jelly (e.g., 235 mg each), so the operator removes the frame and extracts the royal jelly from the cups in it by aspiration. Fresh larvae are grafted into them, and the frame is reinserted in the hive, thus restarting the cycle.

The extracted royal jelly is strained to remove any wax and refrigerated as soon as possible; it can be kept for up to a year at 2°C.

PROPERTIES AND USES

Royal jelly shows wide-spectrum activity against bacteria (although none against fungi). It has been reported in at least one study to have bactericidal activity on *Bacillus metiens; Escherichia coli; Mycobacterium tuberculosis,* which causes tuberculosis; *Proteus vulgaris; Staphylococcus aureus;* and a *Streptococcus* species.

Royal jelly achieved, and has maintained, a place on the world market as a specialized dietary supplement and in cosmetics; by 1990 the annual production was probably between 500 and 600 tonnes. Some scientists have documented a reported feeling of general well-being in humans after consumption of royal jelly, thus justifying its use in anorexia, emaciation, or loss of muscular strength; it has also been recommended for older people. Many of the reported effects of royal jelly on humans and other mammals may, however, be produced much less expensively by other substances.

A queen honey bee differs from a worker in her length of life (several years instead of several weeks or months) and in her reproductive ability. Some people therefore believe that royal jelly can also prolong the life of humans, as well as improving their vigor and sex life. However, these properties have not been substantiated.

WORLD TRADE

Royal jelly production was already proving profitable in France in 1953, and 1.5 tonnes a year was being produced by 1958. Production was established in Cuba before 1957, and in the early 1960s 12 other countries were reported to produce it, including France and Japan (1.5 tonnes each), Canada, Israel, Taiwan, and Korea. The price at that time was from U.S. $220 to $500 per kilogram. By 1984 China and Taiwan were producing 400 and 234 tonnes a year, respectively, and the world price had dropped to U.S. $70 per kilogram. Eastern Asia is still the main center of the world's royal jelly production, and Japan is both a large producer and a large importer and consumer.

See Also the Following Articles
Apis Species • Commercial Products from Insects

Further Reading
Crane, E. (1990). "Bees and Beekeeping: Science, Practice and World Resources," Chap. 14. Heinemann Newnes, Oxford.
Mizrahi, A., and Lensky, Y. (1997). "Bee Products: Properties, Applications, and Apitherapy." Plenum, New York.
Schmidt, J. O., and Buchmann, S. L. (1992). Other products of the hive. *In* "The Hive and the Honey Bee" (J. Graham, ed.), Chap. 22. Dadant, Hamilton, IL.

Salivary Glands

Gregory P. Walker
University of California, Riverside

Salivary glands are glands associated with the mouth or oral cavity and produce secretions (saliva) that are mixed with the food during feeding and are ingested along with the food. Among insects, there are four pairs of glands associated with the mouth or oral cavity, although all four generally are not present in the same insect. Each of these is associated with and is named after its associated mouthpart: the mandibular, maxillary, hypopharyngeal, and labial glands. The presence or absence of each of these glands varies among insect species, and among holometabolous insects, a given gland may be present in only one life stage. For example, in Lepidoptera, mandibular glands occur in the larval stage, but not in the adult. The four glands serve a variety of functions, and the same gland may serve different functions in different species or even in different life stages of the same species. Within a given species, one or more of these four pairs of glands usually function as salivary glands.

GLANDS ASSOCIATED WITH THE MOUTH

Mandibular glands, found in many insects, function as the main salivary glands in Lepidoptera larvae. A common function of mandibular glands in social Hymenoptera is to produce pheromones such as alarm pheromones in ants and honey bees, and the queen substance in queen honey bees. Function can vary with age. For example, the mandibular glands of older honey bee workers (which perform mostly foraging tasks) produce an alarm pheromone, whereas the mandibular glands of young workers (which perform mostly nursing tasks) produce secretions, called royal jelly, that are fed to larvae in differential quantities to control whether a given larva will develop into a queen or worker. Mandibular glands of stingless bees serve in defense and produce a burning sensation when ejected onto the victim.

Maxillary glands occur in Protura, Collembola, some Heteroptera, and some larval Neuroptera and Hymenoptera. They are believed to provide secretions to lubricate the mouthparts, and thus serve one of the salivary functions.

Hypopharyngeal glands occur in the Hymenoptera, and in honey bee workers they produce secretions that are fed to the larvae (a substance different from royal jelly). Hypopharyngeal glands are vestigial in honey bee queens and absent in males. The hypopharyngeal glands of honey bee workers also produce invertase, a common salivary enzyme that hydrolyzes sucrose, and another enzyme that oxidizes glucose to an acid, which is believed to serve as a preservative in honey.

Labial glands occur in the great majority of insect orders (an important exception is the Coleoptera), and most commonly function as salivary glands, but in the larval stage of some groups of silk-producing insects, such as Lepidoptera, Trichoptera, and Hymenoptera, labial glands are the silk-producing organs either throughout the larval stage or at its end, just before pupation. In the Psocoptera, adults have two pairs of labial glands, one pair functioning as silk glands and the other pair functioning as salivary glands. In the primitive orders Collembola and Thysanura, which lack Malpighian tubules (the usual excretory organs of insects), the labial glands function as excretory organs.

STRUCTURE AND FUNCTION OF SALIVARY GLANDS

The salivary glands of most insects are labial glands, which are the focus of this section. Labial salivary glands have been examined in detail in relatively few insect species, and there is great variation among the species examined (Fig. 1). This is not surprising, considering the great variation in mode of feeding (e.g., chewing, piercing-sucking, non-piercing-sucking, sponging, etc.) and types of food consumed by different insect species.

General Description

Several aspects of structure and function are common to most or all variations of insect labial salivary glands. The glands occur in pairs, and the ducts from each gland usually join to form a single common duct that opens to the oral cavity at a single orifice (Fig. 1). Even though the glands originate in the labial segment, the orifice usually occurs just behind or on the hypopharynx, and the glands often extend back into the thorax and even as far back as the abdomen (e.g., Fig. 1A,B).

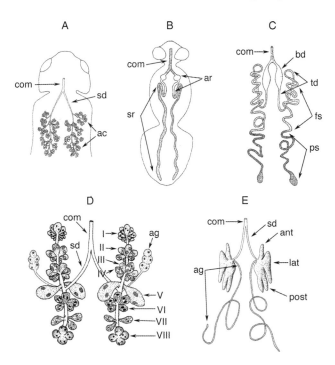

FIGURE 1 Salivary glands of representative insects. (A) The locust, *Locusta*. [After Chapman R. F. (1998). "The Insects: Structure and Function." 4th ed., Fig. 2.16. With permission of Cambridge University Press.] (B) Adult blowfly, *Calliphora*. [After Berridge, M. J., and Prince, W. T. (1971). The electrical response of isolated salivary glands during stimulation wtih 5-hydroxytryptamine and cyclic AMP. *Philos. Trans. R. Soc. Lond. B* **262**, 111–120. Courtesy of the Royal Society of London.] (C) The tobacco horn-worm, moth *M. sexta*. [After Leslie. R. A., and Robertson, H. A. (1973). The structure of the salivary gland of the moth *(Manduca sexta). Z. Zellforsch. Mikrosk. Anat.* **146**, 553–564. Copyright Springer-Verlag GmbH & Co. KG. Used with permission.] (D) The beet leafhopper, *C. tenellus,* showing cell types I through VIII. [Modified after Wayadande, A. C., Baker, G. R., and Fletcher, J. (1997). Comparative ultrastructure of the salivary glands of two phyopathogen vectors, the beet leafhopper, *Circulifer tenellus* (Baker) and the corn leafhopper, *Dalbulus maidis* DeLong and Wolcott (Homoptera: Cicadellidae). *Int. J. Insect Morphol. Embryol.* **26**, 113–120.] (E) The large milkweed bug, *O. fasciatus*. [Modified after Miles, P. W. (1967). The physiological division of labour in the salivary glands of *Oncopeltus fasciatus* (Dall.) (Heteroptera; Lygaeidae). *Aust. J. Biol. Sci.* **20**, 785–797.] Abbreviations: ac, acini; ag, accessory gland; ant, anterior lobe; ar, absorptive region; bd, bulbous duct region; com, common salivary duct; fs, fluid secretion region; lat, lateral lobe; post, posterior lobe; ps, protein secretion region; sd, salivary duct; sr, secretory region; td, thin duct region.

The glands are suspended in the hemocoel and are constantly bathed in hemolymph. The glands generally have at least two regions: a secretory region and a reabsorptive region. Generally, the lumen of the salivary duct is lined with cuticle, at least at the end closest to its opening.

The secretory region produces the primary saliva. The major component of saliva is water. Water is transported from the hemolymph across cells of the salivary gland and into the lumen of the gland. Movement of water from the blood to the gland lumen is accomplished by active transport of potassium or sodium ions from the hemolymph to the lumen, causing water to move from the hemolymph to the lumen down an osmotic gradient. Cells responsible for water transport generally have deep infoldings of the cell membrane and/or dense microvilli on the side of the cell adjacent to the lumen of the gland. This serves to greatly increase the cell's luminal surface area, and also serves to enclose very narrow extracellular spaces into which ions are pumped. The enclosed nature of the spaces helps contain the ions to keep their concentration high, thus facilitating the osmotic movement of water from the cell into the space. The infoldings and microvilli usually are associated with abundant mitochondria to provide the energy for the ion pumps. The secretory region of the gland also synthesizes proteins, such as salivary enzymes, and other organic components of the saliva. Cells responsible for secretion of these components generally possess extensive endoplasmic reticulum, Golgi bodies, and secretory granules that synthesize and transport (intracellularly) the secretions. There may be one or several different types of cell in the secretory region. It should be noted that salivary components are not necessarily produced by the salivary glands themselves but may be produced elsewhere in the body and transported to the salivary glands via the hemolymph.

The reabsorptive region of the salivary gland reabsorbs potassium or sodium ions from the saliva and transports them back into the hemolymph. As a result, potassium and sodium ions are conserved, and the saliva is usually hypotonic to the hemolymph. Reabsorptive cells often have infoldings, especially on their basal side (hemolymph side), to increase surface area. These infoldings, however, tend not to be tightly enclosed (unlike the lumen side of water-secreting cells in the secretory region), to facilitate movement of secreted ions into the hemolymph and away from the cells, thus reducing the osmotic gradient, which would cause the cells to lose water. Reabsorptive cells also have abundant mitochondria to power the active transport of ions from saliva to hemolymph.

Acinuous Salivary Glands

The salivary glands of many insects are composed of clusters of acini, or saclike glandular structures (Fig. 1A) comprising at least two types of cell. Each acinus empties into a cuticle-lined duct, and the ducts of different acini fuse, eventually forming a common duct that leads to an opening just behind or on the hypopharynx. The anatomy of acinuous salivary glands varies greatly among different insects. The acinuous glands of locusts and cockroaches have been particularly well studied and are described next.

LOCUSTS AND COCKROACHES The gross anatomy of locust and cockroach salivary glands is illustrated in Fig. 1A. The acini are the secretory region of the salivary gland and consist of peripheral cells and central cells. The peripheral cells have deep, microvilli-lined invaginations that are contiguous with extracellular canaliculi that open into the gland's duct. The peripheral cells transport water from the hemolymph to the canaliculi, and from there, the water empties into the duct. Additionally in locusts, the peripheral cells seem to synthesize and secrete other salivary components. The central cells of both locusts and cockroaches synthesize salivary enzymes and other salivary components, and secrete them into the salivary duct. The walls of the salivary ducts of locusts and cockroaches are one cell thick, and the lumen is lined with a thin cuticle. The salivary duct contains the reabsorptive region of the gland, and in some cockroaches, a small part of the duct adjacent to the acinus is secretory.

Tubular Salivary Glands

Some groups of insects, such as Lepidoptera, Diptera, and Siphonaptera, possess tubular salivary glands. Generally, the walls of tubular salivary glands are one cell thick and the lumen is lined with thin cuticle. Tubular salivary glands are divided into several regions along the length of the gland. The number of regions and anatomical details of the regions vary among insect groups. Two examples of tubular salivary glands are described.

BLOWFLY ADULTS There are a pair of tubular salivary glands, each with three regions (Fig. 1B). The apical region is the longest and is the secretory region; the shorter middle region is the reabsorptive region; and the short common duct at the proximal end opens to the exterior of the body on the hypopharynx.

The blowfly's secretory region has a long distal section located mostly in the abdomen and a shorter proximal section located in the thorax. Cells in the distal section serve a dual function: they move water from the hemolymph into the lumen of the gland, and they synthesize and secrete salivary enzymes and other salivary components. The cell surface adjacent to the lumen of the gland's central duct encloses extensive canaliculi into which the cells secrete their products. Secretions then move from the canaliculi to the gland's duct. Cells in the proximal section are generally similar in appearance to those in the distal section, but they do not contain secretory granules. Thus, they do not seem to secrete enzymes and probably secrete only water and ions. The reabsorptive region of the blowfly's salivary glands consists of a single cell type that is believed to be responsible for reabsorbing ions from the saliva back to the hemolymph. Finally, the common duct of the blowfly's salivary glands comprises highly flattened cells that seem to play no role in the secretion or reabsorption of any salivary components. This region of the salivary gland is very short and opens on the hypopharynx.

SPHINGID MOTH In the tobacco hornworm, *Manduca sexta*, adults have a pair of tubular salivary glands that are divided into four regions (Fig. 1C). In the apical region, proteinaceous material is synthesized in the extensive rough endoplasmic reticulum and Golgi bodies and is stored in large vacuoles before eventual release into the lumen of the gland. The second region transports water from the hemolymph to the lumen. Cells in this region have the characteristic structure of water-secreting cells, to accommodate what is believed to be the primary function of this region; but in addition, the presence of rough endoplasmic reticulum and Golgi bodies suggests that these cells may secrete more than just water. The third and fourth regions, called the thin duct and bulbous duct, both seem to have a reabsorptive function, moving ions back from the saliva and into the hemolymph. Cells in these two regions differ in the structure of their surface adjacent to the lumen, but the reason for the difference is unknown. After the fourth region, the right and left glands fuse forming the common duct. Cells of the common duct are unspecialized and probably play no role in saliva production.

Salivary Glands of Hemiptera

The most complex insect salivary glands that have been studied occur in the Hemiptera. This complexity is undoubtedly related to the piercing-sucking mode of feeding in this taxon, where saliva is injected into the food substrate via a specialized salivary canal in the elongate maxillary stylets. In this mode of feeding, solid substrates must be pierced and then the food, which is often initially solid, must be liquefied before ingestion through the maxillary food canal. These processes depend greatly on a multitude of salivary components that serve different functions. Consequently, there are usually many distinct types of secretory cells in the same gland, each producing different salivary components. Furthermore, many phytophagous hemipterans produce two distinct types of saliva at different times in the feeding process: sheath saliva and watery saliva. Sheath saliva consists mostly of lipoprotein and is secreted incrementally as the stylets advance through the plant tissue. It gels shortly after secretion and forms a continuous solid sheath around the stylets. As a consequence, only the stylet tips come in direct contact with the plant tissue; the shaft of the stylet bundle is encased by the sheath. Watery saliva, as the name implies, is dilute and does not gel. It contains mostly water and various enzymes.

In general, salivary glands of the Hemiptera are divided into two main parts, the principal gland and the accessory gland. The principal gland is often subdivided into two or more lobes. The principal and accessory glands are served by their respective ducts, and the ducts from each fuse to form a lateral salivary duct. The lateral ducts from each side fuse to form a median duct that leads to the salivary pump (described later in this article). There is considerable variation in the salivary glands among the Hemiptera, even within the same family. Two examples are described.

BEET LEAFHOPPER   The principal salivary glands of the beet leafhopper, *Circulifer tenellus,* are divided into an anterior and posterior lobe, each served by its own duct (Fig. 1D). The ducts from each lobe fuse to form the lateral salivary duct. The accessory gland is not subdivided, and its duct joins the others near the point where they fuse to form the lateral duct. The principal glands have eight different cell types that are arranged in rings around the duct of each lobe, five in the anterior lobe and three in the posterior lobe. All eight cell types are secretory, possessing abundant endoplasmic reticulum and/or secretory granules, and all have intracellular canaliculi that come in contact with the salivary duct. Details of the cells' fine structure and the staining properties of the secretory granules differ sufficiently to indicate that each cell type produces different components of the saliva. Accessory gland cells have some features typical of water-secreting cells but also have abundant endoplasmic reticulum, Golgi bodies, and secretory granules, suggesting that they secrete water and other salivary components.

MILKWEED BUG   As in most hemipterans in the suborder Prosorrhyncha, the principal salivary glands of the large milkweed bug, *Oncopeltus fasciatus,* (Fig. 1E) are divided into discrete lobes that are much more compact than the lobes just described for the beet leafhopper. The lobes comprise a mass of cells with a distinct glandular lumen. The accessory glands and the three lobes of the principal glands each secrete their own salivary components. The anterior and lateral lobes secrete two different components that mix to form the salivary stylet sheath (described earlier). The lateral lobe secretes the bulk of the sheath saliva protein, including most of the components that form the hydrogen bonds that solidify the salivary sheath, while the anterior lobe secretes the components that form most of the disulfide bonds. The posterior lobe produces salivary digestive enzymes, such as amylase and esterase. The accessory gland supplies the bulk of the water in the saliva, as well as polyphenoloxidase and possibly mucoid substances.

CONTROL OF SECRETION, INNERVATION

Secretion of saliva generally is stimulated either by direct innervation or by neurohormonal factors that are released into the blood by secretory neurons. Innervation of the salivary glands varies considerably among different insects. Innervation can come from the subesophageal ganglion, thoracic ganglia, the stomatogastric nervous system, the median–transverse nervous system, or a combination of these.

Two neurotransmitters, serotonin and dopamine, are commonly found in neurons innervating the salivary glands, and each may stimulate different aspects of salivation. For example, in the American cockroach, serotonin induces secretion of proteinaceous saliva, whereas dopamine induces secretion of nonproteinaceous saliva. Other neurotransmitters occur in neurons innervating salivary glands, but their roles are not understood. The diversity of neurotransmitters suggests that control of salivation is a complex process.

The salivary glands of some insects such as blow flies lack direct innervation, and salivation is induced by one or more neurohormonal factors released into the blood. One of these factors seems to be serotonin.

STRUCTURES ASSOCIATED WITH SALIVARY GLANDS

Salivary Reservoirs

Some insects, such as cockroaches, have a pair of distensible salivary reservoirs for storage of saliva. In cockroaches, each reservoir has its own duct, which joins with the duct of its associated salivary gland. The combined gland/reservoir ducts from each side fuse to form the common salivary duct leading to an opening on the hypopharynx. A valve near the orifice of the common duct on the hypopharynx opens and closes as the hypopharynx is raised and lowered during feeding, thus controlling release of saliva. When the insect is not feeding, the hypopharynx is in a lowered position, closing the valve, and saliva produced by the salivary glands then backs up into the reservoirs where it is stored.

Salivary Pumps

Many insects with piercing-sucking mouthparts inject saliva into their food for various purposes. Often a pumping mechanism is used to inject the saliva through elongate hypodermic needle-like mouthparts. In many Hemiptera, the pump is a hollow chamber near the hypopharynx and is referred to as a salivary pump or salivary syringe.

FUNCTIONS OF SALIVA

General

Perhaps the most fundamental and ubiquitous function of saliva in insects is lubrication of the mouthparts and lubrication of the food bolus to assist its transport through the foregut. Lubrication can be achieved primarily by water, the most abundant constituent in saliva. Water in the saliva also can dissolve components in the food, such as sugars, which then become detectable by chemoreceptors on the mouthparts. Thus, saliva also can aid in food recognition.

The most common class of organic constituents of saliva consists of digestive enzymes, such as amylase, invertase, various proteases, and lipases. In many insects with chewing mouthparts, salivary enzymes are mixed with the food during chewing and swallowing, and initiate digestion. However, the midgut usually is the main site of production and secretion of digestive enzymes, and salivary enzymes in these insects generally play only a secondary role in digestion. In other insects, salivary digestive enzymes provide the main digestive function. This is especially common in piercing-sucking insects, such as many Heteroptera, in which digestive enzymes are injected into the food. The enzymes then break down and liquefy the food, and the digested, liquefied food is sucked up through the mouthparts.

Predators

Predaceous insects that have piercing-sucking mouthparts often capture and eat prey that are as big or even bigger than themselves. Large prey size is not nearly as common in predators with chewing mouthparts. This is because many predators with piercing-sucking mouthparts use the mouthparts to inject a salivary toxin into the prey, which enables them to subdue prey without having to be large and strong enough to physically overpower them. Thus, in these insects, saliva assists prey capture and gives the insects a potentially larger range of prey than their chewing mouthpart counterparts. Venoms in predaceous piercing-sucking insects often are accompanied by salivary hyaluronidase, which breaks down hyaluronic acid, an important "intercellular cement" in insects. Hyaluronidase is believed to serve as a spreading agent for toxins (as well as for digestive enzymes), assisting their penetration between cells by breaking down the intercellular cement.

Blood Feeders

Salivary components of blood-feeding insects serve several functions. These have been best studied in mosquitoes, other biting flies, and kissing bugs. Blood feeding is the most dangerous time in the lives of these insects, and survival is greatly enhanced by the ability to complete the task quickly and escape before being detected by the host. Salivary enzymes help to shorten the time it takes to acquire a blood meal. Blood vessels occupy only a small volume of skin tissue, and thus locating blood with the mouthparts can take considerable time. During probing, many blood feeders damage capillaries or tiny blood vessels by random movement of the stylets, and small subcutaneous pools of blood (hematomas) form in the vicinity of the damaged vessels. This increases the volume of the blood in the skin, and thus increases the probability that the stylets will locate a blood source, reducing the time required to locate blood. Successful formation of hematomas is greatly facilitated by factors that inhibit blood clotting, and consequently, the most common salivary components in these insects are factors that inhibit clotting. Clotting comprises two general processes, platelet aggregation and coagulation. In the small vessels used by most blood-feeding insects, platelet

aggregation is the more important of the two, being capable of plugging a damaged vessel in a matter of seconds. One of the compounds that initiates platelet aggregation is ADP, and the saliva of many different blood-feeding insects contains apyrase, which is an enzyme that breaks down ADP. The importance of apyrase is demonstrated in mosquitoes, where the time required to complete a blood meal is directly dependent on the amount of apyrase injected by the mosquito. In addition to salivary components that interfere with platelet aggregation, some blood-feeding insects have salivary components such as antithrombins that inhibit coagulation. Also, substances that inhibit vasoconstriction occur in some blood-feeding insects, thus inhibiting the host's attempt to restrict the flow of blood to the feeding site.

Anticoagulants also serve another function: they prevent blood from coagulating in the food canal. Coagulation would clog the food canal and lead to starvation and death. Finally, antihistamines in the saliva of some blood feeders may act as anti-inflammatory agents and reduce the probability that the feeding insect will be detected by the host.

Herbivorous Hemipterans

Many or most phytophagous hemipterans produce two kinds of saliva, sheath saliva and watery saliva, which were described briefly earlier. Sheath saliva forms a continuous solid sheath around the stylets, and several functions for it have been proposed. One likely function is to reduce friction between the stylets and plant tissue, facilitating advancement and withdrawal of the stylets. Another function may be to shield the moving stylets from plant cells, and thus avoid triggering a defensive response by the plant that could include hypersensitive reactions or release of plant defensive chemicals. This may be especially important in sternorrhynchan hemipterans like aphids and whiteflies, which feed primarily on sap from phloem sieve elements that lie deep in plant tissue. These insects carefully weave their stylets between and around plant cells from the plant surface to the sieve elements, and successful extraction of sap from the sieve elements may be dependent on avoiding the triggering of plant defensive responses during penetration to the sieve elements. In addition to mechanically shielding the stylets from plant cells to avoid plant defenses, salivary sheaths contain the enzyme polyphenoloxidase, which has been proposed to serve the function of oxidizing plant defensive chemicals and converting them to more harmless forms.

Watery saliva contains assorted enzymes and other components that vary among species. Some, like proteases and amylases, serve a digestive function, breaking down insoluble plant constituents into soluble forms that can be ingested through the stylet food canal. Others, like pectinases, break down pectin, which is the "intercellular cement" in plants, and loosen the adhesion between adjacent cells. For hemipterans like aphids and whiteflies, whose stylets penetrate between cells until they reach their actual ingestion site (phloem sieve elements), pectinase facilitates the penetration of stylets between cells by loosening the intercellular cement. After penetrating a phloem sieve element, aphids and whiteflies inject saliva into it, an action that is believed to make the sieve element more suitable for ingestion and to prevent the sieve element from clogging.

The water component of watery saliva also can have a critical role other than simply serving as a carrier for enzymes. Many phytophagous hemipterans (especially phytophagous Heteroptera) feed by a method known as "lacerate and flush." In this feeding method, the stylets are inserted into the plant, and a pocket of cells beneath the surface is liquefied by the combined action of digestive enzymes in the saliva and mechanical laceration by repeated thrusts of the stylets. Once the pocket of cells has been liquefied, the nutrient-rich liquid is flushed out of the pocket by copious secretion of watery saliva and sucked up through the stylet food canal.

Gall Formers

Plant galls are produced by several groups of phytophagous insects that occur especially, but not exclusively, in the dipteran family Cecidomyiidae and the hymenopteran family Cynipidae, as well as many species of sternorrhynchan Hemiptera and some Thysanoptera. The abnormal tissue and cell growth characteristic of plant galls is caused by secretions from the insect, usually salivary secretions, that mimic plant growth hormones or serve as molecular signals that redirect plant cell growth from its normal course to an abnormal form that serves the needs of the gall maker.

Construction of Shelters and Webs

Silk is produced by many insects for a variety of functions such as construction of pupal cocoons in many Lepidoptera, Hymenoptera, and Siphonaptera, and construction of larval retreats or food-gathering nets in most Trichoptera larvae and some chironomid larvae. Many Psocoptera use silk to construct sheetlike shelters under which they aggregate and also use silk to attach their eggs to the substrate. Weaver ants use silk to tie together leaves to construct their arboreal nests, but interestingly, only larvae produce silk; so to weave leaves together with silk, the workers hold the larvae in their jaws and use them as silk dispensers. Not all insects that produce silk do so with their labial glands, but all the examples just cited do, representing some of the diverse uses of these specialized salivary secretions.

Mucoid secretions are produced by the salivary glands of several groups of Diptera. Some Diptera, such as *Drosophila,* produce salivary mucopolysaccharides that glue the puparium to the substrate. In some sciarids and fungus gnats (Mycetophilidae), mucoid secretions are used as a "slime trail" to facilitate larval locomotion, much like terrestrial snails. Larvae of some predaceous fungus gnats use these sticky mucoid salivary secretions to capture prey. The most fascinating of these are the New Zealand glow worms, which construct a silken retreat from which they dangle silk threads that are covered with sticky mucoproteins to trap prey. The larvae reside in their silken retreats, and at night, they produce light by bioluminescence to attract nocturnal flying insects to their traplines.

Trophallaxis

Trophallaxis is the exchange of food between two individuals. The food exchanged may be salivary secretions or regurgitated gut contents. Larvae of many ants and wasps are dependent on adults to feed them. In exchange for being given food by the adult, the larvae of many species secrete a salivary fluid that is greedily consumed by the adult that provided the food. This stimulus for adults to give up food to a larva may have played a role in the evolution of eusociality in the Hymenoptera, in which adult females readily feed larvae that are not even their own offspring.

Others

Larval warble flies bore their way through the subcutaneous tissues of their mammalian hosts. To facilitate movement through the host's tissues, they secrete a salivary collagenase, which breaks down collagen, a main component of connective tissue.

The saliva of some moths contains an enzyme called cocoonase that weakens the silk cocoon. It is produced by the newly eclosed moth to assist its escape from the pupal cocoon.

See Also the Following Articles

Blood Sucking • Digestion • Feeding Behavior • Gallmaking • Mouthparts • Silk Production

Further Reading

Ali, D. W. (1997). The aminergic and peptidergic innervation of insect salivary glands. *J. Exp. Biol.* **200,** 1941–1949.

Berridge, M. J., and Prince, W. T. (1971). The electrical response of isolated salivary glands during stimulation with 5-hydroxytryptamine and cyclic AMP. *Philos. Trans. R. Soc. Lond. B* **262,** 111–120.

Chapman, R. F. (1998). "The Insects. Structure and Function." 4th ed., pp. 30–34. Cambridge University Press, Cambridge, U.K.

House, C. R., and Ginsborg, B. L. (1985). Salivary gland. *In* "Comprehensive Insect Physiology, Biochemistry, and Pharmacology," Vol. 11, "Pharmacology" (G. A. Kerkut and L. I. Gilbert, eds.), pp. 195–224. Pergamon Press, Oxford, U.K.

Kendall, M. D. (1969). The fine structure of the salivary glands of the desert locust, *Schistocerca gregaria* Forskål. *Z. Zellforsch. Mikrosk. Anat.* **98,** 399–420.

Lehane, M. J. (1991). "Biology of Blood-Sucking Insects." HarperCollins/ Academic Press, London.

Leslie, R. A., and Robertson, H. A. (1973). The structure of the salivary gland of the moth *(Manduca sexta). Z. Zellforsch. Mikrosk. Anat.* **146,** 553–564.

Miles, P. W. (1967). The physiological division of labour in the salivary glands of *Oncopeltus fasciatus* (Dall.) (Heteroptera: Lygaeidae). *Aust. J. Biol. Sci.* **20,** 785–797.

Miles, P. W. (1972). The saliva of Hemiptera. *Adv. Insect Physiol.* **9,** 183–255.

Miles, P. W. (1999). Aphid saliva. *Biol. Rev.* **74,** 41–85.

Oschman, J. L., and Berridge, M. J. (1970). Structural and functional aspects of salivary fluid secretion in *Calliphora*. *Tissue Cell* **2**, 281–310.

Ribeiro, J. M. C. (1987). Role of saliva in blood-feeding by arthropods. *Annu. Rev. Entomol.* **32**, 463–478.

Sutherland, D. J., and Chillseyzn, J. M. (1968). Function and operation of the cockroach salivary reservoir. *J. Insect Physiol.* **14**, 21–31.

Wayadande, A. C., Baker, G. R., and Fletcher, J. (1997). Comparative ultrastructure of the salivary glands of two phytopathogen vectors, the beet leafhopper, *Circulifer tenellus* (Baker), and the corn leafhopper, *Dalbulus maidis* DeLong and Wolcott (Homoptera: Cicadellidae). *Int. J. Insect Morphol. Embryol.* **26**, 113–120.

Scale Insect

see *Sternorrhyncha*

Scales and Setae

Shaun L. Winterton

North Carolina State University

Setae are multicellular protuberances on the arthropod cuticle used primarily for mechanoreception. In all groups of arthropods and especially insects, the role of the setae has evolved from simple mechanoreception to various other functions, including defense, locomotion, prey capture, pheromone dispersal, sexual display, preening, and camouflage. Setae are often highly modified, with one common modification being that they may be flattened into a broad, plate-like scale. This article specifically examines the tremendous diversity in shape, structure, and function of setae and scales used by different insect groups and at different life stages. This diversity in setae type is then characterized according to their four major functions in the biology of the insect: mechanoreception, camouflage, defense, and pheromone dispersal.

STRUCTURAL MORPHOLOGY

The terminology applied to scales and setae has historically been confused, with numerous, often interchangeable, terms used by researchers to describe the various types of structures (e.g., hairs, bristles, trichiae, aculei, chetae). Each seta, or trichoid sensillium, is a multicellular protuberance with specifically differentiated cells, with the most diagnostic landmark being its socket. The term trichoid sensillum applies to chemoreceptors as well as mechanoreceptors. The term hair is incorrect when applied to insects because hairs are morphologically different from setae and are considered a characteristic of mammals and not arthropods. Trichobothria is a term

occasionally used to describe large, nontapering seta often associated with the mouthparts and genitalia. Other types of cuticular protuberances not considered here include the various minute projections of the cuticle (e.g., pruinescence, velutum, pollen), collectively termed microtrichia, and large sclerotized extensions of the cuticle (e.g., spines, tubercles).

Most setae form the covering of the body surface, which in very high densities is called a pile. Usually a mechanoreceptor, a typical seta, is composed of four cells: (1) A sensory cell, which innervates the seta, is surrounded by a (2) thecogen cell, which acts as an auxiliary cell by secreting the dendrite sheath. During seta development the (3) trichogen cell secretes the tapered or scale-like protuberance, whereas the (4) tormogen cell is responsible for formation of the socket. While most setae develop as mechanoreceptors, some will become secondarily noninnervated and lose their sensory function. These setae may take on a different role (such as aerodynamics, sexual display, preening) in the functioning of the insect.

Scales are modified setae that have a flattened blade with longitudinal ridges, sometimes with serrate edges (Figs. 1A and 1B). Scales are usually inclined relative to the cuticle, overlapping each other when present in sufficient densities. Each scale has a narrow pedicel, and the enlarged blade may be either gradually attenuated to an apex or truncated.

Scales are found in most groups of insects and serve a variety of functions. In true flies (Diptera), scales are not common but in certain species can be found on the wings (e.g., mosquito, *Anopheles annulipes*—Culicidae), legs (e.g., *Trichopoda* spp.—Tachinidae), or body (e.g., *Metatrichia* spp.—Scenopinidae). Scales are also found on the elytra of various beetle (Coleoptera) families (e.g., Elateridae, Curculeonidae, Buprestidae), but the function of such scales in Coleoptera and in Diptera is unknown. Scales on the body and wings of Lepidoptera have been studied in greater detail and are known to function in cryptic coloration, thermal regulation (e.g., butterfly, *Colias* spp.), and aposematism (warning coloration). In butterflies, wing scales contribute to lift during flight but not to drag, thus enabling them to glide for longer periods.

Lepidopteran wing scales have a specialized structure and are divided into two types. Primitive-type scales, found in nonditrysian Lepidoptera, are solid with longitudinal ridges (Fig. 1C). Normal-type scales, found in the ditrysian Lepidoptera, are composed of superior and inferior lamellae, with an internal lumen subdivided by internal supports called trabeculae (Fig. 1D). Whereas the inferior lamella is smooth, the superior lamella is usually covered with longitudinal ridges, interconnecting transverse ridges (flutes), and/or perforations (windows). Some groups of nonditrysian Lepidoptera have a mixture of the two types, with normal-type scales layered over the primitive-type scales.

Wing color in Lepidoptera is produced by wing scales. Individual scales are usually a single color which may be generated by any one or a combination of pterins (red, yellow, white), melanins (black), flavonoids (white), carotenoids (blue, yellow), papilliochromes (cream-yellow), and ommachromes

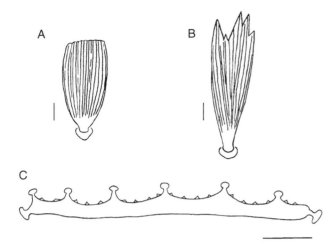

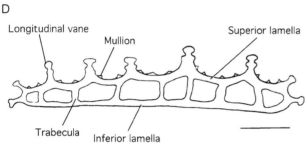

FIGURE 1 Lepidopteran scales. (A) Truncated wing scale, (B) serrate body scale, (C) cross section of primitive-type wing scale, (D) cross section of normal-type wing scale. Scale line length, 0.1 mm.

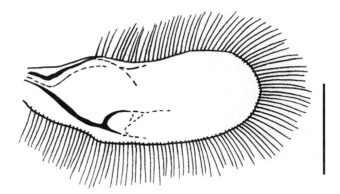

FIGURE 2 Right forewing of *Acrotrichis* sp. (Ptilidae: Coleoptera). Scale line, 0.5 mm.

(red). Scales also produce color by diffraction; regular longitudinal ridge spacing produces various diffraction colors, whereas irregular spacing produces overlap in color spectra and is perceived as white light. Interference colors, from which light is reflected from a series of superimposed surfaces separated by distances equivalent to light wavelengths, result in selective phasing in and out of various colors. This type of color is important for producing bright color hues in adult butterflies (e.g., *Morpho* spp.) and is produced from a series of longitudinal vanes along the scale blade. Each vane is supported by a series of thickenings, or mullions, which act in concert with other vanes to form a series of reflecting surfaces and produce interference colors.

FUNCTIONAL TYPES

The functions of setae are sensory (e.g., touch, taste, and smell), but in many insects the original sensory function of some setae has been discontinued though cell death that creates a noninnervated seta. These setae often have more specialized roles, sometimes with great morphological modification. There is incredible diversity in the shape, structure, and function of both the seta and its morphological derivative, the scale. A detailed examination of almost any insect will reveal a myriad of seta or scale types, often with vastly differing roles in the biological functioning of the organism. The following classification is clearly artificial and categorizes the diversity of setae and scales into functional types for ease of discussion.

Mechanical

LOCOMOTION Elongate setae are used by numerous insect groups for flight and swimming. Elongation of marginal setae is an efficient mechanism to increase the surface area of locomotory structures (e.g., wing, leg), with only a slight increase in weight or developmental investment (Fig. 2). Such elongation of setae along wing margins is common only in very small sized insects, as the degree of elongation is finite and insufficient to generate adequate lift for larger insects. Groups of insects that use this mechanism for flight include all Thysanoptera and various microhymenopterans (e.g., *Mymar*—Mymaridae), microlepidoptera (e.g., *Macarostola*—Gracillaridae), and Coleoptera (e.g., *Acrotrichis*—Ptilidae) (Fig. 2). Aquatic insects that are active swimmers often have paddle-shaped leg segments fringed with elongated setae for added propulsion through water. Examples include various families of Heteroptera (e.g., Naucoridae, Corixidae, and Notonectidae) and Coleoptera (e.g., Hygrobiidae, Haliplidae, Dytiscidae, and Gyrinidae).

PLASTRON To extend their duration under water, some aquatic insects use a plastron, or air bubble trapped by fine hydrofuge setae. The setae are used to hold a large air bubble by surface tension. The plastron enables them to live permanently submerged in water because oxygen passes by diffusion from the water into the plastron. Examples of plastron use by this method include the beetle families Elmidae and Hydrophilidae.

FOOD GATHERING AND PREY CAPTURE In predatory insects, enlarged setae are an economical substitute for teeth or spines for holding prey. Stiff enlarged setae are present along the inner margin of mandibles of some larval Nymphidae and Myrmeleontidae (Neuroptera). Similar setae on the inner margin of fore femora of adult Leptopodidae (Heteroptera) are used for the same purpose. Aquatic insects may use elongated rows of setae around the mouth parts as filters to trap food particles in flowing water (e.g., *Coloburiscoides* sp.—Coloburiscidae: Ephemeroptera) or use setae-fringed foretarsi to sieve for prey in detrital ooze (e.g., *Agraptocorixa*—Corixidae: Hemiptera). Bees (Hymenoptera) commonly have enlarged hind tibiae and basitarsi covered with brush-like setae (scopa), which are used to carry pollen.

WING COUPLING   Some insect groups use specialized setae to couple the fore- and hind wings together, so ensuring synchronous wing beats during flight. The basic pattern, which occurs in Choristidae (Mecoptera), is composed of retinacular setae along the jugal margin of the forewing interlocking with frenular setae along the basicostal margin of the hind wing. More advanced forms are found in Lepidoptera, in which the frenulum may be a large single seta or multiple slender setae. In Trichoptera, a row of large setae is present along the costal margin or the subcostal vein of the hind wing, which engages either the jugal lobe or a ventral ridge in the anal field.

DIGGING   Fossorial insects often have stout setae arranged in rows on the leg segments for digging. A comb of long setae (pecten) is commonly present on the foretarsi of ground-nesting aculeate wasps (e.g., *Bembix*—Sphecidae). Enlarged and thickened setae-like structures borne on acanthophorites are used for digging during oviposition by females of various asiloid Diptera (e.g., Therevidae, Apioceridae).

Camouflage Aids

Highly modified setae are commonly used by insect larvae that camouflage themselves by carrying soil, feces, and/or trash particles on their body. Some of the most elaborate of these may be found on the bodies of myrmeleontoid Neuroptera (i.e., Ascalaphidae, Nemopteridae, Psychopsidae, Nymphidae, and Myrmeleontidae). Other than entangling camouflage materials, the biological function of these highly modified setae, called dolichasters, is unknown. Dolichasters may be simple, scale shaped or highly ornate, star or cup shaped or recurved hooks (Fig. 3). Ascalaphid and nymphid lacewing larvae may also have abdominal extensions called scoli (Fig. 4), on which rows of enlarged scales or setae are used to entangle camouflaging materials.

Defensive

Lepidopteran larvae of several families (e.g., Arctiidae, Notodontidae, Thaumetopoeidae) use specialized urticating (irritating) setae as a defense against potential predators. Urticating setae are modified setae with a poison cell associated with the trichogen cell; the former discharges venom when the tip of the seta is broken off. The urticating setae may be very long and scattered over the body surface or short and positioned on the apex of a scoli. In Limacodidae, the stinging

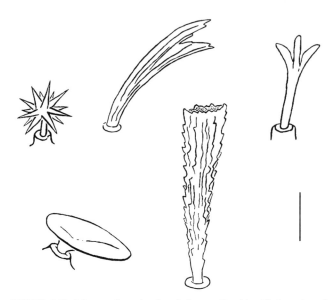

FIGURE 3   Dolichasters from head and thorax of unidentified species of Ascalaphidae (Neuroptera). Scale line, 0.2 mm.

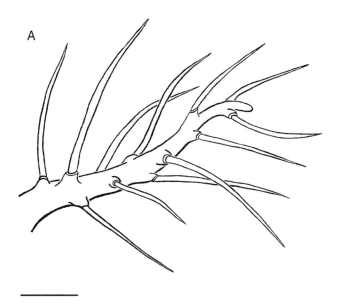

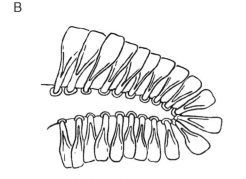

FIGURE 4   Lateral scoli on abdomen of neuropteran larvae. (A) *Osmylops* sp. (Nymphidae), (B) unidentified ascalaphid. Scale line, 1.0 mm.

setae are short and positioned at the apices of often brightly colored, eversible tufts that are extended when the individual is disturbed. Some adults and/or larvae of the coleopteran families Cleridae, Trogossitidae, and Dermestidae are often clothed in long fine setae, which may be recurved. In addition to a sensory function, these setae may be used defensively to obstruct or entangle potential predators or parasitoids.

Secretory Substrate

Setae are commonly used by insects as a substrate of high surface area from which a pheromone is dispersed by evaporation. Adult males of some nemopterid (Neuroptera) genera use distinctive tufts of fine setae (bulla) on the hind margins of the fore- or hind wings to disperse pheromone. Males of the antlion tribe Acanthoclisinae (Myrmeleontidae) have eversible sacs (hair pencils) covered with fine setae for a similar purpose, and many male butterflies use specialized setae (androconia) located in basal depressions along the forewing veins called androconial organs. During courtship pheromone is dispersed from the hairs, often with elaborate "calling" behavior, which can consist of hovering and diving at the prospective female and fanning pheromone in her direction.

See Also the Following Articles

Antennae • Chemoreception • Coloration • Flight • Mechanoreception

Further Reading

Chapman, R. F. (1998). "The Insects: Structure and Function," 4th ed. Cambridge University Press, Cambridge, UK.
Crouau, Y. (1997). Comparison of crustacean and insect mechanoreceptive setae. *Int. J. Insect Morphol. Embryol.* **26,** 181–190.
CSIRO (ed.) (1991). "The Insects of Australia: A Textbook for Students and Research Workers." Melbourne University Press, Cartlon, Australia.
Ferris, G. F. (1934). Setae. *Can. Entomol.* **66,** 145–150.
Keil, T. A., and Steinbrecht, R. A. (1984). Mechanosensitive and olfactory sensilla of insects. *In* "Insect Ultrastructure" (R. C. King and H. Akai, eds), Vol. 2. Plenum, New York.
Richards, A. G. (1951). "The Integument of Arthropods." University of Minnesota Press, Minneapolis.
Richards, A. G., and Richards, P. A. (1979). The cuticular protuberances of insects. *Int. J. Insect Morphol. Embryol.* **8,** 143–158.
Snodgrass, R. E. (1935). "Principles of Insect Morphology." McGraw–Hill, New York.

Scorpions

Stanley C. Williams
San Francisco State University

Scorpions are among the most recognizable groups of arthropods. Their highly segmented body plan is uniquely subdivided into a leg-bearing prosoma, a broad seven-segmented preabdomen (mesosoma), and a narrow, five-segmented postabdomen (metasoma) that terminates in a bulbous stinging organ, the telson. The fossil record indicates scorpions were the first arthropods to occupy the terrestrial environment. A prominent member of the class Arachnida, scorpions survived periods of mass extinctions and today occupy a prominent position in arthropod communities. This gives them the distinction of being the oldest surviving group of terrestrial arthropods.

EVOLUTION

Scorpions date back to the Silurian, some 400 mya. The oldest taxa, *Palaeophonus* spp., are strikingly similar in structure to modern forms in their extensive, characteristic segmentation, appendages, and general body form. Their body was organized into four regions: a prosoma (cephalothorax); a broad mesosoma ("preabdomen" containing digestive and reproductive organs); a taillike metasoma (slender "postabdomen" composed of five ringlike segments); and a terminal segment, the telson. *Palaeophonus* differed from modern scorpions in not having pretarsal claws on walking legs and in having a blunt ending of the telson. The absorptive membranes of the book lungs might have been eversible, forming a gill-like respiratory structure. This suggests that *Palaeophonus* might have existed in aquatic and land habitats.

The fossil record indicates that scorpions underwent a rapid adaptive radiation that led to structural diversity, especially in the segmentation of the ventral prosoma, the development of pretarsal claws, and the formation of a sharp venom-delivering structure on the telson.

MORPHOLOGICAL CHARACTERISTICS

Scorpions are readily distinguished by their unique morphology (Fig. 1). Their prosoma consists of six body segments that are covered dorsally by an unsegmented carapace. Each of these segments bears a pair of characteristic segmented appendages. The first pair, the chelicerae, are small, chelate or scissorlike, and serve as the mouthparts. The second appendages are large, conspicuous pedipalps that terminate in a strong scissorlike chela used to capture and immobilize prey. The next four pairs of appendages are the walking legs, each of which terminates in a pair of distinctive pretarsal claws, the ungues. Along with a single pair of median ocelli, there are two groups of lateral ocelli located at the anterolateral corners of the carapace. Each group of lateral ocelli consists of zero to four facets, evidently derived from a primitive compound eye. The mesosoma is attached broadly to the prosoma, lacks appendages other than the sensory pair of comblike pectines, and houses the digestive and reproductive organs. Expansion of intersegmental membranes permits meal engorgement during feeding and allows for increase in body volume to accommodate internal development of their embryos. The metasoma is composed of five narrow segments forming a freely articulating, taillike

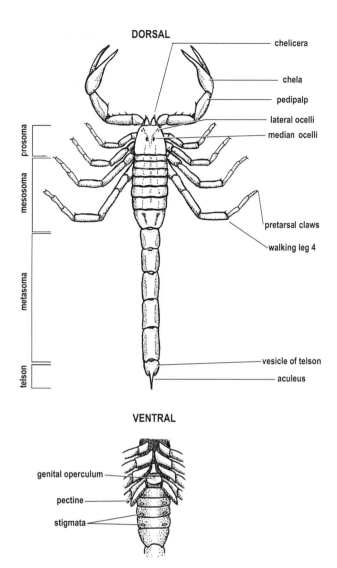

FIGURE 2 A giant hairy scorpion *(Hadrurus concolorous)* in alert posture. Notice the pedipalps positioned forward with their chelae open. The metasoma is in flexed position, ready to strike. The body is balanced over and supported by the walking legs.

FIGURE 1 Dorsal and partial ventral views of scorpion showing the distinctive body segmentation, tagmosis, and morphological features. (Drawing by K. A. Justus.)

body region. A single unsegmented telson freely articulates with the metasoma and is composed of a swollen vesicle that terminates in a sharply pointed aculeus or "sting." Paired venom glands are housed in the vesicle, and the associated aculeus injects the venom into the victim (Fig. 2). Scorpions all over the world have the same basic body structure, which is surprising considering the antiquity of the extant families.

ANATOMICAL ADAPTATIONS

Scorpions demonstrate basically the same organ systems characteristic of most arachnids. These systems do show adaptations that support their successful lifestyle as opportunistic ambush predators. The digestive system begins with a tiny mouth located below the base of the chelicerae. Its small diameter allows intake of food only in fluid form. Preoral enzymes are secreted by salivary glands directly into the body of the prey, and the liquefied, preorally digested tissues are ingested through a sucking action. This is facilitated by a pressing action of the gnathobases at the inner bases of the pedipalps.

Large, expansive midgut ceca facilitate rapid intake and storage of nutrients. Because particulate matter is not ingested with a meal, fecal wastes are minimal. Nitrogenous wastes are excreted by means of Malpighian tubules that deposit wastes in the form of crystalline guanine directly into the digestive tube for subsequent elimination. Some species have also been found to excrete nitrogenous wastes in the form of uric acid and xanthine. Scorpions have no tracheal system but instead use four pairs of book lungs to respire. Air passes through four pairs of tiny stigmata, which are opened and closed by membranous opercula, in the mesosomal sterna. Inspired air then enters the lamellar spaces of the membranous book lungs. This limiting mode of oxygen uptake is correlated with the sedentary lifestyle and metabolic simplicity characteristic of most scorpions. A ventral nerve cord with highly metameric ganglia characterizes the nervous system. At the head end is found the subesophageal ganglion and supraesophageal brain. The circulatory system consists of a pulsing muscular tube, called the dorsal vessel, and colorless blood called hemolymph.

The reproductive system has organs arranged in a somewhat segmented, fishnetlike form within the mesosoma. The male has a paraxial organ that contains a pair of hemispermatophores that fuse to form a sperm-carrying spermatophore used for indirect insemination of a female during the courtship dance. The female system has an inner lumen that exits the body through the ventrally located genital aperture, which is covered by a valvelike genital operculum on the mesosoma. Multiple oocytes in different stages of maturation may be seen on the surface of the reproductive tube. Fertilization is internal, and the female carries the developing embryos until birthing as larvae.

SYSTEMATICS

Substantial numbers of undescribed species are still being found. This largely results from the nocturnal and secretive habits of scorpions, which make them difficult to find. Approximately 1286 species of living scorpions are currently recognized and are assigned to 17 families and 157 genera (Table I). The higher classification of scorpions has been the subject of periodic reevaluation over the past two decades. The trend has been to recognize more families and genera and to base the resulting classification on phylogenetic relationships as suggested by cladistic analyses. The numbers of extant scorpion families and species are surprisingly low considering their success as a group, their broad distribution, their diverse habitats, and their antiquity.

DISTRIBUTION AND BIODIVERSITY

Scorpions are conspicuously common in tropical and subtropical regions throughout the world. They also range into the more temperate regions of both the Northern and Southern Hemispheres. In North America they range as far north as British Columbia, Alberta, and Saskatchewan, Canada (approximately 52° N). In South America, scorpions range south into Chile and Argentina (approximately 50° S).

Scorpions show wide-ranging adaptations to different elevations. In Baja California, Mexico, scorpions are found in the intertidal and beach habitats. In the White Mountains of California they are well established in elevations up to 2130 m. Scorpion species show definite habitat preferences and are normally found in patchy distributions. Although they are often thought of as tropical, scorpions seem to reach their highest biotic diversity in the arid regions of more temperate latitudes. Most scorpion habitats are characterized by a modest diversity ranging from one to five sympatric species. The greatest regional diversity of scorpions is reported from Baja California, where 61 species and 11 genera are found. The higher latitudes and higher elevations have more limited diversity, often with only a single species represented in a habitat.

BIOLOGY

General Life Cycle Strategy

Scorpions often are not noticed, even when abundant. They are mainly nocturnal, secretive animals and usually remain inactive, hidden in their retreats, except when feeding, mating, or disturbed. For many scorpions, such as *Hadrurus* and *Vaejovis*, adults are the stage usually encountered; juveniles tend to minimize their exposure in the environment. This pattern is most conspicuous in scorpions found in more temperate areas and at higher elevations. In other species, such as *Centruroides* spp., and some *Paruroctonus* spp. and *Smeringerus* spp., however, juvenile instars are commonly encountered.

Life Cycle

The scorpion life cycle is a simple one: the first instar larva, five nymphal instars, and the adult. The larva differs morphologically from other instars in lacking the typical unguicular claws at the tips of the walking legs, lack of a sharp aculeus or "sting," lack of dark pigmentation and sclerotization of the exoskeleton, and lack of effective locomotor ability. Behaviorally, the larva is unique in that it remains on the mother's back and does not travel alone or feed. The sedentary larva molts to the second instar nymph on the mother's back at which time it

TABLE I   Higher Classification of Contemporary Scorpions

Family	Number of genera	Number of species	Common genera
Bothriuridae	12	90	*Bothriurus, Brachistostermis, Centromachetes, Cercophonius, Lisposoma, Orobothriurus, Urophonius*
Buthidae	72	528	*Ananteris, Androctonus, Babycurus, Buthacus, Buthus, Centruroides, Compsobuthus, Hottentotta, Isometrus, Leiurus, Lychas, Mesobuthus, Orthochirus, Parabuthus, Rhopalurus, Tityus, Uroplectes*
Chactidae	11	132	*Broteochactas, Brotheas, Chactas, Nullibrotheas, Teuthraustes*
Chaerilidae	1	21	*Chaerilus*
Diplocentridae	8	76	*Didymocentrus, Diplocentrus, Heteronebo, Nebo, Oiclus*
Euscorpiidae	4	14	*Euscorpius, Megacormus, Plesiochactas, Troglocormus*
Hemiscorpiidae	2	7	*Habibiella, Hemiscorpius*
Heteroscorpionidae	1	2	*Heteroscorpion*
Ischnuridae	8	56	*Cheloctonus, Hadogenes, Iomachus, Liocheles, Opisthacanthus*
Iuridae	6	21	*Anuroctonus, Carboctonus, Hadruroides, Hadrurus, Iurus*
Microcharmidae	2	6	*Akentrobuthus, Microcharmus*
Scorpionidae	7	130	*Heterometrus, Opistophthalmus, Pandinus, Scorpio*
Scorpiopidae	6	27	*Alloscorpiops, Euscorpiops, Neoscorpiops, Scorpiops*
Superstitioniidae	4	9	*Alacran, Sotanochactas, Superstitionia, Typhlochactas*
Troglotayosicidae	2	2	*Belisarius, Troglotayosicus*
Urodacidae	1	19	*Urodacus*
Vaejovidae	10	146	*Paruroctonus, Pseudouroctonus, Serradigitus, Uroctonus, Vaejovis*

shows increase in linear dimensions, has pretarsal claws and a sharp aculeus, and becomes increasingly physically active. At this time the second instar normally leaves the mother's back, disperses from the mother's shelter, locates a new shelter, and begins its own independent life. After periods of feeding and growth, nymphs periodically molt to their successive instars, until adulthood is reached. Adults differ from the nymphs in being sexually mature, but they often show little morphological difference except in their larger size and sometimes in the proportion of body parts such as the larger pedipalp chela and elongation of the metasoma. Adults may also show distinctive sexual dimorphism in body proportions.

Feeding

Ambush predation is the main means of prey capture. For example, some burrowing species such as *Anuroctonus phaiodactylus* quietly wait inside their burrow entrance and ambush prey that enters or passes by their burrow. Many species of scorpions, when hungry, leave their protective shelter during nocturnal hours and take a stationary position in their environment, which may be on the substrate surface or in vegetation. Here they will remain motionless until an unsuspecting prey ventures close, at which time it is grasped with the strong pedipalp chela and stung, if necessary. Other scorpions such as *Centruroides* spp. often range great distances from their diurnal shelters in search for prey. At such times they may actively stalk their prey. If a scorpion is unsuccessful it will return to its shelter and resume its predation behavior the next night, and the night after, until successful.

After successful prey capture, a scorpion draws the immobilized prey close to its oral cavity and exudes preoral digestive fluids that digest the prey's tissues internally. During ingestion the chelicerae also shred prey tissues. The resultant fluid is sucked into the oral region, and any remaining fluids are pressed from the prey remains by the gnathobases of the inner pedipalp coxae. Up to 97% of the body mass of a prey may be ingested. A satiated scorpion then returns to its shelter and will normally not be seen until hungry again, which may be from 2 weeks to a month or two. Its effective mode of predation, habit of engorgement feeding, and low metabolic rate result in a lifestyle that requires minimal prey, minimal exposure of the scorpion, and survival when prey are not readily available. Scorpions in the laboratory may go for as long as 12 months between meals. However, in nature, one feeding every 2 to 6 weeks is probably more common. Almost any animal that a scorpion can catch and immobilize is a potential prey, so scorpions feed on a variety of insects and spiders. Predation on centipedes, other scorpions, lizards, and small snakes has also been observed.

Reproduction

The scorpions' reproductive mode is one of the most ancient ones used by terrestrial arthropods. It involves sexual reproduction, with insemination of the female by the male following a ritualistic courtship dance. Internal fertilization is complicated by the lack of an intromittent organ in the male. Fertilization takes place indirectly by a spermatophore deposited by the male.

Pheromones bring potential mates together and initiate courtship. When two potential mates encounter each other, they grasp each other by their pedipalpal chelae and undergo a forward and backward "dancelike" behavior, the *"promenade à deux,"* which may be over in few minutes or last for hours. At the end of this dance, the male emits a sclerotized spermatophore onto the substrate and draws the female over it, at which time she takes up the sperm packet in her gonopore. The two then part and the female normally returns to her shelter. Depending on the taxa, gestation takes from approximately 2 or 3 months to nearly a year. Development is internal and results in the birth of a dozen to over a hundred larvae, which quickly ascend to the mother's back and remain there for the remainder of their first stadium. Parturition takes from an hour to about 3 days, depending on the species and number of offspring. Some species seem to have several birth cycles each year, whereas others seem to have only one per year. Parthenogenesis has been reported in *Tityus serrulatus* and *Liocheles australasiae*. Field observations of *Liocheles* suggest that some natural populations might have parthenogenesis as their primary means of reproduction.

Habitats

Scorpions have been roughly categorized as "ground" versus "bark" dwellers. Ground dwellers construct burrows in the ground, seek shelter in rocky substrate, invade the burrows of other organisms, or occupy protective spaces under or within ground surface debris. Bark dwellers are commonly found sheltered under bark, in bromeliads, in vegetation, in residential thatch, or in other plant material. They are often climbers and may primarily occupy forest canopies. During some seasons, forest-dwelling forms may migrate to the ground to seek shelter under rocks or surface debris. This group has members that invade human habitations and can be seen climbing walls and ceilings at night. They may reside in thatching and in the mortar spaces between bricks and rocks of walls. Most ground scorpions are solitary, although some bark scorpions such as *Centruroides* may be found in aggregations during certain times of the year. *Anuroctonus phaiodactylus* constructs a burrow as a young nymph and occupies it throughout its life, which may last for several years. It seldom ventures far from its burrow and is intolerant of other scorpions in its burrow. Many of the ground-dwelling species probably spend most of their life in their permanent burrow. At time of courtship, however, mature males abandon their burrows or other territory and become more nomadic in search of mates. Scorpions are thigmotactic, seeking closely fitting shelters. The availability of suitable shelters is believed to limit the population size of some species.

Scorpions are found in a variety of habitats from desert to mesic, and they thrive in highly xeric environments. They are common in tropical and subtropical regions, and many species extend into the temperate latitudes. Most scorpions prefer well-drained habitats, but some such as *Centruroides* species thrive in riparian and tropical rain forest environments, and seem to survive drowning by climbing trees during times of flooding. Burrowers vary greatly in their choice of habitats and behavior: *A. phaiodactylus* requires soil with good drainage and adequate compaction to support permanent burrows; *Vejovoidus longiunguis* constructs burrows in unconsolidated soil of sand dunes; species of *Serradigitus* do not burrow but live in crevices, talus slopes, and rock fractures; *Superstitionia donensis* may occupy a burrow or may be found in a simple cell constructed under a rock, animal dung, or vegetation; *Centruroides thorellii* resides in bromeliads within tropical forests; *Vaejovis littoralis* is found in the intertidal wrack zone, where it continually migrates ahead of the changing tides; *Centruroides exilicauda* seeks shelter under bark of trees, but may also be found under surface debris or in rock crevices, and is a common invader of human habitations; and *Uroctonus mordax* is commonly found in wet habitats, where it may reside under mosses and rotting logs, and in soil crevices. Almost any habitat that provides adequate shelter and prey is suitable for some kind of scorpion. As a result, these arthropods are globally widely distributed.

Predators and Parasites

Although scorpions do not have many natural enemies, they are an attractive prey for a few predators. Their rich nutrient content and numerical abundance contribute to their vulnerability as prey. Predators that hunt during the nocturnal hours are particularly effective in feeding on scorpions. Most capture scorpions while they are exposed on the substrate surface, but some, such as coyotes, will dig them up. Some 124 vertebrate and 26 invertebrate predators have been reported to feed on scorpions. Elf owls *(Micrathene)*, burrowing owls *(Speotyto)*, barn owls *(Tyto)*, grasshopper mice *(Onychomys)*, coyotes, bats, desert shrews *(Noteosorex)*, and a variety of lizards are among the most conspicuous scorpion predators. Parasitism of scorpions is not common; reported endoparasites include larvae of a tachinid fly *(Spilochaetosoma californicum)*, mermithid nematode larvae, and the larvae of a sarcophagid fly *(Sarcodexia sternodontis)*. About eight species of mites have been reported as ectoparasites of scorpions, but mite parasitism is not commonly observed.

VENOMS

Scorpions are universally recognized because of their venoms and conspicuous stinging apparatus. All species are venomous, but only a few are harmful to humans. The venom is produced in a pair of glands located in the vesicle of the telson and collects in the lumen of each gland, flowing through a simple duct that terminates near the sharp tip of the aculeus. The venom is delivered during a rapid thrust of the metasoma that results in the penetration of the victim's skin by the sharp aculeus. Contraction of muscles surrounding each venom gland discharges the venom.

The venoms are complex substances composed of water, a number of low molecular weight proteins (neurotoxins), and various organic compounds such as oligopeptides, nucleotides, amino acids, mucus, and cellular debris. A common fraction of venoms function as simple irritants, often causing a sharp burning sensation. Other fractions may cause inflammatory responses resulting in edema. In highly toxic venoms, systemic components may result in neurological symptoms such as convulsion followed by death. The neurotoxic effects seem to be the result of multiple interactions of certain toxins with voltage-dependent ion channels of excitable cell membranes, such as those of neurons. Venom interactions may include membrane depolarization, repetitive firing, prolonged action potentials, and massive release of neurotransmitters, most importantly from the adrenal medulla.

The cumulative effects of a venom in envenomated animals are complex and varied. Envenomation particularly may affect skeletal muscles, the cardiovascular system, lungs, visceral smooth muscle, uterus, and glands. The severe convulsions that may be associated with the sting of more toxic species (e.g., *Centruroides* spp.) are of particular concern. The venom of each species has a unique composition that explains the many varied venom reactions observed. Curiously, each scorpion venom contains a number of different toxic components.

Scorpion venoms are toxic to a variety of organisms, including arthropods and vertebrates, but are often nontoxic to other scorpions. The hemolymph may have the capacity for neutralizing the toxic components in the venom from other scorpions.

The number of reported human deaths from scorpion stings ranges from several hundred to several thousand annually. Stings are considered to be more serious when they happen to children, the elderly, and those in poor health. The number of reported deaths has been declining because of better treatments, development of antivenins, the reduction of human contact with scorpions in critical areas, and public health awareness. Because of unreliable reporting in developing countries, however, the reported incidence of scorpion stings and mortality is probably greatly underestimated.

Venomous scorpions of concern to humans are found throughout the world, but some of the more severe problems are in the more arid regions. Of the 1286 known species, only a dozen or so cause significant health problems for humans. The species of greatest medical concern are members of the family Buthidae and belong to the genera *Androctonus, Buthacus, Buthus, Centruroides, Leiurus, Mesobuthus, Parabuthus,* and *Tityus.*

METHODS OF OBSERVATION AND STUDY

Sampling and Collecting

Scorpion assessments and collections were traditionally made by inspecting trees and by looking under rocks, trash piles, and other surface debris. The discovery that scorpions fluoresce under ultraviolet light [with black light blue (BLB) bulb] led to the use of the "ultraviolet method" of scorpion detection and study. Equipped with a portable ultraviolet light, an investigator can walk through a habitat at night, counting and observing scorpions undisturbed in their natural state. With such a lamp a scorpion can be readily detected on a sand dune at a distance of about 15 m.

Care and Maintenance of Captive Specimens

Scorpions are easily kept in captivity and require minimal care. They may be housed in any closed container such as a terrarium, plastic box, jar, or plastic bag. If soil is provided, a rock or other surface cover should be added for shelter because scorpions are thigmotactic and seek close-fitting shelters. A small amount of water needs to be provided periodically. Scorpions can be maintained at room temperature and should not be exposed to freezing temperatures, especially the tropical forms. Most live insects may be supplied for food, but crickets are particularly well accepted. Generally one cricket every 2 to 4 weeks is adequate feeding for a moderate-sized scorpion.

As a rule, scorpions remain inactive in their shelters during the day and will show normal activity (if any) at night. Lack of physical activity is common and typical of most scorpions. Excessive activity often indicates a lack of food or water, excess light, excessive temperature, or other disturbance. Covering the rearing chamber with red cellophane readily simulates nocturnal environmental conditions. Most scorpions will live for 1 to 5 years in captivity.

See Also the Following Articles

Arthropoda and Related Groups • Spiders • Venom

Further Reading

Baerg, W. J. (1961). Scorpions: biology and effect of their venom. Agricultural Experimental Station Bulletin **649.** University of Arkansas, Fayetteville.
Brownell, P., and Polis, G. (eds.). (2001). "Scorpion Biology and Research." Oxford University Press, Oxford, U.K.
Fet, V., Sissom, W. D., Lowe, G., and Braunwalder, M. E. (2000). "Catalog of the Scorpions of the World (1758–1998)." New York Entomological Society, New York.
Keegan, H. L. (1980). "Scorpions of medical importance." University Press of Mississippi, Jackson.
Polis, G. A. (ed.). (1990). "The Biology of Scorpions." Stanford University Press, Stanford, CA.
Prendini, L. (2000). Phylogeny and classification of the superfamily Scorpionoidea Latreille 1802 (Chelicerata, Scorpiones): an exemplar approach. *Cladistics* **16**, 1–78.
Rubio, M. (2000). "Scorpions: A Complete Pet Owner's Manual." Barron's Educational Series, Hauppauge, NY.
Stoops, E. D., and Martin, J. L. (1995). "Scorpions and Venomous Insects of the Southwest." Golden West Publishers, Phoenix, AZ.
Vosjoli, P. de. (1991). "Arachnomania: The General Care and Maintenance of Tarantulas and Scorpions." Advanced Vivarium Systems, Lakeside, CA.
Williams, S. C. (1980). Scorpions of Baja California, Mexico, and adjacent islands. *Occas. Pap. Calif. Acad. Sci.* **135**, 1–127.
Williams, S. C. (1987). Scorpion bionomics. *Annu. Rev. Entomol.* **32**, 275–295.

Segmentation

Nipam H. Patel
University of Chicago

Segmentation (the repetition of body units along the anterior–posterior axis) is a fundamental property of all insects; indeed, it is an obvious character of all arthropods. Insect segments are clearly visible as reiterated patterns visible in the exoskeleton, but repeating patterns are present in internal structures such as muscles, neurons, and tracheae, as well. Through genetic and molecular approaches in the dipteran fruit fly, *Drosophila melanogaster*, the mechanisms of segmentation in this insect are now understood in great detail. Additional experiments indicate that some aspects of the *Drosophila* mechanisms are conserved in all insects, and others have undergone extensive evolutionary changes.

PATTERN OF SEGMENTS

In virtually all insect embryos, larvae (where present), and adults, the pattern of segmentation in the thorax and abdomen is clearly visible. Segments are usually separated by grooves, called segmental grooves, that lie at the boundary between each pair of segments. The configuration of segments is also often characterized by repeating patterns of pigmentation and elaborations, such as denticles or hairs on the exoskeleton. Internally, segmentation is reflected in repeating patterns within the nervous system, musculature, and tracheal system. Segments of the head and the terminal abdominal regions are sometimes more difficult to recognize. It appears that all insects are composed of six head segments, called the antennal, ocular, intercalary, mandibular, maxillary, and labial segments (going progressively from anterior to posterior), although some authors have suggested the existence of a seventh segment at the anterior of this pattern. The thorax is always composed of three segments (T1–T3), and the abdomen is generally composed of 11 segments (A1–A11) (Fig. 1), the most posterior abdominal segments (A10 and A11) are often fused during later development but usually can be detected separately during embryogenesis.

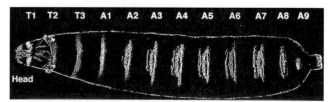

FIGURE 1 The larval segments of *Drosophila*. The pattern of segments is clearly revealed by the pattern of denticles (hairlike projections) on the ventral surface of the larvae. The segments of the head are involuted inside, and the final abdominal segments are not visible on the surface.

MECHANISMS OF SEGMENTATION IN *DROSOPHILA*

Studies of insect segmentation through experimental manipulations, such as ligation, ablation, centrifugation, and transplantation, have a rich history and continue to provide important insights into the mechanisms of segmentation. Beginning about 30 years ago, however, understanding of these mechanisms was rapidly accelerated by the genetic mutant screens in the fruit fly. Indeed, these screens uncovered such fundamental principles of biological pattern formation that the geneticists who carried them out, Edward Lewis, Eric Wieschaus, and Christiane Nüsslein-Volhard, were awarded the Nobel Prize in medicine and physiology for their contributions to the genetic analysis of tagmosis (regionalization) and segmentation in *Drosophila*.

The genetic analysis of segmentation was quickly supplemented with molecular and biochemical studies that have provided detailed knowledge of how segments are generated along the anterior–posterior axis during *Drosophila* embryogenesis. As it turns out, the mutants that effect this process can be grouped into several specific categories that act in a hierarchical manner to sequentially subdivide the embryo into smaller and smaller units, ultimately establishing the pattern of larval segments we see in *Drosophila* (Fig. 1). The mutant classes include maternal effect mutations, which disrupt the anterior or posterior halves of the embryo; gap mutations, which eliminate several contiguous segments; pair-rule mutations, which delete regions in a two-segment periodicity; and segment polarity mutations, which cause deletions and duplications in every segment.

Studies of the maternal effect mutations show that the process of segmentation actually begins during oogenesis, when the female localizes specific messenger RNAs (mRNAs) at either the posterior or anterior end of the developing egg. For example, bicoid mRNA is localized to the anterior end of the egg and forms a gradient of protein in the egg once it has been fertilized (with the highest concentration of bicoid protein at the anterior end). Mothers lacking functional bicoid gene form embryos in which the anterior segments are missing. A reciprocal gradient of the nanos protein is also formed, and *nanos* mutants are missing the more posterior regions of their body. The formation of these gradients is possible because the early development of *Drosophila* is syncytial, with no cell membranes between the nuclei of the early embryo. These gradients of information act to control the expression of the various zygotic gap genes, which come on in individual, well-defined broad regions along the anterior–posterior axis of the embryo.

The interaction of the gap genes (which are all transcription factors) generates the first periodic patterns in the embryo. These periodic patterns are stripes of the pair-rule genes. In *Drosophila*, most of the pair-rule genes display a pattern of seven stripes in the so-called cellular blastoderm stage of the embryo (at about 2.5 h after fertilization, when cell membranes form between the nuclei transforming the embryo from a syncytial to

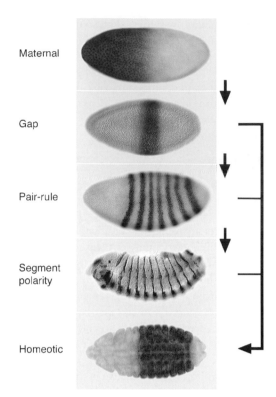

FIGURE 2 The segmentation hierarchy in *Drosophila*. The hierarchy is composed of the sequential expression of maternal, gap, pair-rule, and segment polarity genes. An example of the expression pattern of a single member of each class is shown here. The homeotic genes act to give regionalization to the segments and are primarily controlled by the gap genes, with some input from the pair-rule and segment polarity genes.

cellular blastoderm). The pair-rule genes are again generally transcription factors that regulate the final level of the hierarchy, the segment polarity genes. These genes are expressed in patterns of segmental stripes and include not just transcription factors, but also various receptors, ligands, and enzymes that are used in cell–cell communication and to refine and maintain the pattern of segments that has been elaborated.

A final category of genes, the homeotic genes, give identity to the segments. Mutations in homeotic genes result in the transformation of one or more segments into the identity of adjacent segments, and the homeotic genes are primarily regulated directly by the gap genes, although pair-rule and segment polarity genes also have some control on the precise boundaries of homeotic gene expression. All the homeotic genes encode a family of closely related transcription factors and are organized into two complexes on one of the chromosomes of *Drosophila*. Figure 2 shows the expression pattern of representatives of each of the foregoing classes of genes.

RELATIONSHIP OF *DROSOPHILA* SEGMENTATION TO SEGMENTATION IN OTHER INSECTS

Although the genetic analysis of segmentation in *Drosophila* provided an invaluable insight into the mechanisms of pattern formation, earlier manipulative studies in a variety of insects suggested that some aspects of segmentation differ among the various insects and that indeed *Drosophila* might be somewhat unusual in its mechanisms of segmentation. *Drosophila* is classified as a long-germ insect because various manipulative experiments showed that pattern formation was achieved very rapidly across the entire length of the embryo all at once and without the need for growth. At the molecular level, this is reflected in the nearly simultaneous appearance of pair-rule gene stripes in the *Drosophila* blastoderm. Many other insects, however, display a short or intermediate germ mode of development. In these insect embryos, only the most anterior segments are present in the blastoderm (prior to gastrulation), and more posterior segments are added only as the embryo elongates at later stages. Insects such as the red flour beetle, *Tribolium castaneum,* and the grasshopper *Schistocerca americana* display this type of development.

This progressive addition of segments is supported by comparative studies of gene expression. In both *Tribolium* and *Schistocerca,* the homologues of *Drosophila* segment polarity genes are expressed in stripes as in *Drosophila,* but the stripes appear sequentially as the embryos grow. In *Tribolium,* all the homologues of *Drosophila* pair-rule genes studied so far are expressed in the same pattern as in *Drosophila* (and prior to the expression of segment polarity genes), but again the stripes appear sequentially over time as the embryo elongates. In the more evolutionarily distant embryos of *Schistocerca,* however, the expression pattern of pair-rule gene homologues differs; indeed, some of them are not even expressed in a pattern of stripes. This suggests that extensive evolutionary alterations have occurred at this step of the segmentation hierarchy, although these changes still result in a conserved output of segment polarity gene expression.

The earliest steps of pattern formation are probably even more labile during insect evolution. This is not surprising, given how variable early embryogenesis can be in insects. For example, it is difficult to imagine how a gradient of bicoid protein can form in grasshopper embryos, given that the entire thorax and abdomen arises as a result of cell proliferation well after the blastoderm stage. Recent studies suggest that the bicoid gene, a key component of the maternal gradient *Drosophila* system, evolved somewhere within the dipteran lineage. Apparently extensive modifications have occurred in the segmentation system in different insect lineages, and these changes may reflect adaptive changes in the speed and patterns of oogenesis and early embryogenesis in different insect groups. Nevertheless, the overall logic of the *Drosophila* segmentation hierarchy has been conserved, not just in all insects, but possibly in all arthropods.

RELATIONSHIP TO SEGMENTATION IN OTHER ANIMAL PHYLA

Remarkably, many of the genes involved in *Drosophila* segmentation and regionalization are well conserved throughout animal evolution. In particular, the homeotic genes, which control segment identity, are conserved in both structure and function

between flies and vertebrates. Some changes in the expression patterns of these homeotic genes within insects, however, seem to be responsible for some evolutionary alterations in segment morphology within this group. Homologues of *Drosophila* segment polarity and pair-rule genes are also well conserved, and usually these proteins still play similar biochemical roles, but in different developmental contexts. For example, the segment polarity gene *hedgehog* is used in many pattern formation steps in vertebrates, such as patterning the dorsal–ventral axis of the neural tube, but it has no known function in vertebrate segmentation. Recently, however, it has been found that the mouse and chicken homologue of a *Drosophila* pair-rule gene, *hairy,* is involved in vertebrate segmentation. Thus, there is still considerable debate about the evolutionary origins of segmentation in arthropods, annelids, and vertebrates, with some believing that segmentation is homologous between these groups and other believing that segmentation has evolved independently in these different animal lineages.

See Also the Following Articles
Drosophila melanogaster • *Embryogenesis*

Further Reading
Akam, M. (1987). The molecular basis for metameric pattern in the *Drosophila* embryo. *Development* **101,** 1–22.
Brown, S., Fellers, J., Shippy, T., Denell, R., Stauber, M., and Schmidt-Ott, U. (2001). A strategy for mapping bicoid on the phylogenetic tree. *Curr. Biol.* **11,** R43–R44.
Davis, G. K., and Patel, N. H. (1999). The origin and evolution of segmentation. *Trends Genet.* **15,** M68–M72.
Nüsslein-Volhard, C., and Wieschaus, E. (1980). Mutations affecting segment number and polarity in *Drosophila. Nature* **287,** 795–801.
Palmeirim, I., Henrique, D., Ish-Horowicz, D., and Pourquie, O. (1997). Avian *hairy* gene expression identifies a molecular clock linked to vertebrate segmentation and somitogenesis. *Cell,* **91,** 639–648.
Patel, N. H. (1994). Developmental evolution: Insights from studies of insect segmentation. *Science* **266,** 581–590.
Sander, K. (1976). Specification of the basic body pattern in insect embryogenesis. *Adv. Insect Physiol.* **12,** 125–238.

Sericulture

Satoshi Takeda
National Institute of Agrobiological Sciences, Ibaraki, Japan

Sericulture is an industry that is characterized by a two-step process, the cultivation of mulberry trees and the rearing of silkworms on mulberry leaves to produce cocoons. A cocoon is an oval- to football-shaped object made by a mature silkworm larva by spinning silk proteins; the silkworm larva develops into a pupa inside it. Silkworms are monophagous insects, feeding only on mulberry leaves (Moraceae, genus *Morus*). Because the mulberry leaves must be fresh, it is difficult to transport them over long distances or store them for long periods. This has resulted in the rearing of silkworms and cultivation of mulberry trees generally forming a single enterprise. Mulberry tree cultivation starts with the production of mulberry seedlings, followed by mulberry tree training, cultivation, harvesting, and insect pest control. Silkworm rearing includes preservation of silkworm eggs, management of rearing rooms, handling of rearing equipment, prevention of silkworm diseases, supplying mulberry leaves, and collecting mature larvae to transfer to the cocooning frame.

ORIGINS OF SERICULTURE

It is impossible to document when sericulture began. The silkworm, *Bombyx mori,* now has no wild populations; it is a completely domesticated insect. The oldest written record of sericulture is the Chinese silkworm book *Can-jing,* which states that the queen of the Huang-Di empire started silkworm rearing. The Huang-Di era was around 2650 B.C., but sericulture must have been carried out in China in even earlier times. From China, sericulture spread via the "Silk Road." In the East, it was introduced into the Korean Peninsula and from there to Japan in about the 3rd century B.C. In the West, it spread to Central Asia and India and from India to Persia. Sericulture is thought to have reached Europe in 550 A.D., when silkworm eggs were presented to the Roman Emperor of the East. Silk was an important trade item along the Silk Road, where it was exchanged for its weight in gold. Commerce in silk along the Silk Road also made a major contribution to the exchange of Eastern and Western culture.

THE WORLD'S SERICULTURE INDUSTRY: PAST AND PRESENT

The state of cocoon production worldwide in 1997 compared to 1930 indicates that total global cocoon production, 617,910 tons in 1930 and 620,000 tons in 1997, was almost the same; the cocoon-producing countries, however, have changed considerably. In 1930, Japan ranked first, with a yield of 382,850 tons, and accounted for 62% of global production. China (then the Republic of China), was second with 129,528 tons, 21% of the world total; Italy ranked third with 53,348 tons (8.6%) and the Soviet Union fourth with 15,300 tons (2.5%). At that time cocoons were produced almost everywhere in the world where the mulberry could be cultivated.

In contrast, in 1997 China had by far the greatest production, 423,000 tons, with 68.2% of global production. India ranked second with 127,000 tons (20.5%), followed by Uzbekistan, Brazil, Thailand, Vietnam, and North Korea. Cocoon production in Japan, which accounted for 62% of world production in 1930, dropped precipitously after World War II and today is 2500 tons, a mere 0.4% of the world total. The dramatic fall in cocoon production in Japan was the result of soaring labor and production costs and low cocoon prices compared with other agricultural products. Although an aging population of sericulture workers and a shortage of replacements were factors, the primary cause of the decline in recent years has been the development of large differences in cocoon prices between Japan and other cocoon-producing countries, such as China and Brazil.

CURRENT STATUS OF SERICULTURE IN DIFFERENT PARTS OF THE WORLD

From the 1980s to the 1990s the sericulture industry became concentrated in Asia. Brazil is the sole country outside Asia in which export-quality cocoons are produced. Although sericulture thrived in Europe during the first half of the 20th century, particularly in Italy and France, only a very small scale production remains in Eastern European countries, such as Bulgaria and Romania. The current status of sericulture in the principal cocoon-producing countries is briefly described below.

In China, the provinces of Szechuan, Jiangsu/Chekiang, and Goangdong are the three great sericulture regions. Cocoons of the wild Chinese oak silkworm, *Antheraea pernyi,* add to the production of silk from *B. mori.* About 50,000 tons a year of wild silkworm cocoons are produced by outdoor rearing in mountains and forests, chiefly in the northeast.

Sericulture suited to each of its regions is carried out nationwide in India, the second largest cocoon-producing country in the world, in which production has increased sharply in recent years. In the north of India, temperate-region sericulture is conducted with bivoltine varieties of silkworm, and in the south it is being carried out with polyvoltine varieties or hybrids between polyvoltine and bivoltine varieties. Many species of wild silkworm, including the Tassar silkworm *(Antheraea mylytta)* and the Muga silkworm *(Antheraea assamensis),* are used for sericulture in the northern regions of India. However, sericulture of wild silkworm species remains a manual industry from rearing to reeling, not like that of *B. mori.*

In Southeast Asia, traditional sericulture industries in Thailand, Vietnam, and Laos use tropical polyvoltine varieties suitable for clothing. In central Asia, Uzbekistan is the major sericulture country ranking third among cocoon-producing countries. There, univoltine silkworm varieties are used, and sericulture is conducted as a sideline industry to cotton growing. The Brazilian silk-reeling industry was started by Italian immigrants, and introduction of technology by Japanese silk-reeling companies has enabled the production of high-quality raw silk thread and silk, and Brazil is now a raw silk thread and silk exporter on par with China. In Japan, cocoon production amounts to less than 1%; however, Japan consumes about 25% of global production, ranking second to China.

MANAGEMENT OF SILKWORM EGGS

Management of silkworm eggs is one of the most important sericultural techniques. In tropical regions where the mulberry leaves are available all year, the management of silkworm eggs is not particularly important. Polyvoltine strains used in the tropics hatch year round and can be reared anytime. However, because univoltine and bivoltine silkworms are usually reared in temperate and subtropical regions, larval hatch must be coordinated with the season when mulberry leaves are available. Once the univoltine silkworms are reared, the larvae generally do not hatch from the eggs until early spring of the next year. Such eggs are called "hibernating eggs." Artificial hibernation activates larvae to hatch by incubating hibernating eggs at about 10°C. Hydrochloric acid treatment can be used to induce hatching on demand.

SILKWORM REARING

Rearing silkworms consists of a series of tasks that includes harvesting and transporting mulberry leaves, supplying the mulberry leaf, cleaning the rearing beds, mounting the larvae so they can spin cocoons, and collecting and shipping the cocoons. The goal of silkworm rearing is to produce many high-quality cocoons while economizing on labor and material. The temperature range in which silkworms can be reared is 7 to 40°C, but practical rearing occurs in the 20–30°C range. Larval period rearing is roughly divided into the young silkworm period (first to third instar) and the grown silkworm period (fourth to fifth instar), and the fundamentals of rearing of each of them are different.

Young Silkworm Period

The growth rate is very high during the young silkworm period, especially during the first instar, but this period is characterized by higher susceptibility to bacterial pathogens and malnutrition. The young silkworm period is also physiologically characterized by a low quantity of food ingested, but a high rate of digestion. The amount of mulberry leaves ingested by young silkworms, i.e., first to third instar, is only about 2% of the amount ingested during the entire larval period. The mulberry leaves provided in this period are finely minced.

The basic principles of young silkworm rearing are to provide relatively high temperatures and high humidity. A temperature range of 26 to 28°C is ideal, and as the instars proceed, temperature is gradually reduced. The ideal humidity is 75 to 90% and is lowered as the instars progress. In advanced sericulture countries, rearing of the young silkworms is generally carried out in cooperative rearing facilities.

Grown Silkworm Period

During the fourth instar silkworms ingest 10% of the total amount of mulberry leaves taken, and during the fifth instar they ingest approximately 88%. Because the fifth instar is the period when the silk proteins for cocoons are actively biosynthesized, an adequate supply of mulberry leaves must be provided. When silkworms are reared in temperate regions, at high temperatures of 30°C or more, disease or failure to spin silk can occur. When fifth instars are mature they start to spin. Silkworms discharge a great deal of fluid outside their bodies

FIGURE 1 Silkworm larvae (fifth instar) reared on artificial diet.

with the silk proteins, in addition to that from defecation and urination, during the cocoon production, or mounting, period.

REARING ON AN ARTIFICIAL DIET

In 1960, an artificial diet for silkworms was developed that could successfully sustain them from the first to the fifth instar (Fig. 1). The growth of the silkworms and the cocoon size obtained, however, were considerably inferior to those of silkworms reared on mulberry leaves and therefore this diet was inadequate for actual use in sericulture. Following many improvements the use of artificial diets during the young silkworm period has become widespread, and in 1990 the rate of young silkworm rearing on artificial diets in Japan exceeded 40%. Japan is is the only country in which artificial diets are used for cooperative young silkworm rearing.

See Also the Following Articles

Bombyx mori • *Commercial Products from Insects* • *Rearing of Insects* • *Silk Production*

Further Reading

Hazama, K. (1995). "Sericulture in the Tropics." Assoc. for Int. Cooperation in Agriculture and Forestry, Tokyo.
Japanese Society of Sericultural Science (ed) (1992). "Fundamental Sericulture." Dainihon Sanshikai, Tokyo. [In Japanese]
Kuribayashi, S. (1998). Production and utilization of silkworms. *In* "Insect Resources in Asia" (Japan Int. Res. Center for Agricultural Science, ed.). Assoc. Agric. and Forestry Services, Tokyo. [In Japanese.]
Tajima, Y. (1978). "The Silkworm: An Important Laboratory Tool." Kodansha, Tokyo.

Sex Determination

Michael F. Antolin and Adam D. Henk
Colorado State University

Sex determination depends upon molecular switches that signal whether the male or the female sex-differentiating pathway will be followed during development; it can be triggered by genetic, epigenetic, or environmental cues. Insects display sexual dimorphism, in which males and female differ in form, behavior, and/or physiology. Although sex determination and developmental pathways leading to two distinct sexes are universal within insects, the primary signals that trigger sex determination are highly diverse and differ between groups. The primary signal for sex determination can entail genetic signaling, epigenetic signaling via maternally expressed genes or genomic imprinting of genes, or cytoplasmic factors like B chromosomes and bacterial infections. The primary signals can act alone or in combination. Molecular genetic details of sex determination and sex differentiation are known mainly from *Drosophila,* and these are described below. Comparison of sex determination of insects, nematode worms, and mammals points to a similar genetic mechanism that underlies sex determination: a cascade of gene expression, with alternative splicing of key genes and intermediate genes, leading to alternative splicing of a double-switch gene that ultimately controls differentiation of males and females. In insects, it appears that key genes and intermediate genes early in the cascade are unique to each group of insects, but genetic pathways for controlling sex differentiation after the double switch appear to be the same in all insects. Overall, sex determination in insects is highly variable among groups, and in some groups the mechanisms of sex determination differ between populations of a single species (e.g., the house fly *Musca* and the midge *Chironomus*).

SEX-DETERMINING SIGNALS

Patterns of sex determination have been explored since chromosomes were first described in the late 1800s. By the early 1900s, it was discovered that many insects have distinct chromosomes in males and females; the work is best described in the comprehensive 1973 book by M. J. D. White, *Animal Cytology and Evolution.* A generality about sex-determining mechanisms is that they are tremendously variable among insect orders, although considerable variation also can be found within genera or even within species. Regardless, some broad categories of the primary signal for sex determination can be identified.

The most common pattern of chromosomal sex determination is for females to be the homogametic sex, with two copies of one sex chromosome (XX), whereas males are the heterogametic sex, with two different sex chromosomes (XY). Some insect groups (e.g., Lepidoptera, Trichoptera) have the opposite pattern, in which males are the homogametic sex (ZZ) and females are heterogametic (ZW). Y chromosomes are usually much smaller than X and do not successfully recombine with the X. Unlike mammals, few sex-determining genes are located on Y chromosomes. Thus, the Y chromosome has been lost entirely in numerous orders of insects, leading to a system in which females are XX but males are haploid for sex chromosomes (denoted XO). A molecular genetic mechanism interacting with chromosomal sex determination has been described in *Drosophila,* and it entails a "molecular counting"

mechanism that assesses the ratio of X chromosomes to autosomes early in development. Thus, chromosomal sex determination may generally depend on this type of genic balance, but whether this mechanism is a general feature of insects or is unique to *Drosophila* is unknown.

Sex chromosomes vary greatly in number and size in insects. Translocations and duplications of sex chromosomes are common and fall into two general categories. First, in groups with XO males, small parts of the X chromosomes may be duplicated and form what are called "neo-Y" chromosomes. Second, entire X and Y chromosomes may be duplicated, so that some species have sexes that are homogametic or heterogametic, but contain multiple copies of the sex chromosomes. For instance, in the oriental rat flea, *Xenopsylla cheopis,* males are X_1X_2Y, whereas females are $X_1X_1X_2X_2$, although in other insect groups it is not uncommon for there to be up to five copies of the X chromosome. Multiple sex chromosomes like these would change the balance between sex chromosomes and autosomes and thus bring into doubt the generality of genetic balance as a sex-determining signal.

Many insects have no identifiable sex chromosomes, but still have specific genetic loci encoding genes that act as key sex-determining signals. A common pattern is to have a dominant sex-determining factor that specifies male development for individuals with an *M/+* genotype, whereas females have a *+/+* genotype. In this regard, *M* represents a key gene in a system similar to heterogametic sex determination, but lacking sex chromosomes that differ in size. House flies (*Musca domestica*) demonstrate further variations of the same theme, in which intermediate genes change the sexual phenotype. Another dominant genetic factor, *F,* interacts with rare *M/M* genotypes to produce females with an *M/M F/+* genotype. Individuals with an *M/M +/+* genotype develop as males. Finally, some house fly populations have epigenetic sex determination via a maternally expressed gene. Here, a gene found in females determines whether they will produce only female or only male progeny, regardless of the genotype of their male mates.

Haplodiploid sex determination (sometimes called arrhenotoky—the production of males) is found in all Hymenoptera and Thysanoptera and in some Homoptera and Coleoptera. It provides an interesting study of combinations of epigenetic and genetic sex determination. In haplodiploidy, females arise from fertilized eggs and therefore have the diploid complement of chromosomes, whereas males parthenogenetically develop from unfertilized eggs and are haploid. Recent evidence from the wasp *Nasonia* indicates that haplodiploidy can be explained by an epigenetic switch: genomic imprinting of sex-determining factors during oogenesis. Imprinting renders some female-determining genes inactive in eggs during development. If the genomic imprint is removed after fertilization, fertilized eggs develop as females. Unfertilized eggs would still have inactive female-determining genes and would become male. Haplodiploidy represents a case of a key sex-determining signal under epigenetic, rather than purely genetic, control.

Many Hymenoptera have an additional mechanism that overlays the epigenetic control of haplodiploidy: a single-locus genetic sex-determining system called complementary sex determination (CSD). CSD is found in sawflies, and in most apocritan groups (bees, ants, vespid wasps, ichneumonid wasps, braconid wasps), but not in the Parasitica. In this system, sex is determined by an individual's genotype at a single genetic locus, and the locus is known to have a large number of complementary alleles. As in all haplodiploids, males arise from unfertilized eggs and are haploid. If an egg is fertilized, however, it will develop as a female only if egg and sperm carry two different (complementary) alleles of the sex-determining locus. Thus, CSD of females depends upon heterozygosity at the sex locus. If egg and sperm have the same allele, the resulting diploid individual is homozygous at the sex locus and develops as a diploid male. Diploid males in Hymenoptera are usually inviable or sterile, so that inbreeding in species with complementary sex determination has severe consequences. The genetic mechanism underlying complementary sex determination remains to be discovered.

Haplodiploidy (arrhenotoky) differs from the case of paternal genome loss in some Homoptera (scale insects) that have haploid males. Under paternal genome loss, both males and females arise from fertilized eggs. However, eggs that lose the paternal complement of chromosomes early in development become male and are haploid.

Sex determination in insects also can be influenced by factors passed vertically through an egg's cytoplasm. The most common appear to be by infection with the bacterium *Wolbachia,* an intracellular parasite that is widespread in arthropods. In many cases, *Wolbachia* infections selectively kill one sex early in development and skew sex ratios. *Wolbachia* also induce unisexual parthenogenesis (theletoky) in some Hymenoptera. One case of sex determination being modified by *Wolbachia* is known from insects, that of the Asian corn borer *Ostrinia furnacalis.* Supernumerary chromosomes also can be passed via the cytoplasm. In the wasp *Nasonia vitripennis,* a B chromosome called paternal sex ratio *(psr)* causes condensation of chromosomes shortly after fertilization of eggs. Eggs fertilized by *psr* sperm develop as haploid males rather than as females, and subsequently these males continue to transmit the *psr* B chromosome to future generations.

Environmental cues affect sex determination, especially in reptiles and fish, although few cases are currently known from insects. A temperature-sensitive allele of the intermediate sex-determining gene *transformer* (see below) has been described in *Drosophila,* in which high temperature leads to development of males.

GENE CASCADES IN SEX-DETERMINING PATHWAYS

Despite the variety of sex-determining mechanisms described above, they have in common that both sex determination and sex differentiation ultimately depend on a hierarchical

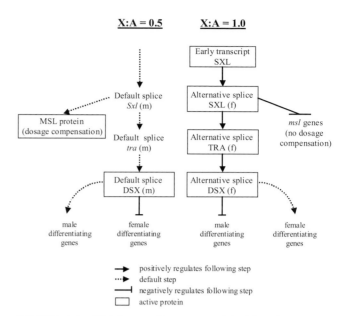

FIGURE 1 A simplified diagram of heterogametic (XX:XY) sex determination in *D. melanogaster,* showing the main genes and proteins contributing to the cascade of gene expression leading to development of males and females. This is the best understood and most fully explored genetic system of sex determination in insects, but is probably unique to *D. melanogaster* and closely related species. Names of genes are in lowercase italics; names of proteins are in uppercase.

cascade of gene expression during development, with numerous switches in pathways that eventually lead to somatic sexual differentiation and germ-line differentiation. Here we describe the hierarchy as it is known from the careful work done on *Drosophila.* The sex-determining "trigger" for male development (XY) is an imbalance between genes on the X chromosome and one of the autosomes (X:A = 0.5), whereas females (XX) have an equal ratio of genes on the X chromosome and autosomes (X:A = 1.0). A molecular counting mechanism controls early transcription of *sexlethal (Sxl),* the key gene in the sex-determining cascade in *Drosophila* that controls sex determination and dosage compensation. Intermediate genes, especially *transformer (tra),* interact with one another to control expression of the double-switch gene *doublesex (dsx).* This pathway controls somatic sexual differentiation of male and female morphology and behavior, in addition to germ-line differentiation leading to mature ovaries and testes.

In brief, the cascades work as follows (Fig. 1). In females (X:A = 1.0), the molecular counting mechanism activates the early transcription of *Sxl* mRNA. SXL protein produced by this mRNA regulates the splicing of subsequent *Sxl* mRNAs to produce the active female-specific form of SXL (f) protein. Active SXL (f) protein regulates splicing of *tra* mRNA to produce female-specific TRA (f) protein. In turn, TRA (f) protein interacts with other proteins (not shown) to regulate the splicing of *dsx* mRNA, generating the DSX (f) form. DSX (f) protein negatively regulates the genes responsible for male somatic sexual differentiation, but allows genes responsible for somatic sexual differentiation to produce a morphologically female fly.

In male *Drosophila* (X:A = 0.5), the molecular counting mechanism prevents transcription of the *Sxl* gene early in development. In the next step, the absence of SXL protein results in unspliced *Sxl (m)* mRNA. This form includes an exon with stop codons embedded in its sequence, so that *Sxl (m)* transcripts make no functional protein. In turn, transcripts of the *tra* gene also are not spliced, again resulting in nonfunctional protein because of internal stop codons. Without TRA protein, there is no alterative splicing of *dsx*, and DSX (m) protein is made. This pathway is the default, because of the absence of functional proteins from key and intermediate genes early in the pathway. Without these proteins, transcription of the double-switch gene *dsx* results in male-specific DSX (m) protein. DSX (m) negatively regulates the genes responsible for female somatic sexual differentiation, but allows genes responsible for somatic sexual differentiation to produce a morphologically male fly.

Because male and female *D. melanogaster* have different numbers of X chromosomes (males have one X and females have two), a mechanism is required to compensate for the difference in the number of genes present in the two genotypes. This mechanism is known as dosage compensation. In male *Drosophila,* a group of genes collectively known as male-specific lethal *(msl)* genes are active and increase transcription/translation of genes on the single X chromosome. Thus, genes on the single X chromosome in males generate approximately twice the gene products, which are approximately equal to those from genes on the two X chromosomes in the female fly. In female *Drosophila* the SXL (f) protein prevents *msl* genes from acting, and there is no subsequent increase in transcription and translation of genes on the X chromosome in females.

A fairly large number of genes also control germ-line sex differentiation and the production of egg and sperm. The regulation of the genes responsible for germ-line sex differentiation is poorly understood. Although different forms of DSX protein appear to affect the morphology of the sex cells in *Drosophila,* which gene products comprise the "switch" that determines the fate of germ cells as either egg or sperm remain to be identified.

Insects other than *Drosophila* do not use exactly the same genetic pathways for sex determination. Not only are sex-determining triggers different, but for many insects that have been studied some or all of the genes described above either do not exist at all or do not perform the same function. In addition to *Drosophila,* the gene *Sxl* has been isolated from four other Diptera (the fruit fly *Bactrocrera,* Mediterranean fruit fly *Ceratitis,* phorid fly *Megaselia,* house fly *Musca*), but in none of these does SXL protein differ between males and females. Thus *Sxl* is not involved in sex determination in these insects, and since no homolog of *tra* has been found in any other insects outside of *Drosophila,* it is unlikely that the same genetic pathway for sex determination exists in other

insects. On the other hand, the gene *dsx* appears to be conserved in structure and function in *Bactrocera, Ceratitis, Megaselia,* and the silkworm *Bombyx mori* (Lepidoptera). Apparent homologs of *dsx* can also be found in the nematode *Caenorhabditis (mab-3)* and in mammals *(DMT1),* and in each they have sex-specific functions in somatic sexual differentiation and/or germ-line differentiation. The control of sex determination is highly variable in insects, but control of sex differentiation leading to somatic sexual differentiation and germ-line differentiation in males and females appears to be conserved, with the possibility that genes within pathways of sex differentiation of all metazoans have common origins.

See Also the Following Articles

Chromosomes • *Drosophila melanogaster* • Sociality • Wolbachia

Further Reading

Baker, B. S. (1989). Sex in flies: The splice of life. *Nature* **340,** 521–524.

Beukeboom, L. W. (1995). Sex determination in Hymenoptera: A need for genetic and molecular studies. *BioEssays* **17,** 813–817.

Blackman, R. L. (1995). Sex determination in insects. *In* "Insect Reproduction" (S.R. Leather and J. Hardie, eds.). CRC Press, Boca Raton, FL.

Bull, J. J. (1983). "Evolution of Sex Determining Mechanisms." Benjamin–Cummings, Menlo Park, CA.

Cline, T. W., and Meyer, B. J. (1996). Vive la difference: Males vs females in flies vs worms. *Annu. Rev. Genet.* **30,** 637–702.

Hodgkin, J. (1990). Sex determination compared in *Drosophila* and *Caenorhabditis. Nature* **344,** 721–728.

Marin, I., and Baker, B. S. (1998). The evolutionary dynamics of sex determination. *Science* **281,** 1990–1994.

Schütt, C., and Nöthiger, R. (2000). Structure, function, and evolution of sex-determining systems in Dipteran insects. *Development* **127,** 667–677.

Stouthamer, R., Breeuwer, J. A. J., and Hurst, G. D. D. (1999). *Wolbachia pipientis:* Microbial manipulator of arthropod reproduction. *Annu. Rev. Microbiol.* **53,** 71–102.

Werren, J. H., and Beukeboom, L. W. (1998). Sex determination, sex ratios, and genetic conflict. *Annu. Rev. Ecol. Syst.* **29,** 233–261.

White, M. J. D. (1973). "Animal Cytology and Evolution." Cambridge University Press, London.

Sexual Selection

Kenneth Y. Kaneshiro

University of Hawaii

S *exual selection depends on the success of certain individuals over others of the same sex, in relation to the propagation of the species; while natural selection depends on the success of both sexes, at all ages, in relation to the general conditions of life. Sexual selection is a struggle between individuals of one sex, generally the males, for the possession of the other sex. The result is not death to the unsuccessful competitor, but few or no offspring....*

—*Charles Darwin, 1859*

HISTORICAL PERSPECTIVES

Darwin proposed the concept of sexual selection to explain the sexually dimorphic characters he observed in a wide diversity of organisms. He introduced the concept in *On the Origin of Species* in 1859 and further expanded on his theory in *The Descent of Man and Selection in Relation to Sex* in 1871. The development of the theory has been rich in controversies and so continues to the present, primarily because of the uncertainty over how much of the sexual dimorphism present in animals is the result of natural selection as opposed to sexual selection.

Notable theoreticians and evolutionary biologists severely criticized Darwin's theory of sexual selection and argued that sexual selection was less important than natural selection in bringing about evolutionary changes. Until recently, most evolutionary biologists accepted the notion that natural selection is the most dominant force in the evolutionary history of the species. The ubiquity of this view is well understood when one considers the notion that populations could ill afford the incapacity to adjust to the effects of their external environment. Even Darwin stated that "sexual selection will also be dominated by natural selection tending towards the general welfare of the species." Nevertheless, Darwin stood firm in his ideas and stated that his "conviction of the power of sexual selection remains unshaken."

THEORETICAL CONSIDERATIONS

Anisogamy and Parental Investment

The evolutionary origins of differences between the male and female sexes lie in what is referred to as anisogamy, which is the difference in gamete size between the two sexes. Females generally produce a few but large gametes, whereas males produce large numbers of much smaller gametes. Furthermore, such sex bias in gamete size is also characteristic of the initial stages of the evolutionary differences in parental investment, because the production of large gametes requires greater energetic investment by the female than the smaller male gametes. In most species, the male's investment ends with fertilization, while that of females may continue throughout the developmental period of the zygote, embryo, and other immature stages of the organism. Consequently, the limited number of gametes and the greater parental investment of the females result in an asymmetry in the degree of sexual selection between the two sexes. One might expect females to be generally less sexually selected but more selective in their choice of males (but see later), with males being under stronger sexual selection, hence attempting to mate with as many females as possible. It is thought that the males with the best territories, the most elaborate structures, the most attractive displays (i.e., secondary sexual traits), and so on will mate more frequently and therefore will leave more offspring. In terms of "investment" then, since it costs females much more to produce fewer large gametes than it does males, which produce lots of small gametes, females must "choose" mates that will produce the highest quality offspring (genetically).

Competition for Mates

It has been suggested that Darwin's concept of sexual selection may be subdivided into two aspects: "intrasexual selection," which involves competition between members of the same sex for individuals of the opposite sex, and "epigamic selection," which is the preferential choice of mates by one sex relative to the other. Intrasexual selection is most evident in males and perhaps can be best illustrated in species that display lek mating behavior. In such species, males compete among themselves for territories to which receptive females are attracted strictly for the purpose of mating (i.e., the territories do not provide other resources such as food or egg-laying substrates). In some species, a dominance hierarchy is established among the males that may be participants in a lek system and as a rule, the males able to occupy and defend the "best" territories will have access to most of the females that arrive at the lek. Epigamic selection, which is also referred to as "intersexual selection," is primarily an attribute of females, since the female sex ultimately discriminates among males and exercises choice of mating partners. Curiously, in most lek species, even though the most dominant males are successful in defending what seem to be the best territories in the lek system (as evidenced by the number of females that arrive at the territories of such males), males nevertheless perform an elaborate courtship ritual before mating occurs. Sometimes a female indeed rejects a dominant male's courtship and mating attempts, moves to an adjacent territory, and eventually mates with a subordinate male. Such occurrences illustrate that although fierce intrasexual selection among males can occur in competition for mating territories, epigamic selection based on female choice is the ultimate determinant of mating success among males, and thus intrasexual selection is not necessarily linked or correlated with epigamic selection. That is, a dominant male that is able to occupy and defend a prime mating territory, must, nevertheless perform courtship displays, which may or may not be acceptable to receptive females.

Runaway Sexual Selection Model

Darwin himself recognized a serious gap in his ideas of sexual selection and stated: "Our difficulty in regard to sexual selection lies in understanding how it is that the males which conquer other males, or those which prove the most attractive to the females, leave a greater number of offspring to inherit their superiority than their beaten and less attractive rivals. Unless this result does follow, the characters which give to certain males an advantage over others could not be perfected and augmented through sexual selection." Critics ridiculed Darwin's ideas and sarcastically implied that Darwin would not be able to prove his theory.

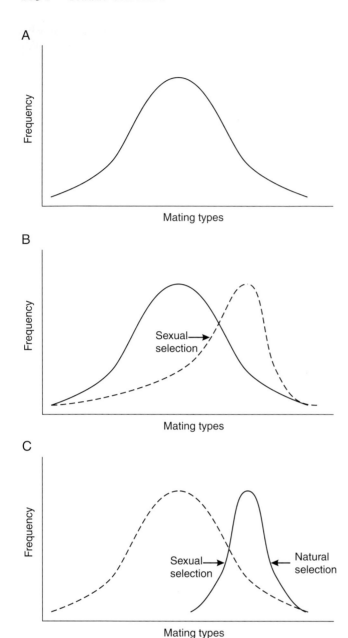

FIGURE 1 The runaway sexual selection model: (A) distribution of mating types segregating in the population, (B) strong sexual selection and elaboration of male trait that confer mating success, and (C) opposing forces of natural selection, which select for optimum phenotype for a particular environment.

The oftentimes sarcastic attacks against Darwin's sexual selection theory challenged theoreticians to develop alternative hypotheses for the evolution of female preference for, and the evolution of, secondary sexual traits in males. The most notable of these hypotheses is the "runaway selection" model, according to which it was inferred that the evolution of a sexually dimorphic character in males could result in a correlated response in the female's preference for that character. The model predicted that female choice and male characteristics influenced by sexual selection would coevolve very rapidly in an interbreeding population. The runaway selection model presumed that two

forces act to counterbalance the runaway process of sexual selection. That is, female preference for a certain male character, whether morphological or behavioral, tends to select for extreme forms of that character until natural selection exerts its forces to maintain the optimum male phenotype that is best able to survive in its particular environment. In an article published in 1972, Ernst Mayr stated that "natural selection will surely come into play as soon as sexual selection leads to the production of excesses that significantly lower the fitness of the species."

The paradox of the runaway sexual selection model is that the opposing forces of sexual selection for elaboration of conspicuous male traits and natural selection, which maintains the optimum condition for the particular environment, result in reduced genetic variation for such male characters (Fig. 1). However, without genetic variation, selection can no longer occur; and unless secondary sexual characters either are linked to or are pleiotropic effects of other components of fitness, such conspicuous characters would be energetically costly to produce, and individuals possessing such traits would be in greater danger of predation.

The Differential Selection Model

In 1989, in an attempt to provide an alternative explanation for the maintenance of secondary sexual characters observed in males, Kaneshiro proposed a model based on his research on the mating behavior of Hawaiian *Drosophila* species. He suggested that there is a range of male mating types segregating within an interbreeding population. That is, some males in the population are highly successful in mating and often accomplish the majority of the matings in the population. Other males are less successful and in fact may not mate at all, despite being given the opportunity to do so with several receptive females. Similarly, among females, there are those that exhibit higher levels of mating receptivity thresholds and strongly discriminate against most of the males in the population. In the same population, there are also females that exhibit lower receptivity thresholds and seem to accept the courtship overtures of most of the males in the population.

Observations of mating experiments of Hawaiian *Drosophila* both in the laboratory and in the field indicate that the most likely matings are between males that are most successful in satisfying the courtship requirements of females and females that are not extremely choosy in selecting mating partners. The genetic correlation between these two behavioral phenotypes in the two sexes (i.e., highly successful males and less choosy females) generates the entire range of mating types in both sexes in subsequent generations (Fig. 2). In this model then, there is *differential selection* for opposite ends of the mating distributions in the two sexes and therefore, sexual selection itself serves as the stabilizing force in maintaining a balanced polymorphism in the mating system of the population. Hence, it is not necessary to invoke the forces of natural selection to counterbalance the elaboration of secondary sexual characters as required in the runaway sexual selection model.

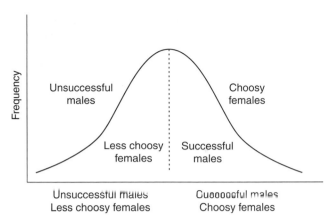

FIGURE 2 Differential sexual selection model, which shows that matings between less choosy females and successful males generate the entire range of mating types in each successive generation.

FIGURE 3 Males of the native Hawaiian *Drosophila, D. heteroneura,* in head-to-head, wingtip-to-wingtip posture, defending a territory to which females are attracted solely for the purpose of mating. The "hammerhead" shape of the males appears to be a result of sexual selection for this behavioral display among males in a lek mating system.

SEXUAL SELECTION IN INSECTS

Different patterns of sexual selection that have been observed can be illustrated by descriptions of sexual selection in a few insect species. This is certainly not a comprehensive review, since space constraints allow only a very few examples to be discussed. Several books present a more in-depth discussion of sexual selection systems in insects, and the reader is referred to such treatments.

Lek Behavior in Hawaiian *Drosophila* and Other Dipteran Species

In an article published in 1997, Shelly and Whittier list the following criteria in defining a lek species: "(1) the absence of male parental care, with males contributing nothing but gametes to the female; (2) the existence of a mating arena in which most matings occur; (3) male territories that contain no resources vital to females; and (4) the opportunity for females to freely select a mate in the arena." These authors also recognize two types of male mating aggregations involved in lek mating species: *substrate-based,* where males spend most of the time perched at stations or territories to which females are attracted for mating (although mating may occur in the air), and *aerial aggregations* or *swarms,* in which males are usually in continuous flight and matings are typically initiated in the air.

In the mating system of the Hawaiian Drosophilidae, males take up station either on the underside of leaves, one male to a leaf, or on branches of understory shrubs. Resident males aggressively defend territories from intruding males with fierce battles (Fig. 3), which in some species resemble sumo wrestling bouts. In the lek mating system of the Mediterranean fruit fly, *Ceratitis capitata* (Diptera: Tephritidae), males occupy and aggressively defend the underside of leaves of host plants as mating territories to which females are attracted solely for the purpose of mating. Once a female has arrived at the mating arena, the males of both the Mediterranean fruit fly and the drosophilids display complex courtship behaviors, which may

or may not result in acceptance by the female. This means that successful aggressive behavior is not necessarily linked to or correlated with successful mating behavior. That is, a male that is successful in occupying and defending prime mating territories still must perform a courtship sequence that is acceptable by the females to ensure mating success. In some cases, the alpha male (i.e., the male that is most successful in defending a prime mating territory) is not successful in mating despite numerous visits by receptive females. Thus, at least for the drosophilids and tephritid fruit flies, although success in intrasexual selection among males may mean increased opportunity to encounter receptive females, the ultimate criterion for successful mating is epigamic selection based on female choice (Fig. 4).

Resource-Based Sexual Selection in the Scorpionfly

The mating system of the scorpionfly, *Hylobittacus apicalis* (Mecoptera: Bittacidae), provides another excellent example of intrasexual and epigamic sexual selection. Here intrasexual selection is evidenced by the competition among males for arthropod prey, which are offered to females as nuptial gifts. Males able to acquire such resources will have greater access to receptive females. It has also been shown that the size of the nuptial prey plays an important role in epigamic selection. Females apparently have the ability to evaluate the males based on the size of the nuptial gift, on which they feed during copulation. Females prefer males with a large prey versus males that offer smaller prey. Such preference results in an increased number of eggs oviposited as well as the survivorship of the female. Essentially, size of prey is correlated with the length of the copulation. It was shown that at least 5 min of mating is required before any sperm is transferred to the female and that copulation duration ranging between 5 to 20 min is directly correlated with the number of sperm transferred. Thus, when a male offers and the female accepts a large nuptial gift,

FIGURE 4 Courtship display of a native Hawaiian *Drosophila, D. clavisetae.* While in a "scorpion-like" posture, the male produces a pheromone bubble and wafts (a row of apically flattened bristles near the terminal end of the abdomen serves as a "fan") the chemical stimulant toward the female by vibrating his abdomen in an up-and-down motion toward the female. The courtship "dance" also involves visual cues (striking white face and wing markings are sexually dimorphic characters), tactile cues (female "kisses" the extended proboscis of the male), as well as acoustical cues. Despite the complex courtship displays and considerable effort on the part of the male, sexual selection via female choice (i.e., epigamic selection) plays a powerful role in determining mating success.

the male is provided an opportunity to transfer a complete quantity of sperm. In some cases, if the prey is large enough, the males may end the mating with the first female to be able to use the prey as a gift for a second female. Furthermore, a male that mates for 20 min or longer introduces a substance into the female's reproductive tract that renders the female sexually nonreceptive and stimulates oviposition. Females that mate for less than 20 min because of small prey size will continue to seek males with larger prey.

Thus, *H. apicalis* females may either reject males with small prey or mate with them for only a brief period, feeding on the prey but receiving few or no sperm. Females may then seek males that can offer larger prey, which will increase the likelihood of a sperm transfer adequate to fertilize her full complement of eggs. Such female preference for males that can offer larger, nutrient-rich prey, which influences the number of offspring produced, has clearly shaped the evolution of male behavioral components in this species. Since insect prey is a limited resource, as evidenced by the small number of males (~ 2–0%) that are carrying prey at any one moment, there is competition among males in securing prey. Males that are successful in securing a gift are not guaranteed to reproduce, since females discriminate against prey-bearing males that offer small prey. On the other hand, males that are able to secure sufficiently large prey may be able to mate with more than one female.

Female choice for prey-bearing males has apparently resulted in the evolution of other male behaviors. For example, it has been shown that males foraging for prey are more prone to predation and that males are 2.3 times more likely to be trapped in spider webs than females. Because hunting for prey can be hazardous, *H. apicalis* males can employ an alternative tactic of securing prey, namely, thievery. Some

males fly to calling males and mimic female behavior—for example, perching next to a prey-bearing male and waiting for the gift to be presented, as if to a real female. If the ruse works, the thief will snatch the prey and use it to attract and feed a female without having to spend energy in a risky search of prey of his own. Thus, it would seem that sexual selection has played an important role not only in the evolution of nuptial gift giving in males but also in the development of female-mimicking behavior and prey stealing as an alternative strategy to minimize predation pressure.

THE ROLE OF SEXUAL SELECTION IN SPECIATION

Two decades ago, Mayr remarked: "Speciation…now appears as the key problem of evolution. It is remarkable how many problems of evolution cannot be fully understood until speciation is understood.…" Over the past 20 years, there has been renewed interest in the process of speciation, initially to address arguments advanced by the proponents of the punctuated equilibrium model of macroevolution but also to revisit the definition of a species. At least two books address questions of speciation and the evolutionary processes of species as populational units of biological diversity. It is not the intent of this chapter to discuss the various models of speciation. *Speciation and Its Consequences* edited by Otte and Endler, published in 1989, and *Speciation and the Recognition Concept* edited by Lambert and Spencer, published in 1995, offer comprehensive reviews of this topic.

In 1997 Hampton Carson suggested that when demographic circumstances force populations to pass through a size bottleneck, the sexual selection system is temporarily disorganized by genetic recombination and the consequences of small population effects. As the population builds back up to restore efficient mate choice, novel genetic recombinants may be generated that in turn may lead to "novel selective change over a relatively small number of generations…" and that "the new population may thus be driven to undergo progressive genetic change under sexual selection."

Based on results of mating experiments on Hawaiian *Drosophila* species, Kaneshiro formulated a hypothesis, that may provide an intuitive explanation for what Carson referred to as the "driver of genetic change." Kaneshiro suggested that under conditions of drastic population reduction, there is even stronger selection for less choosy females in the population, since highly discriminant females may never encounter males that are able to satisfy their courtship requirements. If the bottleneck condition persists over a few generations, there will be a shift in the distribution of mating types in the population until a significant increase in frequency of less choosy females in the population has occurred. (Fig. 5). Such a shift in the distribution of mating types may be accompanied by a corresponding shift in the gene frequencies in the population, resulting in the destabilization of the coadapted genetic systems that had evolved in the population. The resulting destabilized

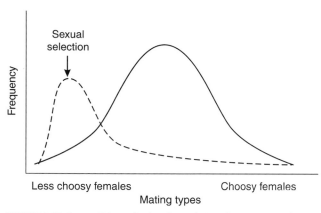

FIGURE 5 Under conditions of reduced population, there is strong selection for less choosy females, since choosy females may never encounter males able to satisfy their courtship requirements.

genetic environment thus sets the stage for genetic changes conducive to the speciation process. That is, the breakup of coadapted sets of genes allows the generation of novel genetic recombinants, some of which may be better adapted to the environmental conditions that led to population reduction. Thus, the dynamics of the sexual selection system can play an extremely important role in the speciation process as well as in the maintenance of genetic variability in the population on which the forces of natural selection can operate.

In the evolution of island biota such as that of the Hawaiian Islands, the most likely mode of speciation is what evolutionary biologists call founder event speciation. For insect groups such as the Hawaiian Drosophilidae, the most probable scenario is that a single fertilized female may be blown to an adjacent island, there to establish a new colony. For a number of generations, during the initial stages of colonization when the population size is small, there would be strong selection for females that are less choosy in selecting mates, once again, because females that are too choosy may never encounter males able to satisfy their courtship requirements. Therefore, within a few generations, there would be a shift in the distribution of mating types toward an increased frequency of less choosy females, resulting in a significant shift in gene frequencies followed by a destabilization of the coadapted genetic system. Genetic recombinants better adapted to the new habitat may be generated and strongly selected, especially if these genotypes are linked or correlated with the genotypes of the less choosy females. Thus, at least during the initial stages of colonization immediately following the founder event, the dynamics of sexual selection may play a significant role in generating a genetic environment in the population that is conducive to the formation of new species.

THE ROLE OF SEXUAL SELECTION AND NATURAL HYBRIDIZATION

It has been suggested that the dynamics of the sexual selection process also provides the opportunity for occasional natural hybridization and permits the "leakage" of small amounts of genetic material from a related species without destroying the integrity of the separate gene pools. When the population is small and there is a shift in the distribution of mating types toward an increase in frequency of less choosy females, such females in the population may occasionally accept males of a related species, which may be less susceptible to the environmental stress conditions. Genes that may be resistant to the conditions responsible for reducing the population size may be strongly selected, especially if such genes are linked to the genotypes of the less choosy females. Therefore, natural hybridization as permitted by the sexual selection process may do more than play an important role in replenishing genetic variability that may be lost owing to drift when population size is small; it may also provide a process by which resistant genes are transmitted between populations.

CONCLUDING REMARKS

The differential sexual selection model discussed here and its role in the processes of speciation may provide a possible explanation of how genetic variability may be not only maintained but actually enhanced. The generation of novel genetic recombinants and the selection for genotypes that are better adapted to changing environmental conditions are enhanced by the sexual selection system in the population, especially when subjected to population bottlenecks. It is suggested that sexual selection is a dynamic process, influenced by density-dependent factors, that enables populations to overcome environmental conditions that result in drastic reduction in population size. Shifts in the distribution of mating types when population sizes are small can generate a genetic environment producing novel recombinants able to respond to the environmental stress imposed on the population. Sexual selection influences the levels of genetic variability generated via novel genetic recombinants resulting from a reorganization of the genome, and via natural hybridization that is permitted by the sexual selection model. Thus, the biology of small populations and the dynamics of the sexual selection process are important aspects of evolutionary biology and the biology of insect populations in general. A better understanding of the role of sexual selection would provide important insights into the mechanisms of species formation as well as enabling us to develop more effective management programs, not only for pest populations but also for the conservation of rare and endangered insects.

See Also the Following Article
Mating Behaviors

Further Reading
Blum, M. S. and Blum, N. A. (1979). "Sexual Selection and Reproductive Competition in Insects." Academic Press, New York.
Carson, H. L. (1997). Sexual selection: A driver of genetic change in Hawaiian *Drosophila. J. Hered.* **88**, 343–352.
Choe, J. C., and Crespi, B. J. (1997). "Mating Systems in Insects and Arachnids." Cambridge University Press, Cambridge, U. K.

Darwin, C. (1859). "On the Origin of Species." Murray, London.

Darwin, C. (1871). "The Descent of Man and Selection in Relation to Sex." Murray, London.

Kaneshiro, K. Y. (1989). The dynamics of sexual selection and founder effects in species formation. *In* "Genetics, Speciation and the Founder Principle" (L. V. Giddings, K. Y. Kaneshiro, and W. W. Anderson, eds.), pp. 279–296. Oxford University Press, New York.

Kaneshiro, K. Y. (1995). Evolution, speciation, and the genetic structure of island populations. *In* "Islands: Biological Diversity and Ecosystem Function" (P. Vitousek, H. Adsersen, and L. Loope, eds.), pp. 23–33. Springer-Verlag, New York.

Lambert, D. L., and Spencer, H. G. (1995). "Speciation and the Recognition Concept." Johns Hopkins University Press, Baltimore.

Mayr, E. (1942). "Systematics and the Origin of Species." Columbia University Press, New York.

Mayr, E. (1972). Sexual selection and natural selection. *In* "Sexual Selection and the Descent of Man, 1871–1971" (B. Campbell, ed.), pp. 87–104. Aldine Press, Chicago.

Mayr, E. (1982). Processes of speciation in animals. *In* "Mechanisms of Speciation" (C. Barigozzi, ed.), pp. 1–19. Liss, New York.

Otte, D., and Endler, J. A. (1989). "Speciation and Its Consequences." Sinauer Sunderland, MA.

Shelly, T. E., and Whittier, T. S. (1997). Lek behavior in insects. *In* "Mating Systems in Insects and Arachnids" (J. C. Choe and B. J. Crespi, eds.), pp. 273–293. Cambridge University Press, Cambridge, U.K.

Thornhill, R., and Alcock, J. (1983). "The Evolution of Insect Mating Systems." Harvard University Press, Cambridge, MA.

Silk Moth

see *Bombyx mori*

Silk Production

Catherine L. Craig

Harvard University/Tufts University

Silks are highly expressed proteins produced by insects and spiders. The ability to produce silk proteins has evolved multiple times in the arthropods via two different pathways. Silks secreted via cuticular glands are produced by either immature or adult insects regardless of developmental mode; silk secretion via systemic and dedicated silk glands is present only in larval insects whose development is holometabolous. The first use of silks by insects was for reproduction; later uses were for protection and foraging.

THE MULTIPLE EVOLUTIONARY PATHWAYS OF SILKS

Phylogenetic comparison across insect taxa suggests that the ability to secrete fibrous proteins is a primitive feature of the

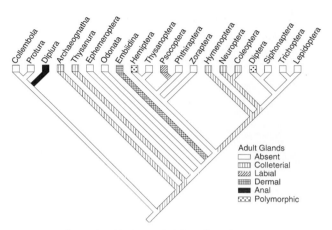

FIGURE 1 Silks produced by hexapod adults. Silk production among hexapod adults is less common than among larvae and all silks are secreted from glands that evolved for some other primary purpose. Most frequently, adult insects produce fibrous proteins in colleterial glands. (Reprinted, with permission, from the *Annual Review of Entomology*, Vol. 45, © 2000, by Annual Reviews, www.annualreviews.org.)

hexapods, hence ancestral to all insect taxa. More specifically, silk production evolved first among adult hexapods, and silks were used during reproduction (Fig. 1). Among larvae, silk production evolved sporadically (Fig. 2). All the derived, holometabolous insect larvae produce silk proteins, which they use for protection.

The evolution of silk-producing glands is equally complex. At least five taxa of adult insects produce silks in colleterial glands, and colleterial silk production has been lost and regained at least once. Among immatures, silk production in Malpighian tubules evolved early in the Ephemeroptera, then was lost and later reappeared sporadically (Neuroptera, Coleoptera, Hymenoptera) across the insect orders. Labial silk glands evolved only twice, once in the Psocoptera and later in the derived, holometabolous insect larvae (Fig. 2).

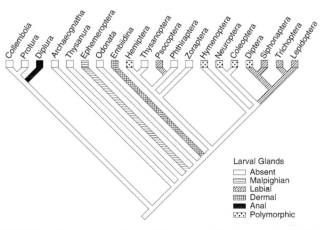

FIGURE 2 Phylogenetic analysis of silks produced by hexapod larvae. Silk production is common among hexapod larvae. The most prolific silk producers, the Embiidina and Lepidoptera, have evolved dedicated silk glands. (Reprinted, with permission from the *Annual Review of Entomology*, Vol. 45, © 2000, by Annual Reviews, www.annualreviews.org.)

Only the Embiidina and the Diplura have the ability to produce silks throughout their life. The Diplura produce silks in their anal glands, and the Embiidina draw silks from specialized dermal glands.

SILK PROTEINS AS SECONDARY PRODUCTS OF OTHER SECRETORY SYSTEMS

For most insect groups, silks evolved as secondary products of insect reproductive and excretory systems and were first used for reproductive purposes. For example, males in the primitive Apterygote order Thysanura (the bristletails) produce protein secretions in accessory genital glands. The secretions are drawn into threads used to restrict females during mating. Female mantids (Mantoidea: Mantidae) produce colleterial gland silks they use to protect their eggs, as does the beetle *Hydrophilus piceus* (Hydrophilidae). Green lacewings, *Chrysopa flava* (Neuroptera: Chrysopidae), produce colleterial gland silks from which they construct an egg sac stalk. Colleterial gland silks are characterized as having either a cross-β or α-helical configuration (Figs. 3 and 4). Except for those produced by male Thysanura, none of the colleterial gland silks are delivered through a specialized spinning organ. Furthermore, only adult insects produce silks in their colleterial or accessory genital glands, perhaps a reflection of their uses and their link to reproductive physiology.

Silks as secondary secretions are also produced in some insect excretory systems, primarily by larval insects in the Hymenoptera, Coleoptera, and Neuroptera. In most cases their silks are used for protection. Larvae of the phylogenetically primitive stingless bee *Plebeia droryana* and the bumble bee, *Bombus atratus,* bind cocoon silks (produced in modified labial glands) with Malpighian tubule secretions. In contrast, *Hypera postica* (Coleoptera) produce cocoons from Malpighian silks (cross-β conformation) that are characterized by loosely fabricated networks of coarse brown fibers. Larvae in the phylogenetically primitive neuropteran taxa Megaloptera and

Raphidioidea produce Malpighian silks as well, but their conformations have not been identified. Insects in the Neuroptera produce silks both as larvae and adult but they are products of different organs. For example, Malpighian silks are produced by larval chrysopids while adults produce colleterial gland silks. Like most colleterial gland silks (except for those of the Thysanura), none of the silks produced in insect Malpighian systems are delivered through a specialized spinning organ.

Orthopterans in the superfamily Gryllacridoidea (not included in the final phylogenetic analysis because of polytomy) are unique among in the Orthoptera in that they produce labial gland secretions throughout their lives. Although, like many fibrous proteins, their molecular structure has not been determined, they have been identified as silks on the basis of their general appearance. *Camptonotus carolinensis* uses silks to roll and bind leaves together; *Cnemotettix* sp. constructs a vertical, cylindrical burrow in which it "sews" small packets of sand grains that it fits into the burrow ceiling or wall; and *Bothriogryllacris brevicauda* uses silks to tie a pebble or soil cap to seal off its burrow and protect it against desiccation.

SILK PROTEINS AS PRIMARY PRODUCTS OF SILK-SPECIFIC GLANDS

Silks produced in dedicated silk glands are used mainly for protection. Character reconstruction shows that except for the silks produced by Siphonaptera, all labial gland silks are some type of parallel-β fibroin. Furthermore, except for the Embiidina and a few Psocoptera, the only insects that produce labial gland silks are insects whose development is holometabolous. Labial gland silks have evolved in some of the hemipteroid insects and insects in the Psocoptera have silk glands similar to those of caterpillars. In particular, Psocidae cover their eggs with silks, Philotarsidae spin webs singly or gregariously in small groups, and Archipsocidae spin extensive webs covering trunks and branches of large trees. Representatives of all endopterygote insect orders (Hymenoptera, Diptera, Siphonaptera, Trichoptera, and Lepidoptera), are characterized by the ability to produce labial gland silks.

SUMMARY

All silk proteins are produced in cells whose lineage is ectodermal. Most insects produce silk proteins in glands that have been modified from some other function. Silks evolved first for reproductive purposes and were produced in the anal and colleterial glands. Later, silks were used for protection and were derived from larval Malpighian tubules and labial glands. Although more ancestral insects produce silks in colleterial glands and Malpighian tubules that are characterized by an α-helical or cross-β conformation, holometabolous larvae produce labial gland silks characterized by parallel-β conformations. Only the paurometabolous Embiidina are able to produce silks characterized by a parallel-β conformation throughout their life (Fig. 4).

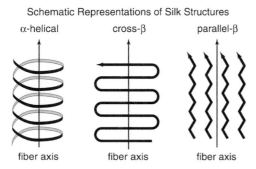

FIGURE 3 Schematic representations of silk structures. (A) The primary unit of an α-helical protein is a polypeptide chain that assumes the conformation of a rodlike coil stabilized by hydrogen bonds that form within the silk backbone. (B) β-Pleated silks are composed of polypeptide chains that are aligned side by side in a parallel or antiparallel configuration along the fiber axis. (C) Cross-β pleated sheet silks are a subset of the parallel fibroins whose polypeptide strands are oriented perpendicular to the fiber axis.

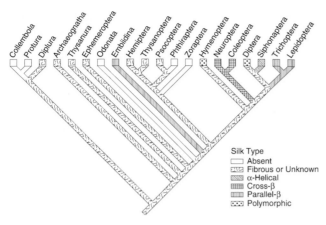

FIGURE 4 Structural configuration of silks produced by hexapods. In most cases the structural configuration of silks produced by hexapods has not been determined. The Hymenoptera produce the greatest diversity of silks. Parallel-β silks evolve only in hexapods that are able to produce both their peculiar molecular sequence and also a means of shearing the protein upon secretion.

All of the parallel-β silks are pulled (spun), in contrast to the α-helical and cross-β silks, which are deposited, secreted, or ejected. Among insects, labial gland silks evolved late in the hexapod clade, and these glands are found only among the larvae of insects characterized by holometabolous development. The tarsal silk glands of the Embiidina represent a unique evolutionary event in the history of the hexapods. All hexapod silk delivery systems or "spinnerets" are simple cuticular hardenings or pores.

This article was adapted, with permission, from the *Annual Review of Entomology*, Volume 42, © 1997, by Annual Reviews: www.annualreviews.org.

See Also the Following Articles

Accessory Glands • Bombyx mori • Embiidina • Salivary Glands • Sericulture • Spiders

Further Reading

Craig, Catherine L. (1997). Evolution of arthropod silks. *Annu. Rev. Entomol.* **42,** 231–267.
Lucas, F., and Rudall, K. M. (1968). Extracellular fibrous proteins: the silks. *In* "Comprehensive Biochemistry." (M. Florkin and E. H. Stotz, eds.), pp. 475–558. Elsevier, Amsterdam.
Rudall, K. M., and Kenchington, W. (1971). Arthropod silks: the problem of fibrous proteins in animal tissues. *Annu. Rev. Entomol.* **16,** 73–96.
Sehnal, F., and Akai, H. (1990). Insect silk glands: their types, development and function, and effects of environmental factors and morphogenetic hormones on them. *Int. J. Insect Morphol. Embryol.* **19,** 79–132.
Stehr, F. W. (1987). "Immature Insects." Kendall/Hunt, Dubuque, IA.

Silverfish

see *Zygentoma*

Siphonaptera
(Fleas)

Michael W. Hastriter and Michael F. Whiting
Brigham Young University

The Siphonaptera are laterally compressed, wingless, holometabolous insects. The order contains approximately 2575 species. All species are parasitic in the adult stage and possess mouthparts modified for piercing and sucking, highly modified combs and setae on their body and legs, and legs that are modified for jumping. Some species are vectors of disease, and current research is providing important insights into flea phylogeny and evolution.

EARLY RESEARCH ON FLEAS

Fewer than 100 species of fleas had been described prior to the end of the 19th century. In 1898 it was demonstrated that fleas were capable of transmitting plague organisms and, later, murine typhus, which stimulated a frenzy of flea studies. In the early 20th century, A. C. Oudemans, Julius Wagner, Nathan C. Rothschild, and Karl Jordan made major contributions. Karl Jordan is recognized as "the father of flea systematics." From 1953 to 1971, G. H. E. Hopkins and M. Rothschild published a comprehensive five-volume series on flea systematics, and three additional companion volumes were published for the remaining families by Mardon in 1981; Traub, Rothschild, and Haddow in 1983; and Smit in 1987. Robert Traub described 153 flea taxa and provided critical insights into flea phylogeny, convergent evolution, and zoogeography. Today there are only a few flea specialists with knowledge of the global flea fauna.

FOSSIL RECORD

Fossils of fleas are extremely rare. Three extinct species *(Palaeopsylla baltica, Palaeopsylla dissimilis, Palaeopsylla klebsiana)* have been reported from Baltic amber and a single specimen *(Pulex larimerius)* has been reported from Dominican amber. The former three species belong to an extant Palaearctic genus, whereas the latter belongs to an extant genus in which five species are confined to the Nearctic and Neotropical regions and a sixth is cosmopolitan *(Pulex irritans).*

ZOOGEOGRAPHY

Current distribution of flea families suggests an ancient origin, possibly in the Jurassic period (>140 mya). In general, families endemic to the boreal continents (North America, Europe, Asia) are predominantly Ceratophyllidae, Hystrichopsyllidae, Leptopsyllidae, Vemipsyllidae (Holoarctic), Coptopsyllidae

(Palaearctic), and Ancistropsyllidae (Oriental), whereas families endemic to the southern continents (Australia, South America, Africa, Antarctica) include Malacopsyllidae, Rhopalopsyllidae (Neotropical), Stephanocircidae, Pygiopsyllidae (Australian and Neotropical), and Xiphiopsyllidae and Chimaeropsyllidae (Ethiopian). The families Ctenophthalmidae, Ischnopsyllidae, and Pulicidae are more broadly distributed in both the Northern and the Southern Hemispheres.

CLASSIFICATION

The approximately 2575 species recognized in the order belong to 244 genera. The most recent review of the order by Lewis in 1998 recognizes 15 families. Several classifications published in recent years have substantially conflicting treatments of superfamilial relationships.

PHYLOGENY

Siphonaptera is a neglected holometabolous insect order from a phylogenetic standpoint. The major obstacle in deciphering flea phylogeny has been their extreme morphological specializations associated with ectoparasitism, which hamper the systematist from homologizing characters across taxa. The majority of characters used for species diagnoses are of limited use for phylogenetic reconstruction. Siphonaptera appears to have many instances of parallel reductions and modifications associated with multiple invasions of similar hosts, further obscuring homology. Recent molecular analyses have provided some insight into flea phylogeny. Ceratophylloidea (Cerato-phyllidae + Leptopsyllidae + Ischnopsyllidae) is a mono-phyletic group, and the families Pulicidae, Ceratophyllidae, and Ischnopsyllidae are also each monophyletic. However, Leptopsyllidae and Ctenophthalmidae are paraphyletic assemblages. It is still unclear which flea group is most basal.

EVOLUTION

Traditional phylogenetic hypotheses favored an origin of fleas from the Diptera or Mecoptera but did not pinpoint the actual sister group. Current evidence suggests that the closest living relative to the fleas is a single mecopteran family, the snow fleas (Boreidae). This relationship is supported by the presence of unusual proventricular spines, a suite of ovariole characters (shared only by fleas and boreids), DNA sequence data, and other morphological and molecular characters. This new phylogenetic hypothesis presents a novel scenario for the evolution of fleas from a mecopteran ancestor. In Boreidae, females are wingless and males have the forewing reduced and modified into hooks, used for grasping the female during copulation. In addition, boreids have the ability to jump in a manner that appears similar to that of fleas. Boreids emerge as adults only during winter months and are closely associated with mosses, and, like many other winter insects, the reduction and the loss of wings reduce the body surface area and are probably an adaptation to the cold. The ability of boreids to jump facilitates movement on soft, fluffy snow and is also probably an adaptation to this extreme environment. When the boreid–flea ancestor shifted from a snowy, mossy habitat to the nest of a host animal, it would have lost its wings and acquired the ability to jump. These adaptations allowed fleas to move efficiently through the host's fur and between hosts. As further adaptations to a parasitic life, modifications to the primitive flea include lateral flattening, the development of sucking mouthparts, and the development of elaborate combs and setae.

MEDICAL AND VETERINARY SIGNIFICANCE

Fleas are capable of transmitting disease organisms to humans, including plague *(Yersinia pestis)* and murine typhus *(Rickettsia typhi)*. They have also been implicated in the transmission of Q fever *(Coxiella burnetti)*, tularemia *(Franciscella tularensis)*, listeriosis *(Listeria monocytogenes)*, salmonellosis, cat scratch fever *(Bartonella henselae)*, and possibly Carrion's disease *(Bartonella bacilliformis)*. Several tapeworms occasionally infect humans who accidentally ingest fleas containing the intermediate stage (cysticercoid) of the tapeworms, including the double-pored dog tapeworm *(Dipylidium caninum)* and several tapeworms of rats *(Hymenolepis diminuta, H. nana)*. Fleas also transmit a protozoan blood parasite *(Trypanosoma lewisi)* among rodents and a filarial worm *(Dipetalonema reconditum)* among dogs. Fleas are notorious for their biting and burrowing habits that cause intense itching and annoyance.

BIOLOGY

Detailed bionomic studies are available for only a small number of known flea taxa, so the following concepts are based on generalities found within the order.

Life Cycle

Fleas display four stadia (egg, larva, pupa, and adult) (Figs. 1A–1D). The number and size of eggs and where they are laid differ from species to species, as does the time interval the larvae require to hatch. There are generally three instars. The exceptional neosomic *Tunga monositus* has only two larval stages. Larvae are ca. 1.5 to 10 mm in length, vermiform (worm-like), eyeless, legless, and without prominent features. Prior to pupation, the third instar ceases feeding and clears its gut contents by defecation. Silk glands present in this last instar are used to spin a cocoon prior to transformation into a quiescent exarate (appendages not "glued" to the body) pupa. The emergence of adults from the cocoon requires specific stimuli (e.g., temperature, vibrations) dependent on the adaptations of the individual species. Fleas may be univoltine or multivoltine. The optimum interval for development from egg to adult ranges from 3 weeks to more than a year. Adults may live for only a few weeks to more than 3 years.

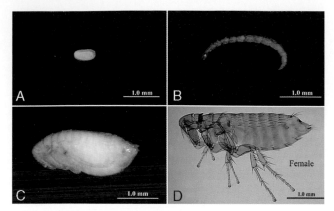

FIGURE 1 Life stages of a flea: (A) egg, (B) larva, (C) pupa, and (D) adult (a female is shown).

Reproduction

Males are smaller than females, have longer antennae, frequently have a dorsal groove on the occipital area of head, have less reticular sculpturing on sternum II, and possess an elaborate aedeagus and modified terminal segments. The interantennal groove or falx frequently divides the preantennal and postantennal areas of the head in males, while this division is often absent in females of the same species. The details of copulation are understood in only a few species of fleas. In general, the male assumes a position under the female, clasps the ventral sternites of the female with his antennae (equipped with suction cup-line organelles), then clasps the female terminal segments (sternum VII) with the tergum IX (clasper), inserts the aedeagus, and transfers sperm by extrusion of long paired penis rods that penetrate the bursa copulatrix and duct of the spermatheca into the spermatheca. Spermatozoa may be stored for long periods of time in the bulga (head) of the spermatheca. The hilla (tail) of the spermatheca of many species has an external muscle extended from its apex to the bulga. When an egg passes though the oviduct, the muscle contracts, pumping the sperm from the bulga through the duct of the spermatheca to the oviduct, where fertilization occurs through a micropyle in the surface of the egg. Depending on the species, the eggs may be laid on the host, in the nest, or randomly in the environment. The reproductive cycles of *Cediopsylla* and *Spilopsyllus* are linked to the hormonal estrous cycle of their rabbit hosts. Other species, such as *Glaciopsyllus antarcticus,* have their cycle synchronized with the seasonal migratory and nesting patterns of their avian hosts.

Host Specificity

Mammals harbor about 95% of the known flea species, with the remaining species living on birds. The seasonal and geographical distribution and host specificity of fleas are largely determined by specific requirements of the larvae. Each species has evolved a narrow range of tolerance for temperature and humidity, enabling exploitation of virtually every environment on the globe where avian or mammalian hosts flourish. These include the Antarctic *(Glaciopsyllus);* extreme xeric conditions of the world's greatest deserts *(Nosopsyllus, Hopkinipsylla, Xenopsylla* spp.); altitudinal extremes of the Andean, Himalayan, and Rocky mountains *(Amphalius, Ctenophyllus, Geusibia, Ochotonobius, Craneopsylla, Plocopsylla, Cleopsylla, Stephanocircus);* tropical rainforests *(Polygenis, Rhopalopsyllus,* Pygiopsyllidae*);* and coastal marine environments *(Notiopsylla, Paraspsyllus).* Although many species are catholic in their host preferences, the larval development is usually limited by the microenvironment of the individual host's nest. Those demonstrating the least host specificity include those free-living taxa whose immature stages can tolerate wide ranges of temperature, humidity, and nutritional requirements *(Ctenocephalides, Pulex).*

Nutritional Requirements

Most flea species require a blood meal prior to oogenesis and spermatogenesis. With the exception of some species whose reproductive cycle is synchronized with that of their host, there are few studies on the nutritional requirements of adult fleas. Although the larvae of several fleas have been noted to burrow into dead host tissues, those of *Uropsylla tasmanica* are the only truly parasitic flea larvae that burrow into and feed on the tissues of their live hosts. Some larvae have been documented to feed directly on fresh blood as it is excreted from the anus of feeding adult fleas, but most scavenge on the dried blood residues or animal dandruff and excreta. A few species are predaceous on other small organisms within the nest.

Neosomy

Neosomy is the formation or enlargement of a morphological structure from secretion of new cuticle during an active stadium of a group that normally changes form by molting. Additional cuticle forms in neosomes, accommodating expansion during enlargement (1000-fold) of gravid females that are capable of laying more than 1000 eggs. Neosomy occurs only among females because the male's principal function is to mate with embedded females *in situ.* Neosomatic growth occurs in the burrowing fleas *Neotunga euloidea* and all species of *Tunga* (Pulicidae). Fleas that do not become embedded, but attach for long periods of time, and also display neosomatic growth include members of *Chaetopsylla, Dorcadia, Vermipsylla* (Vermipsyllidae), *Malacopsylla* (Malacopsyllidae), and *Hectopsylla* (Pulicidae). Neosomatic fleas attach only once and do not feed repeatedly.

ANATOMY OF ADULT FLEAS

The morphological diversity demonstrated among families, and even among species within the same genus, is extraordinary. A glossary of terms published in 1971 by Rothschild and

Traub provides the standard for the interpretations of structures by systematists. The male aedeagus and the female seventh sternite and spermatheca are important anatomical features used in determination of species and subspecies. Study of morphological structures with the aid of a compound microscope requires mounting of fleas on glass microscope slides. Traditional techniques include puncturing the area of abdominal sterna II and III with a minuten pin, soaking for 24 h in 10% potassium hydroxide, transferring to distilled water and gently compressing the flea's abdomen to evaluate dissolved soft tissues, dehydrating in a series of ethanol solutions (70%, 80%, 95%, absolute) for 30 min each, clarifying the chitin for 15 to 20 min in methyl salicylate, transferring to xylene for a minimum of 1 h, and mounting in Canada balsam using a coverslip (No. 1 thickness). Fleas should be arranged with their heads to the right and their legs away from the technician. Thus, when viewed with a compound microscope, specimens will appear in an acceptable anatomical position with head to the left and legs toward the observer. Dissections are frequently required to permit observation of otherwise hidden structures.

Head

The preantennal and postantennal regions of the head capsule may or may not be divided by a groove or ridge that connects the antennal fossae (depressions) that lay on each side of the head. The antennae are usually longer in males than in females and the males of many species are capable of extending both antennae vertically above the dorsal margin of the head. The eye (when present) is always simple (single fascicle) and located anterior to the antennal fossa on each side of the head; it may or may not be pigmented and is often vestigial or absent. Setae arrangements on the head frequently provide important features in distinguishing taxa.

Thorax

The extraordinary jumping ability of fleas is reflected in the development of the pleural arch, which is formed from a fusion of the ventral end of the notal ridge and the transverse ridge of the metanotal area. The notal ridge is homologous with the ridge that divides the wing-bearing notum into an anterior scutal region and a posterior scutellar region of winged insects. The pleural arch is considered homologous with the pleural wing hinge of winged insects, which is composed of a rubber-like protein called resilin. The chitinous pleural arch of fleas (also composed of resilin) has shifted laterally in fleas from the more dorsally located pleural wing hinges of flying insects. This unique elastic material enhances the jumping ability of fleas. Strong jumpers have a hyperdeveloped pleural arch atop the pleural rod of the metathorax. Fleas that live their lives in the nest have little need to jump because they crawl onto the host in the nest, feed, and crawl off. The pleural arch is poorly developed or vestigial in nest fleas and they are infrequently

collected on hosts outside the nest. Many taxa have no ctenidial combs on the dorsal thoracic segments, some possess pronotal combs that extend over the mesothorax, and a few genera have marginal ctenidial combs or spinelets on the metathorax.

Legs

The two front legs are more elongate than the more robust mid- and hind pairs and are articulated beneath the genal process (frequently adorned with teeth) to facilitate forward locomotion in the pelage or feathers of the host. The lengths of individual tarsal segments and the arrangements of bristles on the ventral and lateral aspects of the terminal tarsal segments are frequently important taxonomic characters.

Abdomen

The abdomen comprises 10 tergites and 9 sternites. In the genitalia of the male (Figs. 2A and 2C), the highly modified terga VIII and IX and sterna VIII and IX are used as the principal structures to distinguish many taxa. In general, the 9th tergite

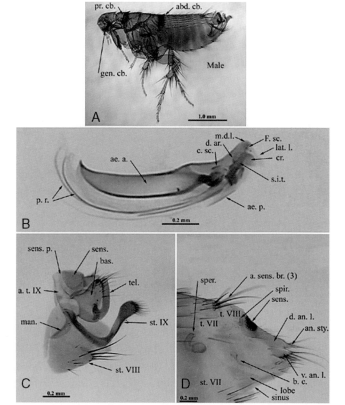

FIGURE 2 (A) Adult male. (B) Male aedeagus. (C) Male terminalia. (D) Female terminalia. Abbreviations: abd.cb., abdominal comb; ae.a., aedeagal apodeme; ae.p., aedeagal pouch; an.sty., anal stylet; bas., basimere; b.c., bursa copulatrix; c.sc., crescent sclerite; cr., crochet; d.an.l., dorsal anal lobe; d.ar., dorsal armature; F.sc., Ford's sclerite; gen.cb., genal comb; lat.l., lateral lobe; man., manubrium; m.d.l., median dorsal lobe; p.r., penis rods; pr.cb., pronotal comb; sens., sensilium; sens.p., sensilial plate; s.i.t., sclerotized inner tube; tel., telomere; v.an.l., ventral anal lobe.

is divided into an immovable process, or basimere; an articulated movable process, or telomere; and a manubrium. Various arrangements of lobes and processes adorn the basimere and are important features to distinguish taxonomic groupings. The 8th tergite is either small and obscure or greatly enlarged to ensheath the 9th tergite and the aedeagus. The 8th sternite is greatly expanded in some groups and reduced or pencil-like in others. An important taxonomic feature of most species is the 9th sternite. The aedeagus is a highly complex structure used extensively in the systematics of the order. The taxonomic significance of the terminal lobes (median dorsal lobe, lateral lobe, ventral lobe), crochet, sclerotized inner tube, crescent sclerite, penis rods, aedeagal apodeme, dorsal armature, and Ford's sclerite are of particular importance. The female (Figs. 1D and 2D) 8th tergite is variably shaped and frequently enlarged to cover the less developed ventral terminal segments. The caudal margin of the 7th sternite is modified in many species and is an important taxonomic character. The various lobes and sinuses of the 7th sternite are thought to be important anatomical features used during copulation. Nine genera possess two spermathecae, each with a duct leading to the perula of the bursa copulatrix. The remainder have but one spermatheca, although a remnant blind duct leading from the bursa copulatrix can be seen in most of them. The bursa copulatrix is often sclerotized and it and the spermatheca distinguish many species. The anal opening passes between the dorsal and the ventral anal lobes, each considered the 10th tergite and sternite, respectively. The 10th tergite usually bears an anal stylet with a variable number and arrangement of apical setae, a character used to distinguish species.

ANATOMY OF LARVAE

The diversity of morphological characters parallels those of adult fleas. Important structures include the shape of the egg tooth (first instar only); chaetotaxy and sensory structures of the head capsule, antenna, and antennal mound; cuticular sensilla of the dorsal plates of the thorax and abdomen; number and length of body setae; sculpturing of the integument; and arrangement of the anal comb and other setae on the 10th abdominal segment.

CONTROL OF FLEAS

Control is often necessary when fleas parasitize livestock and pets, become potential vectors of disease, or are biting people. Household sanitation such as thorough vacuuming of indoor areas and washing of pet bedding removes many of the immature stages (eggs, larvae, and pupae). Insect growth regulators (methoprene and pyriproxyfen) that inhibit the development of immature stages are used in conjunction with organophosphates (chlorpyrifos and diazinon) to interrupt the developmental cycle. Dog and cat flea collars and the use of systemic insecticides add additional preventive measures. Control of potential plague vectors in rural or sylvatic environments requires more drastic measures. Wild mammals such as rats, prairie dogs, ground squirrels, and chipmunks must be controlled around human habitations and recreation sites frequented by people. In such situations it is necessary to apply Carbamates (carbaryl) to burrows and rodent harborages prior to control of animal populations. Failure to control fleas prior to implementing animal control endangers humans that may come into contact with these potentially disease-infected fleas.

See Also the Following Articles
Bubonic Plague • Cat Fleas • Veterinary Entomology

Further Reading
Hopkins, G. H. E., and Rothschild, M. (1953–1971). "An Illustrated Catalogue of the Rothschild Collection of Fleas (Siphonaptera) in the British Museum (Natural History)," Vol. I–V. British Museum (Natural History), London.

Lewis. (1998). Resume of the Siphonaptera (Insecta) of the World. *J. Med. Entomol.* **35**, 377–389.

Mardon, D. K. (1981). Pygiopsyllidae, Vol. VI of "An Illustrated Catalogue of the Rothschild Collection of Fleas (Siphonaptera) in the British Museum (Natural History)." British Museum (Natural History), London.

Rothschild, M., and Traub, R. (1971). "A Revised Glossary of Terms Used in the Taxonomy and Morphology of Fleas." British Museum (Natural History), London.

Smit, F. G. A. M. (1987). Malacopsyllidae and Rhopalopsyllidae, Vol. VII of "An Illustrated Catalogue of the Rothschild Collection of Fleas (Siphonaptera) in the British Museum (Natural History)." British Museum (Natural History), London.

Traub, R., Rothschild, M., and Haddow, J. (1983). "The Rothschild Collection of Fleas. The Ceratophyllidae: Key to the Genera and Host Relationships." University Press, Cambridge, UK.

Whiting, M. F. (2001). Mecoptera is paraphyletic: Multiple genes and a phylogeny for Mecoptera and Siphonaptera. *Zool. Scripta.* **31**, 93–104.

Snakefly

see *Raphidioptera*

Sociality

James E. Zablotny
U. S. Department of Agriculture

I nsect sociality refers to populations of insects that use some form of social behavior. Eusocial insects, for example, are true social species that possess the following three characteristics: cooperative brood care, overlap of two or more generations with offspring assisting with brood care, and reproductive division of labor. In a broad sense, the term "social" is often used for pre- and subsocial insects that have fewer than three of the characteristics of eusociality. In a strict sense, a social insect is usually understood to mean a representative of the ants, termites, vespoid wasps, and social bees, the true eusocial insects. Social insects are an important component of Earth's biological diversity. E. O. Wilson has noted that eusocial insects (including termites, ants, bees, and wasps) make up a significant proportion (75%) of the world's insect biomass. The evolutionary and ecological success of social insects is evident in both tropical and temperate ecosystems.

Sociality is a life history strategy that has evolved multiple times within and among diverse and distantly related insect taxa. Because of large population sizes, social insects collectively consume more energy and biomass than vertebrate animals. In tropical ecosystems, ants and termites play a major role in nutrient cycling and soil turnover. In 1990 B. Hölldobler and Wilson noted that in tropical and temperate habitats, the amount of soil turnover attributable to ants and termites far outweighs the cycling done by earthworms. In the region of the Brazilian Amazon, ants release a significant quantity of atmospheric formic acid, which contributes to natural acidification of rainwater and weathering of limestone-based rock formations.

In eusocial insects, the ability to reproduce has been lost in worker and guard castes. Sterile workers care for the brood, provide nourishment, maintain environmental conditions (e.g., by thermoregulation) inside the nest, and remove waste. The female reproductives or gynes generally outlive several generations of the offspring and are protected and tended by workers in the confines of the nest. The presence of reproductive division of labor is vividly portrayed in social insects, with reproduction restricted to the queen, and nest or colony support being provided by the sterile worker and guard castes. With these three characteristics operating, an insect society is a highly organized, cooperative group of organisms that functions like a superorganism. Social and physiological homeostasis of the insect society is maintained through the integration of caste members through pheromone, tactile, auditory, and sometimes visual communication.

The evolutionary patterns within the Hymenoptera are complex, and eusociality has evolved over eight times within the social bees. The Isoptera (termites) is the only insect order that is exclusively eusocial. Eusociality is rare but present in some aphids (Heteroptera: Cerataphidini) and thrips (Thysanoptera). In aculeate Hymenoptera, many examples of eusocial species are present and include all ants (Formicidae), one group of sphecoid wasps (Crabronini), vespoid wasps (Vespidae, Masaridae, and Eumenidae), and the Apoidea or social bees (Apidae: Bombinae, Apinae, Meliponinae, and Euglossinae).

DIVERSITY IN INSECT SOCIALITY

In 1923 W. M. Wheeler published one of the first contemporary treatments of insect sociality. In the 1960s and 1970s,

TABLE I Synoptic Overview of the Degrees and Characteristics of Sociality Derived from Wilson

Degree of sociality	Characteristics[a,b]		
	Overlapping generations	Reproductive castes	Cooperative brood care
Quasi-social	–	–	+
Semisocial	–	+	+
Eusocial	+	+	+

[a]Plus sign (+) denotes presence and minus sign (–) indicates an absence of a characteristic.

[b]Solitary, subsocial, and communal insects do not have overlapping generations, reproductive castes, or cooperative brood care.

usable classification schemes for insect sociality evolved from the works of C. D. Michener and E. O. Wilson. Wilson's 1971 treatment of the social insects streamlined and established a standard vocabulary for describing the degrees of insect sociality, ranging from solitary to eusocial, with several different grades or intermediate forms (Table I).

Presocial Insects: Subsociality

Most insects lead solitary lives, with many species forming simple aggregations during mating or at commonly used food sources. Sociality appears in insects as a departure from a solitary life history strategy. Presocial insects have one to two characteristics of sociality. One subgroup of presocial insects includes insects that provide limited parental care to their offspring. Labeled as subsocial, these species protect or shelter their eggs, nymphs, or larvae; eggs and larvae may be given parental care in an enclosed nest.

Most subsocial insects provide care of offspring in the absence of a nest or shelter. Many examples of subsociality are known from the true bugs (Heteroptera). For example, in some belostomatid genera *(Belostoma, Abedus)* the female lays her eggs on the male's back, and the male carries the eggs until nymphs hatch. Males of the belostomatid *Lethocerus americanus* assist the female with active guarding of the eggs that are laid on aquatic vegetation. Parental care of eggs is also known in water striders (Gerridae) and leaf-footed bugs (Coreidae). Active guarding of eggs and newly hatched nymphs is known in the Acanthosomatidae, Pentatomidae, Tingidae, Reduviidae, and Scutellaridae. Usually, the female guards the clutch and covers the first instars when a predator approaches. The nymphs often orient themselves toward the parent to facilitate concealment.

More examples of behaviorally advanced stages of offspring care can be found in the orders Heteroptera and Coleoptera. Treehoppers (Heteroptera, suborder Auchenorryncha; Membracidae) also exhibit parental behaviors toward early instars. R. Cocroft, for example, observed female thornbugs *(Umbonia crassicornis)* straddling egg clutches to prevent parasitism and creating feeding slits on plant stems for hatchlings. Additionally, the mother actively herds wandering nymphs to keep them near or around the feeding slits and emits alarm calls to alert offspring to a predator's presence.

Within the Coleoptera, highly developed forms of parental behavior and nest making have been discovered in the Silphidae and in the dung-feeding Scarabaeidae. Known as burying beetles, species of *Nicrophorus* (Coleoptera: Silphidae) provide provisions and actively feed newly hatched offspring through regurgitation. After discovering a small animal carcass, *Nicrophorus* parents process the carcass into a congealed ball of putrefying flesh. After burying the provision, the female chews a shallow depression into the mass, where she deposits her eggs. To solicit parental feeding, the larvae raise the anterior portions of their bodies and wave their thoracic legs. The female parent then regurgitates a small droplet of food material, which is devoured by a larva. In some *Nicrophorus* species, males also participate in larval feeding but less frequently than the female. Parental feeding in *Nicrophorus* is essential for development of the young: removal of the female parent inhibits metamorphosis to the adult.

G. Eickwort and Wilson categorize most bark and ambrosia beetles as subsocial or colonial insects. Members of the Scolytidae and Curculionidae (Platypodinae) have elaborate life history strategies that enable them to evade a host tree's defense mechanism. The scolytid beetle *Xyleborus* produces communal brood chambers where all life stages reside. In similar manner to Hymenoptera, these xyleborines also have haplodiploid sex determination. Another similarity between xyleborines and eusocial Hymenoptera is the presence of a biased sex ratio in favor of females. Female offspring assist in cultivating ambrosia fungi, which provide the cellulase necessary for digestion of wood fibers.

Parasocial Insects: Communal Societies

"Parasocial" refers to species that exhibit communal, quasi-social, and semisocial behavioral patterns with some form of generation interaction between generations. Communal species include insects of the same generation that share a nest site without any form of collective offspring care. A communal life history strategy is not just an aggregation: it may involve complex forms of communication and offer some distinct advantages over a solitary life history strategy. Communal aggregations enhance accessibility to food materials that are difficult to process, facilitate thermoregulation, and offer protection from predation. Examples of communal social behavior in bees can be found in the Andrenidae, Megachilidae, and several subfamilies of the Halictidae. Multiple females construct a large, composite nest, but each female tends only to her own nest cells. This produces a nest having a single nest opening that is more easily defended by multiple female bees. Many different species of beetles have communal life history strategies. Some species of scolytid bark beetles are exemplary communal insects.

Communal behaviors in larval insects can be found in tent-making caterpillars (Lepidoptera: Lasiocampidae) and some of the sulfur butterflies (Lepidoptera: Pieridae), silk moths (Lepidoptera: Saturniidae), nymphalid butterflies (Lepidoptera: Nymphalidae), and tussock moths (Lepidoptera: Lymantriidae). Batch laying of eggs predisposes some larval insects for communal life history strategies, but other species disperse from a common oviposition site to lead solitary lives. Although complex social interactions between group members are limited in many species of caterpillars, some species utilize central-place foraging that requires an extensive communication network within the caterpillar community. Central-place foraging is not tied universally to web making, but instead is scattered throughout the lineages of Lepidoptera. In 1925 F. Balfour-Brown described a model for the evolution of elaborate caterpillar web dwellings from a mat of silk threads or carpet web. This model suggests that carpet webs, the ancestral condition, further evolved with caterpillars that used feeding webs and later to those that used domicile or home webs.

In the jack pine sawfly, *Neodiprion pratti banksinae* (Hymenoptera: Neodiprionidae), communal feeding facilitates the procurement of nutrients from tough waxy pine needles. The collective efforts of many larvae chewing over the same area on a pine needle reduces the amount of time needed to break through the cuticle. Other species of larval sawflies and Lepidoptera use group defense postures in which the aggregation moves in a synchronized manner to thwart a predator. The sawflies *Neodiprion sertifer* and *Diprion pini* use synchronized head jerking and twitching to repel both egg-laying ichneumonid wasps and birds. To be effective, these protective behaviors require a coordinated group response. Similar defense strategies are also known in the woolly or eriosomatid aphids in which a coordinated waving of waxy filaments may intimidate predators.

Parasocial Insects: Quasi-Social Societies

Unlike communal insects, quasi-social insects have some cooperative brood care while sharing a common nest site. Female quasi-social bees participate in nest guarding and assist with building and stocking nest cells with provisions. In 1967, R. B. Roberts and C. H. Dodson discovered that multiple female orchid bees (Apidae: Euglossinae) often occupy a shared nest site, with the number of females present usually exceeding the cells available for egg laying. This finding suggests that euglossine bee societies have a form of cooperative brood care. Temporary quasi-sociality is also known in a few species of polistine wasps with age polyethism. Young females begin their adult lives as foragers and assist with nest provisioning. Later on, these individuals become egg layers in the common nest site.

Parasocial Insects: Semisocial Societies

C. D. Michener has stated that semisocial insects have an additional characteristic of sociality over the quasi-social species in that the former also have some form of reproductive caste that is cared for by a worker caste. Variation in ovary maturation in spring halictine bee populations suggests that the semisocial condition can arise spontaneously within a mixed assemblage of reproductives of different stages of maturity. Similar patterns of seasonal semisocial behavior occur in some *Polistes* wasps.

Eusocial Insects

The fundamental difference between semisocial and eusocial insects is that eusocial insects have overlapping generations that feature cooperative brood care by the offspring. Eusociality has evolved multiple times within insects and first appeared in the termites. Eusociality in lower termites (Hodotermitidae, Termopsidae, Kalotermitidae, Mastotermitidae, Rhinotermitidae) is associated with trophallaxis and enables all individuals of the nest to be inoculated with gut symbionts necessary for the digestion of cellulose. Trophallaxis also serves to transfer chemical communication in lower and higher termites. Higher termites such as the Termitidae consume sources of cellulose other than wood, lack flagellate endosymbionts, and use anal trophallaxis for transferring chemical cues.

N. Lin and C. D. Michener proposed two distinct pathways of character evolution in social insects. Most bees are believed to have evolved eusociality through a parasocial sequence. Ancestral social bees used a communal social strategy with cooperative nest construction. Later generations switched from a communal to a quasi-social strategy incorporating cooperative brood care within an annual life cycle. The semisocial stage followed the quasi-social stage, with a well-defined worker caste providing cooperative brood care. Fully evolved eusocial species then acquired the characteristic of overlapping generations.

Ants, termites, vespid wasps, and some social bees are believed to have used a subsocial sequence for the evolution of eusociality. The subsocial sequence for the evolution of eusociality is based primarily on the degree of brood care by the female with respect to the age of the offspring. In basal or ancestral taxa, brood care was given only to early instar offspring, with the mother abandoning the young well before maturity. In stage two, brood care is extended to the time during which the young reach maturity. In the final sequence of events, some of the brood become permanent, sterile workers and provide cooperative brood care to the female parent.

Eusociality has also evolved in some gall-forming thrips and aphids. These insects have recently been classified as eusocial because of the presence of a sterile guard caste. However, this form of restricted eusociality differs in important ways from the eusocial condition of Hymenoptera. The guard caste is the only nonreproductive caste in aphid and thrips societies, and the guards do not care for the young. In aphids, this is partly because aphid offspring are precocious like small versions of the adult, they are motile and able to feed themselves.

B. Crespi discovered eusociality in several species of phlaeothripine gall-forming thrips. In these eusocial thrips, the first generations of adults develop into sterile soldiers and are used for defending the gall against invasive species such as ants. These Australian thrips reside in *Acacia* leaf galls or leaf mines and utilize social behaviors for defense of the domicile. The guards also assist with domicile construction. Morphological specialization associated with the guard caste in eusocial thrips includes enlarged prothoracic legs that are used to attack invaders. The prothoracic femora may be expanded and adorned with robust spines. Mortality in the soldier caste is high.

S. Aoki along with Y. Ito have shown that soldier castes are also present in the Cerataphidini or bamboo aphids (Hormaphididae) and in the gall-forming pemphigid aphids. Ito found that some of the Cerataphidini produce a sterile guard caste that defends the offspring. In *Ceratovacuna* and *Pseudoregma,* members of the soldier caste are adorned with sharply pointed spines on the head and are used for colony defense on secondary hosts. In *Pemphigus obesinymphae,* sterile guards are also used for protecting the domicile. The guards are larger than the nymphs and kick at invading predators with their long legs. In social aphids, the guard caste never assists with brood care but serves only to protect the colony from predation. Parthenogenesis is a reproductive mode employed by aphids, but how the soldier caste is determined is unknown.

In 1992 D. S. Kent and J. A. Simpson discovered a eusocial ambrosia beetle (Coleoptera: Curculionidae). A member of the Platypodinae, *Austroplatypus incompertus* lives within heartwood galleries of *Eucalyptus.* Platypodine beetles like *A. incomperus* feed on ambrosia fungi, and nonreproductive females tend the galleries and process fungal mycelia. Kent and Simpson found that established colonies older than 2 years were composed of a single fertilized female, and up to five unfertilized females, which helped maintain the gallery structure and provide protection from predators. The presence of overlapping generations, reproductive/nonreproductive castes, and cooperative brood care within a shared domicile provide evidence for eusociality in these ambrosia beetles.

Within the Hymenoptera, eusociality is found in all ants (Formicidae), vespid wasps (Vespidae), and some bees (Apidae). Members of the subfamilies Bombinae, Meliponinae, and Apinae are eusocial. Sex determination in Hymenoptera follows Dzierzon's rule, first shown by J. Dzierzon in 1845 with honey bees and now termed haplodiploidy. Male progeny are produced from unfertilized eggs through a process called arrhenotoky. In many species of social Hymenoptera, males can be produced at the appropriate time from unmated females within the colony. Workers will sometimes undergo facultative thelytoky to produce diploid eggs. Automictic parthenogenesis produces the diploid eggs in queenless hives of *A. mellifera capensis.* This form of thelytoky is uncommon and is restricted to certain subspecies of the honey bee.

COLONY SIZE AND LONGEVITY IN EUSOCIAL INSECTS

Insect societies may range in size from fewer than 10 to over 3 million adults. Some species of termites (families Termitidae,

Mastotermitidae, and Rhinotermitidae) have colonies with millions of members. Colonies of vespid wasps can have 100 to over 5000 members; colonies of the Allegheny mound ant, *Formica exsectoides,* can exceed 40,000; and nests of *A. mellifera* usually contain between 20,000 and 80,000 individuals.

Species that build nests located on or in the ground tend to have more colony members than species that use more restrictive nest sites. Ground-nesting vespid wasps, for example, usually have larger colony sizes than wasps that build arboreal carton nests. Species that have a large colony size generally tend to have more complex social behaviors than species with small colonies. The correlation of increased colony size with highly evolved social behavior suggests that complex behavioral integration is necessary for colony survival.

Although most data gathered on longevity have been derived from laboratory populations, the female reproductive or queen of some eusocial insects can live for more than 10 years. Well-protected inside the nest, most queens are tended by a court of workers and insulated from the hazards of everyday life. Workers have significantly shorter life spans and are quickly replaced with newer offspring. Attrition is high among workers, and approximately 1% of the worker caste of the army ant *Eciton hamatum* is lost daily. Unlike the queen, workers are exposed to many hazards, and activity outside the nest contributes greatly to worker mortality.

ROLE OF ALTRUISM IN INSECT SOCIALITY

Kin selection is the form of natural selection essential for explaining how the characteristics of eusociality evolved. In 1964 W. D. Hamilton published a mathematical model explaining the economic basis for self-sacrifice as a result of altruism. Hamilton's rule, $C/B < r$, suggests that the ratio of the cost C of sacrifice between two closely related individuals to the benefit B of the receiver r must be smaller than the degree of relatedness between the two individuals. Thus, the inclusive fitness of a group of related individuals is the amount of individual and kin selection. The application of Hamilton's rule to social Hymenoptera, which follow Dzierzon's rule with haploid males and diploid females, is illustrated by the following example. A female offspring inherits 50% of its genome from its father and 25% of its genome from its mother. Hamilton's rule predicts that sisters produced from the same mating should tend to sacrifice themselves for each other because they share 75% of their genetic material with each other and 50% of their genetic material with their mother. Thus, the benefit for an altruistic behavior does not have to be high to spread throughout a population.

In 1977 H. E. Evans proposed that selection factors other than kin selection could explain the evolution of altruism and social behaviors in Hymenoptera. Extrinsic factors such as nest parasitism are just as plausible as kin selection for the evolution of altruism. Evans noted that in general, polygynous (multiple queens per nest) wasp species are more common than monogynous (single egg-laying queen per nest) species.

Evans hypothesizes that polygynous nesting strategies are ubiquitous because they are necessary for protecting the progeny and provisions of the extended family group from nest parasitism. Likewise, polygynous nests can provide many foragers for provisioning the nest as well.

Although altruism can be detrimental to the individual, advantages to the group or colony are many. Most social insects utilize group defensive strategies when the nest or colony is under attack. Aggressive defense strategies observed in social Hymenoptera, social aphids, and termites demonstrate that defensive behaviors are usually detrimental to the defenders but facilitate survival of the reproductives and brood within the nest. Self-sacrifice occurs frequently in social insects. In honey bees, for example, the barbed sting of the worker bee eviscerates and kills the worker after stinging, suggesting that the cost of altruism to the individual is high but trivial to entire colony. In termites, guards will position themselves closest to the source of danger when provoked, whereas other colony stimuli are generally ignored.

Although queen adoption is most often associated with colony multiplication, Wilson believes that the genetic advantages for this strategy are clear. Increasing the number of queens in an isolated colony serves to increase the effective population size while incorporating more altruistic traits into the colony. However, polygyny in ants is not restricted to rare, isolated populations, but can also be found among ubiquitous ant populations as well.

CASTES IN SOCIAL INSECTS

The caste system and polyethism determine the division of labor in insect societies. Castes divide colony members into distinct functional roles within the insect society. Specialized caste members conduct such tasks as defense, brood care, nest construction, thermoregulation, provisioning, and egg laying. Morphological specialization or caste polyethism is a key feature of caste structure. The role of a caste member may also change with age (age polyethism).

Morphological specialization associated with caste polyethism is often the result of allometry and usually is tightly coupled with behavioral changes. Allometry or differential growth rates can provide dramatic changes in morphology with simple change in relative growth rates (Fig. 1). Allometric growth provides termite or ant guard castes with hypertrophied mandibles that are used for defense. The head capsule, which may also be enlarged relative to other body parts, accommodates the mandibular muscles used for biting and dismembering enemies.

Overall change in body size also can be important in defining the queen caste. In most eusocial insects, the queen is larger than nonreproductives. The abdomen, which houses the ovaries, may be grotesquely enlarged in mated termite queens and some ant queens. Such enlargement (physogastry) also is known in replete ants, which gather and store honeydew in the crop.

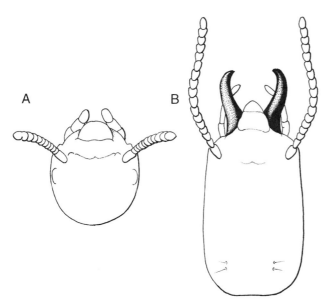

FIGURE 1 Cranial morphological differences in worker and soldier or guard termite castes. In *Reticulotermes flavipes,* workers (A) have short chewing mandibles concealed beneath the clypeus and soldiers (B) have an elongated, enlarged head with prominent mandibles.

The castes of typical social insects can be categorized according to the following classification scheme. Within social Hymenoptera, guards serve as nest defenders and are equipped with well-developed mandibles and potent stings. Workers, the most ubiquitous caste, help with nest construction and repair, thermoregulation, provisioning, and brood and queen care. The queen makes up the reproductive caste, and her egg-laying activity provides the fecundity for maintaining the colony. Most of the members of an insect society are workers, and high numbers of this caste are necessary for maintaining the physiological homeostasis of the domicile.

Many differences are evident between the caste system of termites and social Hymenoptera. In lower termites, the second instars function like workers and are actively involved in nest maintenance. Because, at least in part, of their hemimetabolous development, larval termites are active and also take care of themselves as well. Unless destined to become guards, most apterous termite larvae remain as workers throughout their lives. Termite castes differ from social Hymenoptera castes in

that they contain both sexes, with reproduction restricted to the fertilized queen (Table II). Once established, termite colonies can replenish reproductives from several distinct supplementary reproductive castes. Within the Termitidae, adultoid reproductives can become functional reproductives and replace lost reproductives. Nymphoid reproductives (larvae that possess short wing pads) can become fully functional supplementary reproductives. Ergatoid reproductives are neotenic, apterous individuals with pigmented exoskeletons. Inhibitory pheromones released by functional reproductives prevent the supplementary reproductives from attaining functionality.

In honey bees, morphological and behavioral differences in castes are regulated by endocrine control of development, dietary differences of larval workers and queens, and queen pheromonal conrol. Hormonal developmental control is associated with the activity of juvenile hormone (JH) on target tissue sensitivity to ecdysteroids. This modulates or controls reproductive tissue differentiation in developing larvae. In addition, JH may inhibit the release of ecdysteroid and determine the duration of the last instar feeding stage. In developing queens, JH may inhibit ovariole cell death, a feature common in developing workers. Other hormones also influence ovariole development in larval queens. Makisterone A, an ecdysteroid secreted by the prothoracic glands, promotes ovariole differentiation.

Dietary differences are correlated with reproductive caste development in honey bees. Honey bee worker larvae are fed pollen and honey, whereas queen larvae are fed royal jelly. The hypopharyngeal and mandibular glands of workers produce royal jelly. The production and feeding of royal jelly to select larvae is associated with lowered amounts of queen substance, reduced queen vitality, or queen death; the construction of queen cells is also debilitating to the queen. Different proportions of proteins, lipids, carbohydrates, and trace compounds are apparent in honey and royal jelly, but the active ingredient in royal jelly responsible for queen development is still unknown.

Pheromonal control is also evident in caste formation in honey bees. Queen mandibular gland secretion reduces the amount of JH in developing worker larvae, stimulates worker foraging, promotes queen tending, and inhibits swarming

TABLE II   **Generalized Classification Scheme of Caste Structure in Termites Derived from Wilson**

Caste	Morphological features	Roles
Larvae[a]	Usually apterous (wingless) and lack reproductive and guard caste structures	May develop into workers or guards
	Brachypterous larvae (wing pads present) are known as nymphs	May develop into functional reproductives
Worker	Absence of compound eyes and ocelli; apterous, with reduced pterothorax, well-developed jaws	Nest construction, sanitation, provisioning, brood care
Pseudergate	Regressed nymph, found only in lower termites	May develop into secondary reproductives, soldiers
Soldier/guard	Enlarged mandibles, hypertrophied head glands, head	Defense
Primary reproductives[b]	Queen or male derived from alate adults	Colony founders

[a]Immature termites known as larvae are further grouped by the presence or absence of distinct wing pads.
[b]Primary reproductives become dealate reproductives after shedding their wings.

behavior. Through modulation of JH titers in developing workers, queen substance inhibits ovariole development.

Examples of age polyethism are frequent in social Hymenoptera. In worker honey bees and ants, temporal changes in maintenance behaviors are associated with age. Young workers take care of the queen and feed the larvae, whereas older workers conduct most of their activities outside the nest. Behavioral changes in honey bees also correlate with the thickness of the mandibular, hypopharyngeal, postcerebral, wax, and thoracic glands. The nursing period of worker bees coincides with the maximum development of mandibular and hypopharyngeal glands at between 5 and 10 days of age.

COMMUNICATION IN EUSOCIAL INSECTS

The functional integrity of an insect society is established through complex forms of communication. Communication pathways, either chemical, visual, or acoustic, can be used to convey alarm, simple and multiple attraction for assembly, recruitment to a new domicile or food source, grooming, trophallaxis and exchange of food material, facilitation or inhibition of group or colony effects, nestmate recognition, and caste determination. Trophallaxis and exchange of food material also function as mechanisms for the transfer of chemical information between members of a eusocial insect colony. Sensory cues of many different types are used for communication in social insects, and chemical and acoustic cues have been discovered in termites and social Hymenoptera. Visual communication is limited but, according to Karl von Frisch, plays an important role in the waggle dance of honey bees, in which the bees indicate the location of a food resource through a series of stereotypical movements. Chemical communication is believed to be important for maintaining caste structure, nestmate recognition, and establishment of the queen in an insect society. Chemical signals have an immediate effect that typically ends immediately after the chemical has diffused away from the sensory receptors. Some chemical and acoustic cues have also been implicated for initiating defensive and flight behaviors in social insects.

Queen Dominance and Pheromones

Queen dominance in higher social insects is essential for controlling nestmates and is maintained with a variety of chemical cues in social Hymenoptera. In honey bees such as *A. mellifera,* the mandibular glands and Koschevnikov's gland produce the components of queen substance. In *A. mellifera,* mandibular gland secretion is a pheromone blend composed mostly of 8-hydroxyoctanoic, *(E)*-9-oxo-2-decanoic (9-ODA), *(E)*-9-hydroxy-2-decanoic, 10-hydroxydecanoic, and *(E)*-10-hydroxy-2-decanoic acids. As a newly emerged queen ages and her attractiveness to workers increases, the relative concentrations of these organic acids change. Worker attractiveness is maximized after the queen becomes mated. Although 9-ODA alone does not cause worker court formation, when 9-ODA

is supplied with mandibular gland secretion, court formation occurs. The pheromone 9-ODA is also associated with the inhibition of queen-rearing behaviors and gonad development in worker bees. A possible mode of action of 9-ODA is to reduce the quantity of gonadotrophic hormone so that ovary development is inhibited in worker bees.

Court attraction, in which is a group of workers gather around and tend the queen, is controlled via glandular secretion. Koschevnikov's gland secretions from mated queens are correlated with court attraction. This gland plus mandibular gland secretions may serve to provide an olfactory stimulus for court formation from workers at short distances from the queen. Virgin queens do not form worker court aggregations as readily as do mated queens, which suggests that Koschevnikov gland secretions alone may not be responsible for court formation. Abdominal tergite gland secretions also work as a contact pheromone and facilitate court formation after workers touch the queen's abdominal tergites with their antennae.

Colony Odor

Nest odor is believed to be important for nestmate recognition in social insects. Sources of nest odor are not clearly defined, but a combination of genetic and environmental factors may be important sources of colony odor. Environmental odors from shared foods like nectar or pollen may contribute to the uniqueness of an individual colony's odor. Likewise, food transfer in honey bees may facilitate the spread of environmental odors to all members of the hive or nest. The waxy cuticle of honey bees may also trap environmental odors from the atmosphere of the nest. Cuticular hydrocarbons have been implicated as a means for nestmate recognition in *Polistes* wasps (Hymenoptera: Vespidae).

Queen odors are genetic factors that are believed to be a major component of colony odor. M. D. Breed provided some evidence to support this hypothesis. In 1981 Breed published some research findings suggesting that small groups of worker honey bees are more likely to accept a sister queen over an unrelated queen. Further research has demonstrated that resident bees are able to detect and recognize foreign queens raised under the same environmental conditions. This provides additional support for the presence of queen odor as a component of colony odor.

Trail Pheromones

Trail pheromones are used for recruitment, for marking pathways to resources, and for indicating resource richness. Often released with alarm pheromones, trail pheromones enable guards to aggregate around a nest invader. Furthermore, short-lived and persistent trail pheromones are known. In ants, trail pheromones are produced in the hindgut, Dufour's gland, the venom gland, and the tibial, tarsal, and abdominal glands. In some myrmecine and ponerine ants, the venom gland is the source of trail odors. Odor trails produced by termites originate from the abdominal sternal glands. In *Zootermopsis nevadensis,* sternal gland secretions deposited on a trail from a breached or damaged nest wall to the interior of the nest draws workers toward the area for nest repair.

Odor trails emitted by forest-inhabiting bees provide three-dimensional information on resource location. Some large stingless bees of the genus *Trigona* produce strong odor chemicals from enlarged mandibular glands. *Trigona* forage in tropical forests and utilize mandibular gland secretions to cue in other foragers to nectar sources. Foraging *Trigona* will release more trail pheromone near a food source, thus indicating proximity to the food source for other foragers. The trail pheromone components of *T. subterranea* include both *E-* and *Z*-citrals. Differences in pheromone composition between species may reduce interspecific trail following in stingless bees.

Alarm Pheromones

Alarm pheromones are volatile organic molecules of low molecular weight that diffuse rapidly, forming a concentration gradient away from the signal source. Alarm pheromones initiate arousal, defensive, and assembly behaviors. Alarm pheromones are often not species specific in social Hymenoptera and termites, but the same pheromone often initiates different behaviors in different species. For example, the mandibular gland secretion 4-methyl-3-heptanone is associated with digging behavior in *Pogonomyrmex* species, but in high concentration causes repelling behaviors in the Texas leafcutting ant *Atta texana.*

Contextual responses to alarm pheromones are correlated with the proximity of the nest to the signal's source. Flight occurs when alarm pheromones are perceived far from the nearest nest opening. When these substances are released near the nest, colony members become defensive. Aggregation, defensive postures, and frenzied excitement are often observed in agitated social insects.

Alarm pheromones are often coupled with other glandular secretions such as trail pheromones. Most alarm pheromones originate from the exocrine mandibular gland. Mandibular gland secretions produce a variety of responses that include excitement, attraction, and threat posturing in different species of ants. Other glands associated with alarm signals in ants include Dufour's gland and the anal gland. In honey bees, isoamyl acetate is secreted by the lining of the sting pouch and generates attraction and investigation behaviors. After stinging, honey bees leave the sting and associated glandular components in the adversary's skin, which releases more volatile alarm pheromones for attracting more defenders to the vicinity. The mandibular gland in honey bees secretes 2-heptanone, which initiates alarm behaviors.

The cephalic gland secretions found in termite soldiers are believed to initiate aggressive excitement behaviors in other soldiers and alarm in nonsoldier nest members. Known organic compounds produced by termite cephalic glands include limonene, terpinolene, and α- and β-pinenes.

Acoustic Communication

Termites and ants also use some acoustic alarm signals, and some vespid wasps employ sound to alert offspring to the presence of provisions. Acoustic communication in social insects is conveyed through the substrate, not through the air. Acousticosensory organs present on the legs of social insects detect vibrations transmitted through the substrate.

Ants use subgenual organs and campaniform sensilla located on the legs to detect acoustic signals. Among carpenter ants, the subgenual organ is sensitive to frequencies ranging from 1.5 to 3 kHz. In the leafcutting ant, *Atta cephalotes,* campaniform sensilla function as a sound-detecting organ. Sensilla campaniformae are located near the distal end of the trochanter in *A. cephalotes.* The sensitivity of campaniform sensilla to sound waves varies from the anterior to posterior legs, with the anterior legs being the most sensitive to vibration.

Acoustic signals in ants are generated by stridulation or by tapping the gastral segments on the substrate. Stridulatory files are found universally in pseudomyrmecines, in over 80% of all studied myrmecines, and in nearly half of all studied ponerine ants. Stridulatory files when present are located at the junction of the abdomen (fourth abdominal tergite) with the postpetiolus. Buried or confined ants use stridulatory files to generate acoustic signals through the substrate for initiating digging behaviors in other workers.

Formicine ants are believed to have secondarily lost their stridulatory files and none use different forms of auditory communication. *Campanotus* (carpenter ants) use a combination of mandibular and gastral tapping when disturbed. Response to these alarm cues depends on the state of arousal in the signal receiver. In some cases, the tapping behavior causes some ants to lie motionless, presumably to make them less visible to predators, while more agitated ants orient toward the source of the signal and approach it. Outside the nest, tapping enhances the stimuli associated with disturbance and functions as a danger alarm.

Similar tapping behaviors have also evolved in some vespid wasps but not as an alarm cue. Gastral vibration is believed to signal food or provisioning to larvae. The abdomen or gaster is rapidly tapped on the surface of the comb when provisions are brought into the nest. In *Vespa tropica,* foundresses tap their legs on the nest comb to signal the presence of provisions to hungry larvae. In this species, the sounds produced by leg tapping are loud and audible to humans standing at least 1 m from the nest.

In termites, sounds are generated by rapidly tapping the head or abdomen against the substrate. Head banging in *Zootermopsis* is audible as a rustling sound and is believed to function either as a substrate-transmitted alarm call or as a defensive behavior meant to scare intruders away from the nest. Research on the response of termite subgenual organs to the vibrations generated by head tapping behavior (1 kHz) suggest that these acousticosensory organs are highly sensitive to that frequency.

INQUILINES AND NEST PARASITES OF SOCIAL INSECTS

Eusocial insect nests often harbor guests that take advantage of the food resources and microclimates inside the nest chambers. Arthropod ectosymbionts of social insects are numerous and include some species of mites, spiders, millipedes, isopods, and insects. Most orders of insects contain some ectosymbiotic species. Some nest invaders, such as wax moths (Lepidoptera: Pyralidae), feed on the nest structure of honey bees, whereas others may feed on the young, stored provisions, or scavenge from the debris.

Some nest owners receive a benefit (mutualism) for harboring another species. Trophobionts like aphids (Heteroptera: Aphididae), scale insects (Heteroptera: Coccidae), and mealybugs (Heteroptera: Pseudococcidae) are often protected and sheltered by ants. In return, ants receive from these insects honeydew, a sugary fluid, while providing protection and shelter to these guests. Some have suggested that the posterior morphology of aphids mimics the head region of an ant, with honeydew secretion simulating the transfer of food between ant workers.

A diverse group of guests is usually encountered with army ant colonies. R. D. Akre and C. W. Rettenmeyer studied the association between ectosymbiotic staphylinid beetles and army ants. These beetles have evolved two distinct behavioral strategies to gain access to the resources provided by army ant colonies. Specialized species such as *Ecitomorpha* and *Ecitiosus* are mimetic forms that live within the bivouacs of army ants. They are highly integrated into army ant societies and die if removed from the confines of the colony. Often associated with the larvae or booty, these staphylinid beetles tend to move together within an emigrating column of ants. Phoresy is common, with army ant workers transporting and moving these staphylinids to new bivouacs. *Microdonia* uses a generalized strategy for exploiting army ants. The association between generalists and their host ants is less integrated than specialized species. They usually frequent the periphery of an army ant colony and will often attack wounded or dying ants. The ants will sometimes attack and attempt to drive these insects away from the nest area.

Evolved mechanisms that enable symbiotic associations between symphiles and their hosts may include behavioral and chemical mimicry. Many symphiles can attract hosts and through appeasement gain access to the nest. B. Hölldobler was one of the first researchers to discover that scents are important for attraction of worker ants to larval Staphylinidiae. Some researchers believe that trichome secretions have an intoxicating effect and cause disorientation in host species. The reduviid bug *Ptilocerus ochraceus*, an ant predator, uses a similarly acting substance to attract and paralyze *Dolichoderus* ants. This bug presents its abdominal trichomes to an ant, which begins to lick the trichome hairs leading to intoxication.

The variety of interactions between different species of ants is extensive. Interactions range from coexistence to true parasitic relationships. In plesiobiosis, distantly related ant species can coexist with minimal interactions between closely neighboring nests. Some ant species rely on cleptobiosis or thievery to acquire food or refuse from nests of ants of other species. Cleptobiotic species reside in separate domiciles from their hosts. Lestobiosis differs from cleptobiosis in that one species will actually invade the host's nest to prey on the young or food cache. For example, ants of the genus *Carebara* nest within the walls of termite nests and are believed to prey on termites.

Mixed species colonies usually involve some form of social parasitism. Temporary social parasites were first recognized by Wheeler in 1904. Ants in the *Formica microgyna* group utilize this type of life history strategy. A newly fertilized queen enters a host colony and coerces the host workers to take over the nest through assassination of the host queen. Gradually, the parasitic queen's offspring replace the host workers through attrition.

Slave-making ants invade a host nest and steal pupae for incorporation in their own nests. After eclosion, the slaves work as foragers and as nest constructors, and conduct brood care for the slave makers. Slave makers parasitize other ant species that are close relatives. Some species of *Polyergus* use *Formica* species for slaves, *Formica* species of the *sanguinea* group typically use other *Formica* species as slaves.

Permanent parasitism or inquilinism occurs entirely inside the host's nest. In some species, workers are present but have limited behavioral roles within the host's nest. For example, *Teleutomyrmex schneideri* no longer has a worker caste and represents a highly refined form of inquilinism. *T. schneideri* lives its entire life within the nest of *Tetramorium caespitum,* a close relative of *T. schneideri*. The queens of *T. schneideri* are smaller than the queens of *T. caespitum* and are morphologically patterned to ride on the backs of the host queen. *Teleutomyrmex* queens release attractants, from cuticular glands of the thorax and petiole and are eagerly tended by the host's workers.

See Also the Following Articles

Caste • *Colonies* • *Division of Labor* • *Hymenoptera* • *Isoptera* • *Parental Care* • *Pheromones* • *Sex Determination*

Further Reading

Akre, R. D., and Rettenmeyer, C. W. (1966). Behavior of Staphylinidae associated with army ants (Formicidae: Ecitoninii.) *J. Kans. Entomol. Soc.* **39(4)**, 745–782.

Aoki, S. (1987). Evolution of sterile soldiers in aphids. *In* "Animal Societies: Theories and Facts" (Y. Ito, J. L. Brown, and J. Kikkawa, eds.), pp. 53–65. Japan Science Society Press, Tokyo.

Breed, M. D. (1981). Individual recognition and learning of queen odors by worker honey bees. *Proc. Natl. Acad. Sci. USA* **78**, 2635–2637.

Cocroft, R. B. (1999). Offspring–parent communication in a subsocial treehopper (Hemiptera: Membracidae: *Umbonia crassicornis*). *Behaviour* **136(1)**, 1–21.

Crespi, B. (1992). Eusociality in Australian gall thrips. *Nature* **359**, 724.

Eickwort, G. C. (1981). Presocial insects. *In* "Social Insects," Vol. II. (H. R. Hermann, ed.), pp. 199–280. New York, Academic Press.

Evans, H. E. (1977). Extrinsic versus intrinsic factors in the evolution of insect society. *BioScience* **27**, 613–617.

Frisch, K. V. (1967). Honey bees: Do they use direction and distance information provided by their dancers? *Science* **158**, 1072–1076.

Hamilton, W. D. (1964). The genetical evolution of social behavior, I and II. *J. Theor. Biol.* **7**, 1–52.

Hölldobler, B., and Wilson, E. O. (1990). "The Ants." Belknap Press of Harvard University Press, Cambridge, MA.

Ito, Y. (1989). The evolutionary biology of sterile soldiers in aphids. *Trends Ecol. Evol.* **4**, 69–73.

Kent, D. S., and Simpson, J. A. (1992). Eusociality in the bark beetle *Australoplatypus incompertus* (Coleoptera; Curculionidae). *Naturwissenschaften* **79**, 86–87.

Lin, N., and Michener, C. D. (1972). Evolution of sociality in insects. *Q. Rev. Biol.* **47**, 131–159.

Michener, C. D. (1974). "The Social Behavior of Bees: A Comparative Study." Harvard University Press, Cambridge, MA.

Roberts, R. B., and Dodson, C. H. (1967). Nesting biology of two communal bees, *Euglossa imperialis* and *Euglossa ignita* (Hymenoptera: Apidae), including description of larvae. *Ann. Entomol. Soc. Am.* **60(5),** 1007–1014.

Wheeler, W. M. (1904). A new type of social parasitism among ants. *Bull. Am. Mus. Nat. His.* **20(30),** 347–375.

Wheeler, W. M. (1923). "Social Life Among the Insects." Harcourt Brace, New York.

Wilson, E. O. (1971). "The Insect Societies." Belknap Press of Harvard University Press, Cambridge, MA.

Soil Habitats

Patricia J. Vittum
University of Massachusetts

Many basic and applied studies in insect ecology have considered responses of insect populations to their physical or chemical environment. For insects that live above ground, the mechanisms of behavioral responses to environmental factors often are directly observable. However, behavioral responses of soil-inhabiting insects are much more difficult to observe and quantify. Soil texture and structure can have a direct impact on arthropod behavior and adaptations. Field studies of soil insects often quantify the consequences of behavior, whereas the actual behaviors are only inferred. There have been studies of ecological and physiological adaptations of several soil arthropods that are not considered to be agricultural pests but have certain life history features that make them amenable to investigation. However, most insects that are agricultural pests as soil-inhabiting immature forms (e.g., wireworms or rootworms in corn, scarab grubs in turf grass) often have been studied in detail only in their more accessible, or observable, adult form. This is in part because of the logistical challenge of observing the movements and other behaviors of the soil-bound larvae.

CHARACTERIZATION OF THE SOIL ENVIRONMENT

Soil Solids

Most soils have a complex structure, consisting of solids, liquids, and atmospheric gases. The solid components (e.g., sand or clay) constitute much of the soil matrix in bulk. Inorganic, or mineral, particles range in size from clays (< 0.002 mm in diameter) to silts (0.002–0.05 mm) to sands (0.05–2 mm) to pieces of gravel (> 2 mm). The proportion of different sized particles, or soil texture, determines to a large extent the physical properties and appearance of a soil, as well as its ability to supply chemical nutrients to plants. The soil texture also has an impact on the soil arthropod population. For example, small soil arthropods sometimes find it difficult to move in heavy clay or tightly compacted soils.

Soil organic matter consists of an accumulation of partially disintegrated and decomposed plant and animal residues that have been broken down and resynthesized by microorganisms in the soil. Although organic matter usually does not constitute more than 6% by weight in topsoil (and even less in subsoils), it binds mineral particles into slightly larger granules that produce loose, crumbly soils that can hold more water than their mineral counterparts. Organic matter also is a primary source of energy for a variety of soil organisms, including many arthropods.

Pore Spaces

A typical loam soil will consist of roughly 50% soil solids (a combination of sand, silt, and clay) and 50% pore spaces and water. The size and distribution of pore spaces will depend on the size and shape of the mineral particles, as well as the activity of microorganisms. A predominantly clay soil will usually have very small pore spaces because the clay particles are very small and can pack together effectively. A sandy soil will tend to have much larger pore spaces because the sand grains are more irregular in shape and do not compact as readily. Atmospheric gases (most notably, oxygen and carbon dioxide) also occupy pore spaces and can move passively through the soil profile, depending on surface conditions.

Pore spaces are further characterized as micropores (< 0.06 mm) and macropores (> 0.06 mm). Macropores tend to allow movement of air and percolating water very readily, whereas micropores are the first to be filled with water in a moist field soil and do not permit much air movement into or out of the pores.

Convection can enhance the exchange of gases within a soil or between a soil and the atmosphere above. Aeration (movement of oxygen and other gases) near the surface of a soil occurs most readily in the presence of large, interconnected pore networks or channels. The rate of aeration is influenced by changes in barometric pressure, temperature gradients, and wind gusts. Temperature, relative humidity, surface texture, and continuity of soil pores affect the diffusion of gases into and out of soil.

Soil Water

Water (and dissolved minerals) accumulates in pore spaces and moves vertically through the soil profile if the surface input (e.g., rain or irrigation) exceeds the rate at which any vegetation absorbs the water. Water moves most readily through soils that have well-spaced, interconnecting macropores, but it also fills many of the micropores closest to the surface. Since water displaces air in the pore spaces, soils that are saturated with water cannot retain atmospheric gases that are critical to plant growth.

Soil moisture tension determines how much water remains in the soil at equilibrium and is a function of the sizes and volumes of pore spaces (matric potential), the presence of solutes in the soil (osmotic potential), and gravity. When the surface of a soil dries following an extended period of dry weather (further enhanced by low humidity or steady winds), water can also be drawn back toward the surface through capillary action.

The relative concentrations of oxygen, carbon dioxide, and water vapor often are considerably different in soil pore spaces (soil atmosphere) and in the open air. Oxygen concentrations tend to be lower and CO_2 concentrations tend to be higher in the soil atmosphere as a result of plant and animal respiration and biochemical soil processes. Relative humidity tends to be relatively high, particularly in soils used to produce agricultural crops.

Soil Temperature

Surface temperatures of soils often fluctuate at least as much as ambient air temperatures, but the difference between the daily maximum and minimum temperature decreases as depth in the soil increases. There is a delay of maximum and minimum soil temperatures compared with the overlying surface temperatures, correlated with depth. Seasonally the maximum and minimum temperatures occur in the warmest and coolest seasons, respectively, in the upper soil layers. In contrast, the highest temperatures occur in early winter and the lowest temperatures occur in midsummer, at 7 m depth. Soil insects and other arthropods often move vertically in direct response to soil temperature, moving downward in late autumn to avoid freezing temperatures on the surface and returning to the surface in the spring to resume feeding.

FORMS OF LIFE IN THE SOIL

A vertical cross section of a soil profile reveals several distinct layers. The lowest layer, or horizon, is the C horizon, which consists of unweathered rock. The B horizon contains weathered, rough mineral soil with small deposits of humus. The A horizon normally has fine mineral soil interspersed with organic matter. The O horizon is a layer of plant debris lying on the surface of the mineral soil. The thickness of this layer depends on the amount of vegetation that is deposited seasonally and annually, and on the amount of degradation that occurs as a result of soil organisms. Within the O horizon are several layers, including (from the top) leaf litter, a fermentation layer, and a humus layer. The humus layer often merges into humus-enriched topsoil.

Soils and the overlying organic layer are not homogeneous, but rather are stratified. Similarly, the arthropods that live in these regions are grouped into different life-forms that have adapted to the various conditions that exist in the soil.

Euedaphons

Euedaphic soil arthropods inhabit the lowest soil layers, generally moving within the soil pore system. These arthropods usually are small and are characterized by a round or wormlike body form. The body size corresponds to that of the pore system, and extremities are often reduced. Because they cannot escape predators, many euedaphons generate and release toxic or defensive substances. Most euedaphons are photophobic and either lack eyes or have eyes that have degenerated considerably. They tend to have well-developed mechano- or chemosensitive organs, which compensate for their poor or nonexistent vision. Arthropods occurring in the euedaphon include several species of proturans, diplurans, and symphylans, as well as a few oribatid mites.

Epedaphons

Epedaphic arthropods live on the soil surface and in leaf litter. They are not well adapted to the conditions found in the soil pore system (e.g., high relative humidity, restricted gas exchange, restricted mobility). They are represented by many different body forms, usually are strongly pigmented, and often are dorsoventrally flattened. They have well-developed sensory organs and are highly mobile. Arthropods occurring in the epedaphon include oribatid mites (Oribatei), springtails (Collembola), ectobiid cockroaches, several cricket species, and several predatory beetles, including rove beetles (Staphylinidae) and ground beetles (Carabidae).

Hemiedaphons

The hemiedaphon group represents a transitory form of life, enabling some epedaphic or atmobiotic arthropods to occupy burrows in the soil. Hemiedaphic arthropods have the ability to excavate through soil by means of modified mouthparts or fossorial legs, and often they can enlarge existing cracks and pores. Several insect taxa have adopted a hemiedaphic life for a variety of reasons: to dig channels and then wait for surface prey to fall into the pit, to burrow through the soil hunting for small epedaphic arthropods, to avoid temperature or moisture extremes on the surface, or to feed on roots of plants. Hemiedaphic arthropods include earwigs (Dermaptera), field crickets and mole crickets, tiger beetles (Cicandellidae), and white grubs (scarab larvae).

ADAPTIVE STRATEGIES OF SOIL ARTHROPODS

Response to Soil Texture

Organisms in the soil (and leaf litter) community play very different roles, based in part on their size. Organisms (e.g., protozoa, bacteria, and some nematodes) that exist in water films, often in soil micropores, have resource requirements and defense needs that differ from those of organisms able to move in and out of soil pores independently. Similarly, a soil

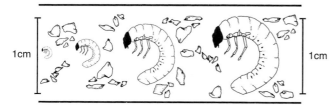

FIGURE 1 Schematic representation of the impact of soil particle size on movement of a soil insect. Japanese beetle larvae (neonate, late first instar, mid second instar, and mid third instar) shown in a square centimeter of typical loamy sand soil. [Drawing by Stephen L. Thomas, University of Massachusetts, adapted from Villani, M. G., and Wright, R. J. (1990). Environmental influences on soil macroarthropod behavior in agricultural systems. *Annu. Rev. Entomol.* **35**, 249–269.]

macroarthropod, as it grows, will perceive the soil matrix differently (Fig. 1). A neonate exists functionally as a microarthropod, able to move only within existing pores in the soil. Thus its ability to move is a function of the porosity of the soil (including the size of the pores and their continuity). As the arthropod grows, less of the pore space is available for free movement. At this point pore space is less important in impeding movement than gross soil structure (impacting insect movement among soil aggregates) and aggregate density (movement through the aggregates). Plant root activity, surface cover, traffic and other sources of compaction, and density of soil organisms all impact aggregate formation. As the insect grows, less of the soil pore space is available to the insect for free movement, but it may take advantage of preexisting soil channels created by soil macrofauna such as earthworms, other arthropods, or small vertebrates. Water-filled soil pores can inhibit movement.

Response to Temperature

Several soil insect species demonstrate a seasonal pattern of vertical movement associated with soil temperature. In temperate climates, many soil macroarthropods move downward in late autumn to avoid freezing and return to the upper soil layers in spring. Some species move away from the soil surface in the middle of summer, in part to avoid high soil temperatures. Species-specific responses to soil temperatures influence periods of feeding activity and may enable similar species to occupy slightly different niches (e.g., wireworms, scarab grubs).

Response to Moisture

Some hemiedaphic arthropods have developmental stages that cannot tolerate extremes in moisture, and many of those stages are very sensitive to soil moisture levels. For example, grubs of several scarab species that had been held in dry soils moved upward almost immediately after moisture was applied to the surface. Several macroarthropods, including some wireworm species, alter the soil environment by creating semipermanent earthen cells. These temporary cavities enable the insects to create nearly saturated chambers in the soil, greatly reducing moisture loss from evaporation. Numerous desert arthropods

create earthen cells in which they can aestivate when soil moisture is extremely low (or temperature is extremely high).

Usually, eggs and pupae of insects are most resistant to moisture loss and least able to escape undesirable conditions. Larvae and adults are often mobile stages and may be able to move away or alter their behavior to minimize the impact of adverse moisture extremes. For example, scarab larvae can move downward in the soil profile to seek moister (and cooler) conditions during periods of heat or drought stress. Heavily sclerotized soil arthropods may be less vulnerable to cuticular moisture loss than are less sclerotized forms, such as grubs or maggots.

Locomotion

Edaphic arthropods move through soil to locate food, to escape predators, or to escape unfavorable abiotic conditions. Adaptations for movement in soil depend on soil type, particle size, pore size, and soil density, among other things. Although many epedaphic species use their bodies as wedges, euedaphic species tend to dig through soil, using their legs and mandibles as shovels. Legs of edaphic insects often are highly modified to facilitate digging or burrowing through soil. At least one leg segment is likely to be enlarged, specially shaped, and edged with spines or lobes to create functional spades. Expanded tibiae or femora provide increased surface area, providing improved leverage when the insect is moving soil. For example, mole crickets (Orthoptera: Gryllotalpidae) use their greatly enhanced fossorial forelegs to burrow through soils very rapidly.

Many euedaphic arthropods lack obvious modifications for digging, but are able to move through soil by fitting between particles or by moving in existing burrows or crevices. Body shapes may be flattened or cylindrical, but in general antennae and legs tend to be reduced or absent. Millipedes serve as an example of the variety of adaptations that have arisen over time. Some species act as bulldozers with long bodies, many legs (for purchase against soil particles), and broad heads; other species have shorter bodies, fewer but longer legs, and tapered heads that allow the millipedes to wedge through small spaces in the leaf litter or upper layers of soil. Still other species of millipedes have very tapered anteriors and compressible bodies that can be used to widen crevices. The heavily sclerotized elytra and terga on adult beetles can provide a very effective protective shield when an insect is pushing aside soil particles or layers of leaf litter.

Host Finding

Much of the behavioral research that has been conducted over the past 50 years has centered on atmobiotic insects, in part because until recently it was virtually impossible to observe soil insects *in situ* without disturbing them. Destructive sampling techniques enabled researchers to make quantitative assessments but revealed very little information about how or why soil insects exhibited certain behaviors.

For many hemiedaphic insects, host finding begins with choices made by the mobile adult female as she seeks oviposition sites. For example, a female corn rootworm beetle (*Diabrotica*) may oviposit in a field where corn is growing, but if the field is planted to a different (nonhost) crop the following growing season, the neonates that emerge in the spring may have to move relatively large distances to reach a suitable host. Similarly, emerging scarab larvae must locate suitable roots and begin feeding within 24 to 48 h of eclosion, and overwintering grubs must relocate to suitable host plants when returning to the upper soil layers in the spring.

There are many plant-derived chemicals that elicit responses in insects in general, including host plant extracts that initiate host-searching behaviors or avoidance mechanisms. Many of the most intensively studied phytochemicals are produced in leaves or stems, but soil insects are more likely to respond to chemicals produced in the root zone. Some of these compounds are quite specific and elicit responses (e.g., host seeking) from limited taxa. Others, like carbon dioxide, are not species specific and influence a wide range of soil insects. Chemicals produced in the soil (typically as root extracts) may diffuse over relatively large distances, but the diffusion rate depends on soil moisture, texture, and compaction.

Defensive Adaptations

Edaphic arthropods produce a variety of compounds that can function as contact toxins, repellents, or irritants. Many different taxa use similar biosynthetic pathways to produce closely related compounds. For example, several soil arthropod groups, including some millipedes and centipedes, as well as some chrysomelid larvae, produce hydrogen cyanide. The glands that produce this nonselective toxin apparently are not homologous, suggesting that the capability has evolved more than once.

Some edaphic arthropods have developed chemical defenses against deep-soil predators that use mechanoreceptors and chemoreceptors to provide cues to locate their prey. Others evade predators by running or jumping. Epedaphic Collembola (springtails) that live in leaf litter are cryptically colored and have a highly evolved mechanism that enables the insect to jump away from a disturbance virtually instantaneously. In contrast, Collembola that live wholly within the soil (and thus are constrained by soil particles) tend to be smaller and paler, and to have less well developed jumping mechanisms than epedaphic species. Instead, the euedaphic springtails secrete noxious fluids that protect them against many predators.

Interactions of Soil Insects with Chemical Control Agents

Edaphic insects that are considered to be pests in production agriculture or the green industry are often much more difficult to "control" than their atmobiotic counterparts. One of the greatest challenges is achieving adequate contact of an insecticide with the target insect. Many insecticides

dissipate or degrade relatively quickly (before they reach the soil) or are adsorbed to soil particles. Most soil insecticides (and other pesticides) remain in the top 5 to 10 cm of the soil, where they are subject to microbial degradation. Soil factors such as pH, organic matter, moisture, temperature, and microbial community diversity will have a direct impact on the mobility and persistence of a soil insecticide. Insecticides that are highly mobile in soil may be ineffective because they move beyond the target zone too rapidly.

Many soil insects can detect the presence of insecticides and other chemicals and will initiate avoidance behavior (e.g., moving away from the soil zone in which the chemical is detected). Abiotic factors, such as soil moisture or temperature, that induce a target insect to move as little as 1 cm further into the soil profile may place target insects beyond the effective "range" of some chemical control agents. Often the manipulation of irrigation apparatus or the use of application equipment that incorporates a control agent directly into the soil at the desired depth can enhance the efficacy of a soil insecticide.

IMPACT OF SOIL ON PATHOGENS, PREDATORS, AND PARASITOIDS

Pathogens

The upper layers of most soils, as well as leaf litter, support active microbial communities. Many of these organisms are natural decomposers, breaking down plant and animal tissues that ultimately become part of the organic matter in the underlying soil. Interestingly, many edaphic arthropod populations are nearly free of disease. Several studies have demonstrated that secretions from many different edaphic arthropods inhibit vegetative growth or suppress the germination of pathogenic organisms.

Edaphic arthropods that are highly susceptible to pathogenic organisms in laboratory tests are rarely infected in the field, suggesting that there may be a critical behavioral component. For example, several soil insects, including earwigs (Dermaptera), mole crickets, and ants, tend and lick eggs. This behavior may remove fungal spores or bacteria or inhibit their germination. In addition, some laboratory studies utilize artificially dense populations of the target arthropod, which may enhance the spread of pathogens from one organism to another.

Nevertheless, several insect pathogens occur naturally in soils. For these pathogens to induce an epizootic, three conditions must be met. A susceptible host must be present, with host susceptibility being governed by population density, species composition, presence or absence of other stress factors, and behavioral responses to the pathogen. A pathogen must be present, with a suitable level of virulence and persistence. Finally, the environment must support both the host and the pathogen. Soil conditions, such as low temperature or high moisture levels, may stress the target insect population, predisposing individuals to infection and ultimately leading to population declines.

Several microbial insecticides were identified and developed during the last half of the 20th century. All were found in naturally occurring epizootics in the soil and were subsequently commercialized. Microbial insecticides are passively mobile because an insect that comes in contact with the microbial product may move some distance from the initial point of contact before it dies. The behavior of the target insect can be important to the spread of the pathogen, particularly when normal (or pathogen-induced) behaviors result in movement of individuals beyond their normal range. Most microbial insecticides are able to replicate within the host.

In some instances, edaphic arthropods are able to detect and avoid fungal pathogens in soil. In 1994 Villani *et al.* conducted a series of microcosm studies proving that when incorporated into soil, mycelial formulations of *Metarhizium anisopliae,* a naturally occurring soil fungus, repel third-instar Japanese beetle *(Popillia japonica)* grubs for as long as 3 weeks. Similar responses have been observed with tawny mole crickets and subterranean termites.

Examples of microbial insect pathogens include bacteria (e.g., *Bacillus popilliae, B. thuringiensis, Serratia marcescens, S. entomophilas*), fungi (e.g., *M. anisopliae, Beauveria bassiana, Fusarisum* spp., *Penicillium* spp., and *Aspergillus* spp.), protozoa (e.g., *Ovavesicula popilliae*), and various rickettsia (bacteria-like) organisms. Many of these have been developed commercially, with varying degrees of success.

External factors relating to soil condition may modify arthropod behavior. Localized flooding may force edaphic insects to move to unsaturated soils, whereas after several weeks without rain these insects may seek moister (usually lower) ground. These localized migrations may result in contacts between populations of edaphic insects and pockets of pathogen activity that otherwise might not have occurred. Disruptions in social behavior may increase or decrease infection rates. For example, studies have demonstrated that applications of sublethal rates of certain soil insecticides, such as imidacloprid, greatly increase the pathogenicity of some pathogens, including *B. bassiana* and some entomopathogenic nematodes (see next section). Apparently the sublethal exposure of the insecticide disrupts normal social behavior in the termite colony, including grooming, trophallaxis (food exchange), and construction of tunnels.

Predators and Parasitoids

Insect predators are mobile and self-replicating. Their effectiveness depends on the interaction of the soil environment with both the agent and the host. The initial contact and subsequent spread through the population depend on temporal and spatial overlap. The predator must be well adapted to the soil conditions. In particular, the predator must be able to move through the soil and to respond to host cues. Some predators are edaphic arthropods, such as predatory beetles or spiders.

Entomopathogenic nematodes can also be considered to be predators, since they move through the soil in search of host insects. Some species move actively, whereas others are more passive, ambushing prey as it moves nearby. Entomopathogenic nematodes enter the host through natural openings, such as the mouth, spiracles, or anus. Nematodes in the genera *Heterorhabditis* and *Steinernema* carry a pathogenic bacterium, which is released in the body cavity of the host. The bacteria multiply within the host and produce toxins that kill the host rapidly. Although several nematodes species are available commercially, their efficacy in field conditions has been inconsistent, in part because the nematodes are extremely sensitive to temperature and soil moisture levels.

Several insects have evolved as parasitoids of edaphic arthropods. Host finding is presumed to be much more complex than for atmobiotic hosts because any host volatiles must move through the soil matrix. The parasitoid also must be able to move through the soil structure to reach the host. Studies attempting to evaluate the specific host-finding behaviors of these parasitoids must include consideration of soil texture, moisture, and temperature, as well as root zone exudates and sensitivity to movement-induced vibrations.

SAMPLING TECHNIQUES FOR SOIL INSECTS

One of the greatest challenges facing soil insect ecologists is the need to develop techniques of following insect movement and feeding behavior *in situ* while minimizing disturbance of the soil system. Until recently, researchers relied on the "snapshot" approach of collecting soil samples and sampling destructively, to determine how many soil insects might be present in the sample. This approach provided limited quantitative assessments but could not provide information over an extended period of time.

One technique that seems to be ideally suited to observing soil insect behavior is radiography, which has been used to trace the movement of scarab grubs and mole crickets in turfgrass, wireworms in corn, and onion maggots in onions, among other arenas. Plastic boxes of varying dimensions are filled with soil, and virtually any edaphic arthropod can be introduced to the microcosm and observed without disturbance. The technique has been expanded since Villani and Wright first described some of the applications in 1988. It has been used to investigate the response of soil insects to the presence of pathogens, to study the movement of two different species in a confined space, and to conduct preliminary basic observations of species behavior in soil. Understanding of insect movement in soil has expanded tremendously as a result of radiographic observations.

See Also the Following Articles

Collembola • Japanese Beetle • June Beetles • Isoptera

Further Reading

Brady, N. C. (1990). "The Nature and Properties of Soils." 9th ed. Macmillan, New York.
Eisenbas, G., and Wichard, W. (1987). "Atlas on the Biology of Soil Arthropods." Springer-Verlag, Berlin.

Hillel, D. (1980). "Fundamentals of Soil Physics." Academic Press, New York.

Villani, M. G., and Wright, R. J. (1988). Use of radiography in behavioral studies of turfgrass-infesting scarab grub species (Coleoptera: Scarabaeidae). *Bull. Entomol. Soc. Am.* **34,** 132–144.

Villani, M. G., and Wright, R. J. (1990). Environmental influences on soil macroarthropod behavior in agricultural systems. *Annu. Rev. Entomol.* **35,** 249–269.

Villani, M. G., Krueger, S. R., Schroeder, P. C., Consolie, F., and Consolie, N. H. (1994). Soil application effects of *Metarhizium anisopliae* on Japanese beetle (Coleoptera: Scarabaeidae) behavior and survival in turfgrass microcosms. *Environ. Entomol.* **23,** 502–513.

Villani, M. G., Allee, L. L., Diaz, A., and Robbins, P. S. (1999). Adaptive strategies of edaphic arthropods. *Annu. Rev. Entomol.* **44,** 233–256.

Sound Production

see *Hearing*

Spermatheca

Marc J. Klowden
University of Idaho

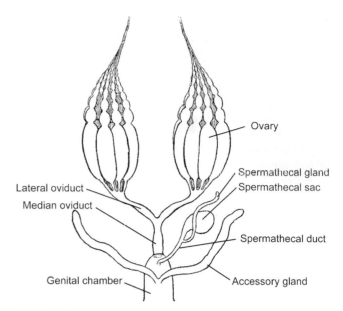

FIGURE 1 The generalized structure of the female reproductive tract, showing the location of the spermatheca. [After Snodgrass, R. E. (1935) "Principles of Insect Morphology." McGraw-Hill, New York.]

The spermatheca is a special pouch in the female in which spermatozoa are stored and maintained after mating. In more advanced insects, the male deposits seminal fluid or a spermatophore in the bursa copulatrix, and the sperm move from the bursa into the spermatheca. The sperm often remain viable within the female's spermatheca for her entire life, which may be as long as 3 to 4 years for social insects such as the honey bee queen. The sperm are released from the spermatheca only when eggs pass down the oviduct so fertilization may occur just before the eggs are laid.

SPERMATHECAL STRUCTURE

A female insect typically has only one spermatheca, but some coleopterans have two, and there are three spermathecae in some dipterans. The spermathecae vary considerably in their overall structure from insect to insect, but as outgrowths of the median oviduct, they are all ectodermal in origin and produce a cuticular lining. They generally arise from the median oviduct near or on the genital chamber (Fig. 1). The spermathecal sac (or receptaculum seminis) is connected to the genital chamber by a secretory duct (or ductus seminalis) through which the sperm are discharged. There also may be specialized secretory cells associated with the capsule that produce secretions that maintain the viability of the sperm and activate them. The sperm move quite actively within the spermathecal sac, and the secretions of the secretory cells are

assumed to provide nourishment to them. Secretions of the spermatheca may also be directed into the hemocoel. In the blood-sucking bug, *Rhodnius prolixus,* a factor produced by the sperm-filled spermatheca is responsible for the increased rate of egg development that is associated with mating.

UTILIZATION OF SPERM

The spermatheca may be innervated by nerves that branch from the terminal abdominal ganglion and control the release of sperm. When sensory receptors on the oviduct are activated by the presence of an egg, a reflex arc through the terminal abdominal ganglion activates a motor neuron that triggers the contraction of muscles surrounding the spermathecal sac. Sperm are squeezed through the spermathecal duct to the oviduct, where the egg is fertilized before it passes out of the body.

There may not always be a random utilization of sperm present in the female spermatheca. When sequential matings occur, although the sperm from different males may be mixed within the storage organ, the sperm from one mating may be used preferentially. Most often there is precedence in the utilization of sperm so that the last sperm to enter are the first ones to leave, increasing the probability that a second mating will be used to fertilize the eggs. To ensure paternity in second matings to an even greater degree, male insects have evolved some novel mechanisms of sperm displacement. For example, the male of the damselfly, *Calopteryx maculata,* uses its scooplike penis both to transfer its own sperm to the female and also to remove from the spermatheca sperm from a previous mating. An unusual method of insemination and sperm storage occurs in some members of the heteropteran superfamily Cimicoidea, which includes the bedbugs. Males puncture the cuticle of the female, injecting the sperm into a specialized tissue mass there. The sperm may then move to the ovaries or remain within the tissue mass for storage, as if in a spermatheca.

See Also the Following Articles
Accessory Glands • Reproduction, Female

Further Reading
Clark, J., and Lange, A. B. (2000). The neural control of spermathecal contractions in the locust, *Locusta migratoria. J. Insect Physiol.* **46**, 191–201.

Davey, K. G. (1985). The female reproductive tract. *In* "Comprehensive Insect Physiology, Biochemistry and Pharmacology," Vol. 1. (G. A. Kerkut and L. I. Gilbert, eds.), pp. 15–36. Pergamon Press, Oxford, U.K.

Okelo, O. (1979). Mechanisms of sperm release from the receptaculum seminis of *Schistocerca vaga* Scudder (Orthoptera: Acrididae). *Int. J. Invertebr. Reprod.* **1**, 121–131.

Spermatophore

Marc J. Klowden
University of Idaho

In the more primitive insect groups, males synthesize specialized sperm carriers called spermatophores to protect the sperm during their transfer to the female genital tract. Insects are believed to have descended from aquatic ancestors whose males released their sperm directly into the water. In adaptation to a terrestrial existence, however, a more precise method of mating was required to prevent the sperm from desiccating. The spermatophore thus represents an initial adaptation for life on land that protects the male gametes until they are within the female reproductive tract. In more advanced insects, the sperm are transferred directly in seminal fluid and the spermatophore is no longer produced.

SPERMATOPHORES IN APTERYGOTE INSECTS

The insect spermatophore is synthesized by the male accessory glands. It consists of a viscous secretion that is shaped by the internal structures of either the male or the female once it has been inserted into the female reproductive tract. A change in pH may account for the transition from a liquid secretion by the male accessory glands to a more solid, gel-like mass. The sperm are contained within the saclike ampulla of the spermatophore before it is transferred to the female.

The structure of the spermatophore and the mechanism of transfer differ among the insect orders. In many primitive apterygote insects, such as the collembolans, the spermatophore consists of a drop of sperm placed in a simple sac at the end of a stalk on the ground. Females must find the spermatophore themselves and actively take it up into the genital tract. The behavior is more complex in other species of Collembola, where the male may actively manipulate the female and direct her to the spermatophore he has already deposited. These males have not developed the modifications necessary to grasp the female for conventional mating. Similarly, some Thysanura males deposit an unstalked spermatophore on the ground, but place a wall of silken threads around it to signal to the female that the spermatophore is nearby. Because all these insects inhabit humid soil environments, there is little chance for the sperm to dry out before the females find them and take them up.

SPERMATOPHORES IN PTERYGOTE INSECTS

In pterygote insects that produce a spermatophore, transfer to the female is more direct. Odonate males deposit their spermatophore from their genital opening at the tip of the abdomen to the secondary genitalia on the anterior segments of the abdomen. During copulation, the male grasps the female with the tip of his abdomen while her abdomen loops forward to receive the sperm from his secondary genitalia. In many orthopterans, the spermatophore is inserted into the female genital tract with a long spermatophore bulb that protrudes from the female genitalia. The female may eat the exposed bulb once the sperm have left, deriving the considerable nutritional investment that the male has made. Some males present a nuptial gift during and shortly after copulation to distract the female and prevent her from eating the spermatophore before the sperm have left it. In the cricket *Aceta domesticus,* the physical coupling of the male and female is followed by the insertion of the spermatophore into the female genital tract and the attachment of its long tube to the female's ovipositor. The sperm travel into the spermatheca of the female, and after 30 to 40 min, the flasklike spermatophore is dislodged.

In the more advanced pterygotes, the males form the spermatophore in a more recently evolved structure in the female genital tract, the bursa copulatrix, and the sperm then move to the spermatheca(e) of the female for ultimate storage and utilization. In some Lepidoptera, the bursa contains spines that rupture the spermatophore and allow the sperm to escape. Once the sperm have escaped, the spermatophore may be digested by enzymes secreted into the bursa, and its raw materials exploited by the female for egg production. Male *Anopheles* mosquitoes produce a modified spermatophore that serves as a mating plug to temporarily block the genital tract and prevent the female from mating with other males.

The metabolic costs of synthesizing a spermatophore may be one reason for its ultimate disappearance in many of the higher insect orders. Accompanying the loss of the spermatophore has often been the development of a male intromittent organ that is capable of placing the sperm directly into the bursa copulatrix or the spermatheca(e), making the spermatophore unnecessary.

See Also the Following Articles
Mating Behaviors • Reproduction, Male

Further Reading

Davey, K. G. (1960). The evolution of spermatophores in insects. *Proc. R. Entomol. Soc. Lond.* **35A**, 107–113.

Davey, K. G. (1985). The male reproductive tract. *In* "Comprehensive Insect Physiology, Biochemistry and Pharmacology" (G. A. Kerkut and L. I. Gilbert, eds.), pp 1–14. Pergamon Press, Oxford, U.K.

Khalifa, A. (1949). The mechanism of insemination and the mode of action of the spermatophore in *Gryllus domesticus. Q. J. Microsc. Sci.* **90**, 281–292.

Proctor, H. C. (1998). Indirect sperm transfer in arthropods: Behavioral and evolutionary trends. *Annu. Rev. Entomol.* **43**, 153–174.

Schaller, F. (1971). Indirect sperm transfer by soil arthropods. *Annu. Rev. Entomol.* **16**, 407–446.

Spiders

Rosemary G. Gillespie and Joseph C. Spagna

University of California, Berkeley

For many, spiders are a cause of fear and a source of revulsion. Even to entomologists, spiders have often been thought of as a mere annoyance, filling nets and pitfall traps meant for insect quarry. It is therefore surprising to learn that spiders have held a prominent role in traditional cultures for centuries. Indeed, the terms "arachnid" and "arachnology" come from Greek mythology: A young maiden, Arachne, dared to challenge the great goddess Athena to a spinning contest. Athena wove a remarkable tapestry, yet that spun by Arachne was far superior in its intricacy and perfection. Infuriated that a mere mortal could spin such a masterpiece, Athena turned Arachne into a spider, condemned to a life of perpetual spinning. In native American culture, it was a spider that brought fire to the Cherokee people; the Navajo told of Spider Woman (Na ashje'ii 'Asdzáá), who came from the "first world" and taught the women how to weave; according to Pueblo legend, Spider Woman was at the core of creation; and the Sioux Indians use the "dream catcher," spun by a spider, to capture the good dreams of life. Spiders are also prominent in African culture, as illustrated in the well-known stories of "Anansi, the Trickster." Rather more recently, Scottish legend tells of King Robert the Bruce, whose observations of a spider inspired him to persevere, going on to conquer the English. The reasoning behind some of the tales can be obscure. For example, in the far-flung Micronesian island of Palau, the ability of women to perform natural childbirth is attributed to a spider. With such a diversity of lore, it is clear that spiders have been held in high regard across a global spectrum of cultures and for a very long time.

Spiders, like insects, belong to the phylum Arthropoda, but they are in the class Arachnida, which includes the orders Acari (ticks and mites), Scorpiones, Pseudoscorpiones, Opiliones (harvestmen or daddy-long-legs), and several less common

orders. Arachnids are only distantly related to the other major terrestrial arthropod group, the insects, and represent a separate evolutionary transition from marine to terrestrial living, because their closest living relatives are thought to be the marine horseshoe crabs (xiphosurans) and sea spiders (pycnogonids). Together, these marine and terrestrial orders are called the chelicerate orders based on the structure of their mouthparts, in contrast to the orders to which insects and crustaceans belong. Recent molecular and morphological evidence points to the Ambylpygi, or tailless whip scorpions, as the group sharing the most recent common ancestor with spiders, with the other arachnid orders more distantly related. Spiders can easily be distinguished from other arachnids by their lack of visible segmentation and the marked constriction between the prosoma and the opisthosoma, dividing the body into cephalothorax and abdomen, respectively.

Although much less diverse than insects in habits and morphology, spiders, which are in the order Araneae, nonetheless occupy nearly all terrestrial environments and can be found wherever there are other terrestrial arthropods to prey upon. Research into spider biology, particularly the diversity of silks, webs, and venoms, together with the associated ecology and behaviors, has increased greatly in recent decades. Moreover, phylogenetic advances are beginning to provide the context for comparisons between spider taxa and between spiders, other arachnids, and other terrestrial arthropods.

EXTERNAL ANATOMY

The body of all arachnids is divided into an anterior prosoma and posterior opisthosoma, and in spiders these major divisions are referred to as the cephalothorax and abdomen (Fig. 1). These are connected by a narrow stalk, the pedicel. The cephalothorax is covered dorsally by the carapace and ventrally by the sternum and labium. Unlike in insects and most other arachnids, the segmentation of the spider body is not visible externally (except in two primitive lineages) and the cuticle is relatively soft, particularly on the abdomen.

At the front end of the carapace are the simple eyes, or ocelli, usually in four pairs, but in some groups one or more pairs may be reduced or absent. The eight simple eyes are usually arranged in two rows of four, though each of these rows may be curved in such a way that individual eye pairs seem to form their own rows, and in some species one, two, three, or even all four pairs of eyes may be lost. The anterior eye row is closer to the chelicerae, whereas the posterior eye row is farther back on the cephalothorax. Each eye row consists of median and lateral eye pairs, so that each pair can be identified both by row and by position within that row. For example, the anterior median eyes (AMEs) are the central pair, closest to the chelicerae. Below the front margin of the ocular area is the clypeus, from which the two chelicerae extend downward. Each chelicera consists of a stout basal section, from the outer corner of which articulates a narrower distal section, the fang. Behind the mouth ventrally is a second pair of mouthparts,

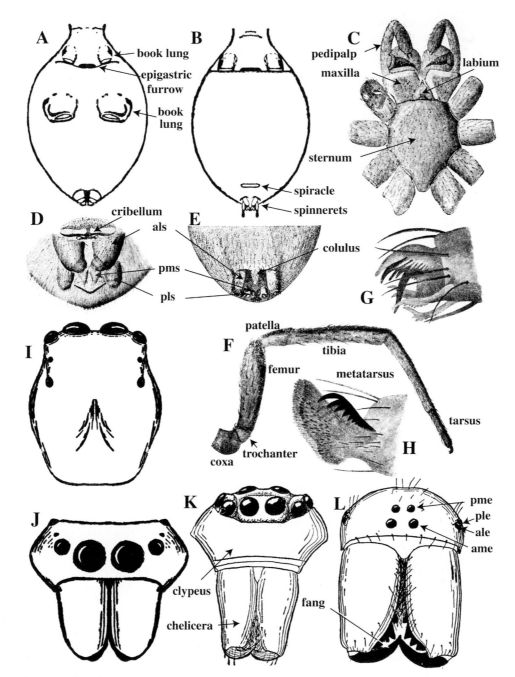

FIGURE 1 External anatomy of a spider. (A–B) Ventral view of abdomen, showing (A) four pairs of book lungs, and (B) two pairs of book lungs. (C) Ventral view of prosoma. (D) Ventral view of female cribellate spider, showing cribellum. (E) Ventral view of female ecribellate spider, showing colulus. (F) Lateral view of leg. (G) Tarsus of leg of three-clawed spider. (H) Tarsus of leg of two-clawed spider and claw tuft. (I) Dorsal view of prosoma (cephalothorax) of jumping spider. (J–L) Anterior view of carapace and chelicerae of (J) jumping spider, (K) tangle-web spider, and (L) orb-web spider. ale, anterior lateral eyes; als, anterior lateral spinnerets; ame, anterior median eyes; ple, posterior lateral eyes; pls, posterior lateral spinnerets; pme, posterior median eyes; pms, posterior median spinnerets.

leg-like in appearance, the pedipalps. In mature males, these are modified for secondary sperm transfer and appear swollen, facilitating recognition of the spider's sex.

Extending laterally from the ventral cephalothorax are four pairs of legs, each consisting of seven segments: coxa, trochanter, femur, patella, tibia, metatarsus, and tarsus. The legs are generally covered with hairs and often have a great diversity of spines, bristles, and scales. These outgrowths serve a variety of functions, including mechanical and chemical sensory functions (see below). Each leg terminates in two or three tarsal claws. In three-clawed spiders, there are usually two larger paired claws and a smaller unpaired median claw, whereas in two-clawed spiders, the median claw is often replaced by a tuft of dense, stiff hairs, called the claw tuft.

A spider's abdomen is carried behind the cephalothorax. The abdomen may be globose or elongate in appearance and is sometimes covered in hairs or scales similar to those found on the legs. The respiratory book lungs (one or two pairs) open at the anterior end of the abdominal venter. They can be seen externally as patches of hairless cuticle on the venter of the spider, adjacent to the genital opening. In mature female spiders, the genital area is found nestled between the book lungs, the form varying from a simple slit or pair of holes to a complex sclerotized copulatory plate, the epigynum.

At the posterior end of the abdomen are the spinnerets—usually in three pairs, the anterior, posterior, and median

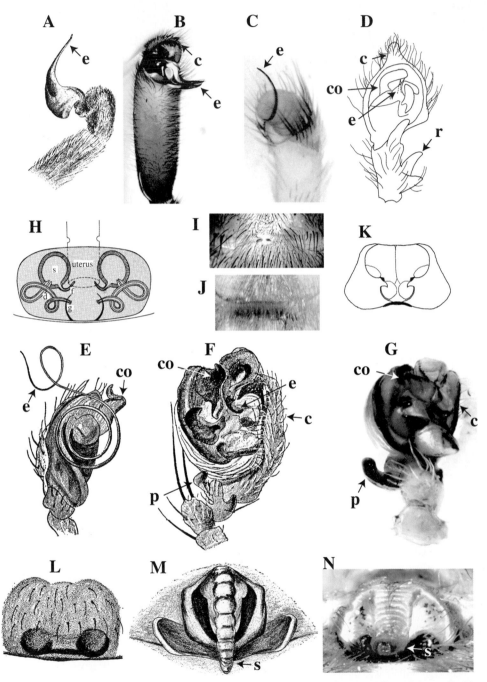

FIGURE 2 Spider genitalia. (A–G) Comparison of male palpal structures showing the change in complexity in different lineages of spiders (see Fig. 5 for major lineages). Lineages represented are (A) *Eurypelma* sp. (Mygalomorphae), (B) *Atypoides* sp. (Mygalomorphae), (C) *Kukulcania* sp. (Haplogynae), (D) *Hololena* sp. (RTA clade), (E) *Theridion spirale* (Orbicularia), (F) *Araneus gigas* (Orbicularia), (G) *Araneus gemma* (Orbicularia). Abbreviations of structures: c, cymbium; co, conductor; e, embolus, p, paracymbium; r, retrolateral tibial apophysis. (H) Diagram of epigynum, ventral view. Abbreviations of structures: d, coiled ducts; f, fertilization duct; g, genital opening; s, seminal receptacle. (Reprinted, with permission, from Fœlix, 1996, Fig. 135, © Oxford University Press.) (I–N) Comparison of female genital structures. (I) *Antrodiatus* sp. (Mygalomorphae), (J) *Kukulcania* sp. (Haplogynae) showing simple slit opening of seminal receptacles (no epigynum), (K) *Hololena* sp. (RTA clade) showing sclerotized epigynum, (L) *Theridion spirale* (Orbicularia), (M) *Araneus gigas* (Orbicularia), (N) *Araneus* sp. (Orbicularia) showing elaborate sclerotized epigynum and scape(s). Panels A, E, F, L, and M, adapted from J. H. Comstock, (1912), "The Spider Book," Doubleday, Page, Garden City, NY. Panels B, C, G, I, J, and N, photographs (microscopic, automontage) taken by the authors. Panels D and K drawn by the authors.

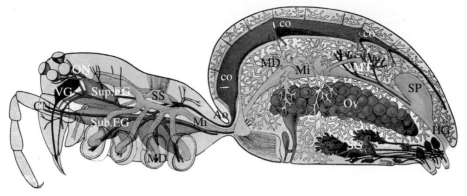

FIGURE 3 Internal anatomy of a female spider (adapted from J. H. Comstock, (1912), "The Spider Book"). Ao, aorta; BL, book lung; co, cardiac ostia of the heart; CN, cheliceral nerve; (P), pharynx (behind the esophageal ganglion); (E), esophagus (behind the esophageal ganglion); HG, hindgut; Mi, midgut; MD, midgut diverticula; MT, Malpighian tubules, Oc, ocular area; ON, optic nerves; Ov, ovaries; SG, silk glands; SS, sucking stomach; Sup. EG, supraesophageal ganglion; Sub. EG, subesophageal ganglion; SP, stercoral pocket; VG, venom gland.

pairs—and the anal tubercle. The median spinneret pair is often obscured by the larger anterior and posterior pairs, and the entire complex of spinnerets may be surrounded by a sclerotized ring. Each spinneret pair has its own complement of silk spigots that extrude silks for specific functions. In addition, many spiders have a cribellum, or spinning plate, found adjacent to the standard spinneret complement that they use to produce an ultrafine looped silk for prey capture. In others, the cribellum has been reduced to a small vestigial lobe, called the colulus, or lost completely.

The sex organs of male spiders are the palps (Figs. 2A–2G). The cymbium, or modified tarsus of the mature male palp, is hollowed out to contain the copulatory organ. The basal appendage of the cymbium is called the paracymbium. The bulb of the male's palp opens through a spine-shaped apparatus, which ends in a fine tube, the embolus. The male palp is generally less complex in primitive spiders (haplogynae) relative to more advanced spiders (entelegynae). More complex palps have hard sclerotized parts (sclerites, which include the conductor, embolus, tegulum, etc.) and soft parts (hematodochae), both of which can carry prominences, or apophyses. At rest, most structures are folded away, with the delicate sclerites protected. During mating, the hematodochae are filled using hydraulic pressure and the sclerites extended, as the male inserts his palp into the female epigynum. The intromittent portion of the palp must navigate through the complex ducts of the female to achieve sperm transfer, while other sclerotized projections on the palp help to lever the embolus into position during copulation. The actual copulation has been referred to as a "lock and key" mechanism because of the often complementary species-specific form of the genital structures.

The female genitalia (Figs. 2H–2N) of primitive spiders comprise a ventral fold within which a hidden opening leads into a single duct, or egg tube. This tube serves for sperm insemination and egg laying. In more advanced spiders (entelegynae), there are two openings, the egg tube and the genital tube, both hidden by an external structure, the epigynum. The epigynum is sclerotized and becomes visible upon maturity. The complex structures on the female epigynum guide the male palpal bulb through a convoluted set of internal ducts into the female's genital tube, which leads to the sperm pocket where the sperm is stored.

INTERNAL ANATOMY

The central nervous system of the spider is located in the cephalothorax and consists of two main ganglia, the larger subesophageal ganglion and the smaller, more anterior supraesophageal ganglion, sometimes referred to as the "brain" of the spider. The two ganglia are divided horizontally by the esophagus, and nerves radiate from both, forming the peripheral nervous system. The supraesophageal ganglion connects to the cheliceral and optic nerves, while the subesophageal ganglion connects to the peripheral nerves of the palps, legs, and abdomen (Fig. 3).

Paired venom glands occupy the upper portion of the chelicera in all spiders except the Uloboridae and, in many spiders, extend well into the cephalothorax about midway between the eyes and the supraesophageal ganglion. The digestive tract consists of a pharynx, esophagus, sucking stomach, and the beginning of the midgut in the cephalothorax, with the rest of the midgut, hindgut, and anus located in the abdomen.

The major respiratory organs are in the lower abdomen and are called "book lungs" because they resemble stacked sheets of paper. More primitive spiders have two pairs of book lungs, others have one pair and may also have a set of tubular tracheae. Within spiders, there has been a sequence of replacement of book lungs by tracheae, apparently in response to problems of circulation, with more active spiders having variably elaborate tracheal systems (Table I). The heart is a tube-like organ, suspended by muscles and ligaments along the dorsal midline of the abdomen, with multiple ostia that serve as valves to keep the blood flowing in one direction. The heart pumps the hemolymph forward through the aorta into the cephalothorax. Silk glands are numerous and may fill up to a third of the volume of the abdomen. In more advanced spiders they can have varied functions (Table II). The gonads consist

TABLE I   Relative Compositions of Book Lungs versus Tracheae in Different Groups of Spiders

Second abdominal segment	Third abdominal segment	Example
Book lungs	Book lungs	Mesothelae, Orthognatha
Book lungs	Short tube tracheae	*Filistata*
Book lungs	Long tube tracheae	*Dysdera*
Book lungs	Long entapophyseal tracheae	*Argyroneta*
Book lungs	Short tube tracheae and short entapophyseal tracheae	Araneidae, Lycosidae
Book lungs	Short tube tracheae and long entapophyseal tracheae	*Cryphoeca*
Book lungs	No respiratory organs	Pholcidae
Sieve tracheae	Long tube tracheae	Caponiidae
Sieve tracheae	No respiratory organs	Symphytognathidae
Tube tracheae	Long tube tracheae	*Telema*

Note. The more primitive groups (Mesothelae, Orthognatha) retain book lungs on both second and third abdominal segments. More advanced groups show loss of book lungs and/or replacement by tracheae.

of paired, coiled, tubular testes in males and paired ovaries in which the follicles may appear grape-like in females.

PHYSIOLOGY

Feeding and Venoms

Upon capture, a spider sinks its fangs into the body of its prey. At this point, the spider must paralyze, or otherwise restrict the movement, of the prey rapidly before it can eat it. Nearly all spiders use venom to incapacitate their prey. When envenomating prey, muscles surrounding the venom glands contract, forcing venom out of the glands and through ducts that carry the venom to the tips of the fangs and into the prey. Spider venoms are toxic cocktails of polypeptides and proteolytic enzymes that are quite effective for paralyzing the spider's (usually insect) prey. Because most of the polypeptides have evolved to act on nervous systems of arthropods, which use glutamate as a neurotransmitter, most spider venoms have little effect on vertebrates, but there are exceptions. Some venoms, like those of the widow spiders *(Latrodectus)*, contain components that broadly affect vertebrate nervous systems. In humans, a black widow bite, which may go unnoticed when it happens, can result in several days of pain, muscle spasms, abdominal cramping, and weakness. More serious symptoms may include respiratory

TABLE II   Silk Gland Types, Silk Uses, and Location of Spigots within Spinnerets

Gland type	Silk uses	Spigot locations
Major ampullate glands	Dragline, web frame	Anterior
Minor ampullate glands	Dragline reinforcement	Median
Aciniform glands	Swathing silk, sperm web, egg sac outer wall	Median, posterior
Cylindrical (or tubuliform) glands	Cocoon silk	Median, posterior
Aggregate glands	Glue for sticky spiral	Posterior
Flagelliform glands	Core of sticky spiral	Posterior
Piriform glands	Attachment disc silk	Anterior

difficulty and hypertension. Deaths are rare, however, and antivenins are available in emergency situations. Other species, such as the brown recluse spider *(Loxosceles reclusa)*, produce venoms that cause tissue death (necrotism) at the site of the wound. While these may not cause systemic effects like the black widow toxins, the wounds produced may take weeks or months to heal, and infection is a serious risk.

Digestion and Excretion

Spider digestion begins outside its body. Once the spider has disabled its prey with venom, silk, or both, it extrudes digestive enzymes into the prey and then, using negative pressure from its sucking stomach, reingests the soup of digestive enzymes and partially digested food. This is repeated until all of the prey's soft tissue has been consumed. Once the liquefied food has passed through the sucking stomach, it enters the midgut where nutrient absorption takes place, with secretory cells producing digestive enzymes and resorptive cells absorbing food into vacuoles.

Waste is concentrated in the cytoplasm as the nearly insoluble products guanine, adenine, hypoxanthine, and uric acid. These products are collected via the Malpighian tubules and moved into the stercoral pocket, which empties through the hindgut and anus. Other excretory tissues in spiders include the coxal glands, which appear to be involved in water balance, and the large nephrocyte cells that concentrate metabolites.

Circulation and Respiration

Although spiders have well-defined blood vessels, they lack capillaries and have few veins; thus, their circulatory system is basically open. The heart is suspended in the pericardial sinus, and blood enters the heart through paired slits called ostia, which open when the heart is at rest. The heart primarily pumps hemolymph from the abdomen forward through the aorta into the cephalothorax, supplying oxygen to the central nervous system and the skeletal muscles. Once depleted of oxygen, the fluid passes into two sinuses, which lead to the base of the abdomen where the fluid is reoxygenated by the book lungs and (if present) by tubular tracheae before pressure pulls it through pulmonary veins back into the pericardial sinus.

Unlike most insects, spider hemolymph has an oxygen-carrying pigment, hemocyanin. Hemocyanin is structurally similar to hemoglobin, but instead of iron it uses copper as its oxygen-binding metal, which can make spider hemolymph appear bluish green. Compared to hemoglobin, hemocyanin is less efficient (~5%) in oxygen transportation and is not concentrated in specialized cells.

NEUROBIOLOGY

The detection of touch and sound, mediated by vibrations, is the primary sense of spiders, although other senses may be well developed in some groups. These senses are discussed in detail in a 1985 book by F. G. Barth.

Touch and Sound

Spiders are notably hairy, and most of the various hairs on their bodies function as mechanoreceptors, sensing movement and vibrations from both the spider's substrate (which may be the web, the ground, or vegetation on which the spider is situated) and the surrounding air. They also serve as touch receptors. The many stout hairs on the legs, cephalothorax, and abdomen, as well as the finer, more upright trichobothria, found only on the legs, are triply innervated and are involved in the localization and identification of potential prey. Additional mechanoreceptors are the slit sensilla and a variety of other proprioceptors, which respond to stresses in the exoskeleton caused by external vibration or by the spider's own movements. The slit sensilla are found all over the exoskeleton, but may be arranged in groups to form "lyriform organs." Together with the trichobothria, the slit sensilla/lyriform organs may be the functional equivalent of spider "ears." Other proprioceptors include internal ganglia at leg joints and hairs that respond to joint movement.

Taste and Smell

Chemoreceptive hairs are localized mainly on the tarsi of the front legs, and in the palps, although some chemoreception may take place at the mouthparts. The hairs involved in chemoreception resemble tactile hairs, but are open-ended and S-shaped. Spiders respond to sex pheromones of conspecifics, and it is also likely that they use chemoreception to identify potential prey, enemies, and environmental change.

Sight

The "simple" eyes, or ocelli, have a single cuticular lens. The back of the eye contains the retina, which consists of visual cells (including the light-sensitive rhabdomeres) and pigment cells. The main eyes (always the AMEs) consist of a lens, vitreous body, and retina (which contains the visual cells), while the secondary eyes may also have a light-reflecting surface, the tapetum, behind the retina, which causes the well-known eye shine of spiders at night. The visual acuity of spiders depends on the shape of the lens and the number of rhabdomeres. For most species vision is poorly developed. Most web-building species rely almost exclusively on touch, with vision used to detect light and dark and (in a few species) direction of polarized light. However, vision is well developed in jumping spiders, with the AMEs having high visual acuity (acting like the fovea of the human eye), the remaining eyes having lower acuity but a broader field of vision and functioning for peripheral vision.

SILKS AND SPIDER WEBS

Spiders are unique among arthropods in their use of silks at all stages of their lives. Silks are produced in the abdomen in specialized silk glands, each of which yields a different kind of silk. The general structure of a silk gland is a tail area that secretes the liquid silk proteins into a sac-like lumen, or storage area. The lumen empties into a duct leading out to the spinneret spigots. The ducts are important in silk production because their tapered shape helps to orient the molecules relative to the axis of the thread to maximize strength, whereas the cells surrounding the duct draw off water from the oriented protein and turn them into solid silk fibers. The ducts of the different silk glands terminate at specific spigots in the spinnerets (Table II).

The various silks produced by spiders show different combinations of remarkable material properties, including high tensile strength, extensibility reminiscent of rubber, and resistance to decay. These properties have led to research in the pursuit of silk genes and techniques for spinning the translated products of these genes; genetically engineered spider silk may soon be found in parachutes, bulletproof vests, car bumpers, artificial ligaments, etc. At the molecular level, spider silks are a family of proteins made up primarily of a subset of amino acids (alanine, glycine, serine, proline, glutamine, and tyrosine make up over 75% of the composition of characterized spider silks) arranged in a highly repetitive manner. The smallest repeating units of two to six amino acids are strung together into larger "tandem repeats." This arrangement is thought to form secondary structures that determine the kind of silk produced. The overall structure can be compared to a composite material with stiff, stress-resistant crystals interspersed in an extensible, energy-absorbing matrix.

The most obvious use of spider silk is in web construction, but it is employed for many other purposes in the life of a spider, including the following:

Ballooning

Spiders are capable of producing silk as soon as they emerge from the egg, and two behaviors, the manufacture of a brood web and ballooning, are characteristic of newly hatched juveniles, or spiderlings. A brood web is made by some spiderlings upon emergence and for the earliest instars can serve as a communal nest for the young to catch prey together. Ballooning generally occurs once dispersal in the spiderlings has been initiated by developmental or environmental cues.

The spiderlings let out a silk thread, which produces enough drag to catch the wind and carry the spider off, sometimes for great distances.

Dragline

Nearly all spiders trail a dragline behind them during locomotion. The dragline is extruded from the ampullate glands as the spider walks along or moves in its web, stopping occasionally to anchor the line with an attachment disc secreted by the piriform glands. The jumping spiders (Salticidae) may make seemingly reckless jumps, but are almost invariably anchored with a dragline, so a missed leap is seldom fatal. Dragline diameters vary from several hundred nanometers to several micrometers, depending on the size of the animal being supported. Draglines are stronger than any other known natural fiber, with the added ability to stretch 15 to 30% beyond their original length before breaking.

Egg Sacs

All spiders use silk to protect their eggs. The silk covering of the egg mass may range from a few strands to a thick covering with multiple silks. The resulting silken cocoon can take on a huge variety of forms depending on the spider and the habitat.

Sperm Webs

Silk is also used to make the male sperm web. Shortly after the final molt to maturity, a male spider makes a small web (sometimes just one or a few strands of silk), upon which he deposits sperm from the abdomen. He then places the tip of his palp into the sperm, which is drawn through the palp's opening into the sperm duct, where it is stored.

Capture Webs

The most remarkable use of spider silk is in the construction of snares for catching prey. In webs, silks can function both directly as a mechanical trap, to stop or slow potential prey, and indirectly as an extension of the spider's sensory apparatus, alerting it to the trapped prey. A number of spiders can produce sticky capture webs, which come in two forms. The "hackled" or woolly silk of cribellate spiders is sticky (akin to Velcro) because of its fine fibers. The silk is formed by rapid combing (by the calamistrum on the rear leg) of a silk produced from the cribellum, a field of fine openings in front of the spinnerets. In contrast, the web silk of ecribellate spiders usually has sticky globules, secreted by the aggregate glands, arranged along its length.

The type of web that spiders produce is often characteristic of a family, with forms including the two-dimensional orb, tangle (cobweb), sheet, and funnel webs (Fig. 4). Each of these webs has its variants as well; for example, orb webs can be oriented either vertically or horizontally or may be reduced, sometimes to a single section of a complete orb. The

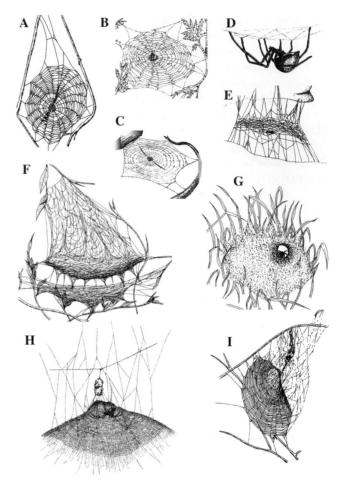

FIGURE 4 Web structures. (A) Typical orb web (Araneidae). (B) Orb web of species of Tetragnathidae. (C) Cribellate orb web (Uloboriodae). (D) Tangle web of black widow (Theridiidae). (E) Tangle web of *Steatoda* sp. (Theridiidae). (F) Sheet web of *Frontinella* sp. (Linyphiidae). (G) Funnel web of *Agelenopsis* sp. (Agelenidae). (Reproduced, with permission, from Wise, 1993; illustrations by Gilbert H. Wise.)

impressive body of literature on the form, function, and evolution of spider webs was reviewed by Eberhard in 1990.

Despite the fact that all webs serve the same basic function and despite the common characterization of spiders as generalist predators, the form of the silk of certain webs can be specialized for capture of a small subset of potential prey types. For example, many comb-footed spiders (family Theridiidae) make tangle webs with viscid threads extended to the substrate below, the last centimeter or two of which are coated with a sticky substance ("gum-foot lines") and serve to snare cursorial prey. More specialized is the bolas spider, an orb weaver whose "web" consists of a single strand of silk with a sticky droplet (bolas) at the end, which it uses to catch moths. In addition to these differences in silk form between web types, silks often have different functions within webs. For example, silk from the flagelliform glands makes up the core fibers of the capture spiral in orb webs and has properties quite different from those of the silks that form the web frame and radial elements. The remarkable property of this capture-spiral silk is its ability to stretch; it may more than double its length while absorbing the kinetic energy of a flying insect. In addition, the capture spiral recovers tension slowly, which prevents prey items from being flung back out of the web.

Web Decorations

One of the most conspicuous features of the webs of some orb spinners is the presence of a stabilimentum, a prominent silk line, cross, and/or spiral, at the center of the orb. The function of the stabilimentum is not clear, though it may serve to camouflage the spider, startle predators, or protect the integrity of the web from accidental damage.

RELATIONSHIPS AND TAXONOMY

Understanding of the relationships among different spider groups and between spiders and other arthropods has increased dramatically in the past 2 decades. Some portions of the spider tree remain unresolved, and additional morphological and molecular study will be needed to settle these uncertainties. Statements of phylogenetic relatedness between spider groups are based on a review of morphological data by Coddington and Levi in 1991, and all groups mentioned can be found on the summary cladogram (Fig. 5), unless otherwise noted.

Spiders are divided into three suborders, the Mesothelae, Mygalomorphae, and Araneomophae, and 106 families. The Mesothelae (1 family, Liphistiidae, two genera, 40 species) are considered the most primitive of all living spiders, based on

their external visible segmentation and location and number of spinnerets (all four pairs present, without a cribellum). The Mygalomorphae, recognized by the articulation of the chelicerae parallel to the body (paraxial), are also considered primitive; they are generally stout-bodied and include the trap-door spiders and the impressively large tarantulas (family Theraphosidae) among their 15 families. Araneomorphs, the "true" spiders, represent over 90% of spider diversity. They can be distinguished from the more primitive spiders by the sideways (diaxial) articulation of their chelicerae. Paraxial and diaxial cheliceral orientations are also referred to as orthognath and labidognath, respectively. The most basal lineage of the Araneomorphae is the Paleocribellatae, consisting of 1 family, the Hypochilidae, in which the body plan is a mosaic of primitive and derived characters.

The rest of the araneomorphs belong to one of two clades, the haplogynae and the entelegynae, distinguished on the basis of the complexity of the female genitalia (Fig. 2). In the haplogynae, the smaller group, the external female genitalia consist of a simple opening, or gonopore, tucked between the book lungs, with a single duct serving both the copulatory and the fertilization functions. Two well-known haplogyne families are the spindly legged Pholcidae (or daddy-long-legs spiders), common around and inside houses, and the Sicariidae, which may also occur in and around houses, and include the much-feared brown recluse spiders *(L. reclusa)*, whose venom may cause necrotic wounds in humans. In the entelegynae, the female genitalia are more elaborate, with separate fertilization and copulatory ducts and sclerotized epigynum. The most diverse and well-known entelegyne spider families belong to

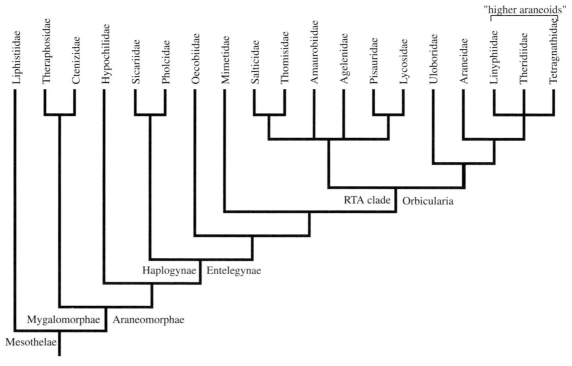

FIGURE 5 Phylogenetic relationships of major spider lineages, with exemplar families. (After Coddington and Levi, 1991.)

the RTA clade, so named because of a retrolateral tibial apophysis on the male's palp, and the orbicularia, which includes the orb-web builders and their relatives. The RTA clade is a huge, ecologically diverse group of spiders and includes about one-third of all described spider species. Many are web builders, for example the funnel-web spiders (Agelenidae). However, some of the most successful lineages have shifted to non-web-building hunting strategies, including the jumping spiders (Salticidae), in which the median tarsal claw is replaced by a tuft of tarsal hairs that facilitates a cursorial lifestyle. Salticidae also have greatly enlarged anterior median eyes, giving them the best vision of all spiders. They use this vision to track prey and carry out unique jumping attacks with great accuracy and in mate choice when faced with flamboyant courtship displays. The crab spiders (Thomisidae) are another family of two-clawed spiders with an interesting variation on the typical spider sit-and-wait game. Thomisids hide in the petals of flowers, where their bright colors often disguise them, and use their raptorial forelegs to capture visiting pollinators. Wolf spiders and their relatives (Lycosoidea) are also cursorial hunters, though they retain all three tarsal claws.

The Orbicularia includes some of the best known web spinners. In particular, the orb web is often considered to be the "classic" spiderweb—a two-dimensional sphere consisting of a frame, with radii projecting outward, and a spiral of sticky silk wrapped surrounding the center. There are three families primarily responsible for these orb webs, the cribellate Uloboridae and the Araneidae and Tetragnathidae, which have lost the cribellum. The orb design, once thought to be the "pinnacle" of web evolution, is now thought to have served as a point of departure for some successful groups (the "higher araneoids") that make webs of quite different designs. The comb-footed spiders (Theridiidae) make seemingly disorganized cobwebs, often with viscid gum-foot lines as mentioned above. On the other hand, "bowl-and-doily" or money spiders (Linyphiidae) make dome-like sheets from which they hang on the lower surface; the viscid silk in these spiders dries up after being produced and serves to cement together the different layers of the sheet.

COURTSHIP, REPRODUCTIVE BEHAVIOR, AND GROWTH

Courtship

The first step in courtship involves maturation of the male spider, at which stage the palps have become modified and swollen for sperm storage and transfer. Once the palps are charged with sperm, the male sets out to find a receptive female. However, spiders are assiduous predators, and the male must overcome this propensity of the female if he is to mate. There is a huge array of courtship strategies that allow the male to approach. Most spiders employ some kind of vibratory communication during courtship. Among web-building spiders, the male often locates himself on the edge of the female's web

and gently plucks. It often takes hours, even days, until the female becomes receptive. Wolf spiders (Lycosidae), although they do not have webs, make use of vibrational cues during courtship. Vibrations generated from stridulating organs are usually transmitted through the leaf litter. Visual cues, although used by wolf spiders, are most complex among jumping spiders (Salticidae). Male jumping spiders generally communicate courtship by performing a variably elaborate dance for the female, waving their often brightly colored legs and body to show off iridescent plumes. If successful, the male can extend his forelegs to touch the female before climbing on top of her.

In a number of species the female must adopt a state of complete immobility before the male can initiate copulation. In other species, the male waits until the final molt and can mate with the female immediately. In certain species the female is very much larger than the male and appears hardly to notice that the male is either approaching or copulating with her. Other males secure the female by wrapping her in silk prior to copulation. Species in the genus *Tetragnatha* have an unusual way of mating whereby the cheliceral fang of the female becomes wedged against a dorsal notch on the cheliceral surface of the male. The male then locks the female in place by closing his fangs over hers.

Copulation

Copulation involves the injection of sperm from the male's palps into the seminal receptacles of the female, with the palps being inserted alternately into the epigynum. In some species, the palp breaks off and seals the epigynum. Once copulation has been completed, the male must escape from the female before her brief period of receptivity ends. If the male fails to escape in time, he can be caught and consumed by the female, although more commonly he escapes to mate again. However, the life span of a mature male is generally short, and many do not eat at all. Female spiders, on the other hand, are able to store sperm and so can produce fertile eggs long after copulation. Accordingly, females may survive for some time before egg laying is complete, and in many species they survive longer, to care for their offspring.

Egg Sacs

Some time after mating, the female will deposit an egg sac. The eggs are always laid within a cocoon of silk. However, like courtship, there is a diverse array of egg sac types and behaviors that spiders use to protect their eggs. In most spiders, the female spins a layer of silk into which she deposits her eggs. She then covers the eggs with more silk. The covering may be scant, as in the daddy-long-legs spiders (Pholcidae). In other spiders the sac is very thick, with multiple silks, soft inside but tough and water resistant outside, as in many spiders of the family Theridiidae (e.g., *Latrodectus*). Egg sacs are also produced in a variety of colors, from white to yellow, green, or black, in pastel or vivid shades. Textures may vary from papery smooth to tufted and furry and shapes from flat to round or angular, with connections to the substrate ranging from a tight fixture to loosely suspended or pendulous. These colors and shades generally match the substrate on which the egg sac is laid, serving to camouflage the developing eggs as protection against predation and parasitism.

Parental Care

Parental care varies tremendously among spiders. Some females abandon their egg sac immediately after it has been laid although, even in these species, the female selects specific sites for deposition of the egg sac. For example, it can be plastered on a twig (e.g., *Tetragnatha*), placed under a leaf or stone, wrapped in a leaf, or suspended in the web or retreat, either with a stalk (e.g., *Argyrodes*) or without. However, extended parental care has been documented in a number of spiders and has broad implications with regard to the origin of social behavior.

EGG SAC DEFENSE A number of female spiders guard their eggs closely until (or after) hatching. Such guarding is common among jumping, crab, sac, and ground spiders, and a variety of others, and can be critical for egg survival in, for example, the green lynx spider *Peucetia viridans* (Oxyopidae) and the Hawaiian happy face spider *Theridion grallator* (Theridiidae). In several groups of spiders, the females carry their eggs with them wherever they go. In particular, wolf spiders (Lycosidae) carry their egg sac attached to their spinnerets; fishing (Pisauridae) and giant crab (Heteropodidae) spiders carry their egg sac under the sternum; daddy-long-legs spiders (Pholcidae) carry their egg sac held in the chelicerae.

CARE OF SPIDERLINGS Maternal care in spiders, when present, often terminates after hatching, and the young disperse. However, female wolf spiders carry their young on their abdomen once they have hatched from the egg sac; fishing spiders build a "nursery web," a large tent-like structure in which the spiderlings live while the mother stands guard. Among a few comb-footed and other spiders, care of the emergent spiderlings can be developed to an extraordinary degree. This is true of the communal comb-footed spiders *Theridion saxatile, T. sisyphium, T. grallator,* and *Anelosimus studiosus.* Females of the latter two species defend the egg sac aggressively and then capture prey for, and even feed, the young, which are unable to capture prey on their own. Providing the young with food appears to be the primary function of brood care once the spiderlings emerge from the egg sac. In addition to simply securing prey, a mother may feed her offspring by regurgitation or by laying "trophic eggs." In some species she may even feed herself to her offspring, a process known as "matriphagy," and in at least one species, *Amaurobius ferox,* the mother is known to expedite this process herself.

Dispersal

One of the most intriguing aspects of spiders is their dispersal. When spiderlings hatch, they are generally aggregated. However, on the first day with suitable wind speeds, they will frequently move up to the highest point they can find (e.g., the tip of a grass stalk) and let out silk which catches in the wind. As the spider lets out more silk, the pull of the wind on the silk becomes sufficient to allow the spiderlings to become airborne. Spiderlings can travel tremendous distances by ballooning and are frequently the dominant component of aerial plankton, although the family groups represented vary with locality: Linyphiidae comprise much of the fauna above land areas that have been examined, whereas Tetragnathidae dominate over oceanic areas. Because of their capacity to balloon, spiders are often the first to colonize unoccupied land masses, whether cleared land or new islands in the middle of the ocean.

Growth

Spiders, like insects, have a rigid exoskeleton and must molt to grow, with three to nine instars (stages between molts) before reaching maturity. Spiders do not metamorphose: A first-instar spiderling looks similar to, though smaller than, an adult. Most spiders live for about a year, though mygalomorphs may live for 30 years.

SOCIALITY

Although most spiders are solitary and highly intolerant of others, there are several species that exhibit some form of social behavior, ranging from aggregations at a certain life stage to prolonged maternal care and even quasisociality or true sociality. Considerable controversy surrounds the origins of quasisocial and true social behavior in spiders, discussed in chapters of a book by Choe and Crespi published in 1997. It may have evolved through coloniality and the development of aggregations around an abundant resource. Alternatively, it may have arisen through extension of brood care into later instars. Many studies have examined the biology of social theridiids, including the formation of colonies and sex ratios in true social species.

Colonial Spiders

Some spiders (e.g., several species in the families Araneidae, Uloboridae, and Tetragnathidae) live in colonies, usually around an abundant food source. The benefits from such behavior include sharing silk support lines for prey capture, sharing prey that is not overall in short supply, and communication.

Subsocial Spiders

In subsocial spiders, there is both maternal brood care beyond the first few developmental instars that is typical of most spider species and an extended phase of tolerance among young within the maternal nest. However, these species live solitarily as adults. True sociality in spiders is thought to have evolved via the "subsocial pathway" by a prolongation of an early tolerance phase without dispersal. The subsocial route to sociality entails decreasing the genetic variance within breeding groups as a result of families staying in proximity.

True Sociality

True sociality appears to have arisen independently multiple times in spiders. Social spiders, unlike eusocial insects, have no castes that are morphologically different or sterile and most individuals within colonies reproduce. In addition, social spiders show a female-biased sex ratio and high population turnover and inbreeding. Their breeding colonies are closed and new colonies are formed by splitting an existing colony, by a swarm of related females, or by single gravid females. High levels of inbreeding and relatedness among females bias the sex ratio toward the dispersing sex. So, unlike most spiders, males do not appear to disperse large distances in social species. Gene flow between established colonies is rare, and colonies show a high degree of genetic structure. It seems likely that sociality may be maintained by behavioral preadaptations that lead to tolerance and cooperation among colony members on the one hand and population structure on the other.

CAMOUFLAGE AND MIMICRY

Crypsis

Spiders have many enemies. In addition to other spiders, perhaps the most important predators are birds and spider-hunting wasps, both of which possess high visual acuity and color vision. Spiders can have specific adaptations for matching background colors (crypsis) such as flowers/leaves, grass/twigs, bark, underleaf surfaces, and the ground. In some spiders, the crypsis is extraordinarily close and can vary between individuals on different backgrounds. There are, for example, several species that are variable in color but always seem to match the bark on which they are living. Selection for crypsis in similar types of habitat has led to the repeated evolution of similar colorations in unrelated species. For example, spiders that live under leaves in the tropics are generally translucent yellow, often with dark leg joints that disrupt the outline of the legs. During the day these spiders lie flat against the underside of the leaf, thereby reducing shadows and becoming highly cryptic against the light transmitted through the leaf. Five unrelated spider species living in the same Hawaiian forests exhibit these adaptations for underleaf crypsis, as do at least four species from the forests of Panama. Within a single lineage, repeated evolution of similar cryptic coloration in different species has been found in a radiation of *Tetragnatha* spiders in Hawaii.

Disruptive Coloration

In some spiders the characteristic shape of the body is concealed, for example by bold, juxtaposed colors. These kinds of colors tend to be found where the propensity to wander over different backgrounds while searching for prey might preclude true crypsis. Crypsis *per se* may also be difficult in diurnal orb-spinning spiders. Some of these have developed mimetic resemblances to dead leaves and sticks but others have apparently adopted disruptive coloration.

Mimicry

Most studies of mimicry in spiders have been concerned with the imitation of ants, thought to be a form of Batesian mimicry. The spiders may gain some protection from predators through their resemblance to aggressive or unpalatable ants. The topic of ant mimicry has recently been reviewed by Cushing, in 1997. Spiders may also mimic a range of other organisms, alive or dead, and inanimate objects. For example, many *Cyclosa* spp. (Araneidae) build vertical "sticks" of prey remains within the web but leave a gap in the center, which is filled by the spider itself. Crab spiders (Thomisidae) mimic the color of flower heads very precisely and prey on pollinators that approach. Some spiders resemble bird droppings, which are attractive to insects. In the garden spider *Argiope argentata,* the visibility of both the contrastingly colored ventral and the UV-reflecting dorsal side of the opisthosoma may increase insect prey caught in the web.

Apostatic Coloration

At least some predators can develop search images, concentrating their search effort on more common forms of individuals of any species. This can result in polymorphism within a prey species. One of the best examples of this is found in *Theridion grallator* (Theridiidae) (Fig. 6). Within this species, there is a remarkable diversity of color forms, yet the frequency of color forms is similar in different populations, apparently maintained by bird predation.

HABITAT SELECTION

How do spiders select a site in which to live? Many studies have demonstrated that there are clear associations between spider abundance and the structural diversity of the habitat, climatic regime, and prey availability. Most spiders are considered to be sit-and-wait predators, spending much time in locating a suitable site in which to wait and remaining there until its quality deteriorates. The time they remain at a site depends on their investment at the site. For species that do not spin webs, the investment is only the time spent finding the site. Ecribellate orb-web spiders can regain most of their resource investment at a site by ingesting the web before abandoning the site. Among other web-building spiders, however, the

FIGURE 6 Some representative color morphs of the Hawaiian happy face spider *Theridion grallator.*

to the vibration of an insect caught in the threads by shaking the web, causing the insect to become more firmly stuck and allowing the spider to locate the position of the insect. The spider may then approach the insect, often doing so very rapidly, and wrap it in silk of various amounts depending on the size and strength of the insect. This approach and attack behavior varies, with some web-building species leaping upon and biting the prey even before it becomes firmly secured in the web. More specialized strategies have evolved in a number of web builders. For example, the ogre-faced spider *(Deinopis)* holds its cribellate web, orb-like in form, between the two front pairs of legs; it uses the web much like a net, hurling it over prey that come close. Comb-footed spiders may use their gum-foot lines to catch crawling prey; the unsuspecting prey may either dislodge the tensed lines, which springs them into the tangle of the web, or they get reeled in by the alerted spider.

For spiders that do not build webs, prey may be detected by air- or substrate-borne vibrational cues or, alternatively, at least in the case of jumping spiders, by visual cues. Cursorial spiders often jump on their prey without wrapping it in silk. However, certain groups have elaborate mechanisms for securing their prey. In particular, the spitting spiders (Scytodidae) have a high-domed cephalothorax, which contains venom glands connected to a posterior gland that secretes a sticky silk. The spider creeps up on its prey and, by rapid contractions of the muscles in the prosoma, ejects sticky silk and venom over the prey, which is thus immobilized. One species of long-jawed spider in Hawaii has long tarsal claws that it uses to impale insects directly from the air. Other specialized mechanisms of prey capture are discussed below.

Diet

Spiders are exclusively carnivorous, although they are usually regarded as generalist predators, taking prey as they are encountered. However, their diet is dictated by the habitat in which they select to live, and habitat specialization can greatly restrict the dietary repertoire of many species. Even within a given habitat, species may specialize on prey, either through choice or because they exhibit specialized predatory behaviors. For example, species in the genera *Dipoena* and *Euryopis* (Theridiidae) are specialized for feeding on ants, while those in *Dysdera* (Dysderidae) appear to specialize on isopods, and pirate spiders (Mimetidae) and some *Argyrodes* (Theridiidae) feed on other spiders.

Kleptoparasitism

Kleptoparasitic spiders steal food from the webs of other, unrelated and usually larger web-building, spiders, which are also potential predators. They may form groups of up to 50 individuals around the web of a "host" spider and glean small insects from the periphery of the host spider's web, eat the host's silk, steal food bundles previously wrapped and left in the web by the hosts, or even approach a feeding host spider

investment at a site cannot be regained; it can only be decreased through reduction in web size. Movement from a site may be dictated by disturbance or web destruction, microclimate change, growth of the spider relative to structural requirements for web construction, and/or prey capture success.

In selecting a site, it is often not clear what aspect of the environment it is to which the spider is responding. It is possible for hunting spiders to move directly to their feeding site, at least when they can perceive air- or substrate-borne vibrations. Web-building spiders are generally confined to their web for prey detection so that movement from a web site is generally not directed, and to select a site they must "sample" a location by building a web. However, the main vibration receptors of web spiders (trichobothria) are basically the same as those of wandering spiders. Accordingly, web-building spiders may also be capable of perceiving vibrations mediated by the air or the substrate.

PREDATION

Prey Capture

For web-building spiders, prey are usually captured upon being trapped in the sticky threads. The spider will generally respond and then feed, undetected, next to the host spider. The most diverse collection of obligate kleptoparasites is found in the spider genus *Argyrodes* (Theridiidae), which contains at least 200 species, about 100 kleptoparasitic.

Aggressive Mimicry

Aggressive mimicry is mimicry that enhances the access of spiders to their prey. Perhaps the most interesting case of aggressive mimicry is shown by certain jumping spiders that can mimic the vibratory signals of a fly by plucking with their legs and palps on the web of other spiders. This may attract the occupant to the jumping spider close enough to be attacked. The best-studied species in this regard is *Portia fimbriata* (Salticidae) from Queensland. This spider preys on web builders with poor eyesight, which allows the araneophagic *Portia* to drum out a pattern of signals on the web that imitates the vibrations of the normal prey of the web spider. In contrast to the vibrational mimicry of *Portia,* the bolas spider (Araneidae) emits odors that mimic the pheromone of female moths to attract male moths to the vicinity. As soon as a moth flies within range, the spider launches its sticky bolas on a silk thread; if successful, the bolas hits and sticks to the moth, and the spider reels in its prey.

DIVERSITY AND CONSERVATION

There are about 34,000 described species of spiders, though the actual number of species has been estimated at around 170,000. In common with insects, species diversity is highest in tropical regions, where knowledge of the biota is least. Also like other arthropod groups, spiders exhibit spectacular examples of adaptive radiation and local endemism in isolated situations, for example the genus *Habronattus* (Salticidae) on the North American mountaintop "sky islands" and the genus *Tetragnatha* (Tetragnathidae) in the Hawaiian Islands. However, in almost every ecosystem, the number and diversity of spiders make them important in the management of ecosystems, whether natural or disturbed, agricultural or urban.

Importance and Conservation in Agricultural Systems

Spiders have been known to be important agents of insect pest control for hundreds of years. It is also well known now that chemical pesticides kill not only the insect pests, but also their major predators, in particular spiders. A review by Riechert and Lockley in 1984 brought attention to spiders as potential agents of biological pest control. Manipulations of habitat structure have resulted in increased spider densities and concomitant decrease in insect pests. The effectiveness of spiders is often enhanced by the ability of the web to kill prey even in the absence of the spider and because many species practice wasteful killing (i.e., they kill more insects than they consume). Control of insect pests by spiders appears to be achieved most efficiently by using an assemblage of spiders

(rather than specific species) and incorporating natural refuges in and around the crops. The particular methods that are best for enhancing the role of spiders in pest control in agriculture are: (i) conservation tillage and retention of stubble, preserving weeds and mulching; (ii) intercropping, provision of native vegetation within crops; and (iii) agroforestry, planting a combination of trees and food crops.

Importance and Conservation in Natural Ecosystems

We have little understanding of the precise role that spiders play in a given ecosystem. However, we do know that spiders are the dominant invertebrate predator in most terrestrial ecosystems and are also important as a food resource for other predators. Accordingly, they must inevitably play a key role in community dynamics. Studies on birds in natural areas have shown that spiders can comprise a significant proportion of the diet of many species, including a number of endangered birds. Spiders also have utilitarian value and have been used as model organisms for research in ecology, behavior, and communication. Moreover, there has been much recent attention on their potential for providing silk for materials science and supplying venom for medical and insecticide research.

Clearly, there is a huge need to augment taxonomic information on spiders, particularly in tropical areas, and to understand patterns of biodiversity. Pursuing this knowledge is imperative as we recognize the obvious importance of spiders in ecosystems, both as prey and as predators. Moreover, given the intrigue that spiders have long held in human society, such an undertaking can only lead to greater heights of fascination.

See Also the Following Articles
Daddy-Long-Legs • Mimicry • Predation • Silk Production • Venom

Further Reading
Barth, F. G. (1985). "Neurobiology of the Arachnids." Springer-Verlag, Berlin.
Choe, J., and Crespi, B. (1997). "Evolution of Sociality," Vol. II of "Social Competition and Cooperation among Insects and Arachnids." Cambridge University Press, Cambridge, UK.
Coddington, J. A., and Levi, H. W. (1991). Systematics and evolution of spiders Araneae. *Annu. Rev. Ecol. Syst.* **22,** 565–592.
Cushing, P. E. (1997). Myrmecomorphy and myrmecophily in spiders: A review. *Fla. Entomol.* **80(2),** 165–193.
Eberhard, W. G. (1990). Function and phylogeny of spider webs. *Annu. Rev. Ecol. Syst.* **21,** 341–372.
Foelix, R. F. (1996). "Biology of Spiders." Oxford University Press, Oxford.
Jackson, R. R., and Pollard, S. D. (1996). Predatory behavior of jumping spiders. *Annu. Rev. Entomol.* **41,** 287–308.
Kaston, B. J. (1978). "How to Know the Spiders." Brown, Dubuque, IA.
Nentwig, W. (ed.) (1987). "Ecophysiology of Spiders." Springer-Verlag, Berlin.
Oxford, G. S., and Gillespie, R. G. (1998). Color polymorphisms in spiders. *Annu. Rev. Entomol.* **43,** 619–643.
Oxford, G. S., and Gillespie, R. G. (2001). Portraits of evolution: Studies of coloration in Hawaiian spiders. *Bioscience* **51,** 521–528.
Riechert, S. E., and Harp, J. M. (1986). Nutritional ecology of spiders. *In* "Nutritional Ecology of Insects, Mites, Spiders and Related Invertebrates" (F. Slansky and J. G. Rodriguez, eds.), pp. 645–672. Wiley, New York.

Riechert, S. E., and Lockley, T. (1984). Spiders as biological control agents. *Annu. Rev. Entomol.* **29,** 299–320.

Roth, V. D. (1993). "Spider Genera of North America," 3rd ed. Am. Arachnol. Soc., New York.

Shear, W. A. (ed.) (1986). "Spiders, Webs, Behavior, and Evolution." Stanford University Press, Stanford, CA.

Skerl, K. L., and Gillespie, R. G. (1999). Spiders in conservation: Tools, targets and other topics. *J. Insect Conservat.* **3(4),** 249–250.

Wise, D. H. (1993). "Spiders in Ecological Webs." Cambridge University Press, Cambridge.

Witt, P. N., and Rovner, J. S. (eds.) (1982). "Spider Communication—Mechanisms and Ecological Significance." Princeton University Press, Princeton, New Jersey.

Springtail

see *Collembola*

Stamps, Insects and

Charles V. Covell Jr.
University of Louisville

One of the major recent developments in philately (stamp collecting) is the popularity of "topical" or "thematic" collecting. The collector looks for anything that relates to the chosen theme. Popular examples are flowers, railroads, birds, sports, stamps-on-stamps, space, and insects. In a recent survey, butterflies and other insects on stamps ranked as the sixth most popular topic.

As of 1991, over 5000 stamps were issued by over 300 countries depicting over 1300 insect species (Fig. 1). Therefore specialization in types of insect being collected might be desirable, as it is with insect collecting itself. Subsets that could be selected as a specialty include butterflies and/or moths, apiculture, sericulture, malaria and other aspects of medical entomology, applied entomology, and noninsect arthropods. A "taxonomic" collection (in which the insect is the only real design) might be restricted to identifiable and named insects. Some, however, prefer to collect stamps with fanciful depictions, such as butterfly carnival costumes, a butterfly in the movie *Bambi* (on Bambi's nose), or a "fly speck" insect in the beak of a bird. Still other philatelists specialize in mint (unused) stamps or canceled (used) ones, including "CTOs," or stamps "canceled to order." The CTOs are stamps with neat cancellations applied before the stamps are released, and the cancels are normally in one corner to avoid obscuring much of the stamp design. Many philatelists frown on CTOs because they are obviously prepared for philatelists and are not used in carrying the mail.

FIGURE 1 This pane of 20 different 33-cent stamps showing U.S. insects and spiders was issued on October 1, 1999 (Scott Catalog item 3351a-1).

PHILATELIC ELEMENTS AND TERMINOLOGY

Philately consists of a number of "elements" (a term used in exhibiting). Mint stamps, used stamps, and CTOs are just the beginning. Other related postal items that can be collected and exhibited are as follows:

Cancel The marking applied in the post office to show the stamp has been used.

Cover An envelope with stamps and cancel attached.

First-day cover (FDC) Cover canceled on the first day of the stamp's issue, often with special cancels and a cachet, (or illustration on envelope created to enhance interest and value) (Fig. 2).

Favor cancel Cancel applied for a collector, not to carry mail.

Piece Fragment of a cover, especially a large one such as from a package, but including stamp and cancel.

FIGURE 2 First-day cover for U.S. 4-cent malaria eradication stamp issued on March 30, 1962. Design of envelope (cachet) and special first-day cancel make the "FDC" a favorite collectible (Scott Catalog item 1194).

Meter cancel   A machine-applied indication of postage paid, often accompanied to the left by a picture or slogan and used in place of stamps.

Proof   A special impression of the stamp's design used to check quality of design and color, with die proofs usually one color and often signed by the artist.

Essay   Original artwork submitted to the stamp-issuing authority for consideration for adoption (these are rare and usually unique; essays for designs that were never printed as stamps are especially rare).

Block   Symmetrical grouping of stamps, still attached to each other, with a block of four being the most common.

Plate block   Block with the selvedge (unprinted border) containing a plate serial number printed in the original pane of stamps.

Revenues   Stamps issued for fiscal purposes (postage stamps are sometimes applied to receipts in place of revenue stamps—an uncommon element to find).

Sheet   A full pane of stamps, usually surrounded by a border (selvedge), which is the way stamps usually arrive at a post office.

Miniature sheet   A special sheet of usually two to several stamps; not a normal full pane; it may have special related pictures in the selvedge.

Postal stationery   Postal cards and letter sheets with stamps printed on them.

Souvenir sheet   One or more stamps inside a printed border that may be a continuation of the vignette (picture) of the stamp itself.

Unified design sheet   A pane in which each stamp is a small part of an overall picture depicted by the sheet.

GETTING STARTED

The first thing a beginner should do is get some help from someone who has experience in collecting stamps. The rudiments of philately, however, do not take long to learn. Knowledge of the terms just defined, and how to handle the different items is a good starting place. There are books and periodicals available as well. In North America, *Linn's Stamp News,* a weekly newspaper, has pages of ads as well as columns on how to "do" philately. People at a local stamp club and/or stamp shop also can be helpful.

Knowledge about stamps and other postal subjects can be found in specialized books. Stamp values (mint and used) in North America usually follow the *Scott Standard Postage Stamp Catalog,* illustrated and revised yearly.

The APS (American Philatelic Society), which publishes a monthly magazine, *American Philatelist,* has departments including library and expertization services. The ATA (American Topical Association) is for thematic collectors and publishes the bimonthly *Topical Time.* This organization is divided into specialty groups. The Biology Unit, which publishes a newsletter called "Biophilately," serves, among others, insect philatelists. The Philatelic Lepidopterists' Association

publishes a newsletter called "The Philatelic Aurelian" (an aurelian is a butterfly collector), which features updates on new butterfly and moth issues (identifiable ones only). Finally, by using key words "stamps" or "philately," one can find many sources of material and information via any search engine on the World Wide Web.

A collector of insect stamps might wish to subscribe to a new issues service run by a dealer who mails new issues to subscribers each month. Such services are listed in the ad section of *Linn's,* and many philatelists are glad to recommend their favorites. New issue dealers very likely can supply older stamps from their stocks of previous issues. There are many stamp club shows and occasional bourses (shows without exhibits, with dealers only), where collectors can fill checklists from dealers' stocks and buy supplies—and meet with other philatelists.

Insect philately keeps alive the thrill of the hunt and the love of the beauty and order of an insect collection. However, for many participants this kind of collection takes less storage space, needs no fumigating, and can keep giving enjoyment when entomologists' more active years of insect collecting are behind them.

See Also the Following Articles
Commercial Products from Insects • Cultural Entomology

Further Reading
American Philatelic Society, P.O. Box 8000, College Station, PA 16803. Web site: www.stamps.org.

American Topical Association, P.O. Box 50820, Albuquerque, NM 87181–0820. Email: atastamps@juno.com.

Coles, A. and Phipps, T. (1991). "Collect Butterflies and Other Insects on Stamps." Stanley Gibbons Thematic Catalogue. Stanley Gibbons Ltd., London.

Hamel, D. R. (1991). "Atlas of Insects on Stamps of the World." Tico Press, Falls Church, VA.

"Scott Standard Postage Stamp Catalog." (revised annually). Scott Publishing, Sidney, OH.

Williams, L. N. (1990). "Fundamentals of Philately." American Philatelic Society, College Station, PA.

Sterile Insect Technique

Jorge Hendrichs
Joint Food and Agriculture Organization/International Atomic Energy Division, Vienna Division, Vienna

Alan Robinson
International Atomic Energy Agency Laboratories, Seibersdorf, Austria

The sterile insect technique (SIT) is a biologically based method for the control of insect pests. Female insects inseminated by released, radiation-sterilized males do not reproduce, and repeated releases of the sterilized insects lead to a reduction in the pest population. Effective control using SIT is achieved when it is applied systematically on an area-wide basis over multiple generations. SIT is species specific, nonpolluting, and resistance free. Since the original concept was developed in the United States in the 1940s and 1950s, SIT has been used successfully for screwworm flies, tsetse flies, fruit flies, and moths. Technical developments in behavioral ecology, mass rearing, strain improvement, global information systems, aerial release, and monitoring systems, combined with economies of scale and a growing demand for low-pesticide agricultural products in local and international trade, have increased the use of SIT in area-wide integrated pest management programs. These programs have decreased insecticide use against some major pest insects and thereby facilitated the efficacy of biological control agents against secondary insect pests.

HISTORY OF SIT

In the late 1930s E. F. Knipling, working with *Cochliomyia hominivorax,* a screwworm fly causing myasis in livestock and humans, developed the concept of using insects to control themselves (i.e., autocidal control) by reducing their reproductive capacity. Two developments helped to translate this concept into reality. The first was a breakthrough in the development of mass rearing procedures for this insect and the second was the identification of an efficient method to sterilize insects. In 1916 G. A. Runner demonstrated that X rays could sterilize the cigarette beetle, *Lasioderma serricorne,* and in the 1920s, H. J. Muller showed that *Drosophila* fruit flies could be sexually sterilized by radiation-induced dominant lethal mutations. Knipling and colleagues confirmed that low

radiation doses given to mature pupae and adults of both sexes of the screwworm sterilized them without excessive damage to their mating ability or diminishing their longevity. The first decisive test that confirmed the feasibility of SIT was implemented in 1954 on the island of Curacao, 64 km off the coast of Venezuela. The success of the experiment so impressed the cattle industry in Florida that in 1957 members persuaded the U.S. Congress to fund a statewide eradication program. A rearing facility with a capacity of producing 50 to 75 million sterile flies per week was established in 1958 at Sebring, Florida, with the result that the last endemic screwworm case was recorded in Florida in 1959. Based on this success, the program was expanded, with new rearing facilities at Mission, Texas, and eventually in Tuxtla Gutierrez, Chiapas, Mexico. Release of sterile screwworms eradicated this pest from North America and more recently from Central America (Fig. 1). Currently, a permanent "sterile fly barrier" consisting of the aerial release of 40 million sterile flies per week in eastern Panama protects Central and North America from reinvasion.

PRINCIPLE OF SIT

The SIT involves industrial-scale mass production of insects in large biofactories. Insects for release are exposed to a precise dose of γ-radiation that is sufficient to induce sterility without significantly impairing the ability of the treated insects to fly, mate, and transfer sperm. The treated insects are released over the target area on a sustained basis and in sufficient numbers to achieve appropriate sterile-to-wild insect overflooding ratios. This tactic substantially reduces the proportion of fertile matings and results in the decline of

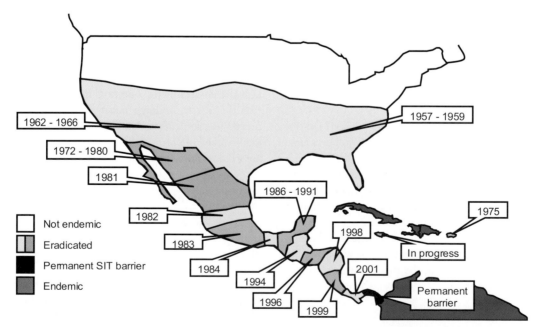

FIGURE 1 Progressive shift of eradication zones over time in the screwworm SIT program in North and Central America. [Reprinted from Robinson, A. S. (2002). *Mutation Res.* **511,** 113–132, with permission.]

the field population. The SIT stands or falls on the ability of the released sterile males to transfer functional sperm to the females that succeeds in fertilizing the eggs. The number of times a female mates is irrelevant, providing that the sperm from the sterile male is competitive with the sperm from the wild male. Knipling devised a simple mathematical model to describe the SIT. Despite some limitations of the model, it provides a valuable set of equations to demonstrate the principles of SIT. As the wild population declines and the number of released insects remains constant, both the proportion of sterile matings and the rate of suppression increase, i.e. the efficiency of SIT is inversely density dependent, and once a wild population begins to decline, the rate at which this occurs increases with time. Also, compared with classical biological control, which involves the introduction of exotic control agents, SIT has important advantages: (1) autocidal control is species specific, (2) no exotic species are involved, and (3) no new strains of the target pest can become established because the released insects are sterile.

REQUIREMENTS FOR SIT APPLICATION

Not Stand-Alone Technology

The use of any single technique in isolation is unlikely to be successful in an insect control program and integrated approaches are now favored. This caveat applies especially to the use of the SIT because insect populations in the field are often of such magnitude that it is impractical to consider releasing sufficient sterile insects to cause the initial population decline. Because the strength of the SIT is at low population densities, to be effective it relies on a variety of presuppression activities (such as application of insecticides), depending on the particular insect species targeted. An alternative, and frequently used, strategy is to apply the SIT when the natural population is at low density because of adverse climatic conditions.

Area-Wide

The effectiveness of integrating compatible pest control methods is increased by coordinated implementation over large contiguous areas that targets the total insect population rather than the protection of individual crops or animal herds. This approach is critical to the SIT because already-mated females that move into an area under treatment are largely unaffected by the presence of sterile males. The area-wide concept has to be seen, not just in terms of the individual farmer or even of the agricultural community, but in terms of the whole ecosystem, including human activity, in the treatment area.

Selection of Target Species

Most insect pest species reproduce by sexual reproduction and therefore, in theory, are amenable to this type of control strat-

egy. However, in reality the technique is used on a selective basis. Probably the first criterion is that the target species is a key pest in its particular ecosystem. It also makes very little sense to develop SIT for a species for which effective, acceptable conventional controls are available. As the technique relies on the release of large numbers of adult sterile insects into the environment, it is important that this stage not be the damaging one. For example, use of SIT against house flies or cockroaches would probably not be acceptable.

Knowledge of Population Dynamics, Ecology, and Mating System

SIT compromises the natural reproduction system of the target species, and it is therefore very important that the relevant population dynamics are well understood in terms of density dependence, life-table parameters, and so on. In addition, the spatial and temporal ecological parameters of the target population need to be defined and the population size must be estimated. It is also essential that the natural mating system is well understood, so that the mating behavior of the released males can be monitored to ensure that no significant changes in the behavior of released insects accumulate during mass rearing.

Colonization, Mass Rearing, and Sterilization

An efficient, economical, and biologically acceptable method for the mass production of competitive insects is essential. Mass rearing of large numbers of insects over many generations in large biofactories imposes major selection and adaptation forces on all aspects of the behavior of the insect. These forces inevitably work to the detriment of the competitiveness of the insect in the field. The art of mass rearing is to find an acceptable compromise between the efficiency of mass production and the overall quality of the mass-produced individuals. Beyond a certain point it is not possible to compensate for the poor quality of insects by simply releasing more of them into the environment. The final and most critical treatment that the insects undergo before release is sterilization. This process uses ionizing radiation and must be subject to stringent quality control protocols to ensure that all the released insects are irradiated and that the required dose is applied.

Quality Control

As in any industrial system, quality control in terms of both the process (mass rearing) and the product (sterile insects) plays a major role in determining the success of an SIT program. For most programs a detailed set of standard operating procedures is in place, which provide some confidence as to the quality of the insects being produced. The use of mating tests under seminatural conditions to ensure continued compatibility of released insects with the wild population is essential.

STRATEGIC APPLICATIONS OF SIT

Eradication

SIT has the unique capability of eradicating populations of pest insects providing that certain ecological and regulatory conditions are met. The inverse density-dependent action of the SIT is very effective in removing the last individuals from the target population. Eradication is often assumed not to be an option if the target population is not isolated. To some extent this is true, but in many cases isolation can be created by making use of natural barriers to pest distribution and by applying the technique on an area-wide basis. The screwworm eradication program relied on this approach with the population being eradicated in a "rolling carpet" approach (Fig. 1). The maintenance of the pest-free status following eradication requires adequate quarantine measures and continued surveillance. SIT is also one of the most powerful tools to eliminate incipient outbreaks of introduced exotic invasive species.

Containment

The very benign environmental impact of SIT means that sterile insects can be released preventively into an area on a continuous basis to avoid the movement of a pest into a sensitive agricultural area. Sterile insects can also be used in a barrier approach to protect an eradicated area from reinvasion.

Suppression

The concept of using sterile insects as a tool in integrated suppression programs to replace the use of insecticides is gaining increasing acceptance as the cost of sterile insects is decreasing and concerns about the environment and food safety are increasing. The use of SIT for suppression removes the need for expensive and difficult to maintain quarantine regulations and the continuous nature of a suppression program will stimulate the commercialization of the SIT.

OPERATIONAL PROGRAMS

In the past decades, SIT has been applied in over 20 countries on five continents to control some 15 insect species that account for an important part of agricultural losses worldwide. Table I provides an overview of the mass rearing facilities currently in operation in support of SIT programs against four main groups of pest insects: fruit flies, screwworms, tsetse, and moths.

TABLE I Sterile Insect Mass Rearing Facilities in the World

Country	Location	Species	Weekly capacity (millions)
		Fruit flies	
Argentina	Mendoza	*Ceratitis capitata*	100–250[a]
Australia	Perth	*C. capitata*	5–15[a]
Chile	Arica	*C. capitata*	50–75[a]
Guatemala	El Pino	*C. capitata*	1000–1500[a]
Mexico	Metapa	*C. capitata*	500–600
Portugal	Madeira	*C. capitata*	35–60[a]
South Africa	Stellenbosch	*C. capitata*	5–10[a]
United States	Hawaii	*C. capitata*	210–450
Australia	Campden	*Bactrocera tryoni*	15–25
Japan	Okinoawa	*B. cucurbitae*	70–200
Mexico	Metapa	*Anastrepha ludens*	150–250
Mexico	Metapa	*A. obliqua*	25–50
Mexico	Metapa	*A. serpentina*	5–10
Philippines	Quezon City	*B. philippinensis*	10–20
Thailand	Pathumtanee	*B. dorsalis*	15–40
United States	Gainesville	*A. suspensa*	10–50
United States	Mission	*A. ludens*	15–40
		Screwworms	
Mexico	Tuxtla	*Cochliomyia hominivorax*	120–500
		Moths	
Canada	Okanagan	*Cydia pomonella*	12–15
United States	Phoenix	*Pectinophora gossypiella*	70–90
		Tsetse (1000s)	
Tanzania	Tanga	*Glossina austeni*	60–100[a]
Burkina Faso	Bobo-Dioulasso	*G. palpalis gambiensis*	40–50[a]

[a]Sterile male-only production, otherwise 50% sterile males and 50% sterile females. [Modified from Hendrichs, J. (1995). *J. Appl. Entomol.* **119**, 371–377.]

Fruit Flies

Few insects have a greater impact on international marketing and world trade in agricultural produce than tephritid fruit flies. Maintaining or achieving a pest-free status has therefore been the goal of a number of large operational programs. Mediterranean fruit fly, *Ceratitis capitata,* ("medfly"), was removed from areas it had invaded in southern Mexico, and it is now close to 20 years that a sterile fly barrier across Guatemala has maintained its medfly-free status. Similarly, Chile eliminated medfly from northern Chile and since the mid-1990s has supported a sterile fly barrier in Peru's two southernmost valleys. The melon fly, *Bactrocera cucurbitae,* was eradicated from southern Japan and the Queensland fruit fly, *B. tryoni,* from Western Australia and some *Anastrepha* species from northern Mexico. Rather than having to eradicate outbreaks of exotic medflies, often on an annual basis, California and Florida maintain permanent medfly preventive release programs in high-risk areas, which is less costly and politically more acceptable than aerial insecticide applications. Increasingly, SIT is also being applied for area-wide suppression purposes without quarantines, with pilot programs in progress against medfly in Israel/Jordan, Portugal, South Africa, and Tunisia and against *Bactrocera* fruit flies in Philippines and Thailand.

Screwworms

In addition to the current barrier against *C. hominivorax* across the Darien region in eastern Panama, an eradication project is under way in Jamaica, and others are under assessment for Cuba and Hispaniola. As part of an Australian contingency planning to maintain Australia free of the Old World screwworm, *Chrysomya bezziana* (OWS), a pilot mass rearing facility was recently established in Malaysia. Application of SIT against OWS is also being considered in the Persian Gulf region, to where it has spread over the past decade, causing large outbreaks in Iraq, and from where it is threatening to invade the Near East and Mediterranean regions.

Moths

Field programs are in progress for prevention or suppression of pink bollworm, *Pectinophora gossypiella,* in various cotton areas in California and for codling moth, *Cydia pomonella,* suppression in apple production areas in British Columbia, Canada. In the northeastern United States, a number of successful pilot tests were carried out against the gypsy moth, *Lymantria dispar.* In addition, the SIT package is in various stages of development against other moth pests in various parts of the world, targeting diamondback moth, false codling moth, cotton bollworm, date moth, sugarcane borer, and peach borer. For moth pests there is considerable potential for expanding the implementation of inherited sterility suppression programs, which have shown a significant synergistic interaction between this type of SIT and biological control.

Inherited sterility refers to the release of moths which have been given a substerilizing dose of radiation.

Tsetse

Tsetse flies and the diseases they transmit (sleeping sickness in humans and Nagana or animal trypanosomosis in livestock) prevent the establishment of sustainable agricultural systems in many areas of great agricultural potential in sub-Saharan Africa. Following pilot SIT projects in Burkina Faso and Nigeria, and breakthroughs in mass rearing and aerial release technology, an area-wide program succeeded in 1996 in eradicating *Glossina austeni* from the island of Zanzibar. This success has led to the consideration of similar programs for selected areas in mainland Africa.

ECONOMIC BENEFITS

Direct benefits of screwworm eradication to the North and Central American livestock industries are estimated to be over $1,500,000,000/year, compared with a total investment over half a century of close to $1 billion. Mexico protects a fruit and vegetable export market of over $1 billion/year through the annual investment of ca. $10 million. Achieving the medfly-free status has been estimated to have opened markets for Chile's fruit exports of up to $500 million. When implemented on an area-wide basis and with economies of scale in the mass rearing process, the application of SIT for suppression purposes is cost competitive with conventional control, in addition to its environmental benefits.

FUTURE TRENDS

The increasing constraints being applied to the use of insecticides, especially for food crops, will enhance the need for alternative control strategies for both native and introduced pest insects. SIT can be an increasingly important component of these strategies, especially when used in pest suppression programs. The expanding use of the SIT will be sustainable only when improvements in the efficiency of the technique continue to be made. This requires substantial investment in research and development related to all components of the technology, from improved insect diets to more sensitive and informative pest monitoring systems. The expansion of the use of the SIT to include suppression will be a major focus in the future, as it will provide a great opportunity for the commercialization of parts of the technology. This could range from customers simply purchasing sterile flies and developing their own release and monitoring systems to the purchase of a complete SIT package. For the medfly, these developments are imminent. Along with any commercialization of the technology, quality control procedures of both the process and the product will need to be reinforced. For the product, i.e., the sterile insect, routine protocols will have to be developed to measure the mating success of the males with females in the field. SIT programs will also benefit tremendously if methods can be developed that enable only male insects to be reared as required. For the medfly this genetic sexing technology is now available, leading to major improvements in program efficiency. As far as the process is concerned, more appropriate artificial diets for insect adults and larvae could make a major contribution through both improved economies and improved insect quality.

In many insect species, transgenic technology has been developed and there is great interest in using it to improve the SIT. A first application could be in replacing the use of fluorescent dyes by a transgenic marker. Suitable markers are available and are currently undergoing evaluation in several pest species. A second application is the use of transgenic technology to develop molecular methods for genetic sexing. A caveat to the use of any transgenic technology will be regulatory approval; however, because the SIT uses sterile insects there is no risk of vertical transmission of the transgene.

See Also the Following Articles

Agricultural Entomology • *Biotechnology and Insects* • *Integrated Pest Management* • *Rearing of Insects* • *Tsetse Fly*

Further Reading

Hendrichs, J. (2000). Use of the sterile insect technique against key insect pests. *Sustainable Dev. Int.* **2**, 75–79.

Knipling, E. F. (1979). Use of insects for self-destruction. *In* "The Basic Principles of Insect Population Suppression and Management." USDA, Washington, DC. [Agricultural Handbook No. 512]

Krafsur, E. S. (1998). Sterile insect technique for suppressing and eradicating insect populations: 55 years and counting. *J. Agric. Entomol.* **15**, 303–317.

Tan, K.-H. (2000). "Area-Wide Control of Fruit Flies and Other Insect Pests." Pernerbit Universiti Sains Malaysia, Paulau, Pinang.

USDA National Agricultural Library. (2000). "The Screwworm Eradication Collection." Special Collections. www.nal.usda.gov/speccoll/screwworm/.

Vreysen, M. J. B., Saleh, K. M., Ali, M. Y., Abdulla, A. M., Zhu, Z.-R., Juma, K. G., Dyck, A., Msangi, A. R., Mkonyi, P. A., and Feldmann, H. U. (2000). *Glossina austeni* (Diptera: Glossinidae) eradicated on the island of Unguja, Zanzibar, using the sterile insect technique. *J. Econ. Entomol.* **93**, 123–135.

Sternorrhyncha
(Jumping Plant Lice, Whiteflies, Aphids, and Scale Insects)

Penny J. Gullan
University of California, Davis

Jon H. Martin
The Natural History Museum, London

The Sternorrhyncha is one of the suborders of the order Hemiptera; it comprises some 16,000 described species. It contains four major groups, all entirely phytophagous and usually recognized as superfamilies: the Psylloidea (psylloids or jumping plant lice), Aleyrodoidea (whiteflies), Aphidoidea (aphids), and Coccoidea (scale insects or coccoids). The name Sternorrhyncha (from the Greek *sternon,* meaning chest, and *rhynchos,* meaning nose or snout) refers to the location of the mouthparts on the underside of the insect, between the bases of the front legs, although sometimes the mouthparts are lacking in the adult.

OVERVIEW OF STERNORRHYNCHA

Insects belonging to the Hemiptera are unique in having their mouthparts forming a rostrum that comprises mandibles and maxillae modified as needle- or thread-like stylets lying in a grooved labium. Two pairs of stylets interlock to form two canals, one delivering saliva and the other taking up plant or animal fluid. The Sternorrhyncha is widely accepted as the sister group to the rest of the Hemiptera. It is a well-defined group diagnosed by the position of the rostrum and the presence of only one or two segments on each tarsus (the part of the leg most distant from the body and bearing one or two claws); in most other hemipterans there are three tarsal segments.

Sternorrhynchans use their stylets to probe plant tissues intracellularly (into or through plant cells) or intercellularly (between plant cells). The tips of the stylets always enter cells at the site of ingestion, which is often phloem-sieve elements. Generally, stylet penetration is accompanied by secretion of a solidifying saliva that forms a sheath around the stylets. Other hemipterans mostly probe intracellularly, may or may not secrete salivary sheaths, and ingest from a wider variety of plant or animal tissues. Most sternorrhynchans are phloem feeders, and thus have a diet rich in carbohydrates (sugars) and deficient in amino acids and other nitrogenous compounds. Generally, there is an intimate association with intracellular bacteria, called endosymbionts, which are housed in special tissue (bacteriomes or mycetomes) and contribute nutrition to the insect host.

Sternorrhynchan excreta is often a sticky, sugary liquid called honeydew that often contaminates foliage; it serves as a substrate for the growth of black sooty mold fungi that can impede photosynthesis and reduce plant vigor. Honeydew often attracts ants that may protect the sternorrhynchans from their natural enemies, especially predatory and parasitic insects.

Adult jumping plant lice, whiteflies, and many adult aphids have two pairs of wings. All adult female scale insects and most adult female aphids are wingless (apterous). Adult male scale insects usually resemble small flies (order Diptera) in having the hind wings reduced to small balancers (halteres), although sometimes the halteres are absent; in a few scale insect species males are apterous. The absence of wings in adult females of scale insects and many aphids means that the immature stages (nymphs) of these groups can be difficult to distinguish from their adults.

Evolutionary interpretations of morphology have suggested either that whiteflies are sister to aphids + scale insects, with

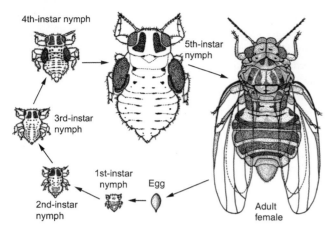

FIGURE 1 The life cycle of *Cardiaspina fiscella* (Psylloidea: Psyllidae). [Adapted with permission from K. L. Taylor, 1962, *Aust. J. Zool.* **10(2)**, 310.]

psylloids sister to all three, or that whiteflies + psylloids are sister to aphids + scale insects. Phylogenetic analysis of nucleotide sequences (of small subunit rDNA, also called 18S rDNA) supports the former hypothesis that Psylloidea is the sister group to the other sternorrhynchans. Some of the morphological traits shared by Aleyrodoidea and Psylloidea are ancestral features (plesiomorphies), such as two, similar-sized, tarsal segments (the first tarsal segment is reduced or absent in aphids and coccoids) and good jumping ability. Also both groups have pedunculate eggs, i.e., one end of the egg has a short stalk or narrow extension (Fig. 1). Whiteflies, aphids, and scale insects all have Malpighian tubules (filamentous excretory organs) reduced in number or absent and their gut often has a well-developed filter chamber that allows most of the water in ingested sap to bypass the absorptive part of the midgut. In contrast, psylloids have four Malpighian tubules (an ancestral feature found in other Hemiptera) and possess only a rudimentary filter chamber.

PSYLLOIDEA (JUMPING PLANT LICE)

Worldwide, there are about 2500 described species of psylloids or jumping plant lice. Sometimes the common name "psyllids" is applied to all jumping plant lice but this name may cause confusion because it is also the common name for one family of psylloids, the Psyllidae. The greatest abundance and species richness of psylloids are found in tropical and south temperate regions, where many new species await discovery. In the Northern Hemisphere, for example, fewer than 100 species occur in Britain and about 300 species are known from North America. Adult psylloids are small, ranging in length from 0.2 to 8.0 mm, and superficially resemble miniature cicadas or small leafhoppers, but are distinguished by their multiseg-mented antennae. Their common name derives from the ability of adults to jump backward when disturbed. All psylloids feed by sucking sap, mostly of woody dicotyledonous plants. They ingest sap from a variety of tissue sources, including phloem, xylem, and mesophyll parenchyma, and thus are not phloem

specialists like most other sternorrhynchans. Psylloid nymphs are either free living, gall inducing, or lerp forming; lerps are manufactured sugary and starchy covers or scales under which the nymphs live. Some species of jumping plant lice are pests of cultivated plants, with damage resulting from loss of plant sap, toxins in the saliva, and/or disease caused by transmitted microorganisms.

Evolution and Classification

Historically, the classification of psylloids was based on the well-known, but relatively species-poor, Holarctic fauna. Expanding knowledge of the diverse tropical and south temperate faunas has led to better understanding of relationships and hence a more natural classification. The system used here is based on that of White and Hodkinson presented in 1985, with emendations suggested by D. Burckhardt and D. Hollis. In this system, six families of extant psylloid species are recognized (with each of the first four families listed here having 10 or fewer genera): Calophyidae, Carsidaridae, Homotomidae, Phacopteronidae, Psyllidae (includes the Aphalaridae, Liviidae, and Spondyliaspididae, recognized by some authors; about 120 genera), and Triozidae (over 40 genera).

Many species of the predominantly Neotropical and tropical/subtropical Asian family Calophyidae feed on plants in the Anacardiaceae. Carsidaridae is almost entirely pantropical in distribution and is restricted to the related plant families Bombacaceae, Malvaceae, and Sterculiaceae. The Homotomidae is predominantly from the Old World tropics and feeds almost exclusively on figs (Moraceae: *Ficus*). The Phacopteronidae occurs in tropical Asia, Africa, and the Americas and many species feed on plants in the Meliaceae. The Psyllidae is the largest family, but it is heterogeneous morphologically. A large number of psyllids are associated with legumes (Fabaceae), although the speciose subfamily Spondyliaspidinae specializes on *Eucalyptus* (Myrtaceae). Many species of Spondyliaspidinae produce lerps or induce leaf galls that they plug with lerp substance. The Triozidae is cosmopolitan and has wide-ranging host preferences.

The fossil record of the Psylloidea extends back into the Permian (>270 mya). The families Cicadopsyllidae, Liadopsyllidae, and Protopsyllidiidae are known only as fossils, and the extant families probably diversified in the Cretaceous. Relationships among psylloid higher groups (families and subfamilies) are largely unresolved.

Life History

Reproduction is always sexual. There are seven life stages (Fig. 1): the egg, which is stalked; five instars; and the adult, with the sexes in about equal numbers. Eggs are either completely inserted into plant tissue or inserted by just the stalk (peduncle) through which water is absorbed. Depending on species, each female may lay as few as 20 to more than 1000 eggs. In cold temperate regions, psylloids frequently have only

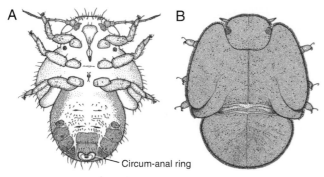

FIGURE 2 Nymphs of jumping plant lice (Psylloidea). (A) The underside, or venter, of a nymph of *Padaukia kino* (Psyllidae). (Illustration by J. H. Martin; reproduced, with permission, from D. Hollis and J. H. Martin, 1993, *Bull. Entomol. Res.* **83**, 207.) (B) The upper side, or dorsum, of a nymph of *Trioza remota* (Triozidae). [Illustration by J. H. Martin; reproduced, with permission, from W. R. Dolling, 1991, "The Hemiptera," p. 171, British Museum (Natural History), London.]

a single generation per year (univoltine). In warmer climates, development continues year round, often with several to many generations per year. Species with multiple generations per year (polyvoltine) are more likely to have population outbreaks than univoltine or bivoltine species.

Morphology

Identification is based primarily on adult characteristics, with males being most important for species-level determinations, although features of the female terminalia are sometimes used. The fifth instar can materially aid identification in many groups. Adult males and females are morphologically similar, except that males are often smaller and have different genitalia and thus different abdominal shapes. In a few species the sexes are color dimorphic, and there is often a seasonal color dimorphism. Body coloration or the pattern of markings can be diagnostic of species. The size and shape of the head and its appendages provide useful taxonomic characters. Immature psylloids (Fig. 2) are usually characteristically flattened dorsoventrally and many secrete wax.

Behavior and Ecology

Adult psylloids disperse over short distances by jumping or flying, but many species may travel long distances on air currents. The adults of a number of species are known to stridulate as a method of mate attraction. Mating behavior is often rudimentary, but precopulatory darting movements and wing flicking have been observed in some species of Spondylaspidinae.

The habits of nymphs vary among the different taxonomic groups. Nymphs of many species are naked, although often covered in waxy body secretions, and live exposed on the shoots of their host plant. Others induce galls of various complexity, from simple pits in leaves to round or elongate structures that totally enclose the nymphs. A third lifestyle is for the nymphs

to develop on foliage but protected under scale-like covers called lerps. Each lerp is formed of substances eliminated from the insect's gut (a special kind of liquid "feces" that is different from honeydew) or of a mixture of "feces" intermingled with wax filaments (exuded from pores in the anal plate). The "feces" of at least some species are composed of starch, especially amylose and amylopectin, synthesized in the insect's gut. The insect builds the lerp by moving its abdomen back and forth while squeezing out the liquid excreta (like glue from a tube), which rapidly hardens to form a lacework of filaments. Lerps have various shapes and sizes, from simple cones to fans or lacy shell-like structures, and often are of diagnostic use. They vary in color from white to cream or brown and may be opaque or transparent. Some lerp-formers (e.g., *Glycaspis* spp.) produce copious honeydew, whereas others (e.g., *Cardiaspina* spp.) produce very little or none.

Most species are specialist feeders as nymphs, with many species restricted to one plant genus or even a single species and often to particular host structures (leaves, new shoots, etc.) or growth stages (either young or mature foliage). Almost all psylloids develop on dicots, a few on monocots or conifers, but none on ferns. In cool-temperate climates, some species overwinter as adults on evergreens, returning to their true hosts in spring, when eggs are laid. Host selection appears to involve taste rather than olfaction. Adults frequently are less discriminating than nymphs and thus host plants are defined as those on which nymphs can complete their development to adulthood. Nymphal feeding frequently damages the host plant (e.g., premature leaf senescence), whereas adult feeding usually causes little injury. Large natural increases, or outbreaks, in psylloid populations appear to result from changes in the suitability of their food supply, rather than from reduction in numbers of predators or parasitoids. Many psylloid species have a mutualistic association with honeydew-seeking ants.

Notable Pest Species and Their Control

Some species of jumping plant lice have become pests of plantation trees. For example, since the mid-1990s, several species of Australian Psyllidae have become pests of South American plantations of *Eucalyptus*. In particular, the eucalypt shoot psyllid *Blastopsylla occidentalis,* the blue gum psyllid *Ctenarytaina eucalypti,* and *Ctenarytaina spatulata* have caused considerable damage. In western North America, Australian psylloid pests appeared in the early 1980s, with *B. occidentalis, C. eucalypti,* the tristania psyllid *Ctenarytaina longicauda,* the red gum lerp psyllid *Glycaspis brimblecombei,* and the lemon gum lerp psyllid *Eucalyptolyma maideni* all established in California.

Two of the world's most serious citrus pests are the Asian citrus psylla (or psyllid), *Diaphorina citri* (Psyllidae), and the African citrus psylla, *Trioza erytreae* (Triozidae). Damage to citrus derives from direct feeding (removing sap), development of sooty mold on honeydew, and, most importantly, from transmission of phloem-limited bacteria (*Liberobacter africanum* and *L. asiaticum*) that cause a disease called citrus greening.

The most characteristic symptom is a yellow-green mottling associated with dieback of shoots and leaves, followed by subsequent deterioration of fruit quality. Adults and nymphs of both psylloid species can transmit both diseases.

Other than the citrus pests, the most economically important psylloids of temperate and subtropical areas feed on cultivated apple and pear trees. *Cacopsylla mali* (Psyllidae) and related species causes various kinds of damage to apples. Pears also host a suite of *Cacopsylla* species, especially *Ca. pyricola,* which can transmit the virus that causes pear decline.

Trioza species (Triozidae) occur as pests on persimmons, cultivated fig, and carrots. *Paratrioza cockerelli* (Triozidae) feeds on potatoes, tomatoes, and other Solanaceae; nymphal saliva contains phytotoxins that cause "psyllid yellows" in potato (reducing tuber growth) and sometimes in tomato. The leucaena louse, *Heteropsylla cubana* (Psyllidae), which is native to the Caribbean, has become a pest of the tropical leguminous tree *Leucaena leucocephala* in several parts of the world where it is variously planted for timber, fuel, animal fodder, or shade for other crops and or because it fixes nitrogen and tolerates drought.

Recorded parasitoids are mostly psylloid specialists, but there is little evidence of parasitoid–host specificity. Worldwide a number of wasp species, especially from the families Encrytidae (e.g., *Psyllaephagus* spp.) and Eulophidae (e.g., *Tamarixia* spp.), are parasitic on psylloids. In Europe, cecidomyiid flies (*Endopsylla*) are parasitic on some adult psylloid pests (e.g., *Cacopsylla* spp.).

ALEYRODOIDEA (WHITEFLIES)

Members of the Aleyrodoidea derive their common name from the powdery, white waxy secretion preened over the body and wings of most adults. The Greek root *aleuro-,* found in many whitefly names, means flour. Adults of both sexes are tiny, delicate, and free flying. They have a wingspan of up to 4 mm, but usually about 2 mm, and resemble minute moths. They are a familiar sight to many home gardeners because, when infestations are heavy, the adults will fly up en masse if disturbed from the foliage of favored host plants. The group occurs worldwide, although few whiteflies occur in the cooler temperate regions. About 1450 species are described, with perhaps two or three times as many species awaiting collection and formal taxonomic study. All adults and nymphs feed by ingesting phloem sap and several species are serious plant pests.

Evolution and Classification

All whitefly species belong to one family, Aleyrodidae, which is divided into just two extant subfamilies, the Aleurodicinae and Aleyrodinae. The Aleurodicinae contains 17 genera and just over 100 described species, mostly from the Neotropics (Central and South America and the Caribbean). All other species (in more than 110 genera) belong to the Aleyrodinae, which has a mostly pantropical distribution. These two sub-

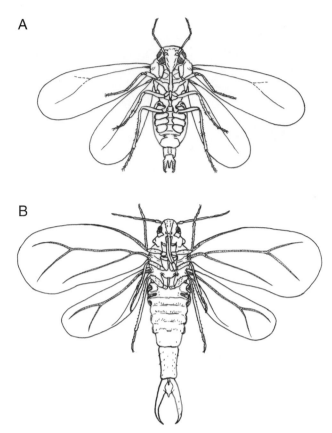

FIGURE 3 Adult male whiteflies (Aleyrodoidea: Aleyrodidae), ventral views. (A) *Trialeurodes vittata,* subfamily Aleyrodinae. (Unpublished illustration by R. J. Gill, adapted with permission.) (B) *Aleurodicus* sp., subfamily Aleurodicinae. [Adapted with permission from R. J. Gill, 1990, "Whiteflies: Their Bionomics, Pest Status and Control" (D. Gerling, ed.), p. 31, Intercept Ltd., Andover, UK.]

families are defined on both adult and nymphal features, whereas species and genera are diagnosed mainly or entirely on characters of the fourth instar, usually known as a puparium. The Aleurodicinae generally have larger adults with more complex wing venation than those of the Aleyrodinae (Fig. 3).

Whiteflies are rarely fossilized because they have small and delicate bodies. The known fossils are preserved mostly in Cretaceous and Tertiary amber with one record from the Upper Jurassic. The two modern subfamilies probably diverged during the Cretaceous, whereas the Aleyrodoidea may have originated in the Late Permian or even earlier.

Life History

Reproduction is usually sexual, with fertilized eggs producing females and unfertilized eggs producing males (arrhenotoky or arrhenotokous parthenogenesis); unmated females lay only haploid eggs. A few species produce only females (thelotoky or thelotokous parthenogenesis). There are six life stages (Fig. 4): the egg stage; the first instar, in which the nymph is often called a crawler; the second and third instars, in which the nymphs are sessile; the fourth instar or last nymphal stage, at the end

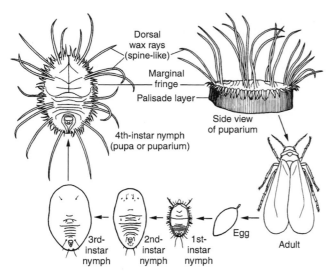

FIGURE 4 Life cycle of the greenhouse whitefly *T. vaporariorum* (Aleyrodoidea: Aleyrodidae). [Adapted with permission from R. J. Gill, 1990, "Whiteflies: Their Bionomics, Pest Status and Control" (D. Gerling, ed.), p. 15, Intercept Ltd., Andover, UK.]

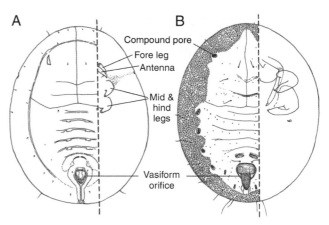

FIGURE 5 Whitefly puparia, the last nymphal stage (Aleyrodoidea: Aleyrodidae), with dorsum drawn to left and venter to right. (A) *Aleurolobus marlatti*, subfamily Aleyrodinae. [Reproduced, with permission, from J. H. Martin, 1999, "The Whitefly Fauna of Australia (*Sternorrhyncha*: Aleyrodidae): A Taxonomic Account and Identification Guide," p. 172, CSIRO, Canberra.] (B) *Aleurodicus cocois*, subfamily Aleurodicinae. (Adapted with permission from J. H. Martin, 1997, *Mem. Mus. Victoria* **56**, 126.)

of which the insect is usually referred to as a "pupa," or a "puparium," (Fig. 5) although it is not a true pupa as seen in holometabolous insects as there is no molt to a totally non-feeding, "resting" stage; and, finally, the adult (or imaginal) stage. After the emergence of the adult, the empty cuticle (exuviae) is often called a "pupal case."

Whitefly eggs are usually attached to the underside of leaves by a short stalk, the pedicel, through which water is absorbed from the plant. Eggs are often laid in circles or arcs, and egg batches are frequently conspicuous because of a dusting of white wax. First instars usually settle adjacent to the eggs from which they hatched. In most species, development occurs on the underside of leaves. After the first molt the nymphs become immobile and cannot move to a new site if food quality deteriorates. During the fourth instar, feeding ceases and the insect transforms into the winged adult.

Species such as the greenhouse whitefly (*Trialeurodes vaporariorum*) can breed continually in greenhouses or indoors. Those whitefly species that have economic impact have several to many generations per year, but many species are thought to have only one or two generations annually.

Morphology

Historically, puparial rather than adult characteristics have been used to recognize species and genera, although adults do display many taxonomically useful characteristics. Seasonal dimorphism, as seen in many aphid species, is uncommon but occurs in the puparia of a few temperate species. Male (Fig. 3) and female whiteflies are similar in appearance except for their genitalia, but males are often smaller than females of their own species. The body and wings are dusted with powdery wax emanating from large, paired wax plates on the underside of the abdomen and thus adults most commonly

appear whitish or grayish even if the body color is yellow, brown, or red or if they are darkly sclerotized. The male genitalia (the claspers or parameres and the intromittent organ or aedeagus) are more informative taxonomically than the female terminalia (ovipositor and associated structures).

Nymphs of all stages (Fig. 4) superficially resemble scale insects (Coccoidea) and frequently have ornate wax secretions in later instars. First instars have functional legs, but nymphs of the subsequent three instars have vestigial appendages and are immobile. The anus of all nymphal stages (and adults) opens in a special pit called the vasiform orifice (Fig. 5), located dorsally near the posterior end of the abdomen. It comprises a depression with a sclerotized border, a dorsal flap (the operculum), and a tongue-shaped structure (the lingula). The anus opens under the operculum, which covers at least the base of the lingula; each droplet of anal excreta (honeydew) that accumulates is catapulted away by an upward flick of the lingula, which is the only mobile body part in instars II to IV.

Puparia are mostly oval or elongate-oval, 0.5 to 2.0 mm in length, with the body margin either smooth or variously sculptured. Color varies from transparent to white, brown, or black. Waxes also contribute to the puparial appearance and may be transparent, translucent, iridescent, or opaque, usually colored white or grayish; mechanical color (iridescence) may be blue, turquoise, bottle-green, or pearly white. Transparent wax occurs in a thin layer over the body or as a marginal fringe and/or dorsal "spines." Opaque wax may occur in powdery deposits, loose mats, or filaments, or be defined as tufts or rays, and is exuded from a variety of pores and tubes in the cuticle.

Behavior and Ecology

Adult males of a few whitefly species, especially *Bemisia tabaci* and *T. vaporariorum,* have been observed to display courtship behavior prior to mating, including abdominal oscillations that result in acoustic signals caused by substrate-borne vibrations. Females may produce sex pheromones to attract males. Adult whiteflies will fly short distances, if disturbed from their host plant. They also undertake longer migratory flights, which are air-current dependent because whiteflies are weak fliers. They have complex host-finding and host-orientation behaviors, involving at least attraction to particular colors, especially yellow or yellow-green.

Whiteflies feed from plant vascular tissue and all feeding stages produce copious quantities of honeydew, yet relatively few species are ant attended. Most whitefly species appear to be oligophagous, with few known to be monophagous; however, polyphagous species are the best documented because they are most likely to be pests. Whitefly host plants are almost entirely flowering plants (angiosperms), especially woody dicots; relatively few species of whiteflies are found on herbs, grasses, ferns, or palms.

The morphology of the puparia of some species has been shown to vary depending on the species of plant that is acting as host, probably because of differences in the nature of the plant surfaces. For example, the size, number, and position of setae of the puparia can be correlated with leaf hairiness in some species (e.g., *B. tabaci*). These phenotypes of the same species often look very different from each other and this phenomenon can confound identification. This variation is especially a problem among polyphagous species in the genera *Bemisia* and *Trialeurodes*.

Notable Pest Species and Their Control

The feeding activity of many whiteflies damages their host plants and some species are known or believed to transmit diseases, especially those caused by viruses. Feeding can cause excessive sap loss as well as physiological changes in host tissues, such as leaf discoloration, wilting and premature shedding, and fouling by honeydew. The most serious aleyrodid pests are those of orchards (especially citrus), greenhouse crops, and vegetable field crops, and a few pest species are almost cosmopolitan in distribution. The past 2 decades have seen an increase in the pest status of certain whiteflies, probably because of the development of aggressive strains that are more fecund, are more efficient virus vectors, and/or have broader host ranges. Successful whitefly control depends upon an appropriate integrated pest management program in which natural enemies are protected or augmented. A number of groups of parasitic wasps (particularly Eulophidae and Aphelindiae) specialize on whiteflies, and useful predators include some ladybird beetles (Coccinellidae). Whiteflies are attacked also by pathogenic fungi, especially *Aschersonia* species.

One of the most important pests is the polyphagous tobacco or sweetpotato whitefly, *B. tabaci,* which occurs predominantly in tropical and subtropical regions and on protected crops elsewhere. It feeds on numerous fiber, food, and ornamental plants, and damage is exacerbated by its ability to vector more than 70 different viruses. *B. tabaci* exists in many biotypes or strains, and one virulent form, biotype B, sometimes is recognized controversially as a separate species, named *Bemisia argentifolii* (the silverleaf whitefly). Effective biological control of *Bemisia* whiteflies is possible using host-specific parasitoids, such as wasps of *Encarsia* and *Eretmocerus* species (Aphelinidae).

The polyphagous greenhouse whitefly, *T. vaporariorum,* occurs worldwide in greenhouses and on indoor plants; the aphelinid wasp parasitoid *Encarsia formosa* is an important biological control agent. *Aleurodicus destructor* and spiraling whitefly, *A. dispersus,* are polyphagous pests with abundant wax secretions covering the puparia. Spiraling whiteflies are parasitized by wasps of *Aleuroctonus vittatus* (Eulophidae) and *Encarsia* species, including *E. guadeloupae* (Aphelindiae), which can reduce the whitefly populations. *Siphoninus phillyreae,* perhaps of Middle Eastern origin, and *Parabemisia myricae* from Japan are pests of fruit trees and have extended their ranges into the Mediterranean region and parts of the United States. The Indian species *Vasdavidius* (formerly *Aleurocybotus*) *indicus* now attacks rice in West Africa. The woolly whitefly, *Aleurothrixus floccosus,* which is endemic to the Americas, has been introduced accidentally into Africa and Southeast Asia, where it is a pest, especially of citrus. *Dialeurodes citri* and *Singhiella* (formerly *Dialeurodes*) *citrifolii* are widespread pests of citrus, with *D. citri* attacking a wider range of other plants than does *S. citrifolii*.

APHIDOIDEA (APHIDS)

Aphids are small soft-bodied insects, ranging from 1 to 8 mm in length. They are usually found living in aggregations on rapidly growing parts of their host plants and occur predominantly in the northern temperate regions of the world. More than 4400 species are known. The life cycles of aphids are frequently complex and usually include parthenogenetic (or asexual), but often also sexual, reproduction; many species display host alternation in which cyclical parthenogenesis is combined with the obligate use of two unrelated host plants. Also many aphids produce either eggs or living young at different parts of the cycle. Adults may be winged (alatae) or wingless (apterae). Most aphids can increase their population size rapidly because of parthenogenetic production of live young combined with "generational telescoping," whereby a mother aphid carries both her daughters and their daughters (i.e., embryos within embryos). Aphid feeding activities can have deleterious effects on their host plants mainly via sap removal and/or virus transmission.

Evolution and Classification

The Aphidoidea contains three families: Phylloxeridae, Adelgidae, and Aphididae. Strictly speaking, the common name of the entire group should be "aphidoids," but "aphids" is used almost universally as a collective name. Very occasionally the name "aphids" has been applied to just members of the Aphididae, which should more properly be referred to as "aphidids."

The Phylloxeridae contains about 75 species, in eight genera, divided between two tribes. The Phylloxerini has seven genera, feeding mostly on oaks (Fagaceae: *Quercus*), or hickory and pecans (Juglandaceae: *Carya*), but with one pest species on grape vines (Vitaceae: *Vitis*). The Phylloxerinini has a single genus associated with the willow family (Salicaceae). The Phylloxeridae is probably the oldest extant family, although only one fossil (*Palaeophylloxera* from the Lower Miocene) is known. Many phylloxerid species induce galls on their host plants; only a few species are host alternating.

Conifer woolly aphids, the Adelgidae, belong to two genera and about 50 species and are entirely Holarctic in distribution. Adelgids induce galls on their primary host, spruce (*Picea*), and move to other conifers as alternate hosts.

The largest family, Aphididae, has been divided into a number of subfamilies and tribes, but different authors frequently use different classifications. The 10 subfamilies listed here are those recognized in the monographs by R. L. Blackman and V. F. Eastop: Anoeciinae (34 species), Aphidinae (over 2700 species), Calaphidinae (including the Drepanosiphinae, Thelaxinae, and several other groups treated as subfamilies by some other authors; 400 species), Chaitophorinae (164 species), Eriostomatinae (formerly Pemphiginae; 319 species), Greenideinae (151 species), Hormaphidinae (176 species), Lachninae (355 species), Mindarinae (5 species), and Phloeomyzinae (1 species).

The oldest aphid fossils are from the Triassic (at ca. 230 mya) but aphids may have originated in the Permian. Phylogenetic analysis of molecular data suggests that aphids underwent a rapid radiation into the current tribes after shifting from gymnosperms to angiosperms some time during the Upper Cretaceous. Furthermore, the ancestral aphid probably had a simple life cycle with host alternation evolving independently in each of the families and perhaps several times within the Aphididae.

Life History

There are six stages in the life history of an individual aphid: the egg or embryonic stage, four instars, and the adult. Aphids have evolved a range of annual or biennial life cycles and other adaptive strategies that often vary within as well as among species. This complexity can complicate the study of aphid biology. Furthermore, aphidologists have developed a special nomenclature for the different life stages and types. Life cycles are called holocyclic (Fig. 6) if a sexually reproducing generation is present or anholocyclic if the sexual generation is absent. A typical complete cycle (holocycle) consists of a single generation of sexual morphs (sexuales) and several to many generations of only parthenogenetic females. In the sexual generation, males mate with females to produce fertilized eggs, which are all female. Each egg gives rise to a wingless "foundress" or "stem mother" (fundatrix) that gives rise

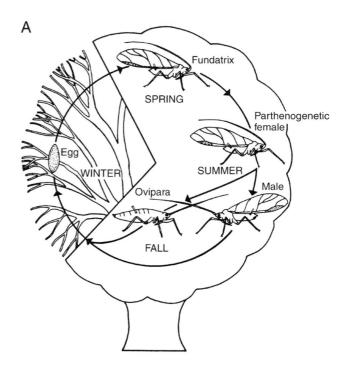

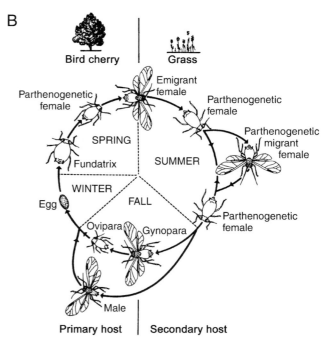

FIGURE 6 Life cycles of aphids. (A) The holocyclic life cycle of the monoecious sycamore aphid, *Drepanosiphum platanoidis*. (B) The holocyclic life cycle of the heteroecious bird cherry–oat aphid, *Rhopalosiphum padi*. Fundatrix, foundress, a wingless parthenogenetic female that produces live offspring rather than eggs; gynopara, migrant that flies back to the primary host where she produces oviparae; ovipara, wingless, sexual egg-laying female that mates with a male. (Reproduced with permission from A. F. G. Dixon, 1973, "Biology of Aphids," pp. 8 and 9, Edward Arnold Ltd., London.)

to a lineage of parthenogenetic females. The parthenogenetic descendants of a single foundress are called collectively a clone and are identical genetically but may be of different phenotypes (morphs). The adults may be apterae or alatae, depending on environmental conditions. In Aphididae, parthenogenetic females (viviparae) give birth to live young, whereas the sexual females (oviparae) lay eggs. In Adelgidae and Phylloxeridae, females lay eggs during both the sexual and the asexual phases of reproduction. The phenomenon of cyclical parthenogenesis is a key feature of aphid biology. Some aphids have lost, or can facultatively lose, the sexual part of the life cycle (a condition then called anholocycly). It is common for aphids to have 15 to 20 generations per year, with up to 40 in tropical climates.

Most aphids are monoecious (Fig. 6A), i.e., they undergo all phases of their life cycle on one host plant or on a small number of closely related hosts. About 10% of aphids are heteroecious (Fig. 6B), having more complex life cycles involving host alternation (heteroecy). The sexual morphs mate and the oviparae lay eggs on the primary host (often a woody plant); however, a regular move occurs to another, unrelated plant, the secondary host (which may be either herbaceous or woody), on which the parthenogenetic generations live. The aphids must return to their primary host for the next sexual generation. Host alternation may occur as part of a 1-year (annual) life cycle, as in many Aphididae, or as a 2-year (biennial) life cycle, as in Adelgidae. In the Aphididae, in temperate climes, alternation is frequently obligate, overcoming the twin problems of poor sap flow in woody primary hosts in summer and the death of many herbs in winter.

Morphology

Aphids are highly polymorphic, with most species occurring in several different forms or morphs. In individuals destined to be winged as adults, the wing buds are usually apparent after the second nymphal molt. Most common aphid species are soft-bodied and green in color, but dark and brightly colored species and a few hard-bodied species also occur. Nymphs superficially resemble adults except that they are smaller in size and never have wings. Aphids usually have two prominent, tube-like structures on the posterior dorsum of the abdomen, called siphunculi or cornicles (Fig. 7), which can discharge defensive lipids and alarm pheromones. In some species, these structures are reduced to pores or are absent. Another structure unique to the aphids is a posterior projection on the tip of the abdomen called the cauda, equivalent to the lingula in white-flies. Protective wax secretions are common.

Behavior and Ecology

Both adults and nymphs can travel short distances by walking, but dispersal of alatae occurs by both active flight and passive long-distance movement on air currents. Apterae and nymphs are frequently transported by attendant ants. In sexual morphs, males are attracted to females by sex pheromones released from glands on the female's legs. Altruistic behavior has been reported in the few aphid species that have a sterile soldier "caste,"

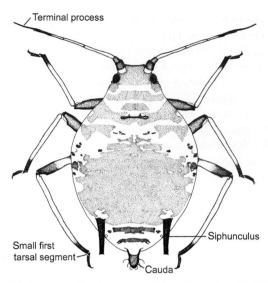

FIGURE 7 An apterous female cowpea aphid, *Aphis craccivora* (Aphidoidea: Aphididae). [Illustration by J. H. Martin; reproduced with permission from W. R. Dolling, 1991, "The Hemiptera," p. 193, British Museum (Natural History), London.]

which is a special kind of first or second instar that defends the family group against competitors and natural enemies.

Aphids generally live in aggregations on the buds, stems, and/or leaves of their host plants. Some species induce galls and a few others live underground on roots. Almost all aphids are phloem feeders, producing copious quantities of honeydew, and thus many species are attended by ants. Some aphids feed from both phloem and parenchyma tissue, and adelgids are largely parenchyma feeders. In general, aphids are quite host specific, with each genus being associated with a particular host-plant family and each species with one genus or species of host plant or at least closely related plant genera. In host-alternating species, however, the primary and secondary hosts are usually unrelated and the specificity to the primary host is generally higher than to the secondary host(s). Pest aphids, especially those of agricultural crops, often feed on plants in a number of unrelated families.

Notable Pest Species and Their Control

Infestations of aphids can grow rapidly to enormous size and cause plant debilitation through nutrient deprivation, but even small numbers of aphids can transmit viruses to uninfected plants. Pest aphids often are more polyphagous than nonpest species and frequently are species exotic to the area where they are of most economic concern. Cosmopolitan pest species are frequently anholocyclic, i.e., reproducing continually by parthenogenesis on their crop hosts, and thus are particular pests in the tropics.

One of the most notorious aphidoid pests is the grape phylloxera, *Daktulosphaira vitifoliae* (Phylloxeridae), which induces galls on *Vitis* species, including the viticulturally important *V. vinifera*. Damage in vineyards results mainly from the rotting of roots when infestations are heavy, rather than sap loss per se. The main method of control is the use of resistant grape rootstocks.

Genera of Aphididae that contain significant pests of crops include *Aphis, Brevicoryne, Macrosiphum, Myzus,* and *Rhopalosiphum;* the species listed below all transmit plant viruses. There are a number of serious pest species of *Aphis,* including the cowpea aphid *A. craccivora* (Fig. 7), bean aphid *A. fabae,* and melon or cotton aphid *A. gossypii.* The cabbage aphid, *B. brassicae,* attacks members of the Cruciferae. The potato aphid *Ma. euphorbiae* and the rose aphid *Ma. rosae* both use roses as primary hosts. The polyphagous shallot aphid *My. ascalonicus* appears to be exclusively anholocyclic. The primary hosts of black cherry aphid *My. cerasi* and green peach aphid *My. persicae* are *Prunus* fruit trees; secondary hosts include a range of economically important plants. The corn leaf aphid *R. maidis* is a worldwide pest of cereal crops, particularly maize, sorghum, and barley, and is habitually anholocyclic in most places. Important tree pests include *Pineus* species (Adelgidae) on pines; *Cerataphis ftransseni* (Hormaphidinae) on coconut and other palms; *Cinara* species (Lachninae) on a wide range of conifers; *Dysaphis, Myzus,* and *Brachycaudus* species (Aphidinae), many of which utilize rosaceous trees as their primary hosts, causing unsightly leaf curling/galling in spring; and *Eriosoma lanigerum* (Eriosomatinae), which develops woolly colonies on apple branches and trunks.

The most common aphidophagous predators are ladybird beetles (Coccinellidae), lacewings (Neuroptera), some hover flies (Syrphidae), a few gall midges (Cecidomyiidae, especially the widespread *Aphidoletes aphidimyza*), and certain predatory bugs (Anthocoridae). Wasps of the large family Aphidiidae and several genera of Aphelinidae are endophagous parasitoids of Aphididae; Adelgidae and Phylloxeridae are not parasitized.

COCCOIDEA (SCALE INSECTS)

The scale insects (also called coccoids) occur worldwide. They are mostly small (less than 5 mm in length) and often cryptic in habit. There are estimated to be about 7500 described species. Many scale insects are economically important pests of agriculture, horticulture, and forestry. Male scale insects display complete metamorphosis, whereas female development is neotenous (adults resemble nymphs). A number of taxa display remarkable diversity in their genetic systems (e.g., parthenogenesis, hermaphroditism, and paternal genome elimination) as well as in chromosome number, sperm structure, and types of endosymbioses. The name "scale insect" derives both from the frequent presence of a protective covering or "scale" and from the appearance of many of the female insects themselves. Most species produce a waxy secretion that covers the body either as a structure detached from the body (a scale or test) or as a substance that adheres to the body surface. Some coccoid species have been used as sources of candle wax, lacquers such as shellac, or dyes. Scale insects are more diverse in terms of major evolutionary lineages (families), species richness, and morphology than any of the other sternorrhynchan groups.

Evolution and Classification

Scale insects have been assigned variously to 20 or more families, but the family-level classification is controversial. Often the Coccoidea is divided into two major, informal groups, the archaeococcoids (or archaeococcids) and the neococcoids (or neococcids). The extant archaeococcoids comprise the Margarodidae *sensu lato* (with over 400 species and sometimes treated as several families), Ortheziidae (ensign scales; about 155 species), Carayonemidae (4 species), Phenacoleachiidae (2 species), and perhaps the genus *Puto* (about 60 species, sometimes placed in their own family, Putoidae, or otherwise in Pseudococcidae). Collectively, the above archaeococcoid families number only approximately 80 genera and 600 species. Some of the morphological features that define the archaeococcoids occur more widely in the Hemiptera, and monophyly of archaeococcoids is uncertain.

The neococcoids, which comprise all of the other extant families (usually 17 are recognized) and most of the species of scale insects (about 7000), are a monophyletic group characterized by derived features, including a chromosome system involving paternal genome elimination, needle-like apical setae on the labium, and loss of abdominal spiracles. Among neococcoids, most families (with the exception of the Eriococcidae) are well characterized morphologically. In contrast, relationships among families are largely unknown or not supported well by available data. The three largest families of neococcoids, in order of size, are the Diaspididae (armored scales; about 2400 species), Pseudococcidae (mealybugs; about 2000 species), and Coccidae (soft scales; over 1000 species). The other neococcoid families are the Eriococcidae (felt scales; about 550 species); Asterolecaniidae (over 200 species), Lecanodiaspididae (about 80 species), and Cerococcidae (about 70 species) (these 3 families collectively are called pit scales); Kerriidae (Tachardiidae) (lac insects; about 100 species); Kermesidae (gall-like scales; about 90 species); Aclerdidae (flat grass scales; about 50 species); Conchaspididae (about 30 species); Halimococcidae (about 20 species); Stictococcidae (16 species); Beesoniidae (9 species); Dactylopiidae (cochineal insects; 9 species); Micrococcidae (8 species); and Phoenicococcidae (1 species).

The oldest fossil scale insects are from the Lower Cretaceous but the group is at least of Triassic and probably of Permian age. The earliest radiation involved the archaeococcoids, with the neococcoids apparently diversifying in conjunction with flowering plants.

Life History

The reproductive repertoire of scale insects includes hermaphroditism, seven kinds of parthenogenesis, and six major types of sexual chromosome systems. In the vast majority of coccoid

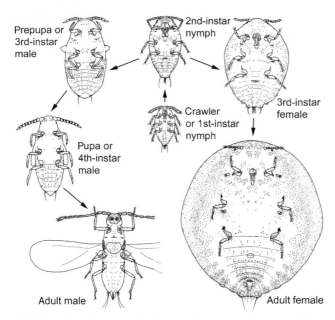

FIGURE 8 The life cycle of the Australian *Acacia*-feeding mealybug *Melanococcus albizziae* (Coccoidea: Pseudococcidae). (Partially adapted, with permission, from G. S. Farrell, 1990, *Mem. Mus. Victoria* **51**, 49–64.)

species, the males are functionally (because of inactivation or elimination of paternal chromosomes), and sometimes actually, haploid.

Each individual female scale insect has four or five growth stages (Fig. 8): the egg, two or three immature (nymphal) instars, and the adult (imaginal stage). The female lays eggs (ovipary) either in a cavity under her body or in a waxy covering (ovisac; Fig. 9E) that may be attached to her body, or she

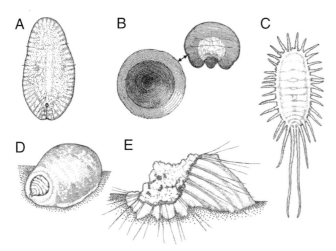

FIGURE 9 The morphological diversity of scale insects (Coccoidea). (A) An adult female of the soft brown scale, *Coccus hesperidum* (Coccidae). (B) Adult female (right) and scale cover of a California red scale, *Aonidiella aurantii* (Diaspididae). (C) An adult female of the longtailed mealybug, *Pseudococcus longispinus* (Pseudococcidae). (D) Adult female of a gum tree scale, *Eriococcus coriaceus* (Eriococcidae), inside her waxy test with just her posterior abdomen visible. (E) The hermaphroditic adult of the cottony cushion scale *Icerya purchasi* with ovisac (Margarodidae). (Illustration by P. J. Gullan.)

retains the eggs in her reproductive tract until the young are ready to hatch (ovovivipary). The mobile first instars, called crawlers, are the main dispersal agents for Coccoidea; other immature instars generally are sessile. Adult females may live for an extended period (months to several years).

After hatching from the egg, male scale insects have a total of four immature or preimaginal instars including their own specially derived form of a complete metamorphosis (holometaboly) involving one or two pupa-like stages (Fig. 8). These are called the prepupa and pupa and develop either under a scale cover or test or inside a waxy cocoon or test that is produced by the second instar. Neither the pupa-like instars nor the adult males feed; adult males are short-lived (at most a few days) and have limited time to seek out the sedentary females for mating.

Host-alternating life cycles, as seen in some aphids, are unknown. The number of annual generations varies among and often within species and ranges from fewer than one to up to seven or eight per year. Annual life cycles are common in cool-temperate regions and numerous annual generations, typical of many aphids, do not occur in scale insects.

Morphology

Morphologically, scale insects are among the most unusual of insects. There is a very marked dimorphism between the adult male and the adult female and identification is based almost entirely on the long-lived adult female. Adult females (Fig. 9) are sac-like without a well-defined head, thorax, or abdomen, and they are neotenic, that is, the adult resembles an immature individual. Adult females are wingless, may or may not have legs, but usually have well-developed mouthparts. The anus opens posteriorly and usually is surrounded by a chitinized ring and often anal plates (Fig. 9A) or anal lobes flank the opening. Species that have a scale cover (e.g., armored scales; Fig. 9B) usually incorporate in it the dorsal part of the cast-off cuticle from the previous growth stage. If the body wax is of the adhering type, then it may vary from a thin translucent sheet (many soft scales) to a thick, wet mass (e.g., wax scales), a cottony secretion (e.g., some margarodids) or a powdery, white dusting (many mealybugs; Fig. 9C). Species that produce waxy tests (e.g., felt scales; Fig. 9D) often also have a dusting of powdery body wax. Wax is produced by epidermal glands and is exuded from a great variety of cuticular pores and ducts and sometimes also from glandular setae. Typically the body also is covered in setae that may be hair-like, spine-like, or other shapes, often with several types present on each species.

Adult males resemble small, delicate flies (Fig. 8) and have a distinct head, thorax, and abdomen. Most have a pair of membranous forewings and a pair of vestigial hind wings (balancers or halteres), although adult males of some species are wingless. The mouthparts of the prepupa, pupa, and adult males are reduced or absent.

Crawlers (Fig. 8) are usually ovoid or elongate ovoid and flattened dorsoventrally. Their antennae, legs, and mouthparts are well developed and various kinds of pores, ducts, and setae are present on the body and its appendages; often a fringe of setae surrounds the body margin.

Behavior and Ecology

The first instars either seek suitable feeding sites on the natal host plant or disperse on the wind; some crawlers display behaviors that increase their chances of becoming airborne. Adult males probably locate their sessile conspecific females using sex pheromones but the presence of these chemicals has been demonstrated experimentally for very few species.

Scale insects primarily feed from either the phloem or the parenchyma, and their host associations range from monophagous to polyphagous. Sap removal is the main cause of plant damage, but a few species (especially of mealybugs and armored scales) also transmit plant pathogens and/or toxins that may further reduce plant vigor and eventually kill the host. Furthermore, most scale insects (except armored scales and a few others) produce honeydew and are ant attended; associations are usually facultative but a number of obligate ant–coccoid relationships have been described.

Cochineal scales (*Dactylopius* spp.) and certain mealybugs and lac insects have been used as biological control agents for particular noxious weeds; e.g., cochineal insects can assist with the control of prickly pear cacti, *Opuntia* species.

Notable Pest Species and Their Control

Some scale insects are serious plant pests, especially of perennial agricultural plants. They can cause damage to nut and fruit trees, forest or plantation trees, glasshouse plants, woody ornamentals, house plants, and sometimes sugar cane and even grass in lawns. Pests are usually either polyphagous (e.g., certain wax scales, *Ceroplastes* species; the pink hibiscus mealybug *Maconellicoccus hirsutus;* and the cottony cushion scale *Icerya purchasi*) or oligophagous (e.g., the beech scale *Cryptococcus fagisuga* on beech trees or *Matsucoccus* species on pines). The cryptic habits and small size of most scale insects mean that they may not be detected until plant damage is substantial. Also, if populations on plants are low, they can be notoriously difficult to detect during quarantine inspections. Most pest scales belong to the Diaspididae (e.g., California red scale *Aonidiella aurantii* on citrus, Boisduval scale *Diaspis boisduvalii* on orchids, euonymus scale *Unaspis euonymi* on ornamentals, and black pineleaf scale *Nuculaspis californica* on pines), Coccidae (e.g., certain polyphagous *Ceroplastes, Coccus, Pulvinaria,* and *Saissetia* species), and Pseudococcidae (e.g., many species of *Dysmicoccus, Pseudococcus,* and *Planococcus* on a wide range of plants), but a few significant pests belong to other families, such as the Margarodidae (especially polyphagous *Icerya* species), Eriococcidae (e.g., certain *Eriococcus* species on ornamental plants), and Asterolecaniidae (e.g., *Bambusaspis bambusae* on bamboo).

The most important predators of scale insects are ladybird beetles (Coccinellidae; especially *Rodolia, Chilocorus,* and *Cryptolaemus*). The main parasitoids of scale insects are chalcidoid wasps, especially species of Aphelinidae (e.g., species of *Aphytis, Encarsia,* and *Coccophagus*) and Encrytidae (e.g., species of *Anagyrus, Leptomastix,* and *Metaphycus*), although some scale insects are attacked by flies that may be either parasitic (e.g., Cryptochetidae) or egg predators (e.g., a few Cecidomyiidae).

See Also the Following Articles

Auchenorryncha • Cicadas • Honeydew • Parthenogenesis • Prosorrhyncha • Rostrum • Symbionts Aiding Digestion

Further Reading

Ben-Dov, Y., Miller, D. R., and Gibson, G. A. P. (2002). ScaleNet. http://www.sel.barc.usda.gov/scalenet/scalenet.htm.

Blackman, R. L., and Eastop, V. F. (1994). "Aphids on the World's Trees: An Identification and Information Guide." CAB Int., Wallingford, UK.

Blackman, R. L., and Eastop, V. F. (2000). "Aphids on the World's Crops: An Identification and Information Guide," 2nd ed. Wiley, Chichester.

Burckhardt, D. (1994). Psylloid pests of temperate and subtropical crop and ornamental plants (Hemiptera, Psylloidea): A review. *Entomologia* **2,** 173–186.

Byrne, D. N., and Bellows, T. S., Jr. (1991). Whitefly biology. *Annu. Rev. Entomol.* **36,** 431–457.

Carver, M., Gross, G. F., and Woodward, T. E. (1991). Hemiptera. *In* "The Insects of Australia: A Textbook for Students and Research Workers" (CSIRO, ed.), 2nd ed., Vol. I, pp. 429–509. Melbourne University Press, Carlton.

Gerling, D. (ed.) (1990). "Whiteflies: Their Bionomics, Pest Status and Management." Intercept, Andover, UK.

Gullan, P. J., and Kosztarab, M. (1997). Adaptations in scale insects. *Annu. Rev. Entomol.* **42,** 23–50.

Minks, A. K., and Harrewijn P. (eds.) (1987–1989). "Aphids: Their Biology, Natural Enemies and Control," Vols. 2A, 2B, and 2C of "World Crop Pests." Elsevier, Amsterdam.

White, I. M., and Hodkinson, I. D. (1985). Nymphal taxonomy and systematics of the Psylloidea (Homoptera). *Bull. Br. Mus. (Nat. Hist.), Entomol.* **50(2),** 153–301.

Stonefly

see *Plecoptera*

Stored Products as Habitats

Rudy Plarre
Federal German Institute for Materials Research and Testing

Wendell E. Burkholder
University of Wisconsin

The protection of stored food against attack by insects and molds has been a necessity since the dawn of agriculture,

FIGURE 1 Traditional storage facility for grain in China.

or perhaps since humans began to manufacture primitive clothing and tools for daily use from plant or animal material. Protection of stored products, thus, has a tradition of tens of thousands of years and is documented in the earliest records of history.

Stored food and other products of plant or animal origin provide a habitat with shelter, food, and breeding sites for insects (Fig. 1). Worldwide losses from these postharvest insects and other food pests are estimated to be from a few percent in developed countries to 20% in an average year in some developing countries. Under certain storage conditions, 50 to 100% of food is lost.

Several hundred insect species are associated with stored products. Most are beetles (Coleoptera) or moths (Lepidoptera), with their parasitoids being wasps (Hymenoptera). However, there are only about 100 species of economic importance. These can be divided into two main categories: species infesting dry botanicals, mainly but not exclusively grain seeds and their products, and scavenger species infesting animal products (meat, furs, wool, feathers, etc.). Very few pest insects can feed and breed successfully on both plant and animal material.

ECOLOGY OF STORED-PRODUCT INSECTS

The storage environment can be regarded as a human-made, simply structured, unstable ecosystem. All trophic levels of a typical ecosystem are represented. However, the system is not self-sustaining, and there is a unidirectional flow of energy and biomass.

Trophic Levels in a Storage Environment

The harvested and stored parts of the plants or the processed material are regarded as the producers (or the first trophic level) and contain the energy that was captured by the crops in

FIGURE 2 *Sitophilus zeamais,* maize weevil, adults on maize. (Photograph from Agriculture and Agri-Food Canada.)

the field. The primary consumers of the second trophic level in stored cereals are insects that infest the intact grain kernel. Most numerically dominant are the grain weevils of the genus *Sitophilus* (Curculionidae) (Fig. 2), the grain borers *Rhyzopertha dominica* (Fig. 3) and *Prostephanus truncatus* (Bostrychidae), and the grain moth *Sitotroga cereallela* (Gelechiidae). Primary consumers provide routes of entry for secondary consumers, such as *Tribolium* spp., that feed on the waste products or on debris, the frass, provided by the primary consumers. However, because damaged grain and grain dust are always present, secondary consumers can also exist without primary consumers.

The oxidation processes of the stored commodity and the

FIGURE 3 *Rhyzopertha dominica,* lesser grain borer, adults on wheat. (Photograph from Agriculture and Agri-Food Canada.)

respiratory activity of primary and secondary consumers generate heat and moisture that affect the microclimate. "Hot spots" may occur, which favor microbial growth. Certain microbial fungi further enhance conditions favorable to primary and secondary consumers. At this stage secondary and tertiary consumers become difficult to separate. Finally, the way is paved for the third trophic level of decomposing fungi, yeasts, and bacteria.

Predators and parasitoids of stored product insects are frequently encountered in highly infested commodities. Some of them are very closely associated with prey or host. Others are less specific: most parasitic wasps, for example, parasitize a number of different host species.

Carnivorous or scavenger species are typically found on stored commodities of animal origin such as dry meat or fish, wool, hair, and furs or products made out of these such as clothing or carpets. For example, dermestid beetles of the genera *Dermestes, Anthrenus, Attagenus,* or *Trogoderma* and the tineid moths *Tineola bisselliella* and *Tinea pellionella* are the most common insects in households. Scavengers feed on dead insect parts as well, which is why they are often found in stored products of plant origin that are infested with phytophagous insects.

Because of the weblike interactions of the different trophic levels, it is difficult to place insect species in specific categories. Their roles may be granivore, carnivore, fungivore, predator, parasite, and scavenger. These roles, together with the microflora and the environmental factors temperature and humidity, result in ongoing fluctuations of physical parameters, species succession, changes in population densities, and continuous degradation of the stored commodity.

Feeding Habits and Damage

Primary and secondary insect pests of stored foodstuff are polyphagous, with a general preference for either dry plant material or material of animal origin. Under a different perspective, these pests can also be classified into internal and external feeders. Internal feeders are species that penetrate the intact outer coating of seeds either as feeding or ovipositing adults or as freshly hatched larvae. External feeders graze on the surface of stored food stuff, on already damaged material, and on debris. Nevertheless, they almost all have a cryptic feeding habit that is probably a result of their negative phototaxis. The larvae of the Indian meal moth, *Plodia interpunctella,* the most common stored-product moth worldwide and also frequently encountered in households, prefers undisturbed and darker places wherever suitable food is available.

The most obvious damage caused by stored-product insects is the destruction of food destined for human consumption. In addition to their capability to cause tremendous losses, insects contaminate the commodity with frass, feces, exuviae, and corpses. The accumulation of filth may lead to the rejection of the stored material. Except for some tenebrionid grain beetles infected with a certain tapeworm, stored-product insects have not been shown to transmit human pathogens or diseases. However, some of their by-products can be highly allergenic to humans.

Environmental Factors

Temperature and relative humidity usually determine the limits of insect infestations in stored products. The range of temperature in which stored-product insects are active lies between 8 and 41°C. Reproduction and larval development above or below these limits are not possible, and the individuals can survive extremes only in diapausing or hibernation stages. Some species are more cold-hardy or heat tolerant than others. Relative humidity also is important because it directly influences the moisture content of the foodstuff. The relative humidities to which pest insects have adapted range from near 0 to close to 100%, with typical optimal conditions being from 60 to 75%. Temperature and relative humidity of the microclimate can be actively influenced through the metabolic activity of aggregated individuals. Light is less important than temperature and humidity as a factor influencing the survival of stored-product insects. However, most larvae have a negative phototaxis and therefore are most likely to be found in dark, less disturbed areas.

BEHAVIOR OF STORED-PRODUCT INSECTS

In addition to physical parameters, the behavior of stored-product insects is greatly influenced by semiochemicals. Living organisms, grain, microflora, insects, and mites, as well as decaying matter, produce an array of volatiles that combine with the oxygen, nitrogen, and carbon dioxide of the system. For example, enzymatic reactions in stored seeds produce unsaturated fatty acids that are highly attractive to a number of pests. Catabolism of decaying plant and animal matter enables pest insects to locate suitable feeding and breeding sites. Fungi have a distinct odor that is attractive to fungivorous species but repellent to others at high concentrations.

Intra- and interspecific communication is directed by a number of pheromones and allelochemicals. For example, spacing and oviposition pheromones are found among bruchids marking already infested seeds. These compounds are repellent to conspecifics. Spacing pheromones of certain species not only act on conspecifics but also repel nonconspecifics and thus act as allomones.

Sex pheromones, nearly always produced by the female, attract mates. These pheromones comprise specific compounds or a highly specific combination of compounds. Aggregation pheromones, typically found in species with long-lived adults, are produced by males and attract conspecifics of both sexes. Certain compounds of some aggregation pheromones are not species specific, and there is cross-attraction, especially among closely related species. In most cases, foods synergize the effect of aggregation pheromones. Although aggregation pheromones increase the likelihood that an individual will find a mate among aggregated conspecifics, the role of these substances as mating pheromones is not necessarily obvious.

Several parasitoids and predators make use of volatiles to locate hosts and/or prey, and under certain circumstances pheromone components may act as allelochemicals. Specialists that parasitize or prey on a single host species, a very narrow range of related host species, or specific developmental stages are attracted to specific kairomones emitted by their host or prey.

ORIGIN OF STORED-PRODUCT INSECTS

The origin of stored-product insects can be examined from three viewpoints: first, from a practical or physical point of view, asking for the natural reservoirs from which the insects spread into the storage system and seeking to learn how they invade a particular commodity; second, from a phylogenetic point of view, examining the ecological and behavioral preadaptations that made it possible for insects to shift from an original or primary habitat to a relatively recent, human-made habitat; and finally, from a historical point of view, searching for the geographic origins of the various insect species that infest stored commodities, which today are considered to be distributed worldwide.

Preadaptations and Natural Habitats

Insects species capable of surviving in stored commodities generally have some biological strategies in common. Species infesting stored products are not monophagous; rather, stored-product insects are food opportunists. For example, the cigarette beetle, *Lasioderma serricorne,* and the drugstore beetle, *Stegobium paniceum,* are perhaps the most ubiquitous of all stored-product pests, but all others also feed and breed on a variety of products. Almost all species found in human-made storage systems also can be found outdoors.

Natural breeding sites of stored-product insects can be sheltered niches where seeds and nesting material were gathered by rodents, birds, or other insects. Dermestids, ptinids (spider beetles), and tineids (clothes moths) are frequently found in empty rodent or bird nests where they feed on hairs, feathers, feces, pellets regurgitated by birds of prey, and food debris brought in or left by the previous occupants. Carrion is a food source as well for some species.

Stored-product insects typically have the ability to survive a long time without food. When environmental conditions are favorable and resources are available, stored-product insects reproduce over prolonged periods and have a rapid population growth rate, exploiting resources maximally. Dispersal is regulated by population density, and most stored-product insects are good fliers, which is essential for locating patchily distributed food and breeding habitats. Therefore, the following preadaptations help insect species to exploit the resources of stored-product systems: a range of natural reservoirs, a range of food habits, ability to survive in the absence of food, a wide range of tolerance for physical factors in the environment, and a high reproductive rate at optimal conditions.

Additionally important for consumers of dry foodstuffs is the ability to conserve water. Pest insects have to be either tolerant of low humidity and moisture content, able to recycle metabolic water, or able to influence and alter microclimatic conditions. Several species, in which the adults feed, tend to aggregate at spots suitable for feeding and breeding. Such aggregation is not incidental but a result of male-produced pheromones that are attractive to conspecific males and females. This behavior raises the question of the evolutionary advantage for males producing pheromone that attracts not only potential mates (conspecific females) but also potential competitors (conspecific males). It is thought that male-produced aggregation pheromones originally served to attract a mate. If this is correct, the use of the pheromone by other males of the same species then evolved secondarily, presumably allowing them to locate resources and assembled females. This would result in a fitness gain only for the intruding males.

An alternative explanation for the production of aggregation pheromones could be that the primary role of the male-produced aggregation pheromone was to attract and congregate conspecifics to overcome host defenses or unsuitable climatic conditions. Because males and females can serve equally well in this respect, the pheromone emitter benefits from attracting beetles of either sex.

To infest dry foodstuffs or stored commodities successfully, stored-product beetles certainly do not have to overcome host defense systems. However, in natural habitats, such as under bark or in wood, stored-product beetles most likely benefit from, and rely on, aggregation pheromones to weaken the host and make the habitat more suitable. This behavior can be observed in non-stored-product species closely related to postharvest pests in the Tenebrionidae, Curculionidae, and Bostrychidae, which partly feed and breed on physiologically active host plant parts. Where seeds and other harvested products are usually stored under low temperatures and rather dry conditions, the aggregation pheromones may serve to congregate groups that alter the microclimate. Such areas of higher temperatures and humidity may provide favorable conditions for insect development.

Not all insects that infest dry foodstuff have aggregation pheromones. In those that do not, there is a tendency of the larvae to aggregate. For many species, development is maximized at an optimal density of individuals. Individual fitness declines as the population shrinks or enlarges. For example, the eggs of many bruchids are deposited in clutches. Ovipositing females in the pyralids are guided to breeding sites by mandibular gland secretions produced by conspecific larvae. (A larval pheromone from mandibular glands is responsible for an optimal spacing, thus regulating population density.) Many aggregation pheromones also serve as guides to ovipositing females.

Temporary Changes in Habitats

The evolutionary origin of most stored-product insects remains speculative. However, autecological studies indicate a more or less xylophilous origin, under bark or in decaying wood, for most species. Omnivorous and partly predatory feeding habits could have allowed those species to invade animal nests in the same habitat or close by. These nests and the materials within could have provided good sources of otherwise scarce carbohydrate- and protein-enriched resources. In the same way, at the beginning of agriculture more than 10,000 years ago, insects could have gained access to storage bins in granaries constructed of infested timber. This kind of behavior can be observed in *Prostephanus truncatus,* the larger grain borer, which can survive in a variety of wooden building materials often used for simple granaries in tropical and subtropical countries. When the new harvest is brought in, this beetle readily invades the stored commodity, returning to its xylophilous habit when the alternative enriched food source declines. Other insect pests associated with stored products also frequently commute between the stored commodity and their cryptic life in cracks and crevices of the storage facility, where they feed on debris.

Geographic Origin and Worldwide Distribution

Today most stored-product insects are worldwide in distribution, and it is difficult to trace species to specific geographic origins with certainty. Global trade of food and other commodities, and the past lack of quarantine regulations, dispersed most storage pests with the products they had infested. Despite their cosmopolitan distribution, not all storage insects are established worldwide or occur outside warehouses or granaries. This is especially true in the colder parts of the world. *Callosobruchus maculatus,* for example, does not survive winters in the Northern Hemisphere unless inside heated storage facilities. For some species it is impossible for climatic reasons to successfully survive as nonsynanthropic populations.

The association of certain insects with humans most likely began in prehistoric times, probably contemporaneous with intensified agriculture and the domestication of animals. Phylogenetic, ecological, and historical evidence suggests that the majority of stored-product insects originated in the tropical and subtropical parts of the world. These centers of origin were also the centers of Neolithic agriculture.

PROTECTING STORED PRODUCTS AGAINST PEST INSECTS

Some of the most modern approaches to keeping stored food from spoiling are based on historical and traditional methods modified by today's technology. The storage of commodities under low oxygen atmospheres, to slow down the biological decay mechanisms, was in principle already used in ancient underground and hermetically sealed storage facilities of the Old World. The same effect occurs in modern grain facilities that are fumigated with modified atmospheres enriched with nitrogen and carbon dioxide. The ancient Egyptians intermixed ash with grain, which acted as abrasives on insects' cuticle, causing desiccation. Today, inert dusts on a base of diatoms are applied, and by a slightly different mode of action these additives cause the insects to desiccate. The Egyptians added specific herbs and spices to the commodities to repel pest insects. Sulphur was one of the first fumigants for stored product protection. The insecticidal property of burning sulfur was known in the New Kingdom around 1,000 B.C. Toxic fumigants have the great advantage over contact insecticides of fully penetrating the commodity and therefore also killing internal infestations. However, the application of fumigants and pesticides is not risk free. Environmental concerns and the potential danger to human health have led to a steady reduction of registered fumigants and contact insecticides. Incorrect handling of pesticides can lead to control failures and resistance.

In the past, eradication of pests was the primary objective, and increasingly high food quality standards allow no, or only little, contamination of stored commodities. On the other hand, consumers demand foodstuff that is free of chemical residues. A zero pest tolerance, however, can rarely be achieved without the application of insecticides. Today's pest control programs must therefore address health-related, ecological, and economic concerns. Modern pest control strategies in stored product protection focus on integrated pest management (IPM). The concept of IPM is based on using threshold levels to make control decisions and on the combination of physical, biological, and chemical pest control measures. One threshold level, the economic injury level, is defined as the pest infestation level at which the loss due to pests is equal to the costs of available control measures. The threshold is not a fixed parameter, but varies with the control options, the infested commodity, and the infesting pest species. Fundamental requirements for a successful application of an IPM program are detailed understanding of the mode of action of any control measure and specific knowledge of the pest's biology.

See Also the Following Articles

Agricultural Entomology • Integrated Pest Management

Further Reading

Anonymous. (1979). "Proceedings of the Second International Working Conference on Stored-Product Entomology," Ibadan, Nigeria, 1978. Technical Assistance Bureau Agency for International Development, Washington DC.

Anonymous. (1984). "Proceedings of the Third International Working Conference on Stored-Product Entomology," Manhattan, KS, 1983. Kansas State University, Manhattan, KS.

Donahaye, E., and Navarro, S. (eds.). (1987). "Proceedings of the Fourth International Working Conference on Stored-Product Protection," Tel Aviv, Israel, 1986. Moar–Wallach Press, Caspit, Jerusalem.

Fleurat-Lesssard, F., and Ducom, P. (eds.). (1991). "Proceedings of the Fifth International Working Conference on Stored-Product Protection," Bordeaux, France 1990, Vols. I–III.

Gorham, J. R. (1991). "Ecology and Management of Food-Industry Pests." Association of Official Analytical Chemists, Arlington, VA.

Highley, E., Wright, E. J., Banks, H. J., and Champ, B. R. (eds.) (1994). "Proceedings of the Sixth International Working Conference on Stored-Product Protection," Canberra, Australia, 1994. CAB International.

Hinton, H. E. (1945). "A Monograph of the Beetles Associated with Stored Products," Vol. I. British Museum, London.

Jayas, D. S., White, N. D. G., and Muir, W. E. (1995). "Stored-Grain Ecosystems." Dekker, New York.

Subramanyam, B., and Hagstrum, D. W. (1996). "Integrated Management of Insects in Stored Products." Dekker, New York.

Subramanyam, B., and Hagstrum, D. W. (2000). "Alternatives to Pesticides in Stored-Product IPM." Kluwer Academic Publishers, Dordrecht, Netherlands.

Zuxun, J., Quan, L., Yongsheng, L., Xianchang, T., and Lianghua, G. (eds.) (1999). "Proceedings of the Seventh International Working Conference on Stored-Product Protection," Beijing, China, 1998. Sichuan Publishing House of Science and Technology, Chengdu.

Strepsiptera

Michael F. Whiting
Brigham Young University

Strepsiptera (twisted-winged parasites) is a cosmopolitan order of small insects (males, 1–7 mm; females, 2–30 mm) that are obligate insect endoparasites. The order comprises ~550 species placed within eight extant and one extinct family. Strepsiptera derive their common name from the male front wing, which is haltere-like, and early workers considered it to be twisted in appearance when dried specimens were examined. All members of this group spend the majority of their life cycle as internal parasites of other insects and consequently have a highly specialized morphology, extreme sexual dimorphism, and a unique biology.

BIOLOGY

The adult male strepsipteran is free living and winged, whereas the adult female is entirely parasitic within the host, with the exception of one family (Mengenillidae) in which the female last instar leaves the host to pupate externally. Strepsiptera parasitize species from seven insect orders: Zygentoma, Orthoptera, Blattodea, Mantodea, Heteroptera, Hymenoptera, and Diptera. In one family (Myrmecolacidae) the males are known to parasitize ant hosts, whereas the females are parasites of Orthoptera. The life cycle of most strepsipteran species is unknown, and only a few species have been studied in detail.

The first instar is free living, emerges from the female brood canal, and disperses to the surrounding vegetation in search of a new insect host. In some species, the larvae have long abdominal setae that are used to propel them on to a new host. These larvae may be carried back to the nest of a social host to begin parasitizing the host's larvae or may simply spring onto the early instars of a nonholometabolous host. In *Stylops pacificus,* the abdominal setae are short and the larvae are transported to the nest inside of the crop of the bee *Andrena complex,* where it will begin parasitizing a single egg of the bee.

Once a first instar enters a new host, it molts into an apodous larva and feeds transcutaneously from the host's hemolymph. *Elenchus tenuicornis* undergoes two additional molts prior to pupation; apolysis is not followed by ecdysis in strepsipteran larvae, such that the cuticle of the second to the fourth instar forms the puparium. In Mengenillidae, both sexes leave the insect host to pupate externally. The remaining Strepsiptera pupate in the host with a portion of the head and thorax extruded from the host's cuticle (Fig. 1C).

Males will emerge from the puparium and seek out a female for mating. Virgin females release a pheromone to attract males, and in Stylopidia the male copulates by rupturing the female's brood canal opening between her head and prothorax in a process referred to as hypodermic insemination. Adult males live only a few hours, which makes them difficult to collect, although they are occasionally found at lights or may be lured to a virgin female kept in a cage in the field. Adult males do not feed and their mouthparts are partly modified into sensory structures. After insemination, females may live for a few weeks to a few months. In *Xenos peckii,* the female is typically inseminated in the fall and she remains in the wasp host until spring, when the planidial larvae emerge. The female may be an enormous endoparasite relative to the size of its host. For example, *X. peckii* may fill up to 90% of the abdominal cavity of the wasp host. It is relatively common to find a host that has been parasitized by multiple males and/or females.

MORPHOLOGY

First instars bear stemmata (simple, single-lens eyes), mandibles, maxillae, and a labium. Antennae are absent. Abdominal tergites are typically smooth, the abdominal sternites are serrate, and the legs are filiform with slender tarsi (Fig. 1A). In Mengenillidae, the pupae are free living and have legs, mouth, and a segmented abdomen. In the remaining Strepsiptera, the puparium is tanned and male pupae possess compound eyes, abdominal and thoracic segmentation, and developing wings, legs, and antennae (Fig. 1B).

The adult male has large compound eyes, with ommatidia separated by cuticle and/or setae, giving the eye a blackberry appearance. Antennae are flabellate and the mandibles are conical, except in Corioxenidae, which entirely lack mandibles (Fig. 1F). The first pair of wings is reduced, clubbed, and morphologically similar to the halteres of Diptera. The hind wings are enlarged and venation is reduced, and these wings are used for flight (Fig. 1G). The prothorax is small and saddle-shaped and the metathorax is large and bears the principal flight muscles. The trochanter is fused with the femora in the fore- and middle legs, the hind legs lack free

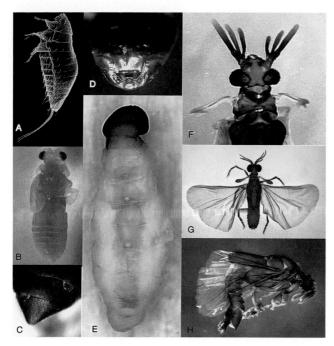

FIGURE 1 (A) First instar of *Mengenilla chobauti*. (B) Male pupa of *Xenos peckii*. (C) Cephalothorax of female *X. peckii* extruded from the abdominal tergites of the wasp host *Polistes fuscatus*. (D) Close-up of cephalothorax of female *X. peckii*. (E) Female *Xenos vesparum*. (F) Head and thorax of *M. chobauti*. (G) Male *Triozocera bedfordiensis*. (H) Male *Coriophagus rieki*.

coxae, and the tarsi have two to five segments and may lack a tarsal claw (Fig. 1H).

The adult female lacks wings, legs, and external genitalia, and only rudiments of the mouthparts, antennae, and eyes remain (Figs. 1D and 1E). The thorax and head are fused to form a heavily sclerotized cephalothorax, and this is the portion of the female that is extruded from between the abdominal segments of its host. The abdomen is large, segmentation is reduced, and the cuticle is unsclerotized, which allows nutrients to pass from the host to the developing embryos. The tracheation and the nervous system are reduced, and the reproductive system is highly modified, allowing the production of vast quantities of eggs that are freely distributed throughout the abdominal cavity.

CLASSIFICATION

The Strepsiptera are divided into two major groups: the Stylopidia and the Mengenillidia. Mengenillidia comprises one extinct family (Mengeidae) and one extant family (Mengenillidae). Mengenillidia is considered to be the most primitive strepsipteran group because females are free living, with rudimentary legs, antennae, and a single genital opening, and the males have robust mandibles and a hind wing with an elongated and sturdy MA vein.

Stylopidia comprises seven families: Corioxenidae, Halictophagidae, Callipharixenidae, Bohartillidae, Elenchidae, Myrmecolacidae, and Stylopidae. The monophyly of Stylopidia is supported by the fact that females are found in the host, the female has multiple genital openings, and the hind wing in the male has only residues of the MA$_1$ vein. Recent research on the phylogenetic relationships among strepsipteran families, based on the first instars, suggests that the Corioxenidae is the most basal family within the Stylopidia and that Stylopidae is paraphyletic with the Xeninae and Stylopinae placed in different groups.

PHYLOGENY AND EVOLUTION

The phylogenetic position of Strepsiptera relative to other insect orders has been one of the most contentious issues in insect systematics, and controversy still exists as to their phylogenetic affinity. There have been four phylogenetic hypotheses presented: (1) placement as a group somewhere within Coleoptera, as suggested by the superficial similarities in their parasitic lifestyle and morphology with certain beetle groups; (2) placement as sister group to Coleoptera, as suggested by the powering of flight via the hind wing; (3) placement outside of Holometabola, as suggested by the presence of larval stemmata; and (4) placement as sister group to Diptera, as suggested by DNA sequence data. The latter hypothesis has been the most controversial in recent years and different interpretations of the molecular data have been presented. However, if Strepsiptera are indeed sister group to Diptera, then this gives rise to some intriguing possibilities in insect evolution. Functionally, the halteres of Diptera and Strepsiptera appear identical, both highly specialized to serve as a gyroscopic balancing mechanism during flight. Therefore, it is possible that the mesothoracic halteres of Strepsiptera were derived via a specialized mutation in the genes controlling the development of the metathoracic halteres in Diptera, essentially causing the wings of the dipteran to switch positions to produce the type of wings as observed in Strepsiptera.

See Also the Following Article
Fossil Record

Further Reading
Crowson, R. A. (1960). The phylogeny of Coleoptera. *Annu. Rev. Entomol.* **5**, 111–134.
Kathirithamby, J. (1989). Review of the order Strepsiptera. *Syst. Entomol.* **14**, 41–92.
Kinzelbach, R. K. (1971). Morphologische Befunde and Facherfluglern und ihre phylogenetische bedeutung (Insecta: Strepsiptera). *Zoologica* **119**, 1–256.
Kristensen, N. P. (1991). Phylogeny of extant hexapods. *In* "The Insects of Australia: A Textbook for Students and Research Workers" (P. B. C. I. D. Naumann, J. F. Lawrence, E. S. Nielsen, J. P. Spradberry, R. W. Taylor, M. J. Whitten, and M. J. Littlejohn, eds.), 2nd ed., pp. 125–190. CSIRO, Melbourne University Press, Melbourne.
Linsley, E. G., and Macswain, J. W. (1957). Observations on the habits of *Stylops pacifica* Bohart. *Entomology* **11**, 395–430.
Pix, W., Nalbach, G., and Zeil, J. (1993). Strepsipteran forewings are haltere-like organs of equilibrium. *Naturwissenschaften* **80**, 371–374.
Pohl, H. (2002). Phylogeny of the Strepsiptera based on morphological data of the first instar. *Zool. Scripta* **31**, 123–134.

Whiting, M. F., Carpenter, J. C., Wheeler, Q. D., and Wheeler, W. C. (1997). The Strepsiptera problem: Phylogeny of the Holometabolous insect orders inferred from 18S and 28S ribosomal DNA sequences and morphology. *Syst. Biol.* **46**, 1–68.

Young, G. R. (1987). Notes on the life history of *Stichotrema dallatorreanum* Hofeneder (Strepsiptera: Myrmecolacidae), a parasite of *Segestes decoratus* Redtenbacher (Orthoptera: Tettigoniidae) from Papua New Guinea. *Gen. Appl. Entomol.* **19**, 57–64.

Swimming, Lake Insects

Werner Nachtigall

Universität der Saarlandes

I nsects of many types, such as beetles (Coleoptera), true bugs (Heteroptera), and fly larvae (Diptera), can be observed swimming in ponds and lakes. Of these, the water beetles in the family Dytiscidae are reputed to be the best swimmers. The trunks of their bodies are well adapted to flow, and they generate thrust by executing synchronous power strokes with their flattened rear legs, which bear two rows of "swimming" hairs.

FLOW ADAPTATION

Measurements on flow adaptation have been carried out on trunks of several Dytiscidae, especially with the large European water beetle *Dytiscus marginalis.*

The technical term coefficient of drag, or c_d, is used as an indicator of flow adaptation. A small coefficient indicates that a beetle with a given frontal area and at a given swimming speed generates little drag; that is, it is well adapted to flow. However, one cannot readily compare the drag created by objects moving at different speeds because the coefficient of drag is a function of the Reynolds number (i.e., of its swimming speed in water and its body length). For certain ranges of Reynolds numbers, there are sufficient data available, ranging from the most streamlined bodies (e.g., drop-shaped trunks; lowest c_d values) to the objects that produce the greatest drag (e.g., parachute forms; highest c_d values). Figure 1A shows the trunks of four *Dytiscus* species viewed from above, from the side, and from the front (the front projection shows the frontal area *A*). Figure 1B shows the measured c_d within the spectrum of coefficients that are possible for the range of Reynolds number (Re) obtained from swimming water beetles (10^3 < Re < 10^4). The coefficients of the beetle fluctuate between 0.38 for *D. latissimus* and 0.43 for *D. pisanus,* with the smallest possible value being 0.2 or slightly less, and the highest possible value approximately 1.4. Compared with the possible spectrum, the beetles' trunks seem to be well adapted to flow, although they do not reach the extreme values technically possible. Low-drag technical constructions are unstable; in the presence of oblique flow

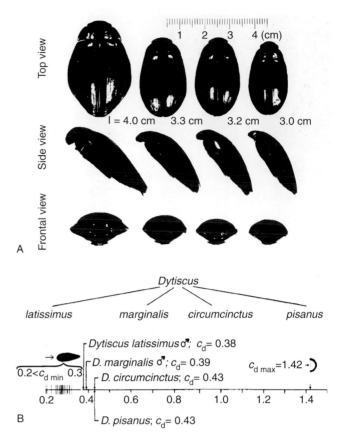

FIGURE 1 (A) Trunks from four large species of *Dytiscus* viewed from above, from the side, and from the front. (B) Classifying the four c_d values of these beetles at zero angle of attack (cf. Fig. 2) within the spectrum of possible coefficients of drag in the range 10^3 < Re < 10^4. [After Nachtigall, W. (1986).]

they turn immediately broadside, and stabilizing surfaces preventing this behavior would increase drag. Enlarged prothoracic and elytra edges in large *Dytiscus* beetles (extreme in *D. latissimus*) serve as stabilizing surfaces for swimming by damping oscillations around the longitudinal and lateral axes and by creating stabilizing moments. If one reduces these stabilizing surfaces along the edges, drag at almost all angles of attack between trunk and flow is perceptibly reduced. But the trunk then is distinctly unstable. These two contradictory demands have led to the evolution of an optimal shape with good swimming stability and good flow adaption.

Geometrically similar, 10-times-enlarged models of large *Dytiscus* beetles have been used to measure the pressure distribution along the dorsal midline of these trunks, with numerous pressure holes. A positive pressure dent appears on the head–prothoracic region (and also on the abdomen), whereas the remaining upper abdominal side is under negative pressure (Fig. 2). Thus, a *Dytiscus* trunk is very similar to a small-span wing profile and should therefore create a certain dynamic lift during fast swimming.

Because the drag coefficient of bodies increases with smaller Reynolds numbers, one may expect that the smallest, most slowly moving Dytiscidae (e.g., *Hyphydrus, Bidessus*) are characterized by relatively high drag coefficient values. Thus,

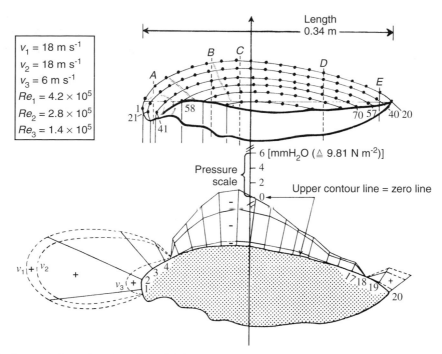

FIGURE 2 Pressure distribution in the median plane of a model of *D. marginalis* (male) enlarged 10 times, measured in a wind tunnel: dashed lines, pressure above atmospheric ("pressure"); solid lines, subatmospheric pressure ("suction"). [After Nachtigall, W. (1986).]

they seem to be less well adapted to flow simply by their smallness (i.e., by their low Reynolds numbers). From flow mechanics, we know that streamlining is less effective at lower Reynolds numbers. Indeed, these small beetles are characterized by blunt body forms that may have biological advantages (more effective space for the organs). Because these beetles are energetically better equipped, this inherent physically disadvantage may not be of primary importance.

FUNCTIONAL MORPHOLOGY

Leg Flattening and Position of Swimming Appendages

The functional formation of a swimming leg is easy to understand because of the simple mechanical demands made upon it. A swimming leg should produce strong thrust (i.e., a force component in the swimming direction) during a rowing stroke (i.e., a power stroke), and little counterthrust while being drawn forward (i.e., a recovery stroke). This action is aided by morphological and kinematic adaptations.

At a given speed the hydrodynamic drag force F_d is proportional to the leg surface A. Therefore, A must be as large as possible during rowing stroke, and as small as possible when drawn forward. This is achieved by flattening the leg segments and by means of thrust-inducing processes such as swimming hairs (e.g., *Dytiscus, Corixa*) or swimming platelets *(Gyrinus)*. During the rowing stroke, the flat broadside of the leg is brought perpendicular to flow, and the hairs and platelets spread out automatically under the pressure of flow, thus increasing the rowing surface. When the leg is drawn forward, it is turned so that the flattened surface lies parallel to flow, and the rowing appendages are folded together and pressed against the leg by the flow; their additional surface disappears completely (Fig. 3).

Leg oscillation is angular movement. Therefore, the drag, which is created by an element of surface, is proportional to the square of its rotating radius r, or its distance from the coxa–trochanter joint. A component of the

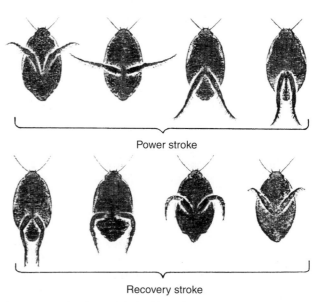

FIGURE 3 Stroke phases of the hind legs of *A. sulcatus* during power stroke and recovery stroke. [After Nachtigall, W. (1986).]

counterforce of this drag, parallel to the median, is the thrust force. The thrust-creating area during a rowing stroke should lie as far out as possible (*r* large) during the power stroke, and as far in as possible (*r* small) during the recovery stroke. Thus, the broad area and its hairs can be called a thrust-creating area. This is achieved principally because the distal leg segments, the tibia and the tarsus, are greatly enlarged and carry thrust-creating swimming hairs or swimming platelets, and because the legs are outstretched during stroke and then bent and drawn forward as close as possible to the median.

Stroke Areas and Radii of Rotation

The leg segments in *Gyrinus* have been flattened so extensively that their surface area is approximately five times that of a round leg; because of the simultaneously increasing coefficient of drag, a hydrodynamic force roughly eight times larger is created. During the power stroke, the leg segments spread out like playing cards, thus increasing their size by roughly 1.6 times over their projection area when not spread. The supplementary surface areas of the swimming platelets increase the stroke surface in *Gyrinus* even more: by roughly 230% in the tibia and roughly 130% in the tarsus.

In *Acilius,* the flattened tarsus positioned very distally occupies 80% of the total stroke area. Its mean rotational radius is also roughly 80% of the leg length. In *Gyrinus,* the corresponding values are approximately 40 and 80% for the tarsus and approximately 50 and 60% for the tibia.

During the recovery stroke, *Acilius* tilts the broad side of its leg parallel to the flow and presents the narrow edge of its leg to the flow. At the same time, the swimming hairs and swimming platelets are flattened and lie close to the leg. Because the exterior leg sections lie more or less parallel to the median during the first phase of drawing forward, its area of projection in the direction of flow is smaller yet. In *Gyrinus,* this projection area in proportion to the area of the power stroke is reduced to 1/13 for the middle legs and 1/16 for the hind legs. The tarsus from *Gyrinus* collapses completely and disappears partly into a groove in the tibia, which in turn slips partly into a hollow in the femur, so that the leg surface area is dramatically reduced by roughly 70% of the leg surface during the recovery stroke.

By folding the legs when they are drawn forward nestled against the body, *Gyrinus* reduces the mean rotational distance relative to that of the power stroke from 60% to approximately 35% of total leg length. The values for the tarsus alone are 80 and 50%.

Two additional phenomena should augment swimming efficiency, although their precise contribution remains to be measured. First, pressure drag is created mainly during the power stroke, and frictional drag is probably when the legs are drawn forward. Second, the leg segments, insofar as they nestle against the ventral surface with their broad side when drawn forward, move within the boundary layer of the trunk, thus creating less drag than when moving through free water.

All results presented here are for "standing water," that is, for the very first strokes after starting. During steady swimming, the relations are more complicated because the leg interferes with flowing water so that the speed of the leg's rowing surface relative to the surrounding water changes. Small fishes are propelled by the rowing motions of their pectoral (or breast) fins in almost the same manner as the water beetles are propelled by rowing with their hind legs. It has been shown in these small fishes that interference of the rowing surfaces with flowing water does not play a major role.

See Also the Following Articles

Aquatic Habitats • Walking and Jumping

Further Reading

Blake, R. W. (1983). "Fish Locomotion." Cambridge University Press, Cambridge, U.K.
Hughes, G. (1958). The co-ordination of insect movements. III. Swimming in *Dytiscus, Hydrophilus* and a dragonfly nymph. *J. Exp. Biol. London* **35,** 567–583.
Nachtigall, W. (1986). Bewegungsphysiologie. Laufen, Schwimmen, Fliegen. Handbuch der Zoologie, Bd. 4, Arthropoda: Insecta. Gruyter, Berlin.

Swimming, Stream Insects

Bernhard Statzner
CNRS-Université Lyon 1, France

The physical world of an insect living in water or air is ruled by the fluid velocity relative to its body. Thus, in principle, it should not matter whether an insect swims fast in still water or slowly against the current of flowing water. However, the flow in streams is physically so harsh that it often constrains the movements of stream insects. Imagine a mountain stream with its typical current velocity of 50 cm s^{-1} and a 1-cm-long mayfly larva swimming upstream. To move upstream relative to the stream bottom, this larva must propel itself at more than 50 body lengths s^{-1}, which is a difficult accomplishment. Therefore, stream insects rarely swim and live predominantly at or in the surface layer of the stream bottom, where friction between the moving water and the solid bottom material generates extremely complex flow patterns. As a consequence, the aquatic stages of stream insects move in a physical world that is much more complicated than that of lake insects.

LIFE IN FLOW NEAR THE STREAM BOTTOM

Streams offer a mosaic of flow conditions, including areas with (almost) still water. Where a stream runs over a rough (e.g., stony) bottom, flow can be laminar or turbulent (or transitional between these two). The difference between the laminar and the turbulent flow regime is easily understood by watching the smoke released by a cigarette in a calm place. Close to the tip, the ascending smoke travels in an orderly manner on a narrow path as a laminar flow, until suddenly the flow becomes turbulent and travels whirling on a wider path. The physical formulas for these regimes are quite different; thus, these two flow regimes constitute different physical worlds.

The Reynolds number *Re* indicates the flow regime for a particular flow situation through $Re = l \times U/\upsilon$, where l is a length dimension (e.g., the body length of an insect), U is the relative velocity to the insect body, and υ is the kinematic viscosity of the water. Low *Re* indicates laminar and high *Re* turbulent regime. Therefore, in general, small and slow (relative to the flow) stream insects (having a low $l \times U$) experience laminar flow in which viscous forces matter, whereas large and fast ones experience turbulent flow, where inertial forces predominate.

However, near the bottom of a stream things are a bit more complicated. When water runs over a stone, the water in contact with the stone surface does not slip relative to the surface (the no-slip condition) and a velocity gradient develops above the surface (e.g., as in Fig. 1c). Near to the surface, the flow can be laminar ("laminar sublayer"). If the stone is embedded among other stones of similar size (Fig. 1a) and the water is shallow relative to the stone size, turbulent eddies may extend to the stone surface and disrupt the formation of the laminar sublayer. As a consequence, velocity varies tremendously over short periods near the stone surface (Fig. 1b), as well as around insects living at that surface. To maneuver in that type of flow, a 1-cm-long insect must deal with velocity changes from 5 to more than 40 body lengths s^{-1}, and this within a fraction of a second. The body vibrations of black flies or elongated caddisflies staying at the surface of a stone in a shallow stream nicely illustrate that these insects have a very "shaky" life.

In natural streams, isolated larger bottom elements that rise far into the water column (Fig. 1a) are often very abundant. Above the surfaces of these elements, the no-slip condition produces a steep velocity gradient (Fig. 1c). Flow is generally laminar nearest to the surface (low standard deviation of the velocity in the laminar sublayer: Fig. 1c), whereas it is turbulent in the remaining part of the steep velocity gradient (elevated standard deviations in Fig. 1c). At a given location, the geometry (e.g., the thickness of the laminar sublayer) of this steep gradient changes with changing flow. That is, discharge variation in a stream causes temporal variability of the physical conditions at that location. At a given flow, the geometry of this gradient varies spatially along the slightly curved bottom element. The changes of the isovel (line of equal velocity) pattern along the curvature (Fig. 1d) demonstrate considerable spatial variability of the physical conditions over the distance of a few millimeters. As a consequence of these temporal and spatial patterns, a relatively sessile black fly larva must deal with temporal physical variation caused by changing

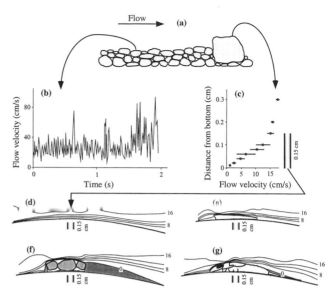

FIGURE 1 Flow patterns near the stream bottom and near three fully grown bottom-dwelling stream insects. (a) Profile of a stream bottom. (b) Short-time velocity variation measured 0.2 cm above the surface of a natural stone that was embedded among other stones of similar size [cf. (a)] in a natural stream riffle. (c)–(g) Flow patterns measured above a curved bottom in a laboratory flume, simulating natural conditions that prevail above bottom elements (large stones, rocks, wooden logs) rising distinctly above neighboring bottom elements [cf. (a)]. (c) Mean and standard deviation of the velocity at different height above the highest point of the curved bottom; the column indicates the height of the layer having a steep velocity gradient. (d) Profile of the curved bottom and isovels [lines of equal velocities, drawn for velocity steps of 4 cm s^{-1} for the same flow as in (c)]; the column indicates the height of the layer having a steep velocity gradient above the highest point of the curved bottom (cf. c). (e)–(g) Changes of the isovel pattern shown in (d) by the cases of two caddisflies [*Micrasema* in (e), *Silo* in (f)] and a dorsoventrally flattened mayfly [*Ecdyonurus* in (g)] [note that (e)–(g) have the same scale as (d) and that velocity fluctuated around 0 cm s^{-1} below the zero isovel]. [Part (b) redrawn and simplified after Hart, D. D., and Finelli, C. M. (1999). Physical–biological coupling in streams: the pervasive effects of flow on benthic organisms. *Annu. Rev. Ecol. Syst.* **30**, 363–395, with permission from Annual Review of Ecology & Systematics, Vol. 30 (1999) by Annual Reviews (www.annualreviews.org); part (g) redrawn and simplified after Statzner, B., and Holm, T. F. (1982). Morphological adaptations of benthic invertebrates to stream flow—an old question studied by means of a new technique (Laser Doppler Anemometry). *Oecologia* **53**, 290–292, and parts (c)–(f) redrawn and simplified after Statzner, B., and Holm, T. F. (1989). Morphological adaptation of shape to flow: microcurrents around lotic macroinvertebrates with known Reynolds number at quasi-natural flow conditions. *Oecologia* **78**, 145–157; (c)–(g) with permission of Springer-Verlag GmbH & Co. KG.]

stream discharge, whereas a relatively mobile mayfly larva must deal with spatial physical variation when moving at the bottom surface.

Whether sessile or mobile, stream insects living in steep velocity gradients can modify the flow patterns to a considerable extent (cf. Fig. 1d and Fig. 1e–g). Frequently, these stream insects are so large (e.g., in Fig. 1f, g) that their lower part is in the laminar flow, their intermediate part is in the layer of elevated turbulence, and their upper part extends beyond the steep velocity gradient. These insects live with different body parts in three different physical "worlds," and the relative importance of these different worlds experienced by their bodies changes when the insects move. Thus, the flow conditions near larger bottom elements are also physically harsh for stream insects because they variably affect factors of physiological relevance for the insects (e.g., facility of respiration, drag, or lift force).

MOVEMENTS IN THE STREAM ENVIRONMENT

The surface of the stream bottom provides essential resources (e.g., food, oxygen-rich water). To exploit these resources, insects have developed many different strategies that relate to movements at and in the stream bottom, or in the free-flowing water column.

Movements at and in the Stream Bottom

A strategy frequently used to deal with the physical harshness at the stream bottom is temporal avoidance. Particularly, mayfly and stonefly larvae move vertically among the interstices of the stream bottom, to the top of the bottom surface and back into deeper bottom layers that offer shelter from the surface flow. Such movement produces daily and/or seasonal patterns in the presence of these larvae at the bottom surface. However, the dissolved oxygen concentration of the water interferes with the use of interstitial flow shelters. If the oxygen in the interstices drops below a species-specific critical value, the physiological means of regulatory oxygen consumption (gill movements or undulatory body movements) become insufficient to meet respiratory needs. Thus, larvae leave the interstices and crawl to current-exposed locations at the front faces or tops of stones, where oxygen concentrations are usually higher and the elevated flow increases the renewal rate of oxygen at respiratory surfaces.

Another means of staying or moving in the flow near the bottom surface is safe fixation. Often long, curved tarsal claws that enable a good attachment to the bottom surface are found in stream insects (e.g., adult riffle beetles). In addition, hooks or claws situated near the end of the abdomen may be used for bottom attachment. Stream-dwelling larvae of caddisflies, moths, and midges use silk for such attachment. Within the true midges and the caddisflies, many species use silk to build tubelike larval retreats that are fixed at the bottom surface. Caseless polycentropodid and hydropsychid caddis larvae crawling over coarse bottom material glue a silken safety thread in a zigzag line on their path. Other caddisfly larvae that build cases temporarily secure these cases with silk. Finally, black fly larvae spin tiny silk carpets, fix them at the bottom surface, and then attach their posterior abdominal hooks to these carpets.

The stream insects that colonize the most extreme physical conditions are the larvae of the net-winged midges. They can live at rock surfaces where velocities exceed 2 m s^{-1}, fixing themselves to the rock with six ventral suckers. The larvae can precisely adapt their locomotion to a given situation because they move by releasing one or more suckers at a time, having at least two suckers firmly attached to the rock at any moment. This highly specialized locomotory system enables the larvae to move straight upflow at a relative velocity (to the rock surface) of ≈ 0.03 body length s^{-1}, which seems to be a rather poor performance. However, the larvae achieve this movement against water velocities of 2 m s^{-1}. Therefore, their physically relevant speed corresponds to ≈ 300 body lengths s^{-1}, which is an extraordinary performance (to achieve equivalent performance, a human would have to swim 100 m in less than 0.2 s).

Claws, silk, suckers, and other means not detailed here enable stream insects to maneuver actively at or in the stream bottom. Daily movements tend to carry the insects upstream because their body (even of dead nymphs and exuviae) is oriented upstream. Thereby, the larvae of mayflies, stoneflies, and caseless caddisflies may move upstream (mostly at night) several meters per day. In comparison, cased caddisfly larvae often move relatively little during periods of normal flow. Other movements at or in the bottom have seasonal patterns. The nymphs of some mayfly or stonefly species crawl bankward prior to emergence, and old larvae of some limnephilid caddisflies even move toward land in early summer, either to feed on semiaquatic plants or to aestivate outside the stream. In streams that freeze down to the bottom, true midges, dance flies, and some caddisflies can overwinter in the frozen habitat, whereas all other insects actively move away when facing an advancing freezing front. Among the mayflies, the nymphs of many species crawl bankward when the water level of the streams rises during springtime. If the adjacent floodplain of a stream becomes inundated, the nymphs may continue their movements toward the floodplain, where most nymphal growth and development can take place. Other mayfly species not only move toward the stream banks but continue to move upstream near the shoreline at a speed of about 100 m day^{-1}. By following the shoreline of the main stream that bends toward the tributaries, the nymphs move into the tributaries and finally reach marshy areas drained by the tributaries. The nymphs complete their development in these marshy areas.

Active movements of stream insects are also caused by occasionally occurring, extreme hydrological events—floods and droughts. During floods, insects may move toward deeper bottom layers to find flow shelter, but evidence for such behavior is equivocal. A giant water bug living in desert streams has a successful response to flooding. During periods of heavy rainfall that often precede flash floods in these streams, adult and juvenile water bugs move toward the banks, leave the stream, and crawl up to about 20 m over land toward sheltered areas (from where they return to the stream after 24 h). Some studies report that stream insects burrow deeper into the humid sediments to avoid effects of drying, whereas other studies do not confirm such movements. With sinking water level in a stream, insects may also crawl away over land (adult beetles), or they crawl up- or downstream to places where water remains. The speed of insects during such movements can be rather high. For example, the larvae of a cased limnephilid caddisfly crawl downstream at about 12 m h^{-1}.

Finally, the aerial females of a few species of stream insects crawl down on solid objects through the water surface and fix their eggs on the submerged surface of the object, or they oviposit in an aquatic host insect. The first behavior has been primarily observed in mayflies, caddisflies, and blackflies, the second in parasitic wasps.

Movements in the Free-Flowing Water Column

Drifting downstream with the flow is the typical movement of stream insects in the water column. Physically, a passively drifting insect barely moves because the water velocity relative to its body is almost zero. However, relative to the stream bottom it travels at the speed of the flow.

The drift of stream insects is perhaps one of the most frequently studied topics in stream ecology. Given that both the diversity of stream insects and the diversity of running water conditions across the continents are very high, there is evidence for almost any conceivable drift response. For example, the following factors may increase, decrease, or have no effect on the natural drift of stream insects: sun- or moonlight; current velocity; stream discharge; type of the bottom substrate; turbidity, oxygen concentration, ion concentration, or temperature of the water; organic matter; food; predators; microbial pathogens; larger parasites; molting process; benthic density; and age or behavior of the drifting species. Thus, it is difficult to identify clear patterns in the drift of stream insects.

Typically, stream insects drift during the night. This drift may be caused by accidental dislodgement of the insects through the current or by deliberate, active entries into the drift. Active entries occur when a habitat patch is overcrowded and resources (e.g., food, space, preferred flow conditions) are lacking. Approximately 10 to 30% of the insect population of a stream reach may drift during one night, and they may travel between 2 and 20 m during one drift movement. However, caddisfly larvae that build heavy cases from gravel typically have distinctly lower drift rates and shorter drift distances. Insects that produce silk may also have shorter drift distances. Hydropsychid and polycentropodid caddisfly larvae, and black fly larvae, can fix a silk thread to the stream bottom, actively enter into the drift and, prolonging the thread by spinning, rope themselves downstream over several centimeters until they resettle at the bottom. Polycentropodid caddis larvae that have lost contact with the bottom often spin "anchor" threads of 30 cm in length. When such a thread adheres to the stream bottom, the larva is able to return to the bottom by climbing down along the thread. Another means of affecting drift distance is used by drifting larvae of mayflies, stoneflies, and caseless caddisflies. They modify their body posture, thereby either decreasing or increasing their sinking speed, and thus the distance traveled in the water column.

The intensity of everyday drift events usually varies across seasons, either because of seasonal changes in factors affecting the insect drift or because of the seasonal occurrence of a particular developmental stage that drifts more than other stages.

A typical example for the first cause is the seasonal change of temperature in temperate streams and the related effects on the insect drift. A typical example for the second cause is found in hydropsychid caddisflies that may have newly hatched first instars that drift at much higher rates than later instars.

Occasional natural events such as floods and droughts, or unnatural ones such as pesticide or pollution spills, also increase drift. During such events and in contrast to the everyday drift, insects also drift during the daytime, and the distances traveled are longer. When discharge increases, and thus near-bottom velocity, many stream insects are dislodged. Among the sessile forms that fix themselves firmly to the stream bottom, those sitting on submerged wood or leaf litter may travel extremely far downstream on dislodged litter pieces that drift for long distances. In contrast, when discharge and thus near-bottom velocity decreases during a drought, many insects release their grasp on the stream bottom and swim upward until they are caught and transported by the free flow in the water column.

Another cause for massive upward swimming of insects from the bottom is relatively low oxygen concentration in the water, a condition typical of streams with organic pollution. Drifting insects may swim, either to avoid sinking so that they stay longer in the water column or to return faster to the stream bottom. Finally, some mayfly larvae swim over short distances when they encounter predators or aggressive conspecifics. These larvae are such good swimmers that they can travel against the current just above the stream bottom, where the velocity is reduced. The larvae of most insect groups swim by body undulations.

An exceptional type of movement among the stream insects is the diving of aerial females to reach submerged oviposition sites. Female black flies can dive through thin water layers flowing over rocks to get foothold at the rock surface, where they affix the eggs. Similarly, female caddisflies dive vertically and swim to oviposition sites below inclined submerged stones.

See Also the Following Articles
Aquatic Habitats • Walking and Jumping

Further Reading
Brittain, J. E., and Eikeland, T. J. (1988). Invertebrate drift—A review. *Hydrobiologia* **166**, 77–93.

Brunke, M., and Gonser, T. (1997). The ecological significance of exchange processes between rivers and groundwater. *Freshwater Biol.* **37**, 1–33.

Frutiger, A. (1998). Walking on suckers—New insights into the locomotory behavior of larval net-winged midges (Diptera: Blephariceridae). *J. N. Am. Benthol. Soc.* **17**, 104–120.

Hart, D. D., and Finelli, C. M. (1999). Physical–biological coupling in streams: The pervasive effects of flow on benthic organisms. *Annu. Rev. Ecol. Syst.* **30**, 363–395.

Koehl, M. A. R. (1996). When does morphology matter? *Annu. Rev. Ecol. Syst.* **27**, 501–542.

Otto, C., and Sjöström, P. (1986). Behaviour of drifting insect larvae. *Hydrobiologia* **131**, 77–86.

Söderström, O. (1987). Upstream movements of invertebrates in running waters—A review. *Arch. Hydrobiol.* **111**, 197–208.

Statzner, B., and Holm, T. F. (1982). Morphological adaptations of benthic invertebrates to stream flow—An old question studied by means of a new technique (Laser Doppler Anemometry). *Oecologia* **53**, 290–292.

Statzner, B., and Holm, T. F. (1989). Morphological adaptation of shape to flow: Microcurrents around lotic macroinvertebrates with known Reynolds numbers at quasi-natural flow conditions. *Oecologia* **78**, 145–157.

Statzner, B., Gore, J. A., and Resh, V. H. (1988). Hydraulic stream ecology: Observed patterns and potential applications. *J. N. Am. Benthol. Soc.* **7**, 307–360.

Vogel, S. (1994). "Life in Moving Fluids." 2nd ed. Princeton University Press, Princeton, NJ.

Williams, D. D., and Williams, N. E. (1993). The upstream/downstream movement paradox of lotic invertebrates: Quantitative evidence from a Welsh mountain stream. *Freshwater Biol.* **30**, 199–218.

Symbionts Aiding Digestion

Andreas Brune

Universität Konstanz, Germany

The interactions of insects with microorganisms range from the cultivation of fungus gardens to intimate associations in which bacteria are housed either within special organs (mycetomes) or intracellularly in dedicated mycetocytes. Many of these associations have nutritional implications; this article will focus on microbial symbionts that colonize the intestinal tract and are directly involved in digestion.

Although insects produce a wide variety of digestive enzymes, many species harbor an intestinal microbiota that converts a substantial portion of the dietary components to fermentation products before they are resorbed by the intestinal epithelia. Interestingly, the occurrence of such associations is always correlated with a dietary specialization, which indicates that the symbiosis provides metabolic capacities that are normally not available to the host.

Digestive symbioses are most common among insects feeding on wood or other lignified plant materials. The most prominent example is that of wood-feeding termites (Isoptera), which represent the only group of insects whose interactions with intestinal microorganisms has been systematically studied. Here, they serve to illustrate principles that most likely govern also other cases of symbiotic associations in which detailed information on the gut microbiota and its function in digestion is lacking.

BIOCHEMICAL BASIS OF LIGNOCELLULOSE DIGESTION

Why would an insect need help in digesting its food, and which benefits would be gained from sharing this resource with an intestinal microbiota? For xylophagous (wood-feeding) insects, the answers are found in the chemical structure and composition of their fiber-rich, low-nutrient diet.

Structure and Composition of Lignocellulose

The major components of plant cell walls are cellulose and hemicelluloses. Although the glycosidic linkages between the sugar subunits of these structural polysaccharides are relatively easy to hydrolyze, the sheer size of the molecules, their primary and secondary structures, and the intimate physical association of the different components, especially in lignified cell walls, impart a considerable recalcitrance to enzymatic digestion.

In cellulose, the linear β-$(1{\rightarrow}4)$-linked polyglucose chains are arranged in microfibrils with a highly ordered, mostly crystalline structure. Cellulose depolymerization is catalyzed by endoglucanases (endo-β-1,4-glucanases), which cleave randomly within the extremely long polyglucose chains, and by exoglucanases (exo-β-1,4-cellobiohydrolases and β-glucosidases), which require free nonreducing ends for their catalytic activity. Because native cellulose contains only one nonreducing end per many thousand glucose units, and because the action of endoglucanases is restricted to the amorphous regions of the microfibrils, an efficient cellulolytic system requires the synergistic action of both types of enzymes.

Enzymatic activity of cellulases is further impeded by the insolubility of cellulose, and depolymerization is usually the kinetically limiting step even in the digestion of pure cellulose. In the cell wall, the cellulose fibrils are intimately associated with hemicelluloses, which comprise a wide variety of homopolymers and heteropolymers of different sugars and sugar acids. Hemicellulose chains are often branched and lack the highly ordered structure of cellulose. The variety of primary structures requires an equal variety of digestive enzymes, and it is not astonishing that the ability to degrade hemicelluloses is a prerequisite for an efficient cellulose degradation.

Sound wood is most difficult to digest because, through lignification, the polysaccharides of the secondary plant cell wall are embedded in an amorphous resin of phenolic polymers, which provide an efficient barrier to enzymatic attack of the polysaccharides. The lignin macromolecule itself is extremely recalcitrant to degradation since the bonds between the subunits cannot be hydrolyzed.

Roles of Symbionts in Digestion

For the reasons outlined above, the degradation of plant cell walls requires the synergistic action of many different enzymes and, in the case of lignified substrates, also a mechanism to break up the lignocellulose complex. The most efficient cellulose and hemicellulose degraders in nature are microorganisms, i.e., bacteria, protozoa, and fungi. Fungi and certain filamentous bacteria (actinomycetes) are also the only organisms that have developed a strategy for the chemical breakdown of lignin. Not surprisingly, insects and other animals have made use of these capacities by using microorganisms as symbionts in the digestion of lignocellulosic food.

In addition to being difficult to degrade, lignocellulose is an extremely nutrient-poor substrate. Whereas nonlignified

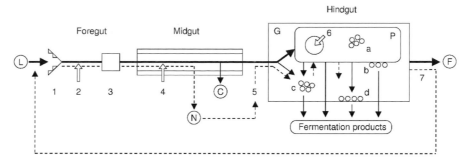

FIGURE 1 Major events in the symbiotic digestion of lignocellulose by wood-feeding lower termites. The bold lines show the path of the insoluble material, the lignin-rich residues of which are released as feces, whereas the thinner lines represent soluble degradation products that are eventually resorbed by the host. The dashed lines indicate the cycling of nitrogenous compounds. Hollow arrows mark the sites where cellulolytic enzymes are secreted. Lowercase letters refer to the different groups of bacteria, which are either endobionts (a) or epibionts (b) of the protozoa, suspended in the gut lumen (c), or attached to the gut wall (d). The scheme has been simplified for the sake of clarity; not all possible interactions are shown. Further details are given in the text. L, lignocellulose; C, carbohydrates; N, nitrogenous compounds; F, fecal matter; G, gut lumen; P, protozoa; 1, mandibles; 2, salivary glands; 3, proventriculus; 4, midgut epithelium; 5, Malpighian tubules; 6, phagosomes; 7, proctodeal feeding.

plant cell walls contain some structural proteins, the C-to-N ratio of sound wood is up to 100-fold higher than that of the insect body. Moreover, a lignocellulosic diet typically lacks most of the essential nutrients required by the animal, such as amino acids, vitamins, and sterols.

In contrast to higher animals, many microorganisms are capable of fixing dinitrogen, assimilating nitrate and ammonia, and synthesizing those amino acids and vitamins essential for the host. Many animals, including insects, have developed means of exploiting these biosynthetic capacities of microorganisms, which include—in the simplest case—the digestion of the intestinal symbionts.

LIGNOCELLULOSE DIGESTION IN TERMITES

The symbiotic digestion of lignocellulose by termites is a complex series of events involving both the host and its gut microbiota, which comprises prokaryotic symbionts (bacteria and archaea) and, at least in lower termites, also protozoa and possibly fungi (yeasts). While the events in foregut and midgut are mainly the result of host activities, the digestive processes in the hindgut are largely controlled by the symbionts (Fig. 1). Many aspects of lignocellulose digestion are common to all termite species, yet there are several noteworthy differences between the phylogenetically lower and higher taxa.

Host-Related Processes

One of the key contributions of the termite to wood digestion is the scraping action of the mandibles and the grinding activity of the proventriculus located at the end of the foregut, which results in the comminution of wood particles to a microscopic size. This is not only a prerequisite for the ingestion of the wood particles by symbiotic protozoa, but it also mechanically destroys the lignin–carbohydrate complexes, which creates an enormous surface area for the digestive enzymes provided by host and symbionts, thereby relieving much of the kinetic limitations of cellulose digestion.

In all insects, the digesta are exposed to a variety of digestive enzymes secreted by the salivary glands and the midgut epithelium. The complement of enzymes released by termites has not been systematically studied, but evidence is accumulating that it comprises enzymes necessary for the digestion of polysaccharides, microbial biomass, and other organic components of the diet and even includes cellulolytic activities contributed by the host (see below). Although experimental data are lacking, one can safely assume that—as in other insects—most of the easily digestible material has been mobilized and resorbed at the end of the midgut.

Insects lacking a pronounced gut microbiota usually have a short hindgut that serves mainly in recovering water and useful electrolytes from the residual material before the feces are voided. The digestive tract of termites, however, is characterized by one or more proctodeal enlargements and may reach enormous dimensions in length and volume (Fig. 2a). The dilatations increase the residence time of the digesta,

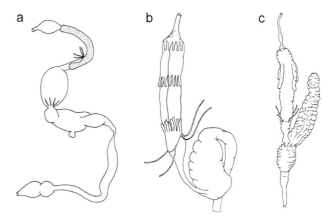

FIGURE 2 Intestinal tracts of insects with fermentation chambers harboring symbiotic microorganisms. (a) *Thoracotermes macrothorax* (Isoptera: Termitidae). (b) *Tipula flaveolineata* (Diptera: Tipulidae). (Reproduced from Buchner, 1928, "Holznahrung und Symbiose," J. Springer, Berlin.) (c) *Potosia cuprea* (Coleoptera: Scarabaeidae). (Reproduced from Werner, 1926, *Z. Morph. Ökol. Tiere* **6,** 150–206. Not drawn to scale.)

thereby prolonging the exposure to the activities of the intestinal microbiota. Moreover, the increased diameter also affects the oxygen status of the "fermentation chambers" because it reduces the proportion of the gut volume rendered oxic by the diffusion of oxygen into the periphery, thus providing a favorable environment for the oxygen-sensitive microbiota and reducing the inevitable loss of fermentation products to aerobic processes (see below).

Fiber-Digesting Symbionts

In lower termites, the solid food particles entering the hindgut are immediately phagocytized by the intestinal protozoa (Fig. 3a). These oxygen-sensitive flagellates, which make up a large fraction of the hindgut volume, are essential for wood digestion and represent a major source of cellulolytic and xylanolytic activities in the hindgut. It appears that the different flagellate species are nutritionally specialized and each species might fill a specific niche in symbiotic digestion.

The hindguts of all families of lower termites are packed with flagellates, whereas the phylogenetically higher termites (family Termitidae) contain a largely prokaryotic microbiota. It was initially assumed that other symbionts took over the cellulolytic function of the protozoa in the course of termite evolution. In the case of the fungus-cultivating termites (subfamily Macrotermitinae), which cultivate basidiomycete fungi (*Termitomyces* spp.) on predigested food in "fungus gardens" located within their nests, the key activities of the symbiotic partner in the fungus combs are extensive delignification and conversion of lignocellulose to fungal biomass. It has been proposed that fungal cellulases ingested by the termites together with the comb material are essential for cellulose digestion in the gut, but this "acquired enzyme hypothesis" is controversial.

All other higher termites do not cultivate fungus gardens, and their intestinal tracts harbor cellulolytic activities that are probably neither produced by symbiotic bacteria nor the result of ingested enzymes. The persisting dogma that higher animals do not produce cellulases has been unequivocally refuted by the demonstration of endoglucanase genes in the termite genome and their expression in the cells of the midgut epithelium and in the salivary glands.

Noncellulolytic Microbiota

Phylogenetic characterization of the prokaryotic microbiota of termite guts, which was boosted by the advent of cultivation-independent methods during the past decade, has revealed an enormous diversity. Although most of these symbionts do not take part in fiber degradation, they seem to play other important roles in digestion and in the nitrogen metabolism of the hindgut.

METABOLIC INTERACTIONS The original concept of the hindgut metabolism in lower termites assumed that the anaerobic flagellates depolymerize the polysaccharides to

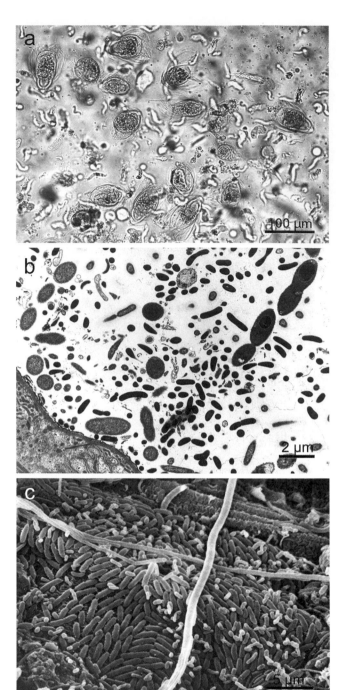

FIGURE 3 Examples of microbial symbionts in the hindgut of *Reticulitermes flavipes* (Isoptera: Rhinotermitidae), a wood-feeding lower termite. (a) Preparation of anaerobic protozoa from the hindgut of a worker larva, showing the large hypermastigote flagellate *Trichonympha agilis,* filled with wood particles, and numerous smaller flagellates (mainly oxymonads, *Dinenympha* spp.). Differential interference contrast photomicrograph taken by U. Stingl. (b) Transverse section through the peripheral hindgut, showing the diverse bacterial microbiota associated with the thin cuticle of the hindgut wall (bottom left). Transmission electron micrograph provided by J. A. Breznak. (c) Preparation of the hindgut wall, showing the dense colonization of the cuticle by numerous rod-shaped and filamentous bacterial morphotypes. Scanning electron micrograph provided by J. A. Breznak.

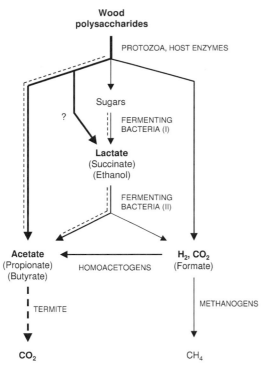

FIGURE 4 Schematic presentation of the metabolic processes involved in the fermentative degradation of polysaccharides in the hindgut of the termite *R. flavipes*. The dashed lines indicate metabolic fluxes that seem to be strongly influenced by the continuous influx of oxygen into the gut periphery.

sugar units, which are then fermented to acetate, hydrogen, and CO_2 as the major products. Recent results indicate that the spectrum of fermentation products released by the diverse protozoa is much wider, giving rise to a variety of intermediates that form the substrates of the endobiotic or epibiotic bacteria colonizing the protozoa. In addition, metabolites released by the protozoa or originating directly from the midgut support numerous bacteria located in the lumen or attached to the wall of the hindgut and are eventually converted to a range of short-chain fatty acids (mostly acetate, propionate, butyrate) that accumulate in the hindgut fluid and that are eventually resorbed by the hindgut epithelium (Fig. 1).

Although it is still difficult to assign functions to individual members of the gut microbial community and to localize exactly the microbial populations involved in these reactions, the major roles of different functional groups are beginning to take shape (Fig. 4). At least two metabolically different groups of prokaryotes are involved in the oxidation of the hydrogen or one-carbon compounds produced in the microbial fermentations: (1) methanogens, which reduce CO_2 to methane, and (2) homoacetogens, which reduce CO_2 to acetate.

Methanogenesis and reductive acetogenesis occur in the hindgut of all termites, although in most wood-feeding species, reductive acetogenesis prevails over methanogenesis as a hydrogen sink and substantially increases the pool of short-chain fatty acids available to the host. In the fungus-cultivating

and soil-feeding termites, however, methanogenesis may amount to a considerable portion of the total respiratory activity of the insect. Although methane production is of little use for the host, methane emission acquires global significance due to the enormous biomass of these groups in tropical rainforests and savannahs.

THE ROLE OF OXYGEN The steep gradients of oxygen between oxic gut epithelium and anoxic gut contents drives a continuous influx of oxygen into the gut. Microsensor measurements have shown that oxygen may penetrate 50 to 200 μm into the periphery of the hindgut, leaving only the central portion of the dilated compartments anoxic. The small insect guts have an enormous surface-to-volume ratio, which lends much greater importance to all surface-related processes than in larger animals. Because oxygen removal in the gut periphery is fueled by the fermentative processes in the hindgut lumen, the maintenance of anoxia is not a trivial issue, and there must be a lower size limit for arthropods with a symbiotic digestion.

The bacteria and protozoa colonizing the gut periphery, especially those directly associated with the gut epithelium (Figs. 3b and 3c), have to be specifically adapted to the presence of oxygen at low levels and may even use oxygen as an electron acceptor. Radiotracer analysis of the *in situ* metabolism in the hindgut of the termite *Reticulitermes flavipes* has demonstrated that the high oxygen fluxes significantly influence the metabolic processes in the hindgut (Fig. 4), but the details of this scenario remain to be studied.

Nitrogen Economy of Termites

Wood-feeding termites have developed several adaptations that help to compensate for the low nitrogen content of their diet, all of which involve the biochemical capacities of their gut microbiota. The most important strategy is a combination of conservation and recycling and is reminiscent of the application of organic fertilizers in agriculture.

Like in any other insect, uric acid and urea, the waste products of nucleic acid and protein metabolism, are secreted into the digesta via the Malpighian tubules at the midgut–hindgut junction. However, they are not voided with the feces, but are readily mineralized by the hindgut microbiota. The resulting ammonia is assimilated into microbial biomass; it remains to be clarified whether the intestinal protozoa can also assimilate ammonia directly or acquire combined nitrogen by phagocytosis and digestion of other microbial symbionts.

The nitrogen cycle is closed by the digestion of microbial cells. Because termites cannot access the microbes in the hindgut directly, worker larvae solicit hindgut contents from their nestmates. This behavior, which has been termed proctodeal trophallaxis and is unique to this group of social insects, increases in frequency with the level of nitrogen limitation. Digestion of the hindgut contents and resorption of the nitrogenous products probably take place in the foregut and midgut (Fig. 1). The efficiency of nitrogen conservation within the colony is increased further by the consumption of exuviae and dead individuals by nestmates.

While nitrogen recycling creates high ammonia concentrations in the hindgut, which allow the maintenance of an active gut microbiota and thus ensure high rates of carbon mineralization, the low nitrogen content of the food still limits the growth of a termite colony severely. However, the presence of bacteria among the hindgut microbiota capable of fixing atmospheric nitrogen may contribute considerably to increasing the nitrogen pool. It has been estimated that the nitrogenase activity in certain *Nasutitermes* termites would be sufficient to double the nitrogen content of a colony within a few years, and stable isotope analysis has revealed that 30 to 60% of the nitrogen in *Neotermes koshunensis* workers is derived via this pathway.

Although most lower termites are strictly xylophagous, the evolutionarily higher termites in the family Termitidae show an enormously successful dietary diversification. The fungus-cultivating Macrotermitinae, which are specialized on degradation of nitrogen-poor plant litter, probably recycle nitrogen within the colony by exploiting the ammonium-assimilating capacities of the fungus. The diet of the other subfamilies (Nasutitermitinae, Termitinae, and Apicotermitinae) ranges from sound wood to lignocellulosic plant materials in different stages of humification, including soil and animal dung. The C-to-N ratio of the diet decreases with increasing humification, which suggests that microbial biomass in the organic matter forms a potential source of nutrition. The extreme gut alkalinity in humivorous termites, which may reach pH 12 in the anterior proctodeal compartment of soil-feeding Termitinae, may have allowed the specialization on nitrogen-rich but recalcitrant diets, because it helps to release fermentable substrates from the ingested humic substances that would otherwise be protected from microbial degradation in soil.

DIGESTIVE SYMBIOSES IN OTHER INSECTS

The occurrence of a specific, autochthonous gut microbiota among insects remains to be systematically studied, but sufficient evidence for the presence of a digestive symbiosis has accumulated for representatives of several insect orders.

Phytophagous insects feeding on protein-rich plant material (e.g., caterpillars) usually have a relatively undifferentiated intestinal tract and, because of the rapid gut passage, digest little or no cellulose. Many xylophagous, detritivorous, and humivorous insect larvae, however, possess hindgut dilations that are missing in closely related species with a different feeding habit. The most prominent examples are among the Coleoptera (e.g., Scarabaeidae) and among the Diptera (e.g., Tipulidae) (Figs. 2b and 2c).

Scarabaeids and tipulids have an actively fermenting gut microbiota, including cellulolytic and hemicellulolytic bacteria and, in the former, also methanogenic archaea, which are attached to brush-like chitinous structures. Also, the guts of omnivorous cockroaches contain a largely prokaryotic microbiota of bacteria and methanogenic archaea, especially when maintained on a fiber-rich diet; many species also harbor anaerobic ciliates with methanogenic endosymbionts in their hindguts. Fiber-degrading protozoa, however, which are so prominent among the Isoptera, are found elsewhere only in *Cryptocercus punctulatus* (Cryptocercidae), which is not astonishing because these wood-feeding cockroaches share a common ancestor with the termites.

Reports on the Orthoptera are somewhat contradictory. The bacteria in the gut of locusts (family Acrididae) have been considered commensals because their absence in germ-free cultures had no obvious effect on the host, although a recent study documented that key components of the pheromone responsible for aggregation of *Schistocerca gregaria* are produced by its gut microbiota. The situation is different in crickets (Gryllidae), which benefit from the presence of a gut microbiota when raised on a fiber-rich diet. In the hindgut of *Achaeta domesticus,* the density of microorganisms is even higher than that in termites, and there are brush-like supports for the attachment of bacteria that resemble those in scarab beetle larvae.

It is very likely that insects other than termites access protein and recycle nitrogen via digestion by microbial symbionts. Proctodeal feeding is a form of social behavior that is restricted to the termites and the wood-feeding cockroach, *C. punctulatus,* but theoretically the consumption of feces would also allow access to the microbial protein. A special adaptation to digestion by symbionts seems to be present in Scarabaeidae, in which a reflux of hindgut contents into the alkaline midgut has been observed. Alkaline midguts are encountered in many humivorous, detritivorous, and coprophagous dipteran and coleopteran larvae and may be an analogy to the extreme alkalinity in the anterior hindgut of soil-feeding termites.

Newly hatched first instars do not possess a gut microbiota, and all larvae have to be reinoculated after each molting because the hindgut intima together with the complete contents is shed in this process. However, many insect larvae eat their exuviae directly after molting, and first instars may pick up their intestinal symbionts together with the food from their environment. Because the oxygen-sensitive flagellates in lower termites do not form resistant cysts, proctodeal trophallaxis is essential for the transfer of the microbial symbionts between nestmates, and it has been speculated that the evolution of sociality and symbiotic digestion in termites might have proceeded hand in hand. This is supported by the strict host specificity of the different flagellates and their apparent coevolution with their hosts.

MUTUALISTS VS COMMENSALS

In the symbiosis between lower termites and fiber-digesting flagellates, both partners are indispensable and the mutual advantage is obvious. However, there are many symbionts in the guts of termites and other insects whose presence—to the best of our knowledge—appears to be of no advantage to the host. When the benefit of the association is unidirectional, a symbiont is classified as a commensal, and the host might even benefit from its elimination.

However, it is also not unlikely that in such cases we are merely lacking insight into the symbiont's role in the symbiosis. An example might be the methanogenic archaea, which are regularly encountered in the hindguts of termites, scarab beetle larvae, and many cockroaches. Although methane, their sole metabolic product, cannot be utilized by the host, methanogenic archaea seem to be an integral part of the gut microbiota. They colonize the hindgut intima or the intestinal protozoa and are often attached to cuticular spines, which apparently represent specific attachment sites for methanogenic symbionts.

Unfortunately, it is often impossible to eliminate a specific member of the intestinal microbial community selectively, and even when possible, it is difficult to distinguish between the direct and the indirect consequences of their elimination. The only way to gain insight into the complicated microbe–microbe and microbe–host interactions in the intestinal tracts of insects is probably in treating the gut microbiota as a whole.

It might also be worth speculating on the problems that would be created by a general absence of specific gut symbionts and on the efforts involved in keeping a gut sterile. Microorganisms are continuously incorporated with the food, and while a rapid passage of the digesta controls microbial growth by washout (as long as attachment is prevented), the slow gut passage, necessary also for the digestion of lignocellulose by host enzymes, would allow the uncontrolled proliferation of foreign microorganisms. Perhaps the promotion of specific symbionts excludes colonization by potential pathogens.

As long as the access of oxygen to the dilated hindgut compartments is limited, the energy expenditure for any microbial symbionts remains relatively small. A fermentative degradation of carbohydrates to acetate releases only about 10% of the free energy contained in the substrates, and a fraction of this energy is returned as nutritionally valuable microbial biomass. Together with the added value of metabolic properties like nitrogen fixation, ammonia assimilation, and the provision of vitamins, the advantages for the host may be well worth the investment.

CONCLUSION

The digestive symbioses of insects with their intestinal microbiota allow the insects to overcome the severe kinetic limitation of lignocellulose digestion and the nutritional restrictions imposed by this special diet. Although the situation is somewhat reminiscent of the digestive symbiosis encountered in ruminants, it is important to consider the small size of insects, which affects especially the oxygen status of the gut. From a microbiological point of view, insect guts are not simply anoxic fermentors, but axially and radially structured environments with physicochemically distinct microhabitats. We are just beginning to understand the complex interactions within the intestinal microbial communities of termites. Further efforts targeted at identifying the microbial symbionts and their metabolic potentials in this and other, hitherto understudied, groups of insects are sorely needed.

See Also the Following Articles
Digestive System • Isoptera • Nutrition

Further Reading
Bignell, D. E. (2000). Introduction to symbiosis. *In* "Termites: Evolution, Sociality, Symbiosis, Ecology" (T. Abe, D. E. Bignell, and M. Higashi, eds.), pp. 189–208. Kluwer Academic, Dordrecht.
Breznak, J. A. (2000). Ecology of prokaryotic microbes in the guts of wood- and litter-feeding termites. *In* "Termites: Evolution, Sociality, Symbiosis, Ecology" (T. Abe, D. E. Bignell, and M. Higashi, eds.), pp. 209–231. Kluwer Academic, Dordrecht.
Breznak, J. A., and Brune, A. (1994). Role of microorganisms in the digestion of lignocellulose by termites. *Annu. Rev. Entomol.* **39**, 453–487.
Brune, A. (1998). Termite guts: The world's smallest bioreactors. *Trends Biotechnol.* **16**, 16–21.
Brune, A., and Friedrich, M. (2000). Microecology of the termite gut: Structure and function on a microscale. *Curr. Opin. Microbiol.* **3**, 263–269.
Hackstein, J. H. P., and Stumm, C. K. (1994). Methane production in terrestrial arthropods. *Proc. Natl. Acad. Sci. USA* **91**, 5441–5445.
Kane, M. D. (1997). Microbial fermentation in insect guts. *In* "Gastrointestinal Microbiology" (R. I. Mackie, and B. A. White, eds.), pp. 231–265. Chapman & Hall, New York.
Radek, R. (1999). Flagellates, bacteria, and fungi associated with termites: Diversity and function in nutrition—A review. *Ecotropica* **5**, 183–196.
Slaytor, M. (2000). Energy metabolism in the termite and its gut microbiota. *In* "Termites: Evolution, Sociality, Symbiosis, Ecology" (T. Abe, D. E. Bignell, and M. Higashi, eds.), pp. 307–332. Kluwer Academic, Dordrecht.
Watanabe, H., and Tokuda, G. (2001). Animal cellulases. *Cell. Mol. Life Sci.* **58**, 1167–1178.

Symbionts, Bacterial

Michael E. N. Majerus
University of Cambridge

Many species of insect are host to symbiotic microorganisms. These symbionts have traditionally been classified as mutualists (i.e., beneficial), parasitic (harmful), or commensal (neutral). A variety of microorganisms among these symbionts are vertically transmitted, i.e., they are inherited. The survival of these heritable symbionts is almost totally dependent on the success of their host. It follows that these symbionts should evolve characteristics that increase host survival. However, some so-called "ultraselfish" symbionts manipulate their hosts to the symbiont's benefit even when this is to the detriment of the host. The strategies used by such microorganisms include sterilization of noncarriers, feminization of genetic males, induction of asexual reproduction, and biasing host sex ratios in favor of females by killing male hosts. The most common of these reproductive parasites are the ultraselfish bacteria.

ULTRASELFISH, MATERNALLY INHERITED BACTERIA

Inherited bacteria that live in the cytoplasm of host cells are transmitted to subsequent generations through the female line. This is because female gametes are heavily resourced with cytoplasm, whereas sperm contain negligible amounts. A maternally inherited bacterium can thus increase its prevalence either by increasing the fitness of infected compared with uninfected hosts or by biasing the sex ratio in favor of females. The phenomenon of cytoplasmic incompatibility (CI), which involves the sterilization of noncarriers, follows the first of these options. Three other strategies, feminization, parthenogenesis induction (PI), and male-killing (M-k), skew populations in favor of females.

CYTOPLASMIC INCOMPATIBILITY

Cytoplasmic incompatibility was first noted when some crosses between *Culex pipiens* mosquitoes from European, Asian, and American populations failed to produce offspring or produced offspring only when the crosses were carried out in one direction. Observed patterns of reproductive successes and failures demonstrated maternal inheritance (usually 95–100% efficiency), indicating the possible involvement of some cytoplasmic factor, hence the name given to the phenomenon. A bacterium of the genus *Wolbachia* was found to be responsible. Subsequently, CI has been reported from many insect orders.

Cytoplasmic incompatibility occurs when a male carrying a CI *Wolbachia* mates with a female that does not bear the same *Wolbachia*. The bacteria allegedly secrete a chemical into the sperm of their host that kills zygotes formed within the female parent if they do not bear the *Wolbachia*. In essence then, an uninfected female that mates with an infected male is rendered sterile thereafter. Other mating combinations are not affected.

CI *Wolbachia* can give rise to two patterns of reproductive failure. If one host population has the *Wolbachia*, whereas the other lacks it, matings fail in only one direction; i.e., the incompatibility is unidirectional. If, however, each population harbors a different strain of CI *Wolbachia*, matings fail in both directions and the incompatibility is said to be bidirectional. For example, in the fruit fly *Drosophila simulans*, both unidirectional and bidirectional incompatibility have been recorded. Indeed, some CI host species carry more than one strain of CI *Wolbachia*, with individuals in some populations hosting two strains concurrently.

Host Range and Identity of the Agents Causing CI

Cytoplasmic incompatibility is commonly reported when crosses of insects from geographically distinct populations are made. It has been reported most often from species of Diptera. However, Coleoptera, Heteroptera, Lepidoptera, and Hymenoptera have also been found to exhibit the phenomenon and other orders will certainly be added to the list in the near future.

The CI phenotype can be cured by treatment with antibiotics or by exposing infected strains to high temperatures (≈37°C) for several days. This suggests that the agent responsible for CI is a bacterium and in every case in which the agent responsible has been identified it is a bacterium of the *Wolbachia pipientis* complex. This complex represents a widespread group of α-Proteobacteria, whose members are known to infect hosts from all the major orders of insects as well as some other arthropods (mites, spiders, isopods) and phyla (nematodes). Not all *Wolbachia* infections result in CI. The bacterium is known to manipulate hosts in a variety of other ways and some appear to have no overt effect on their hosts, whereas some are beneficial. Estimates from molecular surveys using *Wolbachia*-specific probes suggest that between 10 and 20% of insect species harbor this bacterium.

Phylogenetic analysis has shown that *Wolbachia* in taxonomically very different hosts may be more similar to one another than *Wolbachia* in very closely related hosts, suggesting that *Wolbachia* may transfer between host species.

Mechanisms of Incompatibility

The precise mechanics of incompatibility are not well understood, partly because the mechanism of incompatibility appears to vary between host species. In *C. pipiens*, incompatibility results from failure of sperm bearing CI *Wolbachia* to fuse correctly with female gametes that lack the same *Wolbachia*. In fruit flies *(Drosophila)* and moths *(Ephestia kuehniella)*, embryo development appears to be suppressed at an early stage. The situation in the jewel wasp, *Nasonia vitripennis*, is a little more definite. Here *Wolbachia* interferes with condensation of the paternal chromosome set during the first mitotic cell division of the embryo so that the paternal chromosome set is lost. All progeny that get the *Wolbachia* thus carry just the maternal chromosome set, are thus haploid, and develop as males. Here CI results in a male-biased sex ratio.

The Population Dynamics of CI

Wolbachia that cause CI spread through host populations because infected females have an advantage over uninfected females, which are rendered sterile once they have mated with an infected male. The precise dynamics of CI are influenced by several factors, including (1) the vertical transmission of the bacterium, (2) the fitness costs or benefits it imposes on its host and possibly horizontal transmission, and (3) host suppression systems.

In general, the prevalence levels of CI *Wolbachia* in infected populations are high and often approach 100%. Few studies of the population dynamics of CI have been conducted. The most intensive studies involve drosophilid hosts. For example, in *D. simulans,* in California, rapid spread of a CI strain of *Wolbachia* was monitored during the 1980s and the 1990s. A theoretical model, based upon the levels of incompatibility, the vertical transmission efficiency of the bacterium, and the assumption that bearing the symbiont had no direct fitness effect on its bearer, produced theoretical predictions close to the observed rates of increase.

FEMINIZATION

Microbes and Sex Determination

Microorganisms of two types are known to cause feminization of their hosts. One type, microsporidians and other simple protists, as yet are known to infect only crustaceans. The other type, bacteria of the genus *Wolbachia,* causes feminization in both crustaceans and insects. In essence, what both of these groups of microorganisms do is turn genetic males into females.

The best studied cases of feminization involve crustaceans, such as the woodlouse, *Armadillium vulgare.* However, many aspects of the feminization system in crustaceans are probably similar to those in insects, of which two species of moth, *Ostrinia furnacalis* and *Ostrinia scapulalis* harbor feminizing *Wolbachia.*

Sex determination in *A. vulgare* and in the Lepidoptera is normally based upon sex chromosomes; in the Lepidoptera females are the heterogametic sex, carrying two different sex chromosomes (WZ), and males are homogametic.

In *O. furnacalis* all-female strains were first reported in 1998. A second example of feminization has been reported in *O. scapulalis* subsequently. Antibiotic treatment has shown the female biases to be tetracycline sensitive, and molecular analysis has revealed a *Wolbachia.* In both moths this bacterium is taxonomically similar to that causing feminization in *A. vulgare.*

In parallel to *A. vulgare,* cure of feminizer lines with antibiotics led to a strong male bias in progeny of cured females. These moths are ancestrally female heterogametic, with the result that feminized ZZ males, when cured of the bacterium and mated by normal (ZZ) males, can produce only male offspring. It is perhaps no coincidence that in the examples of *Wolbachia*-induced feminization, the host is ancestrally female heterogametic, and further instances of *Wolbachia*-induced feminization in the Lepidoptera and other insect taxa with heterogametic females are likely to be discovered in the near future.

Mechanism of Feminization

The mechanism of feminization in the two lepidopteran examples has not been determined. However, in *A. vulgare,* the difference between developing into a male and developing into a female appears to depend on the activity of a single gene that blocks the expression of one or more genes that cause the differentiation of the androgenic gland.

PARTHENOGENESIS INDUCERS

Microbe-Induced Parthenogenesis in the Hymenoptera

Much of the biotic world indulges in sexual reproduction. However, some organisms reproduce asexually, either ancestrally or secondarily when asexuality has evolved from sexuality. Secondary asexual reproduction, or parthenogenesis, occurs with new individuals arising from unfertilized eggs.

Various types of parthenogenesis exist. Normal haplodiploid species, in which females result from fertilized eggs and males from unfertilized eggs, are called arrhenotokous. In deuterotoky, both males and females arise from unfertilized eggs, whereas thelytoky involves the production of only female offspring without fertilization. The production of only female offspring, without the need for males, raises the possible involvement of cytoplasmic symbionts whose interests are favored by female hosts, which can vertically transmit them. This involvement is realized in a number of haplodiploid Hymenoptera and Thysanoptera.

The first report of bacterially induced thelytoky was in the parasitoid hymenopteran *Trichogramma pretiosum.* Here, administration of various antibiotics or of high temperatures to thelytokous females led to the production of sons among the offspring. This led to the deduction that some sort of cytoplasmic microorganism, probably a bacterium, caused thelytoky in this species. Microscopic examination revealed a bacterium present in eggs of some thelytokous lines, but not in antibiotic-treated, temperature-treated, or arrhenotokous lines. Molecular genetic analysis revealed the presence of a *Wolbachia.*

Evidence of microbe-induced parthenogenesis has been revealed subsequently in over 30 species of parasitoid wasp, in the predatory thrip *Franklinothrips vespiformis,* and in the colembollan *Folsomia candida,* as well as in some species of mite (*Bryobia* spp.). In all cases in which the microbe involved has been identified, it is a *Wolbachia.* The *Wolbachia* involved do not form a single taxonomic group within the *Wolbachia,* but are intermixed with *Wolbachia* that affect hosts in other ways (e.g., CI and male-killing). This suggests that PI has evolved independently several times in the *Wolbachia.* Alternatively, the genes that cause PI may be horizontally transferred between *Wolbachia,* through DNA exchange involving virus-like particles or plasmids. A third possibility is that the same *Wolbachia* causes different effects in different hosts, for example, parthenogenesis in haplodiploids and CI or male-killing in diploids. Phylogenetic analysis of a number of PI *Wolbachia,* particularly in the genus *Trichogramma,* has provided evidence for interspecific horizontal transmission of the symbiont.

Mechanism of Parthenogenesis Induction

Investigations of parthenogenetic organisms have shown variation in the way in which diploid offspring can result from unfertilized eggs. This diversity can be divided into two basic groups: either meiosis is modified with the diploid number of chromosomes being retained or, following meiosis, the diploid chromosome number is restored after the formation of a single pronucleus, usually by the fusion of two haploid mitotic products. In the examples of *Wolbachia*-induced parthenogenesis that have been studied microscopically, the latter of these two routes applies. In infected females, both sets of chromosomes migrate to the same pole during the first mitotic division of the meiotic product. This results in a single diploid mitotic product. The ways in which *Wolbachia* affect the behavior of host chromosomes during the first mitotic division are currently not known.

Dynamics of *Wolbachia* Infection

In most Hymenoptera with PI *Wolbachia,* thelytoky has been fixed. However, in many *Trichogramma* species both thelytokous and arrhenotokous individuals occur together. The reasons for polymorphism in some species and monomorphism in others are as yet not clear, but incomplete vertical transmission, negative-fitness effects of *Wolbachia* on their host, and suppresser genes may all have an influence.

MALE-KILLERS

Types of Male-Killers

Male-killing is perhaps the least sophisticated of the mechanisms of sex ratio distortion practiced by inherited symbionts. The basic mechanism involved is for the symbiont to kill male but not female hosts. Among arthropods, M-k has been reported from five orders of insects (Hemiptera, Coleoptera, Lepidoptera, Hymenoptera, and Diptera) and two species of mite. In addition, perhaps because M-k is easily evolved, M-k's themselves are taxonomically diverse, with α- and γ-Proteobacteria, Mycoplasma, Flavobacteria, and Microsporidia being recorded. Two M-k strategies have been reported. The first strategy, which involves killing male hosts late in their development (typically in the final larval instar), allows symbionts to replicate within their host, so that large numbers of individuals are released from the host's corpse. Here the population dynamics of the M-k depend on both horizontal and vertical transmission. Reported late M-k's are confined to Microsporidia infecting mosquitoes.

The second strategy involves killing males early in their development, typically during embryogenesis. In killing its host, the M-k dies, but in doing so it increases the fitness of copies of itself in female hosts. The death of the M-k in male hosts means that early M-k does not involve significant levels of horizontal transmission. The increase in the fitness of infected females resulting from M-k is called fitness compen-

sation, because for a M-k to invade a host population, infected females must have higher fitness than uninfected females to compensate for the death of the infected males. The nature of cytoplasmic bacterial reproduction means that many or all progeny of an infected mother will carry copies of the symbiont that are virtually identical by descent. If by killing male hosts the symbiont thereby increases the fitness of the dead males' sisters, a fitness benefit will also accrue to "exact" copies of the male-killer within these females.

Two advantages of M-k have been identified. First, the probability of mating with close relations, specifically brothers, will be reduced, thus avoiding the harmful effects of inbreeding. An infected female whose brothers have been killed has no siblings to mate with. The second advantage involves resource reallocation. If the resources that would have been consumed by sons of an infected female become available specifically to female siblings these females will benefit.

Early M-k has been recorded from many insect hosts, and to date, all known early M-k's are bacteria. A diverse array of bacteria is involved, suggesting that M-k has evolved independently many times.

The distribution of M-k's among insects is not random. Certain groups have behavioral traits and ecologies that provide a context in which high resource benefits are obtained by daughters of infected females specifically as a result of the death of males. In these instances, the resources made available to females from the death of males become preferentially available to infected compared with uninfected females.

Hot spots for male-killing are known in the milkweed bugs (Hemiptera), nymphalid butterflies (particularly of the genus *Acraea*), and ladybug beetles (Coccinellidae).

Early Male-Killing in Ladybugs

Female-biased sex ratios were first reported in some families of the ladybug *Adalia bipunctata* in Russia in 1947. The female-biased sex ratio trait was shown to be maternally inherited. More recent work has demonstrated that the trait is curable with both antibiotics and heat treatment, suggesting that a bacterial agent is responsible. Male-killers have been discovered in eight other species of ladybug. In each case, the female sex ratio bias is associated with the death of approximately 50% of eggs before they hatch (Fig. 1). This leads to the interpretation that the eggs of infected females have a normal sex ratio at fertilization, but that the secondary sex ratio is biased toward females because male embryos die before hatching.

Experimental work has shown that female larvae from mothers bearing a M-k gain a considerable nutritional advantage from consuming the contents of their dead brothers' eggs. This means that they are larger when they disperse from their egg clutch than larvae from uninfected mothers, and they have longer to find and subdue their first prey before dying of starvation.

Detailed examination of resource reallocation in coccinellids has provided a set of criteria that are predicted to allow the

FIGURE 1 An egg clutch laid by a male-killer-infected female *Harmonia axyridis*. Approximately half of the larvae (the females) have hatched and the neonates have begun consuming the soma of the dead male eggs.

invasion of M-k's. The ladybug must feed on an ephemeral and often limited food, such as aphids. Neonate larval mortality from starvation must be significant. The ladybug must also lay eggs in clutches and neonate larvae must consume unhatched eggs in their clutch. Lack of one or more of these criteria should make a ladybug immune to M-k invasion. All ladybugs known to harbor M-k's possess all the relevant criteria.

The Mechanism of Male-Killing

Little is known about of the mechanism by which bacteria discriminate between male and female hosts and kill only the former. It is known that the sex determination systems of species that host M-k's vary. So, for example, ladybugs have XX females and XY males, butterflies of the genera *Acraea* and *Danaus* have ZZ males and ZW females, and the hymenopteran parasitoid *N. vitripennis* has typically arrhenotokous haplodiploid sex determination, yet all have M-k's. Thus M-k's do not all detect that they are in males by identifying a specific region of a male-specific chromosome.

Consequences of Population Sex Ratio Distortion

Rather more is known about the evolutionary effects of M-k's on their hosts. The prevalence of M-k's varies from less than 5% to over 95% in different insect groups. In the multicolored Asian ladybug *Harmonia axyridis,* the prevalence of a male-killing *Spiroplasma* is closely correlated to the population sex ratio. At low prevalence, the population sex ratio deviates little from 50% of each sex. However, at high prevalence (50% of females infected) over two-thirds of the population are female. Similarly female-biased populations of *Ad. bipunctata* have been reported. More spectacularly, in the butterflies *Acraea encedon* and *Ac. encedana,* populations in which females comprise over 95% occur. Such highly biased sex ratios have the potential to impact on the evolution of reproductive strategies. This potential may be the cause of a number of unusual features

of the courtship and copulatory behavior of these species.

First, both *Ad. bipunctata* and *H. axyridis* exhibit a variety of genetically determined mate choice patterns, including, unusually, weak male choice of females. Furthermore, in *Ad. bipunctata,* the investment that a male makes in an individual copulation decreases as the ratio of females to males increases. Male *Ad. bipunctata* may transfer one, two, or three spermatophores (sperm packages) during a single copulation. In populations with a 1:1 sex ratio, males transfer, on average, just under two spermatophores. Conversely, in populations approaching two females for each male, males rarely transfer more than a single spermatophore (average 1.2).

In highly female-biased populations of the butterflies *Ac. encedon* and *Ac. encedana,* many females die without mating. This suggests that males are limiting. The result has been the evolution of sex-role-reversed behaviors. The most striking of these is of female lekking behavior. Male leks, which are resource-lacking arenas where males congregate and compete for mates, are known in many taxa. In *Ac. encedon* and *Ac. encedana,* it is virgin females that congregate at resource-lacking sites. Within these arenas females compete for matings. Furthermore, there is some evidence to suggest that males preferentially mate with females that are free from male-killer infection.

IMPLICATIONS AND APPLICATIONS

Ultraselfish inherited bacteria adopt a strategy of harming their host to aid their own persistence and spread. Early reports of such symbionts, which caused abnormalities in host sex ratios or reproductive patterns, were long thought of as evolutionary oddities. Over the past 2 decades, that view has been changed as molecular genetic advances have revealed that many insects are host to these reproductive manipulators. Surveys have shown that 10 to 20% of insect species harbor bacteria of a single genus, *Wolbachia.* It is likely that similar surveys directed toward other clades of inherited bacteria will lead to the conclusion that species of insects that host antagonist bacteria are in the majority.

Invasion by ultraselfish inherited bacteria can have far-reaching effects on the evolution of hosts. It has already been shown that the selection pressures imposed by these cytoplasmic symbionts have led to the evolution of nuclear suppresser genes in some insects. Such genes provide evidence of the intragenomic conflict that results from the maternal inheritance of these microbes. Invasion and spread of any maternally inherited symbiont through a host population will also cause an immediate decrease in the diversity of host cytoplasmic organelles, such as mitochondria, for they are inherited together. The spread of a symbiont, therefore, pulls the organelles of its original female host through the host population with it. Further effects of CI, PI, feminizing, and M-k bacteria undoubtedly await discovery. Given the antagonistic interactions between deleterious symbionts and their hosts, it is difficult to believe that a complete understanding of the biology of any insect species can be achieved without consideration of the symbionts it carries.

Deleterious inherited bacteria, such as *Wolbachia,* currently have a high scientific profile. This is at least partly because they may have a future role in pest control. Several strategies of use can be envisaged. Endosymbionts that cause female biases in host species that are already used as biological control agents, such as parasitoid wasps or aphidophagous coccinellids, may be used to increase the impact of the control agent or reduce costs of a control program. Alternatively, mass release of CI agents could be used to cause sterility in recipient populations, much as mass releases of males sterilized by radiation have been used to control *Cochliomyia hominivorax.* A third strategy would be to use these bacteria to transfer genes. If a symbiont bearing "useful" genes were likely to spread through a target population to fixation, as is likely of a CI agent, the useful genes need be introduced by only a relatively small initial release.

The study of ultraselfish inherited bacteria in insects has blossomed over the past decade. The focus on this group of extraordinary parasites is likely to increase as the breadth of their influence on host evolution becomes more clearly understood.

See Also the Following Articles

Genetic Engineering • Parthenogenesis • Reproduction, Female • Reproduction, Male • Sex Determination • Wolbachia

Further Reading

Hoffmann, A. A., and Turelli, M. (1997). Cytoplasmic incompatibility in insects. *In* "Influential Passengers" (S. L. O'Neill, A. A. Hoffmann, and J. H. Werren, eds.), pp. 42–80. Oxford University Press, Oxford.

Hurst, G. D. D., Hurst, L. D., and Majerus, M. E. N. (1997). Cytoplasmic sex-ratio distorters. *In* "Influential Passengers" (S. L. O'Neill, A. A. Hoffmann, and J. H. Werren, eds.), pp. 125–154. Oxford University Press, Oxford.

Jiggins, F. M., Hurst, G. D. D., and Majerus, M. E. N. (1999). Sex ratio distorting *Wolbachia* causes sex role reversal in its butterfly host. *Proc. R. Soc. London. B.* **267,** 69–73.

Majerus, M. E. N., and Hurst, G. D. D. (1997). Ladybirds as a model system for the study of male-killing symbionts. *Entomophaga* **42,** 13–20.

Rigaud, T. (1997). Inherited microorganisms and sex determination of arthropod hosts. *In* "Influential Passengers" (S. L. O'Neill, A. A. Hoffmann, and J. H. Werren, eds.), pp. 81–101. Oxford University Press, Oxford.

Stouthamer, R. (1997). *Wolbachia*-induced parthenogenesis. *In* "Influential Passengers" (S. L. O'Neill, A. A. Hoffmann, and J. H. Werren, eds.), pp. 102–124. Oxford University Press, Oxford.

Werren, J. H., and O'Neill, S. L. (1997). The evolution of heritable symbionts. *In* "Influential Passengers" (S. L. O'Neill, A. A. Hoffmann, and J. H. Werren, eds.), pp. 1–41. Oxford University Press, Oxford.

Systematics

see *Nomenclature and Classification*

Taste

see *Chemoreception*

Taxonomy

see *Nomenclature and Classification*

Teaching Resources

John H. Acorn and Felix A. H. Sperling
University of Alberta, Canada

In the classroom, insects are perennially popular subjects. Small enough and odd enough to be both simple to manage and intriguing to students, they are sometimes taught about for their own sake and sometimes used as models for the teaching of concepts in biology as a whole. For those who teach about insects in the entomological sense, the best resources have always been a good entomological library, a selection of live insects in culture, a well-maintained collection of preserved specimens, and somewhere to take students to see live insects behaving in a natural fashion. These are still the basics, along with contagious enthusiasm and genuine personal knowledge of the subject. For those who use insects as conceptual models, there are a few standard species (cockroaches, crickets, fruit flies, and others) for which methods of culturing and experiments for demonstrating various principles are well established and generally available.

In recent years, two changes in social attitudes have strongly affected entomological teaching. First, globalization, especially through the Internet, has allowed people to pursue their interests at any time and place, with consequent opportunities to compare information and to communicate on a global stage. Second, localization has also occurred, at least in the sense of a surge in pride in the natural features and biodiversity that can be found in one's own backyard. One might argue that this is a bit of a reaction to globalization as well. The combination of these two seemingly contradictory movements has meant that a deluge of information and resources is available and relevant at varying scales. However, this information must be filtered carefully, because much of what is available now, especially through the Internet, is often not peer reviewed nor generated by trained entomologists.

WEB SITES

Web sites are the most important recent development in entomological teaching. They take various forms, such as clearinghouses to other sites, museum databases, exercises for instruction at various levels, and taxon-based compilations and/or species accounts. The following list of Web sites offers a general overview of what is available, as well as providing links to other sites. Keep in mind, however, that the Web is an ever-changing thing and that sites come and go, as well as change address. In general, institutionally affiliated Web sites provide the most reliable information, but this is not always so. Luckily, entomological Web sites as a group have been less affected by pseudoscience and the fringe element than have many other areas in cyberspace.

- Coleoptera Home Page: http://www.coleoptera.org/. This site is an excellent starting point for Web resources about beetles.
- Electronic Resources on Lepidoptera: http://www.chebucto.ns.ca/Environment/NHR/lepidoptera.html. This site provides many links to Web resources for the study of butterflies and moths.

- Entomology Index of Internet Resources: http://www.ent.iastate.edu/List/. The hub of the virtual entomological wheel, this site has links to other sites on just about every entomological topic imaginable.
- Ephemeroptera Galactica: http://www.famu.org/mayfly/. "Everything you ever wanted to know about mayflies" is at this site.
- Insect Physiology Online: http://lamar.colostate.edu/~insects/. This site is "an interactive, referenced teaching resource" for insect physiology.
- Introduction to the Odonata: http://www.ucmp.berkeley.edu/arthropoda/uniramia/odonatoidea.html. This site is a fine starting place for resources on damselflies and dragonflies.
- NeuroWeb: http://entowww.tamu.edu/research/neuropterida/neuroweb.html. "The neuropterist's home page," with links and information about lacewings, ant lions, and the like, is at this site.
- Orthoptera Species File Online: http://viceroy.eeb.uconn.edu/Orthoptera. This site is "a taxonomic database of the world's orthopteroid insects."
- Crickets in the Classroom: http://members.attcanada.ca/~ecade/cricket_n_classroom.html. This is a good site for those who would like to use crickets to demonstrate various biological principles.
- Social Insects Web: http://research.amnh.org/entomology/social_insects/. A starting point for resources on social hymenopterans, this site has an emphasis on the ants.
- The Diptera Site: http://www.sel.barc.usda.gov/Diptera/. This site is a clearing house for studies of flies, from "young dipterists" to current research.
- The Tree of Life: http://ag.arizona.edu/tree/phylogeny.html. This is a superb site that not only provides information on as many types of living things as possible, but also places this information in an evolutionary context through an explicitly phylogenetic structure. Insects are particularly well covered here.
- UF Book of Insect Records: http://ufbir.ifas.ufl.edu/. This site is a fascinating compilation of subject matter, from "fastest flier" to "least specific sucker of vertebrate blood."

ELECTRONIC MAILING LISTS

Posting a question to a mailing list, or a listserve, is often the quickest way to answer an obscure question. Keep in mind, however, that experts on the list may resent the implication that they are being used to do your work for you. Therefore, before posting a question, make a serious attempt to find the answer in the standard sources first. You may have to subscribe to the list in order to post a question, or you can have a current subscriber post the question on your behalf. Your posting may also be archived, and as a result you may be quoted some time later on. Professionalism is therefore a wise path to follow. A good way to learn how to ask questions appropriately is to visit the Insect Question Page at http://www.ent.iastate.edu/mailinglist/bugnet/question.html.

OUTREACH PROJECTS

A number of outreach projects have entomological themes. A good example is Monarch Watch, a program of the University of Kansas that calls upon schools to help in the monitoring of monarch butterflies *(Danaus plexippus)* and that in return provides educational materials. Other examples of "citizen science" include the Canadian Nature Federation's Ladybug Survey, which may be explored at http://cnf.ca/beetle/guide.html.

Various institutions and organizations combine a Web presence with an opportunity for school visits. One prominent example in the United States is the Young Entomologists Society (http://members.aol.com/YESbugs/mainmenu.html), which offers programs for schools and groups, workshops for educators, a mobile exhibit including live insects, and an education center, located in Michigan. Many universities also have the equivalent of insect hotlines, or extension entomologists whose duties include answering agricultural and urban entomology questions.

BOOKS AND LIBRARIES

The Internet is a wonderful tool, but the library is still the rightful place for scholarly information in entomology. Books and journals are for the most part much more carefully written, reviewed, and designed than their Web-based counterparts, and therefore they are the guardians of mainstream entomological knowledge.

Here we will not cite particular faunal guides, but it is worth noting that there are interesting trends apparent in recently published works. For some groups, formerly obsolete treatments are being updated or replaced on a broad geographic scale. An example is *Dragonflies of North America* by J. G. Needham, Michael L. May, and Minter J. Westfall (2000, Scientific Publishers, Gainesville, FL). In other instances, local faunas are being treated separately and in greater detail, often for the first time (for example *The Butterflies of West Virginia and Their Caterpillars,* Thomas J. Allen, 1997, University of Pittsburgh Press). Still other books are the results of citizen-science survey projects, such as *The Millennium Atlas of Butterflies in Britain and Ireland* (James Asher, Martin Warren, Richard Fox, Paul Harding, Gail Jeffcoate, and Stephen Jeffcoate, 2001, Oxford University Press). One faunal treatment, *The Insects of Australia* (listed below) goes beyond geography and has become a classic of entomology in general.

For the study of most groups of insects, the traditional methods of collecting and preserving specimens are promoted, but for some, and especially for butterflies, many recent books advocate either a hands-off approach to insect study or catching and releasing the insects one finds. This trend seems directly related to geography—dense populations of more-or-less urbanized people, sensitive to conservation issues, are likely to support "nonconsumptive" entomology, despite a lack of evidence that this is a legitimate conservation concern, especially compared with the effects of habitat loss and introduced species.

The following is a list of 25 books, written in English, that are clearly entomological classics. Some are impressive summaries of one aspect of entomological science, and some are masterfully written explorations of particular topics. With them, an instructor should be able to quickly answer most entomological questions. A good entomology library will contain most of them, but the odds are that every entomologist has at least a few of these volumes close at hand.

- *Bugs in the System: Insects and Their Impact on Human Affairs.* M. R. Berenbaum, 1995, Addison–Wesley, Reading, MA.
- *Life on a Little Known Planet,* rev. ed. H. E. Evans, 1993, Lyons & Burford, New York.
- *Man Eating Bugs.* P. Menzel and F. D'Aluisio, 1998, Ten Speed Press, Berkeley, CA.
- *The Forgotten Pollinators.* S. L. Buchmann and G. P. Nabhan, 1996, Island Press, Covelo, CA.
- *The Hot-Blooded Insects: Strategies and Mechanisms of Thermoregulation.* B. Heinrich, 1993, Harvard University Press, Cambridge, MA.
- *The Insect Societies.* E. O. Wilson, 1971, Belknap Press of Harvard University Press, Cambridge, MA.
- *The Evolution of Insect Mating Systems.* R. Thornhill and J. Alcock, 1983, Harvard University Press. Cambridge, MA.
- *Principles of Insect Morphology.* R. E. Snodgrass, reprinted 1993, Cornell University Press, Ithaca, NY.
- *The Insects: Structure and Function,* 4th ed. R. F. Chapman, 1998, Cambridge University Press, Cambridge, UK.
- *Entomology,* 2nd ed. C. Gillott, 1995, Plenum, New York.
- *The Torre-Bueno Glossary of Entomology,* rev. ed. J. R. de la Torre-Bueno, 1989, New York Entomological Society, American Museum of Natural History, New York.
- *The Encyclopedia of Insects.* C. O'Toole, ed., 1995, Facts on File, Inc. (online).
- *Insects of Australia,* 2nd ed., Vols. I and II. CSIRO, 1991, Melbourne University Press.
- *An Introduction to the Study of Insects,* 6th ed. D. J. Borror, C. A. Triplehorn, and N. F. Johnson, 1992, Harcourt Brace, New York.
- *Aquatic Insects of North America,* 3rd ed. R. W. Merritt and K. W. Cummins, 1996, Kendall/Hunt, Dubuque, IA.
- *Immature Insects,* Vols. 1 and 2. F. W. Stehr, 1987 and 1991, Kendall/Hunt, Dubugue, IA.
- *Biology of Spiders.* R. F. Foelix, 1982, Harvard University Press, Cambridge, MA.
- *Dragonflies: Behavior and Ecology of Odonata.* Philip S. Corbet, 1999, Cornell University Press, Ithaca, NY.
- *The Biology of the Coleoptera.* R. A. Crowson, 1981, Academic Press, San Diego.
- *Hymenoptera of the World: An Identification Guide to Families.* Henri Goulet and John T. Huber, eds., 1993, Agriculture Canada Publication 1894/E, Ottawa, Ontario.
- *The Ants.* B. Hölldobler and E. O. Wilson, 1990, Belknap Press of Harvard University Press, Cambridge, MA.
- *The Bees of the World.* C. D. Michener, 2000, Johns Hopkins University Press, Baltimore.
- *Manual of Nearctic Diptera,* Vols. 1 and 2. J. F. McAlpine, ed., 1981 and 1987, Agriculture Canada Monographs 27 and 28, Ottawa, Ontario.
- *The Lepidoptera: Form, Function and Diversity.* Malcolm J. Scoble, 1995, Oxford University Press.
- *Illustrated Encyclopedia of the Butterfly World.* P. Smart, 1996, Salamander Books, London.

TELEVISION AND VIDEO

A number of excellent television programs, as well as regular series, have featured insects as their subject matter. Some of these are available on videotape or DVD. In general, video narration is not peer-reviewed and is notorious for inaccuracies. Nonetheless, the value of video in education is difficult to overestimate, and sometimes a game of "spot the mistakes" can be a learning experience in itself, especially when watching "B movies."

In terms of television series, *Insectia* is a 13-part, half-hour series hosted by Georges Brossard, the founder of the Montreal Insectarium, and it features many exotic locations and extralarge insects. *Alien Empire* is a 3-part series of 1-hour programs produced by the Public Broadcasting System in the United States. It is especially memorable for its use of miniature cameras and close-focusing wide-angle optics that are able to literally track alongside running and flying insects. *Acorn the Nature Nut* is a 91-part, half-hour series, with 26 episodes having entomological themes. It is hosted by Canadian entomologist John Acorn (coauthor of this article) and combines humor, music, and reliable entomological information.

A recent, well-received film on insects is *Microcosmos,* by Claude Nuridsany and Marie Pérennou. It features superb cinematography of the insects of southern France, and has a companion book by the same name (1997, Stewart, Tabori & Chang, New York).

SUPPLIERS

The following suppliers are keenly aware of the needs of entomology teaching. Although they deal with a mix of research customers, hobbyists, and teachers, their catalogs provide a good overview of the sorts of materials available for sale in an entomological context. Field equipment can enhance field-trip experience, whereas models, charts, and properly designed lab apparatus can greatly improve the quality of teaching in the laboratory. This list is incomplete, but the following suppliers have a long history and good reputation in the field.

- Bio Quip Products, 17803 La Salle Avenue, Gardena, CA 90248-3602. Phone: (310) 324-0620. Fax: (310) 324-7931. E-mail: bioquip@aol.com. Web site: http://www.bioquip.com. Bio Quip has been the main supplier of entomological equipment and books in North America for over 55 years. Their selection of educational items is very good.

- Watkins & Doncaster., P.O. Box 5, Cranbrook, Kent TN18 5Ez, England. Phone: +4401580 753133. Fax: +4401580 754054. E-mail: sales@watdon.com. Web site: http://www. watdon.com. The British equivalent of Bio Quip, with similar items in the catalog and a history dating back to 1874.

- Acorn Naturalists, 17821 East 17th Street, No. 103, Tustin, CA 92780-2138. Phone: (800) 422-8886 (Canada and U.S.A.), (714) 838-4888 (international). Fax: (800) 452-2802 (Canada and U.S.A.), (714) 838-5869 (international). E-mail: EMailAcorn@aol.com. Web site: http://www. acornnaturalists.com. Good selection of educational materials, especially for grade school teaching.

- Classey Books, Oxford House, Marlborough Street, Faringdon, Oxfordshire SN7 7JP, England. Phone: +44(0)1367 244700. Fax: +44(0)1367 244800. E-mail: info@classey-books.com. Web site: http://www.classeybooks.com. Since 1949, E. W. Classey Ltd. has carried an excellent selection of entomology books, both new and used.

- Carolina Biological Supply, 2700 York Road, Burlington, NC 27215-3398. Phone: (800) 334-5551 (U.S.A. only), (800) 387-2474 (Canada only), (336) 584-0381 (international), Fax: (800) 222-7112 (U.S.A. only), (800) 374-6714 (Canada only). E-mail: carolina@carolina.com. Web site: http://www.carolina.com. Long-established general source for science and technology teaching materials.

- Ward's Natural Science Establishment, 5100 West Henrietta Road, P.O. Box 92912, Rochester, NY 14692-9012. Canadian address: 397 Vansickle Road, St. Catherine's, Ontario L2S 3T5. Phone: (800) 635-8439 (U.S.A. only), (800) 387-7822 (Canada only), (716) 359-2502 (international). Fax: (800) 635-8439 (U.S.A. only), (905) 984-5952 (Canada only), (716) 334-6174 (international). Web site: http://www. wardsci.com. Another long-standing source for general science teaching materials.

See Also the Following Articles

Collection and Preservation • Entomological Societies • Insect Zoos • Movies, Insects in • Museums and Display Collections • Photography

Temperature, Effects on Development and Growth

František Sehnal, Oldřich Nedvěd, and Vladimír Košťál

Institute of Entomology, Academy of Sciences, Czech Republic

The body temperature of insects, as in other ectothermic organisms, is linked to changes in the ambient temperature. Temperature fluctuations are small in environments such as the tropical rainforest, caves, and some aquatic habitats. In most habitats, however, the seasonal and diurnal temperature oscillations are considerable. For example, insect body temperature can change abruptly by 10°C or more when exposure to direct sunlight is followed by the shade of a cloud. The way of life with fluctuating body temperatures is called heterothermy; this is in contrast to the homeothermy of endothermic organisms, such as birds and mammals, which regulate their body temperature, partly by generating endogenous heat.

DEVELOPMENTAL PARAMETERS

Every organism is adapted to a set temperature range. A general preference for high temperatures is referred to as thermophily, and the inclination to low temperatures is psychrophily. Insects living in warm climates or parasitizing warm-blooded animals are thermophilic, whereas those dwelling in soil are usually psychrophilic. Temperature preferences may change during development; for example, aquatic insects inhabit cold mountain streams during their immature stages but fly in warm air as adults. Temperature fluctuations within the species-specific physiological range determine the rate of development and often exert other physiological effects.

Development and reproduction occur at physiological temperatures that are delimited by an upper and a lower developmental threshold (UDT and LDT, respectively). Within this range, there is an optimal temperature for rapid development. The dependence on temperature can be expressed as a metabolic rate or as a developmental rate. The metabolic rate (MR) reflects the velocity of the energy-supplying biochemical processes and can be measured as oxygen consumption, carbon dioxide production, or heat generation. Many enzymatic reactions and the total body metabolism increase exponentially over a broader temperature range than is the span of physiological temperatures. Metabolic increase is usually two- to threefold with temperature elevation by 10°C and can be expressed by $MR = e^{a+kT}$, where a and k are constants and T is temperature.

Metabolic rate determines the developmental rate (DR), which is a reciprocal value of the developmental time (DT), $DR = 1/DT$. Measuring developmental times, such as duration of larval development, length of the reproductive period, or expanse of the entire life cycle, requires maintenance of defined conditions (notably temperature and nutrition). This is difficult to do for long periods of time and this is why DT and DR values are usually established for individual developmental stages and then recalculated for the entire life cycle. Circadian rhythmicity of some processes, for example, the synchrony of hatching or adult emergence at a certain time of day, complicates DT assessments.

In contrast to the exponential rise of the metabolic rate, the increase in DR within the physiological temperature range is linear. When a series of temperature and corresponding DR values is plotted in a graph, a straight line so obtained crosses the *x* axis at the theoretical LDT point. Close to this point, the relationship between developmental rate and temperature ceases to be linear, and the straight line is bent into a sigmoidal curve. Actual LDT is therefore somewhat lower then predicted. At the upper temperature range, the DR slows down before it reaches a maximum at the optimal temperature. After the maximum, DR sharply drops and at UDT the development is discontinued.

Developmental time depends on the effective temperature, i.e., temperature value above LDT (actual temperature *T* minus LDT). The constant product of effective temperature and developmental time is called the sum of effective temperatures (SET) and represents the heat required for the completion of a particular developmental stage. SET is conveniently expressed as the number of degree days. For example, a SET value of 100 degree days means that development at 5°C above LDT lasts 20 days and at 10°C above LDT 10 days. When the insects develop at fluctuating temperatures, the average temperature above LDT and the length of time when the temperature surpasses LDT are considered for each day, and the number of degree days established in this way is summed. Developmental stage is completed when the summation reaches the SET value.

A temporal drop in temperature below LDT is associated with developmental block and is counted as 0. Usually, there is no "negative developmental rate" at temperatures below LDT, i.e., no delay of development is observed after transfer to an effective temperature. Natural temperature fluctuations, however, may have a signaling effect and influence the SET value in some species, and this possibility must be checked experimentally.

The LDT and SET values are species-specific population characteristics. The LDT values are similar for all developmental stages of a given species, even when they develop in diverse seasons and experience disparate temperature fluctuations. The stability of LDT is manifested as developmental thermal isometry, i.e., the percentage of time spent in a particular stage at any constant physiological temperature is a stable fraction of the entire developmental time. The LDT and SET values established in the laboratory enable prediction of the course of development in the field. Control of many insect pests in agriculture and forestry largely relies on such predictions. For example, on the basis of the LDT and SET data for the codling moth (*Cydia pomonella*) we can predict, on the basis of daily temperature measurements in an orchard, the time of the first egg deposition and time insecticide sprays accordingly.

Both LDT and, especially, SET may vary between geographical populations because of adjustments to local climatic conditions. Insects that have spread to temperate zones from the tropical regions often maintain a high LDT and can reproduce and develop only in the hot season, spending most of the year in a state of dormancy. The survival in cold is made possible by increased cold hardiness, a parameter that seems to be more plastic than LDT.

DEVELOPMENTAL ARRESTS

Most insects must overcome long periods of adverse conditions when food is wanting and temperature remains outside the physiological limits. Insects cease development and reproduction but, if the temperature does not reach lethal extremes, they remain capable of resuming these processes as soon as the conditions become favorable. The state of easily reversible, directly temperature-dependent developmental arrest is known as quiescence. It is typical of insects adapted to relatively short periods of unfavorable, nonlethal circumstances, and it is usually associated with temperature acclimation.

Insects that will be exposed to severely hostile conditions that can last for many months enter a programmed developmental arrest, diapause. Diapause occurs in anticipation of a season in which the insect could not survive in the active state. It is induced by environmental signals acting before the adverse conditions set it, sometimes on a much earlier developmental stage and exceptionally on the parental generation.

Seasonal changes in the environment are specific for each latitude, altitude, and habitat, but always correlate with changes in the photoperiod, i.e., the length of day versus the length of night. Photoperiodic changes therefore provide ideal signal for the advent of unsuitable conditions. However, temperature can shift the diapause-inducing photoperiod response over a broad range. For example, 50% of caterpillars of *Acronycta rumicis* are induced to enter pupal diapause at a daylength of 19 h at 15°C, but at 16 h at 25°C. Low temperature normally enhances the effect of short photoperiod and high temperature enhances the effect of long photoperiod. Daily fluctuations of temperature, the thermoperiod, can induce diapause in a few species kept in constant darkness. On the other hand, high temperature can abolish the diapause-inducing effect of a short photoperiod. For most insects, the diapause is facultative and there is a variation in the critical photoperiod/thermoperiod at which each individual enters diapause.

Once induced, diapause is not terminated immediately after the diapause-inducing conditions disappear. A certain time must elapse, during which neurohormonal regulations return to the pattern supporting development and reproduction. The mechanisms controlling diapause termination are not known. The length of time in diapause depends on its "depth" and on environmental conditions, especially temperature, to which the diapausing insects are exposed. In the overwintering insects, diapause is often shortest at temperatures around 5°C (Fig. 1) and the photoperiod is irrelevant. Due to exposure to low temperatures in late fall, the overwintering insects terminate diapause in early winter, and the resumption of their development or reproduction is then halted only by a direct effect of low temperature: diapause turns into quiescence.

ALTERNATIVE DEVELOPMENTAL PATHWAYS

In certain insects, temperatures within the physiological range affect the course of non-diapause development. For example,

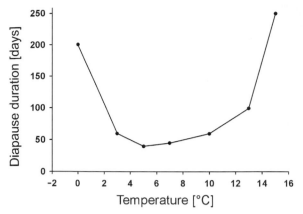

FIGURE 1 The influence of constant temperature on the termination of imaginal diapause in the chrysomelid beetle *Atrachya menetriesi*. The diapause is completed (50% of individuals can resume development) most rapidly at 5°C. Based on data of Ando (1983).

larvae of the yellow mealworm, *Tenebrio molitor,* develop in 11–15 instars at 25°C and in 15–23 instars at 30°C. Less dramatic changes in the number of molts and the growth rate were noted in a number of insects. Abnormal temperatures can also dissociate the onset of metamorphosis from the body size at which it normally occurs. In the wax moth, *Galleria mellonella,* placing newly ecdysed larvae of the last or penultimate instar on melting ice induces an additional larval molt. To cause such anomalies, the temperature must alter the secretion of hormones that control specific developmental events.

Some insect species occur in more than one form and their alternation depends on temperature. For instance, development of young caterpillars of *Colias eurytheme* at 18°C leads to mainly yellow, and development at 27–32°C to orange, butterflies. The spring and summer forms of some other butterflies are well known examples of seasonal dimorphism in which one form is linked to diapause.

LETHALITY AT EXTREME TEMPERATURES

A general response of insects to temperatures just below their LDT or above their UDT is the cessation of development and reproduction while the insects remain active and feed. The larvae may slowly grow and the adults accumulate reserves and to some extent undergo gonadal maturation. These processes are terminated at more extreme temperatures when the insects begin to die.

During cooling, the metabolic rate and motility gradually decrease. At a certain temperature, the neural and muscular activities are impaired and the insect lapses into cold stupor. The metabolic rate of such immobile insects continues to decline with decreasing temperature. The stupor point is as high as 12°C in some tropical insects and the honey bees, around 5°C in many temperate species, near 0°C in most overwintering insects, and below the freezing point in species living in very cold areas.

The nature of chill injuries is little understood. Desiccation and nutrient depletion during the cold-induced starvation are certainly incompatible with long-term survival, but death usually occurs earlier and is probably the result of damaging effects at cellular level. The loss of cell membrane fluidity, imperfect protein functions (enzymatic activities, transport, signaling, etc.), and the resulting asynchrony of the life-supporting processes cause metabolic disorders. For example, the ion pumps in the cell membrane become inefficient and sodium concentration in cytoplasm increases, while the potassium ions flow out into the hemolymph.

The upper temperature extremes are also lethal. Gradual warming past UDT, which is for many species around 35°C but is never sharply delimited, increases the metabolic rate, loss of water, and motility. At a certain temperatures, usually around 40°C, the water loss, and thereby the evaporative body cooling, increases sharply. The spiracles are wide open and the melting of cuticular lipids permits evaporation through the body surface. After some time at such a high temperature, the losses of water and nutrients lead to exhaustion, manifested as a rapid decrease of motility and a drop of transpiration. If this state is brief, it can be reversed. The temperature at which it occurs is the upper lethal threshold. A gradual temperature increase to this threshold may cause heat stupor.

Survival at temperatures above the threshold is a function of temperature and length of exposure. Warming to the absolute upper lethal temperature, which is usually around 50–55°C, causes irreversible tissue damage, and even a short exposure is lethal.

ACCLIMATION AND TEMPERATURE TOLERANCE

The survival at extreme temperatures is improved after an acclimation. Shortening of the photoperiod in late autumn and early winter usually acts synergistically with descending temperature in triggering a seasonal cold acclimation. An exposure to low temperature alone is often insufficient for full cold acclimation and successful winter survival because some physiological adjustments (for instance down-regulation of the ice nucleators and enhancement of cryoprotectant biosynthesis) require a preceding switch to the diapause developmental mode and this is controlled by the photoperiod. In other insects, cold acclimation is associated with temperature-dependent quiescence and the photoperiod is irrelevant.

Cold acclimation is a complex adjustment involving profound changes at the organismic and tissue levels. Accumulation of low-molecular-weight cryoprotective polyols and, in some insects, also synthesis of antifreeze proteins are characteristic features of cold acclimation. Other physiological changes include changes in cuticular lipids, increased fluidity of phospholipids in the cell membranes, conformation changes of some proteins, possibly production of alternative enzymes with activity optima at lower temperatures, and synthesis of heat-shock proteins. Acclimation may also encompass changes in morphology (e.g., cold-acclimated lacewings turn from green to reddish brown) and behavior (e.g., formation of cocoons with higher resistance to desiccation and ice penetration, voiding the gut to get rid of ice nucleators, and seeking dry places to prevent ice inoculation from the surroundings).

The temperature at which insects freeze is relatively low even without cold acclimation. Insects contain only low amounts of nucleators needed for the ice crystal formation and are therefore capable of considerable supercooling. The nonacclimated individuals freeze at −10 to −15°C, while those with accumulated cryoprotectants and antifreeze proteins can be supercooled to temperatures below −20°C. All cryoprotective compounds disappear from the organism after a certain time. The length of this deacclimation process also depends on temperature, but details have not been examined.

Changes associated with heat acclimation are little known. Synthesis of heat-shock proteins, which were discovered in *Drosophila* exposed to elevated temperature, is a very general response not only to heat but also to cold and various other types of stress. Several types of heat-shock proteins are known from organisms ranging from bacteria to plants. It is believed that they are chaperones enabling or protecting functional protein conformations. Temperature or another stressing factor acts as a signal inducing their synthesis at transcriptional level. Accumulation in the cells of "unfolded" proteins is believed to be the common intracellular message triggering this transcription.

See Also the Following Articles

Aestivation • Cold/Heat Protection • Diapause • Fat Body • Greenhouse Gases, Global Warming, and Insects • Growth, Individual • Homeostasis • Thermoregulation

Further Reading

Ando, Y. (1983). Diapause and geographic variation in a leaf beetle. *In* "Diapause and Life Cycle Strategies in Insects" (V. K. Brown and I. Hodek, eds.), Vol. 23 of Series Entomologica, pp. 127–141. Junk, The Hague.

Beck, D. S. (1983). Insect thermoperiodism. *Annu. Rev. Entomol.* **28,** 91–108.

Browse, J., and Xin, Z. (2001). Temperature sensing and cold acclimation. *Curr. Opin. Plant Biol.* **4,** 241–246.

Danilevsky, A. S. (1965). "Photoperiodism and Seasonal Development of Insects." Oliver & Boyd, London.

Danks, H. V. (1987). Insect dormancy: An ecological perspective. *Biol. Surv. Can. Monogr. Ser. 1,* Ottawa.

Lee, R. E., Jr., and Denlinger, D. L. (1991). "Insects at Low Temperature." Chapman & Hall, New York/London.

Leather, S. R., Walters, K. F. A., and Bale, J. S. (1993). "The Ecology of Insect Overwintering." Cambridge University Press, Cambridge, UK.

Lindquist, S., and Craig, E. A. (1988). The heat shock proteins. *Annu. Rev. Genet.* **22,** 631–677.

Lozina-Lozinskii, L. K. (1974). "Studies in Cryobiology." Wiley, New York.

Morimoto, R. I., Tissieres, A., and Georgopoulos, C. (eds.) (1994). "The Biology of Heat Shock Proteins and Molecular Chaperones." Cold Spring Harbor Laboratory Press, Cold Sprimg Harbor, NY.

Smith, A. U. (1961). "Biological Effects of Freezing and Supercooling." Arnold, London.

Vannier, G. (1994). The thermobiological limits of some freezing-intolerant insects: The supercooling and the thermostupor points. *Acta Oecol.* **15,** 31–42.

Termite

see *Isoptera*

Terrestrial Insects

see *Soil Habitats*

Thermoregulation

Bernd Heinrich
University of Vermont

In insects, as in other animals, body temperature strongly affects the rate of energy expenditure, the rate at which food can be located and harvested, growth, the facility with which mates can be acquired and predators avoided, and sometimes also the susceptibility to disease organisms. Thermoregulation refers to the ability to regulate that body temperature which best serves survival and reproduction, and it encompasses numerous conflicting constraints and selective pressures. In insects, major considerations involve body mass and access to either external or internal heat. Thermoregulation operates through behavior, physiology, and morphology. For the most part, insects are too small to be able to appreciably elevate, or regulate, their body temperature by internal heat production, although some are large enough and that, coupled with their high flight metabolism, could easily cause them to overheat. In numerous insects, elaborate mechanisms of thermoregulation have evolved both for heating and for cooling the body that possibly rival those of the typically endothermic vertebrates.

ENDOTHERMY IN FLYING INSECTS

Insects arose on earth at least 350 mya in the Devonian Period of the Paleozoic Era. Little is known about the earliest forms, except that originally they must have been crawlers, not flyers, and their bodies assumed approximately the temperature of the immediate surroundings to which they adapted. This holds true even when the immediate surroundings are quite frigid. The adult form of a flightless midge (*Diamesa* sp.) walks on glacier ice even when its body temperature is chilled to −16°C. It is so sensitive to heat that, when taken from its natural environment and held in one's hand, it is killed by the warmth of one's skin. However, there are insects that maintain quite specific and high body temperatures. Some species of sphinx moths, for example, have thick insulating fur and normally maintain a thoracic temperature near 46°C during flight over a wide range of ambient temperatures. To these moths, our own normal body temperature of 37°C is almost cool. An insect's head and abdominal temperatures are for the most part unregulated.

In the same way that the motor heats up when a car burns fuel, heat is released as an inevitable by-product of cellular metabolism whenever muscle contracts. Close to 94% of the energy expended by muscles during contraction is degraded to heat, while approximately 6% appears as mechanical force on the wings. Insect flight is one of the most energetically demanding activities known, and thus most insects produce more heat per unit muscle mass when they fly than almost any organism on earth. Most insects exist under conditions somewhere in between the cold-blooded crawler and the hot-blooded flyer, but these extremes show us what is possible, and they thus offer us a remarkable window into thermal adaptation from an evolutionary perspective.

The ability of birds and mammals to regulate body temperature at one set point, specifically 37–41°C, has long been considered proof of sophistication and phylogenetic advancement relative to animals whose body temperature varies with that of their environment. Deviation of body temperature from the set point of 37–41°C is, in birds and mammals, often associated with illness and was once thought to be caused by a failure of the thermoregulatory system. We now know that both increases and decreases in body temperature can be and often are adaptive responses. Both responses are often sophisticated physiological mechanisms that involve more thermoregulation rather than less, albeit the body is kept at a more appropriate temperature for specific conditions.

Small insects have much lower body temperatures in flight than large insects, not because they produce less heat—in fact, they may have *higher* rates of heat production than larger insects, but because they have much greater conductance because of their large relative surface area. In bees, for example, only the large species heat up in flight and generate an appreciable elevation of body temperature even though metabolic cost of flight per unit weight declines approximately 230% for a 10-fold increase in mass. A mosquito in flight maintains only a tiny (<1°C) gradient between thoracic and ambient temperature, despite prodigious amounts of heat production. A blow fly *(Calliphora vicia)* may heat up 5°C, and a honey bee heats up its thorax about 15°C. Having a much larger thorax, and hence a smaller relative surface area, means that the internally generated heat during flight is not lost by convection at the same rate that it is produced until a much higher temperature gradient has been generated.

Large insects—those that inevitably generate a high body temperature during continuous flight—must be biochemically adapted to operate their flight muscles at the high temperatures experienced. Temperature is important for mechanical efficiency; at low muscle temperature there is partial overlap in contractions of the up- and downstroke muscles, the

dorsoventral and dorsal longitudinal muscles; the two sets of muscles then work against each other rather than working to move the wings.

Numerous other moths—such as most microlepidoptera and geometrids (inchworms) and some ctenuchids and arctiids—are small or weak flyers that do not heat up but fly at low air temperature. They fly at muscle temperatures much lower even than those at which the large-bodied, small-winged (and hot-blooded) sphinx moths generate zero power. Evolution has acted strongly to tailor the flight motor's capacity for maximum power output for much lower ranges of operating temperatures. For example, the geometrid *Operophtera bruceata* can gain sufficient power to fly at 0°C. (Nevertheless, its capacity to do so is only partially the result of muscle physiology.)

The basis for the evolution of differences between species arising from a common ancestor is variation among individuals. Variation was present in the past, and for many traits, variation is maintained even now. For example, in *Colias* (sulphur) butterflies, the gene locus for phosphoglucose isomerase, one of the enzymes involved in energy metabolism in these butterflies, changes in allele frequency with season and habitat temperature. This suggests that natural selection is occurring even over very short (that is, seasonal) time spans. The different enzyme alleles have different thermal stabilities, and heterozygotes are thought to have an advantage in an environment of rapidly fluctuating temperatures inasmuch as the individuals heterozygotic for this locus fly over a range of temperatures broader than that of individuals of other genotypes.

When a seasonally changing temperature environment, which is the rule, can select for heterozygosity, then one might expect that an environment of constantly high or low temperature, which is the exception, should lead to the fixation of an appropriate genotype. Hence, selection in terms of gene-frequency changes would not normally be present for our inspection in more constant environments, in which appropriate genotypes would already have been selected long ago to adapt to the average temperature. The specific thoracic temperature that is maintained by regulation is "chosen" by evolution probably because it is the temperature most readily regulated for maximum activity over a range of prevailing environmental conditions.

WARM-UP BY SHIVERING

During his classic studies of honey bee communication, Karl von Frisch noted that bees often interrupted flight for a few minutes when they were returning to the hive heavily laden with nectar. He presumed they stopped "to rest," but we now know they were stopping to work: to raise their thoracic temperature. They most likely stopped flight because it was a cold day and they had cooled convectively. Bees are able to raise their thoracic temperature by shivering, which can work their flight muscles harder than flight itself does. Von Frisch could not have known any of this, because shivering and thermoregulation by individual insects was unknown in the

1960s, nor is shivering externally visible in bees even if one looks very closely.

Like the maintenance of an elevated body temperature by internal heat production in flight, physiological warm-up is found in all large, active flyers among the dragonflies (Odonata), moths and butterflies (Lepidoptera), katydids (Orthoptera), cicadas (Clypeorrhyncha or Homoptera), flies (Diptera), beetles (Coleoptera), and wasps and bees (Hymenoptera). That is, it is found from some of the earliest forms, the Odonata, to the most evolutionarily highly derived, the Diptera, Coleoptera, and Hymenoptera. It is not found in the small and therefore nonendothermic members of the same groups. Because no insects shiver except those that then also heat up from flight metabolism, it seems reasonable to conclude that the evolution of shivering behaviors is related to the evolution of flight but is unrelated to the insect's place on the phylogenic tree.

During preflight warm-up, there are synchronous contractions of groups of muscles that normally contract alternately in flight. That is, the main wing-depressor muscles, the dorsal longitudinal muscles, are excited simultaneously—in other words, in synchrony—with the dorsoventral wing elevator muscles. The neural activation pattern of thoracic flight muscles needs to be and is already very labile for flight control, and to add shivering when flight behavior has already evolved is probably a very minor evolutionary step. Physiological warm-up in its most basic form is like the idling of an engine; the engine "evolved" to propel the car, not to warm it up. Once present, the heat-producing flight muscle system required only a slight modification of neuronal activation patterns and, in the more sophisticated models, also the addition of the biological equivalent of a clutch—a mechanism to disengage the wings in the same way that an automotive clutch disengages the car's wheels. Some insects, such as dragonflies and moths, do have visible external wing vibrations that were originally called "wing whirring," these were once thought to pump air into the animal before the true function was elucidated.

The zenith of the shivering response of any hot-blooded animal (vertebrate as well as invertebrate) belongs to some bees. Honey bees and bumble bees have a physiological sophistication either not existing or not yet observed in other insects, and they exploit shivering behavior to an unprecedented extent and in a variety of ways. Like flies and beetles, bees are "myogenic" flyers in which the wing-beat cycle runs in part on automatic; as the downstroke muscles contract they stretch the upstroke muscles. This stretching *by itself* causes the upstroke muscles to contract. The downstroke of the wing therefore automatically causes the upstroke muscles to contract and vice versa, in a repeating cycle that is sparked by neural commands that are at a much lower frequency than the wing beats and are no longer specific to a single wing beat. (This system permits some of the smallest insects, such as midges, to achieve the unprecedented coordination required for wing-stroke cycles of over 1000 beats per second.) But the stretching of opposing muscle groups that maintains the myogenic contraction cycle can occur only if the wings are actually beating—namely,

during flight. When the wings are not in use, as when they are folded back dorsally and the clutch-like wing hinge is engaged, then the muscle groups are in near tetanus. That is the reason for the bees' shivering.

Very vigorous shivering in bees is physically dampened even more by yet another mechanism. One of the two sets of opposite-acting muscles is activated (and hence contracted) slightly more than the other. Because the opposing muscles act like weights forcing down each side of a seesaw, the added force on one set of muscles prevents the "seesaw" from working (and the wings from "vibrating" back and forth).

WARM-UP BY BASKING

Like warm-up by shivering, warm-up by basking occurs in all major orders of insects that have fast flyers large enough to heat up from their flight metabolism. In its simplest form, behavioral warm-up is merely heat-seeking. A basking insect usually takes specific postures that simultaneously maximize solar input and minimize convective heat loss. Heat input is maximized by exposing the maximum surface area to the sun, while convective heat loss is minimized by using body parts (such as the spread wings) as baffles to retard air movement around the body. Orienting the body parallel to the air stream (as a wingless insect might do) would reduce the effect of convective cooling, but orienting the body perpendicular to the sun's rays to facilitate heating should take precedence, because no heat loss can be minimized until heat is first gained. Grasshoppers, beetles, and flies use these basking methods. For some dragonflies and butterflies the wings are especially important during warm-up in their role of reducing convective cooling.

Behaviorally distinct types of basking have been described in butterflies, although some (tropical) butterflies do not bask at all. In one type, called "lateral basking," the butterfly closes its wings dorsally and then tilts to present either the right or the left wing and body surface to the sun. The lower portions of the wings wrap around the body and touch it, and warming the lower wing portions in sunshine then causes heat to be conducted directly through them and into the body.

Many species of small-bodied butterflies, primarily pierids and some lycaenids (which commonly fly in breezy mountain meadows) bask by opening their wings partially in a V so that the body is directly available to the sun's rays at the bottom of the V. The wings then serve as convection baffles to reduce cooling in the breezy environment. "Dorsal baskers" hug a solid substrate, such as the ground, and pull their wings down around them. They thereby expose the dorsal body surface to the sun while simultaneously capturing heat from the sun-heated substrate.

HEAT LOSS MECHANISMS

Insects near the size of a honey bee (approximately 200 mg) or larger may experience body temperatures during forced flight exercise that are potentially lethal to them. Alternately, metabolic heating may inhibit continuous flight at relatively modest ambient thermal conditions (of air temperature and solar radiation), unless one or more of the following mechanisms for heat loss are activated.

Harnessing Convection

The rate of convective heat loss from a body is determined by the conductance of the body, i.e., its intrinsic rate of heat loss (which is a function of body size, shape, and insulation). Conductance, in turn, is a function of the wind speed (a hot body cools more quickly in wind than in still air—meteorologists call this the "wind-chill factor"). However, no convective heat loss is possible, regardless of conductance, if body and ambient temperature are equal, and at any one conductance and wind speed the amount of heat loss is directly proportional to the temperature difference between the body and the ambient surroundings.

Small insects that are endothermic in flight are sufficiently air-cooled such that they almost never reach the potentially dangerous high-temperature ceiling of near 45°C that is common to most animal tissues at normal atmospheric pressures. These small insects thus have no need of a specialized cooling system: they lose sufficient heat passively. Theoretically, larger insects could cool themselves by increasing flight speed and thus increasing convective heat loss, but flying faster would generally increase metabolic heat production, which would cancel out the increased heat loss, unless the internally generated heat is redistributed.

Heat Radiators

A radiator is a device that increases the surface area of a body or object so that more heat can be transmitted to the surrounding environment by convection. In some radiators, a fluid with a high heat capacity (like water, blood, or other liquid) circulates by means of a pump and transfers heat from its source to the radiator site for the heat loss. That is how a car engine is cooled.

Radiators are utilized by many large insects from the very diverse orders Lepidoptera, Odonata, Dipera, and Hymenoptera. The animals have a fluid-transfer cooling mechanism that dissipates heat through an abdominal radiator, while small members of the same groups that are not strong or continuous flyers lack the heat-transfer response. When we humans exercise in the heat, blood is pumped to the skin or extremities to facilitate heat loss, but this is done at the expense of pumping blood and oxygen to the muscles instead. Therefore, work capacity is compromised. Insects, on the other hand, do not need to compromise aerobic work capacity at higher air temperatures because of thermoregulation. In insects, the total separation of respiratory and heat-transfer functions makes it possible for them to continue working, even when the fluid flow is interrupted.

The "radiator tube" that conducts the hemolymph to the abdominal heat radiator in insects also serves as a pump, which operates by peristaltic contractions along its entire length. Although sphinx moths in whom surgery has rendered this "heart" inoperative (by tying it shut) can still fly until reaching near-lethal thoracic temperatures, removal of their insulating layer of thoracic scales makes continuous flight again possible.

Evaporative Cooling

One of the extraordinary examples of an evaporative cooling mechanism specifically for thermoregulation is that found in the workers of honey bees, *Apis mellifera,* and yellowjackets, *Vespula* spp. These insects use the head as a radiator, but they do so with a difference. Nectar-gathering honey bees normally fly with flight-motor temperatures near 15°C above air temperature. They are capable of the astounding feat of flying even at ambient temperatures near 45°C while maintaining the thorax at the same or only slightly lower temperature. They do so by regurgitating nectar from the honeycrop, and while the nectar is held on the mouthparts and the head, water evaporates from it. Because of the physical contact between the head and the thorax, thermoregulation of one effectively results in thermoregulation of the other. Thus, the head is cooled by evaporation of water until there is a large temperature difference between the head and the metabolically heated thorax, at which point heat from the thorax follows the temperature gradient and is transmitted to the head. Head temperature is actively regulated, with thoracic temperature passively following, because artificial heating of the thorax alone does not result in the heat-dissipation response so long as head temperature remains low. However, artificial heating of the head (as with a narrow beam of light from a heat lamp) almost immediately results in nectar regurgitation and evaporative cooling, even while thoracic temperature is still (momentarily) low.

Some insects cool evaporatively from the back. In the hot Australian deserts, the larvae of the sawfly *Perga dorsalis,* in response to solar heat stress, first raise their abdomen to the sun to shade the body and to increase convective heat loss. In an emergency, when this response is insufficient, they also emit rectal fluid and spread it over their ventral surface to cool themselves evaporatively.

Diceroprocta apache of the Sonoran desert of the southwestern United States employs a third evaporative cooling mechanism, this one analogous to sweating. These cicadas are plant-sap feeders, and despite living in a dry environment, they have access to a large fluid supply by inserting their sucking mouthparts into the xylem of deep-rooted shrubs, such as mesquite. They thus indirectly tap water from deep underground stores. Cicadas sing when ambient temperatures in the shade reach 40°C, and the repetitive contractions of their tymbal muscles result in internal heat production that adds to the already considerable external heat load.

Body temperature during this exercise in the heat is reduced to tolerable levels by evaporative cooling from fluid shed through large pores distributed over their dorsal body surfaces. The release of this fluid, and the consequent evaporative cooling, occurs only in response to very high body temperature. Most insects, especially those of desert environments, are instead highly resistant to water loss when alive, and upon death, there results an immediate *increase* in water loss as the spiracles are no longer actively maintained shut. Killing of the cicada, in contrast, immediately stops the sweating response, therefore showing that it is under metabolic control. The cooling response is mediated, ironically enough, by aspirin-like substances produced in their bodies in response to heat stress.

MORPHOLOGY AND THERMOREGULATION

Aside from physiology, various aspects of insects' morphology come into play in their thermoregulating responses.

Insulation

Many hot-blooded insects that regulate their body temperature have, like their endothermic vertebrate counterparts, bodies wholly or at least partially covered with insulation. One type of insulation is derived from air sacs. Insects already have air sacs used for breathing, and still air is, next to a vacuum, the best possible insulator. Many insects of various orders have air sacs between the thorax and the abdomen that greatly retard the leakage of heat into the abdomen. But large-bodied dragonflies have gone one step further: their air sacs surround the thoracic flight motor. The other two types of insulation are derived from exterior cuticular structures.

Lepidopterans are covered with a layer of thin overlapping scales, which are especially noticeable in coloring the wings. Rather than remaining flat and colorful for visual signaling as in butterflies, they have become long and thin to form a thick insulating body (thoracic) pile or fur coat in many moths. This coating of pile is so effective as insulation it more than halves the rate of heat loss, or doubles the temperature excess, hence permitting flight at much lower air temperatures. Endothermic insects with pile now fly in many northern areas and at times of the year at which they would otherwise be excluded. Conversely, insects from tropical environments have no or only sparse pile covering.

A covering of setae, small hair-like projections from the cuticle, is the third source of insect insulation. Setae have various functions and numerous independent evolutionary origins. Within the Hymenoptera, only the northernmost large bees, the bumble bees, have a heavily insulated flight motor. However, even honey bees have a layer of short insulating pile on the thorax, which aids them on cool mornings and at high elevations. Bees inhabiting the tropics and hot deserts do not have a covering of pile dense enough to provide appreciable insulation. But even tropical bees cannot get along totally without setae, because they use these projections to trap pollen from flowers. Some wasps, in contrast, live in the same northern areas that bumble bees do,

but they do not rely on pollen for protein. Instead, many are predators on fast-flying insects, and they are glabrous. Perhaps an advantage of fuel economy in flight, or perhaps the necessity for fast flight, has assumed more importance than drag-inducing insulation for thermoregulatory control.

Color

In a few instances, an insect's color has a functional significance in thermal balance during basking. For example, lateral-basking sulphur butterflies (*Colias* spp., usually yellow or white) found in cool environments (such as mountaintops or high latitudes) or seasons (early spring) tend to have dark wing undersides. These darker individuals are able to heat up the thorax slightly faster than lighter congeners, which buys them additional flight time when basking is needed to prepare for flight.

Although color can have a slight thermal advantage, it is more often subservient to other needs, such as the need to evade predators. Not surprisingly, insects that inhabit open ground often match their background in color and thus are highly camouflaged.

Stilts and Parasols

A beetle walking on hot desert sand might experience temperatures that could kill it in a minute or less. Just a few millimeters above the ground, however, the hot, ground-hugging air layer is disrupted and mixed with cooler air from above. If we were shrunk to Lilliputian size and forced to live where a few millimeters' difference in elevation could mean the difference between life and death, we would find some way to lift ourselves above the searing heat. Numerous ground-dwelling beetles living on hot sands do just that. Tiger beetles (Cicindelidae) begin to stand tall when sand temperatures exceed 40°C. Aside from extending their jointed legs to stand taller, some beetles, like *Stenocara phalangium* (Tenebrionidae) from the Namib Desert of southern Africa, have evolved very long stilt-like legs that allow them to avoid overheating by both avoiding the heat at ground level and losing some of the solar heat by convection through fast running.

Another option is to use one body part to shade another. For example, some beetles reduce their absorption of external heat from direct solar radiation by having an air space beneath the elytra that insulates the abdomen from direct solar radiation.

Countercurrent and Alternating-Current Heat Exchanges

At low air temperatures when the abdomen is cool, the flight motor could cool precipitously if the hemolymph carried heat away from the thorax to be dissipated from the abdomen. Two mechanisms, however, normally prevent this potential problem of thoracic cooling. The first is a temporary reduction or elimination of the circulation: to retard heat loss.

Another mechanism, one more subtle than cardiac arrest, helps some insects prevent heat leakage from thorax to abdomen. Proof of that is seen in honey bee workers and Cuculliinae winter moths, which never show appreciable increases in abdominal temperature, even as the flight motor stays hot. An examination of their circulatory anatomy explains the mystery: they harness countercurrent heat transfer.

A countercurrent implies two separate currents flowing next to each other but in opposite directions, as through the petiole between thorax and abdomen in insects. If the fluid in one current is of a higher temperature than that of the other, then heat (which is not confined by the vessel walls) will passively flow "downhill," from high to low temperature, across these walls. Thus, if the hot blood leaving the thorax flows around the vessel in close proximity to cool blood entering it from the abdomen, as in most bees, then heat exchange is inevitable. At least some of the heat from the thorax will be recycled back into the thorax because the incoming blood is heated by the outgoing blood. In honey bees and winter moths, countercurrent heat exchange is greatly enhanced by prolonging the area for that potential heat exchange to occur, as the aorta in the petiole is lengthened and (in honey bees) convoluted into loops.

Bumble bees and northern vespine wasps have a very much different and seemingly less efficient countercurrent heat exchange circulatory anatomy than honey bees and winter moths (Fig. 1). This situation may seem counterintuitive because they live in cold climates, some species even inhabiting the High Arctic. They might thus be expected to have even better countercurrent heat exchangers than honey bees, which are of temperate and tropical origin. Instead of having more loops for countercurrent heat exchange, they have none! Nevertheless, their anatomy can also be understood in terms of thermal strategy, but as it relates to their social system.

Bumble bee and wasp queens start their colonies very early in the spring; each new queen attempts this task alone, as an individual. To this individual bee or wasp, time is of the essence, for she must complete the whole colony cycle within a single growing season. A queen's first priority, then, is to rear a group of helpers. Temperatures when and where she builds her nest may be near 0°C, and if the brood were left at that temperature it might take years for them to develop to adults—provided they could withstand the freezing temperature. Even in the High Arctic, however, the queens of *Bombus polaris* can produce a batch of workers in about 2 weeks, as can other bumble bees and *Vespula* wasps. Both bees and wasps accomplish these feats by incubating the brood from the egg to the pupal stage. The queens perch upon their brood clump—consisting of eggs, larvae, and/or pupae—and they press their abdomen upon the brood, much as a hen incubates her eggs with her belly. Only the abdomen provides a smooth surface for contact, but only the thorax produces heat by way of intense shivering by the flight muscles. No incubation, and hence social life, would be possible for these insects in a cold environment if, like honey bees, they were incapable of transferring heat from the source of its production into the abdomen that provides the smooth tight contact with the brood.

The bumble bee's aorta is long enough to permit moderate heat exchange and hence retention of heat in the thorax, but it is short and straight enough so that a physiological mechanism can be activated that shunts the fluid and heat through, effectively eliminating countercurrent heat exchange.

Countercurrent heat exchangers in vertebrate animals can be bypassed by rerouting the blood into an alternate (generally external) channel. That is why our own veins seem to pop out when we are active in the heat. Such rerouting of the blood from internal to external channels is not possible, however, in insects with open circulatory systems lacking veins and capillaries. Instead, in bumble bees there is a physiological solution for heat loss in the presence of a countercurrent heat exchange anatomy that serves the same purpose as an alternate blood channel. In the bumble bee, this consists of an *alternating-current* flow of blood. To shunt heat past the heat exchanger and into the abdomen, the bee lifts a small valve that allows a pulse of warm blood to enter the abdomen, and in the fraction of a second after the warm blood enters the abdomen, she then squirts a bolus of cool blood into the thorax. And so it goes back and forth, hot and cold pulses of hemolymph passing alternately through the heat exchange area in the bee's waist. The essential point is that although the blood is not rerouted into a different channel, it is instead temporarily "chopped" into alternating pulses in the same channels. This is the opposite of countercurrent flow because instead of recovering heat from the thorax, the system acts to remove it, in this case into the abdomen.

The pumping of hemolymph by the heart and the ventral diaphragm is also aided by in–out pumping movements of the

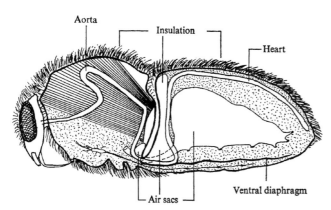

FIGURE 1 Anatomy of a bumble bee *(Bombus)* relevant to thermoregulation. The thorax is insulated with pile that reduces the rate of convective heat loss. The ventor of the abdomen is lightly insulated or uninsulated when the bee presses her abdomen onto brood to be heated. Hemolymph (blood) is pumped anteriorly by the heart, from the abdomen into the thorax. When dissipating heat from the working muscles in the thorax, the blood enters the aorta from the heart in pulses. Each pulse of cool blood from the abdominal heart into the thoracic aorta alternates with a pulse of warm blood entering the abdomen to the thermal window. In this way, countercurrent heat flow (into blood returning to the thorax) is minimized and heat flow (into abdomen) is maximized. [Reproduced, by permission of Oxford University Press, from B. Heinrich, (1976), *J. Exp. Biol.* **64**, 561–585.]

whole abdomen, which otherwise function only for moving gas in and out of the thorax; the in–out telescoping movements of the abdomen are synchronous with the heart beats and the ventral diaphragm beats, and they cause pressure changes that facilitate hemolymph flow in precise alternating currents.

THERMAL ARMS RACES

Against Predators

On a summer day in the Sahara Desert in Algeria, as the sun rises and begins to heat the sands that have been cold at night, an abundance of insect life is forced to retreat to cool underground refuge. Those unfortunate ones caught out in the heat become disoriented and, moving frantically, they heat up even more, then they die. The sun keeps rising, and sand temperatures begin to exceed 46°C. The desert lizards, *Accanthodactylus dumerili,* continue to hunt the incapacitated prey and any ants they can find. But they now dash quickly across the sand, and when they stop and stand, they alternately lift their feet to prevent burning them. Meanwhile, long-legged silver ants, *Cataglyphis bombycina,* avoid the lizards by remaining in their burrows under the sand. However, they are poised to leave, waiting for the sand temperatures to heat up even more, until it reaches about 60°C, when the temperature of the air at ant height is about 46.5°C. Temperature "testers" among them lurk at the nest entrance. At the right moment, they signal the time to come out by releasing pheromones from their mandibular secretions. The rest of the colony then rushes out into the field to forage safely, until they too must retire back to their underground shelters—when they experience air temperatures of 53.6°C, which is just a fraction of a degree below their thermal death point.

In the southwestern deserts of the United States near Phoenix, Arizona, the desert or Apache cicada, *D. apache,* also engages in a thermal arms race against vertebrate predators. These cicadas are active at the hottest time of the year, and even then they wait until the high midday temperatures of 44°C (in the shade) near noon to be most active, when the cicada-killing wasps and birds are forced to retire from the heat.

In the deserts of southern California, the grasshopper *Trimerotropis pallidipennis* endures heat rather than regulating heat loss like the sweating cicada. By blending in with the background of the desert floor, it hides to escape bird and lizard predators. Normally grasshoppers that inhabit the ground stilt high above that substrate when it becomes heated to very high temperatures in sunshine. But to remain camouflaged it is imperative for *T. pallidipennis* to crouch down onto the searing hot ground. When that ground heats to near 60°C in sunshine, the duration of time that a grasshopper can remain hidden is limited by how high a body temperature it can tolerate. *T. pallidipennis* has evolved to tolerate the extraordinary high body temperature of 50°C and can thus escape into the sanctuary of sunlight, where a predator such as a lizard or bird cannot hunt.

It is probably rare that insects escape predators by seeking out low temperatures. Possibly the best candidates are the Cuculliinae, a subfamily of the generally endothermic Noctuidae or owlet moths. The Cuculliinae are a northern circumpolar group of moths, and in northern New England they may fly during any month of the winter when temperatures reach 0–10°C and when most of their bat and bird predators have left. During flight, cuculliinines have flight-motor temperatures near 30–35°C, as do other moths of their size and wing loading. Unlike all other moths, however, the cuculliinines can begin to shiver at the extraordinarily low muscle temperature of 0°C, and they continue shivering to warm up all the way to 35°C.

The giant hornets, *Vespa mandarinia,* attack honey bee colonies. During a typical giant hornet attack, a lone hornet forager first captures bees at the periphery of the bees' nest. After several successful foraging trips to the beehive, the hornet deposits a marking pheromone at the hive entrance from the van der Vecht gland at the tip of the abdomen. This pheromone attracts other hornets from the home nest, and then the slaughter phase of the hornet attack begins: 30,000 bees can be killed in 3 h by a group of 30 to 40 hornets. Subsequently the hornets may occupy the hive itself, and then they carry off the bees' larvae and pupae to feed to their own young.

The above happens when hornets attack colonies of the introduced European honey bee, *A. mellifera,* but the Japanese honey bee, *A. cerana,* has evolved an effective counterstrategy to the hornets' mass invasion. With the latter, those unfortunate hornets that are recruited by the pheromone and then try to enter the hive are met and killed by heat as hundreds of bees envelop each wasp into a tight ball. The interior of these bee balls quickly rises to 47°C, killing the hornet but not the bees, whose upper lethal temperature is 48 to 50°C.

Against Competitors

Contest competition or fighting over food is rare in insects, but at least two species of African dung beetles, *Scarabaeus laevistriatus* and *Kheper nigroaeneus,* engage in combat over dung balls that they make to feed on and/or to serve as sexual attractants. An elevated thoracic temperature plays a crucial role in these contests on the ground. The more a beetle shivers to keep warm (with its flight muscles), the higher the temperature of the leg muscles adjacent to the flight muscles in the thorax and the faster its legs can move and construct the dung into balls and roll it away. Endothermy thus aids in the scramble competition for food, and it reduces the duration of exposure to predators. Additionally, hot beetles have the edge in contest competitions over dung balls made by other beetles; in fights over dung balls, hot beetles almost invariably defeat cooler ones, often despite a large size disadvantage.

Mate Competition

For large insects, endothermic heat production is a requisite for flight, and it is during flight that other activities, including foraging, oviposition, and predator escape, as opportunity and necessity dictate, may occur. Hence, to find a direct effect of body temperature on mating success specifically, one must examine a mating behavior that is not already tightly linked with some other temperature-dependent activity. Singing in some species is a good candidate. Singing is one activity that serves only for mate attraction, and in katydids and cidadas only males sing and the females remain silent. The vigor of this singing activity is associated with and dependent on thermoregulation. Katydids, *Neoconocephalus robustus,* warm up for their ear-shattering mating concerts by shivering, bringing flight-muscle temperatures above 30°C. Males of the Malaysian green bush cricket, *Hexacentrus unicolor,* sing from dusk until well into the night, and before they sing, they prepare themselves by shivering to achieve thoracic temperatures near 37°C. At thoracic temperatures of 37 to 38°C, the males are able to achieve the extraordinarily fast wing movements of up to about 400 vibrations per second.

The dragonfly *Libellula pulchella* demonstrates both the importance of body temperature for mating success and the trade-offs required for maximizing power output as an insect matures. The young, nonreproductive adults of this species are sit-and-wait predators that typically fly with relatively low thoracic temperature. Their flight-muscle performance does not peak at any one temperature; instead, performance is uniformly spread over a wide range of low thoracic temperatures. In contrast, sexually mature males engage in nearly continuous flight in intense territorial contests. At such times, they generate a very high thoracic temperature, and they regulate that thoracic temperature precisely and within only 2.5°C from their upper lethal temperature. Thus, muscle performance of the sexually mature males is narrowly specialized relative to that of young adults that do not engage in strenuous battle.

SOCIAL THERMOREGULATION

Many of the social insects regulate the temperature of their nests in coordinated behavioral and physiological responses involving the adult nest inhabitants. Nest temperature regulation functions primarily to maintain activity and to keep the otherwise thermally labile larvae at the proper temperature for rapid growth. Thermoregulation allows social insects to rapidly build up large nest populations and to inhabit environments where they could not otherwise exist.

Nest Site

One of the first requirements for effective nest temperature regulation is the choice of an appropriate nest site. Typically, northern ants nest in the open, often under solar-heated rocks, or they make solar-heated mounds; many termites also nest to maximize exposure to solar radiation. Honey bees, that live in northern temperate climates require enclosed nest sites such as tree cavities, whereas a variety of other more tropical bees have open and exposed nests.

Nest Construction

Northern vespine wasps enclose their nests in multiple layers of paper that insulate the nest contents. Some termites and ants construct nests so located and constructed as to maximize solar heating in the morning and evening and to minimize overheating at noon. Nests may be constructed so that air circulation and heat transfer are enhanced for thermoregulation.

Behavior and Physiology

Ants regulate the temperature of their brood by carrying it to those parts of the nest with suitable temperatures. Both honey bees and vespine wasps regulate the temperature of the nest, especially near the brood. They fan to circulate air and carry off heat when nest overheating is imminent, and if temperatures continue to increase they carry in water and sprinkle it on the combs for evaporative cooling results. At low temperatures, such as during winter and in swarm clusters outside the hive, the bees crowd together tightly as air temperatures drop, thereby trapping heat inside. As air temperatures rise, the bees on the cluster start to disperse, the cluster loosens, and heat from the interior is released. In hives containing both honeycomb and comb with brood, the bees preferentially cluster around the brood. Brood temperature is maintained near 36°C in hives that may be subjected to air temperatures as low as –50°C and as high as 50°C, provided the bees have access to honey as an energy source for heat production in the cold and to water for evaporative cooling in the heat.

See Also the Following Articles

Cold/Heat Protection • Diapause • Dormancy • Flight • Hibernation • Temperature, Effects on Development and Growth

Further Reading

Esch, H., and Goller, F. (1991). Neural control of honeybee fibrillar muscles during shivering and flight. *J. Exp. Biol.* **159,** 419–431.

Heinrich, B. (1971). Temperature regulation of the sphinx moth, *Manduca sexta.* II. *J. Exp. Biol.* **54,** 141–166.

Heinrich, B. (1993). "The Hot-Blooded Insects: Mechanisms and Evolution of Thermoregulation." Harvard University Press, Cambridge, MA.

Heinrich, B., and Esch, H. (1994). Thermoregulation in bees. *Am. Sci.* **82,** 164–170.

Kammer, A. E., and Heinrich, B. (1972). Neural control of bumblebee fibrillar muscle during shivering. *J. Comp. Physiol.* **78,** 337–345.

May, M. L. (1982). Heat exchange and endothermy in Protodonata. *Evolution* **36,** 1051–1058.

Seeley, T. D., and Heinrich, B. (1981). Regulation in the nests of social insects. *In* "Insect Thermoregulation," (B. Heinrich, ed.). Wiley, New York.

Toolson, F. C. (1987). Water prolifigy as an adaptation to hot deserts: Water loss rats and evaporative cooling in the Sonoran Desert cicada, *Diceroprocta apache* (Homoptera: Cicadidae). *Physiol. Zool.* **60,** 379–385.

Wehner, R., March, A. C., and Wehner, S. (1992). Desert ants on a thermal tightrope. *Nature* **357,** 585–587.

Thrips

see *Thysanoptera*

Thysanoptera

Laurence Mound

CSIRO Entomology, Canberra, Australia

The 5000 described species of Thysanoptera, the thrips, exhibit a wide range of biologies. About 50% feed only on fungi, with most of these feeding on hyphae but some on spores. Of the remainder, approximately equal numbers feed either in flowers or on green leaves; a few are obligate predators on other small arthropods. Several opportunist species are crop pests, causing feeding damage and vectoring tospoviruses, but sometimes acting as beneficials by feeding on other pest arthropods (Fig. 1).

Thrips have unique asymmetric mouthparts involving only one mandible, a life history that is intermediate between those of the hemi- and the holometabola, and a haplodiploid sex control system, and many species exhibit complex behavioral patterns including leking, fighting, and eusociality.

THYSANOPTERA STRUCTURE

Larvae and adults have only a left mandible. Their paired maxillary stylets (lacinia) are coadapted to form a feeding tube with a single central channel and a subterminal aperture. These feeding stylets emerge through a mouth cone that points either downward or backward. The dorsal surface of the head is symmetrical, but the ventral surface is asymmetrical, reflecting the absence of the right mandible. The head bears a pair of antennae commonly with seven or eight segments, although the plesiomorphic (or ancestral) number is presumably nine, and various species have segments fused to produce lower numbers.

Paired compound eyes are usually well developed, although reduced to less than 10 ommatidia in some wingless species. Winged, but not wingless, adults have three ocelli between the compound eyes. The pronotum commonly has a regular number of major setae, five in Phlaeothripidae but usually only two in Thripidae. The legs of adults lack typical insect tarsal claws, but each tarsus has an eversible bladder-like arolium.

Members of the two suborders of Thysanoptera, Terebrantia and Tubulifera, differ from each other considerably in structure. The forewings of adult Tubulifera lack longitudinal veins, have a smooth surface, and bear nonarticulating fringing cilia that insert directly into the wing membrane. The abdominal tergites of these species bear one or two pairs of sigmoid wing-

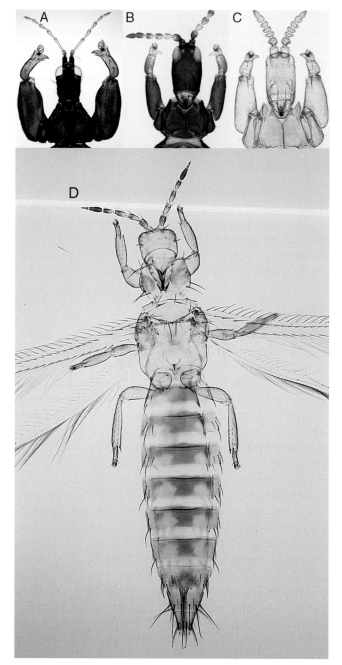

FIGURE 1 Thysanoptera diversity. (A) *Lichanothrips pulchra* female, which creates a domicile by gluing together pairs of Acacia leaves. (B) *Oncothrips waterhousei* foundress, which induces leaf gall on Acacia. (C) *O. waterhousei* first-generation female, which functions as a soldier to defend a gall. (D) Western flower thrips *(Frankliniella occidentalis)*, one of the world's major insect pests.

holding setae, under which the wings lie flat on top of each other when at rest. Moreover, the 10th abdominal segment is tubular, with the anus terminal but the genital opening at the base of the tube, the female's ovipositor being an eversible, chute-like, structure.

In the species of Terebrantia, in contrast, the forewings have two longitudinal veins, the wing surface is covered in microtrichia, and the fringing cilia are inserted into sockets that are figure-8-shaped. The wings lie parallel to each other on the abdomen when at rest, with the cilia of the posterior margins in the midline, but the wing-holding mechanisms vary considerably between species. The 10th abdominal segment of Terebrantia species is incomplete ventrally, and the ovipositor comprises four saw-like blades that are used to insert an egg into plant tissue.

FAMILY CLASSIFICATION

The suborder Tubulifera is usually considered to include a single family of about 3000 species, the Phlaeothripidae, although aberrant species, or groups of species, within this suborder are sometimes segregated into small families. The Terebrantia is a more diverse group of 2000 species and currently includes eight families.

Phlaeothripidae

About 600 of the 3000 species in the family Phlaeothripidae feed by ingesting whole fungus spores, and these species comprise the subfamily Idolothripinae. These are some of the largest thrips, with body sizes up to 15 mm, and the males are commonly considerably larger than the females, with prominent tubercles and large foretarsal teeth. Moreover, these large thrips commonly exhibit patterns of allometry, such that the largest males differ considerably in structure and appearance from the smallest in the same population, and these size differences are reflected in their behavior.

The remaining members of the family constitute the Phlaeothripinae, most species being grouped into three ill-defined lineages. The *Haplothrips* lineage includes those Phlaeothripidae most commonly seen in the north-temperate zone, where they are common in the flowers of Asteraceae and Poaceae, including cereal crops. In the warmer parts of the world, members of the *Liothrips* lineage are more abundant, feeding on the leaves of shrubs and trees and commonly causing them to distort, to roll, or to form discrete galls. Many species and genera have been described in this group, but there are few good studies on biology, host specificity, or structural variation, so that species recognition and generic classifications remain unsatisfactory. Species in the third major group, the *Phlaeothrips* lineage, feed on fungal hyphae on dead branches or in leaf litter. Again, many exhibit complex allometry in males, form colonies, and are subsocial. The complexity of variation in structure within and between sexes is such that males and females can sometimes not be recognized as belonging to the same species unless observed alive within a single colony.

Thripidae

In contrast to the other Terebrantia, adults of all 1700 thripid species have slender emergent, simple or forked, sense cones located on the third and fourth segments of the antennae. The greenhouse thrips, *Heliothrips haemorroidalis,* is placed in the subfamily Panchaetothripinae, together with about 120 related species that have a body and legs heavily reticulate. Similar small numbers of species are placed in two further subfamilies, Dendrothripinae and Sericothripinae, but most thripids are placed in the Thripinae. This subfamily includes most of those flower-living insects that are commonly recognized as thrips, particularly the 450 members of two genera, *Thrips* and *Frankliniella,* many of which are important crop pests (Fig. 1D).

Aeolothripidae

The 200 species in this family all have nine-segmented antennae with at least the last three segments closely joined, and the sensoria on the third and fourth segments are almost always linear along these segments. Aeolothripids are relatively large thrips, with the forewings broadly rounded at the apex and commonly banded black and white. Although they are probably all predatory on other arthropods, the common flower-living members of the genus *Aeolothrips* also feed on plant tissue, the only obligate predators being tropical species in the genera *Franklinothrips, Mymarothrips,* and possibly *Stomatothrips.*

Melanthripidae

The 60 species in this family are sometimes referred to the Aeolothripidae, but they are all flower feeders, not predators. In contrast to aeolothripids, all nine antennal segments are distinct and bear transverse rows of microtrichia, and the sensoria on the third and fourth segments are linear around the apex. Moreover, remnants of an eighth sternite are visible on the abdomen, as in Merothripidae.

Heterothripidae

The 70 species recognized in this family are found only in the New World. The adults all have antennae with nine segments, and the sensoria on the third and fourth segments are continuous around the apex. Little is known about the biology of most of these species, but they probably all breed in flowers, some being known to be host-specific, and a wide range of plant families is involved.

Minor Families

The remaining four families in the Terebrantia include a total of fewer than 30 species. A single, widespread, tropical species is placed in the Uzelothripidae. This is presumed to be fungus-feeding and has a remarkable whip-like terminal antennal segment. The Adiheterothripidae includes two species from the western United States and four species from the flowers of date palms between the eastern Mediterranean and India. The Fauriellidae includes one species from California, two from southeastern Africa, and two from southern Europe. The family Merothripidae includes fewer than 20 species, mostly from

South America, all of which feed on fungus on dead twigs and leaves. These are minute thrips, and they retain structural character states that are presumed to be closest to the ancestral states of all Thysanoptera.

LIFE HISTORY

Reproduction in thrips is haplodiploid, which involves males having one-half the number of chromosomes of females and developing from unfertilized eggs. Despite this, several thrips species can produce females from unfertilized eggs, a process known as thelytoky. In a few common and widespread species, such as the greenhouse thrips *H. haemorroidalis,* males are rare or, as in the introduced basswood thrips, *Thrips calcaratus,* unknown. Larvae usually hatch within a few days, but in some of the larger Idolothripinae eggs develop while still in the abdomen of a female and larvae hatch soon after the eggs are laid.

There are two larval stages, both of which feed actively for about 2 to 5 days. In Terebrantia there are then two pupal stages with the antennal segmentation reduced or absent and the mouth parts nonfunctional. Wing rudiments can be seen in the first, the propupa, but are much longer in the second, the pupa. Tubulifera species are even more remarkable in that they all have three pupal stages.

The site of pupation varies greatly among species. Some Thripidae and Phlaeothripidae pupate on leaves in association with their larvae, but more commonly pupae are found in leaf litter or on the trunks of trees. The pupae of Aeolothripidae, Heterothripidae, and even a few Thripidae are contained within a pupal cocoon that is spun by the second instar. Larvae in these species have stout tubercles near the posterior end of the abdomen that presumably assist burrowing into the soil.

Species that are host-specific within the flowers of particular plants are usually univoltine. Related polyphagous species often breed continuously as long as conditions remain suitable. Overwintering stages are usually pupae, or adults, but the citrus thrips, *Scirtothrips citri,* overwinters as eggs in leaves and shoots.

FEEDING

Little difference is evident between phytophagous, predatory, and fungus-feeding species in the structure of their feeding stylets. The two maxillary stylets fit together along their length with a tongue-and-groove system, and their apices are linked with slender finger-like projections around a subapical feeding aperture. Only in the spore-feeding Idolothripinae are the stylets clearly adapted to the food consumed, being exceptionally broad to facilitate the ingestion of whole spores.

The only obvious difference in the stylets between species is their length. Among the Terebrantia, the maxillary stylets are short and restricted to the mouth cone, but in many Phlaeothripidae the stylets are deeply retracted into the head, sometimes as far as the eyes, and lie close together along the midline. In a few species the stylets are longer than the total body length and are coiled within the head.

When feeding, a thrips initially makes a hole by extruding the single mandible, a solid needle-like structure. The mandible is then withdrawn, and the maxillary stylets are inserted into the food source through the hole, saliva being pumped into the tissues and the resultant fluid being pumped back into the thrips' crop. In contrast to the independent salivary and feeding channels of an aphid or similar bug, a thrips has a feeding tube with only a single channel.

In those thrips species that transmit tospovirus diseases to plants, the virus is acquired only by a first, or early second, instar. The virus is taken up from an infected cell when a larva feeds and passes into the gut and through the gut wall. Ultimately, the virus reaches the salivary glands of the adult, from where it is then reinjected into a plant. Although an adult may take a tospovirus into its gut, the virus is not able to pass through the gut wall and reach the salivary glands, so that no adult can acquire and then transmit a tospovirus. Viruliferous adult thrips must have acquired the virus when they were larvae.

HOST RELATIONSHIPS

Almost nothing is known of host specificity in fungus-feeding thrips. In Europe, certain species of the phlaeothripine genus *Hoplothrips* are associated with *Stereum* fungi on dead branches of trees. Those tropical species of Idolothripinae that have exceptionally broad stylets presumably feed on larger fungal spores than related species with slightly less broad stylets. Such details of thrips natural history are, regrettably, little studied. Even among phytophagous species, precise host relationships have rarely been established. In Europe, the host plants of most species in the genus *Haplothrips* are known—many living only in the flowers of a particular species of Asteraceae—but the plants on which the many tropical species of this genus breed remain unknown. Similarly, the host plant of not 1 of the 43 endemic North American species of the genus *Thrips* has been determined. A major reason for this lack of sound biological information about thrips species lies in the dispersive activity of adults, these sometimes being found in very large numbers on plants on which they cannot breed.

Despite the inadequacy of field studies, some host-plant relationships are well established. For example, the Palearctic species of *Odontothrips* breed only in flowers of the family Papilionaceae, and the Old World tropical species of *Megalurothrips* also breed in such flowers. Similarly, *Dichromothrips* species live on Orchidaceae in the Old World; *Projectothrips* species live only in the flowers of *Pandanus,* the screw pines of the Old World tropics; and all four species of the adiheterothripid genus *Holarthrothrips* live only in the flowers of the date palm, *Phoenix dactylifera.* The Poaceae has a particularly rich fauna of Thripidae, with *Aptinothrips* and *Limothrips* being specific to grasses in the Palearctic; *Stenchaetothrips, Fulmekiola,* and *Bregmatothrips* specific to grasses in the tropics; and *Chirothrips* and *Arorathrips* breeding in grass flowers in many parts of the world.

In contrast to these host associations at genus level, some thrips genera show an entirely different pattern of host exploitation, with each species using as host some entirely unrelated plant. In the New World genus *Echinothrips,* one species lives on a species of *Selaginella* (Lycopsida), another lives on the needles of *Tsuga* (Pinaceae), a third lives on various soft-leaved plants in Euphorbiaceae and Balsaminaceae, and a fourth is polyphagous with no clear host associations.

Host plant data are particularly weak in the Phlaeothripidae, many species being known only from single samples, particularly in the tropics. However, in Australia one suite of about 200 species of Phlaeothripinae is associated only with *Acacia,* and this suite of species appears to represent a single evolutionary lineage. This close association between a thrips lineage and one plant genus is remarkable, particularly so when the richness of the Australian flora and the complete absence of Phlaeothripinae from the leaves of any of the 900 species of *Eucalyptus,* the second largest plant genus in Australia, are considered.

FLIGHT AND DISTRIBUTION PATTERNS

Adults of both phytophagous and mycophagous species can be observed in warm weather climbing to the tops of plants such as grasses or climbing up dead twigs. They spread their wings, and in Terebrantia the marginal cilia are combed from a parked position parallel to the wing margin into a flight position at right angles to the wing. Many species then actively jump into the air and fly vertically upward. Despite this activity, the dispersal of thrips is presumably determined primarily by air currents, and even wingless individuals are dispersed widely by the wind. Wingless species on the mountains of southeastern Australia are recorded dispersing to the northern part of South Island, New Zealand, a distance of more than 1600 km across the Tasman Sea. Such dispersal is probably not entirely fortuitous, but is a function of the behavior of particular species. Some species, such as *Frankliniella schultzei* in Australia, seem to be particularly prone to long-distance migration flights. In contrast, the extensive dispersal of the western flower thrips, *Frankliniella occidentalis,* around the world is probably due mainly to the horticultural trade.

The worldwide distributions of many thrips species are primarily the result of human trading patterns. For example, *Chirothrips* species pupate within the glumes of grass florets and have thus been distributed around the world in commercial grass seed. Other grass thrips were widely distributed in hay used to feed animals on sailing ships during the period of colonial expansion. Thus some members of the European genera *Aptinothrips* and *Limothrips* can now be found in temperate zones all over the world, including mountains in tropical countries. Similarly, orchids, bananas, and sugarcane, all of which are transported and planted from plant parts, not seeds, have been accompanied around the world by their pest thrips species. The vast increase in the use of air transport by the horticultural trade since 1980 is expanding the distributions of pest thrips around the world.

Despite these disrupted patterns of thrips distribution, there remain several natural distribution patterns that are of interest. The family Heterothripidae is confined to the New World, and most species of Merothripidae are restricted to that area. In contrast, the genus *Thrips,* with more than 270 species worldwide, has no species native to the Americas south of Mexico, and the genus *Frankliniella,* with more than 180 species, includes very few that have a natural distribution anywhere outside the New World. Presumably these two advanced genera of Thripinae evolved at about the time that the American continent separated from Europe; the presence of the Heterothripidae only in the New World may also suggest a relatively recent origin for that family.

Within the Melanthripidae, the genus *Dorythrips* has three species in southern South America and two in Western Australia, and *Cranothrips* has one species in South Africa and several in Australia. Within the Aeolothripidae, the Cycadothripinae and Dactuliothripinae are considered sister groups. Species of the first live only in Australia, in the cones of *Macrozamia* cycads, whereas the second is found only along the western side of the American continent in the flowers of various plants. Among the more advanced Aeolothripidae, the two genera *Aeolothrips* and *Desmothrips* are ecological and morphological counterparts of each other, the first restricted to the Holarctic, the second to Australia.

Among the Phlaeothripidae, geographical patterns of distribution are less clear. Some smaller genera are restricted either to the Western or to the Eastern Hemisphere, but larger genera are more widely distributed. The leaf-feeding members of the genus *Liothrips* are found throughout the tropics, including the Pacific region, although they have been little studied in the Neotropics. The large species of the genus *Elaphrothrips,* all of which feed on fungal spores, are found widely throughout the tropics, being replaced east of Wallace's Line by members of the genus *Mecynothrips.*

BEHAVIOR

In Terebrantia, males are usually much smaller than females, whereas in Tubulifera males are commonly much larger than females. These differences are related to different patterns of behavior. In fungus-feeding phlaeothripids, it is not uncommon for a male to defend a female and the egg mass that she produces or, alternatively, for a male to defend a single egg mass to which various females contribute after first mating with him. These strategies lead to competitive behavior between individual males, involving flicking with the abdomen to displace a rival or even stabbing with foretarsal teeth to kill a rival. Moreover, while large males are involved in such competitive activities, smaller males may sneak-mate. Clearly there is a balance of advantages, between developing into a small male on a restricted food supply and requiring more food and developing a larger body.

Competition between males is possibly a plesiotypic behavioral trait in Thysanoptera, because in the basal clades Merothripidae and Aeolothripidae males of some species are polymorphic and presumably competitive. In *Merothrips* and *Cycadothrips* species the largest and smallest males differ considerably in body size and in the strength of their forelegs and abdominal setae. Male competitiveness also occurs in some Thripidae, including the pest species *F. occidentalis*. Males of Kelly's citrus thrips, *Pezothrips kellyanus,* a pest of citrus in Australia and the Mediterranean, form leks on ripe lemons, and females are attracted to these male aggregations for mating.

THRIPS DOMICILES

The term "domicile" is used to include both leaf galls that are induced by thrips and the shelters that many Australian Phlaeothripinae construct by fixing leaves together with glue or silk. Gall induction by thrips, mainly by species of Phlaeothripinae, is widespread in tropical countries although inadequately recorded in the Neotropics. Galls range from simple rolled leaves containing a few thrips to highly contorted masses of leaf tissue enclosing up to 10,000 adults and larvae. In most gall-inducing thrips from the Oriental Region there is little sexual dimorphism, whereas gall thrips on *Acacia* and *Casuarina* trees in Australia exhibit considerable variation both between sexes and between long- and short-wing morphs.

In some phlaeothripines on *Acacia* in Australia, the gall foundress is a fully winged female, but the eggs she lays first develop into short-winged adults of both sexes. These adults sometimes have reduced reproduction and act as soldiers to defend a gall while the foundress produces a second and larger generation that become winged adults. This behavioral strategy falls within the definition of eusociality.

The habit of domicile construction by thrips is recorded only from Australia, in a suite of species on *Acacia*. At least 30 species are now known to form such shelters using a secretion from the anus. In some of these species the *Acacia* leaves, or more precisely phyllodes, are glued together in pairs at an angle, and the thrips breed within the space created by a ring of glue and the two phyllode surfaces. In other species the secretion is more silken in form, and this silk is used to sew together two or more phyllodes enclosing a small space within which the thrips breed. At least two species are known to use this silken material to weave a tent on one surface of a phyllode within which to breed.

These Australian thrips domiciles, whether galls or constructs, are evidently of great value in ecosystems with high temperatures and low humidities. As a result, a range of kleptoparasitic species has evolved, each of which has a different method of trying to usurp a domicile. These methods of driving out the original inhabitants range from a frontal assault with sharp foretarsal teeth to a porcupine-like action of the abdomen which bears many stout setae. A few Australian *Acacia* thrips species seem to have evolved into true inquilines, in that they breed within the colony of a domicile producer without unduly disturbing the original inhabitants.

POLLINATION

Thrips are sometimes abundant in flowers, their bodies often bear large numbers of pollen grains, and they can fly actively between flowers. Despite this, their function as pollinators, even their presence, is frequently overlooked; no less than Charles Darwin complained of thrips interfering with his experiments on larger pollinators! Thrips have been demonstrated to be pollinators in a wide range of flowers: heather plants in the north of the Northern Hemisphere, dipterocarp trees in Malaysia, a rain forest tree in eastern Australia—*Wilkiea huegeliana* (Monimiaceae), the Panama rubber tree *Castilla elastica* (Moraceae), and several *Macrozamia* cycad species in Australia.

Although a thrips individual does not carry a large pollen load, the large number of thrips in each flower, each one carrying between 10 and 50 pollen grains, is enough to deliver pollen to many stigmatic surfaces. Despite this, thrips probably remain the most underestimated of all flower pollinators, the majority of botanists failing to see these small insects, let alone consider their significance.

PEST SPECIES

In general, the only thrips that are noticed by nonspecialists are pest species. Most pest thrips are members of the Thripidae, although particular phlaeothripid species cause leaf damage on a wide range of plants in the warmer parts of the world, including decorative *Ficus* trees, black pepper vines, and olives. In the Northern Hemisphere cereal crops are also damaged by a phlaeothripid species. In contrast there are many thripid species that attack cultivated plants, and some of these cause serious economic losses.

Citrus production in both California and southern Africa can suffer considerable losses because of downgrading of scarred fruit on which thrips have been feeding. The value of a nectarine can be seriously reduced through a single thrips larva feeding on the fruit when it is young. Cucumbers, capsicums, and strawberries are badly distorted at times because of the feeding activity of thrips. Roses, carnations, and chrysanthemum flowers can be devalued through thrips feeding damage, and table grapes burst and become fungal-infected through thrips oviposition scars.

Worldwide, there are four species of thripids that are particularly significant as pests: the onion thrips *(Thrips tabaci),* the melon thrips *(Thrips palmi),* the tomato thrips *(Frankliniella schultzei),* and the western flower thrips *(F. occidentalis).* Feeding by each of these species can cause severe damage on some crops, but the most serious damage associated with them is due to the tospoviruses they can transmit. More than 12 of these viruses have been described, and although they cause damage only to plants it is evident from their molecular structure that they are members of the animal virus family Bunyaviridae. The origin of the plant infections remains unknown, but each tospovirus is dependent for its continued existence on being transmitted from one plant to another by one or more of 10 thrips species.

See Also the Following Articles
Gallmaking • Mouthparts • Plant Diseases and Insects • Rostrum

Further Reading
Crespi, B. J., and Mound, L. A. (1997). Ecology and evolution of social behaviour among Australian gall thrips and their allies. *In* "Evolution of Social Behaviour in Insects and Arachnids" (J. Choe, and B. J. Crespi, eds.), pp. 166–180. Cambridge University Press, Campridge, UK.
Lewis, T. (ed.) (1997). "Thrips as Crop Pests." CAB Int., Wallingford, UK.
Moritz, G., Morris, C. D., and Mound, L. A. (2001). ThripsID—Pest thrips of the world. An interactive identification and information system. CD-ROM published by ACIAR and CSIRO Publishing, Australia.
Mound, L. A., and Marullo, R. (1996). The thrips of Central and South America: An introduction. *Mem. Entomol. Int.* **6**, 1–488.

Thysanura

see *Archaeognatha; Zygentoma*

Ticks

Daniel E. Sonenshine
Old Dominion University

Ticks comprise a distinct group of exclusively blood-feeding ectoparasites familiar to most people in virtually all regions of the world. Ticks transmit a greater variety of disease-causing pathogenic agents than any group of arthropods, including protozoan, viral, bacterial, and even fungal pathogens. An example is Lyme disease (LD), which is now the most important vector-borne disease of humans in the United States, Europe, and Asia. In numerous countries in tropical and subtropical regions of the world, tick-borne diseases of livestock such as babesiosis, theileriosis, and heartwater, have made it difficult or impossible to raise domestic animals for food or animal products. Ticks also can cause irritating or even fatal injury to humans and animals because of paralysis, toxicity, or severe allergic reactions to their bites.

BODY STRUCTURE

The tick body is organized into two major body regions, the anterior capitulum, bearing the mouthparts, and the idiosoma, which bears the four pairs of walking legs (Figs. 1, 2). There is no head, and the highly fused body is not divided into a thorax and an abdomen. The capitulum contains the toothed hypo-

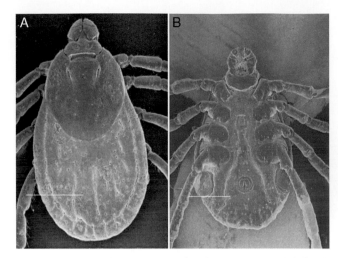

FIGURE 1 Scanning electron micrographs of a representative adult female ixodid tick, *D. variabilis:* (A) dorsal view and (B) ventral view. Scale = 1 mm.

stome, which embeds the tick into the host's skin and also contains the food canal for blood imbibition, the chelicerae, delicate paired appendages that cut into the skin, and the four-segmented paired palps that provide important sensory information for host identification. In argasid ticks, the capitulum is recessed under an anterior extension of the body. The remainder of the body contains the genital pore, anus, and spiracles, which are visible on the ventral surface. In ixodid ticks (so-called hard ticks), a prominent platelike scutum is found on the dorsal surface. Argasid ticks (so-called soft ticks) are similar to the ixodid ticks but lack a scutum, and the body cuticle is leathery.

The interior of the tick body is a simple, open cavity called the hemocoel that is filled with a circulating fluid, the hemolymph, which bathes the internal organs. In ticks, as in other terrestrial arthropods, there is no hemoglobin, and the hemolymph does not function in oxygen transport. Most of the body interior is occupied by the midgut, the largest internal organ of the tick body, which consists of a central saclike stomach and several lateral diverticuli. Other prominent internal organs are the paired salivary glands, which

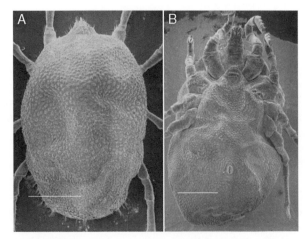

FIGURE 2 Scanning electron micrographs of a representative adult female argasid tick. *O. parkeri:* (A) dorsal view and (B) ventral view. Scale = 1 mm.

appear as white grapelike clusters, and the reproductive organs. In females, these are the ovary, paired oviducts, uterus, and seminal receptacle, and the vagina that connects the system to the genital pore. During feeding, the ovary enlarges and becomes distended with large, brown or amber-colored eggs. In males, the reproductive system consists of the testis, vasa deferentia, and seminal vesicle, and the ejaculatory duct, which is connected to the genital pore. Much of the system is obscured by the large, white multilobed accessory gland. This gland provides the components for the saclike spermatophore that the male tick uses to transfer its sperm to the female. Also present are numerous tracheal tubes, connected to the marginal spiracles, that provide the respiratory system, and the Malpighian tubules and rectal sac, connected to the anal pore, for waste elimination. Argasid ticks have a pair of coxal glands that excrete via the coxal pores excess water and salts accumulated during feeding. The fused central nervous system, the synganglion, is located in the body above the genital pore.

SYSTEMATIC RELATIONSHIPS

Ticks are classified with the class Arachnida, the group that contains the familiar spiders and scorpions. Arachnids have chelicerae, which are appendages with pincerlike or scissorlike cutting edges, instead of mandibles. There is no head or thorax such as occurs in insects. There are no antennae. Ticks are grouped together with the mites in the subclass Acari. Ticks constitute a distinct suborder, the Ixodida, within the acarine order Parasitiformes. The Ixodida contains three families, the Ixodidae, Argasidae, and Nuttalliellidae. The Ixodidae or hard ticks are by far the largest of the different families of ticks, with approximately 650 species. Ixodid ticks have three active life stages, including a single nymphal stage. Hard ticks contain most of the important disease vectors and pest species that plague livestock and wildlife. The Ixodidae are further subdivided into the Prostriata, represented by the single genus *Ixodes*, which is easily recognized by the anterior anal groove, and the Metastriata, which include the remaining 13 genera, in which the anal groove is posterior to the anal aperture. The Ixodidae include the ticks that transmit the agents of LD, Rocky Mountain spotted fever, boutonneuse fever, babesiosis and theileriosis of livestock, and most of the other tick-borne disease-causing agents.

The Argasidae comprise the soft ticks, with their leathery, highly flexible cuticle. There are approximately 170 species divided into five genera (four according to some authorities). Soft ticks also have three active life stages, but most species have multiple nymphal stages before they develop into adults. Ticks of the genus *Ornithodoros* (> 100 species) have a leathery cuticle with innumerable small elevations known as mammillae and a rounded body margin. Ticks of the genus *Argas* (~ 58 species) have a flattened lateral margin marked by a sutural line. The leathery cuticle bears small ridges and folds in a rectangular pattern, each with a pit at the center of these buttonlike enclosures. Except for species

that transmit the spirochetes that cause relapsing fever, most soft ticks are not important in the transmission of disease.

The third family of ticks is the monospecific family Nuttalliellidae, represented by only one species, *Nuttalliella namaqua*, in southern Africa. It contains structures characteristic of both other tick families. Its highly wrinkled cuticle with pits and elevations resembles the cuticle of argasid ticks, but its dorsal pseudoscutum resembles the scutum of the ixodid ticks.

Ticks are an ancient group of specialized acarines that were already well developed during the Mesozoic era (i.e., the era of the dinosaurs). A larval tick found in amber in New Jersey, was dated between 90 and 94 mya (i.e., during the Upper Cretaceous period). Although exhibiting some unusual characteristics in its setal arrangements, the tick was readily characterized as a member of the genus *Carios*, a genus that exists today. This finding suggests that this genus at least (and perhaps other argasid ticks) has not changed very much in many millions of years.

TICK BIOLOGY: LIFE CYCLES, FEEDING BEHAVIOR, DEVELOPMENT, AND REPRODUCTION

The tick life cycle comprises the egg and three active stages, namely, larva, nymph, and adult. There is only a single nymphal instar in the ixodid tick life cycle, but varying numbers of nymphal stages may occur in the argasid tick life cycles. All ticks feed on blood during some or all stages. Most species are three-host ticks; that is, each stage attacks hosts, feeds, and detaches before developing into the next life cycle stage. Adult ticks seek hosts, feed, and, in the case of engorged females, drop off to lay their eggs. Ticks can survive for long periods between blood meals. Consequently, when feeding is delayed, the life cycle may be extended for years or, in the case of some argasids, for a decade or longer. There are major differences between the life cycles of the Ixodidae and the Argasidae, as discussed next.

Life Cycles of Ixodid Ticks

The ixodid ticks feed slowly, from several days to as long as 2 weeks. Immature and adult ticks each take a blood meal, except for the nonfeeding males of some species. After crawling onto their hosts, these ticks embed their mouthparts into the host skin and secrete cement from their salivary glands into and around the wound site to anchor themselves. The cement binds the ticks firmly in place and makes them very difficult to remove. During blood feeding, the ticks secrete potent anticoagulants and anti-inflammatory agents, which suppress host wound healing and facilitate blood flow. As the ticks feed, new cuticle is synthesized to accommodate the enormous blood meals the animals consume, often 10 to 100 times their original body weight. Females feed only once. Mating occurs during feeding, although ticks of the genus *Ixodes* may also mate prior to host attachment. In the metastriate Ixodidae (i.e., ixodids other than ticks of the genus *Ixodes*), mating occurs within a few days after the commencement of feeding and is regulated by sex pheromones, including the volatile 2,6-dichlorophenol and the nonvolatile cholesteryl esters on the body surface. Following mating, females suck blood rapidly (24–48 h) and swell enormously, whereupon the replete females drop from their hosts, find a sheltered location, and lay thousands of eggs. An example is the American dog tick, *Dermacentor variabilis,* which typically lays more than 5000 eggs. Following oviposition, the female dies. In contrast to the females, males swell only slightly during feeding. However, they can mate many times, feeding between matings. In certain species of *Ixodes,* mating may occur either on or off the host (e.g., the blacklegged tick, *Ixodes scapularis,* or the sheep tick, *I. ricinus*). In some nest-inhabiting *Ixodes* species, the males have vestigial hypostomes. These ticks always mate off the host.

Once oviposition has been completed, the larvae hatch in the thousands and begin host-seeking activity. Except for the nest-inhabiting species, the larvae disperse into the vegetation, where they come in contact with passing animals. Once they have attached to a host, the larvae embed themselves into the host skin, form a feeding pool, and engorge in the manner already described. Feeding usually takes 2 to 4 days, whereupon the engorged larvae drop from their hosts to molt on the ground. Molting usually occurs in some sheltered microhabitat such as soil or leaf litter, or in host nests. After molting, nymphal and adult ticks must seek another host and feed. However, more than 90% of the life cycle is spent off the host. When host seeking and feeding occur in all three parasitic stages, the pattern is termed a three-host life cycle.

A few ixodid species exhibit a two-host or one-host life cycle. For example, in the camel tick *Hyalomma dromedarii,* both larvae and nymphs feed on the same host (two-host life cycle), and in the cattle tick *Boophilus annulatus,* all stages feed, molt, and even mate on the same host (one-host life cycle).

Life Cycles of Argasid Ticks

Feeding is very rapid among the argosid ticks. Once they have crawled onto a host, the ticks embed their mouthparts in the same manner as their ixodid relatives, but without secreting cement. Bloodsucking commences quickly and, as feeding progresses, the bloated ticks excrete copious quantities of a clear, colorless coxal fluid (some times this occurs soon after feeding). By eliminating excess water and salts via the coxal fluid, the ticks can concentrate their blood meals and adjust their internal water balance. The ticks expand to about 5 to 10 times their original body weight, depending on the ability of the cuticle to stretch. Following feeding, often completed within as little as 30 to 60 min, the replete ticks drop off to molt or, if female, to lay eggs. Argasid females take repeated small blood meals and lay small batches of eggs (typically < 500 eggs in a batch) after each feeding (multiple gonotrophic cycles). The interval between feedings is typically several

months but may be up to several years, depending on host availability. Mating usually occurs off the host. Because of the multiple nymphal instars (six or even seven in some species), argasid ticks often live for many years. In addition, these ticks are highly resistant to starvation, an advantage that can extend their longevity even further. As a result, the entire life cycle may take from 10 to 20 years.

Following oviposition and hatching, most argasid tick larvae seek hosts, feed rapidly, and molt to the first nymphal instar. These nymphs seek hosts again, feed rapidly, and molt to the second nymphal instar. Subsequently, the life cycle varies considerably, leading to additional nymphal instars or proceeding directly to the adult stage. As a rule, males emerge earlier than females and have fewer nymphal stages. In some argasids, especially bat parasites, the larvae remain attached to their hosts for many days, feeding slowly, just like their ixodid tick relatives, and then molt twice without additional feeding. Thereafter, the life cycle resembles the typical argasid pattern. Another unusual species is *Otobius megnini,* which has only a single nymphal stage. Neither the males nor the females feed, and the females lay eggs without having had a blood meal (i.e., autogeny).

ECOLOGY

Most ticks are exophiles (i.e., nonnidicolous, living exposed in the open environment rather than in shelters). Most ticks live in forests, savannahs, brush, grassy meadows, or under stones, crevices, or even in sand in semidesert environments. Others, however, are nidicoles, surviving in caves, burrows, houses, cracks, and crevices where their hosts obtain shelter. This habit is characteristic of most argasids and many species of the genus *Ixodes.*

Seasonal Activity and Host-Seeking Behavior

Exophilous ticks are active during certain periods of the year when climatic conditions are suitable for development and reproduction. During this seasonal activity period, they attack and feed on suitable animals. This is known as host-seeking behavior. At other times, ticks remain in diapause (i.e., a state of reduced metabolic activity). In temperate and subpolar regions, the seasonal activity period is regulated by ambient temperature, changing photoperiod, and incident solar energy. Tick seasonal activity usually commences with the onset of warmer weather and increasing daylength. In what is termed the ambush strategy, hungry ticks climb on the vegetation to varying heights, depending on life stage (e.g., adults climb the highest) and cling to any passing animals. In some species, the ticks emerge from their shelters and run toward their hosts when they detect animal odors (or, rarely, noise) from animals nearby. This so-called hunter strategy is useful in arid habitats, where there is little vegetation or source of other moist, protective covering. Argasid ticks normally do not exhibit seasonal activity, since they live in proximity to their hosts in nests, burrows, or other shelters. However, in some species specific for migratory birds or bats, host-seeking activity is synchronized with the period of the year when these hosts return to reoccupy their nests.

For many ixodid ticks that occur in temperate or subarctic regions, seasonal activity begins in the spring. In *D. variabilis,* for example, larvae that survived the winter begin to feed on small mammals. Activity accelerates rapidly as increasing numbers of larvae, stimulated by rising soil temperatures and lengthening photoperiods, emerge to attack these animals, reaching the seasonal peak within a few weeks. Feeding by nymphal and adult ticks follows soon afterward, with the adult peak in early summer. In the southern parts of its range, the tick's entire life cycle, from eggs to ovipositing females, is completed in one year. In the northern parts of the *D. variabilis* range, larval emergence is delayed until late spring. Moreover, the cooler soil temperatures and shorter daylengths delay molting of fed ticks. As a result, adults emerge from fed nymphs in late summer or early fall, when soil temperatures and incident solar radiation are declining, and this results in a two-year life cycle. Thus, both a one-year and a two-year life cycle can occur because of variations in climactic conditions within this wide-ranging species.

In *I. scapularis,* larvae and nymphs feed in the spring and summer, as in *D. variabilis,* whereas adults are active in the fall and early spring. However, the order of larval and nymphal feeding is the opposite of the dog tick. Nymphal ticks emerge from their overwintering diapause in spring or early summer, depending on the region of the United States where they occur. Larvae appear next, typically a month or two after the nymphal peak. Meanwhile, fed nymphs molt over the summer, but the newly emerged adults delay host-seeking activity until the cooler months of the fall or early winter. This life cycle pattern enables nymphs infected with *Borellia burgdorferi* to infect mice, providing thereby a reservoir of infected hosts to infect the next generation of larval ticks. The implications of these different tick life cycle patterns for the survival and transmission of zoonotic diseases are discussed further in the sections on specific diseases.

Occasionally, ticks are active only during the winter months. An example is the winter tick, *D. albipictus,* a one-host tick that feeds on horses, deer, elk, moose, and other large ungulates. In this case, larvae commence host-seeking activity in late summer or early fall. Larvae and nymphs feed and molt on the same hosts and the resulting adults reattach, feed, and mate. Feeding and development require many weeks for completion, even though these activities occur on the same host and, as a result, the adults are often found on their ungulate hosts in winter or early spring. In *D. albipictus* declining photoperiod stimulates feeding activity, just the opposite of the pattern seen in *D. variabilis.*

Host Specificity

Host-seeking activity is strongly influenced by the availability of hosts and host selection behavior. All ticks species exhibit varying degrees of host specificity; most (> 85%) exhibit relatively strict host specificity. At one extreme are the argasid ticks that feed exclusively on bats (e.g., ticks of the genus *Antricola* and certain species of *Ornithodoros*). For example, during his graduate student years, Sonenshine colonized the bat tick, *Ornithodoros kelleyi,* in the laboratory. To feed the ticks, he had to maintain a colony of bats collected from limestone caves or attics of old buildings. No other hosts would do. Similarly, the cattle ticks *Boophilus microplus* and *B. annulatus* feed solely on cattle and, when available, on white-tailed deer. Other species exhibit limited host specificity (e.g., *D. variabilis*). Larvae and nymphs of *D. variabilis* feed on a wide range of small mammals (e.g., white-footed mice, meadow voles, etc.), but never on carnivores, ungulates, humans, or other large mammals. In contrast, adults of this species feed on medium-sized and large mammals, including humans (although they can be induced to feed on rodents when confined in capsules). Finally, at the opposite extreme of the specificity spectrum are the opportunistic species that feed on hosts of virtually all types, (e.g., *I. scapularis* and *I. ricinus*). Immatures feed on lizards, birds, and small, medium-sized, and large mammals, including deer and humans. Adults feed on medium-sized and large mammals, including humans. Although the range of confirmed hosts is astounding, even these opportunistic ticks have preferred hosts (e.g., mice for the immatures; deer, sheep, and other mammals for the adults). Host specificity also is strongly influenced by ecological adaptations, so that ticks adapted to a particular habitat in a given region of the world will encounter only vertebrates adapted to the same habitat.

As tick–host associations evolved, ticks gradually developed the ability to facilitate long-term feeding by evading or suppressing host homeostatic systems. For example, *I. scapularis* saliva contains pharmacologically active compounds that suppress edema and inflammation in their hosts while enhancing vasodilation. This leads to greater blood flow into the wound site without the pain and intense itching sensation so characteristic of the bites of mosquitoes or biting flies. These adaptations are most effective for the hosts encountered most frequently by each tick species, so-called preferred hosts, but less effective for uncommon hosts.

Survival between Blood Meals

One of the most remarkable aspects of tick biology is the ability to survive for long periods between blood meals. The tick's midgut serves as a storage organ where the blood meal is digested slowly over long periods. Among the argasids, individuals may survive for several years without feeding while waiting for the occasional wandering hosts that enter their secluded shelters. Among the exophilous ixodids, survival periods are much shorter, but even these ticks may survive for up to one year between blood meals. According to a study by Needham and Teel in 1991, ticks spend more than 90% of their life history off the host.

Ticks also must conserve body water to survive while they wait for hosts. While questing (i.e., perching for attack) on short stems, blades of grass, or other vegetation, ticks are exposed to desiccating conditions that can become life threatening within a few days or weeks, depending on the species. Among the desiccation-intolerant *I. scapularis* and *I. ricinus,* which are adapted to cool, humid forest habitats, desiccated individuals retreat to the forest floor or rotting vegetation at the base of a meadow. In these nearly saturated humid microenvironments, they can restore their water balance by a process known as atmospheric sorption, in which the partially desiccated ticks salivate salt-rich secretions onto their hypostomes. This hygroscopic secretion collects moisture, which is sucked back into the body. Since, however, the process demands a considerable expenditure of energy, the number of cycles of desiccation and sorption is limited as the tick's age. Other species, such as the relatively desiccation-tolerant *Hyalomma asiasticum,* can survive for long periods in the semidesert habitats in central Asia, where it waits for passing camels and other large ungulates. Nidicolous ticks, sheltering in caves, burrows, or other protected microenvironments, are subject to less stressful conditions during the long wait between hosts. These ticks exhibit behavioral patterns that restrict their distribution to these sheltered locations.

REPRESENTATIVE TICK-BORNE DISEASES

In view of the exceptionally large variety of diseases caused by tick-borne pathogens and injurious substances, this section is limited to a brief description of several representative examples, with primary emphasis on development of the infection in the tick and tick vector ecology. For a more extensive review, the reader may wish to consult books by Sonenshine and Strickland, or review articles, for information on the specific diseases. Table I lists some representative tick-borne diseases affecting humans and animals.

Lyme Disease

The most common tick-borne disease affecting human health in the world today, LD occurs throughout most of the United States and southern Canada, Europe and northern Asia. The disease is caused by *B. burgdorferi (sensu latu),* a type of bacterium known as a spirochete. *B. burgdorferi* is the causative agent of LD in most of the United States. A second genospecies, *B. lonestari,* was isolated from lone star ticks in the southeastern United States, but the relationship of these spirochetes to LD in humans is uncertain. In Europe, LD is caused by *B. afzellii* and *B. garinii* as well as *B. burgdorferi.*

In humans, LD results from the bite of an infected tick, either a nymph or adult of the genus *Ixodes.* The tick must have remained attached for several days to allow for the bacteria to travel from the tick's midgut to its salivary glands and into the wound site. Symptoms begin several days to several weeks later. Onset of illness is characterized by mild, flulike fever and, in most patients, a reddish skin rash, known as the erythema migrans (EM). The typical EM rash is a gradually expanding circular or elliptical lesion with a red margin and clear center, at least 5 cm or more in diameter, and often near the site of the tick bite. Some patients show multiple EM rashes. If left untreated in this early stage, the fever abates, the rash fades, and the patient may recover without any further symptoms. Often, however, the bacteria remain in people's bodies and spread into the nervous system and joints, where they cause the long-lasting secondary symptoms of chronic LD. Late manifestations include

TABLE I   Representative Tick-Borne Diseases, Their Causative Agents, Tick Vectors, and Reservoir Hosts

Disease	Causative agent	Primary tick vector species	Affected hosts	Major clinical symptoms
Protozoan				
Human babesiosis	*Babesia microti, B. divergens, B. gibsoni*	*Ixodes scapularis, I. ricinus*	Humans	Malaria-like fevers, myalgia, arthralgia, nausea, sweating
Bovine babesiosis	*B. bigemina*	*Boophilus annulatus, B. microplus,* others	Cattle	Hemoglobinuria (redwater) fever, death
East Coast fever	*Theileria parva*	*Rhipicephalus appendiculatus*	Cattle, buffalo	Fever, lymphadenopathy, pulmonary edema
Tropical theileriosis	*T. annulata*	*Hyalomma anatolicum*	Cattle, horses	Fever, lymphodenopathy, pulmonary edema
Feline cytauxzoonosis	*Cytauxzoon felis*	*Dermacentor variabilis*	Cats	Fever, emaciation, splenomegaly, death
Bacterial, extracellular				
Tularemia	*Francisella tularensis*	*Haemaphysalis leporispalustria,* other tick species	Humans, various other mammals	Fever, headache, pustular, ulcerated papulae; pneumonia, pleuritis, rash; however few deaths
Lyme disease	*Borrelia burgdorferi, B. afzelii, B. garinii, I. persulcatus,* others	*Ixodes scapularis, I. ricinus, I. pacificus,*	Humans, dogs, cats, domestic animals	Initial phase: fever, EM rash Chronic phase: arthritis, neurologic symptoms

(Continues)

TABLE I *(Continued)*

Disease	Causative agent	Primary tick vector species	Affected hosts	Major clinical symptoms
Tick-borne relapsing fever	*Borrelia* spp.	*Ornithodoros* spp.	Humans	Intermittent fevers, chills, fatigue, myalgia, arthralgia; generally mild illness; death rare
Avian spirochetosis	*B. anserina*	*Argas persicus*	Birds	Fever, death
Epizootic bovine abortion abortion	Unknown, possibly *B. coriaceae*	*Ornithodoros coriaceus*	Cattle, deer	Fever, spontaneous abortion
Bacterial, intracellular (Rickettsiales)				
Rocky Mountain fever	*Rickettsia rickettsii*	*Dermacentar variabilis, D. andersoni,* others	Humans	High fevers, spotted whole-body rash
Boutonneuse fever[a]	*Rickettsia conorii*	*R. sanguineus, D. reticulatus,* others	Humans	High fever, rash, ulceration at bite site (eschar)
Human monocytic ehrlichiosis (HME)	*Ehrlichia chaffeensis*	*Amblyomma americanum, D. variabilis*	Humans	Fever, rash (sometimes), muscle aches, joint aches
Human granulocytic ehrlichiosis (HGE)	*Ehrlichia phagocytophilia*	*Ixodes scapularis, I. pacificus, I. ricinus*	Humans	Fever, rash (sometimes), muscle aches, joint aches
Tick-borne fever (sheep pyemia)	*E. phagocytophilia*	*I. ricinus*	Sheep	Fever, weight loss, reduced milk production, abortion
Canine ehrlichiosis	*E. canis, E. ewingli, E. phagocytophilia*	*R. sanguineus, I. ricinus, A. americanum,* others	Dogs	Fever, loss of appetite, weight loss, apathy, death in severe cases
Heartwater	*Cowdria ruminantium*	*Amblyomma hebraeum, A. variegatum,* others	Ruminants	Fever, "pedaling behavior", prostration, coma, death
Anaplasmosis	*Anaplasmosa marginale, A. centrale, A. ovis*	*D. andersoni, D. occidentalis, R. sanguineus,* others	Cattle, sheep, other ruminants	Fever, anemia, death
Q fever	*Coxiella burnettii*	Many tick species	Humans, large domestic livestock	Low-grade fever, sweating, sore throat, pneumonia, severe frontal headache, myalgia, photophobia
Arborviruses				
Tick-borne encephalitis	*Flavivirus[b]*	*I. ricinus, I. persulcatus*	Humans, carnivores	Fever, headache, encephalitis, meningitis, paralysis; death in severe cases
Powassan encephalitis	*Flavivirus[c]*	*Ixodes, Dermacentor, Haemaphysalis* spp.	Rodents, hares, etc.	Fever, headache, encephalitis, neurological symptoms, brain damage, death
Colorado tick fever	*Coltivirus[c]*	*D. andersoni*	Rodents, humans, domestic animals	Biphasic fever, headache, muscle aches, joint pain
Crimean–Congo hemorrhagic fever	*Nairovirus[d]*	*Hyalomma m. marginatum, H. m. rufipes,* others	Hares, humans, small mammals, others	Fever, chills, headache, internal bleeding, rashes; death in severe cases
Louping ill	*Flavivirus[b]*	*Ixodes ricinus*	Sheep	Fever, erratic, louping gait, loss of motor encephalitis, death
African swine fever	*Iridovirus*	*Ornithodoros moubata porcinus, O. erraticus*	Domestic pigs, wild boars, warthogs	Fever, internal damage; death in most cases
Tick-caused diseases				
Tick paralysis	Tick proteins	*I. holocyclus, I. rubicundus, D. variabilis, D. andersoni,* others	Cattle, sheep, humans, other mammals	Ascending paralysis, loss of motor control, no fever; death
Tick bite allergies	Tick proteins	*Argas reflexus, O. cariaceus, I. pacificus,* etc.	Humans	Nausea, vomiting, diarrhea, irregular pulse, shocklike symptoms; rarely death
Sweating sickness (and other tick toxicosis)	Tick proteins	*H. truncatum, O. savignyi, O. lahorensis, A. persicus*	Cattle, sheep, others	Fever, sweating, anorexia, tearing, salivation; high mortality

[a] Also known as Mediterranean spotted fever.
[b] Family Flaviviridae.
[c] Family Reoviridae.
[d] Family Bungaviridae.

arthritis, especially in the knee joints, which frequently spreads to different joints (migratory polyarthritis). Damage to the synovial membranes is a distinctive feature of this type of arthritis because of invasion of the synovial fluid by the spirochetes. Neurological symptoms include nerve pain (peripheral neuropathy), various types of palsy caused by nerve damage (e.g., Bell's palsy), and central nervous system disorders. Many patients suffering from some or all of these symptoms simultaneously also complain of severe fatigue. In Europe, another chronic feature of late LD is acrodermatitis chronica atrophicans, a condition in which the skin atrophies and peels. Chronic LD is greatly feared in regions where it is endemic because, among other reasons, it can persist for many years, despite treatment with antibiotics.

Only a limited variety of ticks of the genus *Ixodes* are competent vectors of LD spirochetes. Other biting arthropods may acquire these bacteria, but they cannot transmit the bacteria when they feed again. In the eastern and central United States, the only proven vector to humans is *I. scapularis.* In the western part of the country the vector is the western blacklegged tick *I. pacificus.* However, other species of the so-called *I. ricinus* complex that feed solely on wildlife contribute to maintaining the disease in nature; examples include *I. spinipalpis (I. neotomae)* in the west and *I. dentatus* in the east. The sheep tick, *I. ricinus,* is the primary vector in Europe and western Asia, while the taiga tick, *I. persulcatus,* is the primary vector from eastern Europe across most of northern Asia. Other species of the *I. ricinus* complex (e.g., *I. ovatus*), also serve as efficient enzootic vectors.

The cycle of *B. burgdorferi* development in the tick begins with ingestion of an infectious blood meal by the larvae as they feed on a mouse or other reservoir host. Spirochetes survive in the midgut diverticula but are not disseminated until the tick molts to the nymphal stage and feeds again. Following the influx of fresh host blood, the spirochetes surviving in the midgut pass between the cells of the gut wall and disseminate to the salivary glands and other internal organs. However, in females, spirochetal invasion of the ovary damages the developing oocytes, which prevents efficient transovarial transmission. As a result, transmission is by the transstadial route. As the tick feeds, spirochetes escape with the saliva and are introduced to the wound site of the new host. Transmission from nymph to adult also is common.

In the eastern United States, the risk of acquiring LD is greatest in the late spring or early summer, because of the peculiar nature of the LD cycle in nature. Nymphal ticks infected with *B. burgdorferi* emerge from their overwintering diapause to attack hosts, especially white-footed mice and other small mammals, as well as humans. This feeding pattern ensures the presence soon afterwards of numerous infected hosts when the larvae emerge, spreading the *B. burgdorferi* infection to vast numbers of these tiny ticks. Fed larvae molt in late summer or early fall, but the resultant nymphs diapause rather than commence feeding. Thus, the infection is perpetuated from one year to the next. Meanwhile, adults that emerge from fed nymphs delay feeding until the fall, whereupon they seek white-tailed deer and other large mammals (including humans). This pattern results in a 2-year cycle throughout most of the tick's geographic range.

Although the vector ticks are opportunistic feeders and will attack most vertebrates, only a limited variety of hosts are competent reservoirs of *B. burgdorferi.* Only competent reservoirs, animals capable of maintaining spirochetes within their tissues for prolonged periods, can infect ticks that feed on them. Examples of competent reservoirs include the white-footed mouse, the dusky-footed wood rat, the California kangaroo rat, and several other rodents in North America, and such animals as the common shrew, bank vole, wood mouse, yellow-necked field mouse, and hares and pheasants in Europe. Several species of birds also are reservoir competent and play an important enzootiologic role by dispersing infected ticks over considerable distances, thereby establishing new foci of infection. In contrast, some of the animals that are excellent hosts of vector ticks, such as white-tailed deer and western fence lizards, destroy invading spirochetes and, therefore, are not reservoirs. Such animals serve as amplifying hosts and are critically important for the expansion and spread of the tick populations, but they play no direct role in the perpetuation of the infection.

In the United States, LD has increased more than 1.7 times since it was first designated a reportable disease in 1991. By 1999, 16,273 cases that met the U.S. Centers for Disease Control and Prevention definition had been reported, for an overall incidence of 6.0 per 100,000. Most cases were reported from the northeastern, mid-Atlantic, and north central United States. Other important foci are in northern Wisconsin, northern Minnesota, and northern California. In Europe, important foci occur in the Scandinavian countries and in Germany, Poland, the Czech Republic, and Russia.

Domestic animals are also susceptible to infection with *Borrelia* pathogens. Dogs, cats, cattle, horses, and possibly other livestock and companion animals were found to be infected with *B. burgdorferi.* High seroprevalence rates occur in hyperendemic areas such as the northeastern United States, where many dogs show typical symptoms of chronic LD, especially lameness in one or more legs, fever, and fatigue.

Rocky Mountain Spotted Fever

Rocky Mountain spotted fever (RMSF) occurs throughout almost the entire United States, southern Canada, and Mexico and, to a lesser extent, in South America. This disease is caused by a tiny intracellular bacterium, the rickettsia *Rickettsia rickettsii.* People become ill with RMSF following the bite of an infected adult tick. Once in the human host, the rickettsia multiply profusely in the epithelial linings of the capillaries, arterioles, and venules. Vessels hemorrhage which, in the dermis of the skin, leads to the characteristic red spots. These innumerable reddish lesions, raised above the skin surface (maculopapular rash), coalesce to form the characteristic spotted rash. Patients also develop high fever, severe headaches, nausea, joint and muscle pain, photophobia, and other symptoms. Unless treated with antibiotics, some patients die and others suffer irreversible injury.

In the United States and southern Canada, the primary vectors of RMSF are the Rocky Mountain wood tick, *Dermacentor andersoni,* in the west and *D. variabilis* in the east. Adults of these tick species readily attack humans. Other tick species (e.g., the rabbit ticks *Haemaphysalis leporispalustris* and *Ixodes dentatus*) transmit the rickettsia when they feed on rabbits and birds, thereby contributing to the maintenance of the zoonosis in nature, but these latter species do not bite humans. The disease survives in the natural environment in overwintering (i.e., diapausing) ticks, but not in the reservoir hosts. When rickettsia-infected larvae emerge from diapause in the spring, they infect mice and other susceptible rodents on which they feed. Other, uninfected larvae feeding on the rickettsemic animals acquire the infection, and the disease spreads rapidly in the tick population. Nymphal ticks spread the infection further as they feed again on other small mammals. Adult ticks seek larger animals (e.g., dogs, raccoons, etc.) as well as humans. Rickettsia are passed to subsequent generations of ticks by transovarial transmission. Ticks also harbor nonpathogenic species such as *R. montana, R. belli, R. rhipicephali, R. parkeri,* and other as yet unnamed rickettsia, complicating attempts to measure the incidence of RMSF in nature.

In the United States, about 600 to 800 cases of RMSF have been reported yearly since 1985, with an estimated annual incidence between 0.24 to 0.32 per 100,000 population. Most cases now occur east of the Mississippi River, with the highest concentration in the south-central and southeastern states, especially along the Atlantic coast. Cases tend to occur in foci in rural areas and suburban communities near major population centers. Since RMSF is a seasonal disease, the frequency of cases follows the seasonal activity pattern of the adults, with highest frequency in July and August in the southern United States, but greater frequency in May and June in the northeastern part of the country.

A closely related disease, boutonneuse fever (Mediterranean spotted fever), caused by *R. conorii,* occurs in southern Europe, North Africa, and Asia. These rickettsia are transmitted to humans by the bite of the brown dog tick, *Rhipicephalus sanguineus.* Although generally similar in its symptoms to RMSF, this illness is distinguished by the formation of a black ulcer, the eschar, at the wound site where the infected tick had attached.

Ehrlichiosis

Another rickettsial disease that has emerged into increasing prominence in recent years is ehrlichiosis. The causative agents are known as ehrlichiae, tiny obligate intracellular organisms that invade the white blood cells of humans and animals. Ehrlichiae develop within cytoplasmic vacuoles in different blood cell types such as monocytes, granulocytes, lymphocytes, or even platelets, depending on the species. Patients develop an acute illness with high fever and severe headaches, as well as aching muscles and joints. In contrast to RMSF, a rash is not common (20–30% of patients) and usually does not involve the palms or soles. Although severe cases can occur and may result in death, most cases are relatively mild. In the southeastern United States, most cases are caused by *Ehrlichia chaffeensis,* which invades the monocytic leukocytes and causes leukopenia (abnormal loss of white blood cells) among other symptoms. This disease is now known as human monocytic ehrlichiosis (HME). The primary vector for *E. chaffeensis* is the lone star tick *Amblyomma americanum.* These ehrlichiae develop primarily in the monocytes. Some cases are caused by *E. ewingii,* also transmitted by lone star ticks. In the northern and western United States, the disease is caused by a related ehrlichia, *E. phagocytophila (E. equi),* which also causes illness in livestock. However, like *E. ewingii,* these ehrlichiae prefer the granulocytic leukocytes and, for this reason, the disease is known as human granulocytic ehrlichiosis (HGE). The primary vector in the northern United States is *I. scapularis.* In California, the primary vector for humans is *I. pacificus.* In Europe, the pathogen is transmitted to humans by the bites of the *I. ricinus. E. phagocytophila,* believed to be the most widespread of the various human-infecting ehrlichiae, has been reported in many countries of Europe and northern Asia. Ehrlichiosis also affects dogs, sheep, and other animals, being caused by different ehrlichiae.

Tick-Borne Encephalitis

Tick-borne encephalitis (TBE) is caused by viruses of the family Flaviviridae. In Europe and northern Asia, the illness in humans is manifested by high, often biphasic, fever and headache, followed soon afterward by inflammation of the brain (encephalitis) and meninges (meningitis). Some patients develop muscle weakness or paralysis, especially in the right shoulder muscles. In the Far East, case fatality rates are relatively high (up to 54%), whereas the disease in Europe is considerably milder. In Europe and the Far East, TBE viruses are transmitted by *I. ricinus* and by the taiga tick, *I. persulcatus.* Other tick species that feed on wild animals may amplify viral infection. Rodents and insectivores are the chief reservoir hosts. The disease is endemic in central, northern, and eastern Europe, where thousands of cases are reported each year. In India, a similar disease known as Kyasanur Forest disease has been responsible for numerous cases of human illness. In North America, a *Flavirus* of the TBE complex causes a disease known Powassan encephalitis, a serious illness that has caused death in some patients and permanent nerve damage in others. The primary vector to humans in the east is a little-known tick, *I. cookei,* a common parasite of woodchucks, foxes, and raccoons; in the west it is transmitted by the *D. andersoni.*

Babesiosis and Theileriosis

Two genera of tick-borne protozoa, *Babesia* and *Theileria,* cause severe illness and even death in domestic livestock and wildlife throughout most of the world. Tick-borne babesiosis also can cause illness in humans. *B. bigemina, B. bovis,* and *B. divergens* are examples of tick-borne protozoans that can cause babesiosis, a serious and often fatal illness in cattle prevalent in Mexico, parts of South America, Africa, and elsewhere. The clinical course of the disease is characterized by high fever and dehydration. Sick animals stop feeding, become lethargic, and show labored breathing. Anemia and bloody urine (so-called redwater fever) is a consequence of the massive destruction of red blood cells. Although some animals recover naturally, most of the diseased individuals gradually become comatose and die. Mortality estimates for infections with *B. bigemina* are 30%; for *B. bovis,* 70 to 80%. The causative agents, which are introduced by the bites of infected ticks, develop as meronts in the erythrocytes and multiply by multiple fission or, in the case of *B. bigemina,* by binary fission. Subsequently, huge numbers of erythrocytes are destroyed, leading to the symptoms noted. Eventually, some parasites invade leukocytes, where they are transformed into piroplasms, the stage that will infect ticks. Piroplasms remain in an arrested state of development until ingested by a suitable tick vector. In the tick's digestive tract, the erythrocytes and leukocytes of the host blood are lysed, releasing the parasites. The meronts are destroyed, but the piroplasms develop into ray-shaped gamonts (strahlenkorpers) that fuse with other, similar gamonts to form the diploid zygotes. These parasites are transformed into cell-penetrating ookinetes that traverse the tick's midgut epithelium and seek out the ovaries. In female *Boophilus* ticks, numerous oocytes are invaded by these ookinetes. When the ticks mate and lay eggs, the ookinetes proliferate in the developing embryos. Eventually, as the embryos mature, the ookinetes move to the salivary glands of the young larvae, where they form the invasive stage known as sporozoites, ready to infect cattle. Thus, a single blood meal in a female tick can result in thousands of disease-infected larval progeny, a phenomenon known as transovarial transmission. As a result, numerous animals or even an entire herd of cattle can be infected in a single tick generation.

Human babesiosis is a malaria-like illness caused by *B. microti, B. gibsoni, B. major,* or *B. divergens.* In the northeastern and midwestern United States, the dominant pathogen afflicting humans is *B. microti,* transmitted by the bite of *I. scapularis.* In Europe, human babesiosis may be caused by *B. microti,* or *B. divergens,* transmitted by *I. ricinus* and/or other ixodids. Although the life cycle in the human body is similar to that of the bovine babesias, development in the tick is quite different. When blood infected with *B. microti* is ingested by immature deer ticks, the invasive stages of the parasite transform into cell-penetrating sporokinetes, which, aided by a cell-piercing arrowhead organelle, traverse the midgut epithelium and find their way to the salivary glands. When the fed ticks molt to the next feeding stage and attack new human or animal hosts, the sporozoites are injected with the tick's saliva.

Babesiosis also afflicts other animals. In dogs, *B. canis,* transmitted by the bites of the *R. sanguineus,* causes severe illness and may be fatal. Other babesias cause illness in horses, sheep, and other animals.

Protozoan parasites closely related to the babesias, *Theileria parva* and *T. annulata* cause serious or fatal illness in cattle and other ungulates. *T. parva,* transmitted by the bites of the African brown ear tick, *Rhipicephalus appendiculatus,* causes a disease in southern Africa known as East Coast fever. *T. annulata,* transmitted by the bites of ticks of the genus *Hyalomma,* causes a disease known as tropical theileriosis. It is much more widespread, ranging from North Africa across western and central Asia to the Indian subcontinent. The life cycle of both protozoan parasites is generally similar to that of the babesias, but with several notable exceptions. *Theileria* parasites multiply in the leukocytes instead of the erythrocytes. Following destruction of the host cells, the *Theileria* progeny infect other leukocytes, thereby damaging the host's immune system, while others invade erythrocytes to form piroplasms. The latter are the stage that will infect ticks. These piroplasms remain in an arrested state of development until ingested by the tick vector. When vector-competent ticks feed on *Theileria*-infected cattle, the piroplasms are released from the disintegrating erythrocytes and transform into gamonts similar to those seen in the babesias. Following gamont fusion and fertilization, the zygotes develop into motile kinetes that cross the midgut epithelium and travel to the salivary glands. Because only the salivary glands are infected, pathogen transmission occurs only when the infected larval or nymphal ticks molt and feed again in the next life stage on susceptible hosts (transstadial transmission).

Tick Paralysis and Tick Bite Allergies

In addition to transmission of infectious microbes, ticks may cause paralysis, allergies, and severe toxic reactions in their hosts. In humans, this affliction is manifested by a gradually ascending paralysis, beginning with the loss of sensation and motor coordination in the legs, abdominal muscles, back muscles and, eventually, the diaphragm and intercostal muscles. Unless the tick responsible for these symptoms is found and removed, the patient will undergo respiratory failure and die. Most patients recover rapidly after tick removal, but those with advanced symptoms may require several weeks for complete recovery.

Tick paralysis also affects animals, especially large ungulates. In Australia, a single *Ixodes holocyclus* female may be sufficient to cause the death of a full-grown cow or bull. Thousands of cattle died in the western United States and Canada because of paralysis induced by the bites of *D. andersoni.* Other than finding and removing all the feeding female ticks, there is no known cure. Some ticks also cause severe allergic and toxic reactions. A notable example is the African tampan *Ornithodoros savignyi,* the bite of which may cause cattle to salivate, tremble, gnash their teeth, become disoriented, and even die. A similar condition, known as sweating sickness, is caused by ticks of the genus *Hyalomma,* especially *H. truncatum.* The illness is characterized by fever, loss of appetite, sweating, lachrymation, salivation and, usually, death. However, there is no paralysis.

CONTROL

Although difficult to estimate in precise monetary figures, the economic importance of ticks is enormous. This is primarily because of losses caused by tick-transmitted diseases affecting livestock. In addition, ticks cause injury from severe blood loss, reduced milk production, and damage to the hides. Worldwide, losses to the cattle industry alone were estimated to be more than $7 billion. Perhaps the most important economically are the cattle ticks, especially *Boophilus annulatus, B. decoloratus,* and *B. microplus,* because of the disease agents they transmit (*Babesia* spp.). Other economically important species are the bont ticks, *Amblyomma variegatum* and *A. hebraeum,* which are vectors of the causative agent of heartwater, *Cowdria ruminantium.*

Tick control is done almost exclusively with acaricides: poisonous compounds such as chlorinated hydrocarbons (e.g., DDT), organophosphorus compounds (e.g., diazinon), carbamates (e.g., carbaryl), and pyrethroids (e.g., permethrin, flumethrin). Animals are treated with acaricides by either bathing or spraying them with mixtures of liquids containing the specific toxicant. Acaricides also can be delivered as pour-ons or spot-ons, highly concentrated mixtures containing surfactants (i.e., spreading agents) that enable the product to disperse naturally throughout the animal's hair coat or as dusts, consisting of mixtures of talc and the acaricide. For long-lasting effects, acaricides are incorporated into plastic collars, ear tags, or even tail tags, which gradually emit the toxic chemicals over a long period of time. Also promising is the tick decoy, in which tail tags are impregnated with the tick's natural pheromone as well as the acaricide. These "tick decoys" demonstrated excellent efficacy (> 95%) for up to 3 months when applied to cattle in Zimbabwe. Other promising innovations for treating tick-infested animals include self-medicating devices, such as the "four-poster" developed by the U.S. Department of Agriculture for treating white-tailed deer, and antitick vaccines. Self-medicating applicators release an acaricide from a reservoir when host animals insert their heads into the device to retrieve corn or other bait. This is perhaps the first practical method that has been developed for controlling ticks on wild animals without killing or injuring them. Antitick vaccines work by immunizing livestock animals such as beef cattle against proteins in the tick's digestive tract, thereby killing the ticks as they attempt to feed. Further scientific advances may make it possible to reduce the current dependence on toxic chemicals to control ticks on domestic animals, pets and even wildlife.

Hikers, campers, and other people entering tick-infested habitats should take appropriate precautions to avoid ticks and the risk of acquiring tick-borne diseases. Each person should wear long trousers, tuck the trouser ends into boots or shoes, and draw the socks over the trousers. The protective barrier should be secured with tape and treated with a repellent—for example, one of the many products that contain diethyl toluamide (DEET). Upon returning from the trip, each person should decontaminate the field clothes and carefully inspect skin and hair for ticks that may have attached unnoticed. Early detection and removal of attached ticks will minimize the risk of transmission of disease-causing agents. Unattached ticks, if any, may be destroyed and discarded. However, attached ticks should be removed carefully with tweezers (forceps) and retained intact if needed for future examination in case of illness.

See Also the Following Articles

Arthropoda and Related Groups • Medical Entomology • Mites • Veterinary Entomology • Zoonoses, Arthropod-Borne

Further Reading

Anonymous. (2001). Lyme disease—United States, 1999. *MMWR* **50,** 181–185.
Hoogstraal, H., and Aeschlimann, A. (1982). Tick–host specificity. *Bull. Entomol. Suisse* **55,** 135–238.
Jaenson, T. G. T. (1991). The epidemiology of Lyme borreliosis. *Parasitol. Today* **7,** 39–45.
Klompen, H., and Grimaldi, D. (2001). First Mesozoic record of a parasitiform mite: A larval argasid tick in Cretaceous amber (Acari: Ixodida: Argasidae). *Ann. Entomol. Soc. Am.* **94(1),** 10–15.
Lane, R. S., Piesman, J., and Burgdorfer, W. (1991). Lyme borreliosis: Relation of its causative agent to its vectors and hosts in North America and Europe. *Annu. Rev. Entomol.* **36,** 587–609.
Needham, G. R., and Teel, P. D. (1991). Off-host physiological ecology of ixodid ticks. *Annu. Rev. Entomol.* **36,** 659–681.
Norval, R. A. I., Sonenshine, D. E., Allan, S. A., and Burridge, M. J. (1996). Efficacy of pheromone-acaricide impregnated tail-tag decoys for control of bont ticks, *Amblyomma hebraeum* on cattle in Zimbabwe. *Exp. Appl. Acarol.* **20,** 31–46.
Pound, J. M., Miller, J. A., George, J. E., and Lemeilleur, C. A. (2000). The '4-poster' passive topical treatment device to apply acaricide for controlling ticks (Acari: Ixodidae) feeding on white-tailed deer. *J. Med. Entomol.* **37,** 588–594.
Ribeiro, J. M. C. (1987). Role of saliva in blood feeding by arthropods. *Annu. Rev. Entomol.* **32,** 463–478.
Sonenshine, D. E. (1993). "Biology of Ticks." Oxford University Press, New York.
Sonenshine, D. E., Hamilton, J. G., Phillips, J. S., and Lusby, W. R. (1991). Mounting sex pheromone: its role in regulation of mate recognition in the ixodidae. *In* "Modern Acarology," Vol. 1 (F. Dusbabek and V. Bukva, eds.), pp. 71–78. Academia, Prague, and SPB Publishing, the Hague.
Strickland, G. T. (ed.). (2000). "Hunter's Tropical Medicine and Emerging Infectious Diseases." Saunders, Philadelphia.
Telford, S. R., III, and Spielman, A. (1989). Competence of a rabbit-feeding *Ixodes* (Acari: Ixodidae) as a vector of the Lyme disease spirochete. *J. Med. Entomol.* **26,** 118–121.

Touch

see *Mechanoreception*

Tracheal System

Jon F. Harrison

Arizona State University

In contrast to most vertebrates, crustaceans, mollusks, and many types of worm, insects do not use respiratory pigments in blood to transport oxygen. Instead, insects have a tracheal respiratory system in which oxygen and carbon dioxide travel primarily through air-filled tubes called tracheae. Usually the tracheal system penetrates the cuticle via closable valves called spiracles and ends near or within the tissues in tiny tubes called tracheoles (Fig. 1). The tracheae primarily serve as pipes that transport gases between the spiracles and the tracheoles, whereas the thin-walled tracheoles are thought to be the main sites of gas exchange with the tissues. In many insects, dilations of the tracheae form thin-walled air sacs or tracheal lungs that serve as bellows for enhancing the flow of gases through the tracheal system or directly deliver oxygen to tissues. The organization of the tracheal system varies dramatically among insects, with spiracle number ranging from zero to 20 and with tracheal branching patterns varying widely across species, between body regions, and during the developmental stages of holometabolous insects.

SPIRACLES

Spiracles are the openings of the tracheal system on the integument of the insect. A few apterygote insects lack valves in their spiracles and therefore have tracheae that are always open to the environment. However, spiracles generally have valvelike structures that allow them to close (Fig. 2). Spiracles

range tremendously in area from 27 mm^2 for the second spiracle of the giant *Petrognatha* beetle to openings of only a few micrometers. Often the outside of the spiracle is fringed with hairs or covered with a filter to help resist entry of dust, water, or parasites.

The valve that closes the spiracle takes various forms but can appear as two lips of cuticle or as a bar of cuticle that pinches the trachea shut (Fig. 2). The movement of the valve usually depends on the action of muscles that insert on or near it, and elastic elements in the cuticle. Some thoracic spiracles have only a closer muscle, and elastic elements tend to open the spiracular valve (Fig 2A). Often the valve is recessed behind the atrium, an enclosed space behind the external opening (Fig. 2B). The most common situation (true for all abdominal spiracles studied to date) is for the spiracle to be recessed behind an atrium and to have both opener and closer muscles (Fig. 2B).

TRACHEAE

Tracheae, the largest tubes of the tracheal system, range in diameter from a few millimeters to about 1 μm. The tracheal wall (Fig. 3A) contains an epithelial cell layer that secretes a

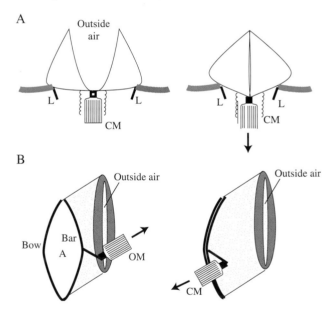

FIGURE 2 Spiracular structure and function. (A) Schematic side view of the locust metathoracic spiracle, a one-muscle spiracle with an external valve. [After Chapman, R. F. (1998). "The Insects: Structure and Function," Fig. 17.11c. Cambridge University Press, Cambridge, U.K., with permission of the publisher and the author.] When the closer muscle (CM) contracts, the lips of the valve come together. When the closer muscle relaxes, ligaments (L) attached to the cuticle passively pull the valve lips apart. (B) Schematic view from the inside of the animal of the abdominal spiracle of a butterfly, a two-muscle spiracle with an internal valve. [After Schmitz, A., and Wasserthal, L. T. (1999). Comparative morphology of the spiracles of the Papilionidae, Sphingidae, and Saturnicdae (Insecta: Lepidoptera). *Int. J. Insect Morphol.* **28**, 13–26, © 1999 with permission of Elsevier Science and the authors.] When the opener muscle (OM) contracts, it pulls on a lever that pulls the bar away from the bow, opening the atrium (A). When the closer muscle (CM) contracts, it pulls the bar against the bow, collapsing the walls of the atrium together.

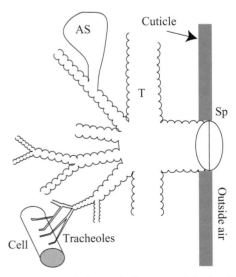

FIGURE 1 Overview of the tracheal system. Spiracles (Sp) in the integument connect to tracheae (T), which branch repeatedly within the insect, eventually leading to the tracheoles, which are located near most cells. In some insects, soft, collapsible air sacs (AS) occur.

A

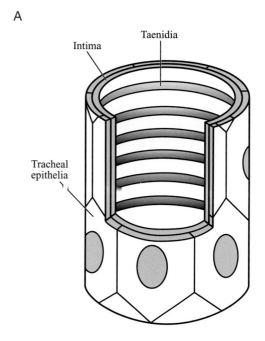

B

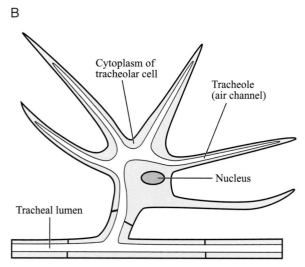

FIGURE 3 (A) Structure of a trachea. The basement membrane that encircles the tracheal epithelia has been omitted. (B) Tracheolar cell connected to a trachea.

FIGURE 4 Electron micrograph showing a tracheole (T) within the flight muscle of an American locust. M, mitochondria; Z, Z line; I, I-band. Scale bar = 1 micron. (Image courtesy of Scott D. Kirkton, Arizona State University.)

basement membrane that forms the outermost layer of the tracheal wall. Internally, the tracheal cells secrete the intima, which contains protein and chitin fibers. The intima forms spiral folds known as taenidia. In general, it is thought that the taenidia prevent the trachea from collapsing inward, yet allow the trachea to flex. However, in the tracheae of some insects, the taenidia are widely spaced within the tracheal wall, and the intima is quite flexible, with the result that such tracheae are compressible. Because tracheal walls are relatively thick, oxygen is not believed to pass out of the larger tracheae in appreciable amounts. Thus, the primary function of the trachea is to transport gases between the tracheoles and the spiracles. However,

because carbon dioxide can diffuse through tissue more easily than oxygen, it is possible that a significant portion of the carbon dioxide emitted may pass through the tracheal walls.

TRACHEOLES

Tracheoles, the small tubes that form the terminal endings of the tracheal system, range in size from 1 to 0.1 μm in diameter. Tracheoles are formed within single tracheolar cells, (Fig. 3B). The tracheolar cells have many branching processes, some of which contain an air-filled channel (the tracheole) that connects to the air-filled lumen of the trachea (Fig. 3B). Tracheole walls are capable of transporting oxygen at high rates by diffusion because they are thin (usually < 0.1 μm), and their ratio of surface area to volume is very large. Thus, it is likely that the tracheoles are the major site of gas exchanges between the tissues and the tracheal system.

Tracheoles are particularly dense in metabolically active tissues such as flight muscle. Most tracheoles occur outside the cells of the insect's body, but sometimes in histological sections they appear to be within cells, particularly in flight muscle (Fig. 4). These tracheoles are believed to enter the flight muscle via infoldings of the muscle plasma membrane, and so they are actually extracellular. The high tracheolar densities and penetration of flight muscle cells by tracheoles allows flying insects to achieve oxygen consumption rates that are among the highest in the animal kingdom.

A

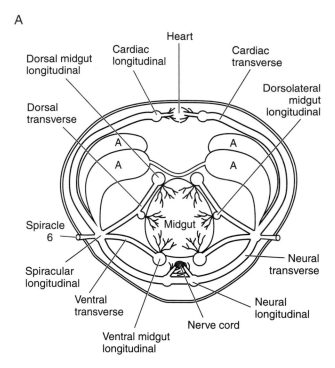

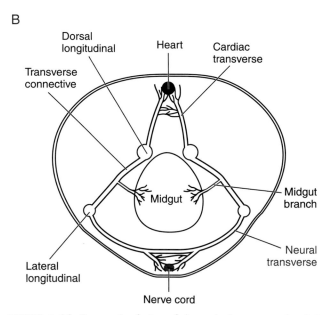

FIGURE 5 (A) Cross-sectional view of the tracheal system at the sixth abdominal segment of an adult grasshopper *(Schistocerca americana),* an insect with air sacs and 10 paired spiracles. (B) Cross-sectional view of the tracheal system at the sixth abdominal segment of an embryonic or larval fruit fly *(Drosophila melanogaster),* an insect without air sacs and with only the first and tenth spiracles functional. In both insects, longitudinal tracheal trunks run the length of the animal, connecting the thoracic and abdominal segments, but longitudinal trunks are much more numerous in the grasshopper.

AIR SACS, AERIFEROUS TRACHEAE, AND TRACHEAL LUNGS

In active insects, particularly those with a rigid cuticle, portions of the tracheal system are enlarged to form thin-walled air sacs. These air sacs lack taenidia and so collapse and expand with variation in hemolymph pressure. Air sacs are common in adult flying insects of many orders, including Diptera, Hymenoptera, Lepidoptera, Odonata, and Orthoptera. These air sacs serve as bellows to enhance the movement of air through the tracheal system. For insects that have them, air sacs often account for the majority of the volume of the tracheal system. Of insects that lack air sacs (such as cockroaches or Dictyoptera), many have slightly dilated, compressible tracheae that may also function as bellows.

Tracheal lungs, or aeriferous tracheae, are similar to air sacs in that they are thin-walled expansions of the tracheal system with few taenidia that float freely in the hemolymph. Their primary function is believed to be oxygen delivery to tissues that lack tracheoles, such as hematocytes, and some ovaries and Malpighian tubules. In the absence of tracheoles, the oxygen must diffuse through the thin wall of the tracheal lung, through the hemolymph, and to the tissues.

ORGANIZATION

The organization of the tracheal system varies substantially among insects, at least partially in correlation with species ecology. One obvious variant is spiracle number. In some aquatic insects, lack of spiracles prevents water entry to the tracheae. Many dipteran larvae have only one pair of spiracles on their abdominal tip, which they extend into the air as they feed head-down in water. In contrast, many terrestrial insects have a full 10 pairs of spiracles. Insects also vary in the number of air sacs, the location and size of tracheal branches, and the number of longitudinal tracheae.

A glimpse of the cross-species variety can be gained by comparing the organization of the major tracheae in the sixth abdominal segment of an adult grasshopper and a larval fruit fly (Fig. 5). Grasshoppers have a spiracle on every segment, whereas fly larvae have only functional spiracles on the head and abdominal tip. In the grasshopper, transverse tracheae lead from the spiracles to longitudinal tracheae that run along the heart, ventral nerve cord, and several positions on the gut. Air sacs are prominent. In contrast, in flies, two large dorsal longitudinal trunks connect the two functional spiracles on each side, with branches extending from the dorsal trunk to the tissues and one other longitudinal trachea. Air sacs are absent.

Tracheal organization also varies with location within the insect. In most insects, the head segments lack spiracles, and so tracheae usually extend forward from the prothoracic spiracles to supply the tissues in the head; similar elaborations of the tracheae occur in the posterior direction in the tenth segment. Structures that extend away from the main body of the insect, such as legs, antennae, and wings, also lack spiracles and usually have long tracheae extending from the spiracle of the local body segment. Antennae, legs, and wings all are known to have accessory pulsatile structures that pump hemolymph and help circulate air through their long tracheae. The tracheal organization in the thorax of flying insects is very different from that in the abdomen, since the thoraces often contain extensive air sacs associated with the flight muscles.

Tracheal organization can also be strongly modified by developmental stage, especially in holometabolous insects. For example, many dipterans transform from aquatic larvae with one or two pairs of spiracles and no air sacs to adults with 10 pairs of spiracles and extensive air sacs, a transition that enables high oxygen delivery and flight. Clearly, flexibility in tracheal structure and function is a key trait in the success, adaptability, and diversity of insects.

See Also the Following Articles

Anatomy • Respiratory System

Further Reading

Chapman, R. F. (1998). "The Insects: Structure and Function." Cambridge University Press, Cambridge, U.K.

Locke, M. (1998). Caterpillars have evolved lungs for hemocyte gas exchange. *Journal of Insect Physiol.* **44,** 1–20.

Manning, G., and Krasnow, M. A. (1993). Development of the *Drosophila* tracheal system. "The Development of *Drosophila melanogaster*" (M. Bate and A. M. Arias, eds.). Cold Springs Harbor Laboratory Press, Plainview, NY.

Nikam, T. B., and Khole, V. V. (1989). "Insect Spiracular Systems." Ellis Horwood Entomology and Nematology, Chichester, U.K.

Pass, G. (2000). Accessory pulsatile organs: Evolutionary innovations in insects. *Annu. Rev. Entomol.* **45,** 495–518.

Schmitz, A., and Perry, S. F. (1999). Stereological determination of tracheal volume and diffusing capacity of the tracheal walls in the stick insect *Carausius morosus* (Phasmatodea, Lonchodidae). *Physiol. Biochem. Zool.* **72,** 205–218.

Schmitz, A., and Wasserthal, L. T. (1999). Comparative morphology of the spiracles of the Papilionidae, Sphingidae, and Saturniidae (Insecta: Lepidoptera). *Int. J. Insect Morphol.* **28,** 13–26.

Snodgrass, R. E. (1993). "Principles of Insect Morphology." Cornell University Press, Ithaca, NY.

Wasserthal, L. T. (1998). The open hemolymph system of Holometabola and its relation to the tracheal space. *In* "Microscopic Anatomy of Invertebrates," Vol. 11b, "Insecta," (F. W. Harrison and M. Locke, eds.), pp. 583–620. Wiley-Liss, New York.

Wigglesworth, V. B. (1983). The physiology of insect tracheoles. *Adv. Insect Physiol.* **17,** 85–149.

Trichoptera (Caddisflies)

John C. Morse
Clemson University

Trichoptera, or caddisflies, are holometabolous insects closely related to Lepidoptera, or moths. However, unlike most moths, their eggs, larvae, and pupae are usually found in or very near fresh water, and adults are aerial, usually not far from their aquatic habitats (Fig. 1). The Trichoptera include

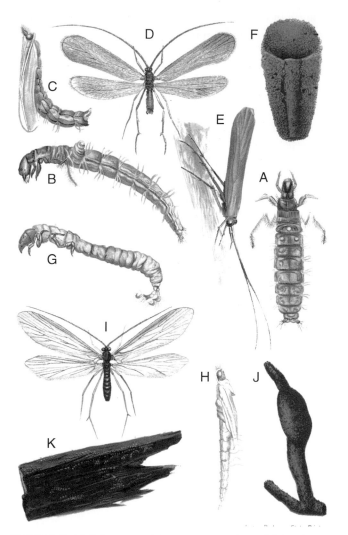

FIGURE 1 Caddisfly larvae, pupal, imagoes, cases, and eggs: (A) dorsal view of larva of *Molanna cinerea,* ×4; (B) lateral view of larva of *M. cinerea,* ×5; (C) lateral view of the pupa of *M. cinerea,* ×4.5; (D) dorsal view of imago of *M. cinerea,* ×4; (E) accustomed resting position of the imago of *M. cinerea;* (F) ventral view of the flat larval case of *M. cinerea,* ×2; (G) lateral view of larva of *Phylocentropus lucidus,* showing the very long anal prolegs and the absence of gill filaments, ×5; (H) lateral view of pupa of *P. lucidus,* ×6; (I) dorsal view of imago of *P. lucidus,* ×3.5; (J) larval case of *P. lucidus,* tube composed of sand and silk, two-layered, enlargement near the end containing the pupa; (K) eggs laid by *P. lucidus* female on a stick protruding from the water in a breeding cage. [From Betten, C. (1901). Order Trichoptera, caddisflies. "Aquatic Insects of the Adirondacks: A Study Conducted at the Entomologic Field Station, Saranac Inn, N.Y, under the Direction of Ephraim Porter Felt D.Sc. State Entomologist. (J. G. Needham and C. Betten, eds.) pp. 561–573 and plates 13,15, 30–34. *Bull. N.Y. State Mus.* **47,** 383–612, plates 1–36.]

more species than any of the other primarily aquatic orders of insects. This high species diversity is correlated with an unusually broad range of ecological specialization. Immature caddisflies occur in almost every type of freshwater habitat and on every continent except Antarctica; they are often one of the most abundant insect orders in streams and ponds. Larvae of many species use silk to construct portable cases of various shapes and materials to serve as physical protection, camou-

TABLE I Extant Families of Trichoptera, with Primary Larval Feeding Strategies, Distributions, and Approximate Numbers of Named Species and Subspecies

Family	Larval feeding strategy[a]	Distribution[b]							Number of species[c]
		AT	AU	EP	NA	NT	OL	WP	
Annulipalpia									
Stenopsychidae	cf	X	X	X		X	X		93
Philopotamidae	cf	X	X	X	X	X	X	X	911
Hydropsychidae	cf, pe	X	X	X	X	X	X	X	1462
Dipseudopsidae	cf	X	X	X	X		X	X	165
Polycentropodidae	cf, pe	X	X	X	X	X	X	X	600
Ecnomidae	cf	X	X	X	X	X	X	X	345
Xiphocentronidae	cg	X		X	X	X	X	X	130
Psychomyiidae	cg	X	X	X	X		X	X	395
"Spicipalpia"									
Rhyacophilidae	pe	X	X	X	X	X	X	X	685
Hydrobiosidae	pe		X	X	X	X	X	X	375
Glossosomatidae	sc	X	X	X	X	X	X	X	522
Hydroptilidae	ph, sc, cg	X	X	X	X	X	X	X	1677
Integripalpia									
Oeconesidae	sd		X						19
Brachycentridae	cf, cg, sh			X	X		X	X	111
Phryganopsychidae	sd			X			X		3
Lepidostomatidae	sd	X	X	X	X	X	X	X	391
Kokiriidae	pe		X			X			8
Plectrotarsidae	sd		X						5
Phryganeidae	sh, pe			X	X		X	X	79
Goeridae	sc, cg	X	X	X	X		X	X	163
Uenoidae	sc, cg			X	X		X	X	79
Apataniidae	sc, cg			X	X		X	X	186
Limnephilidae	sd, cg, sc		X	X	X	X	X	X	873
Tasimiidae	sc		X			X			9
Odontoceridae	sd		X	X	X	X	X	X	106
Atriplectididae	sd	X	X			X			5
Philorheithridae	pe		X			X			21
Molannidae	sc, cg, pe			X	X		X	X	32
Calamoceratidae	sd, sc,	X	X	X	X	X	X	X	124
Leptoceridae	cg, sh, pe	X	X	X	X	X	X	X	1567
Sericostomatidae	sd	X		X	X	X	X	X	97
Beraeidae	cg	X		X	X			X	52
Anomalopsychidae	sc					X			22
Helicopsychidae	sc	X	X	X	X	X	X	X	197
Chathamiidae	sh		X						5
Helicophidae	cg, sh		X			X			21
Calocidae	cg		X						20
Conoesucidae	sc, sd, sh		X						42
Antipodoeciidae	sc		X						1
Barbarochthonidae	sh, sc	X							1
Hydrosalpingidae	sc, sd, sh	X							1
Limnocentropodidae	pe			X			X		15
Petrothrincidae	sc	X							6
Pisuliidae	sd	X							15
Rossianidae	sc, sh				X				2

[a] Abbreviations: cf, collector-filterers; cg, collector-gatherers: pe, predator-engulfers; ph, piercer-herbivores; pp, predator-piercers; sc, scrapers; sd, shredder-detritivores; sh, shredder-herbivores. (From Merritt and Cummins, 1996. Some larval feeding strategy data from M. Winterbourne, University of Canterbury, Christchurch, New Zealand, and F. C. de Moor, Albany Museum, Grahamstown, South Africa.)

[b] Abbreviations: AT, Afrotropical; AU, Australasian; EP, East Palearctic; NA, Nearctic; NT, Neotropical; OL, Oriental; WP, West Palearctic.

[c] The total number of these known species and subspecies is 11,638.

flage, or aids in respiration. Others make stationary retreats of silk for similar purposes or to serve as food-gathering structures. The variety of feeding strategies employed by caddisflies is as great as for the highly diverse freshwater Diptera, or true flies (Table I). Because of their diversity and density in most clean, freshwater ecosystems, the significance of caddisflies for processing nutrients and transferring energy is often great. Their importance as food for predators such as fish is emphasized by fly-fishing enthusiasts who imitate them with artificial lures. The different caddisfly species are variously sensitive to changes in environmental conditions, such that the diversity of the order is commonly used in part as a measure of pollution.

DIVERSITY AND PHYLOGENETIC RELATIONSHIPS OF TRICHOPTERA

More than 11,000 species of caddisflies have been described globally (Table I), with more than 1300 reported for America north of Mexico. The world species are in 504 genera in 45 families. The most species have been described in the micro-caddisfly family, Hydroptilidae, with nearly 1700 species. Other well-represented families globally include the long-horned caddisflies (Leptoceridae, > 1500 species), the common netspinner caddisflies (Hydropsychidae, > 1400 species), and the northern caddisflies (Limnephilidae, almost 900 species). The highest known species diversity and the greatest density of species occurs in the Oriental biogeographical region (> 3700 species, with 1.6 species per kilohectare).

The Trichoptera and Lepidoptera are generally considered to be sister orders, together constituting the Amphiesmenoptera, with earliest fossils dating from the Permian. The families of tube-case-making caddisflies are generally considered to constitute a monophyletic suborder Integripalpia and the net-spinning caddisfly families a monophyletic suborder Annulipalpia (Fig. 2). The relationships of the remaining caddisfly families (Glossosomatidae, Hydrobiosidae, Hydroptilidae, and Rhyacophilidae, collectively the "Spicipalpia") are yet unknown but are apparently near the base of the caddisfly phylogeny. The phylogenetic relationships among genera and species of caddisflies have been studied for most extant families, mostly with morphological characters visible with a light microscope. The phylogeny of Trichoptera species continues to be investigated at all categorical scales by Kjer and others, using both morphological and molecular evidence and computer-managed algorithms. These hypothetical relationships have proven useful in numerous comparative studies, suggesting hypotheses for research of case- and retreat-making behaviors, feeding strategies, historical biogeography, mating behaviors, and other aspects of caddisfly biology.

Triassic fossils of species of Philopotamidae, Prorhyacophilidae, and Necrotauliidae are thought to represent the oldest known Trichoptera. The family Philopotamidae includes modern species, but the other two families are extinct.

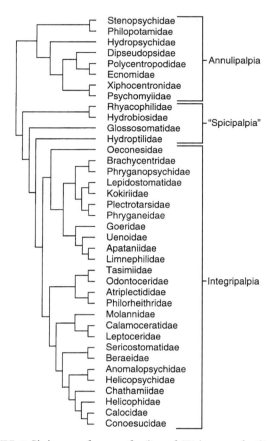

FIGURE 2 Phylogeny of extant families of Trichoptera; families not analyzed by Kjer *et al.* include Antipodoeciidae, Barbarochthonidae, Hydrosalpingidae, Limnocentropodidae, Petrothrincidae, Pisuliidae, and Rossianidae. [After Kjer, K. M., Blahnik, R. J., and Holzental, R. W. (2001). Phylogeny of Tricoptera (caddisflies): characterization of signal and noise within multiple data sets. *Syst. Biol.* **50(6)**, 781–816.]

LIFE HISTORIES OF TRICHOPTERA

Life Cycles

Among the relatively few species of caddisflies for which life histories are known, most exhibit a univoltine, or one-year, life cycle, with some species having more than one generation per year and some with one generation every 2 or 3 years. In general, longer life cycles are found in the higher latitudes and shorter life cycles closer to the equator. A few species have been shown to have more than one cohort in the same locality, with different segments of the same population having their life cycles synchronized apart from the synchrony of one or more other segments. The time of the year during which particular phases of the life cycle occur is distinctive for the species, with most species reaching adulthood in temperate regions during the warmer months of the year.

Eggs

Eggs are embedded in a sticky, gelatinous polysaccharide called spumalin, forming an egg mass. Round or elliptical eggs are generally laid in egg masses of particular shapes with 12 to 636 eggs reported per egg mass and with a single female depositing one to several egg masses. Egg masses may be spherical, pyramidal, butterfly shaped, or some other configuration, or they may be arranged in a flat spiral or as a string of beads. Because of the spumalin, little is known about the surface structure of caddisfly eggs. Eggs may diapause for several months before larvae hatch, especially during winter, but larvae generally hatch within a few days depending on temperature. If the egg mass is out of water, the first instars of some species may remain in the spumalin matrix until it is inundated; for other species whose egg masses are laid on plants or rocks overhanging the water, young larvae drip out of the spumalin into the water.

Larvae

Larvae usually undergo five instars before pupation, although a few species have a generally indeterminate number of instars greater than five. The larval stage usually is the longest stage of the life cycle, with development completed during a period of 2 months to nearly 2 years. The shape of the larva differs usually with the family or genus and depends in part on the habits and feeding strategy peculiar to that taxon. Eruciform larvae, typical of the Integripalpia, are cylindrical, with hypognathous mouthparts oriented at a right angle with the body axis and with the posterior end of the abdomen blunt and having thick anal prolegs fused with segment IX; these live in portable tubular cases and feed generally as shredders or scrapers or collector-gatherers.

Campodeiform larvae, typical of the Annulipalpia and unplaced primitive families, have a tapered shape anteriorly and posteriorly, with prognathous mouthparts nearly aligned with the body axis; the posterior end of the abdomen is slender, and there are slender, independent or semi-independent anal prolegs. These larvae live in silk retreats or roam freely in search of food. Larvae of microcaddisflies (family Hydroptilidae) typically undergo hypermetamorphosis, with the first four instars campodeiform and free-living and the fifth instar with a much-enlarged abdomen and living in a purse case.

Pupae

The last instar builds a shelter or modifies the larval case to serve as a shelter. Modification of a larval case usually means simply sealing the ends with silk closure membranes that are perforated to allow movement of water through the shelter. Larvae may then line the shelter with silk; a few families (Glossosomatidae, Hydrobiosidae, Rhyacophilidae) spin a semipermeable cocoon inside the shelter. The larva may then diapause for a few months as a prepupa inside the shelter, but usually it proceeds immediately to molt into the pupal stage, usually retaining the exuviae of the last instar inside the shelter. The pupal appendages are not fused with the body, as they are in some Lepidoptera and Diptera, but are merely folded against the body, with the wings and antennae wrapped around the sides and venter of the body.

Special structures of the pupa include long mandibles for keeping debris from blocking anterior closure membrane perforations and for cutting the closure membrane or cocoon at the time of emergence, setose caudal processes for keeping debris from blocking posterior closure membrane perforations, and small plates of hooks dorsally on the abdomen for maneuvering the pupa inside the silk-lined shelter or cocoon. Respiration is assisted by undulating movements of the pupa, forcing water through the shelter and cocoon from anterior to posterior ends. The pupa usually completes development to the adult stage in two or three weeks.

Emergence

When the pupa is ready to transform to the adult stage, it uses its long mandibles to cut through the shelter's anterior closure membrane, swims to the water surface, and emerges from the pupal exuviae at the surface in open water or on some emergent object. In the molt from pupa to adult, the period between apolysis and ecdysis is unusually long, allowing body sclerites to complete much of their sclerotization before the adult emerges. In this way, the adult caddisfly is able to fly quickly from the surface of the water, away from predatory fish, usually completing the escape from the shelter and the water surface in less than a minute. The teneral adult then rests on shoreside vegetation or rock until sclerotization is complete.

Adult

Nearly all adult caddisflies are capable of flight, with only a few species having short, flightless wings. They are secretive, hiding among shoreside vegetation or rocks most of the time, with periods of activity in the day or night specific for the different species. Adults of different species live for a few days to several months. Those that have long adult lives have an especially well-developed haustellum for imbibing liquids; apparently they ingest water or nectar for sustenance. Mating behavior for some species is mediated by sexual pheromones, with females emitting odors that attract males. Males of some day-flying species also exhibit distinctive flight patterns or swarms that attract females. Mating is accomplished while standing on shoreside substrate in an end-to-end orientation with tips of the wings overlapping; the male inferior appendages, or "claspers," hold the end of the female abdomen while he inserts his phallus and releases a spermatophore into her spermatheca. The female then lays her egg mass. It may be laid on a rock or vegetation overhanging the water, from which first instars may drip into the water upon hatching. It may be laid on underwater substrates by a female that crawls or swims beneath the water surface. A few species may lay an egg mass in a protected spot in a dry depression that will later be filled with water.

HABITATS AND DISTRIBUTION OF TRICHOPTERA

Habitats

Immature stages of caddisflies may live in a wide range of habitats with fast to slow current speed or in standing water, but the habitat preferences of the individual caddisfly species usually are quite restricted. Most species require relatively clean, cool water with high concentration of dissolved oxygen. In these habitats, larvae may be found in accumulations of dead leaves and sticks, on or in woody debris, on the tops or bottoms or sides of stones, burrowing in sand or silt, in the crevices among gravel, on submerged or floating parts of living plants, or on other relatively stable debris. A few species complete their entire life cycles away from water, with immature stages crawling among leaves on the forest floor or clinging to vertical rock faces. Immature stages of a few species develop in brackish water, and those of species of the family Chathamiidae apparently all develop only in ocean surf on the shores of New Zealand, at least one of them in a complex mutualistic relationship with starfish.

Distribution

Caddisflies are distributed throughout the world, with species known from every continent except Antarctica. The greatest known species diversity is in the Oriental biogeographic region, where the density of species per hectare is nearly twice that of the next most diverse region.

CASE AND RETREAT MAKING BY TRICHOPTERA

Case Making

Caddisfly larvae have long been appreciated for the beautiful and complex cases that many of them build. These usually are not attached to a substrate but are carried by the larva as it crawls or swims in its habitat. Larvae build cases from a wide range of mineral and plant materials, with the type of material and the shape of the case often recognizable in the field for the genus or even for the species. Mineral building materials may include sand grains or small stones, which may be all of one size or may be of two or more sizes organized in a species- or genus-specific pattern, for example with larger "ballast" stones laterally. The larva selects a mineral particle of preferred size and shape and applies it to the anterior edge of the case with silk extruded from the spinneret in its mouth. Examples of plant building materials include living or dead pieces of leaves or wood that have been shaped by the larva before their attachment to the anterior edge of the case with silk. The preferred plant material may be very specific, such as algae, mosses, liverworts, or rootlets or leaves of particular vascular plants growing in the water, or pine needles, sticks, or leaves that fall from shoreside vegetation. These materials may be oriented longitudinally or horizontally; longitudinal materials

may be organized in rings or in a spiral; horizontal materials may be interlocked at corners of a two-, three-, or four-sided case or may be wrapped in a cylindrical case. The larvae of a few species make their cases by hollowing the axis from a stick or piece of wood. Some cases are made entirely from silk or may incorporate pieces of freshwater sponge or abandoned shells of freshwater mollusks. At least one marine species builds its case with bits of coral.

The cases of caddisflies are often classified as tube cases (most Integripalpia), saddle cases or tortoise cases (Glossosomatidae), or purse cases (Hydroptilidae). A tube case is more or less cylindrical, surrounding the larva at approximately the same distance from all sides, although the shape of the case, especially externally, may vary greatly from one species to another. Thus, a tube case may be externally cylindrical or four-sided or three-sided or flattened; it may have lateral and/or anterodorsal flanges; or it may be straight or curved or coiled into a tight spiral like a snail shell. A tube case almost always has recognizable anterior and posterior ends. Pupation occurs within the tube case after it has been fastened to stationary substrate and the ends have been sealed with silk membranes (except for holes left for water circulation).

A saddle case is more or less domelike, oval on its flat side, always of mineral materials, sometimes with larger stones laterally. A transverse strap of fine sand grains connects the longer sides of the oval ventrally, beneath the larva, leaving anterior and posterior openings on the ventral side, from which the head and anus of the larva protrude interchangeably. This ventral strap is removed and the dome is fastened to stable substrate by the larva as it prepares to pupate under the dome.

A purse case typically consists of two flat sheets of silk (often including sand or algae) that have been sewn together at their edges, leaving the ends open for the head and anus to protrude interchangeably. Purse cases sometimes are cemented to the substrate. Pupation occurs within the purse case after the case has been fastened to stationary substrate and the ends have been sealed with silk membranes.

Retreat Making

About half the species of caddisflies do not build cases, but instead spin silken retreats that are stationary, fastened to the substrate. These retreats often are modified to assist with capturing food, usually by filtering it from moving water. The shape of a retreat usually is characteristic of a family or genus and may appear on wood or stone substrate—for example, as a flat sheet over a shallow depression on the substrate; a long, sinuate tunnel or a short covering that incorporates bits of mineral or plant material; a fingerlike net on the undersides of stones; or a bag of silk suspended from substrate in the current. Some species construct a silk-lined vertical tube in sandy soil. Pupation occurs either in the silken retreat or in a special pupation chamber constructed of silk and plant or mineral substrate by the larva.

Free-Living Larvae

Larvae of species of Rhyacophilidae and Hydrobiosidae are free-living predators, roaming about the substrate in search of prey. Pupation occurs under a domelike pupation chamber constructed of silk and plant or mineral substrate by the larva.

FEEDING STRATEGIES OF TRICHOPTERA

Although some long-lived adults may imbibe water or nectar, most nutrients are acquired by larvae. Larvae feed in a variety of ways, with a greater diversity of feeding strategies than any other group of insects having a comparable number of species. Early instars of most species and late instars of many species are collector-gatherers, picking small bits of loose organic material from the substrate. Many other species are collector-filterers, using silken nets (usually) or hairy legs to strain small bits of organic matter from moving water. Shredding herbivores chew pieces of leaves of living plants, often using some of the same leaves in their cases. Shredding detritivores gouge rotting wood or cut pieces of dead leaves that have been "conditioned" by fungi and bacteria, getting most of their nutrition from digestion of the fungi and bacteria. Scrapers graze on the algae, fungi, and associated organic material ("periphyton") that is attached to the surfaces of stones and plant material exposed to sunlight. Predators either chase their prey or lie in wait for it to wander nearby; prey items typically include other insects, microcrustaceans, and annelids; the predators either swallow their prey whole or they chew them in bits. Some caddisfly larvae may eat the flesh of dead animals, such as dead fish, as these become available. Some species of long-horned caddisflies feed facultatively or obligatorily on freshwater sponge, ingesting whole pieces of the sponge, including the spicules, which accumulate in the gut.

IMPORTANCE OF TRICHOPTERA

Caddisflies are one of the major groups of macroinvertebrates in freshwater ecosystems both in terms of species diversity and of density, especially in relatively unpolluted waterways. For this reason, they are significant contributors in the processing of nutrients. On one hand, collecting-gathering and collecting-filtering and scraping caddisflies help concentrate the nutrients of fine particulate organic matter into their own bodies, making the nutrients available to invertebrate and vertebrate predators in the food web. On the other hand, shredding herbivores and shredding detritivores and predators help to break coarse organic materials into small particles, including feces, that can then be used by many animals that are able to ingest only fine bits of organic material.

Food Resources

Because of the many different feeding strategies and habitat preferences of this diverse order, nearly every conceivable food resource is processed by caddisflies. Because their populations are so large, they process significant amounts of these resources for the benefit of the other animals in the ecosystem. There is even evidence that scraping caddisflies help to increase the production rate of the attached algae on which they feed, much as mowing the grass in a lawn or pruning new growth of fruit trees stimulates increased growth and production in those plants. The role of caddisflies in the food web is appreciated very well by sport fishing enthusiasts, who tie imitations of larval, pupal, and adult caddisflies on hooks to entice their game fish to bite a hook. The more those who fish learn about the species diversity, behavior, and biology of caddisflies, the more likely they are to succeed in tricking the fish to take the hook.

Pollution Tolerance

Although caddisflies generally will not tolerate even moderate levels of pollution, the range of tolerances is wide among the various species of caddisflies. For this reason and because of the usual high species diversity and density of caddisflies in unpolluted surface waters, communities of Trichoptera and other macroinvertebrates are often used to detect the presence of pollution. The occurrence of several of the less-tolerant species and high densities of large numbers of species of caddisflies suggest that the water is relatively unpolluted. Pollution may be detected with this technique more reliably and more cheaply than with chemical analyses. Once it has been established that a given waterway is polluted, follow-up analyses may then be attempted to discover the specific polluting substances or microorganisms and their concentrations. Finally, equipped with those data, land managers, engineers, economists, politicians, and other responsible decision makers may be better able to determine appropriate mitigating measures to reduce or eliminate the pollution.

See Also the Following Articles
Aquatic Habitats • Pollution, Insect Response to

Further Reading
Barbour, M. T., Gerritsen, J., Snyder, B. D., and Stribling, J. B. (1999). "Rapid Bioassessment Protocols for Use in Streams and Wadeable Rivers: Periphyton, Benthic Macroinvertebrates, and Fish." 2nd ed. EPA 841-B-99-002. U.S. Environmental Protection Agency, Office of Water, Washington, DC.
Holzenthal, R. W., and Blahnik, R. J. (1997). Tree of Life: Trichoptera. http://phylogeny.arizona.edu/tree/eukaryotes/animals/arthropoda/hexapoda/trichoptera/trichoptera.html, updated 20 February 2001.
Kjer, K. M., Blahnik, R. J., and Holzental, R. W. (2001). Phylogeny of Trichoptera (caddisflies): Characterization of signal and noise within multiple datasets. *Syst. Biol.* **50(6),** 781–816.
LaFontaine, G. (1989). "Caddisflies: A Major Study of One of the Most Important Aquatic Insects—Entomology, Fly Tying, and Proven Fishing Techniques." Lyons Press, New York, NY.
Merritt, R. W., and Cummins, K. W., (eds.) (1996). "An Introduction to the Aquatic Insects of North America." 3rd ed., Kendall/Hunt, Dubuque, IA.
Morse, J. C. (1997). Phylogeny of Trichoptera. *Annu. Rev. Entomol.* **42,** 427–450.

Morse, J. C., (ed.) (1999). Trichoptera World Checklist. http://entweb.clemson.edu/database/trichopt/index.htm, effective 22 May 1999, updated 28 July 2000, 8 January 2001.

Schmid, F. (1998). "The Insects and Arachnids of Canada," Part 7: "Genera of the Trichoptera of Canada and Adjoining or Adjacent United States." NRC Research Press, Ottawa.

Wiggins, G. B. (1996). "Larvae of the North American Caddisfly Genera (Trichoptera)." 2nd ed. University of Toronto Press, Toronto.

Tsetse Fly

Stephen G. A. Leak

International Trypanotolerance Centre, The Gambia

Tsetse flies belong to the single genus *Glossina*, in the family Glossinidae of the order Diptera. They are found only in sub-Saharan Africa, Yemen, and Saudi Arabia, infesting 38 countries and occupying about 11 million square kilometers. As cyclical vectors of protozoan parasites of the genus *Trypanosoma*, they are of major economic and biological importance. Trypanosomosis (sleeping sickness in humans) is a major constraint to livestock production and is a threat to millions of people in Africa. Within the genus, 31 species and subspecies of tsetse have been identified. Fossil *Glossina*, found in Florissant shales in Colorado, date back to the Oligocene, indicating a wider original distribution.

The genus is split into three subgenera or groups: *Austenina*, (*fusca* group), *Nemorhina* (*palpalis* group), and *Glossina* (*morsitans* group). The forest-dwelling *fusca* group are the most primitive. Classification is based largely on morphology of the genitalia; however, the three groups also differ ecologically. Differences in the cuticular alkanes of their sex pheromones can also be used for classification and for estimating "genetic" distance between species.

FIGURE 1 An adult tsetse fly in the process of feeding. (Photograph courtesy of the International Livestock Research Institute, Graphic Arts Unit, Nairobi.)

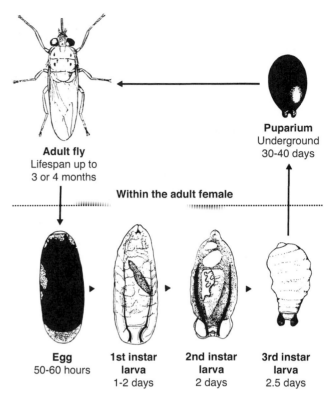

Adult fly
Lifespan up to
3 or 4 months

Puparium
Underground
30-40 days

Within the adult female

Egg
50-60 hours

**1st instar
larva**
1-2 days

**2nd instar
larva**
2 days

**3rd instar
larva**
2.5 days

FIGURE 2 Life cycle of the tsetse fly.

LIFE CYCLE AND REPRODUCTION

An adult female produces one egg at a time, from which a first instar hatches in the uterus, where it is supplied with nutrients. After a period of development and molting in the uterus, a third instar is deposited on the ground. One full-grown larva is produced every 9 or 10 days, which then pupates in light or sandy soil. The adult emerges after a puparial period of around 30 days, depending on the ambient temperature. This process, termed adenotrophic viviparity, results in a very low rate of reproduction.

Spermatogenesis occurs during the puparial stage of males and ceases after eclosion; maturation of adult males takes approximately 3 days. Substances synthesized in the male accessory glands are transferred to the female hemolymph during copulation and stimulate receptivity of female tsetse. Unfed females are unwilling to copulate, and receptivity declines with age or after mating. Most female tsetse are successfully inseminated even at very low population densities and need to mate only once in their lifetime, although multiple mating does occur. Male flies seem to rely on visual attraction to find mates.

Larval Nutrition and Development

In pregnant females, proteins synthesized by highly efficient "milk" glands provide nourishment to the developing larva. Milk is secreted cyclically, with a peak in early pregnancy, which continues from the time of larval hatch until parturition. Some components of the milk probably originate from fat-body secretions, and the milk contains relatively constant proportions of protein and lipid plus phospholipids, cholesterol, triglycerides, and tyrosine. The latter may be necessary for tanning larval and adult cuticles. At eclosion of the first instar, the milk is rich in acidic lipids, which are later replaced by proteins.

The anterodorsal mouth of the first instar is blocked by a chitinized median tooth. This "egg-tooth," lost with the skin of the first instar, is used to puncture the eggshell (chorion), which then splits along the dorsal side, allowing the first instar to hatch.

Larviposition and Pupariation

Unlike most dipteran larvae, which commit to metamorphosis early in the third instar, the third instar tsetse may pupariate only when fully grown. Shortly after larviposition, larvae exhibit a strong photonegative response and attempt to burrow. Uric acid from the anus then spreads over the larva and may protect it from predation. Periodicity of larviposition is affected by temperature, and the tsetse may respond to seasonal temperature cycles such that larviposition occurs in the most favorable conditions. The female fly becomes very active a few hours before parturition, presumably while searching for a suitable larviposition site. Both tactile and visual responses have a role in breeding site location. Females tend to choose darker resting sites as pregnancy progresses and when temperatures increase. The interlarval period depends on maternal nutrition and temperature conditions and varies according to species, localities, and seasons, from 20 to 90 days. Survival of puparia is dependent on relative humidity, which, together with temperature, influences selection of breeding and larviposition sites. Survival of emerging adults depends on the existence of sufficient fat reserves to support them until their first blood meal.

The adult emerges from the puparium using the "ptilinum," a saclike structure folded in a frontal cavity of the head. Rhythmic contractions of thoracic and abdominal muscles help the adult to break out of the puparium and reach the soil surface. Females emerge 1 to 2 days before males. Immediately after eclosion, the wings, abdomen, and thorax are highly compressed, although the legs are already full sized. The newly emerged fly has to expand by about 90%, using the cibarial pump to suck air into the gut, and must then feed to complete growth of the flight muscles and endocuticle.

BLOOD MEAL DIGESTION AND UTILIZATION

During feeding, tsetse discharge saliva containing a powerful anticoagulant antithrombin enzyme. The quantity of saliva secreted increases as tsetse become hungrier and may contain trypanosomes in infected flies. Two platelet aggregation inhibitors identified in saliva may also have immunosuppressive, anti-inflammatory properties. Digestion of blood proteins takes place in the posterior section of the midgut and involves six enzymes, which convert proteins to peptides and free amino acids.

Water Balance and Excretion

After feeding, the increased weight of the fly hampers flight and increases vulnerability to predators. There is therefore a rapid reduction of the water content of the blood meal from 79% to about 55% within 3 h after feeding. A blood meal in early pregnancy provides females with nutrients that are stored for larval development in late pregnancy.

Pheromones and Hormones

Sex recognition pheromones of tsetse are saturated, methyl-branched hydrocarbons found in cuticular waxes and are apparently unique to each tsetse species. They trigger mating behavior in males at ultraclose range or upon contact with baited decoys. The pheromone is released from the cuticle via unicellular glands on the legs and spreads over the external body surface by diffusion and grooming behavior. It is present on the pharate adult female about 2 days before emergence from the puparium and remains throughout life. A larval pheromone may be released from the anal orifice of tsetse larvae. The pheromone seems to affect choice of larviposition site, resulting in aggregation of larvae in breeding sites.

Flight Metabolism

The energy required for flight is produced from proline and is sustained by utilization of lipid reserves. Proline is depleted during flight and cannot rapidly be replenished; therefore, tsetse are limited to short periods of high-speed flight, at between 6.5 and 7.5 m s^{-1} (0.4 km min^{-1}) in open country. Proline is synthesized from lipids in fat bodies and oxidized to alanine in the muscles under hormonal control. Daily flight duration in male flies is about 15 min in the hot season and more than twice as long in the cold season. If fat reserves fall to about 6% of total dry body mass, the fly may die of starvation; such nutritional stress also results in production of small puparia.

Vision and Sense Organs

The compound eyes of both sexes are similar, with a specialized zone of greater visual acuity in the forward-pointing region. Males, however, have a wider region of binocular overlap, which may enhance detection of females. The visual function is consistent with fast flight, detection of drift due to low wind speeds, mating chases, and discrimination of cryptic hosts at high light intensities.

For mating behavior to be initiated, both mechano- and chemoreceptors must be stimulated. The chemoreceptors are thought to be basiconic sensilla on the tibiae and tarsi. Other, hair-shaped chemo- and mechanoreceptor sensilla are found on the front side of the wings.

DISTRIBUTION AND HABITATS

The general distribution of tsetse flies, determined principally by climate, is influenced by altitude, by vegetation, and by the presence of suitable host animals. The limit of distribution is closely correlated with the "tropical savanna (summer rain) climate," which follows the > 508 mm annual rainfall line, whereas the "tropical rain forest (equatorial) climate" controls the habitats of the *fusca* and *palpalis* groups. The surrounding savannas are the habitat of the *morsitans* group. The southern limits of *Glossina* distribution in Africa lie north of a line drawn from Benguela, in Angola, to Durban, in South Africa. The northern limits are roughly a line from Dakar in Senegal across to Ethiopia and Mogadishu in Somalia on the east coast.

The distribution and abundance of *morsitans* group tsetse often correspond to those of wild animals. In West Africa, *G. longipalpis* inhabits forest islands with plenty of shade and moves to forest edges only during seasons of high rainfall. *G. austeni* is confined to the coastal zone of East Africa, from Somalia to South Africa, including the island of Zanzibar until it was eradicated from the island in 1997.

Species of the *palpalis* group occur in drainage systems leading to the Atlantic Ocean or to the Mediterranean, not those draining into the Indian Ocean. *G. palpalis* is found close to water, in gallery forests, and it cannot tolerate the wide range of climatic conditions occurring in the savanna belt, where it is restricted to watercourses or "forest islands."

Species of the *fusca* group are mostly found in forests of West and central Africa. Humans are likely to have an adverse effect on *fusca* group tsetse populations through forest clearing and hunting.

Human activity has an important influence on tsetse distribution and abundance; humans can scare away or kill potential hosts, or destroy the vegetation forming the flies' habitat through agricultural or other development. Most African countries, particularly those in tsetse-infested areas, have low human population densities; however, Nigeria, Africa's most densely populated country, has a population of 89 to 100 million, equivalent to 108 people km^{-2}. Nigeria developed rapidly, and as a result, up to two-thirds of the potential *G. m. submorsitans* population of Nigeria may have been suppressed. *G. m. submorsitans* occurs in areas with human population densities ranging from 0 to 15 km^{-2}, occasionally in areas of 15 to 40 km^{-2}, but never when the population exceeds 40 km^{-2}. Most sub-Saharan African countries have densities below 40 km^{-2}. Flies of the *palpalis* group, particularly *G. tachinoides*, are much less affected by human settlement, possibly because they are able to adapt from a preference for feeding on wild mammals and reptiles, to feeding mainly on humans and their domestic animals.

Tsetse flies can adapt to peridomestic habitats, and *G. tachinoides* commonly follows domestic pigs in villages and utilizes human-made larviposition sites such as clumps of oil palm, cola nut, and banana trees.

HOSTS AND FEEDING BEHAVIOR

Potential hosts are recognized from visual and olfactory stimuli, which, together with mechanical stimulation, activate tsetse and initiate host-oriented responses. Approach to a stationary host is by upwind flight, modulated by olfactory stimuli, flight speed significantly reduced when a fly enters an odor plume. Final orientation toward the host is visual. Heat stimulation after a fly has landed on the host may then cause a probing response and subsequent feeding. Both endogenous and exogenous factors influence host-seeking behavior. Endogenous factors include a circadian rhythm of activity, as well as the level of starvation, age, sex, and pregnancy status of the flies. Exogenous factors include temperature, vapor pressure deficit, and visual, mechanical, and olfactory stimuli.

Olfactory Stimuli

Odor attractants are of three types: those associated with animal breath, such as acetone, octenol, and CO_2; those associated with urine, such as the phenols; and those associated with skin secretions, such as sebum.

Feeding Process

As a tsetse fly lands on a host, the labium is enclosed between the palps. Heat is the prime stimulus leading to probing and subsequent feeding by tsetse. Responsiveness increases with hunger until the final stages of starvation. As the fly starts to probe, the labium moves from the palps, to an angle of 90° to the skin. While the labella rests on the skin, the teeth on the inner surface are everted and penetrate it. At the same time, saliva is excreted from the hypopharynx. Normally, the labellar teeth lacerate capillaries, resulting in a hemorrhage, which is sucked into the labrum. When the fly stops sucking, a small pool of blood forms. If blood is not found, the fly withdraws the labium partially and makes a new penetration. Unmated female tsetse take smaller meals than mated ones either because the gut cannot be distended to the same extent or because of lower metabolic demands.

Host Effects and Preferences

Nonrandom feeding patterns may result from responses of hosts to tsetse attack, such as tail-flicking or skin-rippling reactions. Although normally unable to discriminate between potential hosts at long-range, the upright habit of humans, and lactic acid secretions in the skin, result in visual and olfactory repellency, including inhibited landing responses. Zebras may be protected from being fed on by biting flies, including tsetse, by their striped pattern. Tsetse are less attracted to stripes than to solid colors and seem to avoid horizontally striped objects. Zebras usually have a combination of vertical and horizontal stripes, but the horizontal stripes are on the lower part of the animal, where tsetse normally feeds.

Natural Hosts

There is a close correspondence between ecological niches of common hosts and tsetse habitats, and overlapping habitats are important in determining feeding patterns, in addition to behavioral characteristics of the host.

Tsetse will change to alternate hosts if their usual host(s) become unavailable. For example, *G. longipennis,* which once fed mainly on rhinoceros, adapted to feeding on other large animals as rhinoceros became scarce. Tsetse select host species rather than simply feeding on the most common available host.

Wild pigs (Suidae), particularly warthogs, are important hosts for *morsitans* group tsetse; however, the feeding efficiency of tsetse visiting a single warthog could be as low as 12 to 18%. Between 26 and 31% of tsetse landed on the head region of live adult warthogs, apparently because of a visual response to a dark patch produced by the preorbital glands of mature warthogs. In East Africa Suidae can form about half the food supply of *morsitans* group tsetse in areas with wildlife. Most other feeds are from ruminants, particularly bushbuck.

The *palpalis* group, generally inhabiting riverine or lacustrine vegetation, have a close ecological association with important hosts such as crocodiles and monitor lizards. They are, however, opportunistic feeders, and many mammals, including humans and reptiles, that enter their habitat may be fed upon. In peridomestic situations in the Guinean "forest savanna mosaic" region of West Africa, domestic pigs are an important source of blood meals for *G. palpalis* and *G. tachinoides;* wild Suidae, humans, and small ruminants are the next most common hosts. In the forest zone of Côte d'Ivoire the feeding habits of *G. palpalis* vary with the availability of wild hosts and activities of humans. Around villages, nearly all feeds are from domestic pigs, whereas in plantations, more feeds are from humans. In human trypanosomosis foci outside the edges of villages, *palpalis* group flies may feed equally on humans and antelopes, especially bushbuck. *G. tachinoides* can survive in close association with humans in the virtual absence of wild mammals and reptiles. It will readily feed on domestic pigs or cattle, but feeds have rarely been identified from domestic sheep and goats even when these were common.

RESTING BEHAVIOR

Tsetse rest for most of the day; making use of sites that provide protection from extreme temperatures and from predators. The sites may also provide vantage points from which to seek hosts, and their location can change according to time of day, season, and hunger stage of the fly. Recently engorged tsetse fly poorly and rest low down on tree trunks; the height above

ground level of the resting site increases in relation to the fly's nutritional state.

During the daytime, high temperatures of exposed resting sites result in tsetse moving to rest in sites such as the boles of large trees, where they squeeze into the fissures of the bark, at heights generally less than 0.3 m from the ground. Otherwise, flies hide in rot holes, often quite high up, in big tree trunks.

At dusk, tsetse flies generally move upward, to spend the night on leaves or small twigs, possibly to avoid predators. The choice of nocturnal resting sites is influenced by seasonal effects on the physical condition of the vegetation, such as leaf fall. A reverse migration seems to take place about an hour after sunrise, possibly in response to light intensity falling below a critical level. This rapid migration may occur at different temperatures in different seasons.

Activity

The diurnal pattern of tsetse activity may differ between species and according to sex, hunger, pregnancy, and nutritional state of the population. The pattern may also differ for the same species of tsetse in different localities. Natural activity of tsetse flies can be environmentally stimulated or spontaneous, but it has an underlying endogenous circadian rhythm modified by temperature. Intensive activity occurs for periods of less than a minute. The *morsitans* group tsetse are mostly active early in the morning and/or late in the afternoon.

DEVELOPMENT OF TRYPANOSOMES IN TSETSE

Fly species differ in their capacity to transmit trypanosomes, and individual fly genotypes also vary and affect susceptibility to trypanosome infection. For example, *salmon* mutants of *G. m. morsitans* appear to be better vectors of trypanosomes, perhaps owing the metabolism of tryptophan, which is essential for trypanosomes. Metabolism in *salmon* mutants is affected so that tryptophan accumulates, predisposing the fly to trypanosome infection. Flies of the *morsitans* group are mostly good vectors of all trypanosome species. Species of the *palpalis* group seem to be poor vectors of *Trypanosoma congolense* but efficient vectors of some stocks of *T. vivax* and can be important vectors of human infective trypanosomes. Species of the *fusca* group can be effective vectors of *T. congolense* and *T. vivax* but poor vectors of *Trypanozoon* trypanosomes. Trypanosome infection rates in tsetse are generally low and can depend on the age of the fly at the time of the infective feed. Wild male tsetse can achieve a life span of almost 5 months, but this is probably unusual. Teneral flies taking an infective blood meal on the first day of life, or soon after emergence, are the most easily infected. The prevalence of mature infections appears to increase with age for *Nannomonas* and *Duttonella* group trypanosomes in many tsetse species. Susceptibility to trypanosome infection may be age dependent, but the actual age is less critical than whether the fly has taken a previous, uninfected feed.

SLEEPING SICKNESS

Trypanosomes were first known to infect humans after being detected in the blood of a steamboat captain in the Gambia in 1901. The parasite was identified and named *Trypanosoma gambiense.* Sleeping sickness currently occurs in about 200 distinct disease foci in Africa. These foci place between 35 and 55 million people at risk, although only about 3 million of them are under surveillance and relatively few new cases are diagnosed annually. Human sleeping sickness was largely under control in the 1960s, but its incidence then increased, and in 1994 it was estimated that there were 150,000 cases in the Democratic Republic of the Congo. Other estimates suggest that around 300,000 people are infected in Africa, many of whom will die because of lack of treatment.

Identity and Origins of the Parasite

The two parasites causing human sleeping sickness, *T. brucei gambiense* and *T. b. rhodesiense,* are morphologically indistinguishable, and early diagnosis was based on clinical signs of the disease. Sleeping sickness caused by *T. b. rhodesiense* is an acute disease, with death occurring after only a few months, whereas with *T. b. gambiense* death may not occur for several years. *T. b. gambiense* occurs in West and central Africa and is transmitted predominantly by *palpalis* group tsetse, while *T. b. rhodesiense* occurs in southern and East Africa and is transmitted by *morsitans* group flies.

Although generally accepted that sleeping sickness trypanosomes are derived from *T. b. brucei* (infecting wild and domestic animals and morphologically indistinguishable from the human infective parasites), there are several theories on the evolution of species infecting humans. First, the two types of sleeping sickness may be caused by the same parasite, the disease simply being more acute when the trypanosome becomes adapted to certain species of wild animals. Chronic disease occurs when the trypanosome is adapted to humans as its principal host. Second, *T. b. rhodesiense* may have evolved in the late Miocene or Pliocene, when hominids were exposed to *Trypanozoon* trypanosomes in their savanna habitat. Subsequently, *T. b. gambiense* may have evolved from *T. b. rhodesiense* when hominids invaded forests and became hosts of *palpalis* group tsetse. Third, the two subspecies may have evolved independently, from *T. b. brucei* or a common ancestral species. A fourth theory is that after *T. b. gambiense* spread to savanna areas of southeast Africa, *T. b. rhodesiense* evolved from it and subsequently spread northward. An alternative theory is that *T. b. brucei* gave rise first to *T. b. rhodesiense* and then to *T. b. gambiense* and that the two diseases evolved in humans from an animal infection to an anthropozoonosis *(T. b. rhodesiense),* and then to a pure anthroponosis *(T. b. gambiense).*

Molecular characterization of *T. b. rhodesiense* suggests that it had polyphyletic origins. It is widely accepted that *T. b. gambiense* sleeping sickness spread from West to central Africa. Areas of West Africa where sleeping sickness occurred and the Nile basin are ecologically similar, with a continuous corridor of Sudano-Guinean climate and vegetation. *T. b. rhodesiense* then would have spread northward either from Central Africa through East Africa or from the Zambezi valley in Zambia, increasing in virulence the further north it progressed. Contradictory evidence from biochemical and molecular characterizations suggests that strains of *T. rhodesiense* from Zambia, Kenya, and Uganda have independent origins.

Reservoir Hosts

Until recently, one of the difficulties in proving the existence of reservoir hosts for *T. b. rhodesiense* and *T. b. gambiense* arose from problems in subspecies identification. It is now possible to use molecular genetic and biochemical methods to identify human infective forms. Consequently, there is now evidence of a range of wild and domestic animal host reservoirs. The latter are likely to be less important in the epidemiology of sleeping sickness due to *T. b. gambiense* than for *T. b. rhodesiense.* Many animals can be experimentally infected with either subspecies, and wild and domestic animals can maintain the "*rhodesiense*" disease endemically.

See Also the Following Articles
Medical Entomology • Veterinary Entomology • Zoonoses, Arthropod-Borne

Further Reading
Beard, C. B., O'Neill, S. L., Tesh, R. B., Richards, F. F., and Askoy, S. (1993). Modification of arthropod vector competence via symbiotic bacteria. *Parasitol. Today* **9,** 179–183.

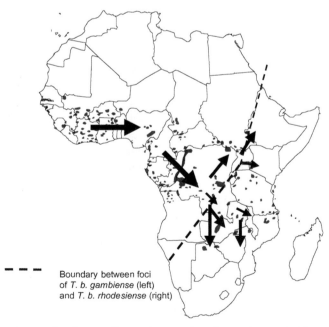

Boundary between foci of *T. b. gambiense* (left) and *T. b. rhodesiense* (right)

FIGURE 3 Distribution of human sleeping sickness foci in Africa. Dashed line indicates boundary between foci of *T. b. gambiense* (west) and *T. b. rhodesiense* (east). Arrows denote the probable direction of spread of the two types.

Brady, J., Packer, M. J., and Gibson, G. (1990). Odour plume shape and host finding by tsetse. *Insect Sci. Appl.* **11,** 377–384.

Carlson, D. A., Milstrey, S. K., and Narang, S. K. (1993). Classification of tsetse flies *Glossina* spp. (Diptera: Glossinidae) by gas chromatographic analysis of cuticular components. *Bull. Entomol. Res.* **83,** 507–515.

Cockerell, T. D. A. (1919). New species of North American fossil beetles, cockroaches, and tsetse flies. *Proc. U.S. Nal. Mus.* **54,** 2337, 301–313.

Colvin, J., and Gibson, G. (1992). Host seeking behavior and management of tsetse. *Annu. Rev. Entomol.* **37,** 21–40.

Gooding, R. H. (1984). Tsetse genetics: A review. *Quaest. Entomol.* **20,** 89–128.

Jackson, C. H. N. (1946). An artificially isolated generation of tsetse-flies (Diptera). *Bull. Entomol. Res.* **32,** 291–299.

Langley, P. A., Pimley, R. W., and Carlson, D. A. (1975). Sex recognition pheromone in tsetse fly *Glossina morsitans. Nature* **254,** 51–53.

Leak, S. G. A. (1998). "Tsetse Biology and Ecology: Their Role in the Epidemiology and Control of Trypanosomosis." CABI, Wallingford, Oxon, Vit.

Maudlin, I. (1991). Transmission of African trypanosomiasis: interactions among tsetse immune system, symbionts, and parasites. *Adv. Dis. Vector Res.* **7,** 117–148.

Mulligan, H. W. (ed.) (1970). "The African Trypanosomiases." Ministry of Overseas Development and George Allen & Unwin, London.

O'Neill, S. L., Gooding, R. H., and Aksoy, S. (1993). Phylogenetically distant symbiotic micro-organisms reside in *Glossina* midgut and ovary tissues. *Med. Vet. Entomol.* **7,** 377–383.

Welburn, S. C., and Maudlin, I. (1992). The nature of the teneral state in *Glossina* and its role in the acquisition of trypanosome infection in tsetse. *Ann. Trop. Med. Parasit.* **86,** 529–536.

Zdárek, J., and Denlinger, D. L. (1993). Metamorphosis behaviour and regulation in tsetse flies (*Glossina* spp.) (Diptera: Glossinidae): A review. *Bull. Entomol. Res.* **83,** 447–461.

Urban Habitats

Michael K. Rust
University of California, Riverside

Urban environments are created from natural or agricultural ecosystems that have been disturbed by human activities, typically by the construction of towns and cities. The urban biotic environment can be divided into plant and animal communities that surround houses and buildings, occur indoors or in close contact with humans, or directly affect the structure. Human activities give rise to special conditions and environments that are frequently exploited by a select group of insects and arthropods. For example, about 25 species of the some 3500 known cockroaches are considered to be domicilary pests. Of these 25 species, only about 5 are considered to be serious urban pests. Similarly, of the approximately 550 species of ants found in the United States, only 30 species commonly infest homes, and only 10 of these are of major importance. Urban landscapes are truly special ecological environments and niches.

Urban environments are similar in developing and developed countries worldwide in that natural systems have been disturbed by human activities. The rate at which urbanization is occurring in undeveloped areas of Africa, South America, and Asia is almost incomprehensible. By the year 2025, 6.3 to 8 billion people will inhabit less developed regions of the world compared with 1.2 to 1.5 billion people that will inhabit the developed regions. Between 1960 and 1990, the number of cities in China, Indonesia, and Brazil with more than 500,000 residents doubled; in India, the number tripled. By the year 2010, about 57% of the world's population will be living in urban centers, and in Brazil, it will be close to 89%. Because of the rapid growth of urban cities in developing countries with tropical climates, urban insect pests, especially those that vector human and animal disease pathogens, are a critically important emerging problem.

DEVELOPMENT OF MEGACITIES

A newly emerging feature of countries with large population centers is the development of mega-urban cities. Megacities typically incorporate two or more large urban centers linked by transportation. Rural zones between such megacenters are characterized by commercialization of agriculture, expansion of transport systems, and employment shift from farming to other activities, accompanied by migration to the cities. The number and size of these megacities in developing countries will continue to grow, and by 2015 it is projected that there will be some 23 megacities with more than 10 million inhabitants (Table I). Tremendous amounts of resources and commerce will be necessary to support the large populations and the waste and garbage they will produce. It is these conditions that are conducive to insect pest problems such as mosquitoes, flies, cockroaches, fleas, termites, and wood-destroying pests. In addition, crowding and waste lead to vertebrate pest problems such as rats, mice, and pigeons.

Urban centers require the movement of goods and food to maintain their large human populations. Human movement and transportation of commerce provide a unique and rapid dissemination of insects and arthropods between centers. At the turn of the 20th century, it took over a week for ships to sail from Europe to North America. Today aircraft make the same journey in less than 7 h. The potential exists for rapidly spreading insects and the diseases they vector from one continent to another. Three recent examples of urban invasive species spread by commerce are presented.

URBAN INVASIVE SPECIES

Asian Tiger Mosquito

TABLE I   Emerging Megacities of the World with More Than 10 Million Inhabitants (population in millions)

1975		2000		2015	
City	Population	City	Population	City	Population
Tokyo	19.8	Tokyo	26.4	Tokyo	26.4
New York	15.4	Mexico City	18.1	Bombay	26.1
Shanghai	11.4	Bombay	18.1	Lagos	23.2
Mexico City	11.2	São Paulo	17.8	Dhaka	21.1
São Paulo	10.0	Shanghai	17.0	São Paulo	20.4
		New York	16.6	Karachi	19.2
		Lagos	13.4	Mexico City	19.2
		Los Angeles	13.1	Shanghai	19.1
		Calcutta	12.9	New York	17.4
		Buenos Aires	12.6	Jakarta	17.3
		Dhaka	12.3	Calcutta	17.3
		Karachi	11.8	Delhi	16.8
		Delhi	11.7	Metro Manila	14.8
		Jakarta	11.0	Los Angeles	14.1
		Osaka	11.0	Buenos Aires	14.1
		Metro Manila	10.9	Cairo	13.8
		Beijing	10.8	Istanbul	12.5
		Rio de Janeiro	10.6	Beijing	12.3
		Cairo	10.6	Rio de Janeiro	11.9
				Osaka	11.0
				Tianjin	10.7
				Hyderabad	10.5
				Bangkok	10.1

Source: United Nations Population Division. "World Urbanization Prospects," 1999 rev. UN, New York.

It is thought that the Asian tiger mosquito, *Aedes albopictus,* entered the United States in Houston, Texas, in 1985 in used automobile tires imported from northern Asia. *Ae. albopictus* is a major human biting pest species throughout its range. In addition, it has been experimentally shown to be a competent vector of a number of viruses including western equine encephalitis, dengue, Japanese encephalitis, and yellow fever. To highlight the impact of commerce on the potential movement of insect pests, over 11 million used tires from 58 different countries entered the United States from 1978 to 1985. *Ae. albopictus* also shows the importance of local transport and the interstate highways. In 1987, of the 92 counties infested with this species, 64 were on an interstate highway, nearly twice the number that would be expected if their spread were not related to these road systems.

Since its introduction, this mosquito has spread to 25 states, mostly throughout the southeastern United States but as far north as Iowa, Illinois, and Ohio (Fig. 1). The female's propensity to lay eggs in bottles, cans, tires, rain gutters, flower pots, watering cans, children's swimming pools, and containers makes this mosquito ideally suited for urban environments.

Formosan Subterranean Termite

The Formosan subterranean termite, *Coptotermes formosanus,* has spread by marine commerce to semitropical and tropical seaports throughout the world. Once ashore, railroad ties provide a rapid method of moving and establishing large colonies of termites. In the 1960s, *C. formosanus* was discovered in Texas, Louisiana, and South Carolina. By 2000, it had spread to nine other southern states, and to California and Hawaii. *C. formosanus* is characterized by extremely large populations of aggressive foragers that can do extensive damage to structures. It is the most import urban pest in Hawaii and has been estimated to cause about $300 million damage a year in New Orleans, threatening nearly all the buildings of the historic French Quarter.

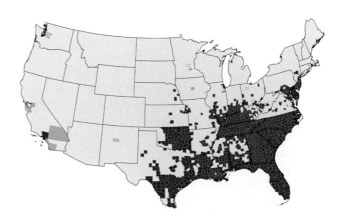

FIGURE 1 Current distribution of *Ae. albopictus* in the United States: red, positive counties; yellow, negative/eradicated/disappeared; blue, intercepted (never established); green, current status unknown. [From Moore, C. G. (1999). *Aedes albopictus* in the United States: Current status and prospects for further spread. *Am. J. Mosq. Control Assoc.* **15(2),** 221–227.]

FIGURE 2 A male and a female *An. glabripennis* on a poplar *(Populus alba)* tree. (Photograph by Baode Wang, taken on July 4, 1999 in Gansu province, China. U.S. Department of Agriculture, Animal and Plant Health Inspection Service, Plant Protection and Quarantines Programs, Otis Plant Protection Laboratory.)

Asian Long-Horned Beetle

The Asian long-horned beetle, *Anoplophora glabripennis* (Fig. 2) has the potential of becoming a serious pest of ornamental trees found in urban landscapes. Infestations of Asian long-horned beetle were first discovered in 1996 in Brooklyn. Another infestation was found in Chicago in 1998. It is believed that the beetle entered the United States in solid wood packing material from China. The beetle has been found in warehouses in some 26 locations thoroughly the United States. This beetle poses a serious threat to many different hardwood trees that are common in urban landscapes, including sugar, silver and red maple, birch, poplar, willow, ash, and black locust. The cost of suppressing the beetle in New York has already reached $4 million.

ADDITIONAL URBAN PESTS

The vast majority of insects, however, are not capable of adapting to the artificial and unnatural conditions created by humans in urban environments. A few pests such as the German cockroach, cat flea, and house fly have readily adapted, have been able to survive in close proximity to humans, and now are cosmopolitan. Many urban insect pests such as the German cockroach, oriental cockroach, confused flour beetle, and granary weevil have lost their ability to fly and rely on human activities for their dispersal. Other species such as the Argentine ant and the pharaoh ant do not swarm and rely on budding for reproduction and commerce for transport. Many of the indoor pests of food and fabric have continuous development and do not require special dormancy or hibernation to complete

development. These insects thrive at temperatures ranging from 18 to 35°C and relative humidity (RH) ranging from 50 to 90%, which are ranges typically found within human dwellings.

In addition to human occupants, domesticated pets may be a source of pests such as fleas, ticks, and mites. The cat flea, *Ctenocephalides felis,* is the most important ectoparasite of domesticated cats and dogs worldwide. From its probable origin in Africa, the cat flea has spread worldwide. In most urban environments, the larval development requirements of *C. felis* (RH > 75%, 21–32°C) are found in restricted microhabitats such as crawl spaces under homes, pet houses and bedding, utility rooms, and bedrooms. Interestingly, surveys of feral animals such as opossums indicate that *C. felis* is encountered only on animals frequenting urban settings.

NEED FOR URBAN PEST MANAGEMENT

The need for urban pest management in and around structures can be directly attributed to the dislike of insects and arthropods by humans. Some pests such as cockroaches, fleas, flies, and mosquitoes represent potential health risks. Others such as termites and wood-destroying pests can cause serious economic damage to structures. Pantry and fabric pests cause damage to foods, fabrics, and items made with furs, skins, and feathers. Many pests such as earwigs (Dermaptera), crickets (Orthoptera), pillbugs (Isopoda), sowbugs (Isopoda), and millipedes (class: Diplopodia) are occasional intruders, but their suppression often requires remedial action. The control of urban pests has created a large and important industry in the United States and Europe; in the United States alone about 1 million structures are treated each year for termite damage, repairs, and treatments at a cost of about $520 million.

The costs of treatment and repair by the pest control industry in the United States are estimated at about $2 billion. Consumers' annual expenditures on over-the-counter pest control remedies probably are at least as much. In developing countries, there will be an increasing need for pest control to protect the health and welfare of the inhabitants. With the increasing levels of international trade and tourism urban pest management will take on special importance in the 21st century.

See Also the Following Articles

Blattodea • Cat Fleas • Introduced Insects • Isoptera • Stored Products as Habitats

Further Reading

Anon. (1992) "Long-Range World Population Projections: Two Centuries of Population Growth, 1950–2150." United Nations Publication ST/ESA/SER.A/125. UN, New York.
Ebeling, W. (1975). "Urban Entomology." University of California Press, Division of Agricultural Sciences, Berkeley.
Moore, C. G. (1999). *Aedes albopictus* in the United States: Current status and prospects for further spread. *J. Am. Mosq. Control Assoc.* **15(2),** 221–227.
Robinson, W. H. (1996). "Urban Entomology. Insect and Mite Pests in the Human Environment." Chapman & Hall, London.
United Nations Population Division. (1999) "World urbanization prospects: The 1999 Revision." UN, New York.

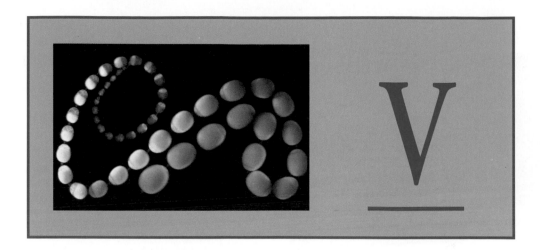

Venom

Justin O. Schmidt
Southwestern Biological Institute, Tucson

Venoms are biologically active liquids delivered into or onto another organism through a piercing structure such as a mouthpart or sting apparatus and often resulting in pain, tissue damage, paralysis, or death in the target organism. Insects contain more venomous species than all other groups of terrestrial animals combined. Within insects, venoms evolved numerous times to occupy positions in groups as diverse as ants and wasps to biting bugs, robber flies, and stinging caterpillars. Venoms empower insects with unique properties that enable them to achieve far greater effects or impacts on target organisms than would be expected for such small animals. For example, a tiny wasp weighing far less than a milligram can rapidly paralyze a caterpillar hundreds of times its own size. Likewise, a single attacking honey bee can send a million times larger human into panicked flight.

Venoms provide the biological means for insects to break free from many ecological restraints, thereby expanding their opportunities to exploit the world around them. One could argue that the omnipresence of ants in today's world is the result of venom combined with a social structure in ancestral ant lineages. Parasitic braconid and ichneumonid wasps, whose species number more than 100,000, with probably several times that number undescribed, owe their successful speciation, in part, to the ability of their venoms to facilitate use of a variety of host organisms. Honey bees (*Apis* spp.) have become dominant pollinators throughout much of the world, in large part because their venomous stings have enabled them to defend against a multitude of large, often intelligent predators. Without their effective defensive venoms, honey bees could not prevent the plunder of their large stores of honey and pollen, which they must maintain to overcome long seasonal periods of harsh conditions. The biological roles, operation, and properties of insect venoms provide dazzling illustrations of the successful adaptation of insects to their environment.

BIOLOGICAL ROLES OF VENOMS

Venoms serve two primary roles—defense and prey capture. In most lineages, the ancestral role, from which a defensive role evolved secondarily, was the use of venom for prey capture. The venoms of many taxa retain primary roles of prey capture. Examples are numerous and include many families or superfamilies of bugs, beetles, flies, neuropterans, and especially parasitoid and solitary aculeate wasps. Many familiar insects have taken venom use another step—to defend against predators. The mere fact that insects such as honey bees, yellowjackets, and hornets are so universally well known attests to the effectiveness of their defensive venoms. Defensive venoms are present in all social wasps, many ant species (including fire and harvester ants), some solitary bees and wasps, a few bugs, and some spiny caterpillars. In many of these species, the role of venom for prey capture has been entirely replaced by a role in defense. In caterpillars, which never used venom for prey capture, the role evolved independently. Some species retain both roles, with the spider wasps (Pompilidae) as conspicuous examples.

EVOLUTION OF VENOMS

Most venoms evolved from preexisting digestive or reproductive systems. Exceptions are the lepidopteran caterpillars, for which the origin of their venomous spines is unclear. The venoms of all other non-hymenopteran insects evolved from salivary or gastric secretions associated with the mouthparts and digestion. In some taxa, the venom toxins are salivary secretions injected into the prey. In others, the toxins are regurgitated from the gut into the prey through piercing mouthparts. Asilid flies use the latter form of venom delivery.

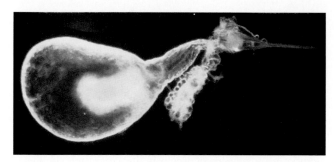

FIGURE 1 Fire ant *(S. invicta)* sting apparatus including the stinger, a derived ovipositor, with attached large venom sac and small Dufour's gland, which are derived accessory reproductive glands. (Photograph by J. O. Schmidt.)

The Hymenoptera evolved an entirely different venom system. In Hymenoptera, the stinger evolved through modification of the female ovipositor, and venom evolved from the accessory reproductive glands (Figs. 1 and 2). Consequently, only female Hymenoptera can sting. The lack of ability to sting and defend a colony against large predators is considered a primary reason for the failure of males, often called drones, of honey bees and social Hymenoptera to meaningfully defend their colonies against predatory assaults or to help the colony via foraging.

VENOMS USED FOR PREY CAPTURE

Two basic types of venom are used for prey capture: those that kill outright and those that only paralyze or alter prey physiologically. Prey paralysis is important for many insects that provide prey for their young. Prey that is paralyzed but still living does not spoil rapidly, giving larvae time to feed and grow. Insects that kill their prey with venom tend to consume it immediately, though exceptions to this rule exist. Wasps that use their venoms to provide a living host for their larvae and oviposit one or more eggs in or on the host are often called parasitoids. The venoms of parasitoids span a paralysis range from no paralysis to temporary paralysis to permanent paralysis. Venoms causing no paralysis, or short-acting paralysis, are found in many species of wasps in which the parasitized host

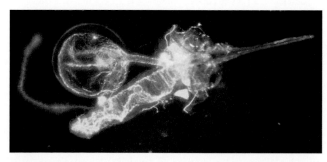

FIGURE 2 *Pogonomyrmex badius* harvester ant sting apparatus including the round venom sac with attached free filament that collects materials from the hemolymph for venom synthesis, and the tubular Dufour's gland. (Photograph by J. O. Schmidt.)

continues feeding and growing until just before it dies, all the while harboring and protecting the parasitoid larvae from environmental and predatory risks. The venoms of these wasps often contain mutualistic viruses and other components that not only help protect the oviposited eggs from host encapsulation and immune defenses, but also alter the hormone balance inside the host to prevent molting to the next stage of development. If the host is in a protected environment, such as exists for wax moth larvae in combs of a dead honey bee colony, an extremely potent venom that causes total and permanent paralysis, like that of a parasitoid wasp such as *Bracon hebetor,* can be beneficial.

Hunting wasps are another group of solitary wasps that use venom for prey capture. These wasps include familiar sphecid, eumenid, and spider wasps plus a variety of other families of aculeate wasps. Like classical parasitoid wasps, hunting wasps possess venoms with a large activity span. For example, the sphecid *Larra analis* uses venom to paralyze its mole cricket hosts temporarily. At the other extreme, represented by the beewolf *Philanthus triangulum* and tarantula hawks (*Pepsis* spp.), the hunting wasps paralyze the honey bee and tarantula prey permanently. No evidence for injection of viruses or factors altering the host's endocrine or immune systems exists for hunting wasps. The majority of parasitoid and hunting wasp venoms are of little value for defense against predators. In many groups, the wasps cannot, or do not, even attempt to sting predators; in others, the pain and activity of the venom, at least to humans, is trivial. Some exceptions exist: most pompilid wasps, some eumenid and bethylid wasps, and a very few sphecid wasps can deliver painful stings; however, none of these venoms is meaningfully toxic to mammals.

DEFENSIVE VENOMS

Because of the extreme size difference between vertebrates and insects, insects are vulnerable to attack by vertebrate predators. Once such a small creature is captured by a large predator, typical defenses such as kicking, scratching, and biting are all but useless. A very few defenses, particularly potent venomous stings and allomones, can neutralize the predator's size advantage and blunt attacks. A key feature promoting the efficacy of defensive venoms is the ability of the sting apparatus to bypass the predator's skin and external defensive barriers and inject venom directly into sensitive tissues. Venoms function as defenses in several ways. First, they cause pain. Second, they can cause body or tissue damage, or death. Beyond these primary activities, venom can have defensive value based on its nasty taste, and, possibly, on its ability to induce allergic reactions. Honey bee venom illustrates the value of a repugnant taste of a venom: presumably birds and other vertebrates reject worker bees while readily devouring drones because of the ability of workers to sting. For example, western king birds selectively prey on drone honey bees, which they manipulate entirely with their hard bill until the prey can be quickly swallowed, head first. The handling and speed of consumption suggested to the author that a king bird could consume worker bees without being stung. Why then were workers rejected? A suspected, but unconfirmed, explanation was revealed upon consumption of whole frozen drone honey bees and abdomens of workers. Drones were generally palatable. In contrast, worker abdomens were noisome, possessing a bitter and "hydrolytic" taste, a taste confirmed (by sampling of an isolated venom sac) to be venom.

No direct evidence confirms a defensive value for venom allergy. However, three lines of reasoning suggest that such a value might exist. First, the 1 to 4% incidence of allergic reaction to insect stings in the general human population is disproportionately high compared with allergy to other proteinaceous substances. Second, observations of social interactions among human groups in which one individual experiences an allergic reaction to a sting reveal that fear of the culprit stinging insect is shared among the whole group. Primates and other intelligent social mammals, major potential predators of social insects, likely also learn to avoid the stinging insects by observation of the plight of a group member. Third, an allergic reaction impairs an individual's ability to avoid its own predators, to obtain food, and to reproduce as effectively as if no reaction had occurred. Consequently, the genes from an individual that attacks stinging insects and suffers a reaction are not as likely to be passed to future generations. Such "genetic learning" would be a consequence of allergic reactions.

Pain, the most notorious property of an insect sting, is key for instant defensive value of venom. Pain is the biological signal that indicates to an organism that bodily damage has occurred, is occurring, or is about to occur (Fig. 3). Consequently, pain is a signal that must be heeded and acted upon if an animal is to enhance its own survival and fitness. Pain acts to stop an attacking predator and to cause it at least momentarily to assess its own risk situation. In the meantime, the stinging insect gains valuable time for escape or further defense. Because pain cannot be measured easily or precisely, exact values to use for comparing the effects of different venoms on potential predators are unobtainable. To provide a means for comparison, Schmidt developed a pain index, based on a scale of 0 to 4. Zero indicates that the stinging insect is too small or otherwise unable to pierce the human skin; 2 is the central value, as represented by the pain of a honey bee sting; and 4, the top of the scale, is typically described as causing immediate, excruciating, totally debilitating pain that completely eliminates the ability of the stung individual to continue to act in a normal fashion. Table I lists some of the pain values for common insects.

Pain, in itself, is not an outright defense—it can be a bluff, mimicking damage and deceiving a would-be predator into believing that it has been injured. Unless confirmation of injury follows, an intelligent animal quickly learns to ignore the pain and continue with its attack. Beekeepers provide a familiar example of the principle—upon learning that bee sting pain does not translate into actual damage or risk, they continue their activities, often receiving scores of stings in a day. If damage did occur to beekeepers, the profession would have become extinct long ago. A minority of beekeepers do experience damage in the form of large local or allergic reactions, and these individuals usually abandon the profession. To avoid the limiting problem associated with pain, venoms of many insects have evolved a step further to the stage of "truth in advertising": pain is the advertising; toxicity is the truth. If the

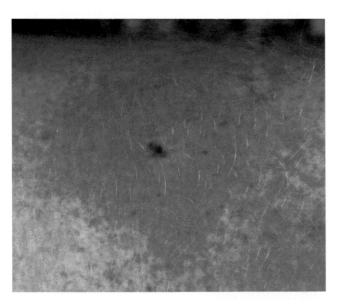

FIGURE 3 Painful local reaction illustrating central tissue damage and surrounding inflammation induced by a honey bee sting. (Photograph by J. O. Schmidt.)

TABLE I Pain Rating and Lethality of Common Insect Venoms[a]

Species	Common name	Pain rating[b]	Lethality[c]
Lasioglossum spp.	Sweat bee	1	n/a
Sceliphron caementarium	Mud dauber wasp	1	n/a
Solenopsis invicta	Fire ant	1	n/a
Sphecius grandis	Cicada killer wasp	1–2	46
Myrmecia gulosa	Bull ant	1–2	0.18
Dinoponera gigantea	Giant ant	2	12
Xylocopa californica	Carpenter bee	2	26
Dolichovespula maculata	Baldfaced hornet	2	6
Bombus impatiens	Bumble bee	2	12
Apis mellifera	Honey bee	2	2.8
Vespula germanica	German yellowjacket	2	2.8
Polistes canadensis	Red paper wasp	2–3	2.4
Dasymutilla klugii	Velvet-ant wasp	3	71
Pogonomyrmex maricopa	Harvester ant	3	0.125
Pepsis formosa	Tarantula hawk wasp	4	65
Paraponera clavata	Bullet ant	4	1.5

[a] For a more complete list, see Schmidt (1990).
[b] Based on a scale ranging from 0 to 4; 4 is the highest score.
[c] Measured as milligrams of venom per kilogram of body weight in mice; the smaller the number value, the more potent the venom.

pain signal is associated with damage, then predators cannot easily bypass the pain signal. Venom toxicity is most easily measured in terms of its ability to cause death in an animal.

Table I also rates the lethality of representative insect venoms. Often, but not always, increased pain parallels increased lethality, hence the tandem evolution of both "advertising" and "truth." Notable venoms on the list are those of the harvester ants *(Pogonomyrmex)*, the most lethal insect venoms in the world, which cause rather nasty intense pain that lasts for 4 to 8 h, and bullet ants *(Paraponera)*, the world's most painful venomous insects, which also have a respectable lethality. Bull ants *(Myrmecia)* and tarantula hawks *(Pepsis)* present curious cases: bull ants are rather lethal, yet they induce relatively little pain; tarantula hawks cause a most intense pain, equaling that of bullet ants, for 2 to 3 min, yet they possess no vertebrate lethality. The explanation for this discrepancy between pain and lethality is unclear for bull ants, but for tarantula hawks the explanation might be that *Pepsis* must permanently paralyze its spider prey without killing them.

ECOLOGY AND EVOLUTION

Insect venoms have played an enormous role in ecology and evolution. For parasitic and hunting wasps, venoms have provided the means to expand the menu of hosts as food sources. This in turn allowed, and continues to allow, isolation of breeding populations and rapid speciation of the wasps. Via their increased impact, wasps alter the foraging and defensive behaviors of their host species, with a ripple effect insofar as other predators and parasitoids are influenced. The influence of defensive venoms is even greater. The defensive potency of venoms provided venomous insects the freedom to enhance further their defensive strengths through evolution of secondary defenses of aposematism and mimicry. Warning colors, sounds, or chemical signals used in conjunction with the honest threats of damage from the sting provide venomous insects with the ability not only to defend against potential predators, but also to reduce risks of being attacked. After one or a few species develop aposematic signals, other venomous insects can become Müllerian mimics of the signals. The system opens to "cheaters"—which are not noxious—to become Batesian mimics of the venomous models.

For most species, freedom from one or more predators, or reduced predatory pressure, provides wonderful behavioral and evolutionary opportunities. Secretive species may expand their niches into more conspicuous environments or to increased periods of time, thereby reaping the benefits of increased quantity and variety of resources. Examples of this are brightly colored bumble bees, social wasps, and spider wasps that can forage during daylight hours in open environments, free from most threats from predatory birds and lizards. Likewise, venomous ants can exploit daylight periods, open areas, or conspicuous areas in the vegetation.

All other effects of venoms pale in comparison to their impact on the evolution and maintenance of sociality in the Hymenoptera. Social wasps, bees, and ants account for a huge percentage of the individuals and biomass of most ecosystems, especially tropical systems. Social wasps are important predators of other arthropods, social bees are responsible for much of the pollination of plants, and ants are major predators and influences on other animals. Insect sociality evolved through kin selection in which related presocial individuals that cooperated had higher inclusive fitness (the successful passing on of their genes to the next generation) than did individuals that did not cooperate. In Hymenoptera, the evolution of sociality, and its maintenance, was possible, in large part, because of defensive venoms. Social insects and their nests full of defenseless brood and/or food stores represent potential bonanzas for predators capable of exploiting them. As colony size increases, larger and more determined predators will attempt to exploit a colony's brood or food cache. Venoms provided the means, both proximate and ultimate, for colony defense against destruction. Once established, sociality in Hymenoptera could progress from small simple social structure to highly organized structures via the evolution of increasingly potent venoms and other defenses. Populous, highly organized social species dominate and rule most environments.

See Also the Following Articles
Bee Products • Chemical Defense • Mimicry • Predation • Sociality • Wasps

Further Reading
Bettini, S. (ed.). (1978). "Arthropod Venoms," Vol. 48 *In* "Handbook of Experimental Pharmacology," (especially Chaps. 18–26, pp. 489–894). Springer-Verlag, New York.
Piek, T. (ed.). (1986). "Venoms of the Hymenoptera." Academic Press, London.
Quicke, D. L. J. (1997). "Parasitic Wasps," pp. 221–242. Chapman & Hall, London.
Schmidt, J. O. (1982). Biochemistry of insect venoms. *Annu. Rev. Entomol.* **27**, 339–368.
Schmidt, J. O. (1990). Hymenoptera venoms: Striving toward the ultimate defense against vertebrates. *In* "Insect Defenses: Adaptive Mechanisms and Strategies of Prey and Predators" (D. L. Evans and J. O. Schmidt, eds.), pp. 387–419. State University of New York Press, Albany.

Veterinary Entomology

Bradley A. Mullens
University of California, Riverside

Veterinary entomology deals with arthropod pests and vectors of disease agents to livestock, poultry, pets, and wildlife. It is allied with the fields of medical entomology, parasitology, animal sciences, veterinary medicine, and epidemiology. The main pests of veterinary concern are sucking and biting lice, biting flies, nonbiting muscoid flies, bot flies, fleas, and Acari (mites and ticks).

ARTHROPOD GROUPS

Arthropods that affect animals can be categorized by the intimacy of their host association, and these range from permanent ectoparasites to pests that contact the vertebrate only briefly once every few days.

Permanent Ectoparasites

Some arthropods, such as lice and many parasitic mites, complete their entire life cycle on the host. All stages of sucking lice (Pthiraptera, suborder Anoplura) are mammal parasites and feed on blood, whereas biting lice (suborders Amblycera and Ischnocera) use either mammal or bird hosts, feeding on skin, hair, and feather debris. Lice tend to be abundant in cool weather or on animals stressed by poor nutrition or overcrowding. Many lice are specific to one or a few closely related hosts and cannot survive more than one to a few days away from the host. Transmission from host to host is mostly by direct contact.

Parasitic mites (Acari, suborders Mesostigmata, Acaridida, and Actinedida) are found on most groups of birds and mammals. Like lice, most are specific to one host species or a small group of related species. Several genera comprise what are commonly known as "mange mites." *Sarcoptes scabei,* which causes sarcoptic mange, burrows at the surface of the dermis and exists in a number of races that generally are host specific to swine, dogs, and so on. *Demodex* mites, causing demodectic mange, live in follicles and can be important especially in immunocompromised hosts. *Chorioptes* and *Psoroptes,* causing chorioptic and psoroptic mange, include species of considerable importance for cattle, and wild and domestic sheep. The latter mites complete their development at the skin surface and do not actually burrow in the skin, although mites often are covered by scabs and are frequently called scab mites. *Ornithonyssus* mites are blood feeders and are especially important for wild and domestic birds, where populations may reach many thousands per host (Fig. 1). They occupy fur and feathers, travelling to the skin surface to feed regularly.

Some of the more advanced flies are also permanent parasites, and a number of species in three families hardly resemble flies at all because they have secondarily lost their wings (apterous). Members of the dipteran families Streblidae and Nycteribiidae live on bats, whereas members of the Hippoboscidae parasitize various birds and mammals. Some hippoboscids are economically important, such as the sheep ked, *Mallophagus ovinus.* In all these families, the adult female nurtures a single larva within her body until it is mature; this is a very unusual pattern for insects. After the mature larva exits the female, it promptly pupates on the host.

Semipermanent Ectoparasites

The semipermanent ectoparasitic arthropods do not complete the entire life cycle on the host, but they do spend at least several

FIGURE 1 (A) Scanning electron micrograph of the northern fowl mite, *O. sylviarum,* a permanent ectoparasite of many birds, including domestic chickens. (B) The vent of a chicken, showing the blackened feathers typical of a heavy infestation. This hen has over 20,000 mites. [Part (A) courtesy of Jeb Owen, University of California, Riverside.]

days at a time on a vertebrate. The hard ticks (Ixodidae) attach to feed for several days in each of the life stages. Although some, like the cattle tick *Boophilus,* complete the entire life cycle on a single host, the most abundant and widespread species tend to use a separate host for each stage. In this case the engorged tick falls off, molts, and then finds a new host by crawling up on vegetation and waiting for a passing vertebrate, to which it attaches. This activity is known as questing. Often the larva hatches from an egg and attaches to a small host such as a rodent, whereas the nymph (the stage after the larval molt) or adult may attach to a larger host such as a deer. Examples of this include the American dog tick *Dermacentor variabilis.* Adult female hard ticks take a large blood meal, produce a single large batch of eggs (typically several thousand), and then die.

Fleas (Siphonaptera) generally are on a host for most or all of the adult stage, and feed on blood. About 94% of flea species live on mammals, and the rest on birds. Flea eggs fall from the host pelage into a nest environment, where the larvae feed on organic debris and sometimes on excess blood produced by the adults. Fleas thus are often lacking on hosts that do not return to long-term bedding areas or nests.

Bot flies include important species in the dipteran families Oestridae, Gasterophilidae, and Cuteribridae. They spend nearly the entire year as immatures within the vertebrate's body. Eggs, often laid on hairs, hatch and enter the host body.

Horse bot larvae *(Gasterophilus)* attach to the wall of the gastrointestinal tract for several months before they pass from the host with feces to pupate in soil. Cattle grubs *(Hypoderma)* migrate through the lining of the esophagus or spinal cord, depending on the species, and eventually form a cyst in the back. After a final period of maturation, larvae exit the cyst and fall into the soil to pupate. Bot fly adults lack functional mouthparts and depend entirely on reserves from the larval stage to sustain them for several days; in this brief time they must find mates and hosts. The invasion of vertebrate tissues by fly larvae is known as myiasis.

Certain other flies also are semipermanent parasites. Among muscoid Diptera, the horn fly, *Haematobia irritans,* is on cattle most of its adult life, where adults of both sexes feed 20 to 30 times daily. They leave the host to disperse and to lay eggs in very fresh dung, and then return to the host.

Occasional Parasites

The broad category of occasional parasites includes a range of arthropods. The most intimate host associations in this group include the soft ticks (Argasidae) and some blood-feeding mites such as the bird parasite *Dermanyssus gallinae.* In both cases, nymphs or adults hide in or near nest areas, sheltered in cracks and crevices or under debris; thus, they are closely associated with animals, although they contact them only periodically. They leave the hiding places, often at night, to feed for periods of 5 to 30 min, returning to the nest to digest the blood meal. Some soft ticks can withstand dry conditions and years without feeding.

Many serious biting fly pests spend the immature period away from the host, exploiting an entirely different resource base. It is common for larvae to feed on detritus in wet habitats, whereas the adults use plant nectar for energy and blood for egg development. For example, larvae of stable flies live in rotting vegetation, blackflies in running water, horse flies and biting midges in swampy mud, or mosquitoes in ponded or slowly moving water. When the adults emerge, they take blood meals at intervals of 1 to 4 days, but they typically are in contact with the host for only a few minutes at a time.

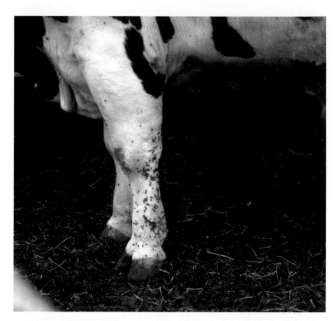

FIGURE 2 The stable fly, *Stomoxys calcitrans,* may attack animals in high numbers: there are approximately 200 flies feeding on the front leg of this bull. This species is illustrative of the occasional parasites because the adult flies blood-feed for only a few minutes every day or two, and larvae are in rotting vegetation away from the host animal.

They leave the host to digest the blood in some sheltered resting location. For the higher Diptera (suborder Cyclorrhapha), such as stable flies (Fig. 2) or tsetse flies, both sexes feed on blood, and multiple blood meals usually are needed to develop a batch of eggs (or, for tsetse, a single mature larva). In the lower Diptera (suborders Brachycera and Nematocera) such as horse flies, blackflies, or mosquitoes, only females take blood, and most species require only a single large blood meal to develop an entire batch of eggs numbering 50 to 300. Some other pests in this general category do not feed on blood but visit the host to take meals of tears or other protein-rich secretions that are also used to develop eggs. A good example is the face fly, *Musca autumnalis.*

HOW ARTHROPODS CAUSE DAMAGE TO ANIMALS

There are several basic ways in which arthropods cause damage to animals, and the different mechanisms interact to impact agricultural production. Arthropod damage to plant crops often is evident to consumers, who react with disgust to cabbage leaves damaged by loopers, corn earworm larvae on an ear of corn, or scale insects on citrus fruit. In contrast, arthropod damage to animal production tends to be hidden from the consumer because the product is purchased in the form of jugs of milk or wrapped packages of butchered meat. Nevertheless, losses are serious for producers, and costs are passed on to consumers.

Loss of Blood and Tissue Fluids

Many arthropods ingest blood, usually for egg development. Impact of blood loss on the animal reflects the style of feeding and the number of arthropods. For example, mosquitoes canulate a vessel and hosts generally lose only the blood mosquitoes ingest. In contrast, not only do biting flies that macerate the capillary beds of the skin to feed from blood pools tend to inflict more painful bites, but the bites themselves lead to a larger quantity of blood loss per feeding insect. For example, horse flies (Tabanidae) may directly remove over 200 ml day^{-1} from a host in a pasture, but blood continues to run from the wound for a period of time, often being fed on by other flies. Many hard ticks (Ixodidae) increase in weight by 100-fold or so as they feed for 7 to 10 days. Females of very large species of ticks may contain over a milliliter of blood at a time, and hundreds may be attached to a single host animal. Blood or fluids such as lymph are metabolically "expensive" for a vertebrate to produce. Such loss is reflected in reduced feed conversion efficiency, which occurs when animals eat more for a given yield of meat or eggs. Insect feeding also causes significantly lower weight gains or milk production. Losses of 10 to 20% feed conversion, 0.1 to 0.2 kg day^{-1} weight gain (cattle), and 5 to 10% loss in milk yield are not uncommon for animals under heavy attack.

Pain and Interference with Activities

Pain and irritation caused by arthropod attack force animals to alter their feeding or activity patterns and to engage in a number of sometimes vigorous behaviors to defend themselves. There may be economic loss as well, since animals are not feeding normally and must expend energy that might otherwise be directed toward growth or reproduction. For example, stable flies *(Stomoxys)* and face flies, as well as other biting flies, can cause animals to retreat into groups for refuge. In groups,

FIGURE 3 "Gadding" behavior (note the tail held up in the air) by a calf being attacked by cattle grub flies *(Hypoderma* spp.). Vertebrate host behavior can be altered by parasites. (Photograph courtesy of Dr. Jerry Weintraub, Agriculture Canada.)

insect attack rates usually are lower per host (the herd dilution effect), particularly for the animals that occupy the interior of an aggregation. Animals also may enter woods or bodies of water in an apparent attempt to escape insects. Cattle pursued by adults of cattle grubs *(Hypoderma)* experience no immediate pain from the flies, which cannot bite. The female flies are trying merely to deposit eggs on the cattle hair at the base of the legs. Still, cattle exhibit an interesting, stereotypical behavior known as "gadding" (Fig. 3). The animals run at full speed with tails raised straight into the air, which expends energy and may cause accidental injury.

Although pests such as house flies may not cause direct losses to the animals, they are produced near animal operations and thus are a veterinary entomology problem. Excessive numbers ("excessive" admittedly is a subjective term) of nuisance arthropods cause great annoyance to people living nearby, and thus constitute serious public relations and legal problems for producers. Public health agencies can close facilities unable or unwilling to mitigate such a problem.

Allergic Responses to Saliva

Blood-feeding arthropods possess a potent arsenal of chemicals in their saliva to maintain blood flow (vasodilators, anticoagulants) and sometimes have anesthetics to reduce host defensive response. Like humans, animals can develop allergies to these compounds. Horses commonly react to biting midge *(Culicoides)* feeding with an allergic reaction called Queensland itch or sweet itch, resulting in skin inflammation and hair loss. Mass emergences of the blackflies *Simulium arcticum* in Canada and *Cnephia pecuarum* in the valley of the southern Mississippi River have resulted in the deaths of livestock, probably from allergic responses as well as blood loss. Larvae of sheep blowfly *(Lucilia)* feed near the skin surface, especially where the wool is wet, and can contribute to a toxic shock–type syndrome fatal to infested sheep. Pets may develop serious allergies to fleas, with resulting hair loss and other symptoms.

Product Damage

Arthropods sometimes cause direct damage to parts of the animal desired by people. For example, cattle grub *(Hypoderma)* larvae form large cysts in the backs of cattle. They cut a hole in the skin to breathe, and this skin is the thickest on the animal. Although the holes heal after the larva exits, the scarred skin is less valuable for leather. The presence of larvae also can affect the quality of the meat in this area of the animal, which is the part where steaks come from, and damaged meat sometimes must be trimmed at the slaughterhouse. Mites such as *Psoroptes* and *Sarcoptes,* as well as many lice, often result in irritation, rubbing, and gross loss or damage to hair and wool.

Cosmetic damage, including rashes or minor hair loss, can be predictably serious to the owner of a pet or a show animal. However, cosmetic damage also can cause losses in animal agriculture out of proportion to actual damage. An example of this is the condition "gotch ear" in cattle caused by the Gulf Coast tick *Amblyomma maculatum* in the southern United States. Damaged ear cartilage is cosmetic, but causes the animals to be placed in an "odd lot," with per-pound prices of $0.05 to 0.10 less than undamaged cattle.

Restricted Trade

Many pests have distinctive distributions, and preventing movement or dispersal into new areas is of paramount importance. Cattle ticks *(Boophilus)* and screwworm *(Cochliomyia hominivorax)* were eradicated from the southern United States in the 20th century, but they persist in Central and South America. The U.S. habitat obviously is still suitable. Without complex systems of animal quarantine, treatment, and examination, it is certain they would reestablish in the United States.

Exotic arthropods pose a great threat either as direct pests or vectors of disease agents such as those that cause heartwater or African swine fever. The Office International des Epizooties lists diseases of risk for animals worldwide, and one of those on List A (greatest risk) is bluetongue. This viral disease of ruminants, such as cattle and sheep, is transmitted by biting midges, and is endemic in the United States. Trade restrictions from bluetongue cost the U.S. cattle industry many millions of dollars annually, even though cattle themselves do not usually develop obvious disease. The reason is that some major trading partners (e.g., western Europe) lack bluetongue, and their agricultural authorities fear an impact on their sheep industries from accidental importation.

Diseases

A number of serious animal disease agents are transmitted by arthropods. The worst of these are tropical, and they cause death and heavy production losses in the affected countries. African trypanosomiasis causes a wasting-type disease known as nagana in animals and sleeping sickness in humans, and *Theileria parva,* called East Coast fever, can cause 90 to 100% mortality in affected cattle in eastern Africa. People in developed countries tend to underestimate the true value of animals in the developing world. Animals are vital there not only for protein-rich food, but for draft and transportation purposes, and as wealth. They are the basis of many pastoral peoples' economies, and the economic impact of some of these animal diseases can far exceed the impact of similar, serious human pathogens.

Temperate zones also have some rather important arthropod-transmitted animal disease agents, including *Anaplasma,* dog heartworm, and equine infectious anemia virus. In the United States and Europe, the direct effects of arthropods on animal production generally exceed losses caused by arthropod-transmitted diseases. However, the role of wild animals as natural reservoirs of pathogens that incidentally infect people is very important in both temperate and tropical zones. Diseases that cycle naturally in animal populations and occasionally infect people are called zoonoses. Zoonoses comprise some of the more notorious arthropod-related human health problems. They include plague (maintained in rodents and transmitted by fleas), Lyme disease (maintained in rodents and transmitted by ticks), and St. Louis encephalitis (maintained in birds and transmitted by mosquitoes). Previously unknown tick-borne ehrlichioses (caused by intracellular bacteria-like organisms in the genus *Ehrlichia*) have been recently discovered infecting humans in the United States. They are zoonotic in origin and typify a category of "emerging" human diseases that is now of great interest in the medical community.

See Also the Following Articles

Bubonic Plague • Dog Heartworm • Medical Entomology • Mites • Tsetse Fly • Zoonoses, Arthropod-Borne

Further Reading

Harwood, R. F., and James, M. T. (1979). "Entomology in Human and Animal Health." Macmillan, New York.

Kettle, D. S. (1995). "Medical and Veterinary Entomology." 2nd ed. CAB International, Wallingford, Dxon, U.K.

Lancaster, J. L., and Meisch, M. V. (1986). "Arthropods in Livestock and Poultry Production." Ellis Horwood, Chichester, U.K.

Mullen, G., and Durden, L. (eds.). (2002). "Medical and Veterinary Entomology." Academic Press, San Diego, CA.

Wall, R. and Shearer, D. (1997). "Veterinary Entomology." Chapman & Hall, London.

Williams, R. E., Hall, R. D., Broce, A. B., and Scholl, P. J. (1985). "Livestock Entomology." Wiley, New York.

Vibrational Communication

Andrej Čokl and Meta Virant-Doberlet
National Institute of Biology, Slovenia

Many insects communicate by vibratory signals that are produced directly by a body part or, indirectly, as a by-product of some other activity. Insects emit vibratory signals in connection with aggression, distress, calling, courtship, rivalry, and other specific behaviors, among other functions enabling mate location and recognition. This article presents mechanisms, structures, and signals of vibratory communication in relation to behavior.

MECHANISMS OF VIBRATION PRODUCTION

Insects produce vibratory signals by percussion, vibration of a body part, tymbal mechanisms, or stridulation. Percussion is a very common because of the hard exoskeleton, which enables either percussion of two body parts or striking against a substrate. The percussive structures are in most cases relatively simple. Book lice (Psocoptera) and stoneflies (Plecoptera) tap their abdomens against the substrate, Orthoptera use their legs, and termites and beetles drum with their heads. Low-frequency signals also can be emitted by vibration of some body part(s). For example, male and female chloropid flies *Lipara* communicate over distances of more than 2 m on a reed with signals produced by vibration of the abdomen. Lacewings (Chrysopidae) oscillate the abdomen without touching the substrate, and in this way shake the stem or leaf on which they are standing with the low-frequency component of the broadband signal being used for communication. Abdominal muscles are involved in song production also in the Hawaiian fly *Drosophila sylvestris.* Bees produce vibratory signals by thorax vibrations; "begging" signals are transmitted to the substrate through the legs and "tooting" and "quacking" signals directly by pressing the thorax to the substrate.

Tymbal-like mechanisms are used to produce vibratory signals in most cicadas, small plant bugs, and hoppers. Because there are no special resonant air sacs behind the tymbal, the carrier frequency remains low and suitable for transmission through a substrate. Many insects produce vibratory signals by friction of two body parts moved one across another. For example, ants stridulate by means of a file on the dorsal surface of the first abdominal segment and a plectrum on the posterior edge of the metathorax or postpetiole. The substrate-borne component of the audible signal is used for communication. Burrower bugs (Cydnidae) communicate with substrate-borne signals produced with the tergal plate, which functions simultaneously as tymbal, as a file with its latero-frontal surface, or as a plectrum for the alary stridulitrum on the postcubital vein.

Vibratory signals are produced by different insect activities. An insect singing on or close to a substrate produces signals with airborne and substrate-borne components, and many of them use both simultaneously or alternatively. Fruit flies communicate with near-field sound radiated from wings and with substrate-borne signals, which are simultaneously produced by muscles driving the wings. Soil buzz vibrations and sounds produced by solitary bees and wasps are used as cues for localization.

TRANSMISSION OF VIBRATIONS

An insect standing on a substrate may be represented by a model of a mass on six springs. The bodies of treehoppers, spiders, and even fiddler crabs respond to substrate vibration with resonance at lower frequencies and attenuation at higher frequencies. Legs transmit body vibrations to the substrate and are the seat of most sensitive vibrational receptors. Transmission over the legs depends on their stiffness, and the response properties of the body–legs oscillating system varies in different species with posture changes. Signal amplitude decreases during transmission to the substrate, mainly because the body also vibrates during singing in a horizontal plane.

Plants are widely used as transmission channels for insect vibrational songs. Although quasi-longitudinal waves and waves of some other types cannot be excluded, communication signals are carried by bending or flexural waves with only a little frictional loss in energy and particle motion, both in the longitu-

dinal direction and in a plane perpendicular to the direction of wave propagation and to the surface. Their propagation velocity varies little with the structure's mechanical properties and is proportional to the square root of the structure's radius and the square root of the frequency. Vibratory signals reflect mainly at the root and top of the plant and may travel back and forth several times. In standing wave conditions with a complicated and frequency-dependent pattern, it is a better strategy for communication to use broadband and not pure tone or narrowband signals.

There are many examples that demonstrate how well the spectra of vibratory signals fit the filtering properties of host substrates. In plants, lowest attenuation with distance occurs with signals of frequencies around 100 Hz. Spectra of the songs of the southern green stink-bug, *Nezara viridula,* have a narrow dominant frequency peak between 80 and 160 Hz (Fig. 1) and the main energy of the broadband "small cicadas" and cydnid signals is emitted at lower frequencies. Carpenter ants, *Camponotus herculeanum,* live in tree trunks from which the soft spring wood has been eaten out, leaving thin, lignified lamellae. The alarm signals, produced by drumming the head and gaster against the substrate, are particularly suited for transmission through the nest but not through to the outside. "Begging signals," emitted by bees following a dancer, are used to solicit food samples. The 1-μm peak amplitude of the signals lies slightly above the threshold for the bee "freezing response," and the signal dominant frequency around 320 Hz lies in the range of the best signal-to-noise ratio of the bee comb.

RECEPTOR MECHANISMS

The most sensitive and specialized receptor for substrate vibrations, the subgenual organ, is derived evolutionarily from chordotonal organs in the body and appendage joints. The organ lies in insects in all the six legs close to the main tracheal trunk and almost completely occludes the dorsal blood sinus of the proximal tibia. The organ is supported proximally by the nerves and distally by accessory cells that form a ligament attached most often to the cuticle of the leg. The structural unit of the organ is the scolopidium, each of which is composed of one or more bipolar sensory cell(s) and two accessory cells, the scolopale and the cap (attachment) cell.

In Orthoptera, the subgenual organ is accompanied by other scolopal organs of which the crista acoustica, supported by sound-transmitting structures, represents a very sensitive receptor organ for airborne sound. In the American cockroach, *Periplaneta americana,* the subgenual and the neighboring Nebenorgan (Fig. 2) are most sensitive to substrate vibrations. The distal organ, which surrounds the hemolymph, responds to variations in hemolymph pressure. Scolopidia of the subgenual organ are organized in a fan-shaped manner and converge radially toward a single insertion point so that vibrations of the substrate cause tension changes. The organ is most sensitive in the frequency range between 1000 and 5000 Hz, with threshold amplitudes around 0.1 nm. Scolopidia of the Nebenorgan

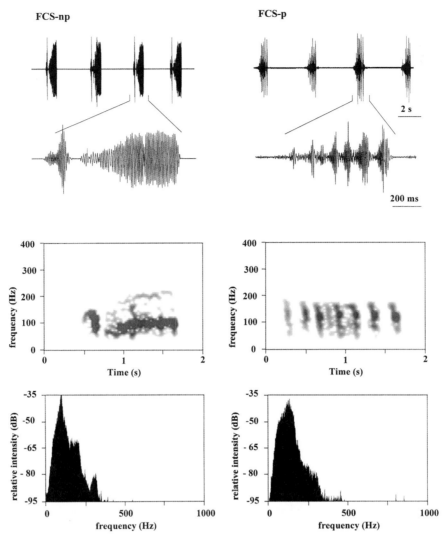

FIGURE 1 Oscillograms (top), sonograms (middle), and power spectra (bottom) of the nonpulsed (left) and of the pulsed (right) type of *N. viridula* female calling song. [After Čokl, A., Virant-Doberlet, M., and Stritih, N. (2000). The structure and function of songs emitted by southern green stinkbugs from Brazil, Florida, Italy, and Slovenia. *Physiol. Entomol.* **25,** 196–205, Blackwell Science Ltd., with permission.]

are stretched perpendicular to the long axis of the tibia so that the organ reacts to tension changes within the scolopale caused by vibration of the cuticular walls.

Subgenual organs of different morphology have been described in other insect groups. In the southern green stinkbug, the subgenual organ is proximally attached to the epithelium of the tibial wall, whereas the two scolopidia with the cilia and the flat and thin ligament are stretched out in the hemolymph of the blood channel. The subgenual organ in the legs of the lacewing *Chrysoperla carnea* (Fig. 3) is composed of only three scolopidia. The cell bodies of the three cap cells form a lenslike part of the organ, the velum, which distally divides the blood channel in two separate parts. Scolopidia are attached to the middle of the velum and extend to the dorsal leg wall. The hollow cone-shaped bee subgenual organ with approximately 40 scolopidia is connected at two points to the cuticle and at two points to the membrane bag surrounding the

organ and the membrane lining the tracheal wall. The organ oscillates with the hemolymph, and the sensory cells respond to displacements of the organ relative to the leg.

Mechanoreceptors like campaniform sensilla, joint chordotonal organs, and Johnston's organ respond to substrate vibrations with lower sensitivity, preferentially in the frequency range below 100 Hz.

VIBRATORY SIGNALS AND INSECT BEHAVIOR

Insects emit vibratory signals in connection with aggression, distress, calling, courting, rivalry, and many other behaviors. In many habitats, vibrational communication represents the only useful way to exchange information. Social insects like bees, ants, and termites live in nests where the possibility of communication by signals of other modalities (other than airborne chemicals) is limited. Insects, whose small body size

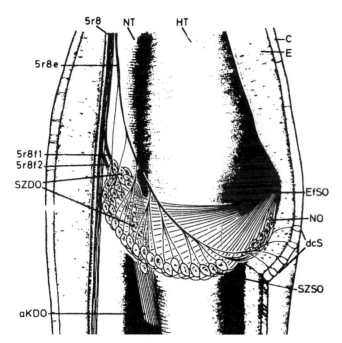

FIGURE 2 Subgenual organs of the cockroach *P. americana.* The left tibia is opened to show the spatial arrangement of the scolopidial organs. aKDO, accessory cap cells of the distal organ; C, cuticle; dcS, distal campaniform sensilla; E, epidermis; EfSO, terminal fila of the subgenual organ; HT, main trachea; NO, Nebenorgan; NT, small trachea; SZDO, sensory cells of the distal organ; SZSO, sensory cells of the subgenual organ; 5r8e, nerve innervating the subgenual organs and the campaniform sensilla. [After Schnorbus, H. (1971). The subgenual organs of *Periplaneta Americana:* histology and thresholds for vibration stimuli. *Vgl. Physiol.* **71**, 16. Springer-Verlag GmbH & Co. KG, with permission.]

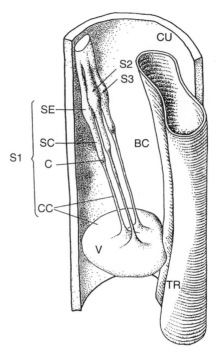

FIGURE 3 Three-dimensional reconstruction of the subgenual organ of the left middle tibia of the lacewing *C. carnea:* BC, blood channel; C, cap; CC, cap cell; CU, cuticle; S1, S2, S3, three scolopidia; SC, scolopale cell; SE, sensory cell, TR, trachea; V, velum. [After Devetak, D., and Pabst M. A. (1994). Structure of the subgenual organ in the green lacewing, *Chrysoperla carnea. Tissue and Cell* **26(2)**, 249–257, Harcourt Inc., with permission.]

does not allow efficient low-frequency sound radiation, can use vibrational communication without attracting predators or parasites as they would by singing. Many insects use the vibratory component of the emitted airborne signals to improve signal discrimination and recognition.

On the other hand, vibrational communication through plants is disturbed by environmental factors such as wind and raindrops. For example, wind induces low-frequency vibrations that are accompanied in apple (but not banana) leaves with broad-band vibrations of spectra up to 25 kHz. Raindrops falling on banana leaves produce vibrations up to 1000 Hz; those on apple leaves cause vibrations composed of an irregular high-frequency phase and a regular low-frequency phase. Signal-to-noise ratio is enhanced by the highly ordered temporal structure of songs with narrower spectra with dominant frequencies, usually above the low-frequency noise level, as well as by band-pass filtering properties of the subgenual organ.

The time pattern of a vibration pulse series usually carries more specificity information than its spectral structure. Generally, a rapid signal divergence occurs in sympatric taxa, and song similarity is expected only in allopatric or allochronic species. In *Chrysoperla,* tremulation songs represent the best cue for species identification within the *carnea* group. Within

the same region, the songs of different cryptic species differ structurally and functionally more than in pairs of species from North America and Eurasia. The vibratory song repertoire of sympatric Californian species *N. viridula* and *Acrosternum hilare* are different, and interspecific mating has not been observed. In Japan, interspecific mating was described between the sympatric species *N. viridula* and *N. antennata,* although songs, which correspond to definite behavioral features, are different.

Vibratory signals are used by insects also as cues for mate, prey, or enemy localization. In *N. viridula,* male directional movement on a plant results from female calling song signals. Leafcutting ant, *Atta cephalotes,* workers stridulate when they cut leaves, and nearby ants respond by orienting toward the source of the vibrations and join in the leafcutting. The response of the heteropteran predator *Podisus maculiventris* to vibrational signals produced by a common prey species demonstrates that these predators are capable of using substrate-borne vibrations as cues for prey location. The male katydid, *Conocephalus nigropleurum,* shakes its body to produce vibratory signals that attract females during courtship and mating.

Central nervous resolution of stimulation time differences and amplitude gradients, which occur when vibration travels through a substrate from one leg to another, is the mechanism that enables directionality in larger insects. For example, the response of the body of the treehopper *Umbonia crassicornis* relative to the substrate reveals resonance at lower frequencies and attenuation at higher frequencies. The transfer functions measured on the body differ substantially depending on whether the stimulus originates in front or behind, indicating that directional information is available in the mechanical response of the body to substrate vibration. Comparison of signal amplitudes between receptors of the front and back legs might enable vibration localization in smaller insects.

See Also the Following Articles
Hearing • Mechanoreception

Further Reading
Bailey W. J. (1991). "Acoustic Behaviour of Insects. An Evolutionary Perspective." Chapman & Hall, London.
Barth, F. G. (1998). The vibrational sense of spiders. *In* "Comparative Hearing: Insects (R. R. Hoy, A. N. Popper, and R. R. Fay, eds.), pp. 00–00. Springer-Verlag, New York.
Devetak, D. (1998). Detection of substrate vibration in Neuropteroidea: a review. *Acta Zool Fennica* **209**, 87–94.
Dusenbery, D. B. (1992). "Sensory Ecology: How Organisms Acquire and Respond to Information." Freeman, New York.
Ewing, A. W. (1989). "Arthropod Bioacoustics: Neurobiology and Behaviour." Edinburgh University Press, Edinburgh, U.K.
Gogala, M. (1985). Vibration producing structures and songs of terrestrial Heteroptera as systematic character. *Biol. Vesn.* **32**, 19–36.
Henry, C. S. (1984). The sexual behavior of green lacewings. *In* "Biology of Chrysopidae." (M. Canard, Y. Semeria, and T. R. New, eds.), pp. 00–00 Junk Publishers, The Hague.
Kalmring, K., and Elsner, N. (1985). "Acoustic and Vibrational Communication in Insects." Verlag Paul Parey, Berlin.
Markl, H. (1983). Vibrational communication. *In* "Neuroethology and Behavioral Physiology. Roots and Growing Points." (F. Huber, and H. Markl, eds.), pp. 00–00. Springer-Verlag, Berlin.
Michelsen, A., Fink, F., Gogala, M., and Traue, D. (1982). Plants as transmission channels for insect vibrational songs. *Behav. Ecol. Sociobiol.* **11**, 269–281.
Shaw, S. R. (1994). Detection of airborne sound by a cockroach "vibration detector": a possible missing link in insect auditory evolution. *J. Exp. Biol.* **193**, 13–47.

Vision

see *Eyes and Vision*

Vitellogenesis

William H. Telfer
University of Pennsylvania

Vitellogenesis is the process by which yolk accumulates in the cytoplasm of an ovarian oocyte. It is one of the final stages of egg formation, occurring just prior to deposition of the chorion. Studies on vitellogenesis have focused during the last 50 years on the major protein yolk precursor, vitellogenin, its synthesis in the fat body, its transport to the oocyte, its sequestration by receptor-mediated endocytosis, and the developmental and hormonal mechanisms that control these processes.

VITELLOGENIN SECRETION BY THE FAT BODY

Before vitellogenesis can begin, the female fat body must transform from a tissue that supports molting, metamorphosis, and intermolt metabolism to one that can secrete vitellogenin and other yolk precursors. When the fat body has reached a requisite stage of maturity and hormonal stimulation, vitellogenin genes are transcribed that encode a polypeptide chain whose molecular weight in many insects is over 200,000. During transport through the endoplasmic reticulum and Golgi bodies, this polypeptide is clipped by endoproteases that remove a secretory signal from its N-terminal end and divide it into shorter polypeptide subunits. The latter differ in number from one in some Hymenoptera to nine in some Hemimetabola. As it moves through the secretory pathway, the complex of subunits is conjugated at sequence-specific sites with high-mannose oligosaccharides, phosphate, and lipids.

Amino acid sequences determined for vitellogenins from at least seven orders of insects are sufficiently similar to indicate a common ancestry. They are members of a protein superfamily that also includes the yolk proteins of vertebrates and nematodes, lipid transport proteins in the blood of both vertebrates and insects, and receptor proteins involved in the endocytosis of lipoproteins.

The yolk proteins of cyclorrhaphan Diptera are exceptional in that their amino acid sequences resemble those of a family of digestive enzymes, the vertebrate lipases. They nevertheless behave like conventional vitellogenins in being synthesized and conjugated in the female fat body, and deposited in cytoplasmic vesicles by the oocyte after endocytosis. A notable difference from conventional vitellogenins is that the several yolk polypeptides in a species are encoded by separate genes, rather than being proteolytic fragments of one gene product.

Other fat body products may supplement vitellogenin in the yolk. The eggs of several Lepidoptera contain lipophorin, a hemolymph protein whose functions include the delivery of lipids from the fat body to other tissues. In the yellow fever mosquito, *Aedes aegypti,* two proproteases are synthesized in the fat body and deposited in the eggs, where they are converted to active proteases during embryogenesis. Endocytosis of hemolymph proteins that bind iron, calcium, heme groups, or biliverdin concentrates these ligands in the yolk of select species. Some of these supplementary proteins are secreted in synchrony with and under the same hormonal control as vitellogenin. Others, like lipophorin, serve somatic functions that require their presence in the hemolymph of males as well.

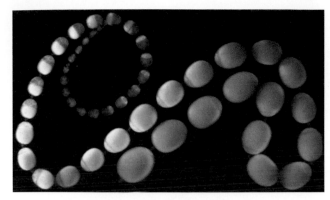

FIGURE 1 Reflected light micrograph of a vitellogenic ovariole from the saturniid moth *H. cecropia*. The largest 36 follicles are vitellogenic. The yellow mass in each follicle is a yolk-filled oocyte, whose opacity is due to light scattering by yolk particles. The yellow color is due to carotenoids carried by vitellogenin and lipophorin, the two most abundant yolk proteins in saturniids. Nurse cells form a transparent cap at one end of each follicle. The epithelium of follicle cells surrounding each oocyte and its nurse cells is too thin to be readily visible, except where it forms a connection between successive follicles. Scale: largest follicle is about 1.9 mm long.

VITELLOGENIC FUNCTIONS IN OVARIAN FOLLICLES

In the ovaries, the morphological unit of vitellogenesis is a follicle—a single oocyte surrounded by an epithelium of somatic cells (the follicle cells) and associated in many insects with a set of modified germ cells (the nurse cells) (Fig. 1). All three cell types are necessary for vitellogenesis, but they contribute to it in very different ways.

Follicles begin to form during or shortly after metamorphosis. They are produced in linear chains termed ovarioles. Within each ovariole is a developmental gradient of follicles (Fig. 1), with the most mature one lying close to the beginning of the oviduct. A common pattern among cyclic egg producers is for only one follicle at a time in each ovariole to form yolk. The penultimate follicle is retarded in its development until the next reproductive cycle. In noncyclic insects, many follicles in each ovariole may simultaneously form yolk (Fig. 1).

Follicle Cells, Patency, and Secondary Yolk Proteins

At the onset of vitellogenesis, the follicle cells develop a system of intercellular spaces that give proteins from the hemolymph access to the surface of the oocyte. In the bug *Rhodnius prolixus,* patency has been attributed to cellular protrusions whose cytoskeletal elongation pushes neighboring follicle cells apart. In the cecropia moth, *Hyalophora cecropia,* osmotic shrinkage of the follicle cells is crucial. Whichever mechanism applies, the intercellular spaces are under tight developmental control: they arise at the onset of vitellogenesis and close when it terminates.

In addition, the follicle cells of many insects secrete proteins that are endocytosed by the oocyte along with vitellogenin.

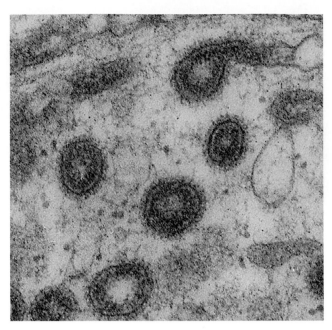

FIGURE 2 Electron micrograph of several endocytotic vesicles near the surface of a vitellogenic oocyte from *H. cecropia*. The bristlelike outer coat is the clathrin lattice that generates the force used to bend coated surface membranes into cytoplasmic vesicles. Receptor-bound vitellogenin is included in the thick layer of granular material that lines each vesicle. An incipient vesicle on the upper right is still attached to an infolding of the cell membrane. The average diameter of coated endocytotic vesicles is about 0.15 mm.

These products may resemble vitellogenin in size, antigenicity, and amino acid sequence; examples occur in Thysanura, Heteroptera, and Coleoptera. A similar relationship holds for the yolk proteins of the cyclorrhaphan Diptera. In Lepidoptera, follicle cell products are instead lipaselike sequences reminiscent of the yolk proteins of the Cyclorrhapha. A few exceptions are known in which either fat body or follicle cells but not both secrete precursors for the protein yolk.

Finally, the follicle cells connect to the oocyte during vitellogenesis via gap junctions that permit direct cytoplasm-to-cytoplasm transfer of ion currents and small organic molecules. These junctions have the potential to function in intercellular exchange of signaling substances such as cyclic nucleotides and calcium ions.

Nurse Cells and the Origin of Egg Cytoplasm

In all holometabolous and a few hemimetabolous orders, the oocytes connect via cytoplasmic bridges to nurse cells (Fig. 1). The bridges are wide enough to permit passage of ribosomes and membranous organelles such as mitochondria. Once believed to be the site of yolk production, nurse cells are now known to be the primary source of egg cytoplasm. They support vitellogenesis by providing the oocyte with the ribosomes, transcripts, and metabolic machinery needed to synthesize the receptors, structural proteins, and enzymes necessary for yolk deposition. In Orthoptera, Blattodea, and other Hemimetabola that lack nurse cells, the requisite transcripts are produced instead within the oocyte's own nucleus by amplified nucleoli and lampbrush chromosomes.

Receptor-Mediated Endocytosis in the Oocyte

The surface of the vitellogenic oocyte contains receptors that can selectively bind vitellogenin and other yolk precursors after they have penetrated the spaces between the follicle cells. The receptors form transmembrane associations with clathrin lattices on the cytoplasmic side of the membrane. The membranes containing these complexes then fold inward to form endocytotic vesicles (Fig. 2). In subsequent processing steps, vitellogenin dissociates from its receptors within the vesicles, and the clathrin is released from the lattices on the outside. The denuded vesicles transfer their cargo of yolk precursors to neighboring yolk bodies by membrane fusion. Receptors, clathrin lattices, and extra membrane are recycled to the oocyte surface for more rounds of endocytosis. In some insects, vitellogenin is later modified in the yolk to a less soluble storage form known as vitellin.

In the other two classes of yolk particles, glycogen is synthesized inside the oocyte from hemolymph-derived sugars, whereas lipid droplets are assembled from precursors carried from the fat body to the oocyte by lipophorin. Because the lipid droplets contain primarily triacylglycerols and lipophorin transports primarily diacylglycerols, enzymatic conversions must take place during the transfer.

CONTROL BY JUVENILE HORMONE AND ECDYSONE

In many insects, juvenile hormone secreted by the corpora allata stimulates the fat body of adult females to initiate the synthesis of vitellogenin. It may also promote patency of the follicle cells, hence making vitellogenin in the hemolymph available to the oocyte for endocytosis. Juvenile hormone, whose power as an inhibitor of metamorphosis requires that its secretion be reduced during pupation and adult development, has thus evolved new kinds of role in adults: it has become an effector of the neuroendocrine networks that synchronize vitellogenesis with feeding, the photoperiod, and mating.

Variations occur on this theme. In species such as stick insects, and native and domestic species of silk moths that complete egg formation during adult development, juvenile hormone is not required for vitellogenesis. In its classical role as an inhibitor of metamorphosis, the hormone may even prevent the fat body and ovaries from completing their essential previtellogenic development.

Another kind of variation occurs among Diptera, whose yolk protein synthesis by fat body is triggered by ecdysone. Juvenile hormone may still be required, but here it promotes posteclosion development of the fat body to a stage capable of responding to ecdysone. In the yellow fever mosquito, the ovaries themselves were shown to secrete ecdysone in response to EDNH (egg development neurosecretory hormone), a brain hormone released from the corpora cardiaca following a blood meal.

See Also the Following Articles

Eggs • Egg Coverings • Reproduction, Female: Hormonal Control of

Further Reading

Hagedorn, H., Maddison, D., and Tu, Z. (1998). The evolution of vitellogenins, cyclorrhaphan yolk proteins and related molecules. *Adv. Insect Physiol.* **27**, 335–384.

Raikhel, A., and Dhadialla, T. (1992). Accumulation of yolk proteins in insect oocytes. *Annu. Rev. Entomol.* **37**, 217–251.

Sappington, T., and Raikhel, A. (1998). Molecular characteristics of insect vitellogenins and vitellogenin receptors. *Insect Biochem. Mol. Biol.* **28**, 277–300.

Telfer, W., Huebner, E., and Smith, D. (1982). The cell biology of vitellogenic follicles in *Hyalophora* and *Rhodnius*. *In* "Insect Ultrastructure" (R. King and H. Akai, eds.), pp. 118–149. Plenum Press, New York.

Telfer, W (2002). Insect yolk proteins: a progress report. *In* "Reproductive Biology of Invertebrates," Vol. 12 of "Recent Progess in Vitellogenesis" (A. Raikhel and T. Sappington, eds), pp. 33–71. Wiley, Chichester, U.K.

Wyatt, G. (1991) Gene regulation in insect reproduction. *Invert. Reprod. Dev.* **20**, 1–35.

Wyatt, G., and Davey, K. (1996) Cellular and molecular actions of juvenile hormone. II: Roles of juvenile hormones in adult insects. *Adv. Insect Physiol.* **26**, 1–155.

Walking and Jumping

Roy E. Ritzmann
Case Western Reserve University

Sasha N. Zill
Marshall University

T he agility of insects certainly contributes to their reign as the most successful creatures in the animal kingdom. Few if any forms of terrain present an insurmountable barrier to all insects. By evolving variations on a basic body plan, they have achieved remarkable dexterity in a wide range of environmental niches. We find insects that walk slowly over floors, scurry under rocks, climb up walls and over ceilings, or jump over barriers that if scaled to human dimensions would represent achievements unattainable by the most accomplished athletes. These abilities have attracted the attention of engineers who study insect locomotion as inspiration for legged robotic devices.

Although these creatures are often described as "simple systems," a close examination of their abilities reveals mechanisms that are elegant and not really simplistic. Their capabilities represent remarkable combinations of mechanical principles, neural control, and sensory input leading to efficient movement of leg joints. Although many aspects of these systems are economical in design and have been studied for many decades, they are only now beginning to be understood.

The problems inherent in insect walking and jumping encompass issues ranging from biomechanics to both central and peripheral neurobiological factors, as well as force development in muscle. To illustrate these points, we will examine the leg structures and neural circuits that produce walking and running in the cockroach. However, a variety of other insects have also been studied extensively by neurobiologists and engineers, including stick insects, grasshoppers, and crickets.

To understand how an insect walks, we must subdivide the process. First, we must understand the movements of the legs and their constituent joints, then we can begin to look at how muscles generate these movements and the circuits within the central nervous system that control motor activity. Over most walking speeds, sense organs of the limbs contribute to motor control by providing detailed information about the position, velocity, and the forces occurring in each leg. Therefore, we must also investigate the role of sensory input in the control of movement. Even with all this information, we will understand only how an animal can walk on a horizontal surface. But the aspects of walking in insects that are truly remarkable, and attract them to robotics engineers, are the abilities to run over and around obstacles, climb up walls, and jump over barriers. Often these adaptations occur through subtle changes in the basic pattern of locomotory behavior.

LEG MOVEMENTS

The walking movements of insects were accurately described in 1887 by Morgan who, in an era in which galloping of horses held the public attention, wrote a letter to *Nature* entitled "The Beetle in Motion" that recounted the coordination seen in a tripod gait. In more detailed accounts in the mid-20th century, Hughes and Wilson noted that insects walk by moving their six legs in reproducible patterns. Each leg alternates between a stance phase, when the tarsus (foot) is on the ground and the animal is pushed forward, and a swing phase, when the tarsus is moved forward through the air. At slow speeds, the legs follow a metachronal pattern, moving from the hind legs to middle and then to front legs on either side. However, to move more rapidly, some legs must be moved at the same time and therefore the insect shifts into a modification of the metachronal pattern called the tripod gait. Here the front and rear legs on one side of the animal move as a unit with the middle leg on the opposite side (Fig. 1A, B). This tripod alternates between swing and stance with the tripod made up of

A

B

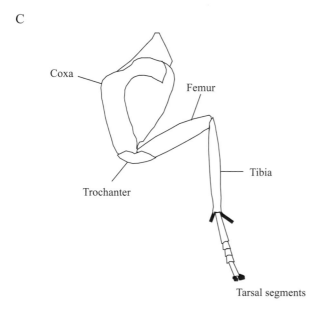

C

Coxa

Femur

Trochanter

Tibia

Tarsal segments

FIGURE 1 Description of leg segments and movements in tripod gait. (A and B) Pictures from a high-speed video record of a cockroach walking on a lightly oiled plate. These images are taken at the beginning and end of one leg cycle. The legs forming one tripod (animal's right front and rear legs and left middle leg) are indicated with triangles; the legs forming the other tripod are designated with circles. Lines connect the triangles to clearly indicate the tripod. Note that in (A) the left front and middle tarsi are very close together, almost overlapping. (C) Diagram showing the segments that make up a typical cockroach leg.

the remaining legs. The tripod gait is very stable, because at most speeds the animal's center of mass remains within the base of support. However, at very high speeds, many insects make dynamic postural adjustments to stay upright. Remarkably, at exceptionally high speeds the American cockroach *Periplaneta americana* has been seen to rise up on its hind legs and run with a bipedal gait. This gait is not statically stable (the cockroach would fall if it stopped), but represents a balance of forces that produces dynamic stability.

Each leg is made up of segments that are similar from leg to leg but differ in dimensions. From the most proximal to distal location, the leg segments are the coxa, trochanter, femur, and tibia, and a series of tarsal segments ending in a retractable claw (Fig. 1C). In the cockroach, the most important joints for walking are the coxa–trochanter (CTr) joint and the femur–tibia (FTi) joint. The CTr joint actually moves the femur relative to the coxa because the trochanter–femur (TrF) joint makes only small movements. Although flexion of the TrF joint can effectively rotate the tarsus, during many movements it acts mechanically as a fused joint. In other insects, relative proportions of these leg segments are changed to match the needs of specialized forms of locomotion. Thus, the hind leg of a locust has a very short coxa and a long muscular femur, making a powerful jumping leg.

Although the legs within a tripod move their feet as a unit, the joint movements and resulting forces are unique for each pair of legs. In cockroaches, the hind legs are specialized to propel the animal forward in walking. To accomplish this the CTr joint and the FT joint move in near synchrony (Fig. 1A). This action allows these rotary joints to direct the movements of the tarsi (feet) in a line nearly parallel to the long axis of the animal's body. The middle legs make similar movements, but with smaller joint [CTr] excursions. The orientation of this leg causes it to first brake and then accelerate forward movements of the animal.

The front legs make very different movements. Unlike the other two pairs of legs, the front legs make much greater use of the body–coxa (BC) joint, which attaches the leg to the thorax. This joint has three degrees of freedom, similar to the ball-and-socket joint in a human shoulder. Movement of the BC joint swings the front legs far forward much like an extending human arm. The CTr and FT joints then move out of phase with each other as the foot is drawn back toward the body and then is pushed backward. The resulting ground reaction forces slow the forward movement of the body and keep the animal from losing control. Clearly, the neural control of this leg is much different from control of the other two pairs of legs. Nevertheless, the sum of the ground reaction forces of all three pairs of legs is similar to that seen in the bipedal leg movements of a human.

Cockroaches can run extremely rapidly (up to 25 steps per second), ranking them as the fastest terrestrial animal (in steps/per second). Running, which follows wind stimulation of the cerci (abdominal appendages) or tactile stimulation of antennae or body cuticle, is used to escape from predators. Although fast running shows a number of similarities to walking, it occurs so rapidly that it may actually be a separate behavior. Passive elastic structures in the legs can play an important role in generating such rapid locomotion, prompting cockroach running to be modeled as a mass on a spring.

MOTOR CONTROL OF LEG MUSCLE

The muscular anatomy of insect legs follows a proximal-to-distal arrangement that makes very good biomechanical sense. The largest and most powerful muscles are proximal or closer to the body. The muscles are smaller in more distal segments of the leg. With the most powerful muscles placed near the body, inertial effects in the distal part of the limb are reduced, allowing for more controlled movement. Also, the movements of the proximal joints act on the tarsus or foot through the relatively large lever arm of the intervening leg segments to the tarsus or foot. Thus, for example, relatively small movements at the BC joint can greatly alter the orientation of the tarsal end point.

The most distal segment, the tarsus, is actually made up of a series of segments ending in a retractable claw that have again remarkable mobility. However, as with human fingers, the muscle for these segments is found in more proximal segments and imparts its movements via a long apodeme that serves the role of a tendon. Indeed, although the claw can be engaged by this muscle, there is no antagonistic muscle. Rather the claw is disengaged and lifted from the substrate by a remarkably efficient elastic protein (resilin) found in the joints of the tarsal segments. The resilin acts like a spring on a screen door; is stretched when the claw is engaged and causes the tarsus to be lifted automatically when the muscle is relaxed.

NEUROMUSCULAR SYSTEM

The neuromuscular arrangement of insects provides distinct advantages for motor analysis. Unlike vertebrate systems in which muscle cells fire action potentials, in insects most muscle fibers produce graded potentials when motorneurons fire action potentials. Variation in tension in a vertebrate muscle requires recruitment of more or fewer motor neurons. However, in arthropods, simply altering the frequency of action potentials in a single motor neuron can control tension. Furthermore, insect muscles are innervated by very few motor neurons. Often, especially in stance phase muscles (the muscles that extend the leg while the tarsus contacts the ground), there are only two motor neurons serving a range of muscles. These can be readily distinguished as either fast or slow motor neurons depending on the types of muscle contraction they produce. Slow motor neurons need a series of action potentials to generate significant movement, whereas fast muscle generates a reasonable twitch with a single muscle potential. Typically, extracellular recordings indicate that fast motor neurons generate larger action potentials. This neuromuscular arrangement allows one to use electromyogram (EMG) electrodes to record muscle activity extracellularly and often to know exactly which motor neuron is observed.

The leg movements that occur during walking on a horizontal surface are associated with typical patterns of motor activity. In both the middle and hind legs of cockroaches, CTr extension is generated by a burst of activity in the slow depressor motor neuron (Ds). At faster speeds, one or more muscle potentials from the fast depressor motor neuron (Df) occurs at specific times in the leg cycle. These actions generate the stance phase of the leg cycle and alternate with activity from several flexor motor neurons that produce the swing phase. The simultaneous extension at the FT joint is generated by the slow extensor of the tibia (SETi) motor neuron and the fast extensor of the tibia (FETi), and is again opposed by activity in several flexor motor neurons that generate the swing phase of that joint. As expected, the motor patterns of the front legs are more complicated, matching the joint movements described earlier.

The cockroach controls speed of walking by altering the frequency of motor action potentials. Motor frequency is positively correlated with stepping frequency and joint velocity. Thus, the animal can move faster by increasing the frequency of Ds and SETi in each leg. It can turn by increasing frequency in one or more legs of the tripod while decreasing the frequency in the leg or legs located on the opposite side. This change creates stronger forces on the outer leg of the turn and weaker lateral forces on the inner leg, thereby turning the animal.

PATTERN GENERATION

Although both stance and swing phase motor neurons increase burst duration with increasing walking speed, the change in stance phase motor activity is much greater. This observation led to a model for insect walking referred to as the flexor burst generator. In this model, a set of interneurons within the central nervous system called a central pattern generator controls the flexor motor neurons directly and the stance phase motor neurons indirectly through inhibition from swing phase interneurons.

Investigators have recorded traces from oscillatory neurons in cockroaches and stick insects that could be part of such a central pattern generation circuit. For example, in cockroach, neurons have been found that undergo membrane potential oscillations in time with levator motor activity. Moreover, stimulation of these cells activates the flexor motor neurons. Thus, these cells could be part of the flexor burst pattern generator. However, they by no means represent the entire circuit. Other work has strongly suggested that there are separate burst generators that activate motorneurons for each joint in the leg. This would provide for considerable flexibility. The circuits that produce walking could be used in other rhythmic behaviors, such as righting (when an animal falls over and uses its legs to regain an upright posture) or grooming, by changing the coupling between the oscillators.

LEG SENSORS

Even where central pattern generation circuits contribute to control of locomotion, sensory systems still play important roles in generating behaviorally meaningful motor activity. This concept was particularly well demonstrated in the locust flight system. Locust preparations that have been isolated from sensory structures in the periphery (deafferented) are capable of generating bursts of activity that could move the wings in an appropriate manner. However, the frequency of the "wing-beat" cycle is much lower than normal and the pattern quickly dies out, whereas in the intact animal, flight can go on for very long periods of time. If appropriately timed activity from wing stretch receptors is added in, a more normal flight pattern occurs that is maintained for long periods of time. Thus, the central circuits are capable of generating part of the cyclical pattern but not a complete, behaviorally relevant one.

Similarly, normal walking behaviors in all insects that have been studied require input from sensory structures on the legs. However, in rapid running movements associated with escape, feedback from sensory inputs probably is too slow to influence walking within a single step, and the animal may be running "on autopilot."

The sensors that play an important role in walking provide precise bits of feedback information to the walking system through structures that are most often found in the same locations in all three pairs of legs (Fig. 2). Internal sensors

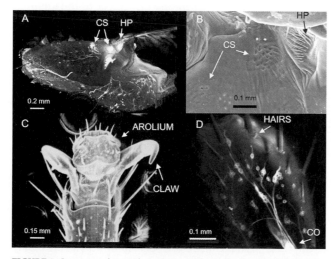

FIGURE 2 Sensory and cuticular structures of cockroach leg (A) View of the trochanter in a confocal microscope in which the main leg nerve was infiltrated with fluorescent dye (diI) that diffuses in membranes and stains sensory neurons when it is applied to peripheral nerves. The sensory neurons of campaniform sensilla (CS) and the hair plate (HP) are filled with the dye and fluoresce brightly. (B) Scanning electron micrograph of the region containing the sense organs showing cuticular caps of campaniform sensilla and long hairs of hair plate. (C) Scanning electron micrograph of the distal end of the tarsus (foot), showing the hooklike claws and the arolium (an adhesive pad between the claws) (D) View of the dorsal side of the claw shows numerous sensory neurons stained with diI that innervate hairs as well as a portion of the chordotonal organ (CO). [Photomicrographs (A) and (D) by Faith Frazier, (B) by David Neff, and (C) by Laura Quimby, Marshall University School of Medicine.]

such as the chordotonal organs span the joints and monitor joint angle. Hair plates (HP) are positioned near joints where adjacent segments will touch during maximal flexion. Campaniform sensilla (CS) are located in strategic locations near joints and muscle insertions and detect strain in the cuticle. These structures are made up of sensory nerve endings positioned in sockets with a flexible dome. As strain increases in the surrounding cuticle, the dome is deformed and activates the sensory neuron.

Each peripheral sensor provides information that is used at discrete phases of the step cycle as the insect walks. Although joint angle detectors such as the femoral chordotonal organ are active throughout the ranges of joint movement, they often reach their highest levels of firing in positions of maximum joint flexion or extension. In walking of stick insects, these receptors trigger a change from stance to swing phase when joint angles reach the extreme joint position. Hair plates monitor the resulting joint movements in swing and can aid in the initiation of stance. Campaniform sensilla are activated during stance as forces are exerted upon the leg and the load of the insect's body is supported. They then act to increase the activity in the extensor motor neurons to rapidly generate forces in the leg. The greater extensor activation in each stance leg propels the animal forward more efficiently and also allows other legs to be lifted in swing. The elevated stance activity continues until the chordotonal organs trigger the switch to swing phase.

Beyond helping to set the normal locomotor pattern, these sensors are, of course, readily available to make rapid adjustments to the motor pattern as the animal moves away from horizontal surfaces and climbs inclines or vertical surfaces, or even walks on ceilings and negotiates obstacles in its path. The increases in load experienced on each leg as the animal climbs a surface ultimately translates into an altered gait. For example, as a cockroach walks on a ceiling, legs hold on to the surface for greater periods of time. Metachronal gaits are used when the animal is inverted, and individual legs make very rapid swing movements before reattaching to the substrate.

Arrays of neurons with diverse properties receive sensory inputs and could mediate these effects. Within each thoracic ganglion, the sensory neurons from these leg sensors project to populations of spiking local interneurons (interneurons that are isolated to a single ganglion). The projections of many receptors follow a topographic pattern that reflects the location of the sensors in the leg. The spiking interneurons project both to motor neurons and to another set of local interneurons called nonspiking interneurons. These interneurons never generate action potentials. Rather, they act through graded potentials to control and coordinate motor activity to various leg muscles within a thoracic segment. Coordination among joints of each leg (intraleg coordination) occurs through interactions in these local circuits.

Finally, spiking interganglionic interneurons project between thoracic ganglia, influencing local circuits for adjacent legs. Through these circuits the sensory activity directs appropriate adjustments in individual muscles, coordinates segments within a single leg, and influences the coordination between legs.

INTERLEG COORDINATION

To generate an effective tripod gait, insects must coordinate movements of joints not only within a single leg but also between legs. This interleg coordination on the surface appears to be a daunting task. The tarsi of the three legs that make up an effective tripod must move in synchrony. However, as described earlier, the legs of each segment make different movements. The patterns of the motor neurons that control those movements also differ in each leg. Thus, the animal must coordinate the tarsi through joint movements and neural activities that are specific to each pair of legs.

Observations on stick insects suggest that these potentially complex connections can be functionally formulated as a set of fairly straightforward rules indicating that events in one leg that are readily detected by leg sensors influence the actions of other legs in discrete ways. For example, one rule is that when one leg is in swing, the adjacent anterior leg cannot enter into swing phase. Alternatively, the start of the stance phase promotes the start of swing in adjacent anterior legs and in the contralateral leg of the same segment. More subtle influences also exist to account for transient changes. For example, an increased load on one leg prolongs the power stroke on the contralateral leg of the same segment. This would result in coactivation of legs when the animal is encountering increased resistance, such as during climbing. The specific neural connections that underlie these rules are an exciting new area of research.

NEGOTIATING OBSTACLES

The remarkable agility that attracts interest from roboticists is seen when insects are faced with an obstacle. Most insects effortlessly negotiate barriers that would pose considerable control problems in robots. This problem is only now beginning to be investigated, since climbing over obstacles is a more transient event than horizontal walking and there is considerable variability in the strategies used. However, some common themes are evident. In climbing over a block, for example, insects must move their center of mass (CoM) upward to surmount the obstacle. For cockroaches, the strategies used to accomplish this task vary with the size of the obstacle and the speed that the insect is moving at when the block is encountered. The front legs are normally lifted up fairly high during normal walking. If the block that is encountered is lower than the normal front leg trajectory, the insect hardly needs to change its movement at all. The front leg will be placed on top of the obstacle and pushed down to move the animal's CoM upward as a natural consequence of the encounter with the block. Larger blocks can be negotiated in similar ways when the animal is moving at faster speeds by taking advantage of the inertia developed by body movements.

At slower speeds, encounters with larger blocks require the animal to first rear upward to get its front legs on top of the block and initiate the climb. Locusts negotiate such blocks by using an elevator reflex, in which the front leg bumps into the front surface of the block repeatedly while moving upward until the top of the block is reached. However, in cockroaches, the front leg typically reaches the top of the block in a single movement, without even touching the front surface.

For very large obstacles, the insect must actually climb up the front face of the object. Here the insect must attach itself as if the face of the block were a wall. This problem is solved by a combination of claws and pads that adhere to various substrates. Cockroach claws, which are remarkably similar to the claws of a cat, are normally held up during horizontal walking. However, when a cockroach is walking up a vertical surface, the claw is pulled down by a muscle in the tibia, which attaches to the claw through a long apodeme. There is no return muscle. Rather, as we have described, the claw pops back up passively with the aid of strategically placed resilin ligaments. For smoother surfaces, including those as polished as glass, some cockroaches have cuticular pads between their claws that can adhere to the substrate for stability but be readily lifted when needed in walking.

JUMPING

Many insects have developed efficient jumping behaviors to move long distances with a single jump. The jump requires a powerful and rapidly accelerating movement of the jumping leg. Orthopterans such as grasshoppers and locusts have very large hind legs that can generate powerful jumping movements. However, even the large femur extensor muscles of these jumping legs cannot generate the quick extension necessary for an efficient jump without some mechanical modification within the leg structure for storing energy. In locusts, the tendon of the tibial flexor muscle moves over a stop that allows the extensor to contract without moving the leg when the muscles are coactivated. As a result, a considerable amount of energy is stored in the mechanical distortion of the femur, tibia, and extensor tendon (like the bow of an archer). Inhibition of the tibial flexor muscle releases locking mechanisms and produces a very rapid and powerful extension that propels the animal upward. The timing of these events is critical, and neural circuits have been identified in association with kicking movements that provide the appropriate control.

Fleas generate remarkable jumps relative to their tiny size. Again, a proportionally very large hind leg is used to generate the jumping movement, with a hook on the hind leg preventing movement until a large isometric force has been achieved. The isometric force is stored in strategically located resilin ligaments and released during the jump.

One of the most remarkable jumping strategies in insects does not even involve legs. Click beetles can jump from a standstill to four times their body length by rapidly accelerating the joint between the prothoracic and mesothoracic segments of the body. Here again a mechanical stop prevents movement until large isometric force has been achieved, this time in thoracic muscles, and then is released suddenly to shoot the animal upward.

ROBOTIC DEVICES INSPIRED BY INSECT WALKING

The efficiency and agility of walking insects as well as specialized strategies such as jumping have not been lost on robotics engineers. Insects have provided popular models for legged robots because the hexapod gait is statically stable. That is, the CoM remains within the base of support of at least one tripod at all times. Thus, the control for a robot need not actively maintain balance during horizontal walking. However, to capture the agility of the insects, postural adjustments, reflexes, efficient leg design, and sensors must be incorporated into the robotic designs. This realization has generated collaborations between engineers and biologists to study insects and then incorporate newly discovered control and mechanical properties into more efficient robots. The efforts to develop and control these new robots in turn provide new insights into how the insect controls its own movements.

See Also the Following Articles
Anatomy • Flight • Legs • Wings

Further Reading

Bässler, U., and Büschges, A. (1998). Pattern generation for stick insect walking movements—Multisensory control of a locomotor program. *Brain Res.* **27**, 65–88.
Burrows, M. (1996). "The Neurobiology of an Insect Brain." Oxford University Press, Oxford, U.K.
Camhi, J. M. (1980). The escape system of the cockroach. *Sci. Am.* **243**(6), 158–172.
Cruse, H. (1990). What mechanisms coordinate leg movement in walking arthropods? *Trends Neurosci.* **13**, 15–21.
Delcomyn, F. (1985). Walking and running. *In* "Comprehensive Insect Physiology, Biochemistry and Pharmacology: Nervous System: Structure and Motor Function." (G. A. Kerkut and L. I. Gilbert, eds.), pp. 439–466. Pergamon Press, Oxford, U.K.
Heitler, W. J., and Burrows, M. (1977). The locust jump. I: The motor programme. *J. Exp. Biol.* **66**, 203–219.
Hughes, G. M. (1952). The co-ordination of insect movements. I: The walking movements of insects. *J. Exp. Biol.* **29**, 267–284.
Morgan, C. L. (1887). The beetle in motion. *Nature* **35**, 7.
Pearson, K. (1976). The control of walking. *Sci. Am.* **235**, 72–86.
Ritzmann, R. E., Quinn, R. D., Watson, J. T. and Zill, S. N. (2000). Insect walking and biorobotics: A relationship with mutual benefits. *BioScience* **50**(1), 23–33.
Watson, J. T., and Ritzmann, R. E. (1998). Leg kinematics and muscle activity during treadmill running in the cockroach, *Blaberus discoidalis*. I: Slow running. *J. Comp. Physiol. A* **182**, 11–22.
Wilson, D. M. (1966). Insect walking. *Annu. Rev. Entomol.* **11**, 103–123.
Zill, S. N. (1990). Mechanoreceptors: Exteroceptors and proprioceptors. *In* "Cockroaches as Models for Neurobiology: Applications in Biomedical Research" (I. Huber, E. P. Masler, and B. R. Rao, eds.), pp. 247–267. CRC Press, Boca Raton, FL.
Zill, S. N., and Seyfarth, E.-A. (1996). Exoskeletal sensors for walking. *Sci. Am.* **275**(1), 86–90.

Walking Sticks
see *Phasmida*

Wasps

Justin O. Schmidt
Southwestern Biological Institute, Tucson

Few entomological words evoke a more vivid image and response than the word "wasp." Wasps live on all continents inhabited by people, and nearly everyone is familiar with their colorful presence and habits. The root of their name is the word "webh" (Low German)—to weave, a reference to the nests constructed by these insects. In entomological literature, "wasp" is often applied to all members of the Apocrita suborder of Hymenoptera except ants and bees. However, in popular usage a wasp is any social species in the family Vespidae, especially in the genera *Vespula, Dolichovespula*, and *Vespa*, which are viewed as irritating, plus a few conspicuous solitary hunting wasps. In this article, emphasis is according to popular usage.

TAXONOMY

In the broad sense, wasps include an enormous number of parasitoid species in the families Ichneumonidae, Braconidae, Chalcididae, and many other insects referred to as "parasitic wasps," plus solitary or hunting wasps, including sphecid wasps (Sphecidae), spider wasps (Pompilidae), potter wasps (Vespidae: Eumeninae and others), scoliid wasps (Scoliidae), velvet ants (Mutillidae), and others. In terms of sheer numbers of species, these groups far outnumber those of the social wasps. In terms of success, however, social wasps have earned due respect as major influences and elements of most ecosystems.

The total number of social wasps in the world is slightly over 900 species divided into three subfamilies: the Stenogastrinae, or hover wasps; the Polistinae, or paper wasps; and the Vespinae, or wasps, hornets, and yellowjackets (Fig. 1). Stenogastrines live in the South Indo-Pacific region, and comprise just over 60 species of long slender adults in six genera; their small, inconspicuous nests often are attached to roots or twigs under overhangs of banks. They provide their larvae with insects, often taken from the webs of spiders, hence, the term "hover wasp." The Polistinae are by far the most speciose of the three groups of social wasps: approximately 800 species are divided into 27 genera, with a cosmopolitan distribution except for the colder regions of Eurasia and North and South America. The Vespinae contains just over 60 species broken into four genera. *Vespa*, the true hornets, inhabit temperate and tropical Eurasia

FIGURE 1 Wasps. Clockwise from upper left: *Polistes exclamans,* larvae and capped pupae of *Vespa mandarinia, Vespula maculifrons* workers stinging a leather target—note the loss of the sting apparatus (sting autotomy) in the leather, *Agelaia myrmecophila* carving "meatballs" from a dead mammal. (Photographs by J. O. Schmidt.)

and northern Africa, number about 23 species and are the largest of the wasps. *Vespula,* called yellowjackets or the true wasps, are a Northern Hemisphere genus of about 22 species that are present from the Arctic to the northwestern fringe of Africa, to India, and into Central America. The 15 species of *Dolichovespula,* sometimes called aerial yellowjackets because many species construct round paper nests above ground, occupy the same range as *Vespula* but are absent in extreme Southeast Asia and Central America plus most of Mexico. The smallest genus of vespines, *Provespa,* consists of three pale nocturnal species limited to forested areas in tropical Southeast Asia.

BIOLOGY AND LIFE HISTORY STRATEGIES

Social wasps are hunters and scavengers that prey on a variety of arthropod and animal protein sources. They are well equipped for this role, possessing large, powerful cutting mandibles for capturing, subduing, and processing prey, large eyes for detecting potential prey, and the ability to fly and hover in pursuit of food. Although well known for their stinging ability, wasps do not capture or subdue prey with the sting. Their venom is used entirely for defense, primarily against vertebrate potential predators. Indeed, all investigated wasp venoms are toxic, painful, and effective against vertebrates, but rather inactive and slow in affecting insects unless the sting is delivered near a ganglionic center of the nervous system. Moreover, wasps have little need to sting prey because they are equipped with powerful mandibles that can be used to chew wood fibers from dead trees or cut through tough insect nets in a matter of minutes.

Prey of wasps is varied. Paper wasps *(Polistes)* are specialists on caterpillars, which they locate by visually and olfactorally searching vegetation likely to harbor them. Once located, the prey is quickly subdued, cut into manageable pieces that are chewed into a "meatball," and carried back to the nest to be fed to the larvae. Yellowjackets and hornets tend to have a broader diet than paper wasps and will capture a variety of arthropod prey including flies, spiders, caterpillars, and an assortment of other groups. House flies *(Musca domestica)* and other flies comprise major prey items of some species. Some species also scavenge for prey, removing insects captured in spiderwebs, carving flesh from dead animals such as rodents, and even removing insects freshly smashed on radiators and grilles of cars. The scavenging habits of some species have earned them the distinction of being considered to be pests at picnics, outdoor events, and around garbage cans.

Prey of the swarm-founding epiponine wasps in the subfamily Polistinae is less well characterized than prey of other wasps. Like other wasps, the epiponines often remove wings, legs, and the head of prey and chew it into a pulp before returning to the nest. Epiponines take a variety of prey, including flies, caterpillars, leafhoppers, and other insects, and many species are likely to specialize on certain taxa, or to prefer particular parts of the habitat, such as open areas, forest canopies, or thick vegetation. Some species in the genus *Agelaia* are well known for removing pieces of flesh from large dead animals and have been given the common name of "vulture wasps."

Adult wasps usually obtain energy for flight and general metabolism from sugar sources. Nectar from flowers, honeydew from aphids and other homopterous insects, and sweet fluids from fruit all can be food sources. In most species, these sugars are supplemented by the sugary trophallactic secretions produced by larvae in response to solicitation by adults.

All wasps construct multicell nests from plant fibers (Fig. 2). The most common materials for making nests are wood fibers scraped from weathered dead wood from trees or twigs, but other materials including rotten wood, fibers from living plant leaves, and stems are sometimes used. In most cases, these fibers are strengthened with salivary secretions during preparation and application to the nest. After application of the fibers, abdominal secretions may be added to further strengthen the material and help repel ants.

Wasp life cycles consist of colony initiation, growth and expansion, production of reproductives, reproductive dispersal, and colony decline. In most taxa the cycle is determinate and lasts less than a year, with complete dissolution of the society at the end. Some taxa have an indeterminate cycle in which the colony population simply decreases after a major reproductive event before again beginning the growth and expansion cycle in the same nest, often with new queens. Some indeterminate species can remain in a nest for years, with a record of 30 years.

Colony founding can be independent, or by swarm founding. In independent founding, the colony is initiated by one or more queens without the aid of workers; in swarm founding, colony initiation is accomplished by a swarm of many workers plus reproductive queens. Independent founding can be by a single queen (haplometrosis) or by several queens joining their efforts to initiate the colony (pleometrosis). Most yellowjackets and hornets are haplometrotic, although *Vespa affinis* sometimes is pleometrotic with multiple foundresses in a colony, and *Provespa,* a greater departure from the rule, repro-

FIGURE 2 Wasp nests. Clockwise from upper left: *Polistes snelleni, Vespa simillima, Apoica pallens*—note aposematic color and alignment of abdomens of adults on nest, *Parachartergus fraternus*, cryptic nest of *Metapolybia aztecoides* attached to a *Bursera simaruba* tree trunk, *Polybia simillima*—workers are in the defensive attack position on nest. (Photographs by J. O. Schmidt.)

duces by swarming. *Polistes* and *Mischocyttarus* independently found colonies with a mix of both pleometrotic and haplometrotic colony founding. Single founding queens have the advantages of absence of competition for egg laying from other queens and a potentially greater number of reproductive offspring per foundress. Disadvantages include greater risks from ants and other predators or parasites attacking the unattended nest when the foundress is foraging, and often a much lower success rate in establishing a nest. In addition to better protection of the nest, advantages for multiple foundresses include faster construction of the nest, more reliable prey capture for the larvae, and increased colony survival in the event of the loss of a foundress. These trade-offs of better survival and growth often prove advantageous for foundresses to join others. Both within individual species and across the genus as a whole, pleometrotic founding tends to be more prevalent at lower latitudes than haplometrotic founding, perhaps in part because of the more intense pressure from predators and parasites in warmer environments.

Reproduction and colony founding by swarms occurs in all epiponine wasps plus *Provespa* and some *Ropalidia*. During swarm founding, hundreds (or even thousands) of workers with young queens leave the parent nest and move to a new location to construct a new nest and reproductive unit. This form of reproduction, though similar in many ways to that of honey bees, differs in several details including the general presence of many more queens in wasp swarms. Clear advantages of swarm founding over independent founding are the

availability of a large worker force to construct the nest quickly and the opportunity for task specialization by individuals within the swarm. Reproductive individuals in the swarm need not forage as must their independent founding counterparts, and individual workers can specialize in different tasks such as collecting nectar, prey, fiber for nest construction, or water for cooling and mixing with fiber for nest construction. Multiple workers also ensure that a large defending force is always available, should intruders appear.

WASPS AND PEOPLE

Nuisance

The general public is not aware of the enormous beneficial role wasps play in controlling the populations of flies, caterpillars, and other often unwanted insects. It could be argued that even wasp scavenging on dead animals might provide competition for flies. These beneficial aspects of wasps are overshadowed by the view that they are nuisances, pests, and threats. Only a very few species of *Vespula*, particularly *V. germanica, V. vulgaris, V. maculifrons, V. squamosa,* and a few others, are attracted to human food and activities. Particularly in the fall, these species will readily forage around outdoor dining areas, areas where food is processed or openly available, or where garbage is present. At these locations, their bright yellow- or white-on-black color patterns make them conspicuous, as does their buzzing and foraging action. At least in many western cultures, an almost innate swatting or flapping reflex is exhibited by people when they detect a buzzing insect, irrespective of whether the insect is a fly or a wasp. The origin of this reaction is uncertain: Is it fear of wasp stings, dislike of flies, or a general dislike of any insect that approaches one's body? In any event, wasps, by their very nature, are perhaps the insect world's masters at stimulating human aversive swatting, in the process gaining a solid reputation for being an irritating nuisance.

Wasp nests themselves are often viewed as an encroachment on human space and aesthetics. Were it not for the habit of many species of *Polistes* to build their exposed comb nests under eaves or inside man-made structures, their presence likely would never be noticed. These nests with their exposed adult wasps, especially when near entrances to buildings, tend to generate fear and dislike on the part of the humans passing through the area. This often results in an unfortunate consequence for the wasps—the destruction of their nests and sometimes the adults. Aerial yellowjackets, especially *Dolichovespula arenaria* and *D. maculata* in North America, build large conspicuous gray, nearly spherical nests attached under eaves or to the sides of buildings. Their large size and busy activities tend to elicit an even less favorable reaction by people for these wasps than for *Polistes*.

The mere existence of wasps is seen as a nuisance by much of the human populace. As long as neither the wasps nor their nests are apparent, they do not come into conflict with people. However, should a hidden *Polistes* nest in vegetation, or a *Vespula*, *Dolichovespula*, or *Vespa* nest in the ground, a tree, or within a building be discovered, the nest is either avoided or destroyed. Such discoveries are usually beneficial neither to the wasps nor to the people involved.

Tramp and Pest Status

Tramp species are those that tend to be able to disperse via the intentional or inadvertent help of humans to areas where they are not native, and to establish large successful populations in the new areas. Numerous examples can be found among the cockroaches, flies, rodents, domestic animals, and others. Among the wasps, only a relatively few species are successful tramps: for example, the German yellowjacket *(Vespula germanica)*, the common wasp *(Vespula vulgaris)*, the western yellowjacket *(Vespula pennsylvanica)*, and the European paper wasp *Polistes dominulus*. *Vespa crabro*, the European hornet, has also been introduced into eastern North America, but it seems to be neither particularly abundant nor a pest in its new location. The others, once released from many of the controlling constraints placed on their populations by predators, parasites, competitors, and possibly pathogens and diseases, tended to expand their populations to enormous numbers and become pests.

The German yellowjacket is classic in this regard: it is successfully established in North America, New Zealand, Australia, South Africa, and several other locations. In New Zealand, where no other social wasps previously existed, it expanded its niche and population to such densities that it constitutes

a major pest and threat to much of the native fauna of the country. In addition to its dense populations, *V. germanica* evolved secondary polygyny in which some colonies do not disintegrate at the end of the season, but instead many mated queens remain in (or return to) the nest and the brood rearing cycle resumes—although this time the colony can become huge, with many tens of thousands of individuals. In North America, *V. germanica* rapidly expanded its population and range to include most of the northern part of the United States and the populated areas of California. Its success at the apparent expense of some native species of *Vespula* is likely a result of its aggressive foraging behavior and its affinity for nesting in human structures. The walls of buildings provide many benefits for the wasps: warmth during the late fall and early winter, shelter from moisture and many predators, and proximity to garbage and other food sources. This proximity to people and their food allows colonies to remain active longer during the year, to grow larger, and to produce more reproductives. It also brings them into direct conflict with people and makes them a major pest in some areas.

P. dominulus, unlike *V. germanica*, is not particularly defensive to people near its nest and tends to be less threatening in appearance. In recent years, it successfully expanded its range in North America to include much of the Midwest, where it can be present in enormous numbers. Also, its tendency to be less choosy in nest site locations causes annoyance beyond what was traditionally elicited by a *Polistes* species.

Stings, Venom, and Medical Risks

At the sight or the mention of a wasp, simultaneous thoughts enter the minds of many people—dislike and sting. The immediate human association of wasp with "sting" demonstrates the success of the wasp sting as a defense against large predators. Wasp stings hurt; and wasp stings provide reinforcement to the pain in the form of local swelling, redness, itching, and tenderness. Social wasps, like many other social insects, live in immobile colonies with many edible and nutritious young, offering an incentive for large predators to attempt to overpower the wasps and consume the brood. The sting with its associated venom is a nearly perfect defensive system to counter attacks by large predators: the sting can penetrate the tough skin of the predator, and the venom is then delivered into richly innervated living tissue, where it causes immediate pain and tissue damage. The consequence of wasps possessing their venomous sting is that few vertebrate predators can successfully attack wasp nests: the exceptions are a few specialists that have secondarily evolved means to exploit wasps and their nests. The effectiveness of wasp stings is not lost on people and forms much of the basis for human aversion to wasps. Until recently, this worked to the benefit of wasps, but now with modern protective clothing, devices, insecticides, and other means to attack wasps, the defense is less effective against people. Nevertheless, general human fear and dislike of wasps continue undiminished.

Wasps can pose a minor medical risk to people hypersensitive to their stings. Overall estimates of the incidence of allergic reactions to insect stings range from about 0.5 to 2% of the population, with slightly over half the individuals being sensitive to wasp stings (the rest are primarily sensitive to honey bee and fire ant stings). Of this wasp-sensitive population of approximately 3 million in the United States, about 20 actually die each year. The remaining 99.999+% of the hypersensitive population suffer at most generalized reactions involving the skin or the respiratory or cardiovascular systems. For people who suffer severe allergic reactions, particularly those in which breathing becomes difficult, or fainting or other signs of blood pressure drop occur, preventive approaches in the form of personal epinephrine injectors or immunotherapy are available. For the rest of the populace, wasps present little risk beyond the affront they cause with their stinging.

See Also the Following Articles

Caste • Colonies • Defensive Behavior • Hymenoptera • Nest Building • Venom

Further Reading

Itô, Y. (1993). "Behaviour and Social Evolution of Wasps." Oxford University Press, Oxford, UK.

O'Neill, K. M. (2001). "Solitary Wasps: Behavior and Natural History." Cornell University Press, Ithaca, NY.

Ross, K. G. and Matthews, R. W. (eds.). (1991). "The Social Biology of Wasps." Cornell University Press, Ithaca, NY.

Schmidt, J. O. (1990). Hymenoptera venoms: striving toward the ultimate defense against vertebrates. *In* "Insect Defenses: Adaptive Mechanisms and Strategies of Prey and Predators" (D. L. Evans and J. O. Schmidt, eds.), pp. 387–419. State University of New York Press, Albany.

Turillazzi, S., and West-Eberhard, M. J. (1996). "Natural History and Evolution of Paper-Wasps." Oxford University Press, Oxford, U.K.

Water and Ion Balance, Hormonal Control of

Thomas M. Clark

Indiana University, South Bend

Insects are found in habitats ranging from the driest deserts to aquatic habitats of diverse ionic composition. Their survival depends on the ability to keep their tissues and cells moist, and to regulate the composition of their body fluids. Body fluid composition is challenged whenever exchanges of materials with the environment take place, and such exchanges are unavoidable. Routes of gain of materials include ingestion through the mouth, osmotic gain of water, active uptake of ions, and diffusion across the body surface. Materials are lost through the excretory system and in the feces, as well as by osmotic, evaporative, or diffusive losses across the body surface. During feeding or drinking, an insect may rapidly ingest a large volume of material that is significantly different in composition from the body fluids. Insects respond to perturbations in body water composition by using a combination of hormonal and autonomous mechanisms to vary the amount and composition of fluid excreted by the excretory system.

STRUCTURE AND FUNCTION OF THE EXCRETORY SYSTEM

The excretory system consists of the Malpighian tubules and the hindgut (Fig. 1). The Malpighian tubules form the primary urine by active secretion of ions into the lumen. Water follows passively by osmosis, drawing with it small hemolymph solutes. This primary urine is then modified as it passes through more proximal regions of the Malpighian tubules and through the hindgut. The Malpighian tubules and the hindgut each frequently consists of several functionally distinct regions and/or distinct cell types, and each region contributes by specific mechanisms to the overall formation and processing of the urine.

Water balance can be adjusted by altering the rate of formation and/or composition of primary urine, and by changing the activity of mechanisms that modify the primary urine. Fluid secreted by Malpighian tubules can be passed forward into the midgut and reabsorbed, or passed back through the hindgut for modification prior to elimination. Evidence suggests that Malpighian tubule, midgut, and hindgut transport processes are each regulated by hormones.

REGULATION OF THE EXCRETORY SYSTEM

A number of chemical factors have been isolated from neural tissue of many insect species that alter fluid and/or ion transport rates of particular regions of the excretory system. Only rarely have these factors been demonstrated to act as hormones, and

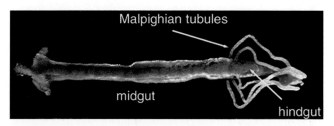

FIGURE 1 The larval mosquito gut and excretory system. The five Malpighian tubules form the primary urine by active secretion of ions, with water following the resulting osmotic gradient. Fluid formed by the Malpighian tubules enters the alimentary canal at the midgut–hindgut junction. This primary urine can be either shunted forward into the midgut and reabsorbed, or passed backward into the hindgut for modification and eventual elimination. Evidence suggests that all three regions, the midgut, Malpighian tubules, and hindgut, contribute to water and ion balance and are under hormonal control. [Modified from Clark, T. M. *et al.* (1999). The anterior and posterior "stomach" regions of the larval *Aedes aegypti* midgut: regional specialization of ion transport and stimulation by 5-hydroxytryptamine. *J. Exp. Biol.* **202**, 247–252. With permission of the Company of Biologists Limited.]

their specific roles in regulation of water and ion balance remain largely speculative. Nevertheless, available evidence suggests that regulation of water and ion balance involves several different hormonal factors, each with a specific mechanism of action and acting on only a portion of the excretory system or even a specific group of cells within a region. At least some regions seem to be regulated by multiple hormones, with combinations of hormones producing responses that are often distinct from the actions of the individual hormones acting alone. The actions of the distinct regions and cell types within the excretory system are apparently coordinated by hormones to produce final excreta of appropriate composition. Evidence to date suggests that hormones may regulate several related parameters, including hemolymph volume, hemolymph ionic composition, and clearance of metabolic wastes and toxins from the hemolymph.

Hormonal Regulation of Malpighian Tubule Transport

Factors hypothesized or demonstrated to play a hormonal role in regulation of Malpighian tubule function include peptides and the biogenic amine serotonin. The best studied of the peptides are the kinins and the diuretic peptides that are related to corticotropin-releasing factor (CRF).

Kinins are small (6- to 15-amino-acid) peptides with a C-terminal amide. Those identified to date have similar C-terminal sequences of Phe-Xxx-Yyy-Trp-Gly-amide, where Xxx can be Asn, Ser, His, Phe, or Tyr and Yyy can be Pro, Ser, or Ala. The kinins have been isolated from several insect species but have not been detected in animals other than insects. A single species may contain several kinins. To date, a hormonal role for a kinin has been demonstrated in only the cricket *Acheta domesticus* (see later).

The CRF-related diuretic peptides range from 30 to 46 amino acids in length and show structural homology to a family of peptides found in vertebrates that includes corticotropin releasing factor (hence the name), sauvagine, and urotensin. Available evidence suggests that a single species produces more than one CRF-related diuretic peptide. To date, a hormonal role for a CRF-related diuretic peptide has been demonstrated conclusively in only one species, *Locusta migratoria* (see later).

Serotonin stimulates Malpighian tubule secretion in a number of insect species and is known to hormonally regulate excretory function in the bloodsucking bug *Rhodnius prolixus* and in larval mosquitoes *(Aedes aegypti)*. In *Rhodnius* and most likely in mosquitoes as well, serotonin stimulates both uptake of fluid across the midgut into the hemolymph and removal of that fluid from the hemolymph by the Malpighian tubules. Thus the neurotransmitter acts to coordinate midgut and excretory function to remove excess fluid and ions from the body following ingestion.

Malpighian tubules of at least some insect species appear to be further regulated by unidentified antidiuretic factors. In addition, short-term hormonal adjustments in Malpighian tubule transport rates in response to immediate challenges can be superimposed on longer term increases and decreases in basal and maximal transport capacities. This process may be mediated at least in part by prostaglandins.

Hormonal Regulation of Hindgut Transport

Hormones acting on the hindgut appear to stimulate recovery of fluid and ions from the primary urine, leading to cycling of hemolymph through the excretory system and clearance of wastes from the hemolymph. The majority of our information about regulation of hindgut transport comes from studies on the locust. The locust hindgut consists of two distinct regions, a more anterior ileum and a more posterior rectum. Two peptide factors isolated from the corpus cardiacum are ion transport peptide (ITP), acting on the ileum, and chloride transport stimulating hormone (CTSH), acting on the rectum. The amino acid sequence of these peptides has been determined. Other factors that stimulate water reabsorption by the hindgut include neuroparsins A and B. To date, none of the factors acting on hindgut transport have been demonstrated in hemolymph.

Hormonal Regulation of Midgut Transport

Regulation of midgut transport has received little attention. Hemolymph serotonin stimulates midgut transport mechanisms in *Rhodnius* and in larval mosquitoes. Generation by larval lepidopterans of a highly alkaline midgut lumen, creating conditions favorable for digestion and nutrient absorption, seems to be regulated by an unknown endocrine or paracrine signal. Several peptides inhibit ion transport across the lepidopteran midgut. Evidence for regulation of more than one midgut transport mechanism in diverse insect species suggests that the regulation of midgut transport may be more important in the control of water and ion balance than has been assumed.

BIOLOGICAL ROLES OF HORMONAL REGULATION OF EXCRETORY FUNCTION

Orthoptera: Grasshoppers

When food is available, locusts ingest considerable volumes of fluid with their food and produce copious, moist feces. When deprived of food and water, the feces produced are very dry. During times of limited water availability, Malpighian tubules secrete fluid at a low rate, and much of the secreted water is recovered in the hindgut. This recovery of ions and water seems to be mediated by a hormone or hormones (ITP and CTSH, mentioned earlier). Upon feeding, a CRF-related diuretic hormone appears in the hemolymph, stimulating fluid secretion by the Malpighian tubules. Presumably, antidiuretic factors acting on the hindgut are no longer released under these conditions, and thus elimination of excess water and ions (especially KCl) becomes possible. Hormones acting on the hindgut also seem to be involved in regulation of hemolymph pH.

Orthoptera: Crickets

Crickets possess Malpighian tubules consisting of distinct distal, mid, and proximal regions. The distal and midtubule regions at least may be regulated independently, and apparently secretion rates of each region can be either stimulated or inhibited. Little is known about the biological roles of the majority of the factors that alter tubule function *in vitro*. Achetakinin (a member of the kinin family of peptides) appears in hemolymph at levels sufficient to stimulate Malpighian tubule secretion in response to starvation, at a time when the insect is conserving water, rather than during feeding or water loading. These data suggest that achetakinin may act to stimulate fluid cycling through the excretory system, leading to clearance of materials from the hemolymph, rather than stimulating diuresis.

Heteroptera: Rhodnius

Most of our information concerning regulation of water balance in Heteroptera comes from studies on the bloodsucking bug *Rhodnius*. The physiology of this animal is unusual in that *Rhodnius* seldom feeds, and when it does it may ingest a volume of blood up to 12 times its own body mass. At this time, serotonin released into the hemolymph causes the cuticle to become more plastic (allowing accommodation of the meal), stimulates anterior midgut fluid and ion absorption, and acts synergistically with a diuretic peptide to cause a 1000-fold increase in Malpighian tubule secretion rates. This leads to rapid removal from the midgut lumen and hemolymph of the unusable fluid and ion load ingested with the meal.

Diptera: Mosquitoes

Much of our information about hormonal regulation of water and ion balance among Diptera comes from studies on larval and adult mosquitoes. Larval mosquitoes are aquatic, but the adults are terrestrial. The adult female of most species seeks out and consumes large blood meals to provide proteins for successful egg production. Larvae and adults thus have very different challenges to water and ion balance.

Larval *A. aegypti* ingest the medium more rapidly when food is available and increase their drinking rates as the ambient salinity increases. Serotonin levels in hemolymph increase in response to increased salinity but do not change with feeding status. As drinking rates increase with increasing ambient salinity, serotonin seems to act to stimulate absorption of ingested fluid from the midgut and secretion of fluid by the Malpighian tubules, leading to clearance of material from the hemolymph. For this strategy to work, at least some of the secreted fluid must be recovered in the hindgut, suggesting that hindgut transport is also regulated.

Adult females of most mosquito species survive for days at a time, feeding only on plant juices and other fluids, then consume within a few minutes a blood meal that may triple their body mass. Even while feeding, they begin to eliminate the bulk of the fluid and ionic load ingested with the meal. This response appears to be hormonally mediated, with CRF-related diuretic peptides, kinins, and serotonin all stimulating Malpighian tubule secretion. In part because of the small size and limited hemolymph volume of these insects, it is not known which of these potential hormones are involved in the response to feeding.

Lepidoptera: Butterflies and Moths

Lepidopteran larvae are able to adjust to diets of different water content by regulating the amount of water lost in the feces. This regulation seems to be under hormonal control. Both CRF-related diuretic peptides and kinins stimulate Malpighian tubule secretion *in vitro*, but they seem to have very different biological roles. When injected into caterpillars, CRF-related diuretic peptide (Mas-DH) has an antidiuretic effect: it stimulates fluid (and presumably ion) recovery from the cryptonephric complex. In contrast, kinins act to increase the fluid content of the feces, leading to marked reduction in weight gain by injected larvae. Allatotropins and a group of small peptides with an amino terminus amino acid sequence of Phe-Leu-Arg-Phe-NH$_2$ (FLRFamides) have been found to inhibit ion transport across the midgut epithelium, but the physiological significance of this response in terms of water and ion balance is not known.

Upon emerging from the chrysalis, butterflies discharge a considerable volume of fluid containing metabolic wastes accumulated during the pupal stage. This diuresis is regulated by unidentified hormones.

Desert Beetles

Beetles of the family Tenebrionidae are generally very tolerant of dry conditions, to the extent that larvae of at least some species can remove water vapor from unsaturated air across the rectum. Despite this, factors that stimulate Malpighian tubule fluid secretion to high rates have been found in some of the most drought-tolerant species of this family. When injected into the animals, these factors stimulate fluid secretion by Malpighian tubules; but rather than being passed back through the hindgut and eliminated, the secreted fluid is passed forward into the midgut. The water and valuable hemolymph components are recovered, leaving toxins and metabolic wastes in the gut where they can be eliminated. These actions have led to the suggestion that factors regulating excretory function often are more properly called "clearance hormones" instead of "diuretic hormones."

See Also the Following Articles

Excretion • Hemolymph

Further Reading

Chung, J. S., Goldsworthy, G. J., and Coast, G. M. (1994). Haemolymph and tissue titres of achetakinins in the house cricket *Acheta domesticus:* Effect of starvation and dehydration. *J. Exp. Biol.* **193**, 307–319.

Clark, T. M., and Bradley, T. J. (1997). Malpighian tubules of larval *Aedes aegypti* are hormonally stimulated by 5-hydroxytryptamine in response to increased salinity. *Arch. Insect Biochem. Physiol.* **34**, 123–141.

Coast, G. M. (1998). Insect diuretic peptides: Structures, evolution and actions. *Am. Zool.* **38(3)**, 442–449.

Nicolson, S. W. (1991). Diuresis or clearance: Is there a physiological role for the "diuretic hormone" of the desert beetle *Onymacris? J. Insect Physiol.* **37(6)**, 447–452.

Patel, M., Hayes, T. K., and Coast, G. M. (1995). Evidence for the hormonal function of a CRF-related diuretic peptide (Locusta-DP) in *Locusta migratoria. J. Exp. Biol.* **198**, 793–804.

Petzel, D. H., and Stanley-Samuelson, D. W. (1992). Inhibition of eicosanoid biosynthesis modulates basal fluid secretion in the Malpighian tubules of the yellow fever mosquito (*Aedes aegypti*). *J. Insect Physiol.* **38(1)**, 1–8.

Phillips, J. E. (1983). Endocrine control of salt and water balance: Excretion. *In* "Endocrinology of Insects" (R. G. H. Downer and H. Laufer, eds.), pp. 411–425. A. R. Liss, New York.

Spring, J. H. (1990). Endocrine regulation of diuresis in insects. *J. Insect Physiol.* **36(1)**, 13–22.

Weevil

see *Boll Weevil*

Wings

Robin J. Wootton
University of Exeter, United Kingdom

Insects are the only invertebrates to have developed the power of flight, and their wings are unique. Unlike those of birds and bats, they are not modified limbs and so have no internal muscles; they consist predominantly of an extracellular material—cuticle. Typical wings nonetheless need to perform—semiautomatically—the complex movements and cyclic changes in attitude and shape associated with flapping flight, and these necessities strongly influence their structural design. The latter shows huge variation within the class Insecta, and this is made greater by the wide range of secondary functions that the wings have evolved to perform, sometimes instead of flight, sometimes in addition to it. While great progress has been made in interpreting wing structures in functional terms, many questions still remain to be answered.

ORIGIN

We have no fossil record of the earliest winged insects nor of the structures from which wings evolved. Majority opinion currently favors their origin from articulated lateral outgrowths—"exites"—of an obsolete basal segment of the legs, long since incorporated in the thoracic wall. Similar, though not homologous, structures are found on the coxae and abdominal segments of some apterygote insects. The articulated gill plates of juvenile Ephemeroptera may actually be wing homologues. Those of some fossil mayfly nymphs indeed resemble tiny wings, though this may simply reflect the fact that both are adapted for accelerating fluid.

The origin of flight is still controversial. Two conflicting theories both have significant support; in 2001 they, and their implications on the later development of flight, were reviewed and compared in a chapter by Wootton in *Insect Movement: Mechanisms and Consequences*. The first, and more traditional, theory is supported by some experimental evidence and theoretical modeling and proposes that powered flight arose in tree-dwelling insects, through parachuting and gliding stages. These "protopterygotes" would have had winglets, probably movable, on most body segments. Selection would favor enlargement and improvement of winglets close to the center of mass of the body, and reduction and loss of the remainder, together with the evolution of an effective mechanism for flapping and twisting the wings. The second hypothesis suggests that flight arose in semiaquatic insects, initially using their winglets to skim on the surface film of water bodies, as do many modern adult Plecoptera and Ephemeroptera. Selection would favor enlargement and improvement of the thoracic winglets to the point at which they could generate sufficient upward force for the insects to leave the surface and fly.

STRUCTURE

Although some Carboniferous and Permian insects had short, apparently movable winglets on the first thoracic segment, the wings of modern insects are borne on the second and third segments only. They articulate with the back (tergum) and sides (pleura) of the thoracic segments via the axillae: complex, three-dimensional mechanisms of stiff and compliant cuticle through which the muscular forces of the thorax are transmitted to the wings and which control the relative movements of the wings' basal components. The axillary structures vary greatly in detail, but those of most insects can be referred to a common plan of up to four so-called axillary sclerites and three other sclerotized plates, the humeral and the proximal and distal median plates, linked together by hinges and broader expanses of soft cuticle (Fig. 1). The axillary structures of the Ephemeroptera and Odonata, orders which appear primitively incapable of folding their wings back over the abdomen, differ from the typical pattern of wing-folding (neopterous) insects and also from each other. Various attempts have been made to homologize these with the neopterous pattern, which is clearly influenced by the need to fold; but no consensus has been reached.

Orthodox wings themselves consist of cuticular membrane supported by, and continuous with, a framework of veins. Both veins and membrane are double structures, formed by juxtaposition of the cuticle of the dorsal and ventral sides as the

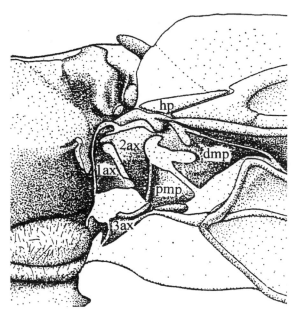

FIGURE 1 Forewing axilla and wing base of the desert locust, *Schistocerca gregaria.* 1ax, first axillary sclerite; 2ax, second axillary sclerite; 3ax, third axillary sclerite; dmp, distal median plate; hp, humeral plate; pmp, proximal median plate. (Modified with permission from Wootton, 1979.)

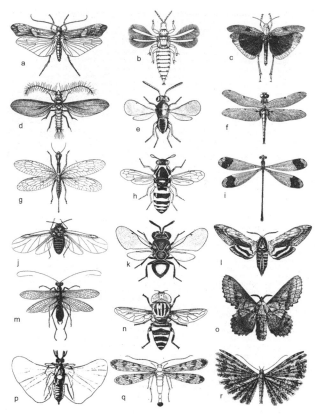

FIGURE 2 The variety of insect wings. The wing spans extend over two orders of magnitude, from ca. 1 mm in the thrips (b) to 100 mm in the damselfly (i). (a) Caddis (Trichoptera), *Limnophilus rhombicus.* (b) Thrips (Thysanoptera), *Liothrips oleae.* (c) Grasshopper (Orthoptera), *Dissosteira carolina.* (d) Male coccid (Homoptera), *Icerya purchasi.* (e) Parasitoid wasp (Hymenoptera), *Coccophagus tschirchii.* (f) Dragonfly (Odonata), *Libellula quadrimaculata.* (g) Snake fly (Megaloptera), *Raphidia adanata.* (h) Wasp (Hymenoptera), *Celonites abbreviatus.* (i) Damselfly (Odonata), *Megaloprepus coerulatus.* (j) Aphid (Homoptera), *Eriosoma lanigerum.* (k) Perilampid wasp (Hymenoptera), *Perilampus chrysopae.* (l) Hawk moth (Lepidoptera), *Hyloicus ligustri.* (m) Mantis (Mantodea), *Mantoida brunneriana.* (n) Hover fly (Diptera) *Lathyrophthalmus quibquelineatus.* (o) Lasiocampid moth (Lepidoptera), *Gastropacha quercifolia.* (p) Male stylopid (Strepsiptera), *Eoxeonos laboulbenei.* (q) Scorpionfly (Mecoptera), *Panorpa communis.* (r) Plume moth (Lepidoptera), *Orneodes cymodactyla.* (Reproduced with permission from W. Nachtigall, 1974.)

wing increases in area, thins, and stiffens after the final molt. The veins are typically tubular and contain hemolymph, and major vein branches usually also carry branches of the tracheal system. The hemolymph circulates or moves tidally within the veins and serves to maintain the moisture content of the cuticle, which would otherwise become stiff and brittle. Little or no epidermal material appears to remain in the expanded, sclerotized adult wing.

The longitudinal veins run distally from the wing base and many branch as the wing broadens along the span. They are usually to some extent linked by crossveins. Together they form a supporting and conducting framework, which may be structurally very complex (Figs. 2c, 2f, 2i, and 2m) or relatively simple (Figs. 2d, 2e, 2j, 2k, and 2p).

Even the simplest wings are far more complex than conventional illustrations indicate. Most have considerable relief, which adds substantially to their rigidity. The primitive condition may well have been a radially pleated structure, with the longitudinal veins alternately occupying the ridges and troughs of the pleats. This is found in many early fossil insects and is still the case in Odonata and Ephemeroptera. Some pleating indeed persists in parts of most insect wings, though it may be absent elsewhere. Many wings show an arched, or cambered, cross section, which again enhances rigidity to bending and introduces other important properties, which will be discussed later.

Further features, often omitted from illustrations, but with great mechanical significance, are lines and bands of flexibility within the wing and the axilla. One can conveniently distinguish between flexion lines, whose primary function is the facilitation and control of wing deformations in flight, and fold lines, whose primary role is in folding the wings to and from the resting position. The distinction is not absolute: there may well be some limited deformation along flexion lines during wing folding and along fold lines in flight. Some other lines of flexible cuticle, for example in the distal parts of some hymenopteran and dipteran wings, seem to fit neither category and are best interpreted as adaptations to allow the wing tip to crumple reversibly on impact with obstacles. The veins themselves vary considerably in diameter, wall thickness, and shape of their cross section, all factors that strongly influence their local resistance to bending and help determine the pattern of deformation in flight. Figure 3 shows some details of contrasting vein morphology. Many crossveins have a characteristic annulated form (Figs. 3f and 3g), a means of reconciling flexibility with the need to maintain an open

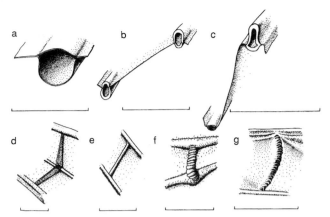

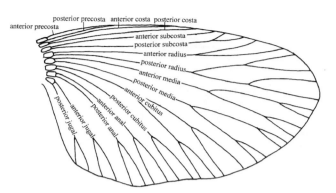

FIGURE 4 Wing vein nomenclature. The system of Kukalová-Peck. (Modified from Kukalová-Peck, 1991.)

FIGURE 3 Details of veins. (a–c) Transverse sections of longitudinal veins. (a) Forewing subcosta of a butterfly, *Papilio rumanzowia* (Lepidoptera). (b) Adjacent veins in the radial sector field of *Calopteryx splendens* (Odonata). (c) Ridge and trough veins from the hind wing fan of *Schistocerca gregaria* (Orthoptera). (d–g) Crossveins. (d) Angle-bracket-like crossveins from the leading edge spar of *C. splendens* (Odonata). (e) Crossvein from the remigium of the wing of *C. splendens*. (f) Flexible crossvein from *Eristalis tenax* (Diptera). (g) Flexible crossvein from the hind wing fan of *S. gregaria*. Scale bars, 0.5 mm. (Modified with permission from R. Wootton, 1992.)

section, as in flexible drinking straws and some hoses and flexible pipes. Elsewhere, crossveins can be stout and rigid and can even form high-relief angle brackets, as in the leading edge spar of dragonfly wings (Fig. 3d). Where longitudinal or crossveins are crossed by flexion lines and fold lines they may be locally interrupted, thinned, or annulated.

The membrane itself can be thickened and sclerotized, either over substantial areas, as in the protective or semiprotective forewings of Coleoptera, Dermaptera, Heteroptera, Orthoptera, and Blattodea, or in more restricted lines or zones. In the latter case it may not always be easy to distinguish the result from veins, especially since some true veins are not tubular, but are simply thickened, sclerotized grooves.

The texture of the membrane is also variable. Under the microscope, areas that undergo appreciable stretching in flight may appear crumpled like crêpe paper, as in the hind wing fan of locusts, or as an expanse of hollow papillae, as in those of some Heteroptera. Microtrichia (small unarticulated hairs) are widespread, and the wing membrane of many insects, particularly among Holometabola, bears articulated hairs (e.g., Trichoptera) or scales (e.g., Lepidoptera, Psocoptera, mosquitoes). The scales are sometimes extraordinarily complex in structure, providing both insulation and color, including some spectacular physical colors. This complexity is particularly remarkable when one considers that each scale is produced by a single cell.

The veins too may bear scales, hairs, and other ornamentation, including airflow- and vibration-sensitive hairs and sense organs that detect distortions of the cuticle in flight.

In several orders the fore- and hind (meso- and metathoracic) wings are in physical contact during the wing stroke, the posterior part of the forewing overlying the anterior part of the

hind and in many cases physically linked to it by a coupling mechanism. Some mechanisms may in fact provide sensory information about the relative positions of the wings rather than rigid mechanical coupling, but in Hymenoptera the hind wing leading edge bears a row of hooks that firmly clasp a fold of the forewing; in Heteroptera the hind wing edge is held by hairs in a groove in the forewing, and in many Homoptera the wings are held together by interlinking grooves.

HOMOLOGY AND TERMINOLOGY

Despite occasional claims to the contrary, there is no good evidence that wings arose more than once within the insects, and it is generally accepted that homologous wing characters should be recognizable throughout the class. In fact this is anything but straightforward. In the course of evolution some veins have degenerated, or their identities have become obscured by fusion. Vein branches have been gained or lost, have become realigned, have come to resemble crossveins. Linear, vein-like membrane thickenings have arisen and are readily mistaken for true veins. The homologies of flexion lines and fold lines, and the extent to which they can be used as landmarks in vein identification, are also far from obvious.

Vein Nomenclature

Unfortunately, several conflicting systems are in use. All derive principally from a scheme established in the late 19th century by J. H. Comstock and J. G. Needham after a series of comparative studies of wing tracheal supply, but the situation has become confused by later additions and alterations. Two widely used schemes are illustrated here. Figure 4, adapted from Kukalová-Peck, shows all the longitudinal veins of which any trace remains in extinct, as well as living, insects and is an attempt at an overall ground plan for the winged insects. It proposes that veins were initially paired, the anterior member of each pair being convex and situated on a ridge of a pleated wing, the posterior member concave and situated in a trough. They are the anterior and posterior precosta, costa, subcosta, radius, media, cubitus, anal, and jugal. This scheme is consis-

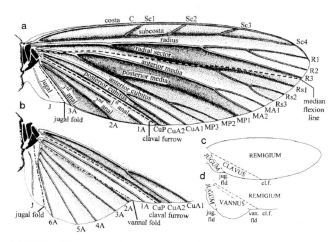

FIGURE 5 Wing nomenclature. A more traditional system, after Wootton, showing veins, some flexion and fold lines, and the main areas of the wing. (a, c) A forewing or a hind wing lacking an expanded anojugal area. (b, d) Cubital and anojugal region of hind wing with an expanded anojugal area. (Reproduced with permission from Wootton, 1979).

tent and logical and is increasingly widely followed. In most modern insects, however, several of the most anterior veins are obscure or absent, and many entomologists tend to ignore these, or indeed to question their existence, and to follow a scheme closer to that of Comstock and Needham. Figure 5 illustrates one such system, showing (Fig. 5a) a diagrammatic wing, (Fig. 5b) the expanded anal fan (vannus) of some hind wings, and (Figs. 5c, and 5d) the names in common use for specific regions of the wing. Kukalová-Peck's precostal veins are omitted, as are her posterior costa and anterior subcosta. The anteriormost veins are hence the convex, unpaired costa and the concave subcosta. Kukalová-Peck's posterior radius is called by the older name of radial sector. The number of anal veins is regarded as unfixed.

Both diagrams show most veins to be branched, as is often the case, but we do not in fact know what was the primitive number of branches for any vein. Crossveins are omitted from the diagrams. Where comparatively few are present in a wing, these are often named by the longitudinal veins that they connect, with the symbols for these veins given in lowercase type.

Regions of the Wing

In the wings of a neopterous (wing-folding) insect distinct areas can be recognized, delimited by flexion lines or fold lines. The main, anterior area, between the leading edge and the claval furrow, is the remigium. In wings without an expanded anal fan, a narrower area between the claval furrow and the jugal fold is the clavus. The remaining, usually small area, which lies posterior to and inboard of the jugal fold, is the jugum. In the hind wings of many forms the area between the claval furrow and the jugal fold is expanded into a broad anal fan, or vannus, which sometimes, notably in Orthoptera, Phasmida, and Blattodea, has fold lines within it and folds into a pleated fan.

Flexion Lines and Fold Lines

Flexion lines and fold lines have received comparatively little attention and indeed are often omitted from diagrams. In neopterous insects, one flexion line and one fold line are particularly widespread. The claval furrow is a flexion line that typically lies posterior to the posterior cubitus and allows the remigium to hinge upward or downward relative to the clavus. The jugal fold is continuous with the basal hinge of the wing and is involved in both flapping and folding.

Two other flexion lines are common, but their positions relative to other wing landmarks are not constant, and they may have arisen independently several times. The median, or remigial flexion line, runs longitudinally from the wing base, often between the radial and the median vein systems. It allows the profile of the wing to be altered in flight and may influence its resistance to transverse bending. Transverse flexion lines, by contrast, facilitate ventral bending across the span. A transverse flexion line occurs in several orders, crossing the wing along a curved or irregular, sometimes oblique, path from the costa to the posterior margin.

DIVERSITY

The wings of insects are very diverse indeed—far more so than those of birds or bats. They vary in proportions; the relative sizes of fore- and hind wings; relief, texture, and ornamentation; thickness; venational richness; the details of veins, fold lines, and flexion lines; and the extent and manner in which they are coupled in flight. Figure 2 illustrates some of this diversity.

It is widely assumed, though without direct evidence, that fore- and hind wings were originally similar in size and shape. This is true today of zygopterous Odonata (Fig. 2i); of many Neuroptera, Megaloptera (Fig. 2g), Mecoptera (Fig. 2q), Trichoptera, and primitive Lepidoptera; of Thysanoptera (Fig. 2b), Embioptera, most Isoptera, and a few Plecoptera (Chloroperlidae). In the last two orders at least the condition is clearly secondary. Some Permian fossil Ephemeroptera and Heteroptera also had similar fore- and hind wings, as did some members of several extinct orders.

However, some of the earliest insect fossils already show relative enlargement or reduction of one wing pair, and one or the other applies to most insects today. In anisopterous Odonata (Fig. 2f); Orthoptera (Fig. 2c) and winged Phasmida; Blattodea; one family of Isoptera (Mastotermitidae) and Mantodea (Fig. 2m); most Plecoptera, Dermaptera, Trichoptera (Fig. 2a), and Coleoptera; and some Heteroptera and male Strepsiptera (Fig. 2p), the hind wing is significantly broader than the fore and has become the principal lifting surface. In Ephemeroptera, Psocoptera, some Heteroptera (Fig. 2j) and Neuroptera, many Lepidoptera (Figs. 2l, 2o, and 2r), most Hymenoptera (Figs. 2e, 2h, and 2k), and all Diptera (Fig. 2n) the converse is true; the forewings have greater area and usually greater length than the hind. In all but the Odonata, which do not fold their wings, it is necessary for the hind wings to fold up along two or more radiating fold lines to lie beneath the forewings at rest. The forewings themselves are often shorter than the hind, and in some cases have become thickened, providing some protection for the abdomen and the more fragile hind wings. When moderately thickened, as in many Orthoptera, Blattodea, Mantodea, and some Homoptera, the forewings are often referred to as tegmina (singular tegmen). When thickened to the extent that their flight function is almost lost or wholly so, as in Coleoptera and Dermaptera, they are known as elytra (singular elytron). The typical forewings of Heteroptera, with the basal part thickened but a membranous tip, are known as hemielytra.

Forewings adapted for protection tend to be less effective in flight and are usually accompanied by disproportionately large, elaborately folding hind wings with expanded anal fans or else by total loss of flight capability. Where the forewings are significantly shorter than the hind, e.g., in some Orthoptera, winged Phasmida, Dermaptera, and Coleoptera, two different solutions have been adopted to meet the problem of stowing the hind wings at rest. In Orthoptera, e.g., Gryllotalpidae, Tridactylidae, and Tetrigidae, and in Phasmida, the longitudinally folded hind wings project from beneath the tegmina along the back of the abdomen. In Phasmida and Tridactylidae, the narrow remigium of the hind wing itself is thickened and protects the more delicate fan. However, in Dermaptera and Coleoptera and in several genera of Blattodea the hind wings fold transversely as well as longitudinally, in a variety of complex and mechanically fascinating ways.

Some reductions and modifications are extreme. Among Ephemeroptera, the Caenidae and some Baetidae have lost the hind wings entirely and fly with the forewings alone; the same is true of male Coccidae (Homoptera, Fig. 2d). The hind wings of Diptera are modified as gyroscopic sense organs, the halteres, and the forewings of male Strepsiptera (Fig. 2p) are apparently similarly adapted. The hind wings of nemopterid Neuroptera are elongate plumes or filaments. In the nemopterid species that have been investigated they do not flap, but are spread out behind the insect, apparently acting as physical stabilizers. Similar long, more or less slender hind wings occur in several families of butterflies and moths, but in some species at least they flap in phase with the forewings, sending ripple-like waves along their "tails."

The wings of many small moths, flies, beetles, and wasps are fringed with long hairs, which appear to operate aerodynamically as if they were a continuous membrane. This is taken to extremes in the tiny feathery wings of Thysanoptera (Fig. 2b), ptiliid beetles, nymphomyiid Diptera, and mymarid Hymenoptera, of which most of the lifting surface consists of long hairs, arising from a rod-like rachis. The emarginated wings of orneodid moths (Fig. 2r) consist of a radiating fan of similar feather-like structures, and the pterophorid moths approach this condition, though with fewer plumes.

The forewings of many male Orthoptera in the suborder Ensifera are adapted for sound production and amplification, with distorted venation enclosing large areas of membrane—the "mirrors." These regions, and their function, are often retained in species whose wings are otherwise reduced. The forewings of some noctuid moths are also adapted for sound production.

Reduction (brachyptery) and loss (aptery) of both wing pairs have happened in many orders. It is particularly frequent in Orthoptera, Blattodea, and Hemiptera, and sometimes occurs polymorphically within a species. The adaptive significance of brachyptery is seldom clear. Short forewings are sometimes retained, apparently as protective structures, as are the elytra of many flightless beetles that lack hind wings.

FUNCTIONS AND DESIGN

The primary function of wings is flight. The wings operate as flexible aerofoils, oscillating as first-order levers over a process of the pleuron, which acts as a fulcrum. They need to be rigid enough to resist unwanted deformation under the forces they receive from the air and the inertial forces from the repeated accelerations and decelerations of flapping, but locally flexible enough to allow the cyclic twisting and bending deformations that are necessary to generate the forces to propel the insect and support its weight in air. Because they have no internal muscles, their shape from instant to instant needs to be controlled partly remotely though the axillae, partly automatically by their own structure: the distribution of rigid and flexible elements, of relief, veins and flexion lines within the wings themselves.

The Wing Beat

To interpret these aspects of wing morphology it is necessary to understand the nature of the wing beat. Simply to beat wings symmetrically up and down would be aerodynamically ineffective, because the vertical forces would tend to cancel out. It necessary to introduce some asymmetry to the wing beat, in the path of the wings and/or their shape and attitude. An abundance of high-speed still and cine photographs of a wide range of insects shows that the downstroke has many features common to all. The wings beat downward, and often rather forward relative to the insect's body axis, and are usually relatively straight along their length, with a rather curved section and a slight nose-down twist along the span. In the upstroke, however, the shape of the wings varies greatly, both between species and according to how the insect is at that moment performing.

In virtually all species the wings twist backward (supinate) to some extent at the bottom of the downstroke and continue supinated throughout the upstroke, twisting forward (pronating) again at the top. The extent to which they can twist varies greatly, as was illustrated by Wootton in 2000. The wings of most Diptera, and of zygopterous Odonata, which have slender bases and operate in individual pairs, are able to twist almost onto their backs. The camber becomes reversed, and they are then capable of generating upward, weight-supporting forces on the upstroke as well as the downstroke. This allows them to fly slowly and to hover. At the other extreme, locusts, and probably most other flying Orthoptera, can twist their wings only slightly and get the necessary stroke asymmetry by limited forewing twisting, but mainly by pulling the hind wings fully forward in the downstroke and partly retracting them for the upstroke. Pulling the hind wing forward extends the fan, and the radiating veins within it become compressed and curved like the spokes of an umbrella, giving the wing an aerodynamically effective cambered section, which is lost as the wing relaxes into the upstroke. Only the downstroke generates significant weight support, and the insects can fly only forward, at a relatively high speed.

Between these two extremes fall many insects whose wings are capable of varying amounts of torsion and which hence display a range of versatility and maneuverability in flight. Broad wings and fore- and hind-wing coupling both tend to limit active twisting, but many wings show a degree of passive torsion along their span, so that the distal part is significantly supinated for the upstroke and can generate useful lift. This is in many insects—Plecoptera, sialid Megaloptera, Mecoptera, many Hemiptera and Hymenoptera, some Trichoptera and Lepidoptera—enhanced by a degree of ventral bending along a more or less marked transverse flexion line. Where such a flexion line runs obliquely across the wing, ventral bending serves also to twist the distal part of the wing, enhancing upstroke lift generation and increasing flight versatility.

The Importance of Wing Camber

A cambered cross section makes a wing rigid to both bending and torsion, but asymmetrically so. A thin, dorsally convex wing is far more resistant to a bending force applied from below (Fig. 6a) than to one from above (Fig. 6b), as the latter tends to make the edges buckle outward and flatten the section. Furthermore if the force is applied behind the axis along which the wing naturally twists, the latter twists far more easily if the force is from above (Fig. 6d) than from below (Fig. 6c). These effects are exploited by many insects to favor ventral bending and supinatory twisting for the upstroke, whether or not a transverse flexion line is present. There is evidence, moreover, that some insects can actively alter the height of the camber, and hence the wing's rigidity, by longitudinal bending along the median flexion line, and this may prove to be the latter's principal function.

Automatic Mechanisms

The mechanisms so far described illustrate that the control of wing shape in flight is to a considerable extent automatic: determined by the local distribution of flexibility and rigidity within the structure, moderated remotely by the controlling muscles at the wing base. Nowhere is this automation better demonstrated than in the wings of dragonflies (Odonata, Suborder Anisoptera).

These agile aerial predators have long complex wings (Figs. 2f and 2i), extensively corrugated, with several richly

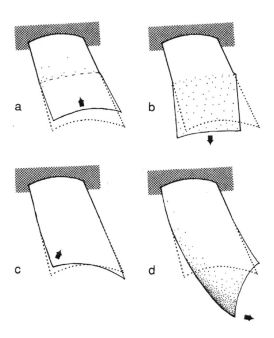

FIGURE 6 Mechanical behavior of cambered plates. (a, b) Cambered plates, fixed at one end, are more easily bent by a force from the convex side than from the concave side. (c, d) Such plates are more readily twisted by a force applied from the convex side behind the torsional axis than from the concave side.

branched longitudinal veins and an abundance of crossveins. The principal support is provided by the costa, subcosta, and radius, which are linked together in the basal part of the wing by high-relief, angle-bracket-shaped crossveins (Fig. 3d) into a spar with a V-shaped cross section, rigid to both bending and torsion. About halfway along the wing the subcosta terminates at the "nodus," a high-relief structure consisting of a strong transverse strut and a patch of flexible cuticle that interrupts the costa. Beyond the nodus the costa and radius form with the first branch of the radial sector a more shallow spar with an inverted V-shaped section, supporting the distal part of the wing but compliant to supinatory twisting. The crossveins here, and elsewhere in the wing, are slender, allowing considerable deformation, so that the flapping wing twists forward and backward about its leading edge spar, like the sail of a tacking dinghy. The nodus serves to reinforce the wing and to minimize stress concentrations at the transition point between the two, mechanically very different, sections of the leading edge spar. The extreme twisting, essential to its versatile, high-performance flight, is assisted by an unusually flexible trailing edge; but it is essential that this latter should not be allowed to swing up like a flag to flutter in the airflow, which would be aerodynamically useless and potentially damaging to the wing. This is automatically prevented by two features: a dense spot called the pterostigma, near the end of the wing tip, which is believed to act as a regulating counterweight to the swinging wing, and a curious, three-dimensional formation of veins near the wing base—the triangle and supratriangle—which automatically levers the trailing edge downward in direct response to the aerodynamic forces that the wing generates. This degree of automation in a propulsive appendage is unique in the animal kingdom and has no obvious parallels in technology.

Wing Folding

All modern insects except the Odonata and Ephemeroptera have the ability to fold their wings back over the abdomen, an adaptation that increases their ability to move freely over a substrate and to penetrate small spaces. Simple folding back is achieved mainly by muscular rotation of the third axillary sclerite and folding of the wing base along the basal hinge between the sclerite and the side of the thorax and along the jugal fold line. In hind wings with a large vannus (anal fan) there may be many extra fold lines lying parallel to the radiating veins, so that the wing folds like a pleated fan. The hind wings of Dermaptera, Coleoptera, and the cockroach *Diploptera,* however, fold up transversely as well as longitudinally; in some cases with extraordinary complexity. Because the wings have no internal muscles, folding and unfolding need to be achieved remotely, by mechanisms operated from the wing base and by the action of other body parts, combined with elasticity within the wing. Hydraulic mechanisms have sometimes been postulated, but recent work suggests that these are unnecessary. The unfolding of beetle wings appears to be achieved by active, scissor-like movements of the bases of the main supporting veins, which set in action a cascade of interconnected, simple opening mechanisms similar to those used in pop-up books and three-dimensional greeting cards; assisted in places by elasticity in the cuticle. Refolding the wings involves the reversal of these mechanisms, assisted by movements of the abdomen. Dermaptera, by contrast, use their abdomens to unfold the hind wings by means of a complex array of similar mechanisms, and refolding is largely elastic.

Secondary Functions

Wings have assumed a range of secondary functions. The sound-producing and -amplifying forewings of many Orthoptera, the protective elytra and tegmina of several orders, and the sensory halteres of Diptera and Strepsiptera have already been mentioned. The wings of Odonata and many Lepidoptera assist in thermoregulation. Members of several orders—Orthoptera, Phasmida, Mantodea, Odonata, Diptera, Lepidoptera—use theirs in active signaling: defensive, territorial, and sexual. Wings with warning patterns and colors are found particularly among Coleoptera, Lepidoptera, Orthoptera, and Heteroptera. The use of wings in camouflage is especially widespread in Orthoptera, Mantodea, Heteroptera, and Lepidoptera and may be reflected in their shape as well as color and pattern. Those of leaf mimics may be unusually broad and may bear petiole-like protuberances and emarginations resembling insect damage. It is often hard to determine whether particular wing characters are flight adaptations or have only secondary functions.

Unsolved Problems

Many wing characters are still poorly understood. We know far too little about the aerodynamic significance of the great variety of wing shapes in, for example, Lepidoptera, and of surface structures like hairs, spines, and scales. While the principles underlying venation pattern and the layout and orientation of flexion lines are broadly clear, the details are often quite obscure. Much work, linking aerodynamics, functional morphology and structural mechanics, remains to be done.

See Also the Following Articles
Anatomy • Exoskeleton • Flight • Fossil Record • Integument • Tracheal System

Further Reading
Brodsky, A. K. (1994). "The Evolution of Insect Flight." Oxford University Press, Oxford/New York.
Dudley, R. (2000). "The Biomechanics of Insect Flight." Princeton University. Press, Princeton, NJ.
Kukalová-Peck, J. (1991). Fossil history and the evolution of hexapod structures. *In* "The Insects of Australia" (CSIRO, eds.), 2nd ed., Vol. 1. Cornell University Press, Ithaca, NY.
Grodnitsky, D. L. (1999). "Form and Function of Insect Wings." Johns Hopkins University Press, Baltimore/London.
Nachtigall, W. (1974). "Insects in Flight." Allen & Unwin, London.
Wootton, R. J. (1979). Function, homology and terminology in insect wings. *Syst. Entomol.* **4,** 81–93.
Wootton, R. J. (1981). Support and deformability in insect wings. *J. Zool. London* **193,** 447–468.
Wootton, R. J. (1991). The functional morphology of the wings of Odonata. *Adv. Odonatol.* **5,** 153–169.
Wootton, R. J. (1992). Functional morphology of insect wings. *Annu. Rev. Entomol.* **37,** 113–140.
Wootton, R. J. (2001). How insect wings evolved. *In* "Insect Movement: Mechanisms and Consequences" (I. Woiwod, D. R. Reynolds, and C. D. Thomas, eds.). CAB Int., Wallingford/New York.
Wootton, R. J., Evans, K. E., Herbert, R., and Smith, C. W. (2000). The hind wing of the desert locust (*Schistocerca gregaria* Forskål). 1. Functional morphology and mode of operation. *J. Exp. Biol.* **203,** 2933–2943.

Wolbachia

Richard Stouthamer
University of California, Riverside

In the heyday of light-microscopy studies, intracellular bacteria were discovered in many organisms. In 1924 M. Hertig and S. B. Wolbach observed bacteria in the cells of a mosquito *(Culex pipiens).* Hertig formally described them in 1936 as *Wolbachia pipientis* in honor of his collaborator in the earlier work. *Wolbachia* was placed in the Rickkettsiales, which belong to the α-Proteobacteria. Its closest relatives are several pathogenetic bacteria (*Cowdria* and *Ehrlichia*) that are transmitted by arthropods to vertebrate hosts. After their discovery little was reported on these bacteria until Yen and Barr in 1971 discovered that they were involved in causing the death of mosquito eggs when sperm of infected males fertilized the eggs of *Wolbachia*-free females. Since that time the interest in *Wolbachia* has increased rapidly because *Wolbachia* appears to be very common in arthropods (up to 70% of all insects) and causes a number of unusual manipulations of their host's reproduction. These manipulations are interesting both from an academic point of view (evolution of sex, speciation, conflict between different sets of genes within one organism) and from an applied point of view (Can these bacteria be used to manipulate their host for our benefit?).

WOLBACHIA'S HOST MANIPULATIONS

Wolbachia is generally passed on only from an infected female to her offspring; infected males do not pass on their infection to their offspring. This can be understood because insect eggs are large and can harbor bacteria, whereas sperm are small and carry very little cytoplasm in which bacteria can be transported. Such asymmetry in the transmission ability of their host explains many of the effects that *Wolbachia* has. An extreme case of this host manipulation is the induction of parthenogenesis (a type of asexual reproduction) in wasps by *Wolbachia*. Infected female wasps produce only daughters; in addition, they do not have to mate to produce daughters. Some wasp species consist entirely of infected females. In this case *Wolbachia* has manipulated its host into producing only females that can transmit the bacterium. In addition no eggs are "wasted" on the production of males, which are a dead end for *Wolbachia*. Other manipulations will be discussed below.

Cytoplasmic Incompatibility

One of the best known and possibly the most common effect of *Wolbachia* is cytoplasmic incompatibility (CI). Several different forms of CI are known. In its simplest form, CI causes the death of embryos when a sperm from an infected male fertilizes the egg of an uninfected female. Consequently, in populations in which infected and uninfected individuals occur together, all possible crosses of infected females result in infected offspring. Uninfected offspring are produced only when an uninfected male mates with an uninfected female, whereas all the offspring of uninfected females that have mated with infected males die. This difference in offspring production causes the CI infection to rapidly spread through the population. In the fruit fly *Drosophila simulans,* the infection in California spread at a rate of approximately 100 km per year.

The underlying mechanism of CI is not yet understood, but it is believed that sperm of infected males are somehow modified during spermatogenesis so that only in eggs infected with the same type of *Wolbachia* can this sperm function properly or be "rescued." This system of modification (mod) of the sperm and rescue (res) in the eggs is used to understand the possible CI types. For instance if in one species two different *Wolbachia* infections exist, this could lead to two-way incompatibility because the modified sperm is not rescued by the eggs infected with the other type of *Wolbachia*. This leads to the death of all the embryos resulting from intertype crosses. Therefore two infections within one species can cause genetic isolation of the two different subpopulations. This isolation may speed up speciation.

Not only is it possible that different populations within a species are infected with a different *Wolbachia,* but superinfections with several *Wolbachia* types are also common. Individuals infected with two different *Wolbachia* may be incompatible with individuals that are infected with only one of the two types. Sperm from doubly infected males may not be rescued in eggs that are singly infected. In these situations the double infection should spread through the population, displacing the single infection.

The types of CI *Wolbachia* that have been discussed so far can all be classified as modifiers (mod+) and rescuers (res+). If we assume that both these functions are energy costly to maintain, one can imagine that once a complete population is infected with the mod+, res+ variant, *Wolbachia* variants that do not modify the sperm any longer (mod–, res+) will spread relative to the modifying form (mod+, res+). Finally, once all of the individuals carry the mod–, res+ variant, the population becomes vulnerable to invasion by mod–, res– variants. Mod–, res– host populations either may lose their infection or can be invaded by new mod+, res+ variants. Examples of mod+, res+, mod–, res+, and mod–, res– are all known. Mod+, res– *Wolbachia* do not exist because the males would be incompatible with the females of their own population.

CI *Wolbachia* is thought to be promising in the spread of beneficial traits through a population. The idea is to genetically modify a *Wolbachia* with genes that inhibit the transmission of disease-causing organisms (malaria, dengue fever, etc.) through the insects that vector the disease. If this gene were inserted into a CI *Wolbachia*, the cytoplasmic incompatibility caused by the *Wolbachia* would lead to the spread of the bacterium through the vector population and all infected insects would be unable to transmit the disease. *Wolbachia* appears to be present in many different parts of the insect body. Consequently the gene products inhibiting the disease organisms may also be delivered to various insect tissues.

Parthenogenesis-Inducing *Wolbachia*

Wolbachia inducing parthenogenesis in their hosts are known from many wasps and some thrips (Fig. 1). Infected females can produce daughters both from fertilized and from unfertilized eggs. The cytogenetic mechanism through which *Wolbachia* cause unfertilized eggs to grow into females is known for a number of parasitoid wasps. *Wolbachia* exploits the normal sex determination in wasps. In this system, males arise from unfertilized eggs, which contain only one set of chromosomes, whereas females arise from fertilized eggs, containing two sets

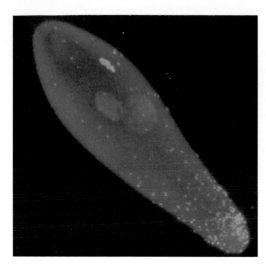

FIGURE 1 PI *Wolbachia* visible as light dots in a 4,6-diamidino-2-phenylindole-stained egg of the parasitoid wasp *Trichogramma kaykai*. The *Wolbachia* are concentrated around the area of the egg from which the future eggs of the wasp develop. Photograph by Merijn Salverda.

of chromosomes. In unfertilized, infected eggs containing only a single set of chromosomes, *Wolbachia* causes a doubling in the number of chromosomes. This is accomplished through a modification of the first mitotic division in the egg, causing the egg to develop into a female.

In most populations in which parthenogenesis-inducing (PI) *Wolbachia* are known, the entire population consists of females. Feeding antibiotics to these females causes them to produce male offspring. In most instances, these males are not able to successfully inseminate the females of their own line. It is assumed that the populations have been parthenogenetic for such a long time that mutations have accumulated in the part of the genome involved with sexual reproduction. An exception to this situation is found in the tiny parasitoid wasps of the genus *Trichogramma* in which, in many populations, both infected and uninfected individuals coexist. In these populations, males that are the offspring of antibiotic-treated mothers can successfully mate with infected females.

PI *Wolbachia* are of interest from the applied point of view as well. Many parasitoid wasps are reared on a large scale in insectaries for release in the field to control pest insects. Only female wasps are effective biological control agents because (1) they lay their eggs in or on the pest insect, (2) the wasp larvae will subsequently eat up the host insect, and (3) new wasps emerge from the remains of the pest insect. PI-*Wolbachia*-infected females are cheaper to produce than normal sexual wasps. No production effort is wasted on rearing of males that are worthless for biocontrol.

Feminizing *Wolbachia*

In many woodlice species (pill bugs, Crustacea, Isopoda), a *Wolbachia* is found that causes infected genetic males to develop into functional females. The feminizing *Wolbachia* accomplishes this through the infection of the male-hormone-producing gland. Without this hormone, genetic males develop into females. The spread of such an infection through a population can result in a severe shortage of males. Because males are needed for reproduction, this has led to counteradaptations by the nuclear genes in such a way that males will be produced even if the individual is infected. The effects of the *Wolbachia* and the counteradaptations against the feminization have led to a dynamic system of sex determination in some species of woodlice.

Feminizing *Wolbachia* have also been discovered in a moth (*Ostrinia scapularis:* Noctuidae). Infected individuals grow into females, but how the *Wolbachia* accomplishes this transition is not known.

Male-Killing *Wolbachia*

Many different microorganisms have evolved methods of killing only male offspring of infected mothers. *Wolbachia* have this ability in several lady beetles and butterflies. The advantage of killing male offspring to *Wolbachia* requires that the infected daughters somehow benefit from the death of their brothers. In the case of the lady beetles this is most easily explained. Lady beetles generally lay their eggs in clusters, and the survival of the larvae is greatly enhanced if they find food soon after hatching. If an infected mother has laid a clutch of eggs the daughters emerge and then consume the eggs containing their dead brothers as their first meal. Because males cannot pass on the *Wolbachia,* the enhanced survival of the infected females benefits the *Wolbachia.* The presence of the male-killing *Wolbachia* can lead to extreme sex ratios in the population and just like in the case of the feminizing *Wolbachia* counteradaptations are expected.

Male-killing *Wolbachia* occur with a high frequency in several species of the butterflies of the genus *Acraea* occurring in East Africa. This has led in some populations to extremely female-biased sex ratios and a concurrent change in behavior of the butterflies. Normally males form leks where females would "choose" the male to mate with, but in these female-biased populations females form leks where males "choose" their partners. Little is known about the mechanism that allows these *Wolbachia* to kill exclusively infected male eggs.

PHYLOGENY OF *WOLBACHIA*

The phylogeny of *Wolbachia* has been studied extensively using several *Wolbachia* genes. Initially the 16S ribosomal DNA sequence was used to derive the relationship between the different *Wolbachia*. From this work it became clear that within the insect *Wolbachia* two groups exist, later named A and B, that had diverged from each other approximately 60 mya. Later several other genes were used to get a finer scale picture of the relationship of the different *Wolbachia* to each other. From all this work two main conclusions can be drawn: closely related *Wolbachia* can have quite different effects on their host and closely related hosts can carry quite different *Wolbachia.* To understand the distribution of the *Wolbachia* over the different hosts it is clear that transfer of *Wolbachia* between species must happen quite frequently over evolutionary time. The fact that the different host manipulations of *Wolbachia* can all be found in closely related *Wolbachia* must be the result of a relatively simple evolution of one effect into the other, for instance from CI to PI, or a more likely hypothesis is that the phylogenies based on different *Wolbachia* genes do not reflect the phylogeny of the genes involved in causing the host manipulation. If the genes causing the effects are found on phages that can jump from one *Wolbachia* to the next, the observed phylogenetic pattern can easily be explained. A potential candidate is the phage WO discovered by Masui *et al.* in 2000, which shows evidence of frequent horizontal transfer between different *Wolbachia.* More information on the genes involved in the different effects of *Wolbachia* will become available once its genome is published.

See Also the Following Articles
Genetic Engineering • *Sex Determination* • *Symbionts, Bacterial*

Further Reading
Dobson, S. L., Bourtzis, K., Braig, H. R., Jones, B. F., Zhou, W., Rousset, F., and O'Neill, S. L. (1999). *Wolbachia* infections are distributed throughout insect somatic and germline tissues. *Insect Biochem. Mol. Biol.* **29,** 153–160.
Hertig, M. (1936). The rickettsia, *Wolbachia pipiens* (gen. et sp.n.) and associated inclusions of the mosquito, *Culex pipiens. Parasitology* **28,** 453–486.
Jiggins, F. M., Hurst, G. D. D., and Majerus, M. E. N. (2000). Sex-ratio-distorting *Wolbachia* causes sex-role reversal in its butterfly host. *Proc. R. Soc. Biol. Sci. B* **267,** 69–73.
Masui, S., Kamoda, S., Sasaki, T., and Ishikawa, H. (2000). Distribution and evolution of bacteriophage WO in *Wolbachia,* the endosymbiont causing sexual alterations in arthropods. *J. Mol. Evol.* **51,** 491–497.
O'Neill, S. L., Hoffman, A. A., and Werren, J. H. (eds.) (1997). "Influential Passengers: Inherited Microorganisms and Arthropod Reproduction." Oxford University Press, London.
Stouthamer, R., Breeuwer, J. A. J., and Hurst, G. D. D. (1999). *Wolbachia pipientis:* Manipulators of invertebrate reproduction. *Annu. Rev. Microbiol.* **53,** 71–102.
Turelli, M., and Hoffmann, A. A. (1991). Rapid spread of an inherited incompatibility factor in California *Drosophila. Nature* **353,** 440–442.
Werren, J. H., Zhang, W., and Guo, L. R. (1995). Evolution and phylogeny of *Wolbachia:* Reproductive parasites of arthropods. *Proc. R. Soc. London B* **261,** 55–63.
Yen, J. H., and Barr, A. R. (1971). New hypothesis of the cause of cytoplasmic incompatibility in *Culex pipiens. Nature* **232,** 657–658.

Worker

see *Caste*

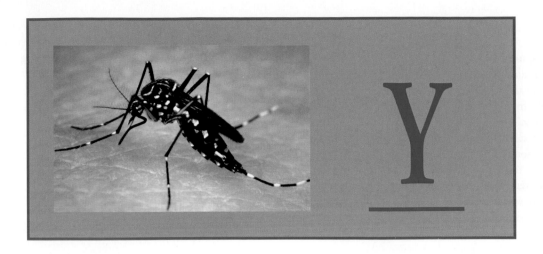

Yellow Fever

Thomas P. Monath

Acambis Inc., Cambridge, Massachusetts

Yellow fever virus is the prototype of the family Flaviviridae, which includes approximately 70 single-stranded RNA viruses, the majority of which are transmitted by mosquitoes or ticks. Yellow fever virus is transmitted principally by insects (mosquitoes), but ticks *(Amblyomma variegatum)* may play a secondary and minor role in Africa.

DISEASE AND MEDICAL IMPACT

The disease caused by the yellow fever virus is a severe hemorrhagic fever, characterized by high viremia (virus in the blood), hepatic, renal, and myocardial injury, hemorrhage, and case-fatality rates of 20 to 50%. Today, the disease occurs in tropical Africa and South America. It is estimated that up to 200,000 cases occur annually but far fewer are reported officially. Between 1990 and 1999, 11,297 cases and 2648 deaths were reported; 83% of these cases occurred in Africa. The highest incidence was in Nigeria, where epidemics occurred between 1986 and 1994. Recent epidemics have affected Cameroon (1990), Kenya (1992), Ghana (1993–1994, 1996), Liberia (1995, 1998, 2000), Guinea (2000), Gabon (1994), Senegal (1995, 1996), and Benin (1996). In South America, 1939 cases and 941 deaths were reported between 1990 and 1999, for an average of about 200 cases per year. Bolivia and Peru had the highest incidence, reflecting low vaccination coverage. In Brazil, an increase in yellow fever activity occurred over a wide area between 1998 and 2000.

ROLE OF INSECTS IN TRANSMISSION

The enzootic transmission cycle involves tree-hole-breeding mosquitoes such as *Aemagogus janthinomys* (South America), *Aedes africanus* (Africa), and nonhuman primates. Infection of mosquitoes begins after ingestion of blood containing a threshold concentration of virus (~ 3.5 $\log_{10}$ ml^{-1}), resulting in infection of the midgut epithelium. The virus is released from the midgut into the hemolymph and spreads to other tissues, notably the reproductive tract and salivary glands. Seven to 10 days is required between ingestion of virus and virus secretion in saliva (the extrinsic incubation period), after which the female mosquito is capable of transmitting virus to a susceptible host. Vertical transmission of virus occurs from the female mosquito to her progeny and from congenitally infected males to females during copulation. Virus in the egg stage provides a mechanism for virus survival over the dry season when adult mosquito activity and horizontal transmission abate.

In tropical South America, the virus is transmitted by *Haemagogus* mosquitoes (principally *Hg. janthinomys*) between monkeys in the rain forest canopy and from monkeys to humans. Transmission peaks during months of high rainfall, humidity, and temperature (January–March), corresponding to the activity of *Haemagogus* mosquitoes. It is during occupational activities, such as forest clearing, lumbering, and road construction that humans acquire the infection ("jungle" yellow fever) from mosquitoes that previously had fed on viremic monkeys; most patients are young adult males. The density of *Haemagogus* and risk of human exposure are relatively low, and human cases occur in a sporadic fashion. Other mosquitoes involved in jungle yellow fever in South America include *Hg. leucocalaenus* and *Sabethes chloropterus*. In the early 1950s an epizootic extended to Central America, where other species were implicated *(Hg. lucifer, Hg. equinus, Hg.*

iridicolor, Hg. mesodentatus). Prior to the 1940s epidemics of yellow fever due to bites from *Ae. aegypti* occurred in the Americas. *Ae. aegypti* breeds in and around houses, principally in artificial containers, and is responsible for direct inter-human transmission of virus ("urban" yellow fever). Age and sex distributions are different from jungle yellow fever, because children and females are affected. Most areas of South America had effectively eliminated *Ae. aegypti* during eradication efforts (1930s–1960s), but the continent was reinvaded in the 1970s, and the vector is now widespread. The first outbreak of urban yellow fever since 1954 occurred in Santa Cruz, Bolivia, in 1997–1998. The wide distribution of *Ae. aegypti* (including areas juxtaposed to the yellow fever endemic zone) and the rise in air travel have increased the risk of reemergence of urban yellow fever in the Americas.

In Africa, the enzootic cycle involves tree-hole-breeding *Aedes* mosquitoes, which transmit virus between monkeys, from monkeys to humans, and between humans. In West Africa, cases appear during the middle of the rainy season (August) and peak during the early dry season (October), the period of maximum vector longevity. The principal enzootic vector is *Ae. africanus;* this species is responsible for transmission between monkeys in the forest zone and in gallery forests. During the rainy season, *Ae. africanus* and other vector species [*Ae. furcifer, Ae. taylori, Ae. luteocephalus, Ae. metallicus, Ae. vittatus,* and (in many parts of East Africa) *Ae. simpsoni*] reach high densities in the humid savanna zone (the "zone of emergence" of yellow fever). The prevalence of immunity in the human population accumulates rapidly with age, and children are at highest risk of infection and illness. The domestic vector *Ae. aegypti* breeds in receptacles widely used by humans for water storage where piped water supplies are unavailable. Where this mosquito is involved in virus transmission, the disease may occur in the dry season in both rural and urban areas. *Ae. aegypti* has been the principal vector in many recent human epidemics in West Africa. Because the density and biting rates of *Ae. aegypti* and sylvatic *Aedes* vectors are high, human infection rates during epidemics may reach 20–30%.

Rainfall and temperature influence vector abundance and transmission rates and the potential for yellow fever epidemics. Prolongation of the rainy season, reflected by vegetational indices in satellite images, predict a risk of yellow fever epidemics. Increasing temperature shortens the extrinsic incubation period of yellow fever virus in the mosquito, resulting in a significantly increased rate of transmission. Even brief exposure to high temperatures (e.g., in a sunlit forest clearing) can have this effect. Warm temperature also increases biting and reproductive rates of *Ae. aegypti.*

Ae. aegypti populations differ genetically in vector competence for yellow fever virus. High vector densities may be required for virus transmission by relatively insusceptible populations, such as occur in West Africa.

CONTROL STRATEGIES

The principal method for prevention and control is human vaccination. A safe and effective live, attenuated vaccine (yellow fever 17D vaccine) has been in wide use since World War II. The vaccine is incorporated into routine childhood vaccination programs in South America and some parts of Africa. The occurrence of human disease reflects incomplete coverage of the population, because vaccine immunity is probably lifelong after a single dose.

In the Americas, control of *Ae. aegypti* through larval source reduction was successful in reducing urban yellow fever in the early decades of the 20th century. The increasing size and complexity of urban areas, the proliferation of detritus and objects such as automobile tires that collect rainwater and breed mosquitoes, and increasing costs and competition with other health care priorities have limited the impact of mosquito control efforts. Emergency vector control using space sprays to kill infected female vectors could undoubtedly be used to contain urban yellow fever outbreaks, but the measure has never been evaluated specifically for this purpose. Spraying of large areas of forest and around villages by air- or ground-operated equipment to contain sylvatic yellow fever has been evaluated, with mixed success.

See Also the Following Articles
Medical Entomology • Mosquitoes • Zoonoses, Arthropod-Borne

Further Reading
Digoutte, J.-P., Cornet, M., Deubel, V., and Downs, W. G. (1995). Yellow fever. *In* "Exotic Virus Infections." (J. S. Porterfield, ed.), pp. 67–102. Chapman & Hall, London.
Miller, B. R., Monath, T. P., Tabachnick, W. J., *et al.* (1989). Epidemic yellow fever caused by an incompetent mosquito vector. *Trop. Med. Parasitol.* **40,** 396–399.
Monath, T. P. (1988). Yellow fever. *In* Monath, T. P. (ed.). "The Arboviruses: Ecology and Epidemiology," Vol. V (T. P. Monath, ed.), pp. 139–231. CRC Press, Boca Raton, FL.
Monath, T. P., Craven, R. B., Adjukiewicz, A., *et al.* (1980). Yellow fever in the Gambia, 1978–1979: epidemiologic aspects with observations on the occurrence of Orungo virus infections. *Am. J. Trop. Med. Hyg.* **29,** 912–928.
Robertson, S. E., Hull, B. P., Tomori, O., *et al.* (1996). Yellow fever. A decade of reemergence. Reprinted in *J. Am. Med. Assoc.* **276,** 11562.
Tabachnik, W. J., Wallis, G. P., Aitken, T. H. G., *et al.* (1985). Oral infection of *Aedes aegypti* with yellow fever virus: geographic variation and genetic considerations. *Am. J. Trop. Med. Hyg.* **34,** 1219–1227.
Trapido, H., and Galindo, P. (1957). Mosquitoes associated with sylvan yellow fever near Almirante, Panama. *Am. J. Trop. Med. Hyg.* **6,** 114–135.

Yellowjacket
see *Wasps*

Zoonoses, Arthropod-Borne

Robert S. Lane
University of California, Berkeley

The term zoonosis (plural, zoonoses) is derived from the Greek roots for animal *(zoon)* and disease *(osis)* and was coined initially to characterize any infection of lower animals that could be transmitted to humans. The Food and Agriculture Organization and the World Health Organization (FAO/WHO) broadened the definition during the 1950s to include infections that are transmitted in either direction between humans and other vertebrate animals. Thus, zoonoses are those diseases and infections for which the agents are naturally transmitted between vertebrate animals and humans. Zoonoses can be transmitted to humans by direct contact with an infectious vertebrate host or a fomite (i.e., an inanimate object such as an article of clothing that can harbor an agent); by ingestion of contaminated water, food, or other organic matter; by inhalation; and by arthropod vectors. This article is devoted exclusively to zoonotic agents that are transmitted from other vertebrates to humans by arthropods. It does not include nonzoonotic disease agents that are transmitted solely from person-to-person by arthropods (e.g., malaria). The terms arthropod vector versus vector, and arthropod-borne versus vector-borne, are used interchangeably.

CHARACTERISTICS

Viruses, prions, and bacteria, which include the rickettsiae, and fungi, protozoa, and helminths can serve as zoonotic disease agents. Taylor and Woolhouse reported in 2000 that 832 (49%) of 1709 infectious organisms known to be pathogenic for humans are zoonotic in origin. These include 507 viruses and prions (30%), 541 bacteria (32%), 309 fungi (18%), 66 protozoa (4%), and 286 (17%) helminths. Furthermore, 156 of these agents produce diseases that are considered to be emerging, and 73% of the emerging pathogens are zoonotic.

Many zoonotic agents are transmitted to humans by means of arthropods, particularly insects or ticks. Indeed, two of the three contemporary internationally quarantinable diseases are vector-borne zoonoses transmitted either by mosquitoes (yellow fever) or fleas (plague). Plague also is transmitted by other routes, but the bacterial agent *(Yersinia pestis)* is maintained primarily in natural cycles involving rodents and their associated fleas. In the tropics and subtropics, the most important arthropod-borne zoonotic disease agents are transmitted by flies (Diptera) such as mosquitoes, phlebotomine sand flies, and tsetse flies, whereas in temperate regions ticks transmit over 90% of the vector-borne human pathogens.

Humans are "dead-end" hosts for most zoonotic agents, i.e., secondary cases do not occur following human infection. Notable exceptions are yellow fever, plague, and Chagas disease. Humans infected with yellow fever virus can serve as a source of infection for uninfected mosquitoes that feed on them, individuals manifesting primary or secondary pneumonic plague potentially can transmit *Y. pestis* to other persons by the respiratory route, and people infected with the trypanosome causing Chagas disease can infect triatomine bugs that bite them.

Several factors that affect the ability of zoonotic agents to produce disease in people are the route of transmission (portal of entry), the genetically determined invasiveness (pathogenicity) of the infecting organism, and the age and immune status of an exposed individual. Two North American mosquito-borne viral encephalitides tend to produce more severe disease in disparate age groups, i.e., children infected with western equine encephalomyelitis (WEE) virus experience graver illness than adults, whereas the reverse is true for individuals infected with St. Louis encephalitis virus. Zoonoses may cause comparable harm to humans and some lower animals, or they may adversely affect one group or the other in a disproportionate manner. Thus, certain of the mosquito-borne encephalitic viruses (e.g., eastern equine encephalitis, venezuelan equine encephalitis, WEE) can be highly pathogenic for both humans and horses.

Most arthropod-borne zoonotic infections can be characterized by their focal distribution within particular geographic landscapes. A natural focus, also known as a nidus, is an area where the complex interplay of various biotic and abiotic environmental factors ensures the temporal persistence of a zoonotic agent. This concept, referred to as "landscape epidemiology," was developed into a doctrine by the Russian scientist E. N. Pavlovsky in 1939. Accordingly, foci exist under certain conditions of macro- and microclimate, vegetation, and soil in localities in which suitable arthropod vectors and reservoir hosts of the disease agent also exist.

Furthermore, foci can be as distinct and limited geographically as the burrow system of a small mammal or a tick-infested cottage, or they can be of the diffuse variety in which case they encompass portions of much broader territories such as coniferous or mixed hardwood forests. For instance, individual cases or localized outbreaks of relapsing fever in California usually are associated with rodent- and tick-infested cabins at higher elevations. In such foci, the spirochete *Borrelia hermsii* occasionally is transmitted to sleeping individuals by the bite of the rapid-feeding, nocturnally active soft-tick vector, *Ornithodoros hermsi*. In contrast, northern Californian foci of the Lyme disease spirochete *Borrelia burgdorferi,* although present in several major habitat types, seem to be most intense in certain leaf-litter areas within mixed hardwood forests where infection prevalences in nymphal *Ixodes pacificus* ticks sometimes reach 20 to 40%.

VECTORS

The phylum Arthropoda contains six classes of varying medical importance but only two of them, the Insecta and Arachnida, are of paramount importance as transmitters of zoonotic agents. Mites, ticks, spiders, and scorpions comprise the Arachnida, but only certain species of mites and particularly ticks are capable of transmitting infectious organisms to humans.

Insects that transmit zoonotic agents belong principally to three orders, namely, the Siphonaptera (fleas), Heteroptera (true bugs), and especially the Diptera (flies). Important families of flies that transmit zoonotic agents are the Culicidae (mosquitoes), Glossinidae (tsetse flies), Psychodidae (phlebotomine sand flies), Ceratopogonidae (biting midges), and Tabanidae (deer flies and horse flies). Other dipteran families, like those comprising the filth flies generally and the house fly in particular (Muscidae), may mechanically contaminate human foodstuffs with bacteria, protozoans, or other agents that can cause gastrointestinal illnesses.

Among the arachnids, ticks overwhelmingly outweigh the mites in importance as vectors of zoonotic agents because of their universal blood-sucking habit. With rare exceptions, all three trophic stages (larva, nymph, adult) of ticks must ingest

a blood meal to complete development and to ensure reproductive success. By comparison, the majority of mites are free living and nonparasitic. In the United States, approximately 10% of the 83 described tick species transmit one or more viruses, bacteria, or protozoan parasites to people with any regularity. One commonality shared by many arachnid and insectan vectors is that they feed intermittently upon their hosts and therefore spend more than about 95% of their entire life cycle off the host.

CLASSIFICATION

Zoonoses were classified by the FAO/WHO in 1967 according to whether their primary reservoir hosts are humans or lower animals and with regard to the type of life cycle of the infecting organism. The latter classification, which is based upon shared epidemiologic features, is more instructive than one based solely on reservoir hosts. In it, zoonoses have been categorized as direct zoonoses, cyclozoonoses, metazoonoses, or saprozoonoses. However, only two of these categories involve transmission by arthropods. Direct zoonoses are transmitted, in part, from an infected to a susceptible vertebrate host mechanically by a vector, whereas metazoonoses are transmitted biologically by vectors.

TRANSMISSION CYCLES

Arthropod-borne zoonotic agents are maintained in transmission cycles of variable complexity. Nevertheless, four components are evident in all such cycles: the agent itself, one or more efficient arthropod vectors and primary reservoir hosts, and a permissive environment. A reservoir host is a vertebrate that is readily infected with the agent, is capable of maintaining the agent within its tissues for an extended period, and can serve as a source of infection for uninfected vectors that feed on it while it is in an infectious state. In reality, arthropods also contribute to the maintenance of zoonotic agents to various degrees; some, like certain hard-tick vectors of spotted fever group rickettsiae or mosquito vectors of some viruses, may be considered reservoirs as well. That is, they can maintain the agent in their tissues for months or even years, efficiently pass it from one trophic stage to the next (transstadial transmission), and eventually transmit it from one generation to the next via the eggs of infected females (transovarial transmission). In such cases, the reservoir of infection may best be considered polyhostal because both the vertebrate host(s) and the vector(s) help to maintain the cycle of infection.

Arthropods may transmit zoonotic agents from an infected vertebrate to a susceptible one either mechanically or biologically. Of these, biological transmission is much more prevalent than mechanical transmission among most groups of arthropod vectors. Mechanical transmission occurs when an agent adheres externally on the mouthparts, legs, or other bodily regions of a vector and then is transported directly or indirectly to a vertebrate, e.g., by means of contaminated foodstuffs or by specific inoculation into the skin or bloodstream by the bite of contaminated mouthparts. In mechanical transmission, the agent does not require the arthropod to complete an essential part of its life cycle, and the arthropod/agent relationship is of an accidental nature.

In biological transmission, the vector plays an indispensable role in the life cycle of the agent. Biological transmission takes three forms: cyclodevelopmental, cyclopropagative, and propagative. In cyclodevelopmental transmission, the agent undergoes cyclical changes within the internal tissues of the vector but does not multiply (e.g., mosquito transmission of the filariid nematode *Wuchereria bancrofti*). In cyclopropagative transmission, the agent undergoes cyclical development and multiplication in the arthropod's body (e.g., malaria plasmodia in anopheline mosquitoes or babesial piroplasms in their tick vectors). In propagative transmission, the agent multiplies but undergoes no cyclical development within the vector's body (e.g., most bacteria and viruses).

Regardless of the specific mode of biological transmission, the vector usually transmits the agent anteriorly via its salivary secretions while ingesting a blood meal from a vertebrate host. In a few instances, the agents are transmitted posteriorly when infectious feces are deposited on the host near the bite wound (e.g., feces of triatomine bugs infected with *Trypanosoma cruzi*, the causative agent of Chagas disease).

PUBLIC HEALTH IMPORTANCE

The public health and economic impact of the arthropod-borne zoonoses is immense, particularly in developing countries in tropical or subtropical regions. Losses include morbidity and mortality among humans and livestock and the resultant direct and indirect economic effects to affected individuals and to society at large. For instance, the African trypanosomiases still are among the most devastating diseases afflicting humans and livestock. African sleeping sickness is estimated to have killed up to 500,000 people between 1896 and 1906, and one epidemic near Lake Victoria in Uganda claimed about 200,000 lives. Today, it is estimated that over 300,000 new cases of the disease are contracted annually and that 60 million people in 36 countries of sub-Saharan Africa and an even greater number of livestock are at risk. In addition to African trypanosomiasis, other zoonotic and nonzoonotic vector-borne diseases such as malaria, dengue, yellow fever, filariasis, leishmaniasis, plague, and louse-borne typhus caused more human morbidity and mortality than all other causes from the 17th to the early 20th centuries, especially in the tropics. Since the 1970s, there has been a resurgence of many of these diseases and the emergence of others, notably tick-borne diseases like Lyme disease, ehrlichiosis, and babesiosis in temperate regions.

Certain population groups traditionally have been at elevated risk of acquiring vector-borne zoonotic infections. Agricultural and forestry workers, hunters, field naturalists, park rangers, wildlife biologists, woodcutters, and tourists rep-

resent just a few of the many groups that are vulnerable to such infections. In the northeastern United States, suburbanites may be at considerable risk for contracting Lyme disease in the peridomestic environment because of the presence of host-seeking, spirochete-laden *Ixodes scapularis* ticks on lawns and in adjacent shrubbery or forested areas. Similarly, in the western United States, the risk of human plague increased in peridomestic settings during the late 20th century as human populations encroached into formerly unpopulated foci of *Y. pestis*.

Apart from the human misery and loss of life attributable to zoonoses, the significant costs associated with the occurrence of individual cases or outbreaks of a few specific arthropod-borne diseases in the United States illustrate their economic impact. Following a single human plague case at Plumas–Eureka State Park in Plumas County, California, in 1976, attendance at that park declined markedly for the remainder of the year. Moreover, a benefit–cost analysis of bubonic plague surveillance and control at that park and another nearby campground revealed that the costs incurred when human plague was contracted at either recreational area averaged $52,000. In 1966, an outbreak of St. Louis encephalitis virus, a mosquito-borne flavivirus, in Dallas, Texas, was estimated to cost nearly $800,000 in terms of morbidity, mortality, patient treatment, and control activities. In total, 172 presumptive or serologically confirmed cases with 20 fatalities were reported during that outbreak. In Massachuetts, the average total cost per case of eastern encephalitis, a mosquito-borne alphavirus, was determined to be $21,000 for a transient case versus a lifetime cost of nearly $3 million for persons suffering severe residual sequelae.

Finally, in contrast to the tropics where dipterans (flies) rank first as transmitters of vector-borne disease agents, ticks are unsurpassed in importance as primary vectors in temperate regions. Lyme disease alone currently accounts for over 95% of all cases of vector-borne diseases reported annually in the United States. In 1999, Lyme disease and Rocky Mountain spotted fever composed 99.5% of the total number of indigenously acquired vector-borne disease cases reported that year; mosquito- and flea-borne infections accounted for the remaining 0.5%. These computations for 1999 exclude numerous cases (probably a few hundred) of several tick-borne diseases (e.g., Colorado tick fever, human granulocytic and monocytic ehrlichioses, and relapsing fever) and 62 cases of the newly introduced mosquito-borne West Nile virus because none of these zoonoses is a nationally notifiable disease. Likewise, tick-borne diseases predominate throughout much of Europe and in parts of Asia, especially in the former Soviet Union.

In conclusion, zoonoses comprise about one-half of all infectious organisms known to cause disease in humans. Knowledge of the specific environmental factors and vectors that maintain and distribute arthropod-borne zoonotic agents is essential before locally effective control methods can be developed and implemented. This information is especially valuable because relatively few vaccines are available for personal protection against vector-borne infections.

See Also the Following Articles
Bubonic Plague • Dengue • Medical Entomology • Mosquitoes
• Ticks • Tsetse Fly • Veterinary Entomology • Yellow Fever

Further Reading
Acha, P. N., and Szyfres, B. (1987). "Zoonoses and Communicable Diseases Common to Man and Animals," 2nd ed. Pan Am. Health Org., Washington, DC. [Scientific Pub. No. 503]
Centers for Disease Control and Prevention (1999). Recommendations for the use of Lyme disease vaccine: Recommendations of the Advisory Committee on Immunization Practices (ACIP). *MMWR Suppl.* **48(RR-7),** 1–25.
Centers for Disease Control and Prevention (2001). Summary of notifiable diseases, United States, 1999. *MMWR* **10(53),** 1–101.
Gubler, D. J. (1998). Resurgent vector-borne diseases as a global health problem. *Emerg. Infect. Dis.* **4,** 442–450.
Harwood, R. F., and James, M. T. (1979). "Entomology in Human and Animal Health," 7th ed. Macmillan Co., New York.
Hugh-Jones, M. E., Hubbert, W. T., and Hagstad, H. V. (eds.) (2000). "Zoonoses: Recognition, Control, and Prevention." Iowa State University Press, Ames.
Kimsey, S. W., Carpenter, T. E., Pappaioanou, M., and Lusk, E. (1985). Benefit–cost analysis of bubonic plague surveillance and control at two campgrounds in California, USA. *J. Med. Entomol.* **22,** 499–506.
Lawyer, P. G., and Perkins, P. V. (2000). Leishmaniasis and trypanosomiasis. *In* "Medical Entomology: A Textbook on Public Health and Veterinary Problems Caused by Arthropods" (B. F. Eldridge and J. D. Edman, eds.), Chap. 8. Kluwer Academic, Dordrecht.
Pavlovsky, E. N. (1966). "Natural Nidality of Transmissible Diseases with Special Reference to the Landscape Epidemiology of Zooanthroponoses." University of Illinois Press, Urbana.
Reeves, W. C. (1990). Epidemiology and control of mosquito-borne arboviruses in California, 1943–1987. Calif. Mosquito Vector Control Assoc., Sacramento.
Schwab, P. M. (1968). Economic cost of St. Louis encephalitis epidemic in Dallas, Texas, 1966. *Public Health Rep.* **83,** 860–866.
Tälleklint-Eisen, L., and Lane, R. S. (1999). Variation in the density of questing *Ixodes pacificus* (Acari: Ixodidae) nymphs infected with *Borrelia burgdorferi* at different spatial scales in California. *J. Parasitol.* **85,** 824–831.
Taylor, L. H., and Woolhouse, M. E. J. (2000). Zoonoses and the risk of disease emergence. *In* "International Conference on Emerging Infectious Diseases, Atlanta, Georgia, Program and Abstracts Book," late breaker poster session, Board 122, p. 14.
Villari, P., Spielman, A., Komar, N., McDowell, M., and Timperi, R. J. (1995). The economic burden imposed by a residual case of eastern encephalitis. *Am. J. Trop. Med. Hyg.* **52,** 8–13.
World Health Organization (1967). Third Report of a Joint WHO/FAO Expert Committee on Zoonoses. *WHO Tech. Rep. Ser.* **378.**

Zoraptera

Michael S. Engel
University of Kansas

The Zoraptera ("zorapterans," "angel insects," or less appropriately "soil lice") are one of the smallest insect orders with only 32 living species and 6 more known from fossils. Zorapterans are rarely observed; indeed many entomologists

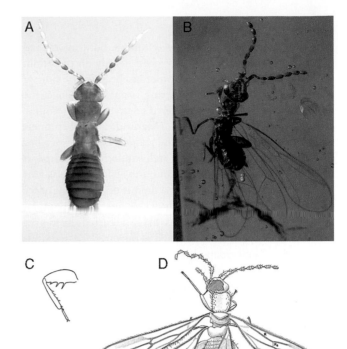

FIGURE 1 (A) Adult female of *Zorotypus huxleyi* (most legs curled under body), blind–apterous morph. (B) Alate female of †*Z. goeleti* in Miocene amber from the Dominican Republic. (C) Hind leg of *Z. huxleyi* depicting femoral and tibial spines. (D) Alate of †*Z. nascimbenei* in Cretaceous amber from Myanmar, with wing veins labeled. (Reproduced, courtesy of the American Museum of Natural History, from Engel and Grimaldi, 2002.)

have never seen them alive. There is a single family (Zorotypidae) and only two genera, *Zorotypus* and the fossil genus †*Xenozorotypus* from the Middle/Late Cretaceous. Some authors have at times attempted to divide living species of *Zorotypus* into multiple genera or subgenera, but none of these have successfully characterized natural (i.e., monophyletic) groups. Owing to the paucity of species and the great morphological homogeneity of the order, the recognition of numerous living genera is currently unwarranted.

Zorapterans superficially resemble termites (Isoptera) and bark lice (Psocoptera) (Figs. 1A, 1B, and 1D) and they have at times been classified near or within these groups. Individuals are typically less than 4 mm in body length (excluding their antennae) and live in small colonies. The most readily noticeable traits of the order are the expanded hind femora that bear stout spines on their ventral surface (Fig. 1C), the presence of only two tarsomeres (the basalmost of which is greatly reduced in size; Fig. 1C), and paddle-shaped wings with reduced venation (Figs. 1B and 1D) that are shed by an ill-defined basal fracture.

---

† Extinct group.

ORIGIN AND EVOLUTION

Although zorapterans are relatively minute both in size and in species diversity, they have weighed heavily on the minds of evolutionary biologists. The order has been at one time or another classified in a bewildering array of phylogenetic positions, at times being considered relatives of the termites, roaches (Blattaria), mantises (Mantodea), webspinners (Embiidina), or bark lice; sister to Paraneoptera; or even allied to the Holometabola. The most well supported hypothesis is that they are polyneopterans and the living sister to the webspinners, both orders sharing a unique development of the musculature in the hind legs, among several other important traits. The Zoraptera–Embiidina likely occurred during the early Mesozoic, perhaps as long ago as the early Triassic, while the ancestor of both groups diverged from stoneflies (Plecoptera) perhaps sometime in the Carboniferous. The Zoraptera are clearly ancient as evidenced by the presence of extinct species of the living genus *Zorotypus* in Middle/Late Cretaceous amber (e.g., Fig. 1D). These extinct Mesozoic species already show the development of the dual-morphs, with blind–apterous and eyed–winged individuals. Only the extinct genus †*Xenozorotypus* shows a departure in morphology (albeit not nearly as dramatically as the anatomical diversity found in other orders) from all other zorapterans by the primitive retention of additional wing veins and the structure of spines on the hind tibia.

DISTRIBUTION

Species principally occur in tropical habitats throughout the world although four species occur north of the Tropic of Cancer (i.e., north of 23.5°N): two in North America and two in Tibet. Similarly, the known fossil zorapterans have all been discovered in ambers that were formed in warm-tropical paleoclimates. The northernmost areas of distribution for *Z. hubbardi* in North America are likely a result of human activity. Most northerly colonies are formed in sawdust piles (rather than natural logs) where warmth from the decaying material can sustain a colony for a few seasons. Thus, once a pile is no longer suitable dispersal to new sawdust piles or into natural logs is required or, conversely, those northernmost populations repeatedly become extinct and are maintained over the long term by perpetual reintroductions from more southern populations. Once believed to be highly endemic as a result of an inability to disperse, several *Zorotypus* species are increasingly being recognized over larger geographic ranges (e.g., *Z. hubbardi* in south-central and southeastern United States), evidence of an ability to disperse to some degree so as to maintain specific integrity.

GENERAL BIOLOGY

The biology of zorapterans is poorly documented. Only three species have been studied and only one, *Z. hubbardi,* generated a sizeable literature. Species are gregarious with colonies consisting of 15 to 120 individuals in relatively ephemeral, subcortical habitats. Colonies occur in logs (or man-made sawdust piles) but only after wood decay is well progressed. Zorapterans feed principally on fungal hyphae and spores, although they can also be predatory, victimizing nematodes, mites, and other minute arthropods such as springtails (Collembola). The number of nymphal instars is estimated to be either four or five. Captive individuals live up to 110 days.

Individuals occur in two morphs. Blind, wingless individuals predominate during the general life of a colony, whereas dispersive individuals (with eyes and wings) are generally rare and are produced as resources become depleted and the colony becomes crowded. Winged females are more common than winged males, suggesting that females perhaps mate prior to dispersal. After arriving at a new log, individuals shed their wings and dealated individuals are often found in young colonies. Wide distributions for some species as well as the presence of species endemic to distantly isolated islands such as Christmas Island, Fiji, and Hawaii are evidence for their dispersal abilities.

BEHAVIOR

Although *Zorotypus* species are gregarious and live in small colonies, they evidently do not distinguish between individuals from the same colony and introduced vagrants. Isolated individuals do not survive.

Zorapterans spend much of their time grooming either themselves or other individuals. Grooming may be a way of removing fungal pathogens and this may be important for their colonial lifestyle. Most of the movements in the grooming repertoire are found in other insect orders but some are unique to zorapterans. Most notable are the movements associated with the cleaning of the posterior sterna, the cerci, and the genital–anal area using their mouthparts. This complex action involves the raising of the body on a four-point stance, with both anterior and posterior ends of the body bent downward (perpendicular with the substrate) to meet under the insect and between the middle and the hind legs.

Females mate multiple times (with up to eight mates) and with multiple males (up to three multiple matings). Prior to copulation, zorapteran species have complex courtship behaviors which appear to be species specific. The process is begun by the male who strokes the female with his antennae. If the female reciprocates with similar antennal signals, the male initiates courtship displays during which his head and a portion of his neck are extended. If the female approaches, the male secretes fluids from a cephalic gland, which she ingests while the male and female bend their abdomens toward one another. Copulation is initiated by the linking of the male to the female via a hook located medially on his 10th abdominal tergum (females have a small groove into which the hook is placed on their eighth sternum). Once the hook is coupled with the female, the male inverts himself, assuming an upside-down position and facing away from the female.

ECONOMIC IMPORTANCE

Although zorapteran colonies at times can be found in wood or sawdust piles at lumber yards, owing to their relatively minute colony sizes, general scarcity, and apparent minimal ecological impact, it is unlikely that they will ever be considered of any economic importance. If the fragmentary reports of wood ingestion by zorapterans are confirmed as well as studies of their gut fauna (bacteria, protozoa, etc.) undertaken, then the economic role of *Zorotypus* might be reevaluated.

See Also the Following Articles

Isoptera • Psocoptera

Further Reading

Choe, J. C. (1992). Zoraptera of Panama, with a review of the morphology, systematics, and biology of the order. *In* "Insects of Panama and Mesoamerica: Selected Studies" (D. Quintero and A. Aiello, eds.). Oxford University Press, Oxford.

Choe, J. C. (1995). Courtship feeding and repeated mating in *Zorotypus barberi* (Insecta: Zoraptera). *Anim. Behav.* **49**, 1511–1520.

Choe, J. C. (1997). The evolution of mating systems in the Zoraptera: Mating variations and sexual conflicts. *In* "The Evolution of Mating Systems in Insects and Arachnids" (J. C. Choe and B. J. Crespi, eds.). Cambridge University Press, Cambridge, UK.

Engel, M. S. (2000). A new *Zorotypus* from Peru, with notes on related neotropical species (Zoraptera: Zorotypidae). *J. Kansas Entomol. Soc.* **73**, 11–20.

Engel, M. S., and Grimaldi, D. A. (2000). A winged *Zorotypus* in Miocene amber from the Dominican Republic (Zoraptera: Zorotypidae), with discussion on relationships of and within the order. *Acta Geol. Hisp.* **35**, 149–164.

Engel, M. S., and Grimaldi, D. A. (2002). The first Mesozoic Zoraptera (Insecta). *Am. Mus. Novitates* **3362**, 1–20.

Gurney, A. B. (1938). A synopsis of the order Zoraptera, with notes on the biology of *Zorotypus hubbardi* Caudell. *Proc. Entomol. Soc. Wash.* **40**, 57–87.

Silvestri, F. (1913). Descrizione di un nuovo ordine di insetti. *Bol. Lab. Zool. Gen. Agric. Portici* **7**, 193–209.

Zygentoma
(Thysanura, Silverfish)

Helmut Sturm
University of Hildesheim, Germany

The Zygentoma (Thysanura *sensu stricto*) are medium-sized apterygote insects with a body length from 5 to 30 mm. Their body is flattened and the eyes are small or absent. There are no ocelli except for the Lepidothrichidae, which has three ocelli. The flagellate (whiplike) antennae can be short (e.g., in Ateluridae) or much longer than the body (Fig. 1). The mouthparts are ectognathous. The mandibles are dicondylic (have two rotation points) and the maxillary palps have five segments that are of normal length. Nearly all parts of the body and appendices bear bristles of different length and

FIGURE 1 Male (above) and female of *Thermobia domestica* during premating (passing-by behavior), vertical view. In this phase the antennae and the cerci are in contact; body length about 12 mm.

structure. Scales are present on the thoracic segments of Lepismatidae and Ateluridae, but lacking in the three other families.

All Zygentoma are able to run quickly, but they cannot jump as do the Archaeognatha. The coxae of all legs are large and flattened. The penis of males is situated on abdominal segment 9; on the hind end, three caudal appendages are present as in the Archaeognatha. Molting continues in adults. Some species of the family Lepismatidae (e.g., *Lepisma saccharina*, *Thermobia domestica*, *Acrotelsa collaris*) prefer to live in human houses.

FOSSIL RECORD, SYSTEMATICS, BIOGEOGRAPHY

The oldest fossils that are clearly Zygentoma were found in Mesozoic layers (Lower Cretaceous) of Brazil. From the Cenozoic on, there are many Zygentoma fossils, and most of them are amber inclusions. Nearly all of them are similar to extant taxa (e.g., *Ctenolepisma*, *Nicoletia*, Lepidothrichidae). However there are representatives of apterygote Hexapoda from the Paleozoic (e.g., Monura and Cercopodata) that cannot be included in extant orders.

The Zygentoma comprise five families and some 400 species in about 90 genera.

1. **Lepismatidae** This largest family includes the well-known silverfishes, some of which occupy human houses. This family is the richest in genera (more than 21) and species (> 200).

2. **Nicoletiidae** Relatively small and slim, and without dark pigment, eyes, and scales, they live in the upper layers of soils within the humus layer or beneath stones. Some species or populations reproduce parthenogenetically. More than 10 genera and more than 30 species are known.

3. **Ateluridae** Small, blind, and of droplike form, apparently all species in this family are commensal with ants or termites. There are more than 10 genera and more than 30 species.

4. **Lepidothrichidae** There are two species: *Lepidothrix pilifera*, which is known from Baltic amber, and the extant

FIGURE 2 Female of *T. domestica,* dorsal view. The female is picking up the spermatophore with her ovipositor. The spermatophore was deposited on secreted threads spun, a short time before, by the male.

Tricholepidion gertschi, which occurs in forests of northern California (Fig. 3).

5. **Maindroniidae**   There is one genus and three species from the Middle East and Chile; their habitat is somewhat similar to that of the Nicoletiidae.

The Lepismatidae and Nicoletiidae are distributed worldwide. Some Lepismatidae are household pests: *L. saccharina* in northern Europe; *A. collaris* in the tropics; and *T. domestica* in the United States (Figs. 1 and 2). In southern Europe, *L. saccharina* and other lepismatid species also live outside houses. In Africa, Lepismatidae occur in areas with sand dunes. Nicoletiidae are found up to 3000 m elevation in Colombia and enter the northern parts of Europe via transport of potted plants to greenhouses. The distribution of Ateluridae coincides fairly well with that of ants and termites. The distribution of Lepidothrichidae and Maindroniidae appears quite restricted.

FIGURE 3 Male of *T. gertschi* (Lepidothrichidae), vertical view, body length about 14 mm. This species is probably a relic and the most archaic within the Zygentoma. It is found only in the forests of Northern California. The animal is seen here on rotting wood, its natural environment.

Their relatively flat body form enables the Zygentoma to enter small fissures. The Nicoletiidae are especially well adapted to the upper layers of the soil, whereas Lepismatidae generally rest in fissures during the day and are active during the night. If the relative humidity is above 50 to 60%, lepismatidae can absorb atmospheric water by means of the anal sac (part of the hindgut).

PHYSIOLOGY AND DEVELOPMENT

As in Archaeognatha and in other orders of apterygote insects, the cuticle of Zygentoma is mainly thin and flexible. The mandibles present an intermediate form between monocondylic and a typical dicondylic structure. The Lepismatidae eat plant material and detritus; *L. saccharina* is able to digest cellulose by the cellulase produced in its midgut.

Development has been studied mainly on Lepismatidae (especially *L. saccharina, T. domestica,* and *Ctenolepisma* spp.). The eggs of *T. domestica* are of oval form with a largest diameter of about 1.2 mm, and are laid underground. At a temperature of 37°C, they need 12 to 16 days until eclosion. In *L. saccharina,* egg development lasts about 35 days at temperatures near 20°C. The first free-living developmental stage possesses an egg tooth on the front of the head, and the first three stages lack scales. The molting intervals are dependent on age, temperature, loss of scales and/or appendages, species, and even population. At temperatures of about 40°C *T. domestica* molts in 9- to 11-day intervals. The life span of *L. saccharina* is up to 3 years, whereas that of *Thermobia* is about 2 years.

EVOLUTIONARY ASPECTS

With flattened form, the Zygentoma are adapted to small fissures in soil, detritus, or in houses. The Lepismatidae leave those places during night to find food, favorable conditions (e.g., high relative humidity), and members of the same species for mating. In contrast with the Archaeognatha, their food use is broad. Lepismatidae have been spread widely by humans. For example, of the 13 lepismatid species in the United States, only three are believed to be native. In comparison to the Archaeognatha, the Zygentoma show a great adaptability in terms of the different habitats they live in and the astonishing adaptation of the Ateluridae, which live with ants and termites. They possess some morphological structures that correspond to the Pterygota: a special form of mandibles; maxillary palps smaller than thoracic legs; an abdomen with tracheal commisures and connections; and an ovipositor base with well-developed gonangulum. Apparently, however, they are not the ancestors of the Pterygota. Because of the crosswise joining of abdominal terga with correspondent coxites (a unique apomorphy within the Hexapoda), the possession of 11 abdominal segments, styli on abdominal segments, eversible vesicles, three abdominal appendages (filum terminale and two cerci), and the structure of the ovipositor, the Archaeognatha and Zygentoma are obviously linked. However the "jump" from Apterygota to Pterygota probably was from a line separate from the Thysanura.

See Also the Following Article
Archaeognatha

Further Reading

Smith, G. B. (1995). Thysanura (Zygentoma). *In* "CSIRO, Zoological Catalog of Australia," Vol. 23, pp. 6–18. Commonwealth Scientific and Industrial Research Organization, Canberra.

Sturm, H. (1997). The mating behavior of *Tricholepidion gertschi* Wygodzinsky, 1961 (Lepidotrichidae, Zygentoma) and its comparison with the behaviour of other "Apterygota." *Pedobiologia* **41,** 44–49.

Wygodzinsky, P. W. (1972). A review of the silverfish (Lepismatidae, Thysanura) of the United States and the Caribbean area. *Am. Mus. Novit.* **2481,** 1–26.

GLOSSARY

The following glossary defines approximately 800 specialized terms that are found in the articles in the *Encyclopedia of Insects*. The authors of the Encyclopedia have defined these words as they are used in their articles, for the benefit of readers who may not be familiar with the technical vocabulary of insect study. As the name indicates, the Glossary is a compendium of the terminology of this particular encyclopedia. It is not a general dictionary and therefore does not include all possible entomological terms that a reader might encounter in other publications.

A

accessory gland Any secondary gland of a glandular system.

accessory pulsatile organs (APOs) Very small muscular pumps and veins that direct hemolymph into appendages such as wings and antennae.

acclimation A process of adjustment to survival at temperatures that are normally lethal.

acetylcholinesterase (AChE) The enzyme responsible for breaking down the neurotransmitter **acetylcholine (ACh)** at nerve synapses, thereby preventing hyperexcitation of cholinergic pathways in the nervous system. This enzyme is the major target of organophosphate and carbamate insecticides.

acrosome The anterior portion of the differentiated spermatozoa that assists in the penetration of the egg membrane.

across-fiber pattern code An often complex code in which each of several cell types sends a message about a single stimulus to higher processing centers. The stimulus may contain one or more than one type of chemical.

aculeate Any member of a group of families that includes the familiar stinging ants, bees, and social and hunting wasps (e.g., sphecids, eumenids, scoliids, pompilid spider wasps). Aculeates are largely characterized by their stinging apparatus, which is derived from the ovipositor of their ancestors but is no longer used for laying eggs.

acylurea A class of insect growth regulators, including diflubenzuron.

adaptive immunity A host defense mechanism that is specific for an epitope; it involves B and T cells and exhibits memory.

adaptive radiation Diversification of species within a lineage associated with ecological change.

adipocytes (trophocytes) The major cell type of the insect fat body that accumulates and stores reserve nutrients such as lipids, glycogen, and proteins used primarily during molting, metamorphosis, flight, and oocyte formation. These cells are actively involved in intermediary metabolism and play a crucial role in regulating blood proteins (enzymes), lipids, and carbohydrates. A fundamental function of adult adipocytes is the synthesis and release of vitellogenins (yolk proteins) taken up from the hemolymph by developing oocytes.

adipokinetic hormone (ADK) A hormone involved in promoting fat metabolism.

aedeagus The sclerotized terminal portion of the male genital tract that is inserted into the female during insemination.

aerobic Using molecular oxygen as an electron acceptor; pertains to organisms or metabolic processes.

agent An organism that can infect a larger organism in which it may cause a specific disease.

alary muscles Muscles of the dorsal diaphragm that sometimes participate in hemolymph movement in and around the pericardial sinus of the abdomen.

alata The parthenogenetic winged morph of viviparae, specialized for migration; among dioecious aphids, alatae termed emigrants fly from the primary to the secondary host each spring.

alate 1. Winged; having wings. 2. An adult individual with fully developed wings.

allele Form of a gene. Genes are considered **allelic** when they occur in the same position on members of a chromosome pair.

allelochemical A nonnutritional chemical produced by one species that affects the growth, health, behavior, or reproduction of another species, e.g., phytochemicals that function in plants as defenses against herbivores or pathogens.

allergen An antigen that induces an allergic reaction.

allometric growth A pattern of growth in which a body part grows at a rate different from that of other body parts or the body as a whole. **Allometry** can be measured within an individual over the course of its development, and also between individuals. **Isometry** is a special case of allometry in which the relative sizes of two or more structures are constant within or among individuals.

allomone An allelochemical that benefits the producing organism and harms the receiving organism, such as chemicals produced by a stink bug to protect it against natural enemies.

alloparent An individual that assists the parents in care of the young.

allozyme A particular variant of an enzyme.

alternative splicing The ability to create two or more types of proteins from the same (eukaryotic) gene, by selective inclusion or exclusion of exons from mature messenger RNA, whereby regulatory interactions between genes result in great diversity of proteins from the same group of genes and alternative developmental pathways.

altruism An organism's behavior that does not directly benefit itself but contributes to the welfare of others within the family group. Associated with kin selection in eusocial insects.

ametabolous Characterized by no significant metamorphosis. The immature body form does not differ from the adult, except for external genitalia, and the number of stages is variable.

anaerobic Not using molecular oxygen as an electron acceptor; pertains to organisms or metabolic processes.

anaerobic flagellates The heterogeneous assembly of phylogenetically ancient protozoa (archaezoa) that is found exclusively in the hindguts of termites and wood-feeding cockroaches.

anchialine Relating to surface and underground habitats associated with coastal ponds that have only underground connections to the sea.

anemotaxis Directed orientation with respect to wind, usually involving walking or flying in an upwind direction.

anholocycly The asexual reproduction of parthenogenetic viviparous aphids, wherein oviparae, males, eggs, and fundatrices are omitted in a nonrecombinant lifestyle.

anisogamy A condition in which the female sex cells (ova) are larger than the male sex cells (sperm).

Anopheles A genus of mosquitoes (order Diptera, family Culicidae) that contains 422 species, of which 40 frequently may be involved in the transmission of human malaria.

anopheline mosquito A mosquito classified in the family Culicidae, subfamily Anophelinae. The important genus *Anopheles* is in this subfamily.

anoxic Devoid of molecular oxygen; describes the oxygen status of an environment.

antifeedant A chemical, often toxic, that prevents or reduces feeding.

ant-tending The activity of ants resulting in the consumption of honeydew produced by the tended insects. Also, **ant attendance**.

aphidophagous Feeding on aphids.

apitherapy Medicinal use of the honey bee or its products.

Apocrita The Hymenoptera with a wasp waist (i.e., the sister group to the Orussidae), which comprises the majority of species in the order.

apolysis Separation of new and old cuticle caused by release of ecdysteroids at the beginning of each molt.

apomixis Parthenogenesis in which eggs are produced without meiosis.

apomorphous Describing a novel character evolved from a preexisting character.

apomorphy (-ic) A feature of an organism in the derived state, contrasted with an alternative one in the ancestral (primitive) state—a plesiomorphy (-ic). For example, with the character of forewing development, the sclerotized elytron is an apomorphy for Coleoptera, and the alternative, a conventional flying forewing, is a plesiomorphy at this level of comparison.

aposematic signals Conspicuous visual, acoustic, olfactory, or behavioral warnings displayed by an animal to advertise its noxious or dangerous properties. Aposematic colors are contrasting patterns of reds, whites, oranges, yellows, or blacks; sounds are usually generalized hisses or stridulatory squeaks that have low complexity and a wide range of frequency emissions; odors are strongly apparent to most organisms and are distinct from common odors; behaviors are obvious jerks, wing flicks, or accentuated movements.

apostatic selection Evolutionary selection force that confers protective advantage to individuals within a population that have rare color or pattern relative to the others.

aptera The parthenogenetic wingless morph of viviparae, specializing in reproduction.

apterous 1. Wingless; without wings. 2. Describing an individual lacking wings altogether.

aptery A state of winglessness.

apterygote 1. Primitively wingless. 2. Insect orders without wings.

arboreal nest A nest built in a tree or shrub.

archedictyon The primitive original vein network characterizing the wings of many of the most recent insect fossils.

arolium Inflatable pad on the tarsus of thrips.

arrhenotoky The process of producing males from unfertilized eggs.

arthropod vector An arthropod, such as an insect or tick, that can biologically or mechanically transmit an animal disease agent from one vertebrate to another.

artificial honeydew A sugar-containing solution sprayed onto plants with the aim of attracting natural enemies of homopteran insect pests.

artificial selection Selective breeding under laboratory conditions to change a trait. Such selection may be unidirectional or bidirectional; in bidirectional selection some metric is both increased and decreased by appropriate breeding.

arylphorins Storage proteins, rich in aromatic amino acids, that are synthesized in the fat body, stored in the hemolymph, and utilized for cuticle synthesis in connection with a molt.

atmobiotic Living above ground.

ATPase An enzyme that cleaves adenosine triphosphate (ATP) and uses the energy derived from this cleavage to drive further chemical reactions. With respect to excretion, ATPase refers to proteins that drive ion transport across cell membranes.

autocidal control The use of insects to control wild-type populations of the same species.

autoecious See MONOECIOUS.

autogeny The ability to produce eggs without feeding as an adult. Species that can reproduce only in this way have **obligate autogeny**; species that can produce eggs with or without feeding have **facultative autogeny**. The opposite condition is called **anautogeny**.

automixis Parthenogenesis in which eggs undergo meiosis.

B

bacillus Bacterium with a rod shape.

balancer chromosome A chromosome with multiple inversions that suppresses recombination.

ballooning Aeronautical dispersal by means of air currents acting on strands of silk.

Batesian mimic A palatable insect that gains protection by resembling another species that is unpalatable to predators.

Batesian mimicry A parasitic relationship, named after Henry Walter Bates (1862), in which an undefended (e.g., palatable) species copies the signal of poisonous or otherwise defended species in order to deceive the receiver.

behavioral ecology The study of movements and activities of animals as they relate to interrelations with other organisms and the environment.

behavioral fixity Principle stating that the behavior, ecology, and climatic preferences of fossil organisms will be similar to what is found in their present-day descendants at the generic level; used in reconstructing the behavior and ecological preferences of extinct organisms.

benthos Bottom-living organisms in streams or lakes, including biota that ranges from bacteria to algae to invertebrates to fish.

Bergmann's rule The principle that individuals within a species should grow larger in size at higher latitudes (and altitudes) to reduce heat losses by increasing the ratio of surface area to volume.

bioassay (biological assay) The measurement of a biological response (e.g., mortality) to different concentrations of a biologically active compound, using standardized conditions in the laboratory or field.

biodiversity (biological diversity) The variation in all living organisms from gene to ecosystems.

biological amplification The increase in concentration of chemical compounds in higher trophic levels (e.g. predators) of a food web.

biological control The use of living organisms, parasitoids, predators, pathogens, and competitors to suppress pest populations below levels that would occur naturally. Biological control programs are intended to suppress invasive species through importation of specialized natural enemies that have evolved with a pest in its native range.

bioluminescence The production of light by living organisms.

biomonitoring The systematic use of living organisms or their responses to determine the quality of the environment.

bionomics Study of the life history and habits of a species. Often used as a synonym of *ecology*.

biotype A distinct strain of a species.

bivoltine Describing a life cycle with two generations a year.

blastoderm A one-cell-thick layer of cells surrounding the yolk.

blastomeres The embryonic cells produced during the process of cleavage.

book lungs Respiratory pouches of arachnids, filled with closely packed sheets of tissue to provide maximum surface for aeration.

Brachycera The so-called "short-horned" flies, characterized generally by short, three-segmented antennae, the last segment of which is either stylate or aristate, and by larvae that usually have a reduced head capsule.

brachyptery (-ous) The condition of having shortened wings, whose function in flight is lost.

brood The immature members of a colony collectively, including eggs, nymphs, larvae, and pupae.

brood cell A chamber or pocket built to house one or more immature individuals.

bursa copulatrix An accessory structure in the female reproductive tract of some insects that temporarily accepts the spermatophore or sperm from the male.

C

cambium The thin layer of cells beneath the bark of woody plants that gives rise to new cells (phloem and xylem) and is therefore responsible for diameter growth.

cAMP Adenosine 3′,5′-cyclic monophosphate. A compound produced by the enzyme adenylate cyclase upon binding of a hormone to certain receptors located on the cell surface. cAMP carries the information from the hormone into the cell where the response occurs.

campodeiform An insect larva with an elongate body shape.

canaliculus *plural,* **canaliculi.** A small channel.

carapace A hard dorsal covering of the cephalothorax in Arachnida.

carboxylesterase A hydrolytic enzyme that can cleave compounds containing a carboxylester group (COOR).

carrying capacity The maximum population size or density of a species that can be sustained indefinitely in a given environment.

Ca²⁺ signaling Changes in the intracellular calcium (Ca^{2+}) concentration, through the release of Ca^{2+} ions from intracellular stores by the opening of ligand-gated ion channels or the entry of Ca^{2+} ions into the cell through different types of Ca^{2+}-selective channels located in the plasma membrane, that play a role in the regulation of various cellular processes, including cell metabolism, gene expression, cytoskeletal dynamics, and neurotransmission.

caste Any set of individuals in a colony that are morphologically distinct and/or that perform a specialized task.

cauda A specialized process on the tip of the abdomen; used to flick away honeydew waste.

cavernicole An organism that lives in caves and other cave-like subterranean habitats.

ceca Outpocketings of the midgut epithelium. The cecal cells are often differentiated from the other midgut cells both histologically and functionally.

cecidium *plural,* **cecidia.** Technical term for plant gall, an abnormal growth of plant tissue under the influence of a parasitic organism; thousands of insect species induce these growths on a variety of plant species and plant parts.

cecidogenesis The mechanism of instigating physiological changes in host plant tissue to form a gall; this begins with a chemical stimulus passed from insect to plant either at the time of oviposition or as the growing larvae feed upon the plant.

cecidology The study of plant galls and gallmaking.

center of mass The point in a body or system of bodies about which mass is evenly distributed.

center place foraging The process whereby an animal searches for a resource (usually food) starting from a more or less permanent location to which the animal returns after searching is complete.

cephalothorax United head and thorax of Arachnida and Crustacea.

cercus *plural,* **cerci.** Simple or segmented appendages of the 11th abdominal segment of insects that usually act as sensory organs; often called "forceps."

chaetotaxy The arrangement and nomenclature of the setae or bristles on the exoskeleton of an insect.

chelicerae *singular,* **chelicera**. Jaws, each comprising a large basal part and fang.

chemoautotrophic Relating to the ability of an organism to obtain its nourishment through the breakdown of inorganic chemical compounds rather than through photosynthesis.

chemoreceptor Either the chemosensillum (preferably called the "sensillum") or the receptor protein on the sensory cell membrane.

chemotransduction The process leading from a chemical recognition signal (molecular interaction) to a depolarization of a sensory dendrite. This involves an amplification step that allows a small number of stimulus molecules to yield a large cellular response, thus increasing sensitivity.

chitin A polysaccharide that serves as the major fibrous component in insect cuticle.

chitin synthase A membrane-bound enzyme that polymerizes N-acetyl-D-glucosamine units into a straight-chain amino sugar homopolymer. The protein has a large number of transmembrane segments that may be involved in translocation of the chitin polymer across the cell membrane.

chlorinated hydrocarbons Organic compounds containing chlorine, hydrogen, and occasionally oxygen and sulfur. The first widely used synthetic organic insecticides.

chorion A complex set of eggshell layers produced by follicle cells toward the end of egg maturation.

clade A hypothesized monophyletic group, i.e., a group of related taxa that includes all descendants of a common ancestor.

cladogram A diagrammatic illustration of the branching sequence of purported relationships of organisms, based on distribution of shared derived features (synapomorphies).

clathrin lattice A honeycomblike structure that adheres to the cytoplasmic face of a cellular membrane, where it folds inward and forms a vesicle.

clavus In a neopterous wing without an expanded anal fan, the area between the claval furrow and the jugal fold.

cleptoparasite An insect that feeds on the food resources or uses the nest of another species.

coadapted genetic system The entire set of interacting genetic factors of an organism.

coelopulse system A part of the nervous system that drives the extracardiac pulsations.

cold hardening An increase in cold tolerance, usually achieved by exposure to a moderately low temperature.

cold hardiness The ability to survive subzero temperatures.

colleterial glands In females, accessory glands that secrete fibrous proteins to fasten eggs to a support; in males, accessory genital glands.

colulus A slender or pointed appendage immediately in front of the spinnerets of some spiders; in others it is greatly reduced or seemingly missing; a homologue of anterior median spinnerets or cribellum.

comb A layer of more or less regularly arranged, closely packed brood cells.

commensalism A form of intimate interaction between members of two different species. One partner of a commensal relationship benefits from the association but the other partner neither benefits nor is harmed during the interaction.

community Populations of different species that coexist and interact with each other in a defined area.

competition An interaction between two or more organisms or species that utilize the same resources, in which the presence of one reduces the birthrate or increases the death rate of the other.

competitive exclusion The elimination of one species from an area of habitat due to competitive interaction with another species.

complementary foods Foods that individually are nutritionally unbalanced, but when combined can provide a balanced diet.

complete (holoblastic) cleavage A coordinated series of mitotic divisions wherein the entire volume of the egg cytoplasm is divided into numerous smaller cells that are fully separated from one another and that together comprise a morula.

compound eye A photoreceptive unit functioning through combined action of two or more ommatidia, each consisting of a convex corneal lens, variably shaped crystalline cone, distal pigment-screening cells, composite rhabdome, and retinular cells.

conductor Semimembranous structure in male palp that supports and guides the embolus in insemination.

condyle Any process by means of which an appendage is articulated into a pan or cavity.

convection The bulk movement of a fluid (either gas or liquid) driven by a physical pressure gradient.

copal Semifossilized (subfossil) resin ranging from several years to thousands of years old with a melting point under 150°C, a hardness of 1 to 2 on the Mohs scale, and a surface soluble to organic solvents.

copulation The period or process during which an intromittent organ (almost always of the male) introduces gametes into a reproductive tract of the opposite sex.

cordon sanitaire Literally, sanitary rope; a line of guards stationed to prevent communication with an infected district.

cornicles See SIPHUNCULI.

corpora allata *singular,* **corpus allatum**. The pair of endocrine glands that make juvenile hormone.

corpora cardiaca *singular,* **corpus cardiacum**. A pair of endocrine glands closely associated with the aorta just behind the brain and connected to it by nerves.

coupling A term for processes and structures that convey mechanical events in the environment to the sensitive plasma membrane of a mechanosensory neuron.

courtship Traditionally viewed as communication between the sexes that brings about successful copulation. However, given that communication during copulation may influence fertilization success, courtship can in theory occur anytime during mating.

coxites Flat appendages on the abdominal sterna, often bearing styli and exsertile vesicles.

critical daylength The daylength marking the transition between a daylength that elicits diapause and one that promotes nondiapause development.

critically endangered Synonymous with endangered; a designation used by the International Union for Conservation of Nature and Natural Resources.

critically imperiled Synonymous with endangered; a designation used by the U.S. Natural Heritage Program.

cryoprotectants Low-molecular-weight polyols and sugars that act as classic antifreezes and thus function to depress the insect's supercooling point.

crypsis Camouflage of color and pattern or of odor such that the individual blends into the background environment and is not readily detected by predators.

cryptobiosis Complete cessation of active life processes that gives hardiness to extreme conditions of heat, cold, or drought.

cryptonephric complex A structure in which the distal ends of the Malpighian tubules are closely associated with the rectum, and the structure is enclosed within a perinephric space by a perinephric membrane.

culicine mosquito In general terminology, a mosquito classified in the family Culicidae, subfamily Culicinae. In strict usage, a mosquito classified in the tribe Culicini of the subgenus.

cursorial Adapted for running.

cuticle The outer covering of the insect that serves as its skeleton to which the muscles attach. It is secreted by the epidermis and must be shed periodically to allow for continued growth and for the change of form at metamorphosis.

cyclic AMP A cyclic derivative of adenosine monophosphate that is synthesized from ATP by adenylyl cyclase. An intracellular second messenger involved in the regulation or modulation of ion channels, protein kinase activity, and gene expression.

Cyclorrhapha A subgroup of Brachycera characterized by a flagellum of composite first flagellomere and three aristomeres, pupation within the cuticle of the last instar (i.e., puparium), and a life cycle including three instars.

cystoblast Egg precursor cell produced mitotically by division of a germ stem cell into another stem cell and the cystoblast.

cytochrome P450 monoxygenases A ubiquitous group of enzymes involved in the NAPDH-mediated oxidation and metabolism of a broad range of endogenous and exogenous substrates.

cytokine A chemical mediator, secreted by blood cells, that acts at close range on other immunocompetent cells.

cytopathology Pathology of disease exhibited at the cellular level.

D

daytime dark firefly A member of the family Lampyridae that is sexually active in the daytime and is without light organs as an adult (e.g., *Ellychnia*, *Pyropyga*, some *Lucidota*).

dealate 1. Describing individuals that have shed their wings in termite and ant reproductive castes. **2.** An adult individual that has shed its wings.

debridement Removal of dead or contaminated tissue from a wound.

decomposer An organism that feeds on dead organic material.

dendrite The part of a chemosensitive cell that bears the receptor proteins. It is in close communication with the environment via pores in the sensillum cuticle.

density The number of individuals of a species per unit of habitat.

density dependence The tendency for the birthrates or death rates to change as density of a population increases or decreases.

density independence Factors operating in population regulation that are not related to population density (e.g., climate).

desmosomes Special contact zones between the cytoplasmic membranes of epidermal cells.

determinate growth A pattern of growth in which both growth and molting cease upon reaching the adult stage.

detoxification The fact or process of reducing or removing toxicity. Once insecticides have entered the body, they may be metabolically altered by enzymes in insects as well as other organisms. Through this process of detoxification, such alterations generally produce less toxic products called metabolites.

detritus Organic matter of unidentifiable origin.

detritus feeder An organism that feeds by eating soil or soil particles, from which nutrients are extracted by its digestive system.

diapause A programmed state of dormancy or arrest of development mediated by the neuroendocrine system, usually occurring in a species-specific stage of the life cycle. It may be induced obligatorily or facultatively in response to seasonal cues (day length, temperature, food quality, etc.).

dicondylic Having two condyles.

diffusion The movement of a molecule from high to low concentration driven by random thermal motion.

diflubenzuron An acylurea insect growth regulator that disrupts formation of normal cuticle via inhibition of chitin synthesis.

digestibility coefficients Coefficients that measure an organism's food input and waste output and thus allow one to estimate how much food was digested.

digestion The biochemical process by which food is dissolved and converted into compounds that can be absorbed by the intestinal epithelia.

dioecious A host-alternating aphid life cycle requiring primary and secondary hosts; synonym: *heteroecious*.

diploid Having two complete sets of chromosomes (like a typical adult animal).

disruptive coloration Patterns of coloration that can blur the outlines of a potential prey's body.

dominance hierarchy A social order of dominance established among individuals of one sex, usually males, by aggressive or other behavioral displays.

dominant (genetic) The stronger of a pair of alleles, expressed as fully when in single dose (i.e., heterozygous) as it is when present in double dose (i.e., homozygous). The opposite of recessive.

dorsal diaphragm A complete or fenestrated membrane separating the dorsal vessel in its pericardial sinus from the alimentary canal in its perivisceral sinus.

dorsal vessel The principal hemolymph pump in the body, consisting of a heart in the abdomen and an aorta in the thorax and head regions.

dosage compensation Differential regulation of genes found in different numbers (dosages) within cells of males and females, usually because of heterogametic (XY) sex determination.

ductus seminalis A duct connecting the spermathecal sac with the median oviduct.

Dyar's law An empirical observation that indicates a geometric progression in head width in successive instars of most holometabolous larvae, as proposed by H. G. Dyar in 1890.

Dzierzon's rule The original term for haplodiploid sex determination in honey bees, as defined by J. Dzierzon in 1845.

E

ecdysis The emergence of an insect from its old cuticle during the final stages of a molt.

ecdysone   In the context of hormonal control of female reproduction, one of a group of hormones called ecdysteroids. Synthesized by the prothoracic gland, it is the precursor to 20-hydroxyecdysone.

ecdysteroids   Steroid molting hormones of insects that are synthesized in prothoracic glands of larvae and in the reproductive organs of adults.

eclosion   The molting of a new adult insect from the cuticle of its preceding stage.

ecological potential   The ability of a species to adjust to environmental factors.

economic loss   For an agricultural crop, the fraction of the realizable yield that is lost to the combined effect of all pests and the physical forces of the environment.

ecosystem   A community of organisms and their interaction with the physical environment.

ectognathous   Having exserted mouthparts.

ectoparasite   A parasite that lives on the surface or within the skin of its host.

ectophagous hyperparasitoid   Hyperparasitoid having larval offspring that feed externally on the primary parasitoid host.

ectotherm   An organism whose body temperature is strongly influenced by the ambient temperature.

ectothermy   The fact of having a body temperature roughly equal to the ambient temperature.

edaphon   An organism living in soils.

effective (physiological) temperature   Temperature permitting development; the sum of effective temperatures equal to the number of degree days needed for completing a certain developmental period.

elateriform   An insect larva with a cylindrical body shape.

elytra   *singular,* **elytron**.   The anterior leathery or chitinous wings of beetles, serving as coverings to the hind wings, commonly meeting in a straight line down the middle of the dorsum in repose.

embolus   A structure in the male palp containing the terminal part of the ejaculatory duct and its opening; it may be small, long, whip-like, or coiled; sometimes divided into several structures.

embryonic diapause   The cessation of development in the embryo stage with metabolism markedly reduced.

encapsulation   The response of host blood cells, known as plasmatocytes, to the presence of a parasitoid egg or larvae, which results in the formation of a multilayered capsule that causes the death of the parasitoid through asphyxiation.

encoding   The creation of a train of action potentials from a sensory receptor potential.

endangered   Referring to species that face a high risk of extinction in the near future unless action is taken to protect them; a designation used by the U.S. Fish and Wildlife Service and International Union for Conservation of Nature and Natural Resources. (A formal listing as endangered or threatened by the U.S. Fish and Wildlife Service is the only designation that provides legal protection to species that are at risk of extinction.)

endemic   Restricted to a certain region or part of a region; describing a species found naturally in one area and nowhere else; such species can be **neoendemics**, frequently formed through adaptive radiation, or **paleoendemics**, formed through genetic isolation from source populations.

endocuticle   The flexible inner layer of procuticle. The endocuticle is not stiffened or cross-linked and is usually resorbed, and its components are recycled with each molt.

endocytosis   The formation of vacuoles that transport fluids and solutes into the cytoplasm of a cell.

endoparasite   A parasite that lives within the body of its host.

endophagous hyperparasitoid   A hyperparasitoid having larval offspring that feed internally inside the primary parasitoid microwasp host.

endoplasmic reticulum   A system of cytoplasmic vesicles whose functions include collecting newly synthesized proteins destined for secretion from the cell.

endopterygote   A complete metamorphosis with egg, larval, pupal, and adult stages.

endosymbiosis   An association of organism in which one organism lives within the other; ants, termites, and other wood-feeding insects contain **endosymbiotic** bacteria that digest cellulose and make it available as food.

endotherm   An organism able to keep its body temperature at a level that is more or less independent of the ambient temperature.

endothermy   The fact of having a body temperature that is primarily controlled by endogenous mechanisms rather than by the ambient temperature.

endoxyly   The fact or condition of living within wood.

enhancer   A DNA sequence that recognizes certain transcription factors to stimulate transcription of nearby genes.

enhancer trapping   The identification and cloning of enhancer regions through the nearby insertion of a transposable element containing a reporter gene that becomes regulated by the enhancer.

entomopathogenic   Causing disease or death in insects.

entomophobia   An unnatural and unreasonable fear of arthropods.

envelope   A sheath or carton surrounding the combs of a social wasp nest.

epedaphon   An organism living on the soil surface and in leaf litter.

epicuticle   The thin (but chemically complex) outermost layer of cuticle. The epicuticle is the first layer to be formed and thereafter delimits the space into which the inner procuticle may be secreted.

epigynum   A more or less complicated apparatus for storing spermatozoa, immediately in front of the opening of internal reproductive organs of female spiders.

epitope   A site on an antigen that is recognized by an antibody or T-cell receptor.

epizootic   Describing an outbreak of disease in an animal population.

epoxide hydrolase   An enzyme that inactivates juvenile hormone through hydrolysis of the epoxide substituent.

eradication   The application of phytosanitary measures to eliminate a pest from an area.

eruciform   An insect larva having a caterpillar-like shape.

essential amino acids   Amino acids that contribute to protein and cannot be synthesized de novo by most animals, usually resulting in a dietary requirement for such amino acids.

etiological agent   The causative agent of a disease, such as a pathogen, a chemical, or a genetic mutation.

euedaphon   An organism living wholly within the soil, often well below the surface.

eusocial   Fully social; a term applied to some species of insects and other animals that live in groups for at least part of their life cycle. Eusocial insects exhibit cooperative offspring care and an overlap of adult generations in their groups, and some individuals (workers) forgo reproduction to help others reproduce.

eutrophic Describing a pond, lake, or other body of water containing a rich supply of plant nutrients.

exocrine secretion An exudate discharged to the exterior from a specialized or exocrine gland that functions as an interspecific (allomone) or intraspecific agent (pheromone).

exocuticle The rigid and sclerotized (chemically stiffened) outer layer of some procuticles. Because of its chemical cross-linking, the exocuticle cannot be resorbed during the molting cycle, but its stability makes it a good candidate for structural color production.

exopterygote A condition in which wings develop externally and increase in size with each molt.

exoskeleton The external skeletal structure typically composed of chitin and often complexed with protein and strengthened with calcium carbonate.

extracardiac pulsations Minute contractions of the abdominal body wall muscles that cause pressure pulses in the hemolymph, which moves hemolymph around organs and aids in ventilation.

F

facultative diapause A form of diapause that occurs in response to specific environmental cues that have been received.

facultative hyperparasitoid Hyperparasitoid having progeny that can develop as either a primary parasitoid or a secondary parasitoid.

fang A claw-like part of each chelicera, usually with a groove down which venom and digestive juices can flow into prey.

fat body A structure within the cockroach body that contains stored nutrients, uric acid, and endosymbiotic bacteria, each housed in a different cell type.

feeding guild A group of organisms that exploit a resource in a similar manner (e.g., sap-feeding, leaf-chewing, leaf-mining).

filarial nematode Characteristically long, thin unsegmented roundworms that live everywhere in vertebrates, with the exception of the lumen of the digestive tract.

filter chamber Anastomoses of the foregut and proximal rectum in many (but not all) honeydew-producing insects.

filter feeding The process of acquiring food by straining small particles from the air or water column.

flagellum The third (most distal) segment of an insect antenna.

flash pattern The species-typical unit of light emission of "lightning bug" fireflies that is repeated at somewhat regular time intervals by advertising, mate-seeking males; commonly it is the flashed entity that stimulates the female response flash.

flexion lines Lines or bands of flexible cuticle within the wing, primarily adapted for deformability in flight.

focus (nidus) A place in the environment where ecological conditions ensure the perpetuation of a zoonotic disease agent for an extended period of time.

fold lines Lines or bands of flexible cuticle, primarily adapted for folding the wing at rest.

follicle cells Cells not derived from germ cells that cover oocytes and contribute to their development.

follicles Sperm tubes in the testes in which spermatogenesis occurs.

formulation The form in which an insecticide is applied; the most common formulations are sprays, dusts, and baits.

fossorial Of a limb, body part, etc., adapted for digging or burrowing through soil.

founder event speciation Mode of speciation generally associated with island biota: a small number of individuals, perhaps even a single fertilized female, colonize and become established on an adjacent island or in a geographically isolated habitat.

fragment island An island formed as a result of separation from a larger landmass.

frass Solid excrement from an insect, often containing undigested plant or wood parts.

frequency-dependent selection Selection that is positively or negatively dependent on the frequency of phenotypes in the population. It is negative when phenotypes that are rare are favored (leading to balanced polymorphisms) and positive when phenotypes that are rare are disfavored (leading to stabilizing selection and monomorphism).

fumigant A volatile chemical that acts as a poisonous gas in a confined area.

functional response The relationship between the quantity of food available to an organism and the quantity actually eaten.

fundatrix The parthenogenetic morph that hatches in the spring from the recombinant egg and gives rise to the viviparae; synonym *stem mother*.

G

GABA γ-Aminobutyric acid, a naturally produced inhibitory neurotransmitter that acts on its specific receptor, the **GABA receptor**, to suppress excitation by opening a chloride channel.

gallmaker The organism that induces the host plant to form the gall structure, which includes an excess of protein-rich parenchyma upon which the gallmaker can feed.

gametocytes Infective haploid blood stage of the malaria parasite for mosquitoes acquired during blood feeding. Sexual union of micro- and macrogametes occurs in the midgut of the definitive mosquito host.

gap genes Genes expressed in broad domains in the *Drosophila* embryo; mutations in these genes cause large deletions of the body plan.

gene The unit of heredity.

gene tagging The identification and cloning of a gene based on the insertion of a transposable element that inactivates it or alters its expression.

genetic marker A gene that enables one to detect a transgenic individual. The marker gene is located within the gene vector, usually a transposable element, and its expression indicates that the vector has inserted it into the target DNA.

genetic polymorphism The occurrence together of two or more discontinuous and heritable forms of a species at such frequencies that the rarest of them may not be maintained merely by recurrent mutation.

genetic sexing The selective production of one sex over the other as a direct consequence of a genetic difference between them.

genetic transformation A stable, heritable change in genotype caused by the incorporation of foreign DNA into the genome.

genome The entire nucleotide sequence of an organism, including the entire set of genes.

genomics   The comparative study of genomes, the entire ensemble of an organism's genetic material.

genotype   All or part of the genetic composition of an individual or population.

geomagnetic field   A dipolar magnetic field surrounding the earth with poles in the vicinity of the earth's poles. The field's lines of force are approximately horizontal in the vicinity of the equator and tilt increasingly upward toward the South Pole and increasingly downward toward the North Pole.

geomagnetic vector   The description of a local geomagnetic field comprising the local total strength of the field, the dip angle (inclination) of its lines of magnetic force relative to the horizontal and the usually small deviation (declination) of these lines of force from true north.

germ anlage   The cluster of cells that forms the embryo proper; the germ anlage represents the ventral surface of the embryo.

germarium   The distal region of an ovariole, consisting of germ stem cells, young cystoblasts, and precursor to follicle tissue.

germ band   In the blastula, a thickened band of cells that forms the future embryo. The germ band consists of a head, three gnathal, three thoracic, and eight to ten abdominal segments; it has been referred to as the "phylotypic" stage for insects.

germ cells   Cells that are ancestors of oocytes.

germ-line differentiation   Specification early in the development of testes (spermatogenesis) or ovaries (oogenesis), from a small group of cells in embryos, and subsequent production of mature sperm and eggs.

glowworm   A term historically and currently applied to the north European glowworm firefly, *Lampyris noctiluca*, but for more than a century used as a general term for worm-like or larviform glowing organisms, including lampyrid and phengodid larvae and larviform females and fungus gnat larvae.

glowworm firefly   A "primitive" member of the family Lampyridae that uses glow signals for sexual communication, typically with glowing females attracting nonluminescent males (e.g., *Microphotus*, *Pleotomus*).

gluconeogenesis   A metabolic pathway responsible for the synthesis of glucose from amino acids, lactate, and glycerol.

glutathione S-transferases (GSTs)   Enzymes that catalyze the metabolism of a range of substrates after their conjugation with the endogenous tripeptide glutathione.

glycolysis   The principal metabolic process responsible for oxidation of glucose to pyruvate during cellular respiration.

gnathal   Referring to the feeding appendages, including the mandible and maxillae.

Golgi bodies   Membrane-bound cellular organelles associated with the endoplasmic reticulum; cellular products produced by the endoplasmic reticulum are bound by Golgi bodies possibly for further modification and/or transport.

Gondwanan distribution   Having a geographic distribution on more than one of the southern continents that were once united to form Gondwanaland (a supercontinent precursor to current landmasses).

G protein-coupled receptors   Integral membrane proteins that constitute a large family of neurotransmitter, hormone, or olfactory receptors. Characterized by seven transmembrane regions. When agonists bind to these receptors, trimeric GTP-binding (G) proteins are activated that then regulate the activity of intracellular secondary effectors, which change intracellular concentrations of second messengers or ion channel activity.

gregarious   Liking to be in a crowd. In locusts, this implies that the insects tend to group together spontaneously. Locusts of the **gregarious phase** tend to be very active and to have conspicuous coloration—black and yellow or black and orange.

gross pathology   Pathology of disease exhibited at the level of the whole animal, generally observed externally.

gyne   Female reproductives in an insect society.

gynopara   The specialized winged vivipara that returns to the primary host in the fall and produces the ovipara.

H

haltere   The modified metathoracic wings of the dipteran adult, which are sense organs concerned with the maintenance of stability in flight.

hamulus *plural,* **hamuli**. Hook-like setae that can allow wings to be temporarily attached.

haplodiploid   Describing organisms that have one sex haploid and the other diploid. **Haplodiploidy** is based upon a sex-determining mechanism by which females develop from fertilized eggs and functional males from unfertilized eggs.

haploid   Having one complete set of chromosomes (like a typical egg or sperm cell).

heartbeat reversal   During immobile stages or behaviors (e.g., pupal stages), contraction of the dorsal vessel in characteristic peristaltic waves that travel in the anterograde (to the front) or retrograde direction.

heat-shock proteins   Stress proteins that are highly expressed in response to heat and other forms of stress and contribute to both high- and low-temperature tolerance.

hematophagous   Feeding on vertebrate blood as a food source.

hemidesmosomes   Structures that connect the basal cytoplasmic membrane to the basal lamina.

hemiedaphon   An organism that completes part of its development in the soil but spends the remainder of its life above ground.

hemimetabolous   Describing immature forms that closely resemble the adult form and pass through a fixed number of juvenile instars; the last instar exhibits wing buds and incomplete metamorphosis.

hemocoel(e)   The open body cavity of insects and other arthropods through which the hemolymph (blood) circulates.

hemocytes   Insect blood cells. The main classes are plasmatocytes, granulocytes, spherule cells, lamellocytes, oenocytoids, and crystal cells.

hemoglobin   An oxygen-carrying pigment of the blood possessing a high oxygen affinity and occurring in some aquatic insects inhabiting low oxygen environments.

hemolymph   The blood of insects that bathes all tissues via an open circulatory system powered by an open tubular heart. It transports all the nutrients and hormones to the cells and removes the cellular wastes.

heterochrony   The precocious or delayed appearance of certain features in the ontogenic development.

heteroecious   See DIOECIOUS.

heterogametic sex   The sex that carries sex chromosomes of different types and thus produces gametes of two types with respect to the sex chromosome they carry. In humans, males are the heterogametic sex, carrying one X and one Y chromosome.

heterogeny Alternation of generations in a single insect species, wherein one generation includes both sexes while the other generation includes only females; among gallmaking insects, these alternate generations induce very different galls.

heterothermy (poikilothermy) The condition of having fluctuating body temperature.

heterotrophic Referring to a life condition in which an organism obtains organic chemicals in a preformed state by consuming other organisms or their byproducts.

heterozygote advantage A situation in which the fitness of individuals carrying two different alleles of a particular gene exceeds that of those carrying two alleles of the same type.

hexamerins Abundant proteins in insect plasma, whose main function is storage of amino acids for later use.

Hexapoda A superclass of arthropods containing the Insecta, Collembola, Protura, and Diplura.

higher termites Descriptive term for a family (Termitidae) of Isoptera that lack gut protozoa, instead having bacteria.

histopathology Characteristics of disease observed in specific tissues.

holoblastic cleavage Cleavage in which the entire egg is divided.

holocycly The seasonally cyclic sexual and asexual reproduction of aphids, in which sexual morphs produce a genetically recombinant overwintering egg.

holometabolous Having complete metamorphosis; having four life stages consisting of egg, larva, pupa, and adult.

homeostasis The maintenance of a steady physiological state by means of self-regulation through feedback.

homeotic genes Genes whose mutations cause transformation of segment identity.

homogametic sex The sex that carries sex chromosomes that are the same. In humans, females are the homogametic sex, carrying two X chromosomes.

homologous Describing or referring to structures or organs derived by evolution from the same ancestral structure.

homology The existence of structures that have a common origin.

homozygote An individual who bears two copies of the same allele at a given locus.

horizon (soil) A layer of soil, roughly parallel to the soil surface, that differs in properties and characteristics from adjacent layers above or below it.

horizontal transmission The transmission of a gene or genetic element from one organism to another by any mechanism other than that normally involved in the inheritance of genetic material by offspring from parents.

host plant The organism upon which a gall grows and from the tissues of which the gall is formed.

humoral immunity Immunity conferred by non-cell-mediated defense mechanisms, with the active molecules being plasma-borne.

humus The relatively stable portion of soil organic matter that remains after the major portions of plant or animal residues have decomposed.

20-hydroxyecdysone The most common biologically active steroid in insects.

hypermetamorphosis Development involving more than one larval stage.

hyperparasitoid A parasitoid that attacks another species that is itself a parasitoid, usually when the latter is feeding with its host. Such species may be detrimental to insect biological control programs.

hypognathous Having mouthparts directed ventrally.

I

ice nucleator An agent that facilitates the organization of water molecules into crystals and thus promotes ice formation.

idiobiont A parasitoid whose host is not allowed to develop further after parasitization (e.g., egg parasitoids that complete their entire development within a host egg, larval parasitoids that paralyze their hosts permanently at the time of oviposition).

imaginal discs Clusters of undifferentiated embryonic cells in holometabolous insects that proliferate during larval stages and then differentiate during the pupal stage upon induction by ecdysteroids in the absence of juvenile hormone.

immature Referring to an egg, larval, or pupal stage.

imperiled Synonymous with threatened; a designation used by the U.S. Natural Heritage Program.

indeterminate growth A pattern of growth in which an insect continues to molt after reaching the adult stage.

indirect fertilization The transfer of sperm by mean of an externally laid spermatophore.

induced chemical defense Increased production or mobilization of plant allelochemicals in response to some cue (commonly leaf damage) that indicates a high probability of future attack and that is usually absent in unstressed and undamaged plants.

infective juvenile (J3) The third-stage juvenile or filariform larva found within the biting mouthparts of the arthropod intermediate host.

inflammation A localized tissue response to an antigen that is characterized by pain, redness, swelling, and the influx of white blood cells.

injury In this context, the mechanical or physiological effect on a plant of an insect's feeding, oviposition, excretion, or nesting and sheltering behavior.

innate immunity A nonspecific host defense reaction, such as inflammation, to an infectious agent.

inoculative biological control A form of augmentative biological control in which the goal is to establish reproducing populations of natural enemies at the start of the crop.

inquilinism A replacement name for permanent parasitism. **Inquilines** are social parasites that reside entirely within the nest of their host.

insectaries Rearing facilities in which parasitoids and predators are mass-produced, usually for sale as agents for augmentative biological control programs.

insecticide targets Physiologically or biochemically important molecules (such as those comprising ion channels, receptors, enzymes, proteins) and structural molecules (such as chitin) that, when disrupted by insecticide chemicals, cause damage or death to pest insects.

in situ hybridization A technique in which small fragments of DNA are labeled with radioactive or chemiluminescent compounds and used as probes to localize genes or other segments of DNA.

instar The growth stage of an insect between two successive molts.

intake target The optimal nutrient intake over a given period in development that can be represented as a point in multidimensional nutrient space.

integrated Coordinated or harmonious use of multiple approaches to control single or multiple pests.

intensity Intensity of a vibratory signal can be expressed as a displacement (i.e., a vector quantity that specifies the change of position of a body), as velocity (i.e., a vector quantity that specifies rate of change of displacement), or as acceleration (i.e., a vector quantity that specifies rate of change of velocity).

interkernel space An air-filled space between grain kernels.

intermolt A period of feeding and growth; it begins with ecdysis from previous stage and ends with apolysis.

Inter-Tropical Convergence A zone extending across Africa and Arabia along which northerly winds meet (converge) with those blowing from the southwest. Rising air currents along the convergence lead to rain if the converging winds carry moisture, and, if rain occurs at all in the arid regions of North Africa, it is most likely to be along the convergence.

intragenomic conflict A situation in which the action of one gene, by increasing its chances of transmission, is in conflict with the interests of other parts of the genome.

inundative biological control A form of augmentative biological control in which natural enemies are released in large numbers throughout the cropping period, with no expectation that the released biological control agents will establish; pest control is expected from the individuals actually released, not their progeny.

involucrum A sheath of resin and wax surrounding the brood chamber in the nest of most stingless honey bees.

isometric growth A pattern of growth in which a body part grows at the same rate as other body parts or the body as a whole.

ivermectin An oral antiparasitic drug also effective against some ectoparasites.

J

Johnston's organ A mechanosensory organ located in the pedicel that responds to changes in antennal position.

jugum The wing area posterior to the jugal fold.

juvenile hormone (JH) One of the two major insect developmental hormones (the other is ecdysteroid or molting hormone). It is produced by the corpora allata under control of the brain. During larval stages JH maintains larval characters. In the adult it influences egg devel-

opment and other traits, including migration.

juvenile hormone esterase An enzyme that inactivates juvenile hormone through cleavage of the methyl ester bond, yielding juvenile hormone acid.

K

kairomone A chemical that is produced by one organism conveying information to another organism of a different species; it is advantageous to the recipient but detrimental to the producer of the chemical.

karst A landform created by solution of a substrate, usually limestone and other carbonates; characterized by subsurface drainage and fissures, sinkholes, underground streams, and caverns.

key-factor analysis An analytic procedure for identifying the main causes of observed fluctuations in population density.

keystone predator A predator whose activities (or lack of same) determine the species composition of an entire community.

kin selection The principle that reproduction by an individual's relatives can increase that individual's genetic representation in future generations, above and beyond its own specific reproductive output. Therefore, adaptations can evolve that favor relatives' survival and reproduction, as well as personal fitness.

kleptoparasite An individual or species that steals the food resource secured by another. This may be by taking the food away, as with some sphecid wasps that enter unguarded nest burrows of conspecifics and remove the prey items another individual has stowed there; or it may involve laying an egg on the food in the nest of another wasp, not necessarily a conspecific, which the kleptoparasite's larva then consumes.

klinokinesis An undirected change in the turning movements of an organism resulting from the intensity of a stimulus.

koinobiont A lifestyle in which a parasitoid allows its host to continue to feed and/or develop after oviposition, such that its larvae feed on an active host that is killed at a later stage. For example, many endoparasitic larval parasitoids lay their egg into a young host larva (or even egg) but do not complete development until the host has grown, and maybe even pupated.

L

labeled-line code A simple sensory code by which one cell or a single cell type sends a unique message to the higher processing centers.

labellum The apical part of the adult proboscis, comprising the modified labial palpi.

labium The "lower lip" of the insect head.

labrum The "upper lip" of the insect head.

lacinia The distal median lobe of the insect maxilla.

Lamarckian evolution A theory proposed by Jean Baptiste Lamarck, describing the passing on of an organism's characteristics to succeeding generations as the result of environmental influence on the organism during its lifetime; a predecessor to the Darwinian theory of evolution.

larviposition The deposition of living larvae that have already hatched inside the female.

latent period The time interval between when a vector acquires a pathogen and when the vector is able to inoculate the pathogen into a susceptible host.

lek behavior A mating system in which assemblages of males defend territories to which females are attracted solely for the purpose of mating.

lerp Sugary and waxy formations made by larval psyllids for their protection.

life cycle The sequence of events in the life of an insect from hatching; immature development up to adult emergence.

life table A summary of the survival rates of individuals in a population to each life stage or age category.

lightning bug firefly A member of the family Lampyridae whose adults use flashes, flickers, or other rapidly controlled bioluminescent emissions for sexual signaling (e.g., *Photinus*, *Photuris*).

lignocellulosic diet A diet consisting of wood or other lignified plant materials, whether sound or in different stages of humification.

lineage A clade, or monophyletic group; the group is defined on the basis of definitive derived traits.

lipid A chemically diverse group of molecules that are insoluble in water and other polar solvents.

lipophorin A plasma protein in insect hemolymph that transports lipids between tissues.

locus A site on a chromosome; the term is sometimes also used to refer to the gene itself.

logistic growth Population growth that is influenced by the carrying capacity of the environment.

lower termites A descriptive term for families (Mastotermitidae, Kalotermitidae, Termopsidae, Hodotermitidae, Rhinotermitidae, Serritermitidae) of Isoptera that have symbiotic intestinal gut protozoa.

luciferase The generic name given to certain biological catalysts of bioluminescence, which are enzymes that facilitate a light-emitting reaction.

luciferin The generic name given to substrate molecules that are oxidized in many light-emitting reactions.

lumen The open center of a duct.

luminescence The emission of light energy by molecules involving shifting of subatomic particles, contrasting with incandescence.

M

macroinvertebrate A collective term for aquatic insects and other arthropods without backbones and generally visible to the unaided eye.

macrophyte A large aquatic plant that can occur above, below, or on the water surface.

maggot therapy Therapeutic myiasis; introducing live fly larvae into wounds to treat them.

magnetite A common crystallized ferromagnetic mineral of iron oxide (Fe_3O_4). Submicroscopic particles of magnetite (<100 nm) have been localized in cells of various organisms, from bacteria to pigeons. These particles may implement the sensing of local geomagnetic vectors.

major/minor morphs Forms that result from dimorphism in the size of beetles and the horns and other ornaments on the head and thorax of males.

malaria paroxysm A clinical attack of malaria in a human host; associated with the liberation of parasites from the red blood cells, featuring cold (shivering, lasting <1 h), hot (fever as high as 41°C, 2–6 h), and sweating (fever breaks, temperature drops rapidly to or below normal) stages.

male accessory glands Secretory glands associated with the male reproductive tract that produce seminal fluid and the structural components of the spermatophore.

Malpighian tubules Long, narrow outpocketings of the gut; the site of primary urine formation in most insects.

mandibulate Having mandibles adapted for chewing or biting.

mating disruption The application of a formulated pheromone to a crop in order to interfere with mate finding by a pest insect.

matric potential The portion of the total soil water potential due to the attractive forces between water and soil solids.

meconium An accumulation of waste products from the larval stage.

melanin A black pigment, toxic to parasites and pathogens; formed from the precursors tyrosine and DOPA via the phenoloxidase reaction.

meristem Undifferentiated tissue capable of active cell division and differentiation into specialized tissues.

meroblastic cleavage Cleavage in which only the nuclear material of the egg is divided.

metabolic rate The rate at which an organism converts chemical energy (fuels) into heat. Increased metabolic rate is correlated with greater oxygen consumption and with greater production of carbon dioxide, water, and heat.

metamere (somite) A true body segment derived during embryonic development.

metapopulation A set of local populations linked together by dispersal.

microfilaria The first-stage juvenile in the developmental life history of dog heartworm found in the host's blood.

microtubules Proteinaceous tubular cytoplasmic structures that are part of the cytoskeleton and are involved in force transmission and movements of cell organelles.

microvilli Fingerlike projections of cells, bound by the cell membrane and extending outward into extracellular space; usually associated with absorptive or secretory regions to increase the surface area of the cell.

microwasp A wasp that is 4 to 5 mm in length or smaller.

midgut A region of the insect gut that lies between the foregut and the hindgut. The midgut is derived from embryonic endoderm. The midgut is the site of food digestion and nutrient absorption in insects.

mimicry The close resemblance of insects to different species with the resultant gain of protection by being mistaken for something else that is dangerous, unpalatable, toxic, etc.

mimicry ring A group of coexisting species that use the same signal, for example, a group of unrelated butterflies bearing the same warning coloration. A mimicry ring typically involves one or several models that evolved the signal first and one or several mimics that were later selected to use the same signal as the models.

mirror A term for the membranous area whose border is formed by the posterior cubitus vein on the right forewing, serving as a sounding board during stridulation.

mitochondria Cellular organelles that, among other functions, produce adenosine triphosphate (ATP), the primary energy source for cellular function.

mode of action A term for the mechanism by which an insecticide affects an insect.

mola Thickened grinding and crushing region at the inner edge of the mandibular base.

mollicute A member of a class of bacteria that lacks a cell wall.

molt The period during which an insect synthesizes new cuticle and other structures appropriate for the next developmental stage; it begins with ecdysteroid-induced apolysis and ends with ecdysis of old cuticle.

monocondylic Having one condyle (especially on the mandible).

monoecious A nonhost-alternating aphid life cycle that requires a single host; synonym: *autoecious*.

monogyny The presence in the nest of a single functional queen.

monomorphic/polymorphic Describing the distribution of body shapes and/or sizes among the members of a social group. If body size variation clusters around a single average value, typically approximating a normal or bell-shaped distribution, and body shapes are isometric, then group members are said to be monomorphic. In polymorphism there are two or more discrete (nonoverlapping) size classes of individuals, or individuals exhibiting differences in body shape.

monophagous Specialized to feed and develop on one host plant.

monophyletic Referring to a taxonomic group (called a clade) that contains all descendants derived from a single ancestor and recognized by the possession of a shared derived feature(s). For example, the clade Diptera is monophyletic, recognized by shared derived development of the hind wing as a haltere (balancing organ).

monophyly   The classification or status of being monophyletic (see above).

morphogenesis   The phase of embryogenesis in which specific tissues, organs, and structures develop.

Müllerian mimicry   A mutualistic relationship, named after Fritz Müller (1879), in which several prey species, or co-mimics, with anti-predator defenses (e.g., bad taste, toxicity, venom) direct the same signal to the receiver. Despite being mutualistic, Müllerian mimicry is not necessarily coevolutionary and co-mimics do not necessarily benefit equally from the mimicry.

multivoltine   Describing species that have two or more generations per year. Those that have a single generation per year are called *univoltine*.

mutation   Any change in genetic material, though the term usually is used to refer to an error in replication of a nucleotide sequence.

mutualism   An interspecific relationship in which both partners (**symbionts**) benefit from the interaction.

mya   Millions of years ago.

mycangium   An invagination of the cuticle within which microorganisms are held, facilitating their transfer to uninoculated host substrate.

mycetocytes   A specialized form of fat body cells harboring symbiontic bacteria that might contribute essential nutrients. Mycetocytes are in proximity to urocytes in the fat body lobe, suggesting a biochemical interaction between the two types of cell.

mycetophagous   Describing an organism that is fungus feeding.

myiasis   The invasion of living tissues by fly larvae (maggots).

myrmecophiles   Organisms, generally insects other than ants, that live in or around ant nests and that exploit resources from the ants by begging, scavenging, or predation.

N

naiad   A term for the larva of the aquatic, hemimetabolous insect orders.

nasute   A soldier in the family Termitidae that has a well-developed horn-like median projection (**nasus**) and reduced mandibles. A defensive fluid may be produced and ejected. Soldiers for some species in the family Rhiontermitidae may also produce a defensive fluid from a fontanelle gland on the top of the head, but do not have the horn-like median projection.

natural control   The level of control produced by natural enemies that occur normally without any active management by humans.

Nematocera   The paraphyletic or grade-level group of so-called "long-horned" flies or "lower Diptera," characterized generally by long, multisegmented antennae and by larvae with a well-developed, sclerotized head capsule.

neonates   Newly hatched young.

neopterous   Having a structure that permits folding the wings at rest down to the sides of the body.

neosome   An organism altered by neosomy.

neosomule   A new external structure resulting from neosomy.

neosomy   The ability to produce new cuticle without molting.

neotenic   An individual that retains juvenile morphological features after having reached sexual maturity.

neoteny   A condition in which nymphal or larval structures are carried over into the adult and the corresponding adult structures are suppressed.

nest   Any modification of the environment by adult insects that provides shelter for the rearing of their offspring.

neurohormone   A small organic or peptidergic substance that is produced in neurosecretory cells; released into the hemolymph at special regions called neurohemal organs and transported to target tissues with the hemolymph.

neuromodulator   A neuroactive substance that is released by synaptic terminals. It simultaneously acts on large numbers of cells in the proximity of the releasing cell and modifies the properties of synaptic transmission and the properties of target cells.

neurosecretory cell   A neuron that produces one or more peptides or small proteins (neuropeptides), stores the peptide(s) at the axon terminal, and then releases it into the hemolymph in response to excitatory neural input.

neurotransmitter   A chemical substance that is released from the presynaptic endings of a neuron. It transmits information across the synaptic cleft to specific receptors located on the surface of postsynaptic cells.

niche   The environmental limits within which a species can survive and reproduce.

nit   Informal term for a louse egg.

nontarget impact   Mortality or injury that may occur to neutral or beneficial species as a consequence of the use of natural enemies in biological control programs or insecticide applications.

nuclear receptors   A family of soluble proteins that are mobilized by steroid hormones to coordinate gene expression through direct binding to DNA.

numerical response   An increase in the number of predators or parasitoids in response to an increase in prey density.

nutritional rail   A means of representing a food in a nutrient space as a line which moves from the origin into the space.

nymph   The larval stage of hemimetabolous insect orders.

O

obligate hyperparasitoid   A hyperparasitoid that is always a secondary parasitoid; that is, its progeny can develop only on or in a primary parasitoid wasp host.

obligate parasite   A parasite that cannot complete development or reproduce without obtaining nourishment from a host animal.

obligatory diapause   A form of diapause that is genetically programmed to occur at a specific developmental stage regardless of the environmental conditions that prevail.

ocellus *plural,* **ocelli**.   A simple eye consisting of photosensitive cells and sometimes a single, beadlike lens; multiple ocelli may be present, but their action is not coordinated.

odor binding protein (OBP)   A protein that binds and transports an odor molecule to the receptor protein on the sensory cell dendrite. It is needed to convey the water-insoluble odor molecule across the watery lumen of the sensillum.

oenocytes   Special cells associated with epidermis and tracheae. They are involved in synthesis of waxes, such as beeswax and waxes used for waterproofing cuticle and eggshells.

official control   A term for the suppression, containment, or eradication of a pest population by a plant protection organization.

oligogyny   The presence in a nest of two or more functional queens.

oligophagous   Feeding on a few, often related plant taxa, such as different species in one genus or a few genera in one family.

oligosaccharide A molecule composed of 4 to 20 monosaccharide sugar units.

oligotrophic Describing a pond, lake, or other body of water that is deficient in plant nutrients.

ommatidium *plural,* **ommatidia.** An individual unit of the compound eye.

oocyte A female gamete or developing egg.

oogenesis The formation and development of the eggs in the female. It includes egg maturation and yolk formation (vitellogenesis).

ootheca A case or other covering of an egg mass that has been laid.

optimal foraging theory A theory that attempts to document the most efficient ways (in terms of evolutionary fitness) for an organism to find food or other resources.

optimal temperature The temperature at which any body performance or activity is most efficient.

orthokinesis An undirected change in speed of movement of an organism resulting from the intensity of a stimulus.

Orthorrhapha A probable paraphyletic subgroup of Brachycera in which the pupae are not enclosed in puparium and in which the sclerotized portions of the larval head capsule are exposed externally.

osmoregulation Control of osmotic pressure or salt and water balance.

osmotic potential The portion of the total soil water potential due to the presence of solutes in soil water.

ostia *singular,* **ostium.** Valves that can be permanently open or flapped to allow uni- or bidirectional flow of hemolymph into and out of dorsal vessel and accessory pulsatile organs.

ovariole An element of the insect ovary that consists of a tapering tubule in which oocytes are produced.

ovipara The sexual female produced by the viviparae (or gynoparae) in the fall; oviparae mate with males on the primary host, producing the overwintering genetically recombinant egg.

oviparity Egg laying as the normal means of reproduction.

ovipary Alternate term for oviparity (egg laying).

oviposition The act of laying eggs; the passage of an egg from the median oviduct to outside the insect's body.

ovipositor The organ by which eggs are deposited.

ovoviviparity The production of well-developed eggs that hatch inside the mother's body.

ovulation The passage of an egg from the ovariole into the oviduct.

oxic Containing a certain amount of molecular oxygen; describes the oxygen status of an environment.

P

pair formation The stage before mating during which the sexually receptive male and female are attracted to one another.

pair-rule genes Genes expressed in seven stripes in the *Drosophila* embryo; mutations in these genes cause deletions in a two-segment periodicity.

Paleoptera Insect orders that have direct flight muscles and lack the ability to fold their wings over their back while at rest.

paleosymbiosis Fossil evidence of associations between two different species; this includes **paleoinquilinism** (two organisms in the same niche but neither benefits nor is harmed by the other), **paleocommensalism** (one organism benefits, neither is harmed), **paleomutalism** (both organisms benefit, neither is harmed), and **paleoparasitism** (one organism benefits by taking nourishment at the expense of the other, which is harmed, often killed).

palp/palpal organs Variably complex structures found in the terminal part of adult male palp.

pandemic A disease that affects populations in many countries.

Pangaea The single continent that existed about 220 mya; about 135 mya, this protocontinent broke up and drifted apart into two pieces, Gondwana to the south and Laurasia to the north.

paracellular permeability The ability of substances to travel from one side of an epithelium to the other by moving through the spaces between cells.

paracrine Describing chemical messengers that act on cells in the immediate area without entering the blood.

parameres pairs of annulated gonapophyses on abdominal segments VIII and IX or IX.

Paraneoptera A superordinal group of insects related to the Holometabola and consisting of the orders Hemiptera, Thysanoptera, Psocoptera, and Phthiraptera.

paraphyletic Referring to a taxonomic group containing a single ancestor and some, but not all, of its descendants.

paraphyly The status of a group that includes some, but not all, the descendants of a single common ancestor.

parasite An organism living in a relationship with another organism in which it benefits but the other does not (and typically is harmed).

parasitoid A specialized form of parasite that develops in or on a host, eventually killing it. Larval parasitoids require one host to complete development. In addition to parasitism, adult female parasitoids can kill hosts through host feeding. Most hosts of parasitoids are other insects, and most parasitoids are in the orders Hymenoptera and Diptera

paratransgenesis The process of genetically modifying endosymbionts, which results in the malnutrition and death of the insect that requires them.

parental care Behavior undertaken by a parent organism that is beneficial to offspring; e.g., in some beetle species the mother remains underground tending the brood mass until development of her offspring is complete.

parental investment Any behavioral or metabolic investment by a parent that increases the individual offspring's survival at the expense of the parent's ability to benefit other offspring.

parthenogenesis Reproduction in which eggs develop without fertilization by a male gamete.

pathogen An organism that causes disease. Viruses, bacteria, microsporidia, and fungi can be pathogens.

pattern formation The process by which embryonic cells form ordered spatial arrangements of differentiated tissues; during insect embryogenesis, patterning begins with specification of the anterior–posterior and dorsal–ventral axes and ends with the formation of the characteristic structures associated with each segment of the head, thorax, and abdomen.

pattern generation Control of repeating movements such as walking, wing flapping, swimming, and breathing. The neural circuits that control these behaviors include neurons within the central nervous system that are often influenced by sensory inputs from the periphery.

pedicel The second segment of an insect antenna.

pedogenesis Reproduction by an anatomically immature life stage.

pedomorphosis An evolutionary change in which adults of a later species retain some of the characteristics of juveniles of the ancestral forms.

perimicrovillar membranes In hemipteran midgut cells, lipoprotein membranes that ensheath the microvillar membranes like glove fingers and extend toward the luminal compartment; they are apparently involved in amino acid absorption from dilute diets.

peritrophic membrane A filmlike anatomical structure made of protein and chitin that separates food from midgut cells; it is believed to have evolved from an ancestral mucus and combines the protection function of this mucus with several roles in digestion associated with midgut compartmentalization.

permissive host A host insect species in which a parasite or pathogenic organism establishes a successful infection. Species or host strains in which the development of the invader is thwarted by hostimmunological defenses are **nonpermissive**.

pest A general term for any species whose activities interfere in some way with human health, comfort, or profit. In **integrated pest management (IPM)**, a pest can be an insect or other animal, a plant, or a microbial pathogen.

petiole The slender abdominal segment of Hymenoptera.

phagocyte A cell that is able to engulf foreign bodies.

phagocytosis The process of "cell eating," in which small particles or microorganisms are engulfed by blood cells.

phagostimulant A chemical that induces and maintains feeding.

phagostimulate To promote (stimulate) feeding, usually by a nutrient.

pharate The early period of an instar in which a new cuticle has been formed, and the insect is still enclosed in the cuticle of the former instar.

phenology The periodicity of life cycle stages in relationship to seasonal occurrences in a given region.

phenotype The outward appearance of an individual, being determined by its genotype and the influence of the environment on the expression of this genotype.

pheromone Chemical message released by an individual that induces either a behavioral reaction or a developmental process in other individuals of the same species.

phloem Active, food-conducting tissue of the inner bark of trees and other woody plants, which with age becomes outer

bark. Food is stored in this tissue and is also conducted to the roots for storage.

phosphorylation Transient, reversible posttranslational modification of proteins in which the terminal phosphate group of ATP is transferred to specific residues of a polypeptide by kinases and often alters the properties of the protein.

photoperiod The length of "day," i.e., **photophase**, and "night," i.e., **scotophase**, usually during a 24-h period.

phreatic Of or relating to groundwater, specifically the saturated zone below the water table.

phylogenetic Incorporating the evolutionary history and expressed in ancestor-descendant relationships of a species or taxonomic group.

physiogenesis A physiological (still unknown) process in the diapause stage that leads to the termination of diapause; sometimes termed *diapause development.*

phytosanitary regulations Official rules to prevent introduction and/or spread of pests by regulating the production, movement, or existence of commodities or other articles, or the normal activities of persons.

planidium The first instar of some hymenopteran parasitoids. **Planidial** larvae can move about actively because they have hardened integuments and spines.

plasma The liquid portion of hemolymph, consisting of water and dissolved solutes.

plasmid An independent, stable, self-replicating piece of DNA in bacterial cells that is not a part of the normal cell genome. Plasmids are commonly used for cloning.

Plasmodium A genus (order Coccidiida, family Plasmodiidae) that contains 100 species of blood parasites of vertebrates, of which *Plasmodium malariae, P. vivax, P. falciparum,* and *P. ovale* cause malaria in humans.

plastron A structure by which a layer of air is held close to the body and by which gas exchange occurs with surrounding water.

plesiomorphic Referring to the ancestral condition of a character state.

pneumonic Spread from the lungs by coughing and spitting.

poikilothermy See HETEROTHERMY.

polydnavirus A viruslike entity comprising multiple circular DNA molecules in a protein coat. Polydnaviruses are encoded by the genomes of some parasitic wasps (some ichneumonids and

braconids), produced in the female reproductive tract, and injected into hosts, where they are expressed and help overcome the host's immune defenses.

polyethism Caste designation or division of labor in social insects. In **caste polyethism**, morphological specialization involves different functional castes, whereas **age polyethism** refers to a change in behavior related to age.

polygenic character (trait) A trait controlled by the integrated action of multiple independent genes.

polygyny The presence in a nest of multiple, functional queens.

polymerization A process by which small molecules (**monomers**) combine chemically to produce a large network of molecules, or **polymers**; one of the main processes by which resin becomes amber.

polymorphism The existence of two or more genotypes for a given trait within a population.

Polyneoptera A superordinal group of insects consisting of the Plecoptera, Embiidina, Zoraptera, Orthoptera, Phasmida, Dermaptera, Grylloblattodea, Isoptera, Mantodea, and Blattaria.

polyphagous Feeding on many plant species from a range of families.

polyphenism The occurrence of alternate phenotypes caused by environmental cues and usually mediated through hormonal signaling. Polyphenisms can occur sequentially as a part of a developmental sequence (larva, pupa, and adult) or alternatively as either seasonal morphs or castes in social insects.

polyphyletic Referring to a taxonomic group derived from more than one ancestor and recognized by the possession of one or more features evolved convergently. For example, if the primitively wingless silverfish were united with secondarily wingless grasshoppers, beetles, and flies, the resulting group would be polyphyletic.

polyploid Having more than two complete sets of chromosomes (e.g., triploid, having three sets; tetraploid, having four sets).

polytene chromosomes Giant chromosomes arising from replication without mitosis, allowing visualization of banding patterns and puffs indicative of transcriptional activity.

population A particular group of interbreeding individuals more or less separated in time or space from other groups of the same species.

preadaptation The existence of a pre-existing anatomical structure, physiological process, or behavior pattern that makes new forms of evolutionary adaptations more likely.

predator An organism that attacks and eats another; a true predator is distinct from a parasitoid in that it does not develop on or within the host and usually kills and consumes many prey during its lifetime.

primary host A term for an aphid's overwintering host, where egg deposition occurs in a dioecious life cycle; usually a deciduous woody angiosperm that was the ancestral host before host alternation was adopted.

primary sex ratio The sex ratio at fertilization or zygote formation.

proctodeal trophallaxis The feeding on microbe-rich hindgut fluid solicited from nestmates; found only in termites and wood-feeding cockroaches.

procuticle The inner layer of cuticle, beneath the epicuticle. In stiffened (sclerotized) cuticle, it typically consists of exo- and endocuticle, but in flexible cuticle, the exocuticle is not present.

prognathous Having mouthparts directed anteriorly.

prokaryotes A collective name for all microorganisms without a true nucleus, comprising the Archaea (formerly Archaebacteria) and the Bacteria (formerly Eubacteria), which form two phylogenetically independent lineages next to the Eukarya (also, eukaryotes).

prolegs Unsegmented, false legs, as found in many larval insects.

pronation A classification of wings; i.e., twisting, anterior margin downward, posterior margin upward.

propagule The initial group of individuals that found an introduced population.

propolis Resins collected from plants and used in the nest by honey bees.

prostheca Any of a variety of structures—rigid or flexible, simple or complex, lobelike or setose—arising from the mesal surface of the mandible just distal of the mola.

prothoracic gland A diffuse endocrine organ in the thoracic area of insects that produces ecdysone, the precursor of 20-hydroxyecdysone.

pseudergate A caste in some species in the families Termopsidae, Kalotermitidae, and Rhiontermitidae; a nonreproducing and nonsoldier caste that is developmen-tally very plastic and can do "worker" tasks within colonies.

pseudogamy A form of sperm-dependent parthenogenesis in which eggs require activation by entry of sperm, but only maternal chromosomes are expressed and passed on.

pseudokarst A landform mimicking karst in form but created by volcanism or erosion; like karst, drainage is primarily subterranean.

pterygote Describing insect orders with wings.

puparium The sclerotized exuvium of the last instar, within which the pupa (intermediate developmental stage between larva and adult) is formed.

pyrethroid An organic synthetic insecticide with a structure based on that of pyrethrum, which is a natural botanical insecticide.

Q

quarantine A secure importation facility designed to prevent the unintentional release of natural enemy species into new geographic regions after their discovery during foreign exploration.

quarantine pest A pest of potential economic importance to a given area endangered thereby and not yet present there, or present but not widely distributed and being officially controlled.

quiescence A forced state of dormancy imposed by an immediate effect of environmental conditions that prevent the normal activity.

R

receptaculum seminis The spermathecal sac in which sperm are stored.

recessive referring to an allele that is not expressed phenotypically when present in a heterozygote, but only when in a homozygote. The opposite of dominant.

rectal pads Thickened regions of the rectal epithelium in which the cells are specialized for active transport of solutes across the epithelial wall.

reference condition The environmental condition that is representative of a group of minimally disturbed sites organized by selected physical, chemical, and biological characteristics.

reflex bleeding The utilization of blood (hemolymph), sometimes fortified with deterrent allomones, as a defensive secretion in adversarial contexts. **Reflex**
bleeders exude hemolymph from weak sutures in the chitin (e.g., at femoral joints). Used by beetles to defend themselves against attack from ants and other predators.

refuge A location in or near crop fields in which natural enemies are protected from pesticides and provided with resources necessary for their survival and reproduction.

remigium The area of the wing of a neopterous insect lying anterior to the claval furrow.

reproductive success An estimate of evolutionary fitness used to measure sexual selection. Measures can include success in obtaining matings with different numbers of the opposite sex as well as fertilization success (by males) when females mate and store sperm from several males.

required nutrients Nutrients, including essential nutrients, that enhance or optimize growth, development, and/or reproduction.

reservoir A term for an animal, usually a vertebrate, that is infected with an animal disease agent, can maintain the agent in its tissues for a prolonged period, is fed upon abundantly by one or more efficient vectors, and is capable of serving as a source of infection for such arthropods. Some vectors that transmit agents transstadially, transovarially, or by both mechanisms also serve as reservoirs.

resilin A very elastic protein found in insects that is particularly associated with joints. Resilin is rubberlike and can store energy and can recoil to produce efficient movement.

resin A water-insoluble exudate emitted from parenchymal cells of various plants, both angiosperms and gymnosperms; it is composed mainly of a mixture of terpenoids, acids, alcohols, and carbohydrates and can be molded by hand.

resonance A condition such that the frequency of an applied force is equal to the natural frequency of a system; the frequency of vibration of a system when the amplitude of vibration is a maximum.

respiration The collection of metabolic pathways responsible for the oxidation of glucose and fatty acids, with the production of energy involving an electron transport chain.

rhabdome The rod-like structures that collectively comprise the "retina" of an ommatidium.

RNA interference The ability of double-stranded RNA to inactivate the expression of the homologous cellular gene.

S

salivary glands In mosquitoes and other blood-feeding dipterans, glands located in the thorax of both immature and adult stages. In adult females, the glands are associated with blood feeding. Certain pathogens causing diseases in vertebrates are injected into feeding wounds during blood feeding, thus resulting in a type of transmission known as salivarian.

sapwood The soft wood just beneath the inner bark of a tree.

scabreiform An insect larva that is grub shaped.

scape The first (most proximal) segment of an insect antenna.

sclerotization A chemical process in which insect cuticle and other materials are rendered hard, insoluble, dehydrated, and resistant to degradation by the oxidative incorporation of phenolic compounds.

scolopale A "sense rod," or minute rod-like structure around the distal end of a sensory neuron.

search imagery The ability of predators to detect cryptic prey more efficiently by learning to search for potential prey of a specific visual appearance based on color, pattern, size, movement, or location within the environment, often with concurrent emphasis on search in specific habitats and reduction of the search rate.

secondary compound A toxic or digestibility-reducing compounds synthesized primarily for defense against herbivores. Not involved in the primary metabolic pathways of plants.

secondary host A term for an aphid's summer and usually herbaceous host in dioecious life cycles; in secondarily monoecious aphids it may be the host that remained after the aphid's ancestors abandoned the primary host.

secondary metabolite A plant chemical product that is not essential to the basic metabolism of the plant, e.g., phenolic glycosides. Contrast with primary metabolites such as nucleic acids, lipids, proteins, carbohydrates, or ribulose diphosphate. Secondary metabolites may or may not function as allelochemicals.

secondary sex ratio A change in the sex ratio of a family or population that occurs after fertilization but prior to reproductive maturity of the individuals involved.

secondary sexual traits Morphological and behavioral differences between the sexes that relate to structures other than the reproductive organs and gametes.

second messenger An intracellular substance, such as Ca^{2+}, cyclic AMP, inositol-1,4,5-trisphosphate, that modifies or modulates cellular responses. Its concentration changes in response to activation of G protein-coupled receptors.

segment polarity genes Genes expressed in segmental stripes in the *Drosophila* embryo; mutations in these genes cause deletions and duplications in every segment.

selectivity The property of given insecticides to selectively affect the target pest species, sparing other organisms collectively called nontarget species.

semiochemical A chemical produced by an organism that has a communication effect on another individual of the same or a different species.

semivoltine Describing a life cycle with one generation every two years.

sensillum *plural,* **sensilla.** An organ used for sensing. For olfactory and gustatory (taste) sensilla, cuticular modifications, accessory cells, and sensory cells combine to make a functional sensillum.

sensitive Referring to species for which population viability is a concern; a designation used by the U.S. Forest Service.

sentinel organism An organism that accumulates pollutants from its surroundings and/or food and is used in tissue analysis to provide an indirect estimate of prevailing environmental concentrations of these substances.

septicemic Characterized by the presence and growth of pathogenic infectious agents in the blood.

serial homology Similarities between repeated structures in different segments of an organism, caused by shared origin in development.

sericulture The industry in which mulberry plants are cultivated and used to raise silkworms, the cocoons of which are used to produce silk.

serotinous cones Cones that remain closed and on the tree for one or more years after seed maturation. High temperatures melt the resin that holds the cones closed, causing them to open rapidly.

sesquiterpenoid A substituted 15-carbon terpene derived from three isoprene units.

setae *singular,* **seta.** Slender, rigid, bristle-like hairs extending from the cuticle.

sex differentiation The genetic pathway that leads to male or female development, including regulation of somatic sexual differentiation, germ-line differentiation, and (possibly) dosage compensation.

sex role reversal A situation that occurs in species in which the sex ratio of individuals available for reproduction is female biased, such that males are the limiting sex. This leads to female competition for males and male choice of females.

sexual dimorphism The development of drastically different morphology in the male and female of a species.

sexually dimorphic Having morphological differences between males and females of the same species.

sexual selection Darwinian selection for traits—including behavioral traits—that promote success in competition for mates or the best mates (operating through discrimination of mates or direct competition). For males—the sex typically subject to the more intense sexual selection—the competition is ultimately for fertilizations.

signal Any trait borne by some individual organism, called a sender, and perceived as information by another organism, called a receiver. The receiver usually displays a signal-dependent response (otherwise the signal is meaningless).

signal system The program/format of signaling by which males and females identify and reach each other; basic systems are system I, a stationary individual broadcasts a signal, the other sex approaches, and system II, one individual broadcasts a signal, the other responds with a signal, the first approaches the second, which typically remains stationary.

silk protein A class of proteins constituting the cocoons produced by the silkworm family (Bombycidae). Silk protein is usually composed of two proteins, fibroin and sericin.

siphunculi A pair of specialized pore-bearing structures on the posterior third of the abdomen; involved in production of aphid alarm pheromones; synonym: cornicles.

sister groups Species or monophyletic groups that arose from the stem species of a monophyletic group by a singular, identical splitting event. For example, the Lepidoptera and Trichoptera are sister groups; they shared a common ancestor that gave rise to no other lineage.

social Living in association and/or cooperation with others of the same species. The term **social insect** is applied broadly to include presocial and eusocial insect species.

solitarious In isolation; not grouped. Locusts in the **solitarious phase** do not swarm. The opposite of *gregarious*.

solubility A measure of the amount of a gas that dissolves in a fluid when exposed to a defined pressure of that gas in the air. Carbon dioxide is approximately 1000 times more soluble in water than oxygen.

somatic cells A collective term for all cells except gametes.

somatic sexual differentiation Specification during development of body parts, including male and female sex organs, and other sex-specific morphological and/or behavioral differences.

spermatheca A sperm storage organ in female insects that dispenses sperm as the eggs pass through the oviduct.

spermatogenesis The production of spermatozoa from germinal cells.

spermatophore An encapsulated package of sperm that is passed during mating and can include proteins that are eaten or absorbed by the female.

sperm competition Competition between sperm from two or more males, within a single female, for fertilization of the eggs.

sperm precedence The increased likelihood that sperm from a particular mating will be used for fertilizing an egg.

spinnerets Finger-like abdominal appendages of spiders containing numerous spigots through which silk is extruded.

spiracle An opening in the insect integument that connects the tracheal system to the air.

sporozoite Infective stage of the malaria parasite for the human host, inoculated with saliva by *Anopheles* females during blood feeding.

stadium The interval between molts.

stance The phase of the leg cycle movement in which the foot, or tarsus, is in contact with the ground and extension of the leg joints acts on the movement of the body relative to the substrate.

station-keeping Activities and movements that keep an animal within its home range or enable it to return to its home range.

stem mother See FUNDATRIX.

stenoxenic phoresy Relationship restricted to only one host genus or species.

sterile insect technique A process in which insects are reared in massive numbers, sterilized, and released to prevent normal mating in target populations.

steroid A lipid containing a 17-carbon nucleus of fused rings: three cyclohexane rings and one cyclopentane ring.

stridulating organ A file and scraper structure for sound production; may be variously located on chelicerae, palps, legs, abdomen, and carapace.

stygobite An aquatic troglobite.

stylet bundle A group of stylets (see below) that penetrates the food substrate as a functional unit.

stylets Thin, elongated mouthparts of piercing–sucking insects that penetrate into the food substrate.

subimago The winged, terrestrial instar of Ephemeroptera that molts to the reproductively mature adult.

subsocial Describing a social system in which adults protect and/or feed their own offspring for some period of time after birth, although typically the parent leaves or dies before the offspring become adults.

subterranean nest A nest below the surface of the ground.

sun compass The ability of an organism to use the sun to maintain a constant direction. This requires a "biological clock" to allow compensation for the passage of the sun across the sky.

supercooling The absence of freezing at or below the normal freezing point of water.

supercooling point The subzero temperature at which body fluids freeze, also known as the **temperature of crystallization**.

supergene A group of tightly linked genes on a chromosome, functioning as an integrated unit and segregating like a single gene. The genetic subunits usually have a more restricted phenotypic effect than the integrated supergene.

supination A classification of wings; i.e., twisting, posterior margin downward, anterior margin upward.

swarm founding The initiation and founding of a colony by a group of workers along with one or more queens.

swing The phase of the leg movement cycle in which the foot, or tarsus, is raised from the ground and repositioned for the next stance phase.

sylvatic Living or found in forests or woods.

symbiont An organism living in intimate association with another dissimilar organism. Symbiotic associations can be mutualistic, neutral, or parasitic.

symbiosis A condition in which a plant or animal is intimately associated with another organism of a different species; the living together of two taxonomically distinct organisms.

sympatric Describing two or more species inhabiting the same or overlapping geographic areas.

synanthropic Associated with humans or their dwellings; living in close association with humans.

synapomorphy (-ic) A derived state shared among the members of a monophyletic group, in contrast to a **symplesiomorphy**, a shared ancestral (plesiomorphic) state from which phylogenetic relationships cannot be inferred.

syncytial cleavage The condition of all the cleavage nuclei containing a common cytoplasm.

syndrome In genetics, an association of traits functioning together. To be adaptive and acted upon by natural selection, the traits must be influenced by the same genes (i.e., they must be genetically correlated).

synomone A chemical signal with a beneficial effect to the sender and to the receiver of the message, as in a floral scent which indicates a nectar source to a pollinator and thereby facilitates pollination.

synovigenic A form of reproduction in which an adult female continues to produce and to mature eggs throughout adult life.

systematics The study of the relationships and classifications and naming of sorts of organisms. The terms *taxonomy* and *systematics* are often used interchangeably.

T

tachygenesis Abbreviation of the developmental cycle, by suppression of one or more instars or their feeding requirements.

tagma *plural,* **tagmata**. A body region consisting of metameres grouped or fused to perform similar functions (e.g., head, thorax, abdomen).

taxa *singular,* **taxon**. A collective term for all the taxonomic groups of organisms within a higher taxonomic group (a genus, family, tribe, etc.).

taxonomic Having to do with the classification and naming of organisms.

taxonomic rank A classificatory level in the taxonomic hierarchy, e.g., species, genus, family, order. No rank is absolute, and comparisons between ranks of different organisms are inexact or even misleading; despite this, traditional ranks used for insects—notably orders and families—have useful didactic and synoptic value.

taxonomy The procedure or discipline of classifying organisms.

tegmen *plural,* **tegmina**. A parchment-like forewing characteristic of the Orthoptera and related orders. The cuticle is somewhat thickened, in association with protection.

temperature threshold The limiting (minimal or maximal) temperature at which growth and development will occur.

teneral Describing a condition in which the cuticle is soft and usually pale immediately after molting.

tentorium A bracing structure inside the head formed by invaginations of cuticle.

testis The organ in the male that produces sperm.

thelytoky Parthenogenesis in which only females are produced; the production of females from unfertilized eggs.

thermoperiod Daily temperature fluctuations.

thermosensitivity A decrease in tolerance to high temperature resulting from prior exposure to high temperature.

thermotolerance An increase in tolerance to high temperature, usually achieved by exposure to a moderately high temperature.

thigmotactic Contact orientation, as in insects that inhabit crevices or cracks.

threatened Referring to species that face a high risk of becoming endangered in the near future unless action is taken to protect them; a designation used by the U.S. Fish and Wildlife Service and the International Union for Conservation of Nature and Natural Resources.

tidal flow (of hemolymph) The reciprocal exchange of oxygen and hemolymph in some adult insects, especially where hemolymph volume is low.

trachea An air-filled branching tube that transports gases between tissues and the environment, formed by a cylinder of cells (the tracheal epithelium).

tracheal gill A body appendage with a thin cuticle and containing many trachea, used for gas exchange by many aquatic insects.

tracheal lung Thin-walled, expanded tracheae, thought to be specialized for oxygen delivery to cells lacking tracheoles such as hematocytes.

tracheole A small, blind-ended, terminal tube of the tracheal system. Each cell receives oxygen and expels carbon dioxide by means of its own tracheole.

transduction The conversion of mechanical displacement at a nerve cell membrane into a receptor current that changes the membrane potential.

transformation A process by which the genetic makeup of an organism is altered by the incorporation of foreign DNA.

transgenic Describing an organism with foreign genes incorporated into its genome by recombinant DNA techniques.

transovarial transmission The transmission of an animal disease agent from an infected female arthropod to its progeny via the egg stage.

transposable elements (TEs) Mobile segments of DNA that are able to replicate themselves and insert copies into new locations in the genome.

transstadial transmission Passage of an animal disease agent from one life stage of an infected arthropod to the next stage during the transstadial molt, e.g., from an infected larval tick to the nymphal stage.

trichomes Specialized tufts of hair on some insects that serve to dispense chemical secretions from underlying glands.

tripod gait A pattern of leg movements used by insects in which the front and rear legs on one side move in conjunction with the middle leg on the opposite side, forming a stable tripod of support. The remaining three legs also form a tripod, and motion occurs when the two sets alternate their motions.

triungulin The campodeiform first instar of Meloidae, Rhipiphoridae, and Micromalthidae that molts to a less active, feeding second instar. Unlike later instars, triungulins have well-developed legs and move around actively.

troglobite A species obligately adapted to live only in subterranean habitats and unable to survive in surface environments.

troglomorphic Relating to the suite of morphological, physiological, and behavioral adaptations that are characteristic of troglobites.

troglophile A species able to live and reproduce in caves but also able to survive in surface habitats.

trogloxene A species habitually roosting or regularly visiting caves for food or shelter but unable to complete its life cycle underground.

trophallaxis The exchange by oral transfer of alimentary liquid among members of a colony of social insects. Typically, this liquid serves to feed and unite members of a colony.

trophic cascade An interaction of trophic levels in a community, such that changes in density of a species at one level have multiple influences on the density of species at other trophic levels.

trophic cells Specialized cells containing a reserve of fat or other nutritive substance.

trophic egg An egg, usually degenerate in form and inviable, that is fed to members of the colony.

trophic level A functional classification of organisms in a community based on what they feed on. The first trophic level consists of primary producers, mostly green plants that obtain energy from the sun. The second trophic level consists of herbivores, organisms that feed on plants. The third trophic level consists of carnivores that feed on the herbivores, and so on.

trypanosome A type of protozoan that has a leaf-like motile stage.

tubular body A microtubule-based structure found in the distal sensory dendrites of many mechanoreceptor neurons.

tychoparthenogenesis The rare or occasional production of eggs that begin developing without having been fertilized.

U

ultraselfish Describing a strategy used by symbionts whose spread and maintenance occur despite and because they cause damage to the individual in which they occur.

univoltine Describing species that have a single generation per year. Those that have two or more are called *multivoltine*.

unprofitability The character of a prey that gives no net reward to the predator once consumed, leading to learned or evolved avoidance. Examples are toxicity, toughness, difficult handling, and difficult/costly capture.

urocytes Cells that accumulate uric acid either as a form of nitrogen waste product or as reserves to be used as a nitrogen source by the mycetocytes.

urogomphi Sclerotized, paired dorsal processes that project from the posterior margin of the ninth larval tergite.

USP Ultraspiracle, a nuclear receptor that forms a dimer with the ecdysteroid receptor to regulate gene expression.

V

vannus The expanded fan-like area of some hind wings, posterior to the claval furrow.

vas deferens The sperm duct leading away from the testis.

vector Any agent that transports a microorganism from one host to another.

vectorial capacity The epidemiological efficiency of *Anopheles* host species in transmitting malaria parasites, expressed as new infections per infection per day, based on mathematical relationships among the mosquito biological characteristics of daily survival, blood meal host-feeding pattern and frequency, and susceptibility to parasite infection.

veins The cuticular, usually tubular, rods that support the wing.

venation The pattern of veins within a wing.

ventral diaphragm A complete or fenestrated membrane always associated with the ventral nerve cord of insects and defining a perivisceral sinus when present.

veriform Describing an insect larva that is worm shaped.

vertical transmission The transmission of genetic elements from parents to progeny.

vesicle An extensible organ.

vitelline membrane The innermost layer of eggshell secreted by the follicle cells; sometimes considered to be the first layer of the chorion.

vitellogenic cycle A cycle of oocyte development associated with the synthesis of **vitellogenin**, a protein that contributes most of the protein in yolk. In some insects, such as mosquitoes, **vitellogenesis** (vitellogenin production) is limited by lack of protein and occurs only following a highly proteinaceous meal, such as blood.

vivipara The parthenogenetic winged or wingless morphs that bear live young; synonym *virginopara*.

viviparity Reproduction by giving birth to live offspring rather than eggs.

voltage-gated sodium channel A large transmembrane protein that regulates the flow of sodium ions across axonal membranes and mediates the rising phase of action potentials.

vulnerable Facing a risk of extinction in the medium-term future; a designation used by the International Union for Conservation of Nature and Natural Resources.

W

Wallace's Line A term for the junction between the Oriental and the Australian zoogeographical regions at the straits between Bali and Lombok.

wasp waist The strong constriction between the first and second abdominal segments, present in virtually all members of the Apocrita. The thorax and first abdominal segment in these insects comprise the **mesosoma**, and the rest of the abdomen, the **metasoma**.

wild type An organism that has no visible mutant phenotype.

X

xylem A plant conductive tissue interior to the cambial layer, composed of several cell types, which transports water from the roots to other sites in the plant and also serves as structural support.

xylophilous Preferring wood.

Y

yolk An accumulation of protein, lipid, and glycogen particles in the cytoplasm of an egg. The particles are assembled during vitellogenesis and consumed during embryonic development.

Z

zoonosis *plural,* **zoonoses.** A disease of vertebrates communicable from animal to animal and from animal to human.

zoonotic Transmitting disease between vertebrate animals and between animals and humans.

SUBJECT INDEX

A